100 YEARS OF VEHICLE SAFETY DEVELOPMENTS

Other related resources from SAE International:

2005 SAE Accident Reconstruction Technology Collection on CD-ROM
Order No. ARCD2005CD

2005 SAE Occupant Protection & Crashworthiness Technology Collection on CD-ROM
Order No. OP2005

Stapp Car Crash Conference Proceedings Collection on CD-ROM
Order No. STAPPCD2004

Automotive Safety Handbook
By Ulrich W. Seiffert and Lothar Wech
Order No. R-325

Vehicle Accident Analysis and Reconstruction Methods
By Raymond M. Brach and Matthew Brach
Order No. R-311

Recent Developments in Automotive Safety Technology
Edited by Daniel J. Holt
Order No. PT-119

Advances in Side Airbag Systems
Edited by Donald E. Struble
Order No. PT-120

The SAE Story: One Hundred Years of Mobility
By Robert Post
Order No. R-360

For more information, or to order a publication, contact SAE Customer Service at:
400 Commonwealth Drive, Warrendale, PA 15096-0001
Web site: http://store.sae.org
E-mail: CustomerService@sae.org
Phone: 1-877-606-7323 (U.S. or Canada) or 1-724-776-4970

100 Years of Vehicle Safety Developments

PT-116

Edited by
Daniel J. Holt

Published by
Society of Automotive Engineers, Inc.
400 Commonwealth Drive
Warrendale, PA 15096-0001
U.S.A.
Phone: (724) 776-4841
Fax: (724) 776-5760
www.sae.org
April 2005

For permission and licensing requests contact:

SAE Permissions
400 Commonwealth Drive
Warrendale, PA 15096-0001-USA
Email: permissions@sae.org
Fax: 724-772-4891
Tel: 724-772-4028

Global Mobility Database®

'rs, standards, and selected
stracted and indexed in the
ity Database.

2005 10 25

For multiple print copies contact:

SAE Customer Service
Tel: 877-606-7323 (inside USA and Canada)
Tel: 724-776-4970 (outside USA)
Fax: 724-776-1615
Email: CustomerService@sae.org

ISBN 0-7680-1499-9
Library of Congress Catalog Card Number: 2005921720
SAE/PT-116
Copyright © 2005 SAE International

Positions and opinions advanced in this publication are those of the author(s) and not necessarily those of SAE. The author is solely responsible for the content of the book.

SAE Order No. PT-116

Printed in USA

Preface

The transformation of the horseless carriage into the automobile

The term "horseless carriage" came about because the early automotive pioneers spent much of their time adapting the newly developed internal combustion engines, electric motors, and steam engines to replace the power of the horse. They took the carriages of the day, added the appropriate powerplant and a way to move the wheels, and added a mechanism to maneuver the wheels left and right because the horse was no longer required. Maneuverability, performance, and braking were not items high on their list of important functions.

When the early automotive pioneers attached their engines to some type of wheeled vehicle, they probably were not concerned with the possibility that these newly developed means of transportation would run into objects and other vehicles. They may not have realized that during the next 100 years, the numbers of these vehicles would grow to the extent that they have today. These pioneers were faced with making sure that the vehicles could traverse the rutted mud paths that had been created by the horses and horse-drawn wagons. In the early days, they had to share the road with these other means of transportation, and one wonders how many times these newly developed automobiles actually had some accidental encounter with a wagon or horse. Horses have some level of intelligence and were less likely to run into another horse, even with the rider incapacitated. However, the early automobile had no "built-in" intelligence, and the operator had total control of the vehicle! With the poor braking and maneuverability of the vehicles, one would imagine others had to try to stay out of the way of many horseless carriages and their drivers.

As automobiles and trucks replaced the wagon and horse as the principal means of personal transportation, a whole new set of challenges faced the designers of these self-propelled vehicles. In 1796, Cugnot is credited with the first accident when his steam-powered tractor lost control and wrecked due to its lack of maneuverability. The inventors had to face the fact that these vehicles had to be improved in many areas. Safety was beginning to be a term that would help drive the design of motor vehicles during the next 100 years and beyond.

In 1893, inventors such as the Duryea brothers were outfitting carriages with the gasoline engine that basically replaced the horse. By the start of the twentieth century, approximately 10,000 automobiles were on the road. In five years, the number of cars more than doubled, Henry Ford was mass producing automobiles, and the Society of Automotive Engineers was formed. The automobile was quickly on its way to affecting the lives of many people. The world, and especially America and Europe, were on a path that has provided citizens with freedom of mobility that is still expanding to this day. By the start of the twenty-first century, car production is recorded in the millions; unfortunately, so are the number of accidents and the number of vehicles involved in those accidents.

This Progress in Technology (PT) book is a compilation of the changes that automotive engineers have made to the automobile since its inception to achieve a level of occupant protection. For the first 100 years or so, motor vehicle safety has been achieved by adding devices and systems that would protect the occupants during an impact. As we enter the first decade of the twenty-first century, much of the new development is occurring in crash avoidance—how the vehicle can help the driver to avoid being involved in an accident. Under crash avoidance, the developers are adding systems to the vehicle that either help the driver avoid the accident by giving him or her better control of the vehicle, or at least reduce the severity of the accident by enabling better stability and maneuverability, better and faster braking activation, and advance warning of possible dangers.

This PT and the 96 SAE technical papers in its 11 chapters allow the reader to get a feel for the changes and extensive work that the developers went through to raise the level of occupant protection to what it is today. Please note that there are two very comprehensive, excellent PTs on seat belts and airbags: PT-92, *Seat Belts: The Development of an Essential Safety Feature*, edited by David C. Viano, and PT-88, *Air Bag Development and Performance*, edited by Richard Kent.

The various chapters in this book take a look at the progression that the developers made to bring the current automobile to what it is today. Chapter One deals with some of the basic changes made to the vehicle design and basic safety items that were added to the vehicles. In the early days of the automobile, the material used changed from wood to metal. Metal bodies allowed for a more rigid structure and had frames constructed of angle iron. Early papers noted that these new bodes were "practically indestructible." Little did these early developers realize that these indestructible bodies would eventually be replaced with collapsible, energy-absorbing/dissipating structures.

Chapter Two is a quick look at basic enhancements made to the lighting of the early vehicles. Electric bulbs were developed, and headlamps were added to vehicles to aid night driving.

The adoption of the hydraulic brake as covered in Chapter Three gave the drivers of the early automobiles much better stopping ability than the mechanical brakes they replaced. Later during the century, power assist was added. Today, vehicles have four-channel anti-lock disk brakes, with additional power boost systems if an accident is sensed.

One of the biggest occupant safety items added to the vehicle to reduce injuries was the adoption of laminated safety windshields, as described in Chapter Four. Early automobiles did not have any type of windshield, and occupants were assaulted by insects and debris from the road and the weather, so they adopted the use of goggles. In the very early days, the speeds of the early automobiles were slow, but as speed increased, so did the need for some type of windscreen. Eventually in the early 1900s, glass windshields were added to give the occupants some protection, and this feature went hand in hand with the closed body. However, the developers soon realized that these glass windshields would shatter during an impact, and occupants would be injured or killed. The French had been working on laminating two pieces of glass with a cellulose layer between two glass panes. Eventually, the safety glass windshield was created by using a high-strength vinyl known as polyvinyl butyral (PVB) between two pieces of glass. This laminated glass was adopted for use in the windshield, and some current-day vehicles are using it in side glass.

Chapters Five and Six look at the changes that were made to the steering columns and the interior of the vehicles. As more vehicles were produced and more accidents occurred, engineers began to do crash testing to see how they could reduce injuries. Early investigations showed that drivers were being impaled by the rigid steering columns. The collapsible or energy-absorbing steering column was developed, which collapsed upon impact. Chapter Five examines the work done on the steel dashes that were used in the vehicles. Various interior padding systems were examined to help reduce the impact of the human body on the steel structure. Many of the changes to the vehicles were due to information collected during crash/impact tests using test dummies. Even today, dummies are used to develop and improve safety systems.

Chapters Seven and Eight deal with changes in door guard beams and door latches, which were made to improve occupant protection. It seems that the more testing and development that were done on occupant protection, more areas were uncovered that needed some attention. Safety systems were evolving into an ever increasing challenge. Side impact crashes and other types of crashes pointed to the need for better door strength and stronger door latches. Side impacts have become a key issue in vehicle safety, and current-day vehicles have side impact airbags and side curtains and are moving toward better rollover and side impact protection.

The development of the energy-absorbing bumpers discussed in Chapter Nine examines the changes that occurred from the time when bumpers were nonexistent, through the progression of the spring steel bumper, through the bumpers equipped with "shock absorbers." There is still much discussion of bumpers for the modern-day automobile, especially with reference to their height above the ground and how they interface with larger vehicles, particularly large sport utility vehicles.

Chapter Ten highlights the seat belt, which even today is scorned by some, with less than 100% use by the occupants of the modern motor vehicle. Seat belts were used in aircraft and in various other industries early in the twentieth century, but they were not readily accepted by the driving public. There has been a progression of variations on seat belt systems. The early belts were lap straps that were single belts that went across one's lap. Over the years, the shoulder belt was developed, with early versions using two separate belts that allowed the driver to choose to use only the lap belt or both the lap and shoulder belts. Eventually, the single-belt, lap/shoulder belt system was adopted. However, mounting/anchoring points have varied—some on the car pillar, some on the seats, and even one automatic system anchored to the door. Systems were developed to get more consumers to use seat belts. One such system used a seat belt interlock that required the seat belt to be fastened for the vehicle to be started. These forced interlock systems did not survive, but today most vehicles do use some type of buzzer/warning light system to encourage the use of seat belts. Much work is still being done on seat belts to make them more efficient. Various systems have been tried and are under development, which employ various belt tensioning systems and seat belt control before and during an impact situation.

Chapter Eleven rounds out the key occupant protection systems that have been developed during the past century. The restraint system known by all as the airbag is highlighted in this chapter. In the early days of vehicle development, researchers looked at air cushions as a method to reduce injuries during vehicle impacts. The eventual system that became the air bag was a bag inflated through a small chemical reaction that deflated quickly after the impact. Current-day vehicles have driver airbags located in the steering wheel, passenger airbags in the dash, and side airbags in various locations including the seat, doors, and pillars. Air curtains for rollover protection are being added to many vehicles, especially minivans and sport utility vehicles. Changes have been made to airbag systems to account for the size, weight, and age (child or adult) of the occupant. Systems are being developed using various sensors and cameras to detect if a seat is occupied, who or what is in the seat, and how much force the airbag must exert to do the best job. Certain types of vehicles also have driver-controlled, passenger airbag deactivation systems. The best occupant protection is achieved when the occupant is using the seat belt in conjunction with an airbag system.

The road to the current level of occupant protection has not been easy, and many would argue that government standards and regulations have forced many of the safety systems. But when all is said and done, motor vehicle safety is a key area in vehicle design. The ultimate goal is to someday reduce to zero the number of deaths and injuries due to crashes. Much progress is being made, and with the current trend toward crash avoidance, vehicle manufacturers are making progress. The following National Highway Traffic Safety Administration (NHTSA) statistics show that we are making progress but have a long way to go:
- Frontal airbag systems have saved 13,967 lives from 1987 through 2003.
- Eighty percent of Americans wear their seat belts, according to an NHTSA study.
- The 80% safety belt usage will save 15,200 lives in the United States.
- Across the world, every year 1.2 million people die as a result of road crashes, and 23–35 million people are injured.

Table of Contents

Chapter Seven -- Door guard beams

Chapter Eight -- Door latches

Chapter Nine -- Energy absorbing bumpers

Chapter Ten -- Seat belts

Chapter Eleven -- Airbags

CHAPTER ONE

Early changes

AUTOMOBILE WARNING SIGNALS

By Alden L. McMurtry

(Member of the Society)

There is opportunity for the exercise of considerable ingenuity and mechanical skill in developing automobile signaling apparatus, but the basis of any system or appliance of this nature that is destined to meet with lasting success is quite as much a matter of psychology as mechanics. No signaling apparatus can wholly succeed of its purpose unless two phases of the human equation have been properly considered. In order to serve its purpose as a warning, the signal must penetrate the wall of partial insensibility with which every human being unconsciously surrounds himself by directing his thoughts, along some particular line. Were the pedestrian fully aware of the dangers that beset him in the street, he would require no reminder of his peril. But his thoughts are elsewhere, and, for the moment, he is unconscious with respect to his surroundings. As there is a variation in the speed of perception and reaction in different individuals, the signal must be designed for its effect upon the least responsive of those who may still be termed normal.

spoke of the average wire wheel of modern construction standing up for the life of the car. I believe that wire wheels have not been on the market long enough for us to accept that statement as a positive fact. We know that wood wheels have been in service thirteen or fourteen years and are still in good condition, but so far as I know no wire wheels have been out long enough to prove that they will stand up for the full life of the car like a well-made wood wheel does.

The wire wheel has several advantages and desirable points. One is the quick-detachable feature. There are many cases in which it is desirable to have a wheel that is more quickly detachable than the average demountable rim which we have to use with the wood wheel. Another important feature is the advantage of enameling the wire wheels; giving a finish which will last longer than the average paint and varnish. Still another advantage is the saving in weight accomplished by the use of wire wheels.

HERBERT CHASE:—I have heard it said by some engineers that they found it impossible to drive with a set of wire wheels over certain very rough road, about two hundred miles in length, averaging, perhaps, thirty miles an hour, without the majority of the spokes in the wheels coming loose. Their tests were made, I believe, with different makes of very well constructed European wheels. In the tests which the A. C. A. conducted we had no trouble whatever with the wire wheels, which might have been considered due to inherent defects. When traveling over a comparatively large proportion of good roads such as were encountered in the tests, wire wheels will, I believe, give very little difficulty. Where the roads are very bad, as for example, where the sand is deep or the mud ruts are severe, there is a liability to serious injury in the use of the wire wheel.

The question of cleaning also is an important one. While it is doubtless true, as Mr. Houk says, that with proper facilities the wire wheel can be cleaned with reasonable ease, the fact remains that the average garage is not equipped with those facilities, and for that reason they charge more in many cases for the cleaning of a car equipped with wire wheels.

A MEMBER:—We have been unable to find any one willing to guarantee wire wheels to give service when equipped with solid tires, although added resiliency, additional strength and more reliability are claimed for the wheels. Using a wire wheel with a solid tire gives about the same condition as a pneumatic tire pumped up to 80 or 90 pounds.

E. R. HALL:—We have not been able to use solid rubber tires on wire wheels, even of the cushion type (that is, with undercut sidewalls to make the tire more resilient and more yielding and more like a pneumatic tire in its effect).

CHAIRMAN WALL:—Is there any further discussion? If not, Mr. Mudge will close the discussion on his paper, if he cares to.

R. B. MUDGE:—I have very little to say other than that the main

4

WARNING SIGNALS

There are times when the mere creation of an unusual sound will not serve to warn. In the congested traffic of the city, dangers arise suddenly; a fraction of a second may decide the issue of life or death. The signal in this case must be distinctive in character so as to assert itself positively above all of the other incidental noises of traffic or industry.

Because the subject is one involving public safety very largely, and also because it has never been reviewed in detail, it is believed that the following treatment is warranted. The history and general requirements of audible warnings have been studied, leading up to an investigation of the subject that has come as a later development of the motor-driven diaphragm horn, which is the highest type of signaling device thus far employed for automobile use.

HISTORICAL

"Horseless carriages," as the first motor vehicles were termed, were universally equipped with some form of warning signal long before they had demonstrated their ability to move under their own power. The addition of a large trip gong on the order of fire apparatus conveyed the impression that the carriage would run successfully. Such appliances were usually superfluous, for in most cases the early vehicles themselves made sufficient noise for all warning purposes. The earlier types of American automobiles were fitted with a gong actuated by the foot, as the hands were required to operate all control levers, including, in some cases the brake.

During the year 1899 foreign-made cars were equipped with small reed horns attached to the steering wheel or column. This was a logical signal from the foreigners' point of view, because horns were used abroad for signaling purposes long before the advent of the motor vehicle, and as the feet were required to operate this type of car, signaling naturally had to be done by hand.

REED HORNS

In 1903 the reed horn was the universal signaling device for motor vehicles. In reed instruments the sound emitted by the reed is generally strengthened by resonance. Reed horns all have resonators but in numerous cases little or no attention is paid to resonance between reed and resonator, or the horn proper. The horn was generally selected with an eye to the size of resonator, irrespective of its efficiency. With improved conditions the field of travel of the motor vehicle was enlarged, and the reed horn became ineffective as a safety signal in the country. It was hardly heard around blind curves and corners. The rumble of a wagon on the road drowned out the sound of the reed horn, even at short range. Reed horns are affected by low temperature to such an extent that in extremely cold weather some horns cannot be sounded.

SIRENS

The year 1905 saw the introduction of the siren. This device was operated generally by a friction pulley which was pressed against the flywheel of the engine. The power was transmitted by means of a flexible shaft or belt. The sirens had an average of eight apertures and required a speed of at least 4,000 r.p.m. before the note became effective. As the rotor of the siren was a considerable mass to accelerate to such a speed, the signal required an interval of at least five seconds before it was at all effective. Even at this rate of acceleration the strain on the flexible shaft often caused it to break.

The note emitted was of a musical character, and, unless the acceleration was sudden, involving abrupt change of pitch, it was an agreeable note to the ear. The great advantage of this device, namely, the ability to sound a continuous signal, made it popular from the start. For the same reason it constitutes an ideal signal for fire apparatus. With the great cost of installation on a car, time required for the sound to become effective and its almost universal use by fire apparatus the siren was, however, soon discarded by car owners.

Various forms or adaptations of the siren principle appeared from time to time. The most popular was the Sireno, which was driven by a small electric motor and fitted with a magnetically operated brake to overcome the flywheel effect of the rotor at the end of a blast. Hand-operated sirens were made but soon discarded, except for fire apparatus. A modification of the steam-driven siren was arranged to be operated from the exhaust of the motor. This, however, required so much pressure that it was ineffective as a signal.

An example of the restrictions now quite generally surrounding the use of the siren may be found in the Connecticut State Law governing motor vehicles. "No person shall use on any vehicle upon the public highways of this State, except upon fire apparatus or upon an ambulance, any siren horn for a signal." A peculiar circumstance in this connection is that such a restriction, in a measure, defeats its own ends. As in the case of sound itself, even scientists have at times difficulty in reaching an agreement as to whether it is the note, or the instrument producing it, to which the designation siren really applies. The sliding pitch tone of the siren instrument can be produced by a variety of other means, by a piston-whistle, for example.

MUSICAL HORNS

A modification of the wind horn known as a musical horn was made in France about 1908. It consisted of four or five reed horns arranged in a musical scale on the order of the bugle. The various reed chambers were opened or closed by a rotary valve operated by the air pressure. Every time the bulb was pressed a different reed was connected, so that the horn gave a variety of notes.

CONTINUOUS WIND HORNS

A reed horn known as the Autovox is made abroad to overcome one of the drawbacks of the wind horn, by giving continuous, instead of intermittent, blasts. This device consists of a large reed horn with an air reservoir fitted with a safety valve and connected by flexible tubing to a rotary pump operated by the flywheel. The pump was controlled through Bowden wire to a lever placed on the steering-wheel. The advantage of a long blast did not overcome the cost of installation, which in a number of cases was excessive.

COMPRESSED AIR WHISTLES

The sound of a small compressed air whistle mounted on the front of a car is a good close range signal. The Watres device, brought out in 1905, consisted of a compression check-valve which permitted part of the explosion of one or more cylinders to pass into a storage tank. An ordinary low-pressure chime whistle constituted the alarm. About six blasts of the whistle exhausted the tank and until it was fully charged, the cylinder that supplied the pressure developed little or no power to drive the car.

Another form of high-pressure whistle was arranged to be connected directly to one cylinder of the automobile motor. A small poppet valve was used to control the signal. An adjusting screw limited the opening of this valve in order to decrease the loss of power of the cylinder. This whistle was very small and gave a shrill note of very high pitch. Its location under the bonnet of the car, however, reduced the effectiveness of the signal, while the heat of the gas soon drew the temper of the spring and allowed the whistle to blow continuously.

EXHAUST WHISTLES

Exhaust whistles appeared in an experimented way about 1902. Various forms of whistle were made, generally arranged to be connected to the exhaust pipe before entering the muffler. A two-way valve closed the passage to the muffler and opened the passage to the whistle. The effectiveness of the whistle depended upon the position of the throttle of the motor and the location of the whistle on the car. These whistles were quite popular at one time, due mainly to the fact that they permitted a continuous blast and that they were quite musical.

A later form of exhaust whistle was arranged to be connected to the exhaust pipe after leaving the muffler. This whistle was easily applied and gave quite a shrill note. The location of the whistle in the rear of the car not only made it ineffective as a warning signal, but a nuisance as well.

The great majority of exhaust whistles were operated by the foot, thereby limiting their usefulness.

THE ELECTRIC HORN

The electric horn is a diaphragm horn in which the diaphragm is caused to vibrate by the influence of electric magnets. The dia-

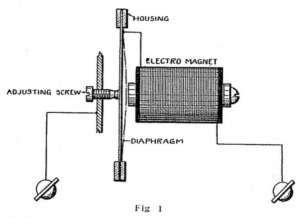

Fig 1

phragm itself being of soft iron is attracted by the electro-magnets and carrying one of the contact points in the electric circuit, serves to make and break the electric current. When the circuit is closed at the push button, a magnetic pull draws the diaphragm toward the magnets. At a certain point on the swing toward the magnets the current is broken at the contact points and the diaphragm is freed. Elasticity carries the diaphragm back to its original position where the electrical circuit is again closed by the diaphragm itself. This type of horn is not new, having been made in France about 1895. At that time it was made to be used in place of electric bells.

A modification of the electric horn employed an armature rigidly fastened to the diaphragm and a contact point carried upon a spring attached to the armature. The addition of this mass of iron lessened the period of vibration while the amplitude was not greatly increased. Various modifications in the principle of the electric horn appeared from time to time, the most noteworthy of which was one in which the diaphragm of a telephone receiver was vibrated by the periodic impulses of current from an induction coil (Fig. 2). The trembler of the induction coil was loaded so as to give a period of vibration or pitch about the same as the reed horn. The secondary winding of the coil was connected to the telephone receiver which was arranged to replace the reed of an ordinary wind horn. While the horn gave a good note, its loudness was not sufficient for practical purposes. Another form of electric horn was made to be operated by the alternating current generated by the Ford type of magneto. However, the tremendous variation in frequency due to the variation in motor speed made this horn inefficient except at certain frequencies or motor speeds.

WARNING SIGNALS

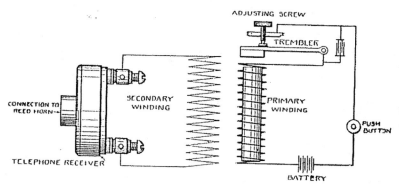

Fig 2

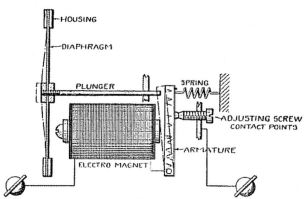

Fig 3

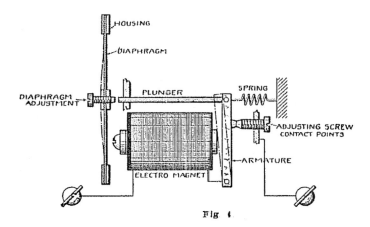

Fig 4

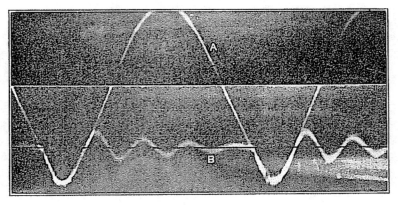

Fig 5

The defects of the electric horn, namely, the constant adjustment that was necessary and its limitations with respect to the volume of the signal, soon condemned it for use on automobiles. The advantage of control by a push button which could be located at any part of the car, soon created a demand for a more effective electrically controlled warning signal. About this time the dry cells used for ignition purposes began to be replaced by large capacity storage batteries, so that the current consumption of such an appliance became negligible. Nevertheless, with this obstacle removed, the electric or magnetic horn was still found inefficient as a warning signal and was soon discarded.

ELECTRIC BUZZER HORN

The principle of the electric buzzer horn is the same as that of an ordinary electric bell except that the clapper hits a diaphragm instead of the bell (Figs. 3 and 4). Striking the diaphragm a hammer-blow causes a greater displacement than in the electric horn, but the pitch or frequency is considerably lower. The striking mechanism, such as plunger, armature, spring and contact points, is not connected mechanically to the diaphragm, and is, therefore, not limited by the diaphragm in regard to amplitude of movement or frequency of vibration. The characteristic of the diaphragm (B, Fig. 5) in regard to period of vibration is of little importance in this type of horn, as the sound is produced by the sudden displacement caused by the hammer or plunger. (A, Fig. 5). The pitch of the sound depends entirely on the frequency of the actuating mechanism. The diaphragm, when freed of the hammer, vibrates at its period, but the sound from these free vibrations is not heard because the sensation to the ear of the blow is greater than that of the symmetrical movement of lesser amplitude, that follows. The dimensions of the diaphragm in regard to diameter of the unclamped portion and thickness are important, as they govern not only the action of the diaphragm under the effects of the hammer-blow, but its useful life as well. A heavy diaphragm will not be displaced to as great an extent as a light one, while on the other hand a light diaphragm will show gradual distortion at the point of contact with the hammer until it becomes useless. As the period of the diaphragm and its normal position while at rest are not essential in the electro-mechanical horn, almost any method of clamping may be employed.

The actuating mechanism of an electric buzzer horn must have weight or mass in order to impart a blow to the diaphragm with sufficient energy to cause a maximum displacement. The greater this mass the greater will be the displacement of the diaphragm, but at a lower frequency of vibration. Therefore, the electric buzzer horn, like its predecessor, the electric horn, is limited to two types, namely, that giving a loud sound at a low pitch, and that giving a soft sound at a high pitch. To increase the current in an effort to increase the loudness of the sound would increase the blow on the diaphragm and slightly increase the pitch due to a quicker response of the

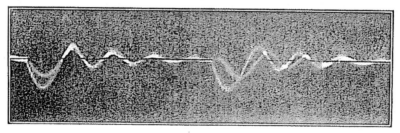

Fig 6

armature to the increased magnetic force (Fig 6). The return of the armature would be at the same speed as before the current was increased. A heavier spring applied to effect a quicker return of the armature would, on the other hand, only tend to offset the increased magnetic force. In one type of electro-mechanical horn a two-contact push button was used to give two different sounds. A slight pressure on the button closed the circuit through a resistance, while a heavier pressure cut out the resistance. The difference in loudness was noticeable, while the increase in pitch was hardly perceptible.

In order to vibrate the diaphragm of an electro-mechanical horn at a higher pitch than is possible under the conditions of the ordinary automobile installation, it would be necessary to use a device on the order of the oscillator developed by Professor Fessenden for use in submarine signaling.* This device requires an alternating current which for automobile signals is manifestly impracticable.

*Proceedings American Institute of Electrical Engineers Vol XXXIII. page 1576, 1914

GENERAL REQUIREMENTS OF SIGNAL

Considering the psychological effect of the automobile signal, namely its audibility, two conditions must be fulfilled: It must satisfy the requirements of an emergency alarm, and also those of the more distant cautionary warning. To satisfy the first requirement, that is, to compel the attention of a person at no great distance, it is essential that the sound possess some quality differentiating it from all other sounds. To this end the warning signal may be loud, abrupt, disagreeable or otherwise violent and unexpected in its nature. While careful driving and proper use of signals should make the alarm signal only an emergency requirement, it is obvious that every properly qualified signal should possess that character in some degree. If one man sees another about to be run over by a truck, he does not stop to sing to him, but yells at him in his ugliest and loudest voice.

The second requirement demands only that the sound have carrying power. A further requirement, applicable to both warning and emergency signals, is that the sound shall be as nearly directional as possible and sufficiently concentrated that its zone of influence shall not be unnecessarily wide.

GENERAL REQUIREMENTS OF APPARATUS

In addition to its sound-producing qualities, which will be dealt with in detail elsewhere in this paper, a warning signal must possess certain other characteristics, such as adaptability to a suitable location on the car, with respect to both the direction of the signal and the convenient location of the operating means. It must operate through a wide range of temperature and be absolutely water-proof. Within reasonable limits, it must be economical. It must also possess durability to an extent seldom required of mechanical apparatus.

At least 50 per cent. of all automobile accessories in use are virtually undergoing a breakdown test, for the reason that they usually receive no attention by the motorist until they fail in some way. Signaling devices are no exception; which accounts for so many effective devices operating in an "ineffective" manner. Instruction books are studied only to the extent of learning how to install the apparatus, while if the system happens to be included as regular equipment on the car the instruction book is rarely opened. It is a familiar fact that the less attention required to keep an appliance in good condition, the more it is apt to be neglected. Here the manufacturer of such an appliance is again brought face to face with the bothersome human equation. The practical conclusion, therefore, is that so soon as it is attempted to render a device in any degree automatic or self-sustaining, its future success is practically measured by the length of time it will continue to perform satisfactorily with no attention whatsoever.

PRACTICAL APPLICATIONS

Having reviewed the requirements of the warning signal and touched upon several points with respect to current practice in the use of such signals, I will now pass to a more detailed consideration of the mechanical signal in order to show how far it goes toward meeting the theoretical requirements of the ideal signal itself, and the practical requirements of every-day service.

THE MECHANICAL HORN

In the mechanical horn the vibration of the diaphragm is accomplished by purely mechanical means. The source of power may be an electric motor, the engine of the automobile or the hand or foot of the operator. In fact, the first mechanical horns used on automobiles were operated by a friction disk pressed against the periphery of the flywheel of the engine.

There are two types of mechanical horn which may be distinguished by the method in which the diaphragm is caused to vibrate. The first is a percussion instrument somewhat on the order of the buzzer horn. In this type a spring is employed to vibrate the diaphragm by means of a blow or "snap." A common form of this type is known as a campaign rattle, and is used more or less as a toy. The diaphragm is pressed from a sheet of tin, in the center of which is attached a small bar of steel. A ratchet wheel is mounted on a shaft made of spring wire (Fig. 7) which permits the ratchet-wheel to spring away from the steel bar on the diaphragm (B, Fig. 8) far enough to free the engaged tooth and allow the next tooth to come into engagement with a snap (C, Fig. 8). The heavier the pressure of the ratchet-wheel against the diaphragm, the more pronounced will be the blow. This type of horn is a noise maker of the simplest order, and was invented to be used (to use the inventor's words) "at such time as noisy demonstration by the people is in vogue."

Another form of horn built on this principle was invented for use on bicycles. It consisted of a diaphragm with a pointer fixed at the center, which pressed against a disk having a roughened or transversely corrugated surface. The disk was revolved by placing it in contact with the rubber tire of the bicycle wheel. This device was never practical. The pointer which pressed against the roughened disk would suffer from excessive wear, as do the steel needles used in the present flat disk gramophones.

In the foregoing described instruments the diaphragm when at rest is pressing against actuating disks and is displaced at its center, which limits the amplitude of vibration. The second type of mechanical horn, which at present is used on about four hundred thousand automobiles, is an instrument the diaphragm of which is caused to vibrate by mechanical impulses applied in harmony with its natural frequency of vibration, acting on the principle of resonance. In this type of mechanical horn, the frequency with which the dia-

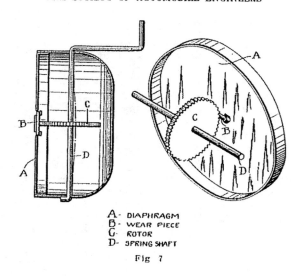

A - DIAPHRAGM
B - WEAR PIECE
C - ROTOR
D - SPRING SHAFT

Fig 7

A - DIAPHRAGM
B - WEAR PIECE
C - ROTOR
D - SPRING SHAFT

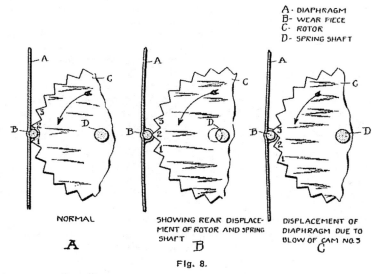

NORMAL

A

SHOWING REAR DISPLACE-
MENT OF ROTOR AND SPRING
SHAFT

B

DISPLACEMENT OF
DIAPHRAGM DUE TO
BLOW OF CAM NO.3

C

Fig. 8.

phragm will freely vibrate, otherwise known as its natural frequency, is of the greatest importance. It depends upon the material of which the diaphragm is made, its size and thickness and the method of clamping the diaphragm to the case of the horn. The action of the diaphragm in the mechanical horn is not in any way similar to that of telephone or phonograph diaphragms, as is popularly supposed. Telephone and phonograph diaphragms are kept under a buckling stress,

14

in the telephone by magnetism, and in the phonograph by spring pressure. Damping is necessary in both cases, because the diaphragms must reproduce correctly sounds varying in pitch up to 4,000 vibrations per second. If these diaphragms were permitted to vibrate freely, the reproduction of sound would be mutilated by the constant effort of the diaphragms to vibrate at their natural frequency. In the mechanical horn every restriction to the free bodily movement of the diaphragm, or damping effect, is eliminated as it is this natural movement which gives the sound, now recognized throughout the civilized world as a warning signal.

In the center of the diaphragm (Fig. 9) is secured a projection or wear piece arranged to receive impulses from a cam wheel, otherwise known as a rotor. The rotor is so placed in relation to the wear piece that the peaks of the cams barely touch it. The movement of the diaphragm when a cam slowly passes the wear piece is not more than two to five thousandths of an inch (0.0508 to 0.1270 mm.)

When power is applied to the rotor, the diaphragm is given a push stresses opposed to the inertia of the diaphragm, the diaphragm

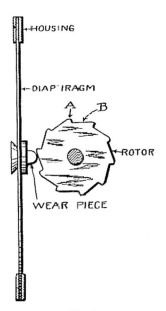

Fig 9

by a cam, so that the diaphragm acquires momentum. The inertia of the diaphragm will carry the diaphragm further than the cam itself goes, so that the outward movement is part forced and part free. When the outward movement ends, by reason of the increasing elastic

15

elasticity plus its inertia, sends the diaphragm back again to the rear side of its normal position at rest. While the rotor is speeding up, impacts are given at various intervals, governed largely by chance, according to the accidental times at which the cams strike the wear piece. During this period these accidental collisions produce severe impacts, giving rise to a very irregular and disagreeable noise on the order of the campaign rattle.

When the speed of the rotor has risen so much that the successive cam impacts are closely timed to the natural period of bodily vibrations of the diaphragm the impact of the cams with the wear piece produces a rapid increase in the amplitude of the diaphragm movement. When the speed of the rotor or the number of cam pushes is properly harmonized to the vibration of the diaphragm, the ear perceives a recognizable musical note. The push of the cam is of a different velocity from the natural sine-wave rate of change with which the diaphragm wishes to move. Otherwise, there would not be a forcing of the diaphragm. The forced part of the movement of the diaphragm must be slightly faster than its natural rate in order to impart any power to the diaphragm movement. This acceleration of one part of the otherwise natural movement shortens the total time, but the forced movements and the natural movement are of the same frequency, even though their velocities and duration are not the same. This represents the normal operating speed of the rotor. If, however, more power is applied to the rotor, it will transfer more power to the diaphragm rather than increase its speed of revolution. This action is known as the governing effect of the diaphragm on the speed of the rotor.

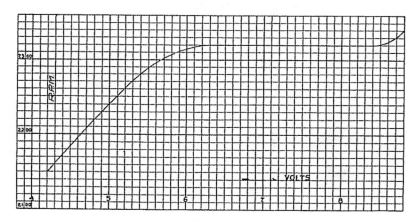

Fig 10

In Fig. 10 is shown a curve indicating that increase of current has no effect on the speed of the motor. The increase in current causes the cam to overtake the diaphragm a little sooner (B, Fig. 9), resulting in wider amplitude of diaphragm swing and a slight increase in the pitch by reason of the increased speed of the outward forcing part of the movement. At the normal operating speed, a cam of the rotor pushes the diaphragm *after* it has *started* on the outward swing, so that if the cam pushes the diaphragm any sooner it will meet it at the end of the inward swing at rest. A greater increase in current will cause the cam to *stop* the diaphragm on the inward swing and suddenly start it on the outward swing.

The governing effect of the diaphragm on the rotor acts on the order of the pendulum governing the speed of the clock movement.

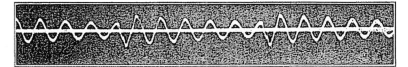

Fig. 11

Fig. 11 shows the movement of a diaphragm caused by a rotor with but one cam. It shows that the diaphragm is forced outward by the cam and then permitted to vibrate freely until again pushed by the cam. This occurs at a time when the diaphragm has just completed the inward swing and is starting on the outward swing. If the cam push was slightly earlier, the force of the push would be so great as to reduce the speed of the rotor so that the next cam push would be properly timed. Fig. 12 shows the normal action

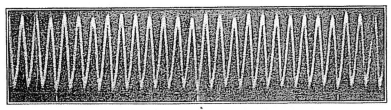

Fig. 12

of a diaphragm with a ten-tooth rotor. This photograph was made with an extremely low film speed in order to compare the action of the different cams of the rotor. It will be noticed that cam No. 1 gives a slightly greater push than the others.

Under the conditions of increased current or overload, the diaphragm does not vibrate symmetrically as the portion surrounding the wear piece is greatly distorted, due to the terrific blows of the cams. (Fig. 13.) This distortion lessens considerably the life of the diaphragm and in extreme cases will destroy it within a short time.

Fig 13.

In some types of motor-driven mechanical horns the diaphragms are purposely overloaded with rotor power in order to get a louder and more distinct sound. Under these conditions the diaphragm will not survive three to five thousand signals.

When the power applied to the rotor is shut off, the vibrations die away rapidly, the irregular collisions and impacts begin again, producing an irregular noise, as when the power was first applied, until the rotor comes to rest. If the power supplied to the rotor is insufficient the collisions occurring during its acceleration will *not be so pronounced*, and there will be a slight interval of time between the final collision stage and the point of resonance between diaphragm and rotor. The sound under these conditions will lack to a certain extent the characteristic note of the mechanical horn as the push of the cam against the wear piece of the diaphragm is reduced. If the adjustment of rotor in relation to diaphragm is greater than 0.005 inch (0.1270 mm.) the effect of the collisions will be greater and the acceleration to note will be slower. Too heavy an adjustment will not permit the rotor to attain full speed, and therefore puts the horn in the class of percussion instruments.

Cracks in the diaphragm will naturally impair its elasticity and result in absolute lack of collisions. The note of a cracked diaphragm resembles the siren in the gradual rise of pitch to the point of resonance when the sound is not unlike that of a cracked bell. Cracks in the diaphragm of a well made horn are the result of fatigue and appear gradually but the destruction of a diaphragm caused by excessive rotor power is sudden and complete.

MECHANICAL CONSTRUCTION

While the principle of operation of the mechanical horn is comparatively simple, its construction is an entirely different matter.

Of greatest importance is the diaphragm and its relation to the rotor, which determines the usefulness of the instrument in terms of the number of effective signals before destruction. Diaphragms covering a wide range of physical qualities will apparently behave alike until subjected to a breakdown test to determine the number of signals. Some diaphragms crystallize and crack in less than one thousand signals while others have withstood two hundred thousand signals before showing defects. There are other factors than the material from which the diaphragms are made that determine their life, the most important of which are method of manufacture, clamping, form, location and method of attaching wear piece. Another very important point regarding the behavior of the diaphragms is their normal position of rest. The over-lap of the cams of the rotor in relation to the wear piece, as previously stated, is approximately 0.002—0.005 inch. This light adjustment permits the rotor to quickly accelerate to normal speed. If there is a variation in the normal position of rest a heavier adjustment is required to assure cam contact with the wear piece of the diaphragm which in itself imposes a high starting torque on the motor. The same is true if irregularities exist in the heights of the cams of the rotor. The size and shape of the wear piece with respect to the characteristics of the diaphragm are of great importance in order to get a full bodily displacement without reverse bending or warping.

There are additional details of refinement in the construction of the mechanical horn which affect to a great extent its efficiency and life, but which for obvious reasons cannot be recorded in this paper. Suffice it to say that one type of mechanical horn alone has received as much detailed study and laboratory research work as any accessory of the modern automobile.

The electric motor which drives the rotor must be designed with a speed-torque curve that will best meet the conditions imposed by the diaphragm and perform its duty for a long period without any attention or care. In some cases such motors have actually given two to three years of service without receiving the slightest attention.

RESONANCE

The term resonance signifies the production of a natural swing or vibration by the action of a periodic force. The ideal case of resonance is where both sounding and responsive elements possess the same natural period of vibration. Remove the reed from a reed horn and blow it by the mouth and the sound is entirely different from that produced when the reed is placed in the horn or resonator. If water be poured into the resonator it will alter the pitch and destroy the resonance. From this it follows that the resonator fulfills a most important function in "re-inforcing the tone" and conversely, that if the resonator is not properly tuned to the sounding element the resulting tone will be produced inefficiently. In a number of cases, the resonators of warning signals are designed merely to improve the appearance of the instrument.

In the mechanical horn the resonator should have a pronounced natural frequency. The air column within the resonator will control and determine the natural frequency of the diaphragm to some extent but obviously the effectiveness of the signal will be enormously greater where the natural frequency of the air column is selected so as to be the same as the normal full speed note of the bare diaphragm.

CONTROL OF THE ELECTRICAL SIGNAL

Various attempts have been made to effect a control of the signaling device so that two different sounds would be obtained. The efforts were generally centered in the electrical push button, the amount of current being regulated by pressure on the button. Other types were made in the form of time limit contacts. A light pressure on the button would close the electrical circuit only for a fraction of a second while a heavy pressure would keep it closed. All these devices were complicated and limited the reliability of the signal to a great extent. The simplest form of push switch is the best because it must be reliable and close the circuit with the least electrical resistance while withstanding in some instances very severe blows. The operation of the button, like that of the accelerator or clutch pedal should be a sub-conscious effort on the part of the driver. For this reason the button should be located in a fixed position and not, for example, on the rim of the steering-wheel, where its location is subject to change. The modern method of mounting the button in the center of the steering wheel, on the other hand, is ideal, since it involves a permanent installation in a location equally accessible to either hand.

HAND-OPERATED DIAPHRAGM HORNS

The hand horn, as it is commonly known, is a mechanical horn operated by pressure of the hand on a plunger or lever, although some hand horns are made to be operated by turning a crank or wheel. The advantage of the hand horn is that it is a unit in itself, the same as the bulb horn. The duration of blast is approximately the same as that of the bulb horn. The characteristic note of the hand horn varies with the make of the horn. The overlap or adjustment between rotor and diaphragm in the majority of cases is very heavy, which classifies the horn as a percussion instrument. One and possibly two makes employ a light adjustment in order to get the full bodily swing of the diaphragm and characteristic note of the motor-driven mechanical horn.

While the hand-operated horn has the advantage of not being dependent upon any source of external energy, its position on the

Fig 14

car is limited. It must be placed within effective reach of the driver, so that it can be operated with the same facility as the electrical push button. It is also necessary that the operator be able to deliver a powerful push on the plunger or other operating device in whatever position it may be placed. At the same time it should be located in a position so that the sound is unobstructed by the windshield or other parts of the car.

In some truck installations a horn of this type has been arranged to be operated by foot pressure. This has the disadvantage of limiting the signaling to moments when the feet are not engaged in operating the controls. Another point against this location is that the horn mechanism is not made strong enough for foot operation and three or four powerful kicks on the plunger will strip the train of multiplying gears in the horn. Hand horns, like all other forms of mechanical signal, suffer from lack of lubrication.

TESTS OF SIGNALS

It is absolutely necessary to subject various signaling devices to breakdown tests in order to establish their reliability. Practically nothing is determined by a test consisting of one continuous blast; while on the other hand, a test of the number of short blasts a signal will give without adjustment produces most valuable data. In the motor-driven type of mechanical horn the motor itself will be subjected to greater effort in repeated starting than in one continuous run. The commutator and brushes are subjected to a series of overloads. The greatest strain on the diaphragm wear piece and rotor occurs in the starting of a blast. Obviously in a continuous test of a horn of this type practically the only data obtained would be in regard to the wearing qualities of the commutator and bearings. It can be readily seen, therefore, that it is the number of short blasts a signaling device will give without adjustment that determines its usefulness in practical operation. Or, to put it another way, unlike certain other classes of mechanical appliances the most effective service implies maximum demand upon the mechanism. Fig. 14 is a photograph of a testing box which is arranged to reduce the sound of a horn under test so that it is barely audible. Mounted above the box is the necessary apparatus for recording the number of blasts, current consumption and other details regarding test.

POSITION OF SIGNAL ON CAR

During the last year for the first time consideration has been given by the automobile engineer to the position of the signaling device on an automobile. It was only competition that caused cars to be sold with full accessory equipment. It then devolved upon the automobile engineer to install this possibly inefficient equipment in the most capable manner possible. The position of the signaling device was determined by the limitation of the device itself. Recently accessory devices have been made with particular makes of cars in view. This refinement has not only insured the permanency of the location but the more efficient operation of the instrument. Of late there has been a tendency toward placing the signaling device under the hood in order to give the exterior of the car as clean an appearance as possible. In this position the sound from the instrument is restricted to a large extent, and in some cases loses its abruptness. Another point regarding this location is that in the case of a closed car the sound is more offensive to the passengers.

As an apparatus, this form of horn answers the requirements of being subject to any location desired. It should preferably be

Fig. 15.

mounted in front of the radiator (Fig. 15) where its sound is most effective and least objectionable to the passengers. The possibility of locating the horn button at any desired point has further advantages.

SIGNAL ABUSE

There has been considerable agitation in the trade papers in England lately regarding loud signals, especially within town limits during the night hours. This is probably caused by the injudicious use of mechanical horns, which, in a number of cases, are used as excuses for reckless driving. In this country some of the States have laws which specify that "every motor vehicle which shall produce a suitable bell, horn, or other signaling device which shall produce an abrupt sound sufficiently loud to serve as an adequate warning of danger, but no person operating any motor vehicle shall make, or cause to be made, any unnecessary noise with such bell, horn or signaling device or use the same except as a warning of danger." It is seldom necessary to operate a loud signal for a longer period than two seconds. There are some types of buzzer horns, the sound of which is not distinctive as a warning. They require, therefore, a series of continued blasts before becoming effective.

CONCLUSION

It will thus appear that in nearly every respect the mechanically operated diaphragm horn is an ideal form of automobile warning signal. Due to a happy combination of circumstances its sound approaches more nearly the ideal than that of any other instrument of the same kind. During the period of action in which the rotor is coming up to speed and getting into phase with the natural period of the diaphragm, there is produced a disturbance that is almost entirely inharmonic, and which, because of its peculiarly rough and abrupt character, is both distinctive and alarming if heard close at hand. This portion of the operation of such a horn is admirably suited to the purposes of the emergency signal. If the circuit be kept closed long enough for the rotor and diaphragm to get into phase, however, there is produced a powerful fundamental wave that is of great carrying power.

It is possible to construct such a horn in a manner comparable with that of any other part of the car, so that with a minimum of attention it should have a life expectation equal to that of the car as a whole. Where a source of electrical energy is available, its cost of operation is negligible. Its only disadvantage is in first cost, which, considering its advantages in other respects is low, when spread over the total term of its usefulness.

In conclusion, I desire to emphasize again the point that the abuse of signaling devices is exceedingly prevalent, and that it results not only in annoyance to the public, but breeds confusion, tends to cause signals to be ignored, and often defeats the purpose of the cautious driver whose whole effort is in the interest of safety. This is due mainly to the indiscriminate use of long blasts, which, as has already been shown, are of less value for signaling than short blasts. Such confusion can be eliminated in only one way, namely, by universal custom or habit. The value of a warning signal is seriously impaired by reason of the fact that it is frequently used for the purpose of a door bell when the car is standing by the curb, as an alarm clock for sleeping garage attendants, and as a plaything for street urchins.

In a certain sense, the warning signal may be said to constitute the only bond of common interest between the motorist and the public. It is the means whereby the motorist in an emergency may impart a warning of danger. In its peculiar function in guarding the public safety, therefore, it is deserving of more attention than it has received heretofore.

DISCUSSION

WILLIAM SPARKS:—The front of a car is a very poor location for a signal, as it will accumulate rain, snow and mud there. Mr. McMurtry states "Of late there has been a tendency toward placing the signaling device under the hood, in order to give the exterior of the car as clean an appearance as possible." That is the proper place for it. The sides of the car should be clean. He states further "In this position the sound from the instrument is restricted to a large extent, and in some cases loses its abruptness." That is true only if the type of bell shown in Fig. 15 of his paper is used. Quoting further, "In a number of cases the resonators of warning signals are designed merely to improve the appearance of the instrument." That applies to the one that you see on the front of the car in Fig. 15. If a straight tapered bell resonator is used instead of an oval bell, 25 to 33 per cent. greater volume, that is, carrying capacity, can be secured.

Therefore, although the sound is somewhat restricted by placing the horn under the hood, as much volume can be secured as with the oval bell on the outside of the car.

ALDEN L. McMURTRY:—The signal shown in Fig. 15, referred to by Mr. Sparks as having an inefficient resonator, is of the same type as the signal used to demonstrate the efficiency of a resonator during the presentation of the paper. The increase of 25 to 33 per cent. of volume is simply a matter of opinion and cannot be substantiated by actual test.

R. H. MANSON:—I have had made a hasty survey of the cars on exhibit at the Palace and find that out of 217 examined, 189 had warning signals mounted under the hood, 22 on the side of the car, and 6 at the front in a position adjacent to the lamp or the front of the radiator. I found also that about 138 cars had the button located in the center of the steering-wheel, 48 at the side or on the door inside of the body, 13 on the steering-column (most of these being located immediately under the steering-wheel), 10 on the instrument board (in most cases very close to the steering-wheel). There were about 8 miscellaneous button locations, some below the seat cushion. I found that there were 211 electric warning signals, 5 reed horns and 1 hand-operated mechanical horn. Sixty-seven different makes of cars were examined.

The car makers having large production exhibited the vibrating type of warning signal exclusively. Inasmuch as many manufacturers are using the vibrating type of horn, I would like to say a few words in regard to it, especially the modern vibrating horn. There was a time when the vibrating horn was nothing more than a door-bell mechanism located in an enclosure of some kind, with a projector placed in the front to give it good appearance. At the present time in certain designs of vibrator horn resonance is taken into consideration, as well as the natural period of vibration of the diaphragm and various other elements which help in producing an instrument giving a signal with a small expenditure of energy and at the same time very uniform in action. I have made a good many experiments to determine the best methods of testing these signals in the factory to get the same results as when mounted on cars. In these tests I find that the voltage of the modern car battery varies from about 5.5 to 7.6 approximately. The old type of electric signal was designed for a six-volt battery, but we have to contend now with a considerable variation of voltage. When a vibrating type of signal is properly designed it is found that if the diaphragm-actuating element and the column of air in the resonator are properly attuned, the only difference in the character of the signal for the range of voltage mentioned will be in the loudness. Other points in connection with the modern type of vibrator horn are that it can be produced so as to be reliable, require only occasional adjustment, and stand more abuse, possibly, than an instrument which has bearings and cams and a few other moving parts.

Three points should be brought up in connection with warning signals, and some sort of standard which can be used by the different manufacturers worked out.

First, the location of the operating button should be standardized so that a driver or an owner purchasing different makes of cars will not have to learn where to reach for the signal button. This is very important.

Second, the location of the horn should be standardized. I think it has been demonstrated that the underhood horn is the correct type to use.

The third and most important point to standardize is the type, shape and drilling of the mounting bracket. A person purchasing a moderate-priced car equipped with a horn not up to the particular requirements, should be able to replace the signal readily. At the present time it is an extremely difficult job to do this, and usually requires a mechanic.

AUTOMOBILE BODIES

By H. Jay Hayes

(Member of the Society)

Automobile bodies cover a multitude of sins. The method of manufacture and materials used have changed considerably, coming gradually to the use of metal. At present, all-metal bodies are being manufactured successfully, the metal being enameled and finished ready to assemble.

Up to 1899, nothing but wood entered into the major portion of body construction. Letters patent were granted in 1900 for an all-metal body. This body construction was used on an electric automobile and proved very satisfactory. Another body was made for a steam runabout and exhibited at the Washington Park Automobile Show, Chicago, in 1899. The body was very rigid, being made of a three-sided integral base frame constructed of angle iron extending from the front around the rear. The base frame supported all the mechanism, this being prior to the use of chassis frames. The upper portion was a skelton frame with sheet steel panels lined with asbestos at the sides and rear to act as insulation, etc. The finish was enamel, baked on at a high temperature. This body was practically indestructible and very desirable in the days when steam was used as motive power. One of these bodies was sold to a machinist in Warren, Ohio, to be used on a car he expected to build but never completed. The body was stored in the basement of his shop for twelve years, after which he wrote the manufacturer of the body with regard to selling it. The body was shipped uncrated and arrived in good shape, the enamel being in especially good condition. I mention this to show that if metal is properly treated before painting rust will not affect it. This body is still in excellent condition, having been enameled fifteen years ago. There is no question that sheet steel rusts more quickly now than formerly, because of the increased use of carbon and manganese to obtain a smooth surface for finishing.

Metal bodies were more expensive to build in the small quantities used about 1900, and did not appeal to buyers for several reasons, all of which have been eliminated. Later, when gasoline engines came into use, composite bodies with wood frame-work, covered with sheet aluminum or steel, were developed, the aluminum type predominating in high-priced cars. In some cases cast aluminum was used instead of sheet aluminum. In fact, one manufacturer of very high-grade cars still uses cast aluminum panels, which, by the way, are prohibitive for medium-priced cars from the standpoint of cost.

Composite bodies have an advantage not possessed by wood bodies. High-grade lumber necessary for body panels has become scarce. This fact, together with the tendency to check or split on account of extremes of heat and cold, made a very serious situation, especially in large production. Many a manufacturer has spent several days in finishing and varnishing wooden bodies in rooms of high temperature, only to see panels split or check upon being exposed to the cold when loading on freight cars for shipment.

Sheet-metal bodies do not require one-third the paint a wooden body does to obtain the same finish, and the cost of painting is much less. Fewer coats are necessary on metal on account of a filler not being required as on wood. Some body manufacturers are enameling the sheet metal before applying it to the body, baking the enamel at a high temperature, making a very durable and satisfactory finish for medium-priced cars.

The fight for supremacy among the manufacturers of automobiles has imposed upon the manufacturers of automobile body sheets a most perplexing problem. Naturally methodical and conservative, the steel manufacturers have been spurred on to a pace that fairly makes them dizzy. No one will deny that the requirements of automobile body sheets are vastly different from and more exacting than the requirements of any other grade of sheet steel. The steel sheets now being made for the automobile industry are the finest sheets in every way that have ever been manufactured, either in the United States or abroad. The steel employed is selected with the greatest care, after being made from materials specially selected for the purpose. It is essential that the chemical composition and the physical structure be such as to permit of developing the high finish necessary, providing at the same time ductility, strength and durability. The carbon content should not exceed 10 per cent. to provide ductility, and the manganese content must not exceed .40 per cent.

to prevent deterioration. During the process of manufacture in each department it is necessary to exercise much skill and employ extra care to prevent the development of irregularities which may later on cause the stock to become unsuitable for the purpose intended. The precautions take the form of additional labor and result in materially decreasing the output. The sheets are subjected to a very accurate heat treatment to establish uniform temper and to relieve any internal strain that may have developed during the course of manufacture.

The perfect automobile body sheet must be "hard" and "soft" at the same time. It must possess high tensile strength and great ductility. Only a mirror-like surface is accepted and oftentimes that surface is subjected to microscopic inspection. Five years ago a full-pickled cold-rolled sheet answered all purposes. Today special heats, analysis, carefully supervised bar treatment, hot-rolling, cold-rolling and annealing must be given to meet each individual requirement of a discriminating customer. Body, fender and radiator sheets no longer belong to the same family. They may be of the same genus and look alike, but the basic ingredients and qualities are radically different. Steel to be used for pressing or stamping must be soft enough for drawing without cracking or straining the metal too severely. Usually the stampings for tonneau backs are made in three pieces, as also for the shroud or cowl, and then joined together by spot-welding or acetylene welding and afterwards soldered and smoothed off to make a good joint. Invariably it is necessary to bump or hammer these parts, which naturally hardens the metal, causing crystallization later. While this has apparently been satisfactory, a great deal of care is necessary in preparing the metal for painting. Painters do not like anything that looks like solder. Occasionally trouble arises after the body has been used a short time, on account of opening at the joint, or the paint coming off, attributable possibly to imperfect cleaning of the metal. Some wonderful press operations are now being conducted, eliminating practically all the above difficulties and crystallization, etc. One-piece stampings of back seats and cowls are now being made very satisfactorily in one operation. This allows the metal to remain ductile with little, if any, tendency to crystallize. The one-piece stampings make a complete body shell containing practically four pieces, as follows:

Tonneau.
Cowl.
Two side panels.
Against eight pieces with the other construction.

Again, less wood frame-work is necessary, as the metal has more rigidity and less weight. The metal is also less expensive to finish.

The gages of steel sheets are usually 18, 20 and 22, according to the strength required. The metal must have clean surfaces before painting. Sand blasting is used to eradicate all irregularities. If metal is to be exposed to the elements for any length of time, it is better to coat it with a primer of red lead, which, being a double oxide of lead, is an excellent anti-corrosive primer.

Metal garnish rails have replaced wooden ones. In most cases the sheet forming the door panel extends over the door opening, eliminating the need of molding and overcoming the former difficulty of paint cracking around the molding.

White ash or maple is usually used for sills, posts, etc. When properly designed, bodies can be made very light and very strong with under-frames strong enough to support them. The lumber is usually dried in the air for three to six months and then exposed to a current of hot air in dry kilns for six to ten days, the temperature depending on the kind and the dimensions of the stock.

Gradually metal is taking the place of wood in body building, as in railroad car construction. Some of the largest automobile manufacturers are using all-metal bodies. It will be only a short time before the majority of medium-priced cars will be equipped with all-metal bodies finished with enamel, baked on at a high temperature, saving time and producing a much more durable finish.

DISCUSSION

J. A. ANGLADA:—I would like to ask Mr. Hayes why wood frames for bodies are apparently preferred to metal framework.

H. J. HAYES:—The standard dies of all bodies are practically made to dimensions, and die work seems to be the most expensive and slower process of producing all-metal bodies. Two or three manufacturers are building all-metal bodies. Ninety days to six months is required to finish a set of dies. Dies for back seats and cowls are made in sixty to ninety days, but those for the seat risers and other things take considerable time.

CHAIRMAN ZIMMERSCHIED:—There are few large assemblies of the automobile, the construction of which is more completely in the hands of specialists than is that of the body. Engineers and designers have their ideas about the lines which the body should have on the outside, and the disposition of the space on the inside, but when it comes to the lay-out of the details of construction I think this is to an unusual extent in the hands of the body-builders themselves.

To me this subject is a very interesting one from another angle, because it is one of the most important elements which determine the peculiar economic situation which the automobile industry occupies. I refer to its status as a semi "fashion industry." In weighing the desirability of one car against another we engineers are prone to lay most stress on the relative efficiencies of these cars as transportation apparatus and their performance as machines primarily; and in this connection it is often pointed out that concentration on and a continuation of one model is a golden road to the crest of this sort of cost-per-ton-mile efficiency. Many buyers purchase automobiles from this same viewpoint, and are satisfied to support a long continued and unchanged design. There is another class, however, which wants style with efficiency or even style at the sacrifice of efficiency. For such people last year's car is not a desirable object, no matter how well it performs; however completely the chassis has been standardized, the outward appearance at least must announce that the car is of the latest model. It is this situation that will always demand a frequent partial change of design and which, on account of the very evident results of his work, will furnish the strongest justification and encouragement for the body-builder's effort.

J. A. ANGLADA:—I think it might be interesting to hear from Mr. Hayes as to the much mooted question of the combined body and frame; that is, practically an integral construction of the body and frame.

H. J. HAYES:—We completed last year 2,500 to 3,000 bodies entirely of metal attached to a sub-frame containing the machinery. The rear seat, the front seat and the shroud were detachable in three separate pieces. That seemed to be very satisfactory if heavy enough metal were used. They were a little harder to fit, because of the ease of shimming up and bolting down a metal body with a wood frame, giving leeway for defects in the chassis frame. The door bumper and trim rails for upholstery were the only parts of wood. The difficulty with the metal body is that the outside body man is unable to get blueprints and working drawings enough in advance to produce all the dies. I believe it is coming.

J. A. ANGLADA:—From your remarks, Mr. Hayes, I judge that you figure the combined metal body and frame is a good thing; that it would save weight, perhaps, and might cost a little more. Are there really any big problems to be overcome before it can be introduced as a general thing?

H. J. HAYES:—The frame I referred to was a very deep one. The frame, running boards, step hangers, body and all, weighed less than the ordinary touring car body. A man could drive up on a curb and lift one wheel to some height, and all the doors would open and close very freely, which we have not seen done with the wood body ordinarily under those conditions. I think the body is not as light as could be constructed with wood framework. But nearly all the manufacturers want large rapid production, bodies produced in one day instead of four or five days. We believe that this will come, which would be very satisfactory for medium-priced car production.

(Vice-president Wall assumed the chair.)

THE IMPORTANCE OF THE IMPACT TEST

By H. A. Elliott

From observations made by the writer it would seem that the lack of appreciation of the impact test in this country is probably due to the fact that automobile engineers have too little data on the subject. In Europe, on the contrary, its value in the selection of steels is becoming more and more evident. I will endeavor to show why it is just as important to consider the resilience (the converse of brittleness) of steels as it is to consider their tensile strength.

The consideration in the selection of steels for parts subjected to sudden shocks and vibrations, such as axles and steering-knuckles, of only the results of the tensile test may be in some measure responsible for the sudden unexpected breaking of parts, at least for the larger sections used here as compared with the lighter European construction. If we knew the resilience of the steels (i. e. their resistance to impact), it is quite likely that different steels would be selected. For instance, let us take R and G. R is an 0.12 per cent. carbon steel having an elastic limit of 46-54,000 pounds per square inch; G is a 0.22 per cent. carbon steel having an elastic limit of 54-60,000 pounds, the reduction in area of both steels being the same. If no reference is made to the results of the impact test, one would in all probability select steel G owing to its greater strength, but if it were known that this steel only resists 153-255 ft.-lbs., whereas steel R resists 306-416 ft.-lbs., surely steel R would be chosen for parts subjected to shocks and vibration.

Such an authority as Professor Guillet states: "Brittleness is not defined by tensile test. An experienced operator can, from the examination of the test piece, express an opinion as to the brittleness, but he cannot assess any definite value. The impact test, which approaches the phenomena of the shocks to which the parts are exposed in practice, gives a definite value for the resilience, a

conditions, and all that I have to say on the subject is that they cannot live. They have been absolute failures in every direction. The only fit place for them is in a belt-driven motor bicycle. Their vibration is terrific. Practically speaking, they do not work out.

SECRETARY H. L. CONNELL:—You are going to four-cylinder water-cooled engines?

W. R. MORRIS:—Yes. If you are going to use the two-cylinder, use the ordinary type, not the V type. I think anything under four cylinders must die a natural death, in England, at any rate.

SECRETARY H. L. CONNELL:—What about the other units standing up on what we call real cyclecars? How have the frames shown up, even in the four-cylinder type?

W. R. MORRIS:—The cyclecar of the past always strikes one as having been built by a novice, shedding bolts and nuts all the way along the road, and I am rather afraid you will find the same thing in these belt-driven cars you have over here. The belt-driven type with a V engine is to my way of thinking an absolute failure on the other side.

SIX VERSUS FOUR ABROAD

C. C. HINKLEY:—Could we get some information on the growth of the six and the growth of the four? The impression is that the six is dying out on the other side. It is a little on the ascendency, if anything, with us.

W. R. MORRIS:—I think the four is coming more and more to the front. The tendency at the present time in England is to get to a smaller car. The 80 x 130, somewhere around 3⅛" by 5⅛", is the motor which is selling in England at the present time in the biggest quantities.

C. C. HINKLEY:—Is that simply due to the greater economy?

W. R. MORRIS:—Over there the tax on everything over fifty horse-power is forty-two guineas a year. That is around five dollars a week. That is what, to my way of thinking, is killing the six more than anything. I think that no one who has driven a six would want to go back to a four. Then, also, we are paying three times as much for gasoline as you are paying here.

C. C. HINKLEY:—Should we understand that the prominence of the small four is due to economy?

W. R. MORRIS:—Certainly.

C. C. HINKLEY:—It is not a matter of choice?

W. R. MORRIS:—No; a matter of economy.

C. C. HINKLEY:—You do not believe the American trade should be governed by the same line of thought as is prevalent in England?

W. R. MORRIS:—Over here the six should be more in demand than the four, considering the cost of gasoline and the tax on the car.

quality whose importance we can no longer misjudge, and which the most elementary prudence must exact in all cases, where the parts in question are subjected to dynamic effects."

To cite a case in point, the crosshead of a 600-horsepower reciprocating steam engine broke suddenly without any apparent reason while the engine was running. Examination of the fracture did not reveal any pre-existing defect in the metal. The tensile test gave normal results, but the impact test on notched bars showed the resilience to be extremely low, which was probably due to the high phosphorus content of the metal.

The impact test has a further value, as has been clearly demonstrated in a paper* read by Mr. Derihon, managing director of the G. Derihon Works, Loncin, Belgium, at the summer meeting of the International Society for Testing Metals. According to Mr. Derihon, the impact test reveals whether the correct heat treatment has been applied. An abbreviated translation of his paper follows:

"In working hard and semi-hard high-grade alloy steels, one quickly realizes that the resistance to shock of these steels depends

RESULTS OF IMPACT TESTS
34,080 Pieces Tested. 2,166 Rejected. Percentage, 6.35.

Ft.-lbs	NUMBER OF PIECES TESTED								
	June	July	Aug	Sept	Oct	Nov	Dec	Jan	Feb
36	5	0	0	0	0	0	0	0	1
72	30	49	5	0	0	0	0	0	0
108	258	56	57	5	0	0	0	0	0
Limit 159	396	486	145	352	178	32	80	30	4
180	22	49	60	25	16	44	11	6	9
216	71	185	110	66	40	186	120	34	1
252	172	347	115	146	131	195	166	41	3
288	451	620	187	310	426	625	630	246	11
324	774	560	102	595	646	912	672	166	70
360	839	570	49	386	719	1179	1211	590	140
360	2490	1084	393	665	1126	2036	2793	2530	3246
Total	5380	3946	1219	2350	3282	5259	5683	3648	3485
Rejected	689	591	207	357	178	39	80	30	5
Percent	13	15	17	15	5	0 6	0 4	0 8	0 14
Max. percent	47	26	32	28	35	38	49	70	93

Fig. 1

solely on the success of the thermic treatment. Although there can be no question as to the good quality, it often happens that a considerable number of manufactured parts is rejected on account of brittleness.

"We soon noticed that these rejections came by groups. Certain

*Determining the correct application of the Thermic Treatment

34

lots were very good, others very bad, yet often they were made from the same metal. We finally succeeded in determining the exact treatment to give the steel not only to avoid rejections but to obtain the maximum resilience so far as possible for all the parts. In reviewing all the tests made, we arrived at a remarkable conclusion.

"When the impact tests are uniform they all give the maximum figures; this shows that the thermic treatment has been properly applied. When the treatment, on the contrary, has been incorrectly given, the whole scale of resilience figures is represented by the results, and the larger this variation the greater is the percentage of rejections. An important question now arose: Since we were certain that the metal was of good quality, should we rely entirely on the results of the impact test on a small test piece and reject parts on the strength of that test? After having made a very careful study of this question, I reply most emphatically, 'Yes, always.'"

Mr. Derihon produced records of results of impact tests on two different kinds of steels made during twelve months. It will suffice for our purpose to show the results of tests made on one steel during nine months, as this steel was difficult to treat uniformly and the influence of the thermic treatments is clearly brought out. (Fig. 1.) The number of foot-pounds has reference in each case to the square centimeter. The Fremont test piece, 8 x 10 x 30 mm. with a 1 x 1 mm. notch, was used. It will be noticed that the tests are assembled by groups of 36 foot-pounds, except on the fourth line, which represents tests giving 108-159 foot-pounds, as all parts are rejected which do not give at least 159 foot-pounds. During the nine months 34,080 pieces were tested, 2,166 being rejected, giving a percentage of 6.35, but it will be noticed that starting in November the number of rejections drops considerably, reaching 0.8 per cent. in January and 0.14 per cent. in February. During this month 93 per cent. of the tests gave more than 360 foot-pounds, whereas during the months of July, August and September, where the rejections reached 15, 17 and 15 per cent., the percentage of pieces attaining the maximum was only 26, 32 and 28 per cent., which shows that as the treatment was improved there was a corresponding decrease in the number of rejections.

From the examination of these figures the conclusion was drawn that the impact test determines a physical condition of the steel, which is more or less perfect, according to whether or not the heat treatment has been correctly applied; also, that when the larger proportion of the test pieces do not attain the maximum results, one should conclude that the treatment has not been absolutely successful.

The brittleness is independent of the quality; a good steel perfectly sound will become brittle if badly treated, but a steel of medium quality can be rendered non-brittle if given an appropriate treatment.

It is interesting to know that the Fremont hammer was used in conducting these tests. It is quick in operation, 700 tests being made daily at the Derihon plant; it does not, therefore, incur great expense.

To the uninitiated the variation in the results in foot-pounds per square centimeter of the impact test has long been a stumbling block, and this test has often on this account not been given the rank it deserves. This is a subject in itself, which formed the basis of a paper read by Mr. Eugene Nusbaumer at the last meeting (December, 1913) of the International Society for Testing Metals (held

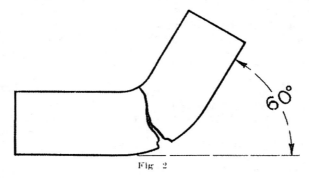

Fig 2

at Paris), in which he recommended that the results of the impact test should be measured by the angle of fracture (Fig. 2) instead of by foot-pounds per square centimeter.

In addition to considering the number of foot-pounds registered, the nature of the fracture is also examined. If the piece has torn apart it is good, provided, of course, it registered sufficient foot-pounds, but if the fracture is clean and crystalline, or if there is a pipe, the piece is rejected even if it registers 216 or 252 foot-pounds.

Another interesting thing was noticed and that is segregation of the steel does not necessarily render the test piece brittle. This is so true in fact that when inspecting certain open-hearth steels of which segregation is suspected, each bar is subjected to an acid test independent of the impact test.

In conclusion it is worth mentioning, however, that during the past year all tests made by the Derihon company were measured by the angle of fracture. The results were so satisfactory that this method has now been definitely adopted.

DISCUSSION

C. E. Cox:—Can one get an idea of the character of the steel after treatment by putting a separate test piece through the same treatment at the same time as the parts under examination. The parts you have shown have had the test pieces forged on them. Is it necessary to test each individual part?

H. A. ELLIOTT:—If you test just a few out of a lot there is no indication that the others are good, because in the heat treatment it is possible that some parts will not have the same heat as others. Mr. Nusbaumer recommends that you test all the pieces subject to shocks and vibrations, such as knuckles, axles, etc.

FACTORS OF SAFETY

By Russell Huff

(President of the Society)

Abstract

The author has selected fourteen automobiles on which to make a study of the factors of safety used in their design. He considers specifically the front axles, front-wheel spindles, propeller-shafts, clutch-shafts, transmission drive-shafts and rear-axle drive-shafts. The method of calculating the stresses is outlined; compositions of the steels used are given; and complete data are presented showing the factors of safety of the various parts, together with the intermediate figures used in obtaining the factors.

Formulas for guiding the engineer in designing the vital parts of automobiles or motor trucks have been published from time to time, but data on the actual conditions of design as they exist in the average automobile of to-day are not available in a convenient, condensed form. The author, having made an intimate study of a number of different models of successful cars, presents in this paper comparative data on designs, materials, stresses and factors of safety, as found to exist in the several cars examined, in the front axle, front-wheel spindle, rear-axle drive-shaft, clutch-shaft, transmission drive-shaft and propeller-shaft. The fact that so much of automobile engineering practice is based on experience, rather than on theory, has prompted the author to make the study of the stresses and factors of safety given in this paper in order to see if there is any approximate uniformity in the designs commonly used by successful engineers and manufacturers.

The data submitted have been collected from a variety of cars ranging from a small 1900-lb. four-cylinder car to a 3-ton truck. Included in the list are cars with front axles of I-beam and tubular section, engines of four, six and twelve cylinders, transmissions located on the rear axle and forward in the frame, and rear axles of the semi-, three-quarter- and full-floating construction. Two 3-ton trucks, one of the chain-driven and the other of the worm-driven type, are covered in the list. The parts relating to the data submitted have been carefully measured and, with few exceptions, analyzed chemically.

Actual physical tests could not conveniently be made on many of the parts examined, but as physical properties of different kinds of

TABLE I—GASOLINE ENGINE DIMENSIONS AND CHARACTERISTICS

Car No.	Car Weight Loaded, Lb.	Car Wheel-base, In.	Engine Bore, In.	Engine Stroke, In.	No Cyl	Piston Disp., Cu. In.	Estimated Maximum Torque, Lb. In.
1	3,315	118	4⅛	4½	4	240.5	1550
2	3,175	112	3¾	5	4	220.9	1450
3	2,523	100	3⅝	5	4	206.4	1350
4	2,828	105	3⅛	5	4	153.4	1010
5	4,360	116	3¼	4½	6	224.0	1510
6	2,932	110	3⅞	4½	4	212.3	1400
7	2,850	110	2¹³⁄₁₆	4¾	6	177.0	1160
8	2,500	105	3¼	5	4	165.9	1095
9	5,200	129	5	5½	4	431.9	2800
10	5,445	139	4½	5½	6	524.0	3360
11	13,870	126	4½	5½	4	349.9	2240
12	5,937	144	4	5½	6	414.7	2700
13	5,635	125	3	5	12	424.0	2750
14	14,070	156	4½	5½	4	349.9	2240

steels, with certain heat treatments, have been so well established by metallurgists, the author has assumed, in submitting the figures herein, that each manufacturer is now actually heat-treating the different steel parts of his automobiles according to the most up-to-date methods. Such parts therefore should show physical properties equal to those given in the S. A. E. Standard specifications

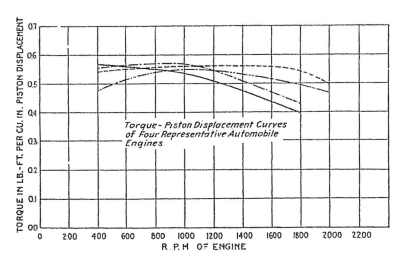

FIG. 1—TORQUE CURVES OF FOUR AUTOMOBILE ENGINES

In the S. A. E. Handbook under specifications for steels, curves are presented showing the variation in the elastic limits to be expected from different steels, depending on the Brinell hardness. Inasmuch as these curves are drawn for the minimum average of results obtained and cover only plain carbon steels, the author found it necessary, in order to preserve uniformity throughout this paper, to select values from similar curves in his possession, covering plain carbon as well as alloy steels and giving average results from extended tests. A hardness numeral of 210 Brinell has been adopted for all except case-hardened parts, because machining operations after heat treatment to this hardness value are satisfactory. The figures on expected elastic limits submitted in the different tables throughout this paper are based on the assumption that the steel has been heat treated to give the same hardness.

In the study of front axles the actual dead load resulting from a full complement of passengers, full gasoline, oil and water systems

TABLE II—COMPOSITION (IN PER CENT) OF FRONT-AXLE STEELS

Car No.	S A E. Steel No	Carbon	Manganese	Phosphorus	Sulphur	Nickel	Chromium	Vanadium
1	1045	0.417	0.82	0.020	0.038	0	0	0
2	1035	0.305	0.82	0.018	0.031	0	0	0
3	1035	0.334	0.78	0.016	0.035	0	0	0
4	1025	0.250	0.46	0.014	0.021	0	0	0
5	1035	0.321	0.60	0.018	0.039	0	0	0
6	6125	0.280	0.70	0.018	0.035	0	0.90	0.18
7	1035	0.349	0.72	0.018	0.035	0	0	0
8	1035	0.364	0.48	0.013	0.045	0	0	0
9	2330	0.250	0.50	0.020	0.036	3.50	0	0
10	3140	0.400	0.60	0.010	0.030	1.20	0.35	0
11	1025	0.250	0.50	0.015	0.035	0	0	0
12	3140	0.400	0.60	0.010	0.030	1.20	0.35	0
13	3140	0.400	0.60	0.010	0.030	1.20	0.35	0
14	3140	0.400	0.60	0.010	0.030	1.20	0.35	0

and spare tire equipment was used in making the calculations. In the study of the power-transmission elements, such as the clutch-shaft and rear-axle drive-shaft, the maximum engine torque, direct or geared-up, irrespective of friction or gear losses, was used in all the calculations. The lowest possible ratio was used in the consideration of all elements affected by a speed reduction through gears. Many of the engines were tested on the dynamometer to determine a fair average torque for each engine. By reducing the data thus obtained to pound-feet torque per cubic inch piston displacement, the figure of 0.55 was obtained and has been used throughout in this paper as being a reasonable average for all engines in the cars examined. (Fig. 1 shows torque curves from four representative engines.) Table I gives the loaded car weights, wheelbases, engine data, and estimated maximum engine torque for all the cars under consideration.

FRONT AXLES

Of the fourteen front axles examined, one had the tubular, one the solid rectangular and all the others the I-beam section. Chemical analyses of the axle materials show that nickel, chrome vanadium, chrome nickel and plain carbon steels are used.

In studying the stresses to which front axles are subjected, it is customary to consider only the bending moments in the vertical plane, due to the dead weight on the axle. There are of course unusual

TABLE III—DIMENSIONS OF FRONT AXLES AT CENTER SECTION

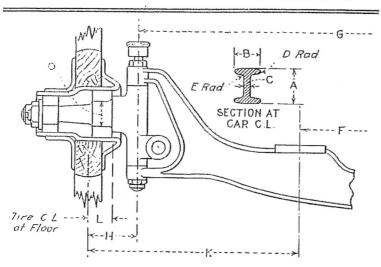

Car No.	A (=2×c)	B	C	D	E	Area in Sq In.
1	$2\frac{3}{16}$	$1\frac{7}{16}$	$\frac{1}{4}$	$\frac{1}{8}$	$\frac{5}{16}$	1.08
2	$2\frac{3}{8}$	$1\frac{3}{4}$	$\frac{3}{16}$	$\frac{3}{32}$	$\frac{3}{8}$	1.44
3	$1\frac{15}{16}$	$1\frac{1}{4}$	$\frac{1}{4}$	$\frac{3}{32}$	$\frac{3}{8}$	0.96
4	2	$1\frac{7}{16}$	$\frac{1}{4}$	$\frac{3}{32}$	$\frac{7}{16}$	1.12
5	$2\frac{1}{8}$	$1\frac{1}{2}$	$\frac{1}{4}$	$\frac{3}{32}$	$\frac{3}{8}$	1.00
6	2	$1\frac{3}{8}$	$\frac{9}{32}$	$\frac{7}{64}$	$\frac{5}{16}$	1.20
7	$1\frac{3}{4}$	$1\frac{1}{4}$	$\frac{1}{4}$	$\frac{3}{32}$	$\frac{3}{8}$	1.00
8	$1\frac{7}{8}$	$1\frac{1}{2}$	$\frac{1}{4}$	$\frac{3}{32}$	$\frac{1}{16}$	0.90
9	Tubular	Axle—I	D 2—O.	D $2\frac{1}{2}$ in		1.75
10	$2\frac{13}{16}$	2	$\frac{1}{2}$	$\frac{3}{16}$	. . . . .	2.53
11	$3\frac{7}{16}$	$2\frac{1}{4}$	. . . . .	. . . . .	. . . . .	7.72
12	$2\frac{1}{4}$	$1\frac{7}{8}$	$\frac{7}{16}$	$\frac{5}{32}$	. . . . .	1.88
13	$2\frac{3}{16}$	$1\frac{7}{8}$	$\frac{7}{16}$	$\frac{5}{32}$	. . . . .	1.86
14	$3\frac{1}{4}$	$2\frac{1}{2}$	$\frac{7}{16}$	$\frac{3}{16}$	. . . . .	2.95

TABLE IV—FRONT AXLE DIMENSIONS AND STRESSES

Car No.	Weight on Frt. Axle, Lb.	F*	G*	H*	K*	Bending Moment on Frt. Axle, Lb. In.	Moment of Inertia of Front Axle	Section Modulus Relative to		Ratio Vert. to Hor. Sect. Mod., Per Cent	Max. Fiber Stress, Lb. per Sq. In.	S.A.E. Steel No.	Expected Elastic Limit, Lb. per Sq. In.	Factor of Safety
								Horizontal Axis	Vertical Axis					
1	1375	31 1/2	51 3/4	2 1/8	12 1/4	8,425	0.777	0.710	0.139	19.6	11.950	1045	70,000	5.9
2	1335	28 1/2	52 1/4	2 1/2	14 3/8	9,600	1.190	1.005	0.229	22.6	9,600	1035	70,000	7.3
3	1065	27 1/16	51 7/8	2 1/2	14 29/32	7,950	0.510	0.526	0.157	29.8	15,100	1035	70,000	4.6
4	1300	28	51 3/8	2 5/16	14	9,100	0.672	0.672	0.167	24.8	13,550	1025	55,000	4.0
5	1550	28 1/2	50 3/4	2 5/8	13 3/4	10,650	0.667	0.627	0.157	125.1	17,000	1035	70.000	4.1
6	1235	26 1/4	50 3/4	3 9/64	15 19/64	9,430	0.897	0.897	0.178	19.8	10,500	6125	90,000	8.6
7	1185	26 3/8	51 1/2	2 3/8	14 15/16	8,850	0.460	0.526	0.128	24.3	17,000	1035	70,000	4.1
8	1255	26 3/4	51 1/4	2 3/16	14 7/16	9,020	0.475	0.506	0.140	27.7	17,900	1035	70,000	3.9
9	2085	29 5/8	48 5/8	3 11/16	13 3/16	13,800	1.130	0.906	0.906	100.0	15,250	2330	90,000	5.9
10	2300	26 5/8	50	3	14 11/16	16,860	1.892	1.345	0.525	39.0	12,500	3140	85,000	6.8
11	3870	27	54 1/2	4 11/16	18 7/16	35,700	7.610	4.310	2.900	67.0	8,050	1025	55,000	6.9
12	2520	26 5/8	50 13/16	2 19/32	14 11/16	18,480	1.430	1.270	0.489	38.5	14,520	3140	85,000	5.9
13	2400	26 5/8	50 13/16	2 19/32	14 11/16	17,600	1.490	1.290	0.494	38.3	13,650	3140	85,000	6.2
14	4020	36	62 1/2	4 1/2	17 3/4	35,700	4.680	2.880	0.878	30.5	12,400	3140	85,000	6.9
									Average	..31.0			Average	..5.8

:3 drawing of axle in Table III for key to reference letters.

loads to be sustained by the axle in both the vertical and horizontal planes when the wheels drop into chuck-holes or strike obstacles in the road at high speed. These special stresses cannot be determined in a practical manner because the conditions are variable. However, accumulated experience has taught the engineer that, if a certain factor of safety based upon the vertical stresses due to the straight dead load is provided, the axle will stand up successfully. Experience has also taught the engineer that certain proportions between the vertical and the horizontal resisting moment of the axle section must be maintained.

Tables II, III and IV cover the essential data relating to front axles. Table II gives the chemical analyses and the equivalent S. A. E. steels; Table III the important dimensions of the center sections of all the

TABLE V—COMPOSITION (IN PER CENT) OF FRONT-AXLE SPINDLE STEELS

Car No.	S. A. E. Steel No.	Car-bon	Man-ganese	Phos-phorus	Sul-phur	Nickel	Chro-mium	Vana-dium
1	1035	0.386	0.67	0.017	0.027	0	0	0
2	5140	0.335	0.71	0.014	0.030	0	0.92	0
3	3140	0.417	0.46	0.031	0.025	1.15	0.5S	0
4	1025	0.250	0.46	0.014	0.021	0	0	0
5	1045	0.435	0.66	0.014	0.033	0	0	0
6	6125	0.280	0.70	0.01S	0.035	0	0.90	0.1S
7	1035	0.300	0.68	0.010	0.035	0	0	0
S	3335	0.370	0.53	0.040	0.036	3.02	0.57	0
9	1035	0.400	0.55	0.015	0.025	0	0	0
10	1035	0.400	0.55	0.015	0.025	0	0	0
11	1035	0.400	0.55	0.015	0.025	0	0	0
12	3140	0.400	0.60	0.010	0.030	1.20	0.55	0
13	3140	0.400	0.60	0.010	0.030	1.20	0.55	0
14	3140	0.400	0.60	0.010	0.030	1.20	0.55	0

axles; and Table IV the data on the vertical bending moments, horizontal moments of inerta, fiber stresses, expected elastic limits of the steels used and factors of safety found in the axles.

The moments of inertia of front axle sections in Table IV were obtained by the graphical method in order to get them exactly, instead of by the approximate method commonly used in estimating moments of inertia of irregular sections.

To obtain the one most interesting value of all, namely, factor of safety, computations were made as follows: Taking the case of Car No. 1, we have in Table IV the weight on the front axle as 1375 lb. The weight then on each spring-pad would be one-half of this, or 687.5 lb. The bending moment resulting from this load would be this weight times the lever arm (the distance between the center of the front wheel and the center of the front spring), in this case 12¼ in., and which, when multiplied by the weight, gives a bending moment of 8425 lb. in. This is the maximum bending moment on the axle and is the same at both spring-pad centers and at any point between these centers.

It is the author's observation that most front axles have a slightly weaker section at the center than at any other point, and therefore, in order to simplify the tables and calculations, measurements at the center only were used.

The maximum fiber stress was obtained by the common flexure formula: $M = sI \div c$, in which M is the bending moment, s the fiber stress, I the moment of inertia and c the distance from the center line of gravity of the section to the extreme fiber. In this formula the

TABLE VI—FRONT-AXLE SPINDLE DIMENSIONS AND STRESSES

Car No.	Load on Front Axle, Lb.	L*	0* (=2×c)	Spindle Bending Moment, Lb. In.	Moment of Inertia of Spindle at Bearing Shoulder	Max. Fiber Stress, Lb. per Sq. In.	S.A.E. Steel No.	Expected Elastic Limit, Lb. per Sq. In.	Factor of Safety
1	1375	1¼	1⅜	860	0.176	3360	1035	70,000	20.9
2	1335	1¼	1⁵⁄₁₆	848	0.146	3810	5140	80,000	21.0
3	1065	1³⁄₁₆	1³⁄₁₆	634	0.098	3840	3140	85,000	22.2
4	1300	1¼	1⅛	812	0.077	5940	1025	55,000	9.3
5	1550	1⅜	1⁷⁄₁₆	1260	0.209	4325	1045	70,000	16.2
6	1235	1¹³⁄₃₂	1⁵⁄₁₆	870	0.146	3930	6125	90,000	22.9
7	1185	⅞	1¼	517	0.120	2690	1035	70,000	26.1
8	1255	1⅜	1⅛	865	0.077	6300	3335	90,000	14.3
9	2085	2¼	1³¹⁄₃₂	2350	0.735	3140	1035	70,000	22.3
10	2300	1⅛	1³¹⁄₃₂	1294	0.735	1735	1035	70,000	40.4
11	3870	2¾	2⅝	5320	2.34	2980	1035	70,000	23.5
12	2520	1	1¹³⁄₁₆	1260	0.532	2150	3140	85,000	39.6
13	2400	1	1¹³⁄₁₆	1200	0.532	2045	3140	85,000	41.6
14	4020	1⁹⁄₁₆	2⁹⁄₁₆	3145	2.13	1895	3140	85,000	44.8
								Average	26.1

*See Table III for key to reference letters.

values for M, I and c are known, and solving s gives us the maximum unit fiber stress in the steel. Illustrating this with Car No. 1; $M =$ 8425 lb. in., $I = 0.777$ and $c = 1\ 3/32$ in. Substituting these values in the formula, we have

$$8425 = \frac{s \times 0.777}{1.09375} \quad \text{or} \quad s = \frac{8425 \times 1.09375}{0.777} = 11{,}950 \text{ lb. per sq. in.}$$

The factor of safety is obtained by dividing the expected elastic limit of the steel by the unit fiber stress. In the axle for Car No. 1

$$FS = \frac{70{,}000}{11{,}950} = 5.9$$

The average of all the factors of safety for front axles, as shown in the right-hand column of Table IV, is 5.8, and, since all of the cars examined in this study are recognized as successful, it is reasonable to believe that any front axle having this factor of safety in the vertical plane will successfully carry the dead-weight load and resist all usual shocks resulting from the wheels striking obstacles in the road.

As previously stated, practical experience extending over a period of years has taught the engineer that certain proportions between the vertical and horizontal resisting moments must be maintained. In order to find an average of this proportion, the modulus of each axle section, relative to both horizontal and vertical axes, has been determined. These values are given in Table IV, as is also the ratio (in per cent) of the vertical to horizontal modulus. The average of this ratio, with Car No. 9 omitted, is 31 per cent, showing the engineer

TABLE VII—COMPOSITION (IN PER CENT) OF CLUTCH-SHAFT STEELS

Car No.	S A. E. Steel No.	Carbon	Manganese	Phosphorus	Sulphur	Nickel	Chromium	Vanadium
1	1020	0 21	0 34	0 009	0 032	0	0	0
2	Malleable Iron	. . .	. . .	. . .	. . .	. . .	. . .	. . .
3	3125	0 25	0 51	0 011	0 023	1 25	0 61	0
4	1020	0 19	0 32	0 013	0 042	0	0	0
5	2330	0 25	0 63	0 015	0 028	3 01	0 20	0
6	6115	0 12	0 33	0 030	0 025	0	0 28	0 14
7	3540	0 44	0 51	0 032	0 036	2 85	1 01	0
8	6115	0 15	0 32	0 031	0 026	0	0 29	0 13
9	Special	0 43	0 45	0 025	0 040	4 5	0	0
10	Special	0 43	0 45	0.025	0 040	4 5	0	0
11	Special	0 43	0 45	0 025	0 040	4 5	0	0
12	Special	0 43	0 45	0 025	0 040	4 5	0	0
13	Special	0 43	0 45	0 025	0 040	4 5	0	0
14	Special	0.43	0.45	0.025	0.040	4.5	0	0

has found it practicable to make front axles about three times stronger in the vertical than in the horizontal plane.

FRONT AXLE SPINDLES

Examination of steering spindles shows a marked similarity in construction. Later models have the spoke center-line as near the inner bearing shoulder as practicable. This is good design, because it reduces the bending moment produced on the spindle by the vertical load and by horizontal shocks. It also reduces the friction in the knuckle thrust and pivot bearings, thus tending to easier steering. A third advantage lies in the shorter turning radius obtained by this construction. The value of ample fillets at the spindle-bearing shoulder is recognized by all designers, as shown by the large fillets found in the cars examined.

In order not to make this study too long, the only analysis submitted is that of the stresses and factors of safety found at the spindle-bearing shoulder. In each case the steel was analyzed in order to determine its equivalent S. A. E. number and the elastic limit expected. These analyses, with the equivalent S. A. E. steel numbers, are given in detail in Table V. The data on spindle sizes, loads, bending moments and stresses, are given in a condensed form in Table VI.

The same formula used for determining the unit fiber stress in the front axles, namely, $M = sI \div c$, was employed in obtaining the spindle fiber stresses given in Table VI. The factors of safety were also obtained in a manner similar to the method used in the front axle calculations.

TABLE VIII—CLUTCH-SHAFT DIMENSIONS AND STRESSES

Car No.	Max. Torque on Clutch Shaft, Lb. In.	Length, In.	Least Diameter, In.	Polar Moment of Inertia, Weakest Section	Max. Fiber Stress, Lb. per Sq. In.	S.A.E. Steel No.	Expected Elastic Limit, Lb. per Sq. In.	Factor of Safety
1	1550	7½	1 3/32	0.125	7,000	1020	40,500	5.8
2	1450	3 5/16	2⅛	5.600	370	Mall. Iron	30,000	81.0
3	1350	7⅜	1½	0.187	5,420	3125	100,000	18.5
4	1010	8 29/32	1¼	0.240	2,630	1020	50,000	19.0
5	1510	11¼	1⅜	0.350	3,000	2330	71,000	23.6
6	1400	9 13/32	1⅛ sq.	0.221	4,250	6115	40,000	9.4
7	1160	7 5/16	1⅞	0.980	1,110	3540	155,000	139.5
8	1095	4⅛	1	0.194	3,700	6115	40,000	10.8
9	2800	16	1.1811	0.193	8,250	Special	100,000	12.1
10	3360	16	1.1811	0.193	10,280	Special	100,000	9.8
11	2240	16	1.1811	0.193	6,850	Special	100,000	14.6
12	2700	15 11/32	1.5750	0.604	3,500	Special	100,000	28.6
13	2750	13	1.3125	0.294	6,150	Special	100,000	16.3
14	2240	16	1.1811	0.193	6,850	Special	100,000	14.6
							Average..	28.8

Illustrating these computations in the case of Car No. 1, by substituting in the formula the values in the table for M, I, and c, we have the following:

$$860 = \frac{s \times 0.176}{0.6875}$$

$$\text{or}\quad s = \frac{860 \times 0.6875}{0.176} = 3360$$

Solving for the factor of safety, we have

$$FS = \frac{EL}{s} = \frac{70{,}000}{3360} = 20.9$$

FACTORS OF SAFETY

The factors of safety found in the spindles average much higher than those found in the axles. According to this it would be expected that less breakage would occur here than in the axle, but it has been the author's observation that this is not the case. The author believes this is due to extreme stresses concentrated at the shoulder, caused by the excessive side strains that occur when the wheel strikes a rut or curb in skidding. These strains are also high when steering through deep ruts that do not quite fit the front-wheel gage.

The most common form of side strain, but one that is not of high value, is the centrifugal-force side-thrust on the front wheels when turning corners in ordinary driving. This thrust can be calculated by knowing the weight on the tires, the speed of the car and the coefficient of friction between the tire and ground. The centrifugal force also introduces an added load upon the outside spindle, which can be computed when the height of the center of gravity of the car is known in addition to the data mentioned. The stresses from this cause are less than the vertical load bending moments, so that they have been ignored in this analysis.

CLUTCH-SHAFT

In studying the stresses and factors of safety in the clutch-shaft, main transmission shaft and propeller-shaft, only the weakest section against torsion was considered. As stated before, the value of 0.55 lb.ft. torque per cubic inch piston displacement was adopted as a standard for estimating the maximum torque from each engine on its respective transmission line. In Table VII the analysis and the equivalent S. A. E. steel specification for each clutch-shaft are submitted. In Table VIII are given data on the maximum torque, length, least diameter, polar moment of inertia, maximum fiber stress, expected elastic limit of the steel and factor of safety of each shaft.

The polar moments of inertia of the weakest section in each shaft have been carefully calculated. These values, together with the known values for torque and shaft diameters, make it an easy matter to compute the maximum fiber stress by the common torsion formula: torque $= sJ \div c$, in which s is fiber stress; J, polar moment of inertia; and c the distance of the most remote fiber from the axis in the cross section under consideration.

Again illustrating this with Car No. 1, we have a maximum engine torque of 1550 lb. in., a polar moment of inertia of 0.125 and a value for c of one-half the diameter of the weakest section, or 0.5468 in. Substituting these values in the formula, we have:

$$s = \frac{1550 \times 0.5468}{0.125} = 7000$$

Using the usual formula for factor of safety, namely, $FS = EL \div s$, and substituting the values in Table VIII for E, L, and s we have

$$FS = \frac{40,500}{7000} = 5.8$$

Some extremely high factors of safety will be noted in this table. The engineers who designed these shafts either never figured them out carefully or else originally designed them for a low-grade steel and later on substituted a high-grade steel without reducing the dimensions accordingly. The average of the entire lot, which is 28.8, is much higher than the factors of safety found in front axles. It is

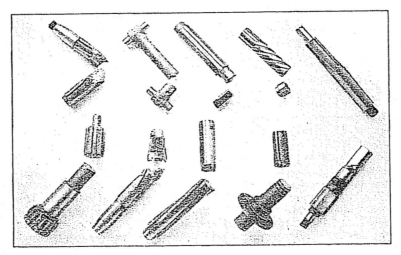

FIG. 2—TRANSMISSION DRIVE-SHAFTS TESTED TO DESTRUCTION

most unusual to hear of a broken clutch-shaft, and it would seem to the author that this one element in the average American car could be reduced in size and weight in many cars without risking any serious danger of breakage.

It is true that shocks on the shafts are high where savage clutches exist, but in the modern smooth-acting disk clutch and the well designed cone clutch, smooth action is the general thing, whereas a few years ago the reverse was common.

The study showed that a greater diversity of engineering practice exists in clutch-shafts than in any other element examined. This is probably caused by the innumerable combinations possible in clutch construction, where so many detail elements, such as bearings, shifting devices, universal joints, flywheels and pedals, are used. Of the fourteen clutches under consideration, seven were of the dry-plate and all the others of the cone type. The almost universal practice of attaching the transmission unit, together with the clutch housing, directly to the engine base, and the importance of making these two units as short and compact as possible, to reduce the unsupported overhang, is no doubt the cause of the prolific clutch-design combinations.

TRANSMISSION DRIVE-SHAFT

In examining the transmission drive-shafts, a much greater uniformity in engineering practice is observed. The four-spline shaft is rapidly superseding the old style square shaft. Many of the shafts examined were case-hardened while others were tempered by heat treatment. The physical properties of most of these hardened shafts

TABLE IX—COMPOSITION (IN PER CENT) OF TRANSMISSION-SHAFT STEELS

Car No.	S. A. E. Steel No.	Carbon	Manganese	Phosphorus	Sulphur	Nickel	Chromium	Vanadium
1	1045	0 492	0 42	0 038	0 035	0	0	0
2	3220	0 239	0 53	0 033	0 035	1 74	0 .081	0
3	3130	0 293	0 .3S	0 009	0 027	1 15	0 .05S	0
4	1020	0 201	0 41	0 01S	0 021	0	0	0
5	1045	0 4S7	0 65	0 015	0 032	0	0	0
6	6115	0 120	0 33	0 020	0 020	0	0 27	0 18
7	1030	0 295	0 42	0 014	0 053	0	0	0
8	1030	0 295	0 42	0 014	0 053	0	0	0
13	Special	0 430	0 45	0.025	0 040	4 5	0	0
9	Special	0 430	0 .45	0 025	0 040	4 5	0	0
10	Special	0 430	0 45	0 025	0 040	4 5	0	0
11	Special	0 430	0 45	0 025	0 040	4 5	0	0
12	Special	0 .430	0 45	0 025	0 040	4 5	0	0
14	Special	0.430	0.45	0.025	0.040	4.5	0	0

were obtained by testing them to destruction in a torsion machine. By this method the effect of the case-hardening and heat treatment on them was obtained. The shafts thus tested are shown in Fig. 2.

The figures on expected elastic limit submitted in Table X are based on the average results secured in these torsion tests. The data on chemical analyses of these shafts submitted in Table IX were secured from samples taken from the soft cores. The values for maximum fiber stress and factor of safety submitted in Table X were obtained by means of the same formulas used in the consideration of clutch-shafts.

In studying the value of factors of safety submitted, the author wishes to call attention to the fact that the first nine cars listed in Table X had transmissions designed with the constant mesh gears at the forward end, and the other five cars had transmissions with the constant mesh gears at the rear. The difference in these two gear layouts may explain why the factors of safety in the first nine are considerably lower than those in the remaining five. In the case of Car No. 9, the factor is unusually high since the shaft is exceptionally large for the size of the engine in the car. An average of the factors of safety of the first nine cars listed in Table X is 5.0, a value the author would feel justified in recommending as a safe figure to employ because all of these nine have a good reputation and so far as the author can find out have had no trouble with their transmission shafts. The factors of safety found in the last five cars average higher than practical experience would seem to indicate as necessary.

TABLE X—TRANSMISSION-SHAFT DIMENSIONS AND STRESSES

Car No.	Length of Shaft, In.	Least Diameter, In	Lowest Gear Reduction	Maximum Torque, Lb. In.	Polar Moment of Inertia, Weakest Section	Max. Fiber Stress, Lb. per Sq. In.	S.A.E. Steel No.	Expected Elastic Limit, Lb. per Sq. In.	Factor of Safety
1	13²⁷⁄₃₂	1¼	5:1	7,750	0 240	20,200	1045	85,000	4 2
2	12⅜	1¼	4 76:1	6,900	0 458	12,120	3220	60,000	5 0
3	8⅜	1¹⁄₃₂	3 99:1	5,375	0 186	17,900	3130	80,000	4 5
4	10¹¹⁄₁₆	1³⁄₁₆	3.24:1	3,275	0 280	8,400	1020	70,000	8 4
5	13¼	1⅛	4.3:1	6,500	0 238	18,000	1045	100,000	5 5
6	10¹⁵⁄₃₂	1⅛	4.66:1	6,520	0 205	21,850	6115	87,000	4 0
7	10¹⁵⁄₁₆	1³⁄₁₆	3.9:1	4,625	0 071	26,500	1030	115,000	4 4
8	9⁹⁄₁₆	1	3 63:1	3,970	0 098	20,300	1030	85,000	4 2
13	12⁷⁄₃₂	1¹¹⁄₃₂	3 42:1	9,400	0 322	19,600	Special	100,000	5 1
9	12²³⁄₃₂	1⁵⁄₁₆	1:1	2,800	0 248	7,400	Special	100,000	13 5
10	11²³⁄₃₂	1⁵⁄₁₆	1:1	3,360	0 248	8,900	Special	100,000	11 2
11	12²³⁄₃₂	1⁵⁄₁₆	1:1	2,240	0 248	5,950	Special	100,000	16 8
12	11²³⁄₃₂	1⁵⁄₁₆	1:1	2,700	0 248	7,150	Special	100,000	14 0
14	11¹³⁄₆₄	1⁵⁄₁₆	1:1	2,240	0 248	5,950	Special	100,000	16 8
								Average	8 4

PROPELLER-SHAFTS

Propeller-shaft construction has had the critical study of the designing engineer during recent years because of the whipping and consequent buckling tendencies introduced by the higher engine speeds and the increase in shaft length made necessary by the longer wheelbases. It is not the object of this paper to go into the theoretical study of the strength of these shafts in relation to whipping tendencies, but rather to present comparative data, taken from practice, and submit figures on the strength of these shafts with relation to torsional resistance only. As a section taken through the central portion of the shaft is usually of the most interest to the engineer, on account of the combined twisting and whipping tendency, an analysis of these middle sections only is submitted. Table XI gives the chemical analyses of the steels and their S. A. E. equivalents. Table XII gives the length, diameter of center section, polar moment of inertia, maximum fiber stress, and factor of safety in each case. The last two values were obtained by calculations similar to those used in studying the clutch and transmission shafts.

Some of these shafts have a solid, while the others have a tubular section. There does not seem to be any standard practice in regard to this point. In some of the designs examined, the propeller-shafts run inside the torque tubes, which might possibly prevent excessive whipping and buckling, while the other shafts run in the open with no protection. In analyzing the figures, it should also be noted that the

TABLE XI—COMPOSITION (IN PER CENT) OF PROPELLER-SHAFT STEELS

Car No.	S A E Steel No	Carbon	Manganese	Phosphorus	Sulphur	Nickel	Chromium	Vanadium
1	1045	0.431	0 590	0 017	0 027	0	0	0
3	3130	0.294	0 540	0 023	0 016	1 20	0 59	0
5	1035	0 405	0 360	0 022	0 031	0	0	0
6	6130	0 270	0 700	0 020	0 035	0	0 90	0 18
7	1035	0 352	0 530	0 008	0 036	0	0	0
8	1045	0 430	0 630	0 015	0 040	0	0	0
13	1035	0 350	0 500	0 015	0 030	0	0	0
14	1035	0 350	0 500	0 015	0 030	0	0	0
2	3130	0.250	0 400	0 020	0 025	1 28	0 56	0
4	1035	0 380	0 510	0 015	0 027	0	0	0
9	1035	0 350	0.500	0 015	0 030	0	0	0
10	1035	0 350	0 500	0 015	0.030	0	0	0
11	1035	0 350	0 500	0 015	0 030	0	0	0
12	1035	0 350	0 500	0 015	0 030	0	0	0

propeller-shafts in the first eight cars listed in Table XII are located to the rear of the transmission and take the maximum geared-up engine torque caused by the lowest transmission-gear reduction, whereas the others withstand the stress of direct engine torque only. The average of the factors of safety in the former case (3.6) is much lower than that (12.3) for the latter.

REAR AXLE DRIVE-SHAFTS

It has been the author's observation that failures in rear-axle drive-shafts are unusually common. It has therefore been interesting in this study to examine the different cars with special attention directed to these shafts to secure their chemical analyses, physical properties and to compute their factors of safety.

In arriving at the figures, it was assumed that the differentials in each rear axle worked sufficiently well to divide the geared-up engine torque equally between the right- and left-hand axles. Torsional resistance only was considered in the computations, although in several cars, where the semi-floating construction was found, the axles are subjected to bending as well as torsional stresses.

In these cases, however, the axle shafts were found to be specially designed with enlarged portions directly under the main outer bearing, gradually tapering toward the differential, to withstand the bending stresses. This form of shaft usually has its weakest section directly adjacent the differential or in the splined portion that engages the differential gears. These inner ends of the axles appeared in all cases to be proportioned to take the torsional load only. Therefore, sections at this point, in the case of semi-floating axles, were measured and used in securing the data. It was found that axles

TABLE XII—PROPELLER-SHAFT DIMENSIONS AND STRESSES

Car No.	Lowest Gear Reduction	Max. Torque Transmitted Lb. In.	Length, In.	Least Dia., In.		Polar Moment Inertia, Center Section	Max. Fiber Stress, Lb. per Sq. In.	S.A.E Steel No.	Expected Elastic Limit, Lb. per Sq. In.	Factor of Safety
				O.D.	I.D					
1	5:1	7750	28²⁹⁄₃₂	1¼	. .	0 240	20,200	1045	55,000	2 7
3	3.99:1	5375	54⁹⁄₁₆	1	. .	0 098	27,500	3130	65,000	2 4
5	4.3:1	6500	. . . . .	1¼	. .	0 240	16,900	1035	55,000	3 3
6	4.66:1	6520	57¹⁵⁄₁₆	1¹⁄₁₆	. .	0 124	27,800	6130	65,000	2 3
7	3.9:1	4625	47⅜	1¼	1	0 141	20,600	1035	55,000	2 7
8	3 63:1	3970	50¾	1⁵⁄₃₂	. .	0 174	13,200	1045	55,000	4 2
13	3.42:1	9400	44¾	1⅜	. .	0 354	18,250	1035	55,000	3 0
14	3 3:1	7400	81¾	1¾	. .	0 920	7,030	1035	55,000	7 8
2	1:1	1450	. . . . .	1¼	. .	0 157	5,200	3130	65,000	12 5
4	1:1	1010	51½	1	. .	0 098	5,155	1035	55,000	10 7
9	1:1	2800	44⅛	1⁹⁄₁₆	1	0 491	4,450	1035	55,000	12 4
10	1:1	3360	49⁷⁄₁₆	1³⁵⁄₆₄	1	0 462	5,630	1035	55,000	9 8
11	1:1	2240	54⅝	1⁹⁄₁₆	1	0 491	3,560	1035	55,000	15 5
12	1:1	2700	52⁹⁄₁₆	1⁹⁄₁₆	1	0 491	4,300	1035	55,000	12 8
									Average. . 7 3	

of the three-quarter- and full-floating constructions were generally made of uniform section between the splined or squared ends engaging the differential gear on the inside and the wheel hub on the outside. In securing the data on these, the smallest diameters were chosen and used in the calculations.

In considering the torsional loads on these axles, it should not be forgotten that the twisting effect of the engine is multiplied many times by the low-speed gears in the transmission and again by the reduction in the driving bevel gears. Take for instance Car No. 1, which has an engine torque of 1550 lb. in. It has a reduction of 5 to 1 in the low ratio gears of the transmission and a reduction of 4 to 1 in the rear-axle bevel gears, or a total of 20 to 1. This total reduction causes a final torque in the axle drive-shafts of 31,000 lb. in., with one-half of this, or 15,500 lb. in. on each section. In the engineer's effort to keep the total rear-axle weights down to a minimum, the smallest possible shafts are used. High-grade alloy steels must therefore be employed to meet the high-duty stresses imposed by the large gear reductions, but even this practice fails to show up a healthy average for factors of safety.

It can be said that great inconsistency prevails in engineering practice when one compares clutch-shaft to rear-axle sizes. Consider, for instance, the clutch-shaft in Car No. 7. Its smallest diameter is 1⅞ in. to transmit 1160 lb. in. With a total gear reduction of 16.85 to 1, the twisting moment on each section of the rear-axle

TABLE XIII—COMPOSITION (IN PER CENT) OF REAR-AXLE SHAFT STEELS

Car No.	S. A. E. Steel No.	Carbon	Manganese	Phosphorus	Sulphur	Nickel	Chromium	Vanadium
1	3135	0.359	0.540	0.018	0.038	1.56	0.81	0
2	3125	0.240	0.360	0.012	0.020	1.56	0.52	0
3	3125	0.259	0.460	0.009	0.015	1.11	0.56	0
4	2330	0.310	0.720	0.020	0.020	3.52	0	0
5	3125	0.228	0.680	0.019	0.023	1.10	0.84	0
6	6125	0.270	0.700	0.020	0.035	0	0.90	0.18
7	1035	0.353	0.530	0.008	0.036	0	0	0
8	3135	0.350	0.360	0.008	0.021	1.10	0.52	0
9	Special	0.430	0.550	0.010	0.030	4.50	0	0
10	Special	0.430	0.550	0.010	0.030	4.50	0	0
11	Special	0.430	0.550	0.010	0.030	4.50	0	0
12	3140	0.400	0.600	0.010	0.030	1.20	0.55	0
13	3140	0.400	0.600	0.010	0.030	1.20	0.55	0
14	Special	0.530	0.600	0.016	0.022	1.47	1.11	0

shafts is 9800 lb. in. If the same steel were used in the rear-axle members that was used in the clutch-shaft in this case, and the same factor of safety desired, the axle would have to be of 3.55 in. diameter. This axle would be prohibitive not only on account of its own weight and expense, but also on account of the increased weight and cost of the other parts entering into the complete rear-axle assembly. These conditions have driven the designer, even with the finest of steels, to the use of a low factor of safety. The average for all the cars is 2.6, but several are in the neighborhood of 1.5.

While considering the stresses in rear-axle shafts, the author wishes to point out the fact that the transmission service brake is now practically obsolete, but in the few cases where it is used, the twisting moments on the axle shafts due to the braking loads are greater than those due to the engine torque. It is customary in careful engineering practice to design, in the case of a transmission brake, to meet the maximum twisting moment caused by the brakes locking the rear wheels and sliding the tires on the road surface.

Car No. 14 has a transmission brake and thus illustrates the possible difference between engine torque and braking torque on the rear-axle shafts. We find from the table that the maximum gear reduction of 32.9 to 1 produces a twisting moment of 36,750 lb. in., with a factor of safety of 4.0 in each axle section. In computing for the values that obtain due to maximum braking conditions, it is necessary to know the greatest load on the rear wheels, the coefficient

TABLE XIV—REAR-AXLE SHAFT STRESSES AND DIMENSIONS

Car No.	Length of Shaft, In.	Least Dia., In. $(2 \times c)$	Lowest Gear Reduction	Max Torque Transmitted, Lb In	Polar Moment Inertia, Weakest Section	Max. Fiber Stress, Lb. per Sq In	S.A.E. Steel No.	Expected Elastic Limit, Lb. per Sq In.	Factor of Safety
1	31¾₁₆	1⅞₁₆	20:1	15,500	0 420	26,500	3135	65,000	2 5
2	....	1⅛	18:1	13,050	0 157	46,600	3125	60,000	1 3
3	31¼	1¹⁄₁₆	14 5:1	9,750	0 124	41,800	3125	60,000	1 4
4	30¾	1¹⁄₁₆	14.5:1	7,325	0.124	31,400	2330	75,000	2 4
5	31	1⅜	16 35:1	12,350	0 352	24,200	3125	60,000	2 5
6	31⁷⁄₃₂	1¹⁄₁₆	16 87:1	11,800	0 124	50,500	6125	80,000	1 6
7	32⅞₃₂	1¹⁄₁₆	16.85:1	9,800	0 124	42,000	1035	55,000	1 3
8	31½	1¹⁄₁₆	14.8:1	8,100	0 124	34,700	3135	65,000	1 9
9	30⅞₃₂	1 574	14.65:1	20,500	0 605	26,700	Special	88,000	3 3
10	30¹⁵⁄₃₂	1 574	11 5:1	19,300	0 605	25,100	Special	88,000	3 5
11	26⅛	1 574	12 5:1	14,000	0 605	18,250	Special	88,000	4 8
12	30⅞₃₂	1 574	11.7:1	15,650	0.605	20,400	3140	70,000	3 4
13	29¾	1.574	14 85:1	20,450	0 605	26,500	3140	70,000	2 6
14	39¹⁵⁄₁₆	1¹⁵⁄₁₆	32 9 :1	36,750	1.380	25,800	Special	104,000	4 0
								Average. 2.6	

of friction between the tire and pavement and the diameter of the rear wheels. In the case of Car No. 14, which was a 3-ton worm-driven truck, the greatest allowable load on each rear wheel is 5025 lb. and the diameter of the wheels is 36 in. A coefficient of friction of 0.6 is commonly used for rubber tires sliding on pavement, but owing to the extended use of block tires, and also the common use of anti-skid chains on the rear wheels, a coefficient of friction of 1.0 is a safer assumption.

With a load of 5025 lb. on the tires of each rear wheel and a wheel radius of 18 in. as a lever arm, a twisting moment of 5025 × 18 × 1.00, or 90,500 lb. in., results in each axle section when the wheels are locked by the transmission brake. Compare this with the geared-up engine torque of 36,750 lb. in. and we see that the twisting moment from the brake is 2.46 times greater and the factor of safety is reduced to 1.6. In order to secure even this margin of safety with the shaft diameter used, an exceptionally fine alloy steel with an elastic limit of 104,000 lb. per sq. in. was employed.

AUTHOR'S CONCLUSION

Table XV was prepared to sum up the data compiled in this paper and to show at a glance all the factors of safety obtained. While the table exhibits several extreme values, the author considers the averages represent values that could be safely adopted as standards for these particular elements in all future engineering practice.

TABLE XV—SUMMARY OF FACTORS OF SAFETY

Car No.	Front Axle	Front-Axle Spindle	Clutch Shaft	Transmission Shaft	Propeller Shaft	Rear-Axle Shaft
1	5 9	20 9	5 8	4 2	2.7	2 5
2	7 3	21 0	81 0	5 0	12.5	1 3
3	4 6	22 2	18 5	4 5	2 4	1 4
4	4 0	9 3	19 0	8 4	10 7	2 4
5	4 1	16 2	23 6	5 5	3 3	2 5
6	8 6	22 9	9 4	4 0	2 3	1 6
7	4 1	26 1	139 5	4 4	2 7	1 3
8	3 9	14 3	10 8	4 2	4 2	1 9
9	5 9	22 3	12 1	13 5	12 4	3 3
10	6 8	40 4	9 8	11 2	9 8	3 5
11	6 9	23 5	14 6	16 8	15 5	4 8
12	5 9	39 6	28 6	14 0	12 8	3 4
13	6 2	41 6	16 3	5 1	3 0	2 6
14	6 9	44 8	14 6	16 8	7 8	4 0
Average	5 8	26 1	28 8	8 4	7 3	2 6

In studying Table XV, attention is directed to the interesting fact that the factors of safety decrease step by step from the clutch-shaft down through the transmission line to the rear-axle drive-shafts. The listed values of all the members affected by the gear reductions are actually lower than would exist in practice because they are estimated on the basis of 100 per cent efficiency for transmission gearing and bearings. A condition of 100 per cent efficiency in any mechanical device is impossible, but the uncertainty of the values of friction and other losses made this assumption imperative. The author believes the theoretical figures for geared-up torque are safe to follow even though they are a trifle high, because the variation in the strength of the parts due to improper heat-treatment could easily exceed the loss due to friction of gearing and bearings.

CHAPTER TWO

Electric headlamps

ELECTRIC BULBS FOR AUTOMOBILES

By Henry Schroeder

(Non-Member)

The tungsten filament used in bulbs has several interesting characteristics. The metal tungsten has a very positive temperature coefficient of resistance, that is, increased temperatures give increased

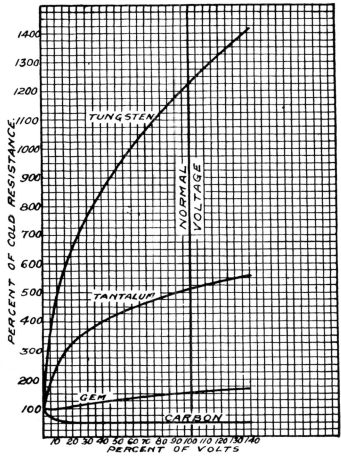

Fig. 1.—Resistance Characteristics of Various Filaments.

electrical resistance, which is common with metals. The resistance of the filament at its normal operating temperature is about twelve and a half times its cold resistance. This is graphically shown in the curve in Fig. 1, which also shows that other filaments do not vary so greatly. Carbon, not being a metal, has a negative temperature coefficient, but the Gem filament has a positive temperature coefficient, and, being a special form of carbon, is therefore often called a metallized carbon filament.

On account of the variation in resistance of filaments at different temperatures, varying electrical pressures (voltages) do not produce corresponding variations in current (amperes) and energy (watts). The greater the increase in resistance with increase in voltage, the less will be the increase in amperes and watts. This gives the tungsten filament an inherent advantage over the other filaments.

The candlepower of a filament is very sensitive to changes in voltage, as slight changes in temperature produce great changes in

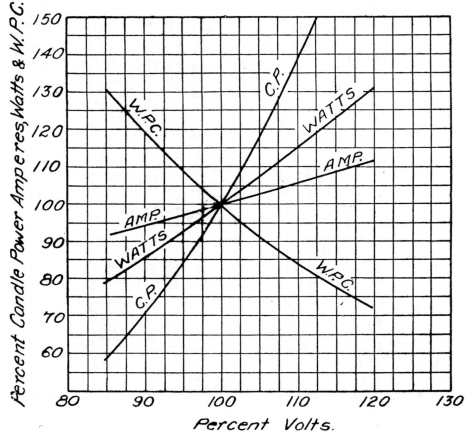

Fig. 2.—Characteristic Curves—Tungsten Filament.

candlepower. On account of the lesser variation in the wattage consumption with voltage changes of the tungsten as compared with other filaments, the tungsten filament varies less than other filaments in candlepower on varying voltage. The efficiency of a filament is measured in watts per candlepower, being the watts it consumes divided by the candlepower given by the filament, which is somewhat misleading, for the lower (numerically) the watts per candle, the higher (better) the efficiency. The characteristic curves of the tungsten filament on varying voltage are shown in Fig. 2.

Between the conditions of a fully discharged battery (1.8 volts per cell) and a fully charged battery (2.6 volts per cell) the variation in volts, amperes, watts, candlepower and watts per candle of a tungsten bulb, assuming 6.5 volts to be the average voltage on the bulb in three-cell systems, is as shown in Table I.

This table indicates that the lighting load in amperes may vary from 21 per cent below to 11 per cent above the normal amperes taken by the bulbs. The candlepower of the bulbs may vary from 77 per cent below to 90 per cent above the normal, and the wattage from 46 per cent below to 33 per cent above normal. This wattage variation is not a large amount compared to the horsepower of the engine, for a lighting load of two 2½-ampere headlight gas-filled bulbs (approximately 21 candlepower each), two .84 ampere sidelights (approximately 4 candlepower each), one rear and one speedometer light

TABLE I.—BULB CHARACTERISTICS WITH BATTERY CHARGED AND DISCHARGED

Battery Condition	Per Cent Normal Volts	Per Cent Normal Amperes	Per Cent Normal Watts	Per Cent Normal C.P.	Per Cent Normal W.P.C.
Fully discharged (1.8 volts per cell)	68	79	54	23	233
Fully charged (2.6 volts per cell)	120	111	133	190	70

(.42 ampere each, approximately 2 candlepower) is less than 50 watts or about one sixteenth of a horsepower, as 746 watts equal one horsepower. The variation in candlepower is, however, most undesirable.

The life of a filament is very sensitive to a variation in voltage, and it is impossible to obtain an exact relation between life and efficiency. In view of the sensitiveness of bulbs to variation in candlepower and life on varying voltage, it is very important that the voltage on the bulbs be kept constant. Increasing the voltage, therefore, shortens the life of a bulb materially and decreasing the voltage materially reduces the candlepower. Tests are now being made of different automobile lighting systems under various conditions of service to ascertain the extent of the voltage variations in practice.

The candlepower of filaments can be measured in two general ways, the simpler being measuring it in a horizontal direction. This does not, however, give the average candlepower in all directions, so that candlepower is often given as the "mean spherical," the former being called the "mean horizontal" candlepower. The light-giving qualities of a headlight are dependent not alone upon the rated candlepower of the bulb, but largely on the position of the filament and the condition of the reflector. As stated, in service the candle-power of a bulb is very variable, depending on the voltage, so that instead of rating a bulb by its candlepower, which is purely nominal, it can be rated more accurately by the amperes it takes. For example, a headlight bulb can be rated better as a 2.5-ampere (not 21 candle-power) 6-8 volt bulb. It is the endeavor of bulb manufacturers, therefore, to have customers order headlight bulbs by amperes instead of candlepower.

There have been put on the market recently headlight bulbs filled with an inert gas, the object being to put a slight pressure on the fila-ment so that it can be raised to a higher temperature without undue

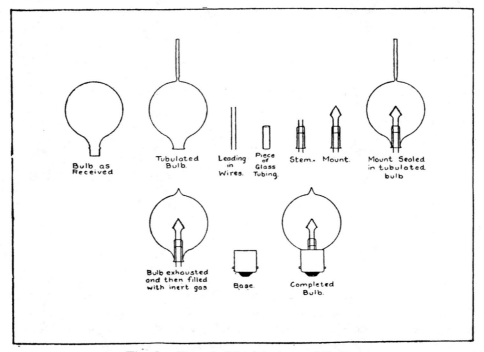

Fig. 3.—Steps in Manufacture of Bulbs.

filament evaporation, one of the factors determining the life of a bulb. This is similar to the steam pressure in a boiler raising the boiling point of water. This gas pressure raises the temperature of rapid evaporation of the filament. The limiting temperature at which a filament can be operated is not its melting temperature, but the high-est temperature that can be obtained without rapid evaporation, which so reduces the cross-section of the filament at its thinnest point that it finally burns out. Tungsten has a lower melting point than carbon, but carbon evaporates readily at temperatures much below that at which tungsten evaporates. The pressure of the gas inside the bulb when lighted is about the same as the atmospheric pressure on the outside of the bulb.

These bulbs are impractical if made with filaments of small cross-section, as the cooling effect of the circulating gas on the filament requires a great amount of electrical energy to maintain the temperature. In other words, unless the gain in candlepower (due to the higher temperature possible) is greater than the extra energy required on account of the cooling effect of the gas, there is no advantage in putting gas in a bulb, as otherwise the efficiency would be no better than with a vacuum. In the present state of the art filaments taking one ampere are the smallest practicable, and side and rear lights are therefore made only in the vacuum type. The filaments are accordingly coiled in order to present as little of the surface as possible to the cooling effect of the circulating gas.

The various steps in the manufacture of bulbs are shown in Fig. 3. The glass bulb as received has a neck which is cut off at the flare. This piece of the neck which is cut off contains the internal strains in the glass. A hole is then melted in the glass bulb and a piece of glass tubing welded to it for the subsequent purpose of exhausting the air. For the stem a larger piece of tubing is used and one end flared out, which is afterward welded to the flare in the neck of the bulb. Two pieces of special wire are then inserted in the stem tube and the end of the tube opposite the flare welded together with the wires imbedded. These wires are called the "leading-in wires." They connect the filament with the base, and are of such a nature that their coefficient of expansion is the same as that of glass. This is necessary, as otherwise when the bulb is lighted and the glass expands, unless the leading-in wires expand at the same rate, air will leak into the bulb. Platinum was used formerly, but is used rarely at present; so there is practically no scrap value in a burned-out bulb, except for the slight amount of brass in the base.

The tungsten filament is made of pure tungsten powder formed into a slug by hydraulic pressure, holding the tungsten particles together by cohesion. This fragile slug is then put under a hood containing an inert gas so the tungsten will not oxidize and a very heavy electric current is passed through it, heating it to a high temperature and electrically welding the tungsten powder together. The slug is then put through a rotary hammer, and formed out hot into a wire about the size of the lead in a pencil. This wire is then drawn out hot by means of diamond dies until it is of the proper cross-section. Diamond dies have to be used as tungsten is such a hard substance that it will soon wear out a die of any other material. The wire is then coiled into its proper form and cut to the right length for the filament desired.

The ends of the filament are then either electrically welded or pinched to the ends of the leading-in wires, making the completed mount. The mount is then put in the glass bulb and the flare on the stem welded to the flare on the neck of the bulb, making an air-tight joint. The bulb is then exhausted and in the case of a gas-filled bulb

63

the gas put in and the glass tube melted off, leaving a tip on the bulb. The base is then fastened to the neck of the bulb by a waterproof cement, and the ends of the leading-in wires cut off and soldered to the base contacts, the bulb then being ready for shipment.

The best filament is one that is comparatively thick because it can be operated at a better efficiency than a thin one. The thicker a filament is, the more current it takes; a six-volt bulb has a much thicker filament than a twelve-volt bulb of the same wattage. On the other hand, the loss in energy due to the heat conducted away by the leading-in wires becomes a smaller percentage of the total the higher the voltage is. From the standpoint of efficiency, therefore, there is practically no choice in bulbs for three or six-cell systems. Bulbs for nine-cell systems are not so good in quality, however, as for either three- or six-cell, and are somewhat more expensive to manufacture.

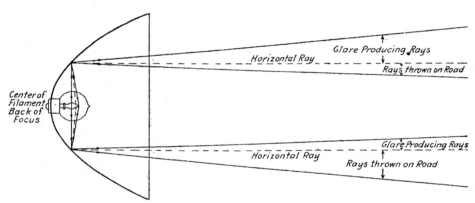

Fig. 4—Diagrams of Light Rays from Headlight.

In any system the wiring must be large enough to carry the current without undue voltage drop, and as the voltage drop is proportional to the square of the current, very large wires must be used for low voltages. For three-cell systems, therefore, the single-wire ground return seems preferable from the bulb standpoint.

For headlight purposes the filament must be concentrated to a certain extent so as to bring it in the focus of the reflector. It is obviously impossible to make the filament a single-point source of light. The result is that every point on the surface of the reflector sends out a cone of light. This is illustrated in Fig. 4.

In order to prevent glare, which is the intense amount of light thrown upward into the eyes of persons, the rays of each cone of light which are thrown upward must be bent down. On the other hand, too much filament concentration will produce too intense and narrow a beam with a parabolic reflector, a certain amount of spread being required to light up the sides of the road.

The focal length for the usual headlight bulb is 13/16 in. The focal length of the reflector in which these bulbs are used should be greater than 13/16 in. in order to allow for adjustment of the bulb. The center of the filament should preferably be back, instead of forward, of the focus of the reflector to get a spread of the beam of light, for the reason that when the center is back of the focus a greater amount of the light rays is utilized and redirected by the reflector, producing a higher beam candlepower. In this case the majority of the reflected rays diverge directly from the headlight instead of crossing each other. If the center of the filament is back of the focus, the upper part of the reflector is producing most of the glare. If, in this case, the upper part of the front glass of a headlight is covered with a translucent material, the glare will be largely reduced, but

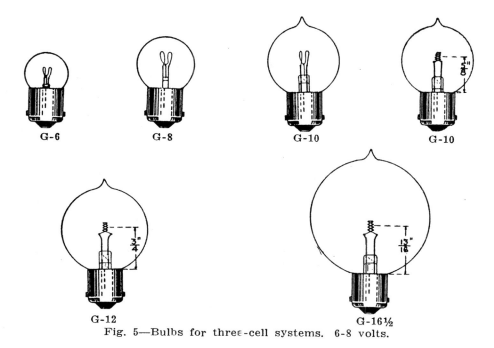

G-6   G-8   G-10   G-10

G-12   G-16½

Fig. 5—Bulbs for three-cell systems. 6-8 volts.

should the center of the filament be forward of the focus, this translucent material will not do the work it was intended for, as most of the glare will come from the lower part of the reflector. Similarly, devices put on a bulb to reflect the light, should be on the upper part of the bulb when the filament center is back of the focus. With gas-filled bulbs, however, this is the hottest part of the bulb.

Headlight bulbs for three-cell systems are now made for 6½ volts, which is believed to be the general average voltage at the socket on all cars having such systems and will operate satisfactorily when the circuit varies within 6-8 volts, the bulbs being so marked. Side, rear and speedometer bulbs are made for 6¾ volts on account of the lesser drop in voltage on their circuits. All bulbs for six-cell systems are made for 15 volts and for nine-cell systems for 21 volts, being designed for satisfactory operation with a voltage variation of 12-16 volts and 18-24 volts respectively. Heretofore bulbs have been made exactly to the voltage a customer's fancy might dictate, but for renewal purposes it is impracticable to have so many voltage bulbs in stock in garages and automobile supply stores. It is also desirable, for self-evident reasons, that there be one standard base; the single-contact bayonet candelabra, style 1100, for use on single-wire ground-return systems seems to be the most popular at present.

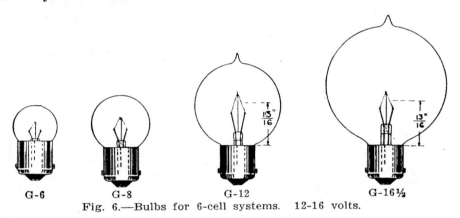

G-6   G-8   G-12   G-16½

Fig. 6.—Bulbs for 6-cell systems. 12-16 volts.

Taking into consideration only the standard sizes now listed for use on three-, six- and nine-cell systems and the different bases regularly used, there are about twenty-four different bulbs that should be kept in stock by a garage. In addition about forty other sizes are generally used, and if individual voltages had to be supplied, taking into consideration the different standard bases, a stock of over two hundred different bulb sizes would be required. Bulb manufacturers

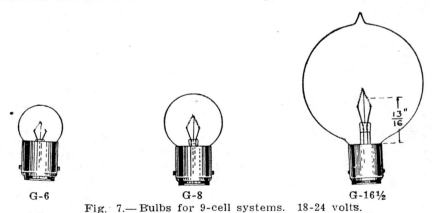

G-6   G-8   G-16½

Fig. 7.—Bulbs for 9-cell systems. 18-24 volts.

would certainly welcome one voltage system, one base and as few sizes of bulbs as possible. Garages and supply dealers cannot possibly carry a large variety of stock and as their knowledge of bulbs is very limited, they now have difficulty in supplying proper bulbs.

The diameter of the glass bulb is expressed in eighths of an inch and its shape by a prefix letter, *G* for round (globular), *T* for tubular, *S* for straight-side, etc. Thus *G*-6 means a round bulb 6/8 or ¾ in. in diameter. The present standard vacuum bulbs as listed by the manufacturers are illustrated in Figs. 5, 6 and 7, and the gas-filled headlight bulbs in Fig. 8. The various bases used are illustrated in Fig. 9.

The total number of bulbs used in the United States for automobile lighting during 1915 is estimated at about ten millions, as given in Table II.

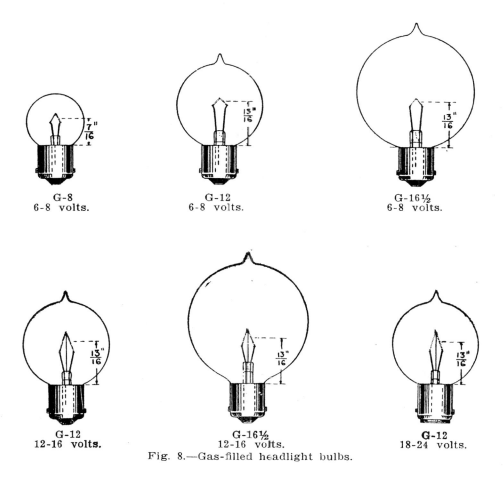

Fig. 8.—Gas-filled headlight bulbs.

Information received from eighty-five car manufacturers regarding their 1916 Model cars indicates the following:

Bulbs for three-cell systems are increasing in favor, for six-cell systems decreasing in favor and for nine-cell sytems remaining about constant. The single-contact base is gaining and the double-contact losing in favor. About three-quarters of the bulbs used for three-cell systems are fitted with single-contact base, and this amount is increasing. Of the bulbs used for six-cell systems, up to the present time little less than one-half of them are fitted with the single-contact base. Next year nearly three-quarters of them will be so fitted. Practically all of the bulbs used for nine-cell systems have been and will be fitted with the double-contact base. This information is indicated in Table III.

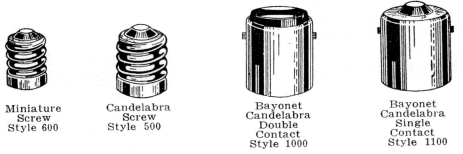

Miniature	Candelabra	Bayonet	Bayonet
Screw	Screw	Candelabra	Candelabra
Style 600	Style 500	Double	Single
		Contact	Contact
		Style 1000	Style 1100

Fig. 9.—Bases for incandescent bulbs.

The special voltage headlight bulbs used two in series on magneto, which requires the use of a double-contact base, unless the fixtures are insulated from each other (in which case the single contact base can be used)—have not been included in either of the tables given.

TABLE II.—BULBS USED IN THE UNITED STATES DURING 1915

	Fitted with Single-Contact Base	Fitted with Double-Contact Base	Total
Three-cell systems............	6,000,000	2,000,000	8,000,000
Six-cell systems..............	700,000	800,000	1,500,000
Nine-cell systems.............	....	500,000	500,000
Grand Total..............	6,700,000	3,300,000	10,000,000

For electric vehicles three sizes of bulbs, Fig. 10, are made: 8, 15 and 25 watts, for 32, 42, 62 and 82 volts, designed to operate on voltages of 30-34, 40-44, 60-64 and 80-84 volts respectively.

The 8-watt bulb is used as a rear light and the 15- and 25-watt for sidelights and headlights. On account of the high voltage the filament cannot be easily concentrated to produce an intense beam in a small parabolic reflector, but electric vehicles do not need such a powerful beam headlight on account of their slower speed.

TABLE III—BULBS USED IN 1915 AND 1916*

Battery System	PERCENTAGE, 1915 GRAND TOTAL			*PERCENTAGE, 1916 GRAND TOTAL		
	Single-Contact Base	Double-Contact Base	Single- and Double-Contact	Single-Contact Base	Double-Contact Base	Single- and Double-Contact
Three-cell..	60	20	80	70	15	85
Six-cell....	7	8	15	7	3	10
Nine-cell...	0	5	5	0	5	5
Total....	67	33	100	77	23	100

*Model cars only.

DISCUSSION

H. W. SLAUSON:—What is the relative depreciation of the carbon, the gas-filled and the vacuum tungsten filaments, due to engine vibration and road shocks?

HENRY SCHROEDER:—The first tungsten filaments made were very fragile. Bulbs were tried out in railway Pullman lighting and stood

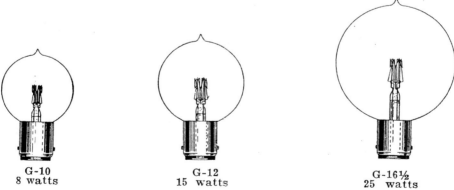

G-10
8 watts

G-12
15 watts

G-16½
25 watts

Fig. 10. Electric Vehicle Bulbs.

up satisfactorily. With the lower voltages and thicker filaments allowable for automobile use we found they stood up very well. The filament is ductile now when it is new. After it has burned awhile it becomes brittle when cold, but is ductile when hot. Even when cold and brittle, the filament seems to withstand all road shocks. A carbon filament, having less than one-third the efficiency of a tungsten filament, requires a generator and battery of more than three times the capacity. Also the carbon filament cannot be concentrated readily into the desired form. The tungsten filament is therefore decidedly better than the carbon.

A. L. McMURTRY:—Mean spherical candlepower is the accepted standard for the rating of incandescent lamps. It is rather an indefinite rating for headlight bulbs unless we know the percentage of light incident upon a reflector or within a certain zone.

In order to show the importance of the form of filament I submit in Table IV the results of tests made of two automobile headlight reflectors and ten different types, Fig. 11, of bulbs. These were purchased at supply stores in New York City.

The bulbs were subjected to a rating test for mean spherical candlepower at marked volts. Various light distributions were made

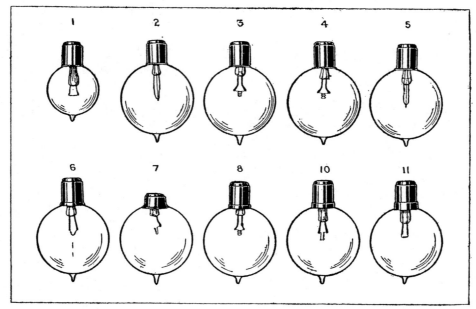

Fig. 11—Group of Ten Incandescent Bulbs Used in Tests. (Drawn from Photograph.)

to determine the light flux incident upon the reflectors. Readings were also taken of the light distribution around the lamps at 90 and 120 deg. from the tips, thus showing the distribution upon the reflectors at these angles.

Fig. 12 shows the comparatively uniform distribution of light of bulb No. 3. Fig. 13 shows the distribution of light of bulb No. 6, where the "leading-in" wires cast a partial shadow at 120 deg. from the tip of the bulb.

The form of bulb filament is of great importance where the light projected by an automobile headlight is restricted by law. Figs. 14 and 15 are photographs showing the effect of different forms of

TABLE IV—RESULTS OF TESTS OF ELEVEN BULBS AND TWO REFLECTORS

BULB NUMBER	1	2	3	4	5	6	7	8	10	11
Rating										
Volts	6-7	....	5	6-8	12-16	6-8	6-7	6	6-7	6-7
Candlepower	....	50	12	18	15	....	32	15	32	50
Amperes	2.5	3	....	....	....	4	4	....	4	....
Measured Values										
Volts	6.0	12.0	5.0	6.0	16.0	5.0	7.0	6.0	6.0	7.0
Amperes	2.45	3.00	2.47	3.01	1.38	3.35	4.20	2.54	3.79	4.44
Mean hor. cp	13.4	41.1	11.2	18.8	23.6	20.0	39.6	15.5	24.6	42.6
Mean spher. cp	13.1	37.3	10.7	19.4	23.6	17.7		14.6	24.5	46.5
Lumens	164	468	135	244	296	222		183	308	584
Watts per mean spher. cp	1.12	0.965	1.15	0.93	0.93	0.945	0.74	1.04	0.93	0.67
Lumens										
On Vesta refl.	121	371	100	182	232	170	*	126	244	444
Outside Vesta refl.	43	97	35	62	64	52	*	57	64	140
On Owen refl.	111	344	92	167	216	157	*	114	227	416
Outside Owen refl.	53	124	43	77	80	65	*	69	81	168
On Owen complete unit	115	356	95	174	223	163	*	119	234	428
Per cent Lumens										
On Vesta refl.	74	79	74	74.5	78.5	76.5	*	69	79	76
Outside Vesta refl.	26	21	26	25.5	21.5	23.5	*	31	31	24
On Owen refl.	67.5	73.5	68	68.5	73	70.5	*	62	73.5	71
Outside Owen refl.	32.5	26.5	32	31.5	27	29.5	*	38	26.5	29
Dimensions—inches										
Top base to center filament	56/64	56/64	56/64	57/64	58/64	57/64	1	55/64	49/64	59/64
Filament, length	.../...	20/64	12/64	18/64	22/64	12/64	18/64	9/64	12/64	.../...
Max. width	18/64	10/64	16/64	18/64	6/64	14/64	14/64	18/64	14/64	9/64

*The base of No. 7 bulb was loose, and the bulb burned out before these readings were taken.

filaments in the same reflector. It was almost impossible properly to focus bulb No. 11 (Fig. 11) without shadows or black spots appearing on the road.

The focal length of the usual headlight bulb is $1\frac{3}{16}$ in. This is the distance from the top of the base to the center of the filament. But

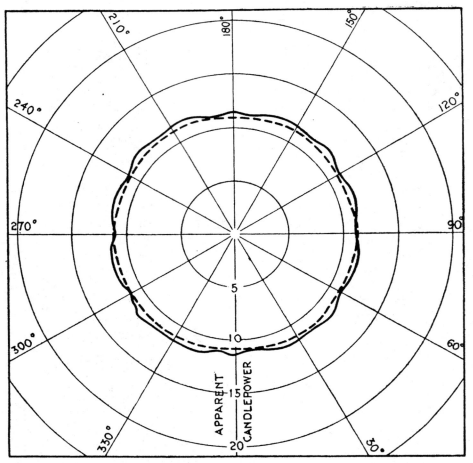

Fig. 12—Distribution of Light from Bulb No. 3. Solid Line—Normal to Bulb Axis. Dotted Line—120 Deg. from Tip of Bulb.

in the lamps, Fig. 11, this distance varies $\frac{5}{32}$ in. This variation is of little importance if the headlight is fitted with a proper bulb focusing device. In some headlights it is almost impossible to focus properly a bulb of the usual focal length.

It is imperative that the Society formulate a standard for headlight bulbs. When headlights are set to comply with present "nonglare" regulations any variation in filament volume will mean a readjustment of headlight setting. A standard form of headlight bulb made within reasonable limits will require little focal adjustment and no alteration of headlight position. I suggest the following be standardized: Voltage (now S. A. E. standard); amperes; maximum length of filament; maximum width of filament; and limit of variation in focal distance. No refinement in bulb manufacture can compensate, however, for the imperfect design and cheap construction of automobile headlights.

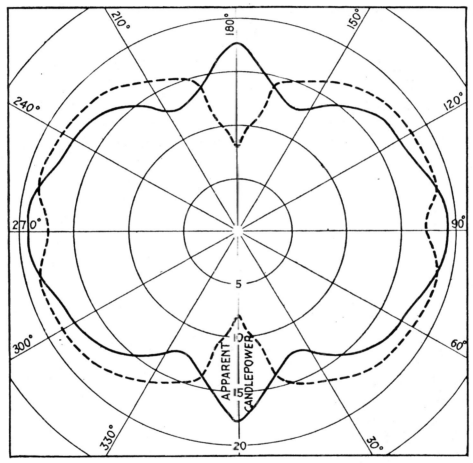

Fig. 13.   Distribution of Light from Bulb No. 6.   Solid Line—Normal to Bulb
Axis.   Dotted Line—12C Deg. from Tip of Bulb.

T. R. COOK:—Some extensive experiments on which I was engaged brought out the fact that the reflected light from a headlight follows the law of inverse squares. To determine this we took a headlight and obtained its distribution and candlepower at a distance of 50 ft. Then we took the same angle on the axis at a distance of 200, 300 and 500 and so on to 1000 ft. The candle-foot rating at those points checked closely with that determined at 50 ft. With a concentrated filament it was difficult to set the filament in the reflector so as to

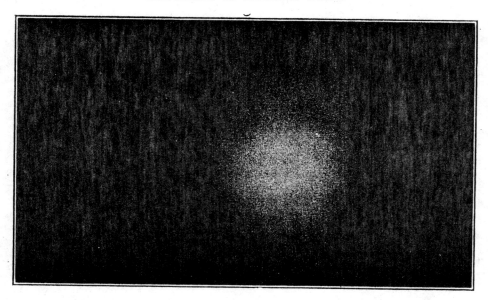

Fig. 14—Light Thrown by Bulb No. 4 and Reflector on White Screen 40 Ft. Distant.

repeat tests. A slight variation changed the candlepower considerably.

HENRY SCHROEDER:—A slight variation in the neck of the bulb, the length of the neck, or a difference in the length of the base and the cement used will vary the focal length. Bulb manufacturers therefore desire as much leeway as possible in filament dimensions.

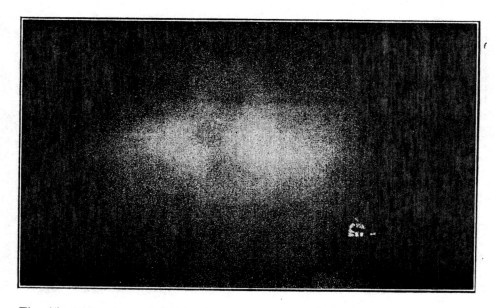

Fig. 15—Light from Bulb No. 2 in Same Reflector Used for Fig. 14 and on Screen 40 Ft. Distant.

HEAD-LAMP LIGHT-CHARACTERIS-TICS AND DISTRIBUTION[1]

BY WALTER D'ARCY RYAN[2]

ABSTRACT

Subsequent to reviewing the circumstances responsible for the present complicated situation existing with respect to satisfactory automobile-headlighting, the author says that headlights glare if they are adjusted for range and that, when adjusted for nonglare, they have no range; hence, careful tests were made on a number of the best types of approved headlight and lens in use. The units were set-up in pairs, in operating position as to height and interval and were tested at a range of 100 ft. with a 1-m. (39.37-in) hemisphere having an aperture 21 in. square, corresponding to 1 deg. at 100 ft. All the lamps were held at proper current-value throughout the tests, and it was demonstrated by the tests that the reflectors of the parabolic type and others of similar characteristics that have proved to be unsatisfactory during many years past must be abandoned.

Regarding the so-called "hot-spot" of light on the road to facilitate high-speed driving and the tendency toward such practice lately evidenced, the author sees no reason that anything beyond 20,000 cp., maximum, is required, and says that half this candlepower would be sufficient in a well-balanced light. The main light-beam should have a good lateral spread so that it will cover for a considerable distance not only the road but also the ditches, with reasonable depth so as to carry a gradually diminishing illumination to within say 20 ft. of the car, and should illuminate the wheels and the running-board of the approaching car clearly at a distance of at least 200 ft. without projecting any direct rays into the eyes of the approaching driver.

Commenting upon fog penetration, the practice of "dimming," tilting reflectors, and two-filament lamps, the author says that headlights of the future should have range without glare and should not require tilting, dimming or any other form of manual operation. He believes that the two-filament lamp will do much to

---

[1] Annual Meeting paper.

[2] Director, illuminating engineering laboratories, General Electric Co., Schenectady, N.Y.

improve the present situation, states specifications for improved headlights and discusses the tests and the illustrations.

Designers of the early electric automobile-headlights had the one thought in mind of producing the maximum amount of light ahead to satisfy the man behind the steering-wheel; they had no thought of the effect of such headlights on the approaching driver, a feature apparently considered only recently. Their effort to outshine everything on the road resulted in an intolerable and dangerous situation which increased with the automobile traffic until, with millions of automobiles on the roads, it has become necessary to enact laws to protect the automobile drivers from each other. Exhaustive tests, which resulted in the so-called "I.E.S. Specifications" that many States have adopted, were made by the Illuminating Engineering Society. They were not advocated as being ideal specifications but were formulated on the basis of a compromise between glare and range of sufficient light on the roadway for safe driving, based on the best available units in existence at the time of the tests in question and also based on the automobile-traffic situation then existing. These specifications, which have stood for so many years with only one addition, appear to be the best basis so far offered as a first line of elimination, provided that we are to continue the use of headlights having the same general light-distribution characteristics as we are using today. The specifications of the Society of Automotive Engineers are mainly based on those of the Illuminating Engineering Society but show higher values at certain points.

Accident statistics were studied and conclusions were drawn to the effect that although glare is disagreeable and dangerous it is not the primary cause of night accidents, a greater number being attributed to insufficient light on the roadway. Further analysis probably would have shown that improper distribution of light on the roadway, irrespective of the number of lumens delivered, was responsible for a large percentage of the accidents. Glare is the only thing that ever put me in a ditch or caused me to turn a car over by side-swiping it on account of misjudging the passing space available. I believe that the public feels glare to be a greater menace than anything else we have to contend with at night. As a result, the automobile adjusters, irrespective of any specifications issued, are depressing the headlights to accomplish minimum glare. Such adjustments mean that the most intense light strikes the road comparatively near the car, thereby diminishing the visibility at long range and resulting in a most unsatisfactory and dangerous driving-light.

HEADLIGHT ADJUSTMENT

It has been demonstrated that headlights glare if they are adjusted for range and that when adjusted for nonglare they have no range. In a paper presented on April 2, 1925, at a meeting of the German Illuminating Engineering Society, Dr. Otto Reeb analyzed the German and the American automobile-headlight practices, stating that the American illuminating engineers have tried to solve the difficult problem of automobile headlighting by combining contradictory requirements; first, to give the driver as much light as possible and, second, to avoid glare for the occupants of approaching vehicles. This is precisely what we have been trying to do. Dr. Reeb concludes that the German method is better; adding:

> It was found, for instance, that electric headlights of 20 watts (40 Hefner units or 36 international cp.), having moderate power, do not cause glare when the axis of the projector is inclined in such a manner that the center ray intersects the road within 30 m. (98.5 ft.) of the car. In the case of the stronger lamps, that is, 35 watts (70 Hefner units or 63 international cp.), the beam must meet the road within 15 m. (49.2 ft.) of the car.

Therefore, it is evident that the Germans are using lamps of higher candlepower but are doing precisely what our automobile adjusters are doing; that is, deflecting the beams to avoid glare until they have no range, without regard to the allowable limits of the specifications of the Illuminating Engineering Society or those of this Society.

Ever since the first automobile headlight was placed on a car we have adhered very closely to the parabolic reflector or reflectors having similar vertical characteristics. Various types of lens and of reflector have been designed, but they have mainly to do with spreading the light in the horizontal plane with little change in the vertical distribution, making it necessary to set a rather high value at the C point, which is on the center line 1 deg. above the horizontal, to bring the most powerful part of the beam a reasonable distance ahead of the machine. We are allowed from 800 to 2400 cp. at point *C*. I consider even the low limits, 800 cp., within the range of glare and, with the best lamps available set at the low limit, the maximum light strikes the road surface between 40 and 60 ft. ahead of the car; when set at the high limit, it strikes between 80 and 100 ft. ahead. This produces a brilliantly lighted area in the foreground which reduces visibility at long range.

Recently, I had a series of careful tests made on a number of the best types of approved headlight and lens in use today. The units were set-up in pairs, in operating position as to height and interval, and were tested at a range of 100 ft. with a carefully calibrated

FIG. 1—SEARCHLIGHT-TESTING RANGE

Extensive Automobile-Headlight Investigations Have Been Conducted in the Last 3 Years in This 155-Ft. Searchlight-Testing Range Located in the Illuminating Engineering Laboratory of the General Electric Co

FIG. 2—ONE-METER HEMISPHERE HAVING PHOTOMETER ATTACHED
This Equipment, Including the Franz, Schmidt & Haensch Photometer, Was Used in the Tests

1-m. (39.37-in.) hemisphere having an aperture 21 in.
square, corresponding to 1 deg. at the distance mentioned. All the lamps were calibrated carefully and
held at proper current-value during the tests. The
curves represent the results on one pair of each type of
unit only, and some variation naturally would exist one
way or the other on the average of a number of pairs,
but the vertical characteristics remain essentially the
same.

The identity of the headlights and lenses used in the
tests is designated merely by letter. Some pairs of the
types have laboratory ratings around 40,000 or 50,000
cp., maximum. While it is possible, no doubt, to obtain
points of these high values under laboratory conditions
with selected lenses of the first run and super-parabolic
testing-reflectors, I believe these values are not attained
in practice with the regular commercial equipment; so,
apparently, we have been misleading ourselves somewhat
as to the candlepower of the automobile headlights in

FIG. 3—HEADLIGHT TESTING-STAND

This Apparatus Is Self-Contained and Has Batteries, Instruments
and Rheostatic Control It Enables Very Fine Horizontal and
Vertical Adjustments To Be Made

service. These tests demonstrate clearly that, if we are to have range without glare, we must abandon the reflectors of the parabolic type and others of similar characteristic which, after a quarter century of juggling, have proved to be unsatisfactory. The characteristic features will be analyzed later.

"Hot-Spot" of Light for High-Speed Driving

In the last few years a tendency has existed to produce a so-called "hot-spot"; that is, a small central area of very high candlepower for high-speed driving. Unless this hot-spot is projected near the horizontal, where it would produce blinding glare, it is of no great value and, if turned down so as to minimize the glare and come within the legal requirements, it produces a new source of dangerous glare from the oil-polished road-surfaces and is especially pernicious on wet nights. It becomes a secondary source of discomfort, in some cases more dangerous than the direct glare from the lamp, owing to the lower angle from which it is projected into the approaching driver's eyes.

In his excellent and constructive paper on Improved Automobile Headlighting[2], Alfred W. Devine, engineer in charge of equipment section, Massachusetts Registry of Motor Vehicles, brings out the following points:

> In attempting to solve the headlighting problem we must have in mind, at each step, the two primary objects which must be attained insofar as it may be practicable, sufficient illumination ahead for safe operation and freedom from dangerous headlight glare for other users of the highway.

> The perfection of the beam-distribution characteristics of automobile head-lamps for average use has been handicapped by the rapid change which has taken place in traffic conditions during the last few years and also by a lack of proper understanding on the part of the manufacturer as to what a head-lamp should be called upon to do. *Modern traffic conditions, except in infrequent cases, do not permit excessively high speed on our State roads. The necessity then for great beam-penetration is not present and even though it were desirable to have strong penetration, the usefulness of the head-lamp on congested State highways should not*

---

[2] Presented at the 19th Annual Convention of the Illuminating Engineering Society, Detroit, Sept. 15, 1925.

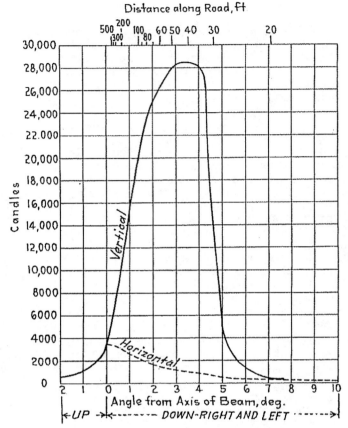

FIG. 4—VERTICAL AND HORIZONTAL CHARACTERISTICS

On the Pair of Units, Letter *A*, It Will Be Noted That They Give
Approximately 28,000 Cp., Maximum, As Indicated by the Solid
Line. With These Lights Adjusted for 1000 Cp. and Projecting to
the *C* Point on the Center, 1 Deg. above the Horizontal, the Maximum-Candlepower Beam Will Intercept the Road Surface Approximately 40 Ft. Ahead of the Car

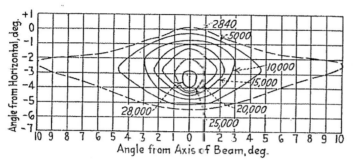

FIG 5—ISOMETRIC CHART OF THE A UNITS

The Maximum Spot of 28,000 Cp. Subtends an Angle of 1 Deg., Horizontally. This Will Produce a Spot on the Road Only 21 In. Wide at 100 Ft or 42 In. at 200 Ft. and Is Altogether Too Small To Be of Any Great Value

be handicapped by the distribution of flux in the beam for a special purpose.

One of the greatest objections to head-lamps that project high intensities in the center of the beam is that the greatest advantage is not taken of them unless the lamp is aimed up so as to bring this intensity close to the horizontal. Under these conditions intolerable glare is bound to result. There is an ever-present temptation for the operator to tilt such a lamp upward. This temptation is reduced or eliminated when a reasonable limit is placed on the maximum.

I see no reason that anything beyond 20,000 cp., maximum, is required, and half this candlepower would be sufficient in a well-balanced light.

SIDE ILLUMINATION AND DEPTH OF BEAM

It is desirable that the main beam have a good lateral spread so that it will cover for a considerable distance not only the road but also the ditches. The beam should have reasonable depth so as to carry a gradually diminishing illumination well back to within say 20 ft. of the car. It should be of such a character that it will illuminate the wheels and the running-board of the approaching car clearly at a distance of at least 200 ft. without projecting any direct rays into the eyes of the approaching driver. The majority of the headlights have a sharp horizontal cut-off ranging from 6 to 15 deg. on either side of the center line, which leaves a dark area on both sides of the car for a considerable distance, making it difficult to see the ditches or to judge the

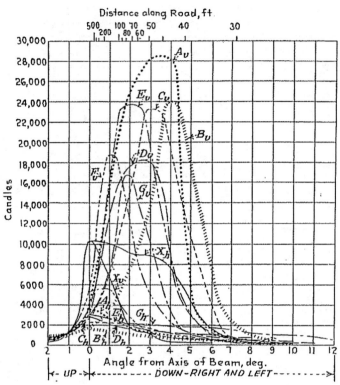

FIG. 6—VERTICAL AND HORIZONTAL CURVES COMBINED

This Drawing Is a Combination of All the Vertical and Horizontal Curves in the Comparison with the Different Units Set for 1000 Cp. at the C Point, Which Is Only 200 Cp. above the Maximum of 800 Cp. Allowed at Point C by the Specifications. With This Setting, All the Lamps Equipped with Parabolic Reflectors A, B, C, D, and E Project Their Maximum Beams at an Angle That Will Intercept the Road Surface Anywhere from 35 to 75 Ft. Ahead of the Car, Which Is Altogether Too Close for a Good Driving-Light. Units F and G, Which Are Not of the Parabolic Type, Show Improved Range with This Setting. Units X Have, if Anything, More Range Than Is Desirable

passing space available for other cars or to see pedestrians who step into the line of travel at short range. Light in this area is fully as important as light in any other section, and should be of sufficient intensity so that the side of the road or near-by ditch is clearly visible, and there should be ample light for rounding sharp curves or turning corners independent of the light from the main beam. It also should make the reading of signs on either side of the road possible without the use of spotlights or other auxiliary units. Many motorists today are equipping the right-hand fender with what might be called a ditch light or are adding a spotlight, mounted below the radiator and inclined to the right, showing that there is a strong appreciation of the necessity for good side-light. To compensate further for the defects in the present headlights, a tendency also exists to add hub-lights, courtesy lights and one or more spotlights mounted on the windshield to perform the functions which should be taken care of by a single pair of headlights so as to reduce the amount of lighting equipment on the car and the current-drain on the battery.

SIMULTANEOUS CONTRAST

Sufficient light should be provided on the front of the machine—the fenders, the wheels and the radiator—so that in case one light fails an automobile cannot be mistaken for a motorcycle. In conjunction with the light at the sides and immediately in front of the car, this illumination reduces, by simultaneous contrast, the glare effect. Again quoting Dr. Otto Reeb:

> Measurements made by Gehlhoff and Schering dealt with the influence of the general illumination on the glare phenomena. One should expect that the luminous intensity necessary to cause glare would increase with an increase in the general illumination. With no general illumination at a distance of 100 m. (328 ft.), 2000 Hefner units (1800 cp.) will cause unbearable glare, while with a general illumination of 1 lux (0.1 foot-candle) the same effect is obtained only at 20,000 Hefner units (18,000 cp.) . . . . .

In other words, if two bright spots shine toward a driver out of darkness, approximately 10 times as much glare effect exists as if these lights shine out of an illuminated field of relatively low intensity. A careful study of this phenomenon undoubtedly would be of great value.

FOG PENETRATION

Just so long as strong reflected rays that illuminate the fog particles intervening between the driver and the point he is trying to see are thrown above the horizontal, difficulty is bound to be encountered and the best the driver can do is to feel his way along, having no idea as to his position on the road. The remedy is to eliminate all high-power reflected-rays between the observer's point of vision and the distant point he is trying to see and, at the same time, to supply sufficient side and local illumination so that, no matter how dense the fog is, he will always know his exact position on the road.

DIMMING, TILTING REFLECTORS AND TWO-FILAMENT LAMPS

I am inclined to believe that all those who have studied the dimming problem carefully are convinced that dimming is a most unsatisfactory and dangerous practice. The light is cut-off at the critical moment and the eye has no chance to recover before passing the approaching car and into the danger zone. Manually operated tilting reflectors have been tried without much success. This may have been due partly to the fact that so few of them were in use, and the relief from glare offered the approaching driver by depressing the beam rarely could be answered, because the majority of machines had no means of responding except by dimming, which many consider too hazardous a chance to take. I believe the two-filament lamp, if put into use rapidly and extensively, will do much to improve the situation. With the millions of headlights in use and the millions that will go into service before any radical changes in headlights of improved light-characteristics can be put into production on a large scale, it is necessary to do something and that without further loss of time. The success of the two-filament lamp will depend, in a great measure, upon the reasonable response of the public in its proper use. As with the tilting-reflectors, we cannot expect 100 per cent relief but, if we persuade say one-third of the operators to use them with discretion it will make a vast difference in the comforts of night driving. The results obtained with the two-filament lamps are vastly superior to dimming and most likely will result in bringing about the repeal of dimming laws in the few States in which they are now in force.

The headlights of the future should have range without dangerous glare and should not require tilting, dimming or any other form of manual operation; but, until such units are widely in use, we should use every possible means to bring about the equipping of new headlights and those in service with any meritorious feature that will assist in taking the curse out of night driving. I wish to repeat the conclusions of my discussions at the meetings of the Metropolitan and the Detroit Sections of this Society. It is not difficult to design a headlight to suit the man behind the steering-wheel, but it becomes a difficult problem when the other driver who faces the lights is taken into consideration. It is right here that the majority of the headlights fail today and, before an automobile headlight can take its place as a decided improvement over the existing equipment, it must meet the following specifications:

SPECIFICATIONS FOR IMPROVED HEADLIGHTS

(1) It must be a non-glare unit having a range between 200 and 300 ft. on a level road; it should be non-focusing and capable of operating with lamps of any form of concentrated filament or candlepower without change of focal adjustment

(2) The light distribution should be of fairly wide characteristic with reasonable depth and should be homogeneous, with a gradual increasing intensity from a point near the machine to the most distant point, and the reflected beam should not rise above the horizontal. Sufficient light from even a macadam road-surface always exists to take care of softening the cut-off above the horizontal at long range. The area of greatest intensity should not be concentrated in a small spot of high candlepower, but should have a reasonable lateral-divergence. It is important to bear in mind that a very intense spot, particularly on a wet road-surface, introduces a new element of glare, reflected, which should be avoided

(3) A reasonable amount of light should be projected at right angles to the plane of the main beam and even a few degrees to the rear, so as to light-up the gutters and curves and make turns in difficult places, and make possible as well the reading of road signs on either side without the use of spotlights or other auxiliary lamps

(4) Sufficient light should be thrown on the front of

the machine—the radiator, the forward wheels and the bumper—so that they are clearly visible, and if one light fails, there should be no chance of mistaking an automobile for a motorcycle. The cut-off of the beam should be such that there will be no upward high-candlepower rays to scatter in the fog and reduce visibility

(5) A general dispersion of unreflected light should be provided to illuminate trees and telegraph poles and give general vista without glare, so that distance can be judged at night as well as in daylight driving. If the non-glare feature of the unit is improved further by lighting-up the front of the machine and the general surroundings, the intensity of the source becomes less brilliant by simultaneous contrast. Further, the main beam should be of a nature such that it will become even less brilliant as the car is approached, which, in turn, will improve the ability of the approaching driver to see beyond the approaching car at the critical moment, and in other ways add a sense of comfort or feeling of security in driving

(6) The lights should be focused definitely for city and for country driving so that there is no necessity for dimming, tilting or other manual operation which, in the majority of cases, with the present increased automobile traffic, is impracticable unless operated at the low point most of the time, except in country driving

Mechanical Headlight Specifications

(1) The lamps should be adaptable to modification of design to meet the aesthetic lines of the car and embody the elements of true art which, at a glance, suggest that the unit is primarily a functioning light-source rather than a decoration

(2) A lamp must be sufficiently rigid in construction so that it cannot get out of adjustment easily

(3) It should be dust and rain-proof and a simple means of opening the door should be provided so that the replacement of the lamp or the cleaning of the reflector can be done without the use of tools, unusual exertion or disturbing the adjustment

(4) A simple means for adjustment of the beam should be provided which will not require the bending of forks, difficult manipulations or technical knowledge; in fact, the means should be so simple that anyone can make the adjustment and

that little excuse will exist for failure to comply with State or with police regulations

(5) The headlights must be produced at a cost that will not make them prohibitive, even for the low-priced cars, and should not be subject to wide variation in production

I believe the above specifications are well within the range of possibility and can be met without insurmountable mechanical or optical difficulties and at reasonable manufacturing cost.

Fig. 1 shows a section of the 155-ft. searchlight-testing range in the illuminating engineering laboratory of the General Electric Co., Schenectady, N. Y., where extensive automobile-headlight investigations have been conducted in the last 3 years. Fig. 2 illustrates a 1-m. (39.37-in.) hemisphere, having a Franz, Schmidt & Haensch photometer attached, which was used in the tests as described. Fig. 3 is a self-contained headlight testing-stand, having batteries, instruments and rheostatic control. This apparatus enables very fine horizontal and vertical adjustments to be made.

Five pairs of the standard headlights now in general use, equipped with parabolic reflectors, were selected and tested for this paper and they are designated as units A, B, C, D, and E. Two pairs of head-lamps now on the market having reflectors differing from the parabolic type are included under letters F and G. These are compared with a pair of experimental lamps embodying improved characteristics designed to increase the range and, at the same time, to reduce the glare and provide adequate side-light. Letter X has been assigned to these units.

Analyzing first the vertical and horizontal characteristics on the pair of units, letter A, Fig. 4, it will be noted that they give approximately 28,000 cp., maximum, as indicated by the solid line. With these lights adjusted for 1000 cp. and projecting to the C point on the center, 1 deg. above the horizontal, the maximum-candlepower beam will intercept the road surface approximately 40 ft. ahead of the car. It will be observed further that with this setting the maximum candlepower on the horizontal is approximately 3500 cp., falling off rapidly to the right and to the left. This low intensity on the horizontal is made less effective by the brilliantly illuminated patch of light striking the road surface a short distance ahead of the car.

Fig. 5 is an isometric chart of units *A*. The curves have not been carried beyond 10 per cent of the maximum, which is the vanishing point of the line of demarcation one observes in looking at a projected beam. As this is sufficient to show the general characteristic, it did not seem necessary to project the curves any farther. The maximum spot of 28,000 cp. subtends an angle of 1 deg. horizontally. This will produce a spot on the road only 21 in. wide at 100 ft. or 42 in. at 200 ft. and is altogether too small to be of any great value, also having the objections of causing too sharp a contrast to the rest of the beam. Even if this so-called hot-

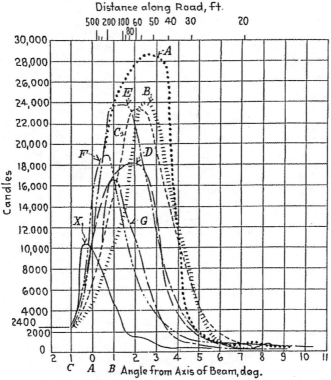

FIG. 8—COMPARISON OF VERTICAL CURVES

This Was Made with the Units All Set at 2400 Cp. at the *C* Point, Which Is the Upper Limit of the Specifications. With This Excessive Permissible Glare. the Range of the Lamps Using Parabolic Reflectors Is Improved, Ranging from 50 to 150 Ft., Which, in the Main. Is Still Too Short, Especially for High-Speed Driving

spot were wider, it is situated with reference to the vertical so that it could not be brought up to the long-range position, which would require an elevation of about 3 deg., without projecting intolerable glare at the C point.

Referring to units B, C, D, and E (Fig. 6), considerable variation in the general distribution of the different units is evident and some modification in the vertical characteristics but not so much but that the previous remarks apply to all. Units F show a step in the right direction by departing from the parabolic reflector, which results in an increase in range and some modification in the horizontal distribution. Units G contain reflectors having a slight modification in the vertical and a wide modification in the general distribution, with resultant increase in range and improvement in cut-off. These are the most efficient units in beam lumens so far tested. It will be noted, however, from the isometric chart, that a considerable amount of this light is thrown far beyond the sides of the road, where it is wasted. It would, therefore, be better if this stray light were utilized in giving greater depth of beam from the maximum down, so as to illuminate farther back toward the car.

Units X show a considerable departure in light characteristics from standard headlight-practice, not only in the vertical and in the horizontal distribution but in the general character of the field. For a setting of 1000 cp. at point C, the maximum light can be projected on the horizontal. It will be observed further, by studying the line X_h, that the horizontal candlepower which coincides with the maximum is retained at relatively high value for a considerable distance to the right and to the left from the center in place of starting at low candlepower and falling-off rapidly, as shown in the curves of the other units. A study of the isometric chart in the lower right corner of Fig. 7 reveals that the general distribution takes the form of a curve so as to follow the vanishing point of the road, thereby utilizing what otherwise would be stray light for the illumination of the sides of the road and ditches well up to the car. This also avoids flashing light of high candlepower into the eyes of the approaching driver in rounding curves. Another feature of importance is the wide dispersion of the maximum-candlepower area. This makes it possible to cover the entire road at long range with the strongest light and permits a reasonably sharp horizontal cut-off following, as near as possible, the crown of the road.

Fig. 6 is a combination of all the vertical and horizontal curves in the comparison with the different units set for 1000 cp. at the C point, which is only 200 cp. above the minimum of 800 cp. allowed at point C by the specifications of the Illuminating Engineering Society and those of this Society. With this setting, it will be noted that all the lamps equipped with parabolic reflectors A, B, C, D, and E project their maximum beams at an angle that will intercept the road surface anywhere from 35 to 75 ft. ahead of the car, which is altogether too close for a good driving-light. Units F and G, which are not of the parabolic type, show improved range with this setting. Units X have, if anything, more range than is desirable.

Fig. 7 is a combination of the isometric diagrams of

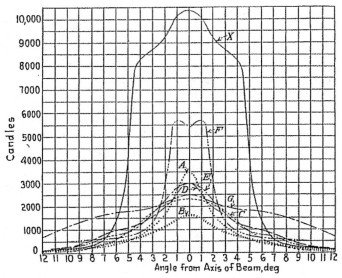

FIG 10—INITIAL-CANDLEPOWER DISTRIBUTION IN A HORIZONTAL
PLANE

With All the Lamps Set for 1000 Cp. at Point C, Which Is within the Present Permissible Limits, the Maximum Illuminations on the Horizontal Ranges, for Units with Parabolic Reflectors, Vary Approximately from 1500 to 3500 Cp., Running As High As 5500 Cp. for Units F. With This Setting, Units X Project their Maximum Intensity of 10,000 Cp. on the Horizontal and Maintain a Relatively High Value for Several Degrees to the Right and to the Left from the Center

the different units in order that their characteristics can be studied simultaneously.

Fig. 8 is a comparison of the vertical curves with the units all set at 2400 cp. at the C point, which is the upper limit of the specifications of the Illuminating Engineering Society and those of this Society. With this excessive permissible glare we find the range of the lamps using parabolic reflectors improved, ranging from 50 to 150 ft. which, in the main, is still too short, especially for high-speed driving. Units F, with this setting, show a range of approximately 250 ft., and units G about 160 ft., while units X project the light altogether too high. With an initial setting of 2400 cp. at point C, the deflection up of 1 or 2 deg., due to road inequalities, car loading or other reasons, would cause the projection of light in line with the approaching driver's eyes of anywhere from 10,000 cp. for units X to 24,000 cp. for units A. Under these conditions the elimination of glare, no matter what type of unit is used, is physically impossible. Ultimately, the candlepower limits at point C must be reduced. It can readily be seen that to establish these limits today would eliminate practically all the headlights on the market. As it is, they all give glare without satisfactory range when working within the present allowable limits.

Fig. 9 shows clearly that, with a setting of 500 cp. at point C, the lamps equipped with parabolic reflectors would have a range varying from 22 to 55 ft. to the point of interception of maximum beam on the road ahead of the car. Units G show 55 ft.; units F, 85 ft.; and X units, 220 ft. I think that almost anyone will agree that headlights having a range of less than 100 ft. for the maximum illumination would be the cause of just as many accidents as glare would cause.

Fig. 10 shows that with all the lamps set for 1000 cp. at point C, which is within the present permissible limits, the maximum illuminations on the horizontal ranges, for units with parabolic reflectors, vary from approximately 1500 to 3500 cp., running as high as 5500 cp. for units F. Units X, with this setting, project their maximum intensity of about 10,000 cp. on the horizontal and maintain a relatively high value for several degrees to the right and to the left from the center.

Fig. 11 shows the efficiency in side illumination of all existing headlights, units A to G inclusive. *If a headlight is to perform all the functions required for safety, side*

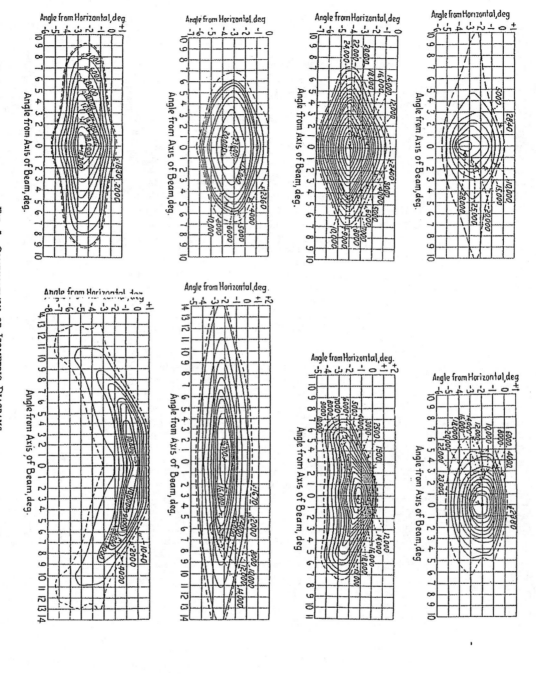

FIG. 7—COMBINATION OF ISOMETRIC DIAGRAMS

This Grouping Was Made in Order That the Characteristics of the Different Units Can Be Studied Simultaneously. From the Top Down the Order of the Units at the Left Is A, B, C, and D. The Units at the Right, Beginning at the Top, Are E, F, G, and X.

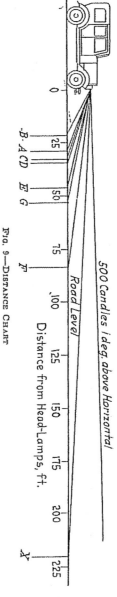

his Shows the Distance from the Car at Which the Beams of Maximum Candlepower Strike the Road Surface. With a itting of 500 Cp. at Point C, the Lamps Equipped with Parabolic Reflectors Would Have a Range Varying from 22 to 55 t. to the Point of Interception of Maximum Beam on the Road Ahead of the Car. Units X, and Units G Show 55 Ft. Units F, 85 Ft., and Units X, 220 Ft.

Fig. 9—Distance Chart

500 Candles i deg. above Horizontal

Road Level

Distance from Head-Lamps, ft.

B-ACD E G   F   X

0 25 50 75 100 125 150 175 200 225

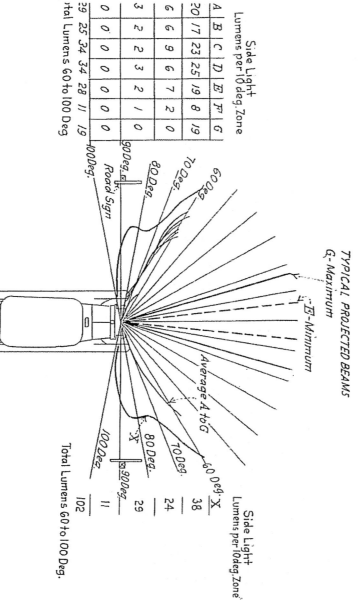

TYPICAL PROJECTED BEAMS

G - Maximum

E - Minimum

Average A to G

Side Light Lumens per 10 deg. Zone	
60 Deg. X	38
70 Deg.	24
80 Deg.	29
90 Deg.	11
100 Deg.	102

Total Lumens 60 to 100 Deg.

Road Sign

100 Deg.   90 Deg.   80 Deg.   70 Deg.   60 Deg.

Side Light Lumens per 10 deg. Zone

	A	B	C	D	E	F	G
	20	17	23	25	19	8	19
	6	6	9	6	7	2	0
	3	2	2	3	2	1	0
	0	0	0	0	0	0	0
	0	0	0	0	0	0	0
	29	25	34	34	28	11	19

Total Lumens 60 to 100 Deg

Fig. 11—Side-Light Comparison. If a Headlight Is To Perform All the Functions Required for Safety, Side Illumination Must Be Given As Much Consideration As Range and the Elimination of Glare the Efficiency in Side Illumination of Units A to G Inclusive Is Shown.

illumination must be given as much consideration as range and the elimination of dangerous glare.

In making the foregoing comparisons it is not my intention to assume that the experimental headlights, units X, are the final solution of the problem. They are probably too far in advance of the times and may have to be led up to by degrees. Further improvements undoubtedly can be made; nevertheless, the various points herein covered indicate a general direction in which we can work to relieve the present automobile-headlight situation, and while some of the present candlepower limitations stand in the way of improved design, it is to be hoped that, when improvements are demonstrated clearly, the laws will be modified as to the point-C values and even to the extent of allowing lamps of higher candlepower, if necessary, to bring about the desired result.

WHAT HAPPENS WHEN A HEAD-LIGHT IS OUT OF FOCUS[1]

By L C Porter[2] and G F Prideaux[3]

ABSTRACT

Since the layman and not the engineer buys and drives most of the automobiles produced and because the literature on automobile headlighting presents too technical a picture of what happens when the light source of an automobile head-lamp is out of focus, the authors planned and executed an extensive study of the subject in an endeavor to clarify the technicalities by presenting them in the forms of photographs and simple charts, the chief object being to obtain data that emphasize the necessity of accurate control of the size and location of the light source with respect to the focal point of parabolic headlight-reflectors. A great difference in the resultant beam of light is produced by a very small displacement of the light source, either through poorly constructed lamps or due to lack of proper adjustment, and the tests made evaluate how

---

[1] Annual Meeting paper.

[2] M.S.A.E.—Commercial engineer, Edison Lamp Works of the General Electric Co., Harrison, N. J.

[3] Engineering department, Edison Lamp Works of the General Electric Co., Harrison, N. J.

small these displacements and how great these differences are.

After considering the difficulties of locating the light source or filament of the 21-cp. headlight-bulb at the exact focal point of the reflector and of taking account of the practice of certain manufacturers who measure tolerances in sixty-fourths of an inch and of trouble due to wabbly sockets, distorted reflectors and the like, the authors constructed a device that enabled the lamp socket to be moved in any direction by micrometer screws having 32 threads per in., thus causing a one-half revolution of the screws to move the light source exactly 1/64 in., backlash being compensated for by springs. A test reflector made as perfectly as possible was used in connection with the device. Variations due to filament size, shape or relative position in the bulb were eliminated so far as possible by selecting lamps exactly correct as to light-center length, axial alignment and bulb image, and the same lamp was used for all the light-center length and axial-alignment tests, these being made by moving the one lamp with the aforesaid accurate focusing-device. Three types of lamp were chosen to give beams of wide, medium and narrow spread, and, with this equipment, the authors set out to ascertain the effects of various specified changes in equipment arrangement and to record them by photographs, as well as by photometric and linear measurements.

In making the measurements, the test reflector with the universal focusing-device was mounted on a rotatable table, graduated in degrees, located 25 ft. from a screen having the Illuminating Engineering Society's headlight-specification test-points plotted upon it. Photometric readings were taken with a Macbeth illuminometer. The headlight-lamps were operated at exactly 21 mean spherical cp. A single reflector was used. The tests brought out clearly the need for accurately made lamps and equipment and for accurate focusing if the headlight situation is ever to be brought under control.

Anyone who has studied automobile lighting, and many who have not, know in a general way what happens when the light source is out of focus. Search of the literature on the subject, however, fails to reveal information that can be clearly understood by the layman and, after all, it is the layman and not the engineer who buys and drives most of the cars. Plenty of information is available in technical or semi-technical journals describing the lumen output of a headlight having a paraboloidal reflecting-surface. The brilliancy of the road surface in milli-lamberts can be calculated and the beam candlepower can be determined 7 deg. to the left or 4 deg. above the lamp axis and the like. Photometric distribution-curves are available, headlight laws have been passed and ideal beams have been described, but what have we in actual operation on our roads and what do the results *look* like? In an endeavor to reduce some of these technical data to plain English and to tell the story by photographs and simple charts, we planned and executed a rather extensive study that is described in the following condensed form.

The chief object of this study was to secure data that would emphasize the necessity of accurate control of the size and location of the light source with respect to the focal point of parabolic headlight-reflectors. Few automotive engineers realize how small a displacement of

FIG. 1—MICROMETER FOCUSING-DEVICE AND TESTING REFLECTOR

The Device Permits Movement of the Socket in Any Direction by Micrometer Screws Having 32 Threads Per In. A One-Half Revolution of the Screws Moves the Light Source Exactly 1/64 In., Any Backlash Being Taken-Up by Springs

the light source, either through poorly constructed lamps or lack of proper adjustment, will make a very great difference in the resultant beam. Just how small and how great these differences are is clearly shown by the accompanying results of tests.

TESTING EQUIPMENT

Certain special types of reflector and of lens are not as sensitive to focal adjustment as is the true paraboloid. Obviously, these data do not hold for such equipment. We had heard and read much about the difficulties of locating the light source of filament of the 21-cp. headlight-bulb at the exact focal point of the reflector. We also had learned of certain manufacturing tolerances measured in sixty-fourths of an inch, which the lamp manufacturers claimed were necessary in making the bulbs. We had heard also of wabbly sockets, distorted reflectors and the like; so, to eliminate all these troubles, we had a focusing device such as that shown in Fig. 1 made-up in our machine-shop, which enabled us to move the socket in any direction by micrometer screws having 32 threads per in. With this device, a one-half revolution of the screws moves the light source exactly 1/64 in., no more and no less, any backlash being taken-up by springs. A test reflector, made as true and as perfect as the best reflector-manufacturer in the Country could construct it, was used with the focusing device.

To eliminate any variations due to filament size, shape or relative position in the bulb, we selected lamps that were exactly correct as to light-center, length or distance from the locking pin to the center of gravity of the filament, axial alignment and bulb image, and the same lamp was used for all the light-center-length and axial-alignment tests; these being made by moving the one lamp with our accurate focusing-device. Three types of lens were chosen to give beams of wide, of medium and of narrow spread. With this equipment we set out to ascertain the following effects and to record them by photographs, as well as by photometric and linear measurements; namely the effect on

(1) Headlight beams of variations in light-center length in the lamp bulbs, or improper location in the reflector

(2) Headlight beams of variations in axial alignment in the lamp bulbs, or improper location in the reflector

(3) Headlight beams of a combination of variation in light-center length and axial alignment of the headlight bulbs, or their location in the reflector

(4) A spotlight beam of variation in filament position in a headlight bulb, or location in the reflector

(5) Headlight beams of mandrel size or coil diameter of headlight filament

(6) Spotlight beams of mandrel size

(7) Spotlight beams of auxiliary bulb image

In making the measurements, the test reflector with the universal focusing-device was mounted on a rotatable table, graduated in degrees, 25 ft. from the screen having the Illuminating Engineering Society's headlight-specification test-points plotted on it. Photometric readings were taken with a Macbeth illuminometer. The headlight lamps were operated at exactly 21 mean spherical cp.

A single reflector was used; hence, in comparing the actual readings obtained with the Illuminating Engineering Society's specifications, the Illuminating Engineering Society's readings should be divided by 2. This has been done in the charts to make the figures more readily comparable. We realize that this method does not give the exact figure that would be obtained from a pair of headlamps but, for purposes of comparison, it was considered to offer less variation than might be expected if two head-lamps were to be adjusted each time. While the exact figures are interesting to the engineer who is making a detailed study, we believe the real value of the work, to most people, lies in the photographs. The pictures bring out clearly the need for accurately made lamps and equipment and for accurate focusing, if we are to get the headlight situation under control.

TESTING PROCEDURE

The first tests were made to determine the effect of light-center length and axial alignment on three well-known types of lens, A, Bausch & Lomb; B, Patterson; and C, Osgood, giving the wide, medium and narrow-spread type of distribution respectively. With the lamp filament located exactly at the focal point of the reflector, we obtained the readings at the Illuminating Engineering Society's test-points which are stated in Table 1, the distribution of the light being shown in Fig. 2 for the respective lenses.

The figures are shown graphically on each chart by the wide solid-columns. At the side of these columns are shown one-half the Illuminating Engineering Society's specifications in cross-hatched columns, the maximum and the minimum values being shown by the top and the bottom of the columns respectively. Where no maximum value is specified, the top of the column is filled-in solid. The lamp was then moved back of the focal point, in steps of 1/64 in. at a time, and readings were taken at the test-points. These results are shown in Table 2 and in the charts reproduced in Fig. 3 for the respective lenses.

It is evident from the charts that a very slight movement of the light source, back of the focus, causes the beam to exceed the maximum limits at points B, C and A and, with lens B, the minimum value required at point Qr is just barely met. The lamp was then returned to the focal point and moved ahead of it 1/64 in. at a time, with the results given in Table 3.

The charts reproduced as Fig. 4 show that moving the light source ahead of the focus also exceeds the maximum values at points B, C and A, but the source has to be moved somewhat farther ahead than behind the focus before this results. Apparently, lens B is the least sensitive of the three in this respect, as the maximum is exceeded only at point D and here only when the source is 4/64 in. ahead of the focal point. Returning again to the focal point, the lamp was moved vertically above the focus, with the results shown in Table 4 and Fig. 5, for the respective lenses.

Moving above the focus, of course, throws the beam down, and the converse is true when the filament is moved below the focus. The latter movement, however, is more serious, as it increases the candlepower at the glare point very rapidly. The charts of Fig. 5 show

TABLE 1—LAMP IN FOCUS

Test-Point	Bausch & Lomb Lens	Patterson Lens	Osgood Lens
D	393	393	393
C	437	525	812
A	2,187	1,437	1,563
B	8,125	7,187	7,500
Pl	14,680	10,810	3,875
Pr	15,437	10,810	3,437
Ql	4,062	2,125	3,125
Qr	4,375	1,500	3,500

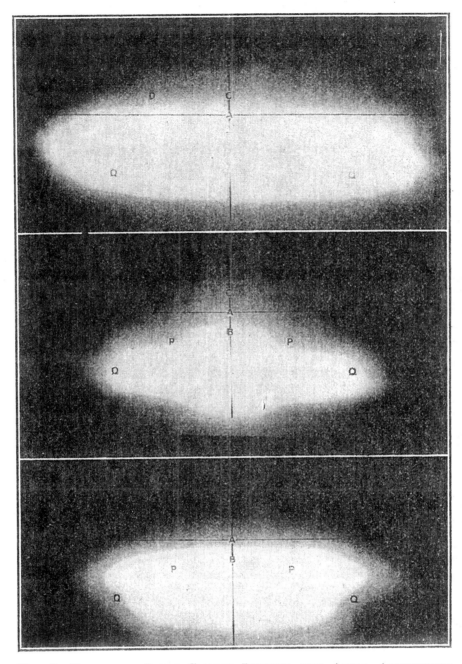

Fig. 2—Effect of Light-Center Length and Axial Alignment
on Three Types of Lens

Wide, Medium and Narrow-Spread Types of Light Distribution Are
Provided Respectively by the Bausch & Lomb Lens Used To Obtain
the Top View, the Patterson Lens Employed for the Middle View
and the Osgood Lens for the View at the Bottom. With the Lamp
Filament Located Exactly at the Focal Point of the Reflector, the
Readings Stated in Table 1 Were Obtained at the Illuminating
Engineering Society's Test-Points

TABLE 2—LAMP MOVED BACK OF THE FOCAL POINT IN 1/64-IN. STEPS

Test-Point	1/64 In.	2/64 In.	3/64 In.	4/64 In.	5/64 In.
Bausch & Lomb Lens					
D	606	1,062	1,750	2,375	2,875
C	606	1,187	1,750	2,437	3,125
A	2,625	3,750	5,000	5,375	5,875
B	10,000	8,750	9,375	7,812	6,875
Pl	12,350	11,562	9,375	7,500	5,625
Pr	11,562	12,350	9,687	7,812	5,625
Ql	4,000	4,625	4,625	4,375	3,750
Qr	3,875	4,125	4,062	3,875	3,687
Patterson Lens					
D	482	581	875	1,562	2,125
C	625	706	1,188	2,062	3,062
A	2,375	3,062	4,500	5,562	4,687
B	7,500	9,375	10,000	10,000	9,375
Pl	10,000	11,562	9,250	8,750	5,687
Pr	10,800	10,000	10,000	9,062	6,000
Ql	1,625	1,437	1,375	1,312	1,312
Qr	1,375	1,312	1,188	937	1,125
Osgood Lens					
D	412	462	481	537	631
C	937	875	1,062	1,500	1,875
A	2,187	2,500	3,187	3,875	4,375
B	5,625	6,562	6,062	6,250	6,062
Pl	2,750	3,250	2,812	3,125	3,062
Pr	2,625	3,000	2,687	3,062	3,062
Ql	3,750	3,875	2,875	2,687	2,062
Qr	3,250	3,125	2,687	2,437	1,875

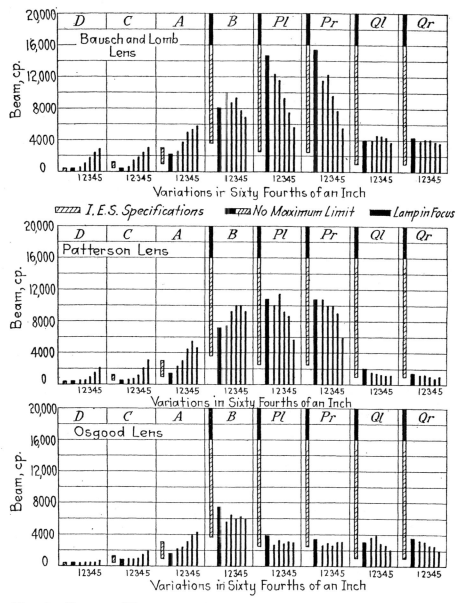

FIG. 3—EFFECTS WHEN THE LAMP IS MOVED BACK OF THE FOCAL
POINT

Moving the Lamp Back of the Focal Point in Steps of 1/64 In.
Produced the Readings Illustrated in the Above Charts for the
Respective Lenses, the Results Being Stated Also in Table 2

TABLE 3—LAMP MOVED AHEAD OF THE FOCAL POINT IN 1/64-IN. STEPS

Test-Point	1/64 In.	2/64 In.	3/64 In.	4/64 In.	5/64 In.
			Bausch & Lomb Lens		
D	406	506	937	1,187	1,562
C	531	687	1,000	1,312	1,875
A	2,125	2,062	3,937	3,750	4,062
B	7,812	8,125	8,750	8,750	7,812
Pl	15,437	13,125	16,187	13,125	10,812
Pr	16,187	13,900	13,125	13,900	12,350
Ql	4,625	3,750	4,000	3,750	4,000
Qr	4,125	4,625	4,375	4,000	4,625
			Patterson Lens		
D	368	425	387	462	687
C	493	581	593	687	1,062
A	1,187	1,312	1,625	2,187	2,937
B	6,562	5,562	6,437	6,250	6,125
Pl	8,500	9,250	8,125	7,812	9,062
Pr	9,375	10,000	7,812	6,437	7,500
Ql	2,875	2,875	2,937	3,687	4,000
Qr	1,937	2,437	3,000	3,187	3,437
			Osgood Lens		
D	306	375	419	494	556
C	750	937	1,187	1,625	2,250
A	1,687	2,312	2,812	4,437	5,625
B	5,125	7,187	7,500	8,750	8,437
Pl	2,375	3,500	3,812	4,000	4,625
Pr	2,812	3,187	2,937	3,875	4,125
Ql	2,500	2,875	2,625	2,437	2,187
Qr	3,437	3,312	3,125	2,750	2,750

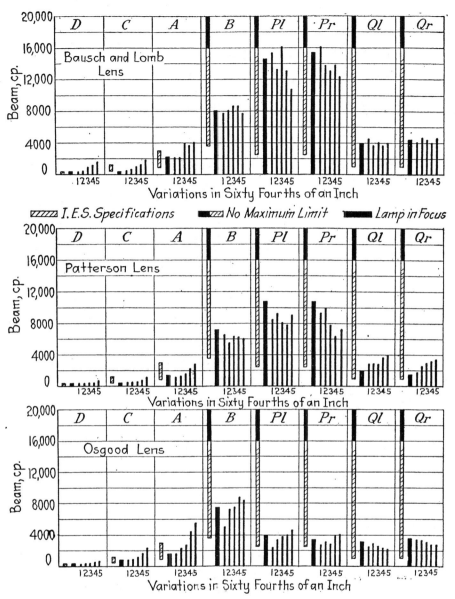

FIG. 4—EFFECTS WHEN THE LAMP IS MOVED AHEAD OF THE FOCAL
POINT

The Charts Show That Moving the Light Source Ahead of the Focal
Point Causes the Maximum Value at the Test-Points To Be Ex-
ceeded, But That the Light Source Must Be Moved Somewhat
Farther Ahead of Than Behind the Focus before This Results

TABLE 4—LAMP MOVED VERTICALLY ABOVE THE FOCAL POINT IN 1/64-IN. STEPS

Test-Point	1/64 In.	2/64 In.	3/64 In.	4/64 In.	5/64 In.
		Bausch & Lomb Lens			
D	375	344	312	269	281
C	425	387	375	325	368
A	1,625	1,625	1,437	1,125	1,125
B	6,250	4,125	3,937	3,000	2,812
Pl	8,750	6,250	5,000	4,062	3,437
Pr	9,375	6,875	4,875	4,375	3,875
Ql	5,625	7,500	8,437	6,875	6,125
Qr	5,750	7,500	8,125	6,875	6,125
		Patterson Lens			
D	375	306	281	275	268
C	544	406	369	375	331
A	1,187	875	750	800	681
B	5,062	2,937	3,250	3,062	2,500
Pl	6,750	5,187	4,375	3,812	3,125
Pr	8,062	5,625	4,750	3,937	3,312
Ql	2,937	3,375	3,500	3,062	2,437
Qr	2,750	2,812	2,937	2,312	2,062
		Osgood Lens			
D	344	294	250	225	225
C	812	687	606	556	531
A	937	1,125	1,062	875	875
B	4,875	3,000	3,125	2,500	2,687
Pl	2,125	1,750	1,500	1,375	1,312
Pr	1,937	1,625	1,500	1,125	1,312
Ql	2,312	2,312	2,312	1,375	1,187
Qr	2,812	2,312	1,625	1,250	1,250

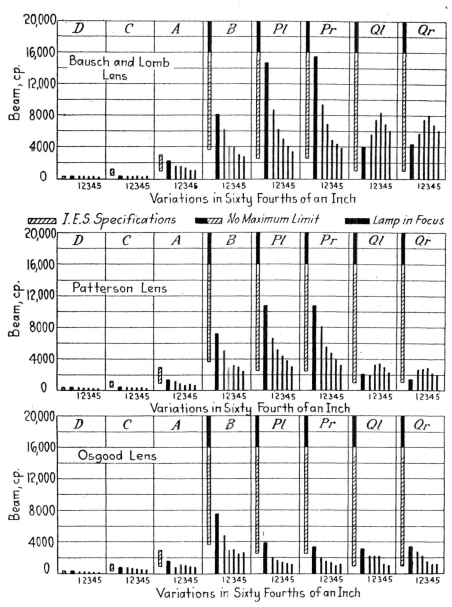

FIG. 5—EFFECTS WHEN THE LAMP IS MOVED VERTICALLY ABOVE THE
FOCAL POINT

Movement of the Light Source Above the Focus in Steps of 1/64
In. Produced the Results Shown in the Above Charts

TABLE 5—LAMP MOVED VERTICALLY BELOW THE FOCAL POINT IN 1/64-IN. STEPS

Test-Point	1/64 In.	2/64 In.	3/64 In.	4/64 In.	5/64 In.
Bausch & Lomb Lens					
D	531	937	1,750	2,812	5,375
C	606	1,312	2,010	3,750	6,250
A	3,125	5,937	9,375	12,350	12,350
B	12,500	18,615	17,750	13,900	10,812
Pl	12,294	17,000	12,350	9,250	6,875
Pr	20,170	17,000	10,000	10,000	6,250
Ql	2,625	2,437	2,125	1,875	1,875
Qr	3,437	2,375	2,187	1,750	1,750
Patterson Lens					
D	519	750	1,062	2,125	4,000
C	606	1,062	1,687	3,125	6,000
A	1,937	4,375	6,437	10,312	10,625
B	10,000	14,600	15,437	14,662	11,575
Pl	13,900	13,125	11,562	11,062	10,000
Pr	13,900	14,600	13,125	10,812	9,250
Ql	1,875	1,312	1,062	1,187	1,000
Qr	1,062	750	1,062	750	937
Osgood Lens					
D	462	531	644	713	937
C	1,187	1,312	2,062	2,812	3,687
A	2,625	3,937	5,000	5,812	7,750
B	9,625	10,375	13,125	17,000	15,437
Pl	5,312	6,875	9,250	9,437	9,250
Pr	3,812	6,437	8,750	10,437	9,062
Ql	3,312	2,687	1,562	937	937
Qr	3,875	2,437	1,437	937	1,000

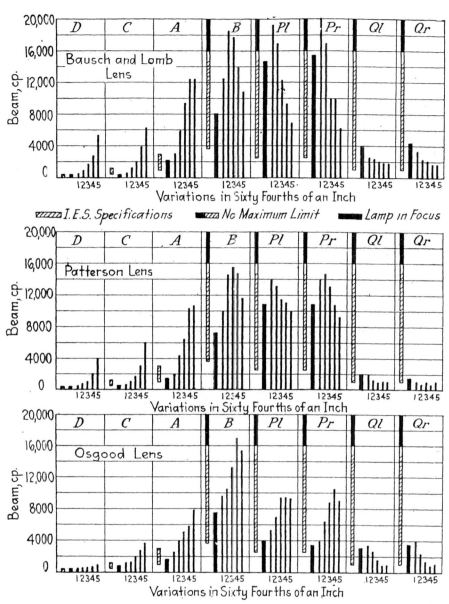

FIG. 6—EFFECTS WHEN THE LAMP IS MOVED VERTICALLY BELOW THE FOCAL POINT

Movement of the Light Source Below the Focus in Steps of 1/64 In. Produced the Results Shown in the Above Charts for the Respective Lenses

the values as the filament goes up. Once again the lamp was returned to the focus, and then moved vertically downward to obtain the data given in Table 5 and Fig. 6 for the respective lenses.

The charts of Fig. 6 show that if the filament is moved but 1/64 in. below the focus, the glare value at point D is exceeded. Again returning to the focus, the lamp was moved sideways from the focal point in 1/64-in. steps. The results are shown in Table 6.

Moving the filament sideways does not affect either the glare-point values or those at A and B materially, but it does shift the entire beam sideways.

To determine what happened with a lamp that just came within the lamp manufacturers' present maximum-variation tolerance of 3/64 in., as to both light-center length and axial alignment, the lamp was returned to the focal point. Then it was moved 3/64 in. behind the focal point, and then 3/64 in. vertically below that position.

TABLE 6—LAMP MOVED SIDEWAYS FROM THE FOCAL POINT
IN 1/64-IN. STEPS

Test-Point	1/64 In.	2/64 In.	3/64 In.	4/64 In.	5/64 In.
		Bausch & Lomb Lens			
D	363	406	406	363	438
C	469	469	456	531	556
A	1,875	1,875	1,937	2,125	2,812
B	7,812	7,500	8,125	8,437	7,812
Pl	11,575	11,575	10,800	9,275	9,375
Pr	11,575	11,575	10,800	11,575	10,000
Ql	3,437	4,125	4,000	3,187	3,250
Qr	4,875	6,250	4,750	6,250	6,250
		Patterson Lens			
D	363	369	363	325	306
C	525	400	469	531	512
A	1,375	1,562	1,562	1,875	2,010
B	5,750	7,312	7,187	7,875	6,375
Pl	9,062	9,625	9,125	7,687	6,250
Pr	10,500	11,575	11,250	10,812	10,625
Ql	1,250	1,125	875	812	656
Qr	2,375	3,500	5,375	6,437	7,187
		Osgood Lens			
D	419	319	356	312	294
C	937	937	1,062	937	875
A	2,062	2,250	2,010	1,937	2,010
B	6,625	7,187	6,562	5,312	5,062
Pl	2,875	2,750	3,437	2,812	2,562
Pr	3,187	4,375	6,562	8,187	9,187
Ql	3,437	2,687	2,812	2,062	1,562
Qr	3,625	4,687	5,500	5,562	5,312

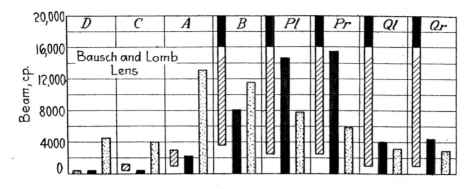

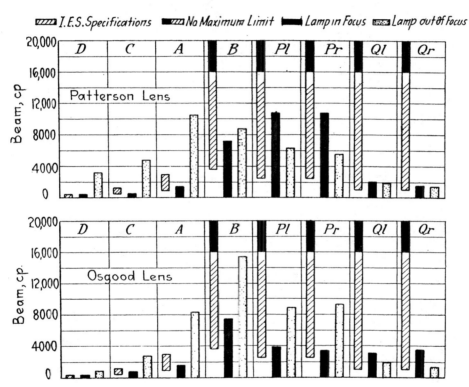

FIG. 7—EFFECTS OF MOVING THE LAMP BACK OF AND BELOW THE
FOCAL POINT

When the Lamp Was Moved 3/64 In. Behind the Focal Point and
Then Was Moved 3/64 In. Below That Position, Not Only Were the
Glare Limits Greatly Exceeded But the Road Illumination Was
Reduced Seriously. This Is Illustrated in Fig. 8 and the Other
Data Are Presented in Table 7 for the Respective Lenses

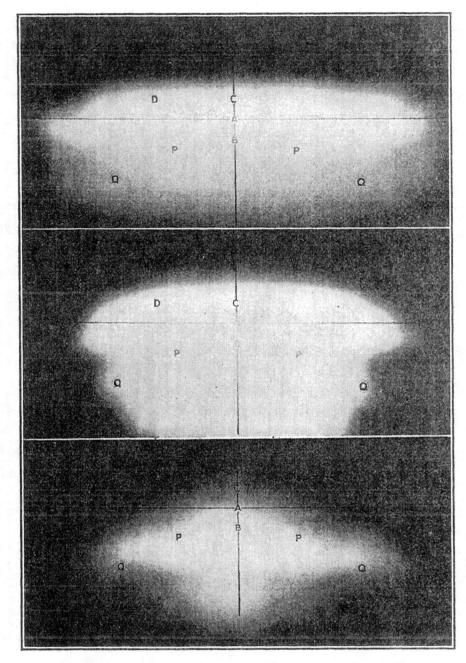

FIG. 8—ILLUMINATION PRODUCED BY MOVING THE LAMP BACK OF AND BELOW THE FOCAL POINT

Movement of the Light Source 3/64 In. Behind and 3/64 In. Below the Focal Point Gave the Results Shown in the Top View for the Bausch & Lomb Lens, in the Middle View for the Patterson Lens and in the Bottom View for the Osgood Lens. Other Information Pertaining to This Procedure Is Presented in Fig. 7 and in Table 7

TABLE 7—LAMP MOVED 3/64 IN. BEHIND THE FOCAL POINT
AND THEN 3/64 IN. BELOW THAT POSITION

Test-Point	Lens Bausch & Lomb	Patterson	Osgood
D	4,500	3,187	937
C	4,000	4,750	2,812
A	13,125	10,562	8,250
B	11,575	8,750	15,437
Pl	7,812	6,250	8,875
Pr	5,875	5,625	9,312
Ql	3,250	1,875	1,875
Qr	2,875	1,437	1,250

The results are shown in chart form in Fig. 7 for the respective lenses. Not only are the glare limits greatly exceeded, but the road illumination is seriously reduced as shown in Fig. 8. The results are presented in Table 7 for the respective lenses.

We often had heard it stated that if a lamp were out of focus, all that needed to be done to make a good driving light of it was to tilt the head-lamp until the beam ceased to be excessively glaring, taking the Illuminating Engineering Society's limit of 800 cp. as that condition. To determine this, the lamp was again returned to focus and moved back of the focal point 1/64 in. at a time, but at each setting of the beam it was tilted down until it came within the Illuminating Engineering Society's specification at point D, or rather one-half of it, as only one lamp was used. The measurements taken are shown in Table 8 and Fig. 9 for the respective lenses.

The charts show that, by this procedure, legal beams can be obtained, but at considerable loss in road illumination, except in the case of lens C, by which the intensity on the center of the road is increased somewhat. Next, the same procedure was repeated, but by moving the lamp below the focus instead of back of it. The results are shown in Table 9. In this case, the results are fairly good for lenses A and B, but poor for lens C, the minimum values at points B and P not being complied with.

At the same time that the photometric readings were taken, measurements were made of the height and width of the beams, and their relative positions, or shift of positions from the normal, obtained with the lamp at the focus. This was done by making a horizontal reference-line on the screen at the same height as the center of the headlight, and a vertical line passing through the axis

TABLE 8—LAMP MOVED BACK OF THE FOCAL POINT IN 1/64-IN. STEPS, BUT TILTED TO COMPLY WITH ILLUMINATING ENGINEERING SOCIETY'S SPECIFICATIONS AT POINT D

Test-Point	1/64 In.	2/64 In.	3/64 In.	4/64 In.	5/64 In.
Bausch & Lomb Lens					
D	394	394	394	394	394
C	531	406	406	394	406
A	2,187	1,875	1,562	1,562	1,562
B	6,875	6,250	4,750	4,125	3,750
Pl	10,812	7,500	6,187	5,937	4,625
Pr	10,000	7,187	5,625	5,562	5,000
Ql	6,875	6,875	4,875	3,875	3,750
Qr	6,500	6,875	5,562	4,500	3,687
Patterson Lens					
D	394	394	394	394	394
C	500	513	475	506	519
A	1,625	1,687	1,562	2,187	2,010
B	6,875	5,937	6,250	5,437	5,625
Pl	10,000	9,562	7,500	6,875	6,187
Pr	10,000	9,187	8,625	6,625	6,500
Ql	1,812	2,062	2,187	2,312	2,250
Qr	1,750	1,812	2,375	2,750	2,875
Osgood Lens					
D	394	394	394	394	394
C	875	1,000	1,000	1,125	1,250
A	1,750	1,937	2,562	3,062	3,000
B	4,687	5,125	6,375	6,875	7,500
Pl	2,687	3,062	3,250	3,125	3,562
Pr	1,875	2,437	2,562	2,937	3,250
Ql	4,000	4,125	3,312	2,750	2,125
Qr	3,312	3,375	2,500	2,312	1,750

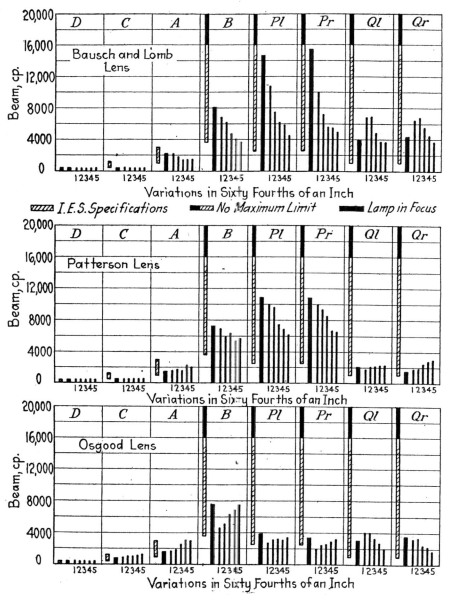

FIG. 9—EFFECTS OF MOVEMENT OF THE LAMP WHEN TILTED

The Lamp Was Moved Back of the Focal Point in Steps of 1/64 In. But Was Tilted To Comply with the Illuminating Engineering Society's Specification at Test-Point *D*

TABLE 9—LAMP MOVED BELOW THE FOCAL POINT IN 1/64-IN. STEPS, BUT TILTED TO COMPLY WITH THE ILLUMINATING ENGINEERING SOCIETY'S SPECIFICATIONS

Test-Point	1/64 In.	2/64 In.	3/64 In.	4/64 In.	5/64 In.
		Bausch & Lomb Lens			
D	394	394	394	394	394
C	413	469	506	438	456
A	1,812	2,125	2,125	1,562	1,687
B	8,750	9,687	10,800	8,437	6,500
Pl	16,200	15,125	17,000	12,350	10,000
Pr	17,000	14,650	15,437	13,125	9,275
Ql	4,375	5,500	3,750	5,250	5,875
Qr	4,500	4,125	5,375	5,750	6,562
		Patterson Lens			
D	394	394	394	394	394
C	512	506	531	531	525
A	1,437	1,250	1,312	1,500	1,625
B	8,750	5,687	6,187	5,312	4,937
Pl	11,575	10,000	9,375	6,500	6,562
Pr	13,900	9,687	8,437	8,125	6,125
Ql	1,500	2,062	2,375	2,187	3,062
Qr	1,562	1,750	2,312	2,687	2,687
		Osgood Lens			
D	394	394	394	394	394
C	937	937	937	750	812
A	1,562	1,375	1,437	1,187	1,187
B	4,250	3,125	3,437	1,937	2,250
Pl	2,062	1,937	1,937	1,375	1,562
Pr	1,875	1,562	1,562	1,000	1,250
Ql	4,500	3,500	4,187	2,437	2,500
Qr	3,750	3,500	5,000	1,187	1,562

of the head-lamp. Linear measurements were then made to the top, bottom and edge of the beams from these reference lines. The results are given in Table 10.

FILAMENT LOCATION IN SPOTLIGHTS

A study was then made to determine the effect of filament location on spotlights. For this purpose, the standard test-reflector was masked-down to an opening of 4½ in. to be comparable with spotlights. We realize that some spotlights have shorter focal-lengths than the 1¼-in. focal-length of our test reflector. It was felt, however, that this would have very little bearing on the test results, distortion of beam and the like, although it might add slightly to the spread of the beam. It was observed that a dark area appeared in the beam as the lamp was thrown out of focus sufficiently; however, these areas were only relatively dark, having about 20,-000 cp. directed to them. Candlepower measurements were taken in this area, as well as in the brightest portion of the beam, wherever that happened to come. As the filament was moved sideways from the axis, the beam became elliptical; hence, both maximum and minimum diameters were measured, as well as the distance the center of the beam was thrown off from the axis at 25 ft. A test was made also to show the effect of a lamp having its filament at the limits of present manufacturing tolerances; that is, 3/64 in. off the axis and 3/64 in. short in light-center length. Short, rather than long light-center length was chosen, because the distortion is greater from short light-center length than from long light-center length. The results of these tests are shown in Table 11.

To study the effect on the beam of a spotlight having a light source made with a filament wound on a large mandrel, No. 15, outside coil-diameter 0.029 in. and length 0.085 in., as compared to one wound on a small mandrel, No. 10, outside coil-diameter 0.024 in. and length 0.098 in., horizontal and vertical distribution-curves were obtained using a lamp of each type. These showed a beam diameter at 25 ft. of 21 in. for the large mandrel and 26 in. for the small mandrel. The maximum beam intensities were 88,000 and 75,500 cp. respectively. Figs. 10 and 11 respectively show the results obtained. The two humps in the horizontal curve are caused by the two filament-coils, each located close to the focal point of the reflector.

TABLE 11—SPOTLIGHT WITH LARGE-MANDREL LAMP-FILAMENT

Position	In.	Maximum Reading, Cp.	Dark Spot Reading, Cp.	Maximum Diameter at 25 Ft., In.	Minimum Diameter at 25 Ft., In.	Distance Center of Beam Is Thrown Sideways, In.
Lamp in Focus		86,906	.....	21	21	0
Ahead of Focus	1/64	72,562	.....	23	23	0
	2/64	54,800	34,731	29	29	0
	3/64	42,450	21,594	32	32	0
	4/64	26,250	7,500	37	37	0
	5/64	16,206	3,125	41	41	0
Back of Focus	1/64	63,837	.....	26	26	0
	2/64	42,450	.....	31	31	0
	3/64	20,837	5,875	36½	36½	0
	4/64	12,812	1,625	40	40	0
	5/64	11,875	.....	47	47	0
Sideways from Focus	1/64	77,187	.....	22	20	1½
	2/64	80,275	41,684	26	21½	3½
	3/64	77,187	.....	29	24	6
	4/64	61,750	.....	31	26	7½
	5/64	43,225	.....	34	26	10
3/64 In. Sideways from and 3/64 In. Back of Focus		34,733	5,500	42	37	5

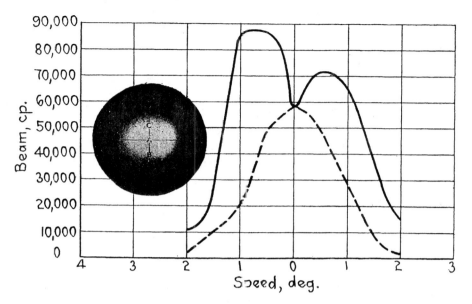

FIG. 10—EFFECTS OF FILAMENT SIZE ON SPOTLIGHT BEAMS
Curves and the Light Spot Obtained from the Large-Mandrel Lamp-
Filament Are Shown

Some claims had been made that a material difference
in the height or cut-off of the beam resulted from putting
the plane of the filament in the headlight reflector vertically, rather than in the usual horizontal position. This
was tested-out with the large and with the small-mandrel
filaments. The tests showed that the mandrel of larger
diameter producing a shorter stockier coil gives the more

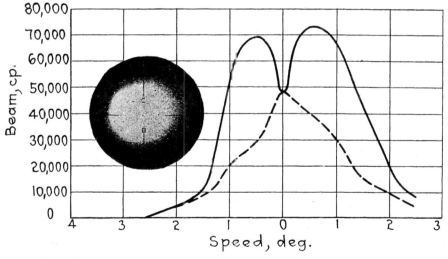

FIG. 11—EFFECTS OF FILAMENT SIZE ON SPOTLIGHT BEAMS
Curves and the Light Spot Obtained from the Small-Mandrel Lamp-
Filament Are Shown

concentrated beam. As a matter of interest, the beam diameters were also measured with the lenses removed. Table 12 gives the results of these tests.

In some cases Table 12 shows a greater spread for the horizontal than for the vertical filament-position. This does not seem logical, and our general conclusions are that, due to the difficulty of determining the exact edge of the beam, we can, for all practical purposes, consider that the difference is little or none and certainly wide variations of the relation of the plane of the base pins to that of the filament will not affect the resultant beam materially.

EFFECT OF AUXILIARY BULB IMAGE

To show clearly the effect of reflected or auxiliary bulb image, and how this is broken-up by the new corrugated bulbs, the photographs reproduced in Figs. 12 and 13 were made. In Fig. 12, the view at the extreme left shows a smooth bulb with the reflected image to one

TABLE 12—HEADLIGHT-BEAM MEASUREMENTS

Type of Filament	Filament Position	Distance from Top of Beam to Reference Line, In.	Distance from Bottom of Beam to Reference Line, In.	Total Vertical Spread at 25 Ft., In.
Bausch & Lomb Lens				
Large Mandrel	Horizontal	4½	24	28½
	Vertical	2	20	22
Small Mandrel	Horizontal	4	25	29
	Vertical	4	25	29
Patterson Lens				
Large Mandrel	Horizontal	1	25	26
	Vertical	4½	24	28½
Small Mandrel	Horizontal	8	25	33
	Vertical	2½	24½	27
Osgood Lens				
Large Mandrel	Horizontal	4½	24	28½
	Vertical	4½	27	31½
Small Mandrel	Horizontal	3	24	27
	Vertical	3	28½	31½

Size of Spots without Lenses
Large Mandrel, 22½ In.
Small Mandrel, 24½ In.

side of the filament, and the adjacent left-central view shows the resultant spot. The central view shows a smooth bulb with the reflected image ahead of the filament, and the adjacent right-central view shows the resultant spot. The view at the extreme right shows the bulb image superimposed on the filament. In Fig. 13, the left view shows the spot when the corrugations on the bulb are too shallow and the strong light from the image is not completely diffused. The middle view shows the spot from a bulb in which the corrugations are too deep, each acting as a small lens. The view at the right shows the spot with a bulb image diffused by corrugations of correct proportions.

SUMMARY

Summarizing all the tests, we can say that they show clearly that the control of the flood of light from a headlamp can be likened to the control of a gallon of water. It can be concentrated into a narrow, powerful beam, corresponding to putting the gallon of water into a deep vessel of small diameter or it can be spread out as if the water were poured into a large shallow pan. When light is taken from one point, it appears somewhere else. The area that it is to illuminate and the intensity of light on that area are controllable as are the depth of the water and the area it will cover. The control of the light depends upon accurately made equipment with light source, reflector and lens held rigidly in *exactly* the correct relation with one another. The tests indicate that a precision of $\pm 2/64$ in. in the position of the lamp filament, with respect to the focal point of the reflector, must be maintained to secure good all-round illumination. This is not only the bulb-manufacturers' problem in producing filaments located more accurately with respect to the lamp bases, but also the problem of manufacturers of lamps in producing sockets, focusing devices and reflectors that will "stay put" within $\pm 2/64$ in. The car-builders' problem is to mount the head-lamps accurately and the lens-manufacturers' job is to maintain accurate lenses. All this will be costly, but if the driving public is asked if it will pay \$5 more for a car having really good head-lamps, the answer "Yes" will be shouted back.

In conclusion, we wish to express our thanks to F. W. Brehme, who did the photometric work.

THE DISCUSSION

R. N. FALGE[4]:—A comment that strikes me as being pertinent with reference to Mr. Porter's interesting and well-arranged data is that they apply to devices which do not include representatives of the modern trend of headlight design. In other words, the art has already advanced well beyond the conditions he suggests.

In another discussion[5], I have pointed out several principles of design, incorporated in a variety of modern equipments, which render the beam pattern far less sensitive to commercial variations in the optical elements of the head-lamp. The chief one is that of forming the upper part of the beam from the middle transverse section of the unit and tilting the light from the upper and the lower sections downward. The value of this compensating feature in minimizing the effects of variations ahead of and back of focus is indicated by a comparison of typical data from Mr. Porter's Tables 1, 2, 3, and 8 with the data on a compensated device, as shown in Table 13 herewith. The difference is most pronounced. Sidewise variations, as given in Mr. Porter's Table 6, are relatively unimportant in practice and hence are not included here.

It appears to be the general opinion of those who have studied the servicing of headlighting that, if focusing could be eliminated so that aiming would be the sole operation in adjustment, to bring about the cooperation of the industry, the trade, and the motorist in keeping headlamps adjusted generally would become relatively easy. Especially is this true with the mountings now used by many car builders that make aiming the lamps a very simple matter. The data of Table 13 point clearly toward the possibility of eliminating focusing when equipments are suitably designed and accurately made. The headlamp manufacturers have learned how to make and assemble reflectors and sockets accurately, and the filaments of incandescent lamps are practically all well within the 3/64-in. positioning tolerance of the S.A.E. Standard, with promise of further progress. The trend of the art with reference to the focusing matter is unmistakable.

Provision for the simple aiming of the head-lamps is, in any event, necessitated by the possibility of the dis-

---

[4] M.S.A.E.—Engineering department, in charge of automotive lighting, National Lamp Works of the General Electric Co., Cleveland.
[5] See p. 560.

TABLE 13—EFFECT OF FILAMENT VARIATIONS AHEAD AND BACK OF THE FOCUS ON (a) A TYPICAL LENS INCLUDED BY MR. PORTER AND ON (b) A COMPENSATED DEVICE

S.A.E. Test-Point	At Focus		Light-Source Position 4/64 In. behind Focus			4/64 In. ahead of Focus			S.A.E. Recommended Practice
	Device No. 1 Aimed	Device No. 2 Aimed	Device No. 1 Not Reaimed	Device No. 1 Reaimed	Device No. 2 Not Reaimed	Device No. 1 Not Reaimed	Device No. 1 Reaimed	Device No. 2 Not Reaimed	
D	786	460	4,750	788	575	2,374		750	0— 800
C	874	870	4,874	788	830	2,624	No	1,000	800—2,400
A	4,374	5,000	10,750	3,124	6,000	7,500	data	5,900	2,000—6,000
B	16,250	31,000	15,624	8,250	26,000	17,500	given	29,000	25,000+
P_l	29,360	10,800	15,000	11,874	17,300	26,250	by	11,200	10,000+
P_r	30,871	9,100	15,624	11,124	15,000	27,800	Mr.	10,600	10,000+
Q_l	8,124	7,300	8,750	7,750	5,200	7,500	Porter	5,550	4,000—8,000
Q_r	8,750	8,800	7,750	9,000	3,500	8,000		4,400	4,000—8,000

TABLE 14—EFFECT OF FILAMENT VARIATIONS ABOVE AND BELOW THE FOCUS ON (a) A TYPICAL LENS INCLUDED BY MR. PORTER AND ON (b) A COMPENSATED DEVICE

Light-Source Position

S.A.E. Test-Point	At Focus		3/64 In. below Focus		5/64 In. below Focus		3/64 In. above Focus		5/64 In. above Focus	
	Device No. 1 Aimed	Device No. 2 Aimed	Device No. 1 Reaimed	Device No. 2 Reaimed	Device No. 1 Reaimed	Device No. 2 Reaimed	Device No. 1 Reaimed	Device No. 2 Reaimed	Device No. 1 Reaimed	Device No. 2 Reaimed
D	786	500	788	600	798	600	No data given by Mr. Porter	700	No data given by Mr. Porter	600
C	874	1,000	1,012	1,200	912	1,400		1,400		1,300
A	4,374	6,000	4,250	6,000	3,374	6,000		6,000		5,000
B	16,250	27,700	21,600	25,000	13,000	21,400		24,400		15,400
P_l	29,360	13,950	34,000	14,400	20,000	15,700		12,900		11,000
P_r	30,874	15,100	30,874	16,200	18,550	15,700		11,700		11,600
Q_l	8,124	5,000	7,500	3,800	11,750	3,800		5,100		5,400
Q_r	8,750	5,000	10,750	4,200	13,124	4,500		5,600		5,800

placement of the complete head-lamps through force. Aiming alone also compensates for variation of the light source above and below the focus. The results for equipments of both the older and newer types are shown in Table 14. Obviously, the beam pattern remains acceptable over the range of modern manufacturing variations. As indicated above, in a well-made equipment to contemplate total variations of 5/64 in. is no longer necessary. On the other hand, it is also apparent that, with suitably designed equipments, the total variation in the device and within the bulb may safely be somewhat greater than the 2/64 in. indicated by Mr. Porter.

TABLE 10—HEIGHT, WIDTH AND RELATIVE POSITIONS OF BEAMS WITH LAMP AT THE FOCUS

Bausch & Lomb Lens

Position	In.	Edge of Beam Left of Reference Line Ft.	In.	Right of Reference Line Ft.	In.	Total Width at 25 Ft. Ft.	In.	Top of Reference Line Ft.	In.	Bottom of Reference Line Ft.	In.	Total Height at 25 Ft. Ft.	In.
Lamp in Focus		4	7	4	4	8	11	..	5	1	10	2	3
Back of Focus	1/64	4	8	4	4	9	0	..	5	2	½	2	5½
	2/64	4	11½	4	4	9	3½	..	6½	2	1½	2	8
	3/64	5	1	4	5	9	6	..	7½	2	4½	3	0
	4/64	5	5	4	5½	9	10½	..	11	2	7½	3	6½
	5/64	5	6	4	6½	10	½	1	3	2	10½	4	1½
Ahead of Focus	1/64	5	0	4	7	9	7	..	4	1	11½	2	3½
	2/64	5	3	4	9	10	0	..	7	2	2	2	9
	3/64	5	3	4	10	10	1	..	7½	2	4½	3	0
	4/64	5	3	5	½	10	5½	..	9½	2	5	3	2½
	5/64	5	8	5	½	10	6½	1	0	2	7½	3	7½
Above Focus	1/64	4	9½	4	5	9	2½	..	3	2	1	2	4
	2/64	5	½	4	5	9	5½	..	3	2	2	2	5
	3/64	5	2	4	5	9	7	..	4	2	3	2	7
	4/64	5	2	4	6	9	8	..	5	2	7	3	0
	5/64	5	2	4	6½	9	8½	..	5	2	11	3	4
Below Focus	1/64	4	9	4	4	9	1	..	4	2	1	2	5
	2/64	4	10	4	3½	9	1½	..	7	2	0	2	7
	3/64	5	1	4	3	9	4	..	9½	2	½	2	10
	4/64	5	1½	4	3	9	4½	1	1	2	1	3	2
	5/64	5	0	4	3	9	3	1	3½	2	2½	3	6
Sideways from Focus	1/64	4	10	4	5	9	3	..	3½	2	0	2	3½
	2/64	4	8	4	7	9	3	..	5	1	11	2	4
	3/64	4	6	4	9	9	3	..	6	1	11	2	5
	4/64	4	6	5	1	9	7	..	7	2	½	2	7½
	5/64	4	6	5	4	9	10	..	8	2	1	2	9
3/64 In. Back of and 3/64 In. Below Focus		4	10	4	6	9	4	..	11	2	5	3	4

Patterson Lens

Position	In.	Left of Reference Line Ft.	In.	Right of Reference Line Ft.	In.	Total Width at 25 Ft. Ft.	In.	Top of Reference Line Ft.	In.	Bottom of Reference Line Ft.	In.	Total Height at 25 Ft. Ft.	In.
Lamp in Focus		3	11	4	5	8	4	..	1½	2	6	2	7½
Back of Focus	1/64	3	11	4	3	8	2	..	3	2	7½	2	10½
	2/64	4	0	4	½	8	½	..	5	2	8	3	1
	3/64	4	4	4	7	8	11	..	7	2	10	3	5
	4/64	4	5½	4	7½	9	1	..	8	3	2	3	10
	5/64	4	5½	4	9	9	2½	1	0	3	4	4	4
Ahead of Focus	1/64	3	10	4	5	8	3	..	1½	2	5½	2	7
	2/64	3	11	4	5	8	4	..	0	2	4	2	4
	3/64	3	11	4	4	8	3	..	1½	2	4½	2	6
	4/64	3	10½	4	4½	8	3	..	4½	2	7½	3	0
	5/64	4	2	4	6	5	8	..	6	2	8	3	2
Above Focus	1/64	3	7	3	11	7	6	..	2	2	5½	2	7½
	2/64	3	9½	4	1	7	10½	..	2	2	4	2	6
	3/64	3	11	3	9½	7	8½	..	2	2	6	2	8
	4/64	3	11	3	6	7	5	..	2	2	9	2	11
	5/64	4	0	3	9	7	9	..	1	2	11	3	0
Below Focus	1/64	3	10	4	1	7	11	..	3	2	7	2	10
	2/64	4	3	4	2	8	5	..	5	2	8	3	1
	3/64	3	4	3	7	6	11	..	7	2	6	3	1
	4/64	3	5	3	6	6	11	..	8	2	5	3	1
	5/64	3	9	3	7	7	4	..	11	2	5	3	4
Sideways from Focus	1/64	3	7	4	2	7	9	..	2	2	5½	2	7½
	2/64	3	6	4	4	7	10	..	2½	2	6	2	8½
	3/64	3	4	4	6	7	10	..	3	2	6½	2	9½
	4/64	3	2	4	9	7	11	..	3	2	7	2	10
	5/64	3	0	5	1	8	1	..	4	2	11	3	3
3/64 In. Back of and 3/64 In. Below Focus		4	4	4	6	8	10	1	1	2	11	4	0

128

Osgood Lens

Lamp in Focus		3	2	3	2	6	4	..	3	2	2	2	5
Back of Focus	1/64	3	3	3	3	6	6	..	4	2	2	2	6
	2/64	3	5	3	4	6	9	..	6	2	3	2	9
	3/64	3	11	3	6	7	5	..	9	2	6	3	3
	4/64	3	7	3	7½	7	2½	..	9½	2	7½	3	5
	5/64	3	9	3	8	7	5	1	1	2	9	3	10
Ahead of Focus	1/64	4	4	3	9	8	1	..	6	2	6	2	10
	2/64	4	4	3	11	8	3	1	0	2	3	3	3
	3/64	4	1	3	10	7	11	..	9	2	9	3	6
	4/64	4	2	3	10	8	0	..	11	2	10	3	9
	5/64	4	8	4	2	8	10	..	11½	2	10½	3	10
Above Focus	1/64	4	0	3	5	7	5	..	5½	2	3½	2	9
	2/64	3	11	3	5	7	5	..	7	2	5	2	10
	3/64	4	3	3	7½	7	10½	..	8	2	10	3	6
	4/64	4	2	3	7	7	9	..	10	2	11	3	9
	5/64	3	11	3	7	7	6	..	9	2	10	3	7
Below Focus	1/64	3	10	3	6	7	4	..	4½	2	1½	2	6
	2/64	3	10	3	6	7	4	..	9	2	1	2	10
	3/64	4	3	3	4	7	7	..	9½	2	½	2	10
	4/64	4	3	3	7	7	10	..	10	2	1	2	11
	5/64	3	11	3	6	7	5	1	1½	2	3½	3	5
Sideways from Focus	1/64	3	4½	3	6	6	10½	..	3	2	4	2	7
	2/64	3	2½	3	8	6	10½	..	3½	2	4½	2	8
	3/64	3	0	3	10½	6	10½	..	4½	2	5½	2	10
	4/64	2	11	3	9	6	8	..	8	2	2	2	10
	5/64	2	10	3	10	6	8	..	8½	2	4½	3	2
3/64 In. Back of and 3/64 In. Below Focus		2	7	3	10½	6	5½	..	11	2	8	3	7

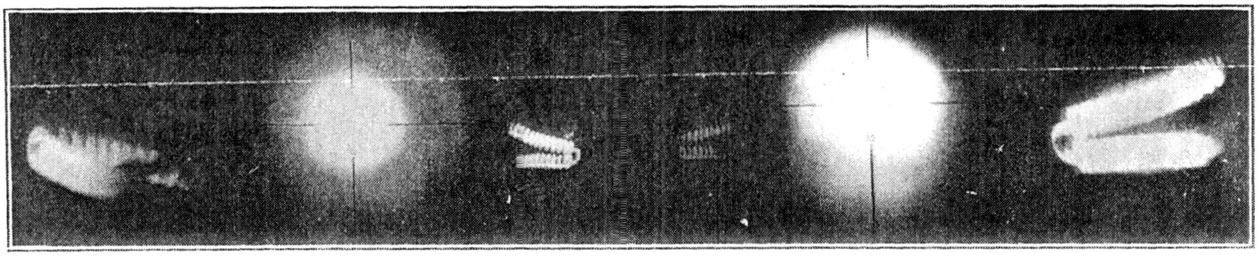

FIG. 12—EFFECTS OF AUXILIARY BULB IMAGE
The View at the Extreme Left Shows a Smooth Bulb with the Reflected Image to One Side of the Filament and the Adjacent Left Central View Shows the Resultant Spot. The Central View Shows a Smooth Bulb with the Reflected Image Ahead of the Filament and the Adjacent Right Central View Shows the Resultant Spot. The View at the Extreme Right Shows the Bulb Image Superimposed on the Filament

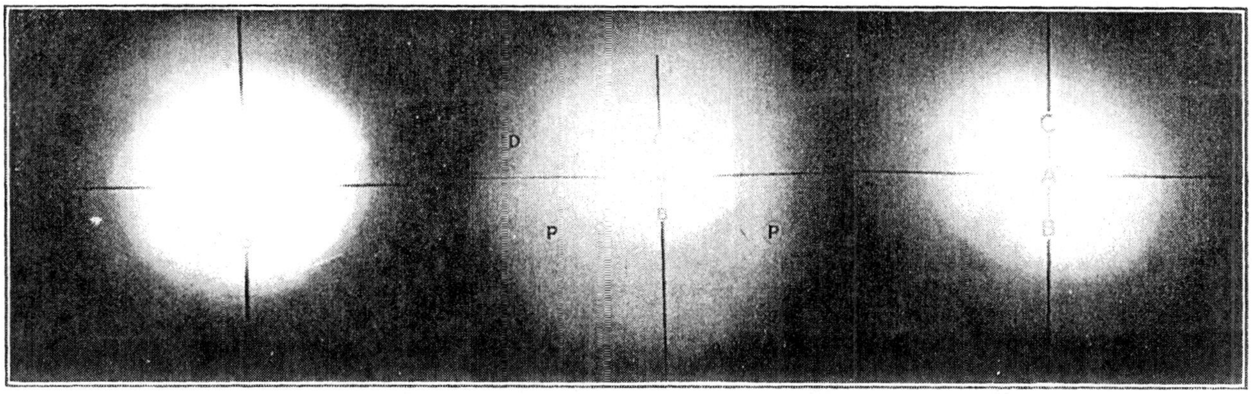

FIG. 13—EFFECTS PRODUCED BY CORRUGATED BULBS
The Left View Shows the Spot When the Corrugations on the Bulb Are Too Shallow and the Strong Light from the Image Is Not Completely Diffused. The Middle View Shows the Spot from a Bulb in Which the Corrugations Are Too Deep, Each Acting As a Small Lens. The View at the Right Shows the Spot with a Bulb Image Diffused by Corrugations of Correct Proportions

CHAPTER THREE

Hydraulic brakes

HYDRAULIC FOUR-WHEEL BRAKES FOR AUTOMOTIVE VEHICLES[1]

By Malcolm Loughead[2]

ABSTRACT

The author answers citations in disfavor of four-wheel brakes and compares external with internal brakes in favor of the external type. The hydraulic four-wheel brake system is illustrated and described in detail, and a statement is made of four-wheel-brake design requirements. These are: Reliability, equalization, stopping ability, control, simple adjustments, minimized service requirements and provision against wear. Each requirement is then sub-divided and given separate consideration, claims being made in regard to how well the hydraulic system meets each requirement, with reasons therefor. A summary of the main advantages of the hydraulic compared with the mechanical type of four-wheel brake is made in conclusion.

Whether automotive vehicles should or should not be equipped with brakes on the front as well as on the rear wheels is a subject of vital importance to the entire industry. The number of deaths caused by automotive vehicles in the United States alone is appalling, especially when we acknowledge that a very high percentage of these accidents could be eliminated if all automotive vehicles were provided with adequate and properly designed braking facilities. Most of our tests of four-wheel brakes have been made at a car speed of 30 m.p.h. but, at this speed, only a small percentage of present-day automobiles equipped with conventional two-wheel brakes can be stopped within a distance of 130 ft., even when these brakes are adjusted and lubricated properly.

DISADVANTAGEOUS FEATURES

Two points commonly cited against the usage of four-wheel brakes are the possibilities of locking the front wheels and of rear-end collisions caused by a too sudden stoppage of the car, with the failure of the car behind to

---

[1] Detroit Section paper

[2] Chief engineer, Hydraulic Brake Co., Detroit

328

stop likewise. If front-wheel brakes are not designed properly, they may lock the front wheels and cause a serious accident due, among other possibilities, to radical interference with the steering ability. My experience is that it is advisable to make the front and the rear brakes of the same design and size; otherwise, if the rear brakes are larger than those in front, there will be a sacrifice of stopping ability under dry-pavement conditions. If the front brakes are larger or more efficient than those at the rear, they will lock the front wheels, especially when the coefficient of friction of the pavement is low. No accidents due to the locking of front wheels of cars equipped with our four-wheel brakes have been reported.

Danger from rear-end collisions lies more in theory than in actuality. They have occurred in a few instances, but the resultant damage has been slight. Damage to lamps and radiators is more expensive than that done to rear tires and fenders, but the car that follows another usually runs 15 to 30 ft. behind and has that much additional space within which to stop.

EXTERNAL VERSUS INTERNAL BRAKES

With regard to the expansion of brake-drums on account of heat generated by friction, the external has a decided advantage over the internal type of brake. When the internal type is used on a long grade, when several successive and severe applications of it are made and especially when the foot-pedal adjustment is very close to the toe-board, the expansion due to heat may reach an amount such that the car is virtually without brakes although, with the same adjustment, the brakes may function perfectly when cold. The opposite condition obtains with the external type; the hotter the brake-drums become, the greater the available pedal-travel will be, a point often overlooked in emergency-brake design and one that can cause a serious accident if a car having hot brake-drums is left standing on a hill.

Oil leakage from the rear axle and over-greasing of front wheels commonly cause internal brake-shoes or bands to become saturated with oil and grease so that the car is practically without brakes, but this is much less likely to occur with the external type of brake. Distortion as well as expansion must be provided against for brakes of the internal-shoe type, and it is very difficult to attain as high a ratio of pedal movement to

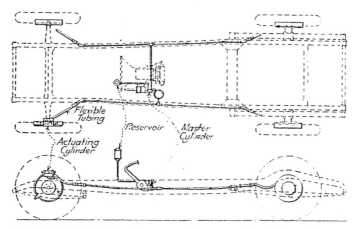

FIG. 1—LAYOUT OF THE LOCKHEAD HYDRAULIC FOUR-WHEEL BRAK-
ING SYSTEM SHOWING THE FLEXIBLE TUBING THROUGH WHICH OIL
IS FED TO THE ACTUATING CYLINDERS

brake-shoe movement for a given pedal pressure as can
be realized with the external type.

To design a brake-drum that will not distort is largely
a matter of weight and cost. My experience is that the
cost of a drum built rigidly enough to withstand the
pressure of internal shoes without becoming distorted
will be triple the cost of a pressed drum suitable for external brakes.

THE HYDRAULIC SYSTEM

Pressure in the hydraulic four-wheel-brake system is
built-up in a cylinder mounted on the transmission case
near the fulcrum of the brake-pedal. Fig. 1 shows the
layout, including the master cylinder, the piping and
the reservoir tank. A piston in this cylinder is connected directly to the lower end of the brake-pedal forg-
ing, as shown in Fig. 2, and is forced into the cylinder
by the depression of the brake-pedal itself. The ratio
of the movement of the pedal to that of the piston is
about 4 to 1. Four copper pressure lines or tubes lead
from the master cylinder to points on the chassis frame
adjacent to each of the four wheels, as indicated in Fig. 1.

Mounted rigidly on the dust shield or anchor bracket
at each one of the four wheels is a brake-band-actuating
cylinder that is connected to the pressure line on the
chassis frame by a suitable length of seven-ply rubber

hose capable of resisting a pressure of 2000 to 2500 lb. per sq. in. before bursting. A section of the hose assembly is shown in the lower portion of Fig. 3. The hose has a close-wound coil-spring inserted while the hose is under a pressure of 1200 to 1400 lb. per sq. in. There is practically no expansion loss up to a pressure of 1000 lb. per sq. in., this being a very important factor with hydraulic brakes because two movements, those of steering and of spring action, must be provided for by using flexible connections. An expansion loss in the hydraulic system causes a lessening of the amount of pedal travel; to compensate for this loss it is necessary to reduce the amount of pedal leverage and this would require a greater pedal pressure.

The copper tubing used for the pressure line on the chassis is supported by clips inside the channel-iron of the frame and terminates opposite each wheel in a standard union nut or coupling. We have never had a failure of brake action due to copper tubing when it was used as a non-flexible member. In some of our first installations we used a coil of copper tubing at each of the brake-bands to take-up the movement, but leaks developed due to the flexing of the tube. The present tubing is not allowed to flex, and I attribute the absence of tubing

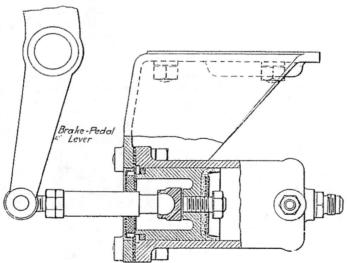

Brake-Pedal Lever

FIG 2—MASTER CYLINDER ASSEMBLY

trouble also to the support of the tubing by the union nut instead of by the flared end of the tube. I have had three failures of the gasoline pipe-line, and each showed a spiral crack at the base of the flare of the copper tubing.

BRAKE-ACTUATING CYLINDERS

Each of the individual brake-actuating cylinders on the four brake-supports contains a pair of opposed pistons, one at either end of the cylinder. The brake-actuating-cylinder assembly is shown in Fig. 3 also. These pistons act against two levers that are, in turn, connected to the brake-band ends, the brake-bands being of the external contracting type. Liquid is admitted to these brake-actuating cylinders between their opposed pistons through an opening at the cylinder centers. When the brake pedal is depressed, the opposed pistons in each cylinder are forced apart by the hydraulic pressure set-up in the system by the master cylinder; one end of each of the levers in each brake-actuating cylinder bears on the head of one of its opposed pistons and the other end exerts a pull on the brake-band end, drawing it tangent to the drum. Smooth brake action is assured by drawing the brake-band ends in a line tangent to the brake-drum and not allowing the ends to "snub." Since the opposed pistons are forced apart by hydraulic pressure, it follows that the pressure on both pistons in all four brake-actuating cylinders must be equal. The brake-actuating-cylinder group-assembly is shown in detail in Fig. 3.

LIQUID COMPOSITION

A mixture of alcohol and glycerine, 40 per cent alcohol and 60 per cent glycerine, constitutes the liquid used in the hydraulic braking system. Any leakage of liquid from the system is compensated for from a small reservoir mounted on the dash that replenishes the master cylinder automatically. When the master-cylinder piston is in the "off" position, it uncovers several small portholes, each about 0.02 in. in diameter, that communicate directly with the reservoir. This allows a free flow of liquid from the tank to the master cylinder and replenishes any loss of liquid that may occur during the application of the brakes. If there is any loss of liquid, it is very slight; in fact, we have run cars for several months with a closed line and have lost practically none of the fluid. Rawhide is used to pack the pistons and prevent any leakage of the fluid from the hydraulic cylinders.

FOUR-WHEEL-BRAKE DESIGN

In the design of a thoroughly satisfactory four-wheel braking-system, the following six factors are of prime importance:

(1) Reliability
(2) Equalization
(3) Stopping ability
(4) Control
(5) Simple adjustments and minimized service requirements
(6) Wear and replacement of parts

For reliability, three conditions are essential; (a) a high factor of safety throughout the system, (b) that all movable parts function as designed and that the least possible number of orifices for their lubrication be included and (c) that satisfactory performance be attained during all kinds of weather conditions.

A high factor of safety is provided in the flexible hose,

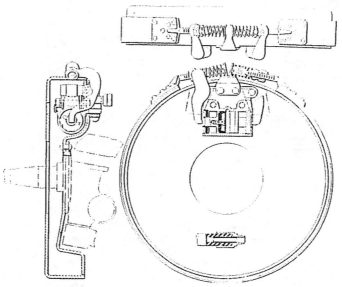

FIG. 3—WHEEL CYLINDER GROUP ASSEMBLY WITH A SECTION OF THE HOSE END SHOWN IN THE LOWER PORTION

the weakest part of the system. Its bursting pressure is 2000 to 2500 lb. per sq. in. and the maximum pressure to which it is subjected in an emergency stop is 300 lb. per sq. in. The pedal clevis-pin is the only movable part that requires lubrication; it is well protected under the floorboards and is accessible. The alcohol content of the alcohol-glycerine fluid can be increased for low-temperature conditions; hence, the system is independent of severe low temperatures.

EQUALIZATION

Proper equalization of four-wheel brakes demands (a) that the application of the brakes should have no effect on steering ability; (b) that it should prevent skidding; (c) that it should prevent locking of the wheels, especially the front wheels, and (d) that it should not reduce the braking effectiveness under varying road conditions.

Steering ability is not affected in the hydraulic system because the pressure of the liquid is the same in all parts of the system and perfect equalization obtains up to the brake-band ends. Variations in the coefficient of friction of the brake-linings, such as may occur when one front brake becomes wet, tend to make that brake more effective than another. By locating the center-line of the knuckle-pin so that its projection is ½ in., or less, from the center-line of the tire on the ground, the rise in the coefficient of friction of the saturated brake is not noticeable and no effect is transmitted to the steering-wheel even at high car-speeds.

As a means of preventing skidding, the four-wheel brake is efficient only to the degree of equalization attained. Positive equalization is secured up to the brake-band ends in the hydraulic system. This claim of positive equalization has been questioned by engineers who say that any great variation in the coefficient of friction of the brake-lining would offset the admitted equalization up to the brake-band ends. For the most part, such variations in the coefficient of friction result from heat generated during a long-continued brake-application and would affect each brake proportionately. Therefore, the change in the coefficient of friction cannot be said to affect equalization at each of the wheels but, rather, would result in requiring a greater pedal effort to accomplish a given result. Assuming that a change in the coefficient of friction is not distributed equally to the four wheels, it is none the less desirable and necessary

TABLE 1—TESTS OF STOPPING ABILITY OF HYDRAULIC
FOUR-WHEEL BRAKES

Model 57, Cadillac Phaeton Car—Speed, 30 m.p.h.	
Total Weight, with Driver, lb.	4,400
Weight on Front Wheels, lb.	2,400
Weight on Rear Wheels, lb.	2,000
Tires { Make	Mason Cord
Size, in.	35 x 5
Pressure, lb. per sq. in.	65
Road Surface	Dry Concrete

Stopping Distances; Average of Five Stops.	Ft.
Brakes { Rear Wheels Only	89.2
Applied { Front Wheels Only	67.6
on { Front and Rear Wheels	36.8

to maintain equalization in the operating mechanism, unless one argues that two wrongs make a right.

Some means to release the brakes somewhat, at the time the wheels stop rotating, is necessary to prevent locking of the front wheels without reducing braking effectiveness under varying road conditions. Some experiments have been made by others, and we have followed European attempts to accomplish this release of the brakes, one being to have each wheel equipped with one or more weights, the centrifugal force of which tends to release the brakes as the car speed decreases. But this is not commercially practicable and we have found no other way of incorporating the principle with a successful brake.

STOPPING ABILITY

Some engineers have advocated a longer rather than the quickest possible stop and have contended that abrupt stops will result in frequent rear-end collisions and much property damage, but a questionnaire answered by 1000 users of the hydraulic four-wheel brake reports very little damage of this character and proves it to be a matter of education. A car driver appreciates the security afforded by this type of brake but will not subject himself and his passengers unnecessarily to the inconvenience resulting from a very sudden stop.

Tests of the stopping ability of the hydraulic four-wheel braking-system were made recently, the results being stated in Table 1. A pistol was mounted rigidly on the chassis of the car; its trigger was connected to the brake pedal and set to fire a bullet into the road surface when the pedal had moved 1 in. After the car had been brought to a stop, the distance from the muzzle of the pistol on the car to the imprint of the bullet in the road surface was measured with a steel tape. The stopping distances stated in Table 1 are an average of five stops in each case, and all stops were made from a car speed of 30 m.p.h.

With a car fully loaded, our tests show that it is necessary to provide 1 sq. in. of brake lining for each 12.5 lb. of car weight. If the ratio of the car weight to the area of the brake-lining is increased to more than 16 lb., the surface of the lining may burn and cause a decrease in the coefficient of friction.

The ratio of 12.5 lb. of car weight to 1 sq. in. of brake-lining conforms fairly well with standard practice. Considering a car that carries seven passengers, weighs 5360 lb. and has 3200 lb. of this weight on its rear wheels, we find the average area of the brake-lining to be 225 sq. in., approximately; this gives 14.5 lb. of car weight on the rear wheels for 1 sq. in. of brake-lining. Owing to a shift of the center of gravity of the car when brakes are applied, this car weight per square inch of brake-lining area will be somewhat greater on the front wheels when the brakes apply to all four wheels or to the front wheels only.

CONTROL

Adequate brake control requires (a) light pedal-pressures, (b) positive release, (c) the minimum length of pedal stroke and (d) a constant pedal-pressure for any given stop.

Because of the low friction-loss in the hydraulic means of brake application, the high friction-loss of the mechanical hook-up is avoided and it is therefore possible to use only a moderate pedal-pressure without employing any servo-mechanism. A positive brake release is accomplished because there are no mechanical joints or bearings to freeze or seize. The brake members are returned positively to their normal position after the release of pressure by an adequate return spring. A minimum length of pedal stroke is permitted because there are no built-up losses of motion due to a multiplicity of mechanical joints. Each inch of pedal stroke is productive of work at the brake-band ends. The pedal pressure remains constant for any given stop, except as it may be affected by variations in the uniformity of the brake-lining and this, in our experience, is not a large factor, because of the effectiveness of the hydraulic actuating means.

SIMPLICITY OF ADJUSTMENT

Desirable adjustment features that minimize service requirements are (a) simplicity, (b) infrequency and (c) accessibility.

No adjustment for brake equalization is required by the hydraulic system; the inherent equalization in the actuating means is complete throughout so long as the friction members are functioning. Adjustments for wear will be infrequent if a liberal area of brake-lining is provided by the design, say 1 sq. in. of lining for each 12.5 lb. of car weight. Cars equipped with hydraulic brakes have been driven 8000 miles without adjustment.

Regardless of the simplicity of adjustment that is provided, many users will neglect adjustments unless they are made accessible, and adjustments should be of a character that do not require skilled service. The average owner-driver should be able to make them quickly and with certainty, and he can do this for hydraulic brakes.

WEAR AND REPLACEMENT

Other desirable factors of satisfactory brake performance are (a) an infrequent replacement of the brake-lining and the consequent removal of the vehicle from service and (b) that wear in the actuating mechanism be reduced to a minimum.

We have records of many cars in service that have hydraulic four-wheel brakes and have run from 25,000 to 48,000 miles without requiring a relining of their brakes. The wear in the hydraulic system can be said to be negligible and the mechanism should outlive the car. It is obvious that removing all necessity for sliding the rear wheels to make an emergency stop, which is accomplished by the use of four-wheel brakes, will result in saving the rear-wheel tires from wear.

ADVANTAGES OF THE HYDRAULIC SYSTEM

Inherent equalization is secured in the hydraulic system of braking. Exactly the same pressure is applied to each of the four brake-bands by utilizing a column of liquid under pressure against pistons acting in cylinders adjacent to the brake-band ends, and equalization is thus maintained regardless of brake-band adjustment.

Since there are fewer mechanical joints, the friction is lowered considerably. This advantage becomes pronounced after several months of service when, due to a lack of proper lubrication, brakes of the mechanical type show wear in some joints and seizing or binding in others.

The flexible connection between the brakes and the chassis prevents trouble due to the improperly related movement of a mechanical brake-linkage.

Brake adjustment is not affected by a change of loading in the car. The movement of the car body in relation to the axles when traveling over rough roads does not react against the brake pedal. This reaction does exist in some forms of mechanical brake and is particularly noticeable when coasting down a rough grade because the brakes then have a tendency to lock and release alternately, due to the movement of the car springs. The operation of the hydraulic system is very smooth on the roughest of roads.

Pedal-pressure requirements are moderate in the hydraulic system and, with a pedal stroke of 5 to 6 in. and a moderate pressure, the necessity of utilizing any servo-mechanism is eliminated by taking advantage of the low friction-loss, absence of lost motion and external-band brakes anchored to give approximately 200 deg. of brake-lining in contact with the brake-drum on the wrapping end and in the forward direction of rotation.

THE DISCUSSION

QUESTION:—How does the cost of hydraulic brakes compare with that of mechanical four-wheel brakes?

ANSWER:—That depends somewhat on the design of the mechanical brakes. To date we have seen no mechanical brakes that are equal in cost to hydraulic. We can build the toggle-joint system as used on some foreign cars for a little less.

QUESTION:—What about the design of front axles to which four-wheel brakes are applied, as regards the effect on the steering?

ANSWER:—Placing the intersecting lines of the knuckle ½ in. or less from the center-line of the tire does not affect the steering much. We have a car that is running now in which the line is ⅝ in. from the center of the tire; the car when traveling at a speed of from 65 to

100 m.p.h. could be stopped with one wheel completely shut off, practically without moving the steering-wheel.

QUESTION:—In case the operating fluid congeals or freezes in cold weather, what means of operating are provided?

ANSWER:—In extremely cold weather we run the alcohol content up the same as we should with radiators.

QUESTION:—When city ordinances require two sets of brakes, are independent mechanical means of operation provided?

ANSWER:—In that case we recommend an emergency brake, if for no other reason than locking the car when it is standing.

QUESTION:—Is there no emergency brake otherwise?

ANSWER:—None, except either the drive-shaft brake or the internal brake used in conjunction with the four outside brakes, which are operated by the hydraulic system.

QUESTION:—Are the rear brakes designed to take effect slightly ahead of the front brakes, on account of the greater weight at the rear end of the car?

ANSWER:—No; because even though there is more weight on the rear axle, in the ratio of 60 to 40 per cent, as, of course, we all know is the case, in stopping with four-wheel brakes the shifting of the center of gravity of the car makes the braking of the front wheels more effective than that of the rear wheels.

As an illustration of the shifting of the center of gravity, braking the front wheels of a car traveling at 30 m.p.h. will stop the car in 67 ft. while braking the rear wheels will stop it in 89 ft., which means a decided shifting of the weight; the rear wheels tend to slide before the front wheels even though more weight is at the rear.

QUESTION:—What would happen if a connection or a pipe should leak or break?

ANSWER:—The same thing that would happen if the threaded end of a brake-rod should break, a contingency that is not uncommon. Our system of piping has shown no breakage, and I should say that the percentage of failures would be no higher than that of the threaded ends of brake-rods. You would have to use a brake in either case, whether mechanical or hydraulic.

QUESTION:—Would there be an indicator on the reservoir to indicate that a leak had occurred in the system?

ANSWER:—No; if there were a leak in the system sufficient to cause a failure, the pedal would not remain in position but would drop to the floor-board when it was applied.

QUESTION:—Are the torsional strains on the axles greater or less when 32 x 7-in. tires are replaced by 35 x 5-in. tires?

ANSWER:—The coefficient of friction is the same whether it be distributed over 10 or over 2 sq. in.

QUESTION:—What effect has front-wheel camber on front-wheel brakes?

ANSWER:—It is taken into account only in designing to make the intersecting line of the knuckle-pin come within ½ in. of the center of the tire.

QUESTION:—Can you give the weight that was added to the Cadillac front axle in the test referred to?

ANSWER:—The axle itself was designed to take the torque load as well as the bending load of the weight of the car. The axle is 12 oz. lighter than the former Cadillac axle; I have no exact figures as to the weight of the drums and the rest of the mechanism. The drums weigh approximately 20 lb. each; the bands and the rest of the mechanism would weigh possibly 15 lb.; between 70 and 100 lb. on a car of that weight.

QUESTION:—Do you find that front axles at present are ample in most cars to withstand the torque of front-axle braking?

ANSWER:—No; they have not a sufficiently high factor of safety. We recommend a better torque-section from the spring pad to the outer wheel or the knuckle.

QUESTION:—Can an internal brake be designed to wrap internally?

ANSWER:—Yes; but not with the same efficiency as an external band. We have shown on tests that when an external band is operated by a straight pull between the gap ends, that is, in a straight line, and an internal brake is used on the same drum and expanded in a straight line, the external brake develops 35 lb. and the internal brake 25 lb. In other words, the efficiency of the two is as 35 to 25.

W. C. KEYS:—Were the bands anchored at the middle in each case?

MALCOLM LOUGHEAD:—They were anchored 180 deg. from the center of the gap.

QUESTION:—How much stronger, approximately, in per cent must the front axle be with front-wheel brakes?

ANSWER:—To give a torsional section. The bending section, of course, is still there and is sufficiently strong.

QUESTION:—Would air in the system have any effect on the action of the brakes?

ANSWER:—It would mean a loss of pedal travel; if there were sufficient air the pedal would go clear to the floor-board without actuating the brakes, so that we take practically all the air out, that is, within a very small fraction of 1 per cent.

QUESTION:—What effect have very cold and very hot weather on operation?

ANSWER:—There is slightly sluggish action with the fluid that we use when the weather gets below zero. This can be counteracted by increasing the alcohol content, which will bring the action back to normal.

QUESTION:—Is tire mileage increased by using four-wheel brakes?

ANSWER:—Yes, considerably. There is practically no necessity for ever sliding the tires with four-wheel brakes; this happens many times a day with two-wheel brakes in traffic, without taking into consideration the conditions in mountainous districts.

QUESTION:—In the ratio you gave of 12½ lb. of car weight to 1 sq. in. of brake surface, is the brake surface the effective area or the total area of the brake-lining?

ANSWER:—The effective area; that is, the area after deducting approximately 4 in. of lining at the anchor-pin point on a 15-in. drum.

QUESTION:—What change, if any, is made in the front springs?

ANSWER:—None. We have never had a spring break. The spring makers have recommended that no change be made when four-wheel brakes are used. The front springs are as strong as the rear springs that take the torque loading in the Hotchkiss drive and the brake loading also.

QUESTION:—Are there two points of adjustment on each wheel?

ANSWER:—There are three on each wheel. One on each end of the band and one at the anchor-pin.

QUESTION:—What is the percentage of loss of liquid?

ANSWER:—That question is very hard to answer because the percentage varies considerably. We have never had a loss, have never had to refill the 2-pint tank in less than 6 months. The car I have mentioned, which has been running for over 20,000 miles, has shown no loss at all in a year's running.

QUESTION:—Is there churning of the oil? If so, what is the effect?

ANSWER:—There is some churning of the liquid, but it has no ill effect on the brakes.

QUESTION:—Is there oil leakage through the coiled wire tubing? If so, does this not rot the rubber casing and cause excessive loss of oil?

ANSWER:—The glycerine and alcohol has a preservative effect, if any, on the rubber.

QUESTION:—What is the minimum weight of car that can be economically equipped with four-wheel brakes? Would a Ford be a suitable car for four-wheel brakes?

ANSWER:—All cars, no matter what their weight may be, whether they are Fords or 6-ton trucks, can be stopped in the same distance; one can be stopped in as short a distance as the other, provided that the torque is sufficient to reach the point just before the wheels begin to slide. All cars should be equipped with four-wheel brakes. A Ford car that hits a child will do as much damage as a 6-ton truck could do. Neither could be stopped in a shorter distance than the other.

QUESTION:—What is the comparative efficiency of four-wheel hydraulic brakes and the commercial two-wheel type of brake on wet pavements?

ANSWER:—It is very hard to give any figures for efficiency, but four-wheel brakes almost totally eliminate the tendency to skid. When I say tendency to skid I mean the tendency of the car to pivot. The four wheels naturally will slide, but as a rule, they will hold to a straight line.

QUESTION:—Is it true that equalized pressure assures equalized braking effect?

ANSWER:—Any hydraulic system must remain in equalization up to the end of the bands. The only variable to consider is that of the coefficient of friction of the brake-lining itself.

QUESTION:—What is the maximum change in the coefficient of friction in the lining throughout its life?

ANSWER:—That will depend to a great extent on the lining. The best lining we have found, in practical use and on the test stand, has shown a variation in coefficient of friction of approximately 0.20 to about 0.46; and this variation usually takes place during wet and dry conditions of the lining.

QUESTION:—Will the brake-drum cool as well with external as with internal brakes?

ANSWER:—Yes, with a pressed-steel drum. Of course, the ribbed drum, whether cast or forged, will cool more rapidly with internal brakes, but with a pressed-steel drum it is about the same in either case. I might add that the drum radiates through almost double the surface that it does with two-wheel brakes. That question really has reference to four-wheel brakes.

QUESTION:—Water on the surface of the lining usually affects the operation of brakes detrimentally. Does not this show more with external than with internal brakes?

ANSWER:—In our experience, water, mud, or anything of that sort gets into the lining of the external-type brake in less time than it would get into the lining of the internal, but there is one point that offsets this difficulty; it gets out quickly, too. The internal brake will trap grease or oil or mud and takes a long time to work it out in some cases. I should say that these characteristics will about balance.

QUESTION:—What is the difference between the action of four-wheel brakes and that of the conventional type when a front tire goes flat at high speed, or on a turn, especially with the great drop due to the size of balloon tires?

ANSWER:—If a tire goes flat on a car equipped with four-wheel brakes you can stop in a much shorter distance than you could stop with two-wheel brakes. As a flat tire has a decided tendency toward pulling the car over into the ditch, and as you can stop in a shorter distance, I should say four-wheel brakes were safer.

QUESTION:—How is leakage prevented with hydraulic pistons?

ANSWER:—The cup packing that we are using now is of rawhide construction. The rawhide is the only form of hide-leather cup that will readily hold the liquid without leaking. When I say "leaking" I mean that no seepage gets past it. In the process of tanning the rawhide the fat cells are not broken down, but the rawhide is cured with them, so that it is not porous. Usually, when leather is tanned all the fat cells are broken, the gelatine is taken out and the leather is impregnated with wax, which can only be put into the leather at a temperature of less than 135 deg. fahr. Temperatures higher than this would injure the leather. In cold weather the ordinary leather cup gives good service, but in summer the heat forces the plugs of wax out of the leather, causing a gradual seepage at all times. We have absolutely no trouble when we use all rawhide packing.

We are making a search now and expect in a very short while to have a cup packing manufactured from fabric, rubber composition and the like, which would assure a uniform grade and a larger production than will any leather we might obtain. This packing has better possibilities from a production standpoint, though our present packing is giving us no trouble.

QUESTION:—What about the car behind that has rear-wheel brakes only?

ANSWER:—I think that is the business of the driver.

QUESTION:—Some builders of foreign cars are advocating that front-wheel brakes should be operated slightly in advance of rear-wheel brakes. What is the reason for that?

ANSWER:—Our reason for not doing it is that the shifting of the center of gravity always throws more weight on the front wheels, so that the rear wheels have a tendency to slide before the front ones, though the efficiency of the front wheels has not been increased. Many foreign cars with toggle operation of the shoe make this more necessary than does a balanced system of the hydraulic type.

QUESTION:—What minimum clearance do you allow between the lining and the drum?

ANSWER:—A minimum of 0.025 in. This will allow for expansion of the drum up to 650 deg. fahr.

QUESTION:—Does the brake-lining wear better with four-wheel brakes, or worse?

ANSWER:—Very much better. I might cite an instance of a Model-66 Pierce Arrow that weighs approximately 7000 lb. and carries 18 passengers, over the Ridge Road in California, which has a 7000-ft. rise. The rear brakes were relined after 48,000 miles, the front brakes after 32,000 miles of service. This shows the difference in efficiency between the front and the rear brakes on account of the shifting of the center of gravity.

QUESTION:—Is this particular hydraulic brake the type used by the Duesenberg car that won the Grand Prix?

ANSWER:—In principle, yes; but it is not the same in detail.

QUESTION:—What would the braking action be on the front wheels while turning a corner, as compared with straight running?

ANSWER:—If the brakes are properly equalized there will be no difference.

QUESTION:—How is the pressure of the brakes regulated to prevent sliding of the tires?

ANSWER:—That is another condition that varies so greatly with road conditions that it is almost impossible to put on a safety device that will keep the wheels from sliding. If a person puts too much pressure on the pedal it will cause the wheels to slide at times.

QUESTION:—Are the brake adjustments each independent?

ANSWER:—Yes, independent as regards taking up wear, but not as regards taking up equalization. They are inherently equalized at all times.

QUESTION:—How are the adjustments made?

ANSWER:—By two nuts, one on the head of each band extension, and by the conventional anchor-pin adjustment that is used on external brakes now.

QUESTION:—If the front wheels lock, do you not lose steering control?

ANSWER:—Absolutely. If a person puts his foot on the brake-pedal sufficiently hard to lock the wheels, he will lose steering control until he releases it enough to let the wheels revolve again. Cars that we have equipped have run an aggregate distance of 7,000,000 to 10,000,000 miles, with all kinds of average drivers, and have never had an accident on this account. There is only one thing to do, and that is to release the pedal slightly when you feel the wheels sliding.

QUESTION:—Does that mean that the ordinary driver must be specially coached in the use of four-wheel brakes?

ANSWER:—If he gets on a slippery pavement he must always be careful. He should work the brakes intermittently, as he would two-wheel brakes on ice. If he does not work the brake-pedal intermittently he will swerve round; we recommend that the pedal be applied in the same way with four-wheel brakes.

QUESTION:—The public would probably have to be re-educated in the use of four-wheel brakes?

ANSWER:—I should not think it would take much education; in the cars we have put out we have done no more than give warning of that condition and have never had trouble from that cause.

QUESTION:—Do you consider the use of chains on front wheels advisable?

ANSWER:—We have really forgotten what skid-chains are. We never use them any more. I have had one occasion in the last 7 years to use them and that was coming over the Tuscarora Mountains last winter on the ice. I have never used them on front wheels with four-wheel brakes. That includes two trips across the United States, back and forth, under very varied conditions.

QUESTION:—Do you have a reservoir for excess liquid to replace that lost by leakage, and how does it function?

ANSWER:—The tank is mounted slightly above the master cylinder so as to have a gravity feed; it feeds through several port-holes in the master cylinder and replenishes any loss due to the application of the brakes. The loss, however, is very slight.

QUESTION:—Is the band of the brake-lining continuous throughout its length or is a gap left at the rear support?

ANSWER:—We always leave a gap at the rear support, for two reasons: (a) it tends to cut out any dirt or foreign matter that may get into the band; (b) it is almost a dead point in the lining.

QUESTION:—Would the churning of the liquid cause any chemical change?

ANSWER:—None that we have found to date. I know of no reason for it.

QUESTION:—In making a turn on a wet pavement with the brakes set, is there any tendency to slide off at a tangent?

ANSWER:—Yes, if the wheels are locked. If the brakes are not applied to such an extent as to lock them the steering is normal.

QUESTION:—How often must the packing be replaced?

ANSWER:—We have never had to replace any packing on account of wear.

QUESTION:—Have you tried riveting the lining to the inside of the drum and using a plain steel band of the internal expanding type to obviate the expansion of the drum?

ANSWER:—No, we have not. To get back to the question of wear, on cups, we continue making tests at the factory; during the last 6 months we have had at least 50 cups that have withstood from 300,000 to 700,000 applications of the brake at a higher pressure than would be used in braking. This is equivalent to a life of 7 or 8 years in a car.

QUESTION:—Does the system become inoperative if a pipe line is broken or plugged up?

ANSWER:—Yes.

QUESTION:—Will not the additional margin of safety provided by four-wheel brakes be absorbed in practice by faster driving?

ANSWER:—It certainly should be.

QUESTION:—If the front wheels lock will the car skid, that is, revolve, or will it merely slide ahead in a straight line?

ANSWER:—Unless there is a decided crown or slope to the road, the car will slide in a straight line. There will be a slight side-swerve; if the car is on a crown it will tend to slide sidewise, but not pivot; the car as a whole will tend to go sidewise slightly.

QUESTION:—Is there an advantage in braking front and rear wheels diagonally in two stages?

ANSWER:—To do so would be to acknowledge that the equalization is poor. If the brakes are equalized I think it would be much better to brake all four wheels and not pay special attention to any two. There can be no advantage in braking diagonally unless the equalization is poor.

CLARENCE CARSON:—Mr. Loughead has said that he found the variation in brake-lining coefficient of friction to be from 0.20 to 0.46, I believe. It is rather hard to reconcile that statement with his other statement that four-wheel brakes have uniform equalization. I do not see how, even if he has uniform pressure per square inch on the pistons and the actuating mechanism, he can have uniform retarding of the vehicle on all four wheels, or even on two wheels, if there is that much variation in friction between lining and drum.

MR. LOUGHEAD:—I think that I supplemented the latter statement by saying that we equalized up to the band ends, and that the variable which we could not control was in the brake-lining itself.

MR. CARSON:—It seems to me that, with the greater part of the braking effort on the front wheels of the car and with such a variation in the coefficient of friction, there would be likelihood of bad "slewing" of the car, particularly if the brakes were applied on a curve. I have found that to be so on four-wheel-brake cars that I have tried. Though the cars hold to a straight course when the brakes are applied on a straight road, if the brakes are applied in making a turn, the car is liable to go wrong end first instead of head first.

MR. LOUGHEAD:—That condition does exist to a certain extent, but is within the control of the driver.

MR. CARSON:—It seems to be a problem for the brake-lining manufacturers to produce a lining with a coefficient of friction as uniform as it should have in order to get the results that are sought with this type of brake. At the present time I do not think that any of them can guarantee it.

MR. LOUGHEAD:—As a rule, a decided change in the coefficient of friction is caused by a severe case of heating. That is what I had in mind when I mentioned the extreme drop down Pike's Peak, where the brakes got very hot on account of the 7000-ft. drop in 18 miles. That did not occur with one wheel nor with another; as a rule, it occurs in all four wheels more or less.

MR. CARSON:—The Society has been conducting experimental work during the last 3 years on friction materials, in conjunction with the Bureau of Standards at Washington; and more recently I think there have been several independent investigations. To date I do not think any brake-lining has been found, regardless of price or make, that will hold to a very close variation of the coefficient. Many of the brake troubles that car builders are criticized for having, are unjust, because from the very nature of the lining and the way it is made, it is almost impossible to produce a uniform coefficient of friction. The problem is to get a satisfactory lining first.

MR. LOUGHEAD:—That is something that is beyond our control in any type of brake, until a better standard of brake-lining is developed. Probably 50 per cent of the poor equalization in mechanical brakes is due to linkage; if that can be overcome, the variables will be reduced from two to one.

MR. CARSON:—I take issue with you a little on that, because these tests have been made after overcoming as far as possible the variation due to brake linkage. They have been made on the drum with the brake-lining applied as directly as possible, and most of them have been on testing apparatus. It has been found that the coefficient of friction of the lining is extremely variable; it will usually start out fairly high on woven types when the lining is new, but the frictional heat developed when the brakes are applied causes the coefficient of friction to fall very quickly. It will recuperate when the drum is allowed to cool. This variation will continue throughout the life of the lining and is caused by different conditions.

Another thing is the possibility of the swelling of the lining. On brakes designed with a very small clearance frictional heat might cause some types of lining to swell and grip. Is it not true that if a front-wheel brake should grip on one side it would be very disastrous?

MR. LOUGHEAD:—No; we have run cars around with one front wheel completely shut off, and there has been no ill effect from it at all. I mean that a three-wheel brake is absolutely practicable from all standpoints in handling the car.

MR. CARSON:—The condition I referred to was of one wheel's suddenly locking.

MR. LOUGHEAD:—Sudden locking is within the control of the driver and depends on the pedal pressure he uses.

MR. CARSON:—The average car-operator will not graduate the first application carefully, but will lock the wheels.

MR. LOUGHEAD:—As I have stated before, I can answer that in only one way; we have 1000 drivers, who have had no trouble whatever in that respect. Of course, 1000 is not a large enough number on which to render final judgment, but we have 1000 drivers who have covered more than 7,000,000 miles with absolutely no trouble from that cause.

MR. CARSON:—We are very much interested on account of its being a frictional material problem, but I think I can see the possibilities of trouble for the brake-lining manufacturers until such time as a lining has been developed that either will be of a sufficient uniformity of coefficient to give the results we are after, or the ratio of braking between the front and rear wheels has been so modified that the "slewing" effect on curves will be overcome, and the public has become so used to four-wheel brakes that the driver behind, in order to save his car, will come to the conclusion that every car ahead of him has four-wheel brakes and will act accordingly. Most of the accidents occur from trying to stop instead of keeping on running. With four-wheel brakes making sudden stops, I can see the possibilities of smashing rear-end collisions, and it will take much slow educational work to offset that condition.

MR. LOUGHEAD:—We expected more of that than has actually occurred. We have had very few accidents. I know of no accident that has resulted in more than a dented fender; when the number of miles that I have stated is taken into consideration, I think the record is remarkable.

MR. CARSON:—A car was brought to me some time last fall, not fitted with hydraulic brakes, however, but equipped with four-wheel brakes, and the statement was made that it could not be made to skid. We succeeded in landing on a man's front lawn rear-end first while steering round a curve; so apparently they do sometimes skid.

MR. LOUGHEAD:—I should say that that was a mechanical-type brake, because it is typical of mechanical four-wheel brakes, but not so in any respect of the hydraulic brake.

MR. CARSON:—Have you had any trouble with the brake-lining's losing its gripping power on account of the lubricant's seeping by the pistons?

MR. LOUGHEAD:—We have no seepage of the lubricant at all; absolutely none.

MR. CARSON:—Have you any cars in operation around Saskatchewan, where the temperature is about 40 deg. below zero fahr.?

MR. LOUGHEAD:—I know of none. The lowest that I know of is approximately 20 below zero fahr. but the fluid we use will not solidify above 60 deg. below zero fahr. if a high enough alcohol-content, which we recommend in that extreme cold weather, is used. We have run cars under test with pure alcohol in the system and have had no trouble on account of it. Of course, it is a little harder to confine and there is more evaporation from the refilling tank under the hood.

MR. CARSON:—Does the high alcohol-content of the fluid affect the rawhide?

MR. LOUGHEAD:—Alcohol has no effect on rawhide at all.

G. L. MCCAIN:—When I wrote my question as to what economical installation could be made on a light car, what I had in mind was that the Ford stops too quickly sometimes. There must be a point at which the light car can be stopped as well as need be with brakes on two wheels. There must be some minimum weight at which it is economical to install four-wheel brakes.

MR. LOUGHEAD:—A car weighing 10 lb. and a truck weighing 6 tons can be stopped in the same distance because the coefficient of friction on the wheels retards them by a given force, so that the weight of a car has nothing at all to do with the distance in which it can be stopped, provided that there is an efficient means of braking the wheels up to the point of sliding; the light car requires it as much as the heavy one.

A. L. CLAYDEN:—Mr. Loughead said he did not know any reason for connecting brakes diagonally except to compensate for poor equalization. That was not the original idea of diagonal braking at all. The original idea of the diagonal connection was this: So long as we had a wheel on each axle rolling free, we would have either end of the car under steering control because it is only when a wheel ceases to roll that steering ability is lost. Consequently if we lock the right front and the left rear wheels, and leave the left front and right rear wheels free to roll, we shall have the locked condition on one wheel on each axle and can still steer the car almost as freely as if no wheels were locked at all.

I was able to experiment several years ago with some of the Argyll cars that were linked up diagonally or linked upon all four wheels. I also experimented a little with a hydraulic system very much like the present system but not so well carried out in detail. I also tried four-wheel brakes with diagonal connections and all four together.

The conclusion I reached at that time was that the increased stopping power to be obtained by applying brakes on all four wheels was of much greater practical value than the theoretical advantage of the diagonal connection.

Henri Perrot, the chief engineer of the Argyll Co., is the owner of a number of patents on the mechanical application of brakes. In 1911 or 1912 he was very strongly in favor of the diagonal connection and the Argyll car was built on that principle, with a rather complicated mechanical connection for the brakes.

I had a letter from M. Perrot only recently in which he said he had entirely given up the idea of a diagonal connection and had come round to my original view; that the plus stopping power of all four wheels was worth more than the gain from diagonal arrangement.

L. C. FREEMAN:—Mr. Loughead, I believe, made the statement that the efficiency of his transmission system is higher than that of the usual mechanical linkage. Has he any specific data to support that statement?

MR. LOUGHEAD:—The efficiency of mechanical linkage, properly lubricated, is undoubtedly in the neighborhood of 70 per cent; that is, it transmits approximately 70 per cent of the energy from the foot-pedal to the band ends but that is seldom the case after a car has been on the road from 6 to 8 months. Very few persons lubricate the brake-rod joints before something freezes or sticks or the rods do not operate; then they begin to drive the pins out, put kerosene on them and so forth. Very few of the cars running today have properly lubricated brakes. They are very inefficient, often not more than 50 per cent of the energy being transmitted from the pedal to the band ends. The efficiency of a fluid column, on the other hand, unless there is seepage, remains constant. The liquid acts as a lubricant on the pistons and the cup packings and is in the line at all times; the lubricant must be there if the brakes are to operate.

MR. FREEMAN:—Constant as approximately what figure? You mentioned 70 per cent for mechanical transmission.

MR. LOUGHEAD:—That is, if well lubricated. I have seen very few transmissions that would average over 60 per cent after they have been in service. An efficient hydraulic system is balanced in all four or five points, whereas with mechanical linkage, by an equalizer bar, each one is added to the other. The efficiency of the hydraulic system is about 82 per cent and remains constant.

MR. CLAYDEN:—The fact has been mentioned that several braking systems developed in England many years ago had been abandoned. Anybody who has had an opportunity of looking at them will have no difficulty in understanding why they have been abandoned.

G. W. HARPER:—Do you find that steering requires more effort at the wheel in making short turns when laterally inclined steering-pivots are used to bring the intersection of the tire and the pivot near together at the ground, than when the steering-pivots are vertical?

MR. LOUGHEAD:—When inclined steering-pivots are used the effort required to turn the steering-wheel is a little greater at very low speeds or when the car is at rest. This is due to the "scrubbing" action of the tire and the slight raising of the car. For several months I have driven a car that has an axle the stub ends of which are set at an angle of 8 deg. from the vertical, the projected knuckle-pin center-line being ⅜ in. from the center-line of the tire. This car has never shown any tendency to "shimmy" and the slight increase of steering effort at low speeds is more than offset by the greater ease of steering over rough roads and at high speeds, for most of the road shocks are absorbed by the knuckle-pin and are not transmitted through the steering-knuckle to the steering-wheel.

MR. HARPER:—How do you determine the theoretically correct position of the steering-knuckle tie-rod ball when inclined pivots are used, and the consequent length, center to center, of tie-rod balls?

MR. LOUGHEAD:—The only difference between determining the position of the steering-knuckle tie-rod balls with inclined pivots and those of the conventional vertical type is that the pivot center-line is on an angle, and the location of the ball must be taken from a line parallel to this center-line. As the location of the ball is taken from the inclined knuckle-pin center-line, the distance, center to center, between these balls will depend on their vertical location.

MR. HARPER:—If the coefficient of friction between the lining and the brake-drum is subject to such wide variations under various conditions as has been stated, what is the advantage of perfect equalization in applying the operating power to the four brakes from the foot-pedal?

MR. LOUGHEAD:—If two wrongs could make a right then we should disregard the equalization of the brake-operating mechanism. As an illustration, take two cars, each of which has the same brake-lining, but in one the operating mechanism is equalized, and the other has very poor equalization. As the braking loss due to the variation in the coefficient of friction is the same in both cases, the difference in stopping ability will be proportionate to the efficiency and the equalization of the operating mechanism. In the design of brakes the variation in the coefficient of friction must be accepted until such time as there shall be a more uniform grade of lining available. When we consider that this extreme change in the coefficient of friction does not occur during more than 1 per cent of the miles traveled, and that when it does occur, the change is more or less uniform at each of the wheels, such change in the coefficient should be considered as a variation in foot-pedal pressure, rather than as a loss of equalization at the wheels.

HYDRAULIC BRAKE ACTUATION

By
Burns Dick
Wagner Electric Corporation

For presentation to this Society, the conventional historical background has been omitted,   t of you are already familiar with it.

Consistent brake performance becomes more and more essential as average speeds increase. There are conditions under which an emergency application of powerful brakes, not properly balanced, may lead to serious consequences. Many accidents are caused by grabby or out of balance brakes, but this is seldom brought to light because the operator is either ignorant of the cause or ashamed to admit it. In other cases, the operator may hesitate too long to apply his brakes when an emergency occurs, knowing that they are out of balance and fearing the consequences of a pivot skid.

Hydraulic actuation tends to maintain a balanced performance.

In any brake actuating mechanism, the force applied to the brake pedal must be distributed, in predetermined proportions, to the eight brake shoes. This predetermined input must be maintained in the face of uneven wear of brake linings. The torque output is not the same on the two shoes of a given brake assembly.

The total torque requirements of the front axle brake assemblies are not the same as the torque requirements for the rear axle brakes, and lining wear is roughly proportional to the torque output.

Therefore, with different rates of wear, the distance the shoes must move from "off" position to "on" position will vary between adjustments. The problem is how to insure the application of a uniform force to the toe of the shoe, in spite of variable travel due to wear and drum distortion.

A two piston hydraulically actuated wheel cylinder gives a close approximation to a perfect floating cam. It automatically accommodates itself to long or short travel, as the shoe clearance demands. At the same time it has an inherent tendency to maintain the correct pressure, irrespective of the travel.

All this is difficult to accomplish and maintain by mechanical means.

Tests indicate that there is a greater tendency to swerve from the in-

tended line of travel if front wheel brakes are out of balance than if the rear
wheel brakes are out of balance. This is probably the reason why many early four
wheel brakes carried on the old tradition of greater braking effect on the rear
axle than on the front.

For the average passenger car, it is now generally recognized that it
is desirable to proportion the braking effect in such a manner that more than 50%
is applied to the front wheels.

Correct proportioning; i.e., more torque on the front wheels than on the
rear wheels; also corresponds with the more favorable cooling conditions encountered on the forward axle. Tire diameters and wheel sizes are now such, that the brake
drums, which should be regarded as "brake heat radiators", are largely hidden in the
"shadow" of the tires, instead of being fully exposed to the air stream. The front
brake drums, however, are in a much more favorable position for cooling than those
on the rear. This is especially true on streamline jobs where tin pants are brought
down close to the ground and the body partly shields the rear drums from the air
flowing under the chassis.

High temperature is probably the greatest cause of changes in brake torque
output with present day friction materials; therefore, since excess heat dissipation
can be better accommodated on the front wheels than on the rear wheels, greater uniformity in operation is to be expected when the maximum possible brake torque is
applied there.

Rear brake assemblies gather more dirt than the front brake assemblies,
being subjected to the stream of mud, dust and water raised by the front tires.
Service reports indicate more frequent cutting and wear from abrasives on the rear
drums and linings due to the sand blast effect set up by the front wheels. Rear
brakes require more careful protection against this bombardment.

Protection against grit entering brake assemblies in the space between
the drums and the backing plates requires to be very carefully worked out on the
rear end, particularly where ribs or bulges on the brake drums may present such an
angle to the flying particles that they are deflected underneath the curl of the
backing plate and into the brake assemblies. An example of how this may happen is
shown in the sketch, Figure 1, wherein the arrows indicate the deflection of the
particles off the rib of the drum.

With hydraulic actuation, the percentage of brake torque on front and
rear axles can be accurately and boldly proportioned, with the knowledge that it will
maintain the predetermined fore and aft ratio, say 40% on the rear and 60% on the
front. This is done by varying the diameters of the wheel cylinders on front and
rear axles. Confidence in the constancy of this relation relieves the driver of
timidness and allows the maximum possible braking effect to be used when needed.

Because of the high efficiency of hydraulic actuation in transmitting the
input from the brake pedal to the brake shoe, it becomes practical to use a shoe
mechanism having a low energizing factor. This, in turn, makes possible smooth deceleration at an approximately constant rate.

If, on the other hand, due to inefficiency of the mechanism between the
brake pedal and the shoe, it becomes necessary to make use of a high energizing
factor in the brake shoes, experience and theory both indicate that more frequent

and delicate adjustments will be required to maintain equal effectiveness on all four wheels. A multiplier appears which magnifies many unavoidable variables.

The higher the overall, every day efficiency from pedal to shoe, the lower the energizing factor in the brake mechanism can be, for a given pedal pressure to torque ratio. The lower the energizing factor the more stable the brake.

Hydraulic actuation plus a shoe mechanism with low energizing factor reduces body stresses and passenger discomfort. This combination also gives increased safety in the cause of emergency application because of its stability.

Efficiency curves of typical small units are shown in Figures 2, 3 and 4.

Hydraulic actuation combined with a brake assembly with a low energizing factor does not give the impression of rapid deceleration, even though measurements may indicate the same deceleration rate or distance to stop, but it is easy to bring the car to rest at the exact point intended without over-shooting or under-shooting the mark.

Affecting any type of car and brake, there is what, for want of a better name, we have called a "driver inertia effect", which we have attempted to measure. This was obtained as shown in diagram, Figure 5.

A person was seated beside the driver in the front seat and provided with a dummy steering wheel and dummy pedal. The pedal was equipped with a weighing scale calibrated in pounds.

The first surprising thing noted was that a "foot" or leg weighs from twelve to twenty pounds, varying with the person's weight, height, and the location of the front seat.

A car, on which the brake performance previously had been plotted, was then operated at various rates of deceleration, and the readings of the scale on the dummy pedal noted.

These readings plotted against rate of retardation are shown at A Figure 6.

The curve B, Figure 7, shows pedal pressure vs retardation performance of the car on which one test was run.

Assuming the values shown at A, Figure 6, are an involuntary "driver inertia effect", we deducted pressures A from pedal pressures B, producing curve C, which may then be regarded as the voluntary physical effort required to obtain the rates of deceleration indicated for that particular car and brake.

In some tests we found that the "driver inertia" pressures were almost as great as the required brake operating pressures, when checked at high rates of deceleration.

If, under emergency conditions, the "driver inertia effect" is excessive, it may indicate a ratio of pedal pressure to retardation unsafe for general use.

In other words, it might indicate that the driver would have to make a positive effort to release part of the total pressure, or run the risk of being thrown on to the pedal with sufficient force to lock the wheels.

The automatic nervous reaction of the muscles may make it difficult to effect this release when a sudden brake application is made, due, for instance, to someone stepping in the path of a fast moving car.

These curves, and the opinions based thereon, are given here with full appreciation of the fact that such tests show wide variations; they may be affected by the personal element; and after all, brake pedal "feel" is largely a matter of opinion.

The apparent "feel" of any brake is quite distinctly affected by the type upholstery, or rather, the softness or hardness of the padding and springs in the seat and back. Perhaps there will some day be an S.A.E. standard established for "feel".

There is a distinct difference in the feel of a hydraulically actuated brake and a mechanically actuated one. The hydraulic appears to harden or get more solid as the pressure is increased; the mechanical appears to get spongy or softer at high pressures. Personal preference seems usually to be for those things to which one has been accustomed.

Mechanical linkage from pedal to brake shoe is liable to be seriously affected by the following variable relations of brake assemblies and chassis.

1. Vertical movement.
2. Horizontal or fore and aft movement.
3. Sidewise movement.
4. Combinations of the above, including rotation due to braking torque and drive reaction.
5. Deflection of springs and changes in wheel to chassis location due to variations in pay load.
6. Steering.

Hydraulic actuation is unaffected by any of them, including those necessary for steering.

The flexible fluid carrying hose, used with Lockheed hydraulics to transmit the power from the chassis to the brake assemblies, is not affected by the movements of the road wheel assemblies relative to the chassis; that is to say, the cubic contents of the flexible hose are not changed by such movements. If the brake shoes, linings, and drum do not vary, and the factory setting is correct, no equalization adjustment is necessary throughout its life. On the other hand, the hydraulically actuated brake will not correct variations in torque output due to imperfect contact between the linings and the drum, or due to variations in lining friction. But hydraulic actuation will give freedom from frequent equalization adjustments and will give fool-proof operation, which produces a feeling of dependability, and I think you will admit "dependability" sells more cars than any other one item.

The hydraulic components on the chassis, conveying the fluid from the master cylinder, such as the metal pipe, fittings, hose, etcetera, as shown in Figure 8, become a part of the chassis itself, and in practice introduce no variations in brake operation.

Lockheed hydraulic components as now in general use are designed for a
maximum of 1,000 pounds per square inch line pressure. This is not a working
pressure, but a maximum which should not be exceeded. In fact, the installation
should be so laid out that it is impossible to exceed this maximum under any circumstances, including the combined full effect of booster and manual effort where
power equipment is used, such as on commercial vehicles.

The flexible hose used to operate with the above mentioned maximum of
1,000 pounds per square inch line pressure actually carries an average working
pressure of perhaps only 300 pounds per square inch on passenger cars and perhaps
500 to 600 pounds on heavy commercial vehicles. This hose has a factor of safety
of five to six, the bursting pressure being between 5,000 and 6,000 pounds per
square inch.

Soft springing is possible on cars with hydraulic actuation because the
amplitude of spring deflection in no way affects the practical operating conditions,
and the balance is not disturbed.

Figure 8 shows in plan view, and in elevation, the hydraulic lines between
the compensator and the wheel assemblies.

The movement of the axles and wheels relative to the chassis is taken care
of by the use of three flexible hose, one on each front wheel, and one for the rear
axle.

The metal piping should be so laid out and clamped to the frame that it
will not be affected by vibration. It should not be pulled tight over sharp edges
or allowed to rest on the corners of belts or other points where vibration may wear
through the wall of the pipe.

If the compensator is attached to a rubber mounted engine, it is necessary to provide an additional flexible hose between the compensator and frame, but
if mounted on the chassis itself, metal piping can be used for the compensator to
frame connection.

Metal piping in exposed positions, such as on the rear axle, where it is
liable to be damaged by flying rocks, must be given extra protection. Ordinary loom
is usually satisfactory for this purpose. Exposed metal pipe lines under the front
fenders should be protected also, as indicated in Figure 8.

It is essential to provide a rigid mounting for the compensator to prevent movement and loss of pedal travel; also called "spongy" pedal.

In Figure 9 are shown two types of compensator and pedal mountings. The
one on the right hand side is bad. The brake pedal shaft tends to move in the
direction indicated by the arrow at A, and the compensator tends to move around the
radius indicated at B. The forces acting at this point are sufficient to sometimes
distort the frame and exaggerate these movements, with consequent loss of pedal
travel.

A more desirable mounting is that indicated at the left hand side, Figure
9, wherein the compensator and brake shaft are rigidly tied together.

For commercial vehicles with power operation, a heavier box type compensator is usually employed, as illustrated in Figure 10. In this case the brake pedal is rigidly mounted in the box itself, which also forms a supply tank, the compensator cylinder being submerged in the fluid.

Figure 11 shows the internal of a typical two piston wheel cylinder. The casting A is of semi steel. The piston E is of aluminum, and piston cups F are of a specially compounded rubber. Note: Figures 11 - 15, inclusive (photographs) are not included with this mimeographed copy.

Figure 12 shows a single ended wheel cylinder intended to be attached by the bolt holes and flange to one brake shoe, the piston D operating through a connecting link against the opposite shoe. This, of course, involves a slight movement of the wheel cylinder casting itself with the brake shoe each time a brake application is made. This calls for a separate flexible hose connection between the two rear brake assemblies and the frame, as well as one for each of the front wheels. This is usually quite easily accomplished, and if brakes for four independently sprung wheels are required, a separate hose for each wheel would, of course, be necessary with any type of wheel cylinder.

Figure 13 shows a compound type wheel cylinder with two different bores. These compound wheel cylinders are used where it is required to apply more pressure to one shoe than to the other. Should the differential between the two bores be great, it is, of course, necessary to provide sufficiently stable mounting of the cylinder on the backing plate to offset the reaction thrust on the cylinder casting itself, so that there will be no movement of the cylinder.

With the conventional straight bore cylinder, as shown in Figure 11, in which the areas of both the pistons are the same, the cylinder casting itself is in perfect balance and produces no reaction upon its mounting. In this case the means for attaching the cylinder to the backing plate can be quite light.

The hydraulic fluid carries a lubricant which makes the system, from compensator to wheel cylinder, self lubricating.

When the compensator, pipe lines, wheel cylinders, etcetera, have been mounted on the chassis, it is necessary to make provision for removing all the air from the system. A solid column of fluid between the compensator piston and the wheel cylinder pistons is essential. This is done by what is known as bleeding the line. Assuming that we have a passenger car combination type compensator, as shown in Figure 14, the supply tank must be kept filled with hydraulic brake fluid, and a certain amount of the fluid pumped through the lines to the wheel cylinders out through the wheel cylinder bleeder valve, to expel the air. This can be done by operating the brake pedal throughout its full stroke, which, due to the arrangement of ports and valves, will cause it to act as a pump and force a fresh charge of fluid into the lines at each stroke. This is continued until all the air is expelled through the wheel cylinder bleeder valve.

The check valve at the base of the master cylinder, Figure 14, prevents air being sucked back into the system on the return stroke of the brake pedal.

Figure 15 shows a set-up for bleeding at the wheel cylinder, and the location of the bleeder valve, which is held open until all air has been expelled, and then closed. By inserting the bleeder hose down to the bottom of the transparent container provided to catch the over-flow, it is easily possible to observe when air-bubbles cease to rise, indicating that the line has been fully bled.

The bleeding system above described is satisfactory for individual cars, but where a production line must be taken care of, a faster method is necessary.

Figure 16 shows the bleeder equipment for a factory assembly line.

Great care must be taken to avoid any possibility of grit, rust or metal particles being introduced into the car system, and removable screens and filters are provided to prevent this.

Pressure is maintained on the production bleeder line by means of a motor driven pump. A flexible hose, conveniently located to reach the compensator filler cap opening on the chassis assembly line, is provided, and the fluid is forced through the line and out the wheel cylinder bleeder valves by pressure, instead of by pumping the brake pedal as with individual cars.

Figure 17 shows a plan view of the chassis assembly line and bleeding equipment suitable for an up to date assembly line.

With all the air eliminated from the lines, assuming the brake shoe clearances have previously been set, no adjustment of the individual assemblies for equalization is required. The brake dynamometer check at the end of the line is unnecessary, because the pressures exerted on the various pistons are governed by fundamental laws which insure uniformity of operation throughout their life.

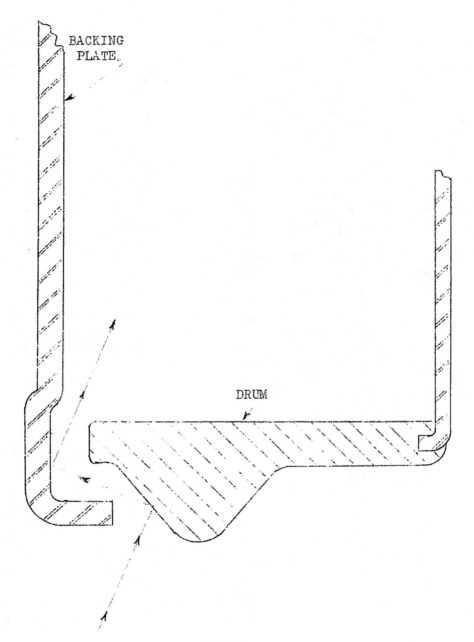

FIG. 1

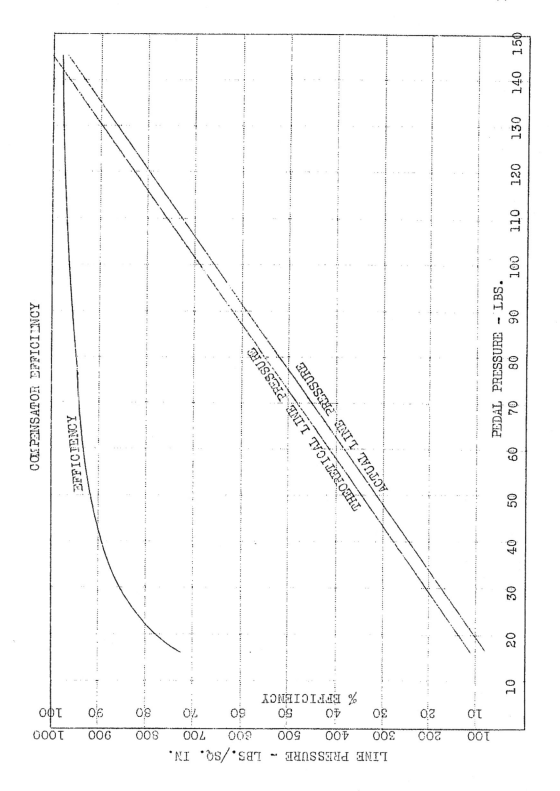

COMPENSATOR EFFICIENCY

FIG. 2

167

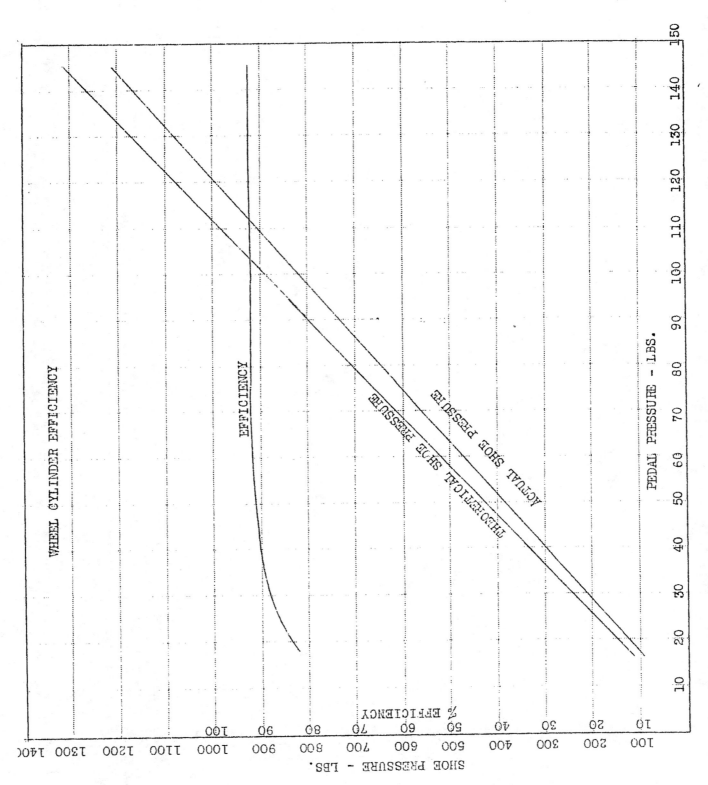

WHEEL CYLINDER EFFICIENCY

EFFICIENCY

THEORETICAL SHOE PRESSURE

ACTUAL SHOE PRESSURE

PEDAL PRESSURE - LBS.

% EFFICIENCY

SHOE PRESSURE - LBS.

FIG. 3

168

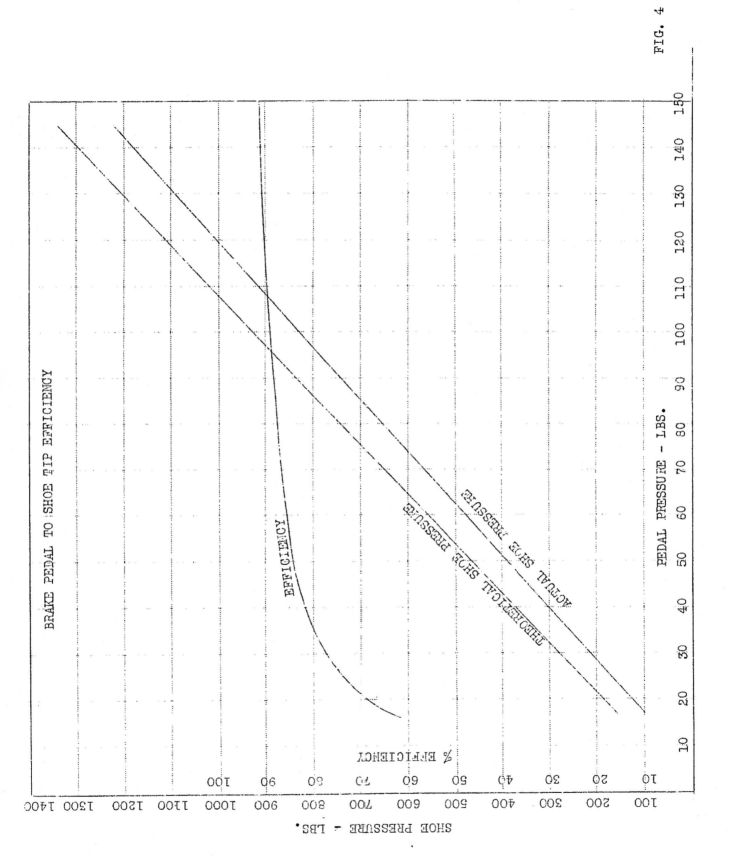

BRAKE PEDAL TO SHOE TIP EFFICIENCY

FIG. 4

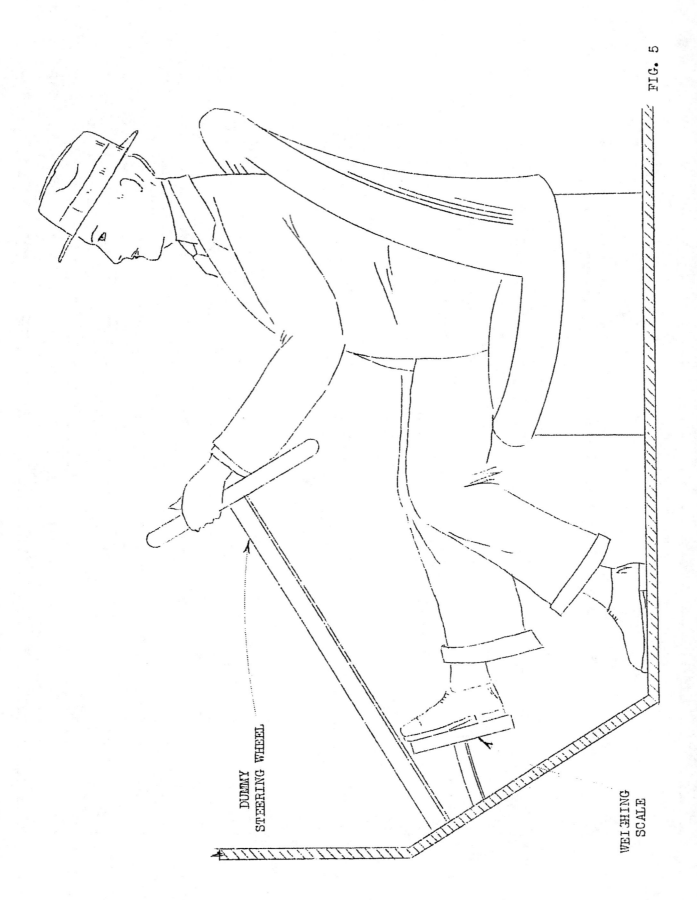

DUMMY
STEERING WHEEL

WEIGHING
SCALE

FIG. 5

170

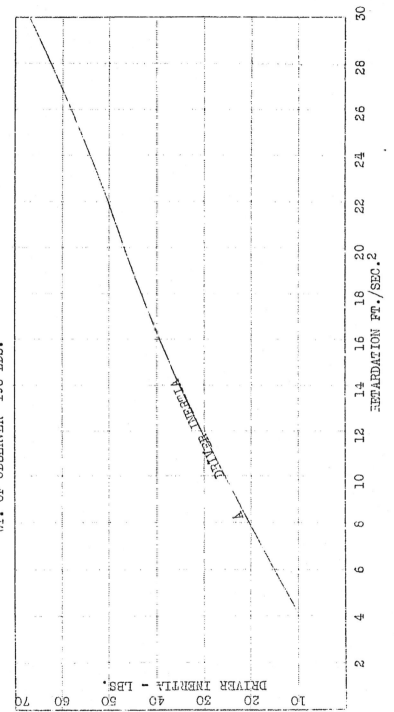

EFFECT OF INERTIA OF DRIVER

WT. OF OBSERVER 196 LBS.

FIG. 6

171

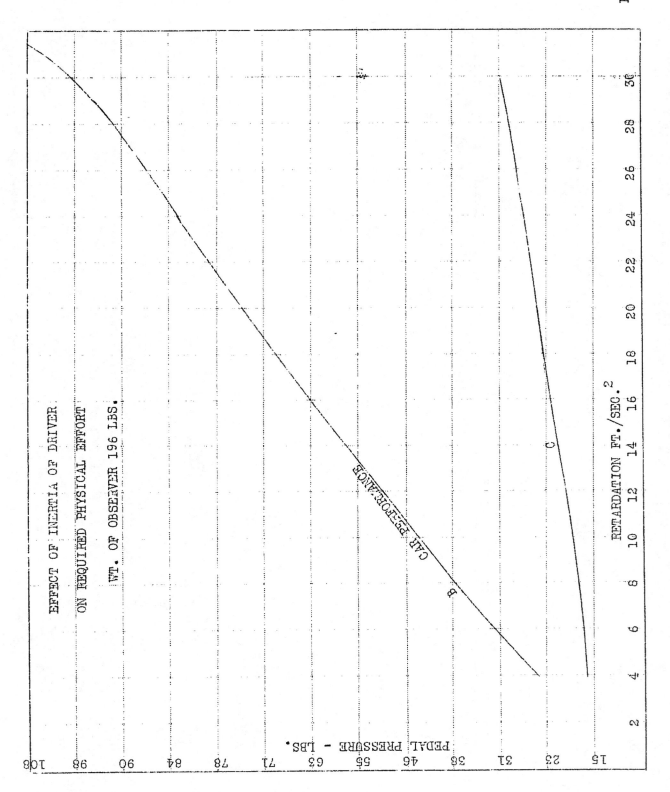

EFFECT OF INERTIA OF DRIVER.

ON REQUIRED PHYSICAL EFFORT

WT. OF OBSERVER 196 LBS.

CAR PERFORMANCE

RETARDATION FT./SEC.^{2}

PEDAL PRESSURE - LBS.

FIG. 7

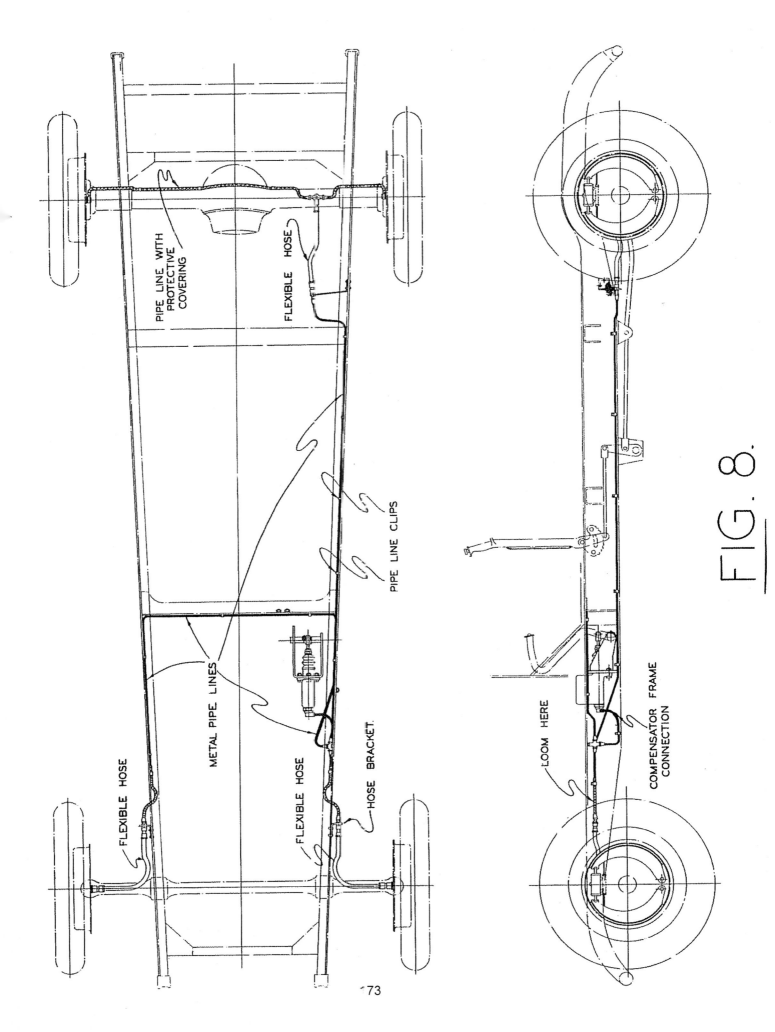

PIPE LINE WITH PROTECTIVE COVERING

FLEXIBLE HOSE

PIPE LINE CLIPS

FLEXIBLE HOSE

METAL PIPE LINES

FLEXIBLE HOSE

HOSE BRACKET.

LOOM HERE

COMPENSATOR FRAME CONNECTION

FIG. 8.

73

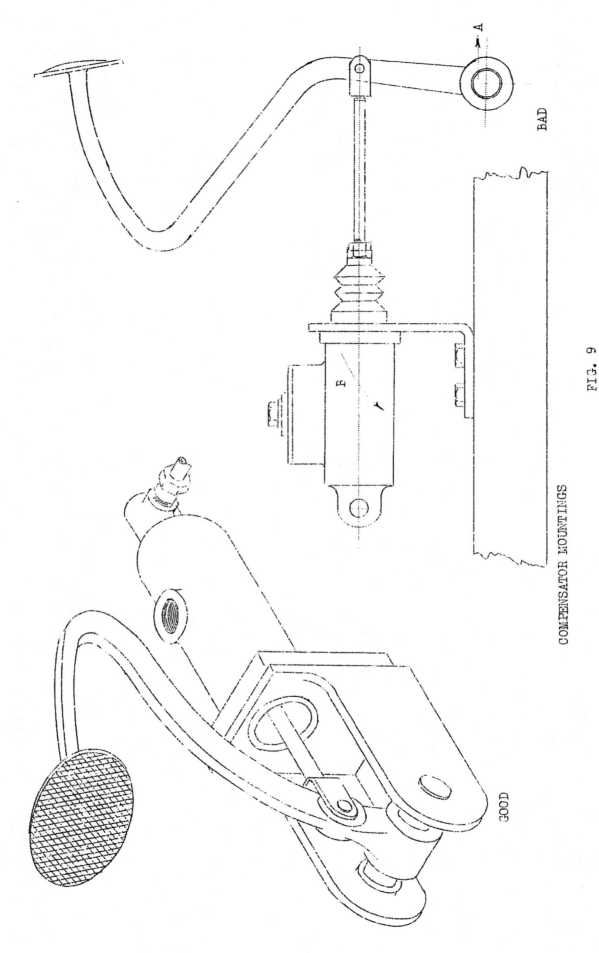

BAD

GOOD

COMPENSATOR MOUNTINGS

FIG. 9

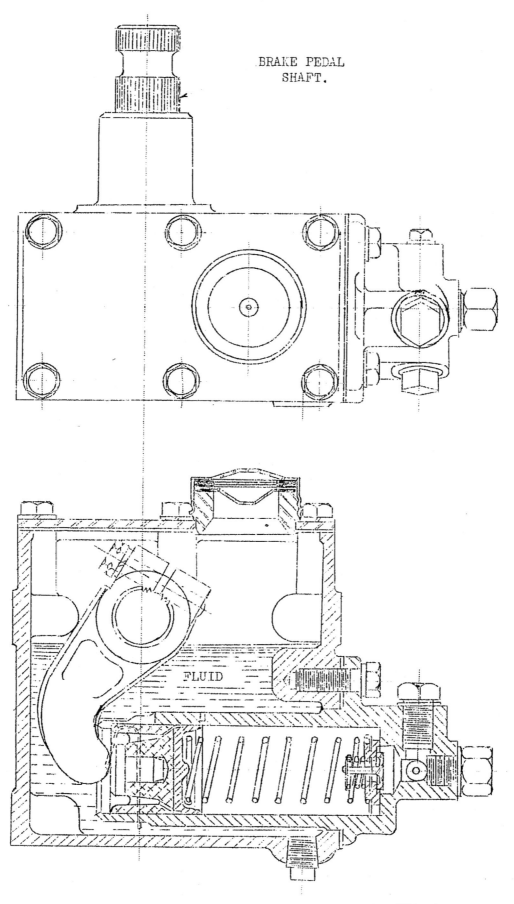

BRAKE PEDAL
SHAFT.

FLUID

FIG. 10

BOX TYPE COMPENSATOR
FOR COMMERCIAL VEHICLES.

175

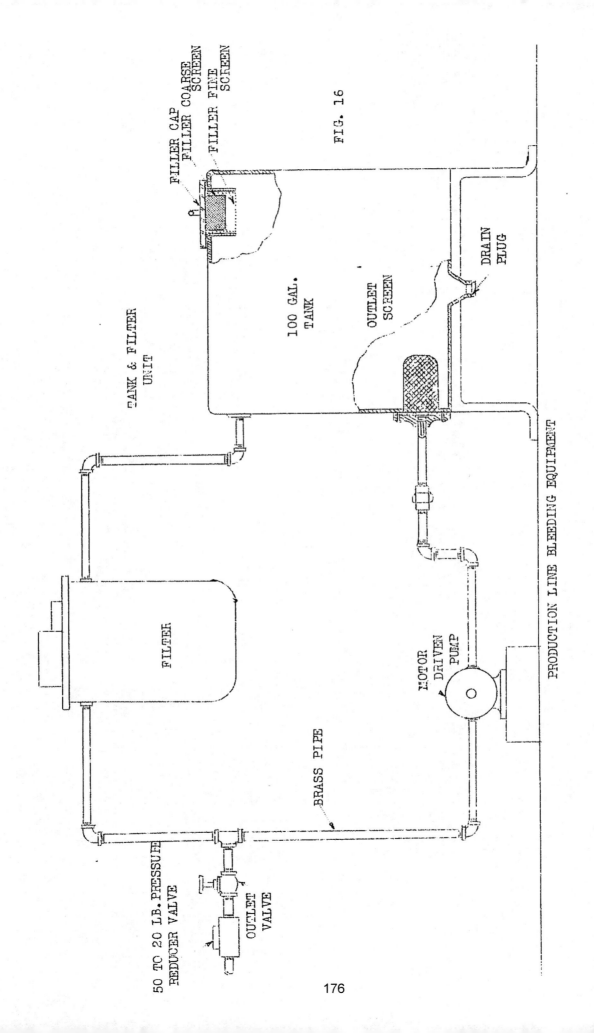

FILLER CAP
FILLER COARSE SCREEN
FILLER FINE SCREEN

FIG. 16

TANK & FILTER UNIT

100 GAL. TANK

OUTLET SCREEN

DRAIN PLUG

FILTER

MOTOR DRIVEN PUMP

BRASS PIPE

50 TO 20 LB. PRESSURE REDUCER VALVE

OUTLET VALVE

PRODUCTION LINE BLEEDING EQUIPMENT

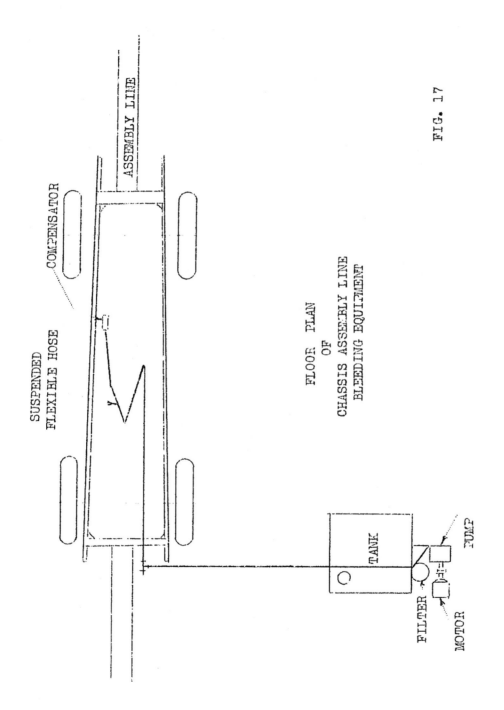

SUSPENDED
FLEXIBLE HOSE

COMPENSATOR

ASSEMBLY LINE

TANK

FILTER

MOTOR

PUMP

FLOOR PLAN
OF
CHASSIS ASSEMBLY LINE
BLEEDING EQUIPMENT

FIG. 17

ONE PHASE OF THE HYDRAULIC BRAKE PROBLEM

By

John C. Cox, Wagner Electric Corp.

Tonight I am going to discuss only that phase of hydraulic brakes which properly includes hydraulics. In other words, all brakes will be considered hydraulic where friction surfaces are held in contact by hydraulic pressure. I propose to acquaint you with some of the problems facing the designer whose work is the successful application of hydraulic actuation to motor vehicles.

Of the many means for transmitting motion, pressure tight cylinders and pistons moved by liquid pressure have the direct appeal of simplicity and reliability. To make these parts pressure-tight has been a work of major importance. That reliability has been established is daily demonstrated by delivery of thousands of cars with hydraulic brakes from the works of all major manufacturers.

But the problem of holding pressure is only one phase of hydraulic actuation, for it is also necessary to transmit motion. The obvious fundamental motion is that required to move the shoes into contact with the drum. This calls for pedal travel to first build up sufficient pressure to overcome the shoe return springs, and friction losses.

Gtom yhr moment the shoes are brought into contact with the drum, further pedal travel is required to build up pressure between the friction surfaces, and additional travel to overcome elasticity effects in the hydraulic lines and brake drums.

To these travel requirements is added further movement for the purpose of following the heated drum as it expands with absorbed heat.

Thus it is quite clear that the same hydraulic system cannot be used with larger brakes as actuate the small ones. Quite apart from the fact that large brakes require more operating force to get the most from their greater lining area, more travel is demanded to overcome the effects previously mentioned. Travel demands are proportionate to increased size in a given series of brake equipment.

To reduce travel by careful design is therefore equally as important as obtaining higher friction linings. And it is of interest to know that much of the improvement in modern brakes is due to effort spent in reducing travel requirements, minimizing losses, if you please.

Primarily, experience has shown that means must be provided to permit unobstructed change in length of the fluid column with temperature variation, otherwise excess travel may result. Furthermore, sealing cups must be accurately fitted, and definitely held in position. Hose and piping must be as free of "spring" as they can be made. Systems not possessing these characteristics have been tried and discarded.

It has been assumed that appreciable lost motion can be traced to compression of the rubber seals at high pressure. Experiment has demonstrated that the loss due to this cause is very slight indeed when proper materials are used.

If then, travel is limited by custom - that is, we habitually apply our brakes with but a single motion, and on the other hand the comfortable pedal pressure is certainly limited; so we are restricted to a definite field by human characteristics.

We do not design for lady, or lady-like truck drivers, and therefore it is customary to provide more travel and to expect increased pedal effort from paid operators than from passenger car owners. For this reason, we can get more from a given size brake in commercial service, provided the pedal action does not produce fatigue.

But where passenger car weights run in the vicinity of two and a half tons, or for trucks over five tons gross weight, it is likely that the driver will be helped by auxiliary power. In the interest of economy, it is usual to proportion the additional power input so that the driver supplies as much effort as practicable. In larger trucks full power equipment is used, for the percentage of driver input available to the total requirements is so low that the expense of adding extra pedal linkages is unnecessary and certainly not warranted.

Now equipment is very carefully checked to assure that all of the pedal linkage is stiff as required and that the master cylinder mountings are well supported. Proper precautions taken to assure rigid mountings may permit greater leverages, and consequently improved brake performance. In fact, the chief difference between an old and a new model with better brakes may be in the pedal group.

Excellent results can sometimes be obtained from older equipment by reducing travel losses inherent in the master cylinder mounting, although it is realized that there are other factors affecting brake output. I desire only to indicate a place where it is sometimes possible to effect improved brake performance where early models are not quite up to standard.

There are now in regular production some eight master cylinder sizes, ranging from one cubic inch displacement to eight, to take care of the entire range of vehicles. That it is customary to fit smaller cylinders than formerly required is proof that constant improvement in design is being accomplished.

With better machine tools came closer production limits and better finishes. To assure satisfactory operation it is essential that these limits be maintained. Repair operations involving cylinders must be carried out with discretion, for here both the surface finish and diameter are involved.

Today rubber is handled by engineers as an ordinary structural material, just as steel and aluminum. Provided care is exercised in the control of basic rubber compounds they can be moulded to exceptionally close limits and over-all performance predicted with certainty. That is one of the very important factors in the successful application of hydraulic brakes.

Equally important as proper rubber compounding is fluid selection. These two are the Siamese Twins of the hydraulic brake - inseparably related, and fortunately they are now on the best of terms.

More output from the brakes themselves and great cooling restrictions demanded fluid of improved heat resistance. Simultaneously, it was desired to offer a single fluid for northern climates to obviate the necessity for changing fluid in northern climates. This was accomplished after a long period of development work. Modern fluid is a really remarkable commodity in that one grade will handle the cars of Alaska or Panama with equally good performance.

Oftentimes it is asked how long will the components of the hydraulic system last. The answer naturally depends on the class of service, but every precaution is taken to assure satisfactory service for an indefinite period. Although we advocate periodic inspection and replacement of certain parts, an illustration comes to my mind. Three hose from a car in service four years were tested and found to pass production tests for burst and other checks. The hose is probably the most involved production problem in the entire system. It is certainly the target for more criticism than any other part. Perhaps, that is the reason it is so good!

It is hoped the high lights given herein will lead to a better understanding of the inner workings of the hydraulic brake. Almost every part is a development in itself, and the history of each would be a lengthy but interesting subject to one concerned with such matters. But to most of us, the hydraylic brake is just a simple, convenient and satisfactory way of stopping a vehicle.

AUTOMOTIVE BRAKE PROBLEMS
by
Paul J. Reese
Automotive Application Engineer, Wagner Electric Corp.
Presented at SAE Section Meetings in March, 1949:
Seattle, Washington
Portland, Oregon
Vancouver, B.C.

The importance of adequate brake performance in the safe operation of motor vehicles has been recognized by all interests directly or indirectly connected with Automotive transportation. The manufacturers of vehicles and braking equipment and the operators of motor vehicles, on the one side, and the officials responsible for the design of highways, the regulation of motor vehicles, and the control of traffic, on the other, have been and are cooperating in a broad program of research on motor vehicle brakes.

One phase of the program has been to obtain additional information concerning the brake performance of vehicles as they operate in everyday traffic.

Public Roads Commissioner Thomas H. MacDonald in his 1948 Beecroft Memorial lecture made this statement:

"Safe vehicle performance depends to a greater degree upon adequate well maintained brakes than on any other single element of vehicle design."

Improvements in roads, tires, motive power and vehicle design have permitted increased speeds and greater payloads for commercial automotive vehicles. Operating under varying conditions - creeping through congested traffic, rolling along the highway, going down long grades - these modern heavy vehicles need brakes with complete and flexible controls which the driver can operate with confidence in their stability and effectiveness and without excessive physical exertion and resulting fatigue.

The functions of the brake system are to reduce the speed, to stop the vehicle, and to hold it at rest and there are three basic elements involved in the performance of these functions.

> 1. The vehicle
> 2. The driver
> 3. The road

Each of these has certain definite limitations for safe and desirable performance insofar as brakes are concerned.

The first problem is that the brakes must have adequate capacity for the weight and speed of the vehicle.

The driver problem involves very definite physical limitations which must not be exceeded. In passenger car brakes, the woman driver must be considered and the effort required to operate the brakes must be kept as low as possible.

The third problem is the road. The character of the surface of the road and the tires are the fundamental factors that dictate the design possibilities and limit the braking performance of the vehicle.

THE COEFFICIENT OF FRICTION OF TIRES ON ROAD SURFACES

Skidding of wheels is the factor which limits the maximum torque that the brakes can deliver for the maximum anticipated weight on the wheels and thus limits the stopping ability of the vehicle.

The Skidding Coefficient of various combinations of tire and road surface conditions has been investigated by Iowa State College and Ohio State University and the results are available in bulletins issued by the U.S. Highways Research Board. The S.A.E. Highways Committee has prepared instructions for a method of measuring skidding coefficient of tire and road surfaces under the title "Determination of Friction between Pnuematic Rubber Tire and Road Using A Self-Propelled Vehicle and Measuring the Straight Ahead Braking-to-Stop Distance with All Wheels Locked."

A report on a Preliminary investigation of the Effect of Front Wheel Brakes, Fifth Wheel location and Tire Chains on the Stopping Ability and Steering Control of Commercial motor vehicles by Joint Committee on 1948 Winter Traction Tests has been published by the National Safety Council.

Mr. T. P. Chase of General Motors Research Laboratories presented a very fine paper on "Passenger Car Brake Performance; Limitations and Future Requirements" at the S.A.E. National Passenger Car and Production meeting in Detroit, March 3-5, 1948, in which the results of the above tests were analyzed and translated into effect on brake performance. Mr. Chase very ably covers his subject and we highly recommend his paper to anyone interested in brakes. I have made liberal use of his paper in preparing this one.

These tests show that the skidding Coefficient may be as high as unity at low speeds, decreasing with speed to 0.6 at highway driving speeds which limits the maximum rate of deceleration available without sliding wheels to 20 ft/sec/sec.

During deceleration there is a transfer of weight from the rear to the front wheels, due to the force of inertia that is equal to the product of the braking force at the ground level and the ratio of the height of the center of gravity to the wheel base. At high rates of deceleration this weight transfer is large enough to materially reduce the braking ability without sliding rear wheels. Mr. Chase cites approximately 500 pounds of dynamic weight being transferred from the rear to the front wheels of a typical modern car for the deceleration available at 0.6 road coefficient.

A perfect braking system should, therefore, vary the brake effort distribution with the dynamic weight distribution changes. Since this is not feasible (yet) and as all wheels can develop their maximum friction only at one rate of deceleration, the solution to the problem of distributing brake effort to suit various body types and possible load distribution must be a compromise.

The percentage of braking on the front wheels of the first four wheel brake installations was limited to 35%, whereas on some late model cars this has been increased to 60%. With present trend of design toward increased weight on the front wheels, this ratio of brake distribution will be required to maintain present performance. The result of this increased percentage of braking on the

182

front wheels has been that the most commonly used rates of deceleration are dictated by the comfort factor of riders rather than the ability of the brake. It has been found by experience that decelerations greater than about 16 ft/sec/sec tend to throw passengers out of their seats. For standing bus passengers rates greater than 12 ft/sec/sec are apt to cause complaints.

TIRES:

Smaller wheels, wider rims and larger section tires increase cooling problems and limit brake drum diameters.

There have been some test results published showing lower skidding coefficient of friction for synthetic than for natural rubber tires.

FOUNDATION BRAKES:

Consist of the brake shoes and related parts, such as backing plates or spiders and means of expanding the shoes against the drum. The two means most commonly used are:

1. A cam mounted on a shaft which is rotated by pressure applied to an external lever.

2. Hydraulic Actuating cylinders with pistons acting directly on the brake shoes.

The best performance is obtained from a brake when the brake shoes do an equal amount of work with resulting equal lining wear on the shoes. This has been accomplished in recent brakes by the use of two cylinders to obtain symmetrical action of two identical floating-type self-centering shoes. Two types are in production by our company at the present time.

The type "F" has single-end cylinders so the shoes are self-energized in forward rotation only and the type "FR" in which double-end cylinders are used so the shoes are self energized in both forward and reverse rotation.

The self-energizing factor causes the type "F" brake to be approximately three times as effective during forward rotation as it is during reverse rotation so its use is generally confined to the front axle of vehicles in conjunction with rear-axle brakes which provide adequate stopping ability in reverse as well as in forward rotation.

The floating type shoes are held against the radially aligned anchor blocks by the shoe-retracting springs. Upon brake application the wheel-cylinder piston forces the shoe into contact with the brake drum. As the shoes roll upon the anchor block to contact the drum, their heels may also slide radially upon the anchor blocks thus automatically (self) centering the shoes in relation to the drum.

The type "FR" brakes are also made in a dual design with two shoes side by side and dual double-end cylinders. This design has the advantage of distributing the applying force at two points across the total brake width and permits a certain amount of shoe follow-up to meet the condition of drum distortion or bell-mouthing. It also provides for the application of adequate force to the brake shoes to meet the requirements for stopping ability.

The type "V" brake is a self-centering, floating-shoe type with a double-end cylinder and one forward and on reverse shoe. The anchor is a solid "V" shaped block with its sides aligned upon the radius of the axle.

The type"V" brake is also produced with a self-adjusting device which automatically compensates for brake-lining wear. Adjustment is necessary only when new lining in installed. It is usually installed on the forward shoes only as the reverse lining wears slowly and the reverse shoe seldom requires adjustment. Drum distortion or expansion has no effect on the adjusting device as the contact-plug movement depends entirely upon lining wear and it does not alter position if the shoe is forced into a distorted or expanded drum.

The type "V" brakes are used on passenger cars and light trucks, the type "F" brakes on front axles of both passenger cars and trucks and the type "FR" brakes on the rear axles of trucks and also on the front axles of trucks in the heavier duty class.

DRUMS:

During braking action, the kinetic energy of the moving vehicle is converted, thru friction between the brake lining and the drum, into heat energy, and brake drums have been known to develop temperatures as high as 1400-1500°F momentarily on the braking surface in severe service.

It is important to remember that the temperature rise is very rapid and it is only the metal on the inner surface of the brake drum that is raised to this very high temperature. If the drum has a heavy section, its outer side may remain relatively cool (in the neighborhood of 300-400°F) at the same time that its surface has reached maximum temperature. It is quite evident that the expansion of the surface metal due to heat is very much greater than that of the metal below which is considerably cooler.

The internal strains produced by this unequal expansion tends to cause the surface metal to "check". The amount of checking will depend on the physical properties of the metal, especailly at high temperatures. A brittle metal will check very readily, and each repetition of the same condition after a check is started will tend to deepen it until eventually it becomes a crack that may go all the way thru the drum section

The outer side of a thin drum will be hotter than the outer side of a heavy section drum for any given surface temperature. Consequently, the average temperatures of a heavy drum will be less than that of a thin drum during the critical period of brake application.

The tensile strength of 35,000 lbs. per sq.in. of a brake drum iron at atmospheric temperature drops to 2000 lbs/sq.in. when its temperature is raised to 1200°F. (The actual loss of tensile strength or resistance to distortion due to heat will depend on the material from which the drum is made, its thickness and design.) (Location and dimensions of ribs.)

The brake drum must withstand the pressure of the brake shoes which is great enough to flex the drum several thousandths of an inch out of round at ordinary temperatures so that, in effect, the brake drum behaves somewhat like a belt running over pulleys. Every point on its circumference is alternately stressed in tension and compression several times in each revolution.

This flexing or distortion of the drum, within certain limits, may increase the torque output of a brake over that obtained with a more rigid drum.

The ability of a brake drum to dissipate heat is increased by adding cooling ribs which permit more drum area to effectively contact the surrounding air or by causing more air to circulate over the drum surface by coring ventilating holes in the brake drum dish.

There is always the hope that some new materials may be found for brake drums that will extend their life and usefulness, and experiments are being conducted with:

> Hardened Alloy Iron Drums.   (Brinnell 230-250)
> Aluminum Drums with Cast Iron Liners.
> Aluminum Drums with Sprayed Steel Liner.
> Cast Iron Drums with Copper Cooling Fins.
> Cast Iron Drums with Aluminum Cooling Fins.

BRAKE LINING:

Uniform and dependable response of the brakes is essential for safety and the output of the brakes must be stable in order to attain this result.

The brake output depends entirely on the friction between the brake lining and the drum, which changes kinetic energy into heat energy, so anything that affects this friction will affect the stability of the brake.

Grease, water, dust, etc., will change this friction and it is important to guard against them since they usually occur in one brake and, therefore, have a pronounced affect on the controlability of the vehicle.

"Another cause of changes in the friction between the brake lining and the drum are variations in the coefficient of friction of the lining and changes in the dimensional relation of the shoe and drum. These are primarily the result of the heat developed during conversion of kinetic energy into heat energy."

"The kinetic energy possessed by a car in the smaller lower priced group traveling at maximum speed is approximately 800,000 ft. lbs. and the heat equivalent is 1050 B.T.U. For the larger and faster cars the kinetic energy possessed at maximum speed is over 1,500,000 foot pounds, and the heat equivalent is 1925 B.T.U.'s."

Metallurgists estimate that 1050 B.T.U.'s will melt a cube of iron 1-3/8 inches square and weighing 1.9 pounds. Similarly, 1925 B.T.U.'s will melt a cube 2-3/8 inches square and weighing 3-1/2 pounds.

Another significant fact is that 75% of this heat is developed during the first half of any deceleration. This is the reason, plus the fact that the brakes must also overcome the force of gravity, why extremely high temperatures are attained on winding, mountain grades.

The drums expand and surface temperatures reach very high values during the brake application. The first result is that the coefficient of friction of most linings decreases to a lower value at elevated temperatures which reduces the torque exponentailly. A further result not generally appreciated is that the shoes heat more slowly than the drum so that the radius of the shoe is less than the radius of the drum. This causes an additional reduction of torque because of reduced shoe contact.

If severe braking is continued any considerable length of time with the brakes hot, the surface of the lining will be burned and wear at a very rapid rate to the radius of the enlarged drum. When the drums have cooled, the radius of the shoes will be larger than normal, and the pressure will be higher than normal at the toe and heel, which makes the self-actuation abnormally high until the linings have worn to the radius of the cool drum.

If high-coefficient lining is used to increase the percentage of actuation and thus obtain greater output, one must expect greater sensitivity or, in other words, less stability of performance.

Practically all modern automotive brakes are fitted with molded asbestos linings. Among these there is a wide difference of characteristics. Some provide greater friction than others for the same pressure against the drum; some suffer more "fading" or loss in friction than others during a long (high speed) application; some show more tendency than others to glaze and lose their frictional values under successive light applications such as are encountered in city driving.

Solving the problem of selecting the most suitable brake lining requires a knowledge of its properties or characteristics. These are determined by dynamometer and road tests, conducted according to a definite procedure which permits a direct comparison of the following:

1. Friction – throughout the life of the lining, hot and cold.
2. Fade - a general description of the reduction of brake effectiveness caused usually by the development of critical temperatures in the brakes.
3. Build up - a general description of a temporary increase in brake effectiveness caused by an increase of temperature in the brake lining or resulting from a reduction in rubbing speed.
4. Glazing tendency
5. Drum wear
6. Life

HYDRAULIC SYSTEM

WHEEL CYLINDERS:

CASTINGS:

A large percentage of steel is included to make a semi-steel, rather than a gray iron casting. This semi-steel material gives the casting a non-porous density that makes it leak-proof and strong enough to easily withstand the pressure to which it is subjected. The casting is diamond bored and rolled within limits of plus .003 inches or minus .000 inches.

Corrosion is a problem and tests are being conducted continuously in an effort to find better materials and finishes without increasing the cost.

PISTONS:

Are made of aluminum, brass, or chrome-plated steel, and are machined within the limits of plus or minus .0005 inches.

HYDRAULIC ACTUATING CYLINDER CUPS:

Are sliding seals that must function through a temperature range from minus 40°F to plus 340°F. Extensive tests have been and are being conducted to improve these parts and extend the range of operating temperature. They are molded to exacting specifications from accurately formulated compounds to ensure their meeting SAE standard test requirements for:

1. Resistance to fluids at elevated temperatures
2. Heat pressure test
3. Stroking test
4. Low temperature test
5. Hardness test
6. Bend test
7. Leakage test
8. Aging test

BOOTS:

Must prevent the entrance of water, dirt and brake lining dust into the wheel cylinder. They are in addition exposed to grease and to sparks thrown off by the brake lining. They are made from heat-resistant synthetic rubber and search for better materials is continuous.

HYDRAULIC BRAKE HOSE:

Are fabricated to rigid specifications and the following SAE standard test requirements must be met:

1. Constriction
2. Expansion
3. Burst Strength
4. Whip test
5. Tensile test
6. Cold test
7. Salt spray test
8. Pressure test

MASTER CYLINDER:

CASTINGS:

Have generally been semi-steel but for the past year aluminum master cylinders have been used on certain passenger cars. They are made in various sizes to provide the required displacement for the wheel cylinders and aside from high temperatures, the master cylinder problems are essentially the same as for wheel cylinders.

HYDRAULIC BRAKE FLUID:

Is of the non-mineral oil type and SAE standard specifications have been prepared for "Heavy Duty" and "Moderate Duty" types covering the following properties or characteristics:

1. Corrosion
2. Stability (295°F BP-HD)
3. Rubber swelling test
4. Compatibility
5. Lubrication
6. Residue
7. Evaporation
8. Viscosity
9. Cold test (-40°F)
10. Flash point
11. Water tolerance
12. Neutrality

GENERAL:

When hydraulic brake systems were first used on Automotive vehicles about 25 years ago, the pressure was limited to about 1,000 psi, as the burst strength of the hose was only about 2,500 psi. Today the burst strength of the hose is at least 6,000 psi by specification, and the other parts of the system will easily withstand the 1,500 psi pressures which are employed on present day commercial vehicles.

MASTER CYLINDER SIZE, PEDAL PRESSURE AND PEDAL RATIO

The master cylinder diameter and stroke required for the vehicle are calculated by a rather complicated formula which takes into consideration original brake shoe clearance, thermal drum expansion, the setting of parts in and the expansion of wheel and master cylinders, hose and tubing expansion, and which provides a reserve for lining wear.

The master cylinder force is found by multiplying its area by the required line pressure. The pedal ratio is than selected to give desirable values of pedal travel and pedal pressures. Practical maximum allowable values are 9 inches pedal travel and 240 pounds pedal pressure. If the brake system does not permit of values below these limits, power braking probably should be employed.

Several types of power braking are at present being used; namely:

1. Vacuum
2. Compressed Air
3. Electric
4. Hydraulic

The type of power employed is determined by economic considerations and by the requirements of the service for which the vehicle is designed.

From the foregoing discussion, it will be apparent that the solution of Automotive Brake problems is dependent upon the cooperation of the manufacturers, the road builders, and the operators.

W-2876-107

BENDIX

TREADLE-VAC

A MODERN POWER BRAKE FOR PASSENGER CARS

Since the invention of the automobile, its history has been one of constant improvement directed toward speedier transportation, increased riding comfort and greater driving ease. Each year has seen such improvements as the self-starter, automatic windshield wipers, hydraulic brakes, automatic transmissions and power steering. The automatic transmission, now almost universally used, has reduced the necessity for physical movement to the point where use of the left foot in driving has become almost unnecessary.

And now, although use of the right foot for operation of accelerator and brake is still essential, improvements have been incorporated which further minimize the physical effort necessary to drive a car. Now it is possible to operate accelerator and brake pedals without even lifting the heel of the foot from the floor. Brake pedal height, and travel - has been reduced so that the pad is at approximately the same level as is that of the accelerator. This improvement has been made possible by the use of vacuum power to assist in the application of the vehicle brakes.

Now, power brakes on motor vehicles are not new. They have been in use for many years on commercial vehicles and have been applied to heavy passenger cars, when necessary, at intervals through the years. Today, however, they are used to provide reduced pedal travel and to contribute a degree of brake control far beyond that available with previous applications.

The device by which this has been accomplished is called the Treadle-Vac. It is a development and a product of Bendix and was introduced last year on the 1952 Packard under the name - "Easamatic Power Brake". This year it is also in use by Oldsmobile (Pedal Ease), Lincoln and Mercury.

This Treadle-Vac is a combined vacuum and hydraulic unit which utilizes intake manifold vacuum and atmos-

pheric pressure as a power source. It is a self-contained unit which requires no exposed linkage since actuation is by directly applied force. As mentioned before, its use offers decreased pedal height, - and travel - and a considerable reduction in necessary leg movement. Slightly lighter pedal effort is also gained.

Normal use of power brakes has been confined, primarily, to commercial and comparatively heavy passenger vehicles. In brief, power is required on such heavy vehicles because of mechanical - and physical - limitations which make it impractical and often impossible to produce sufficient braking with physical force alone. The use of power, as a supplement under the control of driver effort, has been necessary to provide adequate control for these vehicles.

However, in the case of the conventional passenger car these limitations have not been too great a factor. On most cars today vehicle weight brake size and brake effectiveness are such that physical effort required to stop is not excessive and pedal effort, both force and travel, remain in a comfortable range. With introduction of automatic transmissions and consequent elimination of the clutch pedal, the brake pedal became obvious, - like a "sore thumb", - and the advantage of providing reduced pedal travel for driving ease became more apparent. The desirability of such an improvement has been the inspiration for development of the "Treadle-Vac".

Now, actually the Treadle-Vac is the product of a long history in the application of power brakes to all types of vehicles. With specific regard to passenger cars, power brakes found production use as early as 1928 on Pierce-Arrow and Stutz motor cars. They were quite generally used through the 30's by Buick, Cadillac, Lincoln, Chrysler, Packard, Studebaker and others.

189

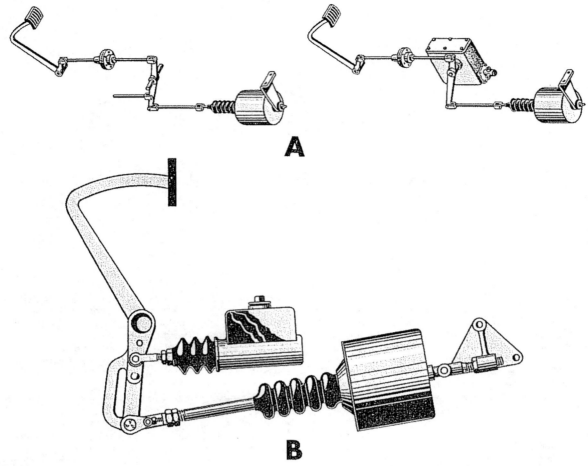

FIGURE 1

These early power brake systems, which were adapted to use with either mechanical or hydraulic brakes, once consisted of a power cylinder actuated and controlled by an external valve mounted in the pedal linkage as shown by section A in Figure 1. Section B illustrates a later method incorporating an internal valve power cylinder controlled through reactionary linkage.

In 1940 a revolution in power braking took place with the introduction of the Bendix Hydrovac to make possible the elimination of these linkage type systems: - and their limitations. This basic type unit, a self-contained vacuum over hydraulic power package as representatively illustrated in Figure 2 has been universally applied to all types of hydraulic brake systems ever since.

Experimentally, the use of power to permit use of a "low" Pedal has been pursued since as early as 1941 when a Hydrovac was used in an installation made on a Hydromatic Oldsmobile. Early in 1945 power was also used to provide a "low" pedal in the "Negotiation B-1" Buick, a futuristic low slung beauty built in 1941 for the G. M. styling section. Later in 1945 the direct acting principle, in use today, was somewhat exemplified by a reactionary linkage power brake system, installed as illustrated by Figure 3.

These are but some of the highlights of the progress that has been made to lead from the power brakes of "necessity" used in former years to those of "convenience" in use today.

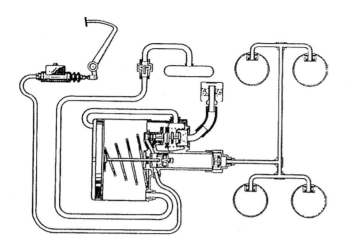

FIGURE 2

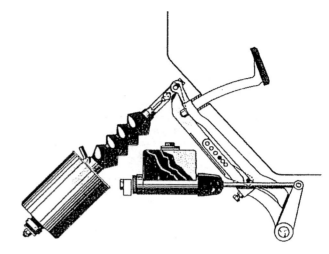

FIGURE 3

To understand the reason for use of the Treadle-Vac to permit low pedal brakes, let us briefly review the requirements fulfilled by the standard hydraulic brake system.

To provide adequate stopping ability in a specific hydraulic brake system, a certain total volume of fluid must be available which can be elevated to pressures sufficient to provide the required thrust at the brake wheel cylinder. This must be done within a comfortable range of physical effort and travel at the foot pedal. These requirements determine the pedal ratio and pedal travel needed in a full physical hydraulic brake system. Fundamentally, two basic hydraulic elements, - displacement and pressure - must always be considered if the hydraulic brake system is to be satisfactory.

Here, in Figure 4, is shown, diagrammatically, a conventional brake system. The pedal and master cylinder combination is such that sufficient hydraulic line pressure can be raised to apply the brakes and adequate fluid can be displaced to provide for brake fade and lining wear. The position of the pedal pad at drum contact, wheel slide, and during seven fade stops from 70 MPH are those that would occur with an average set of brakes adjusted to recommended minimum drum clearance. It can clearly be seen here that it would be impractical merely to lower the pedal by half. There would be insufficient pedal travel without even con-

sidering that required to compensate for lining wear. The basic hydraulic element of displacement would be sacrificed and the system would be unsatisfactory.

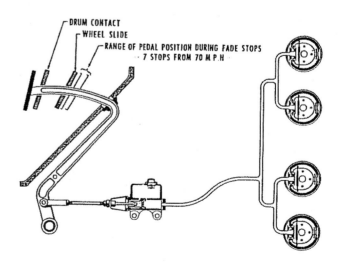

FIGURE 4

If we were to decrease the pedal travel to half that normally available, as in Figure 5, it would be necessary to change the point at which the master cylinder push rod is attached to the pedal lever in order to retain the original master cylinder travel and fluid displacement. Such a step is illustrated by Figure 5. By doing this, however, the mechanical advantage of the pedal lever would be cut in half. It would then be necessary to push twice as hard as before to raise the same line pressure. And, likewise, for the same pedal pressure, we would now produce only half the line pressure. Since the hydraulic element of pressure is thus decreased, such a system would also be unsatisfactory.

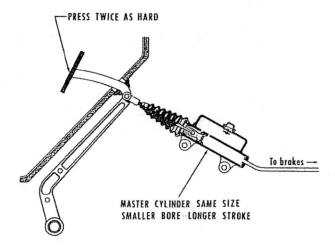

FIGURE 6

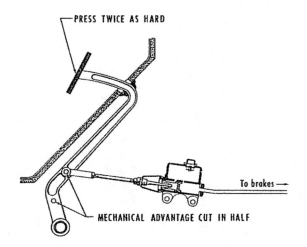

FIGURE 5

that it is possible to raise the same line pressure without pushing any harder than is necessary with the "old long stroke pedal." Figure 7 illustrates the method of power application. In this system is retained the full displacement of the original, and, though the pedal travel is reduced, pedal effort required is no higher than with the old style long stroke pedal because of the assistance of the power brake device. Essentially this is the "Treadle-Vac" low pedal power brake system.

This same result may also be obtained by using a special master cylinder attached to the pedal lever in line with the pedal pad as shown by Figure 6. Pedal pressure is thus applied as a direct force to the master cylinder. The master cylinder used is the same size as that of the standard system. That is, the available displacement is the same. However, it is gained by using a smaller bore than standard since the stroke must now match that of the pedal.

Since we have not changed the size of the master cylinder and the short stroke pedal is retained it is still necessary to push twice as hard, as in the standard system, to provide the same line pressures. Thus, the system is no more satisfactory than the one before.

However, by placing a power brake device between the pedal and the master cylinder, as in Figure 7, a force can be provided to supplement that of physical effort so

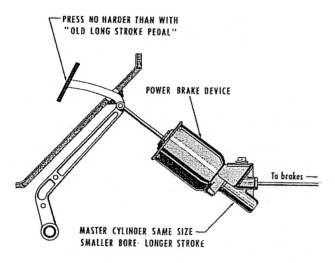

FIGURE 7

Actually the Treadle-Vac system is a bit more refined than shown by Figure 7. In Figure 8, is shown more clearly the form of an actual installation. The Treadle-Vac is directly attached to the underside of the toe board so that actuation is direct without intermediate linkage. Size is such that required pedal effort is actually less than that needed with the standard system. The power unit does approximately two-thirds of the work in applying the brakes so that only one-third is necessary by physical effort. Since the application of force is direct, a lever is no longer required, and a relatively light extension member is needed only to provide location for the pedal pad. Either a suspended pedal as shown, or a treadle like that of the conventional accelerator, can be used.

Although this method of direct action against the unit mounted on the toe board provides the ultimate in simplicity and advantage over the conventional brake system other types of installation have been successfully used. Some of these methods are illustrated in Figure 9.

Section A shows a linkage type as used by Lincoln and Mercury. This method provides the most satisfactory protection of linkage, required when remote unit location is necessary.

Section B presents the most simple means by which remote location of the unit can be provided. Basically, this is still a direct acting type, but with "extended" actuation through a pedal arm segment and long push rod.

Section C is of another method by which remote unit

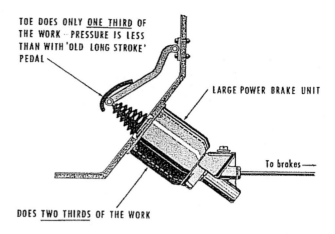

TOE DOES ONLY <u>ONE THIRD</u> OF THE WORK — PRESSURE IS LESS THAN WITH 'OLD LONG STROKE' PEDAL

LARGE POWER BRAKE UNIT

To brakes —

DOES <u>TWO THIRDS</u> OF THE WORK

FIGURE 8

location has been obtained without undue linkage exposure. However, this method (and its variations) is not one of greatest economy.

Section D shows a method of linkage actuation which permits use of conventional pedal geometry. One variation of this design has been devised to provide maximum production interchangeability between power and conventional installations.

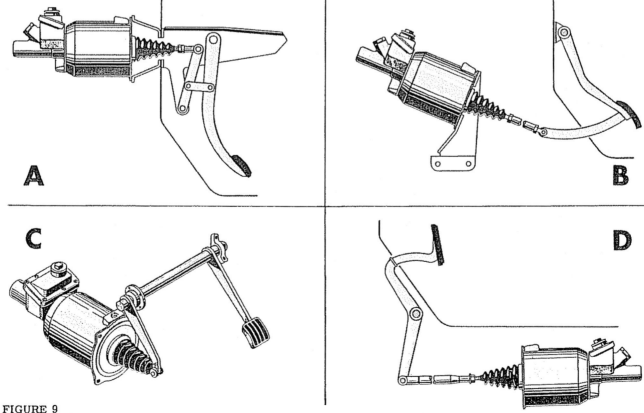

A

B

C

D

FIGURE 9

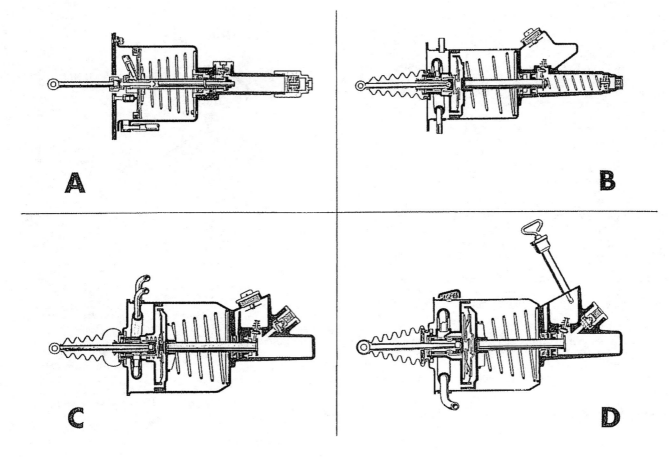

A

B

C

D

FIGURE 10

With the introduction of this low pedal braking by the use of vacuum power, "foot and leg" brake operation has been replaced with "toe and ankle" control. As a result, driving fatigue has been greatly reduced by minimizing leg and foot movements in applying the brakes. By virtue of this reduction in leg movement, reaction time to apply the brakes is decreased and stopping distance is thereby improved. More sensitive control of the brake is attained by better muscle manipulation afforded by use of the heel as a rest. Thus, the Treadle-Vac contributes much to greater safety and driving comfort.

The device which has been developed to provide this sort of low pedal braking is a self-contained vacuum hydraulic power system not unlike some forms of commercial vehicle units, such as the Bendix Hydrovac and Air-Pak. Both types do a good share of the work required during application of the brakes. They do this in much the same manner although the basic design is quite different.

While the commercial type power units are designed to be used as a supplement to the existing brake system, the Treadle-Vac is different in that it replaces the conventional master cylinder and brake pedal.

As it is manufactured today, the Treadle-Vac is the evolution of a basic design created approximately 5 years ago. The development program has carried the design through many stages: - through those illustrated by A, B, and C of Figure 10 to that of the modern unit as presented in section D.

The Treadle-Vac embodies the master cylinder within itself and is caused to operate by a force transmitted to it directly from the foot through the pedal. Actuation is by mechanical force rather than by an input of hydraulic pressure. You have seen, in the diagrammatic form in the previous figures, one method by which the Treadle-Vac unit may be applied to a passenger car brake system. Let us now look at the internal construction of the unit and the way in which it operates during a cycle of brake operation.

Figure 11 illustrates diagrammatically a typical Treadle-Vac installation. The unit, in section, is shown in released position as would be the case when the brakes are not being applied. As you can see the installation is quite simple and the connections are few. A mechanical connection is made at the treadle or pedal pad. A vacuum connection is made to the intake manifold of the engine through a vacuum check valve. A vacuum reserve tank is provided in this line to prevent immediate power failure in case the engine stops running. Air enters the unit through an integrally mounted air cleaner and no connection for this purpose is therefore required. Hydraulic connection is made to the basic brake system in the conventional manner at the output port of the master cylinder.

The Treadle-Vac is composed of three basic sections: - The hydraulic section, the vacuum power cylinder and the control valve assembly.

The hydraulic portion is at the extreme right and is filled with fluid as indicated. Note the construction of the master cylinder. It is unique in that fluid displacement and pressure is created by movement of a hydraulic plunger into the fluid system thereby eliminating the conventional moving piston and cup, and the necessity for close control of diameter and finish of the master cylinder bore. Movement of the plunger toward the right permits the tilting valve, located between the

barrel and the reservoir, to close and fluid is forced to the brakes through the hydraulic lines. Since there is no cup to pass over the usual compensating port, a common ailment of the conventional master cylinder is eliminated.

The second major portion of the unit is the power cylinder itself. With the unit in released position the complete cylinder is open to atmospheric pressure and the vacuum piston within it, which provides the force necessary to move the plunger into the hydraulic portion, has no tendency to move under the condition of balanced pressure on either side of it, - as shown.

Within the vacuum piston is embodied the control valve mechanism and reaction elements which determine the amount of brake application to be made in response to the force of the foot upon the pedal. In the released position, the valve plunger is to the left of the valve sleeve under the force of the valve return spring. Communication of the unit to vacuum source is thus cut off and the power cylinder piston is balanced under atmospheric pressure. The power cylinder return spring forces the hydraulic plunger, and power cylinder piston, to the left against the power piston release stop. In this position the tilting valve in the master cylinder is held open by contact of the lower stem with the washer at the end of the hydraulic plunger. The fluid in the barrel of the master cylinder is thus vented to atmospheric pressure in the master cylinder fluid reservoir.

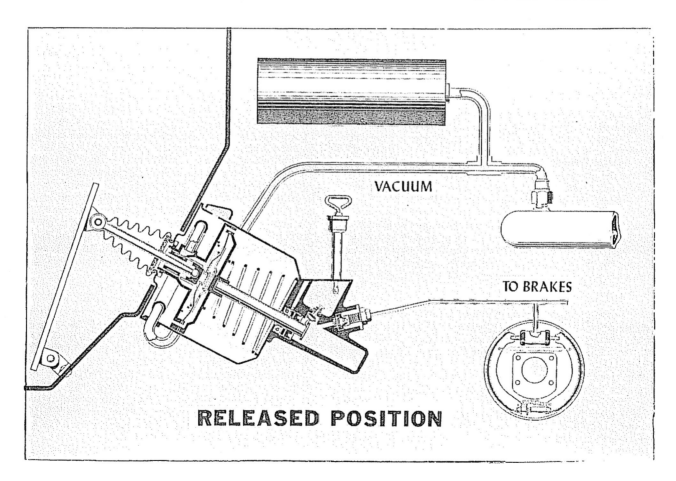

VACUUM

TO BRAKES

RELEASED POSITION

FIGURE 11

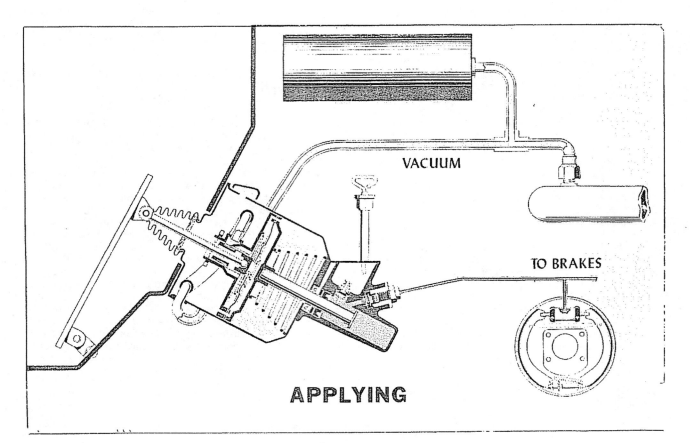

VACUUM

TO BRAKES

APPLYING

FIGURE 12

In Figure 12 the unit is shown in applying position while a brake application is being made. The valve piston has been pushed to the right, against the action of the valve return spring, and the connection of the right side of the power cylinder piston to atmosphere has been cut off while communication to the manifold has been established. The left side of the power cylinder piston remains in constant communication with air at atmospheric pressure. A pressure differential is thus created across the power piston which causes it to move to the right forcing the hydraulic plunger into the master cylinder bore. The tilting valve is permitted to close, by movement of the hydraulic plunger, and the fluid, under pressure, is transmitted to the brake system.

The pressure differential which is in force across the power piston also exists across the reaction diaphragm but in opposite direction. The right side of the diaphragm is in constant communication with atmosphere to the left of the power piston while the opposite side is subject to control vacuum. Thus there is a force exerted back against the driver's foot through the control valve which is always in proportion to the force exerted by the power piston against the fluid in the hydraulic chamber. This force provides feel at the brake pedal which is always in direct relationship to the amount that the brakes are being applied. The total force against the hydraulic fluid is a summation of physical effort and that created by the power cylinder piston.

Figure 13 shows the Treadle-Vac in holding position which occurs when the brakes have been applied and are maintained without increase or decrease in brake

pedal effort. The force of the reaction diaphragm has balanced that of the foot against the brake pedal. This causes the valve to close and cut off further change in pressure differential across the reaction diaphragm and the power piston. In this state, the right side of the power piston is no longer in communication with either atmosphere or vacuum source so that the degree of pressure differential is held and no further action will take place. Further increase in pedal pressure will again connect the control chamber to vacuum and fluid pressure in the master cylinder will be raised. If pedal pressure is relieved, a connection will be established to atmosphere and a decrease in hydraulic pressure will take place.

In figure 14 the unit is now shown releasing. Pedal pressure has been removed and the control valve is to the left against its stop under the load of the valve return spring. This again creates a connection between the right and left sides of the power piston permitting the piston to return to a condition of balance with atmospheric pressure on both sides. The force of the power piston return spring, and existing hydraulic pressure against the plunger, forces the power cylinder piston and hydraulic plunger to the left until full release position against the piston release stop has been reached as was shown in Figure 11.

The tilting valve is opened by contact with the release washer on the end of the hydraulic plunger and communication is again established between the master cylinder pressure chamber and the fluid reservoir. Hydraulic line pressure is relieved and the brakes are again released.

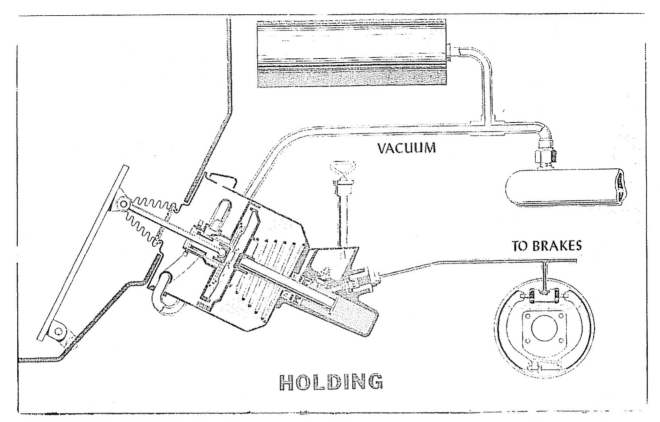

FIGURE 13

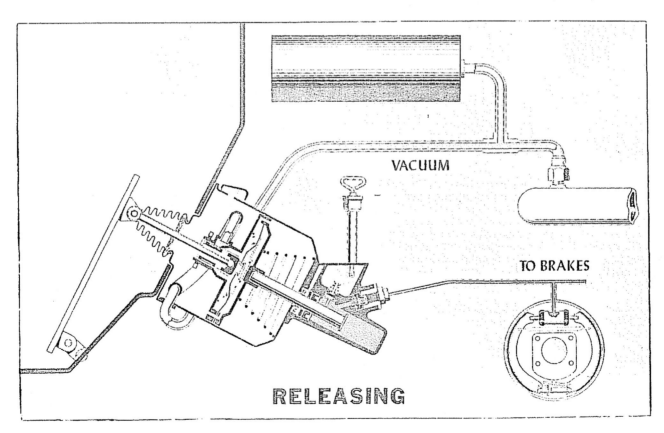

FIGURE 14

This review of the operation of the Treadle-Vac has been rather brief. I would like to continue with a more detailed explanation of the construction and operation of the reaction diaphragm which provides so much of the refinement in control that has been attained with the Treadle-Vac.

A major problem in the development of low pedal power braking has been that of providing a light and "silky" initial pedal feel. It is desirable that the point of initial deceleration be reached with little more pedal effort than that provided by the weight of the toe on the pedal pad and, at the same time, to retain graduated reaction between the point of full pedal release and that at which the brakes just begin to do their work. To accomplish this result, two stages of reaction have been incorporated in the Treadle-Vac design. To explain why, let me show you an evolution of the reaction system.

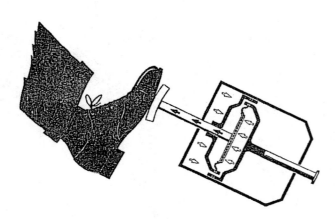

SIMPLE REACTION

FIGURE 15

In Figure 15, is shown diaphragm reaction in its simplest form. A single diaphragm member exerts its force, created by differential pressure, against the foot through the push rod and intermediate control valve. (This valve, of course, is normally interposed between the diaphragm and the push rod as displayed in preceding figures. In these, such detail construction has been eliminated to more clearly show the reaction pattern.)

This form of construction produces reaction - and pedal feel - which directly corresponds to the degree of differential pressure raised across the power piston. However, it does not accurately reflect the degree of brake application, in terms of deceleration, because of inherent and designed losses (return spring loads and seal frictions) in the Treadle-Vac and in the brakes themselves.

The function of the diaphragm is linear with that of pressure differential while the end product of deceleration is not. As a result, with a diaphragm selected to provide adequate reaction for safe and comfortable control of the deceleration range the pedal effort required to produce the sensation of initial deceleration is too high and definition at shoe to drum contact is lost.

To produce a more accurate reproduction of the braking effect it is desirable to obtain lighter initial with more definite indication of shoe contact. This can be accomplished by slight modification of the simple diaphragm form.

In Figure 16, we have a reaction system very similar to that of the previous figure. However, the push rod and diaphragm are no longer connected and a spring has been added to act against the diaphragm. This spring is an anti - or counter reaction device which serves to prevent the force of the diaphragm from being transmitted to the foot until a desired degree of pressure differential is reached. The force of the power piston may be used to overcome system losses without transfer of the normal rate of reaction to the foot. Any desirable value of pedal effort at the point of initial brake effectiveness thus may be obtained by proper selection of counter reaction spring load.

Although the anti-reaction method permits low initial pedal pressure it lacks good control of initial pedal travel. Because there is no change in the load needed to move the pedal from full release to approximately the point of drum contact intermediate positioning of the pedal is difficult and the pedal seems to "fall away" as an application is started.

To prevent this sensation a primary stage of reaction is needed.

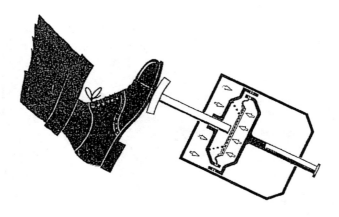

ANTI-REACTION

FIGURE 16

Figure 17 represents the actual form of reaction elements used in the Treadle-Vac. Two stages of reaction are provided to permit light, but progressive control of pedal initial with heavier reaction in the working range of the brake application.

The diaphragm assembly actually consists of two sections. A small inner portion works against the push rod from the beginning of the actuation cycle and provides low rate graduated reaction in the first stage of application. The outer portion is held away by a counter reaction spring as in the anti-reaction form, and is arranged to pick up the inner diaphragm as the application is increased so that the two then act as one to develop the higher rate of reaction desired in the deceleration range.

This is the reaction system that produces the accurate and responsive control available with the Treadle-Vac low-pedal power brake system. A true reflection of brake position and the intensity of application is secured.

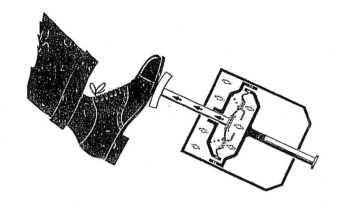

TWO STAGE REACTION

FIGURE 17

This has been but a quick review of the highlights of Treadle-Vac operation and design as it exists today. There are, of course, many more aspects of the present design and of the developments for the future that have not been covered.

The modern power brake to provide "convenience" in driving has been given birth. Improvements we now are developing will contribute to even more general acceptances and universal application.

Proportioning Valve to Skid Control — A Logical Progression

Frederick E. Lueck, William A. Gartland,
and Michael J. Denholm
Borg and Beck Div., Borg-Warner Corp.

UP TO THIS TIME, most braking systems have been designed with a fixed front-to-rear brake ratio. This provides a fundamentally simple system with a given level of brake performance at some arbitrary loading condition.

With continuing emphasis on improved brake performance and improved driver control, this approach to brake system design is becoming outmoded. For the purposes of this discussion, brake controls will be defined as any device which intelligently modifies the driver-controlled brake system input.

It is the objective of any brake control system to maximize the stability and controllability of the vehicle while minimizing stopping distances. Brake control systems attempt to attain these objectives by making the most of the available force at the tire-to-road interface.

While skid control, hold-off valves, fixed-ratio proportioning valves, and load-sensitive proportioning valve systems attain these ends to different degrees and by different methods, the objectives of stability, controllability, and improved stopping distances are the same. For this discussion, we will confine our comments to the areas of skid control and load-sensitive proportioning valves.

LOAD-SENSITIVE PROPORTIONING VALVE

As shown in Fig. 1, as a vehicle increases its deceleration, the ratio of front-axle-to-rear-axle loading changes as a function of the vehicle deceleration, wheel base, and height of the center of gravity of the vehicle.

If we superimpose a fixed front-to-rear braking ratio over this vehicle system, it is obvious that the brake balance can only be correct for one specific vehicle deceleration. At all other coefficients of friction, either premature front-wheel or rear-wheel slide will occur.

To provide the maximum stability and minimum stopping distance, we would prefer that all four wheels lock simultaneously. When we add the variable of changing vehicle load, as shown in Fig. 2, the problem becomes even more severe in that the imbalance between the fixed brake ratio and the impressed axle load becomes even greater.

ABSTRACT

This paper discusses the development of a family of brake control devices capable of handling all vehicles from passenger cars through air-braked heavy trucks. These devices consist of: Hydraulic load-sensitive proportioning valves for passenger cars, light trucks, and vans; hydraulic load-sensitive proportioning valves for medium trucks; pneumatic load-sensitive proportioning valves for air-braked heavy trucks and tractor/trailer combinations; skid control systems for passenger cars; and skid control systems for trucks and tractor/trailer vehicles.

The paper explains how this broad approach to brake controls allows the selection of a system which can be tailored to the particular brake control needs of a specific vehicle and its duty cycle. Except for the passenger car skid control, these devices are capable of being retro-fitted on existing vehicles.

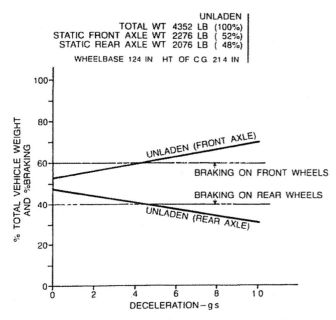

Fig. 1 - Weight transfer of a typical sedan - unladen

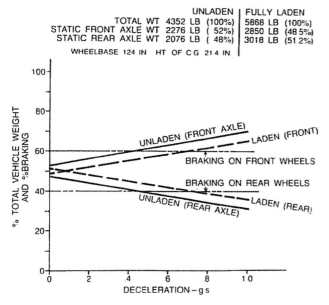

Fig. 2 - Weight transfer of a typical sedan - laden and unladen

It is the aim, then, of the load-sensitive proportioning valve to match the applied brake torque to the wheel torque generated at the tire patch.

WHAT IT THEORETICALLY ACCOMPLISHES

If we look at this same data in a different fashion, we can observe both the effect of vehicle loading and the effect of changing coefficient of friction between the tire and road. Fig. 3 shows a plotted curve of the braking efficiency versus

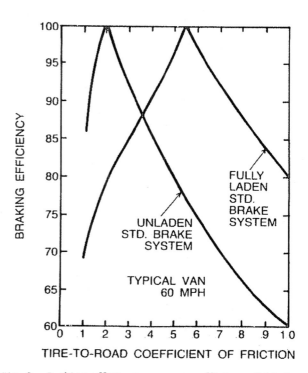

Fig. 3 - Braking efficiency versus coefficient of friction - laden and unladen

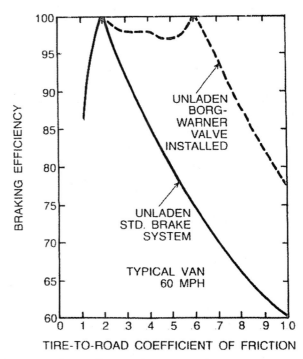

Fig. 4 - Braking efficiency versus coefficient of friction - unladen with and without load-sensitive proportioning valve

the coefficient of friction between tire and road. Braking efficiency is defined as the ratio of actual vehicle deceleration to available tire-to-road coefficient of friction.

In other words, if a vehicle is capable of attaining 0.8 g deceleration at first wheel lock on a 0.8 u surface, the braking efficiency is defined as 100%. It is obviously the goal of any braking control system to provide maximum braking efficiency across the widest range of load and road surface.

Fig. 3 shows the brake efficiency of a typical short wheelbase van in the unladen and laden condition. Note the shift in peak efficiency between laden and unladen conditions. Rear wheel slide, which occurs to the right of the peak. exists over a wider range of coefficient of friction in the unladen condition.

Now, over the unladen curve we will overlay the performance change that can be attained by the use of a load-sensitive proportioning valve. This is shown in Fig. 4. Note that in the unladen condition, the braking efficiency has been increased over a wide range of tire-to-road friction.

ACTUAL PERFORMANCE GAINS

To demonstrate the actual performance improvement that is attainable with a system of this type, a series of tests were conducted on this same van.

Fig. 5 shows the deceleration capability at first wheel slide at 30 and 60 mph in the unladen and laden condition. As would be expected, the major improvements are in the unladen condition, where the brake balance is most compromised in a system without a load-sensitive proportioning valve.

Fig. 6 depicts the performance of a medium truck (24,000 lb gross) under various load conditions. Here, we have plotted stopping distance with first wheel lock. You will note again that the stopping distance is significantly reduced in the unladen and half-laden conditions. This indicates that the basic brake balance was set up for the maximum load condition.

Fig. 7 depicts the same performance characteristics with load-sensitive proportioning valves applied to the driving wheels and the trailer wheels of an air-braked tractor/trailer rig (32,000 lb gross combination weight). Again, since the normal combination rig has its brake balance optimized for the loaded condition, there is little difference with and without the valve when laden. Major performance gains occur at no load or partial load.

HOW PROPORTIONING VALVES ACCOMPLISH THIS

Because of the wide variety of displacement requirements for various brake systems and different operating media, a family of valves is required to cover the entire line of vehicle applications.

Fig. 8 shows a basic installation of the hydraulic load-sensitive proportioning valve for passenger car, van, and light truck. One end of the torsion bar linkage is attached to the rear axle of the vehicle. The other end is connected to the proportioning valve, which is fastened to the chassis. The torsion bar twists in relation to the load applied on the axle. The torsional reaction, which is taken by a piston on the valve, is then the load-sensitive input.

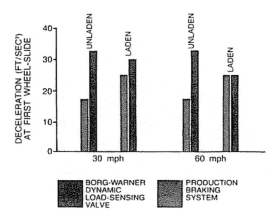

Fig. 5 - Deceleration versus speed and loading - production brake system versus load-sensitive proportioning valve system, 1969 production van

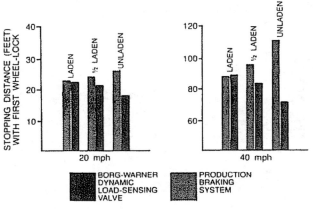

Fig. 6 - Stopping distance versus speed and loading - production brake system versus load-sensitive proportioning valve, 2 1/2-ton truck

Fig. 9 shows a schematic of the valve. The torsion bar input to the valve applies a load to the main valve piston. The torsion bar load plus the return spring load tend to keep the valve piston away from the sealing ring. When sufficient inlet pressure is generated, the inlet pressure on the large diameter of the valve is sufficient to cause the main piston of the valve to move to the left against the combined spring load.

At this point, the valve shuts off against the sealing ring; and the split point of the valve is reached. Above this point, proportioning at a fixed ratio occurs. This ratio is a function of the relative areas of the piston head and piston stem.

A typical performance curve for this valve is shown in

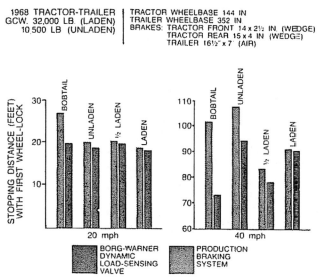

1968 TRACTOR-TRAILER | TRACTOR WHEELBASE 144 IN
GCW. 32,000 LB. (LADEN) | TRAILER WHEELBASE 352 IN
10,500 LB (UNLADEN) | BRAKES: TRACTOR FRONT 14 x 2½ IN. (WEDGE)
   TRACTOR REAR 15 x 4 IN (WEDGE)
   TRAILER 16½" x 7" (AIR)

Fig. 7 – Stopping distance versus speed and loading – production brake system versus load-sensitive proportioning valve, tractor/trailer

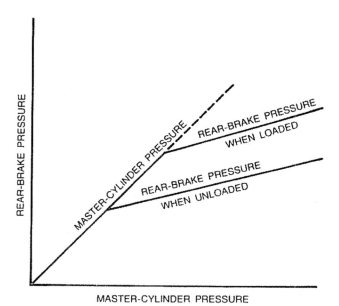

Fig. 10 – Dynamic load-sensitive proportioning valve performance

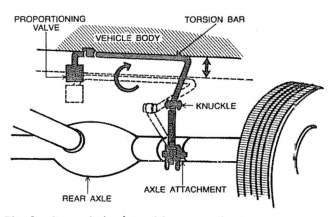

Fig. 8 – Dynamic load-sensitive proportioning valve installation

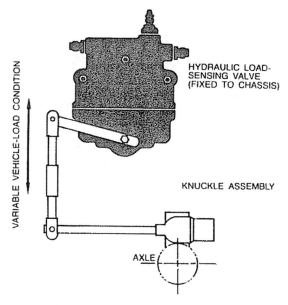

Fig. 11 – Hydraulic load-sensitive proportioning valve installation

Fig. 10. By adjusting the internal ratio of the valve and the rate of the torsion bar, the performance of the valve can be adjusted to the requirements of a particular type of vehicle.

While providing the same function, the hydraulic valve for a medium truck, as shown in Fig. 11, accomplishes this by a different means. This unit, which is fixed to the bed or chassis of the truck, is linked to the axle with a lost motion device or knuckle assembly to prevent over-travel in the valve. In this way, the vehicle axle load is sensed and transmitted to the valve.

As shown in Fig. 12, hydraulic braking pressure from the master cylinder is applied to the load-sensing valve and

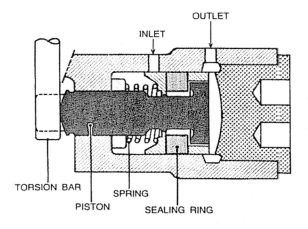

Fig. 9 – Dynamic load-sensitive proportioning valve

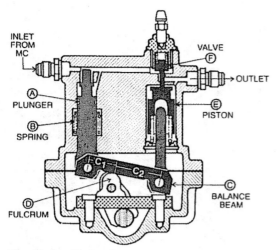

Fig. 12 - Hydraulic load-sensitive proportioning valve

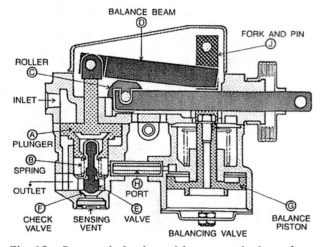

Fig. 13 - Pneumatic load-sensitive proportioning valve

acts upon plunger A, forcing it down against the load in spring B. The plunger then acts on one arm of the balance beam C, which pivots on fulcrum D. The position of fulcrum D is determined by the position of the frame relative to the axle; that is, it is a function of the axle load.

On the other arm of beam C is piston E, which moves valve F to open or close the passage between the inlet port and the outlet port. Valve F can isolate the inlet from the outlet. Balancing of this system occurs when the outlet pressure on piston E multiplied by the lever arm C_2 equals the inlet pressure on plunger A multiplied by the lever arm C_1.

Since the length of the arms is governed by the fulcrum position, which is dependent on axle load, the brake proportioning is determined by axle load. The pneumatic valve shown in Fig. 13 is similar in installation and in principle to the hydraulic valve, except that valve E acts both as an isolation valve and an exhaust valve to release the compressed air to atmosphere on brake release.

SUMMARY

In summary, the load-sensitive proportioning valve has the ability to increase vehicle deceleration before first wheel lock to improve vehicle stability by preventing premature wheel lock. In articulated vehicles, it can minimize the possibility of jackknifing by preventing premature locking of the tractor drive wheels (1) *.

Devices of this type cannot eliminate skidding, but do significantly reduce the hazard by providing greater latitude and margin for driver error.

SKID CONTROL

As mentioned earlier, proportioning valves -- regardless of their level of sophistication -- can only minimize the possibility of skidding by providing greater latitude for driver error.

The true skid control system seeks to prevent skidding by eliminating protracted locking of the wheels. The ideal skid control system, by preventing wheel lock, can improve vehicle stability, vehicle controllability, and stopping distances under almost all driving and road conditions.

At the present "state-of-the-art," this theoretical performance level is not attainable.

REAR-WHEEL VERSUS FOUR-WHEEL SKID CONTROL

The ultimate skid control system must understandably be a four-wheel system if all theoretical advantages are to be obtained. However, before a choice between rear-wheel and four-wheel systems can be made, performance and economic considerations must be taken into account.

It is obvious that the application of a rear-wheel skid control system cannot provide the ultimate objective of vehicle controllability since, by its very nature, the rear-wheel system allows the front wheels to lock. In this condition, no steering forces can be generated (2). For this reason, four-wheel skid control systems are actively under development.

The rear-wheel system, however, can and does provide significant improvement in vehicle stability by assuring straight-line stops under almost all conditions. It also offers significant improvement in vehicle stopping distance. This can be attained at a price which is less than half that required for a four-wheel system of the same concept.

PRINCIPLES OF SKID CONTROL

The basic phenomena which permits a skid control system to function is best described by the family of μ-slip curves shown in Fig. 14 (3). These curves demonstrate the principle that the maximum retarding force is generated when the tire

---

*Numbers in parentheses designate References at end of paper.

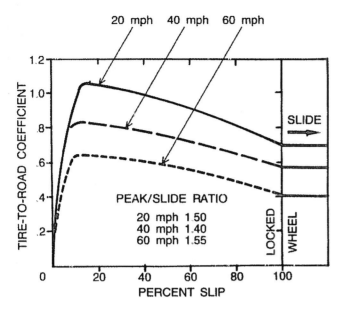

Fig. 14A - Typical μ-slip curves - wet concrete

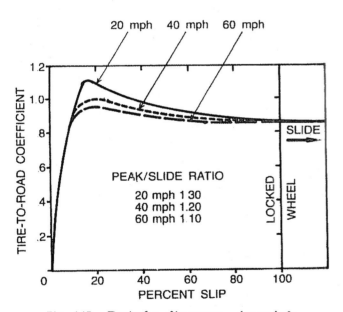

Fig. 14B - Typical μ-slip curves - dry asphalt

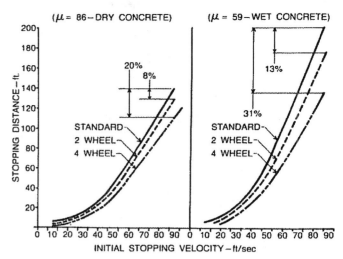

Fig. 15 - Theoretical stopping distance versus velocity for a standard brake system, rear-wheel skid control system, and four-wheel skid control system

is rotating with a peripheral velocity less than the corresponding translational velocity of the vehicle.

The coefficient of friction at 100% slip (or locked wheel) represents the maximum deceleration that can be obtained on a locked-wheel stop. Therefore, the ratio of the peak coefficient of friction to the locked-wheel coefficient of friction represents the improvement in stopping distance that can be attained with an optimum skid control system.

Fig. 15 plots the theoretical improvement in stopping distances for two-wheel and four-wheel skid control systems on dry pavement and on a wet concrete surface. It is readily apparent from these curves that significant improvement can be obtained on lower coefficient surfaces, while less

improvement (on a percentage basis) can be attained on dry pavement. These improvements are a direct result of the ratio of peak coefficient to locked-wheel coefficient.

REAR-WHEEL PASSENGER
CAR SKID CONTROL

Since the present "state-of-the-art" has made the rear-wheel skid control system attainable, we would like to describe a workable rear-wheel skid control system.

From a functional standpoint, today's skid control systems can be divided into three basic elements:

1. The sensor, which monitors the velocity or deceleration of the wheels.

2. A logic device, which processes this information into a usable signal.

3. An actuator, which modulates the brake pressure in some fashion on command from the logic device.

A typical system schematic is shown in Fig. 16.

To move from the theoretical system to a real system, the sensor shown in Fig. 17 is mounted in each of the rear wheels and monitors the deceleration of the rear wheels. When a preselected triggering level is reached, a signal is sent to the logic module (Fig. 18), which processes the signal and provides a command function to the actuator (Fig. 19).

The actuator then reduces the pressure to the rear brakes by increasing the volume of the rear brake system. This reduction in pressure causes a reduction in brake torque and allows the wheel to accelerate. The pressure is then reapplied by the actuator, and the cycle is repeated five to eight times per second.

Typical performance of such a system is shown in Table 1. Note that as predicted, major performance improvements occur on the lower coefficient surfaces where the higher ratio of peak-to-sliding coefficients between the tire and road provide greater potential for improvement.

Table 1 - Performance Summary of the Passenger-Car
Skid Control System (1968 Sedan)

Vehicle Speed (mph)	Apparent Coefficient of Friction	Stopping Distance (Feet)		Percent Improvement
		Standard Brake System	Borg-Warner Skid Control	
10	.09	41.6	39.3	6.0
10	.21	15.6	15.2	3.0
10	.24	14.4	13.5	6.0
20	.19	71.8	62.2	15.0
20	.63	21.6	21.4	--
30	.66	46.8	44.2	6.0
40	.71	75.2	68.7	10.0
40	.76	72.1	68.6	5.0
60	.71	170.2	164.0	3.5
60	.93	129.3	123.4	5.0

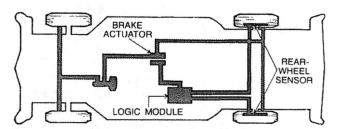

Fig. 16 - Schematic of a rear-wheel skid control system

Fig. 17 - Rear-wheel skid control system sensor

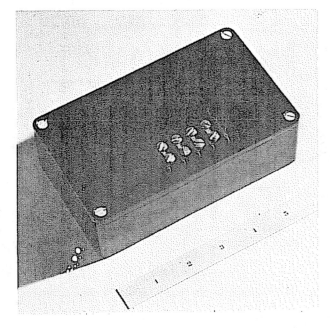

Fig. 18 - Rear-wheel skid control system logic module

SKID CONTROL FOR COMMERCIAL VEHICLES

For trucks and tractor/trailers, skid control can provide the same advantages as for passenger cars. In addition, on articulated vehicles jackknifing can be minimized by preventing locking of the driving wheels on the tractor (4).

A schematic of a typical air-operated skid control system is shown in Fig. 20. Three components make up this system:

206

Fig. 19 - Rear-wheel skid control system actuator

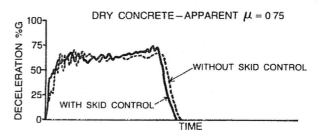

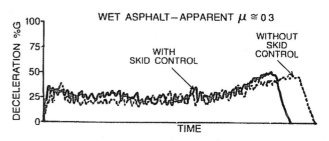

Fig. 21 - Vehicle deceleration versus time - with and without skid control - 30 mph, 32,000 lb GCW, tractor/trailer

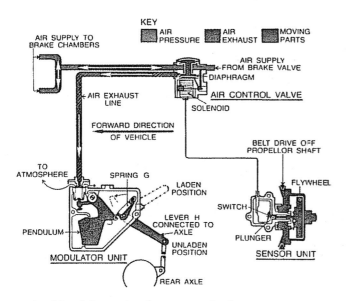

Fig. 20 - Schematic of a pneumatic skid control system

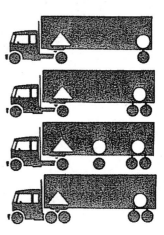

 BORG-WARNER PNEUMATIC SKID-CONTROL SYSTEM

 BORG-WARNER PNEUMATIC LOAD-SENSING VALVE

Fig. 22 - Typical schematics of combination skid control and proportioning valve systems

1. The sensor, which is normally installed on the driving axle.

2. The air control valve.

3. The modulator.

It should be noted that in spite of the difference in names, the three basic skid control elements are present. Whenever the controlled wheels reach incipient skid, the sensor actuates the air control valve. This allows the air pressure in the brakes to vent through the exhaust line to the modulator.

The atmospheric valve in the modulator, which is controlled by the pendulum and the load-biasing spring G, adjusts the pressure to which the brakes are released. In this fashion, the amount of pressure drop is determined both by the coefficient of friction between the tire and road (which is determined by the pendulum, which measures vehicle deceleration) and the load on the bias spring G, which measures the load on the axle.

The improvement in performance of this system is shown in Fig. 21. Again, it should be noted that the performance gains are greatest at the lower coefficients because of the higher ratio of peak-to-sliding coefficients.

COMBINATION SYSTEMS FOR ARTICULATED VEHICLES

To provide maximum cost effectiveness, a combination of skid control on the driving axle of the tractor coupled with load-sensitive proportioning valves on the trailer axles can be utilized. A schematic of this system is shown in Fig. 22. This provides a system with excellent cost effectiveness and minimum complexity.

An advantage of this approach to the problem is that jackknifing can be minimized significantly, even on interchange trailers which do not have skid control or proportioning valves.

SUMMARY

In summary, by approaching the problems of brake controls with a broad viewpoint, this family of brake control systems allows the controls to be selected on the basis of the specific vehicle requirements. It is apparent that no single device can provide the answer for all vehicles under all conditions while maintaining maximum cost effectiveness.

REFERENCES

1. H. A. Wilkins, "Jackknifing." Motor Transport Magazine, March 29, 1968.

2. T. A. Byrdsong, "Investigation of the Effect of Wheel Braking on Side-Force Capability of a Pneumatic Tire." NASA TND-4602, June, 1968.

3. J. L. Harned, L. E. Johnston, and G. Scharpf, "Measurement of Tire Brake Force Characteristics as Related to Wheel Slip (Antilock) Control System Design." Paper 690214 presented at SAE International Automotive Engineering Congress, Detroit January, 1969.

4. "The Jackknife Terror." Precision Magazine, March/April, 1968, pp. 21-25.

5. W. C. Eaton and I. J. Schreur, "Brake Proportioning Valve." Paper 660400 presented at SAE Mid-Year Meeting, Detroit, June 1966.

6. G. Putnam, "Brake Balancing Under All Conditions." Fleet Owner Magazine, July, 1968, pp. 53-56.

7. H. J. H. Starks, "Loss of Directional Control in Accidents Involving Commercial Vehicles." Presented at the Symposium on Control of Vehicles during Braking and Cornering, London, England, June 11, 1963.

This paper is subject to revision. Statements and opinions advanced in papers or discussion are the author's and are his responsibility, not the Society's; however, the paper has been edited by SAE for uniform styling and format. Discussion will be printed with the paper if it is published in SAE Transactions. For permission to publish this paper in full or in part, contact the SAE Publications Division and the authors.

Society of Automotive Engineers, Inc.

12 page booklet.   Printed in U.S.A.

Hydraulic Brake Actuation Systems under Consideration of Antilock Systems and Disc Brakes

Otto Depenheuer and Hans Strien
Alfred Teves GmbH

THE DEVELOPMENT of hydraulic power supply systems for passenger car and light truck brake actuation appears necessary and important for three main reasons:

1. The increasing application of fuel injection and the exhaust emission regulations have led to an inlet manifold vacuum of only 40-55% instead of the previous 75-85%. In order to maintain the same degree of servo assistance to the driver's pedal effort while retaining the vacuum booster (and the trend is to even greater assistance), the booster diameter must be increased 15-30% versus today's units. Thus, the already existing packaging problems will become more acute, and can be alleviated only by boosters operating at a higher pressure level.

2. Investigations by the H.S.R.I. of the University of Michigan and the industry have shown that the application time for brakes (that is, the time elapsed between the driver pressing the pedal and a buildup of brake torque) increases with decreasing pressure differential. This means unwanted and even dangerous lengthening of the stopping distance.

3. Antilock systems require in all cases an energy source at least equivalent to that of a servo brake system, and as a rule even greater, because the assisted pedal effort of the driver is not available during the antilock operation. Logically, the same energy source can be used for service brake and antilock system. The installation space for an antilock system with hydraulic control, including the energy source (pump, reservoir, accumulator), is only a fraction of that for a vacuum-actuated antilock system. When establishing parameters for a new braking system, a hydraulic energy source should at least be planned as an alternative, with the possibility of easily installing an antilock system at a later date.

SYSTEMS

CONTINUOUS-FLOW AND ACCUMULATOR SYSTEMS - For the actuation of a braking system, continuous-flow systems as well as accumulator layouts can be considered. The basic designs of both systems are diagrammatically shown in

ABSTRACT

Hydraulic power braking systems for use in passenger cars and light trucks are attracting considerable automotive design attention. This is due to their compactness, smaller space requirements, and better operating "feel," as well as their more direct control over the braking function, which has extremely short application and release times. Moreover, they are readily adaptable to the energy source and controls of an antilock system, and they contribute to (or even form the basis of) a central power-supply system that would provide servo assistance to other vehicle systems. This paper describes and explains ways of creating these brake systems and gives design calculations of brake layouts based on standard values and comparative judgment criteria.

Figs. 1 and 2. The continuous-flow system may at first seem to be more attractive, considering the smaller number of its elements. Instead of the pressure-controlled suction valve in the accumulator system, a suction orifice is all that is needed, and the accumulator and nonreturn valve are eliminated completely. Also, the pressure regulator valve is simpler in the continuous-flow system compared to other systems. On closer examination, two factors speak against the continuous-flow system.

Acceleration - With currently available friction brakes, the volume requirements for maximum deceleration with 0.10 to 0.15s application time is 30-40 cm^3/s/ 1000 kg vehicle weight (0.83-1.13 in^3/s/ 1000 lb vehicle weight). This would have to be delivered at idling engine speeds (600-800 rpm) by a full-flow pump. In other words, for a car weighing 2000 kg (4400 lb) the pump would have to deliver 6-7 cm^3/rev. The energy absorption of such a constantly running pump at higher engine speeds is not inconsiderable, in spite of the suction orifice.

For accumulator systems, the same vehicle would use a pump with only 0.9-1.1 cm^3/rev, which operates only when charging the accumulator.

Stall or Pump Failure - Should the engine stall or the pump be damaged, the continuous-flow system would fail without warning. Even if this system were used only to augment the driver's pedal effort, its failure would be critical when the driver had to apply more than two or three times the normally assisted effort to obtain the same retardation. Experience

shows that most drivers are at such a time psychologically and physiologically overstressed. Therefore, dependent on the vehicle weight, continuous-flow and power-assisted systems (next section), might require an extra pump, either electrically or transmission driven, which in the event of a damaged main pump or stalled engine would be automatically actuated.

Brake Efficiency Deterioration - The sudden failure of an accumulator is less probable, for with a correct design volume a predetermined number of brake applications with a gradual diminishing efficiency can be made, should the pump be damaged or the engine stall.

With these points in mind, the accumulator system and the development of its associated hardware was given preference.

POWER-ASSISTED AND FULL-POWER SYSTEMS

POWER-ASSISTED SYSTEM - A system that permits the driver's pedal effort to be assisted by servopower can (according to the present European regulations) be applied to a vehicle whose gross weight is such that, despite loss of power assistance, 30% deceleration with max 50 kp* (110 lb) pedal effort is still available.

Efficiency Factor - We shall now introduce a dimensionless value for the comparison between various brake systems, called the "efficiency factor" (W^X). This shows the relationship between the total brake force of a vehicle and the pedal effort to generate this force, and is expressed as

$$W^X = \frac{\text{total brake force}}{\text{pedal effort}} = \frac{aG}{P_F}$$

It is readily established that for current vehicles without boosters (index 0), the efficiency factor W_0^X = 12-15. Therefore, without power assistance and in fulfillment of the given legal requirements, vehicles of 2000-2500 kg (4400-5500 lb)** can use a power-assisted brake system.

With all brakes intact, decelerations of 85-90% with pedal efforts of 25-30 kp (55-66 lb) can be achieved. This means that for vehicles in the 2000-2500 kg (4400-5500 lb) range,

$$W^X = 60\text{-}90$$

and there is no doubt that even higher values can and will be reached.

If we consider the fully working system as a basis (100%) then the "effectiveness coefficient"

$$w = \frac{W_0^X}{W^X} \ 100(\%)$$

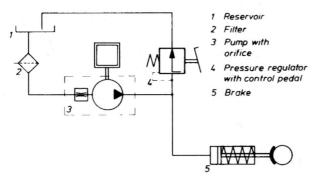

1 Reservoir
2 Filter
3 Pump with orifice
4 Pressure regulator with control pedal
5 Brake

Fig. 1 - Hydraulic circuit plan of continuous-flow system

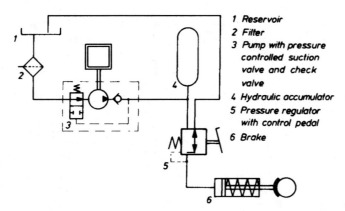

1 Reservoir
2 Filter
3 Pump with pressure controlled suction valve and check valve
4 Hydraulic accumulator
5 Pressure regulator with control pedal
6 Brake

Fig. 2 - Hydraulic circuit plan of accumulator system

*According to FMVSS 105a, pedal effort of 68 kg (150 lb) is specified.
**In the United States this weight would be 2700-3400 kg (6000-7500 lb).

shows by what percentage the system efficiency suffers for the same pedal effort compared with intact brakes. For the aforementioned vehicle weights the usual value for the effectiveness coefficient lies in the range 18-25%. Although such figures have been legally acceptable up to the present, they are not adequate. If the assistance to the driver's pedal effort suddenly fails, the value for "w" should not be less than 35-40%.

Backup System - If the 30% deceleration at maximum 50 kp (110 lb) pedal effort with failed power assistance cannot be achieved, European legislation calls for a reservoir of energy that must be large enough to enable 6-10 full brake applications at 30% deceleration to be made without exhausting the accumulators. (The number of stops varies from country to country.) Heavier vehicles, requiring in any case a larger energy supply, will probably be equipped with full-power braking systems, with efficiency factors W^x of 200 or more.

As already pointed out, there is a clear trend toward further reduction of the driver's effort to effect a stop. This will have the effect of pulling down the vehicle weight class for which full-power braking can be applied. Therefore, such systems for passenger cars must be granted the same degree of importance as power-assisted systems.

HYDRAULIC POWER-ASSISTED BRAKE SYSTEMS

APPLICATION WITH INTERPOSED BOOSTER (ZHS 2.0) - Fig. 3 shows a system where the vacuum booster has been simply replaced by its hydraulic counterpart, the original tandem master cylinder being retained. The energy supply is covered by the "hydraulic module" (pump and electro motor), which consists of the following units:

1. Electric motor with 3800 rpm nominal.
2. Radial piston pump, with a delivery of only 0.1 cm^3/rev.
3. Accumulator, gas loaded, whose design volume can be 250,500 or 750 cm^3 (15,32, or 48 in^3) depending on the brake system. Fig. 4 shows the relationship of fluid reserve, V_l, to nominal volume, V_N, available for braking at: maxi-

mum accumulator pressure (170 bar = 2400 psi); charging pressure (150 bar = 2150 psi); and the pressure at which the driver is warned in case no recharging occurs (130 bar = 1850 psi).

4. Pressure switch, by means of which the electric motor is switched off at 170 bar (2400 psi) accumulator pressure and switched on at a pressure of 150 bar (2150 psi).
5. Warning switch, giving an optical or acoustic signal if the pressure falls below 130 bar (1850 psi).
6. Pressure relief valve, adjusted to open at approximately 190 bar (2700 psi).

The pump is connected to a special reservoir, item 2 in Fig. 3. The pump could also be engine driven (Figs. 2 and 10), in which case the reservoir for the pump could then supply the brake actuation as well (Fig. 7); this will be discussed later.

Fig. 5 shows a cutaway drawing of a tandem master cylinder with an interposed pressure regulator valve. As such a valve arrangement always exhibits a certain leakage rate when modulating high pressures, a check valve is installed between accumulator feed and the regulator valve housing. This valve opens only when the brake pedal is pressed, and thus prevents the accumulator emptying itself when the vehicle is parked or when the brakes are not operated for long periods.

As the pedal is depressed, the control piston advances to

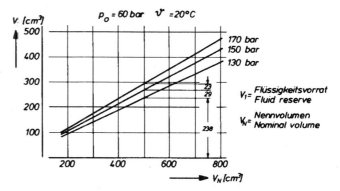

Fig. 4 - Fluid reserve in hydraulic accumulator

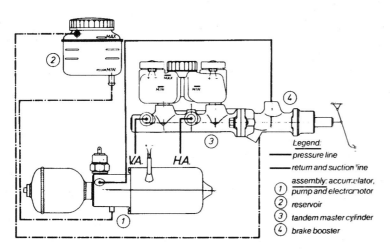

Legend:
——— pressure line
——— return and suction line
① assembly: accumulator, pump and electromotor
② reservoir
③ tandem master cylinder
④ brake booster

Fig. 3 - Power-assisted brake system with interposed booster

contact the booster piston (area F_1) and accumulator pressure (Sp) is simultaneously admitted. Once the control piston porting overlaps the feed ports in the booster piston, pressure is applied to the full area (F_1) as well as to the cross-sectional area of the control piston (d_S). Thus, the value of the regulated pressure is directly proportional to the pedal effort and can be exactly controlled with a high degree of response.

With such an actuation, the boosting need be only single-circuit, if 50 kg (110 lb) pedal effort can achieve 30% deceleration with a failed booster (so-called auxiliary brake effect); that is, if the efficiency factor W_0^x is equal to or greater than six times the maximum vehicle weight (in 1000 kg). Because the chance of an accumulator failure is extremely remote, the efficiency factor W^x may be large, and the effectiveness coefficient "w" can be made small because the driver will be warned that the accumulator is not being recharged and he can gradually adapt to a reduced performance of the brake system.

APPLICATION WITH INTERGRAL BOOSTER (ZHS 2.1) - Means of improving the cost and installation factors of the interposed booster are shown in Figs. 6 and 7.

As in the interposed booster, the energy supply is provided by a "hydraulic module." item 1 in Fig. 6. A brakefluid reservoir, item 2, with a dividing wall supplies the dynamic brake circuit, consisting of pump, accumulator, and pressure regulator valve, as well as the hydrostatic brake circuit comprising master cylinder and wheel brakes. According to regulations, a fluid-level warning device must also be provided.

The ZHS 2.1 booster, item 3 in Fig. 6, combines the spool-valve controlled booster and a single-circuit master cylinder in one unit; Fig. 7 shows its operation.

When depressing the brake pedal, the control piston advances to contact the master cylinder piston so that the primary seal lip passes the recuperation port (at B in Fig. 7)

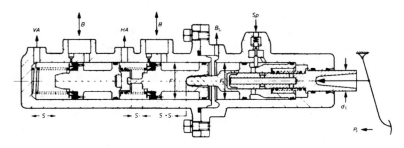

Fig. 5 - Interposed booster with tandem master cylinder (ZHS 2.0)

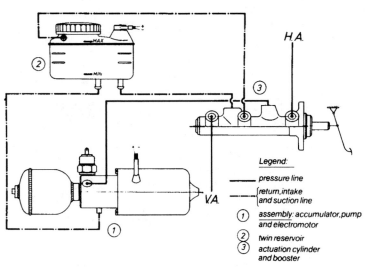

Legend:

――― pressure line

{ return, intake and suction line

(1) assembly: accumulator, pump and electromotor
(2) twin reservoir
(3) actuation cylinder and booster

Fig. 6 - Power-assisted brake system with integral booster

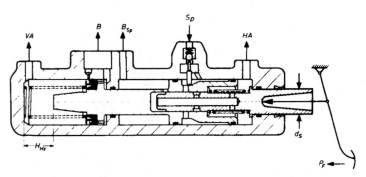

Fig. 7 - Integral booster and single-circuit master cylinder (ZHS 2.1)

and the check valve is lifted open by the ramp on the piston. Accumulator pressure then enters the annular control valve chamber and, passing through the overlapping porting, is applied to the full cross-sectional area at the rear of the master cylinder piston, passes to the control piston cross section (diameter d_S), and via the outlet marked HA enters the brake circuit. The pressure generated in the hydrostatic circuit passes via outlet VA to the other brake circuit. Due to seal drag and differences in the two spring loads, this pressure is slightly lower than that in the boost chamber, the latter value being governed by the input load P_F and the area d_S.

On failure of the accumulator pressure, or a leak in the HA brake circuit, the master cylinder and the VA circuit are mechanically operated. Should the VA circuit fail, the master cylinder piston advances to the bottom of the bore, at which stage the accumulator pressure is directed to the HA circuit.

With the ZHS 2.1 booster, the cross-sectional area of the master cylinder can be considerably less than that of the ZHS 2.0 because it feeds only one brake circuit; in the case of the common front-rear split, this circuit would be for the front brakes, and by a diagonal split, for one rear and one front brake. With the same pedal ratio and pedal effort a higher pressure can thus be generated in the intact hydrostatic circuit than could be with the ZHS 2.0 layout, achieving the same brake efficiency as with a lower pressure in a fully intact system.

Contrary to ZHS 2.0, a sudden failure of the HA circuit would rapidly drain the accumulator. Therefore, when using ZHS 2.1, the value of "w" should (where possible) not fall below 35-4%.

APPLICATION WITH INTEGRAL BOOSTER (ZHS 2.2) - On failure of the power assistance or of a brake circuit, the regulations call for 30% deceleration; that is, only a third of the maximum available with an intact system. Consequently, the fluid volume, even with the "auxiliary system," can be reduced and the unboosted mechanical master cylinder stroke need not be so large as with ZHS 2.0 or ZHS 2.1 systems.

The following example will serve to illustrate the case for a standard front-rear split: With all brakes intact, the brake force for the front axle is

$$B_{VA} = (1 - \phi)a_{max} \times G = P_{max}K_{VA} \qquad (1)$$

where :

ϕ = brake force proportion of the rear axle
$\quad = B_{HA}/aG = B_{HA}/(B_{VA} + B_{HA})$
$1 - \phi$ = brake force proportion of the front axle
$\quad = B_{VA}/aG = B_{VA}/(B_{VA} + B_{HA})$
a_{max} = maximum deceleration
G = vehicle weight, kp
P_{max} = max. line pressure at a_{max}, bar

K_{VA} = a constant, incorporating wheel cylinder diameter, brake factor, etc., kp/bar
(See Nomenclature for definitions of symbols.)

After failure of the boosting and the rear brakes, the front brakes alone must create a brake force of

$$B'_{VA} = a'G = p'K_{VA} \qquad (2)$$

where :

a' = deceleration with failed circuit
p' = line pressure at a'

Combining Eqs. 1 and 2, the required line pressure must be

$$p' = P_{max} \frac{a'}{a_{max}(1 - \phi)} \qquad (3)$$

If a maximum deceleration of $a_{max} = 0.75$-0.85, requiring a line pressure $p_{max} = 100$-120 bar (1420- 1700 psi), then the pressure p' needed to achieve the deceleration $a' = 0.3$ is only $[40/(1 - \phi)]$ to $[4.25/(1 - \phi)]$, or 580-600 psi. If $1 - \phi = 0.6$ to 0.7, then p' is only 56-71 bar (800-1000 psi) or 55-60% of the maximum pressure.

Fig. 8 shows the essential relationship between volume requirement and line pressure for disc brakes (curve "a") and drum brakes (curve "b"). If for the maximum theoretical line pressure a volume of 100% is required for both brake types, it can be seen that at 55-65% of maximum pressure, the fluid volume required for disc brakes is 26-33% below the maximum (ΔV_a), whereas the volume for drum brakes is only 8-11% below maximum (ΔV_b). Thus, in the case of booster failure and mechanical operation of the master cylinder, the stroke needs to be approximately 70% of the maximum available for disc brakes and a full 90% when drum brakes are used.

The other well-known features, such as insensitivity to friction changes and fade resistance, also make disc brakes highly desirable whenever they can be possibly used.

On pressing the brake pedal, with accumulator intact, the preloaded outer simulator spring (which is captive) initially

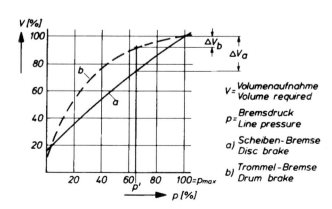

Fig. 8 - Fluid consumption characteristics

213

moves the spool control value and effects three actions; see Fig. 9:

1. Isolates the booster chamber from the fluid reservoir, B_{S_p}.

2. By means of the two sliding pins in the control valve bushing, moves the master cylinder primary seal lip past the recuperation port.

3. Lifts the ball check-valve in the accumulator feed port (S_p).

Further travel causes the rear edge of the control valve to overlap the rearmost annular recess in the control valve bushing. This causes the accumulator pressure to build up in the booster chamber and (via the lower transfer passage) in the chamber at the rear of the master cylinder piston, thus feeding the brake circuit marked HA. Pressure level in the latter is controlled by the relationship between the driver's pedal effort and the cross-sectional area of the control valve piston F_1.

The stroke of the control valve piston is governed by the travel/load characteristics of the simulator and can be a maximum s_1. On the other hand, the stroke of the master cylinder piston is governed by the volume/pressure characteristic of the VA brake circuit, and can be s_3, independent of the control valve stroke. (If, for reasons of poor bleeding, etc., in the VA circuit, 80% of the maximum available stroke is attained, the driver is given a warning signal).

If the accumulator fails, the master cylinder piston can be mechanically operated to a maximum stroke s_2 via the two pins; at the same time, play "s" is overcome and the central valve bushing is urged forward without compressing the strong inner simulator spring. By a correct design of the simulator springs, the control valve piston and thus the pedal travel remain virtually the same for intact brakes or for failure of one circuit. This shows a clear improvement of ZHS 2.2 over ZHS 2.0 and ZHS 2.1.

Dependent upon application or vehicle design, the ZHS 2.2 offers a choice of advantages: application to higher vehicle weights and lower pedal efforts, or shorter pedal travel, or an improvement of the effectiveness coefficient "w" (see formulas in Appendix A).

HYDRAULIC FULL-POWER BRAKE SYSTEM (ZHS 3)

The conditions described under "Backup System" can be satisfied by the system shown in Fig. 10. The engine driven pump 2 is fed by the reservoir 1 and charges both accumulators 3. The pressure level is either controlled from the charging valve 4 or from a pressure-controlled suction valve integral with the pump (see also Fig. 2). Generally, the lower cut-in pressure is 150 bar (2150 psi) and the upper, 170 bar (2400 psi).

The accumulator unit also incorporates a nonreturn valve that prevents a transfer of fluid from one accumulator to the other when the pressure falls below 135 bar (1900 psi). Each accumulator has a warning switch set at 130 bar (1850 psi) and a pressure relief valve set at 190 bar (2700 psi). When depressing the brake pedal, the dual-circuit pressure control

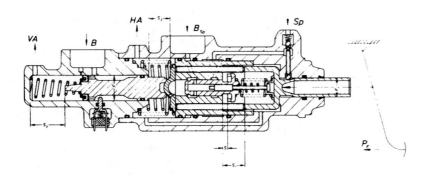

Fig. 9 Modified integral booster and single circuit master cylinder (ZHS 2.2)

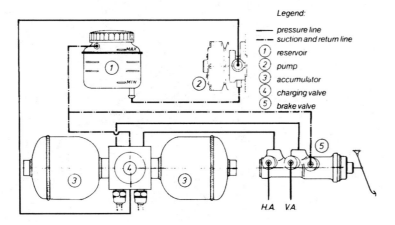

Legend:
— pressure line
—·— suction and return line
1 reservoir
2 pump
3 accumulator
4 charging valve
5 brake valve

Fig. 10 - Hydraulic full-power brake system

valve 5 allows pressure to flow to VA and HA brake circuits, each circuit being fed by its own accumulator.

As in ZHS.2 range, check valves are fitted between accumulators and the control valve, and a load-travel simulator is provided.

On failure of the pump, a number of brake applications, dependent on accumulator volume, charge pressure, and gas pressure, are possible. If the charge pressure falls to that of the gas pressure in the accumulator, no more assistance is possible. Therefore, the gas-filling pressure is chosen sufficiently high to retain-even in the final brake operation-an auxiliary braking effect of a = 30%.

For such a brake system it is important to know how the accumulator pressure behaves when its capacity is repeatedly tapped of volume, and how many brake applications can be made in a certain pressure range. Fig. 11 presents these factors in a diagrammatic form, applied to one brake circuit (approximately 50%) of a 4000 kg (9000 lb) vehicle.

The heavy curve shows the volume usage, and the left coordinate shows the pressure change "p" in a 500 cm^3 (32 in^3) accumulator: Line 1 shows the number of braking operations, item Z, for a deceleration of 20-25%; line 2 shows the number of braking operations, item Z, for a deceleration of 82-87%.

When the warning lamp operates at 130 bar (1850 psi), the volume for 12 applications has already been used (or 6.5 brake applications when considering line 2). After 20 applications more, the pressure remaining is still 92 bar (1300 psi) for line 1 and 70 bar (1000 psi) for line 2. This system thus has plenty in reserve to cope with the severest legal requirements.

The basic design of the full-power control valve is shown in Fig. 12. In the "brake-off" position, the check valves located between each of the two pressure control valves and the accumulators are closed, and both brake circuits VA and HA are in contact with the fluid reservoir, B$_{sp}$. On operating the brakes, both circuits are initially isolated from the fluid return; then both check valves open and, with a fractional advance, pressure is built up in the VA circuit. This pressure also actuates the control valve of the HA circuit, thus equalizing the pressure buildup in both circuits.

The stepped-diameter input piston is influenced by the reaction force from the VA-circuit control valve, F$_2$, transmitted rearward by the simulator springs, and by the HA-circuit line pressure acting on the ring area F$_1$. Thus, despite a control valve diameter of only 0.85 cm (0.355 in), the reaction load from the driver is such that this ZHS 3 control valve can be used with existing pedal layouts.

Should the HA circuit fail after opening the check valve, the forward control valve moves to its stop; if the VA circuit fails, the HA control valve is mechanically (not hydraulically) brought into play.

HYDRAULIC CONTROL FOR USE IN ANTILOCK SYSTEMS

For antilock control, the system of actuation adopted for the service brake plays a decisive role. Some solutions are considered below.

CONTROL WITH A PLUNGER UNIT - If the brake actuation relies on a master cylinder with limited volume

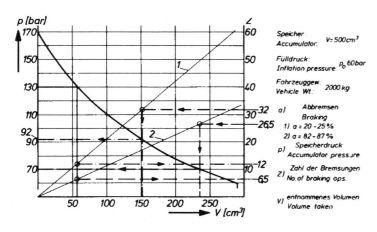

Fig. 11 - Accumulator pressure related to number of brake applications and volume used

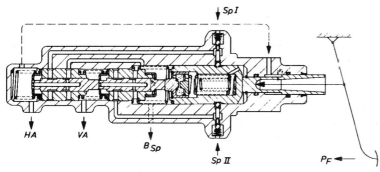

Fig. 12 - Dual-circuit full-power control valve (ZHS 3)

215

output, the fluid can be retained in the brake circuit and the antilock operation carried out by depressurizing and repressurizing the fluid already in the brake's wheel cylinders.

Fig. 13 shows that for each brake supplied by the master cylinder, a plunger unit with a reliable sliding seal is provided. When the antilock function is triggered, the inlet valve 3a opens and outlet valve 3b closes, the spring-loaded stepped piston is moved to the left, and the cutoff valve 4 closes so that fluid from the wheel cylinders flows into the space vacated by the plunger (pressure drop). When the "repressurize" signal is given, valve 3a closes and the spring pressure forces the plunger to the right, reapplying the brakes and opening the cutoff valve to complete the normal brake circuit again.

CONTROL WITH A VALVE BLOCK - If the service brake is directly supplied from an accumulator, then the fluid from the wheel cylinder can be led back to the reservoir, and when the "repressurize" signal is given, the brakes are reapplied, using the energy source or accumulator.

Fig. 14 shows a simplified layout of a valve block without plunger. When the antilock signal is given, the inlet valve 3b closes and the outlet valve 3a opens, releasing fluid from the brakes. At the end of the antilock control, the valve functions return to that shown and the wheel cylinders are repressurized.

THE COMPACT-BREL - For a vehicle equipped with power-assisted brakes (Fig. 6) using ZHS 2.1 or ZHS 2.2 boosters (Figs. 7 and 9) and an antilock system, it is practical to use a plunger unit (Fig. 13) for the one circuit and a valve block for the other (Fig. 14). Thereby the spring loading of the plunger

can be replaced by accumulator pressure, which is reduced or increased during the antiskid operation.

Furthermore, to save space, hydraulic connections, and pipework, the energy source and hydraulic actuation can be conveniently integrated into one BREL unit, diagrammatically shown in Fig. 15. ("BREL" is short for brake-force regulator-electronic.) Apart from this unit, only the ZHS 2 type booster 14 with integral reservoir 9 and a 500 cm^3 (32 in^3) diaphragm accumulator 12 are required to complete the brake system.

The electric motor driven pump "M" charges the membrane accumulator 12 as well as the piston accumulator 4, which is preloaded by the powerful spring 3. With this latter pressure flowing via the open solenoid valves 5b and passage 6, the plungers 1 and 13 are held against their stops, whereby the cutoff valves 7 remain open. In this position the service brake functions quite normally.

When an antilock function is triggered by the computer, valve 5b closes and 5a opens, cutting off the pressurized chamber 8 from the piston accumulator 4 and allowing the fluid to return to the reservoir 9. The line pressure in the upper wheel cylinders urges the plunger 1 to the left and the cutoff valve 7 closes off the return to the ZHS booster.

Thus, the space vacated by the plunger "s" is occupied by the wheel cylinder volume "s" and the brake force in the one or more regulated brakes reduced.

Repressurization occurs after the solenoid valves have returned to their original position and accumulator pressure is reintroduced to chamber 8. If the accumulator pressure fails, one circuit remains intact. The vehicle can be braked with the hydrostatic circuit because the spring 3 and its cup 10 urge the tappets 11 against the plunger 1 and mechanically keep the cutoff valve open all the time.

Because the plungers 1 are located concentrically around the central piston accumulator 2, their number can be increased according to the number of wheels to be regulated. With more than three or four plungers, the spring 3 might have to be strengthened.

Finally, Fig. 16 shows the convincing simplicity and compactness of a total vehicle installation with power-assisted brakes and antilock with the hydraulic and electrical connections. In the vehicle illustrated, both front brakes are individually controlled while the rear brakes use the "select-low"

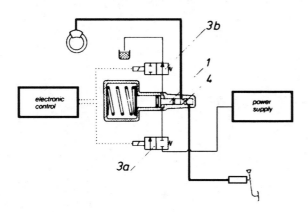

Fig. 13 - Plunger type hydraulic antilock control

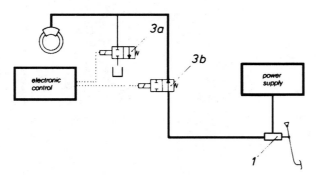

Fig. 14 - Valve block hydraulic antilock control

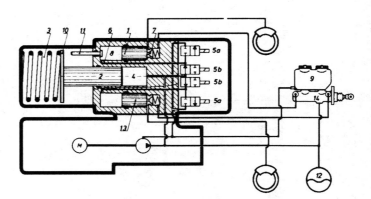

Fig. 15 - Compact BREL with power-assisted brake system

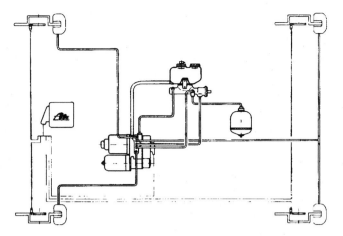

Fig. 16 - Vehicle installation of hydraulic service brakes and antilock system

principle. Dependent on the space available, the separately mounted accumulator can be integrated with the BREL or with the hydraulic booster.

SUMMARY

Three variations of hydraulic boosters for power-assisted brake systems and a dual-circuit control valve for hydraulic full-power brake systems are described. The units all make use of the accumulator, for this offers the driver more security than the continuous-flow system.

It has been suggested that brake systems be classified by means of an "efficiency factor" (total brake force/driver's pedal effort) and that an "effectiveness coefficient" be introduced as a comparison value to correlate the efficiency factor of a brake system having failed boosting with that of an intact system. Applying these values to brake systems using the described boosters helps judge the advantages and disadvantages; and emphasizes the importance of a new ZHS 2.2 booster, especially for heavier vehicles with disc brakes.

A hydraulic actuation system is especially attractive when a vehicle is also to be equipped with antilock. By means of the compact BREL unit, hydraulic actuation of the antilock function and the energy source for service and antilock braking can be integrated into one unit so that a relatively simple installation with a minimum of connections and piping is now possible.

NOMENCLATURE

a	= % deceleration
B	= brake force; kg; wheel cylinder
C*	= brake factor
d	= diameter, cm
F	= cross-sectional area, cm^2
G	= weight, kg
HA	= rear axle
i_p	= pedal ratio
K	= constant
M	= master cylinder
p	= hydraulic pressure, bar
P_F	= pedal effort, kp
r	= effective radius, cm
R	= rolling radius, cm
s	= stroke, cm
S	= control piston
S_F	= pedal travel, cm
VA	= front axle
w	= effectiveness coefficient
W^x	= efficiency factor
W_0^x	= efficiency factor, unboosted
η	= efficiency
ϑ	= temperature, °C
λ	= max. wheel cylinder travel, cm
$\dot{\psi}$	= rear-axle brake force proportion

APPENDIX A
CALCULATION OF EFFICIENCY FACTOR AND
EFFECTIVENESS COEFFICIENT

Brake force on a vehicle with four hydraulically operated disc brakes is expressed as

$$B = p(F_{VA} + F_{HA})\eta_B \left(C * \frac{r}{R} \right) 2 \qquad kp \qquad (A-1)$$

if p, $C*$, r/R on the front and rear axle are equal.

Given

$$K = \eta_B C * \left(\frac{r}{R} \right) 2 \qquad (A-1b)$$

and

$$p = \frac{P_F i_p}{F_M} \eta_M \qquad bar \qquad (A-1b)$$

and substituting Eqs. A-la and A-lb into Eq. A-1,

$$B = aG = P_F i_p \eta_M K \frac{F_{VA} + F_{HA}}{F_M} \qquad kp \qquad (A-2)$$

Balancing volumes,

$$S_M F_M = \left(\frac{S_F}{i_p} \right) F_M = 4\lambda (F_{VA} + F_{HA}) \qquad cm^3 \qquad (A-3)$$

The efficiency factor of the unboosted brake system is

$$W_0{}^X = \frac{aG}{P_F} \qquad (A-4)$$

Combining Eqs. A-2, A-3, A-4 gives

$$W_0{}^X = i_p K \eta_M \frac{F_{VA} + F_{HA}}{F_M} = \frac{S_F}{4\lambda} K \eta_M \qquad (A-5)$$

With an interposed, intact booster ZHS 2.0, as in Fig. 5,

$$p = \frac{P_F i_p}{F_S} \eta_S \qquad bar \qquad (A-6)$$

and since Eq. A-3 still applies, the efficiency factor with booster is

$$W^X = i_p K \eta_S \frac{F_{VA} + F_{HA}}{F_S} = \left(\frac{S_F}{4\lambda} \right) K \eta_M \left(\frac{F_M}{F_S} \right) \qquad (A-7)$$

The percentage effectiveness of the brake system with failed

booster compared with that for an intact system is given by the effectiveness coefficient

$$w = \frac{W_0{}^X}{W^X} (100) \qquad \% \qquad (A-8)$$

For a system with ZHS 2.0,

$$w = \frac{F_S}{F_M} (100) \qquad \% \qquad (A-9)$$

For ZHS 2.1 (Fig. 7), Eq. A-6 still applies, but instead of Eq. A-3, since the VA circuit only is supplied from the master cylinder,

$$S_M F_M = \left(\frac{S_F}{i_p} \right) F_M = 4\lambda F_{VA} \qquad cm^3 \qquad (A-10)$$

and since

$$F_{VA} = (1 - \phi)(F_{VA} + F_{HA}) \qquad cm^2 \qquad (A-11)$$

the efficiency factor of the brake system with intact ZHS 2.1 booster is

$$W^X = i_p K \eta_S \frac{F_{VA} + F_{HA}}{F_S} = \frac{S_F}{4\lambda} K \eta_S \frac{F_M}{F_S(1 - \phi)} \qquad (A-12)$$

If the accumulator pressure fails, only the VA circuit will be operated; thus

$$B = p F_{VA} K = aG \qquad kp \qquad (A-13)$$

and combining Eqs. 1b, 10, 11,

$$W_0{}^X = i_p K \eta_M (1 - \phi) \frac{F_{VA} + F_{HA}}{F_M} = \left(\frac{S_F}{4\lambda} \right) K \eta_M \qquad (A-14)$$

Now, dividing Eq. A-14 by Eq. A-12 gives the effectiveness coefficient

$$w = \frac{F_S(1 - \phi)}{F_M} (100) \qquad \% \qquad (A-15)$$

For ZHS 2.2 (Fig. 9), the pedal travel causes a shorter travel λ ' at the wheel cylinders. Thus, the area F'_M can be smaller than F_M; in this case, instead of Eq. A-10, the following applies:

$$\left(\frac{S_F}{i_p}\right) F'_M = 4\lambda' F_{VA} = 4\lambda'(1-\phi)(F_{VA} + F_{HA}) \qquad cm^3$$

(A-16)

and the efficiency factor for brake systems with a ZHS 2.2 booster will be

$$W^X = i_p K \eta_S \frac{F_{VA} + F_{HA}}{F_S} = \frac{S_F}{4\lambda'} K \lambda_S \frac{F'_M}{F_S(1-\phi)}$$

(A-17)

If the accumulator fails, the efficiency factor is similar to Eq. A-14:

$$W_0^X = i_p K \eta_M (1-\phi) \frac{F_{VA} + F_{HA}}{F'_M} = \frac{S_F}{4\lambda'} K \eta_M$$

(A-18)

and thus the effectiveness coefficient is

$$w = F_S \frac{(1-\phi)}{F'_M}(100) \qquad \% \qquad (A-19)$$

Table A-l - Summary of Most Important Equations

	ZHS 2.0	ZHS 2.1	ZHS 2.2
$W_o{}^X$	A-5	A-14	A-18
w^x	A-7	A-12	A-17
w	A-9	A-15	A-19

NOTES:
If F_M for ZHS 2.1 = $(1-\phi)$ (F_M for ZHS 2.0), then equation applications are as follows:

A-5 = A-14; A-7 = A-12; A-9 = A-15

If $l' < $ hand $F'_M/l' = F_M/l$, then

A-18 > A-14; A-17 = A-12; A-19 > A-15

CHAPTER FOUR

Safety glass

Laminated Safety Glass

By R. H. McCarroll
Ford Motor Co.

AFTER tracing the development of laminated safety glass, this paper describes the manufacture of this material as made at the new glass works of the Ford Motor Co.

In 1927 the Ford Motor Co. introduced it as standard equipment in the windshields of its cars.

Improvement followed shortly after the development became a cooperative effort.

AS an introduction we believe we can do no better than to quote from an article published in *Industrial & Engineering Chemistry*[1] by George B. Watkins and William D. Harkins on this subject:

"The manufacture of laminated safety glass may well be classed as one of the outstanding modern industries since its development has given what is probably the most important single contribution to safety in modern transportation.

"Evidence of public recognition of the merits of the product of this industry is best illustrated by its widespread use as standard equipment by the automobile manufacturers and by the definite legislative steps that have been taken in several States to the effect that all motor vehicles for public and private conveyance must be equipped with laminated safety glass.

"The principle of laminated glass as such is old, dating back to the latter part of the nineteenth century but, like many other industries, during its early stages little money or well-directed scientific and engineering effort were expended by those closely associated with it, with the result that four or five years ago the industry was still in its infancy and required corresponding treatment.

"For the idea of laminated safety glass as we know it today, as far as public records are concerned, the honors go to an Englishman, Wood, who in 1905 obtained a British patent which describes a method for safety-glass manufacture by the use of Canada balsam for cementing a sheet of transparent celluloid between two sheets or plates of glass. The Safety Motor Screen Co., Ltd., made samples of safety glass in this manner and exhibited them at the spring motor show in England in 1906. Because of the high cost of materials, the general unsatisfactoriness of this product, and the small demand, Wood's venture was without success and the patent was allowed to lapse.

"The first man to capitalize on the idea of laminated safety glass was a Frenchman, Benedictus, who obtained French and British patents in 1910. Benedictus named his product "Triplex" and employed the same general principle as Wood, except that he proposed gelatin instead of Canada balsam as the bonding adhesive for the glass plates and celluloid. Benedictus introduced the manufacture of Triplex safety glass in 1912 in England where production started in 1913. The new industry received an enormous impetus during the World War when laminated glass was used for the manufacture of gas-mask lenses and goggles, and for automobiles and airplanes.

"Although the manufacture of laminated safety glass was established as an industry during the war, for some years following it was at a standstill if not in the waning class because the producers of safety glass suddenly found that the high-priced commodities and low standards of quality acceptable during the rage of battle failed to meet the approval of the close-range scrutinizing public in time of peace. However, the merits of safety glass had been demonstrated beyond question."

It was at this time, more than eight years ago, that the Ford Motor Co. entered the picture. It secured the manufacturing rights to the Triplex process and put this process on a large-scale-manufacturing basis, developing equipment and methods – taking it from the stage of "dish-pan and tea-kettle" equipment.

When Ford in 1927 introduced safety glass as standard equipment in the windshields of his motor cars, the glass produced was, of necessity, somewhat experimental. It was a bold move and was attended with many tribulations for the chemical engineer, for there were admittedly obstacles to be overcome before it became possible to produce in large quantities safety glass of a quality capable of resisting the ravages of the sun and the weather. But the results, in the reduction of human damage suffered in motor-car accidents, were well worth all the cost.

The early safety glass had two major faults. Safety glass is in reality a "sandwich" in which two outer plates of glass are cemented to a middle layer of a plastic substance. Use soon revealed that early safety glass was subject to discoloration, and that "rainbows" and even opaqueness soon developed.

A cellulose nitrate first was employed for this transparent middle layer of the sandwich. It soon was discovered, however, that this cellulose compound in many cases deteriorated seriously as a result of the destructive effect of the actinic rays of the sun. This deterioration produced the rainbows and discoloration. It was necessary for the chemical engineer to explore the other available substances and to find a substitute that would be better fitted for the purpose and that would resist to as great a degree as possible the effect of the sun's rays. The chemists finally hit upon cellulose acetate.

Another difficulty experienced with the early forms of safety

[This paper was presented at the Production Meeting of the Society, Detroit, April 24, 1936]
[1] See *Industrial and Engineering Chemistry*, Vol 25, November, 1933, pp. 1187-1192; "Laminated Safety Glass", by George B. Watkins and William D. Harkins

glass was that it sooner or later became somewhat fogged. It was found that this difficulty resulted from the failure of the cement bond between the middle layer and the glass. Inquiry eventually determined that this failure was due to the effects of air and the weather. It led to the development of better bonding substances and to the practice of sealing the edges of the safety-glass sandwich to prevent the entrance of air or moisture between the layers.

In this manner the chemical engineers improved substantially the quality of safety glass and contributed measurably to an increase in the safety of motor-car operation.

The advance in quality of safety glass later became a cooperative work and was helped by the contribution of many companies such as the Triplex Safety Glass Corp. of North America, Libbey-Owens-Ford Glass Co., Pittsburgh Plate Glass Co. (Duplate Division), Eastman Kodak Co., The Fiberloid Corp. and others.

Since that time the making of laminated safety glass has grown to tremendous proportions. Advances have been made in manufacturing technique until today all Ford cars are equipped throughout at no extra cost with clear and colorless safety glass.

We will now describe the present practice of manufacture as it can be seen today in the glass plant at the Rouge plant of the Ford Motor Co.:

Plate glass ground and polished to 1/8 in. thickness is used for the two outer layers of the sandwich.

The materials that go into the making of the glass are silica sand, cullett or salvage glass, soda ash, limestone, salt cake, charcoal, and arsenic trioxide. These materials are all fused together at a high temperature, 2600 deg. fahr.

This glass is held at a high temperature until it has become clear and completely free from bubbles, when it is allowed to flow out of the furnace in a continuous stream onto rollers which convert it into a ribbon about 1/4 in. thick and 36 to 72 in. wide and of indefinite length. This ribbon is run through an annealing oven or lehr 350 ft. long. The temperature of this lehr is controlled very accurately, so that the glass emerges uniformly cooled and free from strain.

The annealed glass is inspected and cut into lengths of convenient size for handling. The cut sheets are set in plaster of Paris on iron tables and run under a long series of grinding, smoothing, and polishing wheels which produce the clear brilliant surface of polished plate glass.

When both sides have been ground and polished, the thickness of the sheet has been reduced to about 1/8 in. These polished sheets are carefully washed, inspected, and cut into rectangular sections or "brackets" from which the finished lights of glass can be cut economically. The washing, inspection, cutting, and sorting are all done while the glass moves along on a continuous conveyor.

The bracket pieces are picked off the conveyor and cut to a templet pattern which gives the size and shape of a finished part, either windshield or body light.

After this plate glass has been cut to the proper windshield shape, it is placed upon the conveyor and first enters a washing chamber where it is scrubbed thoroughly. In an adjoining chamber the glass is dried and then continues down the conveyor.

After the plate glass has been washed and dried, it is inspected and, if found to be clean, it enters a cementing chamber. In an enclosure on the conveyor, the surface of the glass is coated thoroughly with a cement mixture. After be-

ing cemented properly, the glass passes through a drying oven where all the low-boiling solvent or liquid in the cement mixture is driven off.

The operators next place a plastic piece of cellulose acetate on the first piece of glass. The acetate adheres to the glass by means of the cement.

This cellulose-acetate plastic is produced by the manufacturers in the form of a ribbon 0.025 in. thick, 24 to 48 in. wide, and of indefinite length. Brought to the laminated-glass plant in rolls or cut to templet sizes, it is all carefully washed, dried, inspected, and transported to the assembly room. The greatest care is taken to make sure that no dust or dirt gets on the plastic before it is assembled in the sandwich.

Now a second piece of glass is placed upon the acetate, forming what is called in the industry a sandwich – a glass sandwich consisting of two layers of glass with plastic cellulose-acetate material between them. This so-called sandwich continues down the conveyor, and at the end of it, goes through a roller press where it is firmly tacked together so that it may be handled without the plies separating.

These sandwiches are loaded in trays and subjected to heat and pressure in large autoclaves at the end of the assembly line. This operation does the final bonding and renders the sandwich clear and transparent.

If moisture comes in contact with the cellulose acetate, it becomes discolored. Therefore, we seal the edges against the entrance of moisture. The first step is the removal of sufficient plastic around the edges, about 1/8 in., to allow the application of the waterproof sealing material.

This removal is done by the action of an acid mixture that will attack the plastic layer but not the glass. When the acid has removed the plastic to the proper depth, the glass is withdrawn, washed free from acid, and dried carefully.

After the groove has been formed around the edges, it is filled with a moisture-resistant material applied on continuous automatic machines. This sealing material, which is soft like putty when used, is applied by a series of wheels to the edges of the glass. The pieces, fed into one end of the sealing machine, are conveyed along automatically, indexed so that each edge is filled in succession, washed, dried, and presented for inspection as they emerge from the other end of the sealer.

Windshields or other lights set in complete frames have their edges rounded off only, but body glass must have all exposed edges carefully ground and polished. This operation is done on automatic machines which convey the pieces along in contact with grinding, smoothing, and polishing wheels, indexing automatically so that all edges are presented in rotation for finishing. There is no hand labor, the attendant having merely to load and unload the conveyors.

After edge-grinding and cleaning we are ready for the last operation. This is an interesting step done on a little sand-blasting machine. The Ford insignia is sand-blasted clearly into one corner of the glass. This insignia reads "Ford Safety Glass" and also gives the month and year of manufacture. The Ford safety glass is now ready for installation in automobile bodies.

Ford safety glass as manufactured easily meets all the requirements of the American Standards Association Code for Safety Glass. We make only one quality of glass, the quality that is approved not only for body glass, but also for installation in windshields and wings.

Safety Glass Breakage by Motorists During Collisions

DERWYN M. SEVERY[1]

and

HARRISON M. BRINK[1]

ABSTRACT

Five intersection-type collision experiments were conducted at 40 mph to provide data on several categories of collision performance. This paper presents the interactions of passenger heads with car windshields and side-glass. Instrumentation included 60 channels of force and acceleration data, supported by the photographic coverage of 40 cameras. Tri-axial accelerometers, mounted in anthropometric dummy heads and chests, and strain gauges bonded to windshields facilitated data collection on the relative collision performances of different types of safety glass.

I. PURPOSE

Full scale collision tests provide realistic performance data on conventional and experimental configurations of car safety glass. The kinematic behavior of motorists pitched against their car glass is observed by slow motion photography while other instrumentation simultaneously record the forces these motorists sustain during their complex interactions with the glass. These experiments have provided valuable insight for researchers conducting more limited laboratory studies primarily by assisting them to avoid over-simplifications that often invalidate findings.

In addition to the supporting of safety glass studies, this series of experiments was organized to provide detailed information on many other aspects of motoring safety. These include child harness performance, door force systems, occupant ejection kinematics, vehicle dynamics for intersection col-

---

[1] Research Engineer, University of California (UCLA) Institute of Transportation and Traffic Engineering

lisions with one car stationary, and related observations. These specific data categories will be covered in other publications based on this series of five experiments. They are mentioned here only to show the *multiple purpose* of these experiments and to demonstrate that the experimental design was structured to accommodate many special studies. This paper presents only those findings relating to safety glass performance.

II. METHODS OF PROCEDURE

Full scale comprehensive collision experiments are difficult to conduct but they provide the foundation for many separate areas of inquiry. Each collision experiment represents a collection of separate but interrelated studies and, when conducted in a scientific manner, these studies provide answers accepted with confidence by others. In the U.C.L.A. studies, each car is controlled directionally by a front bumper-mounted shoe that slides in a monorail guide track secured to the asphalt pavement. Car speeds and their relative positions of impact are accomplished by means of a tow car, cable and sheave arrangement, Figure 2. Operational procedures for the controller, the instrumentation pace vehicle, and the tow vehicle are also described and illustrated. For the 600-foot guide track used in this series, the required acceleration was supplied by a 360-H.P. Dodge used as the tow vehicle.

III. EXPERIMENTAL PLAN

Five intersection-type collisions were conducted in this series. The striking car in each experiment was traveling 40 mph at impact. The struck car in each

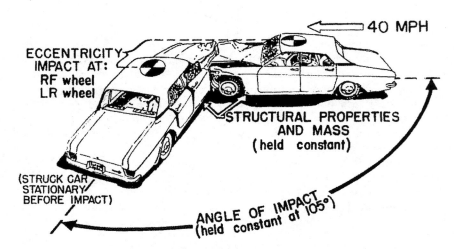

FIGURE 1. Independent Variables of an Intersection Collision.

226

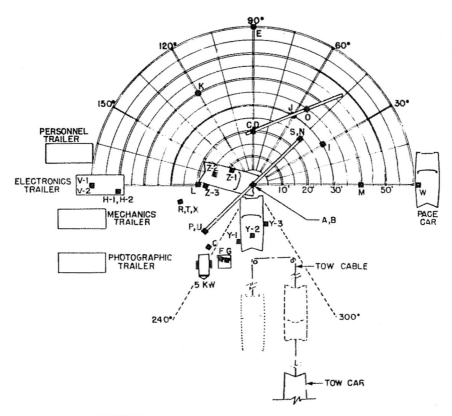

FIGURE 2. Vehicle Control and Photographic Systems

experiment was stationary, as though its driver were poised in the intersection awaiting opportunity to complete his left turn. In experiments 78 and 79, the struck car was positioned for impact at its left rear fender. In addition, the struck car's rear was angled toward the striking car 15 degrees from a broadside exposure. In experiments 80, 81 and 82, where the right front fender of the struck car was impacted, the struck car's front end was positioned 15 degrees toward the striking car. This manner of contact was selected to provide the highest collision forces that could be delivered to the car doors without directly impacting the doors. This impact also provided a realistic exposure for the simulated motorists being pitched against the safety glass.

In addition, the steering wheel of the striking car was removed so that all front seat occupants would have similar exposures for impacts with the windshields.

Type of safety glass in each experiment is shown in Figure 3.

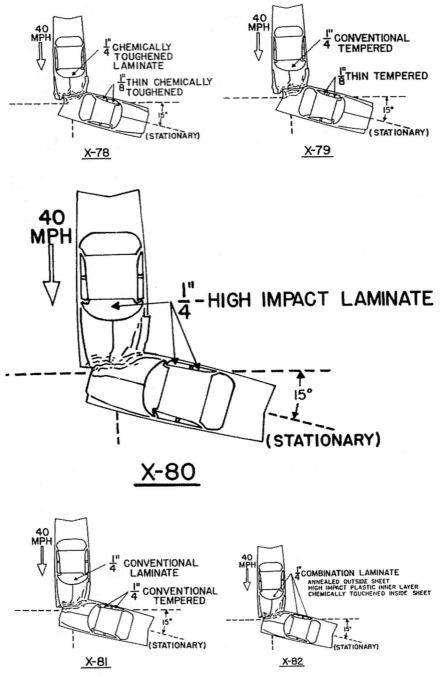

FIGURE 3a to e. Impact Configurations and Glass Assignments in the UCLA-ITTE Experiments.

IV. FINDINGS (SUMMARIZED)

Two important indicators of the relative safety provided for head impacts with glass are: (1) the head's peak acceleration, and (2) extent of cuts and laceration. Acceleration values attained are summarized by Figures 4 and 5. Findings pertain only to the one speed of impact and, therefore, should not be generalized. Additionally, owing to the unpredictable performance variations,

HEAD IMPACTS TO WINDSHIELD
(STRIKING CAR)

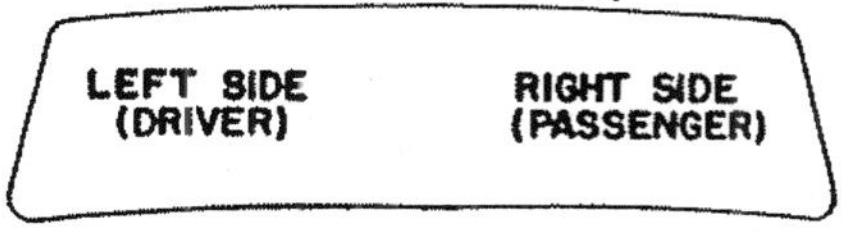

LEFT SIDE (DRIVER)
RIGHT SIDE (PASSENGER)

EXP NO	STRIKING AT	TYPE OF GLASS	HEAD ACCELERATION G's/ms		CHEST ACCELERATION G's/ms	
			DRIVER	PASSENGER	DRIVER	PASSENGER
X78	REAR WHEEL	1/4" CHEMICAL LAMINATE	28/142	26/170	NO INSTRUMENTATION	8/115
X79	REAR WHEEL	1/4" TEMPERED	44/100	36/154		14/130
X80	FRONT WHEEL	1/4" HIGH IMPACT LAMINATE	36/148	47/163		16/146
X81	FRONT WHEEL	1/4" CONVENTIONAL LAMINATE	49/124	23/123		14/142
X82	FRONT WHEEL	1/4" COMBINATION LAMINATE	51/100	63/122		22/128

FIGURE 4. Head Impacts to Windshields of Striking Cars.

HEAD IMPACTS TO SIDE WINDOW GLASS
(STRUCK CAR)

REAR
FRONT

EXP NO	STRUCK AT	TYPE OF GLASS	HEAD ACCELERATION G's/ms		CHEST ACCELERATION G's/ms	
			REAR PASS.	FRONT PASS.	REAR PASS.	FRONT PASS.
X78	REAR WHEEL	1/8" CHEMICAL	106/88	56/103	41/100	14/103
X79	REAR WHEEL	1/8" TEMPERED	98/71	56/127	28/100	13/87
X80	FRONT WHEEL	1/4" HIGH IMPACT LAMINATE	42/125	20/115	8/120	22/110
X81	FRONT WHEEL	1/4" CONVENTIONAL TEMPERED	57/140	22/137	9/125	10/144
X82	FRONT WHEEL	1/4" COMBINATION LAMINATE	40/168	16/128	11/144	8/75

FIGURE 5. Head Impacts to Side Window Glass of Struck Cars.

these findings should be regarded as preliminary even with respect to their conditions of impact

Data within the shaded areas of Figures 4 and 5 are accelerations that were substantially influenced by dummy occupant interaction with non-glass parts of the vehicle compartment and, therefore, represent at best only partial acceleration values for head impacts with automotive window glass. A detailed accounting of head interaction with the car interior is provided in the next section.

Although Figures 4 and 5 are self-explanatory, correct interpretation requires additional data qualification. In Experiment X-78, the driver and passenger of the striking car impacted the "chemically toughened" laminated windshield with a glancing blow after striking and deforming the sheetmetal header above the windshield. This softer impact associated with deforming light gauge sheetmetal was attributed to the dummy's six foot height. To remedy this condition, the dummy height was redesigned to 5 feet 9 inches to make it more representative of the average motorist and to permit head-to-glass impacts.

The windshield values in Experiment X-82, although realistic, cannot be compared with values obtained from other experiments. In X-82, the striking car's front seat anchorage partially failed causing the front seat occupants to shift posture before they impacted the windshield. As a result their heads flailed against the glass.

This change in contact dynamics, evident from viewing slow motion photography, is further substantiated by the higher (22 G) chest deceleration for the passenger. The seat rotational action provided a greater chest-to-dash contact than was true in other experiments. In direct contrast, the chest deceleration of only 8 G's for X-78 is attributed to chest attenuation as a result of the head striking sheetmetal above the windshield header and the hips being snubbed by knee impacts with the instrument panel.

Head acceleration values obtained for side-glass impacts for the struck car were influenced by the torso-inertial force applied to the door by the dummy's shoulder. In each instance, the rear seat passenger sitting adjacent to the impacted door sustained head impacts with the side-panel sheetmetal of the roof. Their heads did not strike the rear side-glass even though the rear seat had been repositioned six inches forward to increase probability of glass impact. This non-glass contact for the rear windows was attributed to the swept back roof panel design and to spin-out dynamics of the struck car.

In Experiment X-80, the front seat dummy's shoulder crushed against the door and pre-cracked the glass before his head contacted it resulting in a reduced head acceleration of 22 G's. This condition was even more apparent in Experiment X-81 where the conventional ¼ inch tempered glass became completely fractured before the dummy's head made contact. Increased door deformation and reduction in door rigidity following fracture of the glass accounted for the lower (10 G) chest acceleration.

In Experiment X-82, conventional door latches were installed in the struck car as contrasted with prior experiments using instrumentation to replace the latches The front door latch failed in X-82 and permitted ejection of two front seat occupants The head and chest accelerations were the lowest recorded for this seated position and represent only the limited containment strength the door latch could provide before the latch failed

Although this study concerns a limited scope of exposures, it should be evident that side window glass performance will attain even greater prominence as improvements are made in the strength and general collision performance of door latch systems The natural head-level location of side windows and the glass performance requirement for passenger containment during collisions suggests that improved glass designs will have to be evaluated for structural integrity and collision injury performance

Experiment 78

A stationary 1964 Plymouth four-door sedan was impacted at the left rear fender by an identical car striking at 40 mph The striking car was equipped with a ¼ inch *chemically toughened laminated windshield* and the stationary car with ⅛ inch *chemically toughened side glass*

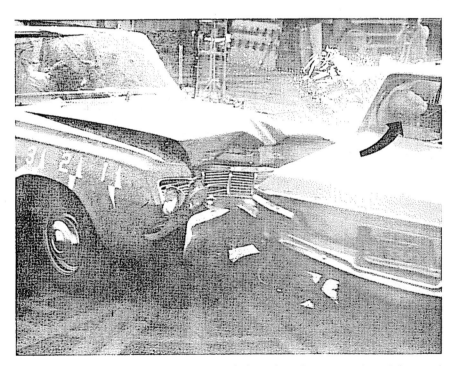

FIGURE 6 Experiment 78. Showers of Glass Fly from the Left Rear Window of the Struck Car. Dummy's Head (Arrow) Slammed into Side-Panel and Window Frame cs His Shoulder Punched into Door Panel.

231

The left rear seat passenger of the impacted car, sitting on the collision side, was forced laterally against the door. His head struck the top side-panel sheetmetal and side-glass frame at about the same time his shoulder punched into the side panel of the door, Figure 6. The door remained closed because of the special strain gauge instrumentation installed for door latch force studies. The thin, chemically toughened side glass was fractured by shoulder inertial forces at 62 milliseconds (ms). The head reached a peak acceleration of 108 G, 85 ms after the cars collided. In the struck car, the left front seat passenger's head was pitched against the thin side glass near the window's top support. The head cracked the glass and sustained a 56 G blow at 100 ms. Laboratory studies are underway to evaluate comparable-condition head impacts at different locations on side glass.

In the striking car, the front seat passenger's head impacted the sheetmetal header above the laminated windshield and sustained 26 G, 170 ms after the cars initially collided. His six-foot height positioned him too high for his head to impact the windshield.

The top of the driver's head grazed the header but his principal head impact was directed to the windshield (steering wheel removed). The driver sustained a 26 G blow at 143 ms after the vehicles contacted each other. These

FIGURE 7. Experiment 79. Left Rear Side-Glass Fractures as Passenger's Head Impacts Top of Quarter Panel and His Shoulders Deform Door Panel.

head impacts did not fracture the windshield. As with the passenger, the driver dummy was six feet tall. Both occupants impacted the sheetmetal header and the top of the windshield in a manner not representative of the average motorist. To achieve a more nearly average size the dummies were redesigned to 5 feet 9 inches for the remaining four experiments.

Experiment 79

A stationary 1964 Plymouth four-door sedan was impacted at the left rear fender by an identical car traveling at 40 mph.

The side-glass was ⅛ inch tempered and the windshield was solid tempered. (Tempered glass windshields are not lawful in this country even though their use is prevalent abroad.)

The passenger seated next to the left rear side-glass of the struck car sustained a head impact primarily involving the sheetmetal supporting the top at the left rear quarter panel, Figure 7.

The thin tempered left rear side-glass fractured from impact at 55 ms owing to distortion clearly evident from the shoulders loading the side panel of the door. The dummy's 98 G head blow at 71 ms indicated the relatively

FIGURE 8. Experiment 79 Driver's Head Passes Through the Already Cracked Window and Then Slams Down Against His Shoulder and the Window Sill.

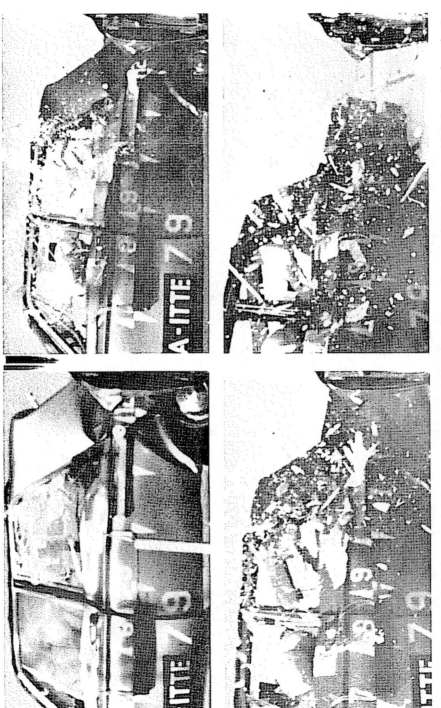

FIGURE 9a to d. Experiment 79. These Stop Motion Enlargements From Motion Pictures Show Initial Disintegration of Left Rear Window; the Beginning Outward Flight of Shards, Chunks and Pebbles, Figure 9b; Expanding Movement of the Pieces as the Flight Enlarges, Figure 9c; and the Closeup Blizzard of the Missiles, Figure 9d.

high impact level sustained from contacting the more rigid sheetmetal rather than the less rigidly supported side-glass

In the struck car, the dummy seated in the driver's position (steering wheel removed) punched his head centrally against the glass. The chest sustained a 13 G blow at 87 ms and the glass broke at 90 ms. After fracture, the dummy's head sustained a 56 G blow at 127 ms. The glass fractured 37 ms before the head sustained its peak G. Film analysis showed that fracture of the left front window occurred before the driver's head contacted the glass. After passing through the broken window, the driver's head slammed down against his left shoulder, contacting the window sill at the same instant to account for the peak G. Figure 8. This was a result of the mechanical limit for head excursion as well as head contact with the sill, rather than head impact with side-glass. Examination of the transducer curves for the 90 ms position, where glass failure occurred, indicates that the chest peak G had been reached and suggests that the sideways shift of the torso contributed to window breakage. At that instant, the head's lateral acceleration was only 10 G and it increased as the head continued in a leftward rotation through the shattered window to develop the maximum 56 G as the head reached its limit of excursion after the glass fractured.

The motorist interaction with ⅛ inch tempered side-glass is shown in Figures 9a through 9d. Large sections of fractured glass are seen being hurled toward the camera. Each of these larger pieces consist of many pebble-sized pieces and smaller particles mechanically interlocked until they strike other objects or the pavement. When propelled into a car, these large sections of glass may cause serious injury.

FIGURE 10. Experiment 79. Driver's Face Is Flat Against the Windshield of the Striking Car, a Position He Reached 54 ms Before the Passenger's Face Struck. Head of the Child Dummy, Who Was Standing Next to the Driver, Can Be Seen Impacting the Header Above the Windshield

FIGURE 11 Experiment 80. In Struck Car, Left, the Front Seat Passenger's Shoulder Bowed the Door Outward and Cracked the Glass His Head Also Impacted the Glass But Only After It Was Already Cracked.

FIGURE 12 Experiment 80 Struck Car Showing Fracture Pattern of High Impact (H-I) Laminated Glass and Permanent Deformation of Window Frame.

On impact, the driver of the striking car (steering wheel removed) is thrown forward with the upper portion of his forehead striking close to the top of the windshield. As this occurred, the head rotated backward to bring the chin forward and up and position the face flat against the windshield with the body continuing forward, Figure 10. The tempered glass windshield does not break although it is partially pushed from its mount at the top of the windshield. A peak of 44 G, equivalent to a 500-pound head blow, occurred at 100 ms for the driver, and 36 G, or a 400-pound head blow, for the passenger at 154 ms. A seat cushion positioned the driver 6 inches closer to the windshield and accounted for his head impact 54 ms before the passenger's head struck the windshield.

The driver's impact displaced the windshield from its mount and it probably accounted for the passenger's consequently lower head impact. Since he was farther from the windshield, the passenger jackknifed and the top of his head contacted the windshield. Both adult heads impacted the windshield directly without striking the header or the instrument panel. The child standing on the front seat between the adults was thrown forward and, because of his standing position, contacted the header or sheetmetal above the windshield.

Experiment 80

The stationary struck car, a 1954 Plymouth four-door sedan, was impacted at the right front wheel by an identical car traveling 40 mph. Both vehicles were equipped with ¼ inch high impact (H-I) laminated windshields.

The struck car was accelerated to the left throwing the front-seat passenger toward the right door setting off the following sequence of events: The passenger's shoulder struck the door centrally, bowed it outwards and cracked the side-glass 95 ms after onset of collision, Figure 11.

In this sideward movement, the dummy developed a peak chest acceleration of 22 G at 110 ms, followed 5 ms later by a peak head acceleration of 20 G. The right front passenger's head struck the side-glass approximately 4 inches from the top, central to the front-to-rear direction of the glass. Since this glass had already been fractured by the earlier shoulder impact, head impact acceleration was greatly reduced. Even though the shoulder action bowed the window sill and frame support, the door remained closed because of its strain gauge instrumentation. Outward bow of the door during impact was approximately 3 inches at the top of the frame.

The right rear seat passenger also struck the door beside him; however, his seated posture and the design of the roof structure side-panel kept his head from striking the side window. Torso inertial forces (8 G chest at 120 ms) bent the door sill sufficiently to fracture the H-I laminated side glass, Figure 12. His head struck the roof side panel registering 42 G at 125 ms. The door stayed closed because of the latch replacement instrumentation.

The striking car in Experiment 80 impacted the frontside of the stationary car, Figure 11. The driver of the striking car was pitched forward into the laminated windshield (bonded with a H-I plastic interlayer). The driver sustained a 36 G head blow as compared with the passenger's 47 G blow. Driver impact occurred 146 ms after the initial car-to-car contact. The driver fractured the windshield with his head and came to rest 20 ms after fracture. His head impacted the windshield with a completely horizontal thrust and contacted neither the instrument panel nor the windshield supports, Figure 13. The head's 36 G appears to be completely attributable to the head's pocketing into the interlayer—an action which abruptly stopped its forward movement, Figure 14. In terms of its total size, the head's pocketing action was very good, moving halfway through the windshield without rupturing the interlayer.

Front seat occupants in the striking car (no steering wheel) generally conformed to these kinematics: As their heads contacted the windshield, knee pressure against the instrument panel elevated the hips. Head-to-windshield impact was accompanied by forward and upward shoulder rotation. This bowed the head downward and exposed the top to impact as it penetrated the windshield.

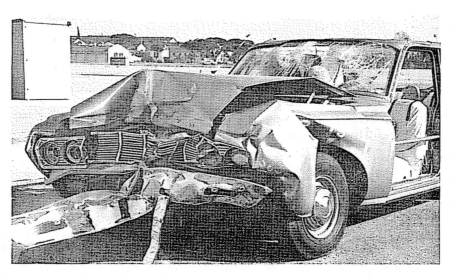

FIGURE 13. Experiment 80. Striking Car Showing Windshield Fracture Pattern of H-I Laminated Glass

The child occupant standing on the front seat between the adults struck the windshield almost as soon as the driver and the right front seat passenger and, about 17 ms later, reached a peak of 47 G at 163 ms, Figure 15.

The collision, viewed from a camera positioned on the rear window deck looking forward at the three front seat occupants, showed that the heads were pocketed by the interlayer after they penetrated the windshield. In all probability, the restraining action of the knees striking the instrument panel developed the upheaval-like action which developed rotation at the hips and caused the head to move upward vertically while still penetrating the windshield. For windshields of ordinary or annealed glass this upward movement against shards of glass will cause serious lacerative action.

Experiment 81

A stationary 1964 Plymouth four-door sedan was impacted at the right front wheel by an identical car traveling 40 mph. Figure 16. Both vehicles were

FIGURE 14. Experiment 80. Striking Car Showing Pocketing Action of High Impact Plastic Interlayer.

FIGURE 15 Experiment 80 Striking Car Showing Front Seat Occupant's Head Contact with Windshield

equipped with standard glass; *the windshield of conventional laminated construction and the side-glass of conventional tempered construction*

In the struck car, the right front seat passenger contacts the door with his shoulder first A split second later the driver follows as the car is spun leftwards. Their combined inertial impact distorts the door near the door latch position causing the glass at the lower rear-side to fracture before the head contacts the glass, Figure 17 This standard 7/32 inch tempered glass fractures 37 ms before the head reaches its peak G. The right front seat passenger's head reaches a peak acceleration of 10 G at 144 ms The head continues its rightward movement and penetrates the pre-shattered tempered side-glass to a position 5 inches beyond the plane of the glass

The right rear seat passenger in the struck car responded in a similar manner as the inertia of the body applied force to the center of the right door. Figure 11 His shoulder penetrated the side of the door substantially before his head contacted the inside upper edge of the rearward member of the window frame The glass broke at 130 ms, and the head contacted the window frame and associated side-glass at 140 ms sustaining a 57 G impact Finally, the chest peaked at 8 G. 215 ms after the cars collided.

FIGURE 16. Experiment 81. Striking Car (40 mph) has Conventional Laminated Windshield. Stationary Struck Car Was Equipped with Conventional Solid Tempered Side Glass

FIGURE 17. Experiment 81. Stationary Struck Car Showing Door Deformation Caused by Dummy's Shoulder Penetration.

The dummy seated on the driver's side (steering wheel removed) thrust his head completely through the windshield, reaching a peak of 49 G at 124 ms after contact; the passenger sustained a 23 G peak at 123 ms.

Although both impacts were essentially simultaneous, their peak values differed markedly. Large showers of glass were punched forward from the windshield during these head impacts. Residual glass adhering to the windshield frame consisted of large sections of a highly lacerative nature, Figure 18.

In the striking car, the driver's body was stopped primarily by his head penetrating the windshield and by his knees striking the lower instrument panel. During contact with the windshield, his head rotated backward about five degrees as it penetrated. Then the movement of the lower torso forced the head upward, rotating it in the opposite direction and then dropped it downward 30°. This rotation, considered in connection with the 35° plane of the windshield, provided a condition for his head to impact the plane of the windshield directly. In contrast, the passenger's head pocketed without rotation, had less of a scrubbing action and avoided direct contact with the intact edges of the glass.

Differing kinematics between driver and passenger are a result of the seat cushion which positioned the driver six inches closer to the windshield.

Experiment 82

This experiment involved two 1964 Plymouth four-door sedans with the 40-mph-car striking the stationary car at its right front wheel. Both vehicles

FIGURE 18 Experiment 81. This View of the Striking Car (40 mph) Demonstrates the Fracture Pattern and "Collaring" Effect of Conventional Laminated Windshields

were equipped with ¼ inch combination laminate (annealed outside sheet, H-I plastic interlayer, and chemically toughened, thin inside sheet) No seat belts were used and no special door latch instrumentation was installed. As a result, the impact had just started when the right front seat passenger began to be ejected, Figure 19. (Ejection of passengers did not occur in the other experiments because of the special instrumentation installed for collecting door latch load data—a modification not included in Experiment 82.)

Failure of the right front door latch is attributed to a combination of passenger and door inertial forces and to collision distortion forces. The right front passenger's immediate ejection began with his head directed downward as he was violently spun out of his car. A split second later the left rear wheel of the striking car ran over and crushed the middle portion of his head.

The middle seat passenger was also ejected, with his feet remaining in the car and his head and shoulders dragging along the pavement as the car spun left.

No side-glass breakage occurred because door latch failure prevented side-glass loadings from reaching failure values. The right rear dummy's head became wedged between the seatback and the top's side panel to reduce loads

FIGURE 19. Experiment 82. Impact Had Just Started When the Right Front Seat Passenger of the Struck Car Was Ejected. Striking Car (40 mph) Was Equipped with a ¼ Inch Combination Laminated Windshield. Stationary (Struck) Car Had ¼ Inch Combination Laminated Side Glass

FIGURE 20 Experiment 82. Striking Car's Front Seat Occupants Contacted Windshield and Partially Displaced It from Its Mount. Door Latch Failure Caused Ejection of Right Front Seat Passenger in Struck Car.

applied to the door. He made no contact with side-glass and, even though his shoulder hit the rear portion of the door, no glass breakage occurred.

In the striking car, the three front seat occupants contacted the windshield and partially displaced the glass from its mount, Figure 20. The outer layer of glass in the striking car's windshield was broken on impact, but the layer on the passenger side remained intact. This inside layer of glass consisted of ⅛ inch chemically toughened glass and the outside was annealed non-safety glass; these two sheets of glass were sandwiched with a H-I plastic interlayer.

Owing to the small amount of deformation in this glass, compared to prior impacts, the driver struck the windshield and sustained a 51 G head blow at 100 ms followed by the passenger sustaining a 63 G head impact at 122 ms. The top of the windshield was displaced from its mount and the outer layer of glass was fractured. The minor overall glass deformation accounts, at least in part, for the higher peak deceleration in comparison with the other two types of laminated glass.

During impact, the front seat occupants' heads could be seen pressed against the inside layer of toughened glass with a smooth pressure area; fracture did not occur in the inside sheet and, therefore, there could be no laceration.

If this rather ideal non-lacerative situation could be accompanied by increased glass yielding, even if it included a pebbled fracture as occurred in prior impacts, the peak G's would be reduced without the attending serious laceration. Such a combination should provide greatly improved collision performance for safety glass because it would bring the H-I interlayer into an active working role.

HEAD-MASS AUGMENTATION BY TORSO

For motorist exposures to head impact with side-glass and windshields, one of the design considerations concerns the strength of the glass for resisting impacts. Head sizes range from four pounds for the small child to twelve pounds for the average adult. On a passenger mile basis, children are infrequent passengers in motor vehicles and, owing to their smaller size, can be restrained more effectively from forceable contact with automotive glass during impacts. For these reasons, the twelve pound head appears to represent the average motorist exposure best.

The question next to be considered is: "To what extent is the average adult head augmented by body mass during head-to-glass impact?"

This question concerns those exposure conditions for which the human head, during some phase of impact with glass, is being pushed partially or directly by its torso mass. Collisions to the front end and sides of vehicles may align the head and torso to allow the torso mass to influence head-to-glass impacts. Full-scale collision experiments using anthropometric dummies and accident collision data (where the data includes examination of the patient and car damages), provide evidence that torso augmentation of head mass is not a major contributor to injury. The front seat occupant during a front-end collision is abruptly displaced forward still maintaining a normal seated posture until his knees crush into the instrument panel. This knee deceleration slows hip movement but does not initially keep the head and upper torso from continuing forward. As the hip deceleration approaches maximum, the chest angle has shifted from vertical to a leaning-forward type posture. For the average exposure condition, however, the head and chest are by this time impacting the windshield and instrument panel. Consequently, for this exposure condition, the head sustains little modification of impact alignment with the windshield as a result of the torso being decelerated at the hips. Head-mass augmentation for this exposure condition may be lessened for the motorist adequately restrained by a seat belt.

For motorists exposed to impacts directed to the side of the vehicle (intersection type), full-scale collision experiments show that an unbelted motorist, sitting alone on a seat struck at the side most remote from his seated position, will be propelled toward the impacted side. Seat drag and leg interference tend to allow the head and upper torso to lead the hips and lower limbs so that head contact with the side-glass, the windshield or doorpost may occur with the head and torso positioned in a reclined posture. For certain exposures, the torso mass may, in great measure, add to the inertial force of the head during impact.

Head-mass augmentation may occur by reason of torso inertial forces being directionally aligned to pass through or close to the head during impact. This is the mechanism of such an event: The head strikes an object, for example, the opposite side-window glass, and is abruptly decelerated by the impact, reaches a peak value of deceleration, followed perhaps 3 to 5 milliseconds later by the inertial force component transmitted from the torso through the neck. The delay will vary according to the relative velocity of the human to the impacted surface, the degree of alignment of the torso vector with the head, and, to a lesser extent, the nature of the compression deformation and shearing action taking place in the neck, torso and head.

Head-mass augmentation will increase the force the head applies to the impacted surface, assuming the surface structure is capable of resisting still higher forces. To this extent, some of the torso inertial force is transmitted to the head through the neck and the peak deceleration of the head mass is reduced even though the reaction force of the impacted surface may be increased. Possible exception to this principle would occur where the variables described above operate to extend the interval between the initial head impact and the torso augmentation force applied to the head. Head-mass augmentation ordinarily will not increase the peak deceleration sustained by the head because this condition augments the inertial force applied by the head, and thereby increases the yield or deformation of the car interior surface being impacted by the head.

This description of head-mass augmentation demonstrates that peak deceleration values recorded for simulated human head impacts, when translated in terms of pounds force for the twelve-pound head, represent minimal values. Any contribution by the torso serves only to increase these force values. Although this mechanism may tend to decrease peak decelerations, it does not mean that injuries are correspondingly reduced because increased force values of these augmented exposures may generate brain stem trauma, spinal cord separation and related critical injuries.

ANALYSIS OF VEHICLE DYNAMICS

These calculations evaluate the influence of occupant seating positions and vehicle positions of impact on occupant acceleration during intersection collisions.

A commonly accepted observation concerning automobile accident victims is that within a given car some passengers fare better than others. It is also generally conceded that compartment surfaces against which the motorists are violently hurled often serve to differentiate injury patterns. It should be apparent that a motorist sitting beside a door that is directly impacted by a striking car, typified by the intersection collision, will in all probability sustain more serious injuries than a motorist seated more remotely from the direct impact. But what about occupants seated in a car struck at its front wheel as compared to its rear wheel? Acceleration levels obtained for passenger chest exposures of this nature, where door latches do not fail, require analytical evaluation to differentiate between front seat and rear seat positions.

This differential in peak G's is illustrated by Figure 21 presenting data on X-79's rear side and X-80's front-side impacts. As may be seen when the motorist, in the struck car, is closer to the position of contact between the vehicles, he sustains higher injury producing forces. For these experiments, the simulated passengers were the same size and weight. However, in addition to differences in seated positions relative to location of car contact, there is another significant difference: collapse characteristics for the front door versus the rear door, as occupants are pitched against the door. Thus, the differences in values noted in Figure 21 for passenger chest accelerations concern a double variable.

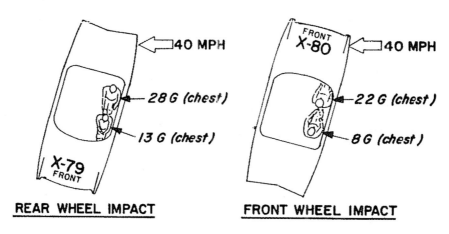

FIGURE 21. Differential in Peak G's.

The influence on the relative exposure hazard for passenger location alone can be calculated. During impact, the struck car's center of gravity causes its center of rotation to shift from the fulcrum (front wheels for X-70 and rear wheels for X-80) by an amount dependent on tire friction and distance of the c.g. from fulcrum wheels. This offset distance was determined by micro-motion analysis to be one foot for X-79 and one-half foot for X-80. Micro-motion analysis also showed at the time of peak occupant loading, that the struck car had only rotational movement and that linear translation was negligible. The composite force acting on passenger position "A" is the sum of force components owing to the distance of "A" from the center of rotation and the linear acceleration of the center of gravity. The composite force acting on passenger position "B" may be similarly described.

For EXPERIMENT X-79 the struck car had the following forces acting at positions A and B 80 milliseconds after contact, Figure 22.

(1) Struck car, X-79, angular velocity and acceleration by micro-motion analysis

$$\omega_1 = \left(\frac{2\pi}{360°}\right)\left(\frac{2.90°}{.012 \text{ sec}}\right) = +20 \text{ rad/sec}$$

$$\omega_2 = \left(\frac{2\pi}{360°}\right)\left(\frac{3.0°}{.012 \text{ sec}}\right) = +35 \text{ rad/sec}$$

$$\omega \text{ avg} = \left(\frac{+20 + +35}{2}\right) = +28 \text{ rad/sec}$$

$$\alpha = \frac{+35 - +20}{.012 \text{ sec}} = 12.5 \text{ rad/sec}^2$$

A....Front seat occupant

B. Rear seat occupant

C...Center of rotation

R_a .Radius arm from center of rotation to point "A" in feet

R_b Radius arm from center of rotation to point "B" in feet

θ. ∠ACP of point "A"

ϕ... ∠BCP of point "B"

μ .75° included ∠ for impacting cars

ω...Angular velocity of car, rad/sec

α. Angular acceleration of car, rad/sec²

ωr Tangential velocity of points "A" or "B", ft/sec

$\omega^2 r$ Centripetal acceleration of points "A" or "B", ft/sec²

αr Tangential acceleration of points "A" or "B", ft/sec²

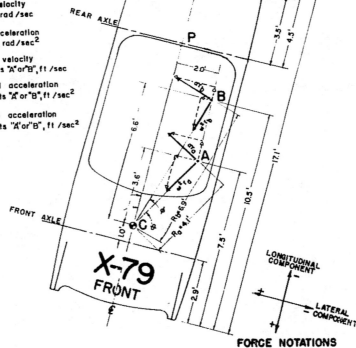

FIGURE 22 Struck Car 80 Milliseconds After Contact.

248

(2) Position "A" X-79 struck car, tangential velocity, tangential acceleration, and centripetal acceleration.

$$Ra = 4.1 \text{ feet from Figure 22}$$

$$\omega r = (4.28)(4.1') = 17.6 \text{ ft/sec}$$

$$ar = (12.5)(4.1') = 52 \text{ ft/sec}^2 = 1.6g$$

$$\omega^2 r = (4.28)(4.1') = 76 \text{ ft/sec} = 2.4g$$

(3) Position "b", X-79 struck car, tangential velocity, tangential acceleration, and centripetal acceleration

$$Rb = 6.9 \text{ feet from Figure 22}$$

$$\omega r = (4.28)(6.9') = 29.5 \text{ ft/sec}$$

$$ar = (12.5)(6.9') = 86 \text{ ft/sec} = 2.7g$$

$$\omega^2 r = (4.28)^2(6.9') = 126 \text{ ft/sec} = 3.9g$$

(4) Resolution to lateral and longitudinal components for position "A" X-79

Tangential velocity resolution at "A"

$$\text{lat } \omega r = \omega r \cos \theta = (17.5)\left(\frac{3.6'}{4.1'}\right) = 15.4 \text{ ft sec}$$

$$\text{long } \omega r = \omega r \sin \theta = (17.5)\left(\frac{2'}{4.1'}\right) = -8.6 \text{ ft/sec}$$

Tangential acceleration resolution at "A"

$$\text{lat } ar = (ar)\cos \theta = (1.6g)\left(\frac{3.6'}{4.1'}\right) = 1.4g$$

$$\text{long } ar = (ar)\sin \theta = (1.6g)\left(\frac{2.0'}{4.1'}\right) = -0.8g$$

Centripetal acceleration resolution at "A"

$$\text{lat } \omega^2 r = (\omega^2 r)\sin \theta = (2.4g)\left(\frac{2.0'}{4.1'}\right) = 1.2g$$

$$\text{long } \omega^2 r = (\omega^2 r)\cos \theta = (2.4g)\left(\frac{3.6'}{4.1'}\right) = 2.1g$$

249

(5) Resolution to lateral and longitudinal components for position "B", X-79

Tangential velocity resolution at "B"

$$\text{lat. } \omega r = \omega r \cos \phi = (29.5)\left(\frac{6.6'}{6.9'}\right) = 28.2 \text{ ft/sec}$$

$$\text{long. } \omega r = \omega r \sin \phi = (29.5)\left(\frac{2.0'}{6.9'}\right) = -8.6 \text{ ft/sec}$$

Tangential acceleration resolution at "B"

$$\text{lat. } \alpha r = (\alpha r) \cos \phi = (2.7g)\left(\frac{6.6'}{6.9'}\right) = 2.6g$$

$$\text{long. } \alpha r = (\alpha r) \sin \phi = (2.7g)\left(\frac{2.0'}{6.9'}\right) = 0.8g$$

Centripetal acceleration resolution at "B"

$$\text{lat. } \omega^2 r = (\omega^2 r) \sin \phi = (3.9g)\left(\frac{2.0'}{6.9'}\right) = 1.1g$$

$$\text{long. } \omega^2 r = (\omega^2 r) \cos \phi = (3.9g)\left(\frac{6.6'}{6.9'}\right) = 3.7g$$

Analysis was made in a similar manner for EXPERIMENT 80, Figure 23. This provided the acceleration data shown in Figure 24 for X-79 and X-80. In X-80, the force of collision was directed to the side of the car near its front-wheel as contrasted with a rear-wheel impact for X-79.

The Impact Velocity (door to occupant) for positions A and B of X-79 and X-80 are shown diagrammatically in Figure 24 and provide the following:

(a) Occupant "B" is accelerated most abruptly for X-79 and least abruptly for X-80, according to its closer position to impact (X-79) and its more remote position from impact (X-80). In a similar manner, impact velocities are a function of the distances of positions "B" from their respective centers of rotation and correlated closely with the lateral peak G of the occupant.

$$\frac{\text{Lat. } \omega r \; 79}{\text{Lat. } \omega r \; 80} = \frac{28.2}{7.6} = 3.7 \qquad \frac{\text{Lat. } G_{79} \text{ For "B"}}{\text{Lat. } G_{80} \text{ For "B"}} = \frac{27.6}{7.3} = 3.8$$

Position A functions similarly.

(b) The force of impact differs for X-79 and X-80 owing to the different structural properties of the impacted rear-end Experiment 79 versus the impacted front-end (X-80). However, the property of degree of remoteness from impact identified in (a) is maintained even for these differing force levels, as shown by:

$$\frac{B^R_{79}}{A^R_{79}} = \frac{8.6'}{5.6'} \;\; 1.5 \quad \text{and} \quad \frac{B^R_{80}}{A^R_{80}} = \frac{9.5}{6.5} = 1.5$$

(c) Item (b) indicates the relative exposure hazard for the seat position most remote from the point of impact compared with the seat closest to the impact.

The ratios indicate a person is safer when seated more remotely. Although the experimental data for dummy-chest accelerations actually sustained for these respective experiments carries the added factor of collapse characteristics of object struck, it is interesting to note that these collapse characteristics do not differ markedly for the chest level impacts as shown by the calculations of (a)

(d) A similar correspondence was not obtained for head impacts of dummies seated in these positions owing to the radically different rates of collapse for head-impacted structures according to whether it was glass, window frame, the side panel of the car-top, etc.

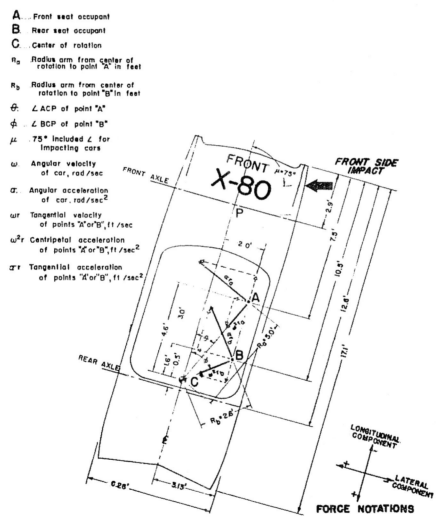

A ....Front seat occupant
B ....Rear seat occupant
C ....Center of rotation
R_a ....Radius arm from center of rotation to point "A" in feet
R_b ....Radius arm from center of rotation to point "B" in feet
θ ....$\angle$ ACP of point "A"
ϕ ....$\angle$ BCP of point "B"
μ ....75° included $\angle$ for impacting cars
ω ....Angular velocity of car, rad/sec
σ ....Angular acceleration of car, rad/sec^2
ωr ....Tangential velocity of points "A" or "B", ft/sec
$\omega^2 r$ ....Centripetal acceleration of points "A" or "B", ft/sec^2
σr ....Tangential acceleration of points "A" or "B", ft/sec^2

FIGURE 23. Analysis for Experiment 80.

251

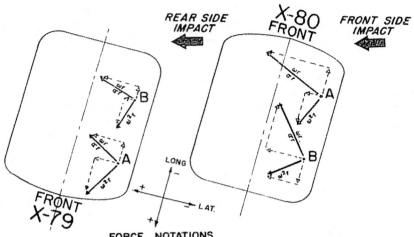

FORCE NOTATIONS

EXPERIMENT NO AND POSITIONS		RECORDED PEAK CHEST ACCEL-ERATION (G)	MICRO- MOTION ANALYSIS (CRASH FILM)			
			IMPACT VELOCITY OF DOOR ωr (ft/sec)	TANGENTIAL ACCELERATION αr (G)	CENTRIPETAL ACCELERATION $\omega^2 r$ (G)	COMBINED ACCELERATION (G)
X-79 "A"	LAT.*	12.1	15.4	1.4	1.2	2.6
	LONG	2.0	-8.6	-0.8	2.1	1.3
	RES.	13.0	17.6	1.6	2.4	2.9
X-79 "B"	LAT.	27.6	28.2	2.6	1.1	3.7
	LONG	3.3	-8.6	-0.8	3.7	2.9
	RES.	28.0	29.5	2.7	3.9	4.7
X-80 "A"	LAT.	21.4	22.0	10.7	1.4	12.1
	LONG	5.0	-9.6	-4.7	3.3	-1.4
	RES.	22.0	23.9	11.6	3.6	12.2
X-80 "B"	LAT.	7.3	7.6	3.8	1.4	5.2
	LONG	3.2	-9.6	-4.7	1.0	-3.7
	RES.	8.0	12.4	6.1	1.8	6.4

* LATERAL, LONGITUDINAL, RESULTANT

FIGURE 24. Acceleration Data for Experiments 79 and 80.

These varying exposure values explain how one person may be killed in a collision where others are not seriously injured, according to where their heads (or other parts of the body) happen to strike the car interior.

CONCLUSIONS

1. Side-glass breakage during intersection-type collisions may occur from: (a) Direct contact by the opposing car's structure. (b) Collision distortion not

involving direct contact by the opposing car. (c) Inertial force applied by motorists from within. (d) A combination of the foregoing factors.

2. Where side-glass breakage is attributed primarily to inertial force from an adjacent motorist, his torso impact with the door precedes head impact by enough time to pre-stress the glass greatly. This pre-stressing is generally of sufficient magnitude to induce glass breakage before the head contacts the side-glass. Exception to this condition relates to more remotely positioned motorists who tend to be pitched more nearly head first against an opposite side-glass panel rather than an adjacent one. Use of seat belts reduces or completely eliminates exposure to this type of head impact.

3. Further research is needed to verify observations described in Item 2. As a preliminary conclusion, however, higher performance side-glass, now receiving consideration by the United States auto manufacturers, appears to be a desirable change for a number of reasons. High strength glass is less likely to break and, therefore, occupant containment during collisions would be more consistently obtainable. In addition, some measure of increased roof strength during upsets would be realized from use of such glass. These important contributions to motorist safety will not, it now appears, be offset to any significant extent by the problem of increased motorist concussion and skull fracture as a result of use of a more impact-resistant side-glass.

This tentative conclusion is based on the observation that: (a) Side-glass is not rigidly mounted; impacts directed to side glass readily deflect the glass outward. Such deflections serve to reduce peak head accelerations. (b) The stronger chemically toughened glass can be made thinner and, therefore, presents a reduced inertia force to the impacting head. (c) The head is less likely to fracture the glass than is the direct fracturing action of the shoulder pitched against the side panel of the door.

4. Side-glass breakage is less likely to occur in cars having *inadequate* door latch mechanisms because these mechanisms fail before the passenger and door inertial forces reach their full potential. (Door latch failure exposes motorists, including those wearing seat belts, to increased hazards. Stronger door latches and door hinges are needed for all cars, domestic and particularly foreign.)

5. In general, head impacts with side window frames represent more dangerous exposures than head impacts with side window glass.

6. Based on observations from this and from prior series of collision experiments, the following tentative conclusions relate to the possible influence the torso may have on head impact characteristics:

a. During impact of car interior and human head, head-mass augmentation may occur when torso inertial forces happen to be directionally aligned to pass through or close to the head. This does not occur frequently.

253

b. Head-mass augmentation cannot be coincident with the initial head impact owing to the elastic nature of the human body. The extent of A-synchrony depends on the velocity of the human subject relative to its impacting surface and to a lesser extent on the torso force direction.

c. Head-mass augmentation increases the force the head may apply to the impacting surface where the structure of such surface is capable of sustaining still higher forces than the head alone provides.

d. Where some of the potential torso inertial force is transmitted to the head through the motorist's neck, the peak deceleration of the head mass is reduced even though the reaction force of the impacting surface may have increased. Possible exception to this principle would occur where the relative speed and torso vector operate to extend the interval between the initial head impact and the torso augmentation.

e. Head-mass augmentation (HMA) is not the same as head inertia increase even when the force the head may apply for each of these conditions is identical. The mechanism of HMA supplies an inertial force, acting through the neck, to the head. In this context, the head tends to be caught in a vise-like action with the torso pushing from one direction and the impacted surface resisting head movement with a force of comparable magnitude. Increases in head inertia alone may allow the same head impact force as attained with HMA, but such forces relate to the inertia of the head alone so that the forces applied to the head from the torso are zero. Where the impacted surface fails, acceleration of the brain mass is actually reduced as a result of a still higher level of force transmitted to the impacted surface, whether by increased head inertia or by increased HMA.

f. Accordingly, where peak accelerations for the head-to-glass impacts were measured, these values, when multiplied by the weight of the head alone, represent conservative force estimates of the head blows sustained; any HMA by the torso that may have been overlooked would result in actually higher force values than estimated because the effective head mass is larger by the amount contributed from the torso.

g. HMA, while tending to reduce peak deceleration, does not necessarily reduce injuries sustained. The reduction in potential exposure to concussion may be more than offset by the increased exposure to brain stem trauma, or even spinal cord separation or related critical injuries.

ACKNOWLEDGMENTS

The authors extend their sincere appreciation to the U.S. Public Health Service for the substantial financial support of this research through a grant to the University of California; to the Chrysler Corporation for donating the passenger vehicles; to the Commanding Officer, U.S. Naval Base, Long Beach,

California, for making the test site facilities available. This research was materially assisted by the cooperation and technical assistance of the following:

Corning Glass Works
Libby-Owens-Ford Glass Company
Permaglass, Incorporated
Pittsburgh Plate Glass Company

Acknowledgment also is made to Wendell Severy, M.D., for medical evaluation of injuries, and in equal measure, to the regular staff members: Jack Baird and Hans Jakob, engineering assistance; Amir Karimi and Roger Seiler, motion picture photography and production; David Blaisdell and William Archibald, electronic instrumentation and data reduction; Ross Sater and Virgil Sibell, mechanician and related services; George Newnam, mathematical analyses; Jack Kerkhoff, engineering illustrations; Louise Kirkpatrick and Adele Fishgold, publication and secretarial assistance.

APPENDIX 1

Instrumentation

The extensive photographic coverage used for UCLA full scale collision studies has been described in detail in prior publications. All such publications are listed in ITTE (UCLA) Bulletin No. 34. Photographic coverage was expanded by 30% to provide saturation coverage of motorist head interactions with car safety glass during collision. Additional high speed cameras were mounted on the cars in special protective steel boxes; these cameras were equipped with wide angle lenses allowing complete coverage of exposure conditions from a close range. The schedule of camera types, locations, and operating speeds are given in Appendix 3. Reference should also be made to Figure 2 for purposes of identifying specific locations of cameras at time of impact. Three recording oscillographs provided in excess of 60 data channels for transducers from both cars and from each anthropometric dummy passenger.

Several categories of new instrumentation were included in this series of experiments. Type SR-4 strain gauges were cemented to the glass for measuring relative impact forces, impact duration, and precise instant of glass breakage. Linear potentiometers were installed in the struck car between each car door and the car body to record elastic plastic distortions.

Tension calibrated strain links were installed, replacing the door latches, to measure door force variations to prevent door openings during excessive loadings.

APPENDIX 2

"Safe on Impact", a documentary motion picture film:

Investigation of the favorable—as well as the hazardous—performance features of various kinds of automotive safety glass is reported by the Institute of

255

Transportation and Traffic Engineering, University of California (U.C.L.A.). This 12 minute documentary motion picture film shows laboratory studies using a head-form drop test stand, supplemented by actual automobile collisions. High speed photography and stop-action film techniques detail what happens when anthropometric dummies collide with various types of windshields and side windows. Practical information of value to motorists is presented as well as demonstrations of the exceptional properties of new forms of safety glass.

This motion picture, 12 minutes long and in color, is priced at $110.00; distribution is handled by its producer, Charles Cahill and Associates, 5746 Sunset Blvd., Hollywood, California.

APPENDIX 3

Camera Desig.	Camera and Lens	Distance, Elevation and Position	Shutter Speed
A	16mm Photo-Sonics 13mm lens	24' overhead, Tower	200 f/s
B	4"x5" Super Speed Graphic 135 mm lens	24' overhead, Tower	1/1000 s
C	16mm GSAP 12mm lens	40' overhead, High Boom	64 f/s
D	35mm Bell & Howell 14.5mm lens	40' overhead, High Boom	48 f/s
E	16mm Photo-Sonics 25mm lens	60' x 12' High Slant Boom	200 f/s
F	16mm Eastman High Speed 63mm lens	30' x 5' High @ 250°	600 f/s
G	16mm Photo-Sonics 13mm lens	30' x 5' High @ 250°	200 f/s
H-1	16mm Fastax 50mm lens	50' x 5' High @ 185°	800 f/s
H-2	16mm Fastax 50mm lens	50' x 5' High @ 185°	800 f/s
I	16mm GSAP 17mm lens	30' on Ground @ 30°	64 f/s
J	16mm GSAP 17mm lens	30' on Ground @ 60°	64 f/s
K	16mm GSAP 17mm lens	40' on Ground @ 120°	64 f/s
L	16mm GSAP II 17mm lens	20' x 4' High @ 180°	200 f/s
M	16mm Kodak K-100 25mm lens	40' x 5' High @ 360°	64 f/s
N	16mm Photo-Sonics 13mm lens	25' x 5' High @ 45°	200 f/s
O	16mm Photo-Sonics 50mm lens	35' x 10' High @ 55°	200 f/s
P	16mm Photo-Sonics 13mm lens	25' x 5' High @ 225°	200 f/s
Q	70mm Hulcher 75mm lens	30' x 5' High @ 235°	20 f/s
R	70mm Hulcher 75mm lens	27' x 5' High @ 195°	20 f/s
S	4" x 5" Super Speed Graphic 135mm lens	25' x 6' High @ 45°	1/1000s
T	4" x 5" Speed Graphic 135mm lens	27' x 3' High @ 195°	1/1000s
U	4" x 5" Super Speed Graphic 135mm lens	25' x 6' High @ 225°	1/1000s
V-1	16mm Arriflex 17mm lens (panning)	60' x 12' High @ 180°	24 f/s
V-2	16mm Photo-Sonics 25mm lens (panning)	60' x 12' High @ 180°	200 f/s
W	16mm Kodak K-100 25mm lens	65' x 4' High on Moving Pace Car	64 f/s
X	35mm Foton Sequence 50mm lens	27' x 4' High @ 195°	1/1000 s
Y-1	16mm Photo-Sonics 13mm lens	Striking Rear Fender	200 f/s
Y-2	16mm Photo-Sonics 8mm lens	Striking Hatrack	200 f/s
Y-3	16mm Photo-Sonics 5.3mm lens	Striking Outrigger Mount	200 f/s
Z-1	16mm Photo-Sonics 8mm lens	Struck Roof	200 f/s
Z-2	16mm Photo-Sonics 5.3mm lens	Struck Trunk Deck	200 f/s
Z-3	16mm Photo-Sonics 13mm lens	Struck Rear Fender	200 f/s

Factors in the Development and Evaluation of Safer Glazing

R. G. RIESER[1] and G. E. MICHAELS[1]

Glass as a component of automobiles has become a subject of widespread study to improve its safety attributes when impacted by vehicle occupants. Windshields have been cited (Schwimmer and Wolf, 1962) as one of the first four major sources of motor vehicle occupant injury. Thus effort is being expended by many groups and laboratories to make the windshield, in particular, less destructive regarding lacerative injury, severe neck ruffle and fatal physical damage. With the availability of advanced materials, these studies have assumed a new scope and bench scale experimentation has expanded into full-scale automobile crash sequences (Severy, Mathewson and Siegel, 1962) and cadaver impact evaluations (Patrick and Daniel, 1964).

In recent years extensive study has been undertaken to improve windshield safety performance by evolutionary and innovational developments. All the information and data reported in this paper except for the cadaver impact studies were developed at the Glass Research Center, Pittsburgh Plate Glass Company. The aim of our research and development effort on safer transparencies is to seek the ultimate in windshield safety as well as develop data and technology for the automotive industry's consideration. This paper represents an interim progress report covering our laboratory data on the impact performance of laminated safety glass as influenced by many component and evaluation variables. New candidate materials are under study and the progress of our investigation will be reported from time to time as research progresses.

In order to simulate on a laboratory scale a more meaningful impact evaluation related to human impact performance, the firms that manufacture and laminate glass and the plastic manufacturers have examined extensively the impact characteristics of experimental panels with a 22 lb headform vertical drop. This is the tentative test proposed by the Glazing Safety Study Group of the SAE Body Engineering Committee as a development test. A 5 lb ball vertical drop test is proposed by the same committee as a SAE "Recommended Practice" for evaluating laminated safety glass for use in automotive windshields.

---

[1] Glass Research Center, Pittsburgh Plate Glass Company, Pittsburgh, Pennsylvania

The details of these two test procedures are presented in another paper at this Stapp Conference.

The instrumented 22 lb headform is dropped vertically onto a 24" x 36" glass panel and the deceleration measured by the accelerometer within the ball. The recorded force-time curves are compared with the "Human Tolerance to Head Impact" curve (Patrick, 1965) developed at Wayne State University to judge the impact forces in terms of dangerous or non-dangerous to life. The proposed 5 lb steel ball vertical drop test procedure involves a 12" x 12" test panel with a specified go-no go drop height. In the vertical free-fall ball tests reported herein, the panel sizes for the 5 lb ball and the 22 lb headform are 12" x 12" and 24" x 36", respectively, unless otherwise specified.

The configurations covered in this report include two types of laminated safety glass interlayer. The first of these is the plasticized polyvinyl butyral used for many years, in 0.015" thickness, as the standard automotive laminated safety glass interlayer. This interlayer composition is referred to as Std in the tables and figures of this report. The second interlayer is plasticized polyvinyl butyral compounded by the manufacturer to control the adhesion to glass when the moisture content is optimized (0.3 to 0.6 wt %). This interlayer, normally used in 0.030" thickness, is referred to as high penetration resistance (HPR) interlayer.

The Std and HPR interlayers are produced by Monsanto Company and E. I. de Pont de Nemours, Inc. The lower adhesion of the HPR product permits the interlayer at the glass fractures to "neck-down" rather than shear thus providing more elongation and energy absorption before interlayer rupture occurs. This recent modification of polyvinyl butyral interlayer followed the work by Rodloff (Rodloff, 1962) in Germany which showed that high moisture content (1 to 1.25 wt %) in the standard interlayer lowered interlayer-glass adhesion permitting higher elongation before rupture.

It is interesting to note a comparison between the ball impact data and Wayne State University cadaver tests. In Table 1 are shown data for the mini-

TABLE 1 Comparison of Ball and Cadaver Impacts on Laminated Safety Glass

			Minimum Velocity for Penetration (mph)				
			Steel Ball (5 lb)**		Headform (22 lb)***		Cadaver*
Configuration (Inches)							Penetration to Ears
Glass	Interlayer	Glass	45°	90°	45°	90°	
1/8	0.015 Std	1/8	12.7	8.5	14	10	14-15
1/8	0.030 HPR	1/8	25.8	25.9	28	28	29
	0.015 Std			29.5			
	0.030 HPR			>38.3			

* Wayne State University—windshield tests with impact sled
** 12" x 12" test panels   (Patrick and Daniel, 1964)
*** 24" x 36" test panels
All glass—annealed plate glass

258

mum penetration velocity with the 5 lb ball and the 22 lb headform and the cadaver impact velocities against windshields to achieve penetration "to the ears" for the two windshield configuration currently in use. The penetration velocities for the cadaver and the 45° and 90° impacts with the dropping ball show quite good correlation.

Additional impact data in Table 1 show that the free polyvinyl butyral interlayer, unrestricted in diaphragmatic action, has very high impact absorption characteristics, being much greater than the same interlayer in a glass laminate. The interlayer alone, if it were possible to use it this way, would be an excellent impact energy absorber.

In order to provide a meaningful velocity performance value for the penetration characteristics of the various configurations tested, the mean penetration velocity (MPV) has been selected for most of the curves and tables of this report. The mean penetration velocity is illustrated on Figure 1 and is defined as follows: For any particular safety glass configuration, the impact velocity can be either low enough or high enough so that the impacting object will not penetrate any of the samples tested or will penetrate all samples tested. There is a transition range where at any set velocity the impacting object will penetrate a percentage of the samples tested. Mean penetration velocity is the midpoint of this transition range and is the velocity at which 50% of a group of samples will be penetrated.

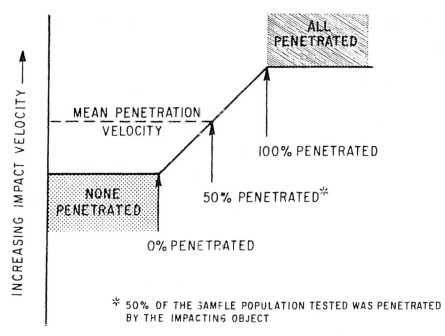

FIGURE 1. Illustration of Mean Penetration Velocity (MPV)

Investigation of the 22 lb headform impact performance of 24″ x 36″ laminates in comparison with the 5 lb steel ball vertical drop on 12″ x 12″ test panels has shown very close performance in terms of mean penetration velocity. The velocities for 90° impacts as well as 45° impacts show excellent correlation by the curve in Figure 2 (tabulated data in Table A, Appendix).

On the basis of these data, it appears acceptable to use the 22 lb headform and 5 lb steel ball interchangeably in the penetration velocity range investigated at the same impact angle.

Neither the 5 lb ball nor the 22 lb headform showed good correlation between the 45° and 90° impact angles as shown in Figure 3 (tabulated data in Table B, Appendix). The tendency was toward a higher mean penetration velocity at the 45° impact angle. This tendency was especially pronounced for configurations with strengthened glass components and for impact of the laminate configuration of two plies of ⅛″ annealed glass with a .015″ standard

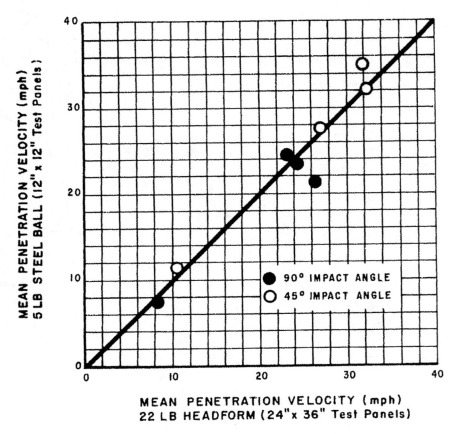

FIGURE 2. Correlation of Ball Impact Tests (5 lb. Steel Ball vs. 22 lb. Headform at 45° and 90° Impact Angles) (Data Tabulated in Table A, Appendix)

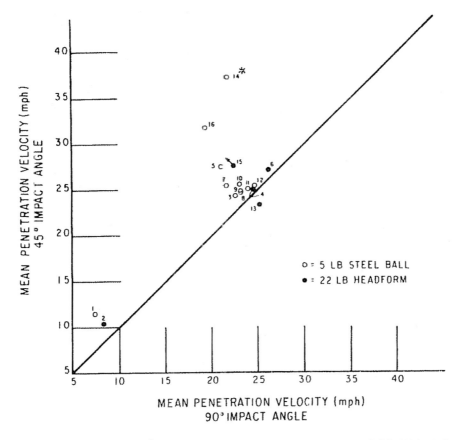

(* NUMBERS REFER TO CONFIGURATION - SEE TABLE B, APPENDIX)

FIGURE 3. Correlation of 45° and 90° Impact Angles (Data Tabulated in Table B, Appendix)

polyvinyl butyral interlayer In both cases, the greater MPV for the 45° impact angle was attributed to the effect of glass strength. Although the annealed glass in the latter configuration was not specifically strengthened, its normal resistance to impact breakage was a relatively large factor because of the low impact velocity and the type of laminate configuration. The same glass with a .030″ HPR interlayer showed no difference in mean penetration velocity between 45° and 90° impact angle at a 25 mph level because glass strength energy absorption was a small factor compared to the impact energy absorbed by the plastic interlayer after the glass was fractured.

Obvious factors which can be used to alter the impact characteristics of a glass laminate are (1) interlayer thickness, (2) inboard glass thickness, (3) total glass thickness and (4) glass strength. The effect of interlayer thickness on penetration is illustrated in Figure 4 (tabulated data in Table C, Appendix). The 22 lb headform impacts at 90° angle for various thicknesses of standard and high penetration resistant interlayer-glass laminates show: (1) the MPV to be a straight line function of the interlayer thickness, and (2) the higher impact penetration resistance of the new controlled adhesion interlayer. For the same thickness in the 0.015" to 0.060" range, the HPR interlayer provides a 75-80% increase in MPV as measured by the 22 lb headform on 24" x 36" panels. The combined effect of doubling the interlayer thickness and the use of HPR interlayer gives triple the MPV in this test.

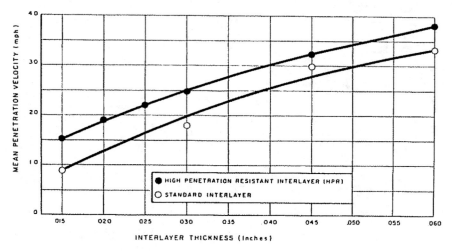

FIGURE 4. Effect of Interlayer Thickness on Mean Penetration Velocity (22 lb. Headform at 90° Impact Angle) (24" x 36" Test Panels, ⅛" Annealed Glass Plies) (Data Tabulated in Table C, Appendix)

Impact tests on safety glass laminates having a ⅛" thick outboard glass ply and various inboard glass ply thicknesses (Table 2) demonstrated that the variation in thickness of impacted ply from ⅛" to 0.050" had no significant effect on MPV. The interlayer film, free of any glass, shows much higher MPV than the interlayer with one or two glass layers regardless of which surface receives the impact. The mean penetration velocity of 0.030" HPR interlayer was found to be greater than 38.3 mph (drop height of 49 feet, the limit of our tower) while the same type interlayer laminated with one ply of ⅛" glass had a MPV of 22-24 mph. The influence of the glass ply in restricting the diaphragm behavior of the interlayer is thus very clearly demonstrated. The new high penetration resistance windshield configuration, i.e., 0.030" HPR interlayer with ⅛" inboard and outboard glass plies shows essentially the same MPV as the structure with only one ⅛" glass layer (24.0 mph for the complete windshield configuration vs. 22.1 to 24.3 mph for the partial laminate). As the penetration

TABLE 2. Effect of Inboard Glass Thickness on Penetration

(5 pound ball at 45° and 90° Impact Angles)
(Test Panel: 12" x 12")

Configuration (Inches)			Impact Angle	MPV[a] (mph)
Glass[b]	Interlayer	Glass		
	0.030 HPR		90	>38.3
	0.015 Std		90	29.5
⅛ P	0.030 HPR	⅛ P	90	24.0
⅛ P	0.015 Std	⅛ P	90	7.4
	Same		45	11.4
0.090 P	0.030 HPR	⅛ P	90	23.2
	Same		45	25.8
0.067S	0.030 HPR	⅛ P	90	23.2
	Same		45	24.9
0.050 S	0.030 HPR	⅛ P	90	23.2
	Same		45	24.2
	0.030 HPR	⅛ P	90	24.3
	Same		45	25.2
⅛ P	0.030 HPR		90	22.1
	Same		45	25.9

[a] Mean Penetration Velocity   P—Annealed Plate Glass
[b] Impacted Surface   S—Annealed Sheet Glass

resistance of the interlayer increases, the influence of ⅛" annealed glass on the impact performance is essentially eliminated whether the impact is at 45° or 90° impact angle.

In the evaluation of total glass thickness on the impact penetration resistance of glass laminates, several configurations having the same total glass thickness and laminates with thin glass plies were studied. The standard safety glass laminate having ⅛" glass plies has a total glass thickness of approximately 0.250". This structure along with combinations of 0.090" + 0.150", 0.050" + 0.195", 0.067" + 0.067", and 0.050" – 0.050", all with 0.030" HPR interlayer, have been examined for MPV using the 5 lb ball at 45° and 90° impact angles. Table 3 shows that total glass thickness of approximately 0.250" regardless of the thickness of the component glass plies within the limits of 0.050" to 0.195" had no significant influence on MPV whether impacted at 45° or 90°. With the thinner glasses, 0.050" and 0.067", two conclusions are apparent; namely, decreased MPV at a 90° impact angle and increased MPV at a 45° impact angle for these laminates compared with the thicker glass structures. The advantage, if any, for thin glass over ⅛" and thicker glasses as regards MPV is not greater than 15% at a 45° impact angle in the thickness range given in Table 3.

Several possibilities exist for controlling the strength of glass used in safety glass laminates. Annealed glass as currently used could be replaced with higher

TABLE 3. Total Glass Thickness vs Mean Penetration Velocity

(5 pound ball at 45° and 90° impact angles)
(Test Panel: 12" x 12")

Configuration (Inches)			Impact Angle	MPV[a] (mph)
Glass[b]	Interlayer	Glass		
⅛ P	0.030 HPR	⅛ P	90	24.0
	Same		45	24.3
0.090 P	0.030 HPR	0.150 P	90	24.5
	Same		45	25.5
0.050 S	0.030 HPR	0.195 P	90	24.0
	Same		45	25.0
0.067 S	0.030 HPR	0.067 S	90	21.8
	Same		45	25.4
0.050 S	0.030 HPR	0.050 S	90	21.2
	Same		45	27.6

[a] Mean Penetration Velocity   P—Annealed Plate Glass
[b] Impacted Surface   S—Annealed Sheet Glass

TABLE 4. Effect of Glass Strength on Mean Penetration Velocity

(5 pound ball at 45° impact angle)
(Test Panel: 12" x 12")

Configuration (Inches)			MPV[a] (mph)
Glass[b]	Interlayer	Glass	
⅛ P	0.015 HPR	⅛ P	15.0
⅛ TT	0.015 HPR	⅛ TT	26.9
⅛ P	0.030 HPR	⅛ P	25.0
⅛ TT	0.030 HPR	⅛ TT	37.5
⅛ P	0.030 HPR	⅛ TT	28.5
⅛ TT	0.030 HPR	⅛ P	30.0
⅛ P	0.030 Std	⅛ P	14.4
⅛ TT	0.030 Std	⅛ P	19.5
0.040 H II	0.015 HPR	0.040 H II	23.0-24.5
0.076 H II	0.015 HPR	0.076 H II	23.0-24.5
0.010 H II	0.015 HPR	0.10 H II	21.4
0.10 H II	0.030 HPR	0.10 H II	32.0

[a] Mean Penetration Velocity
[b] Impacted Surface
P—Annealed Plate Glass
TT—Thermally Strengthened Plate Glass—fully tempered
H II—HERCULITE II, Pittsburgh Plate Glass Company's chemically Strengthened glass

strength products such as thermally tempered and chemically strengthened glasses. The extent of strengthening can be controlled in the manufacturing process for these glasses and thus much freedom exists in the selection of the optimum strength. Various thermally tempered and chemically strengthened glasses in laminate structures have been examined for MPV with the 5 lb ball and 22 lb headform. The data for 45° impact angle have been summarized in Tables 4 and 5.

Impacts with the 5 lb ball as summarized in Table 4 show that the use of ⅛" tempered glass plies has the same effect on MPV as doubling the interlayer thickness from 0.015" to 0.030". With 0.030" HPR interlayer and two plies of ⅛" thermally tempered glass, the MPV increases to 37.5 mph; this is approximately a 50% increase in velocity for penetration over the new high penetration resistance windshield configuration. Penetration resistance is, of course, not the only criterion for safety and other aspects of this type configuration will be discussed later in this report. It is also apparent in Table 4 that an annealed glass ply as either the inboard or outboard component in combination with a ⅛" tempered glass ply, both plies being adhered directly to the interlayer, is the penetration velocity controlling component. It is apparent that the tempered ply in a tempered glass-annealed glass laminate should be the inboard glass to reach the highest MPV for this combination. In this series of tests, symmetrical laminates of thin Herculite II (Pittsburgh Plate Glass Company's chemically strengthened glass) do not show penetration resistance superior to the thermally strengthened glass.

The 22 lb headform penetration is influenced significantly by the size of the impacted test panel. Mean penetration velocities with the 22 lb headform

TABLE 5 Effect of Glass Strength on Mean Penetration Velocity

(22 pound headform at 45° impact angle)

(Configuration (inches)			Size	MPV[a]
Glass[b]	Interlayer	Glass	(Inches)	(mph)
⅛ P	0.030 HPR	⅛ P	18 x 30	21.5
⅛ TT	0.030 HPR	⅛ TT	18 x 30	27.3-29.9
⅛ TT	0.015 HPR	⅛ TT	18 x 30	17.3-21.2
⅛ P	0.030 HPR	⅛ P	24 x 36	25.0
⅛ ST	0.030 HPR	⅛ ST	24 x 36	27.9
0.1 H II	0.030 HPR	0.1 H II	24 x 36	32.5

[a] Mean Penetration Velocity
[b] Impacted Surface
P—Annealed Plate Glass
TT—Thermally Strengthened Plate Glass—fully tempered
ST—Thermally Strengthened Plate Glass—semi-tempered
H II—HERCULITE II, Pittsburgh Plate Glass Company's chemically strengthened glass

have been determined for 18″ x 30″ and 24″ x 36″ size laminates. The slightly lower MPV for 18″ x 30″ panels shown in Table 5 indicate that the increased stiffness associated with the smaller panel lowers the penetration resistance.

The penetration resistances of several more complicated laminate structures considered to have safety potential beyond the new safer windshield are summarized in Table 6.

TABLE 6 Multi-Ply Interlayer Laminates
(5 pound steel ball at 45° impact angle)
(Test Panel: 12″ x 12″)

Configuration (Inches)			MPV[a]
Glass[b]	Interlayer	Glass	(mph)
⅛ P	0 045 Std	⅛ P	32 4
⅛ P	0 015 Std + PL + 0 015 Std + PL + 0 015 Std	⅛ P	>38 3
⅛ TT	Same	⅛ P	37 9
⅛ TT	0 025 HPR + PL + 0 015 Std	⅛ P	29 1
⅛ TT	0 030 HPR + PL + 0 007 Std	⅛ P	>38 3
0 1 H.II	Same	⅛ P	33 3[c]

[a] Mean Penetration Velocity
[b] Impacted Surface
[c] 22 Pound Headform on 24″ x 36″
PL—Parting Layer
P—Annealed Plate Glass
TT—Thermally Strengthened Plate Glass—fully tempered
H.II—HERCULITE II, Pittsburgh Plate Glass Company's chemically strengthened glass

All laminates in Table 6 have ⅛″ annealed plate glass as the outboard ply combined with several different interlayer configurations and strengthened as well as annealed inboard glass. The use of 0.045″ polyvinyl butyral interlayer increases the MPV to 32.4 mph, and, if this same interlayer thickness is separated into three 0 015″ thick layers, the MPV goes to greater than 38 mph. An inboard ply of ⅛″ thermally tempered glass at this impact speed with this same interlayer assembly has no beneficial effect on penetration resistance. A parting layer in these laminates permits the central plastic layer to act independent of any adhesion with the glass plies and this centrally located interlayer is protected from the sharp glass edges of the fractured panel.

The last two configurations of Table 6 hold promise as safer windshield structures. These configurations are but two of many composites and materials being evaluated. This multi-ply interlayer laminated with a strengthened inboard glass ply and a pushaway outboard glass is one of many possibilities with some possible faults along with new safety performance concepts. In these structures, the outboard glass ply along with its thin adhered plastic layer is pushed away by a high energy impact against the inboard glass. Penetration of the inboard thermally tempered glass or the Herculite II-0 030″ plastic composite results in a flexible edge with freedom from jagged glass edges in the "neck ruffle." The lacerative injury potential with this type break pattern should be reduced over the standard annealed glass laminate. In addition, there would not be the rigid glass edge exposed which the Wayne State University tests so vividly demonstrate to be the source of severe and massive laceration and facial bone fractures.

As mentioned previously the use of thermally tempered or chemically strengthened glass plies with 0.030" HPR interlayer results in high mean penetration velocities. However, the resistance to penetration even after fracture is not sufficient to characterize a safer glazing structure. Glass break patterns are important both for impact energies insufficient to penetrate the interlayer as well as those occurring when the head penetrates the interlayer ending with a "neck ruffle." The broken inboard tempered or chemically strengthened glass should contribute less lacerative hazard because the accident victim contacts small glass particles not large jagged splines and shards.

The outboard high strength glass ply of such a two ply strengthened glass laminate shows long thin glass splines located in a radial array from the impact location. Figure 5 shows a picture of the splines ringing the neck of the dummy. The particular panel shown here consists of 0.1" Herculite II — 0.30" HPR — 0.1" Herculite II. The spline pattern developed in this outboard chemically strengthened glass ply is typical also of the splines produced in an outboard thermally strengthened (full temper) glass layer under the same impact conditions. It is obvious that a breakage pattern such as shown here cannot be a characteristic of a safe glazing.

The multi-ply interlayer configuration with the outboard ply that breaks away when the inboard strengthened glass is impacted does not present this serious "neck ruffle" hazard. Figure 6 demonstrates the situation when head penetration occurs. In this case, the rolled-edge "neck ruffle" consists of small particles from the strengthened inboard glass and the plastic interlayer. The broken outboard glass with its integral thin plastic layer has been pushed away.

It has been shown that concussive injuries are prevalent in automobile accidents involving windshield contact (Campbell and Hopens, 1963). Severity of a concussive injury is, of course, proportional to the resultant force on the head upon impact. Many of the safety glass configurations tested for penetration resistance were also tested for deceleration characteristics by recording the force-time curves from an accelerometer mounted inside the 22 lb headform. These curves were then compared to the "Human Tolerance Curve" developed at Wayne State University. "G's" of effective acceleration and the duration of the acceleration are used as coordinates to determine if the deceleration pulse produced by an impact is dangerous or non-dangerous to life.

Typical force-time curves for two automotive windshield configurations currently in use in this country are shown in Figure 7 in comparison with the "Human Tolerance Curve." These two pulse traces are for the laminated automotive glass used for many years (two ⅛" annealed plate glass plies with 0.015" thick standard polyvinyl butyral interlayer) and the new windshield configuration used in 1966 model automobiles (two ⅛" annealed plate glass plies with

267

FIGURE 5 Illustration of Splines in Outboard Ply of Laminated High Strength Glass
(Arrow Indicates Base of Splines)

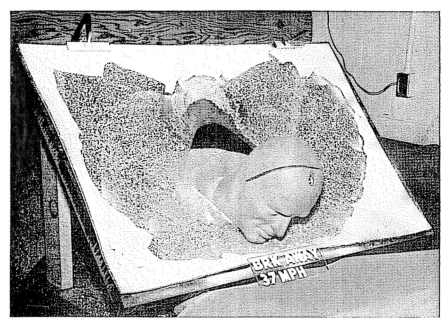

FIGURE 6 Illustration of Soft Neck Ruffle of Penetrated Multi-Ply Interlayer Laminate.

0 030" thick high penetration resistance (HPR) polyvinyl butyral interlayer). This new laminate structure requires approximately three times the impact velocity for penetration that is normal for the 0 015" thick standard interlayer (Std) laminate.

These curves were obtained with the accelerometer equipped 22 lb headform at a 90° impact angle. Neither laminate structure shows impact severity above the tolerance curve even though the impact velocity for the 0 015" Std interlayer laminate was 8 7 mph and the HPR interlayer laminate was 23 2 mph The panels fractured in both cases but the headform was held by the interlayer.

In an impact situation sufficiently severe to cause glass breakage, the first spike of the force-time curve represents deceleration before glass breakage which occurs within 1 to 2 milliseconds after impact. The next portion of the curve represents the plastic interlayer retarding the ball. This latter portion of the curve accounts for an increasing percentage of the total energy absorbed (area under the curve) with increasing impact velocity.

Penetration resistance can be achieved through an increase in thickness of the polyvinyl butyral interlayer; however, deceleration characteristics approach the dangerous to life level at the higher impact velocities for such laminates. Typical deceleration pulses are shown in Figure 8 for configurations of two plies of ⅛" annealed plate glass with 0 030", 0 45" and 0 060" HPR interlayers, a 045" Std interlayer laminate and a single sheet of .030" HPR polyvinyl butyral Impact velocities were at 22.5 or 28.4 miles per hour, the latter being

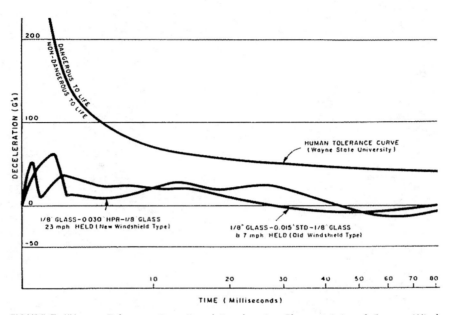

FIGURE 7. "Human Tolerance Curve" and Deceleration Characteristics of Current Wind-shield Configurations (22 lb. Headform, 90° Impact Angle, 24" x 36" Test Panels)

the upper limit of the equipment used. Although none of the pulses was rated dangerous to life, a trend toward this point can be seen with increasing inter-layer thickness. It is expected that use of .060" and possibly .045" interlayers would result in dangerous to life characteristics at higher velocities, somewhere in the 30 to 40 mph range. The pulse for the laminate with a .045" Std inter-layer indicates a greater force-time curve than for a laminate with .045" HPR interlayer tested under the same conditions. This is to be expected because of the greater restriction on interlayer stretching of the standard interlayer because of excellent adhesion to glass.

Penetration resistance testing has shown that the .030" polyvinyl butyral can withstand much higher impact velocities when it is not used for lamination of two plies of annealed glass, i.e., when it is tested as a single sheet (see Table 2). A typical deceleration pulse (Figure 8) for a 28.4 mph impact of a single sheet of .030" HPR interlayer shows good deceleration characteristics. Thus, a better approach than increasing interlayer thickness to achieve high velocity penetration resistance is to design a laminate that will allow the HPR inter-layer to approach its single sheet strength.

Deceleration pulses of laminate configurations which include tempered glass are of particular interest because of the high impact force needed to fracture the glass. Shown in Figure 9 are series of pulses for three different configura-tions. The samples were impacted at a 90° angle with the 22 lb headform at successively higher impact velocities until the tempered glass fractured. Pulses at or near the penetration point are also included.

The shaded areas are those which are dangerous to life as rated by the "Human Tolerance Curve." When the recorded force was large enough to be dangerous to life, it occurred before or at fracture. At higher impact velocities, near the penetration point, the resultant force was not dangerous to life except for one case (Figure 9, third series), the reason being that the glass was fractured very quickly with a high peak force of a time duration too short to be dangerous to life. The fracture cracks in tempered glass propagate at speeds greater than the speed of sound, and therefore after fracture, the many cracks formed so quickly allow the completely broken laminate to act very nearly like a single sheet of .030″ HPR polyvinyl butyral, the interlayer plastic for all laminates tested

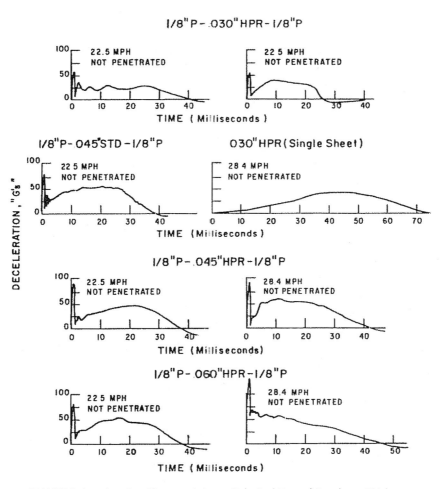

FIGURE 8. Deceleration Characteristics vs. Polyvinyl Butyral Interlayer Thickness.

271

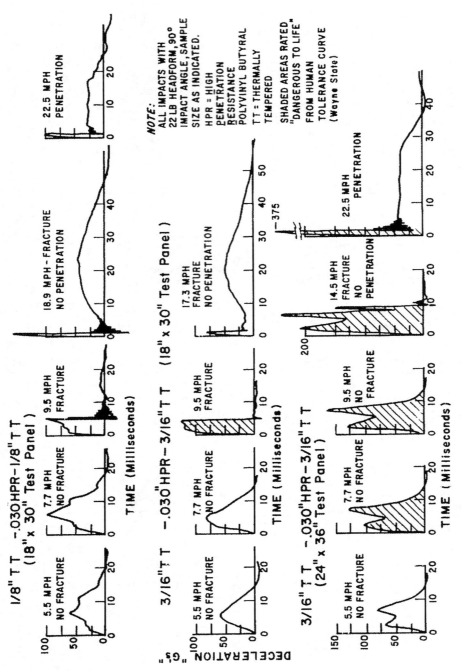

FIGURE 9. Deceleration Pulses of Laminates with Thermally Tempered Glass Plies.

272

The first series in Figure 9 is for an 18" x 30" laminate configuration with two plies of ⅛" tempered plate glass. The second series is for an 18" x 30" laminate configuration with two plies of 3/16" tempered glass. Comparison of these two series shows the thinner tempered glass to be more desirable since it did not yield any force-time curve that could be rated dangerous to life. The third series is also for the configuration with two 3/16" tempered glass plies but for a 24" x 36" panel. These data show (1) a higher impact velocity is needed to fracture the 24" x 36" panels, (2) the deceleration is dangerous to life at a lower impact velocity for the 24" x 36" panels than for the 18" x 30" panels and (3) this high strength laminate is dangerous to life at the penetration velocity (high velocity level).

The differences in the shapes of the various deceleration pulses and the areas under the curves are attributed to the properties of the impact system, i.e., the physical properties of the test panels, the mounting fixture and the 22 lb headform. The deceleration pulses for a human head-automobile windshield impact system may be somewhat different because of expected different physical properties; however, the magnitudes of the pulses are not expected to be greatly different from those determined by the 22 lb headform impacts.

The study of laminates with tempered glass was extended to include thinner glass and a combination configuration of one ply of tempered glass and one ply of ⅛" unstrengthened plate glass (Figure 10). As in the previous samples, all plastic interlayers were .030" high penetration resistance polyvinyl butyral. Deceleration data on a combination configuration with a 3/16" tempered glass inboard ply and a ⅛" plate glass outboard ply show the impact velocity for the dangerous to life situation is improved over that for the configuration with two plies of 3/16" tempered glass. Also, fracture occurred at a lower impact velocity. Similar improvement was observed through the use of two plies of 0.1" Herculite II chemically strengthened glass. The best deceleration characteristics of this group were obtained with a combination of 0.1" Herculite II inboard ply and a ⅛" annealed outboard ply. Neither of the pulses recorded was dangerous to life.

Three other materials tested are worthy of mention. Laminates with two types of rigid plastic used for inboard plies with a ⅛" plate glass outboard ply and .030" HPR interlayer were tested for deceleration characteristics and penetration resistance. The first two curves are for laminates with a 1/16" acrylic plastic inboard ply. Both pulses are very small and the 22 lb headform penetrated the laminates at both 17.3 and 22.5 mph impact velocities. Although deceleration characteristics were good, penetration resistance was poor. Also, laceration hazard was rated high for this type of configuration because the fractured acrylic had long sharp spikes around the area of impact. Similar results were obtained for 1/32" and ⅛" acrylic inboard plies.

The second two curves in Figure 11 are for laminates with inboard plies of ⅛" polycarbonate, a tough rigid plastic. The plastic did not fracture at either 17.3 or 22.5 mph impact velocities; however, the deceleration pulses showed dangerous to life characteristics for each impact. Polycarbonate inboard plies in 1/32" and 1/16" thicknesses were also tested for penetration resistance,

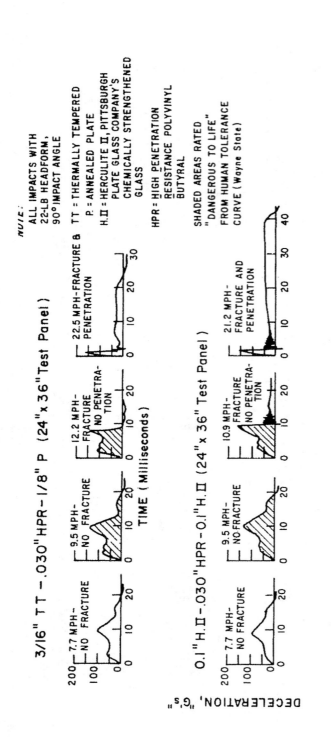

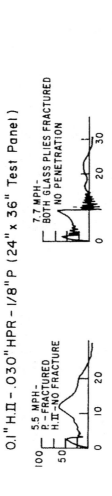

FIGURE 10. Deceleration Pulses of Laminates with Annealed and Chemically Strengthened Outboard Glass.

274

but not for deceleration characteristics. These tests were limited, but results indicated good penetration resistance with a penetration level of about 32 mph impact velocity. However, the thinner polycarbonate plies fractured at a low impact velocity in a manner similar to the acrylic panels. The fracture pattern in these cases where larger plastic pieces developed in the inboard ply would present a serious laceration hazard in the event of a human head impact.

The deceleration characteristics of ⅛" soft aluminum were determined and are shown in Figure 11 in an attempt to simulate a glass that would not break. The soft aluminum has several physical properties similar to that of glass without the brittleness of glass. The deceleration pulses become dangerous to

1/16"ACRYLIC—.030"HPR — /8"ANNEALED PLATE GLASS

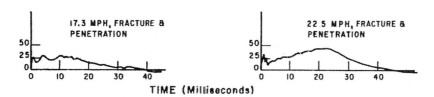

1/8" POLYCARBONATE—.030"HPR—1/8"ANNEALED PLATE GLASS

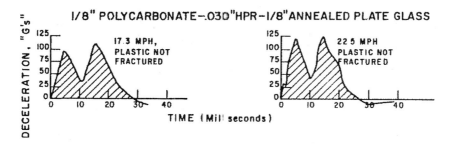

1/8" SOFT ALUMINUM

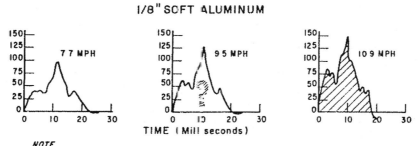

NOTE
ALL IMPACTS WITH 22 LB HEADFORM. 90°IMPACT ANGLE, 24"x 36" SAMPLE SIZE
HPR = HIGH PENETRATION RESISTANCE POLYVINYL BUTYRAL
SHADED AREA RATED "DANGEROUS TO LIFE" FROM HUMAN TOLERANCE CURVE (Wayne State)

FIGURE 11. Deceleration Pulses of Laminates with Rigid Plastic Inboard Ply and of Soft Aluminum Sheet.

life at levels similar to those recorded for several of the tempered glass configurations which did not fracture. Data of this type are valuable for use in basic studies.

Several conclusions can be drawn, based on deceleration pulses as related to the "Human Tolerance Curve":

1. Concussive injuries from impact to windshields of a ⅛" plate—.015" standard interlayer—⅛" plate configuration and a ⅛" plate—.030" HPR—⅛" plate configuration are not dangerous to life.

2. Laminate configurations with 3/16" thick tempered glass plies or a combination of 3/16" thick tempered-unstrengthened glass plies are potentially dangerous to life. The most critical impact velocities are not those at the laminate penetration level (high velocities) but are those at which the tempered glass does not fracture or at the fracture level (low to moderate impact velocities).

3. Thicker tempered glass is more likely to cause dangerous to life deceleration characteristics.

4. Tough, rigid plastic (polycarbonate) as a safety glass laminate component is likely to cause dangerous to life deceleration characteristics.

5. In general, the most desirable safety glass laminate appears to be one in which the rigid members fracture into a multitude of cracks with a relatively low force, thus allowing the flexible plastic interlayer or interlayer components to arrest the motion of the impacting object.

LITERATURE CITED

Campbell, B. J. and Hopens, T., "Report to the American Standards Association Z26 Committee, Number Four: Injuries from Windshield Glass," ACIR, November 1963.

Patrick, L. M., "Human Tolerance to Impact—Basis for Safety Design," SAE Paper 1003B, January 1965.

Patrick, L. M., "Human Tolerance to Impact Conditions as Related to Motor Vehicle Design—SAE J885," SAE Handbook Supplement, SAE Information Report, issued 1964.

Patrick, L. M. and Daniel, R., "Comparison of Standard and Experimental Windshields," Proceedings, Eighth Stapp Car Crash Conference, Wayne State University, Detroit, Michigan, October 1964. In press.

Rodloff, G., Witten-Ruhr, "More Recent Investigations on Compound Safety Glass for Windshields," ATZ, 64, No. 6, 1962.

Schwimmer, S. and Wolf, R. A., "Leading Causes of Injuries in Automobile Accidents," ACIR, 1962.

Severy, D. M., Mathewson, J. H. and Siegel, D. W., "Automobile Side-Impact Collisions, Series II," SAE Journal 1962, SAE Preprint SP232.

APPENDIX

TABLE A Correlation of Ball Impact Tests (Tabulated Data for Figure 2)

(5 pound steel ball vs. 22 pound headform at 45° and 90° impact angles)

Configuration (inches)			Mean Penetration Velocity (mph)			
			45° Impact		90° Impact	
Glass[a]	Interlayer	Glass	5 lb Ball	22 lb H/F[b]	5 lb Ball	22 lb H/F
⅛ P	0 015 Std	⅛ P	11 4	10 5	7 4	8 4
⅛ P	0 030 Std	⅛ P			18 1	14 4
⅛ P	0 045 Std	⅛ P	32 5	29 9		
⅛ P	0 015 HPR	⅛ P			16 2	15 2
⅛ P	0 030 HPR	⅛ P	24 3	25 0	24 0	24 6
⅛ P	0 045 HPR	⅛ P	35 0	32 1		
0 090 P	0 030 HPR	0 150 P	25 5	23 3	24 5	25 4
0 050 S	0 030 HPR	0 050 S	27 6	27 1	21 2	26 5
0 10 H II[c]	0 030 HPR	0 10 H II	32 0	32 5		

[a] Impacted Surface
[b] headform
[c] HERCULITE II, Pittsburgh Plate Glass Company's chemically strengthened glass
P—annealed plate glass
S—annealed sheet glass

TABLE B. Correlation of Impact Angles (Tabulated Data for Figure 3)

(45° vs. 90° impact angle, 5 lb steel ball and 22-lb headform,
various laminate configurations)

	Configuration (inches)				Mean Penetration Velocity (mph)	
	Glass[a]	Interlayer	Glass	Ball	45°	90°
1.[b]	⅛ P	.015 Std	⅛ P	5[c]	11.4	7.4
2.	⅛ P	.015 Std	⅛ P	22[c]	10.5	8.4
3.	⅛ P	.030 HPR	⅛ P	5	24.3	24.0
4.	⅛ P	.030 HPR	⅛ P	22	25.0	24.6
5.	.050 S	.030 HPR	.050 S	5	27.6	21.2
6.	.050 S	.030 HPR	.050 S	22	27.1	26.5
7.	.067 S	.030 HPR	.067 S	5	25.4	21.8
8.	.050 S	.030 HPR	⅛ P	5	24.8	23.2
9.	.067 S	.030 HPR	⅛ P	5	24.9	23.2
10.	.090 P	.030 HPR	⅛ P	5	25.8	23.2
11.	.050 S	.030 HPR	.195 P	5	25.0	24.0
12.	.090 P	.030 HPR	.150 P	5	25.5	24.5
13.	.090 P	.030 HPR	.150 P	22	23.3	25.4
14.	⅛ TT	.030 HPR	⅛ TT	5	37.5	21.9
15.	3/16 TT	0.30 HPR	3/16 TT	22	>27.9	<22.5
16.	0.1 H II	.030 HPR	0.1 H II	5	32.0	19.4

[a] Impacted surface
[b] configuration number used in Figure 3
[c] Test Panels: 12″ x 12″ for 5 lb steel ball
                   24″ x 36″ for 22 lb headform
P—annealed plate glass
S—annealed sheet glass
TT—thermally strengthened plate glass—fully tempered
H II—HERCULITE II, Pittsburgh Plate Glass Company's chemically strengthened glass

TABLE C. Effect of Interlayer Thickness on Penetration
(Tabulated Data for Figure 4)

(22 pound headform at 90° impact angle)
(Test Panel: 24" x 36")

Interlayer*		Mean Penetration Velocity
Type	Thickness (Inches)	(mph)
Std	0.015	8.4
Std	0.030	18.2
Std	0.045	29.9
Std	0.060	33.1
HPR	0.015	15.1
HPR	0.020	19.0
HPR	0.025	22.0
HPR	0.030	24.6
HPR	0.045	32.1
HPR	0.060	38.0

* Conventional laminate construction consisting of ⅛" glass—interlayer —⅛" glass; annealed plate glass was used in all panels

Safety of Windshield Against Flying Stones

G. Rodloff and G. Breitenburger
Deutsche Tafelglas A G

This study deals with the question: what are the various thresholds of safety or degrees of risk of accident when a road stone ricochets and hits the windshield. Will the flying stone penetrate the windshield? Is visibility along the stopping distance guaranteed? The objective of these researches was to determine the type of windshield that offers a maximum of safety to the passengers when the windshield is struck.

The following types of glass were tested:

Tempered glass, 0.197 in. (5 mm) thickness

Chemically-strengthened glass, 0.098 in. (2.5 mm) thickness

Laminate construction with chemically-strengthened glass

Laminated safety glass.

Besides "internal safety", i.e., safety against penetration of the windshield by pas-sengers and safety against head or cerebral injuries (see SAE paper 670191), the external safety of the windshield against stone impact has to be considered. With reference to its external safety, the windshield has to satisfy two conditions:

A. The windshield must not be penetrated by a flying or richocheting stone which hits the windshield (because the passengers would be struck).

B. The visibility through the windshield after the impact of a stone must be preserved at least over the stopping distance (for driver to maintain control).

In this study, "safe speed" designates the speed at which these two conditions are satisfied.

RESULTS

STRENGTHENED GLASS - regardless of thickness or processing method (thermally or

ABSTRACT

The subject of this report is a study of the problem of external safety of windshields under impact from flying stones. Laminated safety glass is resistant to stone impact; tempered glass (for which no road experience is available in the U.S. because it is not used) can only be classified as offering limited safety. More than 1000 road driving trials were made to determine some of the limits of safety offered by the various types of "safety" glass: 1. tempered glass; 2. chemically-strengthened glass; 3. laminated construction with chemically-strengthened glass; 4. laminated safety glass with different thicknesses of the inter-layer.

It was found that windshields made of tempered glass could suddenly break during the study, and this can be considered a potential cause of an accident. The visibility along the stopping distance is not guaranteed. This also applies to laminates made with strengthened glass -- a development which is being pressed in the United States. Such strengthened constructions are not penetrated by stones. However, the risk involved in covering the stopping distance with limited visibility is increased, since some drivers in the face of this danger who have no "second of fear" (delayed reaction) could punch out a hole in an overall-crazed windshield when it is made of a single piece of strengthened glass.

chemically-strengthened) is not sufficiently resistant to impact. Many trials showed that the stone flies into the car with nearly un-diminished speed and thus directly endangers the passengers. Moreover in such circumstances the pieces of flying glass are another and not insignificant source of eye danger. The visibility over the stopping distance after the break is not sufficient and can be considered a potential cause of accident.

LAMINATE CONSTRUCTION WITH CHEMICALLY STRENGTHENED GLASS - offers adequate safety against penetration. However, the visibility over the stopping distance is not guaranteed (see Table 1).

Standard LAMINATED SAFETY GLASS provides sufficient safety against penetration (see Table 1 values). The visibility after break is guaranteed in any case.

TEST PROCEDURE AND DISCUSSION OF LIMIT OF ERROR

It was clear from the beginning that it would not be possible to run the tests with standard test panels of 12 x 12 in. size such as used for the dropping ball impact test or dart impact test. Therefore, the question was which type of windshield should be chosen for the series of tests: curved or flat windshields? As preliminary trials showed, shaped windows give very different performances according to their curvature and the flexure at the site from which the stone is rejected. Consequently, the tests were made on flat windshields to establish a constant set of test conditions. Only with flat windshields could the function be studied that would make it possible to predicate inferences for practical application. Such a limitation of the test series to flat windshields is logical because the performance of curved windshields is better, the resistance is higher rather than lower, which means that the more difficult case is being tested. The main series of tests were conducted on VW windscreens installed at an angle of 58° to the horizontal which is one of the most steeply sloped windshield positions. Other tests were made with an angle of 45°. Most passenger vehicles on the German market are equipped with windshields whose installation angle lies within these values. For completeness sake, the stone impact resistance at an angle of 90° was also determined. For clarity in the following, instead of the angle of incidence of the stone, the installation angle of the windshield to the horizontal is used; this is the same angle with respect to the road. This angle is identical with the angle of incidence of the stone relative to the windshield.

Table 1 - Safe speed in mph at impact of stones (pebbles) at 45° angle of incidence

| Type of glass | Safety against penetration up to a speed of | | | | Visibility after fracture is guaranteed up to a speed of | | | |
| | Stoneweight in ounces | | | | Stoneweight in ounces | | | |
	0.705	1.764	3.527	7.05	0.705	1.764	3.527	7.05
Thermally strengthened glass; 0 197 in thickness	44	31	31	31	44	31	31	31
Chemically strengthened glass; 0 098 in thickness		47	47	40		47	47	40
Laminate configuration of chem. strength glass; 0 118 in norm + 0 098 in chem strength glass; 0 03 in HI interlayer					71	53		31
Laminated safety glass 0 118 + 0 07 in glass; 0 015 in interlayer	124 *	109	84	68	unlimited	unlimited	unlimited	unlimited
Laminated safety glass 0 118 + 0 07 in glass; 0 02 in interlayer	168 *	137 *	106	93	unlimited	unlimited	unlimited	unlimited
Laminated safety glass 0 118 + 0 07 in glass; 0 03 in HI interlayer	249 *	211 *	162 *	121	unlimited	unlimited	unlimited	unlimited

* These values have been extrapolated from figure 10

The tests were conducted with these four types of safety glass:

Tempered glass:
0. 197 in. (5 mm) thickness, plate glass finish with sight zone

Chemically-strengthened glass:
0. 098 in. (2,5 mm) Chemcor and Herculite II

Laminate construction:
0. 098 in. (2,5 mm) thick chemically-strengthened glass laminated with 0.12 in. (3,0 mm) normal glass and 0.03 in. (0,76 mm) interliner; chemically-strengthened glass to the inside of the car

Standard laminated safety glass:
0. 12 in. + 0.07 in. (3 + 1,8 mm) glass with various thicknesses of the interlayer; the thinner glass sheet to the inside of the car.

Laminated safety glass with this construction has previously given the best performance for internal safety (see head-form dropping test, ATZ 6/1962). Therefore, the same glass construction was tested in this study but with opposite direction of impact; the 0.12 in. (3 mm) layer being the struck face, since the present research study deals with external safety. The interliner that was used throughout the tests was Saflex sheets (standard and HI-grade) both products of the Monsanto Company.

Gravel and stones are the most common flying objects which are flung against the car during driving and occasionally against the windshield. Preliminary tests that imitated impact with gravel and stones gave the following results: impacts by a piece of gravel weighing 0.529 oz. (15 g) on a windshield of laminated safety glass cause an insignificant shell-like sink mark to appear on the outside of the laminated construction. If a windshield of tempered glass is struck by the same gravel particle, and if a point happens to strike the windshield, it breaks. But since a point does not always strike, it would be unrealistic to conduct the main series of the test with pointed stones. In contrast to a pointed object, a round one was chosen, a steel ball. These tests accordingly were conducted with steel balls to obtain a correlation with the standard dropping ball impact test. During this study, it was revealed that the values with this steel ball were different when compared to those obtained by the stone test (see Table 2). It is apparent that with an impact test the material properties of the impacting body also have an influence on the result. As for example: the performance of tempered glass under impact tests made with several materials varies greatly. If the weight of impact is not enough to destroy the windshield, impact with a pointed stone produces scratches or point-like surface damages; rounded stones produce hardly visible surface damages. In contrast, steel balls of the same weight and at the same speed cause a small circular fracture at the site of the impact (as shown in Fig. 3). Thus, the destruction of tempered glass by a steel ball impact takes place in quite a different manner and gives other results accordingly. The "performance-values" of laminated safety glass upon impact with steel balls are significantly lower than those obtained by pebble or gravel impact. The reasons are: 1. a higher specific surface load results form the steel ball's higher specific gravity; 2. a steel ball will slip through any small interlayer rupture because of its smooth surface. Thus, the performance of laminated safety glass under steel ball impact is therefore about 18.645 mph

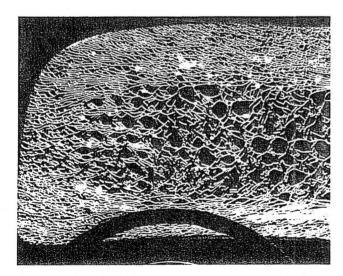

Fig. 1 - A typical dicing-pattern on tempered glass

Fig. 2 - A typical dicing-pattern on laminated-glass

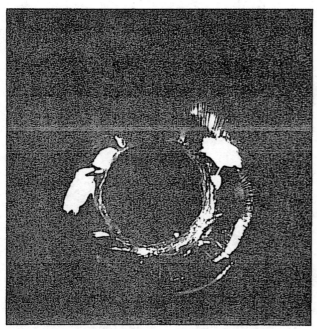

Fig. 3 - shows a circular fissure resulting from the impact of a steel-ball upon thermally strengthened glass. Measure of enlargement 1:8

(30 km/h) lower than its performance when compared with stone impacts. Since steel balls are rarely encountered lying on the road, stones were chosen as the impact medium for a series of tests of the present report. The stones with a weight of 0.529 - 21.162 oz. (15 - 600 g) were selected so that they were as equal as possible in material and form. The use of slag metal particles for the test series can hardly be questioned; they belong, because of their sharp points, to the group of gravel tests.

It was considered pointless to determine the resistance of the windshields against stone impact by shot test, because consistent comparison could be obtained only with steel balls, which is unrealistic because of the above mentioned reasons. A stone shot from a gun barrel leaves the barrel spinning; it attains its breaking efficiency with another power component. To obtain exactly the same impact conditions for a stone, it must not be moved against the windshield, but the windshield has to be moved against the suspended stone in a definite position. The tests were made with a driven car as the moving body. In this way,

Table 2 - Safe speed in mph for different types of safety glass at impact of stones and steel-balls

As "safe speed" the value is named at which the windshield is neither penetrated, nor the visibility after fracture is inadmissibly prevented All values were determined at defined angles of incidence this means at flat windshields

Type of safety glass: glass configuration thickness in in.	interlayer thickness in in.	Angle	\multicolumn{13}{c}{Pebbles weight in oz}													Crushed stone w.i oz		Steel-balls weight in oz				
			21.2	15.5	12.3	8.7	7.2	5.2	3.7	3.0	2.5	2.1	1.8	1.1	0.7	0.81	0.55	22.6	15.4	7.9	1.15	0.48
Thermally strengthened glass 0.197 in glass thickness		90°					18							24	30							
		58°	16	19	25	22	17	19	24	19	28			31	31	33	22	16		19	26	38
		45°			25	22	31	37		25	37		31			44						
Chemically strengthened glass 0.098 in glass thickness		58°						28	40		40											
		45°				37		47					47									
Laminate config.of chem.strength.gl 0.118 N/0.098 C	0.03 HI	58°				31			40	50		47										
0.118 N/0.098 C	0.03 HI	45°		31	36	31		42					56	65	71							
Laminated safety glass 0.118 / 0.07	0.015	90°		44		50	56									x	x					
0.118 / 0.07	0.015	58°	40	44		50	55		71	75	75			90	x	x	x	x	x	25	30	48
0.118 / 0.07	0.015	45°	44	50	56	65	68		87	87	87		109			x	x	x	x			
0.118 / 0.07	0.02	58°		62		68	75		87		99			x	x	x	x	32				
0.118 / 0.07	0.02	45°	62	64		93	92		106	x	x	x	x	x	x	x	x					
0.118 / 0.07	0.03 HI	58°		87		99	121	x	x	x	x	x	x	x	x	x	x	x	x			
0.118 / 0.07	0.03 HI	45°		109		x	x	x	x	x	x	x	x	x	x	x	x					
0.13 / 0.13	0.015	58°		44			68			99			x	x	x	x	x			30		
0.13 / 0.13	0.015	45°		47			75			106			x	x	x	x	x					

Key to symbols used. In columne "glass configuration" the following figures indicate the direction of the impact for instance. 0.118 / 0.07 = 0.118 in struck glass and 0.07 in glass
x = the treshold value for these points could not be determined because only a speed up to 109 mph was driven normally
Angle = angle of incidence related to the windshield area
N = normal glass
C = chemically strengthened glass

excessive space requirements and expensive equipment were avoided. Because of the scope, these tests could not be done in the laboratory, using for instance, a carousel turning with an adjustable speed. For the tests, a VW 1200 windshield was used with a frame support with an adjustable installation angle. The windshield structure was mounted on top of a passenger car and then driven against the suspended stone. Comparison tests in which the normally mounted VW 1200 windshield was used gave the same results.

Different methods of supporting the stone were tested. The support had to work in such a way that the stone was set free at the instant of first impact. On the other hand, the stone must not fall down because of the driving vibration of the car. The portion of the tests in which the speed of impact was increased to higher speeds than 62.15 mph (100 km/h) had to be carried out with two cars driving in opposite direction each other, since there was no high-speed car or sufficiently long test road available. Accordingly the tests were conducted as follows: the test stones were drilled and suspended on 0.098 in. (2,5 mm) thick glass rods which were slightly thicker at the bottom. At the slightest shock, such as caused by contact of the stone against the windshield, the glass rod breaks. The described procedure is very close to reality. The windshield support device was mounted on a luggage-rack. (Fig. 4). At speeds up to 62.15 mph (100 km/h) the car with the windshield to be tested drove against the stones which were suspended from a stationary standing support device. At higher speeds, two cars passed each other to obtain the test speeds in excess of 62.15 mph (100 km/h). For support on the moving car, small stones were suspended by a very thin thread attached by Tesafilm (Scotch Tape) in a way that the stone would just be carried but would fall down at the slightest additional stress. This procedure

with the passing cars was necessary to avoid error in measurement due to the weight of the glass rod. In all cases, the point of impact was the center of the sight zone of the driver's passenger. Strikes at the edges gave for all types of glass slightly lower values, however, the test conditions were no longer controlled so that no comparison could be obtained. The safe speeds (stone impact resistance) were determined step-wise with the driving speed increased each time about 3.108 mph (5 km/h) up to the point of penetration.

SOURCES OF ERROR - The sources of error which result from such testing procedure are:

1. _Speed indicating accuracy of the speedometer_ - This error is present in all the tests and therefore of no consequence.

2. _Read-out accuracy of the speedometer during the test driving_ - Because a great many of the trials have been filmed, it was possible at least to check the test cars. It was found that the statements of the drivers corresponded exactly with the indications of the filmed speedometer.

3. _Changes in temperature_ - Because the tests were made outside and lasted until the beginning of the winter, it was impossible to work at a constant temperature of 64.4° F (18°C). The temperatures ranged between 46.4°F and 64.4°F (8°C and 18°C).

4. _Stone support and attaching_ - Both kinds of support (glass and threads) were checked by drop-tests, as far as this was possible. It was shown that the attachment of the stone was made in such a way that the results were not affected by it.

5. _Differences in the thickness of the interliner_ - The allowed tolerance for interliner thickness is 0.002 in. (0,05 mm). In Fig. 10 you can determine from the differences in the interliner thickness the deviation of performance for stones weighing from 21.162 to 1.058 oz. (600 - 30 g) with the theoretical value of 4.972 - 13.052 mph (8 - 21 km/h) for laminated safety glass with 0.015 in. (0,38 mm) interlayer. However, in practice the variation of the interlayer thickness is such that the deviation resulting from it will hardly exceed 6.215 mph (10 km/h).

TEST RESULTS

TEMPERED GLASS, 0.197 in. (5 mm) THICKNESS - The main series of tests was conducted using pebbles. The results, which are summerized in Table 2 and in Fig. 5 and 8 are surprising: It makes no difference

Fig. 4 - shows the test arrangement for driving trials

what strikes the tempered glass; whether a round or a flat stone, or even a particle of a gravel. The windshields will break immediately by impact at velocities from 24.86 to 31.075 mph (40 - 50 km/h). For this reason all tests were repeated and confirmed by a drop test. The slope of the pebbles graph is such, that one branch stands nearly vertical on the X-axis. Upon impact of dull stones, the windshield will break from the bending stress. With pointed stones or gravel, the break of the windshield is effected in such a way that the outside compression stress is penetrated by a point that joins it with inside compression stress. If this happens, the tempered glass will rupture explosively. The tests conducted showed that a gravel particle of 0.554 oz. (15,7 g) can cause this fracture of the windshield at a speed of only 24.86 mph (40 km/h). The weight of the pointed stone is of no importance, much more so is the exact impact of a sharp point.

In Germany, where 90% of the windshields are made of tempered glass, it happens frequently that windshields shatter without any perceptible reason. The press reports this with headlines such as: "Spook", "Earth Radiation", "Invisible Shots", and "Improper Tension". It is possible that this is a question of damages to the outside surface's layer of compression that are produced by small stones, not deep enough at the time of occurence to cause immediately the fracture of the windshield. However, at a time when an additional bending stress is applied to a pre-damaged windshield, this stress can cause the fracture. The explosion-like bang and the abrupt "crazing-over" of the windshield can be considered a source of accident in a difficult traffic situation, for instance, such as in passing. In summarizing the results obtained with tempered glass, it can be seen that <u>pebbles</u> penetrate a windshield of tempered glass practically regardless of the stone's weight and the installation angle at a speed of 24.86 - 31.075 mph (40 - 50 km/h). The stone penetrates into the inside of the car with almost undiminished speed and can injure the passengers. <u>Pointed stones</u> (gravel) may cause fracture of the tempered glass at even a low speed. The diminished visibility after fracture can be considered as a potential cause of accidents.

CHEMICALLY-STRENGTHENED GLASS, 0.098 in. (2,5 mm) THICKNESS - At 0.098 in. (2,5mm) thick, chemically-strengthened glass shows values for stone impact resistance that are slightly higher than 0.197 in. (5 mm) thick tempered glass. The values are shown in Table 2 and Fig. 5 and 8. For the build-up of the dicing-pattern, what was shown in the "Tempered glass" section also applies. Since with thin glass, the area of bending stress is

significantly larger if a dull object impacts, the build-up of the knife-like structure of the dicing pattern extends also to a larger area than with 0.197 in. (5 mm) thick tempered glass. Researches were only made on windshields that would develop the usual dicing pattern. As Fig. 6 and 7 show, however, the same appearance should result with larger glass interstice-sized dicing patterns.

As "safe speed" is named the value at which the windshield is neither penetrated, nor the visibility after fracture is inadmissibly prevented

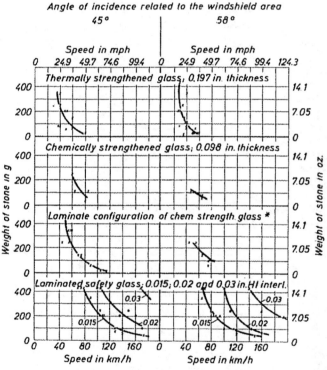

Fig. 5 - Safe speed for different types of safety glass at impact of stones

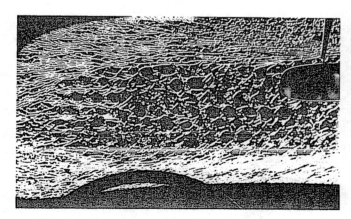

Fig. 6 - Dicing-pattern by pointed steel impact

LAMINATE CONSTRUCTION WITH CHEMICALLY-STRENGTHENED GLASS -

The following glass construction was tested: 0.118 in. (3 mm) normal glass + 0.098 in. (2,5 mm) chemically-strengthened glass laminated with 0.030 in. (0,76 mm) HI-interliner. Since only a restricted number of test windshields were available, the penetration values could not be determined. These may be considered to be of the same order of magnitude as for the standard laminated safety glass with the 0.030 in. (0,76 mm) HI-interliner. In contrast to standard laminated safety glass, with this construction the visibility after fracture is no longer guaranteed when the chemically-strengthened glass on the inside breaks due to the excess bending stress (Fig. 9). These fracture values can be obtained from Tables 1 and 2.

It must be said that for these tests only chemically-strengthened glass that would develop the usual dicing pattern was used. However, it can be assumed that even with the use of other constructions, in which dicing pattern can be expected with bigger interstice-size or semi-tempered glass that the build-up of the knife-like fracture pattern will appear if the windshields break under the bending stress.

LAMINATED SAFETY GLASS -

The resistance to stone impact depends on the following factors:

1. Stone weight
2. Interliner Thickness
3. Glass thickness
4. Installation angle of the windshield.

Fig. 7 - Dicing-pattern by bending-stress

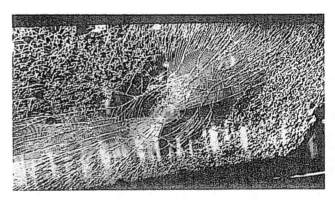

Fig. 9 - Laminate configuration: 0.118 in. (3 mm) norm. glass + 0.098 in. (2,5 mm) chemically strengthened glass with 0.03 in. (0.76 mm) HI-interlayer. Fracture through rounded stone impact at 49.72 mph (80 km/h) - stone-weight 3.739 oz. (106 g)

As "safe speed" is named the value at which neither the windshield is penetrated, nor the visibility after fracture is inadmissibly prevented.

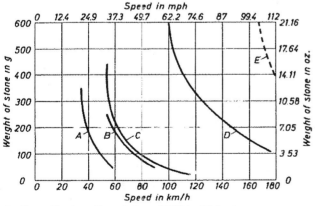

A = thermally strengthened glass; 0 197 in thickness
B = chemically strengthened glass; 0 098 in thickness
C = laminate configuration of chem strength glass;
   0 118 norm + 0 098 in chem strength glass; 0 03 in HI interlayer
D = laminated safety glass; 0 118 + 0 07 in glass; 0 02 in interlayer
E = laminated safety glass; 0 118 + 0 07 in glass; 0 03 in HI interlayer

Fig. 8 - Safe speed for different types of safety glass at impact of stones at 45 deg angle of incidence

In this curve all test values of the trial serie 0 118 + 0 07 in glass with 0 015; 0 02 and 0 03 in interlayer-thickness are indicated The points show the speed at which the stones do not yet penetrate the windshield.

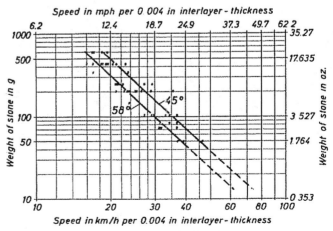

Fig. 10 - Safe speed for windshields made of laminated safety glass per 0.004 in. interlayer thickness at impact of stones at 58 and 45 angles of incidence

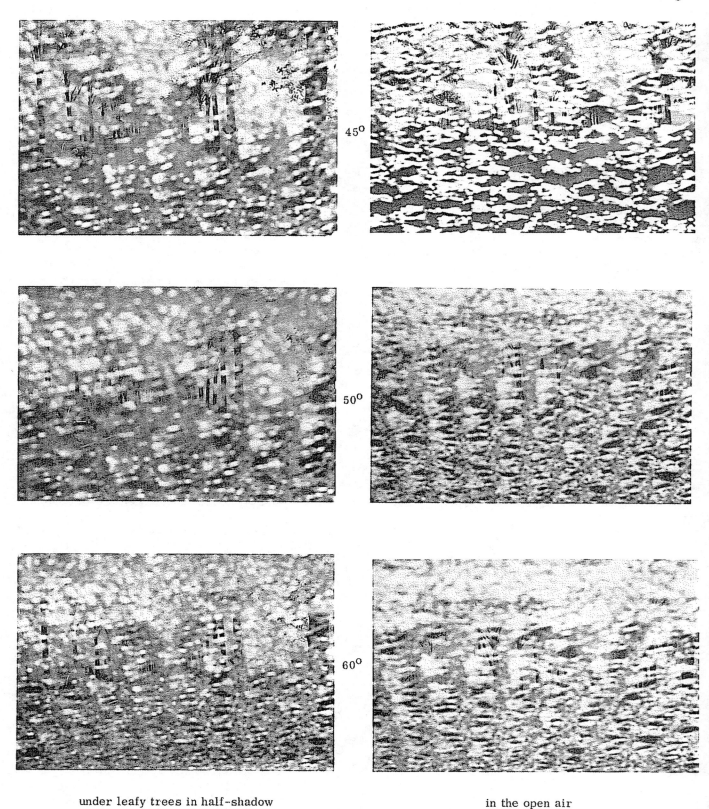

45°

50°

60°

under leafy trees in half-shadow

in the open air

Fig. 11 - shows the visibility after fracture at strengthened glass with dicing-pattern with bigger particle size.
At all pictures the following data were constantly kept:
diffuse light at overcast sky, windshield, distance of
camera, camera adjustment. Variations have been
made on the diagonal position of the windshield (45°,
50° and 60°) and on the incidence of light from above
(in the open air and under leafy trees in half-shadow)

288

Safety Performance of
Laminated Glass Configurations

R. G. Rieser and J. Chabal

Glass Research Center, Pittsburgh Plate Glass Co

EVEN THOUGH RECOGNIZED improvement in the performance of automotive laminated safety glass windshields has been achieved in the past several years, additional expansion of performance attributes is needed. Domestic passenger cars starting with the 1966 models are equipped with the new high penetration resistance laminated safety glass windshields, and limited preliminary reports of accidents involving occupant contact with the new windshield indicate excellent performance[1]. Thus far these windshields show markedly reduced penetration frequency and multiple shallow lacerations instead of the former deep lacerations with facial bone fractures.

This development in laminated safety glass followed by translation to utilization in our domestic passenger cars was achieved through the effective cooperative efforts of the automobile, plastic and glass manufacturers and fabricators. This represents the first major advance in laminated safety glass performance in many years.

Further improvements are desired and the Glass Division Research Center of PPG INDUSTRIES is examining in depth all aspects

requisite of a safer glazing structure for automobile windshield use. Much of the information reported herein is preliminary in a program directed toward basic and applied studies of safety glass performance. Our aim in this research effort is to seek the ultimate in windshield safety and develop data and technology for the automotive industry's consideration.

Over the past 30 years, laminated safety glass has been evaluated in accordance with the ASA Z26.1 American Standard Safety Code (now USA Standards Institute Z26.1 - 1966 Code)[2] by impact of free falling objects such as the 1/2 lb steel ball and 7 oz dart. A go -- no go criterion based on the breakage and penetration of a group of test panels was used to judge the strength and quality of the manufactured product. Over the years a great number of tests of this type has been run as a production quality control evaluation. Recently, however, free falling ball impact tests correlating more closely with automotive laminated safety glass performance[3] have been utilized in production quality control and as a test procedure for development work. These procedures have

ABSTRACT

Impact performance of a large variety of interlayer combinations laminated with annealed plate, heat strengthened, and fully tempered glasses has been analyzed for acceleration pulses, "Severity Index" and penetration. These tests, conducted at 0°F, 70°F, and 120°F, indicate that small interlayer thickness changes are most effective in upgrading performance. Severity Index

numbers well below 1000 were calculated for laminated safety glass with interlayer thickness as high as 0.120". Highly localized bending of glass panels during impact with the 22 lb headform has been recorded through high speed photography as reflected grid board distortion. Lacerative potential of laminated safety glass is being evaluated by using a simulated skin developed through studies of human tissue.

Franke because the method of attachment prevented an inflection curve through the driven point. The variation in resonant frequencies around the head as recorded in our test correspond somewhat to the ranges of frequencies given for the first and third mode by Franke.

The method of determining mechanical impedance of the embalmed human cadaver skull by direct measurement as outlined herein has provided us with a much better understanding of the impact response of the head. We have excited modes which correspond closely to those detected in the living human by Franke[7]. It has been shown that mechanical impedance, apparent weight and dynamic stiffness vary with frequency and location on the skull. Largest rates of change with frequency occur between just below anti-resonance and resonance, which is the range in which bending modes are excited and therefore fracture occurs. At frequencies below anti-resonance the head is accelerated by a force like a rigid body. When a linear change in motion is imparted to the head by a force, an acceleration pulse rich in frequency components between anti-resonance and resonance represents a local acceleration measurement and does not generally indicate the acceleration or deceleration of the center of mass.

1. Von Gierke, H.E.: Measurement of the Acoustical Impedance and the Acoustic Absorbtion Coefficient of the Surface of the Human Body. U.S.A.F. Technical Report No. 6010, March, 1950.

2. Franke, E.K.: Mechanical Impedance Measurements fo the Human Body Surface. U.S.A.F. Technical Report No. 6469, April, 1951.

3. Simonson, E., Snowden, A., Keys, A., and Brozek, J.: Measurement of Elastic Properties of Skeletal Muscle in Situ. Journal of Applied Physiology 1:512, 1949.

4. Oestreicher, H.L.: Fiedl and Impedance of an Oscillating Sphere in a Viscoelastic Medium with an Application to Biophysics. Journal of the Acoustical Society of America, 23,707 (1951).

5. Franke, E.K.: The Impedance of the Human Mastoid. Biophysics Branch Aero Medical Laboratory, Wright Air Development, Wright-Patterson Air Force Base, Dayton, Ohio.

6. Von Békésy, G.: The Mechanical Properties of the Ear. Handbook of Experimental Psychology, S.S. Stevens, editor (John Wiley and Sons, Inc.,) New York, 1951.

7. Corliss, L.R. and Koidan, W.: Mechanical Impedance of the Forehead and Mastoid. Journal of the Acoustical Society of America, 27, 1164, Nov., 1955.

8. Franke, E.K.: Response of the Human Skull to Mechanical Vibrations. Journal of the Acoustical Society of America, 28, 1277 (1956).

9. Bradley, Wilson, Jr.: Mechanical Impedance Testing. TP 202, Endevco Corporation, 801 S. Arroyo Parkway, Pasadena, California 91109.

10. McGregor, H.M.: Impedance Measurement Techniques. Environmental Quarterly, March, 1967.

11. Bouche, R.R.: Instruments and Methods for Measuring Mechanical Impedance. TP 203, Endevco Corporation, 801 S. Arroyo Parkway, Pasadena, California 91109.

12. Wright, D.V.: Bureau of Ships letter to SSN Noise Advisory Committee Structural Impedance Panel, July 13, 1961.

13. Von Kármán, T. and Biot, M.A.: Mathematical Methods in Engineering, p. 376, McGraw Hill Book Company, Incorporated, New York, 1940.

APPENDIX

Post-Mortum Data on Cadavers

Cad. No.	1012	963
Age	66	73
Date of Death	6/7/66	3/25/66
Cause of Death	Cancer of Maxilla	Pneumonia and carcinoma of liver
Date of Embalming	6/13/66	6/30/66
Date of Test	8/24/66 to 9/24/66	5/24/67 to 7/13/67

resulted from the cooperation of the automotive, plastics, and glass industries
working under the auspices of the Society of
Automotive Engineers. These new techniques
are the 5 lb steel ball test and the 22 lb
headform test using 12" x 12" and 24" x 36"
test panels, respectively.

Continued effort directed toward improvement of automotive windshields is justified
because the automotive windshield is rated
as one of the first four sources of motor
vehicle occupant injury and fatality[4].

Our test facilities have been modified and
expanded to provide new evaluation techniques
and higher velocity impacts. Several years
ago, work was directed toward a 2-3 fold
increase in velocity necessary to penetrate
automotive laminated safety glass. Thus
impact performance testing with the 5 lb
steel ball and 22 lb headform had to be increased in velocity from approximately 10
mph to 25-30 mph. Further improvement in
performance over the new high penetration
resistance laminates now requires impact
capability beyond 30 mph velocity. Our
tower for free falling impact tests is limited
to 48 ft. or 37.9 mph. In view of the need
for greater velocities in our development
program, devices have been constructed to
accelerate the 5 lb ball and 22 lb headform
to 85 mph and 65 mph, respectively. Our
test facilities for impact evaluation are
summarized in Table 1.

The photographs exhibited in Figures 1,
2, 3 and 4 show several views of each test
device listed in Table 1. In all cases, a one-
inch wide shoulder covered with 1/8" thick
cork is utilized to support and clamp the test
panels for high velocity testing. Speed and
accelerometer equipment as well as extensive photographic equipment monitor the
performance and provide records for
analysis. In this work, we continue to favor
the 5 lb ball and 22 lb headform principally
because good correlation has been demonstrated for these devices in comparison with
cadaver impacts against windshields[5].

The holding frames for 90° impacts of
12" x 12" and 24" x 36" panels are shown in
Figure 1. Similar jigs are used for 45°
impact tests. A heavy angle iron frame
clamped in place holds the 24" x 36" size
test specimen and small air cylinders apply
the force to hold the 12" x 12" samples.

A pneumatic cannon fires the 5 lb steel
missile (3-1/4" diameter ball or bullet shape
having the same nose radius as the ball)
horizontally into the 12" x 12" test panel
held at 45° or 90° angle. The speed is
measured with photoelectric screen pick-ups
electronically modified to low sensitivity to
prevent air blast triggering of the system.
The cannon and missiles are shown in
Figure 2.

The horizontal impact machine shown in
Figure 3 propels on 2" diameter rods a 22
lb projectile having a 6-1/2" diameter nose
duplicating the weight and size of the dropping
ball headform. In order to achieve the de-

Fig. 2 - Pneumatic cannon and 5 lb missiles for high
velocity impacts

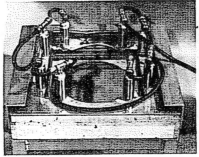

Fig. 1 - Equipment used for vertical free fall impacts (tower)

Fig. 3 - Horizontal impactor for 22 lb headform studies

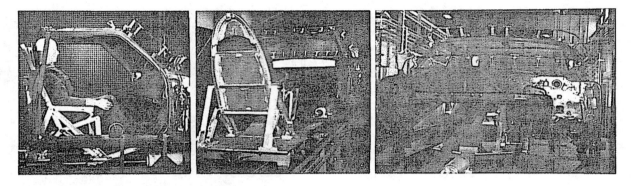

Fig. 4 - Full scale horizontal accelerator (sled) for windshield evaluations

Table 1 - Devices for Impact Studies of Laminated Safety Glass

Facility	Propelling System	Missile	Impact Angle	Sample Size	Maximum Speed(mph)
Tower	Gravity	Ball Headform Dart Shot Bag	45 & 90	6" x 6" to W/S*	38
Cannon	Compressed Air	5 lb Steel Ball	45 & 90	12" x 12"	85
Horizontal Impactor	Rubber Shock Absorber Cords**	22 lb Headform	20° to 90°	24" x 36"	65
Sled	Compressed Air	50th Percentile Dummy	W/S installation angle	W/S in automobile body	65

* Windshield
**Bungee Cords

sired 22 lb weight, this missile is constructed entirely of magnesium metal; the propelling technique is stretched rubber shock absorber cords (six 3/8" diameter bungees stretched 75% provides 65 mph missile speed). Speed and accelerometer data are automatically recorded. The test panel which can be positioned for a 20° to 90° impact angle is held in the frame by air cylinders which provide a means for evaluating clamping force. This horizontal impact machine has been used for the study of glass deflection during impact which is discussed and illustrated later in this report.

The full scale automotive windshield test machine (Figure 4) pneumatically fires a sled (gross load approximates 500 lbs including the 50th percentile Alderson dummy) to 65 mph. The sled and the power cylinder are stopped hydraulically. The air of the drive cylinder is released through one-inch diameter holes in the cylinder wall just prior to contact of the piston with the arresting shock absorber. One view in Figure 4 shows the arrangement of air cylinders to clamp the windshield in the body opening.

In previously reported work on heavy interlayer laminates, it was indicated that the force-time pulses for interlayers to 0.060" thickness at room temperature approached the Wayne State University developed Human Tolerance Curve (HTC)[6]. It appeared questionable that the plasticized polyvinyl butyral interlayers beyond 0.060" thickness would be satisfactory in terms of the HTC[5]. It has been found that in thickness to 0.060" over the temperature range of 0°F to 120°F, the force-time pulses (decelerations) are well below the HTC and thus would not be expected to cause concussive injury. At 0.090", high penetration resistance (HPR) or standard (Std.) interlayers, while closely approaching the HTC curve particularly at 70°F, show "Severity Index" numbers well below the injury hazard threshold for head impacts. The use of a "Severity Index" has been proposed[7,8] as an assessment of head impact pulses relative to interpreting injury hazard. On the basis of published injury tolerance data for frontal impacts to the head, an exponential weighting factor of 2.5 was selected by these authors with a "Severity Index" (S.I.) of 1000 for the threshold to serious or dangerous to life internal head injury. The S.I. numbers for laminates impacted with the 22 lb headform in free fall at 0°F, 70°F and 120°F are given in Tables 2, 3 and 4. It is apparent that these 1/8" annealed plate glass laminates

with interlayer thicknesses in the HPR and Std. types to 0.120" do not exceed the S.I. threshold of 1000 whether or not the 22 lb headform penetrates the test panel.

Sample force-time pulses compared with the HTC for the three temperatures are shown in Figure 5 for the interlayer thickness range of 0.030 to 0.120".

The peak deceleration values listed in Tables 2, 3 and 4 and as shown by the spike in the force-time pulses represents the force developed in fracturing the glass of the test panel. This high load exists for a time interval lasting only 1-2 milliseconds. Since the glass was the same in all these panels, the expected similarity in peak G's (spike) was found in this test series with no significant influence apparent from interlayer thickness or panel temperature.

The photographs of Figures 6 and 7 show the type of rupture produced by the 22 lb headform when penetration occurs for various interlayer thicknesses over the 0 - 70 - 120°F temperature range. Included with each photograph is its deceleration trace compared with the HTC curve. Examination of these impacted 24" x 36" test panels along with the previous three tables leads to the following general conclusions:

1. Except at 70°F, high penetration resistance interlayer shows no significant advantage over the standard interlayer of the same thickness at 0°F and 120°F. At 70°F the mean penetration velocity (MPV)[5] for 0.030" HPR laminates is approximately 24 mph and for 0.030" Standard laminates is approximately 18 mph.

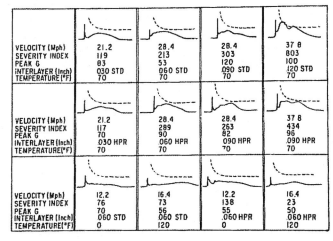

VELOCITY (Mph)	21.2	28.4	28.4	37.8
SEVERITY INDEX	119	213	303	803
PEAK G	83	53	120	100
INTERLAYER (Inch)	.030 STD	.060 STD	.090 STD	.120 STD
TEMPERATURE (°F)	70	70	70	70

VELOCITY (Mph)	21.2	28.4	28.4	37.8
SEVERITY INDEX	117	289	263	434
PEAK G	70	90	82	96
INTERLAYER (Inch)	.030 HPR	.060 HPR	.090 HPR	.090 HPR
TEMPERATURE (°F)	70	70	70	70

VELOCITY (Mph)	12.2	16.4	12.2	16.4
SEVERITY INDEX	76	73	138	23
PEAK G	70	56	55	50
INTERLAYER (Inch)	.060 STD	.060 STD	.060 HPR	.060 HPR
TEMPERATURE (°F)	0	120	0	120

Fig. 5 - Force-time pulses for laminated safety glass compared with the human tolerance curve (22 lb headform - no penetration, 1/8 in. annealed plate glass, 24 in. x 36 in. panels, 90 deg impact)

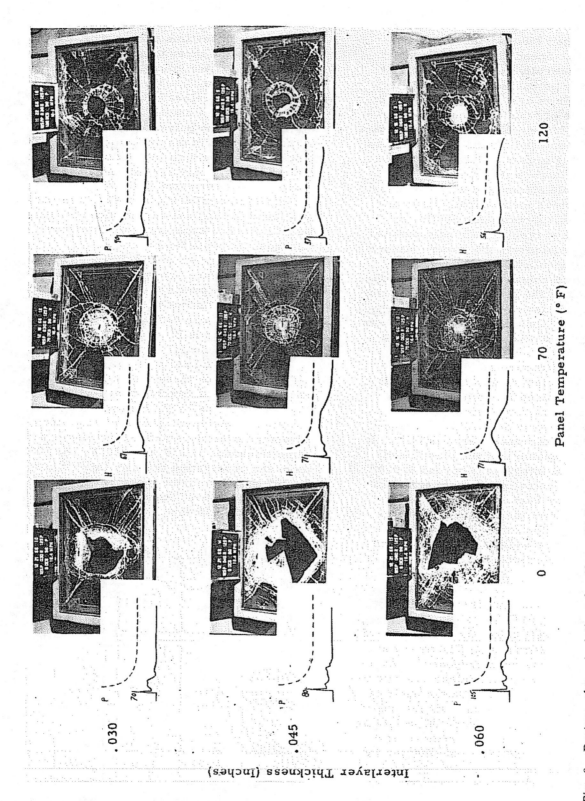

Fig. 6 – Rupture characteristics: headform (22 lb) impact performance – interlayer thickness versus panel temperature – constant impact velocity = 16.4 mph (test panel: 24 in. x 36 in. with 1/8 in. annealed plate glass and standard interlayer) (H = Held   P = Penetrated)

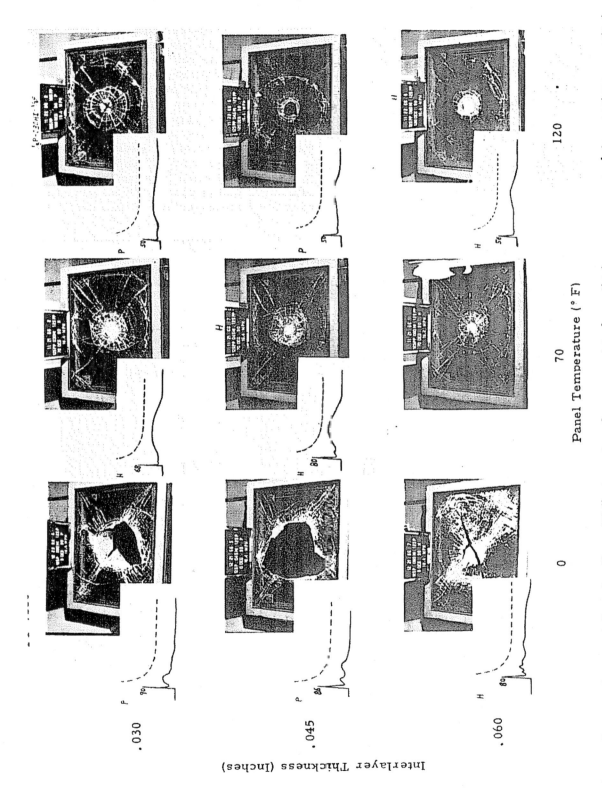

Fig. 7 – Rupture characteristics: headform (22 lb) impact performance – interlayer thickness versus panel temperature – constant impact velocity = 16.4 mph (test panel: 24 in. x 36 in. with 1/8 in. annealed plate glass and high penetration resistance interlayer) (H = Held P = Penetrated)

2. At 0°F and 120°F, resistance to penetration is low for both types of inter-layers. The MPV of 0.030" thick HPR or Standard interlayer with 1/8" plate glass is approximately 10 mph at 0°F and 120°F.

3. The force-time pulses are essentially the same for both types of interlayers over the 0° to 120°F temperature range.

At 0°F, laminates of Standard and HPR interlayers show similar performance. Large gaping holes develop at relatively low impact velocity (12-16 mph). The interlayer behaves as a brittle material and fractures much like glass. At 70°F, much glass fracturing occurs and the penetrated panels exhibit multiple radial interlayer ruptures tending to present a "rolled edge" to the penetrating object. At 120°F, these photographs clearly show the small hole produced which can be viewed as a possible dangerous "neck ruffle" upon head penetration (8-12 mph). This fracture and penetration pattern demonstrates the low tensile strength of the interlayer at the 120°F temperature.

The MPV data over the 0-120°F temperature range for 45° and 90° impacts of 12" x 12" test panels for 0.030" HPR and 0.030" Std. interlayers are given in Table 5. These data indicate the small but significant increase in MPV resulting from change in contact angle from 90° to 45°. In addition to the fact that when impacted at 90° HPR interlayer shows essentially no advantage over Std. interlayer at 0°F and 120°F and a

Table 2 - Deceleration of Headform by Laminated Safety Glass at 0°F
(22 lb headform; 90° impact angle; 24" x 36" test panel.
Configuration: 1/8"P* - Interlayer - 1/8"P)

Interlayer		Impact Speed (mph)	Peak G	Held (H) or Penetrated (P)	Severity Index
Type	Thickness (ins.)				
Std.	.030	12.2	74	H	94
HPR	.030	12.2	60	H	79
Std.	.030	16.4	70	P	30
HPR	.030	16.4	90	P	69
Std.	.030	21.2	60	P	13
HPR	.030	21.2	86	P	64
Std.	.045	12.2	85	H	85
HPR	.045	12.2	80	H	101
Std.	.045	16.4	80	P	49
HPR	.045	16.4	86	P	46
Std.	.045	21.2	96	P	55
HPR	.045	21.2	90	P	53
Std.	.045	28.4	86	P	49
HPR	.045	28.4	87	P	30
Std.	.060	12.2	70	H	76
HPR	.060	12.2	55	H	138
Std.	.060	16.4	105	P	67
HPR	.060	16.4	80	H	103
Std.	.060	21.2	88	P	72
HPR	.060	21.2	86	P	81
Std.	.060	28.4	93	P	135
HPR	.060	28.4	105	P	70

*P = Annealed polished plate glass

1/3 improvement in MPV at 70°F, no difference can be noted between HPR and Std. at the individual temperatures when impacted at 45°.

During the course of evaluating safety glass structures, a large variety of inter-layer configurations has been examined. Included in this particular series of samples are annealed plate, thermally tempered (full temper) and heat strengthened 1/8" thick glasses as the inboard or impact side with 1/8" thick annealed plate glass as the

Table 3 - Deceleration of Headform by Laminated Safety Glass at 70°F
(22 lb headform; 90° impact angle; 24" x 36" test panel.
Configuration: 1/8"P* - Interlayer - 1/8"P)

| | Interlayer | | | | Held(H) | |
Type	Thickness (ins.)	Impact Speed (mph)	Peak G	or Penetrated(P)	Severity Index
Std.	.030	16.4	67	H	81
HPR	.030	16.4	68	H	99
Std.	.030	21.2	83	H	119
HPR	.030	21.2	70	H	117
Std.	.030	28.4	135	P	98
HPR	.030	28.4	80	P	115
Std.	.030	33.7	68	P	39
HPR	.030	33.7	92	P	131
Std.	.030	37.9			
HPR	.030	37.9	90	P	114
Std.	.045	16.4	71	H	109
HPR	.045	16.4	80	H	130
Std.	.045	21.2	80	P	106
HPR	.045	21.2	87	H	159
Std.	.045	28.4	77	P	49
HPR	.045	28.4	92	P	63
Std.	.045	33.7	90	P	72
HPR	.045	33.7		P	
Std.	.045	37.9	88	P	37
HPR	.045	37.9	86	P	196
Std.	.060	16.4	71	H	133
HPR	.060	16.4	76	H	78
Std.	.060	21.2	72	H	186
HPR	.060	21.2	70	H	225
Std.	.060	28.4	53	H	213
HPR	.060	28.4	90	H	289
Std.	.060	33.7	100	P	88
HPR	.060	33.7	96	H	91
Std.	.060	37.9		P	
HPR	.060	37.9	92	H	404
Std.	.090	28.4	120	H	303
HPR	.090	28.4	82	H	263
Std.	.120	28.4	92	H	481
HPR	.120	28.4	100	H	561
Std.	.120	37.8	100	H	803
HPR	.120	37.8	96	H	434

*P = Annealed polished plate glass

outboard ply. The 5 lb ball impacts of 12" x 12" laminates at 0°F, 70°F and 120°F for this wide variety of laminate structures are summarized in Appendix 1 Table A. Each MPV registered in this table has been determined statistically with 95% confidence limits. The various configurations have been grouped relative to the thickness of interlayer. In order to simplify this information a chart of the Table A data is presented in Figure 8.

It is apparent that a small increase in interlayer thickness is as effective as any other change in laminate structure in this test series at room temperature. This includes such modifications as heat strengthened or fully tempered glass combinations of standard and high penetration resistance interlayers, combinations of these automotive interlayer types with aircraft windshield interlayer, and the use of parting layers in the interlayer sandwich. At 0°F, however, regardless of the thickness and configuration modification made in the laminate structure very little improvement in penetration resistance results. At 120°F, a significant improvement in penetration resistance is apparent only when interlayer thicknesses of 0.060" and greater are used.

Several of the laminate structures included in Figure 8 are plotted in Figures 9, 10 and 11 at the three test temperatures 0°F, 70°F and 120°F. The statements made above with respect to the many laminate structures of Figure 8 and Appendix 1 Table A are more apparent from these curves of

Table 4 - Deceleration of Headform by Laminated Safety Glass at 120°F
(22 lb headform; 90° impact angle; 24" x 36" test panel.
Configuration: 1/8"P* - Interlayer - 1/8"P)

Interlayer		Impact Speed (mph)	Peak G	Held (H) or Penetrated (P)	Severity Index
Type	Thickness (ins.)				
Std.	.030	7.7	38	H	23
HPR	.030	7.7	25	H	29
Std.	.030	12.2	40	P	18
HPR	.030	12.2	52	P	11
Std.	.030	16.4	90	P	46
HPR	.030	16.4	50	P	37
Std.	.030	21.2	60	P	23
HPR	.030	21.2	65	P	31
Std.	.045	12.2	48	H	79
HPR	.045	12.2	50	P	15
Std.	.045	16.4	57	P	85
HPR	.045	16.4	50	P	31
Std.	.045	21.2	58	P	45
HPR	.045	21.2	70	P	45
Std.	.045	28.4	55	P	15
HPR	.045	28.4	50	P	11
Std.	.060	12.2	38	H	32
HPR	.060	12.2	40	H	51
Std.	.060	16.4	56	H	73
HPR	.060	16.4	50	H	23
Std.	.060	21.2	115	P	98
HPR	.060	21.2	40	P	123
Std.	.060	28.4	80	P	149
HPR	.060	28.4	82	P	25

*P = Annealed polished plate glass

the standard interlayer (1/8"P - Std. - 1/8"P), high resistance interlayer (1/8"P - HPR - 1/8"P), this same type laminate with a parting layer (1/8"P - HPR - PL2-1/8"P) and the HPR laminates with an inboard ply of heat strengthened (HS) or fully heat tempered (TT) glass (1/8" HS or TT-HPR-1/8"P).

The numerical data for Figures 9, 10 and 11 are tabulated in Appendix 1 Table B. It is apparent from these figures that fully heat tempered or heat strengthened (partial temper) glass in 1/8" thickness in combination with 1/8" annealed plate glass contributes no advantage in penetration resistance. An increase in interlayer thickness greatly improves penetration resistance at 70°F with much less but definite improvement being noted at 120°F. At 0°F, however, the spread in MPV values at 0.030" interlayer thickness tightens up at greater thicknesses with little overall gain in penetration resistance over the interlayer thickness range of 0.030" - 0.090".

A proprietary composition coded PPG-1 (see Appendix 1 Table A) has been included in our studies as a safety glass interlayer and a direct substitute for the plasticized polyvinyl butyral in laminated safety glass. This composition, which is chemically different from the plasticized polyvinyl butyral species, does not include plasticizer to achieve the balance of impact properties over the 0°F to 120°F range. This proprietary composition can be chemically modified to provide optimum performance over the desired temperature range. On the basis of our impact data, the performance of

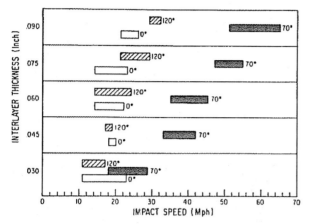

Fig. 8 - Mean penetration velocity (MPV) range* versus interlayer thickness at 0, 70, 120 F (12 in. x 12 in. panels with 1/8 in. thick glass, 90 deg impact angle, 5 lb steel ball)

*MPV range is obtained from penetration performance of many types of panel configuration listed in Appendix I - Table A

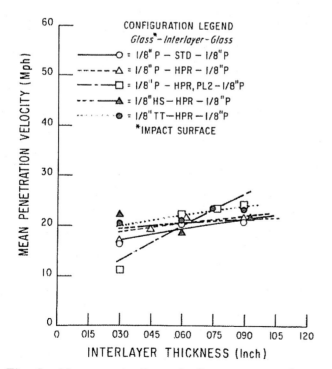

Fig. 9 - Mean penetration velocity versus interlayer thickness (0 F panel temperature; 90 deg impact angle, 12 in. x 12 in. test panel; 5 lb steel ball)

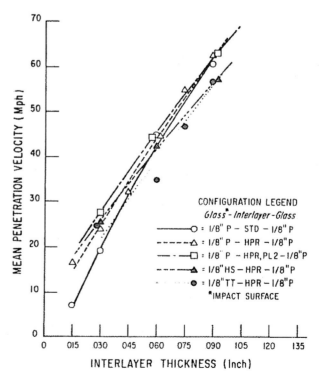

Fig. 10 - Mean penetration velocity versus interlayer thickness (70 F panel temperature; 90 deg impact angle; 12 in. x 12 in. test panel; 5 lb steel ball)

this particular composition evaluated in this series of tests indicates equivalent impact resistance at 0°F and improved performance at 70°F and 120°F. The deceleration values for the PPG-1 material show no significant difference from HPR interlayer of the same thickness. This composition is undergoing

extensive testing and so far meets all the performance requirements of a safety glass interlayer including edge stability and color stability to accelerated and outdoor weathering.

Little has been done to study or evaluate the bending of glass during impact. An evaluation of the dynamics of impact has value in the development of safer automobile glazing for the behavior of glass relative to localized bending and the resultant rapid changes in stress distribution are fundamental to the fracturing and penetration of a safety glass structure. Clues to improvement in lacerative injury potential are also

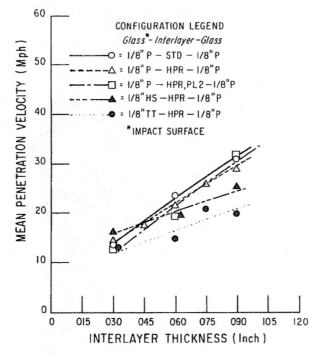

Fig. 11 - Mean penetration velocity versus interlayer thickness (120 F panel temperature; 90 deg impact angle; 12 in. x 12 in. test panel; 5 lb steel ball)

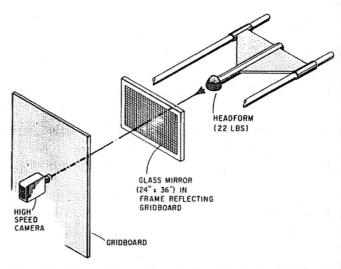

Fig. 12 - Impact dynamics of glass by optical deflection

Table 5 - Mean Penetration Velocity vs Temperature
(5 lb ball at 45° and 90° impact angles)
(Test panel: 12" x 12")

| Configuration (ins.) | | | Impact | Mean Penetration Velocity(mph) | | |
Glass	Interlayer	Glass	Angle	0°F	70°F	120°F
1/8P	0.030 HPR	1/8P	90°	17.3	24.0	12.2
	Same		45°	21.5	24.3	16.4
1/8P	0.030 Std.	1/8P	90°	16.6	18.9	13.7
	Same		45°	21.2	23.8	16.0

P = Annealed plate glass
HPR = High penetration resistance interlayer
Std. = Standard interlayer

anticipated from a study of impact dynamics. Our 22 lb headform horizontal impactor is used in this program.

Figure 12 shows a sketch of the impactor, glass, grid (stringboard) and high speed camera arrangement. The test panel is mirrored prior to clamping in the holding frame and the grid is positioned accurately at the desired distance from the test panel and parallel with it. The high speed camera located behind the grid is placed so that its lens is at the plane of the grid and in line with the impact point of the headform against the test panel. The mirrored glass surface reflects the grid and the camera records bending of the glass as grid distortion. The magnification of the bending is determined by the distance between the grid and the mirrored test panel and is actually double this distance (see Appendix 2).

$$dy = 2 \times \alpha$$

Where dy is the actual shift in grid pattern as measured from the photographic record, x is the distance from test panel to grid and α is the angle of deviation in the glass from its original plane.

Figure 13 shows the localized deflection

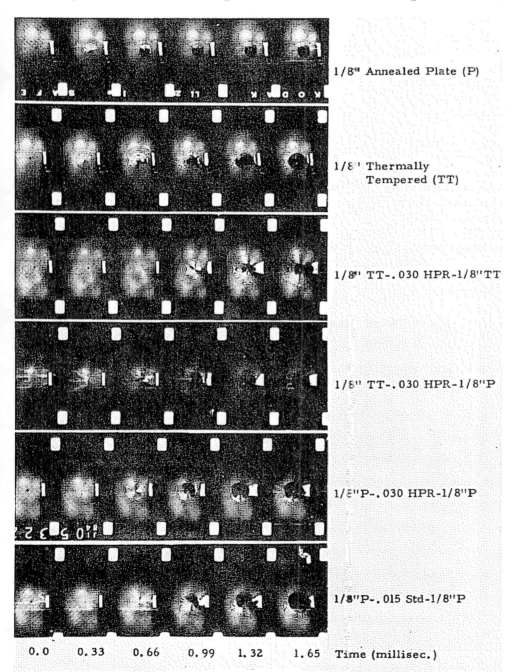

1/8" Annealed Plate (P)

1/8" Thermally Tempered (TT)

1/8" TT-.030 HPR-1/8"TT

1/8" TT-.030 HPR-1/8"P

1/8"P-.030 HPR-1/8"P

1/8"P-.015 Std-1/8"P

0.0   0.33   0.66   0.99   1.32   1.65   Time (millisec.)

Fig. 13 - Localized bending of glass during impact with 22 lb headform at 32 mph

of glass during impact with the 22 lb head-form (90° angle) at 32 mph.

These photographs taken from high speed film show the localized bending of laminated, tempered and annealed glasses. The highly localized bending noted in these pictures shows that fracture occurs prior to distortion of the panel close to the holding frame in these 24" x 36" test samples. They also provide an explanation for (1) the concentric break pattern typical of fractured and penetrated laminated safety glass and (2) the glass spline formation when strengthened glass fractures from high energy impact. These are preliminary results, and, in future reports, mathematical analysis of stress attendant with type of clamping, strength and thickness of glass and speed of impact will be presented. We view this approach to analysis of impact performance as significant and feel that this fundamental study will provide another avenue to improved safety glass performance.

In the study of performance of safety glass configurations, an anthropometric dummy has been used in simulated automotive crashes. The gyrations of the dummy generated by the sudden deceleration akin to a crash situation causes continuous shift in the dummy's center of gravity (CG). The magnitude of this shift for the generated linkage angles and the influence of the locus of CG upon impact performance are significant to analysis of safety performance. In

order to compare test configurations in these full scale sled tests, it is important that the energy of impact and effective mass locus of the dummy be as closely the same as possible. Once the sled is propelled, there is no present physical way in which to control exactly the motions of the dummy as it leaves the sled and flies into the test windshield. However, knowledge of the location of CG for the body contortions photographically recorded permits an assessment of the forces in comparative tests at the moment the dummy contacts the windshield.

Figure 14 shows the swing, linkage angles and reference point used in measuring the CG of our 50th percentile anthropometric dummy (Alderson Model P-1, 50 AU). The sprocket-chain system is used to provide any desired tilt in order to generate several plumb lines for a particular body arrangement on the charts attached to the swing on each side of the dummy. To minimize extraneous mass of the swing a chair was eliminated in favor of two aluminum bars attached between the ends of the chain. By locking the various dummy linkages at the desired angles, the CG can be easily determined with this equipment. In practice, three swing tilt positions were used to obtain each CG. In all the measurements, the CG's were located in space in front of and toward the bottom of the chest cage. The reference point used for these readings is the centerline intersection with the bottom

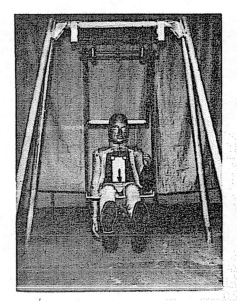 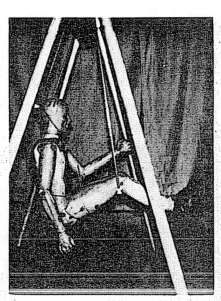 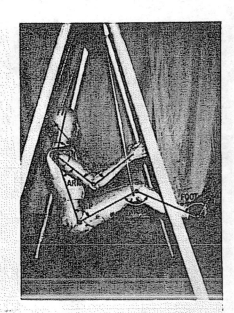

Reference Point = Intersection   **Location of Center of Gravity**   **Linkage Angles**
of Centerline with Bottom Edge   **from Reference Point**

Fig. 14 - Swing reference point and linkage angles used in center of gravity measurements

edge of the breast plate as shown in Figures 14 and 15. An analysis of the segments of a dummy of this type has been reported[9].

Figure 15 summarizes the CG data obtained for changes in hip and knee angles (points 1 through 12). The two curves in Figure 20 were generated from (1) the CG locations for various hip angles while keeping the knees at 125° and (2) the CG locations for changes in knee angle while maintaining the hips at 90°. For comparative purposes and to properly position the dummy at the beginning of a run, CG data have been obtained for the average front seated position (1966 Models) for the major domestic automobile manufacturers (points 13 through 16). Also shown in Figure 15 are two

curves tracing the change in CG of the dummy for sled impacts starting at approximately the same speed (20.6 mph vs 23.8 mph). The difference in the dummy's seated position for these two runs obviously influenced the location of CG at the moment the dummy contacted the windshield. The recommended dummy seated position at the start of a run should be in the region of points 13-16.

One of the areas of safety performance not extensively studied or defined thus far is that of lacerative injury. To those involved in performance studies of laminated safety glass, this is recognized as an important safety aspect because lacerative injury, if not fatal, leads to horrendous physical and psychological damage. Lacerative injury

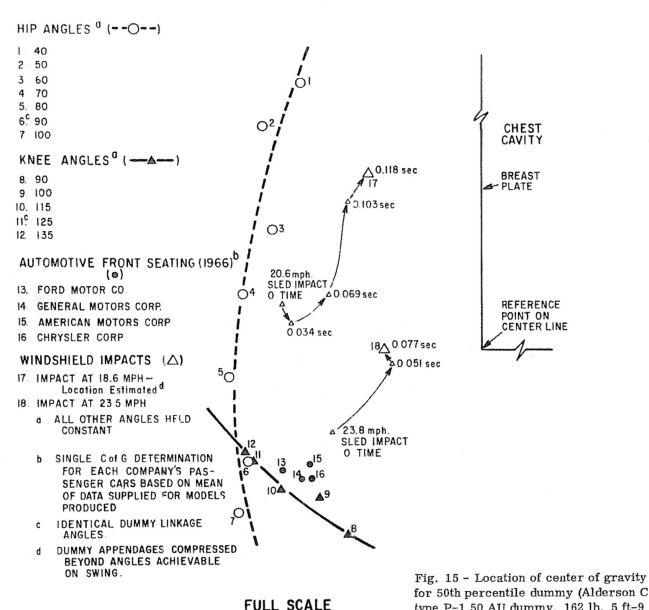

HIP ANGLES [a] (--O--)

1.	40
2.	50
3.	60
4.	70
5.	80
6.[c]	90
7.	100

KNEE ANGLES [a] (—△—)

8.	90
9.	100
10.	115
11.[c]	125
12.	135

AUTOMOTIVE FRONT SEATING (1966)[b] (⊙)

13. FORD MOTOR CO.
14. GENERAL MOTORS CORP.
15. AMERICAN MOTORS CORP
16. CHRYSLER CORP.

WINDSHIELD IMPACTS (△)

17. IMPACT AT 18.6 MPH—
    Location Estimated[d]
18. IMPACT AT 23.5 MPH

a . ALL OTHER ANGLES HELD CONSTANT

b . SINGLE C of G DETERMINATION FOR EACH COMPANY'S PASSENGER CARS BASED ON MEAN OF DATA SUPPLIED FOR MODELS PRODUCED

c . IDENTICAL DUMMY LINKAGE ANGLES.

d . DUMMY APPENDAGES COMPRESSED BEYOND ANGLES ACHIEVABLE ON SWING.

FULL SCALE

Fig. 15 - Location of center of gravity for 50th percentile dummy (Alderson C type P-1 50 AU dummy, 162 lb, 5 ft-9 in.)

potential is one of the three main impact performance characteristics of laminated safety glass, the other two being deceleration and penetration resistance.

Search of the literature shows little information regarding cutting human tissue. Little, if anything, is available on the force necessary to cause incisional wounds or the effect of mechanical violence on human epithelial tissue. Gadd et al. in a recent publication[10] state the "belief that skin fails more often in a tensile or cohesive manner than from a shearing or cutting action." For this reason, the study of skin strength by these authors has been directed principally toward an indenter type test using either a pointed or guillotine type device. Hog tissue was judged to be similar to human skin in mechanical properties as evaluated by these techniques.

In order to study in depth the lacerative injury potential of laminated safety glass, material cutting devices were developed at the Glass Research Center and a variety of materials examined to simulate human skin.

Two machines are currently used in this study; namely, a linear cutting apparatus drawn by an Instron and a swinging pendulum. These two machines are shown in Figures 16 and 17. The Instron is operated at 20" per minute cutting speed while the pendulum (modified Izod machine) travels at 11 ft. per second for a speed ratio of approximately 1 to 400. The type blade chosen from a large

variety that was examined in this program is a scimitar shaped surgical blade (No. 12 rib-back non-sterile carbon blade).

The No. 12 blades are used in both devices and are shown in the blade holders in Figures 16 and 17. An array of 3 blades spaced 1/4" apart is used to cancel some of the variation in blades, and new blades are used for each cut. The scimitar shape of this blade is particularly advantageous for two reasons; first, the point of the blade must penetrate the test sample at the be-

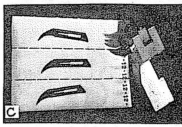
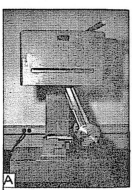

Fig. 17 - Modified izod machine for material cutting (pendulum velocity = 11 ft/sec); A - modified izod, B - vacuum platen and simulated tissue sample (3 x 6 in.), C - blades in holder used on pendulum

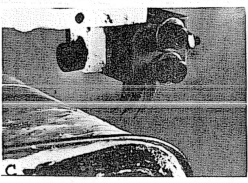
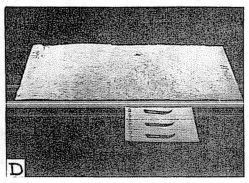

Fig. 16 - Material cutting apparatus used with instron tensile machine; A - arrangement for vertical travel, B - apparatus with simulated tissue on vacuum platen, C - close-up of blade away, D - simulated tissue (6 x 12 in.) with blades and holder

ginning of a run and second, this curvature does not force the surface of the test piece away from the cutting edge.

The test pieces (6" x 12" for the Instron and 3" x 6" for the modified Izod) whether natural or synthetic are held in position on either machine by vacuum on a waffle pattern metal platen. The three-blade array is arranged in the holder to give a cut depth of 3/16". The standard thickness selected for the synthetic samples has been 3/8" but proper adjustment of the blades to give the 3/16" cut depth permits use of other thicknesses as well. As shown in Figure 16 the lead edge of the sample for the Instron de-

vice is clamped below the points of the blades. Thus the points must penetrate the surface of the sample at the beginning of the run (total length of cut = 9"). In the case of the modified Izod machine, the blades enter and exit through the sample surface during the pendulum swing.

Through the use of these machines a series of synthetic compositions has been compared with similar data obtained with fresh human and hog tissues. Some pertinent test results are summarized in Table 6. In this series, the commonly used materials such as chamois and thin polyvinyl chloride show cutting forces considerably greater

Table 6 - Human Skin Simulation
(Cut depth = 3/16"; 3 No. 12 surgical blades)
(Test temperature = 70°F)

Coating[a]		Base	Instron	Mod. Izod
Type	Reinforcement	5/16" Sponge Rubber[c]	(lbs)	(inch lbs)
Silicone 11	0.5 wt. % human hair[b]	yes	4.28	30.5
Silicone 77	do	do	3.83	31.5
Silicone 116	do	do	4.34	36.3
Silicone 504	do	do	5.61	40.6
Silicone 560	do	do	4.26	26.0
Silicone 577	do	do	4.20	32.3
Silicone 580	do	do	4.11	33.5
Silicone 589	do	do	3.87	34.2
Silicone 616	do	do	4.28	--
Silicone 630	do	do	6.39	--
Chamois(dry)	--	do	5.92	32.6
Chamois(oiled)	--	do	5.76	28.8
PVC (1/16")[d]	--	do	6.64	36.0
PVC (1/8")[d]	--	do	9.90	>50
PVC (1/8")[d]	cloth	1/4" polyurethane foam	>50	--
Paper (0.003")	--	3/8" polyurethane foam	1.48	37.2
Silicone 11	0.5 wt. % human hair	1/4" polyurethane foam	1.31	--
Silicone 616	do	do	1.84	--
Hog (3/8")[e]	--	--	3.18-3.39	--
Human (1/8")[f]	--	1/4" rubber sponge	4.7	--
Human (1/4"-3/8")[f]	--	--	2.3-4.8	8-27

a. Coating is 1/16" thick unless otherwise noted
b. Silicon tetrachloride treated hair; wt. % based on wt. of coating
c. Commercial black open cell sponge rubber - Shields Rubber Co., Pittsburgh, Pa.
d. Polyvinyl chloride (Geon 121)　38.9 pph ⎫
   DOP plasticizer　58.3 pph ⎬ Cured at 300°F
   Stabilizer (Ferro 12V63)　1.1 pph ⎪ for 1-1/2 hours
   Paraplex G12　1.7 pph ⎭
e. Abdominal skin
f. Human skin from the calf of leg

than the range for human skin. Human skin removed from the calf of the leg varied over a range of 2.3 pounds to 4.8 pounds in the Instron test and 8-27 inch pounds in the modified Izod apparatus depending upon the age and health of the individual. The data recorded are for fresh tissue samples evaluated at 70°F within a few hours after amputation and kept under refrigeration prior to testing.

A large number of resins mounted on a sponge rubber base was examined in these two devices for similarity to human skin with underlying fatty tissue and from this the silicone rubber system was selected for several reasons. First, the very low sensitivity of silicone rubbers to temperature and second, availability in a wide variety of types that can be cured either at room conditions or at slightly elevated temperature. However, the notch sensitivity of silicone rubbers is normally low and therefore human hair was incorporated (0.5 wt. % based on the weight of silicone rubber) to provide tear resistance. The primers normally recommended for use with silicone rubber to improve bonding and several used with fiber glass were not effective in upgrading adhesion of human hair. It was found that the adhesion between hair and silicone rubber could be satisfactorily increased through pretreatment of the hair with silicon tetrachloride.

The simulated skin system selected for our laceration studies consists of a 1/16" thick layer of human hair reinforced silicone rubber, preferably General Electric RTV 577 (white) or RTV 560 (red), on 5/16" thick open cell sponge rubber (Shields Rubber Company, commercial black open cell). The hair reinforced RTV silicone rubber, after deposition on the sponge rubber, is cured by heating at 200°F for 2 hours.

This simulated skin using the RTV 577 coating has been applied to the 6-1/2" diameter 22 lb headforms (horizontal impactor and dropping ball) and impacted at speeds slightly above the penetration velocity so that the effect of glass fracture as well as penetration could be ascertained. Essentially duplicate performance was obtained with the horizontal impactor and the dropping headform even though the dropping headform is free to roll while the headform of the horizontal impactor travels a defined path. Impact angles of 45° and 90° have been studied with 24" x 36" glass panels. At the 90° impact angle, only superficial surface abrasions were noted on the headforms even though

severe fracturing occurred and the head completely penetrated the test panels.

The behavior at 45° impact angle is quite different because the headform moves forward against the edges of the fractured glass producing severe cutting by those panels which tear into large shards. Some of the simulated skin tests are shown in Figure 18 where the paper flags inserted into the cuts demonstrate both the length and angle of the cuts made at 45° impact angle. Also shown in Figure 18 is the nose of the accelerometer equipped horizontal impactor with type RTV 577 simulated skin cover. The three simulated skin headform covers in Figure 18 show in these preliminary results considerable variation in the extent and severity of damage depending upon the panel tested.

The 1/8"P - 0.015 Std. - 1/8"P panel (old windshield type) penetrated at 8.2 mph caused more lacerative damage than a 1/8" TT - 0.030 HPR - 1/8"P panel penetrated at 29.3 mph. The 1/8"P - 0.030 HPR - 1/8"P (new windshield structure) caused heavy damage to the simulated skin when penetrated at 29.3 mph. While not exhibited in this figure, it has been found that the old windshield type laminate penetrated at 45° angle and 29.3 mph caused slightly more

Laceration Caused by 45 deg Impacts

1/8 P- 015 Std-1/8 P	1/8 P-.030 HPR-1/8 P	1/8 TT-.030 HPR-1/8 P
8.2 mph	29.3 mph	29.3 mph

Accelerator Equipped Headform With Simulated Skin Cover

Fig. 18 - Simulated skin (RTV-577 on sponge rubber) used on 22 lb headform

extensive and deeper damage than the new windshield type structure.

The simulated skin will be used as the face covering of our full size dummy for windshield impact studies at velocities above and below head penetration speeds.

REFERENCES

1. American Glass Review
   May 1967, p. 28
2. USA Standard Z26.1 - 1966
   "Safety Glazing Materials for Glazing Motor Vehicles Operating on Land Highways"
3. Widman, J. C.
   "Recent Developments in Penetration Resistance of Windshield Glass",
   SAE Publication 650474, May 1965
4. Schwimmer, S. and R. A. Wolf
   "Leading Causes of Injuries in Automobile Accidents", ACIR, 1962
5. Rieser, R. G. and G. E. Michaels
   "Factors in the Development and Evaluation of Safer Glazing",
   Proceedings, Ninth Stapp Car Crash Conference, University of Minnesota, October 1965
6. Patrick, L. M.
   "Human Tolerance to Impact - Basis for Safety Design",
   SAE Publication 1003B, January 1965
7. Danforth, J. P. and C. W. Gadd
   "Use of Weighted Impulse Criterion for Estimating Injury Hazard",
   Proceedings, Tenth Stapp Car Crash Conference, Holloman Air Force Base, New Mexico, November 1966
8. Danforth, J. P.
   "Computation of GMR Injury Severity Index",
   General Motors Corporation, Research Publication GMR 574, July 5, 1966
9. Naab, K.
   "Measurement of Detailed Inertial Properties and Dimensions of a 50th Percentile Anthropometric Dummy",
   Proceedings, Tenth Stapp Car Crash Conference, Holloman Air Force Base, New Mexico, November 1966
10. Gadd, C. W., Lange, W. A., and F. J. Peterson
    "Strength of Skin and Its Measurement"
    ASME Publication 65 WA/HUF8 November 1965

Appendix 1

Table A - Mean Penetration Velocity vs Interlayer Thickness
(All glass 1/8" thick; 90° impact angle; 12" x 12" test panel; 5 lb steel ball)

Glass*	Configuration (ins.) Interlayer	Glass	Mean Penetration Velocity (mph) 0°F	70°F	120°F
	Interlayer Thickness 0.030"				
P	.030 Std.	P	16.6	18.9	13.7
P	.015 HPR-.015 Std.	P	18.2	18.3	13.0
P	.015 A-.015 Std.	P		19.2	
P	.015 HPR-PL2-.015 R	P		19.2	
P	.030 Std.-10	P	18.6	20.0	13.1
TT	.015 HPR-PL2-.015 Std.	P	19.2	19.9	
P	.015 HPR-PL1-.015 Std.	P		20.3	
P	.030 HPR-PL1	P	13.5	20.8	
TT	.015 A-.015 Std.	P		20.9	
HS	.015 A-.015 Std.	P		23.3	
P	.015 A-.015 HPR	P		22.5	
P	.030 HPR	P	17.3	24.0	12.7
TT	.030 HPR-PL2	P	18.8	24.3	10.7
TT	.015 HPR-.015 Std.-PL2	P		24.2	
P	.015 HPR-.015 Std.-PL1	P	13.9	25.0	
TT	.015 A-.015 HPR	P	17.4	24.4	
TT	.030 HPR	P	20.5	24.9	12.2
P	.015 HPR-.015 Std.-PL2	P	11.2	25.4	14.7
HS	.030 HPR	P	22.6	25.5	16.1
HS	.030 HPR-PL2	P		26.5	16.5

(cont'd)

307

Table A - (Continued)

Glass*	Configuration (ins.) Interlayer	Glass	Mean Penetration Velocity (mph) 0°F	70°F	120°F
HS	.015 A-.015 HPR	P		26.8	
TT	PL2-.030 HPR-PL2	P		26.7	
TT	PL1-.030 HPR-PL1	P		29.2	
P	.030 HPR-PL2	P	11.6	27.8	12.8
P	.030 PPG-1	P	12.2	26.2	19.0

Interlayer Thickness 0.037"

Glass*	Interlayer	Glass	0°F	70°F	120°F
TT	.030 HPR-PL2-PL1-.007A	P		27.0	
TT	.030 HPR-PL2-.007A	P		27.4	
TT	.030 HPR-PL1-.007A	P		28.8	
HS	.030 HPR-PL2-.007A	P		29.1	
HS	.030 HPR-PL1-.007A	P		33.6	

Interlayer Thickness 0.045"

Glass*	Interlayer	Glass	0°F	70°F	120°F
P	.045 HPR	P	19.4	32.8	17.4
TT	PL2-.007A-PL1-.030 HPR-PL1-.007A-PL2	P		38.3	
HS	.015 Std.-PL2-.015 HPR-PL2-.015 Std.	P		39.5	
TT	PL2-.007A-.030 HPR-.007A-PL2	P		41.6	

Interlayer Thickness 0.060"

Glass*	Interlayer	Glass	0°F	70°F	120°F
TT(a)	.060 HPR	P	21.2	33.6	14.4
TT	.060 HPR	P		35.8	
HS	.060 HPR	P	18.7	42.1	19.4
P	.060 HPR	P	21.2	44.2	21.3
P	.060 Std.	P	20.4	44.9	23.5
P	.060 HPR-PL2	P	22.3	44.7	19.1
P	.045 HPR-.015 Std.-PL2	P	13.0	45.0	23.6
P	.060 HPR-.075	P		45.7	
P	.060 PPG-1	P	19.0	43.0	32.0

Interlayer Thickness 0.075"

Glass*	Interlayer	Glass	0°F	70°F	120°F
TT	.060 HPR-PL1-.015 Std.	P		47.0	
TT	.075 HPR	P	23.9	46.8	20.8
P	.060 HPR-.015 Std.-PL2	P	13.8	52.1	21.9
TT	.060 HPR-.015 Std.-PL2	P		53.2	
P	.060 HPR-PL1-.015 Std.	P		52.8	29.1
P	.060 HPR-.015 Std.-PL1	P		54.4	27.6
P	.075 HPR	P	22.8	55.3	25.8

Interlayer Thickness 0.090"

Glass*	Interlayer	Glass	0°F	70°F	120°F
TT(a)	.090 HPR	P		50.8	
TT	.090 HPR	P	23.4	56.8	19.8
HS	.090 HPR	P	21.8	57.2	25.1
P	.090 Std.	P	20.9	60.9	31.1
P	.090 HPR	P	21.4	62.6	28.6
P	.090 HPR-PL2	P	24.5	62.9	31.8
P	.060 HPR-.030 Std.-PL2	P		64.7	30.0
P	.090 PPG-1	P		65.0	

P - Annealed plate glass
HS - Heat strengthened plate glass
TT - Full thermally tempered plate glass
PL1 - Parting layer No. 1
PL2 - Parting layer No. 2

HPR - High penetration resistance plastic interlayer
A - Aircraft type plastic interlayer
Std. - Standard plastic interlayer
a - Abraded
* - Impact side

Table B - Mean Penetration Velocity vs Interlayer Thickness at 0°, 70°, 120°F
(All glass 1/8" thick; 90° impact angle; 12" x 12" test panel)

Glass*	Configuration (ins.) Interlayer	Glass	Mean Penetration Velocity (mph) 0°F	70°F	120°F
1/8P	.015 Std.	1/8P		7.4	
1/8P	.030 Std.	1/8P	16.6	18.9	13.7
1/8P	.060 Std.	1/8P	20.4	44.9	23.5
1/8P	.090 Std.	1/8P	20.9	60.9	31.1
1/8P	.015 HPR	1/8P		16.2	
1/8P	.030 HPR	1/8P	17.3	24.0	12.2
1/8P	.045 HPR	1/8P	19.4	32.2	17.4
1/8P	.060 HPR	1/8P	21.2	44.2	21.3
1/8P	.075 HPR	1/8P		55.3	25.8
1/8P	.090 HPR	1/8P	21.4	62.6	28.6
1/8P	.030 HPR-PL2	1/8P	11.6	27.8	12.8
1/8P	.060 HPR-PL2	1/8P	22.3	44.7	19.1
1/8P	.075 HPR-PL2	1/8P	22.8		
1/8P	.090 HPR-PL2	1/8P	24.5	62.9	31.8
1/8HS	.030 HPR	1/8P	22.6	25.5	16.1
1/8HS	.060 HPR	1/8P	18.7	42.1	19.4
1/8HS	.090 HPR	1/8P	21.8	57.2	25.1
1/8TT	.030 HPR	1/8P	20.5	24.9	12.2
1/8TT	.060 HPR	1/8P	21.2	34.6	14.4
1/8TT	.075 HPR	1/8P	23.9	46.8	20.8
1/8TT	.090 HPR	1/8P	23.4	56.8	19.8

*Impact surface

Appendix 2 - Measurement of optical deflection

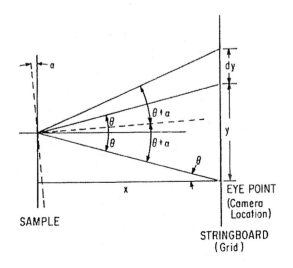

$$\text{TAN } \theta = \frac{1/2y}{x} = \frac{y}{2x}$$

$$\text{TAN } (\theta + 2\alpha) = \frac{dy}{x}$$

$$y + dy = 2 \times \text{TAN } \theta + x \text{ TAN } (\theta + 2\alpha)$$

$$dy = x \text{ TAN } (\theta + 2\alpha) - x \text{ TAN } \theta$$

$$dy = x (\theta + 2\alpha) - x\theta \text{ WHEN } \alpha \text{ IS SMALL}$$

$$dy = 2x\alpha$$

WHERE x IS THE DISTANCE BETWEEN
STRINGBOARD AND SAMPLE

α IS THE BENDING OF SAMPLE

dy IS THE DISPLACEMENT OF GRID IMAGE

309

700481

Safety Performance of Laminated Glass Structures

R. G. Rieser and J. Chabal
Glass Research Center, PPG Industries, Inc.

INTRODUCTION OF THE higher penetration resistance windshield started with the 1966 Model domestic automobiles. Highway crash information indicates this windshield to be performing beyond expectations (1)* relative to lowered lacerative hazard and very low frequency of interlayer rupture. This significant improvement in safety performance has been achieved without any apparent increase in neck or intercranial injury hazard from the windshield (2). Highway crash information indicates that this modification has made the windshield safer in the two major safety aspects; penetration resistance, and laceration, and without evidence of increase in concussive hazard.

While highway crash experience has been excellent thus far, there has been limited evidence that the windshield could be further improved in the area of less severe lacerations occurring when the head or face slides along the broken inner glass surface. In the Glass Research Center of PPG Industries, Inc., there is a continuing effort to evaluate in detail the safety performance of windshield configurations.

The PPG program on safety performance of laminated automotive windshields has been wide ranging in scope and has been aimed at providing the automotive industry with research data for its study and consideration. The program has included field studies as well as laboratory experimentation. In

*Numbers in parentheses designate References at end of paper.

view of our interest in the full scope of windshield safety, we have not limited our studies to those configurations considered economically attractive or even feasible from the large scale fabrication point of view. We have considered from the inception of our intensive safety program some six years ago, a broad search into all potential windshield configurations in the search for possible improvements in safety.

In addition to conventional float (F) and plate (P) glasses, our studies have included our newly developed Vertiglas (V); varied thicknesses of each of these three basic glasses have been used, including some of purely academic interest at this time.

The term Vertiglas is applied to a new type of vertically drawn glass with entirely new optical qualities, quite unlike the characteristic waves found in even the best sheet glass but comparable with plate or float glasses. Vertiglas makes possible consideration of thinner glass plies in windshield fabrication than has hitherto been feasible with either plate or float glasses.

This paper is a continuation of our overall program and presents pertinent information on the influence of two variables on safety performance; namely, glass and interlayer thicknesses.

EXPERIMENTAL PROCEDURE

In order to minimize the gap between laboratory experimentation and actual crash conditions, we have utilized full

—————— ABSTRACT

Full scale laboratory studies using curved laminated safety glass windshields, a horizontal accelerator, and a 50th percentile dummy to provide impact performance are presented. Lacerative damage based on moist chamois face covering and Severity Indices for annealed glass windshield configurations were obtained on the current domestic automobile windshield, experimental thin annealed glass windshields in symmetrical and asymmetrical glass configurations, and windshields with 0.045 in. HPR interlayer.

Symmetrical thin glass windshields, such as, 0.050 in. glass, gave excellent performance to 38 mph. Lacerative damage decreases as the inboard glass thickness decreases for windshields having 0.030 in. HPR interlayer and 1/8 in. float glass outboard ply. Windshields with 0.045 in. HPR interlayer, designed to improve penetration resistance, have somewhat higher Severity Indices. Such windshields will require further study.

scale testing in addition to flat panel tests using a headform. Our horizontal accelerator propels a sled and a 50th percentile dummy into a shock absorber which stops the sled and permits the dummy to move forward into the windshield (3, 4). The windshield, in a frame, is positioned at a typical automobile installation angle or can be adjusted to any other desired angle from 20-90 deg off horizontal.

Early studies with this device demonstrated that the normal automobile instrument panel/dashboard assembly interfered erratically with the movement of the dummy into the windshield (4). Reproducible dummy movement during windshield impact is extremely important in conducting tests for comparative performance. It was found that the normal instrument panel of the body buck used in our early experiments frequently caused the legs to wedge under the panel thus interfering with forward movement of the torso. The magnitude of this interference increased directly with velocity.

In view of this difficulty, the body buck was replaced by a windshield opening constructed of angle iron bent to conform with the windshield curvature. This frame is designed so that adjustment in windshield installation angle can be easily accomplished. The attachments at the ends of the frame, consisting of a shaft and bearing, permit easy adjustment to the desired angle, and then a locking device is utilized to hold this selected installation angle as shown on Fig. 1. A heavily reinforced knee barrier, located below the windshield frame and mounted independently, is the device which stops forward movement of the dummy as the head contacts the windshield. Head contact is arranged to occur at 5-6 in. from the top of the windshield measured in the plane of the glass. Impact of the knees stops forward movement and localizes the expenditure of kinetic energy in windshield and knee barrier. In this arrangement, forward movement of the dummy into the windshield is followed by downward sliding of the head over the fractured glass surface with minimal restriction from the torso and lower extremities.

This dummy-windshield impact arrangement, while obviously different from the automobile in the instrument panel/dashboard area, is considered to be more severe than an automobile crash at the same speed. This is judged to be a preferred situation for the comparative analysis of windshield safety performance in full scale laboratory testing.

The curved windshield employed in the studies reported herein was the NAGS** No. W660 windshield, which is installed at 35 deg off horizontal in the vehicle; this same installation angle was used in our laboratory tests. The windshields were adhered to contoured metal frames with polysulfide adhesive (4). After curing the polysulfide (4 hr at 120 F), the frame with adhered windshield is then bolted to the angle iron opening. On the basis of sequential durometer tests of coupon samples, this cure schedule gives at least 90% of the total cure achieved by long time room temperature storage and thus corresponds closely with the adhesive mounted windshields in service on our domestic automobiles.

The deceleration forces are recorded by a biaxial system mounted in the forward portion of the head behind the bridge of the nose. The vertical (superior-inferior) and horizontal (anterior-posterior) force components are recorded; the lateral force is neglected since tests have shown the force in this plane to be very small in our test procedure. A typical force-time record of a dummy-impact with the windshield is shown in Fig. 2.

In general, it must be said that laboratory full scale evaluation of safety glass performance cannot be on an individual windshield test basis in providing the answer to actual behavior in a highway crash situation. Full scale testing creates the least problem of translating laboratory performance to actual service. It is our belief, based on many full scale impact tests,

----

**National Auto Glass Specifications.

Fig. 1 - Full scale windshield impact evaluation apparatus showing adjustable windshield installation angle arrangement

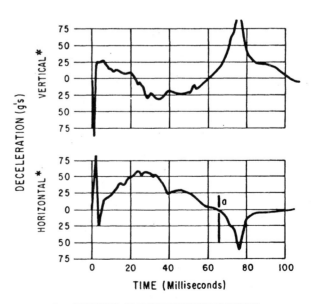

* — DIRECTION OF ACCELEROMETER MOUNTING
□ — EXTENT OF TIME INTERVAL USED FOR CALCULATION OF SEVERITY INDEX

Fig. 2 - Head deceleration during windshield/dummy impact

that the most desirable system is to run many tests on the windshield configurations and examine the data for trends, instead of making comparisons on an individual impact test basis. Approximately 1000 windshield-dummy impacts have been run thus far in our program on windshield performance.

In our study of windshield configurations, it has been our "modus operandi" to run a series of full scale dummy-windshield impacts over a broad speed range. Data recorded from a test series includes lacerative injury rating, penetration resistance, extent of interlayer rupture, and Severity Index from the biaxial accelerometer curves.

All windshields were impacted at room temperature. In addition, in order to more closely simulate the windshields in service, the exterior surface of the outboard glass ply of each windshield was abraded prior to impact. This abrasion was done by pressing a 240 grit silicon carbide abrasive fabric sheet against the glass surface using a 50 lb load on a 6 in. long × 2 in. dia hard rubber roller.

The lacerative injury potential of the windshield is judged by means of a double layer of moist chamois. This double layer of chamois is fastened to the face over 3/8 in. thickness sponge rubber. The demerit scale utilized is based on the severity of damage to the top and second layers as outlined in Table 1.

In our evaluation of lacerative damage as recorded by moist chamois, chin damage is not included in our judgments. The reasons for this are:

1. Without instrument panel and dashborad in our full scale testing, the chin area contacts the glass to a much greater extent than indicated by actual clinical damage.

2. Excessive stretching of the chamois over the chin during the downward slide of the dummy's head over the windshield.

3. Potential for misjudgment of damage resulting from chin contact with the windshield mounting frame.

TEST RESULTS

The test results for our full scale impact studies are based on the impact of a 50th percentile dummy and a stationary windshield, mounted at 35 deg off horizontal—the automobile installation angle for this windshield. In the lacerative injury charts of this report, the damage rating at the velocity of impact is supplied as a bar whose height records the extent of

chamois damage. This lacerative damage demerit scale as discussed above is on a 1-5 basis.

In Fig. 3, is shown the performance of the currently used windshield configurations constructed of nominal 1/8 in. thickness plate and float glasses. Automotive windshields manufactured in this country at this time might have either type of glass; float glass is now used more extensively than plate glass in domestic windshield manufacturing. It can be noted in Fig. 3 that the float glass windshield shows less severe lacerative damage than the plate glass counterpart up to approximately 30 mph impact. Above 30 mph impact velocity, both the current float and plate glass windshields show the highest lacerative damage rating of 5.

Vertiglas, in 1/8 in. thickness plies, on the basis of the limited data presented in Fig. 3 also indicates less lacerative damage than plate glass in windshield impacts.

Comparison of 0.090 in. thickness glass plies with the 0.030 HPR interlayer, as recorded also in Fig. 3 in Vertiglas and float glass demonstrates that the lacerative damage is essentially equivalent for windshields with these two types of glass. It is also evident by comparing the laceration ratings of Figs. 4-6 that the thinner glass windshields are less lacerative than the 1/8 in. thickness glass configurations.

The tests recorded in Fig. 4 represent an extension of the data in Fig. 3. Fig. 4 summarizes impact laceration for thin Vertiglas windshields in balanced glass thicknesses and in asymmetric configurations with 1/8 in. thickness float glass outboard plies. With the symmetrical laminate structures, as glass thickness decreases, lacerative damage decreases. The 0.050 in. Vertiglas tested in this program, constituting both plies of the windshield, yielded low lacerative damage, receiving no ratings above 3 at impact speeds up to 38 mph. This windshield configuration surpassed all other laminates tested

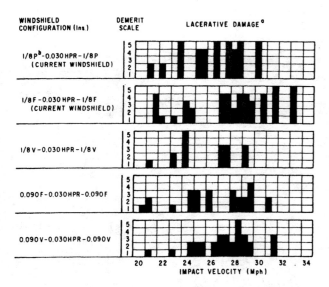

o — ASSESSMENT BASED ON DAMAGE TO DOUBLE LAYER OF MOIST CHAMOIS SKIN

b — INBOARD GLASS GIVEN FIRST FOR ALL CONFIGURATIONS; THIS SURFACE CONTACTED BY THE DUMMY.

HPR = HIGH PENETRATION RESISTANCE INTERLAYER

P = ANNEALED PLATE GLASS

F = ANNEALED FLOAT GLASS

V = ANNEALED VERTIGLAS, A PPG INDUSTRIES GLASS

Fig. 3 - Effect of glass thickness and type on lacerative injury

Table 1 - Demerit Scale Based On Severity
Damage to the Top and Second Layers

Rating	Damage to Moist Chamois
1	Shallow scratches
2	Few shallow cuts—top layer only
3	Cuts and gouges of top layer
4	Many cuts and gouges—few through both layers
5	Extensive cuts and gouges through both layers

in total performance; that is, low laceration, excellent diaphragm action by the interlayer, high penetration resistance, and Severity Index at 38 mph indicate the force-time energy expenditure to be excellent. At the present time, a windshield with two plies of 0.050 in. thickness glass is not considered a "practical" safer windshield because of many unevaluated factors with this type windshield. In particular, windshields with annealed glass as thin as 0.050 in. have not been evaluated in vehicles to ascertain the service life under highway conditions.

The other symmetrical annealed glass windshields, recorded in Fig. 4 in 0.070 and 0.090 in. glass thicknesses, show greater lacerative damage as the glass thickness is increased. Change in thickness from 0.050 in. glass plies to 0.070 in. thickness causes an increase in lacerative damage; the difference in laceration between 0.070 and 0.090 in. glass plies is less significant but, nonetheless, favors the thinner glass.

When these thin glasses are utilized as the inboard glass ply laminated to the currently used 1/8 in. outboard glass, the same trend in lacerative damage is noted. That is, a definite advantage exists for the thinner inboard glass, and this improved performance is noticeably greater in the 0.050-0.070 in. than in the 0.070-0.090 in. thicknesses when laminated with 1/8 in. float glass as the outboard ply. Based on these tests of windshields consisting of thin inboard glass with thicker outboard glass, additional experiments were conducted with inboard glass between 0.05-0.07 in. Such a windshield was considered commercially feasible, and windshields of this design installed in vehicles have indicated no problems relative to actual service on the highway. An intermediate glass thickness of 0.065 in. was examined in depth to ascertain the contribution such a windshield would make to safety performance.

Fig. 5 shows the lacerative damage of several thin inboard glass thickness windshield configurations which have 1/8 in. float outboard glass. In this series, the lacerative damage for the 0.065 in. Vertiglas inboard ply windshield with 1/8 in. float glass is superior to the windshields having heavier inboard glass. While not considered superior to the 0.050 in. V inner ply windshield, this configuration is of importance because it constitutes a commercially feasible structure, and has the attractive low lacerative benefits of thin annealed inboard glass.

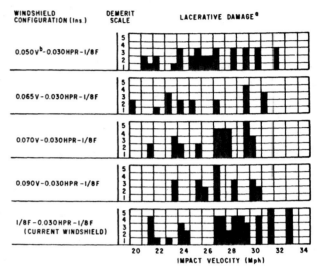

a — ASSESSMENT BASED ON DAMAGE TO DOUBLE LAYER OF MOIST CHAMOIS SKIN.
b — INBOARD GLASS GIVEN FIRST FOR ALL CONFIGURATIONS; THIS SURFACE CONTACTED BY THE DUMMY.
HPR = HIGH PENETRATION RESISTANCE INTERLAYER
F = ANNEALED FLOAT GLASS
V = ANNEALED VERTIGLAS, A PPG INDUSTRIES GLASS

Fig. 5 - Lacerative damage: asymmetric glass windshield configurations

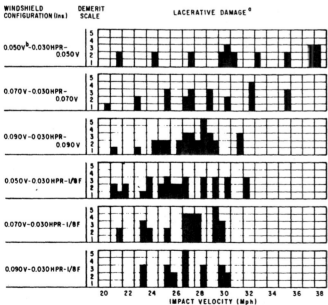

a — ASSESSMENT BASED ON DAMAGE TO DOUBLE LAYER OF MOIST CHAMOIS SKIN.
b — INBOARD GLASS GIVEN FIRST FOR ALL CONFIGURATIONS; THIS SURFACE CONTACTED BY THE DUMMY.
HPR = HIGH PENETRATION RESISTANCE INTERLAYER
F = ANNEALED FLOAT GLASS
V = ANNEALED VERTIGLAS, A PPG INDUSTRIES GLASS

Fig. 4 - Effect of glass on lacerative injury

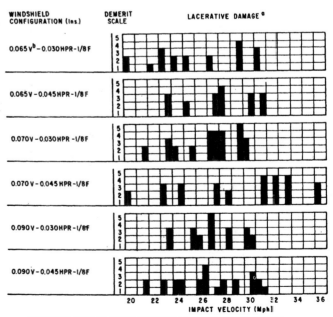

a — ASSESSMENT BASED ON DAMAGE TO DOUBLE LAYER OF MOIST CHAMOIS SKIN.
b — INBOARD GLASS GIVEN FIRST FOR ALL CONFIGURATIONS; THIS SURFACE CONTACTED BY THE DUMMY.
HPR = HIGH PENETRATION RESISTANCE INTERLAYER
F = ANNEALED FLOAT GLASS
V = ANNEALED VERTIGLAS, A PPG INDUSTRIES GLASS

Fig. 6 - Effect of interlayer thickness on lacerative injury

Comparison of this windshield with the current windshield in either plate or float glass clearly shows the lower lacerative damage to be expected with the thin annealed inboard glass.

Several asymmetrical annealed glass windshield configurations were impact evaluated with 0.045 in. HPR interlayer. The lacerative injury ratings in Fig. 6 are for duplicate configurations with 0.030 and 0.045 in. HPR interlayers at dummy impact velocities of 19.5-35.6 mph. With 1/8 in. glass outboard and 0.065, 0.070, and 0.090 in. glass thicknesses inboard, the 0.045 in. HPR interlayer appears to have greater benefit in lowering lacerative damage as the inboard glass is increased to the 0.090 in. thickness. At 0.065 in. inboard glass thickness, there appears to be little if any benefit for 0.045 in. HPR over 0.030 in. HPR; at 0.090 in. inboard thickness, the lacerative damage is significantly less with the 0.045 in. HPR interlayer.

SEVERITY INDEX

In the evaluation of windshields using a dummy and horizontal accelerator, force-time pulses are monitored by instrumentation located in the head of the dummy to permit comparison of energy dissipation characteristics of the individual windshields under impact conditions. The instrumentation used is fully described in a previous publication (4, Appendix B) and consists of a biaxial accelerometer system using piezoelectric quartz crystals monitoring anterior-posterior and superior-inferior forces. A triaxial system was not utilized, because the left-right component in this type of full scale testing is at most only a small force component and is considered insignificant.

A typical force-time trace for the biaxial accelerometer system as cited earlier in Fig. 2 is used to calculate the Severity Index (S.I.). While the S.I. was originally proposed for flat laminated glass panel tests with the 22 lb headform and not for dummy-windshield impacts, it is a recognized term and is the only available calculation utilizing a weighting factor based on laboratory biomechanical data. Our experimental procedure differs from techniques used by others, and thus the force-time responses (S.I.) are not directly interchangeable with those obtained by other investigators. In this calculation, the force in "g's" is taken from the curves over the time interval beginning with the moment of head impact, to the time in milliseconds when there is a sudden change in deceleration rate (recorded by the horizontal accelerometer) that crosses the original zero base line (point 'a' in Fig. 2). The two curves as shown in Fig. 2 are read in "g's" at 0.001 sec intervals for this particular time duration; the resultant "g" values are used in the Danforth-Gadd (5) equation to calculate S.I. as:

$$S.I. = \int_i g^{2.5}\, dt$$

In our operation, the actual calculation of Severity Index value is by means of a computer program.

This section of this report summarizes the Severity Index

values obtained on those windshield configurations discussed earlier with regard to lacerative injury potential. While each sled run is instrumented to record the force-time pulses from the biaxial accelerometers in the head of the dummy, not all recordings obtained have been satisfactory for analysis. Difficulties are occasionally encountered from the Visicorder recording of the input pulse, from whipping of the instrumentation extension cables, and impact causing loosening of electrical contacts and/or loosening of the accelerometer mounting.

An improvement in yield of acceptable recordings has been achieved, by potting the accelerometers and electrical connections in the head with a room temperature cure polymer such as Sylgard 188 (Dow Corning product). The recording instrumentation is now a Visicorder Oscillagraph Model 1912 (Honeywell, Inc.) which has decreased our losses in recording the force-time pulses.

The Severity Index values have been summarized in bar charts for each windshield configuration. The numerical values are presented as tables in the Appendix. A bar located over an impact velocity gives the magnitude of the Severity Index for that impact, and solid or broken designates no penetration or head penetration, respectively. By penetration in these tests is meant extension of any portion of the head through the laminate during the impact event. Whether or not head penetration occurs in each test is ascertained by examination of the high speed photographic record of the impact, and both front and side views are utilized in making this judgment. We employ 400 fps high speed motion pictures for this purpose.

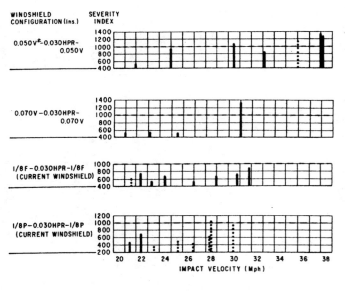

*INBOARD GLASS GIVEN FIRST FOR ALL CONFIGURATIONS; THIS SURFACE CONTACTED BY THE DUMMY.
HPR = HIGH PENETRATION RESISTANCE INTERLAYER
P = ANNEALED PLATE GLASS
F = ANNEALED FLOAT GLASS
V = ANNEALED VERTIGLAS, A PPG INDUSTRIES GLASS.
NUMERICAL VALUES RECORDED IN TABLE I, APPENDIX A

KEY:
| NO HEAD PENETRATION
: HEAD PENETRATION

Fig. 7 - Effect of glass thickness and type on Severity Index

314

In general, these bar charts show a tendency to higher Severity Index values with increase in impact velocity; it also is apparent that rupture of the interlayer sufficient to permit head penetration does not necessarily mean a low Severity Index. Also the windshields with 0.045 in. thickness HPR interlayer show consistently high S.I. values, thus indicating the need for extensive investigation of the concussive potential of this windshield configuration. This windshield shows excellent penetration resistance with no head penetration occurring in full scale tests to 35.6 mph.

Fig. 7 presents the Severity Index data for balanced windshield structrues, wherein the inboard and outboard glass plies are the same thickness and the same type glass. These bar charts show much more head penetration for the current windshield with 1/8 in. plate glass plies compared with the current 1/8 in. float glass windshields. Both these 1/8 in. glass structures show low deceleration forces at impact velocities of 20-32 mph. In this same figure, thinner glass laminates show considerable variation in deceleration forces. In the case of the 0.050 in. V–0.050 in. V glass system, the Severity Index values are higher but still below the 1000 value up to speeds of 30 mph. In fact, the excellent diaphragm behavior of the 0.030 HPR interlayer with this glass thickness was particularly evident even in two impacts at 38 mph. Head penetration did not occur in impacts at 38.0 and 38.2 mph; several inches of interlayer rupture occurred at 38.0 mph and no rupture was found in the windshield impacted at 38.2 mph.

Combinations of thin inboard glass with 1/8 in. float glass outboard in windshield configurations are summarized in Fig. 8. The data show a wide scatter, but it can be noted that those windshields with the 0.050 and 0.065 in. V exhibit lower Severity Indices than the duplicate configuration having .090 in. V inboard glass ply.

The data in Fig. 9 are limited to 0.045 in. thickness interlayer windshields with 1/8 in. float glass outboard and several different glass thicknesses as the inboard ply. In no case up to impact speeds of 35.6 mph did head penetration of these test windshields occur. It can also be noted when comparing the Severity Index values of Fig. 9 with the results in Figs. 7 and 8, that the possibility of concussive injury appears greater with the 0.045 in. thickness interlayer. In view of these results, HPR interlayer thicknesses less than 0.045 in. should be considered in lessening the concussive hazard. The effect of HPR interlayer thicknesses between 0.030 in., the currently used thickness, and 0.045 in. is now under study. A judicious choice of HPR interlayer thickness to provide maximum penetration resistance without increasing concussive danger should result from such studies.

CONCLUSIONS

The following conclusions are based on full-scale dummy-windshield impacts at room temperature.

1. Thin annealed inboard glass reduces lacerative injury.

2. In asymmetrical glass configurations, the thinner the inboard glass, holding outboard glass thickness constant, the lower will be the laceration hazard.

3. Of the windshield configurations studied, the thinnest glass, 0.050 in., as both inboard and outboard glasses was the safest based on laceration, penetration resistance, and force-time head loading during impact. Such a windshield is not yet considered to be "practical."

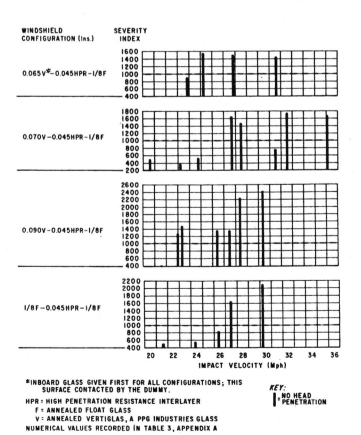

*INBOARD GLASS GIVEN FIRST FOR ALL CONFIGURATIONS; THIS SURFACE CONTACTED BY THE DUMMY.

HPR = HIGH PENETRATION RESISTANCE INTERLAYER
F = ANNEALED FLOAT GLASS
V = ANNEALED VERTIGLAS, A PPG INDUSTRIES GLASS
NUMERICAL VALUES RECORDED IN TABLE 3, APPENDIX A

KEY:
▌= NO HEAD PENETRATION

Fig. 9 - Effect of glass and interlayer thicknesses on Severity Index

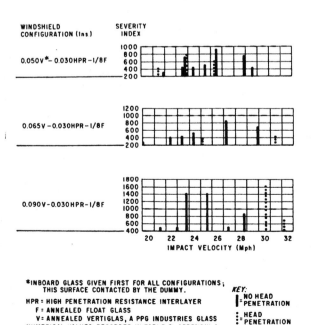

*INBOARD GLASS GIVEN FIRST FOR ALL CONFIGURATIONS; THIS SURFACE CONTACTED BY THE DUMMY.

HPR = HIGH PENETRATION RESISTANCE INTERLAYER
F = ANNEALED FLOAT GLASS
V = ANNEALED VERTIGLAS, A PPG INDUSTRIES GLASS
NUMERICAL VALUES RECORDED IN TABLE 2, APPENDIX A

KEY:
▌= NO HEAD PENETRATION
⋮= HEAD PENETRATION

Fig. 8 - Effect of inboard glass ply thickness on Severity Index

4. Data indicate that a "practical" windshield with 0.065 in. Vertiglas, inboard, 0.030 in. HPR and 1/8 in. float, outboard, has considerable merit in reduced lacerations. Similarly, a symmetrical configuration of the same total thickness has essentially the same improved characteristics.

5. Thicker interlayer (0.045 in.) windshields show higher decelerations in dummy impacts and may be more hazardous for concussion. All 0.030 in. HPR interlayer windshields with annealed glass up to 1/8 in. thickness appear to be about the same in decelerative loading of the head during impact.

REFERENCES

1. A. W. Siegel and A. M. Nahum, "Automobile Collision and the Effect of the New U.S.A. Standards." Proceedings of the Conference on Road Safety, Brussels, Belgium, 1968.

2. D. F. Huelke, W. C. Grabb, and R. O. Dingman, "Automobile Occupant Injuries from Striking the Windshield." Highway Safety Research Institute, The University of Michigan, 1967.

3. R. G. Rieser and J. Chabal, "Safety Performance of Laminated Glass Configurations." Proceedings of Eleventh Stapp Car Crash Conference, 1967, paper 670912. New York: SAE, 1967.

4. R. G. Rieser and J. Chabal, "Laboratory Studies on Laminated Safety Glass and Installations on Performance." Proceedings of Thirteenth Stapp Car Crash Conference, 1969, paper 690799. New York: SAE, 1969.

5. J. P. Danforth and C. W. Gadd, "Use of a Weighted-Impulse Criterion for Estimating Injury Hazard." Proceedings of Tenth Stapp Car Crash Conference, 1966, paper 660793. New York: SAE, 1966.

APPENDIX

Table 1 - Effect of Glass Thickness and Type on Severity Index

(Full-scale tests; NAGS No. W660 windshield at 35 deg installation angle; 50th percentile dummy; room temperature)

(see Fig. 7)

Windshield Configuration, in.									
0.050V[c]-0.030HPR-0.050V	Velocity, mph	21.5	24.5	29.8	32.7	35.6	38.0	38.2	
	S.I.[a]	558	936	1070	860	1381	1374	1365	
	H/P[b]	H	H	H	H	P	H	H	
0.070V-0.030HPR-0.070V	Velocity, mph	20.5	22.7	25.0	30.5				
	S.I.	520	558	532	1336				
	H/P	H	H	H	H				
1/8F-0.030HPR-1/8F	Velocity, mph	21.2	22.9	24.0	26.5	28.5	30.3	31.3	
	S.I.	631	470	667	531	620	784	912	
	H/P	P	H	H	H	H	H	H	
1/8P-0.030HPR-1/8P	Velocity, mph	20.9	21.8	23.1	25.2	26.4	27.8	28.0	29.9
	S.I.	471	680	374	572	477	617	1076	971
	H/P	H	H	P	P	P	P	P	P

[a]S.I. = Severity Index.

[b]H = No head penetration.
P = Penetration of head.

[c]Inboard glass given first for all configurations; this surface contacted by the dummy.

Table 2 - Effect of Inboard Ply Glass Thickness on Severity Index

(Full-scale tests; NAGS No. W660 windshield at 35 deg installation angle; 50th percentile dummy; room temperature)

(see Fig. 8)

Windshield Configuration, in.																
0.050V[c]-0.030HPR-1/8F	Velocity, mph	20.9	21.4	21.9	23.1	23.3	23.4	23.9	24.7	25.0	25.0	25.8	25.9	28.4	29.2	31.9
	S.I.[a]	429	356	489	418	749	893	412	523	295	570	618	906	735	429	890
	H/P[b]	P	H	H	H	H	P	H	H	H	H	P	H	H	H	H
0.065V-0.030HPR-1/8F	Velocity, mph	19.5	21.9	22.8	23.8	24.6	26.7	29.4	30.9							
	S.I.	251	408	419	524	377	805	671	444							
	H/P	H	H	H	H	P	H	H	P							
0.090V-0.030HPR-1/8F	Velocity, mph	20.9	22.5	23.4	25.2	28.3	30.1	31.7								
	S.I.	472	480	1443	1422	882	1667	702								
	H/P	H	H	H	H	H	P	P								

[a]Severity Index.

[b]H = No head penetration.
  P = Penetration of head.

[c]Inboard glass given first for all configurations; this surface contacted by the dummy.

Table 3 - Effect of Glass and Interlayer Thicknesses on Severity Index

(Full-scale tests; NAGS No. W660 windshield at 35 deg installation angle;
50th percentile dummy; room temperature)

(see Fig. 9)

Windshield Configuration, in.									
0.065V[c]-0.045HPR-1/8F	Velocity, mph	23.4	24.7	27.4	31.2				
	S.I.[a]	916	1568	1528	1419				
	H/P[b]	H	H	H	H				
0.070V-0.045HPR-1/8F	Velocity, mph	20.2	22.7	24.3	27.4	28.1	31.0	32.0	35.6
	S.I.	457	373	494	1628	1473	798	1755	1678
	H/P	H	H	H	H	H	H	H	H
0.090V-0.045HPR-1/8F	Velocity, mph	21.5	22.5	23.0	25.9	27.0	27.9	29.8	
	S.I.	402	1231	1473	1382	1377	2230	2406	
	H/P	H	H	H	H	H	H	H	
1/8F-0.045HPR-1/8F	Velocity, mph	21.2	23.8	25.8	27.0	29.6			
	S.I.	464	551	801	1630	2177			
	H/P	H	H	H	H	H			

[a]Severity Index.

[b]H = No head penetration.
  P = Penetration of head.

[c]Inboard glass given first for all configurations; this surface contacted by the dummy.

LAND
SEA
AIR
SPACE

This paper is subject to revision. Statements and opinions advanced in papers or discussion are the author's and are his responsibility, not the Society's; however, the paper has been edited by SAE for uniform styling and format. Discussion will be printed with the paper if it is published in SAE Transactions. For permission to publish this paper in full or in part, contact the SAE Publications Division and the authors.

Society of Automotive Engineers, Inc.
TWO PENNSYLVANIA PLAZA. NEW YORK. N.Y. 10001

12 page booklet.

Printed in U.S.A.

NONLACERATING GLASS WINDSHIELDS –
A NEW IMPROVED APPROACH

E. R. Plumat, R. Van Laethem, and P. Baudin
Laboratoire Central, Glaverbel S.A. (Belgium)

ABSTRACT

This study is to be considered as a part of a research program which aims to develop a laminated glass windshield whose laceration potential is very weak and even almost non-existent.

Two types of new safety windshields are tested. They only differ by the strengthening level of the V.H.R.glass which is used.

In both cases, the very thin 0.050 in. (1,2 mm) thick inner sheet and the 0.110 in. (2,8 mm) thick outer sheet are made of V.H.R. glass. Nevertheless, the glass of the outer ply has a deliberately limited tensile strength.

The polyvinylbutyral plastic interlayer is 0.030 in. (0,76 mm) thick.

The evaluation of the biomechanical behaviour of these windshields is made in different ways among which a laboratory study, described here, and including impact tests with a headform free falling on positioned samples.

During the impact, all measures defining the main safety performances are recorded or filmed at high speed: the deceleration peak along two orthogonal axes, the resultant severity index, relating to the initial impact and to the plow-in, the tearing length of the plastic interlayer and finally the laceration potential.

The latter is evaluated on basis of the laceration rating scale used by Prof. Patrick at the Wayne State University : a laceration index is given following the number and size of the cuts measured on the two superposed chamois leathers covering the headform.

The experimental parameters whose influence is more particularly studied are : temperature, impact velocity, impact location, increase of the mechanical strength of the sheets.

The researches are systematically carried out in order to compare the new windshield safety performances with the conventional laminated ones.

All the results of the measurements are analysed after the statistical method : parameters of distribution, lines of regression, analysis of correlations, signification tests, etc.

The new safety reinforced laminated windshield, whose laceration potential is very low and even non-existent at very high impact energy, might be used as a true passive restraint system if the tensile strengths of the reinforced glass are adjusted.

This double performance will have to be developed in a later series of tests about simulated crashes with anaesthetized primates, anthropomorphic dummies and, if possible, human cadavers.

IT IS CLEAR and well admitted that the present conventional laminated glass windshield with a thick plastic interlayer has much improved occupants' safety in a car when they undergo the "second collision".

And yet, the conclusions of official reports dealing with the whole of the crashes that have been analysed as well as the publications of specialists of automobile safety show that its laceration potential is still a severe disadvantage that ought to be lessened and if possible suppressed.

We come to the same conclusion when we consider the other glazings like the side and rear windows.

Several formulas have been suggested to reduce the risks of cuts for occupants who hit the windshield and whose heads and more particularly faces enter into brutal contact with the bits and broken fragments of glass.

None of the solutions that have been considered up to now seem to be a satisfactory compromise to meet with the requirements of the manufacturers as far as mass-production windshields are concerned.

It is very important to underline that the windshield whose laceration potential would have become very low and even non-existent should also have all the other performances that are required, especially a good optical quality that does not impair too much the visibility of the passengers and does not obstruct the view either because of an unforeseen fracture or by progressive natural aging.

Considerable progress has been achieved in manufacturing very thin glass of good optical quality in view of using it, strengthened as a part of high safety glazings (1)* (2).

However, it has been shown that the withstand against hard flying stones impact is improved when increasing the thickness of the outer sheet and the windshield rigidity : this results in preventing the propagation of fracture after indentations.

To find the best compromise, we have established the program of research described here, with three aims :

1) to study experimentally the numerous parameters which state the laceration potential of a windshield

2) to define, for increasing impact velocities, the best compromise between two antagonistic requirements :
- high resistance to head penetration on one side
- fairly limited severity index with a reasonable peak of deceleration on the other side

3) to set off the factors that could turn the windshield into a real passive restraint system.

This program is to be carried out in three successive phases. The first of them is explained here.

1°) Tests about the impact of the headform free falling at impact angle on flat samples and full scale curved windshields. Measurement of the mechanical factors and estimate of the biomechanical characteristics of the impact, more especially of the laceration index as a function of the different parameters such as impact velocity, temperature, increase of its mechanical resistance, etc. The results of the first phase will be used as a basis and orientation for the other two.

2°) Tests about the impact of primates against the types of windshields selected during the first phase. Estimate of the lacerations together with an X-ray study of the fractures and an attempt of diagnosis of the cerebral and cervical traumatisms.

---

* Numbers in parentheses designate References at end of paper

319

3°) Tests about the impact of anthropomorphic dummies and cadavers against the same windshields in the conditions of simulated crashes following the method used for instance at the Wayne State University.

The estimate of the results gathered then can be made but by direct and systematic comparison with the experimental data related to the conventional glazing undergoing the same tests under the same conditions.

This compared experimentation has two parts : one carried out on flat samples (size : 24 X 36 in) adopted by the S.A.E. Glazing Study Group; and one carried out on large windshields fixed on automobile body openings.

The increase of safety of the new windshield can be roughly rated by extrapolation based on numerous investigations about the behaviour of the post 1966 current windshield during crashes.

LAMINATE CONFIGURATIONS TESTED

All the samples used in this comparative study of performances to headform impact can be classified in three types that differ as far as the kind and thickness of the glass are concerned, as shown by the diagrams of figure 1 and table 1.

The thickness of the outer sheet (0.110 in) is higher than that of the other very thin V.H.R. windshields that have been dealt with in former papers (1) and (2).

Table II gives the mean values and the statistical distribution of the tensile strengths and of the various types of V.H.R. used as well as for ordinary glass.

The coefficient of variation as defined by the comparison of the standard variation with the mean value ranges about 10% in all cases.

The values of the tensile strengths of the ordinary glass and of the V.H.R. glass are obtained by centrally bending circular test on disks of 110 mm diameter loosely displayed on a 100 mm diameter ring. The experimental method together with the mode of calculation have already been described (2).

Table III shows the relative values of stored up energy (per volume unit) to reach rupture (mean value) in each type of glass.

These two types of V.H.R. glass windshields are conceived in order to meet not only with the safety requirements but also with all the criterias imposed by the automobile manufacturers for the mass production of windshields - without the biomechanical considerations - like for instance the optical quality, the easiness to mount...

As far as the so-called conventional windshield is concerned, it distinguishes itself from the current windshield used in U.S. by the thickness of the sheets - both in annealed float - 0.110 in instead of 0.125 in.

All the samples have the same polyvinylbutyral plastic interlayer of a thickness of 0.030 in.

The adhesion to the glass corresponds to an amount of residual moisture ranging between 0.5 and 0.6 % in weight.

FLAT PANELS (Type S.A.E. 24 X 36 in) - IMPACT STUDIES

Before starting the study of the performances of the large curved windshields, it has been thought useful to make a preparatory experimental step with cant flat sam-

320

ples. This was done in order to :

1°) rate the relative level of the biomechanical performances of the two V.H.R. glazings and to compare it with those of the standard glazing, especially from the angle of the danger of laceration;

2°) rate the influence of the variations of the temperature on the resistance to head penetration and on the resulting effects of laceration;

3°) try to discover the effect of the purely mechanical factors like impact localisation point, on the characteristics of the headform impact on the conventional glazing or the V.H.R. ones.

SAMPLES, EXPERIMENTAL DEVICE AND MEASURING EQUIPMENT, CONDITIONS OF THE TESTS - The samples are flat and appear under three configurations showed on figure 1 : conventional laminated glazing, V.H.R. 54 A Glazing, V.H.R. 54 B Glazing. Their size (24 X 36 in) is that of the headform impact test as recommended by the S.A.E. Glazing Study Group since 1963 - 1964 (3).

Before undergoing the impact test at low or high temperature, they stay during several hours in a fridge or in a closet with a temperature regulator at + 2°C.

The experimental device is shown in figure 2.

Its consists of a engine foundation on which a rigid metallic frame, which can be positioned at angle, is mounted.

The glass sample is only fixed on its rim between the two gripping jaws of the frame fastened by ten bolts distributed as shown on figure 3.

The clamping moment of each of them equals 2 mkg in order to counteract the effect of slipping on the edges which will heighten artificially the threshold of tearing and penetration and alter the comparisons made between the tests.

The tightening pressure on the rim of the sample is about 10 kg/cm2.

The headform (figure 4) is rigid and weighs 10 kg. It is similar to the one described in a preceding communication to the 12th Stapp Conference (1) but is equiped with two decelerometers, an axial and a transversal one in order to measure the component forces of the impact in these two directions and to draw its resultant. It is covered with a double layer of wet chamois leather (figure 5) which is renewed each time to enable us to rate the laceration index.

The headform falls freely from various heights necessary to reach the impact velocities. It is possible to reach a maximum velocity of 35 m.p.h.

As shown by figure 6, the headform is guided when it is lifted and when it falls by its outer parts that slide freely along two vertical ropes; the travel of the headform is stopped when the phases of the shock concerning the glass - initial impact and plow-in-are over.

When the headform impacts the glazing, it rolls around its extremities and tilts more and more as it slides on the glass (figure 6).

The measuring instruments comply with the scheme of principe of figure 7.

They also consist of two decelerometers (Kistler model 808 A) and their charge amplifiers, a southern oscillographic recording device with luminous spot and paper sensitized to U.V. on which are recorded (figure 8):
- the deceleration curve in each of the orthogonal

directions

- the evolution of the severity index (after Gadd's method) caused by the total resulting deceleration.

The numerical characteristics of the measurement elements are given in appendix I.

The conditions in which the tests have been run are :

- the impact velocity of the headform is steady and fixed at 20 mph, which corresponds to an impact energy of 40 kgm

- the samples are positioned at 45° off the vertical

- the successive values for the temperature of the samples are - 5°, 0°, 20°, 40°, 50° and 70°C.

MEASUREMENT OF THE LACERATION INDEX -

Measurement method - Several methods are used to pass judgment on the laceration potential of the windshields as they undergo the impact of rigid or articulated dummies.

The most widely spread consists in covering the hard ball simulating the skull with a layer of synthetic material and to stretch above it one or several chamois leathers.

The choice of the materials that can be used to simulate best the subcutaneous tissues is widely discussed and criteria of choice have not yet been defined (4).

The laceration potential is measured by means of the slight scratches, abrasions and cuts caused on the successive layers by the contact and skidding of the headform face against the inner ply.

The method used during this study is a well-tried one. The layer that is in direct contact with the wood is made of a 4 mm thick felt (thus fairly thin). The chamois used to measure the laceration are selected out of a series of skins of 45 X 35 cm (18 X 14 in) the thickness of which ranges between 0.020 in and 0.032 in (0.50 to 0.60 mm) as shown by the histogram of figure 9. Prior to the test, each chamois is dipped into water, then dried by hand.

The top of the headform is covered with two chamois that are stretched so that no wrinkle can appear; these chamois are fastened onto the head by means of a rubber band.

After the impact test, the severity of the laceration is estimated from a laceration index that been defined in terms of the alterations measured on each of the layers.

Laceration rating scale - Several rating scales are used in the researches about laceration measured by means of chamois.

This study will adopt a rating scale where the value of the index ranging from 0 to 10 is linked with the number, depth and length of the deteriorations.

As far as its qualitative definition is concerned, it is identical with the one proposed by other authors (5) and (6).

However, in order to get free from a more or less subjective evaluation, the index is given for each of the chamois on the basis of the total of the measurements of the lengths of the cuts in accordance with the quantitative scale whose details are defined in table IV.

The total index of laceration is equal to the half of the sum of the indexes given to each of the chamois.

By these means, the laceration index measured is the mean value obtained from the sum of the measurement made on each chamois. So that the same importance is devoted to each chamois.

322

RESULTS. (Clamped flat samples 24 X 36 in. Impact velocity 20mph). Impact angle : 45° - The results of the different measurements are shown in figures 10 to 15 and concern the following items :

- deceleration peak versus the temperature: (figure 10)
- severity index versus the temperature (initial impact phase) : (figure 11)
- total severity index versus the temperature (initial impact + plow-in) : (figure 12)
- tearing of the interlayer versus the temperature : (figure 13)
- laceration index versus the temperature:(figure 14)
- laceration index versus the tearing of the interlayer : (figure 15)

The same values are expressed versus the impact location in figures 16 to 20.

On each diagram tendency lines have been drawn that have been adjusted after the method of least squares and based on the estimate of the coefficients of correlation.

On each side of the lines of regression we have drawn the lines at more or less one standard deviation ($\pm$ 1σ) that cover the 2/3 of the distribution that can be supposed normal.

Table V gives the list of all the numerical values recorded and calculated for each impact.

Table VI summarizes the statistical results of the data shown in table V.

The fracture pattern of typical samples in each case together with the state of the two chamois and the corresponding laceration index can be seen on the photographs of figures 21 to 24 next to which an oscillographic record of the impact values can be found : longitudinal deceleration, transversal deceleration and severity index caused by the resultant of both decelerations.

DISCUSSION OF THE RESULTS -

Laceration and rating scale - The rating scale that has been adopted has two limits : a lower one : index 0, an upper one : index 10. This difference needs be taken into account.

When the inner sheet of glass does not rupture, the impact is given a laceration index equal to 0 without any problem of measurement or interpretation. This means the lower limit which is identical with the minimum under which we may not go. In other words, a negative index, for instance - 1, would not have any meaning.

This is however not the case for the upper index (equal to 10), which is not necessarily identical with a maximum of laceration.

When the index is established, as it is the case here, on the base of the measurement of the number, length and depth of the cuts, a figure superior to 10 might be necessary to show the real result.

We must furthermore notice that the upper limit of the scale does'nt state the level of the danger to which one is submitted as far as disfiguration is concerned. It does not give any precision about the degree of reversibility of the damage suffered.

Laceration and changes of temperature - The V.H.R.54 B Glazing distinguishes itself from the conventional glazing and the V.H.R. 54 A Glazing, the two sheets of which break when impacted, by the fact that, about one time out of two, only the outer sheet will rupture whereas the thin inner sheet remains undamaged after having been elastically deformed (values shown in the diagrams by a circled triangle).

This very clear and noticeable difference of behaviour of the V.H.R. 54 B and on occasion only of the V.H.R. 54 A (which recurs with the windshields, as will be seen later) is to be put in relation with the levels of tensile strength and storeable energy as they are mentioned in tables II and III.

Both V.H.R. Glazings prove to be almost indifferent to changes in the temperature ranging from - 5°C (23°F) to 50°C (122°F) at an impact velocity of 20 mph.

In the conditions we have adopted for our tests, the impact of the headform never causes the interlayer of V.H.R. glazings to tear, at any temperature (figure 13). The laceration index is very low, of a mean value ranging between 2 and 3 (figure 14).

On the other hand, the temperature influences in a considerable way the tearing of the interlayer (figure 14) of the conventional glazing, which becomes the more cutting as the tear is larger (figure 15).

It goes without saying that the laceration index is equal to 0 when the inner V.H.R. sheet remains undamaged. When it ruptures, the laceration effect may be very weak but appears even if there is no tearing of the interlayer (figure 15).

When the interlayer tears, the laceration index (as will be seen later in the study of the impact of windshields) is practically independent from the length of the cut.

Figure 10 shows that the deceleration peak is higher at a low temperature and that it tends to decrease as the temperature rises in the case of conventional and V.H.R. Glazings.

At low temperature and in the case of sudden loading like that of an impact, the laminate tends to behave like a monolith. At higher temperature, the layers can slide over one another, which will reduce the rigidity modulus of the whole.

This change does not occur with the V.H.R. 54 B (horizontal tendency line) partly because the data concerning the glazings with unbroken inner sheet have been included here.

Figures 11 and 12 prove that the changes of temperature do not affect in a significant way the severity index during the initial impact and the plow-in for any type of glazing.

The question of compared levels of peak G and severity index will be dealt with in a following part. It brings in another factor than temperature, i.e. the location of the impact.

Its effect, practically equal to 0 for ordinary glass, grows more and more important as the glass is stronger and as the sheets can, consequently, deform more before being broken.

Laceration and mode of fragmentation - As it has just been seen, the level of tensile strength of the glass influences clearly but indirectly the laceration index when the fraction of the initial energy absorbed by the glass is such that tearing does or does not occur in the range of the temperatures used.

The very same factor has a direct influence on the decrease of the lacerative potential since it creates the mode of breakage of the glass. The latter can be characterized by two elements : the size of the fragments (mean value) and the spreading and distribution of the fracture lines in the area covered by the impact and the sliding of the head over the inner sheet.

The fraction of the impact energy absorbed by the glass is directly proportional to the square of the tensile strength.

When rupture occurs, that mass energy is freed and transforms itself into a surface fracture energy.
The fragments are therefore more numerous and tinier as the quantity of the energy absorbed by elastic deformation is more important. It is what happens with V.H.R. glazings (figures 22, 23, 24), whereas the opposite happens with conventional glazings (figure 21).

Next to the difference of fragmentation, the photographs of figure 25 show the appearence of the fracture lines starting from the impact and rupture point, which, incidentally, are always coinciding.

The pattern remains unchanged whether tearing of the interlayer occurs or not. The latter never starts from the impact point but lower, at a variable distance of about 4 or 5 inches.

In the V.H.R. the fracture radiates in very close lines starting from the point of rupture and spreading practically down to the rims of the glazing.

In the ordinary laminated glass, the radial lines are few, relatively distant from each other and intersected by nets of circular fractures originating at a short distance from the breaking point and appearing all along the propagation to the edges of the glazing as can be seen by the high speed movie sequence.

While they are sliding along the glass, just after the initial impact, the head and the face come across these nets of circular fractures at a higher speed as the initial impact velocity is higher and the fraction of energy absorbed by the fracturing of the glass smaller.

This state of things will increase the number and severity of the cuts, namely for all the protruding parts of the face.

If we add that the edges of the fragments of the broken strengthened V.H.R. are relatively blunt compared with those of ordinary glass which are very sharp, it is understandable that this double net of fractures helps to increase the lacerative potential of the ordinary standard windshield.

Laceration, peak G and severity index – The maximum value of deceleration recorded by the headform during impact is not due only to an effect of inertia between the masses of the missile and the target.

This effect is to be seen on the rate of onset which is steeper for thicker ordinary laminated glass as shown by the oscillograms of figures 21 to 24.

The height of the deceleration peak varies accordingly with the ability of deformation of the panel before rupture occurs.

The elastic line results from the interaction of four factors :
- the rigidity to flexion, in the meaning defined by the theory of elasticity, which is proportional to the cube of the thickness

- the location of the impact
- the mounting system
- the tensile strength of the face of the plate opposite to the impact and undergoing a maximum expansion.

As it has been seen in table II, the mean value of tensile strength of the ordinary glass is 14 kg/mm2 and that of the V.H.R. X 28 and V.H.R. X 35 used for the outboard respectively 28 and 35 kg/mm2 for V.H.R. 54 A and V.H.R. 54 B glazings.

Moreover, if we compare the thicknesses of the ordinary laminated and V.H.R. glazings taking into account the thickness of a fictive mechanically equivalent monolith (80% of the real geometrical thickness after disc tests) we can see that the V.H.R. glazings are more than twice less rigid in bending.

Both factors : high tensile strength and low bending rigidity cause an important elastic deflection which can expend a big part of the whole impact energy of 40 kgm developed by the impact of the 10 kg headform at 20 mph.

This explains the higher level of the peak of the V.H.R. glazing and the longer duration of the impact in its initial phase (tables V and VI b, and oscillograms of figures 21 to 24).

When peak G is expressed versus impact location (figure 16) it is noticeable that it increases when the strain moves away from the rim toward the center of the sample.

This evolution occurs with V.H.R. glazing whose deflection calls in more the diaphragm effect next to the simple effect to bending.

The evaluated values of the peak (table VI b) for the mean value of location of the impact at 23 cm off the rim are 82 g with a standard deviation of 14 for ordinary glass and 106 g with a standard deviation of 16 for V.H.R 54 and 54 B together with those whose inner sheet remains unbroken.

Table IV gives the efficient values of the deceleration over a duration of 1 millisecond on each side of the peak.

These values comply with the present data about the tolerance of head impact.

Figures 17 and 18 show that the severity index of the V.H.R. undergoes a greater change than that of peak G with location of the impact point. The value evaluated for initial impact (23 cm off the rim) is 74 (σ = 46) for ordinary glazing and 302 (σ = 87) for V.H.R.

The effect of the plow-in is important in neither of the cases.

The interlayer of the laminated windshield does not tear in dependence with the impact point (figure 19) whereas the laceration index increases considerably as the impact becomes more central.

ANGLE IMPACTS ON CURVED WINDSHIELDS *

The comparative study conducted on curved windshields positioned as they are really in car body openings when they are impacted is aiming at :
1°) evaluating the relative level of the biomechani-

---

* The windshield used is these tests is a cylindrically curved windshield manufactured to equip an American car (A Body Chrysler)

326

cal performances of the two variants of V.H.R. glazing
and compare them with conventional once, namely as far as
the lacerative potential is concerned;

2°) evaluating the influence of the variations of the
impact velocity on the resistance to penetration, the laceration effects and the biomechanical values linked with
the impact deceleration;

3°) discovering the effect due to purely mechanical
factors such as the conditions of the peripheral sealing
on the characteristics of the shock.

SAMPLES, EXPERIMENTAL DEVICE AND MEASURING INSTRU-
MENTS - The means and methods of measuring are identical
to those used for the study of the performances of flat
samples, except for the geometry of the supports and their
attachment.

Two different systems have been used to fix the sample
into the body opening.

In the first case (figure 26) the rim of the curved
windshield is clamped in place by two grupping jaws made
integral by means of a set of 8 bolts similarly to the
flat samples. The clamping moment of each nut is 2 mkg
and the peripheral clamping pressure is about 10 kg/mm2.

In the second ease (figure 27) the curved windshield
sticks to the body opening which is fastened on a rigid
metallic frame that can be positioned at any angle.

Solbit 1107 from Bostik has been used as a sealant.
It is an 8 mm Ø polychloroprene tape containing a vulca-
nizer and a built-in electrical resistance. Before sealing the system, an adherization primer is spread on both
supports.

When the bonding which lasts a minute and consists
in exerting a peripheral pressure to squash the tape during the passage of an electric current, is over, the
adhesion is characterized by a resistance of 15 kg/cm2.

RESULTS - Tables VII, VIII and IX give the list of
all the numerical values recorded and calculated for each
impact on each of the families of conventional, V.H.R.
54 A, V.H.R. 54 B glazings.

In tables VII and VIII the tests have been divided
into two groups in order to stress the peculiar cases
where the impact does not cause any sheet to rupture or
causes only the outer sheet of V.H.R. glazings to break.

The results of the measurements made on the windshields whose both sheets break are shown in figures 28
to 33 and concern the following values.

- Deceleration peak versus velocity (figure 28)
- Severity index versus velocity (initial impact
phase) (figure 29)
- Total severity index versus velocity (initial
impact phase + plow-in) (figure 30)
- Tearing of the interlayer versus velocity (figure
31)
- Laceration index versus velocity (figure 32)
- Laceration index versus tearing of the interlayer
(figure 33)

On each diagram, tendency lines adjusted after the
method of least quares and based on the evaluation of the
correlation coefficients have been drawn.

On each side of the lines or regression are the lines
standing at more or less one standard deviation ($\pm 1\sigma$)
which cover the 2/3 of the distribution that is supposed
to be the normal one.

Table X sums up the statistical results on which the

327

tendencies shown in figures 28 to 33 have been drawn; the different families have been grouped where the statistical tests do not enable us to distinguish them (6th column).

The pattern of the fracture of standard samples and the state of the two chamois together with the corresponding laceration index can be seen on the photographs of figures 34 to 40. Opposite is the oscillographic record of the impact characteristics : axial deceleration, transversal deceleration and severity index caused by the resultant of both.

DISCUSSION OF THE RESULTS -

Influence of the windshield installation system -
The statistical study reveals that the clamped and bonded windshields behave in a fairly identical way, whether they be conventional of V.H.R. ones.

The peripheral pressures (at least equal to 10 kg/cm2) exerted in the test with clamped windshields give a similar result to that obtained from windshields sealed with the Bostik tape.

The different systems cause but a small peripheral dislodging when impacted at velocities up to 30 mph (impact energy : 90 kgm).

Influence of the mechanical resistance of glass in the performances of the V.H.R. variants at increasing impact velocities -
1. Tables VIII and IX show a higher frequency of cases where one of the two sheets remains unbroken for variant V.H.R. 54 B, i.e. for the more resisting V.H.R. as well as with flat samples.

The excessive values of the peak of resultant deceleration, the efficient deceleration during 2 and 3 milliseconds, or the severity index have only been obtainable in tests 14 and 102. These values seem to be excessive for human tolerance. Tests 14 and 102 demonstrate that two sheets remain unbroken at high velocities, i.e. at high impact energies (63 kg.m).

The mechanical resistance of the outer ply of variant V.H.R. 54 B must be limited at a lower level than the adepted one (35 kg/mm2). A reducing of about 7 kg/mm2 to reach the level of the outer sheet of the V.H.R. 54 A (28 kg/mm2) would thus be advisable if we consider that the outer sheet of the V.H.R. 54 A systematically breaks at 20 mph.

2. If we compare the cases where both sheets rupture when impacted, we see
- that the deceleration values and the severity index are always tolerable and only a bit higher for the V.H.R. 54 B than for the V.H.R. 54 A
- that the risks of penetration and laceration are almost similar at high velocities (figures 31-32)
- that the rupture of the inner sheet always occurs some milliseconds after the outer sheet has ruptured.

Comparison of the behaviours of conventional and V.H.R. windshields versus velocity -
a) Effects of the velocity on the laceration created.
Figure 32 demonstrate that the laceration index increases with the velocity for conventional windshields as for V.H.R. ones; this is due to the fact that the headform scrapes the windshield at a higher velocity.

The laceration levels are clearly different for V.H.R and conventional windshields because of the mode of fragmentation as was explained in the study of flat samples.

Figures 34, 35, 38 and 39 show what generally happens at speeds of 20 and 25 mph respectively.

Figures 36, 37 and 40 show what happens when one of the sheets at least remains unbroken.

b) Influence of the velocity on the deceleration peak and on the severity index.

The increase of velocity does not influence greatly the deceleration peak or the severity index of conventional windshields.

A slight tendency upwards appears for the deceleration peak when V.H.R. windshields are tested. The severity index, on the other side, is not influenced.

The evaluated levels of the deceleration peaks are rated at 114 (σ = 25) for conventional and at 141 (σ = 20) for V.H.R. windshields.

The severity index of the initial impact is 144 (σ = 88) for conventional and 389 (σ = 121) for V.H.R. windshields.

The total severity index is 239 (σ = 90) against 470 (σ = 123).

The coefficient of variation is twice higher for conventional windshields.

Figures 34, 35, 38 and 39 show typical examples.

CONCLUSIONS

This study has enabled us to make clearer the influence of the different factors that condition the lacerative injury potential of a laminated windshield.

The characteristics of a new type of windshield have been defined to minimize considerably and even avoid the danger of laceration at pretty high impact energies while at the same time maintaining severity index values below the injury threshold level.

The tensile strength of the very thin inner sheet of the windshield must be very high.

The level and dispersion of the tensile strength of the glass used for the outer ply can be adjusted to limit the impact energy and the severity index to values that have been established at the start.

When evaluated on the basis of currently accepted safety parameters and within the confines of the experimental program considered in this report, this windshield appears as a consistent approach to the best compromise which has to meet the safety requirements as well as numerous other demands of a mass-production windshield.

More data are needed to assess the ability of the new windshield to be considered as a part of a passive restraint system.

REFERENCES

1. R. Van Laethem, "A new high safety glazing for automobile and other vehicles". Proceedings twelfth Stapp Conference Detroit, Michigan, October 1968. Paper 680789.

2. E. Plumat, P. Eloy, L. Leger, F. Toussaint and R. Van Laethem, "Safety improvement of the new laminated V.H.R. glazing for cars." 1970 International Automobile Safety Conference Compendium. New-York : Society of automotive engineers. Inc. 1970. Paper n° 700429, pp. 1152-1170.

3. Proposed S.A.E. recommended practice impact test

for use in development of glazing materials.

4. J. Brinn, "Two anthropometric test forms. The frontal bone of the skull and a typical facial bone." Proceedings of the thirteenth Stapp Conference; New-York: Society of automotive engineers. Inc 1969. pp.381-399.

5. J. R. Blizard and J. S. Howitt, "Development of a safer non lacerating automobile windshield." Paper 690484 presented at SAE Mid-Year meeting, Chicago Illinois May 1969.

6. L. M. Patrick, K. R. Trosien and F. T. Dupont, "Safety performance of a chemically strengthened windshield." Paper 690485, presented at SAE mid-year meeting Chicago,Illinois. May 1969.

APPENDIX 1

Instrumentation for deceleration measurement

1. Accelerometers. Kistler model 808 A (piezoelectric crystal)

a) Resonant frequency : 40.000 Hz

b) Frequency response : near DC to 8.000 Hz

c) Transverse sensitivity (max) : 5%

2. Accelerometer cable : Kistler model 1603 SP Low noise (oil damped)

a) Insulation resistance : 10^{15} ohms x m.

b) Capacity : 70 pf/m.

3. Amplifier : Kistler model 568

a) Input impedance : 10^{14} ohms

b) Linearity : ± 0,1 %

c) Frequency range : 0 - 150.000 Hz

d) Sensitivity : 25 g/Volt

4. Recording device : Ultraviolet oscillograph, Southern model M 1330 AB

a) Acceleration curves : SMI/M Southern galvanometers, frequency response of 0 - 1000 Hz, sensitivity : 1 V/cm

b) Severity index : SMI/N Southern galvanometer, frequency response of 0 - 500 Hz, sensitivity : 1 V/cm.

TABLE I -

Configurations

Kind of glazing	Inner sheet (receiving the impact)	Interlaying in polyvinylbutyral	Outer sheet
Conventional laminated glazing used as reference Total thickness : 0.250 in - 6,4 mm	Regular float 0.110 in (2,8 mm)	0.030 in (0,76 mm)	Regular float 0.110 in (2,8 mm)
VHR 54 A Glazing Thin laminated glass with VHR glass Total thickness : 0.190 in - 4,8 mm	VHR . X 40.. 0.050 in (1,3 mm)	0.030 in (0,76 mm)	Float VHR. X 28 0.110 in (2,8 mm)
VHR 54 B Glazing Thin laminated glass with VHR glass Total thickness : 0.190 in - 4,8 mm	VHR . X 50 0.050 in (1,3 mm)	0.030 in (0,76 mm)	Float VHR. X 35 0.110 in (2,8 mm)

TABLE II - Tensile strengths of glasses (in kg/mm2)

	Conventional Glazing Annealed Float Glass	V.H.R. 54 A Glazing	V.H.R. 54 B Glazing
Glass for the inner sheet	mean : 14 $11.2 < 95\%$ of the distribution < 16.8	mean : 40 $32 < 95\% < 48$	mean : 50 $40 < 95\% < 60$
Glass for the outer sheet	id.	mean : 28 $22.4 < 95\% < 33.6$	mean : 35 $28 < 95\% < 42$

Table III : Energy that can be stored up by volume unit (relative values, arbitrary units).

	Conventional Glazing	VHR 54 A Glazing	VHR 54 B Glazing
Glass for the inner sheet	1	8	12
Glass for the outer sheet	1	4	6

TABLE IV : Rating scale of the chamois lacerative damage.

Laceration Index (1)	Degree (2)	Outer Chamois		Inner Chamois		Cuts of inner-layer material and headform face (7)
		Type of cuts (3)	Total length of cuts (through only) (4)	Types of cuts (5)	Total length of cuts (through only) (6)	
0	-	None	-	None	-	None
1	Minimal	Abrasions cuts to 3/4 in, none through	-	None	-	None
2	Minor	Abrasions cuts over 3/4 in, none through	-	None	-	None
3	Minor	As (2) above, but one 3/4 in cut through	0 - 25 mm	Abrasions	Abrasions	None
4	Moderate	2 or 3 3/4 in cuts through	25 - 75 mm	Cuts but not through	Abrasions	None
5	Moderate	Numerous cuts *	75 - 150 mm	Only one cut through to 3/4 in	0 - 25 mm	None
6	Severe	Numerous cuts	150 - 225 mm	2 or 3 cuts through to 1½ in	25 - 75 mm	None
7	Severe	Numerous cuts	225 - 300 mm	Numerous cuts *	75 - 125 mm	Abrasions
8	Severe	Numerous cuts	300 - 375 mm	Numerous cuts	125 - 175 mm	Cuts up to 1/32 in deep and 3/4 in long
9	Very severe	Numerous cuts	375 - 450 mm	Numerous cuts	175 - 250 mm	One cut deeper or longer than (8)
10	Very severe	Numerous cuts	> 450 mm	Numerous cuts	> 250 mm	Numerous cuts worse than (9)

* "Unlimited cuts" as mentioned in the original scale (5) has been replaced here by "numerous cuts" because of the quantitative evaluation defined in col (4) and (6)

TABLE V - Results of the measurements of the impacts on clamped flat 24 X 36 in samples

Impact velocity of the 10 kg headform : 20 mph

Test n°	Conditions of impact				Initial impact					
	Temp. °C	Temp. °F	Localisation of impact cm	Pull out cm	Axial Peak g	Trans Peak g	Result Peak g	Efficient decel. (2 msec)* g	Total Duration msec	Severity Index
Conventional windshield O. 1	-5	33	24,5	0	60	58	84	42	2	60
2	-5	23	25,0	0	60	53	80	40	2	77
4	0	32	18,0	0	75	70	103	54	3,5	150
5	20	68	18,5	0	65	50	82	60	1,5	40
6	20	68	19,0	0	69	58	90	67	1,5	35
7	20	68	24,0	0	68	60	85	42	2,7	85
8	20	68	22,0	0	72	61	94	43	2,3	50
9	40	104	21,5	0	60	56	82	64	3,5	135
10	40	104	16,0	0	69	55	89	57	3,2	130
11	50	122	22,0	0	43	40	59	44	1,5	35
12	70	158	23,0	0	40	36	54	41	1,5	17
V.H.R. 54 A Glazing I. 3	0	32	24,5	0	80	80	113	82	9,5	340
4	0	32	22	0	80	72	113	78	9,0	290
5	20	68	25,5	0	77	68	102	80	9,6	335
6	20	68	26,5	0	84	84	118	89	9,5	480
7	20	68	23	0	81	76	112	94	10,0	325
8	20	68	21,5	0	72	63	96	68	7,7	200
9	40	104	23,5	0	72	66	98	86	8,5	265
10	40	104	26,0	0	74	72	103	87	8,0	275
11	50	122	22,5	0	68	58	89	85	7,5	215
V.H.R. 54 B Glazing II. 1	-5	23	27,0	0	73	70	103	93	9	417
3	0	32	23,0	0	90	90	147	100	12	550
4	0	32	27,0	0	88	82	120	102	10	515
5	20	68	22,5	0	82	73	110	92	16,2	210
6	20	68	25	0	75	69	102	86	13,4	320
7	20	68	20,5	0	-	-	-	-	-	-
8	40	104	23,5	0	72	68	99	95	8,5	300
10	50	122	32	0	112	119	163	87	5	545

* Efficient deceleration calculated for 1 msec. on each side of the max peak.

(cont'd)

333

TABLE V (continued)

Plow-In					Total				
Axial Peak	Trans Peak	Result Peak	Total Duration	Severity Index	Total Duration of shock	Total Severity Index	Tear of inter-layer	Lace-ration Indexes	Remark
g	g	g	mm		msec	-	cm	-	
25	22	34	19	38	21	98	71	10	
17	15	23	23	20	25	97	80	10	
30	20	36	22	50	25,5	200	58	8	
25	12	28	30	36	31,5	78	54	7	
27	12	30	28	42	29,5	77	44	6	
30	15	33	30	57	32,7	142	40	7	
25	10	27	35	40	37,3	90	51	6	
25	10	27	42	50	45,3	185	60	7	
22	10	25	38	38	41,2	168	64	7	
-	-	-	-	0	15	35	71	9	
-	-	-	-	0	15	17	94	10	
25	15	29	30	45	39,5	385	0	3	
25	15	29	32	48	41	338	0	4	
20	12	24	28	35	37,6	370-	0	1	
25	15	29	28	40	37,5	520	0	2	
20	12	24	34	68	44,0	393	0	3	
25	7	26	34	55	41,7	255	0	4	
20	10	23	37	30	45,5	295	0	3	
10	10	15	30	25	38	300	0	2	
10	10	15	30	25	37,5	240	0	1	
20	13	24	35	30	44	447	0	2	inner sheet unbroken
20	15	25	30	25	42	575	0	0	
15	10	18	30	12	40	527	0	2	inner sheet unbroken
-	-	-	-	0	16,2	210	0	0	inner sheet unbroken
-	-	-	-	0	13,4	320	0	0	
-	-	-	-	-	-	-	16	4	
15	10	18	36	15	44,5	315	0	3	
20	10	23	19	15	24	560	0	4	

TABLE VI - Statistical results of the data given in Table V. Flat samples 24 X 36 in (Clamped).

Impact speed 20 mpn. Impact angle 45°.

(a)

$Y = f(x)$	Conventional			VHR 54 A Glazing			VHR 54 B Glazing		
	(1)	(2)	(3) b	Y	(2)	b	Y	(2)	b
Peak G = f (Temperature)	84	11	− 0.42	106	7	− 0.41	121	25	0
Initial impact S I = f (T)	74	46	0	303	83	0	408	135	0
Total S.I. = f (T)	108	59	0	344	86	0	422	142	0
Tearing [x = −5° to x = 20°] = f (T)	47	7.3	− 1	0		0			
Tearing [x = 20° to x = 70°] = f (T)	46	5.1	0.89	0			1.0	0.2	0
Laceration I [x = −5° to x = 20°] = f (T)	6.45	0.68	− 0.13	2.56	1.13	0	2	0.83	0
Laceration I [x = 20° to x = 70°] = f (T)	6.33	0.64	0.07						
L.I. = f (Tear)	7.69	0.91	0.09				no tear		

(b)

$Y = f(x)$	Conventional			VHR 54 A Glazing			VHR 54 B Glazing		
	(1)	(2)	(3) b	Y	(2)	b	Y	(2)	b
Peak G = f (local. of impact)	82	14	0	106		0	16	4.12	
S.I. (Init. impact) = f (L.I.)	74	46	0	302		0	87	31.3	
Total S.T. = f (local.)	108	59	0	334		0	90	29.3	
Tearing = f (L)	62	16	0	1.0		0	0.2	0	
L.I. = f (L)	8.44	1.38	0.30	2.29		0	0		

(1) estimate Y for the given value of x (T = 20°C)
(2) standard deviation (L = 23cm)
(3) slope of the line of regression
(4) x = tear of 60 cm

335

TABLE VII – Conventional windshields (*)

Temperature 20°C – Impact angle 40°

Test n°	Mode of fixation	Conditions of impact			Initial impact						
		Velocity	Pull out	Localisation of impact	Axial Peak	Trans Peak	Result Peak	Efficient decel.		Total Duration	Severity Index
								*	**		
		mph	cm	cm	g	g	g	g	g	msec	-
3	Cl	20	0	15	110	85	139	61	48	2.2	315
4	Cl	20	0	15	111	80	136	60	41	1.8	170
5	Cl	20	0	14	66	69	95	45	35	2.2	110
6	Cl	20	0	20	65	68	94	45	33	2.1	77
11	Cl	25	0	17	63	65	91	40	22	1.4	50
12	Cl	25	0	20	79	75	103	41	22	1.3	50
13	Cl	25	0	18	90	99	134	60	50	2.3	200
14	Cl	25	0	20	103	95	140	85	68	2.9	315
18	Cl	30	0	15	104	105	148	57	40	1.6	210
19	Cl	30	0	18	118	112	162	65	43	1.6	250
101	Bo	20	0	26	84	78	115	49	33	1.7	130
102	Bo	20	0	13	68	73	105	46	32	1.8	150
103	Bo	20	0	17	74	83	112	53	35	1.9	180
104	Bo	25	0	14	47	54	72	20	13	1.1	35
105	Bo	25	0	13	60	57	85	24	17	1.2	42
106	Bo	25	0	11	67	70	97	37	26	1.6	125
107	Bo	30	20	20	76	63	100	59	27	1.6	125
108	Bo	30	14	19	92	89	128	58	24	1.2	67

(*) A Body Chrysler Valiant

* Efficient deceleration calculated for 1 msec on each side of the max peak

** Efficient deceleration calculated for 1,5 msec on each side of the max peak

(cont'd)

TABLE VII (continued)

Plow-In					Total			
Axial Peak	Trans Peak	Result Peak	Total Duration	Severity Index	Total Duration of shock	Total Severity index	Tear of interlayer	Laceration Index
g	g	g	msec	-	msec	-	cm	-
37	6	38	42	115	44.2	430	1	6
35	5	36	44	102	45.8	272	0	6
25	7	26	37	38	39.2	148	2	8
20	15	25	38	41	40.1	118	5	6
37	15	40	33	100	34.4	150	0	7
37	12	40	28	83	29.3	133	14	9
35	15	39	28	80	30.3	280	6	7
35	8	37	32	80	34.9	395	24	9
30	17	35	28	65	29.6	275	63	10
37	15	41	28	95	29.6	345	52	10
28	12	30	32	48	33.7	178	7	6
23	13	27	63	75	64.8	225	2	9
20	15	26	55	56	56.9	236	4	8
30	20	36	37	85	38.1	120	30	9
35	17	39	65	185	66.2	227	35	9
30	15	34	75	150	76.6	275	36	10
39	15	43	32	125	33.6	250	43	9
50	15	56	28	185	29.2	252	45	9

337

TABLE VIII - V.H.R. 54 A Windshields

Temperature 20°C - Impact angle 40°

Test n°	Mode of fixation	Velocity mph	Pull out cm	Localisation of impact cm	Axial Peak g	Trans Peak g	Result Peak g	Efficient decel. g *	Efficient decel. g **	Total Duration msec	Severity Index -
								Initial impact			
4 a	Cl	20	0	18	58	62	86	60	49	3.5	265
3 b	Cl	20	0	14.5	100	92	140	103	80	5.2	400
8 b	Cl	20	0	18	99	99	140	103	87	5.4	465
11 b	Cl	20	0	17	94	84	126	83	65	5.0	335
12 b	Cl	20	0	12.5	100	99	141	112	90	5.4	465
13 b	Cl	20	0	18.5	103	91	137	105	92	5.5	485
8 a	Cl	25	0	18	89	99	133	81	68	3.8	385
9 a	Cl	25	0	18	114	119	165	100	85	4.3	440
9 b	Cl	25	0	12	121	90	150	85	66	2.7	310
10 b	Cl	25	0	10	138	97	169	112	93	3.8	460
16 b	Cl	25	0	12	109	94	144	95	72	3.1	295
14 a	Cl	30	0	17	85	98	129	75	49	2.4	310
16 a	Cl	30	0	20	126	113	169	117	103	3.9	450
101	Bo	20	20	16	83	73	110	68	40	4.6	450
108	Bo	25	0	10	106	93	141	93	70	3.1	475
104	Bo	25	0	14	118	82	144	83	58	2.8	290
110	Bo	30	0	12.5	102	70	124	60	35	1.7	215
111	Bo	30	0	12.5	120	94	148	85	72	3.0	365
102	Bo	20	0	-	80	105	132	106	94	9.2	260
103	Bo	20	0	-	105	111	154	135	124	10.0	445
105	Bo	25	0	-	101	141	172	145	121	8.5	565

Conditions of impact — Initial impact

* Efficient deceleration calculated for 1 msec
on each side of the max peak
** Efficient deceleration calculated for 1,5 msec
on each side of the max peak

(cont'd)

TABLE VIII (continued)

Plow-In

Axial Peak G	Trans Peak g	Result Peak	Total Duration msec	Severity Index	Total Duration of shock msec	Total Severity Index	Total Tear of interlayer cm	Laceration Index	Remark
25	11	28	47.0	58	50.5	323	0	4	
19	5	20	45.0	24	50.2	424	0	6	
23	10	25	38.0	39	43.4	504	0	1	
30	8	31	46.5	84	51.5	419	0	1	
25	10	27	49.0	41	54.4	506	0	3	
28	10	30	37.0	55	42.5	540	0	3	
37	20	42	32.0	108	35.8	493	0	5	
49	10	52	23.0	140	27.3	580	0	5	
39	13	41	45.0	149	47.7	459	36	5	
43	10	44	46.0	171	49.8	631	14	3	
38	14	41	46.5	240	49.6	535	11	3	
45	20	49	28.0	139	30.4	449	17	5	
12	5	13	18.0	8	21.9	458	40	5	
30	0	30	16.0	25	20.6	475	0	4	
37	15	41	37.0	120	40.1	595	20	5	
47	9	48	51.0	235	53.8	525	10	3	
40	10	41	13.0	40	14.7	255	57	5	
40	10	41	28.0	95	31.0	460	60	5	
-	-	-	-	-	9.2	260	0	0	2 sheets unbroken
-	-	-	-	-	10.0	445	0	0	2 sheets unbroken
-	-	-	-	-	8.5	565	0	0	inner sheet unbroken

TABLE IX - V.H.R. 54 B Windshields

Temperature 20°C - Impact angle 40°

Test n°	Conditions of impact				Initial impact						
	Mode of fixation	Velocity	Pull out	Localisation of impact	Axial Peak	Trans Peak	Result Peak	Efficient decel.		Total Duration	Severity Index
		mph	cm	cm	g	g	g	g *	**	msec	-
4	Cl	20	0	19	83	86	120	73	55	3.8	250
6	Cl	20	0	18	91	119	150	130	15	5.8	595
13	Cl	25	0	15.5	112	101	150	90	75	3.5	345
17	Cl	30	0	17	89	69	113	49	33	1.5	165
18	Cl	30	0	16	125	100	159	103	55	4.2	450
21	Cl	30	0	18.5	144	126	188	137	113	5.2	610
1002	Bo	20	0	16.5	78	76	104	83	52	4.6	415
1003	Bo	25	0	16.5	91	114	146	102	80	4.8	450
106	Bo	25	45	18	89	94	129	95	52	9.2	150
108	Bo	30	0	19	145	112	180	98	82	2.7	600
1	Cl	20	0	-	97	116	150	142	125	9.6	635
2	Cl	20	0	-	93	105	141	138	120	10.0	625
5	Cl	20	0	-	84	112	140	138	110	9.2	325
7	Cl	20	0	-	85	114	141	137	125	9.2	350
8	Cl	20	0	-	86	115	144	138	120	8.8	345
9	Cl	20	0	-	95	116	150	142	122	10.0	400
14	Cl	25	0	-	119	175	211	203	170	8.1	775
101	Bo	20	0	-	93	112	146	133	120	10.0	385
103	Bo	20	0	-	84	104	134	110	94	10.0	310
105	Bo	25	87	-	109	93	143	104	71	3.0	200
102	Bo	20	0	-	121	139	184	170	125	8.5	570

* Efficient deceleration calculated for 1 msec
  on each side of the max peak

** Efficient deceleration calculated for 1,5 msec
  on each side of the max peak

(cont'd)

TABLE IX (continued)

	Plow-In					Total				
Axial Peak	Trans Peak	Result Peak	Total Duration	Severity Index		Total Duration of shock	Total Severity Index	Tear of inter-layer	Lace-ration Index	Remark
g	g	g	msec	-		msec	-	cm	-	
20	10	23	39	32		42.8	282	0	2	
17	7	18	20	10		25.8	605	0	3	
27	15	32	31	53		34.5	398	48	5	
37	22	43	13	46		14.5	213	64	-	
35	10	37	25	68		29.2	518	13	5	
15	0	15	14	5		19.2	615	38	6	
28	10	30	44	65		48.6	480	0	2	
40	10	41	39	145		43.8	585	32	5	
25	10	27	40	50		49.2	200	33	5	
32	12	34	21	42		23.7	642	68	5	
-	-	-	-	-		9.6	635	0	0	2 sheets unbroken
-	-	-	-	-		10.0	625	0	0	"
-	-	-	-	-		9.2	325	0	0	"
-	-	-	-	-		9.2	350	0	0	"
-	-	-	-	-		8.8	345	0	0	"
-	-	-	-	-		10.0	400	0	0	"
-	-	-	-	-		8.1	775	0	0	"
-	-	-	-	-		10.0	385	0	0	"
-	-	-	-	-		10.0	310	0	0	2 sheets unbroken
-	-	-	-	-		3.0	200	0	0	inner sheet unbroken
-	-	-	-	-		8.5	570	0	0	2 sheets unbroken

TABLE X - Statistical results of the data given in Tables VII - VIII - IX. Clamped and bonded windshields - Temperature 20°C

Y = f (X)	Conventional Clamped + Bonded			VHR 54 A Clamped (2 sheets broken)			VHR 54 A Bonded (2 sheets broken)			VHR 54 B Clamped (2 sheets broken)			VHR 54 B Bonded (2 sheets broken)			VHR 54 A + B Clamped + Bonded (2 sheets broken)		
	$Y_{(1)}$	$\sigma_{(2)}$	$b_{(3)}$	Y	σ	b	Y	σ	b	Y	σ	b	Y	σ	b	Y	σ	b
Y = Peak G (g units) X = Impact velocity = 25 mph	114	25	0	141	22	0	133	16	0	147	27	0	140	9	3.80	141	20	1.33
Y = SI (Initial impact) X = Impact velocity = 25 mph	144	88	0	390	77	0	359	109	0	402	182	0	404	187	0	389	121	0
Y = Total SI X = Impact velocity = 25 mph	239	90	0	486	78	0	462	127	0	438	169	0	477	196	0	470	123	0
Y = Tearing (cm) X = Impact velocity = 25 mph	24	10	2.32	14	10	1.38	23	10	3.13	27	27	0	33	1	3.40	22	14	2.22
Y = Laceration Index X = Impact velocity = 25 mph	8.38	1.1	0.13	4.05	1.3	0.12	4.40	0.9	0	4.20	0.8	0.15	4.25	1.1	0.15	4.13	1.0	0.12
Y = Laceration index X₁ = tear < 10 cm	7.00	1.1	0	3.25	1.9	0	(*)	-	-	(*)	-	-	(*)	-	-	3.17	1.4	0
X₂ = tear ≥ 10 cm	9.33	0.5	0	4.20	1.0	0	4.50	1.0	0	5.00	0	0	5.00	0	0	4.60	0.8	0

(1) Estimate Y for the given value of X.
(2) Standard deviation
(3) Slope of the line of regression

(*) N < 3 : no calculation

342

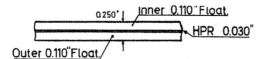

Inner 0.110" Float.
0.250'
Outer 0.110"Float.
HPR 0.030"

Conventional GLAZING

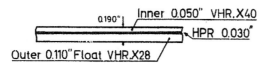

Inner 0.050" VHR.X40
0.190"
Outer 0.110"Float VHR.X28
HPR 0.030"

VHR 54A-GLAZING

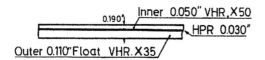

Inner 0.050" VHR.X50
0.190'
Outer 0.110"Float VHR.X35
HPR 0.030"

VHR 54B-GLAZING

Fig. 1 - Cross section of laminate con-figurations

Fig. 2 - Drop test facility and re-cording instrumentation. Flat sample clamped

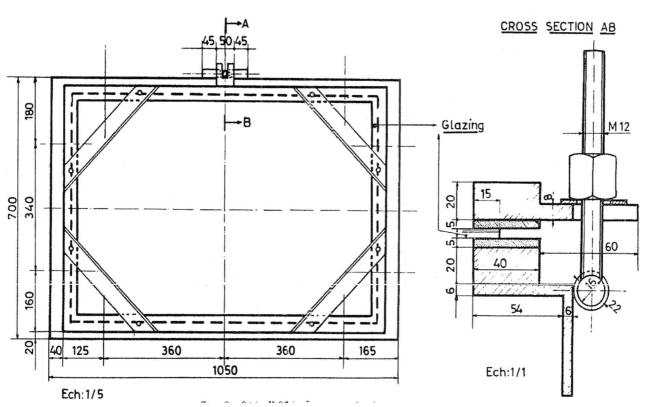

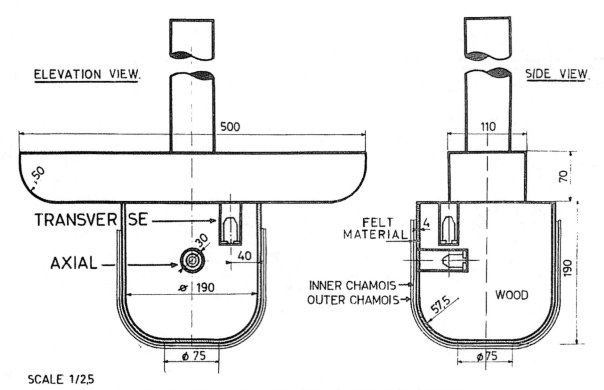

ELEVATION VIEW.

SIDE VIEW.

500

110

50

TRANSVERSE

FELT MATERIAL

4

70

AXIAL

30

40

INNER CHAMOIS
OUTER CHAMOIS

190

ϕ 190

57,5

WOOD

ϕ 75

ϕ 75

SCALE 1/2,5

Fig. 4 - 10 kg headform with axial and transverse decelerometers

Fig. 5 - Headform covered by 2 chamois skins

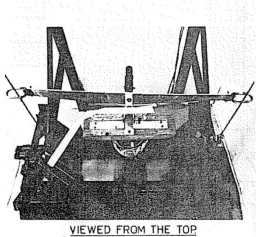

VIEWED FROM THE TOP

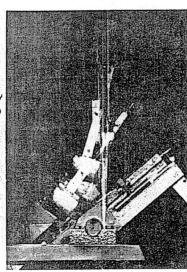

Fig. 6 - Successive headform positions during impact

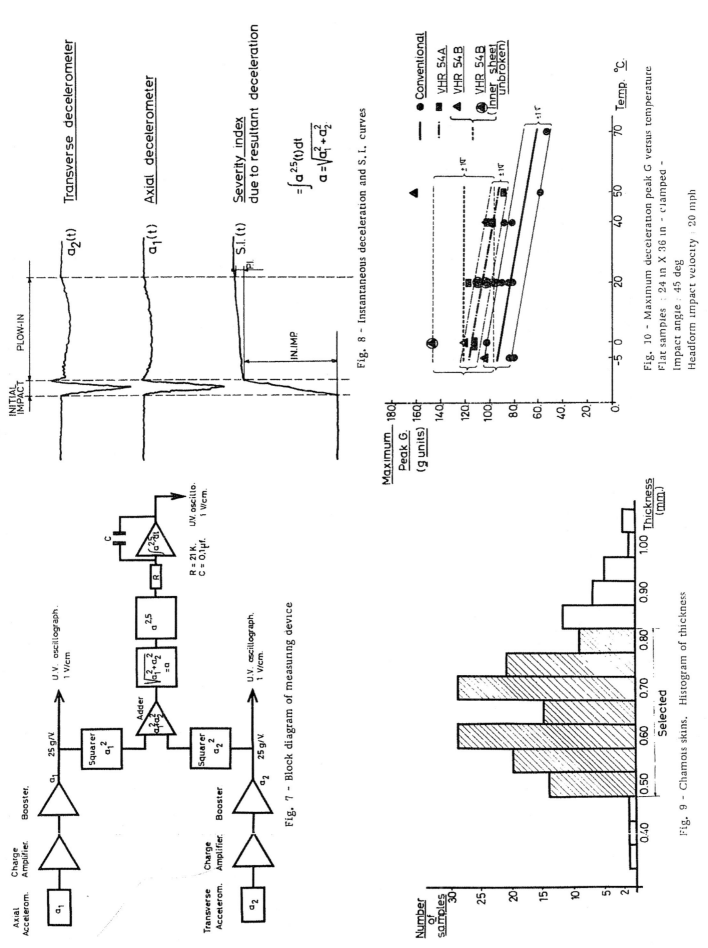

Transverse decelerometer

$a_2(t)$

Axial decelerometer

$a_1(t)$

Severity index due to resultant deceleration

S.I.(t)

$= \int a^{2.5}(t)dt$

$a = \sqrt{a_1^2 + a_2^2}$

Fig. 8 - Instantaneous deceleration and S.I. curves

Axial Accelerom. a_1 — Charge Amplifier. — Booster. — a_1 — 25 g/V — U.V. oscillograph. 1 V/cm.

Squarer a_1^2

Adder $a_1^2 a_2^2$ — $\sqrt{a_1^2 + a_2^2} = a$ — $a^{2.5}$ — R. — $\int a^{2.5} dt$ — U.V. oscillo. 1 V/cm.

C

R = 21 K.
C = 0.1 μf.

Squarer a_2^2

Transverse Accelerom. a_2 — Charge Amplifier. — Booster — a_2 — 25 g/V — U.V. oscillograph. 1 V/cm.

Fig. 7 - Block diagram of measuring device

Maximum Peak G. (g units)

Fig. 10 - Maximum deceleration peak G versus temperature

Flat samples : 24 in X 36 in - clamped -
Impact angle : 45 deg
Headform impact velocity : 20 mph

● Conventional
■ VHR 54A
▲ VHR 54 B
◉ VHR 54 B (Inner sheet unbroken)

Number of samples

Fig. 9 - Chamois skins. Histogram of thickness

345

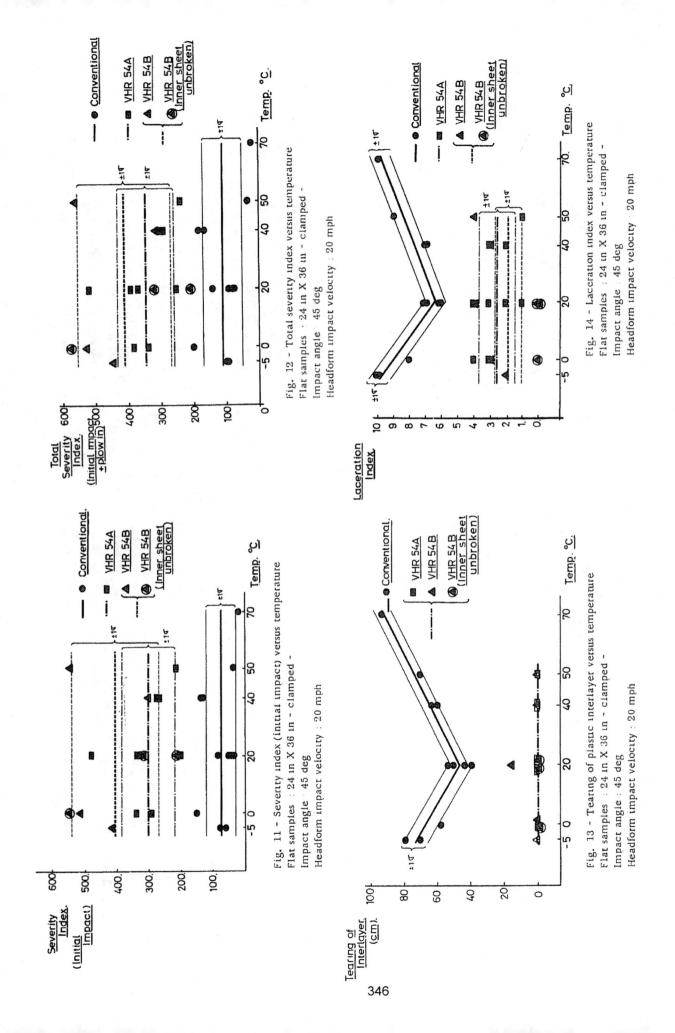

Fig. 11 - Severity index (initial impact) versus temperature
Flat samples : 24 in X 36 in - clamped -
Impact angle : 45 deg
Headform impact velocity : 20 mph

Fig. 12 - Total severity index versus temperature
Flat samples : 24 in X 36 in - clamped -
Impact angle : 45 deg
Headform impact velocity : 20 mph

Fig. 13 - Tearing of plastic interlayer versus temperature
Flat samples : 24 in X 36 in - clamped -
Impact angle : 45 deg
Headform impact velocity : 20 mph

Fig. 14 - Laceration index versus temperature
Flat samples : 24 in X 36 in - clamped -
Impact angle : 45 deg
Headform impact velocity : 20 mph

346

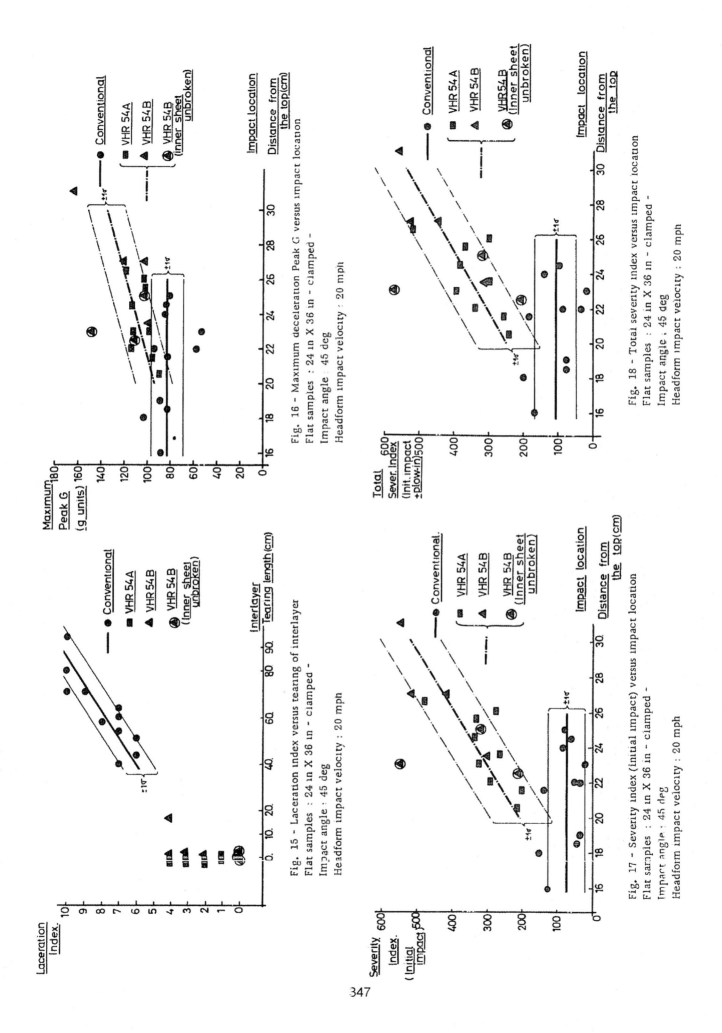

Fig. 15 - Laceration index versus tearing of interlayer
Flat samples : 24 in X 36 in - clamped -
Impact angle : 45 deg
Headform impact velocity : 20 mph

Fig. 16 - Maximum deceleration Peak G versus impact location
Flat samples : 24 in X 36 in - clamped -
Impact angle : 45 deg
Headform impact velocity : 20 mph

Fig. 17 - Severity index (initial impact) versus impact location
Flat samples : 24 in X 36 in - clamped -
Impact angle : 45 deg
Headform impact velocity : 20 mph

Fig. 18 - Total severity index versus impact location
Flat samples : 24 in X 36 in - clamped -
Impact angle : 45 deg
Headform impact velocity : 20 mph

347

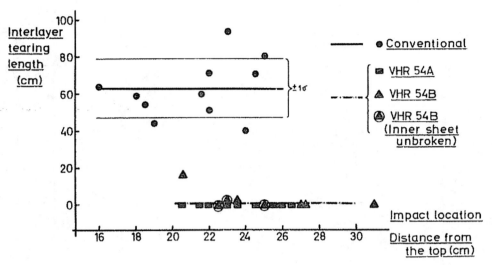

Fig. 19 - Interlayer tearing length versus impact location
Flat samples : 24 in X 36 in - clamped -
Impact angle : 45 deg
Headform impact velocity : 20 mph

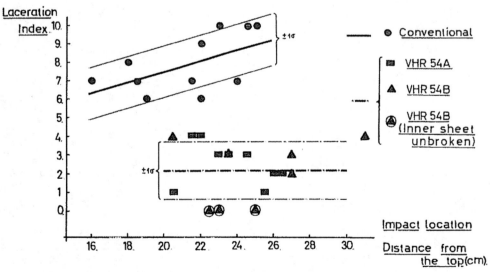

Fig. 20 - Laceration index versus impact location
Flat samples : 24 in X 36 in - clamped -
Impact angle : 45 deg
Headform impact velocity : 20 mph

348

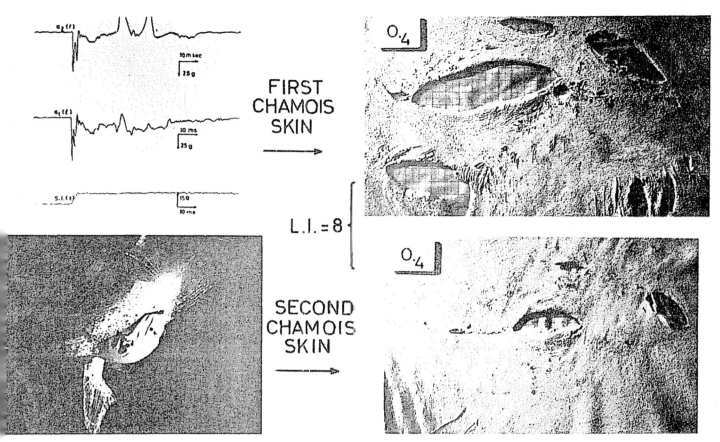

Fig. 21 - Conventional glazing - typical oscillograph record, fracture pattern, and chamois lacerative damage
24 in X 36 in   Temp. : 0 C   20 mph

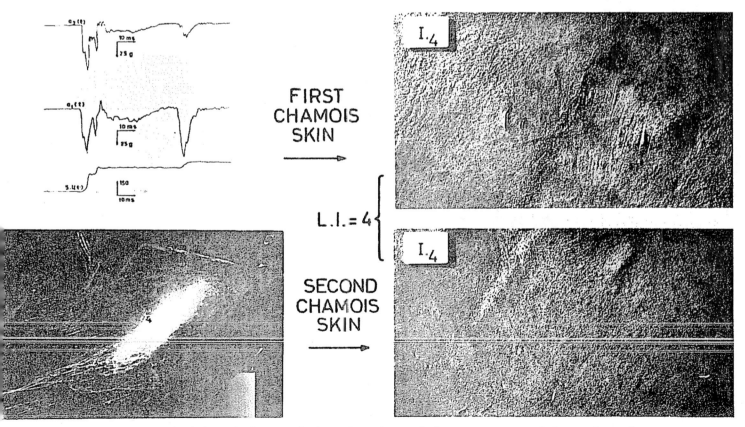

Fig. 22 - VHR 54 A - both sheets broken - typical oscillograph record, fracture pattern, and chamois lacerative damage
24 in X 36 in   Temp. : 0 C   20 mph

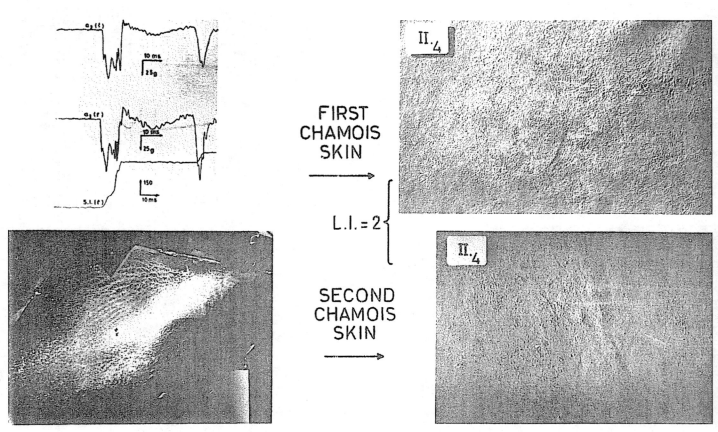

Fig. 23 - VHR 54 B - both sheets broken - typical oscillograph record, fracture pattern, and chamois lacerative damage

24 in X 36 in   Temp. : O C   20 mph

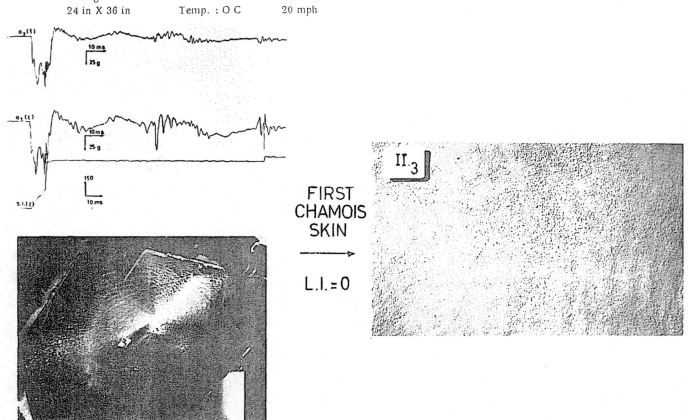

Fig. 24 - VHR 54 B - inner sheet unbroken - typical oscillograph record, fracture pattern, and chamois lacerative damage

24 in X 36 in   Temp. : O C   20 mph

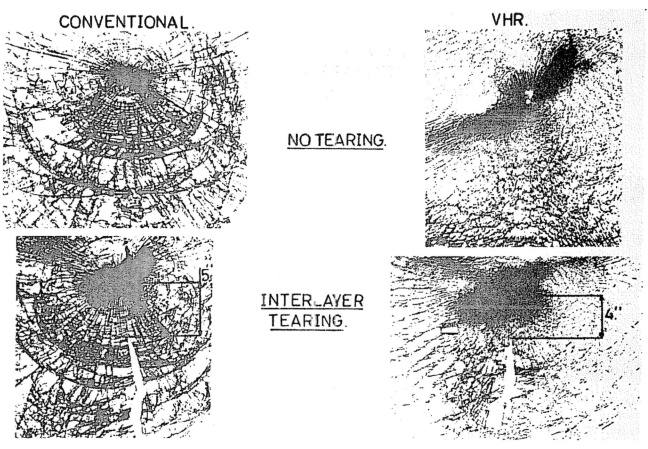

CONVENTIONAL.

VHR.

NO TEARING.

INTERLAYER
TEARING.

Fig. 25 - Typical fracture pattern of conventional and VHR glazing

Fig. 26 - Drop test facility and recording instrumentation

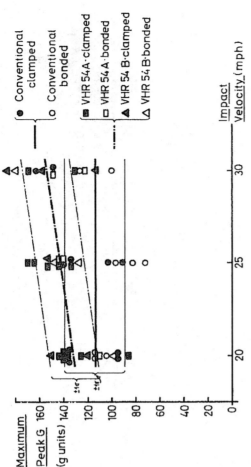

Fig. 28 - Maximum deceleration peak G versus impact velocity. Windshields - clamped and bonded - impact angle ; 40 deg   Temp. : 20 C

Legend:
- Conventional clamped
- Conventional bonded
- VHR 54A-clamped
- VHR 54A-bonded
- VHR 54 B-clamped
- VHR 54 B-bonded

Maximum Peak G (g units)

Impact Velocity (mph)

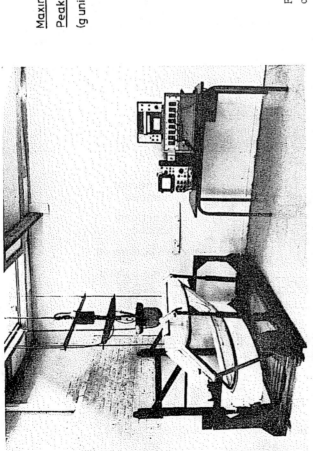

Fig. 27 - Drop test facility and recording instrumentation bonded windshield

352

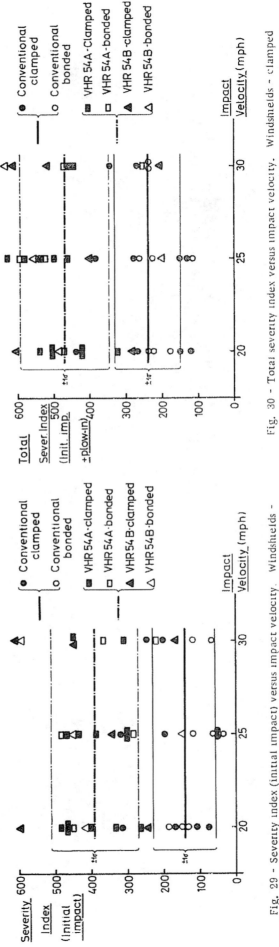

Fig. 30 - Total severity index versus impact velocity. Windshields - clamped and bonded - impact angle : 40 deg   Temp. : 20 C

Legend:
- Conventional clamped
- Conventional bonded
- VHR 54A-Clamped
- VHR 54A-bonded
- VHR 54B-clamped
- VHR 54B-bonded

Total Sever. Index (Init. imp. + plow-in)

Impact Velocity (mph)

Fig. 29 - Severity index (initial impact) versus impact velocity.   windshields - clamped and bonded - impact angle   40 deg   Temp. : 20 C

Legend:
- Conventional clamped
- Conventional bonded
- VHR 54A-clamped
- VHR 54A-bonded
- VHR 54B-clamped
- VHR 54B-bonded

Severity Index (Initial impact)

Impact Velocity (mph)

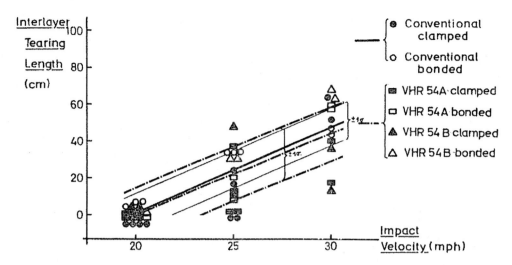

Fig. 31 - Interlayer tearing length versus impact velocity. Windshields - clamped and bonded - impact angle : 40 deg   Temp. : 20 C

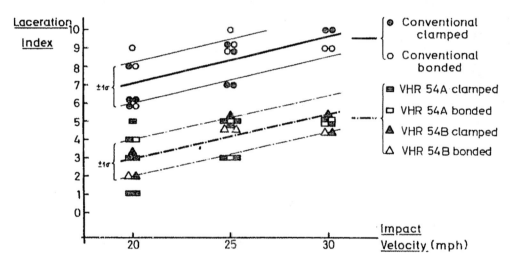

Fig. 32 - Laceration index versus impact velocity. Windshields - clamped and bonded - impact angle : 40 deg   Temp. : 20 C

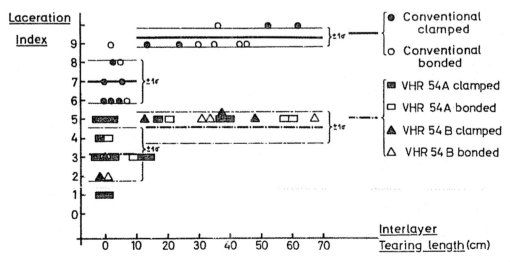

Fig. 33 - Laceration index versus interlayer tearing length. Windshields - clamped and bonded - impact angle : 40 deg   Temp. : 20 C

353

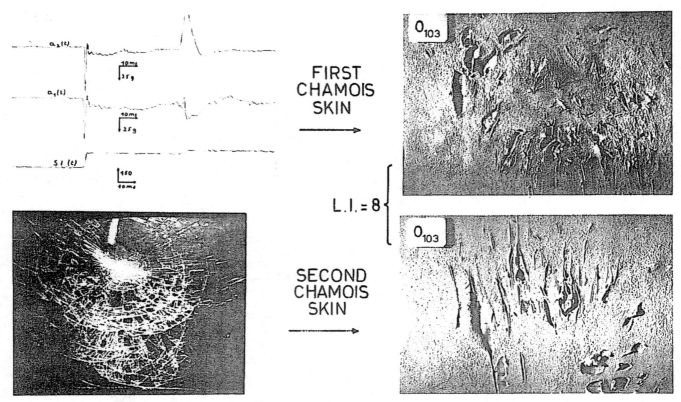

Fig 34 - Conventional windshield - typical oscillograph record, fracture pattern and chamois lacerative damage; 20 mph

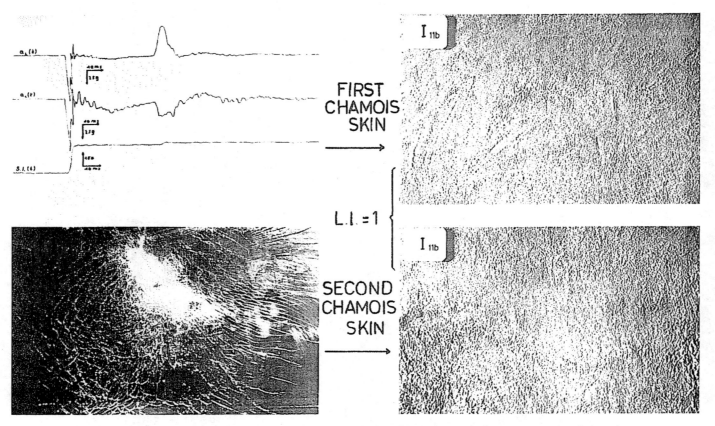

Fig. 35 - VHR 54 A windshield - both sheets broken - typical oscillograph record, fracture pattern. and chamois lacerative damage; 20 mph

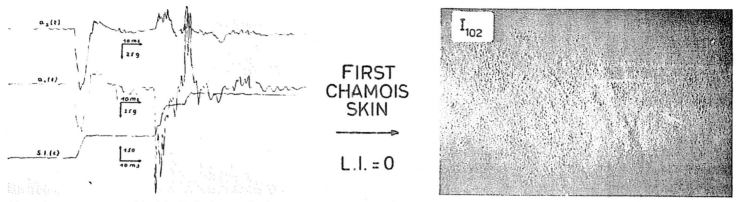

Fig. 36 - VHR 54 A windshield - both sheets unbroken - typical oscillograph record - no fracture and no chamois lacerative damage; 20 mph

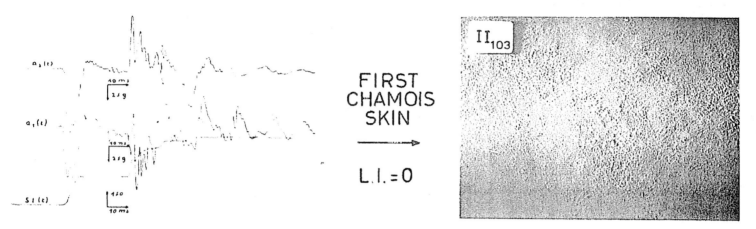

Fig 37 - VHR 54 B windshield - both sheets unbroken - typical oscillograph record - no fracture and no chamois lacerative damage; 20 mph

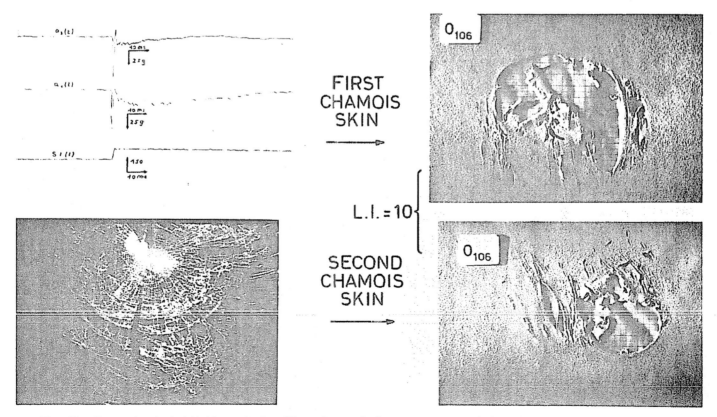

Fig. 38 - Conventional windshield - typical oscillograph record, fracture pattern, and chamois lacerative damage; 25 mph

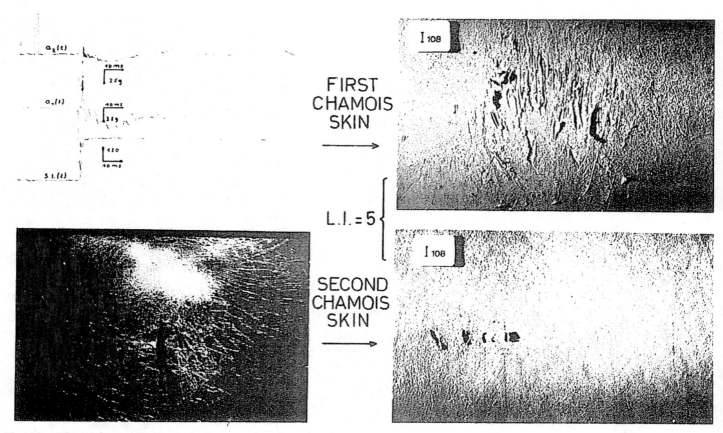

Fig. 39 - VHR 54 A windshield - both sheets broken - typical oscillograph record, fracture pattern, and chamois lacerative damage; 25 mph

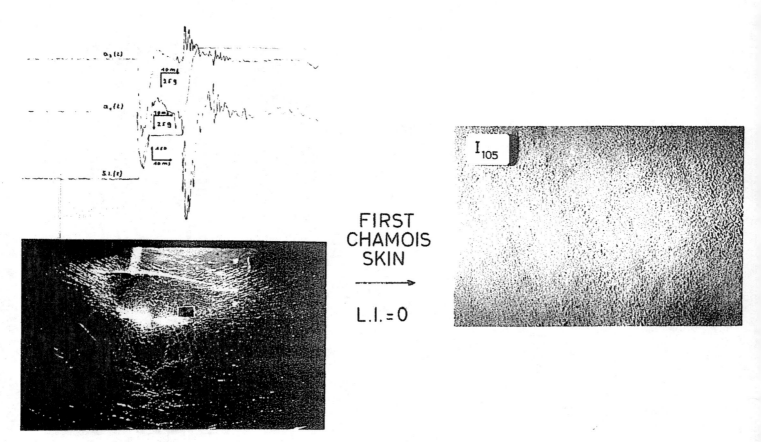

Fig. 40 - VHR 54 A windshield - inner sheet unbroken - typical oscillograph record, fracture pattern, and chamois lacerative damage; 25 mph

356

A New High Safety Glazing for Automobiles and Other Vehicles

Mr. R. van Laethem
Chief of Central Laboratory
Glaverbel, Gilly, Belgium

Summary

It is clear and well admitted that the present conventional laminated glass windshield with a thick plastic interlayer has much improved occupants' safety in a car when they undergo the "second crash."

And yet, the conclusions of official reports dealing with the whole of the crashes that have been analysed as well as the publications of specialists of automobile safety show that its laceration potential is still a severe disadvantage that ought to be lessened and if possible suppressed.

We come to the same conclusion when we consider the other glazings like the side and rear windows.

Several formulas have been suggested to reduce the risks of cuts for occupants who hit the windshield and whose heads and more particularly faces enter into brutal contact with the bits and broken fragments of glass.

None of the solutions that has been considered up to now seems to be a satisfactory compromise to meet with the requirements of the manufacturers as far as mass-production windshields are concerned.

It is very important to underline that the windshield whose laceration potential would have become very low and even non-existent should also have all the other performances that are required, especially a good optical quality that does not impair too much the visibility of the passengers and does not obstruct the view either because of an unforeseen fracture or by progressive natural aging.

Considerable progress has been achieved in manufacturing very thin glass of good optical quality in view of using it, strengthened as a part of high safety glazings[1][2].

However, it has been shown that the withstand against hard flying stones impact is improved when increasing the thickness of the outer sheet and the windshield rigidity; this results in preventing the propagation of fracture after indentations.

To find the best compromise, we have established the program of research described here, with three aims:

1. To study experimentally the numerous parameters which state the laceration potential of a windshield
2. To define, for increasing impact velocities, the best compromise between two antagonistic requirements:
   - high resistance to head penetration on one side
   - fairly limited severity index with a reasonable peak of deceleration on the other side
3. To set off the factors that could turn the windshield into a real passive restraint system.

This program is to be carried out in three successive phases. The first of them is explained here.

1) Tests about the impact of the headform free falling at impact angle on flat samples and full scale curved windshields. Measurement of the mechanical factors and estimate of the biomechanical characteristics of the impact, more especially of the laceration index as a function of the different parameters such as impact velocity, temperature, increase of its mechanical resistance, etc. The results of the first phase will be used as a basis and orientation for the other two.
2) Tests about the impact of primates against the types of windshields selected during the first phase. Estimate of the lacerations together with an X-ray study of the fractures and an attempt of diagnosis of the cerebral and cervial traumatisms.
3) Tests about the impact of anthropomorphic dummies and cadavers against the same windshields in the conditions of simulated crashes following the method used for instance at the Wayne State University.

The estimate of the results gathered then can be made but by direct and systematic comparison with the experiemental data related to the conventional glazing undergoing the same tests under the same conditions.

This compared experimentation has two parts: one carried out on flat samples (size: 24 x 36 in.) adopted by the SAE Glazing Study Group; and one carried out on large windshields fixed on automobile body openings.

The increase of safety of the new windshield can be roughly rated by extrapolation based on numerous investigations about the behaviour of the post 1966 current windshield during crashes.

The studies here considered are a part of this research program whcih aims to develop a laminated glass windshield whose laceration potential is very weak and even almost non-existent.

Two basic types of new safety windshields are tested. They only differ by the strengthening level of the V.H.R. glass which is used.

In both cases, the very thin 0.050 in. (1.2 mm) thick inner sheet and the 0.110 in. (2.8 mm) thick outer sheet are made of V.H.R. glass. Nevertheless, the glass of the outer ply has a voluntarily limited tensile strength.

The polyvinylbutyral plastic interlayer is 0.030 in. (0.76 mm) thick.

The evaluation of the biomechanical behaviour of these windshields is made in different ways among which laboratory studies, described here, and including impact tests with a headform free falling on positioned samples.

During the impact, all measures defining the main safety performances are recorded or filmed at high speed; the deceleration peak along two orthogonal axes, the resultant severity index, relating to the initial impact and to the plow in, the tearing length of the plastic interlayer and finally the laceration potential.

The latter is evaluated on basis of the laceration rating scale used by Prof. Patrick at the Wayne State University; a lateration index is given following the number and size of the cuts measured on the two superposed chamois leathers covering the headform.

The experimental parameters whose influence is more particularly studied are: temperature, impact velocity, impact location, increase of the mechanical strength of the sheets, decrease of the inner sheet thickness combined with increase of the interlayer thickness.

The researches are systematically carried out in order to compare the new windshield safety performances with the conventional laminated ones.

All the results of the measurements are analysed after the statistical method: parameters of distribution, lines of regression, analysis of correlations, signification tests, etc.

The new safety reinforced laminated windshield, whose laceration potential is very low and even non-existent at very high impact energy, might be used as a true passive restraint system if the tensile strengths of the reinforced glass are adjusted.

This double performance will have to be developed in a later series of tests about simulated crashes with anaesthetized primates, anthropomorphic dummies and, if possible, human cadavers.

References

1. R. van Laethem, "A New High-Safety Glazing for Automobiles and other Vehicles," Twelfth STAPP Car Crash Conference Proceedings, New York: Society of Automobile Engineers, Inc., 1968, pp. 360-386.
2. E. Plumat, P. Eloy, L. Leger, F. Toussaint, and R. Van Laethem, "Safety Improvement of the New Laminated V.H.H. Glazing for Cars," FISITA STAPP Car Crash Conference, Proceedings, Detroit et Bruxelles: Society of Automobile Engineers, Inc., 1970, pp. 1152-1170.

Improved Laminated Windshield with Reduced Laceration Properties

S. E. Kay and J. Pickard
Triplex Safety Glass Co. Ltd (England)

L. M. Patrick
Wayne State University

Abstract

A new laminated automobile windshield called Triplex "Ten-Twenty," fabricated from two thermally stressed glass plies of 2.3 mm soda-lime float glass laminated with a 0.76 mm HPR polyvinyl butyral interlayer, has been biomechanically evaluated by Triplex Safety Glass Co., Ltd., using a dropping headform and a skull impactor, and by Wayne State University, using a 50th percentile anthropomorphic dummy on the WHAM III sled test facility. The results of these evaluations at velocities up to 60 km/h are expressed in terms of Gadd index, head injury criterion, and various laceration scales including the new Triplex laceration index (TLI). Some details are also given of other properties of the windshield. The results of the evaluations indicate that the Ten-Twenty windshield offers a reduction of about two units on the TLI scale equivalent to one of the following:

1. A 99% reduction in the number of cuts when the length and depth of cuts remain unaltered

2. A 90% reduction in the length of cuts when the number and depth of cuts remain unaltered

3. A change in depth of cuts from one layer of skin simulation to another, but in particular a 78.5% reduction in the depth of cuts into the polyvinyl chloride base layer when the number and length of cuts remain unaltered

In practice, the length, number, and depth of cuts all change together so that one typical example taken from the test program of a two-unit reduction in TLI is:

1. A 62% reduction in the average depth of cuts into the polyvinyl chloride base

2. A 27% reduction in the average length of cuts.

3. No increase in the total number of cuts.

On the basis of these results, Ten-Twenty is a much safer laminated automobile windshield than those now commercially available due to decreased laceration to the occupants during a collision.

DETAILS of the biomechanical evaluation of a new laminated windshield, called Ten-Twenty, which establishes its performance in terms of head deceleration and facial laceration, are presented in two parts. The first part comprises evaluations by Triplex using drop-test and skull impactor methods. The second part reports the evaluation conducted by Wayne State University (WSU) using a 50th percentile anthropomorphic dummy and the WHAM III sled test facility. Other properties are also discussed.

Desirable Windshield Properties

Automobile safety authorities and the international safety-glass industry generally agree that an improved car windshield should have satisfactory performance in the product properties listed below:

1. Head deceleration performance.
2. Lacerative potential.
3. Resistance to penetration by head impact.
4. Penetration resistance, retention of vision, and protection of occupants against hostile climatic conditions in the event of stone impact.
5. Insensitivity of product properties to the effect of normal handling and abrasion.
6. Insensitivity of the product properties for extremes of temperature.
7. Optical quality.

Clearly, it is important that an improvement in one property should not be offset by a significant worsening of another when compared to existing commercial products and standards.

With the introduction of high penetration resistant (HPR) polyvinyl butyral interlayers, the risk of penetration of the windshield during head impact has been reduced without exceeding the accepted standards for head deceleration performance. The head deceleration performance has been further improved to some extent by the use of thinner glass in laminated windshields (1)* Due to the importance of laceration from windshields, the most significant further improvement in windshield performance would be to reduce the degree of facial laceration in the event of head impact. The work reported in this paper demonstrates that the Ten-Twenty windshield offers a substantial improvement in laceration without penalty or offset to the other six main properties.

Present Product Specification

The Ten-Twenty windshield consists of two pieces of 2.3 mm soda-lime float glass laminated with a standard 0.76 mm HPR polyvinyl butyral interlayer.

---

*Numbers in parentheses designate References at end of paper.

IMPROVED LAMINATED WINDSHIELD

The work reported here refers to laminates in which the inner glass ply is thermally tempered to a center tensile stress of 47 ± 3 MPa. The outer glass ply is lightly thermally tempered to a center tensile stress of about 7 MPa; typical stress profiles for these glasses are shown in Fig. 1.

The windshield performance has been evaluated by comparison with the two main types of symmetrical, laminated windshields now in commercial use:

1 The conventional 2.5 mm windshield consisting of two plies of 2.5 mm float glass laminated together using a standard HPR 0.76 mm interlayer.

2 The conventional 3 mm windshield consisting of two plies of 3 mm float glass laminated together using the same standard 0.76 mm interlayer.

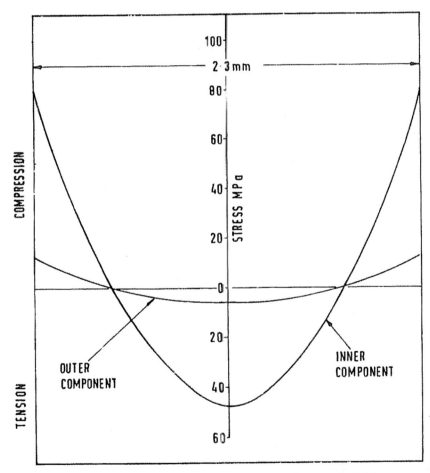

Fig 1—Stress profiles for two components of Ten-Twenty windshield

With all the windshields in these comparative evaluations, the interlayer and adhesion characteristics were kept identical

Head Deceleration and Laceration Testing

All the available test methods were studied and the following chosen as being the most suitable for preliminary determination of the safety characteristics of a windshield:

1. For head deceleration comparisons--a simple dropping headform
2. For facial laceration comparisons--a skull impactor

These have the advantage of being simple, reproducible, and rapid, permitting large numbers of windshields to be impacted with a minimum of scatter in the experimental results This is important in the development of a new product where a large number of variables have to be optimized and new constructions tried as quickly as possible if the development program is not to be delayed by the testing procedures

Drop Test--Head Acceleration Evaluation--A 10 kg (2) headform was dropped onto a windshield mounted normally to the direction of the impact The test was used to determine deceleration characteristics, Gadd severity index (SI), and head injury criteria (HIC) at velocities up to 60 km/h Table 1 sets out results of normal, central impacts on Ten-Twenty windshields of the Volvo 142 shape For comparison, results are also shown for similar impacts made in the same test series on the same shape of conventional annealed glass windshields.

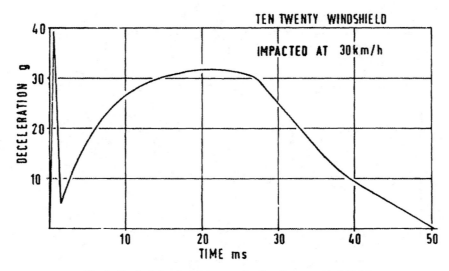

Fig 2--Typical deceleration trace for Ten-Twenty windshield

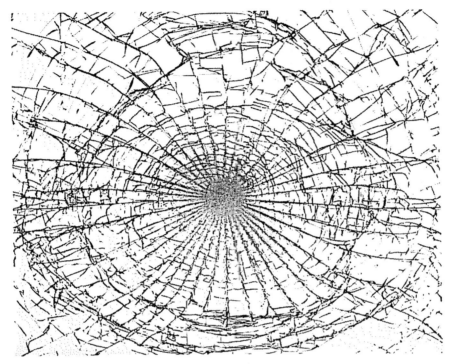

Fig. 3—Typical fracture of conventional 2 5 mm laminated windshield

These impacts were all performed at 20°C on windshields clamped into a wooden buck The headform was equipped with piezoresistive accelerometers The acceleration output was recorded on an ultraviolet recorder with a 1 kHz pass-band A typical deceleration pattern is shown in Fig 2 Photographs of typical fractured windshields are given in Figs 3-5

Graphical presentation of the results in terms of SI against headform impact velocity is presented in Fig 6, where the lines plotted are the least squares fit of straight lines to the data up to penetration of the interlayer.

It can be seen that the results for Ten-Twenty are superior to those for the conventional 2 5 and 3 mm windshields Statistical testing of these results shows that the improvement in SI of Ten-Twenty over that of the conventional 3 mm laminates is significant at the 1% level The improvement over the conventional 2 5 mm is significant at the 5% level When the head acceleration results were plotted in terms of HIC (Fig 7), the results were in a similar order, but the differences were not statistically significant HIC and SI were, for all the versions

Table 1—Results of 10 kg Headform Normal Drop-Tests on Ten-Twenty, 2.5 mm, and 3.0 mm Windshields

Test No.	Windshield Composition	Velocity, km/h	Deceleration			Total Duration of Impact, ms	Gadd SI	HIC	Total Interlayer Tears, mm	Penetration
			Max Deceleration Due to Glass, g	Duration of Deceleration Due to Glass, ms	Max Deceleration Due to Interlayer, g					
1965	3 mm	38.9	56	2.2	45	111	240	187	230	No
1966	3 mm	38.9	114	3.9	41	82	293	173	–	Yes
1967	2.5 mm	38.9	55	2.2	39	57	234	193	180	No
1968	2.5 mm	41.0	59	1.9	38	119	213	179	254	No
1969	2.5 mm	43.0	63	1.2	41	114	212	172	252	No
1970	2.5 mm	46.0	98	1.5	40	43	181	132	–	Yes
1971	Ten-Twenty	38.9	37	1.4	40	55	233	198	130	No
1972	Ten-Twenty	41.0	40	1.2	40	60	209	176	150	No
1973	Ten-Twenty	43.0	42	0.9	38	148	172	141	–	Yes
1974	3 mm	34.9	53	2.2	40	51	210	175	95	No
1975	2.5 mm	34.9	58	1.7	35	61	193	168	105	No
1976	2.5 mm	34.9	75	1.4	38	51	220	181	100	No
1977	2.5 mm	34.9	38	1.9	35	120	167	140	262	No
1978	3 mm	32.0	61	1.7	38	51	183	153	0	No
1979	2.5 mm	32.0	57	1.2	40	51	234	192	100	No
1980	Ten-Twenty	32.0	40	0.9	34	55	175	153	0	No
1981	3 mm	27.8	45	1.7	31	51	115	95	15	No
1982	2.5 mm	27.8	85	2.4	34	51	207	163	5	No
1983	Ten-Twenty	27.8	39	1.2	33	50	131	113	55	No
1984	3 mm	24.0	60	2.4	28	51	104	84	0	No
1985	2.5 mm	24.0	53	1.9	30	51	117	98	30	No

1987	Ten-Twenty	24.0	51	1.2	28	50	98	85	40	No
1990	2.5 mm	20.0	32	1.4	23	51	64	57	0	No
1991	3 mm	20.0	94	3.9	21	53	124	78	0	No
1992	Ten-Twenty	20.0	49	2.2	24	51	68	59	0	No
1993	3 mm	43.0	52	2.7	43	52	161	137	—	Yes
1994	Ten-Twenty	43.0	45	1.4	41	60	232	193	—	No
1995	2.5 mm	43.0	51	2.9	45	71	227	170	—	Yes
1996	3 mm	41.0	43	2.2	43	71	160	141	—	Yes
1998	Ten-Twenty	29.9	39	1.4	33	50	138	121	50	No
1999	2.5 mm	29.9	86	2.4	36	49	176	138	50	No
2000	3 mm	29.9	84	2.2	32	61	185	140	0	No
2001	2.5 mm	20.0	77	2.9	23	52	95	56	0	No
2002	2.5 mm	24.0	63	1.4	30	55	112	89	0	No
2003	2.5 mm	27.8	52	1.4	33	48	123	103	50	No
2004	2.5 mm	32.0	43	1.4	36	50	166	143	80	No
2005	2.5 mm	38.9	65	2.2	45	56	273	215	190	No
2006	2.5 mm	43.0	46	1.2	43	109	208	154	—	Yes
2007	3 mm	20.0	112	3.9	18	51	172	122	0	No
2008	3 mm	24.0	90	3.2	33	43	208	136	0	No
2009	3 mm	27.8	88	2.9	34	47	179	136	0	No
2010	3 mm	32.0	96	2.2	43	46	264	202	0	No
2011	3 mm	38.9	65	2.4	48	47	292	244	100	No
2015	3 mm	20.0	90	3.7	21	53	116	57	0	No
2017	Ten-Twenty	20.0	31	1.7	27	52	76	63	0	No
2018	Ten-Twenty	27.8	66	1.0	34	51	175	145	0	No
2019	Ten-Twenty	24.0	29	1.7	29	50	91	78	0	No
2020	Ten-Twenty	32.0	37	1.0	35	75	153	128	150	No
2021	Ten-Twenty	38.9	36	1.2	38	42	93	72	—	Yes
2022	3 mm	43.0	81	2.0	38	43	139	97	—	Yes

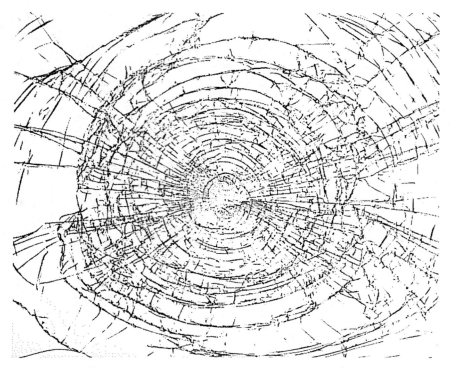

Fig. 4—Typical fracture of conventional 3 mm laminated windshield

of windshields tested, below 300 units at the onset of penetration; that is, well below the limiting value of 1000 units. Analysis of the results in Table 1 shows that for Ten-Twenty, the peak accelerations due to the breaking of the glass plies were somewhat lower than those for the conventional annealed laminates. This advantage is a consequence of using thinner glass which has a moderate modulus of rupture. A similar effect was reported by Alexander, Mattimoe, and Hoffman (1). No significant difference was detected in the penetration velocities of the three types of windshields.

Skull Impact Rig—Facial Laceration Evaluation—This is similar to the type of equipment used by Corning, which has already been reported to give good agreement with the more complex anthropomorphic dummy impacts (3). The rig used in this work was designed to give improved repeatability and also to give scope for possible future upgrading. The rig is shown diagrammatically in Fig. 8. A more detailed drawing of the trolley and pivoting arm is given in Fig. 9. It was made very robust, and the mass of the trolley carrying the pivoted headform was increased to 45 kg to represent more closely the inertia of a dummy torso. The stopping distance used for each of the impact tests was 25 mm, so that the total

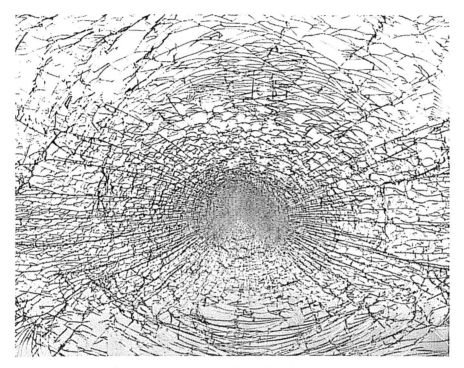

Fig 5—Typical fracture cf Ten-Twenty windshield

movement of the pivot carrying the headform and arm was in each case less than 25 mm for the duration of the impact and the plow-in of the headform. This rig was used principally to measure laceration.

Measurement of Laceration—Two layers of carefully selected moist chamois leather, each of 1 ± 0 15 mm thickness, were used to simulate facial skin. A replaceable skullcap of moldable polyvinyl chloride (PVC) material was used to simulate subcutaneous tissue

During the course of this work, the inadequacy of existing laceration scales and, in particular, the high degree of subjectivity involved in using them became evident. In order to quantify laceration more reliably and much more objectively, a new laceration index was developed, which is known as Triplex laceration index (TLI). This is a nonsubjective, continuous, unbounded scale that relies solely upon accurate measurements of the cuts produced in two layers of moist chamois leather and the underlying layer of soft PVC. It does not depend upon the idiosyncrasies of the assessor. The scale itself and its derivation will not be described in detail here since this is the subject of a separate paper (4). Numerically it is based on, and is in good agreement with, the scale devised by

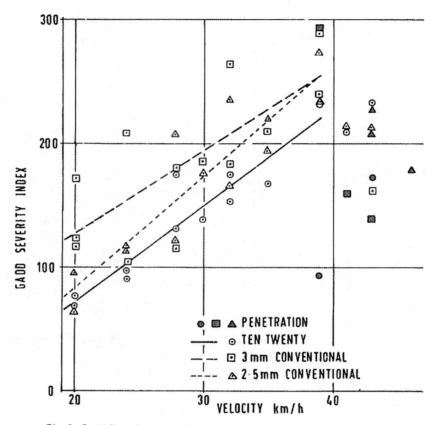

Fig. 6—Gadd SI as function of impact velocity by dropping headform test

Corning (3) between the numbers 0 and 9, while possessing substantial advantages over previous methods in its application. Leathers and PVC were chosen to be as close as possible to the SAE recommendations (5).

Mathematical analysis of the experimental results led to the derivation from first principles of a logarithmic formula for this laceration index. Thus,

$$TLI = 2 + \log_{10} (1 + 1.16 D_1 + 50.8 D_2 + 16,500 D_3) \qquad (1)$$

where:

D_1, D_2, D_3 = lacerative damage in three layers of tissue simulation

$$D_1 = \sum_s n_{1s} s^2 \qquad (2)$$

368

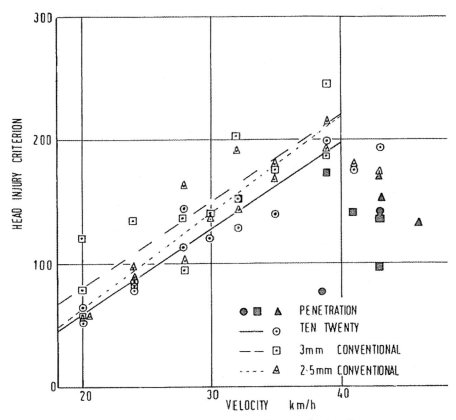

Fig. 7—HIC as function of impact velocity by dropping headform test

$$D_2 = \sum_s n_{2s} \, s^2 \tag{3}$$

$$D_3 = \sum_d \sum_s n_{3sd} \, s^2 \, d^3 \tag{4}$$

where:

s = length of cut, mm
d = depth of cut into PVC, mm
n_{1s} = number of cuts of length s in first (outer) leather
n_{2s} = number of cuts of length s in second (inner) leather
n_{3sd} = number of cuts of length s and depth d into PVC

369

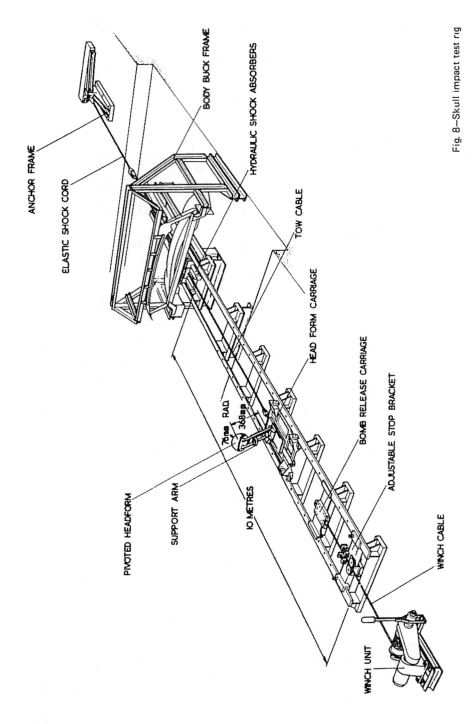

ANCHOR FRAME

ELASTIC SHOCK CORD

BODY BUCK FRAME

HYDRAULIC SHOCK ABSORBERS

TOW CABLE

HEAD FORM CARRIAGE

BOMB RELEASE CARRIAGE

ADJUSTABLE STOP BRACKET

WINCH CABLE

WINCH UNIT

PIVOTED HEADFORM

SUPPORT ARM

IO METRES

76mm RAD

368mm

Fig. 8—Skull impact test rig

IMPROVED LAMINATED WINDSHIELD

LENGTH OF ARM   368 mm
WEIGHT OF HEADFORM & ARM   6·8 kg
RADIUS OF HEADFORM   76 mm
WEIGHT OF TROLLEY   45 kg
ANGLE OF HEADFORM TO ARM   45°

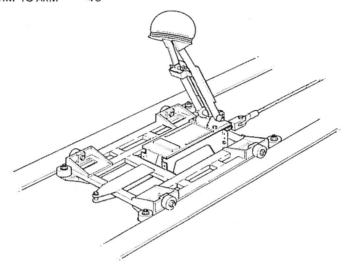

Fig. 9—Carriage and headform assembly of skull impact test rig

Fig. 10—Facial tissue simulation used on skull impact test rig. TLI = 6

Fig. 11—Facial tissue simulation used on skull impact test rig, TLI = 9

Fig. 12—Facial tissue simulation used on skull impact test rig, TLI = 12

It can be seen that the quantity $1.16 D_1 + 50.8 D_2 + 16,500 D_3$ represents a measure of the total lacerative damage. It can also be seen that because TLI is a logarithmic unit to the base 10, a change of one unit in TLI represents a tenfold change in the total lacerative damage. A change of one unit can be represented as a tenfold change in the number of cuts, or a threefold change in the length of cuts, or about a twofold change in the depth of deep cuts into the PVC. Examples of lacerative damage corresponding to TLI values 6, 9, and 12 are given in Figs. 10-12. A conversion table from TLI to total lacerative damage is given in Table 2.

The TLI gives reasonably good straight-line plots of laceration injury against impact velocity, and this greatly assists the statistical analysis of results.

Laceration Results—Table 3 sets out the laceration results obtained on the skull impactor rig on windshields similar to those used on the drop rig. In this case, the windshields were clamped by a continuous rubber-lined metal strip into the glazing aperture of a Volvo car. Again, all these tests were done at 20°C. The velocity indicated is the trolley velocity immediately before impact. The laceration results are quoted in terms of TLI and are shown graphically in Fig. 13. The lines plotted show the least-squares regression of TLI on velocity for all impacts in which the screens were fractured. It can be seen that the average improvement of the Ten-Twenty windshield over the conventional 2.5 mm annealed glass windshield is 2.0 units, and the average improvement over the 3 mm annealed glass windshield is 2.7 units. These results represent a reduction in the quantity "total lacerative damage" of about 100 times and 500 times, respectively. No attempt has been made to give a medical interpretation to these results. The easiest interpretation can be made from the formula for TLI by the fact that an improvement of two units can be brought about by:

1. A decrease in the number of cuts by 100 to 1 (99%), the lengths and depths of all cuts remaining unchanged.

Table 2—Conversion from TLI to
Lacerative Damage

Change in Level of TLI Units	Corresponding Change in Level of Total Lacerative Damage
0	X 1
1	X 10
1.2	X 16
1.4	X 25
1.6	X 40
1.8	X 63
2.0	X 100

Table 3—Data from Skull Impact Rig at 20°C
(Triplex Safety Glass)

Test No.	Type of Windshield	Velocity of Impact, km/h	TLI	Total Interlayer Tears, mm
80	2 5 mm	57 1	10 3	250
81	Ten-Twenty	55 3	9 8	290
82	3 mm	33 8	8 9	20
84	3 mm	50 8	12 1	250
87	3 mm	27.4	5 7	—
88	Ten-Twenty	52.9	8 7	30
89	2 5 mm	27.5	5 9	—
90	3 mm	51.5	12 5	390
91	2 5 mm	41 5	9.4	112
93	2 5 mm	28 3	4 8	—
94	Ten-Twenty	28	2 9	—
95	2 5 mm	50 8	10 8	250
97	3 mm	41.5	9 7	230
98	3 mm	40.7	10	—
100	3 mm	33.6	9 8	—
102	2 5 mm	33.8	9 8	—
104	Ten-Twenty	27.4	0	—
106	3 mm	26.7	8.4	—
107	2 5 mm	58 1	12.9	350
108	Ten-Twenty	57 1	11.9	207
109	2 5 mm	33 1	9 0	3
110	Ten-Twenty	40 4	7.6	134
113	Ten-Twenty	33 1	6	—
114	2 5 mm	50 7	12 7	235
115	Ten-Twenty	40.4	8 1	14
120	3 mm	54.5	12 3	190
122	Ten-Twenty	31.5	9 2	—
123	2 5 mm	37.8	10 1	52
124	Ten-Twenty	46 8	8.2	50
125	3 mm	54 5	12.1	500
127	Ten-Twenty	26 1	6	—
134	Ten-Twenty	48 8	10	148
142	2 5 mm	58 1	12.2	221
143	Ten-Twenty	26 1	5 4	—
145	Ten-Twenty	48.8	10 5	61
147	2 5 mm	49.4	12 4	400
148	Ten-Twenty	54.2	12 8	—
149	Ten-Twenty	32.7	3 1	22
151	Ten-Twenty	55	8 7	161
152	Ten-Twenty	32 3	5 5	—
153	Ten-Twenty	41 0	5 8	12
154	2 5 mm	54 9	12 2	190
157	2 5 mm	26 1	5.7	—
159	Ten-Twenty	39.3	5.7	—

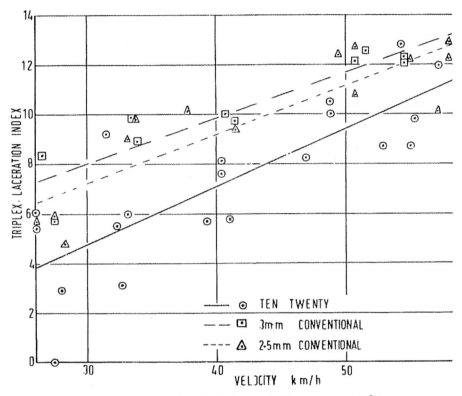

Fig 13—TLI as function of velocity by skull impact test at 20°C

2. A decrease in the length of cuts by 10 to 1 (90%), the numbers and depths of all cuts remaining unchanged.

3. A change in the depth of cuts from one layer to another but in particular in the PVC A decrease in depth by 4 6 to 1 (78.5%), the numbers and lengths of all cuts remaining unchanged.

In any actual case the improvement is brought about by contributions from all three sources.

For example, a decrease in TLI from 12 7 to 10.3 was brought about by (see Table 4):

1. A 62% reduction in average depth of cut into the PVC.
2. A 27% reduction in average length of all cuts.
3. No change in actual total number of cuts.

A reduction from 6.8 to 4.6 on the TLI was brought about by:

1. A reduction of depth of cuts into the PVC by 100%.
2. A reduction of average length of cuts by 30%.
3. A reduction in numbers of cuts by 41%.

Table 4—Laceration Measurements from Skull Impact Machine

TLI	Outer Chamois		Inner Chamois		Face	
	No. Cuts	Length, mm	No. Cuts	Length, mm	No. Cuts	Length, mm
4.6	1	19	1	1	0	0
	5	1				
6.8	1	24				
	1	12				
	2	5	1	2	1	2
	1	4				
	1	3				
	4	1				
10.3	1	50				
	1	45				
	1	40				
	1	35				
	1	27				
	1	18	1	25		
	2	12	1	16	1	20
	2	10	1	5	1	20
	1	9	2	2		
	1	8	4	1		
	3	7				
	1	15				
	4	5				
	3	4				
	4	3				
	8	2				
	9	1				
12.7	1	140	1	80		
	1	95	1	22		
	1	20	2	14		
	1	16	1	10	1	55
	3	7	2	7	1	9
	1	6	1	6		
	8	3	2	5		
	5	2	7	3		
	6	1	2	2		
			4	1		

Illustrating Effects of Length and Depth of Cut on TLI

Mask

Depth, mm	Total Chamois Cuts	Total Length Chamois Cuts, mm	Avg Length Chamois Cuts, mm	Total Depth Face Mask Cuts, mm
0	7	25	3 6	0
1	12	61	5 1	1
3 1.5	55	467	8 5	4.5
10 2	55	467	8 5	4 5

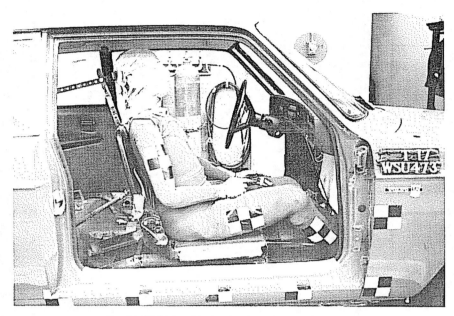

Fig. 14—Vehicle with dummy installed on WHAM III

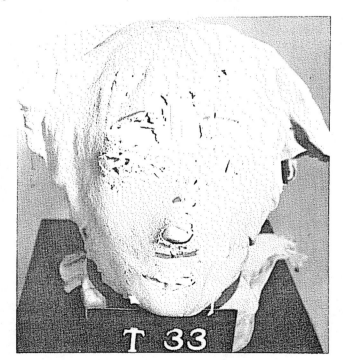

Fig. 15—Chamois on dummy after run T-33 illustrating laceration of TLI 7.4

IMPROVED LAMINATED WINDSHIELD

The improvement of 2 and 2.7 units for the Ten-Twenty windshield over the conventional windshields is significant at the 1% level and has been achieved with no penalty in head deceleration performance.

Full-Scale Vehicle Evaluation at Wayne

A comparison of the safety performance of the experimental Ten-Twenty windshield with conventional annealed glass windshields was made on WHAM III at WSU. The experiments were conducted in a modified Volvo Model 142 automobile with the windshield installed with butyl tape, which is one of the conventional methods used in the United States. The occupant was a Sierra 1050 anthropomorphic dummy located in the passenger seat of the modified vehicle.

Modifications to the vehicle included the removal of the engine, transmission, and other concentrated masses that do not contribute to the interior dynamics of the vehicle, and the addition of strengthening members to the frame of the vehicle so as to permit it to be subjected to deceleration pulses similar to those encountered in collisions without being destroyed. The modified vehicle and occupant were accelerated over 72 ft with uniform acceleration to the predetermined experimental velocity, and then decelerated with a deceleration pulse and stopping distance corresponding to that of the collision under consideration.

A steel seat construction was used in place of the conventional seat to minimize the variations in the test procedure and the effect of degradation of the conventional seat during the program. Nonresilient energy-absorbing foam was placed under the instrument panel in the knee impact area to provide a uniform decelerating medium for the knees. The foam was replaced after each run. The instrument panel and pad were also replaced when damaged. The doors were removed from the vehicle to permit high-speed cinematography of the event. Fig. 14 shows the vehicle with the dummy installed ready for one of the tests on WHAM III. The temperature of the windshields and the laboratory was maintained between 21-23°C.

Laceration from the windshield impact was evaluated by placing two layers of moist chamois over the head of the dummy. The chamois can be seen in Fig 15 after impact. To permit use of the TLI it was necessary to modify the head of the dummy by removing the rubberlike skin normally on the dummy head, and replacing it with a PVC mold of identical exterior contour. The PVC covering was replaced each time it was lacerated in a windshield impact. After placing the PVC covering over the facial area of the dummy, the head was covered with two layers of 1 mm thick chamois. The chamois were moistened and wrung out until they changed to a light color prior to being placed on the dummy head. They were fastened to the head with tape around the neck and in the occipital area of the head. Potential injury to the brain was measured by a triaxial accelerometer mounted at the c.g. of the dummy head. The triaxial acceleration was

Table 5—Summary of Data Analysis for Triplex Windshield Laceration Study at WSU

Run No.	Type of Windshield	Velocity, km/h	Stopping Distance, mm	Head Index		Laceration Index*			Pummel	Remarks
				Gadd	HIC	WSU	Corning Scale	TLI		
T- 4	2.5 mm	34.1	410	253	220	3	4	4.9	3	Abrasion on forehead
T- 5	Ten-Twenty	33.5	410	200	164	1	4	5.3	3	Light abrasion on forehead
T- 6	2.5 mm	34.1	410	248	202	3	5	6.2	4	Windshield slightly pulled out, abrasion on forehead
T- 7	Ten-Twenty	32.8	410	249	196	1	3	3.8	5	Abrasion on forehead, one cut through outer layer on nose
T-10	2.5 mm	16.3	300	11	9	0	0	0	5	Head broke outer glass, no cuts
T-11	Ten-Twenty	16.9	200	13	11	0	0	0	5	Head broke outer glass, no cuts
T-14	2.5 mm	24.8	300	67	55	1	4	5.2	4	Light abrasion and few small cuts on forehead and nose
T-15	Ten-Twenty	24.9	300	131	82	1	3	3.2	3	Abrasion on forehead
T-16	2.5 mm	36.7	500	576	482	3	4	5.4	3	Severe abrasion and a few cuts on forehead, buttocks slid off seat, legs jammed under instrument panel
T-17	Ten-Twenty	41.5	500	677	423	1	3	4.1	4	Severe abrasion on forehead and nose, no SI acceleration
T-18	2.5 mm	42.2	500	465	393	3	4	5.9	3	Lacerations and severe abrasion on forehead, abrasions on nose
T-19	Ten-Twenty	27.5	330	371	121	0	0	0	5	—
T-20	2.5 mm	28.3	330	172	113	0	0	0	3	—
T-21	2.5 mm	33.1	410	174	143	3	4	5.4	3	Abrasions and cuts on forehead

T-22	Ten-Twenty	32.0	410	278	222	1	2	3.4	4	Scratches on forehead
T-23	2.5 mm	33.0	410	229	161	1	3	3.6	4	Scratches on forehead
T-24	Ten-Twenty	32.3	410	175	131	1	1	1	4	Scratches on forehead
T-25	Ten-Twenty	43.9	610	787	606	3	4	4.9	3	Severe abrasions on forehead, dummy slid off seat
T-27	Ten-Twenty	45.9	610	582	494	3	4	5.2	3	127 mm tear (vertical), severe abrasions with few cuts on forehead and nose
T-28	2.5 mm	46.0	610	570	482	6	5	6.2	5	50 mm tear, several cuts on forehead and near right eye
T-29	2.5 mm	45.9	610	593	463	6	6	6.8	3	More severe abrasions on forehead, cuts on forehead, eyes, and nose
T-30	Ten-Twenty	54.1	710	871	523	6	6	7.1	3	267 mm jagged tear, extensive lacerations through outer chamois and couple of cuts through inner chamois
T-31	2.5 mm	55.0	710	747	584	6	7	8.3	3	50 x 152 mm tear, severe lacerations through both layers of chamois, 11 mm cut on chin in rubber flesh
T-32	Ten-Twenty	55.0	710	798	643	6	6	7.3	3	216 mm jagged tear, severe lacerations through both layers of chamois, no cuts in rubber flesh
T-33	2.5 mm	55.0	710	905	673	6	6	7.4	3	Severe lacerations through layers of chamois, no cut in rubber flesh

(Continued)

381

Table 5—Summary of Data Analysis for Triplex Windshield Laceration Study at WSU—Continued

Run No.	Type of Windshield	Velocity, km/h	Stopping Distance mm	Head Index		Laceration Index*			Pummel	Remarks
				Gadd	HIC	WSU	Corning Scale	TLI		
T-35	Ten-Twenty	22.5	330	258	167	0	0	0	3	No fracture
T-36	Ten-Twenty	32.2	410	1380	811	0	0	0	3	No fracture
T-37	Ten-Twenty	39.3	500	836	543	1	2	3.1	3	—
T-38	2.5 mm	23.3	330	114	68	1/3	3	4.4	4	—
T-45	Ten-Twenty	32.2	410	1069	568	0	0	0	4	No lacerations
T-47	2.5 mm	33.1	410	329	268	3/6	5	5.8	4	—
T-48	Ten-Twenty	25.4	330	330	185	1/3	3	2.7	4	—
T-50	2.5 mm	40.1	500	487	369	3	4	5.4	4	—
T-51	Ten-Twenty	39.4	500	534	381	1	2	3.1	4	—
T-54	2.5 mm	47.0	610	689	581	10	9	10.0	3	Severe lacerations, cuts (15 and 4 mm) in flesh of dummy
T-55	Ten-Twenty	45.9	610	1054	585	1	2	2.3	3	—
T-58	2.5 mm	55.0	710	629	528	6	5	6.0	4	63 and 140 mm tears
T-59	Ten-Twenty	54.9	710	1106	689	3	3	4.5	4	—

*See Tables 6 and 7 for criteria.

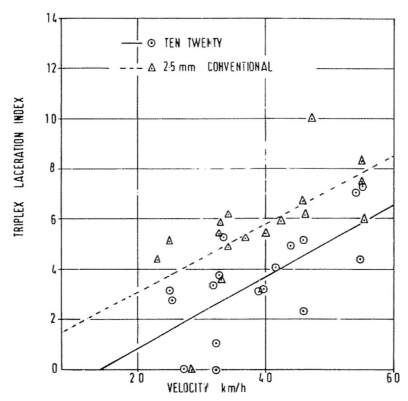

Fig 16—TLI as function of velocity by WHAM III test rig

used to calculate the HIC and SI. Signals from the accelerometers were transmitted through trailing cables to the electronic conditioning equipment on the control console The conditioned signals including the calibrations for each run were recorded on a Sangamo Model 35C0 tape recorder The tape records were displayed on a lightbeam galvanometer oscillograph (Honeywell Model 1508 Visicorder) After each run, a sample was cut from the corner of the windshield and used for a pummel test (6) to check on glass-to-interlayer adhesion levels.

Experimental Results—The experimental results are shown in Table 5, which is a summary of the data including the velocity, stopping distance, type of windshield, head injury indexes, three different laceration indexes, the pummel number for each windshield, and comments on the particular run. Fig. 16 is a plot of the TLI as a function of velocity with the best fit, first-order regression curve for the two types of windshields shown The HIC indexes as functions of velocity are shown in Fig. 17 with best fit, first-order curves included

In Table 5, the velocity varies from 16-56 km/h, the stopping distance from 203 to 710 mm, and the pummel range from 2 to 5 In the "remarks" column,

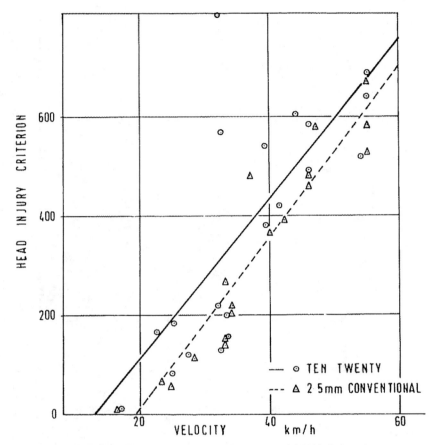

Fig. 17—HIC as function of impact velocity by WHAM III test rig

abnormal conditions are noted. Some runs have been excluded from this table; for example, in runs T-1 and T-2, the head hit the visor or the header invalidating the test. Any run in which there was substantial windshield pull-out due to poor bonding was also eliminated from consideration in the experimental data.

A comparison of three methods of rating laceration can be made from the data in Table 5, which includes the WSU, Corning, and TLI scales. The first two are primarily subjective scales in which the degree of laceration in each of the chamois and the PVC simulated tissue are rated subjectively. The TLI scale is an objective rating based upon physical measurements for the range above 2. The subjective WSU scale explained in Table 6 is the lowest of the three, with the Corning scale described in Table 7 between the WSU and the TLI in most of the ranges. Fairly close correlation prevails between the TLI and the Corning scale. This is not surprising, since the TLI is based upon the Corning scale.

Table 6—WSU Criteria for Rating Laceration Severity

Laceration Index	Description
0	No lacerations
1	Scratches and abrasions in outer chamois with none completely through it
3	More severe abrasions over large areas of face with small cuts through outer chamois, but not through inner one; abrasions on inner chamois
6	Extensive lacerations through outer chamois with limited abrasions or cuts through inner chamois
10	Severe lacerations through both layers of chamois and cuts in rubber flesh of dummy over extensive facial areas

Table 7—Corning Scale

Degree	Outer Chamois	Inner Chamois	Rubber Dummy Face
0	None	None	None
1—Minimal	Abrasions, cuts to ¾ in, none through	None	None
2—Minor	Abrasions, cuts over ¾ in, none through	None	None
3—Minor	As (2) above, but one ¾ in cut through	Abrasions	None
4—Moderate	Two or three ¾ in cuts through	Cuts, but not through	None
5—Moderate	Unlimited cuts	Only one cut through to ¾ in	None
6—Severe	Unlimited cuts	Two or three cuts through to 1-½ in	None
7—Severe	Unlimited cuts	Unlimited cuts	Abrasions
8—Severe	Unlimited cuts	Unlimited cuts	Cuts up to 1/32 in deep and ¾ in long
9—Very severe	Unlimited cuts	Unlimited cuts	One cut deeper or longer than (8)
10—Very severe	Unlimited cuts	Unlimited cuts	Numerous cuts worse than (9)

385

Comparison of the laceration of the two types of windshields in Fig. 16 shows the Ten-Twenty windshield to be approximately two units on the TLI scale less than the conventional 2.5 mm annealed windshield over the entire velocity range. Statistical testing shows that this result is significant at the 1% level. Two units on the TLI scale represent a 100-fold decrease in the number of lacerations of the same depth and length and indicate a substantial improvement in the safety from laceration injury of the windshield.

Alternatively, from Fig. 16, it can be seen that the improvement is equivalent to about 15 km/h impact velocity. That is, in a barrier crash, the conventional windshield gives a laceration value that would only be achieved by a Ten-Twenty windshield at an impact velocity some 15 km/h greater. The WHAM III simulations represent barrier collisions in which the severity is equivalent to a head-on collision or a moving-to-stationary car collision with a closing velocity twice that of the WHAM III velocity. For example, the 15 km/h difference of Fig. 16 between the Ten-Twenty and the 2.5 mm conventional windshields is the same as a difference of 30 km/h in the closing velocity of two identical cars in a head-on collision. Specifically, an occupant involved in a head-on collision between two identical cars equipped with the Ten-Twenty windshield on the average would be expected to receive the same laceration injury as measured on the TLI at a closing velocity of 100 km/h (both cars traveling 50 km/h) as an occupant in the same car equipped with the 2.5 mm conventional windshield would receive at a closing velocity of 70 km/h (both cars traveling 35 km/h).

While lacerations are often disfiguring and sometimes present severe cosmetic effects, they are seldom dangerous to life. Consequently, the improvement in laceration must not be achieved at the expense of an increased potential for brain injury. The brain injury potential from these experiments is shown in Fig. 17, which is a graph of the HIC index as a function of vehicle velocity. The best-fit, first-order curves for the two types of windshields show that the HIC index for the Ten-Twenty windshield is only approximately 100 units higher at low velocities, and only about 50 units higher at 56 km/h than the 2.5 mm annealed glass windshield. However, it is important to note that at 56 km/h, both are under 700 while the allowable limit is 1000. The average difference of 75 units is small and is not statistically significant, so it is concluded that the brain injury potential has not been increased measurably by the Ten-Twenty windshield.

There is one point on the HIC index graph of Fig. 17 for the Ten-Twenty windshield at 32.2 km/h that needs explaining. The HIC value is 811 for run 36 at 32.2 km/h. It appears that this was an extra-strong windshield since it did not fracture at 25.4, 22.5, and 32.2 km/h in runs T-34, T-35, and T-36, respectively. The same windshield did fracture at 39.3 km/h in run T-37 with an HIC index of 543. Even the anomalous HIC value of 811 in run T-36 is well below the allowable 1000 level. The HIC indexes for runs T-34–T-37, all from the same windshield, were above the best-fit curve by an appreciable amount. Consequently, this one windshield influenced the position of the HIC curve by an undue amount.

Since the greatest laceration and HIC values occur at the higher velocities, comparisons of records at nominally 47 and 56 km/h will be examined. Figs. 18–21 are photographs of the oscillograph records for runs T-54, T-55, T-59 and T-58, respectively, for the experimental Ten-Twenty windshield and the 2.5 mm annealed windshield at approximately 47 and 55 km/h. The major difference between the two windshields occurs in the initial deceleration spike.

Examination of the Gadd SI for head injury shows that the initial peak acceleration attributed to the glass is a major contributor to the SI at the 48 and 56 km/h velocities. The initial spike in the case of the Ten-Twenty experimental windshield contributes from seven to ten times as much to the SI as does that of the 2.5 mm. The initial acceleration spike for the Ten-Twenty windshield probably also contributed substantially to the HIC index. However, it should be noted that in all cases, the HIC index, which is the primary indicator of brain injury, is well below the allowable value of 1000, so there appears to be no problem of brain injury with either of the windshields examined.

Conclusions from WSU Program—The conclusions drawn from the experimental investigation of the two types of windshields studied in the WSU program are:

1. The major improvement of the Ten-Twenty experimental windshield is in the decrease of lacerations.

2. The TLI index is two units lower for the experimental windshield than for the production 2.5 mm annealed glass windshield. The magnitude of this improvement is most easily appreciated by the statement that the improvement is the equivalent of a reduction in cuts of a given length and constant depth by 100 to 1.

3. The brain injury potential as measured by the HIC index is not significantly different for the experimental Ten-Twenty windshield than for the production 2.5 mm annealed glass windshields. It is less than 700 at 56 km/h with an allowable level of 1000 according to MVSS 205.

4. The initial anterior-posterior acceleration spike is higher in magnitude and duration for the experimental Ten-Twenty windshield than for the 2.5 mm annealed glass windshield. However, this does not result in excessive HIC numbers.

Comparison of Results from Triplex and WSU Evaluations

The evaluations carried out by Triplex and by WSU show some differences in absolute values, but they arrive at the same broad conclusions:

1. The Ten-Twenty windshield is less lacerative than the conventional annealed glass windshields by about two units on the TLI scale.

2. The lacerative improvement is not counterbalanced by a significant worsening of potential brain injury.

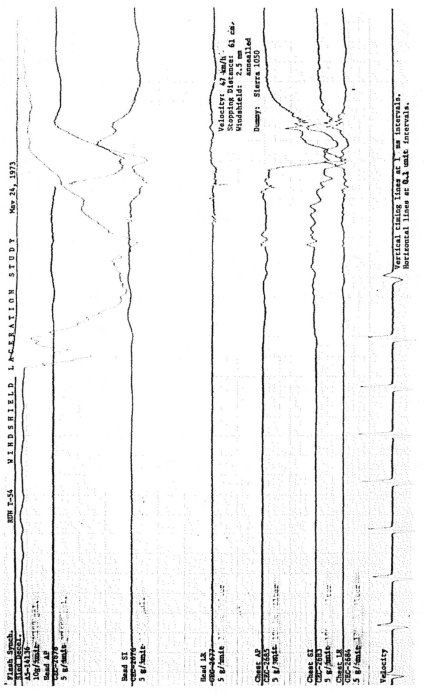

Fig. 18A—Complete record of run T-54, 2.5 mm annealed windshield at 47.0 km/h

IMPROVED LAMINATED WINDSHIELD

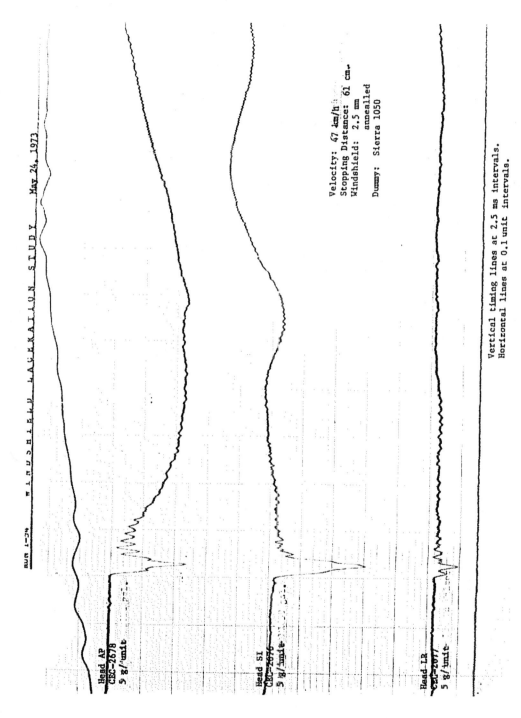

Velocity: 47 km/h
Stopping Distance: 61 cm.
Windshield: 2.5 mm
        annealed
Dummy: Sierra 1050

Vertical timing lines at 2.5 ms intervals.
Horizontal lines at 0.1 unit intervals.

Fig. 18B—Expanded record of head impact portion of run T-54

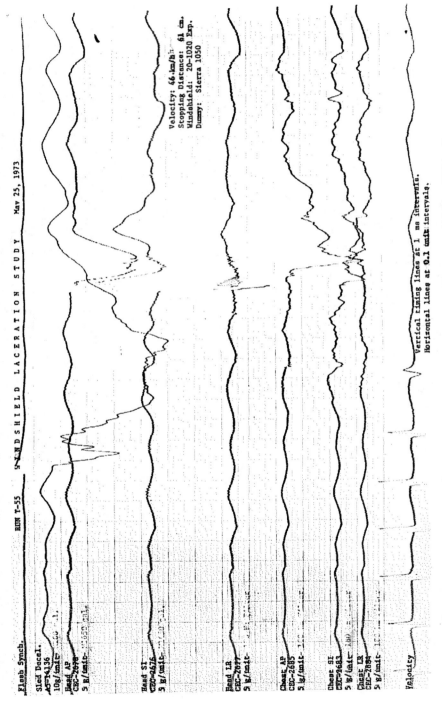

Fig. 19A—Complete record of run T-55 experimental Ten-Twenty windshield at 45.9 km/h

390

IMPROVED LAMINATED WINDSHIELD

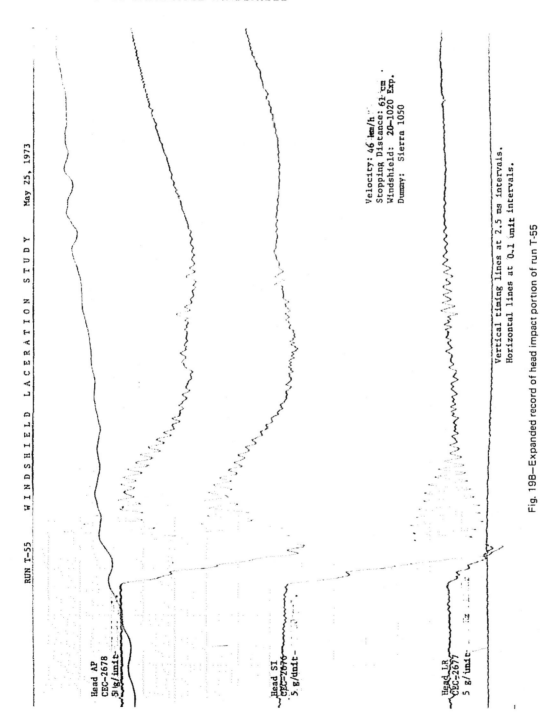

RUN T-55   W I N D S H I E L D   L A C E R A T I O N   S T U D Y   May 25, 1973

Head AP
CEC-2678
5 g/unit

Head SI
CEC-2676
5 g/unit

Head LR
CEC-2677
5 g/unit

Velocity: 46 km/h
Stopping Distance: 61 cm
Windshield: 20-1020 Exp.
Dummy: Sierra 1050

Vertical timing lines at 2.5 ms intervals.
Horizontal lines at 0.1 unit intervals.

Fig. 19B—Expanded record of head impact portion of run T-55

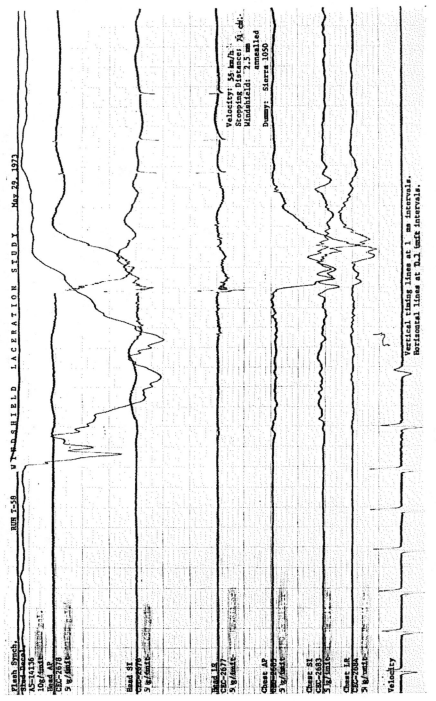

Fig. 20A—Complete record ot run T-59, experimental Ten-Twenty windshield at 54.9 km/h

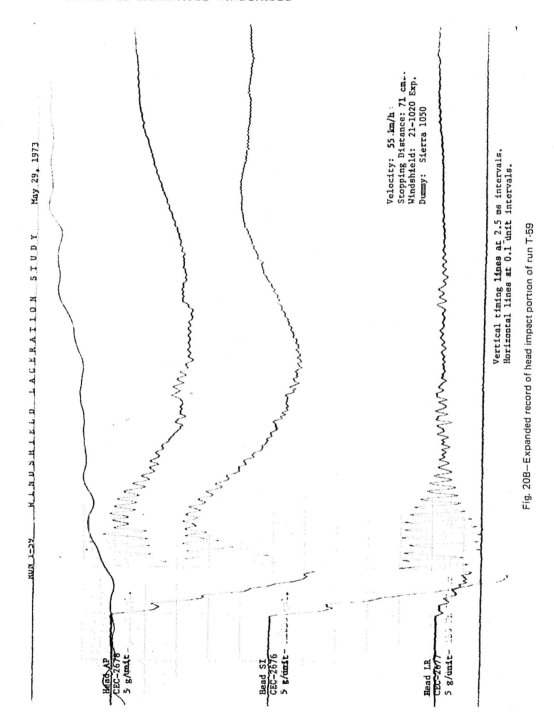

Fig. 20B—Expanded record of head impact portion of run T-59

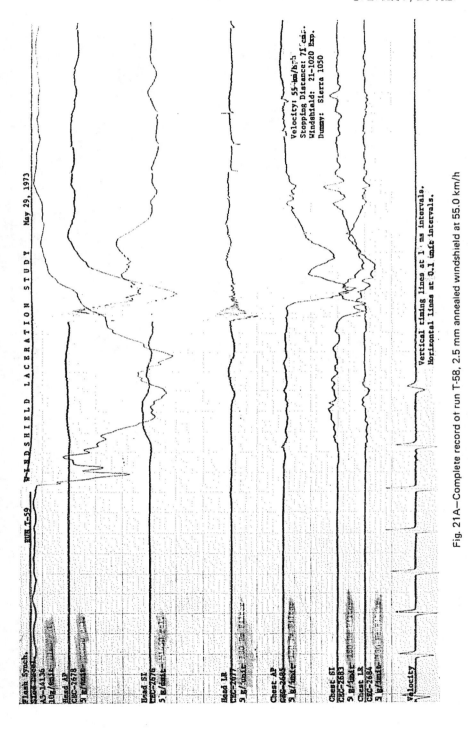

Fig. 21A—Complete record of run T-58, 2.5 mm annealed windshield at 55.0 km/h

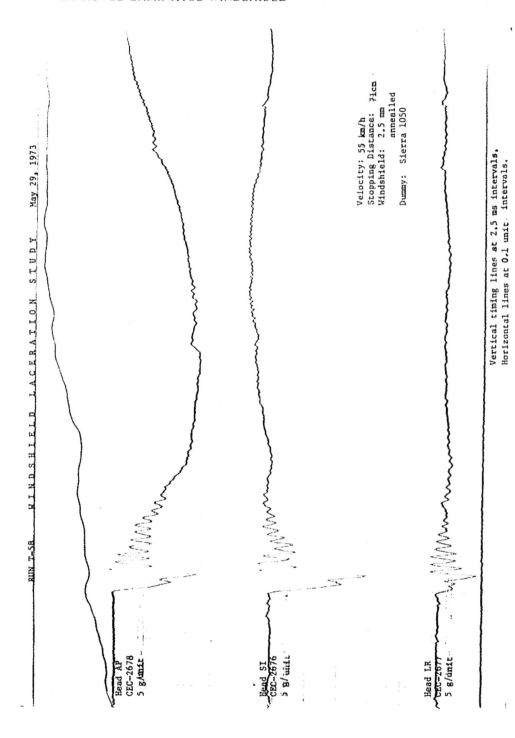

Velocity: 55 km/h
Stopping Distance: 71cm
Windshield: 2.5 mm
          annealed
Dummy: Sierra 1050

Vertical timing lines at 2.5 ms intervals.
Horizontal lines at 0.1 unit intervals.

Fig. 21B—Expanded record of head impact portion of run T-58

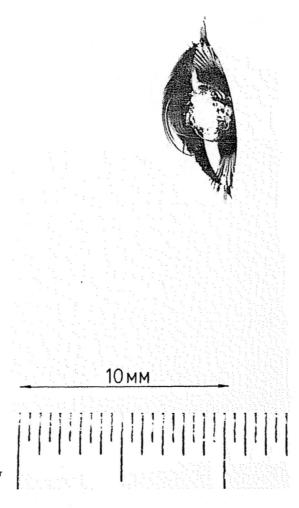

10 MM

Fig. 22—Cone fracture of outer
ply of Ten-Twenty windshield

3. At any given impact velocity, the skull impactor gives a higher overall measure of laceration than does the anthropomorphic dummy test. This is not surprising in view of the obvious differences between the testing methods. There is, however, a very close correspondence between the two methods in establishing the difference in lacerative performance of different products.

4. The HIC values from the drop-test are lower than those obtained from the anthropomorphic dummy. This is partly due to the difference in mass between the dummy head (4.6 kg) and the dropping headform (10 kg), and partly due to the difference in the impact conditions; for example, normal to the glass for the dropping headform, inclined for the anthropomorphic dummy.

10 MM

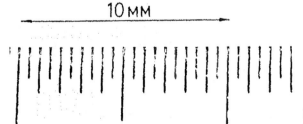

Fig. 23—Star fracture of outer
ply of Ten-Twenty windshield

5. The head acceleration due to the glass is greater in relation to the acceleration due to the interlayer plow-in for the anthropomorphic dummy evaluations This might be due to the glass being broken by the head before the mass of the dummy torso has been effectively coupled to that of the dummy head by appreciable neck flexion. In the simple dropping headform test, the effective mass of the impactor is constant during the impact

Stone Impact Performance, Climatic Protection, and Retention of Vision— Being of a laminated construction, the Ten-Twenty windshield shares the ad-

397

vantages of other laminated windshields in giving protection to the occupant against hostile climatic conditions in the event of glass breakage. The new windshield has several additional advantageous features.

Outer Glass—The stress in the outer glass of the windshield will to some extent increase its resistance to damage by stone impact. For example, in tests conducted on full-scale windshields using the 198 g Z26 dart to simulate the impact of a large, fairly pointed stone on the outer ply, it was found that the Ten-Twenty windshield gave a mean dropping height of 1.2 m to produce fracture of the outer glass compared with 0.7 m for the conventional 3 mm windshield.

The level of the stress in the outer ply is such that in the event of its fracture by a stone impact:

1. The glass will not "craze."
2. The cracks that may be produced will not be self-propagating.

Clarity of vision will therefore be maintained. Typical examples of cone and star fractures caused by external impact on the outer glass are given in Figs. 22 and 23.

Inner Glass—The temper of the inner glass gives a good protection against penetration of pebbles through the windshield. The level of stress of the inner glass is also set so as to minimize crazing in the unlikely event of the inner pane being broken by a heavy stone impact. An example of such an impact (1.4 kg brick at 30 km/h) in which both glasses are fractured is given in Fig. 24. It is evident from the above that even if both plies are fractured, the windshield will still protect satisfactorily against all hostile climatic conditions, provided the interlayer is not impaired and adequate visibility will be maintained.

In total performance against stone impact, the critical test is how actual windscreens perform on the road. An intensive road testing program of the new windshield is being performed at present.

Fig. 24—Fracture of both plies of Ten-Twenty windshield by house brick

IMPROVED LAMINATED WINDSHIELD

Abrasion and Damage in Handling—The thick compressive surface layers of the Ten-Twenty product shown in Fig. 1 make the potentially sensitive glass surfaces relatively insensitive to handling damage during the laminating process. A useful consequence of the stressed condition of each of the glass plies, both over their major surfaces and also at their edges, is that the finished windshield is robust and is an easier product for packaging, transport, or final fitting on the car assembly track. Typical edge compression stresses of the inner and outer plies are 70 and 25 MPa, respectively, as compared with about 5-15 MPa for existing commercial products.

Sensitivity to Temperature—The properties of the glass plies are hardly affected over the range of temperature experienced in field conditions. It is well known, however, that the temperature of the interlayer will markedly affect its performance. The bulk of our testing has, for convenience, been carried out in the region of 20°C. When less temperature-sensitive interlayer materials become available, the advantages these confer can be directly applied to the Ten-Twenty windshield. The improvement in laceration safety offered by the new windshield is not restricted to the use of existing interlayers.

Performance Against Standards and Optical Quality—In-house testing of the Ten-Twenty windshield has shown that it can be produced to meet all existing national and the proposed international standards. Actual approvals have been obtained at the time of writing for American Standard Z26.2-1973 and British Standard 857. Applications are pending for German and French approval.

Satisfactory optical performance of a windshield depends upon two factors:

1. The quality of the raw glass.
2. The control of the processing conditions used in making the product.

The Ten-Twenty windshield has already been produced to a good commercial standard in which the distortions produced during processing are of the same order subjectively as those present in the raw glass. In-house assessments and those of a few automobile manufacturers who have received trial supplies indicate that the Ten-Twenty windshield can meet all commercial standards of performance.

Mechanism of Ten-Twenty Improvement—The improved laceration characteristics of the Ten-Twenty windshield are a consequence of the use of thin-tempered glass for the inner ply, which, under head impact conditions, fractures to produce small glass fragments. This fine fracture is due to the release of the energy stored in the glass at the instant of fracture. Some of this energy is stored in the glass by virtue of the tempering and some is the strain energy stored transiently in the glass by the bending of the glass ply under the impact. The fine glass fracture reduces the size of the cutting particles and hence tends to reduce the depth to which cutting is liable to occur. The individual particles are also less sharp and therefore less liable to cut—a familiar property of tempered glass.

The fine fracture across the full depth of the windshield ensures that a greater area of the windshield is able to yield flexibly as the head impacts and passes down the windshield. In the case of conventional windshields, only a very small area of the glass is finely fractured by the head impact, and the head then has to traverse larger pieces, which have sharp cutting edges and which more rigidly oppose the motion of the head.

The good performance in terms of SI and HIC is a consequence of using a thin glass tempered in such a way that it has a large amount of strain energy but still with only a moderate modulus of rupture. The moderate increase in modulus of rupture compared with annealed glass constructions is offset by the reduced thickness of glass used in the construction of the new windshield.

Summary and Conclusions

Impact performance of an improved laminated windshield known as Ten-Twenty has been described. It is comprised of two layers of 2.3 mm soda-lime float glass, the inner ply thermally stressed to a center tensile stress of 47 ± 3 MPa, and a standard 0.76 mm HPR PVB interlayer. It is a much safer automobile windshield than those now in commercial use due to decreased laceration to the occupants during a collision and without penalty to other product properties such as concussion, optics, stone damage, etc.

The following conclusions are based upon the evaluation programs at Triplex and WSU:

1. The TLI is 2 and 2.7 units less for the Ten-Twenty windshield than for the 2.5 and 3.0 mm conventional windshields, respectively, on the skull impact rig.

2. The TLI is 2 units less for the Ten-Twenty windshield than for the 2.5 mm conventional windshield on the WHAM III simulator. This represents a decrease in laceration equivalent to:

    (a) A 99% reduction in the number of cuts when the length and depth of cuts remain unaltered.

    (b) A 90% reduction in the length of cuts when the number and depth of cuts remain unaltered.

    (c) A change in depth of cuts from one layer of skin simulation to another, but in particular a 78.5% reduction in the depth of cuts into the PVC base layer when the number and length of cuts remain unaltered.

In a severe impact, all three types of damage occur and the improvement is made up of contributions from all three sources. In terms of impact velocity for the barrier crash simulation, a conventional windshield gives a laceration value that is only achieved by a Ten-Twenty windshield at a barrier impact velocity some 15 km/h greater. In terms of actual road accidents, this may be interpreted as an advantage of about 30 km/h closing speed in car-to-car collisions.

3. The TLI is higher for the skull impactor test than for the anthropomorphic dummy on WHAM III test, but the difference of two units between the Ten-Twenty and the 2.5 mm commercial windshield is obtained with both test methods.

4. The initial or glass anterior-posterior acceleration spike is higher in magnitude for the Ten-Twenty windshield than for the 2.5 mm commercial windshield, but the HIC is well below the allowable value of 1000 for both types of windshields with approximately the same HIC values.

5. The Ten-Twenty windshield has greater resistance to stone breakage or stone penetration than the commercial windshields.

6. Stone impact does not cause loss of visibility through excessive crazing.

7. The optical properties of the Ten-Twenty windshield meet the current and proposed commercial, national, and international standards.

8. The thick compressive surface and edge stress of the Ten-Twenty windshield should reduce breakage during fabrication, shipping, and installation.

9. The improvement of laceration characteristics of the Ten-Twenty windshield is achieved without degradation of other required features.

References

1. H. M. Alexander, P. T. Mattimoe, and J. J. Hoffman, "An Improved Windshield." SAE Transactions, Vol. 79 (1970), paper 700482.

2. SAE Recommended Practice J938, "Drop Test for Evaluating Laminated Safety Glass for Use in Automotive Windshields." October 1965.

3. J. R. Blizard and J. S. Howitt, "Development of a Safer Nonlacerating Automobile Windshield." Paper 690484 presented at SAE Mid-Year Meeting, Chicago, May 1969.

4. J. Pickard, P. A. Brereton, and A. Hewson, "An Objective Method of Assessing Laceration Damage to Simulated Facial Tissues." Paper presented at 17th Conference of American Association of Automotive Medicine, Oklahoma, November 1973.

5. SAE Recommended Practice J984, "Bodyforms for Laboratory Impact Testing." March 1967.

6. United States Patent No. 3,434,915, Glass Laminate, William E. Garrison. Patented March 25, 1969.

REDUCED LACERATION FROM A NEW LAMINATED WINDSHIELD

JOHN PICKARD
Research and Development Manager
Triplex Safety Glass Co., Ltd.

η46041

SUMMARY — Accept···

A new car windshield has been developed by the Triplex Safety Glass Co., Ltd. of England which promises a substantial reduction in facial laceration without adverse consequences in other respects. The new windshield is called Triplex Ten Twenty.

Of laminated construction, using two pieces of 2.3 mm thick float glass and a standard HPR polyvinyl butyral interlayer of 0.76 mm thickness, the two glass components are thermally stressed to different levels. The outer glass is lightly stressed to about 7 MN.m^{-2} for optimum stone impact resistance and the inner glass to 47 ± 3 MN.m^{-2} (center tensile stress) for major reduction in laceration.

Test programs have been carried out at Triplex using a dropping head form and using a skull impactor sled and by Wayne State University in Detroit, U.S.A., using a 50th percentile anthropomorphic dummy on their Wham III crash simulator.

The results of these evaluations are given in terms of Head Injury Criterion, Gadd Index and various laceration scales including a newly developed objective scale — the Triplex Laceration Index, or TLI.

The TLI is a formula based on the number and the measured length and depth of cuts in the dummy head coverings. When compared with conventional laminated windshields Triplex Ten Twenty has been demonstrated to give a reduction in laceration of 2 units on this logarithmic scale.

A reduction of TLI units is equivalent to a reduction of:

99% in the number of cuts of constant size,

or 90% in the length of cuts of constant depth and number,

or 78% in depth of cuts of constant length and number,

if the differently weighted attributes (depth, length and number) are considered separately.

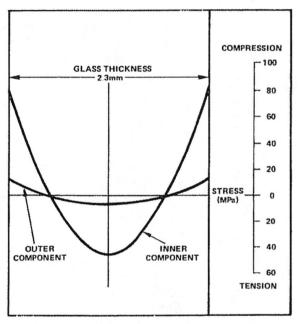

Figure 1. Stress Profiles for Two Components of Ten Twenty Windshield

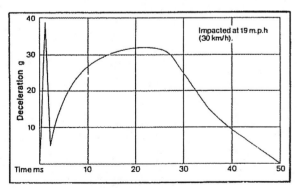

Figure 2. Typical Deceleration Trace for Ten Twenty Windshield

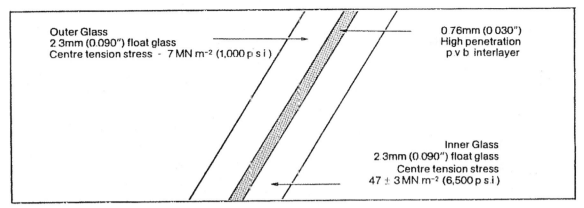

Outer Glass
2.3mm (0.090") float glass
Centre tension stress - 7 MN m⁻² (1,000 p s i)

0.76mm (0.030")
High penetration
p v b interlayer

Inner Glass
2.3mm (0.090") float glass
Centre tension stress
47 ± 3 MN m⁻² (6,500 p s i)

Figure 3. Construction of the Ten Twenty Windshield

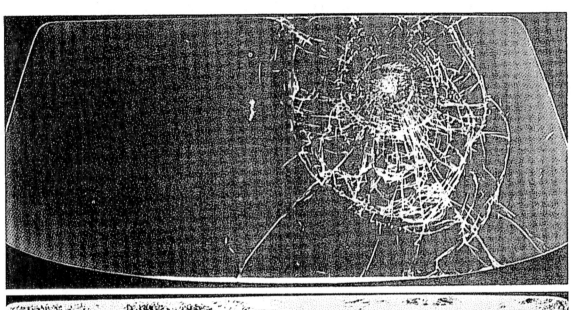

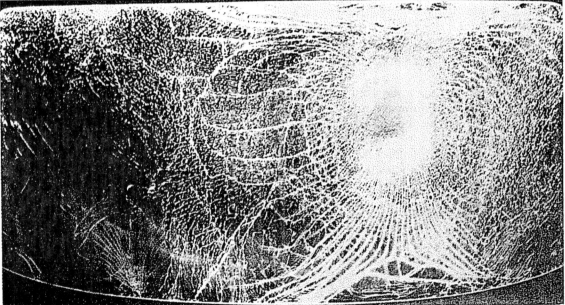

Figure 4. 3 mm + 3 mm Laminated Glass and Ten Twenty after 19 mph (30 km/h) Sled Impact.

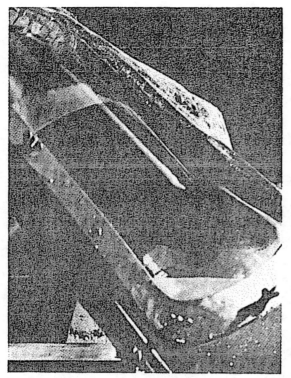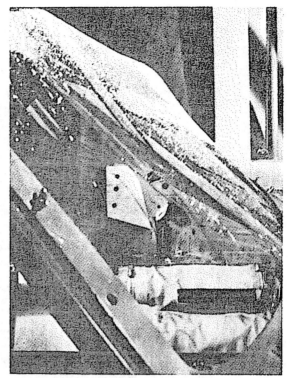

Figure 5. *3 mm + 3 mm Laminated and Ten Twenty Side View after 19 mph (30 km/h) Sled Impact*

In practice all three attributes contribute simultaneously and a case where half the average length, half the average depth and one third of the number of cuts occurred would rate 2 units TLI.

THE OBJECTIVE

A car windshield is required ideally to possess a number of performance characteristics apart from its normal transparency. It is expected:

1. To retain its transparency after normal road stone impact.
2. To retain weather protection after such an impact.
3. To retain an occupant thrown against the glass at an acceptable deceleration rate to avoid permanent brain damage.
4. To retain the occupant within the car.
5. To retain the occupant while causing a *minimum of lacerative injury*

The test program described briefly here (Reference 1) shows that the first four features are met by Triplex Ten Twenty in the same way as by conventional HPR laminated windshields, while providing the important additional advantage of a major reduction in laceration.

TRIPLEX TEST PROGRAM

For deceleration measurement a 10 kg dropping headform was used falling onto windshields mounted normally to the direction of impact. Measurements were made in terms of Gadd Severity Index (SI) and Head Injury Criterion (HIC) at speeds up to 38 mph (60 km/h).

The results are shown in Table 1 comparing Triplex Ten Twenty windshields with conventional 3 mm + 3 mm and with conventional 2.5 mm + 2.5 mm laminated windshields at 20°C.

The results are plotted on Figure 6 and show superior performance of Triplex Ten Twenty in terms of Gadd Severity Index. No significant difference was found in terms of HIC (Figure 7) nor in penetration velocities between the three types.

For laceration measurement a skull impactor, similar to that developed by Corning and shown in Figures 8 and 9, was used.

Windshields were clamped in an aperture set at the usual car mounting angle. The headform was covered first with a 6 mm thick cap of PVC to represent flesh and to enable the depth of "flesh" lacerations to be measured. Over this were placed two 1 mm layers of moist chamois leather to represent skin.

Table 1

RESULTS OF 10 kg HEADFORM NORMAL DROP-TEST ON TEN TWENTY 2.5 mm AND 3 mm WINDSHIELD

			Deceleration							
Test No	Windscreen Composition	Velocity km/h	Max Deceleration Due to Glass g	Duration of Deceleration Due to Glass ms	Max Deceleration Due to Interlayer g	Total Duration of Impact ms	Gadd SI	HIC	Total Interlayer Tears mm	Pene-tration
1965	3mm	38 9	56	2 2	45	111	240	187	230	No
1966	3mm	38 9	114	3 9	41	82	293	173	–	Yes
1967	2.5mm	38.9	55	2.2	39	57	234	193	180	No
1968	2 5mm	41 0	59	1 9	38	119	213	179	254	No
1969	2 5mm	43 0	63	1 2	41	114	212	172	252	No
1970	2 5mm	46 0	98	1 5	40	43	181	132	–	Yes
1971	Ten Twenty	38 9	37	1 4	40	55	233	198	130	No
1972	Ten Twenty	41 0	40	1 2	40	60	209	176	150	No
1973	Ten Twenty	43 0	42	0 9	38	148	172	141	–	Yes
1974	3mm	34 9	53	2 2	40	51	210	175	95	No
1975	2 5mm	34 9	58	1 7	35	61	193	168	105	No
1976	2 5mm	34 9	75	1 4	38	51	220	181	100	No
1977	2 5mm	34 9	38	1 9	35	120	167	140	262	No
1978	3mm	32 0	61	1 7	38	51	183	153	0	No
1979	2 5mm	32 0	57	1 2	40	51	234	192	100	No
1980	Ten Twenty	32 0	40	0 9	34	55	175	153	0	No
1981	3mm	27 8	45	1 7	31	51	115	95	15	No
1982	2 5mm	27 8	85	2 4	34	51	207	163	5	No
1983	Ten Twenty	27 8	39	1 2	33	50	131	113	55	No
1984	3mm	24 0	60	2 4	28	51	104	84	0	No
1985	2 5mm	24 0	53	1 9	30	51	117	98	30	No
1987	Ten Twenty	24 0	51	1 2	28	50	98	85	40	No
1990	2 5mm	20 0	32	1 4	23	51	64	57	0	No
1991	3mm	20 0	94	3 9	21	53	124	78	0	No
1992	Ten Twenty	20 0	49	2 2	24	51	68	59	0	No
1993	3mm	43 0	52	2 7	43	52	161	137	–	Yes
1994	Ten Twenty	43 0	45	1 4	41	60	232	193	–	No
1995	2 5mm	43 0	51	2 9	45	71	227	170	–	Yes
1996	3mm	41 0	43	2 2	43	71	160	141	–	Yes
1998	Ten Twenty	29 9	39	1 4	33	50	138	121	50	No
1999	2 5mm	29 9	86	2 4	36	49	176	138	50	No
2000	3mm	29 9	84	2 2	32	61	185	140	0	No
2001	2 5mm	20 0	77	2 9	23	52	95	56	0	No
2002	2 5mm	24 0	63	1 4	30	55	112	89	0	No
2003	2 5mm	27 8	52	1 4	33	48	123	103	50	No
2004	2 5mm	32 0	43	1 4	36	50	166	143	80	No
2005	2 5mm	38 9	65	2 2	45	56	273	215	190	No
2006	2 5mm	43 0	46	1 2	43	109	208	154	–	Yes
2007	3mm	20 0	112	3 9	18	51	172	122	0	No
2008	3mm	24 0	90	3 2	33	43	208	136	0	No
2009	3mm	27 8	88	2 9	34	47	179	136	0	No
2010	3mm	32 0	96	2 2	43	46	264	202	0	No
2011	3mm	38 9	65	2 4	48	47	292	244	100	No
2015	3mm	20 0	90	3 7	21	53	116	57	0	No
2017	Ten Twenty	20 0	31	1 7	27	52	76	63	0	No
2018	Ten Twenty	27 8	66	1 0	34	51	175	145	0	No
2019	Ten Twenty	24 0	29	1 7	29	50	91	78	0	No
2020	Ten Twenty	32 0	37	1 0	35	75	153	128	150	No
2021	Ten Twenty	38 9	36	1 2	38	42	93	72	–	Yes
2022	3mm	43 0	81	2 0	38	43	139	97	–	Yes

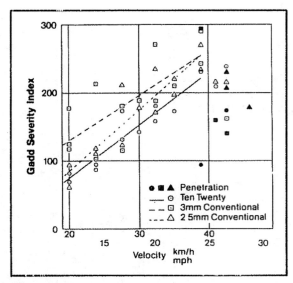

Figure 6. Gadd SI as Function of Impact Velocity by Dropping Headform Test

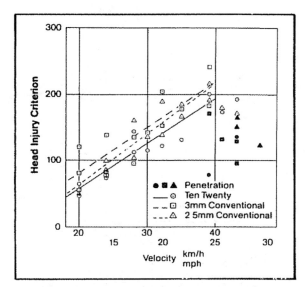

Figure 7. H.I.C. as Function of Impact Velocity by Dropping Headform Test

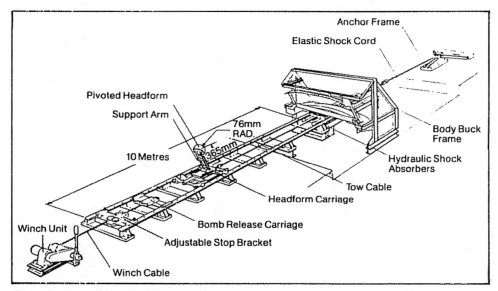

Figure 8. Skull Impact Test Rig

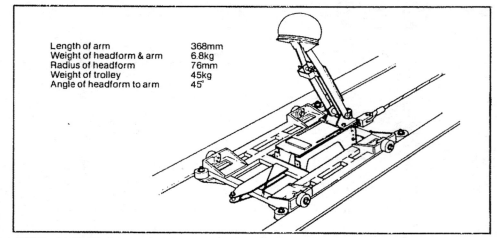

Length of arm	368mm
Weight of headform & arm	6.8kg
Radius of headform	76mm
Weight of trolley	45kg
Angle of headform to arm	45°

Figure 9. Carriage and Headform Assembly of Skill Impact Test Rig

A high degree of subjective assessment was found to be the weakness of existing laceration indices. A new objective index was therefore evolved (The Triplex Laceration Index, or TLI) based solely on measurements of cuts produced in the leathers and PVC facemasks. Full details of this index were the subject of a separate paper presented to the American Association of Automotive Medicine (Reference 2).

Mathematical analysis of experimental results led to the logarithmic formula:

$$TLI = 2 + \log_{10}(1 + 1.16\,D_1 + 50.8\,D_2 + 16{,}500\,D_3).$$

Where $D_1\,D_2$ and D_3 are the lacerative damage in the two layers of chamois and PVC cap respectively. The lacerative damage in each layer is computed as a function of the length, depth and number of cuts.

The results given in Table 2 and plotted on Figure 10 show a 2.7 (logarithmic) unit improvement over conventional 3 mm + 3 mm annealed laminated and a 2.0 (logarithmic) unit improvement in TLI over conventional 2.5 mm + 2.5 mm.

These results indicate reduced laceration by Triplex Ten Twenty of 5 X 100 and of 100 times respectively without penalty in other directions.

WAYNE STATE UNIVERSITY TESTS

The Deceleration and Lacerative performance of Triplex Ten Twenty and conventional annealed 2.5 mm + 2.5 mm laminates was compared on the Wham III crash simulator at Wayne State University using a Sierra 1050 anthropomorphic dummy in a modified Volvo car.

Comparisons were made at speeds between 16 and 56 km/h head impact speeds and the results are shown in Table 3 and on Figures 11 (TLI) and 13 (HIC).

Table 2

DATA FROM SKULL IMPACT RIG AT 20°C (TRIPLEX SAFETY GLASS)

Test No.	Type of Windscreen	Velocity of Impact. km/h	TLI	Total Interlayer Tears. mm
80	2.5mm	57.1	10.3	250
81	Ten Twenty	55.3	9.8	290
82	3mm	33.8	8.9	20
84	3mm	50.8	12.1	250
87	3mm	27.4	5.7	–
88	Ten Twenty	52.9	8.7	30
89	2.5mm	27.5	5.9	–
90	3mm	51.5	12.5	390
91	2.5mm	41.5	9.4	112
93	2.5mm	28.3	4.8	–
94	Ten Twenty	28.0	2.9	–
95	2.5mm	50.8	10.8	250
97	3mm	41.5	9.7	230
98	3mm	40.7	10	–
100	3mm	33.6	9.8	–
102	2.5mm	33.8	9.8	–
104	Ten Twenty	27.4	0	–
106	3mm	26.7	8.4	–
107	2.5mm	58.1	12.9	350
108	Ten Twenty	57.1	11.9	207
109	2.5mm	33.1	9.0	3
110	Ten Twenty	40.4	7.6	134
113	Ten Twenty	33.1	6	–
114	2.5mm	50.7	12.7	235
115	Ten Twenty	40.4	8.1	14
120	3mm	54.5	12.3	190
122	Ten Twenty	31.5	9.2	–
123	2.5mm	37.8	10.1	52
124	Ten Twenty	46.8	8.2	50
125	3mm	54.5	12.1	500
127	Ten Twenty	26.1	6	–
134	Ten Twenty	48.8	10	148
142	2.5mm	58.1	12.2	221
143	Ten Twenty	26.1	5.4	–
145	Ten Twenty	48.8	10.5	61
147	2.5mm	49.4	12.4	400
148	Ten Twenty	54.2	12.8	–
149	Ten Twenty	32.7	3.1	22
151	Ten Twenty	55.0	8.7	161
152	Ten Twenty	32.3	5.5	–
153	Ten Twenty	41.0	5.8	12
154	2.5mm	54.9	12.2	190
157	2.5mm	26.1	5.7	–
159	Ten Twenty	39.3	5.7	–

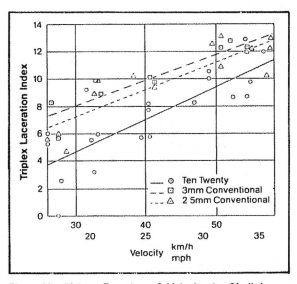

Figure 10. TLI as Function of Velocity by Skull Impact Test at 20°C

Though the Wham III and Triplex laceration tests showed a difference in absolute levels of TLI the Wayne Test program confirmed the improvement of 2 logarithmic units TLI for Triplex Ten Twenty compared with 2.5 mm + 2.5 mm annealed laminate.

At the same time Wayne found that HIC was not significantly increased and remained well below the accepted danger level of 1,000 (U.S. MVSS 208).

MECHANISM

The very large reduction in laceration is attributable to the fine fragmentation on head impact of the highly stressed inner glass of Triplex Ten Twenty.

When the head strikes a Triplex Ten Twenty windshield the energy of the impact plus the stored stress in the inner glass component causes the inner glass to break into a toughened type of fragmentation over the depth of the glass down which the head moves after initial impact. With the characteristic finer and more blunted fragments this causes much less laceration damage than when the head has to pass over the coarser and sharper fragments of a conventional annealed laminated windshield (see Figures 4 and, particularly, 5).

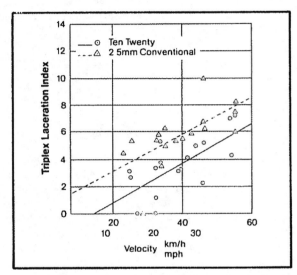

Figure 11. TLI as Function of Velocity by WHAM III Test Rig

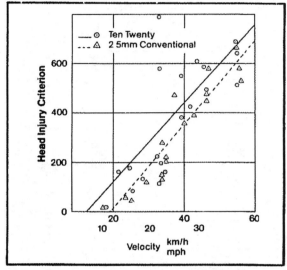

Figure 13. HIC as Function of Impact Velocity by WHAM III Test Rig

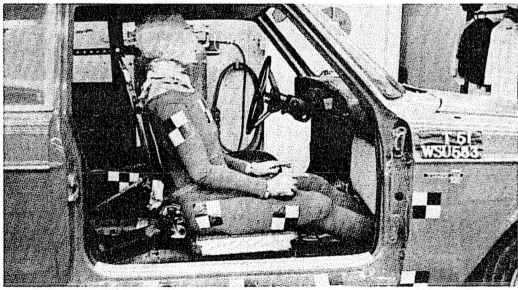

Figure 12. Vehicle with Dummy Installed on WHAM III

408

Table 3

SUMMARY OF DATA ANALYSIS
FOR TRIPLEX WINDSHIELD
LACERATION STUDY AT WSU

Run No.	Type of Windscreen	Velocity km/h	Stopping Distance mm	Head Index Gadd	Head Index HIC	Laceration Index WSU	Laceration Index Corning Scale	TLI	Pummel	Remarks
T- 4	2.5mm	34.1	410	253	220	3	4	4 9	3	Abrasion on forehead
T- 5	Ten Twenty	33.5	410	200	164	1	4	5 3	3	Light abrasion on forehead
T- 6	2.5mm	34.1	410	248	202	3	5	6 2	4	Windshield slightly pulled out abrasion on forehead
T- 7	Ten Twenty	32.8	410	249	196	1	3	3 8	5	Abrasion on forehead one cut through outer layer on nose
T-10	2.5mm	16.3	300	11	9	0	8	0	5	Head broke outer glass no cuts
T-11	Ten Twenty	16.9	200	13	11	0	8	0	5	Head broke outer glass no cuts
T-14	2.5mm	24.8	300	67	55	1	4	5 2	4	Light abrasion and few small cuts on forehead and nose
T-15	Ten Twenty	24.9	300	131	82	1	3	3 2	3	Abrasion on forehead
T-16	2.5mm	36.7	500	576	482	3	4	5 4	3	Severe abrasion and a few cuts on forehead buttocks slid off seat legs jammed under instrument panel
T-17	Ten Twenty	41.5	500	677	423	1	3	4 1	4	Severe abrasion on forehead and nose no SI acceleration
T-18	2.5mm	42.2	500	465	293	3	4	5 9	3	Lacerations and severe abrasion on forehead abrasions on nose
T-19	Ten Twenty	27.5	330	371	121	0	0	0	5	—
T-20	2.5mm	28.3	330	172	113	0	0	0	3	—
T-21	2.5mm	33.1	410	174	143	3	4	5 4	3	Abrasions and cuts on forehead
T-22	Ten Twenty	32.0	410	278	222	1	2	3 4	4	Scratches on forehead
T-23	2.5mm	.30	410	229	161	1	3	3 6	4	Scratches on forehead
T-24	Ten Twenty	32.3	410	175	131	1	1	1	4	Scratches on forehead
T-25	Ten Twenty	43.9	610	787	606	3	4	4 9	3	Severe abrasions on forehead dummy slid off seat
T-27	Ten Twenty	45.9	610	582	491	3	4	5 2	3	127mm tear (vertical) severe abrasions with few cuts on forehead and nose
T-28	2.5mm	46.0	610	570	482	3	3	6 2	5	50mm tear several cuts on forehead and near right eye
T-29	2.5mm	45.9	610	593	463	6	5	6 8	3	More severe abrasions on forehead cuts on forehead, eyes, and nose
T-30	Ten Twenty	54.1	710	871	623	6	6	7 1	3	267mm jagged tear, extensive lacerations through outer chamois and couple of cuts through inner chamois
T-31	2.5mm	55.0	710	747	584	6	7	8 3	3	50 x 152mm tear, severe lacerations through both layers of chamois 11mm cut on chin in rubber flesh
T-32	Ten Twenty	55.0	710	798	643	6	5	7 3	3	216mm jagged tear, severe lacerations through both layers of chamois no cuts in rubber flesh
T-33	2.5mm	55.0	710	905	673	6	5	7 4	3	Severe lacerations through layers of chamois no cut in rubber flesh
T-35	Ten Twenty	22.5	330	258	167	0	0	0	3	No fracture
T-36	Ten Twenty	32.2	410	1380	811	0	0	0	3	No fracture
T-37	Ten Twenty	39.3	500	836	543	1	2	3 1	3	—
T-38	2.5mm	23.3	330	114	68	1/3	3	4 4	4	—
T-45	Ten Twenty	32.2	410	1069	568	0	0	0	4	No lacerations
T-47	2.5mm	33.1	410	329	268	3/6	5	5 8	4	—
T-49	Ten Twenty	25.4	330	330	185	1/3	3	2 7	4	—
T-50	2.5mm	40.1	500	487	369	3	4	5 4	4	—
T-51	Ten Twenty	39.4	500	534	381	1	2	3 1	4	—
T-54	2.5mm	47.0	610	689	581	10	9	10 0	3	Severe lacerations cuts (15 and 4mm) in flesh of dummy
T-55	Ten Twenty	45.9	610	1054	585	1	2	2 3	3	—
T-58	2.5mm	55.0	710	629	528	6	5	6 0	4	63 and 140mm tears
T-59	Ten Twenty	54.9	710	1106	689	3	3	4 5	4	—

REFERENCES

1. Improved Laminated Windshield with Reduced Laceration Properties, S. E. Kay and J. Pickard (Triplex) and Professor L. M. Patrick (Wayne State University), presented at 17th Stapp Conference, Oklahoma, Nov. 1973.

2. An Objective Method of Assessing Laceration Damage to Simulated Facial Tissues, The Triplex Laceration Index, J. Pickard, P. Brereton and A. Hewson (Triplex), presented at AAAM Conference, Oklahoma, Nov. 1973.

Car Crash Tests of Ejection Reduction by Glass-Plastic Side Glazing

Carl C. Clark
Office of Vehicle Research
National Highway Traffic Safety Administration
Peter Sursi
Vehicle Research and Test Center
National Highway Traffic Safety Administration
Currently with: Safety Testing, Inc
Transportation Research Center of Ohio
East Liberty. OH

ABSTRACT

1983 ejection statistics are reviewed: half of the passenger car ejections, some 36,000 people of whom 5,346 died, are through glazing areas. Previous work has shown the remarkable strength of thin plastic coatings, developed for windshield anti-laceration applications, when applied to the inside of tempered glass side windows, in reducing ejection. In the present work, two tests were made, each with the NHTSA Moving Deformable Barrier (MDB) at 39 mph and all four wheels turned at 26 degrees, striking a stationary Volkswagen Rabbit in a perpendicular impact. The Alfred I. DuPont de Nemours Company provided the plastic coating on tempered glass side windows. The plastic layer extended beyond the sides and top of the glass to be wrapped around steel strips bolted to the window frame. On vehicle impact, the tempered glass broke, but the pieces were held in place by the plastic layer, which then deformed outward as a "safety net" with head contact. In the first test, with a low and short striking vehicle hood line simulation, the top of the door bent out as the bottom was pushed in, and the head bulged the glazing out to hit the elevated base plate of the MDB, although the "safety net" held. In the second test, with a more typical vehicle hood line effect simulation, the "safety net" again held, giving a HIC of 616. A window design to allow window up and down motion with the plastic layer movably secured with a "T edge" in the window channel is under construction. An eight inch sphere Glazing Test Device is being developed, with a skin simulation / chamois coat on the lower hemisphere for laceration measurement, accelerometers for Head Injury Criterion (HIC) measurement, and a weight variable from 10 to 20 to 40 pounds, for ejection reduction measurement.

This paper presents the views of the authors, and not necessarily those of the National Highway Traffic Safety Administration (NHTSA). The numbers in parentheses are the references, listed at the end of the paper.

This paper is dedicated to the memory of Henry H. Wakeland, late of the National Transportation Board, who was instrumental through his work in New York in establishing government efforts to build experimental safety cars, and to the memory of William Haddon, Jr., M.D., late of the Insurance Institute of Highway Safety, who taught the world that automobile safety is one of the most important areas of preventive medicine, and that cars could be built, passively, not to harm. Both of these giants of automobile safety research died in early 1985.

EJECTION STATISTICS

EJECTION IS A MAJOR CONTRIBUTION to death in motor vehicle crashes (Table 1), with ejection of the whole body or partial ejection, usually of the head, being involved in about a quarter of the passenger car deaths or on average of all motor vehicle deaths, rising to 38 percent for light trucks and vans. Our earlier paper (1) reviewed the 1981 statistics; this paper presents 1983 data from the Fatal Accident Reporting System (FARS 83) (2) and from the National Accident Sampling System (NASS 83) (3). Hopefully, with the expected increase in the use of occupant restraints in response to mandatory child and adult crash protection use laws by the states, occupant crash deaths will further decrease. Note also that medical technology is succeeding in keeping more injured people alive that previously would have died, including many that are comatose or quadruplegic. A person who survives a motor vehicle crash more than 30

0148-7191/85/0520-1203$02.50

days, and then dies, is not counted as a motor vehicle death by NASS or FARS.

With the decrease in total motor vehicle deaths from 49,268 in 1981 to 42,584 in 1983, and passenger car occupant deaths from 26,545 in 1981 to 22,975 in 1983, the latter a 13.5 percent reduction, reported passenger car ejections have decreased from 43,200 in 1981 to 36,000 in 1983, a 17 percent reduction. However, passenger car deaths with ejection decreased from 5,907 (FARS) in 1981 to 5,346 in 1983, a 9.5 percent decrease. Malliaris, Hitchcock, and Hansen (4) similarly report an increase of risk of injury from 1980 to 1983 in spite of a decrease in fatalities. The decrease in the number of people involved in accidents is greater than the decrease in injuries, so that the risk of injury for those involved in crashes is increasing.

The relative importance of ejection in automotive safety concern may be increasing, although we stress the possible judgemental errors of the investigators in concluding whether an ejection has occurred or not. Often, bodies are moved before the police arrive. Partial ejection as an injury cause is judged by investigators by blood on the hood or dirt in the hair, etc., and hence may well be under-reported. In crash simulation work, which admittedly deal with only certain of the road crash conditions, we see partial ejection with dummy contact, against the hood of a striking car in side crashes for example, far more frequently than we see whole body ejection. From FARS 83 (2) only 18 percent of the reported ejections with death are partial ejections. (See Table 1 below).

For the cases with data for crash change of velocity analysis, ejection occurs most frequently in the more violent crashes. Malliaris et al. (4) report a bimodal distribution of ejectees with change of velocity in the crash, with 42 percent of the ejections occurring with a change of velocity in the crash of under 30 mph, and 57 percent of the ejections occurring with a change of velocity of more than 40 mph. They point out that this distribution does not include crashes for which the change in velocity cannot be determined, including many rollover crashes in which ejections frequently occur, often well before the cars come to rest. Including rollovers, we surmise that more than half of the ejections occur with a change of velocity up to the time of ejection of less than 30 mph, and perhaps even less than 20 mph.

Table 2 data, analyzed in Table 3 below, show that the risk of serious or greater injury (Abbreviated Injury Scale or AIS equal to 3 or greater, including death) in NASS towaway passenger car crashes increases from 0.02 for crashes without ejection or rollover, to 0.04 without ejection but with rollover, to 0.26 with ejection but no rollover, to 0.311 with ejection and rollover: almost a third of the people in passenger car towaway crashes with ejection and rollover will have serious or greater injuries. Table 4 below gives the overall picture of deaths and injuries in selected motor vehicle crashes. In spite of present safety standards and automotive industry safety countermeasures, much more remains to be done in increasing motor vehicle safety.

As an example, from Table 2 below, 18,000 people are ejected of 3,032,000 passenger car occupants in towaway crashes without rollover, whereas 18,000 additional people are ejected of the 228,000 passenger car occupants in towaway crashes with rollover. Note that 35,000 of these 36,000 passenger car ejectees are injured, with 10,200 of them having AIS 3 or above injuries (Table 2), and 5,346 of them being killed (Table 1). Malliaris et al. (4) report that about half of the primary causes of injuries of ejectees occur with contact against interior structures. Kahane (5) reports that about 40 percent of fatally injured ejectees suffered life-threatening lesions due to contacts against interior structures. To reduce ejection deaths, we must not only prevent the glazing area ejection in a way that itself does not produce severe injury but we must also soften the other interior contacts that preceed ejection.

Tables 5 and 6, using computer accessed data from the 1977-79 National Crash Severity Study (NCSS), give the routes of ejection of those ejected and killed, or of those ejected with serious or greater injuries (AIS>=3). Malliaris et al. (4) give moderately similar values for NASS 79-83. Industry staff have suggested to Clark that in this period door latches have been further improved, but this effect is not yet apparent in the studied road crash data.

Note that about half of the ejectees go through the side doors and half go through glazing, with the side windows being the most significant glazing ejection route. The great majority go through glazing and doors which have been loosened or opened by crash forces prior to body contact. For example, Kahane (5) reports that 99 percent of windshield area ejections are with significant bond separation rather than High Penetration Resistant (HPR) glazing penetration.

The NCSS data indicate that in cars of the mid '70's about half of the ejections with death through side windows occur with the windows open. We question these data, for when viewing cars on the road one sees less than half of the driver windows open, particularly with the newer cars with air conditioning and the public's desire for a quiet ride. Tempered glass side windows are frequently broken in fatal crashes. Without care in studying the window crank and broken glass positions, one might mis-report that the window was open.

Table 1: Number of Occupant Fatalities by Ejection and Vehicle Type (from FARS 83, Table II-14)

	Passenger Cars	Light Trucks	All Motor Vehicles Except Mototcycles
Not ejected but killed	16,993	3,243	21,431
Totally ejected and killed	4,371	1,739	6,961
Partially ejected and killed	975	313	1,406
Ejected and killed	5,346	2,052	8,367
Unknown if ejected, but killed	636	84	778
Total occupants killed	22,975*	5,379	30,576**
Percent of killed that are ejected	23%	38%	27%

* This includes 222 deaths in convertibles.
** This does not include 4,264 fatally injured motorcycle and other motorized cycle riders.

Table 2: Ejection and Rollover Injuries in Passenger Car Towaway Crashes (from NASS 83, Table III-30)

	No rollover	Rollover or not	Rollover
All occupants	3,032,000 a	3,260,000 g	228,000 m
not ejected	3,014,000 b	3,224,000 h	210,000 n
ejected	18,000 c	36,000 i	18,000 o
Injured occupants	1,474,000	1,623,000	149,000
not ejected	1,457,000	1,588,000	131,000
ejected	17,000	35,000	18,000
AIS >=3 occupants	65,400 d	80,000 j	14,600 p
not ejected	60,800 e	69,800 k	9,000 q
ejected	4,600 f	10,200 l	5,600 r

Table 3: The Chance of Serious or Greater Injury (AIS >=3) if in Passenger Car Towaway Crashes (derived as shown from Table 2 above)

	No rollover	Rollover or not	Rollover
No ejection	e/b= 0.0202	k/h= 0.0217	q/n= 0.0429
Ejection or not	d/a= 0.0216	j/g= 0.0245	p/m= 0.0640
Ejection	f/c= 0.256	l/i= 0.283	r/o= 0.311

Table 4: Deaths, Injuries, and Vehicles in Police Reported Accidents (from FARS 83 and NASS 83)

	Motorcycles	Passenger Cars	All Motor Vehicles
Deaths			
Riders or occupants	4,264	22,975	34,840
All involved			42,584
People injured seriously of greater (AIS >=3)			
Riders or occupants	29,000	89,000	140,000
All involved			166,000
All injured (AIS >=1)			
Riders or occupants -a	182,000	2,535,000	3,203,000
All involved			3,414,000
Total people in police reported accidents			
Riders or occupants	209,000	11,514,000	14,578,000
All involved			14,852,000
Total number of vehicles in police reported accidents -b	174,000	7,710,000	9,869,000
Occupants injured/vehicle (a/b)	1.05	0.329	0.325
Total registered vehicles	5,585,000	126,728,000	169,446,000

Table 5: Routes of ejection with fatality from passenger cars for those occupants with identified routes of ejection (NCSS).

Killed	917	Ejected and killed, known route:	119
Not ejected	593	Ejected through doors	52%
Unknown if ejected	112	Ejected through side windows	34%
Known ejected	212	Ejected through windshield	8%
Unknown route	88	Ejected through "roof"	3%
Unlisted	5	Ejected through rear window	3%

Percent ejected of those killed with known ejection status: 27%
Percent ejected through glazing area of those killed: 47%
(This assumes that two thirds of the "roof" ejections are through glazing areas).

Table 6: Routes of ejection with serious or greater injury from passenger cars for occupants with identified routes of ejection (NCSS).

Seriously or greater injured	2906	Ejected and seriously or greater injured, known route:	209
Not ejected	2412	Ejected through doors	44%
Unknown if ejected	178	Ejected through side windows	32%
Known ejected	316	Ejected through windshield	11%
Unknown route	103	Ejected through roof	11%
Unlisted	4	Ejected through rear window	2%

Percent ejected of those with serious or greater injury with known ejection status: 12%
Percent ejected through glazing areas of those with serious or greater injury: 52%

It is interesting to compare Tables 5 and 6. With less severe crashes included in Table 6, there is the suggestion that less door ejection occurs, and more glazing ejection, including "roof" glazing ejections, although indeed the numbers of cases observed are so small that this is a suggestion and not a definitive conclusion. In these analyses, the "roof" ejections include ejections through roof glazing or through other "T top" closures, and ejections from open top vehicles such as convertibles. Note that from Table 1, 222 of the 22,975 occupant deaths in passenger cars were in convertibles. Further studies of FARS, NASS, and NCSS of passenger cars excluding convertibles or open top passenger cars would be appropriate, to understand the conditions of "roof" ejection. A study of the "hard copy" details of the crash reports to distinguish "open top" from roof opening or roof glazing ejections would also be appropriate. For estimation purposes in Tables 5 and 6, we assume that two thirds of the "roof" ejections are through glazing areas.

Kahane (5) has estimated that some 100 lives have been saved in changing from windshields with rubber gaskets to bonded windshields. Since 99 percent of the fatal ejections through the windshield area are still with bond separation, a similar order of gain might be expected with a glass-plastic windshield with a "safety net" attachment of the inner plastic layer, and a greater gain from the other more important glazing ejection routes (Table 5).

EARLIER EXPERIMENTAL CRASH STUDIES OF EJECTION

A six year old child dummy (44 pounds) moving with its spine parallel to the car long axis and striking the windshield at 21.3 mph (a 185 inch drop) broke but did not penetrate through the standard High Penetration Resistant (HPR) windshield (1). Children eject through windshields that have already partially separated. The seated unrestrained adult will have head penetration (partial ejection) through the windshield at 25-30 mph (6). But if the adult has been thrown into the air so that the spine is parallel to the car long axis, he ejects totally through the bonded HPR windshield, with glazing penetration rather than bond separation, at about 14 mph (1).

A child, striking the tempered glass side window with some body weight behind the head, can break the window at 14 mph (7), and eject. Likewise, the seated adult head striking the tempered glass side window in a side impact will break the glass and have head partial ejection at about 14 mph (1).

Reducing window ejection is a significant child crash restraint means. We have shown (1) that a 6 year old (44 pound) child dummy moving at 20 mph with the head and spine aligned perpendicular to the glazing can be stopped by the DuPont glass-plastic side glazing, with the plastic layer attached as a "safety net" at the window margins, with the plastic stretching and holding after the glass breaks.

CAR CRASH TESTS OF GLASS-PLASTIC SIDE GLAZING

We have conducted the first two public car crash tests of vehicles with glass-plastic side glazing. (The automotive and glazing industries are conducting developmental tests of glass-plastic windshields, but we know of no public report of cars crash tests with glass-plastic side glazing). These were carried out with the cooperation of the NHTSA Vehicle Research and Test Center "Safety Research Laboratory Project 92 - Side Impact Aggressiveness Attributes" staff, with the further cooperation of the State of Ohio Transportation Research Center staff, all in East Liberty, Ohio.

Both tests were 90 degree impacts of the 3000 pound Moving Deformable Barrier (MDB) (8) moving at 39 mph into the driver side of stationary two door Volkswagen Rabbits of 2500 pounds test weight. The Rabbits had no special padding or structures. Driver and left rear passenger instrumented HSRI Side Impact Dummies were in position. The Alfred I. DuPont de Nemours Company, Polymer Products Department, Wilmington, DE 19898, provided the plastic coatings on the Volkswagen tempered glass windows, consisting of 0.015 inches of polyvinyl butyral and 0.04 inches of polyethylene pterthalate (to provide the inside abrasion resistant surface). The plastic layers extended beyond the sides and top of the glass, and were wrapped around steel strips that were bolted to the window frame. The MDB wheels were all turned at 26 degrees, so that the body long axis made an angle of 26 degrees to the track along which the MDB was drawn. The Volkswagens were at 64 degrees to the track, so that the simulated car front face of the MDB was parallel to the Volkswagen door plane. This dynamics and geometry simulates a crash with the MDB moving at 35 mph and the Volkswagen moving at 17.5 mph. The left front contact of the MDB was 37 inches forward of the midpoint of the Volkswagen wheelbase, just behind the front wheel.

In the first test, DOT 0771 in the NHTSA Office of Vehicle Research Data Center "Vehicle Crash Test Data Base," on September 19, 1984, reported by (9), figure 1 post-crash, the MDB had an "Altered Profile" lower than average hood elevation, striking the Volkswagen door several inches below the usual car to car door contact level. Also, the crushable honeycomb front end simulation is only 19 inches from the bumper front plane to the honeycomb back plate, which rises above the honeycomb to about the level of the bottom of the Volkswagen window.

This MDB "hood" did not simulate the depth of an average car hood. With side penetration in the Volkswagen of 18.4 inches at the level of the H point (dummy hip), the door bent in at the bottom and out at the top, making an angle after the crash of the front of the glazing frame of the door of about 45 degrees from the original door plane. The tempered glass broke due to crash deformation of the door, but the pieces were held in place by the plastic coating. The head hit the plastic "safety net", but because the window frame was bent out and went farther dynamically as the head loaded the plastic layer, the head hit the top of the back plate of the moving barrier through the plastic, making a small tear in the plastic, with a 340 G head resultant acceleration before going off scale at 69 milliseconds, and a Head Injury Criterion (HIC) greater than 3027, the value if the acceleration had not gone off scale. This test illustrated that the glass-plastic glazing would stay intact in a 39 mph side crash, but the window frame should not bend out so far to allow the head to hit outside structures even through the plastic.

The second car crash test, DOT 0780 in the data base, was on October 26, 1984 (10). The Moving Deformable Barrier (MDB) had a "lowered bumper," to engage the side sill of the Volkswagen and so reduce penetration, and a higher than average hood elevation simulation, although the honeycomb base plate was still at 19 inches from the plane at the front of the bumper. The maximum penetration at the level of the H-point was 14.5 inches, four inches less than with the standard bumper. This difference had negligible influence on the driver thorax loads (11).

Figures 2, 3, and 4 show the post-crash conditions. Again, the tempered glass broke, but the pieces were held by the plastic "safety net." There was some separation of the plastic at the back of the window (figure 3). The door, pushed in by the striking MDB, hit the seat, hip, and thorax of the driver dummy. The driver bucket seat was rotated rapidly, rotating the dummy, which ended up in the passenger seat, partially facing the driver window. During this time, the hip was pushed vigorously (pelvis +220 Gy max. at 26 ms), lifting the body off the seat. From the films, the body is seen to slide across the seat toward the driver door and the head to displace laterally toward the left shoulder, then rotate into the plastic "safety net," with a +57 Gy peak of the head at 41 ms.

Figures 5-8 show the driver dummy head accelerations. From figure 5, the head in the X axis is first accelerated rearward (-31 Gx at 41 ms) then is accelerated forward as the turning seat and deforming door pushes the chest forward (head +10 Gx at 58 ms), then is accelerated rearward again (-27 Gx at 63 ms), partially due to the head / body yaw rotation toward the window and perhaps some "plow down" in the plastic layer. From figure 7, head Z axis, the head is first accelerated upward (+15 Gx at 27 ms), then "downward" along the spine (-85 Gx at 60 ms), which is increasingly toward the right side of the car as the body rolls left with the tremendous but brief load on the pelvis, and continuing load on the thorax.

This complex reversal of motions, seen particularly in the X axis, as the body is first accelerated backward, then forward by the penetrating door and rotating seat motion, produces a bimodal head resultant acceleration curve, figure 8, with a head injury criterion (HIC) of 616. If the head, through the deforming plastic "safety net," had hit outside structures, a HIC over 1000 would be expected. Note that with an open window but no head contact, a HIC of under 1000 could occur. Neck loads were not measured in these car crash tests, but remain of concern (1). It is clear that in this very severe side impact (39 mph), protection of more than the head is required to further reduce occupant injury, but glass-plastic glazing can significantly reduce adult head partial ejection and child ejection injuries.

The rear left passenger dummy head, hitting the rear window header (+155 Gy peak at 51 ms) had a HIC of 1583. The rear dummy shoulder indented through the rear window ledge (figure 2) (+64 Gy of the upper spine at 48 ms), with a partial ejection of the shoulder and later the head. This dummy did not receive as extensive door intrusion loads as the driver and remained in the left rear passenger seat.

In future work, we will explore means to prevent excessive bending out of the window frame when the glass-plastic glazing is loaded. Additional door latches around the periphery, and particularly at the window top, will be one method to explore.

CAN A GLASS-PLASTIC "SAFETY NET" WINDOW STILL BE OPENED?

In our preliminary work, we have attached the plastic layer extending beyond the window glass by wrapping it around steel strips which are then bolted to the window frame, preventing the window from opening. We have designed and are building a prototype of the "T edge" structure at the window edge to which the plastic layer is attached, which can then be movably secured within the window channels (figures 9 and 10). This plastic T edge structure would be bonded to the edge of the glass-plastic glazing, possibly with a plastic cap structure for the top edge, although this may not be necessary. The window channels would require the channel frame stiffeners (figure 9) to prevent the T edge from coming out of the channel, and the top of the window frame would have to be openable, to allow replacement of the window, together cost items in mass production of under $10., we estimate.

PRACTICABLE LABORATORY TESTS FOR ADEQUATE EJECTION AND LACERATION REDUCTION GLASS-PLASTIC GLAZING

A review has been made of the American National Standards Institute and other tests of automotive glazing (12, 13), with emphasis on attaining a more direct simulation of road use conditions, and automotive glazing effects, as we contemplate a growing use of glass-plastic glazing, now a standard item for the windshields of 1985 Cadillac Sevilles, for example, although without the "safety net" attachment of the inner plastic layer. The present ANSI Z-26 standard (14), and Federal Motor Vehicle Safety Standard 205 (15) which was historically derived from the ANSI standard, does not measure the head injury criterion (HIC), nor the laceration aspects of the glazing, nor the ejection reduction implications. Practicable laboratory tests should be as inexpensive and simple as possible yet provide the key information of the expected glazing performance. Hence we are in the early stages of exploring an additional set of tests using the "Glazing Testing Device," figure 11.

The prototype is an eight inch sphere made from a plastic bowling ball, hollowed out to allow lead weights and a center mounted triaxial accelerometer. At ball weights of 10 and 20 pounds, the Device will expectedly provide HIC estimations correlatable with heads when seated upright or when tipped forward, with some body weight behind the head. A metal plate, for electromagnetic drop release of the Device, is fashioned on the top hemisphere.

The lower hemisphere is covered with a skin simulation of plastic and chamois layers, classically used to evaluate lacerations. With the indications that the present glass-plastic glazings eliminate glass skin lacerations, the measurement method of the lacerations, a significant cause of the variation of results with the method, could be greatly simplified.

A body is a series of elastically connected masses whose effective mass varies with the time after the start of an impact. A body striking and deforming but not penetrating a glass-plastic glazing collapses, often off the glazing, with a complex time variation of the effective mass at the contact points. None the less, it is felt that a unit mass with a skin simulation coat can be correlated with the ejection and laceration reduction properties of glass-plastic glazing for people. Since a 44 pound child dummy falling at 20 mph with the spine perpendicular to the glazing has been stopped without penetration by the DuPont glass-plastic glazing (1), experiments are beginning with a Glazing Test Device which will be at 10, 20, and 40 pounds. It is expected

that most adults, hitting glazing in ejection also are hitting surrounding structures. Therefore, the glass-plastic glazing does not have to absorb all of the energy of the body, but only enough to prevent the ejection. It is our estimate that a glass-plastic glazing that stops a 40 pound unit mass moving at 20 mph (not yet carried out, although we think within "safety net" glass-plastic capability) would also prevent and probably without lethality the whole body ejection through glazing areas of more than a quarter of the small children and more than an eighth of the adults who are now totally ejected and killed, and more than half of the partial ejections through glazing areas of children and more than a quarter of the partial ejections through glazing areas of adults who are now partially ejected and killed.

Note that these tests require the use of full size automotive glazing in their standard mounts, i.e., test bucks of glazing frames or larger automobile pieces. In the Agency review that led to permitting the use of glass-plastic glazing (17), several groups recommended in the Docket that glazing tests should use the full size glazing rather than the small "coupons" presently used. Particularly with ejection, edge effects of the glazing in its mounting, and bond separation effects, become very important. Kahane (5) also discusses unrealistic results for road glazing separation conditions which may develop from laboratory studies with small "coupons" of the glazing.

The commercial ejection reduction glass-plastic glazing, still to be developed, may have a different plastic thickness than the samples used thus far, which were developed for the anti-laceration application alone. Because of possible neck load problems (1), the stopping distance of the plastic "safety net" deformation cannot be too short, nor indeed too long if the body is not to be hit in rollover, or the head not to hit penetrating structures. The practicable ejection reduction test might be that the glazing in its automotive mounting will stop a defined Glazing Test Device mass moving at a defined speed with a deceleration within defined limits.

FUTURE WORK WITH GLASS-PLASTIC GLAZING

Is it possible to safely prevent ejection through glazing? From the work we and cooperating groups have done in making the glass-plastic glazing into "safety nets," and showing by tests that even the anti-laceration version of the glass-plastic glazing will stop a 44 pound child moving at 20 mph, and that this "safety net" will stay intact through violent car crashes, we conclude that it would be possible with glass-plastic glazing to prevent the deaths of many people who with present glazing would eject and be killed by external impacts. We will continue our study of the forces and conditions (18) involved in ejection through glazing areas, including the contact forces on the surroundings.

In one drop test at 12 mph of a seated adult hitting the side door (7), with tempered glass, the glass broke out as the door was bent out, and whole body ejection occurred. With the 3M glass-plastic glazing and a new door, the glass still broke and the door still bent out but the inner plastic "safety net" held, reducing the head and shoulders excursion out of the window, and preventing the ejection. In the most hazardous situations (Table 3) with rollover and ejection (other than the truck crushing situations), the load initiating rollover and body contact with the glazing preceeds ground contact with sufficient time that the "safety net" may generally return the body or head to within the occupant compartment frame structure before a lethal ground contact, particularly when an elastomeric polymer is used for the plastic layer.

We hope to continue our work with the glazing industry, testing whether an attached plastic layer "safety net" of a glass-plastic windshield can reduce windshield area ejection deaths, and testing improvements of their other glass-plastic products, including glass-plastic windshields with only one layer of glass. If the weight of automotive glass can be further reduced with acceptable road noise isolation and safety aspects, the desired evolution to the use of glass-plastic glazing throughout the car will be furthered.

We will continue to study our present estimate that such glass-plastic glazing in mass production for windshields, and side, rear, and top windows, with the necessary mounting means, will add less than $100. to the cost of the car, and hopefully less if glazing weight reductions can be effected. For the windshield alone, the estimated cost for a change to a glass-plastic windshield was $38 to $45 (17). We plan additional car crash tests and drop tests to advance the development of what can become the proven technology of glass-plastic glazing with both anti-laceration and reduced ejection properties.

CONCLUSIONS

We have shown that glass-plastic glazing can reduce ejections as well as reduce lacerations. An early estimate from NCSS 1977-79 data was that there were some 300,000 lacerations per year occurring with automobile glazing (16). More recent analyses using NASS (5) indicate more than 400,000 lacerations per year are still occurring with HPR windshields. Indications are that most of these would be prevented by glass-plastic glazing (17).

As we discuss, there are also some 18,000 people who eject through glazing areas of passenger cars each year, with 17,500 of them being injured and some 2600 of them being killed. We believe that glass-plastic glazing can save many of these lives, if the glazing is mounted in the "safety net" fashion.

###

REFERENCES

1. Carl C. Clark and Peter Sursi, "The Ejection Reduction Possibilities of Glass-Plastic Glazing," Report SAE 840390, Society of Automotive Engineers (SAE), Warrendale, PA 15096, March, 1984.

2. National Center for Statistics and Analysis, "Fatal Accident Reporting System 1983," FARS 83, National Highway Traffic Safety Administration (NHTSA), Washington, DC 20590, 1985.

3. National Center for Statistics and Analysis, "National Accident Sampling System 1983," NASS 83, NHTSA, 1985.

4. A.C. Malliaris, Ralph Hitchcock, and Maria Hansen, "Harm Causation and Ranking in Car Crashes," Report SAE 850090, SAE, February, 1985.

5. Charles J. Kahane, "An Evaluation of Windshield Glazing Installation Methods for Passenger Cars," Report DOT HS-806 693, NHTSA, February, 1985; available from the National Technical Information Service (NTIS), Springfield Virginia 22161.

6. R.G. Rieser and G.E. Michaels, "Factors in the Development and Evaluation of Safer Glazing," Proceedings of the Ninth Stapp Car Crash Conference (1965), University of Minnesota Printing Department, 1966.

7. Carl C. Clark, "Learning from Child Protection Devices and Concept from Outside of the United States," Report SAE 831666, SAE, October, 1983.

8. Sol Davis and Carl Ragland, "Development of a Deformable Side Impact Moving Barrier," pages 646 - 677 in the Eighth International Technical Conference on Experimental Safety Vehicles, October, 1980, Report DOT HS 805 555, NHTSA, October, 1981.

9. Laura J. Bell, "Side Impact Aggressiveness Attributes, MDB-to-Car Side Impact Test of a 26 Degrees Crabbed Moving Deformable Barrier to a 1981 Volkswagen Rabbit at 38.9 mph," Vehicle Research and Test Center Report, SRL Project 92, Vehicle Crash Test Data Base Test DOT 0771, NHTSA, October, 1984.

10. Laura J. Bell, "Side Impact Aggressiveness Attributes, MDB-to-Car Impact Test of a 26 Degrees Crabbed Moving Deformable Barrier to a 1981 Volkswagen Rabbit at 39.1 mph," Vehicle Research and Test Center Report, SRL Project 92, Vehicle Crash Test Data Base Test DOT 0780, NHTSA, December, 1984.

11. Michael Monk and Donald Willke, "Side Impact Aggressiveness Attributes," Event Report, a report of Vehicle Research and Test Center SRL Project 92, National Highway Traffic Safety Administration, Washington, DC 20590, (in preparation for April, 1985). 12. Harold Wakeley, Linda Wolf, and Steven Godin, (IIT Research Institute), "Optical Engineering Research on New and Used Automotive Glazing," Final Contract Report, Report DOT HS-806 690, NHTSA, April, 1983. Available from NTIS.

13. Edward Jettner, Carl C. Clark, and Harold Wakeley, "A Review of Glazing Road Use Conditions and Laboratory Simulations," Report SAE 840387, Society of Automotive Engineers.

14. American National Standards Institute (ANSI), "ANS Z-26.1-1983, Safety Glazing Materials for Glazing Motor Vehicles Operating on Land Highways – Safety Code," American National Standards Institute, Inc., 1430 Broadway, New York, NY 10018, 1983.

15. National Highway Traffic Safety Administration, "Federal Motor Vehicle Safety Standard 205, Glazing Materials," Code of Federal Regulations, Volume 49 (Transportation), Chapter 5 (National Highway Traffic Safety Administration), Part 571.205. (This is abbreviated as 49 CFR 571.205). The Code is printed annually by the Superintendent of Documents, U.S. Government Printing Office, Washington, DC 20402. Revisions appear in the Federal Register, also available from the Superintendent of Documents, often with background material which is not included in the next printing of the Code.

16. Susan Partyka, "Glass Related Injuries on NCSS," in Accident Data Analysis of Occupant Injuries and Crash Characteristics – Eight Papers, National Center for Statistics and Analysis Collected Technical Studies, Volume 2, Report DOT HS-805 884, NHTSA, April, 1981. Available from NTIS.

17. National Highway Traffic Safety Administration, "Final Rule – Amendment of FMVSS 205, Glazing Materials, To Permit the Installation of Glass-Plastic Glazing as Windshields and Windows in Motor Vehicles," Federal Register, Volume 48, pages 52061 – 52066, November 16, 1983. (This is abbreviated as 48 FR 52061, Nov. 16, 1983). Superintendent of Documents, U.S. Government Printing Office, Washington, DC 20402. In the revised standard, "glass-plastic glazing material" is defined as "a laminate of one or more layers of glass and one or more layers of plastic in which a plastic surface of the glazing faces inward when the glazing is installed in a vehicle."

18. Arnold K. Johnson and David A. Knapton, "Occupant Motion During a Rollover Crash," Report DOT HS 806 646, National Highway Traffic Safety Administration, November, 1984; available from NTIS.

###

Figure 1. Side impact, Test DOT 0771, on September 19, 1984, Moving Deformable Barrier (MDB) with a low hood elevation into a Volkswagen Rabbit, at 39 mph. With the low door impact, the door pivoted in at the bottom and out at the top. The DuPont plastic layer of the driver door glass-plastic glazing, attached at the window edges as a "safety net," held, but with the window almost horizontal dynamically, the driver head through the plastic hit the back plate of the MDB. Note the partial ejection of the passenger.

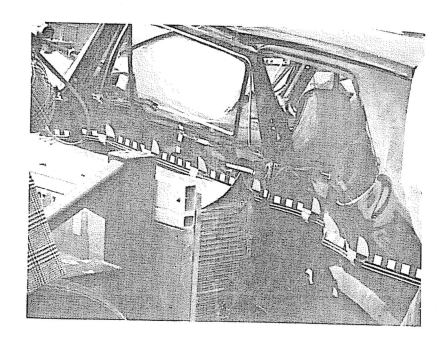

Figure 2. Side impact, Test DOT 0780, on October 26, 1984
(841026), MDB with a low bumper and higher than average hood
elevation into a Rabbit, at 39 mph. The glass-plastic driver
window "safety net" again held, preventing the head partial
ejection and contact with the MDB. Note the rear window ledge
deformation, made by the rear dummy shoulder partial ejection.

Figure 3. Side impact, Test DOT 0780. In this 39 mph crash, there
was some separation of the plastic "safety net." The plastic layer
of this glass-plastic glazing extended beyond the glass edge, and
was attached by steel strips to the window edge.

Figure 4. Side impact, Test DOT 0780. An inside view of the glass-plastic glazing after the crash. The plastic layer, on the inside, holds the broken pieces of the tempered glass, forming a "safety net." The hip and chest loads on the driver dummy, due to penetration at the door, threw the dummy into the passenger seat, partially rotated toward the left window, after the glass-plastic "safety net" had prevented the head partial ejection.

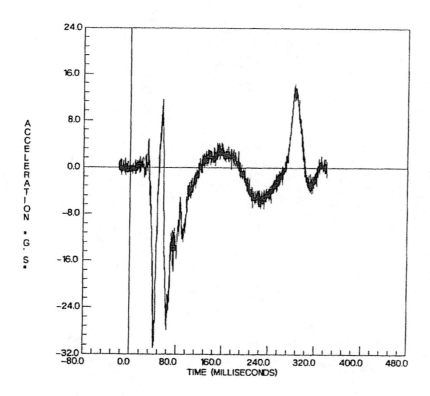

V0780AA00.001

TRC OF OHIO   841026

81 VOLKSWAGEN   RABBIT
VEH 2
OCCUPANT LOC 1   P572DM
HEAD CG
AS MEASURED
FILTER CUTOFF: 1650Hz

39.10 mph   ITV

THREE DOOR HATCH

XG AXIS
YMIN = -30.94000 at 413.7500 msec
YMAX = 14.28000 at 284.1250 msec

Figure 5. Side impact Test DOT 0780, Head Gx acceleration (posterior-anterior). Positive is with a displacement acceleration forward.

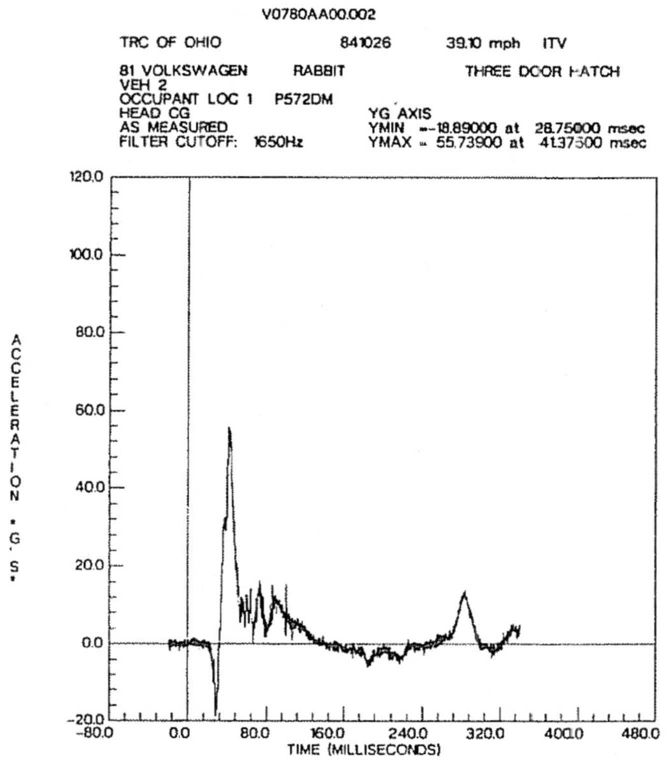

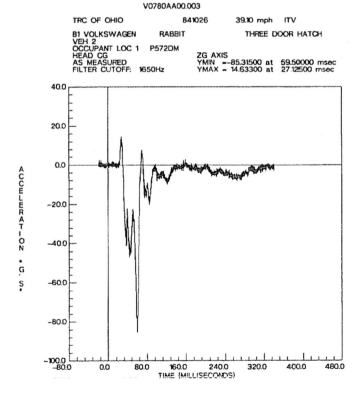

V0780AA00.002
TRC OF OHIO 841026 39.10 mph ITV
81 VOLKSWAGEN RABBIT THREE DOOR HATCH
VEH 2
OCCUPANT LOC 1 P572DM
HEAD CG
AS MEASURED
FILTER CUTOFF: 1650Hz
YG AXIS
YMIN = -18.89000 at 28.75000 msec
YMAX = 55.73900 at 41.37500 msec
ACCELERATION G'S
TIME (MILLISECONDS)

Figure 6. Head Gy acceleration (left to right). Positive is with a displacement acceleration to the right.

V0780AA00.003
TRC OF OHIO 841026 39.10 mph ITV
81 VOLKSWAGEN RABBIT THREE DOOR HATCH
VEH 2
OCCUPANT LOC 1 P572DM
HEAD CG
AS MEASURED
FILTER CUTOFF: 1650Hz
ZG AXIS
YMIN = -85.31500 at 59.50000 msec
YMAX = 14.63300 at 27.12500 msec
ACCELERATION G'S
TIME (MILLISECONDS)

Figure 7. Head Gz acceleration (inferior-superior). Positive is with a displacement acceleration upward.

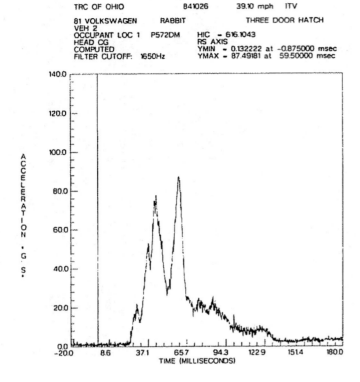

V0780AC00.R1H

TRC OF OHIO　　　　841026　　　39.10 mph　ITV

81 VOLKSWAGEN　　RABBIT　　　　THREE DOOR HATCH
VEH 2
OCCUPANT LOC 1　P572DM　　　HIC － 616.1043
HEAD CG　　　　　　　　　　RS AXIS
COMPUTED　　　　　　　　　YMIN － 0.132222 at −0.875000 msec
FILTER CUTOFF:　1650Hz　　YMAX － 87.49181 at　59.50000 msec

Figure 8.　Head Gr
acceleration (resultant).

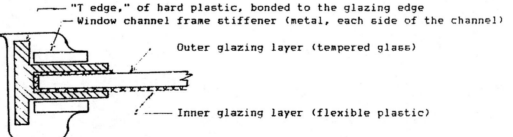

Figure 9.　Top view of the edge of the glass-plastic "safety net"
window which is able to be opened.　The　"T edge" bonded to the
glazing, with the attached flexible plastic "safety net", is
prevented by the channel frame stiffeners from pulling out of the
window channels during crash loads, after the tempered glass is
broken.　This provides head and body support, reducing ejection.

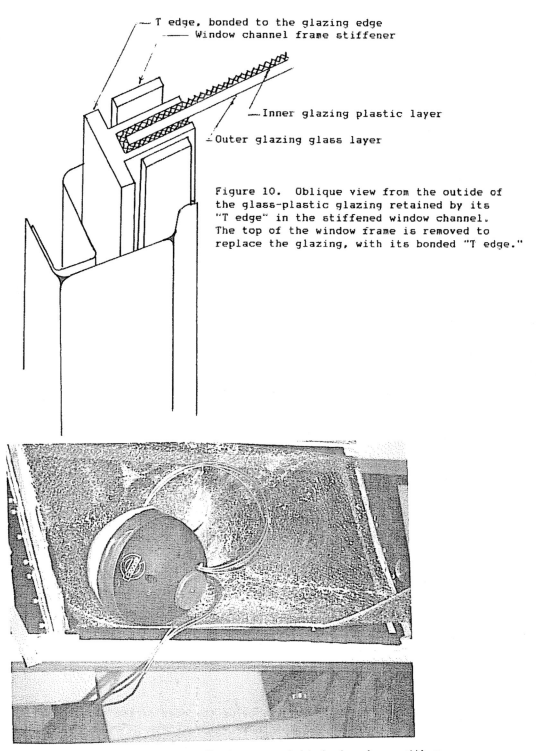

— T edge, bonded to the glazing edge
—— Window channel frame stiffener

——Inner glazing plastic layer

⸺Outer glazing glass layer

Figure 10. Oblique view from the outide of the glass-plastic glazing retained by its "T edge" in the stiffened window channel. The top of the window frame is removed to replace the glazing, with its bonded "T edge."

Figure 11. The Glazing Test Device, an eight inch sphere with a center triaxial accelerometer, weighted to 10 or 20 pounds for Head Injury Criterion (HIC) estimation, coated with plastic and chamois layers on the bottom hemisphere for laceration estimation, and weighted to 40 pounds for ejection reduction estimation. The Device is dropped electromagnetically onto full size automotive glazing mounted in the standard method in the actual automobile window frame or windshield mount.

CHAPTER FIVE

Interior changes

Forces on the Human Body
in Simulated Crashes

LAWRENCE M. PATRICK[1]

CHARLES K. KROELL,[2] and HAROLD J. MERTZ, JR[3]

ABSTRACT

Details of a new crash simulator and preliminary results from a series of cadaver knee impact experiments were presented at the Eighth Stapp Conference. During the past year additional data concerning injury to the knee-thigh-hip complex have been obtained, and the studies have been extended to consider impact to the chest. Results to date indicate that for knee impacts against a moderately padded surface it is not possible to predict whether failure of the patella, femur or pelvis will occur first, although in these studies femoral fractures occurred most frequently. A force of 1400 lb is recommended at this time as a reasonably conservative value for the over-all injury threshold level. Volunteers tolerated impact loads to the knee of 800-1000 lb.

For loads applied over the sternum through a 25-30 in^2 padded surface, static and dynamic thoracic stiffness characteristics were determined for a limited number of cadavers. In two cases the dynamic force-deflection relationship was linear with a slope of approximately 1000 lb/in up to 900 lb., where indications of skeletal damage were manifested. Volunteers tolerated similarly applied static loads of 300-400 lb.

INTRODUCTION

In a paper presented at the Eighth Stapp Conference (Kroell and Patrick, 1964) a new crash simulator and an associated biomechanics research program

---

[1] Professor, Biomechanics Research Center, Wayne State University

[2] Senior Research Engineer. Electro-Mechanics Department, General Motors Research Laboratories

[3] Research Assistant, Biomechanics Research Center, Wayne State University

on human tolerance to impact forces were described. Initial test results were given at that time. In the current paper a presentation is made of additional experimental findings which have been obtained during the past year.

In order to establish human tolerance to impacts indirect methods must be employed since it is obviously not practical to subject live humans to impacts which will cause injury. Volunteers can be used at sub-injury impact levels, but there is usually no way to determine how close to injury the impact is except by comparison with animal, cadaver, or clinical evidence. Nevertheless, such data from volunteers provide a valuable supplement to knowledge derived from other sources; and voluntary studies are conducted to a limited extent as a part of the present program. Mainly, however, the investigations are concerned with the levels of impact required to produce significant trauma, and embalmed cadavers are used as test subjects.

Of the many types of blunt impact trauma observed from automobile accidents, skeletal damage is the most amenable to study with cadavers, since embalmed bones have physical properties similar to those of fresh bones and the skeletal structure and weight distribution of the cadaver approximate those of the living human. Consequently, the effects of impact forces upon the living skeleton are inferred from measurements of forces applied to cadavers using the crash sled and from radiological examinations made before and after ex-

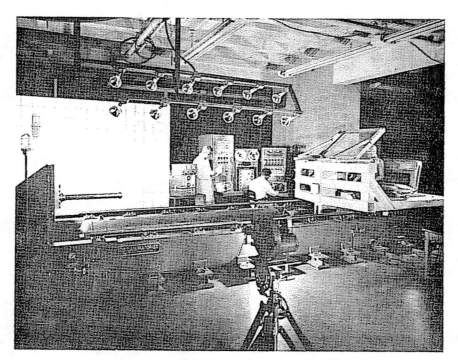

FIGURE 1 Crash Simulator

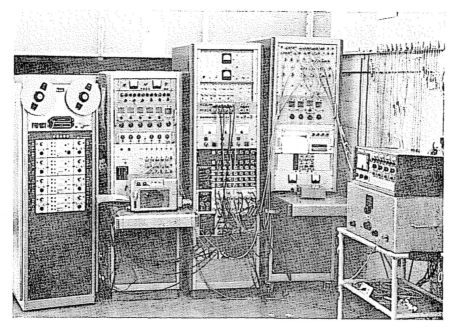

FIGURE 2. Electronic Conditioning Equipment.

posure to impact. Finally, the body is dissected and detailed studies are made of the skeleton after the series of experiments is complete.

For reference the crash simulator used is shown in Figure 1. The sled is propelled up to the desired speed by a pneumatic cylinder, disengages and free wheels for several feet while the velocity is measured, and then is stopped by an energy absorbing device in a predetermined manner. Load cell supported impact targets are positioned to intercept the occupant and to measure the loads developed upon the knees, chest and head. Varying the position of the targets changes the relative times at which the knees, chest and head strike. For most of the runs these were positioned so that the knees struck first, followed by the chest and, finally, the head, with each successive impact starting after the peak load of the preceding one. The forces on the cadaver together with other pertinent test data are recorded with the electronic conditioning equipment shown in Figure 2.

The remainder of this paper will discuss experimental results and findings from investigations concerning impact to the knees and chest.

KNEE-THIGH-HIP COMPLEX SKELETAL INJURY INVESTIGATIONS

One phase of this research program concerns skeletal injury to the knee-thigh-hip complex produced by knee impact forces directed essentially along

429

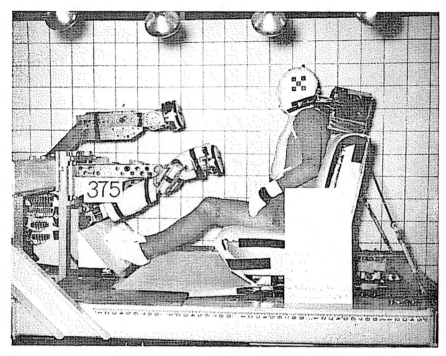

FIGURE 3 Typical Pre-Impact Test Setup.

the femoral axis Figure 3 displays a pre-impact test setup typifying the pertinent experiments During deceleration of the sled, the cadaver, being unrestrained, moves forward over the seat cushion and collides with the impact targets. The feet remain fixed upon the incline while the lower legs pivot about the ankles, forward and upward into the knee targets, kinematically representing impact against an instrument panel lower surface in a frontal automobile collision Except in one case, all of the knee impacts were applied through a padded, conforming target surface which was used to prevent highly localized forces at the site of impact.

The knee-thigh-hip complex, as considered in these studies, consists of three skeletal components—the knee cap, or patella; the thigh bone, or femur; and the hip bone, or pelvis Ten cadavers have been tested to date, and the complete results are summarized in three tabulations—one pertaining to each of the skeletal members involved. Table 1 is a tabulation for the femur Although a fracture was produced at a load as low as 950 lb ,[4] a bone defect was suspected in this specimen, and the data point is not considered representative. This fracture occurred, as did the majority of the femoral fractures at the

[4] All of the force values comprising the knee-thigh-hip data have been rounded off to the nearest 50 lb

TABLE 1

CADAVER SKELETAL TOLERANCE TO KNEE IMPACT - FEMUR

Cadaver No.	RIGHT THIGH Max Force Applied	Result	LEFT THIGH Max Force Applied	Result
1	950	Supracondylar fracture (bone defect suspected)	1400	Intertrochanteric fracture (fractured through bone screw)
2	1500	Mid-shaft fracture	1600	No fractures
3	1500	No fractures	1600	No fractures
4	1650	Supracondylar fracture	1650	Supracondylar fracture
5	2150	No fractures	2100	No fractures
6	2250	No fractures	2250	Supracondylar fracture
7	1900	No fractures	1850	Supracondylar fracture
8	2800	No fractures	1750	No fractures
9	3850	No fractures	2650	Dislocated intertrochanteric fracture
10	2400	Comminuted fracture of distal third of shaft and intercondylar notch	1800	No fractures

Human volunteers tolerated 800-1000 lbs. with only minor pain in knee

distal end of the bone, in the proximity of the condyles. Also, the intertrochanteric fracture produced on the left side for this same specimen occurred through a bone screw which had been used to mount an accelerometer and must likewise be regarded as unrealistically low.

The lowest load producing a normal fracture was 1500 lb. for the right mid-shaft fracture in cadaver No. 2. The results from cadaver No. 4 were characterized by a high degree of symmetry. Very similar right and left supracondylar fractures were produced simultaneously at essentially the same load level of 1650 lb. Supracondylar fractures were also produced at loads of 2250 lb., 1850 lb., and 2400 lb. for cadavers 6, 7, and 10, respectively; and finally a dislocated, intertrochanteric fracture occurred at 2650 lb. in cadaver No. 9. Several of the fractures described as supracondylar (cadavers No. 1, 4, 6 and 10) also extended into the condylar region of the bone.

In addition, several loads in excess of 2000 lb. (one as high as 3850 lb.) failed to produce femoral fractures. Thus it is seen that the data are characterized by a high degree of scatter as would be expected from the variation in size, weight, age and physical conditions represented by the cadavers. Figures 4A, 4B and 4C present a composite of roentgenograms revealing each of the femoral fractures listed in Table 1.[5]

[5] Grateful acknowledgment is made to Drs. J. David Harris and James E. Lofstrom, Detroit, who gratuitously contributed to this program by conducting the radiological examinations.

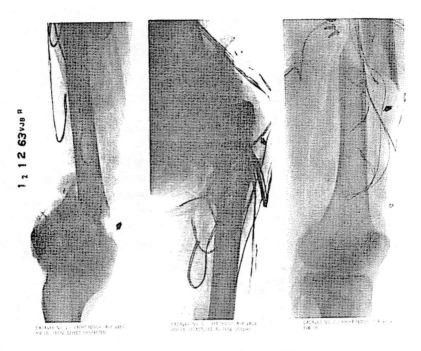

FIGURE 4A Roentgenograms of Femoral Fractures.

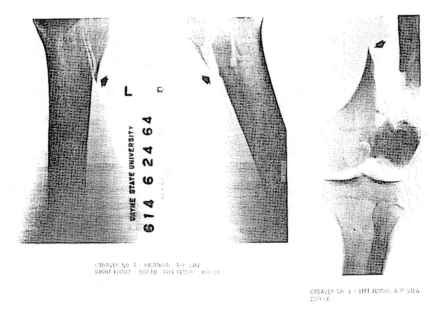

FIGURE 4B Roentgenograms of Femoral Fractures.

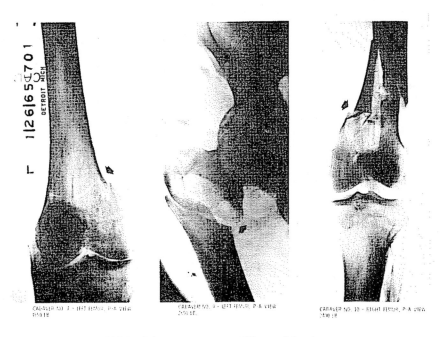

FIGURE 4C Roentgenograms of Femoral Fractures

Table 2 is a tabulation for the patella, or knee cap. There were no fractures for the first four cadavers for loads not exceeding 1650 lb. For cadaver No. 5 the radiological diagnosis was "no definite fractures" for both knees after maximum loads of 2050 lb. (rt.) and 1550 lb. (lt.) had been applied in the usual manner through padded targets.[a] Subsequently, additional loads up to 2000 lb. (on each side) were applied in this fashion. Then the padding was removed and a sequence of loads as tabulated was applied through the bare metal target plates. The final results were comminuted fractures of both patellae, as revealed by both the final roentgenograms and the post-test autopsy. It is assumed that the damage occurred during the unpadded loading sequence.

Bilateral comminuted patellar fractures were also produced in cadaver No. 6 by loads of approximately 2000 lb. (padded targets). In cadaver No. 8 comminution of the right patella occurred at a load of 2550 lb., but it is interesting to note that after additional successive loadings to 2800 lb. and 2400 lb. the fracture pattern was observed not to have changed appreciably. Finally, at the highest knee impact load yet recorded, 3850 lb., a linear patellar fracture was produced in the right knee of cadaver No. 9. Figure 5 exhibits roentgenograms revealing the six tabulated fractures.

---

[a] During the testing sequence, changes in the force-time wave forms occurring at approximately the 1400 lb. level suggested the possibility of some skeletal damage having been produced; however, this was not confirmed by the radiological examination.

433

TABLE 2

CADAVER SKELETAL TOLERANCE TO KNEE IMPACT - PATELLA

Cadaver No.	RIGHT KNEE Max Force Applied	Result	LEFT KNEE Max Force Applied	Result
1	950	No fractures	1400	No fractures
2	1500	No fractures	1600	No fractures
3	1200	No fractures	1600	Abnormality-but not adequately identified as fracture
4	1650	No fractures	1650	No fractures
5	2050	No fractures (padded)	1550	No fractures (padded)
	1550 1800 1950 2150	Heavy abrasion and fracture of patella (unpadded)	1500 1800 2000 2100	Complete fracture of patella, damage to articular cartilage (unpadded)
6	1700	No fractures	1950	No fractures
	2050 2250	Comminuted fracture of patella	2000	Comminuted fracture of patella
7	1900	No fractures	1850	No fractures
8	2550	Comminuted fracture of patella	1750	No fractures
9	3850	Linear fracture of patella	2650	No fractures
10	2400	No fractures	1800	No fractures

Human volunteers tolerated 800-1000 lbs with only minor pain in knee

The third tabulation, Table 3, is for the pelvis. The lowest damaging load was 1600 lb., fracturing the left pubic rami in cadaver No. 3. For cadavers No. 7 and 9 loads ranging from 1900 to 3850 lb. produced multiple pelvic fractures. For cadaver No. 8, a roentgenogram taken subsequent to a 1400 lb. loading of the right knee appeared to the radiologist "vaguely suggestive of possible injury to the right ischium." The succeeding test loaded the knee to 2550 lb. and resulted in multiple pelvic fractures. Again it will be observed that several loads in excess of 2000 lb. were not damaging. Each of the five fracture conditions produced is illustrated in Figures 6A and 6B.

The facts that all of the subjects used exceeded fifty years of age and that all were embalmed should render the foregoing data somewhat conservative, at least as applied to average male adults.

This study is not yet completed, but at this point an approximate threshold load for skeletal damage to the knee-thigh-hip complex produced by knee impact forces applied through a moderately padded target can be given. The results thus far indicate that, under conditions similar to those employed in these experiments, it does not appear possible to predict with certainty whether the patella, the femur or the pelvis will fail first. On the whole, however, the femur appears to be slightly the most vulnerable in that more failures occurred and at a lower average load level than was the case for the patella or the pelvis. On the basis of the data presently available, a load of 1400 lb. should certainly represent a reasonably conservative value for the over-all injury threshold level. It is planned that this threshold will be modified, if warranted, as additional experimental data are collected. Table 4 provides a tabulation of pertinent specifications for the ten cadaver specimens used.

An interesting supplement to the foregoing data is provided by the results obtained from a series of tests on human volunteers. Knee impacts were produced by the volunteer by driving his leg with muscle action against one of the knee impact targets mounted on the stationary sled in approximately the same kinematical manner that the cadaver knees impacted during the dynamic tests. The results of these tests for several volunteers are shown in Table 5; two representative force-time histories are shown in Figure 7A. For comparison, two force-time histories for cadaver impacts of similar peak loads are displayed in Figure 7B. The maximum values of these forces tolerated by volunteers, which ranged from 400 to 1050 lb., were not limited by mild injury or even pain, but only by the inability of the volunteer to generate higher forces by

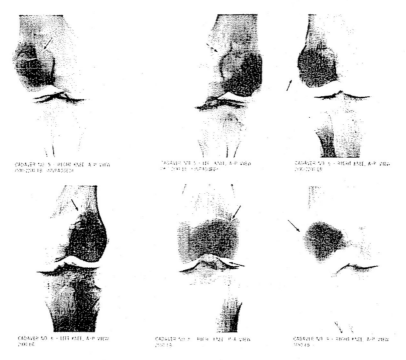

FIGURE 5 Roentgenograms of Patellar Fractures

435

TABLE 3

CADAVER SKELETAL TOLERANCE TO KNEE IMPACT - PELVIS

Cadaver No.	RIGHT HIP Max Force Applied	Result	LEFT HIP Max Force Applied	Result
1	950	No fractures	1400	No fractures
2	1500	No fractures	1600	No fractures
3	1200	No fractures	1600	Fractures of superior and inferior rami of the pubis
4	1650	No fractures	1650	No fractures
5	2150	No fractures	2100	No fractures
6	2250	No fractures	2250	No fractures
7	1900	Severe multiple fractures	1850	No fractures
8	1400	Possible mild fracture of ischium	1750	No fractures
	2550	Severe multiple fractures		
9	2750	No fractures	1950	Possible mild fracture of transverse ramus
	3850	Severe multiple fractures	2650	Severe multiple fractures
10	2400	No fractures	1800	No fractures

Human volunteers tolerated 800-1000 lbs. with only minor pain in knee

the method employed. The maximum volunteer load of 1050 lb. compared to the suggested threshold load of 1400 lb. indicates that this tolerance load may indeed be conservative. However, this is regarded as desirable in view of the variation in skeletal strength among individuals to be expected due to difference in age, size, state of health, etc.

CHEST IMPACT INVESTIGATIONS

Human tolerance to impact loading of the chest represents a most important problem area, yet one concerning which little knowledge has yet become available. A second phase of the research program described in this paper involves an investigation of this problem to the extent which is possible using cadaver and volunteer subjects. Several aspects have been studied in a preliminary manner to date.

Cadaver subjects were exposed to chest impacts applied through a 6 inch diameter padded target, approximately over the sternum and directed approximately perpendicular to the longitudinal body axis. Certain static measurements were also taken and were supplemented by similar data from other cadavers not tested dynamically. Determinations of static and dynamic thoracic stiffness

(A-P, or front-to-back, load-deflection characteristics), force and deflection time histories for the chest during impact, and resultant skeletal damage were made In addition, volunteers were subjected to low level chest loadings for purposes of comparison.

A summarization of findings in the aforesaid areas of investigation follows, although it is emphasized that these findings are yet preliminary and limited and represent specific rather than generalized data However, a continuation of

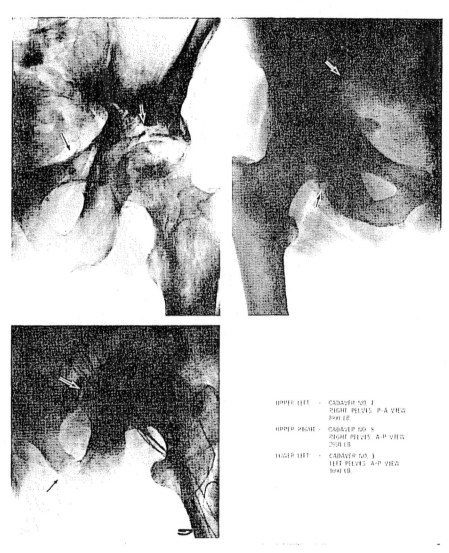

UPPER LEFT   -   CADAVER NO. 7
                 RIGHT PELVIS P-A VIEW
                 1900 LB.

UPPER RIGHT -   CADAVER NO. 8
                 RIGHT PELVIS A-P VIEW
                 2550 LB

LOWER LEFT -   CADAVER NO. 3
                 LEFT PELVIS A-P VIEW
                 1900 LB.

FIGURE 6A   Roentgenograms of Pelvic Fractures

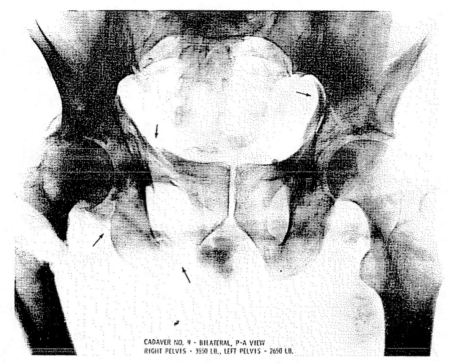

CADAVER NO. 9 - BILATERAL, P-A VIEW
RIGHT PELVIS - 3550 LB., LEFT PELVIS - 2650 LB.

FIGURE 6B. Roentgenogram of Bilateral Pelvic Fractures

this experimental program is planned until sufficient data have been collected to enable a more generalized interpretation.

Static Stiffness Measurements for the Thorax

Figure 8 exhibits plots of the thoracic A-P force-deflection characteristics for three cadavers as measured statically at relatively low load levels. The solid curves designated 7B, 658B and 8B in Figures 8A, 8B and 8C, respectively, refer to specimens which were measured as received with undamaged thoracic skeletons and prior to loosening of the joints as required for dynamic testing. For these cases loads were applied through a weighted, 4 inch wide structural channel section (25-30 $\overline{in}$.^{2} contact area), suspended from a spring scale and lowered onto the chest with a chain fall. Deflection was monitored with a dial indicator, the arrangement being illustrated in Figure 9. Similar measurements made subsequent to the production of multiple rib fractures revealed a definite, although not radical, reduction in stiffness within the low load range investigated. The broken lines in Figures 8A, 8B and 8C define the approximate range bracketing these post-damage measurements. For any given stiffness determination several repeated loadings were applied. In general, the stiffness

TABLE 4

CADAVER DESCRIPTION
(All Male)

No.	Age	Height	Weight	Cause of Death
1	-	-	110	Natural Causes
2	75	--	150	Pneumonia
3	64	-	205	Coronary Occlusion
4	62	5'9"	160	Septicemia
5	52	5'9"	140	Pneumonia
6	60	5'5"	185	Arterial Sclerosis
7	71	6'1"	160	Rheumatoid Arthritis Artereosclerotic Heart
8	60	5'8½"	170	Coronary Occlusion
9	59	5'8"	195	Artereosclerotic Heart
10	75	5'6"	160	Pneumonia

associated with the first load application was observed to be somewhat less than that determined from successive loadings This was especially true for the case of prior skeletal damage and apparently reflected an initial permanent set of the body (visceral shifts, skeletal hysteresis, movement of bone fragments, etc.) occurring at low force levels when first loaded subsequent to handling or testing The stiffness characteristics shown in Figure 8 refer only to the stabilized measurements made following three or four previous loadings

A fourth specimen was loaded in a conventional hydraulic testing machine, the force being applied through a 6 inch diameter (28 in² contact area) metal disc, as shown in Figure 10 The subject was of large stature and barrel chested However, prior to this stiffness measurement it had undergone previous impact exposure; and roentgenograms revealed that ribs 1-6 on the right side had already been broken Figure 11, curve 572A, exhibits the resulting stiffness characteristic As noted on the diagram, additional fracturing occurred during loading

TABLE 5

KNEE IMPACT FORCES ON VOLUNTEERS

Volunteer (All Male)	Age	Weight lbs.	Height in.	Maximum Force Developed, lbs.*	
				Right Knee	Left Knee
MB	20	140	67	450	400
MP	19	141	66	850	850
JB	18	162	70	800	1050
JD	18	155	68	700	550
KT	25	175	70½	450	550
SP	21	213	58	800	500
BM	28	205	72	750	550
LP	45	172	70	750	-
CK	31	165	70½	800	750

* Note: Force maxima were limited by an inability to generate higher values by the muscular loading method employed – not by pain or injury

The foregoing data demonstrate a wide variation in static thoracic stiffness characteristics for the three undamaged cadavers measured. At an applied load of 100 lb., the range extended from 185 lb/in. to 400 lb/in. This variation is greater than that found to exist between the damaged versus the undamaged conditions (stabilized measurements only) for any of the three subjects tested.

Dynamic Thoracic Stiffness Measurements

When considered as a parameter relevant to the dynamic behavior and susceptibility to injury of the chest under impact conditions, the thoracic stiffness must likewise be regarded in the dynamic sense. There is a fundamental difference between loading the chest by a blow delivered in a matter of milliseconds and compressing the chest less rapidly between rigid surfaces In the latter case a force is applied to the front of the body and an equal and opposite reaction is developed at the back. The resulting deformation will be determined by the static stiffness and fracture characteristics of the body. On the other hand, when the chest is subjected to an impact loading on the sled, there is no force reaction developed upon the back. Rather, an inertial force gradient extends across the thickness of the body, diminishing from a maximum value

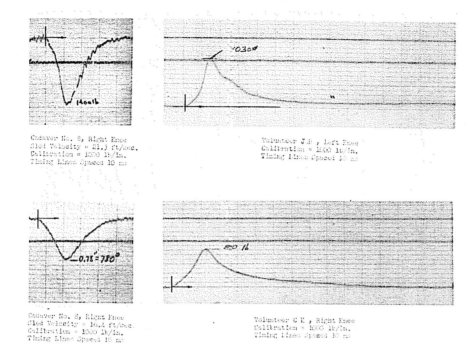

Cadaver No. 6, Right Knee
Sled Velocity = 21.3 ft/sec.
Calibration = 1000 lb/in.
Timing Lines Spaced 10 ms

Volunteer J.B., Left Knee
Calibration = 1000 lb/in.
Timing Lines Spaced 10 ms

Cadaver No. 6, Right Knee
Sled Velocity = 16.4 ft/sec.
Calibration = 1000 lb/in.
Timing Lines Spaced 10 ms

Volunteer C.K., Right Knee
Calibration = 1000 lb/in.
Timing Lines Spaced 10 ms

FIGURE 7A and 7B. Comparative Knee Force-Time Histories for Cadaver and Human Volunteer.

at the site of loading to zero at the surface of the back. Owing to a complex body mass distribution and the presence of flexibly attached thoracic viscera, this reaction gradient will not be a simple relationship. Also, it is to be expected that the chest force-deflection characteristics would be further influenced under dynamic loading conditions by the viscous behavior (sensitivity to rate of loading) of fluid filled body tissues. Furthermore, in vivo, the instantaneous state of respiration may well be of significance, but animal studies will probably be required to evaluate this factor.

Dynamic thoracic stiffness characteristics for two cadavers have been determined by measuring simultaneously the force-time and A-P deflection-time histories for the chest during sled impacts. These data were then crossplotted to obtain the force-deflection characteristics. This procedure is illustrated in Figure 12.

In both cases the impact forces were applied approximately over the sternum and were developed against a 6 inch diameter, padded target. This target was supported by the chest load cell, which provided the necessary force-time histories. To measure the chest deflection a deflectometer rod was inserted, front to back, through the thoracic cavity. Anteriorly, the rod terminated in a swivel ball joint which was sutured to the skin adjacent to the sternum. Posteriorly,

441

the rod passed adjacent to the vertebral column, between ribs and through a metal bushing on the back, also sutured in place. The rod was terminated with a marker "flag" outside the body. During impact, movement of the "flag" end of the rod is monitored with high speed photography, and the deflection-time history is then plotted from the filmstrip (Figure 12).

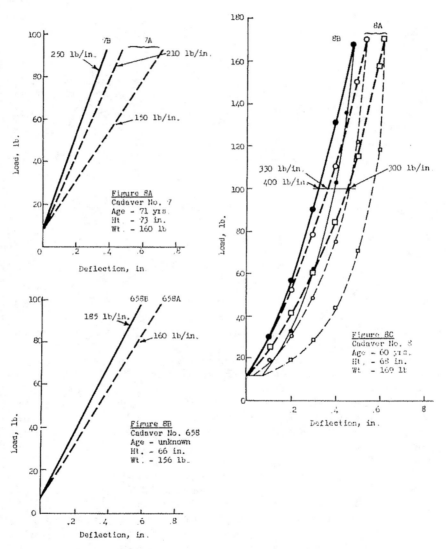

FIGURE 8. Static Force-Deflection Characteristics for Cadavers No. 7, 8, and 658—Weight Loading Method.

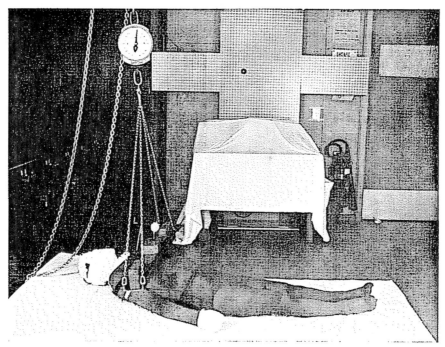

FIGURE 9. Thoracic Static Load-Deflection Device Using Chain Fall.

FIGURE 10 Thoracic Static Load-Deflection Device Using Hydraulic Testing Machine.

443

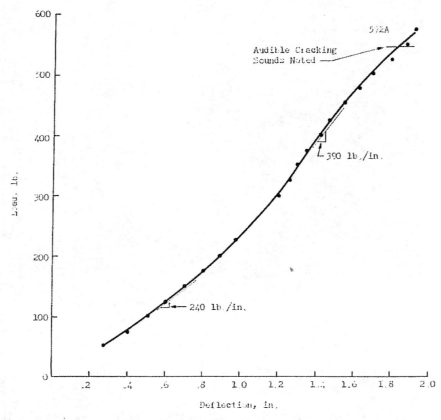

FIGURE 11. Static Force-Deflection Characteristic for Cadaver No. 572—Press Loading Method.

Figures 13A, 13B and 13C exhibit dynamic thoracic A-P force-deflection characteristics as determined from three successive test runs for cadaver No. 9 (195 lb.). Figure 13A represents the first impact, which was delivered against the initially undamaged thoracic skeleton. Roentgenograms taken subsequent to this test disclosed to the radiologist "strong suspicion of incomplete fractures of the right second, third and fourth ribs just distal to the costochondral junction." Figures 13B and 13C were derived from successive impacts. Following the test corresponding to Figure 13B, only slight aggravation of the skeletal damage was read out by the radiologist. Both impacts were at sled velocities of approximately 16½ mph and resulted in similar peak load and maximum deflection values (within 13%). The third test, corresponding to Figure 13C, was at a sled velocity of 22½ mph and resulted in substantially higher values for both parameters. Subsequent to this test, both the results of palpation and radiological examination disclosed massive skeletal damage.

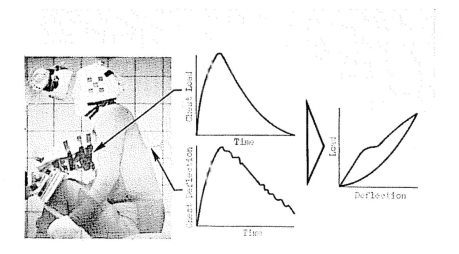

FIGURE 12. Method of Determining Dynamic Thoracic Stiffness by Crossplotting the Chest Load-Time and the Chest Deflection-Time Parameters to Obtain a Load-Deflection Plot.

The relationship demonstrated by Figure 13A is approximately linear with a slope of 1000 lb/in. up to a load of about 900 lb. At this level, a marked reduction in stiffness is observed until the slope once again increases and assumes an approximately linear value of 500 lb/in., which is maintained up to the peak load of 1600 lb. It is conjectured that the moderate skeletal damage diagnosed probably occurred during the "plateau" portion of the curve at approximately the 900 lb level. Then, as the remaining undamaged bone structure underwent further deformation and compression of the thoracic viscera began, the stiffness increased again.

An A-P permanent set of approximately ½ inch remained following this impact. This represents a new reference for zero deflection corresponding to the second impact. The latter is described by the curve of Figure 13B. The basically smooth, slightly concave upward appearance suggests that little, if any, additional skeletal damage was produced. This is consistent with the findings of the radiologist. A second permanent set is observed and must be borne in mind when considering the stiffness characteristics associated with the third impact (Figure 13C). The latter was at a higher severity level, involving peak values for force and deflection of approximately 3000 lb. and 3¼ inches, respectively. This would probably correspond to a 4-4½ inch deflection if referred to the initial zero reference. The curve demonstrates an early increase in stiffness up to approximately 2000 lb/in., and then remains essentially linear between 250 and 1000 lb. Between 1000 and 1700 lb., a region of much re-

445

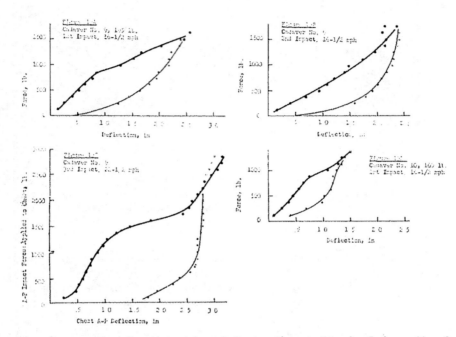

FIGURE 13. Dynamically Obtained Force-Deflection Characteristics for Cadavers Nos. 9 and 10.

duced stiffness is manifested, beyond which the slope again rapidly approaches its earlier value.

Again, it is assumed that the massive skeletal fractures subsequently diagnosed were produced during the flattened portion of the curve, between 1300 and 1700 lb. Beyond this, compression of the viscera probably accounts for the rapid increase in stiffness.

Figure 13D refers to the first loading applied to cadaver No. 10 (160 lb.) As was the case in Figure 13A (cadaver No. 9), the skeleton was undamaged prior to impact and the associated sled velocity was 16½ mph. Although they are by no means identical, these two curves do demonstrate a marked similarity. In both cases the initial slope is approximately 1000 lb./in. up to around 900 lb., where a plateau occurs before the slope increases again. Moderate skeletal damage (fractures of the right third, fourth, fifth and possibly sixth ribs in the anterior axillary line and a questionable fracture of the left fourth rib in the A-A line) was diagnosed and is assumed to have occurred at the 900-1000 level. It is probable that the difference in weight between the two subjects accounts for the difference in maximum force developed at the same test velocity—1600-1800 lb. for cadaver No. 9 and 1400 lb. for cadaver No. 10.

Additional data will be necessary to adequately specify the range of thoracic dynamic stiffness characteristics for the human cadaver. This is one of the pro-

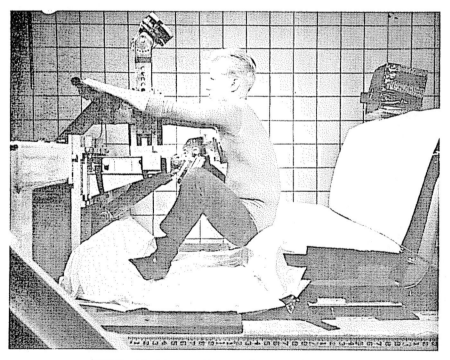

FIGURE 14 Volunteer Chest Loading Bar Pull Method.

gram objectives. Such knowledge will help to better define levels of human impact tolerance and should provide a valuable contribution towards the design of superior, more representative anthropomorphic test dummies. The ultimate aim, of course, is correlation with the in vivo condition.

The chest loading tests on volunteers were conducted by two different methods. In each case the subject used his own muscle action to apply the load, and did so until either he found himself physically incapable of exerting more muscular effort or until he had reached the limit of voluntary discomfort.

In one method the volunteer was required to assume a seated position upon the stationary sled, place his body in contact with the 6 inch diameter chest impact target and, through a crossbar handle affixed to the sled in front of him, pull himself against the target in the manner illustrated in Figure 14.

For the second method, a self-loading test fixture was fabricated as shown in Figure 15. A volunteer applied the load to his chest through a lever arrangement. The force was measured by a load cell supporting the 6 inch diameter loading plate.

Although these two methods of loading are fundamentally different in terms of the mechanism of load transfer through the body, the volunteers did not consistently develop higher forces with one technique or the other. This was true also when the bar pull method was used with both a padded and an

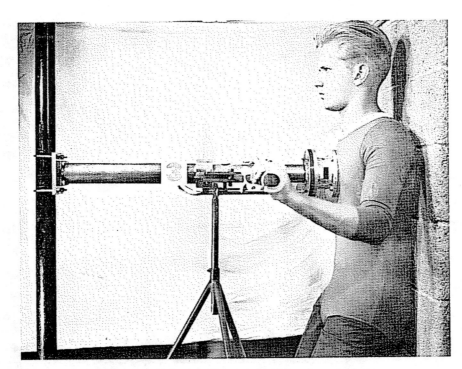

FIGURE 15. Volunteer Chest Loading Lever Arrangement.

unpadded loading plate. The maximum force developed using the padded bar pull method was 400 lb. For the unpadded bar pull this value was 300 lb., and for the self-loading fixture 290 lb. These maxima represent different subjects, however, and do not constitute a general ranking of the methods themselves. Additional, more definitive tests using these procedures are planned.

LITERATURE CITED

Kroell, C. K. and Patrick, L. M. "A New Crash Simulator and Biomechanics Research Program," *Eighth Stapp Car Crash Conference Proceedings*, Detroit, Michigan, October 1964. In press.

A Bio-Engineering Approach
to Crash Padding

Roger P. Daniel

Engineering Staff, Ford Motor Co.

CRASH PADDING has been used within the vehicle for over a decade to reduce injury. However, the evolution of padding into the quality product presently available has not been generally based on human performance data. Present day padding is the product of automobile company and material supplier cooperation to obtain a stable and durable product that absorbs energy under various impact tests. However, there has been insufficient human performance data available to insure that padding selections have always been the optimum for reducing head and face injury. This paper is written to relate the type and performance of padding used in specific vehicle areas to human tolerances.

BACKGROUND DATA - Head injuries are received by 72% of injured occupants in automobile injury-producing accidents and account for about one-half of all vehicle accident fatalities (1)*. Due to the importance of head injuries, this report will be limited to research on padding for head and face protection.

Review of test track full-scale collision data (2) and data from highway investigators (3) indicates that cranial vault**

---

*Numbers in parentheses designate References at end of paper.

**The phrase "cranial vault" will be used throughout this paper to denote the head structure above the eyebrows, in contrast to the face.

impacts are usually with the windshield header, visors, and roof side rails and pillars; whereas, facial impacts are usually with the instrument panel, the seat back, and the steering wheel rim.

Medical data show that the strength of facial bone under localized impact (about 1.0 sq in. area) varies from 15-60% of the frontal skull bone strength, depending on the facial area contacted (4, 5). These relatively low strengths reflect the thinness of many facial bones, some of which are as thin as 0 04 in , compared to 0 20-0 25 in for the cranial vault bone (6). However, when an impact is fully distributed, the face acts as a unit and withstands very high forces without structural damage (4). Thus, human tolerance for distributed facial impacts is nearly as high as for cranial vault impacts (7).

PAD-HEAD RELATIONSHIPS - The above data indicate that padding to protect the face must be different from padding to protect the cranial vault. Since the cranial vault is resistant to localized damage, load distribution is the secondary rather than primary consideration. Of course, small, rigid protrusions must be avoided in areas likely to be impacted by the cranial vault. However, to avoid brain damage, the vehicle area struck by the cranial vault must deform at a load below the brain tolerance level and thus absorb the head energy. Thus, energy absorption is the primary concern in cranial vault impact. On the other hand, a poorly distributed load, below the brain damage threshold,

---

───────────────── ABSTRACT ─────────────────

The injury-reducing functions of crash padding are discussed as they relate to head impact. The bony structure of the cranial vault (above eyebrows) is strong under localized impact compared with the face. Padding used to protect the cranial vault from impact has the primary function of absorbing energy to reduce the possibility of brain damage. On the other hand, padding for facial protection has the primary function of providing uniform load distribution on the face. The pad understructure then supplies the needed energy absorbing capacity. Test procedures to measure both energy absorption and load distribution are described, and evaluation criteria are shown. Other factors that affect padding, such as temperature and cover stock material, are discussed.

Table 1 - Head, Vehicle, and Crash Pad Relationship

Portion of Human Head	Vehicle Areas Most Likely to be Contacted	Function of Crash Pad	
		Primary	Secondary
Cranial vault (above eyebrows)	Windshield header, visors, roof side rails, roof pillars	Energy absorption	Load distribution
Face (below eyebrows)	Instrument panel, front seat back, and steering wheel rim	Load distribution	Energy absorption (understructure must then yield to absorb the energy of the head)

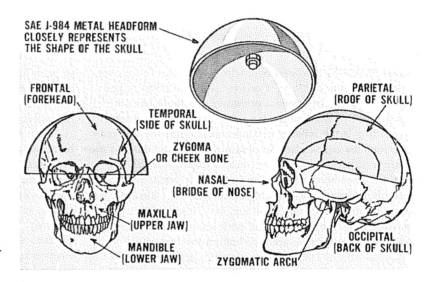

Fig. 1 - Headform for testing energy absorbing materials

applied to the face would likely cause localized bone fracture and facial collapse. Thus, for facial impacts, load distribution is of primary concern, with energy absorption second. These considerations are summarized in Table 1. Cornell Automotive Crash Injury Research (ACIR) data (8) show a 400% increase in frequency of head injury from the instrument panel when a seat belt is worn. A seat belt tends to direct the face against the panel. Of course, without a belt, even more severe injuries are likely to occur from striking other portions of the vehicle.

PADDING FOR ENERGY ABSORPTION

GENERAL DEFINITION - Energy absorbing crash padding, when impacted, deforms for nearly its entire thickness at a uniform load just below the human tolerance value for the body area contacting it.

DISCUSSION - Energy absorbing padding should be used in those vehicle areas where cranial vault impact without facial contact is likely to occur. Padding materials effective for energy absorption are firm to the touch, to the point

of being termed "rigid." In fact, thin (0.028-0.036 in.), "soft" sheet steel, carefully contoured for energy absorption, is one of the most efficient energy absorbers known. Such structure can collapse at a nearly uniform load to about 95% of its original depth, compared to 50-80% for foam padding. However, where complicated shapes and small radii cannot be avoided in some vehicle design areas, foam padding is dictated as the energy absorber.

TEST PROCEDURE FOR ENERGY ABSORPTION - Fig. 1 shows that the cranial vault is very close to a hemisphere in shape, being slightly longer than wide. Anthropometric data (9) give the 50 percentile head length and breadth (assuming half male, half female population) as 7.4 and 5.8 in., respectively. This averages 6.6 in., very close to the dimensions of the SAE 6.5 in. diameter hemispherical metal headform (10). Therefore, this headform was chosen as the impactor for testing energy absorbing materials in this study.

Headform deceleration is an effective criterion for measuring energy absorption for any given thickness of material, at any speed, because it describes how efficiently the mate-

Table 2 - Factorial Analysis - One Inch Thick Energy Absorbing Material Impact Tests

MATERIAL DESIGNATION	FLEXIBLE POLYURETHANE															POLYURETHANE		
SAMPLE DENSITY, LB. FT.³	4			6			9			12.5			15.5			9		
DROP HEIGHT IN FEET *	2	4	6	2	4	6	2	4	6	2	4	6	2	4	6	2	4	6
CORRECTED PEAK DEC IN G S	133	290	-	80	230	350	71	122	220	69	142	220	75	148	247	75	197	340
SAMPLE NUMBER	264	264	-	265	265	265	266	267	266	268	269	268	271	271	272	306	307	308

* 2 FT DROP HEIGHT = 7 75 MPH (11 4 FPS) = 30 FT –LB OF ENERGY
  4 FT DROP HEIGHT = 10 9 MPH (16 FPS) = 60 FT –LB OF ENERGY
  6 FT DROP HEIGHT = 13 4 MPH (19 6 FPS) = 90 FT –LB OF ENERGY

TEST CONDITIONS:
  6.5 IN DIA. METAL HEADFORM (S A E J-984)
  15 0 LB EFFECTIVE WEIGHT
  MATERIAL RIGIDLY BACKED

FLEXIBLE POLYURETHANE									FLEXIBLE LATEX												FLEX. POLYURETHANE					
6			8			12			15 5			25-A			24-B			27-C			7			8		
2	4	6	2	4	6	2	4	6	2	4	6	2	4	6	2	4	6	2	4	6	2	4	6	2	4	6
105	313	-	103	247	430	88	185	295	160	357	-	72	130	200	85	183	293	81	150	257	65	140	270	60	128	220
207	207	-	208	210	208	209	211	209	356	357	-	363	363	363	361	362	361	358	359	360	500	500	500	402	402	403

FLEX. POLYETHYLENE						STYROFOAM						MOLDED POLYSTYRENE RIGID											
2.5			9			2.2			3			2.5			3.5			4.5			6		
2	4	6	2	4	6	2	4	6	2	4	6	2	4	6	2	4	6	2	4	6	2	4	6
125	330	-	72	125	200	66	115	200	92	120	155	68	105	230	75	120	140	92	140	160	115	160	200
553	554	-	555	556	557	606	607	608	610	611	612	613	613	613	614	615	614	618	617	616	619	620	621

FIBERGLASS									FLEXIBLE POLYURETHANE									RIGID POLYURETHANE								
12			15			18												2.1			4.2			5.7		
2	4	6	2	4	6	2	4	6	2	4	6	2	4	6	2	4	6	2	4	6	2	4	6	2	4	6
103	185	250	115	171	238	135	205	240	115	325	-	95	215	440	74	195	370	70	211	380	75	118	165	95	133	175
450	451	450	452	453	453	454	455	454	658	659	-	656	657	656	660	661	660	600	600	600	601	601	601	604	604	604

rial stops the headform. Mathematically, any material thickness has a theoretical minimum deceleration value for each impact speed, given by

$$V^2 = 2as$$

where:

V = Impact speed, fps

a = Acceleration, fps^2

s = The stopping distance, ft

A common variation of this formula is

$$"G" = 0 4 (MPH)^2/S$$

where:

  G = Acceleration in gravity units
  MPH = Miles per hour
  S = Stopping distance, in

Lower decelerations, that is, those closer to the theoretical minimum, indicate good energy absorbing materials.

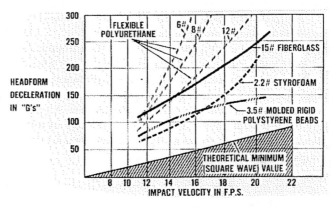

Fig. 2 - Deceleration characteristics of padding material

TEST SET-UP - The SAE metal headform was mounted to a pendulum so that the effective weight at impact was 15 lb. One-inch thick test pads were mounted to a rigid, steel-backed sheet of plywood and impacted at 11.4, 16.0, and 19.6 fps. Ninety-one impacts were conducted. The test data (Table 2) were set in a factorial arrangement to facilitate comparison and analysis.

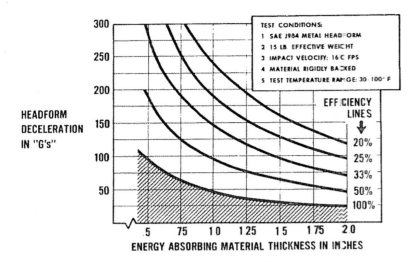

Fig. 3 - Evaluation criteria for energy absorbing material

RESULTS - Pad density is an inconsistent measure of energy absorption outside the bounds of any one material formulation. Pads of the same generic material, supplied by different manufacturers, exhibited considerable variations in density for equal energy absorption. On a density basis, rigid foams are far better energy absorbers than flexible foams. However, none of the materials tested was an efficient energy absorber in the sense of having deceleration near the theoretical minimum. The best material tested had an efficiency of only 45-50%. Fig. 2 is a plot of headform deceleration versus impact speed for representative padding samples, comparing the test data with the 100% efficiency value.

SUGGESTED EVALUATION CRITERIA - Ideally, any material used to cushion cranial vault impact should absorb energy at the brain tolerance level of 80-100 g's (7), in a 100% efficient manner. However Fig. 2 shows that ideal efficiencies do not occur.

Fig. 3 shows evaluation criteria for energy-absorbing materials. Shown are efficiency curves from 20 to 100%. An efficiency curve could be chosen as the minimum allowable efficiency and thus, lower efficiency materials would be rejected. Since the data show that efficiencies greater than 50% rarely occur with present day materials, an efficiency less than 50% would have to be chosen. A test velocity of 16 fps was chosen as being representative of relative head velocity in highway accidents, where cranial vault impacts occur.

LOAD DISTRIBUTING CRASH PADDING

GENERAL DEFINITION - Load distributing crash padding readily conforms to the contours of the human face upon initial impact, and then cushions and uniformly supports the face during collapse of the padded vehicle component.

DISCUSSION - Padding applied to vehicle surfaces likely to be impacted by the relatively fragile human face has the primary function of load distribution and thus should be termed "Load Distributing Crash Padding." The commonly used term "energy absorbing padding" is a misnomer for facial padding since it directs attention to the secondary aspect of padding for face protection (see Table 1). A load-distributing pad does absorb some energy, but in a very inefficient (about 15%) manner. Therefore, when considering the instrument panel and seat back upper surfaces, the padding should be the load distributor and the understructure the energy absorber, and each should be designed to best fulfill its function. However, there is an interrelation between the two, so the final safety determination must be based on a system (pad, cover, and understructure) approach.

Facial contact with a padded component should proceed as follows. The pad (and its cover stock) should offer little resistance to initial penetration of protrusions such as the nose, zygoma (cheek bone), and chin. However, once the entire face is in contact with the pad and penetrates into the material, the resistance of the pad to deformation should build until the understructure begins to yield. This understructure yielding should occur before any portion of the face bottoms against the understructure. The understructure should continue to yield until the energy of the head is dissipated, with the padding providing a resilient cushion between the face and understructure during the entire impact.

TEST PROCEDURE - Fig. 4 shows that the bony structure of the human face is better represented by a slightly irregular pear-shaped plane, than by a hemisphere. The neck of the pear-shaped plane represents the chin. The most reasonable regular geometric approximation is a circular plane, about 5 in. in diameter.

TEST EQUIPMENT AND PROCEDURE - Fig. 5 shows a face-form fabricated to meet the above criteria. The flat surface has an area of 20 sq in. (5.05 in. dia.).

A total of 175 impact tests were conducted with the flat face form to obtain static and dynamic pad load-deflection characteristics. Impact speeds were 4, 10,500, and 15,800 in. per minute (0, 10, 15 mph). The "static" tests were conducted on a Riehle tensile test machine. The 10 and 15 mph tests utilized the face-form on a pendulum. In each case, the padding thickness was 2 in.

The effective weight of the face-form at impact was 35.7 lb. This heavy weight, compared with the human head,

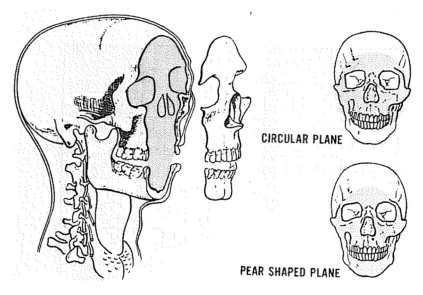

CIRCULAR PLANE

PEAR SHAPED PLANE

Fig. 4 - Face contours compared to flat plane

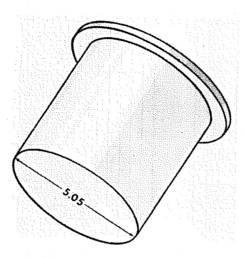

Fig 5 - Face-form for testing load distributing crash padding

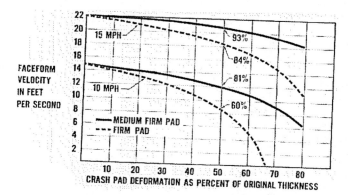

Fig 6 - Typical face form velocity-displacement curves for load-distributing crash padding

was chosen to insure a face-form velocity at 50% pad deflection, nearly equal to the initial velocity, which helped insure accurate dynamic results. Fig 6 shows that the desired results of nearly constant face-form velocity through the first half of the pad were obtained, except for very firm pads impacted at 10 mph. This loss of velocity for firm pads was not serious as such pads were too firm to be good load distributors.

Two types of samples were prepared for each material tested -- a circular section the same diameter as the face-form, and a 12 × 12 × 2 in flat sample. The former measures the load-deflection properties without edge effects and is called a "Compression-Load-Deflection" (CLD) test. The square samples were used in "Indenting-Load-Deflection" (ILD) tests, which take into account edge effects. In some cases, ILD tests were also conducted on samples covered with current production ABS plastic cover stock material

Twin accelerometers were mounted to the back of the face-form for the impact tests. Very careful calibrations were conducted, and the recording paper was run at high speed to give as large a trace as possible An initial contact marker was placed on the pad to obtain a time-zero blip on the record.

The test data were first analyzed on an "Oscar-K" to digitize analog acceleration-time records at 0.001 sec intervals for computer analysis. The computer calculates and prints in tabular form, the time and corresponding acceleration, velocity, penetration, and force. Other information is also printed and graphs plotted as desired. as shown in Fig 7.

RESULTS - The test data, in a factorial arrangement for ease of comparison and analysis, are shown in Table 3. At 15 mph and 50% pad deflection, the average increase in load for flexible foams from the CLD to the ILD tests was 69%, showing that edge effects are important in padding tests Less dense (softer) foams showed somewhat less increase in load than the average, and very firm foams tended

454

Table 3 - Factorial Analysis -- Load-Distributing Crash Padding

*CLD = COMPRESSION – LOAD – DEFLECTION (SAMPLE SIZE EQUAL TO, OR SMALLER IN SURFACE AREA, TO THE AREA OF INDENTOR)
 ILD = INDENTING – LOAD – DEFLECTION (SURFACE AREA OF INDENTOR SMALLER THAN SURFACE AREA OF SAMPLE)

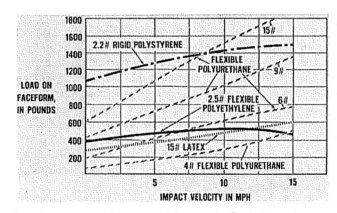

Fig. 8 - Typical velocity sensitivities of padding materials

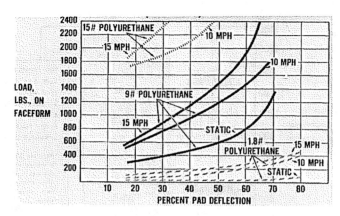

Fig. 9 - Typical deflection-sensitivities of padding materials (ILD tests)

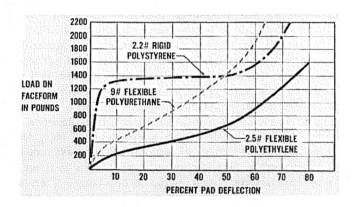

Fig. 10 - Comparison of load-deflection characteristics of flexible and rigid foams

TEST NO. 070667.72   PENDULUM NO. 2B
SEVERITY INDEX = 1207.139
VELOCITY RATIO =   -0.452
EFFECTIVE DECELERATION = 34.215
MAXIMUM PENETRATION = 1.765 INCHES AT 9.233 MILLISECONDS
MAXIMUM DECELERATION = 164.773 G'S AT 8.500 MILLISECONDS
VO = 22.000   NCUR = 2   PCTVO = 100.000   MINVEL = 0.
SCF1 = 125.000   SCF2 = 125.000   SCF3 = 1.000

TIME MILLISECONDS	ACCELERATION G'S	VELOCITY FT/SECOND	PENETRATION INCHES	FORCE IN LBS
0	0	22.0	0	0
1.00	13.55	21.78	0.131	483.16
2.00	20.98	21.23	0.389	748.11
3.00	25.35	20.48	0.639	903.97
4.00	31.47	19.57	0.879	1122.17
5.00	44.14	18.35	1.107	1574.15
6.00	62.50	16.63	1.317	2228.75
7.00	98.34	14.04	1.501	3506.77
8.00	155.59	9.95	1.645	5548.50
8.50	164.77	7.37	1.697	5875.80
9.00	157.78	4.78	1.733	5626.42
10.00	102.27	0.59	1.765	3647.05
11.00	55.51	-1.95	1.757	1979.30
12.00	35.84	-3.42	1.725	1278.02
13.00	24.91	-4.40	1.678	888.38
14.00	15.30	-5.05	1.621	545.50
15.00	9.62	-5.45	1.558	342.88
16.00	9.62	-5.76	1.491	342.88
17.00	9.18	-6.06	1.420	327.30
18.00	10.05	-6.37	1.346	358.47
19.00	12.24	-6.73	1.267	436.40
20.00	13.11	-7.14	1.184	467.57
21.00	12.67	-7.55	1.096	451.98
22.00	13.11	-7.97	1.003	467.57
23.00	11.80	-8.37	0.905	420.81
24.00	12.24	-8.75	0.802	436.40
25.00	10.93	-9.13	0.695	389.64
26.00	7.87	-9.43	0.583	280.54
27.00	6.56	-9.66	0.469	233.78
28.00	4.81	-9.84	0.352	171.44
29.00	1.75	-9.95	0.233	62.34

KINETIC ENERGY = 208.151 FT/LBS

Fig. 7 - Typical computer data print-out sheet

to show greater increases. Most foams showed a definite speed sensitivity (visco-elasticity), with the load increasing somewhat uniformly as the impact speed increased. Load-velocity curves are shown in Fig. 8 for representative foam samples, with the load measured at 50% pad deflection.

Fig. 9 shows that the pad loading increases somewhat linearly as the pad deflects, to about 50-70% deflection, at which point the pad tends to become solid, causing an abrupt increase in loading. Fig. 9 also shows that these load-deflection characteristics are markedly affected by pad density and impact speed.

Rigid foam is not as readily indented as flexible foam, and thus, is a poorer load distributor. The resistance of rigid foams builds to a high value on the face-form immediately upon contact. Fig. 10 compares the dynamic (15 mph) load-deflection curves for a rigid foam with two varieties of flexible foams. The advantage of the flexible foam in permitting easy initial entry of the face into the foam surface is evident.

SUGGESTED EVALUATION CRITERIA - The suggested evaluation criteria are based on the following assumptions, previously discussed.

1. A load distributing pad should permit the face to penetrate its surface relatively easily and then maintain a cushioning layer of foam between the face and the underlying structure during collapse of the understructure.

2. The understructure should deform at close to the 80g

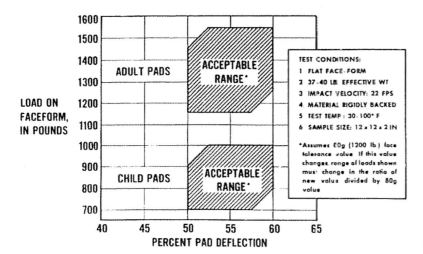

Fig. 11 - Load-distributing material evaluation criteria

(1200 lb) face tolerance level, expressed in both SAE J 885a (7) and Federal Motor Vehicle Safety Specification 201.*

A crash pad for facial load distribution should permit the face to penetrate 50-60% of its thickness to insure full facial contact. At that percentage deformation, the resisting load should be about 1000 lb and the understructure should begin to collapse and absorb energy. However, the face-form (Fig 5) does not fully represent the face. Its flat surface produces higher padding test loads than an actual face, due to the more rounded edges of facial bones. The exact ratio of pressure on the face to pressure on the face-form is not known, but is assumed to be in the 65-85% range. Therefore, a pad sample, tested with the Fig. 5 face-form, at 50-60% deflection, should give a face-form loading of about 1150-1550 lb. These data are plotted in Fig. 11.

The head of a 3-4 year old child weighs about 1/3 less than an adult head (11). Therefore, to maintain the same deceleration on a child's head during an impact as for an adult, padding and understructure designed for child protection should have at least 1/3 less resistance to deformation than when designing for an adult. Fig. 11 also shows an acceptable padding range for child head protection.

GENERAL CONSIDERATIONS AND REQUIREMENTS

ENERGY ABSORBING AND LOAD DISTRIBUTING PADS -
Temperature - Most padding and cover stock materials are sensitive to temperature, becoming stiff at low temperatures and soft at elevated temperatures. Since 1956, Ford Motor Co. has impact-tested foam products at temperatures of 0-120 F. However, the recent increase in sales of factory installed air conditioning units (29% in 1966 (12) and 37%

* Most padded structures are currently designed to yield in the range of 1000 lb to insure meeting all requirements. Should further research indicate that human tolerance values for the face should be increased or decreased, the load values given in this paper should be multiplied by the ratio of the new value to the present 80g value.

for first half of 1967 (13) plus accessory air conditioning sales, and the widespread use of heat-absorbing glass (61% installation in 1966) (12), more efficient heaters, and better vehicle insulation, would indicate that fewer miles are now driven with the vehicle interior below freezing or above about 100 F. Although no data on vehicle interior temperature versus miles driven could be found, a temperature range of 30-100 F is suggested as a practical padding test temperature range. This range should not be confused with the greater temperature range used for padding checks for other purposes since pad temperatures from -40 F to 220 F have been observed in the United States (14).

Density - The resistance to deformation of any given padding formulation is proportional to density. However, two different formulations of the same density may have considerable differences in load-deflection properties. For instance, a 15.5 lb density latex (test 351) gave a face form load of 730 lb at 40% deformation and 15 mph, compared to 2730 lb for a 15.5 lb density polyurethane (test 259). One 6.0 lb density polyurethane produced a 900 lb face form load at 50% deformation and 15 mph (test 203) compared to 1490 lb (test 256) for another 6.0 lb density polyurethane. The 15.5 lb latex and polyurethane gave head-form decelerations of 357 and 148 g's respectively during energy absorbing tests at 16 fps. This shows that the polyurethane is a much better energy absorber. The two samples of 6.0 lb polyurethane gave decelerations of 230 and 313 g's, showing the variation possible with equal densities of the same type of material.

A high density material becomes solid at less percentage of deflection than a low density material. A 15.5 lb density polyurethane is essentially solid at 55% deformation, whereas a 4 lb density polyurethane does not become solid until 85-90% deformation. Therefore, the lowest density material that will satisfy the impact requirements should be chosen. This reduces the weight of the part and the cost, if purchased by the pound.

LOAD DISTRIBUTING CRASH PADS -
Thickness - Although much more research is needed in this area, the very limited data available (15) indicate that

457

a minimum thickness of about 1.00-1.25 in. of padding is required on smoothly shaped understructures to cushion the somewhat irregular shape of the human face. A greater pad thickness than required to maintain a resilient layer between the face and understructure is not needed from a safety standpoint although it may be desirable for appearance.

Cover Stock - A semi-rigid padding cover stock reduces the ability of padding to conform readily to the intricate shape of the human face and also increases the apparent firmness of the pad. The cover stock should be pliant, even at low temperature, and should have minimum effect on the indenting properties of the foam. ILD tests on identical padding samples with and without a 0.030 in. thick ABS cover stock, at 50% deflection, produced 51% higher average static loads. Dynamic tests at 15 mph gave 25% higher average loads for the covered samples.

Tests were then conducted on pad samples with a Chatillon 25 lb push-pull gage. A 0.2 sq in. flat-faced indentor was forced into covered and uncovered ILD test samples (12 x 12 x 2) for 0.25 in. to compare the effects of three types of cover stock material. A 0.005 in. thick polyethylene plastic sheet increased the loads about 10%, compared with uncovered material. Cloth-backed vinyl, 0.035 in. thick such as used in seating applications, increased the load 20-30%. The 0.030 in. thick ABS cover stock increased the load more than 100%.

Since an important property of load distributing crash padding is to permit easy entry of facial protuberances into its surface, high resistance to local penetration caused by the cover stock should be avoided.

Understructure Shape - Understructures with equal collapse properties as far as total force is concerned, but having different contours at the location of facial contact, will require different thicknesses of padding for effective facial load distribution. A smoothly-contoured (flat or large radius) understructure requires the least thickness of padding since it provides a maximum area for facial impact. On the other hand, an understructure with a small radius on the rear edge or with a right-angle bend or corner requires considerably more padding to reduce the high pressure band across the face caused by the reduced impact area. However, since padding has poor shear strength and requires surface support, it is very difficult for padding alone, regardless of thickness, to protect against understructure edge effects (15).

CONCLUSIONS

1. Head injuries are received by 72% of injured occupants during injury producing accidents, and these injuries account for about half of all motor vehicle accident fatalities.

2. The face is considerably weaker than the cranial vault under concentrated loads, but nearly equal to the cranial vault in strength when the load is well distributed.

3. The face and cranial vault require different types of padding. Energy absorbing padding should be used for cranial vault impact and load distributing padding should be used for face impact.

4. The shape of the cranial vault closely approximates a 6.5 in. diameter hemisphere, so the SAE J-984 metal headform can be used to test padding for energy absorption.

5. Headform deceleration is an effective measure of energy absorption.

6. Density is an inconsistent measure of padding energy absorption, but for the same density, rigid foams are considerably better energy absorbers than flexible foams.

7. None of the 32 padding types tested was more than 50% efficient as an energy absorber; however, the author knows of no other padding material with significantly greater efficiency.

8. The bony structure of the face is better represented by a flat-faced, circular plane than by a hemisphere.

9. A load distributing crash pad should allow easy entry of facial protuberances into its surface. The load should then build as the face penetrates into the pad to where, at 50-60% pad deflection, the understructure begins to yield and absorb energy.

10. Most flexible foams exhibit visco-elastic characteristics, with the load increasing with the impact speed.

11. Flexible foams show somewhat uniform increases in load up to 40-80% deflection. The load then increases rapidly as the pad becomes solid.

12. Rigid foam does not allow easy facial entry into the pad. Resisting load builds up to a high sustained value almost immediately upon facial contact.

13. Since the head of a 3-4 year old child weighs about 1/3 less than that of an adult, pad loads for child head protection should be reduced by at least 1/3.

14. Padding and cover stock materials are affected by temperature. A practical test temperature range is 30-100 F.

15. Cover stock should not excessively restrict local penetration into the pad. Common cover stocks increase the apparent padding firmness by 51% statically and 25% dynamically during face-form tests. Some cover stock materials increase the resistance of the pad to localized (0.50 in. diameter) indenting by over 100%.

16. The understructure shape has a marked influence on the ability of padding to distribute load effectively.

REFERENCES

1. The Injury Producing Accident: A Primer Cornell Automotive Crash Injury Research, 1961.

2. Over 700 full-scale vehicle crashes conducted at Ford Motor Co. 1955-1967.

3. Various papers of Cornell A.C.I.R., D. F. Huelke, and others, 1955-1967.

4. J. J. Swearington, "Tolerances of the Human Face to Crash Impact." Federal Aviation Agency, July 1965.

5. V. R. Hodgson, G. S. Nakamura, R. K. Talwalker, "Response of the Facial Structure to Impact." Proceedings of the 8th Stapp Car Crash Conference, 1965.

6. Romanes, Cunningham's Textbook of Anatomy, Oxford Press.

7. "Human Tolerance to Impact Conditions as Related to Motor Vehicle Design," SAE J-885a, November 1966.

8. J. K. Kihlberg, "The Driver and His Right Front Passenger in Automobile Accidents." Cornell University Automotive Crash Injury Research, November 1965.

9. Chapanis, Cook, Lund, Morgan, "Human Engineering Guide to Equipment Design," McGraw-Hill, 1963.

10. "Body Forms for Laboratory Impact Testing," SAE J-984, April 1967

11. E. H. Watson, G. H. Lowery, "Growth and Development of Children." Chicago: The Year Book Publishers, Inc., 1964.

12. Automotive Industries, p. 98, March 15, 1967

13. Automotive News, p. 2, July 10, 1967

14. Automotive News, July 24, 1961

15. R. Daniel, L. Patrick, "Instrument Panel Impact Study." Proceedings of the 9th Stapp Car Crash Conference, 1965.

Society of Automotive Engineers, Inc.

This paper is subject to revision. Statements and opinions advanced in papers or discussion are the author's and are his responsibility, not the Society's; however, the paper has been edited by SAE for uniform styling and format. Discussion will be printed with the paper if it is published in SAE Transactions. For permission to publish this paper in full or in part, contact the SAE Publications Division and the authors

12 _page booklet.   Printed in U.S.A.

Factors Influencing Knee Restraint

Clyde C. Culver and
David C. Viano
Biomedical Science Dept,
General Motors Research Laboratories

OVERVIEW

Restraint of the lower torso via a knee bolster is expected to play a role in occupant restraint systems. The primary functions of a knee bolster are: 1) to control occupant kinematics during a frontal crash, thus minimizing the potential for an occupant to slide under the upper torso restraint; 2) to distribute lower extremity contact loads; and 3) to absorb occupant energy through a body region that is more capable of accepting these kinds of restraining forces. It is, therefore, important to understand the possible range of knee contacts and the mechanisms of potential lower extremity injuries and to interpret them in terms of the variability in occupant size, seating posture, and injury tolerance. Recent laboratory experiments demonstrated [1]*that skeletal fractures or knee joint ligament ruptures can result from contacts which predominantly load the lower leg. The data particularly indicated the

importance of the proper alignment and deformation characteristics of the knee bolster if the potential for lower extremity injury is to be minimized. Further studies investigated various factors which influence the performance of a knee bolster during a frontal crash. A planar model of lower torso kinematics was developed to provide a simple method for studying the interaction with the various parameters on lower extremity loadings. Other crash victim simulation programs (e.g., MVMA-2D [2] and/or CALSPAN-3D [3]) can also be used to study lower extremity kinematics and mechanics; however, the model described in this paper is conceptually less complicated and includes anatomical relationships between the knee joint surface and interior.

Particular attention is given in this paper to the kinematics of various sized occupants (5% F, 50% M, 95% M) since it

---

*Numbers in parentheses designate References at end of paper.

ABSTRACT

A planar mathematical model was developed to provide means of studying factors which can influence the function of lower torso restraint via a padded lower instrument panel or knee bolster. The following factors were judged to play the most significant role: 1) initial fore-and-aft position of the seated occupant relative to the knee restraint; 2) location of the knee-to-bolster contact; 3) angular orientation of the bolster face; 4) primary axis of the bolster resisting force, 5) variations in vehicle crash parameters

(e.g., toepan rotation and displacement and seat deflection); and 6) deformation characteristics of the bolster. The model of a seated occupant included radiographic and empirical data on the anatomy of the links and joints in the lower extremity. Emphasis was also placed on determining a range of reference location, orientation and primary axis of resistance of the knee bolster so that an effective restraint may be provided for the 5th percentile female, 50th percentile male, and 95th percentile male occupant.

highly relates to the anticipated location of lower extremity contacts with the knee bolster. In addition, the locations of the contact of the middle and top of the knee, identified from radiographs, have been included in the model by referencing these anatomical landmarks to the joint or rotational pivot of the knee. The position, orientation, and reaction angle of the knee bolster have also been considered in the development of the model so that the effects of compressive lower leg (tibia) and shearing knee joint loads can be considered in the alignment of a knee bolster. The effects of five different factors on lower extremity restraint with a knee bolster have been specifically investigated: 1) size and initial position (knee clearance) of the seated occupant in the vehicle relative to the knee bolster or lower instrument panel; 2) the trajectory of the knee and the location of knee bolster contact; 3) the angular orientation of the bolster face to loading plane of the instrument panel at the point of contact; 4) the primary axis of the instrument panel resistance or restraining force during contact with the lower extremity; and 5) the changes in vehicle parameters which alter the trajectory of the knee contact point, such as toepan displacement and rotation, or vertical H point motion due to that seat deflection. The effects of knee bolster deformation characteristics have not been addressed in this study.

Although the results of this study can assist in improving occupant restraint systems, additional laboratory tests are needed to investigate further interacting factors which may influence occupant and restraint system performance in the real-world crash environment.

MATHEMATICAL MODEL RATIONALE

The mathematical model (Appendix A) is a planar kinematic model of the basic pin-connected, three-link system shown in Figure 1. These three links represent simplified skeletal connections between the occupant H point (hip joint) and the heel point. Femur, tibia, and foot links are as shown on the lateral seated occupant profile of Figure 2. Simple trigonometric relationships of the system are programmed in Fortran on an IBM 370 computer to provide position and angular accountability of the joints and links under specified conditions. In a simulated frontal crash the H point of a vehicle occupant is assumed to move horizontally toward the instrument panel from an initial position as shown in Figure 3, to a more forward position as shown in Figure 4. Kinematics of the lower extremity are assumed to involve a rotation of the tibia around the ankle joint as the H point moves forward.

In "real-world" crashes, the occupant compartment and seat cushion can deform. An option in the program provides for toepan displacement and rotation and vertical seat deflection during H point motion in the crash simulation. Linear changes of these parameters are made relative to forward H point motion and can be started and stopped at the users option. These optional motions are shown in Figure 4.

Figure 3 shows a lateral profile of the seated occupant relative to a hypothetical vehicle interior profile consisting of a toepan, instrument panel/knee bolster, and seat. Origin of the vehicle coordinate system is assumed to be at the heel point. Toepan angle, length of a line from the occupant's heel to the ankle, and the angle between this line and the bottom of the foot are used to locate the ankle rotation point. Computations for tibia and femur angles, knee

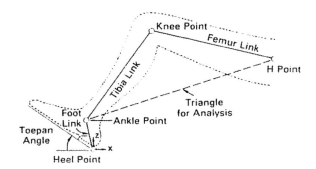

Fig. 1 - Three-link geometrical configuration and basic triangle of the lower extremity used in the planar mathematical model

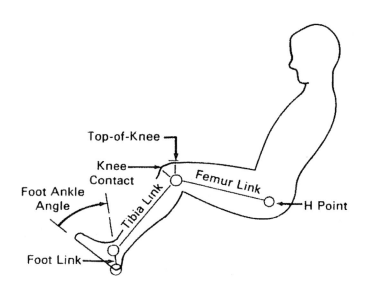

Fig. 2 - Location of principal points, lengths and angles of the lower extremity of a seated occupant

461

joint, knee contact, and knee height locations are made relative to the triangle formed by the femur link, tibia link, and a line between the ankle joint and the H point, as referenced to the heel point.

Since the forward movement of the H point is assumed to be the primary independent variable, the model theoretically accepts additional modes of upper torso restraint as long as the trajectory of the H point follows the basic assumption of this simplified model.

MODEL PARAMETERS DERIVED FROM HUMAN ANATOMY

DISTANCE FROM KNEE JOINT TO KNEE CONTACT POINT - In this model, the distance from the assumed knee joint pivot to the knee contact point, as shown in Figure 2, is specified as a line perpendicular to the tibial axis originating from the knee joint. This distance has been approximated from evaluations of lateral knee radiographs of six random human subjects (with the leg flexed at 90°). Mean value of the distance was 4.3 cm (S.D. = 0.3 cm, coefficient of variation = 6.4%).

DISTANCE FROM KNEE JOINT TO TOP OF KNEE - The top-of-knee point shown in Figure 2 is assumed to be directly above the knee joint for all leg flexion angles. Distance between the assumed knee joint and top-of-knee point was also approximated from radiological measurements (with the leg flexed at 90°). Mean value for the six subjects was 3.8 cm (S.D. = 0.3 cm, coefficient of variation =

7.9%). Single length values of the knee joint to knee contact and knee joint to top of knee distance have been employed in this program and by other researchers [4] for all sizes of occupants. These parameters vary with occupant size. Unfortunately, insufficient data are available at this time to quantify size dependence.

KNEE BOLSTER OR INSTRUMENT PANEL FACE ANGLE - A reference knee bolster face angle is judged to be one that will minimize lower leg and knee joint loads and direct the reaction forces through the femur and pelvis. The authors were guided by previous laboratory findings on potential impact responses and injuries in the lower extremity [1, 5-10], particularly, by the fact that excessive tibia loading could result in fractures of the lower leg or ruptures of the posterior cruciate ligament in the knee joint. Previous experiments indicated that lower extremity loads should be directed primarily through the femur minimizing the tibial and knee joint ligament loads. A series of lateral knee radiographs with the knee flexed at various angles from 30° to 120° were evaluated in three random human subjects. For each radiograph a reference bolster face angle was selected by placing a simulated bolster face on individual radiographs in a position to assure that the patella-femoral interface would be loaded by a deforming bolster face. The subjective bolster angle selection was based on a series of lateral knee radiographs made in human subjects under quasi-static

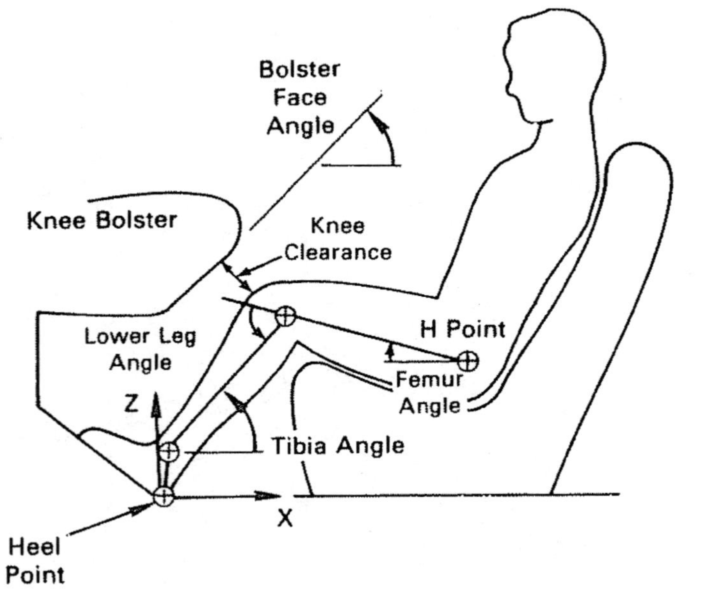

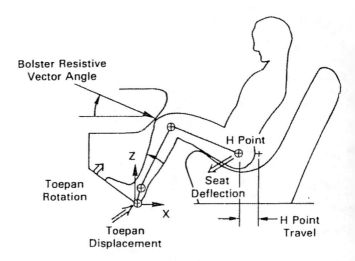

Fig. 3 - Location of additional principal points, lengths and angles of the lower extremity of a seated occupant in a vehicle

Fig. 4 - Occupant position at knee bolster contact assuming horizontal H point motion a rotation about the ankle joint. Also shown are optional vehicle parameters available in the mathematical model to study the effects of toepan rotation and displacement, and seat deflection on occupant kinematics

bolster loading conditions. For each lower leg orientation, the angle of the bolster face relative to the tibia axis was measured, recorded, and correlated with the leg flexion angle. A power fit relationship was then empirically determined:

$$\Theta_1 = 0.10\ \Theta_2^{2.36},\ r^2 = 0.95$$

Where: Θ_1 is the bolster face angle relative to the tibia in radians; Θ_2 is the leg flexion angle in radians; and r^2 is the coefficient of determination between the empirical data.

This equation was incorporated in the mathematical model to relate the reference bolster face angle (bolster angle shown in Figure 3) to the vehicle coordinates and occupant kinematic parameters by the relationship:

$$\Theta_3 = \Theta_4 - \Theta_1$$

Where: Θ_3 is the bolster face angle relative to the vehicle coordinate system in radians; Θ_4 is the tibia angle relative to the vehicle coordinate system in radians; and Θ_1 is the bolster face angle relative to tibia in radians.

Precautionary constraints are placed on the bolster face angle to insure that the bolster face angle is never greater than the tibia angle since this would place the primary restraining loads directly on the tibia.

FACTORS INFLUENCING KNEE RESTRAINT

BOLSTER FACE LOCATION - The mathematical model computes, at 1 cm H point travel increments, the knee contact trajectory (X-Z coordinates) relative to the heel point for various sized occupants. These trajectories identify the midpoint location of the projected knee contact on a hypothetical bolster profile. Some additional bolster face should be considered above and below the contact midpoint to insure load distribution during lower torso restraint. The effects of the design parameters, e.g., initial occupant position and knee-to-bolster clearance, as well as occupant size, can be explored for a given vehicle configuration using the mathematical model.

BOLSTER FACE ANGLE - A reference bolster face angle is computed from the particular lower extremity flexion during the simulated crash based on previously discussed empirical relationships. This reference angle has particular meaning only at the point of knee-to-bolster contact. However, the effect of various bolster face design profiles can be investigated from computed knee contact position and the corresponding reference bolster face angle.

REFERENCE BOLSTER RESISTIVE FORCE VECTOR ANGLE - Bolster resistive forces and tibia compressive load provide the primary equilibrium resistance to the femur force vector during knee impingement on the bolster. The angle of the bolster resistive force, as shown in Figure 4, may greatly influence the quality of restraint. The resistance force can be oriented to minimize tibia compressive forces, mitigate lower leg "jack knifing," and efficiently equilibrate lower torso loads directed through the femur. Because it is a reaction force, a reference favorable orientation can only be obtained through proper deformation of the bolster and supporting structure.

TOEPAN DISPLACEMENT AND SEAT DEFLECTION - Toepan displacement and rotation will cause a movement of the ankle point, which will directly affect the knee contact trajectories and leg flexion angle and under some conditions, increase tibia compressive loading. Downward H point motion by seat deflection can result in a greater leg flexion angle and an increased H point excursion.

EXECUTION OF MATHEMATICAL MODEL

MATHEMATICAL MODEL INPUT - Anatomical dimensions included in the occupant portion of the model are based on SAE templates (SAE J826b) and values obtained from radiographs of human subjects. Incorporated in the model from SAE templates are: 1) tibia length; 2) femur length; 3) heel to ankle joint length; and 4) the foot-ankle angle. Information included from human radiographs included: 1) knee joint to knee contact point; and 2) knee joint to top of knee distances. Pertinent occupant parameters, vehicle parameters, and interaction parameters are listed in Table 1. Anatomical data sets have been established for the 5% F, 50% M, and 95% M occupant as listed in Table 2. For actual studies of knee restraint involving various surrogates, anatomical dimensions from physical measurements or radiographs can be entered into the model.

MATHEMATICAL MODEL DEMONSTRATION - Only minimal model input is required for computer program operation. To execute the program, one of three occupant sizes (5% F, 50% M, or 95% M), initial position of the H point, toepan angle, and details of the toepan displacement and seat deflection are given as program input, e.g., 50.0, 80.0, 26.0, 45.0, etc. Vehicle parameters which are summarized in Table 3 have been used in a hypothetical model demonstration. These data consist of published research finding [11] or hypothetical optional parameters.

COMPUTER MODEL OUTPUT: KNEE CONTACT TRAJECTORIES - Knee contact trajectories for 5% F, 50% M, and 95% M were determined by the

Table 1: Mathematical model parameters

Occupant Parameters	Vehicle Parameters	Interaction Parameters
tibia length (+)	H pt height and distance (+)	knee clearance
femur length (+)	toepan angle (+)	bolster position
heel hard-to-ankle length (+)	optional:	bolster face angle
knee joint-to-top of knee (+)	seat deflection (·)	bolster resistive force
knee joint-to-knee contact (+)	toepan rotation (·)	angle
ankle angle (·)	toepan displacement (·)	

Computer input parameters (+), and optional input parameters (·)

Table 2: Occupant parameters contained in mathematical model

	5th P. Female	50th P.[1] Male	95th P.[1] Male
tibia length (cm) (lower leg)	34.0	41.7	46.0
femur length (cm) (upper leg)	36.0	43.2	45.5
heel hard-to-ankle length (cm)	8.6	13.3	13.5
knee joint-to-top of knee (cm)[2]	3.8	3.8	3.8
knee joint-to-knee contact (cm)[2]	4.3	4.3	4.3
ankle angle (deg)	67.0[3]	53.0	53.0

[1] Approximate anthropometry data obtained from SAE templates. SAE J826B, January 1974 and J267A, January 1961.

[2] From human radiographs

[3] Incorporates higher heel shoe for female occupant.

Table 3: Vehicle input parameters used for demonstration

H pt.   Height = 26.0 cm

H pt.   Distance[1]:

                5th P. Female = 68.0 cm
                50th P. Male = 80.0 cm
                95th P. Male = 86.0 cm

Toepan Angle = 45.0°

User Selected Values:

Toepan rotation = 1°/1cm H point travel

(x) Heel point displacement = 0.25 cm/1cm H point travel

(z) Heel point displacement = 0.1 cm/1 cm H point travel

(z) H point displacement = -0.1 cm/1cm H point travel

[1] Assumed forward, mid, and rear seating position
for the 5%F, 50%M and 95%M, respectively [11].

mathematical model as shown plotted in Figure
5 for the demonstration values. For this
initial demonstration, no toepan rotation,
displacement, or seat deflection was entered
in the model. Each knee contact point
trajectory is a partial arc about the appropriate ankle joint. The computed "top-of-
knee" values for the various occupant sizes
could be plotted as an arc approximately 4 cm
above the knee contact trajectory to serve as
a guide for the upper boundary for a knee
bolster. An arrow was plotted to indicate
the 90° leg flexion angle for each knee
contact point trajectory for referencing
convenience. When toepan displacement,
rotation and H point depression, as defined
in Table 3 were entered into the model, the
knee contact point trajectories were affected
only slightly as shown by Figure 6. For this
example the 90° leg flexion angle is attained
more rearward in the knee trajectory.

COMPUTER MODEL OUTPUT: BOLSTER FACE
ANGLE - Reference bolster face angles for the
three occupant sizes are shown plotted in
Figure 7 versus the knee contact distance
behind the heel point. Again, note that
these angles have meaning only during knee
impingement on the bolster. Arrows again
indicate the 90° leg flexion angle. A bolster
face angle in the range of 35° to 50° is
noted from the position of the three open
circles to encompass bolster face angle for
this spectrum of occupant sizes, initial

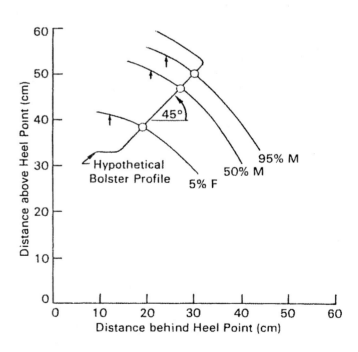

Fig. 5 - Knee contact trajectories for three
occupant sizes assuming a horizontal H point
motion at a vertical height of 26 cm above the
heel point (Arrows indicate 90° leg angle; top
of knee is approximately 4 cm above the knee
contact trajectory; open circles indicate
intersection of knee contact trajectory with
the hypothetical bolster profile)

465

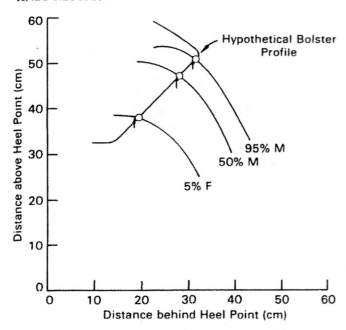

Fig. 6 - Knee contact trajectories assuming a hypothetical heel point displacement, toepan rotation, and depression of H point. (Arrows indicate 90° leg angle; open circles indicate intersection of knee contact trajectory with the hypothetical bolster profile)

seating positions and knee contact trajectories. A hypothetical bolster within this range has been sketched on Figure 5 (i.e., 45° face angle). Toepan displacement and seat deformation modifies the knee contact profiles as shown in Figure 8 where again the hypothetical bolster from Figure 5 is also plotted.

COMPUTER MODEL OUTPUT: LEG ANGLE VERSUS KNEE CONTACT LOCATION - Computed leg flexion angles are shown plotted in Figure 9 versus knee contact distance behind the heel point. This relationship is important since at small leg angles a large amount of femur force can be directed into the tibia and at leg angles greater than 90° femur loading stability is dependent on the bolster resistance force vector. Designers studying new knee restraint concept layouts can relate knee contact or trajectory positions (e.g. Figure 5) to occupant leg and knee positions for various occupant sizes. For this demonstration, the range of lower leg orientations at contact with the hypothetical bolster varies from 71°-76°, whereas displacement of the toepan and seat deflection increase the range of leg flexion angles at contact to 84°-92° as shown in Figure 10. Toepan displacement and seat deflection of the type assumed in this demonstration cause leg flexion to substantially increase for a similar forward knee travel.

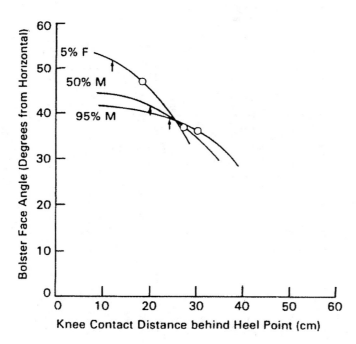

Fig. 7 - Reference bolster face orientation as a function of knee contact location for various size occupants assuming a horizontal H point motion at a vertical height of 26 cm above the heel point (Arrows indicate 90° leg angle; open circles indicate intersection of knee contact trajectory with the hypothetical bolster profile shown in Fig. 5 for each occupant size)

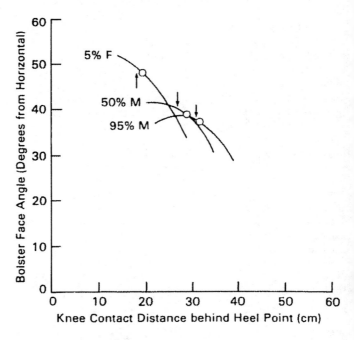

Fig. 8 - Reference bolster orientation as a function of knee contact location for various size occupants assuming heel point displacement, toepan rotation, and H point depression (Arrows indicate 90° leg angle; open circles indicate intersection of knee contact trajectory with the hypothetical bolster profile shown in Fig. 6 for each occupant size)

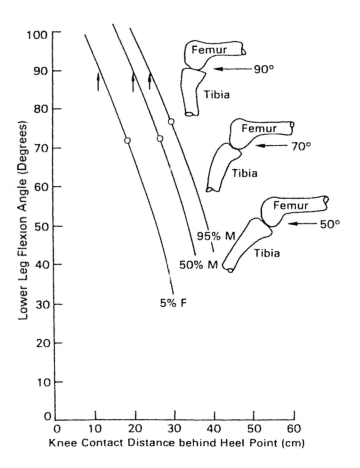

Fig. 9 - Leg flexion angle versus knee contact distance behind heel point as computed by the mathematical model for demonstration parameters (Arrows indicate 90° leg angle; open circles indicate intersection of knee contact trajectory with the hypothetical bolster profile shown in Fig. 5 for each occupant size)

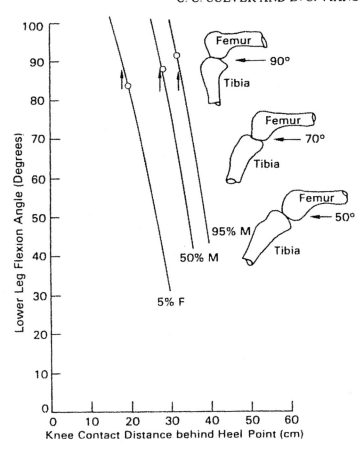

Fig. 10 - Leg flexion angle versus knee contact distance behind heel point as computed by the mathematical model assuming heel point displacement, toepan rotation, and H point depression (Arrows indicate 90° leg angle; open circles indicate intersection of knee contact trajectory with the hypothetical bolster profile shown in Fig. 6 for each occupant size)

COMPUTER MODEL OUTPUT: BOLSTER RESISTIVE VECTOR ANGLE - An understanding of the fundamentals for determining reference bolster resistive vector angles may be an aid to understanding basic bolster design criteria. For example in a crash simulation, two principal modes of bolster resistance were judged to equilibrate lower torso loads depending on the lower leg orientation during knee impingement on the bolster, as illustrated in Figure 11.

In the predominant configuration Mode I, the lower leg angle is less than 90°. Some compressive loading of the tibia is believed to be desirable to minimize the potential for sliding of the knee contact on the bolster face. A bolster resistive vector such as "A" could relieve the tibia from virtually all compressive load. However, the knee contact would be lifted upward during deformation of the panel. The upward action would increase

the forward motion of the H point. This could affect upper torso restraint performance. A bolster vector such as "B," coaxial with the femur force vector, would provide a minimal angle to prevent significant upward movement of the knee contact point and resulting potential for the occupant to underride the upper torso restraint. Resistive vector "C" is tangential to the rotational trajectory of the knee joint about the ankle point and is collinear with bolster deformation under this knee trajectory. A restraint force vector position between "B" and "C" is judged to provide a stable lower torso restraint. A resistive vector angle significantly greater than "C," e.g., position of vector "D" would drive the knee contact reaction downward increasing compressive loads in the tibia and ankle joint as well as leading to potential underriding of the bolster. Thus a bolster resistance angle within the envelope defined by vector angles "B" and "C" should provide a stable equilibrium of the lower torso loads during knee impingement on the bolster.

In Mode II, the lower leg angle is greater than 90°, a situation which is expected to occur much less frequently. A bolster resistive force vector collinear with the femur axis, vector "B" seems to provide a

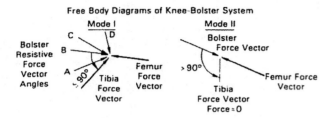

Fig. 11 - The bolster resistive force vector diagrams showing two principal modes of possible equilibrium; one when the lower leg angle is less than 90° (Mode I), and another when the lower leg angle is greater than 90° (Mode II)

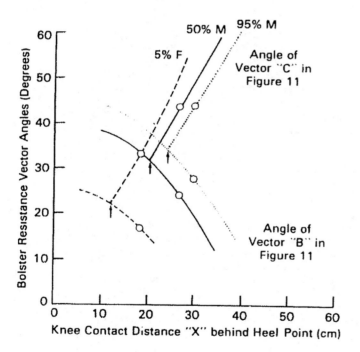

Fig. 12 - Reference bolster resistance vector angles defined by femur angle "B" and knee contact trajectory tangent angle "C" versus knee contact distance behind heel point for the demonstration parameters (Arrows indicate 90° leg angle; again open circles indicate intersection of knee contact trajectory with the hypothetical bolster profile shown in Fig. 5. For each occupant size the two open circles define a reference range of bolster resistance vector angles, e.g. 22.5°-45° for the 50%M)

desirable resistance vector angle. Resistive vector angles significantly less than the collinear angle could be expected to cause an upward lift during compression of the bolster, as vector "A" in Mode I, with increase in forward movement of the H point and a potential for underriding of the upper torso restraint. Resistive vector angles significantly greater than the collinear angle would drive the knee contact reaction downward (as vector "D" in Mode I), at the expense of potentially excessive compressive loads on the tibia and ankle and potential underriding of the bolster.

A range of bolster resistance angles defined by the femur angle (i.e., vector "B", Figure 11) and knee contact trajectory tangent angle (i.e., vector "C," Figure 11) for the demonstration are shown plotted versus knee contact distance behind the heel point in Figure 12 when the lower leg angle is less than 90°. Otherwise the femur angle defines the reference resistance vector angle. Note that the vector "B" and "C" curves intersect at the 90° leg flexion angle for each occupant size. A range of reference bolster vector angles can be defined within this envelope for a lower leg angle less than 90° (Mode I) for a particular knee contact distance behind the heel point and occupant size. For this demonstration the range of overlapping reference bolster resistance vector angles is between 28°-32° for all occupant sizes. When the lower leg angle is greater than 90° (Mode II), a single curve defined by vector Angle "B", is obtained as shown for each occupant size. Figure 13 demonstrates the effects of toepan displacement and seat deflection on the range of bolster resistance vector angles and further indicates no overlapping range of resistance angles for all occupants.

CONFOUNDING FACTORS - Bolster design based on the fundamental methods and simplified mathematical model presented in this paper must be cognizant of confounding influences from:

1. This is a planar model of occupant kinematics which assumes a simple pin joint linkage as representing the lower extremity,

2. Knee dimensions and bolster face angle relationships are based on judgemental evaluations from a limited number of human radiographs,

3. Inertial reactions are not accounted for in the model,

4. Splaying of the legs and other 3D effects are not accounted for in this planar model, and

5. Heel movement, toepan rotation and H point height changes are programmed only as simple linear functions of H point motion.

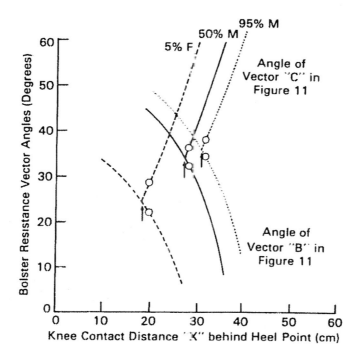

Fig. 13 - Reference bolster resistance vector angles defined by femur angles "B" and knee contact trajectory tangent angle "C" versus knee contact distance behind heel point for the demonstration parameters including toepan rotation and displacement and seat vertical depression (Arrows indicate 90° leg angle; open circles indicate intersection of knee contact trajectory with the hypothetical bolster profile shown in Fig. 6. For each occupant size the two open circles define a reference range of bolster resistance vector angles, e.g. 32°-35° for the 50%M)

ONCLUSIONS

. Use of this mathematical model may aid in the understanding of the fundamentals of lower extremity kinematics and some knee bolster characteristics that influence lower torso restraint by knee bolsters.

. Methodology is presented for determining the possible effects of bolster face angle on loading the human knee and lower extremities.

. Knee contact trajectories, calculated by the model, may aid the designer in the study of the effects of knee-to-bolster clearance and location of bolster impact for a range of occupant sizes and vehicle characteristics.

. Basic data on the range of bolster resistance vector angle may aid in understanding the influence of the bolster supportive structure and crush characteristics.

5. The potential effect of toepan displacement and rotation, and seat deflection on lower torso kinematics can be investigated.

ACKNOWLEDGEMENTS

Mr. Max Bender and Mr. Roger Culver of Highway Safety Research Institute, University of Michigan assisted in the preparation and interpretation of lower extremity radiographs.

REFERENCES

1. D. C. Viano, C. C. Culver, et al, "Bolster impacts to the knee and tibia of human cadavers and a Hybrid III dummy." Proceedings of the 22nd Stapp Car Crash Conference, Society of Automotive Engineers: SAE 780896, 1978.

2. B. M. Bowman, R. O. Bennette, and D. H. Robbins, "MVMA two-dimensional crash victim simulation." UM-HSRI-BI-74-1.

Highway Safety Research Institute, University of Michigan, 1974.

3. J. T. Fleck, F. E. Butler, and S. L. Vogel, "An improved three dimensional computer simulation of motor vehicle crash victims." Calspan Corporation, Department of Transportation Report No. ZQ-5180L1, 1974.

4. J. A. Roebuck, K.H.E. Kramer, and W. G. Thomson, Engineering Anthropometry Methods. John Wiley and Sons, 1975.

5. J. W. Melvin, R. L. Stalnaker, et al, "Impact response and tolerance of the lower extremities." Proceedings of the 19th Stapp Car Crash Conference, Society of Automotive Engineers: SAE 751159, 1975.

6. J. W. Melvin and R. L. Stalnaker, "Tolerance and response of the knee-femur-pelvis complex to axial impact." Highway Safety Research Institute Publication: UM-HSRI-76-33, University of Michigan, 1976.

7. W. R. Powell, S. J. Ojala, et al, "Cadaver femur responses to longitudinal impacts." Proceedings of the 19th Stapp Car Crash Conference, Society of Automotive Engineers: SAE 751160, 1975.

8. M. Kramer and A. Heger, "Severity indices for thoracic and leg injuries." Second International Conference on the Biomechanics of Serious Trauma, IRCOBI, 1975.

9. R. D. Lister and J. G. Wall, "Determination of injury threshold levels of car occupants involved in road accidents." Compendium of the 1970 International Automobile Safety Conference, Society of Automotive Engineers; SAE 700402, 1970.

10. C. K. Kroell, D. C. Schneider, and A. M. Nahum, "Comparative knee impact response of Part 572 dummy and cadaver subjects." Proceedings of the 20th Stapp Car Crash Conference, Society of Automotive Engineers: SAE 760817, 1976.

11. R. W. Roe, "Describing the driver's work space: eye, head, knee, and seat positions." Automotive Engineering Congress and Exposition: SAE 750356, 1975.

APPENDIX A

Computer Program

ANGA	angle of toeboard plus angle of ankle to bottom of foot
ANGBOL	calculated bolster face angle to horizontal
ANGBY	angle of displacement vector of knee relative to horizontal
ANGCHK	check of foot-to-tibia angle > 23° up; > 19° down (user selected)
ANGFEM	angle of femur relative to horizontal
ANGFT	angle between line from heel to ankle and bottom of foot
ANGLEG	angle of leg flexion at knee
ANGTIB	angle of tibia to horizontal
ANGTOE	angle of toepan to horizontal
ANKDX	ankle "X" position relative to heel point on shoe
ANKDZ	ankle "Z" position relative to heel point on shoe
APOSX	ankle "X" position relative to initial heel point
APOSZ	ankle "Z" position relative to initial heel point
B	distance of H point above ankle point
BETA	angle formed by line from H point to ankle point and knee to ankle point
CHECK	check value of arc cos for BETA
CHECKI	check value of arc sin for ANGFEM
DELPX	knee contact-to-knee joint "X" direction
DELPZ	knee contact-to-knee joint "Z" direction
DISTFT	distance from heel to ankle
DISTK	distance from knee joint-to-knee contact; perp to tibia axis
DISTKT	distance from knee joint-to-top of knee
DTR	resultant incremental knee travel
DTX	"X" displacement of knee for step
DTZ	"Z" displacement of knee for step
FEM	length of femur
HEPOSX	"X" position of heel relative to initial heel hard point
HEPOSZ	"Z" position of heel relative to initial heel point
HX	"X" position of H point to initial heel point
HZ	"Z" position of H point to initial heel point
LIM	step at which toepan rotation and displacement, and seat deflection start
LIMI	step at which toepan rotation stops
LIMII	step at which heel "X" displacement stops
LIMIII	step at which heel "Z" displacement stops
N	step number
PAHTAN	"new" knee height above ankle
PAHTAN	"new" knee height above H point
PATCCX	knee contact "X" relative to heel
PATCCZ	knee contact "Z" relative to heel
PATCX	"new" knee "X" contact relative to ankle
PATCZ	"new" knee "Z" contact relative to ankle
PATCXN	"new" knee "X" contact relative to heel
PATCZN	"new" knee "Z" contact relative to heel
PATHTA	knee height above ankle
PATHTA	knee height above H point
PATTCX	top of knee position relative to heel "X"

PATTCZ	top of knee position relative to heel "Z"
PATXA	knee "X" position relative to ankle
PATXAN	"new" knee "X" position relative to ankle
PATXAT	"new" knee "X" position relative to ankle
PATZAT	top of knee "Z" position relative to ankle
PER	percentile of dummy selected
PHI	angle of TIBFEM to horizontal
PHIRR	incremental knee movement
SDTR	summation of knee travel
TIB	length of tibia
TIBFEM	length of line from H point to ankle

```
 2    WRITE(16,4)
 4    FORMAT(' INPUT: PER,HX,HZ,ANGTOE,LIM,LIMI,LIMII,LIMIII')   00000010
      READ(5,*,END=114)PER,HX,HZ,ANGTOE,LIM,LIMI,LIMII,LIMIII     00000020
      N=0                                                          00000030
      SDTR=0.0                                                     00000040
      DTR=0.0                                                      00000050
      HEPOSX=0.0                                                   00000060
      HEPOSZ=0.0                                                   00000070
      IF(PER.EQ.5.) GO TO 6                                        00000080
      IF(PER.EQ.50.) GO TO 8                                       00000090
      TIB=46.0                                                     00000100
      FEM=45.5                                                     00000110
      ANGFT=53.0                                                   00000120
      DISTFT=13.5                                                  00000130
      DISTK=4.3                                                    00000140
      DISTKT=3.8                                                   00000150
      GO TO 10                                                     00000160
 6    TIB=33.97                                                    00000170
      FEM=36.0                                                     00000180
      ANGFT=67.0                                                   00000190
      DISTFT=8.6                                                   00000200
      DISTK=4.3                                                    00000210
      DISTKT=3.8                                                   00000220
      GO TO 10                                                     00000230
 8    TIB=41.7                                                     00000240
      FEM=43.2                                                     00000250
      ANGFT=53.0                                                   00000260
      DISTFT=13.3                                                  00000270
      DISTK=4.3                                                    00000280
      DISTKT=3.8                                                   00000290
 10   ANGA=ANGTOE+ANGFT                                            00000300
      IF(ANGA.EQ.90.) GO TO 12                                     00000310
      IF(ANGA.GT.90.) GO TO 14                                     00000320
      ANKDX=DISTFT*(COS(ANGA/57.3))                                00000330
      ANKDZ=DISTFT*(SIN(ANGA/57.3))                                00000340
      GO TO 16                                                     00000350
 12   ANKDX=0.0                                                    00000360
      ANKDZ=DISTFT                                                 00000370
      GO TO 16                                                     00000380
 14   ANKDX=((-1)*DISTFT)*(COS((180.0-ANGA)/57.3))                 00000390
      ANKDZ=DISTFT*(SIN((180.-ANGA)/57.3))                         00000400
 16   APOSX=HEPOSX-ANKDX                                           00000410
      APOSZ=HEPOSZ+ANKDZ                                           00000420
      B=HZ-APOSZ                                                   00000430
      IF(B.LE.0) GO TO 92                                          00000440
      TIBFEM=(((HX-APOSX)**2)+((HZ-APOSZ)**2))**0.5                00000450
      DISTTF=0.99*(TIB+FEM)                                        00000460
      IF(TIBFEM.GT.DISTTF) GO TO 100                               00000470
      PHI=ATAN((HZ-APOSZ)/(HX-APOSX))                              00000480
      BETA=ARCOS(((TIBFEM**2)+(TIB**2)-(FEM**2))/(2.*TIBFEM*TIB))  00000490
      ANGTIB=PHI+BETA                                              00000500
      ANGCHK=180.0-((ANGTIB*57.3)+ANGTOE)                          00000510
      IF(ANGCHK.GT.109.0) GO TO 84                                 00000520
      IF(ANGCHK.LT.67.0) GO TO 88                                  00000530
      PATHTA=TIB*(SIN(ANGTIB))                                     00000540
      PATHTH=PATHTA-(HZ-APOSZ)                                     00000550
      ANGFEM=ASIN(PATHTH/FEM)                                      00000560
      IF(PATHTH.LE.0.0) GO TO 96                                   00000570
      DELPX=DISTK*(SIN(ANGTIB))                                    00000580
                                                                   00000590
```

472

```
                                                                00000600
      DELPZ=DISTK*(COS(ANGTIB))                                 00000610
      PATXA=TIB*(COS(ANGTIB))                                   00000620
      PATXAT=(TIB*(COS(ANGTIB)))                                00000630
      PATCCX=PATXAT + APOSX                                     00000640
      PATZAT=(TIB*(SIN(ANGTIB))+DISTKT)                         00000650
      PATCCZ=PATZAT+APOSZ                                       00000660
      PATCX=PATXA-DELPX                                         00000670
      PATCZ=PATHTA+DELPZ                                        00000680
      PATCCX=PATCX+APOSX                                        00000690
      PATCCZ=PATCZ+APOSZ                                        00000700
      ANGLEG=ANGTIB+ANGFEM                                      00000710
      WRITE(6,18)PER,HX,HZ,ANGTOE                               00000720
   18 FORMAT(4G10.4)                                            00000730
      WRITE(6,20)                                               00000740
   20 FORMAT(1X,'PATCCX',4X,'PATCCZ',4X,'HEPOSX',4X,            00000750
     +'HEPOSZ',4X,'HX',8X,'HZ',8X,'PHTRR',4X,'SDTR')            00000760
      WRITE(6,22)                                               00000770
   22 FORMAT(1X,'ANGLEG',4X,'ANGFEM',4X,'ANGTIB',               00000780
     +4X,'ANGBOL',4X,'ANGBV')                                  00000790
      WRITE(16,24)                                              00000800
   24 FORMAT(1X,'PATCCX',4X,'PATCCZ',4X,'HEPOSX',4X,            00000810
     +'HEPOSZ',4X,'HX',8X,'HZ',8X,'PHTRR',4X,'SDTR')            00000820
      WRITE(16,26)                                              00000830
   26 FORMAT(1X,'ANGLEG',4X,'ANGFEM',4X,'ANGTIB',               00000840
     +4X,'ANGBOL',4X,'ANGBV')                                  00000850
      WRITE(16,28)                                              00000860
   28 FORMAT(1X,'PATTCX',4X,'PATTCZ')                           00000870
      WRITE(16,30)PATCCX,PATCCZ,HEPOSX,HEPOSZ,IIX,IIZ           00000680
   30 FORMAT(6G10.4)                                            00000890
      WRITE(16,32)ANGLEG,ANGFEM,ANGTIB                          00000900
   32 FORMAT(3G10.4)                                            00000910
      WRITE(16,34)PATTCX,PATTCZ                                 00000920
   34 FORMAT(2G10.4)                                            00000930
      WRITE(6,36)                                               00000940
   36 FORMAT(1X,'PATTCX',4X,'PATTCZ')                           00000950
      WRITE(6,38)                                               00000960
   38 FORMAT(80X)                                               00000970
      WRITE(6,40)PATCCX,PATCCZ,HEPOSX,HEPOSZ,HX,HZ              00000980
   40 FORMAT(6G10.4)                                            00000990
      WRITE(6,42)ANGLEG,ANGFEM,ANGTIB                           00001000
   42 FORMAT(3G10.4)                                            00001010
      WRITE(6,44) PATTCX,PATTCZ                                 00001020
   44 FORMAT(2G10.4)                                            00001030
      WRITE(6,46)                                               00001040
   46 FORMAT(1X)                                                00001050
   48 N=N+1                                                     00001060
      HX=HX-1.0                                                 00001070
      HZ=HZ-0.1                                                 00001080
      IF(HZ.LE.APOSZ) GO TO 92                                 00001090
      IF(N.GE.LIM) GO TO 50                                    00001100
      GO TO 62                                                 00001110
   50 IF(N.GE.LIMI) GO TO 52                                   00001120
      ANGTOE=ANGTOE+1.0                                        00001130
   52 IF(N.GE.LIMII) GO TO 54                                  00001140
      HEPOSX=HEPOSX + 0.25                                     00001150
   54 IF(N.GE.LIMIII) GO TO 56                                 00001160
      HEPOSZ=HEPOSZ-0.1                                        00001170
   56 ANGA=ANGTOE+ANGFT                                        00001180
      IF(ANGA.EQ.90.) GO TO 58
```

473

```
                                                                00001190
                                                                00001200
                                                                00001210
                                                                00001220
                                                                00001230
                                                                00001240
                                                                00001250
         IF(ANGA.GT.90.) GO TO 60                                00001260
         ANKDX=DISTFT*(COS((180.0-ANGA)/57.3))                   00001270
         ANKDZ=DISTFT*(SIN(180.-ANGA)/57.3))                     00001280
         GO TO 62                                                00001290
      58 ANKDX=0.0                                               00001300
         ANKDZ=DISTFT                                            00001310
         GO TO 62                                                00001320
      60 ANKDX=(-1)*DISTFT)*(COS((180.0-ANGA)/57.3))             00001330
         ANKDZ=DISTFT*(SIN(180.-ANGA)/57.3))                     00001340
      62 APOSX=HEPOSX-ANKDX                                      00001350
         APOSZ=HEPOSZ+ANKDZ                                      00001360
         TIBFEM=((HX-APOSX)**2)+((HZ-APOSZ)**2))**0.5            00001370
         IF(TIBFEM.GT.DISTTF) GO TO 100                          00001380
         PHI=ATN((HZ-APOSZ)/(HX-APOSX))                          00001390
         CHECK=((TIBFEM**2)+(TIB**2)-(FEM**2))/(2.0*TIBFEM*TIB)) 00001400
         IF(CHECK.GE.1.0) GO TO 104                              00001410
         IF(CHECK.LE.(-1.0)) GO TO 104                           00001420
         BETA=ARCOS(CHECK)                                       00001430
         ANGTIB=PHI+BETA                                         00001440
         PAHTAN=TIB*(SIN(ANGTIB))                                00001450
         PAHTHN=PAHTAN-(HZ-APOSZ)                                00001460
         PATXAT=TIB*(COS(ANGTIB))                                00001470
         CHECKI=PAHTHN/FEM                                       00001480
         IF(CHECKI.GE.1.0) GO TO 108                             00001490
         IF(CHECKI.LE.(-1.0)) GO TO 108                          00001500
         ANGFEM=ASIN(CHECKI)                                     00001510
         DELPX=DISTK*(SIN(ANGTIB))                               00001520
         DELPZ=DISTK*(COS(ANGTIB))                               00001530
         PATXAN=TIB*(COS(ANGTIB))                                00001540
         PATCX=PATXAN-DELPX                                      00001550
         PATZAT=TIB*(SIN(ANGTIB))+DISTKT                         00001560
         PATCZ=PAHTAN+DELPZ                                      00001570
         PATCXN=PATCX+APOSX                                      00001580
         PATCZN=PATCZ+APOSZ                                      00001590
         PATTCX=PATXAT+APOSX                                     00001600
         DTX=PATXA-PATX.AN                                       00001610
         DTZ=PAHTAT+.POSZ                                        00001620
         PATTCZ=PAHTHN-PAHTH                                     00001630
         DTR=((DTX**2)+(DTZ**2))**0.5                            00001640
         SDTR=SDTR+DTR                                           00001650
         PATHTH=PAHTHN                                           00001660
         PATHTA=PAHTAN                                           00001670
         PATXA=PATXAN                                            00001680
         ANGBV=1.57008-ANGTIB                                    00001690
         ANGLEG=ANGTIB+ANGFEM                                    00001700
         IF(ANGLEG.GT.1.57) GO TO 112                            00001710
      64 ANGBOL=ANGTIB-(0.0953*(ANGLEG**2.4683))                 00001720
         IF(ANGBOL.GT.ANGTIB) GO TO 66                           00001730
         GO TO 68                                                00001740
      66 ANGBOL=ANGTIB                                           00001750
      68 PHTRR=DTR                                               00001760
         WRITE(6,72)PATCXN,PATCZN,HEPOSX,HEPOSZ,HX,HZ,PHTRR,SDTR 00001770
         WRITE(16,70)PATCXN,PATCZN,HEPOSX,HEPOSZ,HX,HZ,PHTRR,SDTR
      70 FORMAT(8G10.4)
      72 FORMAT(8G10.4)
         WRITE(6,78)ANGLEG,ANGFEM,ANGTIB,ANGBOL,ANGBV
         WRITE(16,74)ANGLEG,ANGFEM,ANGTIB,ANGBOL,ANGBV
      74 FORMAT(5G10.4)
         WRITE(16,76)PATTCX,PATTCZ
```

```
   76 FORMAT(2G10.4)                                             00001780
   78 FORMAT(5G10.4)                                             00001790
      WRITE(6,80) PATTCX,PATTCZ                                  00001800
   80 FORMAT(2G10.4)                                             00001810
      WRITE(6,82)                                                00001820
   82 FORMAT(1X)                                                 00001830
      IF(ANGTIB.GT.1.3) GO TO 2                                  00001840
      GO TO 48                                                   00001850
   84 WRITE(16,86)                                               00001860
   86 FORMAT(' TOEBOARD ANGLE IS TOO SMALL')                     00001870
      GO TO 2                                                    00001880
   88 WRITE(16,90)                                               00001890
   90 FORMAT(' TOEBOARD ANGLE IS TOO GREAT')                     00001900
      GO TO 2                                                    00001910
   92 WRITE(16,94)                                               00001920
   94 FORMAT(' SELECT LARGER VALUE OF HZ')                       00001930
      GO TO 2                                                    00001940
   96 WRITE(16,98)                                               00001950
   98 FORMAT(' INCREASE TIBIA ANGLE OR REDUCE HZ')               00001960
  100 WRITE(16,102)                                              00001970
  102 FORMAT(' "H" PT IS TOO FAR FROM ANKLE PT')                 00001980
      GO TO 2                                                    00001990
  104 WRITE(16,106)                                              00002000
  106 FORMAT(' ARGUMENT IS OUTSIDE OF RANGE FOR ARCOS')          00002010
      GO TO 2                                                    00002020
  108 WRITE(16,110)                                              00002030
  110 FORMAT(' ARGUMENT IS OUTSIDE OF RANGE FOR ARSIN')          00002040
  112 ANGBV=ANGFEM                                               00002050
      GO TO 64                                                   00002060
  114 WRITE(16,116)                                              00002070
  116 FORMAT(1G12.5)                                             00002080
      RETURN                                                     00002090
      END                                                        00002100
                                                                 00002110
                                                                 00002120
```

Integrated Panel and Skeleton Automotive Structural Optimization

H. Miura
NASA Ames Research Center
Moffett Field, CA

R. V. Lust
and J. A. Bennett
Engineering Mechanics Dept
General Motors Research Labs

ABSTRACT

Previous work in structural optimization for the automotive structure has been limited to beam models of the major load-carrying structure. This was primarily done to reduce the amount of computer resources required to minimize the mass. In this study, techniques necessary to include a moderately complex representation of the panels are developed in which some compromises between model fidelity and solution time must be accepted.

As an example, plate elements have been included in a vehicle structural optimization model to represent the roof, floor, dash, motor compartment, and rear quarter. Minimum mass designs subjected to stress, displacement, and frequency constraints are obtained by structural optimization. It was found that most panels were at minimum gage in the optimum design. This suggests that these panels are designed by local criteria as opposed to being controlled by global load and stiffness criteria. A few panels in the highly stressed areas are affected by the global load conditions and should be included in this portion of the optimization.

IN AN EARLIER PAPER (1)* a structural optimization capability for automotive structures was discussed in some detail. The thrust of that work was to develop an optimization capability that would be appropriate for designing the global vehicle structure. However little attention was devoted to the identification of an appropriate structural model. Current finite element models of the global vehicle structure (Fig. 1) are quite

complex, containing both beam and plate finite elements and having several thousand degrees of freedom. Since formal structural optimization methods require many evaluations of a finite element model as well as many derivative calculations, an optimization model of this complexity may be prohibitively expensive from both a computer cost and a total computer residence time standpoint. In addition by the time a detailed structural model can be completed, it may be too late to make more than minor changes in the design. In (1) it was suggested that a simple beam optimization model of the major load carrying member of the structure (Fig. 2) could pro-

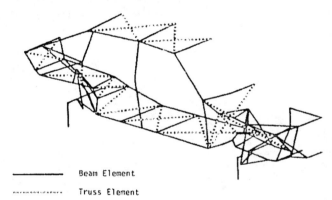

--------- Beam Element

··············· Truss Element

Fig. 2 - Optimization skeleton model

vide valuable information about the primary structure. In that model the shear carrying capability of the various panels was modeled by crossed truss elements. However those elements could not be used to design the panels. The crucial question to be determined then is how accurately that simple optimization model reflects the optimum solution of a more complex beam and plate optimization model. This paper will attempt to address this question by first introducing plate elements into the structural optimization method and then applying the method to a moderately complex automotive structural model representative of the full body.

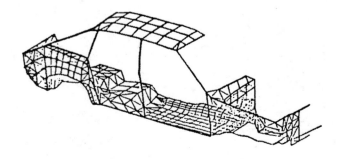

Fig. 1 - Detailed model of automobile structure

* Numbers in parentheses designate References at end of paper.

OPTIMIZATON WITH PLATE ELEMENTS

While there is extensive literature on structural optimization with plate elements (2), most of these efforts have been limited to simple components and have not addressed the problems of complex structures. In addition, much of the work on complex structures has been limited to simple shear elements or constant strain elements (3).

The basic optimization process is usually stated in the following form:

$$\text{minimize } M(\vec{x}) \qquad (1)$$

$$\text{subject to } g_i(\vec{x}) < 0 \qquad (2)$$

where M is the mass, $\vec{x}$ is a vector of design variables and the g_i are constraints on stresses, displacements and frequencies. During the optimization process, derivatives of the constraints with respect to the design variables will be required. These are typically obtained from implicitly differentiating the equilibrium equation to obtain relationships of the following form (1) for displacements

$$\frac{\partial u}{\partial x_i} = [K]^{-1}\left[\frac{\partial K}{\partial x_i}\right]\{u\} \qquad (3)$$

where K is the global stiffness matrix, and x_i is the ith design variable. The terms $\frac{\partial K}{\partial x_i}$ may either be assembled by differencing the elemental stiffness matrix $[k_e]$ or by realizing that the thickness can usually be extracted from the stiffness matrix in such a way that analytical differentiation can be performed. For instance, for the plate element used in this optimization study (a triangular element with constant strain membrane behavior superimposed on a nonconforming, incomplete cubic transverse displacement bending element (4)) the elemental stiffness matrix can be written as

$$[k_e] = [k_{e1}]t + [k_{e2}]t^3 \qquad (4)$$

For the thickness design variable, then, the analytic derivative takes the form

$$\frac{\partial[k_e]}{\partial t} = [k_{e1}] + 3[k_{e2}]t^2 \qquad (5)$$

where k_{e1} and k_{e2} depend only on material properties.

Most of the common methods for solving large scale structural optimization problems rely on approximations based on first order Taylor series expansions of the constraints (3).

$$\frac{\partial g_i}{\partial x} = g_i\text{o} + \sum_{j=1}^{n} \frac{\partial g_i}{\partial x_j} \Delta x_j . \qquad (6)$$

If such expansions are accurate, significant reductions in the number of finite element analyses required for optimization can be made. The accuracy of the approximations is controlled by many things, including the relative importance of plate bending and membrane stiffness to the global stiffness as well as the choice of expansion variable. Clearly if the problem can be adequately formulated by constant strain or simple shear elements, displacement or stress constraints will be almost linear in the thickness. For the purposes of this example we have chosen to retain both membrane and bending behavior and to attempt to use the linear constraint approximation concept expanding the stress, displacement, and frequency constraints in the plate thickness.

OPTIMIZATION METHOD

The ODYSSEY (Optimum DYnamic and Static Structural Efficient SYstems) structural optimization code developed at GM Research is a general purpose code which allows both approximation methods and full mathematical programming methods with exact constraint evaluation to be used. A feasible direction algorithm is used as the optimizer in both cases. A design library of thin-walled beam elements and plate elements (bending and membrane) is available. Multiple load conditions and multiple boundary conditions may be applied and frequency, displacement, and stress conditions may be used. The philosophy and development of this code are described in more detail in (1).

MODEL

The beam and plate model chosen for this study is shown in Fig. 3. As was discussed in the introduction, it was necessary to limit the size of the model for computational cost reasons. In addition computer storage limitations restricted the model to the present size of approximately 1200 degrees of freedom. The design of the model was dictated by the desire to represent the complex curved sheet metal areas as accurately as possible. This meant that the flatter areas

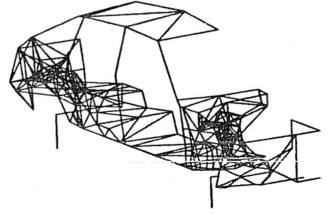

Fig. 3 - Optimization beam/plate model

such as the roof and floor would be modeled
by a minimum number of elements. The load
conditions and constraints, given in Fig. 3b,
are the critical subset of those initially
used to optimize the skeleton shown in
Fig. 2. The full set of load conditions is
given in (5). Note that the constraints are
all global constraints and that no constraints which describe the local conditions
on the panels (such as dent resistance and
oil canning) are included. Therefore, this
optimization model will only design the
structure for the global load conditions.

Frequency Constraints

1. Symmetric
   First mode > 21.4 Hz

2. Asymmetric
   First mode > 30.5 Hz

Symmetric Loading Conditions

1. Front bumper
   3.1×10^4 N rearward at each front E.A. attachment

2. Rear bumper
   2.9×10^4 N forward at each rear E.A. attachment

3. Front barrier
   1.2×10^4 N rearward at front end of front upper rails
   1.4×10^4 N rearward at front end of midrails
   2.6×10^4 N rearward at cradle at midrail attachments
   2.6×10^4 N rearward at front upper rail to cowl attachments
   6.8×10^4 N rearward at A-pillar to rocker attachments
   2.8×10^4 N rearward at cradle to one-bar attachments

4. Rear barrier
   1.2×10^4 N forward at rear end of rear compartment side rails
   1.8×10 N forward at rear end of rear hatch side rails

Non-symmetric Loading Conditions

1. Front chuckhole
   2.9×10^4 N vertical at right front wheel center

Fig. 3b - Load conditions and constraints
for beam and plate model

   More detailed views of two of the more
complex component structures are shown in
Figs. 4 and 5 as well as the companion
detailed finite element models. For the
front wheelhouse inner panel the optimization
model is almost as refined as the detailed
finite element model; however, the rear
wheelhouse inner is much courser than the
detailed finite element model. The stress
contours for the front wheelhouse inner are
shown in Fig. 6. While it is clear that the
model is accurate enough to identify the high
stress areas around the load points at the
shock tower, it is equally clear that the
model is not refined enough to perform a
detailed stress analysis. However, the model
should be accurate enough to identify the
appropriate thickness for a constant
thickness panel. Local reinforcements would
then be determined by a more detailed model.
   In order to reduce the number of design
variables, the structure is divided into
panel areas such as roof, floor, dash, shock

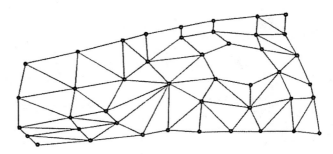

Detailed Model

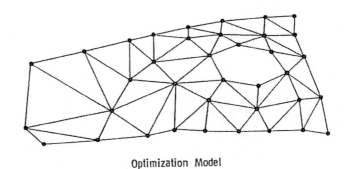

Optimization Model

Fig. 4 - Front wheelhouse inner panel

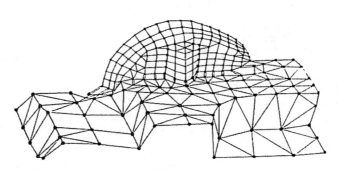

Detailed Model

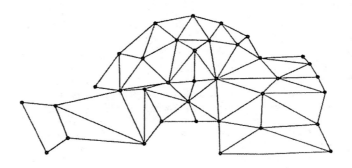

Optimization Model

Fig. 5 - Rear wheelhouse inner panel

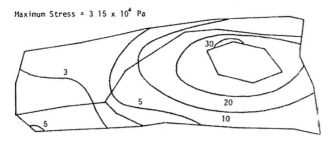

Maximum Stress = 3.15 x 10⁴ Pa

Coarse Mesh (19 Elements)

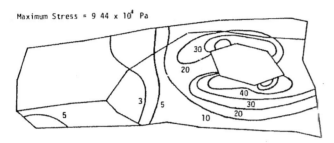

Maximum Stress = 9.44 x 10⁴ Pa

Finer Mesh (47 Elements)

Fig. 6 - Stress contours - front wheelhouse inner

tower, rear quarter, etc. In each panel all thicknesses are linked to a single design variable. The only stress constraint that is tracked is the maximum stress in each panel rather than each element -- this reduces the number of constraints in the problem.

A skeleton version of this structure (Fig. 2) has been optimized in (5) and will be used as a baseline for comparison with the model of Fig. 3. In (5) only the thicknesses of the beams were designed with the widths and heights fixed. This limitation will be continued in the present study.

Since the models shown in Figs. 2 and 3 are different, they will never predict exactly identical stresses or frequencies, and thus we can never expect that they will give identical answers. However, by comparing their behavior we can deduce general trends that the modeling techniques will exhibit. Since each baseline model attempted to simulate the baseline detailed model, we have selected the frequency goals to be approximately the value of the first bending and torsion frequencies of each model when roughly equivalent beam sections and plate thicknesses were used. Thus each frequency constraint is scaled to the particular model. This data is summarized in Fig. 7.

	Initial Frequencies		Frequency Design Constraints	
	1st Bending	1st Torsion	1st Bending	1st Torsion
Beam Model	17.9 hz	27.6 hz	18.0 hz	27.5 hz
Beam and Plate Model	21.4 hz	30.5 hz	21.4 hz	30.5 hz

Fig. 7 - Baseline frequencies and design constraints

INITIAL RESULTS

Three different cases were run: I. Beam and Truss model; II. Beam and Plate model, beams designed; and III. Beam and Plate model, beams and plates designed. The optimized masses for these structures are shown in Fig. 8. The optimum mass difference between the beam only model (Case I) and the more complex models is 7.9%. This difference is indicative that the skeleton model is reasonably accurate. However it is necessary to examine the mass distribution in more detail. This information will be represented in the form of a mass distribution plot.

		Beam Mass (kg)	Plate Mass (kg)
CASE I	Beam & Truss Model	106.0	36.9
CASE II	Beam & Plate Model (beams designed)	97.6	36.9
CASE III	Beam & Plate Model (beams & plates designed)	98.8	34.4

Fig. 8 - Optimum mass summary

In these plots the mass is represented in a type of polar plot. Each vertex on the plot represents a beam or group of beams in the structure as is shown in Fig. 9. The points are then connected to form the plot. Note that the area enclosed by the plot is not proportional to the total mass of the structure since the zero for each beam is not the center of the circle but the inner circle labeled 0 in Fig. 9. Each plot shown in this report will have the same orientation with the front upper section of the vehicle in the upper right or first quadrant of the plot. The major purpose of these plots is to give a rough visual display of the mass distribution among the structures.

The mass distribution plots for Case I and Case II are superimposed in Fig. 10. In general the mass distibutions are quite similar. The major difference is in the rear structure area where the beam model was least representative of the detailed model. The difference in the front upper structure is due to a modeling difference in the radiator support area. Thus it can be seen that the mass distribution errors are localized in the areas of the complex panel structures. In

479

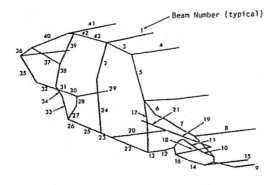

Beam Number (typical)

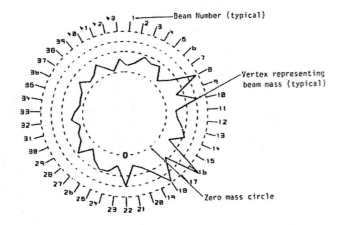

Beam Number (typical)

Vertex representing beam mass (typical)

Zero mass circle

Fig. 9 – Sample mass distribution plot

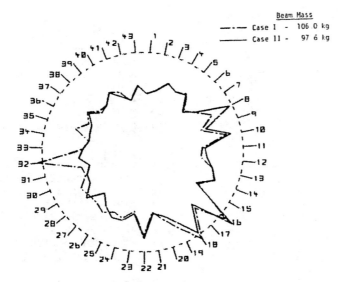

Beam Mass
----- Case I - 106.0 kg
——— Case II - 97.6 kg

Fig. 10 – Mass distribution comparison – case I vs. case II

the Case II design the plates were fixed at an initial value which was their value in the original vehicle. In Case III the plate thicknesses were considered as design variables. Mass distribution plots comparing the Case II and Case III designs are shown in Fig. 11. It is clear that these are virtually identical designs.

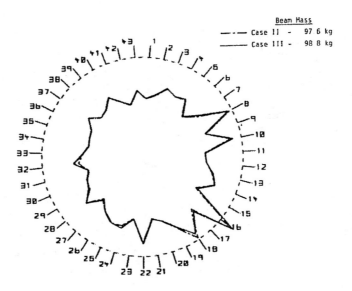

Beam Mass
-----· Case II - 97.6 kg
——— Case III - 98.8 kg

Fig. 11 – Mass distribution comparison – case II vs. case III

The remaining quantities to be examined are the panel thicknesses. These are shown in Fig. 12. There are two points of interest. First, when the panels were designed, all of the panels decreased in mass. Second, most of the panels were driven to minimum gage. This very strongly implies that most of the panels are not designed by the global constraints because the optimization chose to take mass out of the panels without violating any global constraints. Note that there was a slight increase in the skeleton mass (Fig. 8) to compensate for this. It appears, therefore, that most panels should be designed by local conditions and then fixed in the global model.

| | Panel Thickness (cm) | | | | |
	Case II	Case III	Case IV	Case V	Case VI
Front Wheelhouse Inner	1300	1159	1180	1203	.1723
Front Shock Tower	1700	0765	0762	0762	.1491
Dash	0800	0764	0773	0773	.1238
Floor Pan	0762	0762	0769	0762	0772
Roof	0762	0762	0762	0762	.0762
Rear Compartment Floor	0800	0762	0762	0762	0762
Rear Wheelhouse Inner	0762	0762	0762	0762	0762
Rear Wheelhouse Outer	0762	0762	0762	0762	0769
Rear Shock Tower	1700	0765	0763	0762	.0771
Rear Quarter Panel Outer	0800	0762	0762	.0762	0762
Rear Quarter Panel Inner	0762	0762	0762	0762	0765
Rear End Panel	0762	0762	0762	0762	0762

Fig. 12 – Summary of panel thicknesses

In order to confirm these results some additional cases were run.

CASE IV - BEAM AND PLATE MODEL, PLATES DESIGNED

DIFFERENT INITIAL CONDITIONS ON THE PLATES - In order to verify that the solution to Case III was not a local optimum, the initial thicknesses of the plates were doubled. As Figs. 12 and 13 show, the results are virtually identical to those of Case III.

CASE V - BEAM AND PLATE MODEL, PLATES DESIGNED

DIFFERENT INITIAL CONDITIONS ON THE SKELETON - In this case, the design was started with the final solution to Case I, the skeleton alone. As shown by Figs. 12 and 14, the design is very similar to that of Case III and, within the accuracy of the program, can be considered to be the same design. Thus, Cases IV and V led to the conclusion that Case III is most likely a global minimum.

CASE VI - BEAM AND PLATE MODEL, ALL ALUMINUM

Previous cases have duplicated the material distribution of Case I. For this design, the entire structure, including beams and plates, was assumed to be aluminum, roughly attempting to duplicate the all-aluminum structure of (5). The plate thicknesses are shown in Fig. 12; the beam mass distribution is compared to the all-aluminum skeleton in Fig. 15 and to the steel structure in Fig. 16. The mass is virtually identical to that of the all-aluminum skeleton of (5).

Also of interest is the interplay between the steel and aluminum front structures as shown in Fig. 16. Note that the aluminum midrail is actually heavier than its steel counterpart. One possible explanation for

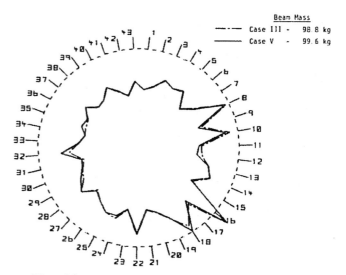

Fig. 14 - Mass distribution comparison - case III vs. case V

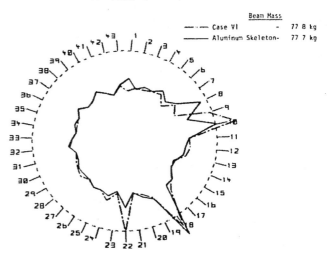

Fig. 15 - Mass distribution comparison - case VI vs. all aluminum skeleton

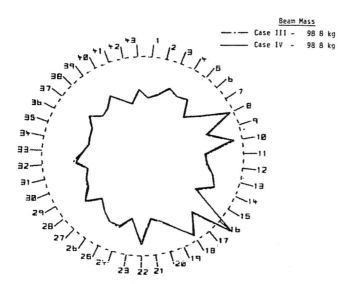

Fig. 13 - Mass distribution comparison - case III vs. case IV

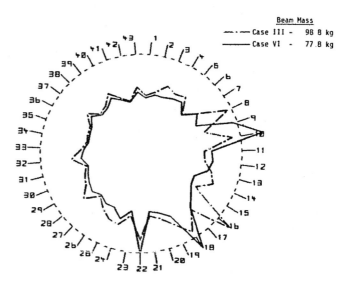

Fig. 16 - Mass distribution comparison - case III vs. case VI

this is that the cradle in the steel design is both stress constrained and an inefficient place to provide overall stiffness. Therefore, when the material is switched to aluminum with a higher specific strength, the optimization removes material from the cradle area and increases the mass of the midrail, which is a more efficient way of providing global stiffness. This is an excellent example of why material substitution schemes based on single components may not be accurate.

A second point of interest is related to the comparison between the skeleton alone and the skin and skeleton models. For the all-steel design, recall that the skin and skeleton design was 8% lighter than the skeleton-only design, whereas they were identical for the all-aluminum design. One plausible explanation for this is that, in the steel design, many parts were stress as well as frequency constrained and, therefore, the modeling of the skin in the heavily stressed areas affects the mass distribution. This is, of course, confirmed by the final mass distribution (Fig. 10). On the other hand, the aluminum design is, in general, not stress constrained and, therefore, the two aluminum designs are extremely close.

DISCUSSION

The optimization results presented in this study have been obtained using first-order Taylor series approximation for the constraints and move limits of 25% in the design variables. The success of this approach implies that the bending effects are not dominating the global constraint behavior. The optimum designs were achieved in 10 finite element solutions or less. Run times are shown in Fig. 17. These results indicate that it is feasible to optimize beam and plate models of automotive structures. Since the run times are strongly dependent on the size of both the analyses problem (number of degrees of freedom) and the design problem (number of design variables), there is a reasonable limit to the size of problem that could be handled. At present it appears that a model of the complexity of that shown in Fig. 1 would require significant resources and would certainly be inappropriate for preliminary design studies.

Time in CPU* Minutes

Skeleton Model	30
Skeleton and Panel Model (beams designed)	45
Skeleton and Panel Model (beams and plates designed)	74

* IBM 3033

Fig. 17 - Approximate computer times

The results of the optimization studies have also shown that the skeleton optimization model predicts the primary structure mass and mass distribution given by the more complex beam and plate model. In addition it is clear that the majority of the panel components are not designed by the global load conditions and should be designed subject to local load conditions. It should be emphasized that these results are based on only one sample structure. However the optimum design for this structure is not dominated by any one constraint. In fact both frequency constraints are active and about 50% of the beams have an active stress constraint. Thus it is quite likely that the behavior of most automotive global structures will be similar.

The results of this study suggest the following approach to optimal structure design (Fig. 18). First a simple skeleton

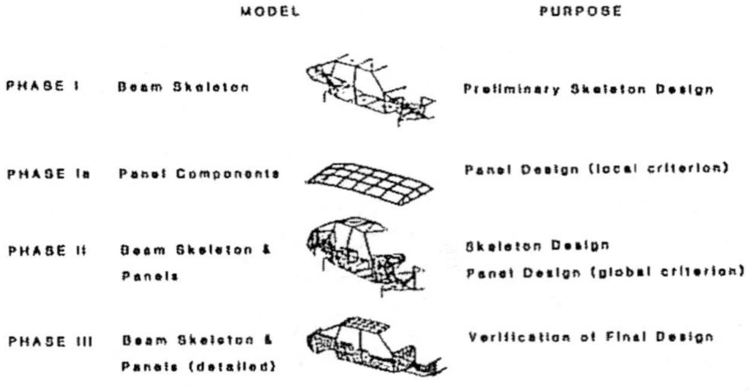

Fig. 18 - Automated design procedure

model is used to make major configuration decisions and to identify areas where structural efficiency would require space which must be obtained at the expense of other design constraints. Subsequently, or perhaps concurrently, panels which are designed by nonglobal criteria can be optimized separately. Examples of these types of problems are given in (6). As a second step, an intermediate model of the global structure is used which contains the basic skeleton structure as well as simple models of the panels designed by local criteria and more complex models of the globally designed panels. This model would be used for more detailed tuning of the beam elements and design of any globally controlled panels. Finally, a detailed finite element model would be built for final tuning of the design. This orderly progression of models would ensure early use of accurate optimization information.

REFERENCES

1. J. A. Bennett and M. F. Nelson, "An Optimization Capability for Automotive Structres," SAE Transactions, Vol. 88, pp.

2. R. T. Hafka and B. Prasad, "Optimum Structural Design with Plate Bending Elements - A Survey," AIAA Journal, Vol. 9, No. 4, pp. 517-522.

3. L. A. Schmit and H. Miura, "An Advanced Structural Analysis/Synthesis Capability — ACESS-2," Int. J. for Numerical Methods in Engineering, Vol. 12, pp. 353-377, 1978.

4. J. Connor and G. Will, "A Triangular Plate Bending Element," TR 68-3, Department of Civil Engineering, M.I.T., Cambridge, Mass., 1968.

5. P. A. Fenyes, "Structural Optimization with Alternate Materials: Minimum Mass Design of the Primary Structure," SAE paper 810228.

6. R. V. Lust and J. A. Bennett, "Structural Optimization in the Design Environment," Proceedings of 4th SAE International Conference on Vehicle Structural Mechanics," November, 1981, Detroit, MI.

840866

Crash Padding Mechanical Properties and Impact Response Analysis

Oscar Orringer and Kevin T. Knadle
U.S. DOT Transportation Systems Center
Cambridge, MA
John F. Mandell
Department of Materials Science and Engineering
Massachusetts Institute of Technology
Cambridge, MA

ABSTRACT

Highly deformable materials are used to pad automobile interior surfaces for occupant crash protection. Thick energy-absorbent pads usually provide better protection than thin pads nonabsorbent pads. Material selection tends to be the primary design variable because interior space is limited. Padding design requires analysis of the effects of typical crashes on the force-time history that an occupant will experience when he strikes an interior surface. To be practical the analysis must be physically realistic but must avoid excess numerical detail that might obscure basic trends. A practical method based on simplified occupant and vehicle models and empirical description of material properties is presented. The models focus on essential geometry and inertia parameters. A constitutive equation for an actual foam rubber is derived from the results of laboratory tests.

CRASH PAD DESIGN is an art based on test experience. The experience is relatively easy to gain for the packaging of inert objects, but the art of designing automobile interior padding is complicated by coupling with the occupant's biomechanical response. In this case, experience can be gained only indirectly via time-consuming and expensive laboratory tests with pads and cadavers and full-scale crash tests of automobiles containing instrumented dummies. Realistic but simple mathematical simulations are needed to explore the full range of design possibilities.

The National Highway Traffic Safety Administration (NHTSA) has done and continues to do extensive research on the mathematical modelling of biomechanical response. A recent report summarizes several human response models and shows the complexity of as well as the progress made in this area[1]*. Two general features have emerged. First, different body parts require different kinds of models to correlate medical in-

*Numbers in parentheses designate references at end of paper.

jury levels with the physical parameters of a crash. Second, those models which deal with injuries other than bone fracture must often treat the affected body part as a deformable system to obtain good results.

The evolution of head-injury models illustrates these points. Early work based on cadaver and animal tests led to an empirical correlation between the onset of concussion or other serious head injury and the Head Injury Criterion (HIC) number $H = 1,000$ seconds. The HIC number for a given crash event is calculated as:

$$H = \max\left[(t_2-t_1)^{-1.5}\left|\int_{t_1}^{t_2} n(t)dt\right|^{2.5}\right] \quad (1)$$

where $n(t)$ is the head acceleration-time history in dimensionless "g"s, t_1 and $t_2>t_1$ are any time limits within the crash event, and the notation $\max[..]$ means that the limits are chosen to give the maximum possible result for H.

The Mean Strain Criterion (MSC) is a recent development based on consideration of the biomechanical interaction between the brain and the skull. The MSC model is a damped second-order two-mass system (Figure 1), and the criterion is based on the maximum relative displacement Z between the brain and the skull. Finite-difference analysis of the MSC model with prescribed base-motion input to the skull mass M_1 can be used to compute Z for a given crash.

A hypothetical case (constant acceleration) illustrates how the MSC gives a better response prediction than the HIC. Eq. (1) reduces to $H = n^{2.5}\Delta t$ for this case, where Δt is the total time of the event. The HIC thus predicts that any level of acceleration will cause injury if applied for a long enough time. Calculations show, e.g., that the HIC predicts concussions for such non-injurious events as railroad and automotive vehicle crash "ridedown" by restrained occupants or maneuver or launch accelerations routinely imposed fighter pilots or astronauts. Conversely,

484

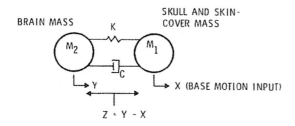

BRAIN MASS   SKULL AND SKIN-
              COVER MASS

M_2 K M_1

Y   X (BASE MOTION INPUT)

$Z = Y - X$

TYPICAL HEAD-IMPACT PARAMETERS

CASE	W_1 (LB.)°	W_2 (LB.)°	K (LB./IN)	C (LB SEC/IN)
HUMAN (FRONT)	0.60	10.00	50,000	2.0
HUMAN (LAT)	0.40	9.00	26,000	2.4
CHIMPANZEE	0.08	4.75	35,000	2.4
RHESUS MONKEY	0.07	1.95	40,000	1.7
°M = W/G; G = 386.4 IN/SEC²				

INJURY CRITERION

PEAK CRUSH, Z (IN.)	AIS INDEX
0.015	1
0.025	2
0.035	3
0.045	4
0.055	5

Figure 1. Occupant Model for MSC

the MSC model predicts $Z \sim n\Delta t$ if the event is short or $Z \sim n$ if the event is long with respect to the characteristic response time of the brain in the skull. This behavior is in better accord with experience because it avoids spurious predictions for routine accelerations but preserves an inverse relation between duration and injurious acceleration level for impulsive events.

The foregoing example oversimplifies events such as a head striking a padded dashboard. Such events involve time-dependent accelerations that must be found by solving the coupled dynamic-material problem. The acceleration-time history must be found before the HIC number can be calculated, although Z for the MSC can be computed with compare-and-save logic during the dynamic analysis. In either case, both the engagement and rebound phases of the head motion relative to the dashboard must be analyzed to perform the injury assessment. The head-injury models pose the most extensive requirements for a crash padding design simulation.

The NHTSA is interested in such simulations for the purpose of assessing the extent to which reasonable improvement of interior padding might mitigate unrestrained occupant injuries in typical crashes. The present work on impact simulations and crash padding mechanical properties started in 1982. This paper reports the results of the first phase of the work(2,3).

COLLISION ANALYSIS OF DEFORMABLE BODIES

If masses M_1, M_2 with initial velocities V_1

and V_2 collide, and if the coefficient of restitution C between the bodies is given, then the laws of conservation of energy and momentum can be used to derive the post-collision velocities:

$$V_1' = [(M_1 - M_2\sqrt{C})V_1 + (1+\sqrt{C})M_2 V_2]/(M_1+M_2) \quad (2a)$$

$$V_2' = [(1+\sqrt{C})M_1 V_1 + (M_2 - M_1\sqrt{C})V_2]/(M_1+M_2) \quad (2b)$$

The quantity 1−C is by definition the fraction of available collision energy that is inelastically absorbed:

$$1-C = E_{AB}/E_{AV} \; ; \; E_{AV} = M_1 M_2 (V_1-V_2)^2/[2(M_1+M_2)] \quad (3)$$

If the bodies are an occupant M_1 and a vehicle structure $M_2 \gg M_1$, and if the collision is described in terms of occupant motion relative to the vehicle, then Eqs. (2) reduce to $V_2' \cong V_2 = 0$ and $V_1' \cong -\sqrt{C} V_1$, i.e. the coefficient of restitution defines the occupant's rebound speed.

Coefficients of restitution are worthwhile to find per se because rebound velocities have immediate use as inputs to computer programs such as the NHTSA Crash Victim Simulator(4), which predicts the path of an unrestrained occupant engaging in multiple collisions with different parts of the passenger compartment. To find C requires constitutive equations for the deformable components and a complete analysis of the collision, i.e. the same basic simulation as is needed for injury assessment. Suppose, e.g.,

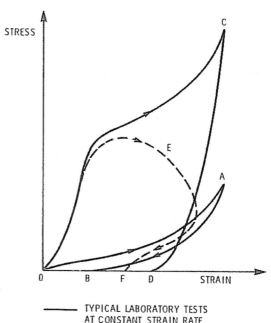

STRESS

C

E

A

0   B   F   D   STRAIN

——— TYPICAL LABORATORY TESTS AT CONSTANT STRAIN RATE

— — — COLLISION WITH TIME-DEPENDENT STRAIN RATE

ENERGY ABSORBED PER UNIT VOLUME = AREA ENCLOSED BY STRESS-STRAIN CURVE

Figure 2. Laboratory and Collision Stress-Strain Diagrams for an Inelastic Material

485

that a material is nearly elastic at low strain rate (curve OAB) but strongly inelastic at high strain rate (curve OCD), as shown in Figure 2. A colliding body will then cause the material to trace an intermediate path on the stress-strain diagram, say curve OEF. The shape of the path, the value of C, the stress-time history, etc. all depend on the mass and initial speed of the colliding body.

Figure 3 illustrates a model of the coupled dynamic-material problem in which three simplifying assumptions have been made. First, the occupant is assumed to be rigid. Second, the mass of the structure is much larger than the mass of the occupant. Third, the deformable component is struck uniformly, i.e. its constitutive equation can be written directly in terms of force F and deflection X instead of stress and strain. With the convention that positive F and X correspond to compression of the deformable component, the model response can be formulated as the coupled differential equations and auxiliary conditions:

$$M\ddot{X} = -F \qquad \text{(dynamical equation)} \quad (4a)$$

$$\dot{F} = f(X,\dot{X},F) \qquad \text{(constitutive equation)} \quad (4b)$$

$$X(0)=F(0)=0 \; ; \; \dot{X}(0)=V \qquad \text{(initial conditions)} \quad (4c)$$

$$F(T)\leq 0 \text{ for } T>0 \qquad \text{(termination condition)} \quad (4d)$$

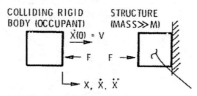

X = t = 0 WHEN BODY FIRST TOUCHES STRUCTURE

Figure 3. Simplified Model of Coupled Dynamic-Material Collision Problem

Eqs.(4a) and (4b) are easily solved with a forward first difference algorithm. Eq. (4c) provides the starting conditions, and Eq. (4d) is a test that is to be applied at each step to find the time T when the occupant loses contact with the structure. The time histories can be saved for post-simulation analysis (e.g. for calculating the HIC number), and $\dot{X}(T)$ is the rebound velocity.

It is useful in some cases to divide a deformable structure into linear and nonlinear components. For example, it is reasonable to treat metal dashboards as linearly elastic over the range of forces and deflections of interest to injury assessment, but padding materials are nonlinear and rate-dependent over this range. One can then replace Eq. (4b) with:

$$\dot{F} = f(X',\dot{X}',F) \; ; \; X' = X - F/K \qquad (4e)$$

where K is the dashboard stiffness and X' is the deflection of the padding.

Eqs. (4) contain the essential ingredients for collision analysis. The next step is to examine the simplifying assumptions and to elaborate the formulation where the assumptions appear to be unrealistic.

The assumption of a rigid occupant may seem questionable at first glance but is actually a reasonable approximation in some cases. For example, stiffnesses of typical auto dashboards lie in the range of 200-600 lb./in.(3,5), i.e. two orders of magnitude less than the MSC model stiffnesses (Figure 1). It is reasonable in such cases to neglect the effect of occupant deformation on the collision analysis and to reserve a calculation of Z for post-simulation analysis.

Conversely, typical auto A-posts are much stiffer than the MSC model, and the thinness of A-post padding tends to limit the structure deformation to the order of Z. Cases of this type require a two-degree-of-freedom formulation in which the MSC model is coupled with the constitutive relation for the padding.

Suppose for example that the A-post padding is assumed to be linearly elastic up to a given deflection X_p and rigid thereafter (Figure 4). It is then easy to show that the collision analysis can be carried out in two phases using nondimensional equations(3). The first phase covers engagement up to the deflection limit:

$$\ddot{Z}+(1+\mu)(2\zeta\omega\dot{Z}+\omega^2 Z)-\chi\mu\omega^2 X = 0 \qquad (5a)$$

$$\ddot{X}+\ddot{Z}+2\zeta\omega\dot{Z}+\omega^2 Z = 0 \qquad (5b)$$

$$X(0)=Z(0)=\dot{Z}(0)=0 \; ; \; \dot{X}(0)=V \text{ at } t=0 \qquad (5c)$$

$$\chi = K_p/K \; ; \; \mu = M_2/M_1 \; ; \; \omega = \sqrt{K/M_2} \; ; \; \zeta = C/2\sqrt{KM_2} \qquad (5d)$$

Eqs. (5a) and (5b) can be solved by finite differences. The second phase begins when the deflection limit is attained, say at t=T, and Eqs. (5) are replaced by the following equations on a shifted time scale $t'=t-T$:

$$\ddot{Z}+2\zeta\omega\dot{Z}+\omega^2 Z = 0 \qquad (6a)$$

$$Z(t'=0)=Z(t=T) \; ; \; \dot{Z}(t'=0)=\dot{X}(T)+\dot{Z}(T) \qquad (6b)$$

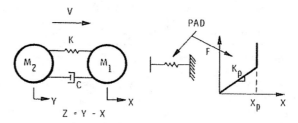

Figure 4. MSC Model Striking Idealized A-Post

Eq. (6a) can be solved analytically. The peak value of Z can occur in either phase. The entire analysis has been programmed on a pocket calculator (see Appendix). Figure 5 presents some example results to show that the least injurious outcomes are associated with cases in which the pad deflects to approximately the limit value.

Another useful variation on the occupant is to account for effects of distributed body mass. Figure 6 shows a simple example of a rigid two-mass model, where only one mass is directly involved in the impact. Because of the rotational effect introduced, Eq. (4a) must be replaced by two nonlinear dynamical equations:

$$\ddot{X} = (1 - \mu F \sin^2 \theta - M_2 D \dot{\theta}^2 \cos \theta)/(M_1 + M_2) \qquad (7a)$$

$$\ddot{\theta} = F \sin \theta / M_1 D \qquad (7b)$$

The rigid-body assumption is not reasonable for

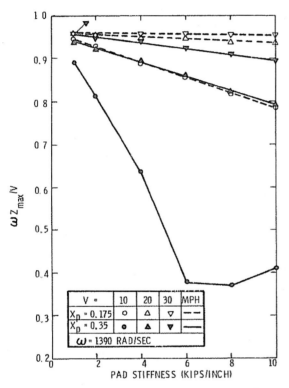

Figure 5. Crush Curves for Human Head (Frontal)

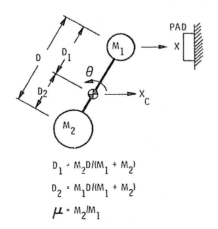

$$D_1 = M_2 D/(M_1 + M_2)$$

$$D_2 = M_1 D/(M_1 + M_2)$$

$$\mu = M_2/M_1$$

Figure 6. Occupant Model with Rotational Ability

whole-body models, to which one should add some flexibility between the masses. In either case, however, it is still a routine task to solve the coupled dynamic-material problem by finite differences.

The second assumption was that the mass of the structure is much larger than the occupant's mass. The basic formulation, in effect, assigns infinite mass to the structure. The error thus introduced can be estimated by considering the result of an elastic collision. The significant quantity is the occupant's velocity change, $\Delta V = V_1 - V_1'$, which can be calculated by taking C=1 in Eq. (2a). For the case of a 180-lb. occupant in an 1,800-lb. subcompact car, the simplifying assumption leads to $\Delta V = 2V_1$. For the exact case, $\Delta V = 1.82V_1$ if the entire occupant mass is effective or $\Delta V = 1.96V_1$ if half the occupant's mass is effective. The simplified model is therefore conservative, certainly less than 20 percent in error, and probably less than 5 percent in error for practical cases.

The third assumption, uniform engagement, was convenient for the purpose of introducing the basic formulation but is not reasonable for practical situations. Most components of the human body have singly or doubly curved surfaces which tend to locally indent padded structures. The following section discusses some possible approaches to the simulation of such behavior.

NONUNIFORM CONTACT

Full treatment of nonuniform contact involves three-dimensional constitutive relations, material nonlinearity, and geometric nonlinearity. Such analyses are unwieldy, as will be shown by a later example formulation. Therefore, the development of nonuniform contact analysis has concentrated on approximate models which simulate some of the significant features of local contact.

One can begin by assuming that occupant deformation can be neglected (e.g., as in a head-dashboard collision). Such situations can be approximated by a rigid sphere which imposes its shape on an initially flat pad. The approximation is also useful in that it allows one to isolate the validation of a pad material-contact model from biomechanical effects by controlled experiments. The sphere radius R and pad thickness L are characteristic dimensions for this type of model.

Lockett et al.(6) first proposed the sphere contact model in connection with their research on several foam padding materials. They observed that the so-called "rigid" foams, which crush without much recovery when compressed, have extremely low Poisson's ratios. They then proposed that a local contact be treated as if the pad could be divided into concentric cylinders, each of which reacts to the sphere penetration as an independent column. A one-dimensional constitutive equation for the material is then applied to the compression of each cylinder, and the stress distribution is integrated to obtain the contact force. Lockett et al. restricted their formulation to shallow penetration (R>>L) but applied it to recoverable as well as rigid foams.

Figure 7 illustrates the geometrical relations for a similar sphere contact model that allows deep penetration (R>L). The sphere displacement X for the dynamical equation corresponds to the maximum penetration depth at the center of contact. The penetration, strain, and strain-rate fields can then be defined as functions of the radial distance from the center of contact:

$$w(r) = X - R[1 - \sqrt{1 - (r/R)^2}] \quad \text{(penetration)} \quad (8a)$$

$$\epsilon(r) = w(r)/L \quad \text{(strain)} \quad (8b)$$

$$\dot{\epsilon} = \dot{X}/L \quad \text{(strain rate)} \quad (8c)$$

for $0 \leq r \leq r_{max}$, where

$$r_{max} = \sqrt{2RX - X^2} \quad (8d)$$

The stress field is then obtained from the constitutive relation,

$$\dot{\sigma} = f(\epsilon, \dot{\epsilon}, \sigma) \quad (9)$$

and the contact force is formally given by:

$$F = 2\pi \int_0^{r_{max}} \sigma(r) r \, dr \quad (10)$$

but is actually computed by means of the numerical algorithm suggested in Figure 7.

Figure 8 illustrates a generalization of the model to ellipsoidal contact, restricted to cases in which a principal axis of the ellipsoid lies along the line of collision. The kinematic equations are modified as shown, and the right hand side of Eq. (10) must be multiplied by C/B.

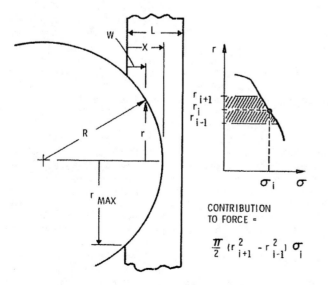

Figure 7. Spherical Contact Geometry

The sphere and ellipsoid contact models can be added to the basic collision analysis by substituting Eqs. (9) and (10) for Eq. (4b) or (4e). An application will be presented later.

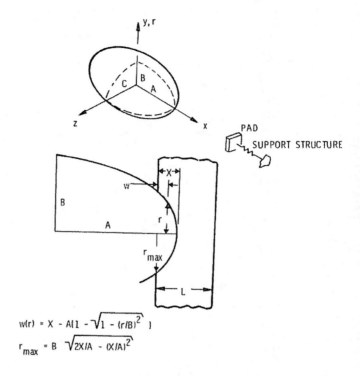

$$w(r) = X - A[1 - \sqrt{1 - (r/B)^2}]$$

$$r_{max} = B \sqrt{2X/A - (X/A)^2}$$

Figure 8. Ellipsoidal Contact Geometry

Recoverable foams and solid elastomers which might be used for padding tend to have Poisson's ratios of about 0.35 to 0.40. It is thus natural to infer that such materials have considerable ability to spread a reaction to compression by means of load transfer in shear. Simple compression experiments with flat-ended cylindrical indenters confirm the expectation: more force is needed to produce a given penetration in a specimen whose diameter exceeds the indenter diameter than when the two diameters are equal.

One should also expect pads made of such materials to behave more stiffly when indented by a curved object. Cylindrical indenter tests can be used to determine an empirical adjustment factor for the sphere contact model. This factor probably overestimates the pad stiffness, however, since the radial distribution of shear is more diffuse under sphere contact than under flat-ended cylinder contact. The material may also have different rate effects in shear and compression, i.e. pad responses to static and dynamic contact may not be homologous. A full simulation would thus require axially symmetric analysis of the padding continuum using three-dimensional constitutive relations for dilatational and distortional behavior.

The load-spreading effect of shear can also make the contact radius an unknown quantity, and the model then has a geometric nonlinearity. One can formulate a simple example by assuming that the pad consists of two linearly elastic components: a thin skin with non-negligible Poisson's ratio bonded to a thick foundation which behaves like the idealized material in the basic sphere contact model (Figure 9). The Poisson's ratio of the skin does not appear directly in the model, but the skin itself introduces a load-spreading effect via membrane action.

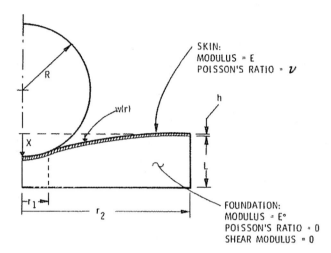

Figure 9. Linearly Elastic Skinned-Pad Model

The field equations of the model can be derived from the theory of a plate on a continuous elastic foundation[7] by treating the skin as an ideal membrane with in-plane stiffness but no bending stiffness:

$$\nabla^4 \phi + (E/2r)(w'w')' = 0 \qquad (11a)$$

$$(h/r)(w'\phi')' - (E*/L)w = 0 \qquad (11b)$$

where the $(E*/L)w$ term in Eq. (11b) is the foundation compressive stress, $(..)' = d(..)/dr$,

$$\nabla^2 \phi = (1/r)(r\phi')' \; ; \; \nabla^4 \phi = (1/r)[r(\nabla^2 \phi)']' \qquad (12)$$

and ϕ is a stress function from which the skin membrane stresses can be calculated as:

$$\sigma_r = \phi'/r \; ; \; \sigma_\theta = \phi''/h \qquad (13)$$

Variational mechanics[8] can be applied to Eqs. (11) to recast the formulation in terms of the energy integral:

$$\Pi(w,\phi) = \int_0^{r_2} [(E*E/L)w^2 - h(\nabla^2 \phi)^2 + (Eh/r)(w')^2 \phi'] r\, dr \qquad (14)$$

where r_2 is the outside radius of the pad. The functional $\Pi(w,\phi)$ is extremized by exact solutions of the field equations. Approximate solutions can be found by expressing w and ϕ as two series of assumed functions with unknown coef-

ficients, integrating Π, and then solving the resulting algebraic extremum problem. Assumed functions with suitable asymptotic behavior can be used to obtain solutions for generic cases in which there is no boundary influence on the contact process $(r_2 \to \infty)$.

Geometric nonlinearity can be embodied in the variational formulation by dividing Eq. (14) into contact and outer region components:

$$\Pi(w,\phi,r_1,C) = \Pi_1 + \Pi_2 \qquad (15a)$$

$$\Pi_1 = \int_0^{r_1} [(E*E/L)\bar{w}^2 - h(\nabla^2 \bar{\phi})^2 + (Eh/r)(\bar{w}')^2 \bar{\phi}'] r\, dr \qquad (15b)$$

$$\Pi_2 = \int_0^{r_2} [(E*E/L)w^2 - h(\nabla^2 \phi)^2 + (Eh/r)(w')^2 \phi'] r\, dr \qquad (15c)$$

where $\bar{w}(r)$ is the prescribed pad deflection in the contact region, given by Eq. (8a),

$$\nabla^2 \bar{\phi} = -(E/2)\int (1/r)(\bar{w}')^2 dr + 2C \qquad (16a)$$

$$\bar{\phi}' = (1/r)\int r \nabla^2 \bar{\phi}\, dr + D/r \qquad (16b)$$

are found from Eq. (11b), and C,D are constants of integration with D chosen to supress singular behavior from $\bar{\phi}'$ at $r = 0$. The constant C is an additional unknown which is to be determined as part of the solution. The outer region variables w and ϕ can be assumed as series of functions to satisfy the inter-region continuity conditions:

$$w(r_1) = \bar{w}(r_1); \; \phi'(r_1) = \bar{\phi}'(r_1) \qquad (17)$$

and the outer boundary conditions term by term. Integration of Eqs. (15) then leads to a nonlinear algebraic extremum problem for r_1, C, and the coefficients of the series for w and ϕ. Note also that the Leibniz rule for differentiating at the limits of an integral must be applied to properly account for the terms which result from the variation of r_1.

Numerical experiments with the skinned pad model have shown that the extremum problem needs a predictor-corrector iteration method to avoid instability. One must also anticipate similar difficulties for problems with nonhomologous material behavior. Hence, it remains an open question whether the benefit of such refinement over the basic sphere contact model is worth the extra computational effort for design analysis.

MATERIAL CHARACTERIZATION

Laboratory tests were run[2,3] to measure the dynamic mechanical properties of Uniroyal Ensolite AAC, a recoverable closed-cell foam used in NHTSA investigations of injury reduction concepts for automobile passenger compartments. The main purpose of the testing was to develop an empirical constitutive equation for uniform uniaxial compression over realistic ranges of strain and strain rate. Unrestrained occupants colliding with padded structures can be expected to subject the pad to peak compression strains of at least 0.8 and initial strain rates as high as 2,000/second. Strain-control tests covered strains up to 0.9 and strain rates up to 73/sec, and a few uncontrolled tests investigated strain rates up to 10,000/second.

Cylindrical samples 0.14 and 1.2 in. thick with 1-in. diameter were subjected to stress-relaxation and stress-strain tests under strain control on electromechanical (EM) and servohydraulic (SH) machines at temperatures between 66 and -40°C. Crosshead motion and piezoelectric load cell signals recorded the strain input and stress output in a Nicolet digital oscilloscope. The EM machine was used for most of the tests because of its ability to accurately control strain, but the strain rate was limited to about 1 per second. Hence, the time-temperature superposition principle(9) was applied with graphical construction techniques(10) to organize the test results at different temperatures into master curves to cover higher strain rates at a 25°C reference temperature. The SH tests and some uncontrolled tests were then used to confirm the master curves at higher strain rates.

The stress-relaxation tests were performed by rapidly imposing and holding a fixed strain on the samples while measuring the stress-time decay curve. The prescribed strain could be applied in a time as short as one millisecond (ms) on the SH machine, but transient ringing of the sample and fixture prevented accurate measurement of stress decay until 20 ms had elapsed. The controlled stress-strain tests were run by commanding a triangular strain-time signal (i.e. constant strain-rate magnitude) and a prescribed maximum strain. Both the loading and unloading stress-strain curves were measured, but some inaccuracy was observed in the unloading curves at high values of strain and strain rate.

Figure 10 shows EM master curves for three strains. The data points at low temperature and high strain depart from the curves, as indicated by the dashed lines. Microscopic examination of these samples revealed cell wall crushing which was not observed at other combinations of strain and temperature. Consequently these results were discarded from the master curve constructions. Figure 11 shows that the curves agree well with the SH tests where the stress decay was measured as early as 20 ms after strain application. The high-strain data scatter in Fig. 11 is an artifact of strain-control error in the SH machine. Stress decay was also measured at times up to 2,000 seconds, and the curves for temperatures above 0°C were found to have increased rates of decay at about 300 seconds. This observation suggests that the results of conventional (i.e. long-time) rheological tests should not be extrapolated to the impact time range.

Figure 12 plots EM stress-strain curves obtained at low strain rate (0.12 per second) and

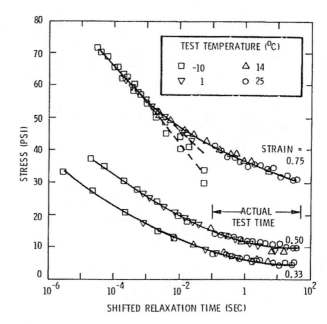

Figure 10. Master 25°C Stress-Relaxation Curves

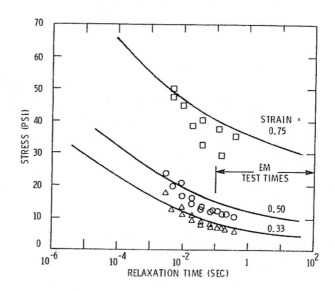

Figure 11. Comparison of SH Stress-Relaxation Data with EM Master Curves

various temperatures. A dramatic increase in hysteresis is evident as the test temperature decreases. The time-temperature superposition concept then leads one to expect a similar increase in hysteresis for tests at fixed temperature as the strain rate increases, i.e. the material should have highly rate-sensitive mechanical properties. A better way to organize the stress-strain test data is to cross-plot master curves of stress during loading versus strain rate for selected values of strain. Figure 13 compares the 25°C master curves constructed from EM tests with test results at higher rates. The data includes some drop-weight impact tests that achieved initial strain rates up to 10,000/sec.,

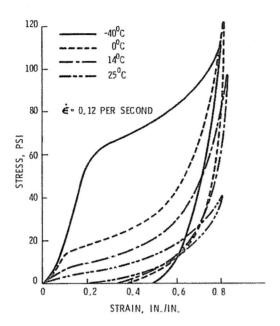

Figure 12. Typical Low-Rate Stress-Strain Curves

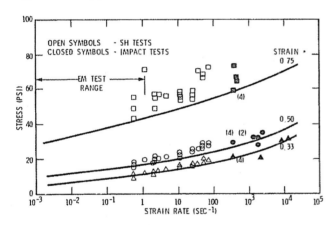

Figure 13. Comparison of SH Stress-Strain Test
Data with EM Master Curves

stiffness of 1.8 to 2.7 times its uniform compression stiffness (Figure 15). The last result suggests that Ensolite AAC foam has significant ability to spread compression loads via shear.

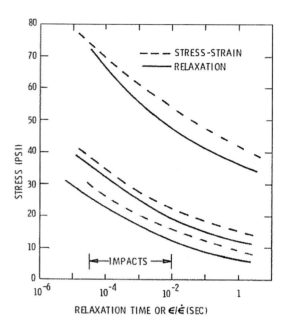

Figure 14. Comparison of Stress-Relaxation and
Stress-Strain Test Master Curves

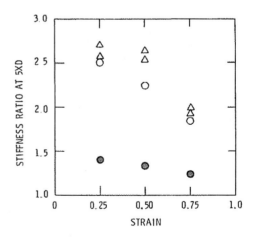

STRAIN RATE(SEC^{-1})	THICK	THIN
0.48		●
0.55	○	
2.20	△	

Figure 15. Results of Indenter Tests

but the strain rate varied during those tests. The stress-strain test results can also be plotted in terms of stress versus the time required to reach the selected strain ($\epsilon/\dot{\epsilon}$) to allow for comparison with the results of the stress-relaxation tests. Figure 14 compares the two sets of master curves to show that the results agree.

Lateral extensometry was also used in some of the low-rate compression tests at, and other samples were subjected to the cylindrical indenter test mentioned earlier. The extensometry showed that the 1.2-inch thick foam is anisotropic, and its major Poisson's ratios were found to vary from 0.32 to 0.48, generally increasing with strain. The indenter tests were performed on both 0.14- and 1.2-inch thick material at strain rates from 0.5 to 2.2 per second. There was only a small shear effect in the thin material, but the thick material showed an apparent

491

CONSTITUTIVE EQUATION

The laboratory tests revealed the strong dependence of Ensolite mechanical properties on strain rate and the nonlinear characteristic of its stress-strain curve. The next step was to model the material with a one-dimensional constitutive equation for uniform compression.

The first modelling efforts were based on construction of simple mechanical analogs consisting of discrete spring and damper elements. This is a well established and popular method for modeling the multiple relaxation mechanisms observed in solid polymers[9]. Figure 16 illustrates the basic linear viscoelastic model that is often used as the starting point. The basic model consists of two springs and a damper, E_0, E_1, C_1, and its constitutive equation is:

$$\dot{\sigma} = E_\infty \epsilon / \tau + E_0 \dot{\epsilon} - \sigma / \tau \qquad (18)$$

where E_0 can be viewed as the material's asymptotic modulus in the limit of infinite strain rate, $E_\infty = E_0 E_1 / (E_0 + E_1)$ is the asymptotic modulus for static loading, and the characteristic decay time for stress-relaxation is $\tau = C_1 / (E_0 + E_1)$. The general solution of Eq. (18) is given by the Boltzmann integral:

$$\sigma(t) = [E_\infty + (E_0 - E_\infty)e^{-t/\tau}]\, \epsilon(0) +$$

$$+ \int_0^t [E_\infty + (E_0 - E_\infty)e^{-(t-t')/\tau}]\, \dot{\epsilon}(t')dt' \qquad (19)$$

where it is understood that $\sigma(0) = E_0 \epsilon(0)$. The particular solution:

$$\sigma(t) = [E_\infty + (E_0 - E_\infty)e^{-t/\tau}]\epsilon \qquad (20)$$

describes the response to an ideal stress-relaxation test in which strain is instantaneously imposed at t=0.

The mechanical analog method was applied in the present case by substituting a nonlinear element for a basic model element. It was easy to match either a stiffening or a softening type of stress-strain curve in this way, but such models produced an unrealistic relation between hysteresis and strain rate. For example, the consti-

tutive equation for the softening-type model in Figure 17 is easily found to be:

$$\dot{\sigma} = \frac{E_\infty \epsilon / \tau + E_0 \dot{\epsilon} - [1 + (E_\infty / B)(\sigma/B)^{n-1}](\sigma/\tau)}{1 + n(E_0/B)(\sigma/B)^{n-1}} \qquad (21)$$

Figure 18 illustrates a typical numerical experiment which shows that this model always has a narrow hysteresis loop and that the maximum loop width occurs at an intermediate strain rate. The behavior differs from that of Ensolite (compare with Figure 12); the mechanical analog approach was consequently abandoned.

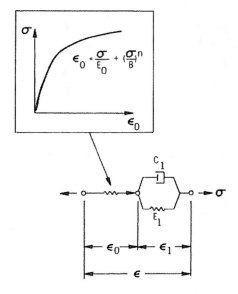

Figure 17. Model with Softening Characteristic

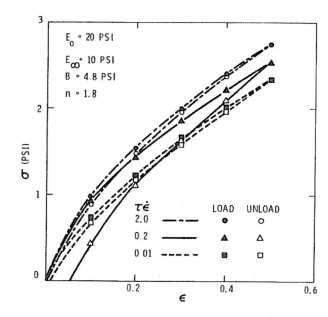

Figure 18. Computed Softening Model Response to Stress-Strain Test at Constant Strain Rate Magnitude

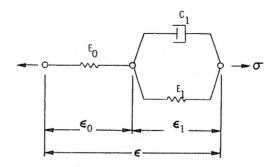

Figure 16. Linear Viscoelastic Solid

A second method involves models of the type proposed by Lockett et al.(6), who fitted empirical equations of the form:

$$\sigma = f(\epsilon) \dot{\epsilon}^r \qquad (22)$$

to the results of stress-strain tests. The function $f(\epsilon)$ was chosen to match the nonlinear behavior of the loading stress-strain curve, but the authors did not attempt to model the unloading curve. Consequently, this type of model is unable to carry a collision analysis into the rebound phase, although such ability is essential for injury assessment. Two further shortcomings of Eq. (22) are its asymptotic behavior for static loading and its inability to model relaxation effects because it lacks a term in stress rate.

Consideration of the problems posed by the foregoing methods led one of us (Orringer) to formulate the following conditions which a constitutive equation must satisfy to be consistent with stress-relaxation and stress-strain test results:

$$\underset{t \to \infty}{\text{Lim}}[\sigma_1(t,\epsilon)] = \underset{\dot{\epsilon} \to 0}{\text{Lim}}[\sigma_2(\epsilon,\dot{\epsilon})] =$$

$$= \underset{\dot{\epsilon} \to 0}{\text{Lim}}[\sigma_3(\epsilon,\dot{\epsilon},\epsilon*)] = \sigma_\infty(\epsilon) \qquad (23)$$

where σ_1 is the stress-relaxation behavior, σ_2 is the stress response to loading ($\dot{\epsilon} > 0$) at constant strain rate, σ_3 is the stress response to unloading ($\dot{\epsilon} < 0$) at constant strain rate after the material has reached a maximum strain $\epsilon*$, and σ_∞ is the static (in practice, very low strain rate) stress-strain curve. A limitless variety of functional forms including higher time-derivatives of stress and strain can be chosen, but in practice models of the general form $\dot{\sigma} = f(\epsilon,\dot{\epsilon},\sigma)$ are sufficient to satisfy the conditions of consistency.

We first developed such a model with nine parameters, eight of which appear in the constitutive equation:

$$\dot{\sigma} = S(\epsilon,\dot{\epsilon}) - \sigma/\tau \qquad (\dot{\epsilon} > 0) \qquad (24a)$$

$$\dot{\sigma} = \frac{\epsilon^{m-2} \epsilon*(\tau*\dot{\epsilon})^p (E_\infty + E_r \dot{\epsilon}^r)}{(1-\epsilon)^n [1-(\tau*\dot{\epsilon})^p]} +$$

$$+ \frac{\epsilon - \epsilon*(\tau*\dot{\epsilon})^p}{\epsilon[1-(\tau*\dot{\epsilon})^p]} S(\epsilon,\dot{\epsilon}) - \sigma/\tau \quad (\dot{\epsilon} < 0) \qquad (24b)$$

$$S(\epsilon,\dot{\epsilon}) = [\epsilon^m/(1-\epsilon)^n][E_\infty +$$

$$+ E_r \dot{\epsilon}^r][m\dot{\epsilon}/\epsilon + n\dot{\epsilon}/(1-\epsilon) + 1/\tau] \qquad (24c)$$

where it is understood that $\dot{\epsilon}$ is to be used in all calculations including Eq. (24b). The parameters E_∞, m, n model a static stress-strain curve which softens ($m < 1$) at moderate strain and stiffens at high strain such that $\sigma \to \infty$ as $\epsilon \to 1$. The parameter τ is a characteristic decay time, while E_r and r scale the loading stress-strain curve nonlinearly with strain rate. The quantity $(\tau*\dot{\epsilon})^p$ should be viewed as an asymptotic residual strain which would be present in the material at the instant it returned to the state of zero stress after having reached the asymptotic state $\epsilon = 1$ in a stress-strain test. The parameters $\tau*$ and p nonlinearly scale the hysteresis loop width according to the working residual strain, which is linearly proportional to the actual maximum strain $\epsilon*$. Note also that Eq. (24b) defines the unloading stress-strain curve shape as a linear mapping of the loading curve from the strain range $[0, \epsilon*]$ onto the strain range $[\epsilon*(\tau*\dot{\epsilon})^p, \epsilon*]$.

The particular conditions corresponding to ideal stress-strain and stress-relaxation tests can be imposed on Eqs. (24) to see how the model represents the experimental results. The static stress-strain curve corresponds to $\dot{\sigma} = \dot{\epsilon} = 0$:

$$\sigma_\infty(\epsilon) = E_\infty \epsilon^m/(1-\epsilon)^n \qquad (25a)$$

The stress-relaxation curve can be found by taking $\epsilon = \text{const.}$, $\dot{\epsilon} = 0$, integrating Eq. (24a), and defining a ninth parameter, $E_0 = E_\infty \sigma_1(0,\epsilon)/\sigma_\infty(\epsilon)$ by analogy to the linear case:

$$\sigma_1(t,\epsilon) = [1+(E_0/E_\infty -1)e^{-t/\tau}]\sigma_\infty(\epsilon) \qquad (25b)$$

The stress-strain curves are obtained by taking $\dot{\epsilon} = \text{const.}$, noting that $\dot{\sigma} = \dot{\epsilon}(d\sigma/d\epsilon)$, and integrating to find:

$$\sigma_2(\epsilon,\dot{\epsilon}) = [1+(E_r/E_\infty)\dot{\epsilon}^r]\sigma_\infty(\epsilon) \qquad (25c)$$

$$\sigma_3(\epsilon,\dot{\epsilon},\epsilon*) = \frac{\epsilon - \epsilon*(\tau*\dot{\epsilon})^p}{1-(\tau*\dot{\epsilon})^p} \sigma_2(\epsilon,\dot{\epsilon}) \qquad (25d)$$

It is then easy to show that Eqs. (25) satisfy the conditions of consistency.

In practice, such models are developed in reverse order. A convenient procedure is to formulate σ_2 and σ_3, derive the respective branches of the constitutive equation from the quantity $\dot{\epsilon}(d\sigma/d\epsilon) + \sigma/\tau$, and then find σ_1 and σ_∞.

The model parameters were fitted to the Ensolite test results by least-squares regression of the logarithms of Eqs. (25). Stress-strain data taken at 0.0012/sec. was used to calculate E_∞, m, and n. Points were read from the stress-relaxation master curves (Figure 10) to provide data for E_0 and τ. The high-strain-rate data points shown in Figure 13 were averaged at each strain rate to provide data for the calculation of E_r and r. The master curves were not used in this case because the high-rate data was considered to better represent impact behavior. High-rate data should also have been used to find $\tau*$

and p, but measurements of residual strain were found to be inaccurate at strain rates exceeding 73/sec. Lower strain-rate data was used, and in consequence this part of the model is only an extrapolation for impact rates.

Table 1 summarizes the parameter values and Figures 19 through 22 compare the model with the data. The static stress-strain curve, the available residual strain data, and high-rate stress-strain data up to 0.5 strain are well matched. However, the stress response at 0.75 strain and stress-relaxation behavior are poorly matched. Thus, while the nine-parameter model fits the Ensolite data better than the mechanical-analog model or Eq. (22), it is not flexible enough to match all aspects of the material behavior.

Table 1. Ensolite AAC Nine-Parameter Model

Parameter	Value	Standard Error	Eq.	Fig.
E_∞ (psi)	7.96	0.86 psi	25a	19
m	0.675			
n	1.09			
E_0 (psi)	22.12	10.80 psi	25b	20
τ_0 (ms)	8.81			
E_r (psi)	5.45	5.90 psi	25c	21
r	0.166			
τ^* (ms)	0.00976	0.00948 in/in	25d	22
p	0.0987			

The apparent shortcomings of the nine-parameter model led us to modify Eqs. (25b,c) to the forms:

$$\sigma_1'(t,\epsilon)=\sigma_\infty'(\epsilon) +$$
$$+ \sigma_0' \exp(-t^A/BC^\epsilon)\{\exp(\epsilon^a/b)-\exp(-\epsilon^c/d)\} \quad (26a)$$

$$\sigma_2'(\epsilon,\dot\epsilon)=\sigma_\infty'(\epsilon) +$$
$$+ \sigma_r'\dot\epsilon^{r}\cdot\exp(-\epsilon^u/v)\{\exp(\epsilon^w/x)-\exp(-\epsilon^y/z)\} \quad (26b)$$

while retaining Eqs. (25a,d). The constitutive equation of this model contains 21 parameters:

$$\dot\sigma=S(\epsilon,\dot\epsilon)-(At^{A-1}/BC^\epsilon) \qquad (\dot\epsilon>0) \quad (27a)$$

$$\dot\sigma = \frac{\epsilon^*(\tau^*\dot\epsilon)^p}{\epsilon^2[1-(\tau^*\dot\epsilon)^p]} \text{ X}$$

$$X[\sigma_\infty'(\epsilon)+\sigma_r'\dot\epsilon^{r}\cdot\exp(-\epsilon^u/v)\{\exp(\epsilon^w/x)-\exp(-\epsilon^y/z)\}]+$$

$$+ \frac{\epsilon-\epsilon^*(\tau^*\dot\epsilon)^p}{1-(\tau^*\dot\epsilon)^p} S(\epsilon,\dot\epsilon)-(At^{A-1}/BC^\epsilon) \qquad (\dot\epsilon<0) \quad (27b)$$

$$S(\epsilon,\dot\epsilon)=\sigma_\infty'(\epsilon)[m\dot\epsilon/\epsilon + n\dot\epsilon/(1-\epsilon) + At^{A-1}/BC^\epsilon] +$$
$$+\sigma_r'\dot\epsilon^{r}\cdot\exp(-\epsilon^u/v)[(w/x)\epsilon^w\exp(\epsilon^w/x) +$$
$$+ (y/z)\epsilon^y\exp(-\epsilon^y/z)+\{\exp(\epsilon^w/x)-\exp(-\epsilon^y/z)\}X$$
$$X\{At^{A-1}/BC^\epsilon-(ry/z)\epsilon^y\ln(\dot\epsilon)\}] \qquad (27c)$$

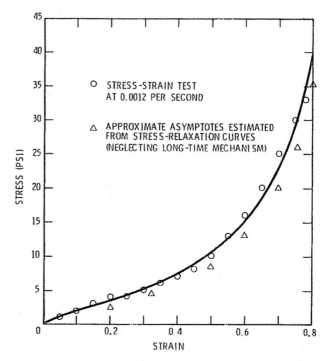

Figure 19. Model Static Stress-Strain Curve

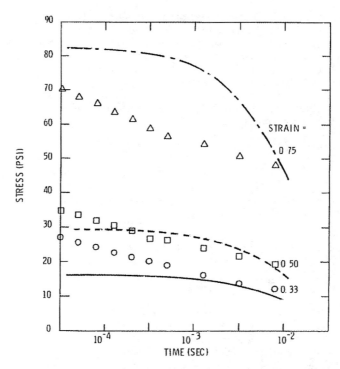

Figure 20. Model Stress-Relaxation Curves

494

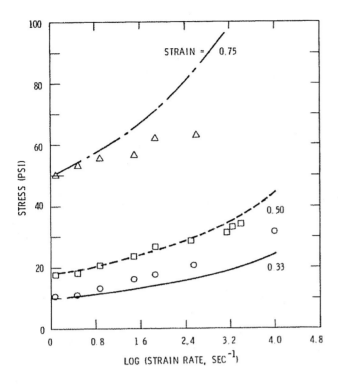

Figure 21. Model Dynamic Stress-Strain Response

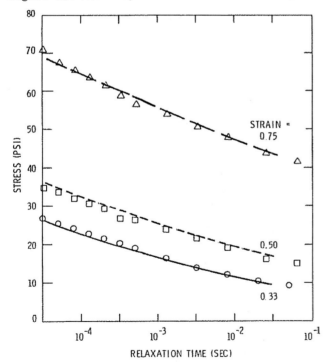

Figure 23. Stress-Relaxation Response
of 21-Parameter Model

Table 2 summarizes the parameter values for Ensolite, and Figures 23 and 24 compare the revised parts of the model with the data. The decreased standard errors and comparison of Figs.
23 and 24 with Figures 20 and 21 demonstrate the

improvement in the fit.

Both the nine- and 21-parameter models were
exercised in a finite-difference simulation of a

Table 2. Ensolite AAC 21-Parameter Model

Parameter	Value	Std. Error	Eq.	Fig.	Note
E_∞ (psi)	7.96	0.86 psi	25a	19	(a)
m	0.675				
n	1.09				
σ_0 (psi)	66.0	0.96 psi	26a	23	(b)
A	0.130				
B	0.156				
C	3.08				
a	10.12				
b	0.279				
c	1.94				
d	0.029				
σ_r (psi)	4.153	1.25 psi	26b	24	(c)
r	0.190				
u	3.17				
v	0.580				
w	2.57				
x	0.296				
y	2.23				
z	0.00164				
τ^* (ms)	0.00976	0.00948	25d	22	(a)
p	0.0987				

(a) Same as nine-parameter model.
(b) Parameter A allows nonlinear time scaling;
    BC^ϵ scales decay time with strain; a,b,c,d
    scale $\sigma(0)$ with strain.
(c) Parameters u,v scale rate exponent with
strain; w,x,y,z scale the loading stress-
strain curve with strain.

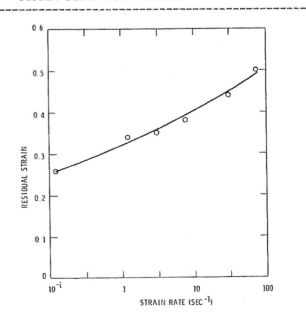

Figure 22. Model Residual Strain After Unloading

495

collision using the sphere-contact model. Figure 25 illustrates a case in which a 10-lb. sphere with 3-in. radius and 20-mph initial velocity strikes a 1-in. thick Ensolite pad supported by an elastic structure with 600 lb./in. stiffness. These parameters represent a typical head-dashboard collision. The time histories of force and deflection for the two models are nearly identical, a result which suggests that the simpler nine-parameter model may represent the material well enough for collision analysis.

CONCLUDING REMARKS

Transient analysis of collisions between occupants and padded structures in automobile passenger compartments is an essential part of crashworthiness assessment. However, the collision event per se is only a small detail in a large picture that includes vehicle structure collapse, crash pulse transmission, biomechanical effects, criteria for human injury, and the dynamics of unrestrained occupants who may repeatedly collide with passenger compartment surfaces. Hence, the collision analysis method must be both accurate and simple.

The collision analysis has been formulated with only a few degrees of freedom by focusing on the most important variables and behavior aspects. It is even possible to deal with local contact effects in such models by making a simplified kinematic assumption and concentrating on the one-dimensional compression characteristics

of padding materials. Experiments with a foam rubber have shown that a well fitted one-dimensional constitutive equation can be developed from standard materials tests, provided that the tests are at impact strain rates, and that the complete relations needed for injury assessment and rebound analysis can be derived from general conditions of consistency.

This engineering approach meets the criterion of simplicity, but it remains to be seen if it possesses sufficient accuracy. The next phase of the research program, now in progress, is intended to answer the question of accuracy. Steel projectiles with hemispherical noses will be allowed to collide with elastically supported foam rubber pads at initial speeds from 15 to 30 mph. These tests will produce transient data and rebound velocity measurements against which the engineering model predictions can be checked.

REFERENCES

(1) R.H. Eppinger, "Proposed Injury Criteria and Mathematical Analogs for Selected Body Areas," National Highway Traffic Safety Administration (NRD-12), Washington, DC, memorandum to A.C. Malliaris, July 6, 1982.
(2) K.T. Knadle, "Impact Rate Properties of an Auto Safety Foam," SB thesis, Dept. of Materials Science and Engineering, MIT, Cambridge, MA, June 1983.
(3) O. Orringer, J.F. Mandell, D.Y. Jeong, and K.I. Knadle, "Research on Crash Padding: Mechanical Properties and Impact Analysis Methods,"DOT

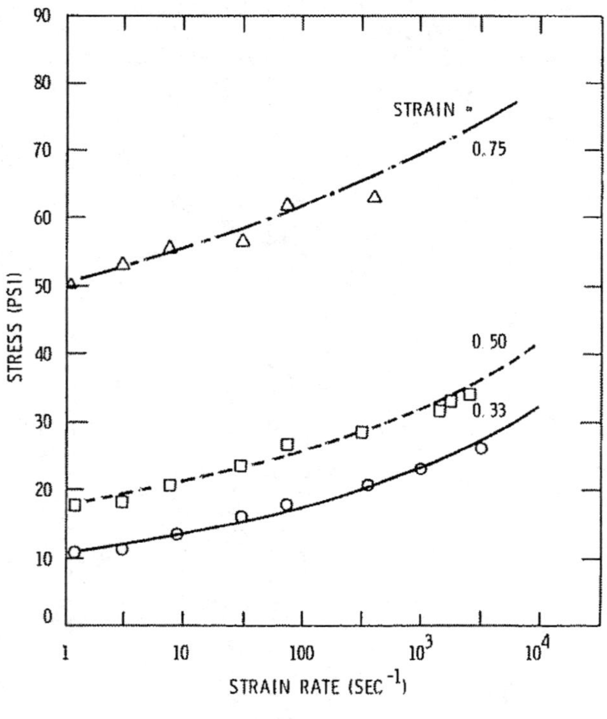

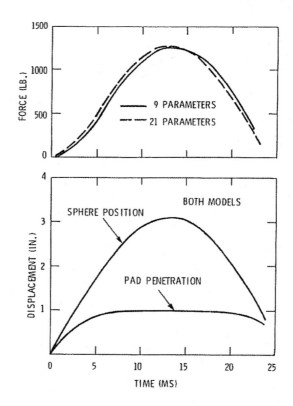

Figure 24. Dynamic Stress-Strain Response of 21-Parameter Model

Figure 25. Results of Example Collision Analysis

Transportation Systems Center, Cambridge, MA,
Project HS-376 interim report, June 1983.
(4) J.T. Fleck and F.E. Butler, "Validation of
the Crash Victim Simulator,"CALSPAN Corporation,
Buffalo, NY, Report No. ZS5881, final report to
NHTSA, Contract DOT-HS-6-01300, 1982 (4 vols).
(5) R.W. Carr, "Automobile Consumer Information
Study Crash Test Program — Volume II — Technical
Report," Ultrasystems, Inc., Phoenix, AZ, Report
No. 8268-75-189, December 1975.
(6) F.J. Lockett, R.R. Cousins, and D. Dawson,
"Engineering Basis for Selection and Use of
Crash Padding Materials," Plastics and Rubber
Processing and Applications, Vol. 1 No. 1(1981),
25-37.
(7) S. Timoshenko and S. Woinowsky-Krieger,
Theory of Plates and Shells, McGraw-Hill, New
York, 1959.
(8) K. Washizu, Variational Methods in Elastic-
ity and Plasticity, Pergamon Press, New York,
3rd ed., 1982.
(9) A.V. Tobolsky, Properties and Structure of
Polymers, Wiley, New York, 1967.
(10) I.M. Ward, Mechanical Properties of Solid
Polymers, Wiley-Interscience, London, 1971.

APPENDIX

This appendix outlines an HP-67 program to
perform the calculations outlined in Eqs. (5)
and (6). If the maximum deflection of the padding is reached, the structure is then treated
as rigid, and the minor mass of the MSC model
instantaneously comes to rest with respect to
the structure.

The time integration is performed by the
forward difference method in specified time
steps t. The value $t = 10^{-5}$ second was found
to be adequate for the MSC parameters given in
Figure 1 and pad stiffnesses between 1 and 10
kips/inch. However, the user should experiment
with the time step size to establish numerical
convergence.

The pre-"bottoming" phase of the computation includes compare-and-save logic to update
the peak pre-bottoming MSC model crush Z. If
the occupant rebounds without bottoming the pad
or if the peak pre-bottom crush exceeds the peak
post-bottom crush, the saved value becomes the
peak crush for the collision. (The displayed
"time to peak" t_1 will be incorrect in the second case, but this is unlikely in practice.) If
the occupant bottoms the pad, the integration
stops at the approximate bottoming point, the
initial conditions are reset, and the peak crush
is computed analytically.

To key in the program the user must set the
keyboard in radian mode. To initialize the program the user must calculate and store the non-
dimensional parameters:

$$\mu = W_2/W_1; \quad \omega = \sqrt{386.4K/W_2}; \quad \zeta = (C/2)\sqrt{386.4/KW_2}$$

where the weights are in lb., K is in lb./in.,
and C is in lb.sec./in. The user must also

store K and Δt. The program is then run in two
steps. First, enter the pad stiffness K_p lb./in.
and maximum deflection X_p in., followed by keystroke $\boxed{A}$. The program computes, stores, and
displays $\varkappa = K_p/K$. Then enter V (mph) and follow
with keystroke $\boxed{B}$. The program internally converts V to in./sec. before starting the time integration. Step B can be repeated as often as
desired until the occupant or pad properties are
to be changed.

If the occupant bottoms the pad, the program gives flashing displays of the time from
first contact to bottoming (t_1), the computed
deflection ($X_k \cong X_p$), and the crush value at
"bottom" (Z_k). A large difference between the
X-values means that a smaller time step should
be used. The program continues briefly to a
flashing display of the time from bottoming to
peak post-bottom crush (t_2) and stops with Z_{max}
in the display. If the occupant does not
bottom the pad, the first three display items
are omitted, 0.0 is flashed, and the correct
Z_{max} from the pre-bottoming phase is displayed.
Table 3 presents the input for an example
analysis. Table 4 lists the program.

Table 3. Test Example

User:

9	ENTER	.4	÷	STO	A										
386.4	ENTER	26	EEX	3	X	9	÷	f√x	STO	B					
386.4	ENTER	26	EEX	3	÷	9	÷	f√x	2.4	X	2	÷	STO	C	
26	EEX	3	STO	D											
1	EEX	5	CHS	STO	E										
6	EEX	3	ENTER	.35	A										

Program: 0.23077

User:

$\boxed{10}$ $\boxed{B}$

Program: 0.00000 (Flash) t_2
             0.07146 Z_{max}

User:

$\boxed{20}$ $\boxed{B}$

Program: 0.00126 (Flash) t_1
             0.35134 (Flash) x_k
             0.07197 (Flash) z_k
             0.00121 (Flash) t_2
             0.27664 Z_{max}

Table 4. Program Listing*

f LBL A	RCL C	STO+1	RCL_2 C	STO 5
STO 9	RCL B	f x<0	g x^2	RCL 7
h v	X	GTO 2	X	RCL 6
RCL D	2	RCL 4	RCL_2 B	X
÷	X	RCL E	g x^2	f sin
STO 8	RCL 4	X	X	STOX5
h RIN	X	STO+3	STO+6	RCL 7
f LBL B	RCL 6	RCL 5	RCL 3	RCL 6
1	RCL 3	STO 4	STOX6	X
7	X	RCL 6	RCL 2	f cos
.	+	STO 2	RCL 4	RCL 3
6	RCL E	h RCI	+	X
4	X	RCL 3	STO÷6	STO+5
X	STO 6	g x>y	RCL B	RCL B
STO 2	RCL A	h STI	RCL C	RCL C
0	1	RCL 1	X	X
STO 0	+	RCL 9	RCL 7	RCL 6
STO 1	RCL 6	g x>y	÷	X
STO 3	X	GTO 1	STO+6	g e^x
STO 4	CHS	RCL 0	RCL 6	STO÷5
h STI	RCL 7	f −x−	h 1/x$_{-1}$	RCL 5
f LBL 1	+	RCL 1	g $\tan^{-1}$	h RCI
RCL E	RCL 4	f −x−	RCL 7	g x>y
STO+0	+	RCL 3	÷	STO 5
RCL_2 B	STO 5	f −x−	STO 6	RCL 6
g x^2	CHS	RCL_2 C	RCL B	f −x−
STO 6	RCL 2	g x^2	RCL C	RCL 5
RCL 8	+	CHS	X	h RIN
X	RCL 4	1	RCL 3	f LBL 2
RCL A	+	+	X	0
X	RCL 6	f √x	RCL 4	STO 6
RCL 1	−	RCL B	+	f −x−
X	STO 6	X	RCL 2	h RCI
RCL E	RCL 2	STO 7	+	STO 5
X	RCL E	STO 6	RCL 7	h RTN
STO 7	X	h 1/x	÷	

*Read down columns left to right. Symbol "v" means roll down stack.

498

860654

Improvements in the Simulation of Unrestrained Passengers in Frontal Crashes Using Vehicle Test Data

Kennerly H. Digges
National Highway Traffic Safety Administration

ABSTRACT

The absence of data on the load deflection and energy absorption characteristics of vehicle interiors has been a factor which limits the accuracy of crash victim simulations.

A recent test program conducted for the National Highway Traffic Safety Administration has developed data on the interactions of dashboards and knee panels with chests and knees. This paper summarizes the test results for several vehicles and shows how these results are used in simulating vehicle crash tests.

Comparisons between crash tests and computer reconstruction using the 3-Dimensional Crash Victim Simulator (CVS-3D) for a late model car are included. The simulation shows good agreement with test and illustrates the application of available static and dynamic test data to improve occupant simulations.

ANALYTICAL MODELS OF HUMAN OCCUPANTS are extremely useful in studying occupant kinematics during a crash. The Three-Dimensional Crash Victim Simulator (CVS-3D) is a well documented model which has been validated for belt restrained occupants in frontal collisions. (1)* The model has been used in studying how design variations affect occupant protection (2, 3) and for reconstructing actual vehicle crashes which involve unrestrained occupants. (4, 5, 6) Authors of previous studies have stated that the literature contains little data on the stiffness and energy absorbing properties of vehicle interior surfaces. In particular, data on the mechanical properties of dashboards, headers, and knee panels of current automobiles has been virtually nonexistent in the literature. In order to assist in occupant modeling, several research projects are underway at the National Highway Traffic Safety Administration. These programs involve static and dynamic tests which are producing useful information on the properties of vehicle interior surfaces.

The purpose of this paper is to provide a preview of recently developed experimental test data on the characteristics of knee panels and dashboards, and to illustrate how this data can be used to improve predictions using the CVS-3D model.

COMPONENT TEST PROGRAM

In order to evaluate the characteristics of vehicle interior components when impacted by an occupant, a program of static and dynamic component testing has been initiated. (7)

The vehicles selected for testing represent a range of vehicle sizes and model years. A listing of the vehicles for which test data is being developed is shown in Table 1.

The static tests were conducted by loading dummy segments against vehicle interior components and measuring the load and deflection during the test.

For tests of the lower dashboard and knee panel, a pair of 572 dummy legs were used to apply the load. The upper legs were attached to the loading mechanism, and the feet were placed against the toe board. Load cells were located in each upper leg to measure the left and right femur loads separately. The test was performed by simultaneously loading both knees against the knee panel with the force applied parallel to the femur of a seated dummy.

Typical force deflection measurements of

---

*Numbers in parentheses designate references at end of paper.

Table 1

Component Test Vehicles

Year	Make	Model
1983	Chev.	Celebrity
1980	Chev.	Citation
1979	Chev.	Chevette
1978	Buick	Le Sabre
1976	Chev.	Monza
1979	Ford	LTD
1979	Ford	Mustang
1978	Ford	Fairmont
1977	Ford	Pinto
1980	Dodge	Omni
1977	Plym.	Volare

passenger side knee panels are shown in Figure 1. The data presented in the figure is the combined load on the two knees measured in a static test.

Testing of the mid-dashboard was conducted by pressing the chest section of a Part 572 dummy into the dashboard. This test was conducted after the knee panel test, so that the dashboard had been previously deformed by the knees. Typical test results of passenger dashboards are shown in Figure 2.

In the dynamic tests, rigid body forms representative of sections of a 572 dummy were propelled into the vehicle component by a pneumatic device. The speed of impact for the forms was 15 mph for the knees and 20 mph for the chest. Force and deflections were calculated from accelerometer readings.

Typical dynamic test results for knee panels and dashboards for the passenger side of a Chevrolet Celebrity are shown in Figures 3 and 4. The static test curves are also shown for comparison.

Tests are underway to characterize windshields, headers, and other interior components. In addition, dynamic tests to simulate impacts over a range of severities are planned. The complete data on all test results will be reported in the final report for contract No. DTRS-57-84-C-00003, scheduled for completion during 1986.

SLED TESTS OF UNRESTRAINED DUMMIES

Dynamic crash tests of unrestrained dummies provide a basis for comparing computer predictions with experimental results.

A series of dynamic tests using a Chevrolet Celebrity compartment on a Hyge acceleration sled have been conducted by NHTSA's Vehicle Research and Test Center. The impacts simulated a frontal crash with severity of approximately 30 mph. (8) Unrestrained Hybrid III dummies were used in the test series.

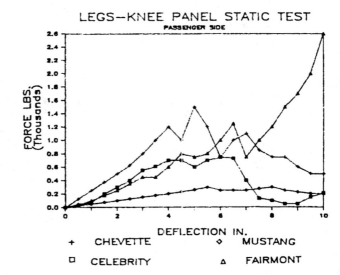

Figure 1

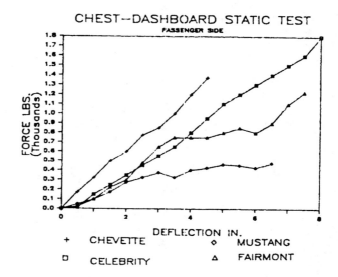

Figure 2

The impact of the unrestrained dummies on the vehicle interior surfaces caused extensive damage. Therefore, it was necessary to replace interior components after each test. However, some of the ducting and other nonstructural components between the knee panel and firewall were not replaced. Instead, these components were simulated by substituting blocks of foam for the damaged components. (9)

For the purpose of illustration, Test 10 from Reference 8 has been selected for simulating, using the CVS-3D model.

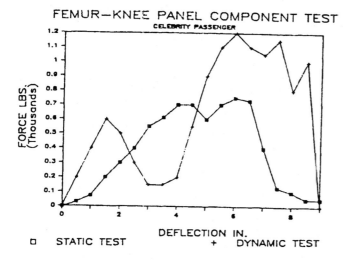

FEMUR—KNEE PANEL COMPONENT TEST
CELEBRITY PASSENGER

Figure 3

□ STATIC TEST   + DYNAMIC TEST

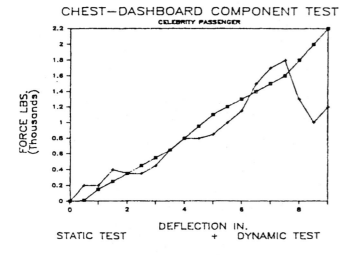

CHEST—DASHBOARD COMPONENT TEST
CELEBRITY PASSENGER

Figure 4

STATIC TEST   + DYNAMIC TEST

Test 10 simulated a 31.3 mph barrier test. The test results were: HIC: 685; maximum head acceleration: 101 G; maximum chest acceleration: 59 G; femur loads: left 1,534; right 1,311; neck extension moment: 85 foot pounds.

COMPUTER SIMULATION

A simulation of the Celebrity sled test using the CVS-3D model provides insights into the performance of the model and the utility of component test data.

The crash pulse used in the simulation was the same as that used on the sled.

The occupant data set was the same as that reported in Reference 1. This occupant data set has been validated by tests of the

Part 572 dummy. This data set was used because no validated data set for the Hybrid III dummy is currently available. The Hybrid III and Part 572 dummies have different lumbar spines and different neck extension characteristics. Therefore, minor differences between the Hybrid III sled test and the Part 572 model simulation might be expected.

The data for simulating the vehicle interior geometry was based upon measurement of an undamaged Celebrity. A review of the test films showing occupant impact revealed that the mid-dash was moved upward and rearward by knee penetration, prior to the impact of the thorax. Therefore, an additional panel was included in the simulation to represent the mid-dashboard in its displaced position.

The windshield properties were based upon the data set presented in Reference 1. This data was used because reliable windshield test data for the Celebrity was not available.

The knee panel properties were based upon the static test data in the range of penetration from zero to 7 inches. Beyond 7 inches, the static test force dropped to virtually zero whereas the sled test produced femur forces in excess of 3,000 pounds. Likewise, the dynamic test produced maximum forces less than half those recorded in the sled test.

It was evident that neither test adequately described the "end-of-stroke" characteristics of the knee panel in the sled test at the energy level of a 31.3 mph crash. An inspection of the vehicle and the test data resulted in estimates of 8 and 10 inches for the available penetration of the left and right knees, respectively, prior to end-of-stroke. The stiffness beyond the end-of-stroke point was assumed to be approximately 500 pounds per inch, based upon the maximum slope of the dynamic test curve, shown in Figure 3. The force level at 9 inches was assumed to be equal to the maximum force produced in the dynamic test.

The "end-of-stroke" characteristics were also introduced into the dashboard simulation. The use of static and dynamic component test data for the CVS model simulation permitted much greater penetration of the dummy into the dash panel than observed in the sled test. An examination of the static test data in Figure 4 shows that the dashboard has a stiffness of about 200 pounds per inch up to 3 inches of penetration. The 20 mph dynamic component test shows an increase in stiffness at approximately 5 inches, and a stiffness of about 500 pounds per inch beyond 6 inches.

For the simulation, a stiffness of 1,400 pounds per inch was used for values greater than the inflection point at 5 inches, which was observed in the dynamic test data.

TEST AND SIMULATION RESULTS

A comparison of the chest acceleration history for the simulation and the sled test is shown in Figure 5. The maximum value of chest acceleration for the test was 59 G, compared to 54 G for the simulation.

Figures 6 and 7 show the femur loads for the right and left knee panels, respectively. For the simulation, the femur load was not calculated directly, but was assumed to equal the resultant force between the knees and the knee panel.

Figure 8 shows the neck extension moment measured in the Hybrid III dummy and the resultant head torque produced at the head to neck joint in the simulation.

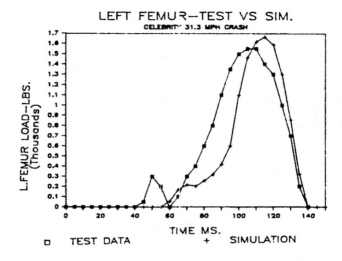

Figure 7

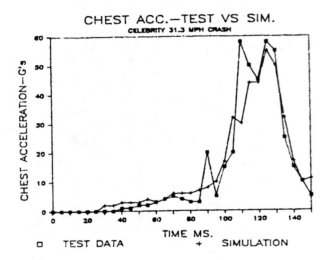

Figure 5

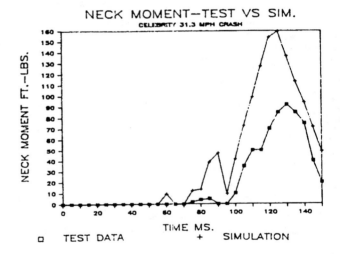

Figure 8

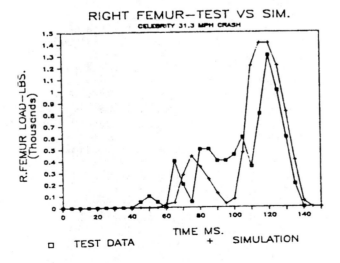

Figure 6

Figures 9 through 14 show a comparison between the kinematics of the dummy in the sled test and the simulation.

Figures 9A and 9B show the initial position of the occupant.

Figures 10A and 10B show the occupant at 90 milliseconds (ms) after the beginning of the crash. At this point in time, the knee penetration into the dashboard has begun, and the dashboard is being displaced upward and rearward by the knee penetration. Head impact with the windshield occurred at approximately 80 ms. The maximum head acceleration was 110 G for the simulation vs. 101 G for the test, both produced at windshield contact.

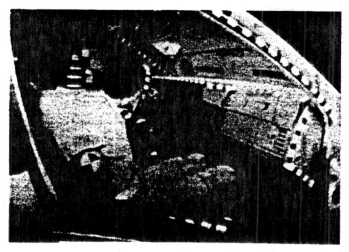

Figure 9A.   SLED TEST: 0 ms

Figure 9B.   SIMULATION: 0 ms

Figure 10A.   SLED TEST: 90 ms

Figure 10B.   SIMULATION: 90 ms

Figure 11A.   SLED TEST: 100 ms

Figure 11B.   SIMULATION: 100 ms

Figure 12A.   SLED TEST: 110 ms

Figure 12B.   SIMULATION: 110 ms

Figure 13A.   SLED TEST: 120 ms

Figure 13B.   SIMULATION: 120 ms

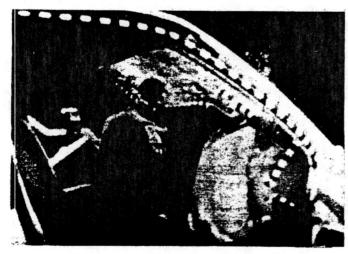

Figure 14A.   SLED TEST: 130 ms

Figure 14B.   SIMULATION: 130 ms

Figures 11A and 11B show the occupant at 100 ms. At this point, the thorax is beginning to contact the displaced dashboard, and the neck is being extended by continued head-windshield contact.

Figures 12A and 12B show the 110 ms time frame. At this time, the occupant thorax has reached the original plane of the dashboard. The interaction between the neck and head continues to extend the neck. The resulting moment reaches a level in excess of 50 foot-pounds.

Figures 13A and 13B show the 120 ms position. At this point in time, the neck is continuing to extend and the chest is beginning to contact hard structure in the dashboard.

The final Figures, 14A and 14B, at 130 ms show the maximum penetration of the occupant.

DISCUSSION OF RESULTS

The static test curves for different vehicles presented in Figures 1 and 2 show large differences between vehicles. For example, the initial stiffness of the Chevette knee panel is 250 pounds per inch while that for the Mustang is 60 pounds per inch.

The knee panel properties can generally be characterized by an initial stiffness, followed by a plateau at fairly constant load. Finally, an increased stiffness occurs at the end of the stroke. The deflection at which the increased stiffness occurs is an important parameter in simulating high severity crashes.

The knee panels in the Celebrity and Fairmont both reach constant force plateaus at 4 inches deflection and at force levels between 600 to 800 pounds. The Chevette has a plateau at around 1,200 pounds, and at the same deflection. The Fairmont plateau appears to end at about 7 inches where the stiffness increases to around 600 pounds per inch.

For the dashboard static tests, the stiffness range is from 300 pounds per inch for the Chevette to about 125 pounds per inch for the Mustang. The test results for the Mustang and Fairmont exhibited force plateaus at 350 and 750 pounds, respectively. No similar plateaus were noted in the Celebrity or Chevette tests.

A comparison of the dynamic and static knee panel test data shown in Figure 3 indicates a dynamic force plateau at approximately 1,300 pounds, as compared with a static force plateau at 600 pounds. The energy of the dynamic test device was dissipated over a stroke of about 9 inches. Consequently, the components which produced the end-of-stroke forces greater than 3,000 pounds in the sled test may not have been engaged in the component test.

The chest to dashboard stiffness characteristics at high deflections appeared to be much higher in the 30.3 mph sled test than those measured in either the static or the dynamic component test. Dynamic component testing at higher energies than those achieved in the 20 mph tests may be required to explain the differences.

A comparison of the chest acceleration curves in Figure 5 shows that the model predicted somewhat lower values of chest acceleration, as compared to the test. The peak values were 54 G and 59 G, respectively. The sled test produced acceleration peaks at 105 and 125 ms. The initial peak was probably due to the inertia spike after initial contact. The model has the capability to simulate the inertia spike. However, this feature was not employed due to lack of input test data. Except for the inertia pulse, the two curves are quite similar in magnitude and duration.

A comparison of test vs. simulation for the right and left femur loads, shown in Figures 6 and 7 indicates interesting similarities and differences. The general shape of the test and simulation curves is similar. In the sled test, the left femur begins end-of-stroke loading approximately 30 ms earlier than the right femur. Data from a 30 mph crash test of a Chevrolet Celebrity also indicated that the right femur loading was significantly lower and later than that of the left femur. (10) These differences in femur loadings were not observed in the static test. This may be partially explained by changes in the dashboard test configuration between the static component test, the sled test, and the vehicle test. The sled test configuration contained padding beneath the dashboard to simulate the nonstructural components. The static test configuration beneath the dashboard may have also differed from the actual vehicle. It is quite probable that these physical differences caused different stiffnesses at high knee penetrations.

A comparison of the neck extension moment, shown in Figure 8, indicates that the duration is similar, but the peak value is much higher for the simulation. It may be noted that the neck for the 572 dummy used for the simulation is stiffer than the neck of the Hybrid III dummy which was used in the test. Consequently, the higher neck moment might be expected. In addition, generic windshield properties were used in the simulation. Properties of the Celebrity windshield, when available, could improve the simulation of head to windshield contact and the resulting neck extension.

An examination of Figures 9 through 14 shows an excellent correspondence between the kinematics of the dummy and those predicted by the model. In the simulation, the dummy sits somewhat higher and strikes the windshield slightly higher. This difference may be due to differences in the spine, the geometry, and the inertial properties of the two dummies.

CONCLUSIONS

The static and dynamic testing of interior dashboards and knee panels currently underway has shown that the stiffness of interior panels varies widely among different car models. The static test data is extremely useful in developing data sets for modeling, particularly with regard to the initial stiffness and deflection characteristics. However, dynamic data and/or sled tests may be required in determining panel displacements and stiffness characteristics for large deflections in high severity impacts of unrestrained occupants.

The accurate representation of under dashboard components is important when conducting sled tests and component tests. These components can influence the loading of the knees, particularly at large penetrations.

The use of the existing model data set for the Part 572 dummy produced reasonable CVS-3D simulations of the Hybrid III dummy kinematics when tested as an unrestrained passenger in a Chevrolet Celebrity in a 31.3 mph crash. However, the neck moment was higher for the simulation than for the test. A validated data set which simulates the Hybrid III neck, spine, and inertial properties would be desirable for future use to minimize differences.

Except for the neck extension, the simulation results which employed end-of-stroke data adjustments generally agreed quite well in magnitude and duration with the test results.

The use of the CVS-3D model offers a useful way of augmenting testing programs to develop and evaluate the built-in safety of vehicle interior components.

ACKNOWLEDGEMENTS

The author wishes to express his thanks to the following who have assisted during the course of the work presented in the paper:

Mr. Dan Cohen, Office of Crashworthiness, NHTSA Research and Development, who was the project manager for research on passenger occupant protection, and who assisted in preparing data sets on interior components.

Dr. Herb Gould, Transportation Systems Center, U.S. Department of Transportation, who is contract manager on programs to collect data on vehicle interiors, and who provided data sets on vehicle interior geometry.

Mr. Roger Saul, Vehicle Research and Test Center, NHTSA Research and Development, who was responsible for sled testing, and who provided information regarding test configurations.

REFERENCES

1. Fleck, J.T. and Butler, F.E., "Validation of the Crash Victim Simulator," NHTSA Report DOT-HS-6-01300, February 1982.

2. Miller, Patrick M. and Green, James E., "Design, Development and Testing of Calspan/Chrysler Research Safety Vehicle - Phase II," Report DOT HS-802 250, Calspan Corporation, February 1977.

3. Romeo, D.J., "Development of a Front Passenger Aspirator Air Bag System for Small Cars," Report No. DOT-HS-802-039, August 1976.

4. Kelleher, B.J. and Walsh, M.J., "Computer Simulations of Occupant Responses in Frontal Crashes Using CVS III," Report No. DOT-HS-6-01470, April 1980.

5. Digges, K., "Reconstruction of Frontal Accidents Using the CVS-3D Model," SAE Paper 840869, May 1984.

6. Ommaya, A.K. and Digges, K., "A Study of Head and Neck Injury Mechanisms by Reconstruction of Automobile Accidents," 1985 International IRCOBI/AAAM Conference on the Biomechanics of Impacts, Goteborg, Sweden, June 1985.

7. MGA research under contract with the Transportation Systems Center, Contract No. DTRS-57-84-C-00003, "Instrument Panel Static/Dynamic Force-Deflection Characteristics."

8. Bell. L., "Frontal Occupant Sled Simulation Correlation, 1983 Chevrolet Celebrity Sled Buck," NHTSA Report DOT HS 806 728, February 1985.

9. Private communication with Roger Saul, NHTSA, Vehicle Research and Test Center, East Liberty, Ohio, December 1985.

10. Beebe, M. and Wade B., "Frontal Crash Responses, A 1983 Chevrolet Celebrity into a Fixed Rigid Barrier at 30.0 MPH," NHTSA Report 840926, December 1984.

CHAPTER SIX

Steering columns

Accident Investigations of the Performance Characteristics of Energy Absorbing Steering Columns

Donald F. Huelke
Dept. of Anatomy, Medical School, The University of Michigan

William A. Chewning
The Highway Safety Research Institute, The University of Michigan

DATA FROM FIELD investigations of automobile collisions have previously indicated that the nonenergy absorbing steering assembly -- the steering column and wheel -- was one of the leading causes of serious and fatal injuries to the driver (1, 2)*. In the Michigan study, ejection was the leading cause of all occupant deaths, and driver deaths from ejection or steering assembly impacts were equal in number. More recently, investigations of accidents involving new model cars have been conducted in the Ann Arbor, Mich. area through the Highway Safety Research Institute (HSRI). For this report, 87 accidents involving 1967 or 1968 model cars were reviewed. Of these, the majority (78) had the General Motors energy absorbing column and nine were 1968 model Ford automobiles.

The rationale, design, and construction details of the energy absorbing steering column have been presented elsewhere and need not be detailed here (3).

The principal area of impact to each vehicle is pre-

---

*Numbers in parentheses designate References at end of paper.

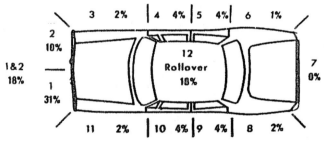

Fig. 1 - Impact locations - 1967 and 1968 model cars (87 cases)

sented in Fig. 1. Twelve areas were plotted as follows: area 1 - left front; area 2 - right front; area 3 - right front fender; area 4 - right front door; area 5 - right rear door

---

Note: This research was supported by USPHS Grant Number UI-00021 from the Injury Control Program, National Center for Urban and Industrial Health, and by funds from the Automobile Manufacturers Association through the Highway Safety Research Institute of The University of Michigan.

ABSTRACT

Investigations of 1967 and 1968 model cars indicate that the injuries sustained by driver impacts to the steering assembly are markedly reduced because of the energy absorbing steering column. Drivers, however, are sustaining facial injuries from impact to the steering wheel rim even in low speed crashes. In more severe head-on collisions, the driver is compressing the energy absorbing column and is striking his face on the upper padded instrument panel in front of the steering wheel. Relatively severe facial fractures are sustained by impacting this portion of the panel.

Table 1 - Summary of Energy Absorbing Steering Column Cases

Impact Location	Case No.	Vehicle(s), Impact Object, Accident Type	E. A. Mesh	Column Capsules	Driver Sex/Age	Seat Belt	Injuries	Injury Source	Police Traveling Speed	HSRI Estimated Impact Speed
1	156	1967 Chevrolet Chevelle hit rt. side of 1965 Ford MINOR	0 25* *underhood shaft telescoped 3 15 in.	?	M 29	No	None		50 5	30 ?
2	174	1967 Plymouth Fury hit trees MODERATE	0.5	?	M 22	No	Abrasions of knees	Lower instrument panel	45	25
1 + 2	182	1967 Chevrolet Impala convert. hit side of 1966 Ford Van MINOR	0.5	?	M 18	No	1-1/2 in. lac. lt. eyebrow; Rt. knee abrasion; Rt. forearm bruise	Wheel rim; Lower instrument panel; Instrument panel	30 16	20 5
3	198	1967 Chevrolet Corvette hit tree MODERATE	0	0	M 24	No	Abrasions and contusions of face	Instrument panel	70	?
1	214	1967 Chevy II hit by 1960 Chrysler MINOR	0	0	M 61	No	Contusion of rib cage	Wheel rim	45	10
1	219	1967 Chevrolet Sports Coupe hit 1967 Ford in rt. side MINOR	3.0	?	M 52	No	Scalp lac. above lt. eyebrow	Windshield	55 5	25 5
1 + 2	227	1967 Chevrolet Camaro hit 1962 Chrysler in rt. side MODERATE	4 0	?	M 16 5'7" 200 lb	No	1-1/2 in. lac. chin; Bruised lt. knee; Sore lt. elbow; Small scratches at hairline; Sore back of neck	Wheel rim; Instrument panel; Unknown; Windshield; Front seat back	45 5	35 5
3	234	1967 Oldsmobile Cutlass hit 1967 Firebird (front) MODERATE	0.15	?	M 24	No	Depressed skull fx. intracranial injury - fatal	Rt. "A" pillar	70	35
1 + 2	234	1967 Pontiac Firebird (front) MODERATE	5.0	2.0+	M 31 6'3" 170 lb	Yes	Scalp lac. small; Small lac. lt. elbow; Chest sore; Seat belt bruises; Abrasions of knees	Header; Unknown; Wheel; Seat belt; Lower instrument panel	30	25
1	235	1967 Pontiac Catalina hit rear of parked 1960 Mercury MINOR	0.5	0	F 28 5'6" 143 lb 7 mos. pregnant	No	Lac. lower lip; Bruise - rt. knee	Wheel rim; Lower instrument panel	20	15-20
1	238	1967 Oldsmobile Toronado hit tree MINOR	0 25	?	F 54 5'7" 145 lb	No	Bruise - Chin; Contusion - rt. arm; Bruise below lt. knee	Wheel; Unknown; Lower instrument panel	30	15
1 + 2	247	1967 Pontiac Grand Prix hit tree head-on MODERATE	8.0	Free	M 40 5'10" 170 lb	Yes	Lac. - forehead; Chest bruises; Bruise - lt. leg; Fx. rt. leg above ankle	Wheel rim; Wheel; Lower instrument panel; Floor pan	95	40-45
1 + 2	254	1967 Oldsmobile 442 hit trees MINOR	0.75	?	M 21	No	Lac. - rt. side of nose	Wheel rim	85	15

(cont'd)

(Table 1 cont'd)

Impact Location	Case No.	Vehicle(s), Impact Object, Accident Type	E. A Mesh	Column Capsules	Driver, Sex/Age	Seat Belt	Injuries	Injury Source	Police Traveling Speed	HSRI Estimated Impact Speed
1	279	1967 Chevrolet Camaro hit tree MINOR - MODERATE	2.0	?	M 24 5'11" 150 lb	No	Pain - lower ribs Bruise - behind rt. knee Lac. lt. side of chin	Wheel Floor shift Wheel rim	?	20-25
1	UM-03-68	1967 Pontiac Firebird head-on into 1966 Mustang - 50% overlap SEVERE	8.1	Free	M 18	No	Fatal: internal head, thoracic and abdominal injuries	Wheel, column and others (?)	? 50	40-50 40-50
1	UM-04-68	1967 Oldsmobile Delta 88 hit in rear - then struck rear of 1967 Dodge MINOR	?	0.75	M 48	No	Multiple muscular strains	Impact	0	0 rear 5-10 front
1	UM-05-68	1968 Oldsmobile Delmont 88 hit tree MODERATE	2.7	1.0	M 28	No	Lac. at hairline Lac. lt. eyelid and beneath lt. eye Lacs. behind both ears	Windshield Windshield Windshield	65	20-25
1 + 2	UM-06-68	1967 Dodge Dart hit in rear by 1964 Buick. Dodge hit rear of 1963 Pontiac Tempest MINOR	0	0	F 23 5'6" 110 lb	Yes	Whiplash Bruise - rt. knee	Impact Ignition key	0 10	0 rear front 5-10
1	UM-07-68	1967 Dodge Dart hit 1967 Mustang head-on MINOR	0	0	F 27 5'5" 115 lb	Yes	Abrasions of knees Seat belt bruises	Lower instrument panel Seat Belt	35 0	5-10 0
1	UM-14-68	1968 Chevrolet Camaro hit 1962 Plymouth head-on SEVERE	7.1	4.0 (?)	M 19 6'0" 140 lb	No	Lac. of liver Fx. lt. 12th rib. Abrasion - lt. lower thorax Lac. chin Abrasions of knees Abrasions rt. chest	Wheel column Wheel Wheel rim Lower instrument panel Wheel	40 40 5	35 35 0-5
1	UM-17-68	1968 Ford Station Wagon hit by 1967 Volkswagon MINOR	0	0	M 22 5'10" 180 lb	Yes	None		15	5-10
1	UM-20-67	1967 Chevrolet Camaro hit pole MODERATE	7.2	0.9	M 20 5'10" 180 lb	Yes	Bruised forehead (unconscious 5 min.) Bruised ribs Pain - low back Pain - abdominal wall Seat belt bruises across hips Abrasions - both legs below knees Sprained rt. ankle Abrasions lt. elbow and arm	Wheel rim Wheel Seat belt (?) Wheel Seat belt Lower instrument panel Floor pan Door panel	?	25

(cont'd)

511

(Table 1 cont'd)

Impact Location	Case No.	Vehicle(s), Impact Object, Accident Type	E. A. Mesh	Column Capsules	Driver, Sex/Age	Seat Belt	Injuries	Injury Source	Police Traveling Speed	HSRI Estimated Impact Speed
1	UM-21-68	1968 Ford Fairlane hit cement drain wall MODERATE	1.1	0.5	M 21 5'9" 160 lb	No	Lacs. eyelid, forehead, and nose Concussion Bruised knees	Windshield and rear view mirror Windshield (?) Lower instrument panel	70	30
1 + 2	UM-29-68	1967 Pontiac Catalina hit pole MODERATE	4.7	2.9	M 32 5'6" 110 lb	No	Fxs. - midface and mandible Fx. - lt. femur	Upper instrument panel Floor pan	40	40
1 + 2	UM-30-68	1968 Ford Falcon hit rear of 1965 Chevrolet MINOR	0	0	M 18 5'4" 120 lb	No	Bruise over lt. eye Bruise - rt. knee	Wheel rim Lower instrument panel (?)	45	10
1	UM-36-68	1968 Pontiac Catalina hit guard rail MODERATE	1.0	.75	M 25 5'9" 140 lb	No	Lac. upper lip Contusion - top of head Pain - all ribs mostly on rt. front Abrasions - lt. knee	Wheel rim Sun visor Wheel Lower instrument panel	55-60	25-30
1 + 2	UM-39-68	1967 Chevrolet Impala hit 1966 Chevrolet head-on MODERATE	3.5	3.5	M 33 6'0" 153 lb	No	Pain - lt. chest Abrasions - rt. knee Fxs. - midface	Wheel column Lower instrument panel Upper panel	50 45	35-40 35-40
1 + 2	UM-40-68	1967 Oldsmobile Cutlass hit rear of 1967 Ford MINOR	0	0	F 24 5'10" 150 lb	Yes	Lac. chin Lac. lower lip Lt. knee abrasion	Wheel Wheel rim Lower instrument panel	25 0	10 0
1	UM-49-68	1968 Chrysler intersection; hit 1966 Pontiac in lt. side MINOR	0.5	0.5	F 47	No	Lt. shoulder and arm bruised Contusions - knees	Wheel (?) Lower instrument panel	5 30	25 5
2	UM-51-68	1968 Ford Fairlane hit tree MODERATE	0	0	M 54	No	Lac. on scrotum Contusion - rt. knee Fx. - pelvis Neck sore	Unknown Ash tray Rt. door Impact	70	30
1	UM-57-68	1968 Buick Le Sabre hit embankment MINOR	0.3	0.3	F 73	No	Nasal lac. Fx. - lumbar vertebra	Wheel rim Unknown	50	5-10
2	UM-59-68	1968 Ford Galaxie 500 down embankment MINOR	0	0	M 42	No	Fx. - rt. 5th rib Bruise - rt. elbow and forearm Bruise - rt. anterior chest Small puncture wounds of lt. knee	Rt. door Rt. door Rt. door Lower instrument panel	70	20-25
11	UM-60-68	1967 Chevrolet Caprice down embankment MODERATE	1.0	0.5	F 43	No	1/2 in. lac. over lt. eye Sore lt. shoulder Pain between shoulders Both knees sore Pain - chest and upper abdominal wall	Window glass Lt. door Door (?) Lower instrument panel and brake handle Wheel	50	25

(cont'd)

512

(Table 1 cont'd)

Impact Location	Case No.	Vehicle(s), Impact Object, Accident Type	E. A. Mesh	Column Capsules	Driver Sex/Age	Seat Belt	Injuries	Injury Source	Police Traveling Speed	HSRI Estimated Impact Speed
1	UM-62-68	1957 Pontiac Catalina hit 1964 Chevrolet head-on SEVERE	1.2	Free	M 27	No	Ruptured spleen Lac. forehead Fxs. - lt. ribs (4-7) Pain - lt. knee Abrasions - lower lt. abdominal wall	Wheel Lt. "A" pillar Lt. door Lower instrument panel Wheel	? ?	55 15
1	UM-63-68	1967 Pontiac Firebird intersection; hit 1966 Pontiac in rt. front MODERATE	0.9	0.5	M 43 5' 4" 150 lb	No	Swollen lt. knee and thigh Abrasions - lt. elbow Swollen nose; eyes black and blue "Aches all over"	Lower instrument panel Door sill Wheel rim	35 8	35 5
2	UM-65-68	1968 Mercury Monterey hit tree MINOR-MODERATE	1.6	1.5 Free	M 48 5' 8" 145 lb	No	Lac. lower lip and chin Lacs. inside mouth and throat Pain - chest Larynx sore - can't swallow Bruise - lt. knee Pain - neck "Aches all over"	Instrument panel eyebrow Dentures Wheel Upper instrument panel Lower instrument panel Impact	?	20
1	UM-67-68	1968 Chrysler Newport hit guard rail MODERATE	0.5	0	M 53 5' 3" 185 lb	Yes	Lac. lt. ankle Fx. dislocated lt. humerus	Foot controls Lt. door	70	?
1 + 2	UM-68-68	1968 Dodge Charger intersection; hit 1965 Buick Riviera (lt. front) MINOR-MODERATE	0.5	0.5	M 22 5' 11" 140 lb	No	Contusions and pain - back Bruise - lt. knee Pain - chest	Unknown Lower instrument panel Wheel	25 50	10-15 25
2	UM-69-68	1967 Chevrolet Impala hit rear of 1955 Buick MODERATE	0.4	0.3	M 25	No	Contusion - forehead and neck Pain - rt. arm, hand, and lt. elbow	Wheel and header Unknown	60 35	55-60 35
1 + 2	UM-73-68	1968 Oldsmobile Delmont 88 hit tree MODERATE	5.1	0.8	F 34 5'4" 170 lb	No	Large lacs. - sides of head Pain - lt. lower rib cage Lac. rt. knee and fx. rt. patella Bruises - lt. leg and knee Dislocation - lt. ankle	Eye glasses and lt. "A" pillar Wheel Instrument panel and e. a. column capsule bolt Lt. door Lt. side panel and toe board	70	30-35
1	UM-75-68	1968 Pontiac Tempest hit 1968 Pontiac Tempest (lt. front) MINOR	0	0	M 22 5' 11" 150 lb	No	Cuts on forehead and nose Lac. - inside mouth, four lower teeth loose Pain - neck Pain - rt. shoulder	Windshield Wheel rim Impact Wheel	0 35	0 20
1	UM-81-68	1968 Pontiac Catalina hit sign post MODERATE-SEVERE	3.7	0.1	M 16 5'9" 130 lb	Yes	Seat belt abrasions Fx. - three teeth and fx. - mandible Fx. - T-12 vertebra	Seat belt Wheel rim Flexing over belt (?)	80	30

(cont'd)

513

(Table 1 cont'd)

Impact Location	Case No.	Vehicle(s), Impact Object, Accident Type	E. A. Mesh	Column Capsules	Driver, Sex/Age	Seat Belt	Injuries	Injury Source	Police Traveling Speed	HSRI Estimated Impact Speed
1 + 2	UM-82-68	1968 Chevrolet Chevelle hit sign post MODERATE-SEVERE	9.0* *underhood shaft telescoped 4.5 in.	3.0 Free	F 24 5'2" 105 lb	No	Lac. - forehead	Wheel rim or instrument panel	60	35
							Lac. - rt. upper eyelid	Instrument panel		
							Lac. - side of head temporal area	Instrument panel		
							Contusions - chest	Wheel		
							Bruise - lt. arm	Door (?)		
							Abrasions - lt. abdomen	Wheel		
							Severe lac. - rt. knee	Lower instrument panel		
							Small lac. - lt. knee	Lower instrument panel		
							Fx. - lt. ankle	Floor pan		
							Fx. - rt. heel bone	Floor pan (?)		
							Fx. - mandible	Wheel rim		
							Six month fetus stillborn			
2	UM-86-68	1967 Chevrolet Corvette hit tree MINOR	0	0	M 38 5'8" 170 lb	No	Lac. - forehead	Sun visor and header	50	15
							Bruise - lt. knee	Instrument panel (?)		
							Lac. - inside lt. thigh	Gear shift (?)		
1	UM-90-68	1968 Dodge Charger hit front of 1968 Pontiac Catalina Station Wagon MODERATE-SEVERE	0	0	M 50 5'10" 210 lb	Yes	Fx. nose	Wheel rim	65	35
							Lac. nose and eyelid	Wheel rim		
							Contusion - chest	Wheel		
1	UM-90-68	1968 Pontiac Catalina Station Wagon	0.7	0.6	F 44 5'4" 125 lb	Yes	Fx. nose	Wheel rim	45	25
							Fx. lt. ulna	Unknown		
1	UM-92-68	1968 Plymouth Fury hit head-on by 1961 Chevrolet MODERATE	0.5	0.5	F 47	?	Lacs. - lips	Wheel rim	0	0
							Fx. tooth	Wheel	40	30
2	UM-93-68	1967 Pontiac Firebird hit sign posts MODERATE	0.6	0.5	M 34 5'11" 168 lb	No	Lac. - forehead	Windshield	40+	30
							Lac. lips	Windshield		
							Pain - chest	Wheel		
							Bruise - rt. knee	Lower instrument panel		
							Sprain - rt. foot	Toe board		
							"Aches all over"			
2	UM-94-68	1968 Dodge Dart hit rear of truck MODERATE	1.2	1.0	M 26	?	Pain - neck and hand	Unknown	75 50	65 50
1	UM-95-68	1967 Oldsmobile Delmont 88 hit tree MODERATE	1.0	0.5	F 79 5'7" 160 lb	Yes	Abrasions - forehead	Windshield	30	20
							Sprain - rt. ankle	Toe board		
							Fx. lt. wrist	Unknown		
2	UM-96-68	1967 Dodge Dart hit truck MODERATE	0.5	0.5	F 47 5'3" 130 lb	No	Ruptured duodenum	Wheel	30 0	20 0
1 + 2	UM-100-68	1967 Chevrolet Impala hit 1960 Ford head-on MINOR	1.5	0.7	M 52 5'10" 170 lb	No	Bruise - forehead	Sun visor and header	20	15
							Pain - back of neck	Impact	30	20
							Pain - rt. forearm	Unknown		
							Fx. - lt. rib	Wheel		
							Bruise - lt. thigh	Instrument panel		
1	12/67	1967 Pontiac GTO hit tree SEVERE	6.0* *underhood shaft telescoped 8.4 in.	6.0	M 23	Yes	Fx. - skull and intracranial injury - fatal	Instrument panel		45-50
							Fx. - rt. 5th rib	Wheel		
1 + 2	18/67	1967 Chevrolet Station Wagon hit rear of tractor trailer MODERATE	1.5	0.5	M 52	No	Lac. thoracic aorta - fatal	Wheel	?	
							Fxs. - rt. ribs (6, 7)		50	30-35
							Lt. hemothorax			
							Lacs. - face	Windshield		

514

or panel between the B and C pillars; area 6 - right rear fender; area 7 - rear end; area 8 - left rear fender; area 9 - left rear door or panel between the B and C pillars; area 10 - left front door; area 11 - left front fender; area 12 - rollover. In 18% of the cases, the impact was across the entire front end or directly to the center front of the car and thus the impact location is presented as area 1 and 2 (Fig. 1).

Of the 87 cases, 63% of the impacts were to the car in front of the A pillars. It is in these collisions that driver contact to the wheel-column system may be expected. Details of these "front end" collision cases are presented in Table 1.

When the mesh section of the energy absorbing column is compressed it has been noted that, in general, one-half or more of the compressed distance is from below, and less than half of the compressed distance of the mesh section is produced by driver loading of the column from above. There are cases where all of the compression of the mesh section is from driver impact but, conversely, cases are recorded where almost all of the mesh compression is from below (Table 1).

Comparison of 1967 and 1968 model cars involved in accidents with crashes previously studied (cars without the energy absorbing column) indicates a marked reduction in serious and fatal driver injuries at vehicle impact speeds that were the same or higher than those crashes involving cars without the energy absorbing steering column (4). Previously, internal thoracic and abdominal injuries were the leading causes of deaths of drivers who struck the wheel-column system. Drivers are still striking their chest and abdomen on the steering wheel and energy absorbing column as indicated by bruising, stiffness, or soreness in the thoracic and/or abdominal areas. Yet, in general, the injuries to these body areas are in the minor to moderate category.

The steering wheel rim of the energy absorbing column system is still a source of injury. The driver, with or without a seat belt, strikes his face on the steering wheel rim even in minor to moderate collisions. These facial injuries are not life threatening but are painful, require medical attention, and are sometimes disfiguring. These injuries are usually near the midline of the face and are found between the hairline and the chin. They include forehead or nasal lacerations, lacerations about the nasal openings, lacerations of the lips or chin, fractures of the nasal bones or nasal septum, fractures of the upper or lower jaw, or of the teeth or supporting bone of the teeth. The vertical location of the injuries appears to be related to the seated height of the driver and to impact speed. Steering wheel rim injuries to the side of the face usually occur when the head is turned at the time of impact. All of these various injuries may occur even in individuals who are wearing seat belts. Thus, facial injuries from the steering wheel rim probably can be prevented only by the combined use of a lap seat belt and shoulder belt.

A new injury causation has been identified in cars equipped

with the energy absorbing column. As the driver strikes and compresses the energy absorbing column in a front-end impact, the middle portion of the driver's face has been noted to contact the upper portion of the instrument panel directly in front of the wheel. Here, the padding usually distributes the force over a wide area of the face so that facial lacerations are infrequently seen. However, the relatively thin bony structures of the midface skeleton are weaker than those of the cranial vault -- the forehead area or top of the head. Thus, serious, but not life threatening, facial fractures are produced. These fractures can be of only the upper jaw, or also of the nasal bones, nasal septum, or of the zygomatic (cheek) bones as well. Often, there is a laceration of the nasal lining that produces extensive hemorrhaging. In that this area of the upper panel is within the inboard edge of the steering wheel rim, it is excluded from the Federal Standard Number MVSS201 that indicates a maximum impact tolerance for the head (5). In that the strength of the midface bones is less than that of the cranial vault bones, and that there appears to be a concentrated loading of the midface onto the panel beneath the padding, this specific area of the instrument

Fig. 2 - Head-on impact into dirt embankment 1954 Ford

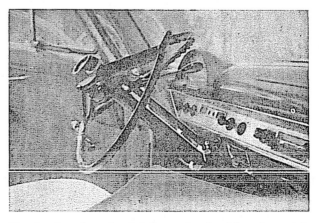

Fig. 3 - Steering wheel deformation by driver impact in vehicle shown in Fig. 2

panel needs to be reevaluated to reduce the severity of facial injury.

To date, three driver fatalities have occurred in cars equipped with the energy absorbing column. In one accident (case UM-03-68), the column was almost completely compressed and was separated from the flexible coupling. The estimated car-to-car impact speed was 40-50 mph. In another (case 12/67), the driver flexed over his seat belt and struck his head on the instrument cluster in front and to the left of the column. In the third crash (case 18/67), the car struck the rear end of a truck and then spun out. In this case, the driver struck the wheel-column system and possibly was loaded from the rear by luggage.

The following clinical case histories were selected to document driver injuries in a variety of front-end crashes.

Case 1. 5/62 - This 1954 Ford went off the right side of a curve and struck an embankment head-on. Slight rearward movement of the column was noted. The 35-year-old male driver sustained a laceration of the thoracic aorta, ruptured right ventricle, and multiple rib fractures with perforations of the lungs. Injuries were due to steering wheel-column impact (Figs. 2 and 3).

Case 2. 26/64 - This 1964 Oldsmobile struck the left rear dual wheel of a dump truck on the expressway. The 28-year-old male driver died of thoracic and abdominal injuries from steering wheel impact (Figs. 4 and 5).

Case 3. 19/65 - This 1964 Ford struck the side of a truck. The 32-year-old female driver sustained multiple rib fractures, bilateral hemothorax, and lacerations of the liver and spleen from steering wheel-column impact (Figs. 6 and 7).

Case 4. 29/65 - This 1964 Chevrolet went off the left side of the road, down a slight incline, and struck a tree head-on. The 52-year-old male driver struck the steering column and sustained multiple rib fractures, hemothorax, and a bruising injury of the heart. He died 8 hr after the accident (Figs. 8 and 9).

Case 5. UM-39-68 - In this accident, a 1967 Chevrolet Impala struck a 1966 Chevrolet Impala directly head-on.

Fig. 6 - On-scene photo of fatal accident (see Fig. 7)

Fig. 4 - 1964 Oldsmobile - collision with dump truck on expressway

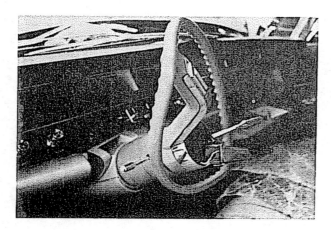

Fig. 5 - Fatal thoracic and abdominal injuries of driver from wheel-spoke impact (see Fig. 4)

Fig. 7 - Deformed steering wheel in 1964 Ford. Wheel-column impact produced multiple thoracic and abdominal injuries

The 33-year-old male driver (6 ft, 150 lb) of the 1967 Chevrolet was not wearing a seat belt. The mesh section was compressed 3.5 in. and the capsules separated by 3.5 in. The driver bent the upper half of the wheel forward. He sustained abrasions to his right knee (lower instrument panel), and had pain in his lower chest from wheel contact. He struck his face on the top of the instrument panel in front of

Fig. 8 - Head-on tree impact. 1964 Chevrolet

Fig. 9 - Chest-column impact producing fatal injuries

Fig. 10 - Head-on, car to car collision. 1966 Chevrolet

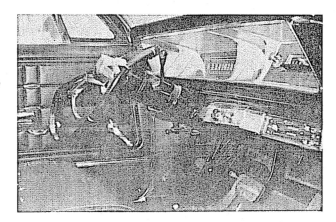

Fig. 11 - Bent steering wheel of car shown in Fig. 10. Driver sustained serious chest injuries

Fig. 12 - 1967 Chevrolet head-on collision with vehicle shown in Fig. 10

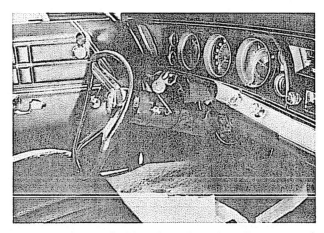

Fig. 13 - Interior of 1967 Chevrolet (Fig. 12). Energy absorbing mesh section compressed 3.5 in. Facial fractures sustained on upper padded panel in front of steering wheel

the wheel sustaining fractures of his upper jaw, nasal bones, and of the right orbital floor. The 43-year-old male driver (5 ft 9 in., 180 lb) of the 1966 Chevrolet (not equipped with an energy absorbing column) also bent the upper half of the wheel forward. He sustained rib fractures on the right side and had blood and air in his right chest cavity (hemopneumothorax) from wheel-column impact, a lacerated area about the left eye (windshield), a laceration of his left elbow (windshield), a left leg fracture (lower instrument panel), and a fracture of his right forearm (source unknown). He was not wearing a seat belt (Figs. 10-13).

Case 6. UM-20-68 - 1967 Camaro went off the roadway and struck a utility pole at an estimated impact speed of 25 mph. The front end deformation was 28 in. The 30-year-old male driver (6 ft, 180 lb) was wearing his seat belt snugly. On impact the driver bent the upper half of the wheel rim forward and separated the capsules 0.5 in. The mesh portion of the column was compressed 8.2 in. The driver had abrasions on both legs from impact to the lower instrument panel, seat belt bruises, pain in the abdomen and an-

terior chest wall (steering wheel). He also bruised his forehead and was unconscious for a short while from striking the top of the instrument panel in front of the steering wheel. (See arrow, Figs. 14-16.)

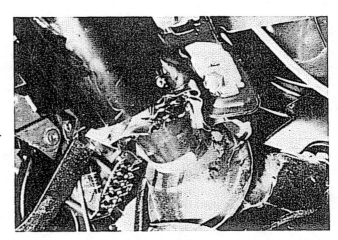

Fig. 16 - Compressed mesh section (8.2 in.) of 1967 Camaro

Fig. 14 - 1967 Camaro hit utility pole - 28 in. of front end deformation

Fig. 17 - Mid-center front end crush of 32 in. in 1967 Pontiac after striking steel pole

Fig. 15 - Bent steering wheel of Camaro (Fig. 14). Seat belted driver struck his forehead on upper panel in front of wheel

Fig. 18 - Bent steering wheel (mesh section compressed 4 7 in.). Driver sustained multiple facial fractures on upper padded panel in front of steering wheel

Case 7. UM-29-68 - This 1967 Pontiac Catalina, driven by a 32-year-old male, struck a steel pole at an estimated impact speed of 40 mph. There was 32 in. of mid-center front end crush. The unrestrained male driver (5 ft 6 in. 110 lb) struck the steering wheel bending the upper half of the wheel forward. The mesh section of the column was compressed 4.7 in. with capsule separation of 2.9 in. The driver flexed over the steering wheel, struck the windshield with his head (no reported injuries), and contacted the top of the panel with his face sustaining multiple facial fractures. He fractured his left hip by jamming his left knee against the steering column and instrument panel and his left hip against the door (Figs. 17 and 18).

Case 8. UM-65-68 - 1968 Mercury Monterey. This vehicle hit a tree adjacent to the road at an estimated speed of 20 mph. There was 25 in. of front end deformation in the right front area. The 48-year-old male driver (5 ft 8 in., 145 lb) was not wearing his seat belt. The energy absorbing section of the column compressed 1.6 in. with 1.5 in. of separation of the capsules. The steering wheel rim was slightly bent. The driver sustained a sore chest (wheel-column impact), a laceration of his lower lip and

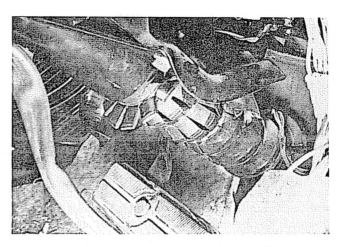

Fig. 21 - Energy absorbing mesh compression of 1.6 in. 1968 Mercury

Fig. 22 - Head-on collision with large tree. 1968 Oldsmobile

Fig. 19 - On-scene, car-tree impact (1968 Mercury)

Fig. 20 - Depression of panel in front of steering wheel from impact of driver's face

Fig. 23 - Undeformed steering wheel with angulated column. Driver struck her knee on one of capsule bolts to sustain severe knee injury

519

chin (instrument panel), lacerations inside mouth and throat (dentures), sore larynx (voice box) from instrument panel contact, a left knee bruise (lower instrument panel), pain in back of neck, and "ached all over" (Figs. 19-21).

Case 9. UM-73-68 - This 1968 Oldsmobile Delmont 88, driven by a 64-year-old female (5 ft 4 in., 170 lb), went out of control and struck a tree at 30-35 mph. There was 23 in. of front center deformation. The steering wheel was not deformed, but the column was angulated upward. The mesh section of the column was compressed 5.1 in. with 0.8 in. separation of the capsules. The driver lacerated and fractured her right knee on one of the capsule bolts; she bruised her left leg and knee on the parking brake release handle and lower instrument panel. She dislocated her left ankle (side wall-toe board area), and had pain in her lower left chest wall (steering wheel and door). Horizontal lacerations of the side of her head were due to broken eyeglass frame when she struck the left A pillar-vent window area. She was not wearing a seat belt (Figs. 22-24).

Fig. 26 - Seat belted driver struck steering wheel, fracturing his mandible. Energy absorbing mesh section compressed 3.7 in.

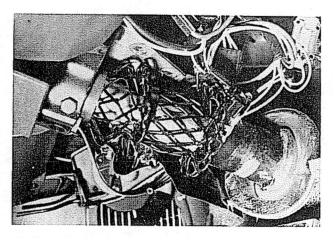

Fig. 24 - Energy absorbing mesh section compression of 5.1 in. (see Figs. 22 and 23)

Fig. 27 - On-scene, head-on collision of 1968 Chevrolet with cement base of expressway sign

Fig. 25 - Head-on impact into cement base of expressway sign

Fig. 28 - Front center deformation of 34 in.

Case 10. UM-81-68 - A 1968 Pontiac Catalina struck the cement base of an expressway sign support at an estimated 30 mph. The 16-year-old male driver (5 ft 9 in., 160 lb) was wearing a seat belt. He sustained a compression fracture of the 12th thoracic vertebra, had seat belt belt bruises, and fractured his lower jaw and three teeth on the steering wheel rim. The mesh section of the column compressed 3.7 in. with a 0.1 in. separation of the capsules. The steering wheel was not bent (Figs. 25 and 26).

Case 11. UM-82-68 - A 1968 Chevrolet Chevelle, driven by a 24-year-old female (5 ft 2 in., 102 lb, six months pregnant), struck the cement base of an expressway sign at an impact speed of 35 mph. The front center deformation was 34 in. The underhood section of the column telescoped 4.5 in., the mesh section compressed 9.0 in., and the capsules were separated and free for approximately 3.0 in. The wheel and spokes were severely deformed. The fetus was stillborn after the accident. The unbelted driver sustained a fracture of her right heel and of her left ankle (toe board), a severe laceration of her right knee, and

Fig. 31 - Lip laceration and chipped tooth from striking steering wheel rim

Fig. 29 - Deformed steering wheel producing abdominal wall abrasions, anterior chest contusions, and fractured mandible

Fig. 32 - Damage to 1968 Dodge Dart from striking rear of truck on expressway

Fig. 30 - Front end damage of 1968 Plymouth

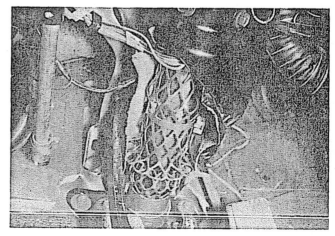

Fig. 33 - Energy absorbing mesh section compression of 1.2 in. Steering wheel slightly deformed. No chest injuries reported

521

a small laceration to her left knee (lower instrument panel). She had anterior chest contusions, abrasions to the left abdominal wall, a bruised left arm, and lower jaw fractures; all these injuries were due to impact to the steering wheel. Multiple scalp lacerations were sustained from either the steering wheel or instrument panel (Figs. 27-29).

Case 12. UM-92-68 - A 1968 Plymouth Fury III was stopped at a stop light when an oncoming car struck it head-on. The Plymouth was pushed backwards into a parked Mustang. Rear damage to the Plymouth was negligible. The impacting car struck the Plymouth at 30 mph. The 47-year-old driver of the Plymouth sustained lip lacerations and a chipped tooth from steering wheel impact. No other injuries were reported. The lower half of the wheel was bent forward, the mesh section compressed 0.5 in., and the capsules were separated by 0.5 in. (Figs. 30 and 31).

Case 13. UM-94-68 - A 1968 Dodge Dart struck the rear of a truck on the expressway. The Dodge was traveling at 70 mph, and the truck at 40 mph (police estimate). The impact speed is estimated at 15-20 mph. The 26-year-old male driver complained of pain in his right hand and back of the neck. The mesh section of the column compressed 1.2 in., and the capsules were separated by 1.0 in. The steering wheel rim was only slightly deformed. No chest injuries were reported. Seat belts were not worn (Figs. 32 and 33).

Case 14. UM-95-68 - A 1967 Oldsmobile Delmont 88, driven by a 79-year-old female (5 ft 7 in., 160 lb), went out of control and struck a tree. Front end deformation measured 17 in.; the impact speed was 20 mph. The seat belted driver slightly deformed the steering wheel, the mesh section compressed 1.0 in., and the capsules were separated 0.5 in. The driver sustained seat belt bruises, a bruised right knee (lower instrument panel), a lacerated lip and a bruised right cheek from steering wheel contact (Figs. 34 and 35).

Case 15. 12/67 - A 1967 Pontiac GTO left the roadway and struck a large tree. The front end crush measured 32 in. The seat belted male driver (6 ft 1 in., 150 lb) deformed the steering wheel, the mesh section compressed 6.0 in., and the capsules were separated and free. The driver sustained a fracture of the right 5th rib, slight hemorrhagic areas of the intestines and mediastinum (wheel), fracture of the left humerus (door area), a horizontal forehead lacera-

Fig. 34 - Head-on collision of 1968 Oldsmobile with tree. Front end damage of 17 in.

Fig. 36 - On-scene, head-on collision of 1967 Pontiac

Fig. 35 - Slight steering wheel deformation by seat belted driver who sustained lacerated lip and bruised right cheek from wheel rim impact

Fig. 37 - Front end deformation of 32 in.

Fig. 38 - Deformed steering wheel (energy absorbing mesh section compression of 6.0 in.) from driver impact. Seat belted driver struck his head on instrument panel face to left of column, sustaining fatal head injuries

tion, and a basilar skull fracture with intracranial injury. He struck the instrument cluster in front of the steering wheel to sustain his fatal head injuries. Impact speed was 45-50 mph. The underhood section of the steering shaft telescoped 8.4 in. (Figs. 36-38).

SUMMARY

Field investigations of automobile collisions involving vehicles equipped with the energy absorbing steering column indicate a significant reduction in serious and fatal injuries to drivers at impact speeds equal to or higher than in accidents where no energy absorbing column was present. Compression of the column in front end collisions allows the driver to strike the padded instrument panel in front of the steering wheel where fractures to the midface area are sustained.

REFERENCES

1. D. F. Huelke and P. W. Gikas, "Causes of Deaths in Automobile Accidents." Final Report to the Office of Research Administration, University of Michigan under Project No. 06749-1-F, April 1966.

2. S. Schwimmer and R. A. Wolf, "Leading Causes of Injury in Automobile Accidents." Automotive Crash Injury Research, Cornell Aeronautical Laboratory, Cornell University, Ithaca, N Y., June 1962.

3. P. C. Skeels, "The General Motors Energy Absorbing Steering Column." Tenth Stapp Car Crash Conference, Holloman AFB, November 1966.

4. D. F. Huelke and W. A. Chewning, "The Energy-Absorbing Steering Column." Highway Safety Research Institute, Report No. Bio-7, 1968.

5. Motor Vehicle Safety Standard No. 201. "Occupant Protection in Interior Impact - Passenger Cars." Federal Register. Vol. 32, Aug. 16, 1967, pp. 11777.

This paper is subject to revision. Statements and opinions advanced in papers or discussion are the author's and are his responsibility, not the Society's; however, the paper has been edited by SAE for uniform styling and format. Discussion will be printed with the paper if it is published in SAE Transactions. For permission to publish this paper in full or in part, contact the SAE Publications Division and the authors.

Society of Automotive Engineers, Inc

16 _page booklet.   Printed in U.S.A.

Correlation of Accident and Laboratory Impacts to Energy-Absorbing Steering Assemblies

L. M. Patrick and D. J. Van Kirk
Biomechanics Research Center, Wayne State University

ENERGY-ABSORBING steering columns were introduced by General Motors Corp., Chrysler Corp., and American Motors Corp. in their 1967 model automobiles. The injury potential of the energy-absorbing column was established by limited cadaver research at Wayne State University prior to the use of the columns which showed the forces were not high enough to cause rib fractures with the wheel-column characteristics used in the investigation. Part of the results of that investigation were reported at the 1968 SAE Automotive Engineering Congress (1)*. The investigation was based on a body buck and steering assembly mounted on a sled in a forward-force simulated impact. The purpose of this paper is to compare the results of the simulated impacts with the injuries observed in accidents by a team of accident investigators at Wayne State University.

---

*Numbers in parentheses designate References at end of paper.

Note: This research was supported in part by a grant from the Automobile Manufacturers Association.

In the reported series of experiments with lap-belted drivers the total force on the upper body varied from 1630 to 1830 lb with the force divided between the hub and the rim. The hub forces, or the forces on the central part of the chest, varied from 550 to 740 lb, while the force on the rim was about 1075 lb. The velocities varied from 24.4 to 29.4 mph in the barrier impact simulation. One of the conclusions drawn from these experiments was that the distribution of the force between the rim and the hub was important in maximizing the force which the body could stand without injury. The three-spoke wheel used in these experiments deformed at static forces in the neighborhood of 800-1000 lb when the force was applied across two spokes and at about 250 lb with a concentrated axial force at the top of the rim (normal face impact area). The type of steering wheel makes a major difference in the injury with the energy-absorbing columns in 1967 and 1968 automobiles. Since the 1967-1968 columns used by the three manufacturers (GM, Chrysler, AMC) are essentially the same, the investigation includes the effectiveness of the steering column when used with dif-

ABSTRACT ————————————————————————————

Data are presented for 19 frontal-force collisions involving vehicles with collapsible steering columns with collision severity rating from minor to very severe (1-7) and an injury severity index from minor to fatal. Injury results are compared to laboratory experiments in which a force of 1800 lb distributed over the rim and hub was measured for a fairly stiff wheel and collapsible column combination.

When the steering wheel did not deform excessively and the force reached the 1800 lb level as evidenced by column collapse, there were no serious thoracic injuries. Gross deformation of the steering wheel with exposed sharp spoke ends or small diameter hub resulted in serious abdominal and thoracic injuries.

Two cases of hood intrusion are presented, each of which resulted in fatalities.

524

ferent steering wheels (two and three spoke, stiff and flexible rims) and with different installations.

A brief review of the forces and displacements available in the wheel-column combination used in the laboratory experiments will provide insight into the theoretical performance of the combination. While the peak force was about 1800 lb, the average force was considerably less. If the 1800 lb force is assumed to act over the available collapse distance of about 8 in., it can absorb 1200 ft-lb of energy. If the average weight of a driver is assumed to be 160 lb and the total kinetic energy is removed by collapsing the steering column, the maximum velocity representing the 1200 ft-lb for a 160 lb individual is 15 mph. Thus, at 15 mph the total kinetic energy could be dissipated by collapsing the column a distance of 8 in. at a force of 1800 lb. At 30 mph the energy dissipation available would represent about 25% of the kinetic energy, and at 60 mph it would represent about 6-2/3% of the kinetic energy.

If a more realistic average force over the total available 8-in. distance is assumed to be 900 lb (600 ft-lb), then the maximum speed for which the total kinetic energy of the individual could be dissipated in the collapsing column is about 10-1/2 mph At 30 mph with the assumed 600 ft-lb dissipation in the column, it can absorb only about 12% of the total kinetic energy.

Since there are many documented cases in which the driver was only slightly injured or not injured at all when accidents occurred at much higher speeds than the 10.5 mph for which the column can dissipate the total energy, it is interesting to study the circumstances permitting the higher speed impacts to occur without injury, especially since the collapse distance is generally appreciably less than the 8 in. available with the available collapse energy correspondingly less. The circumstances include:

1. The energy-absorbing column does not have to dissipate the total energy. The torso deforms the steering wheel and the knees deform the instrument panel, providing energy dissipation, and if the occupant is wearing a lap belt, it takes out part of the energy.

2. The impact or collision is seldom as severe as the barrier type collision simulated in the laboratory. Car-to-car, car-to-pole, and car-to-guard rail collisions, for example, are less severe than car-to-barrier collisions at the same speed. The less severe impact observed in most collisions provides the dissipation of the kinetic energy of the driver through the so-called ride-down (2).

These factors permit a much higher velocity collision without major injury to the occupant than the severe type simulated in the laboratory.

The primary function of the energy-absorbing column is to limit the forces on the body to a sub-injury level during the collision. In most collisions the driver hits the wheel before the vehicle is completely stopped. If, during the collision while the driver is in contact with the wheel, the forces are below the injury level and below the column-wheel crush level, the kinetic energy is dissipated without any relative motion or collapse of the steering column. The

column collapses only when the force level reaches a dangerous level when it limits the maximum force on the body. As soon as the force drops below the collapse or injury level, the column ceases to collapse and additional ride-down is experienced without collapse. Thus, the main purpose of the column is to limit the forces by providing a relative motion between the occupant and the vehicle during the intermittent times in the collision when the force is above the injury level.

One of the earliest documented energy-absorbing column collisions was reported by Hanson in a paper by Skeels (3) in which an American Motors employee was driving an Ambassador at a speed of 55 mph when he struck a stationary truck loaded with 33,000 lb of cargo. At the estimated 55 mph speed the energy-absorbing column could only absorb a small fraction of the total kinetic energy represented by the moving mass of the occupant. Even if only the weight of the torso is considered, the collapsing column still could not dissipate the energy. However, an analysis of the impact shows that the total stopping distance was well over 10 ft. This represents the distance the truck moved forward, the truck-car under-ride, and the car collapse. During the collision the energy-absorbing column collapsed a distance of 5.4 in. (Case S-6, Table 1A). It is hypothesized that the remaining kinetic energy of the occupant was dissipated by ride-down at a sub-injury level. The major injuries were severe facial lacerations, lateral lacerated right knee and thigh, and diabetic shock (probable cause of the accident).

FIELD INVESTIGATION OF COLLISIONS INVOLVING ENERGY-ABSORBING STEERING COLUMNS

The WSU accident investigating team has 19 cases involving 1967 and 1968 model cars with energy-absorbing column: nine standard size automobiles, six intermediate, and four compacts divided among GM, Chrysler, and Ford Motor Co.* The collisions, which are essentially of the forward-force type, include data on the injuries to the driver and the deformation of the energy-absorbing column along with other pertinent information collected during the accident investigation. Table 1 shows the details of the accidents including data on the vehicle, the age of the driver, the type of accident involved, the amount of steering column collapse, the steering wheel deformation, the estimated speed, the injury index of the driver, the equivalent sled speed, and the Vehicle Damage Scale according to the TAD rating (4).

Fig. 1 is a graph showing the relationship between maximum column deformation and sled impact speed for impacts to the energy-absorbing column by seat-belted drivers on the small sled described in a paper by Daniel and Patrick (5) and shown in Fig. 2. The equivalent sled speed shown in Table 1 is based on the speed corresponding to the column deformation shown in the table. It should be pointed

---

*Ford Motor Co. introduced the collapsible column on its 1968 models.

Table 1A - Detailed Data on Forward-Force Collisions of 1967 and 1968 Automobiles
with Energy-Absorbing Steering Columns

Standard Vehicles (Full Size)

Case No.	Vehicle Make & Year	Estimated Speed, mph	Age & Sex	Vehicle Severity[a]	Total S/C Collapse, in.	S/W[b] Deform., in. deg	Driver Injury	Interior Deformation	Injury[c] Index	Equiv. Sled Speed,[d] mph
S-1	1967 Chevy	50	25/M	FL-4	0.0	0.0	Concussion	Visor mounting bracket bloody	B_0	10
S-2	1967 Chev.	1 - 35 2 - 70	18/M	FLO-6 -7/8	3.6	0.4 at 315 SP1 - 0.5 SP2 - 0.5 SP3 - 0.5	3/8 in. knee dent, bruise r. bicep, lac. to cheek, bruise eye		A_0	21
S-3	1967 Pont.	1 - 65 2 - 0	47/M	FD-5	5.0	1.0 at 0 0.0 at spokes	Tender r. ribs	S/W damage, header & visor impact, knee dent 1.	A_0	26
S-4	1967 Ambas.	1 - 45 2 - 10-20	22/M	FR-2-1/2	0.0	0.0	None	No I.D. on driver's side	A_0 - Belted	10
S-5	1967 Ambas.	1 - 35	39/M 5'11" 180 lb	FC-7	6.8	8.0 at 180 SP1 = SP2 = 8 SP3 =	Lac. of liver, torn mesenteric vein, lac. face, fx r. ankle	Seat broke loose injuries due to S/W	D_1	32
S-6	1967 Ambas.	50-60	33/M	FL-7-2/3	5- 3/8	1.1 at 0	Severe facial lac., lat. lac. to r. knee & thigh	W/S and dash	B_1 - Belted	8
S-7	1968 Dodge	30	62/M 6'0" 150 lb	FR-6	0.4	3.9 at 225 SP1 = 1.6 SP2 = -1.1	Fx 1-6 ribs, traumatized myocardium, abr. r. leg		D	10
S-8	1968 Ford	1 - 50 2 - 50	38/M	FR-7-3/4	0.5	5.6 at 0 SP1 = 0.2 SP2 = 0.2	Mult. cont. & abr., fx mandible		C - Seat belt & shoulder harness	--
S-9	1968 Buick LeSabre	1 - 45	60/M	FL-7	3.0 from bottom	1.0 at 0	Brain concus. fx disloc. l. hip, post.	Hood came thru W/S into driver	F - 4 days	18

Table 1B – Detailed Data on Forward-Force Collisions of 1967 and 1968 Automobiles
with Energy-Absorbing Steering Columns

Intermediate Vehicles

Case No.	Vehicle Make & Year	Estimated Speed, mph	Age & Sex	Vehicle Severity [a]	Total S/C Collapse, in.	S/W Deform., in. deg [b]	Driver Injury	Interior Deformation	Injury Index [c]	Equiv. Sled Speed, mph [d]
I-1	1967 Dodge Cor.	1 – 50 1 – 20	23/M	FLO-4-F	5.8	2.5 at 0	Lac. upper gums trauma to maxilla - S/W		A	28
I-2	1967 Chevelle	1 – 40 2 – 35	41/F	FD-5	Col. pushed 1.4 in. at hub, 0 in. collap.	0.5 at 0 1 = 0.7 2 = 1.2 3 = 1.2	Lac. scalp W/S, cont. r. hip (belt)	W/S cracked	A_1 - Belted	10
I-3	1967 Ply.	1 – 45 2 – 45	48/M	FR-6-3/4	0.0 pushed 1.3 in. at top	2.5 at 225 SP1 = 0.9 SP2 = 2.4 SP3 = 1.2	Mild conc. - 3 days, disoriented, shock, abr. head, cont. chest & knee	Header & visor bent, knee dent 1. 3/4 in., s/lever broken off	C_1 - Belted	10
I-4	1967 Chevelle	25-30	34/M	FD-4	5.2	3.2 at 0 SP1 = 2.0 SP2 = 0.1 SP3 = 0.2	Lac. of head, chin & r. hand	W/S cracked, ashtray dent	A_o - Belted	26
I-5	1967 Dodge Cor.	25	22/F	FL-1/ FR-1	0.0	0.0	Bump on head	Horn ring	A_o - Belted	10
I-6	1968 Pont. Temp.	No P. R.	45/M	FR-7	5.8	8.2 at 90 SP1 = 3.5 SP2 = 3.6 SP3 = 8.8	Fx maxilla, fx ribs l., proven myocard. cont., fx r. acet., disloc. femur	1 in. deep l. knee dent, r. knee dent over contr. total S/W collapse	E	28

out that in the laboratory the force is applied essentially along the axis of the column, while in the accident situation there are often lateral or vertical forces which vary the collapse characteristics of the energy-absorbing column. Also, in field accidents the front-end collapse often causes the column to collapse from the bottom or front of the vehicle while in the simulated impacts the total column collapse is from the driver impact (3,6). In most instances the equivalent sled speed based on column collapse is much lower than the estimated accident speed. The only cases where the estimated speed and the sled speed are approximately the same are those involving severe types of accidents such as car to culvert or car to utility pole (Cases I-4 and S-5, Table 1).

A meaningful comparison of injury for each accident can only be made if the severity of the accident is established. Consequently, each collision is rated in severity according to the TAD rating (4). The rating for several of the collisions is given in the following section with the description of the accident case where the deformation of the vehicle

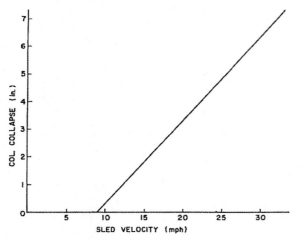

Fig. 1 - Column collapse versus sled speed

shows the severity of the accident. In order to provide uniformity of reporting, it is recommended that the TAD rating or some similar method of establishing severity be adopted by all accident investigators.

A uniform method of reporting degree of occupant injury is also essential. Several different laboratories have proposed injury indices (7-9). The injury index used in this report was developed at Wayne State University to meet the specific requirements of forward-force collisions and is a modification of several other rating systems. It has been reported by Van Kirk and Lange (10) (see Appendix).

With only 19 cases available for this study statistical inferences cannot be drawn. However, since the accidents were investigated in depth and each case analyzed by the same team of investigators, the results are more significant than a similar number of cases with less stringent analysis, and are considered valid for establishing general trends.

COLLISION CASES

Each case listed in Table 1 will be described briefly in order of increasing severity with pertinent photographs of some of the vehicles to illustrate the severity of the collision and the Vehicle Damage Scale. The Vehicle Damage Scale is described in Technical Bulletin #1 published by the Traffic Accident Data Project (4) of the National Safety Council. The Vehicle Damage Scale is described by a series of photographs with different degrees of deformation. Since the cases reported herein are essentially forward-force collisions, the rating is generally indicated by two letters followed by a number, that is, FC-4. The first letter indicates the damaged area -- for example, F for front-end damage -- while the second letter indicates the location of the impact such as C for concentrated, L for left, R for right, and D for distributed. The number rating from 1-7 indicates the severity and is obtained by comparing the vehicle or photographs of the damaged vehicle with photographs corresponding to the different degrees of severity in the Ve-

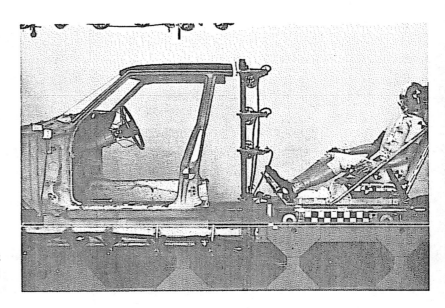

Fig. 2 - Set-up view of body buck mounted on small impact sled for steering assembly impacts

Table 1C - Detailed Data on Forward-Force Collisions of 1967 and 1968 Automobiles with Energy-Absorbing Steering Columns

Compact Vehicles

Case No.	Vehicle Make & Year	Estimated Speed, mph	Age & Sex	Vehicle Severity [a]	Total S/C Collapse, in.	S/w Deform., in. deg [b]	Driver Injury	Interior Deformation	Injury Index [c]	Equiv. Sled Speed, mph [d]
C-1	1967 Val.	20	22/M	FR-1-2/3	0.0	0.0	Lac. upper lip	W/S bulge, l. knee dent 3/8 in.	A$_o$	10
C-2	1967 Firebird	35	30/M	FC-6	5.3 1.5 driver	3.0 at 0 0.3 at 90 0.3 at 270	Lac. to scalp, conc. mild, lac. l. elbow tender chest, belt bruises	Visor & header dent, S/w damage window vent glass	B$_o$	14
C-3	1967 Firebird	40	18/M	FD-4	1.0 driver	0.0	Lac. forehead	W/S cracked 1 in. l. knee dent	A$_o$	12
C-4	1967 Camaro	20/M		FC-5	7.0 vehicle 1.5 driver	6.0 at 0	R. chest cont., sore back, abr. to head and face	S/w only	A$_o$ – Belted	14

[a] Vehicle Severity obtained from Technical Bulletin No. 1 published by the National Safety Council.

[b] Steering wheel deformations are measured in the laboratory and are given in inches of deformation of the rim and degrees from the top, center of the wheel moving in a counterclockwise direction.

[c] Occupant Injury Index developed at Wayne State University, Biomechanics Research Center (see Appendix).

[d] Equivalent sled speed is determined from laboratory experiments and shown in Fig. 1.

hicle Damage Scale report. In some cases a third number will appear in the rating. It is usually a fraction and denotes the amount of area impacted, that is, FR-4-1/3.

Figs. 3 and 4 show Vehicle Damage Ratings of 1; Fig. 5 is a 2 rating; Figs. 6 and 7 are 4 ratings; Fig. 8 is a 5 rating; Figs. 9, 12, 13, and 15 are 6 ratings; and Figs. 18, 21, 24, and 27 are 7 ratings. The position of the impact is obvious in most of the photographs, and the complete rating is listed under vehicle severity in Table 1.

The individual cases are discussed under each case number.

CASE C-1 (Fig. 3) - A 1967 Valiant struck the left front of another vehicle in a 90 deg intersection type of accident resulting in a damage rating of FR-1. There was no steering column collapse and no deformation of the steering wheel. The injury to the driver consisted of a lacerated upper lip, resulting from contact with the steering wheel rim. The windshield was broken, but there was no injury which could be attributed to the windshield impact. The estimated speed at impact was 20 mph, while the equivalent sled speed for no column deformation is under 10 mph.

CASE I-5 (Fig. 4) - A 1967 Dodge Coronet skidded out of control on an icy pavement while making a turn and struck two sign posts resulting in a damage rating of FL-1 and FR-1. The driver, who was wearing a seat belt, suffered a bump on the head from the horn ring for an injury index of A_o.

There was no damage to the column or to the steering wheel.

CASE S-4 (Fig. 5) - A 1967 Ambassador struck another car broadside in an intersection type accident traveling at an estimated speed of 45 mph. There was no driver injury (injury index of A_o), no column collapse, and no steering wheel deformation although the accident is rated as an FR-2 damage.

CASE I-4 (Fig. 6) - The belted driver of the 1967 Chevelle lost control of his car and hit a culvert at an estimated speed of 25-30 mph resulting in a damage rating of FD-4 with a total steering column collapse of 5.2 in. and a maximum steering wheel rim deformation of 3.2 in. He received a laceration of the head, chin, and right hand for an injury index of A_o. The estimated speed of 25-30 mph corresponds to the equivalent sled speed of 26 mph based on the 5.2 in. column collapse. Impact to a culvert or other immovable type barrier is about the same severity as the sled impact. The lap belt worn by the driver probably reduced the severity of his injury.

CASE C-3 (Fig. 7) - A 1967 Firebird struck another car broadside at an estimated speed of 40 mph resulting in a damage rating of FD-4. The steering column collapsed 1 in., and there was no steering wheel deformation. The equivalent sled speed is 10 mph or less. The driver sustained an A_o injury index with a lacerated forehead.

CASE I-2 (Fig. 8) - A 1967 Chevelle struck another car broadside at an intersection resulting in a damage rating

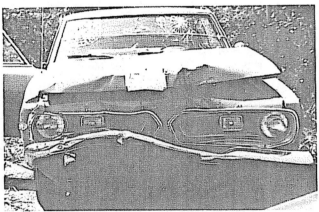

Fig. 3 - Case: C-1 Vehicle damage rating: 1
Vehicle: 1967 Valiant Column collapse: 0 in.
Estimated speed: 20 mph Driver injury index: A_o

Fig. 4 - Case: I-5 Vehicle damage rating: 1
Vehicle: 1967 Dodge Column collapse: 0 in.
Estimated speed: 25 mph Driver injury index:
          A_o - belted

Fig. 5 - Case: S-4 Vehicle damage rating: 2
Vehicle: 1967 Ambassador Column collapse: 0 in.
Estimated speed: 45 mph Driver injury index:
          A_o - belted

Fig. 6A - Case: 1-4 Vehicle damage rating: 4
Vehicle: 1967 Chevelle Column collapse: 5.2 in.
Estimated speed: 25-30 mph Driver injury index: A_o -belted

of FD-5 corresponding to an equivalent sled speed of less than 10 mph when based on column collapse. The steering column did not collapse but the center of the hub was pushed 4 in. to the left. The steering wheel was deformed up to 1.2 in. at the spokes. The belted driver received an injury index of A_o consisting of a laceration to the scalp from the windshield and a contusion of the right hip from the belt.

CASE S-2 (Fig. 9) - A 1967 Chevrolet struck a spinning car at the left rear with the impact to the Chevrolet appearing at the left front producing an FLO-6 damage rating (front left oblique impact of damage rating 6). The estimated speed was 35 mph for the Chevrolet and 70 mph for the other car prior to impact. The steering column of the Chevrolet collapsed 3.6 in. with a maximum steering wheel deformation of 0.5 in. The driver sustained a bruise to the right bicep, laceration to the cheek, and a bruised eye for an injury index of A_o. There was a 3-in. deep knee dent shown in Fig. 10 with the wiper control knob knocked off and the

Fig. 6B - Interior view of Case 1-4 showing deformed steering wheel and knee dent to right of column

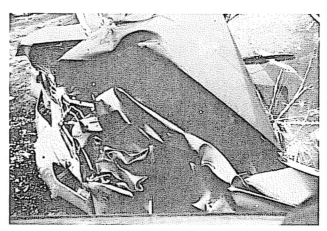

Fig. 8 - Case: 1-2
Vehicle: 1967 Chevelle Vehicle damage rating: 5
(El Camino) Column collapse: 0 in.
Estimated speed: 40 mph Driver injury index: A - belted

Fig. 7 - Case: C-3 Vehicle damage rating: 4
Vehicle: 1967 Firebird Column collapse: 1 in.
Estimated speed: 40 mph Driver injury index: A_o

Fig. 9 - Case: S-2 Vehicle damage rating: 6
Vehicle: 1967 Chevrolet Column collapse: 3.6 in.
Estimated speed: 35 mph Driver injury index: A_o

headlight knob bent approximately 90 deg. Fig. 11 shows the steering column with several collapsed convolutions.

CASE S-7 (Fig. 12) - A 1968 Dodge struck a utility pole in a right center location at an estimated speed of 30 mph

resulting in a damage rating of FR-6. The steering column collapsed 0.4 in. and the steering wheel had a maximum deformation of 3.9 in. with the spokes deformed 1.6 and 1.1 in. respectively. The driver suffered six fractured ribs, traumatized myocardium, and an abrasion of the right leg for an injury index of D_o. The equivalent sled speed for the 0.4 in. column collapse would be about 10 mph.

Fig. 10 - Left knee dent, bent light control knob, and broken off wiper control knob in Case S-2

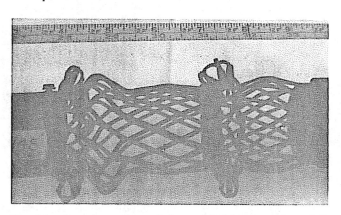

Fig. 11 - Steering column deformed 3.6 in. from Case S-2

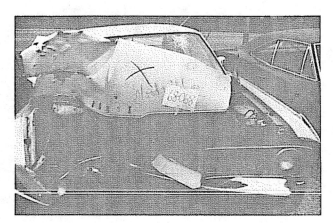

Fig. 12 - Case: S-7   Vehicle damage rating: 6
  Vehicle: 1968 Dodge   Column collapse: 0.4 in.
  Estimated speed: 30 mph   Driver injury index: D

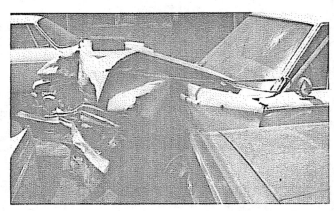

Fig. 13 - Case: I-3   Vehicle damage rating: 6
Vehicle: 1967 Plymouth   Column collapse: 0 in.
Estimated speed: 45 mph   Driver injury index: C_1 - belted

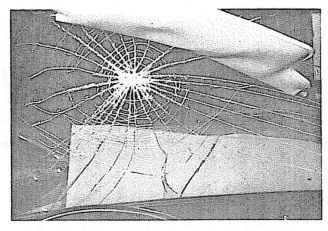

Fig. 14 - Visor and windshield damage in Case I-3

Fig. 15 - Case: C-2   Vehicle damage rating: 6
Vehicle: 1967 Firebird   Column collapse: 6.8 in. total
Estimated speed: 35 mph   Driver injury index: B_o

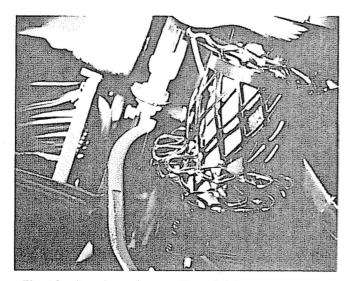

Fig. 16 - Steering column collapsed 6.8 in. in Case C-2

Fig. 17 - Deformed steering wheel and column in Case C-2 with instrument panel pad torn off showing instrument panel controlled deformation design

Fig. 18 - Case: S-5 Vehicle damage rating: 7
Vehicle: 1967 Ambassador Column collapse: 6.8 in.
Estimated speed: 35 mph Driver injury index: D_1

CASE S-5 (Fig. 18) - A 1967 Ambassador struck a utility pole at an estimated speed of 35 mph near the center of front of the vehicle for a damage rating of FC-7. The steering column collapsed 6.8 in. with a maximum steering wheel deformation of 8 in. The driver sustained a laceration of the liver, torn mesenteric vein, laceration of the face, and fracture of the right ankle for an injury rating of D_1. In this collision the seat broke loose and probably contributed to the degree of injury by adding a crushing force between the occupant and the wheel. The grossly deformed wheel is shown in Figs. 19 and 20. During the deformation the spokes tore, presenting a sharp, jagged edge to the driver which probably aggravated the internal injuries. The equivalent sled speed based on column collapse is 32 mph.

CASE I-6 (Fig. 21) - The estimated 45 mph impact to the 1968 Tempest was to the right front of the vehicle, producing an FR-7 damage rating. The total column collapse was 5.8 in. with a maximum rim deflection of 8.8 in. occurring on the right-hand spoke. The injuries consisted of a fractured mandible, fractured ribs, myocardial contusion,

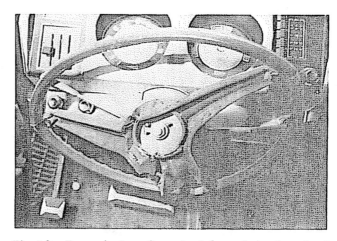

Fig. 19 - Forward view of grossly deformed steering wheel in Case S-5

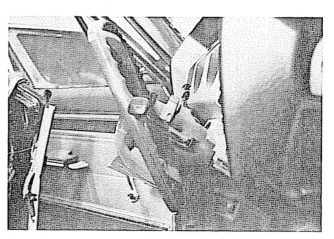

Fig. 20 - Lateral view of grossly deformed steering wheel and seat which tore loose in Case S-5

fractured right acetabulum, and dislocated femur for an injury index of E$_o$. The equivalent sled impact speed is 28 mph. Fig. 22 shows the gross deformation of the steering wheel with the hub and center ends of the spokes which

Fig. 21 - Case: I-6   Vehicle damage rating: 7
     Vehicle: 1968 Tempest   Column collapse: 5.8 in.
     Estimated speed: --   Driver injury index: E

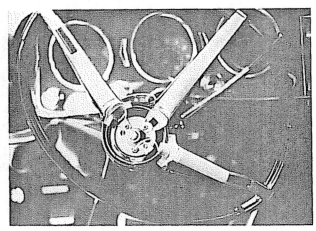

Fig. 22 - Deformed steering wheel in Case I-6

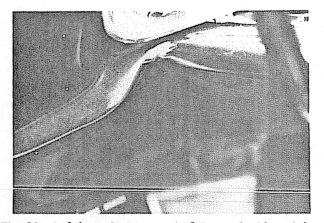

Fig. 23 - Left knee dent in panel of Case I-6 with no injury. Right knee hit stiff section of panel adjacent to column with fractured acetabulum and dislocation of right hip

caused the thoracic injuries. The left knee dent is shown in Fig. 23 which did not result in injuries to the left knee-thigh-hip complex. The small dent adjacent to the steering column on the right-hand side indicates that the deformation at this point was not sufficient to absorb the energy of the right knee-thigh-hip complex and the remaining force caused a fracture of the right acetabulum with dislocation of the right hip.

CASE S-9 (Fig. 24) - A 1968 LeSabre traveling at an estimated speed of 45 mph struck another car head on with

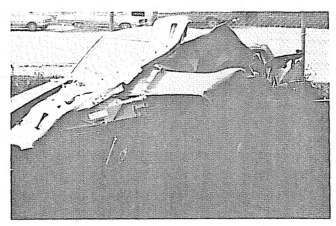

Fig. 24 - Case: S-9   Vehicle damage rating: 7
     Vehicle: 1968 LeSabre   Column collapse: 3 in.
     Estimated speed: 45 mph   Driver injury index: F

Fig. 25 - Left knee dent resulting in fracture dislocation of left hip - Case S-9

about a 50% overlap on the left front of each car for a damage index of FL-7-1/2. The steering column collapsed 3 in. from the bottom and the steering wheel deformation was 1 in. on the top. The injury index is F_1 (fatal) with death occurring four days after the accident. Fig. 25 shows the knee deformation of the instrument panel which produced a fracture dislocation of the left hip. Fig. 26 shows the intrusion of the hood into the windshield area. Contact of the head with the windshield area produced severe head injuries which resulted in the death of the driver.

CASE S-8 (Fig. 27) - This is a car-to-car collision with approximately a 3/4 overlap on the right-hand side. The estimated speed of both vehicles was 50 mph and the damage rating is FR-7-3/4. The column collapse was 0.5 in. and the maximum rim deformation was 5.6 in. at the top. The driver was wearing a seat belt and shoulder harness and received multiple contusions, abrasions, and a fractured mandible for an injury index of C_o. Fig. 28 is a lateral view of the same car showing how the hood penetrated the windshield on the passenger's side, fatally injuring the right

front passenger. There was no penetration of the hood on the driver's side. Fig. 29 is an interior view of the driver's area with the steering column detached from the instrument panel. The bent steering column is shown in Fig. 30 with very little collapse (0.5 in.).

It is interesting to compare Case S-8 and S-9. In S-8 the hood penetrated the passenger's side causing a passenger fatality, while the driver, without hood penetration, was only moderately injured in an extremely severe collision. Case S-9 is another offset front-end car-to-car collision of approximately the same damage rating where the driver was not wearing a restraint system and the hood penetrated the driver's side of the windshield to a degree where the head struck it, causing fatal injuries. These two cases illustrate the importance of maintaining passenger compartment integrity.

RESULTS

The data for the collisions reported herein are summarized in Table 1. A further presentation of the data is shown in Fig. 31, which is a plot of injury index as a function of ve-

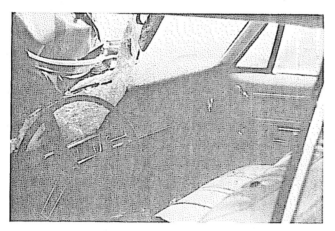

Fig. 26 - View through windshield showing penetration of rear edge of hood through windshield - Case S-9

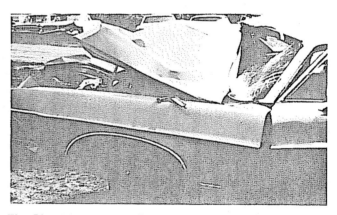

Fig. 28 - Lateral view of Case S-8 showing intrusion of hood into right front passenger compartment

Fig. 27 - Case: S-8   Vehicle damage rating: 7
Vehicle: 1968 Ford   Column collapse: 0.5 in.
Estimated speed: 50 mph   Driver injury index: C - lap and
                                        shoulder

Fig. 29 - Interior view of Case S-8 showing bent steering column broken away from panel attachment

535

e accident severity. In general, the injury increases with accident severity as would be expected. The lap-belted driver has a lower injury index for a given collision severity than the unbelted driver. The one case in which a lap belt and upper torso restraint was used provided a considerably lower injury index than would be expected for the very severe collision. In Fig. 31 each collision is indicated by a symbol representing the type of vehicle and by a number and letter representing the age and sex of the driver. When a lap belt was worn, it is indicated next to the symbol with the letter "B" and the single case of a lap belt and shoulder harness is indicated by the letters "B-SH."

Fig. 32 shows the injury index plotted as a function of total column collapse. The direction of each impact is indicated by letters adjacent to the symbol. When the impact is to the right front portion of the car, the column collapse is less for a given degree of injury than when the impact is on the left or distributed across the front. There is some spread in the collapse of the column as a function of sled speed in the simulated laboratory impacts. Further investigation is needed to determine why the column does not collapse in some cases. One possibility is that the impact is off center, or oblique to the column.

It is important to maintain the passenger compartment integrity as evidenced by Cases S-8 and S-9 in which the hood intruded into the compartment causing a fatal injury. In one case (S-9) where the intrusion was on the driver's side, the driver was fatally injured, while in the other where the intrusion was on the passenger's side (S-8), the driver was only moderately injured when wearing a lap belt and shoulder harness while the passenger, who was not restrained, was fatally injured.

There were no serious thoracic injuries when the wheel was not grossly deformed indicating that the 1800 lb load measured in the laboratory with a fairly rigid three-spoke wheel and a collapsible column is a noninjurious force level. In those cases where the column did not collapse and the wheel deformed grossly, thoratic and abdominal injuries resulted. Special emphasis should be placed on design of the steering wheel so that the spokes will not fracture or partially fracture when deformed, presenting a sharp, jagged edge to the thorax or abdomen (see Case S-5). Case I-6 is another instance where poor deformation characteristics resulted in thoracic injuries. In this case the rim deformed, leaving the hub, center spoke ends, and steering shaft exposed to produce thoracic injuries.

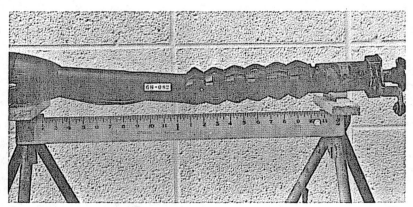

Fig. 30 - Bent steering column from Case S-8 which collapsed 0.5 in.

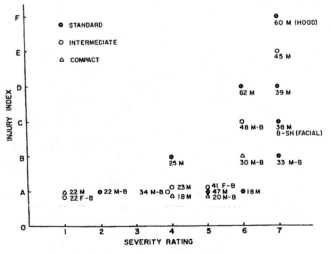

Fig. 31 - Injury Index as a function of Severity Rating (Vehicle Damage Rating)

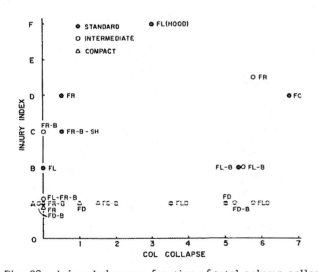

Fig. 32 - Injury Index as a function of total column collapse

Injuries to the knee-thigh-hip complex occur when the knee strikes a portion of the instrument panel over the rigid brace adjacent to the steering column. When the instrument panel deforms without coming in contact with a rigid brace or other stiffening member, the knee-thigh-hip complex is usually not injured. The collapsible column has resulted in more driver knee contacts with the instrument panel than were observed in the cars without collapsible columns, but there has not been a corresponding increase in knee injuries.

CONCLUSIONS

The following conclusions are drawn from the limited number of frontal-force collisions in which collapsible columns were installed in the vehicles.

1. In general, injury increases with collision severity.

2. Injury is reduced by lap belts.

3. The column collapse is only minimal in right-front impacts.

4. Provision should be made to prevent the hood from intruding into the passenger compartment under semi-frontal impacts.

5. An 1800 lb force distributed over the thorax with a stiff steering wheel will not produce serious thoracic injuries.

6. The wheels should be designed so they will not deform in a manner which will result in concentrated loads being applied.

7. An investigation should be conducted to determine why the columns do not collapse in some impacts when the force is obviously above the collapse force as evidenced by gross deformation of the stiff steering wheel.

8. Driver knee impacts are more prevalent with the collapsible column and injuries occur when the impact is near the rigid section of the instrument panel adjacent to the steering column.

REFERENCES

1. C. W. Gadd and L. M. Patrick, "System versus Laboratory Impact Tests for Estimating Injury Hazard." Paper 680053 presented at SAE Automotive Engineering Congress, Detroit, January 1968.

2. D. E. Martin and C. K. Kroell, "Vehicle Crush and Occupant Behavior." SAE Transactions, Vol. 76, paper 670-034.

3. P. C. Skeels, "The General Motors Energy Absorbing Steering Column." H. L. Hanson, "--A Case History." Proceedings of Tenth Stapp Car Crash Conference. New York: SAE, 1967.

4. "Vehicle Damage Scale for Traffic Accident Investigators." TAD Project Technical Bulletin No. 1, National Safety Council, 1968.

5. R. P. Daniel and L. M. Patrick, "Instrument Panel Impact Study." Proceedings of Ninth Stapp Car Crash Conference. Minneapolis: University of Minnesota Press, 1966.

6. D. P. Marquis and T. Rasmussen, "Status of Energy Absorption in Steering Columns." Proceedings of General Motors Corp. Automotive Safety Seminar, July 1968.

7. G. A. Ryan and J. W. Garrett, "A Quantitative Scale of Impact Injury." CAL No. VJ-1823-R34, Cornell Aeronautical Laboratory, Inc., October 1968.

8. A. W. Siegel, A. M. Nahum, and M. R. Appleby, "Injuries to Children in Automobile Collisions." Proceedings of Twelfth Stapp Car Crash Conference, New York: SAE, 1968.

9. G. M. Mackay, "Injury and Collision Severity." Proceedings of Twelfth Stapp Car Crash Conference. New York: SAE, 1968, pp. 207-219.

10. D. J. Van Kirk and W. A. Lange, "A Detailed Injury Scale for Accident Investigation." Proceedings of Twelfth Stapp Car Crash Conference. New York: SAE, 1968, pp. 240-259.

INJURY INDEX CODE

A_0 - Minor - No Hospitalization - abrasions, contusions, fracture of nasal bones, and lacerations without enough loss of blood to cause physiological damage or necessitate transfusion.

A_1 - Minor - Hospitalization - For reasons other than injuries; that is, no means of care in vicinity, at home or otherwise; age of patient; medical history (diabetic shock, insulin shock, and such).

A_2 - Minor - Hospitalization - Considered necessary for observation where patient's obvious injuries are minor, and no additional injuries are found during the observation period.

A_3 - Minor - Fracture of the small bones of the hands and feet.

B_0 - Moderate - Cerebral concussion with loss of consciousness for a short period (usually less than 1 hr) with or without minor injuries.

B_1 - Moderate - Hospitalization for repair of severe facial wounds.

C_0 - Moderately Severe - Cerebral concussion with loss of consciousness for a period greater than 1 hr without residual signs or symptoms and/or: fractured patella, fracture of the long bones of extremities, dislocation of major joints, fractured ribs -- without pneumo-thorax, fractured facial bones -- excluding nasal bones, fracture of frontal sinus, and extensive lacerations. Operative procedure necessary, other than emergency room, and including postoperative hospitalization.

C_1 - Moderately Severe - Same injury as listed above but with complications; that is, pregnancy, peneumo-, or hemothorax.

D_0 - Severe - Extended loss of consciousness; compound fractures of long bones; multiple fractures of extremities, crushing injuries to chest; skull fracture -- X-ray and symptoms, temporary loss of consciousness; multiple facial fractures; fractured vertebra (e); cardiac contusion; fractured sternum; fractured pelvis without bladder involvement.

D_1 - Ruptured spleen, liver, mesenteric tear, perforated colon, and the like.

E_0 - Critical - Shock; multiple, comminuted, compound fractures of long bones; fracture of pelvis with bladder involvement; skull fracture with extended periods of unconsciousness.

E_1 - Critical - Unconsciousness extending to state of coma.

F_0 - Fatal - Immediate.

F_1 - Total - Traceable to accident.

Society of Automotive Engineers, Inc.

This paper is subject to revision Statements and opinions advanced in papers or discussion are the author's and are his responsibility, not the Society's; however, the paper has been edited by SAE for uniform styling and format Discussion will be printed with the paper if it is published in SAE Transactions For permission to publish this paper in full or in part, contact the SAE Publications Division and the authors.

16 page booklet,   Printed in U.S.A.

851724

Mechanism of Abdominal Injury by Steering Wheel Loading

John D. Horsch, Ian V. Lau, David C. Viano, and Dennis V. Andrzejak
GM Research Laboratories
Biomedical Science Dept.
Warren, MI

ABSTRACT

The introduction of energy absorbing steering systems has provided a substantial reduction of occupant injury in car crashes. However, the steering system remains the most important source of occupant injury. Injury associated with steering assembly contact is due to high exposure; energy absorbing steering systems reduce the risk of injury for drivers when compared to the injury risk of right front passengers. Our investigation addressed loading of the upper abdominal region by the steering wheel rim using a physiological model for study of soft tissue injury.

Injury to the liver was related to the abdominal compression response associated with rim loading. Although liver injury correlated somewhat with peak abdominal compression, a better correlation was found when the rate of compression was also considered. Force limiting by the steering wheel, not by column compression, most strongly influenced the outcome of abdominal injury. The force generated by wheel-rim compression of the abdomen was insufficient to cause either column compression or significant whole-body motions. Subjects loaded by "stiff" steering wheels (having rim deformation forces greater than the abdominal compression force) exhibited greater abdominal compression, rate of compression and resulting extent of injury than did subjects loaded by "softer" wheels which deformed more and thus reduced abdominal compression.

ENERGY ABSORBING STEERING SYSTEMS introduced in 1967 were designed with a compressible element as a primary feature for occupant protection. The systems also controlled the amount of steering assembly rearward displacement in the occupant compartment in a frontal collision. Analysis of crash data by NHTSA (1)* has concluded that the energy absorbing steering system reduced the annual fatalities in frontal impacts by 12% and the critical to fatal injuries associated with steering system contact by 38%. Despite the reductions, the steering system remains the most frequent source of injury to car occupants. NHTSA (2,3) attributes more than 25% of the total automobile occupant "Harm" to steering system contact, Fig. 1.

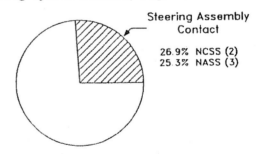

Steering Assembly Contact

26.9% NCSS (2)
25.3% NASS (3)

CAR OCCUPANT "HARM"

Fig. 1 Percent of car occupant "Harm" associated with steering assembly contact (2,3).

The presence of the steering system does not increase and may even reduce the injury risk for drivers (4) when compared to the injury risk of right front passengers. The relatively high incidence of injury associated with steering assembly contact is due to high exposure. Most car occupants are drivers -- many times the only occupant. Due to the driver's proximity with the steering assembly, the driver is more likely to interact with the steering assembly than other vehicle components -- particularly in frontal crashes -- and there is a greater risk of

---

* Numbers in parentheses designate references at end of paper.

injury since frontal crashes tend to be more severe than the general spectrum of crashes.

Driver interaction with the steering system has been the subject of many investigations which have substantially improved our understanding of driver protection by the steering system (1,5-11). With a compressible column system, the column plays the major role in force limitation and energy absorption when the driver is aligned with the steering assembly. The steering wheel also plays a major role since it is the component contacted by the occupant, and it determines the force distribution on the occupant. The steering wheel provides additional force limiting and energy absorption by deformation, particularly in situations where the driver is off-set or off-axis (6). Thus relative deformation of the wheel and compression of the column are highly dependent on the impact alignment of the occupant with the steering assembly (5,6). Crash data indicate that both column compression and steering wheel deformation are important occupant protection aspects (6). Efforts to further improve built-in occupant protection by the steering system must consider both the range of crash situations in which occupants are exposed and injured, and the response and tolerance of the various body regions contacting the steering assembly.

Injury ("Harm") related to steering assembly contact has been identified by body region (2,3), Fig. 2. The thorax has the highest portion of "Harm", followed by the abdomen (24% to 35%), and the head, which primarily involves facial injury. Injury is somewhat uniformly distributed across the range of impact severities as measured by the vehicle change of velocity (ΔV), from below 20 mph to at least 40 mph, (2,3), Fig. 3. Although the "Harm" per exposed occupant is greater at high ΔVs, the exposure frequency is greatest at lower ΔVs (3), resulting in nearly equivalent injury "Harm" over a large ΔV exposure range, Fig. 4. This suggests that high, moderate and low severity crashes

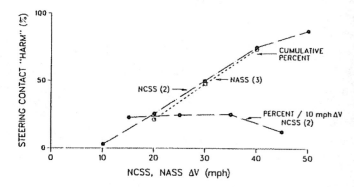

Fig. 3 Distribution of "Harm" associated with steering assembly contact as a function of vehicle velocity change (ΔV (2,3)

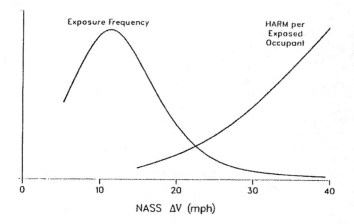

Fig. 4 NASS exposure frequency and "Harm" per exposed occupant (all sources) as a function of ΔV (3). "Harm" for each ΔV is the product of exposure frequency and "Harm" per exposed occupant.

should be considered in analysis of occupant protection systems.

In the laboratory, the response of the steering assembly depends on construction and impact stiffness of the test dummy used for the evaluation (7). Additionally, trends associated with changes in steering assembly characteristics are strongly dependent on which dummy and response is used for the evaluation (7). The Hybrid III dummy was judged to be the best of the current mechanical surrogates to study the steering system in the laboratory based on its more human-like construction and frontal impact response, and its expanded instrumentation capacity -- chest compression being an important response for study of an unrestrained driver interacting with the steering assembly (7).

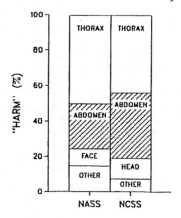

Fig. 2 Distribution of "Harm" associated with steering assembly contact by body region (2,3).

In spite of gains in understanding of driver protection and in laboratory evaluation tools such as the Hybrid III dummy, abdominal injury associated with steering contact is less well understood than thoracic trauma because current anthropomorphic dummies are not instrumented to assess injury potential for frontal abdominal loading. The steering column is elevated from the horizontal and its elevation may increase during vehicle deformation in a crash. The lower rim of the wheel can contact the driver first, depending on occupant kinematics. The liver is only partially protected by the rib cage and is a potential loading site for the lower rim.

We used an anesthetized animal to study the mechanics of upper abdominal injury from steering wheel rim contact in sled tests. 50-Kg swine,* were chosen for these experiments based on the need for a physiological model for soft tissue injury and human-like approximation of the torso and organ mass, and midtorso dimensions of the 50th percentile male. However, because of differences in the head, shoulder and pelvic regions, the swine may not be a representative human surrogate for other studies of the steering system. The swine has been successfully used as a physiological model for the study of injury mechanisms for vital organs in the torso in previous impact studies (12-16).

We investigated force, acceleration and compression response parameters in this study. The compression parameter included a "viscous" response defined as the time varying product of the percent compression [C(t)] and the velocity of compression [V(t)]. The maximum value of the "viscous" response [V(t)*C(t)]max was previously shown to correlate with soft tissue injury for frontal thoracic (17-19), lateral abdominal (20), and frontal abdominal (21) impact, where it biomechanically correlated with liver injury (19,21).

---

* The rationale and experimental protocol for the use of an animal model in this program have been reviewed by the Research Laboratories' Animal Research Committee. The research follows procedures outlined in publications by the U.S. Department of Health, Education, and Welfare, 'Guide for the Care and Use of Laboratory Animals,' or the U.S. DHEW National Institute of Health (NIH), 'Guidelines for the Use of Experimental Animals,' and complies with U.S. Department of Agriculture (USDA) regulations as specified in the Laboratory Animals Welfare Act (PL 89-544), as amended in 1970 and 1976 (PL 91-579 and PL 94-279).

METHODOLOGY

The test fixture consisted of a steel frame mounted on the Hyge sled carriage (Fig. 5). The fully anesthetized subject was supported by a suspension suit attached to a trolley at four corners. The position was adjusted to result in the lower rim loading directly in line with the liver (5 cm below the xyphoid). The trolley system consisted of four wheels suspended on two parallel rails. Slots were machined in the rails for the trolley to drop prior to subject contact with the wheel. During the impact, the surrogate was unrestrained by the tethers. To retain the subject on the sled fixture, the trolley had secondary restraints which came into play following the impact. Kinematics of the lower torso were controlled by belt restraints around each leg and anchored to a force limiter with a yield force of 2 kN. The belts did not interfere with abdominal loading by the wheel.

A sled velocity of 32 km/h was used for all the tests. The sled acceleration pulse and separation between the lower rim and the abdomen allowed the sled to reach test velocity before significant interaction of the subject with the wheel.

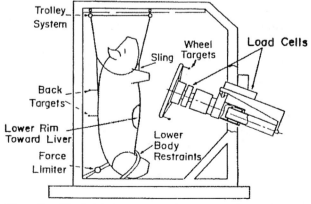

Fig. 5 Sled fixture showing mounting of steering assembly and a schematic of the sled test configuration.

THE STEERING ASSEMBLY - The steering assembly consisted of a standard compressible column and modified wheels. Column angle was 20° or 30° compared to the horizontal. The column was mounted on a bracket which was rigidly attached to the sled fixture by a triaxial force transducer. Instrumentation of the steering assembly consisted of a triaxial force transducer located between the wheel and the column and a displacement transducer for column compression.

Several steering wheels were used in this study. A simulated wheel, Fig. 6a, and two experimental versions of a two-spoke wheel, Fig. 6b, provided several levels of wheel deformation stiffness for rim loading, Fig. 7. The simulated wheel provided a relatively rigid deformation characteristic. A "stiff" version and a "soft" version of the two-spoke wheel provided two deformation stiffnesses without changing other characteristics of the wheel. Additional variations of steering assembly characteristics were studied by alternating between wheel spoke across and vertical positions for tests with the modified two-spoke wheel. Table 1 provides the test matrix for this study.

ANIMAL PREPARATION FOR THE SLED TEST - Seventeen swine weighing 49.5 ± 2.0 Kg, were restrained without excitement by an injection of ketamine (20 mg/Kg, IM) and acepromazine (200 µg, IM) followed by atropine (0.08 mg/Kg, IM). The swine were then induced to a surgical plane of anesthesia with a mixture of fentanyl (40 µg/Kg, IM) and droperidol (2 mg/Kg, IM) and maintained on an equal mixture of oxygen and nitrous oxide with 1-1.5% halothane. They breathed spontaneously in a dorsal recumbency.

A tracheotomy was performed on all subjects, a tube inserted through the tracheal incision, and the mixture of inhalation gases and oxygen was reconnected to the endotracheal tube. The location of the midsternum was determined anatomically (as midpoint between the manubrium and the xyphoid). The midsternum was exposed by cauterizing the skin and the subcutaneous layers. A uniaxial accelerometer was attached to the sternum using bone screws. Catheter tip pressure transducers were inserted into the right femoral artery and vein and guided to the level of upper abdominal viscera. ECG leads (II) were permanently sutured to the limbs for monitor during the test. The swine was then rotated 180° to rest in a ventral recumbency. The thoracic spinous process transverse to the midsternum (T3 and T4) and that transverse to a location 5 cm below the tip of the xyphoid (T11 or T12) were exposed by cauterizing the skin and the subcutaneous layers. Two spinous clamps with mounted accelerometers and photographic targets were fastened to the exposed processes.

Particular adaptational aids were necessary for the quadruped to assure basal

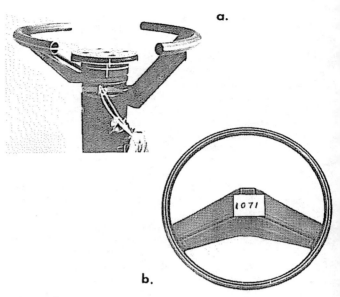

Fig. 6 Steering wheels used in the sled experiments. (a) Simulated wheel. (b) Two-spoke wheel.

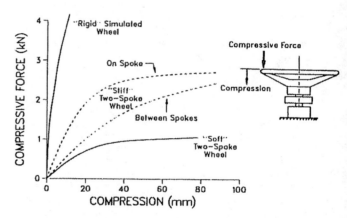

Fig. 7 Wheel quasistatic deformation characteristics for loading as shown in diagram.

physiologic functions in vertical suspension during the sled experiment. Before the suspension, an increase in blood volume was induced by intravenous infusion of 250 mL of isotonic lactated Ringer's solution to mitigate orthostatic hypotension (transient pooling of blood due to change in position). The animal was supported in a suspension suit with belts originating from four tethering points for attachment to the sled fixture (Fig. 5). The suit had adjustable abdominal supports which were securely tightened to minimize blood pooling in the pelvic girdle and the lower extremities. The suit was open at the steering wheel rim impact site. Since no surgical procedures were performed on the swine during the suspension, the concentration of halothane was reduced to maintain a level of 0.5%. With these provisions, the swine retained normal cardio-vascular physiologic function during vertical suspension.

Just before the onset of the test, the swine was disconnected from the inhalation gases and breathed room air for less than two minutes during the sled test. Immediately following the test, the swine was returned to a dorsal recumbency and inhalation gases were reconnected. The animal showed no signs of distress or sensation during or after the experiment. It was observed for 15 minutes, sacrificed by an overdose of sodium pentobarbital (>60 mg/Kg), and a truncal necropsy was performed immediately.

THE NECROPSY - The abdominal viscera were exposed by a midsagittal incision of the linea alba. Hemoperitoneum when present was noted and classified. Hemoperitoneum was considered moderate if peritoneal blood loss was less than 200 mL, hemorrhage stopped spontaneously and bleeding did not resume upon exposing the abdominal cavity to atmospheric pressure. Hemoperitoneum was considered severe if peritoneal blood loss was more than 200 mL and hemorrhage resumed as the abdominal cavity of the sacrificed animal was exposed to atmospheric pressure, indicating that hemorrhage initially stopped because peritoneal pressure equilibrated with central venous pressure. The liver, the spleen, the kidney, the stomach, and the intestines were then examined for lacerations or contusions. The number of independent tears was used as a measure of injury severity for each organ. An AIS score was assigned reflecting the overall severity of abdominal injury, the extent of concomitant hemorrhage, and possible prognosis if resuscitation were attempted.

The right and left diaphragms were incised from the abdominal cavity and possible pneumothorax or hemothorax were noted. The thoracic cavity was exposed by bisecting the left and right ribs and reflection of the sternum. Attention was paid not to interfere with any existing fractures, the number and location of which were noted. The lung, the major vessels and the pericardium were then examined for contusions or lacerations. The heart was excised from the chest. The epicardium and endocardium of all four chambers were examined for possible lacerations, contusions, and petechiae. All heart valves were similarly examined. To differentiate impact injury from possible artifacts resulting from the procedure, the necropsy was also performed on a sham operated swine which received identical treatments as the experimental swine including the vertical suspension, excluding the sled test.

RESULTS

INJURY - The impacts resulted in no cardiac arrhythmia among the test subjects. Only two animals from the "stiff"-wheel tests and the animal from the "rigid" wheel test

died within the 15-minute observation period post impact. The immediate case of death was respiratory arrest, although they also received critical liver injuries. The other animals all breathed spontaneously following the experiments.

Gross necropsy showed that liver laceration was the only abdominal injury. Liver injury ranged from none to extensive laceration. The most frequently observed lacerations were tears of the Glisson's membrane and submembrane tissue (Fig. 8a). A less frequent but more critical injury was laceration of the central venous junction between lobes (Fig. 8b). Junctional lacerations always resulted in severe hemoperitoneum, an immediate potential threat to life.

The most severe thoracic injury was multiple rib fractures. None of the thoracic injury appeared life threatening. Pulmonary injury was restricted to isolated areas of contusion. The only cardiac injury observed were occasional petechiae on the endocardium. No tearing of valves or chordae tendineae were noted. No cases of hemothorax or pneumothorax were observed. The aortic, the pulmonary and the vena caval blood vessels received no injury.

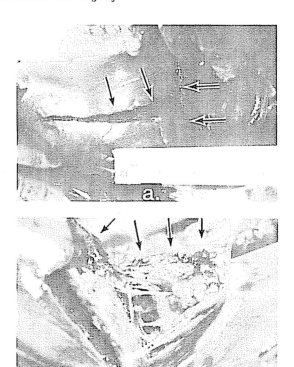

Fig. 8 Examples of liver injury from the "stiff"-wheel tests. (a) Laceration of the posterior surface of the right medial lobe. (b) Junctional laceration between adjacent lobes tearing the central venous junction.

543

Table 1 - Summary of test conditions and responses

Wheel Stiffness[1]	Column Angle	Spoke Position	Number Rib Fractures	Liver Laceration AIS	Liver Laceration Number	Abdominal Compression Maximum Percent[2]	Abdominal Compression Maximum V*C (m/s)[3]	Lower Spine Acceleration 3 ms (g)[4]
"Stiff"	20°	Vertical	12	5	10	42	1.7	41
			6	5	3	43	2.1	28
		Horizontal	8	5	5	36	1.6	43
			7	5	4	50	2.3	47
"Stiff"	30°	Vertical	7	5	10	41	2.0	--
			6	5	5	45	2.0	34
		Horizontal	10	5	7	41	1.5	55
			4	5	9	38	1.2	33
"Soft"	20°	Vertical	0	0	0	32	0.9	47
			6	0	0	34	1.0	43
		Horizontal	7	0	0	36	0.9	43
			6	0	0	33	0.8	46
"Soft"	30°	Vertical	10	4	1	33	0.9	44
			11	4	2	39	1.1	31
		Horizontal	1	0	0	36	0.7	59
			7	4	1	32	0.8	40
"Rigid"	30°		14	5	Extensive	50	2.4	32

[1] See Fig. 7.
[2] Compression normalized by subject thickness at lower rim contact point.
[3] Normalized compression multiplied by velocity of compression.
[4] Accelerometer located at T-11 or T-12.

Table 2 - Correlation of injury with steering assembly parameters. The mean amplitude is given for each level of the listed parameter variation.

Parameter	Rib Fracture Number per Subject	Liver Laceration AIS	Liver Laceration Number per Subject
Spoke position Vertical/Horizontal	7.25/6.25 p 0.57	3.75/3.38 p 0.45	3.88/3.25 p 0.53
Column angle 20°/30°	6.50/7.00 p 0.78	2.88/4.25 p 0.38	2.75/4.38 p 0.12
Wheel stiffness "Stiff"/"Soft"	7.50/6.00 p 0.40	5.63/1.50 p 0.0002	6.63/0.50 p 0.0001

Injury outcome as a function of steering wheel parameters is given in Table 1. Analysis of injury outcome as a function of steering wheel parameters is given in Table 2. Thoracic injury (number of rib fractures) was not influenced by column angle, spoke position, or rim deflection stiffness in these tests. Rib fractures ranged in number from 0 to 14.

Liver laceration was strongly influenced by the wheel deformation characteristic, a possible influence of column angle, but appears unaffected by spoke position. All eight animals in the "stiff"-wheel tests received critical to fatal liver injury, regardless of spoke position or column angle. The livers in five of the test animals were clearly irrepairable and likely fatal. Liver lacerations in these cases were extensive and involved all the major lobes. Liver laceration for the test with the "rigid" rim response of the simulated wheel was greatly increased compared with the worst case for the "stiff" version of the two-spoke steering wheel, and thus only one test was performed because of the severe outcome. By contrast, only three of eight animals in the "soft"-wheel tests received minor liver laceration. The other five animals sustained no abdominal injuries. The difference of liver injury

544

between the "stiff"- and "soft"-wheel groups
was obvious and statistically significant
judged by the assigned AIS values. The difference was even greater based on the actual
number of liver lacerations. Since hemoperitoneum resulted from the liver lacerations,
there was also a parallel difference in the
occurrence and the severity of hemoperitoneum
between the "stiff"- and "soft"-wheel groups.

STEERING INTERACTION MECHANICS - The
interaction mechanics can be viewed from
laboratory (stationary) coordinates (Fig. 9),
steering assembly coordinates (Fig. 10), or
subject coordinates (Fig. 11). The sled was
accelerated to test velocity before significant interaction of the rim with the abdomen,
while the surrogate remained stationary. The
rim would have compressed the abdomen at constant (test) velocity except for possible
whole body motion of the surrogate, or motion
of the rim relative to the sled by column
compression or wheel deformation. However,
the force generated by the rim compressing
the abdomen was insufficient to produce
significant whole body motion of the surrogate or column compression, which were initiated only by the greater forces associated
with hub loading on the thorax. Hub contact
occurred about 15 ms after abdominal contact.
The only mechanism reducing abdominal compression response at test velocity before
thoracic loading was wheel deformation.
Since the "rigid" and "stiff" wheels
exhibited relatively little wheel deformation, most of the sled displacement resulted
in abdominal compression. The "soft" wheel
exhibited relatively large deformation
(Fig. 10) which reduced the velocity and
extent of abdominal compression (Fig. 11).

CORRELATION BETWEEN INJURY AND MEASURED
RESPONSES - Abdominal injury represented by
AIS correlated significantly with the maximum
"viscous" response [V(t)*C(t)]max, and ade-
quately with maximum abdominal compression,
[C(t)] max, (Table 3). There was no correla-

Table 3 - Correlation of AIS liver injury
with response parameters

	Correlation Coefficient (R)	p-Value
3 ms Lower Spine Acceleration	0.46	0.0712
Maximum Abdominal Compression	0.62	0.0077
Maximum [V(t)*C(t)]	0.72	0.0012

***Probability that the value of the
correlation coefficient (R) or greater will
occur when in fact no correlation exists.**

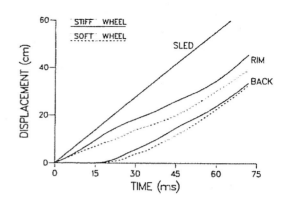

Fig. 9 Movement of the sled, the lower rim
of the wheel, and the back of the
surrogate in laboratory coordinates.
Time zero is contact of the lower rim
with the abdomen. Figures 9, 10 and
11 are from the same matched pair of
tests having a "stiff" and a "soft"
version of the two-spoke wheel.

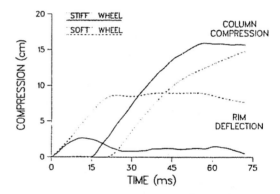

Fig. 10 Column compression and rim
deflection (relative to wheel hub)
measured in the direction of the
column axis as a function of time
from rim contact.

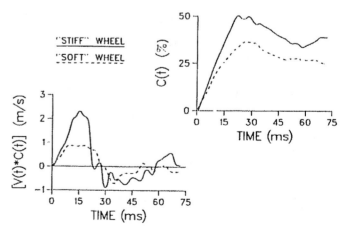

Fig. 11 Abdominal compression normalized by
initial anterior-posterior thickness
at contact location and the derived
"viscous" response [V(t)*C(t)] as a
function of time from rim contact.

tion between abdominal injury and maximum upper or lower spinal acceleration (measured directly opposite to the impact site) nor with maximum sternal acceleration, or arterial or venous pressure. This suggests that dummy injury assessment for abdominal loading by the steering system should be based on measurement of the abdominal deformation as a function of time.

DISCUSSION

TEST ENVIRONMENT - A wide range of occupant alignments and impact severities and directions relative to the steering assembly occur in car crashes. Variations in vehicle design, vehicle crash kinematics, and vehicle crush all affect loading conditions for the occupant. Human factors such as occupant size, seating position, posture and reactions to an imminent crash provide additional variation. The response of the steering assembly strongly depends on impact severity and occupant alignment (5,6). The test environment in this study was chosen to focus on abdominal loading by the lower rim. The torso of the surrogate was upright with slight forward bowing before contact. This alignment of the surrogate to the steering assembly promotes interaction of the lower rim with the abdomen by delaying the contact between the hub and the thorax. Such an impact environment should be considered as one of various possibilities, depending on factors such as the knee impact with the vehicle interior. The test results were independent of the sled acceleration vs. time profiles because test velocity was reached prior to abdominal compression. The test severity provided a range of injury outcomes depending on the steering assembly parameters.

TEST SUBJECT - The 50-Kg swine was chosen for these experiments because we needed a physiological model which offered a good approximation of human function and size. At the midsternum and the impact site, the anterior-posterior dimensions of the experimental model were 27.2 ± 1.7 cm and 26.3 ± 1.5 cm, respectively. That compares favorably with similar dimensions of a 50th percentile male. The subject mass also approximated that of a 50th percentile anthropomorphic dummy interacting with the steering system. Slightly greater column compression occurred with the test subjects than for a seated Hybrid III dummy in a similar sled exposure, suggesting a slightly greater effective mass for the swine. The pelvic structure of the swine differs from that of the human and cannot be restrained similarly by a lap belt. The swine also lacks human-like knees which interact with the instrument panel and help to restrain the occupant in a crash. In a previous study where lap belts were used on swine (22), fatal abdominal injury resulted from belt

loading. In the present study, we used an independent belt around each leg which restrained the lower torso with no abdominal loading. Together with the special adaptational support for the vertically suspended quadruped, the swine appears to be a good injury model for upper-abdominal loading by the rim under these controlled conditions. The swine was successfully used as an injury model for the torso in previous impact studies (12-16). One other study suggested that the swine was an unsuitable surrogate because of inconsistent hardware performance (22), a problem not encountered in our study.

INFLUENCE OF STEERING SYSTEM PARAMETERS - Among the steering wheel parameters, wheel stiffness had the greatest influence on the severity of abdominal injury. The spoke vertical configuration represents a stiffer contact interface than the spoke across configuration (Fig. 3) but no difference in injury was observed. The spoke vertical configuration of the "stiff" wheel had a greater abdominal contact area on impact, abdominal loading was distributed on the rim and the spoke, whereas it was concentrated on the rim in the spoke across configuration. Therefore, the effects of increased wheel stiffness and increased contact area may have compensated for each other in the spoke vertical configuration. The greater interaction force over a larger contact area resulted in no apparent difference in abdominal injury or kinematics. Spoke position did not affect the extent of wheel deformation, another indication of a compensating mechanism.

Column angle had no measurable influence on injury outcome for the "stiff" wheel -- all subjects had an AIS 5 liver injury. However, column angle appears to be a factor for the "soft" wheel tests -- at least on an AIS basis. AIS increases from 0 for no injury to 4 for any liver laceration but can be no greater than 5 for extensive liver laceration. Thus AIS is not a sensitive indicator of the extent of liver laceration. The number of liver lacerations does not suggest a strong correlation of injury with column angle. The loading severities represented by the "stiff" and "soft" wheels are near the injury extremes. Analysis of the influence of other factors might be more meaningful at an intermediate loading severity.

ABDOMINAL PROTECTION MECHANICS - The respective kinematics of the "stiff"- and "soft"-wheel tests explain the reduced abdominal compression and rates of compression and thus the reduced "viscous" response and injury by the "soft" wheel. There was no column compression and rearward displacement of the spine until after 15 ms of abdominal contact by the lower rim of the wheel. For that 15 ms, only the wheel or the abdomen could deform to account for the instantaneous sled displacement. Since the "soft" wheel deformed significantly more than the "stiff" wheel (Fig. 10) the instantaneous abdominal compression and velocity of compression (Fig. 11) were similarly reduced in the "soft"-wheel tests. Even the "stiff" wheel had appreciable deformation at this time; and an increase in wheel stiffness further exacerbates abdominal injury as exhibited by the "rigid" wheel test. We observed that the maximum "viscous" response occurred before column compression initiated, and hypothesize that liver injury is occurring before column compression. Thus abdominal protection must be provided by steering wheel characteristics.

In our test environment, the lower rim contacted the abdomen before the hub contacted the thorax. Column compression and significant whole body motions were associated with the greater force due to thoracic loading. A wheel with a deep dish would delay thoracic contact and might increase the potential for abdominal compression. Occupant alignment with the steering wheel at contact is an important parameter.

In our tests abdominal compression by the rim did not produce sufficient force to cause column compression. If column compression force is reduced to protect the abdomen from rim loading, the column would have a significantly lower energy absorbing capacity than it has today which would affect the higher severity collision protection of drivers well-aligned with the steering wheel. Severe frontal crashes require high energy absorption capacity which implies high column compression forces. Thus the wheel, not the column, is probably the steering system component best suited to limit compression of the abdomen. For our test situation and results, the data suggest that the rim and spoke should deform before sufficient force is developed to produce a high level of "viscous" response and abdominal compression. However, the test environment represents only one of many occupant interaction configurations. Thus, other loading conditions and body regions should be considered in a full analysis of occupant protection provided by the steering system.

ACKNOWLEDGMENTS

Many persons have contributed to the reported study. In particular the authors express appreciation to Donald Barker, Mary Foster, Richard Gasper, Gerald Horn, Edward Jedrzejczak, Joseph McCleary and Kathleen Smiler for their technical support and contributions.

REFERENCES

1. C. J. Kahane, "An evaluation of federal motor vehicle safety standards for passenger car steering assemblies," DOT HS-805 705, National Highway Traffic Safety Administration, Washington, DC, 1981.

2. A. C. Malliaris, R. Hitchcock, and J. Hedlund, "A search for priorities in crash protection," SAE Paper 820242, 1982.

3. A. C. Malliaris, R. Hitchcock, and M. Hansen, "Harm causation and ranking in car crashes," SAE Paper 850090, 1985.

4. S. Parks, "Relative risk of driver and right front passenger injury in frontal crashes," GMR-4802, August 3, 1984, Transportation Research Dept., General Motors Research Laboratories.

5. J. D. Horsch, K. R. Petersen, and D. C. Viano, "Laboratory study of factors influencing the performance of energy absorbing steering systems," SAE Paper 820475 (SP-507), 1982.

6. J. D. Horsch and C. C. Culver, "The role of steering wheel structure in the performance of energy absorbing steering systems," SAE Paper 831607, 1983.

7. J. D. Horsch and D. C. Viano, "Influence of the surrogate in laboratory evaluation of energy-absorbing steering systems, SAE Paper 841660, 1984.

8. J. W. Garrett and D. L. Hendricks, "Factors influencing the performance of energy absorbing steering columns in accidents," Fifth International Technical Conference on Experimental Safety Vehicles, 1974.

9. P. F. Gloyn and G. M. Mackay, "Impact performance of some designs of steering assembly in real accidents and under test conditions," Proceedings of 18th Stapp Car Crash Conference, SAE, 1974.

10. D. F. Huelke, "Steering assembly performance and driver injury severity in frontal crashes," SAE Paper 820474, 1982.

11. C. W. Gadd and L. M. Patrick, "System versus laboratory impact tests for estimating injury hazard," SAE Report 680053, 1968.

12. J. P. Verriest, A. Chapon, and R. Trauchesses, "Cinophotogrammetrical study of procine thoracic response to belt applied load in frontal impact -- Comparison between living and dead subjects," SAE Paper 811015, 1981.

13. D. C. Viano, C. K. Kroell, and C. Y. Warner, "Comparative thoracic impact response of living and sacrificed porcine siblings," SAE Paper 770930, 1977.

14. D. C. Viano and C. Y. Warner, "Thoracic impact response of live porcine subjects," SAE Paper 760823, 1976.

15. M. E. Pope, C. K. Kroell, D. C. Viano, C. Y. Warner, and S. D. Allen, "Postural influences on thoracic impact," SAE Paper 791028, 1979.

16. C. K. Kroell, M. E. Pope, D. C. Viano, C. Y. Warner, and S. D. Allen, "Interrelationship of velocity and chest compression in blunt thoracic impact," SAE Paper 811016, 1981.

17. D. C. Viano and V. K. Lau, "Role of impact velocity and chest compression in thoracic injury," Journal of Aviation, Space and Environmental Medicine, 54(1):16-21, January, 1983.

18. V. K. Lau and D. C. Viano, "Influence of impact velocity and chest compression on experimental pulmonary injury severity in an animal model," Journal of Trauma, 21(12), December, 1981.

19. D. C. Viano and I. V. Lau, "Thoracic impact: A viscous tolerance criteria," 1985 NHTSA Symposium on Experimental Safety Vehicles, Oxford, England, June 1985.

20. S. W. Rouhana, I. V. Lau, and S. A. Ridella, "Influence of velocity and forced compression on the severity of abdominal injury in blunt, nonpenetrating lateral impact," Journal of Trauma, 25(6), 1985.

21. V. K. Lau and D. C. Viano, "Influence of impact velocity on the severity of non-penetrating heptatic injury," Journal of Trauma, 21(2):115-123, February, 1981.

22. R. G. Snyder, J. W. Young, and M. Q. Doyle, "Biomechanical evaluation of steering wheel design," SAE Paper 820478, 1982.

23. V. K. Lau and D. C. Viano, "An experimental study of hepatic injury from belt-restraint loading," Journal of Aviation, Space and Environmental Medicine, 52(10):611-617, October, 1981.

Methodologies and Measuring Devices to Investigate Steering Systems in Crashed Cars

Clyde C. Culver
GM Research Labs.
General Motors Corp.

ABSTRACT

Post-crash conditions of a car's steering system, when properly measured and documented, provide an insight to the interaction between the driver and the steering system that occurs during a frontal car crash. Steering system conditions were investigated in two interrelated phases: 1. Deformation of the wheel rim, spokes, and hub, and 2. Compression resistance force of the steering column. Two devices were developed to document the "crash loading" response of these two segments of the car's steering system. One device was designed to measure the deformations of the steering wheel and the other the force required to further compress the steering column. An initial test series on 19 "crashed" cars "field tested" the devices, developed the test techniques and procedures needed for in-depth studies, and formulated necessary data handling methods and data collection forms. The column loading device described here was primarily developed for columns incorporating an EA element but should function with other types of steering columns as well.

INTRODUCTION

Accident investigators frequently attempt to relate the post-accident condition of the steering system to the the driver's injuries from frontal accidents. Probably, the two most obvious and important factors to be considered by the investigators are the deformation of the steering wheel and the column "compressed" condition. The wheel deformation relates to the kinetic energy the driver's body expends in deforming the wheel structure (1)*, to the nature of the loading interaction surface between the driver's body and the wheel, and to the deformation resistance of the wheel structure to loading. The compressed condition of the column relates to the force exerted on the column by the driver during the crash, the driver's kinetic energy absorbed by the column, the driver's kinematics, the construction of the column, and also the interaction of driver and other components of the vehicle because the compression of the column permitted the driver to move farther forward in the vehicle and form added load paths from the driver to the vehicle.

Currently, in-depth column investigations require the removal of the column from the vehicle and the testing of the column in the laboratory. Laboratory conditions are significantly different from those of the crashed car and the information obtained relates basically to only the column. In-place column loading may reveal column mounting condition as well as column crashed load angles and steering wheel and column interaction with the instrument panel and other components. Data on the crash behavior of the various steering system components are needed for a comprehensive understanding of the driver injury mechanisms.

Some researchers (2) have reported that with the conventional methods of steering system deformation description that they found no correlation between reported driver in jury and reported steering system damage. Earlier

* Numbers in parentheses designate references at end of paper

researchers (3), using a computer analysis of field data, stated that the variable "steering wheel rim damage" significantly related to occupant injury for the one vehicle studied.

Other investigators (4) have reported that steering system peak loading force along with effective load area appear to relate field accident deformations and laboratory tests. The two devices described in this paper obtain values related to these conditions of peak loading and loading area.

The post-crash measurement of the steering column angle has often been difficult to obtain and questionable in previous accident investigations(5). Proper column positioning and a low initial column loading by the column force device tends to position the column in more nearly the final crash position so that more meaningful column angles can be obtained.

RATIONALE

Determining the post-crash conditions of the car's steering system can be performed in several phases and in several sequences. Each phase will provide detailed information on a segment of the performance of the steering system in real world collisions.

It is often informative if, for example, the the wheel, spoke, and hub deformations can be measured with the wheel in the crashed condition, especially if there is interaction between the rim and the instrument panel. Normal procedure is to measure the steering wheel deformation after the wheel has been removed from the vehicle. In some studies, the wheel may remain in place while the column is being loaded and the rim deformation frequently measured at significant load levels. Measurement points on the steering wheel rim are arbitraily designated as the upper central curvature of the face of the rim.

The column also may be loaded in various ways: with the wheel in place, with the wheel removed, or a combination as desired. The mounting of the device permits some variation of the input angle of the force cylinder for a simulation of the driver's loading on the system.

Measurements made post-crash of the deformed steering wheel and of the column force resistance level cannot be directly related to the forces on the wheel and column that occurred during the crash as there are inertial forces involved in the force level during the crash and elastic deformations of the system which cannot be measured post-crash. Measurements made on crashed test cars and comparable measurements on field accident cars may, in the future, permit meaningful relationships to be made between the field accident data and test data.

Although direct correlations may not be made at this time between the steering system's post-crash condition, the dynamic deformation of the system, and the driver's injuries, an in-depth under standing of the steering system interaction with the driver during a frontal crash may permit a better understanding of the injury mechanisms that the driver is subjected to and mechanisms to dissipate the driver's kinetic energy.

TEST DEVICES

WHEEL MEASURING DEVICE - The device to to measure the steering wheel deformation is basically an adjustable framework of three scales that are mounted on an indexing hub on the wheel axis and can be rotated around the wheel to measure the xy points on the rim, spokes, and hub face (Fig.1). The mounting hub contains twelve clock position detents for measuring reference. When the wheel has a

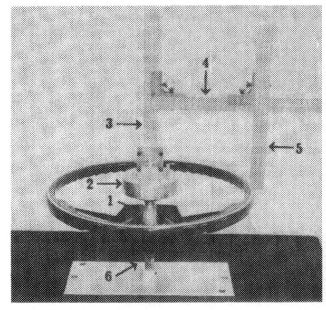

Figure 1 Wheel measuring device in use.
 1. Indexing head shaft/nut
 2. Indexing head
 3. Axial column scale
 4. Radial scale
 5. Vertical rim scale
 6. Short steering column shaft

significantly deformed energy absorbing hub, the mounting hub cannot be mounted in the normal way but must be mounted to an adaptor frame (Fig.2).

Figure 2 Wheel measuring device being used with adaptor for EA hubs.

COLUMN FORCE DEVICE - The device to apply an axial force to the column is basically a 8.9 kN capacity hydraulic cylinder mounted on an adjustable structure which is usually mounted on the vehicle "B" pillars (Fig.3). In some cases where the "B" pillars may not be structurally sound, the adaptors can be mounted on the rear quarter panel. A large diameter aluminum telescoping cross shaft between the "B" pillar adaptors supports a mounting block and a perpendicular arm to support the forcing cylinder and accept the reaction force. The force cylinder is mounted on a parallel linkage on the mounting arm and its reaction force measured with a mechanical proving ring (Fig.4). Usually the device is mounted directly behind the column hub, however some lateral adjustment is permitted for crash load simulation (Fig.5). Considerable force cylinder angle adjustment is available in the verticl plane.

LIMITATIONS OF THE TEST DEVICES - We found in our field proof tests of the devices that some vehicles were so severely crashed that the devices,

Figure 3   Column force device mounted on adaptors clamped to the "B" pillars.
1. Extender rod
2. Forcing cylinder mounting arm
3. Torque arm
4. Cross shaft
5. Cross shaft mounting block
6. "B" pillar adaptors
7. Morehouse proving ring
8. Parallel mounting link for force cylinder
9. Force cylinder

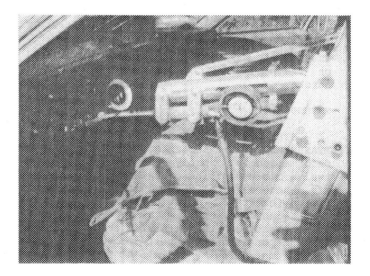

Figure 4   Hydraulic forcing cylinder mounted on parallel links that permit axial force to be measured by Morehouse proving ring.

Figure 5 Cross shaft mounting block permits lateral adjustment. Arm mounting for forcing cylinder can be adjusted up and down the face of the block and torque arm length can be adjusted to provide desired force arm angle. Arrows indicate cross shaft locking pins.

especially the column forcing device, could not be used to give meaningful measurements. Usually the wheel can be removed from the vehicle and measured. Obviously when the column has been completely compressed, force on the column cannot produce added compression. If the column is in a position almost perpendicular to the forcing vector, a force at the top of column can do little to compress the column. If the driver's door cannot be opened or is not open, examination of the shear capsules and the EA unit may be virtually impossible or the readings of a questionable nature.

METHODOLOGY

STEERING WHEEL DEFORMATION MEASUREMENTS - Steering wheel rim deformation is normally measured at the twelve clock positions, at the spoke-to-rim junctures, at maximum and minimum rim deformation positions, and at significant sharp rim bends. Measurements can be taken with the wheel in place or with the wheel removed from the column, See Appendix "A" for the detailed use procedure. Field data are recorded as shown in Table I. Rim and spoke fractures can be documented by footnotes on the data sheet.

STEERING COLUMN FORCE VS.COMPRESSION MEASUREMENTS - Steering column resistive force can be measured with the wheel in place or with the wheel removed. Initial column position measurements should be carefully docummented under light loading prior to the application of significant force. In some cases the column may have detached from the shear capsules and must be lifted into an approximate "crash" position prior to measurements and loading (Fig.6). The steering column forcing device is installed in the vehicle and used as per the procedure detailed in Appendix "B". The column is normally loaded in 0.5 kN increments and the shear capsule readings and energy absorbing device dimensions taken before increasing the column loading to the next loading level. Field measurements are recorded on a data sheet as in Table II.

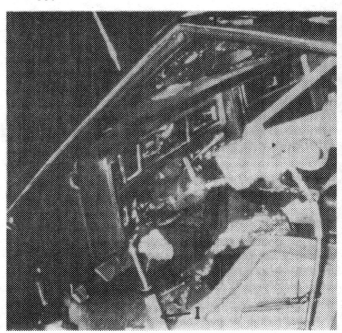

Figure 6 Column lift hydraulic cylinder to position column prior to loading. Arrow (1)

RESULTS

STEERING WHEEL MEASUREMENTS - All steering wheel measurements must be compared to similar measurements made on a new undeformed wheel before the true deformation can be determined. The corrected measurements are then ready for data processing. Sample data sheets and data are shown in Table I and Table III.

STEERING COLUMN MEASUREMENTS - Steering column measurements are basically the force exerted on the column vs. the column travel measurements and the column condition

Table I

SAMPLE DATA SHEET
FOR WHEEL RIM, SPOKE, AND HUB DIMENSIONS

Rim "clock" Position	Measured values				Corrected dimensions*			
	Rim		Hub Face		Rim		Hub	
	Radial (mm)	Axial (mm)	Radial (mm)	Axial (mm)	Radial (mm)	Axial (mm)	Radial (mm)	Axial (mm)
12	120	295	–	–	-54	-165	–	–
1	122	298	–	–	-52	-168	–	–
2	145	280	–	–	-29	-150	–	–
3	177	212	–	–	+3	-82	–	–
4	179	137	–	–	+5	-7	–	–
5	163	101	–	–	-11	+29	–	–
6	155	86	–	–	-19	+44	–	–
7	155	100	–	–	-19	+44	–	–
8	161	102	–	–	-13	+28	–	–
9	171	147	–	–	-3	-17	–	–
10	164	218	–	–	-10	-88	–	–
11	138	274	–	–	-36	-144	–	–

* Corrected by comparison to a new wheel (Table III)

Table II

SAMPLE DATA SHEET FOR COLUMN LOADING

Cylinder force (kN)*	Shear capsule separation distance (mm)	Energy absorption device measure (mm)	Force cylinder angle		Column angle		Total shear capsule separation (mm)	TotalEA device compression (mm)
			xz	xy	zx	xy		
0.0	8	121	13	0	22	0	8	5
0.89	8	121	–	–	–	–	8	5
1.78	8	120	–	–	–	–	8	5
2.67	9	120	–	–	–	–	9	6
3.11	9	120	–	–	–	–	9	6
3.56	9	120	–	–	–	–	9	6
4.00	10	119	–	–	–	–	10	7
4.45	11	118	–	–	–	–	11	8
4.89	12	118	–	–	–	–	12	8
5.38	16	114	–	–	–	–	16	12
5.78	23	108	–	–	–	–	23	23
5.69	27	104	–	–	–	–	27	22
4.48**	38	95	–	–	–	–	36	31
3.20	51	81	12	0	25	0	51	45

Notes: * Cylinder force used to calculate axial column force,
lateral force, and force perpendicular to the column
in the xz plane
** Column out of shear capsules at 4.49 kN load.

Table III

SAMPLE DATA SHEET FOR WHEEL RIM DIMENSIONS
FOR NEW STEERING WHEEL

Rim "clock" Position	Radial dimension (mm)	Measured axial dimension (mm)	Dimension to hub face (mm)
1 2	167	123	77
1	171	125	75
2	176	126	74
3	176	123	77
4	174	129	71
5	170	129	71
6	169	129	71
7	172	127	73
8	176	125	75
9	176	124	76
10	173	122	76
11	166	122	78

Ave. = 172 mm wheel radius

measurements of column angle and in some cases the column mounting movement. Important loading phases such as the point at which the shear capsules release, the initiation of the energy absorbing unit compression, and interaction between attachments on the column and the instrument panel must be carefully documented.

DATA ANALYSIS

Major data analysis is performed on mainframe computers. However, the data must be properly organized initially in data sets.

STEERING WHEEL DATA

Rim deformation measurements - After the measured rim deformation has been corrected to those of a new wheel they are ready to be plotted vs. the clock positions of the wheel. Any "abrupt" rim angle change can be calculated using the "peak" of the bend and the axial dimensions of two known rim points, one an each side of the bend.

Spoke deformation measurements - Spoke deformations are normally described using the spoke-to-rim measurements.

Energy absorbing hub measurements - Energy absorbing hub face measurements

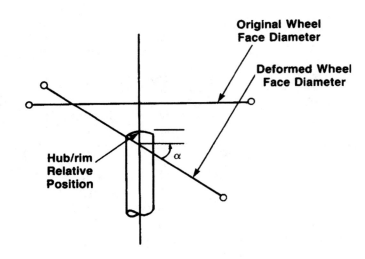

Figure 7 Hub/rim position calculations

can be documented as to slope of the hub face, the direction of the hub face slope, and the offset of the hub face with its direction of offset.

Hub/rim relative position - Hub/rim relative position calculated values are determined from the rim deformation values of diametrically opposite rim readings. Calculations are performed as indicated in Figure 7. The hub face location relative to the rim is required from a new wheel. In some cases, the steering wheel hub cover may be easily deformed under a rather light loading and the dimensions from the rim to the more rigid hub structure can be used to compute an added hub/rim position value.

STEERING COLUMN DATA

STEERING COLUMN FORCE VS. COMPRESSION MEASUREMENTS - Each incremental force applied by the forcing cylinder to the upper end of the steering column is separated into an axial force along the axis of the column and into forces that are perpendicular to the column by calculations based on the cylinder's angle and the column's angle. The data is then in a form that can be plotted for displacements vs. axial column force (Fig.8).

DISCUSSION

The devices have been used in over 19 field accident crashed cars of various types and found to be simple to use and produce unique reliable measurements of the steering wheel deformation and of the force to compress the column under quasi-static conditions. Steering column compression load may serve two purposes for the investigator: to give an indication of the loading forces related to the static force deflection characteristics of the column and to provide the compression at which to start the compression vs. force plotting (Fig.8). The cylinder force at the top of the column must be broken down into the force along the axis of the column, the lateral force on the column, and the force perpendicular to the column and in the vertical plane. Hub/rim relative position values pretain

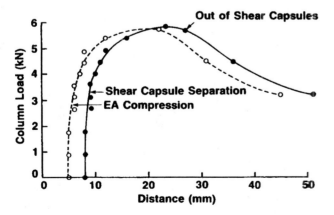

Figure 8 Sample plot of shear capsule separation and EA compression vs. column loads

to the driver/steering wheel interface and provides for a numerical indication of the deformed steering wheel surface presented to the driver. The calculated values are useful as individual values as descriptors of the wheel surface and also, when summed, as a total value for the wheel surface.

If the analyzed data is to be entered in a computer data file to be subsequently examined for interactive relationships, a relatively rigorous analysis procedure must be followed for each case examined.

SUMMARY

1. Both the steering wheel measuring device and the column forcing device have demonstrated an unique ability to numerically identify the degree of steering wheel deformation and the static force necessary to compress the column.
2. Both devices have been found to be simple to use and to produce accurate objective information in a reasonable length of time (1 to 1 1/2 hours).

REFERENCES

l.J.D.Horsch and C.C.Culver, "The Role of the Steering Wheel Structure in the Performance of Energy Absorbing Steering Systems", Society of Automotive Engineers Inc., Twenty-Seventh Stapp Car Crash Conference, San Diego, Ca., SAE paper # 831607, Oct., 1983.
2. D.F.Huelke, "Steering Assembly Performance and Driver Injury Severity in Frontal Crashes", Society of Engineers Inc., International Congress and Exposition, Detroit, Michigan, SAE Paper # 820474, Feb. 22-26, 1982.
3. L.Phillips, A.khadilkar, T.P.Egbert, S.H.Cohen, and R.M.Morgan, "Subcompact Vehicle Energy-Absorbing Steering Assembly Evaluation", Twenty-Second Stapp Car Crash Conference, SAE # 780899.
4. P.F.Gloyns and G.M.Mackay, "Impact Performance of Some Designs of Steering Assembly in Real Accidents and Under Test Conditions", SAE paper # 741176
5. J.W.Garret and D.L.Hendricks, "Factors Influencing the Performance of the Energy Absorbing Steering Column in Accidents", Fifth International Technical Conference on Experimental Safety Vehicles, June 1974.

APPENDIX "A"

INSTRUCTION MANUAL FOR STEERING WHEEL MEASURING DEVICE - Note of caution! Crashed vehicles should be carefully inspected prior to the investigation for fuel leaks, electrical hazards, broken glass, etc.. Safety glasses are a must!

Setup for "in-place" steering wheel measurements

1. Remove steering wheel cover pad.
2. Remove steering wheel nut.
3. Install indexing hub shaft/nut and tighten firmly.
4. For deep hubs or EA hubs it may be necessary to attach an optional 100 mm extension shaft to the indexing hub shaft.
5. Install indexing head and lock in position with the 9 o'clock detent position at approximately 12 o'clock on the wheel.
6. Install axial column scale and lock in position (Fig.1).
7. Attach radial scale to the axial column scale with a 90 degree adaptor. Set it with the lower edge of the scale at 123 mm on the axial column scale. (This positions the lower edge of the radial scale at 200 mm above the wheel hub face.) Inner end of scale is at centerline of the steering column.
8. Attach vertical rim scale to the radial scale with a second 90 degree adaptor.
9. Set the indexing head to the 12 o'clock detent loosen the indexing head lock, position the vertical scale directly to the 12 o'clock position on the wheel rim and lock the indexing head in position.

Measurement procedure

1. With the indexing hub set at 12 o'clock, move the vertical rim scale to position the corner of the scale at the top surface and center of the rim (Fig.1).
2. Read radial and vertical scale values and record.
3. Next, release the vertical rim scale, lift the scale, release the index detent and reposition to the 1 o'clock position. Take readings.
4. Continue readings around the wheel with added readings at maximum and mminimum rim positions and at each spoke-to-rim point.
5. If the wheel has a deformable EA hub, an added fourth scale attached to the vertical rim scale will permit the measurements of the face of the deformed hub to be obtained. When the wheel has a severly deformed hub, it can be mounted on a special fixture for measuring (Fig.2).
6. If there is significant interaction between the wheel rim and the instru-ment panel, the segment of the rim interacting with the panel should be designated and recorded also.

Setup for "out-of the-car" steering wheel measurement - In most cases the steering wheel is removed from the column before it is measured. In these cases a short steering column shaft is inserted into the wheel for the indexing head to be attached to (Fig.1). Steps 3 through 9 as described in the "in-the-car" procedure are followed. Measurements of a new undeformed wheel, necessary for determining actual wheel deformation, is performed in a similar manner.

Maintenance of the wheel deformation measuring device - When not in use the scales, angle brackets, and detent hub should be kept in the storage box. After each usage each component must be carefully cleaned.

APPENDIX "B"

INSTRUCTION MANUAL FOR COLUMN FORCING DEVICE - The installation of the column forcing device can be separated into several distinct phases.
1. Normally the device is attached to the "B" pillars of the vehicle with two clamp-on units (Fig.3).
2. With the "B" pillar adaptors securely installed, the telescoping cross shaft is installed. Before the cross shaft is placed in the vehicle it is necessary to check that the end screws in the cross shaft are turned well into the ends of the shaft (Fig.9). The cross shaft is positioned on the "B" pillar adaptor pins by placing the "captive screw heads" over the pins. Next, the cross

Figure 9 Details of cross shaft mounting-to-"B" pillar adaptor.
  1. Threaded adjustment shaft

shaft end screws are extended to bring the locking pin holes in the cross shaft into alignment for the installation of two locking pins (Fig.5). When the locking pins are in place, the cross shaft end screws are securely tightened outward on the adaptor pins. These should be checked frequently during the loading operation.

3. The cross shaft mounting block for the forcing cylinder mounting arm is attached on the cross shaft directly behind the end of the steering column and securely clamped in place (Fig.5). It is positioned on the cross shaft with the torque arm mount at the top of the block and the forcing cylinder mounting arm to the front of the block.

4. The forcing cylinder mounting arm is attached with a bolt/pin to the cross shaft block (Fig.5). The torque arm positions the forcing cylinder relative to the cross shaft.

5. The forcing cylinder mounting arm, the forcing cylinder with the parallel linkage mounting, and the Morehouse proving ring force measuring unit are normally mounted as an assembly.

6. If the column has compressed sufficiently to permit the column to "drop down" out of the shear capsules, it is lifted into approximately the position of final loading with a hydraulic cylinder (Fig.6) and maintained in that position with as light a load as possible during the test. The lifting cylinder should be as nearly perpendicular to the column as possible to avoid interference with the column as it compresses.

7. A proper length extender rod is selected and the forcing cylinder attached to the column with a threaded "U" joint adaptor (Fig.10). If the

Figure 10 The "U" joint connection between the force cylinder and the top of the column

steering wheel is removed before the test, install the dummy wheel hub in place of the steering wheel.

8. Normally the force cylinder is adjusted to an angle of 10 to 15 degrees to the vehicle level. Two provisions are made for this adjustment. The forcing arm mounting block can be positioned up or down on the face of the mounting block and/or the torque arm can be adjusted for length.

9. Installation must be made with a slight clearance between the back of the forcing cylinder and the load sensing face of the Morehouse proving ring to preclude any preload. Also, the parallel linkage must be checked so there is clearance between the cylinder and the mounting arm for the cylinder to articulate and load the proving ring (Fig.4).

10. After the installation is rechecked, the pump hose is attached to the cylinder and the unit is ready for loading.

Measurement procedure

1. A low level initial load is applied to the column to remove "slack" from the system and the initial readings taken. These include shear capsule separation, EA unit compression, and column and cylinder angles (4 angles - see Table II).

2. Loadings are made on the column in approximately 0.5 kN increments and measurements made at each loading. Any changes in the column mounting, column-to-instrument panel interaction, and shear capsule status must be carefully documented. (At higher loads, > 2.5 kN, the loading should be reduced before entering the vehicle to perform "under the instrument panel" measurements.)

3. Column loading is continued until a specified loading or column travel has been reached.

Maintenance of column forcing device
Relatively little maintenance is required on this unit. Occasionally a small amount of hydraulic fluid may be required in the hydraulic pump. A small quantity of grease should be applied to the adjustment screws on the ends of the cross shaft and to the "U" joint attachment to the column. When tests are completed each component should be cleaned of sand, broken glass, etc. before it is returned to the storage box. A simple dead weight calibration should be performed on the proving ring periodically depending on usage.

557

912890

History of Safety Research and Development on the General Motors Energy-Absorbing Steering System

John D. Horsch, David C. Viano, and James DeCou
General Motors Research Labs
Biomedical Science Dept.
Warren, MI

This paper covers the development of the General Motors Energy Absorbing Steering System beginning with the work of the early crash injury pioneers Hugh DeHaven and Colonel John P. Stapp through developments and introduction of the General Motors energy absorbing steering system in 1966, evaluations of crash performance of the system, and further improvement in protective function of the steering assembly. The contributions of GM Research Laboratories are highlighted, including its safety research program, Safety Car, Invertube, the biomechanic projects at Wayne State University, and the thoracic and abdominal tolerance studies that lead to the development of the Viscous Injury Criterion and self-aligning steering wheel. Also discussed are engineering efforts of the Saginaw Steering Gear and Oldsmobile Divisions, the extensive testing program at the GM Proving Ground, government interactions and regulations, and the field accident reports of the many lives saved by Energy Absorbing Steering Systems.

THE AUTOMOBILE INDUSTRY HAD BEEN INTERESTED in safety long before the introduction of the Energy Absorbing (EA) Steering System in 1966. General Motors developed such early safety features as the speedometers first installed on 1901 Oldsmobiles and the first electric headlamps on their 1908 models. In the 20's and 30's, dashboard lights, electric windshield wipers, dual tail lamps, electric turn signals and steel body structures were also introduced. The auto industry's first proving ground was opened by GM at Milford, Michigan in 1924, and the first barrier impact tests and rollover tests were conducted there in 1934 to evaluate the structural crashworthiness of their automobiles (Yanik 1983).

During these early days of automotive safety engineering and through the 1950's, the emphasis was on accident avoidance, handling and braking, but to a lesser degree on crash impact engineering of vehicle interiors. Highway safety advocates, the courts, and the general public were convinced that if cars were designed for safe normal operation, drivers were well trained and licensed, and roadside hazards were removed, then few if any accidents would occur (Eastman 1973). However, accidents were occurring and people were being killed and injured by impacting interior structures such as the instrument panel, windshield and steering assembly. At least a half dozen patents for compressible steering systems appeared in the late 20's and 30's (Figure 1), but these ideas were not developed at the time because of the lack of measures for impact protection and the common attitude that the best way to avoid injuries was to eliminate accidents.

The effect of interior structural design on accident pathology was first studied in small aircraft crashes. When airplanes first got off the ground, most pilots believed that accidents were a part of operating these machines and that injuries sustained in the accidents were attributed to either bad luck or acts of God. In 1917, when a young American cadet named Hugh DeHaven (1969) in the Canadian Royal Flying Corps suggested to his Commanding Officer that "the luck could be changed by better engineering and design," the officer told him that "the hand had been dealt as it was and that changing anything would amount to stacking the deck." He also told the cadet "that flying was dangerous and the best way to prevent injuries was to stay on the ground." DeHaven had recently sustained serious abdominal injuries in a crash that killed the other three cadets involved. He realized that he survived because his was the only cockpit that remained relatively intact (Hasbrook 1964). He also identified the narrow pointed buckle on his safety belt as the cause of his near-fatal injuries (DeHaven 1969). It was then that DeHaven first considered the concept of designing the plane for crash survival and proposed a crash injury research program.

DeHaven's ideas were rejected by a) the pilots who thought that accidents were all due to fate, b) the safety groups who said that money used for research would save more lives if it were put into extra pilot training, and c) those who considered the human body to be a fragile object that would be cut in two by a safety belt. DeHaven (1969) could not easily disprove these first two points, but he had witnessed enough airplane and automobile crashes to know that the body could withstand tremendous forces under certain conditions.

To demonstrate the strength of the human body, in 1938 DeHaven began an engineering and pathological study of survival in falls from heights of 50 to 150 feet. He presented eight of these cases in 1942 in the American Medical Association's journal, War Medicine. In this paper, DeHaven (1980) stated that "the fact that these survivals occurred when the necessary factors were accidentally contributed is strong evidence of the large increase in safety which can be provided by design." This study was the catalyst that gained him the support of the Civil Aeronautics Board and the National Research Council in setting up the Crash Injury Research (CIR) project at Cornell University Medical College in 1942 (DeHaven 1969).

DeHaven's first work at CIR was a field study of injuries in non-fatal small plane crashes. He reported his initial findings to the American Society of Mechanical Engineers in 1943 in which he stated two main conclusions: "(a) The force of many accidents now fatal is well within physiological limits of survival" and "(b) Needless injuries-- both serious and fatal-- are caused by the unfortunate placement and design of certain objects and structures." He identified the control wheel as a structure causing intrathoracic lesions in crashes, and also called for the support of physicians in assisting accident investigators in a 1946 editorial to The Journal of the American Medical Association (DeHaven 1946). By 1950, the effect of DeHaven's work could be seen in the numerous crash survival design features in new airplanes (Hasbrook 1964).

Several physicians were becoming interested in this idea of injury reduction through design,

especially in automobiles. Detroit plastic surgeon Dr. Claire L. Straith was well aware of what the car interior could do to the human body in an accident and added seat belts and crash padding to his own car as early as the 1930's. He achieved only limited success, however, with automobile manufacturers in implementing his ideas (Eastman 1973). Dr. Fletcher D. Woodward of the University of Virginia Hospital called for specific design changes in his "Medical Criticism of Modern Automotive Engineering" presented at the annual meeting of the American Medical Association in 1948. Woodward (1948) demonstrated among other things that "crushing injuries to the chest of the driver occur when he is driven against the steering wheel," and recommended the installation of a hydraulic steering column that would move forward on impact. By the mid-50's, more physicians had begun discussing ways to combat automobile accident injuries, including Drs. Jacob Kulowski (1955) of St. Joseph, Missouri, C. Hunter Shelden (1955) of Pasadena, California, and Horace E. Campbell (1954) of Denver, Colorado.

Campbell was critical of automobile manufacturers for not following the example of the aircraft industry, although several of the changes made in cockpits were also seen in car interiors by the early 50's, such as better positioning of instrument panel control knobs (Gandelot 1951). In his paper "Deceleration, Highway Mortality, and the Motorcar," Campbell (1954) alleged that "it is possibly about time that we build the motorcar to crash," although GM had been running barrier impact and rollover tests for 20 years to study the crashworthiness of its cars. He also stated: "Motorcar manufacturers advertise at great cost the thrills and satisfactions of increased acceleration. It is time that they point out in their advertising the

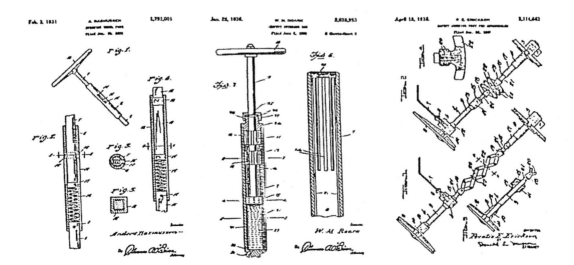

Figure 1: Three examples of compressible steering assembly patents from the 1930's.

even more abiding satisfaction of controlled deceleration." In 1955, however, the widespread idea that "safety doesn't sell" was emerging, as the safety-promoting campaign of Ford Motor Company's Robert McNamara was dropped after only three months, when Ford sales fell further behind those of its' stylish competitors (Eastman 1973). At that time, consumers wanted style and performance, and as sales showed, safety was not an important selling point.

Another physician who was interested in crash injury research and the effects of mechanical force on living tissues was Air Force Colonel John P. Stapp (Figure 2), an M.D. with a Ph.D. in biophysics. From 1946 to 1958, he studied the effects of deceleration on both humans and animals at two facilities he established for the Air Force. Stapp exposed belt restrained volunteers including himself to deceleration forces of up to 40 times the force of gravity (40 g) using rocket sleds at the Aeromedical Facility at Edwards Air Force Base in California (Stapp 1966) and the Aeromedical Field Laboratory at Holloman Air Force Base in New Mexico (Snyder 1970). This work produced invaluable data for design engineers and also aroused tremendous interest in impact research (Hasbrook 1964). Stapp became involved in automobile crash research at the Holloman facility in 1953 and two years later, he invited the seat belt standards committee of the Society of Automotive Engineers and others involved in crash research to view car crash experiments and other demonstrations. This was the first Stapp Car Crash Conference, which is now an annual event (Eastman 1973).

In 1948, Corporal Elmer C. Paul of the Indiana State Police, in talking with a representative of Hugh DeHaven's Crash Injury Research about small plane crashes, became interested in undertaking a similar project on automobile accidents. By 1951, with the help of CIR, Paul had set up a statewide study of rural fatal accidents. In a 1952 report on 495 of these accidents, 50% were classed as survivable and the steering wheel assembly was identified as a major cause of fatal injuries (Eastman 1973). Although it had its weaknesses, DeHaven was impressed with the Indiana project, and together with Paul, representatives of the automobile manufacturers, and officials of several states, set up a nationwide study of injury-producing accidents. This would be the first large scale acquisition of automobile field accident statistics. In March, 1954, this project became Automotive Crash Injury Research (ACIR) located at the Cornell Aeronautical Laboratory in Buffalo, New York and headed by John O. Moore (Eastman 1973).

Two developments within ACIR were instrumental in arousing public interest in automobile safety and convincing the manufacturers to design and build protective interiors. In 1952, Liberty Mutual Insurance Company of Boston contracted with Edward R. Dye of the Cornell Aeronautical Laboratory to develop several automobile safety concepts and combine them into a crash-proof car capable of withstanding a 50 mph impact. Of the 88 safety features on the "Cornell-Liberty Safety Car" (Figure 3), there was a hydraulic steering mechanism controlled by levers and covered with a large chest pad. The car's design was announced in 1956 and a prototype was built and put on display, receiving much publicity (Cornell, 1957). A second car, "Survival Car II", was designed and built by Cornell for Liberty Mutual in 1961 and featured a deformable steering assembly (Figure 4).

In 1955, ACIR made its first attempt to rank injury causes. The steering assembly was found to be the leading cause of injury based on exposure, frequency of injury, and degree of injury (Schwimmer, Wolf 1962). This initial report and others that followed from ACIR (1961) that similarly identified the steering assembly helped focus the priorities of automobile manufacturers. Once the auto industry had sufficient understanding of the issues and had public support, work was begun to introduce an energy-absorbing steering system and other safety items into their products.

DESIGN QUESTIONS

In designing an energy absorbing steering system, many questions had to be answered. The first and most important was whether or not a compressible steering system would improve driver safety. It was known from ACIR (1961) studies and other reports in the mid-50's (White 1955) that the number of injuries and fatalities experienced by drivers was significantly less than by right front passengers, especially in mild to moderate accidents. This was supported by an ACIR report at the Ninth Stapp Conference in 1965 that specifically compared the injuries sustained by the driver and the right front passenger in injury-producing accidents in which both occupants were present (Kihlberg 1965). The conventional steering system was therefore acting as a safety device by providing restraint to the driver in certain crashes. If the steering system were to simply collapse, the driver would respond like the right front passenger since his head would be free to interact with the windshield and instrument panel.

Many studies, however, including one by ACIR in 1961 and another by Donald F. Huelke and Paul W. Gikas (1965) of the University of Michigan Medical School, have shown that in more serious accidents, the driver was sustaining very serious and often fatal injuries to the thorax due to impacting the steering system. The ideal steering system therefore should stay intact in mild to moderate accidents in which the load to the chest is below a certain injury threshold in order to restrain the driver, and in severe crashes it should compress and absorb the impact energy while limiting the load to a level just below that which causes injury.

An important question immediately arose -- What was this load limit for the chest? In 1959, when GM began its work on EA steering systems, there was a general lack of knowledge on the survival parameters of the human body. Important work had been done by DeHaven and Stapp and a few others, but there was much more that needed

Figure 2:   Colonel John P. Stapp.

Figure 3:   Cornell-Liberty Safety Car.

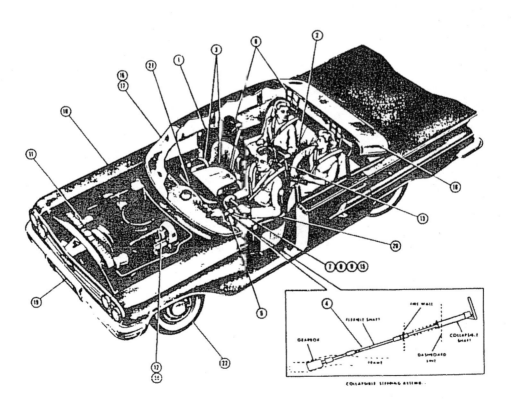

Figure 4:   Liberty Mutual "Survival Car II."

to be learned about the tolerance of specific body regions for impact and restraint loadings. In order to obtain this needed data, the General Motors Research Laboratories (GMR) became involved in biomechanics research.

Another question at the time concerned the rearward axial displacement of the steering column relative to the passenger compartment during severe frontal collisions. In an impact to a rigid barrier, if the vehicle front end was crushed sufficiently to deform the frame, the steering gear and attached shaft would come to a stop before the passenger compartment, thus causing relative movement of the shaft towards the driver. The question concerned whether or not this was detrimental to the driver. Some people argued that the collision of the car with the barrier, the so-called first collision, occurred before the second collision between the driver and the steering wheel, and thus the relative rearward displacement of the column aided in restraining the driver (Abersfeller 1965). However, Snyder (1953) reported that "axial movement of the steering column as a result of crash impact is not unusual, and such movement is potentially dangerous." This question was not even fully answered when Federal Motor Vehicle Safety Standard Number 204 limiting steering control rearward displacement was issued (NTSA 1967).

Besides these questions of safety and biomechanics, there were also many engineering questions that had to be answered. The most important consideration for the engineers was that the steering system still had to fulfill its main function -- that of steering the car. Keeping this in mind, they could then think about the problems of designing the system to absorb energy in a crash. First, the EA element could be put in the wheel, column, or vehicle structures, and if the column were used, various column components could function as the energy absorber. The steering column is made up of the shaft, which connects the steering wheel to the steering gear, the shift tube in some columns, which connects the gear shift lever to the transmission, and the mast jacket, which gives structural support to the column and attaches it to the instrument panel.

If, for example, the EA element were in the mast jacket, the shaft and the shift tube would have to telescope and the instrument panel attachment would have to release to allow the assembly to compress. Also, there are many different mechanisms for absorbing energy, each having unique force-displacement relationships, vibrational and inertial properties, torsional strengths, and many other parameters. These and numerous other engineering questions had to be answered during the developmental stages of this project.

INITIAL GM INVOLVEMENT

In the early 1950's, Charles W. Gadd of the General Motors Research Laboratories became convinced of the need for General Motors to develop a crash injury research program and that GMR was the corporate unit where this should be carried out. In 1955, the Special Problems Department utilized its experience with transient vibration theory, its rapid computation facilities, and its strain gage and vibration instrumentation to carry out a project for improving the design of dash pads and other features for increasing highway crash protection. This was among the first scientific investigation within GM to design the car interior to protect occupants in the event of a crash.

Safety research activities intensified at GMR in 1959 including methods of controlled energy absorption, a means of absorbing energy while limiting the load involved to some maximum value, and maintaining this value throughout the range of displacement. This was considered an ideal load-displacement curve from the standpoint of efficiency, because for a given maximum load, energy absorption, which is the area under the curve, would be maximized for a given displacement. This led to the invention of the Invertube, a ductile, thin-walled aluminum or steel tube which is progressively turned inside out when loaded axially (Figure 5). Three different configurations for gripping and inverting these tubes were developed (Figure 6). All three types of Invertubes (Kroell 1962) were made up in various dimensions and tested on a new gravity drop impact testing facility (Figure 7).

The potential usefulness of the Invertube as an energy absorbing device was immediately recognized and incorporated into an experimental energy absorbing steering wheel (Figure 8). The assembly incorporated an Invertube in the wheel hub to minimize the inertia forces the driver had to overcome to initiate compression, and included a flat wheel with a large rigid padded hub zone to provide maximum contact area for distributing the load. Prototypes were tested on a newly designed drop test apparatus, in which an angled simulated body block was dropped on a vertically mounted wheel/column assembly (Figure 9).

The compressible wheel proved to be markedly superior to a conventional wheel also tested on the facility, based on acceleration-time histories recorded by transducers in the body block. The test was unrealistic, however, because the momentum vector of the falling body block was directed along the column axis. Tests were then set up at the Proving Ground in which an anthropomorphic dummy was dropped on an experimental wheel/column assembly inclined to the dummy's momentum vector (Figure 10). The assembly did not compress axially but instead experienced bending failure (Figure 11). The wheel was redesigned for added bending strength and in subsequent tests, axial compression was achieved without bending.

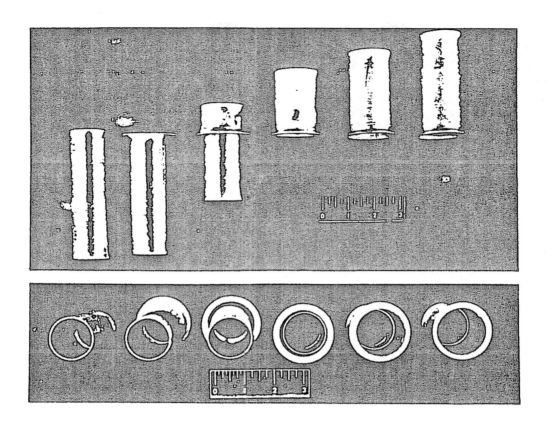

Figure 5: Progressive stages of Invertube formation and inversion.

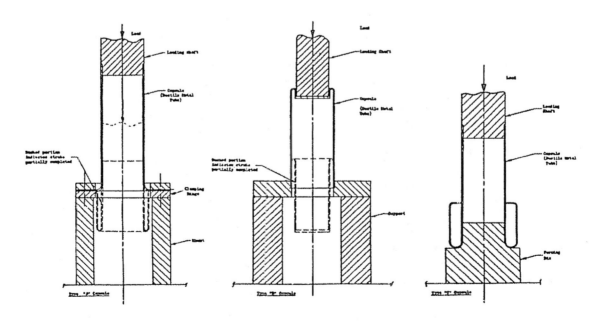

Figure 6: Three Invertube configurations developed by Kroell.

563

Figure 7:   Invertube drop test facility.

Figure 8:   Invertube EA steering wheel.

Figure 9:   Drop test of Invertube wheel.

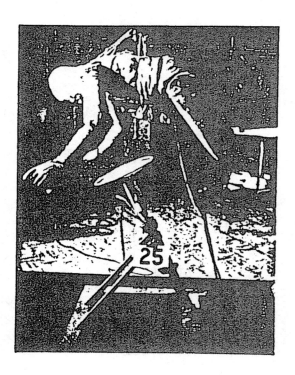

Figure 10:   Dummy drop test on Invertube wheel.

The assembly was then tested at the Proving Ground Snubber Facility to establish its in-vehicle performance (Figures 12 and 13). The Snubber test was originated at the Proving Ground in 1955 as a means of testing one vehicle component at a time without the costly destruction of an entire car in a barrier impact test (Stonex and Skeel 1963). In this test, a stripped-down car with a reinforced frame, equipped with the experimental steering assembly and a dummy driver, had its rear end connected by a long cable to a hydraulic energy absorbing device called a Snubber (Figure 14). This test vehicle was brought up to speed by a tow car connected by a breakaway link. When it got to the end of the cable, the test car was quickly decelerated by the Snubber with a deceleration pulse that could be regulated between 3 g and 35 g. The deceleration-time curves were very similar to those measured in barrier tests. This in-vehicle testing was found to be more realistic compared with drop tests for steering system impact evaluation.

During 1959, to demonstrate the feasibility of possible safety improvements and promote their crash injury research program, the Special Problems Department designed and built a concept car from a 1959 Buick that incorporated 24 safety features,

Figure 12:   Installation of Invertube wheel into Snubber car.

Figure 11:   Failed Invertube wheel after dummy drop test.

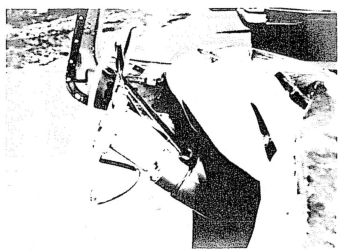

Figure 13:   Deformed wheel after Snubber test.

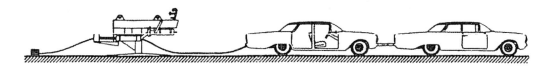

Figure 14:   Proving Ground Snubber test facility.

including an Invertube compressible steering wheel (Figure 15). Stylistically, the wheel was designed by the GM Styling Section with the large flat hub area specified by Special Problems (Figure 16). The car was displayed in late 1959 along with demonstrations of the Invertube and the compressible wheel (Figures 17 and 18). GM executives in attendance were impressed with these innovative concepts and expressed interest in the further development of several of the ideas, especially the compressible steering system.

With the support of the Corporation and after having viewed the GMR display themselves, Saginaw Steering Gear engineers began work to incorporate the Invertube as an energy absorbing element in the steering column. Whereas Special Problems had incorporated the Invertube in the steering wheel, Saginaw moved it into the column as part of the steering shaft for two reasons. The first was simply that Saginaw made columns, not wheels. Second, by locating the compressible element in the shaft below the instrument panel, the relative rearward displacement of the steering column during severe frontal crush could be controlled while still absorbing the energy of the driver. Such placement also was a means for controlling bending loads applied to the invertube element.

GM Research continued the testing of its EA steering wheel design in 1960 using both drop tests and Snubber tests. However, after Saginaw Steering Gear became involved, the Special Problems Department helped them get started running Snubber tests and then relinquished further engineering work to them because of their greater manpower and unmatched experience with the steering system. Special Problems, in a joint venture with Wayne State University's Biomechanics Laboratory, then began work on a task of equal importance to the development of the EA Steering System -- that of determining the tolerance of the human chest to frontal (anteroposterior) impact loading.

ENGINEERING DEVELOPMENT

Saginaw Steering Gear was the General Motors division responsible for most of the engineering work on the Energy Absorbing Steering Column. After seeing the Invertube with its nearly constant load-displacement characteristics and its application in GMR's compressible steering wheel, Saginaw engineers decided to incorporate this safety feature into their product. Their first idea in early 1960 for attaching the steering shaft to an Invertube is shown in Figure 19. However, this shaft had much too large a diameter to be practical in a steering column. It was redesigned to fit inside a column assembly (Figure 20), and in mid 1960, prototypes of this shaft were built and tested by Saginaw.

The Invertube steering shafts were not tested for their dynamic axial compression and energy absorption characteristics until they were tested for the more important function of steering the car. The shafts were tested statically for axial load and torsional yield and dynamically for torsional fatigue.

A major problem with the early designs was the attachment of the steel shaft to the aluminum Invertube. In 1960, Saginaw attempted to make a solid aluminum shaft that would be continuous with the hollow Invertube (Figure 21). They first tried a soft aluminum shaft but it failed the static torsion tests. A shaft of harder aluminum passed the static tests but failed the torsional fatigue tests. These designs were never snubber tested because they were not considered functionally adequate for steering the car.

A steel shaft and aluminum Invertube design was finally found that performed well in the torsion tests and also exhibited desirable energy absorption properties in dynamic tests. The next step was to design the other components of the column to allow the shaft to carry out its energy absorption function. The shift tube and mast jacket were both designed as two overlapping parts that would telescope under relatively low axial loads. The instrument panel mounting bracket was designed as a simple breakaway element between the column and panel. These components were also tested for their performance under both normal and impact conditions. By late summer, 1960, a complete steering column assembly was drawn up (Figure 22), ready for in-vehicle testing.

Toward the end of 1960 with the help of GMR's Special Problems Department, Saginaw ran the first snubber test of the Invertube Column prototype at the Proving Ground. It was designed to compress up to 6 inches at a load of 650 pounds. A 1959 Oldsmobile snubber car carried an unrestrained Sierra 165 pound dummy driver containing two accelerometers in the head and four in the chest. The car was snubbed at 26.5 mph causing a 19 g deceleration peak for the car. The dummy chest experienced a 38.6 g peak. The impact of the dummy bent the column and caused it to jam on the instrument panel (Figure 23) after only 1.25 inches of compression.

The instrument panel mounting was reworked to reduce interference for the second test, but the column again only compressed 1.25 inches as it bent and jammed this time on the dash bracket. For the third snubber test, the column mount was modified by adding a track and roller guide system (Figure 24) which allowed up to 5 inches of compression. This system solved the problems of bending and dash interference, and the column compressed the maximum distance available. This iterative process of testing and redesigning continued as many more experimental Invertube Columns were snubber tested at the Proving Ground during 1961 and the first half of 1962.

By early 1962, the Invertube Column was performing fairly well in the snubber tests and appeared to be the EA steering system that would eventually go into production. In anticipation of this demand for Invertubes, cold extrusion dies and aluminum tubing were ordered from Alcoa, Aluminum Company of America. These Invertube-forming dies had to be specially made and took about a year to produce, slowing the development process.

Figure 15:   1959 GMR Safety Car.

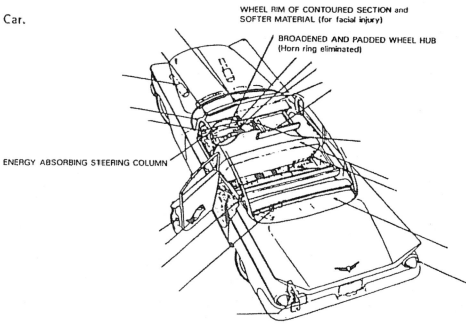

WHEEL RIM OF CONTOURED SECTION and
SOFTER MATERIAL (for facial injury)

BROADENED AND PADDED WHEEL HUB
(Horn ring eliminated)

ENERGY ABSORBING STEERING COLUMN

Figure 16:   Invertube steering wheel installed in
GMR Safety Car.

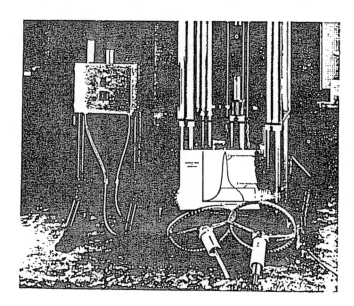

Figure 18:   1959 Invertube wheel display.

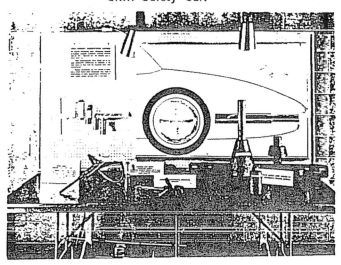

Figure 17:   1959 Display.

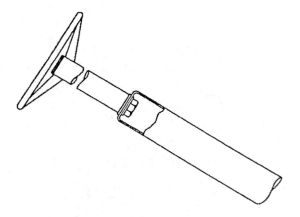

Figure 19:   Invertube steering shaft.

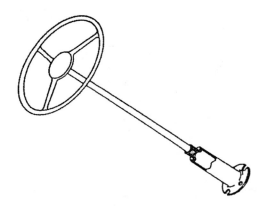

Figure 20:   Redesigned Invertube shaft.

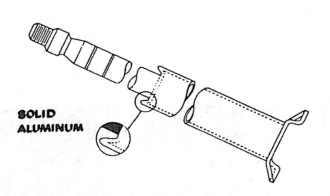

SOLID
ALUMINUM

Figure 21:   Solid aluminum shaft continuous
with Invertube.

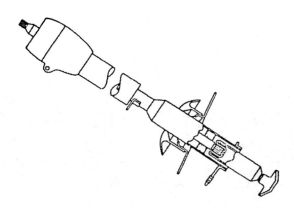

Figure 22:   Complete steering column with
Invertube shaft.

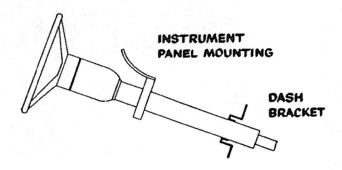

INSTRUMENT
PANEL MOUNTING

DASH
BRACKET

Figure 23:   Steering column mountings.

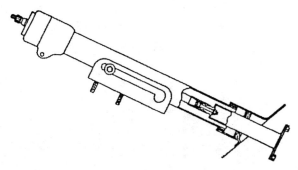

Figure 24:   Column mounted with track and
roller guide system.

In April of 1962, an Invertube Column was installed in a 1962 Oldsmobile both for road testing and for demonstrating the feasibility of the installation to corporate executives and the car divisions. The car was driven and evaluated by GM engineers for six months, after which the column was removed and examined for fatigue damage. GM executives and the carlines were impressed with the installation and encouraged Saginaw to continue their development work. Cadillac Motor Division became so interested in the project that they ordered some columns previously tested by Saginaw to be rebuilt for testing in their own laboratory.

A new method for analyzing steering system impact began to be used in 1963 as testing and design modification of the Invertube Column continued. In 1962, the Proving Ground installed the Impact Sled (Figure 25) originally to test seat belts and later expanded to test other safety features (Cichowski 1963). This vehicle crash simulator, the first of its kind in the auto industry, consisted of a car body buck containing anthropomorphic dummies mounted to a sled on a track (Figure 26). The sled was accelerated backwards by firing a high pressure air piston, causing the stationary dummy driver to be impacted by the moving steering assembly. This simulated a barrier crash by exposing the dummies to the same dynamic loading conditions as would have occurred in such a collision having a deceleration-time pulse and velocity change of equal amplitude but opposite direction to those of the sled test.

The Impact Sled was more repeatable and much less costly than the barrier impact, although barrier tests were still necessary to analyze rearward relative column displacement due to front end crush. Being an indoor facility illuminated by 64 floodlights, the Impact Sled was superior to the Snubber because it could be operated independent of outside weather and lighting conditions. It also made possible more precise instrumentation for data accumulation. During the development of the EA Column, such a large number of sled tests had to be run that a sled test simulator had to be constructed (Cichowski 1963). The simulator consisted of a hammer, divided into seven load areas to represent different regions of the torso, dropped on a steering system aligned to a proper driver/column angle.

By 1963, Saginaw engineers were aiming for a compression load of 1000 pounds, which the Invertube was capable of achieving. However, when it was dimensioned to the proper wall thickness and diameter to reach this load, the torsional strength was reduced to a level well below anything that had ever been used to steer a full-sized car. Other problems were also appearing. In some impact tests, the tube wall tore causing the column to simply collapse without absorbing energy. In other tests, if the dummy struck the wheel sufficiently off center, there was a tendency for the telescoping jacket to bind up and not compress. Also, there was concern about a potential danger of the Invertube separating from the shaft, possibly due to electrolytic corrosion between the aluminum and steel parts, thus leaving the driver without steering.

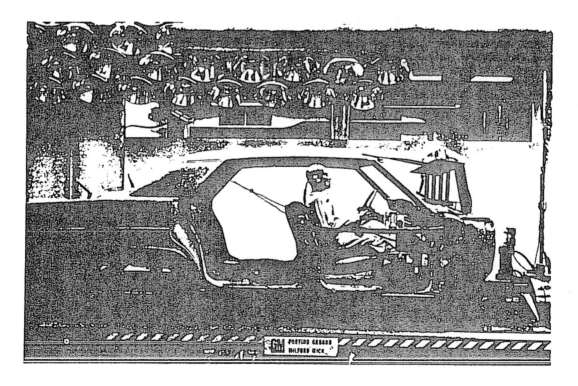

Figure 25:   Proving Ground Impact Sled.

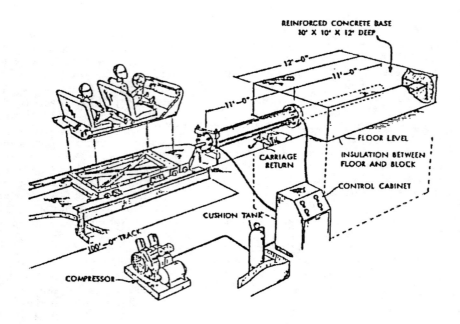

Figure 26:   Impact Sled set-up.

For these reasons, the Invertube steering shaft design was dropped in early 1964, and all the dies received from Alcoa were scrapped.

Although the Invertube Column never made it past the experimental stage, it was very important in the development of the EA Column that eventually did go into production, because it set up certain technical design considerations: 1) provide approximately 6 inches of compression distance; 2) eliminate binding of telescoping components; 3) allow passage of the column under the instrument panel; 4) preserve steering function after moderate frontal impact; and 5) provide a strong mounting to minimize bending effects influencing axial compression (Marquis 1967). Saginaw engineers also decided from the work on the Invertube shaft that the energy absorbing element should be in the mast jacket, because the shaft and shift tube already performed specific moving functions whereas the mast jacket only had a static support function under normal conditions.

Saginaw had investigated a few other methods of energy absorption during their involvement with the Invertube System, so when that project was dropped, they already had some knowledge of alternative mechanisms. In early 1960, they tried pressing a hexagonal shaft into a round tube (Figure 27) which showed good torque capacity, but its compression load of 3000 pounds was too high to be used in an EA steering column. A second design attempt was a jacket with slits cut longitudinally, resulting in local expansion of the diameter when compressed axially (Figure 28).

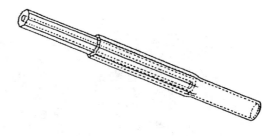

Figure 27:   Hexagonal shaft into round tube.

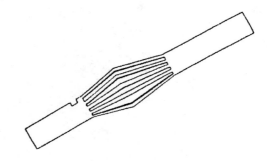

Figure 28:   Japanese Lantern jacket.

Called the Japanese Lantern, the design was first conceived in 1960 by a Saginaw engineer originally assigned to the project. It was not tested however, until August, 1963 when a technician at the Proving Ground asked for permission to work on an EA column design during his spare time. It was then that he ran the first tests of the Japanese Lantern on the Impact Sled. Another design was tried in late 1964 in which a mast jacket of two concentric interfering tubes would absorb energy when it telescoped (Figure 29). By this time, however, most of the testing at the Proving Ground and design work at Saginaw was going into development of the Japanese Lantern concept.

By early 1965, Saginaw had defined what they thought were the necessary parameters for implementing the Japanese Lantern mast jacket in an EA steering column (Figure 30). In early 1965, Saginaw Steering Gear presented a Japanese Lantern design to the Corporation that would compress up to 6 inches under a 700 pound load. A week later, an Impact Sled and Barrier Crash testing program was set up for this column at the Proving Ground. These tests were proposed for mid 1965 but before they could be run, a telescoping shift tube and steering shaft had to be designed.

The first shift tube design consisted of two concentric tubes joined by a press fit hexagonal overlap for torsional stability during normal function (Figure 31). This had problems with space and durability so other designs were attempted (Figure 32). Finally, a telescoping design containing two key-hole stamped recesses (Figure 33) injected with acetal resin plastic for torsional and axial integrity was settled upon. An axial force of approximately 100 pounds was required to shear the plastic keys and allow telescoping to occur (Marquis 1967).

The proposed steering shaft, known as the "Double D", was made up of two rods with cold-formed, half-rounded ends connected by a tubular element that was magnetically shrunk in place (Figure 34). This design allowed the shaft to telescope when loaded from either end, and maintained steering function after compression had occurred. However, excessive shaking problems were discovered in testing, prompting the development of other designs (Figure 35). Although not ready for the mid 1965 tests, a flattened upper shaft in a flattened lower tubular shaft design was settled on in the Fall of 1965 (Figure 36). Plastic was injected into the annular space to eliminate noise caused by looseness at the connection (Marquis 1967).

After the testing of the Japanese Lantern Column was completed, the results were carefully examined. A compressed column assembly from one of these tests is shown in Figure 37. The column performed well in reducing steering column load on the dummy and limiting relative rearward displacement, but it was not known if 6 inches was enough compression distance when load was applied to both ends by occupant contact and front-end deformation. In a July meeting it was decided that more compression was better, so the 6 inch

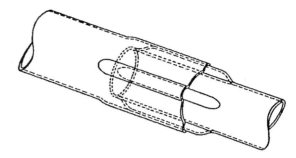

Figure 29: Telescoping EA jacket.

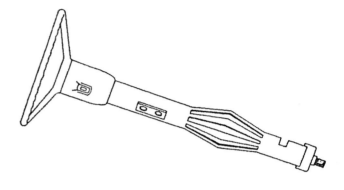

Figure 30: Complete column with Japanese Lantern.

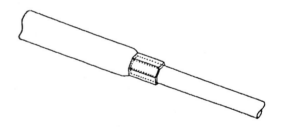

Figure 31: Telescoping shift tube with hexagonal overlap.

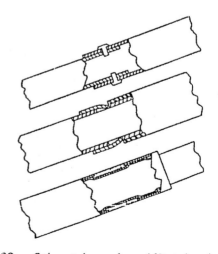

Figure 32: Other telescoping shift tube designs.

distance was increased to 8 inches (Figure 38). It was also decided that jacket diameters of 1-1/2 inches, 2 inches, and 2-3/8 inches accommodated all GM models. A meeting between engineers from Saginaw Steering Gear and other parts of the Corporation defined that as much as 11 inches of compression distance might be needed to prevent rearward displacement and provide energy absorption for all GM cars and light trucks. It was then decided that they would simultaneously pursue programs of 8 inch and 11 inch compression distances.

In order to make the EA Steering System available on 1967 models, which came out in September of 1966, preparation had to be made immediately in order to tool the plant for production. The Corporation suggested that since the Japanese Lantern performed reasonably well in tests, that Saginaw begin tooling for that design with a 6 inch compression distance. However, the engineers argued convincingly that more development was needed, so the design and testing iteration continued.

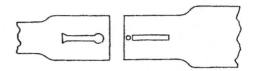

Figure 33:   Key-hole stamped recesses in shift tube sections.

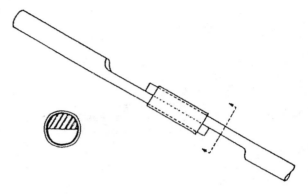

Figure 34:   "Double D" steering shaft.

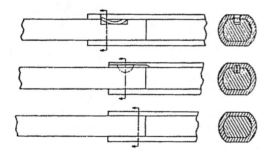

Figure 35:   Other steering shaft designs.

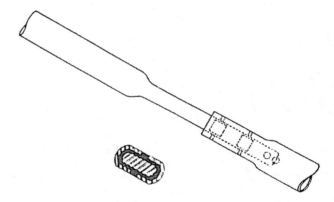

Figure 36:   Flattened tubular steering shaft design.

Figure 37:   Compressed Japanese Lantern jacket.

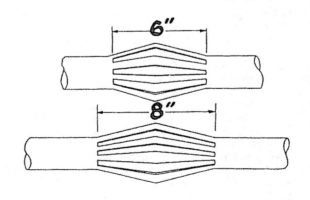

Figure 38:   6" and 8" Japanese Lantern jackets.

572

Increasing the compression distance intensified and aggravated certain problems with the Japanese Lantern: 1) Non-uniform load characteristics; 2) Lack of space for the compressed column; and 3) Decreased structural rigidity. This prompted revised designs of the Japanese Lantern idea (Figure 39), which eventually led to the development of the diamond-shaped Mesh Column jacket in the Fall of 1965 (Figure 40). The Mesh Column with its multiple convolutions was an improvement over the Japanese Lantern in that the compressed diameter was not as large, and also that another convolution could be added if more compression distance was found to be needed.

In late 1965, the Corporation approved Saginaw Steering Gear tooling for a 10-1/2 inch compressing Mesh Column with six convolutions (Figure 41). Two months later, however, a need for more structural stability led to the reduction from six to five convolutions, which lowered the compression distance to 8-1/4 inches (Figure 42). Not only did the jacket have to be redesigned, but so did the steering shaft and the shift tube. Also, a substantial amount of production equipment had to be scrapped.

The design was finally completed by the end of 1965 for the first Energy Absorbing Steering Column, to be produced by General Motors for its 1967 model cars and also those of Chrysler and American Motors. The final design comprised: 1) the Mesh Column mast jacket (Figure 43), which would compress up to 8-1/4 inches at loads of approximately 500 pounds; 2) the three-piece telescoping shift tube (Figure 44), held together by injected plastic keys that would shear under a 150 pound axial load; and 3) the telescoping steering shaft (Figure 45), which was similar to the design settled on earlier (Figure 36). The instrument panel mounting bracket (Figure 46) was held in place at three attachment points by injected plastic keys, which would shear under relatively small loads, allowing displacement in the forward direction only.

A design issue for the system was that the load to initiate compression would be the sum of the compression load of the mast jacket, the load to break the plastic keys of the shift tube, the shaft, and the bracket, and the inertia load needed to accelerate the upper mass of the wheel/column assembly. This summed load was higher than the compressive force of the column, but it became possible to sequence the breakaway and yield so the maximum load was significantly less than the sum. A static load-compression diagram for the Mesh Column (Figure 47) demonstrates this sequence as it relates to column displacement. Dynamic loading averages the peaks and valleys to give more of a square curve, but the relative spacing of the events does not change (Marquis 1967).

Another major issue with the column was the large number of different column assemblies that had to be designed, tooled and produced. Since the 1967 models did not undergo a major body change, the columns had to be fitted to available structures, which precluded standardization. There

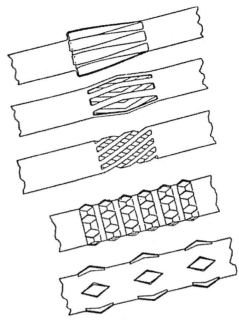

Figure 39: Revised Jacket designs.

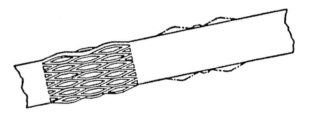

Figure 40: Mesh column jacket.

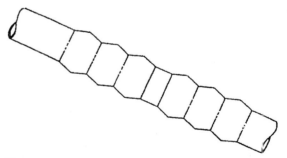

Figure 41: 10-1/2" compressing mesh jacket with six convolutions.

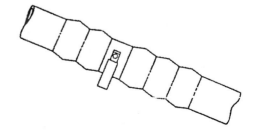

Figure 42: 8-1/2" compressing mesh jacket with five convolutions.

were also many special features that added to the number of necessary assemblies. Whereas some steering assemblies were fixed and not adjustable, there were others that could tilt up and down for driver comfort, and others still that could not only tilt but also telescope axially.

Most cars had the gear-shift lever on the steering column and a shift tube inside the mast jacket, but some sportier models with manual transmissions had the gear-shift mounted on the floor and thus had no shift tube. Most cars had rearward mounted gearboxes and crushed as shown in Figure 48, but others had forward mounted gearboxes and required an additional telescoping section in the shaft, as shown in Figure 49 (Skeels 1966). These various factors combined with the numerous different models that had to be equipped led to the large number of assemblies that had to be designed, tooled, and produced.

The task of tooling a plant for production of the new EA Steering Column was a tremendous one, especially with the large number of different assemblies and parts and the short amount of time available. For all of the cars using the new column, including those of Chrysler and American Motors, 101 column assemblies comprised of over 1100 different parts were needed. To produce all of these assemblies, a completely new 570,000 square foot plant facility employing 1400 workers was constructed in Saginaw, and more than 300 pieces of major capital equipment were purchased.

Dies, molds, fixtures, and other special assembly equipment had to be purchased for 600 different parts. Another 400 parts were produced by and purchased from over 100 vendors. Fulfilling these needs required a major effort from construction workers, machine tool builders, vendors, and various skilled trade shops, especially considering their already heavy workload during this period of a booming economy. Coordinating this effort was truly remarkable in that it was mostly accomplished within the six month period between the finalization of the design at the end of 1965 and the beginning of production in June, 1966. The 1967 model year automobiles with the new General Motors EA Steering System (Figure 50) were introduced in September, 1966.

EARLY BIOMECHANICS AT GMR

When GM Research became involved with the development of their Invertube Steering Wheel Assembly in 1959, they recognized the need for basic biomechanics data on the human body, especially for the chest. This information was needed by engineers working on EA steering systems in order to set the load at which the system should yield and deform. The concept was adopted that the steering system should remain intact at relatively low loads that do not cause injury to the chest, but compress and absorb impact energy to keep chest loads below injury tolerance. Once this injury tolerance was known, the steering system could be engineered to compress at a slightly lower load.

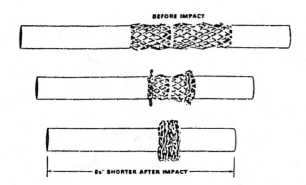

Figure 43: Compression sequence of mesh jacket.

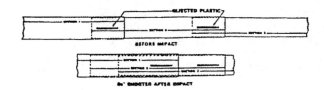

Figure 44: Telescoping of shift tube.

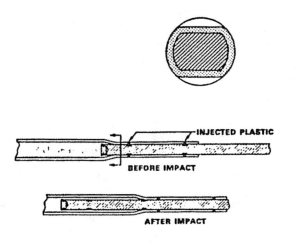

Figure 45: Telescoping of steering shaft.

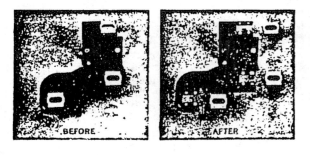

Figure 46: Instrument panel mounting bracket.

In 1959, it was proposed that GMR expand its automotive crash injury research program, by sponsoring a series of human cadaver drop tests at Wayne State University in Detroit. The tests would be carried out at the Biomechanics Research Center in cooperation with the Wayne State Medical School. The Wayne State Biomechanics Research Center was started in 1939 when Drs. E.S. Gurdjian and J.E. Webster, neurosurgeons at the Medical School, teamed up with Professor H.R. Lissner of the College of Engineering to perform concussion experiments on animals (Stapp 1964).

Professors Lissner and Lawrence M. Patrick, then directors of the Center, were approached about GMR's proposed drop tests. They were quite interested in the project and subsequently became biomechanics consultants to GMR. A decision was made for GMR to design and build a full-scale crash simulator that would be used at Wayne State to expose cadavers to actual collision conditions. The design and construction process took many months to complete, but in early 1963, after some trial tests at GMR, the crash simulator was moved to Wayne State and installed at the College of Engineering.

The crash simulator was composed of a test sled riding on rails mounted to a heavy, rigid base (Figure 51). The sled could be used for either a cadaver or dummy and was equipped with an automobile bench seat, an inclined toe board, and load cell supported impact targets for the head, chest, and knees -- all of which were adjustable (Figure 52). A large air cylinder, supplied through a quick-acting valve from an accumulator, was used to accelerate the sled through a ten foot stroke. The piston and rod assembly of the cylinder was then stopped by means of an Invertube, allowing the sled to coast at constant velocity for several feet. Either a large Invertube or a hydraulic piston was impacted by the sled at the end of the track, bringing it to a stop within three feet or less. Design capacities for the simulator were a maximum sled velocity of 40 miles per hour and a maximum deceleration rate of 35 g for a 1600 pound sled (Kroell, Patrick 1964).

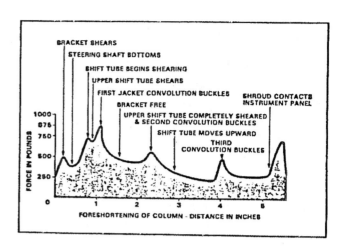

Figure 47:   Column breakaway and yield sequence under static loading.

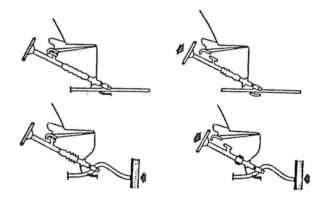

Figure 48:   Column operation with rearward mounted gearbox.

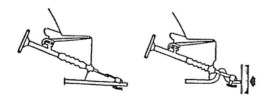

Figure 49:   Operation of lower telescoping section with forward mounted gearbox.

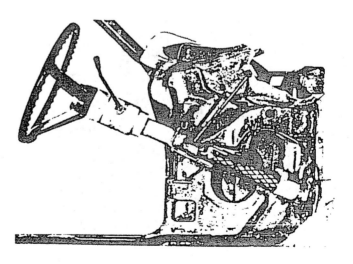

Figure 50:   Installed mesh column assembly.

During 1963, several problems had to be worked out with the simulator. The tri-axial load cells supporting the impact targets had to be specially designed, fabricated, and tested. The many transducers for measuring the displacement, velocity, deceleration, and decelerative force of the sled and the accelerations, stresses, and strains at numerous points on the cadaver or dummy all had to be wired into the main instrumentation consoles (Figure 53). Providing the proper transducer responses and working out the bugs in this extensive instrumentation proved to be quite a large task. Also, the hydraulic decelerator was not giving the desired square wave pulse and had to be reworked by the manufacturer, during which time the Invertube decelerator was used. When the simulator seemed to be functioning properly, dummy tests were run as a final check of the completed test equipment.

The first full-scale impact tests of human cadavers were run at Wayne State on the crash simulator in early 1964. New problems were encountered with cadaver subjects. These included providing articulation of the normally stiff joints in the anatomically embalmed subjects. Physical manipulation was necessary to obtain realistic movement. Also, a special chest deflectometer was developed to measure the compression of the rib cage (Figure 54). Initial studies concentrated on skeletal damage to the knee-thigh-hip complex. When Charles Kroell of GMR and Professor Patrick of Wayne State first reported on the crash simulator at the Eighth Stapp Conference in October, 1964, they presented some of their early results on the leg.

Figure 52:   Cadaver positioned on sled.

Figure 51:   Crash simulator at Wayne State.

Figure 53:   Instrumentation consoles for crash simulator.

Figure 54:   Cadaver fitted with chest deflectometer.

By the end of 1964, the biomechanic studies at Wayne State were concentrating on the more complicated and important matter of chest impact. The first studies looked at static loading of the chest using both cadavers (Figure 55) and human volunteers (Figure 56). Dynamic loading of the chest was then analyzed by testing cadavers on the crash simulator at impact speeds up to 23 miles per hour. Figure 57 shows a photographic time history of one of these chest impact runs which also included targets for measuring knee and head impact loads. The test shown was at 16.8 mph sled velocity and produced a maximum chest load of 1350 pounds and a maximum chest deflection of 1.7 inches. Preliminary results on the tolerance of the chest were presented in October, 1965 at the Ninth Stapp Conference (Patrick et al 1965). From this and subsequent testing, guidelines for the Energy Absorbing Steering System were given to Saginaw Steering Gear -- a maximum chest load of 1000 pounds distributed over 30 square inches of contact area.

After the Impact Sled testing of the Japanese Lantern Column at the Proving Ground in June 1965, it was believed that the chest was experiencing high loads due to inertia forces needed to accelerate the 19 pound moving portion of the column assembly. A dummy with its chest plate mounted on load cells was completed by the first of September, and in tests that followed using this dummy, inertial force spikes of approximately 2000 pounds and 3 milliseconds duration were measured. It was hoped that this spike was due to the stiffness of the dummy chest and that the more compliant human chest would not show this high load. In late 1965, a prototype of the new Mesh

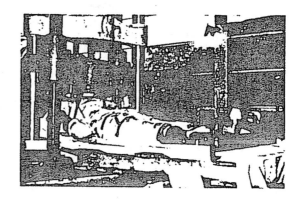

Figure 55: Static loading of cadaver chest.

Figure 56: Static loading of human volunteer.

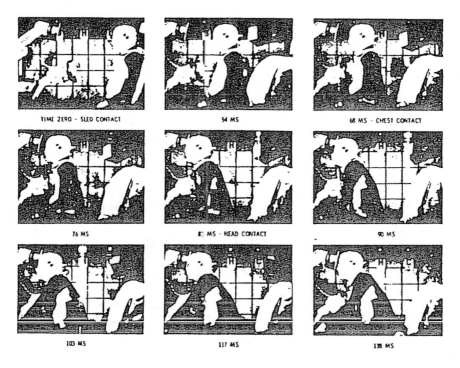

TIME ZERO - SLED CONTACT   54 MS   68 MS - CHEST CONTACT

76 MS   8? MS - HEAD CONTACT   90 MS

103 MS   117 MS   138 MS

Figure 57: Time history of cadaver tolerance test.

Column was installed on the crash simulator at Wayne State, and a cadaver test was run (Figure 58). In this simulated 25 mph crash, four inches of column compression was achieved, and minimal chest deflection and no skeletal damage were observed.

Extensive dummy testing of the Mesh Column over the next few months still indicated these high chest forces, so in 1966, two additional cadaver tests were run with EA Columns using more sophisticated instrumentation. Results of these tests showed that the direct load to the chest was in the range of 500 and 700 pounds -- much less than that registered by the hard-chested dummies and well below the limit of 1000 pounds determined earlier. It was also determined that the shoulders and other body areas were taking an additional 1000 pound load through the wheel rim and outer spokes.

The distribution of load over other parts of the body was very important in limiting the force directly to the chest, primarily the function of the steering wheel. The most important result of this testing was that after the first test, no injuries were observed by x-ray examination of the cadaver. The same subject was used in the second test, following which 3-4 rib fractures and a sternal fracture were noted, but these were judged not to be life threatening. This work experimentally certified the collision performance of the new EA Steering Column, and helped demonstrate the critical need for human-like dummies to design and evaluate safety hardware and components. This understanding led to extensive programs in GM to develop improved test dummies resulting in Hybrid's I, II and III (Foster et al, 1977).

EA STEERING SYSTEM INTRODUCTION

The General Motors Energy Absorbing Steering System was introduced in September, 1966 on the 1967 model passenger cars built by GM, Chrysler, and American Motors. GM used this and other new safety items such as the dual master brake system, the four-way hazard warning flasher, and the lane change signal as selling points in many of their advertisements. An ad featuring the EA steering system for the new '67 Chevrolets is shown in Figure 59. Although the EA column was a primary feature, anti-intrusion and improved steering wheel load distribution were additional aspects of the EA steering system.

The first known real-life accident involving a 1967 car equipped with an EA Column occurred in Detroit in October, 1966. A 33 year old man was driving a 1967 American Motors Ambassador Hardtop (Figure 60) at an estimated speed of 50-60 mph when he ran into the rear end of a stationary tractor-semitrailer (Figure 61). The car suffered major front end damage and had its roof nearly ripped off, as shown in Figure 62. In the impact, the 3,334 pound Ambassador moved the tractor-semitrailer and its 32,000 pound cargo load over eight feet. Impact by the torso of the 130 pound driver compressed the steering column of the Ambassador 5-3/4 inches (Figure 63). The only injuries sustained by the driver were shallow lacerations on the face due to broken glass particles and a bruised left hip from wearing his seat belt. He suffered no cracked ribs or other broken bones and did not even receive a bruised chest. Several of the doctors who attended the patient were of the opinion that the Energy Absorbing Steering Column and the use of a seat belt were responsible for saving his life (Hanson 1966).

During late 1966, when these first accident reports were coming in on the field performance of the EA steering system, the Federal Government was busy preparing its initial safety standards for passenger cars, which would include a standard for the steering assembly. The Ambassador-semitrailer accident and two similar crashes involving the EA Column were all that were known about by early December. More data was needed quickly in order to write effective standards and meet the January 31, 1967 deadline. In early January, Dr. Alan M. Nahum, Director of the Vehicle Trauma Research Group and Assistant Professor of Surgery at the UCLA School of Medicine, reported on 17 additional crashes of 1967 models equipped with the EA steering system in which none of the drivers had sustained significant chest injuries (NTSA 1967). These cases convinced government officials that the EA steering system was a major safety improvement and was a motivation for them to write steering system standards based on the GM Energy Absorbing Steering system that was already in use.

Dr. Nahum, along with Drs. Irving I. Lasky and Arnold W. Siegel, also of UCLA, presented a comparison of the collision performance of different steering system designs at the Automotive Engineering Congress of the Society of Automotive

Figure 58:   Cadaver test of mesh column on crash simulator.

Engineers (Lasky et al 1968). No driver fatalities were reported in frontal impact accidents involving cars with the EA steering system, and the injury severity was rated as dangerous in only 2 of the 27 cases. They stated that the EA steering system "represents an advance in the safety design of wheel-column configurations and as such deserves special commendation."

Other accident investigators were similarly documenting the life-saving effects of the EA

Steering System. A study by Dr. Donald F. Huelke and William A. Chewning (1968) of the Highway Safety Research Institute of the University of Michigan analyzed EA Column performance in fatal and non-fatal accidents. Twenty-six cases were presented in which the column compressed and the driver escaped death and received minimal injuries to the chest. There were only two cases in which the driver died of chest injuries from contacting the steering assembly. Similar studies with promising

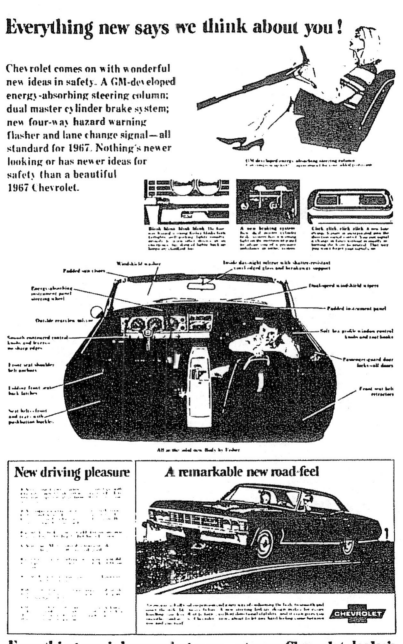

Figure 59:   Advertisement of the 1957 Chevrolets featuring the EA steering column.

results were being performed by Dr. B.J. Campbell of the University of North Carolina Highway Safety Research Center (UNC 1967) and by Professor Patrick (1968) of Wayne State University.

These early reviews on the Energy Absorbing Steering System were all quite positive, but a few possible side effects appeared after a larger number of cases was investigated and the data was more closely analyzed. The study by Patrick (1968) at Wayne State suggested that knee impacts were more prevalent from interaction with the instrument panel, although most of the impacts did not produce injury. It was also observed that the column did not compress in some crashes, when the steering wheel deformed, sometimes significantly. This observation led to an early misunderstanding about column compression performance largely because the analysis did not include a consideration of driver injury or the level of force needed to deform the steering wheel.

An analysis by General Motors of accident data from Cornell's ACIR and reports filed by Motors Insurance Corporation concluded that the EA Steering System significantly reduced injury severity in front end collisions, but that more head injuries were occurring (Lundstrom et al 1969). The increased head injuries were accounted for as mostly being fractures of the facial bones due to contact with the padded instrument panel ahead of the steering wheel and were not life threatening. The GM report also noted that the column might not have compressed in a few cases when it possibly should have.

These early investigations of the performance of the EA steering system indicated a significant advance in automotive safety. The protection the column gave the driver in a crash due to its energy absorption properties was becoming widely known, and its application to other situations of impact protection was considered. In April, 1968, the U.S. Navy began to use the Mesh mast jacket of the EA Column for shock protection chairs and deck platforms on its river patrol boats used by troops in Vietnam (Hawkins, Hirsch 1968). Saginaw Steering Gear and the Navy Ship Research and Development Center in Washington, D.C. jointly developed the chairs and platforms, each of which was equipped with eight of the Energy Absorbing Jackets.

Figure 60:   Typical 1967 Ambassador Hardtop.

Figure 61:   Semi-trailer involved in accident (photo courtesy of Carriers Insurance Company).

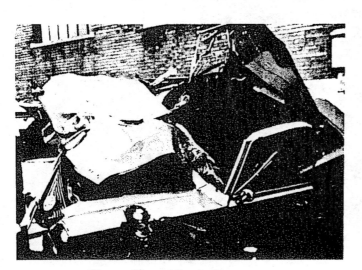

Figure 62:   Wrecked car.

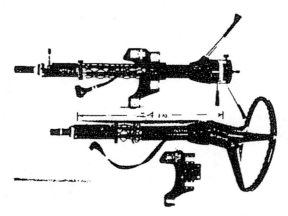

Figure 63:   New column compared with collapsed column from Ambassador.

GOVERNMENT REGULATION

The Federal Government first became involved in safety regulation of the auto industry through the efforts of Congressman Kenneth A. Roberts of Alabama, chairman of the House Subcommittee on Public Health and Safety. In 1956, Roberts set up a Special Subcommittee on Traffic Safety to undertake an extensive study of the highway accident problem. During the first session of the 86th Congress in 1959, Roberts (1963) introduced bill H.R. 1341 which required passenger-carrying motor vehicles purchased by the Federal Government to meet certain safety standards. The bill failed to pass the 86th and 87th Congresses, but after years of work by Roberts, it finally became Public Law 88-515 on August 30, 1964 (Abersfeller 1965).

These standards were prescribed by the General Services Administration (GSA), the agency responsible for purchasing cars for the U.S. Government. The GSA was assisted in writing their standards by representatives of the auto industry, the Department of Commerce, the U.S. Public Health Service, the medical profession, engineers, and safety experts. Federal Standard 515, composed of 17 detailed standards (Figure 64) was published in the June 30 Federal Register (1965) and took effect September 28, 1966. An important consideration was that these were performance rather than design standards, meaning that they specified what was to be accomplished but not the mechanism to be used. This gave design freedom to the manufacturers and thus kept automobile technology from stagnating (Abersfeller 1965).

Standard 515/4 dealt with the steering system. At the Ninth Stapp Conference, October, 1965, H.A. Abersfeller of the GSA stated that 515/4 "is the most controversial subject we have included" and noted particular disagreement over the limit of rearward displacement. Some experts felt that absolutely no displacement should be permitted whereas others advised "that a certain amount of displacement is beneficial and will actually help cushion the body and prevent the build up of forces." It was decided after much analysis and debate to limit the maximum load to the driver from the steering assembly to 2500 pounds and the rearward displacement to 5 inches.

The standard specified that the force developed in a load cell mounted directly behind the wheel could not exceed 2500 pounds when impacted by a body block at 22 feet per second (15 mph). Dimensions and materials of the body block, known as a "blak tuffy", were given in the standard, as shown in Figure 64. GM used a pendulum impact facility for these tests (Figure 65). The 5 inch limit on rearward displacement was to be evaluated in a standard SAE J850 barrier impact test.

GM had been working on its Energy Absorbing Steering Column since 1959 and therefore was able to meet standard 515/4 by the deadline of September 28, 1966. Not only were the requirements for its GSA government contracts met, but also for all of its 1967 model year passenger cars. Chrysler and American Motors did not have systems developed by this time so they purchased EA Columns from GM to install in their cars.

In 1966, the traffic safety issue was a focus in Washington. Amidst the urgings of safety activist Ralph Nader and Senators Gaylord Nelson of Wisconsin, Abraham Ribicoff of Connecticut, and Warren Magnuson of Washington, President Lyndon B. Johnson worked up a transportation program which included a traffic safety bill (Drew 1972). On September 9, President Johnson signed into law the National Traffic and Motor Vehicle Safety Act of 1966 (Public Law 89-563). This law provided for the creation of the National Traffic Safety Agency (NTSA) and the National Highway Safety Agency (NHSA), which later joined to become the National Highway Traffic Safety Administration (NHTSA). NTSA was set up to prescribe safety standards applying to all passenger cars sold in the U.S., not just those bought by the government (NTSA 1967).

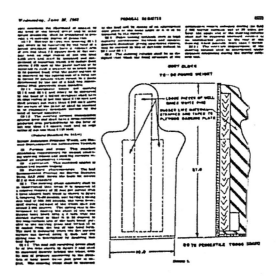

Figure 64: Federal Register of June 30, 1965 with standard 51 5/4.

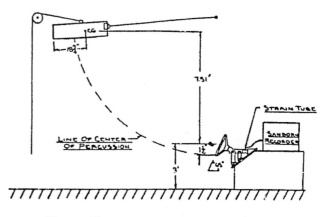

Figure 65: Pendulum impact facility.

From October, 1966 to January, 1967, NTSA consulted with representatives of the auto manufacturers and numerous safety experts and accident investigators, specifically contacting Professor Patrick of Wayne State and Dr. Nahum of UCLA about steering systems. NTSA issued their 20 initial Federal Motor Vehicle Safety Standards on January 31, 1967. The only change made from GSA steering system standard 515/4 was dividing it into number 203 for chest load and number 204 for rearward displacement. The success stories that NTSA had received from the field on the new Energy Absorbing Steering System certainly influenced their decision not to change the 2500 pound steering wheel load and 5 inch rearward displacement limits. These standards went into effect January 1, 1968, nearly 16 months after GM introduced the EA Steering System on its 1967 model year passenger cars (NTSA 1967).

2ND GENERATION EA COLUMN

During the development of the Invertube, Japanese Lantern, and Mesh Columns, General Motors engineers investigated other methods of energy absorption. Oldsmobile was the GM division with lead responsibility for the steering system, and they did design and testing of EA Columns before the engineering work was completely shifted to Saginaw Steering Gear. Beginning in late 1964, at the same time Saginaw was investigating the Japanese Lantern, Oldsmobile designed a mast jacket that absorbed energy by tearing a longitudinally scored metal strip from the jacket -- similar to the action of opening a pull top soda-pop can or an old style key-operated coffee can.

In an Impact Sled test in early 1965, the assembly telescoped but the metal strip broke off and the energy absorption properties were lost. Further tests showed that the tearing force was difficult to control, and therefore this idea was dropped. Oldsmobile also did some work at this time on a flexible steering shaft designed to limit rearward displacement by buckling into an "S" shape when loaded axially. However, this too had problems in early tests, and the idea was dropped when a new concept came along.

During 1965, a new EA device was conceived and tested by Oldsmobile. In this design, the mast jacket was composed of two telescoping tubes, in which the upper tube had two rows of three dimples each that formed an interference with the lower tube (Figure 66). The jacket was assembled by pressing the two tubes together. It absorbed energy when loaded axially by deforming the metal in the lower tube with the dimples in the upper tube. Laboratory drop tests showed that under dynamic conditions, the resulting load-deflection curve had a nearly constant force characteristic (Figure 67). Oldsmobile then combined this jacket with their own designs of a telescoping steering shaft and shift tube to create a complete column assembly.

Impact Sled tests at the Proving Ground indicated that the new type of EA column was feasible, and after changes were made in the compression load, the tube overlap, and the instrument panel mounting bracket, barrier impact tests were run. These tests showed that the column performed well in preventing rearward displacement due to front end crush while still absorbing the energy of the driver. Oldsmobile appeared to have a competing design for an EA column, but late in 1965, the Corporation decided that undivided responsibility should be given to the EA steering system program, and consequently transferred Oldsmobile's development work over to Saginaw Steering Gear.

Saginaw engineers knew that the Mesh Column could have insufficient rigidity and thus, problems with column shake. This led to the addition of structural reinforcement in order to provide sufficient bending strength, which corrected these problems, but at the expense of increased bulk, weight, and number of manufacturing procedures. Even with this reinforcement, there might be potential fatigue life problems with the Mesh section or a potential need to change the compression load of the column, depending on the results of accident field studies and possible changes in future Federal regulations. For the Mesh Column, a change in compressive load would require a major design change and a large and costly retooling job. If the load were reduced, this change would only exacerbate the existing bending strength problems. For these reasons, a new energy absorbing method was being looked for, even before the Mesh Column was introduced (Marquis 1970).

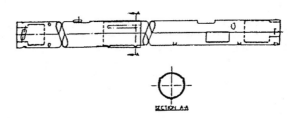

Figure 66: Telescoping jacket with interfering dimples.

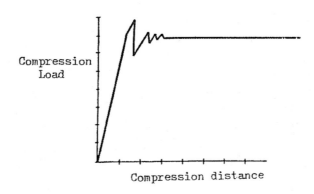

Figure 67: Load-displacement curve of dimpled column.

A Saginaw engineer learned from a NASA publication about an energy absorption mechanism similar to Oldsmobile's dimpled tube column, but instead of using sliding friction caused by the dimples, it had a rubber torus fit between the tubes that absorbed energy on compression by rolling deformation. However, the machining needed to produce the necessary tolerances for this design proved to be too difficult and costly. This led to an idea of placing metal balls in an interference fit between the two tubes (Figure 68).

This new design, known as the Ball Jacket or Ball Column, used balls that were 0.0075 inches larger in diameter than the space into which they were fitted. Energy was absorbed in the deformation of the metal tubes by the balls, which occurred with axial compression. The first jackets that went into production for the 1969 model year had 16 balls arranged in two circles of eight, spaced two inches apart, with each ball operating in its own individual longitudinal track. A static compression load in the range of 450-600 pounds and a maximum available compression distance of 8-1/4 inches were also specified (Marquis 1970).

The Ball Jacket provided improvements compared with the Mesh design. Its greatest attribute was its design flexibility and adaptability to changing needs. If the compression load needed to be decreased, the tubes simply had to be fitted with smaller-sized balls. The measurement of compression force during assembly resulted in a 100% inspection of the column's EA characteristic. Tests showed that for each 0.001 inch decrease in interference, the average compression load decreased by 70 pounds. Increased load could be accomplished by providing greater interference and/or more balls and more longitudinal ball tracks.

To increase bending strength and decrease column shake, a heavy concentration of balls was placed at the six and twelve o'clock positions, since most of the loads that cause bending and shake are in the vertical plane. If more bending resistance was desired without increasing the compression load, individual ball positions could be reinforced by placing a second ball close to the first and in the same track. Changing ball sizes, numbers, and positions provided a relatively simple means of altering column parameters without extensive redesign and testing or a large retooling job.

Besides the adaptability advantage over the Mesh Column, tests of the Ball Column showed it to have greatly improved structural characteristics. Laboratory test stand and in-vehicle analyses of rigidity and natural frequency along with various tests of bending strength and shake gave results for the Ball Column that were equivalent to those for the solid 1966 non-EA columns. Long-term vibration testing at natural frequency indicated a two to three times increase in fatigue life over the Mesh design. This improved stability allowed the removal of the bulky structural reinforcements that were added for the Mesh Column, meaning less weight, less cost, and more available space. This extra space facilitated the addition of the column-mounted anti-theft locking system on the new 1969

model cars, which permitted the locking of ignition, steering, and gearshift all with one key (Marquis 1970).

There were still other areas in which the Ball Column was found to be superior to the Mesh. Whereas the Mesh Jacket had a somewhat sawtooth load-displacement curve due to the staggered compression of the different sections, the Ball Jacket exhibited a nearly ideal, square curve (Figure 69). Also, the Ball Column reduced the number of assembly operations, making its production and installation more efficient and less costly, and lessening the chance of misassembly. For example, the instrument panel mounting bracket had two attachment points rather than three, requiring less tooling and the installation of one less breakaway capsule.

Another advantage of the Ball Column was that the load characteristics could be checked at assembly for all columns, since the load to press the balls between the tubes was the same as that to compress it upon impact. One potential disadvantage with the Ball Jacket was that it could only telescope to one half of the length of the column that was clear of attachments, whereas the Mesh Jacket could compress by essentially this entire length. The Ball Column did provide 8-1/4 inches of compression distance, however, which was indicated by a large quantity of Mesh Column accident data to be sufficient for system performance (Marquis 1970).

The testing program for the Ball Column far exceeded that conducted on the Mesh prior to its production release. Hundreds of Ball Columns were

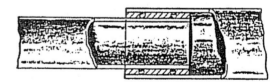

Figure 68: Cut-away view of ball jacket.

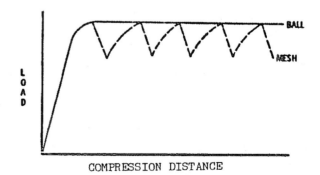

Figure 69: Comparison of load-displacement curves for ball and mesh columns.

tested in barrier impact, Proving Ground Impact Sled, Blak Tuffy mini-sled, and laboratory drop tests -- all with collision performances equal to or better than the Mesh design. Barrier tests on the Ball Column indicated a slightly better compression distance and rearward displacement and chest loading values in the same range as the Mesh Column. Sled tests were run with both belted and unbelted dummy drivers in straight-on, angular and offset impacts. Various torsion, tension, compression, and salt spray tests were performed in the laboratory on over 500 Ball Jackets and Column assemblies. No problems developed in six cars driven on Saginaw Steering Gear's annual cold weather trip during the winter of 1967 or in a car driven in a 5000 mile road test on the Proving Ground's tough Belgian Block course. This list of tests along with several others indicate the very thorough testing job that was performed on the Ball Column before it was released for production.

Although the initial field studies indicated that the Mesh Column was performing well in reducing driver crash injuries, the Ball Jacket had several engineering advantages and also appeared to be slightly better from a safety standpoint. A rigorous testing program verified these claims, and in September, 1968, the second generation Energy Absorbing Steering Column, featuring the Ball Jacket, was introduced on General Motors 1969 model automobiles. The EA Columns on today's GM cars feature a Ball Jacket compressible element. Front wheel drive cars do not mount the lower end of the column to the front of dash and therefore do not require as much available column compression or ball spacing in the EA element.

FURTHER BIOMECHANICAL RESEARCH

During the development and testing of the Ball Column, Oldsmobile initiated and directed a major study of the overall performance of energy absorbing steering systems. This was a multi-faceted, corporate effort and included further cadaver testing at Wayne State. The main objectives of this project were: 1) to aquire additional human tolerance data for the chest and torso; 2) to correlate this human cadaver data with that from dummy testing in order to establish a more realistic and meaningful laboratory steering system test; 3) to obtain a performance comparison between Ball and Mesh Columns using cadavers; and 4) to establish design recommendations for steering system components involved in energy absorption.

In October, 1967, the major part of this study was contracted through GM Research to Professor Patrick at the Wayne State Biomechanics Research Center. From January, 1968 to May, 1969, 38 tests were run of cadavers and Sierra Stan dummies impacting 1968 Mesh Columns and two Ball Column designs at velocities of 16, 22, and 30 mph. A total of six different steering wheel designs were used on the columns. Wayne State used their own impact sled, in which a dummy or cadaver in a moving seat was propelled into a stationary vehicle body section with mounted steering assembly.

The results of this study showed no significant differences between the impact performances of the Mesh and Ball designs based on peak upper torso load. The Sierra Stan dummy chest was found to give different results than the cadaver chest, but test values correlated well enough to develop guidelines for testing with the dummy. Also determined by this study were peak load tolerances for the frontal thorax and the upper torso, both of which were found to be dependent on the steering wheel used. Understanding for the role of the steering wheel were developed in this study.

Work on designing a compatible wheel for the EA Column had actually been going on for years, mostly through the efforts of Inland Division, GM's manufacturer of steering wheels, along with Oldsmobile, the Proving Ground, and GM Research. In designing a wheel from a crash injury point of view, some changes were obvious, but in other cases, a trade-off had to be made. Pieces that could potentially break off and others with sharp edges clearly should be removed. However, there were many arguments over how stiff to make the wheel rim. On one hand, the rim needed to be very stiff to support the shoulders and protect the chest, but on the other hand a soft rim was needed to minimize abdominal and facial injuries.

Many tests were performed on wheels of various sizes, spoke numbers, and rim stiffness, and on others with built-in EA elements. Deep dish wheels with a recessed hub, which were used on cars for several years before the EA Column came out, were also tested with the EA Column. The understanding that was developed by 1969 was therefore the result of years of testing, and attempted to make the steering wheel function with the EA Column as a system to minimize injuries.

The tolerance data acquired from these Wayne State cadaver tests were used by GM engineers in the development of a new dummy. This dummy, called the Hybrid II, was an improvement over previous dummies in that it better simulated human biomechanics. It also had improved repeatability, durability, and serviceability. The Hybrid II was evaluated by NHTSA and subsequently adopted as the standard dummy for crash testing. However, work on development of a more human-like dummy than represented by the Hybrid II was already in progress, leading to the Hybrid III dummy.

At about the same time that GMR became involved in these latest cadaver tests at Wayne State, they also initiated a cadaver testing program at the University of California. When the first Federal Motor Vehicle Safety Standards were being written during late 1966 and early 1967, and Dr. Alan M. Nahum of UCLA was convincing the NTSA of the safety value of GM's Mesh Column via his accident studies, Nahum became interested in the biomechanics program at GMR. In early 1968, a collaborative biomechanics research program at UCLA was initiated to further study impact tolerances of human cadavers. Their first work concerned impact tolerance of the skull and facial bones (Gadd et al 1968), and during the ensuing twelve years, a variety of biomechanics studies were carried out. The main thrust of the GMR/U of C program, however, concerned blunt impact to the thorax; and unembalmed subjects were now used for enhanced biofidelity.

The thoracic tolerance studies were begun at the University of California at Los Angeles (UCLA) in 1969 then transferred to the San Diego campus (UCSD) in 1971. The tests were run using an elastic cord propelled horizontal impactor with a six inch diameter, flat, rigid impact face to apply a midsternal blunt loading to the cadaver chest (Figure 70). In the first phase of this study (Nahum et al 1970), which consisted of static tests of four embalmed and six unembalmed cadavers, the force-deflection characteristics showed a significantly higher chest compliance for the unembalmed subjects (Figure 71). For this reason, only unembalmed cadavers were used to acquire chest tolerance data for the remainder of the UCSD studies. The next phase of the study consisted of an additional fourteen cadaver tests and seven volunteer tests (Kroell et al 1971, Lobdell et al 1972). Live human volunteers were used to analyze the effects of muscle tensing on chest compliance as it was believed that this occurred during many crashes.

From this data, thoracic force-deflection response corridors were developed (Kroell et al 1974) for two combinations of impactor mass and velocity (Figure 72). Combining this data with that from twenty-three more cadaver tests at UCSD, Raymond F. Neathery (1974) of GMR's newly formed Biomedical Science Department then developed scaled response corridors for 5th, 50th and 95th percentile dummies. Prediction equations for maximum force and normalized compression were also determined, using dimensional analysis. Neathery's response corridors for the 50th percentile male dummy chest were used in the design of the Hybrid III (Figure 73), which is a biomechanically improved frontal impact dummy introduced in 1977 by GM's Safety Research and Development Lab (Foster et al 1977).

Neathery, Kroell, and Mertz (1975) also used the UCSD data to develop a normalized chest deflection criterion for predicting injuries from dummy responses, which they showed to be more appropriate than the current criterion based on spinal acceleration. In re-examining the UCSD data, GMR's David C. Viano (1978a) showed that a good correlation also existed between injury level and the initial kinetic energy of impact. In a further analysis, he also proposed that the difference between moderate and life-threatening injury was associated with a specific limit of normalized chest deflection (Viano 1978b).

In order to assess the physiological responses to impact trauma that could not be measured on cadavers, GMR's Biomedical Science Department has conducted many other thoracic loading studies on live subjects, including anesthetized laboratory animals and human volunteers. In 1971, the thoracic tolerance to whole-body deceleration was analyzed by instrumenting professional high diver Ross Collins with accelerometers and having him dive from heights of up to 57 feet and land supinely into a foam pad (Figure 74). He experienced no discomfort in withstanding a 49.2 g

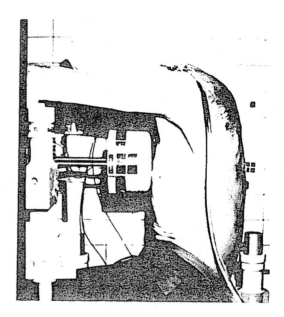

Figure 70:   UCSD cadaver test set-up.

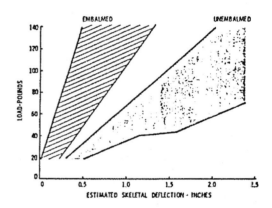

Figure 71:   Embalmed vs. unembalmed cadaver chest load-deflection characteristics.

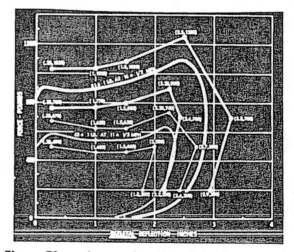

Figure 72:   Averaged, adjusted load-deflection curves and recommended response corridors.

peak thoracic deceleration (Mertz, Gadd 1971). Another volunteer study involved 70 year old stunt diver, Henri LaMothe, who dove from heights of up to 15 feet while "wearing" accelerometers and 34.5 feet without instruments, landing "belly flopper" fashion into 13 inches of water and providing further data (Viano et al 1975) on human response and tolerance to dynamic loadings of the torso (Figure 75). The first thoracic impact studies of animals at GMR were done by Schreck and Viano (1973) on fresh pig carcasses.

Another GMR collaborative research program was initiated — this time with Professor Charles Y. Warner of the Mechanical Engineering Department at Brigham Young University (BYU) in Provo, Utah. Live, anesthetized pigs were used in a series of blunt, frontal thoracic impact tests. This work was done at BYU over a seven year period using experimental set-ups such as shown in Figures 76 and 77. These studies included: 1) a determination of the thoracic impact response of live pigs (Viano, Warner 1976); 2) a comparison between the responses of living and postmortem sibling pigs (Viano et al 1977); 3) a comparison of various thoracic impact sites (Viano et al 1978); 4) a comparison of impact responses in the spine vertical (Figure 76) and spine horizontal (Figure 77) postures (Pope et al 1979); and 5) an analysis of the interrelationship between impact velocity and chest compression (Kroell et al 1981).

Figure 73:   Hybrid III dummy.

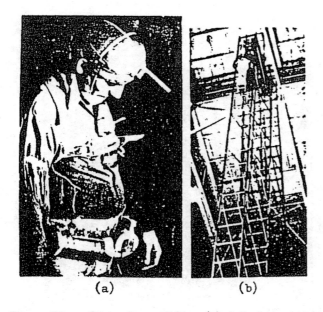

Figure 74:   Diver Ross Collins (a) fully instrumented and (b) preparing to dive.

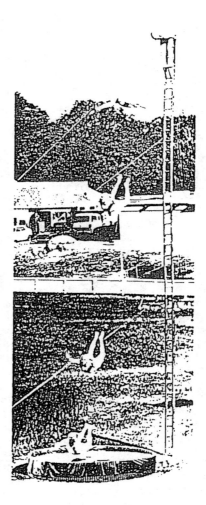

Figure 75:   Sequence of events in Henri LaMothe's high-dive performance.

586

Laboratory research was also conducted at the Biomedical Science Department. Anesthetized rabbits were used to develop a theoretical relationship to assess the severity of thoracic impact based on velocity of impact and normalized chest compression (Viano, Lau 1983). A viscous tolerance criterion represented by the maximum product of velocity and compression was developed by Viano and Lau (1985) as a supplement to the chest compression criterion currently used to measure impact severity in GM's crash testing of dummies.

The main objectives of these chest impact studies of cadavers, volunteers, and anesthetized animals were: 1) to determine the impact tolerance and response of the human chest; 2) to use this data in the development of anthropomorphic dummies; 3) to establish predictive criteria for these dummies to assess the probability of injury to live humans exposed to similar loading conditions; and 4) to use the dummies and test criteria to design safer vehicle interiors and restraint systems for automobiles.

EA STEERING SYSTEM EFFECTIVENESS

Since the introduction of the General Motors EA Steering System in 1966 and of the various designs which have followed, many different accident investigators have been trying to answer the question, "Just how effective are EA Steering Systems in preventing injuries and saving lives?" As already mentioned, the early reports on the Mesh Column were very favorable, although a few questions arose concerning its possible failure to compress when the investigators thought that it should have. The Ball Column provided an engineering improvement on GM's 1969 models. Since that time, many more studies covering larger data bases have been published on the effectiveness of EA Columns, with somewhat mixed reviews.

In 1970, UCLA's Nahum, Siegel, and Brooks (1970) presented a study of 178 injured drivers of 1960-66 pre-EA Column cars and 328 injured drivers of 1967-69 cars equipped with EA Steering Systems. Their data indicated a 61 percent decrease in fatal and dangerous injury rates for drivers striking the steering assembly. A study by Levine and Campbell (1971) at the University of North Carolina showed that EA steering systems reduced serious injuries by 14 percent among drivers not wearing seat belts. They also showed that when drivers wore seat belts in cars equipped with EA steering systems, serious injuries were reduced 52 percent over drivers without either safety feature. A report by Anderson (1974) of Calspan Corporation indicated that from a subset of drivers that wore seat belts, EA systems still reduced serious injuries by 14 percent. Although the effectiveness values given by these and other studies covered a large range, there was a general consensus that compared to previous systems, the EA Steering Columns were protecting many drivers from serious and fatal injuries (Figure 78).

A few reports began coming out in 1973, however, of alleged performance failures with EA steering systems. A majority of the criticism was coming from a British group at the University of Birmingham's Department of Transportation and Environmental Planning, headed by G.M. Mackay and P.F. Gloyns. Mackay first presented their findings in June, 1973 at a conference in Amsterdam (Gloyns, Mackay 1973), in which he stated that axial compression steering systems, such as the EA Column, were not performing as well in real-world crashes as systems with a solid column and a deformable hub element just behind the steering wheel. They blamed this performance difference on the testing procedures outlined in Federal Safety Standards 203 and 204, which tended to favor the axial compression systems. However, the differences in injury level between the two systems were not found to be statistically significant, and also, the small British cars were not similar to GM cars equipped with the GM Ball Column. Gloyns and Mackay (1974) presented the same arguments in December, 1974 at the Eighteenth Stapp Car Crash Conference. Garrett and Hendricks (1975) of Calspan investigated 549 frontal accidents and found that the EA Column compressed less when the impact was at an angle rather than head-on. They incorrectly concluded that the decreased compression was generally associated with increased injuries. They did show, however, that GM's Ball Column achieved the most compression of any design tested.

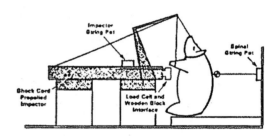

Figure 76:   Set-up for BYU impact studies of anesthetized pigs in spine vertical position.

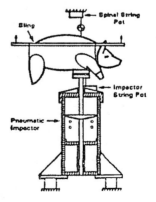

Figure 77:   Spine horizontal impact set-up.

Following the criticism of U.S. Federal Safety Standards 203 and 204 by Gloyns and Mackay, many other investigators made their own evaluations of these standards. Huelke and O'Day (1975) of the University of Michigan classified them as being only "partially effective" and called for the development of new standards. Quality control expert J.M. Juran (1976) labelled the entire set of safety standards "a failure" from a benefits/costs standpoint. In 1981, NHTSA came out with an evaluation of their own standards. This report, by Charles J. Kahane (1981), was the first investigation of its kind that was based on a large number of crashes and with a greatly reduced bias for selection of cases. It found the Steering Assembly standards (MVSS 203 and 204) to be effective in that for frontal crashes, the driver's overall risks of fatality and serious injury were reduced by 12% and 17.5%, respectively, while only

$10 was added to the lifetime cost of owning and operating a car (Figure 79).

At the same time, A.C. Malliaris, Ralph Hitchcock, and James Hedlund (1982) of NHTSA developed a method for establishing priorities in crash protection based on human Harm. Harm was defined as "a cost-weighted sum of the number of people injured or the number of injuries sustained, whether fatal or not," with "cost" based on injury severity. Part of this study focused on the source of injury, and the steering assembly was found to be the area of contact representing the most Harm. Steering assembly contact was associated with 27% of the total Harm to car occupants, with the instrument panel next at 11% (Figure 80). Subsequent analysis demonstrated that the steering assembly is highly represented in injury data due to high exposure and not "negative" benefit. Park (1984) and Horsch et al (1985).

Effectiveness of the Energy Absorbing Steering System

Field Accident	Dates	Injury Severity	Injury Reduction Steering System Contact	Driver Frontal Accidents
Auto. Crash Injury Research (Torso only)	1964-69	AIS ≥ 1	32%	
Multidisciplinary-UCLA	1962-69	AIS ≥ 2	54	
Multidisciplinary Michigan-UCLA		AIS ≥ 3	45	
North Carolina	1966 & 68	K + A		14
North Carolina	1971-72	K + A		20
New York State	1968-69	K + A		24
FARS	1975-79	Fatal		12
NCSS	1977-79	AIS ≥ 3	38	18

Figure 78

Effectiveness of FMVSS 203 and 204 Energy Absorbing
Steering System and Anti-Intrusion Standards

Fatalities

FARS 1975-1979   12% (8-16%) Reduction of Risk in Frontal Crashes

Serious Injury

NCSS 1977-1979   38% (28-48%) Reduction of Injury From Steering System Contact

18% Reduction in Overall Injury Risk in Frontal Crashes

Figure 79

FURTHER DEVELOPMENT OF THE STEERING SYSTEM

Although the ball-sleeve column remained the primary energy absorbing and force limiting device for the steering system, further evaluation of crash performance and refinements of the steering system protective function continued. One result of studies such as the Harm paper was an expansion by General Motors of its on-going efforts to further improve the steering system through a coordinated corporate effort. Several column features, initially developed in air bag programs, were evaluated for unrestrained drivers and were subsequently introduced in appropriate passenger vehicles. These features were: 1) the quick-release shear capsule which reduced the release forces from the anti-intrusion bracket especially in non-axial loadings, and 2) the ramped jacket, which increased the force of compression and energy absorption after the initial part of column compression which has an inertial component related to column acceleration by the forces of driver impact (Figure 81).

GMR's Biomedical Science Department also expanded its activity to provide an improved understanding of the issues and mechanics of the driver interaction with the steering system in car crashes. As part of the critical review of the literature, the "popular" criticisms based on analysis of car crashes and various hypotheses were evaluated in sled tests with dummies and human cadavers. This effort resulted in a series of papers dealing with the steering system impact performance.

GMR addressed the assertion of a poor safety performance of the EA column due to a "frequent" failure of the column to compress. The position in the literature was developed from laboratory tests which observed less column compression in off-axis impact than in axial loadings; and, it was supported by analysis of crash data which involved less column compression in the EA device than the author expected from the general severity of the accident. Because driver injury is highly associated with the steering assembly, the conjecture of a "frequent" failure to compress was embraced by some authors as a flaw of the EA steering system. However, the laboratory studies did not consider the severity of the loading nor did the crash data analysis consider injury outcome of the driver.

The analysis of FMVSS 203 and 204 by Kahane (1981) continued the presumption of a "frequent failure to compress" by concluding that "a failure of the energy absorbing devices to compress" occurred in about half the crashes in which they were heavily impacted by drivers. He speculated that the columns "tend to bind rather than compress when they are exposed to nonaxial loads." Because he was among the first to recognize that the primary reason for noncompression in most cases was insufficient impact force resulting from lower severity crashes, he proposed a criteria to indicate "heavy" driver loading based on steering wheel deformation in which a severely deformed wheel and less than one inch EA element compression was considered a failure, irrespective of injury outcome. However, Kahane failed, as did previous researchers, to evaluate if occupant injuries were associated with their proposed criterion of "failure to compress." This was one reason why GM (1981) stated in its official comments to NHTSA on the Kahane Report that the number of cases of noncompression due to "failure" was greatly overestimated.

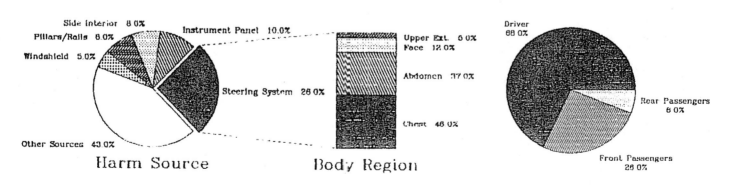

Figure 80: Distribution of crash injury harm by contact source and by body region for steering system impact.

589

GMR's Biomedical Science Department used analysis of accident data and an understanding of impact mechanics to show that no direct correlation existed between wheel deformation or column compression and injury outcome. When various "failure to compress" criterion were applied to a large set of accident data, Horsch et al (1982) found that no greater percentage of steering-related injuries existed in the "failure" group than for the overall data set (Figure 82). In fact, most of the injury was associated with cases in which the column did compress and that these were typically the more severe crashes.

Sled tests were used to evaluate the mechanics of the various non-compression hypotheses and developed an understanding of the impact mechanics. These tests demonstrated that the ball-sleeve column would compress at loadings of more than 50° from the column axis (Figure 83). These studies also confirmed that less column compression was associated with off-axis loading and driver off-set impacts, but most importantly, little or no change in injury risk was associated with this reduction in column compression as energy absorption was shifted to steering wheel deformation. Reduced column compression in off-axis and off-set impacts was shown to be due to geometric effects and shifting force limiting and energy absorption to the steering wheel and not due to increased column compression force for the GM ball column.

A dramatic demonstration of the wheel and column interaction and energy absorbing deformations was accomplished by bracing the dummy's arms on the wheel (Figure 84). When the arms were braced between the spokes, the wheel required less force to deform than for the column to compress. Thus, for this situation, the wheel experienced large deformation and the column had almost no compression. In contrast with the arms braced at the spokes, the wheel was stronger than the column and the column fully compressed. Impact force and dummy responses were similar in the two tests, steering assembly deformation was completely different. Sternal impact from a typical seating position also resulted in full column compression but with greater impact force and dummy response amplitude due to a greater acceleration of column and wheel mass. Locking column compression greatly increased impact force and dummy response amplitude with minimal deformation of the steering wheel. Although the intent of the experiment was not to study "bracing" by the driver, the forces measured with the dummy's arms braced were similar to reported arm forces from a "stunt" driver who braced in a staged crash (Wagner 1979).

The GMR research was able to demonstrate in laboratory tests that the ball-sleeve column would compress in off-axis impacts, that the column would compress less in such impacts, that the wheel would provide force limiting and energy absorbing functions in these situations, and that these facts did not greatly increase the risk of injury based on dummy responses.

The hypothesis had been tested against car

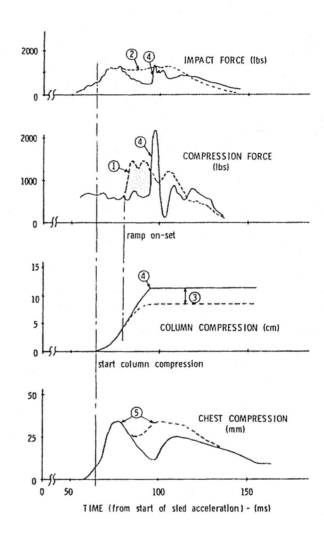

Ramped Force Jacket - Dashed line
Constant Force Jacket - Solid line

① Ramped jacket additional energy absorbing capacity

② Ramped force maintains nearly constant impact force

③ Additional ramped jacket EA results in less compression

④ Non ramped column "bottoms"

⑤ Chest compression is not increased due to jacket ramp force

Figure 81: Ramped jacket increases column energy absorbing capacity without increasing impact force.

UNIVERSITY OF MICHIGAN (HSRI) ACCIDENT DATA

Shear Capsule Separation (mm)	Steering Wheel Deformation Rating	Thoracic Injury* (AIS)				Totals - Wheel Deformation		
		0	1	2	≥3**	None	Minor	Severe
0	none	30	4	0	0	34		
	minor	11	2	0	0		13	
	severe	5	1	1	0			7
1-13	none	9	2	0	0	11		
	minor	16	8	0	1		25	
	severe	9	4	1	0			14
14-25	none	5	3	0	0	8		
	minor	5	1	0	1		7	
	severe	3	0	0	2			5
26-75	none	1	0	0	0	1		
	minor	2	3	0	1		6	
	severe	3	5	0	2			10
>75	none	1	0	0	0	1		
	minor	3	1	1	0		5	
	severe	1	2	1	3			7
	Injury Totals	104	36	4	10			
	Wheel Deformation Totals					55	56	43

U of M (HSRI) Accident Data -
Frontal (11, 12 and 1 o'clock) -
GM Ball-Sleeve Columns -
Unrestrained or Lap-Belted Drivers
154 Cases

* Thoracic injury due to steering system interaction.
**There were no cases of thoracic injury with AIS > 3

"Failure to Compress" Criteria	No. Cases	Injuries (AIS 3+)	Injury Rate
"Frequent Failure to Compress"			
"Failure" < 1 inch column compression	124	4	3%
"Success" > 1 inch column compression	30	6	20%
"Failure to Compress when Heavily Loaded"			
"Failure" < 1 inch column compression and severe wheel deformation	26	2	8%
"Success" > 3 inch column compression	13	3	20%
Injury Rate for Data Set	154	10	7%

Figure 82: "Failure to compress" criteria from the literature applied to a data set where steering contact injuries are known. Clearly injuries are more likely for the "successes" than for the "failures."

Sled Velocity Km/h	Column Angle (degree)	Column Compression (cm)	Wheel Deformation Rating	Interaction Force 3 ms Resultant (kN)	Interaction Force 3 ms Axial Component (kN)	3 ms Thorax Resultant Acceleration (g)	Maximum Chest Compression (mm)
36	20°	13.0	minor	5.96	5.83	44	29
36	35°	12.4	minor	5.74	4.67	45	27
36	43°	7.6	moderate	7.21	5.12	39	29
36	50°	7.1	moderate	9.12	6.05	43	35
25	15°	9.4	negligible	5.29	-	32	20
25	15°*	2.8	moderate	4.33	-	22	17

* Dummy off-set laterally from column axis by 125 mm

Figure 83: Influence of column vertical angle and dummy off-set on test responses. 36 km/h sled velocity.

Test Configuration		Column Compression (mm)	Wheel Deformation	Impact Force (kN)	Resultant Thorax Accel. (3 ms-g)	Chest Compression (mm)
Typical		130	minor	6.0	44	29
Typical (non-compressible column)		none	moderate	14.2	62	45
Arms Braced on Wheel	on spoke	150	moderate	3.4	21	0
	between spoke	6	severe	3.5	24	13

Figure 84: Demonstration of energy management by column and wheel for a range of test situations. 36 km/h sled velocity. Part 572 dummy.

crash data which demonstrated that injury was infrequently associated with little or no EA device compression and that most of the driver injuries were associated with crashes having significant EA device compression. However, this analysis did not explain the high association of injury with steering assembly impact. Determining this cause-and-effect was critical. Park (1984) and Evans and Frick (1987) have shown that the driver has a lower or equal risk of injury compared with a right-front passenger, indicating a possible protective benefit of the steering system. Seventy percent of car occupants are drivers and frontal crashes comprise the most frequent injury category. This indicates that frequent exposure and not defective performance was the reason for the association of injury with steering assembly impact.

Horsch and Viano (1984) evaluated the mechanical surrogates or test dummies available for laboratory testing of the steering system and its safety performance. These studies demonstrated that the Hybrid III dummy was the most appropriate test device based on a more human-like construction, biofidelity in frontal impact responses and an extensive injury assessment capability related to the viscous and compression mechanisms of injury in addition to "traditional" acceleration based criteria. In contrast, the SAE J-944 body block was shown to be an inappropriate test device (Figure 85) based on its non-humanlike construction, a fundamental lack of biofidelity, and no injury assessment capability. The study demonstrated that FMVSS 203 which requires an impact evaluation of the steering system using the SAE body block cannot assess improved occupant protection features of the steering system because of the inherent limitations of the test device responses and methods and that these limitations could result in a counterproductive design.

The available surrogates were also used to determine how various design and test parameters such as steering column mass, steering wheel load distribution and test severity would be rated as affecting injury (Horsch 1982). These tests demonstrated that the choice of surrogate and test response is crucial to identifying which design strategy best protects car occupants (Figure 86). Thoracic stiffness is important for the peak impact

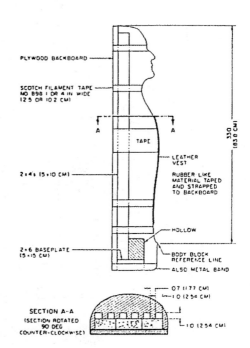

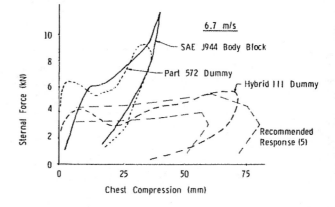

Figure 85: Blunt frontal thoracic impact responses of the SAE J944 body block, the Part 572 dummy, and the Hybrid III dummy for the recommended test method and compared to recommended response envelopes.

force, primarily due to acceleration of the column and wheel mass. The excessively stiff body block caused 57% greater impact force than did the more human-like Hybrid III dummy.

Perhaps more importantly, only the Hybrid III dummy with chest compression biofidelity could distinguish the importance of steering wheel load distribution on the upper torso. A standard steering wheel was compared with a "disk" which applied the impact loading only to the rib cage. The body block and Part 572 dummy indicated minimal difference between these wheels. However, the Hybrid III demonstrated that applying all of the impact force to the rib cage increased chest compression by 80% and injury risk from very low to a high risk of severe thoracic injuries at a test severity well below that of a 50 kmph frontal barrier crash. These tests provided an additional understanding of the importance of load distribution by the steering wheel on the upper body. These are among the reasons that GM developed a "similar" test but substituted a Hybrid III torso at almost three times the impact energy of the 203 test.

In spite of significant advantages in using the Hybrid III dummy, it also has limitations in mimicking an occupant's interaction with the steering system. One limitation was addressed in a GMR sponsored series of dummy and cadaver tests conducted at Wayne State University (Begeman et al 1990). The experiments involved sled tests with a steering wheel mounted on a rigid (non-compressible) steering column and chest impacts by the unrestrained driver at severities sufficient to result in inversion of the steering wheel in cadaver tests, a situation sometimes found in fatal, severe frontal crashes. Sled velocities approaching 50 kph with the full velocity change occurring before steering wheel impact resulted in significant thoracic injuries in cadavers and significant deformations of the steering wheel (Figure 87).

In comparable tests, the Hybrid III and Part 572 dummy produced much lower deformations of the steering wheel in spite of significantly higher impact forces. Although the Hybrid III dummy provided the most representative wheel deformations of the test devices evaluated, it was unable to invert the steering wheel because of the dummies' non-deformable steel spine structure. Fortunately, these types of tests are well beyond the standard evaluation of the steering system and forces would have been limited in real applications by the EA column. The experiments did quantify one deficiency in even state-of-art test dummies to duplicate a fatal injury mechanism by steering wheel impact.

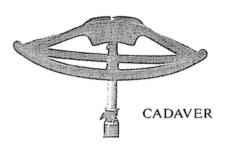

CADAVER

PART 572
DUMMY

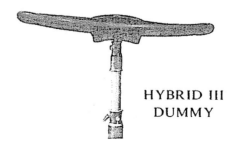

HYBRID III
DUMMY

Ratio of Response	Wheel	Mass
Body Block/Force	1.04	1.08
Part 572/Chest Acceleration	1.09	1.08
Hybrid III/Chest Compression	1.82	1.22

Ratio of Response - Hybrid III	Wheel	Mass
Force	0.84	1.16
Chest Acceleration	0.91	1.08
Chest Compression	1.82	1.22

Figure 86: Influence of the test surrogate and response on the perception of which of two levels of steering wheel load distribution and steering assembly mass provides the best occupant protection.

Figure 87: Deformed steering wheels from WSU tests at 42 km/h with a non-compressible column.

In summing up the GMR work, the studies demonstrated that following recommendations in the literature to improve the consistency of EA device compression was not the key for improvement in driver protection. Additionally, it was demonstrated that the "standard" test could not be used to improve design strategies in preventing severe to fatal flail chest and internal organ trauma. Thus, the GMR studies looked at factors in car crashes that were associated with steering system contact injuries (Horsch, Culver 1983). These investigations found that injury was associated with a wide range of crash directions and severities, driver alignment with the steering system, and injury mechanisms in body regions involved. This new understanding suggested a shift in focus from the EA column to the steering wheel.

A review of fatal crashes involving driver interaction with the steering system showed the possibility of significant steering wheel deformations with the plane of the wheel rim sometimes below that of the hub by occupant loading (Figure 88). This situation would result in a more concentrated sternal loading on the occupant which would reduce the load sharing and distribution between chest and shoulders. This would lower the occupant's tolerance to steering wheel force. One obvious countermeasure to this situation in fatal crashes would be to increase the stiffness of the spokes and rim of the steering wheel. Such increases would reduce a tendency for the wheel to deform during severe occupant loadings and thus maintain the balance of shoulder and chest impact forces.

Although a stiffer steering wheel would likely improve thoracic protection in very severe loadings, it might inadvertently increase injuries for other body contacts. Figure 89 shows the force tolerance for the chest, abdomen and face for steering wheel loading. The higher tolerance of the chest indicates its ability to withstand impact forces on the rib cage that are well above those necessary to cause soft tissue injuries in the unprotective portion of the abdomen. A similar lower force tolerance can be expected for a facial impact which may occur during driver interactions with the steering system. These biomechanical data suggested the need for a systematic improvement in steering wheel design. While stiffening the steering wheel may reduce the risk of crush injury of the chest in severe driver loadings, this design direction did not match the need for lower forces associated with abdominal and facial impact. On this basis, there was a conflict in requirements of the safety performance of the steering wheel depending on the body region in contact.

DEVELOPMENT OF THE SELF-ALIGNING STEERING WHEEL

In frontal and oblique crashes, the driver contacts the steering wheel. Although the column limits the overall impact of force, injury risk for the driver depends on the location, distribution and magnitude of impact forces transmitted by steering wheel contact. In addition, the driver impact with the steering wheel occurs from various positions and

directions in a real-world crashes because of a range in alignments with the steering system prior to and during a crash, belt restraint use being one important factor.

The difficulty in designing an optimum steering system to cover the wide spectrum of impact situations was further compounded by the need to protect various body regions having widely different impact characteristics and tolerances. In addition, some important body regions such as the abdomen were not routinely evaluated in even the state-of-art test devices. Thus, the effect of design changes on the risk of abdominal injury would not be included in the evaluation of system performances and not part of steering wheel criteria for overall safety.

Figure 88: Deformed steering wheel from a severe frontal crash. The driver had a crushed chest.

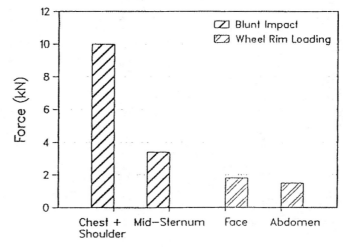

Figure 89: Tolerable forces for various body regions based on recommendations from the literature.

These observations led to a research strategy to deal with load magnitude, distribution and body region contacted by a driver in the range of real-world crash severities. The research focused on the mechanism of blunt impact injury of the chest, abdomen, face and head and considered opportunities to improve the built-in safety of the steering system for both the unrestrained and lap-shoulder belted driver. The research addressed the conflicting requirements for steering wheel deformation which included a "stiff" steering wheel plane to assure sufficient energy management and load sharing between the chest and shoulder thus maintaining tolerable loads on the sternum in severe frontal crashes, and a "soft" wheel rim to reduce the risks of injury to abdominal organs or the face related to the initial contact with the steering wheel. A Hybrid III torso impact test was developed by a corporate task group as a means to assure good upper body load distribution for an unrestrained driver (Figure 90). This test was more relevant to car occupants and real world crashes than the FMVSS 203 body block test due to a more human-like surrogate, a wide range of injury assessment instrumentation, and almost three times the impact energy.

Pivotal to the development of the self-aligning steering wheel was continued GMR research on the mechanisms of internal organ and soft tissue injury in blunt impact. Anesthetized animal experiments provided the only opportunity to determine life-threatening injuries. By the mid 1980's, Viano and Lau (1985) had developed the theory and experimental support for chest and abdominal injury by a viscous mechanism. They subsequently extended their analysis of injury mechanisms by merging in a crushing mechanism due to high forces squeezing organs and tissues against the spine in frontal impacts. Complimentary research was conducted by Kroell et al (1981, 1986) using high velocity

impacts on anesthetized pigs and body region specific studies on mechanisms of liver injury and lung trauma (Lau, Viano 1981 a,b,c,d,), all of which substantiated a rate-dependent mechanism of injury. The final tolerance relationship (Figure 91) emphasized the importance of the amount and velocity of body deformation on injury risk (Viano, Lau 1988).

A reanalysis of earlier UCSD cadaver data (Lau, Viano 1986) provided additional support for the viscous criterion so that by 1988, the viscous injury mechanism was being routinely used in the assessment of safety systems and had an established procedure and recommended tolerance level. The criterion effectively demonstrated not only injury risk but the time of greatest injury potential (Lau, Viano 1988). Knowing the time of injury risk in an impact exposure helped to further focus attention on countermeasures that would be effective in improving driver safety in a crash. In contrast, spinal acceleration continued to show no statistical or practical relation to injury causation or risk. This was most evident in a study of abdominal injury by steering wheel loading which showed that maximum injury risk to abdominal organs occurred prior to sufficient force to initiate column compression and prior to peaks in spinal acceleration. Both of these responses would falsely indicate that injury was associated with events later in the impact (Horsch et al 1985). The data led to an evaluation of steering wheel deformations occurring early in an impact exposure at the time of the peak viscous response (Figure 92); and before the maximum forces are developed by the chest impacting the steering wheel and initiating column compression (Lau et al 1987, Lau, Viano 1988).

Concepts for energy-absorbing steering wheels had been well known and some designs have been used; however, the criteria for an effective EA wheel

Figure 90: Hybrid III torso impact test to evaluate steering assembly EA management and steering wheel load distribution on the upper body.

and hub were not available until analysis of dummy, cadaver, and animal test data. The new understandings provided a set of performance criteria which would be effective in improving driver safety for a range of impact exposures in body regions involved. In 1979, the Inland Division of General Motors developed an energy-absorbing steering wheel by a unique design of a multi-finger hub element which provided a flexible attachment between the steering wheel and column. This EA hub provided a design concept that could resolve the conflict between a "stiff" and "soft" steering wheel for the protection of different body regions. Conceptually, the EA wheel provides the deformation performance given in Figure 93.

A prototype self-aligning steering wheel was evaluated in 1983 using sled tests with the Hybrid III dummy. The experiment successfully showed that the self-aligning steering wheel provided improved protection for a range of initial seating positions and orientations of the driver dummy (Figure 94). Not only were impact forces better distributed in the experiments, the tests showed that one possible negative aspect of the self-aligning feature was, in fact, an advantage. There was a concern that the aligning feature would allow the driver to slip off the steering wheel in oblique impacts and thus increase secondary contacts within the vehicle. However, in oblique and off-axis loading the aligning feature used in this wheel actually helped maintain better contact with the driver and more uniform load distribution on the thorax over the range of exposures. Additional experiments were conducted to address the potential improvement in head impact with the EA hub used in the self-aligning steering wheel. Drop test experiments with the Hybrid III head indicated a reduction in head HIC with the new steering wheel (Figure 95), while further demonstrating the feasibility of limiting force and absorbing impact energy remote from the impact location.

The experiments clearly supported the efficacy of the self-aligning feature and increased wheel stiffness in a range of tests. It also demonstrated the improvement provided by a consistent rim deformation force around the perimeter of the steering wheel, which was controlled by the tipping force of the hub element remote from the rim. In contrast, standard steering wheels could have greater than a 2 to 1 range in rim deflection forces depending on the relative position of rim loading with respect to spoke attachments. Thus, the self-aligning feature reduced the sensitivity of rim loading to the location around the steering wheel.

To further verify the safety potential of the self-aligning steering wheel, Horsch et al (1985) conducted experiments using anesthetized pigs in Hyge sled tests with the self-aligning steering wheel concept compared to the same steering wheel without a self-aligning feature. These tests marked the first time that the Hyge sled installed in the Biomedical Science Department in 1979 was used for animal experiments. A total of 16 well-controlled experiments were performed to assess the safety performance of the self-aligning steering wheel in preventing abdominal injury and reducing risks of

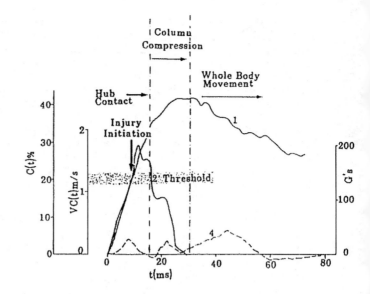

Figure 91: Time-phase of body responses and steering system deformation for blunt impact 1:chest compression (C), 2:Viscous Response (VC) and 4:spinal acceleration (G) (from Lau et al 1988).

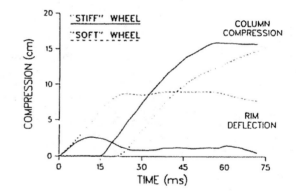

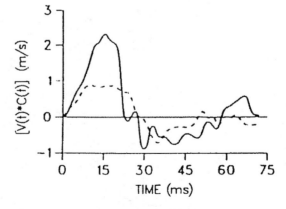

Figure 92: Column compression, wheel deformation, and abdominal viscous response for anesthetized pigs impacting EA steering assemblies in 32 km/h sled tests. "Stiff" wheel tests resulted in severe to critical abdominal injury. "Soft" wheel tests resulted in minimimal or no abdominal injuries.

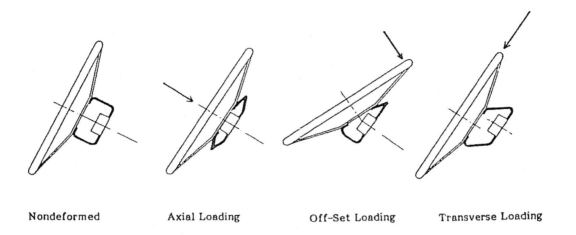

| Nondeformed | Axial Loading | Off-Set Loading | Transverse Loading |

Figure 93: Primary deformation modes of a stiff spoke and rim structure wheel with the deformable hub element.

chest trauma (Lau et al 1987). Using a exposure severity of 9.0 m/s, anesthetized animals impacted by a "stiff" steering wheel experienced severe life-threatening abdominal injuries due to rim loading on the abdomen. Analysis of data showed that injuries were associated with a viscous mechanisms of injury and that injuries were occurring before sufficient forces had been developed by chest impact to initiate column compression (Figure 92). Tests with the self-aligning steering wheel resulted in minimal injury to the abdomen with lower levels of chest injury for a comparable exposure. These data helped validate the self-aligning steering wheel concept and define wheel stiffness guidelines.

The biomechanical data supporting the self-aligning steering wheel concept were presented to the Corporation in early 1985. Its endorsement of the concept subsequently led to a corporate-wide Task Force headed by CPC to consider the broader aspects of production feasibility of the self-aligning steering wheel. Task Force members represented a cross section of General Motors Divisions and Staffs. The Inland Division had the major role for a series of evaluations to refine the concept for the other many product functions, requirements, and manufacturing considerations for the self-aligning steering wheel. The demonstration of a production feasible prototype for the wheel was completed in 1988 and the first production installation of the self-aligning steering wheel manufactured by Inland Division was in the 1989 Cavalier with an accelerated phase-in for steering wheels in non-airbag vehicles.

During the final development of the self-aligning steering wheel, a series of analyses was conducted on the cost-effectiveness of the new safety feature in relation to other safety considerations for vehicle improvement. These comparisons showed a significant advantage for the self-aligning feature over other more costly ideas largely because of its effectiveness in improving safety for a wider range of crash environments and potential body region contacts involving both unrestrained and belt restrained drivers.

Sled Test Conditions Configuration	Velocity (km/h)	Wheel	Maximum Chest Compression (mm)	3-ms Thoracic Spine Resultant Acceleration (g)
15° Column Angle	27	Baseline	42	28
		Self-Aligning	38	24
	33	Baseline	41	34
		Self-Aligning	41	30
125 mm Off-set 15° Column Angle	27	Baseline	33	24
		Self-Aligning	16	28
	33	Baseline	33	31
		Self-Aligning	19	31
30° Column Angle	27	Baseline	39	22
		Self-Aligning	32	25
	33	Baseline	64	32
		Self-Aligning	42	36

Figure 94: Comparison of Hybrid III thoracic responses for a "typical" baseline steering wheel and a prototype self-aligning wheel over a range of alignments with the steering assembly demonstrating improved load distribution (reduced chest compression).

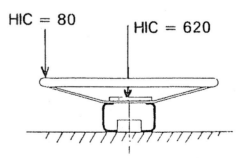

HIC = 80   HIC = 620

Figure 95: Impact locations on the prototype self-aligning steering wheel for 25 km/h free fall head drops. 6 mm thick load distributing pad was used on the hub.

The crash performance of the self-aligning steering wheel was monitored in GM's Motors Insurance Corporation (MIC) cases. The early data has shown success in improving passenger protection while not demonstrating any durability or reliability problems with the new feature. A significant crash occurred shortly after introduction in which an unrestrained driver survived a severe driver-side frontal crash with minimal fractures of chest ribs and extremities. The self-aligning feature performed as designed and helped manage the driver's impact energy. In contrast, the unrestrained right-front passenger was fatally injured by instrument panel and windshield impact.

A SCIENTIFIC BASIS FOR SAFETY ENGINEERING

The role of the steering system for protecting drivers is changing due to increased belt use and wider availability of supplemental inflatable restraints. Work has continued to improve head and facial protection for the belted driver (Viano et al 1986, Melvin and Shee 1989). Much of the technology developed for the energy absorbing steering system remains important for continued advancement of occupant protection. Steering system energy management and intrusion control features developed for unrestrained drivers have been applied to driver air bag systems. Improved understanding of human tolerance and dummy injury assessment which were developed concurrently with steering system designs are keys for further occupant protection.

A scientific basis for safety engineering, which links biomechanical relevant laboratory tests to real world performance has evolved with the many tests and evaluations of the energy absorbing steering system over thirty years. Investigations to determine the tolerable impact force for the torso led to the understanding that tolerance depends on how force is applied to the human. This led to new injury assessment technologies in dummies and injury criteria which could distinguish load distribution and lead to designs with good load application. These needs led to the Hybrid III dummy which has injury assessment based on a biofidelic compressible chest.

The development of the energy absorbing steering system has demonstrated the need for relevant laboratory tests with sufficient injury assessment capability to evaluate primary injury risks to car occupants in a similar exposure. This is critical and provides a foundation in assessing crash safety improvements. However, these tests are not the final link to real world performance. Laboratory tests represent specific, well controlled conditions; the real world consists of large ranges of many parameters such as crash severity and car occupant tolerance to impact. The presumption that a crash test represents real world performance can be highly misleading because it represents the injury risk for only one of many crash situations. Consideration of injury types, frequencies of occurrence, distribution in tolerance and crash situations led to a strategy for steering wheel design which could not have been developed from routine tests.

The analyses of occupant injuries by Malliaris et al (1982, 1985) showed that although "high" injury risk is associated with "severe" crashes, much of the injury is associated with crash severities having "low" risk of injury but a "high" exposure associated with "low" severity crashes (Figure 96). Laboratory tests over a range in crash severities provide more information on the distribution in injury risk. The association of exposure frequency must come from additional analysis. An analysis which considers the range of tolerance in terms of a risk of injury and the exposure frequency associated with particular test or exposure situations provides a better extrapolation to real-world performance (Horsch, 1987, Viano 1987). Research continues on these safety strategies, since they appear to provide an enhanced approach to improve crash protection in real-world crashes.

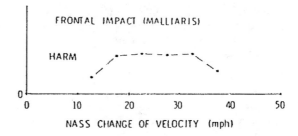

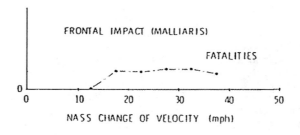

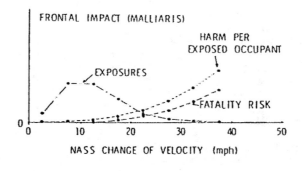

Figure 96: Distribution of the number of exposed occupants as a function of NASS change of velocity showing the strong bias toward low severity impact. Also shown is the risk of injury indicating a strong bias toward high severity impact. Injury distributions are the product of exposure frequency and injury risk.

ACKNOWLEDGEMENTS

The review of the history of the safety development of the energy absorbing steering system was accomplished by James DeCou, MD, while he was a summer student at the Biomedical Science Department. He completed the story of the early years of development through studies of the effectiveness of the EA system in field accidents. The results of more recent studies were added to make the review current, including studies leading to further improvements in safety design and evaluation. This comprehensive history is meant to portray the continuous process of design, evaluation, and review of the built-in safety of General Motors products and the technical and scientific analysis of occupant protection that is an important part of that process. These accomplishments are the result of contributions by many individuals throughout the Corporation.

BIBLIOGRAPHY

1. Abersfeller, H.A., "Federal Program of Automotive Standards. In Proceedings of the Ninth Stapp Car Crash Conference, pp. 287-300. Society for Automotive Engineers, Minneapolis, MN, 1965.
2. ACIR, "Injury-Producing Automobile Accident: A Primer of Facts and Figures," Automotive Crash Injury Research of Cornell University, August, 1961.
3. Anderson, T.E., "The Effects of Automobile Interior Design Changes on Injury Risk," DOT HS-801 239, National Highway Traffic Safety Administration, Washington, D.C., October, 1974.
4. Begeman, P.C., Kopacz, J.M., and King, A.I., "Steering Assembly Impacts Using Cadavers and Dummies." In Proceedings of the Thirty-Fourth Stapp Car Crash Conference (P-236), pp. 123-144, SAE Technical Paper 902316, Society of Automotive Engineers, Warrendale, PA, 1990.
5. Begeman, P.C., Levine, R., King, A.I., and Viano, D.C., "Biodynamic Response of the Musculoskeletal System to Impact Acceleration." SAE Transactions, Volume 89, 1980. In Proceedings of the Twenty-Fourth Stapp Car Crash Conference, pp. 477-510. SAE Technical Paper #801312, Society of Automotive Engineers, October, 1980.
6. Campbell, H.E., "Deceleration, Highway Mortality, and the Motorcar," Surgery, 36(6):1056-1058, 1954.
7. Cichowski, W.G., "A New Laboratory Device for Passenger Car Safety Studies," SAE Paper No. 663A, presented at the National Automobile Meeting, March, 1963.
8. Cornell University, "Cornell-Liberty Safety Car," sponsored by Liberty Mutual Insurance and produced by Cornell Aeronautical Laboratory, 1957.
9. DeHaven, H., "Beginnings of Crash Injury Research." In Proceedings of the Thirteenth Stapp Car Crash Conference, pp. 422-428, Society of Automotive Engineers, Boston, December 2-4, 1969.
10. DeHaven, H., "Mechanical Analysis of Survival in Falls from Heights of Fifty to One Hundred and Fifty Feet." In War Medicine, Volume 2, pp. 586-596, July, 1942, reprinted in Proceedings of the Twenty-Fourth Stapp Car Crash Conference, pp. 3-13, Society of Automotive Engineers, Troy, MI, October 15-17, 1980.
11. DeHaven, H., "Mechanics of Injury Under Force Conditions," Mechanical Engineering, pp. 264-268, April, 1944.
12. DeHaven, H., "Research on Crash Injuries," Journal of the American Medical Association, p. 524, June 8, 1946.
13. Drew, E.B., "The Politics of Auto Safety." In The Atlantic, pp. 95-102, Oct., 1966, reprinted in The World on Wheels, by R.S. Baker and P.L. Van Osdol, Allyn and Bacon, Inc., Boston, pp. 261-275, 1972.
14. Eastman, J.W., "Styling vs. Safety: The American Automobile Industry and the Development of Automotive Safety, 1900-1966," Ph.D. dissertation, University of Florida, 1973.
15. Evans, L. and Frick, M., "Relative Fatality Risk in Different Seating Position Versus Car Model Year." In Proceedings of the Thirty-Second Annual Association for the Advancement of Automotive Medicine Conference, pp. 1-14, 1988.
16. Federal Register, National Archives, Washington, D.C., Volume 31, p. 8323, June 30, 1965.
17. Foster, J.K., Kortge, J.O., and Wolanin, M.J., "Hybrid III -- A Biomechanically-Based Crash Test Dummy." In Proceedings of the Twenty-First Stapp Car Crash Conference, pp. 973-1014, Society of Automotive Engineers, New Orleans, October 19-21, 1977.
18. Gadd, C.W., Nahum, A.M., Gatts, J., and Danforth, J.P., "A Study of Head and Facial Bone Impact Tolerances," presented at General Motors Safety Seminar, GM Proving Ground, Milford, MI, July 11, 1968.
19. General Motors, "Comments of General Motors Corporation Regarding NHTSA Technical Report 'An Evaluation of Federal Motor Vehicle Safety Standards for Passenger Car Steering Assemblies'," Docket 81-03, Notice 1, USG 2062, June 26, 1981.
20. Gandelot, H.K., "Engineering Safety into Automobile Bodies," SAE Paper No. 592, presented at SAE National Passenger Car, Body and Materials Meeting, Detroit, March 6-8, 1951.

21. Garrett, J.W. and Hendricks, D.L., "Factors Influencing the Performance of the Energy Absorbing Steering Column in Accidents," Report on the Fifth International Technical Conference on Experimental Safety Vehicles, Pub. No. 050-003-00210, U.S. Government Printing Office, Washington, D.C., 1975.

22. Gloyns, P.F. and Mackay, G.M., "Impact Performance of Some Designs of Steering Assembly in Real Accidents and Under Test Conditions." In Proceedings of the Eighteenth Stapp Car Crash Conference, pp. 1-27, Society of Automotive Engineers, Ann Arbor, MI, December 4-5, 1974.

23. Gloyns, P.F., Mackay, G.M., Hardy, J.L.G., and Ashtor, S.J., "Field Investigations of the Injury Protection Offered by Some 'Energy Absorbing' Steering Systems." In Proceedings of the International Conference on the Biokinetics of Impacts, pp. 399-410, IRCOBI Secretariat, Amsterdam, June 26-27, 1973.

24. Hanson, H.L., "Energy Absorbing Steering Column: A Case History." In Proceedings of the Tenth Stapp Car Crash Conference, pp. 7-13, Society of Automotive Engineers, Holloman AFB, New Mexico, November 8-9, 1966.

25. Hasbrook, A.H., "The Historical Development of the Crash-Impact Engineering Point of View." Clinical Orthopaedics, Volume 8, pp. 268-274, 1956, reprinted in Accident Research- - Methods and Approaches, by W. Haddon, et al., Harper & Row, New York, pp. 547-554, 1964.

26. Hawkins, J.T.and Hirsch, A.E., "General Motors Energy-Absorbing Steering Column as a Component of Shipboard Personnel Protection," Shock and Vibration Bulletin, no. 37, pt. 4, pp. 79-84, January, 1968.

27. Horsch, J.D., "Evaluation of Occupant Protection from Responses Measured in Laboratory Tests." SAE Technical Paper #870222, Society for Automotive Engineers, Warrendale, PA, 1987.

28. Horsch, J.D. and Culver, C.C., "The Role of Steering Wheel Structure in the Performance of Energy Absorbing Steering Systems." In Proceedings of the Twenty-Seventh Stapp Car Crash Conference, pp. 95-108, Society of Automotive Engineers, San Diego, October 17-19, 1983.

29. Horsch, J.D. and D.C. Viano, "Influence of the Surrogate in Laboratory Evaluation of Energy-Absorbing Steering Systems." SAE Transactions, Volume 93, 1984. In Proceedings of the Twenty-Eighth Stapp Car Crash Conference, pp. 261-274, SAE Technical Paper #841660, Society of Automotive Engineers, Warrendale, PA, November, 1984.

30. Horsch, J.D., Lau, I.V., Andrzejak, D.V., and Viano, D.C., "Mechanism of Abdominal Injury by Steering Wheel Loading," SAE Transactions, Volume 94, 1985. In Proceedings of the Twenty-Ninth Stapp Car Crash Conference, P-167, pp. 69-78, SAE Technical Paper #851724, Society of Automotive Engineers, Warrendale, PA, October, 1985.

31. Horsch, J.D., Petersen, K.R. and Viano, D.C. "Laboratory Study of Factors Influencing the Performance of Energy Absorbing Steering Systems." SAE Transactions, Volume 91, 1982. SAE Technical Paper #820475. In Occupant Interaction with the Energy Absorbing Steering System (SP-507), pp. 51-63, Society of Automotive Engineers, Warrendale, PA, February, 1982.

32. Huelke, D.F. and Chewning, W.A., "The Energy-Absorbing Steering Column: A Study of Collision Performance in Fatal and Nonfatal Accidents," Highway Safety Research Institute of The University of Michigan Report No. Bio-7, 1968.

33. Huelke, D.F. and Gikas, P.W., "How Do They Die? Medical-Engineering Data From On-Scene Investigations of Fatal Automobile Accidents," SAE Paper No. 1003A, presented at International Automotive Engineering Congress, Detroit, January 11-15, 1965.

34. Huelke, D.F. and O'Day, J., "The Federal Motor Vehicle Safety Standards: Recommendations for Increased Occupant Safety." In Proceedings of the Fourth International Congress on Automotive Safety, pp. 275-292, July 14-16, 1975.

35. Juran, J.M., "Automotive Safety Legislation -- Ten Years Later." Presented at the 20th Annual Conference of the European Organization for Quality Control, June 15-17, 1976.

36. Kahane, C.J., "An Evaluation of Federal Motor Vehicle Safety Standards for Passenger Car Steering Assemblies," DOT HS-805 705, National Highway Traffic Safety Administration, Washington, D.C., January, 1981.

37. Kihlberg, J.K., "Driver and His Right Front Passenger in Automobile Accidents." In Proceedings of the Ninth Stapp Car Crash Conference, pp. 335-3534, Minneapolis, MN, October 20-21, 1965.

38. Kroell, C.K., "A Simple, Efficient, One Shot Energy Absorber," Bulletin #30, Shock, Vibration and Associated Environments, Part III, pp. 331-338, February 1962.

39. Kroell, C.K., Allen, S.D., Warner, C.Y., and Perl, T.R., "Interrelationship of Velocity and Chest Compression in Blunt Thoracic Impact to Swine II." SAE Technical Paper #861881, Society of Automotive Engineers, Warrendale, PA, 1986.

40. Kroell, C.K. and Patrick, L.M., "A New Crash Simulator and Biomechanics Research Program." In Proceedings of the Eighth Stapp Car Crash Conference, pp. 185-228, Society of Automotive Engineers, Detroit, October 21-23, 1964.

41. Kroell, C.K., Pope, M.E., Viano, D.C., Warner, C.Y., and Allen, S.D. "Interrelationship of Velocity and Chest Compression in Blunt Thoracic Impact." _SAE Transactions_, Volume 90, 1981. In _Proceedings of the Twenty-Fifth Stapp Car Crash Conference_, pp. 549-582, SAE Technical Paper #811016, Society of Automotive Engineers, September, 1981.

42. Kroell, C.K., Schneider, D.C., and Nahum, A.M., "Impact Tolerance and Response of the Human Thorax." In _Proceedings of the Fifteenth Stapp Car Crash Conference_, pp. 84-134, Society of Automotive Engineers, California, 1971.

43. Kroell, C.K., Schneider, D.C., and Nahum, A.M., "Impact Tolerance and Response of the Human Thorax II." In _Proceedings of the Eighteenth Stapp Car Crash Conference_, pp. 383-457, Society of Automotive Engineers, Ann Arbor, MI, December 4-5, 1974.

44. Kulowski, J., "Auto Crash Injury Research," Medical Aspects of Traffic Accidents." In _Proceedings of the Montreal Conference_, pp. 185-190, 1955.

45. Lasky, I.I., Siegel, A.W., and Nahum, A.M., "Automotive Cardio-Thoracic Injuries: A Medical-Engineering Analysis." SAE Technical Paper #680052. Society of Automotive Engineers, Detroit, January 8-12, 1968.

46. Lau, V.K. and Viano, D.C., "Influence of Impact Velocity on the Severity of Non-penetrating Hepatic Injury." _Journal of Trauma_, 21(2):115-123, February, 1981.

47. Lau, V.K. and Viano, D.C. "An Experimental Study on Hepatic Injury from Belt-Restraint Loading." _Journal of Aviation, Space and Environmental Medicine_, 52(10):611-617, October, 1981.

48. Lau, V.K. and Viano, D.C. "Influence of Impact Velocity and Chest Compression on Experimental Pulmonary Injury Severity in an Animal Model." _Journal of Trauma_, 21(12):1022-1028, December, 1981.

49. Lau, I.V. and Viano, D.C., "The Viscous Criterion: Bases and Applications of an Injury Severity Index for Soft Tissues." _SAE Transactions_, Volume 95, 1986. In _Proceedings of the Thirtieth Stapp Car Crash Conference (P-189)_, pp. 123-142. SAE Technical Paper #861882, Society for Automotive Engineers, Warrendale, PA, 1986.

50. Lau., I.V. and Viano, D.C., "How and When Blunt Injury Occurs: Implications to Frontal and Side Impact Protection." _SAE Transactions_, Volume 97, 1988. In _Proceedings of the Thirty-Second Stapp Car Crash Conference_, pp. 81-100, SAE Technical Paper #881714, Society of Automotive Engineers, Warrendale, PA, October, 1988.

51. Lau, I.V., Horsch, J.D., Andrzejak, and Viano, D.C., "Biomechanics of Liver Injury by Steering Wheel Loading," _Journal of Trauma_, 27(3):225-235, April, 1987.

52. Lau, I.V., Horsch, J.D., Viano, D.C., and Andrzejak, D.V., "Mechanism of Injury by a Deploying Airbag." Submitted to _Accident Analysis and Prevention_, 1990.

53. Lau, V.K., Viano, D.C., and Doty, D.B., "Experimental Cardiac Trauma Ballistics of a Captive Bolt Pistol." _Journal of Trauma_, 21(1):34-41, January, 1981.

54. Levine, D.N. and Campbell, B.J., "Effectiveness of Lap Seat Belts and the Energy Absorbing Steering System in the Reduction of Injuries." University of North Carolina Highway Safety Research Center, Chapel Hill, N.C., November, 1971.

55. Lobdell, T.E., Kroell, C.K., Schneider, D.C., Hering, W.E., and Nahum, A.M., "Impact Response of the Human Thorax." presented at GMR Symposium, October 2-3, 1972.

56. Lundstrom, L.C., et al., "Energy-Absorbing Steering Column Successfully Reduces Injury but New Impact Patterns Emerge." _SAE Journal_, Volume 77, no. 5, pp. 60-64, May, 1969.

57. Malliaris, A.C., Hitchcock, R., and Hedlund, J., "A Search for Priorities in Crash Protection." SAE Technical Paper #820242. Society of Automotive Engineers. Detroit, February 22-26, 1982.

58. Marquis, D.P., "The General Motors Energy Absorbing Column." SAE Technical Paper #670039. Society of Automotive Engineers, Detroit, January 9-13, 1967.

59. Marquis, D.P., "Second Generation Energy Absorbing Column with Locking Feature." SAE Technical Paper #700002. Society of Automotive Engineers, Detroit, January 12-16, 1970.

60. Melvin, J.W. and Shee, T.R., "Facial Injury Assessment Techniques." In _Proceedings of the Wayne State University Bioengineering Center 50th Anniversary Symposium_, pp. 97-108, Detroit, MI, 1989.

61. Mertz, H.J. and Gadd, C.W., "Thoracic Tolerance to Whole-Body Deceleration." In _Proceedings of the Fifteenth Stapp Car Crash Conference_, pp. 135-157, Society of Automotive Engineers, Coronado, California, November 17-19, 1971.

62. NTSA, "Report on the Development of the Initial Federal Motor Vehicle Safety Standards Issued January 31, 1967," National Traffic Safety Agency, United States Department of Commerce, Washington, D.C., March 17, 1967.

63. Nahum, A.M., Gadd, C.W., Schneider, D.C., and Kroell, C.K., "Deflection of the Human Thorax Under Sternal Impact." SAE Technical Paper #700400, In _1970 International Automobile Safety Conference Compendium_, pp. 797-807, Detroit, MI, May 13-15, 1970, Brussels, Belgium, June 8-11, 1970.

64. Nahum, A.M., Siegel, A.W., and Brooks, S., "The Reduction of Collision Injuries: Past, Present, and Future." In *Proceedings of the Fourteenth Stapp Car Crash Conference*, pp. 1-43, Society of Automotive Engineers, Ann Arbor, MI, November 17-18, 1970.

65. Neathery, R.F., "Analysis of Chest Impact Response Data and Scaled Performance Recommendations." In *Proceedings of the Eighteenth Stapp Car Crash Conference*, pp. 459-493, Society of Automotive Engineers, Ann Arbor, MI, December 4-5, 1974.

66. Neathery, R.F., Kroell, C.K., and Mertz, H.J., "Prediction of Thoracic Injury from Dummy Responses." In *Proceedings of the Nineteenth Stapp Car Crash Conference*, pp. 295-316, Society of Automotive Engineers, San Diego, November 17-19, 1975.

67. Park, S., "Relative Risk of Driver and Right Front Passenger Injury in Frontal Crashes," GMR-4802, August 3, 1984.

68. Patrick, L.M., "Analysis of Automobile Accident and Experimental Data -- Steering Assembly Impacts," Wayne State University Biomechanics Research Center, sponsored by Automobile Manufacturers Association, December 4, 1968.

69. Patrick, L.M., Kroell, C.K., and Mertz, H.J., "Forces on the Human Body in Simulated Crashes." In *Proceedings of the Ninth Stapp Car Crash Conference*, pp. 237-259, Society of Automotive Engineers, Minneapolis, MN, October 20-21, 1965.

70. Pope, M.E., Kroell, C.K., Viano, D.C., Warner, C.Y., and Allen, S.D., "Postural Influences on Thoracic Impact." In *Proceedings of the Twenty-Third Stapp Car Crash Conference*, pp. 765-795, Society of Automotive Engineers, San Diego, October 16-19, 1979.

71. Roberts, K.A., "Passenger Safety Standards for Government Motor Vehicles," Report No. 491, House of Representatives, 1st Session 88th Congress, July 1, 1963.

72. Rouhana, S.W., Viano, D.C., Jedrzejczak, E.A., and McCleary, J.D., "Assessing Submarining and Abdominal Injury Risk in the Hybrid III Family of Dummies." In *Proceedings of the Thirtieth Stapp Car Crash Conference*, pp. 257-279. SAE Technical Paper #892440, Society of Automotive Engineers, Warrendale, PA, October, 1989.

73. Schreck, R.M. and Viano, D.C., "Thoracic Impact: New Experimental Approaches Leading to Model Synthesis." In *Proceedings of the Seventeenth Stapp Car Crash Conference*, pp. 437-450, Society of Automotive Engineers, Oklahoma City, November 12-13, 1973.

74. Schwimmer, S. and Wolf, R.A., "Leading Causes of Injury in Automobile Accidents," Automotive Crash Injury Research of Cornell University, June, 1962.

75. Shelden, C.H., "Prevention, the Only Cure for Head Injuries Resulting from Automobile Accidents." *Journal of the American Medical Association*, 159(10):981-986, November 5, 1955.

76. Skeels, P.C., "The General Motors Energy Absorbing Steering Column." In *Proceedings of the Tenth Stapp Car Crash Conference*, pp. 1-7, Society of Automotive Engineers, Holloman AFB, New Mexico, November 8-9, 1966.

77. Snyder, R.G., "Human Impact Tolerance." In *Proceedings of the 1970 International Automobile Safety Conference Compendium*, Detroit, MI, May 13-15, 1970, Brussels, Belgium, June 8-11, 1970.

78. Snyder, W.W., et al., "An Engineering Pilot Study to Determine the Injury Potential of Basic Automotive Interior Design," Human Engineering Laboratory, Aberdeen Proving Ground, Maryland, October 1, 1953.

79. Stapp, J.P., "Past, Present, and Future of Biomechanics at Wayne State University." In *Proceedings of the Eighth Stapp Car Crash Conference*, pp. 391-399, Society of Automotive Engineers, Detroit, MI, October 21-23, 1964.

80. Stapp, J.P., "Biography of the 'Conference Founder.'" In *Proceedings of the Tenth Stapp Car Crash Conference*, pp. v-vii, Society of Automotive Engineers, Holloman AFB, New Mexico, November 8-9, 1966.

81. Stein, P.D., H.N. Sabbah, Viano, D.C. and J.J. Vostal, "Response of the Heart to Nonpenetrating Cardiac Trauma." *Journal of Trauma*, 22(5):364-373, May, 1982.

82. Stonex, K.A. and Skeels, P.C., "A Summary of Crash Research Techniques Developed by the General Motors Proving Ground," *General Motors Engineering Journal*, pp. 7-11, 4th Quarter, 1963.

83. University of North Carolina, "Collapsible Steering Assembly Prevents Injury, N.C. Study Shows." *American Association of Motor Vehicle Administrators*, Volume 32, nos. 5 and 6, p. 5, June-July, 1967.

84. Viano, D.C., "Evaluation of Biomechanical Response and Potential Injury from Thoracic Impact." *Aviation, Space, and Environmental Medicine*, 49(1):125-135, January, 1978a.

85. Viano, D.C., "Thoracic Injury Potential." Presented at the 1978 International Research Committee on the Biokinetics of Impact, Lyon, France, September 12-13, 1978b.

86. Viano, D.C., "Evaluation of the Benefit of Energy-Absorbing Material for Side Impact Protection: Part I." *SAE Transactions*, Volume 96, 1987. In *Proceedings of the Thirty-First Stapp Car Crash Conference* (P-202), pp. 185-204, SAE Technical Paper #872212, Society of Automotive Engineers, Warrendale, PA, November, 1987.

87. Viano, D.C., "Evaluation of the Benefit of Energy-Absorbing Material for Side Impact Protection: Part II." *SAE Transactions*, Volume 96, 1987. In *Proceedings of the Thirty-First Stapp Car Crash Conference* (P-202), pp. 205-224, SAE Technical Paper #872213, Society of Automotive Engineers, Warrendale, PA, November, 1987.

88. Viano, D.C., "Cause and Control of Automotive Trauma." <u>Bulletin of the New York Academy of Medicine</u>, Second Series, 64(5):376-421, June, 1988.

89. Viano, D.C. and Lau, V.K., "Role of Impact Velocity and Chest Compression in Thoracic Injury." <u>Aviation, Space, and Environmental Medicine</u>, 54(1):16-21, January, 1983.

90. Viano, D.C. and I.V. Lau, "Thoracic Impact: A Viscous Tolerance Criterion." In <u>Proceedings of the Tenth Experimental Safety Vehicle Conference</u>, pp. 104-113, National Highway Traffic Safety Administration, Oxford, England, 1985.

91. Viano, D.C., and Lau, I.V., "A Viscous Tolerance Criterion for Soft Tissue Injury Assessment." <u>Journal of Biomechanics</u>, 21(5):387-399, 1988.

92. Viano, D.C. and Warner, C.Y., "Thoracic Impact Response of Live Porcine Subjects." In <u>Proceedings of the Twentieth Stapp Car Crash Conference</u>, pp. 731-765, Society of Automotive Engineers, Dearborn, Michigan, October 18-20, 1976.

93. Viano, D.C., King, A.I., Melvin, J.W., and Weber, K. "Injury Biomechanics Research: An Essential Element in the Prevention of Trauma." <u>Journal of Biomechanics</u>, 22(5):403-417, 1989.

94. Viano, D.C., Kroell, C.K., and Warner, C.Y., "Comparative Thoracic Impact Response of Living and Sacrificed Porcine Siblings." In <u>Proceedings of the Twenty-First Stapp Car Crash Conference</u>, pp. 627-709, Society of Automotive Engineers, New Orleans, October 19-21, 1977.

95. Viano, D.C., Melvin, J.W., McCleary, J.D., Madeira, R.G., Shee, T.R., and Horsch, J.D., "Measurement of Head Dynamics and Facial Contact Forces in the Hybrid III Dummy." <u>SAE Transactions</u>, Volume 95, 1986. In <u>Proceedings of the Thirtieth Stapp Car Crash Conference</u>, P-189, pp. 269-290. SAE Technical Paper #861891, Society of Automotive Engineers, Warrendale, PA, October, 1986.

96. Viano, D.C., Schreck, R.M., and States, J.D., "Dive Impact Tests and Medical Aspects of a 70 Year Old Stunt Diver." In <u>Proceedings of Nineteenth Conference of the American Association for Automotive Medicine</u>, pp. 101-115. Society of Automotive Engineers, San Diego, November 20-22, 1975.

97. Viano, D.C., Warner, C.Y., Hoopes, K., Mortenson, C., White, R., and Artinian, C.G., "Sensitivity of Porcine Thoracic Responses and Injuries to Various Frontal and a Lateral Impact Site." In <u>Proceedings of the Twenty-Second Stapp Car Crash Conference</u>, pp. 167-207, Society of Automotive Engineers, Ann Arbor, MI, October 2-4, 1978.

98. Wagner, R., "A 30 mph Front/Rear Crash with Human Test Persons." In <u>Proceedings of the Twenty-Third Stapp Car Crash Conference</u> (P-87), pp. 825-841, Society for Automotive Engineers, October 1979.

99. White, A.J., "An Engineering Pilot Study to Determine the Comparative Injury Potential of Steering Wheel Assembly Designs," Motor Vehicle Research, Inc., June, 1955.

100. Woodward, F.D., "Medical Criticism of Modern Automotive Engineering," <u>Journal of the American Medical Association</u>, 138(9):627-631, October 30, 1948.

101. Yanik, A.J., "GM Innovations in Safety Engineering, Test Technology and Biomechanical Research," GM Environmental Activities Publication No. A-4266, June 22, 1983.

CHAPTER SEVEN

Door guard beams

LIMITATION OF INTRUSION
DURING SIDE IMPACT

Mr. William J. Wingenbach, AMF

Abstract.

"Limitation of Intrusion
During Side Impact"

One of the more difficult engineering problems encountered in providing vehicle crashworthiness is that of limitation of intrusion during side impact.

The work done on the AMF Experimental Safety Vehicle project has led to the evolution of a design concept which has as its basic element an aluminum honeycomb sandwich door panel. The sequential modes of behavior of the concept under increasing load levels are elastic beam action, plastic beam action, honeycomb crush, and finally, membrane stretching. As these behavioral modes progress, very large resistance to transverse loads develops even though transverse deflection remains small.

The concept has been modeled and analyzed for load deflection characteristics, and several evolutionary models have been built and tested under both static and dynamic loading, including full-scale vehicle crashes. Actual behavior has agreed very well with analytically predicted behavior enabling the side structure system to meet ESV design goals.

Introduction

The Advanced Systems Laboratory of AMF Incorporated at Santa Barbara, California, has been engaged in the development of an Experimental Safety Vehicle for the U.S. Department of Transportation. The scope of this project includes the complete span of activity from original conceptual design through developmental testing and evaluation, and culminates with the delivery of complete vehicles at the end of this year. To the extent possible, the project utilized a systems approach wherein each component was synthesized, analyzed, designed, and tested and evaluated with consideration of the interfaces with other components and of total vehicle objectives.

This paper is concerned specifically with the work associated with resistance of intrusion into the vehicle passenger compartment during side impacts. Included are an enumeration of technical objectives, descriptions of the system and method of analysis and a discussion of the results of developmental testing.

Objectives

The design of the AMF Experimental Safety Vehicle side structure was directed towards a set of objectives derived from explicit and implicit Department of Transportation goals. This set of objectives is as follows:

Passenger compartment intrusion is to be limited to three inches measured from a normal inside surface when struck on the side by the front bumper of a vehicle of equal mass. Impact velocity of 30 mph is normal to the side and at any point along the side. The impacting bumper structure is equivalent to the required ESV front bumper system which has the characteristics of providing override/underride protection over the range of 14 to 20 inches above ground, and a vehicle acceleration force which is velocity dependent. Maximum permissible vehicle acceleration versus impact velocity is shown in Figure 1.

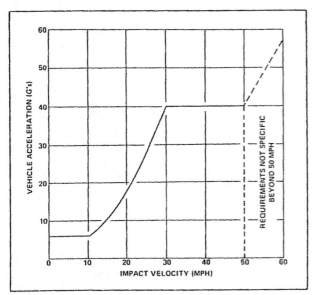

Figure 1

Passenger compartment intrusion is to be limited to three inches at the pillars and four inches at the longitudinal centerline of doors during impact into a fixed 14-inch pole. Impact velocity of 15 mph is normal to the side and at any point along the side.

Passenger compartment doors are to remain closed during any ESV specified crash condition. These conditions include front and rear impacts up to 50 mph and impact angles up to 45 degrees; rollover at 60 mph and side impacts. Impacts may occur with vehicle carrying five restrained or unrestrained occupants.

Passenger compartment doors are to remain operable after impact at any specified ESV crash condition.

Intrusion resistance is to be accomplished with minimum impact on overall vehicle cost and weight. Design concepts employed on the ESV should be susceptible to mass production.

Passenger compartment door systems are to provide ease of opening and closing, and ease of passenger ingress and egress equivalent to current production vehicles.

Side structure is to provide maximum driver field of view. There are to be no more than four pillars of minimum width within the driver's 270 degrees of forward view.

System Description

Major components of the intrusion-resistant side structure are the door panel, door retention hardware, and the passenger compartment side structure. A description of these components and their intended functions follows:

The door panel is an aluminum honeycomb sandwich consisting of an outer sheet, a honeycomb core, an inner sheet and a pair of vertical beams as shown in Figure 2. Door retention hardware along with the passenger compartment structure serve to provide the door panel with non-yielding pin supports fore and aft. Under transverse deflection of the panel, the pin supports develop a longitudinal tensile force which tends to stretch the door panel.

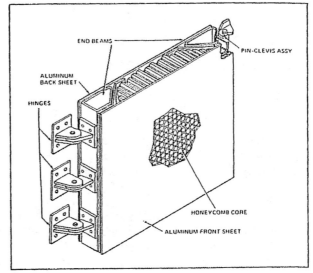

Figure 2

The sequence of behavior of the door panel under transverse loading is initially elastic and then plastic beam action during which the outer sheet is loaded in compression and the inner sheet is in tension. With increasing transverse load, the hoeycomb core begins to crush, decreasing the effectiveness of the panel as a beam. During this action, stress in the outer sheet reverses from compressive to tensile, while the inner sheet increases in tension. When the honeycomb is completely crushed both the inner and outer sheets are plastically stretched as membranes under tensile stress. This tensile stress in the sheets is reacted by the vertical beams which transmit the load to the door retention hardware. During deformation of the door panel, energy is absorbed in crush of the honeycomb and in plastic membrane stretching of the door sheets. In addition, a small amount of energy is stored by elastic deformation of the various structural elements.

Door retention hardware consists of three high-strength steel hinges, three pin clevis assemblies and a conventional door latching assembly. The hinges and pin clevis assemblies serve to provide the non-yielding pin supports for the door panel. As such, they provide the load path to the structure for the applied external transverse force and for the generated longitudinal tensile force. The conventional latch is adequate to resist internal transverse forces generated by an occupant striking the door during impact. The door retention hardware, and door panel along with a molded fiberglass inner and outer panel, window and window actuating mechanism, interior padding, etc., form the complete door assembly shown in Figure 3. Front and rear doors are similar in concept, although slightly different in shape. Rear doors also have fixed glazing.

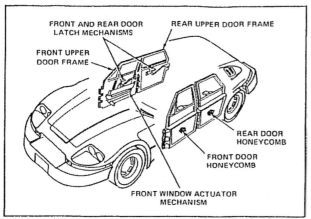

Figure 3

The portion of the passenger compartment structure which is involved in intrusion resistance to side impacts includes the A, B, and C posts and pillars, the perimeter frame, roof rails, and a honeycomb sandwich padding. In addition, there are several auxiliary transverse and longitudinal members which are employed to distribute loads throughout the vehicle. With the exception of the honeycomb sandwich padding, the structure is fabricated from high-strength sheet and tubing. The assembly is an all-welded integral structure shown in Figure 4. During impact by another vehicle, the side structure remains essentially elastic.

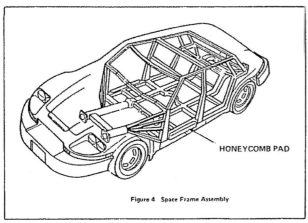

Figure 4 Space Frame Assembly

Figure 4

The aluminum honeycomb sandwich is installed outside of the perimeter frame in the region of the front door. During pole impact in this region, the honeycomb is crushed absorbing energy and distributing the load to the perimeter frame which remains essentially elastic. Pole impacts in the region of the shorter span rear door or at the posts will result in plastic deformation of the structure. A summary of properties of material utilized in the intrusion-resistant side structure is given in Table 1.

TABLE I
Properties of Material Utilized in
Intrusion Resistant Side Structure

	Material	Property
Door Honeycomb sandwich outer sheet	6061 0 AL	17 ksi Ultimate
Door Honeycomb sandwich core	ACG AL	178 psi Crush
Door Honeycomb sandwich inner sheet	7075 T6 AL	76 ksi Ultimate
Door Vertical beams	6061 T6 AL	42 ksi Ultimate
Hinges & Pin Clevis Assembly	AISI 4140 ST	200 ksi Ultimate
Latch Assembly	American Motors	5000 lb Transverse
Passenger Compartment Structure	ASTM 517 ST	100 ksi Yield
Frame Honeycomb sandwich sheets	2024 T3 AL	65 ksi Ultimate
Frame Honeycomb sandwich core	5052 H39 AL	750 ksi Crush

Method of Analysis

The general method of achieving a design is shown in Figure 5, and involves three separate analyses. The first is a deflection analysis to obtain the load-deflection characteristics of components of a structural system. Depending on complexity and the nature of the structure, this is accomplished through hand analysis or by use of a suitable computer model. The second analysis is that of determining the dynamic response loads of a structural system under various crash conditions. This is accomplished using a computer model, and the load deflection data previously obtained. Finally, stress levels in the structure are calculated at the peak dynamic response. A preliminary design may be cycled several times through the analytical loop before acceptable results are obtained.

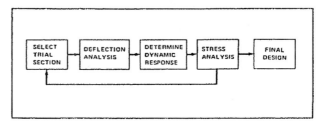

Figure 5

There are currently three mathematical models utilized in the analysis of impact problems. These are:

- SHOCK — a nonlinear, lumped mass, dynamic response program for determining the behavior of systems under impulsive loadings.
- STRESS — a finite element beam program for the solution of space frame type structures.
- SAP — a general finite element program providing versatility in modeling three-dimensional structures.

The manner in which this approach was utilized in the specific problem of designing an intrusion-resistant side structure under the pole impact condition follows. The critical design condition was considered to be pole impact at the center of the front door.

STEP 1 Determine load deflection characteristics of door panel. The load-deflection behavior of the door panel was obtained by superposition of the various behavioral modes of the structure. Sequentially, the behavioral modes involved are: (a) elastic beam action; (b) plastic beam action, slight stretching of inner and outer sheet; (c) crush of honeycomb core, stretching of outer sheet, gradually increased stretching of inner sheet; (d) pure membrane stretching of inner and outer sheet. Membrane behavior was modeled and analyzed for load deflection characteristics using the truss model shown in Figure 6. This model is based on the

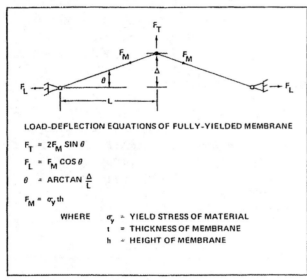

LOAD-DEFLECTION EQUATIONS OF FULLY-YIELDED MEMBRANE

$F_T = 2F_M \sin\theta$

$F_L = F_M \cos\theta$

$\theta = \arctan\dfrac{\Delta}{L}$

$F_M = \sigma_y th$

WHERE σ_y = YIELD STRESS OF MATERIAL
t = THICKNESS OF MEMBRANE
h = HEIGHT OF MEMBRANE

Figure 6

assumption that a plastic hinge is initially formed at the load point and that the membrane behaves as a truss mechanism in post-yield behavior. The equations governing the load-deflection behavior of the mechanism in the fully plastic stage are presented in Figure 6. The accumulative load deflection curve for the panel is shown in Figure 7. Three regions under the curve are defined. Region "A" indicates the energy absorbed through beam action which also provides most of the

panel stiffness at low deflections. Region "B" indicates energy absorbed by crush of the honeycomb core. Region "C", which accounts for the major energy absorption capacity of the panel structure is obtained through membrane stretching of the sheets.

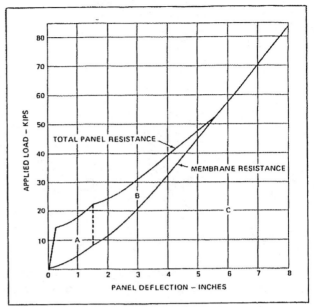

Figure 7

STEP 2 Determine load deflection characteristics of the frame honeycomb sandwich panel. The load deflection characteristics of this panel are shown in Figure 8.

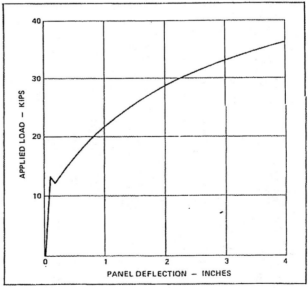

Figure 8

STEP 3 Analysis of the system under dynamic loading. The side structure system was modeled as shown in Figure 9. Included are the load-deflection characteristics of the door panel, frame honeycomb

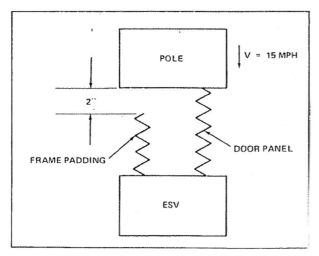

Figure 9

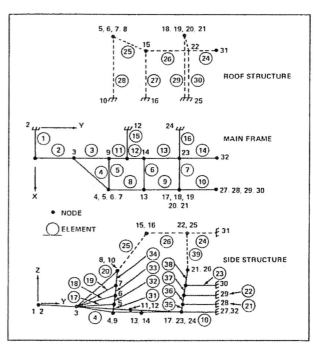

Figure 10

TABLE II
Applied Loads During Pole Impact

Node	Type	Direction	Magnitude
5	Force	X	-12.0 k
5	Force	Y	43.0 k
5	Moment	Z	150 in k
6	Force	X	-12.0 k
6	Force	Y	43.0 k
6	Moment	Z	150 in k
7	Force	X	-12.0 k
7	Force	Y	43.0 k
7	Moment	Z	150 in k
18	Force	X	-12.0 k
18	Force	Y	-43.0 k
18	Moment	Z	-150 in k
19	Force	X	-12.0 k
19	Force	Y	-43.0 k
19	Moment	Z	-150 in k
20	Force	X	-12.0 k
20	Force	Y	-43.0 k
20	Moment	Z	-150 in k
13	Force	X	-100 k

sandwich panel, mass of the vehicle and velocity discontinuity at the point of impact with the pole. A more complete model would include load deflection characteristics of the "A" and "B" posts and the perimeter frame, as well as rotational moments and vehicle moment of inertia. This model was exercised using the SHOCK computer code. The results of the analysis indicated that the pole would intrude into the striking vehicle about 7¼ inches measured from the point of initial contact with the door panel. This would result in an intrusion of the vehicle inside surface of approximately 3¼ inches compared to the allowable 4 inches. Door panel transverse load at this deflection is approximately 72 kips while the longitudinal load applied to the "A" and "B" posts is approximately 129 kips. The analysis indicated that the frame honeycomb sandwich panel would be completely crushed allowing direct pole contact with the perimeter frame resulting in a short duration load spike.

STEP 4 Design of structural elements. The final step in the analysis is that of sizing structural members using the dynamic response loads developed previously. In general, this step is an iterative process in which element dimensions are selected, analyzed and new dimensions chosen until a satisfactory design is achieved.

The passenger compartment side structure was designed using the STRESS computer code. The structure was modeled as shown in Figure 10, and analyzed for the loading conditions given in Table 2. Loads given in Table 2 are derived from the dynamic response analysis while the applied moments are those developed by the door retention hardware. A summary of member cross-sections and calculated peak axial and bending stress in each member is given in Table 3. These results were considered satisfactory, and the structure was fabricated for testing as modeled.

Test Results

The test program supporting the development of the intrusion-resistant side structure was conducted in two phases. The first phase involved component development

611

in which several evolutionary versions of the aluminum honeycomb panel and door retention hardware were built and tested until satisfactory results were obtained. The second phase involved the construction and testing of a complete structural vehicle.

TABLE III
Side Structure Cross Sections & Peak Stresses

Element	Name	Section	Axial	Bending
1	Front Housing Stabilizer	3x3x.120	6 8 (ksi)	63 2 (ksi)
2	Front Housing	3x5x 188	1 3	46 3
3	Front Housing	5x5x 188	7 3	40 2
4	Lower A Post Support	2x2x 062	5 4	133 0
5	A Post Lateral	4x4x 120	22 4	116.2
6	Front Door Lateral	3x3x 100	85 9	40 7
7	B Post Lateral	4x4x 120	13 5	91 5
8	Floor Sill A to Center	5x4 to 3x3x 120	41 5	118 6
9	Floor Sill- Center to B	5x4 to 3x3x 120	66 4	64 3
10	Floor Sill B Aft	3x4x 120	29 3	33 1
11	Front Housing	5x5x 188	1 2	85 9
12	Main Frame	4x4x 120	4 5	131 2
13	Main Frame	4x4x 120	5 0	127 2
14	Main Frame	4x4x 120	9 7	21 2
15	Torsion Bar Frame	3x4x 120	56 4	58 4
16	Center Cross Frame	4x4x 120	32 6	34 3
17	Lower A Hinge Support	1¼x1¼x 250	18 5	20 8
18	Mid A Hinge Support	1¼x1¼x 250	35 8	24 8
19	Upper A Hinge Support	1¼x1¼x 250	18 7	11 4
20	Upper A Post Support	1¼x1¼x 250	14 7	135 3
21	Rear Panel Simulation	1 3/8x 095		
22	Rear Panel Simulation	1 3/8x 095		
23	Rear Panel Simulation	1 3/8x 095		
24	Roof Sill - Aft	3x2x 188	15 5	37 0
25	A Pillar	2½x2½x.120	26 5	115 5
26	Roof Sill -Fwd	3x2x 188	14 9	128 9
27	A Roll Bar	2x2x 120	10 7	109 0
28	A Cross Frame	2½x2½x.120	35 5	122 5
29	B Cross Frame	3x4½x 188	22 9	112 1
30	B Roll Bar	2x2x 120	1 6	38 0
31	A Post	3x8 to 3x3x 120	12 5	101 5
32	A Post	Same as 31	5 2	112 8
33	A Post	Same as 31	11 4	139 5
34	A Post	Same as 31	11 1	139 5
35	B Post	3x8 to 3x3x 120	1 8	57 3
36	B Post	Same as 35	2 3	151 5
37	B Post	Same as 35	2 3	151 5
38	B Post	Same as 35	3 3	129 1
39	B Pillar	2x2x.120	1.3	143.2

The first phase — component development test program — was conducted using the AMF crash simulator. This facility, which has been utilized to support a variety of automobile research programs, has the following features:

- Volumetric capacity in excess of a full size automobile
- Static or dynamic load capacity
- Tri-directional loading capacity
- Load application capacity of 100,000 pounds, 25,000 pounds, and 15,000 pounds along orthogonal directions
- Ability to apply loads separately, sequentially or simultaneously
- Ability to simulate crash load pulses including control of onset rate, pulse magnitude and pulse duration

- Fixturing to mount and provide controlled retention of a wide range of component size, shape and configuration

The result of a dynamic test of the door panel and door retention hardware is shown in Figure 11, along with analytically predicted behavior. Up to the point at which a hinge attachment failed, the behavior was reasonably close to expected behavior and was considered to be satisfactory. The failure in a weld at the hinge attachment was attributed to a manufacturing deficiency and not to the design. Observations made using high-speed photography and strain gages confirm the occur-

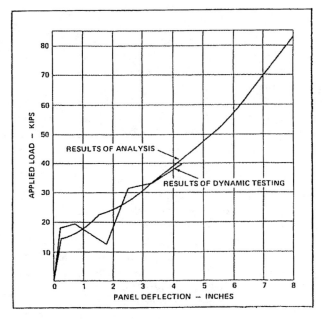

Figure 11

rence of expected sequential modes of behavior. Prior to failure, the panel had exhibited both elastic and plastic beam action, stress reversal in the front sheet and honeycomb core crush through more than half of its thickness. Peak loads were 39.6 kips, transverse, and 70.1 kips, longitudinal at a ram stroke of 4.25 inches. Post-test observation disclosed a vertical break through the front sheet at the point of contact. This was attributed to the occurrence of buckling while stressed in compression and subsequent failure in tension at the crease formed by buckling.

A door panel of the same design was installed on a structural vehicle. The assembly was impacted against a

13 inch rigid pole at 15 mph. Observations made during the event were:

- Maximum deflection of door panel — 7 inches
- Maximum intrusion into passenger compartment —3¼ inches
- Maximum pole load — 106 kips
- Complete crush of honeycomb core in door panel and frame padding

The side structure behavior during the test was considered to be satisfactory and close to analytically predicted behavior. Post-test observations revealed the occurrence of a crack through the outer door panel sheet identical to that observed during component testing. Prototype door panel design was modified to incorporate a slightly thicker and more ductile outer sheet.

Summary and Conclusions

A passenger compartment side structure system capable of resisting intrusion during side impacts has been designed, built, and tested for the Department of Transportation ESV program. The system consists of honeycomb sandwich door and frame panels, high-strength door retention hardware, and high-strength supporting structure.

Observed behavior of the system during component testing and full-scale vehicle crash testing is in close agreement with analytically predicted behavior and satisfies intrusion limits defined in performance objectives. Energy absorption capacity attributed to the various structural behavioral modes of the door panel are isolated and evaluated. This exercise demonstrates the value of the various design features of the panel, and will allow the design to be easily modified to satisfy a change in energy absorption requirement or a change in intrusion limitation.

The design of the side structure system concept discussed in this paper employs state-of-the-art materials and manufacturing techniques, although these materials and techniques may not be currently employed in mass production techniques. The concept is seen to satisfy the immediate objectives, but its ultimate worth is dependent on whether the design can evolve from a hand-crafted laboratory version to an economic, reliable mass production version. To date, no study has been made of the feasibility or cost of this evolution. However, because of the fact that the system is relatively light weight, adaptable to varying requirements, and employs state-of-the-art materials and techniques, there is good potential for successful conversion to mass production.

Development and Analysis of
Door Side-Impact Reinforcements

John S. Haynes
Chrysler Corp.

OUR CURRENT DIRECTION at Chrysler for door side-impact protection involves the use of a rolled section beam in conjunction with reinforcement of the basic body structure. This beam design allows efficient use of material in supporting loads and facilitates economical manufacturing methods. Structural analysis techniques used with the design have proven to be a reliable predictor of static door crush performance. Steps that led up to these results will be covered in this report, along with a detailed description of the analytical techniques that are now used to evaluate door reinforcement beams. Due to the flexible nature of the analysis, it is recommended for general use.

DYNAMIC CRASH PROTECTION

It was first decided that the objective of a side-impact protection system should be the limitation of passenger compartment intrusion under dynamic crash conditions. Initially, no specific goals in terms of loads or deformation were established, other than general agreement that significant penetration reduction in staged car crashes would be desirable. A 25 mph, 45 deg to rear impact at the center of a door was established as a test criterion that would reveal the most information about door structural performance. Several tests of nonreinforced cars were run as a baseline for evaluating design proposals. It was soon found that body pillar, door hinge, and door lock construction played a significant role in protecting passenger compartment security. Fortunately, specific modifications in this area were easily identified and have been incor-

porated into basic body design. This discussion deals only with the door reinforcing structure.

The baseline tests also pointed out a tendency for the striking car to ride up the side of the struck car and cause maximum encroachment to occur at the door belt line. On the basis of this evidence it was concluded that any door reinforcement should be placed as high as possible in the door to trap the encroaching car and force it down into the underframe structure.

DOOR REINFORCEMENT DEVELOPMENT

Traditional structural engineering philosophy was applied to the first proposal for a door side-impact reinforcement, that is to fill up the available space in the door with a welded hat section impact reinforcement. Using these guidelines, the beam for our standard size cars turned out to be a two-cell boxed reinforcement with an average depth of 0.87 in and a metal thickness of 0.075 in (See Fig. 1.)

With confidence that more space would eventually be made available in doors for side-impact reinforcements, we began looking at the efficiencies that could be derived by increasing beam section depth. The analytical tool that was used for evaluation of various proposals was based on section modulus (Z) as an indication of beam strength, expressed as $Z = I/c$ in $S = M/Z$. It was observed, in a series of static bench tests, that section shape had as much of an effect on beam strength as I/c. This outcome was really no big surprise. The neatly squared beam corners that produced a high section modulus

———————————————— ABSTRACT ————————————————

A door side-impact reinforcement beam has been developed that allows efficient use of material in resisting side crush loads. The beam section can be roll formed, thus permitting further economies in fabrication. Analytical techniques have

been developed that evaluate and handle bending, buckling, and crippling in beam design. This paper covers the development that led up to these results and includes a detailed description of how to apply the analytical methods.

were unstable and buckled before the metal reached its yield strength. Large radiused corners and beads in the faces helped a beam hold its shape longer and yield at a higher load. Many basic section shapes were evaluated with static bending tests. It is interesting to observe the ranking of these various section shapes in terms of weight-to-energy ratio, as is shown in Fig. 2. As it turns out, the most efficient of all the proposals was a three-cell constant or corrugated beam section.

Further investigation of this design revealed more advantages and no serious disadvantages. (See Fig. 3.) The main reinforcement member could be fabricated by rolling a continuous

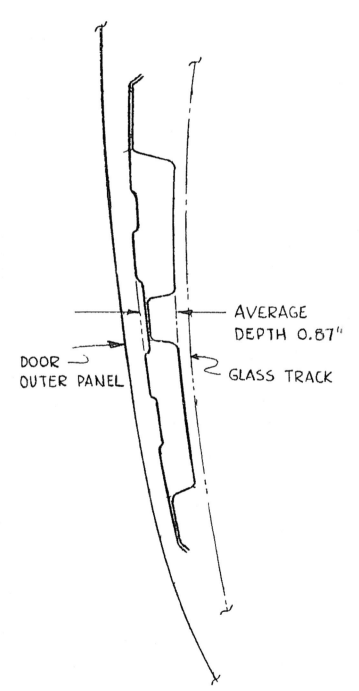

AVERAGE DEPTH 0.87"

DOOR — OUTER PANEL

GLASS TRACK

Fig. 1 - Two-cell boxed beam section—standard size car

section that is cut off at different lengths to fit various doors. Resistance spot-weld attachment of a center plate for additional strength and end plates for attachment to the door completes the reinforcement assembly. Tooling of all these components is simple, resulting in significant cost and lead time reductions. Further refinements in our analytical technique supported the empirically based conclusion that the three-cell rolled section is indeed a very efficient design.

MATERIALS DEVELOPMENT

At about the same time that we were optimizing the beam section, our metallurgy department in conjunction with the Inland Steel Co. developed a new low carbon-nitrogen grade of sheet steel. This steel has about the same forming properties as commercial quality body sheet with a 50-75% increase in yield strength at 7% elongation. Considering actual elongation and buckling characteristics, use of this material for side-impact beams represents at least a 35% increase in effective strength over conventional sheet steel.

FEDERAL DOOR CRUSH STANDARD

Introduction of Federal Motor Vehicle Safety Standard 214 for door side crush strength in October 1970 shifted development priorities to static door strength performance rather than actual crash performance. Since the optimum beam section had already been established at this point, conformance to the standard was solely a matter of selecting the proper metal gage for each application.

SECTION ANALYSIS

When the three-cell rolled beam was selected for use, the section shape variable was eliminated from our comparative analysis. However, even under these conditions, section modulus was not a consistent predictor of beam strength performance. Comparing the bench test results of beams with various metal gages and section depths, it was observed that as the section became deeper and the metal gage was reduced, bending strength performance dropped below the calculated level. To help identify this relationship, the following beam performance function f was established:

$$f = 1 + (\text{actual bending load}$$
$$- \text{calculated bending load})/\text{calculated bending load} \qquad (1)$$

The function f would be 1.0 for beams that performed exactly as calculated. Another function, d/t (beam depth/beam gage), was used to describe the beam shape. A graph of f versus d/t was plotted for all of the corrugated beam bench tests. Although there was some scatter of data, a straight line relationship was established for the effect of beam depth and thickness. By calculating the slope of this curve, the following relationship between f and d/t was established:

$$f = -0.019 (d/t) + 1.51 \qquad (2)$$

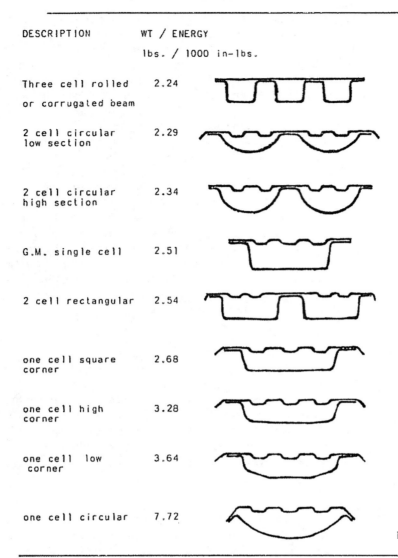

DESCRIPTION	WT / ENERGY
	lbs. / 1000 in-lbs.
Three cell rolled or corrugated beam	2.24
2 cell circular low section	2.29
2 cell circular high section	2.34
G.M. single cell	2.51
2 cell rectangular	2.54
one cell square corner	2.68
one cell high corner	3.28
one cell low corner	3.64
one cell circular	7.72

Fig. 2 - Comparison of impact beam section proposals

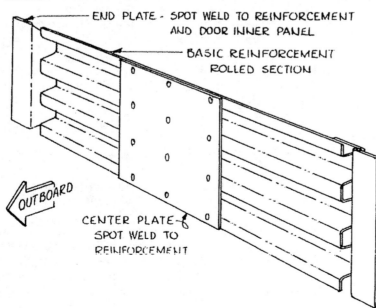

END PLATE - SPOT WELD TO REINFORCEMENT AND DOOR INNER PANEL

BASIC REINFORCEMENT ROLLED SECTION

OUTBOARD

CENTER PLATE SPOT WELD TO REINFORCEMENT

Fig. 3 - Rolled section beam construction

Since the d/t factor of any untested beam section proposal is known, this relationship can be used to find f, which will allow adjustment of bending strength calculations to produce correct results. This method allowed for a significant improvement in our ability to specify correct metal gage for door side-impact beams.

BEAM STRENGTH AND STABILITY ANALYSIS

The d/t relationship was used with full realization that it was an empirical function derived from a limited amount of data. We had a useful analytical tool, but our ultimate goal of finding an absolute method for calculating door crush performance was still unfulfilled. It was then discovered that the uniform rolled-beam section lent itself well to stress analysis methods used in the aerospace industry. These methods consider that simply supported thin walled structural members, subjected to a bending load, can fail either by bending, buckling, or crippling.

BENDING STRESS - In bending failure the beam section holds its shape until the metal yield strength is reached. In analyzing this failure mode, allowable bending moment is increased over that found by the classical $S = Mc/I$ relationship by the addition of a section factor f. This factor is a function of the first area moment of the beam section. A separate value is determined for each side of the neutral axis to account for loading direction. In order to take full advantage of material strength properties, a beam must fail in this mode.

The bending stress allowable (Fig. 4) can be increased by the section factor K, which is a function of the first area moment of the beam above the neutral axis (1)*. A typical example follows:

Section properties of the entire section:

$$I_{na} = 0.316 \, in^4 \quad A = 0.974 \, in^2 \tag{3}$$

First area moment of solid area about the neutral axis:

$$Q = \Sigma A_i d_i \tag{4}$$

For the example section $Q = 0.253 \, in^3$.

---

*Numbers in parentheses designate References at end of paper.

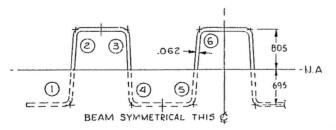

Fig. 4 - Calculation of bending stress allowable

For compression side:
Section factor $K_c = 2 \, Q \, c/I$;

$$K_c = 1.30 \tag{5}$$

For tension side:
Section factor $K_t = 2 \, Q \, c/I$;

$$K_t = 1.10 \tag{6}$$

Assuming that $F_{tu} = 65$ ksi,
compressive ultimate bending allowable $= 1.3 \, (65 \, ksi) = 84.8$ ksi
tension ultimate bending allowable $= 1.1 \, (65 \, ksi) = 71.5$ ksi

BUCKLING STRESS - Local buckling is defined as wall deformation without translational or torsional movement of the beam as a whole (hard corners remain fixed). Allowable buckling stress, σcr, is described by the following relationship:

$$\sigma cr = \frac{K_t \, \pi \, E}{12 \, (1 - \gamma^2)} \left(\frac{t}{b_t} \right) \tag{7}$$

The value for K_t can be obtained from empirically determined curves as a function of section shape, as shown in Fig. 5. Local buckling does not constitute failure but means that the beam wall can no longer carry load. This action transfers the load to hard corners, thus precipitating a crippling failure. A section should be designed so that allowable buckling stress is higher than crippling stress.

CALCULATION OF CRIPPLING STRESS ALLOWABLE

In crippling failure, wall deformation is accompanied by translational or torsional movement of the entire beam section (hard corners distort). Allowable crippling stress, F_{cc}, is described by the following relationship:

$$F_{cc} = \frac{\Sigma F_{cci} A_i}{\Sigma A_i} \tag{8}$$

Definitions of variables and curves for determining F_{cc} are shown in Table 1 and Figs. 6 and 7.

The crippling stress allowable, F_{cc}, can be determined from the curves of Figs. 6 and 7. This method is empirical, using an angle as the basic element. The allowable stress given is for a compressive load with a maximum cutoff stress for each angle element $= F_{cy}$.

For a beam in bending with the top of the hat section in compression (as in the example), a cutoff stress for each angle element can be equal to the compressive bending allowable with bending modulus applied.

Crippling allowable, F_{cc}, for the example section in Fig. 4 can be found by summing the properties of each of the angle elements as shown in Table 1. Since the beam is symmetrical, there will be six angle elements.

COMPRESSIVE CRIPPLING ALLOWABLE FOR SECTION -

$$F_{cc} = \frac{\Sigma F_{cci} A_i}{\Sigma A_i} = \frac{39,540 \text{ lb}}{0.484 \text{ in}^2} = 81.6 \text{ ksi}$$

BEAM ANALYSIS

In evaluating a particular design proposal, the allowable stress for all three failure modes is calculated. The section shape is optimized by producing the highest possible bending stress for a given metal area. An extension of this basic method is used for three-cell rolled section beams. A complete analysis is performed for both the base rolled section and the boxed section in the center plate area. Bending moments are then calculated at the beam center and at the edges of the center plate. A computer program, developed to perform this analysis, finds the optimum center plate length to produce simultaneous failure at both places. Results of this analytical method correlate within 8% of actual static bending strength developed by beams that were tested.

SYSTEM PERFORMANCE

It has been observed, in many door side crush tests, that load deformation performance of a reinforced door can be divided

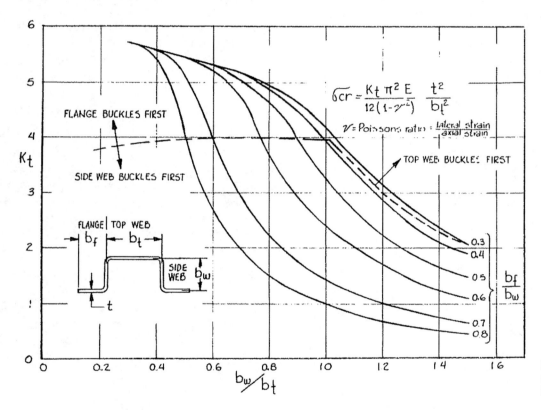

$$\sigma cr = \frac{K_t \pi^2 E}{12(1-\nu^2)} \frac{t^2}{b_t^2}$$

$\nu = $ Poissons ratio $= \frac{\text{lateral strain}}{\text{axial strain}}$

Fig. 5 - Compressive buckling stress σcr for hat section; from NACA-TN

Table 1 - Calculation of Crippling Stress Allowable

No.	a, in	b, in	t, in	$\frac{b'}{t} = \frac{a+b}{2t}$	$\frac{F_{cc}^*}{\sqrt{F_{cy}E}}$	F_{cc}, ksi	F_{cc} cutoff, ksi	Area, in^2	P_{cc}, lb
1	0.82	0.65	0.062	11.9	0.053	74.0	_**	0.091	6730
2	0.78	0.50	0.062	10.3	0.064	89.5	84.5	0.079	6670
3	0.50	0.78	0.062	10.3	0.064	89.5	84.5	0.079	6670
4	0.65	0.60	0.062	10.1	0.065	90.8	84.5	0.078	6590
5	0.60	0.65	0.062	10.1	0.065	90.8	84.5	0.078	6590
6	0.78	0.50	0.062	11.9	0.057	79.6	—	0.079	6290
							Summation	0.484	39,540

*Use F_{cu} (ultimate compressive stress) for rolled section beams; 65 ksi, for the example.

**Cutoff stress is equal to compressive bending allowable with bending modulus (see Fig. 4)

618

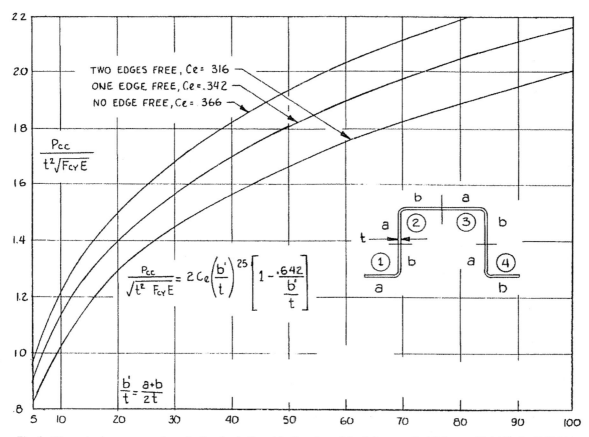

Fig. 6 - Dimensionless compressive crippling load allowable, P_{cc}; from Jrl. of Aeronautical Sciences, Vol. 21, April 1954

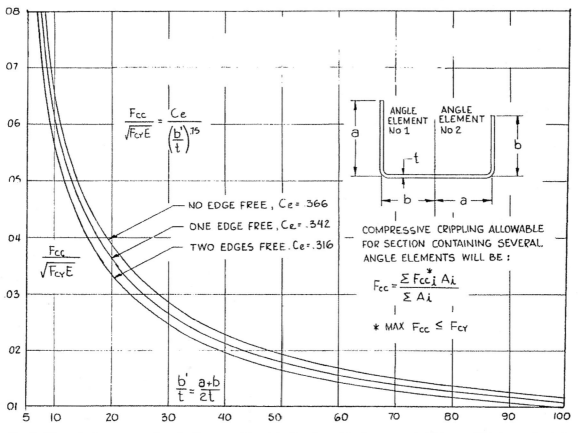

Fig. 7 - Dimensionless compressive crippling stress allowable, F_{cc}; from Jrl. of Aeronautical Sciences, Vol. 21, April 1954

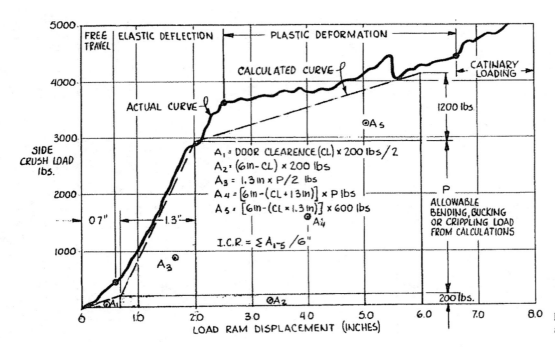

A_1 = DOOR CLEARENCE (CL) × 200 lbs/2
A_2 = (6 in - CL) × 200 lbs
A_3 = 1.3 in × P/2 lbs
A_4 = [6 in - (CL + 1.3 in)] × P lbs
A_5 = [6 in - (CL × 1.3 in)] × 600 lbs

I.C.R. = $\sum A_{1-5}$ /6"

Fig. 8 - Comparison of actual and calculated door crush loads

into four distinct phases: free travel, elastic beam deflection, plastic beam deformation, and catinary loading. (See Fig. 8.)

FREE TRAVEL - Door crush load versus deformation for FMVSS 214 compliance is measured from the point where the loading ram contacts the door outer panel. Load deformation rate follows the characteristics of a nonreinforced door until contact is made with the reinforcement beam. This free travel phase therefore is a function of impact beam-to-door outer panel clearance.

ELASTIC BEAM DEFLECTION - This second phase involves elastic bending of the impact beam, characterized by a marked increase in the slope of the load deformation curve.

PLASTIC BEAM DEFORMATION - The beginning of this third phase is marked by a sharp drop of the load deformation rate that occurs when the impact beam cripples or yields. The slope of this phase is a function of the beam's ability to support load after it initially fails. The three-cell rolled section beam has demonstrated its superiority over other designs in supporting load after failure as a beam.

CATINARY LOADING - After about 6 in of ram travel, the impact reinforcement has lost its ability to support load as a beam. Its resistance to side crush load is now a function of the tensile load developed at the end attachments. This ability is, in turn, a function of door latch and hinge strength. We have found that our current designs, which are a product of the initial side crash performance investigation, are more than adequate to meet the requirements of FMVSS 214.

ABSOLUTE ANALYSIS

It has been found that these four phases occur in a predictable manner that allows overall crush performance, relative to FMVSS 214 requirements, to be calculated. If a door rein-

forcement system meets I.C.R. requirements of the standard (average load to 6 in of 2250 lb), it will also meet or exceed intermediate and peak load requirements. Therefore, our analysis is performed only to a deformation of 6 in.

Load rate during free travel is 286 lb/in. Average load during this phase then is a function of beam-to-door outer panel clearance.

Elastic deflection of a rolled section beam follows a rate that will arrive at a peak load after 1.3 in of ram travel. The value for P is the critical bending, buckling, or crippling load, calculated by the previously described method.

The final phase, applicable during this analysis, is plastic beam deformation and is equal to 50 lb/in.

I.C.R. is calculated by summing the contribution of these three pertinent phases, as shown in Fig. 8. Since the only undefined variable, beam-to-door outer panel clearance, is usually a design variable, it is possible to predict door side crush performance of early design proposals using this analytical method.

ACKNOWLEDGMENTS

Development of a side-impact protection system and related analytical procedures is the product of the efforts of many people in the Chrysler Engineering organization. Of particular importance, however, is the contribution of three individuals in the Computer Aided Design & Development group. F. J. Glasgow directed early investigation of dynamic crash performance and beam section development. D. J. Courtright conducted extensive analysis of the corrugated-type beam and established the d/t relationship. T. L. Treece developed the absolute method for calculating door side crush performance.

OCCUPANT PROTECTION IN SIDE IMPACTS

MR. M. RODGER, *Principal Research Engineer*
Ford Motor Company of Britain

INTRODUCTION

The special problems associated with improved occupant protection in side impacts are well known. By comparison with frontal impacts, the door and side structure of cars offer only limited space in which to provide structural strength and energy absorption. It is also possible for the occupant to be closer to the point of impact than in any other type of accident, except, perhaps, rollover. It should be noted that this applies largely to those occupants on the impact side of the vehicle, for whom seat belts are ineffective. In contrast, a lap and diagonally belted occupant on the far side from impact is fairly well protected, and the universal wearing of seat belts would effect an important reduction in injuries from side impacts.

In terms of overall statistics, Figure 1 shows that side impacts constitute almost 30% of all injury-producing accidents[1], and Figure 2 shows that more than half of these are vehicle to vehicle impacts.[2] It can also be seen from this figure that the breakdown of fatal and serious injuries is very similar to the breakdown of accidents. The greatest potential gain, therefore, lies in reducing the effects of car to car impacts. This approach is also the most cost effective, since it is even more difficult to mitigate the effects of truck and fixed object impacts.

SIDE STIFFNESS AND INTERIOR PADDING

Occupant protection in side impacts is dependent on both the form of restraint employed and the vehicle structural strength. For the occupant on the impact side, the short distance between occupant and impact renders ineffective most of the systems currently used for protection in frontal impacts. Seat belts cannot offer enough lateral restraint to head and torso, and current air bag systems are too slow, both in sensing the impact and inflating the bag. The most

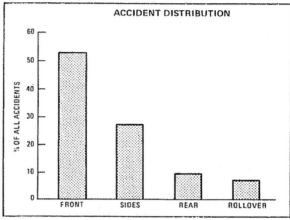

FIGURE 1

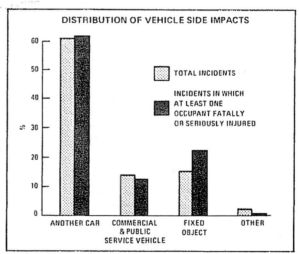

FIGURE 2

effective type of restraint is therefore a form of energy absorbing padding, which, while technically possible, has the disadvantage of permanently taking up space inside the car. This space may have to be paid for in reduced occupant comfort. The alternative is an undesirable increase in the overall widths of cars

2-118

on our already crowded roads. Exactly how much space will be required is at this stage uncertain, because of limited human tolerance data on side impacts. American studies have indicated that lateral tolerance limits may be much lower than longitudinal ones.

The structural requirements are even more difficult to accurately define. While strengthening the vehicle side may seem to be an obvious choice, it can be over-rated. As a general trend it is clearly advantageous, but in those cases where the vehicle slides bodily into a fixed object, the increased side strength brings no benefit, and can in some instances be harmful in increasing the decelerations of the undeformed parts of the vehicle. In cases where a vehicle is hit fair and square in the side, as in Figure 3, the bullet car initially penetrates the target car at the impact velocity, but the strength and inertia of the target car gradually slow it down until a common velocity is reached. During this phase, the deforming side of the target car is effectively moving towards the occupants at the speed of the bullet car. The effect of increased stiffening is to reduce the time and penetration which occur prior to common velocity. This in turn will reduce the velocity of impact of an occupant with the interior if it occurs at this stage. However, no amount of stiffening will reduce the common velocity, and the impact velocity of an unrestrained passenger can never be very much lower than this velocity.

The important point is that side strength should really be considered in conjunction with energy absorbing padding if optimum protection is to be given in really severe impacts. Present efforts should continue to be directed at doors, locks and hinges, which, although they may deform, do not actually fail. This can go a long way to reducing dangerous intrusions and to increasing the tendency for cars to glance apart in the more acute angles of impact. It is also a vital step in reducing the incidence of occupant ejection.

MASS AND AGGRESSIVITY

Both mass and aggressivity are functions of vehicle size and design, and to investigate the effects of mixing the extremes, a trio of 90° vehicle to vehicle side impacts was carried out using Escorts and Mark IV's. These tests are referred to later in this report, but one of the chief conclusions drawn with respect to these particular vehicles was that under side impact conditions, it is safer to be in a Mark IV hit by an Escort than in an Escort hit by a Mark IV. This is due not only to the greater mass of the Mark IV, but

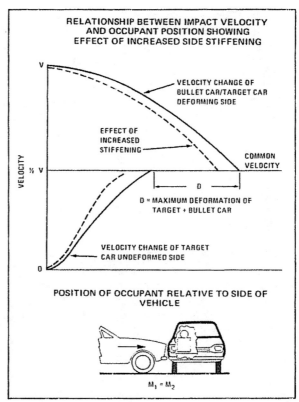

FIGURE 3

also to the strength of its center door pillar and the greater distance between the occupant and the initial impact. Nor is the Mark IV as dangerous a bullet car as might be expected. Its large flat front proved to be an effective load spreader; and it produced less penetration when impacting an Escort than was produced by another Escort.

As a target car, the large car is safer only if its higher mass is effectively utilized for structural strength. The light car can be made to provide adequate protection, but in impacts between vehicles of differing masses, the secondary impact velocities are higher for the occupants of the lighter car, thus increasing the design problem of providing adequate energy absorption in the space available.

It is difficult to generalize on the effects on side impacts of the high impact bumpers which are legislated in America. While they may improve load spreading in a 90° impact, this benefit may well be cancelled out by the increased structural stiffness which is required to sustain a 5 or 10 mph impact. Their effect in angled impacts will be even more acute, and the opener effect of a reinforced bumper corner very difficult to resist. There is no doubt that some of the design trends which made vehicles safer

2-119

622

for their own occupants can make them more dangerous to other road users, and especially to other vehicles not built to the same safety standards.

TEST PROCEDURES

SAE Contoured Mobile Barrier

One of the earliest proposals for a standardized side impact test involved the mobile barrier shown in Figure 4. From any angle, this is a very aggressive piece of equipment, and when used at 45° from the front, the curved-under lower edge, and the almost infinitely stiff corners, combined to take it over the door sill and through the door in a highly effective and unrealistic fashion.

FIGURE 4

Vehicle to Vehicle Angled Impacts

Far more realistic, but still severe, is the vehicle to vehicle impact, and Figure 5 shows the set up for a 45° from the front impact. Figure 6 shows the results of such an impact at 50 km/h on an Escort, while Figure 7 shows the results of a slower speed (33 km/h) impact on a Capri. Figure 8 shows the result of reducing the impact angle to 20° for a 50 km/h impact on a Capri. The effect of reducing either the angle of impact or the speed is to considerably reduce the vehicle damage caused.

Approximately 30% of side impacts are between vehicles whose directions intersect at angles between 0° and 70° from the front.[3] Unfortunately, such data are incomplete, unless the velocities of both vehicles are known in order to determine the true direction of the impact forces. In practice, it is likely that the proportion of actual impacts from the front is more than 30%.

Figure 9 shows some occupant head acceleration data from a 50 km/h 45° vehicle to vehicle impact using standard Mark II Cortinas. As mentioned earlier in this paper, restrained occupants on the far

FIGURE 5

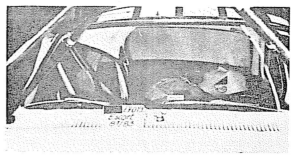

FIGURE 6

FIGURE 7

FIGURE 8

2-120

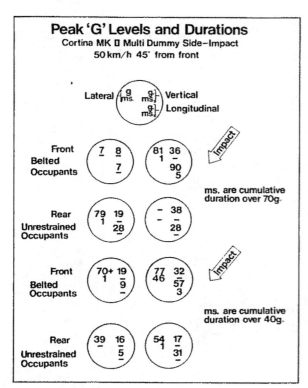

Peak 'G' Levels and Durations
Cortina MK II Multi Dummy Side-Impact
50 km/h 45° from front

Lateral $\left(\frac{g}{ms}\ \frac{g}{ms}\right)$ Vertical
$\frac{g}{ms}$ Longitudinal

Front
Belted
Occupants $\left(\frac{7\quad 8}{7}\right)$ $\left(\frac{81\quad 36}{1\quad -}{90\ \ 5}\right)$

Impact →

ms. are cumulative
duration over 70g.

Rear
Unrestrained
Occupants $\left(\frac{79\quad 19}{1\quad 28}\right)$ $\left(\frac{-\quad 38}{-\quad 28}\right)$

Front
Belted
Occupants $\left(\frac{70+\ 19}{1\quad 9}\right)$ $\left(\frac{77\quad 32}{46\quad 57}{3}\right)$

Impact →

ms. are cumulative
duration over 40g.

Rear
Unrestrained
Occupants $\left(\frac{39\quad 16}{5}\right)$ $\left(\frac{54\quad 17}{1\quad 31}\right)$

FIGURE 9

FIGURE 10

FIGURE 11

side from impact are relatively safe, which is demonstrated by the lower levels of acceleration.

Vehicle to Vehicle 90° Impacts

The 90° impact has a number of advantages from the analytical point of view, especially in the reduction of rotational effects. It is also a representative case in that about half of all side impacts occur at around this angle.[3] With current vehicles, however, the precise angle is less critical than the precise point of impact. In the worst case, the front sidemembers of a small car may fit neatly between the front and center pillars of some large, or two door, vehicles.

The Escort/Mark IV tests have been referred to earlier, and included Escort into Escort, Escort into Mark IV, and Mark IV into Escort. Figures 10 and 11 illustrate the results of these latter two tests. All the tests were 90° side impacts at 50 km/h. The evidence of the dummy head accelerometers indicated that the impact side front occupant was unlikely to survive in the Escort, but would do so in the Mark IV.

Static Crush

At the same time as these vehicle to vehicle impacts were being carried out, investigations were made into the use of a static crush rig of the type shown in Figure 12. This type of rig has many advantages over other forms of dynamic and vehicle to vehicle tests in that it is inherently more controllable, more repeatable and easier to perform. The main criticism is that it does not directly represent actual crash conditions. Our work showed, however, that the results of this test correlated well with both vehicle to vehicle crash tests and real world accidents. At the same time, they yielded valuable information on the sequence and nature of door and pillar distortion and failures of hinges and locks.

In effect, it is not a simulation of a fixed pole impact, but rather of an impact from the corner of a very stiff and aggressive vehicle. While there is no complete substitute for vehicle to vehicle impacts, the static crush test is a valuable development tool which is not subject to the wide range of results scatter found in vehicle to vehicle tests.

Flat Faced Mobile Barrier

Probably because of criticism of the contoured SAE barrier, several later proposals have favored flat faced barriers. An early proposal for testing fuel tank

2-121

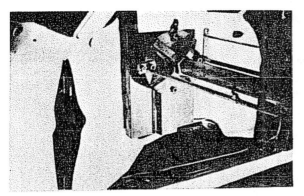

FIGURE 12

integrity in a side impact at 25 km/h is shown in Figure 13. In practice, this is by no means a severe test. A more recent proposal is for the even larger flat face required by FMVSS 208. Figure 14 shows this barrier. It is hard to see what this test represents, as the percentage of injuries caused by vehicles sliding bodily sideways to impact a flat rigid structure at exactly 90° must be extremely small. It has to be considered, therefore, as purely an attempt to assess the protection given to the occupant against secondary impact with the inside of the car.

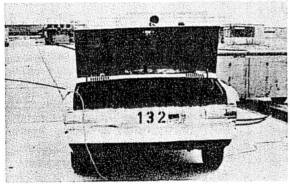

FIGURE 13

FIGURE 14

The face of the barrier is so large that the vehicle side strength is not at all critical. The occupants' heads on the impact side are likely to strike the side pillars or even the face of the barrier. The critical aspects are therefore the thickness of the door, the effectiveness of any interior padding and the positioning of side pillars relative to the occupants. This latter effect is quite arbitrary as a position which is correct for 90° impact could be equally wrong for an angle of 80° or 100°.

Pole Impact

This test basically represents an impact in which the vehicle slides sideways into a fixed pole, although similar results can be obtained using a stationary vehicle, and a pole mounted on a mobile barrier, as in Figure 15. This type of test has featured in several proposals associated with experimental safety vehicles, and undoubtedly much of the reason for this emphasis is the extreme difficulty of protecting an occupant immediately next to the impact, and in the path of the intrusion.

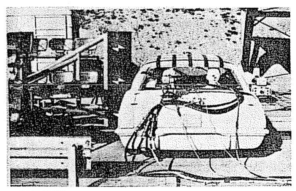

FIGURE 15

However, its frequency of occurrence as a source of serious and fatal injuries is low. Only 20% of side impacts involve fixed objects, and clearly not all of these are pole shaped. Furthermore, because the number of these accidents is small, there is virtually no information on speeds of impact, on which to base a realistic test.

The pole impact can be the hardest of all tests in which to provide protection for the occupant next to the point of impact. In view of its severity and relatively low incidence, a better way of tackling it is through improved design and siting of the roadside furniture involved.

CONCLUSIONS

The work discussed in this paper forms only a small part of the total effort put into safety by Ford of Britain. It was selected to re-emphasize the problems associated with improved occupant protection in side impacts.

The most fortunate aspect of these problems is that at least the most severe cases are also the least frequent. While our work shows that significant improvements can be made with appropriate reinforcement and interior padding, it is also apparent that we must avoid any tendency to isolate the various types of accidents. About half the side impact accidents are also frontal impacts for another car, and the damage and injury caused is dependent on the design of both vehicles. The introduction of stronger and more aggressive front ends for frontal impacts is largely detrimental to side impact performance.

To provide protection in side impacts, the relative merits of increased side strength, and increased interior padding, depend upon the size and weight of the car. Our results suggest that side strength is more critical for the large car, while interior padding is critical for the small car. The real problem for the small car is not strength or weight, but space. Whatever the design, space is required to cushion the violent velocity changes produced by crash conditions. To keep this space to a minimum, we need more real world accident data from which to determine the speeds at which protection should be given. We must also avoid the temptation to be sidetracked by severe and impressive forms of testing, into directing effort into unrewarding areas.

REFERENCES

1. "The Frequency and Severity of Injuries to the Occupants of Cars subjected to Different Types of Impact in Accidents; A Pilot Investigation of Road Accidents in the Winchester Area" G. Grime — University College, London, report.

2. "Protection of Car Occupants against Side Impacts" — R.D. Lister and I.D. Neilson, Road Research Laboratory — 13th Stapp Conference.

3. "Causes and Effects of Road Accidents" — University of Birmingham Department of Transportation and Environmental Planning.

2-123

(This page intentionally left blank)

Development of Lightweight Door Intrusion Beams Utilizing an Ultra High Strength Steel

T. E. Fine
Inland Steel Co.

S. Dinda
Chrysler Corp.

MANY NEW DEVELOPMENTS in the automotive industry are strongly influenced by a desire to increase the degree of passenger protection and a need to reduce the overall vehicle weight while minimizing the cost to the consumer. In certain instances, an initial solution to one of these problems was incompatible with the others. Side door impact protection systems were added to the automobile to provide increased passenger protection during crash conditions (1)* The system which is in use in most current production automobiles consists of door intrusion beams located approximately at each door midline in conjunction with reinforced body pillars, door hinges and door locks. This system limits the amount of passenger compartment intrusion by resisting the impact loading and absorbing the crash energy through controlled deformation. The original versions of side door intrusion beams were fabricated from thick, hot or cold rolled steel and added up to 50 lb (22.7 kg) per automobile. Improved technology and new beam designs have reduced door beam weight slightly, but most commercially produced door beams are not as cost efficient as possible and still add excessively large amounts of weight to the car. To further improve this situation, high strength steels which exhibit high strength to weight and/or cost ratios are being considered in this application. This paper

*Numbers in parentheses designate References at end of paper.

details the development of door intrusion beams fabricated from a low carbon, ultra high strength steel, MartINsite®. These beams meet all the federal protection requirements while adding only 15 lb (6.8 kg) to 20 lb (9.1 kg) per vehicle.

DOOR BEAM SECTION ANALYSIS

The primary function of a door protection system is to resist intrusion of a foreign object into the passenger compartment as a result of a side impact or collision. To satisfy this purpose with the current door beam systems, the beam system must be capable of withstanding significant forces during impact deformation conditions. Failure of a beam system does not occur when the beam yields or cripples but can be defined as the inability of the beam system to sustain standard load values.

The standard average loads as detailed by U.S. Government specification 1973 FMVSS 214(2) are the results of consideration of both peak loads and energy absorption over 18 in (460 mm) of door intrusion. The load-deflection curves obtained in an FMVSS 214 test can have several different shapes and the beam system will still meet all the specifications. An example of this is shown in Fig. 1 where the load-deflection curves from acceptable FMVSS 214 tests of a standard production

®Inland Steel Co. Registered Trademark.

ABSTRACT

Door intrusion beams have been fabricated from an ultra high strength steel resulting in an efficient side impact protection system. Despite the ultra high strength (yield strength ~150 ksi) the steel may be roll formed into beams from thin gauge material (0.035 in) resulting in significant vehicle weight savings while still meeting all federal specifications. This paper covers the development of this door intrusion beam system and includes detailed descriptions of beam design, beam testing and steel properties.

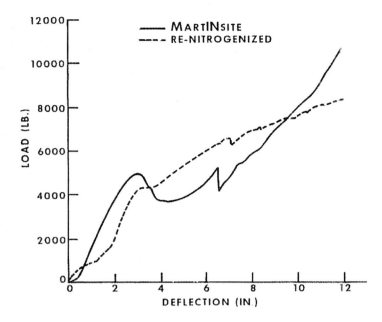

Fig 1 - Load-deflection curves from acceptable FMVSS 214 tests

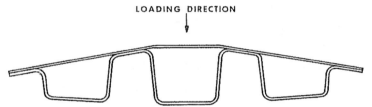

LOADING DIRECTION

Fig 2 - Door beam cross section of beams utilized in FMVSS 214 tests (Fig 1) and three point bend tests (Fig. 3)

beam (0.060 in (1.5 mm) thick, renitrogenized steel) and a corresponding MartINsite beam (0.035 in (0.9 mm) thick, M-160 grade MartINsite steel) are presented. Both beams were fabricated with the beam design shown in Fig. 2.

The shape of the initial (6 in (152 mm) of deflection) crush resistance (ICR) portion of the load-deflection curve is largely determined by the elastic bending behavior of the door beam and by the door beam's ability to support a load after the peak load has been reached and the door beam yields or cripples. The shape of the remainder of the FMVSS 214 load-deflection curve is primarily a function of the tensile loads developed at the door to beam connections, the door latch and the hinges. Since the current door latch and hinge designs are adequate with respect to load-bearing ability, the problem of developing a satisfactory intrusion protection system reverts to that of developing an adequate door beam. In general, it appears that many beam systems which meet the ICR requirements also satisfy the entire FMVSS 214 requirements.

Since FMVSS 214 tests are relatively expensive and difficult to perform, the initial means used to evaluate beam sections is the simple three-point bend test. While there is no generally accepted relationship between simple beam bending and FMVSS 214 test performance, the simple beam bend test results can be used as a relative ranking of the probable effectiveness of various intrusion protection systems by indicating obvious performance differences between various beam designs. However, utilizing the simple beam test to compare beams fabricated with the same configuration but from different materials can be misleading. For example, the load-deflection

curves shown in Fig. 3 are from simple beam tests of MartINsite and renitrogenized beams of the type shown in Fig. 2. Even though these curves show widely differing bend characteristics, both of the intrusion protection systems manufactured with these beams were acceptable during the subsequent FMVSS 214 tests.

The maximum elastic load in three-point bending, P_{max}, can be obtained from the classic flexure formula $\sigma = Mc/I$ and is given approximately by:

$$P_{max} = \left(\frac{4\sigma_{max}}{L}\right)\frac{I}{c} \qquad (1)$$

where σ_{max} is the design strength of the beam material, L is the distance between the beam supports or span, I is the cross sectional moment of inertia and c is the distance from the neutral axis to the outer fiber (3).

It can be seen from Eq. 1 that the maximum elastic load can be increased by designing the beam to maximize the section modulus (I/c), by fabricating the beam from stronger material, that is, increasing σ_{max}, or both. The beam span is limited by the door size. Since the available space between the door outer and inner panels will establish the maximum beam corrugation height, (I/c) can be varied by either changes in material thickness, by changes in the number and type of corrugations, or by the presence or absence of a top plate at the mid-plane cross section. In addition, it can be shown that for a multiple element beam stressed in the three-point bending

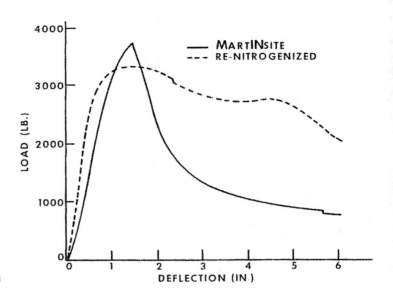

Fig. 3 - Load-deflection curves from three point bend tests

the load-bearing ability is proportional to material thickness and to material yield strength. Therefore, lightweight beams which support high elastic loads can be obtained by utilizing a multiple corrugation design fabricated from thin, high strength material.

For a beam to retain its load-bearing ability after the peak load has been attained, structural integrity must be maintained to as great a degree as possible. There are three ways a beam can distort from its original shape: yielding, buckling, crippling. The point of beam yielding can be determined approximately from Eq. 1 by letting σ_{max} equal the tensile yield strength of the material. Beams which deform solely by yielding usually maintain loads at or near the peak load by a process of continuous yielding and work hardening. For a given steel thickness the peak load in beams of this type is relatively low, since the steel will have low strength. Buckling can be defined as an unstable deflection of a beam element due to a compressive action upon this element. An unrestrained, slender element subject to bending conditions will often fail catastrophically if the load increases past the buckling load. In actual door beams, however, most flat elements, both horizontal and vertical with respect to the applied force, are stiffened in some way, either by properly spaced welds connecting the top plate to the corrugated section or by the physical presence of material adjacent to each flat element in the corrugated section of the beam. When a corrugated door beam with a top plate is loaded in three-point bending, plate buckling occurs first in the top plate and manifests itself as a series of wave shapes.

Under ideal, unrestrained conditions, this plate would buckle and distort at a stress determined by the following modified Euler relationship:

$$\sigma_b = \frac{K\pi^2 E}{12(1 - \nu^2)}\left(\frac{t}{w}\right)^2 \qquad (2)$$

where E is Young's modulus, ν is Poisson's ratio, t is the ma-

terial thickness, w is the plate width, and K is a constant which is a function of the plate supports (4).

In the top plate of an actual beam, the amplitude of the stress waves is limited by the amount of unrestrained material, that is, the spacing of the welds which connect the top plate to the corrugated section. Similar buckling phenomena can occur in both the horizontal and vertical flat areas of the corrugation, but since the width of these areas is generally much less than that of the top plate, their buckling stresses calculated from Eq. 2 will be much higher.

In either case, loss of structural integrity will not occur when the calculated buckling stress is reached due to the phenomenon of "post buckling strength." As the compressive stress level in the flat element increases, the force is transferred to the edges of the element, thereby redistributing the stress pattern to one where the stress is highest at the edges or corners. Eventually, permanent deformation occurs when the materials yield stress is reached at the edges or corners. At this point, the corners will distort causing an attendant loss of load-bearing ability. This type of deformation, termed "crippling," eventually causes total beam collapse since the crippling process will spread from one element to another. Since buckling generally precedes crippling, an ideal beam will have the crippling stress greater than the buckling stress.

Two current empirical theories reach different conclusions on the proper method to calculate a crippling stress (4-6). One theory proposes that as the sum of the lengths of the legs, any a + b, at any corner (Fig. 4B) decreases and the corner radii (R) increase, the tendency to cripple will decrease (5, 6). The second theory proposes that as the web height to sheet thickness ratio, h/t, (Fig. 4A) decreases and the bearing surface to sheet thickness ratio, D/t, (Fig. 4B) increases and the corner radius to sheet thickness ratio, R/t, decreases, the tendency to cripple decreases (4). Independent studies performed at Chrysler (5) and Inland indicate that while corners with sharp radii promote high I/c values they also promote corner crippling.

Upon consideration of all of the factors which can affect a

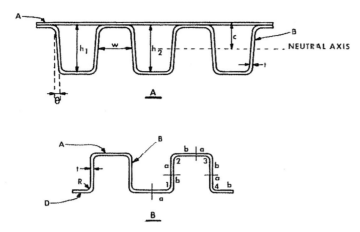

A - TOP PLATE (a.k.a. TOP SHEAR PLATE, COMPRESSION ELEMENT)
B - WEB (a.k.a. SIDE WEB, WALL)
c - DISTANCE FROM NEUTRAL AXIS TO OUTERMOST SURFACE
D - BEARING SURFACE
t - THICKNESS
R - CORNER RADIUS
a,b - LEGS OF CORNER ANGLES 1, 2, 3 AND 4
h_1, h_2 - CELL HEIGHT
w - CELL WIDTH
θ - CELL WALL ANGLE

Fig. 4 - General door beam description

door beam's performance, it becomes obvious that several of the variable parameters work in opposite directions. For example, calculations have shown that for simple MartINsite corrugated beams the tendency for premature failure decreases as the beam cell height and width decrease due to relative changes in both buckling (4) and crippling stresses (6). This tendency has been preliminarily confirmed by laboratory tests. However, these same calculations also show that concurrent with the decrease in tendency for premature failure is a decrease in the section modulus and a decrease in the predicted elastic peak load. This demonstrates that an acceptable door beam design must strike a balance between these two factors.

DOOR BEAM SECTION SELECTION

A cross section through the mid-plane of a generalized corrugated door beam with a face plate is shown in Fig. 4A. By varying the mid-cell height, h_2, the outer cell height, h_1, the cell width, w, and the cell wall angle, θ, and the number of cells, it was possible to construct beams with widely varying values of the section modulus, I/c. The initial series of MartINsite door beams were made with either two, three, or four cells, mid and outer cell heights equal and with values from 0.75 in (19 mm) to 1.5 in (38 mm), cell width varied from 0.75 in (19 mm) to 1.25 in (32 mm) and cell wall angle either 0 deg or 15 deg. Subsequent series of MartINsite beams were fabricated to production designs in current Chrysler models.

GENERAL DESCRIPTION OF MATERIAL

The properties of low carbon, alloy free martensites in general and Inland MartINsite in particular have been characterized in a series of papers (7-11). Basically, MartINsite is a low carbon, fully aluminum-killed steel sheet material currently produced in thicknesses from 0.010 in to 0.035 in

Fig. 5 - Microstructure of automotive grade MartINsite, nital etch, 1,100X magnification

(0.25 mm to 0.89 mm), widths up to 36 in (914 mm) and guaranteed minimum tensile strength up to 220 ksi (1517 MPa). The total elongation in 2 in (50.8 mm) is generally between 3 and 4%, and the 0.2% offset yield strength is in the range of 88-97% of the ultimate tensile strength. Minimum bend radii of approximately 7T when the steel is stressed transversely to the rolling direction and 4T when the stress is parallel to the rolling direction (where T is the sheet thickness) are recommended for production fabrication operations. Tensile strength is a direct function of the carbon content and the relationship, tensile strength (ksi) = 119 + 560 (w/o C) or TS (MPa) = 821 + 3860 (w/o C), has been found to be a reasonable predictor of the strength. Fig. 5 shows a photomicrograph of a typical MartINsite microstructure. The structure shown in this figure is virtually 100% martensite, the normal structure for MartINsite.

Automotive grade MartINsite for door beam applications is

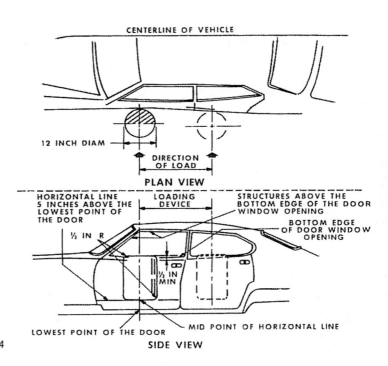

CENTERLINE OF VEHICLE

12 INCH DIAM

DIRECTION OF LOAD

PLAN VIEW

HORIZONTAL LINE 5 INCHES ABOVE THE LOWEST POINT OF THE DOOR

LOADING DEVICE

STRUCTURES ABOVE THE BOTTOM EDGE OF THE DOOR WINDOW OPENING

BOTTOM EDGE OF DOOR WINDOW OPENING

½ IN R

½ IN MIN

LOWEST POINT OF THE DOOR

MID POINT OF HORIZONTAL LINE

SIDE VIEW

Fig. 6 - Schematic of federal door crush test, FMVSS 214

Table 1 - Strength per Unit Cost for Various Door Beam Steels

Steel Type	I
MartINsite M-160	7.1
Low carbon cold rolled	2.1/1.7
Low carbon hot rolled	2.3/1.9
Renitrogenized	3.2
Prestrained renitrogenized	3.8

currently produced to a 0.11-0.13 w/o C specification to assure weldability equivalent to standard low carbon cold rolled steel and has a minimum 0.2% offset yield strength of 150 ksi (1030 MPa). Materials currently used in door beam applications include standard low carbon hot or cold rolled steel, yield strength 25 to 30 ksi (172 to 207 MPa); renitrogenized steel, yield strength 50 ksi (310 MPa); and prestrained renitrogenized steel, yield strength 60 ksi (480 MPa).

It can be shown that for automotive beam bending applications there is a linear relationship between yield strength and material thickness, that is:

$$t_s \propto \frac{\sigma_y(M)}{\sigma_y(S)} t_m \qquad (3)$$

where t_m and t_s are the thicknesses of MartINsite and of any other steel, respectively. On this basis, if optimum design door beams were fabricated from 0.035 in (0.89 mm) thick MartINsite, it would require the following approximate thicknesses of the currently used steels to obtain equivalent theoretical beam strength: low carbon cold or hot rolled steel - 0.175 in-

0.210 in (4.45 to 5.3 mm), renitrogenized steel - 0.105 in (2.67 mm), and prestrained renitrogenized steel - 0.0875 in (2.22 mm). Table I shows a comparison of these steels on a basis of strength per current unit cost (I = yield strength/unit cost).

DOOR BEAM FABRICATION

The beams employed in this development were all of a corrugated, multiple cell construction with a flat sheet face plate spot welded to the outer or compression side of the beam. Initially all the beams were fabricated by using a brake press with a V-notch, air form die. Some subsequent beam sections were produced on a standard roll forming line. The load-bearing equivalence of beams fabricated by these two methods was subsequently established by beam bending tests.

Door Beam Testing Methods - Four basic tests are currently performed to evaluate side door protection and door beam characteristics:

1. Vehicle to vehicle crash.
2. Federal Standard 1973 FMVSS 214 door intrusion test (2).
3. Door fixture test, and
4. Static simple beam bending.

The first test has been used as part of vehicle crashworthiness studies (12, 13, 14) or for establishing a data base for later comparison with reinforced protection system tests (1, 13-15). The second test specifies the legal requirements for a protection system. This test establishes minimum average and peak forces to be obtained via a load-deflection curve when an instrumented 12 in (305 mm) diameter semi-cylinder is forced a maximum of 18 in (457 mm) into the door and door beam of a rigidly fixed vehicle at a loading rate of not more than 1/2 in/s (12.7 mm/s). The test is portrayed in Fig. 6. The third test is a modified version of FMVSS 214 and is used as a pre-

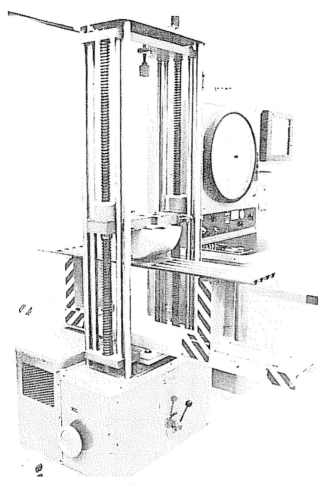

Fig. 7 - Three point beam bending apparatus

car crush ranking test. In this test, a proposed door beam is welded into a full car door and the entire assembly attached to a rigid frame. The same instrumented ram assembly used for FMVSS 214 tests is then employed to crush this door assembly in the manner specified by FMVSS 214 to a total deflection of 6 in (152 mm). This test appears to correlate well with the results for the first 6 in (152 mm) deflection in the FMVSS 214 test. The fourth test is a preliminary test, used to evaluate relative beam strength performance. The test procedure consists of loading a beam in simply supported three-point bending and then ranking the beam configurations by their load-displacement characteristics. The data that appear to relate to FMVSS 214 results are peak load and energy absorbed during deformation. This test apparatus is shown in Fig. 7.

The initial series of beam tests were performed during 1971-72 to evaluate the potential of MartINsite in Chrysler designed beams and to evaluate the effects of variations in corrugation profile on beam performance. Cross sections of these beams fabricated from 0.035 in (0.89 mm) thick M-190 grade MartINsite are shown in Figs. 8 and 9 with the corresponding three-point beam bending data developed at the Inland Research Laboratories presented in Table 2. From these results it can be concluded that increases in peak load are primarily attributable to increasing the cross-sectional moment of inertia by either increasing the number of corrugations or increasing the corrugation depth. The moment of inertia and peak load also increase when a top plate is welded to the corrugated section. However, it was found that for each corrugated section there is an optimum top plate length. If the top plate is too short, a premature failure could occur at the moment of inertia discontinuity. If the top plate is increased in length past that needed to induce bending failure at the

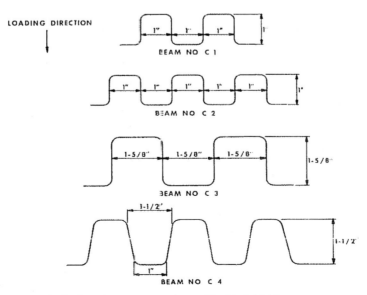

Fig. 8 - Door beam cross sections utilized in initial beam tests

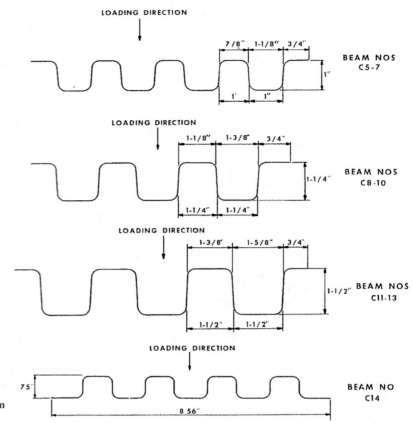

Fig. 9 - Door beam cross sections utilized in initial beam tests

Table 2 - Initial Three Point Beam Bending Tests*

Door Beam Identity Number	Number of Cells	Cell Height, in	Mid-Cell Width, in	Top Plate Length, in	Beam Weight, lb	Cross-Sectional Moment of Inertia in^4	Peak Load, kip	Energy Absorbed, ft-lb	Average Load in 6 in, kip
C1	2	1.0	1.0	None	4.0	4.5×10^{-2}	1.80	540	0.97
C2	3	1.0	1.0	None	6.0	7.0×10^{-2}	2.62	750	1.36
C3	2	1.625	1.625	None	6.5	1.94×10^{-1}	3.30	870	1.6
C4	3	1.5	1.25	None	7.5	1.73×10^{-1}	4.56	1150	1.98
C5	4	1.0	1.06	None	7.9	9.6×10^{-2}	2.87	1060	1.78
C6	4	1.0	1.06	18.0	9.5	1.45×10^{-1}	4.06	1200	2.07
C7	4	1.0	1.06	50.0	12.2	1.45×10^{-1}	4.12	1200	2.03
C8	3	1.25	1.31	None	7.5	1.36×10^{-1}	2.32	870	1.31
C9	3	1.25	1.31	18.0	8.9	2.1×10^{-1}	4.45	1250	2.31
C10	3	1.25	1.31	50.0	11.4	2.1×10^{-1}	4.54	1270	2.23
C11	3	1.5	1.56	None	8.75	2.3×10^{-1}	2.98	1130	1.85
C12	3	1.5	1.56	18.0	10.4	3.5×10^{-1}	>5.0	–	–
C13	3	1.5	1.56	50.0	13.25	3.5×10^{-1}	>5.0	–	–
C14	4	0.75	1.0	None	7.2	5.2×10^{-2}	2.13	870	1.45

*SI conversion factors: mm = 25.4 (in), kg = 4.535924 E-01 (lb-mass), mm^4 = 4.162314 E + 05 (in^4), kN = 4.448222 E + 00 (kip-force), J = 1.355818 E + 00 (ft-lb).

Table 3 - Standard Design Door Beam Tests*

Automobile Model	Steel Type**	Steel Thickness, in	Top Plate Length, in	Beam Weight, lb	Simple Beam Test		Door Fixture Test		FMVSS 214 Test		
					Test Span, in	Peak Load, kip	Beam Length, in	Ave. Load in 6 in, kip	Ave. Load in 6 in, kip	Ave. Load in 12 in, kip	Peak Load in 18 in, kip
Compact size body— 2 door	M-160	0.035	25	7.0	46.3	3.35	46.3	2.95	3.06	3.79	>7.60
	Renitrogenized	0.062	6	12.4	46.3	2.58	46.3	2.85	--	--	--
Compact size body— 4 door front	M-160	0.035	25	5.5	36.8	4.25	36.8	3.50	--	--	--
	Renitrogenized	0.048	6	7.8	36.8	3.16	--	--	--	--	--
Standard size body— 2 door	M-160	0.035	25	7.8	52.3	2.55	52.3	2.45	2.45	4.05	--
	Renitrogenized	0.075	6	16.5	52.3	2.95	52.3	2.80	--	--	--
Standard size body— 4 door front	M-160	0.035	25	5.7	37.8	3.75	37.8	2.90	3.18	5.28	8.4
	Renitrogenized	0.063	6	8.0	37.8	3.35	--	--	3.35	5.20	8.2

*Note: SI conversion factors: mm = 25.4 (in.), kg = 4.535924 E – 01 (lb-mass), kN = 4.448222 E + 00 (kip-force).

**M-160 ⟶ MartINsite, 160 ksi (1103 MPa) minimum tensile strength.

LOADING DIRECTION

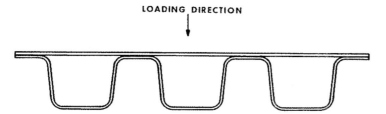

Fig. 10 - Door beam cross section utilized in second beam trial

beam center, the additional material is wasted since no further increases in peak load results.

The significant points to be made by these data are that high peak loads, high average loads in 6 in (152 mm) and high energy absorption can be obtained from a lightweight beam. For example, as shown in Table 2 beam C9, a three-celled, 1.25 in (31.7 mm) deep corrugated beam, 50 in (1270 mm) long, with an 18 in (457 mm) top plate obtained a peak load of 4450 lb (19.8 kN), an average load in 6 in (152 mm) of 2310 lb (10.3 kN), an energy absorption of 1250 ft-lb (1.7 kJ) when deformed on a 39.5 in (1003 mm) span while weighing only 8.9 lb (4.0 kg). A duplicate of beam C14, which is 0.75 in (0.19 mm) deep, with four cells of 1 in (25.4 mm) width, was door crush tested at the Chrysler Laboratories and performed in excess of the FMVSS 214 initial and intermediate crush resistance requirements.

A second series of tests was initiated in mid-1973 to determine if MartINsite could be directly substituted into the pre-existing standard production door beam designs. This beam, shown in Fig. 10, is a roll formed, corrugated section with a flat top plate welded to the corrugation about the beam center. The first phase of this program involved the testing of brake press formed sections of compact and standard body beams at the Chrysler Laboratories. Table 3 contains the results of these tests and presents data for standard production beams fabricated from a renitrogenized steel and for MartINsite beams where both were available.

The simple three-point bending beam test data show that MartINsite beams in standard production model designs develop high peak loads, in most cases higher than that for the renitrogenized steel beams. The door fixture test data, on the same set of section designs, show that ICR values in excess of the FMVSS 214 specifications are obtained in all cases. The FMVSS 214 test data show again that the MartINsite beams obtained average and peak load values in excess of the federal requirements. This again demonstrates that it is possible to obtain the required loads with a part which weighs about half as much as the standard production part.

The second phase of this program involved a roll forming trial of standard body—4 door rear beam section (beam span 33.75 in (857.25 mm)), Fig. 11, and FMVSS 214 testing of both roll formed and brake press formed sections. The results of the FMVSS 214 tests are presented in Table 4.

The MartINsite version of this beam weighs 4.56 lb (2.07 kg) vs 8.19 lb (3.71 kg) for the standard production model. Roll formed MartINsite standard body—4 door rear beams were also mounted in automobiles and subjected to Chrysler's door "slam" test and to a crash test (16 mph (7.15 m/s) at a 45 deg angle). The beam performance in all tests was satisfactory.

As a result of these successful tests, MartINsite door guard beams will be incorporated in the future models. The changeover may be accomplished in a stepwise manner with the substitution of MartINsite in the top plate section first. Conversion of the corrugated beam section roll forming lines, to

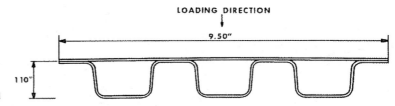

Fig. 11 - Door beam cross section—roll forming trial

Dept. of Transportation, National Highway Safety Bureau, Docket No. 2-6, 49CFR Part 571, January 1, 1973.

3. F. L. Singer, "Strength of Materials." New York: Harper & Bros., Second Edition, 1962.

4. G. Winter, "Cold Formed Steel Design Manual." New York: AISI, 1968.

5. J. S. Haynes, "Development and Analysis of Door Side-Impact Reinforcements." SAE transactions, Vol. 81 (1972), paper 720494.

6. R. A. Needham, "The Ultimate Strength of Aluminum-Alloy Formed Structural Shapes in Compression." Journal of the Aeronautical Sciences, Vol. 21, No. 4, 1954.

7. W. H. McFarland, "Mechanical Properties of Low Carbon-Alloy Free Martensites." Trans. AIME, Vol. 233, 1965, pp. 2028-2035.

8. K. J. Albutt and S. Garber, "Preliminary Assessment of the Properties of Martensitic Low Carbon Steel Sheet." IISI, Vol. 204, 1966, pp. 278-279.

9. J. M. Wallbridge and J. Gordon Parr, "Effect of Rapid Heat Treatment on Mechanical Properties of Low Carbon Steel Sheet." JISI, Vol. 205, 1967, pp. 750-755.

10. W. H. McFarland, "Production and Properties of Martensitic Low Carbon Steel Sheets." Yearbook of AISI, 1968.

11. W. H. McFarland and H. L. Taylor, "Properties and Applications of Low Carbon Martensitic Steel Sheets." Paper 690263 presented at SAE International Automotive Engineering Congress, Detroit, January 1969.

12. M. Tani and R. Emori, "A Study of Automobile Crashworthiness." Paper 700175 presented at SAE Automotive Engineering Congress, Detroit, January 1970.

13. T. Bartol, et al, "Performance Requirements for Doors, Door Retaining Devices and Adjacent Structures." A. M. F. Co., prepared for U.S. Dept. of Transporatation, National Highway Safety Bureau, July 1969.

14. R. P. Mayor and K. N. Naab, "Basic Research in Automobile Crashworthiness—Testing and Evaluation of Modifications for Side Impacts." Cornell Aeronautical Laboratory Report CAN No. YB-2684-V-3, November 1969.

15. Dynamic Science Laboratories, "Side Impact Crashworthiness of Full-Size Hardtop Automobiles." Prepared for U.S. Dept. of Transportation, National Highway Traffic Safety Administration, January 1972.

Table 4 - FMVSS 214 Tests—MartINsite Beams*

Beam Section	Ave. Load in 6 in, kip	Ave. Load in 12 in, kip	Peak Load, kip
Standard size car—4 door rear—roll formed	2.90	**	**
Standard size car—4 door rear—brake press formed	2.90	4.60	>8.40
FMVSS 214 requirements	2.25	3.50	>7.00

*SI conversion factors: mm = 25.4 (in), kN = 4.448222E + 00 (kip-force).

**Test stopped after 6 in to preserve the test car.

accommodate the thinner, high strength MartINsite, will require some roll modification. Substantial weight savings without cost penalty will be achieved by the changeover to MartINsite beams. This will help to alleviate a severe problem in today's material-short and weight-conscious automotive industry.

Extensive testing has established that standard production door beam designs can be fabricated from MartINsite, an ultra high strength steel. The MartINsite door beam develops peak and average loads equivalent to those obtained from the standard production material door beams while weighing approximately half as much. These beams can be roll formed from MartINsite and both roll formed and brake press formed sections passed the FMVSS 214 requirements. MartINsite door beams will be incorporated in the future models with a resultant substantial weight saving.

REFERENCES

1. C. E. Hedeen and D. D. Campbell, "Side Impact Structures." Paper 690003 presented at SAE International Automotive Engineering Congress, Detroit, January 1969.

2. "Side Door Strength—Passenger Cars." FMVSS 214

Improved Test Procedures for Side Impact

I. D. NEILSON, S. PENOYRE, and
S. P. F. PETTY
Transport and Road Research Laboratory
Department of the Environment—
    Department of Transport
United Kingdom

A. K. DYCHE
Motor Industry Research Association

INTRODUCTION

After frontal impact, side impact is the most frequent cause of fatalities to car occupants. Although a majority of these side impacts are car to car, important minorities are heavy goods and other large vehicles into the sides of cars and accidents in which the cars spin and slide sideways into roadside obstacles. Accident studies clearly show that most of the worst injuries occur when the car door or side structure collapses inwards and directly strikes the occupant. This inwards bulge may be caused by a tree, the bumper of a heavy vehicle or the front corners of cars striking at different points and angles of impact. Theoretical studies going as far back as 1969 show that although the change of velocity of the car in a side impact is important, it may be the relative speed at first contact that is more significant. This relative speed can be attenuated as far as the occupants are concerned in two ways. Firstly the side of the car can be strong so that more energy is absorbed by the striking car, thereby reducing the severity of the impact experienced by the occupant in the struck car. Secondly the violence of this blow can be reduced by designing energy absorbing structures and padding on the inner face of the side structure of the car. This means that the peak loadings at the point at which the occupants are struck are reduced to below critical tolerance levels.

Another way of attenuating side impact is by careful design of the object striking the side. This is not possible if it is a tree but the fronts of cars can be redesigned and the advantages of having the front structure and bumper low down were exemplified by the good performance of the SRV Marina displayed at the 5th ESV Conference. What is needed is a strong lower front structure and a weak upper section (if it is included in the design for other purposes such as pedestrian protection). So it is clear that compatibility of design between the fronts and sides of all cars is desirable for this reason as well as for others.

Since the last two ESV Conferences the United Kingdom has been supporting the studies of Working Group 5 of EEVC which has been considering the impact test procedures most suitable for legislative test purposes for front and side impact to cars. For side impact the series of tests described below have been carried out. They show the value of using a suitably shaped small rigid faced mobile barrier propelled perpendicularly into the sides of cars. It had a small flat face with rounded edges which made a bulge in the doors and B post and produced damage very like that seen in accidents although the impacting body and its direction might be different. Some have argued that such an im-

pactor gives a high impact velocity in relation to the change in velocity it gives to the car being struck. This is true, but it does not invalidate the test if the impact speed is somewhat reduced below that of the car to car impact it is representing (35 km/h for 50 km/h in these tests). Possibly the only serious objection to this test is that the impacting face is rigid so that if the door is flattened locally between it and the dummy the loads on the dummy will reach very high levels. However in this respect the barrier face is similar to a tree or heavy vehicle bumper; furthermore this situation will not arise for car door designs with adequate depths of padding on their inside faces.

An advantage of flat rigid faced barriers is that the loadings on small sections of a face can be measured by load cells. So the face receives an imprint of a side of a car as it strikes it, in terms of a pattern of force loadings. Similarly if the barrier is run into the fronts of cars, the fronts of the cars leave imprints as force loading patterns. These patterns give first indications of where the fronts and sides of cars are strong and where they are weak. This paper indicates how these patterns may be used to check the likely compatibility of the side of one car with the fronts or front corners of many other cars.

The paper shows that a specialized dummy for side impacts (a TRRL Mark 1 Side Impact Dummy[1] was used) greatly increases the potential value of full scale side impact testing by measuring the violence of loading on the head, chest and pelvis.

A problem with a single full scale test for legislative purposes is that it can be a demonstration of only one accident situation and in reality there are many. The main shortcoming is that, as is shown later on, the head of the dummy can hit only one point on the side but there are many points along the cant rails and B posts that could be struck. For this reason the separate head impact test, described in the companion paper to this on frontal impact testing,[2] is appropriate for side as well as front impacts.

INITIAL TEST SERIES

Method of Test

The test procedure was based on the 1972 ECE draft regulation[3] which proposed an 1100 kg moving barrier with a 1300 mm wide 600 mm high impacting surface to hit the stationary car in a 90° side impact centered on the driver's R point. The proposed pass criterion was a minimum residual transverse space of 350 mm/seat for each row of seats. For the initial test series the barrier mass of 1100 kg and speed of 35/38 km/h were retained but the impacting face was reduced to a steel plate 1000 mm wide and 500 mm deep faced with a layer of plywood 75 mm thick with a 75 mm radius on all edges. The impact face was rigidly secured to the moving barrier and was adjustable so that the height of its bottom edge above the ground was 175 mm for a low, and 300 mm for a high impact position (see fig. 1). The barrier was moving freely on impact and the impact speeds are given in Table 1.

The intended impact speed of 35 km/h was chosen to reproduce the damage experienced in a 30 mile/h (48 km/h) car to car side impact. The vehicle to be tested was placed at right angles to the direction of impact. The center line of the moving barrier was in line with the front seat H point with the seat in its fully back position. The car fuel tank was filled with water and the handbrake was applied. Two 50th percentile dummies were placed in the seating positions on the impacted side. The rear dummy was uninstrumented and unrestrained. The front dummy was a fully instrumented TRRL Side Impact Dummy restrained with the vehicle's standard 3 point belt. The internal dimensions of the passenger compartment from the inner faces of the sides of the cars were measured both before and after impact and some of the significant changes are given in Table 1. Four types of cars were used in the tests.

- Car A. A 1972 two door Vauxhall Viva HC saloon with the 'B' post to the rear of the front seat so that the door strength was the major factor in protecting the front seat occupant.
- Car B. A 1972 four door B.L. Marina saloon with the 'B' post by the front seat back rest so that the B post was part of the rectangle used to measure residual transverse space in Reference 3. Results were also available for this car from a previous Marina into Marina side impact test which gives a comparison with the barrier results.
- Car C. A 1975 four door Volvo 244 saloon which has a strong door and 'B' post. The car was modified, without affecting the passenger compartment structural integrity, so that its mass, mass distribution fore and aft, H point in relation to the axle positions, and the door sill height were all similar to those of car B. The front seat dummy was also positioned so that its distance from the inside of

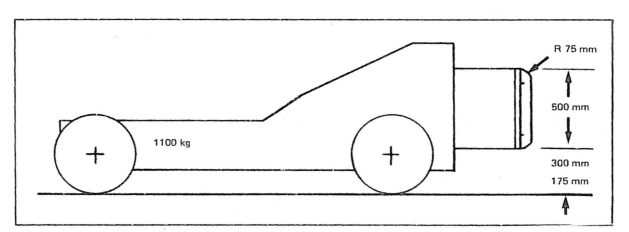

Figure 1. Sketch of a mobile barrier.

Table 1. Data from impact tests.

Barrier height		Tolerance level	Car A			Car B		Car C	Car D
			Low	High	Car to car	Low	High	High	High
Impact speed	km/h		36	32	48	36	36	36	36
Barrier mass	kg		1100	1100	1136	1100	1100	1100	1100
Barrier peak deceleration	g		13	11	17	12	10	14.6	15
	ms		(60)	(44)	(17)	(20)	(25)	(36)	(30)
Car mass	kg		945	984	1052	1082	1056	1068	970
Car peak acceleration	g		22	20	15	26	12	15	14
	ms		(47)	(51)	(17)	(20)	(25)	(23)	(34)
Initial clearance	mm		1315^x	1256^x	*	1188	1202	1282	1162
Final clearance	m	>700	1082	1032	*	958	930	1076	858
Intrusion base A post	mm		83	82	75	97	63	*	94
Intrusion inner door	mm		241	208	279	233	274	224	309
Intrusion top B post	mm		60	65	38	84	56	37	48
Intrusion mid B post	mm		230	237	337	260	288	257	284
Intrusion base B post	mm		218	183	362	227	251	206	248
Head impact			None	None	*	Slight	Slight	Slight	*
Chest acceleration 60 g for	ms	<3	0	0	8.3	7.0	9.5	6.5	13.5
Chest acceleration peak for 3 ms	g	<60	25	23	78	97	119	81	89
	ms		(55)	(28)	(25)	(25)	(24)	(25)	(28)
Peak shoulder load	kN	<7	4.20	3.33	12.49	12.55	11.91	6.05	5.1
	ms		(57)	(28)	(28)	(25)	(23)	(51)	(28)
Load rib 1	kN	<1	1.0	2.0	0.4	1.4	1.4	1.9	1.4
	ms		(57)	(34)	(26)	(37)	(37)	(32)	(33)
Load rib 2	kN	<1	0.6	1.3	0.75	1.5	1.3	2.4	1.8
	ms		(51)	(35)	(45)	(38)	(40)	(33)	(27)
Load rib 3	kN	<1	0.2	0.2	1.6	1.3	1.4	3.0	1.5
	ms		(57)	(34)	(45)	(37)	(39)	(31)	(25)
Load rib 4	kN	<1	*	0.3	1.4	1.3	0.7	1.2	0.7
	ms		*	(31)	(46)	(38)	(37)	(32)	(25)
Load pelvis	kN	<6	7.1	6.5	9.1	5.2	6.7	13.1	22.2$^/$
	ms		(54)	(41)	(43)	(44)	(45)	(32)	(26)
Pelvis peak lateral Acceleration	g		55	55	90	53	60	70	107
	ms		(59)	(43)	(38)	(32)	(43)	(31)	(30)
Pelvis peak lateral Deceleration	g		22	11	*	34	44	19	32
	ms		(107)	(110)	*	(86)	(89)	(88)	(70)
Pelvis velocity	km/h		34	30	*	33	31	32	25

x Different door handles
* Data not available
/ Peak load greater

the door trim was the same as that in the Car B impacts.

- Car D. A 1975 two door Datsun Sunny 120 Y saloon chosen because it appeared to be representative of cars with a light side structure. Though a two door car, its B post was approximately in line with the front seat dummy.

RESULTS

The results from the tests are tabulated in Table 1. The figures given in brackets are the times in milliseconds from impact to reach peak loads. The human tolerance figures for the dummy in the front seats are the preliminary TRRL estimates quoted in the Stapp paper.[1] No head criteria figures are given as in most of the tests the head did not make contact with the car.

Car A, Vauxhall Viva

It should be noted that the difference in the internal clearance between the two cars tested was due to a difference in door handle design. Although the handles were within the defined survival rectangle neither was in a position to hit the occupant in this test procedure. The intrusion at the inner door (by the occupant's torso) was greater for the low barrier than for the high. This was probably because the impact speed achieved was only 32 km/h in the high barrier test compared with 36 km/h in the low barrier test. In both impacts the upper part of the door panel collapsed outwards giving a low load on the shoulder and higher loads on the upper ribs than the lower. In both impacts the barrier missed the A post and caused the door to fold inwards down the vertical edge of the barrier. The door sill either collapsed under impact or was pulled inwards by the B post and energy was also absorbed by the seat frame transmitting the impact load to collapse the tunnel.

Car B, B. L. Marina

The major difference between the car-to-car and the barrier impacts is highlighted by the form of the intrusion into the car. The barrier reproduces its shape in the side of the car with structural collapse around the shape. With the car-to-car the damage was spread over a greater area but the maximum intrusion was concentrated between the B post and the H point of the front seat. The two heights of barrier gave generally similar results. Although the higher barrier gave a similar intrusion to that of the car-to-car test at the inner door, the loads on the ribs and pelvis were somewhat different. The vertical face of the barrier gave higher upper rib loads which were reflected in the higher chest acceleration (the shoulder loads are similar, both being high). The reduced intrusion of the high barrier

compared with the car-to-car, at the base of the B post, gives a lower pelvic load.

For the barrier tests the collapse of the door was similar to that of car A and, as the B post did not tear away from the sill, the sill collapse was also similar. Because of the strong seat structure there was a greater load transferred to the tunnel than for car A.

Car C, Modified Volvo

The strong B post and reinforced door gave a collapse of the structure which was radically different from the barrier tests on the other cars. Instead of the door folding round the vertical edge of the barrier, the impact load was transmitted through the A post into the fascia which then crumpled (preventing the measurement of the A post intrusion being taken). Despite the increased strength of the B post, the inner door intrusion was still significant though only adjacent to the front occupant, however it was less than that for car B. The transmission of the high barrier load through to the floor structure of the passenger compartment was also shown by the increased peak of the car acceleration.

A significant difference between cars B and C was in the loads on rib 3 and the pelvis. The high loads for car C were due to the inner skin of the door being very stiff and transmitting the impact forces directly to the dummy. For car B the inner door skin was much softer, permanently deforming during the impact and acting as an energy absorber and hence giving lower force levels. This difference between the cars also highlights the need for a dummy measuring forces at the ribs and shoulder as part of the test, since an examination of the chest acceleration does not show the high rib loads that were present.

Car D, Datsun Sunny

Considering the car mass and its basic design of passenger compartment the car might have been expected to collapse in the same manner as car A, but it did not do so. Although the barrier speed was only slightly greater than for the other tests, the intrusion at the inner door was the greatest found in all the tests. The B post collapsed and although the door lock did not open the metal round the latch tore out of the B post. Also the fascia started to collapse under load from the middle of the A post allowing an increase in the folding of the front of the door around the vertical edge of the barrier. The load transmitted by the seat structure was sufficient to cause the tunnel to collapse. Despite the substantial failure of the car structure, the final clearance in the passenger compartment was still within the proposed ECE Regulation.[3]

The load on the dummy pelvis was extremely high (at least 22.2 kN) and this is reflected in the high lateral acceleration of the pelvis. The high value probably resulted from the complete collapse of the outer door skin so that the barrier loads were directly transmitted to the pelvis. The upper torso loads were not particularly high but the duration of the chest acceleration above 60 g was large. This was due to the shoulder load cell deforming the inner skin of the door until it hit the outer skin which was above the top of the barrier. This gave a pulse of a long duration but without any high peak.

DISCUSSION

Barrier

In order to reproduce the intrusion into the passenger compartment that is found in injury producing accidents, it was considered necessary that the rigid impacting face should not hit either the A post or the rear suspension mountings. This can be achieved on the majority of cars with a face 1000 mm wide whose center line is the H point of the front seats adjusted in their most rearward positions. The depth of the impactor of 500 mm is such that even in the high position the top edge is not over the bottom edge of the side window. The top and side edges of the impactor were radiused to prevent any cutting effects. The bottom edge was also radiused to compensate for variations in car sill height when the bar-

rier was in the high position. The mass of the barrier is similar to that of a typical European family car with two adult occupants and some luggage. A barrier speed of 35 km/h should give an intrusion similar to a car travelling at 48 km/h ($= 34\sqrt{2}$ km/h) if it can be assumed that the energy absorbed in a car front to car side impact is equally distributed between both cars. This was shown to be true in the case of car B where the inner door intrusion by the torso of the front dummy was very similar for the high barrier and the car to car tests. The two heights of the impactor were chosen to determine the effect upon intrusion of hitting the present design of door sill. The results suggest that although the high impactor is better the more critical factor is barrier speed. For legislative purposes, therefore, a tight tolerance must be placed on barrier speed.

Effect of the Barrier on the Car Structure

The impactor puts the A post, front door, B post, rear door and C post linkage into tension, testing not only the strength of the major structural members but also the strength of the intermediate door hinges and locks and their mountings. This was highlighted in Car D where the door lock survived the impact but metal holding it in the B post failed. In all the tests the door sills collapsed either under impact from the low impactor or under tension from the deformation of the B post with the high impactor. The high barrier test thus tests not only the strength of the sill but also the strength of the joint between the B post and sill. Reinforcing the door transfers some of the impact load to the main passenger compartment frame through the A post. Another simple method of reducing intrusion is by transmitting the impact load through the structure of the front seats. This method has the advantage that a strong seat frame will give protection to the pelvis whatever the seat position. Since the seats may react against each other, test requirements should allow that having defined the impact point with reference to the rearmost seat position a testing authority may move the seat to any of its normally adjustable positions. This clause would allow the two front seats to be staggered and would also ensure that any internal door padding would allow for all seat positions and the fact that in accidents there is usually a rearwards component of impact which throws the occupants forwards and sideways into the doors. A measure of intrusion or survival space within the vehicle after impact is a poor indicator of occupant protection in side impacts and should not be the sole criterion for safety legislation. Instead, the forces acting on a suitable design of dummy should be measured.

This is apparent when comparing the results of cars B and C.

Results From the Dummy

The need to use a force measuring dummy as well as measuring accelerations is highlighted in the results of car C where the high rib loads (particularly rib 3) would not be apparent from the chest acceleration measurements. The need for some form of energy absorption to give an even distribution of load is shown by comparing dummy loads from the thick inner door panel of car C which did not give, with those from the thinner panels of cars A and B which gave but did not bottom on to the barrier and with those from car D which did. Should there be insufficient survival space for the occupant, a load measuring dummy will give high forces but its acceleration could be as low as that of the car. The peak pelvic velocity of the dummy in all tests was greater than the final velocity of the car; after this is reached the dummy is partially held by its three point seat belt. In all cases the pelvic deceleration due to the seat belt later in the impact was less than the acceleration due to the impact, so the wearing of a seat belt in a side impact is thought not to increase the chance of injury.

General Comments

The simplicity of the rigid barrier test makes it attractive if legislation is required in the near future. A barrier to car test is likely to be more attractive from the manufacturer's viewpoint than a car to car side impact test because it requires only one car. A rigid mobile barrier also does not raise the problems of a deformable mobile barrier, but these tests suggest that a small rigid mobile barrier does produce damage similar to that found in real road accidents.

The barrier tests described in this paper would encourage the design of energy absorbing inner door skins and reinforcement of the side structure of the car (to transmit the impact loads to the car structure through the B post to the sill and roof and also through the latches and hinges). The variations of results between the cars tested highlights the need for a side impact test and also shows that with minor modifications it should be possible to produce vehicles that pass this proposed test in such a way that they would be safer when impacted in the side.

DEVELOPMENT OF A LOAD MEASURING RIGID BARRIER AND ITS USE IN A PRODUCTION CAR IMPACT TEST PROGRAM

Since the mobile barrier design used in the side impact tests described above seemed very satisfactory, it was decided to use a similar design in the program of tests of 12 current production car models being undertaken for TRRL by MIRA. These tests are intended to investigate (1) the side impact protection offered to occupants of these car models and (2) the problem of achieving compatibility between the designs of car fronts and car sides to minimize side impact injuries.

Side Impact Occupant Protection

Injuries in side impacts are caused both by intrusion and by acceleration, and a worthwhile side impact test must check both these aspects. The 12 current production cars (VW Golf, Colt Lancer, Datsun Sunny, Renault 5, Fiat (SEAT) 133, Vauxhall Chevette, Chrysler Alpine, BL Marina, BL Princess, Ford Fiesta, Ford Popular, Reliant Kitten) are therefore being tested in 35 km/h impacts using an 1100 kg mobile barrier with a rigid wood face 1 meter wide, 600 mm high, and 200 mm above ground with edge radii of 75 mm. The barrier center line strikes the car at the driver's R point, and the front seats are placed in the fully back position. Two TRRL Side Impact Dummies are used on the struck side of the car, the front one being restrained by the vehicle's standard seat belt, and both intrusion and dummy loads and accelerations are measured as in the TRRL tests. In addition, the face of this barrier is made up of eight load cells, each 250 mm wide and 300 mm high, arranged in two rows, to provide information about barrier load distribution during the impact. This barrier and the load cells were developed by MIRA for TRRL.

A limitation of this 90° mobile barrier testing is that in each impact the dummies will only hit one part of the car interior, while in real accidents with different occupant sizes and impact directions, the occupants may strike the car interior anywhere over a wide region. In addition to this barrier impact, therefore, an interior padding test is proposed using the Free Flight Headform Rig developed by MIRA and described in the companion paper on frontal impact testing.[2] This test would be carried out both on the head impact areas for side collisions (B posts, cant rails) and on the rib, pelvis, and femur impact areas on the door.

Front to Side Compatibility

The problems of providing side impact occupant protection cheaply would be considerably eased if the front of the striking car could be designed to be compatible with the car side struck. This compatibility requires that the front absorbs as much energy as possible by crushing, since the distance available for frontal crush before intrusion reaches its passenger compartment is so much greater than the acceptable intrusion into the struck car's side. The stiffness of car sides must therefore be greater than the crushing stiffness of car fronts, up to the levels of frontal deformation experienced in front to side impacts. Fortunately these levels are much lower than those found in head-on impacts and the fixed frontal barrier tests which represent them, both because the speed changes in front to side impacts are typically only half those of front to front impacts and because the struck side crushes while rigid barrier does not. Thus there is no serious conflict between the need for an initially soft front for low side impact aggressivity and the need for a stiff front to limit intrusion in high speed frontal crashes.

To ensure that side stiffness is greater than initial front stiffness, it is clearly necessary to measure both these quantities under conditions which represent impacts as closely as possible. The mobile barrier side impact test already described offers a promising means of doing this for side stiffness since the barrier deceleration is an excellent measure of the force the car side is exerting on it. Adequate side stiffness can thus be ensured by designing the car to provide at least an agreed minimum level of peak barrier deceleration during this test. Table 1 above indicates that current cars produce levels of 10–15 g, although their designs could clearly be easily and cheaply improved to increase stiffness and reduce intrusion, so a value of 15 g appears a reasonable minimum basis for a future requirement. With an 1100 kg barrier this of course corresponds to a force of $1100 \times 9.81 \times 15 = 162$ kN. To make sure that frontal stiffness is not too great, it is proposed to carry out a perpendicular frontal impact against the stationary car with a similar 1100 kg barrier, with face width increased to 2 meters to cover the whole vehicle front, at a speed of 40 km/h. The barrier peak deceleration would again be measured, and should be less than the value agreed for the side impact test, i.e., 15 g.

This simple criterion should achieve compatible overall front and side stiffness but it unfortunately does nothing to ensure that the structural strengths of the fronts and sides occur at the same height above the road. Without a further criterion, therefore, it could still allow gross incompatibility such as that now found between a high-bumpered HGV and a sports car, which of course proves lethal to the car occupants because the strong parts of their vehicle under-run those of the HGV. To solve this height mismatch problem it is proposed to make use of the mobile barrier's load cell measurements, and it is first necessary to agree on the height at which the main car front to car side forces should be developed. Papers at previous ESV Conferences and elsewhere[4,5] have shown the great advantages of choosing a low height for this front to side load path, e.g., (1) it allows the necessary side strength to be provided cheaply and simply by minor modifications to door sills (foam filling, use of door retaining buttons, etc.), rather than by strengthening the doors themselves, (2) it makes the struck car roll toward the impacted side and so reduces occupant head contacts with B post or cant rail, (3) intrusion is more likely to be below the occupant's pelvis and femur, not opposite them, (4) the seat frames can be used to transmit load to the car floor and tunnel, (5) if door beams are not required the space they occupied can be used for deformable padding to reduce occupant impact loads, see paragraph 'Car C, Modified Volvo' above.

These benefits can only be obtained if the main side impact load is applied at not more than about 350 mm above the road surface. There seem to be no technical reasons why the frontal structures of European-sized cars should not have their main load paths at this height, and indeed low positioning of the mass of the engine and the structure carrying it should benefit both vehicle handling and pedestrian safety (by allowing more crush depth before stiff parts are hit). Most fixed objects struck in side impacts (trees, lamp posts, etc.) extend to ground level, so a strong

sill will be as effective as strong doors in protecting occupants in these collisions. HGV front to car side impacts might become a more serious problem if car side strength is concentrated at sill height, but regulations requiring front underrun protection devices at heights of about 400 mm are already being considered. It is likely that lower underrun preventers able to swing upwards if grounded on ramps could easily be provided. The main difficulty with requiring a low front to side load path appears to be the conflict with US bumper height regulations. The unfortunate results of these on car occupant and pedestrian safety have frequently been pointed out, e.g.[4] If American car designs are now moving towards European sizes without excessive overhangs beyond their wheelbases the need for high bumpers to achieve adequate ramp angles should disappear.

If it is agreed that a front to side load path at about door sill height is required, the question of how to use mobile barrier load cell results to achieve this needs to be considered. The aim should be to make any car compatible with any other (rather than just compatible with itself; the proportion of all front to side impacts which occur between similar models will always be very small). As described above, the barrier used in the tests at MIRA has two rows of load cells, the lower covering the band 200 mm to 500 mm above the ground and the upper covering 500 mm to 800 mm. There may be no need to complicate the side impact barrier test by including these load cell force measurements in that test; the requirement to achieve at least a specified peak barrier deceleration of 15 g will guarantee adequate total strength, and as explained above cost and engineering considerations can probably be relied on to encourage the car designer to place this strength low down. For the car front design, however, the load cell measurement must be used to ensure that the main load path is low enough to meet the strong lower part of the car side rather than the weak doors. It is suggested that this can be achieved by specifying that the peak total load cell force in the upper row must be less than some agreed force level.

The division between the load cell rows of the present barrier at 500 mm above the ground may prove too high, and 400 mm may be a better choice. The peak frontal load which may be carried above this dividing line is not determined yet, but the results of the impact tests of the 12 production cars at MIRA are expected to give a good indication of how low this level can be set without requiring unacceptably large redesign of car frontal structures. Limited results from fixed and mobile barrier testing available at present suggest a value of 50 kN, i.e., about 30% of the maximum allowable peak frontal force of 162 kN set by the "under 15 g on an 1100 kg barrier" criterion.

Although only the peak force measured by the top row of load cells would be used in the proposed pass/fail criterion, it appears that a useful check on the barrier load cell calibration could be obtained by recording the loads in the lower row also, to confirm that the peak total measured force on the barrier is compatible with the observed barrier peak deceleration.

It is possible that the readings of individual load cells may also be required for two purposes: (1) to check that the designer has not used strong longitudinal members of very small area, which would give good barrier impact test results but would cut through the structure of a real car in an accident, (2) to check the lateral distribution of strength across the car's front, to ensure that the corners are soft enough not to intrude into car sides in the common front-corner-to-side collision situation. It is not yet clear, though, whether the injury reductions achieved by this requirement would be outweighed by an increase caused by more intrusion in perpendicular front-to-side collisions if the corners no longer carry their fair share of the load.

Although the main purpose of this proposed 40 km/h mobile barrier frontal impact test is to give front to side compatibility, the test may also be a useful check on the protection offered for unrestrained occupants in a gentle impact (approximately 20 km/h change of velocity for an 1100 kg car). Rear seat occupant behavior or provisions for child seat mounting could also be checked. The MIRA tests are using two unrestrained OPAT dummies in the front seats in this 40 km/h mobile barrier frontal impact.

At this time of writing, the results of these production car mobile barrier tests are not yet available. It is hoped to present a supplement to this paper at the Conference giving these results.*

Number of Cars Required for Testing

When the proposed tests are taken in conjunction with those described in the companion paper on frontal impact occupant protection[2], the complete set of new tests proposed is: (1) interior testing with Free Flight Headform of areas struck by occupants in front impacts; (2) similar testing of areas struck in side impacts; (3) high speed (55–60 km/h) 30° angled wood faced fixed barrier impact using restrained front impact dummies (e.g., OPAT, Hybrid 3); (4) 40 km/h perpendicular front impact with 1100 kg rigid faced mobile barrier using load cells, unrestrained front impact dummies; (5) 35 km/h side impact with 1100 kg rigid faced mobile barrier, using force measuring side impact dummies (e.g., TRRL).

It appears that this set of tests can be carried out with only two cars without introducing significant errors. One car would be used first for Test (2), Side Impact Area Interior Padding, followed by Test (3), High Speed Angled Barrier. The other car would be used first for Test (4), Frontal Mobile Barrier, followed by Test (1), Front Impact Area Interior Padding (omitting the areas already deformed by and thus tested by the dummies in the frontal Mobile Barrier test), and finally the Side Impact Mobile Barrier test (5). It is most unlikely that a car model strong enough to pass the 55–60 km/h angled barrier test will be so deformed by the 40 km/h frontal mobile barrier

---

*Results of these tests subsequently became available and appear as a supplemental table at the end of this paper

test that the subsequent side impact test will give invalid results.

As explained in the Frontal Impact paper[2], the high speed angled barrier test is intended to replace the present 48 km/h perpendicular barrier test. If the proposed set of tests were to be adopted it seems likely that several current regulations would prove to be redundant and could be dropped. The suggested tests would therefore cost little more to carry out than the present regulations, but should be considerably more effective in safety terms.

CONCLUSIONS

- A full-scale side-impact of a mobile barrier with the side of a car is a desirable test of the protection provided for car occupants.
- The loadings on the head, torso and pelvis of a specialized dummy are essential parameters of such a test; dummy acceleration levels are not sufficient as they fail to detect crushing injuries.
- The impact may conveniently be perpendicular into a stationary car under test and representative of a vehicle striking at 50 km/h between the A posts and rear wheel arch.
- A small rigid faced barrier 1 meter wide, 600 mm high and 200 mm above ground with edge radii of 75 mm striking the car under test at 35 km/h is a good representation of this situation.
- The test results given in the paper show the need for strengthening the side structures of some cars, particularly to maintain tensile strength from A post to the rear through the door hinges and latches. A high tensile strength for the B post and its ends is particularly important. At the same time the need for the components on the inner faces of doors to be sufficiently soft to dummy impact is also demonstrated and energy absorbing padding may usually be needed. Some current car designs come close to passing the suggested test.
- Auxiliary tests of the padding or suitable design of the side components that might possibly be struck by the head are also nec-

essary. A suitable test is described in the companion paper on front impact testing.[2]

- It would be desirable to improve compatibility between car fronts and car sides of different designs. Testing with a rigid faced mobile barrier is a means of checking this, if load cells are fitted to the barrier face.
- It appears that a suitable procedure for compatibility testing would be (1) impact the stationary car head on using a 2 meter wide 1100 kg mobile barrier at 40 km/h, recording barrier deceleration and load cell forces, (2) impact the same car at 90° into the driver's R point, using a 1 meter wide 1100 kg mobile barrier at 35 km/h, recording barrier deceleration. Requirements for passing this compatibility test would be

  — in the side impact, peak barrier deceleration must exceed a set value, provisionally 15 g,
  — in the front impact, peak barrier deceleration must be less than this chosen value (15 g),
  — in the front impact, the peak force measured on the top half of the barrier (between 500 mm and 800 mm above ground) must be less than an agreed value, provisionally 50 kN.

Results of carrying out this procedure on 12 current production cars will be available shortly.

- It is considered that very similar results would be obtained if the barrier were to be fitted with a deformable front.
- Whether the face of the barrier be rigid or deformable, its shape, positioning and design are critical factors for the designs of cars that would have to meet such a regulatory side impact test. The desirability of coping with tree and heavy vehicle impacts into the sides somewhat reduces the value of testing with anything but a rigid barrier.

ACKNOWLEDGMENT

The work described in this paper forms part of the program of the Transport and Road

Research Laboratory and the paper is published by permission of the Director.

REFERENCES

1. Harris, J. The Design and Use of the TRRL Side Impact Dummy. Twentieth Stapp Car Crash Conference, SAE 1976.
2. Neilson, I. D., S. Penoyre, and S. P. F. Petty. Improved Test Procedures for Front Impact. Report on the Seventh International Technical Conference on Experimental Safety Vehicles, Paris 1979. U.S. Dept. of Transportation, NHTSA.
3. Draft Regulation: Uniform provisions concerning the approval of vehicles with regard to the behavior of the structure of the impacted vehicle in a lateral collision. W/TRANS/W29/REV1/Amend. United Nations Economic and Social Council. 28 Aug/21 Nov 1972 (Unpublished).
4. Martin, J. K. Standardization of the Heights of Vehicle Bumpers. Report No. VE 508, Dept. of the Environment, Vehicle Engineering Division, Oct 1975.
5. Finch, P. M. Vehicle Compatibility in Car-to-Car Side Impacts and Pedestrian-to-Car Frontal Impacts. Report on the Fifth International Technical Conference on Experimental Safety Vehicles, London 1974, p. 710–711. U.S. Dept. of Transportation, NHTSA.

Supplement to Production Car/Mobile Barrier Side Impact Test Results.

Make and Model	Barrier Speed km/h	Barrier Peak Decel g (3ms)	Front Occupant: TRRL Side Impact Dummy							
			Peak Resultant Accel g (3ms)			Peak Loads kN				
			Head	Chest	Pelvis	Rib 1	Rib 2	Rib 3	Rib 4	Pelvis
		Tolerance	80	60		1	1	1	1	6
BL Marina Coupe	34.9	10.0	89	77	57	1.2	2.2	1.0	0.8	2.4
Datsun Sunny	37.0	10.4	53	49	80	2.2	1.6	1.3	1.1	9.1
Ford Escort	38.0	11.0	58	63	62	1.8	2.4	3.1	2.1	3.5
Colt Lancer	37.0	12.0	63	46	59	1.5	0.9	0.2	0.1	6.9
Reliant Kitten	35.9	9.7	55	59	66	1.5	1.9	1.5	2.1	14.0
Vauxhall Chevette	36.6	12.0	67	59	90	3.2	2.8	0.5	0.2	5.5
BL Princess	37.3	9.5	46	57	70	1.6	1.1	0.1	0.2	3.0
Chrysler Alpine	36.5	11.0	59	57	90	0.8	1.9	2.0	0.7	12.6
Ford Fiesta	35.7	9.0	57	60	86	2.2	2.0	1.2	1.7	4.9
Renault 5	35.7	11.0	57	45	73	2.0	1.9	0.8	1.1	4.2
VW Golf	36.7	10.0	50	43	70	2.3	1.9	0.8	1.2	5.5
Fiat 133	35.9	13.0	80	49	85	1.9	2.5	1.0	0.9	9.0

Barrier Mass 1100kg: Rigid Face 1m x 0.6m, Edge Radius 75mm: Ground Clearance 200mm

Development of Door Guard Beams Utilizing Ultra High Strength Steel

Hiroshi Nousho,
Yasuo Sasakura,
Takeshi Miyamoto,
and Hiroshi Sakurai
Nissan Motor Co, Ltd.

WITH TODAY'S GREAT INTERNATIONAL INTEREST in the conservation of natural resources and energy, improvements in the utilization of automotive space and reduction of vehicle weight are more important than ever. Both are urgent technical problems that require innovation and persistence.

Roomier interiors cannot be designed within applicable vehicle size limits unless the conditions that restrict vehicle layout design are overcome and technology to reduce vehicle weight is developed.

Another advantage to be gained from the reduction of body weight is that it will lead to an overall weight reduction in the vehicular system,(1)* and thus conserve fuel. Accordingly, a lightweight body structure offers the key to deriving successful answers to several existing problems. This requires qualitative changes in the technological development of vehicle bodies.

One example of such a change that has been promoted in recent years, is the use of high strength steel sheets in body structures, primarily to strengthen members.(2) Bumpers of high strength steel are typical applications.(3) Door guard beams are another example in which high strength steel was employed comparatively early.

Guard beams which are installed in doors to improve the door's side collision characteristics not only increase the body weight, but also restrict the vehicle's layout. However, weight was reduced considerably and miniaturization achieved when the material used to make the beams was changed from mild steel sheets to high strength steel sheets in the 50-60 kg/mm^2-class.(4)

---

*Numbers in parentheses designate References at end of paper.

ABSTRACT ———————————————————————

Door guard beams have been developed through the utilization of ultra high strength steel (tensile strength>100 kg/mm^2).

At first, the sheet metal gauge was reduced in proportion to the strength of the ultra high strength without changing the shape of the beam section. This caused beam buckling and did not meet guard beam specifications.

Analyzing this phenomena in accordance with the buckling theory of thin plates, a design criteria that makes effective use of the advantages of ultra high strength was developed.

As a result, our newly designed small vehicle door guard beams are 20% lighter and 26% thinner than conventional ones. This makes it possible to reduce door thickness while increasing interior volume.

In recent years, new manufacturing techniques for making ultra high strength steel sheets have been developed both at home and abroad. These ultra high strength steel sheets can provide strength greater than 100 kg/mm^2 in thicknesses that are suitable for automotive use. Attempts to put such sheets to practical use have been made in the U.S.(5)

This paper outlines the experimental study that we conducted in order to promote further miniaturization and weight reduction of door guard beams through the use of ultra high strength steel.

NECESSITY OF GUARD BEAMS AND CURRENT BEAM SPECIFICATIONS

MVSS NO. 214 REQUIREMENTS - At present, safety standards for vehicular side collisions are codified in the U.S. and Australia. The experimental method required by the U.S. standards, or MVSS No. 214, is illustrated generally in Fig. 1, along with the experimental results. Concrete performance requirements are shown in Table 1.

The main characteristic of MVSS No. 214 is that deformations of up to 6 and 12 inches are specified by mean crushing loads, or absorbed energy amounts. The two inclined lines in Fig. 1 serve as indexes that show the mean crushing load that caused a deformation of less than 6 and 12 inches as a triangular area.

Fig. 1 also shows the experimental results obtained from a body without guard beams. These are far below the MVSS requirements.

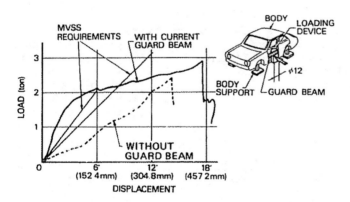

Fig. 1 - Experimental results with MVSS No. 214

CURRENT GUARD BEAM SPECIFICATIONS - In studying the use of ultra high strength steel guard beams, a typical small vehicle intended for export to North America was used. Fig. 2 shows the specifications of the guard beam currently used. It is constructed of a beam and a patch made of roll-formed 60 kg/mm^2-class high strength steel and press-formed brackets of mild steel. Its sectional size is 144 mm x 34 mm and its weight is 3.2 kg per door, including the brackets.

As shown in Fig. 1, the car equipped with this guard beam satisfies MVSS requirements. Compared to the body having no guard beams, it was found that the guard beam increased resistance during the initial compression stage, and effectively improved energy-absorption characteristics in the 6 to 12 inch deformation range

GUARD BEAM PERFORMANCE REQUIREMENTS AND SECTIONAL CONDITIONS

MAXIMUM LOAD AND ENERGY-ABSORPTION CHARACTERISTICS - The condition at which bending force is exerted upon the guard beam is built into a model, as shown in Fig. 3. The maximum load, P_{max}, at this time is calculated as follows:

$$M_{max} = \frac{P_{max} \cdot \ell}{4} = \sigma_y \cdot Z_c$$

Therefore, from the above equation,

$$P_{max} = \frac{4\sigma_y \cdot Z_c}{\ell} \qquad (1$$

where,

σ_y : Yield point of the material
Z_c : Section modulus of guard beam (compression side)

Accordingly, if P_{max} is constant, in a guard beam made of ultra high strength steel, Z_c can made smaller (i.e., miniaturized and weight reduced) in proportion to an increase in σ_y, in accordance with equation (1).

Typical concrete examples of miniaturization and weight reduction are shown in Fig. 3. Roughly speaking, the two types of sectional patterns shown in the figure are considered to be practical. Type (A) is called a straight patch and type (B) is called a stepped patch.

Fig. 4 shows a comparison between experimental values and calculated values with regard to the maximum load, P_{max}. From the figure, it is apparent that P_{max} of the stepped patch type approximates the value derived from equation (1) while the experimental P_{max} value in the straight patch type is considerably lower than the calculated value.

With respect to the guard beam's energy-absorption characteristics, as shown in Fig. 3 (A), a sudden drop in load appears after P_{max} is produced. If this were built into the body, a drop in load would appear at the time that the guard beam was broken, as shown in Fig. 5, so its energy-absorption characteristic would be lower than the current guard beam's.

Fig. 6 shows the deformation modes in the side intrusion test. With the current guard beam, since the beam's reaction force does not decrease even after the crushing load reaching P_{max}, the deformation progresses along the entire body side, as if the center pillar were being dragged along while load is produced. On the other hand, a guard beam made of ultra high strength steel does not produce a reaction force

Table 1 – MVSS No. 214 Requirements

DISPLACEMENT	CRUSHING LOAD
0~6"	THE MEAN CRUSHING LOAD MUST BE GREATER THAN 2250 POUNDS (1021 kg) BEFORE A 6" DEFORMATION OCCURS.
0~12"	THE MEAN CRUSHING LOAD MUST BE GREATER THAN 3500 POUNDS (1588 kg) BEFORE A 12" DEFORMATION OCCURS.
0~18"	THE MAXIMUM CRUSHING LOAD THAT MUST BE APPLIED BEFORE A 18" DEFORMATION OCCURS MUST BE MORE THAN TWICE THE CURB WEIGHT OR 7000 POUNDS (3172 kg) WHICHEVER IS LESS.

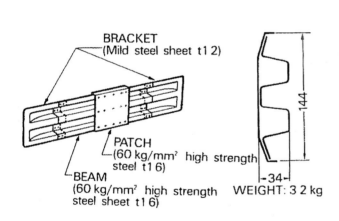

Fig. 2 – Specifications of current guard beam

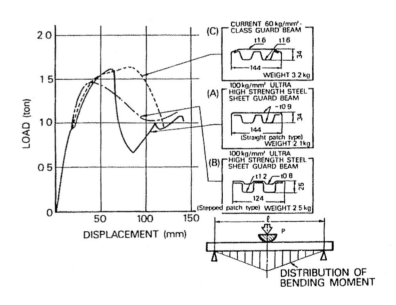

Fig. 3 – Bending test of guard beams

after P_{max} is reached. Thus, the guard beam's deformation proceeds rapidly as if the center pillar were left behind.

For this reason, a consideration of the guard beam's performance must take into account its energy-absorption characteristics, as well as the maximum load, P_{max}.

As noted above, the following two points should be mentioned as being characteristic of the ultra high strength steel guard beam:
(1) The experimental P_{max} value will not necessarily agree with the calculated value.

(2) Its energy-absorption characteristics decline more than the current guard beam's.

INTRODUCTION TO BUCKLING THEORY – The preceding problems were attributed to the buckling phenomenon of the wall plate. This is peculiar to thin gauge structures because ultra high strength steel guard beams use thinner gauges than present beams. Hence, this was taken into consideration.

As shown in Fig. 7, when the guard beam wall plate that received load P is removed individually, as the plate on the compression side and

651

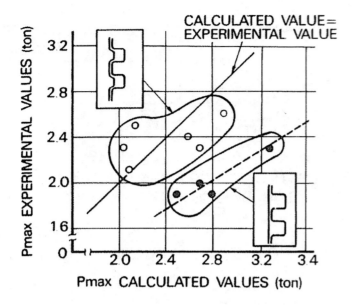

Fig. 4 - Comparison of maximum loads

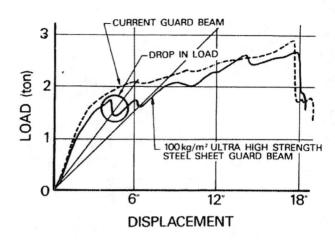

Fig. 5 - Results of side intrusion test

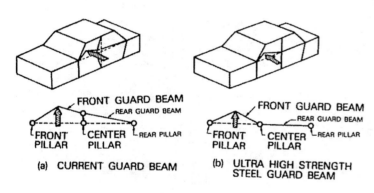

(a) CURRENT GUARD BEAM   (b) ULTRA HIGH STRENGTH STEEL GUARD BEAM

Fig. 6 - Modes of deformation

side plane, the buckling phenomena of the wall plates are built into the respective models in the figure. Their theoretical equations can be expressed as described below:

(1) Compression side plane
Wall-plate buckling stress:

$$\sigma_{cr} = \frac{4\pi^2 E}{12(1-\nu^2)} \cdot \left(\frac{t_p}{h}\right)^2 \qquad (2)$$

where,

E: Young's modulus of material
ν: Poisson's ratio

Here, a wall-plate buckling when σ_{cr} is smaller than the yield strength of the material σ_y, is expressed by the following equation:

$$\sigma_{cr} < \sigma_y \qquad (3)$$

From equations (2) and (3), by substituting numerical values,

$$\frac{t_p}{h} < 3.63 \times 10^{-3} \sqrt{\sigma_y} \qquad (4)$$

(2) Side plane
Proceeding in the same manner as in the compression side plane (1), equation (5) is obtained as follows:

$$\sigma_{cr} = \frac{23.9\pi^2 E}{12(1-\nu^2)} \cdot \left(\frac{t_b}{b}\right)^2$$

$$\frac{t_b}{b} < 1.48 \times 10^{-3} \sqrt{\sigma_y} \qquad (5)$$

As can be seen, the conditions that cause wall-plate buckling can be described as the relationship between the rate of the wall-plate thickness to the plate width and the material's yield point derived from equations (4) and (5). This relationship is graphically demonstrated in Fig. 8, from which the following statements can be derived:

(1) The smaller t_p/h or t_b/b (i.e., the thinner the plate), the more likely it is that wall-plate buckling will occur.

(2) Even when t_p/h or t_b/b is the same, the greater σ_y is, (i.e., the greater the material's strength), the more likely it is that wall-plate buckling will occur. (For example, when Point A shifts to Point B in Fig. 8.)

(3) Wall-plate buckling is likely to occur on the compression side surface.

Hence, the physical significance of wall-plate buckling is recognized once again. The occurrence of wall-plate buckling means that the wall-plate buckling phenomenon will occur before plate stress reaches the yield point.

CONFIRMING PREVENTIVE WALL-PLATE BUCKLING CONDITIONS - In order to prove the preceding analytical results, experimental data on the

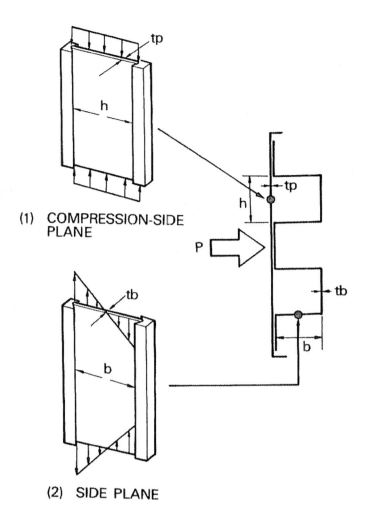

(1) COMPRESSION-SIDE PLANE

(2) SIDE PLANE

Fig. 7 - Wall plate buckling models

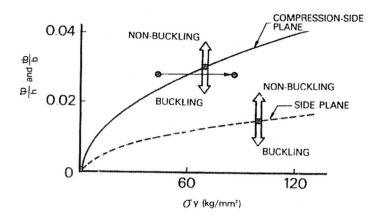

Fig. 8 - Conditions that cause buckling of wall plate

bending of simple beams was obtained by t_p/h and t_b/b, the results of which are mentioned below.

First, Fig. 9 shows the relationship between P_{max} and t_p/h. If t_p/h is large, the P_{max} value that is derived from equation (1) nearly agrees with the experimental value, whether the patch is a straight or a stepped type. However, as t_p/h becomes smaller, the actual P_{max} value will become lower than the calculated value. The boundary is somewhere between 0.025 and 0.03 of t_p/h.

As shown in the figure, t_p/h of the straight type patch fills the space between the spot-welded points, so t_p/h naturally tends to be smaller. On the other hand, the stepped patch type has the advantage of having the stepped part regarded purely as plate width.

Table 2 shows the relationship between the energy-absorption characteristics, t_p/h and t_b/b. According to this table, few straight patch types are satisfactory and are unsuitable for high strength steel guard beams. As

for the stepped patch types, satisfactory results can be obtained only if the range in which plate thickness is reduced is limited.

As can be seen from the facts described above, the following sectional conditions must be satisfied in order to secure the P_{max} and energy-absorption characteristics that will prevent the occurrence of wall-plate buckling:

$$\left.\begin{array}{l} t_p/h \geqq 0.027 \\ t_b/b \geqq 0.040 \end{array}\right\} \qquad (6)$$

RIGIDITY OF THE GUARD BEAM - The third characteristic required of a guard beam is rigidity.

The guard beam's rigidity decreases if it is made small and lightweight through the use of ultra high strength steel. For this reason, the load curve in Fig. 5 displays a small inclination during initial stroke. The amount of energy-absorption decreases in proportion to that small inclination. Although this effect may seem slight, if less energy is absorbed until a crush of 6 inches is produced, the guard beam will eventually fail to satisfy the mean crushing load requirement for 12 inches of deformation. Thus, the difference between the initial inclination must be closely investigated.

The effects of the characteristics of a simple guard beam on the initial gradient are described below.

In 3-point bending as in Fig. 3, the deflection of the guard beam can be expressed by the following equation:

$$\delta = \frac{P\ell^3}{48EI}$$

Therefore,

$$\frac{P}{\delta} = \frac{48E}{\ell^3} I \qquad (7)$$

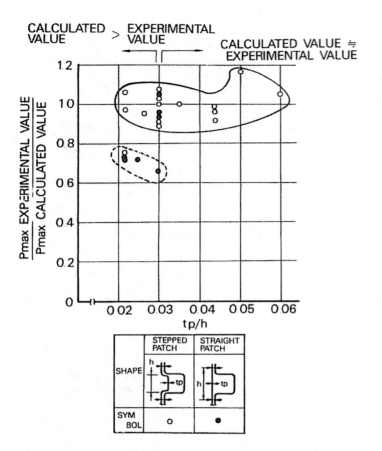

Fig. 9 - Relationship between tp/h and Pmax

Table 2 - Experimental Results of Wall Plate Buckling Condition

tp/h tb/b	STEPPED PATCH TYPE					STRAIGHT PATCH TYPE			
	LARGE ←			0 027				→ SMALL	
LARGE				O	△			×	×
				O				×	
	O	O	O	O	O	×	△	×	×
0 04				O				×	
				×				×	
SMALL	×			×		×		×	×

NOTES : "O" DENOTES THE PATCHES WHICH DID NOT CAUSE WALL PLATE BUCKLING OR A LOAD DROP
"×" DENOTES THE PATCHES WHICH CAUSED WALL PLATE BUCKLING AND A RAPID LOAD DROP
"△" DENOTES CHARACTERISTICS BETWEEN "O" AND "×"

This means that the initial gradient is proportional to the second moment of area I. The experimental value also shows a similar tendency, as shown in Fig. 10.

Table 3 shows the percentage of MVSS requirements to experimental values, from which about 40000 mm⁴ is the minimum second moment of area that is required in order to obtain characteristics that approximate the current guard beam's in terms of the MVSS 12-inch requirement.

$$I \geqq 40000 \ (mm^4) \qquad (8)$$

Incidentally, the 6 inch deformation performance appears to have declined; however, it presents no particular problems because the current guard beam exceeds the MVSS requirements.

CALCULATION OF THE OPTIMUM CROSS SECTION

The door guard beam design criteria that permit effective use of ultra high strength steel have now been established. However, in the actual design business, much time and labor will be required to find the lightest weight guard beam, given the preceding conditions.

In an attempt to solve this problem, a program has been developed which automatically derives the optimum cross section and shape.

Speaking figuratively, when the values that each parameter shown in Fig. 11 can assume are "input," the cross sections that satisfy the preceding sectional conditions are "output" from the combinations in the order of the lightest weight.

Fig. 12 represents a flow chart of the computer program. Sectional conditions are checked with equations (6) and (8) when the combinations are calculated.

EVALUATION OF GUARD BEAM SATISFYING SECTIONAL CONDITIONS

The miniaturized, weight-reduced guard beam made of ultra high strength steel that satisfied the preceding sectional conditions were put to practical use in our vehicles.

First, the experimental results of bending simple guard beams are shown in Fig. 13. Although guard-beam width was decreased from the current 34 mm to 25 or 30 mm and the weight was equal to that of the current one or reduced by 20%, the new guard beams demonstrated even higher crushing loads than the current. No rapid drop in load occurred even after the maximum load was generated.

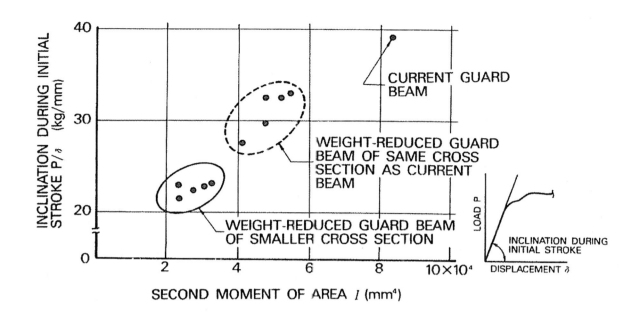

Fig. 10 – Relationship between inclination
during initial stroke and second moment of area

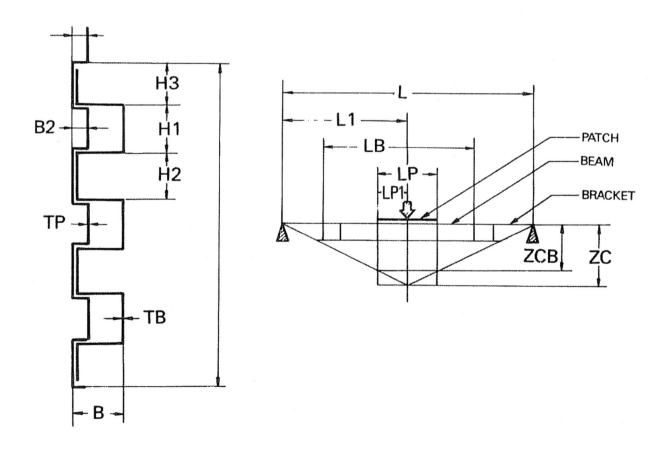

Fig. 11 – Parameters of program for calculating
optimum beam section

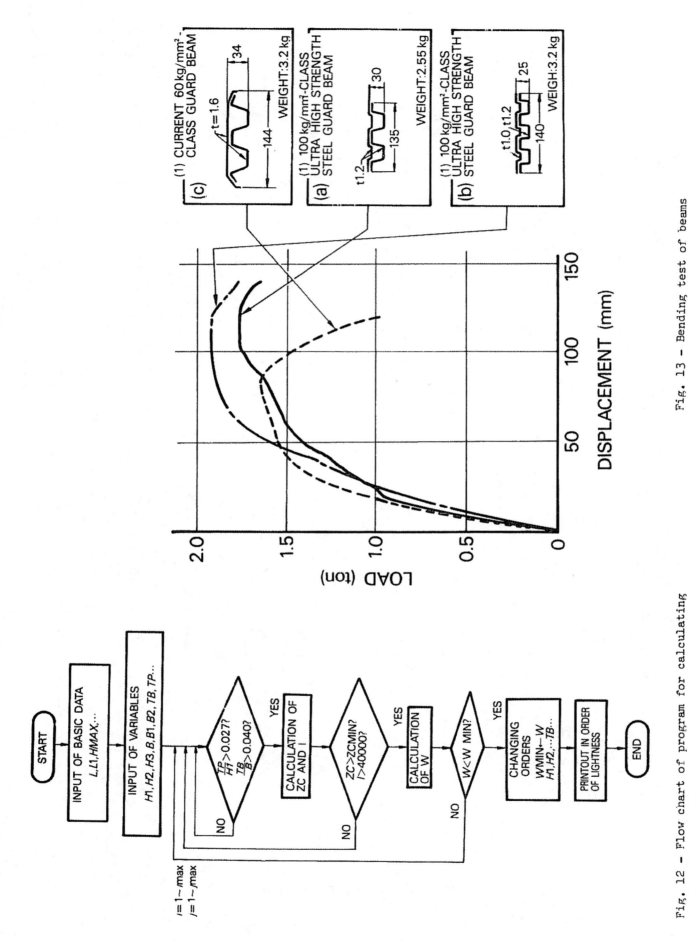

Fig. 13 - Bending test of beams

Fig. 12 - Flow chart of program for calculating optimum beam section

Table 3 - Relationship Between Second Moment of Area and MVSS Requirements

SECOND MOMENT OF AREA	EXPERIMENTAL RESULTS	RATIO TO CONVENTIONAL GUARD BEAM		
		6"	12"	18"
60 kg/mm²-CLASS HIGH STRENGTH STEEL		100 %	100 %	100 %
ULTRA HIGH STRENGTH STEEL	mm⁴ 20000	81	90	88
	40000	89	98	116
	50000	89	99	114
	60000	89	102	122

Table 4 - Spot-Welding Conditions of Ultra High Strength Steel Sheets

		ELECTRIC CURRENT (A)	WELDING PRESSURE (kg)	WELDING TIME (cycle)	CHIP DIAMETER (mm)
ULTRA HIGH STRENGTH STEEL	100 kg/mm²-CLASS + 100 kg/mm²-CLASS	9000 ⌇ 11000	400 ⌇ 600	15	6
	100 kg/mm²-CLASS + MILD STEEL	10000 ⌇ 12000	↑	↑	↑
CURRENT	60 kg/mm²-CLASS + 60 kg/mm²-CLASS	12000 ⌇ 15000	700 ⌇ 800	↑	↑

Experimental results obtained when the guard beams were built into the body are exhibited in Fig. 14. They displayed satisfactory characteristics without rapid load drops and met MVSS requirements.

STUDY ON PRODUCTIVITY AND PRACTICALITY

With a view toward merchandising guard beams made of ultra high strength steels, studies on roll forming characteristics, low-temperature brittleness, etc., were conducted. Satisfactory results were obtained. A detailed discussion of these must be left for another opportunity. We would like to touch on the spot-welding condition here since it is closely related to productivity.

SPOT-WELDING CONDITION - Because the following combinations of materials are spot welded in the manufacture of guard beams, an examination into the feasibility of welding them within the limits of existing production facilities was conducted.
(1) Ultra high strength steel sheet (beam) + Ultra high strength steel sheet (patch)
(2) Ultra high strength steel sheet (beam) + Mild steel sheet (bracket)
As shown in Table 4, ultra high strength steel sheets spot weld easier than the current 60 kg/mm²-class of high strength steel. Existing facilities can be adapted for the spot welding that is required.

The effect of amperage variation on the strength of spot-welds was investigated, as shown in Fig. 15.

CONCLUSIONS

(1) In developing door guard beams that utilize ultra high strength steel sheets, the performance requirements required of guard beams

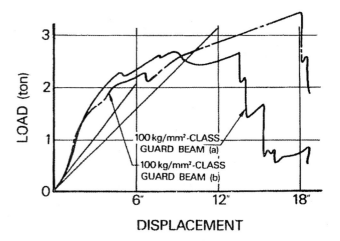

Fig. 14 - Results of side intrusion test

were studied, and reduced to the following three points: (1) Maximum load, (2) Energy-absorption characteristic, and (3) Rigidity. (2) The conditions that secure the above requirements were studied by applying thin plate buckling theory, etc. As a result, the following three concrete sectional conditions were found: (1) $t_p/h \geqq 0.027$, (2) $t_b/b \geqq 0.040$, and (3) $I \geqq 40000$ mm².

(3) A program from which the optimum cross section could be calculated was developed. This program meets the sectional conditions and automatically retrieves the lightest weight guard beam specifications.

(4) An ultra high strength steel was chosen for the manufacture of the guard beams in our small vehicle destined for North America, based on the above findings. As a result, a 20% reduction in weight or a 26% reduction in thickness were achieved.

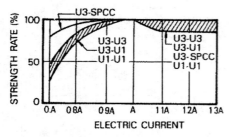
T.S.S. (Tensile shear strength)

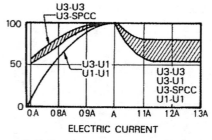
C.T.S. (Cross tension strength)

U3: 130 kg/mm²-CLASS
U1: 100 kg/mm²-CLASS
SPCC: MILD STEEL SHEET

Fig. 15 - Welding current variation and spot strength change

REFERENCES

1. Shimokawa, et al., "Automotive Materials and Weight Reduction: Present and Future," Journal of the Society of Automotive Engineers of Japan, Vol. 33, No. 8 (1979), p. 637.

2. Furubayashi, et al., "Application of High Strength Steel Sheets to Automotive Components," Nissan Technical Review No. 14 (1979), p. 31.

3. Shiokawa, et al., "Thorough Production System of Bumpers by Roll Forming," Nissan Technical Review No. 14 (1979).

4. Sasaki, et al., "Strengthening Side Door in Small Vehicles," Journal of the Society of Automotive Engineers of Japan, Vol 31, No. 3 (1977), p. 190.

5. T.E. Fine, D. Dinda, "Development of Lightweight-Door Intrusion Beams Utilizing An Ultra High Strength Steel," SAE Paper 750222.

The Effectiveness and Performance of Current Door Beams in Side Impact Highway Accidents in the United States.

CHARLES JESSE KAHANE, Ph.D.
National Highway Traffic Safety
Administration

ABSTRACT

Federal Motor Vehicle Safety Standard 214 sets static strength requirements for the doors of passenger cars sold in the United States since January 1, 1973. The requirements have led to the installation of longitudinal reinforcement beams inside the doors.

Standard 214 has reduced the fatality risk in side impacts with fixed objects by approximately 23 percent and has reduced hospitalizations by 25 percent. In collisions with fixed objects, the standard has reduced both ejection and nonejection casualties. The beam enables the car to take a more glancing trajectory, resulting in a wider, shallower damage pattern and a generally less severe collision.

In angle collisions with motor vehicles, Standard 214 has reduced hospitalizations by about 8 percent and has had little or no effect on fatalities. The standard has been effective in reducing door intrusion when the impact is centered on the compartment; in such impacts, the nearside occupant's risk of torso and leg injury is significantly reduced. The standard appears to have little effect on other types of injuries in multivehicle crashes, nor does it enable a car to take a more glancing trajectory in these crashes.

INTRODUCTION

During the 1960's, the American motor vehicle manufacturers conducted side-impact research and crash-testing. They developed a longitudinal beam, installed inside the door, for protection in side impacts. The National Highway Traffic Safety Administration promulgated Federal Motor Vehicle Safety Standard 214, which specifies minimum static strength requirements for the doors of passenger cars sold in the United States after January 1,

1973. The door beams were installed in all passenger cars, accompanied, in some models, by minor local reinforcements of the pillars. Beams were installed in some cars as early as model year 1969, in others as late as January 1973.

It is United States Government policy, as outlined in Executive Order 12291, for Federal agencies to review the benefits, costs and impacts of major existing regulations, based on the actual performance of production vehicles. A comprehensive evaluation of Standard 214 was completed in 1982 (1). This paper summarizes the evaluation's findings on the effectiveness of Standard 214 and its analyses of how beams perform in crashes. It discusses the implications of the evaluation results for current side impact research.

THE SIDE IMPACT SAFETY PROBLEM

A large corpus of accident analysis and crash testing, much of it presented at earlier ESV conferences, has identified the principal injury mechanisms in side impacts: 1. Nearside occupants' torso injuries due to contact with the car's intruding side structure when it is struck in the compartment area by another motor vehicle. 2. Other side structure contact torso injuries (farside occupants, impacts not centered on the compartment, fixed object collisions) where intrusion is less likely to be a major contributing factor. 3. Head injuries due to side structure contact. 4. Ejection. 5. Contact with frontal components (dashboard, steering wheel, etc.).

It is of critical importance to know, in quantitative terms, the relative frequency of the injury mechanisms in highway accidents. Mehta, Pearson and Wilson presented the distribution of fatal and life-threatening injury mechanisms in General Motors accident data at the 8th ESV conference (2). Here are the distributions of fatalities (based on the Fatal Accident Reporting System, FARS) and hospitalizations (based on the National Crash Severity Study, NCSS) that would be occurring in side impacts if Standard 214 had not been promulgated:

659

	Percent of fatalities and hospitalizations	Percent of fatalities
Vehicle-to-vehicle, nearside occupant, compartment-centered damage	27	
Vehicle-to-vehicle, nearside occupant, not compartment-centered	15	} 40
Vehicle-to-vehicle, farside occupant	31	18
Vehicle-to-object, nearside occupant, compartment-centered damage	10	
Vehicle-to-object, nearside occupant, not compartment-centered	4	} 25
Vehicle-to-object, farside occupant	13	17

Throughout this report, a "compartment-centered" impact is defined to be a car in which the midpoint of the damaged area is no more than 45 inches to the front or 15 inches to the rear of the midpoint of the car; this restrictive definition is used to exclude collisions which only peripherally damage the passenger compartment and in which side structure intrusion is relatively unimportant.

The distribution of individual injuries resulting in fatality or hospitalization was the following:

	Percent of serious injuries
Torso, arm or leg injuries due to contacting side structure—nearside occupants, vehicle-to-vehicle crashes centered on compartment	17
Torso, arm or leg injuries due to contacting side structure—other persons	18
Head injuries due to contacting side structure	14
Exterior objects (mostly ejection)	8
Frontal interior components (dashboard, etc.)	30
Other (especially non-contact injuries)	13

The injury mechanism which has been most widely researched—torso injury due to contact with intruding side structures—may be the largest single cause of serious nonfatal injuries but, as shown above, it only accounts for a fraction of all side impact casualties. Moreover, Mehta's accident analysis suggests that this injury mechanism is superseded by head injuries due to side structure contacts and by ejection as a cause of fatal lesions.

EFFECTIVENESS OF STANDARD 214

The primary evaluation objective was to determine the number of lives saved and hospitalizations prevented as a result of Standard 214, based on statistical analysis of highway accident data. An important guideline for the analysis was to isolate the effect of Standard 214 from the effects of other safety standards that are beneficial in side impacts and from the effects of other changes in side structure design (especially the shift from genuine hardtops to pillared hardtops, which took place in the mid-1970's in the United States). Typically, these other safety standards and side structure changes took place 3 or more model years before beams were installed, or 3 or more years afterwards.

Therefore, where possible, the accident analyses were limited to cars of the last model year without beams and the first year with beams or, at most, cars of the last 2 years without them and first 2 years with them. In the analyses that involved a wider range of model years, various statistical techniques were used to identify and remove the effect of factors, other than Standard 214, that affected side impact injury risk.

A second guideline for the analysis was to consider single-vehicle and multivehicle side impacts separately, because Standard 214 could be expected to affect side structure performance quite differently in impacts by objects or vehicles. Furthermore, in the multivehicle crashes, special attention is devoted to nearside occupants in compartment-centered impacts, since they are the type of person most vulnerable to contact with intruding structures.

Fatality Reduction

The Fatal Accident Reporting System (FARS), a census of traffic deaths in the United States since 1975, offers a unique opportunity to assess the fatality reduction of a standard independently of the serious injury reduction. The effectiveness of Standard 214 was estimated by comparing side impact fatalities in cars of the last model year without beams and the first year with them to a control group of frontal impact fatalities in these cars. These estimates were checked by comparisons involving cars of the last 2 years without beams and the first 2 years with them; also by regressions of side impact fatality risk by vehicle age and Standard 214 compliance.

Standard 214 reduced the risk of death in a single-vehicle side impact by a statistically significant 14 percent (confidence bounds: 7 to 21 percent). This amounts to 500 lives saved per year in the United States. The effectiveness was almost equally large for nearside and farside occupants. Moreover, in the preceding estimate, the definition of "single-vehicle side impact" included grade crossing accidents, rollovers with primarily side damage and complex off-road excursions. When the definition was restricted to side impacts with fixed objects, the effectiveness rose to 23 percent.

Standard 214 had no observed effect on fatalities in vehicle-to-vehicle side impacts (confidence bounds: −9 to +7 percent).

Serious Injury Reduction

The National Crash Severity Study (NCSS) is a large, statistically representative sample of towaway accidents involving passenger cars (3, 4). It contains nearly 1600 cases of passenger car occupants who were killed or hospitalized in side impacts; over 500 of them were in cars of the last 2 model years without beams or the first 2 years with beams. The large sample made it possible to apply statistical modeling techniques in a meaningful way and/or restrict the analysis to cars of the last 2 years without beams or first 2 years with them.

In single-vehicle side impacts, the best estimate of Standard 214 effectiveness appeared to be a 25 percent reduction of fatalities and hospitalizations (confidence bounds: 11 to 35 percent). In the NCSS analysis, the single-vehicle side impacts were explicitly defined to exclude rollovers and consisted overwhelmingly of fixed-object collisions. The "best" estimate of effectiveness represents a synthesis of various statistical procedures (multivariate analyses, estimates for restricted model year ranges, comparisons with a frontal control group), all of which yielded positive estimates of Standard 214 effectiveness.

In vehicle-to-vehicle side impacts, the best estimate of Standard 214 effectiveness was 8 percent (confidence bounds: −3 to +17 percent). For the subset of nearside occupants in impacts centered on the compartment, the effectiveness was 25 percent (confidence bounds: 6 to 38 percent). For all other persons in vehicle-to-vehicle side impacts (farside occupants; crashes not centered on the compartment), there was little or no injury reduction.

WHY IS STANDARD 214 EFFECTIVE?

The following 5 hypotheses on why Standard 214 may be effective are stated not as facts but as conjectures. They are tested by examining the effect of Standard 214 on vehicle damage patterns and on specific types of injuries in NCSS and, where possible, by a review of staged

crashes. Hypotheses 1–3 have been mentioned elsewhere in the literature:

1. Crush resistance: beams slow down the rate of door intrusion, at least to some extent, because they increase the door's crush resistance. The post-Standard 214 door dissipates more energy in a shorter distance, causes the frontal structure of the striking vehicle to absorb a larger portion of the energy, allows a more rapid momentum transfer from the striking to the struck car and/or more effectively transmits loads to the vehicle's pillars (5, 6).
2. Deflection: in an oblique side impact, the beam acts like a highway guard rail to help the vehicle partially deflect the striking vehicle or object, resulting in a more glancing collision trajectory, spreading out the damage and reducing the depth, helping maintain the integrity of the door structure and possibly reducing overall collision severity (6, 7, 8, 9).
3. Sill override protection: the beam holds the striking vehicle down, forcing it to engage with the struck car's sill, rather than override it (5).
4. Greenhouse protection: the beam provides a strong horizontal component in the side structure above the sill. It prevents a car from partially "tipping over" into a fixed object and keeps the object away from the car's greenhouse area.
5. Door integrity protection: the beam helps the door maintain its basic shape during a crash, preventing it from being deformed to the point where hinges or latches fail. As a result, fewer occupants are ejected through the door area.

Side Impacts with Fixed Objects

The overwhelming majority of side impacts with fixed objects involve large trees, poles or other rigid, massive, immovable objects that extend from the ground to a point above the car's roof. Hypothesis 3 (sill override protection) does not apply. Hypothesis 1 (crush resistance) would appear to be of limited importance, since momentum transfer is not involved, the fixed object cannot be induced to absorb a significant portion of the energy and the car's sill and roof rails are immediately engaged and absorb most of the energy, not the doors. Thus, the significant reductions of fatalities and serious injuries could only be attributable to hypotheses 2, 4 or 5.

Hypothesis 2 (deflection) is strongly confirmed by the damage patterns in single vehicle crashes on NCSS. Standard 214 decreased the depth of crush by an average of 20 percent while increasing the width of the damaged area by an average of 20 percent. In other words, Standard 214 resulted in significantly shallower and wider damage patterns—evidence of a more glancing collision trajectory. Also, the percentage of cars in which damage was centered on the compartment decreased significantly (from 50 percent to 38 percent) as a consequence of Standard 214: beams help to spread the damage from the compartment area to less vulnerable parts of the car. Finally, as a consequence of Standard 214, the Principal Direction of Force reported by NCSS investigators was, on the average, 9 degrees more oblique—i.e., the effect of beams on the damage pattern made it look as if the collision had been more oblique.

The result of this effect on damage patterns is to reduce not only intrusion but also the overall severity of the collision—e.g., the force levels and, possibly, the velocity change. As a result, the injury reduction need not be limited to nearside occupants' torso injuries. Indeed, on NCSS, nearside occupants' torso injuries due to side structure contact were reduced by 50 percent in post-Standard 214 cars but farside occupants' torso injuries also decreased by 23 percent. Head injuries due to side structure contact decreased by 25 percent and serious injuries due to contact with frontal components, by 27 percent.

Hypothesis 4 (greenhouse protection) is only weakly supported by the NCSS data. There was a nonsignificant 13 percent decrease in the incidence of greenhouse damage in Standard 214 cars; moreover, the reduction might be a consequence of hypothesis 2 (spreading damage away from the compartment area) rather than hypothesis 4. The significant reduction of head injuries for Standard 214, however, could be evidence in favor of hypothesis 4.

On both NCSS and FARS there were large, statistically significant reductions of occupant ejection in fixed object collisions as an immediate consequence of Standard 214. Hypothesis 5 (door integrity protection) cannot be denied. On NCSS, the incidence of occupant ejection through the door area dropped by 40–60 percent as a consequence of Standard 214 (while ejection through other portals was unaffected); the frequency of doors opening in crashes declined by 20–40 percent and the incidence of latch or hinge damage, by 10–20 percent. On FARS, fatal ejections in fixed object crashes decreased by 24 percent because of Standard 214 (while nonejection fatalities decreased by 22 percent). The reduction of ejection was the primary benefit for Standard 214 for farside occupants and a major benefit for nearside occupants.

For these reasons—significantly more favorable damage patterns, door integrity protection and, possibly, greenhouse protection—Standard 214 has been effective in side impacts with fixed objects. Moreover, the benefits are not limited to nearside occupants' torso injuries due to contacting the door but include a reduction of ejections, head injuries and farside occupants' injuries. Fatalities as well as nonfatal serious injuries are prevented.

Side Impacts by Another Vehicle

Standard 214 significantly reduced the depth of crush in the NCSS cases, by an average of 20 percent, when a car was struck in the side by another vehicle and the damage was centered in the compartment area. It had little or no effect on crush when the damage only peripherally included the compartment area. The crush reduction observed in the highway accidents on NCSS was about the same as what had been found by Chrysler and Kitamura et al. in staged crashes (10, 11).

The damage data from NCSS suggest that hypothesis 2 (deflection) was not a major factor in the vehicle-to-vehicle crashes. Unlike the fixed object collisions, the reduction in crush depth was not accompanied by an increase in the width of the damaged area. The percentage of cars with damage centered in the compartment area did not change as a consequence of Standard 214—i.e., the damage was not spread out to other areas. The Principal Direction of Force reported by NCSS investigators was, on the average, about the same for pre- and post-standard cars.

Thus, the significant reduction of intrusion is mainly attributable to hypothesis 1 (crush resistance) or 3 (sill override protection). On NCSS, the incidence of sill override decreased by about 20 percent as a consequence of Standard 214—confirming the evidence for hypothesis 3 that had been obtained by General Motors from staged crashes (5). Nevertheless, the overall incidence of sill override in pre-standard cars was relatively low on NCSS and, given a 20 percent reduction, only one-eighth of Standard 214's intrusion reduction can be attributed to sill override protection. The remainder must be attributed to hypothesis 1—crush resistance.

Since Standard 214 significantly reduced intrusion in vehicle-to-vehicle crashes without otherwise affecting the overall damage pattern or collision severity, it might be expected to mitigate primarily those injuries associated with door intrusion. This is precisely what happened on NCSS. Nearside occupants' torso injuries due to contact with the side structure declined by 33 percent in crashes where damage was centered on the compartment—accounting for almost the entire serious injury reduction for Standard 214 in vehicle-to-vehicle crashes. The standard had little or no effect on head injuries, on farside occupant's injuries, in crashes not centered on the compartment, or on injuries involving contact with frontal components.

The NCSS data do not support a firm conclusion on the effect of Standard 214 on ejection in vehicle-to-vehicle crashes. Ejection through the door area dropped by 10–50 percent and the frequency of doors opening in crashes declined by 20–40 percent—the lower reductions are based on cars of the first 2 years with beams vs. the last 2 years without them and they are not statistically significant. More importantly, ejection is a much less predominant injury source in multivehicle crashes (5 percent of serious casualties) than in fixed-object collisions (15 percent of serious injuries and 24 percent of fatalities). Thus, while Standard 214 appears likely to have reduced ejections in vehicle-to-vehicle crashes, the effect accounts for only a small proportion of the overall benefits in these crashes.

In short, the benefits of Standard 214 in vehicle-to-vehicle crashes are mainly limited to a reduction of nearside occupants' torso injuries due to contact with intruding side structures. The reduction is mainly explained by the increased crush resistance provided by the beams. There has been a significant reduction of nonfatal serious injuries. But since torso injuries are relatively less predominant among fatalities (and head injuries much more predominant) and since the increase in crush resistance due to beams is probably of little consequence on energy dissipation in an extremely severe crash, the effect of Standards 214 on fatalities in vehicle-to-vehicle crashes is negligible.

CONCLUSIONS—AND IMPLICATIONS FOR CURRENT RESEARCH

The results on vehicle-to-vehicle crashes support a number of conclusions that are relevant to current side impact research. The principal conclusion is that existing production door beams have significantly reduced the type of serious injury that they were intended to reduce (nearside occupants' torso injuries due to side structure contact). The beams have performed as intended—reducing the depth of intrusion when the car is directly impacted in the compartment by another vehicle, primarily as a consequence of increased crush resistance. This is evidence that the structural approach to side impact protection will work in highway accidents involving human subjects. What is more remarkable is that existing production beams—which represent a far lower level of structural improvement than what has been furnished in experimental side structures—have nevertheless eliminated 33 percent of these injuries in highway accidents.

Despite the fact that existing beams have eliminated 33 percent of the injuries they were designed to mitigate, they have only reduced the overall injury risk in vehicle-to-vehicle side impacts by 8 percent and have had little or no effect on fatalities. The key point here is that the majority of serious casualties in highway accidents cannot be directly attributed to contact with intruding door structures—especially the large number of fatal head injuries. That point was clearly stated at the 8th ESV Conference by Mehta, Pearson and Wilson (2) and it is confirmed by the NCSS and FARS data. It sets limits on the benefits that can be obtained by structural strengthening unless accompanied by other improvements.

The current Standard 214 appears to have been partially effective in mitigating sill override in highway accidents even though it has only been designed for that purpose to, at most, a limited extent. Perhaps a substantial additional reduction of sill override could be achieved on the highway with relatively simple modifications of existing beams. For example, Hollowell and Pavlick reported at the 8th ESV Conference that considerable reductions of override and intrusion were obtained in crash tests by providing a load path from the beam to the sill (12).

The current door beams did not appear to significantly promote deflection of the striking vehicle in highway accidents—the effect that was prominently displayed in fixed-object crashes and also, to some extent, in General Motors' and Renault's staged crashes involving two vehicles (5, 7). The potential benefits of a deflecting action were not realized in vehicle-to-vehicle crashes. Could the beams be modified so as to deflect the striking vehicle in a highway accident?

Regarding side impacts with fixed objects, the principal conclusion is that the existing Standard 214 has significantly reduced fatalities and serious nonfatal injuries. It has accomplished these benefits because beams have promoted a more glancing, less severe collision and because the likelihood of ejection through the door area has been reduced. The ejection reduction may be an unforeseen consequence of the effect on damage patterns.

The benefits are not limited to a reduction of nearside occupants' torso injuries. Many other categories of injury, including head injury and ejection were mitigated. The annual savings of 500 lives make the current Standard 214 one of the most effective existing safety regulations in the United States. These benefits, and the mechanisms that led to them, should not be overlooked in the development of more sophisticated side impact protection.

REFERENCES

1. C. J. Kahane, "An Evaluation of Side Structure Improvements in Response to Federal Motor Vehicle Safety Standard 214." National Technical Information Service, 1982.
2. R. Mehta, J. L. Pearson and R. A. Wilson, "Side Impact Insights from General Motors Field Accident Data Base." 8th ESV Conference, 1980.
3. C. J. Kahane, R. A. Smith and K. J. Tharp, "The National Crash Severity Study." 6th ESV Conference, 1976.
4. J. Hedlund, "The National Crash Severity Study and its Relationship to ESV Design Criteria." 7th ESV Conference, 1978.
5. C. E. Hedeen and D. D. Campbell, "Side Impact Structures." Society of Automotive Engineers Paper No. 69003, 1969.
6. *Federal Register*, Vol. 35, April 23, 1970, p. 6513.
7. A. Burgett, Letter No. 15 to NHTSA Docket 2-6-GR, August 22, 1975.
8. M. Rodger, "Occupant Protection in Side Impacts." *Toward Safer Road Vehicles*, Transport and Road Research Laboratory, 1972.

9. J. E. Greene, "Occupant Survivability in Lateral Collisions," Vol. 1. National Technical Information Service, 1976.

10. S. L. Terry, Letter No. 13 to NHTSA Docket 2-6-NO2, July 14, 1970.

11. O. Kitamura, K. Watanabe and K. Matsushita, "An Experimental Study of Lateral Impact Protection." 8th ESV Conference, 1980.

12. W. T. Hollowell and M. Pavlick, "Status of the Development of Improved Vehicle Side Structures for the Upgrade of FMVSS 214." 8th ESV Conference, 1980.

Global Trends in Side Impact Occupant Protection

Sherman E. Henson
Ford Motor Company
United States

Stephanie L. Janczak
Ford Motor Company
United States

Paper No. 94-S6-O-03

ABSTRACT

The recent increase in safety belt use and the corresponding phase-in of supplemental air bags has brought about a significant decline in frontal impact injuries and fatalities. This trend will continue worldwide as belt use continues to increase and more cars can be fitted with supplemental dual air bags. As a result of declining front impact trauma, increased attention in the 1990's is being given to side impacts and disabling lower limb injuries. Attention to side impact is not new. Research and evolutionary vehicle enhancements have been made for more than four decades. In the 1950's, Ford offered safety belts and strengthened door latches to help reduce the risk of ejections in side impacts and other collision modes. General Motors developed side door beams and offered them in its products in the late 1960's. The U.S. Government made door beams mandatory for passenger cars as of January 1, 1973. A major inhibitor to progress in side impact safety has been the lack of human-like test dummies and an understanding of injury mechanisms. Amid widespread controversy and criticism of its dummy and injury criteria, the U.S. Government issued the world's first dynamic side impact regulation, which begins phase-in this year and affects all passenger cars beginning September 1, 1996. A regulation using a completely distinct dummy, injury criteria, and barrier is nearing approval in Europe.

In this paper we examine from the perspective of a global manufacturer of cars and trucks some trends in side impact occupant protection and their effects on future products. We look at interactions of side impact protection improvements with other societal goals such as fuel efficiency and the environment. We also discuss emerging technologies such as side air bags, the potentially adverse consequences of international disharmony of side impact regulations, and the benefits of global agreement on injury mechanisms and injury criteria. Finally, we stress the need for a family of anthropomorphic test dummies.

Ten new ways THE BIG M provides you with advanced motoring safety

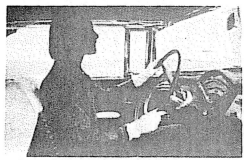

1. New impact-absorbing safety steering wheel, with deeply recessed hub, offers greater protection for the driver during quick stops. The wheel is also positioned for better visibility.

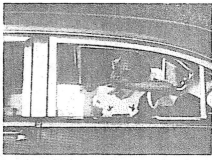

2, 3. New triple-strength safety door locks give extra protection against doors springing open upon sudden impact. Child-proof safety-locking device for rear doors is also available.*

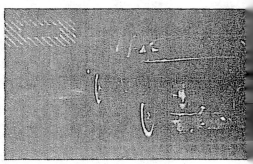

4. New safety-beam head lamps increase effective seeing distance up to 80 feet; improve visibility in fog, dust, rain, sleet or snow; and greatly reduce the glare for oncoming drivers.

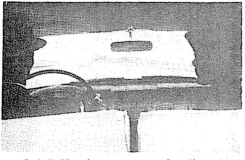

5, 6, 7. New instrument panel pad*, padded sun visors* have a special plastic impact-absorbing filler for your protection. Full-swivel *Safety Rearview Mirror minimizes shattering.

8. New safety seat belts*—bolted to floor supports—are available for driver and passengers. Made of durable high-strength nylon, they are designed for easy one-hand adjustment.

9. New 225- or 210-HP safety-surge V-8 gives you more *usable* power—an extra reserve for greater safety in passing, hill climbing; faster acceleration where you need it most.

10. Improved safety-grip brakes provide quicker, smoother stopping action, last longer. But—besides these new safety features, you'll want to see *all* the big, new things THE BIG M offers this year—in beauty, power, performance. Best place to start looking: your Mercury dealer's.

A MAGNIFICENT VALUE IN THE FORD FAMILY OF FINE CARS

For 1956_the big move is to THE BIG MERCURY

*Optional at extra cost

MERCURY DIVISION • FORD MOTOR COMPANY

Figure 1 1956 Mercury Advertisement

HISTORICAL PERSPECTIVE

Occupant Ejection

In the United States, side impact was part of the increased attention to safety during the 1950's. The growing science of automotive safety was in its infancy and most of the engineering tools needed for advancement were not yet available. Contrary to conventional wisdom, accident investigations and crash tests demonstrated that being "thrown clear" of the accident was not an advantage to the occupant, but a frequent source of death and injury. In side impacts, doors sometimes opened and the unbelted occupants thrown from the vehicle to be injured by objects often more risky than the vehicle interior. Therefore, safety belts and stronger door latches were among the first improvements effective in side impacts. In 1956, Ford offered the "Ford Lifeguard Design" package, that included such items as a collapsible steering column, padded instrument panel and sun visors, lap belts and double-grip door latches. The latter two items were effective in reducing injuries and fatalities from occupant ejections in side impacts.

Door Beams

In 1969 General Motors offered side impact guard beams in most of their large vehicles. The beams were intended to provide protection in oblique vehicle-to-vehicle side impacts by deflecting the striking vehicle away from the struck vehicle's doors in the same way that higway guard rails deflect vehicles away from the roadside. By 1971, General Motors had installed door beams in all of their vehicles and other manufacturers had beams in some of their vehicles as well. In the early 1970's, the National Highway Traffic Safety Administration (NHTSA) promulgated a rule requiring that vehicle doors meet certain force-deflection minimums in a static laboratory push test. The requirement was designed so that the only way to pass the test was to install door beams similar to those offered by General Motors in the late 1960's. NHTSA was later to find[1] that beams were not effective in the two-vehicle collisions for which they were designed, but were effective in collisions with tall, narrow fixed objects such as poles and trees. A puzzling finding of the same NHTSA study was that the beams actually increased the risk of injury in collisions with other types of fixed objects. NHTSA reported the net effectiveness of beams to be the life savings from the pole/tree collisions minus the losses from the other collision types. The unexpected findings, to our knowledge, have never been duplicated in controlled crash tests.

Early Dummies

Many researchers at this time believed that further progress in side impact would require the development of tests which better imitated actual collisions. Dummies had been used for some time in frontal collision research, but the science of side impact dummies lagged far behind. To fill this void, NHTSA issued a contract to the Highway Safety Research Institute (HSRI) at the University of Michigan (now called UMTRI) to develop a Hybrid II-based side impact dummy (SID) for use in crash testing. At the same time the Association Peugeot/Renault (APR) and other European research laboratories were designing their own dummies and Ford Motor Company developed a simpler test device, the Side Impact Body Block (SIBB)[2]. These first generation test devices were based on biomechanical data available at that time and lacked the biofidelity and improved measurement capabilities of second generation dummies such as the EUROSID-1 and BIOSID which were to follow.

Crash Tests

The dynamic side impact tests developed during the 1970's and 1980's were designed to simulate a two-vehicle intersection type side collision representative of the real world. Accident data studies indicated that most serious injuries (AIS 3 or greater) were to the thorax and resulted from contact with the side interior. An analysis of the National Crash Severity Study (NCSS) and the Motors Insurance Company (MIC) data by Partyka of NHTSA and Rezabek[3] of Ford confirmed other studies and further found that the rankings of side impact inury sources depended on collision severity. Below a 32 km/h (20 mi/h) ΔV of the struck car, serious head inury is uncommon relative to thorax and other injuries. Above 32 km/h ΔV, the proportion of serious head injuries increases. Side collisons with struck car ΔV above 32 km/h are very severe, considerably more severe than today's FMVSS 214 and the regulation proposed for Europe.

Therefore, the finite resources available for side impact were first directed at thorax injury, the most frequent serious injury and the one which practicable countermeasures were judged to have the greatest chance of success.

Sub System Tests

Also in the 1970's, General Motors developed component tests for both the head and the thorax. Although these tests were not as all-encompassing as the full-vehicle crash tests, General Motors argued that using these tests could bring about improvements in side impact protection considerably sooner and at lower consumer cost than the full-vehicle tests. Based on the GM work, the Motor Vehicle Manufacturers Association (MVMA) later developed and tested a more biofidelic thorax impactor and offered the device for NHTSA consideration in its forthcoming revision to FMVSS 214.

The idea of a simpler, less costly, and more repeatable test than the full-vehicle test sparked further development in Europe. Volkswagon developed the Composite Test Procedure (CTP), a novel approach which combined exterior and interior vehicle crush tests with a mathematical model to simulate dynamic test results. The Motor Vehicle Manufacturers Association (MVMA, now the American Automobile Manufacturers Association or AAMA), Common Market Automobile Constructors (CCMC, now the Association des Constructeurs Europeans d'Automobile or ACEA) and Japanese Automobile Manufacturers Association (JAMA) combined their efforts to refine the VW procedure and presented it to regulatory authorities in the United States Europe and Japan. Although NHTSA showed little interest in the CTP, until recently the procedure was included as an alternate to the full vehicle test in the European proposal. The strength of the CTP was that it replaced the dummy with a simplified static crush test device and used a math model to reduce test-to-test variability. It was also hypothesized that the math model could be made more human-like than a dummy, because the model is not constrained by the limitations of engineering materials and fabrication processes. On the negative side, the static crush tests did not account for speed-dependent visco-elastic foam properties, and demands for more realism in the interior loading devices led to excessive complexity. In addition, the

CTP was never fairly compared with full-vehicle tests, i.e., the two types of tests have not been graded on their respective representations of real world injuries. It as been assumed that the face validity of the full-vehicle test represents "reality" and that the CTP should give identical results to be acceptable. This "no-win" situation finally led to the abandonment of the CTP.

TODAY'S REGULATIONS

Trends

Over the past ten years, the proportion of fatalities from front impacts has declined while the proportion of fatalities from side impact has increased. This is shown in Figure 2, the results of a Ford study of Fatal Accident Reporting System data for the 1982-1992 time period. To appropriately reflect advances in vehicle safety, the data for each year is restricted to the most recent three model year vehicles. The increasing proportion of side impact fatalities is probably due to a decrease in front fatalities due to such factors as increased safety belt use, less drinking and driving, the installation of airbags, and the proliferation of accident avoidance features. These trends have increased the attention given to side impact regulations.

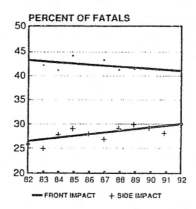

Figure 2. Passenger Car Occupant Fatalitites by Front and Side Impacts.

FMVSS 214

The modest effectiveness of door beams led NHTSA to promulgate a rule requiring the world's first dynamic side impact crash test for passenger cars. This rule, initiated in 1979, was made final in

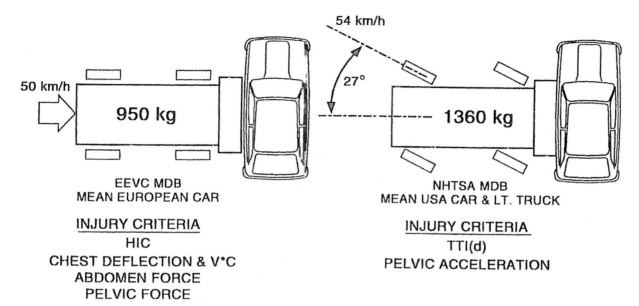

PROPOSED EUROPEAN
EUROSID - 1 FRONT SEAT

50 km/h

950 kg

EEVC MDB
MEAN EUROPEAN CAR

<u>INJURY CRITERIA</u>
HIC
CHEST DEFLECTION & V*C
ABDOMEN FORCE
PELVIC FORCE

USA FMVSS 214
SID - FRONT & REAR SEAT

54 km/h

27°

1360 kg

NHTSA MDB
MEAN USA CAR & LT. TRUCK

<u>INJURY CRITERIA</u>
TTI(d)
PELVIC ACCELERATION

Figure 3. Comparison of the proposed European test procedure and the U.S. FMVSS test procedure.

October 1990 and is to phase-in over a four year period beginning with the 1994 model year. The test procedure requires the use of the NHTSA SID dummy, acceleration-based injury criteria, and the NHTSA moving deformable barrier. There was considerable protest by biomechanics experts worldwide that inadequacies in the SID and its associated Thoracic Trauma Index (TTI(d)) would limit the rule's effectiveness, and some suggested that the rule would lead to padding too stiff for the elderly, potentially increasing their injury risk. There was little effort by authorities in the U.S. or Europe to collaborate and develop a common test based on the latest scientific data.

Regulation in Europe

While NHTSA was developing their dynamic crash test, researchers in Europe were developing a procedure of their own. Although the test looks like the the American test, the similarity ends there. The European test, now a proposed regulation, uses a different moving barrier, a different dummy and different injury criteria. A lighter vehicle population in Europe compared to the United States led to a smaller and lighter moving barrier. The stiffness of

the deformable barrier face are also significantly different than the U.S. barrier face. A 50 km/h (31 mph), 90 degree impact centered on the driver's seating reference point was chosen, again different from the NHTSA test. A single, belted EUROSID-1 dummy is used in the front seat. The NHTSA test has an additional dummy in the rear seat. Because of its later development, the European dummy, EUROSID-1, can measure more modern injury criteria, such as chest deflection and abdomen load, than the NHTSA SID.

The EUROSID-1

The EUROSID was designed to be more biofidelic than the SID and capable of measuring more parameters.[4][5] While establishing a test device for the European requirement, three prototype dummies were created as a part of the EEC Biomechanics Programme (1978-1982). A group of European research laboratories worked together as the European Experimental Vehicle Committee (EEVC) to produce a set of specifications for desirable dummy characteristics. These three prototypes and the U.S. SID were evaluated based on the desirable characteristics and all were found to be unacceptable in their current form. From 1983 to

1985, European researchers worked on designs for a new dummy that used the best features of each original prototype, and EUROSID evaluations began in 1986. The developers continued to make improvements to the dummy and the second phase, EUROSID-1, is currently in use. Due to the EUROSID-1's advanced design, the proposed European regulation limits the thorax deflection, viscous criterion, abdominal force and pubic symphysis force.

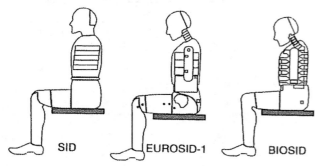

SIDE IMPACT DUMMIES

SID   EUROSID-1   BIOSID

Figure 4 . Profiles of the SID, EUROSID-1, and BIOSID side impact dummies.

The BIOSID

In the mid-1980's, General Motors, working with an SAE Committee, developed a dummy more advanced than SID or EUROSID and offered it for inclusion in FMVSS 214. The BIOSID[6] can measure deflection of the chest and abdomen, and tests to International Standards Organization (ISO) Technical Reports[7] have shown the BIOSID to be much more human-like than the SID and somewhat more humanlike than the EUROSID-1. NHTSA refused to consider the use of the BIOSID or the EUROSID-1 saying that while the SID and injury criteria may not be perfect, a switch to either of the new dummies would delay the NHTSA rule.

Other Countries

It appears that several countries are considering the U.S. or the European test procedure. Those that have a greater market presence in North America may be leaning toward the U.S. procedure.

The ISO has developed a set of specifications for side impact dummy biofidelity[8]. The ISO

procedure is guiding the development of tomorrow's dummies. Also, the ISO 10997 is a compromise test procedure that ISO intends to use as a starting point for a future international standard. The procedure is a hybrid of the U.S. and European test, allowing either a EUROSID-1 or a BIOSID dummy. To date, no country has adopted the ISO procedure.

Consequences

The automobile business is becoming more global and it appears that this trend will continue. Efficiencies gained from world-wide commonality of components result in lower prices and greater choice for automobile buyers. For example, Ford will market the Mondeo and its North American versions, Ford Contour and Mercury Mystique, in 59 countries. Unfortunately, Side Impact and other regulations which differ from country to country will likely require the use of unique components for each country having significantly different regulations. Some regulatory distinctions are based on differences in the traffic environment and may be justified. Others, such as differing specifications for a 50th percentile adult male dummy, are without reason or justification. As shown by Henson, et al, the BIOSID and the EUROSID-1 show similar responses for TTI(d) values and prescribe similar design changes to reduce TTI(d) values.[9] But when the viscous criterion (V*C) is used as the comparator, the BIOSID and the EUROSID-1 show opposite reactions.

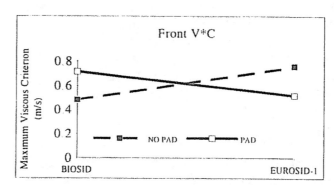

Figure 5. BIOSID vs EUROSID-1 Response to Padding

Without considering which dummy is showing the correct representation of the crash test, the authors concluded that designers would make markedly different design decisions depending on the dummy and injury criteria. Until further research is mpleted, researchers and engineers should use the rib deflection and V*C responses cautiously. The special designs for each country will add cost to cars and trucks, and will act as a brake on trade similar to a tariff. Furthermore, the added engineering and testing cost to homologate a vehicle for multiple markets will limit exports of some low volume models, further hurting trade and narrowing consumer choice. Before more regulations or consumer information tests are instated, the authors ge bioengineering experts representing governments and manufacturers from North America, Europe, and Asia to create a single family of "WORLDSID" dummies, using the latest data and combining the best features of the BIOSID and EUROSID-1. Moreover, this consortium should develop a range of dummy sizes to help study injury patterns in smaller adults and children.

THE FUTURE

Harmonization

The authors predict that Europe and North America will not harmonize side impact rules in the near future. The European rule is not in effect yet and NHTSA has planned extensive biomechanical research to be completed before considering modifications to its side impact dummy. Furthermore, there is no significant effort at harmonization underway. If harmonization is achieved after several more years, product cycles will make it impractical in some cases and impossible in others to change designs which have already been frozen for several more years. Nevertheless, it is urgent that collaboration begin now so that the benefits of harmonization can be realized at the earliest time.

The Occupant Safety Research Partnership (OSRP), established under the United States Council for Automotive Research (USCAR), has taken the first steps in creating a family of dummies. Along with First Technology Safety Systems, USCAR has agreed to support the development and testing of a small occupant side impact dummy. The dummy will approximate an average 12 or 13 year old adolescent and be similar in size to the 5th percentile Hybrid III female. It is expected that the small dummy will be based on existing side impact dummies and will have the capability to record data for the head, upper and lower neck, shoulder,

thorax, abdomen, ilium, acetabulum, and legs. The latest biomechanical information available will be used to develop this dummy.

Consumer Information

NHTSA is asking the U.S. Congress to fund Side Impact Consumer Information crash tests similar to the frontal NCAP tests that have been run since 1979. The Agency is considering running these tests at 61.2 km/h (38 mph), 7.2 km/h (4.5 mph) higher than FMVSS 214. Should market pressures compel manufacturers to meet these requirements, consumers will notice the largest effects on small cars. These cars cannot meet the higher speed test requirements without losing interior hip and shoulder room or increasing exterior width. Thus many small cars will get wider and heavier, making them "bigger" cars. Furthermore, some observers predict that the flaws in the SID dummy and the TTI(d) will lead to interior padding too stiff for the elderly. The overly stiff padding, it is predicted, could further increase the risk of injury to the elderly.

Dummies

Similar to the history of frontal impact dummies, side impact dummies will continue to improve and will multiply into "families" including large, small and child surrogates.

Light Trucks

As directed by the U.S. Congress, NHTSA is now proposing a dynamic side impact requirement for light trucks, vans and multi-purpose vehicles. As can be seen from Figure 6, side impacts account for 30% of all passenger car fatalities but only 13% of all light truck fatalities. With over six times the fatalities (6,437 for car, 1,039 for light truck), NHTSA first directed its resources at passenger cars.

The proposed test procedure is similar to the passenger car procedure except that NHTSA is suggesting the use of a heavier moving barrier with the bumper located higher than for passenger cars. Although trucks are exposed to the same traffic environment as passenger cars, NHTSA argues that side impact injuries in trucks more often result from being hit by another truck. Therefore, the argument goes, a more truck-like barrier is needed to improve

671

side impact performance in light trucks. It is clear from U.S. accident data that light trucks are already safer in side impacts than passenger cars.

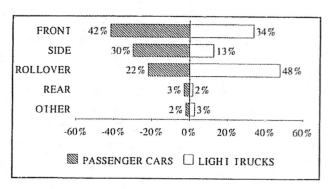

Figure 6. Distribution of Fatalities by Impact Type.

Therefore, there will be little or no benefit from the proposed rule. In addition, a Light Truck Dynamic Side Impact regulation would be yet another unique U.S. requirement the will make American products less competitive in foreign markets.

Interior Head Impact

The revised FMVSS 214 does not include a head impact requirement, due in part to the relatively poor biofidelity of the SID dummy neck. NHTSA had been conducting research and planned to address head impact with a component test. In late 1991, the United States Congress passed legislation effectively requiring NHTSA to begin rulemaking in 1992 and to issue a final rule in 1994.

NHTSA issued a proposal in February 1992 for Interior Head Impact Protection. Manufacturers expressed several objections to the proposal including the lack of objectivity and the impossibly short lead time. The AAMA proposed revisions to the rule that would address these and other issues. The proposal suggests a 15 mph impact into various

To meet the proposed requirement, new interior and structural designs will need to be developed. points along the rails and pillars using a modified Hybrid III head. NHTSA has suggested that 1 to 1 1/2 inches of foam added to the structures would pass the requirements. This solution brings up many issues such as reduced visibility and interior spaciousness. It also calls to question safety belt anchorage locations which tend to be placed on

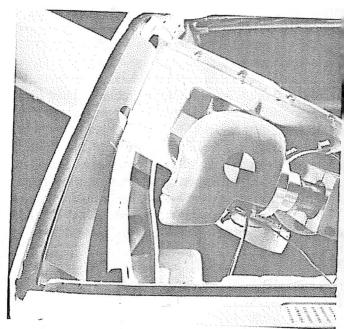

Figure 7 . In-vehicle head impact test using a robotics are to locate and fire the headform.

the B-pillar. Also, NHTSA is considering revising the roof crush requirements. The solutions for increasing roof crush strength run opposite to those for improving interior head impact protection.

Disabling Injuries

More research needs to be done to understand injury mechanisms in side impact accidents. Figure 8 shows data from the Folksam car accident data file indicating that different types of injuries are associated with fatalities than with serious, potentially disabling, injuries.[10]

The commonly used Abbreviated Injury Scale (AIS) shows the likelihood of a fatal injury. The results displayed in Figure 8 show that the most severe injuries in side impact collisions were to the chest or to the abdomen/pelvis region. The researchers also devised the Risk of Serious Consequence (RSC) scale. Injuries on this scale are either fatal or disabling. On the RSC scale, head and neck injuries replace chest and abdomen/pelvis injuries as the most frequent. These data also highlight debilitating leg injuries which are not a significant threat to life using the AIS. The AIS rates a broken ankle as an AIS 2 injury because the

672

likelihood of this injury being fatal is very low, but
the risk of long term disability is high.

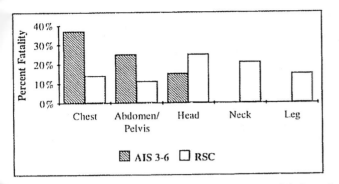

Figure 8. Risk of Fatality versus Risk of
Injury.

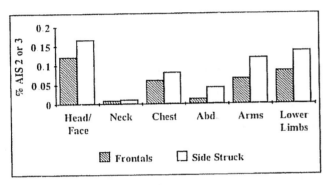

Figure 9. Risk of AIS 2 or 3 Injury in Frontal
and Side Struck Accidents.

Information presented at the 35th Stapp Car
Crash Conference[11] also shows a preponderance of
lower limb injuries in side impacts. Of the skeletal
injuries noted in side impacts, 65% occurred above
the knee; the pelvis accounted for 51% of the
fractures and the thighs for 14%. In most cases the
source of injury was contact with the front door
although intrusion was associated with disabling
skeletal injuries.

The Environment

New safety technologies may have the
potential to harm the environment. Difficult trade-
offs will need to be made in some cases. For
example, increases in structural strength to manage
impact energy will usually add weight, resulting in
increased fuel consumption and emissions. Some
relief may be found in new structural composite
materials which promise high strength and low
weight. However, these materials raise

environmental concerns. Can new materials be
economically recycled? Can environmentally
friendly, high volume production methods be
developed? Some materials may be prohibitively
expensive, forcing more difficult tradeoffs between
mobility and the environment.

ACCIDENT PREVENTION

Another key means of reducing side impact
injuries and fatalities is by reducing the number of
side impact collisions. Several methods of traffic
management, such as single direction highways and
prohibited left turns, are commonly used today. The
American Association of Retired People (AARP)
have already begun instructing their members that it
is safer to make three right turns around a block
rather than one left turn across traffic.

Other means of reducing side impact collisions
will come through technologies related to Intelligent
Vehicle Highway System (IVHS) in the U.S. and
PROMETHEUS in Europe. Two aspects of these
programs which may help to reduce side impact
collisions are intelligent cruise control and proximity
detection. Intelligent radar will either warn the
driver that the brakes should be applied or will apply
the brakes automatically. This will help to avoid the
impact or at least decrease the speed at which the
collision occurs. Proximity detection will become
another source of collision avoidance. Again, the
system will either warn the driver or apply the
brakes when a vehicle or object closes within a predetermined distance or rate.

FUTURE DESIGNS

Computers, air bags and composite materials
will play key roles in the future of Side Impact
Protection. Engineers now use computers to
simulate the crash tests. With rapid advances in
hardware and software predicted, scientists will be
better able to understand the energy dissipation
through the vehicle structure and the forces acting on
the occupants. This understanding will help
engineers develop better vehicles. NHTSA is also
continuing research on a computer model to study
the interactions of safety standards. The scope of
the model will expand beyond a single crash test to
include a simulation of the "real world." The
NHTSA effort is based on programs developed in
the 1970's by Ford as part of the Research Safety

Vehicle effort sponsored by NHTSA, and by Volkswagen. Using today's far more detailed and accurate simulations, the NHTSA effort will account for the vehicle population on the road and historic accident data.

Side impact air bags are being developed in the U.S., Japan and Europe. Volvo and General Motors have already applied for patents for their versions of the side impact air bag. Several key designs predominate research at this point. One design has the air bag stored in the rail above the door. The bag will have a long rectangular shape and will inflate in the downward direction to protect the head from impacting a hard surface. Another option, recently shown at the Frankfurt Auto Show by Audi, has the air bag stored in the occupant head restraint. As the air bag deploys, it deflects against the vehicle window to help prevent the occupant from impacting external objects such as trees, poles, or striking vehicles. This design has the advantage of moving forward and back in the vehicle as the seat position is adjusted. A third concept has the air bag located in the door structure itself. The air bag can then deploy out towards the occupant and up against the window. This air bag design has the added advantage of protecting both the head and the thorax when other designs concentrate on minimizing head contact.

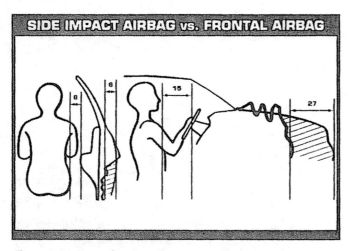

Figure 10 . Comparison of Crush Space for Front and Side Impacts.

There are many issues which still need to be resolved with side impact air bags. As can be seen in Figure 10, there is considerably less crush space available in a door compared to the front end of a vehicle, the impact forces reach the occupant more

quickly. To be effective, side air bags will have to be activated and inflated more quickly than air bags designed for frontal impacts. This may require an advanced sensor not yet developed to detect the impact quickly enough to activate the bag. A sensor which would anticipate the collision would be capable of activating the system prior to contact. This would allow a slower inflation rate and would result in a less aggressive air bag. With the reduced crush space in conjunction with the occupant being located closer to the impact point, the sensor must be highly sensitive to activate so quickly, but must also be selective to assure that the air bag activates only when intended. Injury criteria, test procedures and countermeasures are needed before progress can be made.

There are also concerns due to the wide range of seating positions. With a properly worn safety belt, the occupant is restrained near the center of the air bag during front impacts, but in a side impact the occupant can be located anywhere within the range of travel of the seat track. There is also the possibility that an occupant will be leaning against the door or will have an arm on the window sill causing further complications for inflation. Many companies are trying to determine the technical and manufacturing feasibility of these systems and if there are real world increased safety benefits for the occupants in a side impact.

Another major area of research will be in new composite materials. Scientists will be looking for new methods of energy management. With increasing demands for fuel efficiency and safety, vehicles will have to be made lighter and stronger. Composite materials have the advantage of being lighter in weight but higher in strength than commonly used materials. As mentioned before, there may be issues with recyclability and manufacturing processes.

SUMMARY

Attention to side impact is not new. Efforts to reduce injuries during a collision have been on-going since the 1950's. But the ability to reduce injuries from side impacts has been hindered by the lack of knowledge regarding injury mechanisms and the lack of human-like test devices. As more information is uncovered, more steps can be taken to improve the technology.

As has been shown in this paper, the technology needed to help reduce side impact injuries has not been available until recently. Today, the injury criteria are being reassessed as new information on injury mechanisms becomes available. Research has shown that bones may need different injury criteria than soft tissue.

Although much research has been completed in the area of side impact collisions, much work still needs to be done. At the same time, many governments have either passed or will pass requirements on side impact. With the science still evolving, it is unclear if these requirements will be beneficial to vehicle occupants. New injury mechanisms, injury criteria and human-like test devices are being developed that have the potential to be used to design vehicles which offer better protection in a variety of situations. It is not prudent to pass improper requirements due to political pressures. Time must be allowed for the proper development of knowledge and technology to support the effort.

ACKNOWLEDGEMENTS

The authors would like to acknowledge Mr. James Prybylski and Mr. Jack Edwards for their assistance in compiling the information used in Figure 2 and Figure 6.

REFERENCES

1. Kahane, C. J., "The Effectiveness and Performance of Current Door Beams in Side Impact Highway Accidents in the United States," SAE-826066, Ninth International Technical Conference on Experimental Safety Vehicles, Kyoto, Japan, 1982.

2. Daniel, R. P., P. Prasad, C. Yost, "A Biomechanical Evaluation of the Ford Side Impact Body Block and the SID and APR Side Impact Dummies," SAE-840882, 1984.

3. Partyka, S. C. and Rezabek, S. E., "Occupant Injury Patterns in Side Impacts - A Coordinated Industry/Government Accident Data Analysis," SAE 830459, Detroit, Michigan, February 28- March 4, 1983.

4. "EUROSID - Result of a European Research Programme," TNO Road-Vehicles Research Institute, May 1987.

5. Janssen, E. G. and A. C. M. Vermissen, "Biofidelity of the European Side Impact Dummy - EUROSID," SAE 881716.

6. Beebe, M. B., "What is the BioSID?" SAE-900377, February, 1990.

7. International Standards Organization Test Report 9790, Sections 1 through 6.

8. Ibid.

9. Henson, S., R. Hultman, R. Daniel, A. Spadafora, I. Parekh, "Comparison of BIOSID and EUROSID-1 Dummies in Full-Vehicle Crash Tests," SAE 940563, SAE International Congress & Exposition, Detroit, Michigan, February 28 - March 3, 1994.

10. Holand, Y.; Lovesund, P.; Nygran, O., "Life-Threatening and Disabling Injuries in Car-to-Car Side Impacts - Implications for Development of Protective Systems," Accident Analysis and Prevention, Vol. 25, Issue 2, April 1993.

11. Pattimore, D., E. Ward, P. Thoman, M. Bradford, "The Nature and Cause of Lower Limb Injuries in Car Crashes," SAE 912901, 35th Stapp Car Crash Conference Proceedings, San Diego, California, November 18-20, 1991.

CHAPTER EIGHT

Door latches

An Investigation Into Vehicle Side-Door Opening Under Accident Conditions __

Keith J Dale,
Keith C Clemo,
Motor Industry Research Association
United Kingdom

Abstract

Accident studies in the UK have identified a number of instances where the door latch has been activated following a side impact, allowing the door to open and the occupant to be ejected. The latch activation occurs in cars having a pull rod linkage connecting the latch and interior release handle, through disturbance of the linkage following an impact by the occupant's shoulder or upper arm on the door inner panel.

An investigation into the phenomenon was conducted by MIRA and tests carried out which successfully reproduced this effect in the laboratory. It is proposed that this would be useful in predicting the behaviour of a vehicle door system as a development or legislative tool.

The test comprises an impact on the door's inside surface at 24.1 km/h (15 mile/h) by a resilient bodyform of 36 kg (80 lb) mass, reproducing the shape of the upper torso and shoulder.

In addition, a number of design changes to the door structure and linkage arrangement have been proposed to alleviate the problem.

Introduction

It has long been recognised that preventing the ejection of an occupant in an accident is crucial in minimising his resulting injuries. Accident studies have consistently refuted the popularly-held view that it is beneficial to be "thrown clear" and shown that the risk of injury is greatly increased where ejection occurs.

Although seat-belts are to some extent effective in preventing ejection, it is the vehicle's side door which remains the principal defence against its occurence. If the effectiveness of the door is destroyed by either mechanical failure or inadvertent operation of the latch, a restrained occupant may suffer partial ejection. Similarly, an unrestrained occupant, such as a rear-seat occupant in a four-door car, may suffer complete ejection.

The importance of maintaining an effective barrier between the occupant and the exterior is reflected in the adoption of standards such as FMVSS 206 and ECE 11 as a mandatory requirement in most industrialized western nations. These standards stipulate minimum strength requirements for the latches and hinges, and also require the latch not to release when subject to acceleration. Compliance with the ECE 11 standard has been compulsory for all new cars in the UK for several years, and accident studies have demonstrated its effectiveness in reducing occupant ejection through door bursting(1).

However, a study of UK accidents for the period 1978 to 1980 has identified several instances where occupants have been ejected in accidents following the opening of the car door. In every case, the door appeared to open not as a result of mechanical failure of the door-hinge-latch system but through inadvertent activation of the latch. It is believed that this occurred as a consequence of an impact by the occupant against the inside face of the door disturbing the linkage between the latch and the interior release handle.

Concern over these accidents led to the UK Department of Transport commissioning MIRA to investigate the phenomenon by attempting to reproduce the effect under laboratory conditions. It is this study which forms the basis of this paper.

Accident Data

The Risks Associated with Occupant Ejection

Investigation into cases of accidents involving occupant ejection provide a valuable source of information into the mechanisms of this type of accident.

The incidence of these cases amongst accidents as a whole underwent a significant change as a result of the enforcement of mandatory standards for hinge and latch performance during the 1950's and 1960's. This served to reduce the number of occupant ejections resulting from the mechanical failure of these components(1).

Despite this reduction, ejection remains a very common feature in accidents involving death or serious injury. This is well illustrated in the paper by Carlsson(2) which studied cases of ejection amongst a sample of 10,000 accidents involving Volvo cars. Considered as a proportion of all of these accidents, ejection occurred in only 0.5% of cases. Yet amongst the accidents involving occupant fatality, ejection occurred in 12%. A similar study of the NCSS records(3) put this figure even higher, at 27%. Clearly, despite its low incidence this phenomenon is frequently associated with this most severe category of accidents.

This does not imply, however, that ejection was responsible for the extent of injury in all of these cases. To assess the role of occupant ejection in this respect it is necessary to distinguish between those cases where, on the one hand, ejection occurred as a consequence of the severity of the accident, and on the other where the accident injury was made more severe as a consequence of the ejection. In other words, we need to assess the separate roles of the ejection and the remaining accident conditions as a contribution to the resulting injuries.

It is, however, difficult to measure this directly, but several studies have attempted to assess the role of ejection in causing injuries. O'Day and Scott(4) conducted a "back-to-back" comparison of injuries to selected pairs of occupants, where one of the pair was ejected from the vehicle and the other remained inside. This reported a 22% rate of fatality amongst those ejected, compared with a 9% rate of fatality amongst those not ejected. At the other end of the severity scale, the proportion which received no hospital treatment was 3% for those ejected but 21% for those not ejected. The paper by Carlsson(2) quotes the figures for the risk of severe or fatal injury as being 48% for ejected occupants compared with 8% for belted (presumably non-ejected) occupants.

On the basis of these figures, it is therefore likely that the level of injury is significantly increased by the ejection process. Although crashes, where ejection occurs, appear to be the most severe ones, this does not seem sufficient to explain the high level of injury amongst ejected occupants.

The Mechanism of Occupant Ejection

Before the introduction of mandatory standards for latch and hinge performance typified by FMVSS 206 and ECE 11, a large proportion of ejections were caused by latch failure. The engineering improvements brought about as a result of these regulations have served to reduce the number of cases of this type from the accident figures and give greater prominence to other ejection mechanisms, for example, ejection via the window apertures or through the disruption of the passenger compartment structure. This pattern of distribution has remained essentially unchanged until the present. However, in the UK despite the requirement for all new vehicles to be fitted with latches which meet the ECE 11 requirements for latch strength, hinge strength and the latches' resistance to opening under acceleration, cases of door opening continue to form a recognisable proportion of ejection accidents.

A recent study by Green et al of the Loughborough Institute for Consumer Ergonomics and the Leicester Royal Infirmary(5) illustrates this.

In this study of serious accidents involving ejection in the UK, of 919 vehicles in all, door opening occurred in 129 (12.7%) of the vehicles. This was divided between 79 (7.8%) side door opening and 66 (6.6%) tailgate opening. On 16 cars, both of these opened, and some had openings of more than one side door. These openings were associated with 12

occupant ejections or 12% of the total ejections registered in the study.

The principal reason for non-ejection seems to be that the occupant was not seated next to a door which opened. Amongst exposed occupants, that is, occupants who were unrestrained in a seat next to a door which opened, 30% of such front seat occupants were ejected and 55% of rear seat occupants.

Finally, this study investigated the mechanism of latch release. The greatest proportion of these occurrences, some 51% were considered to be due to activation of the linkage, compared with 28% due to latch failure, 5% due to handle activation and 16% due to other mechanisms.

To obtain an indication of the total number of ejection cases in the UK it is possible to assign the proportion of ejection cases in various detailed studies with comparable totals for the UK as a whole. Such an approach is, of course, subject to the many pitfalls of matching sample and population, and so such figures must necessarily be crude estimates only.

If we apply the 0.5% ejection rate amongst all injury accidents in the Volvo study to the 246,000 such accidents in the UK in 1985, we can estimate a total of 1230 occupant ejections. Alternatively, the figure of 12% given as the proportion of ejections amongst occupant fatalities, when compared to the UK total fatalities of 5165, gives 620 cases. A third approach applied to a wider range of car types is to consider the 40 cases of complete ejection and 16 cases of partial ejection in the Loughborough study. The study was stratified so that serious or fatal injuries represented 30% of the study, giving a total of 526 such cases in the study against 70980 for the UK. On this basis, therefore, we would expect to see 5400 complete ejections and 2160 partial ejections.

Allowing for the disparity in these figures, therefore, we could say that as a minimum figure, 620 ejections occur in the UK per year. If the proportion of cases due to various causes reflects the figures in the Loughborough study, we might expect to see, at least, 140 ejections via the side door. About half of these may be expected to be due to linkage failure, that is some 70 ejections.

The Injury Mechanism-Occupant Trajectory

The mechanism of injury, that is, the circumstances of the accident, the dynamics of the occupant, and their interaction with the door structure and the latch mechanism may be illustrated by reference to six representative accidents chosen from a selection of accident studies collected in the UK.

Table 1 summarizes and illustrates the relevant features of the accidents. The door activation investigation was limited to three specific vehicle models—A, B and C—and two situations have been selected for each of the vehicles concerned.

Table 1. Accidents featuring door latch release.

Accident No.	Vehicle	Collision Configuration	Occupants	Ejection	Severity AIS
1	A		Driver F. Passenger	Yes Possibly	2 to 3 1 to 3
2	A		Driver	Yes	5 (Fatal)
3	B		Driver	Yes	3
4	B		Driver F. Passenger	NO Yes	- 2 to 3
5	C		Driver	No	1
6	C		Driver	No	1

Accident Number 1 concerns a vehicle of type A which was struck on the front right corner from a 5 o'clock direction by a vehicle of similar size, with a severity estimated to be equivalent to an ETS of 30 mph. The unrestrained driver struck the inside of the door, denting the inner panel at the beltline. The driver's door latch released and the door opened. The driver was ejected via the open door and sustained injuries of MAIS 3 in the form of a comminuted fracture of the right ulna, and a fractured scapula. The front seat passenger was reported to have been ejected via the same route, but her injuries are not known.

Accident Number 2 again concerns a vehicle of type A whose driver lost control on a bend and struck a tree. The impact with the tree occurred on the right hand rear door at an angle of 4 o'clock, with a severity estimated to be equivalent to a 35 mph pole impact. The driver, the sole occupant, was unrestrained, he struck the door with his hip and chest, causing moderate distortion of the inner panel. The door released and the occupant was ejected, receiving a fatal injury in the form of contusions of the liver.

Accident Number 3 concerns a vehicle of type B which struck a bridge parapet with the right front corner in a 2 o'clock direction. The impact speed has not been estimated, but there was reported to be minor intrusion of the right footwell. The unrestrained driver struck the right hand door, which opened, allowing the driver to be ejected. The driver suffered a fractured rib.

Accident Number 4 concerns a vehicle of type B which was struck on its front left corner in a 9 o'clock direction by a vehicle of 10% greater mass. The collision speed was estimated to be equivalent to a 30 mph impact. The unrestrained female front seat passenger struck the left front door. The latch released and the door opened, leading to complete ejection and injuries to the head and left scapula of MAIS 3.

Accident Number 5 concerns a vehicle of type C which rolled over and struck a hedge on its right hand side. The impact with the hedge and low bank was distributed over most of the side of the car and was in a 3 o'clock direction. The unrestrained male driver struck the door inner panel with his shoulder, causing the door to unlatch. In this case, the presence of the hedge prevented the door from opening completely and the occupant was not ejected. The resulting injuries were slight and not caused by contact with the door. Despite the insubstantial nature of the hedge at door height, the vehicle suffered moderate intrusion at rocker panel level, presumably from the earth bank or the root system of the hedge. This accounts for the severity of the occupant door contact.

Accident Number 6 concerns a vehicle of type C. The car lost control and struck a tree just forward of the right A-pillar area. This was followed by a roll-over and a further impact with a pole on the rear right quarter panel. Both impacts were estimated to be in a 3 o'clock direction. During the accident the driver struck the right door waist rail with his shoulder and the door released and opened. The driver was restrained; this prevented complete ejection through the door and he suffered only minor injuries (AIS 1).

Although this is only a small sample, it nevertheless represents a wide range of accident conditions in terms of severity, accident type and resulting injuries. Despite the small sample size, however, there are a number of conclusions that we can draw from this study.

a) In each of the accidents, a similar pattern of damage to the door inner panel has been noted and is shown in Figure 1. This was in the form of a dent or crease at the belt-line. Since this is strongly linked with the impact by the occupants shoulder it appears that it occurs before, and not as a result of the door opening. Furthermore, since in most accidents there is very little damage to the door and B-pillar apart from this, it would appear to be the sole cause for the unlatching.

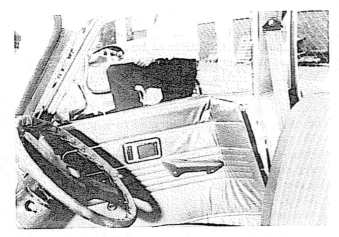

Figure 1. Road accident—door inner panel deformation

b) The recurrence of the same type of vehicle in different accidents suggests that the problem may be related to specific designs of door system.

c) The mechanism of door opening is clear from examination of the door layout. In each type of car involved in the accidents, the linkage between the latch and the interior release handle is in the form of a pull-rod routed along the inner surface of the door, at a height very close to the waist rail. It appears that the impact by the occupant's shoulder has caused the linkage to activate the latch.

d) The absence of damage to the latch, striker pin or surrounding structure suggests that the phenomenon occurs as a result of low energy impacts, less than the energy needed to deform the door structure significantly. This is confirmed by the low level of shoulder injury in those occupants who were not ejected as a result of the door opening. If this is so, it suggests that, firstly, the conditions for door bursting are the same as for ejection, namely an occupant projected outward. In addition to this, if the door bursts under a relatively low force, this implies that an occupant projected against it would experience only a small change in velocity and would exit the open door and strike some external object at a higher velocity than if the door offered more resistance.

Experimental Investigation

MIRA was commissioned by the UK Department of Transport to conduct an experimental investigation of this phenomenon. The terms of reference of this study were to attempt to reproduce this phenomenon in the laboratory, using doors from vehicles known to be prone to this type of occurrence. In addition to this, it was hoped that other features observed in the accident studies, particularly the pattern of door damage, would also be reproduced.

In initial discussions with the sponsors, it was felt that one of two types of test might offer a suitable method of achieving this. The first of these was a quasi-static test, in which an intrusion device, representing the shoulder of the occupants, was forced slowly into the inside face of the door until unlatching occurred. The second was in the form of an impact test using a bodyform representing the torso, shoulder and upper arm of an occupant, projected against the inner panel of the door. However, before the development work on these tests commenced, some additional work was done to gather data on the forces and deflections required to release the latch on undeformed door specimens.

Investigation into Latch Release Parameters

Front doors from two types of car known from the accident studies to be subject to this effect were selected and mounted in a test fixture. A small hole was made in the outer skin of the door opposite the centre of the latch release rod and a thin cable looped around the rod. The cable was then pulled outwards until latch release occurred, the force and deflection of the rod being noted at the time of release. The results of the tests are given in Table 2.

It can be seen from the table that the two specimens from Vehicle A unlatched at a similar force but with some variation in deflection, presumably due to manufacturing and assembly tolerances. The latch from Vehicle B required the same order of force to release but less deflection.

The deflections appear to be rather less than the degree of plastic deformation observed in photographs of the accident vehicles.

It was observed that on one of the latch systems that the locking linkage and latch linkage had some

Table 2. Results of latch rod deflection tests.

	Door Release	
Sample	Force N	Deflection mm
B1	200	15
A1	178	30
A3	182	40

common components. A similar test was conducted on the interior lock linkage rod which was incorporated into the handle system to enable lock release from inside the car.

The result of the test on the lock linkage was interesting, in that it required a similar load and deflection to unlock the latch as it does to unlatch it. The design of latch effectively isolates the latch from the interior release when the lock is activated. However, the result suggests that even if the lock was activated, it might not prevent the door from unlatching, since the lock would be de-activated by the time the door was unlatched.

Development of Quasi-Static Test Method

A quasi-static test was devised similar to a side door intrusion test but from inside the car.

These tests were conducted by mounting a door specimen by its hinges on a fixture, with the latch in engagement with a striker pin. An intrusion device was constructed from 50 mm diameter steel tube to represent the humerus of the occupant. This was mounted on a hydraulic ram which forced it slowly into the inner panel of the door. A typical test arrangement is shown in Figure 2.

The initial test was conducted with the intrusion device axis inclined at 20° to the vertical in a longitudinal plane and aligned with the centre of the longest unsupported section of rod. Later tests used a similar tube mounted vertically, a striker pin mounting allowing it to pivot, a 100 mm diameter intrusion device and a device allowing the intrusion device to pivot in a transverse plane. A total of nine tests were conducted on front doors from vehicles of type A and B.

Despite these variations only one test, the first, where activation was not repeated, resulted in unlatching of the door. All of the others continued until the door failed mechanically or until deformation of the door rendered unlatching impossible. From observa-

tions during the tests, it was felt that the test method was not producing a similar pattern of deformation to that seen in the accident studies. In particular, the test was leading to rotation of the latch about a vertical axis which prevented the striker from clearing the latch.

Development of Dynamic Test Method

It is clear that the closest simulation of the conditions existing at the door on the instant of release, as seen through the accident case studies, would be achieved through some form of impact test. Such a test would need to reproduce the important parameters such as the geometry and mass of the impacting occupant, yet be simple enough to allow the test to be conducted at moderate expense.

It was felt that a suitable test method would be broadly similar to the Bodyform Impact Test specified to measure the effectiveness of the Steering Assembly in Regulations such as FMVSS 203 and ECE Reg 12. Although the bodyform used in this test is not suitable for representing lateral impacts, the use of a common propulsion device would clearly allow the test to be carried out more easily, and provide good repeatability.

As in the previous series of tests, a basic test method was established which was common to all tests, with variations introduced to attempt to achieve repeatable latch release.

The basic test utilised a bodyform constructed from the upper torso and arm of a Sierra fiftieth-percentile male anthropomorphic dummy. This was adapted to enable it to be mounted on a sled-type propulsion device used for ECE Reg 12 tests by fixing a frame to the rigid thoracic spine of the dummy. The whole unit was then ballasted to the ECE Reg 12 bodyblock mass of 36 kg. The arm was fixed at the elbow to constrain articulation and the centre of gravity position adjusted so that it coincided with that of the bodyblock specified by ECE Reg 12. This test arrangement is shown in Figure 3.

Figure 2. Static door latch activation test

Figure 3. Dynamic door latch activation test

To allow information to be gathered which would be useful in the investigation, an accelerometer was mounted on the dummy spine, measuring in a lateral direction.

As in the ECE Reg 12 test, the bodyform was mounted on its propulsion sled, the sled was propelled to impact velocity and arrested a short distance from the door, projecting the dummy forwards to strike the door in free flight.

An impact velocity of 24.1 km/hr was chosen initially as it seemed that this represented the accidents studied earlier in the project. It was felt that the choice of velocity was not important except that the kinetic energy of the bodyform should exceed a certain threshold value necessary to deform the door and that a higher velocity would merely lead to a higher final velocity of the bodyform as it was projected through the door after unlatching.

The initial tests were conducted with a bodyform having an upper arm attached to the torso at the shoulder; in later tests, the upper arm was fixed at the elbow. This change meant that the dummy's ribs no longer contributed to the performance of the bodyform, leading to a simpler device. Other variations were introduced by varying in the impact velocity and the means of mounting the striker pin on the rig. A total of 19 tests were conducted in this series, using front doors from Vehicle A and these are summarised in Table 3.

The most interesting aspect emanating from the analysis of the results from these tests was the effect that realistically representing the surrounding vehicle structure of the door had in allowing latch activation to occur. The door closure seal pressures were reproduced by encorporating actual seals from the vehicle concerned so that normal opening and closing efforts were maintained. In the early tests the striker was attached to a simple box section to represent the

B-pillar. To investigate the forces acting on the latch and striker the components were instrumented with displacement transducers and tri-axial loadcells. The effects of configuring the test rig in these ways tended to lead to poor repeatability. Analysis of high speed cine suggested that the "B" pillar location was too rigid. In an effort to overcome this problem the rig was adapted to encorporate the body "B" pillar fixed at its extreme ends. The fixing position was designed to represent the "B" pillar connection at the cant rail and the rocker panel, thus increasing the degree of representativeness of the door system in the rig. During the second half of the test programme it became clear that a test which very accurately reproduced the circumstances in real world accidents had been achieved. The door was opening on more than 80% of occasions and the damage to the door was virtually identical to that seen in the accidents used to model the test. An example of the damage is shown in Figure 4.

Validation Testing

The investigation had up to this point in time focused on reproducing the circumstances of real-world accident situations with vehicle components that had exhibited the problem. The next stage of the study was to investigate the possibility of the same problem occurring on vehicles that were not currently included in the national accident data file. Trials were conducted to evaluate the performance of a random selection of six models of vehicle readily available on the UK market.

Each test was conducted using the front door taken from cars which had been purchased new for a crashworthiness evaluation programme. Two specimens of each type of door (both left hand) were tested. The complete "B" pillar was removed from the vehicle to mount the striker for the test, and the vehicle was carefully measured before removal of

Table 3. Results of dynamic impact tests.

Test	Test Configuration	Latch Activation Yes	Latch Activation No	Bodyform Data km/hr	Bodyform Data (g)	Test Result
1&2	Striker supported on hollow box section 10 SWG	•		24.8	30	Door Opened
3	Triaxial Loadcell on Striker Rotary Pot on Latch Lever Latch Rod Removed		•	24.7	25	Door Remained Closed
4,5 &6	As 3 but with Latch Rod Fitted. 16, 20, 27 km/hr		•			Door Remained Closed
7	As 3 but 24 km/hr Impact Speed	•		24.8	28	Latch Released but Rig Moved, Trapping Door
8	Striker Supported by B-pillar Secured at Top and Bottom	•		24.5	44	Door Opened
9	As 8	•		24.5	52	Latch Pawl Jammed
10&11	As 8	•		24.3	44	Door Opened
12	As 8		•	24.4	47	Door Remained Closed Latch Mechanism Corrosion
13	As 12. Latch Lubricated	•		24.4	37	Door Opened
14	As 12. But with Interior Trim	•		24.4	37	Door Opened

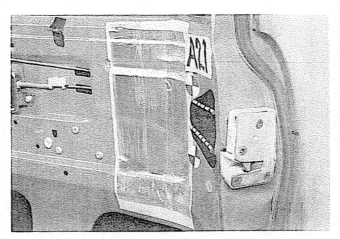

Figure 4. Typical door inner panel distortion

these components so that the relationship of the latch and striker could be accurately reproduced in the test.

The tests were conducted in accordance with the final test method used in the development trials, that is, propelling the bodyform at a 24.1 km/hr impact velocity, with the arm of the bodyform aligned toward with the front seat R-point of the vehicle, to strike the door in free flight.

The results of the tests showed that the door on two of the models activated and released, three failed to activate and one activated but failed to release completely. However, in the case of one of the models where the door opened, the unlatching did not persist long enough to allow the door to pass the intermediate-latched position and the door remained closed.

Engineering Evaluation of Results

The evidence gathered in experimental investigation has suggested that the occurrence of door unlatching is determined to a great extent by features in the design of the door and the latch linkage. In other words, the potential for this type of accident seems to lie in specific types of linkage rather than every linkage of the pull-rod.

It is apparent that in most cases the linkage rod is situated in a vulnerable region of the door and that the degree of outward lateral deformation of the rod and the force to achieve this deformation is small.

The performance of any particular design using a pull-rod appears to be determined to a great extent by the pattern of deformation of the door when struck on the inner face. In particular, it is whether the deformation of the door extends into the end face adjacent to the latch and the behaviour of the rod itself under impact which is crucial in deciding whether the door will open or not.

This suggests that the simplest solution may be to increase the beltline stiffness to reduce intrusion but not enough to cause serious shoulder injury. Alternatively, additional structural stiffness could be incorporated into the corner of the door, in the region of the latch, which will more easily transmit the bending into the end face of the door. However, this will work best where the latch incorporates a fairly restrictive slot, around the head of the striker pin. Doors having a more open slot, or a channel in the striker would not benefit from this.

An alternative solution to the problem, still using the pull-rod linkage, might be to incorporate some free-play into the handle so that it was free to move when the rod is displaced in a direction opposite to its normal travel. If sufficient movement were allowed, it is probable that the rod would not activate the latch under accident conditions.

The problem could also be solved by using other forms of mechanism to release the latch. A "bowden" cable is used in some of the luxury cars produced in Europe. Provided that a generous length of outer casing is incorporated, this would be free from activation problems when routed away from the critical area.

General Observations

The investigation has shown that the opening of car doors as a result of inadvertent unlatching is a possibility in a significant number of accidents and that some action is necessary in order to reduce its occurrence. The choice of an appropriate course of action is a complex one, since there are many considerations which will ultimately determine the practicability of any solution which might be adopted.

In the UK the Department of Transport consider that the necessary changes in vehicle design should be stimulated by amending the existing ECE Regulation concerned with these components. This item has been on the agenda for discussion by the ECE Group of Rapporteurs on General Safety matters (GRSG) on a number of occasions over the last two years. The occurrence of the accident circumstances in question for a variety of traffic environments has clearly to be established. The implications of seat belt use by the occupants of vehicles has a bearing on the efficacy of any proposed changes. In Europe, restraint use is generally quite high for front seat occupants, although wearing rates are variable over a large group of countries[7]. However, restrained occupants can be partially ejected and seriously injured, and the majority of rear seated occupants are not restrained. Therefore a significant risk of injury as a result of latch activation is present.

Engineering solutions which will reduce the incidence of inadvertent door opening are not difficult or expensive to introduce, and this research study has demonstrated that a simple component test procedure can be used to evaluate design changes. The test has all the desirable characteristics for a crashworthiness assessment tool in having the ability to discriminate levels of safety performance for various vehicle safety designs. It has a high degree of repeatability and reproducibility and a low level of test complexity. The research has provided yet a further example of a component test that could provide real benefits for improving vehicle crashworthiness.

Conclusions

1) Passenger car door design, for good crashworthiness, remains an important area for future improvements even though legislation currently exists concerning the performance of hinges, latches and side door intrusion strength.

2) The component (sub-system) test can provide an effective contribution as a development

and legislative tool in reducing the level of road accident injury.

Acknowledgements

The opinions, findings and conclusions expressed in this paper are those of the authors and do not necessarily reflect the views of MIRA.

The authors wish to extend their thanks and appreciation to the United Kingdom Department of Transport for sponsoring the research programme.

References

1. Mackay, G.M. et al—European Vehicle Safety Standards and Their Effectiveness. Proc. 4th International Congress of Automotive Safety 1975, 431-54.

2. Carlsson, G. "Ejection—A Hazard in Traffic Accidents" Int Assoc for Vehicle Design. Vol 4 No 2 March 1983.

3. Huelke, D.F. and Gikas, P.W. "Ejection—The Leading Cause of Death in Car Accidents." 10th Stapp Conference. SAE Paper 660802, 1966.

4. O'Day, J. and Scott, R.E. "Corrected Findings on Injuries Associated with Ejected and Non-Ejected Occupants." UMITRI Review May/June 1984.

5. Green, P.D. et al. "Car Occupant Ejection in 919 Sampled Accidents in the UK 1983-1986." Paper for the International Congress and Exposition, SAE, Detroit USA. 1987. SAE Paper 870323, 1987.

6. Her Majesty's Stationary Office. "Road Accidents Great Britain" (1985).

7. Mackay, G.M. "Seat Belts in Europe—Their Use and Performance in Collisions." Proc International Symposium on Occupant Restraint 1981, pp 39-55 (AAAM, Toronto, Canada).

940562

Collision Performance of Automotive Door Systems

David Blaisdell and **Gregory Stephens**
Collision Research & Analysis, Inc.

Uwe Meissner
Volkswagen of America, Inc.

ABSTRACT

Historically, most safety related improvements to door systems have involved retention of occupants within the vehicle. However, such improvements have not been without some safety trade-offs. The recent update to FMVSS 214 (Side Impact Protection) has focused attention on increased occupant protection in side impacts. The standard essentially increases vehicle side strength requirements in order to reduce intrusion into the occupant space. The safety consequences associated with strengthening vehicle side structure will be evaluated with respect to various impact configurations. Energy management considerations of current as well as conceptual door systems during a collision will also be discussed. Individual latch and hinge component testing as currently required by FMVSS 206 does not completely evaluate the collision performance of the door as a system. From field collision evaluation, it has been seen that doors and surrounding side structure must act as a system to efficiently manage collision forces and distribute occupant loads. Procedures for evaluating current and future door systems by means of revised laboratory testing procedures will be evaluated.

BACKGROUND

Before the advent of enclosed passenger compartments, door and door-latch integrity was of little concern. With introduction of all-steel, fully enclosed passenger compartments however, the value of a door system that could remain closed not only under operational conditions, but also during minor and moderate collisions, began to be seen as a worthy goal. Up until approximately 1950, most cars, whether 2 or 4-door style, had a full-height B-Pillar that served to strengthen the occupant compartment and to provide a relatively rigid door frame. Door latch-to-striker continuity depended on this door frame strength for longitudinal retention. Although this system provided reasonable protection during normal use and even during some moderately severe collisions, many of the more severe crashes resulted in unintended door opening and resulting occupant ejection.

The introduction of the first hardtop "convertible" design in 1949 began a trend that was so successful that by 1955 almost half of the new domestic cars being sold were hardtops. European manufacturers did not follow this trend and continued producing the 2-door and 4-door sedans with full pillars. The hardtop designs resulted in reduced structural rigidity of the passenger compartment and occasionally allowed doors to open, even under non-crash conditions.

In 1952, traffic casualty studies undertaken by the Crash Injury Research Division of the Indiana State Police[1]*, found an early trend in the data which indicated that at least one out of 10, and possibly as many as one out of five persons, were being killed in survivable car accidents because the door latches were inadequate.

A study in 1954, looking at data collected for the Automobile Crash Injury Research (ACIR) at Cornell University Medical College[2], indicated that one or more doors opened in 44% of injury producing accidents. With this collection of data, ACIR for the first time gave the automotive designers some fundamental information identifying injury frequency and the possible causes of injuries. This study concluded:

> "...the development of restraining devices, which will keep occupants inside automobiles, prevent doors from opening in crashes is a matter of the highest importance!"

Others confirmed these conclusions. In November of 1954 the American College of Surgeons[3] made the following statement:

> "...one out of 10 fatalities occurs because doors pop open."

E.R. Dye[4] identified the hazards of door opening during crashes:

> "...one of the hazards of crashing is that of being thrown out of the car through open doors"

---

* Numbers in parenthesis refer to references at the end of the paper.

An article in the Journal of the American Medical Association[5] pointed out that in the experience of physicians, 25% to 35% of all automobile crash fatalities occur as a result of persons being thrown out through doors that have sprung open.

The need for improvements in the retention ability of door and door latch systems had been proven. These improvements primarily focused on a need to reduce the number of door openings and ejections in automobile collisions.

Daimler Benz attempted to incorporate some self-contained longitudinal restraint capability within the latch system with their 1949 introduction of the "cone lock" design, for which they received a patent in 1952. In 1955 and 1956 US manufacturers started to install door latches incorporating additional longitudinal retention capabilities. The incorporation of this additional retention in conjunction with other latch design modifications, served to additionally resist door openings during collision.

These early attempts at including longitudinal restraint capability within the latch system itself did not add a large component of strength to the system. In fact, when tested in a longitudinal direction outside the car, some of these early latch systems could be pulled apart with as little as 445 newtons (100 lb. force). These early latches still relied primarily on the integrity of the door frame for keeping doors within their frames.

The Traffic Safety Hearings in the House of Representatives in 1956 brought together the Government, the American College of Surgeons, the automotive industry and various scientists. These groups started stressing occupant safety as a basic factor in automobile design, including designing doors which would be less likely to open on impact.

By the mid-fifties, research engineers had become aware of the necessity to evaluate door openings by means of crash testing. These tests were designed to evaluate both occupant and vehicle interactions during collision. Some of this early testing began to identify the various mechanisms of door opening during collisions.[6,7]

Pioneering research testing was conducted at UCLA in 1958 with a series of head-on crash tests to evaluate some of the dynamic effects of crash forces on door latch systems. For the first time measurements were made by the use of special instrumentation regarding the inertial effects of the crash on individual latch components. The authors of this study[8] reported that:

> "These experiments disclosed the possibility of doors opening even when the differential movements described above are prevented by the improved types of door latch mechanisms. In certain cases, for example, the forces of collision deceleration are sufficient to open doors as a result of the inertia of door latch mechanisms in a manner comparable to normal operation. Under these circumstances, therefore, a door may open during collision without post-collision evidence of door versus door-frame deformation."

This series of head-on collision experiments which were conducted at closing speeds from 67.6 km to 167.4 km (42 to 104 mph), provided additional observations regarding the performance of the improved latch systems.

> "..the improvements in door structures and latching mechanism for the 1956 cars was evident, even in the higher velocity impacts."

By 1958 it had become widely known that ejected occupants had a considerably higher injury and fatality rate than occupants who remained inside the vehicle throughout the collision event. A study by B. Tourin[9] concluded:

> "..the risk of fatality among the ejected was ... nearly five times as great as that among those not ejected"

and

> "The relationship of automobile design to injury is nowhere more apparent than in the comparisons of the frequency of fatality among ejected and non ejected occupants"

As the door systems began to become strengthened, it became apparent that except for rollovers, the side-impact was the most vulnerable exposure for the system. The first fully instrumented side impact collision experiments having both vehicles in motion at impact was conducted at UCLA in 1959[10]. This series of crash tests dramatically recorded occupant ejection and the resulting injury exposure. The authors noted the following observation:

> "In the case of the impact at the rear wheel.. before the cars had separated, the driver's door had sprung open and the driver was being ejected."

Thus, the effect on doors of occupant inertial loading was documented by both electronic and photographic instrumentation for the first time.

Other results of door strengthening were documented by R. A. Wolfe who stated[11] in the "Discovery and Control of Ejections in Automobile Accidents" that

> "..among the post -1955 cars, the frequency of door opening was reduced about 33% and the frequency of occupant ejections was reduced about 40%.."

However, it was noted in another of Wolfe's studies[12] that in evaluating changes in risk from 1953-55 data to 1956-59 data, it was found that while the risk of being injured by ejection was reduced (-4%) the risk of being injured by the door structure itself was enhanced (+36%). This finding was to be repeated in subsequent years as scientific studies continued to evaluate and monitor door system performance in crashes.

A Cornell report evaluating the newer latch designs came to the following conclusion [13]:

> "For those impacts involving speeds not over 29 miles per hour, the rate of ejection of occupants has been reduced by 64%; for accidents from 30-59 miles per hour the reduction has been 45%"

Continuing research at UCLA which involved static as well as dynamic testing of vehicle structures, doors and latch systems, also found that although newer latch designs had been improved over the earlier designs of the mid-fifties[14], considerable additional improvement was needed.

"..indications are that all door latch mechanisms will have to be improved, strength-wise, several fold before they will remain engaged under severe collisions."

In the same series of studies, it was also found that interactions between occupant, seat and door could act to affect door latch performance in a crash[15]. This study also compiled the following comprehensive list of door opening mechanisms:

1. Longitudinal tensile loads caused by lateral bending of the entire car structure in the horizontal plane

2. Longitudinal compressive loads caused by the inertial loading of the non-impacted portion of the car structure.

3. Occupant loading caused by impact of the occupants against the door.

4. Inertial loading of door-latch mechanism in which the impact forces cause actuation of the door latch mechanism.

5. Seat inertial loading caused by failure of seat anchorage with subsequent seat impact with the door.

6. Seat column loading, or loading caused by transfer of the forces in the area of impact to the opposite door through the seat acting as a column.

7. Latch actuator rod deformation resulting in door (latch) actuation caused by bending of the rod from direct lateral encroachment during impact

8. For the latch mechanism investigated, it was observed in a number of cases that, due to deformation of the end retaining plate of the ratchet housing, the latch tongue was allowed to disengage at a load less than the full potential of the design

9. Inadvertent actuation of inside door handles by occupants during impact.

10. Outside door handle actuation from contact with the ground during rollovers.

We find similar attempts to describe the reasons for door opening and ejections in automobile crashes in other publications of the time[16,17]. Garrett's additional studies[18] continued to confirm the prior findings that door openings and ejections were reduced by the stronger door latches. However, he further concluded that injuries, fatalities and ejections in general clearly had not been eliminated.

"Door openings among 1963 cars with the improved safety door latches has been cut to almost half of that observed among pre-1956 cars"

In the 1965-67 time period, those European automobile manufacturers that had not already done so, started incorporating longitudinal restraints to their latch systems, thus adding to the longitudinal yield resistance provided by the door frame. Between 1961 and 1965 the US automobile manufacturers designed and started to install a new generation of improved door latches. In 1964 W. G. Cichowski discussed "A New Laboratory Device For Passenger Car Safety Studies"[19] which was built to duplicate the longitudinal loading caused during car to car side impacts.

This trend towards latch strengthening continued well into the seventies when latch strength had increased to the point that it matched and often exceeded the strength of its surrounding structure.

TABLE 1: REGULATORY HISTORY OF LATCH SYSTEMS

Date	Source	Item	No.
1961	SAE Committee Meeting	First formal discussion of latch strength regulations	
Jan. 1963	SAE Journal	First public announcement of forthcoming latch strength requirements.	
Dec. 1963	SAE Journal	Announcement that 1964 SAE Handbook would contain latch strength recommendations.	
Jan 1964	SAE Handbook	Recommended practice for latch strength Longitudinal 1500 lbs (Primary) Transverse 1250 lbs (Primary) Transverse 500 lbs. (Secondary)	J839
June 1964	SAE Committee	Consideration of adding inertial requirement to J839 recommended practice	
1965	SAE Handbook	SAE recommended Practice was modified to include the 30G inertia requirement.	J839a
1965	General Service Administration (GSA)	First Federal Regulation relating to door latches Applicable to all 1967 model year passenger cars purchased for use by the Federal Government (effective September 28, 1966) This standard required SAE type static lab tests: Longitudinal 2,500 lb (fully latched), Transverse 1,700lb (primary) 500 lb (secondary) (No specified inertial load test)	515/5
1966	SAE Handbook	Revision to require increased force resistance Longitudinal 2500 lb (Primary) 1000 lb (Secondary) Inertial: 30 G in any direction	J839b
July 1965	SAE Handbook	Recommended Practice for passenger vehicle door hinge systems Longitudinal - 2500 lb. Transverse or lateral - 2000 lb.	J934
Jan 1 1968	Federal Motor Vehicle Safety Standard	FMVSS for door latches and hinges Followed the recommendations of SAE J839 b and SAE J934 Applicable to all passenger cars sold in the US.	FMVSS 206

Table 1 Summarizes the chronology of SAE Recommended Practices, GSA and NHTSA regulations relating to doors and door latch systems and side door structures

Regulatory History of Doors and Latch Systems

Starting in 1961, there was a move within the Society of Automotive Engineers to develop a recommended practice relating to door latch strength requirements. As depicted in

Table 1, this activity was first publicized in January 1963, when the SAE Journal announced a series of forthcoming performance test recommendations for side door latch and striker strength requirements. No specifications or objective data were provided at that time.

Up until this point, the government had played only a small part in evaluation and regulation of vehicle collision performance. This lack of involvement was pointed out in a door latch study reported by A. G. Gross in 1964[20] on the factors relating to accidental door opening. The following observations were made:

"From the information currently available, it appears that no one has yet established either the critical loads that a door latch system must withstand during collisions or developed a positive method of evaluation of door latch performance."

Door Systems Research

A federally funded project by the Advanced Systems Laboratory Industrial Product Group of the American Machine and Foundry Company, tried to develop standards by which automobile door safety could be determined[21]. In this study, recommendations were made to increase dynamic longitudinal load requirements to 7,000 lbs. compressive and 3,500 lbs. tensile, and which would vary with the weight of the vehicle. Other findings were described as follows:

"...it can be said that a door has failed when it hazards the occupants. This condition can result from a deficiency in either the door's operational or its structural performance. Operationally a door should remain closed during and just after an accident, when the car comes to rest, a door should be capable of being readily opened to permit occupant egress. Structurally, a door should limit impact intrusions towards an occupant to 10 inches or less, and prevent the car top from collapsing. Also, it should not be so rigid as to seriously injure an occupant thrown against the inside of the door, (i.e. the door compliance or yielding should match passenger fragility)".

This comprehensive set of goals, although perhaps desirable, had elements that were not only unrealistic, but also conflicted with each other. For example, when sufficiently strengthened doors were involved in severe collisions, substantial distortions would considerably modify the geometry of the system, such that the door would be solidly jammed closed. When jammed, the stronger latches resisted the efforts of rescuers to pry them open. As a result of this difficulty, post-crash conditions such as fire sometimes caused injuries that otherwise could have been prevented. In another area of conflicting design requirements, structure that somehow could limit intrusion to ten inches could itself cause serious injury when occupants were slammed against the strengthened door structure during a collision.

Stronger door and latch system designs of the early seventies, reduced, but by no means eliminated occupant ejections. This was especially true for multiple collision events and

rollovers. Many ejections continued to occur through portals other than an open door. It was observed that when doors with increased resistance to opening were forced outward by occupant loading during a collision, the geometry of the door between the latch and hinges acted to form an outward and upward slope that funneled the occupant toward the window opening, Fig. 1.

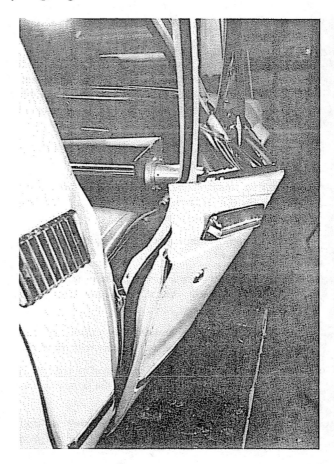

Fig. 1 - Door bowed outward from simulated occupant impact during test.

In October 1974, Calspan's Transportation Safety Department under contract for NHTSA constructed a study looking at "Ejection Risk in Automobile Accidents"[22]. It was found that:

"...the door is the primary avenue of ejection in non-rollover collisions (all "impact areas"). The door window is the second most probable source of ejection for front and side impacts. For roll-over collisions the sequence is reversed." (Window ejection more frequent)

A final report done by NHTSA[23] continued the earlier research done by the Chi Associates[24]. They attempted to determine whether or not a correlation existed between the longitudinal and/or lateral door latch failure loads during FMVSS 206-type testing and the ejection rates for passenger cars. Some of the findings were as follows:

"...ejection rates for 2-door cars were higher than those for 4-door cars. Longitudinal failure loads of

latches were found to be highly dependent upon some of the test parameters, thickness of mounting plate and size of mounting hole. Ejection rates found in this study and that by Chi Associates may not be good indications of ejections caused by latch failure"

It was also found that ejection rate dependency on the number of doors may relate to the vehicle type (sport car, etc.), size of window opening (2-door windows are usually substantially larger than those of 4-door automobiles) or some combination of these factors.

NHTSA's Technical Report of November 1989 "An Evaluation of Door Locks and Roof Crush Resistance of Passenger Cars"[25] (FMVSS 206 and 216) based on FARS, Texas, NCSS, NASS and MDAI data and roof crash tests found that:

"Door latch improvements implemented during 1963-1968 saved an estimated 400 lives per year, reducing the risk of ejection in rollover crashes by 15%,"

and

"...the shift from hardtops to pillared cars, in response to Standard 216 saved on estimated 110 lives per year"

By this time it was becoming clear that ejection was not the only concern related to door and door latch design. Further, it was also apparent that the door system strength had increased at least partially due to the increased attention by the research community. Figure 2 depicts the general strength evolution over the past 30 years. Note the sharp increase in strength in the late sixties and early seventies.

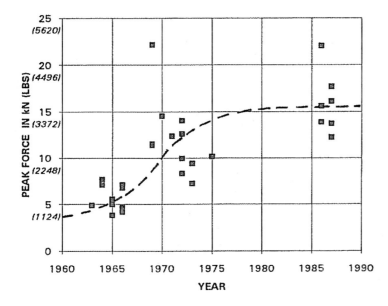

Fig 2 - Door Strength Evolution

History of Side Structure Intrusion Protection

The considerable progress in injury reduction from strengthening of door latches and hinges to decrease the door openings and ejection numbers was accompanied by research to improve the other aspect of protection that doors must attempt to provide, namely the minimization of injuries caused by intrusion. This research led to the Federal Motor Vehicle Safety Standard (FMVSS) 214, which became effective on January 1, 1973. This standard sought to mitigate occupant injuries in side impacts by reducing the extent of intrusion during certain types of collisions.

The side door strength provisions of the standard required each door to resist crush forces applied by a piston pressing a vertical steel cylinder against the middle of the door in a quasi-static manner. Force level requirements increased with depth of penetration. The requirement for the initial six inches of crush resulted in the introduction of additional structure immediately beneath the door skin, usually in the form of a horizontal beam. This beam also provided some additional resistance to intrusion throughout the test procedure.

In 1979, six years after its implementation, a study was constructed to evaluate the statistical results of the implementation of FMVSS 214. This study was the first comprehensive evaluation of a safety standard. However, this preliminary evaluation did not convince the research community that FMVSS 214 was effective. Thus, a second evaluation was made in 1982 entitled "An Evaluation of Side Structure Improvements in Responses to FMVSS 214"[26] that superseded the findings of the preliminary study.

This second study resulted in an estimation that FMVSS 214 eliminated 480 fatalities and 4,500 non-fatal hospitalizations per year in side impacts with fixed objects. It was also estimated that the standard eliminated 4,900 non-fatal hospitalizations per year (no fatality reductions) in vehicle to vehicle side impacts. It was observed that the standard generally brought a significantly shallower and wider damage pattern and that it reduced ejections in side impacts. NCSS, Texas and FARS were the source of statistical data.

On October 30, 1990, NHTSA published a final rule, adding dynamic test procedures and performance requirements to Standard 214, as well as, specification for the side impact dummy to be used in the dynamic crash test and the attributes of the moving deformable barrier to be used in the crash test.

On August 31, 1993, NHTSA announced the Side Impact Standard (modified FMVSS 214) to be the most significant improvement in vehicle safety since its 1984 standard requiring automatic crash protection (air bags or automatic seat belts). It has been estimated that 500 fatalities and 2,600 serious injuries per year will be prevented when the standard is fully implemented.

The problems of ejection from vehicles and latch strength were revisited in the mid to late eighties. In 1984, V. L. Roberts and D. A. Guenther[27] re-evaluated data published in the 50's and 60's and looked at more recent studies based on the NCSS files. They reported that:

"...most injuries are associated with impact with the vehicle interior and occur either before or during the ejection process"

They concluded that:

"…ejection is more a measure of the violence of a collision than a measure of injury causation. Accidents, which involve ejection tend to be high speed, rural events, involving young male drivers, a set of circumstances which are over-involved in fatal and severe accidents, whether ejection occurs or not"

In 1987, W.R.S. Fan[29] of NHTSA reexamined the data produced by the Chi Associates and J. Bartol of NHTSA(see references 23 &24). He concluded that:

"…the ejection rate is strongly related with the ultimate latch load in the longitudinal direction and has a reasonable correlation in the transverse direction. Latch specimens selected from cars with low ejection rates generally can sustain two to three times the 2500 LB longitudinal load specified in FMVSS 206"

Its important to note that these conclusions were drawn off the same data that earlier were reported as non conclusive. Additionally, it should be noted that since structures surrounding the latch and striker cannot withstand these higher longitudinal loads, door systems subjected to such loads would be subject to opening, notwithstanding the stronger door latch system.

TERMINOLOGY

Door system technology progressed from single panels with bars for latches to bulky and complex multi-paneled doors with latches to match. Today door systems have progressed to the point where lightweight doors with advanced latching systems provide more effective retention and energy absorption than the designs of the seventies.

For this improved collision performance to be optimized, the door must act as a system in managing the collision forces. This system includes the following sub-systems:

DOOR   The door is comprised of a number of individual components including outer and inner panels, and the interior trim panel. This inner panel may include some measure of additional energy absorbing material. The housing comprised of the outer and inner panels, which are attached to reinforced structural assemblies at each end, contains the window mechanism, remote actuating levers and rods as well as the side guard beam. This longitudinal structure or beam may be one of many configurations, ranging from a simple cylinder to a more complex W shaped cross section. At one end of this housing is the latching mechanism.

LATCHING MECHANISM   The latching mechanism is comprised of rotors, claws, ratchets, rods and springs that work together to provide easy remote activation and to resist inertial or inadvertent activation. Secondary latched positions are incorporated as a backup for the primary position in the event the door is inadvertently not fully closed.

STRIKER   The striker is typically mounted to the rear portion of the door frame and acts as the interlocking stationary portion of the door latch system. The striker is usually comprised of a U-bolt or a longitudinal mushroom-type stud to which the latching mechanism anchors. Strikers typically have a backing plate or doubler to assist in energy management and load distribution.

DOOR FRAME   The door frame surrounds the door and window and is usually comprised of the pillar structure, roof rail and door sill. The door frame is typically designed to interact with the door in such a way as to mitigate vibration and noise during operation and to resist collision forces.

DESIGN CONSIDERATIONS FOR DOOR SYSTEMS

Notwithstanding the concerns for the safety aspects of the door system, it must be noted that the door serves a number of purposes not related to safety. The automotive designer must consider a number of items when the door system design is considered. Some of these considerations are listed below.

1. Geometry (size, opening angle, ground clearance, window openings for visibility and air flow).
2. Human factors (appearance, ease of locking and unlocking, ease of latching and unlatching as well as ease of ingress and egress once the door is open).
3. Production and efficiency considerations (Weight, cost of parts, tooling considerations, materials & production process requirements, uniformity of parts between various models).
4. Service and reliability considerations (Durability of moving parts, resistance to exterior damage, ease of service, corrosion resistance).

In addition to the practical considerations, safety requirements often conflict with practical considerations and with each other. Various safety related considerations are listed below.

1 Strength of door against intrusion
    a) Compatibility of latch, hinges and door structure.
    b) Side guard beam.
    c) Overlapping pillars, roof rail, door sill
    d) Alignment of door sill and side door beam to opposing structures.
    e) Latch and hinge placement.
    f) Strength of door frame (i.e. roof rail, pillars, door sill)
2 Energy absorbing material and comfort padding.
3. Deformability during impact from occupant
4 Energy dissipation (both inward and outward).

It is important to note that the automotive designer must consider the varying and conflicting requirements for protection during numerous types of accidents. This concept was noted in a report from K. A. Stonex at GM[30], where he stated:

"One of the fundamental principles of safety engineering is to anticipate every possible type of accident which may occur because of machine failure or human failure and then to establish safeguards to minimize the hazards or injury which may result when such a failure occurs."

COLLISION PERFORMANCE - SIDE-IMPACT CONSIDERATIONS

Collision performance of a door system relates to the manner in which a door and its surrounding environment act together to manage collision forces. These forces are primarily either inward or outward with respect to the vehicle, and are transverse to principal direction of force in front or rear-end collisions. As in these front or rear-end collisions however, crush space is a significant factor in properly managing and distributing collision forces so the occupant can be given additional time for an improved "ride-down" or "ride-up". In side impact collisions, space is severely limited by the inherent requirements of vehicle design. Therefore, attempts must be made to utilize the available limited space in an optimum manner.

As door latch designs improved and strength capabilities increased, door openings during collisions were reduced dramatically. To a somewhat lesser extent, modifications to meet intrusion resistance requirements have resulted in additional safety improvements. When the passenger compartment is directly impacted by the opposing vehicle, a more rigidified door system performs better in distributing the intruding forces and maintaining survival space for the occupant. However, as strength requirements continue to increase, concern must be given to energy absorption for impacts to front side and rear side which can cause severe occupant inertial loading to door structures.

More recent activity to meet the new requirements of FMVSS 214 has focused considerable research attention on intrusion reduction. At the onset of FMVSS 214 in 1973, side door beams were introduced into the vehicle environment. As we approached the mid-eighties, more aerodynamic and weight efficient designs were implemented. Examples of these designs were wrap-around window frames and vertically rounded doors. These design changes have modified the space between the outer door skin and the occupant, thus, modifying the potential for "ride up/down" space. These designs have also allowed for better distribution of crash loads to other areas of the vehicle including the roof and rocker panel areas.

In addition to rigidified door systems, some research[31] has focused on doorbags as a possible means for better distributing and controlling collision forces. This research provided a means of optimizing the limited space for energy management. Questions relating to rapid deployment requirements as well as hazards and complications associated

with the close proximity of such a system to the occupant remain to be determined.

Notwithstanding the positive results of reduced intrusion for direct impacts to the passenger compartment, the question remains as to how this increased rigidification will affect the energy moderation requirements of occupants who are not directly in line with the impacting structure and who depend to a great extent on door structure deformation for distribution and moderation of crash forces. Occupant protection and containment in these offset, front or rear side-impacts has special importance when one considers that they account for approximately half of all side impacts[32]. In this type of impact, the occupant is inertially forced against the door. The door must yield outward to better absorb energy while at the same time, reducing the accelerative forces experienced by the torso. This energy management is a key factor when collision performance is considered. It is dependent upon efficient use of space and structure to distribute collision forces applied to the occupants.

Velocity-Time Analysis

One method of evaluating and determining the efficiency of a particular system is to graph the velocity changes for different portions of the structure with respect to the occupant. With regard to side-impacts involving intrusion, it has been shown that bringing the occupant up to speed in a more controlled manner from the onset of impact is highly desirable. This concept has previously been described in the evaluation of seating concepts[33] and is shown in Fig 3.

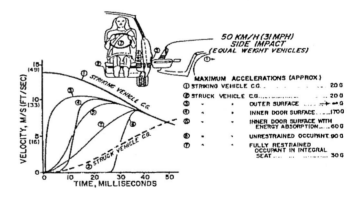

Fig 3 - Velocity-Time analysis - Side impact intrusion

In front or rear side-impacts, the occupant has the potential for considerably more distance to "ride-up" or "ride-down" the collision. A velocity-time analysis, Fig. 4, depicts the differences between the rigidified and yielding environments. The rigidified door system shown subjects the occupant to higher acceleration, thus increasing the potential for injury. Additionally, the occupant has an increased potential for head-torso misalignment that may result in serious cervical injuries. The more yielding environment utilizes energy absorbing deformation in the door system to extend the acceleration pulse. In addition, the differential forces on the head and torso are generally less, resulting in a tendency for the head and torso to remain in better alignment. This type of velocity-time analysis has proven a valuable evaluation tool in determining system performance for particular exposures. This type of evaluation is especially important in the limited space environment of the side impact collision.

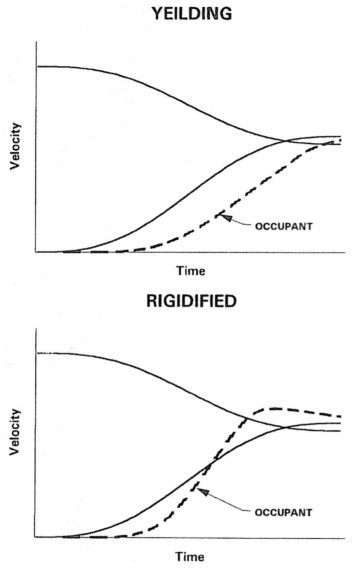

Fig. 4 - Velocity-Time Analysis examples

TESTING AND ANALYSIS

FMVSS 206 currently requires individual latch and hinge testing outside the vehicle environment. This testing generally follows the recommendations of SAE J839b. The individual components are attached to rigid fixtures and are testing to the specified longitudinal and lateral loads. Using this procedure, the individual components are tested in isolation. In the actual collision environment, however, the door must act as a *system* to distribute and moderate the collision forces.

This system involves not only the latch and hinges, but the door, door frame and pillars. Therefore, if the collision performance of a door is to be evaluated, relying on individual component testing is not adequate. A testing

methodology by which the entire door system is evaluated would better approximate the actual collision environment and yield more realistic results. When a door system is tested, the energy absorption capability of the entire door environment can be measured. As can be seen from Fig 5, the strength achieved during FMVSS 206 testing is only a partial measure of occupant protection that may be provided by the door system. Peak forces do not evaluate energy absorption, which is a measure of not only force, but also the distance over which it acts. In these tests, the energy absorption displayed when the entire system was included was several times greater than that of the latch and striker alone.

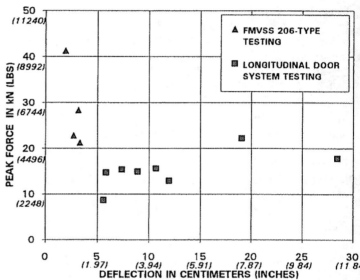

Fig. 5 - Comparison of FMVSS 206 testing to longitudinal door system testing.

A similar situation could exist with respect to FMVSS 214 testing of side structure intrusion resistance. If the structure could be made completely rigid, injuries and fatalities could still occur, notwithstanding the complete absence of intrusion, if occupant forced movements could not be controlled and if available crush space was not sufficient to keep occupant loadings below critical levels.

Door system testing methodologies have been developed over a number of years by the research community. Different variations have been observed, however, the same basic criterion has been met. The entire vehicle or a substantial portion thereof is placed laterally across a testing fixture and a body block or anthropometric dummy is pushed laterally against the door. This type of test setup is depicted in Fig. 6.

Force and deflection are measured in order to determine energy absorption capabilities of the system. In this type of testing, typically the belt line of the door deforms outward with the pillars also rotating outward. The metal surrounding the latch and striker areas deform extensively as the load is increased with eventual tearing of the metal occurring. This deformation and tearing moderates and limits forces applied to the occupant, however, at some point the system capabilities are exceeded and the door opens. It has been observed that the area encompassed within the triangle formed by the two hinges and the latch/striker combination provides most of the energy absorption of the door system.

Fig. 6 - Test setup for door system evaluation

Due to the geometry of the various components, this area is generally substantially below the belt line of the door allowing the upper part of the door to deform outward. An example of this type of deformation was presented in figure 1. A side view of a typical latch/striker and hinge setup is demonstrated in figure 7.

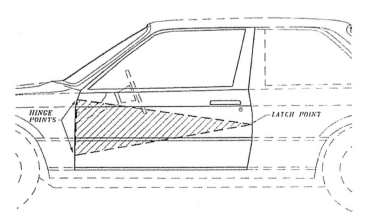

Fig 7. - Depiction of typical force triangle

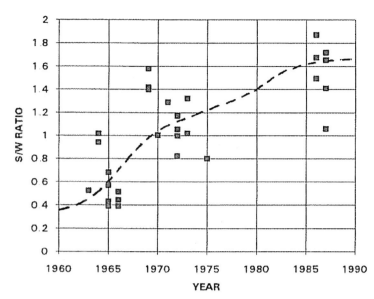

Fig. 8 - Door Strength/Weight Ratio Comparison

Even though vehicle door systems have improved in terms of stronger retention capabilities, the systems have *decreased* substantially in weight. Therefore, the strength to weight ratio has substantially increased, Fig. 8. This is an important consideration when comparing door systems of different eras.

FUTURE CONCEPTS FOR DOOR SYSTEMS

As current systems evolve into stronger, more rigidified systems to meet the new demands of FMVSS 214, the injury exposure of the occupant in front, rear and offset side impact collisions must be considered. Concepts can be explored to increase the intrusion resistance while maintaining the energy absorbing capabilities of current systems. Concepts for possible system improvements include the following:

1. Some means for additional door engagement around the door frame. This concept could assist with the management of intrusion forces by distributing them around the door frame.

2. Increasing the interlocking capability of side beam structures to better engage both pillars. This additional engagement could provide greater intrusion resistance without additionally compromising energy absorption capabilities during the occupant impact with the door structure.

3. Incorporation of a hinge-concept within the door beam. As demonstrated in Fig. 9, this concept of beam design could provide intrusion resistance while maintaining the ability to yield outward from occupant loading. The hinge in the figure is utilized to exhibit the concept. It is possible that beam geometry could be developed to attain the same result without the complexity of an actual hinge mechanism.

695

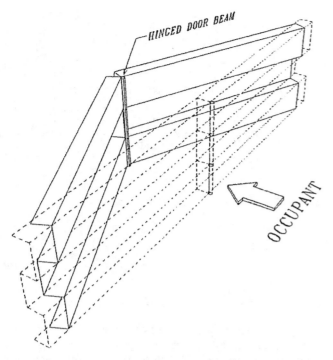

Fig. 9. - Concept for hinged beam to allow additional energy absorption during occupant loading.

4. Relocation of the energy absorbing triangle somewhat higher on the door, Fig. 10. This concept could potentially minimize the funneling effect. However, as shown below, the lower section of the door could be subject to increased intrusion during direct impacts into the passenger compartment.

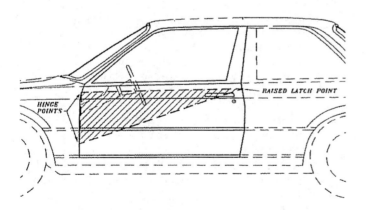

Fig. 10. Raised force triangle concept.

5. Utilization of some type of dual latching door system to increase the energy absorbing area and to better distribute the forces along the sides of the door frame. This concept is shown by Fig. 11. This concept, although appealing, could prove difficult to accomplish without compromising safety in other areas. The introduction of additional latching mechanisms complicate operation, add remote rods and increase the likelihood of jamming during collision.

It is important to note that each of these concepts are suggested, based on a limited safety-related objective. As mentioned earlier, the automotive designer must consider all safety related implications of any design change and also keep in mind the other functions that a door system must provide.

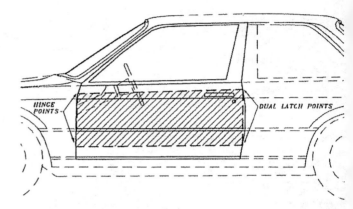

Fig. 11. Dual latch concept could help distribute forces and could act to reduce the funneling effect.

With respect to safety, any substantial modifications to existing designs should be preceded by a thorough analysis and testing to assure that improvements in one area of safety do not result in degradation elsewhere.

CONCLUSIONS

- Early latching systems had little if any longitudinal restraint contained within the latch and striker system itself, relying primarily on the door frame for strength.

- As latch designs progressed, their retention capability was substantially increased, however injuries from occupant impact with the door structure increased.

- Attempts to utilize FMVSS 206 test data to predict door openings does not take into account the strength limitations of the door and door pillar to which the latch and striker are attached. Such procedures may lead to incorrect conclusions regarding the potential effectiveness of increasing latch system strength requirements.

- Latch system strength has increased significantly since the introduction of FMVSS 206 testing requirements. Latch strength currently matches or exceeds the strength of the surrounding structure.

- When correlations between ejection rates and latch component strength were first analyzed, they were reported as "non-conclusive."

- In the mid-eighties studies indicated that ejections were more a measure of the severity of the collision and less a result of door opening.

- Velocity-time analysis for evaluating occupant kinematics in a side impact collision is an invaluable tool in the determination of injury exposure.

- Increased intrusion resistance should not be accompanied by reduced energy absorption for collisions not directly involving the door structure itself.

- Merely increasing door and latch strength without considering the entire door system will not necessarily provide additional occupant protection, and may be counterproductive.

REFERENCES

1. DeHaven, H., "Accidental Survival - Airplane and Passenger Car", SAE, January 1952.

2. Moore, J.O. & Tourin, B., "A Study of Automobile Doors Opening Under Crash Conditions", ACIR at Cornell University Medical College, August 1954.

3. "How to prevent Injuries in Auto Accidents", American College of Surgeons, November 1954.

4. Dye, E.R., et al., "Automobile Crash Safety Research", Cornell Aeronautical Laboratory, March 1955.

5. Sheldon, C.H., MD., "Prevention, The Only Cure For Head Injuries Resulting From Automobile Accidents", Journal of American Medical Association, November 1955.

6. Hanes, A.L., "Design Factors in Automotive Safety", SAE, June 1955.

7. Severy, D.M. & Mathewson, J.H., "Technical Findings From Automobile Impact Studies", SAE Transactions, Vol. 65, 1957.

8. Severy, D.M., Mathewson, J.H. & Siegel, A.W., "Automobile Head-on Collisions, Series II", SAE Transactions, Vol 67, March 1958.

9. Tourin, B., "Ejection in Automobiles Fatalities", Public Health Reports, May 1958.

10. Severy, D.M., Mathewson, J.H. & Siegel, A.W., "Automobile Side Impact Collisions", SAE SP#174, 1960.

11. Wolfe, R.A., "Discovery and Control of Ejections in Automobile Accidents" Journal of American Medical Association, 1961.

12. Schwimmer S. & Wolfe, R.A., "Preliminary Ranking of Injury Causes in Automobile Accidents", Fifth Stapp Automotive Crash and Field Demonstration Conference, September 1961.

13. Garrett, J.W., "Evaluation of Door Lock Effectiveness Pre-1956 vs. Post-1955 Automobiles", ACIR at Cornell University Medical College, July 1961.

14. Severy, D.M. & Siegel, A.W., "Engineered Collisions", Fifth Stapp Automotive Crash and Field Demonstration Conference, September 1961.

15. Severy, D.M., Mathewson, J.H. & Siegel, A.W., "Automobile Side Impact Collisions, Series II", SAE 491A, National Automobile Week, March 1962.

16. Gross, A.G., "Accidental Motorist Ejection and Door Latching Systems", SAE 817A, Automotive Engineering Congress, January 1964

17. Gross, A.G., "Why Doors Spring Open During Crashes", SAE Journal, 1964.

18. Garrett, J.W., "The Safety Performance for 1962-63 Automobile Door Latches and Comparison with Earlier Latch Designs", ACIR at Cornell University Medical College, November 1964.

19. Cichonski, W.G., "A Laboratory Device for Passenger Car Safety Devices", 1965.

20. Refer to reference 16.

21. Bartol, J., Wingerlach, W., Hodosy F., et al., "Performance Requirements for Doors, Door Retaining Devices and Adjacent Structures", DOT #HS-800 254, July 1969.

22. Anderson, T.E., "Ejection Risk in Automobile Accidents", Calspan's Transportation Safety Department - NHTSA, October 1974.

23. Donald T. Wilke, Monk, M.W., et al., "Door Latch Integrity", DOT #HS-807 374, December 1988

24. Shams, T., Nguyen, T.T. & Chi, M., "Side Door Latch/Hinge Assembly Evaluation", DOT #HS-807 039, October 1986.

25. Kahane, C.K., "An Evaluation of Door Locks and Roof Crush Resistance of Passenger Cars", DOT #HS-807 489, November 1989.

26. Kahane, C.K., "An Evaluation of Side Structure Improvements in Responses to FMVSS 214", DOT #HS-806 314, 1982.

27. Roberts, V.L. & Guenther, D.A., "Ejection Versus Injury: A Re-Evaluation", Proceedings of the Canadian Multidisciplinary Road Safety Conference III, May 1989

28. Fan, W. R S., "Two New Areas Concerning Side Protection for Passenger Car Occupants", SAE 871114, May 1987.

29. Stonex, K.A., "Roadside Design for Safety", General Motors, January 1960.

30. Warner, C.Y., Strother, C.E., et al., "Crash Protection in Near-Side Impact - Advantages of a Supplemental Inflatable Restraint", SAE 890602, March 1989.

31. Daniel, R.P., "Biomechanical Design Considerations for Side Impact", SAE 890386, March 1989.

32. Severy, D.M., Blaisdell, D.M., Kerkhoff, J.F., "Automotive Seat Design and Collision Performance", SAE 760810, March 1976.

33. Ruden, T.R., Murty, R. & Ruch W., "Design and Development of a Magnesium/Aluminum Door Frame", SAE 930413, March 1993.

Door Latch Strength in a Car Body Environment

Andrew Gilberg
Teknacon

John Marcosky
Gateway Engineering

Linda Sherman
L. S. Sherman Consulting

Richard Clarke
Clarke Automotive Consulting

ABSTRACT

Federal Motor Safety Standard (FMVSS) 206 regulates the minimum strength of side door latches in passenger carrying vehicles. The purpose of the standard's requirements is "to reduce the likelihood of occupants being ejected from vehicles in real world accidents."

Investigation of unwanted door openings during accidents has revealed various types of latch failures that do not produce latch and/or striker damage consistent with that found in Federal Motor Vehicle Safety Standard compliance testing.

An intersection collision in which a striking vehicle contacts the struck vehicle aft of the affected door has for many years been considered the "most critical to door latch performance" (1). This type of car to car collision will often result in structural separation of the door end panel, "B" post striker panel or latch/striker assembly. These structural failures combine tensile and shear loading on the latch/striker assembly.

This paper defines the state of the industry with regard to vehicle door latch strength using a simple testing method that produces door latch separations more consistent with common intersection side impact collisions. The test methodology simultaneously applies both longitudinal and lateral loads as the vector sum of the FMVSS 206 defined loads. To place the test findings in perspective, results are normalized using the existing 206 requirements.

BACKGROUND

Federal Motor Safety Standard (FMVSS) 206 regulates the minimum strength of side door latches and hinges in passenger carrying vehicles. For side door latches, the essence of FMVSS 206 is four static load requirements and a theoretical calculation for resistance to inertial shock.

The static load tests individually measure the strength of the latch and striker combination in a longitudinal direction (X - parallel to the vehicle fore/aft axis) or lateral direction (Y - perpendicular to the vehicle fore/aft axis and parallel to the ground plane). These lateral and longitudinal tests are done both in the primary (fully latched) and secondary (partially engaged) position of the latch.

To meet the longitudinal load requirements, it is necessary for a latch to withstand at least 11,000 N (2500 lb.) in its primary position and 8,900 N (2000 lb.) in the lateral direction. The load requirements fall to 4,450 N (1000 lb.) for both directions with the latch in its secondary position.

To demonstrate compliance with FMVSS 206, manufacturers typically conduct a series of load tests of the latch and striker combination in laboratory tensile test machines. When mounted in the test machines, the latches and strikers are often bolted to plates (typically 2.5 mm. to 3.2 mm. thick steel) which substantially reinforce the latch frame. Further, the test fixtures rigidly control alignment between the latch and striker throughout the deflection range of the test.

Figure 1. A side impact to the right rear quarter of this 1990 Mitsubishi Eclipse produced a latch failure that resulted in the ejection of a passive torso belted occupant.

The tests, therefore, distort and frequently enhance latch hardware performance. While the 206 test protocol offers the advantage of repeatability, it cannot claim to reproduce failure modes or represent strengths encountered in real world traffic accidents.

Because the latch and striker are both flexibly mounted in passenger vehicles, on (typically .75 mm. thick) sheet metal in a car body, there is a strong tendency for the striker to seek the path of least resistance when it is pulled from the latch. Furthermore, depending upon where an impact takes place, pure tension as in the longitudinal test or pure shear loading as in the lateral requirements almost never occurs.

Most latch failures encountered in the field in which the latch and striker are forcibly separated evidence a combination of tensile and shear loading or, on occasion, compression and shear loading. This combined loading manifests as a prying or levering action between the latch and striker. This prying action may create a mechanical advantage that substantially alters the load level at which separation occurs.

A test, which reproduces the failures encountered in most field accidents, would need to simultaneously apply both longitudinal and lateral force to the latch and permit some self-alignment of the latch and striker. Such a test could combine the separate lateral and longitudinal requirements of the existing FMVSS 206 tests and provide for controlled flexible mounting of the latch system components.

One way to adapt the existing requirements of FMVSS 206 for this purpose would be to sum the longitudinal and lateral load requirements vectorially. This would result in (approximately) a 14,200 N (3200 lb.) load applied at about 39 degrees off the vehicle's longitudinal (X) axis.

Figure 2. Proposed test geometry.

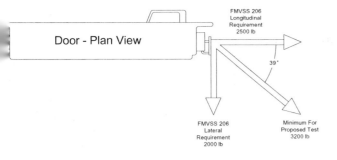

Door - Plan View

FMVSS 206
Longitudinal
Requirement
2500 lb

39°

FMVSS 206
Lateral
Requirement
2000 lb

Minimum For
Proposed Test
3200 lb

Testing the latch in its specific door using its designed mounting provisions could accommodate the flexible-mounting requirement.

Physically, this load geometry could represent a typical intersection collision (with both vehicles moving) in which the striking vehicle contacts the target vehicle just behind the door/striker interface. Although the loading direction would be arbitrarily defined by the existing requirements for FMVSS 206, it is nevertheless a geometry that is encountered in real world collisions.

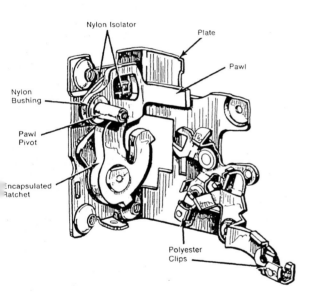

Nylon Isolator

Plate

Pawl

Nylon Bushing

Pawl Pivot

Encapsulated Ratchet

Polyester Clips

Figure 3. Typical latch showing basic mechanical elements. Illustration is from October 1982 Automotive Engineering.

The 14,200 N test load, it could be argued, is consistent with the intent of the existing 206 standard in that it represents the vector sum of perpendicular components for that same standard.

TEST PROGRAM

TERMINOLOGY AND DEFINITIONS

Most contemporary automotive side door latches contain two basic mechanical elements that control its operation. These components are usually referred to as the **fork bolt** and **detent lever**, the **ratchet** and **pawl** or combinations of these terms. For the purposes of this paper, fork bolt and ratchet will be used interchangeably, as will detent lever and pawl.

The latch, of course, is contained within or mounted on the door, which swings on its hinges. The hinges bolt or are welded to the "A" or "B" pillars of the car body. It is the function of the fork bolt to capture the **striker**, which is fixed to the car body on the "B" or "C" pillar. It does this by rotating in an eccentric or cam-like fashion when it contacts the striker.

Most strikers take the form of a cantilevered bolt projecting from the "B" or "C" pillar or a "U" shaped loop which projects from the pillar in an approximately horizontal (X-Y) plane. Some loop type strikers incorporate a wedge shaped base for alignment purposes. Others incorporate a cone shaped receptacle which mates with a pin projecting out from or near the latch.

Figure 4. Striker variations including pin and cone design (phased out beginning 1983) and wedge shaped loop design used by Mercedes-Benz.

The fork bolt term is more descriptive of this component because it contains a slot or channel into which the striker must slide. Once the striker enters the slot and

rotates the fork bolt, it is retained in the slot by the detent lever that is urged, usually by spring pressure, to engage a tooth or similar projection on the fork bolt.

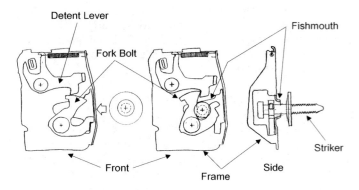

Figure 5. Fork bolt engages striker and is retained by detent lever.

The fork bolt and detent lever are generally arranged to rotate in a common plane (usually the plane of the door end panel - the Y-Z plane) and are usually riveted or bolted to the latch frame. The fasteners act as pivots and, in some designs, as mounting provisions for the latch.

The latch frame, in turn, is screwed or bolted to the sheet metal door end panel. In some vehicles, an extra layer of sheet metal reinforces the mounting area of the latch, which is sometimes an end flange of the side guard or intrusion beam. The Oldsmobile Cutlass is an example of this design feature.

All passenger automobiles have intrusion beams, to meet the requirements of FMVSS 214. However, some vehicle designs attempt to integrate the beams with the hinges and latch thereby enhancing longitudinal crush stiffness of the occupant compartment and reinforcing the mounting provisions for the doors.

Most side door latches are fitted with numerous levers intended to control the locking and remote release functions. Most of these levers are named for their specific role in control or locking with the exception of one multi-function lever sometimes known as the **intermittent lever**. The intermittent lever is typically a rocking link that will cause the remote handle linkages to move without effect or "idle" when the lock is engaged. It triggers the detent lever when the lock is not engaged.

Some latches have an **override** feature that will

automatically move the intermittent lever to its active state when the inside handle is used. This makes the inside handle active whether or not the lock is engaged. The Ford and Mazda latches are examples of designs that include this feature.

Most rear door latches (four door vehicles only) include a child safety feature which, depending upon the position of the associated lever, disables the inside remote handle.

TEST SAMPLE

Seventeen doors were purchased from salvage yards for the tests reported here. Most were from the 1985 through 1987 model years. These years were targeted because, in most cases, the latch designs are still current or one generation removed from current designs. The doors were also substantially less costly than more recent model years.

Driver's side front doors and one passenger front door was used. Only doors with minimal or no evidence of damage were selected. Any accident related damage to the latch, striker or end panel or significant rust was cause for rejection. All latches and strikers were examined prior to testing and determined to be in excellent operating condition.

TEST EQUIPMENT

The test apparatus was designed to operate on a side door latch, whether or not the door is attached to a car body. It consisted of a welded steel tube framework that carried a 44,480 N (10,000 lb.) capacity bi-directional hydraulic cylinder with 25.4 cm. (10 in.) of available travel. The framework was mounted to the doors via horizontal plates that were welded to the upper and lower portions of the door end panels.

These mounting plates were individually shaped to conform to the profiles of the test door but were positioned relative to each other and the engagement center of the latch via a simple jig. The plates were separated by 38.1 cm. (15 in.), well clear of the latch mounting area but sufficiently close together to insure that both mounting brackets usually fall on the door.

The exception to this was the Mercedes-Benz doors that locate the latch high in the end panel, near the belt line.

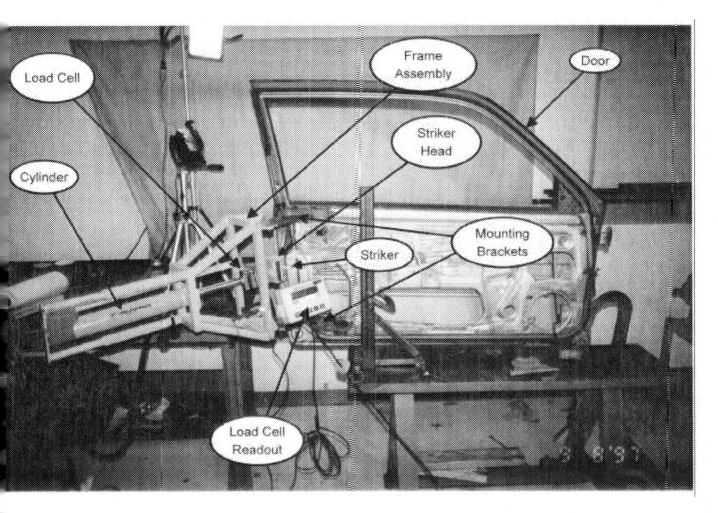

Figure 6. Test apparatus mounted to a door via welded brackets.

To accommodate these doors, additional bracing was welded to the window frame or **sash** to reinforce the upper bracket. Spreading the attachment points of the test fixture minimized the influence of the bracket on the deflection characteristics of the end panel and maximized the bracket stiffness.

A 44,480 N (10,000 lb.) capacity load cell was attached to the threaded piston rod from the cylinder and the striker attached to a fixture, named the striker mounting head, that was located on the opposite end of the load cell. For convenience and simplicity a universal-mounting fixture was created to accept most types of strikers. The strikers bolted to a 7.6 cm. x 8.9 cm. x .64 cm. steel plate drilled with alternative mounting holes for direct mounting of strikers or adapter plates.

The striker mounting plate was pivoted perpendicular to and aligned with the piston rod axis. Positioning of the striker on its mounting plate was adjusted so that the engagement center of the latch and striker also aligned with the axis of the piston rod.

To simulate the flexibility of a striker mounted in the auto body environment, the striker fixture had a provision to allow rotation with controlled resistance. This was achieved by using a second, non-pivoting plate to back up the striker mounting plate.

The rotational stiffness of the striker was adjusted by insertion of high-density elastic pads between the pivoting striker mounting plate and the fixed plate. For tests reported in this paper, the filler material was composed of several layers of high durometer rubber sheet of about 3 to 6 mm. thick.

In the tests reported here, the striker was also permitted to rotate about the axis of the piston rod. No attempt was made to lock the threaded junctions on the load cell or restrict the piston's angular position although in most cases, rotation on this axis was minimal.

The first series of five tests in this program were conducted with a slightly smaller hydraulic cylinder with a 22,240 N (5000 lb.) capacity and 15.2 cm. (6 in.) range

Figure 7. A jig was used to precisely locate the attachment brackets for the test apparatus relative to latch engagement center.

of travel. This range of travel was found to be insufficient for the Plymouth Horizon latch, the fifth door tested. Also, loads in excess of 19,300 N were measured during the test. The latch did not fail at the measured peak loads. For this reason,

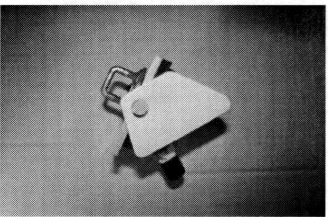

the test rig was modified for increased travel and load capacity.

Figure 8. Striker mounting head.

TEST PROCEDURE

Each door was prepared for testing by fitting and then welding the mounting brackets using the latch center location jig. Once the brackets were properly fixed, the test frame was bolted to the brackets and the striker was engaged at the primary position and centered in the latch. The latch was then locked.

Loads were applied gradually (quasi-static) using a hand operated lever-action hydraulic pump. Readings from the load cell were displayed on a digital meter and recorded real time on videotape.

Three video cameras were used to record the (1) load cell output, (2) an overview of the test and (3) an overhead view of the latch/striker interface. A strobe flash prior to initiation of the test provided a timing synch signal.

The overhead view proved useful for evaluating displacement of the striker and response of the end panel to load. It became apparent, after the test of the Plymouth Horizon door, that door end panel displacement would be significant in some cases. After test number 6, displacement of the piston rod was measured by using a steel tape in the video camera's field of view.

TEST RESULTS

Test number 1

1985 Chevrolet Spectrum (Isuzu)
VIN J81RF69K7F8410166
Date of manufacture: unavailable
Driver's door and latch assembly
Latch housing: within door
Latch fork bolt: descending rotation
Loop type striker assembly
Latch fork bolt in primary position and unlock mode

During force application to the striker, the latch housing and adjacent door panel experienced local deformation, permitting the fork bolt to bend outboard of the housing and release the striker assembly. The fork bolt exited the frame in its primary position. Striker force at latch separation was 9,581 N.

Test number 2

1986 Chevrolet Nova (Toyota)
VIN 1Y1SK1943GZ103994
Date of manufacture: 3/86
Driver's door and latch assembly
Latch housing: within door
Latch fork bolt: descending rotation
Loop type striker assembly
Latch fork bolt in primary position and lock mode

During force application to the striker, the door end panel experienced metal tearing and subsequent latch housing distortion. Latch fork bolt, detent lever, return spring and fork bolt pivot pin separated from the latch housing assembly with resultant striker release. The fork bolt exited the latch housing in its primary position as evidenced by gouges in the latch frame. Striker force at latch separation was 17,205 N.

Test number 3

VIN 1HGBA5435GA065908
Date of manufacture: 3/86
Driver's door and latch assembly
Latch housing: within door
Latch fork bolt: ascending rotation
Loop type striker assembly
Latch fork bolt in primary position and lock mode

During force application to the striker, the fork bolt experienced longitudinal displacement and bending extending outside of the latch case with resultant release of the striker assembly. Post test examination revealed the fork bolt primary position was maintained. Slight door end panel deformation was observed. Striker force at latch separation was 11,716 N.

Test number 4

1985 Nissan Stanza
VIN JN1HT11S5FT308016
Date of manufacture: 10/84
Latch housing: within door
Latch fork bolt: descending rotation
Loop type striker assembly
Latch fork bolt in primary position and lock mode

Minor local door end panel and latch housing deformation was experienced with subsequent fork bolt bypass of the detent lever. Fork bolt rotated to the secondary position and bent outboard releasing the striker assembly. Striker force at latch separation was 5,360 N.

Test number 5

1986 Plymouth Horizon
VIN 1P3BM18C8GD137062
Date of manufacture: unavailable
Driver's door and latch assembly
Latch housing: external
Latch fork bolt: ascending rotation
Loop type striker with end plate assembly
Latch fork bolt in primary position and lock mode

Load application caused extreme door end panel deformation, local separation of the end panel from the external door panel but very little deflection in the striker or latch. Ram bottoming and hydraulic leakage was experienced which prematurely concluded this test. There was extensive buckling in the outer door skin and the outside handle lifted to almost the full limit of its available travel. Maximum force applied to the striker was approximately 19,300 N without striker/latch disengagement. Post test examination revealed the fork bolt remained in the primary position.

Test number 6

1989 Oldsmobile Cutlass Supreme
VIN 1G3WH14T8KD374297

Date of manufacture: unavailable
Right front door and latch assembly
Latch housing: within door
Latch fork bolt: ascending rotation, with plastic coating
Disc head bolt striker assembly

Latch fork bolt in primary position and lock mode. Tip of fork bolt cut through frame in primary position releasing striker. Substantial end panel deflection and separation of the outer skin from the end panel was observed. Striker force at latch separation was 22,658 N.

Test number 7

1987 Plymouth Caravelle
VIN 1P3BJ46D8HC184508
Date of manufacture: 12/86
Driver's door and latch assembly
Latch housing: within door
Latch fork bolt: ascending rotation, with plastic coating
Disc head bolt striker assembly
Latch fork bolt in primary position and lock mode

Slight door end panel and latch housing deformation experienced. Fork bolt bypassed detent lever to secondary position, began to bypass secondary, then bent outside of housing and released striker pin. Post test the fork bolt secondary lobe was found wedged between latch housing and detent. Striker force at separation was 8,224 N. This latch was almost identical in appearance to subject 8 with the exception of plating color and minute deviations in non-structural plastic parts.

Test number 8

1987 Chrysler LeBaron
VIN 1C3BH48D6HN412037
Date of manufacture: unavailable
Driver's door and latch assembly
Latch housing: within door
Latch fork bolt: ascending rotation, with plastic coating
Disc head bolt striker assembly
Latch fork bolt in primary position and lock mode

Latch fork bolt tip fractured and pulled out of latch housing. Striker head remained partially engaged with fork bolt even after fork bolt pulled out of latch housing. Door end panel and latch deformed severely during test. Door latch mounting surface (end panel) was torn and separated from exterior door panel. Striker force at latch separation was 16,724 N. The primary difference

between tests 7 and 8 appeared to lie in the manner of attachment of the latch to its linkages and thereby to the door itself. It is thought that there was slightly less freedom for the LeBaron latch to rotate (about the vertical or Z-axis) due to an additional connection for a power locking solenoid.

Test number 9

1985 Mazda 626 LX
VIN JM1GC3114F1718636
Date of manufacture: 2/85
Driver's door and latch assembly
Latch housing: within door
Latch fork bolt: descending rotation
Loop type striker assembly
Latch fork bolt in primary position and lock mode

Latch fork bolt bypassed detent and rotated to the open position. Latch housing and door end panel experienced deformation. Door panel metal surface was torn adjacent latch opening. Latch housing fractured on interior surface opposite fork bolt. Post test examination of the striker loop revealed a gouge at the interior contact corner with the fork bolt. Striker loop and mounting surface did not exhibit deformation. Striker force at separation was approximately 10,675 N.

Test number 10

1986 Honda Civic
VIN JHMAK5430GS048904
Date of manufacture: 4/86
Driver's door and latch assembly
Latch housing: within door
Latch fork bolt: ascending rotation
Loop type striker assembly
Latch fork bolt in primary position and lock mode

Fork bolt locally deformed latch housing and door end panel permitting fork bolt to extend outside of housing and release striker assembly. Fork bolt remained in primary latch position. Striker loop did not appear deformed. Striker force at separation was approximately 10,222 N.

Test number 11

1987 Hyundai Excel
VIN KMHLD11J3HU068758
Date of manufacture: unavailable
Driver's door and latch assembly
Latch housing: within door
Latch fork bolt: descending rotation
Loop type striker assembly
Latch fork bolt in primary position and lock mode

Latch fork bolt end experienced fracture separation and release of the striker assembly. Local deformation of the latch housing and door end panel adjacent the fork bolt

fracture was observed. The door interior panel revealed metal tears adjacent the latch opening. Post test inspection of the striker mounting-base revealed cupping between the mounting bolts. Striker force at fork bolt fracture was 14,527 N.

Test number 12

1985 Subaru GL 4 door
VIN JF1AN43B0FB446999

Date of manufacture: 8/85
Driver's door and latch assembly
Latch housing: within door
Latch fork bolt: descending rotation, with plastic coating
Loop type striker assembly
Latch fork bolt in primary position and lock mode

Latch fork bolt remained in the primary position and was displaced by the striker loop outside the housing releasing the striker assembly. Door end panel experienced a tear adjacent the latch housing. Striker loop did not appear deformed. Striker force at latch separation was approximately 8,985 N.

Test number 13

1986 Ford Taurus
VIN 1FABP29UXGG198163
Date of manufacture: 6/86
Driver's door and latch assembly
Latch housing: within door
Latch: double-acting jaws (upper and lower)
Disc head bolt striker with anti-jam loop
Latch jaws (2) in primary position and lock mode

Upper and lower latch jaw pivots experienced displacement within housing. Lower jaw rotated to open position, upper jaw rotated to secondary position. Upper and lower jaws were bent outboard consistent with deformed and torn metal in door end panel. Striker bolt and anti-jam loop deformed. Striker force at latch separation was 13,798 N.

Test number 14

1985 Audi 5000
VIN WAUHC0449FN139656
Date of manufacture: unavailable
Driver's door and latch assembly
Latch housing: external
Latch fork bolt: descending rotation
Disc head bolt striker – welded to test fixture plate
Latch fork bolt in primary position and lock mode

The test was incomplete due to premature striker/test fixture attachment failure. Striker weld failure load was 11,596 N.

Test number 15

1986 Mercedes-Benz 190E
VIN WDBDA24DXGF165406
Date of manufacture: unavailable
Driver's door and latch assembly
Latch housing: within door
Latch fork bolt: ascending rotation
Wedge shaped striker loop
Latch fork in primary position and lock mode

ork bolt pulled out of latch housing. Partial fork bolt fracture
t root and complete fracture at tip. Door end and exterior
anel separation at rolled seam. End panel deformation
onsistent with outer door panel seam separation. Striker
orce at latch separation was 21,070 N.

Test number 16

984 Mercedes-Benz 300SD
VIN WDBCB20A1EA063880
Date of manufacture: unavailable
Driver's door and latch assembly
Latch housing: within door
Latch fork bolt: ascending rotation and supplementary conical
ocating pin
Conical receiver for pin in rubber insert contained by slotted
striker housing assembly
Latch fork bolt in primary position and lock mode

Fork bolt was displaced outside the latch housing but
remained in the primary position. Slight deformation of the
upper and lower surfaces of the latch housing and door end
panel adjacent the latch opening was observed. During
separation the fork bolt locally deformed striker pocket (slot),
which engages fork bolt. Striker/load cone rubber insert
compliance permitted striker separation from fork bolt.
Striker force at latch separation was 19,611 N.

Test number 17

1986 Renault Encore
VIN 1XMAC9633GK179064
Date of manufacture: unavailable
Driver's door and latch assembly
Latch housing: external
Latch: descending (single notched-lobe) pawl
Interlocking cam and restrictor tab striker assembly Latch
pawl in primary position and lock mode

Latch pawl remained in primary position. Latch housing
adjacent striker interlock experienced deformation that
released striker assembly. Striker force at latch separation was
8,932 N.

A summary chart of these results is attached as Appendix A.
The peak latch/striker loads are listed as well as a peak load
normalized by the 14,200 N
calculated load derived from 206 requirements. Note that in
some tests, the striker remained engaged with the fork bolt
after the fork bolt left its housing. This sometimes resulted in

a separation load that exceeded the true "failure" load. The
data presented here are the peak loads whether or not they
exceeded the true "failure" load.

FINDINGS

1. Presumably all of the latches in the sample comply
   with FMVSS 206 when tested according to the
   methodologies of the standard. However, over half
   of the latches tested under more realistic conditions
   failed to meet the proposed 14,200 N minimum
   strength derived from the FMVSS 206 requirements.
2. A factor of 4 difference in strength was observed in
   the performance of the strongest door and latch in
   the sample compared to the weakest.
3. A high ultimate strength performance for a
   latch/striker combination often correlated with
   extensive yielding of the door end panel.
4. Bypass of the detent by the fork bolt correlated with
   a relatively low strength. Often, there was much less
   damage present in the door end panel when bypass
   was identified.
5. The Caravelle and LeBaron had essentially the
   same latch but displayed drastically different ultimate
   strength (by a factor of 2) due to different failure
   modes.
6. Within the sample tested, no generalizations could
   be made about the advantages or disadvantages of
   loop versus post type strikers. There were good and
   poor performers in both categories.
7. The geometric details of a latch as well as the
   flexibility of its frame and mounting structure play a
   large role in its performance in these tests and in
   real world crash conditions.
8. The combined loading test, as described herein,
   proved enlightening, required minimal test facilities,
   was simple and relatively inexpensive to perform.
9. When compared to FMVSS 206 tests, failure modes
   observed during this program more accurately reflect
   those found in the real world.
10. The appropriateness of the proposed test as a
    supplement to or substitute for the FMVSS 206 test
    has not been established. Results to date are
    promising, however.

DISCUSSION

There is considerable historical precedence for testing under a combined loading scenario. Arthur Gross, in his January 1964 SAE paper (1), identified the side impact configuration simulated by this test as "most critical to door latch performance."

designed to simulate the type of door opening accident shown in Figure 9.

**HORIZONTAL ROTATION TEST
SIMULATES THIS TYPE OF
DOOR OPENING ACCIDENT**

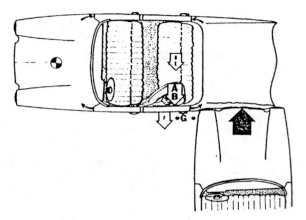

Fig. 37 - Planned impact configuration for forthcoming UCLA experimental collision series relating to door latch loading

Figure 9. Illustration taken from reference (1).

J. Bartol, W. Wingenbach and F. Hodosy, in a July 1969 contract report (2), called for dynamic, in-body testing of door systems in numerous combined loading situations including one similar to that illustrated by Gross.

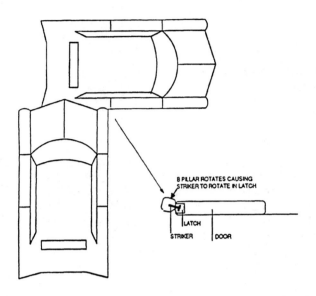

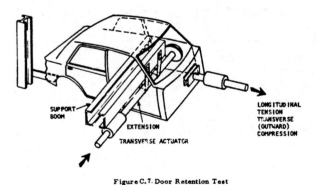

Figure C.7. Door Retention Test

Figure 10. Illustration taken from reference (2).

Figure 11. Illustration taken from reference (3).

The name "horizontal rotation test" made reference to the concept of striker rotation or pry-out. The test was used to reproduce failures seen in field accidents within a laboratory environment. The procedure made use of the routine tensile test machinery and a specialized fixture for holding the latch and striker. As with the 206-test procedure, the latch and striker were rigidly mounted, but to an articulated test fixture.

Although the proposed test configuration was different, the net result, i.e. combined transverse and longitudinal loading on the latch/striker interface including flexible mounting effects, was the same.

In October 1987, General Motors invited NHTSA to the Milford Proving Grounds to familiarize its representatives with a proposed "horizontal rotation test" procedure and related equipment (3). According to GM, this test was

708

HORIZONTAL ROTATION TEST
DOOR LATCH TEST FIXTURE

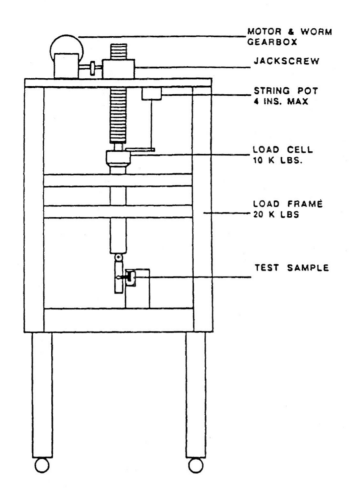

MOTOR & WORM
GEARBOX

JACKSCREW

STRING POT
4 INS. MAX

LOAD CELL
10 K LBS.

LOAD FRAME
20 K LBS

TEST SAMPLE

Figure 12. Illustration taken from reference (3)

NHTSA evaluated the procedure and ultimately rejected it for repeatability problems but was unable to establish whether the difficulties stemmed from the test procedure or variability in the latch hardware that was tested (4).

The lack of availability of identical test subjects makes validation of any door latch test extremely difficult. It is also clear, from the example of the Caravelle and LeBaron tests, that small differences in the structure of the door (or "B" pillar) may result in large changes in the performance of the same latch design.

However, both failure modes of the Chrysler latch have been encountered in real world accidents, so any apparent unpredictability of the test results is by no means an invalidation of the method.

The test scenario described considers only combined tensile and shear load on the latch/striker interface and door end panel. Some collision configurations will place a latch under combined compression and shear. A latch that performs well in this test won't necessarily excel when the loading directions are reversed. Therefore, this test will not answer every question about latch safety.

Figure 13. Chrysler "K" latches from tests 7 and 8. Left side latch experienced bypass of pawl by the ratchet.

Clearly there is also a dynamic (impact) element to latch failure that has traditionally been ignored when seeking a suitable performance test. That criticism applies to the testing methodology suggested here as well.

Also, while this test explores latch strength in the horizontal (X-Y) plane, there is also a third dimension that has implications for the choice of an ascending or descending fork bolt design, the detent arrangement and the latch frame design. It is not unusual to see evidence of relative vertical displacement between the latch and striker in accident vehicles suffering latch failures.

Nevertheless, testing in a combined loading format will probably produce more meaningful results in terms of the stated objective of FMVSS 206, i.e. "to reduce the likelihood of occupants being ejected from vehicles in real world accidents."

CONCLUSION

The car body environment has been shown to have a profound influence on the real world strength of an automotive door latch. A proposed test has been used to demonstrate a large range of performances for production latching systems in a car body environment and accurately reproduce real world latch failure modes. Adoption of a combined loading test represents a needed step forward toward the ultimate goal of safer door latching systems in automobiles.

REFERENCES

1. Gross, A., "Accidental Motorist Ejection and Door Latching Systems," University of California at Los Angeles, SAE Paper No. 640165, originally published in SAE Transactions, Vol. 73 (1965).
2. Bartol, J., Wingenbach, W., Hodosy, F., "Performance Requirements for Doors, Door Retaining Devices and Adjacent Structures," American Machine & Foundry Co., DOT Contract No. FH-11-6891, DOT Report No. DOT-HS-800-254, Publication No. PB193379, July 1969.
3. Letter from Robert A. Rogers – Director of Automotive Safety Engineering, General Motors (signed by Milford Bennett) to Kennerly H. Digges – Deputy Associate Administrator for Research and Development, NHTSA, November 3, 1987.
4. Howe, G.J., Leigh, M., Willke, D.T., "Door Latch Integrity Study: Evaluation of Door Latch Failure Modes," Vehicle Research and Test Center, DOT Report No. DOT-HS-808-188, January 1994.

APPENDIX A

LATCH TEST RESULTS

Rank	Test No.	M. Year	Test Subject	Separation Load	Normalized	Pass
1	6	1989	Olds Cutlass Supreme	22658	1.60	Yes
2	15	1986	Mercedes 190E	21070	1.48	Yes
4	16	1984	Mercedes 300SD	19611	1.38	Yes
3	5	1986	Plymouth Horizon	19300+	1.36	Yes
5	2	1987	Chevy Nova (Toyota)	17205	1.21	Yes
6	8	1987	Chrysler LeBaron	16724	1.18	Yes
7	11	1987	Hyundai Excel	14527	1.02	Yes
8	13	1986	Ford Taurus	13789	0.97	No
9	3	1986	Honda Accord	11716	0.83	No
10*	14	1985	Audi 5000	11596	*	*
11	9	1985	Mazda 626	10675	0.75	No
12	10	1986	Honda Civic	10222	0.72	No
13	1	1986	Chevy Spectrum (Isuzu)	9581	0.67	No
14	12	1985	Subaru	8985	0.70	No
15	17	1986	Renault Encore	8932	0.63	No
16	7	1987	Plymouth Caravelle	8224	0.58	No
17	4	1985	Nissan Stanza	5360	0.38	No

* It is the authors' opinion that this test sample would have exceeded the 14,200 N criterion had the test fixture not failed.

+ Indicates that the test sample did not fail at the peak applied load.

2000-01-1304

Safety and Security Considerations of New Closure Systems

Stephan Schmitz
Robert Bosch Schließsysteme GmbH

Jacek Kruppa
Robert Bosch GmbH

Peter Crowhurst
Robert Bosch Pty

ABSTRACT

A closure system for automotive security and driver comfort has been developed. The system combines a passive entry system and an electronic door latch system.

The passive entry system utilises a single chip transponder for vehicle immobilisation, passive entry and remote control functionality. The form factor free transponder enables the integration into a key fob or a smart card. The system can be activated by either pulling the door handle or by using a push button transponder. Due to the inductive coupling between the transponder and the vehicle mounted antennas, the vehicle door or trunk opens on successful verification as if there were no locks. Additionally, inside the vehicle, the transponder can be used as a far range immobiliser.

The electronic door latch system utilises electronically controlled latches. Symmetrical housing of the electronic latch (E-latch) and the absence of a mechanical connection to the actuators enable the latch to be used not only for the left and right side doors but also for trunk applications. The locking pawl of the E-latch is controlled by an electric motor and the functionality is entirely software dependent.

INTRODUCTION

Vehicle immobiliser systems based on transponder technology are currently used as standard equipment for vehicles. Vehicles equipped with immobilisers have impacted on theft rates, reducing the numbers of stolen vehicles. Due to the demanding security requirements, the immobiliser is gaining more and more importance. With the introduction of passive entry systems additional functionality like remote control, passive entry and vehicle immobilisation is made available to maximise customer comfort. The introduction of electro-mechanical latches provided the customer with new locking functions like central door locking (CDL), double locking (DL) and electronic controlled child safety (CS).

The new closure system described here was developed for automotive security and driver comfort and will replace existing immobilisers, UHF remote controls and mechanical locking systems. To achieve this objective, a vehicle based interrogation system is activated by pulling the door handle and starts a medium range, bi-directional signal transmission (up to 2.5 m) between the vehicle and the transponder carried by the user. Additionally, a special protocol provides long range access to the vehicle (more than 40 m), whenever the user presses a button located on the transponder. A control unit verifies the transmitted data and locks/unlocks the latches on verification.

Closure systems are commonly used in access building applications etc.. However, the practical operation of such a system in vehicle applications requires an important development effort, which has to meet the following requirements:

- The communication area of the system has to be optimised for vehicle applications. Special antennas have to be developed to provide the required operating range.

- A high data rate and a fast communication protocol in combination with a fast latch unlocking time are desired to avoid the utilisation of approach detection sensors.

- A sophisticated cryptographic algorithm has to be developed to guarantee the security of the system for driving permission, vehicle access and remote control under several attacks like statistical attacks or relay station attacks.

- The utilisation of E-latches without any existing mechanical connection and therefore back-up needs special consideration to ensure at least the same safety requirement of latches in series production today.

The paper firstly gives a brief description of the operating principles and functionality. Then we present physical properties such as the operating range and the overall timing of the closure system. In a next chapter we discuss the safety and security of closure systems.

CLOSURE SYSTEM DESCRIPTION

Figure 1 illustrates a smart card based closure system and the components involved. As soon as the user pulls the door handle, the antenna is activated and a LF (125 kHz) signal transmission between the vehicle and the portable transponder is initiated. All transponders within the communication area are identified and one transponder is selected for the security check and transmits a UHF data sequence to the vehicle. On successful verification of the transponder, the E-latch is activated and the door opens.

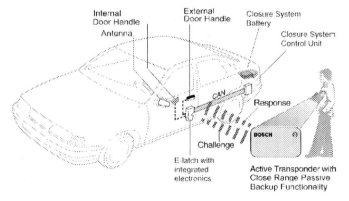

Figure 1.   Closure System with passive entry and E-latches

For long range entry control, the user can have access to vehicle locking or unlocking applications from a greater distance by pushing a transponder mounted button on the key fob. A unidirectional UHF sequence is sent which wakes up the vehicle based UHF receiver. The UHF receiver demodulates the button initiated UHF transmission and sends the bit pattern to the control unit, which validates the code and determines which action is being requested (i.e. locking or unlocking the doors or opening the trunk).

In order to provide higher comfort for the end user, a special transponder memory organisation supplies the user with the ability to use his transponder for multiple application. The unique structure also allows up to three different

applications (e.g. vehicles, garage door opening, building access control). For each of these applications, two different kinds of security protocols exist. Whenever the security should be at a maximum, a three pass challenge response protocol is carried out. In vehicle access applications, where the security is limited by the breaking of a window, a single challenge response protocol is utilised. For vehicle driving applications, where the security should be maximised, a three pass challenge response protocol (mutual authentication) will be carried through. The interrogation unit is able to select between the different security levels by emitting different transponder commands, which can be controlled by software.

The closure system for vehicle applications includes the following components:

CLOSURE SYSTEM BATTERY – A separate redundant closure system battery (CSB) as illustrated in Fig. 2 is used to supply the energy for the E-latches and the passive entry system. The CSB is charged by the main battery and the charge status and the condition can therefore be monitored. In failure of the CSB the vehicle's main battery is directly providing the required energy for the closure system.

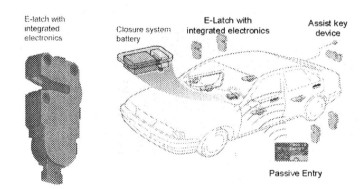

Figure 2.   Closure System Battery and E-latch

E-LATCH – The E-latch as illustrated in Fig.2 with its integrated electronic is purely controlled by software and its unlocking time of 70 ms is ideal for passive entry applications. Due to its symmetric design the E-latch can be used for right side as well as for left side door applications. The integrated electronics establishes the communication to the body work bus via CAN.

TRANSPONDER – As illustrated in Fig. 3, the transponder consists of a hardwired core logic with interfaces for external components such as the antenna coil, battery, UHF transmitter and push buttons. The form factor independent packaging features of the transponder enables embedding of the transponder into a smart card or into a key head/fob.

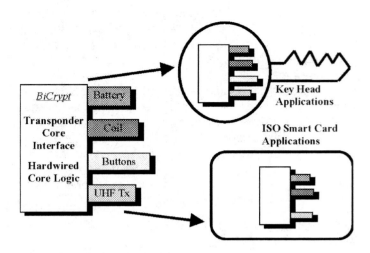

Figure 3. Transponder schematic overview and packaging features

LF- ANTENNA – A wound loop antenna is built into i.e. the driver's door mirror as shown in figure 1 for generating a rotating magnetic field.

UHF RECEIVER – The UHF receiver wakes up when an appropriate UHF sequence is received from the transponder due to an intentional activation by the user. The UHF receiver can be also woken up by the closure system electronics for passive entry applications. The vehicle's door is opened after protocol security verification.

DOOR HANDLE – In order to use existing vehicle components, a door handle without approach detection sensors initiates the passive entry system. An additional push button integrated into the door handle activates the lock process after a successful verification.

CLOSURE SYSTEM CONTROL UNIT – The control unit is connected to the body work bus (e.g. CAN) and controls the functionality of the latches, the passive entry as well as the closure system battery. The closure system control unit provides a secure power source by either selecting the vehicle's main battery or the closure system battery as the closure system energy supplier.

PHYSICAL PROPERTIES

The operating range is a critical issue for any passive entry system. Two requirements conflict. On the one hand, the range should be maximised to provide as much comfort as possible, while on the other, the range should be minimised to ensure anti-theft security. If the communication area is too large, a thief could gain access to the car, while the driver is either approaching or leaving the vehicle.

To discover the best transponder opening positions, many situations were evaluated and this resulted in us

determining that the optimum range is between 2 m and 2.5 m. However, the range should be shaped more rectangularly than quadratically, with the long side in parallel with the vehicle (see figure 4).

To have the best control of communication range, we use inductive coupling between the transponder and the antenna. For very small distances compared to the wavelength, the magnetic field generated by small loop antennas decreases with the cube of the distance. It is therefore well suited for data transmission, where the range has to be controlled.

Another advantage of inductive coupling is the possibility to transfer energy at close distances via the magnetic field. This is necessary in a flat transponder battery scenario. Ideal places for mounting an antenna are i.e. the driver and passenger side mirrors and the rear bumper.

The overall timing for the vehicles' door opening is a critical issue. Mechanical door locks, as used today, provide a door handle which has to be pulled to open the door. Tests with random users have revealed an opening time from as fast as 60 ms to as long as 300 ms. Most car manufactures require, that there is no noticeable hesitation between pulling the door handle and opening of the door. Several tests have revealed that the overall timing from pulling the door handle to opening of the door should be less than 130 ms. Therefore it is essential for any closure system to have fast communication and opening times. With a communication time (anti-collision and challenge response) of 44 ms and an E-latch opening time of 70 ms the overall action time of less than 130 ms as illustrated in Fig. 5 can be achieved. Any action time with more than 130 ms demands a different utilisation concept or the installation of approach detection sensors in order to start the communication protocol before the door handle has been pulled.

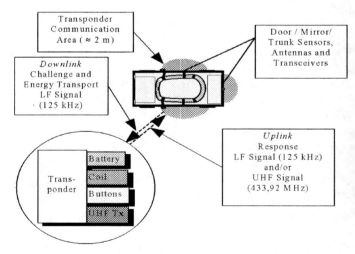

Figure 4. Schematic overview of passive entry system communication features

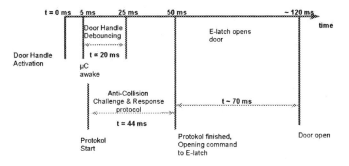

Figure 5. Closure system timing overview

By using the E-latch in combination with the fast protocol customer requirements without utilising approach detection sensors can be fulfilled.

SECURITY CONSIDERATIONS OF A CLOSURE SYSTEM

In a closure system the two subsystems, locking system and the passive entry system, have to be discussed. The security of an E-latch with its integrated electronics provides a new level of security. The sealed housing and the missing mechanical connections between latch and the actuators like door handle prevent any mechanical attack on the latch. Due to the integrated electronics and the CAN communication the closure system control unit and the E-latch have the possibility to communicate via encrypted messages. The security presented by the E-latch reveals new kind of security levels in the vehicle architecture, which have yet not been reached.

The passive entry system can be regarded as a far range immobiliser. It is a fact that the various implemented cryptographic algorithms in the immobiliser function of today have not been the weakest part in the vehicle security chain. Most of the time the security architecture, the secrete key storage and the service concept are prone to a successful attack. It can be assumed that in addition to the above the long distance communication of a passive entry system might reveal new attack possibilities for a non trusted party. A possible threat to a passive entry system might be the so called relay station attack as illustrated in fig. 6.

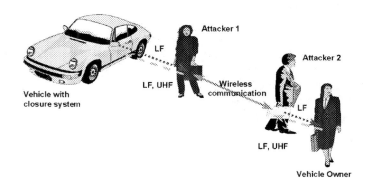

Figure 6. Schematic illustration of the relay station attack

The vehicle with the closure system is under attack by attacker 1, who has initiated the vehicle communication. The attacker 1 receives and demodulates the transmitted data and sends the data by wireless communication to an attacker 2. The attacker 2 demodulates the data and sends the data on the original vehicle frequency to the transponder bearer, the vehicle owner. The transponder processes the received data and sends back his answer. The answer is transferred via attacker 2 and 1 to the vehicle. The transponder's response was only transmitted and not further processed and the vehicle authenticates the data stream and the vehicle opens the door. The same procedure would start, if the attacker would use the relay station attack for passive go. As a consequence the attacker would be authorised to drive the vehicle.

Several counterfeiter possibilities exist. Of course one would think directly of measuring the time of flight of such an signal as this is already well know technology for military and scientific applications (i.e. radio/light detection and ranging). Another solution would be to measure the pulse build up time in a resonance circuit as this is routinely done for laser tuning.

Whereas in principle the time of flight counterfeiter is in principle possible, the application to the automotive market with its cost pressure seems to be quite unlikely, especially when a time of less than a few nano-seconds (i.e. 30 ns) should be measured. The measurement of additional pulse build up times in resonance circuit can be easily achieved with existing technologies. However, the attacker may use low Q-circuits with high power antenna drivers to minimise the additional resonance build up time. As a resume both counterfeiters are not practical for automotive applications.

A feasible counterfeiter would be to use of a non linear approach by the emission of two tones by the transponder of the vehicle owner. The two tones are emitted simultaneously and caused by the low power emission, the communication is limited to less than 5 m. For any distance greater than 5 m the attackers have to amplify the 2 tones with a defined gain. As it is well known that each amplification can be separated into a linear and non linear part we will receive third order distortion as inband lines in the vehicle receiver as illustrated in fig. 7.

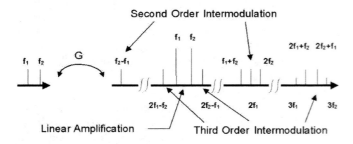

Figure 7. Transformation of two tones by non linear amplification (G) into the linear and non linear terms.

If two tones have been received and emitted by a relay station, the necessary amplification (G) in the relay station causes second order and third order intermodulation. As can be seen from the figure above the third order intermodulation terms are falling into the bandwidth of the receiver by selection of a suitable two tone frequency spacing. The figure 8 presents a typical example of the received spectrum of a two tone emission after passing the amplification (G) of relay station. The distortion in the specturm is caused by third order intermodulation.

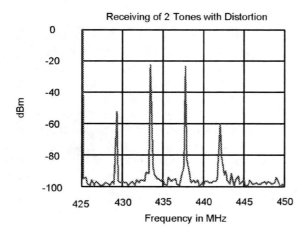

Figure 8. Receiving of 2 tones with distortion after passing a non linear amplification.

Several hardware realisations of a two tone test are possible to implement the relay station counterfeiter which are left to the reader.

SAFETY CONSIDERATIONS OF CLOSURE SYSTEM

As soon as a pure electronic closure system is discussed the reliability of such a system moves into the focus of interest. Especially for the electronic latches the reliability has not only to be considered during the warranty time of the vehicle, but also to be discussed after a life time of 10 to 15 years. An appropriate way to investigate the reliability of a closure system is i.e. a fault tree analysis (FTA).

We are especially interested in the probability that a critical event as described below will occur. The following critical events have been identified.

ENTER – This event describes the malfunction of one of the locked doors, which can not be opened from outside. The consequence of such an event is that the person has to use another door.

EXIT – In analogy to the event enter, this malfunction describes the event that one of the doors can not be opened from inside. As a consequence the person has to use another door.

RESCUE – Much more severe is the event rescue. Here a person in the car shall be rescued i.e. after a crash. The consequence of such an event is that none of the doors can be opened from outside and the person is trapped in the vehicle. This events can result into a danger for the person in the vehicle.

ESCAPE – The event escape is comparable to the event rescue. Here the person wants to leave the car i.e. after a crash and none of the doors can be opened from inside. The person is trapped in the vehicle and the malfunction can result into a danger for the person in the vehicle.

SELF RELEASING – One of the doors releases itself and the door opens. The driver may feel threatened, if the seatbelt is not fastened and this scenario may result into a dangerous situation.

We have laid out the architecture of a closure system with a electro-mechanical latch in series production today. The failure ratios of the components have been derived from the field experience. By undertaking an analysis of the possibilities one can derive the failure tree which will lead to the events described above.

Following this approach we have defined the architecture for an E-latch based closure system and used the failure probabilities of the components of the electro-mechanical latch to derive the failure probability of the critical events.

We have normalised the failure probabilities of the electro-mechanical latch based closure system to 1 and compared the results with the E-latch based closure system as presented in fig. 9. For each closure system the fault tree analysis is carried out to derive the probability of the cirtical events.

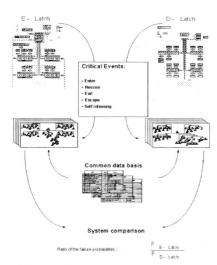

Figure 9.   Schematic description of the comparison between an electro-mechanical latch and E-latch based closure systems.

By studying the fault tree and isolating the critical paths with the failure probabilities of the components involved the relevant components can be optimised to achieve a very low failure probability. By optimising the actuators, sensors and the closure system energy supply one can optimise the system's reliability. In table 1 the electro-mechanical and the E-latch based closure system are compared.

Table 1.   Normalised comparison of the electro-mechanical and the E-latch based closure system failure rates.

Critical Events	Electro-mechanical latch based closure system	E-latch based closure system
Enter	1	0,12
Rescue	1	0,0003
Exit	1	0,71
Escape	1	0,61
Self Release	1	0,79

As a result it can be stated that the safety of the E-latch based closure system is even better than the electro-mechanical latch based closure system.

CONCLUSION

We have described a new closure system. The system combines a electronic E-latch and a passive entry system. The introduction of E-latches offers the automotive manufacturer only one latch for right, left side and trunk applications with the advantage that the functionality is purely controlled by software to minimise application work. The comfort for the end user is increased due to the feature of passive entry and the forceless pull of the door handle due to the missing mechanical connection by the E-latch.

To satisfy the increased security requirements, we have shown that the E-latch and the passive entry system offer a new kind of security level. The sealed E-latch housing and the possible bus connection due to the integrated E-latch electronics even offer the possibility to use encrypted communication between a central ECU and the latches. The utilisation of challenge response protocols in a passive entry system ensures the same security of the protocol as in today's banking system. Possible attacks like the relay station attack are overcome by simply emitting two tones and verifying the third order intermodulation lines, which can be received in the bandwidth of the UHF receiver.

The normalised comparison of a closure system based on electro-mechanical latches in series production today and the utilisation of E-latches have been presented. As a result it can be stated that the safety of the E-latch based closure system is better than an electro-mechanical closure system.

REFERENCES

1. Balanis, C. A., <u>Antenna Theory</u>, 2nd ed., Wiley, New York, 1997.

2. Hirano M., Takeuchi, M. Tomoda, T., Nakano, K.-I., "Keyless Entry System With Radio Transponder", IEEE Trans. Ind. Electr., Vol. 35, No2, 208-216, 1988.

3. Schmitz, St. and C. Roser, "A New State-of-the-Art Keyless Entry System", SAE Technical Paper Series; 980381, 1998.

Door Latch Vulnerability to Rollover Induced Loads

Andrew Gilberg and Jeremy Buckingham
Teknacon

Richard McSwain, Dirk Paulitz and Mark Hood
McSwain Engineering

ABSTRACT

Light truck and SUV rollovers often involve ground contacts at the roof rails or door sills that can induce significant vertical shear loads at the latch/striker interface. These vertical loads are not evaluated in Federal Motor Vehicle Safety Standard testing yet they are known to cause latch failures. Such failures expose both belted and unrestrained occupants to increased injury risk. An example of two such failures can be found in open literature in a single van rollover test.

A simple vertical load test for latches is described by the authors and evaluated for discrimination, suitability and repeatability. This test was applied to an array of current and past generation latches found on many popular SUVs and light trucks. A large range of failure loads was encountered. A review of the structural features of the superior performing test samples suggests simple modifications that could dramatically improve performance of the remaining latches. Improved performance in this vertical load test could be expected to translate to fewer unwanted door openings and ejections in field accidents.

BACKGROUND

Federal Motor Safety Standard 571.206 (FMVSS 206) regulates minimum strength of side and rear door latches and hinges in passenger carrying vehicles. For the side door latches, the essence of FMVSS 206 is four static load requirements and a theoretical calculation for resistance to inertial shock.

For the side door latches, the static load tests individually measure the strength of the latch and striker combination in a longitudinal direction (X axis – parallel to the vehicle fore/aft axis) or lateral direction (Y axis – perpendicular to the vehicle fore/aft axis and parallel to the ground plane). These lateral and longitudinal tests are done both in the primary (fully latched) and secondary (partially engaged) position of the latch.

For rear door latches, two additional static load requirements are imposed (one for the primary and one for the secondary positions) in a direction that is orthogonal to those imposed on the side door configuration. This load test could be thought of as acting in the vertical (Z axis) direction except that the axis system is in the frame of reference of the latch rather than the vehicle. Typically, the Z axis would be oriented along the rear door hinge axis.

Sliding side and back doors have load requirements similar to those of the hinged side doors but are subjected to a *system* test that includes the door. The subtleties of in-door latch testing were addressed in reference (1).

To meet the X axis load requirements, it is necessary for the latch to withstand at least 11,000 N (2500 lb.) in its primary position and 8,900 N (2000 lb.) in the Y axis direction. The Z axis requirement for rear doors also calls for 8,900 N in the primary position. The load requirements fall to 4,450 N (1000 lb.) for all three directions with the latch in its secondary position.

The need to consider the Z axis strength of side doors has been repeatedly demonstrated in field rollover accidents, underride side impacts and other commonly encountered accident events that impose vertical shear loads on side door latch/strikers.

In the well-documented rollover test of a model year 2000 Ford Econoline 15 passenger van conducted to demonstrate the Controlled Rollover Impact System (CRIS) (2) both the driver's and rear cargo doors opened as a result of vertical overload failures of the latches (Figures 1, 2).

Figure 1. CRIS Econoline test vehicle left side showing failed driver's door.

Figure 2. CRIS Econoline showing failed rear doors.

In the case of the driver's door, the latch was displaced downward relative to the striker, whereas in the rear doors the latch was displaced upward relative to the striker.

The design of the latch in the Econoline (Dortec D-21) places the detent lever above the opening into which the striker enters (often called the "fishmouth") and the fork bolt below it. By spreading the fishmouth, the contact depth of the detent pawl is progressively reduced until it disengages (Figure 3). The shearing action of the striker on the latch frame vertically expands the fishmouth opening and causes the detent pawl to disengage from the fork bolt.

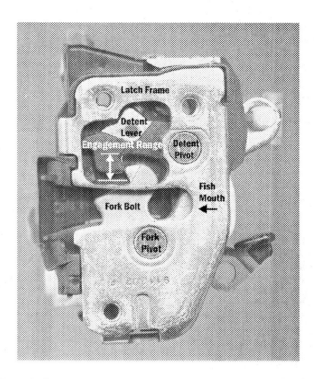

Figure 3. Spreading the fishmouth reduces fork/detent engagement.

In the CRIS test, the striker produced a dent in the fishmouth that indicated the direction of loading on the latch (Figures 4, 5).

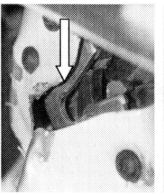 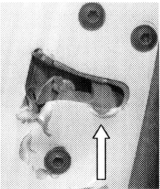

Figure 4. (Left) Witness mark from striker is at top of fishmouth in front door.

Figure 5. (Right) Most prominent witness mark is at bottom of fishmouth in rear door.

A similar failure has been observed in a liftgate using the same type of latch when the vehicle was struck on the side. In the liftgate application, the latch was oriented so that the vertical axis of the latch aligned with the transverse axis of the vehicle (Figures 6, 7).

Figure 6. Liftgate latch failure due to laterally oriented load.

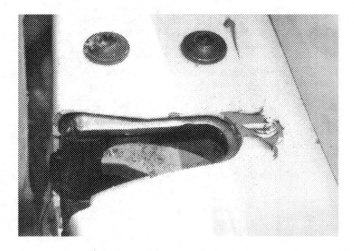

Figure 7. Liftgate latch is nearly identical to side door latch. In an undamaged latch, upper lobe of fork would be partially concealed by top of frame opening, but is now visible due to fishmouth expansion.

The test developed for this paper is similar in concept to that used for the Z axis test of rear latches in FMVSS 206 with two important exceptions. First, the latch is fixed directly to the test machine by its own mounting provisions rather than to an intervening plate that can artificially enhance the strength of the frame. Secondly, the load is applied to the latch through its anchorage points, rather than the striker.

Applying the load directly through the frame changes the points of load application on the latch frame as indicated in Figure 8.

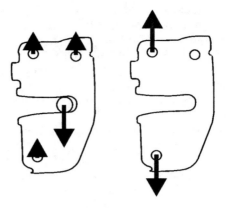

Figure 8. In-door loading is indicated by free body diagram on left; vertical load test is on right. Net effect in both cases is to spread fishmouth.

In spite of this difference, the appearance of the failed latch is very similar (without the striker dent) to field examples. Other justifications for this procedure include that it permits a clear view of the fishmouth area of the latch as it undergoes loading. It also reduces the sources of variability in the test and provides a reliable, repeatable procedure for making A:B:C comparisons among latches.

TEST PROGRAM

Test latches for this program were selected from current and recent model year sport utility vehicle and light truck side doors. All were purchased new from local dealerships for the driver's door of the models shown in the list below.

Several latches were selected because they represented the immediate past generation design. It was anticipated that they would provide benchmarks for measuring improvements in vertical load capacity. Identification of the latch is for the convenience of this discussion and may not represent the manufacturer or supplier's designation.

TEST SAMPLE:

Vehicle	Latch
1994 Chevrolet Blazer	Type III modified
2003 Chevrolet Silverado	Mini Wedge
2003 Dodge Durango	Gecom (version A)
2001 Dodge Ram	Gecom (version B)
2003 Ford Explorer (3 tests)	Dortec D21
2003 Ford Explorer Sport Trac	Ford Mini
2003 Jeep Liberty	Gecom (version C)
cedes Benz ML320 (2)	Mercedes Benz

2003 Nissan Pathfinder	OHI
2002 Toyota 4Runner	Aisin (version A)
2003 Toyota 4Runner	Aisin (version B)

TEST EQUIPMENT – Vertical loads were applied with a computerized laboratory Instron tension/compression test machine. (Figure 9) Latches were linked to the base and crosshead with spade shaped attach plates made of 6 mm (¼ inch) 1018 low carbon steel. These attach plates were clamped into standard tensile test mounting grips.

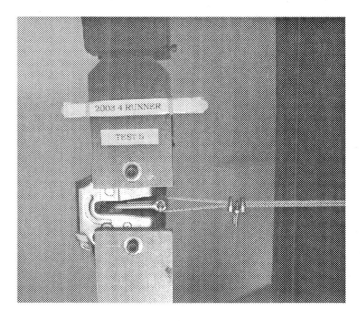

Figure 10. Test sample mounted in attach plates with preload cable in place.

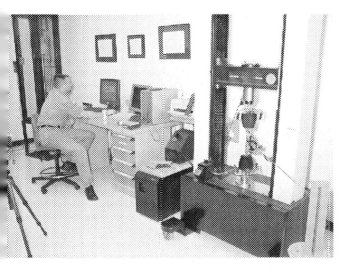

Figure 9. Instron test machine with test sample installed. Preload cable runs to pulley and weight package (not visible).

For latches with counter sunk attachment points, an assortment of conical washers were turned to adapt the fasteners and attach plates to those test subjects. Shallow spot faces were cut into the attach plates to help locate the conical washers.

The latch forks were preloaded with a horizontally oriented braided wire cable terminated by a stainless steel clevis. A consistent horizontal preload of 670 N (150 lbf) was used for all tests based upon pretest analysis of the force required to rotate the fork bolt in exemplar failed latches. (Figure 10)

TEST PROCEDURE – The "vertical" axis of the latch was determined by selecting the two fastening points in the frame that positioned it in the test machine most like it would be installed in a vehicle door.

Each latch was affixed to the attach plates and the tensile test machine was then adjusted so the plates were correctly aligned and clamped in the mounting grips. The preload cable was then inserted into the latch jaws and the latch was placed in the primary position. The lock was then engaged on those latches that could be mechanically locked.

The Instron head was driven at a (quasi-static) constant speed of 5.08 mm/min until the latch jaws released or the sample parted. The force/deflection data was automatically recorded by the machine and displayed in the form of a plot. A video camera recorded the behavior of the sample in real time and still digital photos were used to document details of the failed test samples.

TEST RESULTS

Test number 1

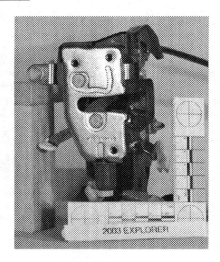

Figure 11. 2003 Explorer (D21-1).

Failure mode:	Fork/detent disengaged
Maximum load:	4772 N (1073 lbf)
Head travel at M/L*:	6.36 mm
Head travel at E/O/T**:	6.36 mm
Energy at E/O/T:	22.18 J
Lock position:	Locked
Fork/detent relationship:	Opp. sides of fishmouth
Fork bolt type:	Ascending
Inside release mech:	Cable
Outside release mech:	Push rod
Frame material:	see test 3
Frame hardness:	see test 3
Frame thickness:	see test 3

*M/L = maximum load
**E/O/T = end of test

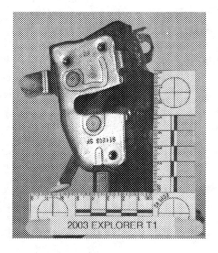

Figure 12. Ford D21 post test (D21-1).

Test number 2

2003 Explorer (D21-2).

Failure mode:	Fork/detent disengaged
Maximum load:	4542 N (1021 lbf)
Head travel at M/L:	6.69 mm
Head travel at E/O/T:	6.69 mm
Energy at E/O/T:	21.85 J
Lock position:	Locked
Fork/detent relationship:	Opp. sides of fishmouth
Fork bolt type:	Ascending
Inside release mech:	Cable
Outside release mech:	Push rod
Frame material:	see test 3
Frame hardness:	see test 3
Frame thickness:	see test 3

Test number 3

2003 Explorer (D21-3)

Failure mode:	Fork/detent disengaged
Maximum load:	4793 N (1078 lbf)
Head travel at M/L:	7.17 mm
Head travel at E/O/T:	7.17 mm
Energy at E/O/T:	22.61 J
Lock position:	Locked
Fork/detent relationship:	Opp. sides of fishmouth
Fork bolt type:	Ascending
Inside release mech:	Cable
Outside release mech:	Push rod
Frame material:	steel (.07C)
Frame hardness:	77 - Rockwell B
Frame thickness:	3.20 mm

est number 4

Test number 4

Figure 13. 2003 Explorer Sport Trac (Ford Mini).

Failure mode:	Upper frame screw hole tore out*
Maximum load:	5290 N (1189 lbf)
Head travel at M/L:	17.36 mm
Head travel at E/O/T:	25.46 mm
Energy at E/O/T:	85.80 J
Lock position:	Locked
Fork/detent relationship:	In line with fishmouth between 2 forks
Fork bolt type:	Ascending and descending
Inside release mech:	Push rod
Outside release mech:	Push rod
Frame material:	steel (.15C)
Frame hardness:	60 - Rockwell B
Frame thickness:	1.73 mm

*Fork and detent remained engaged

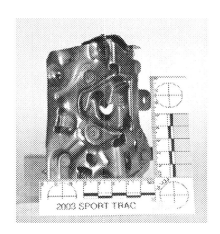

Figure 14. Ford Mini post test.

Test number 5

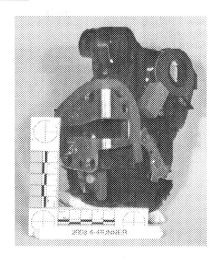

Figure 15. 2003 Toyota 4Runner (Aisin – A).

Failure mode:	Fork/detent disengaged
Maximum load:	9131 N (2053 lbf)
Head travel at M/L:	13.52 mm
Head travel at E/O/T:	14.74 mm
Energy at E/O/T:	94.83 J
Lock position:	Unlocked*
Fork/detent relationship:	Opp. sides of fishmouth
Fork bolt type:	Descending
Inside release mech:	Cable
Outside release mech:	Push rod
Frame material:	steel (.10C)
Frame hardness:	92 - Rockwell B
Frame thickness:	3.20 mm

*Electronic locking only

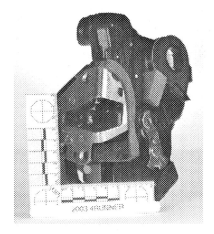

Figure 16. Aisin A post test.

723

Test number 6

Figure 17. 2002 Toyota 4Runner (Aisin – B).

Failure mode:	Fork/detent disengaged
Maximum load:	5839 N (1313 lbf)
Head travel at M/L:	11.35 mm
Head travel at E/O/T:	13.44 mm
Energy at E/O/T:	70.46 J
Lock position:	Locked
Fork/detent relationship:	Opp. sides of fishmouth
Fork bolt type:	Descending
Inside release mech:	Pull rod
Outside release mech:	Push rod
Frame material:	steel (.04C)
Frame hardness:	89 - Rockwell B
Frame thickness:	2.64 mm

Figure 18. Aisin B post test.

Test number 7

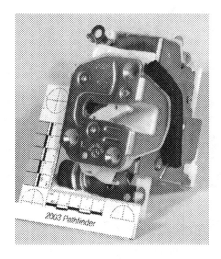

Figure 19. 2003 Nissan Pathfinder (OHI).

Failure mode:	Fork/detent disengaged
Maximum load:	7321 N (1646 lbf)
Head travel at M/L:	11.16 mm
Head travel at E/O/T:	11.16 mm
Energy at E/O/T:	49.70 J
Lock position:	Locked
Fork/detent relationship:	Opp. sides of fishmouth
Fork bolt type:	Ascending
Inside release mech:	Pull rod
Outside release mech:	Push rod
Frame material:	steel (.10C)
Frame hardness:	78 - Rockwell B
Frame thickness:	2.64 mm

Figure 20. OHI post test.

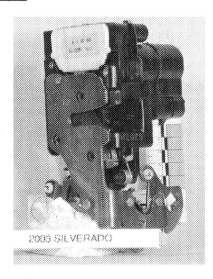

Figure 21. 2003 Chevrolet Silverado (Mini Wedge).

Failure mode: Fork/detent disengaged
Maximum load: 7203 N (1619 lbf)
Head travel at M/L: 9.83 mm
Head travel at E/O/T: 12.37 mm
Energy at E/O/T: 64.81 J
Lock position: Locked
Fork/detent relationship: Opp. sides of fishmouth
Fork bolt type: Descending
Inside release mech: Pull rod
Outside release mech: Push rod
Frame material: steel (.08C)
Frame hardness: 85 - Rockwell B
Frame thickness: 2.11 mm

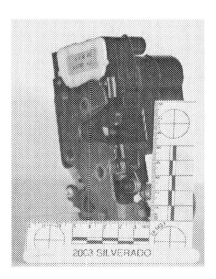

Figure 22. Mini Wedge post test.

Figure 23. 1994 Chevrolet Blazer (Type III modified).

Failure mode: Fork/detent disengaged
Maximum load: 4639 N (1043 lbf)
Head travel at M/L: 10.47 mm
Head travel at E/O/T: 10.47 mm
Energy at E/O/T: 34.76 J
Lock position: Locked
Fork/detent relationship: Opp. sides of fishmouth
Fork bolt type: Ascending
Inside release mech: Pull rod
Outside release mech: Push rod
Frame material: steel (.07C)
Frame hardness: 48 - Rockwell B
Frame thickness: 2.06 mm

Figure 24. Type III modified post test.

Test number 10

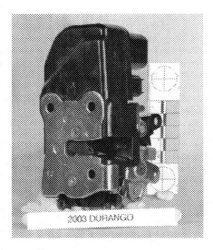

Figure 25. 2003 Dodge Durango (Gecom – A).

Failure mode:	Fork/detent disengaged
Maximum load:	6938 N (1560 lbf)
Head travel at M/L:	14.82 mm
Head travel at E/O/T:	21.37 mm
Energy at E/O/T:	108.78 J
Lock position:	Locked
Fork/detent relationship:	Opp. sides of fishmouth
Fork bolt type:	Descending
Inside release mech:	Pull rod
Outside release mech:	Push rod
Frame material:	steel (.08C)
Frame hardness:	80 - Rockwell B
Frame thickness:	2.59 mm

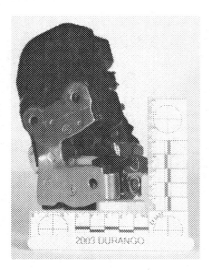

Figure 26. Gecom A post test.

Test number 11

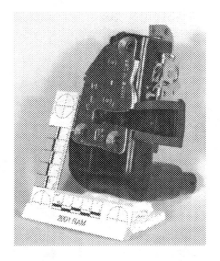

Figure 27. 2001 Dodge Ram (Gecom – B).

Failure mode:	Frame tore through*
Maximum load:	5709 N (1283 lbf)
Head travel at M/L:	11.76 mm
Head travel at E/O/T:	49.83 mm
Energy at E/O/T:	98.20 J
Lock position:	Locked
Fork/detent relationship:	Same side of fishmouth
Fork bolt type:	Descending
Inside release mech:	Pull rod
Outside release mech:	Push rod
Frame material:	steel (.08C)
Frame hardness:	77 - Rockwell B
Frame thickness:	2.87 mm

*Fork and detent remained engaged

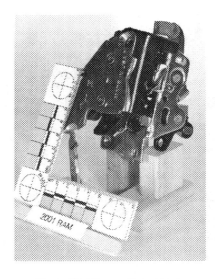

Figure 28. Gecom B post test.

Test number 12

Figure 29. 2003 Jeep Liberty (Gecom – C).

Failure mode:	Fork/detent disengaged
Maximum load:	6271 N (1410 lbf)
Head travel at M/L:	13.02 mm
Head travel at E/O/T:	27.60 mm
Energy at E/O/T:	131.20 J
Lock position:	Locked
Fork/detent relationship:	Opp. sides of fishmouth
Fork bolt type:	Descending
Inside release mech:	Pull rod
Outside release mech:	Push rod
Frame material:	steel (.06C)
Frame hardness:	82 - Rockwell B
Frame thickness:	2.62 mm

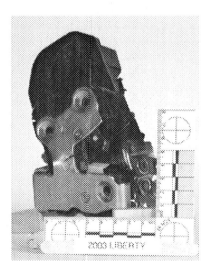

Figure 30. Gecom C post test.

Test number 13

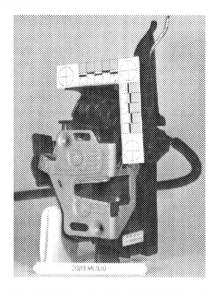

Figure 31. 2003 Mercedes Benz ML320 (Mercedes - 1).

Failure mode:	Mounting screw broke*
Maximum load:	10566 N (2375 lbf)
Head travel at M/L:	8.01 mm
Head travel at E/O/T:	8.01 mm
Energy at E/O/T:	56.54 J
Lock position:	Unlocked**
Fork/detent relationship:	Opp. sides of fishmouth
Fork bolt type:	Ascending
Inside release mech:	Direct drive pull lever
Outside release mech:	Cable
Frame material:	steel (.08C)
Frame hardness:	86 - Rockwell B
Frame thickness:	2.29 mm

*Fork/detent remained engaged; latch was functional following the test
**Electronic locking only

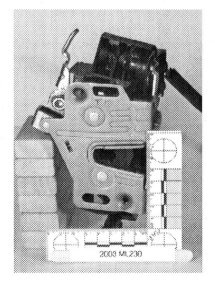

Figure 32. Mercedes 1 post test.

Test number 14

2003 Mercedes Benz ML320 (Mercedes – 2)

Failure mode:	Mounting screw broke*
Maximum load:	9654 N (2170 lbf)
Head travel at M/L:	9.31 mm
Head travel at E/O/T:	12.84 mm
Energy at E/O/T:	93.25 J
Lock position:	Unlocked**
Fork/detent relationship:	Opp. sides of fishmouth
Fork bolt type:	Ascending
Inside release mech:	Direct drive pull lever
Outside release mech:	Cable
Frame material:	see test 13
Frame hardness:	see test 13
Frame thickness:	see test 13

*Special high-strength mounting screws used
**Electronic locking only

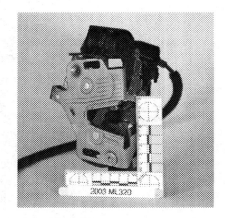

Figure 33. Mercedes 2 post test.

REPEATABILITY AND DISCRIMINATION – For the first three tests, identical 2003 Ford Explorer D21 latches were used to evaluate test repeatability. Figure 34 shows the force/deflection results plotted on the same axes. The ultimate strength of the three samples fell within 5% and the general character of the load curves was very similar.

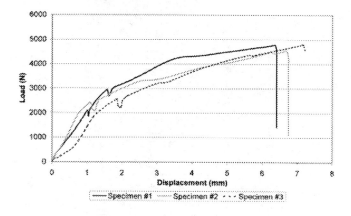

Figure 34. Repeatability plot for three D21 Explorer latches.

The variability encountered in these three tests compares favorably with repeated load testing done for FMVSS 206. In such tests, 3 to 13% variability in the ultimate load is not uncommon. The small variability that does exist appears to reflect production variations among the test subjects.

Figure 35 compares the top and bottom performers on the same axes. The ultimate strength of the strongest latch exceeded the weakest by a factor of 2.33 to 1. In the case of the top performer, the fastening screws, not the latch failed.

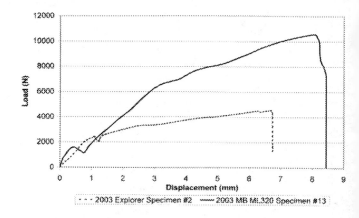

Figure 35. Strongest and weakest test specimens compared.

To place this in perspective with the requirements of FMVSS 206, the peak loads were normalized with the 8,900 N minimum level for the primary position in the transverse direction and the hinge axis direction of rear door latches. The normalized failure load ranged from 0.51 to 1.19.

Following test number 13 in which a mounting screw failed leaving the Mercedes Benz latch with very little damage, it was decided to try to retest this latch using special high-strength fasteners (Test number 14). It was anticipated that this would improve the performance of the top ranked subject.

Instead of improving its performance, however, the ultimate strength declined. The deflection at peak load and at the end of test both increased. Inspection of the test samples revealed that more distortion of the latch frame had occurred with the higher strength (and stiffer) fasteners. There was also less distortion of the failed fastener. The net effect was an increase the energy absorbed by the system but not its ultimate strength. (Figure 36)

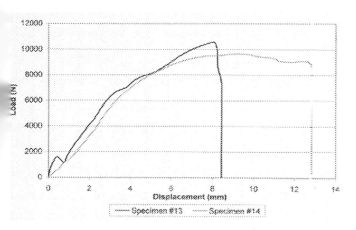

Figure 36. Stronger fasteners transferred deflection into the latch frame increasing energy absorbed but not ultimate strength.

FINDINGS

1. As with all of the testing required by FMVSS 206, the vertical load test does not replicate typical crash loads. It does, however, provide an additional repeatable and discriminating means of comparing latch hardware in a loading direction that can be encountered in rollovers, sideswipes and side impacts.

2. Using the logic underlying FMVSS 206, larger force values for ultimate strength or larger displacements at release/separation in the vertical load test should equate to improved occupant retention in crash events.

3. The "vertical" nomenclature refers to the top/bottom component of the latch axis system, which may be found in other orientations in some applications. Some of the latches tested can be found in liftgates. When mounted in a liftgate, lateral loads on the door can create vertical loads in the latch.

4. When compared to the transverse load requirement in FMVSS 206, most latches tested for vertical strength proved deficient. Eleven of fourteen test subjects failed to match the minimum strength requirement of FMVSS 206.

5. The strongest latch tested was more than 2.32 times stronger than the weakest.

6. Displacement required for latch release or separation varied by a factor of 7.83. (Comparing head travel at E/O/T for test 11 to test 1 generates this ratio.)

7. When only the fork bolt/detent disengagement failure mode was considered, displacement at failure varied by a factor of 4.34. (Comparing head travel at E/O/T for test 12 to test 1 generates this ratio.)

8. Energy absorbed { ∫force) dx} may be a useful performance metric in the vertical load test.

9. In the energy category, the top-performing latch absorbed 6 times more energy than the lowest.

10. With the possible exception of the three latches that exceeded 8900 N, none of the test samples appeared to have been designed with consideration for the potential vertical loads that can be encountered in rollovers, sideswipes and side impacts.

11. Based upon the narrow margins by which the best performers of the test samples exceeded the minimum strength requirements of FMVSS 206 and evidence of vertical loading failures from controlled rollover testing and field accidents, it is clear that some form of vertical load testing should be implemented.

12. The vertical load test proposed here could be swiftly and inexpensively implemented and would fill a large gap in the existing requirements.

DISCUSSION – While rollovers are spectacular and extraordinarily hazardous accident events, they frequently are made up of a series of low or moderate impacts spread out over a relatively long period of time. This means that properly restrained and retained occupants should do better than we have come to expect. Ejection (or partial ejection) with its greatly enhanced risk and its high association with rollover accidents offers a partial explanation for this phenomenon.

The less severe nature of the impacts in rollover accidents also means that door latches should not be stressed as severely as they are in more common planar collisions. This makes the appearance of unwanted door openings in rollovers, both puzzling and intensely compelling to study.

It should be noted that the D21 latch, for example, or its twin the D5, (a D21 minus most of the external levers) can be found in some SUV and minivan liftgates. Although it does not meet the 8,900 N benchmark in this vertical load test, presumably it does meet the requirement when tested according to the protocol of FMVSS 206. A possible explanation for this is that the 3.2 mm mounting plate to which the latch is fastened for the FMVSS test artificially strengthens the latch. For the D21/D5 latch, this mounting plate effectively doubles the thickness of its outer frame.

The thickness of the mounting plate has little importance in A:B comparisons, as long as all latches are tested under equal conditions. But it does obscure the real world significance of the results. Latches that meet all federal requirements still fail in low or moderate impact events such as rollovers.

Analysis of the test samples and a review of their normal structural environment offered some insight into why e failures can occur.

729

A typical door end panel, to which most latches attach, is constructed of sheet steel of about 0.8 mm thickness. In some side doors, the intrusion beam mounting flange is sandwiched between the latch and the door end panel. This could add an additional 1.8 to 2.0 mm thickness to the latch mounting surface. Many multipurpose vehicles, sport utility vehicles and light trucks lack this feature, however. Such reinforcements are exceptionally rare in liftgates and rear doors.

In the vertical load test described here, the latch structure absorbs the entire applied load; therefore, the results more nearly reflect the contribution of the latch to maintenance of the door security.

Physical characteristics of the latch that may influence performance in a vertical load situation include the following:

1. Housing material
   a. Composition
   b. Heat treatment
   c. Thickness
2. Frame geometry
   a. Inner/outer layer separation
   b. Wrap-around features (e.g. Aisin A, OHI)
   c. Surface mount vs. through-bolting (e.g. Ford Mini vs. Mini Wedge)
   d. Material section at maximum offset from vertical load axis
   e. Offset of fishmouth vertex from vertical load axis
3. Frame fixing and attaching provision
   a. Number of mounting screws
   b. Placement of mounting screws
   c. Screw diameter
   d. Screw strength and style
   e. Number of through-frame pivot shafts
   f. End fixing details of pivot shafts
4. Internal configuration
   a. Relationship of fork bolt and detent pivot axes to fishmouth or thinnest section
   b. Number of fork bolts
   c. Orientation of detent lands (vertical, horizontal, inclined)
   d. Depth of engagement of detent pawl(s)
   e. Width of engagement of detent pawl(s)
   f. Coatings on fork/detent interface
   g. Diameters of cross shafts
   h. Dual-function pivot shaft and screw anchorage

These are some of the design parameters that can be manipulated to improve performance in this test. Other, innovative approaches not evaluated in a hardware test could be used to supplement the performance of the latch. Examples of this include the following:

1. Multiple latches
2. Supplemental vertical reaction elements in the door jamb such as bolt-in-pocket or hook-in-loop catches
3. Integration of latch and intrusion beam structure
4. Relocate the latch in "B" pillar and striker in door
5. Items 1 and 4 combined
6. Others - Imagination is the only limit

CONCLUSION

Vertical load testing of automotive door latches can be repeatably and reliably accomplished using techniques described in this paper. Such testing would be useful in reducing the rate of unwanted door opening in various common real world accident scenarios that are not currently addressed by Federal Motor Vehicle Safety Standards.

ACKNOWLEDGMENTS

Thanks are due to Jerry Knight of McSwain Engineering for his enthusiastic assistance in the laboratory during the testing phase of this project.

The authors also wish to acknowledge Edward Caulfield of Packer Engineering whose concept for a static vertical door latch load test inspired the test described in this paper.

REFERENCES

1. Gilberg, A., Marcosky, J., Sherman, L., Clarke, R., "Door Latch Strength in a Car Body Environment," SAE paper No. 980028, published as part of special publication SP-1319.

2. Carter, W.J., Habberstad, J.L., Croteau, J., "A Comparison of the Controlled Rollover Impact System (CRIS) with the J2114 Rollover Dolly," SAE paper No. 2002-01-0694, published as part of special publication SP-1664.

CONTACT

Send Correspondence to:

Dirk Paulitz
McSwain Engineering Inc.
3320 McLemore Dr.
Pensacola, FL 32514

dpaulitz@mcswain-eng.com

APPENDIX A

TEST RESULTS SUMMARY

Rank	Test No.	Latch	Failure Mode	Maximum Load	Normalized	Pass	Energy
1	13	2003 MB ML320 (MB - 1)	mounting screw broke*	10566 N	1.19	Yes	56.54 J
2	14	2003 MB ML320 (MB - 2)	mounting screw broke**	9654 N	1.08	Yes	93.25 J
3	5	2003 4Runner (Aisin - A)	disengagement	9131 N	1.03	Yes	94.83 J
4	7	2003 Pathfinder (OHI)	disengagement	7321 N	0.83	No	49.70 J
5	8	2003 Silverado (Mini Wedge)	disengagement	7203 N	0.81	No	64.81 J
6	10	2003 Durango (Gecom - A)	disengagement	6938 N	0.78	No	108.78 J
7	12	2003 Liberty (Gecom - B)	disengagement	6271 N	0.76	No	(highest) 131.20 J
8	6	2002 4Runner (Aisin - B)	disengagement	5839 N	0.66	No	70.46 J
9	11	2001 Ram (Gecom - C)	frame tore through***	5709 N	0.64	No	98.20 J
10	4	2003 Sport Trac	screw hole tore*	5290 N	0.60	No	85.80 J
11	3	2003 Explorer (D21 - 3)	disengagement	4793 N	0.54	No	22.61 J
12	1	2003 Explorer (D21 - 1)	disengagement	4772 N	0.54	No	22.18 J
13	9	1994 Blazer (T-III modified)	disengagement	4639 N	0.52	No	34.76 J
14	2	2003 Explorer (D21 - 2)	disengagement	4542 N	0.51	No	(lowest) 21.85 J

*Latch remained engaged and functional
**Special high strength screws used
***Latch frame split but latch did not release

CHAPTER NINE

Energy absorbing bumpers

We have concentrated on the latter system as long earlier we had been working on a similar element surrounding the whole car to limit bodywork damage following minor impacts or swiping.

We had noticed that pneumatic elements could provide considerable car protection with negligible penalty on weight and, besides, they could be manufactured so as to ensure good abrasion resistance.

Some pilot solutions were realized and tested during the last four years and were used to study a new type of bumper — to be fitted at car front and rear — which was capable of performances similar to those required by the ESV program low speed collision specifications.

It consists essentially of three parts (see Fig. 1):

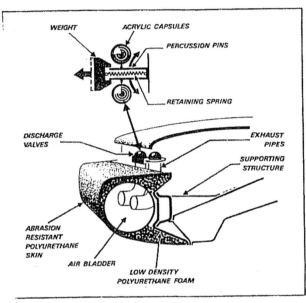

igure 1

- A cylindrical rubber air cushion diagonally and longitudinally reinforced to make its surface inextensible. The two ends are hemispherical and provided with exhaust tubes.
- A polyurethane foam protection, lined with an abrasion resistant skin — enveloping the air cushion.
- Two exhaust tubes whose ends are connected to inertial decelerometer actuated valves. They consist of two thin diaphragms blanking the exhaust tube ends and of two percussion pins, held at a distance from the diaphragms by a hook connected to the decelerometer mass. When deceleration exceeds a pre-fixed limit, the diaphragms are perforated by the pins and air escapes from the cushions.

Design and dimensions of the elements were determined by the following specifications:
- Vehicle weight: 800 kg
- Type of collision: head on against vertical barrier

PNEUMATIC BUMPER PROTECTION SYSTEM FOR PASSENGER CARS

Dr. Richard Sapper, Fiat-Pirelli Consultant

This project, conducted as a joint effort with Pirelli and Fiat, aims at the reduction of the consequences of low speed collisions.

Among the new utilizable systems of energy-absorbing bumpers we examined the features of the following:
1. Elastic metal elements
2. Hydraulic systems
3. Water bumpers
4. Foam structures
5. Pneumatic elements

- Impact speed: 15 km/h
- Maximum tolerable deceleration: 10g
- Maximum diameter of the cushion: 200 mm

The optimum values for diameter, initial pressure, exhaust section and critical deceleration for valve opening were calculated making the following hypotheses:

- The discharge of the gas was considered without losses of any kind; whereas two different equations were used depending on the pressure ratios being higher or lower than the critical state.
- The cylindrical cushion was considered inextensible and infinitely flexible. The influence of the hemispherical ends on the behaviour of the cushion was neglected; it was considered as a cylinder with a horizontal, straight axis whereas in reality its axis is slightly curved with about 25 mm camber.
- The structure of the car was considered non-deformable during collision and deceleration was supposed to result exclusively from the pneumatic element applied forces.
- The opening of the valves and the discharge of the gas was considered instantaneous as soon as an established deceleration value was reached.

Based on these considerations, the equations simulating the vehicle motion during collision against barrier were introduced into Pirelli IBM 360/44 computer and the optimum dimensions investigated by the trial and error method.

Some results of these calculations can be seen on the following diagrams.

They show clearly how the discharge of gas through the exhaust valves lowers considerably the maximum deceleration as compared to an element without valves.

In Diagram A, which shows a collision at 15 km/h, with opening of the valves at a deceleration of 6.3 g, maximum deceleration is lowered from 23 to 9 g.

Diagram B shows the progression of deceleration during the collision versus time, with and without opening of the discharge valves.

Diagram C shows the maximal values of deceleration in function of impact speed, with and without gas discharge.

In all these diagrams, the comparison between the behaviour of the two types of pneumatic elements, with or without valves, has been made not only with equal quantities of energy, but also with the same amount of deflection of the pneumatic element; for this purpose it was necessary to establish different initial pressures: in the case of the element with valves, this initial pressure is considerably higher.

The calculations resulted in the following optimum dimensions:

- Cushion diameter: 200 mm

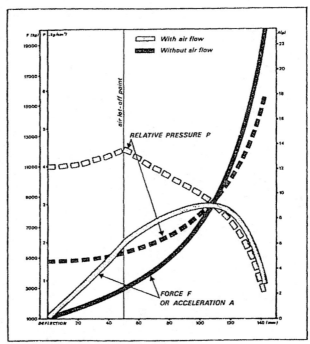

Diagram A

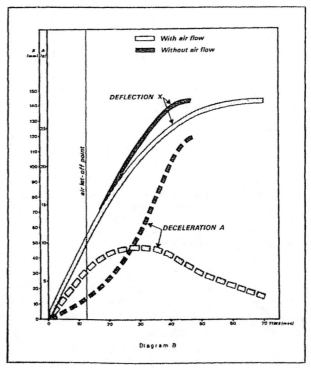

Diagram B

- Initial pressure: 4 atmospheres (gauge)
- Discharge section: 0.0048 sgm
- Critical deflection for valve opening: 50 mm — corresponding to 6.3 g

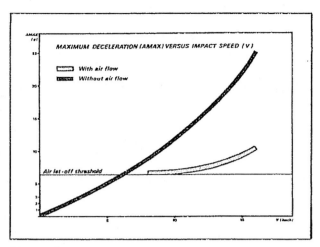

Diagram C

As a next step, a series of prototype bumpers was built and fitted to modified Fiat 128 bodies. This pneumatic bumper protection system, the test results of which have been illustrated in Mr. Franchini's report, represents the present state of the project. We are now analyzing these test results in order to go on with the development.

J. D. Withrow
D. N. Renneker
Chrysler Corporation

The 1973 Chrysler Energy Absorbing Bumper System

1. Introduction

The purpose of this paper is to discuss the 1973 Chrysler Corporation Bumper System, its detailed construction, the energy management scheme used, and the engineering methods applied to the design of the elastomeric bumper guards and the effect of these guards on the overall energy levels transmitted into the vehicle.

The 1973 Chrysler Corporation Bumper System is designed to meet and exceed the 1973 Department of Transportation Exterior Protection Standard. This Federal standard states in part that all vehicles manufactured after September 1, 1972, for sale in the United States must be capable of withstanding impacts into a large immovable concrete barrier, perpendicular to the vehicle centerline, at speeds of 5 m.p.h. in the front and 2-1/2 m.p.h. in the rear, without impairing the function of specified safety related items. These items include lights, brakes, doors, deck lids, etc. The 1973 Chrysler system is designed not only to protect these safety related items, but also to withstand the barrier impacts without significant damage to any part of the vehicle, including bumper or sheet metal.

Prior to the Federal standard, Chrysler Corporation had a corporate bumper standard for many years. This standard required that all bumpers be able to withstand a series of 2 m.p.h. flat face pendulum impacts anywhere along the bumper surface without the bumper or vehicle suffering any significant damage. The pendulum weight was equivalent to that of the car and the pendulum face was 16 inches square. Most of our bumper systems since the late 1950's passed this Chrysler Engineering standard. During the latter part of the 1960's, competitive pressures resulting from consumer buying preferences in the market place, dictated that styling be modified to more closely integrate the bumper with the styling of the sheet metal and ornamentation. This of course reduced the bumper's ability to provide previous levels of protection and in some cases the 2 m.p.h. pendulum standard could not be met. This styling concept, although aesthetically attractive, made it more difficult for the engineer to provide appropriate exterior protecting bumpers because of the close relationship between the body and the bumper and the difficulty in supporting these integrated bumpers.

The D.O.T. Exterior Protection Standard was issued April 14, 1971. By that time all of our 1973 vehicles were well into the design phase. A great deal of engineering work had to be accomplished to analyze the requirements and determine what had to be done to each design to assure its passage. We have 28 unique bumper systems for 1973, and each one had to be analyzed, designed, tested and tooled.

As stated previously every 1973 Chrysler bumper system has been designed to not only pass the federal bumper standard relating to protection of safety items but also to exceed the standard by being capable of impacting a fixed barrier at speeds of 5 m.p.h. in the front and 2-1/2 m.p.h. in the rear without any significant damage to any portion of the vehicle.

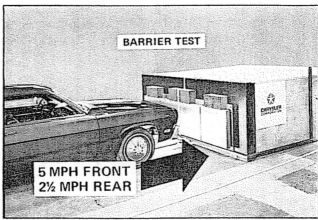

Fig. 1

738

This means that no significant damage will be found on the bumper, bumper supports, fenders, grilles, mouldings or any other portion of the vehicle after impacting the barrier at the speeds previously mentioned. In addition, all 1973 Chrysler Corporation cars have bumper systems designed to withstand, without significant damage to the vehicle, the concentrated load of a 16 x 16 inch flat face pendulum impacting the bumper anywhere along its surface. The impacting pendulum used has a weight equivalent to that of the vehicle which range from approximately 2500 lbs. to 5500 lbs. (A 2500 lbs. car is struck with a 2500 lbs. pendulum while a 5500 lbs. car is struck with a 5500 lbs. pendulum, etc.). The pendulum travels parallel to the longitudinal axis of the vehicle at speeds, for most cars, of 3 m.p.h. in the front and 2 m.p.h. in the rear. This test is designed to simulate the concentrated impacts of such items as poles, trees, etc.

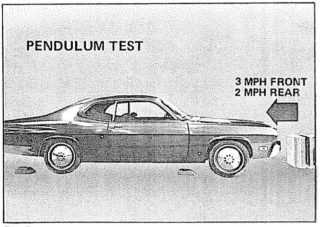

PENDULUM TEST

3 MPH FRONT
2 MPH REAR

Fig. 2

II. The 1973 Chrysler Bumper System Description

All 1973 Chrysler Corporation bumpers have been set out away from the sheet metal as much as 2" to provide better protection to the sheet metal, grilles, lamps, ornamentation and other vehicle equipment during low speed impacts.

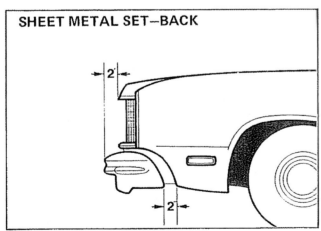

SHEET METAL SET—BACK

2"

2"

Fig. 3

Considerable changes have also been made to the structure of the bumpers and their supporting system. A full width reinforcement has been added behind the bumper surface to provide better beam strength to the bumper during impact. This reinforcement has been nested into the bumper to minimize local denting during impacts with small concentrated objects. In some cases high strength steel has been used where necessary because of space limitations.

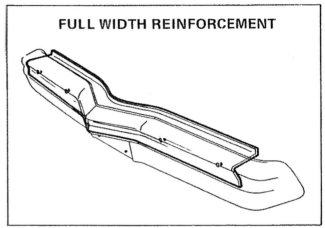

FULL WIDTH REINFORCEMENT

Fig. 4

The bumper supporting structure has also been increased in strength. This was achieved by using heavier gage steel with deeper sections. Where it was required, additional bolts for attaching the supporting structure to the body were also used to increase strength and raise the load carrying capability of the support system.

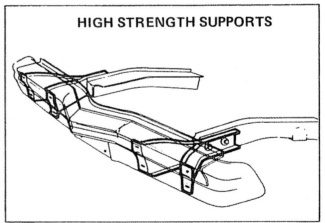

HIGH STRENGTH SUPPORTS

Fig. 5

The last component of the 1973 Chrysler Corporation bumper system is the standard equipment energy absorbing bumper guards. These guards consist of an elastomeric pad attached to a metal supporting structure.

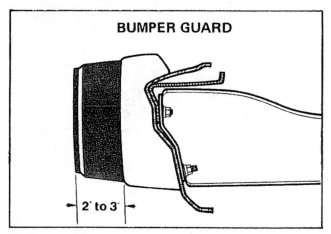

BUMPER GUARD

2" to 3"

Fig. 6

The guards range in depth from 2" to 3" and not only dissipate some of the impact energy but also add significantly to the 2" sheet metal setback previously mentioned. The guards present a large flat vertical surface which reduces the tendency of the bumper to slip over or under other bumpers during low speed collisions.

These components then comprise the 1973 Chrysler Corporation energy absorbing bumper system.
1. Full width bumper reinforcement
2. Heavy gauge bumper supports
3. Standard equipment bumper guards
4. Increased sheet metal setback

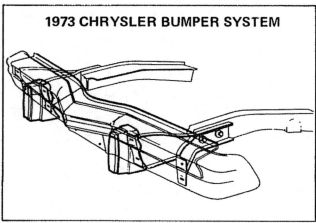

1973 CHRYSLER BUMPER SYSTEM

Fig. 7

Although strength and sheet metal protection are two important features of any good bumper system, it must be recognized that another important factor must also be considered if complete low speed vehicle protection is to be provided. This important factor is, that bumpers must line up with the objects they are impacting. Many collisions of course, are between a vehicle and a wall or a pole or some other object that does not have height constraints.

Fig. 8

However, low speed accidents also occur between cars in which bumper height and face configuration are important factors. If the two vehicles in question do not have bumpers that match, and do not have bumper face shapes that are reasonably compatible then bumper strength will have very little effect in protecting the vehicles.

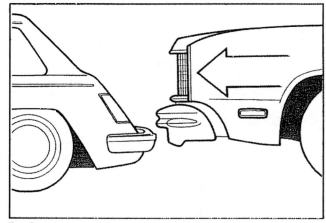

Fig. 9

It has been obvious for some time that a bumper standard was needed to regulate bumper heights. Although impact speeds have been getting most of the publicity lately, we feel that bumper matching is as important if not more important than strength in reducing vehicle damage. The DOT standard regulates only bumper strength on 1973 models. Bumper heights and contours will become standardized on 1974 models.

By the end of 1973, there will be approximately 86,500,000 cars on the road including the more than 10,000,000 1973 cars that will have improved bumper systems. It should not be anticipated that even with the improved bumper systems on the 1973 vehicles, any major nationwide reduction in vehicle damage will be realized since bumper heights will vary greatly in 1973. Even in 1974, with standard bumper heights, no dramatic reduction should be expected because of the small percentage of vehicles that will be on the road with improved bumper systems with matching heights. It will take several years, probably ten, before the full effect of improved bumper matching and strength will be fully felt in reducing vehicle damage and lowering repair costs.

The rest of this paper will be devoted to a description of the energy management approach used in designing the 1973 Chrysler Bumper System which will include a discussion of the elastomeric bumper guards and their contribution to the overall energy management problem.

III. Energy Management

In discussing the subject of energy management, one must first analyze the input energy of the collision that goes into the vehicle. This input energy varies with the type of object that is being struck or is striking. We will therefore discuss impacts with fixed objects such as walls and barriers and non-fixed objects such as other cars.

A. Impact With Fixed Objects

Input energy is a function of the mass, velocity and path of the impact objects. Although a vehicle is not a rigid body, mathematical model studies indicate that most low speed impacts are of sufficient duration that major vehicle components attain the same acceleration (deceleration) level. A vehicle can, therefore, be treated as a rigid body for purposes of input energy calculation. Passengers and objects carried within the vehicle, however, add to the effective mass only if sufficiently restrained by seat belts or other means of restraint.

The kinetic energy of a moving vehicle may be calculated as follows:

$$E_k = \frac{1}{2} \left(\frac{W}{g}\right) V^2 = \frac{WV^2}{29.94} \quad (1)$$

where E_k = Kinetic energy (Ft. Lb.)

W = Vehicle Weight (Lb.)

V = Vehicle Velocity (MPH)

Fig. 10

For a direct impact with a fixed object, the input energy is equal to the total kinetic energy of the vehicle.

B. Direct Impact With Non-Fixed Objects

Impacts with non-fixed objects generate less input energy since some kinetic energy generally exists after impact. The simplest case of this type is a moving object which impacts an unrestrained stationary object directly, such that the line of action of the impact force passes through the C.G. of each object. As the objects collide, deflection takes place at the interface until a point in time when both objects have attained the same velocity. At that point, the end of the compression phase of impact, deflections are at a maximum and the input energy has been stored or absorbed. This input energy can be calculated by subtracting the kinetic energy of both objects at this point in time from the system kinetic energy prior to impact as follows:

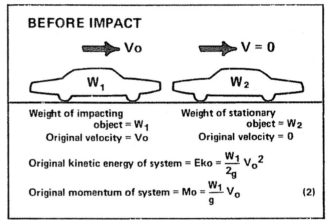

BEFORE IMPACT

Weight of impacting object = W_1

Weight of stationary object = W_2

Original velocity = Vo

Original velocity = 0

Original kinetic energy of system = $E_{ko} = \frac{W_1}{2g} V_o^2$

Original momentum of system = $M_o = \frac{W_1}{g} V_o$ (2)

Fig. 11

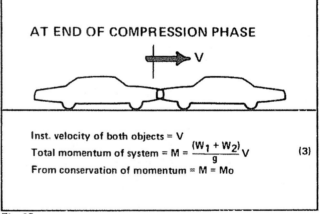

AT END OF COMPRESSION PHASE

Inst. velocity of both objects = V

Total momentum of system = $M = \frac{(W_1 + W_2)}{g} V$ (3)

From conservation of momentum = $M = M_o$

Fig. 12

Therefore: $V = V_o \frac{(W_1)}{(W_1 + W_2)}$

Inst. kinetic energy of system = (4)

$$E_{k2} = \frac{(W_1 + W_2)}{2g} V^2 = \frac{(W_1)^2 V_o^2}{2g (W_1 + W_2)}$$

Input energy = $E = E_{ko} - E_{ki}$

$$E = \frac{W_1}{2g} V_o^2 - \frac{(W_1)^2 V_o^2}{2g (W_1 + W_2)} = \frac{(W_1 \cdot W_2)}{2g (W_1 + W_2)} V_o^2$$

Therefore: $E = \frac{W_2}{W_1 + W_2} E_{ko}$ (5)

Fig. 13

741

Equation (5) gives the total input energy to both colliding objects. The relative energy absorption characteristics of the two objects determine what percentage of the total input energy each absorbs. If, for example, a moving vehicle directly impacts a stationary vehicle of equal weight, the total input energy shared between them is equal to 50% of the striking vehicle's original kinetic energy. Further, if the energy absorption characteristics the two vehicles were identical, each would receive an input energy equal to 25% of the original kinetic energy.

It should be noted that all calculations presented here apply only to the compression phase of impact, when maximum load and deflection values occur. Since component rebound is of no consequence in this phase of impact, the term "energy absorption" will be applied to elastic storage as well as to non-elastic dissipation of energy.

C. Indirect Impact

When an object is hit such that the line of action of the impact force does not pass through the C.G., rotation is introduced, complicating the problem. The effect of this factor on input energy can be approximated by computing an "effective" weight of the object along the line of action involved and then treating the problem as a direct impact. The effective weight of a vehicle for indirect impact can be approximated as follows:

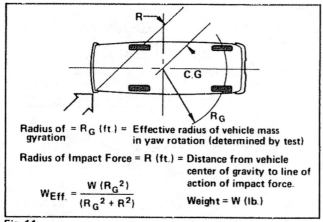

Radius of gyration $= R_G$ (ft.) = Effective radius of vehicle mass in yaw rotation (determined by test)

Radius of Impact Force $= R$ (ft.) = Distance from vehicle center of gravity to line of action of impact force.

$$W_{Eff.} = \frac{W(R_G^2)}{(R_G^2 + R^2)}$$

Weight $= W$ (lb.)

Fig. 14

Practically, R values of one foot or less can be neglected with negligible error.

II. Vehicle Structure For Energy Absorption

The bumper impact loads must be carried without yield by the basic vehicle structural members. These members deflect elastically in the process, providing a significant amount of energy absorption. Even the engine on its mounts contributes to basic vehicle energy absorption capacity as it moves during the impact and thereby absorbs energy. Due to complexities involved, it appears at this time that impact testing is the only way to determine the force and energy capacity of a vehicle structure

Chrysler conducted a series of pendulum tests to determine the energy absorbing capabilities of our vehicles. As shown in Figure 16, the test procedure involved bolting a rigid member similar to a conventional bumper support to the vehicle and impactng it at nominal bumper height with a rigid pendulum. The impact force was recorded at various increments of pendulum velocity up to a level where structural yielding occurred. The amount of energy absorbed by the structure at a given force level was determined by calculation of input energy using equations 1, 2 and 3. For purposes of bumper system performance calculations, it is useful to determine the relationship between impact force and structural energy absorption. If all the many deflections which occur in the vehicle structure were linear, and resulted from forces and moments proportional to the total impact force, then the total energy stored would be expected to be proportional to the square of total impact force. Under this assumption, an energy storage constant (C) has been defined such that

$$E_S = CF_M^2 \qquad (7)$$

Where: E_S = Energy Stored In Vehicle Structure (Ft. Lb.)
F_M = Peak Total Impact Force (Lb.)
C = Structural Energy Storage Constant (Ft./Lb.)

Fig. 15

While this relationship is not exact, test data to date indicates that it is a good approximation.

The implications of equation (7) are important in comparing pendulum and barrier impact performance. It can be reasoned that if a given vehicle can withstand a certain level of force and input energy when struck by a pendulum on one rail, then the same vehicle impacting a flat barrier with the two rails equally loaded might accept twice the pendulum force level and four times the input energy. Inertia loading on every part of the vehicle would, of course, be doubled under this condition. Our testing to date indicates that the maximum dynamic load capability of both front rails can be utilized in a barrier impact and structural energy absorption under these conditions appears to conform to equation (7).

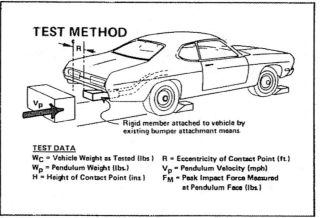

TEST METHOD

Rigid member attached to vehicle by existing bumper attachment means.

TEST DATA

W_C = Vehicle Weight as Tested (lbs.) R = Eccentricity of Contact Point (ft.)
W_P = Pendulum Weight (lbs.) V_P = Pendulum Velocity (mph)
H = Height of Contact Point (ins.) F_M = Peak Impact Force Measured at Pendulum Face (lbs.)

Fig. 16

Structural Constants To Be Determined

1. $F_{max.}$ = Maximum Safe Dynamic Load Per Rail (lb.)
2. C = Structural Energy Storage Constant (FT/Lbs.)

Calculation Procedure

Effective Weight of Car:

$$W_{CE} = W_C \left(\frac{R_G^2}{R_G^2 + R^2} \right) \text{ (Lbs.)}$$

R_G = Radius of Gyration (Ft.)

Pendulum Energy:

$$E_P = \frac{W_P \ V_P^2}{29.94} \text{ (Ft. Lbs.)}$$

Input Energy:

$$E_i = \frac{E_P \ W_{CE}}{(W_P + W_{CE})} \text{ (Ft. Lbs.)}$$

Energy Storage Constant: $C = \dfrac{E_i}{F_M^2} \ \dfrac{(Ft.)}{(Lb.)}$

III. Sample Calculations

The 1973 Chrysler Corporation passenger cars are designed to meet or exceed the requirements of the new federal standard by mounting the bumper to the vehicle with cold formed steel supports and providing molded elastomeric bumper guards as standard equipment. Molded elastomeric guards can be designed to absorb moderate amounts of energy at maximum rail strength levels. A capacity of 500 to 600 ft. lbs. appears to be the practical limit above which size becomes objectionable. A pair of these guards provide sufficient capacity to raise the performance level of cold formed steel support systems to the necessary impact speeds. Considering armature and ineffective material weight, density values of 200 to 250 ft. lbs. of energy per pound of guard weight are attainable. To illustrate how input energy can be calculated and how this energy is absorbed, the following sample calculation has been provided.

Consider an 'A' Body (Valiant or Dart) front bumper system that is mounted to the vehicle with cold formed supports and designed with molded elastomeric energy absorbing pads. The system is to be designed to withstand a 5 m.p.h. barrier impact. The impact should not damage either the bumper system or any other part of the car.

Input Energy:

$$E_i = \frac{W_P \ V_o^2}{29.94}$$

$$E_i = \frac{(3500) \ (5.0)^2}{29.94} = 2900 \text{ Ft. Lb.}$$

Energy Stored In The Structure

$E_S = C \ (F_M)^2$

$E_s = (1.95 \times 10^{-6}) \ (32,000)^2 = 2000$ Ft. Lbs.

F_{Max} = 32,000 — based upon maximum permissible rail loads of 16,000 lbs. per rail

C for this 'A' Body vehicle = 1.95×10^{-6} ft./lb.

Resultant energy to be absorbed by rubber compression pads.

2900 ft. lbs. – 2000 ft. lbs. = 900 ft. lbs.

Energy absorption per pad   450 ft. lbs.

@ 16,000 lb. Max. Force

IV. The Design Of Elastomeric Energy Absorbing Pads

The actual dimensions of the rubber pads can now be considered as an optimization procedure. Sufficient contact area must be provided to reduce compressive stresses within tolerable limits and yet not require excess rubber.

As shown in the preceding example, a typical pad must be designed to absorb a portion of the kinetic energy of the vehicle without exceeding a specified force level. While a vertical pad height of approximately 5 inches is required to assure contact with other vehicle bumpers, it is desirable to minimize pad width and depth to keep vehicle cost and length increases to a minimum. For this reason, extensive testing of various elastomers was carried out to determine the optimum pad material, size and shape. Materials were compared on the basis of efficiency of energy absorption and space utilization.

Testing was performed dynamically by mounting sample pads to a rigid vertical surface and impacting them with an instrumented pendulum at 5 m.p.h. The pendulum weight was varied to produce impact energy levels ranging from 200 to 800 ft. lbs. This energy range covers the pad absorption requirements for all 1973 Chrysler built vehicles. A schematic of the pad test fixture is shown below.

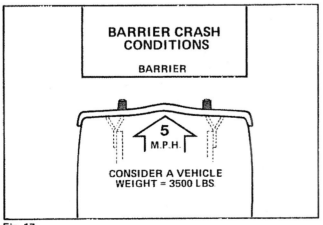

Fig. 17

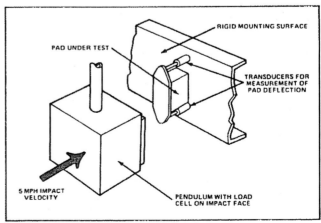

Fig. 18

The following data was then recorded from each test:

A = cross-sectional area of pad
D = Depth of pad in direction of impact
E_i = Input energy (kinetic energy of pendulum less small correction factor for deflection of pad mounting surface)
$F_{max.}$ = Peak force recorded during impact
$C_{max.}$ = Maximum pad deflection recorded during impact

From this data the following calculations were made:

Max. compressive stress $\quad S_{max.} = F_{max.}/A$

Max. percent compression $d_{max} = (C_{max.}/D) \times 100\%$

Depth Efficiency $\quad E_D = \dfrac{E_i}{F_{max.} \times D} \times 100\%$

E_D is an index of the efficiency with which available pad depth is utilized. A pad of 100% depth efficiency would absorb energy at a constant force level and deflect 100% of its original depth.

Energy per unit volume $\quad E_v = \dfrac{E_i}{A \times D} = \dfrac{E_d}{100\%} \times S_{max.}$

Data Summary:
The following graph gives depth efficiency vs. compressive stress for EPDM, butyl and S.B.R. materials over a -20°F. to $+158^\circ$F. temperature range. Although samples of various hardnesses were tested, data is presented here for only optimum hardness values. (Complete data is included in the appendix.)

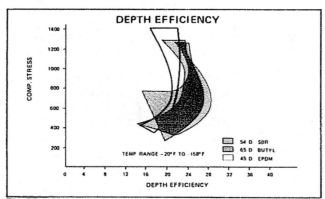
Fig. 20

It can be seen that although the three materials are quite close in their depth efficiency, S.B.R. (Styrene-Butadiene Rubber) is superior through most of the stress range. Depth efficiency is considered important when the plan view or top evaluation of the bumper is very flat and the pad can be deflected without the barrier coming in contact with other parts of the bumper and preventing further deflection.

Althought the efficiency that the guard depth can be utilized is important, the cost efficiency or efficiency of material used is also very important. This is important in the majority of bompers where the guard size is dictated by the bumper configuration. In these cases, volume efficiency is very important. To evaluate this factor a plot of energy absorbed per volume was made vs. compressive stress. From this plot it can also be shown that the S.B.R. rubber is superior, but again by a small margin.

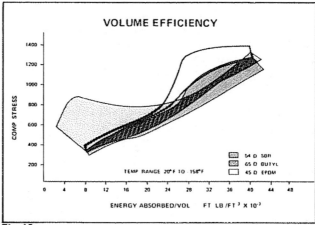
Fig. 19

From this information a selection of S.B.R. rubber was made and the size and shape of each guard was calculated to provide the necessary energy absorption for each bumper shape and energy level requirements.

V. Conclusions
1. Chrysler Corporation has designed its 1973 Bumper Systems to meet and exceed the 1973 DOT exterior protection standards by methodically analyzing the impact event and developing an efficient energy management scheme which takes full advantage of the vehicle as an energy absorber. Additional energy absorption is provided by the efficient use of elastomeric bumper guards. This system has proven to be both theoretically and practically efficient for impact speeds up to 5 m.p.h.

2 There are several materials that can be considered for use as impact pads. These materials have various properties over the full temperature range but show efficiencies ranging from 25% to 40% at the effective stress levels.

3 Although bumper systems will be improved industry wide in 1973 and further improved in 1974, a dramatic drop in property damage should not be expected for several years. This will not be because the government standard is not strict enough, but because it will take some time for vehicles with improved systems to gain a large enough percentage of total vehicle population to make the effects felt in the statistics.

4 A general trend of increased consumer protection benefits in low speed collisions should begin as the 1973 and later model cars take the road and older model cars become phased out of operation.

744

Appendix
Elastomeric Pad Test Data — 5 M.P.H. Impact
Std. sample sizes:
2 in. Pad — D = 2.05 in., A = 12.70 in.^{2}
3 in. Pad — D = 3.10 in., A = 11.85 in.^{2}

| Test Data | | | | | | | Calculated Values | | | |
Material	Hardness Durometer	Sample Size	Temp. ($^\circ$F.)	Input Energy (Ft. Lb.)	Max. Force (Lb.)	Max. Defl. (In.)	Max. Stress (PSI)	Max. Defl. (%)	Depth Efficiency (%)	Energy Per Unit Volume (Ft.Lb./Ft.)
EPDM	45	2 in.	−20	182	6400	.55	504	26.8	16.6	12040
EPDM	62	2 in.	−20	151	8500	.80	669	14.6	10.4	9974
EPDM	74	2 in.	−20	113	10500	.20	827	9.8	6.3	7457
EPDM	45	3 in.	−20	195	5300	.85	447	27.4	14.2	9141
EPDM	61	3 in.	−20	160	7900	.40	667	12.9	7.9	7531
EPDM	71	3 in.	−20	135	9350	.25	789	8.1	5.6	6357
EPDM	45	2 in.	−20	334	9500	.86	748	42.0	20.6	22146
EPDM	62	2 in.	−20	320	10250	.55	807	26.8	18.2	21159
EPDM	74	2 in.	−20	309	10750	.38	847	18.5	16.8	20464
EPDM	45	3 in.	−20	372	7250	1.27	612	41.0	19.9	17465
EPDM	61	3 in.	−20	350	8650	.80	730	25.8	15.7	16423
EPDM	71	3 in.	−20	325	10000	.55	844	17.7	12.6	15239
EPDM	45	2 in.	−20	591	15900	1.13	1252	55.1	21.7	39113
EPDM	62	2 in.	−20	591	15900	.85	1252	41.5	21.7	39113
EPDM	74	2 in.	−20	558	16900	.65	1331	31.7	19.3	36940
EPDM	45	3 in.	−20	673	13050	1.75	1101	56.5	20.0	31601
EPDM	61	3 in.	−20	664	13400	1.25	1131	40.3	19.2	31164
EPDM	71	3 in.	−20	621	14900	.87	1257	28.1	16.1	29174
EPDM	45	2 in.	70	188	5875	1.10	463	53.7	18.8	12470
EPDM	62	2 in.	70	180	6525	.79	514	38.5	16.2	11934
EPDM	74	2 in.	70	172	7125	.56	561	27.3	14.1	11391
EPDM	45	3 in.	70	204	4375	1.64	369	52.9	18.0	9563
EPDM	61	3 in.	70	196	5175	1.20	437	38.7	14.7	9202
EPDM	71	3 in.	70	185	6175	.97	521	31.3	11.6	8671
EPDM	45	2 in.	70	342	9075	1.20	715	38.7	22.1	22662
EPDM	62	2 in.	70	331	9700	.88	764	42.9	19.9	21887
EPDM	74	2 in.	70	325	10000	.67	787	32.7	19.0	21497
EPDM	45	3 in.	70	373	7200	1.86	608	60.0	20.0	17502
EPDM	61	3 in.	70	361	7975	1.40	673	45.2	17.5	16948
EPDM	71	3 in.	70	356	8300	1.02	700	32.9	16.6	16700
EPDM	45	2 in.	70	561	16800	1.35	1323	65.9	19.6	37166
EPDM	62	2 in.	70	534	17600	1.06	1386	51.7	17.7	35378
EPDM	74	2 in.	70	541	17400	.81	1370	39.5	18.2	35801
EPDM	45	3 in.	70	623	14850	2.05	1253	66.1	16.2	29244
EPDM	61	3 in.	70	615	15100	1.61	1274	51.9	15.8	28873
EPDM	71	3 in.	70	603	15500	1.27	1308	41.0	15.1	28319
EPDM	45	2 in.	158	184	6200	1.17	488	57.1	17.4	12212
EPDM	62	2 in.	158	198	5000	.91	394	44.4	23.2	13099
EPDM	74	2 in.	158	185	6150	.80	484	39.0	17.6	12252
EPDM	45	3 in.	158	201	4675	1.75	395	56.5	16.6	9432
EPDM	61	3 in.	158	203	4475	1.45	378	46.8	17.5	9521
EPDM	71	3 in.	158	198	4950	1.13	418	36.5	15.5	9310
EPDM	45	2 in.	158	334	9500	1.25	748	61.0	20.6	22145
EPDM	62	2 in.	158	348	8750	1.15	689	56.1	23.3	23046
EPDD	74	2 in.	158	355	8375	.85	659	41.5	24.8	23477
EPDM	45	3 in.	158	372	7250	1.95	612	62.9	19.9	17465
EPDM	61	3 in.	158	373	7175	1.68	565	54.2	20.1	17516
EPDM	71	3 in.	158	370	7225	1.35	627	43.5	19.3	17347
EPDM	45	2 in.	158	541	17400	1.40	1370	68.3	18.2	35801
EPDM	62	2 in.	158	563	16750	1.25	1319	61.0	19.7	37272
EPDM	74	2 in.	158	558	16900	1.05	1331	51.2	19.3	36940
EPDM	45	3 in.	158	603	15500	2.10	1308	67.7	15.1	28319
EPDM	61	3 in.	158	629	14650	1.90	1296	61.3	16.6	29484
EPDM	71	3 in.	158	623	14850	1.56	1253	50.3	16.2	29244

Appendix
Elastomeric Pad Test Data — 5 M.P.H. Impact

Std. sample sizes:
2 in. Pad — D = 2.05 in., A = 12.70 in.^{2}
3 in. Pad — D = 3.10 in., A = 11.85 in.^{2}

Test Data							Calculated Values			
Material	Hardness Durometer	Sample Size	Temp. (°F)	Input Energy (Ft. Lb.)	Max. Force (Lb.)	Max. Defl. (In.)	Max. Stress (PSI)	Max. Defl. (%)	Depth Efficiency (%)	Energy Per Unit Volume (Ft.Lb./Ft.)
Butyl	43	2 in.	−20	172	7100	.40	559	19.5	14.2	11417
Butyl	55	2 in.	−20	133	9500	.17	748	8.3	17.7	8781
Butyl	65	2 in.	−20	96	11250	.15	886	7.3	5.0	6371
Butyl	43	3 in.	−20	197	5050	.67	426	21.6	15.1	9263
Butyl	54	3 in.	−20	140	9100	.25	768	8.1	6.0	6573
Butyl	60	3 in.	−20	116	10350	.25	873	8.1	4.3	5432
Butyl	43	2 in.	−20	368	7500	.95	591	46.3	48.7	24397
Butyl	55	2 in.	−20	342	9100	.52	717	25.4	22.0	22636
Butyl	65	2 in.	−20	325	10000	.40	787	19.5	19.0	21497
Butyl	43	3 in.	−20	393	5600	1.25	473	40.3	27.2	18460
Butyl	54	3 in.	−20	368	7500	.65	633	21.0	19.0	17296
Butyl	60	3 in.	−20	346	8850	.60	747	19.4	15.1	16258
Butyl	43	2 in.	−20	626	14750	1.20	1161	58.5	24.8	41444
Butyl	55	2 in.	−20	635	14450	.90	1138	43.9	25.7	42026
Butyl	65	2 in.	−20	614	15150	.85	1193	41.5	23.7	40656
Butyl	43	3 in.	−20	720	11100	1.85	937	59.7	25.1	33812
Butyl	54	3 in.	−20	711	11500	1.25	971	40.3	23.9	33390
Butyl	60	3 in.	−20	687	12500	1.00	1055	32.3	21.3	32263
Butyl	43	2 in.	70	177	6750	.87	532	42.4	15.4	11735
Butyl	55	2 in.	70	191	5675	.83	447	40.5	19.7	12623
Butyl	62	2 in.	70	191	5650	.67	445	32.7	19.8	12642
Butyl	43	3 in.	70	203	4475	1.65	378	53.2	17.5	9521
Butyl	54	3 in.	70	204	4300	1.25	363	40.3	18.4	9592
Butyl	60	3 in.	70	203	4475	.97	378	31.2	17.5	9521
Butyl	43	2 in.	70	339	9275	1.22	730	59.5	21.4	22424
Butyl	55	2 in.	70	351	8575	.88	675	42.9	24.0	23252
Butyl	65	2 in.	70	351	860	.78	677	38.0	23.9	23219
Butyl	43	3 in.	70	367	7600	1.78	641	57.4	18.7	17221
Butyl	54	3 in.	70	377	6900	1.40	582	45.2	21.2	17700
Butyl	60	3 in.	70	378	6850	1.22	578	39.4	21.3	17732
Butyl	43	2 in.	70	590	15925	1.21	1254	59.0	21.7	39060
Butyl	55	2 in.	70	578	16300	1.04	1284	50.7	20.7	38259
Butyl	65	2 in.	70	587	16000	1.02	1260	49.8	21.5	38901
Butyl	43	3 in.	70	618	15000	1.92	1266	61.9	16.0	29033
Butyl	54	3 in.	70	668	13250	1.66	1118	53.5	19.5	31352
Butyl	60	3 in.	70	681	12750	1.52	1076	49.0	20.7	31962
Butyl	43	2 in.	158	189	5825	1.20	459	58.5	19.0	12510
Butyl	55	2 in.	158	194	5400	1.05	425	51.2	21.0	12821
Butyl	65	2 in.	158	194	5375	.95	423	46.3	21.1	12841
Butyl	43	3 in.	158	204	4300	1.90	363	61.3	18.4	9592
Butyl	54	3 in.	158	206	4075	1.50	344	48.4	19.6	9681
Butyl	60	3 in.	158	206	4075	1.43	344	46.1	19.6	9681
Butyl	43	2 in.	158	348	8750	1.40	689	68.3	23.3	23046
Butyl	55	2 in.	158	355	8325	1.20	656	58.5	25.0	23530
Butyl	65	2 in.	158	361	7975	1.15	628	56.1	26.5	23907
Butyl	43	3 in.	158	377	6925	2.13	584	68.7	21.1	17681
Butyl	54	3 in.	158	381	6575	1.85	555	59.7	22.5	17906
Butyl	60	3 in.	158	383	6450	1.65	544	53.2	23.0	17981
Butyl	43	2 in.	158	570	16350	1.46	1287	71.2	20.6	38152
Butyl	55	2 in.	158	592	15850	1.35	1248	65.9	21.9	39219
Butyl	65	2 in.	158	594	15800	1.30	1244	63.4	22.0	39325
Butyl	43	3 in.	158	638	14350	2.35	1211	75.8	17.2	29930
Butyl	54	3 in.	158	664	13400	2.10	1131	67.7	19.2	31164
Butyl	60	3 in.	158	665	13350	1.95	1127	62.9	19.3	31230

Appendix
Elastomeric Pad Test Data — 5 M.P.H. Impact

Std. sample sizes:
2 in. Pad — D = 2.05 in., A = 12.70 in.^{2}
3 in. Pad — D = 3.10 in., A = 11.85 in.^{2}

Test Data							Calculated Values			
Material	Hardness Durometer	Sample Size	Temp. (°F.)	Input Energy (Ft. Lb.)	Max. Force (Lb.)	Max. Defl. (In.)	Max. Stress (PSI)	Max. Defl. (%)	Depth Efficiency (%)	Energy Per Unit Volume (Ft.Lb./Ft.)
SBR	54	2 in.	−20	175	6900	.45	543	22.0	14.9	11603
SBR	62	2 in.	−20	172	7100	.40	559	19.5	14.2	11417
SBR	75	2 in.	−20	134	9400	.25	740	12.2	8.4	8901
SBR	54	3 in.	−20	197	5100	.74	430	23.9	14.9	9239
SBR	61	3 in.	−20	191	5650	.70	477	22.6	13.1	8962
SBR	74	3 in.	−20	187	6000	.62	506	20.0	12.1	8770
SBR	54	2 in.	−20	344	9000	.75	709	36.6	22.3	22755
SBR	62	2 in.	−20	344	9000	.68	709	33.2	22.3	22755
SBR	75	2 in.	−20	311	10650	.42	839	20.5	17.1	20609
SBR	54	3 in.	−20	376	7000	1.23	591	39.7	20.8	17634
SBR	61	3 in.	−20	368	7500	1.10	633	35.5	19.0	17296
SBR	74	3 in.	−20	348	8750	.87	738	28.1	15.6	16338
SBR	54	2 in.	−20	608	15350	1.12	1209	54.6	23.2	40252
SBR	62	2 in.	−20	591	15900	1.04	1252	50.7	21.7	39113
SBR	75	2 in.	−20	566	16650	.78	1311	38.0	19.9	37497
SBR	54	3 in.	−20	703	11850	1.74	1000	56.1	23.0	33005
SBR	61	3 in.	−20	690	12400	1.53	1046	49.4	21.5	32376
SBR	74	3 in.	−20	652	13850	1.18	1169	38.1	18.2	30592
SBR	54	2 in.	70	196	5200	.49	409	45.9	22.0	12967
SBR	62	2 in.	70	193	5500	.82	433	40.0	20.5	12755
SBR	75	2 in.	70	183	6275	.62	494	30.2	17.1	12146
SBR	54	3 in.	70	207	3925	1.40	331	45.2	20.5	9737
SBR	61	3 in.	70	204	4375	1.20	369	38.7	18.0	9563
SBR	74	3 in.	70	197	5125	.66	433	21.3	14.8	9225
SBR	54	2 in.	70	365	7750	1.03	610	50.2	27.5	24139
SBR	62	2 in.	70	364	7800	1.00	614	48.8	27.3	24093
SBR	75	2 in.	70	350	8625	.75	679	36.6	23.8	23192
SBR	54	3 in.	70	392	5750	1.70	485	54.8	26.4	18380
SBR	61	3 in.	70	387	6175	1.40	521	45.2	22.2	18146
SBR	74	3 in.	70	376	7000	1.05	591	33.9	20.8	17634
SBR	54	2 in.	70	638	14350	1.30	1130	63.4	26.0	42219
SBR	62	2 in.	70	636	14400	1.15	1134	56.1	25.9	42219
SBR	75	2 in.	70	591	15900	.93	1252	45.4	21.7	39113
SBR	54	3 in.	70	698	12050	2.10	1017	67.7	22.4	32779
SBR	61	3 in.	70	686	12550	1.82	1059	58.7	21.2	32202
SBR	74	3 in.	70	654	13750	1.40	1160	45.2	18.4	30718
SBR	54	2 in.	158	191	5600	1.05	441	51.2	20.0	12675
SBR	62	2 in.	158	192	5575	.94	439	45.9	20.0	12695
SBR	75	2 in.	158	183	6375	.71	502	34.6	16.7	12066
SBR	54	3 in.	158	206	4050	1.65	342	53.2	19.7	9690
SBR	61	3 in.	158	203	4500	1.35	380	43.5	17.4	9512
SBR	74	3 in.	158	196	5200	1.05	439	33.9	14.6	9192
SBR	54	2 in.	158	357	8250	1.25	650	61.0	25.3	23609
SBR	62	2 in.	158	361	8000	1.12	630	54.6	26.4	23881
SBR	75	2 in.	158	351	8600	.85	677	41.5	23.9	23219
SBR	54	3 in.	158	386	6200	1.85	523	59.7	24.1	18131
SBR	61	3 in.	158	382	6500	1.60	549	51.6	22.8	17953
SBR	74	3 in.	158	371	7300	1.25	616	40.3	19.7	17432
SBR	54	2 in.	158	591	15900	1.45	1252	70.7	21.7	39113
SBR	62	2 in.	158	592	15850	1.28	1338	62.4	21.9	39219
SBR	75	2 in.	158	573	16450	1.05	1295	51.2	20.4	37934
SBR	54	3 in.	158	668	13250	2.35	1118	75.8	19.5	31352
SBR	61	3 in.	158	671	13150	1.95	1110	62.9	19.7	31479
SBR	74	3 in.	158	650	13900	1.50	1173	48.4	18.1	30526

A Study of Automotive Energy-Absorbing Bumpers

Hai Wu
Scientific Research Staff, Ford Motor Co.

AUTOMOTIVE BUMPER STANDARDS have been established through government regulations (1).* For the 1973 model year the standards require all new passenger cars to pass barrier tests without damage to safety-related components of the vehicle in a 5 mph front or a 2.5 mph rear impact. In order to meet the standards, the vehicle bumper systems must have the energy-absorbing capability, among other requirements. Various energy-absorbing devices, that is, impact absorbers, are fitted into the automotive bumper systems used by the automobile manufacturers on their 1973 models (2-7)

Previous studies (4-11) were performed mostly either for a particular type of impact absorbers or for a nonbumper application. This paper, however, introduces several general criteria concerning the design and performance of various impact-absorbing units for automotive bumper applications. The analysis also discusses the problem of impact protection for the case of a low-speed, two-car, head-on collision. The objective of the paper is to provide an analytical outline of the parameters involved. It is expected that the data reported here will be useful in evaluating and selecting different types of impact absorbers. However, the paper does not consider problems related to detail design, production, reliability, cost, or similar factors.

---

*Numbers in parentheses designate References at end of paper.

UNIT DESIGN ENERGY

The impact energy of a vehicle collision is simply the kinetic energy of the vehicle at the instant of impact. It is proportional to the vehicle mass but also to the square of the impact velocity. For inch-pound-second units,

$$E_i = \frac{1}{2}\left(\frac{M}{386}\right)\left(\frac{88V_o}{5}\right)^2 \qquad (1)$$

where:

M = vehicle mass, lb
V_o = impact speed, mph

Fig. 1 shows impact energy versus impact velocity for various vehicle weights

In general, the car frame is capable of storing certain amounts of energy, E_f, without damage; the remaining impact energy is necessarily stored by the impact absorber units. The unit design energy is defined as

$$E_d = \frac{(E_i - E_f)/1_1}{N/1_2} \qquad (2)$$

ABSTRACT

This paper presents an analysis of various energy-absorbing units for automotive bumper applications. It also introduces several general criteria concerning their design and performance to provide an analytical outline of the parameters involved. An example case of head-on impact between two vehicles with mass ratio of 1:2 reveals the effect of heavy vehicle absorber design upon lighter vehicle damageability.

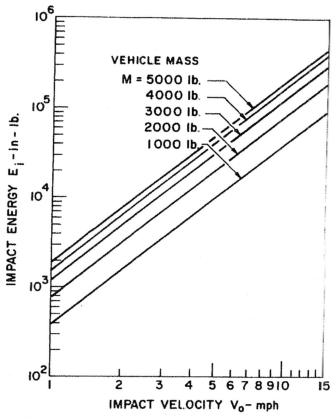

Fig. 1 - Impact energies for different vehicles at various impact velocities

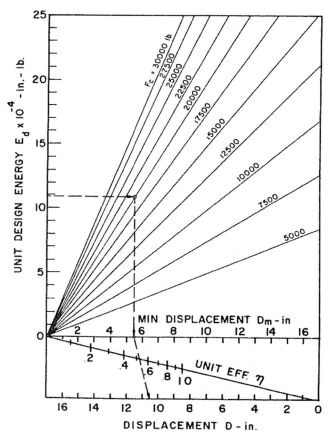

Fig. 2 - Minimum displacements

where:

N = number of units

I_1 = barrier test design constant, one

I_1 = pendulum test design (same mass) constant, two

I_2 = uniformly distributed load design (square hit) constant, one

I_2 = unevenly distributed load design constant, > 1

UNIT SYSTEM EFFECTIVENESS

The impact absorber unit transforms this kinetic energy into other forms, such as potential energy. In general, this process is expressed as

$$E_d = \int F_b \, dX_b \qquad (3)$$

where:

F_b = resisting force

dX_b = displacement

Besides other considerations, there are two overriding requirements in designing the energy absorber units. First, the maximum resisting force of the unit during the impact under design conditions should not at any time exceed a certain

critical load: $F_M \leq F_c$. The critical load F_c can be, for instance, the critical frame load beyond which unacceptable damage will occur. Secondly, the total displacement (stroke) D should be kept in a certain range for styling, stability, and other considerations. It is stipulated in this analysis that the optimal design unit represents one whose resisting force is constant and equal to the critical load of the vehicle:

$$E_d = F_c \cdot D_m \qquad (4)$$

The displacement so determined is called the minimum stroke D_m. It is the limit the absorber designers must have in order to design their units to protect the vehicle from damage during impact under design conditions. The minimum strokes for various systems are shown in Fig. 2, with the unit system effectiveness defined as

$$\eta = \frac{D_m}{D} \qquad (5)$$

Fig 3 shows the unit system effectiveness for several typical impact absorber units. It is seen that not only the shape of the force-displacement curve but also the magnitude of the maximum resisting force F_M determines the unit effectiveness. For units with low unit-design energies (most 2 5 mph units), the

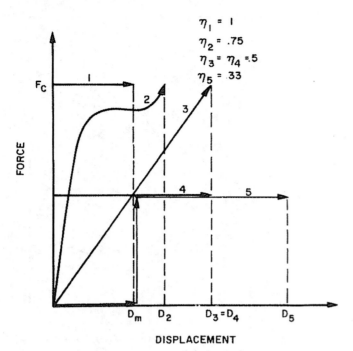

$\eta_1 = 1$
$\eta_2 = .75$
$\eta_3 = \eta_4 = .5$
$\eta_5 = .33$

Fig. 3 - Typical unit system effectiveness

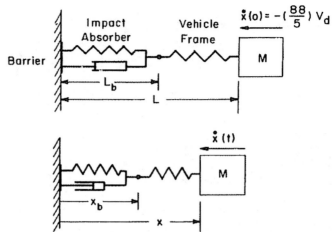

Fig 4 - Coordinates for barrier design

stroke is usually small and does not pose a severe design problem. The unit system effectiveness, consequently, is not a critical design factor. However, as the unit design energy increases because of high speed or of heavier vehicle impact, the significance of the unit system effectiveness increases and becomes critical for units with very large unit-design energies.

For instance, a 10 mph unit for a full-size car ($E_d = 176,000$ in-lb and $F_c = 26,600$ lb) has a minimum stroke $D_m = 7$ in, which is a definite problem in styling, bumper jacking, parking length, and so forth. If this unit has an effectiveness $\eta = 0.5$, and thus needs a stroke $D = 14$ in, one can certainly imagine the difficulty involved.

COEFFICIENT OF RESTITUTION

The unit coefficient of restitution is defined as

$$\xi = 1 - \frac{E_a}{E_d} \qquad (6)$$

where E_a is the amount of energy absorbed and dissipated by the unit during impact.

For the case $\xi = 0$, all stored energy is absorbed and dissipated by the unit, and the unit must be either replaced or restored to its original state by certain external means after each impact (for instance, a friction type absorber). This is not desirable in an automobile bumper because of the inconvenience and the lack of continuous protection. For the case $\xi = 1$, the unit dissipates no energy. The vehicle, at least in theory, will bounce back after a barrier impact at the same speed with which it hit (for instance, linear spring without damping).

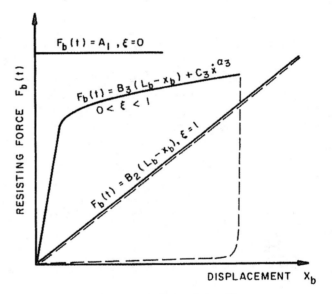

Fig. 5 - Typical force deflection curves

For all the other cases, $0 < \xi < 1$, part of the stored energy is absorbed and dissipated, the unit can be self-restored, and the vehicle will bounce back somewhat after impact (for instance, linear spring with damping and hydraulic shock absorber). For the same reasons mentioned previously, it seems desirable to have a unit that absorbs and dissipates most of the energy and yet preserves the self-restoring property.

For most impact absorber units, the resisting forces (Fig. 4) can be described as

$$F_b(t) = A + B(L_b - X_b) + C\dot{X}_b^{\alpha} \qquad (7)$$

where A, B, C, and α are four constants. The constant-force type of absorber (such as a shock rod) has force characteristics

750

of

$$F_b(t) = A_1 = \text{constant} \tag{8}$$

Linear springs without damping, on the other hand, have

$$F_b(t) = B_2(L_b - X_b) \tag{9}$$

These two types of impact absorber units are not velocity-sensitive. Most hydraulic impact units, however, are velocity-sensitive and have the general force characteristics given in Eq. 7. The force-deflection curves of these three cases are shown in Fig. 5.

DESIGN FOR BARRIER IMPACT (SQUARE HIT)

Since for the same impact speed, the barrier impact is more severe than the pendulum impact (assuming the vehicle is free to move), the design for barrier impact is considered here as the standard design condition. The maximum impact speed a vehicle can withstand without damage* under barrier impact conditions is then called "barrier design speed" V_d (in mph units). For one-dimensional consideration, the process (Fig. 4) can be expressed as

$$\ddot{X}(t) = \frac{F(t)}{(M/386)} \tag{10}$$

$$\dot{X}(t) = -\frac{88}{5} V_d + \int_0^t \ddot{X}(t)\, dt \tag{11}$$

$$X(t) = L - \int_0^t \dot{X}(t)\, dt \tag{12}$$

where $F(t)$ is the resisting force and L is a certain reference equilibrium length (for instance, the length between the vehicle center of gravity and the bumper face). The initial conditions are (Fig. 4)

$$\ddot{X}(0) = 0 \tag{13}$$

$$\dot{X}(0) = -\frac{88}{5} V_d \tag{14}$$

$$X(0) = L \tag{15}$$

The end conditions for the impact stroke are

*Since this analysis is limited to the discussion of impact absorber units, it is assumed here that damage occurs whenever the resisting force exceeds the critical load F_c. The extent of damage is not considered.

$$\ddot{X}(t_1) = \ddot{X}(t_1) \tag{16}$$

$$\dot{X}(t_1) = 0 \tag{17}$$

$$X(t_1) = X(t_1) \tag{18}$$

and those for the recovery stroke are

$$\ddot{X}(t_2) = 0 \tag{19}$$

$$\dot{X}(t_2) = \dot{X}(t_2) \tag{20}$$

$$X(t_2) = L \tag{21}$$

Depending upon the form of resisting force $F(t)$, the problem can be solved either analytically or numerically (Appendices A and B).

Within the yield limit, the car frame may be treated as perfectly elastic, or

$$F_f(t) = K_f[(L - X) - (L_b - X_b)] \tag{22}$$

As a first-order approximation, Eq. 7 may also be written as

$$F_b(t) = \frac{K_f}{N}(L - X), \qquad F_b < A$$
$$F_b(t) = A + B(L_b - X_b) + C\dot{X}^\alpha, \qquad F_b(t) \geq A \tag{23}$$

Since the resisting force for the frame and that for the impact absorber units must be equal, calculation gives

$$F(t) = K_f(L - X), \qquad F(t) < NA$$
$$F(t) = \frac{NK_f}{K_f + NB}[B(L - X) + A + C\dot{X}^\alpha], \qquad F(t) \geq NA \tag{24}$$

$$X_b = L_b - \frac{1}{K_f + NB}[K_f(L - X) - NA - NC\dot{X}^\alpha] \tag{25}$$

As a check, the impact energy can be calculated by a simple integration:

$$E_i = \int_{t=0}^{t=t_1} F(t)\, dX \tag{26}$$

A major portion of the impact energy is absorbed in the impact absorber units (the bumper system), which is given by

$$N \cdot E_d = \int_{t=0}^{t=t_1} F(t)\, dX_b \tag{27}$$

The remaining portion is absorbed by the frame.

751

ENERGY-ABSORBING BUMPERS

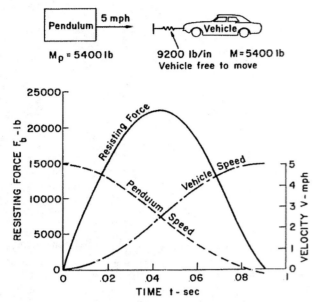

Fig. 6 - Characteristics of a pendulum hit

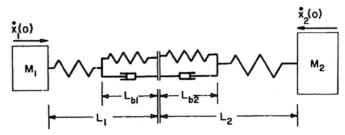

Fig. 7 - Model for low-speed, two-car, head-on collision

PENDULUM TEST (SQUARE HIT)

For a vehicle equipped with the barrier-designed, energy-absorbing bumper system to survive the pendulum test, one of the following three conditions must be met (assuming the vehicle is free to move and the bumper itself is strong enough):

1. Whatever the size of the pendulum mass M_p no damage will occur if the pendulum speed is less than or equal to the barrier design speed V_d as defined by Eq. 11:

$$V_p \leqslant V_d \tag{28}$$

2. For energy-absorbing systems that are not velocity-sensitive ($C = 0$), the speed limit is (2)

$$V_p \leqslant \left(\frac{M + M_p}{M_p}\right)^{1/2} V_d \tag{29}$$

If the pendulum and the vehicle have the same mass, $M_p \doteq M$, Eq. 29 simply reduces to

$$V_p \leqslant \sqrt{2} \; V_d \tag{30}$$

3. For energy-absorbing systems that are velocity-sensitive ($C > 0$), in addition to the conditions of Eqs. 29 and 30, it is also required that the resisting force during pendulum hit will at no time exceed the maximum resisting force of design for barrier impact:

$$F_b(t) \leqslant F_M \tag{31}$$

An example of a 5 mph pendulum hit with $M_p = M$ is illustrated in Fig. 6.

TWO-CAR COLLISION (SQUARE HIT)

When two vehicles collide, there exists initially a relative velocity between them. By exchanging energy through the two bumper systems, this relative velocity is reduced to zero and the two vehicles will move with a common velocity at the end of impact.* The dynamics of the individual vehicle is again described by Eqs. 10-12. In addition, there is the momentum equation:

$$M_1 V_1 + M_2 V_2 = (M_1 + M_2) V_f \tag{32}$$

As sketched in Fig. 7, this coupled problem can be solved only numerically (except certain simple cases). However, if one is interested only in the performance trend rather than in details, the following points should be worth mentioning.

1. Neither vehicle will sustain any damage during collision if the relative impact velocity is less than or equal to the smaller barrier design speeds of the two vehicles.

2. The relative magnitude of the maximum resisting forces of the two energy-absorbing bumper systems involved is the main factor in determining the damageability. In general, the vehicle equipped with a bumper system that has the lower F_M value of the two is likely to be damaged during the crash. In particular, if both vehicles have bumper systems (with $\eta = 1$), then the one with the lower F_c value (critical load) is likely to be damaged.

3. In order for neither vehicle to sustain damage during a two-car crash, it is necessary to have

$$\frac{1}{2}\left(\frac{1}{386}\right)\left(\frac{88}{5}\right)^2 \frac{M_1 M_2 (V_1 - V_2)^2}{M_1 + M_2}$$

$$\leqslant \int_0^{F_{M_1}^+} F_1(t)\, dX_1 + \int_0^{F_{M_1}^+} F_2(t)\, dX_2 \tag{33}$$

provided

$$F_{M_1} \leqslant F_{M_2}$$

_____

*In the case of two vehicles of the same weight traveling at the same speed in opposite directions, the common velocity after impact will be zero.

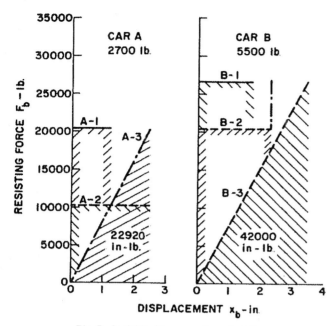

Fig. 8 - Individual bumper characteristics

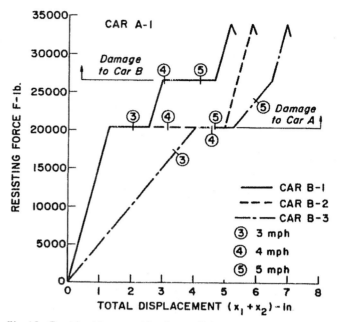

Fig. 10 - Combined low-speed head-on collision characteristics between car A-1 and cars B-1, B-2, and B-3, respectively

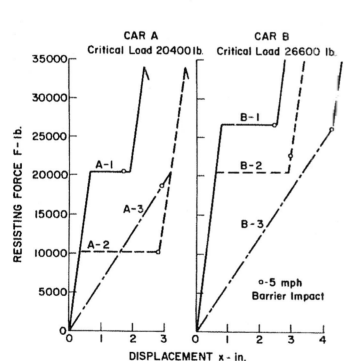

Fig. 9 - Individual vehicle low-speed impact characteristics

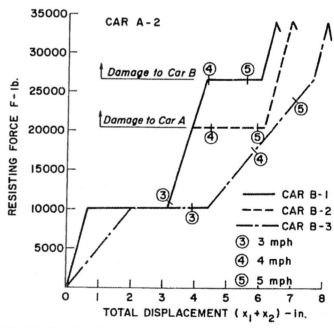

Fig. 11 - Combined low-speed head-on collision characteristics between car A-2 and cars B-1, B-2, and B-3, respectively

Examples of head-on impacts at speeds of 3, 4, and 5 mph between a subcompact car (car A) and a full-sized car (car B) is illustrated in Figs. 8-12. The weight ratio between car A and car B is about 1:2. Fig. 8 shows the individual bumper characteristics for both vehicles. The design energy for the bumper systems on car A is 22,920 in-lb and that on car B is 42,000 in-lb. The other design limits are 20,400 lb maximum force and 2.5 in maximum deflection for car A, and 26,600 lb maximum force and 3.5 in maximum deflection for car B.

Assuming the structures of both vehicles are perfectly elastic, the individual synthesized crash characteristic is shown in Fig. 9.

The results of head-on collisions of 3, 4, and 5 mph between car A and car B are shown in Figs. 10-12. The maximum force and maximum deflection for each case of head-on impact are found by simply integrating the corresponding combined vehicle-crash characteristics such that the area under the curve is equal to the energy to be absorbed. It is seen that during a

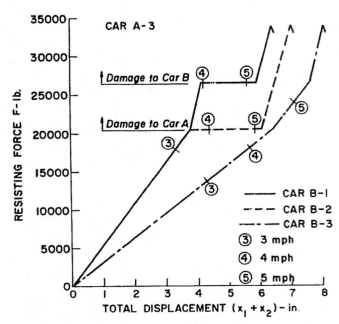

Fig. 12 - Combined low-speed head-on collision characteristics between car A-3 and cars B-1, B-2, and B-3, respectively

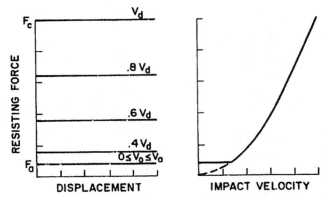

Fig. 13 - Force characteristic of an idealized impact-absorbing bumper

head-on impact at speeds up to 5 mph, car B (the full-sized car) is not damaged in any of the three cases. Car A (the subcompact car) may or may not sustain damage, depending upon the type of bumper system used on car B and, in particular, upon the magnitude of the maximum resisting force of the energy-absorbing units used on car B.

In order to protect both vehicles involved in a head-on collision at speeds up to the barrier design speed, it is necessary to limit the maximum force levels of both vehicles to the smaller design force limit of the two (in this case 20,400 lb for car A) so that Eq. 33 is satisfied. Unfortunately, this usually means lower effectiveness of the energy-absorbing system on the heavy car and results in a longer stroke. However, even though a linear system such as car B-3 has lower effectiveness, it does offer better protection to the weaker vehicle involved in a two-car head-on collision at speeds less than the barrier design speed (Figs. 10-12, curves 3 and 4).

AN IDEALIZED IMPACT-ABSORBING BUMPER

Based on the preceding discussion, one can conclude that a good fixed type of automotive impact-absorbing bumper system should require the smallest stroke possible for a given unit-design energy and unit-design force limit, and should generate the lowest resisting force possible during impact at speeds up to the unit barrier design speed for a given stroke limit. This implies a constant stroke, up to design speed. The characteristic of such an idealized system is shown in Fig. 13. It is seen that the system has the force characteristic of

$$F_b(t) = F_a, \qquad V_o \leqslant V_a \qquad (34)$$

$$F_b(t) = kV_o^2, \qquad V_a \leqslant V_o \leqslant V_d \qquad (35)$$

The cutoff point (F_a and V_a) is necessary to stabilize the system under low-load conditions and to avoid bumper NVH problems. This kind of system needs the smallest possible stroke within the design-force limit, and this should please the stylist. On the other hand, the system produces the lowest possible resisting force at impact up to the barrier design speed within the design stroke limit. This implies that the discomfort to the passengers of the impacting vehicle and the damage to the outside object it hits will be minimized.

However, as mentioned previously, these effects are not important for low impact-energy applications. Only when the impact energy becomes large will these effects become important.

NOMENCLATURE

A, B, C = constants defined in Eq. 7
D = stroke, in
D_m = minimum stroke, in
E_d = unit design energy, in-lb
E_f = energy stored by frame, in-lb
E_i = impact energy, in-lb
F = resisting force, lb
F_a = cutoff force, lb
F_b = resisting force for energy absorber, lb
F_c = vehicle critical load, lb
F_f = vehicle frame resisting force, lb
F_M = maximum resisting force of design for barrier impact, lb
I_1, I_2 = constants defined in Eq. 2
k = constant defined in Eq. 35
K_f = spring constant defined in Eq. 22, lb/in
L = reference length, in
M = vehicle mass, lb
M_p = pendulum mass, lb
t_i = time constant defined in Eq. B-11
V = vehicle velocity, mph
V_a = cutoff speed, mph

V_o = impact speed, mph

V_d = barrier design speed, mph

V_f = final velocity after two-car collision, mph

V_p = pendulum speed, mph

X = displacement, in

$\dot{X}$ = velocity, in/s

$\ddot{X}$ = acceleration, in/s^2

α = constant defined in Eq. 7

η = unit system effectiveness, defined in Eq. 5

ξ = coefficient of restitution, defined in Eq. 6

γ = constant defined in Eq. B-10

ζ = constant defined in Eq. A-4

ACKNOWLEDGMENTS

The author is grateful to A. E. Anderson, G. F. Herbert, J. E. Mayer, Jr., and R. C. Petrof for their suggestions in the preparation of this paper.

REFERENCES

1. "Exterior Protection—Passenger Cars." Federal Motor Vehicle Standard No. 215, Section 517.21 of Title 49, Code of Federal Regulations. U.S. Department of Transportation.

2. D. G. McCullough, "Recent Advancements in Materials." SAE Transactions, Vol. 77 (1968), paper 680239.

3. N. P. Chironis, "Energy-Absorbing Systems Vie for Role Behind Auto Bumpers." Product Engineering (Jan. 18, 1972).

4. G. A. Kendall, "The Menasco Energy Absorbing Unit and Its Application to Bumper Systems." Paper 710536 presented at SAE International Mid-Year Meeting, Montreal, June 1971.

5. D. P. Taylor, "Application of the Hydraulic Shock Absorber to a Vehicle Crash Protection System." Paper 710537 presented at SAE International Mid-Year Meeting, Montreal, June 1971.

6. F. J. Limbert and W. J. Persin, "Impact Testing of High-Density Semirigid Urethane Foam for Automotive Bumper Applications." SAE Transactions, Vol. 81 (1972), paper 720132.

7. L. M. Niebylski and R. J. Fanning, "Metal Foams as Energy Absorbers for Automobile Bumpers." SAE Transactions, Vol. 81 (1972), paper 720490.

8. R. C. Cline, "Are Shock Absorbers Here to Stay?" Paper S106 presented at SAE Atlanta Section, May 1958.

9. G. W. Jackson, "Fundamentals of the Direct Acting Shock Absorber." Paper 37mR presented at SAE Passenger Car Meeting, Detroit, March 1959.

10. W. G. Price, "Cushioning Railroad Impacts with Hydraulics." SAE Transactions, Vol. 71 (1963), pp. 209-217.

11. F. F. Timpner, "Vehicle Impact Analysis." Paper 710540 presented at SAE International Mid-Year Meeting, Montreal, June 1971.

APPENDIX A

VEHICLES WITH LINEAR-SPRING TYPE of ENERGY-ABSORBING BUMPER

For vehicles equipped with the linear-spring type of energy-absorbing bumper, the resisting forces for the vehicle structure and for the bumper system, respectively, are

$$F_f(t) = K_f [(L - X) - (L_b - X_b)] \tag{A-1}$$

$$F_b(t) = B(L_b - X_b) \tag{A-2}$$

By eliminating the $(L_b - X_b)$ term, one obtains

$$\ddot{X}(t) = \zeta^2 (L - X) \tag{A-3}$$

where:

$$\zeta^2 = \frac{386 K_f NB}{M(K_f + NB)} \tag{A-4}$$

With the initial condition given in Eqs. 13-15, Eq. A-3 can be solved. The crash characteristics are

$$X = L - \left(\frac{88}{5}\right)\left(\frac{V_d}{\zeta}\right) \sin \zeta t \tag{A-5}$$

$$X_b = L_b - \left(\frac{88}{5}\right)\left[\frac{K_f V_d}{(K_f + NB)\zeta}\right] \sin \zeta t \tag{A-6}$$

$$\dot{X} = -\frac{88}{5} V_d \cos \zeta t \tag{A-7}$$

$$\dot{X}_b = -\frac{88}{5} \frac{K_f V_d}{K_f + NB} \cos \zeta t \tag{A-8}$$

$$F(t) = NF_b(t) = F_f(t)$$

$$= \left(\frac{M}{386}\right)\left(\frac{88}{5} V_d\right) \zeta \sin \zeta t \tag{A-9}$$

APPENDIX B

VEHICLES WITH CONSTANT-FORCE TYPE of ENERGY-ABSORBING BUMPER

For vehicles equipped with the constant-force type of energy-absorbing bumper, the resisting forces for the vehicle structure and for the bumper system, respectively, are

$$F_f(t) = K_f [(L - X) - (L_b - X_b)] \qquad (B-1)$$

$$F_b(t) = F_f/N \qquad (F_f < NA)$$
$$= A \qquad (F_f \geqslant NA) \qquad (B-2)$$

By eliminating the $(L_b - X_b)$ term, one obtains

$$\ddot{X}(t) = \frac{386 K_f (L - X)}{M} \qquad \left(X > L - \left(\frac{NA}{K_f}\right) \right)$$
$$= \frac{386 NA}{M} \qquad \left(X \leqslant L - \left(\frac{NA}{K_f}\right) \right) \qquad (B-3)$$

With the initial condition given in Eqs. 13-15, Eq. B-3 can be solved. The crash characteristics are given below.

For $X > L - (NA/K_f)$,

$$X = L - \frac{88}{5} \left(\frac{V_d}{\gamma} \right) \sin \gamma t \qquad (B-5)$$

$$X_b = L_b \qquad (B-6)$$

$$\dot{X} = - \frac{88}{5} V_d \cos \gamma t \qquad (B-7)$$

$$\dot{X}_b = 0 \qquad (B-8)$$

$$F(t) = NF_b(t) = F_f(t)$$
$$= \frac{M}{386} \left(\frac{88}{5} V_d \right) \gamma \sin \gamma t \qquad (B-9)$$

where:

$$\gamma^2 = \frac{386 K_f}{M} \qquad (B-10)$$

For $X = L - (NA/K_f)$,

$$t_i = \frac{1}{\gamma} \sin^{-1} \left[\frac{5}{88} \frac{\gamma}{V_d} \frac{NA}{K_f} \right] \qquad (B-11)$$

For $X < L - (NA/K_f)$,

$$X = L - \frac{NA}{K_f} - \frac{88}{5} V_d \left[1 - \left(\frac{5}{88} \frac{\gamma}{V_d} \frac{NA}{K_f} \right)^2 \right]^{1/2} (t - t_i)$$
$$+ \frac{386 NA}{2M} (t - t_i)^2 \qquad (B-12)$$

$$X_b = L_b - \frac{88}{5} V_d \left[1 - \left(\frac{5}{88} \frac{\gamma}{V_d} \frac{NA}{K_f} \right)^2 \right]^{1/2} (t - t_i)$$
$$+ \frac{386 NA}{2M} (t - t_i)^2 \qquad (B-13)$$

$$\dot{X}_b = - \frac{88}{5} V_d \left[1 - \left(\frac{5}{88} \frac{\gamma}{V_d} \frac{NA}{K_f} \right)^2 \right]^{1/2} + \frac{386 NA}{M} (t - t_i) \qquad (B-14)$$

$$\dot{X}_b = \dot{X} \qquad (B-15)$$

$$F(t) = NF_b(t) = F_f(t) = NA \qquad (B-16)$$

A PRACTICAL APPROACH TO THE PROTECTION OF MOTOR VEHICLES BY THE ABSORBTION OF IMPACT ENERGY

G. PERSICKE, F. I. PLANT,
and J. R. CHILD
Road Research Limited

INTRODUCTION

It would seem that today's motorist has never had it so good; his car has air conditioning, a powerful engine, disc brakes, power assisted steering, power operated windows and even a stereophonic radio, record player or tape recorder. He is able to drive in comfort and luxury for hundreds of miles, from one end of the country to another, without even the single interruption of a traffic light.

But there is a paradox here; for although he can drive in luxury he cannot drive in safety.

Motorization has exceeded all expectations, and has led to an incresae in traffic density which is reflected in high accident figures. The suffering, loss of life and high costs of injury arising from accidents on the roads of the world can no longer be tolerated. Ways must be found to reverse the trend and systems must be developed to reduce this morally and economically indefensible toll. We may classify such systems as active, passive, palliative and preventative. Active and preventative systems generally operate to reduce the number of accidents but their cost tends to be much higher than the passive and palliative types which aim to reduce the severity of the resultant damage. Examples of the latter include safety belts, roadsied safety barriers, soft interior padding and burst-proof doors. However, little attention has been paid to reducing the effects of those minor accidents occurring on impact between a vehicle and an obstruction which are rarely reported in statistics, but which cost much in time, money and inconvenience.

We must, therefore, give serious consideration to any device which reduces the initial severity of the impact between a vehicle and an obstruction, at either slow or fast speeds. This will have a dual effect by reducing material damage and lessening or preventing injury to the people involved. If really effective at low speeds, minor accidents immediately become negligible in their effects, thus alleviating a problem which is becoming more frequent both in the increasingly congested cities of wealthy countries, and among the rapidly growing traffic of the developing nations.

The first part of a vehicle to come into contact at the point of collision in most accidents is the front bumper. This is closely followed in terms of frequency of initial contact by the rear bumper. Hence, any system which uses the bumpers and their attachments to the car, to absorb part of the impact energy, offers a valuable route towards the reduction in the cost of accidents. In addition, any such system being both passive and palliatory, is also relatively cheap to use.

DESIGN CRITERIA

The essential requirements of energy absorbing bumpers are as follows:
1. High energy absorbtion.
2. Act without time-lag.
3. Fully effective at very low speeds.
4. Controlled retardation throughout their range.
5. Continued effectiveness at speeds well above the design range.
6. No rebound.
7. Low initial cost.
8. Re-useable after minor impact.
9. Self-resetting after minor impact.
10. Effective at all angles of impact.
11. Low repair cost.
12. Compactness.
   (Not necessarily in order of importance.)

Criteria 1 to 7 can be met to varying degrees of efficiency by any means employing the permanent deformation of a suitable material, even such a simple solution as a stiffer version of the common bumper and dumb-iron design in use today. But, after the slightest impact, the effectiveness of such a system is seriously decreased and it is certain that few, if any, vehicle owners will bother with, much less pay for, repairs to such an apparently unimportant item. The major key to a practical system is, therefore, found in items 9 and 10, provided the other requirements are also adequately met.

EVOLUTION OF A PRACTICAL DESIGN

Road Research Limited accepted the challenge presented by the Transport and Road Research Laboratory in producing a specification for a highly advanced design of energy absorbing bumper system. In doing so, we recognized the well-known principle that a fluid flowing through an orifice can absorb great amounts of energy and when combined with a mechanism to restore the unit to its original state, will produce a system conforming to all the above criteria.

The requirement called for a system capable of being effective at speeds of up to 40 mph (64 km/h) and to prevent all damage at speeds of up to 5 mph (8 km/h). These specifications were fulfilled and subsequent tests confirmed the accuracy of our designs.

The resulting specification consists of a unit functioning in three stages, the design of which is dictated largely by considerations of size and economy. The first stage of action consists of a short stroke hydraulic buffer with a spring return, and this meets all the requirements of the energy absorbing bumpers at low speeds — up to 5 mph (8 km/h). At higher speeds — up to 20 mph (32 km/h) — the conflicting demands of criteria 4 and 12 dictated that the unit could not be self-resetting and that some part of the vehicle would suffer damage. As the vehicle would require repair it was thought permissible to envisage a simple garage repair that would recondition the bumper system in order that it may accept further service. Finally, in high speed impacts of up to 40 mph (64 km/h), where major damage would have been suffered by the car, the bumper system is designed to deform just enough to keep the forces applied to its attachment below the level at which the attachments fail, so absorbing a large amount of energy that would otherwise contribute to increasing the severity of other damage or personal injury. In doing so, the bumper system is irreparably damaged.

MATHEMATICAL TREATMENT

The design problems arose from the need to reconcile the conflicting requirements of the three functional stages and to achieve a smooth transition between them while maintaining a high efficiency at all times. The well-known theory of fluid shock absorbers using variable flow control orifices or valves is, of course, the basis of the design but the conventional solutions of shaped needles, multiple orifices, valves, etc., were all ruled out by the special considerations of the multi-stage action required. The answer was found in the patented combination of a specially designed liner to the cylinder, supported by an elastic member and supplemented by a yielding member which deforms to prevent the cylinder bursting at the highest rated speeds.

The design of the orifice is based on the retardation formula:

$$G = \frac{(V_1{}^2 - V_2{}^2)}{(L_1 - L_2) \times Kg} \qquad (1)$$

derived from $\dfrac{M(V_1{}^2 - V_2{}^2)}{2g} = F(L_1 - L_2) \qquad (2)$

The quantity of oil displaced by the piston movement from position L_1 to L_2 has to pass through the orifice and it is known that the hydraulic resistance of a rectangular orifice gives a flow rate per second of:

$$Q = \frac{P}{\eta} \left(\frac{t^3}{l}\right) \times K_2 \qquad (3)$$

So the displacement of oil is:

$$Q \times \frac{(L_1 - L_2)}{V}$$

units of volume, which (taking V to be the mean of V_1 and V_2) equals the volume of oil to be moved, i.e. $(L_1 - L_2) \times$ Piston area $= (L_1 - L_2)$ A; so Q = AV, which, inserted in formula 3 and rearranged, gives:

$$P = \frac{AV\eta}{K_2} \left(\frac{l}{t^3}\right)$$

But the retarding force 'G' is equal to Pressure P acting on the piston area A, so:

$$G = \frac{A^2V\eta}{K_2} \left(\frac{l}{t^3}\right) \qquad (4)$$

and combining with formula (1),

$$I = \frac{(V_1{}^2 - V_2{}^2)}{(V_1 + V_2)(L_1 - L_2)} \times \frac{t^3}{A^2\eta} \times \frac{2k_2}{K_1} \qquad (5)$$

where,

I	= the variable width of our orifice
t	= the fixed dimension of the orifice
A	= the effective piston area
V_1 and V_2	= the vehicle speeds at distances L_1 and L_2 along the stroke of the shock absorber
η	= fluid viscosity
K_1 and K_2	= constants depending on the units employed

So it is seen that the required width of orifice over any short lengths of stroke is proportional to the term:

$$\frac{V_1{}^2 - V_2{}^2}{V_1 + V_2} \qquad (6)$$

Having chosen suitable values for the terms that do not change during piston travel and a desired value for 'G' for a particular impact speed, 'V_2' can be calculated from formula (1), inserted in formula (5) to give an appropriate orifice width 'I' and the process repeated progressively, using the previous value of 'V_2' as the new value of 'V_1' in the next step, giving a profile of values of "I" which can be smoothed out as desired by adjusting the value adopted for "G". Normally, though, it is desired to keep "G" as near constant as possible for maximum energy absorption efficiency which results in values of "I" approximating to a parabolic curve. Special considerations, particularly those resulting from the inherent elasticity of the vehicle structure as well as the design of a unit to be of high efficiency at both low and high speeds of impact, suggest modifications to the basic profile and the effect of these can be calculated from the formula above.

For any chosen profile, however, the designed value of "G" will be exceeded in impacts of a higher initial speed. "G" is directly proportional to the internal pressure "P" so it is common to provide a pressure sensitive valve to "dump" excessive pressure. While such valves are valuable they do not provide an ideal solution due to a number of factors, of which inertia is a major item, and often allow over – or under – correction of the excess forces.

The orifice design adopted by Road Research Limited is constructed so that it is supported by the metal of the cylinder barrel and if excess pressure expands the barrel, dimension "I" increased by $\delta D \times \pi$. At the breaking stress of steel used, δD can be as much as 9 percent and the normal value of "I" in the order of 1/5 × D, therefore:

$$\frac{\delta I}{I} = \frac{.09 \times \pi D \times 5}{D} \qquad (7)$$

$$= 1.41 \text{ approximately}$$

So a 41 percent increase in orifice size can be obtained which provides an equally large drop in the excess "G" force without bursting the unit which, therefore, continues to absorb energy effectively Even larger deformations can be obtained by interposing an elastic material between the cylinder barrel and the liner tube containing the orifice for special applications. Because the deformation of the barrel is generated directly by the internal pressure and because there are no rapidly moving components, inertia does not affect the response of the orifice which is rapid and precise.

DEVELOPMENT, TESTING, RESULTS

Development of this system was undertaken under contract to the Transport and Road Research Laboratory, the necessary tests being carried out with the help of the Laboratory's impact test rig and recording equipment and with the extensive cooperation of the TRRL and its staff, to whom our thanks are due. Early problems were overcome, including a quite spectacular tendency of the fluid to catch fire under high-speed impact which was entirely eliminated by the rapid development of a suitable fluid by the Castrol Research Laboratory. The result of this work can be seen fitted to the Rover 2000 TC on display in the Exhibition. The results of the impact rig tests have shown that this unit has met all the requirements and these are summarized in Table 1

Table 1.				
Impact Speed		Cylinder Travel		Equivalent "G" Force for 1 ton vehicle
mph	km/h	Inches	Cms.	
5.53	8.85	1.92	4.88	5.2

PERFORMANCE GRAPH OF SHOCK ABSORBER TEST OF TABLE 2

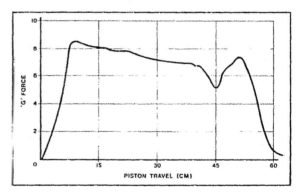

MEETING TODAY'S REQUIREMENTS

Private vehicles at present in production and under development do not, in many cases, have either the space or the strong points that will allow full use of the Road Research limited bumper system. For this reason, a lighter, lower rated and cheaper unit has been designed, using the same basic principle as the larger units but with a reduced specification that offers the same complete protection at impacts up to 5 mph (8km/h) and considerable energy absorbtion at speeds up to 20 mph (32 km/h). Tests made at the TRRL demonstrated its effectiveness (Table 2).

Table 2.				
Impact Speed		Cylinder Travel		Equivalent "G" Force for 1 ton vehicles
mph	km/h	Inches	Cms.	
6.72	10.75	2.5	6.35	8.08

FURTHER DEVELOPMENT

A special case of personal injury exists in the frequent incidents of small cars running under the overhanging rear of commercial vehicles. Where this occurs, even a light impact may be at the occupant's face level and therefore result in serious injury. Many heavy transport vehicles are fitted with rigid rear bumpers which can cause serious damage to cars hitting them, not only because they are too strong to absorb energy, but also because they are vulnerable to damage through incautious reversing or careless loading of the vehicle. Road Research Limited has therefore extended the use of its principle of energy absorbing pumpers to this application and is exhibiting a prototype of a heavy duty unit capable of stopping a one ton car travelling at a relative speed of 35 mph (56 km/h) with minimum damage to the occupants.

While we would not suggest for one moment that effort devoted to the prevention of accidents should be relaxed, we believe that the simple developments outlined in this paper offer an economical and effective means of reducing the loss and suffering caused by those accidents which cannot be prevented.

HYDRAULIC SHOCK ABSORBER AND BUMPER ATTACHMENT

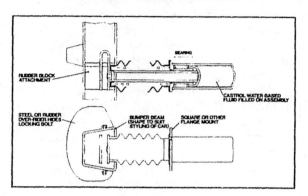

850511

Occupant's Safety by use of Variable Energy Absorbing Bumpers

Katsuya Shibanuma
The Institute of Vocational Training
Hiroaki Tanaka
Unversity of Tokyo
Niichi Nishiwaki
Nishiwaki Laboratory

ABSTRACT

When the gap between the seat belt and occupant's body is considered, it is of vital importance that the seat or the seat belt has sufficient energy absorbing functions. In this conconnection, it was found that the ideal energy absorbing pattern to stop the vehicle was to provide greater deceleration at the first phase of impact and less deceleration at the second phase to be followed by slightly greater deceleration at the last phase of impact. The authors proposed and designed a special bumper capable of simulating this ideal deceleration pattern applicable over a wide range of pre-collision velocities. The performance of the new bumper was verified by a series of experiments using a test vehicle with a belted dummy.

VARIOUS STUDIES HAVE BEEN CONDUCTED on the protection of the occupant during a car crash. These studies include the method of how to mitigate the impact on the occupant by effective absorption of the impact energy by modifying vehicle body structure (1,2)* and research and development of a seat belt system equipped with an energy absorbing mechanism (3). The subject of first concern in this paper is the fact that the occupant, even though restrained by the seat belt, is subject to move forward some distance i.e. equivalent gap s_0 until the seat belt becomes tight. This foward movement results from slackness of the seat belt and the elastic property of the occupant's body. This equivalent gap s_0 causes the occupant to move in a different manner from the vehicle until they stop after collision. During this deceleration process, it is most desirable for the occupant to minimize the maximum deceleration level from the viewpoint of occupant protection. To creat the desirable stopping process, it is very necessary to find a way to absorb the relative kinetic energy of the occupant against the vehicle.

This paper deals with the seat that can move relative to the vehicle (hereinafter this type of seat is referred as seat bumper). The subject of second concern is the necessity of sudden brake application at first phase and gentle deceleration afterwards instead of the constant deceleration to stop bring the vehicle. This necessitates the impact force to have special characteristics in which force-deformation displacement is large at the first phase. This paper discusses how to create the most ideal stopping situation under comparatively simple energy absorbing (EA) force versus force-deformation displacement in consideration of the easiness of bumper fabrication. The authors propose here a new type of bumper which is applicable over a wide range of initial collision velocities having the capability of absorbing the energy by scraping wood stuffs. This bumper can also adjust the EA force by changing the work force while maintaining the same EA characteristics pattern. The latter half of this paper deals with the test results by the use of a full scale test vehicle having a specially designed bumper a belted dummy.

THEORETICAL ANALYSIS

ANALYTICAL MODEL - Fig. 1 shows a model considered. The mass of the vehicle is m_1 and at the front end of the vehicle a vehicle bumper with an EA force of f_1 is equipped. The occupant mass is represented by a point mass m_2 and the occupant is assumed to be supported on the seat by a seat belt with an equivalent front gap s_0 (elastcity of seat belt and occupant are neglected assuming that the seat belt becomes tight at the instance the belt gap s_0 becomes zero). It is considered that the model vehicle running at a velocity of u_0 collides with a barrier fixed to the ground. Under the circumstances, time t, vehicle body displacement x_1 ,

----

*Numbers in parentheses designate References at end of paper

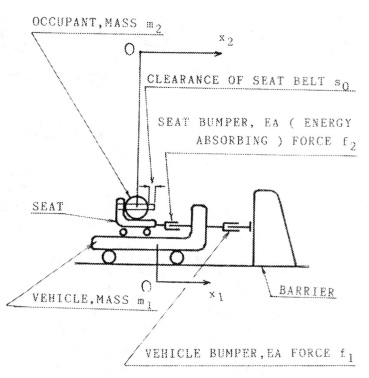

Fig. 1 - Analytical model

and occupant displacement x_2 (in stationary coordinates system) are measured from the moment, the front end bumper begins to contact with the barrier. The motion of the vehicle and occupant can be expressed by the following equations.

$$m_1(d^2x_1/dt^2) = -(f_1 - f_2) \quad (1)$$

$$m_2(d^2x_2/dt^2) = -f_2 \quad (2)$$

when $t=0$, initial conditions are

$$dx_1/dt = dx_2/dt = u_0 \quad , \quad x_1 = x_2 = 0 \quad (3)$$

The functional equations of the EA force of the vehicle bumper and that of the seat bumper are given as $f_1 = f_1(x_1)$ and $f_2 = f_2(x_2-x_1-s_0)$, respectively. These equations are derived from EA characteristics of each bumper as a function of force-deformation displacement x_1 and $x_2-x_1-s_0$. However, $f_2 = 0$, provided that $x_2-x_1-s_0 < 0$, in other words, before the seat belt gap becomes zero. In addition, such a case will never happen in which occupant velocity is lower than vehicle velocity ($dx_2/dt < dx_1/dt$) during deceleration. Whether the vehicle and occupant move as a unified system after the vehicle velocity coincides with the occupant velocity ($dx_2/dt = dx_1/dt$)(In this case, the equation is given by the sum of equation (1) and (2)) or move separately, depends on whether the force exerted on the seat bumper $f_1 m_2/(m_1+m_2)$ exceeds the EA force of the seat bumper or not. Taking this into account, a numerical calculation program based on Runge-

as would be described in the following paragraph.

Because of structural restrictions of the vehicle, it is assumed that the maximum EA stroke is less than s_1 whereas the maximum allowable displacement of the occupant within the vehicle compartment is s_2. Under these restrictions, the problem of authors' concern is how to determine the braking method to stop both vehicle and occupant by minimizing maximum deceleration imposed on the occupant. In other words, the problem is how to determine the EA force-deformation displacement characteristics of both bumpers.

IDEAL STOPPING PROCESS AND PROPOSAL OF BUMPER WITH A PERFORMANCE CLOSE TO THE IDEAL STOPPING PROCESS - Fig. 2 illustrates the process by which the vehicle at an initial velocity of u_0 stops and relationship between u (velocity) and t (time). Fig. 2(a) shows an ideal stopping process which clearly demonstrates the benefit caused by full stroke movement of the vehicle bumper (s_1) and maximum displacement of the occupant (s_2) in the vehicle compartment s_1+s_2. The area surrounded by broken lines represents the distance traveled by the occupant. This figure clearly demonstrates that the optimum deceleration can be accomplished in such a manner that the maximum deceleration imposed

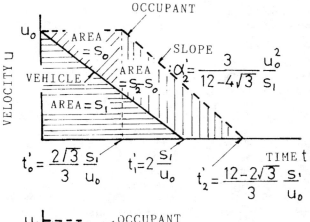

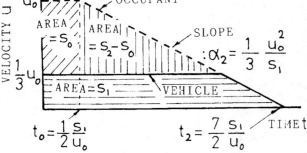

Fig. 2 - Comparison between (a) ideal stopping process and (b) stopping at constant deceleration (where, $s_2 = s_1$, $s_0 = (1/3)s_1$)

on the occupant is minimized, in other words, the maximum slope of the broken line should be as small as possible. For this purpose, it is necessary that the free motion time elapsed by the occupant until he is restrained by the seat belt, t_0 is short as possible, which thereafter should be followed by the constant deceleration on the occupant all the time till the vehicle stops. It is also seen that the best way of vehicle deceleration is to stop the vehicle in the manner as shown by the solid line in Fig. 2(a). The hatched area by the slanted lines is identical to the equivalent gap of seat belt s_0 whereas other hatched areas by both vertical and horizontal lines represent s_0-s_1 and s_1, respectively, as shown in Fig. 2.

For comparison with Fig. 2(a), Fig. 2(b) shows the case in which the vehicle stops by constant deceleration. A comparison between Fig. 2(a) and Fig. 2(b) indicates the fact that free motion time of the occupant (t_0') in Fig. 2(b) is much longer than that of Fig. 2(a). It is assumed that $s_2=s_1$ and $s_0=1/3$ s_1 (actual length in authors calculation and experiment is $s_1=30$ cm, $s_2=30$ cm, $s_0=10$ cm, respectively). Under this assumption, main values of both time and acceleration on the occupant α_2 are indicated in Fig. 2. A wide discrepancy, as much as $\alpha_2'/\alpha_2 =9/(12-4\sqrt{3})=1.77$ times is noted between Fig. 2(a) and Fig. 2(b) in terms of deceleration imposed on the occupant. As shown in Fig 2, velocity difference between occupant and vehicle is much greater at the moment the seat belt becomes tight. This, from the standpoint of occupant protection, clearly explains the necessity and importance of a seat bumper or EA type belt application.

Since the ideal deceleration process in Fig. 2(a) requires infinite EA force when t=0, it is considered unpractical. The designing of a new type bumper is attempted with anticipation to create a deceleration pattern close to the ideal deceleration pattern, while considering the benefit of easy fabrication and possible EA force vs. displacement characteristics as shown in Fig. 3(a). In other words, a standard pattern anticipated is the linear decrease of EA force with respect to force-deformation displacement from a given point to zero and then a quick return to its' original value. In the calculations described below, 30 cm is used as the force-deformation displacement when EA force returns to its' original value in this standard pattern. Actually, the vehicle mass $m_1=675$ kg and occupant mass $m_2=90$ kg (including mass of seat, 20 kg) were used for the experiment performed later. In accordance with the actual test to be conducted later, it is assumed that the allowable force-deformation displacement of the vehicle bumper is $s_1=33.8$ cm and the allowable distance that the occupant can travel in the vehicle compartment is $s_2=30$ cm and the equivalent gap of the seat belt is $s_0=10$ cm. When the initial velocity of $u_0=7$ m/sec=25.2 km/h is given by considering vehicle velocity during the test, the absolute scale of the EA

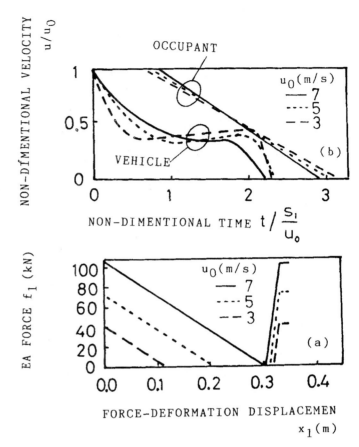

Fig. 3 - Trial calculations for vehicle bumper realizing the ideal stopping conditions: (a)EA force vs. force-distortion displacement characteristics; and (b) velocity to time diagram of stopping conditions

force shall be determined on the ordinate of Fig. 2(a) in order to stop the vehicle while satisfying the value of s_1, s_2 and s_0 under the above-mentioned EA force pattern. The solid line in Fig. 3(b) shows stopping situation on the u-t coordinate where s_1 and u_0 are representative values in Fig. 3(b) and these are put in non-dimensional form by u and t. Under the given condition, it is practical to consider that the seat bumper has a constant EA force just like an ideal stopping process. The EA force in this case is given in form of the product of m_2 and occupant deceleration that is expressed by the slope of the straight line illustrating the occupant's motion as shown in Fig. 3(b).

As will be explained in the subsequent paragraph, by using a bumper capable of absorbing energy by scraping an object, it is possible to change the EA force by changing the amount of work even when the work is done on the same material. Fig. 3(a) shows a case in which EA characteristics can be expressed by dotted and broken lines. In other words, a case is assumed in which the EA force in the standard state of

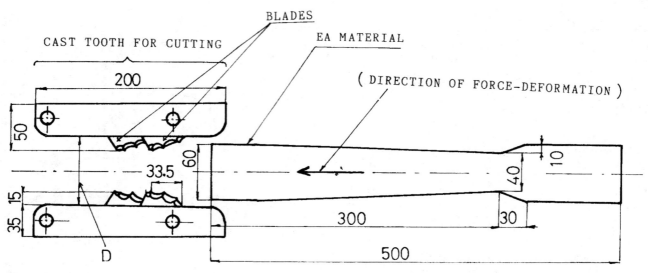

Fig. 4 - Outline of vehicle bumper

Fig. 5 - Photograph of cast tooth

charactristics can be lowered to a certain extent until zero level and any line which goes below zero level can be made zero. This assumption leads to the possibility of fabricating a bumper applicable over a wide range of vehicle velocities by adjusting the absorbing energy by the bumper. Both dotted and broken lines in Fig. 3(a) indicate the EA force level applicable to an initial velocity of 5 m/sec and 3 m/sec. Fig. 3(b) indicates the deceleration pattern corresponding to this applicable stopping situation. This clearly simulates an ideal stopping situation as the EA force is concentrated at the initial phase of impact of low vehicle velocity. Furthermore it is observed that the vehicle is slightly accelerated in a very short span even during the deceleration process. This is attributed to the fact that the occupant pushes the vehicle as the occupant is subject to the relative deceleration against the vehicle during the period when the EA force of the vehicle bumper is reduced to zero.

Dimensional representative values, s_1 and u_0 in the basic equations can be put into non-dimentional form to EA force $m_1 u_0^2/s_1$. The calculation in Fig. 3 is performed in the case of comperatively low velocity range to comply with the expriments conducted as mentioned below. The upper limit of vehicle velocity $\hat{u}$ (7 m/sec in case of Fig. 3) can be increased only by changing the scale of the EA force in proportion to u^2. This creates a situation identical to the low velocity stopping situation.

EXPERIMENTAL BUMPER

A full size exprimental woodworking type bumper was fabricated as shown in a previously presented paper (4). Fig. 4 illustrates the outline of the vehicle bumper and Fig.5 shows the photograph of cast too h. One set of cast

tooth consists of two lines of dull cutting blades. The first line has four and the second line has three cutting blades and both lines facing each other. These two lined cutting blades are designed and fabricated to cut a square test column made of merapi lauan placed inbetween. When this lauan test column of 80 mm wide and 60 mm thick at the base is inserted between these two blades, it is scraped by the dull blades facing each other to produce thin fiber-like shavings and absorbs a large amount of energy. As shown in Fig. 4, the EA material (lauan) is designed so that it has different thickness in longitudinal direction, thus creating different cutting depths on the woodstuff and enabling it to have characteristics close to the required EA characteristics. Futhermore this EA material (lauan) is designed to change the EA force level by changing the cutting depth of the blades. This is accomplished by adjusting gap D between the two cast teeth.

Fig. 6 indicates EA force vs. force-distortion displacement characteristics of test vehicle bumper. This data was obtained as a result of static load tests by the use of a material

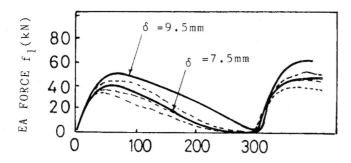

FORCE-DEFORMATION DISPLACEMENT
x_1 (mm)

Fig. 6 - EA force vs. force-distortion displacement characteristics of test vehicle bumper (3 examples of measurements with δ=7.5 mm indicated by dotted lines, and mean values indicated with solid lines)

testing machine. The nominal cutting depth by the cutting blades of δ = 9.5 mm as shown in Fig. 3(a) is aimed at vehicle speed of u_0=7 m/sec, in the same way, δ = 7.5 mm corresponds to u_0 = 5 m/sec. Since the test vehicle is equipped with two sets of this type bumpers, the EA force of each bumper is one-half of the EA force in Fig. 3(a). The test results presented in the previous paper suggest that no distinctive difference was observed in the work force between high speed impact collision test and low speed impact machine test. The seat bumper is designed to work with a single blade as its' EA force is smaller compared with the vehicle bumper.

VEHICLE TEST AND TESTING PROCEDURE

Fig. 7 shows an outline of test vehicle and Fig. 8 shows the photograph of test vehicle and barrier. This test vehicle has an open frame and was fabricated by modifying a small sized truck with a total mass of 675 kg. The masses of the test dummy and the seat were 70 kg and 20 kg, respectively. The five points seat belt system (usually used for racing cars) was used. Pre-collision speed, the vehicle speed just before the collision with the barrier, was measured by two sets of photoelectric tubes and a universal counter. The strain gauge type accelerometer was used for measuring deceleration.

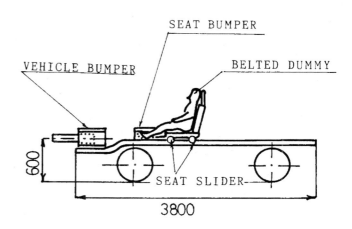

Fig. 7 - Outline of test vehicle

Fig. 8 - Photograph of test vehicle and barrier

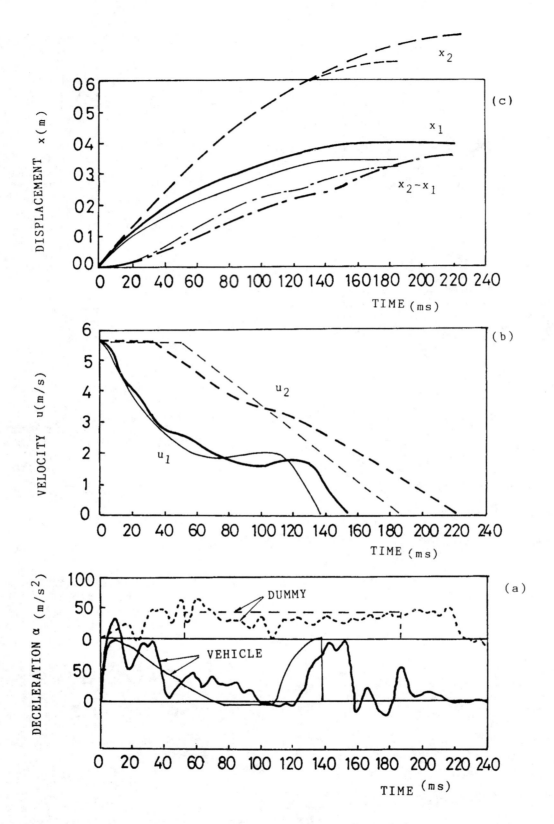

Fig. 9 - Test results, Part 1: Collision at initial velocity of 5.57 m/sec close to optimum initial velocity of 5.41 m/sec : (a) Deceleration; (b) Velocity; and (c) Displacement (heavy lines for measured values and light lines for calculated values)

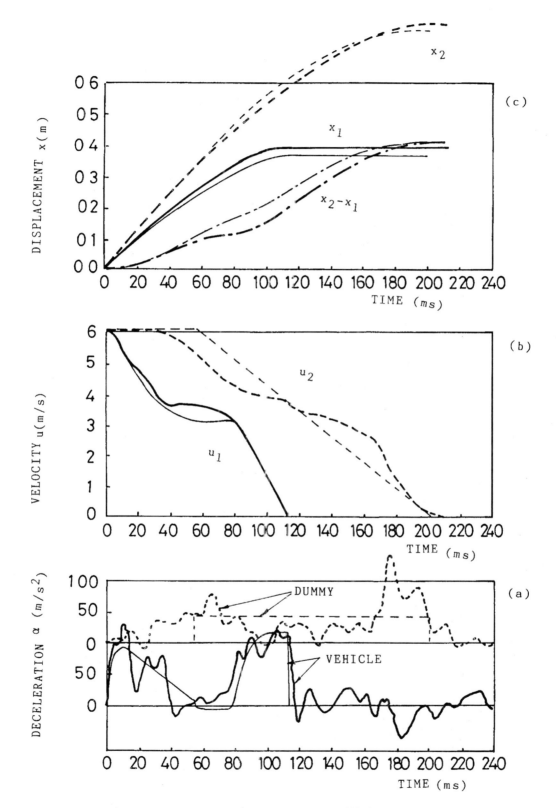

Fig. 10 - Test results, Part 2: Collision at initial velocity of 6.09 m/sec higher than optimum initial velocity of 5.41 m/sec: (a) Decelaration; (b) Velocity; and (c) Displacement (heavy lines for measured values and light lines for calculated values)

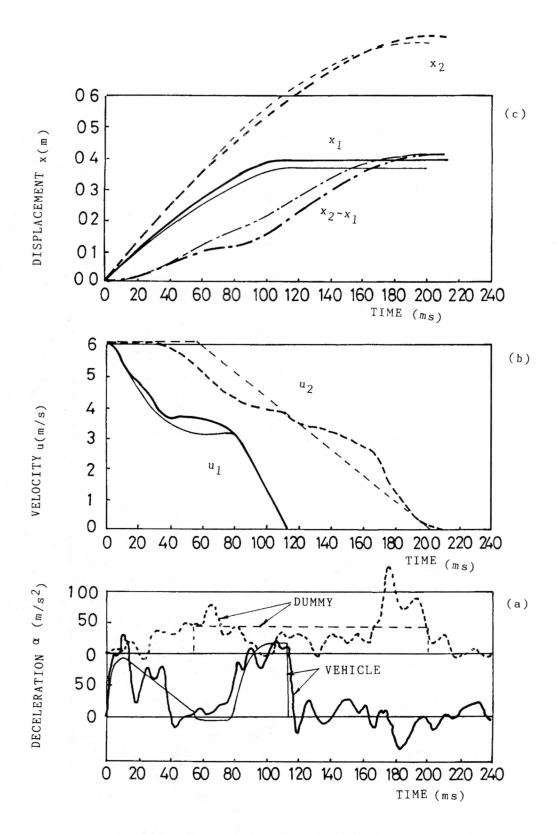

Fig. 11 - Test results, Part 3: Collision at
initial velocity of 5.25 m/sec lower than
optimum initial velocity of 5.41 m/sec : (a)
Deceleration; (b) Velocity; and (c) Displacement
(heavy lines for measured values and light lines
for calculated values)

Output data from the strain gauge was recorded by an oscillograph. Accelerometers were placed on the frame near the seat and on the pelvis of the test dummy. Both force-deformation displacements of seat and vehicle bumpers were measured.

DISCUSSION OF TEST RESULTS

The impact test was conducted by the use of a full size test vehicle equipped with vehicle bumper adjusted to meet one of the required characteristics i.e. at = 7.5 mm, nominal cutting depth of the blade, as shown in Fig. 6. Prior to the test, various initial velocities were input as data necessary for calculating the program explained before. The result of the calculation suggests that the optimum EA effect can be cobtained in the case where the vehicle at initial velocity of 5.41 m/sec (slightly higher than the estimated initial collision velocity of $u_0 = 5$ m/sec) stops causing the final force-deformation displacement of $s_1 = 32.8$ cm. (This deceleration pattern is quite similar to the one nondimentionally indicated by the dotted lines in Fig. 3(b)). This result suggested that the test should be carried out with a pre-collision speed of around 5.41 m/sec (hereinafter this velocity is referred as optimum collision speed and is expressed as u_{op}). In addition, it was also found that the EA force of the seat bumper shoud be $f_2 = 3.72$ kN in order to make the final force-deformation displacement $s_2 - s_0$ to be estimated value of 20 cm at this optimum pre-collision speed. In the test, efforts were made to adjust the cutting depth of the seat bumper so that the EA force would be very close to the estimated EA force of 3.72 kN.

Fig. 9 shows the result of the impact test at pre-collision velocity of $u_0 = 5.57$ m/sec which is close to the optimum pre-collision velocity. The heavy lines in Fig. 9(a) indicates deceleration oscillograms of both vehicle and test dummy. The values on the oscillograms were is read at time intervals of 2 milli-seconds and integrated and subtracted from the initial velocity to get the time history of vehicle velocity and dummy velocity those are shown by the heavy lines in Fig. 9(b). The above-mentioned pre-collision velocity of 5.57 m/sec was actually measured with photoelectric tubes and was input in the calculation program including the full size vehicle bumper characteristics with the EA force of the seat bumper of $f_2 = 3.72$ kN. The results are illustrated by the thin lines in Fig. 9(b) those are integrated with respect to time to get the displacement as shown in Fig. 9(c).

Fig.10 shows the comparison of test results and calculated results at the initial velocity of $u_0 = 6.09$ m/sec which is much higher than the optimum initial velocity of $u_{op} = 5.41$ m/sec. Fig.11 indicates the result of test at an initial velocity of $u_0 = 5.25$ m/sec, much lower than the optimum initial velocity U_{op}. Recorded wave forms on the deceleration

oscillograms in Figs. 9 to 11 seem to include frequencies proper to vehicle frame and dummy. Another integration of these wave forms results in suppressed components and smooth variation of wave forms. Especially good agreement is seen in the vehicle velocity between test and calculattion. The quality of the lauan used for EA material has less variation and similar characteristics can be seen under very high speed working conditions as well as under static load test as shown in Fig. 6. Test results also suggest that deceleration to the dummy began earlier than anticipated. This is explained by the assumption that the racing type 5 points seat belt system with stronger restraining force resulted in a smaller equivalent gap between the seat belt and occupant's body. For some unknown reason, it is realized that the EA force of the seat bumper tends to be smaller than the set value. Compared with the case of the vehicle, a not so satisfactory agreement is observed between test and calculation in case of the dummy. Fig.10 indicates a greater deceleration just before the dummy stops. This is interpreted by the fact that the seat bumper slides its full stroke and comes in contact with the seat stopper.

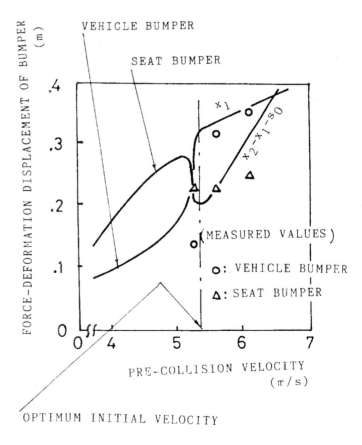

Fig. 12 - Force-deformation displasement vs. precollision velocity of bumper

The solid lines in Fig.12 illustrate the calculated data those show the time history of the final force-deformation displacements of both vehicle bumper and seat bumper by varying pre-collision velocity under given conditions. The force-deformation displacement of each bumper has been measured and plotted with symbols on the basis of the tests shown in Figs. 9 to 11. The noticeable point in Fig.12 is the fact that the force-deformation displacement of the seat bumper becomes mimimum during impact at optimum velocity u_0. A sudden increaces in force-deformation displacement of the seat bumper when the initial velocity is lower than u_{op} can be understood without difficulty by a coincidental sudden decrease in the force-deformation displacement of the vehicle bumper. Since the areas surrounded by the EA force characteristics curve in Fig. 3(a) and Fig. 6 represent EA works, the vehicle stops when the vehicle bumper is force-deformed to the point where the kinetic energy of the vehicle has the same area under the same characteristics line. This explains the reason why force-deformation displacement sharply decreases when it becomes slightly below 30 cm. (This is typical in the characteristics shown by dotted and broken lines in Fig. 3(a)).

CONCLUSIONS

When the gap between the occupant and seat belt is considered, it is very important for the seat or the seat belt to have shock absorbing functions for occupant's safety. The ideal shock absorbing pattern to stop the vehicle is to give higher deceleration at the first phase of collision and lower deceleration again at the last phase. The authors proposed and designed a special bumper capable of simulating this ideal deceleration pattern applicable over a wide range of initial collision speeds. The performance of new bumper was proven by a series of vehicle tests with a belted dummy. The problem still remains for future research, pertaining to the mechanism to detect vehicle velocity so to adjust the bumper to the optimum situation (to adjust gap D between two cutting teeth in Fig. 4).

Nomenclature

f_1 : energy absorbing force (= EA force) of vehicle bumper, N
f_2 : EA force of seat bumper, N
m_1: vehicle mass, kg
m_2: occupant mass, kg
s_0: equivalent front belt-to-dummy gap, m
s_1: allowable full energy absorbing stroke, m
s_2: allowable displacement of occupant in vehicle compartment, m
t : time, sec
u_1: vehicle velocity, m/sec
u_2: occupant velocity, m/sec
x_1: vehicle displacement, m
x_2: occupant displacement, m
α_1: vehicle deceleration, m/sec^2
α_2: occupant deceleration, m/sec^2
δ : nominal cutting depth of work blade, mm

ACKNOWLEDGMENTS

The authors wish to thank Professor Michio Kuroda of Seikei University for his appropriate advice and guidance.

REFERENCES

(1) Yoshiro Okami et al.,"Explanatory Review on RSV Phase 2 - Phase 3 Program." Journal of the Society of Automotive Engineers of Japan, Vol.34(1980),No.1,p.75.
(2) Takashi Sasaki."Body Construction for Higher Speed Protection for Front Collision." Journal of the Society of Automotive Engineers of Japan,Vol.33(1979),No.3,p.162.
(3) Tsuguhiro Fukuda et al.,"The Energy Absorbing Seat System for Occupant Protection." 3rd Automotive Engineering Conference, Paper 18, (1977-Nov).
(4) Katsuya Shibanuma et al.,"Study on Woodworking Shock-Absorer."Transactions of the Society of Automotive Engineers of Japan, No.19,(1979).p.(67).

Persons wishing to submit papers to be considered for presentation or publication through SAE should send the manuscript or a 300 word abstract of a proposed manuscript to: Secretary , Engineering Activity Board, SAE.

Printed in U.S.A.

Dynamics of Low Speed Crash Tests with Energy Absorbing Bumpers

Thomas J. Szabo and Judson Welcher
Biodynamics Engineering, Inc

ABSTRACT

Low speed crash tests with Ford Escorts of model years 1981-1983 were conducted. The response of these vehicles equipped with energy absorbing bumpers in bumper-to-bumper impacts and in an underride of one bumper under another are evaluated.

Bumper displacement, vehicle acceleration, and vehicle velocity time histories are presented for both bullet and target vehicles. Two impacts each at 2.23, 4.47, and 6.71 m/s were conducted. Similar data is presented for a 4.47 m/s impact in which the front bumper of the bullet vehicle underrode the rear bumper of the target vehicle. Results indicate increased damage associated with the underride test, without corresponding increases in vehicle responses.

INTRODUCTION

Interest in low speed vehicular collisions has recently increased. This is largely due to changes in standards governing vehicle bumper performance, corresponding changes in bumper design, and a growing confusion concerning the responses of both the vehicle and occupant to low speed impacts. Application of many principles derived from higher speed impact research to the low speed environment is questionable and unvalidated. Low speed test data is needed in order to better understand the dynamics of these impacts, to enhance future bumper designs, and to facilitate reconstruction efforts.

Much of the previous research in low speed vehicle-to-vehicle impacts used vehicles from model years in which no federal bumper performance standard existed [Severy 1955, Severy 1968]. Consequently, the results of these tests have limited value as an aid to understanding current low speed impacts between vehicles equipped with "energy absorbing bumpers."

Pendulum impacts have been conducted on vehicles equipped with energy absorbing bumpers [Thomson 1990, Romilly, et al., 1988,]. Barrier impacts have also been conducted [Bailey, et al., 1991]. These test results can be extrapolated to vehicle-to-vehicle collisions; however, a paucity of crash test data exists from vehicle-to-vehicle low speed collisions with which to validate such extrapolations.

Emori and Horiguchi [1990] conducted vehicle-to-vehicle collisions with a cushion mounted on the front of the bullet vehicle. The cushion was meant to mimic the compliance of the production energy absorbing bumper, which was removed. Using this technique, it was not necessary to replace the energy absorbing device after each test. The target vehicle was a van which was not equipped with an energy absorbing bumper, and the maximum impact speed reported was 1.2 m/s, which is well below the threshold for damage of most energy absorbing bumpers.

Bailey, et al. [1991] conducted a rear-end impact between two vehicles equipped with energy absorbing bumpers. The impact speed was 1.6 m/s. Maximum velocities and bumper displacements were reported, although no time histories of these responses were published.

A phenomenon frequently encountered is that of "underride", wherein the front bumper of the bullet vehicle travels or protrudes under the rear bumper of the target vehicle. Braking on the part of the bullet vehicle driver precipitates this downward "pitching" of the vehicle. Mismatch in bumper heights between impacting vehicles can also result in "underride." No published data was found for this class of impact.

This test series provides low speed impact responses for bumper-to-bumper collisions in which the bumpers are equipped with energy absorbing displacement pistons. The impact velocities encompass the capacity of the energy absorbing systems. In addition, bumper underride is investigated, in terms of vehicle responses and damage.

TEST PROCEDURES

Seven vehicle-to-vehicle rear end crash tests were conducted, using 1981-1983 Ford Escorts. This vehicle was chosen for several reasons, one of which was the known capacity of the bumper to withstand impact. The 1981 Ford Escort bumper remains one of the few tested by the Insurance Institute for Highway Safety which sustained no damage in all bumper tests. The

same vehicle could thus be used for several tests. Only a change in the bumper system between the low speed tests was anticipated.

Each vehicle was equipped with identical stock bumpers, with shock absorbers for energy absorption. The target vehicle was stationary on a level surface, and the center lines of the vehicles were aligned at impact. Braking was absent for both vehicles. Two impacts were conducted at each of three velocities: 2.23 m/s, 4.47 m.s, and 6.7 m/s, for a total of 6 tests.

An additional 4.47 m/s impact was conducted in which the front bumper of the bullet vehicle was lowered in order to underride the rear bumper of the target vehicle.

INSTRUMENTATION

Impact velocities were measured using a time trap and tape switches affixed to the ground. Both the bullet and target vehicles were instrumented with a uniaxial accelerometer, located at the approximate center of gravity of the vehicle.

The rear of the target vehicle was equipped with two displacement potentiometers extending from the underside of the bumper to the vehicle frame, in line with the bumper shock absorbers.

The sampling rate for data collection was 10,000 Hz. Data was processed according to SAE J211 standard practice. Presented data was filtered at 160 Hz, Class 60. High speed film of each test was obtained, at a rate between 440 and 510 frames per second.

Test	Bullet Vehicle	Mass (kg)	Target Vehicle	Mass (kg)	Impact Velocity (m/s)	
					Target	Measured
1	1982 Escort	930	1981 Escort	966	2.23	2.12
2	1982 Escort	930	1981 Escort	966	2.23	2.11
3	1983 Escort	931	1981 Escort	966	4.47	4.39
4	1983 Escort	931	1981 Escort	966	4.47	4.40
5	1983 Escort	931	1981 Escort	966	6.71	6.75
6	1981 Escort	966	1983 Escort	931	6.71	6.75
7	1982 Escort	930	1981 Escort	966	4.47	4.34

Table 1: Test Parameters

RESULTS

Table 1 summarizes the test parameters

BUMPER-TO-BUMPER IMPACTS

Accelerations - Figures 1-3 show the acceleration-time histories for each vehicle in the 6 tests.

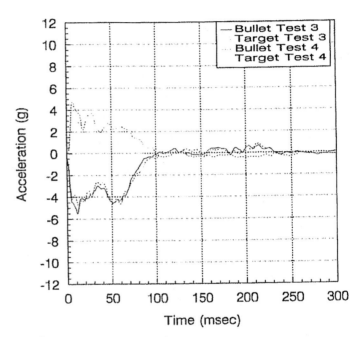

Figure 2: Vehicle Accelerations for 4.47 m/s Impacts

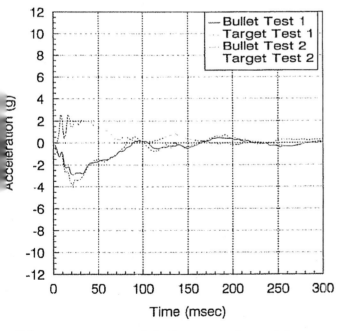

Figure 1: Vehicle Accelerations for 2.23 m/s Impacts

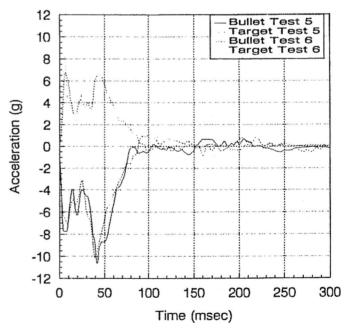

Figure 3: Vehicle Accelerations for 6.71 m/s Impacts

In spite of the relatively complex nature of the interaction between the vehicles and bumpers, a high degree of repeatability of the accelerations for a given test velocity and vehicle is observed. The traces generally exhibit two distinct peaks in acceleration, reflecting initial impact to the bumper, and subsequent increasing stiffness within the bumper shock absorber.

Table 2 contains the peak and average accelerations for each test. The average accelerations were calculated over 100 msec, the approximate duration of each impact. The average accelerations are approximately one-half of the peak accelerations. Both the peak and average accelerations increased with increasing impact velocity, in a relatively linear manner.

Test	Target Acceleration (g)		Bullet Acceleration (g)	
	Peak	Average	Peak	Average
1	2.64	1.30	-2.92	-1.45
2	3.29	1.50	-3.84	-1.74
3	4.78	2.46	-5.58	-3.40
4	5.90	2.37	-4.81	-3.32
5	6.67	3.57	-10.73	-5.14
6	8.23	3.86	-10.14	-4.71

Table 2: Peak and Average Vehicle Accelerations

Velocities - Figures 4-6 show the velocity-time histories for the bumper-to-bumper impacts.

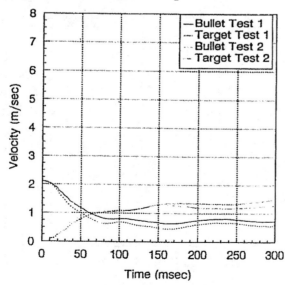

Figure 4: Vehicle Velocities for 2.23 m/s Impacts

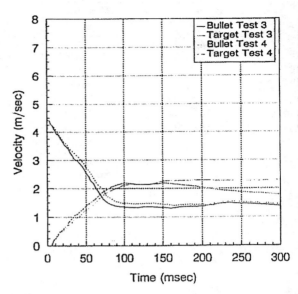

Figure 5: Vehicle Velocities for 4.47 m/s Impacts

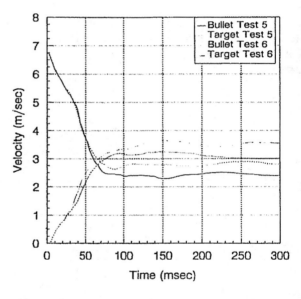

Figure 6: Vehicle Velocities for 6.71 m/s Impacts

Repeatable velocity histories for each impact velocity are observed. The bullet and target vehicle velocity traces are noted to intersect at approximately 70 msec for all impact speeds. This, however, does not represent the time at which the vehicles separate. The true separation times, as determined from film analysis, were 113 msec and 121 msec for Tests 1 and 2, 130 msec and 131 msec for Tests 3 and 4, and 96 msec and 116 msec for Tests 5 and 6, respectively. This phenomenon is ue to the elastic rebound of the umpers during the latter stages of he collision, wherein bumper-to- umper contact exists, but does not ontribute significantly to the cceleration of either vehicle.

Displacements - Figures 7-9 ontain the target vehicle bumper isplacements over time. Repeat- ble bumper displacements were bserved for each impact speed, and nly results from Tests 2, 4 and 6 re presented.

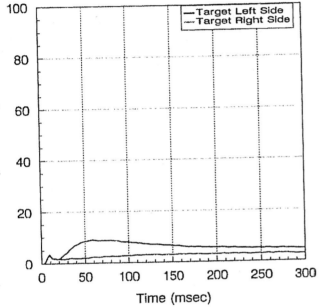

Figure 7: Target Bumper
Displacement for Test 2

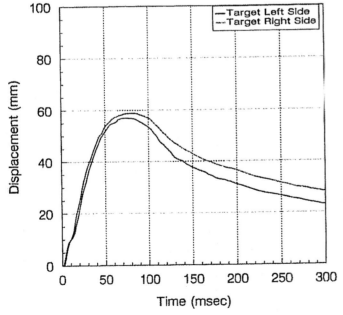

Figure 8: Target Bumper
Displacement for Test 4

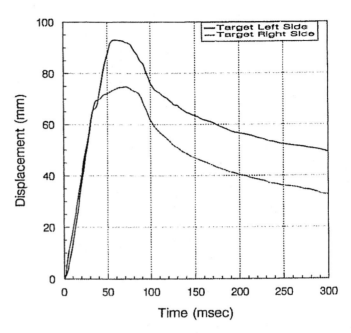

Figure 9: Target Bumper
Displacement for Test 6

Predictably, increased bumper displacement with increased impact velocity was observed. The bumper shock absorbers struck at the two higher impact velocities required in excess of 300 msec to return to their initial positions. Post collision inspection of all shock absorbers revealed each to have returned to its initial length.

As the piston of the shock absorber moved relative to its housing while being stroked, stria- tions along the shaft were created. These striations were measured post-collision, and denoted as a static measure of piston displace- ment. Bumper displacement data is contained in Table 3.

Test	Bumper Displacement (mm)		Piston Displacement (mm)			
	Target Left	Target Right	Bullet Left	Bullet Right	Target Left	Target Right
1	6.9	10.2	10.2	7.6	6.6	5.8
2	8.9	3.3	0.0	3.8	7.4	0.0
3	50.0	37.8	35.6	49.0	49.3	34.8
4	56.4	58.9	44.5	40.9	48.5	53.3
5	60.5	67.1	53.3	55.1	54.6	55.6
6	93.0	74.7	53.6	53.6	54.1	54.9

Table 3: Displacements

The target vehicle bumper and piston displacement measurements compare favorably for the 2.23 m/s and 4.47 m/s impacts. The bumper displacements for the 6.71 m/s impacts are significantly greater than the piston displacement. This indicates that the shock absorber was maximally compressed during the higher speed impacts, and relative motion between the shock absorber and vehicle resulted. This was confirmed by the permanently distorted post-collision bumper position, and the damage sustained by the vehicles.

Vehicle Damage - Vehicle inspections were conducted both before and after each test. Each vehicle showed scuffing on the struck bumper face, and striations along the shock absorber piston, with one exception: one target vehicle shock absorber struck at 2.23 m/s, showed no striations. There was no additional damage to the bullet and target vehicles for the 2.23 m/s and 4.47 m/s impacts.

Both the bullet and target vehicles in the 6.71 m/s impacts sustained damage to the bumper shock absorber mounts on the vehicles, such that the bumper position was permanently distorted, and the shock absorbers partially embedded into a deformed mounting structure.

Underride Test - The front bumper of the bullet vehicle was induced to underride the rear bumper of the target vehicle in one test. The impact speed was 4.34 m/s. The bumper positioning was achieved through the towing mechanism attached to the underside of the front of the bullet vehicle. Time history response data for the bullet and target vehicles for this impact was partially lost due to a delayed triggering of the time zero, and will not be presented. Vehicle responses were obtained from an analysis of the high speed film. Table 4 contains a chronology of the underride collision.

Time	Event
0 msec	Top portion of front face of bullet bumper strikes lower portion of rear face of target bumper
24 msec	Bullet bumper begins to slide under target bumper
37-233 msec	Target bumper contacts and displaces into grille of bullet vehi
61 msec	Target vehicle begins to move forward
233 msec	Target bumper disengages from grille of bullet vehicle
367 msec	Vehicle separation

Table 4: Chronology of Underride Collision

The peak final velocities of the bullet and target vehicles were calculated to be 2.10 m/s and 2.2 m/s, respectively. The duration of the collision was approximately 50 greater than the bumper-to-bumper impacts at the same velocity. The average accelerations for the bullet and target vehicles were 1.55 g and 1.53 g, respectively These are similar to the average accelerations observed i bumper-to-bumper impacts at hal the underride impact speed.

The piston displacement meas urements for the bullet vehicl were 15.0 mm and 14.2 mm, for th left and right shock absorbers respectively. The target vehicl piston displacement measurement were 11.9 mm and 16.5 mm, for th left and right shock absorbers respectively. These were slightl higher than those for bumper-to bumper tests at half the underrid impact velocity, and significantl lower than those recorded in bump er-to-bumper impacts at the same velocity.

Damage - Both target and bullet vehicles sustained bumper scuffing and striations of the shock absorber pistons in the underride impact. There was no additional damage to the target vehicle.

The bullet vehicle sustained significant damage to the grille, grille support, hood, right head- lamp assembly, left headlamp assem- bly, right side marker lamp, right fender, motor mounts, and required frame repair. This additional damage was a result of direct contact between the target vehicle bumper and the grille/headlamp assemblies of the bullet vehicle.

DISCUSSION

BUMPER-TO-BUMPER IMPACTS - Energy in a bumper-to-bumper rear impact is supplied by the bullet vehicle. Upon contact, that energy is transferred to several elements of the system, while some is retained by the bullet vehicle in the form of motion and/or damage. The bumper shock absorbers absorb some of the energy lost by the bullet vehicle as the bumpers of both vehicles displace relative to their respective vehicles. Some of the energy is transferred to the target vehicle, resulting in acceleration and/or damage to that vehicle. In addition, energy is dissipated by friction between the tires and the roadway. The remainder is converted into thermal and acoustic energy.

Assuming the amount of energy converted to thermal and acoustic energy to be minimal, it is possible to estimate the amount of energy absorbed by the energy absorbing bumpers, damage to both vehicles, and that dissipated through friction. This is simply a measure of the difference in the sum of bullet and target vehicle energies before and after the impact. Table 5 contains the estimates of this absorbed energy.

Test	Peak Velocities		At 150 msec	
	Energy Lost (J)	% Initial	Energy Lost (J)	% Initial
1	871	41.5	1216	58.0
2	1189	57.2	1309	63.0
3	5881	65.7	5810	64.9
4	5591	82.0	5822	64.5
5	13953	65.8	13653	64.4
6	12577	57.2	13198	60.0

Table 5: Energy Absorbed

The common convention in calculating energy employs peak pre- and post-collision velocities of the vehicles, in conjunction with the vehicle masses. This implicitly assumes that peak velocities of each vehicle occur simultaneously. Table 5 shows the energy results using this approach, compared to results obtained using velocities at a common time shortly after the collision. It can be seen that the two compare favorably, although the peak velocity approach tends to slightly underestimate the energy absorbed by the bumpers.

Since no damage beyond the bumper system was observed on either vehicle for the 2.23 m/s and 4.47 m/s impacts, the figures in Table 5 reflect energy absorbed primarily by the bumper systems. Increased energy absorption is noted for increased impact velocity, although the percentage of energy absorbed is similar for the two impact speeds (60-65%).

The vehicles in the 6.71 m/s impacts did exhibit damage beyond that to the bumper system. The shock absorber mounting structures were deformed. Thus, for these higher velocity impacts, the energy absorption figures indicated in Table 5 reflect the energy absorbed by both the bumper system, and damage to the vehicles. The percent of energy absorption, however, is similar to that found for the two lower impact speeds, remaining at 60-65%.

The coefficient of restitution for two colliding bodies is defined as the ratio between the separation velocities and the approach velocities. Coefficients of restitution were calculated for the peak velocities, and for velocities at 100 msec. They are presented in Table 6.

Test	Peak Velocities			At 100 msec		
	Post-Impact Velocity (m/s)		e	Post-Impact Velocity (m/s)		e
	Bullet	Target		Bullet	Target	
1	0.63	1.47	0.39	0.80	1.10	0.14
2	0.45	1.28	0.39	0.69	1.06	0.18
3	1.29	2.18	0.20	1.35	2.18	0.19
4	1.39	2.29	0.21	1.48	2.12	0.15
5	2.23	3.20	0.14	2.39	3.18	0.12
6	2.61	3.63	0.15	2.73	3.34	0.09

Table 6: Coefficients of Restitution

The coefficients of restitution for the peak velocities, and those at 100 msec, compare favorably for the two higher impact velocities. The coefficient calculated from the peak velocities for the 2.23 m/s impacts is significantly higher than that obtained using the velocities at 100 msec. It is felt that the velocities obtained immediately post-collision, at 100 msec, are more representative of the actual collision, and yield a more reliable coefficient of restitution.

Results indicate a coefficient of restitution for this series of impacts between 0.10 and 0.20. These values approximate the lowest values reported by Bailey, et al. [1991], who reported coefficients of restitution for vehicle-to-vehicle impacts of between 0.14 and 0.40.

Collins [1979] indicates coefficients of restitution of between 0.97 and 0.67 for vehicle-to-vehicle impacts between 2.23 m/s and 6.71 m/s. In light of the current test series, these coefficients would seem to overestimate the elasticity of low speed impacts between vehicles equipped with energy absorbing bumpers.

UNDERRIDE IMPACT - The impact velocity was 4.34 m/s, while the post-collision velocities of the bullet and target vehicle were 1.46 m/s, and 2.27 m/s, respectively. The amount of energy lost in the collision was 5280 J, or 60% of the initial energy, similar to the results obtained for the bumper-to-bumper collisions at a similar impact velocity.

The coefficient of restitution was calculated to be 0.19, similar to that for the bumper-to-bumper impacts. Given the fact that some initial bumper-to-bumper contact occurred in the underride test, values not grossly different from the bumper-to-bumper impact would be expected. However, in an underride without bumper contact, i.e., in which the target vehicle bumper strikes the grille of the bullet vehicle, a somewhat more plastic collision would result.

CONCLUSIONS

1. A high degree of repeatability in the accelerations and velocities is observed for a given test speed and vehicle.

2. The acceleration traces generally exhibit two distinct peaks.

3. Average accelerations are approximately one-half of the measured peak accelerations.

4. Peak and average accelerations increased relatively linearly with increasing impact velocity.

5. Elastic rebound of the bumpers increased the duration of contact beyond the time point defined by the intersection of the velocity traces.

6. Target vehicle piston and bumper displacements were essentially equal in impacts at 4.47 m/s and below.

7. Post-impact inspection of each vehicle revealed scuffing of the bumper face and striations along the shock absorber shaft.

8. The bullet and target vehicles in the 6.71 m/s impacts sustained damage to the bumper shock absorber mounts on the vehicles, such that the bumper position was permanently distorted, and the shock absorbers partially embedded into a deformed mounting structure.

9. Duration of the underride impact was approximately 50% greater than the duration of the bumper-to-bumper impacts at the same velocity.

10. The average accelerations for the bullet and target vehicles in the underride test were 1.55 g and 1.53 g, respectively. These are similar to the average accelerations observed in the bumper-to-

bumper impacts at half the under-ride impact speed.

11. Following the collision, the bullet vehicle in the underride test required frame repair. It sustained significant damage to the grille, grille support, hood, right headlamp assembly, left headlamp assembly, right side marker lamp, right fender, and motor mounts. This additional damage occurred as a result of direct contact between the target vehicle bumper and the grille/headlamp assemblies of the bullet vehicle.

12. Energy absorption increased with increased impact velocity. The per cent of energy absorbed for the 2.23 m/s and 4.47 m/s was similar (60-65%).

13. Deformation of the shock absorber mounting structures contributed to the total energy absorbed in the 6.71 m/s impacts. The percentage of energy absorbed remained similar to that found for the two lower impact speeds (60-65%).

14. Coefficients of restitution between 0.10 and 0.20 were calculated for the bumper-to-bumper impacts. The underride test resulted in a coefficient of restitution of 0.19.

15. The amount of energy lost in the underride collision was 5280 J, or 60% of the initial energy. This is similar to results obtained for the bumper-to-bumper collisions at approximately the same velocity.

REFERENCES

Severy, D.M., Mathewson, J.H., and Bechtol, C.O., "Controlled Automobile Rear-End Collisions, an Investigation of Related Engineering and Medical Phenomena," Canadian Services and Medical Journal, Nov. 1955.

Severy, D.M., Harrision, M.B., Baird, J.D., "Backrest and Head Restraint Design for Rear-End Collision Protection," SAE Paper 687009, Jan. 1968.

Thomson, R.W., "An Investigation Into Low Speed Rear Impacts of Automobiles," Master's Thesis, University of British Columbia, 1990.

Romilly, D.P., Thomson, R.W., Navin, F., and MacNabb, M.J., "Low Speed Rear Impacts and the Elastic Properties of Automobiles," Proceedings, 12th International Conference of Experimental Safety Vehicles, 1989; Gothenburg, Sweden.

Bailey, M.N., King, D.J., Romilly, D.P., and Thomson, R.W., "Characterization of Automotive Bumper Components for Low Speed Impacts," Proceedings of the Canadian Multidisciplinary Road Safety Conference VII, 1991, Vancouver, British Columbia.

Emori, R.I. and Horiguchi, J., "Whiplash in Low Speed Collisions," SAE Paper 900542.

Collins, J.C., Accident Reconstruction, Charles C. Thomas, Publisher, 1979.

93184

Design, Development, Testing and
Evaluation of Energy Absorbing Bumper

M. K. Chaudhari, S. S. Sandhu, and R. Ragh
The Automotive Research Assn of Ind

ABSTRACT :

At present most of the Indian vehicles are fitted with rigidly mounted bumpers without energy absorbing devices. An alternative to the rigidly mounted bumper is considered by designing and developing various energy absorbing devices and energy attenuators such as elastomers in various shapes and sizes, the collapsible, sheet metal fenders, metal flat springs and the 'U' type energy attenuator. The various factors such as aerodynamics, aesthetic, easy removal and mounting of the bumpers were also considered while designing and fabrication of the bumpers.

The sheet metal and Fibre Reinforced Plastic (FRP) bumpers are designed and fabricated in four different types with various combinations of elastomers, energy absorbing devices and attenuators.

A major emphasis is given on low cost fabrication of three dimensional bumper shell, dies, moulds, collapsible and energy absorbing devices. Various designs of the bumpers are evaluated with the help of pendulum impact test rig by making additional provision to find out the impact load and the deflection transmitted at the mounting point of the vehicles which gives additional information to the vehicle designers.

The paper covers various designs, methodology used for low cost prototype fabrication and the testing and evaluation of bumpers simulating 8 km/hr impact. The data presented also gives the relative weight of various thicknesses of the bumpers and the energy absorbed during impact and transmitted on vehicle mounting point.

1. INTRODUCTION :

The function of the bumper is to absorb the shock during slow speed collision without causing damage to other parts of the vehicle. The bumpers are rated in terms of the speed upto which they can effectively absorb the shock.

The bumper is to be suitably designed for a particular speed. An under-designed bumper deflects more during the impact and damages adjacent parts of the vehicle. On the other hand, an overdesigned bumper acts as a rigid member and directly transmits the shock to other parts of the vehicle. To have a uniformity in their performance, bumper standards were developed and is in practice since 1975. In India this awareness is created very recently with the formulation of the Automotive Research Association of India (ARAI) Safety Standard.

This project is sponsored by the Government of India, to Design and Develop Bumper for the Indian Vehicles. Four different designs are made and their performance is compared with the currently used ones. A test facility is also developed conforming to popular test standards for evaluation of the bumpers.

2. BUMPERS DEVELOPED BY ARAI :

All the bumpers have been designed for a speed of 8 km/hr. The bumpers developed are for a passenger car having an unladen weight of approximately 10,000 N (10kN). Four basic designs are developed keeping the outer shell shape same while changing the internal energy absorbing elements. Bumpers are fabricated out of Cold Rolled Critically Annealed (CRCA) sheet of 1.2 mm thick, 0.7 mm thick and in Fibre Reinforced Plastic (FRP) having average thickness of 3.00 mm.

3. SHELL DESIGN:

The shell is trapezoidal in section. The front part has been strengthened by forming to the shape of a channel.A rubber beading sits over the channel which absorbs part of the energy and also adds to the aesthetics. Since expensive dies are required for producing pressed components, it was decided to fabricate the bumper shell out of sheet metal. For ease of fabrication, cuts were made transversely through out the shell length at intervals. The shape of the shell was designed, keeping in'mind the aerodynamics and aesthetic consideration. The trapezoidal shape has been considered from the point of view of absorbing maximum energy, as the shape factor for this design would be around 1.3 to 1.5. (1 & 2)*

The shell is fabricated in 1.22 mm and 0.71 mm thick Cold Rolled Critically Annealed (CRCA) sheet and 3 mm thick Fibre Reinforced Plastic (FRP). The values for bending moment and safe load of shell are given below :

Table -1

Type of Shell	Thickness in mm	Weight in N	Bending Moment in Nm	Safe Load in KN
CRCA Sheet	1.22	58	7506	30.95
CRCA Sheet	0.71	38	4650	24.50
FRP	2.8	30	-	-

4. FABRICATION:

4.1 SHEET METAL:

The production of bumpers using dies and press would have resulted in higher strength,better aesthetic appeal and uniformity of size. But this was out of the scope of the project,due to limitations in the cost and time. The fabrication was carried out using simple and cheaper techniques.

A mould was fabricated for fabrication of the bumpers. The female part of the mould is fabricated out of steel and the male part was made of wood.The wooden surface was given a protective covering using metal to avoid chipping of the wood. A 400 kN press, used for material testing, was used to press the sheet metal to get the desired shape. For ease of pressing, transverse cuts were made along the length of the bumper. After pressing, the shape was retained using clamps and the cuts were welded together.

4.2 FIBRE REINFORCED PLASTIC (FRP) :

Chopped strand mat having a density of 450 gms is used. This material is not found to be suitable for bumper application because of high modulus of elasticity. The bumper broke into pieces on impact. For similar application, continuous fibre glass composites are ideal. They exhibit very high tensile strength and low modulus of elasticity which is desirable for energy absorbing devices.

5. INTERNAL ENERGY ABSORBING DEVICES :

The combination of bumpers are made with different energy absorbing devices designed and developed, which are given below :

5.1 SPRING TYPE:

The basic shell shape is retained. A spring of 40 x 5 mm cross section bend into the shape of a bow constitute the main energy absorbing element.The spring is fixed to the bumper using bolts. A bellow shaped rubber elastomer is provided between the spring and the mounting structure which holds the bumper to the structure.

On impact,the bumper moves backward compressing the spring to absorb the major part of the impact energy.The remaining part of the energy is absorbed by the rubber elastomer.

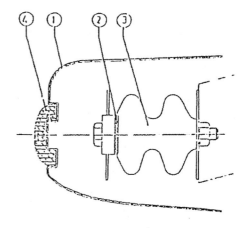

FIG.1 BUMPER (RUBBER WITH SPRING)

1) Bumper   2) Spring   3) Collapsible
4) Rubber Stopper

* Numbers in Parenthesis Designate References at the end of Paper.

5.2 RIB AND BRACKET TYPE :

The shell is reinforced internally throughout the length at intervals with vertical ribs. Due to the impact force, the shell moves backward deforming the bracket, thereby absorbing a part of the energy. The rubber elastomers which are mounted parallel to the bracket absorb the remaining part of the impact energy. The bumper corners are given additional supports to take care of corner impacts.

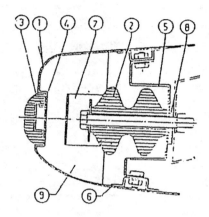

FIG.2 BUMPER (COLLAPSIBLE RUBBER + RIB STRUCTURE)

1) Bumper 2) Collapsible
3) Stopper 4) Bracket 5) Support
6) Bracket 7) Rib Strip 8) Plate
9) Rib

5.3 COLLAPSIBLE TYPE :

A collapsible type of energy absorber in conjunction with rubber elastomers is used to absorb the impact energy. The collapsible type energy absorber is made out of sheet metal of 0.9 mm thickness. The sides are given corrugations to absorb the energy.

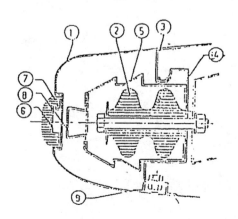

FIG.NO.3 BUMPER (COLLAPSIBLE STRUCTURE + RUBBER)

1) Bumper 2) Collapsible Rubber Bush
3) Bracket 4) Support 5) Collapsible
Bracket 6) Bracket 7) Rubber Stopper
8) Bracket 9) Mounting Bracket

5.4 'U' TYPE ENERGY ABSORBER :

The energy absorber has the shape of the letter 'U'. The two ends of the absorber are fixed to the bumper using bolts. The bumper is held on to the structure by a bolt passing through the center of the absorber. During impact the absorber deflects and absorbs the impact energy. The arms of the absorber are of uneven length to match the inner profile of the bumper.

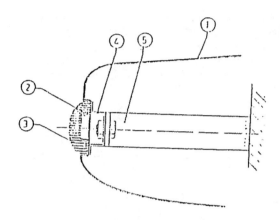

FIG. 4 BUMPER WITH ~U' TYPE ENERGY ATTENUATOR ELASTOMER

1) Bumper Shell 2) Bracket for Rubber Stopper 3) Rubber Stopper 4) Bracket for Energy Attenuator 5) Energy Attenuator

6. THE BUMPER TEST RIG :

The rig is designed to test all bumpers of vehicles with a maximum GVW of 2000 Kg. It incorporates the requirements laid out by standards such as ISO 2958, SAE J 980a, ARAI SS No.36 and FMVSS Part 581. (3),(4),(5),(6)

The pendulum type test set up is preferred over other types for its simplicity, lesser cost and reliability. The main structure is 4 metre high, fabricated out of steel channel to take the shock of impact. The pendulum swings freely and is provided with a maximum swing of 30° corresponding to 10.5 km/hr. A total height adjustment of 200 mm is possible. Weights are to be added to simulate the mass of the vehicle under test. Since this is a developmental type of work, load sensors and associated instrumentations are used for recording the impact and the forces transmitted on the structure. Load sensing elements of suitable capacity are provided on the structure (Bumper mounting points), so as to determine the impact force transmitted to chassis components. As per the standard it is not necessary to measure this force, but from the Research & Development point of view this information is important for vehicle design.

The technical specification of the Bumper Evaluation Test Rig is as given below :

SPECIFICATION :

- Bare Weight of Test Rig – 250 Kg
- Impact Simulation upto – 2000 Kg
- Max Swing of Pendulum – 30 Degree
- Max Centre Height – 500 mm
- Min Centre Height – 400 mm
- Pulling & Releasing
  Mechanism – Mechanical/ Pneumatic
- Floor Space Reqd. – 25 Sq.m. + Parking Space for Vehicle

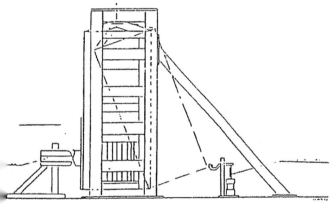

FIG. 5 BUMPER EVALUATION TEST-RIG

MODEL - W 101
SIX CHANNEL
MEMORY PER CHANNEL-8KB
SAMPLING INTERVAL-1 10
MICRO SECOND

NO. OF SAMPLES-UPTO 8000 SAMPLES

CHARGE AMPLIFIER
(C1) (C2) (C3)

PENDULUM TO SIMULATE
UPTO 2 TON GVW
L1 L2 L3-LOAD WASHERS
F1 F2 -FORCE LINK
D1 D2 D3-DISPLACEMENT
TRANSDUCER

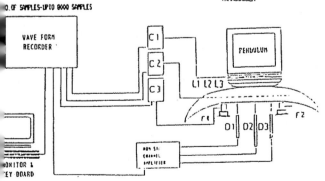

FIG. 6 INSTRUMENTATION SET-UP FOR THE BUMPER TEST-RIG

The various type of energy absorbing devices used in the fabrication of prototype bumpers are given below :

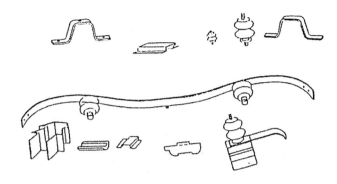

FIG. 7 SPRING,BRACKET,RIB,COLLAPSIBLE ELEMENT, U-TYPE ATTENUATOR & ELASTOMER

7. TESTING AND EVALUATION OF BUMPERS :

Performance evaluation of the bumpers developed is carried out on the test facility described earlier. Four channels of the wave form recorder are used for recording. The signals from the load washers are fed into the first channel; signals from the two force links (Structural Forces) are fed to the second and third channel and the displacement signal from LVDT is connected to the fourth channel. For effective recording, all the channels are to be simultaneously triggered. This is done by using the output signal from the load washers to trigger the channels internally.

8. TEST RESULTS AND EVALUATION :

8.1 TEST CONDITIONS :

Test Rig – Pendulum Type Impact Test Rig

Impact Mass – 12.50 kN – Max upto 16 kN

Impact Velocity – 2.2 m/sec (8 km/hr) Max upto 10.5 km/hr

Impact Energy-(KE) $= 1/2 \, M \times V^2 = 3025$ N-m for 8 km/hr $= 756$ Nm for 4 km/hr

8.2 PARAMETERS MEASURED :

Impact Force – Using Load Washer

Structural Force – Using Force Links

Bumper & mounting   — Using LVDT
points deflection

8.3 INSTRUMENTATION SET UP :

Charge Amplifier   —   To amplify the
                        charge output
                        from load sensors
                        to a measurable
                        voltage

LVDT Amplifier   —   To amplify the
                        signal form the
                        LVDT

Wave Form Recorder —   Facility to store
                        the data and for
                        analysis later.
                        Data stored in
                        floppies on
                        personal computer.

8.4 TEST RESULTS :

The graph of energy absorbed i.e. load
V/s deflection for various prototypes
bumper Designed & Developed by ARAI is
given below. Test results are based on 8
km/hr impact simulation.

SAMPLE NO.1

SHEET METAL — 1.2 mm
SPRING TYPE
ENERGY ABSORBED — 1.26 kNm

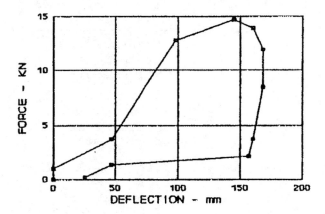

FIG. NO. 8 ENERGY ABSORBED IN BUMPER

SAMPLE NO.2

SHEET METAL — 1.2 mm
COLLAPSIBLE TYPE
ENERGY ABSORBED — 1.4 KN –M

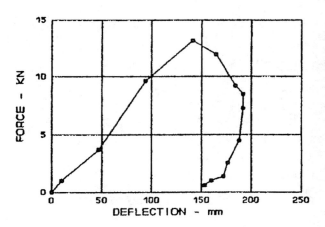

FIG.NO. 9   ENERGY ABSORBING BUMPER

SAMPLE NO.3

SHEET METAL — 1.2 mm
ATTENUATOR TYPE
ENERGY ABSORBED — 1.3 kNm

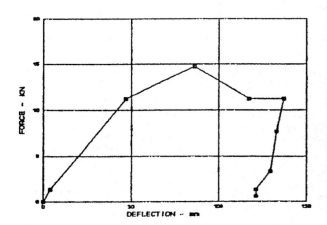

FIG. 10 ENERGY ABSORBING BUMPER

784

SAMPLE NO.4

SHEET METAL   – 0.7 mm
SPRING TYPE
ENERGY ABSORBED – 0.90 kNm

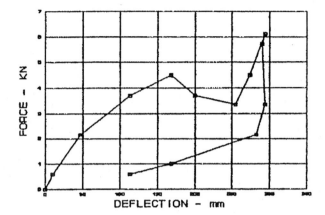

FIG.NO.11 ENERGY ABSORBING BUMPER

SAMPLE NO.6

SHEET METAL – 0.7 mm
BRACKET TYPE
ENERGY ABSORBED – 1.4 kNm

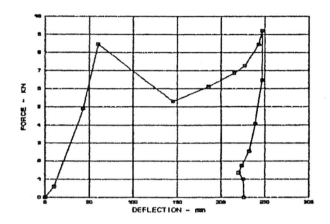

FIG. 13 ENERGY ABSORBING BUMPER

SAMPLE NO.5

SHEET METAL   – 0.7 mm
COLLAPSIBLE TYPE
ENERGY ABSORBED – 1.0 kNm

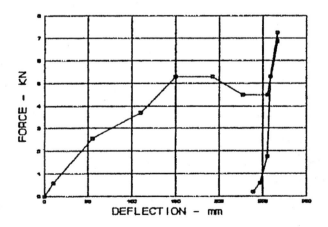

FIG. 12 ENERGY ABSORBING BUMPER

SAMPLE NO.7

FRP – SPRING TYPE
ENERGY ABSORBED – 1.36 kNm

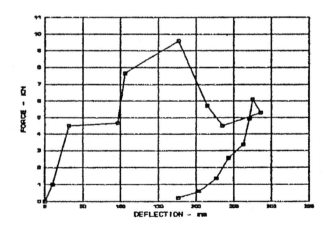

FIG. 14 ENERGY ABSORBING BUMPER

SAMPLE NO.8

FRP - COLLAPSIBLE TYPE
ENERGY ABSORBED - 0.65 kNm

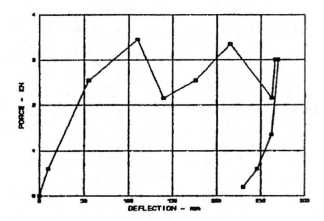

FIG. 15 ENERGY ABSORBING BUMPER

SAMPLE NO.10

ABS PLASTIC
ENERGY ABSORBED - 0.6 kNm

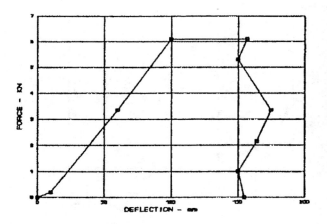

FIG. 17 ENERGY ABSORBING BUMPER

SAMPLE NO.9

POLYURETHENE
ENERGY ABSORBED - 1.96 kNm

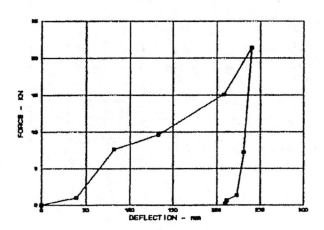

FIG. 16 ENERGY ABSORBING BUMPER

SAMPLE NO.11

SHEET METAL - 1.5 mm
ENERGY ABSORBED - 0.36 kNm

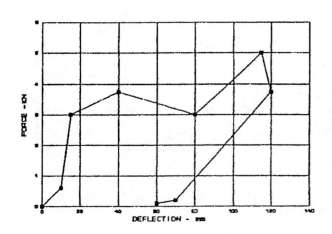

FIG. 18 ENERGY ABSORBING BUMPER

SAMPLE NO.12

SHEET METAL - 3 mm
ENERGY ABSORBED - 2 kNm

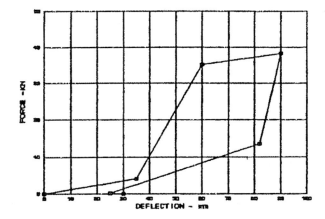

FIG. 19 ENERGY ABSORBING BUMPER

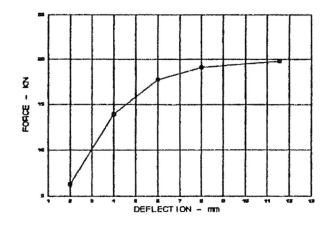

**FIG. 20 'U' TYPE ENERGY ATTENUATOR
FORCE - DEFLECTION CHARACTERISTICS**

The test results have been tabulated and is given in Table - 2.

Table - 2

COMPARISON OF TEST RESULTS OF ARAI DEVELOPED BUMPERS AND CURRENTLY USED BUMPERS

SAMPLE NO.	MATERIAL	INTERNAL DETAILS	MAX. REACTION FORCE BY THE BUMPER (kN)	MAX. FORCE ON STRUCTURE (kN)	DEFLECTION (mm)	PERMANENT SET (mm)	ENERGY ABSORBED (kNm)	EFFICIEN PERCENTA
(1)	(2)	(3)	(4)	(5)	(6)	(7)	(8)	(9)
1	SM (1.2)	Spring	14.7	9.18	168	80	1.26	41
2	SM (1.2)	Collapsible	13.15	9.54	192	90	1.4	46
3	SM (1.2)	Attenuator	14.7	15.92	137	-	1.3	43
4	SM (0.7)	Spring	6.1	20.0	294	160	0.9	30
5	SM (0.7)	Collapsible	7.25	13.40	266	155	1.0	33
6	SM (0.7)	Bracket	9.2	7.46	247	110	1.4	46
7	FRP-3	Spring	9.6	5.26	286	-	1.36	-
8	FRP-3	Collapsible	3.35	20.0	266	-	0.65	21
9	Polyurethene-2 (Currently used)		21.35	20.0	239	-	1.96	65
10	ABS Plastic-4 (Currently used)		6.1	-	172	-	0.6	20
11	SM (1.5) (Currently used)		4.9	20.0	121	-	0.36	12
13	SM 3 (Currently used)		38.2	-	125	125	2.0	66

787

9. OBSERVATIONS AND CONCLUSIONS :

9.1 The efficiencies of all the

bumperslie between 20% and 50% on account of 12 to 15 cuts provided on top and bottom side of the shell throughout the length. These cuts are welded to form the shell.

9.2 The permanent set in all the cases exceeded the stipulated limit of 19 mm. This can be improved by using materials of higher strength, which gives better strength to weight ratio and also byhaving energy attenuators of higher capacity.

9.3 In the present design several cuts are made throughout the length, transversely at intervals for easy bending. These cuts are later welded together to form the shell. The joints are spot welded at intervals which turned out to be the weak spots in the shell. If the cuts are continuously welded this problem can be overcome and the energy absorbing capacity can be increased by approximately 40 to 50%.

If the bumper is fabricated as a pressed or formed component in one piece, then the energy absorbing capacity would have improved by 60 to 70%.

9.4 All the FRP bumpers tested are damaged beyond repair. The shell broke into two or three pieces before the designed deflection is attained. This is inspite of having a shell thickness of 3 to 4 mm. It is felt that the material selected (Mat and Resin) are not suitable for this kind of service. The shell should be able to take more deflection.

9.5 Bumpers fabricated out of (0.7mm) gauge sheets are not able to withstand the impact force. The calculated maximum safe force, the bumper can take (124.5 kN) is roughly 60% of that of the 18 gauge bumper

(139.5 kN). The deflections experienced are between 250 to 290 mm.

9.6 The variation in the permanent set for the same type of bumpers is due to the variation in the strength of the welded joints of the shell. Study of the tested bumpers have revealed that bumpers with weak welding is weak and have larger values of permanent set. This is because, the shell during deflection is pulled apart the welded joints thereby snapping the welding.

9.7 Flexibility in the Design of bumpers can be introduced by keeping the same shell shape and changing the energy absorbing devices in the bumper to fulfill the requirements of individual cases. The shape, size, mounting and energy absorbing devices within the bumper will vary from vehicle to vehicle.

9.8 It is observed that mounting of bumper is equally important from the point of view of supporting the bumper on body shell to enhance the energy absorbing capacity of bumper with less damage to vehicle aggregates.

10. ACKNOWLEDGEMENT :

The authors are grateful to the Director of ARAI for giving permission to publish this work. We are deeply indebted to all our Colleagues at ARAI for their assistance. The authors thank Mr.A.S. Tamhankar, Project Engineer, Mr.D.A. Jagdale for his secretarial work and Mr.A.N. Ghongade and Mr.J.D.Kasar for fabrication of the bumpers. This work was carried out as a part of R & D project financed by the Government of India.

11. REFERENCES :

1. "Basic Design Criteria for Aluminium Bumpers", Reynolds Aluminium Co.

2. Sharp M.L., Peters R.M., Weiss R.B - "Structural Design Considerations for Aluminium Bumpers"

3. ISO-2958 - "Road Vehicles - Exterior Protection for Passenger Cars"

4. SAE J 980a - "Bumper Evaluation Test Procedure - Passenger Cars"

5. ARAI Safety Standard - No.36/1986 - "Bumper"

6. Federal Motor Vehicle Safety Standards, Part 581 - Bumper Standard

7. McDougall M.K., Epel J.N., Wilkinson R.E. - "Bumper Energy Attenuators made from Fibre Reinforced Plastic - SAE 790334"

Low Speed Car Impacts with Different Bumper Systems: Correlation of Analytical Model with Tests

Irving U. Ojalvo, Brian E. Weber and David A. Evensen
Technology Associates

Thomas J. Szabo and Judd B. Welcher
Biomechanical Research & Testing

ABSTRACT

A coordinated test and analysis program was conducted to determine whether a previously proposed, linear, analytical model could be adapted to simulate low speed impacts for vehicles with various combinations of energy absorbing bumpers (EAB). The types of bumper systems impacting one another in our program included, in various combinations; foam, piston and honeycomb systems. Impact speeds varied between 4.2 and 14.4 km/h (2.6 and 9.0 mph) and a total of 16 tests in 6 different combinations were conducted. The results of this study reveal that vehicle accelerations vary approximately linearly with impact velocity for a wide variety of bumper systems and that a linear mass-spring-damping model may be used to efficiently model each vehicle/bumper-system for low speed impacts.

INTRODUCTION

Low speed impact tests have been conducted for over four decades to determine vehicle and occupant response (e.g. see References 1-7). Recently, an efficient linear model was proposed [8,9] to simulate such impacts for piston-type, energy absorbing bumper (EAB) systems. As a result of the success achieved therein, the present effort was undertaken to explore whether the same linear model could also be applied to reasonably approximate other popular bumper systems currently in use, such as honeycomb and foam energy absorbing systems. The specific systems considered herein are: a Mazda/piston-EAB striking a Honda/foam-EAB at 3 different speeds; a Pontiac/honeycomb-EAB striking a Ford/piston-EAB at 2 different speeds; an Hyundai/foam-EAB striking an Audi/piston-EAB at 3 different speeds; a Ford/foam-EAB striking an Hyundai/foam-EAB at 3 different speeds; a Pontiac/piston-EAB striking a Nissan/foam-EAB at 2 different speeds; and a Buick/piston-EAB striking a Pontiac/honeycomb-EAB at 3 different speeds. Impacts were conducted at speeds which varied between 4.2 and

14.4 km/hr (2.6 and 9.0 mph) and a total of 16 tests in 6 different configurations were conducted. Target vehicle Δv's ranged from 2.7 to 9.2 km/hr and bullet vehicle Δv's ranged from 2.5 to 10.2 km/hr.

TEST PROCEDURE & DATA SCALING

Impact configurations were selected to examine the interactive effects of the three predominant types of energy absorbing bumper systems, i.e. those involving piston, foam and honeycomb elements. Bullet and target vehicles were equipped with on-board instrumentation. A tri-axial block of IC Sensor 3031-050 (50 g) accelerometers were mounted at the target and bullet cars' approximate centers of mass. Electronic instrumentation systems employed were in accordance with standard SAE J211, with the X, Y and Z axes directed, in terms of the vehicles, forward, to the right and downward, respectively. Pressure sensitive tape switches were mounted on the ground to record the bullet vehicle impact speeds.

Vehicle accelerometer data were collected with a self contained, on-board data acquisition system. Data collection was initiated by a pressure sensitive tape switch on the target vehicles' rear bumpers. Acceleration data were collected at 1000 samples per second. Prior [10] power spectral density analysis has found this to be a sufficiently high collection frequency to avoid aliasing (i.e. false signal) problems for low speed impacts. Target vehicle accelerations were filtered using an SAE Class 60 filter. Bullet and target vehicle velocity changes were calculated by integration of the vehicle X-axis acceleration (an SAE Class 180 filter was used for change-in-velocity determinations).

The test response results for each struck vehicle within a given test configuration were then scaled linearly to a single impact speed (12 km/h) and plotted on the same scale to permit examination as to whether the responses were linear with impact speed. The method of scaling used was to multiply all acceleration data by the factor of

Based upon our comparisons, which are discussed later in this paper, it was determined that the responses were reasonably linear with impact speed, for a given vehicle combination.

ANALYSIS PROCEDURE

The analytical model used to simulate the in-line impact is depicted in Figure 1, wherein the springs and dashpots shown are assumed to operate linearly. The positions of each vehicle, X_1 and X_2, are measured from each vehicle's center of gravity at the time of impact contact initiation. The governing dynamic equations and initial conditions for this idealized system are given by [5]:

$$m_1 x_1'' + c_{eq}(x_1' - x_2') + k_{eq}(x_1 - x_2) = 0$$
$$m_2 x_2'' + c_{eq}(x_2' - x_1') + k_{eq}(x_2 - x_1) = 0 \qquad \text{(Eq. 1)}$$

where the m's, c's and k's for vehicles 1 and 2 are depicted in Figure 1, the following definitions apply

$$(x)' = d(x)/dt, \quad c_{eq} = c_1 (c_2/(c_1 + c_2)),$$
$$k_{eq} = k_1 (k_2/(k_1 + k_2)) \qquad \text{(Eq. 2)}$$

with initial conditions

$$x_1(0) = x_2(0) = 0, \; x_1'(0) = 0, \; x_2'(0) = V_0 \qquad \text{(Eq. 3)}$$

Eq (1) are only valid when the forces between them are compressive. Thus, the c.g. of the target vehicle can only experience a positive acceleration. Accordingly, these equations, and their corresponding solutions, no longer apply after the initial contact interval and their separation forces reach a value of zero. After this, the (assumed unbraked during impact) vehicles part at their relative speed at this (zero contact-force) instant.

To solve the governing equations, they are cast in modal form using the transformation

$$\{X\} = [\Phi]\{\xi\} \qquad \text{(Eq. 4)}$$

where the matrices are defined as

$$\{X\} = \begin{Bmatrix} x_1 \\ x_2 \end{Bmatrix}, \; [\Phi] = [\phi_1 \; \phi_2], \; \{\xi\} = \begin{Bmatrix} \xi_1 \\ \xi_2 \end{Bmatrix},$$

and

$$[K][\Phi] = [M][\Phi][\Lambda] \text{ and } [\Phi]^T [M][\Phi] = [I] \qquad \text{(Eq. 5)}$$

with

$$\Lambda = \begin{bmatrix} 0 & 0 \\ 0 & \omega_n^2 \end{bmatrix}, \; [K] = k_{eq}[F], \; [F] = \begin{bmatrix} 1 & -1 \\ -1 & 1 \end{bmatrix}$$

and

$$[M] = \begin{bmatrix} m_1 & 0 \\ 0 & m_2 \end{bmatrix}.$$

After pre-multiplication by $[\Phi]^T$, the governing equations become:

$$[\Phi]^T [M][\Phi]\{\xi''\} + [\Phi]^T [C][\Phi]\{\xi'\} + [\Phi]^T [K][\Phi]\{\xi\} = 0 \,\text{(Eq. 6)}$$

where

$$[C] = c_{eq}[F].$$

Solution of Eq (6) then follows the procedure described and set forth in Reference 5.

In the absence of test vehicle data, such as K_1, K_2, C_1 and C_2, the correct value of ω_n is determined from the impact test data by observing the duration of the target vehicle's positive-g pulse. This determines the damped frequency from the relationship $T_d/2 = \pi/\omega_d$. The damping ratio ζ is then selected, using trial and error, so that the peak g response agrees with the impact test data.

NUMERICAL RESULTS AND DISCUSSION

Comparisons of the analytically-based numerical solutions for each configuration were made with each set of scaled test data. These are presented in Figures 2-7, from which it may be seen that the test results are reasonably linear with impact speed and the analytically determined solutions reasonably approximate the normalized test results.

It is noted that the test results are not smooth and show erratic variations. This is to be expected as the detailed response contains many frequencies associated with a complex mechanical system. However, a detailed observation reveals that a dominant lower frequency response (with an early peak acceleration and more prolonged decay) exists, upon which many higher frequencies are superimposed. It is this dominant response which the analytical solution has been designed to capture, as it does reasonably well, considering its comparative simplicity.

With regard to any nonzero accelerations obtained by analysis for the start of the impact, these occur because the idealized car masses have a finite relative velocity at the start of impact and the dashpot forces start acting immediately upon the idealized rigid masses to give them sudden finite accelerations at t = 0.

Table 1 summarizes the vehicles tested, their bumper system elements, impact speeds and the normalized velocity changes experienced. It may be seen that the analytical procedure tends to over-predict the velocity changes by approximately 25%. However, the analytical procedure is far more accurate for predicting the peak impact forces and accelerations.

CONCLUSIONS AND RECOMMENDATIONS

Based upon the results presented and discussed herein, it appears that in-line, low speed (up to 24 km/h or 15 mph), impacts of cars with various types of bumper sys-

osed model has been validated by the favorable comparisons with the various tests cited herein, in that a wide range of different bumper system configurations were considered. Although the simulation equations presented and solved have been described in numerous papers (e.g. see [8, 9]), it is believed that the present work is the first successful general validation of the proposed model.

To employ the proposed analytical model in cases where no crash test data is available, it would be desirable to have single vehicle nondestructive impact data to help determine each vehicle's frequency response and decay rate. This information could then be used to determine each vehicle's stiffness and damping constants. If such test data is not available, it could be easily obtained from current federally mandated tests on vehicle bumper systems by simply requiring that rigid compartment mounted accelerometers be used to record their response with time. Such data could then be used, together with the proposed model, to help analyze low speed impacts better and to design superior vehicle bumper systems, with lower responding g-levels, in the near future.

REFERENCES

1. Severy, D. M., et al, "Controlled Automobile Rear-End Collisions, An Investigation of Related Engineering and Medical Phenomena", *Canadian Services Medical Journal*, Volume XI, No. 10, Nov 1955.

2. Romily, D.P., Thomson, R.W., Navin, F.P.D. and MacNabb, M.J., ALow Speed Impacts and the Elastic Properties of Automobiles@, *12th International Technical Conference on Experimental Safety Vehicles*, pp. 1199-1205, 1989.

3. Chandler, R. F., and Christian, R. A., "Crash Testing of Humans in Automobile Seats", *Society of Automotive Engineers* (SAE No. 700361), 1970.

4. Szabo, T. J. and Welcher, J., "Dynamics of Low Speed Crash Tests with Energy Absorbing Bumpers", *Society of Automotive Engineers* (SAE No. 921573 and SAE SP-925), 1992.

5. King, D.J., Siegmund, G.P., and Bailey, M.N., AAutomobile Bumper Behavior in Low-Speed Impacts@, *Society of Automotive Engineers* (SAE No. 930211), pp 1-18, 1993.

6. Malmsbury, R.N. and Eubanks, J.J., ADamage and/or Impact Absorber (Isolator) Movements Observed in Low Speed Crash Tests Involving Ford Escorts@, *Society of Automotive Engineers* (SAE No. 94012), 1994.

7. Bailey, M.N., Wong, B.C. and Lawrence, J.M., AData and Methods for Estmating the Severity of Minor Impacts@, *Society of Automotive Engineers* (SAE No. 950352), 1995.

8. Thomson, R.W. and Romily, D.P., ASimulation of Bumpers During Low Speed Impacts@, *Proceedings of the Canadian Multidisciplinary Road Safety Conference VIII*, June 14-16, 1993.

9. Ojalvo, I. U. and Cohen, E. C., "An Efficient Model for Low Speed Impact of Vehicles", *Society of Automotive Engineers* (SAE No. 970779 and SP-1226), 1997.

10. Szabo, T. and Welcher J., "Human subject Responses to Various Acceleration Fields", presented at and published for the SAE Low Speed Collision TOPTEC, August 19-20, 1996.

$$x'_2 = v_0 \quad \text{at} \quad t = 0$$

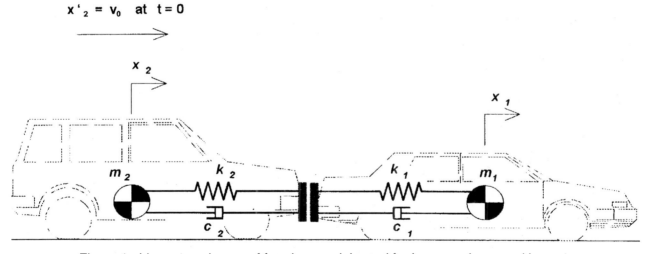

Figure 1. Linear two degree of freedom model, used for low speed rear end impact.

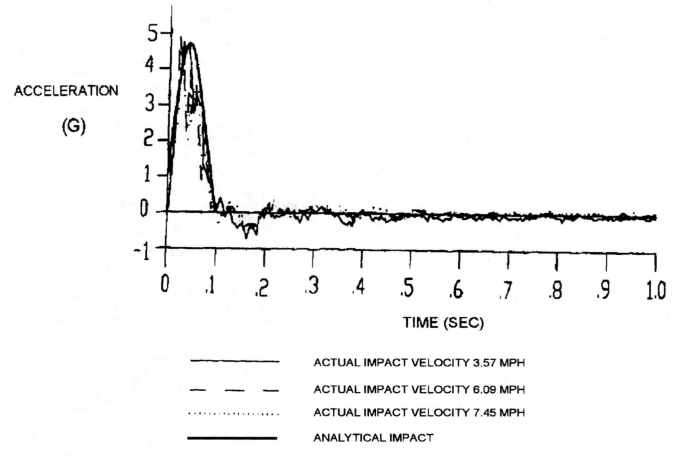

Figure 2. Impact accelerations for '84 Audi 4000S (piston bumper) rearended by '87 Hyundai Excel (foam bumper). Curves normalized to 7.5 mph impact velocity.

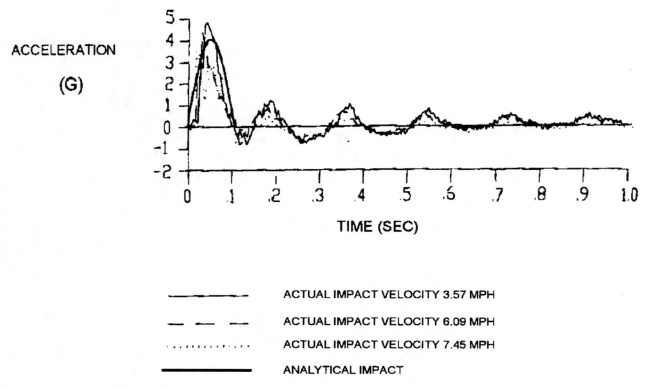

Figure 3. Normalized impact accelerations for '90 Honda Accord (foam bumper) rearended by '79 Mazda RX-7 (piston bumper). Curves normalized to 7.5 mph impact velocity.

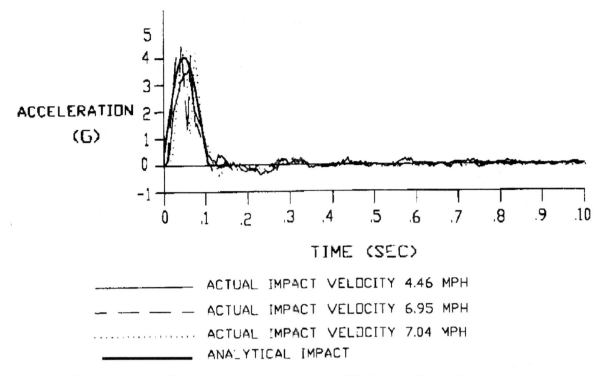

Figure 4. Normalized impact accelerations for '89 Hyundai Excel (foam bumper) rearended by '89 Ford Festiva (foam bumper). Curves normalized to 7.5 mph impact velocity.

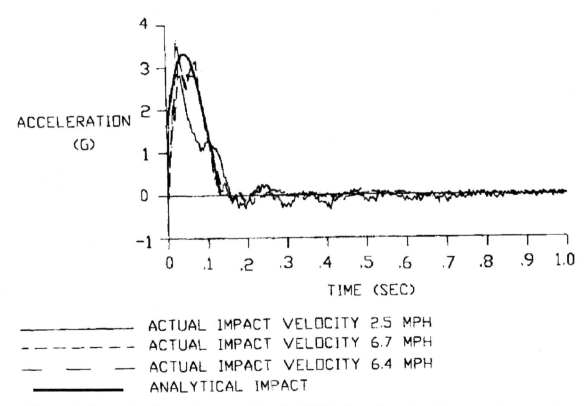

Figure 5. Normalized impact accelerations for '87 Pontiac Grand Am (honeycomb bumper) rearended by '84 Cutlass Supreme (piston bumper). Curves normalized to 7.5 mph impact velocity.

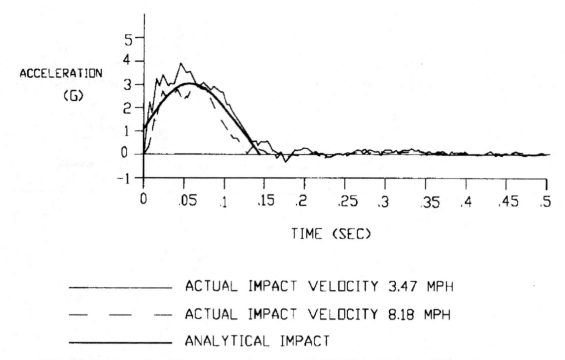

Figure 6. Normalized impact accelerations for '89 Nissan Stanza (piston bumper) rearended by '84 Pontiac 6000 (foam bumper). Curves normalized to 7.5 mph impact velocity.

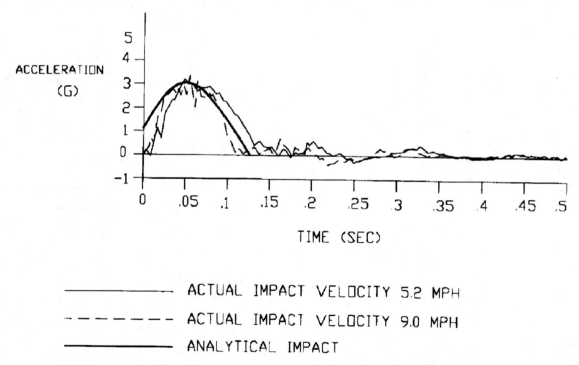

Figure 7. Normalized impact accelerations for '93 Ford Taurus (piston bumper) rearended by '94 Pontiac Grand Am (honeycomb bumper). Curves normalized to 7.5 mph impact velocity.

Table 1. Summary of Test and Analysis Speeds

Test	Assumed Analytical Damping	Bullet Vehicle	Bumper Type*	Test Impact Speed†	Normalized Test ΔV†‡	Analytical ΔV†‡	Target Vehicle	Bumper Type*	Normalized Test ΔV†‡	Analytical ΔV†‡
1	.1	87 Hyundai Excel	F	5.7	9.9	11.3	84 Audi 4000S	P	8.2	11.5
2			F	9.8	8.6			P	7.6	
3			F	11.9	8.3			P	7.2	
4	.05	79 Mazda RX-7	P	4.2	7.0	12	90 Honda Accord	F	not recorded	10.1
5			P	5.9	5.9			F		
6			P	8.3	7.5			F		
7	.05	89 Ford Festiva	F	7.2	9.4	12.1	89 Hyundai Excel	F	8.0	10.0
8			F	11.2	9.3			F	8.2	
9			F	11.2	9.4			F	8.4	
10	.2	84 Cutlass Supreme	P	4	7.7	9.1	87 Pontiac Grand AM	H	8.2	10.4
11			P	10.7	7.5			H	8.7	
12			P	10.2	7.2			H	8.2	
13	.21	84 Pontiac 6000	P	7.8	8.5	7.0	89 Nissan Stanza	F	8.4	9.4
14			P	13.1	8.2			F	8.4	
15	.15	94 Pontiac Grand AM	H	8.3	9.9	10.4	93 Ford Taurus	P	8.8	9.2
16			H	14.4	8.5			P	7.6	

* P = Piston, F = Foam, H = Honeycomb

† All velocities are in km/h

‡ Normalized for impact speed of 12 km/h

CHAPTER TEN

Seat belts

Effectiveness of Safety Belts Under Various Directions of Crashes

D. Cesari, R. Quincy, and Y. Derrien
Organisme National de Securité Routière

THE APPROVED STANDARDS for safety belts, now in force, recommend frontal crash tests. Investigations carried out in France as well as in foreign countries have shown that frontal crashes in a direction parallel to the vehicle axis seldom occurred.

Left and right frontal crashes are more frequent; in France in 1968, the total number of accidents having occurred within a 90 deg (± 45 deg) frontal area represent three-fourths of all the crashes that took place in France, whereas those having occurred within a 24 deg frontal area represent only one-third of all the crashes (1).*

The effectiveness of safety belts in frontal impact has been studied previously (2, 3). The aim of this paper is to complete the results by studying the effectiveness of safety belts in nonfrontal impacts. This paper investigates the effectiveness of safety belts under various crash directions. The investigation was carried out in three stages.

DYNAMIC SLED INVESTIGATION

PURPOSE - The main purpose of the first stage is to study the effectiveness of safety belts during impact tests performed

---

*Numbers in parentheses designate References at end of paper.

under conditions as far as possible similar to those found in actual crashes. We thus simulated impacts in directions frequently occurring in accidents. The selected crash angles ranged from 0-45 deg both ways.

DYNAMIC SLED - The achieve the first part of the research work, we used the catapult of the Organisme National de Securité Routière (ONSER) Crash Test Laboratory in Lyon.

This dynamic sled (Fig. 1) includes three components:

1. The launching system. A low-powered electric motor drives a flywheel (3000 mN). At the required velocity, the flywheel is coupled to a winch through a hydraulic clutch. The winch pulls the sled along by means of a wire rope (Fig. 2).

2. The sled is a welded metal assumbly mounted on rollers that guide it along two rails. Various types of equipment can be mounted on the sled, ranging from a plain seat to a complete car structure. To perform this investigation, we used a Peugeot 204 body mounted on a revolving platform (Figs. 3 and 4). This platform allows the car to be oriented both ways, in relation to the sled, under various angles by 5 deg increments.

3. The stopping system (Fig. 5) is based on the principle of absorption resulting from the deformation of a polyurethane cylinder under the action of a punch forcing its way into it, or rather from the flattening of a plastic material between the

_____ ABSTRACT

Studies of the effectiveness of safety belts were carried out under various directions of crashes, including dynamic sled investigations, destructive barrier tests, and impact tests. The studies showed that three-point belts were effective in frontal impact from 0-30 deg, but that their effectiveness diminished after 45 deg because the belt slips off the chest. The three-point belt did not provide protection for the knees. The studies also pointed out that the anchorage system of the belt is a very important factor in its effectiveness.

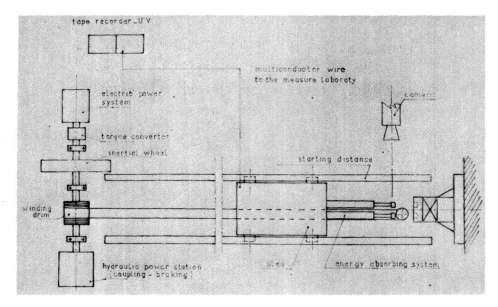

Fig. 1 - Installation of dynamic sled

Fig. 2 - Launching system

Fig. 3 - Platform

punch and the cylinder outer cover. Polyurethane was chosen as a constituent material because it is easy to use and can be reused.

The sled deceleration curve a = f (t) should be inscribed in the cross hatched part of Fig. 6, in compliance with the specifications of the European Economic Community's No. 16 regulations concerning safety belt dynamic tests.

MEASUREMENTS - The parameters being measured are:

1. Dummy head deceleration in three directions.
2. Dummy chest deceleration in three directions.
3. Belt strain at the three anchoring points.

The deceleration sensors are Statham strain-gage accelerometers (50 g for the car and the dummy chest, 100 g for the dummy head). The strain sensors (Fig. 7) are semiconductor strain-gage dynamometric bars.

The signals are amplified by a battery of Honeywell Accu-Data 112 differential input amplifiers, with adjustable ampli-fication factors ranging from 0.005-1000; they are recorded during the impact by a FM Ampex 1300 14-channel recorder, and can be easily visualized by direct recording on ultraviolet sensitive paper with a Honeywell 3508 recorder.

Two high-speed films are taken in the course of the tests, the first one perpendicularly to the car, at 1000 frames/s, the second one perpendicularly to the dynamic sled axis at 4000 frames/s. Analysis of the films permits study of the dummy displacement during the crash.

EXPERIMENTAL PROCEDURE - We utilized a car body (complete passenger compartment portion) mounted on the test sled by means of a revolving platform, allowing it to be oriented in every direction. The car-sled assembly was

Fig. 4 - Car body fitted on sled

Fig. 5 - Stopping system

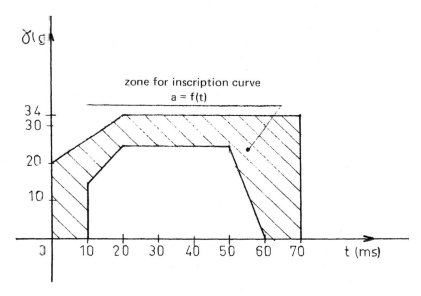

Fig. 6 - Allowed deceleration curve

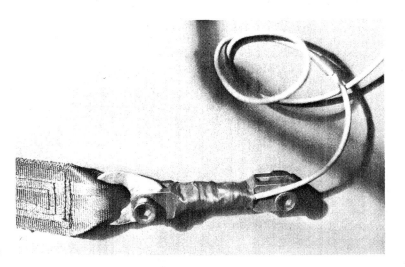

Fig. 7 - Strain sensor

Table 1 - Results of Sled Testing

Code	Direction, deg	Speed, km/h	Duration, ms	Car	Decelerations, g Head	Thorax	Belts, N A	B	C	SI	Observations*
101	0	48.1	110	30	X = 87 Y = 52 Z = 92	X = 44 Y = 22 Z = 27	5350	12700	11400	1970	
102	20L	47.3	105	45	X = 115 Y = 15 Z = 76	X = 40 Y = 15 Z = 34	5200	13000	12500	2900	
103	30L	46.2	100	35	X = 160 Y = 35 Z = 110	X = 45 Y = 20 Z = 27	7000	11200	12000	2645	(A)
104	20R	48.5	120	35	X = 80 Y = 37 Z = 66	X = 30 Y = 30	1000	17000	17000	(1043)	(B)
105	20R	48.2	105	45	Y = 30 Z = 125	X = 36 Y = 25 Z = 16	5500	7800	9700		(C)
106	45L	47.6	100	35	X = 63 Z = 100	X = 40 Y = 38 Z = 30	4500	7000	12000		(D)
107	30R	48.0	100	35	X = 50 Y = 40 Z = 40	X = 30 Y = 35 Z = 25	5700	8300	9900	546	

*Observations:

A—Breaking the twofold strap under a 12,000 N stress is shown in Figs. 9 and 10.

B—Opening of the buckle (probably by the dummy hand) happened when the tension began increasing.

C—The central pillar was bent under a 7800 N strain (Fig. 13). During test 105, a data-transmitting cable was broken, which explains why there is ▮ deceleration level indicated in X.

D—In test 106, the buckle gave way under 12,000 N strain (twofold strap) (Fig. 14). As the dummy was no longer restrained, it was thrown out ov▮ the front left door reinforcing beam (Fig. 15). A measuring cable was then broken under the tensile strain; consequently, we cannot know the decele▮ tion value at the head level in Y.

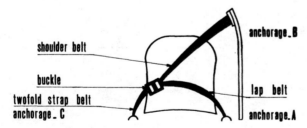

shoulder belt
buckle
twofold strap belt
anchorage_C
anchorage_B
lap belt
anchorage_A

Fig. 8 - Parts of belt

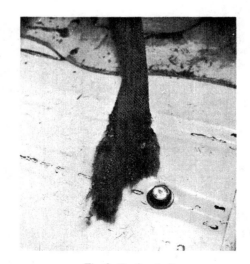

Fig. 9 - Broken belt

launched at the selected velocity, then decelerated by means of a suitable device permitting reproduction of the deceleration to which a vehicle is subjected in frontal barrier impacts. In order to comply with actual conditions, the vehicle was equipped with its initial seat and interior fittings. An Alderson VIP 50 A dummy was placed on the driver's seat and restrained from any movement by a three-point belt composed of separate straps, a nylon strap webbing not breakable at a strain of less than 12% elongation, and a Klippan buckle (Fig. 8). All the belt components have received approval in France.

The following were recorded in every test:

1. Deceleration at vehicle floor level.

2. Deceleration of the c.g. of the dummy head in three directions.

3. Deceleration in the chest in three directions.

4. Strains applied to the belt straps at the three anchoring points.

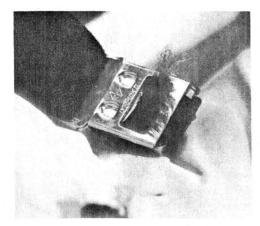

Fig. 10 - Belt broken at buckle

Fig. 13 - View of bent B-pillar

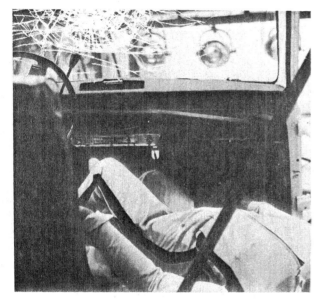

Fig. 11 - Inside view of cracked HPR laminated windshield (test 104)

Fig. 14 - View of broken buckle (test 106)

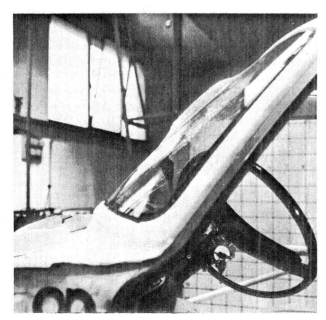

Fig. 12 - Outside view of cracked HPR laminated windshield (test 104)

Fig. 15 - Dummy posture after test 106

803

Fig. 16 - Car before crash barrier test 206

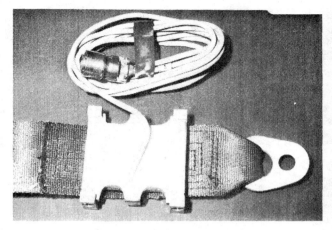

Fig. 18 - New crash sensor

Fig. 17 - Dummy before crash barrier test 206

Fig. 19 - Detail of knee impact

RESULTS - This investigation included a series of seven tests. The directions being studied were; 0, ±20, ±30, and ±45 deg. The results of the tests are indicated in Table 1.

VALUES OF SEVERITY INDEX - The severity index (SI) is defined by

$$SI = \int_{t_o}^{t_f} a^n \, dt \qquad (1)$$

where:

a = acceleration
n = weighting factor (2.5)
t_f = final time
t_o = starting time

For this study, the SI values have been calculated by computer.

The SI has been calculated in every case when three head deceleration curves were available (4). We often found high value. This results from three main causes:

1. The Alderson VIP 50 A dummy utilized in this test was fitted with a neck, the displacement of which was limited by stops offering a hard metal contact surface, which creates acceleration peaks.

2. The stopping system has a severe acceleration law.

3. The seat was not fitted with a hard rest, and so the return motion acceleration increases the value of the SI. The SI value of test 104 (a test in which the belt did not work) is by far lower than those obtained in other tests. We can therefore say that, under such test conditions, the restraint provided by an HPR laminated windscreen gives an SI much lower than that provided by a safety belt.

In test 107, the dummy shoulder broke loose from the strap, and the dummy slowly lay down on the right side, without striking anything. The SI value then obtained is very low.

Table 2 - Destructive Barrier Test

Code	Direction, deg	Speed, km/h	Duration, ms	Decelerations, g			Belts, N			Femur, N	SI	Observations*
				Car	Head	Thorax	A	B	C			
201	0	40.5	90	X = 20	Y = 100 Y = 40 Z = 18	X = 35 Y = 10 Z = 15	3000	7500	12000		637	
202	0	42.1	100	X = 35	X = 110 Y = 10 Z = 70	X = 40 Y = 15 Z = 25	4000		9000	R = 15000	1514	(A)
203	30R	48	100	X = 23 Y = 15 Z = 15	X = 50 Y = 35 Z = 26	X = 27 Y = 20 Z = 16	1000	6000	10000		500	
204	45R	49.3	130	X = 32 Y = 25 Z = 10	X = 40 Y = 50 Z = 60	X = 30 Y = 30 Z = 15	8500	(2750)	1200	R = 4500	524	
205	0	49.3	130	X = 19 Y = 12 Z = 10	X = 170 Y = 65 Z = 73	X = 32 Y = 33 X = 10	3300	6100	3900	R = 3500 L = 1300	3180	(B)
206	30L	48	130	X = 20 Y = 18 Z = 17	X = 113 Y = 36 Z = 58	X = 39 Y = 23 Z = 12	5000	6100	8200	R = 3000	1444	
207	45L	49.2	130	X = 15 Y = 10 Z = 26	X = 29 Y = 18.5 Z = 72		7700		4400	R = 4600 L = 4400	2035	

*Observations:

A—The value of the load applied to the B strap length is wrong. The magnetic recorder became saturated on account of too great a signal amplification.

B—The head deceleration curves and the film evidence a head impact against the steering wheel, which explains the unusually high value of the SI in this test.

Fig. 20 - Car after crash barrier test 205

Fig. 22 - Car after crash from 30 deg

Fig. 21 - Dummy after crash barrier test 205

Fig. 23 - Dummy after crash from 45 deg

LEFT SIDE IMPACTS - In left side impacts, the twofold strap is subjected to high strains. Under 30 and 45 deg angles, the belt was broken at the twofold strap level, and under 20 deg angles, we were approaching breaking point, since the maximum load is slightly higher than the breaking point of the following tests (12,500 N, as compared with 12,000).

RIGHT SIDE IMPACTS - In right side impacts, there occurs a deformation of the central pillar owing to the strain exerted by the shoulder strap toward the inside of the vehicle. The deformation tends to make for better belt effectiveness since it absorbs part of the dissipated energy. Under 30 deg angles, the shoulder does not break loose from the strap, whereas it does so under 45 deg angles. Therefore, the ultimate angle for good operating conditions should be included between these two values.

This effectiveness limit is valid just for the system tested: the car design, the restraining system, the dummy design; certainly, by changing one of these factors, we would not obtain the same effectiveness limit.

DESTRUCTIVE BARRIER TESTS

PURPOSE - The theoretical study of the test device utilized in the first stage shows that the car interior space was not in the car axis. The vehicle being simulated was thus skidding rather than running along normally. The dynamic sled did not enable us to simulate a normally running car, subjected to an oblique crash, while making use of a simple device. So, we deem it interesting to check the results obtained in dynamic sled tests by performing barrier crash tests under actual conditions.

TEST DEVICE - We have performed a series of barrier crash tests (Figs. 16 and 17). Because, owing to the barrier location, it was impossible for us to deflect the impact direction, we had placed one or two heavy angular concrete blocks in front of the barrier. We could thus obtain a barrier front face diverging from the impact direction. We had two blocks at our disposal, with respective angles of 15 and 30 deg. We could therefore obtain either one of these angles by using the blocks separately, or a 45 deg angle by superimposing them.

In every test, the car is radio-controlled from an accompanying car. Remote control operated both the steering gear and

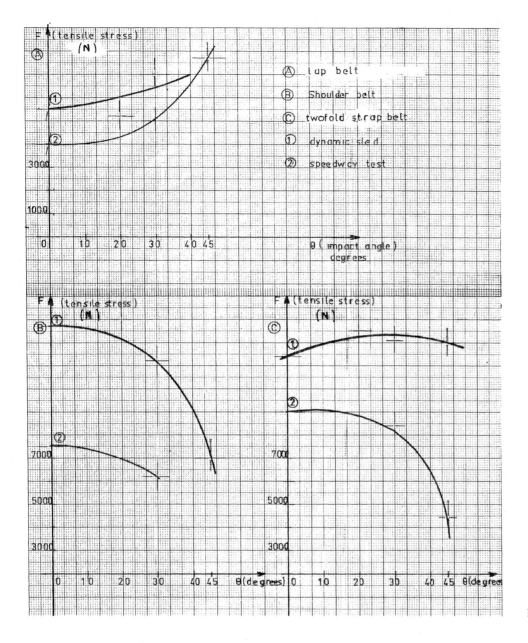

Fig. 24 - F = f (θ) curves

an emergency brake. Speed is regulated by controlling ignition advance.

MEASUREMENTS - The measurements carried out in the course of dynamic sled tests were repeated in these barrier crash tests, under the same conditions. On the preceding tests, a few improvements have been effected:

1. We replaced the dummy statham accelerometers with JPB sensors of equal value, which offer the advantages of having a smaller size, of being easier to position inside the head, and of being more readily available in France.

2. The dynamometric bars utilized to measure the strains sustained by the strap had the disadvantage of shortening the strap's useful length (10 cm at every anchorage). We replaced them by semiconductor strain sensors, which maintain the strap whole length (Fig. 18).

3. As the dynamic sled tests showed that the dummy knees

hit the dashboard underpart (Fig. 19), we fitted the dummy femurs with compression stress sensors.

EXPERIMENTAL PROCEDURE - To permit these tests to be compared with the dynamic sled tests, they were conducted under conditions as similar as possible.

The car utilized was always a Peugeot 204, manufactured between 1966 and 1968, the dummy being placed in the driver's seat and restrained by the same type of belt as that employed in the preceding tests. The seat was identical to the initial one and changed for every test.

RESULTS - The results are shown in Table 2.

SEVERITY INDEX - The value of the SI is at present accepted as a valid comparing element to assess human tolerance of a given impact.

As indicated above, the values are exceptionally high, owing to the use of mechanical type of neck in the tests. So we

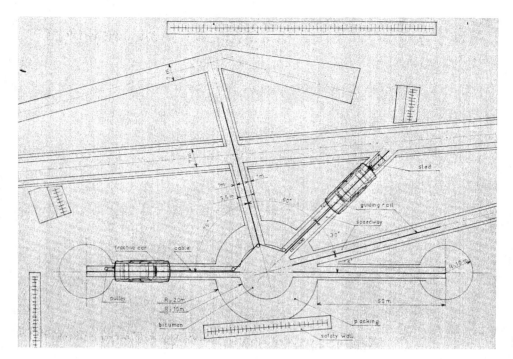

Fig. 25 - Installation of speedway

Fig. 26 - Before car-to-car crash test

Fig. 27 - After test 301

Fig. 28 - Detail of head impact on instrument panel

mainly use these values as comparative data. The analysis of the SI values shows that in speedway tests, the values obtained are always lower.

In these speedway tests, the highest value (frontal test—3180) corresponds to a head impact against the steering wheel. Speaking of the same trials, the lowest values are obtained in right side impacts. This results from the fact that the dummy shoulder breaks loose from the strap during the impact and the head then moves slowly and does not hit anything. In the course of this motion, the head remains in line with the chest, the neck is not bent, and we therefore do not find the

Table 3 - Car-to-Car Crash Test Results

| Code | Angle, deg | Speed, km/h | Pulse Duration, ms | Decelerations, g | | | Femur (N) | Belts | | | SI |
				Car	Head	Thorax		A (N)	B (N)	C (N)	
301	45R	46.5	180	X = 34	47, 4	28, 5		4150	4900		385
				Y = 32	48, 2	33, 3					
				Z = 15, 8	25, 3						
302	15R	38.5	150	X = 28, 8	_	34, 2	R = 1800				
				Y = 8, 3	32	29	=	2000	5850	5900	_
				Z = 13, 2	86	14, 7					
303	0	42	170	X = 25	113	48, 8	R = 3340				
				Y = 6, 5	15, 3	20, 4	L = 1440	3750	7280	5150	1110
				Z = 8, 5	82	18					
304	45R	45	140	X = 27, 2	46, 2	_	R = 1800				
				Y =	54, 5	112	=	4600	5300	8400	342
				Z =	24, 7	29					

peaks corresponding to the mechanical stops, as in the other tests. Using a neck deprived of mechanical stops—a rubber one, for instance—would have lowered the values obtained in the other tests, without increasing those we are considering.

EFFECTIVENESS OF RESTRAINT - In the course of the speedway trials, the operational failures that occurred in the dynamic sled tests did not recur. The behavior of the strap, buckle, and anchorages has been correct in all the tests. The highest strain suffered by the twofold strap was recorded during a frontal test. This stress maximum value was 12,000 N; this is the value for which the belt was broken in the course of two dynamic sled tests (right side impact). However, the strap and the buckle did not show any signs of fatigue in that test.

Comparing the stresses exerted on the three anchorages in relation to the angles shows that the stresses recorded in the dynamic sled tests are stronger than those recorded in the speedway trials and that the curves $F = f(\theta)$ are very much alike in the two series of tests.

In both series of tests, the dummy knees hit the dashboard underpart. According to Patrick (5), the impact tolerance limit of the kneecap under direct impact is 4500 N. In the course of our tests, we recorded one impact that reached the value of the human tolerance limit, and another that was three times as much. In order to give the restraint system full efficiency, protection should therefore be ensured at knee level by means of a shock-absorbing material or a knee strap. The tests are shown in Figs. 20-23.

VIOLENCE OF IMPACT - Comparing the two series of tests shows that the dynamic sled tests are more violent than the speedway tests. The maximum deceleration recorded in the sled tests ranged from 30-45 g; whereas those recorded in speedway trials ranged from 20-35 g. Similarly, impact times range from 63-72 ms, where in speedway trials, they range from 90-130 ms.

The difference between the various deceleration levels and deceleration duration times can be explained by the fact that the vehicles utilized were used cars and not all in the same condition; furthermore, the elements contributing to deformation vary with the crash directions; for example, when the front wheels bear on the side members during the crash, a peak appears on the deceleration curves, but in oblique crashes, the front wheels often break free from the side members.

COMPARING THE CURVES - Interpreting the curves (Fig. 24) is difficult to achieve, because we lack some of the values concerning certain impact angles, a shortcoming that arises when the measuring cable of the safety belt is broken. Consequently, only left side impacts will be studied.

Comparing the values of the stresses applied to the safety belt will show that up to a 30 deg angle, the stresses vary little and the belt maintains the dummy in the initial posture (Fig. 22).

Beyond this value, the stress exerted on the lap strap increases while the strains suffered by the shoulder strap and the twofold strap decreases rapidly. It can thus be inferred that from 30 deg upward, the safety belt does not play its part any more and the dummy has a tendency to tip over toward the right. A human being would probably show the same behavior.

CAR-TO-CAR TESTING

PURPOSE - Approximately half of the accidents in France involve two or more cars. We thus tested the effectiveness of safety belts in these realistic conditions.

TEST DEVICE - Our laboratory studied and realized a test device that allows car-to-car crash tests under different directions of impact, the two cars being at the same speed. The speedway is shown in Fig. 25; tests in Figs. 26-28.

The two cars are guided on a rail and joined by a wire. For this study, we used the engine power of the two cars to obtain the crash speed. In the future, for higher-speed tests, we are building a steam rocket giving a pushing power that allows tests up to 100 km/h.

MEASUREMENTS - The parameters measured during this

test are the same as those recorded in the crash barrier tests. However, with our measures system, we can record a maximum of 14 parameters, so we recorded the parameters just in one of the two cars. In the other, we put an impact-o-graph giving the three curves of the car deceleration.

EXPERIMENTAL PROCEDURE - To be in the same conditions as the other tests, we use the same cars (Peugeot 204 type). We chose to test the following directions of impact; 0, 15 and 45 deg. The crash speed was as near as possible to 48 km/h.

In each car, we put an Alderson VIP 50 A type dummy in the driver's seat. The dummy was restrained by a three-point safety belt, the same as those used on the other tests. On each was fitted an embarked high-speed camera: one on the top of the crashed car on the right side, the other near the right seat of the crashed car on the left side.

RESULTS - At the moment four car-to-car crash tests have been made. The results of these tests are in Table 3. There are significant differences in the results of crash barrier tests and of car-to-car crash tests.

In car-to-car crash tests the pulse duration is longer (140/180 ms compared to 90/130 ms) and the head accelerations values are lower, as are SI values.

In 45 deg tests the high speed films show that car movements after impact are complicated. This movement increases the transverse acceleration values.

These car-to-car crash tests corroborate the other results obtained in sled tests and crash barrier tests: in 45 deg angle test the dummy shoulder slips off the shoulder belt, the femur loads are tolerable, but after impact the dummy feet are often jammed between the pedals and the deformed firewall.

CONCLUSIONS

The results of this study are shown above; however, the main ones are:

1. This study showed that three-point safety belts are effective in frontal impact for directions between 0 and 30 deg.

2. From 45 deg, in the conditions of this study, the risk that the shoulder belt slips off the chest is important, so the effectiveness of the system is diminished.

3. In a small car (such as that used for the tests), the belts have to be associated with protection of knees.

4. The shoulder belt anchorage location has a great effect upon the effectiveness of the system. If this anchorage is located behind the shoulder, the system would work better.

REFERENCES

1. C. Berlioz, "Distribution et Gravité des Collisions en Fonction de la Partie Heurtée du Véhicule et de l'Obstacle Rencontré." Report to the Second International Technical Conference on Experimental Safety Vehicle, Department of Transportation, NHTSA, 1971, pp. 2-131.

2. R. W. Armstrong and H. P. Waters, "Testing Program and Research on Restraint Systems." SAE Transactions, Vol. 78 (1969), paper 690247.

3. R. Coermann and W. Lange, "Beitrag zur Dynamischen Prüfung von Sicherhbeitsgurten." A.T.Z., Vol. 70, No. 8 (1968), pp. 280-287.

4. C. W. Gadd, "Criteria for Injury Potential." National Academy of Sciences—National Research Council, Publ. No. 977, 1962.

5. L. M. Patrick, "Human Tolerance to Impact—Bases for Safety Design." SAE Transactions, Vol. 74 (1966), paper 650171.

Occupant Protection . . .
Back to the Basics

Dr. Philip D. Vrzal
Chrysler Corporation

THE SUBJECT OF this paper deals with what is without a
doubt one of the most critical and emotional aspects of the
automobile as a product of modern technology. The question
of occupant protection with restraint systems is but one of a
multiplicity of problems which sets tasks of the first order
for all of us concerned with this product. Importantly, since
the introduction of the National Traffic and Motor Vehicle
Safety Act in 1966, the scope of this subject has evolved
from the basic engineering of restraint systems into the
complexities of increasing economic, ecological, and social
considerations.

The active search for better restraints on U.S. production
vehicles began with the inclusion of standard front outboard
lap belts on January 1, 1964 and standard lap and shoulder
belts for these positions four years later.

The concept of "passivity" was formally introduced by
the NHTSA eighteen months later in July, 1969, via an
advance notice of proposed rulemaking under Motor Vehicle
Safety Standard 208 (Docket 69-7; Notice 1). The interest in
passivity spawned the widespread use of terms such as
injury criteria, anthropomorphic dummies, multiple crash
modes, and barrier equivalent velocity. Tests were
conducted with fully passive bags, bars, belts, blankets, and
bolsters. Arrays of crash-active and anticipatory sensors,
diagnostics and crash recorders emerged. Existing belt
systems were harnessed with more complex warning systems
culminating with the infamous starter interlock. Further
studies brought the perturbations of semi-passive,
active-passive as well as fully passive belt systems.

Little field experience was available to substantiate or
refute the effectivity of passive restraints and the forum for

discussion in the restraint community seemed to naturally
flow to associated subjects such as anthropomorphic dummy
development and testing, computer simulations and math
modeling, benefit/cost analyses, and restraint effectiveness
extrapolations from unrestrained field accident data. These
studies all have one common denominator . . . increasing
sophistication.

The execution of many of these efforts have yielded
fruitful results and additional insight into the complexities of
occupant protection. However, sophisticated investigations
also tend to induce some preoccupation with analytical
details, data and methods. Oftentimes, in studies the real
purpose and primary goal of occupant protection, the
reduction of injuries and fatalities, appears somewhat
obscure or secondary. Perhaps, it is valuable from time to
time to reassess and reflect on the basics of occupant
protection in order to maintain a favorable perspective in our
ideas and actions, and order in our priorities. This paper is
one attempt to provide: 1) a structured but simple
reorientation to the basics of occupant protection with
restraint systems, and 2) a general assessment of various
kinds of restraints as they relate to those basics at this point
in time.

THE BASICS OF OCCUPANT PROTECTION

The use of the term "basics" here refers to the technical
performance or effectiveness of restraint systems. One
important consideration, system usage, is related to comfort
and convenience. It alone is a subject for separate discussion
and is not included herein. Clearly, the impact of economic,

ABSTRACT

This paper provides a structured but simple reorientation
to the basics of occupant protection with restraint systems.
A general assessment of the various kinds of restraints as
they relate to those basics at this point in time follows. The
evaluation of typical restraints and restraint concepts

include: conventional belts, energy-absorbing pads and
panels, conventional, air cushions, full crash mode air
cushions, passive belts and belt improvements, and inflatable
belts. Expanding the basics to include other than technical
considerations is also briefly discussed.

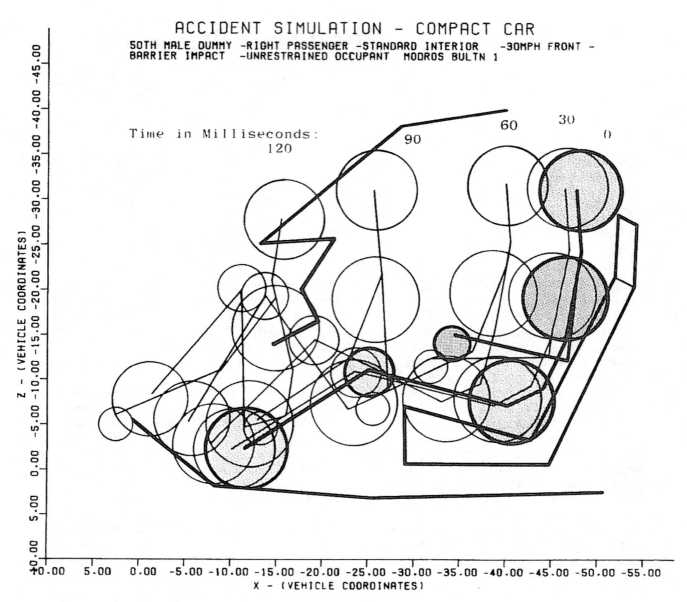

Fig. 1 - Simulated occupant response in an accident

ecological and social considerations must also be weighed and these will be discussed later in this paper.

Simply stated, an automobile accident may be considered as a significant change in a vehicle's motion which causes an unrestrained occupant to impact an interior surface of the passenger compartment with an intensity likely to cause injury or death. Therefore, the starting point for the selection of any restraint system parameters for all car sizes requires an analysis of both the vehicle and unrestrained occupant dynamics during various crash modes.

The crash energy management capability of the vehicle proper is a function of three major items:

1. Shape and magnitude of the crash pulse of the vehicle.

2. Total velocity change of the vehicle along the impact axis.

3. Duration of the crash pulse.

These vehicle related factors place limits on the effectiveness of the restraint system within the passenger compartment.

In our daily oral and written communications, we have learned that to be the most effective, we must know the world of the receiver. The same logic applies to knowing the restraint system environment. The most effective restraints will be those:

1. Within or a logical projection of the state-of-the-art.

2. Within the limits of known passenger compartment integrity.

3. Designed with some capacity for growth with the vehicle.

4. Practicable to implement.

5. Configured with concern for occupant comfort and convenience.

6. Oriented towards all types and sizes of people, not just test dummies.

7. Designed for accidents in the field, not merly fixed barriers. (i.e., for front, angular, sideswipe, lateral, rear, and roll-over mode impacts with other vehicles, poles and other fixed objects.)

Given these general considerations in the restraint system environment and the crash energy management capability of the vehicle proper, the third, and perhaps most important element of the "basics" — occupant dynamics, can be viewed. As an initial point of reference, a brief look at one occupant response in an auto accident is warranted. Figure 1 is derived from the simulation of typical 30 mph frontal barrier impact acceleration functions for a compact size vehicle and the response of an unrestrained 50th percentile male occupant. In the first 25 milliseconds or so, there is little occupant response to the crash management occurring in the forward part of the vehicle. In the next 25 millisecond segment there is significant forward excursion and the beginning of contact by the occupant with the passenger compartment. Above 50 milliseconds, we note the increasing interaction of the occupant with the interior surfaces. In this case, the rebound segment of the simulated crash is not shown but is also very important. Typically (assuming a given impact velocity), as a vehicle gets larger, the impulse duration is longer, the vehicle displacement is greater, and there is more distance within the passenger compartment. These factors generally result in later occupant response to the impact and greater occupant excursion in the passenger compartment prior to contact with the interior surfaces of the passenger compartment of the larger vehicle.

Since there is little opportunity for direct observation of occupant dynamics in actual field crashes, analysts are restricted to film and dummy response data from simulated sled runs and barrier impacts. The results of these analyses in limited crash modes at lease suggest some reasonable rationale for the development of restraint systems. This rationale can be expressed via five favorable factors:

1. Match the acceleration imposed on the occupant by the restraint system with those of the vehicle in all crash modes.

2. Begin occupant deceleration with the smallest possible time delay.

3. Maintain the greatest distance between the occupant and rigid interior surfaces as well as other occupants.

4. Control the kinematics of the occupant.

5. Distribute loads over the body to optimize occupant tolerance.

Although it is doubtful that any one restraint system can totally fulfill these factors on all vehicles, general adherence to these factors should produce positive results.

APPLICATION OF THE FIVE FAVORABLE FACTORS

A short discussion of each of these factors and their relationships is in order. Although some restraint system designs will inherently have more energy absorbing capability than others, thus imposing various acceleration levels on the occupant, the attempt should be made to permit the occupant to ride down the pulse of the vehicle as much as possible, regardless of crash mode. To effect maximum ride down the occupant should be arrested before he begins his forward excursion in the vehicle, ideally within the first twenty milliseconds of the crash event. Early arrest of the occupant helps to maintain the occupant at his original location within the occupant compartment.

Importantly, the occupant should also be isolated within the passenger compartment to reduce or eliminate the possibility of impact with rigid interior surfaces and other occupants. This is the third factor. While the reasoning for isolation from rigid interior surfaces is self-evident, isolation from other occupants may not be. However, recent efforts have included results suggesting that occupant to occupant contact in accidents may cause an injury that otherwise would not have been sustained or causes a more serious injury that would not have happened, if such contact had been avoided. (Reference 1) These injuries are produced not only through direct occupant contact, but also when such contact forces an occupant into some interior part of the passenger compartment. In the referenced study, occupant to occupant collisions caused or aggravated injuries in over 20 percent of the reported crashes where the car contained more than one occupant.

The fourth favorable factor, occupant kinematic control is critical with any restraint system. For example a knee restraint where used, must control occupant motion by limiting lower torso movement to retain the occupant on the seat. It must also absorb the energy of the lower torso while keeping the upper leg loads within tolerable limits. Similarly, a snug lap belt prevents occupant submarining by restricting lower torso movement. Knee restraints or lap belts appear essential to the performance of all types of upper torso energy absorbers such as shoulder belts, air bags, energy-absorbing pads, panels and columns.

The remaining factor, the distribution of loads over the body to optimize occupant tolerance needs little comment except for a special caution note due to the incomplete evidence of human tolerance levels with restraints, especially for the head and upper torso.

The exclusion of supplemental factors from these just discussed, such as maintaining vehicle control in evasive maneuvering or after the first impact in a multiple impact accident, does not downgrade their importance in occupant protection.

Taken as a whole, the five favorable factors suggest that the ideal restraint system be one that secures and isolates the occupant within his seating position. Experience gained from sled runs with a dummy strapped down to a steel seat, as in Figure 2, demonstrates the intent of the five factors nicely. However, the restraint test set up in Figure 2, replete with a bolt through the top of the dummy's head and a collar of belt webbing restraining the neck is hardly feasible. But it does provide a theoretical benchmark.

Fig 2 - Hyge sled run with a dummy strapped down to a steel seat

Indianapolis 500 racing car as well as U S and British fighter aircraft harness configurations provide objective, but highly controlled applications of the five favorable factors. While these have proven to be quite effective in crashes involving those types of craft, they appear highly impractical for passenger car applications for the vast motoring public

EVALUATION OF TYPICAL RESTRAINTS AND RESTRAINT CONCEPTS

Given the five factors as a basic yardstick, typical existing restraints and restraint concepts for front seating positions can be placed in fundamental perspective for automotive applications. Each can be generally rated at five levels of performance, namely, excellent, very good, good, poor and very poor. An additional assumption of impact severity appears reasonable:

- Frontal, angular, and sideswipe modes - 30 mph fixed barrier equivalent
- Lateral mode - 20 mph equivalent from a platform
- Rollover mode - 30 mph equivalent from a platform
- Rear mode - 30 mph moveable barrier equivalent

For ease of comparison. the restraints can be classified by type:

1. Conventional Belts
2. Energy Absorbing Pads and Panels
3. Conventional Air Cushions
4. Full Crash Mode Air Cushions
5. Passive Belts and Belt Improvements
6. Inflatable Belts

CONVENTIONAL BELTS - There is really little need to define conventional belt systems. In short, they are those contained in the models of this year. They are the three-point systems which include anchors, belt webbing. autolock retractors and emergency locking retractors On the whole, this type of restraint, when properly worn has had an enviable record in the field. A comparison with the five factors provides some insight into the whys. Belt systems provide very good performance in matching occupant accelerations with those of the vehicle and importantly in all

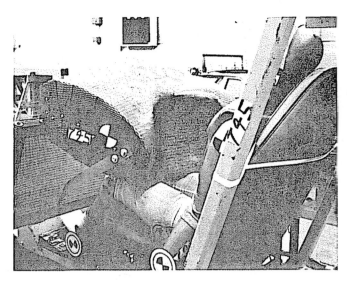

Fig. 3 - Expanded metal panel, post-test condition, 30 mph sled run

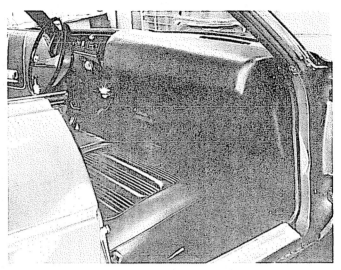

Fig. 4 - Energy-absorbing instrument panel and knee restraint

crash modes. However, occupant deceleration is not accomplished with the smallest possible time delay. Occupant arrest occurs sometime after the first twenty-five milliseconds. Occupant isolation in his seating position is good, except for the head. Occupant containment within the vehicle in all crash modes has been a strong favorable factor. Control of occupant kinematics in relation to the lower torso is very good in spite of some forward translation in the vehicle. Upper torso kinematic control of the chest and head is good with some better means of controlling head motion desirable. The distribution of loads with a belt system to optimize occupant tolerance continues as a subject of strong debate. However, this factor must be rated as good based on the evidence to date.

ENERGY ABSORBING PADS AND PANELS - This type of restraint was addressed largely as a means of providing passive and non-deployable protection. This group usually includes the combination of a knee restraint and a fixed extension of an existing instrument panel. Panel construction and materials range from large expanded metal panels with plastic fillers as shown in Figure 3 to sheet metal panels with a foam and vinyl skin as in Figure 4. Some concepts have included the use of structural foams of varying densities. Panels and pads, when measured against the five factors must be rated as very poor. Occupant to vehicle acceleration matching is difficult. Occupant deceleration occurs very late in the crash event and the occupant must translate forward in the vehicle prior to energy absorption by the panel. Notably, occupant kinematic control appears very difficult in this group. Oftentimes, severe impact of a test dummy's head with the windshield header area has been observed in vehicle crash tests. This is primarily due to a combination of vehicle pitch at the time of impact with the kinematic variations associated with crushable panels and pads. Finally, localized hard spots in many pads and panels have contributed to inordinate load distributions over the upper torso and head of the dummy.

Beyond the five factors considered there are further negatives associated with this group, e.g., occupant confinement and passenger compartment ingress and egress difficulties.

The addition of small air cushions to these fixed panels as shown in Figure 5 do not show any indication of making them more favorable. Similarly, deployable panels and pads increase the complexity of this group since the deployment time budget is critical. Some of the problems associated with the energy absorbing pads and panels have been solved with the introduction of the conventional air cushion.

CONVENTIONAL AIR CUSHIONS - Like conventional belt systems, there is little need to define conventional air cushions. In short, they are those offered currently as an option on standard size cars by one manufacturer. The typical locations are in the steering wheel hub and lower instrument panel area as shown in Figure 6.

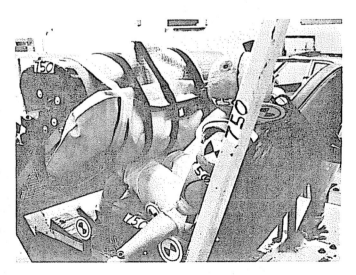

Fig. 5 - Expanded metal panel with air cushion, post-test condition, 30 mph sled run

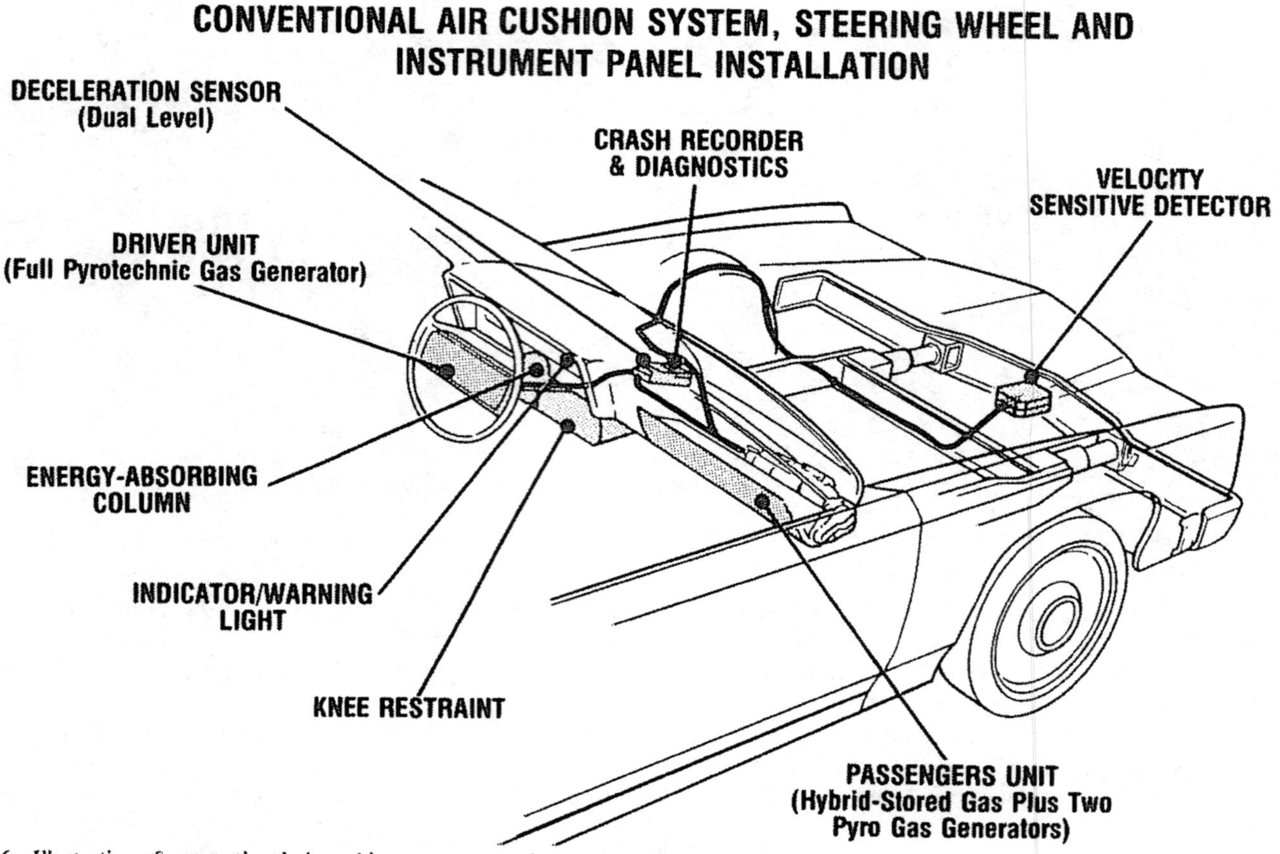

CONVENTIONAL AIR CUSHION SYSTEM, STEERING WHEEL AND INSTRUMENT PANEL INSTALLATION

DECELERATION SENSOR
(Dual Level)

CRASH RECORDER & DIAGNOSTICS

VELOCITY SENSITIVE DETECTOR

DRIVER UNIT
(Full Pyrotechnic Gas Generator)

ENERGY-ABSORBING COLUMN

INDICATOR/WARNING LIGHT

KNEE RESTRAINT

PASSENGERS UNIT
(Hybrid-Stored Gas Plus Two Pyro Gas Generators)

Fig. 6 - Illustration of conventional air cushion system, steering wheel and instrument panel installation

Generally speaking, this type of restraint, when properly deployed has performed very well under limited field experience in frontal and angular crash modes. The system was originally intended to be used without lap belts and is not specifically designed to deploy in other than frontal and angular impact modes. The continued concern and debate over the capability of conventional air cushions to provide protection in all crash modes has been somewhat ameliorated since the inclusion of lap belts into the system. The addition of lap belts has also eased some of the problems common to both conventional air cushions and energy-absorbing pads and panels.

A comparison with the five factors provides additional insight. Air cushions provide very good performance in matching occupant accelerations with those of the vehicle in frontal and angular crashes. However, in other modes of non-deployment they are limited to the effectiveness of the lap belt. Occupant deceleration is not accomplished with the smallest possible time delay. Occupant arrest occurs after the first twenty-five milliseconds and later than belt systems. Unbelted occupant isolation in his seating position is good, except for the driver femurs and a tendency for some occupant to occupant contact. Occupant containment within the vehicle in all crash modes has been a major concern, which is eased when lap belts are used.

Control of unbelted occupant kinematics in relation to the lower torso is good in spite of some forward translation on

the passenger side, but not as good on the driver side. The knee restraint is the critical element in controlling the kinematics of the driver. It affects the load distribution on the entire torso and controls the upper leg forces within specified limits. Use of a lap belt tends to increase lower kinematic control to the very good level. Upper torso kinematic control is very good, especially that of head motion control in frontal and angular impacts. Finally, the distribution of loads on the unbelted occupant is excellent for the head, very good for the chest and sometimes poor for the femurs, especially those of the driver in frontal and angular impacts. The use of a lap belt relieves the femur problem. Other arguments generally advanced against the air cushion fall outside the scope of the five factors established earlier and are not included in this discussion.

FULL CRASH MODE AIR CUSHIONS - This restraint concept was pursued to provide a "totally passive" level of protection equivalent to air cushions and lap belts in the frontal and angular modes. Additionally, lateral and rollover protection at least equal to that of lap and shoulder belts was envisioned. This system features a passenger's cushion (deployed from the header area) a steering wheel mounted driver cushion, a passive driver inflatable belt, and cushions deployed along the roof rails on both sides of the vehicle for lateral impact protection as illustrated in Figure 7. To date this type of restraint has had some success in its advance development phase. The small, higher pressure

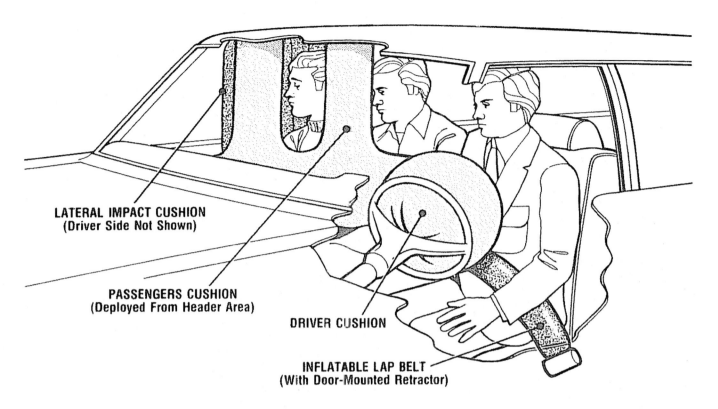

LATERAL IMPACT CUSHION
(Driver Side Not Shown)

PASSENGERS CUSHION
(Deployed From Header Area)

DRIVER CUSHION

INFLATABLE LAP BELT
(With Door-Mounted Retractor)

Fig 7 - Illustration of full crash mode air cushion concept

driver bag (using a blow-off valve) when attended by a passive inflatable belt possibly could provide acceptable performance in all crash modes. For the front seat passengers, the roof (header) mounted system is designed to provide the restraint forces for rollover and side impacts directly from the load carrying structure of the pressurized bags. The side curtain bags are intended to bridge the side glass to provide containment and protect against roof rail and A-pillar impacts. A comparison with the five factors provides some insight into the probably advantages and disadvantages of this system. The driver air cushion and inflatable belt should provide very good performance in matching occupant accelerations with those of the vehicle in all crash modes. The passengers cushion should provide good performance. The air belt begins decelerating the lower toso in a very short time and is later augmented by the driver bag and side curtain as required. However, deceleration via the passengers bag is not accomplished with the smallest possible time delay, with arrest occurring after the first twenty-five milliseconds. Isolation of the driver in his seating position is very good, while there is a tendency for some occupant to occupant contact between the passengers. Occupant containment within the vehicle in lateral and rollover crash mode tests with the side curtain bags is very good.

Control of the driver kinematics and load distribution is very good. The kinematic control of passengers at this point must be considered to be poor from the standpoint of the obvious hazards of deployment from above with forward seated or out of position passengers. Until these deployment

hazards can be solved, inordinate distribution of loads are likely to occur and must be rated as unacceptable. Importantly, sensors for the remaining modes, lateral, rear, and rollover need to be developed. Although this concept appears to have more potential benefit than the conventional air bag system, it also has more complexity and more unresolved problems.

PASSIVE BELTS AND BELT IMPROVEMENTS - Concepts and developmental efforts to make conventional belts passive have been too numerous to single out any one configuration as typical. This restraint group includes elements such as moveable arms, sliders, pivots, rods, pulleys, drive motors and belt pullers. The goal, regardless of system geometry, is to provide a viable belt system requiring no action by the occupant. Those configurations using lap and shoulder belts should at best relate to the five factors in a manner similar to conventional belts, previously discussed.

Those configurations using knee restraints in lieu of lap belts should be less effective, having problems similar to those discussed for knee restraints in the conventional air bag segment in the pages above. There continues to be some debate on the effectivity of the knee restraint and shoulder belt combination. More experience in the field is necessary to settle this issue.

Belt improvements receiving the most attention are force limiting and preloading devices. Belt force limiters do just that and increase absorption of displacement energy thereby reducing local belt loading. This is basically accomplished by lengthening the belts somewhat at the instant of maximum

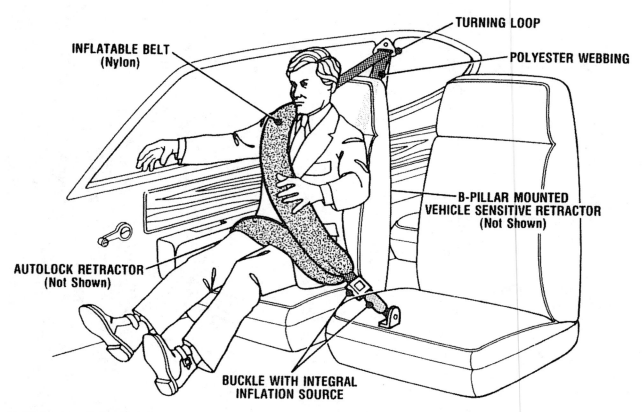

INFLATABLE BELT
(Nylon)

TURNING LOOP

POLYESTER WEBBING

B-PILLAR MOUNTED
VEHICLE SENSITIVE RETRACTOR
(Not Shown)

AUTOLOCK RETRACTOR
(Not Shown)

BUCKLE WITH INTEGRAL
INFLATION SOURCE

Fig 8 - Illustration of inflatable belt system

force limitation. Typical devices include: controlled tearing of the webbing material proper, plastic deformation of torsion bars, dry friction devices and bending flexible plastic materials such as sheet metal. The better load distribution achieved carries the penalty of greater forward displacement of the occupant, a decrease in maintaining the greatest distance between the occupant and rigid interior surfaces (Factor Number 3).

One way to decrease the forward displacement of the belted occupant is to use a preloading device. Preloaders reduce the time delay of occupant deceleration after vehicle deceleration. The forward excursion or displacement of the belted occupant is thereby reduced. Preloader concepts usually include an electrical response from a sensor to ignite a pyrotechnic or stored gas source. The resulting energy causes a piston to stroke and quickly take up the slack in the belt system.

INFLATABLE BELTS - This type of restraint combines many of the energy-absorbing characteristics and elements of both belt systems and air cushions. The inflatable belt system is evolved from conventional belt system designs. It incorporates the use of small columnar-shaped cushions, inflated as required from an integral inflation source as illustrated in Figure 8. The uninflated system is shown in Figure 9, and performs in a manner similar to conventional belt systems. A simple sensor signals the need for energizing the lap and shoulder cushion segments for augmented protection in more severe impacts. A simulated inflation is shown in Figure 10. Limited development testing of the

inflatable belt system to date has yielded very encouraging results.

Some fundamental reasons for the favorable results can be seen by a comparison with the five factors. Because the inflatable belt is pre-tensioned to the occupant it absorbs energy from him as his motion relative to the vehicle begins to occur. In other words, the slack is taken out of the system and there is a very low delta velocity impact by the occupant to the inflatable belt. Therefore, acceleration matching (ride-down potential) is very good. Most importantly, occupant deceleration begins with the smallest possible time delay, prior to occupant forward translation. The system rates as excellent on this point, superior to the air cushion, and better than a belt system. The foreshortening effect and quick reaction time of the inflatable belt keeps the head, chest, and lower torso of the occupant isolated excellently within his original seating position and provides the greatest distance between him and the interior vehicle surfaces and other occupants. This is especially important in more severe crashes where passenger compartment integrity may become compromised and more steering column and instrument panel movement occurs. Control of occupant kinematics is also very good. The lower torso is controlled better than an typical lap belt or knee restraint since the foreshortened (inflated) belt acts as a preloading device prior to occupant movement. The upper torso is controlled better than with a shoulder belt since the system provides energy absorbing and head motion controlling under the chin and across the chest of the

818

Fig 9 - Inflatable belt, uninflated in demonstration vehicle

Fig 10 - Inflatable belt, inflated in demonstration vehicle

occupant. The inflated upper torso cushion appears superior to the head protection of an air cushion system since the protection is provided earlier, requires less crush distance and occurs further rearward in the vehicle.

The distribution of loads to an occupant appears very good with this system. The rounded section of an uninflated inflatable belt is superior to typical belt edges and reduces the "roping" effect. In the inflated mode, the system provides a greater area and rounded section configuration that appears to distribute the load in a better manner than belt systems. Additionally, the inflatable belt system provides the redundancy of equivalent belt system protection for all crash modes in the event of a system malfunction. As in the maturing of any restraint system many development problems exist with inflatable belts, but from the standpoint of the basics, the progression of such systems should remain of high interest.

SUMMARY

Originally in this discussion the use of the term "basics" was restricted to the technical performance or effectiveness of restraint systems. A structured, but simple reorientation to the technical basics of restraints was introduced. Then a general assessment of various types of restraints as they relate to these basics was explored. Having completed these two segments, some fundamental conclusions begin to take shape:

1 A new system doesn't necessarily mean a better system.

2. Increased sophistication or system complexity, while solving some problems often can introduce greater problems, especially if there is inadequate development and test time.

3. Singularity of approach may restrict technological developments that may have yielded greater benefits.

4. A reassessment of the basics from time to time serves to maintain a favorable perspective in our ideas and actions, and order in our priorities.

EXPANDING THE BASICS

The isolation of technical problems is paramount to the development of viable solutions. However, in reality the impact of economic, social, ecological, psychological and political considerations must also be introduced into the final product at some point. Examples of those concerns currently receiving the greatest amount of national attention include:

1 Continued economic recession aggravated by inflation . . increased costs with decreased buying power.

2. Impositions on the freedom of the individual . . . consumer independence.

3. Persistent energy shortages with international implications . . . an imbalance of supply and demand.

4 Pursuit of an ecological balance . . . a better environment for all.

Any or all of these could easily outweigh the technical

considerations along and must be weighed in the final product trade-offs.

The area of occupant protection is not immune to these considerations. Since the products in occupant protection are subject to safety standards, there is a co-responsibility of both those who generate the respective standards as well as those whose products must comply with them. Many relatively recent standards have resulted in improved and more serviceable products for the motoring public. However, some existing standards in the area of safety appear to have resulted in marginal benefit at best while resulting in both tremendous economic and technical opportunity costs. For example, FMVSS 202 made head restraints mandatory equipment for passenger cars sold in the United States after December 31, 1968. The purpose of head restraints is to protect front seat occupants of vehicles which are struck from behind from the possibility of neck or back ("whiplash" type) injuries. Studies to date (References 2 and 3) indicate that while fixed head restraints appear to be more effective than the adjustable type, as a whole, head restraints reduce the frequency of whiplash by a mere 14 percent. This marginal benefit appears attributable in part to the failure of occupants to adjust their head restraints. (About three-fourths of the adjustable head restraints are left in the down position and not adjusted.) A second reason cited is that less than optimal head restraint designs exist, largely due to an inadequate knowledge of human neck kinematics.

The benefits of side guard door beams are also suspect even though they did not become standard equipment on cars sold in the U.S. until January 1, 1973. Some analytical conclusions (Reference 4) suggest no real difference in the frequencies for each level of injury severity between side beam and non-side beam cars for both the driver and right front passenger locations. Others argue (Reference 5) that the side door beam reduced the driver injury levels in left side impacts to those injury levels associated with right side impacts in cars not so equipped. A third study (Reference 3) holds that the effectiveness of side beams is related to car size - they appear to have no effect in intermediate sized cars and may be associated with a detrimental effect in full sized cars. Even the effectiveness of energy absorbing steering columns has been analyzed with mixed results.

More recently, seat belt interlocks in 1974 model cars, praised by some and loathed by many did increase lap and shoulder belt usage notably. However, its rejection by the vast majority of motorists was culminated in the removal of the device via Congressional legislation a few months ago.

The interlock has since been replaced by a driver reminder system of dubious value which features a light and buzzer of 4 to 8 seconds duration.

Finally, a proposed amendment to FMVSS 215, the standard for energy absorbing bumper systems suggests bumper modifications as a trade-off for lower weight and presumably better fuel economy on passenger cars. These are merely a few examples of the action and reaction, hit and miss standards atmosphere that continues to contribute to unprecedented cost and price increases with marginal benefits at best. Accident data and the associated analyses are often misinterpreted, clouded and controversial and add to the total problem.

In conclusion, a few positive steps could be initiated to alleviate these conditions:

1. A system to remove controversy from accident data gathering, storage, and analysis is a must. A National Accident Data Gathering System has been suggested by others and would be of merit.

2. Reasonable periods of development and test time should be allocated prior to the inclusion of mandatory safety equipment.

3. A slowdown on the issuance of what amounts to "trial and error" standards could provide for the establishment of a more realistic and practicable planning base, one that reflects an attitude that is clearly . . . back to the basics.

REFERENCES

1. D. F. Huelke, et al, "The Hazard of the Unrestrained Occupant." A paper presented at the Eighteenth Annual Conference of the the American Association for Automotive Medicine, Toronto, Ontario, September, 1974.

2. J. D. States, et al, "Injury Frequency and Head Restraint Effectiveness in Rear-End Impacts." Proceedings of the Sixteenth Stapp Car Crash Conference, Society of Automotive Engineers 720967, New York, November, 1972.

3. L. I. Griffin III, "Analysis of the Benefits Derived from Certain Presently Existing Motor Vehicle Safety Devices: A Review of the Literature." University of North Carolina, Highway Safety Research Center, Chapel Hill, North Carolina, December, 1973.

4. F. L. Preston and R. M. Shortridge, "An Evaluation of Side-Guard Door Beams." HIT Lab Reports, 1973.

5. A. J. McLean, "Collection and Analysis of Collision Data for Determining the Effectiveness of Some Vehicle Systems." A study prepared for the MVMA, Agreement No. UNC 7301-C19, 1973.

Efficiency Comparison Between Three-Point Belt and Air Bag in a Subcompact Vehicle

M. Dejeammes and R. Quincy
Laboratore des Chocs
Organisme National de Sécurité Routière (Onser) France

Abstract

The purpose of this paper is the comparison of the protection efficiency between three-point belt and air bag systems under various crash conditions.

Dynamic tests have been performed with subcompact vehicles (Renault R 12) in which two dummies were restrained, either by three-point belts with load limiting devices, or by air bags consisting of solid gas generators and bags including porous outlets (the driver's knees were protected by a collapsable structure).

Three types of crashes were chosen :
- frontal barrier crash at 50 km/h (13,9 m/s)
- head-on crash between two vehicles with overlap at 50 km/h (13,9 m/s)
- crash against a guardrail at 80 km/h (22 m/s) with 30° angle of incidence.

The comparison drawn from commonly used biomechanical indices shows that the three-point belt ensures a protection in each analysed crash type but it should be improved in order to reduce head deceleration. The air bag results depend on the crash type and show the problems of adaptation in a subcompact vehicle.

The frontal barrier crash tests conducted with another type of dummy reveal that the results obtained for the two restraint types depend on the dummy, so that the efficiency assessment is difficult.

IN AUSTRALIA, EUROPE OR USA, the efficiency of the three-point belt has been proved and the experience of Australia and the latter one in France indicates that the safety belt may induce a reduction of 50 % of death. However, in spite of instigations or compulsions, it appears that the wearing of seat belts is insufficient and it seems difficult to get higher wearing rates than 60-70 %.

So, in a parallel direction to the researches on the improvements of safety belts, it is important to try to evaluate the effectiveness of passive restraint devices such as air bags. In fact, since 1967, many tests have been performed in order to evaluate the efficiency of air bags comparatively to lap belts or four-point harnesses, in frontal test conditions, using different types of models - dummies (1, 2, 3)*, baboons (4, 5, 6), human volunteers (7, 8, 9).

The purpose of this study is the comparison of protection efficiency between the three-point belt and the air bag. A subcompact vehicle, largely sold in Europe, has been chosen because this car size induces important problems for the adaptation of the air bag system (3).

Moreover, different test conditions, more realistic than the frontal barrier impact, have been looked for, in order to investigate the problems that they induce in the air bag protection, as revealed by some tests of MARTIN (10)

METHODOLOGY

VEHICLE AND RESTRAINT SYSTEM - The test vehicle is a Renault R 12 (weighting 1000 kg) on which some modifications have been introduced (they will be soon provided on the mass produced vehicles) in order to allow a good action of the restraint system. These are :
- reinforcement of the wheel box for preventing wheel intrusion
- lengthening of the frame side rail for strengthening of the passenger compartment.

Special modifications have been provided for the tests with air bags. These are :
- strengthening of the apron to prevent the air bag generators from moving.
- collapsible steering column (possibility of axial collapse at the end of the steering gear).

Besides, each vehicle is equiped with a laminated windscreen, and integrated head-rest seats.

Two types of restraint systems have been used :
- the three-point belt used by the Peugeot - Renault Association, except for the interior common part which is not rigid but is a textile strap. The sash belt is equiped with an absorbing device made of five breakable webbings. A padding is provided on the steering wheel hub in case of head impact.
- the air bag restraint system was chosen after adjustment tests were conducted by Renault on the vehicle Renault R 12. It is a completely passive restraint without lap belt,

_____

* Numbers in parentheses designate References at end of paper.

Fig. 1 - Vehicle with air bag restraint

consisting of solid gas generators (SNPE GSS 60) and IRVIN bags on which the blow-out parts are made of porous tissue.

For the driver, one generator is fixed on the steering wheel hub (delivering 60 liters) with a circular bag. The knee restraint consists of a foam padding (rigid polyurethane foam) (fig. 1).

For the front passenger, three generators inflate a 180 liters bag which provides the restraint of the torso and the knees.

The ignition of the generators was made by a "timer" which gives an impulse 15 ms after the crash start. The crash sensor has not been used because of its lack of reliability in non-frontal impacts.

CRASH TESTS - Some realistic test conditions have resulted from investigation of statistical data. The national statistics (11) show that the most frequently crashed parts are the center front (33 %), the left front (26 %) and the right front (15 %). The bidisciplinary investigation performed by the ONSER in France (12) shows that 38 % of the frontal impacts occurred with an overlap inferior or equal to one third of the vehicle width.

These results led to the choice of the front crash between two vehicles with an overlap of 1/3 width (drivers in line, approximately) (fig 2). The vehicles move at 50 km/h (13,9 m/s), each. This crash induces a rotation after the maximum vehicle collapse.

The front barrier crash test at 50 km/h (13,9 m/s) has been selected as reference test.

Fig. 2 - Head-on crash between two vehicles with overlap — vehicles
before impact

Besides, the crash against a guardrail on highway is
simulated by a crash at 80 km/h (22 m/s) with 30° angle of
incidence. This crash induces a rapid rotation of the vehicle without important deformation.

For these simulated crashes, the vehicles are propelled by a water steam rocket along a guide-rail. For the
crash with two moving vehicles, the vehicles are joined by
a cable, one of them being propelled by the rocket.

MODELS AND MEASUREMENTS - Two vehicle occupants, the
driver and the front passenger, are simulated by dummies.
All the tests have been conducted with Alderson VIP 50 A
dummies, modified with an Alderson rubber neck. For the
front barrier crash test, another test has been performed
for each restraint system with new dummies made by ONSER
which include metallic discs as cervical vertebrae.

All these dummies are equipped with three accelerometers (JPB 100 G) in the head, three accelerometers (JPB 50 G)
in the thorax and two femur axial load cells.

Other measurements are recorded during the impact,
which are three strap axial loads near the anchorage points
of the belt, two vehicle accelerations (horizontal and vertical) with gages fixed under the driver's seat. Conventionally, the axis X, Y, Z will be respectively the longitudinal, lateral and vertical axis. The impact is photographed
by seven high-speed cameras, two of them being located on
the vehicle.

SEVERITY PARAMETERS - Usual severity parameters are
considered and will let to the comparisons of the impact
tests :

EFFICIENCY COMPARISON

- the acceleration of the head and the Gadd Severity
  Index (SI)
- the acceleration of the thorax
- the femur compression load
- the strap loads

Special emphasis will be laid on the kinematics of the occupant during the crash and the possible body impacts on vehicle parts.

COMPARATIVE EFFICIENCY OF AIR BAG AND THREE-POINT BELT

Seven crash tests have been conducted with Alderson dummies so that it is possible to compare the efficiency of the two restraint systems under various test conditions. (Table 1)

FRONTAL BARRIER CRASH — This is the crash used for standard tests, where the vehicle impacts a fixed barrier at 50 km/h (13,9 m/s). It is a symetric, very severe crash with deceleration peaks of 45 Gx et 15 Gz, (fig. 3).

Three-point belt restraint (1 A 11) — For this crash, the initial speed was 52,5 km/h (14,5 m/s). The vehicle was decelerated during 85 ms and its permanent deformation was 50 cm. During this test, the buckle of the driver's belt opened at the beginning of the belt load but it was difficult to determine the exact reason of this failure which could have been a mechanical defect or a bad installation. So it was not possible to analyse the driver's kinematics with the three-point belt.

The passenger deceleration begins at about 20 ms on the thorax and 35 ms on the head. The kinematics of the head and the torso are acceptable, particularly that of the head which has no impact against the vehicle structure but its deceleration is relatively high (86 Gx and 118 Gz with a resultant of 122 G) (Fig. 4). The knees slightly impact the dashboard while the pelvis moves ahead to the seat extremity. This is not really submarining but rather a long displacement due to strap elongation after the breaking of the five load limiting webbings. However, it is interesting to notice that this device working between 35 and 65 ms, gives a good result on the loads, particularly on the sash belt (maximum load of 8600 N) (Fig. 5).

Air bag restraint (2 A 10) — The vehicle used for this crash has some modifications in comparison with the previous vehicle.

- a false gear box tunnel
- higher bumpers
- the steering column was uncoupled from the steering
  box to prevent upward movement.

These modifications were made on behalf of the air bag efficiency for this severe crash type.

Table 1 - Results of Tests

Type of Crash	Restraint / Position	Head Acceleration (G) R	X	Y	Z	Head SI	Displacement (cm)	Torso Acceleration (G) X	Z	Belt Load (DaN) Shoulder	Lap Inboard	Outboard	Remarks
Frontal Barrier	Seat Belt Passenger 1 A 11	122	86		118	3600	54	76	34	860	1270		
Frontal Barrier	Air bag Passenger 2 A 10	118	116	30	104	2300	65	65	69				Head hit upper windscreen rim
Head-on	Seat belt Driver 1 C 6	100	97	15	80	1500		59	35	660	920		Head hit steering wheel rim
	Passenger	72	61	38	62	950		22	45	670	820	640	
Head-on	Air bag Driver 2 C 9	104	50	84	62	775		51	69				Head hit windscreen
	Passenger	85	44	60	68	440		40	27				Head hit windscreen
Guardrail	Seat belt Driver 1 B 5	8	8	10	10			8	7	120	200	100	Head hit side door rim
	Passenger	30	30	50	25			5	20	140	180	100	Head hit driver's shoulder
Guardrail	Air bag Driver 2 B 4	30	30	50	47			15	30				
	Passenger	12	12	14	18			10	14				

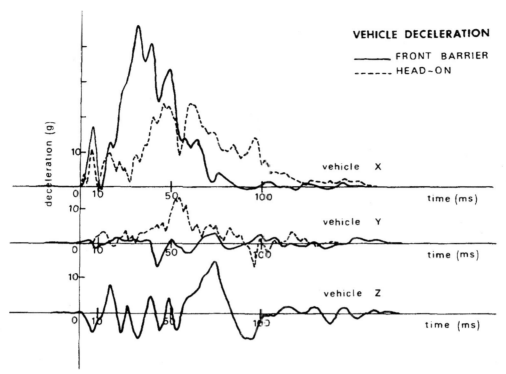

Fig. 3 - Vehicle deceleration

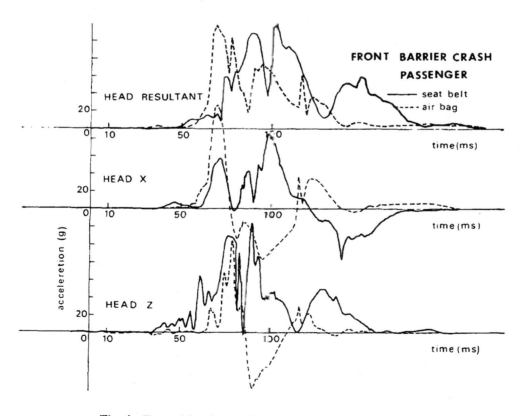

Fig. 4 - Frontal barrier crash — passenger head acceleration

827

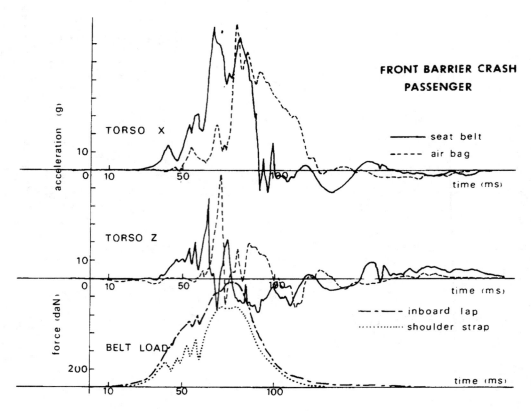

Fig. 5 - Frontal barrier crash — passenger torso acceleration and belt load

During the test with an initial speed of 52 km/h (14,4 m/s), the measurements on the driver dummy were not recorded, so it will be only possible to analyse the kinematics of the dummy. The two bags began to deploy 27 ms after the beginning of the crash and were completely inflated at 55 ms.

The passenger contacted the bag at 50 ms but the bag deployed poorly with its upper part turned downward. The head described a high trajectory so that it impacts the upper windscreen rim with a high deceleration peak (116 Gx, 104 Gz and a resultant of 118 G) (Fig. 4), then it impacts the windscreen. The thorax decelerations are high (65 Gx and 69 Gy) (Fig. 5). With the knees restrained by the air bag, the pelvis reached the seat extremity.

The driver had a high trajectory in the vehicle similar to the passenger. Since the air bag only restrains the thorax, the head impacted the upper windscreen rim with hyperextension of the neck. The movement seemed to predict a high deceleration peak of the head. The right knee impacted the steering column (Fig. 6).

Results comparison - For this type of crash, the air bag restraint seems insufficient particularly for the head protection. In fact, the head moves above the bag. The vertical movement of the vehicle during its deceleration

Fig. 6 - Frontal barrier crash – vehicle after impact

may contribute to this dummy trajectory which does not occur with the three-point belt restraint.

The three-point belt restraint appears efficient for the passenger, even if the head decelerations are a little too high. However, the relative motion of the pelvis should be reduced and it may be assumed that the motion increased by the possibility of lap and sash belt sliding in the buckle combined with belt elongation.

CRASH AGAINST A GUARDRAIL - The crash against a guardrail at 80 km/h (22 m/s) with 30° angle of incidence (standardized conditions in France), was characterized by a long time duration of about 500 ms. During the deformation of the guardrail, the vehicle was decelerated with small deformations on the left side, so that low decelerations were recorded more noticeable in the lateral axis of the car. Then, the vehicle left the guardrail at a low speed with a 10° angle (Fig. 7).

Three-point belt restraint (1 B 5) - The vehicle speed before crash was 77 km/h (21,4 m/s). The driver seated on the impact side, underwent a low amplitude movement with deceleration below 10 G on the head or the thorax. But during this movement, the head went through the side door rim (the window open) and the left femur impacted the side panel door with an axial load of 1200 N.

The kinematics of the passenger shows the influence of the car rotation on the performance of the asymmetric restraint. Indeed, the thorax moved laterally, away from the sash belt, up to the impact of the head against the driver's shoulder which corresponds to a head deceleration peak (30 Gx, 50 Gy, 25 Gz). Then, the passenger moved ahead while restrained by the lap belt only, but the head did not impact the front structure.

Fig. 7 - Crash against a guardrail — vehicle after impact

Air bag restraint (2 B 4) - The vehicle speed is 82 km/
h (22,8 m/s), and the vehicle has a movement similar to the
first crash (1 B 5). The ignition of the air bags started
19 ms after the crash beginning so that the bags are completely inflated after 50 ms.

The driver kinematics are similar to the previous crash
test because he moved laterally with low displacement of the
pelvis, but the head impacted the side door rim with a deceleration peak (resultant acceleration of 69 G). It appears
that the wheel bag has not been impacted by the driver.

The passenger began to move laterally without contact
with the front bag. The left shoulder impacted the shoulder
of the driver, so that the lateral movement was modified
with a good deceleration of the head in the driver's bag
(12 Gx, 14 Gy, 18 Gz). The high load of the right femur
corresponds to the impact on the generators' support.

Results comparison - For this type of crash, with low
car deceleration, it appears that the two restraint systems
did not function properly. The three-point belt ensures a
good restraint of the pelvis and low extremities but does
not avoid head impacts. On the other hand, the air bag does
not protect the driver and its effectiveness is not reliable
for the passenger.

HEAD ON CRASH BETWEEN TWO VEHICLES - For this crash,
each vehicle (R 12 of the same weight) traveled at 50 km/h

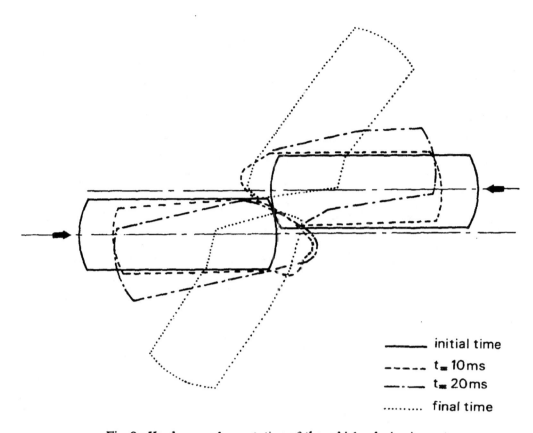

———	initial time
– – – –	t = 10 ms
—·—	t = 20 ms
·········	final time

Fig. 8 - Head-on crash — rotation of the vehicles during impact

(13,9 m/s) and impacted the other one with a 1/3 width overlap (drivers in line approximately). During the first part
of the crash, the two vehicles kept a longitudinal movement
with frontal deformation up to 1 meter. Then, just before the
maximum collapse, the vehicles began to rotate (Fig. 8).
Maximum vehicle decelerations were about 25 Gx and 15 Gy
(Fig. 3).

    <u>Three-point</u> <u>belt restraint (1 C 6)</u> - For this crash,
the initial speed was 48 km/h (13,4 m/s) and nearly 100 ms
after the beginning of the impact, the cars rotated up to
43° from their initial position (Fig. 9).

    The driver had a linear motion in the vehicle compart-
ment, slightly left of center. As four of the breakable
webbings worked, the strap elongation was quite high. The
thorax impacted the lower steering wheel rim, inducing a
high longitudinal deceleration (59 Gx) while the head impacts the upper left steering wheel rim with a resultant
deceleration peak of 100 G (Fig. 10). The strap loads were
not too high (8200 N on the common interior strap, 6700 N
on the sash) (Fig. 11). A slight impact of the knees on the
dashboard was noticed.

    The passenger moved from the seat back as the vehicle
rotated. He slid slightly along the sash belt. During his

Fig. 9 - Head-on crash — vehicles after impact

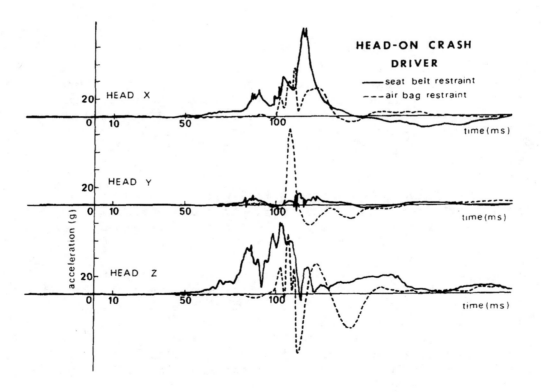

Fig. 10 - Head-on crash — driver head acceleration

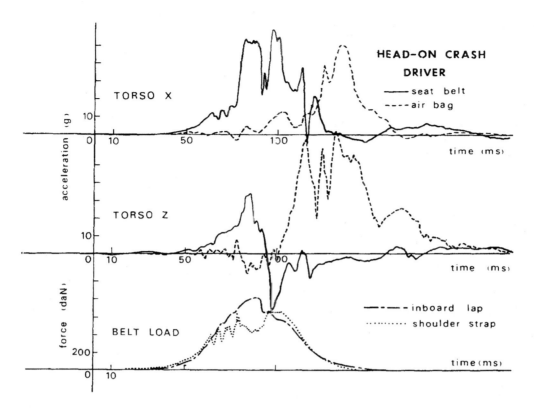

Fig. 11 - Head-on crash — driver torso acceleration and belt load

deceleration, each force limiting webbing broke. The two combined phenomena induced a long displacement of the body and the head impacted the dashboard. But the deceleration peak is not very high (61 Gx, 38 Gy which gives 72 G resultant) (Fig. 12). The thorax deceleration is acceptable and the strap loads are quite distributed (Fig. 13). No impact of the knees was noticed. Besides, the small back motion indicates a good energy absorption by the belt.

Air bag restraint (2 C 9) - The vehicle speed before impact was 48 km/h (13,4 m/s). During the crash the deceleration was low (18 Gx, 10 Gy) but the rotation of the vehicles was more important (90°) than the previous one (43°) and could have been due to a slight divergence of one vehicle after the guide-rails.

The ignition of the air bag occurred 19 ms after the crash beginning, the bags being well inflated 45 ms after the crash beginning.

The driver contacted the bag at about 80 ms. During his movement, the head slid from the bag and impacted the upper left part of the windscreen, inducing a quite high deceleration (56 Gx, 84 Gy, 70 Gz and 104 G resultant) (Fig. 10). The thorax which was not very well decelerated, impacted the steering wheel and broke its rim. The deceleration is also high (51 Gx, 69 Gz) (Fig. 11). The backward

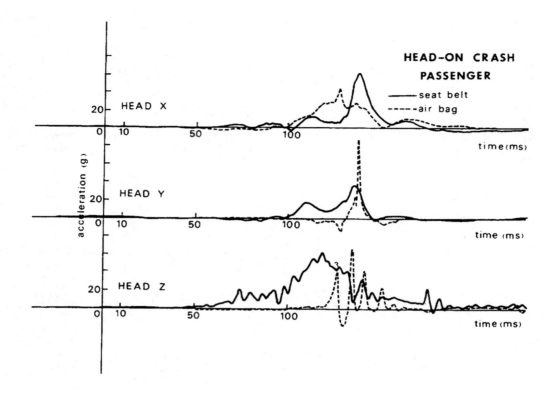

Fig. 12 - Head-on crash − passenger head acceleration

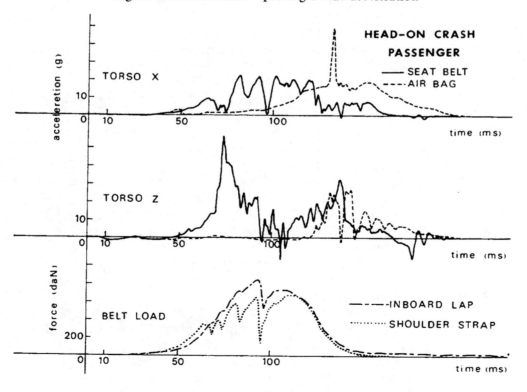

Fig. 13 - Head-on crash − passenger torso acceleration and belt load

movement began rapidly so that the head impacted the upper side door rim.

The passenger contacted the bag nearly at the same time as the driver. Because of the car rotation, the head was off-centered, reducing the protection possibilities of the air bag. The head impacted the windscreen through the bag at 130 ms with a deceleration peak (resultant of 85 G) (Fig. 12). The thoracic deceleration was relatively low (Fig. 13). The quite fast speed during the backward motion indicates poor protection and results in an impact against the "B" pillar.

Case of pre-ignition of the air bag (2 C 8) - For the first test of this type with the air bag, the ignition of the air bag occurred when the vehicle went off the guide-rail (about one second before the crash) so that the bags were not fully inflated during the real crash, because of the porous outlets. The analysis of this test is interesting because such a situation could appear in case of crashes with multiple impacts. The vehicle motions are quite similar to the previous ones with a rotation of about 50°.

The driver's head and thorax contacted the deflated bag at 80 ms, and impacted respectively the sun-shield and the steering wheel with high deceleration peaks (114 Gx, 106 Gz on the head, 85 Gz on the thorax). Moreover, the femurs impacted the dashboard.

The passenger contacted the deflated bag at 70 ms. The head deceleration was severe with impact on the windscreen through the bag while hyperextended. The high femur loads correspond to impact on the generators' supports.

Results comparison - For this type of crash with high vehicle deceleration and rotation after maximum deformation, both restraint systems present disadvantages. The three-point belt gave good protection to the passenger who was seated on the side opposite to the crash, even if the relative displacement was long. But for the driver, the protection was good for the pelvis and low extremities but insufficient for the thorax and the head. It appeared that the load limiting device reduced the passenger deceleration (Gadd Severity Index (SI) lower than 1000), but contributed to the driver's impact on the steering wheel (SI of 1500).

With the air bag, the low deceleration levels and low severity indices seem to indicate a good efficiency. But, the head impact on the windscreen and thorax impact on the steering wheel show the limited protection reached in these test conditions.

This crash test with pre-ignition of the bags reveals that this air bag design would not provide protection for long duration or multiple impacts because it cannot stay inflated for a long time.

Table 2 - Alderson and ONSER Dummies Used — Main Anthropometric Dimensions

Dimensional location (m)	dummies	
	ONSER	ALDERSON
Popliteal height	0.425	0.439
Knee height (sitting)	0.540	0.543
Buttock popliteal length	0.470	0.495
Sitting height (erect)	0.890	0.935
Buttock knee lenght	0.580	0.592
Shoulder breadth	0.480	0.455
Weight (kg)	73.5	74.5

INFLUENCE OF THE DUMMY

Two tests were conducted with ONSER dummies in the standard barrier crash condition, the first one with the three-point belt restraint, the second one with the air bag restraint.

The ONSER dummy, compared to the Alderson's, has a shorter sitting height and femurs articulated on the pelvis. (Table 2)

These tests results show behaviour differences which are interesting to analize.

THREE-POINT BELT RESULTS - The comparison of the passenger movements indicates mainly that the head trajectory is different for the two models (Fig 14). The maximum head displacement was longer with the ONSER dummy which impacted the windscreen (0,70 m instead of 0,54 m in horizontal direction). This fact may be explained by the greater degree of freedom of the ONSER dummy. During the crash, the thorax tended to swivel upon the sash belt.

Besides, it appears that the strap loads (common interior and sash belt) are differently distributed for the two dummies.

AIR-BAG RESULTS - The differences in results which appear for the two crashes are directly connected to the sitting height of the dummies and the vehicle motion (Fig 15 - 16). Thus the trajectories of the driver and passenger indicated a better head protection in the case of ONSER dummy because the Alderson's head did not run into the bag. This phenomenon has certainly been amplified by the vertical rising of the vehicle during this type of crash.

FRONT BARRIER CRASH
PASSENGER
SEAT BELT RESTRAINT

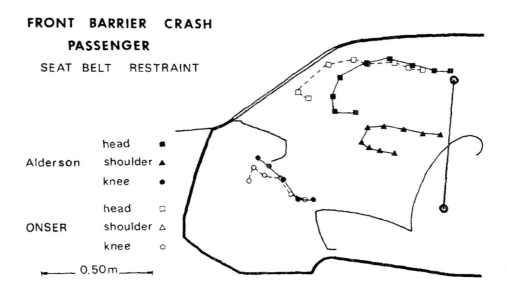

Alderson
- head ■
- shoulder ▲
- knee ●

ONSER
- head □
- shoulder △
- knee ○

0,50m

Fig. 14 - Frontal barrier crash — passenger kinematics three-point belt

FRONT BARRIER CRASH
PASSENGER
AIR BAG RESTRAINT

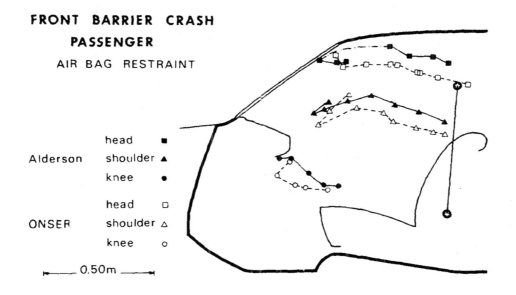

Alderson
- head ■
- shoulder ▲
- knee ●

ONSER
- head □
- shoulder △
- knee ○

0,50m

Fig. 15 - Frontal barrier crash — passenger kinematics air bag restraint

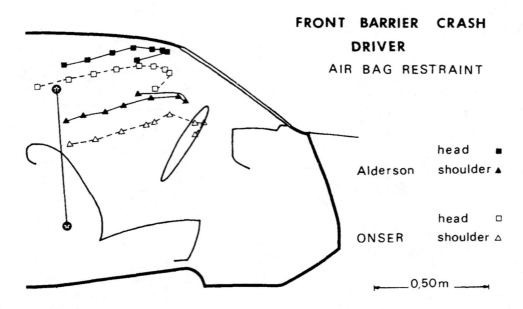

FRONT BARRIER CRASH
DRIVER
AIR BAG RESTRAINT

| | head ■ |
| Alderson | shoulder ▲ |

| | head □ |
| ONSER | shoulder △ |

0,50m

Fig. 16 - Frontal barrier crash — driver kinematics air bag restraint

DUMMY RESULTS COMPARISON - These tests indicate the great influence of the dummy on the assessment of restraint efficiency. This difference is all the more difficult to analyze as it also depends on the type of restraint used. In the case of three-point belt, the mechanical characteristics such as thorax rigidity and articulations, induce differences.

But in the case of air bag, the difference comes from the dimension characteristics of the dummy, particularly the sitting height.

CONCLUSION

This study about two restraint systems, three-point belt and air bag, in real test conditions, leads to the analysis of the problems set by each restraint type in a subcompact vehicle.

It appears that the three-point belt ensures protection in each crash type that has been performed. The pelvis is generally well restrained but it seems that the restraint of the torso is not sufficient and diminishes the head protection in the case of crashes with vehicle rotation (impacts with the vehicle interior are not avoided). This phenomenon has been probably enhanced by the additional slack due to the load limiting device and the advanced position of the anchorage point on the "B" pillar. Moreover, the decelerations and severity indices are too high, particularly on the head, so that the frontal barrier crash test at 50 km/h seems to be the limit protection of this three-point belt.

EFFICIENCY COMPARISON

The air bag appears to give different levels of protection according to the crash type. Good protection is obtained in the case in which two vehicles crash with overlap, wherein the head impact on the windscreen through the bag does not induce high deceleration peaks. However, in the frontal barrier crash, the poor protection comes from the vertical movement of the vehicle so that the head impacts the upper windscreen rim or the windscreen. In addition, the lower extremities protection is not sufficient. For the driver, the polyurethane padding does not absorb enough energy in the area of the steering column. For the passenger, the air bag with only one compartment does not avoid knee impacts in some crash cases.

It is important to notice that the two types of conducted front crashes, the front barrier crash and the head-on crash between two vehicles with overlap, show different characteristics and consequences which should be taken in account into the restraint systems studies.

Thus, the improvement of the used three-point belt must concern with the head deceleration from vehicle contact which should be reduced. The poor results of the used air bag show the difficulty of this restraint adaptation on a subcompact vehicle and the influence of the crash type on its protection.

ACKNOWLEDGMENTS

This study has been supported by the French Highway Administration. The authors acknowledge the Society Renault and the Society SNPE for their technical aids.

REFERENCES

1. L.M. Patrick, G.W. Nyquist, K.R. Trosien, "Safety Performance of Shaped Steering Assembly Air Bag." Proceedings of 16 th Stapp Car Crash Conf. Paper 720976, pp 434/471, November 1972.

2. R. Quincy and Y. Derrien, "Sac Gonflable" Rapport final (Objectif 16). Organisme National de Sécurité Routière, Décembre 1972.

3. U.W. Seiffert, G.H. Borenius, "Development problems with inflatable restraints in small passenger vehicles." Proceedings of 2nd International Conference on Passive Restraints, paper 720409, May 22-25, 1972.

4. T.D. Clarke, J.F. Sprouffske, E.M. Trout, C.D. Gragg, W.H. Muzzy, H.S. Klopfenstein, "Baboon tolerance to linear deceleration (-Gx) : Air Bag Restraint". Proceedings of 14th Stapp Car Crash Conf. Paper 700905, pp 263/278, November 17-18, 1970.

5. T.D. Clarke, J.F. Sprouffske, E.M. Trout, C.D. Gragg, "Baboon Tolerance to Linear Deceleration (-Gx) : Lap Belt Restraint." Proceedings of 14th Stapp Car Crash Conf. Paper 700906, pp 279/298, November 17-18,1970.

6. R.G. Snyder, J.W. Young, C.C. Snow, "Experimental Impacts Protection with Advanced Automative Restraint Systems." Proceedings of 11th Stapp Car Crash Conf. Paper 670922, pp 271/285.

7. T.D. Clarke, C.D. Gragg, J.F. Sprouffske, E.M. Trout, "Human Head Linear and Angular Accelerations during Impact." Proceedings of 15th Stapp Car Crash Conf. Paper 710857, pp 269/286, November 17-19,1971.

8. C.D. Gragg, C.D. Bendixen, T.D. Clarke, H.S. Klopfenstein, J.F. Sprouffske, "Evaluation of the lap Belt, Air Bag and Air Force Restraint Systems during Impact with Living Human Sled Subjects." Proceedings of 14th Stapp Car Crash Conf. Paper 700904, pp 241/262, November 17-18, 1970.

9. G.R. Smith, S.S. Hurite, A.J. Yanik, C.R. Greer, "Human Volunteer Testing of GM Air Cushions." Proceedings of 2nd International Conference on Passive Restraints, Paper 720443, May 22-25, 1972.

10. J.F. Martin, D.J. Romeo, "Preliminary Vehicle Tests Inflatable Occupant Restraint Systems." Proceedings of 15th Stapp Car Crash Conf. Paper 710866, pp 518/551, November 17-19, 1971.

11. C. Berlioz, "Distribution and Gravity of Collisions as a Function of the Damaged Part of the Vehicle and the Obstacle Hit." Proceedings of 2nd International Conf. on ESV, part 2, pp 131/134, October 26-29, 1971.

12. D. Cesari, M. Ramet, "Biomechanical Study of Side Impact Accidents". 5th International Conf. on ESV, pp 511/520, June 4-7, 1974.

Safe and Free

JEAN C. DERAMPE
Safety Pilot
Centre de'Etudes de Paris des Automobiles Peugeot

SUMMARY

Owing to its efficiency, reliability, and most favorable cost/efficiency ratio, the 3-point seatbelt tends to be generalized; in Europe, wearing seatbelts is gradually becoming compulsory for front occupants. Is it reasonable to go further and make the wearing of belts compulsory at rear of vehicles, too? The answer is difficult, for:

- Young children, who in 28 percent of the cases occupy these places, cannot be valuably protected by belts.
- Rear seat occupants may adopt most different positions; their number and size vary greatly according to the travel.

Wearing belts would be particularly constraining at the rear of vehicles and less efficient because they would be less accepted, not to mention the trouble created by fastening belts.

No doubt safety shall be the prior aim provided that the automotive transportation qualities will not be ruined; the aim of this presentation is to propose an alternative to the 3-point seatbelt, which is at least as efficient as the latter and much more comfortable. It is a passive protection system fitted into the passenger compartment.

The main results of an impact testing program are also presented; it involved various occupants (men, women, children) adopting different postures, while ascertaining the system's efficiency. This system provides protection to rear passengers while it does not interfere with the front occupants' protection.

INTRODUCTION[1]

The 3-point seatbelt has experienced widespread utilization. Its efficiency was fully demonstrated in numerous laboratory tests and was further ascertained by the analysis of real life accident consequences, particularly in Sweden, Australia, and France.

Then, while facing the ever-increasing number of accidents, authorities in many countries decided to make the use of seatbelts compulsory.

In 1972 it happened in Australia and Sweden, then in France, and recently in most European countries.

French manufacturers believe that the seatbelt has demonstrated its efficiency in affording protection to front occupants.

Peugeot advocated the compulsory use of seatbelts for front occupants when driving outside cities . . . for the driver's and the front passenger's seating position are well defined by two arm chairs. These allow, together with the minimum inconvenience, being adequately restrained for long travels, especially since the adoption of retractors.

---

[1] Text from the film presented at the "Structural Properties and Occupant Protection" seminar.

Would the compulsory use of seatbelts at rear places be a suitable solution?

The manufacturer believes that making the use of back-seat seatbelts compulsory would impose intolerable discomfort to rear passengers . . . indeed, seatbelts would be a lot of trouble on rear bench occupants, not to mention adults or children may be concerned as well.

One can imagine for example, a mother who drives her children to school every morning. Of course she has to make sure that seatbelts are correctly worn. She checks every time before starting. She spends time applying herself to do it: safety first!

Some of these children are not belted: this is not due to carelessness. Seatbelts are designed to fit grown up people and are not suitable to children of less than 1.30 meters height.

Peugeot is working jointly with the French Institute of Transportation to develop an efficient restraint system for children. The system is not yet available and children are only protected by front seat backs.

The inconvenience of seatbelts at rear places is a fact and not only for children: let's imagine two businessmen in a taxi; their movements are limited while they once again examine their files before the next meeting.

AN ALTERNATIVE TO THE SEATBELT AT REAR PLACES EXISTS

The reader may think this is not quite serious: have we the right to contest an existing means of protection, for reasons belonging to comfort and convenience? Is it possible to equip rear seats with a restraint system as efficient as the seatbelt in front impact situations?

Peugeot has defined its own solution to the problem while developing a vehicle affording protection to rear occupants. The aim of a back-seat restraint system should take account of:

Figure 1. Operating principle of the child restraint system.

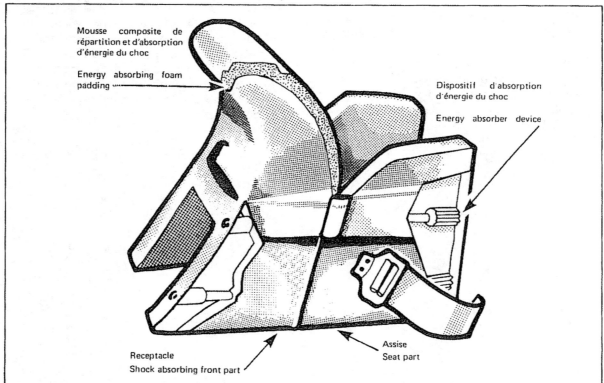

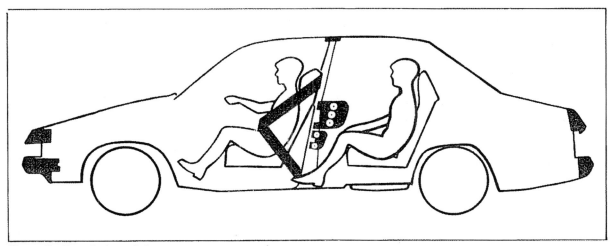

Figure 2. Description of the rear occupant restraint system.

- The occupation rate of seats
- The nature of the occupants
- The severity level of injuries to the rear occupants in accidents

Occupation Rate

From the ONSER inquiry (ONSER is the French National Agency for traffic safety) the occupation rates of vehicles involved in road traffic accidents are as follows:

- In 40 percent of the cases the driver was alone.
- In 28 percent of the cases there was another front occupant.
- There were occupants in the rear seat in 32 percent of cases.

From the Peugeot/Renault Association's multidisciplinary inquiry it appears that the average occupation rate of back seats is 1.7. Assuming there are three places in the back, the utilization rate of a restraint system at any one of these places is at most, 20 percent.

Nature of Occupants

Again, from the multidisciplinary injury, the results are that age classes may be distributed as follows:

- Under 6, 17 percent
- From 7 to 12, 11.5 percent—bearing in mind that the belt is not appropriate as a restraint system for children in these age classes

Figure 3. Deceleration law of the test sled.

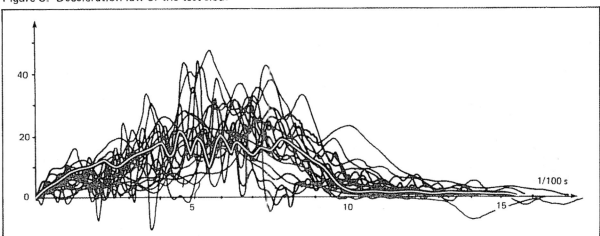

- 8 percent of the young people from 13 to 17 who can wear seatbelts
- 60 percent of the grown up people

To conclude, let us assume that the seatbelt would be adopted as a restraint system for rear occupants, children representing 28 percent of rear occupants, the *actual utilization rate of seatbelts* at rear places would be only 15 percent.

Then, compulsory use of seatbelts in the back seat is seven times less justified than in the front.

Severity Level of Accidents

The restraint system shall be designed on account of the actual risk to rear occupants:

- Out of 100 nonbelted front occupant victims, 7 are killed
- Out of 100 belted front occupant victims, 3.8 are killed
- Out of 100 rear occupants, 2.6 are killed

The risk is therefore lower for the rear occupants than for the front occupants; this may be explained as follows.

The analysis of a front impact shows that rear occupants are restrained by the front seat backs.

This only restraint provides a long stopping distance for rear occupants, resulting in low severity, but implies a hazard to front occupants through excessive loads imposed to their seat backs. Then, the driver's head HIC

rises up to 4 450 and the torso deceleration to 70 g.

SELECTING A NONBELT RESTRAINT SYSTEM

Net Protection System

In limiting the rear occupants, trajectory to the front, a possible solution consists of a net installed between the reinforced front seat backs and the car top.

However, in examining occupants' trajectories, one may notice the very dangerous hyperflexion of the head, though injury criteria are met:

$$\text{Head HIC} = 574$$
$$\text{Torso deceleration} = 40 \, g$$

The restraint system shall not involve any head restraint.

Safety Seat

The protection of rear occupants was also considered in reinforcing the front seat anchorages and hinges and equipping seat backs with a receptacle for the occupant. However, occupants are still thrown over the seat backs: for the seat backs, being very rigid and stopping the knees, entail an upward trajectory.

Locating the receptacle nearer the occupants would be the only way to avoid this but

Figure 4. 50th-percentile dummies - maximum levels measured.

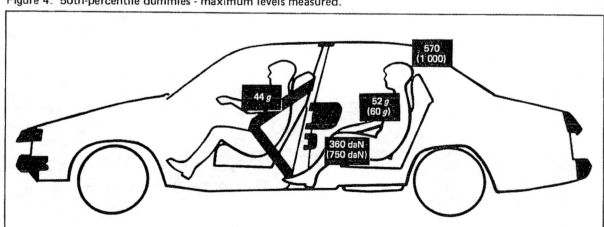

would seriously impair rear occupant space. This makes it clear that modification of the front seat back is not a suitable solution.

PRINCIPLE OF THE PROPOSED RE-STRAINT SYSTEM

Peugeot has experimented with quite a different restraint system for the rear seat. This is a seat especially designed to protect young children. It consists of two parts: the seat itself, attached to the seatbelt anchorage points, and the receptacle located in front of the child, who is thus restrained by a large surface padded with an honeycombed mate-

rial distributing deceleration forces on the whole torso (fig. 1).

The plastic rigidity of the belt is replaced by energy absorbing devices connecting the receptacle to the seat while limiting the forces applied to the torso. The receptacle is de-signed so that the occupant's trajectory avoids any impact to the head.

When this system demonstrated its effi-ciency, Peugeot adapted the principle to adults. The restraint system is attached to the center pillars: it consists of a set of tubes at the torso level (that is, the receptacle) and at the lower member's level (fig. 2).

When an impact occurs, the occupant follows a trajectory at the initial speed of the

Table 1. Four 50th-percentile dummies—measured levels

Item	3-point seatbelts		Static passive	
	Hybrid II 50th (Right front)	Hybrid II 50th (Left front)	Hybrid II 50th (Right rear)	Hybrid II 50th (Left rear)
Head:				
(γ) 3 ms (g)	—	—	62	62
HIC	—	—	563	570
Thorax:				
(γ) 3 ms (g)	42	44	52	52
Femur loads (daN):				
Left	—	—	220	320
Right	—	—	400	360

Table 2. Two 50th-percentile male dummies at front, one 5th-percentile female and one child at rear—measured levels

(V = 52 km/h after braking)

Item	3-point seatbelts		Static passive	
	Hybrid II 50th (Right front)	Hybrid II 50th (Left front)	6-year-old child (Right rear)	Female 5th (Left rear)
Head:				
(γ) 3 ms (g)	—	—	57	85
HIC	—	—	—	712
Thorax:				
(γ) 3 ms (g)	35	31	39	49
Femur loads (daN):				
Left	—	—	—	360
Right	—	—	—	380

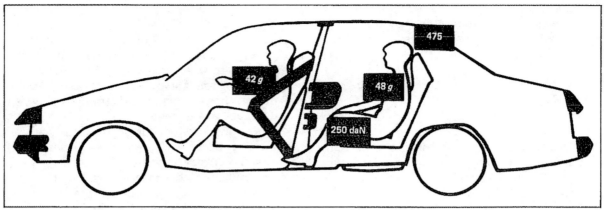

Figure 5. Impact speed 54 km/h with prior braking simulated—maximum levels measured.

vehicle. The distance between the occupant and the receptacle is about 500 mm, which is covered approximately in 0.08 s, thus banning any coupling with the vehicle during the deceleration phase.

Then, the occupant impacts the restraint system at a speed equivalent to the speed variation of the vehicle. The occupant is restrained at the torso level. The tubes are padded with a composite foam material in order to lessen the effects of contacting the receptacle. The energy created in restraining the occupant is dissipated through flexion and distortion of the tubes, their rigidity being calculated to maintain the torso deceleration under 60 g.

The risk of submarining is eliminated by additional tubes located at leg level. The advantage of this solution is to preserve the quality of protection provided by the system for any location of front seats and for every occupant sizes from 6-year-old children on.

Test Results

The quality of protection provided by this system was ascertained through impact testing. The receptacle was placed in a passenger compartment mounted on a test sled. Test conditions were selected as representative of actual impact situations. The deceleration law adopted is an averaging of deceleration laws measured on numerous vehicles in 30-degree front impacts (fig. 3).

Impact velocities are about 50 km/h. We reached 43 km/h at the present state of technology and 54 km/h at the time of the impact with braking simulation beforehand.

Table 3. Four 50th-percentile dummies—measured levels

Item	3-point seatbelts		Static passive	
	Hybrid II 50th (Right front)	Hybrid II 50th (Left front)	Hybrid II 50th (Right rear)	Hybrid II 50th (Left rear)
Head:				
(γ) 3 ms (g)	—	—	48	48
HIC	—	—	475	225
Thorax:				
(γ) 3 ms (g)	42	39	48	46
Femur loads (daN):				
Left	—	—	200	740
Right	—	—	250	550

Figure 6. Vehicle development of the rear occupant protection system.

When four Hybrid II male dummies of the 50th percentile were placed in the vehicle, rear-passenger trajectories were appropriate; their presence did not interfere with front occupant protection: we experienced neither a modification of the torso deceleration nor any submarining. The measured levels are lower than the tolerance levels generally admitted (fig. 4 and table 1).

We also verified the quality of protection of the system with a 5th-percentile woman

and a 6-year-old child, front occupants being 50th-percentile males (table 2).

The protection afforded to the legs avoided submarining effects and the results were very good.

They were in a situation where the woman folded her legs and the man is sitting obliquely. We also performed a test where the speed at the moment of impact was raised to 54 km/h by simulating a prior braking: the results remain satisfactory (fig. 5 and table 3).

Table 4. Synthesis impact test—measured levels

Item	3-point seatbelts		Static passive	
	Hybrid II 50th (Right front)	Hybrid II 50th (Left front)	Hybrid II 50th (Right rear)	Hybrid II 50th (Left rear)
Head:				
(γ) 3 ms (g)	–	–	60	72
HIC	–	–	633	737
Thorax:				
(γ) 3 ms (g)	36	28	46	48
Femur loads (daN):				
Left	–	–	460	380
Right	–	–	480	200

All these results obtained with a test sled were confirmed by a synthesis impact test, performed at 44 km/h. (See table 4.)

CONCLUSION

The results registered with this device demonstrated that the safety belt is neither the only solution nor the best one for back seats. The tests were carried out only at 45 km/h, but the technology may be improved and we reached 54 km/h by simulating a prior braking registered in more than one case out of two.

One may object that we propose a device whose cost is without doubt higher than that of the safety belt. Nevertheless, we have seen that the safety belt cannot be used by 30 percent of the vehicle users, composed of children, so that the cost/efficiency ratio is of the same order.

But it is quite different for the comfort/ efficiency ratio. This device offers the best safety without entailing at the same time uneasiness and discomfort.

It would be advisable that the authorities in charge of safety do not treat the problems only by constraining laws; so much more than the speed limitations increase the time passed in the vehicles.

Any physical discomfort in the occupant compartment may have psychological effects on the safety in vehicle driving. That is why the master coach-builder, Pininfarina, has been charged with adapting this safety system to the occupant compartment. (See fig. 6.)

This research should be the starting point of a new approach to car safety that would involve compulsory belt wearing in the front of vehicles but in the back seats a less constraining solution, better adapted to the comfort of individual transport, should be devised.

791004

Seat Design— A Significant Factor for Safety Belt Effectiveness

Dieter Adomeit
Institute for Automotive Engineering
Technical University Berlin

Abstract

Production seats and specially designed research seats were analyzed with respect to seat-safety belt interaction under frontal crash conditions. The objective was to evaluate seat influences on effectiveness of safety belts, or any other restraint system. We determined that classical measurement techniques alone are insufficient to completely cover problems of seat-safety belt interaction. Supplementary evaluation parameters of dummy kinematics were therefore defined to clarify

- the poor safety design of current production seats

- necessary seat design development for increased safety.

Proposals for possible effective seat design were then derived to satisfy necessary new safety requirements for seating.

849

BACKGROUND: THE CURRENT LEGAL SITUATION AND STAGE OF
RESEARCH WITH RESPECT TO THE INTERACTION OF SEATING AND
THE SAFETY BELT

THE AUTOMOBILE SEAT has, as an important component of the
overall restraint system for vehicle occupants, a decisive
role in the protective effectiveness of the safety belt.
This is well known and has been proved in numerous dummy
crash test series. It has also been evidenced, with the
increasing amount of data and experience gained, that the
seat also plays an important role in the occurrence of
belt-related injury phenomena (1, 3, 8).[*]
    Investigation of the characteristics and of the influ-
ence of vehicle seats in their interworking with the three-
point belt is therefore especially justified in view of the
fact that no national or international safety regulations
or standards exist on the structure or the operation of
vehicle seats under these conditions.
    There exist only strength requirements for the seat
mountings and fittings as prescribed in ECE R 17/FMVSS
207—requirements which, however, are not directly related
to the deformation characteristics of the seat and the
seat frame which are of interest in study of the head-on
crash.
    Up to now, legal safety regulations concerned with
head-on crashes have concentrated essentially on the
safety belt in the over-simplifying assumption that the
quality of this protective system alone dictates the
degree of auto occupant protection.
    The seat, however, plays a decisive role—precisely
for the belt-protected occupant. The seat determines the
nature and the extent of the vertical collapse of the
occupant's body caused by the lap belt, anchored as it is
in the floor of the vehicle. The seat structure furthermore can under certain conditions determine the point in
time and the force of the vertical impact of the hips
onto the lower frame parts of the bottom of the seat
structure. This fact can have critical significance for
the resulting head and cervical spine forces.
    In this manner, the lower seat framework controls the
entire sequence of movement and forces and is critical
with respect to the "biomechanical quality" of the motion-
loading sequence of the strapped-in occupant; i.e., the
manner of introduction of restraining forces into the
occupant's body. The seat characteristics furthermore

---

[*]Numbers in parentheses designate References at end of
paper.

determine varying values for the absolute degree of
occupant loading (accelerations, forces, etc.) with seatbelt systems which are in all other features identical.
The series of head-on crash tests presented here and
utilizing production and research seats were performed in
order to evaluate the effect of the seating structure on
the effectiveness of the three-point automatic belt.

Evaluation criteria utilized were partly the protective criteria according to US FMVSS 208. Utilization
was furthermore made of kinematic variables with the aid
of which the manner of introduction of the safety system
restraining forces into the body of the occupant—i.e.,
the "biomechanical quality" of the system—could be evaluated. Work preparatory to this methodology has already
been presented (1, 2).

TEST APPARATUS; MEASURING AND EVALUATION TECHNIQUE

The Δv values chosen for the crash series were approximately 40 km/h. A deceleration curve lying within the
range of tolerance is shown in Fig. 1 and indicates an
approximate deceleration peak of 28 g as well as a deceleration period of approx. 65—70 milliseconds. Distribution in the impact velocity and deceleration pulse resulted
from slightly differing values for the mass of the seat
and mounting, as well as from the differences in the interaction among the three-point automatic belt, the seat, and
the dummy. These factors, however, do not affect the basic
kinematic test relationships.

Since the impact velocities were lower than under test
conditions specified in the US regulation no. 208, direct
comparison of the dummy acceleration values with FMVSS 208
empirical values is not possible. Despite this, however,
the velocity of impact chosen corresponds more closely
to accident reality (3).

The basis for the series of experiments was a 50%
Hybrid II dummy and a crash sled system at the Institute
of Automotive Engineering at the Technical University of
Berlin. Both dummy and sled were fully equipped with
respect to measurement technology.

The technical evaluation of data gathered was performed in the conventional manner. Mention should be made
here, however, that the high-speed film (500 frames per
second) was subjected to highly sophisticated evaluation
techniques for determination of the trajectories and rotation of the dummy head, chest, pelvis, and knee over the
duration of the crash. The dummy was equipped for marking in the positions shown in Fig. 2: the point S' is
fixed to the upper thorax segment, point B is fixed to

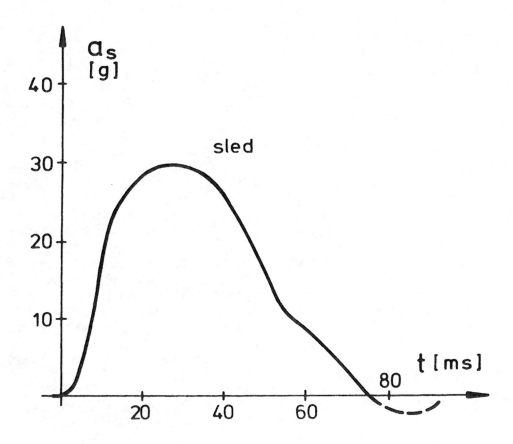

Fig. 1 - Sled deceleration curve

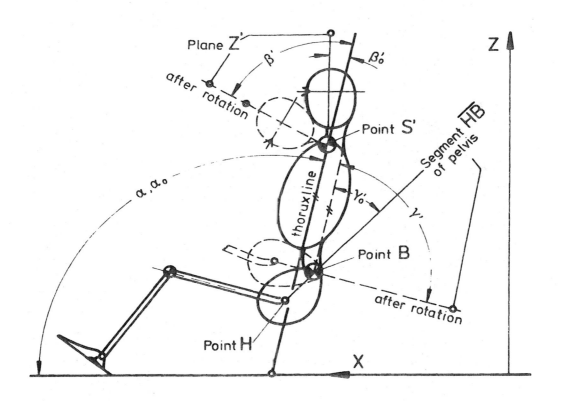

Fig. 2 - Marking scheme and angle definitions

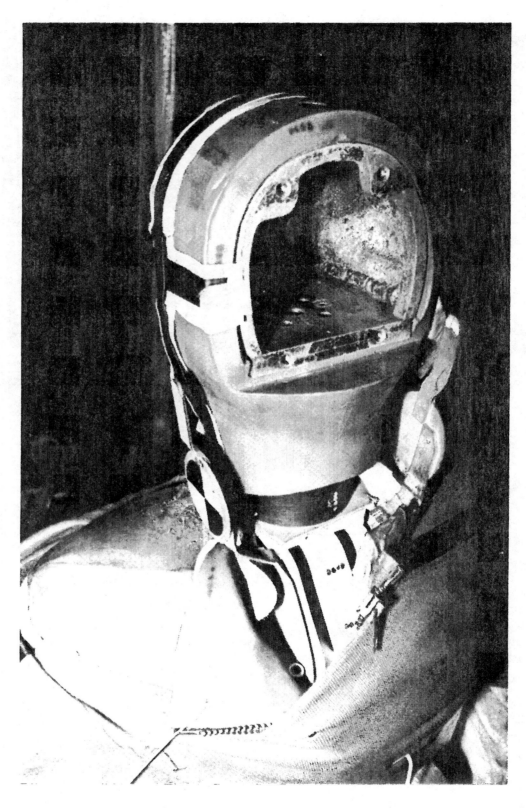

Fig. 2a – Marking and direction arrow at dummy's upper thorax

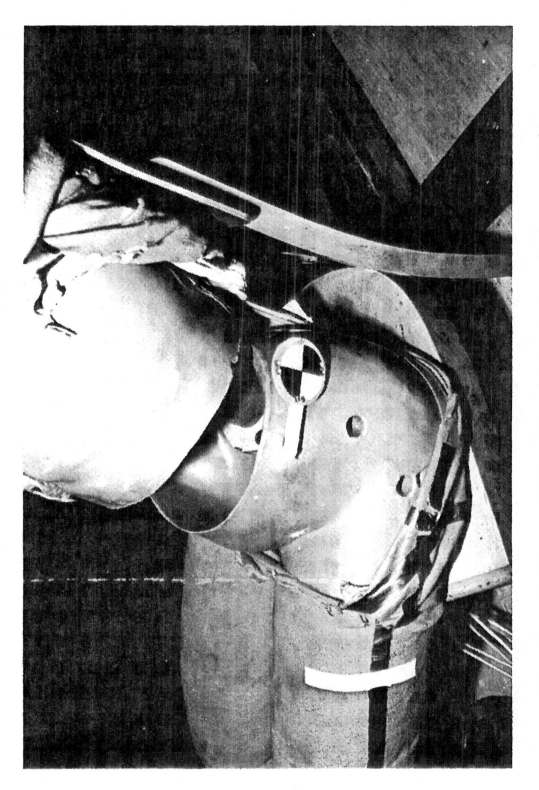

Fig. 2b - Marking and direction arrow at dummy's upper pelvis

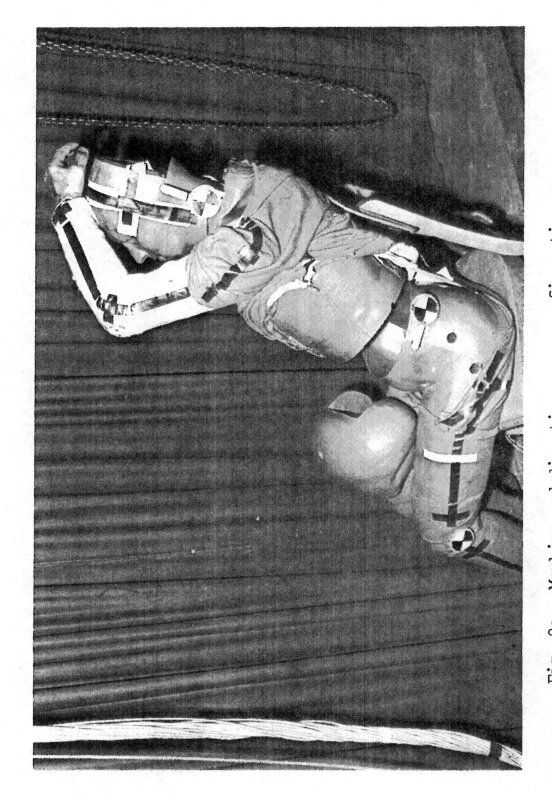

Fig. 2c - Marking and direction arrow configuration on the dummy

the top rear of the iliac crest, and point K is fastened to the knee joint. These special positions of the reference points on the dummy skeleton enable exact measurement of the angular relation of the individual body segments.

Parallax distortions, always present owing to the relative motion of the test object to the camera lens, were eliminated during the evaluation by means of geometric plan view reconstruction.

The following values were either measured, reconstructed, or calculated (see also Tables 1 and 2):

Forces and accelerations:

a_s	= sled acceleration
v_{os}	= sled impact velocity
$a_{H\ x,y,z}$	= dummy head acceleration
$a_{H\ rot}$	= dummy head rotational acceleration
$a_{Th\ x,z}$	= dummy thorax acceleration
$a_{P\ x,z}$	= dummy pelvic acceleration
HIC	= Head Injury Criterion
SI	= Severity Index (chest)

Kinematic variables (dummy):

x_H, z_H	= displacement of the head
x_{Th}, z_{Th}	= displacement of the thorax
x_P, z_P	= displacement of the pelvis
β'	= bending angle of the neck
α	= angular position of the thorax in stationary coordinate system
γ'	= bending angle of the lumbar spine

TEST CONDITIONS

It is well known that, in addition to the influencing variables to be discussed in this paper, there are further structural parameters which have graduated influence on the interaction of the seat and the safety belt. Such factors include the vehicle front end with its deformation characteristics, the rigidity of the interior space structure, the safety belt anchor points, and the interior floor complex. For this reason, care must be taken in formulation of the sled test conditions that excessive simplification does not change, or even reverse, the basic tendencies obtained in the results. In the following, presentation and justification of the sled test conditions are made.

SEAT AND VEHICLE INTERIOR - The production seats were tested in their original mode of slide positioning and mounting, and in their original installation configuration (distance above the vehicle floor, angular position, etc.). The slide mountings were fixed on the sled frame in such a way that any vehicle floor deformations would not be taken into consideration. The test loading values were therefore more severe for the seating console area and the mountings than under identical impact conditions with an actual vehicle.

The research seats were rigidly fixed onto the sled frame so that comparisons with the results with the production seats were possible.

The angle of the seat back for all tests was a constant of 110° to the horizontal, measured as a tangent or as a secant over the support area of the lumbar spine of the seat back curvature in the seat back center line, and referenced from the initial dummy position. A foot support enabled positioning of the legs of the dummy in a realistic manner.

The sled tests were conducted in all cases without an instrument panel. This is of considerable significance, since the lower part of the instrument panel is very often struck by the knees of the strapped-in occupant as he moves forward, and is thereby deformed, as safety-belt accident studies show. Only by disregard of the panel, however, can the interaction of the belt (and therefore the characteristics of the seat) be measured and determined as an isolated, undisturbed phenomenon. The characteristics of the trajectories are, however, not basically modified by the instrument panel, since a knee impact will take place only in late phases of the movements. This is as a result of the free space measured to the panel contour in various European vehicles. Evidence can therefore be presented that such knee impact (uncontrolled in any individual case) can have no tendency-changing effect on the results.

In individual cases, indeed, the knee impact even has a mitigating effect. Recent accident evidence shows namely that this impact has the fortunate result of preventing expected abdominal injuries due to submarining. The frequency of knee impact in reality furthermore shows that even if classical submarining does not occur, a type of trajectory is observed which can be characterized as indeed similar to the submarining phenomenon, as evidenced by the patterns of injuries (1, 3).

The currently valid FMVSS 208 limit for axial forces in the femur is, in this connection, moreover, merely an auxiliary criterion and cannot serve by itself as a basis for optimization, since kinematic requirements there are not sufficiently taken into account.

SEATBELT GEOMETRY AND THE SEAT - Tests for all seats were conducted with the same safety belt geometry, with fixed anchoring points on the sled. The influence of the belt system on the trajectory can therefore be considered constant, with the result than an analysis of the seats for their effectiveness, and in their comparison with each other, was possible.

The safety-belt geometry on the test sleds, when judged in relation to the actual installation conditions on the respective motor vehicles, was, in each of the cases identical, or, in isolated cases, more effective, i.e., steeper lap belt angle, shortened buckle part.

The following represents the belt geometry in figures, in relation to point H, with a constant 50% dummy seating position:

Lap belt angle:

$$\text{Buckle side} = \beta_B = 55 - 60°$$
$$\text{External side} = \beta_E = 60 - 65°$$

Length of the buckle part, with belt force transducer:

$$1 = 220 \text{ mm}$$

The buckle was therefore situated considerably lower than the seat sitting surface level in the conduct of the sled tests. This thereby eliminated the possibility that the shoulder belt could pull the lap belt up at the buckle side over the iliac crest—an occurrence surely leading to submarining.

The upper shoulder belt anchor point was approx. 900 mm above the plane of the sled frame—a position similar to the level as prescribed in ECE R 16. The transversal plane of the upper belt anchor lay approx. 200 mm behind the common transversal plane of the lower anchor points.

SUBJECTED TO TESTING: PRODUCTION SEATS AND RESEARCH SEATS

For the case of the head-on crash, only the structure of the seat cushion and of the lower cushion framework is of significance for our study. Testing revealed that, in the case of production seats, the seat back fittings and seat back structure remained as a rule without significant deformation and/or without effect on the trajectory of the dummy.

The types of production seat structures can be broken down into three structural groups, in accordance with Fig. 3, as follows:

Type SC   Sheet metal frame with spring center in cushion form, and cover: the classical spring—cushion seat

Type SC/SF   Sheet metal frame with steel coil netting or elastic belting in support of a soft foam cushion in seat form

Type SF   Sheet metal pan (self-supporting or mounted on tubular frame) with soft foam cushion in seat form.

Basic differences in the design of the seat cushion structure can be immediately recognized. This is evident both as far as comfort and subjective sitting evaluation are concerned, as well as for the compression characteristics observed for strapped-in passengers in the case of head-on crashes. These differences are of fundamental nature and cannot be explained away by alleged "adaptation" to the frequency response of the vehicle system alone. This fact assumes greater significance once one has become involved with the design of the following research seats and has established a reference to the production seats.

Fig. 4 shows the basic designs of the research seats, described as follows:

Type S   A simple, thickly upholstered soft foam structure (*"soft"*)

Type EA M   A soft foam upholstery, into which a medium-hard energy-absorbing wedge of styrofoam material has been integrated into the forward seating surface region (*"energy-absorbing, medium"*)

Type EA St   A soft foam upholstery, into which a stiff energy-absorbing wedge-formed piece of styrofoam material has been integrated into the forward seating region (*"energy-absorbing, stiff"*).

The basic design and the contour of the frame, of the metal pan, and of the seat upholstery are the same for all three graduated types above. The soft foam layer is installed for comfort and is made of polyurethane with a unit weight of approx. 18 kg/m³. The energy-absorbing styrofoam supporting wedge has the same cross-section in both of the last two EA types. Type EA M has a density of 34 kg/m³,

Seat Design
(production seats)

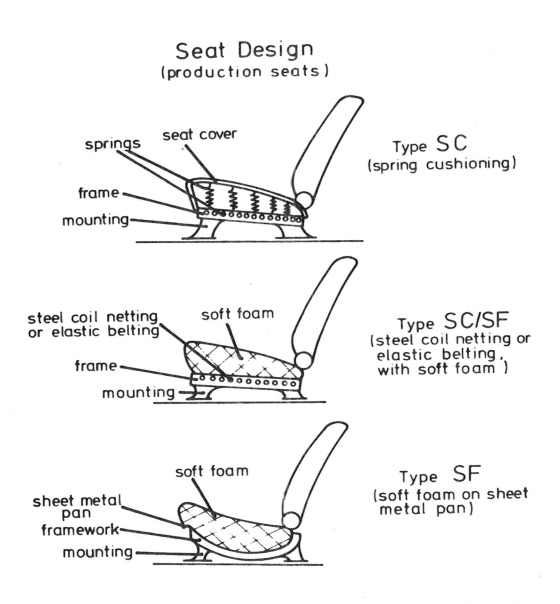

springs

seat cover

frame

mounting

Type SC
(spring cushioning)

steel coil netting
or elastic belting

soft foam

frame

mounting

Type SC/SF
(steel coil netting or
elastic belting,
with soft foam)

soft foam

sheet metal
pan

framework

mounting

Type SF
(soft foam on sheet
metal pan)

Fig. 3 – Three current production seat types used for testing

Seat Design
(research seats)

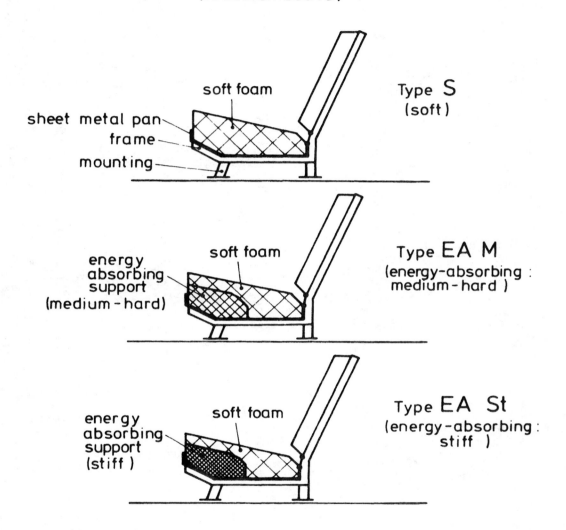

soft foam

sheet metal pan
frame
mounting

Type S
(soft)

energy
absorbing
support
(medium-hard)

soft foam

Type EA M
(energy-absorbing :
medium-hard)

energy
absorbing
support
(stiff)

soft foam

Type EA St
(energy-absorbing :
stiff)

Fig. 4 - Research seat design

whereas type EA St is considerably heavier, with 60 kg/m³. The wedge form of the EA elements is designed in such a way that, with the exception of an overall more rigid seat feeling, there are no basic effects evident on the vertical cushioning characteristics in the more important rear part of the seating region. Subjective seating comfort comparisons do not show marked differences in the three research types from corresponding type groups of production seats.

ANALYTICAL STUDY AND DEFINITION OF THE EVALUATION PARAMETERS

Evaluation parameters were defined for valuation of the loads as well as of the kinematics of the loading process.

LOADING CRITERIA - The FMVSS 208 loading criteria alone are not sufficient for valuation of the influence of the seat on the effectiveness of the safety belt (2). These well-known criteria, nevertheless, were measured, calculated, and listed.

KINEMATIC EVALUATION PARAMETERS - The kinematic evaluation parameters serve the purpose of classifying the manner of occurrence of the loads on the passenger, i.e., to judge the more or less biomechanically favorable points and directions of applications for the restraint forces on the occupant's body.

Fig. 2 represents the dummy skeleton in its initial position, in the stationary x-z coordinate system. The measuring points, rigidly fixed to the skeleton, can be seen at the thorax, pelvis, and knee. The positions of installation for the points S' and B have been chosen such that these points form the instantaneous center in first approximation for the crash-caused relative movements (bending between head and thorax, and between thorax and pelvis).

[Explanation may be inserted here that the measuring points S' and B are positioned approximately at the level of the lower rubber element flange of the cervical rubber cylinder and the lumbar cylinder of the dummy. See Fig. 2a, 2b, and 2c with S' and B supplemented by rigidly fixed white auxiliary arrows for indication of relative rotation of the body segments.]

On the basis of the crash tests, the measured angles further defined in Fig. 2 proved to be especially valuable for the following conclusions:

• The cervical spine bending angle (β') is the difference between the angle of initial head position (α_o) and the angle of final head position with respect to the thorax (α). As can be seen in the above-mentioned figure, it is defined as the angle between the thorax line and the z' plane of the head.

For the evaluation of the results for β', it must be remarked that, although a human has an anteflexion limit of 75 to 80° (threshold of pain) in a static case, the rubber neck of the dummy easily reaches 90° since there is no energy absorption through the contact of cervical vertebrae. Nevertheless, the measured values in relation to each other can be utilized for the drawing of evaluative conclusions.

• The thorax angle α is defined as the angle between the vertical axis of the thorax and the stationary x axis. It is measured in the forward-facing direction and indicates to which angular position the thorax ellipse has turned in the shoulder belt with respect to the horizontal x axis. Final positions with the angles of $\alpha \leq 90°$ are advantageous because, under the assumption of purely horizontal movement, provision is made in this way that the resulting shoulder belt force is applied at the significantly more resistant upper thorax segment in the sternum area (1, 2).

• The angle γ' deserves special attention. It is measured as the angle between the thorax line when displaced parallel up to the point B, and the segment $\overline{HB}$ of the pelvis. The pelvic rotation, determined primarily by the seat characteristics, influences the entire dummy kinematics. The mechanical pelvic model in accordance with Fig. 5 highlights the following fundamentally critical relationships under the influence of the lap belt:

Because the lap belt force F_{LB} always acts above the common center of gravity of pelvis and femur, the force of reaction of the seat $F_{Seating}$ provided via the lower seat framework has the only possible (and therefore decisive) influence on the degree of loading of the lumbar spine, i.e., the internal forces M, N, and Q of the lumbar spine. The crux of the problem is the imbalance of the moments produced, which lead to excessive bending of the lumbar spine. Finally, bending of the lumbar spine has as a result other biomechanically unfavorable and dangerous movement and loading configurations in the abdominal area and in the lower thorax, as well as in the areas of head and neck (5, 6, 8).

On the left side of Fig. 7, these interrelations are demonstrated using the example of a test performed on a soft seat. The dummy is shown in the initial position and at the end of the forward displacement. This is not an especially extreme case: it is merely the tendency generally observed with production seats and soft research seats.

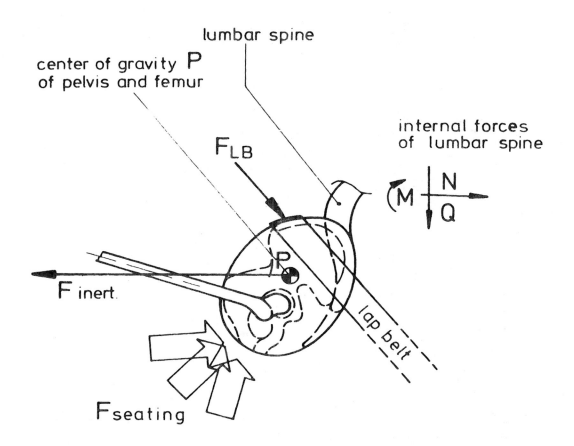

center of gravity P
of pelvis and femur

lumbar spine

F_{LB}

internal forces
of lumbar spine

M $\quad N \atop Q$

$F_{inert.}$

P

lap belt

$F_{seating}$

Fig. 5 - External forces on a lap-belted model of pelvis
and lumbar spine

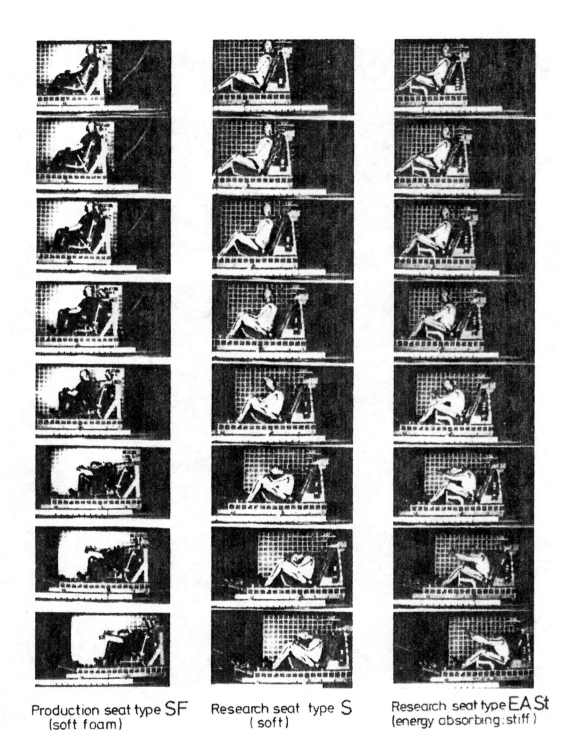

Production seat type SF
(soft foam)

Research seat type S
(soft)

Research seat type EASt
(energy absorbing:stiff)

Fig. 6 – Motion-sequence comparison for seat—safety belt interaction on production seat, and on soft and on stiff research seats

Seat - safetybelt interaction

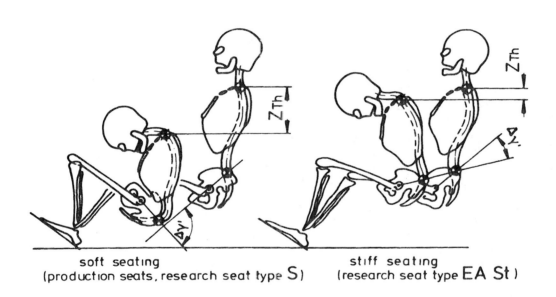

soft seating
(production seats, research seat type S)

stiff seating
(research seat type EA St)

Fig. 7 - Seat—safety belt interaction: final position
demonstrating lumbar spine bending and vertical thorax
travel on soft and on stiff seating

The following purely qualitative determinations and valuations were initially derived:

• The pelvic rotation alone, i.e., lumbar spine bending, causes a vertically downward-directed component of movement of the thorax and of the head.

• The vertically downward-directed components of the upper torso movement are reinforced by compression of the pelvic contour into the seat upholstery.

• The downward vertical thorax displacement leads also to a sinking of the point of application of the shoulderbelt loop force into the lower thorax segment, despite an otherwise favorable thorax angle of $\alpha \leq 90°$. The resulting shoulderbelt loop forces are thereby applied not horizontally via the sternum, but from below via the 5th to 12th ribs.

• Large pelvic rotation angles of $\gamma' \geq 50$ to $60°$ endanger the lumbar spine through excessive bending (fracture of vertebrae, lumbar injuries, etc.) (6, 8).

• Large values of γ', furthermore, lead to averting of the iliac crests, with the danger of submarining or other biomechanically dangerous loading configurations.

• A pelvic rotation angle of $\gamma' \cong 80°$ leads to extension of the femur segment $\overline{HK}$ by the distance $\overline{HB}$ (see Fig. 2). Pelvic rotation alone can therefore be a parameter for knee contact with the lower instrument panel.

Table 1 shows an outline of the kinematic evaluation parameters and their biomechanical significance.

ANALYSIS OF THE RESULTS

Table 2 shows a summary of the numerical results of the individual vehicle seat tests on the basis of the above-explained, constant, and easily reproducible test conditions.

Two representative tests for each type of seat were chosen for this listing. It must be conceded once again at this point that the actual conditions in the interior of any particular vehicle involved could possibly shift the numerical values slightly; after careful analysis, however, there are in our opinion no indications whatsoever of a reversal in the tendencies discussed below on the basis of such possible conditions.

DUMMY LOADING - The numerical results with respect to the loading on the dummy are shown in columns 3 to 7 of Table 2; nothing extraordinary is to be observed here for the group of production seats. The loading values measured are noncritical as judged by FMVSS 208, if consideration is taken of the normal distribution to be found in crash test measurements.

Table 1 - Evaluation Parameters for Kinematics

Evaluation parameter	Description	Responsible for influence on:
	HEAD / NECK	
$x_{H\ max}$	Maximum horizontal head displacement	Head to steering system and dashboard impact
$z_{H\ max}$	Maximum vertical head displacement	Head to steering system and dashboard impact
β'_{max}	Maximum angle of neck anteflexion relative to the thorax line	Neck injuries caused by hyperanteflexion
	THORAX	
$z_{Th\ max}$ $= f\ (x_{Th\ max})$	Maximum vertical thorax displacement at moment of maximum horizontal thorax displacement $x_{Th\ max}$	Loading of lower thorax area (ribs 5 – 12) Increase of vertical upward component of loading direction on lower thorax area
α_{max}	Thorax angle to the horizontal at the moment of maximum upper belt loading	Loading of lower thorax area (ribs 5 – 12) Increase of vertical upward component of loading direction on lower thorax area
	PELVIS	
$x_{P\ max}$	Maximum horizontal displacement of pelvis (point H)	Thorax positioning at the moment of the maximum shoulderbelt force
$z_{P\ max}$ $= f\ (x_{P\ max})$	Maximum vertical displacement of the pelvis (point H) at the moment of maximum horizontal pelvis displacement	Compression and bending of the lumbar spine Vertical impact severity
γ'_{max}	Maximum angle of pelvis rotation relative to the thorax line	Basic guidance of motion; lumbar spine loading; possibility of knee impact

869

Table 2 - Results of Seat Tests: Comparison of Production Seats and Research Seats Under 40 km/h and 30 g Impact Conditions

1	2	3	4	5	6	7	8	9	10	11	12	13	14	15	16
		LOADINGS OF:					KINEMATICS OF:								
		Head	Thorax		Pelvis		Head		Neck	Thorax			Pelvis/lumbar spine		
			Maximum decelerations		Max. decelerations		Maximum displacements		Max. bending	Maximum displacement		Max. angle	Maximum displacement		Max. bending
Type of seat	Test no.	HIC	$a_{Th\,res}$ [g]	SI	$a_{P\,x}$ [g]	$a_{P\,z}$ [g]	$x_{H\,max}$ [mm]	$z_{H\,max}$ [mm]	β'_{max} [°]	$x_{Th\,max}$ [mm]	$z_{Th\,max}$ [mm]	α_{max} [°]	$x_{P\,max}$ [mm]	$z_{P\,max}$ [mm]	Y'_{max} [°]
SC	5	813	47	330	25	33	510	480	104	370	110	80	260	70	100
SC	23	536	49	290	37	37	550	440	96	410	160	90	350	190	82
SC/SF	3	677	51	360	30	41	540	460	97	340	120	72	270	80	99
SC/SF	4	875	50	390	20	38	480	500	110	290	90	74	260	60	87
SF	11	1036	47	270	26	28	550	500	109	390	100	72	290	160	93
SF	20	688	55	320	32	35	580	510	108	420	130	74	310	110	96
				$a_{Th\,x}$: $a_{Th\,z}$:											
S	7'	1114	52	25	28	40	500	440	113	320	140	94	335	50	87
S	8'	1208	52	28	22	48	500	450	114	340	190	98	376	25	91
EA M	13'	517	52	8	38	40	550	410	89	350	60	71	215	0	58
EA M	14'	542	50	7	34	48	500	400	90	340	10	68	220	0	58
EA St	23'	178	50	14	52	36	480	360	72	330	10	68	175	-10	50
EA St	24'	272	48	14	42	38	490	410	75	340	20	77	163	-28	56

PRODUCTION SEATS (rows SC – SF)

RESEARCH SEATS (rows S – EA St)

Legend: SC = spring cushion; SC/SF = spring cushion/soft foam; SF = soft foam; S = soft; EA M = energy-absorbing, medium-hard; EA St = energy-absorbing, stiff

Attention must be called, however, to type SC of the production seats, in test no. 23, to the extraordinarily low HIC value and the small pelvic decelerations. This can be explained as part of an extreme, classical submarining sequence and therefore stands out in comparison to the other production seats. This example is renewed evidence of the well-known and significant phenomenon that, precisely under classic submarining conditions, the FMVSS 208 limits can be especially easily observed—with margin to spare—*with the unfortunate result, however, that the occupant will have it particularly rough.*

Furthermore, it can be easily seen that our research seat type S (see Fig. 4) easily fits into the range of production seats. This seat, moreover, is identical to the future seat provided for the child restraint approval test in the context of an expected ECE regulation. A significant characteristic here—a desirable one within this context as well—is the tendency toward a certain increase of loading, especially in the head and neck region (HIC values). This is caused by the rigid, nondeformable lower seat framework, which leads to a "hard" vertical final impact of the pelvis.

Moreover, determination can on the other hand be made of the good coincidence of the dummy loading values for the type S research seat and for the group of production seats—a phenomenon to be desired within this context. This similarity permits formulation of a fundamental and conclusive correlation between the production seats and the two further-developed stages of the research seat—types EA M and EA St.

With regard to the loading values, types EA M and EA St (the energy-absorbing research seats) show more marked trends than had been expected:
• The HIC decreases markedly with increasing hardness of the energy-absorbing element.
  • The vertical acceleration component of the thorax, a_{Thz}, is clearly reduced.
  • The vertical acceleration component of the pelvis, a_{Pz}, decreases slightly.
  • The horizontal acceleration component of the pelvis, a_{Px}, increases, as expected, but remains noncritical.

This overall positive influence on the scheme of loading has already been discussed in other studies (2).

An analysis of the measured parameters of the associated kinematic processes, however, reveals significant consequences from the standpoint of the biomechanics involved.

It is already clear at this point that the loading

criteria alone do not allow the safety engineer a real picture of the kinematics involved in the loading process. They do not reveal the manner in which the forces measured at the occupant have arisen as a result of the protective restraint system. The following analysis is an attempt to clarify the significance of the evaluation parameters for the kinematics in accordance with Table 1.

KINEMATICS OF THE DUMMY - Columns 8 to 16 of Table 2 allow an overall picture .

In accordance with our experience, one can obtain the first typical impression from the various sequences of movement of the dummy with the aid of the values for the maximum vertical thorax displacement, $z_{Th\ max}$, as given in column 12. These values show just how markedly the thorax has vertically displaced as a result of the components of acceleration and force.

The reasons for this are the following, as also explained by Fig. 6 and Table 1:

• The compression characteristics of the seat upholstery in the forward seat area

• The rotation of the pelvis as a result of the bending of the lumbar spine.

The results associated with the bending of the lumbar spine are listed in column 16 under γ'_{max}. It hardly requires the evaluation of orthopedic specialists to conclude that the relative bending angle of γ'_{max} = 90 - 100° for the production seat group and for type S of the research seats is dangerous and unacceptable. Accident statistics (3, 4, and 5) as well as human cadaver experiments (6, 7, and 8) vertify the above.

An overall evaluation of the kinematic parameters also shows the clear similarity of the research seat, type S, to the production seats. The orders of magnitude of the displacements of the individual body segments of the cervical spine (β'_{max} in column 10) and of the lumbar spine (γ'_{max} in column 16) show no significant deviations with respect to each other.

The data measured for this group of soft seats (lumbar spine flexures of γ'_{max} = 90 - 100°; cervical spine bending of β'_{max} = 96-115°; and x and z displacements of the head, thorax, and pelvis) require that the associated loading values (columns 3 to 7) be subjected to more differentiated valuation. Clearly, a resulting restraint force introduced into the lower thorax half (ribs 5 to 12) should especially be derived from these kinematic measured values, in addition to the absolute bending loads on the

spine. Under consideration of the differentiated strengths
of various parts of the thorax, the loading values as presented may therefore by no means be designated as noncritical.

In contrast to these results, a comparison with the
research seats type EA M and EA St shows the following
favorable biomechanical-anatomic trends:

- Lumbar spine bending is greatly reduced (γ'_{max} in
column 16).
- Vertical thorax displacements decrease significantly
($z_{Th\ max}$ in column 12).
- Vertical head displacements decrease ($z_{H\ max}$ in
column 9).
- Cervical spine bending is reduced (β'_{max} in column
10).
- The thorax shows more favorable positioning under
maximum shoulder belt force (α_{max} in column 13).

- Reduction or complete avoidance of vertical components of movement leads to a fundamental lower vertical
loading on the spine and the thorax.

A summary of the trends mentioned above is reflected
in the following figures, 6 to 8.

Fig. 6 shows the motion sequence in test no. 20 (production seat type SF), test no. 07' (research seat, type
S), and test no. 24' (research seat, type EA St) in adjacent, synchronized representation. The close similarity
between test no. 20 and no. 07' can clearly be seen. The
skeleton configurations during the phase of maximum loading are, on the other hand, not so obvious, because the
dummy clothing prevents a better view. In contrast, however, Fig. 7 gives an idea of the difference in the loading caused by kinematic conditions in the comparison of
the initial and final dummy skeleton positions for the
type S research seat (examples: test no. 20 and no. 07')
and for the type EA St research seat (example: test no.
24'). With the aid of evaluation of the above-explained
dummy markings by high-speed film, Fig. 9 schematically
shows typical trajectories of the head, the thorax, the
pelvis, and the knee on the type S research seat (similar
to the production seats), and on the type EA St research
seat. The vertical axis of the head, thorax, and pelvis
planes has been entered onto the respective trajectories
of these body parts in the time-dependent angular position
in order to represent the relative rotation of the individual body segments.

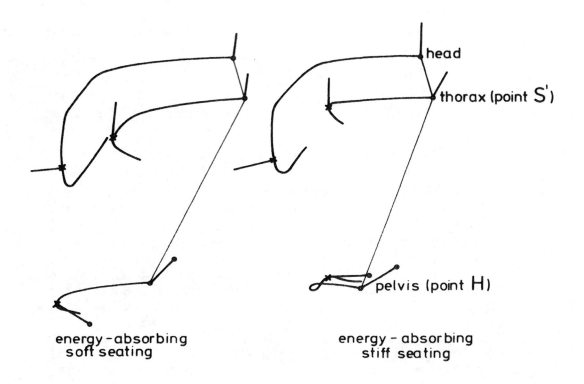

Fig. 8 – Typical trajectories and rotations of head, thorax (point S), and pelvis (point H) on soft and stiff seating

CONCLUSIONS

The seat is the decisive element in the overall protection system to perform functions of guidance of occupant kinematics. In this manner, the seat decisively contributes to the effectiveness of the protective system.

There are, however, no legal requirements for the operational effectiveness of the seating under conditions of head-on crash.

The loading measurements on the dummy in accordance with FMVSS 208 criteria do not provide a clear picture of the interrelations of the kinematics involved, i.e., on the manner in which the restraint system introduces its forces into the occupant's body, and on the manner in which seat and seat belt interact.

The kinematic evaluation parameters defined and introduced here as supplementary aids represent a motivation for further consideration. No effort was made in this stage of the work toward the subsequent and necessary definition of criteria for these kinematic evaluation parameters, as would be required for safety regulations. The trends alone as described here already sufficiently clarify the importance of posing the problems of "Interaction of Seat and Seat Belt," and the topic "Kinematic Control."

Measures taken to optimize the restraint system at any one of its individual elements cannot, therefore, be rendered fully effective before optimization as well of the seat is accomplished.

The first objective for the design of a safe seat should initially, and purely theoretically, be the design of the forward seating area in such a way that—as shown in Fig. 5—a seating reaction force $F_{Seating}$ is properly provided in direction and magnitude, i.e., so that a balance of the moments at the pelvis and lumbar spines is created in interaction with the remaining external forces—the lapbelt forces F_{LB} and the force of inertia F_{inert}.

The energy-absorbing deformation elements outlined here show a practical, constructive direction.

The advantages to be gained would show up in the form of further reduction in the frequency and severity of restraint-specific injuries.

REFERENCES

1. D. Adomeit and A. Heger, "Motion Sequence Criteria and Design Proposals for Restraint Devices to Avoid Unfavourable Biomechanic Conditions and Submarin-

ing." SAE Paper No. 751146.

2. D. Adomeit, "Evaluation Methods for the Biomechanical Quality of Restraint Systems During Frontal
Impact." SAE Paper No. 770936.

3. D. Adomeit, Behrens, and H. Appel, "Bewertung
von typischen Verletzungsmustern gurtgesicherter Insassen
im realen Frontalunfall." Published in DER VERKEHRSUN-
FALL, vol. 9, 1978.

4. Goegler, "Sicherheitsgurt und Mitverschulden."
Presentation at the 16th Convention of the German Traffic
Courts, 1978, held in Goslar, West Germany.

5. D. Havemann and L. Schröder, "Verletzungen unter
Dreipunkt-Sicherheitsgurten: Ergebnisse einer Studie an
Fahrzeuginsassen." Printed in the Report of the 1979
Annual Convention of the Deutsche Gesellschaft für Ver-
kehrsmedizin e.V., held in Cologne, West Germany.

6. G. Heess, "Wirbelsäulenverletzungen menschlicher
Leichen bei simulierten Frontalaufprallen." Doctoral
dissertation, University of Heidelberg, West Germany,
1977.

7. L.M. Patrick and A. Andersson, "Three Point Harness Accident and Laboratory Data Comparison." SAE
Paper No. 741181.

8. Schmidt, Kallieris, Barz, Mattern, and Schulz,
"Untersuchungen zur Ermittlung von Belastbarkeitsgrenze
und Verletzungsmechanik des angegurteten Fahrzeuginsassen."
FAT Research Project No. 3102, Frankfurt, West Germany.

Submarining Injuries of 3 Pt. Belted Occupants in Frontal Collisions — Description, Mechanisms and Protection

Y. C. Leung, C. Tarrière, and D. Lestrelin
Laboratory of Physiology and Biomechanics
Peugeot-Renault Association (France)

J. Hureau
Faculty of Medicine, University of Paris
France

C. Got, F. Guillon, and A. Patel
I.R.B.A. Raymond Poincaré Hospital
Garches, France

ABSTRACT

Accidentological studies show, firstly, what kind of injuries are sustained by seatbelt wearers in frontal collisions, to abdomen, lumbar spine and lower members and, secondly, how to determine their frequencies and severities. Corresponding data are presented.

Then, a synthesis is made, in which the results of extensive cadaver testing - more than 300 human subjects - are examined with particular emphasis on the abdominal injuries, and on the association of injuries, such as lumbar spine injuries. Causation is particularly looked at. This experimental survey is completed by the results of specialized testing in abdominal tolerance when submarining occurs.

These two surveys enable the development of protection.

Finally, former attempts for defining an abdominal protection criterion are reviewed and a final definition for such a criterion is presented and justified.

ABDOMINAL PROTECTION IN AUTOMOBILE ACCIDENTS IS A SUBJECT WHICH HAS NOT BEEN FULLY EXPANDED UPON IN LITTERATURE ON ACCIDENTOLOGY AND BIOMECHANICS.

It is true that the head, thorax and lower members were the areas where those sustaining multiple injuries were the most frequently and most severely injured. Injuries in these areas were often evident (coma, breathing problems and fractures of long bones) whilst the possible share of abdominal injuries is more difficult to ascertain. In fatal accidents in particular, the autopsy is too often incomplete and only indicates the main injuries which are enough to account for the cause of death. Here again, abdominal injuries are often ignored in the same way as spinal injuries which require long and tedious research work before they can be detected.

In the 1970's, the wearing of seatbelts became general practice, especially in Europe. A new approach to regulation from the USA is gradually making headway. It is known as the global approach and sets out to appraise the secondary safety provided by a car, by simulating the most frequent and most deadly accidents and by measuring performance on one or more anthropomorphical dummies.

The first proposal for this type of regulation dates back to 1971. The American Administration wishes to appraise the protection provided by the vehicle in a frontal collision. The performance to be measured on the dummy concerns acceleration of the head and thorax and the load through the main axis of the thigh bone.

It is not surprising that the head, thorax and lower members are privileged areas we wish to protect. Indeed, methodological tools exist which we are not challenging. It is easy to measure the resultant acceleration of almost rigid models (head and thorax) by installing a three directional accelerometer at their centre of gravity.

Difficult research work is being conducted to determine the thresholds beyond which occupant protection cannot be ensured (to eliminate, for example, the risk of irreversible injuries). These thresholds must be based on human impact tolerance.

For 30 years, data have been progressively gathered and refined. Curiously, the abdomen was overlooked. Until 1979, the first tolerance values (load-deflection) of a human abdomen loaded by a seatbelt in static test conditions was presented(1). Because the tolerance could not be measured by accelerometer techniques, a specific methodology was used to define abdominal tolerance and translated into a measurable performance on a dummy. To make the distinction between values which are specific to the human being and their applications to a dummy, we employ the term "tolerance" for the former and "protection criterion" for the latter. The protection criterion may be different if the dummy itself changes definition.

The purpose of this publication is to establish a consolidated version of the most recent work devoted to abdominal protection <u>in frontal</u>

collisions(xx) for belted occupants and to attempt to ascertain whether the protection criterion currently available would be sufficient to avert the great majority of abdominal injuries. Because submarining is the most involved mechanism, the analysis is enlarged to all injuries induced, by submarining that is to say lumbar, lower members, pelvic as well as abdominal injuries.

An analysis is presented for the abdominal injuries observed in a sample of 1542 subjects wearing 3-point seatbelts and involved in frontal impacts. This is followed by a corresponding analysis of abdominal injuries observed in the tested cadavers which were divided-up into two independent samples groups, one composed of 70 subjects, and the other 211 subjects, corresponding to the observations supplied by different research groups.

After having specified the relative importance of the risk of lumbar, lower members and abdominal injuries, the study goes on to analyze and consider the prevention of abdominal injuries caused by the mechanism the most often involved, well-known under the name of "submarining" (factors which influence the occurrence of submarining and in particular the definition of a protection criterion to eliminate the risk of submarining by means of principles of a well-designed vehicle).

By discussing all available information, it is possible to estimate the extent of protection provided by currently available criterion.

(x) Numbers between parentheses are references at the end of paper
(xx) Other publications cover a similar study on abdominal protection in side impacts.

SUBMARINING INJURIES IN REAL-LIFE ACCIDENTS

The accident survey file contains 1017 frontal impacts in which drivers wearing three-point seatbelts were involved.

The impacts are selected based on the following body injury criterion ; at least one of the occupants involved in the accident was taken to the hospital.

SEVERITY OF SUBMARINING INJURIES IN RELATION TO OVERALL RISK - The population of seriously injured among the 1017 drivers (M.AIS 3, 4 or 5) is slightly less than that observed among the 525 passengers. Respective proportions of those killed are not significantly different, (Table 1).(See also Appendix I)

From among the severely injured front occupants, 1 driver out of 5 and 1 passenger out of 3 sustained severe injuries to abdomen and dorso-lumbar spine fractures (Table 2).

Table 2 - Proportion of AIS $\geqslant$ 3 to Abdomen and Dorso-Lumbar Column Among Severely Injured Drivers and Passengers (M.AIS $\geqslant$ 3) wearing 3-pt belt.

Thoracic-lumbar spine and Abdomen AIS

	≤ 3	$\geqslant 3$	Total
Drivers	69 (79%)	18 (21%)	87 (100%)
Passengers	37 (64%)	21 (36%)	58 (100%)
Total	106 (73%)	39 (27%)	145 (100%)

SEVERITY OF ABDOMINAL INJURIES IN RELATION TO OTHER BODY AREAS - The abdomen, with the head and thorax, is, for belted drivers, the body area most exposed to critical injury (AIS $\geqslant$ 4). Injuries to lower members very largely dominate at level 3. (Table 3).

One of the advantages of the adoption of retractor belt systems which are fitted to cars produced since 1978 has been to reduce the slack with which static belts were habitually worn. Further, their geometrical lay-out in the car is a great improvement; the central mounting systems do, in particular, eliminate any further risk of initial contact of the adjusting buckle against the abdomen.

This progress accounts for the very great difference in the risk of abdominal injury between those wearing static and those wearing retractor belts (Table 4).

The improvement to seat belt has, therefore, approximately halved the share of abdominal injuries

However, they are still dangerous and some progress should be made. A precise description of the injuries and their most probable mechanisms may help to throw some light on the subject.

Table 1 - Severity of Frontal Impacts for Front, Belted (3-point) Drivers and Passengers.

	M.AIS			
	0-1-2	3-4-5	$\geqslant 6$	Total
Drivers	892 (87.8%)	87 (8.5%)	38 (3.7%)	1017 100%
Passengers	451 (86%)	58 (11%)	16 (3%)	525 100%

Table 3 - Severity of Injuries by Body Area Sustained by Belted Drivers and Passengers During Frontal Impact (Excluding Those Killed and not Subjected to Autopsy).

DRIVERS   Frontal Impacts (11.12.01 o'clock)

	Head/Face	Neck	Thorax	DL. Spine	Abdomen	Pelvis	L. Members
AIS - 0	651	917	762	968	948	938	670
1	225	75	207	24	31	40	229
2	96	1	16	6	3	6	50
3	16	4	2	1	4	15	50
4	2	1	11	-	11	-	-
5	9	1	1	-	2	-	-
	999	999	999	999	999	999	999

RIGHT FRONT PASSENGERS

	Head/Face	Neck	Thorax	DL. Spine	Abdomen	Pelvis	L. Members
AIS - 0	395	456	326	432	477	498	366
1	90	61	159	27	30	23	123
2	33	4	28	4	1	1	19
3	4	3	7	3	5	4	18
4	3	1	6	-	13	-	-
5	1	1	-	-	-	-	-
	526	526	526	526	526	526	526

Table 4 - Proportions (%) of Occupants Sustaining Severe Abdominal Injuries (AIS $\geqslant$ 3) and/or Dorso-Lumbar Fractures (AIS $\geqslant$ 3) Among the Seriously Injured Drivers and Passengers (M.AIS 3, 4 or 5).

	Static Belt	Retractor Belt
Drivers	22.6%	12 %
Passengers	37.5 %	16.7 %

THE RISK OF SUB-MARINING INJURIES INCREASES WITH THE VIOLENCE OF IMPACT - This relationship, which can be seen in Table 5 allows us to rule out the assumption,occasionally put forward, whereby the risk of injury to the abdomen is primarily governed by malfunctions or by the presence, in the population studied, of dangerous and badly designed systems, irrespective of the violence. The mechanisms which sometimes cause the abdomen to be subjected to excessive loading, re-

sulting in injury, are not easy to describe for the accidentologist. A number of factors do,however, allow us to state, with a good degree of certainty, that each case matches up with one of the following configurations :

1- Restraint is achieved by the most resistant bone segments, including-and more especially- the pelvic region, but the abdominal mass undergoes a deceleration beyond the level of tolerance, It results in tears,wounds or rupture of viscera which certain authors thought could be described by specific features (2).

2 - A variant of these cases on which the seatbelt firmly fastens the pelvis and where we could still observe abdominal injuries, would correspond to another mechanism, that is a very important flexion of the trunk around the pelvis. This flexion can be clearly observed in experimental impacts (Figure 1), and the head can come into contact with the knees. A great kinematics was produced possibly within the certain car models, in which the superior anchorage was very behind that favored the sliding of the shoulder under the shoulder belt. Abdominal injuries would not be the result of the deceleration of the viscera in relation to the belt restrained skeletal frame, but would be induced by very high pressure in the viscera, and even compression between the trunk and thighs. The stout persons are more exposed than others to this type of compression. If here is a submarining process, the lap-belt still increases the pressure in viscera.

3 - The belt is worn in such a way that, when the slack is taken up, the lap-belt presses against the abdomen, penetrating through even as far as the spinal column when deceleration is high (Table 6). This second configuration, where play is a major factor, is frequent when the seat belt anchorage installation is faulty. Its consequences are aggravated when the restraint system on the oldest static belt models includes a buckle for adjusting belt tension, located on the end of a flexible stalk anchored to the centre of the car. This system has significantly deteriorated seat belt performance in numerous accidents which have occurred over the past ten years.

4 - The lap-belt is correctly positioned against the pelvic bone but, when the occupant is displaced forward, loading conditions of the restraint system are such that the lap-belt, under tension, runs up over the iliac crests and compresses the abdomen. This is submarining in the strict sense of the term.

The last two configurations have the most adverse effects on the abdomen. The injuries they cause have no specific typology whereby it would be possible, from a review of the medical file, to determine whether the case in point falls into configuration 3 or 4. Amongst the relevant indications capable of distinguishing between them, we can name the presence of cutaneous erosion in the pelvic region, particularly below or above the anterior superior iliac spines: the distance over which the belt shows signs of friction compared to occupant stopping distance.

In the last two configurations, the pressure exerted on the abdomen is the risk to be avoided. Whether it be at the just beginning of occupant displacement or after slipping from the pelvis, that in no way obviates the need to design restraint systems and their installation in such a way that restraint acts only against the pelvis throughout the whole of the occupant's deceleration phase. The share of the shoulder belt as a direct cause or aggravating factor in injuries to tissues or organs in the upper abdomen region (liver, spleen, in particular) is not negligible but the results of an analysis indicated in Table 7 show that, when it exists, it is most often accompaning the preponderant action from the lap-belt.

Table 5 - Proportion of Occupants With Severe Abdomen Injuries and Dorso-Lumbar Fractures and Proportion of Severely Injured (M. AIS $\geqslant$ 3) in Each ΔV Category.

ΔV (kph)	$\leqslant$ 35	36-45	46-55	56-65	$\geqslant$ 65
AIS abdomen $\geqslant$ 3	0.3%	2.4%	13.3%	13.5%	46.2%
M.AIS $\geqslant$ 3		2% 16.0%	48.0%	59.5%	92%

Likewise, drivers sometimes strike against the steering unit which contributes to the injuries or is the sole cause of them, but this is a rare occurrence, since we have seen (above) that the abdominal risk is lesser for driver than for passenger.

The 26 drivers and 21 passengers who sustained severe submarining injuries (AIS $\geqslant$ 3) were equipped with seatbelts almost half of which had a geometrical defect and were worn too loose. The presence of one or other of these two factors or both combined, was noted in more than two out of three cases (Table 8).

CONCLUSIONS

1. The process described by the term of submarining of the lower part of the trunk (pelvis-abdomen-lumbar-spine and the lower thorax) under the lap-belt is a mechanism fairly frequently encountered among belted occupants in frontal impact.

Submarining is considered to be the mechanism most probably at cause in the vast majority of cases involving the seriously injured (AIS $\geqslant$ 3) sustaining injuries to the abdomen, the lower members, the pelvis or the dorso-lumbar column.

One of the original aspects of this in-depth cases analysis is that it shows how submarining can cause severe injuries to the lower members (legs-knees-femurs), the pelvis and the dorso-lumbar column (fracture-compression of the first lumbar vertebraes or fracture of the transverse processes) as well as to the abdomen. There are besides often combinations of these different injuries.

Other injuries may more exceptionally be combined, such as wounds or injuries to the neck and occasionally rib fractures.

2. The lap-belt is certainly the cause of 68% (32/47) of cases of victims of abdominal (AIS $\geqslant$ 3) and/or dorso-lumbar column (AIS $\geqslant$ 2) injuries.

Table 6 – Dorso-Lumbar Spine Injuries (AIS ≥ 2) – N = 14 Cases

INJURIES' MECHANISMS

LEVEL OF FRACTURES	SUBMARINING			DIRECT REAR IMPACT (rear passenger or object)		
	Sure	Probable	Uncertain	Sure	Probable	Uncertain
D9 – D11			1*			
D 12	1	1	1			
L 1	4					
L 2	2					
Lumbar vertebrae process-transverse:	1	2*	1			
L5 process transverse:				1		

(*) aggravating influence of the rear passenger

881

Table 7

SYNTHESIS OF SUBMARINING - ANALYSIS OF INJURIES' MECHANISMS

PARAMETERS	LAP-BELT			SHOULDER BELT			OTHER DIRECT IMPACTS (steering-system for example)		
	Sure	Probable	Uncertain	Sure	Probable	Uncertain	Sure	Probable	Uncertain
POOR GEOMETRY OF THE RESTRAINT 43%(A)	14	6	1	1	0	2	0	0	0
BELT WORN WITH SLACK 36%	13	4	2	0	0	1	0	0	0
FRONT SEAT TRACK DAMAGE 38%	13	5	1	0	1	2	0	0	0
ADDITIONAL LOADING 21%	7	3	0	0	0	1	0	0	0
NECK CONTUSIONS OR ABRASIONS BY SHOULDER-BELT 19%	6	3	0	0	1	1	0	0	0
LOWER MEMBERS IMPACT AGAINST LOWER PANEL 83%	30	9	5	1	1	4	0	2	1
CRASH VIOLENCE $\Delta V > 50$ km/h mean $\gamma \geq 10$ g 58%	19	6	2	0	1	3	0	1	1
ABDOMINAL INJURIES SUB-MESOCOLIC 64%	25	5	2	0	1	3	0	1	0
ABDOMINAL INJURIES ABOVE MESOCOLIC 21%	8	2	1	1	1	3	0	1	0
DORSO-LUMBAR FRACTURES 23%	8	3	1	0	0	0	1	0	0
LOWER MEMBERS FRACTURES 45%	17	4	2	1	1	2	0	0	0
PELVIC INJURIES 8%	3	1	2	0	0	1	0	0	0
RIB FRACTURES 32%	14	1	1	1	0	1	0	0	0
AGGRAVATING INFLUENCE OF RIB FRACTURES 8%	3	1	0	1	0	0	0	0	0

IMPORTANT REMARKS:

(A) This percentage corresponding to the two first columns means that the parameter involved plays for x% of all cases where lap-belt submarining is sure or the most probable. For example, "Poor geometry..." plays in 20 cases out of 47 where submarining is sure or the most probable.
(B) The total in each ROW could overpass the number of involved people because, for some cases, two mechanisms could play simultaneously (ex.: lap-belt submarining and abdominal injury induced by shoulder-belt (rib fracture).

Its rôle is very probable in 19% of the other cases; only in 13% of cases it is absent or doubtful.

The rôle of the shoulder belt is unquestionable in 1 out of 47 cases as complementing the action of the lap-belt and is possible in one other case. In 4 cases there is some doubt.

3. It is difficult to strictly specify the mechanism causing abdominal injuries as it means taking into consideration all the medical, anthropometric and technical data in the accidentology files. Two analysis grids have been established and they are given in Appendices I and II.

Submarining is evident when contributory factors such as poor geometry of the restraint system and (or) slack in the belts (one or other or both these factors combined are present in 74% of cases) are observed with the following:

a) Seat-tracks damages in 38% of cases

b) Knee or leg impact under the instrument panel with very obvious deformation of the latter in 83% of cases.

c) Violent impact ($\Delta V \geqslant$ 50kph and $\delta \geqslant$ 10g in 58% of cases where submarining is certain or very probable).

4. It should be noted that apart from submarining and the auxiliary rôle played by the shoulder belt, the responsability of the instrument panel and/or the steering unit is exceptional in the occurrence of abdominal or lumbar injuries (one probable case and one doubtful).

5. The additional load from the rear passengers is an aggravating factor (in 21% of cases where submarining is certain or probable).

6. Fractures of the pelvis are seldom but could be the result of submarining (8% of cases where submarining is certain or probable) whether by knee impact or by direct action of the lap-belt.

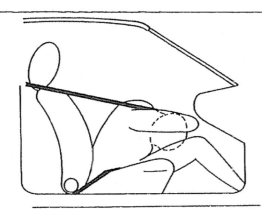

FIG. 1 . DIAGRAM OF A CADAVER TEST.

7. Another Sub-marining consequence already noted in a previous publication (2) is the great frequency (23%) of dorso-lumbar vertebraes fractures (D12-L1-L2) by direct impact of the lap belt or (and) hyperflexion of the trunk around the lap-belt penetrated in the abdomen.

8. The lower members injuries (legs, knees and femurs fractures) are not specific ; however they are observed in 45% of cases of sub-marining sure or most probable and could be considered as another consequence of this process.

ANALYSIS OF ABDOMINAL INJURIES OBSERVED IN EXPERIMENTS CONDUCTED ON HUMAN CADAVERS FOR 281 FRONTAL IMPACT TESTS.

Two test samples with cadavers simulating frontal collisions with three-point seatbelts provide a total of 281 observations. All the subjects were fresh cadavers conserved at a temperature of around 0° and experiments were conducted a few hours or days following their death.

An initial sample known as "APR" groups 70 subjects. A second one, the "Heidelberg" sample groups 211 subjects.

ABDOMINAL INJURIES OBSERVED IN THE "APR" SAMPLE - The most frequent injuries are fractures of the rib cage and litterature already exists on this subject.

In this paper, the injuries indicated are only the abdominal injuries or the lumbar vertebra injuries which can be produced by the same mechanism, in particular submarining.

Out of the 70 frontal impact tests using cadavers, there were 47 cases with injuries, 23 with lap-belt submarining and 24 without lap-belt submarining. The rest were cases where no injury was sustained and there was no evidence of submarining. The appended Table 9 illustrates which of the tested cadavers sustained only the injuries observed in the liver, the spleen, the lumbar vertebra and other abdominal injuries, including the intestines, mesentery, colon, iliac crest and abdominal muscles. Cases with only iliac crest fracture or abdominal muscle injury are also presented in this Table which contains 24 cases, 6 of which are non-submarining and 18 submarining.

Table 8 - Quality of Restraint System Geometry and Belt Wearing Among Victims of Abdominal injuries and Dorso-Lumbar Fractures

		POOR GEOMETRY		
		YES	NO	TOTAL
	YES	8	8	16
	NO	12	13	25
EXCESS SLACK				
	UNCERTAIN	3	3	6
	TOTAL	23	24	47

Table 9 - Abdominal and spine injuries observed in frontal impact tests with cadavers (APR sample).

Test N°	Seat N°	Submarining	Liver	Spleen	Lumbar vertebra	Other abdominal injuries and pelvic bones fractures
2	1	yes	yes		L4 sacrum	colon, mesentery
4	1	no	yes			
6	2	yes				
8	2	no		yes		ilium, diaphragm
10	1	no	yes			ilium
12	2	yes				mesentery
16	2	no		yes		diaphragm
25	2	no				mesentery, ilium *
44	2	no				
127	4	yes			L1	
154	4	yes	yes		L2	iliac crest
170	1	yes			L4,L5	
182	2	yes			L1	
183	1	yes	yes			
184	1	yes	yes			
185	2	yes				mesentery
189	4	yes			L5,S1	abdominal muscle, mesentery
231	2	yes	yes	yes		mesentery
243	4	yes				colon, abdominal muscle
244	4	yes				left ilium crest
245	4	yes				abdominal muscle
246	4	yes			T6,T7	abdominal muscle
247	4	yes				abdominal muscle
255	2	yes		yes		

Seat N°. 1 : driver
Seat N°. 2 : front passenger
Seat N°. 4 : right rear passenger

(*) very light injury

Table 10 - Abdominal injuries observed in Frontal Impact tests with cadavers ("Heidelberg" sample).

	Mean $\gamma \leqslant$ 16 g		Mean $\gamma >$ 16 g	
	Drivers	Passengers	Drivers	Passengers
Size of sample	81	0	46	84
Liver	0	0	2	14
Spleen	0	0	0	5
Kidneys	1	0	1	3
Mesenterium	2	0	4	11
Intestines	1	0	3	4
Vessels	0	0	2	4

The following comments can be made :

1. No lumbar vertebra fracture was found in non-submarining cases but lumbar vertebra fractures occurred frequently in submarining cases (5 cases out of 14).

2. No injuries observed in intestine,colon, and mesentery were found in non-submarining cases (except for N° 25 who sustained a light mesentery injury). These injuries were frequently observed in the submarining cases (6 cases out of 17).

3. 5 drivers sustained liver injuries, 2 of which were non-submarining cases.

4. 2 passengers sustained liver injuries with submarining.

5. All four "spleen injury" cases were observed only on front passengers, 2 of which were non-submarining cases, three were submarining cases

ABDOMINAL INJURIES OBSERVED IN THE "HEIDEL-BERG" SAMPLE - An ISO document based on experimental data from the University of Heidelberg which have appeared in numerous past publications presents the following summary as shown in Table 10. (3)

Except for spleen injuries, there is no preferential distribution of injuries between drivers and passengers. Liver is injured in 16% of cases for the passenger and 3.5% for the driver. The mean sled deceleration was exceeding 16g.

These results confirm the absence of specificity regarding liver injuries which would, as for spleen injuries, seem to be sustained more often by the passenger. This does, therefore, consolidate the findings of the APR sample analysis.

DISCUSSION

1. Out of 23 cases of liver injuries, 7 were drivers. 2 of them were non-submarining cases, 5 were submarining cases. This is not sufficient to ascertain that the shoulder belt is responsible for liver injuries in non-submarining cases, as we could see similar results for the front passenger. 16 cases of liver injuries were observed among front passengers (or right rear passengers),and they were submarining cases. The possible effect of the shoulder belt on liver injuries was relatively small in this group of the sample.

2. All of 5 spleen injury cases were observed in the front passengers. Two of them occurred in submarining cases.

3. As observed for the spleen, or for the liver, any sustained injury cannot be automatically due to the shoulder belt. It would be more suitable to say that the shoulder belt favours submarining on the buckle (interior)side. In certain cases, the shoulder belt could play a role in provoking such injuries without the occurrence of submarining.

4.No intestine,colon,mesentery injuries were found in non sub-marining cases (except one case where a light mesentery injury occured).

5.In our previous submarining studies (4),

it was found that submarining began,in most case on the interior side,for both cadaver and dummy tests.Two cases were found where submarining was limited to the interior side only, This illustrates the statement made in 3. to the effect that "the shoulder belt favours submarining on the buckle (interior) side".

CONCLUSIONS

1. Abdominal injuries occurred rarely in non submarining cases.

2. Dorso-lumbar vertebra fracture was never found in non-submarining cases. But lumbar vertebra fractures are often associated with submarining and, sometimes, without abdominal injury.

3. Spleen and lever injuries are possible consequences of lap-belt sub-marining. Liver injuries are not specific to anyone occupant seat. Spleen injury is found particularly to the right passenger (front or rear).

4. The possible effect of shoulder belt is rather low. The shoulder belt could favour the submarining process on the buckle interior side.

5. All sub-mesocolic injuries (except one) are consequences of lap-belt sub-marining.

DEFINITION OF SUBMARINING (MECHANISM)

A definition problem regarding the word"submarining" was experienced on several occasions these last years, in particular with I.S.O. (ISO TC22/SC12/GT 6).

Corresponding to a personal communication of G.M. NYQUIST, several types of submarining referred to by various authors in the past are :

- Lap-belt submarining
- Shoulder belt submarining
- Air bag submarining
- Instrument panel submarining.

He stressed the fact that abdominal injuries are not the only consequences of a submarining and that knee injuries due to impact following submarining should be considered as submarining injuries. We also agree with this point of view. One should note that, in practice if lap belt submarining is avoided, the risk of knee impact is very much reduced.

For the wearer of a three point safety belt, submarining is reduced to the relative movement of the body in relation with the belts. The submarining can only take place through release of the iliac crests passing under the lap belt. A slanting displacement of the pelvis takes place downwards and frontwards.This movement of the pelvis corresponds to a simultaneous movement of the whole trunk which could be named thoracic submarining relating to the thoracic belt. However, no thoracic syndrome specific to submarining has yet been described. Some cervical injuries could be due to this whole trunk submarining as we have seen in our accidentological sample.

So as not to complicate things, at least as regards the wearer of a three point safety belt, one should avoid talking of submarining regarding the body as a whole. To make things clear one should talk about pelvis submarining relating to the pelvis belt. This usually causes abdominal injuries, injuries of the dorso-lumbar vertebra, of the knee-femur-pelvis axis, all these injuries which can either be isolated or associated.

It is difficult to distinguish, within this category, the difference of the submarining occurred with the initial positioning of the lap-belt on abdomen and with the initial positioning on the pelvis. In real accidents, in most cases, it is impossible to know exactly the initial positioning. This depends on the care taken by the wearer in placing correctly the safety-belt (however does the wearer know that the lap-belt, sometimes called "abdominal", should really restrain the pelvis ?) The placing of the belt depends firstly on the restraint geometry but also of the posture and slack in the belt.

FACTORS INFLUENCING THE OCCURRENCE OF LAP-BELT SUBMARINING

"Submarining" is a complicated problem because its occurrence is associated with many parameters, for example the geometry of the seat-belt system, orientation of the pelvis, dynamic characteristics of the seat and vehicle, impact velocity, etc... These are the external parameters determined by test conditions. An internal parameter is the pelvis shape; the area in question is the upper half of the notch below the Anterior Superior Iliac Spine (A.S.I.S.). Here as in previous publications (4)(5), this pelvic area situated just below the ASIS is given the name "Sartorius" and has a length of A1A2 (see figure 2).

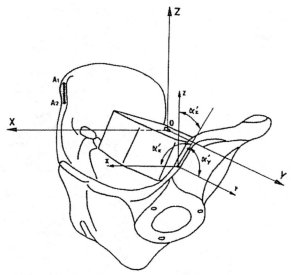

FIG. 2 . DIRECTION OF THE UPPER HALF NOTCH

(A1A2)"SARTORIUS" DEFINED IN THREE DIMENSIONS.

In fact, to avoid any confusion, it is necessary to state that Sartorius is the original name given to the muscle attached to this referred bone area in Gray's Anatomy (pp. 228, 492).

INFLUENCE OF LAP-BELT ANGLES - During frontal impact, the load applied to the "Sartorius" on one side of the pelvis is the resultant of lap-belt tensions on the corresponding side. This load can be resolved into two components: one is parallel to the "Sartorius", the other is perpendicular. The parallel one orients downward and backward along the "Sartorius"; it can be defined as a load to keep the pelvis in a favorable position necessary to prevent the process of submarining. This load is related to the orientation of the "Sartorius" and the geometry of the lap-belt as well as to lap-belt tension. The coefficient of the load can be used as an indicator of the efficiency in reducing the submarining tendency. A brief technical term referring to "anti"-submarining scale was used in a previous paper (4).

If the orientation of the Sartorius is considered as a constant, the anti-submarining scale is only a function of the geometry of the lap-belt defined in three dimensions. Since some experimental results cannot be explained by the traditional geometry definition of the lap-belt given in two dimensions, a complete geometrical definition of the lap-belt determined in three dimensions was developed.

Thanks to the anti-submarining scale curves established on the basis of the present theory and the submarining tendency as a function of lap-belt angles $\beta 1$, $\beta 2$, determined in experimental results, an anti-submarining scale ξ=0.63 is proposed for a limit of submarining risk on lap-belt geometry. Hence, this paper shows that the graphs of the anti-submarining scales can be used for checking the submarining risk in the seat-belt system.

Theoretical study: using a geometrical model of the lap-belt during impact (Fig. 3), the following equations can be found:

F3 = F cos θ
F4 = F sin θ

where load F is the resultant of lap-belt tension applying to the "Sartorius" (A1A2) on one side.

When the magnitude of F is known, F3 and F4 depend only on the angle θ which is formed between the direction of Sartorius and the resultant load F. F3 is a load parallel to the Sartorius, preventing the lap-belt from riding-up over the A.S.I.S. and keeping the pelvis in a correct position. Therefore, it is a load that plays an important role in reducing the submarining tendency. In contrast, F4 is a load which rotates the pelvis backward and downward, i.e. it increases the risk of submarining. F3 could be used as an indicator of the efficiency in reducing the submarining tendency.

By a mathematical study (4), load F3 is given by another expression,

F3 = ξF1

Lap-belt tension in this equation depends on collision velocity, seat and vehicle dynamic characteristics, mass of the belted occupants, etc... If Fl is determined, F3 is a function of the coefficient ξ. Since F3 is defined as an indicator of the efficiency in reducing the submarining tendency, the coefficient ξ can be designated as an anti-submarining scale indicating the efficiency to prevent the process of submarining for a given configuration.

The coefficient ξ depends, on one part, on the direction cosines of the "Sartorius" and, on the other part, on the lap-belt geometry designated by angles $\beta 1$, $\beta 2$. It is presented precisely by the following equation:

$$\xi = \cos\beta 1.\cos\left[\tan^{-1}(\cos\beta 1.\tan\beta 2)\right].\cos\alpha'_x +$$

$$\left[\cos 1/2\tan^{-1}(\cos\beta 1.\tan\beta 2) - \sin 1/2\tan^{-1}(\cos\beta 1.\right.$$

$$\left.\tan\beta 2)\right]^2.\cos\alpha'_y + \sin\beta 1.\cos\left[\tan^{-1}(\cos\beta 1.\tan\beta 2)\right].$$

$$\cos\alpha'_z$$

Since $\alpha'x$, $\alpha'y$ and $\alpha'z$ were three-dimensions defined previously for the "Sartorius" of an occupant within a car (5), anti-submarining scale could be determined as a function of lap-belt angles $\beta 1$, $\beta 2$.

By using above equation, the relationship between the coefficient ξ and the lap-belt angles ($\beta 1$, $\beta 2$) can be obtained, as illustrated in Figure 4 for human subject (male) and for Part 572 dummy (Figure 5).

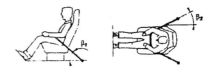

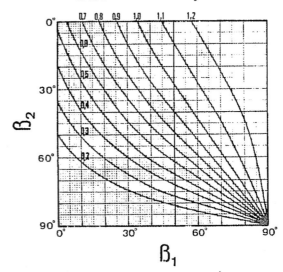

FIG. 4 . ANTI-SUBMARINING SCALE ξ CURVES AS A GUNCTION OF LAP-BELT ANGLES $\beta 1$, $\beta 2$ DERIVED FOR HUMAN SUBJECTS (MALE).

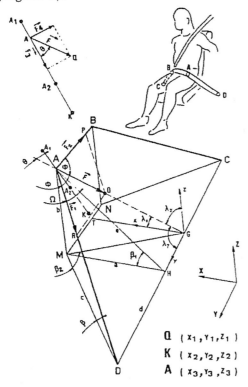

FIG. 3 . GEOMETRICAL MODEL OF THE LAP-BELT DURING IMPACT.

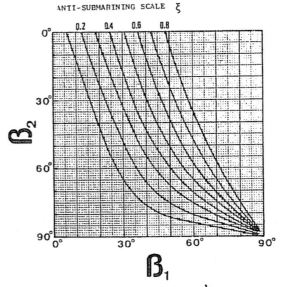

FIG. 5 . ANTI-SUBMARINING SCALE ξ CURVES AS A FUNCTION OF THE LAP-BELT ANGLES $\beta 1$, $\beta 2$ DERIVED FOR PART 572 DUMMY.

Experimental study - In order to put into practice and examine this theoretical study and the results, a series of sled tests was performed with human cadavers, the Part 572 dummy and the modified dummy. Three different bodies of European car models were used and mounted on the sled for the tests.

The lap-belt angles $\beta 1$, $\beta 2$ were measured with the tested subjects in the initial position, the test conditions and the complete experimental results were given in reference (4).

The details of the conservation of human cadavers can be consulted in reference (6). Briefly, the cadavers were fresh and non-embalmed. Death occurred less than 4 days before the test. They were put in a cold box and then taken out a few hours before the test.

Following the experimental results, submarining consequences in relation to the lap-belt angles $\beta 1$, $\beta 2$ are illustrated generally in Fig. 6 for the human subjects, Part 572 dummy and modified dummy.

Based on both the theoretical study and experimental results, the anti-submarining scale ξ can be traced between the groups of submarining data points corresponding to the 0.66 curve on Figure 4. It can be found in this figure: firstly that the submarining data points locate at the area which is formed by the smaller angles $\beta 1$ and the greater angles $\beta 2$ corresponding to the lower anti-submarining scale zone. Secondly, non-submarining data points locate inversely at the other area where $\beta 1$, $\beta 2$ are respectively greater and smaller, corresponding to the higher anti-submarining scale zone. Thirdly, two non-submarining data points for a cadaver test and a modified dummy test are situated in the submarining zone. This may be explained once more by the fact that the cadaver and the modified dummy are less inclined to submarine than the Part 572 dummy.

The figuration and location of the curve indicate that for the same anti-submarining scale, the decrease of $\beta 1$ should be associated with the decrease of $\beta 2$, i.e. if angle $\beta 1$ is favorable to submarining (smaller), it is necessary to associate a good angle $\beta 2$ (smaller) to reduce the submarining tendency if other conditions remain unchanged. This case corresponds to the highest data points (circle). On the other hand, while angle $\beta 1$ is great enough, according to previous beliefs, it seems to be unfavorable to submarining. Unfortunately, the submarining process occurred. In Figure 6, the three or four lowest black data points correspond to this case and it can be explained by the fact that these data points have greater angles $\beta 2$. Experimental results reveal that the geometrical definition of the lap-belt designated only by angle $\beta 1$ is insufficient to define its submarining tendency.

The traditional geometrical definition is used in two dimensions given as the angle formed by the lap-belt and the X-axis from side view, specified as angle $\beta 1$. A wide range of this angle had been proposed as an unfavorable angle to submarining by the previous researchers. There were,

for example, 45-50° proposed by Haley (7), 45° by Patrick (8), 50-70° by Adomeit (9), Hontschik (10), and 69° by Billault (11). These reference angles are represented in Figure 6 for a comparison.

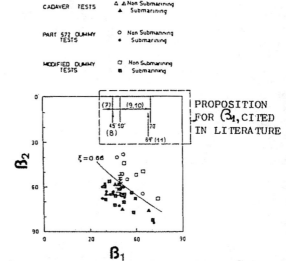

FIG. 6 . ANTI-SUBMARINING SCALE DETERMINED IN CADAVER, PART 572 AND MODIFIED DUMMY TESTS. THE RESULTS GIVEN IN THREE DIMENSIONS (β_1, β_2) ARE COMPARED WITH THOSE GIVEN IN TWO DIMENSIONS (β_1) PROPOSED IN LITERATURE.

Angles $\beta 1$ proposed in the above references cover almost all the submarining and non-submarining data points in the Figure 6. This variation was caused certainly by the variable angles $\beta 2$ which had been used but not observed by the authors of the references. Indeed, it is difficult to state what is the unfavorable angle $\beta 1$ to submarine if angle $\beta 2$ is not given.

Actually, the submarining tendency cannot be indicated when using the traditional definition of the lap-belt geometry as determined by angle $\beta 1$ alone. It is according to our theoretical and experimental study.

INFLUENCE OF SEAT DESIGN - Seat design is another parameter influencing the submarining consequence. Necessary motions of the pelvis are the forward and downward displacements which provoke probably a lap-belt submarining. To limit these two displacements, there is the conception for the creation of an "anti-submarining" seat.

Adomeit and Heger (9) proposed an energy-absorbing seat constructed with the front wall of the seat metal pan. On the forward area of the seat pan, a density of foam of 53 kg/m^3 was used (Figure 7).

Lundell et al. (12) presented a seat which had a contourned floor pan with a pronounced ridge at the front end. The seat cushion had a greater thickness which decreased gradually to the front edge of the cushion (Figure 8). In comparison with their flat seat tests, it showed that the new seat design gave both reduced injury criteria and low risk of submarining.

On the previous study (4), a comparison of submarining tendency was taken for the front occupant in the passenger seat and rear occupant in the bench. It was shown clearly that the rear occupant submarined more easily than the front occupants.

It can be explained by, firstly, a bad fixation system of the lap-belt on the interior side, corresponding to a smaller anti-submarining scale; secondly, no ridge support under the rubber foam at the front end of the bench.

These two parameters played together a role in the greater submarining tendency of rear occupant.

INFLUENCE OF "INTERNAL" PARAMETERS - Seat-Belt geometry, seat design, impact violence are external parameters. Others could be called "internal" parameters, such as anthropomorphic data (size and weight, relative bigness and shape of the pelvic bone, volume and stiffness of soft tissues in front of the "Sartorius", direction in space of "Sartorius" and direction of the pelvis in sitting position.

As indicated in the anti-submarining scale equation, if the lap-belt angles (β_1, β_2) are determined, we can see how the direction of "Sartorius" affects the submarining tendency. A comparison of the anti-submarining scales can be shown in Figure 9 and Figure 10 for human sub-

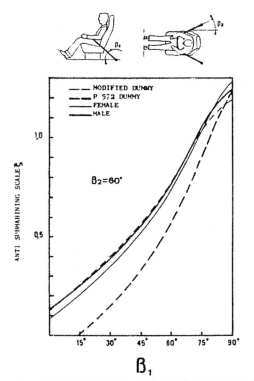

FIG. 9 . COMPARISON OF THE ANTI-SUBMARINING SCALE CURVES BETWEEN THE DIFFERENT SUBJECTS STUDIED AS A FUNCTION OF β_1 WHEN β_2 = 60°.

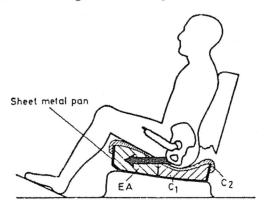

FIG. 7. DIAGRAM OF A SEAT DESIGN GIVEN BY ADOMEIT ET AL. (9).

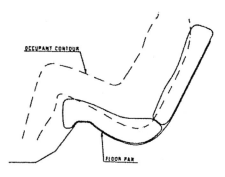

FIG. 8. DIAGRAM OF A SEAT DESIGN GIVEN BY LUNDELL ET AL. (12).

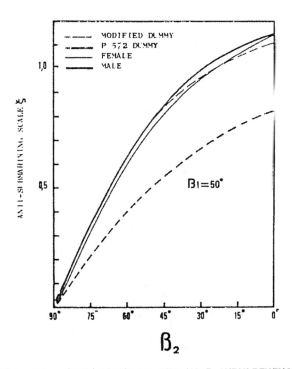

FIG. 10 . COMPARISON OF THE ANTI-SUBMARINING SCALE CURVES BETWEEN THE DIFFERENT SUBJECTS STUDIED AS A FUNCTION OF β_2 WHEN β_1 = 50°.

jects (male and female), Part 572 dummy and modified dummy.

In (5), it was shown that the direction of the Sartorius varies with the sex of the individual and with the different pelvic models. For example, the angle $\alpha'x$ (Fig. 2) is the most critical angle for indicating the tendency to submarine. This angle varies from 85° for males to 79° for females and to 109° for Part 572 and 83° for the FAA-USAF-NHTSA pelvic model. The submarining tendency is stronger when this $\alpha'x$ is higher (4)(5).

Flexibility of the soft abdominal tissue is greater for human subjects than for Part 572 dummy (5). An attempt was made in order to have a more human-like dummy. A Part 572 dummy modified at the level of "Sartorius" and abdominal tissue was defined (Fig. 11) and evaluated (4).

The theoretical approach still applies to these evaluations and experiments already widely described (4) have shown that the frequency of submarining is very similar for the modified dummy and the cadavers (65 % and 62.5 % respectively), smaller than for Part 572 (75 %). It was also observed that submarining occurs in a shorter time for the Part 572 (67.8 ms against 74.5 and 73.5 ms for the human subject and the modified dummy). This constitutes another reference for the justification of submarining tendency.

The direction of the "Sartorius" depends not only on the pelvic shape of the individual but also on the orientation of the pelvis in sitting position. It is known that the lap-belt submarining performance depends significantly on pelvic orientation. The submarining tendency increases with the rearward rotation of the pelvis. This rotation was defined for a sitting position within a car in relation to a standing position. An angle of 36.7° was recommended by Nyquist (13); it was determined an X-ray-radiograms-study on the basis of two volunteers from his study (14) and 5 volunteers from reference (5).

STUDIES BY MATHEMATICAL MODEL OF THE INFLUENCE OF CERTAIN OCCUPANT PARAMETERS ON THE RISK OF SUBMARINING OCCURRING.

A study by mathematical model of occupant related factors was conducted in 1979 with the financial backing of the French Government, as part of the Programmed Thematic Action "Vehicle Safety - Submarining Criterion". The purpose of this study was to evaluate the influence of parameters such as stiffness of the lumbar column, hips and knees, or mass distribution adjacent to the pelvis, on the tendency of a dummy to submarine (15).

MODEL - The model used, "Prakimod", is a two-dimensional one. It has ten degrees of freedom and offers the advantage of being a sophisticated and tried method of simulating a belt restraint system (16).

REFERENCE TEST - An experimental test - 50th Percentile Part 572 dummy in front passenger position, restrained by a 3-point-inertia-reel-belt type on a sled (Renault 5 configuration wi-

thout instrument panel), catapulted at 50 kph against a 30° inclined rigid wall - was reproduced by mathematical modelling, to act as a reference.

OUTPUT PARAMETERS - Actual submarining is not simulated by the model, but it does provide, at each moment, an output of the angle formed by the inboard and outboard webbings of the lap-belt with the pelvis projected along the sagittal plane (Figure 12). The maximum value of this angle during impact, together with its value when maximum forces in the lap-belt occur, make it possible to identify the risk of submarining. Maximum head, thorax and pelvis accelerations are also recorded as well as the maximum tensions of the different webbings and resultant forces applied to the dummy.

SIMULATIONS FOR COMPARISON PURPOSES - After the reference simulation, eleven other simulations were carried out using the same set of data but modidying each time the value of one of the input parameters to be tested.

RESULTS (Table 11) - Stiffening of the lumbar-column would greatly reduce the risk of submarining which is natural. A major increase in head accelerations would be observed simultaneously, along with a significant decrease in the longitudinal displacement of the centre of the head.

Stiffening of the hips generally heightens the risk of submarining, reduces head and pelvis accelerations and increases thorax accelerations. Moderate stiffness of the knees is of very little consequence.

Significant stiffening tends to slightly reduce the risk of submarining, as well as diminishing all maximum accelerations and belt forces. Transferring load from the thigh to the trunk tends to reduce the risk of submarining, especially if the load is transferred to the abdomen and not to the pelvis. Furthermore, a reduction in forward movement of the head and maximum acceleration of the pelvis and an increase in shoulder belt loading were recorded.

FINDINGS OF STUDY - Two possible orientations emerged from this study to reduce the tendency of the Part 572 dummy to submarine, independently of any change to the shape of the pelvis:

–Pending accurate accurate and reliable anthropometric data, displace load from the thighs to the pelvis, or better still, to the abdomen. This would entail modidying the dummy build.

–Or simply modify the adjustment of hip and knee friction torques, by requiring that these torques balance not the weight of the members as specified in Standard 208 (Figure 13A) but the weight of the whole body, realistically positioned in line with the working of the muscles of these members (Figure 13 B). This gives a smaller hip torque (61Nm against 100Nm for both hips), but a knee torque more than seven times greater ! (146Nm against 20Nm).

Two further simulations confirmed the accuracy of this solution regarding the change in output parameters (risk of submarining, accelerations..) but also showed that excessive blocking of the

Table 11 - Part of Mathematical Simulation Results

TEST No.	1 REFERENCE	2 HIP STIFFNESS DIVIDED BY 3	3 MULTIPLIED BY 3	4 MULTIPLIED BY 9	5 KNEE STIFFNESS DIVIDED BY 3	6 MULTIPLIED BY 3	7 MULTIPLIED BY 9	8 LUMBAR STIFFNESS DIVIDED BY 3	9 MULTIPLIED BY 3	10 MASSES DISPLACEMENT -4kg from upper legs +4kg onto pelvis	11 -4kg from upper legs +2kg onto pelvis +2kg onto lumbar segment	12 -4kg from upper legs +4kg onto lumbar segment
MAX. HEAD ACCELERATION (G)	71.00	70.38	57.51	53.59	71.12	70.64	67.85	42.16	75.50	72.04	64.18	50.81
HEAD G.S.I.	511.75	521.09	397.71	371.35	514.13	504.84	473.78	281.06	568.40	535.11	417.17	371.35
H.I.C. (*1)	408.17	417.15	327.41	298.28	410.38	401.64	373.87	242.50	465.93	416.61	380.61	315.55
MAX THORAX ACCELERATION (G)	46.32	47.75	47.14	50.02	46.64	45.37	42.37	49.01	53.10	46.55	49.27	51.35
THORAX G.S.I.	262.28	260.97	269.15	325.80	265.11	253.70	228.97	297.28	354.94	249.76	258.16	276.90
MAX PELVIS ACCELERATION (G)	71.30	74.27	70.36	63.21	71.42	70.88	69.53	69.83	69.27	68.90	67.40	65.97
MAX FORCE 1 (*2) (N)	12813	13266	13117	13509	12833	12750	12523	13082	13529	13251	13669	14025
MAX FORCE 2 (*2) (N)	9589	10163	9532	9937	9624	9482	9133	10974	10550	9961	10522	11148
MAX FORCE 3 (*2) (N)	15713	16177	15753	16329	15809	15423	14484	14569	17848	16089	15876	15590
AT TIME (ms)	50	50	52	63	50	51	51	66	51	50	50	50
θ ANGLE WHEN FORCE 3 IS MAX. (°)(*2)	127.97	124.73	132.67	138.03	128.13	128.93	127.14	158.15	123.30	126.42	125.64	124.88
MAX. θ ANGLE (°)	152.14	149.92	147.44	138.04	152.35	151.47	148.91	165.02	134.59	150.54	149.39	148.16
HEAD x-DISPLACEMENT (cm)	51.78	50.33	51.22	50.18	51.76	51.86	51.78	51.24	43.72	51.41	49.92	48.07

(*1) Not significant because of absence of head impact

(*2) see Figure 12

knees could significantly increase the horizontal displacement of the head.
            -It can be clearly seen that the study to reduce the severity of head impacts cannot be carried out separately from the controlling of submarining.

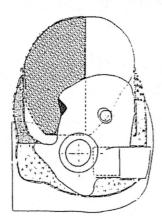

 Silicone foam RTV 5370
Density 0.20 instead of 0.26

 Silicone foam RTV 5370
Density 0.16 instead of 0.20

 Removed portion of the pelvis

To provide a human like three dimentional direction of "Sartorius" in sitting posture

FIG. 11 . PELVIC MODIFICATIONS.

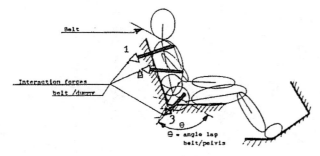

FIG. 12 . INITIAL POSITION OF THE DUMMY.

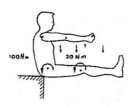

A

According to standard 208.

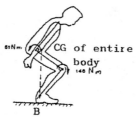

B

According to a more natural position.

FIG. 13 . TORQUES APPLYING TO DUMMY'S LOWER LIMB JOINTS.

DEFINITION OF AN ABDOMINAL PROTECTION CRITERIA

Past publications have described the complexity of the factors influencing the occurence of submarining and concluded that it was impossible to avert it by a single geometrical criterion as exists in Regulation 14,for example.That is why a great deal of work had been carried out with the object of seeking an abdominal protection criterion.
These different approaches have in common the determination to eliminate the risk of abdominal injuries induced by submarining under the lap-belt.

AN ACCOUNT OF THE DIFFERENT ATTEMPTS MADE TO DEFINE AN ABDOMINAL PROTECTION CRITERION

To our knowledge,the first attempt at defining a protection criterion for abdominal organs was published in 1973 (17).It stated that the "protection of abdominal will be satisfactorily guaranteed if any rising of strap -in dynamics- (during test) above anatomical reference marks materializing the limit compatible with a satisfactory rest on pelvis is forbidden"(Figure 14) "Compliance with the criterion is checked by cinematographic observation".
This first method successfully used in research could not be retained for the regulation test as the presence of doors -and unmodified doors-precludes filmed observation of the pelvis.

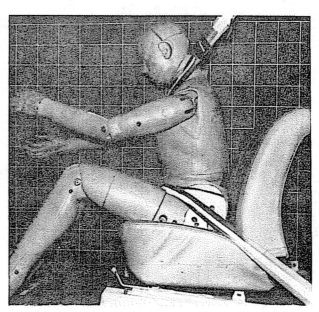

FIGURE 14.

The following year,in June 1974,at the ESV Conference in London,Citroen proposed fitting the dummy with optical transducerslocated on the pelvis beneath the Anterior Superior Iliac Spine (A.S.I.S.).These transducers would allow movement of the lap belt in relation to the A.S.I.S. to be detected. (18)
In the same year,a Ford patent dated 15th October 1974,proposed an almost identical system of detec

tion in which the optical transducer was replaced by force transducers. (19) .This solution was also to be retained by Citroen in 1977 (20).
A criticism which can be levelled against two above solutions is that abdominal danger cannot be stated to exist by the mere fact of the belt running up over the iliac crests.Indeed,the belt may be positioned against the abdomen at the end of impact without there being the risk of the occupant sustaining abdominal injury,if belt load is zero or low.Experimental work on human subject showed that certain cases of submarining proved to be of little danger provided that belt penetra tion into the abdomen as well as belt load –these two variables are obviously interdependent– were lower than a given limit.Moreover,the abdominal tolerance which justifies the abdominal protec–tion criterion presented in following chapter is based on such values.
Three other ideas were formulated which would ena ble the detection of submarining:
    –measure the angle of rotation of the pelvis (21)
    –measure the pressure in the abdominal air bag with which dummy Part 572 is equipped (22)
    –measure the pressure exerted on a set of pressure transducers arranged over the anterior face of the dummy (abdomen and thorax) (23).
These ideas are still in the early proposal stage. With the lack of experiments on dummies and refe–rences to the tolerance of human beings,they have not been able to be transposed into a protection criterion determining a limit which must not be exceeded if the non–occurence of abdominal inju–ries is to be guaranteed.

EXPERIMENTAL WORK CONDUCTED ON CADAVERS TO DEFINE THE ABDOMINAL TOLERANCE IN THE SUBMARINING CASE.

To define this abdominal tolerance in the event of submarining under the lap–portion of a three–point belt,10 cadaver tests were performed. Results have already been published as an ISO document(24).

METHODOLOGY USED IN CADAVER TESTS
Ten human cadaver tests were first and last per–formed with a current Renault 18 and Renault 20 car body mounted on a sled.The maximum sled dece–leration and impact velocity were 21–30g and 50 kph.The fresh cadavers were placed on the front passenger seat (Nb 2) or on the rear bench seat at the right side (Nb 4).Three–point retractor belts were used.Because the car occupant is more like–ly to submarine in the rear seat than in the front seat,owing to different seat–belt geometry (4), most of the fresh human cadavers were placed on the rear bench seat at the right side (seat Nb 4). Eight cameras (velocities 500 or 1000 pictures/ second) were used for this study,The positions of these cameras are shown in figure 15.
For two subjects (Nb 246 and 247),accelerations were recorded for the thorax at D1,D4,D7,D12,the ribs and the pelvis (on sacrum).No head decelera–tion recordings were taken because of recording capacity.Two shoulder belt loads (upper and lower) two lap–belt loads (inboard and outboard) were

also recorded.
Autopsies on the human cadavers were carried out after tests by the specialist of IRBA under the direction of Pr. C.GOT.

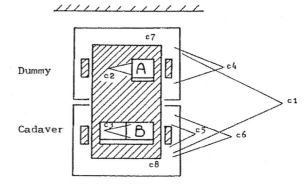

Cameras C1,C4,C5,C6:Placed at the right side of the sled.
Cameras C2,C3:Placed at the left side of the sled.
Cameras C7,C8:Placed at the top of the sled.

FIG. 15 . POSITION OF THE CAMERAS USED IN THE TESTS.

RESULTS AND DISCUSSION
The acceleration recordings are presented in Table 12.The anthropometrical data of the human cadavers and the results of the dissections are shown in Tables 13 and 14.
In five recently conducted tests ,Nb 244 and Nb 245 were normal cases of submarining.This means that the lap–belt firmly restrained the pelvic bone region below the Anterior Superior Iliac Spi–nes.It then rode–up over the iliac crests and pe–netrated into the abdomen.In this case,the ten–sion time curve appeared as a distinct "saddle shape".In precise terms,the first peak of lap–belt tension was always greater than that of the second peak which corresponded to a lower compres sion load penetrating into the abdomen(Fig.16). That is why no dangerous abdominal injuries were found.Heavy submarining occured in the remaining tests.The lap–belt rode–up rapidly over the ASIS and tension increased continuously(Fig.17).The tension was usually greater at the second peak applied to the abdomen than at the first peak, to the pelvic bone region.Therefore,abdomen AIS⩾3 were found in the cadavers.
Penetration of the lap–belt into the abdomen was measured using the same method as described in reference (1).By means of kinematical studies, this penetration was taken into account together with the fact that the lap–belt rode–up over ASIS. Because no significant difference in abdominal injuries between left and right hand sides was observed,either in the autopsies or in the real–life traffic accident,the severity of the abdomi–nal injuries is specified for the whole abdomen as given by the anatomo–pathologists.The average lap–belt tension was taken from both sides. In order to compare the dummy tests,lap–belt ten–

TABLE 12- DECELERATIONS RECORDED - TESTS WITH HUMAN CADAVERS
======== ========================= ================================

Test N°	Seat N°	Sled decel. γ(g)	Sled speed (km/h)	Thorax [γ(g)/SI]				Sacrum γ(g)/SI	Shoulder belt (N)		1st peak lap-belt (N)		2nd peak lap-belt (N)	
				D1	D4	D7	D12		upper	lower	int.	ext.	int.	ext.
127	4	23	50.9	-	-	-	-	86/1501	6800	8500	4000	9100	3900	9000
148	2	21	50.1	-	-	-	-	≐50/563	7700	2250	7600	4100	4200	2000
148	4	21	50	-	-	-	-	53/340	6000	5000	8600	4500	6500	2800
154	4	37	50.7	-	82/620	147/8064	-	71/779	6400	10050	5400	7400	5400	6700
182	2	24	47.5	39/171	36/76	81/555	107/437	45/272	5200	5500	6450	3900	7750	4300
243	4	27	47.4	-	-	-	-	93/471	4200	4600	600	700	4900	5100
244	4	34	50.1	-	-	-	-	63/334	5000	2800	3400	3900	1800	2200
245	4	22	49.8	-	-	-	-	52/859	5000	1800	3000	3700	2800	3700
246	4	30	50.1	56/521	82/678	81/774	79/637	153/1070	6600	4500	1500	4800	6000	7200
247	4	30	50.5	67/622	76/683	79/723	65/494	133/1067	8900	5800	5000	5200	6800	8600

TABLE 13 - NORMALIZED LAP-BELT TENSIONS FOR THE SECOND PEAK (AFTER SUBMARINING)
========= ==

Test N°	Mass Kg	Coefficient for normalized tension	Average 2nd peak tension of 2 sides (N)	Normalized lap-belt tension (N)	Abdomen AIS	Relative bigness x 10^{-6}
127.4	41	1.49	6450	9610,5	3	10.2
148.2	59	1.17	3100	3627	0	13,9
148.4	67	1.08	4650	5022	0	13,2
154.4	42,5	1.46	6050	8833	5	8.5
182.2	62	1.14	6025	6868,5	3	11,4
243.4	74	1.01	5000	5050	4	14,5
244.4	54	1.24	2000	2480	0	12.0
245.4	62	1.14	3250	3705	1	16.0
246.4	52	1.28	6600	8448	3	11.5
247.4	58	1.19	7650	9104	4	13.3

Table 14 - Injuries of the Tested Human Cadavers

Cadaver No.	Sex	Age	Size (m)	Weight (kg)	Injuries and Corresponding AIS	
					Thorax (Rib fractures)	Abdomen
127-4	M	57	1.59	41	28 ribs, AIS 4	L 1 fracture — AIS 3
148-2	F	65	1.62	59	8 ribs, AIS 3	AIS 0
148-4	M	62	1.72	67	12 ribs AIS 3	AIS 0
154-4	M	63	1.71	42.5	40 ribs+sternum AIS 4	liver + L2 fracture — AIS 5
182-2	M	57	1.76	62	8 ribs AIS 3	L 1 fracture — AIS 3
243-4	M	61	1.72	74	7 ribs AIS 3	colon + break of rectus — AIS 4
244-4	F	57	1.65	54	24 ribs AIS 4	fractures of left ilium crest — AIS 3
245-5	M	56	1.57	62	7 ribs.AIS 3	fissure of adipose tissue — AIS 1
246-4	M	62	1.65	52	12 ribs AIS 3/T6-T7 AIS 3	break of rectus — AIS 3
247-4	M	42	1.63	58	8 ribs+sternum AIS 3	break of rectus — AIS 4

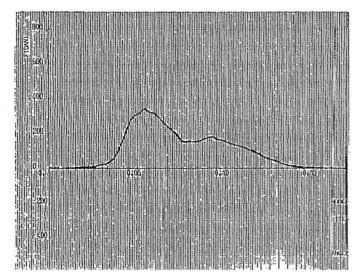

FIG. 16 . LAP-BELT TENSION/TIME HISTORY RECOR-
-DED AT THE INTERIOR SIDE OF CADAVER N° 244

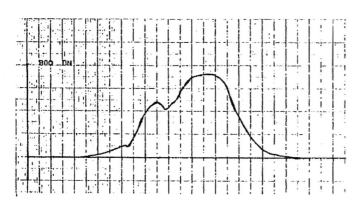

FIG. 17 . LAP-BELT TENSION/TIME HISTORY RECOR-
-DED AT THE EXTERIOR SIDE OF CADAVER N° 246.

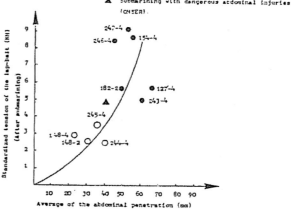

FIG. 18 . ABDOMINAL TOLERANCE TO THE
SUBMARINING.

Test N°	Mass	Size	AIS	
127-4	41kg	152cm	1	Previous results
148-2	55kg	161cm	0	
148-4	67kg	172cm	0	
154-4	42,5kg	171cm	5	
182-2	62kg	176cm	3	
243-4	74kg	172cm	4	Recent results
244-4	41kg	165cm	0	
245-4	62kg	157cm	1	
246-4	53kg	165cm	3	
247-4	58kg	163cm	4	

sion recorded in the cadaver tests should be standardized with respect to the mass of the Part 572
dummy (75kg).A formula given by Eppinger is used
in this study,which has been described in references (24)(25).
In figure 18,a parabolic curve can be drawn as a
relationship between the standardized tension of
the lap-belt and abdominal penetration.The scatter
of the data points is due to the differences in
the anthropometric data of the cadavers.A similar
curve can be found also in figure 19 for the standardized lap-belt tension (after submarining) and
the abdomen AIS.Previous results are also presented in these figures.It can be seen that Nb 243-4
is the lowest point (5KN) in the new group with
dangerous injuries.This value is smaller than that
(5.7KN) of Nb 182-2,the lowest one in the old
group of previous results.One datum point (without
test number) given by ONSER is also presented in
these two figures.
Specific comments on the severity of abdominal injuries (AIS) concerning cadavers Nb244 and 245 are
given below:

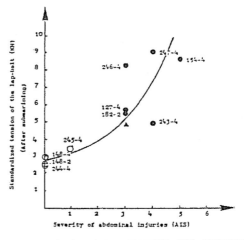

FIG. 19 . RELATIONSHIP BETWEEN THE STANDARDIZED
TENSION OF THE LAP-BELT (AFTER SUBMARINING) AND
THE ABDOMEN AIS.

Nb 244 sustained only pelvic fracture with
an AIS=3 but no abdominal injuries.An abdominal
AIS=0 was given to this case.It implied that if
no pelvic fracture has occured,the lap-belt would
probably have penetrated into the abdomen,with
greater force and caused abdominal injuries,which

in real-life could be a dangerous situation. It is an ambiguous case difficult to interpret and we prefer to avoid underestimating the abdominal risk.

-Nb 245 sustained abdominal AIS=1 and was the only subject situated between the group with dangerous injuries (AIS ⩾ 3) and the group with injuries (AIS=0). Since the severity of the abdominal injuries due to submarining is probably influenced by the "relative bigness" defined as "weight/height³" (kg/cm³) in reference (24), a further study of this factor is described here. The "relative bigness" of the human cadavers versus their abdomen AIS is illustrated in figure 20 where the previous results are also presented. Cadaver Nb 245 had the greatest "relative bigness" in all results. We made the assumption that this subject would sustain a greater AIS if his bigness had been so to speak, smaller. It may be for this reason that the only fissure of the abdominal adipose tissue was found in this subject.

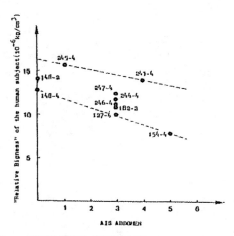

FIG. 20 . "RELATIVE BIGNESS" VS SEVERITY OF THE ABDOMINAL INJURIES (AIS).

The abdomen AIS is certainly influenced by other parameters, for example rib cage strength, which are not discussed in this study. In general, no big difference can be found between previous and recent results. But the limits between the groups with and without injuries are closer due to the recent results; therefore the latest determination of the abdominal tolerance will be more accurate. The determination of critical lap-belt tension should be considered more carefully. As just mentionned above, the case with Nb 244 cadaver produced no injuries, yet it is nevertheless a dangerous situation. On the other hand, Nb 243 (5KN) produced the lowest datum point in the group with injuries in recent results; Nb 182-2 (5.7KN) was the lowest one in previous results. Based on previous and recent available data points presented in figure 18 and figure 19, the critical standardized lap-belt tension was reconsidered. A value of 3 KN can be contemplated, which is smaller than the previous value of 4.4 KN which had been proposed (25,26,27).

EXPERIMENTAL WORK WITH DUMMIES INSTRUMENTED WITH "ILIAC CREST TRANSDUCERS"

EXPERIMENTAL WORK WITH DUMMIES INSTRUMENTED WITH "ILIAC CREST TRANSDUCERS"

SCHEDULE OF RECENT RESEARCH PROGRAMS FOR DUMMY TESTS

-First series - Dummy tests with a Renault 20 car body and seat -
The same car body and sled used in cadaver tests were employed for the dummy tests. The sled deceleration and impact velocity were similar to those registered in cadaver tests. The Part 572 dummy was equipped with short (40mm) or long (55mm) APR submarining transducers and placed in the front passenger seat (Nb 2) or in the rear bench seat on the right side (Nb 4). Three-point retractor seat belts (60mm width) and different positions of lap-belt were used in these tests. Standard recordings were made. Test conditions were similar or more severe with respect to test Nb 5 given in reference (28).

-Second series - Dummy sled tests with Renault 20 seat directly on the sled -
This series of tests was performed by T.N.O. The test conditions were similar to those used in reference (28). Some tests were performed under excessive test conditions (with the seat back set at an angle of about 30° from the vertical for example. Classical recordings were taken.

-Third series - Dummy sled tests with Citroen Visa seat directly on the sled -
This series of tests was the reconstruction of test NB 10 in reference (29). Briefly, the test conditions were:

    -front passenger Citroën Visa seat
    -Part 572 dummy without fore-arms
    -3-pt static seat belts (50mm width)
    -sled deceleration and impact velocity about 20g and 50kph respectively
    -inclination of seat back 28°
    -weakened seat frame
    -slack of the lap-belt and shoulder belt about 25 mm

Recordings were made for pelvic deceleration, shoulder belt tensions (upper and lower) and lap-belt tensions (inboard and outboard). Citroën transducer and APR short or long transducers were used.

RESULTS

-First series - Dummy tests with Renault 20 seat.
The results of this series of tests are presented in Table 15. The first three tests (Nb 1013, 1014, 1015) are cases of non-submarining with the lap-belt positioned correctly. Tests Nb 1018, 1019 and 1021 are submarining cases with the lap-belt positioned 1 cm higher than in the first three tests. The compression loads recorded by the submarining transducer are in the range between 463N and 798N. Tests Nb 1016, 1017 and 1020 are cases of excessive submarining in which the lap-belt rode up over the submarining transducers. Nb 1016 and 1017 are tests in seat Nb 2 with the worst position of the lap-belt. Nb 1020 is a test performed in seat Nb 4 with a correct position of the lap-belt and using short transducers (Figure 21). The recorded loads range from 805N to 1116N and it serves as an exam-

T A B L E   15 - D U M M Y   T E S T S   W I T H   R E N A U L T   2 0   S E A T

Test N°	Seat N°	Submarining transducer	Position of lap-belt with relation to ASIS from side view (cm)	Lap-belt tension after submarining Int (N)	Ext (N)	Transducer (N) Int.	force Ext.	Remarks
1013	2	short	+ 4	-	-	-	-	no submarining
1014	2	"	+ 3	-	-	-	-	"
1015	2	long	+ 2	-	-	-	-	"
1016	2	"	- 1	2200	2240	1030	994	Lap-belt rode-up over transducers
1017	2	"	+ 1	1940	1550	869	805	"
1018	2	"	0	1667	4520	798	614	Transducers worked correctly
1019	2	"	0	1117	1620	638	521	"
1020	4	short	+ 4	5237	-	1116	-	lap-belt rode-up and touched direct submarining trans-ducers
1021	2	"	+ 1	1122	-	463	-	transducers worked correctly

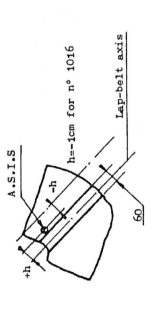

A.S.I.S

h=-1cm for n° 1016

-h

+h

60

Lap-belt axis

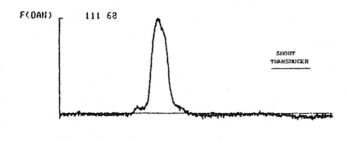

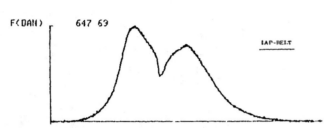

FIG. 21 . TENSION (T) AND FORCE (F) CURVES OF
TEST N° 1020 (INTERIOR SIDE).

ple where the short transducer works correctly and records a greater load under test conditions more favorable to submarining in seat Nb 4 (rear bench) than in seat Nb 2 (front individual seat) (4).
In order to make a comparison with cadaver tests, a new research program is being prepared to perform the dummy tests under these same conditions. It can be recognized from the above analyses that the responses of the transducer to the submarining process were in accordance with the different test conditions.

 -Second series - TNO sled dummy tests -
The results of this series of tests are presented in Table 16.No submarining occurred in the first three out of a total of nine tests,even with the lap-belt located 15 mm higher than the correct position.In the rest of the tests,a submarining phenomenon could only be obtained by placing sheets of double plastic foil between the dummy and the car seat in combination with a 25 mm "higher than correct" position of the belt on the pelvis.Excessive submarining in tests Nb 454,456 and 457 could only be obtained by additionally inclining the seat back to about 30° from the vertical.The excessive submarining cases all resulted in a short peak loading of the transducers (duration 6 ms),followed by an over-riding of the transducers.Consequently,the lap-belt was loaded again by the lumbar spine.In all tests,a reasonable symmetrical loading of left and right lap-belt and transducers was found.

 -Third series - Dummy tests with Citroën Visa seat
In table 17 tests Nb 5363 and 5364 correspond to non-submarining cases.The lap-belts were positioned correctly.Tests Nb 5360,5361 and 5365 are normal submarining cases.Their descriptions are simi

lar to the ones given for the first series of dummy tests.The positions of the lap-belts are relatively worse than in the non-submarining group.Both short and long transducers work correctly in these three tests.The long transducer is used in test Nb 5361;the lap-belt tension and transducer load as function of time for the outboard side are illustrated in Figure 22.

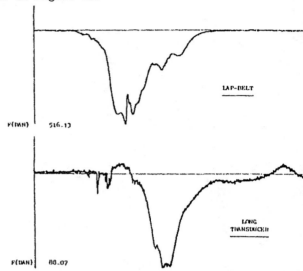

FIG. 22 . TENSION (T) AND FORCE (F) CURVES OF
TEST N° 5361 (EXTERIOR SIDE).

Test Nb 5362 corresponds to a submarining case in which the lap-belt rides-up over the transducer. The position of the lap-belt was the least desirable one observed in this series of tests.However, a significant load of 2531N was recorded by the short transducer.This sample again illustrates how the submarining transducer can provide a correct response:when the test is performed under more severe submarining conditions,a greater compression load can be recorded by the transducer, so indicating that it would be a more dangerous submarining case.The lap-belt tension (t) and transducer load (f) for outboard side in the test Nb 5362 are illustrated in Figure 23.

RELATIONSHIP BETWEEN LAP-BELT TENSION AND THE SUBMARINING TRANSDUCER LOAD

 All data points obtained from the recent results of dummy tests are illustrated in Figure 24 The two parameters are the maximum transducer load and the corresponding lap-belt tension at the same points in time.The previous results are also presented.A linear correlation between lap-belt tension and transducer load can be proposed.
Since the standardized critical lap-belt tension (3000N) is already determined in Figures 18 and 19 a corresponding transducer load of 800N can be found in Figure 24.
Based on previous and recent results obtained from cadaver and dummy tests,a submarining transducer load of 800N is proposed as an abdominal protection criterion in the dummy tests.

CONCLUSIONS

TABLE 16 – T.N.O. SLED TEST EVALUATION OF APR SUBMARINING TRANSDUCERS ON PART 572 DUMMY WITH RENAULT 20 SEAT

Test No.	Submarining transducer	lap-belt position upper ASIS (cm)	Seatback	Max. seat belt tension after submarining (N) Int.	Ext.	Transducer force (N) Int.	Ext.	Remarks
449	long	correct	25°	-	-	-	-	no submarining
450	"	1	25°	-	-	-	-	" "
451	"	1.5	25°	-	-	-	-	" "
452	"	2.5	25°	1500	1300	900	1000	submarining
453	"	2.5	25°	2000	1100	1130	1100	"
454	"	2.5	30°	2200	1000	1100	1000	excessive submarining
455	short	2.5	25°	-	-	-	-	no submarining
456 *	"	2.5	>25°	(?)	3200 (?)	750	450	excessive submarining
457	long	2.5	>25°	3600	2100	1100	820	" "

* In this test, there were the problems found in the recordings.

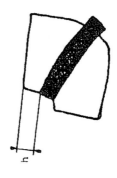

TABLE 17 - DUMMY TESTS WITH CITROEN VISA SEAT

Test No.	Seat No.	Submarining transducer	Position of lap-belt(cm) *	Seat	Lap-belt tension after submarining Int (N)	Ext (N)	Transducer force (N) Int.	Ext.	Remarks
5360	2	short	2	weaken	2627	1564	-	607	transducer worked
5361	2	"	1	"	-	2198	-	880	" "
5362	2	"	0	"	-	4050	-	2531	lap-belt rode-up over transducer
5363	2	"	3	weaken+flexible	-	-	-	-	no submarining
5364	2	long	3	"	-	-	-	-	" "
5365	2	"	3	"	4506	-	1239	-	transducer worked

* Definition is same as given in the diagram presented in Table 2.

1. A standardized critical lap-belt tension of __3000N__ is determined, based on available values from previous and recent submarining cadaver tests.
2. In human cadaver submarining tests, a relationship between abdominal injury (AIS) and "relative bigness" of the subject can be found: abdomen AIS increases as "relative bigness" decreases.

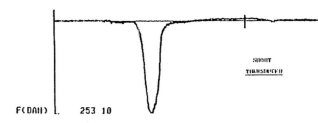

FIG. 23 . TENSION (T) AND FORCE (F) OF TEST N°5362 (EXTERIOR SIDE).

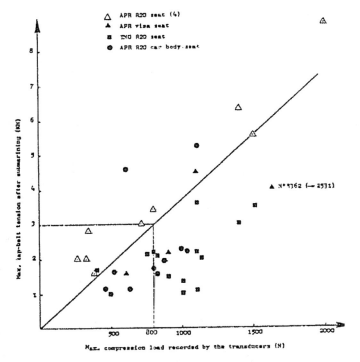

FIG. 24 . RELATIONSHIP BETWEEN THE LAP-BELT TENSION AND THE COMPRESSION LOAD RECORDED BY THE TRANSDUCERS.

3. With a dummy in the front passenger seat (Nb 2) and a correctly positioned lap-belt, non-submarining cases were usually obtained.
4. Submarining cases were obtained at the front passenger seat (Nb 2) by deliberately positioning the lap-belt at an excessive level on the abdomen or by increasing backrest rake, only weakening the seat frame.
5. In all submarining cases, the short (40 mm) and long (55 mm) transducers were able to detect the submarining phenomenon.
6. The values recorded by the short or long submarining transducers corresponded to the different conditions used in the dummy tests.
7. Based on recent critical lap-belt tension determined from the cadaver tests and, also the data points concerning a relationship between the lap-belt tension and the transducer load found in dummy tests, the abdominal protection criterion of __800N__ is proposed for dummy tests.
8. The short transducer can usually be used in submarining tests for the abdominal protection criterion; the long transducer can be used successfully in tests performed particularly under excessive submarining conditions.

GENERAL CONCLUSIONS

1. LAP-BELT SUBMARINING in real-life accidents is a process inducing severe injuries in a relatively high proportion of severe frontal crashes (27%) - but it is too often underestimated in previous studies-. Reasons of underestimation were for example that dorso-lumbar spine fractures, or lower members fractures, or above mesocolic injuries were not often considered as submarining consequences. Present findings constitute an attempt to clarify the whole pattern of such lap-belt submarining.
   From a whole sample of 1423 three-point belted front occupants involved in 958 frontal car crashes, a sub-sample of 45 cars have been selected in which, at least one of the 77 front occupants sustained either a severe abdominal injury (AIS ⩾ 3) or a dorso-lumbar column fracture (AIS ⩾ 2). Among this survey sample, 47 injured people sustained a lap-belt submarining which was sure or most probable (61%). These submarinings induced three main types of injuries:
   —abdominal injuries (sub-mesocolic but also above mesocolic) (AIS ⩾ 3)
   —dorso-lumbar spine fractures (mainly T12, L1, L2) (AIS ⩾ 2)
   —lower members fractures (mainly legs, knees and femurs fractures) (AIS ⩾ 2)

2. Among submarining cases, 68% of cases of abdominal injuries (AIS ⩾ 3) and (or) dorso lumbar spine fractures (AIS ⩾ 2) were SURELY induced by the lap-belt section. The percentage reached 94% if we considered the cases where the lap-belt influence was sure or only MOST PROBABLE.

3. The influence of shoulder-belt plays only as an

903

aggravating factor - complementary to the lap-belt- for less than 10% of cases.

4. Poor geometry and (or) slackly worned belts were present in 74% of submarining cases.

5. Most of the submarining cases were observed in high violence crashes ($\Delta V \geqslant 50$ kph and mean $\gamma \geqslant$ 10g occured in 58% of cases).

6. In real-life accidents, the most frequent severe consequences of lap-belt submarining are according to a decreasing order:
    - sub-mesocolic injuries
    - lower members fractures
    - dorso-lumbar spine fractures
    - above-mesocolic injuries
    - pelvic fractures

7. Taking into account the severity of above-mesocolic injuries (liver and (or) spleen), it is noticeable that 3/5 of such victims sustained a lap-belt submarining which is sure or most probable (of course these lesions are often associated to sub-mesocolic injuries).

8. In cadaver tests (with blood pressure restored at normal level), submarining process is checked by special in-board camera which shows then all these above described injuries which could be induced by lap-belt submarining, even for above-mesocolic injuries (liver or spleen injuries) or dorso-lumbar spine fractures. These experiments also confirm the possible aggravating influence of shoulder-belt section.

9. Abdominal or dorso-lumbar spine TOLERANCE to lap-belt submarining is low. Injuries are observed for lap-belt tension higher than 3000N.

10. A PROTECTION CRITERION - ANTI-LAP-BELT SUBMARINING CRITERION - has been proposed. It consists in the record of lap-belt loading against specific ILIAC-CREST TRANSDUCERS symetrically installed on the pelvis of the dummy. Based on recent critical lap-belt tensions determined from specific cadaver tests and - also - the data points concerning a relationship between the lap-belt tensions and the Iliac-Crest-Transducer loads found in dummy tests, the LAP-BELT-SUBMARINING PROTECTION CRITERION of 800N is proposed for dummy test.

REFERENCES

(1) G. Walfisch, A. Fayon, Y.C. Leung, C. Tarriere; C. Got, A. Patel "Synthesis of Abdominal Injuries in Frontal Collisions with Belt-Wearing cadavers Compared with Injuries Sustained by Real-Life Accident Victims. Problems of Simulation with Dummies and Protection Criteria." in Proceedings of IRCOBI GOETEBORG, Sweden, 7-9 Sept. 1979.

(2) J.S. Dehner, "Seat Belt Injuries of the Spine and Abdomen." American J. Roentgen, VIII, PP833-843, April 1971.

(3) Document ISO/TC 22/SC 12/GT6 N107.

(4) Y.C. Leung, C. Tarriere, A. Fayon P. Mairesse, P. Banzet. "An Anti-Submarining Scale Determined from Theoretical and Experimental Studies Using Three-Dimensional Geometrical Definition of the Lap-Belt " SAE Paper n° 811020, in the Proceedings of 25th Stapp Car Crash Conference, San Francisco, Sept. 28-30, 1981.

(5) Y.C. Leung, C. Tarriere, A. Fayon, P. Mairesse, A. Delmas and P. Banzet, "A Comparaison Between Part 572 Dummy and Human Subject in the Problem of Submarining ." in the Proceedings of 23rd Stapp Car Crash Conference, San Diego Calif., Oct. 17/19, 1979, SAE Transaction Paper N° 791.026.

(6) A. Fayon, C. Tarriere, G. Walfisch C. Got, A. Patel "Thorax of 3-Point Belt Wearers During a Crash (Experiments with Cadavers)" in the Proceedings of 19th Stapp Car Crash Conference, SAE paper 751148, San Diego, Calif., Nov. 17/19, 1975.

(7) J.L. Haley Jr., "Fundamentals of Kinetics and Kinematics as Applied to Injury Reduction "in "Impact Injury and Crash Protection ", C.C. Thomas Publisher 1970.

(8) L.M. Patrick and A. Andersson "Three -Point Harness Accident and Laboratory Data Comparaison". SAE Paper N° 741.181, in the Proceedings of the 18th Stapp Car Crash Conference, Ann Arbor Michigan, Dec. 4th, 1974.

(9) D. Adomeit and A. Heger, "Motion Sequence Criteria and Design Proposals for Restraint Devices in order to Avoid Unfavorable Biomechanic Condition and Submarining". SAE Paper N° 751.146 In the Proceedings of 19th Stapp Car Crash Conference, San Diego, Calif. Nov. 17th 1975.

(10) H. Hontschik, E. Müller and G. Rüter, "Necessities and Possibilities of Improving the Protectice Effect of Three -Point Seat -Belts ", in the Proceedings of the 21st Stapp Car Crash Conference, New Orleans, Louisiana, Oct. 19/21st 1977.

(11) P. Biilaut, C. Tisseron, M. Dejeammes, R. Biard, P. Cord, P. Jenoc, "The Inflatable Diagonal Belt "7th

International Technical Conference on the Experimental Safety Vehicles,Paris June 5/9 ,1979.

(12) B. Lundell,H.Mellander,I.Carlson ,"Safety Performance óf a rear Seat Belt System with Optimized Seat Cushion Design",SAE Paper N°810.796,Passenger Car Meeting Dearborn ,Michigan ,June 8 12,1981

(13) G.W. Nyquist "Comparaison of Vehicule-Seated Volunteer Pelvic Orientations Determined by Leung et al, and by Nyquist et al .Document ISO /TC22/SC12/WG5?April 10,1980.

(14) G.W. Nyquist et al . ,"Lumbar and Pelvic Orientations of the Vehicle Seated Volunteer" SAE 760821 ,20th Stapp Car Crash Conference 1976.

(15) D. Lestrelin "Etude par modele Mathematique de l'Influence de quelques Paramètres du Mannequin Part 572 sur sa Propension au sous-Marinage ".Rapport Interimaire N°2 Contrat N°78043 "Critères de Sous-Marinage" dans le Cadre des Actions Thematiques Programmees Françaises,1979.

(16) D. Lestrelin ,A. Fayon , C. Tarrière "Development and use of a Mathematical Model Simulating a Traffic Accident Victim" Proceedings of 5th International IRCOBI Conference , Birmingham,Sept. 1980.

(17) C. Tarrière "Proposal for a Protection criterion as Regards Abdominal Internal Organs" P371 Proceedings of Conference of A.A.A.M., Oklahoma City ,Oklahoma,Nov.14/17,1973.

(18) M. Clavel "Restraint Systems Improvement " Proceedings of 5th Interna -tional Technical Conference on Experi -mental Safety Vehicles,London ,June 4-7 1974.

(19) R.P. Daniel "Test Dummy Submarining Indicate in United States Patent ,3.841.163 Oct. 15, 1974.

(20) Citroën,"Methode de Detection de Depassement des crêtes Iliaques" Document ISO/TC 22/SC 12/GT6(F3)21F,1977.

(21) D. Adomeit "Seat Design -A Significant Factor for Safety Belt Effec -tiveness " SAE Paper N°791004, in Proceedings of 23rd Stapp Car Crash Conference ,San Diego ,Calif. ,Oct.17-19 1979.

(22) Bröde,Personal Communication Nov.24,1980.

(23) Fiat "Development of a Device to Evaluate the Abdominal Injuries in Submarining",Document ISO /TC 22/SC 12/ GT 6(Italie 1)N 71,Sept.1980.

(24) Y.C. Leung ,C. Tarrière,J. Maltha "A Review for the Abdominal Protection Criterion" Document ISO/TC 22/ SC 12/WG6 N97 October 1981.

(25) "Experimental Elements for the Definition of Abdominal Protection Criterion in a Submarining Possibility" July 27th,1980. ISO/TC 22/SC 12/WG. 6 N°72

(26) "Proposal for an abdominal Protection Criterion ",March 1980. ISO/TC 22/SC 12/WG. 6 N°58

(27) Y.C. Leung ,P. Mairesse,P.Banzet "Submarining Criterion "Sept. 1980. ISO/TC 22/SC 12/WG N°77.

(28) R.L. Stalnaker "Submarining sled Testd Part 572 Pelvis with and without PSA/Renault Submarining Transducers and PSA/Renault modified Part 572 Pelvis with Submarining Transducers" TNO, October 8th,1980.

(29) M.Dejeammes,R.Biard,Y.Derrien "Factors influencing the estimation of Submarining on the Dummy". ISO/TC 22/SC 12/WG.6-96, August 1981.

905

APPENDIX I - CHARACTERISTICS OF SAMPLE - FRONTAL IMPACTS WITH 3-POINT-BELTED FRONT OCCUPANTS ONLY

	WHOLE SAMPLE	SURVEY SAMPLE	VICTIMS WITH ABDOMEN AIS > 3 AND/OR DORSO-LUMBAR SPINE ≥ 2	SUBMARINING PROBABLE OR SURE
N. Front Occupants	1423*	77	47	47
N. cars	958	45	45	39
[1] Abdomen AIS ≥ 3		33	33	30
[2] Lumbar Spine AIS ≥ 2		1	1	1
[3] [1] + [2]		1	1	1
[4] Dorsal Spine AIS ≥ 2		3	3	1
[5] Submarining Without [1] or [2]		6	-	6
[6] Other Cases		24	-	-
Male Drivers	791 (55 %)	37	19	18 (38 %)
Female Drivers	167 (12 %)	8	3	3 (7 %)
Male Right Front Passengers	155 (11 %)	13	9	8 (17 %)
Female Right Front Passengers	310 (22 %)	19	16	18 (38 %)
	100 %	100 %		

(*) 119 cases are not included in this table because there are no available data concerning them about sex

APPENDIX II - SECTION OF THEBELT (LAP OR SHOULDER-BELTS) PRODUCING ABDOMINAL INJURIES
(Analyzis Process Based Upon Medical Data)

LAP-BELT			SHOULDER BELT	
ALMOST SURE	MOST PROBABLE	UNCERTAIN MECHANISM	MOST PROBABLE	ALMOST SURE
Two Above-Mesocolon Injuries At the Same Horizontal Level Exemple: Liver + Spleen PLUS at Least One Sub-Mesocolon Injury	Two Above-Mesocolon Injuries At the Same Horizontal Level Or One Above-Mesocolon Injury PLUS One Sub-Mesocolon Injury	Absence of Typical Pattern of Injuries Or Discordances (Technical Data Can Clear-Up Misunderstandings.)	One and Only One Above-Mesocolon Injury Close to Buckle (Example: Spleen for Right Front Passengers or Liver for Drivers	One and Only One Above-Mesocolon Injury Close to Buckle PLUS Rib Fractures Close to the Injury
Or				
Symmetrical Rupture of Sheath of Rectus Or Horizontal Abrasions Above Antero-Superior Iliac Spine				

Plus, Possibly:

- Wound, or Abrasion of the Neck Due to the Shoulder-Belt

- High Rib Fractures on the Buckle Side

- Lower Members Injuries

- High Lumbar or Low Dorsal Vertebrae Fractures

APPENDIX III - ACCIDENTOLOGICAL SIGNS OTHER THAN MEDICAL ONES WHICH HIGHLIGHT SUBMARINING UNDER LAP-BELT

[1] BELT

1.1. POOR GEOMETRY: . ANGLE $B_1 \leqslant 50°$

.. ANGLE $B_2 \geqslant 100°$

[THE GEOMETRY IS THE WORSE WHEN B_1 IS LITTLE AND WHEN B_2 IS HIGH.]

. BUCKLE ABOVE THE SEAT (ADJUSTABLE BUCKLE ON THE WEBBING AT MIDDLE-ANCHORAG.)
(THE HIGHER THE BUCKLE, THE WORST IT IS).

AS A WHOLE, 20 CASES OF CERTAIN OR PROBABLE SUBMARINING OUT OF 47 (43 %).

1.2. BELT WORN WITH SLACK: 17 CASES OF CERTAIN OR PROBABLE SUBMARINING OUT OF 47 (36 %)

[2] ASSOCIATED FACTORS

2.1. IMPACTS OF GREAT VIOLENCE ($\Delta V \geqslant 50$ KPH AND MEAN $\gamma \geqslant 10$ G): 25 CASES OF CERTAIN OR PROBABLE SUBMARINING OUT OF 47 (53 %).

2.2. IMPACTS OF THE KNEES UNDER THE PANEL: 39 CASES OF PROBABLE OR CERTAIN SUBMARINING (83 %)

2.3. FRONT SEAT TRACK DAMAGE OR FAILURE: 18 CASES OF PROBABLE OR CERTAIN SUBMARINING (38 %)

2.4. REAR OVERLOAD: 10 CASES OF PROBABLE OR CERTAIN SUBMARINING (21 %°

REMARKS (See the Bottom of Appendix I):

(1) Right Front Passengers Are Statistically Significantly More Involved in Submarining Than Drivers

(2) There is a tendency for Women to Be More Involved Than Men But This Difference Is Not Quite
Significant From a Statistical Standpoint.

APPENDIX IV
Details of soft tissus injuries observed in the tested human cadavers.

Cadaver n°   Descriptions of the injuries

2
- Multiple wounds : superior and inferior part of the liver AIS 5 ※
- Perforation of the low part of the descendant colon AIS 5
- Rupture of the mesentry (length of 8 cm) near to posterior AIS 4
- Fracture of L4 transverse process AIS 2
- Fracture of left lateral sacral crest AIS 3

4
- Wound of 2 cm of the right diaphragm AIS 3
- Wound of superior face of the liver AIS 4

8
- Rupture of the left diaphragm AIS 3
- Bruise of the spleen AIS 4
- Fracture of 2 ilium crests AIS 2

10
- Wound of the liver (little fissure of the vascular bad) AIS 4
- Fracture of the anterior part of 2 ilium crests AIS 2

12
- Fissure of the mesentary root without blessure of the artery AIS 3

16
- Rupture of the left part of the diaphragma AIS 3
- Irregular wound of the superior part of the spleen AIS 4

25
- Little fissure of the mesentary root without blessure of the artery AIS 3
- Fracture of the anterior part of 2 ilium crests AIS 2

127-4
- Compression of L1 vertebra body (centrum) AIS 2

154-4
- Wound of the liver (anterior exterior face of right lobe) AIS 4
- L2 fracture with marrows' elongation (no rupture) AIS 5

170
- Compression of L4 and L5 vertebra body AIS 2

183
- Fissure of superior part of the liver AIS 4

185
- Erosion of the skin in front of the A.S.I.S. AIS 1
- Wound of the gashed muscle at the level of abdomen AIS 3
- Rupture of the mesentery (10 cm) AIS 3

189-4
- Dislocation of L5-S1 AIS 4
- Fracture of the right lateral sacral crest AIS 3
- Fissure of the mesentary AIS 3

231
- Depth wound of the liver (anterior board of 2 lobes, superior face of the right lobe) AIS 5
- Wound of the spleen (superior part) AIS 4
- Bruise of the mesentery (without rupture) AIS 3

255
- Wound of the spleen (exterior face) AIS 4

※ AIS numbers are referred from "1980 AIS"

909

840392

Historical Perspective on Seat Belt Restraint Systems

H. George Johannessen
OmniSafe Associates, Inc.

ABSTRACT

Landmarks in the chronology of the development of seat belt restraint systems are highlighted and examined on the bases of the influence of requirements of legislation, product performance, and marketing considerations. Recent developments in component hardware are identified.

SEAT BELT HISTORY will soon be in its hundredth year. As early as 1885 belts were used on horse-drawn vehicles to prevent passengers from being ejected from the vehicles on rough roads. The "incubation" period before seat belts were installed in all seating positions in all new passenger cars was over 80 years - until 1968 - and nearly 100 years will have been passed before seat belts are present in virtually all seating positions in all passenger cars in the total car population in the United States. This long lag is difficult to understand in light of the early use of seat belts in airplanes and racing cars. U.S. Army Plane No. 1 was equipped with a seat belt in 1910. Open cockpit commercial aircraft were required to be equipped with seat belts as of 31 December 1926, and all commercial aircraft had this requirement a few years later. Barney Oldfield equipped his racer with a seat belt in 1922. All race drivers in recognized races in the United States are now required to wear seatbelts. Many passenger car occupants in the United States continue to disregard the seat belts installed in their cars and fail to use them.

Seat belt history may be considered from the standpoint of seat belt geometry and components, seat belt performance requirements, comfort and convenience, usage rates, legislation and regulations affecting seat belts, and litigation relating to the use or non-use of seat belts.

SEAT BELT HISTORY - SYSTEM GEOMETRY AND COMPONENTS

The very early seat belts were nothing more than conventional leather belts secured by primitive means to the vehicle seat or structure. No outstanding improvements were seen until the 1950s and 1960s when a significant amount of development effort began to be directed to improved designs. The webbing-to-metal buckle design (Fig. 1), characteristic of earlier airplane applications, was supplanted quite quickly by the metal-to-metal buckle designs. The non-locking

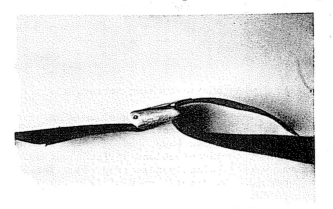

Fig. 1 - Webbing-to-metal buckle

retractor (Fig. 2) was added for improved stowage of the webbing when the belt was not in use. Soon afterward, the automatic-locking retractor (Fig. 3) became available. At about this time, in the mid-1960s, upper torso restraint was receiving increased attention. A limited production application of a Y-yoke shoulder harness with a webbing-sensitive emergency-locking retractor (Fig. 4) appeared in the Shelby-American GT 350 and GT 500 cars. As of 1 January 1968 all seating positions in all passenger cars manufac-

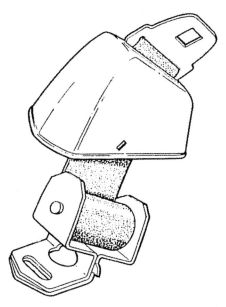

Fig. 2 - Non-locking retractor

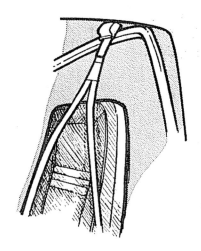

Fig. 4 - Shoulder harness with
emergency-locking retractor

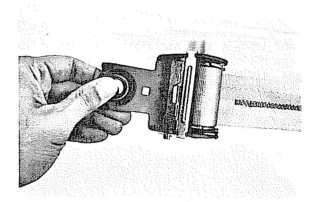

Fig. 3 - Automatic-locking retractor

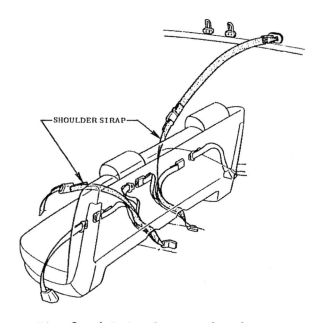

Fig. 5 - 4-Point front outboard
seat belt system

tured for sale in the United States were required
to have seat belts and front seat outboard positions were required to have shoulder straps as
well. The first installations in front outboard
positions were 4-point systems (Fig. 5). These
were superseded by 3-point systems (Fig. 6) in
which the detachable shoulder strap was connected
to the lap belt by means of a terminal fitting on
the shoulder strap that could be attached or detached from the tongue of the lap belt. As of
the start of production of the 1974 Model Year
cars - at the same time that the starter-interlock was introduced - both shoulder strap and lap
belt were required to terminate in a common tongue that engaged the seat belt buckle (Fig. 7).
Shortly thereafter, a modified 3-point system
(the "continuous-loop", or "unibelt" system) was
introduced in which a continuous length of webbing passing through a slot in the tongue comprised the webbing for both lap and upper-torso
portions of the belt system (Fig. 8). The introduction of a new, thinner polyester webbing made

it possible to store the additional webbing required by the continuous-loop system on the retractor without unduly increasing the package
size of the retractor. The polyester webbing
provided strength equal to that of the Nylon webbing. In addition, it provided lower elongation
of the webbing under load, which would result in
decreased excursion of the restrained occupant in
the event of an accident.

SEAT BELT HISTORY - PERFORMANCE REQUIREMENTS

Early model automotive seat belts were based

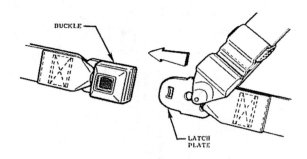

Fig. 6 - Latching arrangement in 3-point seat belt assembly with detachable shoulder strap

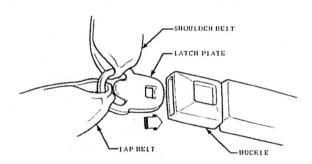

Fig. 7 - Latching arrangement in 3-point seat belt assembly with non-detachable shoulder strap

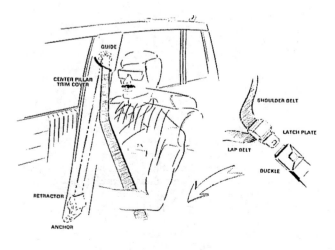

Fig. 8 - "Continuous-loop" seat belt system

on designs for aircraft belts and early standards for these belts issued by government agencies (1)* were influenced by requirements for aircraft belts. In 1961, SAE issued an industry standard, SAE J4, that upgraded the requirements in automotive seat belts as compared with aircraft belts. In 1963 an amendment, SAE J4a, was issued to incorporate additional requirements considered necessary in light of the results of intensive developmental and testing efforts of car manufacturers, seat belt manufacturers, webbing manufacturers, and governmental agencies. Another revision, SAE J4c, issued in 1965 became the basis for a federal standard (2) issued shortly thereafter and the current federal standard, FMVSS 209, issued in 1967 by the National Highway Safety Bureau (the predecessor of the National Highway Traffic Safety Administration).

The performance requirements in the early SAE and the current NHTSA standards appear to have been exceptionally well-conceived, as borne out by the excellent field experience of seat belts in the past twenty five years. Seat belts are not a panacea to prevent all crash fatalities and injuries, but they appear to be able to prevent about half the crash fatalities and two thirds of the serious and severe injuries (3). With further analysis of the substantial number of definitive crash statistics now being acquired in the FARS (4) and NASS (5) data files at NHTSA, it may be found that improved seat belt performance at higher crash velocities maybe not only be desirable but also cost-effective. VW demonstrated in barrier crash tests that an improved belt restraint system incorporating a belt pretensioner would provide occupant protection at 40 mph Barrier Equivalent Velocity (6). The VW system was displayed in 1972 at "Transpo 72" in Washington, D.C. Daimler-Benz has offered an optional advanced supplemental restraint system in Germany since late 1980 having a pretensioner on a 3-point belt restraint in the front passenger seating position (7). Similar systems are now being offered on an optional basis in the United States on Mercedes cars. Field data for cars having these improved systems should soon provide a basis for determining the cost-effectiveness of these systems.

SEAT BELT HISTORY - COMFORT AND CONVENIENCE

Since Model Year 1964, when seat belts began to be factory-installed in all passenger cars, changes in system geometry and components ahve occurred quite rapidly to improve the comfort and convenience of the systems. In 1964 non-locking retractors (Fig. 2) became available to roll up and stow the webbing of the outboard portion of the lapbelt when it was not in use. Non-locking retractors, however, did not provide for automatic adjustment of the lap belt length. Automatic adjustment was first provided in 1965

*Numbers in parentheses designate References at end of paper.

by automatic-locking retractors (ALR) which provided for automatic take-up of paid-out webbing until it was snug on the occupant and automatic lock-up of the webbing at this length, in addition to retracting and stowing the webbing on a covered roller when the seat belt was doffed. With this device on the outboard portion of the seat belt assembly, the inboard portion containing the buckle could be shortened to project only a short distance above the seat. The buckle end could then also be enclosed and supported by a stiffening sleeve which would position the buckle for ready accessibility. Automatic-locking retractors were quite ideally suited for use in lap belts, but they were subject to one criticism. They could "cinch up" - that is, tighten up more than a desirable amount - when the occupant moved downward and compressed the seat when travelling over excessively bumpy roads or terrain. Also, automatic-locking retractors were not useful in upper torso restraints because they did not permit occupant movement.

Emergency-locking retractors (ELR) were developed to overcome the deficiencies of the automatic-locking retractors. They were well-suited for application in shoulder straps since they did not "cinch up" and did provide for free movement of the webbing unless lock-up occurred because of quick evasive movement of the car or an emergency event such as an accident or a panic stop. Emergency-locking retractors may lock up as a result of webbing acceleration or vehicle acceleration. Emergency-locking retractors sensitive to vehicle acceleration (VSR) are basically simple in concept and design (Fig. 9), reliable, and well-suited for applications in passenger cars. (They are not as well-suited for certain specialized applications, such as on the air-suspension seat on a heavy truck where the retractor could be subject to frequent unwanted lock-ups as a result of the bouncing and "chucking" movements of the seat to which it is mounted. A webbing-sensitive ELR is better suited for this application.) Lock-up in webbing-sensitive emergency-locking retractors is initiated by an inertial device which responds to webbing acceleration (Fig. 9). Emergency-locking retractors were applied on a low-production basis in 1967 in the Shelby-American GT 350 and GT 500, and were used in shoulder straps of all new U.S. passenger cars in Model Year 1974 and thereafter.

Most foreign cars sold currently in the United States have emergency-locking retractors that have two inertial sensing modes. One is sensitive to webbing acceleration, and the other to vehicle acceleration. This dual-sensing redundancy has been retained to secure the earlier response and lock-up of the vehicle-acceleration sensing mode, and at the same time permit the user to test the operability of the device by jerking on the webbing to effect lock-up by means of the webbing-acceleration sensing mode.

In 1975, a lap-shoulder belt having only a single emergency-locking retractor appeared in some U.S. passenger cars (Fig. 8). This design

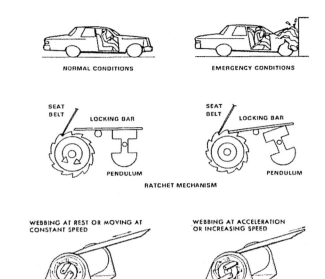

NORMAL CONDITIONS   EMERGENCY CONDITIONS

SEAT BELT   LOCKING BAR   SEAT BELT   LOCKING BAR

PENDULUM   PENDULUM

RATCHET MECHANISM

WEBBING AT REST OR MOVING AT CONSTANT SPEED   WEBBING AT ACCELERATION OR INCREASING SPEED

LOCKING CAMS OUT OF LOCK POSITION   LOCKING CAMS LOCK POSITION

Upper: Vehicle acceleration sensitive mode
Lower: Webbing acceleration sensitive mode

Fig. 9 - Operation of emergency-locking retractors

used a continuous length of webbing for both the lap portion and the upper torso portion of the seat belt assembly. A tongue sliding on the webbing was positioned manually by the user to permit insertion into the buckle. In some versions of this design the tongue incorporated an adjustment feature to permit snug adjustment of the lap-belt portion (Fig. 8). The upper-torso portion was subject to the action of the emergency-locking retractor. This system design provided adequate performance, good stowage when not in use, and lower cost. Surveys conducted for NHTSA have shown that the subjects queried considered this system less convenient to access and don than a two-retractor system. (8)

The "dual-spool" two-retractor system (Fig. 10) was developed to retain the comfort and convenience advantages of the two-retractor system and be more cost-competitive with the "continuous-loop" single-retractor system. This system first appeared in Model Year 1979 in vans and in light trucks in Model Year 1980.

SEAT BELT HISTORY - LEGISLATION AND REGULATIONS

Legislative and regulatory actions have had significant effects on seat belt development and installation. The enactment in 23 states of legislation effective 1 January 1964 requiring installation of seatbelts in front outboard seating positions in all new cars sold in those states prompted all car manufacturers to begin installation of these seat belts in all passenger cars regardless of destination at the start of produc-

Fig. 10 - Dual-spool retractor

tion for Model Year 1964. This action wrought a fundamental change in the seat belt supply scene. Before 1963, most seat selts were produced in relatively small production quantities by a large number of small manufacturers for "after-market" sales. After 1963, most seat belts were produced by "OEM" (original equipment manufacturers) who were generally well-established suppliers of the car manufacturers. Thereafter, seatbelt development benefitted from the broader base of these manufacturers' facilities and funding. Also in 1963, the Congress passed Public Law 88-201 to enable the issuance of standards for seat belts sold or shipped in interstate commerce. The resulting standard (9) was issued by the Department of Commerce in 1965 and a revision (10), incorporating the new requirements of J4c, was issued in 1966.

In 1966, the Congress enacted Public Law 89-593 (11), the "National Highway Traffic Safety Act of 1966", to "reduce traffic accidents and deaths and injuries. . . resulting from traffic accidents". This enabling legislation was implemented by the National Highway Safety Bureau (NHSB), which issued "Initial Federal Motor Vehicle Safety Standards" in 1967. These included MVSS 208, "Seat Belt Installations - Passenger Cars" (12), which designated the number and kind of seat belts to be installed, and MVSS 209, "Seat Belt Assemblies - Passenger Cars, Multi-purpose Vehicles, Trucks, and Buses" (13), which specified the requirements for the seat belts. MVSS 209 became effective in March of 1967 and has continued with minimal changes and interpretations since then. MVSS 208, which included a requirement for seat belt installations in all forward-facing seating positions and shoulder straps in front outboard seating positions, became effective on 1 January 1968 and has had a complicated and colorful history ever since.

Dr. William Haddon, M.D., the original director of the National Highway Safety Bureau, viewed the high incidence of highway traffic deaths and injuries as an "epidemic" and held that the most effective way to bring the "epi-

demic" under control would be by "passive" means of car occupant protection, that is, occupant crash protection requiring no action on the part of the car occupants. The available means for such "passive" or "automatic" protection were air bags and automatic seat belt systems. The direction set by Dr. Haddon was essentially continued by his successors, the Administrators of the National Highway Traffic Safety Administration (the successor agency to the NHSB), until October 1981 when rulemaking was issued during the tenure of Mr. Raymond Peck that amended MVSS 208 to rescind the requirement for automatic restraints for front seat occupants. This rulemaking action was subsequently challenged in the United States Supreme Court and remanded to NHTSA by action of the Court. It is currently undergoing reconsideration. A parallel action by NHTSA is a comprehensive, nationwide effort to increase seat belt usage.

SEAT BELT HISTORY - USAGE RATES

Seat Belt usage rates have been carefully monitored for the past several years in the NHTSA "19-City" program (14). In earlier years, usage rates were determined by numerous local studies which varied considerably in quality and coverage. Many surveys were also conducted to determine the reasons for the low usage rates. Before Model Year 1964, usage of installed lap belts was undoubtedly high because all belt installations resulted from a conscious decision by the car owner to pay extra for the belts as either a retrofit or dealer-installed option. Early statistics from various sources tended to show usage rates after 1964 in the order of 30-35% in the period from 1964 to 1976 (Fig. 11).

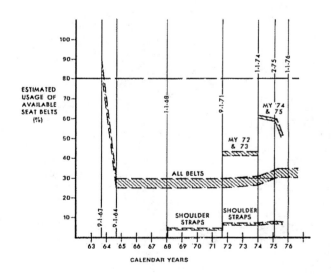

Fig. 11 - Seat belt usage rates from early surveys

Statistics from the "19-City" program indicate a small but significant average increase in usage, from 10.9% to 14.6%, in the period from 1979 to 1983. The range in usage rates from one locality to another is very large, however, and is explained on the basis of the demographics of the human and car populations. In general, for example, usage rates tend to be higher among females, higher socio-economic and better educated groups, users of foreign and small cars, and freeway travellers. Drivers of VW Rabbit cars with automatic restraints are a unique group, with usage rates of about 80% persisting through the years (Fig. 12).

These two special groups taken together, however, constitute only a third of the total vehicle-borne population (Fig. 13). The remaining two-thirds are the target of programs aimed to increase seat belt usage. The intuitively obvious truism that discretionary actions will be done more often when they can be done more easily would support the contention that the perception of an individual in this group as to the comfort and convenience of the seat belt system will significantly influence his dicision to use or not use the seat belt.

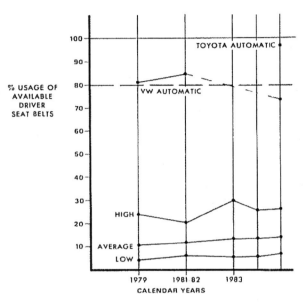

Fig. 12 - Seat belt usage rates from "19-City" program

The possibility for increasing usage rates in defined groups through local programs has been demonstrated by General Motors and others (15). In the General Motors program at the GM Technical Center, the usage rate increased from about 36% to over 70% during the course of the program and maintained a residual rate of about 60% after the termination of the program.

Surveys to analyze seat belt usage rates as affected by public perception of seat belts consistently show that people generally feel that seat belts provide effective protection, but that present seat belt systems are considered uncomfortable and inconvenient. Some investigators have concluded from sophisticated studies (16) that comfort and convenience are not really significant factors in determining an individual's decision to use or not use a seat belt. To a degree this is probably true. A committed user of seat belts (a "hard-core user") will not be deterred by inconvenience or discomfort of the available system. A determined non-user (a "hard-core non-user") will not don the most comfortable and convenient system conceivable.

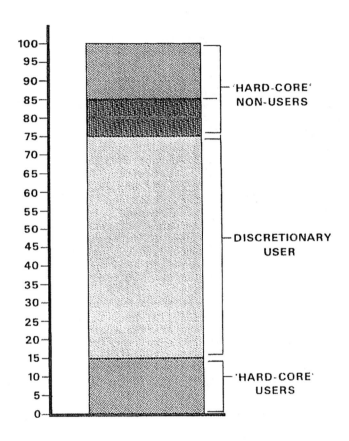

Fig. 13 - Percentage distribution of seat belt user population

Existing statistics provide some measure of usage rates that may be anticipated with certain populations under specified conditions. The "19-City" study, for example, shows that the average usage rate for those nineteen cities is increasing and is now approaching 15% (Fig 12). The range for the individual cities in the study extends from a low of 5-6% for Fargo, North Dakota to a high of about 25% for Seattle. The GM study showed that a well-conceived, well-executed program for a specific, well-defined group could achieve a usage rate of about 70%. A continuing usage rate of about 80% of the VW automatic restraints demonstrates that a satisfactorily high usage rate can be achieved with automatic belt restraints with car owners who have made a conscious decision to order and pay extra for these systems. Experience in foreign countries having mandatory belt usage laws shows that usage rates of 80-85% are achieved with reasonable enforcement of the laws and nominal fines.

SEAT BELT HISTORY - LITIGATION

Litigation involving seat belts is increasing and can be expected to continue to increase. The seat belt involvement in lawsuits may arise from either the use or non-use of the available seat belts.

Injured plaintiffs who were wearing seat belts may have had the erroneous perception that the seat belt would be a panacea to protect them from all harm in all events. They may have had unrealistic expectations as to the degree of protection at excessive speeds, or in certain crash modes (e.g., certain side or rear impacts), or in the event of excessive intrusion into the occupant space. Realistic appraisals of protective capabilities of seat belts in many typical crash modes have been documented in reports presented before the SAE, AAAM, and other technical and professional groups (17) (18).

Injured plaintiffs may allege that they were wearing seat belts that were defective in some way and did not afford the protection that could reasonably be expected. Such allegations are addressed to determine whether the seat belt was indeed used, was operational, and did afford reasonable protection.

Non-use of seat belts is becoming increasingly important as an issue in civil litigation (19). Recent studies undertaken for NHTSA (20) (21) have held that the prudent person can increasingly be expected to avail himself of the protection afforded by the seat belt available in his car and that he exhibits some degree of negligence in not doing so.

SEAT BELT HISTORY - SUMMARY

The chronology of some of the landmark events in the history of seat belts is summarized graphically in Figure 14. It is apparent from the graph that most of the significant developments in seat belts have occurred in the last twenty years in spite of the recognition of the utility and effectiveness of seat belts nearly one hundred years ago.

RECENT DEVELOPMENTS IN SEAT BELT HARDWARE

Developmental efforts with seat belt hardware are proceeding continuously to improve performance, reliability, comfort and convenience, universality in application, and amenability to packaging in the vehicle; and to decrease complexity and cost.

With the continued downsizing of cars and increasing numbers of cars with front wheel drives, seat belt buckles are being designed to sustain the higher loads that may be experienced by the restraint systems in accidents involving these cars. In addition, with the increase in bucket seats in downsized cars, buckle designs and buckle installation arrangements are undergoing intensive review to permit design modifications that will improve convenience in donning and doffing the seat belt in these installations.

The belt pretensioners introduced in production cars by Mercedes (22) may very well prove to provide a new dimension in occupant protection against fatalities and severe injuries in higher-speed accidents.

To reduce weight, complexity, and cost of retractors in two-retractor systems and to increase comfort, convenience, and reliability, the "dual-spool" retractor has been broadly applied in vans and light trucks. An additional feature was added to a "dual-spool" retractor for application on the 1983 Corvette (Fig. 15). To accommodate dedicated drivers who appreciate the secure feeling of a snug lap belt, the emergency-locking retractor on the lap belt portion of the two-retractor, 3-point system incorporated a manual locking mechanism that provides the continuously locked retractor. This same feature could be incorporated into the front passenger location of more conventional cars to provide a convenient, positive, continuously-locked condition in a lap belt securing a child restraint.

Developmental effort in belt restraint systems in the future, when seat belt usage increases - by whatever means - toward more acceptable rates, will continue to be driven by the same objectives as those currently operative: increased performance and reliability, comfort and convenience, and reduced weight, size, complexity, and cost.

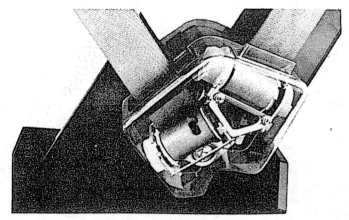

Fig. 15 - Corvette "dual-spool" retractor assembly - Emergency-locking lap belt retractor has manual optional locking provision

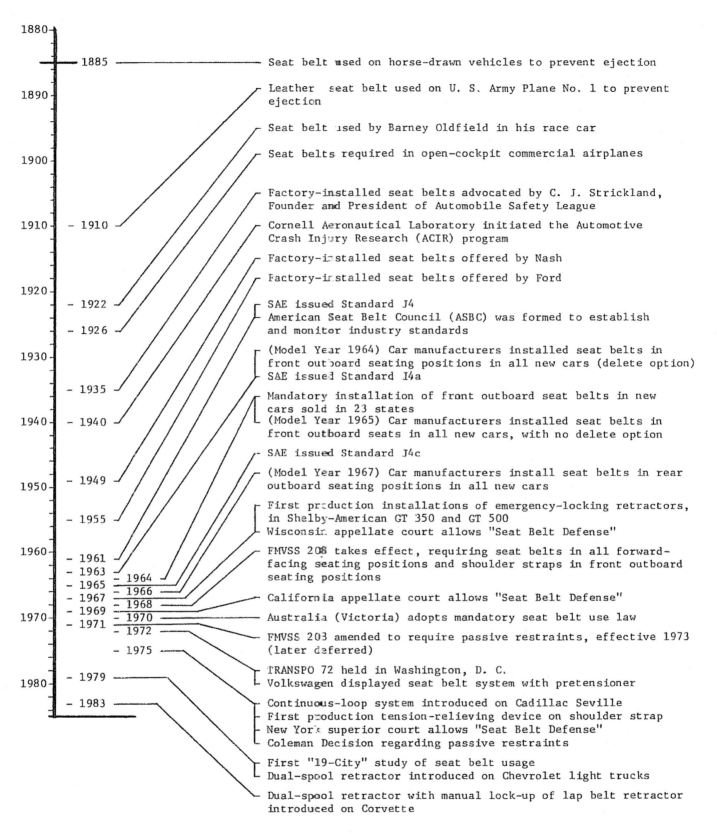

Fig. 14 - Seat belt history

ACKNOWLEDGEMENT

The author acknowledges with thanks the support provided by the American Seat Belt Council in the preparation of this report, and the cooperation of ASBC member companies in providing hardware for the illustrations.

REFERENCES

1. GSA - JJB - 185a

2. "Seat Belts for Use in Motor Vehicles", 15 CFR 9, 30 FR 5132 (July 1965)

3. D.A. Huelke, et al, "Effectiveness of Current and Future Restraint Systems in Fatal and Serious Injury Automobile Crashes" - SAE 790323 (1979)

4. FATAL ACCIDENT REPORTING SYSTEM, administered by National Center for Statistics and Analysis, National Highway Traffic Safety Administration, U.S. Department of Transportation, Washington, D.C. 20590

5. NATIONAL ACCIDENT SAMPLING SYSTEM, administered by National Center for Statistics and Analysis, National Highway Traffic Safety Administration, U.S. Department of Transportation, Washington, D.C. 20590

6. H.P. Willumeit, "Passive Preloaded Energy-Absorbing Seat Belt System", SAE 720433 (1972)

7. W. Reidelbach and H.J. Scholz, "Advanced Restraint System Concepts", SAE 790321 (1979)

8. "Evaluation of the Comfort and Convenience of Safety Belt Systems in 1980 and 1981 Model Vehicles", DOT HS-805 860, (March 1981) (available through NTIS)

9. See 2 above

10. "Seat Belts for Use in Motor Vehicles", 15 CFR 9, 31 FR 11528 (August 1966)

11. 15 USC 1391 et seq., National Traffic and Motor Vehicle Safety Act

12. 49 CFR 571.208, Standard No. 208; Occupant Crash Protection

13. 49 CFR 571.209, Standard No. 209; Seat Belt Assemblies

14. "Restraint System Usage in the Traffic Population", NHTSA Contract DT-NH 22-82-C-07126

15. T.D. Horne and C.T. Terry, "Seat Belt Sweepstakes - An Incentive Program", SAE 830474 (1983)

16. T.J. Kuechenmeister, "A Comparative Analysis of Factors Impacting on Seat Belt Use", SAE 790687 (1979)

17. D.J. Dalmotas and P.M. Keyl, "An Investigation into the Level of Protection Afforded to Fully Restrained Passenger Vehicle Occupants", 22nd Proceedings, American Association for Automotive Medicine (1978)

18. D.J. Dalmotas, "Mechanisms of Injuries to Vehicle Occupants Restrained by Three-Point Seat Belts", SAE 801311 (1980)

19. J.D. States and T.G. Smith, "Use of the Safety Belt Defense - The New York Experience", 27th Proceedings, American Association for Automotive Medicine (1983)

20. Unpublished NHTSA Staff Report, "The Seat Belt Defense - The Education of a Reasonable Person", Summer 1981

21. "Non-Use of Motor Vehicle Safety Belts as an Issue in Civil Litigation", DOT HS-806 443, (August 1983), (available through NTIS)

22. J.E. Mitzkus and H. Eyrainer, "Three-point Belt Improvements for Increased Occupant Protection", SAE 840395 (1984)

Safety Belt Development –

S. Sano
Honda Research & Development

Last summer, NHTSA issued a rule under MVSS 208 that created the regulation to equip automobiles with passive restraints under a phase-in schedule.

Beginning with model year 1987, 10 percent of all automobiles sold will be equipped with passive restraints. That percentage will increase unless enough States representing more than two-thirds of the entire U.S. population pass legislation mandating seatbelt use by drivers and passengers.

I believe the ideal situation would be the use of manual 3-point seatbelts in conjunction with the adoption of seatbelt use laws by all the States. It is my hope that effective regulation will be in place, as I feel this is the most effective way of providing adequate protection to automobile occupants, rather than by mandatory regulation requiring automatic restraints beginning with the 1987 model year.

We are, of course, preparing to comply with the passive restraint requirement and are considering an automatic seatbelt. Before considering what type of system should be used, at this point I would like to discuss the requirements of a system and the utilization of it.

First, I want to discuss the occupant protection performance, especially the performance of the system in restraining a driver or occupant at the moment of impact. Discussions for the passive belt system have involved both the 2- or 3-point types. However, the important issue is not whether the system is 2-point or 3-point, but how efficiently the automatic system works for a particular car; that is, how efficiently any system can be integrated with other crashworthy characteristics of an automobile, such as crush characteristics, impact duration, and the survivability space available for the occupants.

Another consideration surrounds the system performance during a postcrash emergency escape, either by the occupants or by outside rescue personnel. This relates to the selection of a detachable or nondetachable system and to the related issues of decreased seatbelt usage by user tampering or disconnection of the system. I do not think that whether the system is detachable or non-detachable will have a lasting influence on the restraint usage rate, especially in the long term. For example, during model year 1974, there was a starter-interlock device that was adopted for improving seatbelt usage rates. This rate was raised for a short period, but the interlock system was subsequently defeated, and there does not appear to be any significant difference in the usage rate of seatbelts between the 1974 models and any other year model.

It is my opinion that adopting a nondetachable system for the purpose of improving the restraint usage rates might have the same results as the earlier use of the interlock. Furthermore, it is quite possible that users would subvert the restraint system by removing the belts, which would require a subsequent replacement by any future vehicle owners that would want the protection of the system. The most effective way to improve the rate of restraint system usage is to adopt legislation making seatbelt or restraint use mandatory, which would certainly make the question of whether to use the detachable or nondetachable system no longer a point of discussion.

When considering most emergency escapes in accident situations or rescue of an occupant by outside personnel, it is obvious the detachable system is far superior to the nondetachable system. Because of these considerations, I am proceeding toward the adoption of a detachable system as my passive restraint system of choice.

There are two considerations affecting acceptance by consumers. One is that the car equipped with a passive restraint system will be acceptable from a cost/benefit standpoint on the open marketplace, and the second is that the system will be used in an unaltered configuration. In the initial stages of the phase-in program, there will be a mixture of cars with passive restraints and cars with the current active seatbelts on the market. This will allow consumers to choose from among them. During the phase-in period, especially before most of the cars being sold are equipped with the passive system, we can assume that those who decide to purchase the cars with the passive belts will have the intention and desire to use that system.

If a consumer, given the choice, decides to purchase a car with the passive restraint, he has considered the benefits and determined they outweigh any detriments such as increased cost, perceived difficulty in entering and exiting the vehicle, or what we may call the spiderweb appearance of the vehicle's interior.

The degree of perceived benefit of the automatic system will depend on the mind-set of the purchaser. People who presently use seatbelts regularly and like the manual system will probably not care for an automatic system; while people who don't like the inconvenience of manual belts but use them because of the perceived safety benefits will most likely prefer an automatic system. Finally, people who are uninformed or simply don't care about the added safety of seatbelts probably won't like the automatic restraint system and are most likely the ones who will take the opportunity to disengage or defeat the system.

As a consequence, we feel the focus during the phase-in period will be on those consumers who do utilize seatbelts, although they feel seatbelts are difficult or inconvenient to use.

Next, I would like to discuss the motorized system. Certainly a motorized system will improve the application and removal situation when a vehicle occupant enters or exits the automobile; however, it does not eliminate all the problems of putting on and removing a belt. Moreover, certain discomforts will be generated for some occupants, such as the belt rubbing against a user's face or neck since the belt moves automatically regardless of user's intention.

A motorized unit will increase the cost substantially when compared to the cost of the belt itself, either under

the present system or a nonmotorized passive restraint system. Additionally, there is no doubt a system that has simple component parts will have superior reliability when compared to the complexity of the motorized system.

Again it is very difficult to evaluate the cost/benefit of the passive restraint systems. Because no system is a perfect system in every aspect, one approach could be to evaluate various manufacturers' systems over a period of time.

We began to develop the automatic seatbelt system in the early 1970's and exhibited a motorized version 11 years ago at the ESV meeting held here in England. We abandoned that system after an evaluation program because of concern about injury-causing potential from the positioning of the mechanical arm.

Since that time, we have tried many different combinations, such as a motorized arm operating on the door side, a motorized arm operating from the mid-portion of the car interior, a system without a motorized arm, one with a convenience hook, another system with an ELR set in the middle of the automobile interior as well as an ELR on the door, 2-point systems, 3-point systems, and every combination of all the elements mentioned.

From our extensive experience, we concluded no perfect system exists that is acceptable and satisfies every single consideration. The 2-point restraint system and the motorized arm system are not favored by people accustomed to the traditional manual belts, since space is further restricted by a knee bolster and is not considered by those persons as a benefit. System users are influenced apparently more by the discomforts caused by the passive restraint system than by any improvement of the putting on or removal of the system.

Therefore, we consider the simple, nonmotorized 3-point passive restraint system that is similar to the current manual belt system to be the most adequate during the phase-in period.

However, we are well aware that the situation concerning development of automatic restraints keeps changing. We are keeping an open mind as to appropriate forms of automatic restraints that are most desirable from the standpoint of safety, efficiency, and comfort, and will change the system any time substantial and significant advancements are made in accordance with our ongoing research programs.

Figure 4. Complementary Interior

Automatic Belt Restraint Systems for Motor Vehicle Occupants ⎯⎯⎯⎯⎯

H. George Johannessen, P.E.
OmniSafe, Inc.

Abstract

Following early U.S. Government regulatory action requiring passive occupant crash protection for vehicle occupants covered by Motor Vehicle Safety Standard 208, intensive development began on passive seatbelts as promising alternatives to the inflatable systems being considered. The general types of passive seatbelt systems are identified. Passive systems that have been offered to date as optional installations in production car lines are described. Design considerations are discussed. Recent usage rates of installed passive systems are compared with manual belts.

Introduction

The U.S. Congress enacted the Highway Traffic Safety Act of 1966 to address the problem of traffic fatalities and injuries on American roadways. The National Highway Safety Bureau was quickly established to carry out the provisions of the new law. (The National Highway Safety Bureau later became the National Highway Traffic Safety Administration.) The late Dr. William Haddon, Jr., was appointed to head the new agency. He was a medical doctor with a long and distinguished role in public health and traffic safety in the State of New York. He viewed traffic fatalities and injuries as a public health problem of epidemic proportions that could best be overcome by passive means—that is, by means requiring no overt action on the part of the parties affected. He cited past examples of successful passive solutions to major public hazards, such as inoculation of entire populations to overcome dread diseases, pasteurization of milk, and use of automatic sprinkler systems to control fires.

Passive Systems for Passenger Cars

In November 1970, the U.S. Federal Motor Vehicle Safety Standard 208, which addresses vehicle occupant crash protection, was amended to incorporate the original requirements for passive restraints in the front seating positions of all passenger cars manufactured on or after the first day of July 1973 for sale in the U.S. car market. Challenges to these original requirements and original effective date resulted in a series of subsequent actions from 1970 to the present by legislative and judicial branches of the U.S. Government, in addition to the responsible regulatory agency, NHTSA, which reconsidered and revised requirements and effective installation schedules for passive occupant protection. The most recent action calls for passive occupant protection in front outboard seating positions on a phased-in schedule with installations in a minimum of 10 percent of the manufacturers' car build after September 1, 1986, 25

921

percent after September 1, 1987, 40 percent after September 1, 1988, and 100 percent after September 1, 1989. The passive requirement will be rescinded if States representing two-thirds of the Nation's population enact mandatory seatbelt usage laws before April 1, 1989, that will be in effect and enforced by September 1989.

The passive system for vehicle occupant protection that received immediate attention following the issuance of the original rulemaking in 1970 was the inflatable restraint, or airbag. Alternative noninflatable means, however, were considered as well. Many categories of such noninflatable passive candidate systems were identified, as shown in Table 1. These included transparent shields, deployable nets and blankets, cushions, restraining arms and barriers, integrated seat systems, and seatbelt systems.

Table 1. Noninflatable automatic occupant restraints

- Transparent Shields
- Deployable Nets
- Deployable Blankets
- Integrated Seat Systems
- Restraining Arms
- Cushions
- Seatbelt Systems

Developmental Passive Seatbelts

Of these various categories, passive seatbelts showed the greatest promise by far. Passive seatbelts were developed in many variations. Several distinctive classes may be identified. These are listed in Table 2. They may be 2-point, 3-point, or 4-point systems based on the number of load-bearing connections to vehicle structure. They may be all mechanical or may be all or partially motorized. They may have retractors located in either inboard or outboard locations and fixed anchors located either inboard or outboard. Some may require knee bolsters to control lower body movement. All must have some provision to permit disengaging the seatbelt after an accident. Prototypes of all these classes of systems have been developed and installed in production cars.

Table 2. Typical automatic seatbelt systems

2-POINT — Passive Shoulder Belt Only

3-POINT — Passive Lap and Shoulder Belts

4-POINT – Separate Passive Lap and Shoulder Belts

- Mechanical and Motorized Models in All Types

Installations in Production Car Lines

Some of the more promising developmental designs have been offered to the public in production cars. These are described in Table 3. The first production model to appear was a 2-point system, installed as an extra-cost option in a Volkswagen Rabbit in model year 1975 and subsequent years. Figure 1 shows the conceptual design of the Rabbit installation. This system includes a 2-point

Table 3. Production automatic seatbelt systems

CAR LINE	MODEL YEAR INTRODUCED	TYPE
VW RABBIT	1975	2-POINT
GM CHEVETTE	1978 ½	2-POINT
GM CHEVETTE	1980	3-POINT
TOYOTA	1981	2-POINT — MOTORIZED

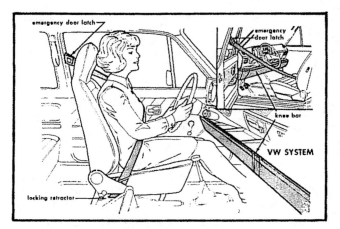

Figure 1. First automatic seatbelt system in production car line

upper torso belt with the retractor inboard and the fixed anchor attached outboard on the door. The door connection incorporates a buckle for emergency disconnection. The buckle has an electrical interlock with the starter system to insure that the buckle is connected in normal driving use. The lower body movement is controlled by a knee bolster and a specially designed seat pan. Figure 2 shows the appearance of the VW system as installed in the Rabbit.

The first production installation in a car produced in the United States appeared in the Chevette manufactured by the Chevrolet Division of General Motors in model year 1978. It was quite similar to the VW installation in the Rabbit, except that it used a seat of conventional design and included a manual lapbelt. Figure 3 shows the conceptual design, with a passive 2-point shoulder belt, a knee bolster, and a lapbelt.

Figure 2. VW automatic seatbelt installation

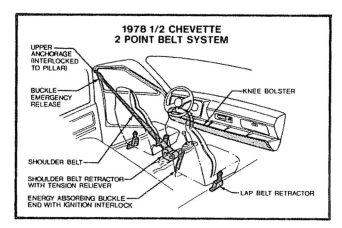

Figure 3. First automatic seatbelt system in U.S. domestic production car line

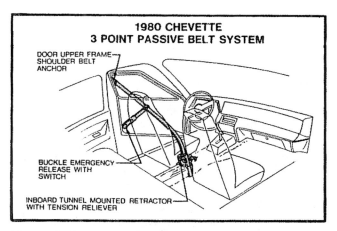

Figure 4. Second generation automatic seatbelt system in U.S. domestic production car line

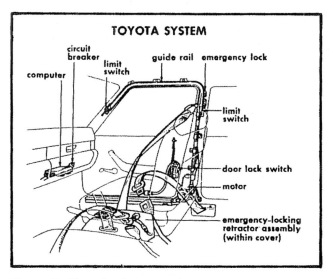

Figure 5. Motorized automatic system

The next production installation appeared in the Chevette in model year 1980. This was a 3-point system with an inboard retractor and two fixed anchor points on the door. Figure 4 shows the conceptual design of this system. A buckle on the lower anchor on the door was provided for emergency release. This buckle also incorporated a switch to provide an electrical connection with a buzzer to insure that the buckle was engaged during normal use.

The most recent automatic system has been installed in the Toyota Cressida since model year 1981. It is a 2-point system with an inboard emergency-locking retractor and an outboard anchor location on the B-pillar. Figure 5 shows the conceptual design of the Toyota system. A manual belt is included to control lower body movement. The lower portion of the instrument panel provides a knee bolster as well. The motorized outboard anchor point is routed along the roof rail and the upper portion of the A-pillar. This action moves the shoulder belt forward out of the way for easy occupant ingress into the vehicle, and then returns it to don the shoulder belt on the occupant after he is seated and the door is closed. The

traverse of the belt has been carefully timed to move the belt quickly, yet not so quickly as to startle the occupant. Figure 6 shows the relative movement and location of the motorized shoulder belt.

Design Considerations for Automatic Seatbelt Systems

The general design considerations for all seatbelts are shown in Table 4. The list includes performance in accidents and evasive maneuvers, simplicity in design and operation, reliability, durability, vulnerability to damage during normal use or in accidents that could render them inoperable, comfort during use, convenience in use, packaging size and shape for accommodation in the vehicle, esthetics, and cost. Automatic seatbelts have the added requirement for automatic donning and doffing.

In any given system design, tradeoffs inevitably are encountered such that optimization of one variable is

Figure 6. Toyota automatic shoulder belt system

accompanied by some sacrifice in optimization of another. Some obvious examples are (1) the possible degradation of performance in emergencies for the sake of increased comfort in normal usage by means of a tension-relieving device on the shoulder belt, and (2) the possible decrease in simplicity and reliability for the sake of increased convenience and improved stowage through the use of retractors.

Table 4. Design considerations

- Performance in Emergency Events
- Reliability
- Durability
- Vulnerability to Damage
- Simplicity in Operation
- Comfort
- Convenience
- Packaging
- Esthetics
- Cost

Usage and Performance

The usage rates and performance of these automatic seatbelts in the field have been very satisfactory. Figure 7 shows the usage rates for automatic systems as compared with the usage rates for manual belts, representing data obtained in the ongoing program sponsored by the National Highway Traffic Safety Administration in which usage rates are monitored in 19 cities across the United States. Average driver usage of manual belts in

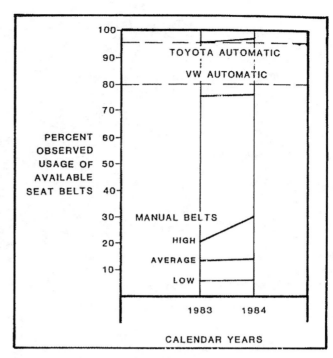

Figure 7. Recent seatbelt usage rates in U.S.A.

these 19 cities increased from 13.3 percent in 1983 to 14.4 percent in 1984 and 15.3 percent in the last quarter of 1984. The upward trend was due mainly to the intensive efforts of NHTSA and others in the safety community to increase seatbelt usage in the United States. Usage rates varied appreciably from one city to another in the 19-city study, with a low rate in 1984 of 7.1 percent and the highest rate of 30.1 percent. The principal predictors of usage, in the absence of mandatory use laws, are age and sex of the driver (with the greatest usage in the age range of 25 to 49, and women consistently being more frequent users than men); size of car (with greater usage in smaller cars); foreign versus domestic U.S. cars (with drivers of foreign cars more than two times as likely to be wearing seatbelts); and the age of the car (with seatbelts more likely to be in use in later model cars).

The usage rate for the VW system has persisted in the range of 80 percent through the years since 1975, when it was introduced, with observed usage rates of 75 percent in 1983 and 76 percent in 1984. The Toyota Cressida system has done even better, with usage rates of more than 95 percent observed in the NHTSA 19-city program. The Chevette system's usage rates are in the same general range as the VW, but somewhat lower.

Performance data for these systems in accidents are limited because of the relatively short time they have been exposed on the roadways and the relatively small number involved in accidents. The data acquired to the present through the National Accident Sampling System operated in the United States by NHTSA, referred to as the NASS data, indicate the performance of these automatic systems in the real world is generally comparable to the performance of manual systems.

The Future of Automatic Seatbelts

The key to the need for automatic belts is the usage rate of seatbelts. In the many countries and jurisdictions around the world having mandatory seatbelt use laws, starting with Australia in 1970 and including England more recently, usage rates of 75 to 95 percent are achieved with manual seatbelt systems when minimal—but continuing—enforcement is observed and nominal fines are imposed. In the State of New York, where a mandatory usage law became fully effective on the first day of January of this year, the usage rate is becoming acceptably high. Seatbelt performance in New York, in terms of reduction in fatalities and serious injuries, has followed the patterns observed in other countries having successful mandatory use laws.

Automatic seatbelts are not inherently better performers than manual belts, but they do appear to induce higher usage rates when installed as extra-cost options. The public acceptance of automatic seatbelts when installed as standard equipment remains to be seen. The degree of success in enacting and enforcing mandatory seatbelt usage laws in the United States will determine whether automatic seatbelt systems will be required in the long term.

Seat-Integrated Safety Belt

Heinz P. Cremer
Keiper Recaro GmbH & Co

ABSTRACT

In the case of fronttal or rear collision, the vehicle seat and safety belt act as a retaining system that is supposed to protect the person and prevent injury as much as possible. A double shoulder belt (harness belt) was integrated into the seat as a 4-point belt in order to examine the possibilities of improving personal protection. Due to the transfer of the test requirements for the safety belt to the seat with integrated safety belt, the load level in the direction of motion increases considerably. The normal test conditions were increased by using a 95% dummy rather than a 50% dummy. According to these requirements, a seat back with seat back adjuster was designed, built, and tested.

The test showed that by proper deformation of the seat back, the load on the person is reduced and that the seat with integrated 4-point belt provides an improvement of the retaining system for the frontal as well as the rear crash.

Seat with Double Shoulder Belt (Harness Belt) - Various examinations showed that a belt system with a double shoulder belt, i.e. a 4-point belt, provides better protection than the currently used diagonal belt.

The goal of this development was to examine:

- whether there are further advantages when a 4-point belt is completely integrated into the vehicle seat.
- whether an acceptable seat back structure with seat back adjuster can be built, that can take the increased load.
- whether the seat back structure can be designed in such a way that the energy will be changed by proper shaping to achieve a lower load level than in other retaining systems with comparable testing conditions.

Not included in this report is the highly resistant seat track and swivel mechanism necessary for belt integration.

Seat integreated safety belt

Since the maximum protection of the person is at hand, it was accepted that:
- up to this day the 4-point belt is only used in motor racing sport vehicles, i.e. not common in assembly line production.
- the request for a one-hand buckling operation of the belt has not been met satisfactorily yet.

Generally, it can be stated that the 4-point belt has proven itself under extreme conditions in the motor sport for years.

The issue was not to determine which 4-point belt system is the most appropriate, since this was already done elsewhere.

<u>Driving and Sitting Comfort of Seat with Belt</u> - In order to provide drivers of various heights with good vision from the vehicle, the vehicle seats are equipped with a height adjuster. The currently used seat height adjusters primarily adjust the entire seat in height.

For the seat-integrated safety belt system, however, it is of advantage to adjust the height of the seat cushion next to the stationary back.

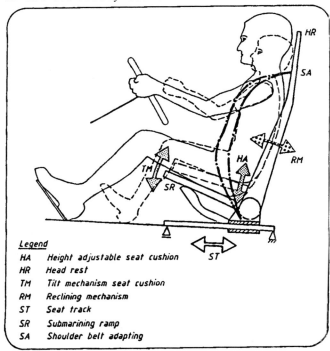

Legend

HA	Height adjustable seat cushion
HR	Head rest
TM	Tilt mechanism seat cushion
RM	Reclining mechanism
ST	Seat track
SR	Submarining ramp
SA	Shoulder belt adapting

Height adjustable seat cushion

The height-adjustable seat cushion achieves:
- the actual purpose of adjusting the visual point of people of various heights.
- an adjustment of the shoulders of people of various heights to the level where the seat belt emerges from the seat back.
- an improved positioning of the head in relationship to the headrest.

- an adjusted location of the seat belt lock, since it is appropriately attached to the upper part of the seat track.

The seat-integrated 4-point safety belt achieves:
- improved wearing comfort provided by better belt location for people of various heights.
- improved wearing comfort for men and women provided by symmetrical belt location.
- decreased and more even pull-out and retraction force of the belt due to smaller change in direction and shorter belt.
- improved comfort when putting on safety belt, since the parts of the belt are always in the same position and always easily reachable, independent from the position of the seat.
- symmetrical and, therefore, natural movement of putting the belt on, however, with both hands.
- improved ease of getting in and out of the vehicle, especially for the rear passengers in 2-door automobiles.
- reduction in the number of anchor points for seat and safety belt, less robot mounting.
- new design of the floor pan for the different force introduction only by the seat.
- more freedom of design in the area of the B-column.
- no problem when used in vehicles without B-column; convertible, sliding doors, etc.

<u>Better Crash Record is Provided by Seat with Belt</u> - Seat and safety belt are the systems that support and hold the person in an accident. Even though the position of the person in a vehicle is determined by the position of the seat, the safety belt in most cases is still attached to the car body. Increasingly, the safety belt lock is attached to the seat or moved back and forth with the seat movement in order to reach a better position of the person, seat and safety belt at this point.

At an accident, the "slack", i.e. the distance a body can move freely before he is caught by the restraining system, seat or safety belt, is of great importance. The greater the slack, the greater the impact on the body. See in this aspect the "Crash Dynamic" illustration on the following page.

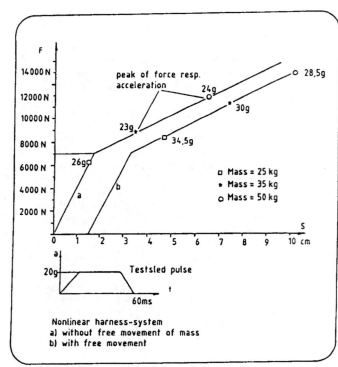

Crash Dynamic

Today, floor pans are designed in a way that they include:
- anchoring points for the seat and
- anchoring points for the safety belt.

For the seat with the integrated safety belt the anchoring points can be reduced to the anchor points for the seat. This, however, due to the greater force, requires a new concept of the floor pan in the area of the front seats.

The following features are characteristics for a seat with a 4-point integrated safety belt:

FRONTAL CRASH
- symmetrical force on the body, therefore, no torsion.
- good position of the shoulder belts for people of different heights.
- good positioning of the hip belt, since the belt lock as well as the belt anchoring point is attached to the upper part of the seat track.
- less surface pressure on the body due to two shoulder belts.
- shorter people sit higher due to the height-adjustable seat cushion, so that the lesser weight acts on a larger lever
- the slack is less due to the closed position of the body, seat and safety belt and can be minimized even more with a belt tightener.
- less elongation of the belt strap due to stretching, since the total length of the belt is shorter due to the integration of the belt roller in the seat back.
- less elongation of the belt strap due to film spool effect, since a lesser belt

length has to be on the spool.
- energy change by the seat back is greater than by the B-column.
- earlier retention of the body by the forward-shaped back during rebound.

REAR CRASH
- shoulder belts resist unwanted upward movement of the body.
- timely retention of the body during rebound.

DIAGONAL OR SIDE IMPACT
- the body is well retained by the closed retaining system seat - safety belt.

The Highly Load-Resistant Seat Back - A larger force acts upon the seat structure when the shoulder belts are attached to the upper edge of the seat back as compared to the force created only by the rear crash (when the belt is attached to the vehicle body). For this reason the construction of the seat structure is to be designed in such a way that as few structural elements as possible are located in the high force flux. This, however, should not impair the functions that belong to a modern seat. Here too, the height-adjustable seat cushion is advantageous, since it is outside of the force flux, that comes from the seat back in a front or rear crash. The seat cushion just has to have a submarining ramp that, together with the hip belt, prevents the plunging of the person. This way the force flux in this concept goes through the seat back, the seat back adjuster and the seat track directly in the vehicle floor. See illustration "Flux of force."

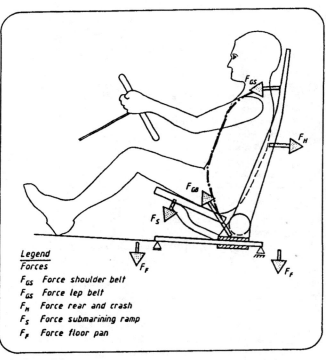

Legend
Forces
F_{GS} Force shoulder belt
F_{GS} Force lep belt
F_H Force rear and crash
F_S Force submarining ramp
F_F Force floor pan

Flux of force

Seat Back Adjuster - The seat back also includes a seat back adjuster that is suitable for the task and the high force. Principally, an adjuster based on a gear (gear type reclining mechanism) is suitable for the task, since the flux of force is not interrupted, even during adjustment. This requirement of being locked constantly seems to be inevitable for the belt that is attached to the seat back and is fulfilled by the TAUMEL seat back adjuster (TAUMEL reclining mechanism).

Calculation of the Seat Back Structure - In the seat back structure with integrated safety belt, the force passes through the belt or the back and exits through the seat back adjuster into the seat track. The shorter the distance of the force, the better the problem solution. In order to find an optimal structure for the seat back with integrated safety belt, calculation procedures for stability calculations, FEM and a crash simulation were used. The symmetrical force on the structure due to the double shoulder belt is advantageous. However, the non-symmetrical force during a diagonal impact and the side stability during the side impact were not neglected.

For series production, the structure of the seat has to be adapted to the deforming characteristics and the delay impulse of the floor pan as well as the available area of the respective vehicle; this achieves a minimum force on the body, while taking advantage of the forward movement of the person in the vehicle.

Vehicle, seat, and person are a dynamic system that can lead to optimal results only when all components are taken into consideration.

For the crash simulation an 8-mass dummy model was used. Random impulses can be entered, as a normal signal or acceleration impulses of floor pans of certain vehicles, as well as the retaining outlines of the seat back and safety belt. See illustration in the next column.

According to the entered force, the following can be calculated:
  - forces of contact to the seat (max. 6)
    - - force on the submarining ramp
  - the belt forces
  - the acceleration of head, chest, and pelvis
  - base forces at the floor pan
The simulation can be shown on the screen in motion so that the deformation of the structural parts as well as the movement of the dummy can be observed. See Crash Simulation illustration in the next column.

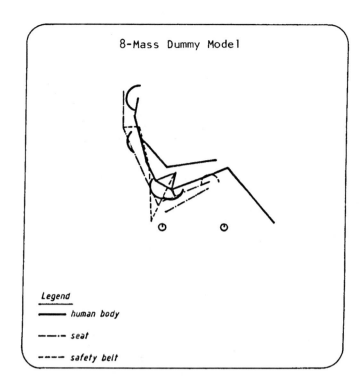

8-Mass Dummy Model

Legend
——— human body
—·— seat
- - - - safety belt

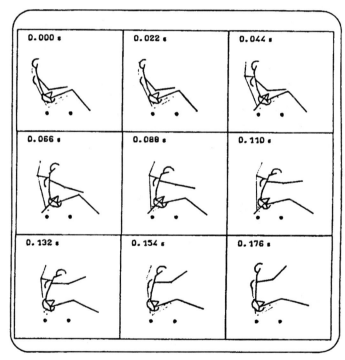

0.000 s	0.022 s	0.044 s
0.066 s	0.088 s	0.110 s
0.132 s	0.154 s	0.176 s

Crash Simulation

Test Results - According to the previous mentioned criteria and requirements, proto-types of the seat back structures with seat back adjusters of the TAUMEL type and seat-integrated safety belt were built and tested. The seat structure was mounted firmly onto the test fixture.

A good conformity between
- theoretical interpretation
- static force and
- dynamic force in the crash test
were achieved.

The following illustrations show the evaluation of high speed films of the crash tests. The tests were conducted with a 95% male dummy (Hybrid II) and with the crash impulses shown in the illustration, i.e. the current legal requirements were surpassed. This allows for possible future developments, e.g. testing with the 95% dummy. The diagrams show the maximum plastic deformation and the especially advantageous reconversion during rebound. The seat backs were evenly deformed and showed controlled deforming on the pull and pull side. The dummy showed no uncontrolled movements during the tests and sat in a normal position after the test. In no test did a seat back or seat back adjuster fail.

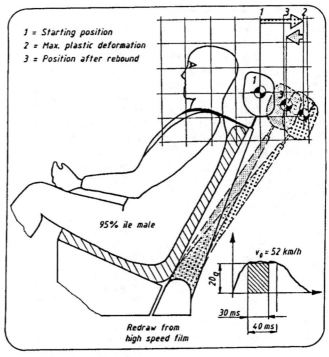

Rearend crash and seat structure

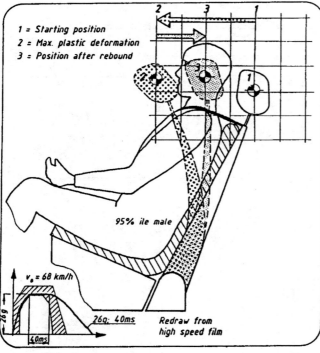

Frontal crash and seat structure

The improved protection of the person becomes clear when observing the process of the deformation of the seat back. From the original position 1) the seat back is deformed forward or backward by the belts or the upper body. At the end of this force the body is moved in the other direction, so that the force is taken off the seat back. The elastic deformation disappears and the seat back takes the position of the maximum plastic deformation 2). In the following instant, the seat back is being deformed into the final position 3) by the remaining energy from the rebound. This timely retention of the body during rebound is only possible with the seat-integrated 4-point safety belt.

The weight of the seat back with seat back adjuster is 25% to 50% higher than a conventional seat back, whereby, however, a 300% to 500% higher load capacitance is achieved. It is to be expected that an acceptable increase in the total weight will be necessary for a uniform layout of seat, safety belt, floor pan and B-column, whereby, however, simultaneously a considerably better protection of the person is achieved by the seat-integrated safety belt. Should only the currently required test conditions be used, the above mentioned weight can be reduced.

Comparison Values - For comparison, the following were tested:
- seat with integrated 3-point belt
- 3-point belt with shoulder belt
  attached to B-column

The illustration "Different seat belt systems" shows the measured results. A distinct advantage for the seat-integrated 4-point safety belt can be recognized.

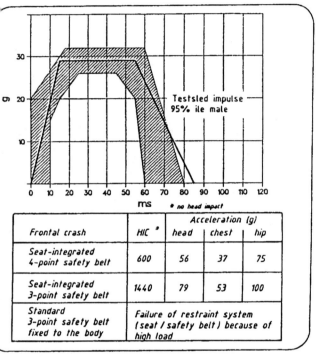

Frontal crash	HIC [a]	Acceleration (g)		
		head	chest	hip
Seat-integrated 4-point safety belt	600	56	37	75
Seat-integrated 3-point safety belt	1440	79	53	100
Standard 3-point safety belt fixed to the body	Failure of restraint system (seat / safety belt) because of high load			

Different seat belt systems

Historical Review of Automatic Seat Belt Restraint Systems

H. George Johannessen
OmniSafe, Inc.

ABSTRACT

Following early rulemaking by the
National Highway Traffic Safety Administration requiring passive occupant crash protection for occupants in vehicles covered by
Motor Vehicle Safety Standard 208, intensive
development began on passive seatbelts as
promising alternatives to the inflatable
systems being considered. The general types
of passive seatbelt systems are identified.
Passive systems that have been offered to
date as optional installations in production
car lines are described, as well as the
production models being introduced in Model
Year 1987 car lines in response to recent
rulemaking. Design considerations are discussed. Recent usage rates of installed
passive systems and manual belt systems are
compared.

THE U.S. CONGRESS ENACTED the Highway
Traffic Safety Act of 1966 (1)# to address
the problem of traffic fatalities and
injuries on American roadways. The first
director of the agency established to
carry out the provisions of the new law --
namely the National Highway Safety Bureau
(NHSB) in the Department of Commerce,
which was later superseded by the National
Highway Traffic Safety Administration
(NHTSA) in the Department of Transportation- was Dr. William Haddon, Jr., a
medical doctor with a long and distinguished role in public health and traffic
safety in the State of New York. He viewed
traffic fatalities and injuries as a public
health problem of epidemic proportions
that could best be overcome by passive
means -that is, by means requiring no overt

---

*Numbers in parentheses designate
references at end of paper.

action on the part of the parties affected.
He cited past examples of successful passive
solutions to major public health problems,
such as inoculation of entire populations to
overcome dread diseases, pasteurization of
milk, and use of automatic sprinkler systems
to control fires. He strongly advocated passive solutions to the problem of highway
fatalities and injuries, including passive
means for occupant crash protection.

PASSIVE SYSTEMS FOR PASSENGER CARS

The early patent literature relating to
passive restraint systems goes back to 1949
when a collision mat was described (2). The
first patent relating to a passive belt system (3) was issued in 1958, in which a system
was described having a safety belt that moved
into position by closing the door and then
automatically released upon opening one or
both doors. The patent described a single
belt used to restrain all the passengers in
the front seat, which could be one, two or
three passengers. Such a conceptual flaw was
symptomatic of the inadequacies of many of
the early concepts for passive restraints
that did not embody performance characteristics consistent with requirements dictated by
the realities of crash dynamics, occupant
kinematics, and limits of human tolerance to
injury. Meaningful development of passive
occupant protection systems required the
concurrent advance in the knowledge and technology relating to the real-world car crash
event.

In November 1970, the U.S. Federal Motor
Vehicle Safety Standard 208 (MVSS 208) (4)
which addresses vehicle occupant crash protection, was amended (5) to incorporate the
original requirements for passive occupant
protection in the front seating positions of
all passenger cars manufactured on or after
the first day of July 1973 for sale in the
U.S. car market. The original requirements

and original effective date embodied in the amendment were challenged and further considered in a series of subsequent actions from 1970 to the present. The most recent action by NHTSA (6) calls for passive occupant protection in front outboard seating positions on a phased-in schedule with installations in a minimum of 10 percent of each manufacturer's car build after September 1, 1986, 25 percent after September 1, 1987, 40 percent after September 1, 1988, and 100 percent after September 1, 1989. The passive requirement will be rescinded if states representing two-thirds of the nation's population enact mandatory seatbelt usage laws before April 1, 1989 that will be in effect and enforced by September 1, 1989.

The passive system for vehicle occupant protection that received substantial attention following the issuance of the original rulemaking in 1970 was the inflatable restraint (or "airbag" or "air cushion"), which was already under development. Alternative noninflatable means, however, were considered as well. Many categories of such noninflatable passive candidate systems were identified and explored (Table 1). These included transparent shields, deployable nets and blankets, cushions, restraining arms and barriers, integrated seat systems, and seatbelt systems.

DEVELOPMENTAL PASSIVE SEATBELTS

Of these various categories, passive seatbelts showed the greatest promise by far. Many conceptual designs of passive seatbelts were investigated. Several distinctive classes may be identified (Table 2). They may be 2-point, 3-point, or 4-point systems based on the number of load-bearing connections to vehicle structure. They may be all-mechanical or may be all- or partially-motorized. They may have retractors located in either inboard or outboard locations and fixed anchors located either inboard or outboard. Some may require knee bolsters to control lower body movement. All must have some provision to permit disengaging the seatbelt after an accident. Prototypes of all these classes of systems have been developed and installed in exemplar production cars.

INSTALLATIONS IN PRODUCTION CAR LINES

In spite of the very substantial amount of effort directed to the automatic belted restraint systems and the large number of prototype systems that were developed and presented to vehicle manufacturers and to NHSTA, the first production-ready automatic seatbelt system was not installed in a production car line until the 1975 Model Year (Table 3).

The first production model to appear was a 2-point system, installed as an extra-cost option in a Volkswagen Rabbit in Model Year 1975 and subsequent years. Figure 1 shows the conceptual design of the VW Rabbit installation. This system includes a 2-point upper torso belt with the retractor inboard and the fixed anchor attached outboard on the door. The door connection incorporates a buckle for emergency disconnection. The buckle has an electrical interlock with the starter system to insure that the buckle is connected in normal driving use. The lower body movement is controlled by a knee bolster and a specially designed seat pan.

The first production automatic seatbelt installation in a car produced in the United States appeared in the Chevette manufactured by the Chevrolet Division of General Motors in Model Year 1978 1/2. It was quite similar to the VW installation in the Rabbit, except that it used a seat of more conventional design and included a manual lap belt. Figure 2 shows the conceptual design, with an automatic 2-point shoulder belt, a knee bolster, and a manual lap belt. Figure 3 shows the system as installed in the Chevette.

The next production installation appeared in the Chevette in Model Year 1980. This was a 3-point system with an inboard retractor and two fixed anchor points on the door. Figure 4 shows the conceptual design of this system. A buckle on the lower anchor on the door was provided for emergency release. This buckle also incorporated a switch to provide an electrical connection with a buzzer to insure that the buckle was engaged during normal use. Figure 5 shows the system as installed.

A more recent automatic system was installed in the Toyota Cressida in Model Year 1981 and subsequent years. It is a motorized 2-point system with an inboard emergency-locking retractor and an outboard anchor location on the B-pillar. Figure 6 shows the conceptual design of the Toyota system. A manual lap belt is included to control lower body movement. The lower portion of the instrument panel provides a knee bolster as well. The motorized outboard anchor point is driven along the upper portion of the B-pillar, along the roof rail, and the upper portion of the A-pillar. This action moves the shoulder belt forward to the A-pillar for easy occupant ingress into the vehicle, and then returns it to the B-pillar to don the shoulder belt on the occupant after he is seated and the door is closed. The traverse of the belt has been carefully timed to move the belt quickly, yet not so quickly as to startle the occupant.

CURRENT INSTALLATIONS OF AUTOMATIC SEATBELTS

In addition to the automatic seatbelt systems installed in cars prior to Model Year 1987, more systems will appear in Model Year 1987 in selected car models from all manufacturers producing cars for sale in the U.S.A. as a result of the federal requirement in MVSS 208 (6). The seatbelt installations include motorized and non-motorized systems (Table 4).

The systems installed by the Ford Motor Company in their Ford Escort and Mercury Lynx car models closely follow the Toyota system concept. Figure 7 shows the installed system.

The systems adopted by General Motors for installation in their Oldsmobile Calais and Delta 88 models are three-point systems having separate emergency-locking retractors for lap belt and shoulder belt mounted within the door and a fixed inboard buckle end, as shown in Figure 8. Figure 9 shows the system as installed in an Oldsmobile Delta 88 Sedan.

DESIGN CONSIDERATIONS FOR AUTOMATIC SEATBELT SYSTEMS

The general design considerations for all seatbelts are shown in Table 5. The list includes performance in accidents and evasive maneuvers, simplicity in design and operation, reliability, durability, vulnerability to damage during normal use or in accidents that could render them inoperable, comfort during use, convenience in use, packaging size and shape for accommodation in the vehicle, esthetics, and cost. Automatic seatbelts have the added requirement for automatic donning and doffing.

In any given system design, tradeoffs inevitably are encountered such that optimization of one variable is accompanied by some sacrifice in optimization of another. Some obvious examples are (1) the possible degradation of system performance in emergencies for the sake of increased occupant comfort in normal usage by the incorporation of a tension-relieving device on the shoulder belt, and (2) the possible decrease in simplicity and reliability for the sake of increased convenience and improved stowage through the use of retractors.

USAGE AND PERFORMANCE

The usage rates and performance of automatic seatbelts already in the field have been quite satisfactory. Figure 10 shows the usage rates for automatic systems as compared with the usage rates for manual belts, representing data obtained in the ongoing program sponsored by NHTSA in which usage rates are monitored in 19 cities ac-ross the United States. Average driver usage of manual belts in these 19 cities increased from 13.3 percent in 1983 to 14.4 percent in 1984 and 21.4 percent in 1985, with 23.3 percent in the last half of 1985. The upward trend was due mainly to the intensive efforts of NHTSA and others in the safety community to increase seatbelt usage in the United States and the increasing number of mandatory use laws adopted by states. Usage rates varied appreciably from one city to another in the 19-city study, with a low rate in 1985 of 11.1 percent and the highest rate of 46.3 percent.

The usage rate for the VW system has persisted in the range of 70 to 80 percent through the years since 1975, when it was introduced, with observed usage rates of 75 percent in 1983 and 76 percent in 1984 and 71 percent in 1985. The Toyota Cressida system has done even better, with observed usage rates of 95 percent in 1983, 97 percent in 1984, and 92 percent in 1985.

Performance data for these systems in accidents are limited because of the relatively short time they have been exposed on the roadways and the relatively small number involved in accidents. The data acquired to the present through the National Accident Sampling System (NASS) operated in the United States by NHTSA indicate the performance of these automatic systems in the real world is generally comparable to the performance of manual systems.

THE FUTURE OF AUTOMATIC SEATBELTS

The key to the need for automatic belts is the usage rate of seatbelts. In the many countries and jurisdictions around the world having mandatory seatbelt use laws, starting with Australia in 1970, usage rates of 75 to 95 percent are achieved with manual seatbelt systems when continuing enforcement is practiced and fines are imposed. It is reasonable to assume that comparable usage rates can be achieved in the states in the U.S. that have mandatory usage laws, provided that appropriate enforcement is practiced and fines are imposed.

Automatic seatbelts are not inherently better performers than manual belts, but they do appear to have induced higher usage rates when installed as extra-cost options. The public acceptance and usage of automatic seatbelts when installed as standard equipment remains to be seen. Automatic seat belts can be defeated, and their presence in the car does not guarantee their being used by car occupants. The degree of success in enacting and enforcing mandatory seatbelt usage laws in the United States will determine whether automatic seatbelt systems will be preferred in the long term.

ACKNOWLEDGEMENT

The author acknowledges with thanks the support provided by the American Seat Belt Council in the preparation of this report, and the cooperation of vehicle manufacturers and ASBC member companies in providing information and illustrations.

REFERENCES

1. 15 USC 1391 et seq., National Traffic and Motor Vehicle Safety Act

2. U.S. Patent 2,477,933 (August 2, 1949) - Collision Mat for Vehicles

3. U.S. Patent 2,858,144 (October 28, 1958) - Safety Belt for Vehicles

4. 49 CFR 571.208 Standard No. 208; Occupant Crash Protection

5. 35 FR 16927 - Docket 69-07; Notice 7 - Occupant Crash Protection

6. 51 FR 9800 - Docket 74-14; Notice 43 - Occupant Crash Protection

INFLATABLE

NON-INFLATABLE

- TRANSPARENT SHIELDS

- DEPLOYABLE NETS

- DEPLOYABLE BLANKETS

- SEAT-INTEGRATED SYSTEMS

- RESTRAINING ARMS

- CUSHIONS AND BOLSTERS

- SEATBELT SYSTEMS

Table 1 - Automatic restraint systems

- 2-POINT* — AUTOMATIC SHOULDER BELT ONLY

- 3-POINT* — AUTOMATIC LAP AND SHOULDER BELTS

- 4-POINT* — SEPARATE AUTOMATIC LAP
AND SHOULDER BELTS

* MECHANICAL AND MOTORIZED MODELS IN ALL TYPES

Table 2 - Automatic seatbelt system types

CAR LINE	MODEL YEAR INTRODUCED	TYPE
VW RABBIT	1975	2-POINT
GM CHEVETTE	1978 1/2	2-POINT
GM CHEVETTE	1980	3-POINT
TOYOTA	1981	2-POINT -MOTORIZED

Table 3 - Pre-1987 automatic seatbelt systems

MANUFACTURER	CAR LINE
MOTORIZED SEATBELTS	
ALFA ROMEO	SPYDER-QUADRIFOGLIO
CHRYSLER-PLYMOUTH	CONQUEST
FORD	FORD ESCORT
	MERCURY LYNX
MAZDA	626 4-DOOR SEDAN
MITSUBISHI	STARION
NISSAN	MAXIMA
SAAB	900 S 3-D
SUBARU	XT COUPE
TOYOTA	CRESSIDA
	CAMRY
SEATBELTS (NOT MOTORIZED)	
AMERICAN MOTORS	ALLIANCE L AND DL
CHRYSLER -PLYMOUTH	LE BARON COUPE
	DODGE DAYTONA
GENERAL MOTORS	BUICK SOMERSET
	BUICK SKYLARK
	BUICK LE SABRE
	PONTIAC GRAND AM
	PONTIAC BONNEVILLE
	OLDSMOBLE CALAIS
	OLDSMOBLE DELTA 88
HONDA	ACCORD HB
VOLKSWAGON	GOLF
	JETTA

Table 4 - Model Year 1987 automatic seatbelt Systems

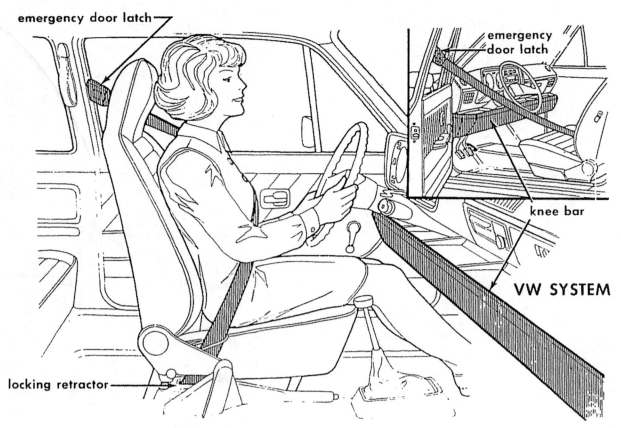

emergency door latch

emergency door latch

locking retractor

knee bar

VW SYSTEM

Fig. 1 – First automatic seatbelt system in production car line

1978 1/2 CHEVETTE 2 POINT BELT SYSTEM

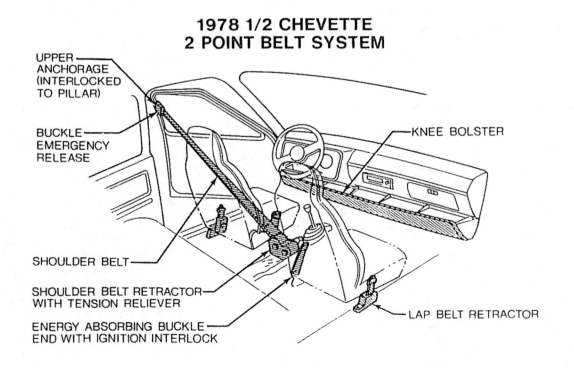

UPPER ANCHORAGE (INTERLOCKED TO PILLAR)

BUCKLE EMERGENCY RELEASE

KNEE BOLSTER

SHOULDER BELT

SHOULDER BELT RETRACTOR WITH TENSION RELIEVER

ENERGY ABSORBING BUCKLE END WITH IGNITION INTERLOCK

LAP BELT RETRACTOR

Fig. 2 – First automatic seatbelt system in U. S. domestic production car line

938

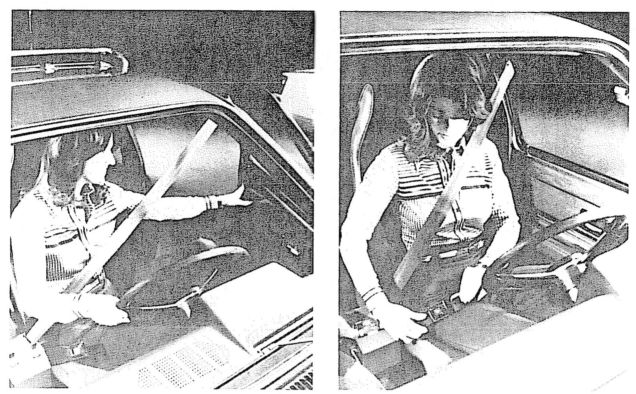

Fig. 3 - 1978 1/2 Chevette automatic seatbelt installation

1980 CHEVETTE
3 POINT PASSIVE BELT SYSTEM

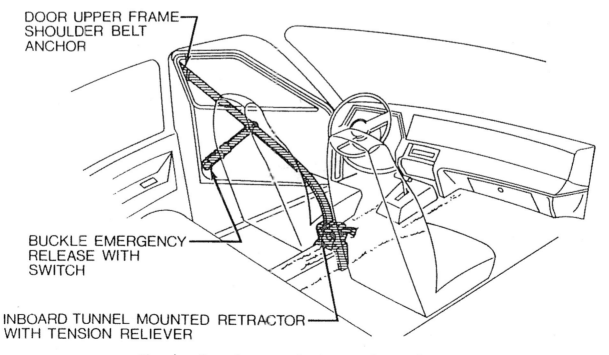

DOOR UPPER FRAME— SHOULDER BELT ANCHOR

BUCKLE EMERGENCY RELEASE WITH SWITCH

INBOARD TUNNEL MOUNTED RETRACTOR— WITH TENSION RELIEVER

Fig. 4 - Second generation automatic seatbelt system in U. S. domestic production car line

Fig. 5 - 1980 Chevette automatic seatbelt
installation

TOYOTA SYSTEM

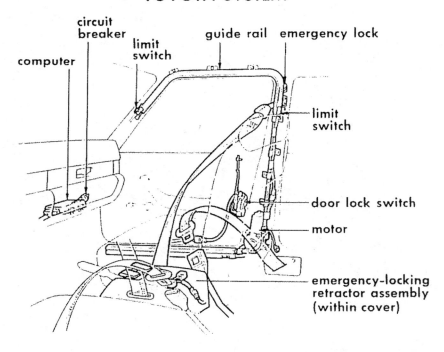

Fig. 6 - Motorized automatic seatbelt system

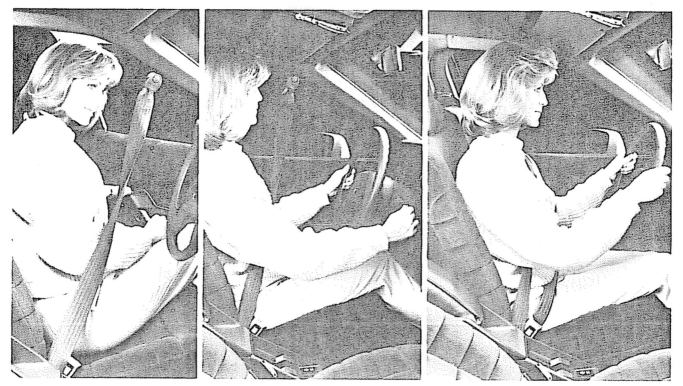

Fig. 7 - Ford automatic seatbelt installation

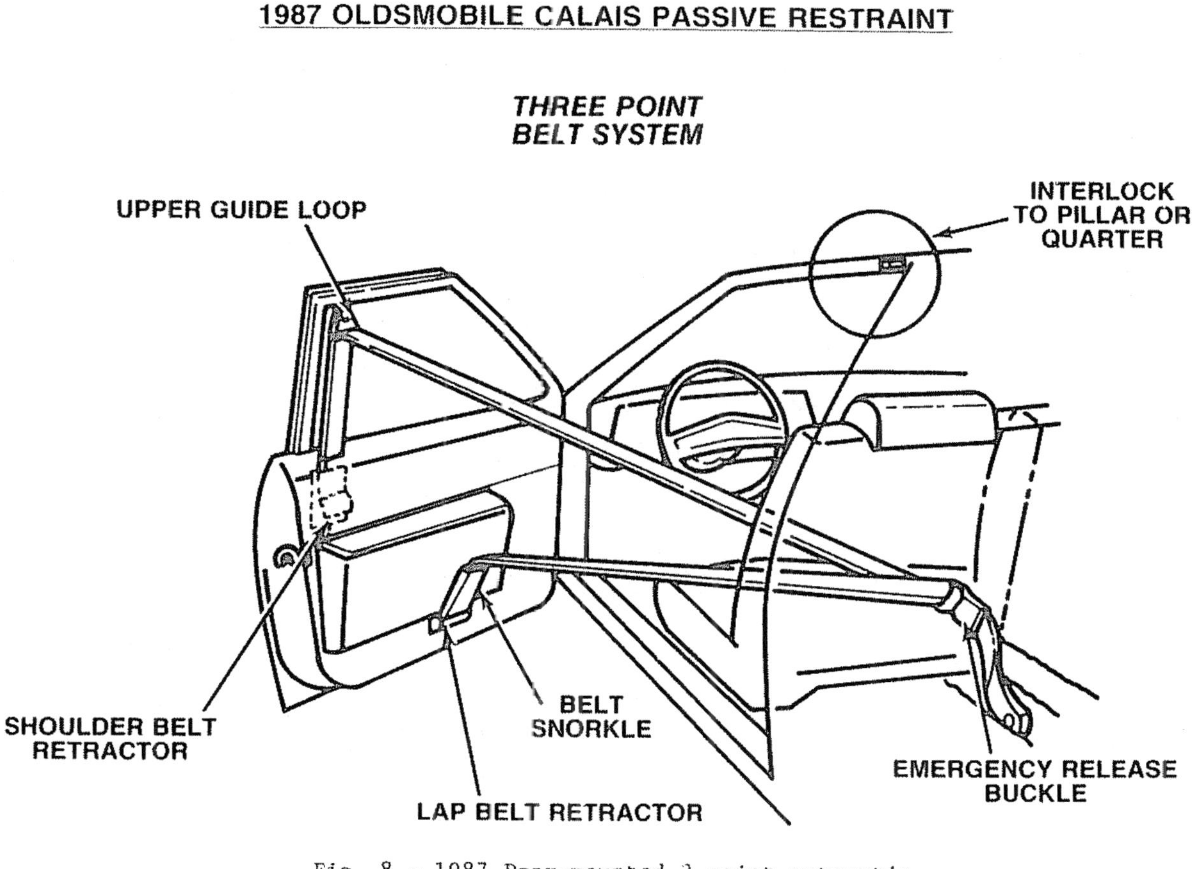

1987 OLDSMOBILE CALAIS PASSIVE RESTRAINT

THREE POINT BELT SYSTEM

UPPER GUIDE LOOP

INTERLOCK
TO PILLAR OR
QUARTER

SHOULDER BELT
RETRACTOR

BELT
SNORKLE

EMERGENCY RELEASE
BUCKLE

LAP BELT RETRACTOR

Fig. 8 - 1987 Door-mounted 3-point automatic
seatbelt system

Fig. 9 - General Motors automatic seatbelt
installation

MANUAL SEATBELT SYSTEMS

- PERFORMANCE IN EMERGENCY EVENTS
- RELIABILITY
- DURABILITY
- VULNERABILITY TO DAMAGE
- VULNERABILITY TO DELIBERATE
  OR INADVERTENT TAMPERING
- SIMPLICITY IN OPERATION
- COMFORT IN NORMAL OPERATION
- CONVENIENCE IN NORMAL OPERATION
- ACCEPTABLE PACKAGING IN CAR
- ESTHETIC CONSIDERATIONS
- COST

AUTOMATIC SEATBELT SYSTEMS

- ALL OF THE CONSIDERATIONS LISTED ABOVE
- AUTOMATIC DONNING AND DOFFING

Table 5 - Seatbelt design considerations

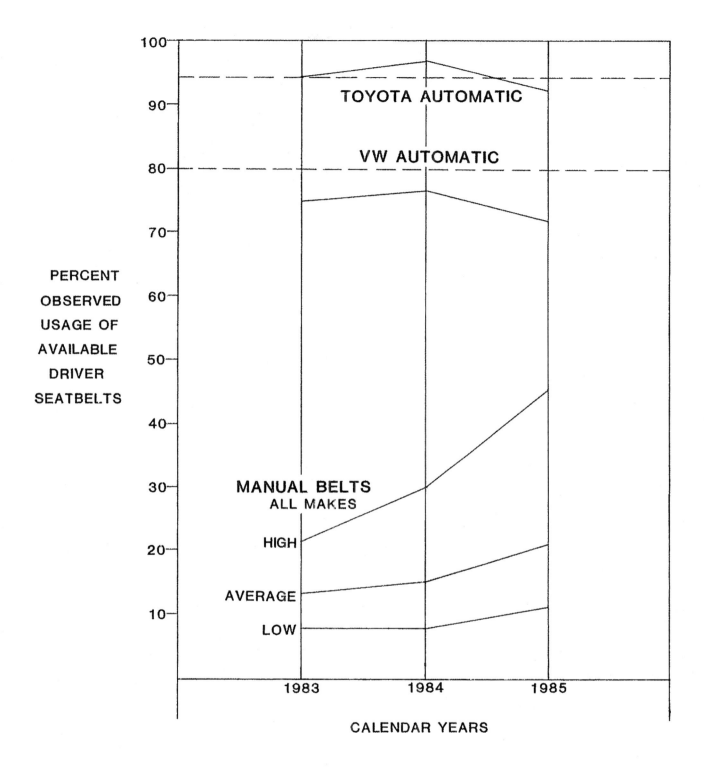

RECENT DRIVER SEATBELT USAGE RATES

Fig. 10 - Recent driver seatbelt usage rates

902328

Assessing the Safety Performance of Occupant Restraint Systems

David C. Viano and Sudhakar Arepally
General Motors Research Laboratories
Biomedical Science Dept.
Warren, MI

ABSTRACT

The purpose of this study was to investigate approaches evaluating the performance of safety systems in crash tests and by analytical simulations. The study was motivated by the need to consider the adequacy of injury criteria and tolerance levels in FMVSS 208 measuring safety performance of restraint systems and supplements. The study also focused on additional biomechanical criteria and performance measures which may augment FMVSS 208 criteria and alternative ways to evaluate dummy responses rather than by comparison to a tolerance level.

Additional analysis was conducted of dummy responses from barrier crash and sled tests to gain further information on the performance of restraint systems. The analysis resulted in a new computer program which determined several motion and velocity criteria from measurements made in crash tests. These data provide new insights on restraint performance including torso angle change, the effectiveness of occupant restraint from velocity buildup -- called the Restraint Quotient -- and forward displacement and rebound of the chest and pelvis. The various approaches are used to discuss Hybrid III responses from 24 barrier crash and 23 Hyge sled tests.

Four additional experiments were critically evaluated to validate the new analysis. They involved a lap-shoulder belted Hybrid III female dummy with the lap-belt positioned low on the pelvis for restraint or pre-positioned on the abdomen for submarining. By comparing calculated and film analyzed data, the Restraint Quotient program was found to predict accurately those test results.

Numbers relating to FMVSS 208 criteria values were reduced 15%-25% by lap-belt submarining over lap-belt restraint. The addition of an abdominal injury criterion and use of a frangible abdomen helped predict injury risk to the abdomen and an overall higher risk of injury with lap-belt submarining. The use of motion and velocity criteria also identified poorer performance with lap-belt submarining.

Several approaches were reviewed to interpret the overall performance and injury risks of safety systems in a crash. The preferred method was found to be injury risk assessment which represents dummy responses as injury risks using Logist probability functions. This enables injury in each body region to be assessed by the maximum risk from applicable biomechanical criteria. Whole body injury is determined by summing the individual risks from each body region.

The motion and velocity criteria determined from the Restraint Quotient program were useful in complementing the biomechanical criteria to assess injury. They may be helpful in an overall determination of the crash performance of restraint systems, supplements and enhancements.

THERE HAVE BEEN RAPID CHANGES in the type of restraint systems in passenger vehicles with the phase-in of passive safety requirements. While each system meets the response limits of FMVSS 208 in a barrier crash test, there is an emerging understanding that this represents only a basic requirement for safety assessment in a crash (1). Based on a review of the methods to assess the effectiveness of safety belt restraint, the following requirements are considered important to the overall quality of occupant restraint:

•   Maintain FMVSS 208 responses below required limits.

- Provide primary restraining forces on the boney structures of the pelvis, upper thorax and shoulder of the belted occupant.

- Minimize loads on the compliant regions of the abdomen and lower thorax, thus limiting abdominal compression.

- Control forward excursion of the lower extremity and minimize rearward rotation of the pelvis.

- Ensure a slightly forward rotation of the torso to load the shoulder and upper thorax.

- Minimize contact HIC and facial loads.

- Minimize neck shear, bending and axial responses.

- Reduce chest compression, Viscous response and lateral shear deformation, and

- Minimize femoral bending, knee shear and tibial-ankle loads.

The above requirements represent a complicated and inter-related set of responses that need to be balanced systematically to ensure "overall" occupant restraint in a crash and minimize injury risks in critical body regions. This is particularly important when the distribution in age, tolerance and seating position for the whole family of occupants is considered in the development of new restraint systems.

For example, the current Part 572 dummy does not assess abdominal injury with lap-belt loading so that a system may adequately satisfy federal requirements while involving submarining and abdominal injury risk. In addition, occupant kinematics can include rearward rotation of the upper torso which may lower HIC and thoracic spinal acceleration but increase lower thoracic loading by the shoulder belt. This may inadvertently result in rib cage fractures, particularly in older occupants. Thus, as further refinements and advances in restraint systems are considered for use in passenger vehicles, a sufficient set of performance requirements should be utilized to help strive for reasonable safety with different systems. This will involve considering individual responses as part of the overall safety performance.

This paper is part of a three-part program to improve the assessment of occupant protection systems and the evaluation of restraint enhancements which may provide complementary protection. The research includes: 1) improvements in dummy injury criteria and assessment procedures to adequately address injury risk to various body regions (2-4), 2) attention to the biomechanical quality of restraints including motion sequence criteria (5-7) and injury biomechanics parameters (8-10), and 3) additional analysis of current test responses to better interpret restraint performance, including the influence of the distribution in crash types and severities, occupant age and tolerances, and injury mechanisms (11,12).

The study provides the rationale and procedures to determine motion sequence properties which are an important aspect in the evaluation of restraint performance. These data can be obtained from current dummy responses in barrier crash or sled tests of safety systems. They provide complementary information on the amount of restraint provided. The relative velocity of the occupant with respect to the interior, rebound, kinetic energy developed, and the forward displacement and posture of the occupant during restraint are evaluated and discussed in terms of potential injuries in a crash. In this study, five parameters of belt restraint quality are analyzed from current safety test data:

- Velocity of the occupant with respect to the interior.

- Rebound velocity as a fraction of the maximum forward velocity.

- Kinetic energy of the occupant with respect to the interior.

- Angle change of the upper torso during restraint.

- Displacement of the pelvis.

It is the intention of this study to show how these values relate to the nine restraint requirements listed above.

The lower the velocity of the occupant with respect to the interior, the greater the restraint and the lower the energy that must be managed. If the relative velocity is zero, the occupant would, in effect, be glued to the vehicle structure and experience the same deceleration as the passenger compartment. In contrast, an unbelted occupant builds up speed with respect to the interior as the vehicle stops by front-end crush. This can result in occupant impact with the interior at speeds approaching the vehicle impact velocity or possibly higher if interior impact occurs during rebound.

In a frontal barrier crash, an unbelted occupant can develop speeds that approach the initial impact speed of 30 mph (13.4 m/s) particularly in the passenger seating position since the steering system provides ridedown for the driver. The potential range in velocity build-up is zero to approximately 13.4 m/s. This produces a greater range in occupant kinetic energy because energy is related to the square of velocity. Thus, a 75 kg occupant may develop 0 to 6,733J of kinetic energy with respect to the passenger compartment.

If the occupant's instantaneous velocity in the vehicle is determined as a fraction of the overall change in velocity of the passenger compartment, the ratio is called the Restraint

Quotient (RQ). It normally varies between 0 and 1. The lower the RQ, the greater the restraint in a crash.

Another factor in occupant restraint is the angle change of the upper body during belt loading. For greater protection, the torso angle (θ) should change so that the spine is upright or slightly forward to ensure that belt load is on the shoulder. This applies force to the strong skeletal regions of the shoulder and upper chest.

At this time, the torso angle is not routinely reported in crash tests; yet, it provides information on the overall quality of restraint performance in loading the shoulder and upper chest. Poor quality restraint can involve a rearward orientation of the upper body with belt loads directed on the lower rib cage and abdomen. This can result in rib fractures and internal injury, and increases the risk of submarining. The torso angle is particularly important in properly restraining older occupants whose rib cage tolerance is lower than younger passengers. However, the need to manage the torso angle must be considered in context with the potential for head impact with the steering assembly.

This study provides an approach to obtaining new information from currently measured responses in dummy tests. It involves additional analysis of test results which may be helpful to making decisions on alternative safety systems and supplemental components. It provides data on the degree of restraint in terms of occupant velocity in the interior. This involves kinetic energy to be managed by occupant loadings. In addition, the angle change of the upper body is an indirect measure of the concentration of belt loading on the shoulder and away from the more compliant regions of the lower chest and abdomen.

METHODOLOGY

The following accelerations are used to determine the Restraint Quotient (RQ) and torso angle change (θ):

1) x- and z-axis of thoracic spinal acceleration;

2) x- and z-axis of pelvic spinal acceleration; and,

3) x-axis sled or passenger compartment acceleration.

The accelerations are routinely measured in barrier and sled tests. Accelerometers in the dummy are fixed to the AP and SI axis and move with the dummy during a crash. A computer analysis of the accelerations is used to generate properties of safety belt restraint.

The following assumptions are made and represent a first-order approximation based on dummy and vehicle responses through maximum occupant restraint in a crash test:

1) the resultant biaxial acceleration of the chest and pelvis is co-linear with the vehicle deceleration axis;

2) the lateral components of chest and pelvic acceleration can be neglected for this analysis;

3) the accelerometers are accurate enough and the responses do not involve sharp impacts so they may be double integrated to approximate occupant displacements; and,

4) the connection between the spinal and pelvic accelerometers is a rigid link.

Obviously, the lumbar spine is flexible and bends during restraint as much as 25°-40° in the Hybrid III dummy and the lap belt causes downward motion of the body but the effects are considered second order factors to the calculation of torso angle change. Using these approximations, the following calculations are made with respect to an inertial reference frame and a moving coordinate frame fixed to the passenger compartment. The calculations determine occupant dynamics and kinematics with respect to the vehicle interior. The resultant accelerations of the chest x_c and pelvis x_p are computed from the x-axis (AP direction on the dummy) and z-axis (SI direction) responses as a function of time:

$$\ddot{x}_c = (\ddot{x}^2_{cx} + \ddot{x}^2_{cz})^{1/2}$$
$$\ddot{x}_p = (\ddot{x}^2_{px} + \ddot{x}^2_{pz})^{1/2} \tag{1}$$

The deceleration of the occupant compartment or sled ($\ddot{x}_v$) represents the average deceleration of the passenger compartment. For this analysis, data are filtered using SAE 180 channel class. The velocity of the dummy and vehicle are determined by standard numerical integration:

$$\dot{x}_c = \int \ddot{x}_c dt,$$
$$\dot{x}_p = \int \ddot{x}_p dt, \tag{2}$$
$$\dot{x}_v = \int \ddot{x}_v dt.$$

These responses are related to the inertial reference frame. The velocity of the occupant with respect to the moving vehicle reference frame are:

$$v_c = \dot{x}_v - \dot{x}_c,$$
$$v_p = \dot{x}_v - \dot{x}_p. \tag{3}$$

Displacements of the dummy chest and pelvis are shown in Figure 1. There are inherent problems in reliably determining velocity and displacement from integration of accelerometers. However, the placement in these tests on the dummy spine and pelvis reduces the potential for sharp spikes in acceleration and rapid changes in position, which are primary factors in increasing errors. Displacements in the vehicle are determined by integration of Equation 3:

$$d_c = \int v_c dt,$$
$$d_p = \int v_p dt. \qquad (4)$$

Occupant restraint during vehicle deceleration is measured as the relative velocity of the occupant in the passenger compartment divided by the maximum velocity change of the vehicle or sled. This parameter is a function of time and is called the Restraint Quotient (RQ). It varies between 0 -- assuming the occupant is glued to the passenger compartment and does not develop velocity -- and 1 -- assuming the occupant attains the total velocity change in the vehicle before impacting the interior. The Restraint Quotient is defined as:

$$RQ_c = \frac{v_c}{(\dot{x}_v)_{max}},$$
$$RQ_p = \frac{v_p}{(\dot{x}_v)_{max}}. \qquad (5)$$

The angular velocity of the upper body of the dummy is calculated using the relative velocity between the chest and pelvis:

$$\omega = \frac{v_c - v_p}{D}, \qquad (6)$$

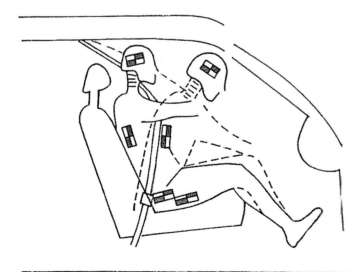

Figure 1: Motion of the occupant during a crash demonstrating the displacement of the chest and pelvis with belt restraint.

where D is the distance between the chest and pelvic accelerometer packages. For the 50th percentile male Hybrid III dummy the distance is D = 35 cm and for the 5th percentile female Hybrid III dummy it is D = 31 cm.

A positive angular velocity implies forward rotation of the upper torso, whereas a negative angular velocity represents rearward rotation of the upper body. The change in angle of the torso (θ) is found by integration of Equation 6:

$$\theta = \int \omega \, dt. \qquad (7)$$

Other parameters can be calculated from these responses. The term ridedown implies that the occupant benefits from restraining forces that act during front-end crush of the vehicle. This relates to an increased stopping distance for the occupant in the inertial reference frame and is associated with lower deceleration.

For the purposes of quantifying ridedown in barrier and sled crash tests, it is defined as starting when the occupant experiences restraining forces sufficient to reduce chest velocity in the interior more than 10% of the maximum velocity change of the vehicle. Ridedown continues until the end of vehicle deceleration. This is the point in which vehicle deceleration passes through zero, which separates front-end crush from rebound.

The duration of vehicle deceleration is defined as the time of positive deceleration of the passenger compartment in a barrier crash. Rebound is experienced by the vehicle and occupant. For the purpose of quantifying rebound for the occupant, it is measured as a ratio of the maximum rearward velocity normalized by the maximum forward velocity of the occupant in the interior:

$$RB_c = \frac{max(-v_c)}{max(+v_c)},$$
$$RB_p = \frac{max(-v_p)}{max(+v_p)}. \qquad (8)$$

Certain other parameters are reported from barrier crash or sled tests because they relate to occupant protection. For example, a relative kinetic energy is calculated using the maximum occupant velocity normalized by a velocity of 5 m/s. This kinetic energy factor (E) is:

$$E_c = \frac{max(+v_c)^2}{25},$$
$$E_p = \frac{max(+v_p)^2}{25}. \qquad (9)$$

The results of the Restraint Quotient computer analysis are presented in a one-page format with five separate plots as shown in Figure 2. This is an example of restraint performance in a 30 mph barrier crash with a lap-shoulder belted Hybrid III dummy in the passenger seating position. The responses include from top to bottom: 1) occupant compartment deceleration ($\ddot{x}_v$) and chest and pelvic acceleration ($\ddot{x}_c$ and $\ddot{x}_p$), 2) vehicle velocity change ($\dot{x}_v$) and occupant velocities in the vehicle (v_c and v_p), 3) displacement of the occupant in the vehicle (d_c and d_p), 4) the Restraint Quotient of the chest and pelvis (RQ_c and RQ_p) and 5) the torso angle change (θ).

Currently, the recline angle of automotive seatbacks is 26°, but the dummy doesn't typically have that great of a recline angle in the seat. Dummies in this study were usually seated with an initial reclining angle of 15°, so a positive torso angle change of 15° brings the torso into an upright (vertical) posture. Angle changes greater than 15° represent a forward rotation of the upper body beyond vertical and more belt loading into the shoulder region.

<u>Validation Tests:</u> The validity of the assumptions in the computer analysis was checked by analysis of a series of sled tests that had been conducted for another study (13). The experiments involved a lap-shoulder belted 5th percentile Hybrid III female dummy on a conventional bucket seat. Using a standard lap-shoulder belt system, two test conditions were simulated: one involved positioning the lap-belt low on the pelvis to ensure proper belt restraint and another pre-positioning the belt on the abdomen resulting in submarining. Each test condition was repeated twice to compare the calculations of occupant dynamics and kinematics against data obtained from high-speed movies of the sled test.

<u>Barrier Crash Tests:</u> The Restraint Quotient computer analysis was further evaluated on a series of 30 mph barrier crash tests conducted at the General Motors Proving Grounds during routine product development of several model passenger cars. The Hybrid III 50th percentile male dummy was used in the driver and right-passenger seating position. The following test conditions were evaluated:

- Unbelted occupants
- Manual lap-shoulder belted occupants
- Automatic lap-shoulder belted occupants
- Manual lap-shoulder belted driver with supplemental airbag

These tests routinely involve a full complement of dummy instrumentation so that the chest and pelvic accelerations were available for processing through the Restraint Quotient program. There was a similar mix of vehicle types in each restraint category so the responses are comparable between restraint type.

<u>Sled Tests:</u> Hyge sled tests were conducted in the Biomedical Science Department using an automotive body buck with standard interior including bucket seat, steering system and instrument panel components. The Hybrid III 50th percentile male dummy was used in various restraint configurations at increasing exposure severities. The combination of test conditions and velocities evaluated in this series of experiments included:

<u>Restraint System</u>

- Unbelted Driver
- Lap-Shoulder Belted Driver
- Lap-Shoulder Belted Driver with Supplemental Airbag
- Unbelted Driver with Supplemental Airbag
- Lap-Shoulder Belted Passenger

<u>Test Velocity</u>

- 14 mph (6.3 m/s)
- 20 mph (8.9 m/s)
- 33 mph (14.7 m/s)
- 38 mph (17.3 m/s)

The dummy was instrumented with sufficient channels to process the data through the Restraint Quotient computer program. In addition, the dummy was instrumented with advanced injury assessment transducers to compare other biomechanical measures of safety belt and restraint component performance. High-speed photographs of the experiments were also taken.

<u>Statistical Analysis:</u> The response data with different types of occupant restraints were compared using standard statistical tests in SAS. The analysis included a t-test to determine the significance of response differences between restraint types.

RESULTS

<u>Validation of Restraint Quotient Calculations with Film Responses:</u> Table 1 provides response data from two tests with lap-belt restraint and submarining using the Hybrid III 5th-percentile female dummy. Three out of four comparisons of chest and pelvic displacement by the Restraint Quotient program and direct film analysis were within 2%, at 100 ms. This degree of accuracy is not expected because of variability in response measurements, high-speed photography and film analysis. Accuracy within 10% would be acceptable for this type of analysis. The chest displacement with lap-belt submarining was 18% lower by film analysis than by calculation. The principal direction of rotation of the upper torso was accurately indicated by the RQ analysis and within 12% for the cases of lap-belt loading. For lap-belt submarining, the rearward angle calculated by RQ was significantly smaller than identified by film analysis.

Figure 2 shows examples of the one-page output from the Restraint Quotient program for lap-belt restraint and submarining. With lap-belt restraint, the torso angle increases as chest displacement is greater than pelvic dis-

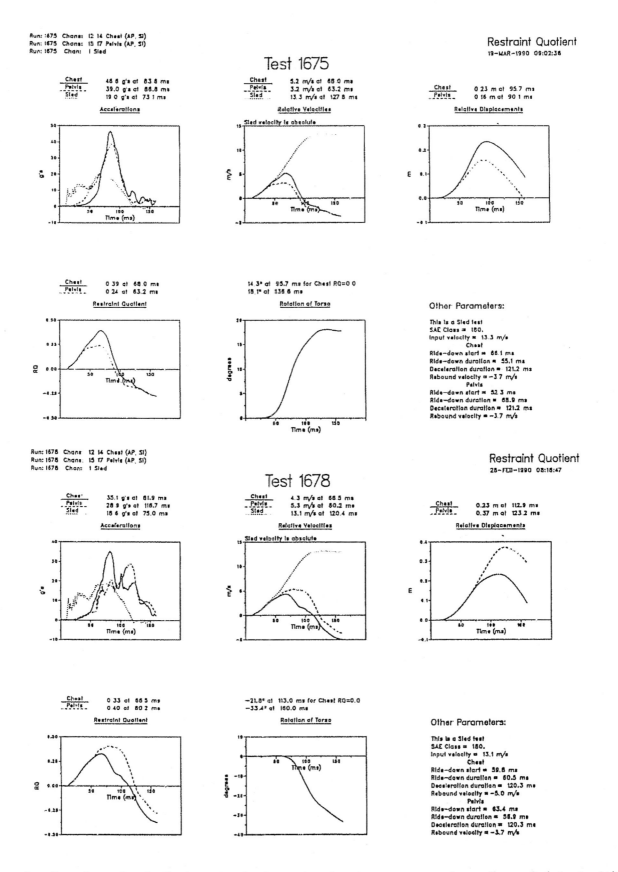

Figure 2: Examples of calculations in the Restraint Quotient program for a Hyge sled test with a 5th percentile female Hybrid III dummy experiencing lap-belt restraint (#1675) or lap-belt submarining (#1678).

placement. In contrast, lap-belt submarining involves a rearward rotation of the upper torso as the pelvis leads the chest in forward displacement. For the case of lap-belt restraint, there is tighter coupling and greater restraint of the pelvis than chest as indicated by a larger RQ_c than RQ_p. In contrast, lap-belt submarining involves a lower chest RQ_c than pelvis RQ_p.

Figure 3 compares the calculated and film analyzed chest and pelvic displacement and torso angle for lap-belt restraint and submarining. The data show acceptable similarity in the predicted and observed responses except for the lag in computed torso angle from the film response with lap-belt submarining.

Figure 4 shows the superimposed plot of Restraint Quotient with the corresponding shoulder or lap-belt load. The responses describe some of the dynamics of safety belt restraint and the Restraint Quotient. For example, with lap-belt restraint, RQ_c increases in magnitude until slightly after the sharp rise in shoulder belt load. This shows that shoulder-belt loading produces upper torso restraint and limits velocity build-up of the chest with respect to the interior.

The peak in RQ_c occurs between the onset and peak in shoulder belt loading and returns to zero slightly after the peak shoulder belt load. The effect of the shoulder belt on chest kinematics are illustrated by a reduction in RQ_c (indicating greater restraint) which coincides with an increasing shoulder belt load. This indicates that the peak shoulder belt load occurs at about the same time as the maximum forward excursion (and velocity) of the chest. Similar dynamics can be observed in the lap-belt submarining responses.

In summary, there is similarity between the computed and film analyzed displacements, in spite of known difficulties in using acceleration to determine velocity and displacement. The greatest differences were observed with the displacement response of the chest and torso angle for lap-belt submarining. For lap-belt restraint, there was excellent agreement between all calculated and film analyzed responses.

Comparing Lap-Belt Restraint and Submarining Responses: Lap-belt submarining reduced head HIC, chest compression, and chest and pelvic acceleration; but, there was an increase in peak head acceleration. This implies that lap-belt submarining may improve FMVSS 208 performance in some situations.

RQ_c was higher than the pelvic response with lap-belt restraint (Table 1: 0.38 v 0.24 or 58% higher). Lap-belt submarining involves a much higher RQ_p (62% increase) indicating a lower resistance to pelvic displacement. Lap-belt submarining also involved lower lap-belt loads (20% and 38% reduction respectively) without significantly affecting the peak shoulder belt tension.

The upper torso angle is positive with lap-belt restraint indicating forward rotation of the upper body. This resulted in a 15°-17° forward rotation which slightly exceeds an upright angle and involves belt loading on the shoulder and upper chest structure of the dummy at the peak in chest RQ_c. In contrast, lap-belt submarining involves a rearward rotation of the upper torso by 15°-20°. This increases the rearward torso angle to greater than -30° with respect to vertical.

Pelvic displacement is significantly larger with lap-belt submarining than restraint. It involves an increase in forward displacement of 14 cm (92% increase) because of belt compression of the abdomen. Chest displacement is slightly reduced as rearward torso rotation counteracts the forward displacement due to shoulder belt restraint. As indicated previously, the FMVSS 208 criteria of head HIC and chest acceleration are reduced by lap-belt submarining. There is also lower chest compression and pelvic acceleration. Lap-belt restraint results in higher chest velocities in the crash (5.1 m/s v 4.5 m/s, or 13% increase) than with lap-belt submarining. This results in slightly lower values of the RQ_c and higher levels of RQ_p with submarining.

With lap-belt submarining the film value of torso rotation is larger than computed with the RQ program. One reason is that the pelvis has a larger forward displacement with lap-belt submarining, whereas the chest is caused to rotate rearward and downward as the upper body pivots about the pelvis. Another assumption in the RQ program is that chest and pelvic accelerometers are attached by a rigid link. Since the RQ calculation assumes a horizontal displacement of the chest and pelvis, and there is a downward motion of the chest, a smaller negative angle change will be calculated by the RQ program due to the assumption of horizontal displacement of both components of displacement.

The peak resultant head acceleration increased by 22% with lap-belt submarining. This increase occurs in contrast to significant reductions in other head and chest responses. Figure 5 compares the resultant, longitudinal (AP) direction, and vertical (SI) direction accelerations for lap-belt restraint and submarining. The higher accelerations with lap-belt submarining are due to a larger longitudinal (AP) component of head acceleration near the end of shoulder belt loading. This is as the head is being arrested from its sweeping motion during shoulder belt restraint and as the head is facing downward. This can be seen by the large AP component of acceleration after the peak in vertical response which is typical of the phase of maximum torso restraint where the head is horizontal, experiencing maximum deceleration through the restraint system, and undergoing a large SI component of acceleration.

Comparing Unbelted Versus Lap-Shoulder Belted Responses in Barrier Crashes: Table 2 provides data from 30 mph barrier crash tests using Hybrid III dummies in the driver and

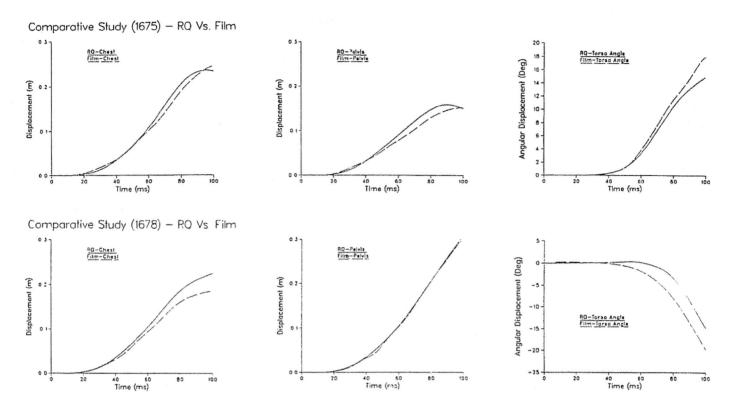

Figure 3: Comparison of calculated and film analyzed displacements of the chest and pelvis and torso angle change for tests #1675 and #1678.

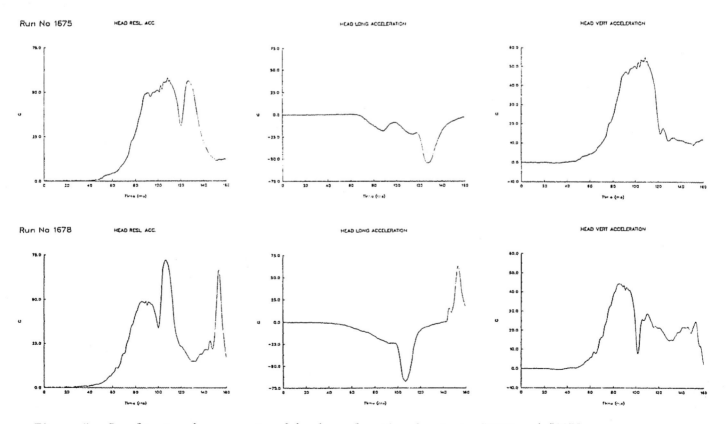

Figure 5: Resultant and components of head acceleration for tests #1675 and #1678.

951

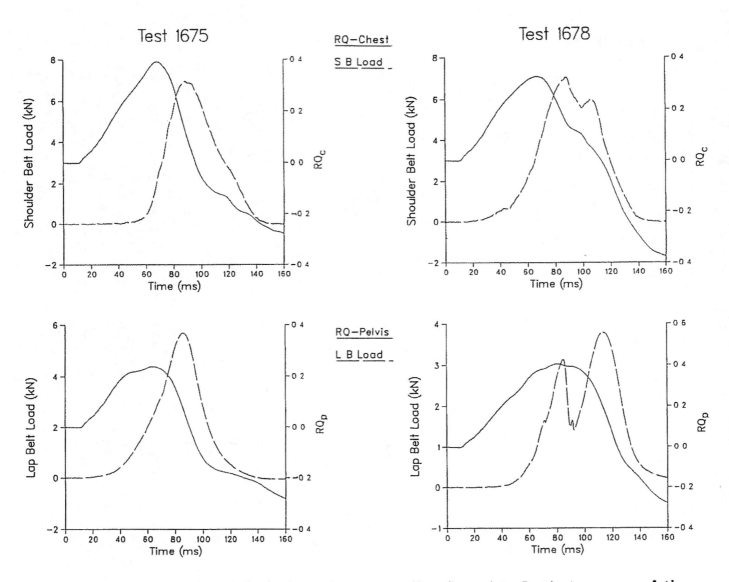

Figure 4: Shoulder and lap-belt loads with corresponding Restraint Quotient response of the chest and pelvis, respectively, for test #1675 and #1678.

Table 1

Hyge Sled Test Results Using a Hybrid III 5th-Percentile Female
Dummy with Frangible Abdomen Experiencing Lap-Belt
Restraint or Submarining in a Lap-Shoulder Belt System

Test Number	Head HIC	Head Acc (g)	Chest Compression (mm)	Acceleration (g) Chest	Acceleration (g) Pelvis	Displacement (cm) Chest RQ	Displacement (cm) Chest Film	Displacement (cm) Pelvis RQ	Displacement (cm) Pelvis Film	Restraint Quotient Chest	Restraint Quotient Pelvis	Torso Angle (°) RQ	Torso Angle (°) Film	Belt Loads (kN) Shoulder	Belt Loads (kN) Inboard Lap	Belt Loads (kN) Outboard Lap
1675	863	68.1	18.9	46.8	39.0	23.0	22.5	16.0	15.4	0.39	0.24	14.6	17.8	6.90	8.52	5.64
1676	889	61.0	23.8	48.4	39.2	22.0	21.8	15.0	15.0	0.37	0.24	15.6	16.0	7.20	8.88	5.74
Avg.	876	69.6	21.4	47.6	39.1	22.5	22.2	15.5	15.2	0.38	0.24	15.1	16.9	7.05	8.70	5.69
SD	18	2.0	3.5	1.1	0.1	0.7	0.5	0.7	0.3	0.01	0	0.7	1.3	.21	.25	.07
1678	706	72.0	17.4	35.1	28.9	23.0	18.5	30.0	30.0	0.33	0.40	-15.0	-19.6	7.00	6.84	3.76
1680	771	73.0	20.4	38.5	28.7	23.0	19.0	27.5	28.5	0.35	0.38	-8.2	-16.2	7.43	7.00	3.32
Avg.	739	72.5	18.9	36.8	28.8	23.0	18.8	28.8	29.3	0.34	0.39	-11.6	-17.9	7.22	6.92	3.54
SD	46	0.7	2.0	2.4	0.5	0.	.4	1.8	1.0	.01	0.01	(4.8)	(2.4)	0.30	.11	.31
% Diff.	-16.6%	+21.6%	-12.1%	-22.7%	-26.9%	+2.2%	-15.3%	+86.6%	+92.8%	-16.5%	+62.5%	-	-	+2.4%	-20.4%	-37.8%

Displacement and torso angle values compared at 100 ms.

Table 2a
30 mph Barrier Crash Tests Using a Hybrid III Driver Dummy

	Test		Head		Chest								Pelvis					
OBS	RUN	RES	G	HIC	G	C	V	E	R	D	RQ	TH	G	V	E	R	D	RQ
1	6386	U	64	810	78	-	7.6	2.3	0.91	29	0.51	-8	73	8.0	2.6	0.24	35	0.54
2	6387	U	67	490	88	-	9.1	3.3	0.74	60	0.83	25	55	9.1	3.3	0.31	46	0.61
3	9411	U	89	1110	108	-	10.3	4.2	0.84	48	0.77	21	46	8.7	3.0	0.46	36	0.65
4	6723	U	82	820	79	4.3	10.5	4.4	0.37	50	0.70	28	46	7.1	2.0	0.49	34	0.47
5	6736	U	98	910	107	5.3	10.7	4.6	0.34	53	0.69	25	53	7.7	2.4	0.35	38	0.50
6	9755	U	125	1870	98	7.8	10.5	4.4	0.65	53	0.70	15	43	9.0	3.2	0.41	45	0.60
7	10023	U	93	1050	109	-	11.8	5.6	0.82	58	0.79	27	68	9.0	3.2	0.51	39	0.60
Mean			88	980	95	5.8	10.1	4.1	0.64	50	0.71	19	55	8.4	2.8	0.40	39	0.57
SD			20	451	14	1.8	1.3	1.0	0.22	10	0.10	12	12	0.8	0.5	0.10	5	0.07
8	6242	M	113	780	40	3.7	5.3	1.1	0.83	24	0.35	12	59	4.6	0.9	0.61	17	0.30
9	6244	M	92	630	37	4.0	4.6	0.9	1.13	21	0.31	10	50	4.0	0.6	0.77	15	0.27
10	6656	M	90	670	46	6.1	5.9	1.4	0.91	30	0.39	3	50	6.4	1.6	0.68	28	0.42
11	6409	M	92	810	40	4.3	5.1	1.0	1.17	26	0.34	-4	51	6.2	1.5	0.71	28	0.42
12	6694	M	54	630	47	-	5.3	1.1	0.51	34	0.33	-9	74	6.9	1.9	1.3	40	0.43
13	6695	M	70	670	42	-	5.3	1.1	0.69	32	0.33	15	55	5.0	1.0	0.34	24	0.31
14	6696	M	78	710	42	-	4.6	0.9	0.98	30	0.28	13	43	4.7	0.9	0.53	23	0.29
Mean			84	700	42	4.5	5.2	1.1	0.89	28	0.33	6	54	5.4	1.2	0.71	25	0.35
SD			19	71	4	1.1	0.5	0.2	0.24	5	0.03	9	10	1.1	0.5	0.30	8	0.07
15	6639	A	71	620	34	-	5.4	1.2	0.65	26	0.35	14	50	5.1	1.0	0.35	20	0.34
16	6640	A	62	470	37	-	5.3	1.1	0.9	27	0.36	8	47	5.5	1.2	0.71	23	0.37
17	6693	A	48	410	52	-	7.2	2.1	0.37	31	0.47	19	72	6.1	1.5	0.47	20	0.40
18	6816	A	72	770	50	4.0	7.8	2.3	0.68	35	0.51	12	61	6.9	1.9	0.87	29	0.46
19	6472	A	85	880	47	-	5.6	1.3	0.69	34	0.35	12	48	5.0	1.0	0.28	27	0.31
20	6470	A	70	690	35	-	6.2	1.6	0.47	41	0.39	33	46	4.8	0.9	0.41	24	0.30
21	6536	A	68	580	40	-	4.6	0.9	0.97	29	0.30	16	43	3.8	0.6	0.74	20	0.26
22	6544	A	69	730	45	-	4.9	1.0	0.94	32	0.31	16	51	4.7	0.9	0.42	23	0.29
23	6555	A	75	800	44	-	5.1	1.0	0.84	31	0.32	12	45	4.1	0.7	0.54	24	0.27
Mean			69	661	43	-	5.8	1.4	0.72	32	0.37	16	51	5.1	1.1	0.53	23	0.33
SD			10	155	7	-	1.0	0.5	0.21	5	0.07	9	9	1.0	0.4	0.20	3	0.07

Table 2b
30 mph Barrier Crash Tests Using a Hybrid III Passenger Dummy

	Head		Chest								Pelvis					
OBS	G	HIC	G	C	V	E	R	D	RQ	TH	G	V	E	R	D	RQ
1	65	440	80	1.4	12.2	6.0	0.57	63	0.82	32	34	7.6	2.3	0.53	44	0.51
2	94	1280	97	3.6	12.9	6.7	0.48	67	0.86	25	32	9.2	3.4	0.38	52	0.61
3	93	1050	104		10.6	4.5	0.81	58	0.83	22	36	8.8	3.1	0.38	45	0.69
4	83	660	79	1.5	11.8	5.6	0.57	62	0.78	20	29	7.9	2.5	0.16	51	0.52
5	76	720	100	2.3	12.0	5.8	0.55	62	0.78	21	42	8.3	2.8	0.24	50	0.54
6	96	920	59	2.9	11.4	5.2	0.57	62	0.82	20	32	8.4	2.8	0.41	50	0.61
7	83	610	87		12.2	6.0	0.71	67	0.89	38	79	9.0	3.2	0.54	44	0.65
Mean	84	811	87	2.3	11.9	5.7	0.61	63	0.83	25	41	8.5	2.9	0.37	48	0.59
SD	11	288	16	0.9	0.7	0.7	0.12	3	0.04	7	17	0.6	0.4	0.14	4	0.07
8	66	970	40	3.8	5.5	1.2	0.98	24	0.36	8	49	5.1	1.0	0.70	20	0.33
9	60	850	40	3.9	5.1	1.0	1.21	23	0.35	12	60	4.9	1.0	1.06	17	0.33
10	52	760	42	5.1	6.0	1.4	1.11	31	0.39	-2	47	6.9	1.9	0.27	33	0.45
11	51	490	37	4.4	4.7	0.9	1.27	25	0.32	-5	49	6.0	1.4	0.78	28	0.40
12	39	470	31		5.2	1.1	0.50	34	0.32	6	40	5.4	1.2	0.31	30	0.34
13	47	600	33		4.8	0.9	0.83	31	0.3	-1	40	5.7	1.3	0.29	32	0.35
14	45	560	35		4.7	0.9	1.0	32	0.29	1	43	6.1	1.5	0.49	31	0.37
Mean	51	671	37	4.3	5.1	1.0	0.98	29	0.33	3	47	5.7	1.3	0.55	27	0.37
SD	9	191	4	0.6	0.5	0.2	0.28	4	0.04	8	7	0.7	0.3	0.30	7	0.04
15	51	520	33		5.3	1.1	0.66	27	0.35	16	56	4.7	0.9	0.69	19	0.31
16	42	450	32		5.6	1.3	0.69	31	0.38	22	59	5.2	1.1	0.92	21	0.35
17	61	520	44		6.3	1.6	0.66	27	0.41	-3	54	7.4	2.2	0.28	29	0.48
18	62	830	44	3.1	5.7	1.3	0.53	25	0.38	1	63	6.2	1.5	0.37	24	0.41
19	53	660	36		5.0	1.0	1.06	31	0.31	2	42	6.1	1.0	0.41	30	0.32
20	50	600	32		5.4	1.2	0.68	36	0.33	3	36	6.2	1.6	0.21	34	0.38
21	43	490	31		4.7	0.9	0.70	31	0.31	14	43	4.1	0.7	0.78	23	0.27
22	48	560	35		5.2	1.1	0.88	35	0.33	8	40	5.7	1.3	0.49	30	0.38
23	46	520	45		4.8	0.9	2.25	23	0.3	1	45	4.9	1.0	0.28	27	0.32
Mean	51	572	38	-	5.3	1.1	0.09	30	0.34	7	49	5.5	1.1	0.48	26	0.36
SD	7	114	6	-	0.5	0.2	0.53	4	0.04	8	10	1.0	0.3	0.24	5	0.06

passenger seating positions. The test data were processed using the Restraint Quotient program and the results are summarized for three conditions for the two seating positions: 1) unbelted (U), 2) manual lap-shoulder belt restrained (M), and 3) automatic (passive) lap-shoulder belt restrained (A). While the Restraint Quotient program was developed to evaluate belt and airbag systems, the analysis can be applied to unbelted occupants for comparative purposes.

The head HIC and chest acceleration were significantly reduced by safety belt use in the driver and passenger seating positions. There was a 31% reduction in HIC (980 v 678, p < .002) in the driver position and a 24% reduction (811 v 616, p < .05) for the passenger. There was a larger reduction in chest acceleration averaging 56% for the driver (95 v 42, p < .001) and 57% for the passenger (86 v 37, p < .001).

The use of safety belts significantly reduced chest velocity with respect to the interior from an average of 10.1 m/s for the unbelted driver to 5.5 m/s for the belted driver (a 45.5% reduction, p < .001). There was a 11.9 m/s velocity for the unbelted passenger, which was reduced to 5.3 m/s with belt use (a 55.5% reduction, p < .0001). These changes resulted in a statistically significant reduction in the Restraint Quotient with safety belt use. It reduced RQ_c from 0.71 for the unbelted driver to 0.36 (a 49.3% reduction, p < .001) and from 0.83 for the passenger to 0.34 with belt use (a 59.0% reduction, p < .0001).

Because of greater pelvic than chest restraint for the unbelted dummies in both seating positions by early knee contact (RQ_p = .56 v .71 for the driver and RQ_p = .59 v .83 for the passenger), the upper torso rotated forward in the crash more in the unbelted test condition than with belt use. There was no statistical difference in torso angle for the unbelted and belted driver (θ = 18.6° v 11°, N.S.). However, the change in torso angle was statistically different in the passenger seating position (θ = 25.2° v 5°, p < .0001).

Manual vs Automatic Lap-Shoulder Belt Responses:
Although there were slight differences in average responses between the manual and automatic lap-shoulder belt tests, there was no statistical significance in the differences in 31 out of 32 comparisons. In the one case where there was a statistical difference, the driver's torso angle was 5.7° with manual safety belts and 15.6° with the automatic belts (p < .05). A similar effect was not seen in torso angle in the passenger seating position. The data indicate comparable performance between the two restraint systems.

All Belted Driver vs Belted Passenger Responses:
In general, the biomechanical and kinematic responses were similar in the two seating positions with safety belt use. There were three response comparisons out of 32 that were statistically different. The peak head acceleration was 76 g for the driver and 51 g for the passenger (p < .0001), the peak chest acceleration was 42 g for the driver and 37 g for the passenger (p < .0005) and the torso angle was greater for the driver than passenger (11° v 5°, p < .05).

Figure 6a-d shows examples of the Restraint Quotient analysis for unbelted and belted occupants in 30 mph barrier crash tests. In Figure 6a, chest velocity increased to 9.7 m/s resulting in an RQ_c = 0.61. Because of greater pelvic restraint by knee contact (RQ_p = 0.37), the peak velocity of the pelvis was 40% lower than the chest. This resulted in a 22° forward rotation of the torso. In contrast, Figure 6b shows a manual lap-shoulder belted driver. The chest velocity increased to 4.6 m/s and resulted in an RQ_c = 0.28 as well as a significantly lower forward displacement of the chest than occurred with the unbelted occupant (30 cm v 64 cm, p < 0.0001).

Figure 6c shows barrier test results for an automatic lap-shoulder belted driver. The chest developed slightly higher velocity than with manual belts and an RQ_c = 0.32 with a similar forward displacement and torso rotation. Figure 6d shows a lap-shoulder belted driver with supplemental air bag. The results of the Restraint Quotient analysis yielded an RQ_c = 0.28, forward displacement of 27 cm, and a lower rebound velocity than observed in other barrier tests analyzed.

Hyge Sled Tests: Table 3 provides Hyge sled test data for various exposure severities and restraint conditions. This matrix of tests provides a rich set of biomechanical responses to evaluate and compare trends in the 17 parameters determined for each experiment. Figures 7a-d shows examples of the Restraint Quotient analysis for a selection of test conditions showing changes in responses with crash severity. At the highest crash severity, there is an increase in head responses for the lap-shoulder belted occupant, but a dissimilar effect with the use of a supplemental airbag. In contrast, chest deflection and Viscous response remain low for belt restrained occupants irrespective of impact severity.

DISCUSSION

One of the key points addressed in this analysis is whether the motion sequence criteria espoused by Adomeit (5-7), velocity criteria presented here and other biomechanical measures (2,8) provide necessary and sufficient conditions to ensure a quality restraint system. It is not possible to answer the question fully in this paper but the criteria are probably not necessary to ensure the adequacy of an occupant safety system. Consider a radically different concept of occupant protection which focuses on, for example, rotation of the occupant's seat to place the individual in a spine horizontal position during a crash. This concept would necessarily violate many of the motion sequence criteria yet may be a viable -- although hypothetical -- system.

Based on the current concepts of occupant restraint, motion and velocity criteria seem to

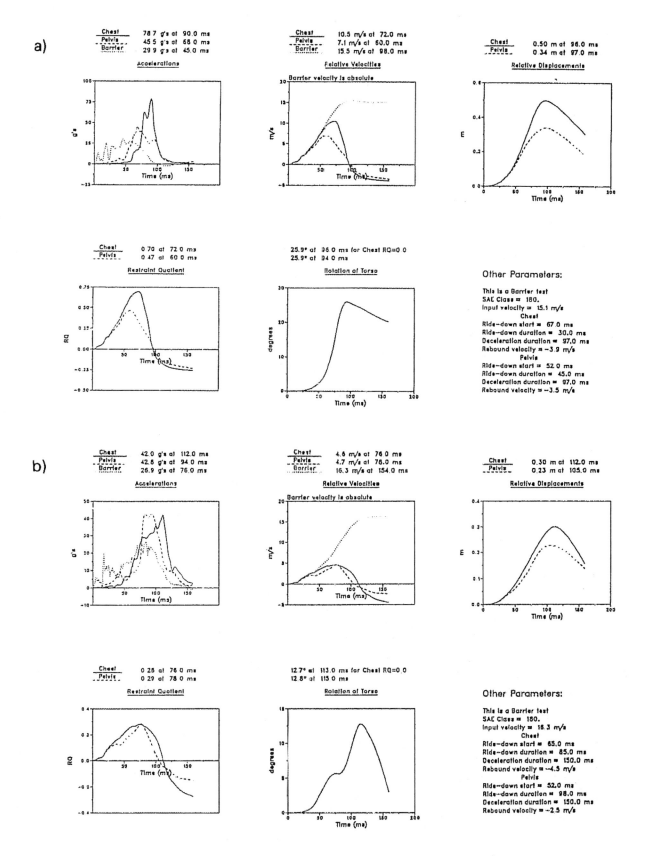

Figure 6: Examples of Restraint Quotient analysis of 30 mph barrier crash data for: a) an unbelted driver, b) a manual lap-shoulder belted driver, c) an automatic lap-shoulder belted driver, and d) a lap-shoulder belted driver with supplemental airbag.

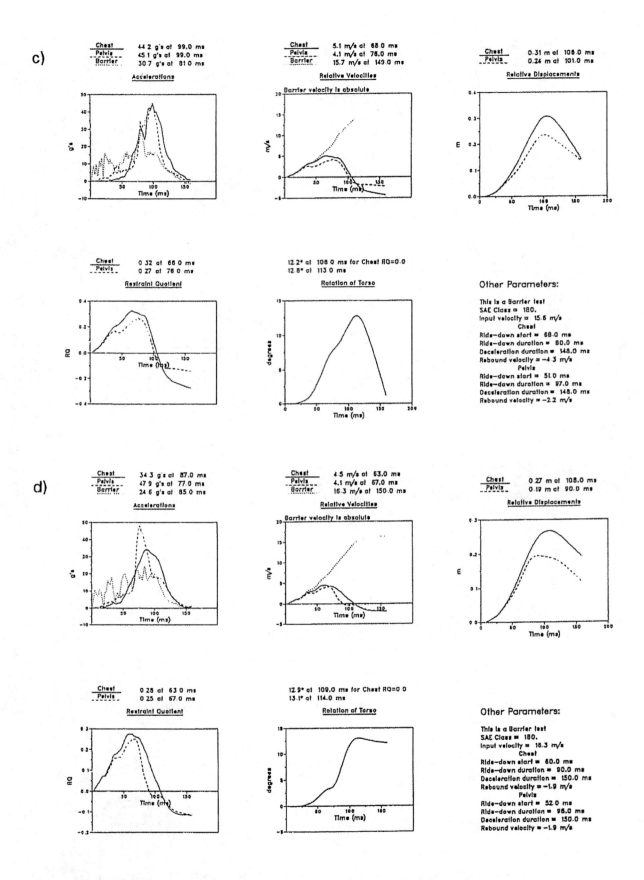

c)

Chest	44.2 g's at 99.0 ms
Pelvis	45.1 g's at 99.0 ms
Barrier	30.7 g's at 81.0 ms

Accelerations

Chest	5.1 m/s at 68.0 ms
Pelvis	4.1 m/s at 76.0 ms
Barrier	15.7 m/s at 149.0 ms

Relative Velocities

Barrier velocity is absolute

Chest	0.31 m at 108.0 ms
Pelvis	0.24 m at 101.0 ms

Relative Displacements

Chest	0.32 at 68.0 ms
Pelvis	0.27 at 76.0 ms

Restraint Quotient

12.2° at 106.0 ms for Chest RQ=0.0
12.6° at 113.0 ms

Rotation of Torso

Other Parameters:

This is a Barrier test
SAE Class = 180.
Input velocity = 15.6 m/s
Chest
Ride-down start = 68.0 ms
Ride-down duration = 80.0 ms
Deceleration duration = 148.0 ms
Rebound velocity = -4.3 m/s
Pelvis
Ride-down start = 51.0 ms
Ride-down duration = 97.0 ms
Deceleration duration = 148.0 ms
Rebound velocity = -2.2 m/s

d)

Chest	34.3 g's at 87.0 ms
Pelvis	47.9 g's at 77.0 ms
Barrier	24.6 g's at 65.0 ms

Accelerations

Chest	4.5 m/s at 63.0 ms
Pelvis	4.1 m/s at 67.0 ms
Barrier	16.3 m/s at 150.0 ms

Relative Velocities

Barrier velocity is absolute

Chest	0.27 m at 108.0 ms
Pelvis	0.19 m at 90.0 ms

Relative Displacements

Chest	0.26 at 63.0 ms
Pelvis	0.25 at 67.0 ms

Restraint Quotient

12.9° at 109.0 ms for Chest RQ=0.0
13.1° at 114.0 ms

Rotation of Torso

Other Parameters:

This is a Barrier test
SAE Class = 180.
Input velocity = 16.3 m/s
Chest
Ride-down start = 60.0 ms
Ride-down duration = 90.0 ms
Deceleration duration = 150.0 ms
Rebound velocity = -1.9 m/s
Pelvis
Ride-down start = 52.0 ms
Ride-down duration = 98.0 ms
Deceleration duration = 150.0 ms
Rebound velocity = -1.9 m/s

be complementary properties of a quality restraint system in comparative tests because they represent fundamentally important aspects of an occupant's kinematic response. The use of other biomechanical responses of the occupant to assess the risk of critical body injury is another important aspect of restraint evaluation. Injury assessment should be based on the best supported criteria and appropriate biofidelity in the test dummy to interpret the potential for occupant injury, and an analysis of the system's performance in managing occupant energy and enhancing protection in a crash.

Based on current procedures to evaluate the performance of restraint systems, the following criteria are considered a comprehensive set of properties determining the relative quality of restraint function. However, specific guidelines may depend on the particular type of safety system and crash test. Restraint criteria involve two types of information, one dealing with motion sequence and velocity criteria, and the other, occupant injury assessment:

Motion Sequence and Velocity Criteria
- Torso angle
- Pelvic rotation
- Pelvic (H-point) displacement
- Restraint Quotient
- Rebound velocity

Injury Assessment
- Contact HIC
- Facial impact force
- Neck bending, shear and extension
- Chest compression, shear, Viscous response, and spinal acceleration
- Abdominal compression
- Femur load
- Knee shear
- Foot-ankle load

Motion and Velocity Criteria: In terms of motion sequence and velocity criteria, there isn't much biomechanical data to validate limits on the parameters. For the purposes of comparing responses to a reference value, the recommendations of Adomeit (5-7) and limits set by judgment are used to normalize observed responses from individual tests. If each motion sequence criteria is equally weighted in determining the quality of belt restraint, a sum of the six criteria can be made determining a single performance index for the Motion Criteria (MC).

Table 3
Hygo Sled Tests Using a Hybrid III Dummy

Test				Head		Chest									Pelvis					
OBS	RUN	Speed	RES	G	HIC	G	C	VC	V	E	R	D	RQ	TH	G	V	E	R	D	RQ
1	1511	8.28	U	20.98	22.16	13.7	1.62	0.05	3.5	0.49	0.34	26	0.58	-12.9	9.3	4.3	0.74	0.00	41	0.71
2	1518	8.94	U	102.07	195.89	27.9	3.37	0.15	5.2	1.08	0.48	38	0.57	17.0	16.0	6.1	1.49	0.00	67	0.67
3	1345	14.22	U	98.97	348.04	41.0	5.70	0.78	7.2	2.07	0.64	47	0.54	44.5	58.2	5.4	1.18	1.04	28	0.41
4	1320	14.22	U	214.35	889.88	83.0	5.93	1.43	8.7	1.79	0.73	42	0.47	41.3	71.4	5.3	1.12	0.83	21	0.37
5	1338	17.17	U	114.13	897.52	96.7	7.20	3.48	9.2	3.38	0.72	46	0.58	39.5	106.4	6.9	1.90	0.98	27	0.42
6	1510	8.28	BD	16.30	37.62	14.9	1.05	0.02	2.2	0.19	1.09	16	0.36	-18.4	6.9	4.3	0.74	0.00	49	0.71
7	1512	8.94	BD	28.33	137.95	27.4	1.52	0.04	4.5	0.81	0.36	29	0.43	-33.2	11.8	7.1	2.01	0.00	90	0.67
8	1310	14.75	BD	38.17	354.59	32.0	2.24	0.05	3.6	0.52	1.03	14	0.26	-1.5	36.2	3.3	0.43	0.64	16	0.24
9	1318	14.75	BD	66.57	398.81	34.7	2.52	0.08	3.8	0.57	0.84	18	0.27	0.8	39.0	3.4	0.46	0.68	17	0.24
10	1319	17.32	BD	148.26	1028.95	41.8	2.76	0.13	4.6	0.84	1.24	17	0.28	-0.3	40.4	4.1	0.67	0.88	17	0.25
11	1340	17.32	BD	131.74	1021.19	49.0	4.08	0.20	4.1	1.16	1.04	23	0.32	2.9	65.1	5.1	1.04	0.69	22	0.31
12	1311	17.32	BD	132.03	648.07	38.0	3.17	0.28	4.2	0.70	0.95	16	0.25	-1.0	42.7	4.0	0.64	0.62	17	0.24
13	1518	8.94	BA	27.25	101.03	25.9	2.43	0.08	2.9	0.34	1.24	18	0.32	-21.3	9.7	6.4	1.64	0.00	75	0.71
14	1316	14.94	BA	44.64	329.96	32.6	3.24	0.15	3.2	0.41	0.93	12	0.22	0.4	33.9	3.2	0.41	0.62	12	0.22
15	1317	17.14	BA	56.89	552.89	40.8	3.16	0.23	3.8	0.57	1.13	15	0.23	2.8	40.9	3.8	0.58	0.89	13	0.23
16	1511	8.28	A	20.98	22.16	13.7	1.62	0.05	3.5	0.49	0.34	26	0.58	-12.9	9.3	4.3	0.74	0.00	41	0.71
17	1517	8.94	A	21.51	62.89	25.1	2.61	0.03	3.9	0.61	0.69	23	0.43	-24.6	10.2	6.0	1.44	0.00	74	0.66
18	1314	14.81	A	47.06	351.99	37.5	3.40	0.55	4.1	0.87	1.02	21	0.29	11.7	32.3	3.9	0.61	0.64	15	0.27
19	1315	17.03	A	132.36	578.99	68.5	-	-	5.7	1.29	1.12	30	0.35	20.1	47.3	4.4	0.77	0.89	19	0.27
20	1530	8.94	BP	28.74	163.87	32.5	2.40	0.08	2.9	0.34	2.17	15	0.31	-2.6	23.6	3.2	0.41	1.47	19	0.34
21	1309	14.97	BP	46.04	347.43	31.2	2.64	0.10	3.8	0.58	0.79	19	0.28	0	32.4	3.5	0.49	0.57	19	0.24
22	1342	17.51	BP	73.38	1254.64	59.7	3.71	0.25	6.5	1.69	0.87	29	0.39	2.9	59.7	6.4	1.64	0.78	28	0.39
23	1308	17.51	BP	56.77	661.93	39.7	3.12	0.18	4.5	0.81	0.80	21	0.28	2.1	47.0	4.3	0.74	0.37	20	0.25

U Unbolted Driver
BD Lap-Shoulder Belted Driver
BA Lap-Shoulder Belted and Supplemental Driver Airbag
A Driver Airbag Only
BP Lap-Shoulder Belted Passenger

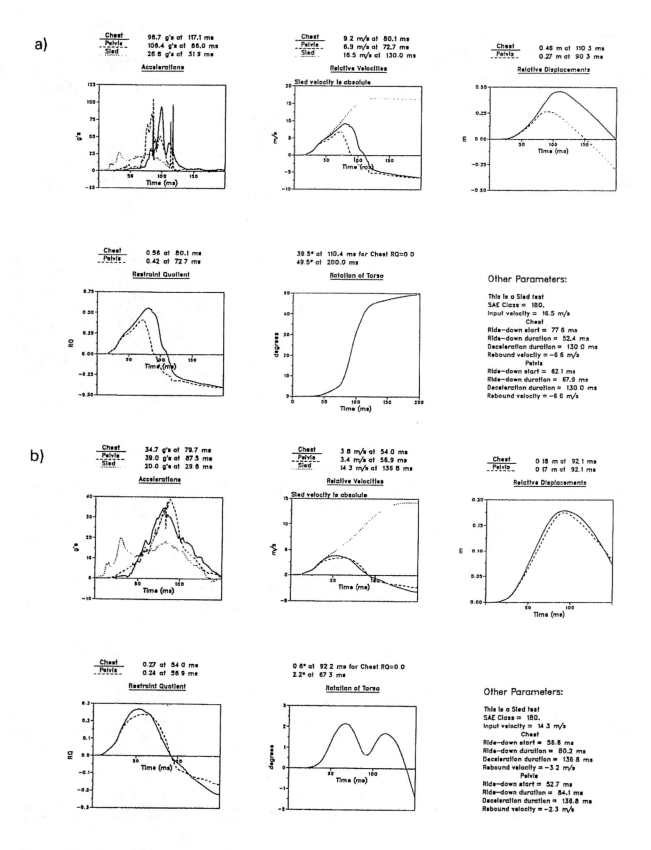

Figure 7: Examples of Restraint Quotient analysis of Hyge sled tests in a body buck using: a) an unbelted driver in a 17.2 m/s test, b) a lap-shoulder belted driver in a 14.8 m/s test, c) a lap-shoulder belted driver in a 17.3 m/s test, and d) a lap-shoulder belted driver with supplemental airbag in a 17.1 m/s test.

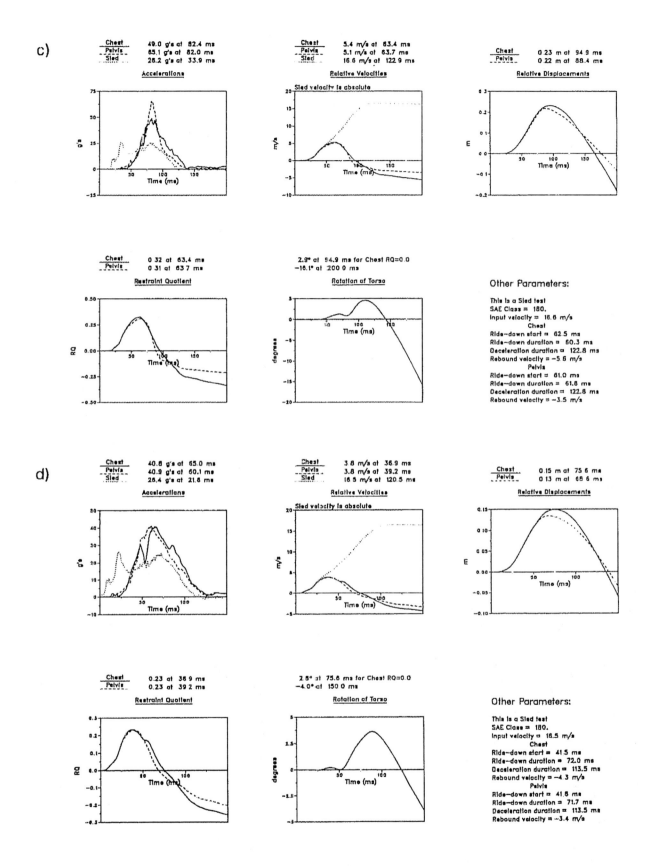

c)

	Chest	49.0 g's at 82.4 ms
	Pelvis	65.1 g's at 82.0 ms
	Sled	28.2 g's at 33.9 ms

Accelerations

	Chest	5.4 m/s at 63.4 ms
	Pelvis	5.1 m/s at 63.7 ms
	Sled	16.6 m/s at 122.9 ms

Relative Velocities

Sled velocity is absolute

	Chest	0.23 m at 94.9 ms
	Pelvis	0.22 m at 88.4 ms

Relative Displacements

	Chest	0.32 at 63.4 ms
	Pelvis	0.31 at 63.7 ms

Restraint Quotient

2.9° at 94.9 ms for Chest RQ=0.0
−16.1° at 200.0 ms

Rotation of Torso

Other Parameters:

This is a Sled test
SAE Class = 180.
Input velocity = 16.6 m/s
    Chest
Ride-down start = 62.5 ms
Ride-down duration = 60.3 ms
Deceleration duration = 122.8 ms
Rebound velocity = −5.6 m/s
    Pelvis
Ride-down start = 61.0 ms
Ride-down duration = 61.8 ms
Deceleration duration = 122.8 ms
Rebound velocity = −3.5 m/s

d)

	Chest	40.6 g's at 65.0 ms
	Pelvis	40.9 g's at 60.1 ms
	Sled	26.4 g's at 21.6 ms

Accelerations

	Chest	3.8 m/s at 36.9 ms
	Pelvis	3.8 m/s at 39.2 ms
	Sled	16.5 m/s at 120.5 ms

Relative Velocities

Sled velocity is absolute

	Chest	0.15 m at 75.6 ms
	Pelvis	0.13 m at 66.6 ms

Relative Displacements

	Chest	0.23 at 36.9 ms
	Pelvis	0.23 at 39.2 ms

Restraint Quotient

2.5° at 75.6 ms for Chest RQ=0.0
−4.0° at 150.0 ms

Rotation of Torso

Other Parameters:

This is a Sled test
SAE Class = 180.
Input velocity = 16.5 m/s
    Chest
Ride-down start = 41.5 ms
Ride-down duration = 72.0 ms
Deceleration duration = 113.5 ms
Rebound velocity = −4.3 m/s
    Pelvis
Ride-down start = 41.8 ms
Ride-down duration = 71.7 ms
Deceleration duration = 113.5 ms
Rebound velocity = −3.4 m/s

959

Since four of the six criteria are specified by the Restraint Quotient program, the following subset of responses is used for comparative purposes in this study:

$$MC_1 = \left[\frac{30° - \theta}{30°}\right] \begin{array}{l}torso \\ angle\end{array} + \left[\frac{RQ_c}{.75}\right] \begin{array}{l}chest \\ restraint \\ quotient\end{array}$$

$$+ \left[\frac{V_R}{5\ m/s}\right] \begin{array}{l}rebound \\ velocity\end{array} + \left[\frac{d_{ph}}{25\ cm}\right] \begin{array}{l}H\text{-}pt \\ hor.\ disp.\end{array} \qquad (10)$$

The two other criteria are:

$$MC_2 = \left[\frac{d_{pv}}{5\ cm}\right] \begin{array}{l}H\text{-}pt \\ ver.\ disp.\end{array} + \left[\frac{30° - \phi}{30°}\right] \begin{array}{l}pelvic \\ rotation\end{array} \qquad (11)$$

so that the combined criteria becomes $MC = MC_1 + MC_2$.

Comparing the results of MC_1, for experiments involving lap-belt restraint and submarining, yield the following calculation for lap-belt restraint (Test 1675, Table 1):

$$MC_1 = \left[\frac{30° - 14.8°}{30°}\right] + \left[\frac{.39}{.75}\right] + \left[\frac{3.7}{5.0}\right] + \left[\frac{16}{25}\right]$$

$$MC_1 = 2.41$$

and for lap-belt submarining (Test 1678):

$$MC_1 = \left[\frac{30° + 22.5°}{30°}\right] + \left[\frac{.33}{.75}\right] + \left[\frac{5.0}{5.0}\right] + \left[\frac{37}{25}\right]$$

$$MC_1 = 4.67$$

Lap-belt submarining increases the Motion Sequence Criteria 94% above the level with lap-belt restraint. This increase is in spite of reduced FMVSS 208 numbers with submarining. The occupant kinematics shown in Figure 8 involve significant lap-belt loading on the abdomen with submarining. The confirmation of potential injury can now be made in such tests by the use of a frangible abdomen in the dummy (2). The average abdominal crush in comparable tests was 8.5 cm with lap-belt submarining and 0.0 cm with lap-belt restraint. The crush can be interpreted as a probability of occupant injury. These submarining tests resulted in a high probability of severe abdominal injury (p > 48%), whereas lap-belt restraint involved no serious injury risk to the abdomen.

Weighted Tolerance Criteria: In terms of injury assessment, each biomechanical criterion has a tolerance level which is based on an analysis of injury from a series of impact exposures. A common approach to setting tolerance has been to select the response at which about 25% of the tested population experience serious injury. This approach is consistent with current tolerance levels in Federal Motor Vehicle Safety Standards and those set for chest compression and Viscous response (9). However, the various criteria do not have the same significance in terms of the actual occurrence of occupant injury in real-world crashes.

It is possible to use the distribution in human injury to weight the relative significance of individual responses from dummy tests. In this way, the importance of head injury is emphasized. This is important because of the risks of permanent disability and death associated with central nervous system trauma. One approach normalizes each response by its tolerance value and then multiplies it by a weighting factor which, when summed over all injury measures, provides a single value for overall injury risk. The approach helps balance the interpretation of various responses based on current occupant injury patterns. The approach can also be extended to injury disability. This helps avoid emphasis on a single response or parameter at the expense of other injury problems.

Table 4 gives the distribution in occupant injury by body region from two sources for unbelted and lap-shoulder belted occupants (14,15). Much of the field injury data isn't collected in a manner that is identical to the available measurements made in test dummies, and there are differences in injury severity and weighting that complicate direct comparison. However, it is possible to provide a reasonable weighting based on the injury data. For simplicity, the weighting is assumed equal for belted and unbelted occupants, and for occupants in the driver and passenger seating position. Obviously, as more refinements are made in the approach, separate procedures can be considered.

Table 5 provides a possible approach to determining occupant protection from dummy responses in crash tests. Since several responses can be measured in a single body region, the approach equally weights each response within a category so that the sum is representative of the overall significance of that body region to the current distribution of human injury in automotive crashes. Recognizing the wide distribution of crash and injury types involving occupants and injury, this is a simple approach; but, it does provide a way to balance interpretations of test data.

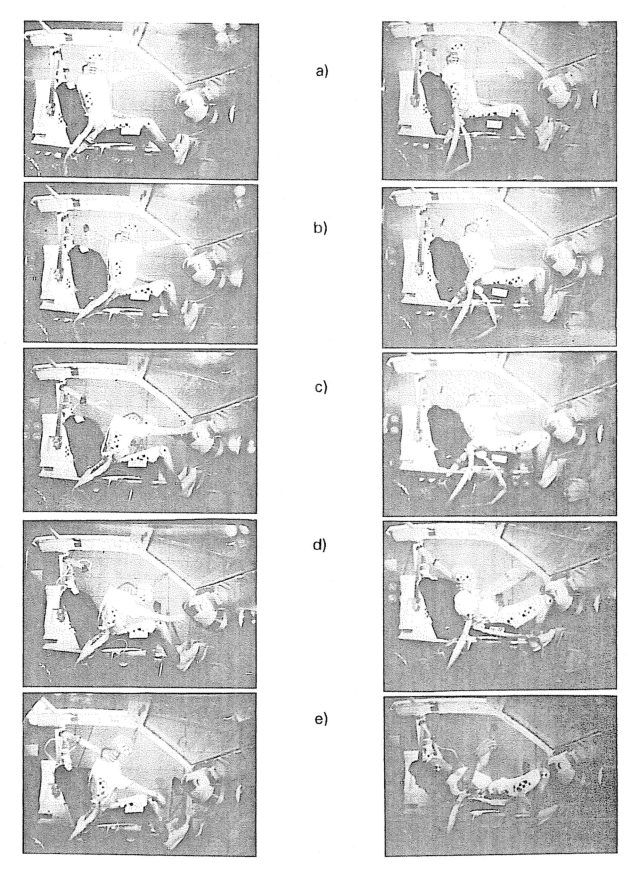

Figure 8: Sequence photographs from the high-speed movies of sled tests with lap-belt restraint (test #1675 - left sequence) and lap-belt submarining (test #1678 - right sequence).

Table 4

Distribution in Occupant Injury for Unrestrained
and Safety Belted Front Seat Occupants

Body Region	Unrestrained		Lap-Shoulder Belted	
	USA[1]	Sweden[2]	USA	Sweden
Head	43.2	31.7/20.7	73.1	22.3/19.7
Face	8.5	9.2/10.3	11.9	15.2/7.5
Neck	9.3	–	1.9	–
Chest	22.1	15.5/21.0	4.1	23.3/27.9
Abdomen	6.6	18.2/27.4	1.7	18.4/26.1
Extremities	6.2	21.2/19.3	7.0	20.2/18.3
Other	4.3	4.2/1.3	0.5	0.7/0.4
	5,553		258	

[1] Fraction of occupant injury harm.

[2] Fraction of injury for occupants with ISS 11+.
Driver/right-front passenger injuries.

Table 5

Injury Criteria, Tolerances and
Logist Parameters for Injury Risk Assessment

Body Region	Injury Significance	FMVSS 208	Criterion	Tolerance	Risk Function α	β
Head	40%	60%	HIC	1000	5.02	0.00351
			A_{3ms}	>100	–	–
Face	10%	–	F	–	–	–
Neck	10%	–	Faxial	1100 N, 45 ms +	–	–
			Fshear	1100 N, 45 ms +	–	–
			Mextension	57 Nm	–	–
			Mflexion	190 Nm	–	–
Chest	20%	35%	A_{3ms}	60 g	5.55	0.0693
			C	34%	10.49	0.277
			VC	1.0 m/s	11.42	11.56
Abdomen	15%	–	C	48%	5.15	0.108
			VC	–	–	–
Extremity	5%	5%	F	10 kN	7.59	0.660
			A_{3ms}	–	–	–

Injury Criteria (IC) can be combined into a single parameter to assess overall safety performance. By using the data in Table 5:

$$IC = \frac{0.40}{2}\left[\frac{HIC}{1000} + \frac{A_{3ms}}{(A_{3ms})_T}\right]\text{Head}$$

$$+ 0.10 \left[\frac{F}{(F)_T}\right]\text{Face}$$

$$+ \frac{0.10}{4}\left[\frac{F_{ax}}{(F_{ax})_T} + \frac{F_{sh}}{(F_{sh})_T}\right]$$

$$+ \frac{M_f}{(M_f)_T} + \frac{M_e}{(M_e)_T}\right]\text{Neck}$$

$$+ \frac{0.20}{3}\left[\frac{C}{(C)_T} + \frac{VC}{(VC)_T} + \frac{A_{3ms}}{60g}\right]\text{Chest}$$

$$+ 0.15\left[\frac{C}{(C)_T}\right]\text{Abdomen}$$

$$+ \frac{0.05}{3}\left[\frac{F_1}{10\ kN} + \frac{F_r}{10\ kN} + \frac{(A_{3ms})}{(A_{3ms})_T}\right]\text{Extremity} \quad (12)$$

In terms of FMVSS 208 criteria, Equation (12) reduces to:

$$IC_{208} = 0.60\left[\frac{HIC}{1000}\right]\text{Head} + 0.35\left[\frac{A_{3ms}}{60\ g}\right]\text{Chest}$$

$$+ \frac{0.05}{2}\left[\frac{F_1}{10\ kN} + \frac{F_r}{10\ kN}\right]\text{Extremity} \quad (13)$$

It is possible to use the terms in Equation (13) for tests 1675 and 1678 involving lap-belt restraint and submarining. This yields a value of $IC_{208} = 0.793$ for Test 1675 and $IC_{208} = 0.629$ for Test 1678 (Table 6). On this basis lap-belt submarining could be interpreted as lowering injury risk.

If the results of frangible abdomen crush are used to reanalyze occupant injury, another term can be added to Equation (13) for abdominal injury assessment. This yields a different interpretation of injury risk, since:

$$IC_{208*} = 0.60\left[\frac{HIC}{1000}\right]\text{Head}$$

$$+ 0.20\left[\frac{A_{3ms}}{60\ g}\right]\text{Chest} + 0.15\left[\frac{C}{48\%}\right]\text{Abdomen} \quad (14)$$

$$+ \frac{0.05}{2}\left[\frac{F_1}{10\ kN} + \frac{F_r}{10\ kN}\right]\text{Extremity}$$

and $IC_{208*} = 0.676$ for Test 1675 and $IC_{208*} = 0.688$ for 1678 (Table 7). Emphasis on abdominal loading now indicates greater injury risk with submarining. An increased injury risk is in agreement with the results of the motion sequence criteria, but the differences are small and don't seem to adequately quantify comparative risks of serious injury.

Injury Risk Assessment: A more informative and the preferred approach for assessing occupant protection involves interpreting dummy responses using an injury risk function which is based on Logist probability analysis of biomechanical data. This enables injury risk to be assessed in each body region.

Table 6 shows the situation using FMVSS 208 responses. Since only one response is measured for the head and chest, injury risk for these body regions can be directly calculated from the probability function in Logist using $p(x) = [1 + \exp(\alpha - \beta x)]^{-1}$. This function relates the probability of injury to a dummy response x using two parameters α and β, which are determined from the best fit using a sigmoidal relationship to approximate biomechanical data on impact injury.

In terms of interpreting risks for the extremities, where two independent leg responses are measured for each test, the overall risk is the sum of the two risks. Using this approach and the FMVSS 208 responses, the lap-belt restraint test results in a risk of p = 21.2% probability of serious injury and lap-belt submarining a lower risk of p = 11.6%.

If abdominal injury assessment is included in the interpretation of performance, the risk of serious injury with lap-belt restraint is identical, at p = 21.2%, whereas there is a 5-fold increase in perceived risk of injury to p = 59.6% with lap-belt submarining (see Table 7). This is due to significant crush of the frangible abdomen.

Augmenting FMVSS 208 responses seems to provide a more relevant interpretation of restraint performance since risk is reported as a probability for critical body regions. The overall risk of injury is determined by the sum of individual risk. Since individual risk varies from zero to one, the overall risk of injury can exceed p = 100%, but that would be consistent with some crash victims experiencing multiple injuries or causes of death. In addition, any discussion of trade-offs between responses is aided by this approach because of a more specific relation to potential injury in specific body regions.

A strength of injury risk assessment is the form of the sigmoidal function that relates a dummy response to injury risk and setting human tolerance at or below a 25% probability of injury. Tolerance levels are typically set in the low-risk region of the sigmoidal response (Figure 9). Thus, maximum dummy responses below tolerance involve a gradual change in risk with responses in the low-risk region of the sigmoidal function.

Table 6
FMVSS 208 Criteria to Assess Lap-Belt Restraint
and Submarining

Body Region	Injury Significance	Criterion	Tolerance	Risk Function α	β
Head	60%	HIC	1000	5.02	0.00351
Chest	35%	A$_{3ms}$	60 g	5.55	0.0693
Femurs	5%	F	10 kN	7.59	0.660

Evaluation of Lap-Belt Restraint Responses
and Injury Risk

Body Region	Criterion	Test Value 1675	Weighted Criterion	Injury Probability
Head	HIC	863	0.520	0.120
Chest	A$_{3ms}$	46.8	0.273	0.091
Femurs	F	0.65/0.82	0.000	0.001
		Total	0.793	0.212

Evaluation of Lap-Belt Submarining Response
and Injury Risk

Body Region	Criterion	Test Value 1678	Weighted Criterion	Injury Probability
Head	HIC	706	0.424	0.073
Chest	A$_{3ms}$	35.1	0.205	0.042
Femurs	F	0.71/0.63	0.0	0.001
		Total	0.629	0.116

Table 7
FMVSS 208 Criteria Augmented by Abdominal Injury Risk
to Assess Lap-Belt Restraint and Submarining

Body Region	Injury Significance	Criterion	Tolerance	Risk Function α	β
Head	60%	HIC	1000	5.02	0.00351
Chest	20%	A$_{3ms}$	60 g	5.55	0.0693
Abdomen	15%	C	48%	5.15	0.108
Femurs	5%	F	10 kN	7.59	0.660

Evaluation of Lap-Belt Restraint Responses and Injury Risk

Body Region	Criterion	Test Value 1675	Weighted Criterion	Injury Probability
Head	HIC	863	0.520	0.120
Chest	A$_{3ms}$	46.8	0.156	0.091
Abdomen	C	0	0.000	0.000
Femurs	F	0.65/0.82	0.000	0.001
		Total	0.676	0.212

Evaluation of Lap-Belt Submarining Responses and Injury Risk

Body Region	Criterion	Test Value 1678	Weighted Criterion	Injury Probability
Head	HIC	706	0.424	0.073
Chest	A$_{3ms}$	35.1	0.117	0.042
Abdomen	C	47%	0.147	0.480
Femurs	F	0.71/0.63	0.0	0.001
		Total	0.688	0.596

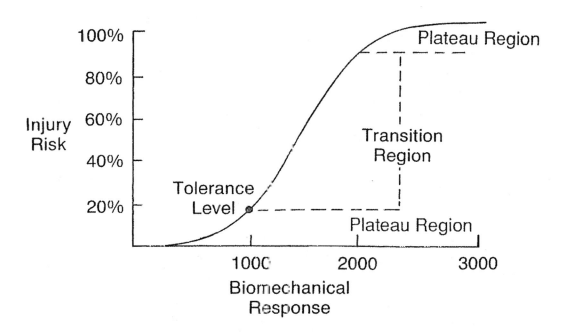

Figure 9: Example of an injury risk function based on HIC and serious head injury. The sigmoidal function is from Logist analysis of biomechanical data.

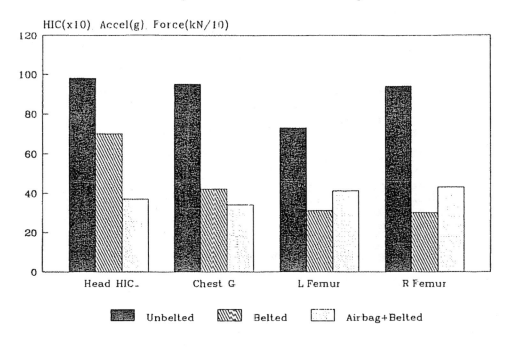

Figure 10: Comparison of FMVSS 208 responses for barrier crash tests with the Hybrid III dummy.

However, if a response approaches or is above tolerance, slight differences in response involve major changes in perceived risk because they occur in the transition region from low to high risk. This increasingly penalizes responses near or above tolerance, until responses are so large that further increases do not significantly modify the nearly absolute probability of injury occurrence. For cases where several responses are made in the same body region, such as in the chest where acceleration, compression and the Viscous response are evaluated, injury risk can be interpreted as the greatest risk from any of the applicable criteria.

Injury risk assessment was used on the average values from the unbelted and manual belted dummies in barrier crash tests, and an additional test involving belt and airbag restraint. Figure 10 shows the FMVSS 208 criteria. Figure 11 shows the overall risk of serious injury based on the analysis in Table 8. There is a 116.8% risk (range 64.6%-209.8% based on one standard deviation in test responses) of serious injury to the head, chest and legs. This is based on a 90.9% (55.7%-138.2%) probability of AIS 4+ injury to the head and chest, and a 28.9% (8.9%-71.6%) probability of AIS 3+ injury to the legs. Safety belts reduce the risk of injury to 14.7% (range 11.1%-19.6%), an 87% reduction from levels with an unbelted driver. The risk is further reduced to 8.0% with the combination of belts and airbag restraint.

Although the above is an informative example of the risk assessment approach, it also shows the uncertainty due to variability in dummy responses. If the confidence interval in the Logist function for injury probability is also considered, the risk range would be larger. However, this seems to be the appropriate way to consider the significance of differences in restraint performance. This approach takes advantage of the sigmoidal form of the injury probability function since dummy responses in the transition region are penalized in comparing occupant protection. Furthermore, the analysis shows that there is injury risk even with belt and airbag restraint.

While injury risk assessment seems to be the most meaningful approach to evaluate occupant restraints, it needs further evaluation, review, and discussion particularly in terms of the Logist parameters in determining the actual distribution of human injury, the adequacy of biomechanical data in quantifying occupant injury, the biofidelity of dummy responses in mimicking human response and validity of measurement approaches. The eventual approach should probably involve a sufficiently instrumented dummy to assess responses in the key body regions and use of appropriate parameters defining the injury probability functions from consensus review of biomechanical data.

Generalizing the Approaches: The Restraint Quotient is different for the chest in barrier and Hyge sled tests with similar restraint conditions (Figure 12). Higher values of RQ_c imply a greater chest velocity with respect to the interior in barrier tests than in comparable Hyge sled experiments. However, both crash types show a 48%-54% reduction in RQ_c with belted versus unbelted occupants. The combination of belts and driver airbag further reduced RQ_c by 15% over belt only restraint. While higher values of RQ_c were observed with airbag only restraint in Hyge sled tests, this type of evaluation involves only frontal crash analysis of restraint performance.

In terms of injury and motion sequence criteria, it is certainly advisable to extend the restraint criteria to consider non-frontal crash conditions and a range in crash severities and circumstances. This would probably involve an envelope of motion limits in the forward and lateral directions of occupant motion. Some work has already been completed on the oblique response of test dummies with safety belt restraints which have demonstrated the value of limiting lateral displacements as an approach to improved occupant protection.

In addition to more global motion sequence criteria, there would be a need to introduce lateral components of injury assessment. This would eventually lead to setting omni-directional restraint criteria which would be compatible with developing procedures and criteria in side impact crashes. This would build upon the work of Horsch (16, 17) and Culver et al (18) for oblique responses of belted dummies and the more recent work on pure lateral impact injury. Obviously, this is a future goal that would require an omni-directional dummy with the interpretation of injury risks and motion sequences based on a range of impact directions and severities occurring in motor vehicle crashes.

The Restraint Quotient concepts lose relevance when the occupant compartment undergoes deformations such that the mean acceleration of the passenger compartment is not a sufficient measure of the impact exposure. This is the case in severe side impacts because the door is rapidly accelerated in the crash and impacts the occupant at high velocity while the passenger compartment experiences a more gradual velocity change. In this situation, the injury risk associated with high-speed impact with the door is based on the Viscous response of the chest and the abdomen. This crash situation also points out the significance of the point of contact and passenger compartment deformation on potential occupant injury (19-21).

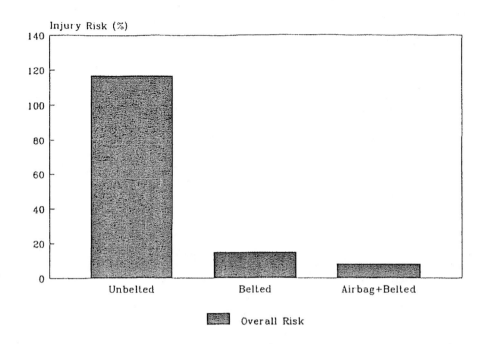

Figure 11: Overall risk of injury based on barrier crash test results.

Table 8

Injury Risk Assessment of Barrier Crash Data

| | HIC | | Injury Risk | |
	Mean	Range	Mean	Range
Unbelted	980	529–1431	17.1	4.1–50.1
Belted	700	629–771	7.2	5.7–9.0
Airbag & Belts	370	–	2.4	–

| | Chest G | | Injury Risk | |
	Mean	Range	Mean	Range
Unbelted	95	81–109	73.8	51.6–88.1
Belted	42	42–46	6.7	5.1–8.6
Airbag & Belts	34	–	3.9	–

| | | Femur kN | | Injury Risk | |
		Mean	Range	Mean	Range
Unbelted	L	7.3	4.4–10.2	5.9	0.9–29.8
	R	9.4	7.8–11.0	20.0	8.0–41.8
Belted	L	3.1	1.8–4.4	0.4	0.2–0.9
	R	3.0	1.3–4.7	0.4	0.1–1.1
Airbag & Belts	L	4.1	–	0.8	–
	R	4.3	–	0.9	–

967

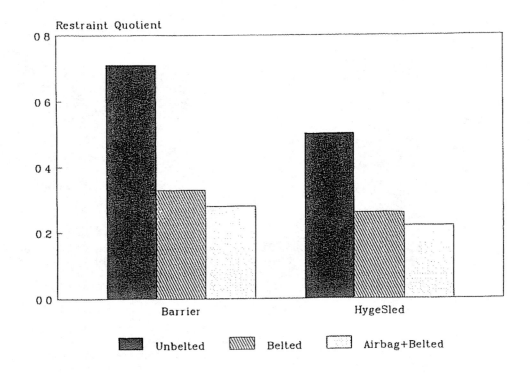

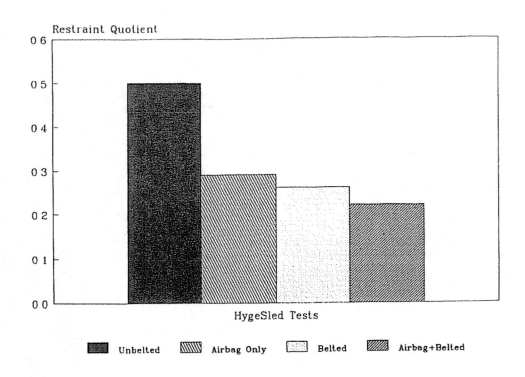

Figure 12: RQ_c for barrier and Hyge sled tests with various restraint conditions (data from Tables 2 and 3).

968

ACKNOWLEDGEMENTS

We would like to recognize the assistance of Kenneth Baron of the Biomedical Science Department for his involvement in writing code and developing graphics for the Restraint Quotient program. The participation of Milan Patel, while a summer employe, is appreciated in the early efforts to develop the method. Special thanks are also due to John Horsch, Stephen Rouhana and the engineering support team in the Biomedical Science Department for conducting the Hyge sled tests analyzed in this study.

REFERENCES

1. Transportation Research Board, "Safety Belts, Airbags and Child Restraints: Research to Address Emerging Policy Questions." Special Report 224, National Research Council, Washington, DC, 1989.

2. Rouhana, S.W., Viano, D.C., Jedrzejczak, E.A., and McCleary, J.D., "Assessing Submarining and Abdominal Injury Risk in the Hybrid III Family of Dummies." In Proceedings of the 33rd Stapp Car Crash Conference, pp. 257-280, SAE Technical Paper #892440, Society of Automotive Engineers, Warrendale, PA, 1989.

3. Rouhana, S.W. and Horsch, J.D., "Assessment of Lap-Shoulder Belt Restraint Performance in Laboratory Testing." In Proceedings of the 33rd Stapp Car Crash Conference, pp. 243-256, SAE Technical Paper #892439, Society of Automotive Engineers, Warrendale, PA, 1989.

4. Horsch, J.D. and Hering, W.E., "A Kinematic Analysis of Lap-Belt Submarining for Test Dummies." In Proceedings of the 33rd Stapp Car Crash Conference, pp. 281-288, SAE Technical Paper #892441, Society of Automotive Engineers, Warrendale, PA, 1989.

5. Adomeit, D., "Seat Design - A Significant Factor for Safety Belt Effectiveness." In Proceedings of the 23rd Stapp Car Crash Conference, pp. 39-68, SAE Technical Paper #791004, Society of Automotive Engineers, Warrendale, PA 1979.

6. Adomeit, D., "Evaluation Methods for the Biomechanical Quality of Restraint Systems During Frontal Impact." In Proceedings of the 21st Stapp Car Crash Conference, pp. 913-932, SAE Technical Paper #770936, Society of Automotive Engineers, Warrendale, PA 1977.

7. Adomeit, D. and Heger, A., "Motion Sequence Criteria and Design Proposals for Restraint Devices in Order to Avoid Unfavorable Biomechanic Conditions and Submarining." In Proceedings of the 19th Stapp Car Crash Conference, pp. 139-165, SAE Technical Paper #751146, Society of Automotive Engineers, Warrendale, PA, 1975.

8. Viano, D.C., A.I. King, J.W. Melvin and K. Weber, "Injury Biomechanics Research: An Essential Element in the Prevention of Trauma." Journal of Biomechanics, 22(5):403-417, 1989.

9. Viano, D.C., and Lau, I.V., "A Viscous Tolerance Criterion for Soft Tissue Injury Assessment." Journal of Biomechanics, 21(5):387-399, 1988.

10. Viano, D.C., "Biomechanics of Head Injury: Toward a Theory Linking Head Dynamics Motion, Brain Tissue Deformation and Neural Trauma." SAE Transactions, vol. 97, 1988, 32nd Stapp Car Crash Conference, SAE Technical Paper #881708, pp. 1-20, October, 1988.

11. Horsch, J.D., "Evaluation of Occupant Protection from Responses Measured in Laboratory Tests." SAE International Congress and Exposition, Detroit, MI. SAE Technical Paper #870222, Society of Automotive Engineers, Warrendale, PA, 1987.

12. Viano, D.C., "Evaluation of the Benefit of Energy-Absorbing Material for Side Impact Protection: Part II." SAE Transactions, vol. 96, 1987, P-202, 31st Stapp Car Crash Conference, SAE Technical paper 872213, pp. 205-224 November, 1987.

13. Rouhana, S.W. Personal communication.

14. Carsten, O., "Relationship of Accident Type to Occupant Injuries." UMTRI-86-15, Final Report, The University of Michigan Transportation Research Institute, Ann Arbor, MI, 1986.

15. Nygren, A., "Injuries to Car Occupants - Some Aspects of the Interior Safety of Cars." Acta Oto-Laryngologica, Supplement 395, ISSN 0365-5237, The Almqvist & Wiksell Periodical Company, Stockholm, Sweden.

16. Horsch, J.D., "Occupant Dynamics as a Function of Impact Angle and Belt Restraint." In Proceedings of the 24th Stapp Car Crash Conference, pp. 417-438, SAE Technical Paper #801310, Society of Automotive Engineers, Warrendale, PA, 1980.

17. Horsch, J.D., Schneider, D.C., Kroell, C.K., and Raasch, F.D., "Response of Belt-Restrained Subjects in Simulated Lateral Impacts." In Proceedings of the 23rd Stapp Car Crash Conference, pp. 71-103, SAE Technical Paper #791005, Society of Automotive Engineers, Warrendale, PA, 1979.

18. Culver, C.C. and Viano, D.C., "Influence of Lateral Restraint on Occupant Interaction with a Shoulder Belt or Preinflated Air Bag in Oblique Impacts." In SAE Transactions, vol. 90, SAE Technical Paper #810370, Society of Automotive Engineers, Warrendale, PA, 1981.

19. Evans, L. and Frick, M., "Seating Position in Cars and Fatality Risk." Am J Pub Health 78:1456-1458, 1988.

20. Thomas, C., Koltchakian, S., Tarriere, C., et al, "Primary Safety Priorities in View to Technical Feasibility Limited to Secondary Automotive Safety." In Proceedings of the 1990 FISITA Conference, Milan, Italy, 1990.

21. Grosch, L., Baumann, K.H., Holtze, H. and Schwede, W., "Safety Performance of Passenger Cars Designed to Accommodate Frontal Impacts with Partial Barrier Overlap." In Automotive Frontal Impacts SP-782, pp. 29-36, SAE Technical Paper #890748, Society of Automotive Engineers, Warrendale, PA, 1989.

905139

Effectiveness of Pretensioners on the Performance of Seat Belts

F. Zuppichini, Verona University Medical School
Y. Håland, Electrolux Autoliv

ABSTRACT

During a frontal impact, a normal three-point seat belt allows forward movements, with possible violent contacts with steering wheel and dashboard; this situation can be improved by pretensioning of the belt. Pretensioners are projected to actively withdraw a portion of belt (about 10–15cm) at the very beginning of impact, in order to keep the car occupant adherent to his seat.

This experimental study was performed on a mechanical pretensioner, on a pyrotechnical pretensioner, on a web-clamping system, and on normal three-point seat belts with and without slack at the moment of impact. Each device has been tested by simulation in a frontal impact at 30 mph with an Hybrid dummy.

The following data were looked for: • Head Injury Criterion (HIC and HIC_{36}); • Head velocity during the restraint action; • Head, chest, and pelvis forward displacement; • Chest acceleration (>3 msec); • Belt loading at the shoulder.

The laboratory results demonstrate a good performance of pretensioners; the reduction in HIC and HIC_{36} vs three-point belts is significant, especially when poor wearing conditions are present (slack).

The dynamic differences of the pyrotechnical and the mechanical pretensioners (mainly in the direction of the retraction force) reflect on the results, with different tendencies in chest displacement and head criteria. The reduction in pelvis displacement is particularly important in order to prevent submarining and its related abdominal lesions.

These characteristics may provide, if correctly linked to the specific deceleration pattern of each car model, better protection to passengers.

IMPACT, DECELERATION AND SEAT BELT PRETENSIONING

An ideal safe vehicle is composed by a rigid core and by progressively deformable periferic structures; its interiors are out of contact (or fully padded), and the occupants are perfectly adherent to their seat. In this motorvehicle, the limiting factor to the effectiveness of restraint systems is the conservation of the structural geometry of the car interior ("survival space").

In the actual car, this dampening effect is always partial and disomogeneous, due to lack of structure and/or to the presence of not deformable mechanical devices (e.g., engine or steering components); each model of car has its own deceleration pattern, characterized by a sequence of spikes that can reach dangerous levels.

Both laboratory and epidemiological research have conclusively proved the effectiveness of seat belts in preventing or ameliorating the lesions caused by car accidents[6,8,13]; nevertheless, it is well known how in frontal impacts the face of the driver can hit some parts of the steering wheel, even at low speed [5,6]. The so produced face lesions play an important role in the described "redistribution of lesions" [9,10] following the use of seat belts.

If the belt system is malpositioned, the driver can move forward against the belt itself, or can submarine, producing lesions of various kind ("seat belt syndrome" [2,3,12]), usually located at the abdomen, thorax or side of neck.

Abdominal lesions usually regard peritoneum and small bowel, with pathogenetic mechanisms involving pressure from the belt, compression of gas-filled ansæ, or sudden deceleration alone [2,6]. Thoracic lesions usually consist in fractures of sternum and/or ribs or clavicula, while neck lesions involve vascular structures like the jugular vein or the carotid artery [7].

During the impact, a portion of the belt is extracted from the system, due to the delay from the first contact and the locking of the retractor, to the narrowing of spires around the retractor ("film spool effect"), and then to the stretching of the belt under the pressure of the body. The sum of these effects is referred to as "web pay-out" (Picture 1).

The "ignition" of a restraint system starts with a defined retard from the initial impact; this histeresis, that is characteristic of each single car model, is around 25–35 msec for standard three-point automatic belts.

A first way to get rid of the spool effect is locking the belt otside the retractor, when the occupant body starts moving forward. The **web clamp** provides one or two branches that grasp the web in case of sudden deceleration [1]; of course, this action does not affect slack. Despite their conceptual ease, web clamps need sophysticated engineering to ensure that additional friction is not added while normally operating the belt.

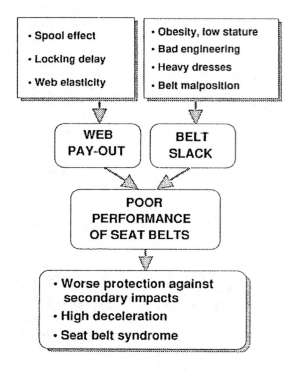

• Spool effect	• Obesity, low stature
• Locking delay	• Bad engineering
• Web elasticity	• Heavy dresses
	• Belt malposition

```
WEB          BELT
PAY-OUT      SLACK
```

```
POOR
PERFORMANCE
OF SEAT BELTS
```

• Worse protection against
  secondary impacts
• High deceleration
• Seat belt syndrome

Picture 1: *Components of web pay-out and factors involved in slack creation.*

The elimination of both web pay-out and slack can be pursued with belt pretensioners [4,5,14]. In its general form, a pretensioner is composed by a crash sensor, a power source and an effector connected to the belt retractor or buckle.

The **pyrotechnical pretensioner** comprises a small explosive charge, that is ignited by an electronic control system when an impact is detected. The electromechanic sensor is based on an inertial mass, that for decelerations over a given threshold switches an electric contact. A central check control detects the changed signal and drives the explosion of the charge, connected with the retractor by means of a steel wire. Pyrotechnical pretensioners were introduced in Mercedes cars.

The **mechanical pretensioner** is powered by a preloaded torsional spring bar, which can be mounted transversally under the seat. Since the torsional bar is hinged as a pendulum, the bar itself is the crash sensor; during the impact, one end of the bar will move forward and an overbent kneejoint mechanism will collapse. This device too has already been adopted and installed in Volvo cars.

Both kinds of pretensioners are activated in frontal impacts, with collision angles within ±30°.

Another kind of mechanical pretensioner, named Procon/ten, is installed as optional on some Audi models. This interesting device is powered by the displacement of the engine during the impact; a steel wire provides both pretensioning of seat belts and retraction of the steering wheel. Procon/ten is active only with severe crashes.

Pretensioners have an intervention time in the range of 10–15 msec; otherwise, the occupant body displacement relative to the car will be too high (>20 mm). The threshold for activation is triggered usually at around 8 g, depending on the specific standards of the car manifacturer.

Adding active components to traditional seat belt systems can be adjusted on the peculiar deceleration pattern of the assigned car model, so to smooth its spikes and furtherly prevent head trauma.

MATERIALS AND METHOD

The experimental study is aimed to evaluate the performance of mechanical and pyrotechnical pretensioners in a simulated car impact, in relation to the performance of a web-clamp and of standard three-point automatic seat belts.

All the tests were performed at the Autoliv Development laboratories in Vårgårda (Sweden), in 1989.

The instrumentation was set as following (Picture 2):
• Crash sled, coupled to a data acquisition system with computer evaluation;
• Soft seat, currently produced, modified and reinforced in its structure;
• 50° percentile Hybrid Dummy, equipped with decelerometers;
• High speed film recorder (1000 photograms/sec).

In each test, one of the following devices were installed on the sled:
• Standard three-point automatic belt, produced by Electrolux-Autoliv in conformity to ECE rules;
• Three-point automatic belt, with a web clamp system produced by Electrolux-Autoliv (KC 1);
• Three-point automatic belt, with a mechanical pretensioner produced by Electrolux-Autoliv (MBP I);
• Three-point automatic belt, with a pyrotechnical pretensioner developed by Electrolux-Autoliv (Type 23 P).

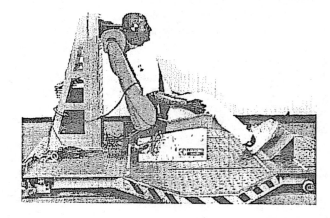

Picture 2: *Crash sled, soft seat and dummy as used in this study.*

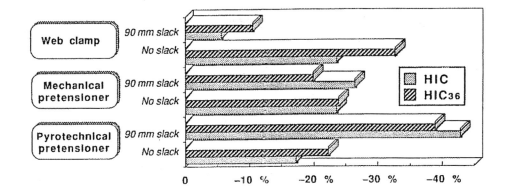

Picture 3: *Reductions in head injury criteria, with and without slack.*

The geometry of the belt was responding to a front installation on an average 4-door sedan with lower anchorage point fitted to the seat; the tests were run without steering wheel or dashboard.

The test conditions were performed both with a perfectly worn belt, and with 90 mm slack, distributed 2/3 at the chest (60 mm) and 1/3 at the pelvis (30 mm). Slack was obtained by inserting foam material between the web and the dummy at the chest and pelvis areas.

The deceleration pulse was set at a maximum of 22 g for the comparative test among web clamp and standard belts, and a maximum of 28 g for the test about mechanical pretensioner and standard belts. For the comparison of the pyrotechnical pretensioner vs standard belts, a bimodal deceleration pulse was choosen, with peaks at 20 and 25 g.

The following data were looked for:
- Head Injury Criterion (HIC and HIC_{36});
- Head velocity during the restraint action;
- Head, chest, and pelvis forward displacement;
- Chest acceleration (>3 msec);
- Belt loading at the shoulder.

In particular, HIC and chest acceleration were reported from the decelerometers, while the forward displacements of the dummy head were measured by a single-photogram computerized review of the high-speed film.

Head forward movement was referred to a fixed point on dummy's head, at the base of temporal squama (head inertial point).

The velocity of impact against a theoretical steering wheel was calculated using head displacement vs time data; this fictitious steering wheel was set responding to an average driver, and was supposed not to be displaced or deformed, by impact forces.

RESULTS

All the results are visualized in pictures 3—10.

Head Injury Criterion (HIC and HIC_{36}) was significantly reduced by the use of pretensioners, with values from -17.3% to -42.8 %. The best target was obtained by a pyrotechnical one with slack, coupled however to a longer head movement (+35 mm).

Web clamps behaved well without slack (-23.5% HIC, -32.6% HIC_{36}), but were less effective with 90 mm slack (only -5.6% HIC, -10.5% HIC_{36}).

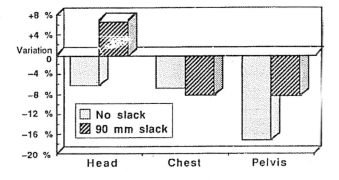

Picture 4: *Variations in head, chest and pelvis displacement with the pyrotechnical pretensioner*

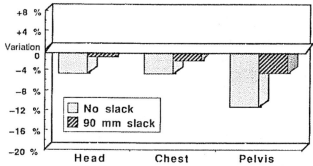

Picture 5: *Variations in head, chest and pelvis displacement with the mechanical pretensioner.*

973

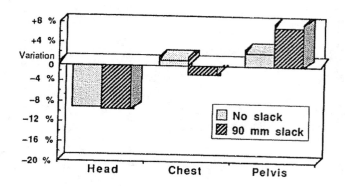

Picture 6: *Variations in head, chest and pelvis displacement with the web-clamp*.

Head displacement was reduced by pretensioners without slack (-5.6% and -4.2%), but not with slack (+6.9% for the pyrotechnical and -0.9% for the mechanical). The web-clamp showed a similar pattern in the two conditions (-7.7% and -8.2%).

Chest displacement reduction was a common finding for all devices; only the web clamp showed a slight increase in absence of slack.

Pelvis displacement was reduced by pretensioners without slack (-16.7% and -10.9%), but not with slack (+7.7% for the pyrotechnical, while -4.1% for the mechanical). The web-clamp increased pelvis displacement in all conditions (+5 mm and +15 mm).

Chest acceleration (>3 msec) was smaller both with the pyrotechnical pretensioner (-11.9%; -2% with slack) and with the mechanical (-8.3%; -15.3% with slack). The reduction was very little for the web clamp (-2% with and without slack).

Shoulder belt force was slightly decreased by the pyrotechnical (-0,2 kN; -0,5 kN with slack), but slightly increased by the mechanical (+0,1 kN; +0,4 kN with slack) and the web-clamp (+0,7 kN; +1 kN with slack).

Head velocity while impacting the steering wheel was reduced by the pyrotechnical only without slack (-3.8%), while with slack it increased by 2.0%. For the mechanical pretensioner, the values were respectively -8.3% and -15.3%. The web clamp (at 22 g) reduced the velocity by 40% without slack, while with 90 mm slack there was no contact against the steering wheel.

The head movement pattern is of difficult interpretation; nevertheless, there is evidence of greater and later deceleration spikes in some impact conditions.

With the mechanical pretensioner, we measured a buckle pulling of 30 mm (with slack, 50 mm) with a deceleration of 28 g; the respective values for the pyrotechnical are 40 mm and 70 mm (at 20–25 g). Of course, the web-clamp does not give any retraction.

CONCLUSIONS

The analysis of the data yields encouraging informations about the performance of the mechanical and pyrotechnical pretensioners.

Both reduction in HIC and in HIC_{36} are statistically significant ($p < 0,01$) and constitute a good predicting factor for the effectiveness of pretensioners in reducing head injuries.

The better performance of the pyrotechnical pretensioner is probably linked to the different and easier deceleration pattern (20–25 g vs 28 g for the mechanical one).

The good performance of pretensioners even in presence of slack can offer great advantages in the prevention of lesions caused by loose seat belts, that are a major contributor to the seat belt syndrome.

Head displacement is not strictly connected to HIC reduction: for instance, the best HIC value was obtained by a pyrotechnical pretensioner with a reduction in chest displacement (-25 mm), but an increase in head displacement (+35 mm).

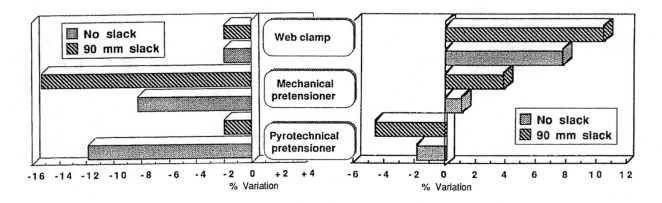

Picture 7: *Variations in chest accelerations (>3 msec)*.

Picture 8: *Variations in belt loading at shoulder*.

974

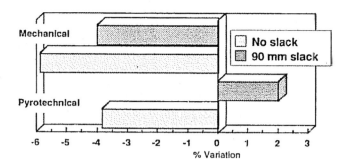

Picture 9: *Variations in the theoretical velocity of impact of head against a steering wheel. Data for mechanical and pyrotechnical pretensioners, with and without slack.*

The high-speed film examination shows that with slack the head tends to "go flat" in the first instants after impact, bending suddenly; this can justify for the high HIC values in impacts with slack and without pretensioning.

Shoulder belt force is slightly decreased by pyrotechnical pretensioners; at the opposite side, the other devices (especially the web-clamp) show meanly an increase of this force.

Chest acceleration is better reduced by the mechanical pretensioner when slack is present.

The retraction force of the mechanical device is directed towards the inner of the car, not towards the B-pillar as for the pyrotechnical one; this feature could also prevent dangerous contacts with side glasses and/or rigid components of car structure.

The retraction is directly performed also on the abdominal tract of the belt, furtherly preventing submarining.

About reducing the velocity of head impact against the steering wheel, the mechanical pretensioner proved to be more effective than the pyrotechnical one. The diagrams show how pretensioning go near to the target of avoiding contacts at all with the steering wheel, even with sharp decelerations.

The mechanical pretensioner has a cost well below the pyrotechnical; it is feasible that this feature could offer pretensioning choice also to lower-class cars, whose passenger compartment is usually smaller. All pretensioners are disposable: once activated, they ought to be replaced with a new spare part.

The mechanical pretensioner overcomes any trouble caused by electric wires and connections, and the possible, even if rare, "decapitation" of the system by cutting away of the battery during a violent crash. The absence of explosive eliminates the caution measures that are necessary while repairing or destroying a pyrotechnical-equipped car.

Pretensioners have no parts in contact with the web, which is a warranty for the integrity of the restraint system in very violent impacts.

Data on performance of pretensioners with child seats and with rear seat belts are not available, although they can be very interesting.

Pretensioning could aggravate the consequences of some patterns of voluntary malposition: for instance, underarm wearing of seat belts [11], or the abnormal slack given by the so-called "comfort clips".

From a practical point of view and beyond the simple numbers, the available data suggest that belt pretensioning is a major issue for everyday use of the restraint system.

The protection of real world car occupants must face a crowd of problems, from a "physiological" amount of slack due to heavy dressing, to the uneasy dissipation of deceleration forces due to comfortable

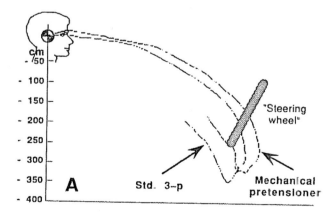
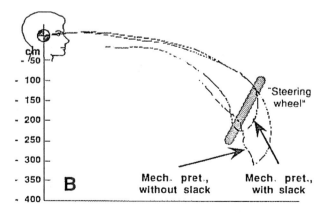

Picture 10. *Head forward movement during impact, referred to head inertial point.*
   A: Pyrotechnical pretensioner vs Standard 3-point belt, with   slack.
   B: Mechanical pretensioner without slack vs Same device with 90 mm slack.

postures or open malposition of belt. Looking for a wide acceptance (and wearing) of the restraint system, it seems more effective to admit a moderate amount of slack, if it can be effectively eliminated in the first milliseconds of impact.

Apart from this behavioural consideration, it could be possible to fit the deceleration pattern of the body to the specific deceleration pattern of the car, with regard to the type of impact and optimizing the filling of the survival space without any secondary impact.

Since a contact lesion is always an "all or nothing" happening, it seems more reasonable to look carefully for protection against secondary impact, but to use all the free space in a car to decelerate the body.

In this concept, the stiffness of the belt is modulated to optimize the deceleration of the body and the flexion of cervical spine.

The available data cannot give a clear sign of relations between pretensioning and changes in whiplash injuries dynamic; further studies are necessary to have a complete understanding of the involved mechanisms.

Since pretensioners seem able to improve the pattern of deceleration for a car occupant, it will be very interesting to investigate their performance in real car accidents.

REFERENCES

1. ADOMEIT D, BALSER W: Items of an engineering program on an advanced web-clamp device. Paper SAE 870328, 119–128, 1987.

2. CHRISTOPHI C, McDERMOTT FT, McVEY I, HUGHES ESR: Seat belt induced trauma to the small bowel. World J Surg, 9:794-797, 1985.

3. GARRETT JW, BRAUNSTEIN PW: The seat belt syndrome. J Trauma, 1962, 2:220-237.

4. HÅLAND Y, SKÅNBERG T: A mechanical buckle pretensioner to improve a three-point seat belt. 12th ESV Conference, Göteborg, Paper 89-5B-0-001, 1989.

5. LOWNE R, ROBERTS A, FENN M, FARWELL A: Potential for improvements in the protection afforded by seat belts. 12th ESV Conference, Göteborg, Paper 89-5B-0-002, 1989.

6. MACKAY GM, HILL J, SMITH M, PARKIN S: The limits of seat belt performance in crashes. 12th ESV Conference, Göteborg, Paper 89-5B-0-003, 1989.

7. NEWMAN RJ: Chest wall injuries and the seat belt syndrome. Injury, 1984, 16(2):110-113.

8. NILSSON G, SPOLANDER K: Seat belts and road safety: some conclusions. Swedish Road and Traffic Research Institute, Paper, Stockholm 1984.

9. NYGREN Å: Injuries to car occupants - Some aspects on the interior safety of cars. Acta Oto-Laryngol (Suppl.), 395-1:164, 1984.

10. OTTE D, SÜDKAMP N, APPEL H: Variations of injury patterns of seat-belt users. Paper SAE 870226: 61–71, 1987.

11. STATES JD, HUELKE DF, DANCE M, GREEN RN: Fatal injuries caused by underarm use of shoulder belts. J Trauma, 27(7):740-745, 1987.

12. TSCHERNE H, OTTE D: Invited commentary. World J Surg, 1985, 9:797.

13. WALZ FH: Unfalluntersuchung Autoinsassen. Eidgenössisches Justiz- und Polizeidepartement, Bern, 1982.

14. ZUPPICHINI F: Effectiveness of a mechanical pretensioner on the performance of seat belts. 1989 Ircobi Congress, Stockholm, Proceedings:93–98, 1989.

912905

Investigation of Inflatable Belt Restraints

John D. Horsch, Gerald Horn, and Joseph D. McCleary
Biomedical Science Dept
General Motors Research Labs.

ABSTRACT

Studies conducted in the 1970's suggested that inflatable belt restraints might provide a high level of occupant protection based on experiments with dummies, cadavers and volunteers. Although inflating the belt was one factor which contributed to achieving these experimental results, much of the reported performance was associated with other features in the restraint system. Exploratory experiments with the Hybrid III dummy indicated similar trends to previous studies, belt inflation reducing dummy response amplitudes by pretensioning and energy absorption while reducing displacement. The potential advantage of an increased loaded area by an inflatable belt could not be objectively demonstrated from previous studies or from dummy responses

Clearly, belt inflation can be one component of a belt restraint system which tends to reduce test response amplitudes. However, other belt system configurations have demonstrated similar test response amplitudes Additionally, packaging and comfort issues for inflatable belts have not been completely resolved Thus inflatable belts do not appear competitive with other approaches to achieve belt restraint system performance

IN THE EARLY 1970's, Allied Chemical developed an inflatable belt restraint to achieve FMVSS 208 performance [1]. They suggested that inflatable belts might be a superior restraint system compared with belts or air bags, by providing lower dummy response amplitudes than conventional belts or air bags in frontal tests. In lateral and oblique crashes, and potential ejection situations the inflatable belt could perform better than air bags and at least as good as belts. They suggested that inflatable belts might cost less than an air bag system and have lower replacement costs, less stringent triggering requirements, and fewer inflations based on a higher inflation threshold because of the protection provided by the uninflated restraint. However the belt would need to be "used" to protect an occupant

Allied Chemical conducted various general and specific development efforts. These efforts included volunteer tests sponsored by the NHTSA and conducted at Southwest Research Institute in which no significant injury was noted in sled tests to 32.5 mph. Sled tests with dummies achieved HIC and chest acceleration well below "208" limits at 32.5 mph. Other efforts chose inflatable belt restraint systems for front or rear seating positions in Research Safety Vehicles with good test results well above 30 mph barrier crash severities [1]

In spite of these favorable demonstration efforts, the system never achieved sufficient development to be placed in a fleet of experimental vehicles. Unresolved design and packaging issues, and the use of other technologies to achieve protection may be among the reasons that inflatable belt design never reached a fully developed product.

Based on the previous test results, the trend of increased belt usage, and the increase of elderly car occupants who generally have lower tolerance to belt loading, a study was undertaken at the General Motors Research Laboratory in the mid-1980's to evaluate inflatable belts. The study's objective was to review the state-of-art for inflatable belts in terms of performance and design by reviewing the literature and conducting exploratory sled tests with current dummies and advanced injury assessment techniques.

TEST RESPONSES

PREVIOUS STUDIES · Studies of inflatable belt restraints had been summarized by Allied Chemical [1]. Digges and Morris provide an overview of previous efforts [2]. All of the studies achieved low dummy response amplitudes for HIC and chest acceleration when compared to "208" requirements and test severity, Table 1. HIC amplitudes of <300 and chest accelerations of <25 g's were observed in 32 mph sled tests conducted by Southwest Research Institute in conjunction with their volunteer program [1]. Factors to be considered in extrapolating these 1970's sled test results to the 1990's include the use of a Sierra 1050 dummy, minimal injury assessment instrumentation, simulation of "older" vehicles, no steering assembly, no belt retractors, and little velocity change at the time of inflation (early inflation). This was one reason for us to conduct exploratory sled tests with inflatable belts

Development efforts generally demonstrated large amplitude reductions in measured responses when comparing the original conventional belt system with the inflatable belt system, Table 2. Analysis of the data provided by Reference [1] indicates that good system test responses and large amplitude reductions from the base conventional belt system were the result of not only belt inflation, but also due to other system modifications and/or enhancements. Estimates of the influence of belt inflation independent of other factors are provided in Figure 1. The estimated ratio of response amplitudes comparing inflated belts to conventional webbing ranges from 0.41 to 0.77 for HIC and from 0.67 to 1.02 for chest acceleration.

The rear seat ESV and the Minicars study appear to provide the least subjective estimate of the influence of belt inflation for HIC and chest acceleration. In the ESV rear seat development conducted at 47 mph [1], the initial conventional belt system tests resulted in HIC's averaging about 2400 while the final inflatable belt system averaged about 1050. Importantly, this improvement was due to many changes and additions including the seat, head rest, anchor locations, allowing "partial submarining", and removing the front seat structure. The belt failed to inflate in four development tests, providing the most direct comparison as to the influence of inflation. On the basis of tests with every thing similar except belt inflation indicates that inflation of the belt provided 27% of the total HIC reduction from 2400 to 1050.

Minicars [1] reported a reduction of HIC from 1720 with conventional belts to 426 with an inflatable belt system in a 54 mph sled test [1]. A major feature of the inflating belt system was a force limiter at all three anchor locations. The force limiters alone resulted in a reduction of HIC from 1720 to 693, 79% of the improvement. Clearly belt inflation provided an important contribution to system HIC performance, but force limiting appears to be a more important factor in these tests.

Inflatable belts were studied as the restraint system for front occupants in two vehicle applications. Minicars applied an inflatable belt system with force limiting anchors to a modified Pinto. A HIC of 302 and chest peak acceleration of 36 g's were reported for a 42 mph frontal barrier crash. These results are for the right-front location, and thus did not have a steering assembly.

A Calspan/Chrysler Research Safety Vehicle used a modified Simca 1308 with a target weight of 2700 pounds. A webbing lock to reduce spooling from the retractors and a force limiting mechanism were part of the inflatable belt system. HIC's of 1422 and 2161, and chest acceleration's of 60 and 50 g's for Hybrid II dummies in the left front and right front respectively, were reported for a 45.8 mph frontal barrier crash. In a frontal head-on crash having a 26.3 mph change of velocity for the RSV, HIC's of 716 (340 not considering B-pillar contact) and 537 and chest accelerations of 37 and 46 for Hybrid II dummies in the left front and right front were reported.

Dummy responses reported in the literature for inflatable belt systems are in most cases representative of an overall system effort to reduce HIC and chest acceleration. Clearly belt inflation was an important feature in obtaining these results. However many other features such as force limiting anchors also played important roles in achieving the reported response magnitudes. As with any restraint system, the performance is that of the system and cannot be based only on single features of that system. Thus adding an inflatable belt to an existing belt restraint will not assure achievement of the reported results.

EXPERIMENTAL STUDY - Exploratory Hyge sled tests were conducted in the mid-1980's at the GM Research Laboratories with inflatable belt restraints to help separate belt inflation from other system features, to evaluate response amplitudes using the Hybrid III dummy and additional injury assessment techniques, and to develop an experimental model for possible future studies. A simplistic fixture containing a bucket seat, "rigid" belt anchors, and a floor/toe-pan surface was used for these tests. Anchor locations and the amount of webbing on the retractors approximated door-mounted lap-shoulder belt systems parameters. Importantly, the overall system did not fully model these restraint systems, differences include the rigid fixture, lack of an instrument panel or steering assembly, and a "severe" generic sled pulse, chosen to provide HIC and chest acceleration near the 208 limits as a "sensitive" region to evaluate the influence of an inflated belt restraint. This differs from previous studies which made system changes to provide low dummy responses. The fixture with a conventional belt webbing restraint are shown in Figure 2. Sled pulses are shown in Figure 3.

Instrumentation included standard head and thorax triaxial accelerometers, axial chest compression, and six-axis top-of-the-neck force and moment transducer. An accelerometer array was used to determine head dynamics and head contact forces [3].

The inflatable belt hardware was obtained from Morton International for the exploratory sled tests, having a gas generator located inside of the inflatable cushion, Figure 4. The inflatable cushion, constructed from a coated material, was attached to conventional webbing by either of two methods. For both configurations conventional webbing carried the tensile loads. For one configuration, the cushion was located between the dummy and the conventional webbing, Figure 5a. The other configuration placed the cushion between two conventional belts, similar to that used by Minicars [1], Figure 5b. A third configuration developed by Allied Chemical [1] was not tested. In this configuration the cushion was the tension member, attached to conventional webbing at each end, Figure 5c. Systems were obtained in four groups of three restraints to allow parameter adjustments based on test results.

TEST RESULTS - The first group of three restraints had a cover over the inflating portion of the belt. One restraint had a cushion on both the lap and shoulder portion, the other two had a cushion only on the shoulder belt portion. The cushions were located between the belt and the dummy as shown in Figure 5a. The first and second tests had "late" cushion inflation, 56 and 33 ms instead of the 10 ms desired. In all three tests, the gas was forced into the free length over the shoulder, compounded by the covers not fully opening, Figure 6. The responses were similar to tests with conventional webbing.

Table 1

Summary of Sled Test Results for Systems
Having Inflatable Belts (1)
Sierra 1050 Dummy, Frontal Impact

Test Series	Sled Velocity (mph)	Number Observations	Mean Response	
			HIC	Chest Accel. (3 ms-g)
Allied Initial Demonstration	33	2	425	--
Rear Seat ESV	27	3	375	33
	34	3	740	50
	47	6	1049	55
Allied Development				
Pinto Fixture	32	4	467	42
1957 Ford Fixture	32	4	560	44
Southwest Research Institute	32	2	235	21
Minicars (Hybrid II Dummy)	54	2	426	47

Table 2

Sled Test Results Comparing Initial Conventional
Belt System with the Final Inflatable Belt System (1)
Sierra 1050 Dummy, Frontal Impact

Test Series	HIC		Chest Acceleration (3 ms-g)	
	Initial System	Final System	Initial System	Final System
Allied Initial Demonstration	676	425	--	--
Rear Seat ESV	2431	1049	67	50
Allied Development				
Pinto Fixture	730	467	63	42
1957 Ford Fixture	1379	560	53	44
Minicars (Hybrid II Dummy)	1720	426	68	47

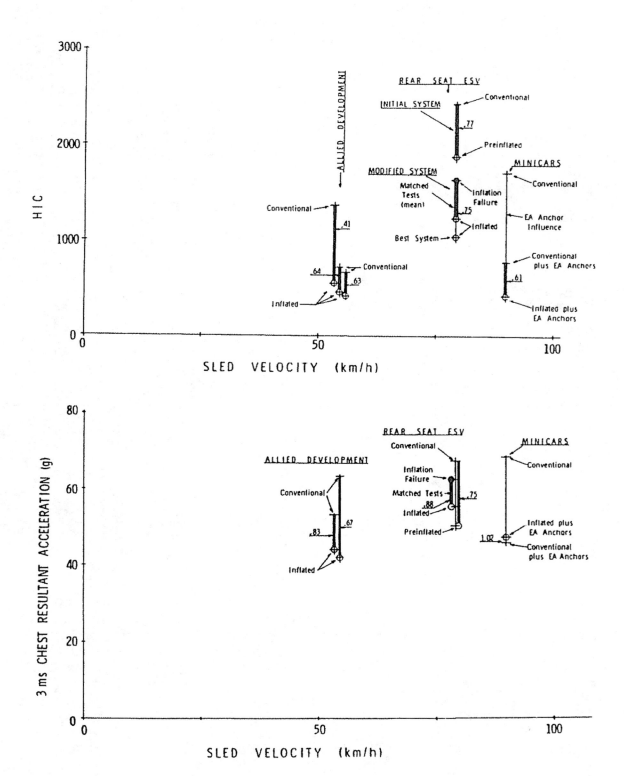

Figure 1 Estimates of the influence of belt inflation from data reported in the literature [1], indicated by heavy bars Ratio of inflated belt system response to the corresponding conventional belt system response is given for each estimate

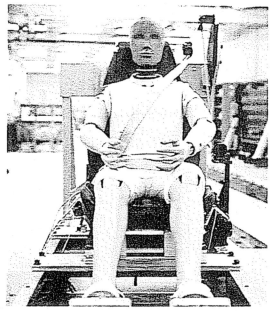

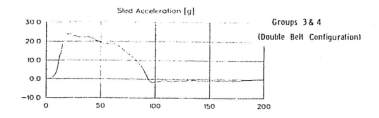

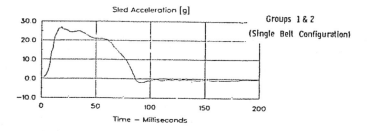

Figure 3 Sled acceleration pulses used for the exploratory tests

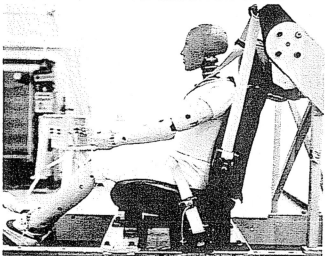

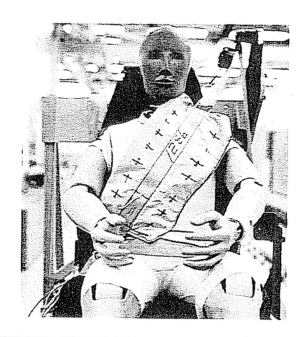

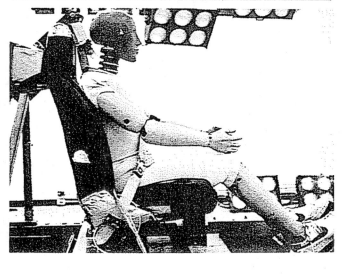

Figure 2 Sled test fixture shown with the conventional webbing restraint

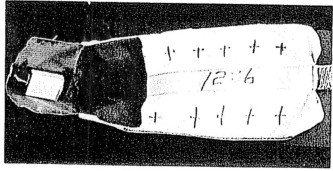

Figure 4 Inflatable belt cushion and inflator as tested for the single belt configuration

The second group of three restraints did not have covers. The cushion was initially spread as shown in Figure 4. The cushion length was reduced over the shoulder. These restraints had a cushion on only the shoulder belt and the cushion was located between the belt and dummy as shown in Figure 5a. One restraint with an inflatable cushion experienced a webbing tear at the inboard anchor attachment, believed to be due to the reuse of the anchor attachment hardware. The mean responses from the other two tests are provided in Table 3. Responses for a similar exposure except with conventional webbing are also provided in Table 3. The high speed movies indicated that the belt rolled to the underside of the cushion during the loading which provided an EA mechanism, Figure 7.

The third group of three restraints had a double belt configuration similar to that used by Minicars [1], Figure 5b to determine if this could prevent "rolling". These restraints maintained the same cushion size and inflation as the previous group. The test severity was reduced to eliminate the potential for webbing tearing with the experimental hardware, the sled acceleration shown in Figure 3. The mean responses from these tests are provided in Table 3. The high speed movies indicated reduced rolling of the cushion with the double belt configuration, Figure 8. Responses for a similar exposure except with conventional webbing are also provided in Table 3.

The final group of three restraints used the double belt configuration, Figure 5b, but with larger circumference cushions of 480 and 430 mm compared with 330 mm in previous restraints. One cushion had a 16 mm diameter vent hole. Responses were similar to those of the previous group of restraints. Cushion diameter and venting are likely important performance parameters. These exploratory experiments might not have adjusted system parameters such as inflator output to best match these parameters.

Due to the difference in test severity between inflatable belt configurations, responses for the "single" and "double" belt configurations should be related by corresponding conventional webbing tests. On this basis, the single belt configuration (Figure 5a) reduced dummy responses more than the double belt configuration (Figure 5b), but allowed a greater displacement of the dummy. Thus the rolling action of the single belt configuration appears to provide an energy absorbing displacement. Which configuration might best reduce injury potential likely depends on the application, such as the available space, and on performance in other crash situations. Further study would be required to determine how belt rolling influences performance over a range of impact severities and directions.

Additional injury assessment responses compared to previous studies were measured in the exploratory experiments including chest compression, neck forces and moments, head angular acceleration and head contact force. These responses provide additional information on occupant loading and indicate a decrease of loading magnitude associated with belt inflation except for the head contact force due to the cushioning of the chin by the inflated belt, Figure 9.

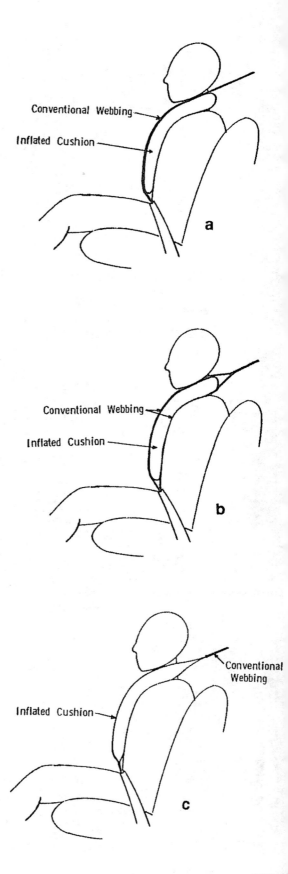

Figure 5 Inflatable belt configurations a) The inflatable cushion located between the subject and the conventional webbing b) The inflatable cushion located between conventional webbing c) The inflatable cushion attached to conventional webbing at each end

982

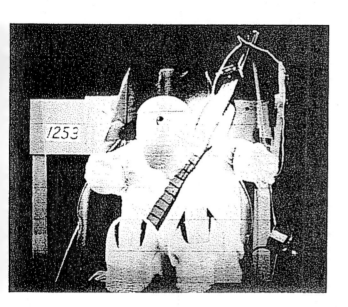

Figure 6. Initial test in which the cushion was inflated late and was filled only behind the shoulder The black "webbing" seen in the side view is a cover used in the initial tests

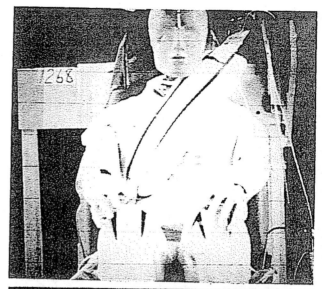

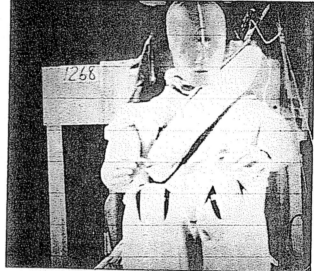

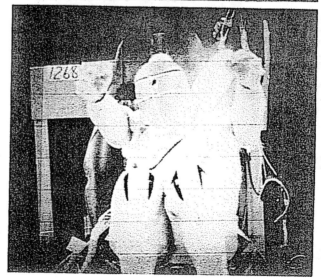

Figure 7. High-speed movie showing "rolling" of the single belt configuration

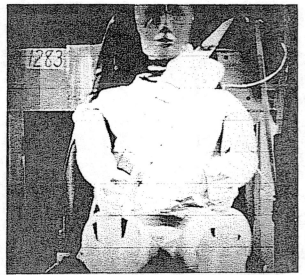
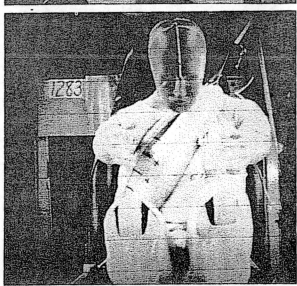
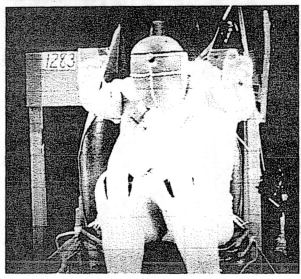

Figure 8 High-speed movies showing reduced "rolling" of the double belt configuration compared with the single belt configuration.

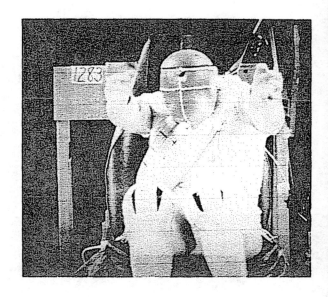

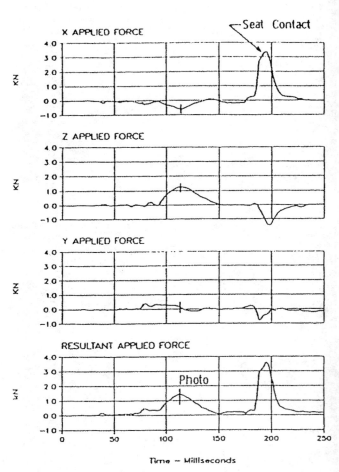

Figure 9 Computed external interaction force acting on the head from the head dynamics instrumentation, illustrating potential for the inflated cushion to restrain the chin

Table 3

Exploratory Sled Test Responses. Hybrid III Dummy.

Response*	Double Belt Configuration (Fig. 5b)			Single Belt Configuration (Fig. 5a)		
	Inflated	Conventional	Ratio	Inflated	Conventional	Ratio
Head						
HIC	796	995	0.80	968	1289	0.75
Max Ang Accel (rad/sec^2)	3350	4050	0.83	3450	4300	0.80
Max Ang Vel (rad/sec)	51	58	0.88	48	61	0.79
Max Chin Force (kN)	1.40	0.67	-	1.29	0.98	-
Max Horizontal Disp (mm)	441	528	0.84	499	554	0.90
Neck						
Max X Force (kN)	1.50	1.78	0.84	1.69	2.09	0.81
Max Z Force (kN)	2.96	3.42	0.87	2.85	3.47	0.82
Max Y Moment (N:m)	71.7	141	0.51	124	158	0.79
Chest						
Max Comp (mm)	56	58	0.97	35	58	0.60
Max Comp VC (m/s)	0.34	0.37	0.92	0.19	0.31	0.61
3 ms Res Accel (g)	37.5	38.4	0.98	36.0	40.7	0.88
Max Shldr Belt Tension (kN)	10.3	11.0	0.94	9.9	11.9	0.83

* Inflated belt data is the mean of two tests. Conventional is a single test.

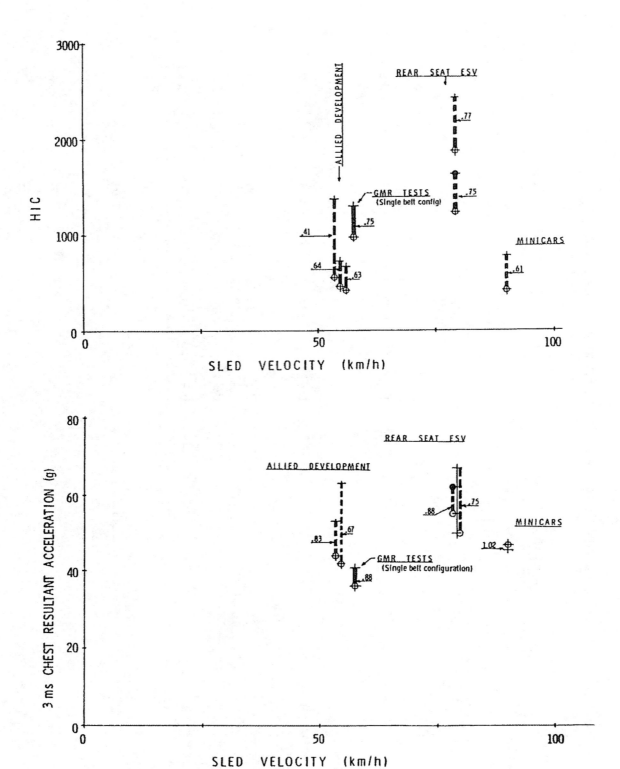

Figure 10 Comparison of the influence of belt inflation observed in the exploratory sled tests for the single belt configuration (Fig. 5a) with estimates made from data provided in the literature (from Fig 1)

COMPARISON OF EXPERIMENTAL RESULTS - The single belt configuration (Figure 5a) provided the greatest reduction of response amplitudes and is used as the basis for comparison with previous studies for the improvement of HIC and chest acceleration When only the influence of inflation is considered in the previous studies, the results of the exploratory experiments fall within the range of previous studies The estimated changes of HIC and chest acceleration due to belt inflation are shown in Figure 10 Clearly such a comparison is at best suggestive of relative performance since different test environments, different dummies, and different restraints were used among the studies

It is unlikely that optimum performance for an inflatable belt was achieved in the exploratory experiments. On this basis, the results may be conservative for predicting the potential for inflatable belts as part of a restraint system. Additionally, performance cannot be fully demonstrated by testing at a single severity [4], or based on only dummy responses of HIC and chest acceleration [5] What might be perceived as best likely depends on the application and on the criteria used to judge performance. In the exploratory tests, all responses indicated a less severe dummy interaction associated with belt inflation compared with conventional webbing, while at the same time reducing dummy displacement. This further suggests that belt inflation might be one means of enhancing belt system test performance

DISCUSSION

Inflatable belt restraints have been previously evaluated as a restraint system component in combination with other components to reduce dummy response amplitudes. Low dummy responses can be achieved by a variety of restraint systems Some production vehicles have response amplitudes less than "208" limits in NCAP 35 mph barrier crashes

The Calspan/Chrysler Research Safety Vehicle provides the best comparison for an inflatable belt restraint relative to the current situation. It is the most recent inflatable belt development and used an FMVSS dummy, the Hybrid II. This experimental vehicle had belt retractors, a steering assembly, and was vehicle based as opposed to a sled fixture. The restraint system combined belt inflation, force limiters, and web grabbers to achieve "low" response amplitudes in high severity tests (45 mph frontal barrier crash) Frontal crash tests which span the NCAP test severity suggest that the RSV test responses would have been close to "208" limits in an NCAP test, Figure 11

A study [6] using conventional webbing with retractors in 50 kmph sled tests achieved a HIC of 280, as low as that from inflatable belt studies which did not use retractors, by the use of seat design, anchor design and location, webbing grabbers, and pretensioning It appears probable that other system configurations can provide equivalent dummy test responses compared with systems that used inflatable belts What is perceived as best is dependent to some extent on the test environment, the overall system, the available space for dummy displacement, what criteria are used to define

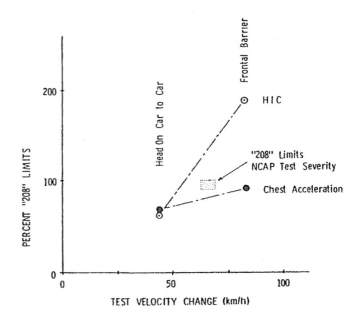

Figure 11 Comparison of test responses obtained for the Calspan/Chrysler Research Safety Vehicle with NCAP test severity and FMVSS 208 limits.

performance, the test goals, and non-safety related issues such as complexity, availability, cost, comfort, etc

Test response amplitudes are dependent on the interaction of vehicle and restraint system components and parameters. System constraints such as space and vehicle crash pulse influence system configuration and restraint system requirements. Thus a variety of effective restraint system components and configurations have evolved Among features which tend to reduce test response amplitudes are pretensioning and force limiting.

Belt inflation provides pretensioning by shortening the belt and/or by expansion of the cushion between the belt and dummy. Increased cushion inflated thickness should increase pretensioning, but will result in increased gas requirements and increased bulk Other less complex pretensioning or other methods to reduce test response amplitudes have been demonstrated Pretensioning might not always decrease test response amplitudes dependent on other factors.

Force limiting or energy absorption is provided by cushion compression and enhanced by venting of the cushion (demonstrated by previous testing [1]) or by rolling of the cushion in the single belt configuration. Other methods of force limiting and energy absorption have been demonstrated that directly limit belt tension. One example is the force limiting belt anchors demonstrated by the Minicars study [1] which appears to provide much greater reduction of HIC in the specific test environment than for any of the reported inflatable belt results in their respective environments. Force limiting results in increased displacement which should be consistent with available space.

LOAD DISTRIBUTION - An inflatable belt appears to provide increased load area on the dummy and on this basis probably a more uniform load distribution on an occupant and a lower potential for lateral loading of the thorax than other belt systems or enhancements due to the inflated belt being able to "roll", Figure 12. The Allied Chemical configuration in which the inflated cushion is also the tension member, Figure 5c, might achieve a better load distribution than would either of the configurations which have conventional webbing as the tension member. The double belt configuration, Figure 5b, would locate conventional webbing against the occupant. The single belt configuration, Figure 5a, could place the conventional webbing against the occupant with sufficient roll.

The hypothesis that improved load distribution would reduce injury potential seems correct. However tests with dummies do not appear to directly demonstrate this possible important feature of inflatable belt restraints. Thus inflatable belts might not appear as competitive as other methods to achieve restraint system responses if the influence of load distribution cannot be objectively demonstrated.

Volunteer Tests - Human volunteer tests conducted at Southwest Research Institute [1] provide a more relevant injury model and no significant injury was found with test velocities to 32.5 mph. In evaluating the significance of these tests relating to load distribution, it would be helpful if results were available for conventional webbing tests having similar restraining forces.

All volunteers were in their 20's, were thoroughly examined to assure that they were not likely to be injury prone, and all had previous sled experience. "Pretest briefing emphasized the importance of coordinated body bracing." The volunteers had peak shoulder belt loads less than 65% of those of the test dummy in spite of the volunteers being heavier, likely due to "coordinated body bracing". Thus injury risk was influenced by the relatively low restraint load due to bracing and good system performance and a probable high tolerance of the volunteers.

The peak shoulder belt tension as a function of the volunteer's age is compared with car occupant injury as a function of shoulder belt tension reported by Foret-Bruno [7], Figure 13. On this basis the belt tension would need to double or the volunteer's age increased to about 60 years to have experienced thoracic injury if a conventional webbing had been used with the same level of restraint loads. The volunteer tests were conducted at too low of a severity to establish the potential benefit of the greater loaded area associated with an inflatable belt.

Cadaver Tests - Sled tests with human cadaver subjects were reported by Calspan [8]. Less severe injury was associated with tests having an inflatable belt restraint system compared with a belt restraint system having conventional webbing, even though the inflatable belt tests were conducted at a greater severity. However there were important age and skeletal differences among the cadaver subjects and important differences among the test situations than inflatable vs conventional webbing.

The conventional webbing tests used a Citroen fixture, the inflatable belts a Pinto fixture. Two of the three tests with the conventional webbing had a steering assembly which was contacted, the third test with conventional webbing and the two tests with the inflatable belt did not have a steering assembly. The tests with the inflatable belt restraint also had force limiting anchors, the conventional webbing tests did not. The mean age for the subjects was 59 years with the conventional webbing and 51 years with the inflatable restraint (65 vs 51 years for tests without a steering assembly). The mean subject weight was 60.9 kg for tests with conventional webbing compared with 73.2 kg for tests with inflatable belts (56.8 kg vs 73.2 kg without a steering assembly).

Shoulder belt tension was chosen from the reported test responses as the best measure of subject loading severity. This estimates the influence of the inflatable belt compared with conventional webbing separated from the other system differences. The mean belt tension for three tests with conventional webbing was 8.3 kN (9.1 kN for the test without steering assembly). This compares with 6.0 kN for the force limited inflatable restraint. The second test with the force limited inflatable restraint did not have shoulder belt tension reported.

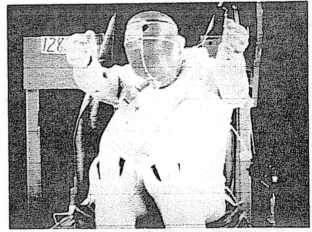

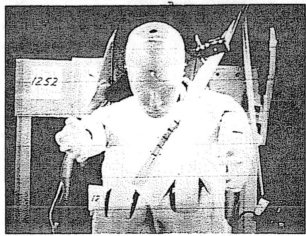

Figure 12 Visual comparison of load distribution for inflated cushion and conventional webbing

Eppinger [9] developed an empirical relationship between the number of rib fractures and the loading severity in terms of shoulder belt tension for belt restrained cadavers. His analysis included the age and weight of the subject. Applying his relationship to the tests without the steering assembly predicted 31 rib fractures compared with 32 observed for the conventional webbing test. The autopsy stated, "generalized osteoporosis of the skeletal structures were noted, this could represent disuse atrophy, however, specific pathologic changes other than the age alone may be involved, such as --"

Eppinger's relationship predicted 13 rib fractures compared with 8 observed for the in the first test with the inflatable belt restraint. Shoulder belt tension was not reported for the second test having force limiting. If it is assumed that shoulder belt tension in this second test was the same as in the first test due to force limiting, Eppinger's relationship predicts 14 rib fractures as compared with only one observed. Although this data could be interpreted to support the hypothesis that inflatable belts provide a beneficial increase in load area, the results are too few and the restraints, environments and subjects have too many differences to consider this a proof of the hypothesis.

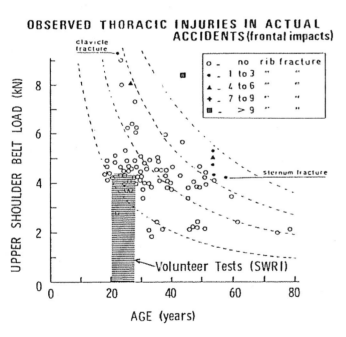

OBSERVED THORACIC INJURIES IN ACTUAL ACCIDENTS (frontal impacts)

Figure 13 Peak shoulder belt load (tension) vs occupant age with thoracic skeletal fractures noted, derived from car crash investigations by Foret-Bruno [7]. Superimposed (shaded area) from the Southwest Research Institute volunteer tests with inflatable belt restraints is the range of peak shoulder belt tensions and age.

DESIGN ISSUES - A variety of system configurations were considered by previous inflatable belt studies including both manual and automatic placing of the belt on the occupant. Both front and rear seat restraint systems were studied. Selection of restraint system features will likely be influenced by dummy responses and other issues such as comfort, and component availability. Inflatable belts will continue at a disadvantage if reduction of injury risk due to increased load area can not be objectively demonstrated and if design issues are not fully resolved.

Several fundamental design issues have not been fully resolved, the most important being the location of the gas generator and possible discomfort from the "bulky" nature of the inflatable portion of the belt. Gas for inflation of the belt was provided either from stored high-pressure gas or generated by pyrotechnics in previous studies. The hardware used in the exploratory experiments had a pyrotechnic inflator (64 mm by 48 mm by 22 mm) located inside the cushion, Figure 4. Location of the inflator inside of the cushion adds to the "bulk" of the inflatable belt with potential comfort implications. Location of the gas source external to the belt requires "piping" from the gas source to the cushion with packaging issues related to buckles and/or retractors. Allied Chemical studied both internal and external location of the inflator. Inflation of the belt was successful for either location; however practical problems were not fully resolved for either location.

The inflatable portion of an air belt restraint would include that portion over the sternum, shoulder, and beside the neck. The extra "bulk" might represent a comfort issue, with decreased use rates if not properly addressed. The attachment of a cushion to conventional webbing as shown in Figures 5a and 5b will likely have greater "bulk" than the use of the cushion as the tension member as shown in Figure 5c. In this configuration, the cushion must be strong enough for both the pressure generated tensions and the restraint tension. Allied Chemical developed cushions which successfully carried restraint loads in severe environments with most design problems relating to the cushion-to-belt attachment. With advances in materials and processes, further advances in inflatable belt design might be possible.

Our study did not evaluate inflation of the lap-belt. This should provide pretensioning, force limiting, and increased load area for lower body restraint. How an inflated belt would influence the belt riding over the pelvic structure and directly load the abdominal region might be an issue.

SUMMARY

Our study clearly indicates that an inflatable belt can be one component of a belt restraint system which tends to reduce test response amplitudes. However, other belt restraint system configurations have demonstrated equivalent test response amplitudes to those having an inflatable belt. Inflatable belt restraints on this basis appear to have a disadvantage because comfort and packaging of inflatable belts have not been fully resolved in spite of significant efforts in the past.

Inflatable belt restraints were developed by Allied Chemical in the 1970's. These and later studies demonstrated "low" dummy response amplitudes in crash severities up to 50 mph and no significant injuries to volunteers up to 32.5 mph. Belt inflation contributed to these improvements. However other restraint system features were also important to obtain the reported system test results.

Exploratory sled tests conducted at the GM Research Laboratories in the mid-1980's produced results with an inflatable belt restraint and Hybrid III dummy that were consistent with previous studies. These tests demonstrated that not only were reductions of HIC and chest acceleration associated with belt inflation, but also neck forces and moments, chest compression, head angular acceleration, and upper body displacement.

Inflatable belt restraints can provide pretensioning, energy absorption, and increased load area. Pretensioning and energy absorption can be achieved by less complex means, the desirability of these features depends on system characteristics and constraints. The potential benefit of increased load area could not be determined from current dummy responses. Because of "system" approaches and insufficient severity, previous volunteer and cadaver test results do not objectively demonstrate the benefit of increased load of an inflatable belt restraint compared with conventional webbing.

Advancement of the state-of-art for inflatable belt restraints might be best served by an objective demonstration of the influence of the increased loaded area to reduce belt related injury risks and by addressing design issues such as packaging and comfort than by another study which achieves "low" dummy HIC and chest acceleration.

ACKNOWLEDGMENT

The inflatable cushions, inflators and attachment to the conventional webbing were obtained from Morton International. David Dahle of Morton International coordinated the adaptation of the test hardware which had been developed for a helicopter crew restraint system. Richard Frantom of the Bendix Safety Restraints Group of Allied Signal Inc. provided helpful discussion and a technical summary of previous inflatable belt studies. David Viano, John Melvin, Steve Rouhana, David Dahle, and Richard Frantom provided technical review of the paper.

REFERENCES

1 "Technical Summary of Inflataband Development," A summary of inflatable belt development efforts made available by the Bendix Safety Restraints Division of Allied Automotive

2 K H Digges and J B Morris, "Opportunities for Frontal Crash Protection at Speeds Greater than 35 MPH", SAE 910807, SAE Congress, 1991.

3 D C Viano, J W Melvin, R Madeira, J D McCleary, R Shee, and J D Horsch, "Measurement of Head Dynamics and Facial Contact Forces in the Hybrid III Dummy," SAE 861891, 30th Stapp Car Crash Conference, 1986

4 J D Horsch, " Evaluation of Occupant Protection from Responses Measured in Laboratory Tests," SAE No 870222, SP-690, SAE Congress, February, 1987

5 J D Horsch, J W Melvin, D C Viano, and H Mertz, "Thoracic Injury Assessment for Belt Loading" In preparation for 1991 Stapp

6 L G Svensson, "Means for Effective Improvement of the Three-Point Seat Belt in Frontal Crashes," SAE 780898, 22ond Stapp Car Crash Conference, 1978

7 J Y Foret-Bruno, F Hartemann, C Thomas, A Fayon, C Tarrier, C Got, and A Patel, "Correlation Between Thoracic Lesions and Force Values Measured at the Shoulder Belt of 92 Belted Occupants Involved in Real Accidents," SAE 780892, 22ond Stapp Car Crash Conference, 1987.

8 M J Walsh, "Sled Tests of Three-Point Systems Including Air Belt Restraints," Final Report, Contract No DOT-HS-5-01017, August 1976

9 R H Eppinger, "Prediction of Thoracic Injury Using Measurable Experimental Parameters". Report 6th International Technical Conference on Experimental Safety Vehicles, pp 770-779. National Highway Traffic Safety Administration, Washington, D.C.

S9-O-13
Frontal Impact Protection Requires a Whole Safety System Integration ———

C. Tarriere, C. Thomas, X. Trosseille
Renault

Abstract

Beyond the generalization of the belt wearing, the improvement of the frontal impact protection is one of the most efficient action to reduce the number of the severe road victims. However, the attempt to evaluate the potential gains shows some important limitations to this efficiency and indicates the necessity of complementary actions. Among them:

- the front-end of the trucks needs to be modified to avoid underide and too severe decelerations of car occupants,
- due to the interaction between the protection in frontal and in lateral impact, the gain in frontal could be lost by an increase of the aggressiveness of the impacting car in side collisions,
- in car-to-car head-on collisions, the gain would be reduced by the increasing aggressiveness of the heavier car.

The author presents the quantification of the expected gains for the most prioritary countermeasures, discusses the major interactions between them, and tries to define the required conditions to optimize the whole safety system.

Introduction

Achieving the most substantial benefits in secondary safety requires a two-pronged approach aimed both at improving the performance of the "structure/restraint systems" combination in asymmetric frontal impacts for a broad-enough range of velocities (delta-V from 55 to 60 km/h), and at reducing frontal aggressiveness against other vehicles, particularly in the case of side-impact crashes.

In addition, such an approach cannot ignore the inescapable fact that cars are getting lighter in order to cope with environmental demands (better fuel economy, less CO_2 and pollution).

In the final analysis, the search for safety benefits will necessarily involve a broad, systemic approach based on the following indicators:

- protection criteria, measured on instrumented dummies,
- criteria for reducing aggressiveness, measured on a dynamometric barrier,
- and criteria for reducing fuel consumption and the production of CO_2, which will need to take into account a reduction in the power/weight ratio in order to control and then reduce the death rate in single-vehicle crashes.

Why Is Improving Protection Against Frontal Impact the Number 1 Priority?

Bureaucratic Truth vs. Scientific Proof

There are already so many regulations on frontal impact, some say, that the new priority should be introducing regulations aimed at improving side impact protection. This line of thinking is purely bureaucratic and simply ignores the scientific facts. According to our own assessment, the benefits to be expected from the side impact regulations that are being contemplated in Europe—and that were recently introduced in the U.S.—would be 6 times less effective than those expected from improvements in frontal impact protection.

Conditions Required for Achieving the Best Gains in Frontal Impact

Selecting a test that matches as closely as possible the deadliest front impact configurations. Such a test will be deemed representative of global frontal impacts if its configuration agrees with the characteristics of real-life accidents, therefore ensuring that test results will eventually translate into better highway safety.

Investigating actual collisions provides a basis for designing a suitable configuration with the most appropriate velocity.

The asymmetric configuration:

This configuration is the one used most widely in studies relating to overlap, deformation and resultant trajectories.

Based on 413 cars involved in front-end crashes (all severities of injuries combined) that were investigated in the region of Hanover (Germany), Otte (1) reports that 79% of them showed no evenly-distributed frontal deformation. The same author notes that in frontal impact the resultant trajectory of the forces is exactly parallel with the longitudinal axis of the car in only 23.7% of the cases.

Zeidler et al. (2) reports that 84% of the 822 cars they analyzed that were involved in injury-inducing frontal

crashes were not symmetrically and evenly deformed along the front end.

Gloyns et al. (3) reports that 90% of the fatal frontal crashes they investigated could not be represented by the full frontal barrier test.

Lastly, the results from a survey conducted by the Peugeot-Renault Association involving 1831 belted front seat occupants confirm these findings. Figure 1 gives the state of these occupants according to the type of frontal overlap with the obstacle and the type of deformation for one sample, all brands and models combined. The number of occupants was reduced to 1000 for easier reading. Only 23 out of the 126 fatally and seriously injured (MAIS 3+)—i.e. 18%—had the front ends of their cars evenly and symmetrically deformed. Before making any comparisons with the full frontal barrier test, it may be useful consider the mean acceleration sustained by these occupants. For 7 out of the 23 seriously injured reported under these conditions, the obstacle crashed against most often was the door of another car. In these cases, the estimated mean acceleration is much below that recorded in tests against an inflexible barrier. These cases appear in the "Symmetric (low deceleration)" category in Figure 1.

		SEVERITY OF INJURIES FOR 1,831 BELTED FRONT OCCUPANTS (N = 1,831 occupants reduced to 1,000 for this table)				
		MAIS 0	MAIS 1-2	MAIS 3-5	Killed	TOTAL
(1/4 track)	All cases	64	79	9	2	154
(1/3 track)	All cases	54	54	9	2	119
(1/2 track)	Rectangular deformation	6	7	1	-	14
	Oblique deformation	59	102	18	6	135
(2/3 track)	Rectangular deformation	9	10	2	1	22
	Oblique deformation	52	87	21	7	167
(distributed)	Rectangular deformation (low mean acceleration)	49	54	6	1	109
	Rectangular deformation (high mean acceleration -impacts related to 0° test)	9	33	13	3	58
	Oblique deformation (impacts related to 30 test)	13	27	10	4	54
(pinpoint)	All cases	10	10	2	-	22
	Overhanging obstacle with passenger compartment intrusion (all types of overlap)	5	9	2	4	20
	Above or below the frame No passenger compartment intrusion (all types of overlap)	38	35	3	-	76
	TOTAL	367	507	96	30	1000

Figure 1. Frontal Real-World Accidents—MAIS of Belted Front Occupants According to Type of Overlap and Type of Deformation

All in all, the full frontal barrier test represents only 13% (16/126) of the belted front seat occupants who were seriously or fatally injured in frontal crashes.

Figure 1 also shows the types of overlap and deformation observed in the other cases. The data show that the deformation equivalent to the 30° angled barrier test resulting from frontal impact with a 50 to 100% overlap covers:

- 41% of the belted front seat occupants, all injuries combined,
- 52% (66/126) of the seriously and fatally injured,
- and 57% (17/30) of the fatalities alone.

Other tests such as the corner impact (1/2 overlap) can also cause asymmetric front-end deformations, but at a right angle. However, this type of deformation is rarely observed in actual crashes.

Selecting a priority-type asymmetric frontal test configuration—such as the 30° angled barrier on the driver's side—also requires estimating the force of the impact sustained by the belted occupants.

Force of the impact:

The force involved in actual frontal crashes is traditionally estimated in terms of the instantaneous velocity variation of the occupant (delta-V) and the mean acceleration of the vehicle (Am).

Figure 2 shows the cumulative delta-V percentages for belted front seat occupants that were not subjected to additional loading from rear seat passengers, according to various severities of injury. A delta-V value of 53 km/h covers half of the 169 severely and fatally injured for whom the force of the impact could be evaluated.

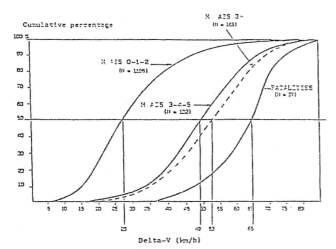

Figure 2. Frontal Impacts—Cumulative Percentages of Belted Front Occupants (Without Rear Occupant Overload) According to Delta-V and Severity of Injuries (N=1,365)

Another way to quantify the most representative velocity consists in examining the overall severity of the lesions in each delta-V category (see Figure 3). The data show that one out of two belted front seat occupants is

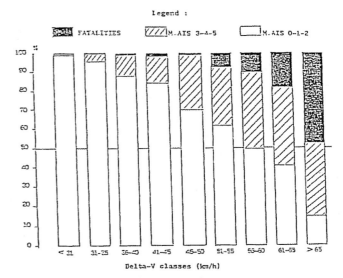

Figure 3. Frontal Impacts—Breakdown of Severity of Injuries for Belted Front Occupants (Without Rear Occupant Over Load) by Delta-V Classes

seriously or fatally injured in the delta-V category ranging between 56 and 60 km/h.

Thomas et al. (4) investigated the estimated mean acceleration in actual frontal crashes equivalent to the 30° angled barrier test with delta-V values on the order of 55 km/h. Based on 320 actual frontal crashes that could be assimilated to that test, the data show that the majority of the mean acceleration values for delta-V values between 50 and 60 km/h fall between 10 and 13 g (4). Most of the mean acceleration values measured in 30° barrier tests for this same range of velocities fall between 11 and 15 g (4). One may conclude that the 30° barrier test simulates rather faithfully the mean acceleration sustained by car occupants involved in crashes between 50 and 60 km/h.

In addition to delta-V and mean acceleration, intrusion is yet another factor that may bear on the risk incurred by belted front seat occupants. Table 1 shows the distribution of 403 such occupants broken down into delta-V categories according to the intrusion level inside the passenger compartment. The data show that for delta-V values ranging from 51 to 60 km/h, intrusion involves one fourth (25/101) of the belted front seat occupants, and most of all, 46% (17/37) of the seriously and fatally injured.

Table 1. Distribution of 403 Belted Occupants by Classes of Delta-V in Relation to the Intrusion

Level of Intrusion (inward displacement of the lower edge of the windshield)	Overall severity	Delta-V (in km/h)			
		40-50	51-60	61-70	TOTAL
Null or moderate (< 250 mm)	M.AIS 0-1-2	208	56	11	275
	M.AIS 3-4-5	26	19	19	64
	Killed	1	2	2	4
	Total	235	32	32	343
Critical (> 250 mm)	M.AIS 0-1-2	7	8	2	17
	M.AIS 3-4-5	7	13	7	27
	Killed	0	4	12	16
	Total	14	25	21	60

Figure 4 provides a distribution of the severities of injury according to the delta-V, mean acceleration and intrusion for actual crashes equivalent to the 30° angled barrier test, although no information is given on the resultant trajectory of the occupant.

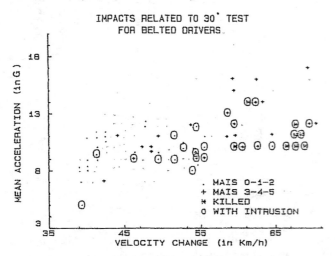

Figure 4. M.AIS of Belted Drivers According to Velocity Change, Mean Acceleration of Cars and Intrusion for Frontal Real-World Accidents Related to the 30° Angled Barrier Test

Otte (1) reports that the mean resultant of the forces is located to the left of the longitudinal axis, forming an angle estimated to be about 5°. The frequency of head impacts against the steering wheel also provides some indication as to the drivers' trajectories.

P. Thomas (5) observes that for values of delta-V between 50 and 59 km/h, 56% (56/99) of the belted drivers experience head/wheel impact. G. Walfisch et al. (6) report a similar frequency in the same delta-V category, based on the crash sample investigated by the PSA/Renault Association.

From an experimental standpoint, these authors note that in the delta-V category from 46 to 60 km/h, the impact of the head of the belted dummy driver is almost systematic in 0° or 50% offset tests, which is not the case in real life.

At any rate, the problem is elsewhere. In frontal crash tests—run as part of car design, regulation compliance or even car rating—the reference to a single head-wheel impact is unacceptable. In real life, there are not one but many conditions under which head-wheel impacts can occur, depending on the size of the occupant, the kinematics of the occupant, the characteristics of the frontal crash, the way the steering column moves, and the exact location of the impact against the steering wheel.

Therefore, a specific test needs to be designed that can measure the protection to the head (face, brain, skull, neck) provided by the steering wheel, and at various locations thereof. Such a test is the necessary comple-

ment to any overall frontal barrier test, whatever that test may be.

In the last analysis, any representative frontal test should restitute the parameters that govern the risks incurred by the maximum number of belted front seat occupants, i.e. it needs to provide for:

- a delta-V of at least 55 km/h,
- a mean acceleration on the order of 13 g,
- an asymmetric deformation of the front end of the car giving rise to a potential risk of intrusion on the driver's side,
- and a slightly oblique occupant trajectory that provides for the possibility that the head of the driver dummy will hit the left hand side of the steering wheel.

It should be specified however that such a test will not, of course, be sufficient to predict how safe a car is in all frontal crash situations. The test described above only seems the one most appropriate at present to cover close to half the seriously and fatally injured belted front seat occupants of cars involved in frontal crashes.

Selecting the tool for predicting the risk of lesions. This means designing a biofaithful dummy that can help scientists avoid making errors when they interpret test results. The base for such a dummy already exists: Hybrid III. It has a thorax, neck and head that are much improved compared with its predecessor's—Hybrid II—and that gives dynamic responses which are more closely related to those of human beings.

But although Hybrid III represents a significant step forward, further improvements are still needed in the following areas:

- collar bone stiffness,
- face biofidelity,
- pelvic behavior in submarining,
- chin/sternum contact,
- and adding asternal rib simulation.

In addition to progress in terms of biofidelity, Hybrid III makes it possible to measure several parameters that can be used to assess the risk of lesions in various body regions. A thorough analysis of all the parameters—both technical and medical—for a large number of actual crashes can help scientists clearly specify the risks facing car occupants: Accidentology can thus determine the types of impact encountered, their force, the body regions that are most at risk, and the types of lesions that are observed.

Based on the data, the role of biomechanics is to translate the physical parameters measured on dummies into protection criteria, and to set the various thresholds beyond which lesions may occur. That is the only way the risk of bodily injuries to the occupants involved in actual crashes can be predicted. Table 2 sums up the criteria proposed for frontal impact.

Table 2. Criteria Synthesis Suggested for the Frontal Test

	US CRITERIA	RENAULT CRITERIA : Proposal
SKULL	HIC < 1000	HIC<1000 If direct head impact To be calculated during head contact
BRAIN		w<25000 rd/s2
FACE		F1<500 daN (forehead on rim) F2<250 daN (below forehead, with deformable face and on a large area)
NECK		My<250 Nm) Fx <250 daN) without head impact with impact : under determination
CHEST	a < 60 g	Deflection < 50 mm a < 60 g
ABDOMEN		F < 150 daN in case submarining is detected
FEMUR	F < 1000 daN	F < 1000 daN

Why Is It Important Both to Improve Protection Against Frontal Impact and to Reduce Frontal Aggressiveness Against Other Vehicles, Particularly in Side-Impact Crashes?

At present, car safety policies vary from country to country, but the one feature they share is that they only address the issue of protection inside the vehicle being certified. they ignore the risks that such a vehicle may pose to other users.

This void is that much more serious that there is no other truly effective means of protecting these users. Accordingly, a car's frontal aggressiveness cannot be effectively compensated by a regulation aimed at making an impact against for instance the doors of another vehicle tolerable by its occupants. Of course, the overall number of injuries can be lowered by making specific design changes to the side of cars, but the effectiveness of such changes is too limited to affect the number of serious injuries and fatalities (7, 8). Effective protective measures—measures that could cut down that number in half—are technically and economically unfeasible, as shown by the evaluation program for the Renault COVER (9 to 11). Naturally, such inability to compensate reaches 100% in the case of road users that are most vulnerable, i.e. pedestrians and two-wheelers.

Charges against the aggressiveness of certain cars were levelled over fifteen years ago (12 to 17). Its components were identified: weight, architecture and stiffness. Spectacular demonstrations were made, such as Philippe VENTRE's to the 3rd International Conference on ESV (18) in a communication entitled: "A Homogeneous Safety in a Heterogeneous Fleet." The author clearly demonstrated that a small Renault 5 weighing 660 kg could offer the same protection as that of a standard-sized American car weighing twice as much, but whose aggressiveness would have been reduced. The case was made using a front-end collision between the two vehicles (Figure 5).

To date, such studies have not prompted any new regulatory developments. Could it be for lack of quantifying the influence on highway fatalities? If so, the void is now filled: we know for instance that the risk is 5 times higher in the most aggressive car category (over

Figure 5. Head-On Crash Test Between Two Cars Weighing Respectively 660 kg and 1320 kg Running Each at 70 km/h (Source: from Ventre (18))

1000 kg) than in the least aggressive one (under 800 kg) (Figure 6). And these are but averages by weight category. The risk runs probably from 1 to 20 between the least and the most aggressive car.

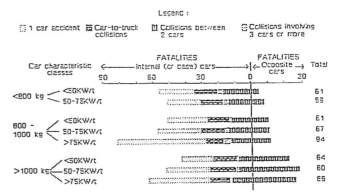

Figure 6. Distribution of Fatalities Among 1,000 Drivers in Each Car Class of Weight and "Power/Weight" Ratio According to Obstacle Type (Source: Sample of 41,944 Drivers, France 1989, Gendarmerie File)

The stakes are high, even higher when pedestrians and two-wheelers are taken into account—and justifiably so—since here again the risk varies in relation with the aggressiveness of the vehicle (Figure 7).

It is often said, particularly in the United States, that safety inside a car drops as the car gets lighter. This indeed complies with the laws of mechanics. The velocity variation of car occupants wearing seat belts anchored to the non deformed section of their cars is proportional to weights ratio. For example comparing two cars involved in a frontal collision, one weighing 700 kg and the other 1400 kg, with a closing speed of 100 km/h, delta-V will be 67 km/h for the lighter car and 33 km/h for the heavier one. Turning now to the statistics on actual crashes, the death rate is 25 per thousand drivers of cars under 800 kg, compared to 10 for cars over 1000 kg. Therefore, the death rate indeed increases as cars get lighter, but the ratio is still below 2.5. Accordingly, the

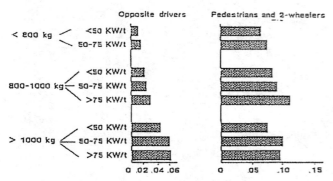

Figure 7. Fatality Rates of Opposite Car Drivers and Pedestrian or Two-Wheelers According to the Striking Car Weight Class and "Power/Weight" Ratio (Source: Sample of 50,710 Car Drivers, France 1989, Gendarmerie File)

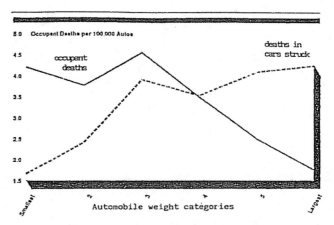

Figure 8. Deaths in Two Cars Accidents—USA, FARS 1975-1988 File (Source: From Chelimsky (19))

increased safety inside the heavier cars does not offset, by far, the risk they pose to the outside since, as indicated above, the corresponding ratio is 1 to 5 when comparing the average for cars under 800 kg and that for cars over 1000 kg.

Figure 6 clearly illustrates the results drawn from a data base on 40,000 drivers involved in injury-inducing traffic accidents which occurred in France in 1989. Our findings are in good agreement with those published recently in the United States (19). The author, E. Chelimsky, points out that " ... the safety provided to the occupants of a large car must be considered together with the risk posed by that same car to passengers in other automobiles.... (I)t is not true that cars become more dangerous simply by getting lighter."

Our findings are in full agreement with C. Thomas's paper (20) presented and discussed 6 months earlier at the AAAM Conference held in Scottsdale (USA) in October 1990. The author shows that although the death rate is higher inside lighter-than-average cars in the French automobile fleet, overall, the heavier cars are more dangerous. The number of victims in the struck cars (only 2-car collisions are studied) increases faster with weight than their number decreases inside these same vehicles.

The results obtained in the US regarding only 2-car accidents relate to crashes that occurred in the 1978-1988 period. These crashes are broken down into six equal-sized categories of car weights, from category 1 (smallest cars) to category 6 (largest cars).

It is noteworthy that the risk inside the cars (as opposed to the risk outside, in cars that are struck) is highest in the 3rd weight category, and not in the two lightest car categories (Figure 8).

The same figure also shows (dashed line) the effect of car size as an initiator of force. The risk of fatalities in the "other" car increases dramatically starting from the 3rd heaviest car category.

It is clear that when you add the two categories of fatalities (inside and outside) in each weight category,

the lightest car category is the least dangerous overall. For the rest, the results differ slightly from those recorded in France, and such for various reasons including the lower seat belt use rate in the US.

Thus, on the basis of two very different samples, the French one representing 90% of all traffic injuries sustained in that country, one can better understand why there is an imperative need to tie in the search for better frontal protection with a limit on the aggressiveness of vehicles. Any improvement to frontal protection alone is likely to cause more heterogeneity in the fleet, particularly between the heaviest cars—that would be made even stiffer to meet the new requirements—and the existing fleet. Accordingly, there is an urgent need to introduce objective means of measuring aggressiveness, such as those proposed hereinbelow.

Analysis of the Influence of the Power/Weight Ratio

In this comparative study of risk in each weight category of cars in the fleet, there is yet another important parameter beside weight: the power/weight ratio.

The above-mentioned US study also investigates one-car accidents (Figure 9). For this type of crashes the sample only includes recent—one to two-year old—passenger cars involved in fatal accidents from 1986 to 1988 grouped in the same six weight categories as above.

Here, the most dangerous cars are in the fourth weight category, where the risk is much higher than that in the three lower weight categories. The risk is lowest in the sixth category, which is comprised of the heaviest cars. These cars also have the lowest risk of rollover.

The author offers no explanation for the fact that cars in the fourth category are more dangerous.

Referring to our own results, we submit that an additional parameter plays an important role in one-car accidents: the power/weight ratio.

The influence of the power/weight ratio is felt across all the weight categories (see Figure 6). For a same category of power to the ton, this influence tops out in the

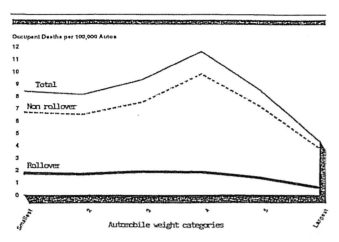

Occupant Deaths per 100,000 Autos

Total

Non rollover

Rollover

Automobile weight categories

Figure 9. Deaths in One Car Accidents—USA, FARS 1975-1988 File (Source: From Chelimsky (19))

cars of average or below average weight. For example in cars weighing between 800 and 1000 kg with a power/weight ratio of 75 kW/ton, there is a sizable excess risk that represents 66% of the "inside" fatalities and 59% of all the fatalities ("inside" + "outside") associated with that category. Loss of control is the major cause, particularly going off the road round a curve and crashing against a fixed obstacle. But the power/weight ratio also translates into more aggressiveness toward the most vulnerable users (Figure 7).

One can see how important it will be to limit the power/weight ratio in the near future, when the average weight of cars tends to drop in order to meet environmental requirements.

Proposals on How to Estimate the Aggressiveness of Car Structures

Several approaches can be imagined to estimate the aggressiveness of cars. But rather than choose a specific test, we suggest that the most cost-effective solution consists in tying in this estimate with the future frontal impact test.

Indeed, would it not suffice to measure the forces applied to the barrier in dynamic tests to get a good description of their maximum values and distribution? The side members and the engine/gearbox unit transmit the brunt of the forces, and it needs to be accurately recorded by the selected frontal impact test configuration. Among the asymmetrical impacts, the so-called "overlap higher than 1/3" impacts do not adequately account for the engine/gearbox unit forces. However, the asymmetrical impact against a 30° angle barrier, shown above to be quite representative, would be acceptable to estimate the aggressiveness involving the quasi instantaneous impact of one side member and the engine/gearbox unit and, at high speed, of the whole front end of the car.

How feasible is estimating a car's aggressiveness using the 30° angle barrier test? Instrumenting the 30° angle barrier using force cells has already been done (21). Measurement platforms divide up the barrier area

into several sectors, each sector corresponding to one of the vehicle's structural components involved in the impact (engine block, side members, wheels, hood, and so on); each platform is made up of a stiff plate mounted onto four piezoelectric sensors with two or three components. The way in which the forces normal to the surface are distributed between the cells of the same platform helps determine the location of the point where the resultant of these forces is applied on that platform. A measurement is therefore technically feasible.

What needs to be agreed on next is how many cells there should be and how they should be distributed. Indeed, the spatial measurement of the forces must be accurate enough so that the authorized limits—which are yet to be determined—depending on the location are compatible with an actual decrease in side impact aggressiveness.

Some may think that running additional tests is not cost effective so long as the full front impact test remains a mandatory requirement.

With respect to this type of impact, we have a picture of the French car fleet in the 80s as given by the moving deformable barrier test proposed by UTAC (22). For impacts at 35 km/h covering 20 representative vehicles, mean overall stiffness values ranging from 500 kN/m to 5000 kN/m (!) and peak forces from 179 kN to 510 kN were recorded. In a first step, it would be interesting to get a new picture of the fleet in 1991. This picture would help define specifications based on the least aggressive vehicle, making sure that the vehicle that is selected offers good protection against frontal impacts.

However, such initial approach to the 0° barrier test should not prevent us from considering the feasibility of estimating aggressiveness using the 30° angle barrier test, a potentially fruitful path for the future.

Conclusion

In the light of these findings, it should be obvious that the problems at hand are highly complex.

Gone are the times when simplistic proposals such as "Improving vehicle safety merely requires better side impact protection" or "Improving vehicle safety merely requires tougher regulations in terms of passive safety, and running tests at a higher speed" could be made.

Before making users pay for implementing such proposals, let's first ascertain what benefit they would derive in terms of a diminished risk on the road.

As concerns protection against side impacts for instance, our estimates show that the benefit would indeed be very small, due to the shape—thinness of the side wall—and to the highly unfavorable ratio between the stiffness of the front end of cars and that of the side wall.

Increasing protection against frontal impacts is a much higher priority since the expected benefits are five times higher than in side impacts. Nevertheless, such benefits should be attained using the "softest" methods, i.e. those that improve protection inside the vehicle without increasing its aggressiveness outside. This means putting a priority on optimizing the effectiveness of protection systems inside the passenger compartment.

Such optimization should be applied to each seat in the vehicle: indeed, the requirements are different for the driver, the front seat passenger and the rear seat passengers depending on whether they are seated next to a door or in the middle, and whether they are adults or children. And for the latter the demands are still different for each age group.

There is still a lot of work to be done in this area, and the "COVER" experimental vehicle presented by RENAULT provides many useful applications.

Progress in the area of frontal impacts can and must be sought by improving the behavior of car structure. This does not mean stiffening, as stiffening fails to deliver better overall protection. Recent examples show that cars that are too stiff transmit forces to the occupants through the restraint system, forces that exceed human tolerance particularly in the thoracic region.

To improve structural behavior means getting better control over this parameter with a view toward maximizing the dissipation of energy while at the same time avoiding unacceptable deformation of the passenger compartment. The criteria measured on instrumented dummies can be used to predict the risk of injury, and hence to assess any improvement to the "structure-restraint system" couple.

It should also be stressed that the damage to a car (its deformation) does not provide—in and by itself—any pertinent measurement of safety.

But there is yet another imperious reason not to increase structural stiffness. In addition to the risk stiffness poses to the occupants in crashes against fixed obstacles—i.e. 39% of all the frontal impacts involving fatalities in France—it also constitutes a considerable danger on the road by increasing the aggressiveness of striking vehicles. Let's not forget that any car on the road is as much a potentially struck vehicle than a potentially striking one.

Limiting front end aggressiveness is an imperious necessity. This objective can only be reached if one acts simultaneously to improve protection against frontal impacts—a desirable goal—and to limit aggressiveness.

The aggressiveness of vehicles is highly dependent on their weight. One can easily conceive how dangerous it would be to let the heterogeneity of cars on the road in terms of weight and stiffness further increase without controlling their aggressiveness.

Studies have also shown that the weight-safety relationship is not a linear one. There is an excess risk related to the power/weight ratio that comes into play independently and that primarily manifests itself in the form of loss of control of the car in a curve (one car accidents).

In the final analysis, it is becoming ever more obvious that safety can only be served by a systemic approach.

Frontal impact protection of course remains the number one priority in passive safety, but it needs to be considered by curbing the evolution of weight and by limiting the aggressiveness of cars against one another.

Overall highway safety will advance as the heterogeneity of the fleet decreases. This goal has to be reached in the context of an inescapable drop in weight in order to lower both fuel consumption and pollution.

The stakes thus appear in an overall perspective where safety imperatives are confronted with those of the environment.

Safety benefits can be expected in spite of—or thanks to—lighter cars, but two conditions need to be fulfilled:

- fleet heterogeneity must be fought against and reduced,
- and power must decrease as cars become lighter. In other words, the power/weight ratio must not increase, it must go down.

Fulfilling these conditions requires a clearly stated political will. Indeed, the natural trend in the past decades has been toward more power and heavier cars, and specifically, more weight in cars that are among the heaviest in the existing fleet.

Finally, it is on purpose that this report has only addressed those issues that concern secondary safety. In fact, safety as a whole is a much more complex matter. A coherent road safety policy cannot avoid comparing the cost effectiveness of the various proposals being made, according to whether they relate to injury prevention (so-called passive or secondary safety) or accident prevention (so-called active or primary safety). The fact that protection against side impacts is not very effective does not mean that nothing should be done. It only shows that lives should be saved by other, more cost effective means, i.e. crash avoidance. Measures of the highest priority need to be taken to improve the road infrastructure. For instance:

- a traffic circle in place of an intersection cuts down the risk of accidents involving serious injuries twenty-fold;
- and putting up guardrails along roads lined with trees would likely cut almost in half the number of killed against trees recorded in FRANCE in 1990 by the state police force.

Accident prevention will account for the great majority of the lives saved in the future (23, 24). Broadening the analysis to all the safety issues and integrating them into a comprehensive system should not be used as an alibi to do nothing. Measures designed to improve frontal impact protection constitute a priority that is that much less arguable that they form an overall system coupled with support measures that give it a clearly established coherence and effectiveness.

References

1. D. Otte, "Comparison and Realism of Crash Simulation Test and Real Accidents Situations for the Biomechanical Movements in Car Collisions" SAE paper 90 23 29.

2. F. Zeidler, H. H. Schreier and R. Stadelmann, "Accident Research and Accident Reconstruction by the EES-Accident Reconstruction Method" Proceedings of SAE Congress, Detroit, Michigan, February 25-March 1, 1985 P. 159. SAE Paper 850256.

3. P.F. Gloyns, S.J. Rattenbury and I.S. Jones, "Characteristics of Fatal Frontal Impacts and Future Countermeasures in Great Britain" Proceedings of 12th ESV Conference, Goteborg-Sweden May 29-June 1, 1989.

4. C. Thomas, S. Koltchakian, C. Tarriere, C. Got and A. Patel, "Inadequacy of A 0° Degree Barrier with Frontal Real-World Accidents" Proceedings of 12th ESV Conference, Goteborg-Sweden, May 29-June 1, 1989.

5. P. Thomas, "Head and Torso Injuries to Restrained Drivers From the Steering System" Proceedings of IRCOBI Conference, Birmingham-U.K., September 8, 9, 10, 1987.

6. G. Walfisch, D. Pouget, C. Thomas and C. Tarriere, "Head Risks in Frontal Impacts: Similarities and Differences Between Tests and Real-Life Situations" Proceedings of 12th ESV Conference, Goteborg-Sweden, May 29-June 1, 1989.

7. D. Viano, "Estimates of Fatal Chest and Abdominal Injury Prevention in Side-Impact Crashes" Journal of Safety Research, Vol. 20, pp 145-152, 1989.

8. C. Henry, C. Thomas and C. Tarriere, "Side Impacts: Expected Benefits of Planned Standards" Proceedings of 12th International ESV Conference, Goteborg, Sweden, May 29-June 1, 1989.

9. N. Casadei, "VSS—Safety Synthesis Vehicle" Proceedings of 13th International ESV Conference, Paris-France, November 4-7, 1991.

10. G. Walfisch, "Renault VSS Safety Vehicle: Occupant Safety in Frontal Impacts" Proceedings of 13th International ESV Conference, Paris-France, November 4-7, 1991.

11. J. Rio, "Renault VSS Safety Vehicle: Occupant Safety in Lateral Impacts" Proceedings of 13th International ESV Conference Paris-France, November 4-7, 1991.

12. G. Chillon, "The Importance of Vehicle Aggressiveness in the Case of a Transversal Impact" Proceedings of 1st International ESV Conference, January 25-27, 1971.

13. H. Appel, "Optimum Deformation Characteristics for Front, Rear and Side Structures of Motor Vehicles in Mixed Traffic."

14. E. Chandler, "Car-to-Car Compatibility" Proceedings of 4th International ESV Conference, Kyoto, Japan, March 13-16, 1973.

15. P. Ventre, "Proposal for Test Evaluation of Compatibility Between Very Different Passenger Cars" Proceedings of 4th International ESV Conference, Kyoto, Japan, March 13-16, 1973.

16. U. Seiffert, "Compatibility on the Road" Proceedings of 5th International ESV Conference, London, England, June 4-7, 1974.

17. P. Ventre, "Compatibility Between Vehicles in Frontal and Semi-Frontal Collisions" Proceedings of 5th International ESV Conference, London, England, June 4-7, 1974.

18. P. Ventre, "Homogeneous Safety Amid Heterogeneous Car Population?" Proceedings of 3rd International ESV Conference, Washington, D.C. May 30-June 2, 1972.

19. E. Chelimsky, "Automobile Weight and Safety" Statement to the Senate Committee on Commerce, Science and Transportation, Subcommittee on Consumer, Washington D.C., April 11, 1991. TECH-91-229.

20. C. Thomas, G. Faverjon, C. Henry, J.Y. Le Coz, C. Got and A. Patel, "The Problem of Compatibility in Car-to-Car" Proceedings of 34th AAAM Conference, Scottsdale, Arizona, October 1-3, 1990.

21. Kistler, Technical information: crash Dynamometer System.

22. E. Chapoux, J.C. Jolys, R. Dargaud, "An Approach for the Design of a Deformable Mobile Barrier to Evaluate the Protection Afforded to Occupants of a Passenger Car Involved in a Side Collision" Proceedings of IXth ESV, Kyoto, Japan, 1982.

23. D.C. Viano, "Limits and Challenges of Crash Protection" In Accident Analysis and Prevention, Vol. 20, No 6, pp 421-429, 1988.

24. C. Thomas, S. Koltchakian, C. Tarriere, B. Tarriere, C. Got and A. Patel "Primary Safety Priorities in View to Technical Feasibility Limits to Secondary Automotive Safety" Proceedings of 23rd FISITA Congress, Torino, Italy, 7-11 May 1990.

The Seat Belt Syndrome in Children

J. C. Lane
Monash Univ.

ABSTRACT

Lap belts, fitted to the centre seats of Australian cars for the past twenty-two years, have come under criticism as being injurious to children. The weight of evidence is that lap belts provide substantial protection, though less than three-point belts. A specific injury, the seat belt syndrome (SBS), to abdominal viscera and/or lumbar spine has been particularly associated with lap belts, an association confirmed by a hospital-based study in Melbourne. Roadside observations of belt use and Transport Accident Commission claims permitted the calculation of the incidence of SBS and the relative risks of SBS by seated position. The centre rear seat (lap belt) carried about twice the risk of SBS as outboard rear seats (three-point belts) which in turn have 2.7 times the risk of the outboard front seat. The number of SBS cases in Victoria has increased with penetration of the car fleet by 1971 and later cars. Suggestions are made for improvements in the restraint system.

FOR THE PAST TWENTY-TWO YEARS, Australian Design Rule 5A has required cars and station wagons to be fitted with three-point seat belts (lap and shoulder belts) in all seat except for the front and rear centre seats, which have lap belts. Lap belts have come under criticism on the ground that, in rears seats, they provide little effective protection and because of an association with a particular injury, the so-called seat belt syndrome (SBS).

In a review of the literature, the great weight of evidence indicates that lap belts provide substantial protection, ranging from 18% to 50% reduction in injuries, though less protection than three-point belts in comparable crashes. Reports from Sweden and the United States show that lap belts provide protection for, in particular, children and the elderly (Kraft et al., 1989; Morris, 1983; Orsay et al., 1989; Partyka, 1988).

The occurrence of specific serious injury presumptively associated with the belt itself was first reported by Kulowski and Rost (1956). An attempt to estimate the incidence of the visceral injury was made in 1962 by Garrett and Braunstein, who coined the term "seat belt syndrome" (SBS). They analysed data from the Cornell Crash Injury Research files of highway accidents in which at least one occupant was wearing a belt - lap belts at that time. Of 3325 belt wearers, 944 were injured and, of these, seven had "reported or possible" abdominal injuries, seven had pelvic injuries and twelve had lumbar spine injuries.

Analysis of two large samples of hospital admissions the U.K., for the year before and the year after the introduction a law requiring belts to be worn in the front seats of cars presumably with mostly three-point belts, demonstrated an increase in injuries to abdominal and pelvic organs in the after period, except for kidney injuries, for which there was a decrease. Changes in the occurrence of lumbar spine injuries were inconsistent (Rutherford et al, 1985; Tunbridge, 1989).

Review of records of car occupant casualties at a trauma centre, over a five year period (1984-88) during which seat belt usage increased substantially, showed that lumbar spine fractures and injuries to the small and large intestine also increased (Anderson et al., 1991). There was high belt usage, of both three point and lap belts, by the casualties.

A comparison of injuries usually associated with belts, a large sample of hospital admissions from motor vehicle crashes showed that gastro-intestinal injuries had a higher incidence in the belted than the unbelted group, but lumbar spine injuries were not significantly different in incidence (Rutledge et al., 1991).

Thus kidney injuries should be excluded from the list injuries associated with belts. Injuries of the spleen, when the only visceral injury, should probably also be excluded because of the propensity of this organ to be damaged in any blunt impact. But, because of practical difficulties of reclassification, the seat belt syndrome (SBS) is defined as injury to the abdominal viscera and/or injury to the lumbar spine.

REPORTED CHILD SBS CASUALTIES

There are many case reports of seat belt injuries in general in the medical literature. As regards children, a search of the English language literature through 1991 yielded single cases or small series in which individual cases are detailed, to a total of 69 children, 19 of whom had visceral injuries only, 11 spinal injuries only, and 39 both visceral and spinal (sources and detail are given by Lane, 1992). These cases cannot be regarded a sample, but it can be noted that the modal age was ten years and 81% were reported to have been wearing lap belts - three quarter f these in the rear seat. It seems likely that selection for reporting was biased towards severity.

A series of papers by Agran et al. (1985 -1990) refers to paediatric patients from car accidents at emergency rooms of hospitals serving a community of 1.9 million people. Two hundred and twenty-nine belt wearers yielded one case of small bowel laceration and one of ruptured spleen (in a three-point belt wearer). There were three cases of bladder contusion but no spine fractures, dislocations or cord injury.

None of the 160 paediatric patients reviewed by Orsay et al. (1989) suffered a visceral or lumbar spine injury. There were no SBS cases in 46 restrained children from 231 crashes in which at least one child less than eight years had been transported by ambulance (Corben and Herbert, 1981). Eight hundred and twenty responses to a newspaper questionnaire referred to 288 unrestrained and 865 restrained children. There were four SBS cases (0.5%) in the restrained and four (1.38%) in the unrestrained children: none had lumbar spine injuries (Langweider and Hümmel, 1989). Clearly the incidence of SBS in children is not high.

Paediatric casualties from motor vehicle accidents admitted to the Royal Children's Hospital in Melbourne for the period 1984 to 1989 have been reviewed by Hoy and Cole (1992). (This hospital admits children up to age 16 directly or on transfer from the whole State of Victoria.) Of 541 casualties, 29 had belt injuries of the abdomen and of these seven had Chance fractures of the spine. One had a cord injury without radiological abnormality. Four of the SBS cases were in the first half of the period and 25 in the second, a nearly four-fold increase. Lap belts were used by 19 of the 28 cases for which the type of restraint was ascertained. The crashes generating the cases were severe: 21 came from vehicles in which at least one occupant was killed.

For rear seated children (a child in a safety seat excluded) 19 were in the centre (lap belt) seat and 5 in outboard seats (with three-point belts). There are more child occupants in outboard than centre seats, so there was a pronounced tendency for SBS injuries to be associated with lap belts.

MASS DATA ANALYSIS

In the State of Victoria, no fault injury compensation of casualties from motor vehicle accidents is the function of the Transport Accident Commission. The TAC has amassed a great volume of data on casualties, to which access was provided to the Monash University Accident Research Centre for mass data analysis.

For estimating the incidence of SBS, a file was used containing casualty information on claims from July 1978, to June 1988 derived from crashes involving 1975 and later vehicles. Total claimants were approximately 77,000, of whom 26,863 were passengers, including 3369 aged 0-14 years.

SBS cases were defined as those car occupants with lumbar spine injuries (ICD 9 codes 805.4, 805.5, 806.4, 806.5 and 952.2) and/or abdominal visceral injuries (ICD 9 codes 863.0 through 866.9 and 868.0 through 869.9) (World Health Organisation, 1975). These rubrics embrace the wider definition of SBS referred to above. The nature of the restraint available could be inferred from the seating position, but information on belt use could not be obtained from this file. This was supplied from the surveys. The case frequencies are shown in Table 1.

Age	L Front	C Front	O/B Rear	C Rear	Total
0-4	1	0	5	2	8
5-9	2	1	11	5	19
10-14	6	1	7	5	19
All children	9	2	23	12	46
>14	312	3	138	30	483
Total	321	5	161	42	529

Table 1 - SBS in Car Occupants in Crashes, July 1978 - June 1988, of Post 1975 Cars

(Australian road traffic is on the left; the driver's seat is the right front)

The number of SBS cases among children (14 years or less) for whom a claim was made to TAC in the ten year period was 46. This compares with 32 (estimate derived from 29 in nine years) at the Royal Children's Hospital. The time periods overlap: TAC 1978-1988; RCH 1980-1989. The RCH casualties may be regarded as a subset containing the more severe SBS injuries in Victoria as a whole.

Although those SBS cases in children reported in the literature represent only an unknown fraction of those which occur, the rather small number of cases listed in 23 years suggests either that the risk of occurrence is low and/or that the exposure has been low. Most of the case reports are from North America where the use of restraints in the rear seat has been low until recent years. In addition, the size of the population from which the cases are derived is generally unknown, with some exception in the series of Agran et al.

The TAC data permit an estimate in relation to the populations of children and of vehicles in the State of Victoria, as shown in Table 2.

Table 2 - Incidence of Child SBS Casualties in Victoria

Cases (1978-1988)	46
Vehicles (1975 and later)	1.5×10^{6}*
Cases/10,000 Vehicles p.a.	0.058*
Population 0-14, Persons	929×10^{3}*
Cases/100,000 p.a.	1.10*
All Child Occupant Casualties #	3369
SBS as % of Child Occupant Casualties #	1.37%

*for the year 1987.   # 1975 and later cars.

The number of child SBS cases 1978 - 1988, 46, is unsuitable for an estimate of incidence, because children's exposure to risk (by being restrained) increased substantially over the ten year period (1978 and 1988 were half years). Australian Design Rules 4 and 5A required belts in rear seats of cars and station wagons (hereafter referred to as "cars") manufactured after January 1971. The Victorian belt wearing law of 1976 required children less than eight to be restrained if riding in a front seat. This is likely to have had the effect of moving children to the back seat, where they were not required to be restrained and where, in any case, there were often no belts available.

The belt-wearing law was changed in December 1981 to require all children, wherever seated, to be restrained (if a restraint was available). In the back seat, this change in the law effectively applied to 1971 and later cars. The number of these cars increased almost linearly with calendar year, from about 940,000 in 1978 to 1,850,000 in 1988. This, together with probable increase in compliance, greatly increased the number of child passengers exposed to risk. In effect, the 1981 law required all children to be restrained and the 1971 design rule provided the means of restraint in the back seat. Consequently the number of SBS cases in children also increased.

Because of this increase, instead of the mean annual number of cases, it is appropriate to consider the expected frequency from the regression of case frequency on calendar year. For 1987 the expected number is 8.7, relating to 1975 and later cars, and must be factored up to account for 1971 to 1974 cars. The factor, from the 1988 vehicle census, is 1.17, and the expected number of child SBS cases in Victoria in 1987 is thus 10.2, rounded to 10.

EXPOSURE - Information on seating position occupancy by age and restraint use was, derived from surveys with matched observation sites, on Melbourne arterial roads in 1985, 1986 and 1988, and for restraint type usage in 1990.

The survey observations can be summarised as follows. Centre front seats: low occupancy for all age groups; belt use moderately high (60%) for 0-7, medium (40%) for 8-13, belt use low for adults.

For the centre rear seat: low occupancy but wearing rate improving from 1985 to 1988 for all age groups. The number of belt wearers in the centre front seat has tended to increase in the period 1985-88, in the centre rear seat it has remained constant.

Child occupants of the centre front seat constitute 2.9% of all child passengers; child occupants of the centre rear seat constitute 18% of all child passengers. Overall, child occupants of the centre seats constitute 4% of all car passengers.

RELATIVE RISK OF CENTRE SEATS - The TAC data (Table 1) provide the cases recorded in the 10 year period mid-1978 to mid-1988. Measures of exposure are provided by the survey data. To make use of he survey data it is necessary to assume that the arterial road samples are reasonably representative of Victoria both at the times of observation and for some years earlier.

The second assumption is, from the viewpoint of risk calculations, conservative, since the restraint use rate is likely to have been lower in the years 1978 to 1984. As only "wearers" are used in the estimation of exposure, the estimate of wearers is likely to be inflated and the estimate of relative rates of SBS correspondingly reduced. The observed frequencies in the various seating positions, factored by the survey wearing rates, are used as measures of relative exposure. In addition, in the youngest age groups, some "wearers" (though perhaps not as many as the 35% shown in the 1990 survey) will have been using the restraints such as child seats, generally regarded as safer than lap belts. They have been counted as lap belt wearers, so the estimate of relative risk is conservative on this count also. The

child group 0-13 years in the survey has been adjusted to conform with 0-14 in the TAC data. The risk calculations are shown in Table 3. The frequencies in the centre front seats are too small for significance calculations.

Table 3 - Relative Risks of SBS in Various Seating Positions with Present Restraints

A. CHILDREN - CENTRE REAR VS OUTBOARD REAR

	Distribution of exposure to risk of SBS*	SBS Cases fo	fe	Relative Rate of SBS (%)
OBR	79.7%	23	27.9	0.77
CR	20.3%	12	7.1	1.57
	100%	35	35	

* based on sample of 3764 wearers; chi-square = 4.24, 0.02<p<0.05

B. CHILDREN - LEFT FRONT VS OUTBOARD REAR

	Distribution of exposure to risk of SBS*	SBS Cases fo	fe	Relative Rate of SBS (%)
LF	30.1%	9	9.64	0.696
OBR	69.9%	23	22.36	0.767
	100%	32	32	

* based on sample of 19232 wearers; The relative rates are not significantly different.

C. ADULTS - CENTRE REAR VS OUTBOARD REAR

	Distribution of exposure to risk of SBS*	SBS Cases fo	fe	Relative Rate of SBS (%)
OBR	93%	138	156.17	4.38
CR	7%	30	11.83	12.56
	100%	168	168	

* based on sample of 22278 wearers; chi-square = 30.02, p<0.001

D. ADULTS - LEFT FRONT VS OUTBOARD REAR

	Distribution of exposure to risk of SBS*	SBS Cases fo	fe	Relative Rate of SBS (%)
LF	85.8%	312	386.29	1.63
OBR	14.2%	138	63.71	4.38
	100%	450	450	

* based on a sample of 22278 wearers; chi-square = 100.92, p<<0.001

E. CHILDREN VS ADULTS (SUMMARY)

relative rate of SBS (%)

	Child	Adult	chi-square	p
CR	1.57	12.56	52.5	<.001
OBR	0.77	4.38	76.5	<.001
LF	0.696	1.63	6.85	<.01

In rear seats, the centre seat is seen to confer a significant increased risk of SBS. The increase is by a factor of two (1.57/0.77) for children and by a factor of almost three for adults (12.56/4.38). The assumptions about exposure referred to above make these estimates conservative, especially for children. Children appear to be less at risk of SBS than adults in the same seating positions.

Adults in outboard rear seats are at greater risk of SBS, by a factor of 2.7 (Table 3D, 4.38/1.63), than occupants of the left front seat. The increase in risk for children is small and non-significant (.767/.696). Since front seats are well known to be less safe than rear seats (for example, Evans and Frick, 1988), this is an unexpected result suggesting that there is a deficiency in the rear seat belt installations.

DISCUSSION

It appears that lap belts were initially accepted uncritically by those concerned with crash protection: more recently they have been perhaps unreasonably condemned. The limitations of the lap belt are, first, it provides insufficient protection, by failing to prevent the head and upper body from contact with unyielding surfaces. Second, it may cause injuries to the abdomen and lumbar spine by direct loading combined with the body motion, "submarining", that the belt induces under impact. The standard Emergency Locking Retractor (ELR) three point combination shares these shortcomings but to a much smaller degree.

Despite much case description, it has been difficult to estimate the numerical size of the SBS problem, through the collections of Agran and associates and casualty series based on accidents suggest that the incidence has been low. The increased child case frequencies noted in recent years in Australia can be related to increased restraint use in child passengers as post 1974, and by inference, post 1970, cars have penetrated the car fleet (Figure 1).

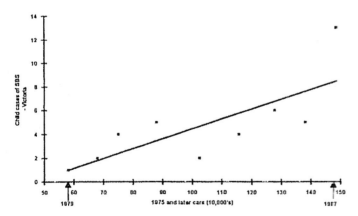

Figure 1 - Relation of child SBS to post-1974 cars

The fatality rate in casualties with SBS derived from published papers is 14.5%, but, as noted earlier, this collection seems selectively biased towards severity and also unduly weighted by the five fatalities from the NTSB series of 26 severe accidents (1986). If these are excluded, the rate is 8.3%. The casualties may have had other serious injuries, as they did in the Royal Children's Hospital series but in which there were no deaths.

SBS is a serious condition, usually requiring emergency surgery and carrying the risk of missed early diagnosis. The characteristic visceral injury is to the gastro-intestinal tract, especially to the small and large intestines. When there is a lumbar spine fracture, there is risk of spinal cord damage and paraplegia. There was 3.5% paraplegia in the Royal Children's Hospital series. A particular type of lumbar spine fracture, the Chance fracture, is especially associated with lap belts.

The substantial case literature indicates a preponderance of lap belt restraint, though the association is confounded with rear seat position in many reports. This association of lap belts with SBS is confirmed in the Royal Children's Hospital series.

From the Victorian mass data the relative risk of incurring SBS from a lap belt is now estimated, for adults, as three times that from a three point belt in the rear seat. For children the relative risk is twice that of a rear-seat three point belt. In addition, rear seat three point installations have, themselves, nearly three times the risk of front seat passenger three point belts.

A deficiency in the published information is any clear estimate of the incidence of SBS. The mass data analysed above indicate a case rate of about ten child cases per annum (for 1987) in the State of Victoria (less confidently, about 14 for 1991). Not all these children were using lap belts: in fact, twice as many children sustained SBS when using adult three point belts, because, despite the greater risk in lap belts, more children are seated in outboard than in centre seats.

Although the focus of this investigation was on children, there are many times more SBS passenger casualties in adults than children. In addition there were about 485 SBS cases in drivers in the years surveyed. Some of the drivers' visceral injuries may come from steering wheel contacts, but the lumbar spine injuries must be ascribed to the belt. Adult car occupants with SBS amount to 159 (the expected number for 1987 in 1975 and later cars), eighteen times the number of child cases. This approximates to 186 adult cases for all cars for 1987. Overall there is a case frequency for all car occupants, in Victoria, of about 196 per annum for the year 1987.

It is to be noted that these totals may contain an uncertain number of occupants who were not wearing belts. They cannot be eliminated from the data because the TAC file does not contain information on belt wearing and because it has been necessary, to use a broad definition of the injuries contributing to the Seat Belt Syndrome.

COUNTERMEASURES

A substantial benefit may be gained by providing upper body restraint in the centre seat positions. More than four fifths of centre-seated occupants are in the rear (87% of children, 77% of adults). Most recent cars do not provide a centre front seat.

The rear centre seat can be provided with a tethered harness for children of appropriate body size. For adults and larger children a three point belt is needed. For new cars, this could become standard practice - a few sedan car models already have three point belts in the centre rear seat. Replacing the lap belt with a lap-sash belt could be expected to eliminate about two thirds of the SBS cases in occupants of the centre rear seat.

Reducing SBS in all seats already required to be fitted with three point belts requires attention to the geometry of installations. In brief, the needed improvements are in making the lap belt angle steeper, appropriate compliance and profile of seat cushions, better access o the buckle in outboard rear seats and vertically adjustable D rings (Fildes et al, 1991). Belt tensioners would be very useful additions, to minimise belt slack (Haland and Nilson, 1991).

For smaller children, for whom adult belts with or without a booster are unsuitable, there is the available child seat, or, best for children up to 9 kg, a backward facing seat or capsule (Turbell, 1990; Lutter, Kramer and Appel, 1991).

Choice, by parents, of restraint type for young children, positioning of belts and the installation of child restraint seats all need to be greatly improved.

CONCLUSIONS

1. Lap belts provide substantial protection to occupants, both adult and child, of both front and rear seats. Lap belts should always be used if no better restraint is available.

2. Both lap belts and three point belts have a disbenefit, being liable to cause a particular type of injury, the seat belt syndrome (SBS) consisting of abdominal visceral injury and/or lumbar spine injury. Lap belts appear to have two to three times the propensity to cause this injury as three point lap sash belts. Children appear to be less at risk than adults.

3. Of adults restrained by three point belts, rear outboard occupants have a greater liability to SBS than left front passengers.

4. Lap and three point belts, as a group, cause about ten cases of SBS in children and 186 cases in adults in Victoria per annum (1987 expected totals). Annual child frequencies are increasing in parallel with increased numbers of post-1970 cars.

5. Two- thirds of the SBS cases in lap belt wearers in the centre rear seat could be expected to be eliminated by replacing the lap belts with three point belts, or, for children, adding a tethered harness. For children of appropriate sizes, a tethered booster with adult lap-sash belt, a tethered booster with lap belt and harness, or child seat are the preferred restraints. For infants and small children backward facing devices are the restraints of choice.

6. Reduction of SBS in general (three quarters of all passenger cases occur in three point wearers) requires improved seat and belt design.

ACKNOWLEDGMENT

The author is indebted to Mr Max Cameron, Dr Brian Fildes, Dr Peter Vulcan, Ms Jennie Oxley, Mr Foong Chee Wai and Mr Raphael Saldana of the Accident Research Centre.

Information on roadside surveys and on the relevant laws was kindly provided by Mrs Pat Rogerson, Ms Sandra Torpey and Ms Pam Francis of Vic Roads. The Transport Accident Commission made available a file of claims arising from crashes to post 1974 cars. Professor W G Cole and Dr Gregory Hoy, of the Royal Children's Hospital generously granted the use of material from a paper in publication.

The project was supported by the Federal Office of Road Safety of the Department of Transport and Communications.

REFERENCES

Agran P F, Dunkle D E, Winn D G, (1985). Motor vehicle Accident trauma and restraint usage patterns in children less than 4 years of age. *Pediatrics, 76*, 382-386.

Agran P F, Dunkle D E, Winn D G, (1987). Injuries to a sample of seat-belted children evaluated and treated in a hospital emergency room. *The Journal of Trauma, 27*, 58-64.

Agran P F, Winn D G, (1987). Traumatic injuries among children using lap belts and lap/shoulder belts in motor vehicle collisions. *31st Proceedings, American Association for Automotive Medicine*, 283-296.

Agran P F, Winn D G, Dunkle D E, (1989). Injuries among 4 to 9 year old restrained motor vehicle occupants by seat location and crash impact site. *American Journal Diseases of Children, 143*, 1317-1321.

Agran P F, Castillo D, Winn D G, (1990). Childhood motor vehicle occupant injuries. *American Journal Diseases of Children, 144*, 653-662.

Anderson P A, Rivara F P, Maier R V, Drake C, (1991). The epidemiology of seat belt-associated injuries. *The Journal of Trauma, 31(1)*, 60-67.

Corben C W, Herbert D C, (1981). Children wearing approved restraints and adult's belts in crashes. Traffic Accident Research Unit Report 1/81. Department of Motor Transport, New South Wales, Australia.

Evans L & Frick M C, (1988). Seating position in cars and fatality risk. *American Journal of Public Health, 78(11)*, 1456-1458.

Fildes B N, Lane J C, Lenard J, Vulcan A P, (1991). Passenger Cars and Occupant Injury. Report No. CR 95, Federal Office of Road Safety, Canberra.

Garrett J W & Braunstein P W, (1962). The seat belt syndrome. *The Journal of Trauma, 2*, 230-238.

Haland Y and Nilson G (1991). Seat belt pretensioners to avoid the risk of submarining - a study of lap-belt slippage factors. Paper 91-S9-0-10, Thirteeenth International Technical Conference on Experimental Safety Vehicles.

Hoy G A & Cole W G, (1992). Concurrent paediatric seat belt injuries of the abdomen and spine. *Pediatric Surgery International, 7*, 376-379.

Kahane C J, (1987). Fatality and injury reducing effectiveness of lap belts for back seat occupants. SAE Paper 870486, *Restraint Technologies: Rear Seat Occupant Protection, SP-691*, 45-52.

Krafft M, Nygren C, Tingvall C, (1989). Rear seat occupant protection. A study of children and adults in the rear seat of cars in relation to restraint use and car characteristics. *12th International Conference on Experimental Safety Vehicles, 2*, 1145-1149. Also in *Journal of Traffic Medicine (1990)*, 18(2), 51-53.

Kulowski J & Rost W B, (1956). Intra-abdominal injury from safety belt in auto accidents. *Archives of Surgery, 73*, 970-971.

Lane, J C, (1992). The child in the center seat. Report CR 107, Federal Office of Road Safety, Canberra.

angwieder K, Hummel T, (1989). Children in cars - their injury risks and the influence of child protection systems. *12th International Conference on Experimental Safety Vehicles* NHTSA, US Department of Transportation, Section 3, 39-49.

utter G, Kramer F, Appel H, (1991). Evaluation of child safety systems on the basis of suitable assessment criteria. *Proceedings IRCOBI Conference*, 157-169.

Morris J B, (1983). Protection for 5-12 year old children. SAE Paper 831654, *SAE Child injury and restraint Conference Proceedings, P-135*, 89-100.

National Transportation Safety Board, (1986). Performance of lap belts in 26 frontal crashes. Report no. NTSB/SS-86/03. United States Government, Washington D C.

Orsay E M, Turnbull T L, Dunne M, Barrett J A, Langenberg P, Orsay C P, (1989). The effect of occupant restraints on children and the elderly in motor vehicle crashes. *12th International Technical Conference on Experimental Safety Vehicles, 2*, 1213-1215.

Partyka S, (1988). Lives saved by child restraints from 1982 through 1987. Technical report HS 807 371, NHTSA, Department of Transportation.

Rutherford W H, Greenfield T, Hayes H R, Nelson J K, (1985). The medical effects of seat belt legislation in the United Kingdom. Research Report No. 13. Department of Health & Social Security, Her Majesty's Stationery Office.

Rutledge R, Thomason M, Oller D, Meredith W, Moylan J, Clancy T, Cunningham P, Baker C, (1991). The spectrum of abdominal injuries associated with the use of seat belts. *The Journal of Trauma, 31(6)*, 820-826.

Tunbridge R J, (1989). The long term effect of seat belt legislation on road user injury patterns. Research Report 239, Transport & Road Research Laboratory, Department of Transport.

Turbell T, (1990). Safety of children in cars - Use of child restraint systems in Sweden. Swedish Road & Traffic Research Institute, VTI.

C498/32/145/95

Smart seat belts – what they offer

M MACKAY
University of Birmingham, UK

SYNOPSIS

Current restraint systems represent one of the most effective engineering developments which have prevented and mitigated car occupant injuries worldwide. However, they are fixed systems, optimized in design terms around a single crash condition with a single occupant in one sitting position. Population issues are addressed only partially through use of the 5th percentile female and 95th percentile male dummies again in a single crash condition. Real world crash injury studies of current seat belts indicate five limitations to occupant protection: head contacts with steering wheels for drivers, intrusion, rear loading by unrestrained objects, mispositioning of the seat belt and injuries from the belt itself. A restraint system of fixed characteristics cannot address the variations in weight, sitting position, biomechanical tolerance and crash severity which occur in the crash populations. Intelligent restraint systems have the potential to address these varying demands so that protection could be optimized for a specific person, in a specific sitting position in a specific crash. The techniques for achieving these aims are variable pretensioners, discretionary web locks and load limiters, position sensing, and airbags with variable inflation rates and volumes.

1. INTRODUCTION

Current seat belts have been shown to be very effective in diminishing the frequency and severity of injuries to car occupants. So much so that high levels of seat belt use are a prime aim of all national transport safety policies in motorized countries. The limitations of the protective abilities of current seat belts have been well documented in many analyses of both field accident data and experimental studies [Bacon, 1989].

Real world accident studies have identified five categories of limitations to the performance of current seat belts. These are:

1) Head and face contacts with the steering wheel by restrained drivers [Rogers et al, 1992] - It is inherent in the kinematics of a restrained occupant that, in a severe collision at a velocity change of around 50 km/hr, the head will arc forwards and downwards, having a horizontal

translation of some 60 to 70 cms. (Figure 1). If a normal steering wheel position is superimposed on such a trajectory, the head and face necessarily will strike the steering wheel. Such contacts usually produce AIS 1 to 3 injuries and are best addressed with the supplementary airbag systems becoming common throughout the new vehicle fleet.

2) Intrusion of Forward Structures - A seat belt requires a zone ahead of the occupant so that the occupant can be decelerated by the compliance of the restraint system. If intrusion compromises that space, then specific localized contacts can occur. The injury risk from such contacts may well be small if they are occurring with structures which have been engineered appropriately. Indeed, in the ultimate condition, it is better for the occupant to be decelerated not just by the seat belt alone but through a combination of belt loads and contact loads. Those contact loads are through the feet at the firewall, through the knees into the lower dash and through the airbag and belt at chest level. In severe collisions, however, major intrusions are destroying the passenger compartment so that exterior objects are actually striking the occupants. This is a feature of restrained fatalities in frontal impacts [Mackay et al, 1990].

3) Rear Loading - Correctly restrained front seat occupants can receive injuries from unrestrained occupants, luggage or animals from the rear seats. Such events contribute to some 5% of restrained front seat fatalities [Griffiths et al, 1976].

4) Misuse of the Seat Belt - Seat belts must be positioned correctly on the human frame to work effectively. Dejeammes (1993) in a survey of belt use in France found that some 1.6% of front seat occupants had the shoulder belt under the arm or behind the back whilst some 3.3% had introduced slack because of the use of some clip or peg to relieve the retraction spring tension. A more important type of misuse relates to the positioning of the lap section. Many occupants, especially the overweight, place the lap section across the stomach instead of low across the pelvis. Indeed for the obese, it is often impossible to position the lap section so that it will engage on the iliac spines of the pelvis in a collision. These problems are reflected in abdominal injuries from the lap section of the seat belt [Gallup, St-Laurent, Newman, 1982].

5) Injuries from the Seat Belt Itself - As with any injury mitigating device there are limits to effectiveness. Those limits are when biomechanical tolerances are exceeded and thus the most vulnerable segment of the population begin to receive injuries. The usual thresholds are sternal and rib fractures occurring, especially in the elderly [Hill et al, 1992].

Current restraint design aims to achieve a compromise in the sense of optimizing protection for the largest number of people exposed in the largest number of injury-producing crashes. The end point, however, is a fixed design with single characteristics optimized around a single crash condition. That crash condition for most manufacturers is usually the 35 mph (56 km/hr) rigid barrier crash test.

The next evolutionary stage in restraint design is to move away from a restraint system with fixed characteristics which need to be considered if the concept of variability is introduced into restraint design.

2. POPULATION CONSIDERATIONS

The ideal restraint system would be tailored to the following variables:
- the specific weight of the occupant,
- the specific sitting position of the occupant,
- the biomechanical tolerances of the occupant,
- the severity of the specific crash which is occurring,
- the chances of specific passenger compartment intrusion occurring which might compromise restraint performance,
- the specifics of the compartment geometry and crush properties of the car.

2.1 Anthropometric Considerations

Current dummies and modeling cover the 5th percentile female to 95th percentile male range. Assuming for simplicity that males and females are exposed equally and that there are few males smaller than the 5th percentile female or females larger than the 95th percentile male, these conventional limits put 2.5% (1 in 40) of the small population and 2.5% of the larger population beyond those limits; 5% or 1 in 20 overall.

Table 1 gives the 1% and 99% ranges for height, sitting height and weight. These data show what would be required if the design parameters were extended to cover this wider range, so that only 1 in 50 of car occupants would be outside the design parameters [Society of Actuaries, 1979].

More importantly, it is implicitly assumed in current designs that height (or sitting height) and hence sitting position are colinear with the weight of the occupant. In fact, there are data available to suggest that the relationship between height and weight are rather complex. For example, the body mass index (i.e., the ratio of weight in kilograms to height in meters squared) varies to a greater degree in women than in men, and particularly at the 75th percentile and above, women have higher BMIs than men. In addition, the prevalence of overweight increases with age, more with females than males [Williamson, 1993].

Therefore to optimize a restraint system it would appear appropriate that sitting position and body weight should be assessed independently if variability is to be introduced into restraint design.

2.2 Population Characteristics by Position in the Car

European data show that some 80% of drivers in injury-producing collisions are male, whilst some 65% of front seat passengers are female [Bull and Mackay, 1978]. Approximately one-third of rear passengers are children of 10 years of age or under [Huelke, 1987]. These simple frequencies suggest that restraint characteristics should not necessarily be the same for all sitting positions in the car.

2.3 **Sitting Positions**

Current design is predicated on the positions established for the three conventional dummies. Observational studies by Parkin et al (1993) have demonstrated that there are substantial differences between those three positions and an actual population of drivers. Passive observations of drivers in the traffic stream have been made using video recording techniques, and drivers classified by sex and general age groups of young (35 years), middle (36-55 years) and elderly (56 years and older). Make and model of car were recorded and measurements made of the following distances:

·   nasion to steering wheel upper rim and hub,
·   top of head to side roof rail,
·   back of head to head restraint, horizontally and vertically,
·   shoulder in relation to 'B' pillar.

Such techniques allow thousands of observations to be made quickly and therefore population contours can be drawn. Figure 2 illustrates how particularly for the 5th percentile female population the actual sitting position is significantly closer than that of the 5th percentile dummy, by some 9.2cm. The 5th percentile, small female population sits some 38cm (15 inches) or closer to the hub of the steering wheel.

2.4 **Biomechanical Variation**

An extensive literature exists concerning human response to impact forces, mostly conducted in an experimental context. A general conclusion from that body of knowledge is that for almost any parameter, there is a variation of at least a factor of 3 for the healthy population exposed to impact trauma in traffic collisions [McElhaney, Roberts, Hilyard, 1976]. That variation applies to variables which are relatively well researched such as the mechanical properties of bone strength, cartilage, ligamentous tissues and skin. It is likely to be even greater when applied to gross anatomical regions such as the thigh in compression, the thoracic cage, the neck or the brain.

How such variability is demonstrated in populations of collisions is less well understood. Data from a ten year in-depth study of European crashes for restrained front seat occupants are given in Figures 3 and 4. The methodology of that work has been described elsewhere [Mackay et al, 1985].

Figure 3 illustrates the effect of age on injury outcome in terms of the frequency of AIS 2 and greater injuries for three age groups. The 60+ age group especially shows greater vulnerability than the younger groups. As a broad generalization one may conclude that for the same injury severity, the younger age groups must have a velocity change of some 10 km/hr more than the elderly. The effect is more marked if a more severe injury level is chosen. Figure 4 illustrates the cumulative frequencies for the three age groups for injuries of AIS 4 and greater.

Figure 5 shows similar frequency curves for crash severity by sex of occupant. Thus at a velocity change of 48 km/hr (30 mph), some 2/3 of male and some 80% of female AIS 2+ injuries have occurred. As a starting point, therefore, as well as specific body weight and sitting position, a combination of age, sex and biomechanical variation could be developed as a predictor of the tolerance of a specific person within the population range.

An intelligent restraint system therefore would perhaps require a smart card, specifying the height, weight, age and sex of the occupant. On entering the card for the first time, the card would be read and the characteristics of the seat belt and airbag adjusted accordingly.

3. SENSING CRASH SEVERITY

Besides assessing the specifics of the occupant's characteristics before impact, protection could be enhanced if the nature and severity of the collision could be assessed early enough during the crash pulse so that the characteristics of the restraint system could be modified. That would require, for example, sensors to discriminate between distributed versus concentrated impacts, and between, for example, three levels of collision severity such as less than 30 km/hr, 30 to 50 km/hr, and greater than 50 km/hr. In addition, conceptually one might have an array of sensors which would detect the early development of compartment intrusion. Such electronic data could then instruct the restraint system to change its characteristics early enough during the crash phase to alter the characteristics of the restraint and thus the loads on and forward excursion of the occupant.

4. VARIABLE RESTRAINT CHARACTERISTICS

The advantages of a variable restraint system are illustrated by considering some examples. A front seat passenger, 70 years of age and female, weighing 45 kg sitting well back, in a 30 km/hr frontal collision with no intrusion, would be best protected by a relatively soft restraint system which would maximize the ride-down distance and minimize the seat belt loads. That would require a low pretensioning force, a long elongation belt characteristic provided by load limiters and a soft airbag.

Such a system is very different from what would be required by a 25 year old, 100 kg male, sitting close to the steering wheel in a 70 km/hr offset frontal collision. He would need a very stiff seat belt, an early deploying stiff airbag and a large amount of pretensioning load.

Consider thirdly a 9 year old girl, weighing 30 kg sitting in a rear seat in a 56 km/hr frontal impact. Maximizing her ride-down distance and minimizing the seat belt loads would require low pretensioning loads and a very soft belt system, but one which would still have a biomechanically satisfactory geometry at the forward limit of excursion. Possible techniques for introducing variability into restraint design are now discussed.

4.1 Variable Pretensioning Force

A retractor pretensioner might be devised which would have a variable stroking distance or perhaps two stages of pretensioning to address the population and crash severity requirements outlined above.

4.2 Combined Retractor Pretensioner and Buckle Pretensioner

Such a system of pretensioners might maintain good seat belt geometry especially for the small end of the population, such as the 9 year old girl in the rear seat, when soft restraint characteristics and hence large amounts of forward excursion are required.

4.3 Discretionary Web Locks

If the seat belt system needs to be stiffened for the heavy occupant with high biomechanical tolerance in a high speed crash, then the switching in of a web lock would be appropriate. Such a device would shorten the active amounts of webbing being loaded and diminish forward excursion at the expense of somewhat higher seat belt loads.

4.4 Discretionary Load Limiting Devices

One way of providing for biomechanical variability would be to have a load limiting mechanism which would be calibrated for the specifics of the occupant's age, sex and weight. Such a device could also be adjusted according to transient sitting position. Belt loads would be limited at the expense of increased forward excursion.

4.5 Variable Sitting Positions

Ultrasonic, infrared or other techniques of sensing might be used to monitor continuously the head position of each occupant. Such information could be used at a minimum to provide a warning that an occupant was sitting too far forward and in particular too close to the steering wheel. At a more advanced level it could be used to tune the seat belt and airbag characteristics to be optimized for that occupant in that specific position by adjusting the other restraint variables.

4.6 Variable Airbag Firing Threshold

The need for an airbag varies according to seated positions in the car and the characteristics and sitting position of the occupant. For most drivers in most sitting positions a supplementary steering wheel airbag becomes desirable in crash severities above 30 km/hr [Rogers et al, 1992]. For a front seat passenger however, particularly one who is towards the top end of the biomechanical tolerance spectrum and sitting well back, an airbag at 30 km/hr is unnecessary. For a child sitting a long way forward in such a crash, it might also be disadvantageous. Hence specific sensing techniques at a minimum could discriminate between the presence or absence of a passenger, and at the next level assess the need for the airbag to inflate or not.

4.7 Variable Airbag Characteristics

In response to the sensing data about the occupant's characteristics and transient sitting position, and the accelerometer data about the nature and severity of the collision which is occurring, the airbag properties could be varied. Specifically, gas volume and inflation rate could be changed. Compressed gas systems instead of chemical gas generators have the potential for providing those characteristics by having time-based adjustable inflation ports. This requires very advanced sensing and control systems but these aims could well be addressed through future research and development.

5. OTHER CRASH CONFIGURATIONS

The discussion so far has focused on frontal collisions which constitute some 50% to 65% of injury producing collisions in most traffic environments. Lateral, rear and rollover crashes also suggest opportunities for optimizing protection through intelligent restraint systems.

5.1 Lateral Collisions

The technology is now developing for side impact airbags with two versions becoming available on 1995 model-year passenger cars. The observational data of Parkin et al (1993) have illustrated the range of driver sitting positions which reflect the requirements of side impact airbag geometry to cover both the door and the B pillar. Because a significant part of the population, tall males, choose to sit as far rearward as possible, in a side impact in many four door vehicles the thorax would be loaded by the B pillar rather than the door.

A practical issue is the nature and position of the sensor for a side impact. Because of the extremely short time available for sensing, around 5 milliseconds, a simple switch system is appropriate [Haland, 1991]. An analysis of a representative sample of AIS 3 plus lateral collisions has demonstrated that if a switch sensor is located in the lower rear quadrant of the front door then approximately 90% of all such side impacts would be sensed appropriately. A set of several sensors would be required to address the remaining few collisions, whilst rear seat occupant protection would also be addressed in large part by a sensor in the same position in the front door as is appropriate for front seat occupants [Hassan et al, 1994].

5.2 Rear Impacts

Occupant protection in rear end collisions is addressed largely through the appropriate load deflection characteristics of seat backs and the provision of correctly positioned head restraints. The real world data of Parkin et al (1993) demonstrates that head restraints are frequently positioned both too low and too far to the rear of the occupant's actual head position. The head position sensors discussed above could also be used for adjusting automatically both the vertical and horizontal position of the head restraint. Such a technology is relatively simple but the costs and reliability, as well as acceptability by the driving population, present serious practical problems.

5.3 <u>Rollover Accidents</u>

Actual mechanisms of injury in rollover accidents have been well researched by Bahling et al (1990) for occupants in current seat belts. Conceptually one can suggest that a buckle pretensioner might have some benefits in rollover circumstances by diminishing the relative vertical motion of an occupant. However, in rollovers current dummies do not have the appropriate soft tissue or thoracic and lumbar spine response characteristics, in comparison to the human frame. The basic clearance of current bodyshell design and packaging limit intrinsically the ability of any restraint system to modify the nature of any roof contacts under the forces of actual rollover circumstances even with no roof deformation taking place. Raising current roof lines leads to many undesirable consequences. Nevertheless it would be of interest to explore occupant kinematics in rollovers using more realistic techniques with volunteer and cadaver subjects in the context of buckle pretensioners and the requirements of a sensor to detect incipient rollover.

6. CONCLUSIONS

This paper only attempts to outline in conceptual form some of the issues which need to be addressed in advancing from today's seat belts and airbags towards some form of intelligent restraint system. Of fundamental importance is to recognize the population issues of size, sitting position, biomechanical variation and changing crash exposures. Beyond these issues lies a larger amount of challenging research and development to actually produce the sensors and hardware to provide variability in a seat belt and airbag system. Proximity sensing has its advocates, and if radar techniques could actually discriminate an impending collision from a near miss or a passing object, then the provision of say 500 milliseconds warning would alter many of the restraint issues reviewed in this paper. However, the basic premise remains; the next generation of restraints must change from having single fixed characteristics towards variable ones which recognize the real world population variables of weight, sitting position, biomechanical tolerance and crash exposure.

REFERENCES

(1)   BACON, D.G.C., The Effect of Restraint Design and Seat Position on the Crash Trajectory of the Hybrid III Dummy, 12th International Technical Conference on Experimental Safety Vehicles, 12:451-457, Göteborg, Sweden, 1989.

(2)   BAHLING, G.S., BUNDORF R.J., KASPZYK G.E., MOFFATT E.A., ORLOWSKI K.R., and STOCKE J.E. Rollover and Drop Tests. The Influence of Roof Strength on Injury Mechanisms Using Belted Dummies, <u>Proceedings 34th Stapp Car Crash Conference</u>, 34:101-112, Society of Automotive Engineers, Warrendale, PA, 1990.

(3)   BULL J.B. and MACKAY G.M., Some Characteristics of Collisions, the Population of Car Occupant Casualties and Their Relevance to Performance Testing, <u>Proceedings, IRCOBI 3rd Conference</u>, pages 13-26, Lyon, France, September 1978.

© IMechE 1996 C498/32/145/

(4)   DEJEAMMES M., ALAUZER A., TRAUCHESSEC R., Comfort of Passive Safety Devices in Cars: Methodology of a Long-Term Follow-Up Survey, SAE Paper No. 905199, Society of Automotive Engineers, Warrendale, PA, 1990.

(5)   GALLUP B.M., St. LAURENT A.M. and NEWMAN J.A., Abdominal Injuries to Restrained Front Seat Occupants in Frontal Collisions, AAAM Proceedings, 26:131-145, October 1982.

(6)   GRIFFITHS D., HAYES M., GLOYNS P.F., RATTENBURY S., and MACKAY M., Car Occupant Fatalities and the Effects of Future Safety Legislation. Proceedings, 20th Stapp Car Crash Conference, 20:335-388, Society of Automotive Engineers, Warrendale, PA, 1976.

(7)   HALAND Y. and PIPKORN B., The Protective Effect of Airbags and Padding in Side Impacts - Evaluation of a New Subsystem Test Method, 13th International Technical Conference on Experimental Safety Vehicles, 13:523-533, Paris, 1991.

(8)   HASSAN A., MORRIS A., and MACKAY M., The Best Place for a Side Impact Airbag Sensor. Proc. AAAM/IRCOBI Conference on Advances in Occupant Restraint Technologies, Lyon, France, September 1994.

(9)   HILL J.R., MACKAY M., MORRIS A.P., SMITH M.T., and LITTLE S. Car Occupant Injury Patterns with Special Reference to Chest and Abdominal Injuries caused by Seat Belt Loading, Proc. IRCOBI Annual Conference, pp. 357-372, Verona, Italy, September 1992.

(10)   HUELKE D.F., The Rear Seat Occupant in Car Crashes, American Assn for Automotive Medicine Journal 9:21-24, 1987.

(11)   HUELKE D.F., OSTROM M., MACKAY M., and MORRIS A. Thoracic and Lumbar Spine Injuries and the Lap/Shoulder Belt. Proceedings SAE Congress, SAE930640, Society of Automotive Engineers, Warrendale, PA, 1993.

(12)   MACKAY M., ASHTON S., GALER M., and THOMAS P. Methodology of In-Depth Studies of Car Crashes in Britain. Proceedings, Accident Investigation Methodologies, SP159, pages 365-390, SAE Paper 850556, Society of Automotive Engineers, Warrendale, PA, 1985.

(13)   MACKAY G.M., CHENG L., SMITH M., and PARKIN S. Restrained Front Seat Car Occupant Fatalities. AAAM Proceedings 34:139-162, October 1990.

(14)   MCELHANEY J.H., ROBERTS V.L. and HILYARD J.F., Handbook of Human Tolerance, Japanese Automobile Research Institute, Tokyo, 1976.

(15)   PARKIN S., MACKAY M., and COOPER A. How Drivers Sit in Cars. Proceedings AAAM 37:375-388, November 1993.

(16) ROGERS S., HILL J., and MACKAY M. Maxillofacial Injuries Following Steering Wheel Contacts by Drivers Using Seat Belts. Brit. J. Oral & Maxillofacial Surgery, 30:24-30, 1992.

(17) Society of Actuaries Build and Blood Pressure Study, London, 1979.

(18) WILLIAMSON D.F., Descriptive Epidemiology of Body Weight and Weight Change in U.S. Adults. Ann Intern Med. Oct 1; 119(7 Pt 2):646-9, 1993.

Adult	Height		Sitting Height		Weight	
	ins	cm	ins	cm	lbs	kg
1%ile female	57	145	28	72	82	37
5%ile female	59	150	29	75	90	41
95% male	73	185	37	93	225	102
99% male	75	190	38	96	236	107

Table 1 - Population Ranges for Height, Sitting Height and Weight

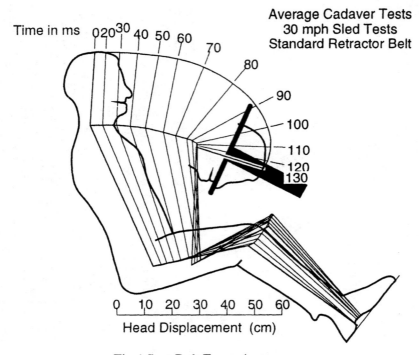

Fig 1 Seat Belt Excursion

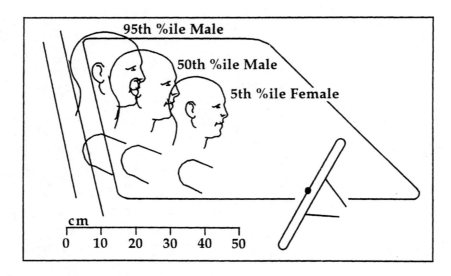

Fig 2 - Drivers' Sitting Positions

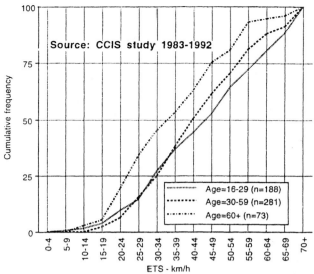

Fig 3 Crash speed distributions for frontal impacts (PDF of 11 to 1 o'clock) to drivers (by age groups) who experienced injuries with a MAIS >= 2

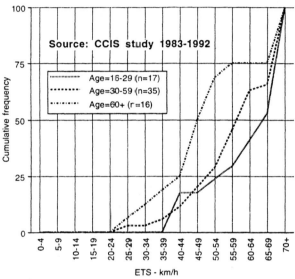

Fig 4 Crash speed distributions for frontal impacts (PDF of 11 to 1 o'clock) to front seat occupants (by age groups) who experienced injuries with a MAIS >= 4

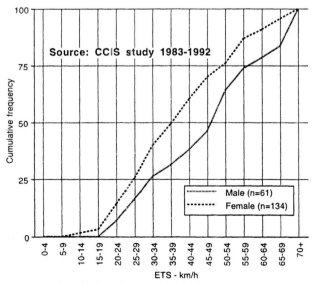

Fig 5 Crash speed distributions for frontal impacts to front seat passengers (by sex) who experienced injuries with a MAIS >= 2

Testing of Seats and Seat Belts for Rollover Protection Systems in Motor Vehicles

Mark W. Arndt
Transportation Safety Technologies Inc.

ABSTRACT

A series of controlled experimental programs were conducted for the purpose of improving the motor vehicle rollover protection system. Test results reported in this paper have been previously presented in SAE Paper No. 980213 [1]. Experiments tested lap belt restraints utilizing a variety of lap belt geometric and webbing slack conditions. Tests utilized in the series include dynamic and static tests and the use of test mannequins and human volunteers. In the first test program, utilizing a rigid seat, human volunteers were subjected to minus 1.0 Gz acceleration and a 95th percentile Hybrid III mannequin was subjected to minus 5.0 Gz acceleration for a variety of lap belt conditions. A second program utilized a rigid mannequin in production vehicle seats for the purpose of measuring and comparing seat belt system effective slack. Finally, the rigid mannequin from the second test and the rigid seat and lap belts from the first test were brought together and tested. The last test program provided information regarding factors affecting occupant displacement and flail in motor vehicle rollover crashes. Test methods and test device's were constructed to study and develop objective understandings of the effect of vehicle seats and seat belt systems for the purpose of improving or anticipating improvements to a motor vehicle rollover protection system. Overall, it is determined that occupant displacement from the seat in rollover conditions is affected by factors associated with a vehicle seat belt restraint system and seat. Compliance of live human occupant is shown to be a significant factor. Change in vehicle seat belt anchorage location and other reductions in effective slack of seat belt systems have the effect of substantially reducing occupant displacement. A variety of rollover injury mechanisms may be favorably influence by the demonstrated changes that reduce occupant rollover induced displacement.

INTRODUCTION

Proper use of seat belt restraint systems has been shown to be one component of a rollover protection system that favorably affects occupant response during rollover. This conclusion is shown retrospectively through numerous in-depth and statistical analyses of motor vehicle crashes [2,3,4,5].

An understanding of seat belt restraint system effects upon occupant response is further substantiated through simulation and experimentation. In belted dummy dynamic dolly rollover tests and inverted vehicle drop tests conducted by Bahling, et al., safety belts were noted to prevent both ejection and projected impacts with the vehicle interior, yet did not result in reduced head and neck loads for dummies in the area of ground contacts [6].

While a historic understanding of seat belt restraint systems in rollover crashes has emphasized their effectiveness in ejection prevention, more recent research demonstrates potential for seat belts in reducing occupant displacement and flail inside the vehicle. Moffat, et al., utilizing a rigid seat, documented cadaver and mannequin flail in a test fixture that produced translational and rotational accelerations [7]. Results of this testing demonstrate changes in occupant flail due to changes in seat belt system anchorage geometry and seat belt system webbing pretension, yet point to practical limitations in the seat belts ability to prevent all potential harm in a rollover collision.

RIGID SEAT TEST

Tests employing a rigid seat were conducted in a static and dynamic mode utilizing human volunteers and a sitting pelvis 95th percentile Hybrid III mannequin [8]. All volunteers and the mannequin were positioned upside down in an unpadded rigid steel seat mounted in a drop cage. Volunteers and the Hybrid III were each tested statically. The Hybrid III underwent minus five Gz accelerations after a .914 m (3 ft) free fall producing impact velocities of approximately 4.25 m/s (9.5 mph).

Each volunteer and surrogate were restrained by lap belts of varying configuration. The varied lap belt configurations included two different nominal lap belt angles, 45 and 90 degrees, and a variety of lap belt adjustment strap tensions. The adjustment strap tension is not the lap belt sion. The lowest adjustment tension, 62 N (14 lb.) was tension chosen to cause a nominal zero lap belt tension, a sufficient adjustment strap tension to remove available slack from the lap belt system. A schematic drawing of the seat and lap belt mounting positions is shown in figure 1. The seat was 508 mm (20 in.) wide and lap belt anchorage's were 559 mm (22 in.) apart.

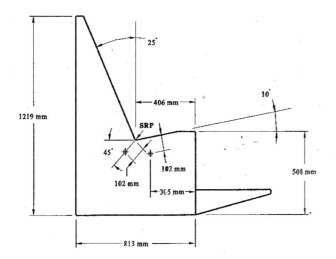

Figure 1. Rigid seat configuration

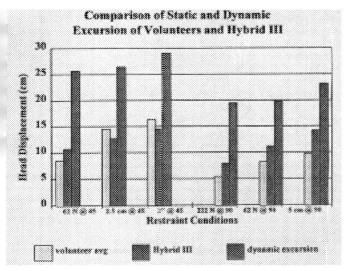

Figure 2. Comparison of displacement

For the rigid seat testing static measurements were recorded on an available population of seven human volunteers. Dimensions of individuals from this population varied with minimum and maximum values measuring: stature -- 170 cm to 188 cm; mass – 61 kg to 104 kg; seated height -- 76.0 cm to 94.6 cm; and age -- 29 yr. to 45 yr. It was thought the measurements from the population of volunteers would provide some basis for comparing the static and dynamic response of the 95th percentile Hybrid III mannequin for the various seat belt configurations in a minus Gz acceleration environment.

For the rigid seat tests, average static head displacement of the volunteer, static head displacement of the Hybrid III and maximum head displacement due to dynamic test of the Hybrid III is compared in figure 2. A summary of the peak dynamic response of the Hybrid III is shown in Table 1. The Hybrid III head did not contact the drop cage during dynamic testing. Neck moment due to dynamic testing of the Hybrid III are not available.

The purpose of the test program utilizing the rigid seat was to provide some objective quantification of the effects of different restraint system variables. The relationship of the variations in lap belt slack to the effective slack that any production vehicle may exhibit was undefined. Despite the lack of any specific vehicle's seat belt characteristic, the test data demonstrates the obvious interference that may occur between an occupant's head and a modern passenger car's roof in rollover conditions. It was clear from the test results that reducing lap belt effective slack by reducing the length of webbing in the seat belt system and providing more efficient anchorage location made substantial reduction in occupant displacements.

RIGID MANNEQUIN TEST

In testing utilizing a rigid test mannequin two different motor vehicles with different seat and seat belt systems were measured to determine occupant displacements associated with effective slack. Arndt and co-authors [9] describes details of the testing method in a previous paper. A force equivalent to 91 kg (200 lb.) theoretically hanging from the seat belt was applied to the mannequin. The actual vertical force exerted on the mannequin with the seat in the upright position was 1045 N (235 lb.) The rigid mannequin had a mass of 16 kg (35 lb.)

While there were numerous differences between the two vehicles and their respective occupant protection systems, they were similar in that both contained bench seats with continuous loop three-point lap/shoulder belts. Effective slack in vehicle 1 with the seat in the aft position resulted in mannequin displacements: X – 1.52 cm (0.60 in), Y – 2.54 cm (1.00 in), Z – 8.64 cm (3.40 in). Displacements associated with vehicle 2 with the seat in the aft position were reported: X – 1.02 cm (0.40 in), Y – 3.81 cm (1.50 in), Z – 11.94 cm (4.70 in).

For vehicle 1, the additional measurements to determine a conceptual equivalent webbing length associated with effective slack was measured. For the tested vehicle the conceptual equivalent seat belt webbing associated with effective slack to produce zero net vertical displacement of the test mannequin under the test condition is 20.6 cm (8.1 inches) and 24.4 cm (9.6 inches) for the seat in the-full rearward and full forward positions respectively. The determination of effective slack is shown in figure 5.

Table 1. Summary of peak dynamic response

		SUMMARY OF PEAK DYNAMIC RESPONSE			
Test No.	Nominal Belt Angle (deg)	Belt Tension /Slack (N) or (mm)	Peak Resultant Neck Load (N)	Peak Belt Load Load (N)	Peak Z Displacement (cm)
B	45	62N	867	8553	25.29
C	45	222N	890	9677	22.51
D	45	25mm	850	8497	26.44
E	45	50mm	841	8408	29.05
F	90	222N	721	5754	19.45
G	90	62N	716	5687	19.77
H	90	444N	672	5498	15.60
I	90	50mm	738	5947	23.13

Figure 3. Rigid Mannequin in Vehicle

In the process of developing the test protocol for determining a seat belt system's effective slack with the rigid mannequin, it was noted that characteristics of the vehicle seats had important effects on mannequin displacement. Compliance of the seat cushion played a roll in the attainment of zero net mannequin vertical displacement. A compliant seat allowed the dummy to compress downward in the cushion when webbing is removed from the seat belt system.

Significant aspects of effective slack include characteristics of the seat belt system. Webbing spool out from the retractor was significantly greater for vehicle number 2. This spool out is associated with considerably greater webbing length in the seat belt system. Seat belt anchorage position and routing geometry also played a significant role in the mannequin displacement. The effect of seat belt anchorage geometry may be noted in measured differences of effective slack for the forward and rearward seat position shown for vehicle 1. It should be obvious that removing webbing from the seat belt system, effectively pretensioning the seat belt webbing, also favorably reduced displacement of the test mannequin.

Figure 4. Rigid Mannequin Lifting Device

The prototype rigid mannequin, by virtue of its weight, rigidity and smooth surface, presented an idealized representation of the human occupant. The idealized characteristics of the mannequin provided the opportunity to consider the seat and seat belt characteristics in isolation. The low weight of the mannequin, 16 kg (35 lb.), inadequately compressed the seat cushion in baseline measurements and certainly has an affect on the determination of zero net vertical displacement of the mannequin in measuring effective slack. Effective slack is an equivalency to a length of seat belt webbing removed from the seat belt system that results in zero net vertical displacement of the mannequin under the test condition. A picture of the test device utilized while testing with the rigid mannequin is shown in figure 3 and figure 4.

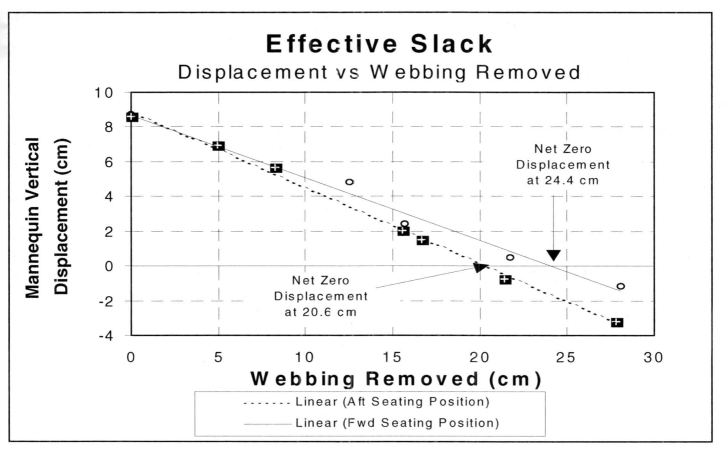

Figure 5. Determining effective slack

RIGID SEAT/RIGID MANNEQUIN TEST

Prior testing had shown results utilizing, in one instance, a rigid seat, and in a second instance, a rigid mannequin. Both test programs had examined the motor vehicle seat belt system and its role in the kinematics response of occupants during roll over collisions. In this third test program the rigid test devices were applied together in a simple test [9].

Similar to the method described in testing utilizing the rigid mannequin [9], the mannequin and lifting device were positioned over the rigid seat test apparatus. The seat was positioned such to provide maximum access. Test configuration is shown in figure 6. The seat geometry is shown in figure 1. A force equivalent to 91 kg (200 lb.) theoretically hanging from the seat belt was applied to the mannequin. Lifting of the mannequin was chosen over inversion in the seat due to the lightweight of the mannequin, 16 kg (35 lb.) Initial positioning of the mannequin resulted in a slightly unnatural position. Because of the rigid mannequin's lightweight, initial positioning of the mannequin is substantially dependent upon its rigid geometry. Figure 7 shows the pitched slightly forward position of the mannequin, with the nominal 90-degree seat belt, relative to the seat. The pretest mannequin position was chosen because of the ability to easily reproducing the initial condition. Testing proceeded as described in the rigid mannequin test program [9].

Seat belt configurations analogous to those in the rigid seat test were utilized [8]. The lap belt configurations included two nominal lap belt angles, 45 and 90 degrees, and a variety of lap belt adjustment strap tensions. The adjustment strap tension is not the lap belt tension. The lowest adjustment tension, 62 N (14 lb) was a tension chosen to cause a nominal zero lap belt tension, a sufficient adjustment strap tension to remove available slack from the lap belt system. The tested seat belt configurations assumed a natural seat belt position with sufficient belt tension to remove all extra webbing from the restraint system. The test configurations are listed in table 2. The column in table 2 labeled "removed" refers to the amount of webbing removed from the baseline system when adjustment strap tension was applied. Negative removed webbing refers to the addition of webbing to the seat belt system. For condition of negative removed webbing the adjustment tension is not applicable (N/A). Since in the original testing webbing tension was employed by applying force to the adjustment strap, that method was duplicated.

Corrected average vertical displacements for the rigid mannequin tested in the rigid seat are shown for each test configuration in table 3. In general three tests were conducted for each seat belt configuration. For the nominal 45-degree seat belt angle configuration, interference between the lower back of the mannequin and the seat back played a role in limiting the vertical movement of the nnequin. This interference account for the reduced vertical displacement associated with test performed in this configuration. In nominal 90-degree seat belt test the mannequin would pitch forward to a greater extent than the nominal 45-degree configuration. This pitching movement was corrected, but is attributed to the short leg of the mannequin and the lift point of the test apparatus.

For both seat belt nominal angle configurations increased webbing in the system allowed increased mannequin displacement. Considering the interference and rigid test device the lower displacement of the nominal 45-degree configuration, compared too nominal 90 degrees, is understandable. As demonstrated in figure 8, a linear relationship between seat belt webbing length and vertical mannequin displacement appears to exist for the conditions tested. This relationship is analogous to that shown in figure 5.

When comparing the nominal 45-degree response to the nominal 90-degree response, the beneficial effect of belt angle can be observed at the higher adjustment force condition. In the higher tension condition the mannequin is held closer to the rigid cushion and is less affected by the interference between the seat back. At the nominal 90-degree seat belt configurations the interference between the mannequin and the seat back did not occur.

While the test conditions are not the same, since the seat is the same and the direction and magnitude of the lifting force applied to the mannequin in the rigid mannequin/ rigid seat test program is similar to the direction and magnitude of the force due to the hanging mass of a 95th percentile Hybrid III dummy, comparison of static test results is presented. Head displacement of the 95th percentile Hybrid III in rigid seat test was significantly greater under the same restraint conditions when compared to the rigid mannequin. The significant greater displacement of the Hybrid III is due to the dummy compliance versus the rigid mannequin. A comparison between results of these tests is shown in table 3. The column labeled "Rigid Seat 95th Hybrid III Vertical Displacement" shows the reported results of prior testing.

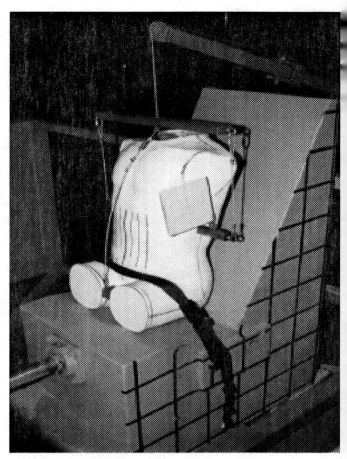

Figure 6.   Rigid Mannequin/Rigid Seat Configuration

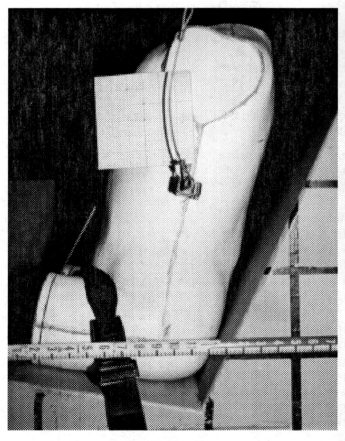

ure 7.   Mannequin Initial Position

Table 2. Test Conditions

Test no.	Nominal Belt Angle (deg)	Adjustment Tension (N)	Removed (cm)
A	45	62.16	0.00
B	45	444.00	1.27
C	45	N/A	-2.54
D	45	N/A	-5.08
E	45	62.16	0.00
F	90	62.16	0.00
G	90	N/A	-5.08
H	90	444.00	0.79

Table 3. Test Results and Comparison

Test no.	Test Result Vertical Displacement (cm)	Rigid Seat 95th Hybrid III Vertical Displacement (cm)
A	4.81	11.00
B	2.67	N/A
C	7.23	14.75
D	8.93	16.10
E	5.05	11.00
F	4.78	11.50
G	11.07	14.00
H	1.50	N/A

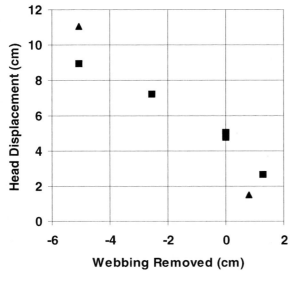

Figure 8. Seat belt webbing length versus vertical mannequin displacement

DISCUSSION

The procedure of applying tension to the seat belt adjustment straps as described in the rigid seat and rigid seat/mannequin testing was problematic. While a consistent methodology produced consistent seat belt webbing tension, the tension in the belt webbing is significantly lower than the tension force applied to the adjustment strap. Do not confuse the seat belt webbing tension with the seat belt adjustment strap tension.

Effective slack has been identified as a seat belt system characteristic that may be expressed as a conceptual equivalent to a length of seat belt webbing removed from the seat belt system that results in zero net vertical displacement of a rigid test mannequin under the test condition [9]. Effective slack quantifies characteristics of the seat belt and seat system, which manifest in the dynamic conditions of crashes. For rollover conditions the manifestation of effective slack is the observation of occupant whole body vertical displacement due to combined seat system and seat belt system characteristics. The characteristics included, among many factors, seat position, seat cushion compliance, seat back compliance, seat geometry, seat belt spool out and/or pretension, seat belt routing and fit, seat belt webbing stiffness, seat belt anchorage geometry and other anchorage compliance.

MOTOR VEHICLE SEATS – Characteristics of the motor vehicle seats that have been identified as important in the control of occupant flail during rollover conditions include, among other possible factors, cushion and seat back compliance and seat geometry. Seat cushion compliance, or the seat's ability to allow occupant whole body movement away from the vehicle roof was a factor in measuring the effective slack of a seat belt system in rollover conditions. The ability of the seat or other structure to allow or cause occupant movement away from the roof prior to the potential harm of a rollover crash may affect the risk of injury.

Because seat belt, lap belt angle forces interaction between the seat back and the occupant lower back or buttocks the geometry and characteristics of the seat back are important in the effectiveness of seat belt systems in rollover condition. As was observed during testing with the relatively stiff Hybrid III mannequin and the rigid mannequin, the effects of the rigid seat back were observable. Human volunteers reported pain due lower back contact with the rigid seat back. In the rigid mannequin to rigid seat test series mannequin displacement was observed to be limited significantly for seat belt angles at the nominal 45 degree condition due to non compliance of the seat back to mannequin lower back interface.

In rollover conditions the seat back has effectiveness as a barrier restricting rearward occupant movement. The effect of the seat back was observed in the rigid seat test and the rigid seat/rigid mannequin test. In the rigid seat t volunteers complained of lower back pain due to seat

1023

back contact in the nominal 45 degree lap belt configuration. Dynamically, in the rigid seat test, lap belt tension was observed to be significantly higher for the nominal 45-degree conditions. This higher lap belt tension is due to the seat back restricting rearward dummy movement. For conditions where the seat back angle is large, or effectively large, occupant movement may be essentially unconstrained in a rearward direction thereby allowing forces due to occupant loading to align over the rear located lap belt anchorage. Occupant movement due to the alignment of forces will increase the effective slack of the seat belt restraint system and increase the possible occupant flail envelope. The potential affect of seat geometry and seat back compliance is greater for lower lap belt angles. Further, possible occupant movement and flail in the rearward direction are limited, under the conditions of a normally upright seat and the wearing of a seat belt.

MOTOR VEHICLE SEAT BELTS – Characteristics of the motor vehicle seat belt systems that have been identified as important in the control of occupant flail during rollover conditions include, seat belt spool out and/or pretension, seat belt routing and fit, seat belt webbing stiffness, seat belt anchorage geometry and other anchorage compliance. These characteristics generally influence the effective slack of a seat belt system and dictate the mechanism through which the seat belt provides a load path between the occupant and vehicle. High seat belt angle in combination with removal of seat belt webbing from the seat belt restraint system (high seat belt webbing pretension loads) has resulted in a demonstrated reduction in occupant displacement and flail in rollover test conditions. Overall a combination of seat and seat belt characteristics brought together in a rollover protection system appear to provide the greatest potential for reducing occupant displacement, flail and injury in rollover crashes.

Testing utilizing, in combination or alone, the rigid test seat and mannequin point to the importance effect of the human condition when considering the flail of motor vehicle occupants in rollover conditions. The human condition affects flail in rollover conditions due to several factors. These factors include kinematic movement of body segments, and viscoelastic compliance, compression and stretch, of body tissues. Flail of body segments in rollover conditions includes movement of the head, upper torso and extremities. To date test mannequins, whether rigid or Hybrid III, lack sufficient biofidelity to replicate expected human response in rollover conditions. These deficiencies do not negate the potential contribution that these important test tools may have in the development of rollover occupant protection systems.

CONCLUSION

Test data demonstrates the opportunity for occupant contact and interference between modern passenger car interiors -- roofs and door structures. For injuries inside the vehicle simple contact or interference inadequately describes potential rollover injury mechanism. Injury in rollover crashes is caused by a more complex series of conditions related to the position and velocity of the occupant and the deformation and velocity of intruding vehicle components. The means to reduce and possibly negate occupant conditions related to many interiors caused rollover injuries through the use of improved seat and seat belt technology has been demonstrated. Possible improvements can be achieved by reducing the seat belt system effective slack.

Design attributes that may favorable affect the performance of seat belt restraint systems for rollover protection include:

1. Mechanism which reduce the length of webbing in the lap belt restraint system,

2. Mechanism which increase the lap belt angle,

3. Seat and seat belt attributes which reduce the effect of occupant compliance

4. Seat or seat cushion attributes which lower the occupants body in the vehicle,

5. For seat belts which are not integral to the seat back, seat backs that are upright and include attributes which restrict occupant movement. The primary mechanism identified would restrict rearward displacement of the occupant torso and there by reduce vertical excursion,

6. Inclusion of passive or deployable passive mechanism in the seat and seat belt system which employee the above described concepts, possible in combinations.

Rollover protection can not occur solely through a seat belt restraint system, a guiding principal in designing for rollover protection would consider all motor vehicle attributes that may produce a rollover protection system. While seat belts have proven to make substantial difference in rates of whole body ejection during rollover, partial ejection remains a primary important injury mode with modern seat belt systems. In addition to improved seat belt systems, occupant ejection and partial ejection could be addressed by other motor vehicle design intervention. The process of research and innovation relative to rollover protection system should be ongoing and occur outside the paradigm of old or existing vehicle components.

REFERENCES

1. Arndt, M. W., "Testing for Occupant Rollover Protection," SAE 980213, February 23-26, 1998

2. Huelke, D. F., Lawson, T. E., Scott, R., and Marsh, J. C., "The Effectiveness of Seat Belt Systems in Frontal and Rollover Crashes." SAE 770148, 1977.

3. Huelke, D. F., Lawson T. E., Marsh, J. C., "Injuries, Restraints and Vehicle Factors in Rollover Car Crashes." Accident Analysis and Prevention, Vol. 9, pp. 93-107, 1977.

4. Evans, L., "Restraint Effectiveness, Occupant Ejection from Cars, and Fatality Reductions." General Motors Research Laboratories, GMR-6398, September 1, 1988.

5. Moffatt, E. A., Padmanaban, J., "The Relationship Between Vehicle Roof Strength and Occupant Injury in Rollover Crash Data." 39th Annual Proceedings, Association for the Advancement of Automotive Medicine, pp. 245-267, October 16-18, 1995.

6. Bahling, G. S., Bundorf, R. T., Kaspzyk, G. S., Moffatt, E. A., Orlowski, K. F., and Stocke, J. E., "Rollover and Drop Tests - The Influence of Roof Strength on Injury Mechanics Using Belted Dummies." Thirty-fourth Stapp Car Crash Conference Proceedings, SAE P-236, pp.101-112, November 4-7, 1990.

7. Moffat, Edward A., Cooper, Eddie R., Croteau, Jeffrey J., Parenteau, Chantal, and Toglia, Angelo, "Head Excursion of Seat Belted Cadaver, Volunteers and Hybrid III ATD in a Dynamic/Static Rollover Fixture," SAE 973347, 41st Stapp Car Crash Conference, 1997.

8. Arndt, M. W., Mowry, G. A., Dickerson, C. P., "Evaluation of Experimental Restraints in Rollover Conditions." 39th Stapp Car Crash Conference Proceedings, SAE P-299, pp. 101-110, November 8-10, 1995.

9. Arndt, M. W., Mowry, G.A., Baray, Pete E., Clark, David A., "The Development of a Method for Determining Effective Slack in Motor vehicle Restraint Systems for Rollover Protection," SAE 970781, February 24-27, 1997.

C524/190/97

Adaptive restraints – their characteristics and benefits

M MACKAY, BSc, PhD, **A M HASSAN** BSc, PhD, and **J R HILL**
Birmingham University, UK

Synopsis

The use of current seat belts has been shown to be effective in reducing deaths and serious injuries to restrained car occupants by 50% compared to unrestrained. Real world accident studies have identified limitations to the performance of set belts. This has led to the next major evolution in restraint design which is the development of the intelligent restraints. The options for intelligent retraints include making the system variable, to take account of occupant age and sex, occupant weight, occupant sitting position (relative to forward structures) and the severity of the collision which is occurring thus changing the characteristics of the seat belt. Data are presented on how a population of drivers and passengers actually sit in cars, and accident analyses will illustrate how injury outcome varies with age and sex for restrained occupants. The implications of the position of the hands on the steering wheel during normal driving and the rotational orientation of the steering wheel during an impact for airbag design are also included.

2 INTRODUCTION

Current seat belts have been shown to be very effective in diminishing the frequency and severity of injuries to car occupants. So much so that high levels of seat belt use are a prime aim of all national transport safety policies in motorized countries. The limitations of the protective abilities of current seat belts have been well documented in many analyses of both field accident data and experimental studies (1).

Real world accident studies have identified five categories of limitations to the performance of current seat belts. These are:

1) Head and face contacts with the steering wheel by restrained drivers (2) - It is inherent in the kinematics of a restrained occupant that, in a severe collision at a velocity change of around 50 km/hr, the head will arc forwards and downwards, having a horizontal translation of some 60 to 70 cms Figure 1. If a normal steering wheel position is superimposed on such a trajectory, the head and face necessarily will strike the steering wheel. Such contacts usually produce AIS 1 to 3 injuries and are best addressed with the supplementary airbag systems becoming common throughout the new vehicle fleet.

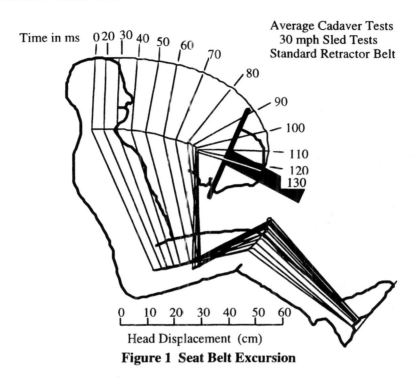

Figure 1 Seat Belt Excursion

2) Intrusion of Forward Structures - A seat belt requires a zone ahead of the occupant so that the occupant can be decelerated by the compliance of the restraint system. If intrusion compromises that space, then specific localized contacts can occur. The injury risk from such contacts may well be small if they are occurring with structures which have been engineered appropriately. Indeed, in the ultimate condition, it is better for the occupant to be decelerated not just by the seat belt alone but through a combination of belt loads and contact loads. Those contact loads are through the feet at the firewall, through the knees into the lower dash and through the airbag and belt at chest level. In severe collisions, however, major intrusions are destroying the passenger compartment so that exterior objects are actually striking the occupants. This is a feature of restrained fatalities in frontal impacts (3).

3) Rear Loading - Correctly restrained front seat occupants can receive injuries from unrestrained occupants, luggage or animals from the rear seats. Such events contribute to some 5% of restrained front seat fatalities (4).

4) Misuse of the Seat Belt - Seat belts must be positioned correctly on the human frame to work effectively. Dejeammes (5) in a survey of belt use in France found that some 1.6% of front seat occupants had the shoulder belt under the arm or behind the back whilst some 3.3% had

introduced slack because of the use of some clip or peg to relieve the retraction spring tension. A more important type of misuse relates to the positioning of the lap section. Many occupants, especially the overweight, place the lap section across the stomach instead of low across the pelvis. Indeed for the obese, it is often impossible to position the lap section so that it will engage on the iliac spines of the pelvis in a collision. These problems are reflected in abdominal injuries from the lap section of the seat belt (6).

5) Injuries from the Seat Belt Itself - As with any injury mitigating device there are limits to effectiveness. Those limits are when biomechanical tolerances are exceeded and thus the most vulnerable segment of the population begin to receive injuries. The usual thresholds are sternal and rib fractures occurring, especially in the elderly (7).

Current restraint design aims to achieve a compromise in the sense of optimizing protection for the largest number of people exposed in the largest number of injury-producing crashes. The end point, however, is a fixed design with single characteristics optimized around a single crash condition. That crash condition for most manufacturers is usually the 35 mph (56 km/hr) rigid barrier crash test.

The next evolutionary stage in restraint design is to move away from a restraint system with fixed characteristics which need to be considered if the concept of variability is introduced into restraint design.

2 POPULATION CONSIDERATIONS

The ideal restraint system would be tailored to the following variables:
- the specific weight of the occupant,
- the specific sitting position of the occupant,
- the biomechanical tolerances of the occupant,
- the severity of the specific crash which is occurring,
- the chances of specific passenger compartment intrusion occurring which might compromise restraint performance
- the specifics of the compartment geometry and crush properties of the car.

3 ANTHROPOMETRIC CONSIDERATIONS

Current dummies and modeling cover the 5th percentile female to 95th percentile male range. Assuming for simplicity that males and females are exposed equally and that there are few males smaller than the 5th percentile female or females larger than the 95th percentile male, these conventional limits put 2.5% (1 in 40) of the small population and 2.5% of the larger population beyond those limits; 5% or 1 in 20 overall.

Table 1 gives the 1% and 99% ranges for height, sitting height and weight. These data show what would be required if the design parameters were extended to cover this wider range, so that only 1 in 50 of car occupants would be outside the design parameters (8).

Table 1 Population Ranges for Height, Sitting Height and Weight

Adult	Height		Sitting Height		Weight	
	ins	cm	ins	cm	lbs	kg
1%ile female	57	145	28	72	82	37
5%ile female	59	150	29	75	90	41
95%ile male	73	185	37	93	225	102
99%ile male	75	190	38	96	236	107

More importantly, it is implicitly assumed in current designs that height (or sitting height) and hence sitting position are colinear with the weight of the occupant. In fact, there are data available to suggest that the relationship between height and weight are rather complex. For example, the body mass index (BMI) (i.e., the ratio of weight in kilograms to height in meters squared) varies to a greater degree in women than in men, and particularly at the 75th percentile and above, women have higher BMIs than men. In addition, the prevalence of overweight increases with age, more with females than males (9).

Therefore to optimize a restraint system it would appear appropriate that sitting position and body weight should be assessed independently if variability is to be introduced into restraint design.

4 POPULATION CHARACTERISTICS BY POSITION IN THE CAR

European data show that some 80% of drivers in injury-producing collisions are male, whilst some 65% of front seat passengers are female (10). Approximately one-third of rear passengers are children of 10 years of age or under (11). These simple frequencies suggest that restraint characteristics should not necessarily be the same for all sitting positions in the car.

4.1 Sitting Positions

Current design is predicated on the positions established for the three conventional dummies. Observational studies by Parkin et al (12) have demonstrated that there are substantial differences between those three positions and an actual population of drivers. Passive observations of drivers in the traffic stream have been made using video recording techniques, and drivers classified by sex and general age groups of young (35 years), middle (36-55 years) and elderly (56 years and older). Make and model of car were recorded and measurements made of the following distances:

- nasion to steering wheel upper rim and hub,
- top of head to side roof rail,
- back of head to head restraint, horizontally and vertically,
- shoulder in relation to 'B' pillar.

1029

Such techniques allow thousands of observations to be made quickly and therefore population contours can be drawn. Figure 2 illustrates how particularly for the 5th percentile female population the actual sitting position is significantly closer than that of the 5th percentile dummy, by some 9.2cm. The 5th percentile, small female population sits some 38cm (15 inches) or closer to the hub of the steering wheel.

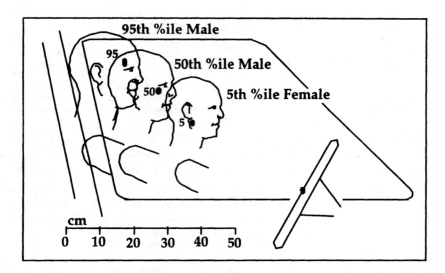

Figure 2   Drivers' Sitting Positions

5th 50th and 95th %ile naison positions are illustrated for "real drivers" (head outlines) and dummies (black spots).

5 BIOMECHANICAL VARIATION

An extensive literature exists concerning human response to impact forces, mostly conducted in an experimental context. A general conclusion from that body of knowledge is that for almost any parameter, there is a variation of at least a factor of 3 for the healthy population exposed to impact trauma in traffic collisions (13). That variation applies to variables which are relatively well researched such as the mechanical properties of bone strength, cartilage, ligamentous tissues and skin. It is likely to be even greater when applied to gross anatomical regions such as the thigh in compression, the thoracic cage, the neck or the brain.

How such variability is demonstrated in populations of collisions is less well understood. Data from a ten year period of the European Co-operative Crash Injury Study (CCIS) for restrained front seat occupants are given in Figures 3 and 4. The methodology of that work has been described elsewhere (14).

Figure 3 illustrates the effect of age on injury outcome in terms of the frequency of AIS 2 and greater injuries for three age groups. Data are presented for frontal impacts involving a principal direction of force (PDF) of 11 to 1 o'clock, controlling for crash severity by equivalent test speed (ETS). Injury severities were rated by Maximum Abbreviated Injury Scale (MAIS;(15).

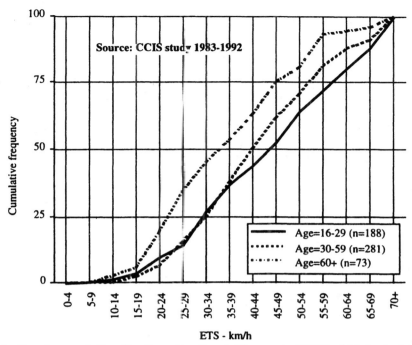

Figure 3 Crash speed distributions for frontal impacts (PDF of 11 to 1 o'clock) to drivers (by age groups) who experienced injuries with a MAIS > = 2

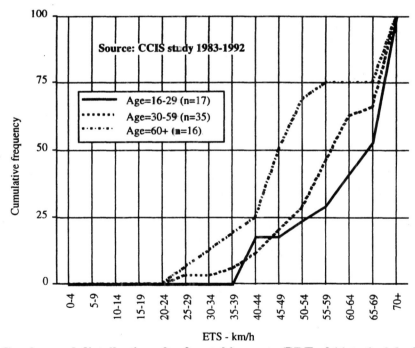

Figure 4 Crash speed distributions for frontal impacts (PDF of 11 to 1 o'clock) to front seat occupants (by age groups) who experienced injuries with a MAIS > = 4

The 60+ age group especially shows greater vulnerability than the younger groups. As a broad generalization one may conclude that for the same injury severity, the younger age groups must have a velocity change of some 10 km/hr more than the elderly. The effect is more marked if a more severe injury level is chosen. Figure 4 illustrates the cumulative frequencies for the three age groups for injuries of AIS 4 and greater.

Figure 5 shows similar frequency curves for crash severity by sex of occupant. Thus at a velocity change of 48 km/hr (30 mph), some 2/3 of male and some 80% of female AIS 2+ injuries have occurred. As a starting point, therefore, as well as specific body weight and sitting position, a combination of age, sex and biomechanical variation could be developed as a predictor of the tolerance of a specific person within the population range.

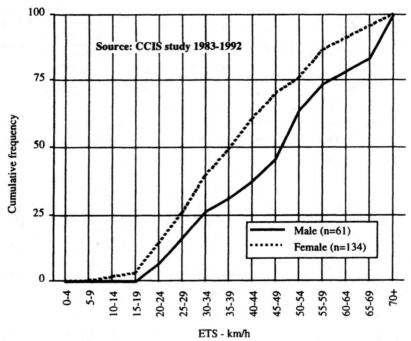

Figure 5 Crash speed distributions for frontal impacts to front seat passengers (by sex) who experienced injuries with a MAIS > = 2

An intelligent restraint system therefore would perhaps require a smart card, specifying the height, weight, age and sex of the occupant. On entering the card for the first time, the card would be read and the characteristics of the seat belt and airbag adjusted accordingly.

6 SENSING CRASH SEVERITY

Besides assessing the specifics of the occupant's characteristics before impact, protection could be enhanced if the nature and severity of the collision could be assessed early enough during the crash pulse so that the characteristics of the restraint system could be modified. That would require, for example, sensors to discriminate between distributed versus concentrated impacts, and between, for example, three levels of collision severity such as less than 30 km/hr, 30 to 50 km/hr, and greater than 50 km/hr. In addition, conceptually one might have an array of sensors which would detect the early development of compartment intrusion. Such electronic data could then instruct the restraint system to change its characteristics early enough during the crash phase to alter the characteristics of the restraint and thus the loads on and forward excursion of the occupant.

7 VARIABLE RESTRAINT CHARACTERISTICS

The advantages of a variable restraint system are illustrated by considering some examples. A front seat passenger, 70 years of age and female, weighing 45 kg sitting well back, in a 30 km/hr frontal collision with no intrusion, would be best protected by a relatively soft restraint system which would maximize the ride-down distance and minimize the seat belt loads. That would require a low pretensioning force, a long elongation belt characteristic provided by load limiters and a soft airbag.

Such a system is very different from what would be required by a 25 year old, 100 kg male, sitting close to the steering wheel in a 70 km/hr offset frontal collision. He would need a very stiff seat belt, an early deploying stiff airbag and a large amount of pretensioning load.

Consider thirdly a 9 year old girl, weighing 30 kg sitting in a rear seat in a 56 km/hr frontal impact. Maximizing her ride-down distance and minimizing the seat belt loads would require low pretensioning loads and a very soft belt system, but one which would still have a biomechanically satisfactory geometry at the forward limit of excursion. Possible techniques for introducing variability into restraint design are now discussed.

7.1 Variable Pretensioning Force
A retractor pretensioner might be devised which would have a variable stroking distance or perhaps two stages of pretensioning to address the population and crash severity requirements outlined above.

7.2 Combined Retractor Pretensioner & Buckle Pretensioner
Such a system of pretensioners might maintain good seat belt geometry especially for the small end of the population, such as the 9 year old girl in the rear seat, when soft restraint characteristics and hence large amounts of forward excursion are required.

7.3 Discretionary Web Locks
If the seat belt system needs to be stiffened for the heavy occupant with high biomechanical tolerance in a high speed crash, then the switching in of a web lock would be appropriate. Such a device would shorten the active amounts of webbing being loaded and diminish forward excursion at the expense of somewhat higher seat belt loads.

7.4 Discretionary Load Limiting Devices
One way of providing for biomechanical variability would be to have a load limiting mechanism which would be calibrated for the specifics of the occupant's age, sex and weight. Such a device could also be adjusted according to transient sitting position. Belt loads would be limited at the expense of increased forward excursion.

7.5 Variable Sitting Positions

Ultrasonic, infrared or other techniques of sensing might be used to monitor continuously the head position of each occupant. Such information could be used at a minimum to provide a warning that an occupant was sitting too far forward and in particular too close to the steering wheel. At a more advanced level it could be used to tune the seat belt and airbag characteristics to be optimized for that occupant in that specific position by adjusting the other restraint variables.

7.6 Variable Airbag Firing Threshold

The need for an airbag varies according to seated positions in the car and the characteristics and sitting position of the occupant. For most drivers in most sitting positions a supplementary steering wheel airbag becomes desirable in crash severities above 30 km/hr (2). For a front seat passenger however, particularly one who is towards the top end of the biomechanical tolerance spectrum and sitting well back, an airbag at 30 km/hr is unnecessary. For a child sitting a long way forward in such a crash, it might also be disadvantageous. Hence specific sensing techniques at a minimum could discriminate between the presence or absence of a passenger, and at the next level assess the need for the airbag to inflate or not.

7.7 Variable Airbag Characteristics

In response to the sensing data about the occupant's characteristics and transient sitting position, and the accelerometer data about the nature and severity of the collision which is occurring, the airbag properties could be varied. Specifically, gas volume and inflation rate could be changed. Compressed gas systems instead of chemical gas generators have the potential for providing those characteristics by having time-based adjustable inflation ports. This requires very advanced sensing and control systems but these aims could well be addressed through future research and development.

8 HAND POSITIONS AND STEERING WHEEL ORIENTATION

In addition to the seating position of the driver before impact, the position of the hands on the wheel and the orientation of the steering wheel at impact need to be considered. These factors may influence airbag characteristics.

8.1 Hand Positions on Steering Wheels

An observational study was carried out which looked at the position of the drivers hands on the steering wheel during normal driving condition on major roads with speed limits of 40 to 60mph (64 - 96km/h)in the UK and US, excluding motoways and freeways. Driving with only one hand on the steering wheel seems to be more common. Fifty eight percent of UK drivers used one hand only, while the proportion was much higher in the US at 70%.. Drivers were more inclined to hold the wheel in the upper semicircle, above the 3 and 9 o'clock positions. When two hand were used both tended to be at same height. The distribution of the positions of the hands on the steering wheel are shown in Figure 6 and Figure 7 where one and two handed positions have been counted together.

A quarter (26%) of the drivers were considered to be at risk of injuries because hands or arms were observed in very close proximity to the airbag module. The risk for US drivers was lower at 16%. Drivers were considered to be at risk of receiving injuries an airbag deployed while hands were (a) at the 1, 11 or 12 o'clock positions, or (b) at the 3 or 9 o'clock positions while resting inside the wheel rim on or near the airbag module. The risk of injury may be increased at junctions where 91% of the drivers in the UK and 98% of the drivers in the US were observed to cross arms while turning the wheel.

Therefore a significant group of the population may be at risk of injuries to the upper extremities if airbags deploy while the steering wheel is held near the top or while turning at a junction. The inclusion of sensors to assess arm position and tight steering manoeuvres at low speeds should be considered with smart restraints.

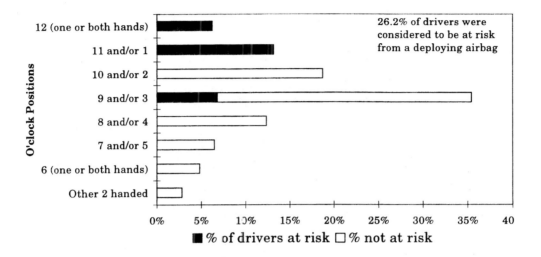

Figure 6 Hand positions on steering wheels observed for 850 U.K. drivers

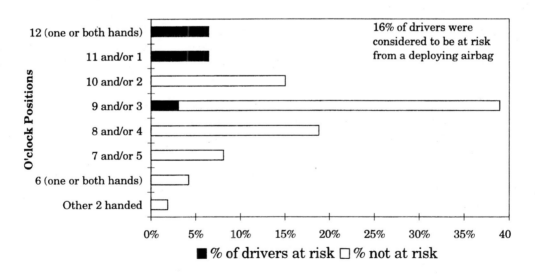

Figure 7 Hand positions on steering wheels observed for 850 US drivers

8.2 Steering Wheel Orientation

Accident records of cars, in the CCIS database, with steering wheels jammed by crushing at the time of impact were examined to determine the rotational orientation of the steering wheel. Only cars involved in single frontal impacts with a principle direction of force between 11 to 1 o'clock and with at least one front road wheel displaced rearwards (strutted) and firmly jammed by crush were included. Steering wheel orientation was assumed not to have changed post impact by considering factors such as degree of strutting, steering wheel damage, steering column damage and orientation of blood stains.

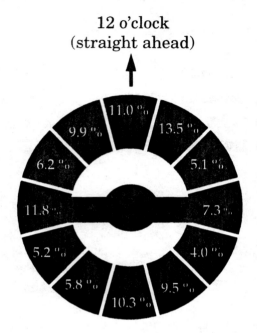

Figure 8 Observed steering wheel orientations grouped into 30° sectors (n = 272)

Wheels were more often orientated in the 11, 12 and 01 o'clock quadrant (Figure 8). However, over 65% of steering wheels were observed at other positions suggesting that steering wheels are in all possible rotational orientations at impact, and it can not be assumed that drivers crash their vehicles with the wheels in the straight ahead position.

There are implications for factors in airbag design including shape of the deploying bag, location of vent holes and port design. The airbag modules and their doors and inflating bags may activated when wheels are at any orientation. Their interaction with occupants may not be as predictable as the current design procedures imply. Symmetry should be incorporated so that components deploy in the appropriate manner and inflate over the wheel orientation.

9 OTHER CRASH CONFIGURATIONS

The discussion so far has focused on frontal collisions which constitute some 50% to 65% of injury producing collisions in most traffic environments. Lateral, rear and rollover crashes also suggest opportunities for optimizing protection through intelligent restraint systems.

9.1 Lateral Collisions
The technology is now developing for side impact airbags with two versions becoming available on 1995 model-year passenger cars. The observational data of Parkin et al (12) have illustrated the range of driver sitting positions which reflect the requirements of side impact airbag geometry to cover both the door and the B pillar. Because a significant part of the population,

tall males, choose to sit as far rearward as possible. in a side impact in many four door vehicles the thorax would be loaded by the B pillar rather than the door.

A practical issue is the nature and position of the sensor for a side impact. Because of the extremely short time available for sensing, around 5 milliseconds, a simple switch system is appropriate (16). An analysis of a representative sample of AIS 3 plus lateral collisions has demonstrated that if a switch sensor is located in the lower rear quadrant of the front door then approximately 90% of all such side impacts would be sensed appropriately. A set of several sensors would be required to address the remaining few collisions, whilst rear seat occupant protection would also be addressed in large part by a sensor in the same position in the front door as is appropriate for front seat occupants (17).

9.2 Rear Impacts
Occupant protection in rear end collisions is addressed largely through the appropriate load deflection characteristics of seat backs and the provision of correctly positioned head restraints. The real world data of Parkin et al (12) demonstrates that head restraints are frequently positioned both too low and too far to the rear of the occupant's actual head position. The head position sensors discussed above could also be used for adjusting automatically both the vertical and horizontal position of the head restraint. Such a technology is relatively simple but the costs and reliability, as well as acceptability by the driving population, present serious practical problems.

9.3 Rollover Accidents
Actual mechanisms of injury in rollover accidents have been well researched by Bahling et al (18) for occupants in current seat belts. Conceptually one can suggest that a buckle pretensioner might have some benefits in rollover circumstances by diminishing the relative vertical motion of an occupant. However, in rollovers current dummies do not have the appropriate soft tissue or thoracic and lumbar spine response characteristics, in comparison to the human frame. The basic clearance of current bodyshell design and packaging limit intrinsically the ability of any restraint system to modify the nature of any roof contacts under the forces of actual rollover circumstances even with no roof deformation taking place. Raising current roof lines leads to many undesirable consequences. Nevertheless it would be of interest to explore occupant kinematics in rollovers using more realistic techniques with volunteer and cadaver subjects in the context of buckle pretensioners and the requirements of a sensor to detect incipient rollover.

10 CONCLUSIONS

This paper only attempts to outline in conceptual form some of the issues which need to be addressed in advancing from today's seat belts and airbags towards some form of intelligent restraint system. Of fundamental importance is to recognize the population issues of size, sitting position, biomechanical variation and changing crash exposures. Beyond these issues lies a larger amount of challenging research and development to actually produce the sensors and hardware to provide variability in a seat belt and airbag system. Proximity sensing has its advocates, and if radar techniques could actually discriminate an impending collision from a near miss or a passing object, then the provision of say 500 milliseconds warning would alter many of the restraint issues reviewed in this paper. However, the basic premise remains; the next generation of restraints must change from having single fixed characteristics towards variable ones which recognize the real world population variables of weight, sitting position, biomechanical tolerance and crash exposure.

11 ACKNOWLEDGEMENTS

This project was undertaken on behalf of the Co-operative Crash Injury Study Consortium. The study is funded by the Department of Transport, Ford Motor Company Limited, Nissan Motor Company, Rover Group, Toyota Motor Company and Honda Motor Company. The Project is managed by the Transport Research Laboratory. Grateful thanks are extended to everyone involved with the CCIS data collection process. The authors would also like to acknowledge the considerable effort made by Paul Kidman and Theo Gillam in the preparation of data used in this study.

12 REFERENCES

(1) BACON, D.G.C., The Effect of Restraint Design and Seat Position on the Crash Trajectory   of the Hybrid III Dummy, 12th International Technical Conference on Experimental Safety Vehicles, 12:451-457, Göteborg, Sweden, 1989.

(2) ROGERS S., HILL J., and MACKAY M. Maxillofacial Injuries Following Steering Wheel Contacts by Drivers Using Seat Belts. Brit. J. Oral & Maxillofacial Surgery, 30:24-30, 1992.

(3) MACKAY G.M., CHENG L., SMITH M., and PARKIN S. Restrained Front Seat Car Occupant Fatalities. AAAM Proceedings 34:139-162, October 1990.

(4) GRIFFITHS D., HAYES M., GLOYNS P.F., RATTENBURY S., and MACKAY M., Car Occupant Fatalities and the Effects of Future Safety Legislation. Proceedings, 20th Stapp Car Crash Conference, 20:335-388, Society of Automotive Engineers, Warrendale, PA, 1976.

(5) DEJEAMMES M., ALAUZER A., TRAUCHESSEC R., Comfort of Passive Safety Devices in Cars: Methodology of a Long-Term Follow-Up Survey, SAE Paper No. 905199, Society of Automotive Engineers, Warrendale, PA, 1990.

(6) GALLUP B.M., St. LAURENT A.M. and NEWMAN J.A., Abdominal Injuries to Restrained Front Seat Occupants in Frontal Collisions, AAAM Proceedings, 26:131-145, October 1982.

(7) HILL J.R., MACKAY M., MORRIS A.P., SMITH M.T., and LITTLE S. Car Occupant Injury Patterns with Special Reference to Chest and Abdominal Injuries caused by Seat Belt Loading, Proc. IRCOBI Annual Conference, pp. 357-372, Verona, Italy, September 1992.

(8) Society of Actuaries Build and Blood Pressure Study, London, 1979.

(9) WILLIAMSON D.F., Descriptive Epidemiology of Body Weight and Weight Change in U.S. Adults. Ann Intern Med. Oct 1; 119(7 Pt 2):646-9, 1993.

(10) BULL J.B. and MACKAY G.M., Some Characteristics of Collisions, the Population of Car Occupant Casualties and Their Relevance to Performance Testing, Proceedings, IRCOBI 3rd Conference, pages 13-26, Lyon, France, September 1978.

(11) HUELKE D.F., The Rear Seat Occupant in Car Crashes, American Assn for Automotive Medicine Journal 9:21-24, 1987.

(12) PARKIN S., MACKAY M., and COOPER A. How Drivers Sit in Cars. Proceedings AAAM 37:375-388, November 1993.

(13) MCELHANEY J.H., ROBERTS V.L. and HILYARD J.F., Handbook of Human Tolerance, Japanese Automobile Research Institute, Tokyo, 1976.

(14) MACKAY M., ASHTON S., GALER M., and THOMAS P. Methodology of In-Depth Studies of Car Crashes in Britain. Proceedings, Accident Investigation Methodologies, SP159, pages 365-390, SAE Paper 850556, Society of Automotive Engineers, Warrendale, PA, 1985.

(15) AAAM, The Abbreviated Injury Scale. Ill, USA. 1985.

(16) HALAND Y. and PIPKORN B., The Protective Effect of Airbags and Padding in Side Impacts - Evaluation of a New Subsystem Test Method, 13th International Technical Conference on Experimental Safety Vehicles, 13:523-533, Paris, 1991.

(17) HASSAN A., MORRIS A., MACKAY M., and HALAND Y., The Best Place for a Side Impact Airbag Sensor. Proc. AAAM/IRCOBI Conference on Advances in Occupant Restraint Technologies, Lyon, France, September 1994.

(18) BAHLING, G.S., BUNDORF R.J., KASPZYK G.E., MOFFATT E.A., ORLOWSKI K.R., and STOCKE J.E. Rollover and Drop Tests. The Influence of Roof Strength on Injury Mechanisms Using Belted Dummies, Proceedings 34th Stapp Car Crash Conference, 34:101-112, Society of Automotive Engineers, Warrendale, PA, 1990.

976007

OPTIMIZING SEAT BELT USAGE BY INTERLOCK SYSTEMS

Thomas Turbell
Swedish Road and Transport Research Institute (VTI)
Torbjörn Andersson
AUTOLIV Development AB
Anders Kullgren
FOLKSAM
Peter Larsson
Swedish National Road Administration
Björn Lundell
VOLVO Car Corp.
Per Lövsund
Chalmers University of Technology (CTH)
Christer Nilsson
SAAB Automobile AB
Claes Tingvall
Swedish National Road Administration
Sweden
Paper Number 96-S1-O-07

ABSTRACT

Seat belts are known to be very effective, reducing the risk of injury by approximately 50% when used. Such high effectiveness is, however, based on the fact that all car occupants use the available belts. In several studies it has been shown that, in severe accidents, the seat belt use was less than 50%.

In order to increase the wearing rate more drastic solutions than information, legislation etc. have to be used. A Swedish group, representing government, research, insurance companies, car and restraint systems industry has approached the problem by proposing a smart system that will force car occupants, that normally are unbelted, to use the seat belts by systems that will interfere with the normal use of the car. Different technical approaches, which not in any way will interfere with the normal belt user, will be put forward and evaluated. The problem will also be discussed from a cost-effectiveness point of view and the potential of saving lives in an international perspective will be analyzed. It is shown that more than 6.000 lives could be saved per year in the European Union if the existing seat belts were used.

INTRODUCTION

For a couple of years there has been a concern within the Swedish road safety community about the fact that the safety potential of the seat belts is not fully used.

One of the first alarms came in a report from 1992 where fatally injured car occupants in Stockholm were studied. (Kamrén, 1992). The belt use in this group was only 40% compared to 80% in the general population.

At the last ESV Conference preliminary thoughts on a seat belt interlock system were presented by the Folksam Research Group. (Kamrén, 1994)

Another Swedish study of the belt use and the injuries showed again that the belt usage rate among severely injured was 50% in rural and 33% in urban accidents. (Bylund 1995)

In a Finnish study (Rathmayer, 1994) a clear pattern of the behavior of non-users was observed. Seat belt users committed one traffic offense in every 13 km on highways while the non-users committed one offense every 5,5 km. In urban traffic the distance between offenses was 9 km for the belted and 2,5 km for the unbelted. Non-users also drove faster and had driving histories with longer violation records than the control drivers.

Seat belt usage rates are generally observed in daylight. It can be assumed the rates are lower in darkness and in other situations where the risk for being caught without a belt is lower.

There are also a number of international research results that confirms these findings regarding the belt use and the non-user.

Being aware of that further campaigns and enforcement could only have a limited effect and that technical solutions like automatic belts were not realistic the Swedish National Road Administration last year formed a group of people representing the administration, research, insurance and car industry in order to analyze the situation and propose solutions. This paper reflects the thoughts of that group so far.

THE US EXPERIENCE

Because of various delays in introducing mandatory automatic protection in the USA in the beginning of the seventies the starter interlock requirement was introduced for the period August 15, 1973 until August 15, 1975 for vehicles without automatic protection produced during that period. These systems were connected to both front seats in such a way that if any front seat belt in an occupied seat was not locked, the starter was disabled. If a buckle was opened later a buzzer-light system was activated. All 1974 model year cars sold in the United States came with this ignition interlock except a few thousand GM models that came with airbags that met the automatic protection requirement.

In March 1974 NHTSA described the public reaction to the ignition interlock as follows:" Public resistance to the belt-starter interlock system currently required has been substantial with current tallies of proper lap-shoulder belt usage at or below the 60% level. Even that figure is probably optimistic as a measure of results to be achieved, in light of the likelihood that as time passes the awareness that the forcing systems can be disabled, and the means for doing so will become more widely disseminated, ..." There were also speeches on the floor of both houses of Congress expressing the public's anger at the interlock system. On October 27, 1974 President Ford signed into law a bill that prohibited any Federal Motor Vehicle Safety Standard from requiring or permitting the use of any seat belt interlock system. NHTSA then deleted the interlock option from October 31, 1994.

Thus the interlock systems were required in the USA for 14½ months instead of the 24 months that were originally intended. (Kratzke. 1995)

LESSONS TO BE LEARNED FROM THE US EXPERIENCE

The failure of the interlock systems in the USA 1974 can be explained by the following factors:

● Many people felt that it was an infringement of personal freedom. This is probably a typical US reaction that may not be valid for e.g. the European market.

● The voluntary seat belt use was very low.

● The belt systems that were used in the USA at that time were usually difficult to use and had a bad fit.

● The interlock system itself was too unsophisticated. It did not allow low speed maneuvers or sitting in the car with the engine idling.

NEW APPROACHES

Basic principles

Some basic principles for a new system have been established:

● The normal seat belt user shall not notice the system.

● It shall be more difficult and cumbersome to cheat on the system than using the belt.

● Permanent disconnection of the system shall be hard to make.

● The system must be very reliable and have a long lifetime.

● All seating positions in the car shall be covered by the system.

● The accident risk must not increase by any malfunctions in the system.

● Retrofit systems for old cars should be available.

Detection and processing

One input to any interaction system is the situation in the car. Which seats are occupied and are the belts properly used on these seats?

The basic sensor for an occupied seat is a contact that will detect a certain load on the seat. This concept can give false signals from e.g. luggage on the seats. Modern techniques with photocells, IR-detectors, inductance, pattern recognition and load measurements on the seatback can be used to overcome most of the problems.

To determine if the belts are properly used is maybe more complicated. The US experience showed that simple systems like a switch in the buckle or measuring the amount of webbing coming out from the retractor could easily be tampered with. There is more information available from the belt system that could be used e.g. angles and forces at anchor points. Also sophisticated systems like pattern recognition or transponders in the webbing could be used.

Information from the doors, the seats and from the belt system can be combined and analyzed in such a way that the proper conclusions can be made.

Other problems that must be considered are how child restraint systems will work in this new environment and how to handle the situation when a passenger disengages his belt during travel. In that case it is probably not possible to influence the behavior of the car other than gradually and after some proper warnings to the driver.

Interaction

Several ideas for interaction could be considered, of which some are presented here.

The **starter-interlock** as used in the USA is the most aggressive solution. As mentioned above there are several shortcomings with this system so it is not on the agenda for the new approaches.

External visual signals is a new concept that is worth considering. By flashing the headlights or the hazard warning flashers the surrounding traffic (and the police) will notice the vehicle with non-belt users. The social pressure and the risk of being caught ought to be a good incentive to use the seat belts.

Internal light and sound warnings are used already in cars today but they can be made more aggressive and more directed to the individual non-user.

Interactions with comfort and audio systems is another approach that is discussed. This is a "soft" countermeasure but by disabling the radio, the air-condition, opening the windows etc. some users may get the message.

Throttle pedal feedback can also be used so that the force on the pedal will increase at a certain speed. This will make it possible but very tiresome to exceed that speed. Another solution may be to introduce severe vibrations in the pedal at a certain speed level.

Maximum gear level makes it impossible to put in any gear than number 1 and reverse. This takes care of one of the main faults with the US starter interlock which made it impossible to garage the car without using the seat belt. It also makes it possible to remove a stuck car from e.g. a railway crossing or a burning garage.

Maximum speed is a similar solution to the maximum gear level. The limit that is discussed so far is 30 km/h.

The final solution may be a combination of these systems i.e. the sequence can start with a visual and audible warning and then increase in intensity and finally reduce the maximum speed.

POTENTIAL EFFECTS

Sweden has got one of the highest seat belt use rates in the world with a front seat use of about 88% in observational studies. Other countries in Europe have a marginally higher use with UK in top with 91%.

In Sweden, the 88% use is to be compared to the less than 50% use among fatalities. The following table on the number of fatalities can be derived from the present situation in Sweden.

Table 1.
Seat belt use among fatally injured car occupants in Sweden 1994, based on a sample of 32 cases, and estimated number of fatalities with 100% belt use

	1994	with 100% belt use
Seat belt used	155	272
Seat belt not used	234	-
Total	389	272

	1994	with 100% belt use
Saved lives in relation to current seat belt use in Sweden (50% effectiveness)		117
Saved lives in relation to 0% seat belt use in Sweden (50% effectiveness)	155	272

The potential number of savings is 272 fatalities per year, but we have only come to a level where we have used 57% of the potential savings. This also shows that we have a higher benefit per user from the last 15% than we have had from the 85% seat belt use that we have today. This is different from other areas where the major benefits comes from the first part of an investment and with a decreasing marginal benefit. The relation between the seat belt usage rate in the population and the potential effect based on the Swedish situation can be described by the curved line in the following curve. A low usage rate gives a very limited effect since these individuals drive very safe anyway. The last 10% probably represents the most accident prone group so this is where we find the largest benefits from the belt use. The straight line describes the common belief that there is a linear correlation between the usage rate and the effect.

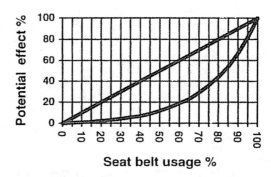

Figure 1. Correlation between seat belt usage and the potential effect on the fatalities.

If we use the Swedish figures and assume that the total European situation is not better, 15.200 unbelted occupants are killed every year in Europe. With a 100% seat belt use and a 50% injury reducing effectiveness, the total number of savings is around 7.600 per year. Given the Swedish situation, this is probably not an overestimation, although the potential savings may vary from country to country. It must be remembered that these figures apply only if the whole vehicle fleet is equipped with an interlock system.

A recent study from the European Transport Safety Council (ETSC) shows similar results with a potential reduction of 5.570 fatalities by a 95% belt usage rate.

Table 2.
ETSC estimations of seat belt use potential

Country	Killed car occupants 1993	Belt usage % Front seat 1991-95	Potential number of saved lives
Austria	747	70	175
Belgium	1050	55	277
Denmark	254	92	58
Finland	274	87	53
France	6168	85	1243
Germany	6128	92	1097
Greece	781	63	199
Ireland	187	53	51
Italy	3931	~55	998
Luxembourg	54	71	14
Netherlands	615	73	139
Portugal	1140	~63	234
Spain	3606	~75	834
Sweden	389	90	69
United Kingdom	1835	91	329
EU Total	27159	80	5570
USA	21987		

(IRTAD 1993) (ETSC 1996)

Only fatalities are discussed in this paper. An interlock system will of course also have a similar effect on the number of severely injured which is about 10 times larger than the number of fatalities.

ALTERNATIVE MEASURES

Preliminary calculations of the cost-effectiveness of an interlock system show that this is a very effective measure compared to some other ones.

By using the Swedish calculations of the willingness-to-pay for risk reductions, it is possible to calculate the possible economic benefits for an interlock system. It can be estimated that the savings from interlock in Sweden is in the region of more than 5 billion SEK/year (~700 million US$/year). With the medium age of cars that we have in Sweden for the moment, the cost that can be spent on each car for an interlock system is therefore approximately 20 000 SEK (~3 000 US$). With an anticipated cost of 200 SEK (~30 US$) per car for an interlock system, the ratio between benefit and cost is 100:1, which by margin is higher than for any other known safety measure. As an example, a 100% fitting of airbags from now on in Sweden would save approximately 50-60 lives annually, but for a cost that is ten times higher than for the interlock, still leaving us with a positive balance between cost and benefit, but serving as an indicator of the extreme benefits of interlock.

ATTITUDES

Preliminary results from a study made by the Swedish National Road Administration in 1995 based on interviews with 5914 persons aged 15-84 years show that there, in general, is a positive attitude for introducing interlock systems

Table 3.
Swedish interviews

Do you agree or disagree that cars should not be able to run faster than 30 km/h if the driver is not using the seat belt?			
%	Male	Female	Total
Strongly agree	24,1	36,6	30,2
Agree	17,7	19,8	18,7
Neither / or	14,1	15,1	14,6
Disagree	19,4	13,7	16,6
Strongly disagree	24,8	14,8	19,9
Total	100,0	100,0	100,0

Table 4.
Swedish interviews

Do you agree or disagree that cars should be equipped with buzzers and lights to warn that someone is not using the seat belt?			
%	Male	Female	Total
Strongly agree	37,1	51,8	44,5
Agree	25,4	26,4	25,9
Neither / or	13,4	9,4	11,4
Disagree	12,5	7,2	9,8
Strongly disagree	11,6	5,3	8,4
Total	100,0	100,0	100,0

These two tables show that women are more positive than men and that the less aggressive buzzer-light system is preferred.

This investigation also shows that older persons are more positive to interlock systems than younger persons.

An alarming fact is that of those who state that they seldom or never buckle up in the front seat on rural roads we can find that 77% disagree or strongly disagree on a 30 km/h speed limiting interlock. 55% of this group are also against the buzzer and light warning system. This is actually our target group so we need to find out how to change their attitudes and how to prevent them from disconnecting the interlock system.

INCENTIVES FOR INSTALLATION

Since a legislation on a national level is difficult or impossible after Sweden has become a member of the European Union, other ways to have these systems installed in new and existing cars have been discussed

• A majority of new cars in Sweden are bought as company cars. There is a possibility to lower the tax liability of the benefit in kind on cars if they are equipped with interlock systems.

- The general vehicle tax can also be moderated depending on the safety equipment of the car.
- Insurance companies are discussing to adopt the premiums along these lines.
- Another proposal is that drivers that are caught without using the belt will be obliged to install an interlock device in their cars.

FUTURE ACTIVITIES

Attitudes

During the spring of 1996 a survey of non-users will be made in Sweden. In cooperation with the police, non-users will be stopped and interviewed at 11 locations spread over Sweden. The interviews will concentrate on seat belts in general and reasons for non-wearing in particular.

System specification

A technical specification, probably in the format of a draft ECE-Regulation will be made during 1996. This draft will allow the car manufacturers to use different options but the goal will be a 99% belt use rate. Also retrofit systems for existing cars will be considered.

An interesting alternative to have detailed technical specifications is to measure the actual belt use rate in traffic for the different cars model years. The tax and other benefits could then be applied with about a one year delay from the introduction of a new car. This alternative will give the vehicle manufacturers free hands to do whatever they want to increase the use in their vehicles. If the coupling between the usage rate and the benefits for the car industry and the car user are strong enough this approach could lead also to a voluntary installation of interlock systems in the existing car fleet.

Implementation

Since this concept has been very positively accepted by the Swedish Ministry of Transport, discussions will go on in order to find the proper tax and other incentives to have some kind of system implemented on the Swedish market as soon as possible.

The car industry may object to this since they do not want to have different equipment on different markets. An international standard would of course be better for everybody but considering the time it will take - and the number of unbelted people killed during that time - our position is that we ought to do something wherever it is possible to get something done quickly in this field.

CONCLUSIONS

At least 6.000 lives can be saved in the European Union annually if the seat belts that are already in the cars are used by 100% of the occupants. The only way to reach this level is to have a technical solution that will make it impossible or very cumbersome to use the car without using the seat belts. There are several technical solutions available that could be implemented in a short time. The main obstacle to reach this goal is probably of a political nature.

REFERENCES

Bylund, Per-Olof; Björnstig, Ulf, Låg bältesanvändning bland allvarligt skadade bilister. Rapport nr 54, Olycksanalysgruppen Umeå, 1995

ETSC Seat belts and child restraints, Brussels, 1995

IRTAD, International Road Traffic and Accident Data, BASt, Cologne 1995

Kamrén, Birgitta; Kullgren, Anders; Lie, Anders; Tingvall, Claes; The Construction of a Seat Belt System Increasing Seat Belt Use, ESV paper 94 S6 W 32, Munich 1994

Kamrén, Birgitta, Seat belt usage among fatally injured in the county of Stockholm 1991-1992, Folksam, Stockholm 1994

Kratzke, Stephen R, "Regulatory History of Automatic Crash Protection in FMVSS 208". NHTSA, SAE Technical Paper Series 950865. February 1995.

Rathmayer, Rita; Mäkinen, Tepani; Piipponen Seppo, Seat belt use and traffic offenses, Poliisiosaston julkaisuja 12/1994.

CHAPTER ELEVEN

Airbags

Steering Wheel Airbag Collision Performance

K. R. Trosien and L. M. Patrick
Mechanical Engineering Sciences Department, Wayne State University

THE EXPOSURE OF THE driver to injury is greater than any other occupant seating position in the automobile, simply because there is always a driver. The steering assembly makes the driver position unique and different from all other positions. While the steering assembly has been maligned because statistically it produces more injuries than any other component in the vehicle, it should be remembered that this is primarily because of greater exposure. Based on the number of injuries, it has been proposed by some safety experts that the steering wheel be removed. This is fallacious in that a careful study (1)* shows that the steering assembly has been beneficial from a safety standpoint for many years with the driver generally injured less than the passenger. In fact, it is the first passive restraint system in the automobile.

While the steering assembly has been an asset for many years, the introduction of the energy-absorbing column (2-4) was a conscious effort to take advantage and improve its passive restraint performance. Analysis of 30 accident cases (5) showed the energy-absorbing column to be remarkably effective when it collapsed if the steering wheel did not deform grossly, resulting in concentrated loads on the torso or head of the occupant. Force distribution (6) is necessary to prevent local injury and also to permit the maximum force to be applied to the torso to decelerate the occupant without injury.

---

*Numbers in parentheses designate References at end of paper.

Addition of an airbag to the steering wheel performs two desirable functions:

1. It distributes the load better than the steering wheel can.
2. It takes advantage of the space between the occupant and the steering wheel.

For maximum protection, it is mandatory that all of the available stopping distance be utilized (7).

BASIC REQUIREMENTS

The steering assembly is both beneficial and detrimental at the same time from a safety standpoint. It requires the driver to maintain a known position to a greater extent than is the case with other occupants which minimizes the problems of the airbag installation. On the other hand, the steering assembly decreases the time before the occupant hits an interior component, making the sensing and deploying of the bag more critical. A successful steering wheel airbag system must:

1. Utilize the space between the occupant and the steering wheel to decelerate the occupant.
2. Distribute the load.
3. Prevent the face from striking the steering wheel rim.
4. Prevent concentrated loads on the torso.
5. Function in conjunction with the EA column.
6. Keep the driver from hitting other interior components.
7. Provide knee restraint.

ABSTRACT

Though the steering wheel has been maligned as a primary cause of injuries in automobile collisions, studies show it is the first passive restraint system in the automobile. Adding an airbag to the steering wheel distributes the energy load better than the wheel alone, and the airbag takes advantage of the space between occupant and steering wheel to protect the driver further.

Specifically, the airbag utilizes space to decelerate the occupant, prevents concentrated loads on the torso, stops the face from hitting the steering wheel rim, and helps distribute impact load over a larger area. The airbag has three major components—the sensor, inflator, and airbag. The functioning of these components, as well as experimental investigations conducted to determine operational capabilities of the system, are discussed.

In order to provide the lowest occupant decelerations, it is essential to utilize all available stopping distance. The airbag fills the space between the occupant and the steering wheel and applies a decelerating force to the driver sooner than would be the case without the airbag. In addition to the distance gained by using the space between the driver and the steering wheel, a further advantage accrues from using part of the crush distance of the front end of the vehicle. The ride-down is dependent upon the crush characteristics of the front end of the vehicle plus the time required to inflate the bag and start decelerating the occupant.

Better distribution of the decelerating force over the driver's body is the primary goal of the airbag. The force distribution permits a larger total force to be applied to the driver without causing injury. The greater the force that can be applied without producing injury, the greater is the impact velocity for which protection can be provided. Force distribution in an airbag impact is optimum since it provides a uniform distribution of the force over a large area of the body with no concentrated forces.

A major advantage of the steering wheel airbag is the minimization of facial injuries. Without the steering wheel airbag, even though the steering wheel and column combination collapses at a subinjury level insofar as the torso is concerned, concentrated forces caused by the face striking the rim can cause rather severe facial injuries. While these are not serious from a danger-to-life standpoint, they are often grievous and can require long convalescent periods. It is necessary to have the rim rigid enough to distribute the impact force over a rather large area of the torso and, when it is that stiff, the facial bones are often fractured.

Concentrated loads from the steering wheel rim on the abdomen and rib cage can cause serious injuries. The steering wheel airbag distributes the load over a greater area and prevents rib fractures and abdominal injuries. One investigation (6) indicates that the allowable force applied to the torso can be approximately doubled by distributing it between the hub and the steering wheel rim. Utilization of the airbag should further increase the total force that can be applied to the torso without injury.

If the airbag is to be successful, it must absorb the impact force and attenuate it without permitting the driver to hit other rigid components in the vehicle. This requires that the airbag be large enough to control the trajectory of the occupant. The geometry of the steering assembly prevents the steering wheel bag from being large enough to cover the driver's body completely if the bag is in the wheel. If the airbag is mounted elsewhere in the vehicle such as in the interior part of the roof or header, there is a major problem in inflating it and getting it between the driver and the steering wheel before he hits the steering wheel. Oblique impacts and driver positions next to the door complicate the problem of keeping the occupant from hitting rigid interior components.

The knees are probably the best point to apply forces to the unrestrained human body to decelerate the lower torso during vehicle collisions. Thus, it is essential that the knee be used for decelerating the driver and also for maintaining his position

to provide the optimum airbag impact. Two obvious methods of providing a knee restraint are:

1. A deformable, energy-absorbing instrument panel in the knee impact area.
2. A separate airbag for knee impacts.

The driver position lends itself to knee impacts to a deformable instrument panel more than the passenger positions. The reason for this is that the driver usually controls the position of the seat and will generally have a more predictable knee-to-instrument panel position than the passenger because of the necessity of operating the foot controls. Also, the driver is less likely to have his knees and legs in an unusual position such as crossed or tucked under him, as is often the case with the passengers who are not required to maintain their position to operate the vehicle.

AIRBAG OPERATION

The airbag under consideration for use in the steering assembly consists of three major components, the sensor, the inflator, and the airbag. The three components must work as a system and requirements for each are dependent upon the particular installation.

The function of the sensor is to determine when a collision has occurred. The collision must be of some predetermined severity determined by an acceleration level and a time. Acceleration level alone is insufficient since the vehicle can be subjected to very high accelerations for a short period of time without damage to the vehicle or without being involved in a collision. For example, slamming the door or hood on the car can cause very short duration, high-amplitude acceleration spikes on interior components to which the sensor might be attached. The current specification for the sensor is that it must not activate the airbag system at collisions of lower severity than a 10 mph barrier impact. Thus, the sensor must be designed to operate when this severity of collision is reached and must not operate at the lower severity. The extremely short duration of the collision puts stringent requirements on the sensor for determining that a collision has occurred in a few milliseconds. For example, in a barrier collision of 30 mph with an intermediate size car, the occupant hits the steering assembly (that is, generates significant forces or decelerations) about 60 ms after the collision is initiated. If the sensor is to determine that a collision has occurred in time to permit the airbag to function before the occupant hits the steering assembly, the sensor must determine that a collision has occurred in 15-20 ms.

The function of the inflator is to provide the gas to inflate the airbag after the sensor has determined that a collision has occurred. Again, time is critical and, if the sensor requires 20 ms to determine that a collision has occurred, the airbag must be inflated in approximately 20 ms which results in inflation in a total of 40 ms after the collision is initiated. This leaves about 20 ms before the occupant would normally hit the steering assembly and does permit the use of some of the stopping distance between the occupant and the steering wheel.

There are three main types of inflators that are currently being considered:

1. The stored-gas generator.
2. The hybrid system.
3. The gas generator system.

The stored-bag system consists of gas stored at high pressure in a tank and supplied to the airbag through a fast-acting valve and manifold. Since the conventional mechanical valve will not operate fast enough, the usual valve is a diaphragm which is ruptured with a squib or other detonator. Gas pressures generally used in the stored-gas system are in the order of 3500 psi. The tank required to store the high-pressure gas is usually heavy and expensive. However, the system is comparatively simple and has been developed to a greater extent than the other systems. It is anticipated that the stored-gas system will be used for the instrument panel installation and, perhaps, for the steering wheel assembly in early installations.

The hybrid system consists of a combined stored gas system and gas generation system. The advantage of the hybrid system is that the gas pressures are lower and the adverse temperature effect is not as great as on the straight stored-gas system. The stored gas in the hybrid system is usually at a pressure of about 2000 psi and the volume required is considerably less, which leads to a smaller storage tank. The disadvantage of the hybrid system is that it has the storage tank and its disadvantages, together with the gas generator and its disadvantages.

The gas generation system used in airbags consists of a propellent or fast-burning powder which generates the gas from burning. It has a disadvantage that many of the gases are toxic, they are at high temperature, and there is real or imagined danger of exploding. The advantages are small size, light weight, and probably less cost than the other systems. While the potential explosion is thought to be a major disadvantage, it should be pointed out that in the other two types of systems, a propellent is used in one and a squib or detonator is used in the other. Thus, there is the possibility of an explosion in the other two systems as well, albeit the amount of energy would be considerably smaller.

The airbag is usually made of a coated nylon and must be strong enough to withstand the forces generated by the pressures and the opening forces. The geometry of the airbag must prevent extreme relative motion of body components such as neck hyperextension which can occur if the force on the head is applied before the force on the rest of the body. Pressure in the bag is determined by the force required to prevent the occupant from penetrating it during the impact and striking the interior components on the far side of the airbag.

The overall operation of the airbag system consists of a sensing function which determines when a collision of sufficient severity has been reached, an inflating action which fills the airbag after the sensor has determined that a collision has occurred, and the airbag which is the cushion struck by the occupant. The entire sequence must take place in about 40 ms for the steering assembly if maximum benefit is to be achieved.

The system must remain operational without any functional check during the average 10 year lifetime of a car. During this time, it can be subjected to temperatures as high as 200 F and as low as -40 F for extended periods. Satisfactory operation is required at temperatures ranging -20 to +160 F. High reliability is essential, and the system must be independent of failure of other vehicle systems insofar as possible.

POTENTIAL INJURY SOURCES

Operation of the airbag is violent and offers some injury potential to vehicle occupants. The primary injury sources are:

1. Ear injury from high noise levels during rapid bag inflation.
2. Impact injury from the bag and/or steering wheel contact.
3. Injury to the head from being struck by the driver's arm if it is between him and the steering wheel at the time the bag inflates.
4. Injury from rebound if the energy dissipation is not adequate.
5. Injury to the driver if he is wearing glasses or smoking when he hits the bag.
6. Burns, if the bag surface temperature is too high in a gas generation system.
7. Injury from inhaling toxic gases if such are present.

When the bag inflates, the velocity of inflation results in a pressure pulse in the vehicle which is potentially dangerous to the ears. Sound measurements have been made at the ear level in the range of 160-170 db based on the 0.0002 dynes/sq cm pressure level. These levels are high enough to produce ear damage under long exposure according to some investigators, and there is some apprehension that even for the short duration of the airbag inflation that some injury can occur. An investigation at a governmental agency (8) in which over 90 volunteers were exposed to the inflation noise in a vehicle during bag inflation did not produce a single ear injury. Ear injury investigations at WSU provided the same result. Injury from the bag or steering wheel contact has been studied extensively and will be discussed in detail in a later section of this report. Primarily, the problem is to prevent the abdomen, thorax, or head from collapsing the airbag to the point where concentrated forces on the steering wheel rim or hub will produce injury to these body components.

If the driver is turning a corner or for some other reason has his arm across the wheel at the instant the bag inflates, there is a danger that it can be accelerated to the point where it will impact his head or face in a manner similar to a blow from a boxer. Preliminary tests with this configuration indicate that the arm and hand is flung back violently, but not enough to cause serious injury from the blow. Also, it was found that unless the arm is positioned in one exact position, it is forced over the driver's head or under the bag rather than hitting him in the face.

Rebound characteristics are important and require that the airbag have a damping valve. If no valve is present, the airbag acts as a spring and, if the occupant impacts it at 30 mph, he would theoretically rebound at 30 mph which could cause injuries from hitting the seat back, roof rail, or other interior components. At 30 mph, it is found that a rebound velocity of 10 or 12 mph is satisfactory. Higher velocities can cause injuries from striking the B-Post with the head or hyperextension injuries from hitting the back of the front seat on rebound.

When the occupant is wearing glasses or smoking a pipe, it is possible that the inflating airbag could cause the glasses to break and be forced into the eyes or the pipe could be forced back into the driver's mouth and neck. Neither of these situations has been investigated thoroughly, but it appears that especially in the case of the glasses there is little likelihood of injury since the eyes are protected by the socket. The alternative of hitting the steering wheel rim or hub when wearing glasses or smoking a pipe is probably much worse than hitting it under the same conditions with the airbag. Considering the serious facial injuries that occur from striking the steering wheel without the airbag, it appears that the trade-off is greatly in favor of the airbag.

If the gas-generating system is used, there is some danger of burns if the surface temperature of the bag is high enough. However, it appears that this problem can be overcome by an adequate design of the generating system with a coolant to prevent excessive temperatures from occurring. Also, the thermal capacity of the lightweight bag is low and the time of contact between the skin and the bag is short in most cases. If the bag is hot and the driver is conscious, he would involuntarily pull away from it before serious burns could occur.

Toxicity of the products of combustion of the gas generator can be a problem which is being overcome by eliminating the toxic gases through adequate choice of materials used in the gas generator and by the addition of chemical components which eliminate toxicity. If the car is adequately ventilated, there is little likelihood that a sufficient amount of toxic gas from the steering wheel airbag would be ingested by the occupants to cause injury. However, if gas-generation type systems were to be installed to protect all passengers, the toxic gas danger would be multiplied many fold.

EXPERIMENTAL INVESTIGATION

An experimental investigation into the performance of the steering wheel airbag has been underway for about two years at WSU, using a 1969 Valiant modified to operate on WHAM II (9). Fig. 1 shows the modified vehicle in position ready for a run on WHAM II.

The vehicle is propelled to the desired speed by a 10 ft long propulsion cylinder and decelerated with an acceleration-time history and stopping distance corresponding to the collision being simulated. Thus, the occupant is subjected to the same acceleration environment that he would be in a collision of the type simulated. Instrumentation on the vehicle includes accelerometers to record the acceleration-time history and a velocity transducer to indicate the velocity at impact.

Instrumentation for measuring the effect of the impact include biaxial accelerometers mounted in the head of the anthropomorphic dummy or cadaver driver, and load cells on the lap belt when worn. High-speed cameras record the dynamics and trajectories of the impact and movement of the steering

Fig. 1 - Modified 1969 Valiant ready for simulated barrier impact on WHAM II

column and wheel after the impact to determine the degree of deformation. A pressure transducer is installed to measure the pressure in the airbag before and during impact.

Experimental studies on the effect on the safety performance of the airbag include:

1. Velocity variation in the range of 15-30 mph.
2. Comparison of results with and without airbag at 30 mph.
3. Position of the driver from centered behind the steering wheel to leaning against the door.
4. Comparison of results with and without lap belt.
5. Potential injury from arm across steering wheel at time of bag inflation.
6. Comparison of results with and without collapse of the steering column.

With only 16 runs and several variables considered, only a limited examination of the effect of each can be achieved. However, the careful laboratory control of each variable permits some conclusions to be drawn and trends to be established.

Much of the research on airbag safety performance has been based upon a carefully controlled, so-called normal position of the occupants. Since much of the time the occupants are not in this predetermined position, it is felt that examination of the nonstandard positions and postures must be taken into consideration. For this reason, the driver has been placed at several locations behind the steering wheel representing those that are known to exist in driving situations. Also, the driver's arm was placed across the steering wheel to determine whether the inflating bag would produce a sufficient velocity of the arm and hand to cause injury if it were to strike the driver in the face. While these nonstandard positions of the driver are difficult to evaluate and represent a large number of variables, it is essential at least to study the trend since many of the collisions in which the airbag will be involved will include the conditions referred to herein. The gross difference in performance when the driver is centered behind a wheel and when he is leaning against the door is evidence of the necessity of studying different occupant positions.

The bags were triggered at different times after the beginning of deceleration to observe the effect on occupant kinematics and kinetics. The collapse of the column was limited on sev-

Table 1 - Steering Wheel Airbag Impact Data

Run No.	Vel. mph	Occupant[a]	Position	Lap Belt Loads		Bag Press, psi	Peak Dorsal Head Accel, g			Col. Crush, in.	Bag Firing Time ms	Comments
				Inboard	Outboard		AP	SI	TOT			
CB-2	15.4	Dummy	Centered	Loads Not Recorded		—	30	20	41	None	12	
CB-3	24.0	Dummy	Centered	980	700	—	22	40	46	4-1/2	18	
CB-4	30.1	Dummy	Centered	1170	900	—	35	65	74	4-7/8	20	
CB-5	30.6	Dummy	Centered	1100	1120	16	25	55	61	6	12	Hand on wheel at 9 o'clock— hand and forearm trapped between bag and lower spoke
CB-6	30.6	Dummy	Centered	Unbelted		—	60	65	89	6-5/8		Bag installed, but not inflated, dummy would have impacted windshield
CB-11	30.2	Dummy	Centered	1460	1060	15	30	65	72	4-5/8	4	Hand at 11 o'clock— blown over head of driver
CB-12	29.7	Dummy	Centered	1350	900	16	45	65	79	1-1/4	4	Restraining collar on column to minimize collapse
CB-13	30.0	Dummy	Centered	Unbelted		—	30	65	72	5/8	4	Restraining collar on column to minimize collapse
CB-14	30.0	Dummy	Next to door[e]	Unbelted		13.5	70	50	86	1/8	4	Restraining collar on column to minimize collapse
CB-15	30.0	Cadaver 1538[b]	Centered	1080	1040	—	25	60	65	3-5/8	4	Impact of bag is high on driver
CB-16	30.0	Dummy	2 in. Left	1300	1120	12.5	18	40	44	4	4	Rebound impact to upper window opening and B-post
CB-17	30.0	Dummy	Next to door	1180	1020	13	20	60	63	3-7/8	4	Impact to A-post on bag rebound, driver would have hit windshield
CB-18	30	Cadaver 1595[c]	Centered	980	1040	13	40	65	76	3-5/8	4	Rebound caused roof impact
CB-19	30	Cadaver 1595[c]	2 in. Left	960	1000	13	40	75	85	1-1/2	4	Rebound to inner roof rail
CB-20	30	Cadaver 1461[d]	Centered	640	560	10	50	70	86	1-7/8	3	22 in. diameter bag. Driver leaning forward at impact. Made early contact with bag
CB-21	30	Cadaver 1461[d]	Centered	680	660	13	40	65	76	1-5/8	2	18 in. diameter bag. Driver leaning forward at impact. Made early contact with bag

[a]Sierra Model 292-750
[b]Weight 140, Height 5'3"
[c]Weight 164, Height 5'8"
[d]Weight 185, Height 5'11"
[e]Four inches left of column centerline

eral runs to determine whether the dummy would "bottom out" and impact the column hub and steering wheel rim.

RESULTS

Sixteen simulated collisions were conducted with 14 of them at 30 mph. The 30 mph velocity was chosen since it is estimated to cover 85% of the injury accident collisions insofar as severity of collision is concerned. One run was conducted at a nominal 15 mph velocity and one at a nominal 25 mph velocity to study the effect of velocity on performance. Five runs were conducted with lap-belted cadavers. There were 11 dummy runs—eight with and three without lap belts. One run conducted without airbag inflation serves as a base for comparison of performance with the inflated airbag. All but one run was conducted with an 18 in. airbag while the other was conducted with a 22 in. airbag. The results of these runs will be discussed under individual headings in which the variation of

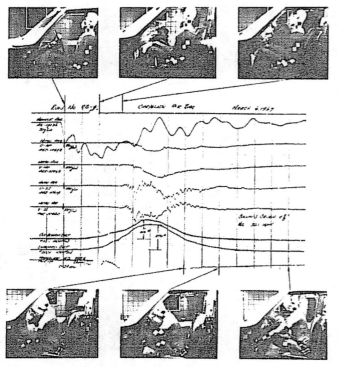

Fig. 2 - Oscillograph record for Run CB-4 with several frames from high-speed movies taken during the run

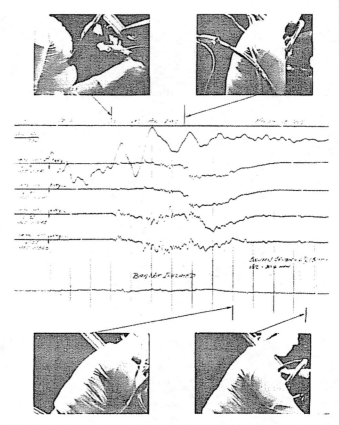

Fig. 3 - Oscillograph record for Run CB-6 with single frames taken from movies from onboard camera

individual parameters will be discussed and the results presented.

A summary of the results is presented in Table 1 which summarizes the relevant data pertaining to each run including the velocity, the head accelerations, and conditions peculiar to each run. Stopping distances for the different velocities were: 12 in. for 15 mph, 17 in. for 24 mph, and 22 in. for the 30 mph runs. The bench-type seat is locked in its midposition for all experiments and the three spoke steering wheel is positioned with the plane of symmetry in a vertical position resulting in the middle spoke being in the normal down position.

CENTRAL IMPACT VELOCITY VARIATION - The "normal" position of the driver centered behind the steering wheel is given the greatest emphasis and was a condition for 12 of the 16 runs. Eight of the 12 runs were conducted with the dummy with the remaining four runs conducted with three cadavers.

Runs CB-2, 3, and 4 illustrate the effect of increasing velocity on the performance of the airbag. The velocities were 15.4, 24, and 30.1 mph. At the 15.4 mph impact, the column did not crush due to insufficient force to cause it to collapse. At 24 and 30.1 mph, the column collapse was 4-1/2 and 4-7/8 in., respectively.

The resultant head acceleration is increased from 41 to 74 g for the change in velocity from 15.4 to 30.1 mph. The AP accelerations were not as greatly different as the resultant accelerations. The blow-out valve limits the pressure in the bag during the impact which would tend to maintain the AP force on the head about constant, providing approximately uniform head acceleration. The difference can be accounted for by the difference in head angle during the impact which results in the

force from the airbag being applied at an angle to the head AP direction.

Loads were not recorded on the lap belts during the 15.4 mph run. The lap load increased from 1680 to 2070 lb going from 24 to 30.1 mph.

In general, the lap belt loads, the resultant head accelerations, and the column crush increased with increasing impact velocity.

CENTRAL IMPACTS—30 MPH WITH AND WITHOUT AIRBAG INFLATED - Comparison of the performance with and without the airbag inflated is illustrated by Runs CB-4 and CB-6 shown in Figs. 2 and 3. Fig. 2 shows several frames taken from the high-speed movies from the lateral, offboard camera from run CB-4, which are shown together with the oscillograph record for the run. The test velocity for CB-4 is 30 mph with a vehicle stopping distance of 22 in. The driver is located directly behind the wheel with his hands initially at his sides. The dummy is also lap belted and the impact should be normal.

The first frame shows the position of the dummy at the beginning of deceleration. At this instant, the dummy torso is about 6 in. from the lower rim. The bag must be inflated before the abdomen impacts the rim, which can cause injury because the contact area is small. The second frame is at 20 ms later and the dummy has translated relative to the car 3 in. and the bag is just beginning to inflate. The outer covering is being blown open and the first folds are visible. At this point, the vehicle is traveling at 22 mph and the dummy at 28 mph. The sensor fired at 20 ms and the seat belt load cells show that the belts have tightened and are just beginning to restrain the lower torso.

In the third frame, the bag is more fully inflated and the dummy has just made contact with the lower rim at 34 ms. Although the lap belt loop load is now 850 lb, the dummy is still upright and no head accelerations are indicated. The next frame at 70 ms shows the dummy position just before the column begins to collapse. The column is a mesh-type and the total load on the dummy can be assumed to be 1200-1800 lb depending on the rate of loadings. The head is leaning forward relative to the torso and is in full contact with the bag. The shoulders are forward as the dummy is wrapped around the bag and wheel and the arms keep him in position as they swing forward. It is imperative that the geometry and pressure of the bag prohibit injurious facial contact with the upper rim. The velocity of the vehicle is now about 3-4 mph and the velocity of the dummy at the head and neck targets is 7-8 mph.

The fifth frame is at the bottom of column crush at 90 ms after deceleration. The vehicle no longer has forward velocity and the dummy is beginning to rebound with a rearward velocity relative to the vehicle. The bag should not have sufficient stored energy to accelerate the occupant with a velocity great enough to cause injury-producing impacts with other interior components.

The final frame shows the dummy at 130 ms or about 40 ms after the beginning of rebound. The velocity of the neck and head targets are about 5 mph, indicating that a bag with collapsible column has absorbed most of the occupant's kinetic energy since the rebound is augmented by the energy stored in the lap belts.

The oscillograph traces in Fig. 2 show the accelerations to be below the injury level. The AP acceleration is probably the most significant since it is a more direct measure of the decelerative action of the bag on the head. The vertical of SI acceleration component is sensitive to angular rotation and tensile forces in the neck. The peak dorsal AP value is 35 g and the peak SI value is 65 g.

The severity of an impact with an installed but not inflated bag was observed in Run CB-6, in which the unbelted driver was seated centrally. Fig. 3 shows a series of frames from the onboard camera. The movies from the offboard lateral camera are not complete because of camera malfunction. The timing of events is similar to Run CB-4 with the dummy contacting the lower rim about 34 ms after impact. At 70 ms (second frame) the face of the dummy has made contact with the upper rim and the AP accelerations rise to 50-60 g with duration of about 20 ms. The third frame shows the column fully collapsed 6 in. The occupant would have impacted the windshield and sustained additional accelerations and possible facial lacerations. The steering wheel in CB-6 is severely deformed and there is a far greater possibility of torso damage than in CB-4 where the forces were distributed and attenuated by the airbag.

Mitigation of facial injury potential with a steering wheel airbag is shown in Fig. 4. The photo from CB-4 shows the head buried to the maximum depth in the bag with adequate cush-

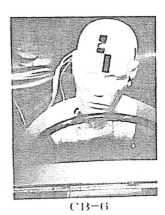

CB-4   CB-6

Fig. 4 - Single frames taken from front camera showing head contact with airbag in CB-4, and upper rim in CB-6

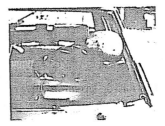

Fig. 5 - Lateral and front view of final position of driver initially seated next to door prior to impact

ioning still available. This is contrasted with CB-6 where the dummy rode the column down with the rim across his chin and mouth and deformed the hard covered wheel as shown in Fig. 3.

OFF-CENTER IMPACTS - Many drivers prefer to sit slightly to the left of the wheel or very close to the door. In the event of a frontal impact, the driver would not be centrally loaded by the bag and column and unusual results are possible. Two runs, CB-14 and CB-17, were conducted at 30 mph with the driver seated next to the door, or 4 in. left of the column centerline. In CB-14, the collapse of the column was prevented by a rigid collar and the dummy was unbelted. The column and bag did not stop the driver. He rotated around the wheel and became wedged between the deformed wheel and the door. His head would have impacted the windshield. Fig. 5 shows the final position of the dummy as seen from inside and in front of the vehicle. The peak accelerations in Table 1 should be higher because they do not include lateral components. Fig. 5 shows the head rotated and illustrates that lateral accelerations occurred but were not measured by biaxial accelerometers in the midsaggital plane. The biaxial head accelerations are higher from striking the A-post. It is doubtful that energy absorbing column would have greatly mitigated the severity of the impact since the wheel probably deflected enough to let the dummy pass without transmitting forces large enough to cause column collapse.

The driver in CB-17 was also seated next to the door but was belted, and the column was permitted to collapse. Because of the restraining of the lower torso and the column movement, the dummy did not get past the wheel but did still impact the

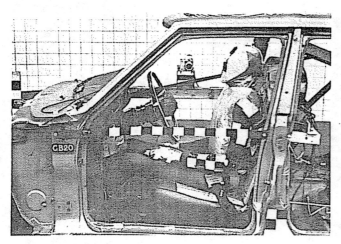

Fig. 6 - Lateral view of test vehicle with cadaver wearing foil and wire vest prior to Run CB-20 with larger, hand-packed airbag

A-post. The peak AP and SI accelerations during bag contact were 20 and 60 g but at A-post contact they were 65 and 35 g, respectively. The dummy also passed through the plane of the windshield. Column collapse was 3-7/8 in. of a possible 6 in.

Run CB-15 at 30 mph was conducted with a belted driver 2 in. to the left of the centerline of the steering wheel. The result was a more ideal impact, but the rebound was to the left. The head of the dummy grazed the upper window opening with AP and SI values of 15 and 10 g and then hit the B-post with peak accelerations of 65 and 45 g, respectively. The impact to the B-post is more severe than that due to the bag which resulted in accelerations of 18 and 40 g. Column crush was 4 in. with a lap belt loop load of 2420 lb.

EFFECT OF ARM OVER BAG - Accidents while turning are not uncommon and the position of the hands varies greatly among drivers. The possibility of a bag inflation with an arm across the wheel is likely. In Run CB-11, the right hand of the dummy was taped in the 11 o'clock position. The driver was belted and the bag was inflated 4 ms after deceleration initiation. The inflating bag forced the arm over the driver's head without striking it.

The right hand was taped in the 9 o'clock position in Run CB-5 and the arm became trapped beneath the bag and was forced against the center spoke by the dummy's chest. The wheel was deformed more in this run than others with normal impact. The forces could be great enough to break an arm in a situation like this. However, the danger is no greater than an impact without an airbag.

Several attempts were made to get the driver's hand to impact the face during the simulated impacts but they were unsuccessful. Inflation of the bag with the vehicle stationary with the dummy and cadavers led to successful hand-to-face impacts. In one experiment, the cadaver's right hand struck its face at approximately 86 mph. However, x-rays of the right hand and facial bones revealed no skeletal damage.

CHEST INJURY POTENTIAL - The dummy used in these experiments is obsolete and did not lend itself to accelerometer installation in the chest. Accelerometer installation in the cadaver chest is difficult at best and, hence, the injury poten-

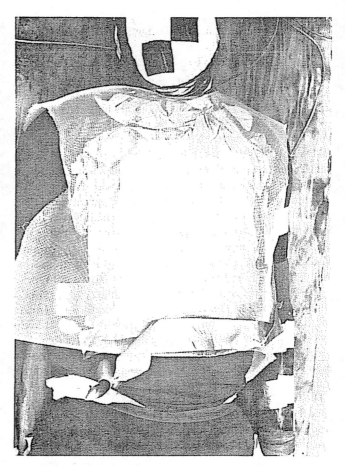

Fig. 7 - Chest of cadaver after Run CB-20 showing impression of lower rim in foil over the chest

tial for the chest must be based on observation and x-ray of cadaver chests after impact. A method was developed on the five cadaver runs to determine the approximate chest-to-bag and possible chest-to-rim contact areas by fashioning a two-layered vest of foil and wire cloth. Fig. 6 shows a cadaver in the vehicle prior to impact with this installation. The foil is taped to the torso and the wire cloth, cut like a vest, is taped over the top. During impact to the airbag and wheel, the wire is pressed into the foil and an imprint of the contact area remains. Fig. 7 shows the chest after the run. Most clearly seen is the imprint of the bottom steering wheel rim.

Three different cadavers were used in the barrier impact simulations. Cadaver 1538 was used in CB-15, seated on-center and lap-belted. This cadaver was only 5 ft 3 in. tall and weighed 140 lb. Consequently, it impacted the bag high on the torso. Fractures of the left fourth and fifth ribs were sustained along with crushing of the illiac crests. This cadaver had been used on previous experiments involving upper torso loading so these fractures are not indicative of normal human tolerance. The size of the cadaver, being about 5th percentile, provided data on the rebound of a small driver. He was thrown straight back and upward with velocities up to 7.5 mph measured on the lateral head target. This rebound velocity is not different from other rebound velocities calculated for other impacts.

Runs CB-18 and CB-19 were conducted with Cadaver 1595, a 5 ft 8 in., 164 lb male, to look at chest damage and rebound with on-center and off-center impacts. This cadaver also had been used on previous chest impact work and is a poor subject for human tolerance determination. The type of rib fractures is of interest, however. In Run CB-18, fractures of the first, second, and seventh ribs in the right axillary line were noted. After CB-19, additional fractures of the fourth through seventh ribs in the left axillary line were noted. With a distributed chest load, it is anticipated that fractures will occur in the axillary lines or near the spinal column. Very little human tolerance data are available for this type of impact.

Rebound in both runs would have caused roof impacts and in CB-19 with the driver off-center, the roof rail was impacted. Roof impacts are not realized in these experiments because a portion of the sheet metal is removed to provide adequate lighting for photography. Rebound velocities were calculated and found not to exceed 9 mph.

Cadaver 1461 used in CB-20 and CB-21 had not been used in previous chest impact experimentation, and no fractures were noted after either run. Cadaver 1461 was 5 ft 11 in. and 185 lb, and was leaning forward at impact in both runs which accounts for the low lap belt loads. Contact with the bag was made about 20-30 ms after beginning of deceleration. The bag was impacted early while still inflating before the lap belts could substantially load the torso. Column collapse was limited because of column bending and, hence, the upper torso loads are higher.

A larger, 22 in. bag (compared to the 18 in. bag in the other runs) was used in CB-20 with the same gas generator which accounts for the lower pressure. No differences in impact behavior were discernible between the two sizes of airbags. Additional experiments with the larger bag will be required before a valid comparison of performance between the two sizes can be made.

DISCUSSION

There were not enough runs to permit a detailed, statistical evaluation of all of the variables which were considered. Therefore, it should be understood that these results are based on a limited number of runs and indicate trends only.

In several of the experimental runs, it was observed that the column bent upward during the impact. It is desirable to minimize this upper deflection since it affects the collapse characteristics of the steering assembly. Furthermore, as the column bends upward the lower part of the rim is exposed to abdominal impact which is undesirable, and the head takes a greater portion of the load because the force applied to the driver is at a higher level on his body. It should be pointed out that it was often desirable to have the column bend upward in the older, rigid column vehicles since the bending absorbed energy and limited the force and, in effect, acted as a collapsible column. With the collapsible column, however, it is undesirable to have bending which inhibits column collapse.

The lap belt was beneficial in controlling the position and trajectory of the driver to provide optimum impact to the steering wheel and airbag. Even in the off-center impacts represented by Run CB-17 with the lap belt and Run CB-14 without the lap belt, it was observed that the lap belt maintained partial control and prevented the driver from moving forward and becoming wedged between the wheel and the door. The lap belt ensures that the trajectory is in a direction which will produce column collapse and, thus, obtain the benefit of the additional distance available from the deformation of the column.

The airbag should distribute the load over the torso and head of the driver to prevent concentrated forces which result when any part of the body strikes the rim or hub of the steering wheel. In the case of the unbelted driver who moves forward in an upright position, the inclination of the steering wheel puts the lower rim in a position to be impacted first by the abdomen. The steering wheel airbag must project beyond the plane of the rim far enough to provide cushioning between the occupant and the rim with a force adequate to cause the driver to rotate about the lower rim and, thus, distribute the force over the torso with a more normal impact to the bag. It must also prevent the head and face from penetrating to the point where concentrated forces result from contact with the upper part of the rim. The pressure in the bag must be high enough so that, when it is distributed over the contact area of the torso, it will cause the column to collapse. By collapsing the column prior to complete deflation of the bag, the maximum distance is utilized with minimum injury. The function of the blow-out valve is to limit the pressure in the airbag. By setting the blow-out valve pressure at the correct level, the column will collapse before the bag is deflated. If the blow-out valve pressure level is too low, the bag will collapse before the column and the driver will be subjected to concentrated forces from the steering wheel rim and hub.

No chest accelerometers were mounted on the dummy or cadavers for these experiments. The injury criterion based on chest acceleration was not established at the time the program was initiated and, consequently, the use of the chest accelerometer was not considered necessary. The bag pressure at which the blow-out valve operates is 10-16 psi as shown in Table 1. By projecting the pressure over the area, a rough estimate of force on the chest is obtained. The force is estimated to be 1200-1800 lb. Chest acceleration and force on the column will be measured in future experiments.

Inflation initiation measured in milliseconds after the deceleration started varies 2-20 ms. The longer time periods occurred with a sensor designed to include a measure of the collision severity while those with the very short duration were initiated the instant deceleration commenced. A study of the high-speed movies shows that even with the 20 ms delay the bag is inflated before the occupant hits it. Thus, there appears to be no insurmountable obstacles in sensing and inflating the bag during a barrier-type collision.

CONCLUSIONS

Based on the experimental program described herein with this vehicle, airbag installation, and a limited experimental program, the following conclusions are reached:

1. The airbag can be sensed, triggered, and inflated after

collision initiation in time (40-50 ms) to provide driver protection.

2. The airbag cushions the face and abdomen at velocities up to 30 mph, and prevents injurious concentrated loads when the impact is centered.

3. The efficacy of the steering wheel airbag installation in mitigating injury is increased by lap belt usage.

4. Head accelerations were below the 80 g/3 ms injury criterion in all cases of impact to the bag.

5. Rebound velocities up to 9 mph appear satisfactory and achievable at impact velocities up to 30 mph barrier equivalent.

6. Velocities up to 86 mph were imparted by the inflating airbag to the arm initially across the steering wheel. No fractures or head accelerations in excess of 80 g/3 ms injury criteria resulted from arm and hand impact to the head under these conditions.

7. The steering column should not rotate (bend) upward during frontal impact to expose the lower rim to abdominal impact. Also, column bending inhibits collapse of the column.

8. Bag shape, size, and pressure blow-out valve should be designed to cause column collapse prior to bag collapse.

9. Bag and column should operate together to achieve maximum driver decelerating distance—the airbag does not eliminate the need for the energy absorbing column.

10. Provision for protection from off-center (and oblique) impacts is necessary and can be attained by using a large bag to prevent the driver from wedging between the steering wheel and the door.

ACKNOWLEDGMENT

This research was supported by Chrysler Corp.

REFERENCES

1. J. K. Kohlberg, "Driver and His Right Front Passenger in Automobile Accidents." CAL Report No. VJ-1823-R16.

2. L. C. Lundstrom and W. G. Cichowski, "Field Experience With the Energy Absorbing Steering Column." Paper 690183 presented SAE Automotive Engineering Congress, Detroit, January 1969.

3. Donald F. Huelke and William A. Chewning, "Accident Investigations of the Performance Characteristics of Energy Absorbing Steering Columns." Paper 690184 presented at SAE Automotive Engineering Congress, Detroit, January 1969.

4. I. I. Lasky, A. W. Siegel, and A. M. Nahum, "Automotive Cardio-Thoracic Injuries: A Medical-Engineering Analysis." Paper 680052 presented at SAE Automotive Engineering Congress, Detroit, January 1968.

5. L. M. Patrick and D. J. Van Kirk, "Correlation of Accident and Laboratory Impacts to Energy-Absorbing Steering Assemblies." Paper 690185 presented at SAE Automotive Engineering Congress, Detroit, January 1969.

6. C. W. Gadd and L. M. Patrick, "System Versus Laboratory Impact Tests for Estimating Injury Hazard." Paper 680053 presented at SAE Automotive Engineering Congress, Detroit, January 1968.

7. Lawrence M. Patrick, "Prevention of Instrument Panel and Windshield Head Injuries." The Prevention of Highway Injury, Symposium Proceedings, Highway Safety Research Institute, University of Michigan, April 19-21, 1967.

710182

EVOLUTION OF THE
AIR CUSHION

W. C. House, W. J. Eggington,
and C. A. Lysdale
Aerojet-General Corporation

THE KEY TECHNOLOGY associated with development of Air Cushion Vehicles (ACV) and Surface Effect Ships (SES) is undoubtedly that associated with the air cushion itself. This supporting air cushion is the unique element of these machines; it is expected to provide the necessary lift and stability for the craft, it controls habitability and ride characteristics, and it contributes to the vehicle power requirements through drag and cushion airflow. The air cushion also defines many aspects of vehicle operability, including mobility in the presence of obstacles and rough terrain, reliability by susceptibility to skirt damage, and availability of the vehicle, as derived from the required frequency of skirt maintenance and the ease with which this maintenance may be accomplished.

Since the early demonstration efforts by Christopher Cockerell in the mid 1950's, air cushion developments and improvements have been achieved in an impressive, continuing manner. This paper will not attempt a comprehensive documentation of these achievements. Rather, it will first discuss the general direction and objectives of these developments; second, review briefly the historical sequence of the developments; and third, indicate the current and projected status and capabilities of air cushion technology.

DEVELOPMENT OBJECTIVES

A major aspect of the potential of air cushion vehicles involves their amphibious capability and their mobility over all sorts of surfaces and terrain, particularly those where low footprint pressure is an advantage. The surface characteristics to be considered for an unlimited cross-country capability are summarized in Figure 1: the surfaces may involve land, water, or ice and snow - plus all of the "in-between" cases. This figure gives some indication of the spectrum of operating surfaces to be tackled by air cushion craft. Overland capability must clearly involve both the ability to accommodate ditches, banks, and obstacles, as well as adequate thrust for slopes and hills. Between land and water, we find the marshes, mud flats, and riverbeds where the low foot pressure and accommodation capability of the air cushion truly excels. Over water, we find that the air cushion must develop reasonable levels of drag and acceptable habitability when operating in a seaway, while amphibious operations demand adequate stability for surf crossing. Ice and snow fields and the numerous ice-water interfaces characteristic of pack ice have been successfully traversed by air cushion craft; the difficulties experienced in these tests have generally been related to limitations peculiar to the specific craft involved. The most demanding basic

ABSTRACT

The characteristics of the supporting air cushion system are key to the cross-country mobility and overwater performance of Air Cushion Vehicles and Surface Effect Ships. The direction of developments in this technology are discussed, and the primary cushion design variables are identified. The historical sequence of air cushion development is reviewed, and it is concluded that current and future advancement will be based on "3rd generation" systems, in which the cushion system design is tailored to the specific application.

2

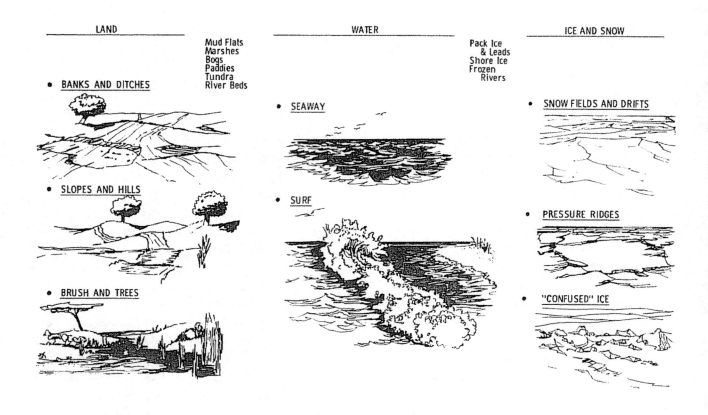

Figure 1. Cross-country operating surfaces

aspects of operation on these surfaces appear to be associated with the crossing of pressure ridges (obstacles) and the tendency of an ACV to "dig itself in" when hovering over loose snow.

Several aspects of the ambient conditions encountered in unlimited cross-country operation also have particular significance on air cushion technology. Temperature extremes (particularly low temperatures) have significant effects on flexible skirt materials. The ingestion of rain, spray, sand, and snow by the air cushion fans can have obviously detrimental effects. Under icing conditions, the spray generated by the air cushion can result in ice buildups capable of significant performance degradation and even severe skirt

damage.

The direction of the air cushion technology developments has, by and large, been pointed toward improving vehicle capabilities over various operating surfaces and under various operating conditions. Continued improvements in these directions can be expected to further extend the capability and applicability of air cushion vehicles in both commercial and military operations.

Certain performance parameters define the effectiveness of a given air cushion system in various cross-country operating circumstances. These performance parameters can be directly related to the design variables in air cushion system design. As illustrated in Table 1, the primary parameters which define

1058

Table 1 - Air Cushion Performance Parameters

PARAMETER	PRIMARY SOURCES & CONTRIBUTORS	CUSHION DESIGN VARIABLES		
		Pressure	Airflow Rate	Configuration
LIFT	Cushion Pressure Cushion Area	P_c/L		L/B
STABILITY	Cushion restoring forces C. G. & Thrust line locations	— Fan P_c-Q —	Air gap	L/B, Cushion ht. Compartmentation Skirt details
DRAG	Skirt contact Wave making (over water) Air momentum	P_c/L	Air gap	Skirt details
CUSHION AIR POWER	Cushion pressure & leakage Stability and control flow Ducting losses	P_c/L	Air gap	Compartmentation Skirt details
HABITABILITY	Cushion compliance & leakage Hard structure contact	P_c/L — Fan P_c-Q —	Air gap	Cushion volume Cushion height Compartmentation Skirt details
OPERABILITY	Skirt contact Hard structure contact Complexity Maintainability		Air gap	Cushion height Compartmentation Skirt details Accessibility

cushion system performance include total lift, pitch and roll stability, drag, cushion air power, habitability or vehicle "ride", and finally operability (which is taken to include obstacle and terrain mobility, reliability and vulnerability to damage, and availability with respect to maintenance and overhaul). For this discussion, the parameters are referenced to the air cushion subsystem only, rather than to the overall vehicle.

The primary sources and contributors to the performance parameters are listed in the center column. For example, the total available lift is simply dependent upon cushion pressure and total cushion area. Alternately, cushion air power is affected by cushion pressure and leakage, by special airflow requirements for stability and control, and by the ducting losses associated with air distribution for compartmented cushion systems. It might be noted that the cushion system actually develops no useful work and can ideally achieve

negligible contact drag - thus no theoretical limit exists for potential improvements in cushion power performance.

The cushion performance sources can be related to the specific design variables which are available for definition of the air cushion system. As indicated in the third column of Table 1, the design variables can also be further grouped quite effectively into just three classes: (1) variables related to design cushion pressure, (2) variables related to design airflow rate, and (3) variables defined by the physical cushion configuration. On this basis, the values of cushion "density" (P_c/L) and planform (L/B) are the key factors which define how the total lift is developed for a given gross weight. These factors are clearly dependent on cushion pressure and configuration, respectively. Similarly, the stability characteristics of the cushion system are defined by the lift fan and duct head-flow characteristic; the cushion length to beam ratio; the cushion height as it affects overall vehicle c.g.; the extent and nature of cushion compartmentation; skirt details, in particular the change in cushion center of lift due to skirt deflection after contact; and finally, the air gap as it affects local leakage and skirt contact conditions. It might be noted that the majority of the primary stability determination factors relate to the physical cushion configuration, rather than to the nominal lift system operating parameters.

The drag associated with the air cushion is dependent upon the air gap (as it affects skirt contact), cushion density, and details of the skirt configuration. The latter two factors are primarily involved for overwater operation, since they relate to wave-making drag and to variations in skirt contact and "scooping" tendencies under certain speed/wave height conditions. The sources of cushion air power demand may be reflected in the variables P_c/L, air gap, and skirt details plus the nature and extent of duct losses associated with cushion compartmentation.

Habitability or ride characteristics of the vehicle will vary with P_c/L, air gap and the associated nominal leakage flow, the nature and extent of cushion compartmentation, and the absolute size of the vehicle as reflected in cushion volume and cushion height. Both cushion volume and P_c/L have significant influence on vehicle habitability, through the mechanism of varying the effective compliance or "springiness" of the contained air cushion. The cushion height parameter alone also affects habitability, due to varying the probability of hard structure contact in large waves or over extremely rough terrain. Finally, the operability of the vehicle is substantially affected by the nominal air gap, as it bears on skirt wear and skirt damage; cushion height and its effect on probability of hard structure contact; skirt details plus the nature and extent of cushion compartmentation, as they reflect on the vulnerability and complexity of the flexible skirt structure and the lift system

ducting and control; and finally, the accessibility of the skirt elements as they influence the ease of inspection and maintenance.

There is one key point which is apparent in the grouping of design variables (Table 1). The various, unrelated cushion performance parameters (left column) are eventually all reflected into a limited number of design variables which really collapse to only three different independent selections; (1) design cushion pressure, (2) design airflow rate, and (3) details of the cushion and skirt physical configuration. Thus potential advancements in air cushion technology really have very few available degrees of freedom. It therefore seems clear that the desired continuing improvements in cushion performance will require both innovative configuration design and careful selection of operating parameters -- both optimized for the specific applications at hand.

HISTORICAL DEVELOPMENT

A brief review of the sequential development of air cushion technology is instructive, as a frame of reference for the "third generation" status of current developments.

BASIC CONCEPTS - If we limit our discussion to aerostatically supported vehicles, the technology grows from two basic concepts, the pure plenum cushion, and the peripheral jet cushion illustrated schematically in Figure 2. The pure plenum cushion relies entirely on the physical restriction of the leakage airflow. Although the concept is extremely simple and acceptable heave stability is achieved, the plenum concept has poor pitch/roll stability and relatively high leakage airflow. Alternately, the peripheral jet cushion makes use of inward momentum of the cushion supply air to reduce the total leakage flow for a given air gap. This principle demonstrates improvement in both pitch/roll stability and heave stability due to jet turning, but does involve increased duct losses and a more

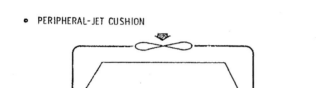

	ADVANTAGES	DISADVANTAGES
• PURE PLENUM CUSHION	Simple construction	Poor pitch/roll stability High air flow
• PERIPHERAL-JET CUSHION	Improved pitch/roll and heave stability Reduced air flow	More difficult for flexible material construction Increased duct losses

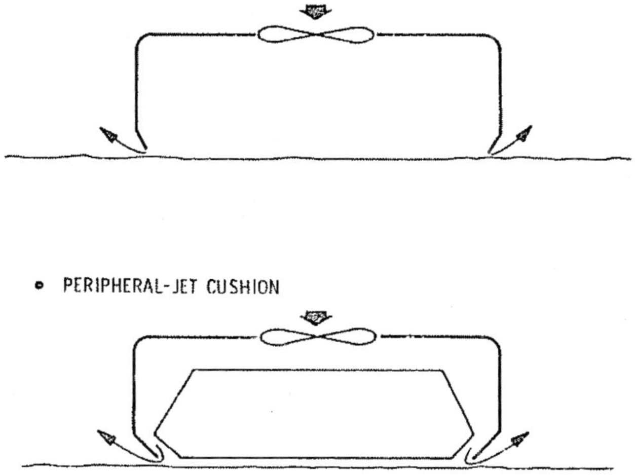

Figure 2. Basic air cushion concepts

difficult configuration, particularly if flexible skirt construction is considered.

1ST GENERATION - Figure 3 indicates the types of R&D craft which might be considered representative of the "first generation" of air cushion technology. These craft and their skirt systems primarily served as research, development, and demonstration tools. For example, the original peripheral jet designs incorporated hard-structure-only cushion systems. However, it was soon recognized that flexible skirts were absolutely necessary to improve obstacle capability and reduce cushion leakage flow, at least for any useful cross-country operational craft.

Craft based on the pure plenum concept generally took the form of multiple cell cushions, due to inadequate pitch/roll stability for a single plenum. A peripheral skirt was soon added to reduce total airflow and improve efficiency for this configuration.

By and large, the R&D craft suffered from high cushion air power demands, excessive spray over water, some stability problems, and poor operability records - which should not be unexpected due to their R&D design objectives.

2ND GENERATION - The air cushion systems of Figure 4 represent what we consider to be the primary examples of "second generation" technology. These "standard" cushion configurations have been well proven and have been applied in a number of operational vehicle systems.

CONFIGURATIONS	ACCOMPLISHMENTS	DISADVANTAGES
• PERIPHERAL JET	Demonstrated amphibious capability and mobility over difficult terrain	High lift air power demand
Initial craft - hard structure only		Excessive spray over water
Flexible nozzle extensions added	Achieved high speeds over land and over smooth water	Inadequate stability under certain conditions
Flexible fingers added	Developed requirements and concepts for improved stability and control	Poor operability record: - Skirt wear and damage - Engine and fan ingestion damage - Excessive maintenance time
• PLENUM/MULTIPLE CELL		
Initial craft - "bare' cells, common air supply	Flexible skirts and gas turbine power adopted	
Peripheral skirt added	Established feasibility for commercial and military applications	

Figure 3.   First generation - R&D craft

	ADVANTAGES	DISADVANTAGES
• MULTIPLE CELLS WITH PERIPHERAL SKIRT	Strong stability by fan characteristic and cell area change	Complexity of ducting and machinery
	Cell redundancy	Vulnerability of peripheral skirt
		High over-water drag for some conditions

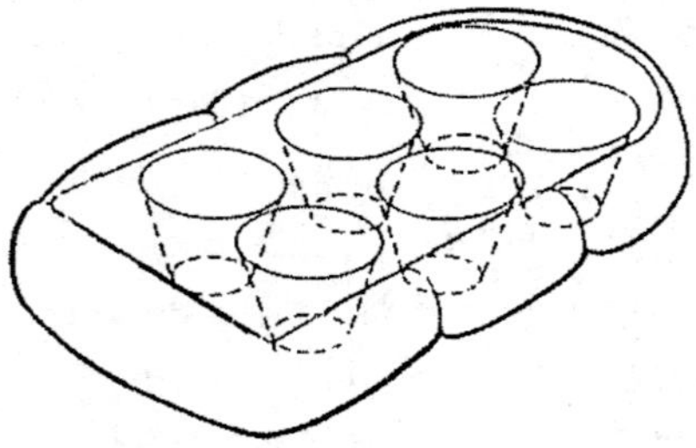

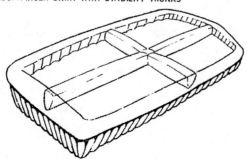

• LOOP/FINGER SKIRT WITH STABILITY TRUNKS	Low drag	Stability limitations over water and rough terrain
	Low skirt wear	
	Individual finger replacement	Complexity of skirt
	Stability by skirt area change plus some jet turning and fan characteristic	Vulnerability of stability trunks

Figure 4.   Second generation - standard configurations

The multiple individual cell system with peripheral skirt has been pioneered in France. It demonstrates strong stability achieved by the separate fan and duct P_c-Q characteristic supplying each cell; the conical cell configuration also provides a continued increase in restoring lift force even after the base of the cell is completely blocked by surface contact. The multiple independent cells do result in increased complexity of ducting and machinery, and the peripheral skirt leaves something to be desired with respect to construction and susceptibility to damage. The individual cells can also incur high over water drag for some conditions, where there is a tendency for the trailing edges to "bucket" or scoop water under adverse wave conditions.

The loop/finger skirt design with stability trunks might well be considered the "standard" cushion configuration pioneered by the British, and has been used in a number of operational craft in series production. This skirt configuration demonstrates both low drag and low skirt wear; in most designs, the fingers can be replaced individually. Stability for this system is achieved mainly by area change of the compartmented cushion as the skirt is deflected, plus minor contributions due to jet turning and the P_c-Q characteristic of the distribution ducting system. The most significant limitation of this system has been in the way of stability shortcomings under certain overwater and rough terrain conditions. Certain wave patterns can render the stability trunks ineffective and result in significant stability reductions, potentially leading to plough-in. Modifications intended to improve this characteristic have included "trip strips" for drag reduction on the bow finger surfaces. The skirt system is relatively complex in design and construction, and the stability trunks have proven vulnerable to damage, particularly over rough terrain.

3RD GENERATION - We believe that air cushion technology has advanced to what might be called a "third generation", in which the air cushion characteristics are tailored to the specific application at hand. Figure 5 is intended to indicate some of the specific design approaches which have been developed to meet specific application requirements.

To achieve strong stability as demanded by certain military applications, we wish to capitalize on a strong fan P_c-Q characteristic plus the changing area effect of tapered cells. If we distribute these cells around the periphery of the craft, the moment arm is increased for a given restoring force, and we achieve good economy with respect to the cushion air power contributing to stability. The peripheral arrangement of cells also provides excellent maintainability, in that all skirt elements can be maintained or replaced without jacking the craft. Note that the substantial number of peripheral cells makes it impractical to serve each cell by a separate fan/duct system. However, the desired effect is achieved by breaking the air distribution loop into four separate quadrants, with each quadrant served by a separate fan/duct system. If the loop bulkheads are designed so that they can be opened, lift system redundancy is available under damage or failure conditions, at some cost in craft stability stiffness.

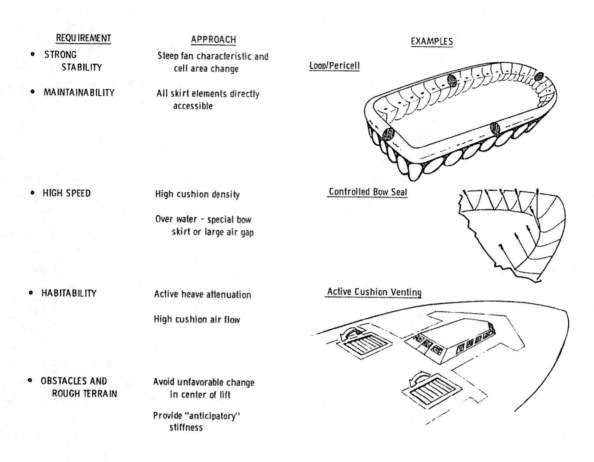

REQUIREMENT	APPROACH	EXAMPLES

- STRONG STABILITY — Steep fan characteristic and cell area change — **Loop/Pericell**
- MAINTAINABILITY — All skirt elements directly accessible
- HIGH SPEED — High cushion density; Over water - special bow skirt or large air gap — **Controlled Bow Seal**
- HABITABILITY — Active heave attenuation; High cushion air flow — **Active Cushion Venting**
- OBSTACLES AND ROUGH TERRAIN — Avoid unfavorable change in center of lift; Provide "anticipatory" stiffness

Figure 5. Third generation - tailored to the application

With respect to habitability, it is clear that the cushion "pumping" effect associated with operation in a seaway or over very rough terrain tends to result in undesirable ride characteristics at high speeds. The ride does naturally improve somewhat with craft size, both due to the obvious reduction in relative wave height, and due to the effect of increasing compliance or "springiness" of the cushion as P_c and cushion volume are increased. Increased cushion airflow also improves the ride characteristics, although in a relatively inefficient manner. Improvements in habitability may be obtained with an active cushion venting system consisting of automatically controlled vent valves. These vent valves will be programmed with various amounts of control "lead", which appears to significantly improve the effectiveness of the vented cushion flow.

With respect to obstacles and rough terrain, work is currently underway which is directed toward tailoring the air cushion configuration to improve obstacle capability, particularly for Arctic vehicle applications. We anticipate that substantial improvements will be possible in this regard, particularly by avoiding unfavorable cushion area changes as the skirt deflects, and also possibly by providing a form of natural "anticipatory" stiffness or lift, to be provided by the forward-most cushion cells.

The series of examples illustrated in Figure 5 was intended to provide some insight into the possibilities which do exist if the air cushion system is tailored to the specific application at hand. It is true that these concepts have not yet been entirely proven at full scale; however, substantial model test work has verified the design performance, and we believe that exciting advances will be demonstrated by current ACV and SES development programs.

SUMMARY & CONCLUSIONS

In summary, it is clear that developments in the technology of air cushion systems have been both rapid and continuous, since the early vehicle demonstrations in the mid-1950's. The basic independent variables in cushion system design are quite limited in number -- thus both innovative configuration design and careful optimization of operating parameters for the specific application will be necessary if continued advancement is to be realized. Air cushion technology has therefore progressed to what might be called "3rd generation" systems, wherein both operating parameters and physical configuration are tailored to the basic vehicle mission. Significant improvements in performance capability are currently being realized by this design philosophy, and anticipated future developments should continue to extend the commercial and military application of air cushion vehicles.

NOMENCLATURE

B = cushion beam - ft

L = cushion length - ft

P_c = cushion pressure - lb/ft^2 gage

P_c/L = cushion "density" - lb/ft^3

Q = cushion air flow rate - ft^3/sec

AIR CUSHION RESTRAINT SYSTEMS
DEVELOPMENT AND VEHICLE APPLICATION

D. D. CAMPBELL
Fisher Body Division
General Motors Corporation

ABSTRACT

The discussion presented in this paper includes: A design review of General Motors air cushion restraint system, performance in dynamic testing, and the General Motors program for production build and fleet tryout. Among the items covered in the design review are: The driver's restraint system, the front passenger's restraint system, and the sensing mechanism. The discussion of air cushion system performance testing includes: The text matrix, two significant crash configurations, the importance of variable inflation levels, and the need to continue baboon and human volunteer testing. The discussion of the General Motors pilot production build and fleet tryout program covers the purpose for gaining "on-road" air cushion system experience.

TABLE OF CONTENTS

INTRODUCTION

General Motors has adopted a position of complete freedom of public release of information, relative to technical advances on the air cushion restraint system. This was done to aid in the development of these systems throughout the world. It is GM's intention to again follow this policy.

In June of 1970, at an opening hearing called by the NHTSA in Washington, GM announced a potential plan and schedule for the development of air cushion systems, pending resolution of remaining problems. GM has no hope of production options for 1973 models and some hope for 1974 production. The delay in the execution of that plan is the result of two problems which occurred in GM's development program at approximately this time one year ago.

PROBLEMS

These two problems, which will be explained in detail in paper entitled, "Special Problems and Considerations in the Development of Air Cushion Restraint Systems (720411)," were: First, occupant rebound and secondly, the deployment hazard of the cushion. The problems were discovered in the first human volunteer testing at Holloman Air Force Base and the baboon testing conducted at Wayne State University.

These two problems are extremely important in the design of GM's latest system. Approximately one year ago, a complete redesign of the system was undertaken to solve these problems.

LOCATION OF THE SYSTEMS

The following relates to the design of the General Motors system. Figure 1 shows a schematic drawing of the sub-systems, components and their location in the vehicle.

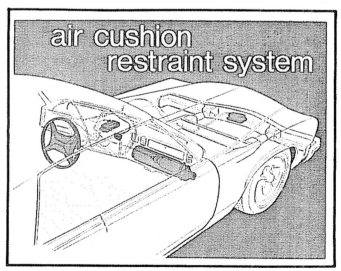

Figure 1

Driver's System

On the driver's side of the front seat, there are three basic components, the steering column and its mounting system, the steering wheel and air cushion assembly and a driver knee restraint pad.

The steering wheel (Figure 2) is new, to provide space for the air cushion module, and also functions as a load reaction member for the inflated air cushion.

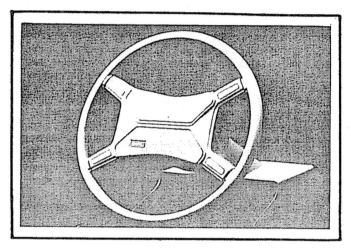

Figure 2

The air cushion module is a self-contained unit and is attached directly to the steering wheel (Figure 3).

When deployed in an accident condition, the cushion is approximately 22 inches in diameter, extends approximately 10 inches rearward of the wheel rim and is about 2.8 cubic feet in volume, as shown in Figure 4. It is neoprene coated woven nylon. A chemical gas generator produces the inflating energy upon an electrical signal through a slip ring arrangement with redundant contacts. The slip ring assembly is housed in the upper part of the steering column and allows the steering wheel to rotate for normal steering needs and yet maintain electrical continuity. The cushion is presurized to approximately 3 psi in free inflation condition in room temperature.

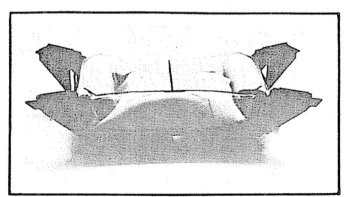

Figure 3

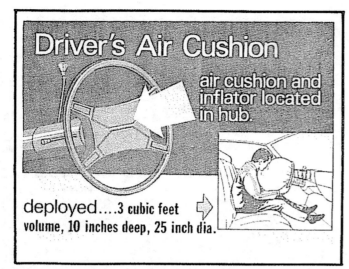

Figure 4

The steering column is new and is the major energy absorbing element in the system.

Several new mounting components (Figure 5) are necessary to provide proper mechanical control of the steering column during impact of the vehicle, and during energy absorption from the driver.

A significant feature of this column is a guide bracket which spans the energy absorbing element in the steering column to isolate it from the vehicle front end deformation. Under most accident conditions, this preserves the entire column travel for absorption of the driver's energy. In addition, this construction is unique in that it allows the resisting force of the column to be staged. An initial low force segment is included to reduce the effect of the column inertia, which is increased by the addition of the inflator weight. After travel begins, a second higher force segment is introduced, this provides a nearly uniform restraining force throughout the column stroke.

Figure 6 shows an exploded view of the instrument panel showing the new structural components required to support the air cushion system, and the energy absorbing driver knee restraint.

The knee restraint (Figure 7) itself is constructed of a formed sheet metal substrat covered by foam and a vinyl skin. This device, in a frontal collision, provides a restraint and kinematic control for the driver's lower torso similar to a lap belt.

The driver's position is well suited to this approach, since he positions himself fore and aft relative to the control pedals; and is, therefore, quite consistently located (Figure 8).

These basic components provide, as a system, torso control, in the knee restraint; force distribution for the head and chest from the inflated air cushion; and energy absorption of the upper torso by means of the energy absorbing column. Although femur load measurements in direct forward impacts are generally acceptable, some concern exists in the angular impact condition where the energy absorbing space available between the knee pad and the steering column is limited. Further development studies and re-evaluation of femur tolerance and design are required.

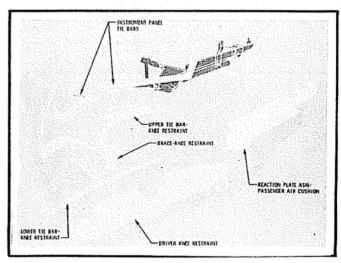

Figure 6

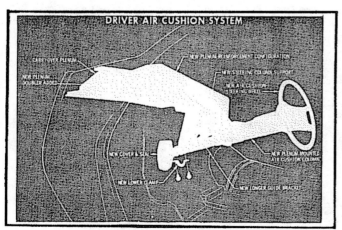

Figure 5

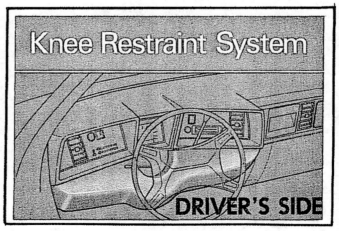

Figure 7

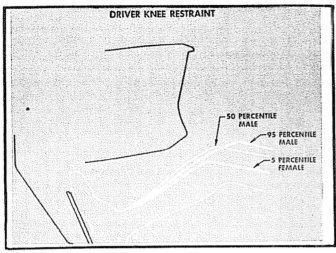

Figure 8

Passenger's System

The passenger's system provides coverage for both the right and center front passengers. The basic system is positioned low on the instrument panel and to the right of centerline of the car (Figure 9). The glove compartment is located above the air cushion package.

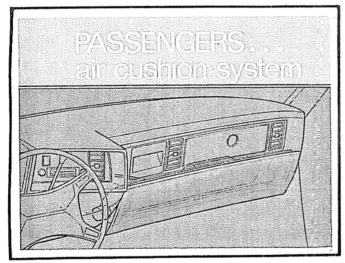

Figure 9

System Module

Figure 10 shows the basic passenger system module, as well as the bumper switch and crash sensor. The module consists of an inflator, duct work to carry the gas to the stored air cushions, a sheet metal housing for the air cushions, the cushions themselves, and the protective cover.

Inflator

The inflator is a chemical gas generator, augmented compressed gas device, sometimes called a hybrid construction. It contains compressed gas, Argon,

and two individual chemical gas generators (Figure 11).

Upon signal from the crash sensing system, the stored gas is released and either one or both of the generators is initiated -- this is very important and will be expanded on later in this paper. The gas is ducted to the two stored air cushions, the pressure build up causes the vinyl-skinned foam appearance cover to open along predetermined lines, the cushions then deploy rearward and upwardly (Figure 12).

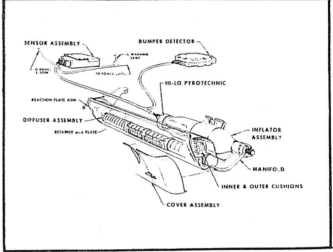

Figure 10

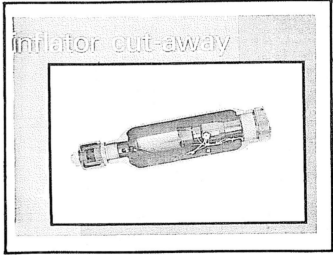

Figure 11

The internal knee restraint cushion is needed to fill the space between the occupant and the instrument panel. Unlike the driver, passenger occupants may be close to, or far away from the instrument panel. The knee restraint bag accommodates these variations. The outer or torso cushion is an uncoated woven nylon construction, approximately 14.5 cubic feet in volume and extends laterally from the door to the steering wheel. In a crash situation, the occupants displace the cushion upwardly until it almost contacts the roof.

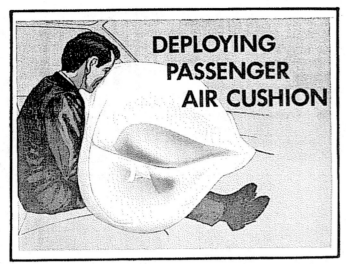

Figure 12

SENSING MECHANISMS

Compartment-Mounted Sensor

The heart of the General Motors system is the crash sensing mechanisms (Figure 13). Two separate devices are utilized: First, a passenger's compartment mounted deceleration sensor; and secondly, a bumper switch. The first device is made up of several pendulum-type magnetic detectors, as well as certain electronic components.

The design of the detector is shown in Figure 14. It consists of a weight hung on a pendulum-type wire and suspended in a permanent magnetic field. The level at which the detector moves can be adjusted by controlling the strength of the magnetic field. The sensor uses several different levels of detector settings. Figure 15 is a schematic diagram of the sensor and shows how the detectors are arranged, as well as a simplified picture of the electrical system.

Figure 13

In addition to crash deceleration detection, this package also performs electronic self-diagnostic functions, records crash levels and records improper maintenance. These areas are covered in more detail in paper entitled, "Special Problems and Considerations in the Development of Air Cushion Restraint Systems (720411)."

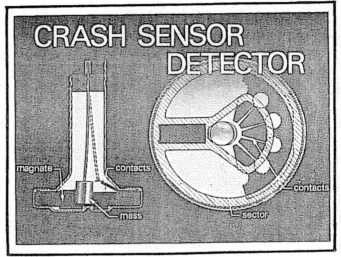

Figure 14

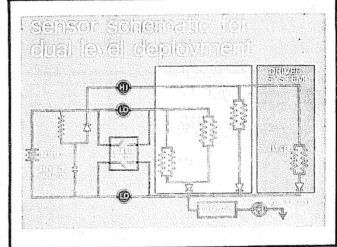

Figure 15

Bumper-Mounted Sensor

The second device is mounted to the bumper and effectively detects velocity change of the bumper early in the accident sequence.

Mechanically, this detector (Figure 16) consists of a mass biased to an at-rest position by a spring and a pair of electrical contacts that are connected to the sensor and passenger system inflator. As the bumper and the bumper switch housing is stopped, the mass continues to move forward until it touches the electrical contacts and a crash signal is then completed.

Testing for the more common types of accidents which occur in the field, such as car-to-car, car-to-pole, or car-to-tree shows that the bumper sensor will greatly improve the performance of the passenger's system by obtaining early deployment of the cushion. This is illustrated in Figures 17 & 18, which show the crash event and important timing relationships for a typical body frame car in two different accident conditions. The first curve is of a 30 mile per hour car-to-barrier impact (Figure 17).

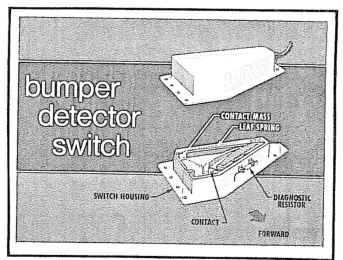

Figure 16

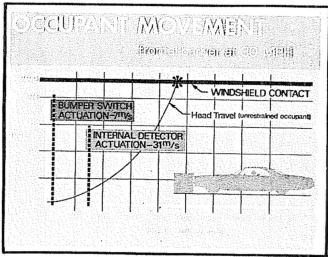

Figure 17

The ordinate of this graph shows unrestrained occupant movement in the body compartment, while the abscissa shows the length of time from vehicle contact with the barrier.

The bumper switch in this case would operate approximately 7 milliseconds after vehicle contact, and the internal body deceleration detector signals approximately 31 milliseconds after the contact. This then shows a time advantage of 24 milliseconds, which is meaningful but not necessarily of major significance.

Figure 18 projects those same timing relationships for a 25 mile per hour car impact into a pole.

The bumper switch actuation occurred at 8 milliseconds, while the internal detector signalled at

89 milliseconds for a difference of 81 milliseconds. This now shows a real advantage in real accidents for a bumper-mounted detector, since the pole collision final high energy signal alone might not provide sufficient time to properly deploy the air cushion.

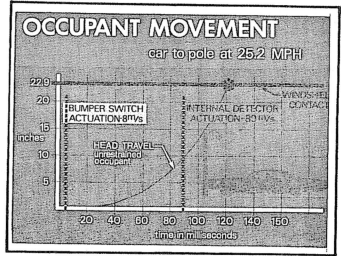

Figure 18

In addition, the combined usage of the bumper and internal deceleration detection devices and the two chemical gas generators in the inflator provides a feature that is called "variable inflation" for the passenger's system. Since the bumper switch senses crashes early, the inflator has been arranged such that on signal from the bumper switch the stored gas is released and one of the solid fuel generators is initiated. This produces a "low level" inflation (Figure 19). That is, the energy of the deploying bag is reduced from that required in higher speed collisions. This level of inflation is provided for restraint of two passenger dummies in an 18 mile per hour barrier equivalent accident. If the crash is more severe than 18 miles per hour, the internal deceleration sensor high level detectors will initiate the second gas generator in the inflator and thereby increase the restraint capability of the air cushion (Figure 20). The time span between initiation of the low level inflation and the high level inflation will vary, depending upon the accident severity. The greater the accident severity, the shorter the interval between actuation of the two generators.

This feature provides restraint proportioned to accident severity and thereby presents lower energy deployment in the lower level of the more frequent accident conditions.

This feature has been added to the system for two basic reasons:

1. To reduce the deployment hazard -- which was one of the major problems mentioned earlier.

2. To improve the restraint capability in real life non-barrier-type collisions.

1071

Figure 19

Figure 20

real occupants must be faced. The work described in this paper relates to tests with dummies which are sometimes misleading. To be sure that such a system is working as well as is required, it is necessary that companion experiments be conducted with living subjects.

This is to be accomplished through repetition of the same experiments at Wayne State University and Holloman Air Force Base, that prompted the afore described modified system. At Wayne State University, additional experiments are being conducted with out-of-position baboons to assure that the question of the deployment hazard to small children has indeed been resolved. And, in cooperation with the National Highway Traffic Safety Administration, the human volunteer program will be continued.

It is also necessary that such a complex system be evaluated in real life road and environmental conditions. This is one of the purposes of GM projected field test program. A second, and perhaps more important purpose, is to gain manufacturing experience. Because of the desire to

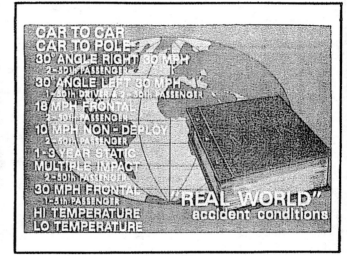

Figure 21

Development and Tests

During the development of this system, many combinations of accident conditions, occupant positions and sizes have been examined. Basically, for the front seat passive requirements, Federal Safety Standard 208 deals with the frontal and 30 degree angular barrier conditions with a full complement of 50th percentile male dummies.

Many additional varieties of test conditions were studied as a part of the air cushion restraint system development program. General Motors has expanded studies beyond the safety standard, to examine the "real world" accident conditions. This "test matrix" is shown in Figures 21 & 22.

GM's Laboratory and Proving Ground programs have progressed to a point that permits us to believe that we are approaching a workable first generation air cushion system for front seat occupants. But the conditions of real life exposure to all types of road and environmental conditions, as well as performance in "real-life" accidents and the effects on

Figure 22

introduce passive restraints at the earliest practicable date, GM is initiating a field test program in advance of completion of the aforementioned volunteer and baboon programs. Current plans should lead to installation in approximately one thousand 1973 model vehicles.

This field test will give actual, although limited, production experience to the several of GM divisions and outside suppliers who would be involved in the manufacturing and assembly of air cushion systems. Another objective will be to determine how the air cushion performs in real traffic conditions. This one thousand car test is expected to yield a limited number of air cushion deployments in the first year, and more importantly, significant information on the question of inadvertent deployments. Also, it should give more information on how the system is affected by temperature, humidity, corrosion, and other factors.

In summary, General Motors believes a workable front seat air cushion restraint system is approaching. It is necessary that the performance of the system be validated through additional animal and volunteer studies which are proceeding. Concurrently, GM is beginning the field test program that is necessary for both final "real world"

appraisal of the system and the development of manufacturing experience.

GM has devoted over three years of intensive effort to the development of this system. It is sophisticated and complex. It is also based on many laboratory test interpretations and engineering judgements.

In the future, there are many difficult design responsibilities yet to come. We must ask ourselves questions, such as:

1. How can the system be simplified?

2. What adjustments will be required by the experience of the "real world" conditions?

3. What is the next design generation in sensors, inflators, and cushion construction?

4. And, what additional reliability can be built into the design of the system?

The search for answers to these questions will challenge all industry and government.

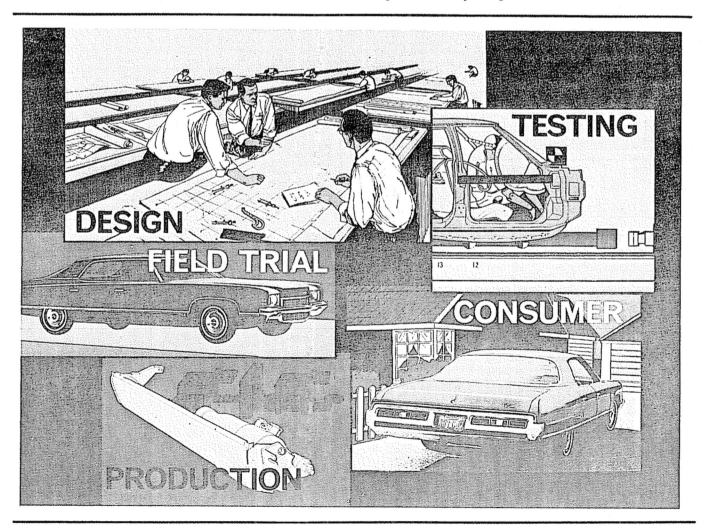

Air Cushion Systems
for Full-Sized Cars

by
J. A. Pflug
FORD MOTOR COMPANY

INTRODUCTION

The purpose of this paper is to relate the technical approach of Ford Motor Company in developing air cushion restraints for possible application to front passenger restraint in full-sized cars. Special considerations in the design of instrument panels and restraint system components are discussed, particularly as influenced by packaging. The differences between restraint designs for full-sized and compact cars are identified and evaluated.

Our efforts to develop the air cushion as a potential means of improving occupant protection and satisfying the goals defined in FMVSS 208 have resolved some technical problems and exposed others. This paper will review our efforts aimed at conformance to the Standard as well as those directed toward satisfying additional performance requirements in the "real world" of occupant protection. Progress in research and development of this restraint concept is reflected in Ford's field test program starting in March, 1972 of several hundred Mercury Monterey's equipped with experimental air cushion/seat belt systems for use by selected fleets (See Figure 1). The air cushions in these vehicles are designed to supplement the seat lap belt in head-on collisions. Use of the lap belt, which we believe is essential to assure proper positioning of the occupant, is enforced by a starter interlock/warning system similar to that planned for 1974 production. We are now monitoring the results of usage in the field.

The sole purpose of this 1972 build program is to provide information related to the following:

- Potential volume production and assembly problems. (Delivery of the first group of these vehicles was delayed by discovery of faulty solder joints in the module circuitry resulting in their replacement.)

- Realistic data on both inadvertent and demand-fire situations.

- Effects of aging, vibration and temperature cycling.

- Effects of everyday vehicle usage.

DESIGN CONSIDERATIONS — FULL-SIZED CARS

The full-sized car was chosen for initial development work at Ford because several dimensional aspects appeared to be inherently more favorable to air cushion development. In the design of the air bag system, the relative position of the module diffuser and the occupant is one of the primary factors affecting performance. Development testing of the so-called low mount system has identified the need for a relatively small tolerance in locating the diffuser side view centerline in the instrument panel package. Closely related with locating the diffuser and of equal importance is the restriction of instrument panel surfaces forward of the path of the inflating bag. Our development guidelines call for 20° side view and 140° plan view angles to define the inflation zone (See Figure 2).

ABSTRACT

The initial air cushion research by Ford and suppliers has primarily involved full-sized cars. Technical approaches for these cars in packaging, kinematics, sensing and deployment are presented. Also, efforts to meet the requirements of the potential restraint standard as well as to resolve "real world" problems are discussed.

Figure 1 — 1972 Mercury Monterey — Air Cushion Restraint Installation

AIR CUSHION DEPLOYMENT GUIDELINES

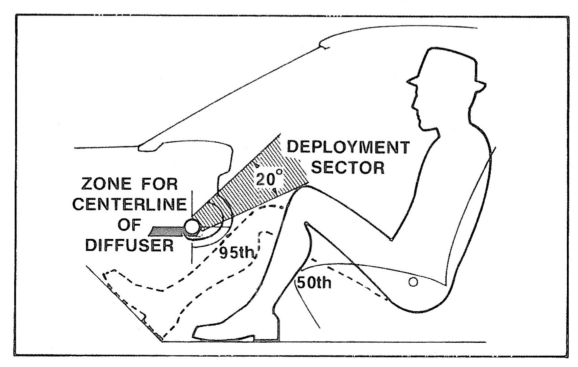

Figure 2 — Air Cushion Deployment Guidelines

This inflation zone must be free of surfaces or objects if bag inflation capability is to be optimized. These guidelines are equally applicable to full-size and compact cars due to the similarities in seating positions of the occupant to be restrained.

Redesigning the traditional instrument panel to accommodate an air bag module leads to some significant changes. A very obvious one is the deletion of the glove box, which conventionally occupies the zone now claimed by the low-mounted module. This loss is typical for both full-sized and compact cars. There may be ways to reclaim part of this zone in the future through higher mounting of the module and/or reduction of its size by use of different gas sources to minimize the total package of the air bag.

The climate control system is also vitally affected by the introduction of the air bag module. A simplistic solution to the interference is to relocate all climate control elements forward of the module. This is seldom a simple thing to do and can lead to encroachment in the engine compartment or possibly an increase in wheelbase. Both are undesirable effects. The full-sized car generally provides more space in this area and thus is less difficult to deal with in this respect. In the 1972 Mercury, even though we were fitting an air bag module into an already existing instrument panel, it was possible to compromise the ducting in a minor way to overcome the packaging problem. Ideally, the packaging problems would be minimized with an all-new design in which climate control, instrument panel and air bag can be designed simultaneously.

Car width also influences the detail of air bag design although the number of passengers to be protected per air bag is of greater importance. For example, the width from steering wheel rim to the right front door trim panel can vary as much as 4 to 10 inches between full-sized and compact cars. The air bag volume may be about 2 cubic feet larger on a full-sized car if it is to protect two passengers instead of one.

Two other package differences of full-sized vs. compact cars have been of concern in air bag design. One is the space available in a full-sized car to allow a longer "ride down" after crash impact for the front occupant. For example, the Mercury measures nearly 2.5 inches or 10% greater than Pinto in the distance from the manikin's torso to the instrument panel with the seat in full rearward position. Development data, however, indicates this difference may be insignificant since a properly tuned air cushion will stop the occupant in a 30 mph head-on barrier crash after 10 to 12 inches of penetration into the filled bag, well before he contacts the panel.

Another package influence to be considered is the position of the windshield and header in relation to the occupant. With the air cushion restraint, this proximity is not considered a significant hazard in most full-sized cars. It may, however, be of greater significance in small vehicles or those with very fast-sloped windshields if occupant head contact occurs during entry into the bag.

The "trigger" or sensor that signals bag inflation responds to the g-load and pulse time experienced in the crash impact. We have found that the "quickness of the trigger" depends more on sensor location and vehicle structure than on vehicle size. A faster response in 30 mph barrier crashes has been achieved by mounting remote sensors on the radiator support. With this system about 18 milliseconds elapse from the first contact until the sensor reacts to "fire" the system in front barrier crashes for either full-sized or compact cars.

In contrast, sensors mounted on the air bag module in the instrument panel/dash area close within 20-24 milliseconds on a compact Pinto but not until 27-34 milliseconds on a full-sized Mercury. This difference in signal time results primarily from the difference in crush characteristics of front sheet metal and the body. Tests show a fast buildup of g-load on the Pinto which is maintained during the collapse of the front end. For the Mercury, the g-load buildup is more gradual during collapse. Both cars, however, show approximately 30 inches of front end collapse in 30 mph barrier crashes.

A new factor requiring evaluation in crash sensor design is the effect of energy absorbing front bumpers which become mandatory in 1973. Our studies are proceeding to accommodate this requirement by keeping the signal within the time limits of the system. The adjustment which will be required is substantially more dependent on the type of energy absorbing device used and its force characteristics than on the size or weight of the vehicle itself.

SAFETY STANDARD COMPLIANCE — FULL-SIZED CARS

The Federal Safety Standard calls for completely passive restraints in a variety of crash modes in all 1976 passenger cars; full-size, intermediate and compact. In our efforts to achieve this goal, Ford has concentrated initial development on full-sized cars rather than attempting to resolve all models simultaneously. We felt it was sufficient challenge initially just to validate the concept, and the challenge was so great that it should be directed at the most favorable environment.

The total requirements of the Standard for 1976 are still beyond our current level of technical capability. Eight recent tests of the present design level air cushion in 30 mph barrier impacts showed that with added lap belts, all injury criteria of the Standard were met with the 50th percentile dummy held in a normal, front passenger position. The same air cushion restraints, when tested without belts, but with the manikin seated in a fully erect position allowed the femur loads to exceed the Standard (1740 lb. vs. 1400 lb.) (See Figure 3.)

In Hyge sled tests at 30 mph with the same air bag designs and the same dummies, chest g-loads exceeded the Standard (61-69g vs. 60g) when unbelted. Head and leg injury criteria were recorded at levels well below the specifications.

In evaluating air cushion protection for the 5th and 95th percentile dummies to cover the entire range of adult occupants, high chest g-loads were also noted in Hyge sled tests. Mean chest loads of 68g's were recorded for the 5th percentile female in a test series of modules from one supplier. The head S.I. was below the limit allowable for the full range of occupant sizes. The test data show that tuning the system to meet the requirements of the FMVSS with a 50th percentile dummy will not give optimum results for 5th and 95th percentile dummies.

Passive protection in side impacts is also required in 1976 by Federal Regulation. Our analysis of this requirement shows that there is insufficient space in full-sized as well as compact cars to stow and deploy an effective air cushion for this type of event. We have, as a result, concluded that air bags will not provide the performance required by FMVSS 208 for side impacts. Some energy absorption can be accomplished in door structure and padding, but corollary effects of padded

door bolsters and padded roof side rails, on things such as package, visibility, appearance, cost and weight, may be totally unacceptable.

To date, we know very little about air cushions in the rear passenger compartment. However, much of the design knowledge gained from higher priority work on front passenger compartments will undoubtedly be applicable. We do know that, in addition to even more difficult packaging limitations, the structural implications are formidable. Whereas in the front compartment the instrument panel which contains the air bag module is backed up by the cowl and dash structure, the rear air bag may be stored either in the roof or the front seat back. The loads imparted during a collision by three rear occupants are very high and must be principally borne by the front seat structure. Loads of 8,000 pounds or more have been projected which will require the addition of significant structural beams from pillar-to-pillar as well as energy absorbing features in the seat back.

"REAL WORLD" REQUIREMENTS — FULL-SIZED CARS

Ford is equally concerned about requirements for air cushion use which are not defined in Federal Regulations. We term these "real world" requirements, and they include the normal protection expected in the everyday use of cars over and above the somewhat

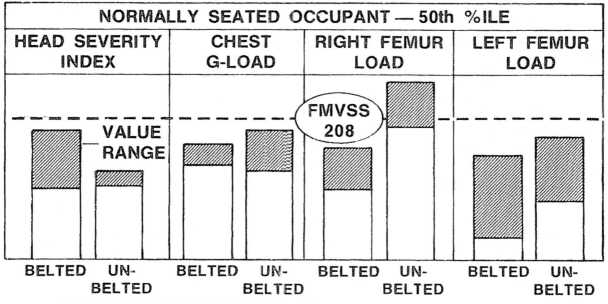

Figure 3 — Compliance Status — 1972 Mercury Air Cushion Restraint.

artificial controlled test conditions of the Standard. Let's take a look at how these affect the full-sized car.

OUT-OF-POSITION OCCUPANTS — People are inclined to relax when riding as a passenger in today's full-sized cars, and frequently you will see them slouched or otherwise out of the ideal seating position. Also, statistics show that panic braking before a crash, which tends to throw the unsuspecting passenger forward, occurs in 36% of all injury producing collisions. Thus consistent use of seat belts is required to reduce the likelihood of being out-of-position.

In developing the air cushion to protect the "out-of-position" occupant, the panic braking situation is simulated by positioning the dummy forward on the seat in a jackknife attitude with the hands/arms on the crash pad. This was determined from human volunteer tests conducted by Ford.

The child who stands up in the front compartment also presents a real problem for protection during a crash or inadvertent bag deployment. Even though the Standard prescribes no requirements for this everyday condition, the complete engineering job dictates that it be considered with equal importance to the task of passing the compliance tests. The magnitude of this problem could grow in severity with incorporation of rear seat air bags. Again though, tuning the air bag design to meet this condition can have a detrimental effect on full-size adult occupant protection.

Our early designs of the air cushion proved to be inadequate in protecting either panic braking or standing child conditions. For example, only the 95th percentile dummy "survived" the 30 mph Hyge sled tests from a panic braking position as indicated by the recorded impact severity index levels versus the injury criteria specified in the Standard. And, static deployment of the air cushion subjected the standing child dummy to impact loads exceeding the criteria of FMVSS 208. (See Figure 4.)

Significant reductions in loading have been achieved by design changes to control the forces and timing of air cushion inflation. For the pure gas system, a flow control valve was introduced. For the hybrid system (partial stored gas and gas generator), the gas generator characteristics were revised.

The slower inflation accompanying these changes led, however, to overall cycle time problems. The bag was coming up too late to provide the intended "ride down" for the normally seated occupant. The effect was higher levels for head S.I. (730 vs. 539) and chest load (38g vs. 35g). (See Figure 4.) But a solution may be the adoption of remote sensors mounted on the radiator support which were mentioned earlier. The faster crash response of these sensors (18 ms. vs. 30 ms. on Mercury) can compensate for the slower bag inflation.

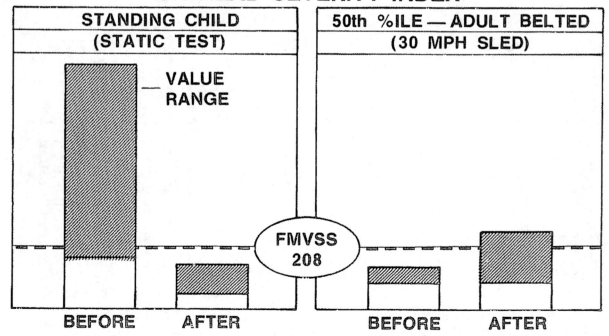

Figure 4 — Effect of Air Bag Flow Control on Head Severity Index

The flow control changes primarily reduce the standing child problem Static tests show both head S.I. and chest g-loads on the child dummy were reduced to well below the Standard's criteria. (See Figure 4.) Further evaluation under dynamic conditions show the head S.I. at permissible levels up to 20 mph. At higher speeds, head injury control requires more air bag development Also, the chest g-loads under dynamic conditions are not acceptable as yet. Our objective for air cushion restraint performance for the standing child in dynamic conditions is to meet the FMVSS 208 criteria.

Hyge sled test results with normally seated manikins show a trend toward higher impact loads with the "flow control" air cushion system In the panic braking simulation, tests made with "flow control" have not shown consistent results thus far and further study is underway. In the next design phase, our objective is to meet FMVSS 208 performance criteria for 50th percentile occupants in both front passenger positions, with and without lap belts The same performance is targeted for the "out-of-position" occupant including the standing child, as well as the range of occupants from 5% female to 95% male. Should this prove to be impossible, we want, at a minimum, to avoid increasing exposure to injury over present levels

EFFECTS OF INADVERTENT FIRING — Human factors relating to the air cushion restraint are still to be determined in some areas. For example, the effect on the driver of being startled by deployment is an unknown His response to inadvertent firing is, of course, of primary interest. Our field test of the 1972 Mercury experimental air bag/seat belt restraint system may furnish the first actual data from publicly driven, air cushion equipped vehicles

POSSIBLE HEARING DAMAGE — Another area of human factors relating to the air cushion which needs research, is the risk of hearing damage from the noise of inflation. Early air bag systems deployed with a sound intensity of 175dB which we felt too nearly approached the threshold level (185dB) for eardrum rupture in normal, healthy ears. Today, the sound level produced by a single bag system is only 1/10 that of the early system, or 165dB vs. 175dB.

Hearing damage at lower levels of intense impulsive noise may be limited to neural impairment. Experimental studies have been documented by Bolt, Beranek and Newman, "Noise and Inflatable Restraint Systems," BBN Report No. 2020, DOT Contract HS-006-1-006, April 3, 1971 to define the criteria for this hearing damage as related to the low and high frequency portions of the pulse. (See Figure 5.) Effort continues to develop reliable systems which will have acceptable sound levels in both categories. There is concern for controlling noise impulses in vehicles equipped with air cushion restraints for occupants in all seating positions, particularly in compact vehicles in which the compartment volume is minimized. In

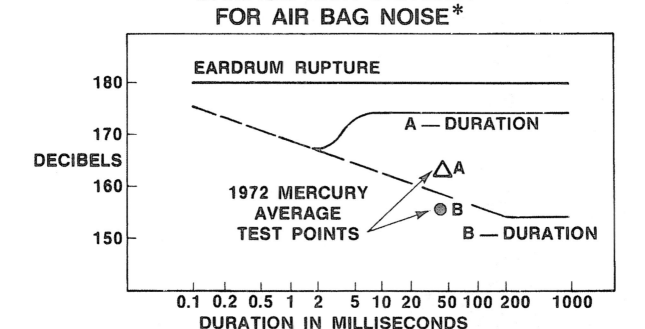

Figure 5 — 1972 Damage Risk Criteria for Air Bag Noise

those cars pressure rise can be equally important to the decibel level. We will not regard the noise issue as resolved until we are convinced that the injury threshold criteria are based on firmly proven medical grounds and systems designed to reduce noise to those levels can also provide acceptable restraint. Such systems have not, as yet, been developed although samples are now available for testing. Again, an adjustment to reduce noise can affect other performance characteristics of the air bag and, in the initial evaluations, the protection capability of the system suffered. This is further evidence of the fact that the air bag system is a finely tuned device and any adjustment to improve one characteristic can have serious consequences for others.

SYSTEM RELIABILITY AND LIFE EXPECT-ANCY — Extremely high reliability has been judged an essential goal for the performance of the air cushion. It could be described as the first "one-shot" system in automotive history, with the sobering responsibility to save lives and prevent injury. Another "one-shot" device is the ejection seat for military aircraft. The Air Force requires that it shall successfully perform all specified functions with a reliability of at least 0.90 at the 90% confidence level. Our reliability goals for the air cushion were even higher than this on the experimental 1972 Mercury system, with our ultimate objective being 0.999 for "demand fire" and 0.99 for compliance to the Federal Standard at 90% confidence. Both component and total system reliability have been computed for the 1972 Mercury air cushion based on tests of appropriate statistical samples. Accordingly, its demand fire reliability has been estimated as low as 0.86 compared to a goal of 0.96 at 90% confidence. With respect to reliability of conformance to the criteria of FMVSS 208 of recent air bag/lap belt systems they averaged 0.66 at 90% confidence although reliability to meet some individual criteria was 0.99. It is obvious that much work is yet to be done in order to meet reasonable levels of reliability for this complicated system.

Thus far all of our efforts have been directed toward simply making a satisfactorily functional air bag system. At the same time, even though the initial objective is not yet accomplished, we recognize that second, and even third, generation systems must be developed in order to improve performance, reduce costs and minimize limitations on styling and packaging. An intensive analysis of the next generation air cushion design is planned to define all potential problems related to proper function, followed by determination of corrective actions to improve the system. Changes in design, processing and handling are anticipated as a result. Equally vital in improving reliability will be the analysis of failure modes and development of a "fault-tree" as further means of modifying system design.

As the new design air cushion is developed, component and system testing programs will allow statistical analysis of the resulting reliability.

A unique aspect of air cushion use is that, after deployment, the entire system must be replaced. Again, the "one-shot" concept is evident. There is no alternate for continued occupant protection. The decision for complete restoration is strictly up to the owner, for he can operate his car without the air cushion.

The effects of prolonged vehicle use on the reliability of the air cushion are still largely unknown. Several approaches are being studied including durability testing of vehicles, laboratory environmental tests, and of course, the field testing of 1972 Mercury air bag cars.

MULTIPLE IMPACTS — Another "real world" situation which affects occupant protection is the "multiple" impact, or a series of crashes involving the same vehicle. Analysis of ACIR accident statistics show that nearly one quarter (23.5%) of the fatalities recorded involved multiple impacts. In a special study of such impacts, it was found that only 8% of these fatalities occur in frontal first collisions; another 19% of the fatalities occur in subsequent frontal impacts. The current concept of air cushion under development has the capability of occupant protection only in the initial frontal impact. In the typical system, deflation follows very quickly after deployment (starting within 0.5 second) leaving an unrestrained occupant vulnerable to the effects of any subsequent impacts. The bag is designed to deflate through ports or by porosity of its material as it restrains the occupant and absorbs his energy in the initial impact. Rapid deflation is a requirement to prevent violent rebound of the occupant.

New air cushion concepts with second inflation capability have not reached the test and hardware stage, and therefore, lack feasibility. Preliminary studies of such a system show that test conditions would be very difficult to control. The unknown manikin position at the time of the second impact and the possibility of losing control of the vehicle in the test area are examples of the test problems. Of greatest concern is the inherent decrease in reliability with the increased complexity of the system. At the expense of repeating once more, the air cushion restraint is a finely tuned system and every adjustment to meet a need can have serious effects on the other performance characteristics as well as total system reliability.

CONCLUSIONS

Based on the extensive development of air cushion restraints already accomplished, it is concluded that full-sized cars, at least for the right front passenger, represent a typical application rather than a substantially more simple one. This is apparent in packaging, occupant kinematics and sensor requirements for protection of the right front passenger. The size and weight differences of the full-sized car versus the compact do not appear to significantly affect the performance of the air cushion restraint.

The passive protection in frontal impacts specifically defined by the Federal Standard may ultimately be achieved with the air cushion restraint. But the air cushion does not provide protection in side impact and rollover in either full-sized or compact cars.

In the "real world" of occupant protection there are still many unresolved problems with the air cushion restraint system. These include the following:

- Inability to accommodate people of all sizes and in all positions.

- Effects of inadvertent firing.

- Possible hearing damage.

- System reliability.

- Survival in multiple impacts.

Some of these problems may be resolved by such programs as the field test of 1972 Mercury air bag cars and also by continued research and development effort.

SPECIAL PROBLEMS AND CONSIDERATIONS IN THE DEVELOPMENT OF AIR CUSHION RESTRAINT SYSTEMS

E. H. KLOVE, Jr.
Fisher Body Division
General Motors Corporation

ROBERT N. OGLESBY
Engineering Staff
General Motors Corporation

ABSTRACT

Presented in this paper is a discussion of the details of the General Motors air cushion restraint system and of specific technical problems of system development and of implementing a production build program. The details of the General Motors system include a description of the components of the driver's and front passenger's systems, crash sensing, and "variable inflation." The discussion of specific technical problems includes performance considerations; such as: Occupant rebound, child-size occupants, out-of-position occupants, non-barrier type crashes, and the function of the appearance cover. Also included is a discussion of the toxicity potential, noise risk, sensor development, reliability considerations, and field service requirements.

TABLE OF CONTENTS

INTRODUCTION

This paper will focus on some of the technical details of problems that have been encountered during the evolution of General Motors front seat air cushion restraint system; the steps taken in dealing with them; some of the existing problems and what is intended to be done about them. However, before entering into those details, a review of the basics of the General Motors front seat system is appropriate.

Figure 1 shows the vital components of this front seat system -- both driver and passenger sides.

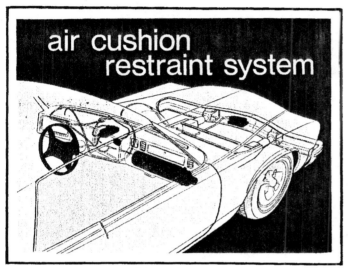

Figure 1

Current GM System

This system features several refinements from the arrangements that have been previously reported. First, in crash sensing two separate devices are utilized to signal the energy source of collision conditions. A bumper-mounted switch, which operates very early in the accident sequence, and if the collision is of a sufficient level, a portion of the deployment-restraint energy for the passenger's system is released. No deployment of the driver's system is planned, to allow maximum vehicle control in low speed collisions. The passenger's compartment mounted deceleration sensor will also provide a signal to the passenger's system accomplishing the same deployment. This is planned for the accident cases not involving the bumper detector, such as an angular collision from the front of the vehicle at the front axle.

If the collision is of sufficient magnitude above the bumper and low level deceleration detector thresholds, another deceleration detector in the passenger's compartment signals a second release of energy to the passenger's system, as well as producing deployment of the driver's system.

This combination of sensing and staged deployment for the passenger's system provides a second feature that is called "variable inflation." The benefit of this feature is that the low level of deployment ener-

gy is released early in an accident, and in a pole collision, the second or high mode is released substantially later. In this way, the hazard of deployment is lessened, and in fact the total deployment is patterned to the severity of the accident.

Occupant Rebound

GM's original passenger air cushion system described in a public presentation in June of 1970 and SAE paper entitled "Safety Air Cushion Systems - GM's Progress to Date," presented in January 1971, utilized a 10 to 11 cubic foot teardrop-shaped cushion, constructed of neoprene coated woven nylon with uncoated or porous end panels. Our sled testing with anthropomorphic dummies showed that the dummy occupant could be restrained from forward movement, yielding severity index results below 1,000. However, upon rebound contact with the seat back and head restraint area, another head acceleration was measured, and the total of both the forward impact and the rebound impact was over 1,000 in some cases of a 30 mph barrier crash simulation (Figure 2).

In addition, kinematic study of the high speed films of these tests showed that the rebound velocity of the dummy was as high as 21 miles per hour.

This system was used in a series of volunteer tests run by the U.S. Air Force under Department of Transportation contract in the Summer of 1971, and similar results were noted with dummies and volunteers from that series.

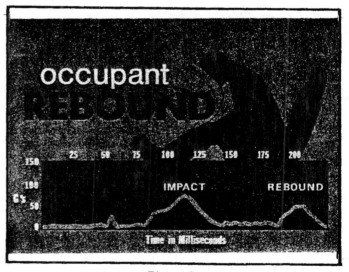

Figure 2

The rebound velocity at seat back contact was analyzed for both volunteers and dummies. Figure 3 compares those measurements -- the dummy and man seem to rebound at similar velocities for the lower collision speeds. At higher speeds, the human rebound velocity is definitely below the dummy's. However, the rebound velocity and accompanying impact were unacceptable. Therefore, the air cushion fabric was modified in construction and shape to reduce this action. Figure 4 shows a schematic comparison between the earlier

"Holloman" system cushion shape and the revised shape. It was also found that the material porosity greatly reduced the rebound action; in fact, eliminated the need for fabric coatings. Controlled porosity has been accomplished, by specifying a tighter weave construction.

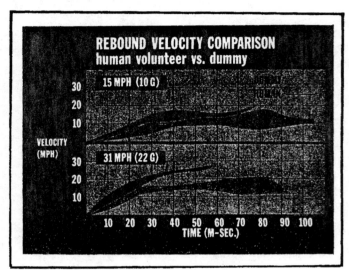

Figure 3

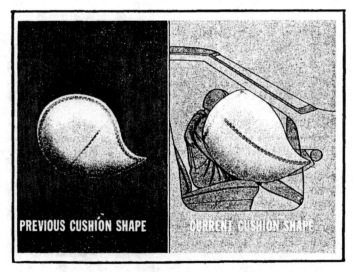

Figure 4

The latest woven, uncoated fabric specified for the system utilizes a rip or tear stop construction. This inhibits small tears from propogating to large openings. The driver's system was found to produce a low rebound velocity inherently, since the steering column absorbs a large porition of the occupant's energy in its normal E.A. stroke.

Child-Size Occupants

Another important consideration in the development of any occupant protection system, is the protection afforded child occupants (Figure 5). The prime reasons for the difficulty in providing a safety system which protects child occupants are:

1. They are small, light, and are believed to be more susceptible to some types of injuries.

2. They are frequently out of the normal seated posture, often standing to see out through the windows.

GM's air cushion development program has included testing with dummies, which simulate a three-year old child placed in both seated and standing positions, with favorable results.

The most significant reasons for these improved child dummy results are:

1. Variable inflation, which lessens the initial deployment surge to ease the action of the cushion on the occupants.

2. The cushion is deployed from the lower instrument panel which produces a less direct impact on the head and upper torso of a standing child.

3. The character of the method of cushion deployment, it unrolls against the occupant and thereby minimizes the deployment shock.

4. The fabric porosity feature that reduced rebound. The greater porosity reduces the thrust of the total system on an occupant.

It is hopeful that the baboon testing with the new system currently in progress at Wayne State University will corroborate these indicated improvements.

Figure 5

Out-of-Position Occupants

It is also important that the air cushion performance be considered in the event that occupants might be out of their normal seated positions.

A common out-of-position occurrence is a result of severe pre-crash braking, which could cause the passengers to be leaning forward when the air cushion deploys (Figure 6).

Test results have indicated that forward leaning dummies benefit from the protection of the air cushion system, with no measurable hazard, resulting from their being "out-of-position."

Additional evidence of the benefit of the variable inflation system was recognized in the comparison of static (non-crash) deployments between the earlier single level inflation and the variable inflation conditions for the forward leaning occupants.

Non-Barrier Collisions

Previously mentioned, the benefit of the multiple sensing and inflation features of this system, relative to non-barrier or real-life conditions, is explained here.

The crash signal, measured by an accelerometer inside the passenger's compartment, produces a rather late firing signal to the air cushion inflator, when the object struck is small in area, or contacts only a small area of the vehicle's front end.

The addition of a sensing device mounted at the bumper of the vehicle produces a signal that is substantially earlier. That signal is used to initiate the low level portion of the passenger's inflator output. In this way, time is gained in the deployment to move the air cushion into a restraint position, as well as minimize the action of the cushion against out-of-position occupants. If the accident is severe enough, the later internal high level sensing system initiates the high level mode of the inflator and additional restraint capability is provided.

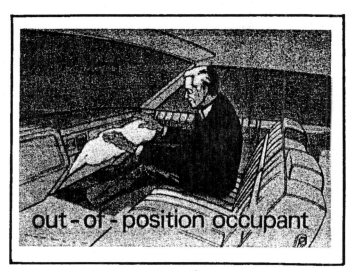

Figure 6

Figure 7 shows the sequence of happenings, relative to the movement of an unrestrained occupant within the vehicle for a typical car-to-pole crash.

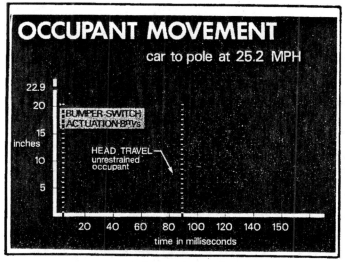

Figure 7

Appearance Cover Function

Figure 8 shows the location of the air cushion system it is hidden from view by a cover that must open upon cushion deployment.

The functional requirements for the cover are simple in nature but difficult to achieve. The cover must withstand abuse; being kicked for instance, yet in a crash situation it must be capable of opening quickly and fully, to permit the air cushion to deploy without appreciable hindrance. And, it must do so without producing injury to the lower leg.

The number of possible cover designs is extensive, but one which has been developed and tested is constructed as shown in Figure 9. The exposed trim surface and the inner backing are expanded vinyl modified A.B.S., approximately .50 inch of polyurethane foam is contained between the two skins. Embossed on the outer skin is a decorative stitch seam, but behind this seam the inner skin is offset so that it contacts the outer skin and provides predetermined split lines at the sides of the cushion deployment opening. The upper section of the cover is forced open by the deploying cushion and pivots around the two horizontal depressions of the cover. The sides and bottom of the opening in the cover is reinforced by a steel insert to provide stability at the attachments, and to add insurance the cover will open properly along the designed split lines.

When the air cushion deploys, the cover opens rapidly, rotating 180 degrees about its lower hinge line in about 20 milliseconds. The cover will contact both legs of the right front occupant (Figure 10), and the right leg of the center front occupant below the knee, which subjects both tibia and fibula to a bending stress.

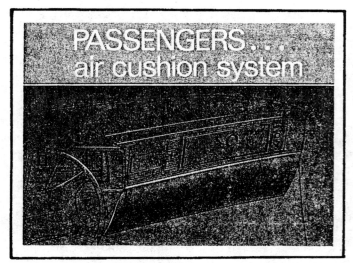

Figure 8

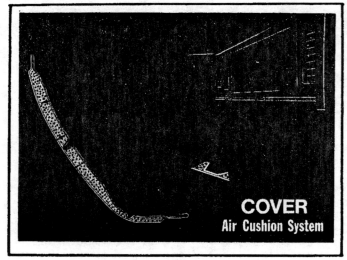

Figure 9

To determine the degree of potential injury that the air cushion cover will produce as it contacts the lower leg during an air cushion deployment, a series of tests were conducted at Wayne State University using cadavers and dummies to measure and observe the effects.

Figure 10

A total of seven tests were conducted using the subject cover. Four of these tests evaluated the injury inducing potential with cadavers, and the other three tests used a Sierra 1050 dummy to establish a correlateable measure of physiological response relative to knee and leg impacts; however, no injury was exposed to either leg of the cadavers.

The cover over the packaged driver's cushion is a molded vinyl skin with relieved areas to provide predetermined opening lines. Upon deployment, this cover opens in a clamshell fashion (Figure 11).

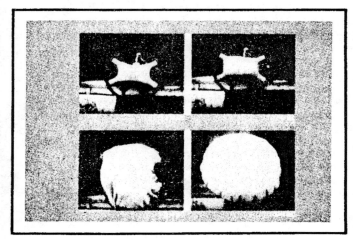

Figure 11

Toxicity

Another special concern regarding the hazards of an air cushion system is the minimization of any possibility that the chemical gas generators in the driver and passenger systems will produce gaseous, toxic products, which would be harmful to the vehicle occupants.

Analysis of the gaseous output of an augmented Argon inflator, indicates that no problems to date were encountered with toxic products interacting with the vehicle occupants. However, as with other aspects of the system, careful studies are continuing to corroborate such analysis. Of particular interest are studies to determine if there might be synergistics effects of the various toxic products with reduced oxygen in the passenger's compartment. It is believed that, for the front seat system alone, this question will be favorably resolved, because of the dilution of the air cushion into the full passenger compartment.
Testing is currently in process, using small primates in a vehicle environment at the Industrial Bio-Test Laboratories, Inc., Northbrook, Illinois.

Noise

A much discussed result of the air cushion deployment is the attendant noise. In the opinion of many people, the risk of impact injury potential in higher speed accidents can be weighed against hearing injury from the sound of deployment. A "trade-off" of sorts.

Although significant studies have been made to reduce the noise pressure of the air cushion deployment, the effect on hearing injury to occupants on the verge of hearing loss, with head colds, aged, etc., is not known.

However, General Motors feels that the application of the variable inflation feature produces a significant step in reducing the frequency of maximum noise exposure. That is, in the more frequent accident cases those in low speed ranges, the benefit of the low level or slower speed inflation inherently reduces the level of noise

Figure 12 shows the damage risk criteria for A and B duration noise pressures, as well as measured pressure for both the high level and low level passenger system deployment modes of the GM system. Under analysis, the air cushion deployment noise, exhibits two distinct characteristics. The first is a single smooth impulse associated with the expansion of the gas and the simple inflation of the bag. This is identified as the "A" pulse and is compared to the low frequency "A" duration criterion. The second characteristic is the high frequency oscillations related to turbulance, bag unfolding, reflections and resonances in the vehicle, and is compared to the high frequency "B" duration criterion.

The solid lines are the risk criteria for exposure to impulse noise, as shown in the Bolt Beranek and Newman Report No. 2020, prepared under Department of Transportation Contract No. DOT-HS-006-1-006. The dotted line for the "A" duration limit is a conservative limit GM is currently considering, because of the unknown effect of loudness on injury. You will note that the measured noise pressures for both the high and low deployment modes are below the risk criteria.

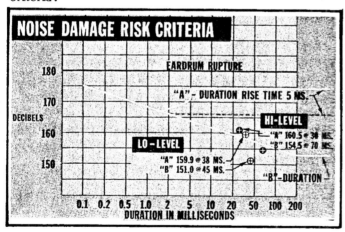

Figure 12

The noise levels of front seat systems have been substantially reduced from earlier systems, and in particular, the more frequent low level accidents are favored by "variable inflation." Since the volume of gas generated for the driver air cushion is less than 3 cubic feet, the accompanying noise level is also low. The combined noise of both systems would be substantially the same as that measured for the passenger's system.

Sensor Development

The basic accident sensing arrangement for the GM air cushion system was described in paper entitled, "Air Cushion Restraint Systems Development and Vehicle Application (720407)." The following expands on that system.

The passenger's compartment-mounted deceleration sensing arrangement (Figure 13) consists of three detectors or inertia switches that provide two levels of sensing. The first is calibrated to a low level collision, approximately 12 miles per hour barrier equivalent. Two of the switches are in series with the low mode of the passenger's system to increase the protection against inadvertent deployment. The second level is calibrated to fire in 18 mph plus barrier equivalent accidents and initiates the passenger's system high level inflator component, as well as the driver's system. Note that the 18 mph level switch is in series with one of the low level switches, thereby providing increased protection against inadvertent oscillation. The bumper sensing arrangement fires only the passenger's low mode inflation and, as previously mentioned, provides additional deployment time. This device also uses the two component series concept for increased reliability (Figure 14).

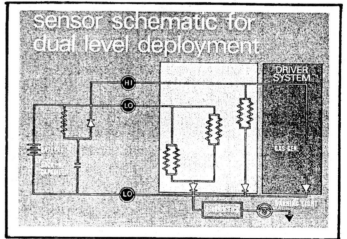

Figure 13

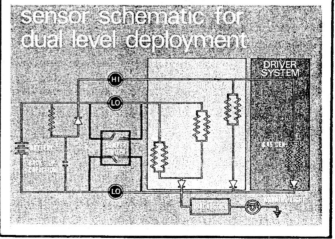

Figure 14

Electronic means for diagnosing electrical continuity and inflator pressure is included in the passenger's compartment-mounted package (Figures 15 & 16). System readiness or malfunction is signaled to the driver with a red light in the instrument cluster.

In addition, a crash recorder is included that will indicate:

1. If a crash occurs before system deployment.

2. If the crash was below the deployment threshold.

3. If the crash was above an accident severity equivalent to a 30 mph barrier collision.

Should the driver ignore a malfunction indication, a "failure to maintain" recorder is also included in this assembly that records the length of time the ignition is on with a "failed" system light.

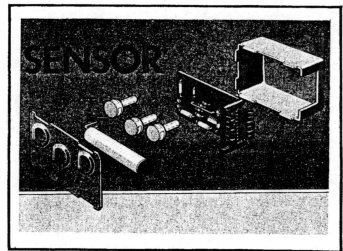

Figure 15

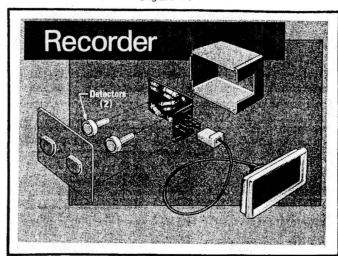

Figure 16

Reliability

A full assessment of reliability must consider inherent or designed in reliability, as well as manufacturing reliability. Inherent reliability seeks to

become final on-the-road reliability, but only with adequate manufacturing processes and control procedures.

That is, total reliability includes both design adequacy (inherent reliability) and build integrity (manufacturing reliability). The reference paper, J.B. Hopkins, et al., "Development of Anticipatory Automotive Crash Sensors," National Highway Safety Administration Report No. DOT-TSC-NHTSA-71-3, July 1972," stated that reliability is primarily among the considerations for public acceptability of passive restraint systems·

> "The ultimate success or failure of a dynamic passive restraint system will probably be determined by considerations relating to reliability."

The number of "on command" firing tests, and the number of road miles in the many variations of the auto driving environment, necessary to demonstrate empirically, high levels of reliability, cannot be realistically accomplished in the time remaining prior to production. Until such data can be accumulated, GM has chose to apply controls in the design, development and production stages of its program that are aimed at best controlling the known variables.

In the design state, reliability is studied utilizing, among other techniques, design check lists and failure mode analyses.

In the development stage, test program techniques are utilized, such as the following:

1. Static testing on both components and systems.

2. Dynamic sled testing including dummies, cadavers, baboons, and human volunteers.

3. Barrier testing consisting of a complete system installed in a vehicle with dummy occupant.

4. Environmental testing on both components and systems.

5. Road testing consisting of a complete system installed in a vehicle with an experienced driver.

6. Field testing consisting of a complete system installed in a vehicle with an inexperienced driver.

For the production stage, GM is currently planning failure prevention analysis, and process and inspection controls.

The purpose of the failure prevention analysis is to consider the effects of potential failure conditions, critical and non-critical to the systems performance in order to incorporate process and inspection controls necessary to minimize the possibility of defects during manufacturing caused by man, machines, or materials.

In manufacturing a part with critical specifications, process controls must be implemented that are automated as far as possible, including in-process inspection and/or test operations with monitoring during the production run.

Lot control inspection is planned and consists of inspections and tests of various product characteristics at specific points in manufacturing, in an effort to control the reliability levels for each of the components, sub-assembly, and shipping assembly in the system for each manufactured lot.

Specifically related to the reliability of the crash sensor, the following facts are of interest, since this device is at the base of two major reliability considerations:

1. The probability of normal operation in a crash situation.

2. The probability of inadvertent operation in a non-crash situation.

Sensor Testing

Through February 1972, the Delco Electronics crash sensor has been durability tested at General Motors Milford and Desert Proving Grounds, and has undergone static exposure testing at the Florida Coastal Test Field. Although not yet complete, a similar evaluation program is in process on the new bumper sensing system. The crash sensor is installed in cabs around the country and has accumulated over two and one-half million miles, and while this mileage is just a small fraction of that necessary to demonstrate that good reliability has been achieved, there have been no inadvertent actuations to date. There have been five real crashes to date, all at speeds below the actuation threshold, and there were no sensor actuations under these conditions. (There were no accidents over 10 mph.)

Prior to field durability testing, sample units were extensively tested in Delco's laboratory under the following environments:

- Random vibration.
- High and low temperature.
- Temperature cycling.
- Humidity.
- Dust.
- Acceleration.

The sensor road shock test program at the Milford Proving Ground consists of 500 miles of driving over six test roads (often with drained shock absorbers to further emphasize the adverse acceleration levels). The rough roads are called:

1. Belgian Block Road.
2. Five-Inch Curb.
3. Hop Road.
4. Tramp Road.
5. Square Block Road.
6. Messoit Bump of Hyne Road.

These test roads provide a good cross-sectional representation of severe rough roads (Figure 17), including changes to the road surface with various weather conditions.

Through February 1972, there have been seven vehicles fully equipped with GM's most recent system configuration that have undergone this qualification testing with no inadvertent actuations nor deployment. Additional testing is planned, as it must be realized that all of these tests constitute only a minute sampling of the billions of actual road miles accumulated each year under all types of driving conditions. Again, it must be stressed that no road testing has been accomplished on the new bumper sensors.

You have read the considerations that are being incorporated into the GM design, development, and production planning, in order to obtain high performance and reliability.

General Motors believes that it is approaching a workable front seat air cushion system, and that the dual level sensing and variable inflation approach offers a significant advance over the simpler single level systems. While the dummy studies described herein require corroboration with human volunteers and baboons, GM is continuing concurrently with the plans for a field and manufacturing test program that will place approximately 1,000 1973 model vehicles on the road this fall.

Figure 17

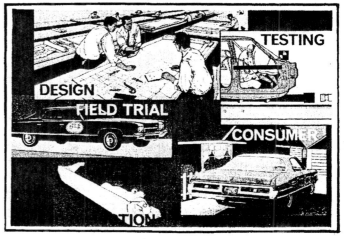

The Anatomy of an Inflator for Air Cushion Occupant Systems

Ronald C. Lawwill
Energy Systems Division
Olin Corporation

There is a great deal of controversy over the Federal Government's introduction of Motor Vehicle Safety Standard No. 208 which requires some form of passive restraint system to be incorporated in passenger vehicles built on or after August 15, 1975.

In order to meet the stringent requirements imposed by MVSS-208, the inflatable occupant restraint system containing a pyrotechnically augmented inflator appears to be the most promising approach to the solution. The power source, or inflator, for this system has been viewed with anxiety and misunderstanding by many people — both in and out of the automotive industry. A basic augmented inflator and its components are discussed herein to provide the knowledge necessary to set aside the questions and anxieties of persons not already familiar with those details.

As one might imagine, the design goals for an air cushion inflator were that it occupy essentially no space because of the already crowded instrument panel area, provide the gas necessary to fill an air cushion as large as the passenger compartment, produce a noise level like the whisper of a summer breeze, and reaction products with the odor and toxicity of orange blossoms. I might begin by admitting that inflators have failed to meet these goals.

THE INFLATOR IN GENERAL

Figure 1 illustrates the basic exploded view of an inflator design which has been coupled with complete passenger restraint packages to provide satisfactory frontal barrier crash protection for front passenger positions.

Figure 1 is an illustration of a dual level inflator design, as seen in the propellant chamber cross-section, where the augmentation can be staged depending upon the severity of the crash. The two squibs ignite separately and independently the two propellant grains shown so that the augmentation can be programmed to provide the air cushion fill rate demanded by the speed of impact. In a low level crash, only the squib igniting the Primary propellant grain is actuated. In a higher speed crash, additional augmentation assist is provided by the second squib and Secondary propellant grain. The signal for the second squib is typically provided by a separate sensor detecting a higher deceleration level with the delay between the first and second signals dependent upon the sensitivity of the two sensors and the severity of the crash. A single level inflator operates similarly with a single squib and propellant grain responding to the full range of pre-selected deceleration values.

ABSTRACT

Many anxieties have been generated concerning the introduction of an air cushion passive restraint system in automobiles for a variety of reasons; but, one of the most prevalent anxieties stems from the common lack of understanding relative to the functioning characteristics and safety of the inflator which is to contain some quantity of pyrotechnic material — this inflator becoming a permanent fixture in the passenger compartment of the vehicle.

This presentation discusses in understandable detail the physical and functional characteristics of the basic pyrotechnically augmented inflator and how each of its major components contribute to form a controllable system which can be varied to meet various systems demands. It further discusses a proposed marriage of inflator and sensor in a system totally restricted from functioning by any other mode than the properly prescribed crash energy signature.

Whether dealing with a single or dual level inflator, the basic functional responses are the same once the propellant has been ignited. Referring again to Figure 1, the ignited propellant grain provides sufficient chamber pressure to burst a machined rupture disc in the propellant chamber and sufficient force to propel the push rod assembly to result in opening of a machined burst disc in the pressure vessel. With this accomplished, the stored gas in the pressure vessel is released with the heat from the burning propellant greatly expanding the stored gas volume thereby "augmenting" the gas flow for air cushion fill.

While each of the major parts of the inflator will be detailed later in this report, the pressure vessel or bottle will be described here. At Olin Corporation, a one-piece aluminum impact extrusion is utilized with the configuration as shown in Figure 2. The output end of the vessel has a machined-in rupture disc which releases the inflation gases upon activation of the system and performs as an over-pressure safety device as well. The opposite end of the vessel is threaded to accept a propellant chamber which also contains a machined-in rupture disc. When the bottle and propellant chamber are put together and welded in the flange area it is easily observable that they form one of the most important attributes of a vessel for use in occupant restraint systems — that of a metallurgically sealed container for the stored gas pressure dependent only upon a single seal. Since leakage of stored gas reduces the inflation level possible from the inflator, it is imperative that the seal integrity of the storage vessel not be interrupted, and that the number of closures be minimized to reduce the number of potential leak paths and also reduce fabrication and inspection problems.

There are two additional features that are a part of the inflator package which do not contribute to the inflation characteristics of the inflator, but provide safety features to permit the safe handling of an inflator. Referring to Figure 3, it can be observed that a "free-volume" shipping cap is utilized so that, should an inflator ignite in the shipping configuration, sufficient volume is added so that all gases can be self-contained while maintaining an internal pressure well below the yield point of the vessel. The cap is designed with bleed down ports which then exhaust the gases without thrust within a short period of time. Again from Figure 3, it can be seen that the transition piece has equally spaced circumferential holes which provide an anti-reaction feature should the inflator be ignited without the presence of the shipping cap.

SQUIBS

Squib, as a term, can be easily replaced with initiator, electric match, actuator, etc.; but, regardless of what its terminology, its function is extremely critical to the successful completion of an inflator's task. Therefore, in going through the major parts of an inflator, it is fitting that one should begin with the squib.

The squib selected must provide closely controlled characteristics to respond properly in the automotive vehicle environment. It must have a low enough activation energy threshold to guarantee performance from the limited power source of the normal passenger vehicle and an activation energy threshold high enough to permit safe handling during assembly operations and protection from transient and desired monitoring signals generated by the vehicle electrical and diagnostic systems.

Electrostatic sensitivity is another important consideration in squib design and selection. Since the unit is designed to activate upon the receipt of a preselected relatively low energy signal, special design considerations are necessary to ensure that the energy capable of being stored by handling personnel or vehicle occupants — which is appreciable — will not cause activation. Surprisingly enough, the human body is a capacitor of up to 500 picofarads capable of storing as much as 25,000 volts. Fortunately, it is also resistive, with the minimum resistance approximately 5000 ohms, so that any discharge dissipates proportionally between the resistance of the human body and the resistance of the discharge path.

Another often overlooked characteristic important in the selection of a squib is its ability to destroy its own electrical activation circuit immediately upon activation. Since each squib will represent only one of many activation circuits in the finalized passive restraint vehicle, its failure to open following activation would potentially jeopardize the success of the remaining activation circuits — since all must be exposed to the same power source.

Figure 4 of this report illustrates the typical squib schematic responsive to the characteristics demanded as detailed above, and lists typical requirements imposed by the vehicle activation circuits and vehicle environment. It is regretted that the schematic shown cannot detail the design configuration of a properly engineered squib, but detail beyond that illustrated infringes upon the proprietary elements of the design.

PROPELLANT

Of course, the basic element of an "augmented" inflator is the propellant. The principal advantage of the propellant augmented inflator is size because the propellant provides hot gas to directly oppose the refrigerative effects, and accompanying reduction in gas volume, caused by the normal bleed-down of a relatively small volume, high pressure tank into a significantly larger "Air Cushion" volume. In a typical augmented inflator approximately 75% of the inflation energy is resultant from the propellant and the remaining 25% from the stored gas.

Another advantage offered by a propellant augmented inflator is tailored air cushion fill rates. While the stored gas inflator permits only moderate tailoring through the use of reasonably sophisticated variable nozzles, the augmented inflator has the same option plus numerous propellant grain configuration options to provide augmentation to meet a wide variety of vehicle needs. Augmented systems largely offset the variability associated with high and low temperature performance of stored gas systems because the propellant energy is essentially unaffected by temperature. In a stored gas system the inflation energy is greatly affected by temperature change due to significant variation in gas pressure with temperature change.

Although propellant provides several advantages to the air cushion inflator, the selection of a propellant for this application has not been an easy task. Although the passenger vehicle environment is extremely familiar to all of us, it posed new and very challenging requirements to the propellant field. The propellant/stored gas combination must produce reaction products which are non-toxic, the effluent gas temperatures must be low enough not to damage the air cushion material or inflict any discomfort to the protected occupant, and the long term storage characteristics must manifest chemical and physical stability during and following exposure to temperature cycling extremes of $-40^\circ F$ and $+220^\circ F$.

Fortunately, although the combination of requirements was unusually severe, it has been possible to use standard state-of-the-art materials in formulating a propellant composition which meets all of the aforementioned requirements. The selected materials also have years of history in manufacturing, storage, and use. Tailoring the resultant grains has been a relatively straightforward task based on Olin's background in propellant applications as complex as accurately controlled course correction rocket grains.

It is therefore well to conclude the propellant grain in a typical, well designed augmented inflator is environmentally and functionally stable, reliably reproducible, with acceptable toxicity and thermal output characteristics.

TOXICITY

In the discussion of an inflator, the gas toxicity is definitely a vital and essential design consideration, and therefore considered appropriate in an anatomical analysis.

Evaluating an overview of augmented inflators, the typical potential reaction products are listed in Table 1 of this report. Of those listed, only three occur in sufficient quantity in the inflator under discussion to be significant. These three are Carbon Monoxide (CO), Nitrogen Dioxide (NO_2), and Nitric Oxide (NO).

A detailed literature survey was conducted by the Olin Research Center Library on the toxicity effects

of CO, NO_2, and NO. This work was summarized in a report dated November 1, 1971, entitled "Toxicity Considerations for Air Cushion Inflation Gases" by Olin's Dr. T. F. McDonnell. Figure 5 and Table 2 of this report entitled "Toxicity of CO" and "Toxic Effects of NO and NO_2" respectively were extracted from Dr. McDonnell's report. A thorough review of all data led to the conclusion that the maximum allowable quantities of CO and NO_X within the air cushion should be limited to 3,200 ppm and 100 ppm, respectively. This conservative conclusion was based on one hour exposure of the occupants in a sealed vehicle assuming a dilution factor of 8:1 which relates favorably to the ratio of the passenger compartment to air cushion volume.

For comparison between the values suggested as acceptable maximums and those actually observed from air cushion inflator tests, Table 3 shows actual test results of the last seven toxicity measurements made on the Olin augmented inflator. As can be observed, the values are significantly below the suggested maximum values.

NOISE

Noise, in the same vein as toxicity, is appropriately a part of the inflator discussion. There appears to be reasonable agreement that the noise produced by the air cushion inflator is a function of pressure and time, but for a given restraint system where air cushion size and inflation time requirements are nonvariable, very little is known about how to effect a true reduction in the resultant sound level through inflator design.

Various sound evaluation techniques have been pursued such as microphone variations, preamplifier and amplifier improvements, selective cycle filtering, and variation in calculation techniques. Although different decibel values are achieved by the various techniques, it is doubtful whether any real sound reduction methods have been discovered — beyond alteration of the pressure/time characteristics of the inflation system, which is clearly a "need" type function.

Based on published, as well as unpublished, data, the noise levels commonly being obtained for augmented inflators range from 157 to 167 db, depending on the frequencies being evaluated as outlined in the Bolt, Beranek and Newman Report No. 2020, entitled "Noise and Inflatable Restraint System." Although this noise level is higher than one would hope, very little is known concerning the effects on the human hearing of one-time noise pulse exposure in this range. It is of interest to note, however, that the April 1972 issue of Consumer's Report had an article entitled "Auto Safety: What Lies Ahead" in which it was stated that toys are not banned from the market until their noise level exceeds 158 db — with this based on repeated exposure. It was concluded in the same Consumer's Report article that "In the judgment of CU's consultants, the rare and near instantaneous exposure to the sound of one or several air bags deploying in a car would not be likely to impair hearing permanently."

THE INTEGRATED SENSOR/INFLATOR

In the inflator discussed thus far, the actuation energy is supplied by the electrical system of the vehicle and triggered by a crash sensor based on the rate of change of acceleration and/or a displacement switch. This approach, although well within the current state-of-the-art, is complex, expensive, and subject to several problems that must be solved through intricate design. Work has been done on the incorporation of mechanical prestressed snapper into an integrated sensor/percussion initiated inflator design. It is felt this approach offers certain potential advantages, and is therefor worthy of mention in this paper.

The mechanical snapper operates somewhat like the toy "cricket" in that a spring blade, when subjected to certain prescribed energy, passes through a point of equilibrium and snaps, thereby delivering sufficient energy to initiate a percussion primer on the inflator. The advantages are detailed below in hopes that interest will be generated to result in active evaluation of this approach to the occupant restraint inflator/sensor system.

1 - Energy itself is sensed rather than a function of energy. The system can only be actuated by the energy of the crash event, with the actuation energy required adjustable to match vehicle dynamics.

2 - No electrical wiring harnesses are necessary.

3 - There are no RF, EMI, or electrostatic problems with which to contend as is the case with the use of electroexplosive devices.

4 - There are no electrical components or solder connections thus reducing the number of potential failure modes and thereby increasing reliability.

5 - There is a significant overall cost reduction.

6 - An excellent modular approach to the entire air cushion system is provided.

The reliability of the integrated sensor/inflator package suggested herein is believed to be at least equivalent, if not actually superior to, the currently used sensor/electrical circuit/inflator system. The inflator would not be significantly different in function when changing from the electrically actuated squib to the percussion primer, and millions of percussion primers are used every year in center fire ammunition with remarkable reliability. The mechanical snapper, although new to the air cushion inflator, has been successfully used in a wide variety of applications which have proven the repeatability and reliability of its performance characteristics.

In conclusion, the augmented inflator combines basic state-of-the-art components resulting in highly controllable and reliably repeatable performance. The design and materials utilized have been based on previously qualified man-rated systems to provide a satisfactory level of guaranteed protection to the occupants of passenger vehicles. It is further the opinion of the author that utilization of the integrated sensor/inflator concept discussed herein would enhance the system performance by increasing potential reliability while reducing cost which is more palatable to the passenger vehicle consumer.

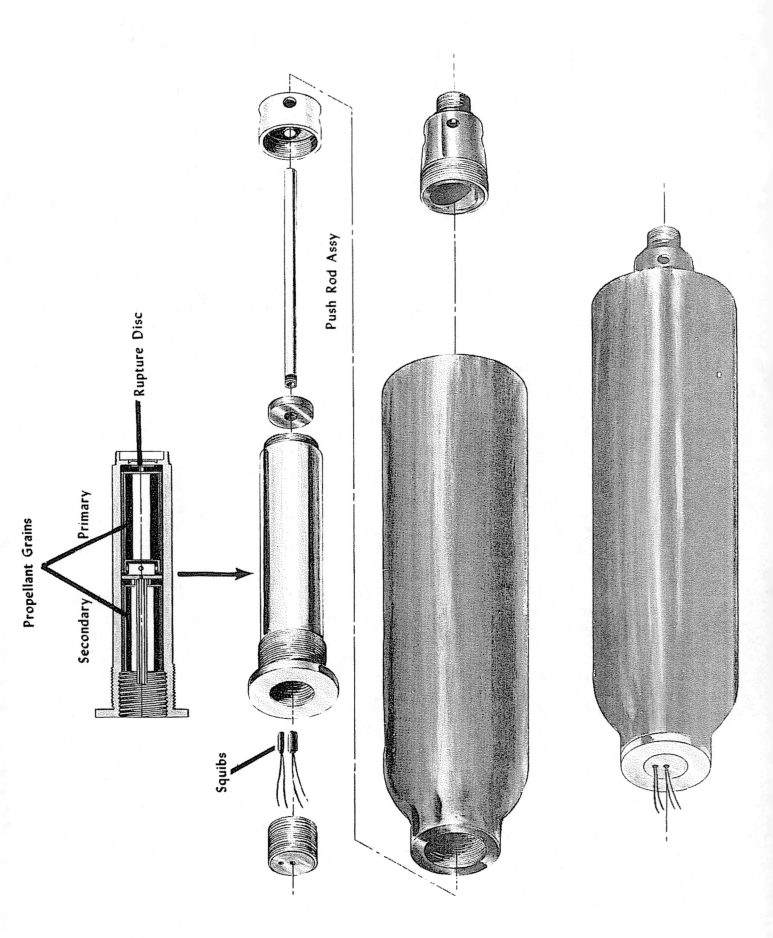

Propellant Grains

Rupture Disc

Primary

Secondary

Push Rod Assy

Squibs

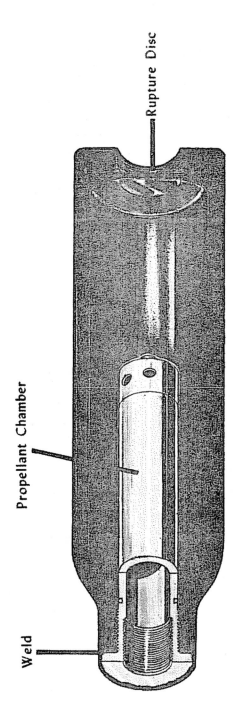

Rupture Disc

Propellant Chamber

Weld

Fig. 2

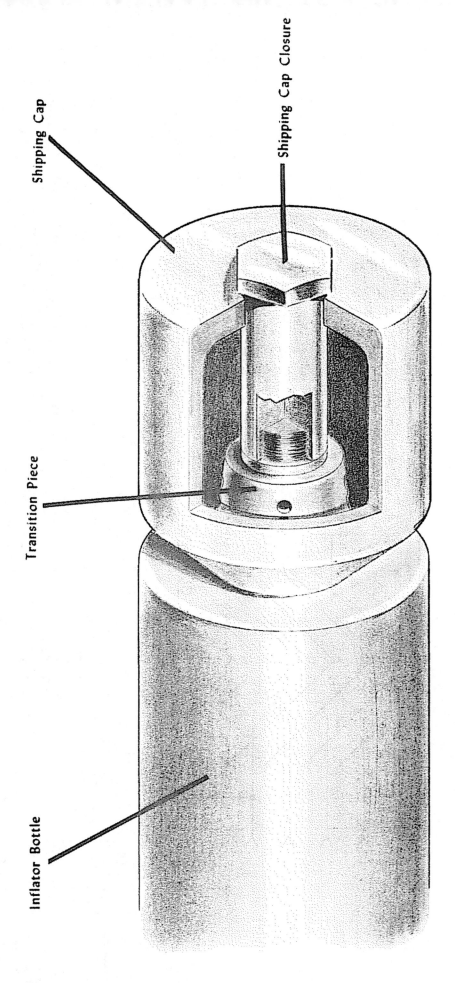

Shipping Cap

Shipping Cap Closure

Transition Piece

Inflator Bottle

1096

Typical Requirements

1. Bridgewire Resistance, R_1 and R_2, 4.5 ± 0.5 ohms.

2. No Fire Current – 5 minutes, 0.100 amperes.

3. Functioning Time at 0.565 amperes, 2.0 milliseconds maximum.

4. Maximum Firing Energy in 2.0 milliseconds or less, 35,000 ergs.

5. Static Discharge Test from pin to case through R_3, R_4, R_5, or R_6 without firing –

   500 picofarad capacitor

   25,000 volts

   5,000 ohms series resistance

Squib Schematic

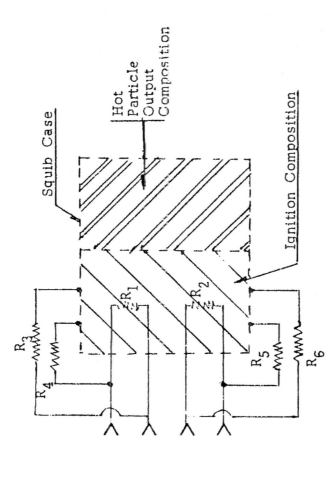

Squib Case

Hot Particle Output Composition

Ignition Composition

Bridge Wire Input

Bridge Wire Input

FIG. 4

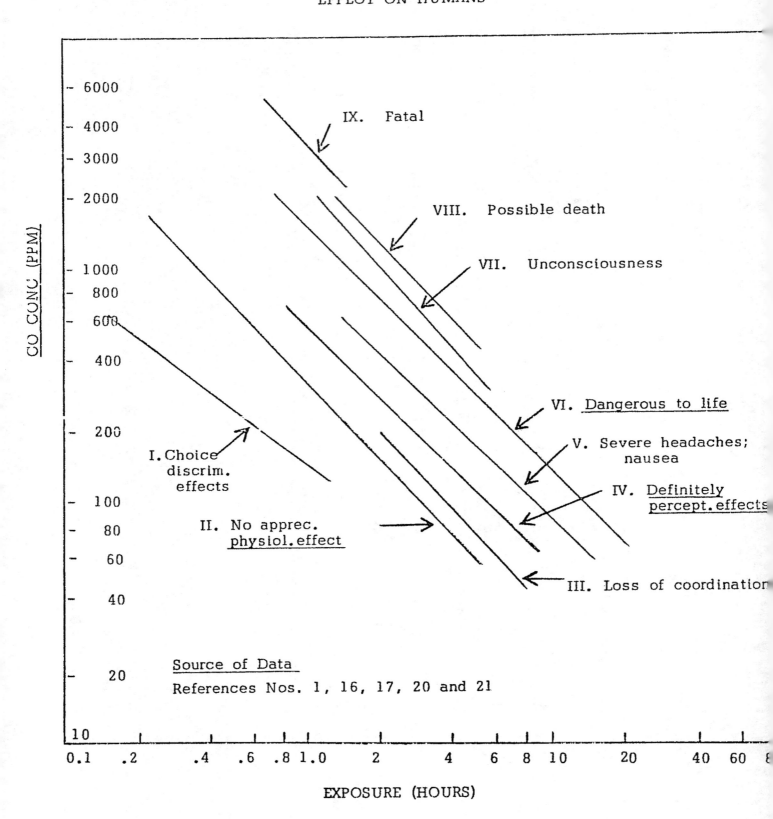

TOXICITY OF CO
TIME/CONCENTRATION
vs.
EFFECT ON HUMANS

FIG. 5

TABLE 1

Gas	Chemical Formula	Gas	Chemical Formula
Carbon Monoxide	CO	Hydrogen Chloride	HCl
Carbon Dioxide	CO_2	Hydrocarbons	
Chlorine	Cl_2	Ammonia	NH_3
Nitrogen Dioxide	NO_2	Water	H_2O
Nitric Oxide	NO	Oxygen	O_2
Nitrous Oxide	N_2O	Potassium Chloride	KCl
Sulphur Oxide	SO	Phosgene	$COCl_2$
Sulphur Dioxide	SO_2	Hydrogen Sulfide	H_2S
Sodium Chloride	$NaCl$	Nitrogen	N_2
Hydrogen	H_2		

TOXIC EFFECTS OF NO & NO$_2$

Subject	Oxide	Conc. ppm	Exposure Condits	Toxic Effects and Notes
Man	NO$_2$	5	10 min.	Respiratory trouble delayed 30 min.; subjects recovered but recovery time not given.
Rats	"	2	Natural Lifetime	Some respiratory system damage; a pre-emphysema-like condition.
Var. Animals	"	5	Contin.	Subjects depressed, low weight gain but low mortality.
Dogs	"	65 28	15 min. 60 min.	These are the "min. effect" conc./time exposures with pulmonary edema as most pronounced symptoms.
Man & Lower Animals	"	0.5-1.0 3.5 3.0	Long,contin. Transient Up to 1 hr.	These are thresholds for no appreciable effects or observed damage.
Man	"	35 10	15 min. 60 min.	These are proposed "Emergency Expos. Limits" for single exposure and no appreciable effect.
Man	"	5 5	8 hr/day & 5 day/week	Army investigation to confirm ACGIH's 5 ppm TLV.
Man	"	5 10-25 100-150	" " 30-60 min.	TLV given in 1963 edition of Sax's book. MAC given in earlier 1951 edition of Sax Dangerous to life
Man	NO	25	8 hr/day & 5 day/week	MAC given in 1951 edition of Sax
Mice	NO " NO$_2$ NO$_2$ NO$_2$	2500 2500 1000 100 30	6-7 min. 12 min. 19 min. 318 min. Extended	Narcosis Death Death Death No deaths occurred
Man	NO$_x$	<10 >25	8 hr/day & 5 day/wk.	Ave. conc. in ship's hold during welding A few air samples showed this conc.

1100

Test No.	NO ppm	NO_2 ppm	CO ppm
2150	85	0	600
2151	77	0	700
2177	34	0	2500
2179	32	0	1650
2222	40	0	700
2223	80	0	1280
2226	47	0	1780

720976

Safety Performance of Shaped Steering Assembly Airbag

Lawrence M. Patrick, Gerald W. Nyquist,*
and Kenneth R. Trosien
Biomechanics Research Center, Wayne State University

VEHICLE OCCUPANT PROTECTION is achieved when the occupant is decelerated during a collision without exceeding his injury tolerance level. Current safety technology indicates that the best way to achieve protection is by restraining the occupant within the vehicle. Controlled occupant deceleration is provided by tailoring the restraint system and/or by utilizing part of the vehicle deceleration to decelerate the occupant (ride down). The major restraint systems currently under development are the airbag and the harness. A detailed description of an airbag system for installation on the steering assembly is presented, together with an analysis of its performance during full-scale simulated and destructive collisions. A comparison of the airbag performance with that of harness systems under identical impact conditions is included.

More effort appears to be directed toward the development of the passenger airbag system than the driver airbag system, as evidenced by the greater number of publications pertaining to the passenger system. Development of the passenger system is not complicated by the steering assembly as is the driver system. Statistics show that the passenger is more vulnerable to injury than the driver, which may be a contributing factor to the greater emphasis placed on the passenger airbag system.

While the passenger is more vulnerable to injury than the driver in a given impact condition, there are a greater number of driver exposures since there is always a driver in the moving vehicle. Design and installation of the driver airbag system are complicated by the steering assembly. Concepts that have been suggested include an airbag system mounted in the header, above the steering wheel, with the airbag deploying downward between the steering wheel and the driver. Another proposed system has the airbag installed in the instrument panel, deploying over the steering wheel and down between the driver and the wheel. These two systems do not appear to be feasible, since the distance the bag must travel is large, and thus a high velocity is required for deployment before the driver moves forward too great a distance. Installation of the airbag system in the steering wheel appears to be the most feasible approach to solving the problem.

A steering wheel-mounted airbag system must be light in weight and small in package size. A heavy system mounted on the steering wheel increases the force on the driver during the impact due to the increased inertia of the moving part of the energy-absorbing steering assembly. The size of the nondeployed system mounted on the steering wheel is also limited by the necessity for the driver to see the instruments and, in some cases, through the steering wheel in order to drive the car. Any steering assembly installation must not interfere with the normal driving task.

In general, the requirements of a successful driver airbag

Note: This paper is based upon the results of DOT Contract FH-11-7607.

*Currently with Research Laboratories, General Motors Corp.

_____ABSTRACT

This paper discusses a program wherein studies were made of forward force simulations of crashes and destructive barrier crashes using a shaped steering assembly airbag. It was shown that the airbag offered the best protection when compared with the performance of lap belts and lap and shoulder belt combinations. This shaped airbag deploys between the abdomen and the steering wheel and between the head and the steering wheel, thus providing protection of these two important areas.

system include deceleration of the head and torso without injury. In a collision without a restraint system, the driver moves forward relative to the vehicle in his precollision posture until part of his body strikes the vehicle interior. In the usual case, the knees strike the instrument panel, the abdomen hits the lower steering wheel rim, the chest often makes contact with the steering wheel hub, and the face or neck strikes the upper steering wheel rim. The function of the steering assembly airbag is to prevent injurious contact of the head and torso with the steering wheel by applying a distributed force over these body areas at a subinjury level. Protection of the knees is accomplished by a crushable instrument panel or, in some cases, a knee airbag. The system described in this paper accomplishes the above requirements by an airbag installed in the steering wheel and a custom instrument panel to provide a controlled crush at subinjury forces to the knees.

BACKGROUND

As a result of the failure of the general public to take advantage of the safety features available (safety belts) and the reluctance of the government to legislate their mandatory use, the present trend is toward a passive restraint system. In this respect, the energy-absorbing steering column is a passive restraint system, as are the instrument panel and windshield to a lesser extent, since they provide protection with no action required on the part of the occupant. These latter passive devices are not nearly as effective as the harness or the inflatable occupant restraint systems, both of which provide much greater stopping distances.

A brief review of the advantages and disadvantages of the airbag system puts the problem in correct perspective. The advantages of the airbag system include:

1. It is always available when needed (passive restraint).
2. It provides optimum load distribution.
3. It usually results in minimum injury parameters.
4. It does not restrict occupant mobility during normal vehicle operation.

The disadvantages of the airbag system are:

1. It is more complex than other restraint systems.
2. There is little highway experience with it.
3. Current designs provide only limited protection in other than forward-force collisions.
4. It does not keep the driver in position to control the vehicle in subinflation collisions.
5. There is a possibility of inadvertent deployment (reliability, sensor design).
6. There is a possibility of nondeployment in a collision (reliability).
7. There is danger of an explosion of the energy source (reliability).
8. Current designs have a high noise level (not expected to be a serious problem).
9. It is considerably more expensive than other restraint systems.

The advantages and disadvantages of the airbag system are based on comparisons with the harness, which is the only other viable system currently available. Probably the most potent argument in favor of the airbag is its availability when needed. No matter how good other systems are, if they are not used they will not provide protection. Since harness systems are only used from 5% (lap and shoulder) to 35% (lap) of the time (1)*, it is obvious that no matter how good they are, they cannot protect a major portion of the driving public. The harness system is designed to distribute the decelerating force over the occupant's body in those areas that can withstand the greatest force with minimum injury. Even with a well-designed harness, however, the distribution is not as good as a well-designed airbag system. Thus, the airbag system can, theoretically, apply greater forces than other restraints to the human body without exceeding the tolerance of the occupant. Stated in another way, the injury criteria as measured from accelerometers mounted in the dummy occupants are usually lower with the airbag than the currently available harness systems.

The complexity of the airbag system contributes to the higher cost, and also to a concern for reliability. The complexity is no greater than that of many other systems used in the automobile; consequently, there is no reason to believe that the reliability and cost cannot be brought into a reasonable range. In the rare case of nondeployment during a major collision, the occupants are no worse off than if they did not have the airbag installed. Even in the extremely unlikely condition of inadvertent deployment, the event may not cause a catastrophic collision, and the incidence of such inadvertent deployment should be as low as, or lower than, other operating systems affecting safety in the vehicle.

The airbag system provides protection primarily in forward-force collisions in its present state of development. Also, it does not provide protection in subinflation collisions. Both of these items favor the harness system if it is utilized.

The biggest disadvantage of the airbag system at the present time is a lack of highway experience. It is imperative that the system be installed in a large sample of automobiles to provide some experience in its performance during normal highway use before it is installed in all vehicles. If the highway experience shows it to be as beneficial as laboratory experiments indicate, and if the harness systems, either passive or mandatory active, are not used, there appears to be little doubt that the airbag system will be grossly beneficial in reducing highway fatalities and injuries.

SYSTEM DESCRIPTION

The major airbag system components are:

1. Crash sensor.
2. Initiator.
4. Inflator or energy source.
4. Manifold or diffuser.
5. Airbag.
6. Housing.

---

*Numbers in parentheses designate References at end of paper.

Each of the major components has been treated in detail by other authors, so their function will be described only briefly as they pertain to the system under consideration.

The sensor is a device that determines that a significant collision has occurred. It must sense the collision in 10-15 ms, with the lower value preferred, and must differentiate between a significant collision and a severe bump or minor collision below the designed threshold limit. Accelerations of 50 g magnitude can be achieved for very short times by slamming the hood or door of the car. Obviously, the airbag must not deploy under these conditions. Therefore, the purpose of the sensor is to determine that a significant collision has occurred and to initiate the airbag inflation cycle. A seismic spring-mass type of crash sensor was used in this research. The sensor, provided by the Eaton Corp., is shown in Fig. 1.

The initiator is usually a squib or other explosive device. The sensor triggers the initiator in most cases by supplying an electrical pulse to the squib. Other types of initiators, used in stored-gas inflators, include a stab type for rupturing a highly stressed metal diaphragm and a prick type for rupturing a highly stressed glass diaphragm. In a gas generator inflator such as the Rocket Research Corp. unit used in this research, the initiator ignites the energy source (propellent).

Three types of inflators are in widespread use. The stored-gas system was the first type to be used. Gas is stored at a pressure of approximately 3500 psi and released through a ruptured diaphragm to inflate the bag. Gas generators of the type used in this system obtain the gas for inflating the bag by burning a propellent. A third inflator, the hybrid system, is a combination of the above two in which less stored gas is required than for a pure stored-gas system with additional gas obtained by burning a propellent.

The gas-generator type of inflator used in this program consists of 18 individual generator units, each with its own initiator and propellent supply. All of the individual generator elements are mounted on a common 5 in diameter nylon base. Each generator consists of a solid-propellent combustion chamber and a liquid Freon coolant chamber located immediately downstream from the combustion chamber. The coolant is contained by means of two burst discs—one located over

the nozzle exit of the combustion chamber and one over the outlet of the gas generator element. The purpose of the Freon is to cool the gases to an acceptable level. Vaporization of the Freon cools the hot gas from the generator, and also increases the volume of the mix to assist in the bag inflation. Details regarding the construction and performance of the inflator are available from the manufacturer (2).

The function of the manifold, or diffuser, is to prevent the high-velocity gas issuing out of the gas source from tearing or perforating the bag. Distribution is often accomplished by a tubular manifold with a series of slots or holes to provide a large number of openings through which the high-pressure gas escapes at a lower pressure. The manifold or diffuser function is accomplished in the system under consideration by the multiplicity of the generators, and also a deflector assembly against which the gases from the generators impinge. The deflector directs the flow radially outward from the steering wheel hub. This tends to inflate the bag more efficiently and also reduces the thrust (the inflator is essentially a rocket). Fig. 2 is a composite photograph showing three views of the gas generator with the deflector assembly included in one view.

The airbag is the envelope containing the gas from the inflator and provides an interface between the occupant and the pressurized gas. Pressure in the bag varies from about 1.5 psi for the large passenger bags to values as high as 6-7 psi for the smaller driver airbag. In addition to the driver and passenger airbag, there is sometimes a knee bag for the driver or the passenger that provides a cushion for the knee. The knee bag usually operates at a higher pressure and can operate up to 30 psi. Since the airbag is of primary interest in this paper, it will be discussed in greater detail in the following section.

The housing for the airbag system must contain the system and protect it from mechanical damage during long periods of vehicle use with the bag in its packed configuration. In addition to the protective function, it must provide an opening through which the bag can deploy upon command. The opening is generally created by pressure buildup in the confined inflating bag. The pressure typically ruptures tear lines (built-in weak seams) in the housing and enables a door to hinge

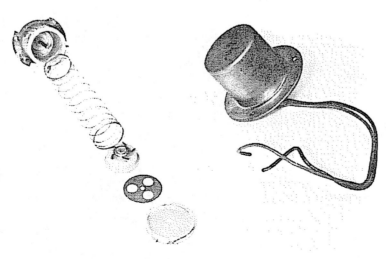

Fig. 1 - Eaton Corp. crash sensor

open, or a cover to split open in a manner similar to the blossoming of a flower bud. A typical cover design used in this research is shown in Fig. 3, which illustrates the geometry of a complete inflator-bag-cover configuration mounted on a steering wheel.

AIRBAG DESIGN

The material available for fabrication of the airbags is generally limited to films and fabrics. For this program, fabric was considered to be superior to the films on the basis of strength, flexibility, low weight, small packed volume, minimal aging affect, small variation of physical characteristics with temperature change, and variable porosity. Nylon fabrics of various weave patterns are widely used for airbags and were chosen for this program. Neoprene or other flexible coatings are used on the fabric if a nonporous material is desired. A good discussion of airbag materials is presented in Ref. 3. The final airbag design in this program utilized J. P. Stevens No. 34219 uncoated nylon fabric. Some material properties are listed in Table 1.

The airbag, when inflated, is essentially an airspring. If the bag has no damping, the object impacting it will rebound with the same speed it had on impact. While there is always some natural damping, it is desirable to control the damping to provide the desired rebound condition. The goal for this program was to limit the rebound to approximately 30% of the impact speed. Damping can be accomplished by valves or by venting through a porous material. Several different valves were designed and tried with varying degrees of success. At the same time, a program was initiated to evaluate the porosity of material to permit a porous bag design without valves with the desired damping characteristics. An advantage of the porous bag over the use of most of the valves is that once the valves are opened, the bag deflates rapidly even though there is no force on it. With a porous bag design, it is possible to provide multiple-impact protection since the initial impact forces part of the air out of the bag, but once the impact is over, the remaining gas tends to stay in the bag, ready for a second impact. It should be pointed out that the bag still deflates rapidly enough so that it does not impede egress after a collision.

A detailed analysis of fabric porosity, introducing a quantity referred to as "flow porosity," was conducted under the program. The results of the analysis (4) can be used in designing airbags to achieve desired porosity and damping characteristics. Flow porosity is defined as the mass rate-of-flow of air at room temperature through a unit surface area of fabric for a

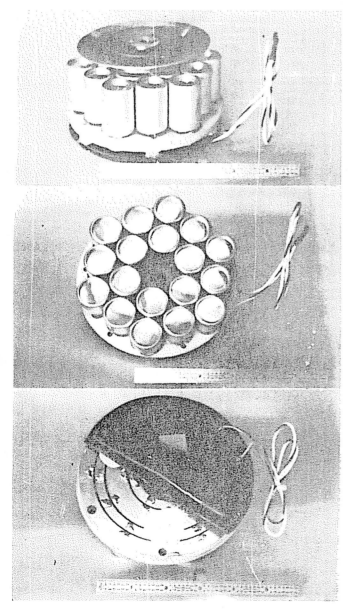

Fig. 2 - Rocket Research Corp. Inflator

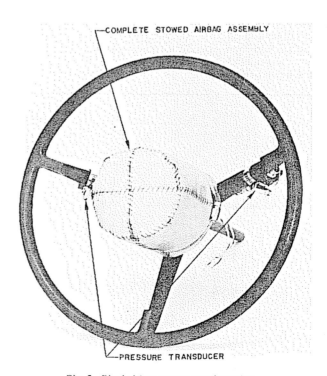

Fig. 3 - Final airbag system configuration

given upstream pressure with atmospheric pressure downstream. A plot of flow porosity versus pressure drop across the fabric is presented in Fig. 4 for the J. P. Stevens No. 34219 fabric used in the final airbag design.

Analytical and experimental investigations of bag shape indicated that a flat circular disc is preferable to a spheroidal shape, which is the easiest to attain. A spheroidal bag mounted on the steering wheel hits the driver high on the chest and is forced upward as the occupant moves forward, due to the inclination of the steering wheel, exposing the lower rim for contact by the abdomen. The flat circular disc design permits the bag to be imposed between the rim and the abdomen. This provides a cushion to keep the concentrated force of the rim from causing injury in the abdominal area.

An analysis of the size of the airbag required included consideration of the possibility of keeping the bag within the diameter of the wheel so the driver could maintain control of the car even after the bag inflated. This was found to be impractical, and the final inflated bag size is 19 in in diameter, while the steering wheel is only 16.2 in in diameter.

Any bag design made from a flat pattern tends to take the form of a sphere. The desired shape was obtained by attach-

ing two concentric cylindrical struts to the top and bottom fabric circles, as shown in Fig. 5. The cylindrical struts have large ports to enable free passage of gas from one region to another. The edges of the two fabric circles are then sewn together to complete the bag envelope, and a concentric sleeve is sewn to the front of the bag to attach to the inflator. The sleeve has a tight fit on the round inflator base and is held in place by a strap type of clamp, which is tightened around the periphery to retain the bag during inflation and impact. A half cross-sectional scale drawing of the inflated bag is shown in Fig. 6. The inflated volume is 1710 in^3, and the surface area is 796 in^2. The diameter of an equivalent (same area) sphere is 15.9 in, whereas the diameter

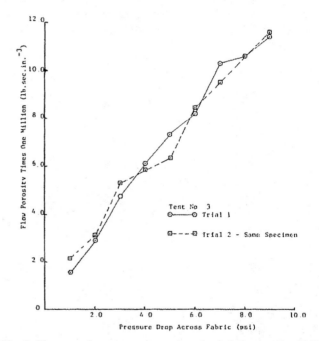

Fig. 4 - Flow porosity versus pressure drop for J. P. Stevens No. 34219 airbag fabric

Table 1 - Airbag Fabric Description

Manufacturer: J. P. Stevens
Manufacturer designation: No. 34219
Material: Nylon, uncoated
Denier: 420 × 420
Weave: Plain
Weight: 6.1 oz/yd^2
Thickness: 0.015 in
Tensile strength (grab test): 447 × 400 lb/in
Percent elongation: 38 × 28
Energy absorption: 269 × 233 in lb
Stiffness (cantilever test): 4.2 × 4.1 in overhang
Count (yarns/in): 56 × 44
Tear strength (tung tear test): 68 × 46 lb

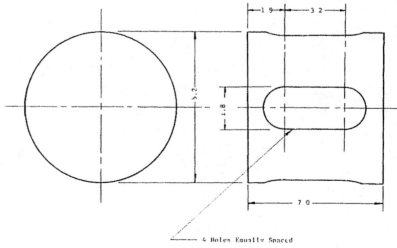

Fig. 5A - Inner shaping strut

Note: No allowance has been made in dimensions for sewing overlap, etc.

of this bag was 19 in. A sphere having the 1710 in³ volume of the airbag would have a diameter of 14.2 in.

A steering wheel that would prevent serious injuries to the thorax and head when impacted at velocities of 15-20 mph without airbag deployment was desired. An attempt was made to find a steering wheel that had a rim with low-durometer energy-absorbing padding, as well as a controlled, uniform load-deflection characteristic. A standard three-spoke wheel with a special soft rim covering was obtained from Sheller Globe Manufacturing Co. The soft steering wheel rim cover was not large enough in cross-sectional diameter and did not have the desired shape, but it was a good compromise considering that the cost for new molds was beyond the range of the funds available. The concentrated force required to deflect the rim axially down the column 1.5 in at different points around the steering wheel varied from 225-290 lb.

VEHICLE MODIFICATION

Modifications of the 1969 Chevrolet Impalas used in this research can be divided into three major categories:

1. Modifications required to adapt vehicles for full-scale dynamic tests on the WHAM II and WHAM III crash simulators.*

2. Preparation of vehicles for destructive barrier crash tests on WHAM III.

3. Modifications required for the installation and correct operation of the air-bag system.

The modifications for use on the crash simulators will be discussed only briefly, while those required to obtain correct operation of the airbag system will be dealt with in detail.

For crash simulations, the test vehicle is subjected to the same deceleration levels prevailing in the crash being simulated and is used repetitively. It is necessary to strengthen the frame and eliminate concentrated masses such as the engine and transmission so that the remaining structure will withstand the accelerations without damage. Structural box section weldments are added to the frame to increase the structural integrity of the vehicle. The engine, transmission, differential, and other masses not required for the study under consideration are removed, resulting in a weight reduction to about 2000 lb. Body-to-frame attachment is reinforced by the addition of bolts and welding. For use on WHAM II, a snubber cable attachment point was welded to the aft end of the car. When the same car was adapted for use on WHAM III, additional members were added to the front of the vehicle to permit the deceleration forces to be

---

*See Ref. 4 for descriptions of WHAM II and WHAM III.

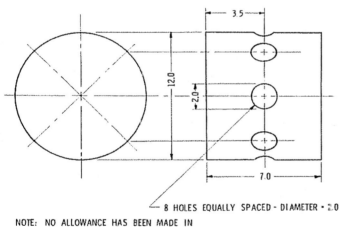

NOTE: NO ALLOWANCE HAS BEEN MADE IN
DIMENSIONS FOR SEWING OVERLAP, ETC.

Fig. 5B - Outer shaping strut

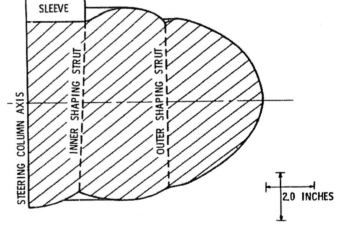

Fig. 6 - Inflated airbag half cross-sectional drawing

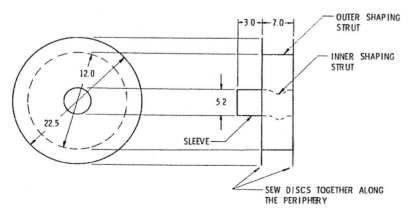

Fig. 5C - Standard airbag

transmitted from the snubber, through the sled, to the front of the vehicle. No vehicle modifications were made for the destructive barrier crash tests except for the attachment of a tow-link beneath the car and interior changes for adapting the airbag system.

In all cases, the seat was reinforced by a cross member between the B-posts, behind the seat, with attachments to the seat. In addition, the seat tracks were welded to prevent the seat from translating forward during deceleration.

Instrumentation added to the vehicle included an accelerometer to record the deceleration history of the occupant compartment and a velocity transducer for measuring the vehicle speed. Instrumentation cables from the dummies were connected to a terminal strip mounted in the rear portion of the car. The leads from the terminal strip were fed through a conduit attached to the vehicle, extending outward from the right rear portion. The cables were dragged along a cable trough and connected to the signal conditioning equipment.

Two crash sensors were mounted on the firewall of the vehicle. They were wired in parallel so that either one would initiate deployment when the vehicle acceleration reached the sensor trigger level. The use of two sensors provided redundacy, and also an opportunity to compare the performance of two sensors under identical vehicle acceleration environments, but on separate locations on the firewall. The time of operation of both sensors was noted on the records.

The major modification to the vehicle for installation of the airbag system consisted of removal of the instrument panel and replacement with a tubular structural frame to which components including the experimental knee restraint/instrument panel can be attached. The modified form is shown in Fig. 7 with the panel removed. The frame was also used to provide an attachment point for a slider-support weldment found necessary to achieve column collapse during impact with the airbag inflated. An adapter

was mounted on the steering wheel hub for mounting the airbag system. Finally, the steering column angle to the horizontal was reduced to 19 deg from the original 25 deg.

COLLAPSIBLE STEERING COLUMN

The energy-absorbing column installed in the 1969 Chevrolet Impala used as the basic vehicle for this program was the ball type of second-generation energy-absorbing column developed by General Motors. A primary crush force is developed through the interference fit of two rings of balls located in the space between concentric tubes. Since the balls are of larger diameter than the clearance between the two concentric tubes, they form grooves in the metal of the tubes as they telescope one over the other. In addition to the force generated by the interference fit of the balls as the column collapses, there is a capsule at the instrument panel that fractures initially at a force on the steering assembly of 200-350 lb from driver impact.

Simulated barrier impacts at velocities up to 40 mph failed to cause column collapse with the standard installation. Analysis of the column installation and the direction of the applied force in the high-speed movies showed that failure of the column to collapse was due to the bending moment applied when the airbag was installed. In order to obtain the desired column collapse, the column angle was decreased from 25 to 19 deg and a slider support assembly was added to the top of the column, as shown in Fig. 8. It consists of a structural attachment to the tubular frame installed for mounting the instrument panel. A frame made from small structural angles is attached to the weldment and terminates in a V-shaped angle guide on top of the steering column. A second angle is clamped to the top of the column to provide a mating, sliding surface. The lubricated surfaces eliminate the large bending moment

Fig. 7 - Tubular weldment for attachment of experimental airbag system components

hat locked the columns under operation without the
slider-support.

WEIGHT ANALYSIS OF STEERING WHEEL AIRBAG SYSTEM

The weight of a steering wheel airbag system is an important
factor if a collapsible steering column is utilized. This
importance stems from the fact that the airbag system is
part of the moving mass that is accelerated by the reaction
force of the airbag as it decelerates the driver and collapses
the column. The force exerted by the airbag (in the direction
of the column axis) is equal to the force necessary to crush
the collapsible element of the column plus the inertia force
associated with accelerating the moving mass of the column
and other components attached thereto. Moving mass information is, therefore, of obvious importance in mathematical
modeling and/or establishing the appropriate quasi-static
column collapse force for a steering column.

A complete weight breakdown for the airbag system
developed in this research program is presented in Table 2.
The pressure pickup and associated parts were present in all
experimental tests. However, they would not be included in
a production type of installation. It can be shown using the
data of Table 2 that the increase in moving mass of the
steering assembly for a vehicle retrofitted with an airbag
system developed in this program would be 3.57 lb, which
represents an increase of approximately 21.4% With the
airbag installed, the acceleration of the column during
collapse will be much lower so the inertial force will be
reduced. There is evidence that the static crush force should
be increased to make better use of the crush distance in
decelerating the driver.

KNEE RESTRAINTS

The knee is one of the best places for applying decelerating
forces to the human body. A conservative value of 1400 lb
for each knee has been established in FMVSS 208. A total
force of 2800 lb is allowed through both knees. The
importance of utilizing this as part of the deceleration
mechanism can be emphasized by realizing that the total
energy represented by the 50th percentile male traveling
at 30 mph is approximately 4800 ft-lb. This total energy, if
it could be dissipated through the knees at a constant force
of 2800 lb, would require only 1.7 ft of stopping distance.
While it is obvious that the total kinetic energy cannot be
dissipated in this manner, the knees can and should be used
as a major point for applying decelerating forces to the body

Table 2- Weight Breakdown—Steering Wheel Airbag Assembly

Component	Weight, lb
Moving mass of collapsing 1969 Impala column	12.71*
Steering wheel (1969 Chrysler design)	4.08
RRC gas generator (charged)	1.97
Gas deflector	0.26
Gas generator steering wheel adaptor plate	0.47
Pressure pickup, tube, and clamp	0.29
Hose clamp retaining airbag	0.08
Airbag (final design)	0.72
Airbag cover (typical design)	0.07
Total weight of moving mass during column collapse (sum of all above)	20.65
Weight added due to airbag system installation	3.86

*Information obtained from General Motors Corp

Fig 8 - Slider support assembly to facilitate column collapse

The general requirements for the knee impact area include a deformable surface area to conform to the knee contour and to eliminate high, concentrated forces on the patella and to avoid forces in excess of the 1400 lb requirement. A sheet-metal panel without sharp edges and small radii in the impact area serves as a suitable load-distributing mechanism. With suitable backing, it can also provide the necessary force-limiting and energy-absorbing capability.

Evaluation of the viable options for design of the knee impact area resulted in the conclusion that either a knee airbag or a deformable instrument panel was a logical solution. The knee impact airbag was discarded as being unnecessarily complex. Previous research in knee impact indicates that a simple energy-absorbing panel can be fabricated at a low cost.

Examination of the knee impact area in the vehicle showed that the rigid bracket for supporting the steering column complicates the instrument panel design, since the bracket is adjacent to the column in the area where the knee would normally hit. In an oblique impact, the knee will hit the steering column and panel intersection at the point where the bracket presents a rigid area. To avoid this dangerous situation, it is necessary to incorporate a fairing or deflector that will guide the knee away from the rigid bracket at impact. In future vehicle design, it is suggested that the support for the column be removed from the knee impact area and placed higher up on the panel and column. The bracket can be incorporated into the slider support.

The experimental knee restraint/instrument panel shown in Fig. 9 has an easily fabricated, semicircular cross section extending farther into the passenger compartment than the standard panel to provide greater crush distance. The advantages of the increased crush distance are twofold: it increases the energy absorption and applies a force before the occupant has moved forward as far as was necessary to strike the original panel. This keeps the occupant on the seat and controls the trajectory of the impact on the steering wheel airbag. If the driver slides forward off the seat, there is danger of injury to his back, either as he slides down or on rebound where contact with the seat causes hyperextension of the spine and possible injury.

TEST DUMMIES AND INSTRUMENTATION

Sierra anthropomorphic test dummies were used in this research, and included the 292-1050 50th percentile male, 292-895 95th percentile male, and 592-805 5th percentile female. Triaxial accelerometers were mounted at the head and upper torso c.g., and force transducers were inserted in the femurs. These transducers were coupled to broadband d-c bridge types of signal conditioning units and amplifiers, and the outputs were recorded on a Sangamo 3500 tape deck and Honeywell 1508 light-beam oscillograph. Sensitivities were adjusted by shunt resistance calibration, and analog filters were used to conform to the requirements of FMVSS 208.

High-speed cinematography was used to observe the airbag system performance during dynamic vehicle tests. A minimum of three 16 mm cameras were used at all times, including a stationary left lateral view, stationary frontal view, and an onboard right lateral view. The stationary left lateral camera film was analyzed quantitatively as indicated in the section entitled, "High-Speed Cinematography Analysis." A typical film sequence is shown in Fig. 10.

OSCILLOGRAPH RECORD ANALYSES

A photographic negative of the light-beam oscillograph record is used for accurate analysis on a Vanguard Film Analyzer, and 8-1/2 × 11 in glossy prints are made for general observation and inclusion in reports. A typical oscillograph record is shown in Fig. 11. The records are identified across the top, and some specific details regarding the test run are included on the right side. Trace identification includes indications of the physical phenomenon measured (acceleration, force, etc.), location and direction of the measurement, transducer identification (including serial number), calibration or sensitivity, and type of oscillograph

Fig. 9 - Experimental knee restraint/instrument panel

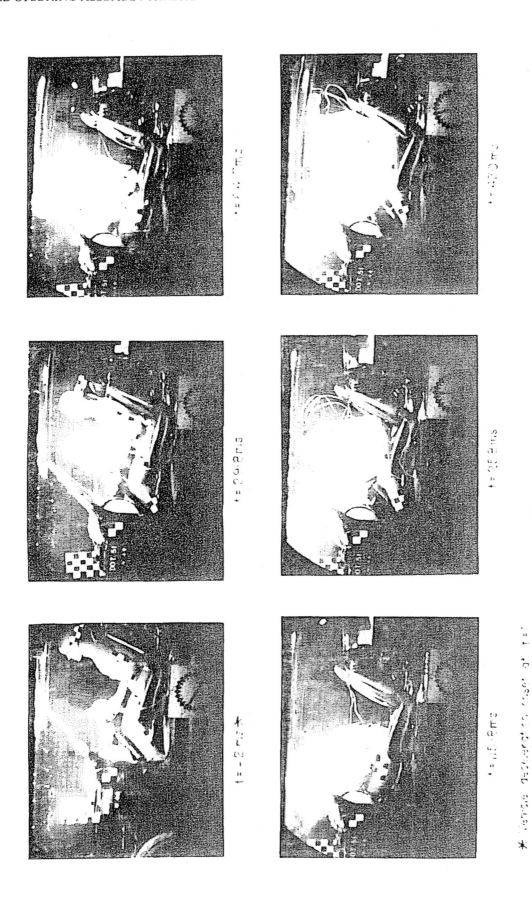

Fig. 10 - High-speed film sequence from run 51 (26.1 mph, 50th percentile dummy, unbelted)

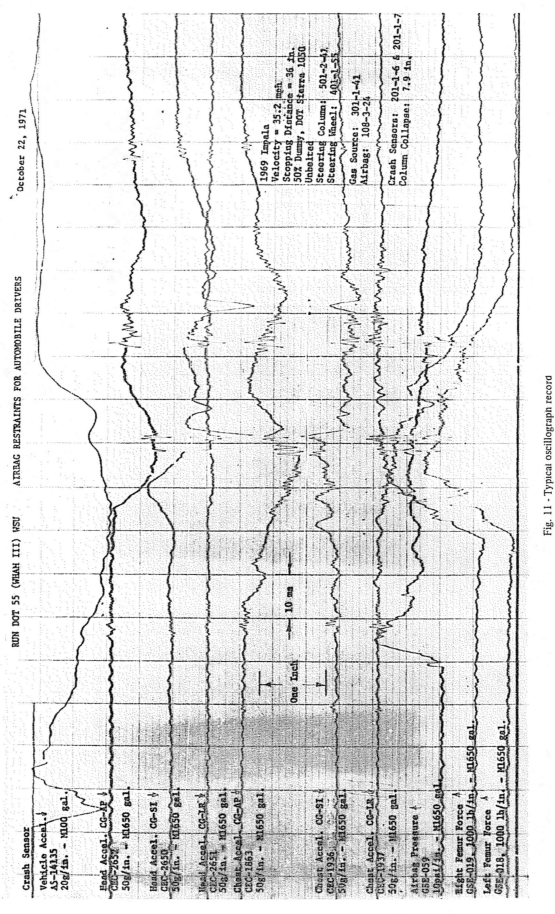

Fig. 11 - Typical oscillograph record

galvanometer used in recording the data. If signal filtering in addition to that provided by the damped galvanometers was used, the filter characteristics are listed on the record.

Computer programs are used to calculate the Gadd Severity Index (SI), the Head Injury Criterion (HIC) index, and the cumulative time duration the chest acceleration is above the 50 g level. These data are calculated for each of the three orthogonal components and the resultant.

HIGH-SPEED CINEMATOGRAPHY ANALYSIS

A computer program is used to analyze the high-speed films to determine vehicle and occupant kinematics during full-scale dynamic vehicle tests of the airbag system.

Input data for the program are obtained using a Vanguard Motion Analyzer that enables one to "track" on a target in the field-of-view and record X, Y, θ position coordinates and angles as a function of film frame number. The analyzer is interfaced with an IBM 526 keypunch machine that prepares data cards for the computer.

Computer input data consist of 14 separate measurements on each film frame, and the output consists of occupant displacements, velocities, and accelerations as a function of time, relative to the vehicle. Typical driver head, shoulder, and H-point displacement versus time curves are shown in Figs. 12 and 13. Displacements are relative to the vehicle.

RESTRAINT SYSTEMS EVALUATED UNDER SIMULATED CONDITIONS

Four modes of driver restraint were evaluated:
1. Steering wheel airbag.
2. Lap and shoulder belts.
3. Lap belt only.
4. Unrestrained.

Vehicle velocities ranged nominally from 15-40 mph. Airbag experiments were conducted at nominal velocities of 20, 25, 30, 35, and 40 mph, and the belted restraint evaluation included nominal velocities of 15, 20, 25, 30, and 35 mph. Unrestrained driver experiments were conducted at nominal velocities of 15, 20, 25, 30, and 40 mph. A packed airbag assembly was mounted in place at the steering wheel hub for all tests.

Seventy-three full-scale laboratory simulated and destructive barrier crash tests were conducted on WHAM II and WHAM III. Runs 1-35 were simulated crashes on WHAM II, and runs 36-73 were simulated crashes on WHAM III, with the exception of runs 60 and 73, which were destructive crashes. Runs 10-19 were 50% dummy belt restraint evaluations on WHAM II, and included a nondeployed airbag packed in place on the steering wheel. The remaining WHAM II runs were airbag developmental tests, as were runs 36-45 on WHAM III enabling evolution of the final design, and will not be discussed further.

DESTRUCTIVE BARRIER CRASH TESTS

In addition to the crash simulations, two full-scale destructive barrier crash experiments were conducted on WHAM III to evaluate the airbag system performance under actual crash conditions. Destructive crashes are performed on WHAM III by accelerating the vehicle in the usual manner, disconnecting it from the propulsion unit, and allowing it to crash into a concrete barrier located down-range from the accelerator. Two destructive crashes were made. The first was a frontal force crash in which the vehicle azimuth was perpendicular to the plane of the crash barrier, while in the second one a 30 deg left oblique barrier impact was achieved by placing a 30 deg structural steel wedge on the front of the barrier.

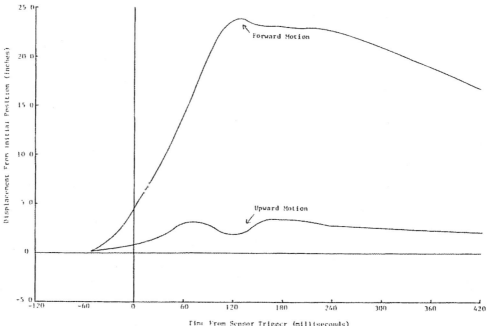

Fig. 12 - Head cg displacement versus time, DOT run 47

In both of the barrier tests, the unbelted driver was protected by the steering wheel airbag and the modified instrument panel. The right front passenger was restrained by a conventional three- or four-point harness system. Both occupants were Sierra 292—1050 50th percentile dummies with triaxial accelerometers in the head and chest and axial load transducers in the femurs. Belt load transducers recorded the forces in the three- or four-point belt components in the passenger restraint harness.

BELT RESTRAINT EXPERIMENTS

The four-point lap and shoulder belt tests of runs 10-16 and the lap belt only tests of runs 17-19 were conducted to establish baselines with which the airbag system performance can be compared. Vehicle velocities and stopping distances on these belt tests ranged from 16.4 mph and 16 in to 35.3 mph and 37.5 in. The head resultant SI ranged from 33-1013 and exceeded 1000 only on run 18 (26.1 mph). The face hit the steering wheel rim on all lap belt only runs, and the head contacted the packed airbag in the higher-velocity lap and shoulder belt tests. The cumulative time duration above 60 g for the anterior-posterior (A-P) component of chest acceleration on runs 18 and 19 (26.1 and 30.6 mph, respectively) was 3.7 and 4.0 ms, respectively. The remaining belt tests all had resultant acceleration durations above 60 g of less than 3 ms. The steering column collapse on runs 18 and 19 was 1.4 and 2.4 in, respectively. Additional details regarding the above discussion are presented in Table 3, including the head to steering assembly impacts when they occurred.

Peak forces in each belt component are listed in Table 4. There is a general increase in harness forces with increasing

velocity, but even at 35.3 mph with lap and shoulder belts, the injury criteria established in FMVSS 208 were not exceeded for the driver in the simulated collisions. At shoulder belt loop loads over approximately 1600 lb (run 13), the driver's head usually hits some part of the steering assembly, indicating a need for a steering assembly design that will minimize injury to the face and neck if the harness is used without the airbag.

The simulated collision run 15 at 30.5 mph and the frontal force barrier crash run 60 at 32.2 mph permit an approximate comparison of results between the following with the same dummy type:

1. A simulated and a destructive barrier crash.
2. Performance of three- and four-point harness.
3. Driver and passenger.

The stopping distance for the simulated collision was 32 in, and the permanent crush of the barrier crash was 29 in. Addition of the elastic deformation of the vehicle should make the total stopping distance about the same for the two experiments. The stopping time for both deceleration pulses was between 110-120 ms. Both pulses were cyclical with 6 or 7 major cycles ranging in amplitude from about 5-25 g except for one peak. The simulated collision had one 30 g peak at 25 ms, and the barrier crash had a 50 g peak at 60 ms.

A crude comparison of total restraint force for the two systems can be made by adding the peak forces in the belts, which results in 5590 lb for the barrier crash and 5580 lb for the simulated collision. These are tensile forces in the belts, which are somewhat different in geometrical configuration, so the comparison is not exact even though the values are essentially the same.

Chest acceleration for the barrier crash was 55 g in the

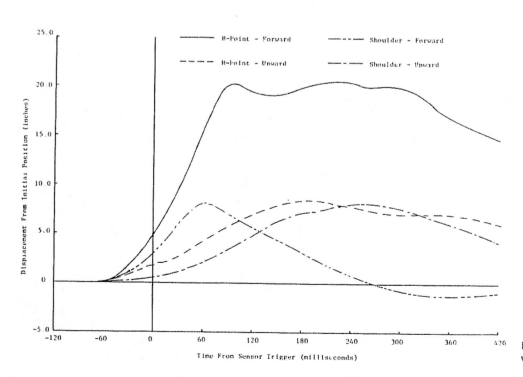

Fig. 13 - Hip and shoulder displacement versus time, DOT run 47

A-P (front to back) direction, 18 g in the S-I (top to bottom) direction, and 43 g in the A-P and 23 g in the S-I direction for the simulated collision. The difference is probably due to the difference in geometry, impact velocity, and the usual variations between runs. The resultant chest acceleration is 56 g for the crash and 49 g for the simulation, both of which are below the maximum prescribed in FMVSS 208.

The greatest difference in results between the two experiments falls in the SI, which for the right front passenger in the crash was 2900 and for the driver in the simulation was 689, both based on the resultant acceleration. It is surprising to find such a great difference with the passenger having a higher value, especially since the driver's head hit the steering assembly and the passenger's head hit nothing except his chest. Examination of the records and the high-speed movies provides an explanation. The SI was calculated for the three orthogonal components in addition to resultant acceleration with values of 398 A-P, 202 S-I, and 15 L-R for the driver in run 15 and 493 A-P, 2027 S-I, and 41 L-R for the passenger in the crash of run 60. The major factor in the difference in the resultant SI is obviously the S-I value. The chin of the dummy passenger contacted the chest at the end of the free rotation of the head, causing high forces and accelerations in the S-I direction and some increase in the A-P direction. Comparison of the acceleration values in the three directions verifies this theory with the S-I peak acceleration being 48 g for the driver of run 15 and 104 g for the passenger of run 60. The A-P accelerations were almost the reverse at 104 g for the driver of run 15, and 57 for the passenger of run 60. Greater time duration for the nonimpact condition of run 60

explains the higher SI for the same peak acceleration in the two directions.

SIMULATED BARRIER CRASH AIRBAG PERFORMANCE

Timing of the airbag cycle is extremely important if correct performance is to be achieved. The sensor, which is designed for a particular acceleration pulse, changes the timing of the system if it is subjected to a grossly different

Table 4 - Belt Loads

Run		Vehicle Velocity, mph	Peak Belt Loads, lb			
			Lap Belt		Shoulder Belt	
No.	Type		Outboard	Inboard	Upper	Lower
10		16.4	660	670	–	270
11		21.0	980	1030	820	430
12	Four-point lap	19.6	830	1220	790	440
13	and shoulder	26.5	1070	1470	1010	630
14	belts	30.8	970	*	1190	870
15		30.5	1660	1970	1250	700
16		35.3	1660	1870	1380	810
17		20.8	1240	1410	–	–
18	Lap belt only	26.1	1590	1760	–	–
19		30.6	1530	1930	–	–
60	**	32.2	1520	2230	1840	–

*Inboard floor anchor failed.

**Right front passenger, destructive barrier crash, three-point harness.

Table 3 - Belt Restraint Test Results

Run		Vehicle Velocity, mph*	Head Resultant Gadd SI	Chest Acceleration				Steering Column Collapse, in	Remarks
				Peak g		Cumulative Duration at 60 g, ms			
No.	Type			A-P	S-I	A-P	S-I		
10	Four-point lap and shoulder belts	16.4	33	12	–	0	–	0	–
11		21.0	124	22	–	0	0	0	–
12		19.6	101	21	12	0	0	0	–
13		26.5	738	33	21	0	0	0	Face hit packed airbag
14		30.8	–	36	27	0	0	0	Inboard floor anchor failed Forehead hit steering wheel upper rim
15		30.5	689	43	23	0	0	0	Face hit packed airbag
16		35.3	789	44	34	0	0	0	
17	Lap belt only	20.8	435	66	16	1.7	0	1.9	Chin and mouth hit steering wheel upper rim
18		26.1	1013	110	95	3.7	–	1.4	Mouth hit steering wheel upper rim
19		30.6	810	126	44	4.0	0	2.4	Mouth hit steering wheel upper rim
60	Three-point	32.2	2900	55	18	0	0	–	Right front passenger, barrier crash

*Stopping distances for nominal velocities of 15, 20, 25, 30, and 35 mph were 16, 21, 26, 32, and 38 in respectively.

1115

pulse. The pulse chosen for WHAM III was estimated to be that of the vehicle for which collision simulation was undertaken. The correct stopping distance was used, and an estimate of the pulse shape was used to establish the snubber condition on WHAM III. After the barrier crash, when a true deceleration profile was available, it was found that the pulse for WHAM III was not optimum. This resulted in a delayed signal to initiate deployment of the airbag. In the vehicle crash, the sensor triggered 16 ms after the onset of vehicle deceleration. The earliest sensing on the simulation runs was 27 ms, which occurred at 35 mph. The increased time to actuate the airbag wastes part of the crush distance and/or part of the available space between the occupant and the steering wheel. At simulated impact velocities greater than 30 mph, the sensing time ranged from 27-40 ms. The true deceleration pulse resulted in faster sensing and consequently improved performance of the airbag. Thus, the values listed are all conservative, and better performance is to be expected in real-world crashes.

Even with slow sensing, the allowable SI and HIC indexes were exceeded only on runs 52, 56, and 66 at 39.5, 34.6, and 34.3 mph, respectively, as shown in Table 5. Run 66 was conducted with the 5th percentile female dummy which always gives a higher SI and HIC than the 50th percentile male dummy. No windshields were present in these simulated collisions. In some cases, where the steering assembly crushed the full distance, the dummy head traveled far enough forward to hit the windshield, had it been present. In these runs, the head sometimes struck the header on the rebound, increasing the acceleration. This is an artifact that would not have been present with a windshield. The SI ranged from 224-528 in the ten 30 mph runs (including developmental runs), which is considerably lower than the 1000 allowed in FMVSS 208.

An interesting comparison between the results of the impacts for the different sized occupants as evidenced by the SI is available in Table 5 and Fig 14, which is a plot of the SI as a function of simulated velocity for the three dummy sizes. The 5th percentile female dummy reaches the maximum allowable SI of 1000 at a velocity of about 32 mph; the 50th percentile male reaches it at about 38 mph, and the 95th percentile male would reach the 1000 level at about 39 mph if extrapolated beyond the 35 mph maximum simulated velocity.

In addition to the quantitative measure of head injury as evidenced by the SI and HIC indexes, a qualitative evaluation of the impact was made from the high-speed movies. The driver's head did not strike the steering wheel rim or hub as it did during lap and shoulder belt runs over 25 mph and all of the lap belt and unbelted restraint experiments. While the force from the facial impact to the steering wheel rim or to the packed airbag is usually below that required to exceed the allowable SI, there would probably have been injury to the facial bones and soft tissue. Consequently, the airbag performance is better than expressed by the SI for the head.

The chest injury criterion established in FMVSS 208 (60 g/3 ms) was exceeded only in runs 52 and 54 at 39.5

and 36.4 mph. The steering column collapse of 2.8 in for 36.4 mph in run 54 was lower than normal, as was the airbag pressure. These two factors probably contributed to the high chest loads. The specified chest criterion was met with no difficulty at the 30 mph specified in FMVSS 208.

FMVSS 208 does not specify an injury criterion for the abdomen. In the unrestrained runs without the airbag, the abdomen struck the lower steering wheel rim and would probably have caused internal injuries. With the airbag, this contact is eliminated, providing better protection than evidenced simply by the evaluation of FMVSS 208 injury criteria.

The greatest number of failures to meet the FMVSS 208 maximums occurred in the femur force, with 11 of the individual knee loads exceeding the allowable 1400 lb, as shown in Table 5. Failure to meet the specified 1400 lb limit on each knee is not considered to be serious from several standpoints. First, the 1400 lb is a very conservative limit and could probably be increased to 1800 lb or higher without causing a substantial number of injuries. Even when injuries occur from knee impact, they are seldom dangerous to life and rarely represent permanent disability. Second, the instrument panel is separate from the airbag system and can be substantially improved by tooling for uniformity and desired deceleration characteristics, which were difficult to achieve with the handmade sheet-metal and cut plastic foam. Finally, the knee loads over the allowable limit of 1400 lb, as shown in Table 5, all occurred at impact speeds greater than the 30 mph specified in FMVSS 208.

The final system performs considerably better than the maximum permitted in FMVSS 208 at 30 mph, and would meet or better the injury criteria at 35 mph with the 50th percentile dummy. The 95th percentile dummy met the criteria maximums at close to 40 mph, while the 5th percentile female met them at velocities up to 32 mph. With further development to refine the design for production, it appears that the injury criteria can be met for all three dummy sizes at 35 mph, and possibly even at 40 mph. Better utilization of the column and instrument panel crush distance by increasing the crush force, or possibly making it velocity-sensitive, would probably improve the performance at the higher speeds.

DESTRUCTIVE BARRIER CRASH
AIRBAG PERFORMANCE

The results of the two barrier crashes are presented in Table 6. The right front passenger in both crashes was restrained by a lap and shoulder belt combination. Results of the harnessed right front passenger in run 60 were included in the discussion of restraint systems. The 30 deg oblique impact sensor actuation time of 35 ms is far too great, indicating a need for an improved sensor. The right femur load was 1600 lb, while the left femur load was 1000 lb. This difference can probably be attributed to the right knee contacting the panel at the juncture of the steering column with higher forces due to the stiffening of the panel by the column

Table 5 - Summary of Results for Experimental Barrier Simulations and Crashes on WHAM III

Run No.	Vehicle Velocity, mph	Steering Column Collapse, in	Peak Resultant Head Acceleration, g	Peak Resultant Chest Acceleration, g	Peak Femur Force, lb		Gadd SI	HIC Index	Remarks*
					Right	Left			
45	30.2	5.7	75	37	1150	840	310	243	1,4
46	34.2	5.2	74	47	1250	1250	646	524	1,4
47	15.4	0	50	14	170	200	143	100	1,5
48	21.1	2.4	76	27	600	450	371	232	1,5
49	25.2	2.9	58	42	880	700	576	503	1,5
50	19.6	0	34	21	400	300	197	181	1,4
51	21.6	4.8	55	37	600	400	205	151	1,4
52	39.5	7.7	101	74	1500	1580	1397	1174	1,4, no head L-R acceleration
53	35.2	2.0	57	59	1650	1600	598	484	1,4, no head L-R acceleration
54	36.4	2.8	–	74	1800	1500	–	–	1,4
55	35.2	7.9	67	44	1380	1340	421	323	1,4
56	34.6	6.9	89	48	1500	1100	1426	1316	1,4, head hit header on rebound
57	35.0	2.2	88	72	1300	1700	997	720	1,4
58	35.0	5.8	82	51	1260	1380	625	524	1,4
59	35.0	7.9	79	46	1450	1360	532	439	1,4
60	32.2	5.2	87	47	1450	1150	460	372	1,4, driver, barrier crash
60	32.2	–	122	56	–	–	2900	2449	1,7, passenger, barrier crash
61	21.2	0.3	79	–	400	350	663	529	2,4
62	20.8	0	66	24	100	310	443	347	2,4
63	25.4	1.4	78	38	940	940	664	573	2,4
64	20.2	0	45	37	180	310	336	303	2,5
65	29.7	1.1	78	46	900	530	708	616	2,4
66	34.3	2.4	102	60	1850	1000	1457	1293	2,4
67	21.4	3.4	–	–	580	500	84	69	3,4, no head L-R acceleration
68	20.8	3.0	–	–	510	500	234	228	3,5, no head L-R acceleration
69	24.8	3.0	33	37	450	680	105	89	3,4
70	29.6	6.3	51	–	700	780	257	225	3,4
71	30.2	5.8	34	–	870	750	173	150	3,4
72	34.2	3.8	42	–	1000	1000	440	406	3,4
73	30.6	5.8	78	36	1600	1000	1274	1029	1,4, driver, 30 deg barrier crash
73	30.6	–	–	–	–	–	–	–	1,6, passenger, 30 deg barrier crash

*Codes: 1. 50th percentile male dummy; 2. 5th percentile female dummy; 3. 95th percentile male dummy; 4. airbag; 5. no restraint; 6. four-point lap and shoulder belts; 7. three-point lap and shoulder belts.

bracket. The SI was 1274 for the oblique impact, compared with 460 for the frontal force impact. The increase in head SI is attributed to the head hitting the windshield, and probably the header, to the left of the normal driver center-line during the impact. The chest accelerations were lower for the 30 deg impact than for the frontal force impact for the airbag-protected driver.

For the frontal force barrier impact, the resultant SI for the driver was 460. The chest accelerations were well under the 60 g/3 ms level, and the left femur force was 1150 lb. The only injury criterion exceeded was the right femur force, which reached a peak force of 1450 lb. This femur force would probably have been less than the allowable 1400 lb if the velocity had been 30 mph. Also, elimination of impact to instrument panel at the steering column attachment bracket, which locally stiffens the panel, would undoubtedly lead to lower femur loads.

ANALYSIS OF 35 MPH BARRIER CRASH SIMULATIONS

Special emphasis was placed on providing airbag protection at velocities greater than 30 mph in this program. Simulations up to 40 mph were made, but the injury criteria could not be met at 40 mph. Therefore, analysis of four runs at a nominal 35 mph will be the basis for evaluating this system at the highest practical velocity. With additional development, it is expected that this velocity can be increased to 40 mph.

Head protection is the most critical item in the system, both from the standpoint of the number and severity of injuries to vehicle occupants and the difficulty in achieving protection. Fig. 15 presents the resultant head acceleration as a function of time during the simulated collisions. Runs 46, 55, 58, and 59 are all plotted to the same scale. The heavy black line represents the average of the four accelerations at any given time. The curves all have the same general shape with some variation in timing (the curves were placed on the scale to provide the same initial starting point), but a greater variation in acceleration for any given time. Varia-

tions greater than 50% of the average are observed at several of the higher acceleration levels.

To compare the A-P accelerations, Fig. 16 was drawn from the same set of runs utilizing the same technique for finding the average. Again, the curve has the same general geometric shape with about the same percentage of spread.

In Fig. 17, the SI is plotted along with the average index as a function of time. The curves all have the same general shape with a substantial deviation from run to run. The average SI for the four runs is 550, with a range from 415-630.

A graphic presentation of the movement of the occupant and steering wheel is presented in Fig. 18 for run 59, which is closest to the average of the four runs based on SI. The data for this curve were obtained from film analysis. The point of reference is taken as the center of the steering wheel, and is a point that remains fixed to the vehicle even when the steering assembly moves. A point at the same vertical height as the reference point at the anterior of the dummy chest is also plotted as a function of time. The difference in these curves represented by the ordinate on the graph is the distance between the chest and the reference point at the steering wheel. When the column collapses, the distance between the chest and the steering wheel is the difference between the curves marked "center of hub" and "anterior of chest." For convenience, the distance between the reference points on the hub and chest is plotted. The total column collapse is the difference between the center of the hub and the reference point at the steering wheel. For this run, the total column collapse was 7.9 in. The time is referenced to initial vehicle deceleration.

The time at which the sensor actuated and the time of column collapse initiation are also indicated. All distances are horizontal components. At time zero, the anterior of the chest is at -13 in, or 13 in behind the reference point at the center, or the rear plane of the steering wheel. The total vehicle crush of 36 in is represented by the maximum horizontal displacement forward of the reference point in the vehicle, and is so indicated on the graph.

COMPARISON OF RESTRAINT SYSTEMS

A comparison of no restraint, lap belt only, lap and shoulder belts, and the airbag gives a good perspective of the currently available safety features. The remarks will be restricted to forward force destructive and simulated barrier crashes, since these were of primary interest in this program.

With no belt or airbag restraint, the injury criteria were not exceeded at impacts up to 25 mph; however, the abdomen and facial impacts would probably have produced serious injuries at this speed. Therefore, the maximum safe speed for no restraint condition with this steering assembly is considered to be 20 mph.

With the lap belt only, the chest accelerations were excessive at 26 mph and marginal at 20.8 mph. In fact, the chest accelerations were higher for the same impact speed with the lap belt only than with no restraint. This is probably

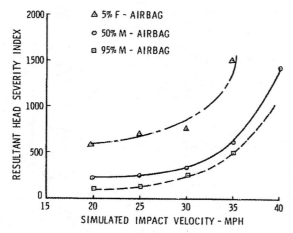

Fig. 14 - SI versus vehicle velocity for 50th percentile and 95th percentile male and 5th percentile female dummies

Table 6 - Summary of Test Conditions and Results For Destructive Barrier Crash Tests

Test Condition, Parameter, or Result	Description or Value	
	First Destructive Barrier Crash— Run 60	Second Destructive Barrier Crash— Run 73
Laboratory test facility	WHAM III	WHAM III
Type of impact	Frontal	30 deg left oblique
Steering column angle from horizontal deg,	19	19
Test dummy		
Driver	DOT Sierra 1050	DOT Sierra 1050
Right front passenger	WSU Sierra 1050*	WSU Sierra 1050**
Vehicle velocity, mph	32 2	30 6
Vehicle stopping distance (dynamic crush), in		
Centerline	29	18 8
Left front	29	37 5
Airbag pressure, psig		
Peak	22	17
Mean	7	6
Peak driver femur force, lb		
Right	1450	1600
Left	1150	1000
Deployment chronology, ms (t_o = time at start of significant vehicle deceleration)		
t_o-to-sensor actuation	16	35
Actuation-to-initial pressure	4	33
t_o-to-chest deceleration onset	35	--
t_o-to-head deceleration onset	67	--
Head peak accelerations, g		
Driver		
A-P	50	62
S-I	73	59
L-R	6	51
Right front passenger		
A-P	57	--
S-I	104	64
L-R	14	13
Gadd Severity Index		
Driver		
A-P	312	677
S-I	93	216
L-R	1	174
Result	460	1274
Right front passenger		
A-P	493	--
S-I	2027	1089
L-R	41	11
Result	2900	--
Steering column collapse, in	5 2	5 8
Maximum steering wheel rim deformations away from driver and toward center, in		
Axial	4 0	4 4
Radial	2 6	1 9
Chest peak accelerations, g		
Driver		
A-P	42	32
S-I	34	23
L-R	8	32
Right front passenger		
A-P	55	33
S-I	18	--
L-R	6	14

*Three-point belt restraint
**Four-point belt restraint

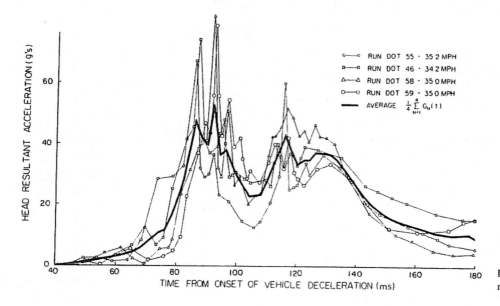

Fig. 15 - Average and individual driver head resultant accelerations for 35 mph airbag runs

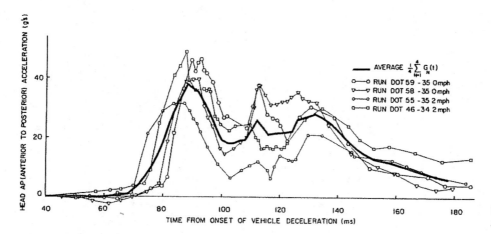

Fig. 16 - Average and individual driver head AP accelerations for 35 mph airbag runs

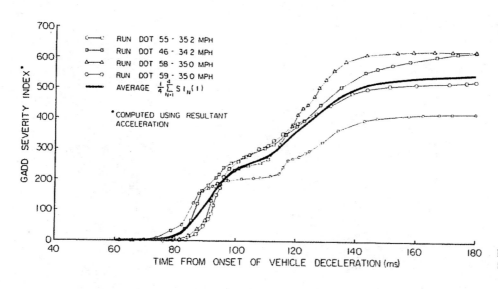

Fig. 17 - Average and individual head SI for 35 mph airbag runs

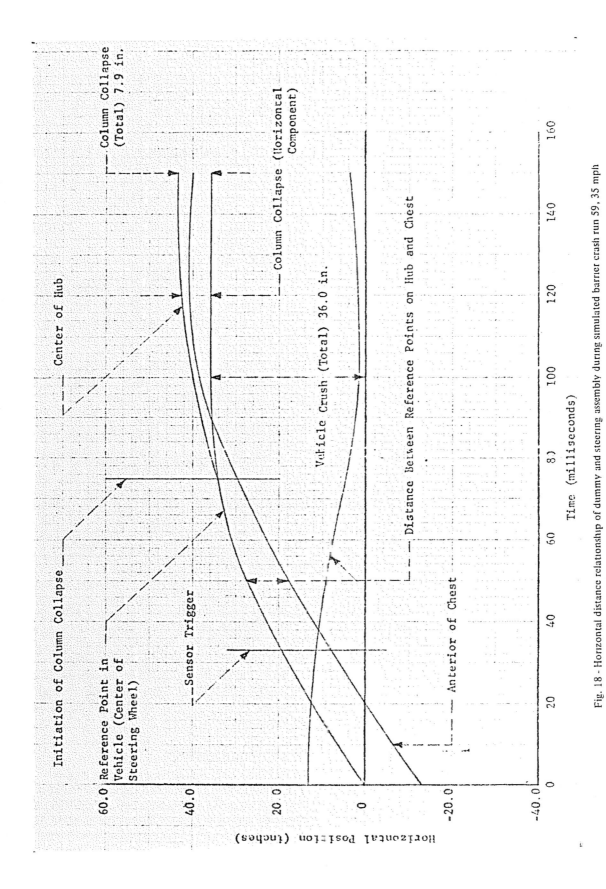

Fig. 18 - Horizontal distance relationship of dummy and steering assembly during simulated barrier crash run 59, 35 mph

due to the guiding effect of the lap belt, which causes the chest to strike the steering wheel hub in a direction perpendicular to the column axis, rather than the abdomen striking first as it does in the unrestrained case. The injury criteria do not consider abdominal and facial impacts; therefore, the unrestrained condition looks better than the lap belt only. Based on these data, the maximum acceptable velocity for the lap belt only is also 20 mph.

The lap and shoulder belt combination was well within the allowable levels at velocities up to 35 mph, except for the barrier crash in which the passenger head SI was 2900. However, at velocities of 25 mph and over, the head did hit the packed airbag. In those cases, the SI was not excessive, indicating that the force was not high enough to cause brain damage. The harness had decelerated the head to a point where injury would probably have been minimal. Also, impact to a padded hub is preferable to impacting the steering wheel rim, which provides concentrated loads to the face with danger of facial bone fracture. According to the injury criteria, the other limiting factor would be the 2900 SI for the right front passenger in the 32.2 mph frontal destructive barrier crash. However, as pointed out earlier, the majority of the SI was due to a large S-I acceleration component, which apparently did not cause injury in the field. Based on these data, the limiting velocity for the lap and shoulder belt would be 25 mph if the SI is considered, or greater than 35 mph if the SI is eliminated when the head does not hit an interior component or hits a noninjury-producing component such as a well-padded and force-controlled steering wheel hub.

The airbag system developed under this program appears to be safe at velocities up to 35 mph. With further development, it could probably meet a 40 mph requirement without exceeding the injury criteria. The airbag has the big advantage of not permitting the face or abdomen to strike the steering wheel rim or hub. Thus, the potential injuries to the abdomen and facial bones are greatly reduced in the case of the airbag, while in the other restraint systems these components present very real dangers.

On the negative side, for the airbag there are little data available on the performance when the occupants are out of position. Also, the harness or lap belt holds the driver in position, thereby lessening likelihood of losing vehicle control during severe operating conditions.

CONCLUSIONS

The following conclusions are based on the forward force simulated and destructive barrier crashes of this program:

1. With the modified steering assembly, the driver can withstand a 20 mph collision unrestrained, without exceeding the FMVSS 208 injury criteria. The abdomen and face hit the steering assembly, but with correct design, the injuries to these areas will probably not be excessive.

2. With the lap belt only, the limiting velocity is also 20 mph. The lap belt causes the occupant to jacknife so that the chest strikes the steering assembly in a normal

direction (perpendicular to the column axis), resulting in a force on the chest higher than without restraint, as evidenced by the higher accelerations.

3. With a lap and shoulder belt, the driver is protected at velocities up to 25 mph without striking the steering assembly, and also, with the head striking the steering assembly, at velocities up to 35 mph, which was the highest velocity used in this program with a lap and shoulder harness. The FMVSS 208 injury criteria were not exceeded for the driver, even though the head struck the packed airbag. The velocity of this head impact was low and the injury criterion was not exceeded. If the head injury criterion is eliminated, except for that portion of the impact during which the head strikes an interior component, the allowable velocity will be at least 35 mph. No experimental results are available for higher velocities.

4. Injury criteria were not exceeded with the airbag for simulated velocities up to 30 mph. With some further development, it is expected that 40 mph barrier equivalent can be withstood without exceeding the injury criteria.

5. For a given impact condition, the airbag appears to offer the best protection based upon established injury criteria and high-speed film analysis. The head and abdomen do not strike the steering wheel with the airbag, while with the other restraints or no restraint, one or both strike the steering assembly with possible injury.

6. Additional injury criteria should be established to include danger of injury or fracture to the facial bones, neck, and abdomen.

7. The knee impact area of the instrument panel is critical and should be considered an important part of any airbag system design. It is desirable to have the instrument panel impact provide force near the allowable 1400 lb.

8. The shaped airbag deploys between the abdomen and the steering wheel and between the head and steering wheel. Careful analysis of the high-speed movies shows that protection for these two areas is provided.

9. It is important that the column collapses to provide the protection afforded by the nearly 8 in. displacement available during the forward motion of the steering assembly.

10. The injury criteria as measured by the head and chest accelerations and femur force vary inversely with the occupant size. With the airbag system developed under this program, the 5th percentile female reaches the maximum injury criterion at about 32 mph, the 50th percentile male reaches it at about 37 mph, and the 95th percentile male reaches it at about 39 mph.

ACKNOWLEDGMENTS

Inflators used in this research were provided by Rocket Research Corp. through Eaton Corp., which also furnished the crash sensors and fabricated custom airbags. The cooperation and advice of their engineers contributed to the success of the program. Gratitude is also expressed to the Detroit Auto Inter-Insurance Exchange (AAA) for donation of the destructive barrier crash vehicles.

REFERENCES

1. Forrest M. Council, "Seat Belts: A Follow-Up Study of Their Use Under Normal Driving Conditions." Jrl. of Safety Research, Volume 1, No. 3 (September 1969), pp. 127-136.

2. Rocket Research Corp. (Redmond, Washington), "Inflatable Restraint System for Automobile Steering Assembly." Report No. 70-P-477 submitted to Biomechanics Research Center, Wayne State University, Aug. 21, 1970.

3. D. Streed and C. B. Rodenbach, "Material, Fabrication, and Packs for Air Cushions." Paper 710018 presented at SAE Automotive Engineering Congress, Detroit, January 1971.

4. L. M. Patrick, "Airbag Restraints for Automobile Drivers." Final Report to Federal Highway Administration, National Highway Traffic Safety Administration, Washington, D.C., for Contract FH-11-7607, September 1972. (HS 800725, National Technical Information Service, Springfield, Va.)

General Motors Driver Air Cushion Restraint System

T. N. Louckes
Oldsmobile Division, General Motors Corp.

R. J. Slifka
Delco Electronics Division, General Motors Corp.

T. C. Powell
Saginaw Steering Gear Division, General Motors Corp.

S. G. Dunford
Inland Manufacturing Division, General Motors Corp.

OLDSMOBILE HAS BEEN INVOLVED in coordinating safety improvements for the driver since the initial development of an energy-absorbing steering column. With the need to provide a passive restraint system to meet the new MVSS No. 208, Oldsmobile began developing a driver air cushion restraint system. This paper is intended to describe how this system is designed to function and give a detailed description of its components.

We will describe the overall system indicating how the components must function together to provide optimum vehicle barrier performance. Following this, three additional discussions are included by Delco Electronics Div. of GM on the sensor system, by Saginaw Steering Gear Div. of GM on the new steering column and mounting system and by Inland Manufacturing Div. of GM on the new steering wheel and wheel-mounted air cushion restraint module.

OVERALL SYSTEM OPERATION

The first thing necessary in understanding how any air cushion restraint system must work is to fully appreciate the time available for sensing and inflation of the system during a 30 mph barrier impact. The total time for both sensing and inflation is approximately 0.04 (40 ms), or less than an eye blink. To give you an idea of some of the significant events that must occur during this time, I would like to describe the events and display them graphically with a series of illustrations.

Fig. 1 shows the vehicle as it first contacts the barrier and is referred to as time zero. At this point no vehicle crush or deceleration exists, thus there is no crash information available. The main elements of the driver air cushion system are shown schematically in this illustration. There is a new energy-absorbing steering column with a revised mounting system, a new steering wheel which houses a chemical gas generator and air cushion, and a new instrument panel which includes a knee restraint to absorb lower torso energy. The sensing system consists of a velocity sensitive bumper switch and a vehicle deceleration sensor mounted inside of the body. This is the same sensing system used to sense and deploy the passenger restraint cushion.

In Fig. 2 we see the vehicle at 10 ms after initial contact with the barrier. The bumper switch has experienced a velocity change of about 14 mph. This change has caused the

ABSTRACT

Presented in this paper is a discussion of the General Motors air cushion restraint system for the driver position. A discussion of the overall system operation and performance is followed by a detailed description of the components including the crash sensors, steering column, and air cushion module.

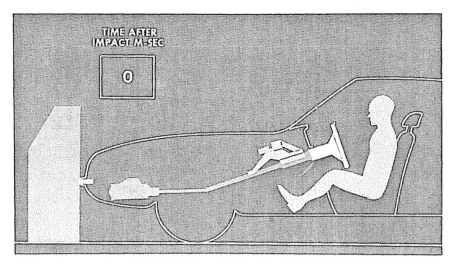

Fig 1 - Driver and air cushion system

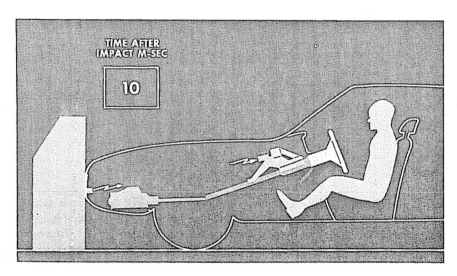

Fig. 2 - Inflation signal

switch contacts to close and send an electrical signal through the sensor to the gas generator located in the steering wheel. If for some reason, such as lack of bumper contact in an accident, this switch did not function, the internal sensor located in the body would actuate the system. This type of actuation results in a deployment signal at approximately 30 ms for this impact, which is adequate for proper system performance.

In Fig 3 we see the vehicle after 40 ms. The cushion is deployed and in position to distribute and absorb the driver's upper torso energy. The cushion is constructed of neoprene-coated nylon and is 23 in in diameter It extends approximately 14 in rearward of the steering wheel rim.

In Fig. 4 we see the vehicle after 50 ms Here the dummy's knees have contacted the lower instrument panel. The relative velocity between the dummy's legs and the vehicle at contact is about 8 mph. The lower instrument panel knee restraint is constructed of a sheet metal substrate covered with 1/2 in of foam and a vinyl cover This knee restraint must perform two very important functions in the driver system. It controls the driver motion or kinematics by limiting the lower torso movement to retain the driver on the seat in an upright position and also absorbs the energy of the lower

torso while keeping the femur loads within the specified limits.

In Fig. 5 we see the vehicle after 80 ms. Here the driver is firmly against the cushion. The pressure in the driver cushion has reached a maximum of about 10 psi and will then start to decrease as the column absorbs energy. It is important to note that the compression of the steering column is the principal means of absorbing energy in the driver system. The cushion is primarily a load distributing element. This is somewhat different than other air cushion systems or the GM cushion used for the front seat passengers, where the main energy absorption is accomplished by the cushion venting gas. Note that the driver has moved forward in an upright position due to the restraint of the knee module.

Fig 6 is a vector diagram of the forces on the steering column and the loading of the air cushion. The load on the cushion is 2300 lb and is basically in a forward direction. This results in a 2000 lb load along the axis of the column combined with a 1200 lb upward load at the base of the steering wheel. This upward force is much greater than that found on non-air-cushioned systems and is the result of the longer effective moment arm and reduced upper torso rotation caused by

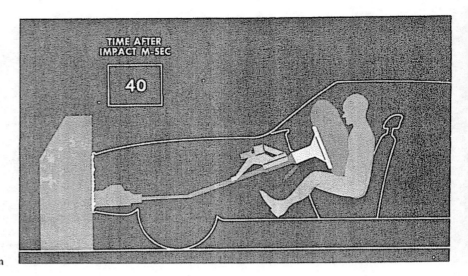

Fig. 3 - Cushion inflation

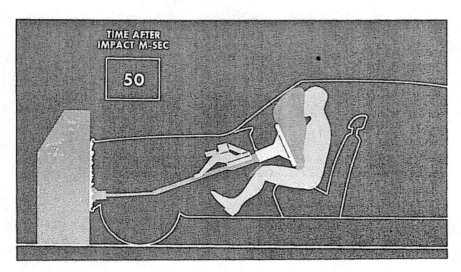

Fig. 4 - Driver loading—cushion and knee restraint

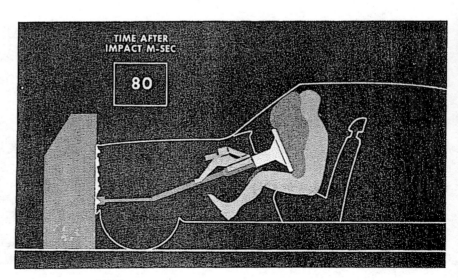

Fig. 5 - Driver energy absorption—column and knee restraint

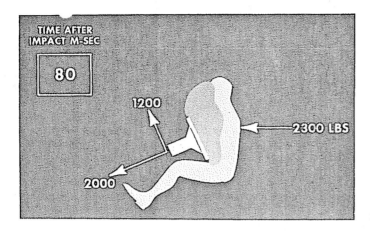

Fig. 6 - Column loading

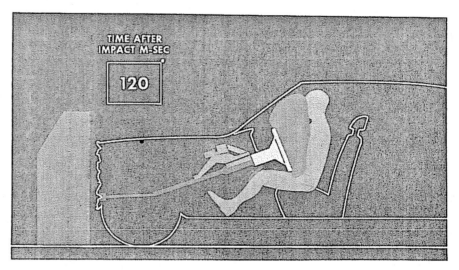

Fig. 7 - Start of rebound

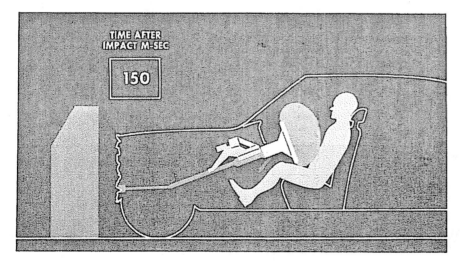

Fig. 8 - Rebound completed

the air cushion. This increased upward force has made necessary several revisions in the column construction and its mounting system which will be covered in more detail in the discussion of the air cushion restraint steering column.

Fig. 7 shows the vehicle after 120 ms. It has come to rest after rebounding away from the barrier. A major portion of the driver's energy has been absorbed by stroking of the energy-absorbing steering column and plastic deformation of

the sheet metal knee restraint. Additional driver energy has also been dissipated through ride-down with the vehicle as it crushes. At this point, more than 85% of the driver energy has been dissipated, with less than 15% remaining in the compressed air cushion. This remaining energy causes the driver to rebound rearward into the seat with a velocity of approximately 10 mph (Fig. 8).

Now, we would like to review the main components in-

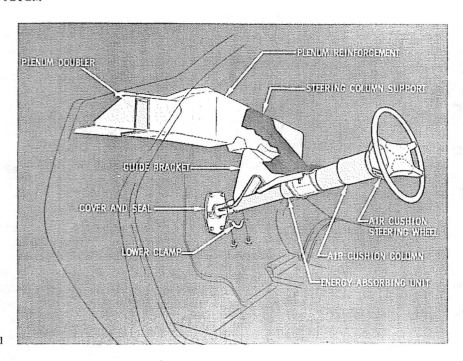

Fig. 9 - Air cushion column and wheel

cluded in the driver system to give you a better understanding of the design features necessary to achieve our performance goals.

STEERING COLUMN SYSTEM - Fig. 9 shows the air cushion restraint column and mounting system. The column support and guide bracket have a deeper section than in our present non-air-cushioned vehicles. These items are attached to a larger plenum with an added doubler. The section modulus of all of these components is much higher than on our current vehicles. This mounting system retains its integrity during 30 mph barrier tests and is an important factor in providing improved performance with the higher loading imposed by the addition of the air cushion.

Also note that the column guide bracket spans the energy-absorbing unit located between the lower attachment clamp and upper shear capsule. This mounting system also uses a floor cover and rubber seal which allows the toe pan to deform without affecting the column position or collapse travel. Full-column stroke is retained for absorbing the driver's energy. It also allows for lowering the initial column collapse load to offset inertial forces. The low initial rate section is followed by an increased load section for greater total energy absorption. This lead-in feature aids in starting the steering column in motion under higher vertical loading conditions and also improves the impact protection at lower speeds when the cushion is not deployed.

Fig. 10 shows the load-versus-time curve of the dual load energy-absorbing column. Note that the load curve is an efficient square wave shape as indicated by the large area under the curve for our new column. Another important element of the new air cushion restraint steering column is the electrical circuit. This circuit required the development of a slip ring assembly at the top of the column for transmitting the deployment signal to the inflator module. The design of this

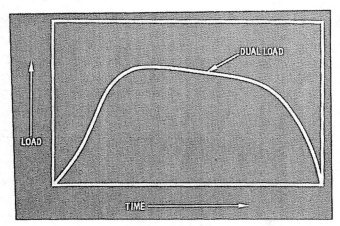

Fig. 10 - Column collapse load

element will be covered in detail in the discussion of the air cushion restraint steering column.

STEERING WHEEL AND CUSHION MODULE - Fig. 11 shows the main components of the steering wheel and air cushion module. These will be discussed in detail in the section on the air cushion steering wheel system, but some of the initial considerations affecting the choice and configuration of these units are worthy of mention. Choice of a chemical gas generator as the inflation element was the result of our need to optimize the size of the air cushion module package. The steering wheel hub area is an extremely important element in providing good visibility to instruments and controls. The size and shape of our module was chosen after many alternatives were examined for visibility, optimum bag stowage, and deployment. The size and shape of the inflated cushion was also a very important factor. Consideration had to be given to providing restraint to the driver on all frontal impacts and yet not make the cushion any larger than necessary. Steering

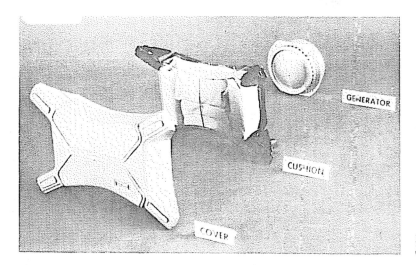

Fig 11 - Main components of the steering wheel and air cushion module

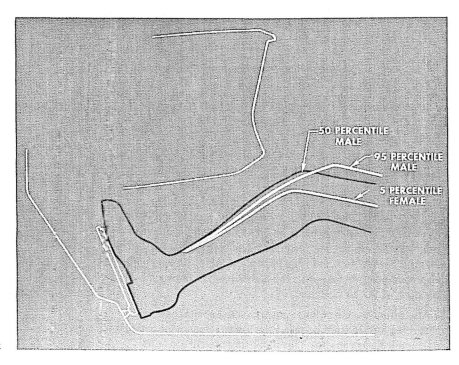

Fig 12 - Driver knee restraint

wheel performance for an air cushion system also required consideration. It must provide conventional wheel and energy-absorbing column performance when the cushion is not deployed and yet provide a stable platform for supporting the air cushion under reasonable off-center loading conditions.

KNEE RESTRAINT - Another main element of the driver air cushion system is the knee restraint. Fig 12 shows a cross section of a typical knee restraint, showing the location of the knees relative to the panel for a 5th, 50th, and 95th percentile driver. Notice that the space between the driver's legs and the panel is very similar for all sizes. This results from the normal adjustment of the seat position, which is forward for small drivers, mid-position for average drivers, and full rearward for large drivers. This consistent spacing makes a fixed type of restraint ideal for use in the driver air cushion restraint system and allowed us to eliminate an inflatable type knee restraint as used in the passenger system. Our studies

showed that an inflatable knee restraint for the driver would be extremely complex. Also it may have undesirable aspects such as trapping the leg against the accelerator or not allowing leg movement from the accelerator to the brake after inflation. Development of the knee restraint was a critical item in the driver air cushion system program, as it is the initial element in controlling kinematics and, therefore, affects the load distribution on the entire torso.

Fig. 13 shows the construction of a typical knee restraint. It is designed so that it provides sufficient restraint to rotate the upper torso and prevent the driver from sliding off the seat. This requires a fast load buildup to a minimum force level of about 1000 lb/leg. In addition, the knee restraint must control the peak forces within the specified limits. We have been successful in producing a humpbacked square wave load curve which is quite good. The shape, construction, and supporting structure of the knee restraint are all important

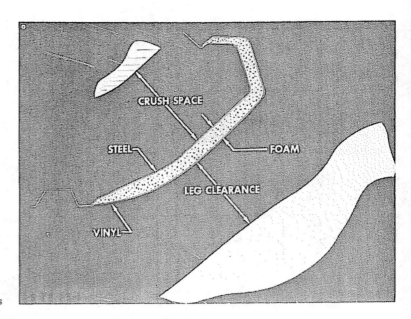

Fig 13 - Knee restraint requirements

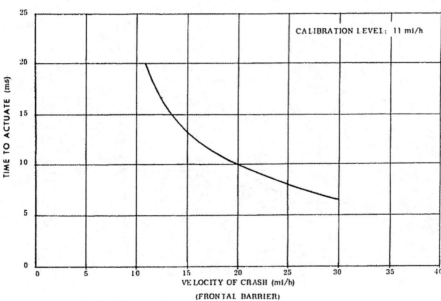

CALIBRATION LEVEL: 11 mi/h

Fig 14 - Nominal actuation time (impulse detector)

elements in achieving this kind of energy management performance. Another important element in the design of a fixed driver knee restraint is control of crush in the area of the column. The knee restraint should not interfere with normal energy-absorbing compression of the steering column. This required the development of local supporting structure in the steering column and knee restraint pass-through area which would minimize hard spots to possible off-target knee impacts.

BUMPER SENSOR SYSTEM - The bumper impulse detector switch itself will be covered in detail in the section on crash sensing. However, there are some considerations with regard to the system and bumper structure that are worthy of mention. It was decided early in the program that it would be desirable to place only one detector on the bumper. Therefore, the best choice for symmetrical impact detection is to place it as near the middle of the bumper as possible.

Also, it is necessary to specify sufficient clearance behind the bumper so that the impulse detector will not be damaged in the early stages of the barrier impact. Routing of the cable to reduce the possibility of it being cut during impact, or being chafed during vehicle operation is another consideration. Also, consideration was given to protect the bumper impulse detector from damage during towing or jacking of the vehicle. The bumper is obviously a very hostile environment for any electrical device; therefore, the sealing and construction of the cable and housing are important. Environmental requirements were developed for laboratory testing of the housing and cable design.

SYSTEM PERFORMANCE

In this section we would like to show how this system performs during the barrier impacts required by MVSS No. 208.

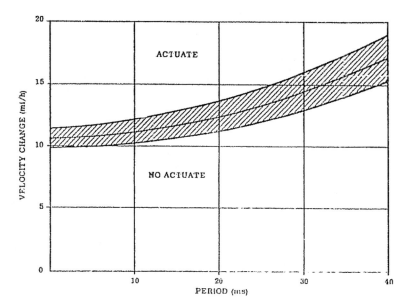

Fig. 15 - Velocity/time characteristics

Table 1 shows typical results from barrier impacts with this system as compared to the requirements of MVSS No. 208. It can be seen that they are met with some margin for variation. These data were generated from the 1973 production-built field trial vehicles and therefore represent the performance of those cars that are on the public roads today. Additional barrier tests conducted at the maximum nondeployed barrier speeds also meet the requirements of MVSS No. 208.

Table 1 - Barrier Results

Test Condition	Head	Chest	Femur
MVSS 208 requirements	1000 HIC	1000 SI	1700 lb
30 mph front	335	280	1530/1050
30 mph left	382	260	1170/760
30 mph right	229	310	1400/1100

SYSTEM COMPONENTS

The following discussions include a detailed description of the operational function and design considerations of the components included in the GM air cushion restraint system. These discussions include the crash sensing system, the steering column, the steering wheel, and air cushion module assembly.

CRASH SENSING - The true heart of the air cushion restraint system is the sensing mechanism. This is where precision is required. In general, the sensing mechanism must determine that the vehicle is in a crash of sufficient severity to require restraint for the occupants. Specifically, it must provide early detection, dual-level sensing, angular as well as frontal sensitivity, high operating reliability, and extremely high reliability against inadvertent deployment.

Early detection is accomplished by placing a detector in the bumper, the foremost part of the vehicle to be involved in a crash. Fig. 14 shows actuation times versus crash velocity. The bumper impulse detector (BID) is calibrated to close an electrical circuit when a predetermined velocity change occurs over a finite period of time. Fig. 15 shows the curve for a BID with frontal barrier threshold of 11 mph. The vertical axis represents velocity change, and the horizontal axis the period of time in which the velocity change occurred. The curve, then, is the threshold of the detector with all points above the curve indicating actuation. Angular crashes, for example, are less severe than frontals. The velocity change occurs over a

longer period of time and, therefore, a greater velocity change is required for actuation. The actuation of the BID closes the circuits for the driver's system and one of the circuits to the passenger's system. Fig. 16 shows this simple device.

Mounted in the vehicle's passenger compartment is another sensing device. It also is set at a low frontal barrier threshold of 11 mph. The detector is a mass on a wire which forms a pendulum. When the car is decelerated, the mass continues forward, making contact and closing a circuit. The mass is held in its rest position by a magnet which also keeps it from moving about while the vehicle is on rough roads or involved in minor accidents which do not require ACRS protection. The magnet also determines the threshold. Fig. 17 shows the detector. Since it is a g-switch and not directly related to vehicle velocity changes, it was necessary to determine the size and shape of the deceleration pulse at its mounting location on the vehicle for the speed of crash actuation required. Once the pulse was defined, the detector is calibrated to close with that input.

The crash pulses observed on our vehicles have the shape of haversines, therefore, input pulses in the shape of haversines are used when testing the detectors.

The dual level detection is accomplished by including an identical detector calibrated at a higher level. As a matter of fact, by varying the magnet's strength and mass, a very large

1131

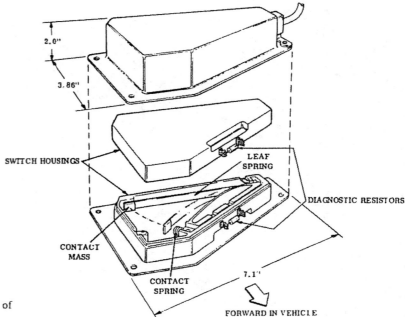

Fig. 16 - Curve of a BID with a frontal barrier threshold of 11 mph

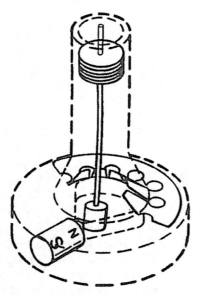

Fig. 17 - SD-1 dectector configuration

calibration range can be obtained. The low level detector is calibrated at 18 g and the high level at 30 g. Fig. 18 shows a response curve for the 18 g detector. Fig. 18 indicates the threshold in g's versus the haversine period. The lower curve is the threshold curve, and the other curves indicate the response time. For example, a 25 g, 25 ms haversine would close the detector in 15 ms, measured from the start of the haversine. Angular sensitivity is obtained by placing the pendulum weight in a sector of 37 deg on either side of the input axis. The interaction of the BID, low and high level, can be seen in Fig. 19 which shows the system requirements. High reliability has been sought by making the detectors simple with minimum parts. In order to minimize the possibility of inadvertent

operation, the sensor uses two detectors placed in series and the BID uses dual contacts and springs in series. If one of the detectors should fail closed, actuation could only occur if the other fails in an identical manner, or a crash occurs.

The schematic of the sensing mechanism and the actuator circuit are shown in Fig. 20.

Battery Backup - Since some vehicles have their batteries near the front of the car and could possibly have them damaged or disconnected during an accident, a backup to the battery is provided in the sensor. It consists of a capacitor which will maintain its charge for at least 100 ms after battery power is removed. Fig. 20 shows the capacitor in the schematic.

Diagnostics - The government standards require that a vehicle operator know that the system is in a ready condition. The diagnostics developed for this system are part of the sensor and do considerably more than indicate readiness.

When the vehicle ignition key is turned to the On position, the fuses, lamp, and diagnostics are checked. Satisfactory operation is indicated by the warning lamp illuminating for 5 s and then going out. This check is made each time the ignition key is turned on. The warning lamp is also turned on any time malfunction occurs and remains on until the malfunction is removed.

The types of malfunctions diagnosed and indicated by a continuous warning lamp-on condition are a shorted or a closed detector, shorted or a closed BID contact, lack of battery power (fuse or open circuit), open circuit in BID harness or inflator harness, low pressure, short-to-ground in BID, inflators or associated harnesses, lack of battery backup power, and diagnostic failures.

The type of malfunction diagnosed and indicated by a no-light condition on turn on are as follows: no lamp power (fuse or open circuit), warning lamp burned out or circuit open, and diagnostic failures.

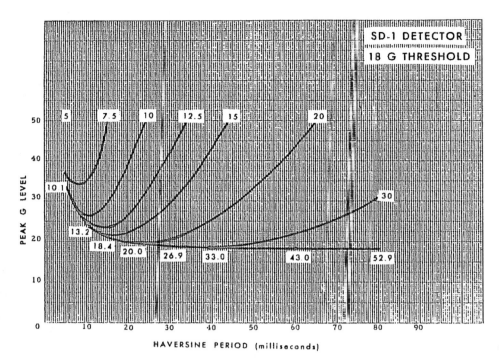

Fig. 18 - Response curve for 18 g detector

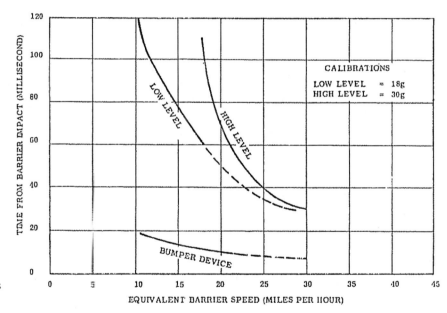

Fig. 19 - Two-stage system requirements (detector upper limits)

The diagnostics consist of a bridge network with monitoring points at strategic locations, essentially one monitoring point each for the passenger inflator, driver inflator, and BID circuit. A 10 mA current, considerably below the no-fire current of squibs, is circulated continuously through these three circuits. If an open or short-to-ground would develop, the voltage would change at the monitoring points. The monitoring points feed into comparators, and a signal less than 3 V or greater than 9 V will develop an output signal which controls the lamp driver transistor. Fig. 21 is a simplified diagnostic schematic of the diagnostic circuits. Fig. 22 shows the assembly of the crash sensor which includes detection, battery backup, and diagnostics.

CRASH MONITORING AND FAILURE RECORDING

To facilitate the reconstruction of what happened after an accident or inadvertent deployment, or both, crash monitoring was included in the system. This circuitry is also included in the sensor. The monitoring provides a record that the crash occurred prior to inflation and within design limits. It records the event sequence of crash/bag inflation and records the length of time the vehicle was operating with a failure diagnosed.

Crash level monitoring is accomplished by a separate set of detectors which when closed, blow fuses. One detector, set slightly above the crash detector, blows a fuse if a low-level

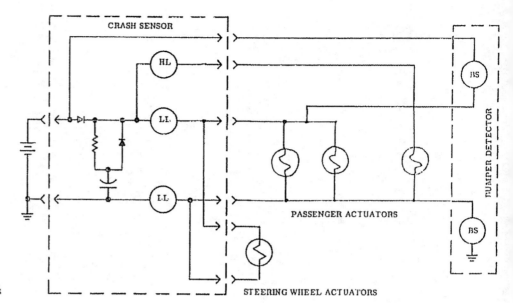

Fig. 20 - Actuation circuits

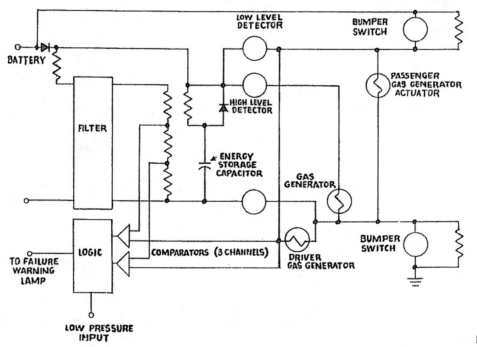

Fig. 21 - Diagnostic circuit

crash occurred which required ACR system protection. The closure also sends power to another fuse which will blow only if the passenger's bag was not deployed. A second detector is calibrated at a level which closes when a crash beyond the capability of the system occurs. This detector blows a fuse if this high-speed crash occurs.

In addition to the fuses there are two actuation signal recorders. One records when the driver's system receives an actuation signal, and the other records when the actuation signal is received from the BID. The signal is not recorded if the driver's bag or passenger's bag is inflated.

The length of time the vehicle has been operated with a malfunction is recorded by a microcoulombmeter. This device plates silver when the warning lamp is on. The time is

determined by replating the silver with test equipment after the event.

A block diagram of the crash monitoring and recording is shown in Fig. 23 and the assembly in Fig. 24.

All of the monitoring and recording circuits have been designed fail-safe. By this we mean any failure will not affect the functioning of the ACRS.

The crash detection and actuation circuits, including the battery backup energy source, the diagnostics and the crash monitoring and recording, are all included in one component shown in Fig. 25.

AIR CUSHION RESTRAINT STEERING COLUMN - The addition of the air cushion to the driver impact protection system and the removal of belt restraints has significantly

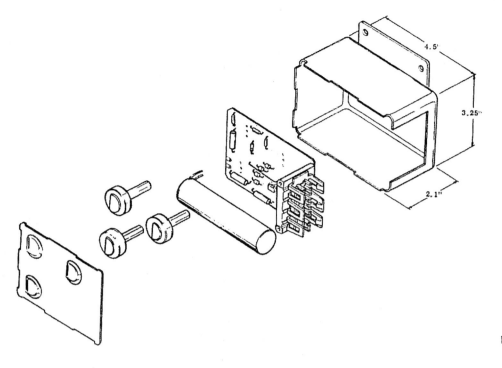

Fig. 22 - Sensor assembly

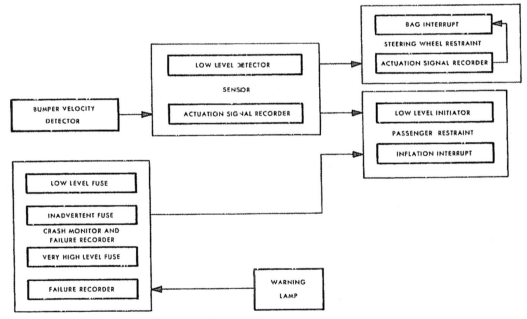

Fig. 23 - Crash monitoring and failure recording block diagram

altered the performance requirements of the GM energy-absorbing steering column assembly.

When the ball energy-absorbing locking column was being designed for introduction on 1969 GM vehicles, one of the prime design goals achieved was the potential flexibility for the unknown requirements of the future. Because of this flexibility, Saginaw Steering Gear Division will be able to meet the new performance requirements of air cushions without major modification of the basic energy-absorbing elements.

The areas of steering column modification are those due to

the following:

1. The vehicle plenum mounting system.
2. The variable energy-absorption rate.
3. Increased strength requirements.
4. The addition of an electrical circuit to fire the wheel air cushion.

Mounting System - As previously stated in the section on overall system operation, we find significant advantages in the new plenum mounting system. Taking another look at this system (Fig. 26), you can see that the column is entirely sup-

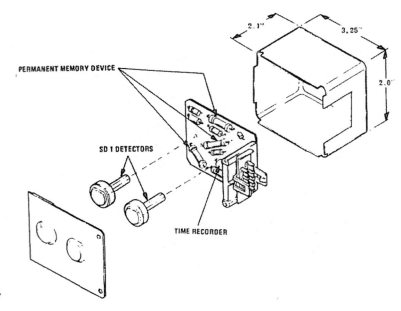

Fig. 24 - Crash monitor assembly

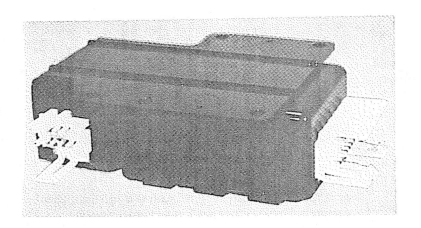

Fig. 25 - Integral crash sensor and motor assembly

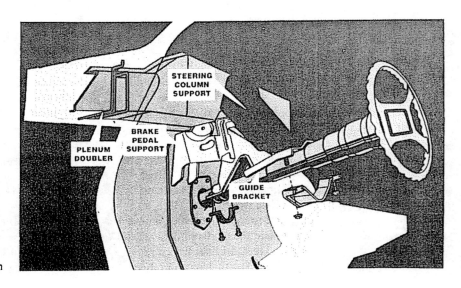

Fig. 26 - Plenum mounting system

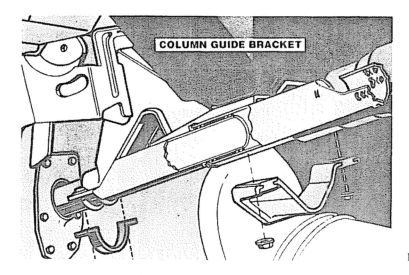

Fig. 27 - Column guide bracket

ported by the guide bracket. The only modification required for the column to fit the new guide bracket is a shortening of the lower jacket and changing the "thumb bumps" on the bottom of the lower clamp. The thumb bump and clamp area will provide the anchor when the driver compresses the column.

This mounting system makes the steering wheel and air cushion an improved target for the driver to react against during a frontal collision. Since the new guide bracket (Fig. 27) spans the energy-absorption area of the steering column, we can now isolate the energy-absorption capability entirely for the driver, with no loss to the frontal collision, and we can tune the absorption rate to provide the capability where and when required.

Energy Absorption - To obtain the desired energy-absorption rate for the new system, we considered two factors. First, the relative velocity of the moving driver contacting the air cushion on the steering wheel produces an inertia load spike of very short duration. The load spike lasts until the wheel and column head mass is accelerated. Thus, an initial low resistance load in the column is desirable. Once past the load spike, however, it is desirable to absorb the maximum amount of the driver's energy. This requires a significantly higher secondary load in the column.

By sled testing, we have determined that the total energy-absorption capability of the column could be raised from the current 4800 in-lb to approximately 7400 in-lb.

This need for a variable absorption rate (Fig. 28) has been accomplished by using a larger ball size than currently used, between the telescoping jackets, to raise the basic jacket status collapse load to a mean of 1200 lb and then "pretracking" the balls for the first inch of travel to reduce the load for that distance to approximately 200 lb. "Pretracking" is an assembly process achieved by pressing the jackets, ball sleeve, and balls together in an axial direction to a given distance shorter than the required overall length, and then extending the jackets to the correct overall length. Thus, the balls have a preformed path to follow for a distance at a low load and then

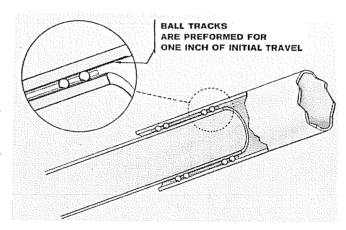

Fig. 28 - Variable rate energy absorber

must form a continuation of that path at a high load, when compressed by the driver in a frontal impact situation.

Structural Revisions - Sled and barrier testing have demonstrated that the vertical load trying to bend or jackknife the column has increased substantially. This is due to the increased energy requirement, the action of the cushion, and the kinematics of an unrestrained driver

There are several areas of the column which must be strengthened to withstand this new load requirement.

The jacket energy-absorption joint (Fig. 29) has been strengthened by increasing the number of balls from 32 to 40 and spreading the balls over a longer span to achieve lower unit ball loading due to bending. A "skid" has been added to limit ball loading.

Stronger bracket weld nuts (Fig. 30) are used to better retain the upper energy-absorbing mounting bracket to the column jacket.

Additional areas requiring strengthening against the high vertical load include the use of four larger screws to attach the head of the column to the jacket and thicker sections in the lock housing die casting.

Electrical Circuit - A new requirement of the steering col-

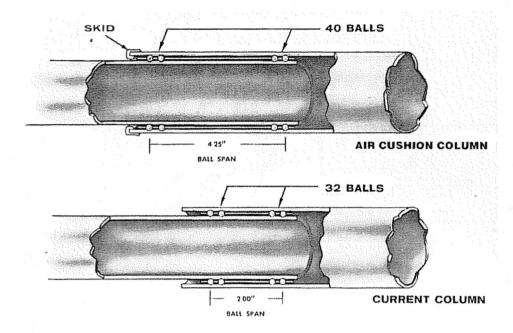

Fig. 29 - Jacket energy absorption joint

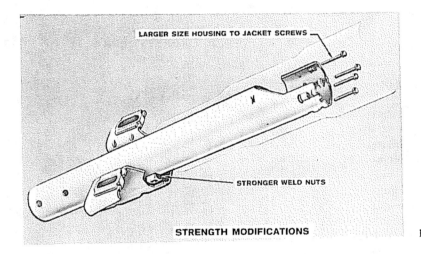

Fig. 30 - Strength modifications

umn is to provide a mechanism to electrically connect the squib in the steering wheel gas generator to the impact sensing mechanism.

The completed electrical circuit (Fig. 31) is designed for a dual function of firing the squib under proper circumstances in frontal collisions, and monitoring the system at all other times to indicate that the system is capable of firing when and if required.

The column circuit is further complicated since it must bridge the gap between the stationary column head and the turning steering wheel. This is the only junction of its type in the entire air cushion wiring system.

The high durability and reliability requirements, coupled with an unusually low electrical resistance allowance, require a mechanism of advanced technology. Saginaw has developed a redundant sliding brush system, utilizing high silver content contacts sliding on silver-plated collector rings, suitable for automotive mass production.

The wires and sliding brushes are shielded against stray radio-frequency radiation, and the mechanism is tuned to resist vehicle vibration frequencies.

The mechanism is housed in the upper lock housing casting, which has been increased in size to contain it and provide for wire passage. The larger housing also has more inherent strength, as previously discussed.

The upper column mounting bracket will maintain its current construction but will be wider to meet the clearance requirements generated in the plenum area by the larger upper column head.

Option Column Program - The same basic design concepts will be extended to the tilt, and tilt and telescoping option columns. Two areas become considerably more complex: strength requirements in both option columns, and wiring in the tilt and telescoping option. In addition, Saginaw has designed an option column head which has a very high degree of standardization between the two types of option columns.

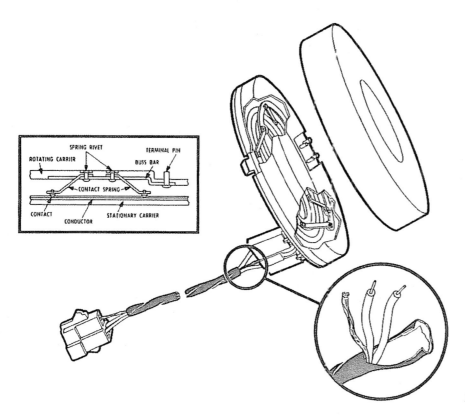

Fig. 31 - Completed electrical circuit

These optional column assemblies are currently being developed for a later introduction date.

In conclusion, the GM family of air cushion restraint steering columns has been specifically designed to meet the exacting standards of the air cushion system and is being subjected to a rigorous test schedule in preparation for production release.

AIR CUSHION STEERING WHEEL SYSTEM - The principal function of the air cushion steering wheel is to provide optimum distribution of impact forces to the driver's head and torso in an impact situation. In accomplishing that function, the steering wheel system has been developed with several additional considerations:

1. Minimum weight added to the steering wheel.
2. Maximum visibility of the car's instruments.
3. Optimum drivability.
4. Accessibility of horn mechanism.
5. Durability and protection of the system components.
6. Compatibility with car assembly operations.
7. Component and assembly reliability.
8. Serviceability of non-air-cushion parts.

Air Cushion Module and Steering Wheel Assembly - The production system for the 1973 air cushion field test program is shown in Fig. 32. The molded vinyl wheel assembly and the air cushion module assembly are attached to the steering column in the same manner as regular production steering wheels and horn shrouds.

Steering Wheel Development - The steering wheel molding assembly, as shown in Fig. 33, is a unique design which took 14 months and 220 sled tests to develop.

Using these sled tests, which encompased a matrix of

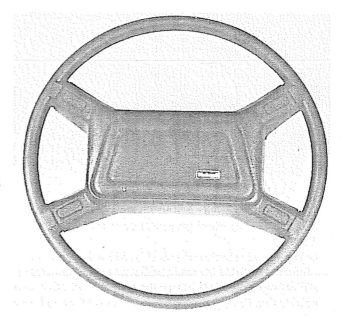

Fig. 32 - Wheel and module

dummies from the 5th to the 95th percentile size in several positions of seating, a specific bending mode and rate of bending was established. Determining this rate was further complicated by the necessity of providing a proper foundation for the air cushion, particularly during angular barrier simulation. It is necessary to achieve a balance of initial conformation of the rim and upper spoke area and maintenance of the wheel integrity in order to transfer the majority

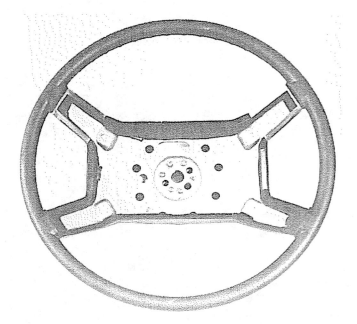

Fig. 33 - Wheel molding assembly

1000 CAR-FULL RIM LOADING

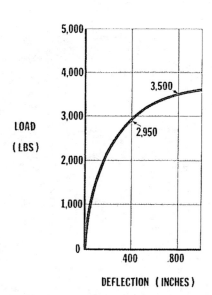

Fig. 34 - Wheel rates, 1000 car-full rim loading

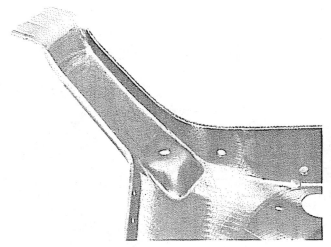

Fig. 35 - Spoke

Fig. 36 - Air cushion module

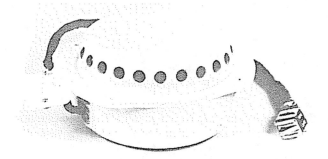

Fig. 37 - Inflator

of the force axially to the steering column. The static rates resolved for this particular system are shown in Fig. 34. It should be pointed out that these rates are tuned to a specific set of conditions, which are partially dictated by the steering column design and load/deflection rate plus the knee restraint position and its crush rate.

The steering wheel starts with a deeply formed steel stamping, which is placed in injection molding machine and covered with polyvinyl chloride material. The heavy channel cross sections in the spokes, as seen in Fig. 35, are necessary to provide the flexural strength required of the total assembly.

During an angular impact, the wheel is subjected to extreme rim loadings owing to the bending moment created by the driver impacting the inflated cushion. Note that the hub area of the molding assembly accommodates the production steering column shaft and horn tower. The larger oval-shaped hole at the bottom of the spoke area is for the connector assembly for the inflator.

Fig. 38 - Fabric of cushion

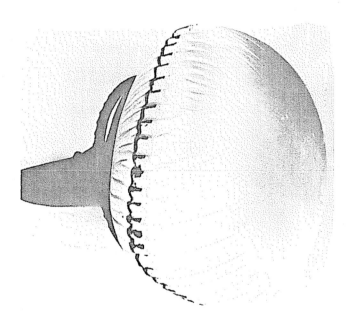

Fig. 39 - Inflated cushion

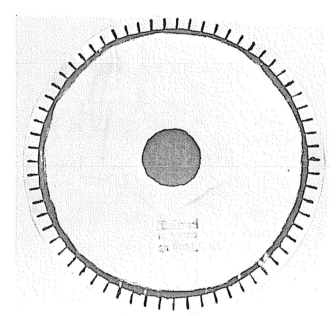

Fig. 40 - Flat cushion

Air Cushion Module Development - The air cushion module is an even more complex development. As viewed in Fig. 36 it has the unique feature of combining the functional components of an air cushion system along with the other essential components of the steering system, such as the horn mechanism, in a consolidated package, which is highly durable, easy to drive, and compatible with the car interior.

In establishing the optimum combination of air cushion size and packaging, several criteria were considered. The air cushion was developed with the objective of meeting the federal MVSS 208 requirements during a 30 mph frontal barrier impact and including tolerance of angular impacts of 30 deg to the left or right. Correspondingly, it is important to consider the overall incidence of driver seating positions and driving attitudes. That is to say, the inflation force

transmitted to the driver must be minimized through design of the cushion. In our system, the cushion was selected by a series of sled and car barrier tests, and it provides the required 30 mph impact performance while acting in combination with the passenger cushion. The steering wheel cushion transfers the force of the driver impact to the rim and spokes of the wheel, as the energy is absorbed by the collapsing steering column.

The module includes a protective plastic container for the air cushion and the inflator. This container also houses the horn mechanism assembly; and during the cushion deployment, a special printed circuit attached to the container registers the inflation in the actuation recorder, which is also located within the module. So, in operation, the module provides a durable environment for the air cushion compo-

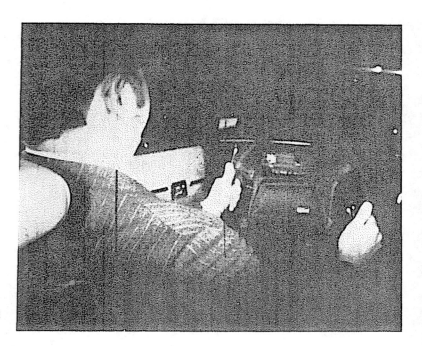

Fig. 41 - Module prior to deployment as electrical signal is acquired

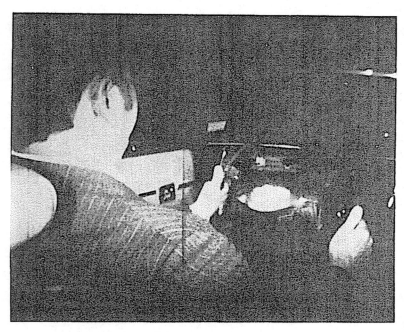

Fig. 42 - Initial opening of vinyl cover as the inflator generates gas and the cushion begins to deploy

nents, it acts as a platform for touch-blow horn mechanism, and it records the air cushion inflation as it occurs. The use of the module concept has also contributed to reproducibility of the cushion's inflation kinematics, while consolidating the system into a package which is compatible with visibility of the car's instrumentation cluster.

The module assembly is composed of a total of 29 different part numbers; and counting all of the internal parts of the inflator and the assembly fasteners, the total number of parts within each module comes to over 100.

Inflator - To make best use of the available space in the steering wheel environment, a chemical gas-generating inflator has been utilized, as shown in Fig. 37. This device contains no

stored gas supply, and it permits us to utilize the rather extraordinary geometry of the available space in the steering wheel without the dimensional limitations imposed by using a high-pressure cylinder.

In developing the inflator, it was necessary to establish a device which was compatible with available cushion materials while the exhaust composition maintained acceptable levels of undesirable gases and particulates. We have achieved these system objectives while using propellents designed and compounded for long-term durability and life.

The system has been subjected to several toxicity exposure tests without adverse consequences. The degree of engineering in developing the inflator has been perhaps the greatest

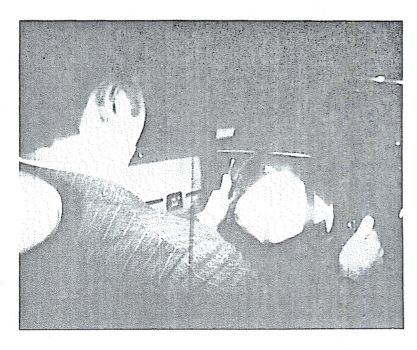

Fig. 43 - Full opening of cover as the cushion begins to open

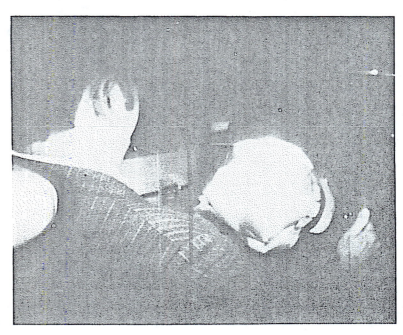

Fig. 44 - Outboard folds of the cushion open

accomplishment in the air cushion steering wheel. The 30 ms conversion of 100 g of propellent into a filled cushion involves reaction temperatures of 2500°F, while the temperature rise of the cushion's surface material is held to 70°F (that is, $\Delta T_c = 70°F$).

The inflator used in the field test program is just over 4 in in diameter, about 2 in deep, and it weighs 4 lb. It is designed with clearance for the steering column shaft, the horn mechanism tower, and it has a mounting bracket for the actuation recorder unit. When the sensor induces the electrical signal, the inflator responds with 2 ft^3 of gas delivered in 30

ms. To date, over 1200 test firings have been conducted in the development of the Inland steering wheel system.

Cushion - In development of the air cushion itself, three parameters were basic: material selection, size and shape, and packaging. Neoprene-coated nylon, as shown in Fig. 38, satisfied the performance requirements of the air cushion system, and it has a background of military and recreational applications.

From 18 months of sled testing came the cushion's inflated dimensions of 23 in diameter and 14 in depth above the rim (Fig. 39). Its size is tuned to the steering column and the knee

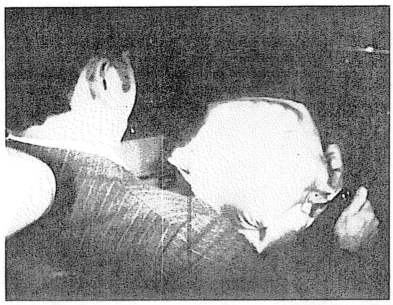

Fig. 45 - Inboard folds of the cushion open

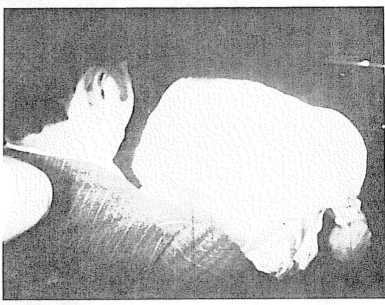

Fig. 46 - Cushion deploys toward driver position

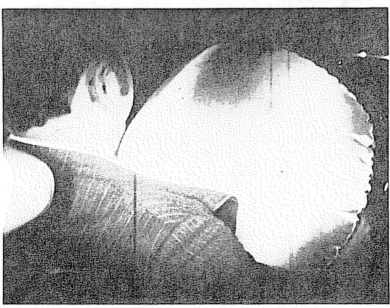

Fig. 47 - Cushion is fully inflated

restraint, and it intersects the inflated envelope of the passenger cushion, thereby providing a continuous span of protection across the car interior.

Packaging is accommodated by the cushion's simplicity. When deflated, it is 28 in in diameter, composed of two flat discs of material (Fig. 40). It is accordion folded along 12 axes into a rectangular package.

The inflator and air cushion function as shown in the following staged inflation sequence are viewed from over the driver's shoulder (Figs. 41 through 47).

NHTSA'S EVALUATION OF AIR CUSHION RESTRAINT SYSTEM EFFECTIVENESS (ACRS)

DONALD F. MELA
Office of Statistics and Analysis
National Highway Traffic Safety Administration

746023

INTRODUCTION

The General Motors Air Cushion Restraint System (ACRS), popularly known as the "air bag," is being installed in several thousand 1974 U.S. General Motors passenger car production models.

There is intense interest in the performance of this passive restraint system because the occupant passive restraint safety standard* is presently scheduled to take effect in two years (September 1976). Therefore, the evaluation of the ACRS effectiveness is a high-priority program for NHTSA. Let me describe how NHTSA is carrying out the evaluation.

The material in the paper developed by a task group of members of NHTSA's Office of Statistics and Analysis (formerly Office of Accident Investigation and Data Analysis). William E. Scott, C. J. Kahane, John Keryeski and Scott Lee were the principal contributors, and should be given credit for the work that is described here.

General Motors Corporation in 1973 announced its aim to produce and sell 50,000 full-size ACRS-equipped cars in the 1974 model year, and 100,000 1975 model cars. The introduction of these cars began about January 1, 1974. Subsequently, there has been a drastic reduction in the expected rate of introduction of ACRS vehicles. It now appears (May 1974) that the number of ACRS cars on the road in the United States by the end of 1974 will not exceed 8,000 cars. So the original NHTSA estimates of crash data availability have also had to be reduced.

OBJECTIVES

The basic objectives of the evaluation program are:
1. Assess the injury-reducing effect of ACRS.
2. Determine operational characteristics of the ACRS.
3. Evaluate public/owner acceptance.

This discussion will be concerned mainly with the first objective, but objectives two and three will also be discussed briefly. As noted above, the drastic reduction that has occurred in the rate of introduction of ACRS vehicles has made it necessary to replan. However, much of the general approach and many of the specifics will either be preserved or will be utilized at a later date, when more ACRS

---

*Standard 208.

vehicles are on the road. Therefore, it is felt that the information should be of interest to participants in this Conference, particularly since it affords the opportunity to comment and to influence the later conduct of the evaluation.

GENERAL CONSIDERATIONS

Based upon the expected sales rate, it had been estimated that there would be about 12,000 ACRS vehicles in crashes by the end of August 1975. This includes all accident types from minor to total demolishment. It also includes crashes in which the ACRS is not deployed (ACRS is designed only to deploy in impacts with a significant frontal component). Of these 12,000 crashes, about 4,000 would damage the vehicle severely enough to require towaway and 1,300 crashes would result in some injury. Table 1 provides a further classification, according to severity level of the injury.

The estimates in Table 3 were developed from previous accident experience for vehicles similar to the ones expected on the road. The injury frequencies presented would be expected if all occupants in the control group were unrestrained. The observed

Table 1
INJURY SUMMARY

ACCIDENT SEVERITY	PARTS OF THE BODY AFFECTED					
	HEAD	NECK	THORAX	ARMS	UPPER LEGS	LOWER LEGS
VDI = 1						
VDI = 2						
VDI = 3						
VDI = 4						
VDI = 5						

Table 2
PREDICTED ACRS CRASH EXPERIENCE
(Predicted in December 1973)

VEHICLES/EVENTS	LEVEL OF INVOLVEMENT
12,000	ALL POLICE REPORTED ACCIDENTS
4,000	TOWAWAYS
1,300	TOWAWAYS WITH MINOR OR GREATER INJURY
360	TOWAWAYS WITH MODERATE OR GREATER INJURY
120	TOWAWAYS WITH SEVERE TO FATAL INJURY
20	TOWAWAYS WITH FATAL INJURY

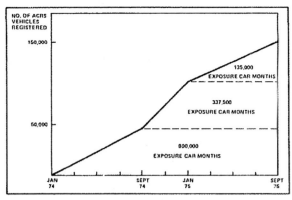

Figure 1. Original Schedule for Planning Accident Investigations (Basis for Table 1)

frequencies of injuries in ACRS crashes should be much lower.

To compare the ACRS with active restraint systems such as safety belt systems, we ask two general questions:

1. Is the overall safety performance of the ACRS better than that of cars equipped with devices such as the ignition interlock or buzzer supplements to safety belt systems?

2. How do vehicle occupant injury rates compare for persons involved in crashes who are protected by
   a. ACRS,
   b. Three-point belt,
   c. Lap belt, or
   d. With no restraints?

This identifies several groups for comparisons with the ACRS fleet.

1. Interlock Fleet — 1974 and 1975 model years.
2. Buzzer Fleet — 1972 and 1973 model year cars.
3. Car occupants wearing lap and shoulder belts, 1968 and later model years.
4. Car occupants wearing lap belts.
5. Unrestrained occupants.

Since the ACRS will be in full-size 1974-75 Cadillacs, Buicks and Oldsmobiles, these control groups should be restricted to cars of the same size and body construction, viz., full-size 1973-75 Cadillacs, Buicks, Oldsmobiles, Pontiacs and Chevrolets.

Rationale for Objective #1

Ideally, it would be best to compare injury performance of the ACRS and control group for the "same" exposure or accidents. There are two well-established approaches to the ideal of "sameness" that is unattainable in real-life:

1. Assume that, overall, ACRS and control-group cars have similar exposure or accident experience and that any difference in gross injury rates is attributable to the restraint system alone.

2. Use some criteria to stratify the exposure units or accidents into sub-groups. The stratification is of a sort that corresponding sub-groups may be compared using approach (1). A statistical method is employed to aggregate the comparison of the sub-groups into a net comparison of ACRS and control group.

Approach (1) is rejected because there is no evidence to support its underlying assumption. ACRS-car owners may well be highly unrepresentative of the automobile population with regard to both exposure and accident experience. In fact, some have conjectured that ACRS buyers would be extremely safety-conscious, while others felt they would be reckless drivers eager to evade the interlock system.

Therefore, Approach (2) was selected. However, the use of "exposure units" had to be rejected. At this time, there is no wide recognized method for stratifying the exposure units — e.g., sub-dividing the miles driven into "dangerous" and "non-dangerous" categories — making it impossible to use fleet exposure as a basis for stratifying until such theory is developed.

On the other hand, there are four well-established means for stratifying accidents into severity classes: accident descriptors used by the police, such as "pre-impact speed"; economic descriptors such as "dollar damage"; damage descriptors that can be measured by looking at post-crash photographs, such as "inches of crush"; and highly sophisticated engineering descriptors such as "velocity-vector change during impact."

Accident and economic descriptors are highly inadequate because past experience show them to be poorly correlated with real accident severity, subject to large errors of measurement, and subject to different methods of measurement in different States — an important consideration since the sparse ACRS data will require collection in more than one State. Dollar damage has year-to-year inconsistencies due to secular economic trends.

The engineering descriptors were considered the best measures of accident severity, but were rejected for three reasons: (1) some of the descriptors, such as "vehicle aggressiveness" have not yet been adequately defined; (2) the methods for measuring them, either crash recorders or the Calspan computerized accident reconstruction program, would not be available in time; and (3) these methods are exceedingly expensive.

There is a damage descriptor presently being used on an international basis — the Vehicle Deformation Index (VDI). It is described in the SAE recommended practice J224a, Reference 3. *With vehicles that are of essentially the same body style and with similar crashworthiness characteristics,* the VDI is considered to give a good comparative measure of the magnitude of impact forces sustained by a vehicle. The VDIs may be readily and inexpensively calculated by looking at photographs of the damaged vehicle.

It was decided to collect photographs and construct VDIs while, left open, was the option to extend data collection at a later time in obtaining engineering descriptors. This procedure was followed just in case the VDIs failed to come up to expectations, or in case there were unforseen advances in the theory and measurement of engineering descriptors.

An examination of the inputs determined the following analytic approach:

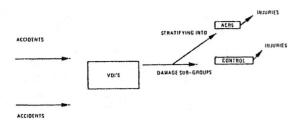

Finally, the expected data on injury production were examined. Usually "injury severity" is coded for each *occupant* of the vehicle, but it can also be construed as a vehicle characteristic by using some composite measure such as "worst injury in vehicle." The latter approach was rejected because, for example, if the control group consists of occupants wearing "lap belts," then "worst injury in vehicle" cannot be fairly defined for a vehicle in which some occupants used belts and others did not.

Thus, injuries per occupant were considered, but here again there were several possibilities: occupant injuries could be stratified according to seated position, or seated position could be ignored. For this study, a middle course was taken. Since the ACRS would have no influence whatever on back-seat occupants, only front-seat occupants would be studied. Moreover, further stratification would dilute the statistical significance of the sparse ACRS data. Hence, a restriction of output "injury production" to "injuries to front-seat occupants" was effected.

The ACRS system is not expected to totally eliminate all injuries, but rather to lower their *severity*. It is, therefore, imperative that the measure of injury production be not a "yes-no" tally, but a severity scale.

The doctor-reported Abbreviated Injury Scale (AIS) defines severity levels in precise medical terms that are fairly consistently interpreted nationwide and thus, would properly distinguish intermediate

levels of injury. It was decided to obtain a medical report on each injured front-seat occupant, and to compare ACRS and control-group injury production at each severity level from minor through fatal, but with special interest in the intermediate levels.

Following the decision to obtain injury rates at each level of severity, analytic methods for comparing the respective rates of ACRS and Control Groups were considered. In addressing the primary question, we queried whether statistical comparison should be a qualitative or quantitative statement. A *qualitative statement* would take the form: "Front-seat occupants of ACRS cars had a significantly lower severe-injury rate than unrestrained occupants"; a quantitative statement would be: "Front-seat occupants of ACRS cars had a 25 ± 5 percent lower severe-injury rate than unrestrained occupants."

The qualitative statements are second-best for a cogent reason: when comparisons are made between the ACRS and belts, it is expected that the ACRS may perform *better* with regard to severe injuries and worse on less severe injuries.* This is expected because the bag, if it deploys, should be superior to the belt; however there will be many less severe injuries in crashes where the bag (through design) fails to deploy.

Therefore, unless cost-benefit analysis is used to determine *how much better* and *how much worse*, it will be impossible to decide which restraint system is superior. (See Tables 2 and 3 for examples of quantitative results.)

Three acceptable statistical methods for comparing injury rates for a population stratified by damage severity were considered:

1. Stratify the ACRS and control-group front-seat occupants into four or five classes, according to the VDI extent and impact direction of the crash involved vehicle. Compare injury rates for corresponding strata. Take an appropriate weighted average over the strata to estimate net injury reduction.

2. For each front-seat occupant of an ACRS car, search the file of control-group cases and find the one which best matches the ACRS car with regard to numerous vehicle damage descriptors and other characteristics. Obtain two groups of vehicles of equal size who are so "similar" that any significant difference in gross injury rate of occupants is entirely attributable to the different restraint systems.

3. For each front-seat occupant of an ACRS car, search the file of control-group cases and find

---

*Unbelted occupants of ACRS cars may suffer minor injuries in crashes below the deployment speed.

about 20 that closely match the ACRS car. Compare the ACRS occupant's injuries to the median injury or to other percentiles of the corresponding 20 or, perhaps, do a Ridit analysis, Reference 4. Some of the ACRS occupants will fare better than the median, some the same, and some worse. An appropriate non-parametric test can be applied to the list of comparisons to determine which system, if any, is significantly better.

Only Methods (1) and (2) give adequate quantitative comparisons. Method (1) is preferred because it requires fewer and less detailed data than for (2) to attain a given level of precision.

In the contingency that the ACRS fleet is much smaller than expected (which now appears to be the case) Method (1) will not yield statistically significant quantitative comparisons. This is precisely where Method (3) is best, that is, when the test group is small and the control group large.

Hence, if things were still to go as originally planned Method (1) would be used to obtain quantitative results, but due to the apparent reduced ACRS fleet contingency, Method (3) will probably be used for qualitative results.

Most of the conceptual model was constructed simply by carefully examining the objectives. Figure 2 displays the model constructed. The details of the operational plans were not described in the rationale presented in the preceding pages. These include:

1. Details of notification, sampling, data collection, and analysis chosen to respond to objectives #2 and #3 (e.g., the collecting of more detailed data on injuries in deployment accidents).

2. Sampling techniques used to optimize statistical efficiency — i.e., maximize statistical precision per data-gathering dollar (e.g., use of stratified rather than simple random sampling).

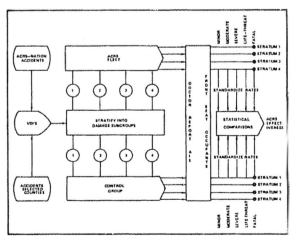

Figure 2. ACRS evaluation program model.

3. Real-world constraints that force the collection of extraneous data or prevent the collecting of data.

Rationale for Objective #2

Evaluate the performance of the ACRS in those accident events for which protection was designed.

The first point which must be clarified is the definition of "Operational Characteristics." This is taken to include the following:

1. The types of crashes which deploy the ACRS.
2. The types of injuries that result from ACRS deployment.
3. The tendency of the ACRS to deploy inadvertently.

The Types of Crashes Which Deploy the ACRS: What one would ideally want would be a polar plot as shown where

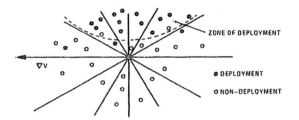

the principal direction of force (θ) and the change in velocity of the vehicle during impact (ΔV) would be plotted for accidents in which the bag both did and did not deploy. The "zone of deployment" would thus characterize those accidents which result in ACRS deployment.

To construct such a plot one would need detailed information concerning a significant number of accidents: "detailed" means either crash recorders on the vehicles or a computerized accident reconstruction program used together with an investigating team to estimate ΔVs and directions of force from physical evidence at the scene. This degree of data-taking cannot be done on every accident and hence no true engineering description of the "zone of deployment" can be obtained.

An inferior but perhaps useful description of the type of accidents that result in ACRS deployment, is a graph of mean or median VDIs and principal direction of force, such as shown in Figure 3.

One would not expect the correlation of deployment with the VDI to be as strong as with ΔV. Both of these data reductions require data on crashes where the bag did not deploy. So it is necessary for NHTSA to investigate a representative sample of ACRS non-deployments.

A third option is merely to summarize those VDIs and θs where ACRS deployment occurred. This could

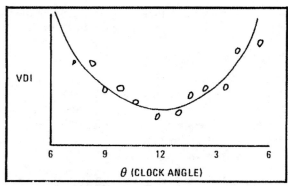

Figure 3. ACRS deployment by VDI and principal direction of force (θ).

be misleading since we could not say whether accidents outside this range would have caused deployment. However, the data might be useful to point out some unexpected deployment modes (e.g., side impacts). This could be done, of course, using the data from the hundreds of deployments that will be investigated.

The Types of Injuries That Result from Crashes Where the ACRS Deploys: Again the ideal is a polar plot showing injury levels (AIS) as a function of ΔV and θ using only deployment data, for example, as shown below:

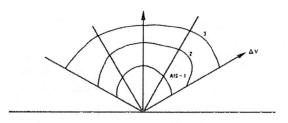

However ΔVs will not be obtained, so we must depend upon the next best indicator of crash severity, VDI, and consider plots like the ones in Figures 4 and 5.

Again, one would not expect the correlation of injury level with VDI to be as strong as with ΔV.

It is expected that among 460 random deployments there would be at most 240 injuries (.52 x 460), of which only about 80 (.175 x 460) will be moderate or worse, 32 severe or worse, 17 life-threatening or fatal, and 8 fatals. Obviously then, constructing the low-injury level contours should be possible (with up to 100 data points). The moderate injury contours (with up to 20-25 data points) might be possible. Clearly, insufficient data will preclude constructing the higher injury severity contours. This justifies additional investigations of accidents with severe and worse injuries.

So far only the severities of the injuries where deployment occurs has been discussed. Certainly we

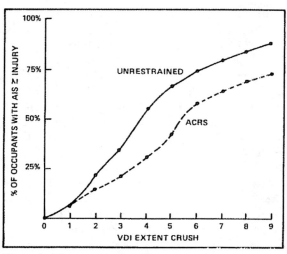

Figure 4. Injury rates and accident severity.

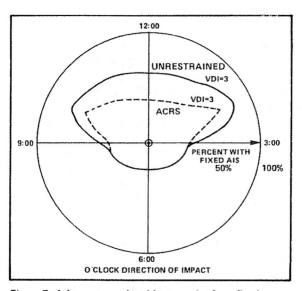

Figure 5. Injury rates and accident severity for a fixed extent of crush.

also would like to know about the nature and mechanism of the injuries.

Lower limb injuries at all levels of accident severity and perhaps some neck injuries in otherwise fatal level crashes can be expected. Fatal injuries will probably be due to ejection, or severe underride/override frontals. Table 2 exemplifies a tabulation summarizing these injuries.

The Tendency of the ACRS to Deploy Inadvertently: What is desired is the inadvertent firings per vehicle mile. It is probable that NHTSA will know of inadvertent firings. It will require more effort to determine the total vehicle mileage accumulated at any point in time. Estimates will probably have to be made based on owner surveys and odometer readings.

1150

To evaluate owner/public acceptance we must determine how the general population of new car buyers will react when they purchase a new car equipped with an ACRS.

The method to achieve this would require owner surveys to determine the reaction to the General Motors ACRS. There are problems with this approach in that those people who will purchase the GM luxury class vehicles with the relatively expensive ACRS option are likely not representative of the new car buyer population. For the purpose of this survey, this and other similar potential biases must be regarded as setting upper bounds for favorable reaction. It is simply anticipated that the biases are not significant and that this group of purchasers will be similar to that of the general population of new car buyers. There would be less concern if the ACRS option was available to all potential new car buyers.

The owner survey function is planned to be carried out under contract. The following indicates the type of information that can be obtained.

DATA TO BE OBTAINED FROM OWNER

1. Demographic Information
   - Age
   - Sex
   - Occupation
   - Education

2. Vehicle Data
   - Model
   - Air bag
   - Air bag with lap belt
   - Date of purchase

3. Use of Vehicle
   - Work
   - Pleasure
   - Other
   - Annual mileage (driven by respondent)
   - Mileage at time questionnaire is answered

4. Restraint System (Previously Owned/Driven Car)
   - Year and model of previous car
   - Use of lap and shoulder belts in this car
   - If 1972-73 car, was warning system defeated or circumvented

5. Restraint System (Air Bag Car)
   - How learned of air bag option
   - Why chose air bag car
   - Understanding of system operation
   - Why or why not chose additional lap belt
   - Use of lap belt
   - Reason for use or nonuse of lap belt
   - Modification or tampering with system(s)

- Satisfaction/security derived from system and basis for such
- Reaction to system by friends, relatives, etc., when they are driven in car

6. Restraint System Defects (Air Bag Car)
   - Nature of defect
   - How it was discovered
   - Was it repaired, by whom, cost

7. Accident Involvement
   - Has ACRS vehicle been involved in an accident
   - Did air bags deploy
   - Was anyone in ACRS vehicle injured
   - No treatment required, treated and released, hospitalized
   - Major damage was sustained to the front, rear, or sides
   - Was vehicle towed from the scene
   - Repair cost

ANALYSIS OF THE DATA

Objective #1. Injury Reduction

From general considerations, it was concluded that the best course of action was to collect ACRS crash data; to obtain or collect control-group crash data; to stratify each data set into four classes according to vehicle damage; to estimate, for various AIS injury levels, the injury rates for front-seat occupants in test and control groups, for each stratum; to estimate the "true" sizes of each stratum for the "total" vehicle population; to obtain the net injury rates by taking the averages weighted by the "true" sizes of the strata; and to give some confidence range on the percentage injury reduction (or increase) due to ACRS with respect to the other restraint system.

For each crash, the Vehicle Damage Index (VDI), including "damage extent" (measured on a 0-9 scale) and "direction of impact" (measured by "clock direction") will need to be obtained. Also required for each injured front-seat occupant will be a medical report, including the overall injury severity as measured using the Abbreviated Injury Scale (AIS).

Data of these kinds must be collected by technicians, who, in turn, must be notified by the police that an accident has occurred. A realistic notification threshold is that either the car had to be towed from the scene or that an occupant had a disabling injury. Non-injury, non-towaway accidents are more elusive.

Table 3 gives the nationwide totals for the 20-month time-frame from January 1, 1974, to August 31, 1975, for towaway-involved, front-seat occupants of ACRS cars, and for front-seat occupants of the 1973-1975 full-sized GM cars equipped with

	ACRS Group	Control Groups		
Cummulative Injury Severity	ACRS 1974-75	Unrestrained 1973-75	Lap/Shoulder 1973-75	Lap 1973
All Towaway-involved front seat occupants	5,300	220,000	130,000	50,000
All AIS ≥ 1	1,660	68,700	40,150	15,600
All AIS ≥ 2	430	17,600	10,700	4,000
All AIS ≥ 3	140	5,760	3,500	1,300
All AIS ≥ 4	46	1,850	1,100	420
All Fatalities	22	900	500	200

TABLE 3

EXPECTED NUMBER OF INJURIES AMONG TOWAWAY-INVOLVED FRONT SEAT OCCUPANTS (20-MONTH PERIOD – FULL SIZED GM CARS – NATIONWIDE)

other restraint systems. It also gives the number of injuries more severe than or equal to a given AIS level, that would be expected if none of the restraint systems were effective.

The reader who has some experience in statistical hypothesis-testing will quickly note that nearly the entire national ACRS occupant population would be needed to detect significant injury reduction at the life-threatening or fatal levels. Each of the control groups, on the other hand, is much larger and need only be sampled.

The four damage strata into which the towaway crashes are likely to be subdivided are defined as follows:

1. Major Frontal Impacts: VDI ≥ 3 and 11:00-1:00; or VDI ≥ 4 and 9:00-3:00; or VDI ≥ 4 and rollover.
2. Nonmajor Frontal Impacts: VDI ≤ 2 and 11:00-1:00.
3. Nonmajor Side Impacts and Rollovers: VDI ≤ 3 and 9:00-3:00; or VDI ≤ 3 and rollover.
4. Rear-end Impacts: All towaways with 4:00-8:00 o'clock angles of impact.

It is anticipated that within each stratum, the ACRS and control groups will be closely matched with regard to accident severity.

The strata were defined this way because it is expected that nearly all of the (non-inadvertent) deployments will be concentrated in one stratum. Essentially, the ACRS occupants are "unrestrained" outside Stratum 1 (or, at worst, Strata 1 and 2). Hence, outside these strata, we may replace or supplement the sparse ACRS-Group data with more easily obtained "unrestrained" data. Further, the other strata would, as a result, contribute no variance when comparing with unrestrained. As a result, the precision of our estimates would greatly increase.

Furthermore, this method is, in a sense, self-checking. VDIs were justified in the foregoing rationale because they were believed to be a good surrogate for engineering descriptors of crash severity for vehicles of similar makes and models. Since the sensor threshold is defined in terms of an engineering descriptor (velocity vector change during impact), most deployments will be in Stratum 1 if and only if this rationale is correct.

The following discussions give the formulas to be used to give net injury rates. A definitive measure of the precision of these rates for various sample sizes is under development.

Let t be the total number of front seat occupants of ACRS towaways in the data collection. Let t_1, , t_4 be the number of occupants in each stratum.

Let t', t'_1, , t'_4 be the corresponding numbers of _unrestrained_ occupants in the various control groups.

Let t'', t''_1, , t''_4 be the lap-shoulder belted and t''', t'''_1, , t'''_4 be the lap-belted.

Let $T = t + t' + t'' + t'''$ be the total number of front-seat occupants in all of the data.

Let $T_i = t_i + t'_i + t''_i + t'''_i$ be the totals for the ith stratum. T_i/T will be used to give an estimate of the "true" size of the ith stratum.

Let x_i be the number of occupants of ACRS cars who are in the ith stratum and who sustained injuries of at least some specified AIS (e.g., the number with severe or worse injuries). Let x'_i, x''_i, x'''_i be corresponding numbers for control groups.

Then x_i/t is the sample estimate of the injury rate for the ith stratum of ACRS occupants and x'_i/t'_i is the estimated rate for unrestrained occupants.

The net rate for injuries at some AIS level to ACRS occupants in towaways is estimated by $R = (x_1/t_1)(T_1/T) + (x_4/t_4)(T_4/T)$.

The estimated net rate for injuries to unrestrained front seat occupants is given by $R' = (x'_1/t'_1)(T_i/T) + (x'_4/t'_4)(T_4/T)$.

If the assumption is correct that most of the deployments fall in Stratum 1, R' may be replaced by $\widetilde{R}$. Then: $\widetilde{R} = (x_1/t_1)(T_i/T) + (x'_2/t'_2)(T_2/T) + (x'_3/t'_3)(T_3/T) + (x'_4/t'_4)(T_4/T)$

Now, $\widetilde{R}$ and R' have a large covariance, and this will improve the precision of $(R' - \widetilde{R})/R'$, i.e., the estimate for the net injury due to ACRS.

Improved Sampling Plan

There appears to be a method for greatly improving statistical efficiency — i.e., getting the same degree of precision with a much smaller number of investigations. In the first approach collection of many thousands of noninjury towaways is only for the purpose of seeing which stratum they were in — i.e., finding t_1, , t_4; t'_1, , t'_4; etc. This is wasteful of data. In fact, after looking at a random sample of only 25-50 percent of the noninjury towaways in ACRS and each control group one may estimate with great precision the distribution of the remaining noninjury towaways among the 4 injury causing strata. Hence, one need only keep a tally of all noninjury towaways and extrapolate their distribution among the strata from the small sample. The disabling injuries, of which there are few, already have to be investigated to find the AIS. It is necessary also to collect VDIs for all of them; there are not enough of them so that they can be precisely distributed among the strata on the basis of a less than 100% sample.

The more efficient sampling plan, then, consists of:

1. Keeping a tally of all towaway crash involved front-seat occupants in ACRS and control groups, and collecting police accident reports.

2. Get the AIS and the vehicle VDI for all occupants for whom the police report stated that they were taken to a hospital — this will include most AIS > 2 injured.

3. Get the VDIs for a random sample of crashed vehicles for which no occupants required treatment. The appropriate sample sizes are now being determined by NHTSA.

As in the first approach, the net injury rates R and R′ are calculated and $(R' - \tilde{R})/R'$ is estimated. The only difference is that t_i, t_i', and T_i are algebraically calculated from a smaller sample, and are subject to more variance than before. Sample sizes will be chosen to optimize the balance of the variance of the t's against the variance of the x's.

The control groups can be obtained by using existing data that classifies VDI and AIS, or by collecting new data. Each control group should be somewhat larger than the ACRS group. In order to obtain the precision desired, we need an ignition interlock crash involvement that will produce 7,500 towaways in the 20-month study period and a 1973 vehicle crash involvement which will have a like number. Fleets of that size will also give us slightly more lap/shoulder belted front seat occupants than ACRS occupants, and twice as many unrestrained occupants as in ACRS vehicles. As indicated in the rationale for Objective #1, these cars should be of the same makes and models as the ACRS cars, i.e., they should be full-sized GM cars, preferably Buicks, Oldsmobiles, and Cadillacs.

Neither Calspan intermediate level data nor the Multi-disciplinary Accident Investigation files which are available have anywhere near the sufficient number of crashes. Motors Insurance Corporation file, which is at this time proprietary to GM, is unusable because of missing data. The VDIs are not coded for noninjury accidents, so it is impossible to determine which of the damage strata they belong to.

Therefore, it will be necessary for NHTSA to collect the needed control group data.

The control group must obviously be collected in some pre-defined areas of the country where either NHTSA or MVMA* accident investigation teams are presently located and where all moderate to fatal injury producing crashes in those regions could be investigated. A systematic sample of the "minor" and "no injury" towaways would also be collected. As for the ACRS group, every moderate to fatal ACRS injury in the country needs to be collected because of the smallness of the population. With regard to the "minor" to "no injury" ACRS towaway crashes, there are two approaches to collecting the sample towaways not involving occupants brought to a medical facility:

1. Collect every ACRS towaway in areas where control group data is collected,** and

2. Collect the towaways across the nation by a systematic sampling procedure.

Approach #1 appears to involve less travel and thus to be less costly, but it is fraught with difficulties:

● It presumes that teams must cover areas large enough to contain the required sample of ACRS towaways.

● The distribution of accidents among the damage strata in the areas may be unrepresentative of the nation. The only clue as to whether they are representative is to compare the percentage of the area towaways resulting in deployment to the national percentage. This clue is worthless if the assumption were incorrect that the deployment population closely resembles Stratum 1 (major frontals). Therefore, Approach (2) is highly recommended.

DATA COLLECTION

Two distinct data collection systems are employed to cover both ACRS equipped vehicles involved in crashes and control group crashes. ACRS data are collected nationally by Multidisciplinary Accident Investigation Teams (MDAI) operating in five regions around the country. Control group crashes are also investigated by the five MDAI teams but in selected counties in each of the team regions.

Notification of the occurrence of a crash involving an ACRS equipped vehicle and the initiation of an investigation is based on a 24-hour-per-day, 7-days-a-week operation of a National Response Center (NRC) located at DOT Headquarters in

---

*Motor Vehicle Manufacturers Association
**Currently, and until more ACRS cars are on the roads, all possible ACRS towaways are being investigated.

Washington, D.C. The control group sampling system depends heavily upon the liaison that has already been established between the MDAI teams and the police agencies in their counties for notification of crashes involving acceptable control group vehicles.

Through the help of the NHTSA regional administrators and the respective governors' highway safety representatives in their regions, it is expected the ACRS evaluation program will receive wide publicity and cooperation. In addition, an explanatory letter containing the NRC toll free telephone number 800/424-8802 was mailed out to all police jurisdictions having a population of 15,000 or greater requesting that the NRC be notified of all traffic crashes in their area involving an ACRS equipped vehicle. General Motors has also requested owners and dealers to provide information. Once the NRC has been contacted, a member of the NHTSA's Accident Investigation staff makes the decision as to what depth the accident will be investigated. The level of investigation is determined by accident severity, whether it involved towing the vehicle from scene or not as well as injury severity based on the Abbreviated Injury Scale (AIS) developed under the auspices of the American Medical Association.

In the case of crashes involving a 1973-1975 GM full size car (control group) the procedure is basically the same except the NRC is not notified. Notification and response are accomplished at the regional level. Local and State police operating in the selected counties will notify the team whenever a crash involving a vehicle meeting control group criteria occurs. If preliminary information provided to the MDAI team indicates that an occupant was taken to a treatment facility, a technician is dispatched to carry out an intermediate level type of investigation. Intermediate level investigations provide police and medical reports plus photographs of the vehicle to provide for developing a VDI.

When the crash did not require an occupant to be transported to a treatment facility, the MDAI staff will apply the systematic sampling procedure specified by NHTSA. If the procedure selects the crash, an intermediate level investigation will be made. The sampling fractions will be determined by the optimization procedure mentioned in "Analysis of the Data." There will probably be different fractions for buzzer and for ignition interlock cars.

REFERENCES

1. J.D. States, H.A. Fenner, Jr., E.E. Flamboe, et al., "Field Application and Research Development of the Abbreviated Injury Scale," SAE Print 710783, New York. Society of Automotive Engineers, 1971.

2. Committee on Medical Aspects of Automotive Safety: "Rating the Severity of Tissue Damage I. The Abbreviated Injury Scale," Journal of American Medical Association, 215:277-280, 1971.

3. Irwin D.J. Bross, "How to Use Ridit Analysis," Biometrics, March, 1958.

4. Collision Deformation Classification — SAE J224a — SAE Recommended Practice, New York: Society of Automotive Engineers, 1972.

A Study of Driver Interactions with an Inflating Air Cushion

John D. Horsch and
Clyde C. Culver
Biomedical Science Department
General Motors Research Laboratories
Warren, MI

Abstract

Conceptually, a steering wheel mounted air cushion is inflated before the upper torso of the driver significantly interacts with the cushion. However, this might not be the case for some seating postures or vehicle crash environments which could cause the driver to significantly interact with an inflating cushion.

These experiments utilized several environments to study the interaction between an inflating driver air cushion and mechanical surrogates. In these laboratory environments, the measured responses of mechanical surrogates increased with diminishing distance between the surrogate's sternum and the steering wheel mounted air cushion.

AIR CUSHION RESTRAINED human volunteer subjects (1,2)*, under well controlled frontal impact conditions and in "normal," upright posture, received no more than minor injuries in impact severities up to a 48 km/hr equivalent barrier crash of a full sized vehicle. The degree of restraint provided for an occupant initially in the space to be filled by the inflated cushion may be different than that for an occupant initially outside of this space.

Studies concerned with child surrogates, postured to interact with the inflating cushion of dash mounted passenger-side air cushions, have been reported by Alderman et al (3), by Patrick et al (4), and by W. Tang et al (5). The investigation reported here was undertaken to study the interaction of an inflating steering wheel mounted air cushion** with variously postured adult driver surrogates.

METHODOLOGIES

The approach of this study was to place a mechanical surrogate at various spacings with respect to the steering column mounted air cushion, independent of the total air cushion system or particular vehicle for which such a cushion might be used. This separation of the steering column mounted air cushion from the total restraint system provides response information that can be applied to a variety of possible system configurations. However, the additional effects of the total air cushion system and the particular vehicle should be considered to relate this study to potential field inflations.

Preliminary testing investigated the mechanism of cushion inflation when a human sized object interacted with an inflating steering column mounted air cushion. These tests were conducted by placing an SAE J944 body block, mounted on force transducers, in front of the cushion module*** to measure interacting forces. Because these body block tests did not address the responses of a human in such an exposure, and because of nonrealistic mechanical constraints, further investigations were made using a GM Hybrid III dummy (7) as

---

*Numbers in parentheses designate References at end of paper.
**The operation and description of a steering wheel mounted air cushion restraint system is discussed by T. Louckes et al (6).
***The term air cushion module refers to the module which fits into the steering wheel hub and contains the gas generator, air cushion, and the packaging covers (see Figure 1).

the driver surrogate. The thorax was chosen as the
body segment to be aligned with respect to the steering
column mounted air cushion. This choice was based upon
the general alignment of the thorax of a seated driver
with the steering assembly and upon the known blunt
frontal thoracic impact response of present dummies.
The GM Hybrid III anthropomorphic dummy was chosen for
these tests due to its significantly improved blunt
frontal thoracic impact response compared to the Part
572 dummy. However, to best utilize the known thoracic
response, the dummy thorax was aligned as close as
practical with the steering column, not necessarily
maintaining complete realism of posturing. Although
the initial seated posture of the dummy was somewhat
typical for a large passenger car, a special seat, a
fixed foot position, and a nonautomotive knee restraint
were used to provide the desired alignment of the
thorax to the steering assembly. The test fixture did
not have a windshield. This assured that an unobstructed
posture of the dummy could be achieved.

Several sled test conditions were used:

1. A static condition with the dummy initially
placed at various spacings from the cushion module, the
only energy input provided by the inflating cushion.

2. A "ridedown" dynamic test condition in which
the dummy was initially placed against the steering
assembly with normal cushion inflation timing.

3. A delayed inflation such that the sled deceleration caused the dummy, initially in an "upright" posture,
to articulate toward the steering assembly. The
desired posture at the initiation of cushion inflation
was achieved by controlling the amount of inflation
delay.

TEST METHOD - Description of Components Tested -
The tests were performed using components from the GM
driver (steering wheel mounted) air cushion restraint
system (ACRS), designated as Type A components for this
discussion. These components, shown in Figure 1, are:

1. The driver air cushion module which mounts in
the steering wheel hub (the module contains the cushion,
the gas generator, and packaging).

2. The ACRS steering wheel.

3. The ACRS energy absorbing steering column.

4. The ACRS steering column mounting brackets.

The air cushion modules were actuated remotely.
Not included were a crash sensor, the ACRS knee restraint,
or components from the passenger side ACRS. An experimental steering wheel mounted air cushion module was
also tested, designated as a Type B module for this
discussion. However, the Type A ACRS steering wheel,
steering column, and mounting brackets were used for
all tests.

Body Block Test Environment - Figure 2 is a photograph of the body block test environment. The SAE J944 body block was mounted on force transducers to a rigid wall to measure horizontal reaction forces. The lower steering column was mounted horizontally on force transducers attached to a large fixed mass to measure the axial lower column reaction force. This configuration allowed relative displacement of the steering wheel and air cushion module with respect to the body block only by stroking of the steering column. An accelerometer was attached to the upper column to estimate upper column inertial forces. High speed movies were obtained. The initial separation between the air cushion module and the body block was set for each test (measured along the steering column axis). This initial separation ranged from zero to 178 mm.

Sled Test Environment - The sled test fixture is shown in Figures 3 and 4, illustrating the two initial postures used for the Hybrid III dummy. The "hard" horizontal seat pan consisted of flat plywood covered with 16 mm of ribbed rubber pad and a layer of vinyl seat material. The seat back was flat plywood angled 25° from the vertical and covered with a rigid foam. The dummy's feet were strapped to a rigid foot rest. The ACRS energy absorbing steering column was mounted on two force transducers such as to measure the axial lower column reaction loading on the sled fixture. The steering column axis was at an angle of 20° from horizontal. The knee restraints were made of a crushable foam, resulting in peak femur loads of about 5,000 Newtons in the 37.4 km/hr tests. The foam was enclosed on the sides to prevent splitting. The fixture was adjusted to have the dummy in an upright seated posture. A loose belt was placed around the dummy's thighs to assure that the dummy remained on the sled during rebound.

Since the Hybrid III thoracic response was developed with midsternal blunt frontal impacts, it was desired to center the dummy's sternum with the steering column axis for the various conditions of the dummy being postured against the steering assembly. However, the Hybrid III dummy's sternum could not be centered vertically on the steering column axis as desired due to the interference of the steering wheel rim with the dummy's thighs. The center of the sternum was about 50 mm low with respect to the steering column axis with the dummy placed against the steering assembly, the wheel rim against the thighs. In the dynamic tests the dummy articulated to this alignment with the steering column from the upright seated posture. It was necessary to place a 100 mm block on the seat pan for both static and dynamic tests in which the dummy was initially placed against the steering assembly to achieve this alignment.

Fig. 1 - Driver air cushion restraint system components
used in tests: a) air cushion module (contains the cushion
and gas generator); b) ACRS steering wheel; c) ACRS
energy absorbing steering column; d) guide bracket;
e) shear capsule; and f) lower clamp

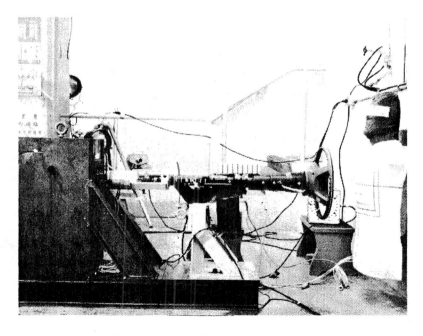

Fig. 2 - Body block test setup

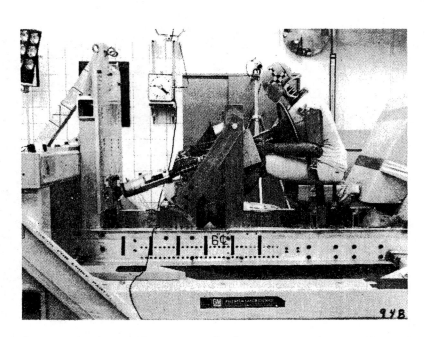

Fig. 3 - Hybrid III dummy "placed against-the-wheel" posture

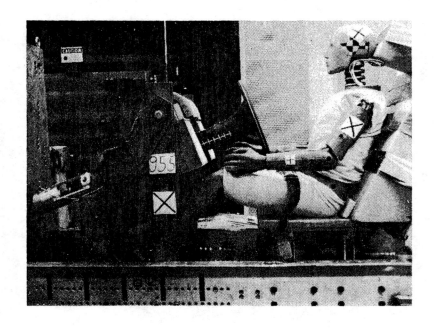

Fig. 4 - Hybrid III dummy in "upright" seated posture

Two sled operating conditions were used:
1. A static test environment for which the sled was stationary.
2. A dynamic test environment having a nominal 37.4 km/hr sled velocity and a nominally constant 14 g acceleration pulse.

The constant level sled acceleration pulse is characteristic of the GMR sled facility and is not an accurate simulation of an actual vehicle barrier impact.

The dummy was initially placed against the wheel (Figure 3) for both the stationary and dynamic sled operating conditions. For the stationary tests, the energy input to the system was provided by the inflating cushion. For the dynamic tests, the "ridedown" of the dummy against the steering assembly provided additional forces to those of the inflating cushion.

Tests having the dummy initially in the upright seated posture (Figure 4) achieved the desired separation from the cushion module by delaying the inflation until the dummy had articulated to the desired posture relative to the steering assembly. This articulation produced a relative velocity between the dummy and steering assembly at the time of cushion inflation, or if with sufficient delay, pre-impact of the steering assembly by the dummy's thorax.

In addition to the delayed inflation tests, tests were conducted with normal inflation timing such that the cushion was fully inflated before the dummy significantly interacted with the cushion. Dynamic sled tests were also conducted, using both initial dummy postures, similar to the previously described tests except that the air cushion was not inflated.

TEST RESULTS - Body Block Tests - Several inflation tests were conducted with Type A air cushion components in the body block test environment. The initial separation, measured between the body block and air cushion module along the steering column axis, ranged from zero to 178 mm. Table 1 lists the initial separation for each test.

The primary system response was the axial force as measured by the force transducers supporting the body block to a rigid wall. Figure 5 shows the time history of the measured body block axial force as a function of time for initial separations of 0 mm, 67 mm, and 178 mm, respectively. In addition, sketches of the cushion were made from the high speed movies at selected times to indicate the state of cushion inflation. The body block forces indicated one or more force peaks. Referring to the sketches shown in Figure 5 for the tests having 0 and 67 mm initial separation, the first large force peak occurred with the cushion still partly folded but appearing to have internal pressure. This force then decreased to low levels for which the cushion fabric appeared to be

TABLE I

BODY BLOCK TEST RESPONSES—TYPE A COMPONENTS

Initial Separation[1] mm	Maximum Body Block Force[2] N	Initial Force Peak[2] N	Steering Column Stroke mm
178	4230	3050	10
67	11570	10450	102
0	20000	20000	152

[1]Air cushion module to body block separation, measured along the steering column axis.

[2]Data frequency response – SAE J211 class 60.

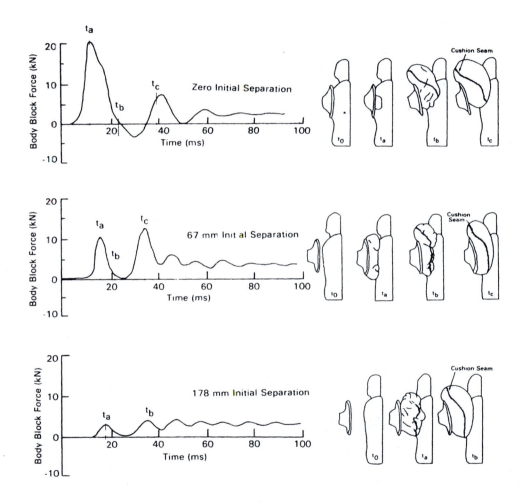

Fig. 5 - Body block force time history with sketches of
the cushion at selected times

loose and not internally pressurized. The cushion then appeared to become completely filled with the fabric having a smooth, tight appearance. For the test having the 178 mm initial separation, Figure 5, the greatest body block force did not develop until the cushion appeared completely inflated. Table 1 lists the magnitudes of the first force peak and the maximum force peak, the first force peak being maximum for zero initial separation. It should be noted that the body block had a beaming vibration mode which interacted with the body block force measurement. The negative force in Figure 5 is an indication of this beaming effect on the body block force response.

The axial force acting on the steering wheel was computed for the test having zero initial separation to compare with the body block force. The force acting on the steering system was determined as the force necessary to stroke the steering column, plus the force necessary to accelerate the upper steering assembly mass. The column stroke force was measured by the force transducers which supported the lower column to the test stand. The upper steering assembly acceleration was measured by an accelerometer mounted on the upper column and multiplied by the upper steering assembly mass of 12.6 kg. These forces are plotted in Figure 6 as a function of time.

The computed steering wheel force is compared to the measured body block force in Figure 7. The computed force acting on the steering wheel assembly is not identical to the measured force acting on the body block. This may be due to the body block or steering assembly structural vibrations and/or inertial effects of the cushion and gas. However, both measurements indicate about the same general magnitude and the same phasing and duration of the primary loading. The agreement of the phasing and duration of the two forces suggest that the primary portion of the forces developed due to this cushion inflation is a pressure force between the steering assembly and body block. Although these body block test responses suggest the characteristic of the forcing function between a human sized object and the inflating cushion, it should be considered that the rigid mounting of the body block provides a nonrealistic mechanical constraint and that the body block has many deficiencies in simulation of a human.

Static Sled Tests - Cushion inflation tests were performed on the stationary sled fixture by initially posturing the Hybrid III dummy near or against the air cushion module. Figure 8 is a plot of various thoracic responses as a function of time for test 948 with the dummy placed initially against the cushion module using Type A components. Table 2 lists the static tests, the air cushion module and steering wheel used for each

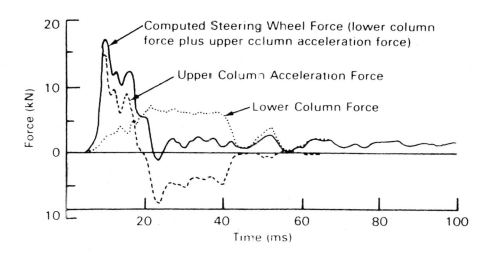

Fig. 6 - Computed steering wheel force for body block test having zero initial separation

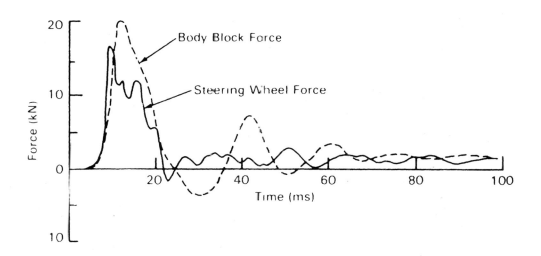

Fig. 7 - Comparison of computed steering wheel force with body block force for test having zero initial separation

1165

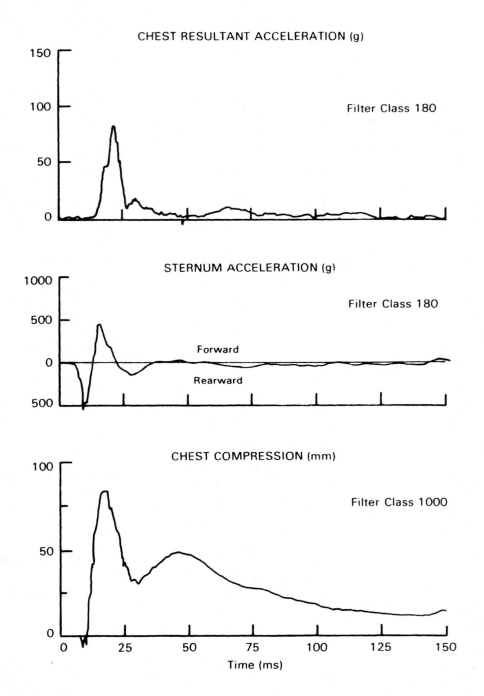

Fig. 8 - Chest acceleration, sternum acceleration, and chest compression time histories for test No. 948--static test--dummy postured against-the-wheel

TABLE II

SUMMARY OF STATIC TESTS -- SLED STATIONARY -- GM HYBRID III DUMMY

Test No	Air Cushion Module	Steering Wheel	Initial Separation (mm)	Thorax Response[2]					Head Resultant Acceleration[3]	Steering Column Stroke mm
				Chest Compression		Chest Acceleration				
				Maximum (mm)	Rate (m/s)	Sternal Peak(g's)	Spine Resultant 3ms level	Spine Resultant GSI	HIC-I	
948	Type A	Type A	0	84	17	538	51	165	137	84
969	Type A	Type A	0	81	17	374	53	175	112	94
963	Type A	Type A	25	> 84	16	304	*	97	133	48
962	Type A	Type A	51	58	11	282	29	39	147	38
964	Type A	Type A	102	25	5	459	22	18	93	43
975	Type A	25 mm dish[1]	12	71	10	240	*	*	*	114
981	Type A	25 mm dish[1]	12	71	12	490	*	*	*	74
974	Type A	51 mm dish[1]	35	76	13	360	*	*	*	86
971	Type A	76 mm dish[1]	60	51	5	450	*	*	*	74
978	Type A	76 mm dish[1]	60	53	5	400	*	*	*	41
984	Type B	Type A	0	43	5	190	*	*	*	33
985	Type B	Type A	0	43	4	190	*	*	*	37

[1] Type A steering wheel modified by welding in spoke extension - Dish measured from straight edge spanning driver side of wheel rim to surface of air cushion module facing driver. Type A wheel has zero dish.

[2] Data frequency response SAE J211 class 180.

[3] Data frequency response SAE J211 class 1000.

*Data not available.

1167

test, and the initial axial separation between the dummy and air cushion module. The responses of peak chest compression, rate of chest compression, peak sternal acceleration, thorax severity index, resultant thoracic spinal 3 ms acceleration level, head resultant HIC-I index, and steering column stroke are given where available.

Dynamic Sled Tests - Cushion inflation tests were performed in the dynamic sled environment by either placing the dummy against the air cushion module or by delaying the cushion inflation until the dummy articulated forward from the upright seated posture. Figures 9, 10, and 11 are plots of various thoracic responses as a function of time for three tests using Type A components. The dummy had an upright initial seated posture for all three tests. Test 958, shown in Figure 9, delayed cushion inflation until the dummy was just contacting the cushion module. Test 955, shown in Figure 10, had normal inflation timing which resulted in a completely inflated cushion before the dummy significantly interacted with the cushion. Test 966, shown in Figure 11, did not inflate the cushion and the dummy impacted the steering assembly. Table III lists the dynamic tests, test conditions, air cushion modules, and the responses of peak chest compression, rate of chest compression, peak sternal acceleration, thorax severity index, resultant thoracic spinal 3 ms acceleration level, head resultant HIC-I index, and steering column stroke where available.

The Hybrid III dummy chest has a compression range of slightly greater than 84 mm at which time the sternum contacts the thoracic spine. Measured values of chest compression which exceeded 84 mm are given as > 84 mm in Table 3.

DISCUSSION OF RESULTS

Hybrid III dummy responses of head triaxial acceleration, thoracic spine triaxial acceleration, femur axial forces, chest compression, and sternum acceleration were made during the experiments. Of these measured dummy responses, Neathery et al (8) suggest that peak chest compression best correlates with thoracic injury when thoracic impact is predominent. Neathery et al (8) have stated that, "Current methods of evaluation of occupant protection (severity index, or 3 ms level of thoracic spinal acceleration or chest load level) should not be used when significant blunt frontal chest impact occurs." The authors recommend that, "Sternal deflection measured in dummy chests with substantial biomechanical fidelity (GM corridors) should be used to evaluate the injury potential of automotive systems where blunt thoracic impact is predominant, e.g., steering column involvement."

Due to the frontal loading of the thorax for these

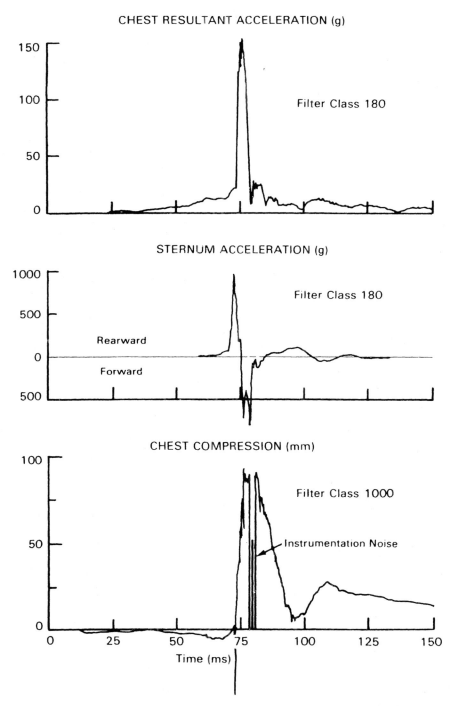

Fig. 9 - Chest acceleration, sternum acceleration, and chest compression time histories for test No. 958--37.4 km sled velocity--65 ms delayed inflation--dummy seated posture initially upright

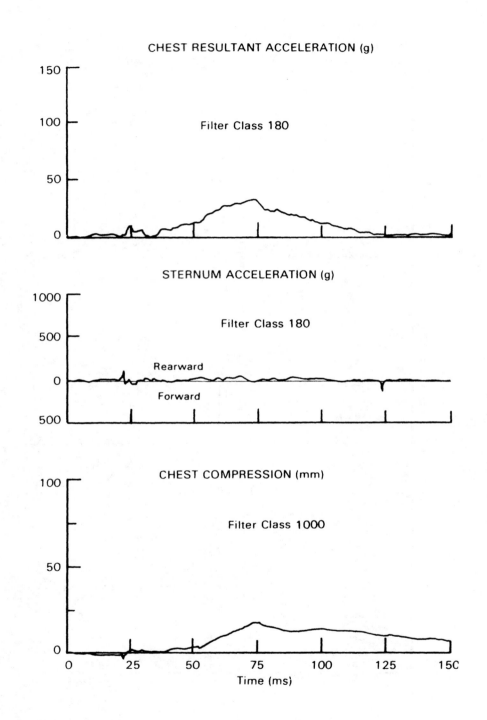

Fig. 10 - Chest acceleration, sternum acceleration, and chest compression time histories for test No. 955--normal air cushion inflation timing--37.4 km/h sled velocity-- dummy seated posture initially upright

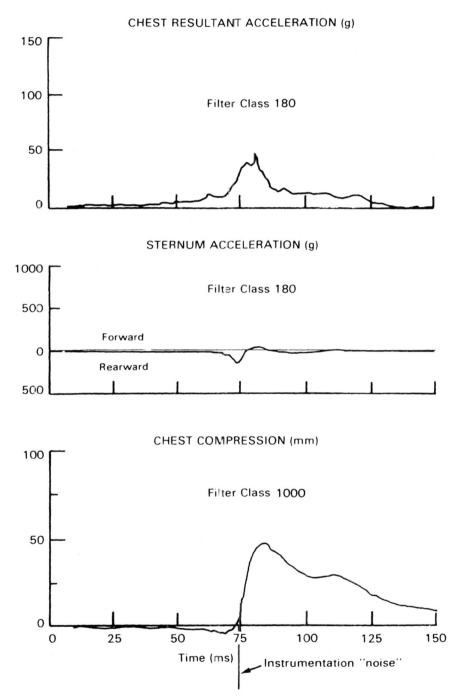

Fig. 11 - Chest acceleration, sternum acceleration, and chest compression time histories for test No. 966-- 37.4 km/h sled velocity--air cushion not inflated--dummy seated posture initially upright

TABLE III

SUMMARY OF DYNAMIC TESTS - 37.4 km/hr SLED VELOCITY - GM HYBRID III DUMMY

| Test No | Air Cushion Module | Dummy Initial Position | Separation at Initiation of Inflation [3] (mm) | Inflation Delay [4] (ms) | Thorax Response[5] | | | | | Head Resultant Acceleration [6] HIC-I | Steering Column Stroke mm |
| | | | | | Chest Compression | | Chest Acceleration | | | | |
					Maximum(mm)	Rate(m/s)	Sternal Peak(g's)	Spine Resultant 3 ms level	GSI		
955	Type A	Upright[1]	267	9	18	1	110	32	151	90	81
965	Type A	Upright[1]	43	55	58	15	558	64	317	412	152[8]
967	Type A	Upright[1]	20	60	79	16	500	75	457	463	152[8]
958	Type A	Upright[1]	0	65	> 84	23	973	113	807	428	152[8]
1032	Type A	Upright[1]	Preimpact[7]	72	> 84	15	386	93	Not Available		152[8]
966	Type A	Upright[1]	No Inflation	No Inflation	46	5	148	39	140	598	79
956	Type A	Against[2]	0	9	> 84	17	506	66	292	203	122
957	Type A	Against[2]	No Inflation	No Inflation	25	2	47	23	92	116	76
986	Type B	Against[2]	0	65	71	11	255	75	Not Available		152[8]

1 See Figure 4; Nominally a 267 mm separation between dummy thorax and cushion module.

2 Thorax positioned against cushion moduel -- See Figure 3.

3 The separation of the thorax from the air cushion module at the initiation of inflation, measured from high speed movies.

4 Inflation delay is that time between the initiation of sled deceleration and the electrical pulse which starts the gas generation.

5 Data frequency response -- SAE J211 class 180.

6 Data frequency response -- SAE J211 class 1000.

7 The thorax impacted the air cushion module before cushion inflation had been initiated.

8 152 mm is the maximum stroke in the sled test fixture.

inflation tests and to the Hybrid III dummy thoracic response having been shown to approximate the recommended corridors for a 50th percentile tensed adult male, Foster et al (7), we have chosen peak sternal deflection (chest compression) as the best of the measured responses to rank order various air cushion modules and test environments. However, differences in impact environment between the blunt frontal impact responses used by Neathery et al (8) to develop their injury criteria and the inflating air cushion interacting with the Hybrid III dummy should be considered. Neathery et al (8) state, "The criteria established will be valid for blunt frontal impact. They may also be a good estimate, perhaps even the best available in the laboratory, for the evaluation of other forms of restraints; but that remains to be demonstrated. Accordingly, great care should be exercised in extrapolating these results to other impact conditions. The usual precautionary statements regarding cadaveric material apply." For example, they found that a test modification appeared to change the relationship between injury and peak chest deflection. They state, "six of the subjects... were tested with the spine rigidly supported. The pattern of injury obtained in these six tests was erratic and did not fit the remaining data."

Effect of Separation From the Module - Figure 12 is a plot of the body block peak forces versus initial separation, the values taken from Table 1. The closer the body block was initially placed to the cushion module, the greater was the magnitude of the resulting body block force.

A similar relationship for chest compression has been found for both static and dynamic tests performed on the sled fixture. Figure 13 is a plot of peak chest compression versus the separation between the dummy thorax and cushion module at the initiation of cushion inflation. The separation was measured from high speed movies for the delayed inflation dynamic tests. There is an increase of the magnitude of chest compression with decreasing separation between the thorax and air cushion module.

In a similar fashion, dished steering wheels (modified Type A air cushion steering wheels) were tested in the static environment, the Hybrid III dummy thorax placed against the wheel. The depth of dish was the distance between the surface of the air cushion module and the wheel rim plane measured along the steering column axis. The results of the dished wheel static tests are also plotted in Figure 13. From Table 2, it should be noted that the thorax to air cushion module separation is less than the wheel dish due to the thorax protruding into the wheel.

Comparison of Responses for Various Test Conditions - Peak chest compression is shown in Figure 14 for the three

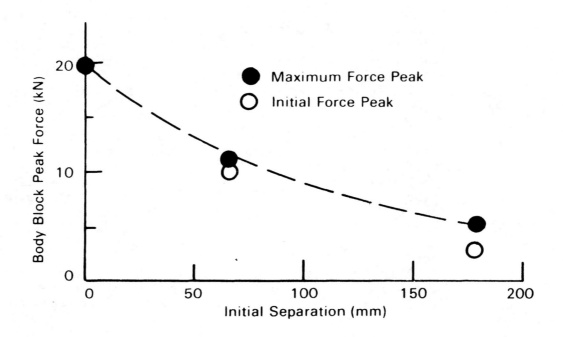

Fig. 12 - Body block peak force response as a function of the initial separation between the air cushion module and body block

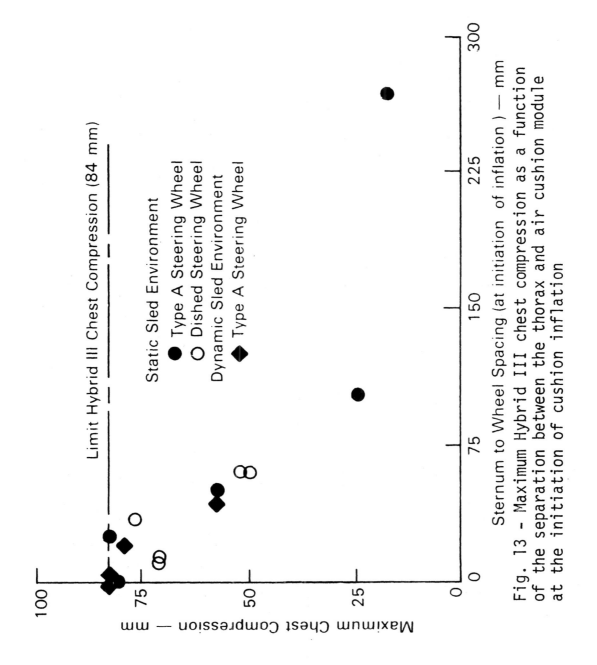

Fig. 13 - Maximum Hybrid III chest compression as a function of the separation between the thorax and air cushion module at the initiation of cushion inflation

conditions with the dummy against the cushion module at the initiation of cushion inflation; preposturing in both the sled static (948) and dynamic modes (956), and articulation forward from the upright seated posture by a delayed inflation in the dynamic sled mode (958). Peak chest compression is also shown for dynamic test modes for which the cushion was not inflated; prepostured against the wheel (957) and articulation forward from the upright seated posture (966). In addition, the peak chest compression from a "normal" cushion inflation with the dummy in the upright seated posture is shown (955). For these test conditions, the highest magnitude of chest compression occurred when the dummy's chest was against the ACRS module at the initiation of cushion inflation (tests 948, 956, and 958). Lower magnitudes of chest compression were observed in similar test environments when the cushion was not inflated (tests 957 and 966). The lowest magnitude of chest compression was observed for test 955 in which the cushion was inflated before the dummy became significantly involved with the cushion from the upright seated posture.

Two types of air cushion modules were comparatively evaluated in several inflation environments. The peak chest compression observed for comparative test environments for each system are presented in Figure 15. The Type B air cushion module produces lower amplitude thoracic responses as compared to the Type A air cushion module for the Hybrid III dummy postured against the cushion module in both static and dynamic environments. Various air cushion modules can be compared for their response with a closely postured driver by these techniques.

CONCLUSIONS

This study of driver surrogate interactions with an inflating steering wheel mounted air cushion involves hypothetical test environments with various initial occupant-air cushion separations or air cushion module modifications. Static tests using a rigidly mounted body block placed in the path of the inflating cushion showed how interaction forces might be developed when the body block interacted with the inflating air cushion. The magnitude of the peak interaction force was an inverse function of the initial separation between the body block and air cushion. When the body block was initially in close proximity of the air cushion, the interaction force was characterized as primarily a pressure existing between the steering assembly and the body block and peaking during the early phases of cushion unfolding. However, the lack of whole body motions and the rigidity of the body block have an effect on the interaction forces and the test responses do not address the responses of a

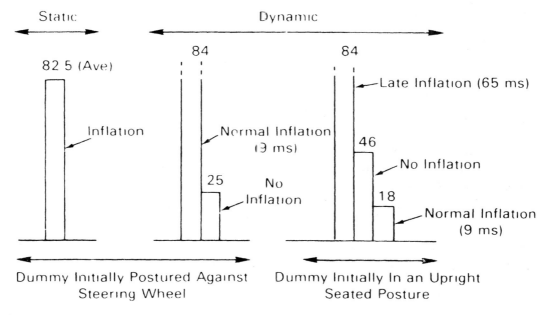

Maximum Hybrid III Chest Compression   mm

Static | Dynamic

82 5 (Ave)

Inflation

84

Normal Inflation
(9 ms)

25   No
Inflation

84

Late Inflation (65 ms)

46

No Inflation

18

Normal Inflation
(9 ms)

Dummy Initially Postured Against
Steering Wheel

Dummy Initially In an Upright
Seated Posture

Fig. 14 - Comparison of maximum Hybrid III chest compression
for various test environments

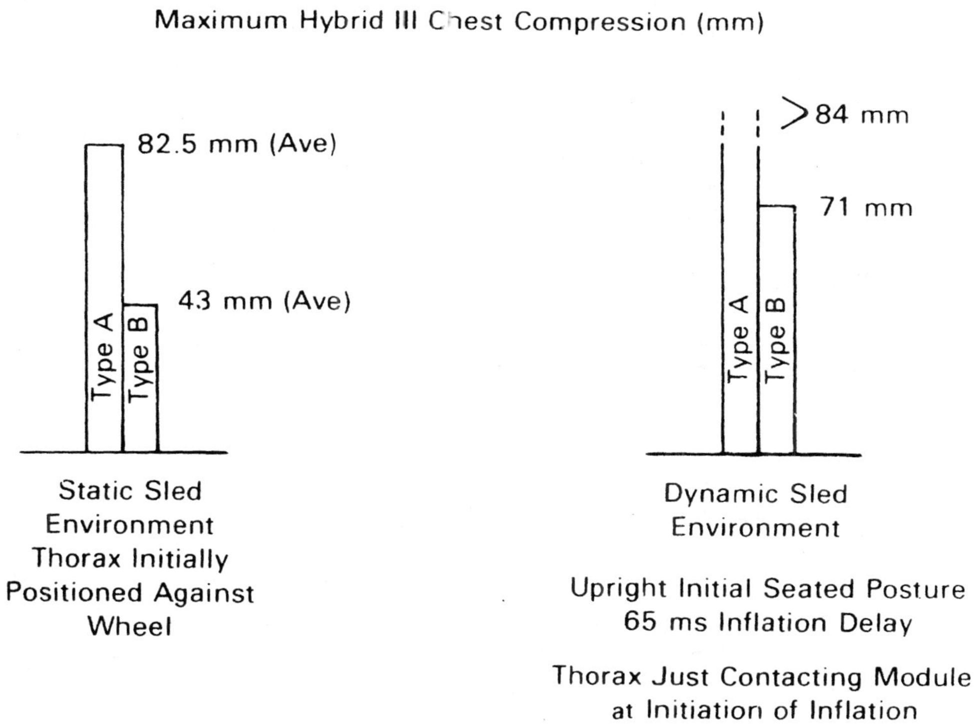

Maximum Hybrid III Chest Compression (mm)

82.5 mm (Ave)

Type A Type B

43 mm (Ave)

Static Sled
Environment
Thorax Initially
Positioned Against
Wheel

>84 mm

71 mm

Type A Type B

Dynamic Sled
Environment

Upright Initial Seated Posture
65 ms Inflation Delay

Thorax Just Contacting Module
at Initiation of Inflation

Fig. 15 - Comparison of maximum Hybrid III chest compression
for type A and type B air cushion modules

human in such an exposure.

Static and dynamic inflation tests performed with the Hybrid III thorax placed in line with the path of the inflating cushion provided a more realistic simulation of a driver's interaction with an inflating air cushion. The Hybrid III dummy allowed whole body motions, body articulation, and chest compression. However, there are many differences between the Hybrid III dummy and car occupants. Peak chest compression was considered the best of the measured responses to compare the interaction of the inflating cushion and dummy for the various test environments, surrogate separations from the air cushion at the time of inflation, or air cushion module modifications.

These test methodologies might be useful during the development of an air cushion restraint system, to allow relative evaluation of air cushion performance for hypothetical situations which place the driver in a posture to interact with the inflating air cushion. Consideration of these hypothetical situations should maintain proper perspective to more typical occupant postures and crash environments for a proper overall evaluation of the restraint system.

It should be considered that these tests may not represent what might happen in actual field performance and what injuries an actual driver might experience. This is, in part, due to:

1. The use of a nonautomotive test environment.

2. The use of only part of a total air cushion restraint system.

3. The nonconsideration of the probability or possibility of occurrence.

4. Limitations of the surrogates and of the injury criteria based upon the measured responses.

ACKNOWLEDGEMENTS

The authors would like to thank personnel from the GM Environmental Activities Staff, GM Inland Division, GM Oldsmobile Division, and Biomedical Science Department, GMR, for their cooperation in furnishing technical assistance.

REFERENCES

1. Smith, G.R., Harite, S. S., Yanick, A. J., General Motors; Greer, C. R., M.D., St. Joseph Hospital, Houston, Texas, "Human Volunteer Testing of GM Air Cushions." 2nd International Conference on Passive Restraints, May 22-25, 1972, SAE No. 720143.

2. Smith, G. R., Gulash, E. C., Baker, R. G., "Human Volunteer and Anthropomorphic Dummy Tests of General Motors Driver Air Cushion System." International Conference on Occupation Protection, 3rd Conference Proceedings, 1974, SAE No. 740578.

3. Aldman, B., M.D., Department of Traffic Safety, Chalmers University of Technology, Anderson, A., M.D., and Saxmark, O., Eng. AB Valvo Goteborg, Sweden, "Possible Effects of Air Bag Inflation on a Standing Child." Proceedings of the 18th Conference of the American Association for Automotive Medicine, 1974.

4. Patrick, L. M. and Nyquist, G. W., "Air Bag Effects on the Out-of-Position Child." 2nd International Conference on Passive Restraints, Detroit, Michigan, May 22-25, 1972, SAE Transactions 720942.

5. Wu, H., Tang, S. C., and Petrof, R. C., "Interaction Dynamics of an Inflating Air Bag and a Standing Child." Automobile Engineering Meeting, Detroit, Michigan, May 14-18, 1973. SAE Paper 730604.

6. T. N. Louckes, et al, "General Motors Driver Air Cushion Restraint System.' SAE Paper No. 730605.

7. Foster, J. K., Kortge, J. O., and Wolanin, M. J., "Hybrid III-A Biomechanically-Based Crash Test Dummy." 21st Stapp Car Crash Conference, October 19-21, 1977, New Orleans, Louisiana. SAE Paper 770938.

8. Neathery, R. F., Kroell, C. K. Mertz, H. J., "Prediction of Thoracic Injury from Dummy Responses." 19th Stapp Car Crash Conference, November, 1975.

Restraint Performance of the 1973-76 GM Air Cushion Restraint System

Harold J. Mertz
Safety & Crashworthiness Systems
General Motors Corp

ABSTRACT

Case reviews are given of deployment accidents of the GM 1973-76 air cushion restraint system where the occupant injury was AIS 3 or greater. Many of these injuries occurred in frontal accidents of minor to moderate collision severity where there was no intrusion or distortion of the occupant compartment. Dummy and animal test results are noted that indicate that these types of injuries could have occurred if the occupant was near the air cushion module at the time of cushion deployment. An analysis is given that indicates that for frontal accidents a restraint effectiveness of 50 percent in mitigating AIS 3 or greater injuries might be achieved if an air cushion system can be designed which would not seriously injure out-of-position occupants while still providing restraint for normally seated occupants.

GENERAL MOTORS EQUIPPED 11,321 vehicles with driver and front seat passenger air cushion restraint systems (ACRS) during the 1973 through 1976 model years. The first cars equipped with the ACRS were 1000 1973 Chevrolet Impalas. This initial effort was called the Field Trial Program and the cars were identified as FTP cars. During the 1974 through 1976 model years, 10,321 Buicks, Oldsmobiles and Cadillacs were produced. For these cars, the ACRS was a customer option available on Buick Le Sabre, Electra and Riviera; the Oldsmobile 88, 98 and Toronado; and the Cadillac Deville, Brougham, Fleetwood and Eldorado. The ACRS option was priced on the basis of a projected high sale volume. This level of customer demand never developed and the ACRS option was cancelled at the end of the 1976 model year.

DESCRIPTION OF THE 1973-76 GM ACRS

The 1973-76 GM ACRS consisted of driver and passenger (center and right front) systems, two crash sensors, a large capacitor to provide ener-gy to deploy the driver and passenger systems in the event that battery power was lost during the collision event, and a diagnostic system to provide the driver with information about the ACRS readiness to deploy. The driver air cushion module was mounted to a specially designed steering wheel and energy absorbing column. A fixed knee bolster was used to provide lower torso and leg restraint. The passenger air cushion module was mounted in the lower part of the instrument panel in front of the right front passenger seating position. It was designed to restrain both the right front and center front passengers, either singularly or together. Lower torso and leg restraint was provided by a separate high pressure, low volume knee cushion which was deployed inside of the larger volume, lower pressure head/torso cushion. Manual lap belts were provided for each front seat occupant position. A crash recorder was installed to provide data to determine (i) if the crash was below the deployment threshold, (ii) if the crash started before the cushion was deployed, and (iii) if the crash severity exceeded the severity of a 30 mph frontal, rigid barrier collision. Detailed descriptions of the various components of the 1973-76 GM ACRS are given in papers by Campbell(1)*, Klove and Oglesby(2), and Louckes, et al(3).

DEVELOPMENT AND VALIDATION OF THE 1973-76 GM ACRS

The 1973-76 GM ACRS was subjected to extensive development and validation test programs. A variety of sled tests was conducted to assure that the ACRS would provide restraint for different occupant sizes (5th percentile female, 50th percentile male and 95 percentile male), different seat positions (full forward, mid and full rear), different combinations of front seat

---

* Numbers in parentheses refer to papers listed under References

passengers (right front, right and center front, and center front) and different inflator environmental temperatures (180° F, 72° F and -30° F).

Human volunteer sled tests were conducted with both the driver system at Southwest Research Institute(4) and the passenger system at Holloman Air Force Base(5). In both test programs, the volunteers were seated in normal occupant posture and the cushions were deployed before the occupants had moved significantly forward relative to the body buck in response to the sled acceleration. Tests were conducted with increasing simulated collision severity with the most severe for each system being the 30 mph frontal, rigid barrier collision simulation. The restraint performance for the center front occupant was not evaluated with human volunteers. No significant injuries occurred. Only minor abrasions were noted.

For an early version of the passenger system (a pre FTP system), a series of static deployment tests was conducted to evaluate the effect of the interaction between the deploying cushion and the occupant. In these tests, a volunteer was seated on the seat and leaned forward, in various degrees, toward the instrument panel. The passenger cushion was deployed without the sled being accelerated. The test series was terminated after the volunteer experienced a slight concussion in a test where he was leaning forward with a torso angle of 29 degrees past vertical(5).

An animal test program to evaluate the effect of the interactions between the deploying passenger cushion and children who may be near the instrument panel at time of deployment was conducted at Wayne State University(6). Anesthetized baboons were placed in various positions near the instrument panel and various passenger system concepts were deployed while the body buck remained stationary. Test results indicated that significant injuries to the animals could be produced if the cushion was inflated too rapidly. Based on these results, the variable rate inflator concept used in the 1973-76 GM ACRS passenger system was developed. Subsequent anesthetized baboon and chimpanzee tests of the 1973-76 GM passenger ACRS produced no significant injuries for the animal positions and gas inflation rates evaluated. (Note: The results of the animal tests of the 1973-76 GM passenger ACRS are not described in the Patrick and Nyquist(6) paper.)

A series of cadaver tests was conducted at Wayne State University to evaluate the effect of the interaction between the deploying trim cover panel of the passenger air cushion module and the legs of front seat occupants. The tests were conducted in a stationary body buck with the cadavers seated in a normal seating posture. Their legs were aligned squared to the passenger module trim cover panel and the cushion was deployed. No lower extremity fractures occurred with the 1973-76 GM passenger ACRS.

To assess the efficacy of the crash sensing system, a variety of car-to-car and car-to-obstacle collisions was staged. These tests and

their results are described in a paper by Wilson and Piepho(7). The tests included 30 mph and 40 mph frontal rigid barrier impacts; right and left 30 mph, 30 degree angle rigid barrier impacts; 30 mph offset rigid pole impacts; 30 mph frontal impact with a bumper underride; and car-to-car impacts involving front of ACRS car to side of car, front of ACRS car to rear of car and a multiple impact scenario. In the majority of these tests, the Hybrid II dummy responses were well below the FMVSS 208 limits. The main exception was the 40 mph, rigid barrier tests where none of the Hybrid II dummies (driver, center front, right front) met all the limits. This test result indicated that in the more severe accidents, the 1973-76 GM ACRS would not mitigate all serious occupant injury.

The results for FMVSS 208, 30 mph, frontal rigid barrier tests are given in Table 1. Note that the 1973-76 GM ACRS values were well below the compliance limits defined by FMVSS 208 for passive restraint systems. These results, plus the results obtained from the human volunteer, animal, cadaver, dummy sled tests and full scale vehicle tests, indicated that the 1973-76 GM ACRS should have provided reasonably effective occupant restraint for a variety of real-world, frontal accident configurations and severities.

RESTRAINT EFFECTIVENESS OF THE 1973-76 GM ACRS

A note was put in the owner's manual of all vehicles equipped with the 1973-76 GM ACRS requesting that GM be notified of any accident involving the vehicle. A toll-free telephone number was given for the purpose of communicating this information. In addition, agents of GM Motors Insurance Corporation (MIC) and GM car dealers were asked to report the occurrence of accidents involving the ACRS vehicles. For each deployment accident reported, an investigator from the GM Field Accident Review group was sent to inspect the vehicle. Photographs were taken of the exterior and interior damage. Police reports were reviewed for accident descriptions and occupant injuries. If agreeable, occupants of the ACRS vehicles were interviewed and medical records reviewed. The severities of the injuries were rated according to the AAAM Abbreviated Injury Scale (AIS). For each reported accident, a report was written summarizing the pertinent observations.

A study was done by Pursel et al(8) to estimate the restraint effectiveness of the 1973-76 GM ACRS in mitigating AIS 2 and greater and AIS 3 and greater injuries. For each reported deployment accident, a search was done of the GM Motors Insurance Company (MIC) files for comparable accidents involving non-air cushion restrained occupants. Comparable MIC accident cases were selected on the basis of collision condition and severity, exterior vehicle damage, and occupant sex, age, and seating position. Percent occurrences of AIS = 0 or 1, AIS = 2, AIS = 3 or 4 and AIS = 5 or 6 were calculated for the ACRS deployment cases that were matched and for the corresponding non-air cushion restraint cases. The ACRS effectiveness was ob-

Table 1 - Results of FMVSS 208, 30 mph,
Frontal Barrier Tests of the 1973-76
GM ACRS Using Hybrid II Dummies(7)

	HEAD HIC	CHEST ACC. (3 ms, G)	LT. FEMUR (lb)	RT. FEMUR (lb)
Test C-3352 (Lap Belts Not Used)				
Driver	340	40	1420	1270
C. Front	320	45	470	1210
Rt. Front	360	44	930	1110
Test C-3353 (Lap Belts Not Used)				
Driver	310	36	1330	1180
C. Front	370	44	540	1190
Rt. Front	490	45	660	1170
Test C-3321 (Lap Belts Not Used)				
Driver	590	50	1250	1230
R. Front	260	40	910	1010
Test C-3094 (Lap Belts Used)				
Driver	370	47	1230	500
R. Front	430	42	660	760

Table 2 - Estimates of Percent Effectiveness
of the 1973-76 GM ACRS in Mitigating
AIS 2 or Greater and AIS 3 or Greater
Injuries in Frontal Deployment
Accidents

	MITIGATING AIS = 2 OR GREATER INJURIES	MITIGATING AIS = 3 OR GREATER INJURIES
1973-76 GM Driver ACRS	18%	21%
1973-76 GM Passenger ACRS	-34%	16%
1973-76 GM Combined ACRS	6%	18%

tained by subtracting the percent of deployment accident occupants that experienced a given AIS level from the percent of non-deployment accident occupants that experienced the same AIS level. This difference was divided by the latter percentage to give the ACRS effectiveness. The combined effectiveness estimates of the driver and passenger system for mitigating AIS 2 or greater injuries and AIS 3 or greater injuries were 6 and 18 percent, respectively. Effectiveness estimates for the driver and passenger systems separately were calculated using the same procedure. All effectiveness estimates of the GM 1973-76 ACRS are given in Table 2. Note these effectiveness estimates are for frontal, deployment type accidents and do not assess the ACRS effectiveness in non-deployment type accidents such as roll overs, side impacts and rear end collisions. If these non-deployment type accidents were included, the overall effectiveness would be lower.

The effectiveness of both the driver and passenger systems in mitigating AIS 3 or greater injuries was less than expected when compared to the "no significant injuries" results obtained in the human volunteer programs. The negative 34 percent effectiveness for mitigating AIS 2 or greater injuries for the passenger system was quite disturbing since the implication is that overall the ACRS passengers experience more AIS 2 or greater injuries than their matched case, non-air cushion restrained counterparts.

MODERATE TO SEVERE DEPLOYMENT ACCIDENTS WHERE THE OCCUPANT WAS PROTECTED

A case-by-case study was done of each ACRS deployment accident. At the time of that review (December 1980), there were 216 deployment accidents in the 1973-76 GM ACRS file (FTP cases through AC-3020 and 1974-76 cases through LA350). These accidents involved 216 drivers, 86 right front occupants and 11 center front occupants. Of the 97 front seat passengers, thirteen (13/97 = 13.4%) were children whose ages were ten years or less.

Based on analyses of collision descriptions and photographs of exterior and interior damage, each deployment accident was classified in terms of collision severity and interior intrusion/-distortion severity. Three collision severity classes were used: minor, moderate and severe. An accident was judged as minor collision severity if the exterior damaged involved only front-end sheet metal (hood and/or fenders), grille and/or the bumper. For an accident to be judged as moderate collision severity, some rearward displacement of the engine and/or wheels had to occur and the estimated speed of the accident had to be greater than 15 mph. If there was extensive damage to the vehicle and the accident speed was estimated to be greater than 25 mph, then the accident was classified as severe. Two classifications of interior intrusion/distortion were used: significant and nonsignificant. If the interior intrusion/distortion occurred in the vicinity of a front seat

occupant and if it was difficult to conceive how any air cushion could provide protection from the intrusion/distortion, then the accident was judged as having significant occupant compartment intrusion and/or distortion.

The majority of the 216 deployment accidents were classified as minor. None of these accidents had significant occupant interior intrusion or distortion. Several of the severe and moderate collision accidents did have significant occupant interior intrusion/distortion. In these accidents it was difficult to conceive how any air cushion system could have mitigated the injuries experienced by the occupants.

There were 14 accidents where the collision severities were classified as moderate or severe, there was no significant occupant interior intrusion or distortion, and the occupant injuries caused by the restraint forces applied by the ACRS were of minor consequence. A summary of the accident conditions, collision severities and occupant injuries for these accidents is given in Table 3. Note that with the exception of one driver injury, the injury severities for these ACRS occupants were AIS 2 or less with the majority being AIS = 1. The exception was a driver who experienced a bruised kidney which was rated as AIS = 3. The kidney injury could have been caused by the vertical loading associated with the car going into a 10 foot deep ravine. Even if the kidney injury occurred during the principal front collision event, the restraint performance of the 1973-76 driver ACRS would be judged as effective because the matched case drivers all experienced AIS = 4, 5 or 6 injuries. These 14 accidents demonstrate the occupant restraint potential of an air cushion system in moderate to severe frontal accidents when there is no significant intrusion or distortion of the interior in the vicinity of the front seat occupants.

1973-76 GM ACRS DEPLOYMENT ACCIDENTS WITH UNEXPECTED OCCUPANT INJURIES

DEPLOYMENT ACCIDENTS WITH AIS 3 OR GREATER INJURIES - Twenty-three front seat occupants (23/313 = 7.3%) experienced AIS 3 or greater injuries: fourteen drivers (14/216 = 6.5%) and nine front seat passengers (9/97 = 9.3%), three of whom were children. Pertinent information concerning the collision conditions and occupant injuries is given in Table 4. There were five fatalities, four drivers and a one-month-old child who was lying on the seat cushion prior to the accident. Four occupants, all right front passengers, experienced AIS = 4 injuries, all related to leg fractures. Ten drivers and four passengers experienced AIS = 3 injuries. These injuries consisted of fractures to the ribs, arms, legs, and lumbar vertebra; a concussion; a persistent neurological impairment and a bruised kidney.

An analysis of the accident data indicated that in ten of these twenty-three cases there was either significant intrusion/distortion of the interior in the vicinity of the occupant

TABLE 3 - SUMMARY OF GM 1973-76 ACRS DEPLOYMENT ACCIDENTS OF MODERATE TO SEVERE
COLLISION SEVERITY, NO SIGNIFICANT OCCUPANT INTERIOR INTRUSION OR
DISTORTION, AND RESTRAINT PROTECTION PROVIDED BY THE ACRS

CASE NO.	COLLISION INFORMATION		OCCUPANT INFORMATION				
	SEVERITY	DESCRIPTION	POS.	AGE	SEX	AIS	INJURY DESCRIPTION
LA 350	SEVERE	HEAD-ON WITH SECOND CAR	DR	16	M	3	BRUISED KIDNEY
AC 356	SEVERE	UNDERBODY TO RAILROAD TRACKS FOLLOWED BY RT. FRONT POLE IMPACT	DR	14	F	1	BROKEN NOSE, CUT LIP
			RF	14	F	1	ABRASION TO HAND
LA 36	SEVERE	PARTIAL HEAD-ON WITH SNOW PLOW, BIASED TO DRIVER'S SIDE	DR	41	M	1	BRUISED NOSE, CHEST, KNEE
			RF	35	F	1	FACIAL BRUISES
LA 250	SEVERE	PARTIAL HEAD-ON WITH SECOND CAR, BIASED TO DRIVER'S SIDE	DR	36	F	1	BRUISED KNEES, FOOT, ELBOW
			RF	21	F	1	BRUISED TIBIAS
LA 96	SEVERE	PARTIAL HEAD-ON WITH BUS	DR	62	M	2	BROKEN RIB, BRUISED TIBIA
AC 3012	MODERATE	UNDERBODY TO RAILROAD TRACKS, LT. FRONT POLE IMPACT	DR	18	M	1	DAZED, FACIAL BRUISES
LA 265	MODERATE	LT. FRONT TREE IMPACT	DR	69	F	2	SPRAINED ELBOW, KNEE BRUISES
LA 307	MODERATE	LT. FRONT IMPACT TO CONCRETE TRAFFIC SIGNAL BASE	DR	21	M	1	BRUISED KNEE, CHIN ABRASION
			RF	20	F	2	SPRAINED ANKLE, BRUISES
AC 1504	MODERATE	PARTIAL HEAD-ON WITH SECOND CAR, BIASED TO DRIVER'S SIDE	DR	18	M	1	CUT ON BACK OF HEAD
			RF	19	M	1	DAZED
LA 85	MODERATE	LT. FRONT IMPACT TO TREE	DR	16	M	1	BRUISED KNEE
			RF	15	M	1	BRUISED KNEES
LA 275	MODERATE	HEAD-ON WITH SECOND CAR	DR	38	F	1	BRUISED CHEST, KNEES, ARM, HAND
			RF	38	F	2	SPRAINED KNEE
LA 174	MODERATE	IMPACT TO SIDE OF CAR	DR	20	M	1	FACIAL BRUISES
			RF	18	M	0	NO INJURIES NOTED
LA 234	MODERATE	PARTIAL HEAD-ON WITH CAR	DR	49	M	1	BRUISED CHEST, CHIN, ARMS, HANDS
LA 235	MODERATE	IMPACT TO SIDE OF CAR FOLLOWED BY POLE IMPACT	DR	68	M	1	SCALP CUT, ARM ABRASIONS

880400

TABLE 4 - SUMMARY OF FRONT SEAT OCCUPANTS WITH AIS 3 OR GREATER INJURIES IN DEPLOYMENT ACCIDENTS INVOLVING THE 1973-76 GM ACRS

CASE NO.	COLLISION DATA		OCCUPANT INFORMATION					INJURY EXPECTED
	SEVERITY	SIGNIFICANT INTRUSION	POS.	AGE	SEX	AIS	MOST SEVERE INJURY	
LA 130	SEVERE	YES	DR	29	M	6	SEVERE HEAD AND CHEST INJURIES	YES
LA 303	SEVERE	YES	DR	29	M	6	SEVERE CHEST INJURIES	YES
LA 343	SEVERE	YES	DR	81	M	6	MASSIVE HEAD INJURIES	YES
LA 324	SEVERE	YES	DR	49	F	3	FRACTURED ARM AND FEMUR	YES
LA 350	SEVERE	NO	DR	16	M	3	BRUISED KIDNEY	YES
AC 1291	MODERATE	YES	DR	39	M	3	FRACTURED RIBS AND ANKLE	YES
LA 23	MODERATE	YES	DR	42	M	3	FRACTURED RIBS	YES
LA 46	MODERATE	NO	DR	58	M	3	FRACTURED LUMBAR VERTEBRA	YES
LA 128	MODERATE	NO	DR	41	M	6	POSSIBLE CHEST/NECK INJURY	NO
LA 8	MODERATE	NO	DR	46	M	3	FRACTURED RIBS	NO
LA 74	MINOR	NO	DR	45	M	3	FRACTURED RIBS	NO
LA 126	MINOR	NO	DR	24	F	3	UNCONSCIOUS FOR 25 MINUTES	NO
LA 152	MINOR	NO	DR	46	F	3	FRACTURED ARM	NO
LA 165	MINOR	NO	DR	56	F	3	FRACTURED CLAVICLE AND RIBS	NO
LA 180	MODERATE	YES	RF	10	M	4	FRACTURED FEMUR	YES
LA 53	MODERATE	NO	RF	65	F	3	FRACTURE OF LUMBAR VERTEBRA	YES
AC 316	MODERATE	NO	RF	1 Mo	M	6	SEVERE HEAD INJURY	NO
LA 173	MODERATE	NO	RF	34	F	4	FRACTURED FEMUR AND ANKLE	NO
LA 195	MODERATE	NO	RF	40	F	4	FRACTURES OF BOTH LEGS	NO
LA 294	MINOR	NO	RF	59	F	4	FRACTURES OF BOTH LEGS	NO
LA 310	MODERATE	NO	RF	21	F	3	FRACTURES OF TIBIA AND FIBULA	NO
LA 246	MODERATE	NO	RF	4	M	3	PERSISTENT NUMBNESS LEFT SIDE	NO
LA 170	MINOR	NO	RF	75	M	3	FRACTURED TIBIA	NO

and/or the injuries were produced by non-frontal collision forces. In these cases it was judged that no air cushion restraint system could have mitigated these injuries. These cases, involving eight drivers and two passengers, are classified as injury expected in Table 4. For the thirteen other cases, the collision severities were minor or moderate. The collision forces were primarily frontal and there was no significant intrusion or distortion of the interior. Since an air cushion system is expected to provide occupant protection in such accidents, these injuries were more severe than expected. These thirteen cases, involving six drivers (6/216 = 2.8%) and seven front seat passengers (7/97 = 7.2%, two of whom were children), are classified as occupant injuries more severe than expected in Table 4. It is these occupant injuries which reduced the effectiveness of the 1973-76 GM ACRS for AIS 3 or greater injuries.

SUMMARY OF AIS 2 OR GREATER INJURIES THAT WERE MORE SEVERE THAN EXPECTED - The remainder of the deployment accident cases were reviewed to identify all occupants who experienced neurological problems; significant thoracic organ injuries including rib fractures; fractures of the hand, arms, legs, pelvis and vertebrae; and significant abdominal organ injuries that were classified as AIS 2. In each case that was identified, the accident data were reviewed in the same manner as was described for the AIS 3 or greater cases and a judgment was made as to whether or not the injury was more severe than expected. A summary of all AIS 2 or greater injuries which were judged as more severe than expected is given in Table 5 including those injuries noted in Table 4. Twenty drivers (20/216 = 9.3%) and seventeen passengers (17/97 = 17.5%) experienced injuries that appeared to be more severe than expected. These injuries were the principal reason why the effectiveness estimates of the 1973-76 GM ACRS noted in Table 2 were lower than expected.

POSSIBLE CAUSES OF UNEXPECTED INJURIES

By definition, the only injuries noted in Table 5 were to occupants involved in accidents of minor to moderate collision severity where the collision force was primarily frontal and where there was no significant intrusion or distortion of the car interior in the vicinity of the occupant. In such accidents, only minor injuries to the occupants were expected provided the occupant was not near the air cushion module at the time of deployment. This expectation is based on the fact that human volunteers experienced either no injuries or only minor abrasions in similar simulated collision environments while being restrained by the ACRS(4, 5). On the other hand, if the occupant was near the air cushion module at the time of deployment, significant forces would be developed between the deploying cushion and the occupant. Human volunteer and animal evaluation of such out-of-position occupant/cushion interactions indicated the potential for significant injuries(5, 6). Analysis of the accident data for cases noted in Table 5 indicated that both the fatally injured driver and the one-month-old passenger may have been close to the air cushion module at the time of deployment. The legs of the passengers with leg fractures were likely close to the passenger air cushion module at the time of deployment since the module is located in the lower part of the instrument panel, directly in front and close to the legs of the passenger. Consequently, it was hypothesized that most of the injuries noted in Table 5 were due to the occupant being close to the air cushion module at the time of deployment and interacting with the deploying cushion. To investigate the nature of these interactions a series of dummy and animal tests was conducted. The following is a summary of these tests.

DRIVER CHEST INJURIES - Horsch and Culver(9) conducted a series of simulated frontal collisions of moderate severity (23 mph, 14 G) sled tests of the 1973-76 driver ACRS using a Hybrid III dummy. The dummy was seated in a normal driving posture and was allowed to slide forward on the seat in response to the simulated collision pulse. The air cushion was deployed when various spacings between the dummy's chest and the cover of the driver's inflator module were reached. These spacings included the condition where the driver's chest had impacted the module cover and was compressing the column when the air cushion was deployed. For comparison purposes, two tests were conducted where the dummy was leaning forward with its chest against the module prior to subjecting it to the sled pulse. In one of these tests, the air cushion was deployed. In the other test, the cushion was not deployed.

Pertinent results from these tests are given in Table 6. When the dummy was initially in a normal driver's posture and the cushion was deployed prior to the dummy moving close to the module (Test 1), the maximum chest deflection experienced by the dummy was only 0.7 inch. However, in those tests where the dummy's chest was against the module at deployment (Tests 4, 5 and 6), the dummy's sternum bottomed out on its spine box and a maximum chest compression of 3.3 inches was recorded in each instance. According to Neathery et al(10), this level of thoracic compression is indicative of life threatening thoracic injury.

The added effect of deploying the air cushion when the chest is against the module is demonstrated by comparing the results of Tests 6 and 7. For these tests, the dummy was leaned forward with its chest against the module prior to subjecting it to the sled acceleration. The air cushion was deployed in Test 6 and the dummy's sternum was bottomed out on its spine (3.3 inches). In Test 7, the cushion was not deployed and the maximum chest deflection was only 1 inch. These results clearly demonstrate that the 1973-76 driver ACRS had the potential to produce the various thoracic lesions noted in Table 5 if the drivers were near the inflator module when the cushion was deployed. Although

Table 5 - Summary of Injuries to 1973-76 GM ACRS Occupants
That Were More Severe Than Expected

Body Region	Driver Injuries	Passenger Injuries
Head/Neck	3 Drivers Concussed 1 AIS = 3, Unconscious 25 M. 2 AIS = 2	1-Month-Old Baby With Subdural Hematoma, AIS = 6 4-Year-Old With Persistent Numbness, AIS = 3
Arm/Hand	9 Drivers With Fractures (9/216 = 4.2%)	6 Passengers With Fractures (6/97 = 6.2%)
Thorax	Possible Fatal Chest Injury, No Autopsy 5 Drivers With Rib Fractures (3 AIS = 3, 2 AIS = 2)	None
Abdomen	None	None
Leg/Pelvis	2 Drivers With Fractures (2/216 = 1.1%)	9 Passengers With Fractures (9/97 = 9.3%)

Horsch and Culver did not evaluate the potential for head injuries, it is not difficult to envision the potential to produce head/neck injuries if the head is close to the module when the cushion is deployed.

CHILD INJURIES - Anesthetized pig tests conducted by Chambers University for Volvo(11) indicated that fatal liver injuries could be produced if the pig was oriented with the region of the abdomen containing the liver aligned with the path of the deploying cushion. These results suggested that the animal positions used by Patrick and Nyquist(6) to evaluate the out-of-position child injury concern for the 1973-76 GM ACRS may not have been the most critical positions. To evaluate this possibility, a second series of animal sled tests was conducted using anesthetized pigs and baboons(12). In these tests, the animals were positioned near the 1973-76 GM ACRS passenger inflator module with either their head, chest or abdomen placed in the path of the deploying cushion. The sled was subjected to a variety of simulated collision pulses and the cushion was deployed. Life threatening injuries to the brain, cervical spine, heart and liver were produced. These results suggest that children who were near the 1973-76 GM ACRS passenger air cushion module when it was deployed could have experienced life threatening injuries as well. These results are entirely consistent with the injuries experienced by the two children noted in Table 5 and suggest that the children may have been close to the air cushion module when the cushion was deployed.

PASSENGER LEG FRACTURES - The relative high frequency of passenger leg fractures (9/97 = 9.3%) which occurred in minor to moderate collisions without significant occupant compartment intrusion was quite unexpected because the axial compressive femur loads that were measured in the 30 mph, frontal barrier tests (Table 1) were quite low, ranging from 470 to 1210 pounds.

Seven of the nine fractures were to the tibia suggesting that axial compressive femur load may not be the most appropriate response measurement for evaluating the restraint potential of the 1973-76 GM ACRS deploying passenger knee cushion. A pair of specially instrumented lower legs for the Hybrid III dummy were developed by Nyquist and Denton(13). Since the majority of the fractures were to the tibial plateau and malleolus, these legs were instrumented to provide measurements of the internal medial and lateral loading of the knee and ankle joints. The shafts of the legs were instrumented to provide measurement of the bending moment at two cross sections and the axial compressive load. Sled tests were conducted using a Hybrid III dummy equipped with these instrumented lower legs. Since none of the passenger leg fractures noted in Table 5 occurred in a se-

1187

Table 6 - Summary of Sled Tests Conducted to Investigate Out-of-Position Driver Interactions With the 1973-76 Driver ACRS. Moderate Collision Severity Pulse (23 mph, 14 G) and Hybrid III Dummy Used(9).

Test No.	Dummy Initial Position	Distance Between Chest and Module at Deployment (in)	Thoracic Responses Chest Compression Max. Defl. (in)	Max. Rate (ft/s)	Spine Acc. (3 ms, G)	Column Stroke (in)
1	Normal	10.5	0.7	3	32	3.2
2	Normal	1.7	2.3	49	64	6.0[5]
3	Normal	0.8	3.1	52	75	6.0[5]
4	Normal	0	3.3[4]	75	113	6.0[5]
5	Normal	Impacting[1]	3.3[4]	49	93	6.0[5]
6	Chest on Module	0[2]	3.3[4]	56	66	4.8
7	Chest on Module	0[3]	1.0	7	23	3.0

Notes:

1. In Test 5, cushion deployed 7 ms after chest contacted module.

2. In Test 6, cushion deployed 9 ms after beginning of sled pulse which is same timing as Test 1.

3. In Test 7, cushion was not deployed.

4. In Tests 4, 5 and 6, Hybrid III sternum bottomed out on spine box.

5. Maximum column stroke available in test fixture was 6 inches.

vere collision environment, a moderate collision severity (18 mph, 10 G) was chosen and only the low level of the passenger inflator was activated. Two different leg positions were evaluated: knees squared to the inflator module cover and knees rotated inboard with the right leg extended and closest to the module. The pertinent leg loads measured in these tests are summarized in Table 7. With the knees squared to the module cover all the leg loads, including the femur loads, were quite low indicating a low potential for leg fracture. In contrast, with the knees rotated inboard, the lateral compressive internal knee load of the right leg (the one closest to the module) was 1607 lb. Cadaver fracture load data of Hirsch and Sullivan(14) which was analyzed by Mertz(15) indicate that internal compressive loads of 900 lb between the tibial plateau and femoral condyle may fracture the tibial plateau if the load is biased medially or laterally. Clearly the results of the inboard facing leg test demonstrate the impor-

tance of leg orientation in evaluating the efficacy of deploying knee cushions and that many of the passenger tibial plateau fractures noted in Table 5 could have been caused by the legs not being squared to the inflator module cover when the air cushion was deployed.

HAND AND ARM FRACTURES - No tests were conducted to investigate possible causes for the hand and arm fractures noted in Table 5. However, it is easy to deduce that many of these fractures may have occurred if the cushion impacted the hand, or if the hand was flung hard against the car interior by the deploying air cushion.

RESTRAINT EFFECTIVENESS OF AN AIR CUSHION SYSTEM DESIGNED TO REDUCE THE SEVERITY OF CUSHION FORCES APPLIED TO OUT-OF-POSITION OCCUPANTS

If a driver and passenger air cushion system can be designed which would not injure out-of-position occupants while still providing re-

Table 7 - Summary of Peak Femur Loads and Lateral and
Medial Internal Knee and Ankle Loads for Two
Positions of the Legs, Moderate Severity Sled
Pulse (18 mph, 13 G), and Low Level Deployment
of the 1973-76 GM Passenger ACRS.

Leg Position	Femur Forces (lb) L/R	Knee Forces (lb)		Ankle Forces (lb)	
		LT. LAT. LT. MED.	RT. LAT. RT. MED.	LT. LAT. LT. MED.	RT. LAT. RT. MED.
Knees Squared To Module	452	259	479	265	319
	470	308	281	324	297
Knees Rotated Inboard, RT Leg Extended and Closest to Module	72	486 T	1607	104	339
	387	517	1310 T	145	299

Note:

1. All loads are peak compressive forces except those followed by a T which indicates tension.

straint for normally seated occupants, then the restraint effectiveness of that system would be greater than the effectiveness of the 1973-76 GM ACRS. For example, if the 1973-76 GM ACRS had been designed so that the unexpected AIS 3 or greater injuries noted in Table 4 were AIS 2 or less, then the restraint effectiveness in frontal deployment accidents of such a system would have been 50% instead of 18%. The individual effectiveness estimates for such a driver system and passenger system would have been 41% and 71% instead of 21% and 16%, respectively. These restraint effectiveness estimates were calculated using the same data and method used to calculate the 1973-76 GM ACRS effectiveness estimates except the injury severity classification for all the AIS 3 or greater unexpected injuries were changed to AIS = 2. The results of this analysis indicate that a sizeable gain in restraint effectiveness can be achieved if air cushion systems are designed to reduce the severity of cushion forces applied to out-of-position occupants.

Some methods to achieve this objective are discussed in papers by Klove and Oglesby(2) and Horsch and Culver(9). Mertz(16) notes that the tradeoff between inflating the cushion fast enough to provide restraint protection in the 30 mph, frontal barrier crash test required by FMVSS 208, but slow enough so as to not to seriously injure out-of-position occupants is the design dilemma of air cushion systems. Much more effort is needed to address this concern (17, 18, 19, 20) if the restraint effectiveness

potential of the air cushion system concept is to be realized.

SUMMARY

General Motors equipped 11,321 vehicles with driver and front seat passenger air cushion systems during the 1973 through 1976 model years. Based on an analysis of field accident data, Pursel et al estimated the restraint effectiveness of the 1973-76 GM ACRS to be 18% in mitigating AIS 3 or greater injuries when compared to unrestrained occupant injuries that occurred in comparable accidents. A case by case review of AIS 3 or greater injuries indicated that many of the injuries occurred in frontal accidents of minor to moderate collision severity where there was no intrusion or distortion of the occupant compartment in the vicinity of the occupant. These injuries were classified as unexpected and appear to be due to the occupant being near the air cushion module at the time of deployment. Dummy and animal tests of the 1973-76 GM ACRS were conducted and confirmed the fact that the types and severities of the unexpected injuries could be produced if the occupant was near the ACRS module at the time of deployment. It was estimated that a restraint effectiveness of 50% in mitigating AIS 3 or greater injuries in frontal deployment accidents could be achieved if an air cushion system could be developed that reduced the cushion forces applied to out-of-position occupants.

REFERENCES

1. Campbell, D. D., "Air Cushion Restraint Systems Development and Vehicle Application", SAE 720407, Second International Conference on Passive Restraints, May, 1972.

2. Klove, E. H. and Oglesby, "Special Problems and Considerations in the Development of Air Cushion Restraint Systems", SAE 720411, Second International Conference on Passive Restraints, May, 1972.

3. Louckes, T. N., Slifka, R. J., Powell, T. C., and Dunford, S. G., "General Motors Driver Air Cushion Restraint System", SAE 730605, May, 1973.

4. Smith, G. R., Gulash, E. C., and Baker, R. G., "Human Volunteer and Anthropomorphic Dummy Tests of General Motors Driver Air Cushion System", SAE 740578, 1974.

5. Smith, G. R., Hurite, S. S., Yanik, A. J., and Greer, C. R., "Human Volunteer Testing of GM Air Cushions", SAE 720443, Second International Conference on Passive Restraints, May, 1972.

6. Patrick, L. M. and Nyquist, G. W., "Airbag Effects on the Out-of-Position Child", SAE 720442, Second International Conference on Passive Restraints, May, 1972.

7. Wilson, R. A. and Piepho, L. L., "Crash Testing the General Motors Air Cushion", Fifth International Technical Conference on Experimental Safety Vehicles, London, England, July, 1974.

8. Pursel, H. D., Bryant R. W., Scheel, J. W. and Yanik, A. J., "Matching Case Methodology for Measuring Restraint Effectiveness", SAE 780415, February, 1978.

9. Horsch, J. D. and Culver, C. C., "A Study of Driver Interactions With an Inflating Air Cushion", SAE 791029, Twenty-Third Stapp Car Crash Conference, October, 1979.

10. Neathery, R. F., Kroell, C. K. and Mertz, H. J., "Prediction of Thoracic Injury From Dummy Responses", SAE 751151, Nineteenth Stapp Car Crash Conference, November, 1975.

11. Aldman, B., Andersson, A., and Saxmark, O., "Possible Effects of Airbag Inflation on a Standing Child", Proceedings of the International Meeting on Biomechanics of Trauma in Children, Lyons, France, 1974.

12. Mertz, H. J., Driscoll, G. D., Lenox, J. B., Nyquist, G. W., and Weber, D. A., "Responses of Animals Exposed to Deployment of Various Passenger Inflatable Restraint System Concepts for a Variety of Collision Severities and Animal Positions", Proceedings of the Ninth International Technical Conference on Experimental Safety Vehicles, Kyoto, Japan, November, 1982.

13. Nyquist, G. W. and Denton, R. A., "Crash Test Dummy Lower Leg Instrumentation for Axial Force and Bending Moment", Transactions of the Instrument Society of America, Vol. 18, No. 3, 1979.

14. Hirsch, G. and Sullivan, L., "Experimental Knee Joint Fractures -- A Preliminary Report", ACTA Orthopaedica Scandinavica, Vol. 36, 1965.

15. Mertz, H. J., "Injury Assessment Values Used to Evaluate Hybrid III Response Measurements", ISO/TC22/SC12/WG5, Document N123, May, 1984.

16. Mertz, H. J., and Marquardt, J. F., "Small Car Air Cushion Performance Considerations", SAE 851199, 1985.

17. Prasad, P. and Daniel, R. P., "A Biomechanical Analysis of Head, Neck, and Torso Injuries to Child Surrogates Due to Sudden Torso Acceleration", SAE 841656, Twenty-Eighth Stapp Car Crash Conference, November, 1984.

18. Montalvo, F., Bryant R. W. and Mertz, H. J., "Possible Positions and Postures of Unrestrained Front-Seat Children at Instant of Collision", Proceedings of the Ninth International Technical Conference on Experimental Safety Vehicles, Kyoto, Japan, November, 1982.

19. Wolanin, M. J., Mertz, H. J., Nyznyk, R. S., and Vincent, J. H., "Description and Basis of a Three-Year-Old Child Dummy for Evaluating Passenger Inflatable Restraint Concepts", Proceedings of the Ninth International Technical Conference on Experimental Safety Vehicles, Kyoto, Japan, November, 1982.

20. Mertz, H. J. and Weber, D. A., "Interpretations of the Impact Responses of a Three-Year-Old Child Dummy Relative to Child Injury Potential", Proceedings of the Ninth International Technical Conference on Experimental Safety Vehicles, Kyoto, Japan, November, 1982.

890602

Crash Protection in Near-Side Impact — Advantages of a Supplemental Inflatable Restraint

Charles Y. Warner, Charles E. Strother, Michael B. James,
Donald E. Struble, and Timothy P. Egbert
Collision Safety Engineering
Orem, UT

ABSTRACT

Collision Safety Engineering, Inc. (CSE), has developed a test prototype system to protect occupants during lateral impacts. It is an inflatable system that offers the potential of improved protection from thoracic, abdominal and pelvic injury by moving an impact pad into the occupant early in the crash. Further, it shows promise for head and neck protection by deployment of a headbag that covers the major target areas of B-pillar, window space, and roofrail before head impact. Preliminary static and full-scale crash tests suggest the possibility of injury reduction in many real-world crashes, although much development work remains before the production viability of this concept can be established. A description of the system and its preliminary testing is preceded by an overview of side impact injury and comments on the recent NHTSA Rule Making notices dealing with side-impact injury.

PROBLEM DEFINITION

Side impacts, according to the National Highway Traffic Safety Administration (NHTSA), account for 30% of all fatalities and 34% of all serious injuries to passenger-car occupants (1,2)*. The problem of improving side-impact protection has received much attention in recent years, leading to NHTSA's issuance of two notices which propose changes to the present Federal Motor Vehicle Safety Standard (FMVSS) 214: its Notice of Proposed Rulemaking (NPRM Jan. 27, 1988) and Advance Notice of Proposed Rulemaking (ANPRM - Aug. 19, 1988). The January NPRM addresses torso and pelvic injuries, while the August ANPRM

---

*Numbers in parentheses indicate references are at end of paper.

addresses the issues of head and neck injuries and ejection (1, 2).

INJURY STATISTICS

Many statistical studies of side-impact accidents are reported in the literature. However, significant variations in data collection and analysis procedures make it difficult to directly compare the results of these studies.

INJURIES - Rouhana and Foster (3) made an excellent compilation of some of the most important side-impact statistical studies. They analyzed the National Crash Severity Study (NCSS) side-impact data and then compared their results (insofar as possible) to other published studies of NCSS and other data. Some of their findings include:

(1) Approximately 40% of all accidents are side impacts.

(2) With regard to occupant seating position, near-side occupants experience three times the incidence of serious or immediately-fatal injuries as do far-side occupants.

(3) Serious injuries are three to ten times more likely if the passenger compartment sustains intrusion.

(4) While thoracic injuries are most prevalent among "serious injuries," head and neck injuries are most prevalent among "immediately-fatal injuries" (3).

Although there has been no detailed, in-depth analysis of the National Accident Severity Study (NASS) data with regard to side impact collisions, Hackney, et al. (4) did compare the available NASS data to the NCSS analysis made by Partyka and Rezabek (5). They found that "upper torso/side surface" injuries were the major injury category in both NASS and NCSS as analyzed by Partyka and Rezabek.

Using the selection criteria of Rouhana and Foster, we looked at the NASS data to see whether they would also show that head and neck injuries are the most prevalent among the immediately-fatal injuries. We found that head and neck injuries made up slightly more than one half of the immediately-fatal injuries, with chest injuries making up the remaining half. In both the NCSS and NASS data a significant number of head injuries in side impacts have unknown contact points. While this lack of data makes it difficult to know the precise mechanisms of these injuries, it is reasonable to expect that some (and maybe even a substantial number) of the side-impact head injuries are attributable to the head passing through the side window opening and contacting either the lower window frame or parts of the oncoming vehicle.

ACCIDENT SEVERITY - It is essential to establish some basis for categorizing injuries according to accident severity, if accident statistics are to be used effectively in designing for injury reduction. Both NASS and NCSS use the CRASH3 program to calculate vehicle center-of-gravity-change in velocity (delta-V), which is used as a measure of accident severity.

Research by CSE and others now in progress on the NASS side impact data indicates there may be important reasons to question the validity of the delta-V's found in both NCSS and NASS, particularly for side impact collisions. Five areas of concern are: 1) missing data, 2) CRASH3 stiffness coefficients, 3) the effect of principal direction of force (PDOF) on crush energy computation, 4) NASS field procedures for measuring vehicle side crush, and 5) the "missing vehicle" algorithm. Each of these five areas are briefly discussed below.

The missing data problem has been acknowledged (6), but its ramifications have never been seriously analyzed. For side-impact collisions the NCSS study has delta-V information for only 55% of the reported cases, while NASS has delta-V values for only 45% of the reported cases. Drawing conclusions on the basis of these minorities of cases is equivalent to assuming that the cases with delta-V data represent a random sample of all cases. The correctness of that assumption needs to be explored. It seems likely that the cases with delta-V information are not randomly distributed through the data base but rather are grouped in some way that skews the overall picture.

The CRASH3 program uses pre-programmed stiffness coefficients to compute the crush energy from vehicle deformation measurements. These stiffness coefficients are selected according to the vehicle's wheelbase and the location of deformation (side, frontal rear). Recent analysis of crash-test data shows that the stiffness coefficients used by CRASH3 significantly overestimate vehicle deformation energy associated with relatively small values of frontal crush (striking vehicle) and

underestimate energy associated with side crush (target vehicle). Since it is the total deformation energy that is used in CRASH3 to calculate the delta-V,s in vehicle-to-vehicle collisions (damage algorithm), these errors may tend to compensate for each other. However, in single vehicle side-impact collisions, the CRASH3 program may consistently under predict the delta-V.

Another confounding effect in the way CRASH3 computes crush energy from vehicle deformation is the effect of the principal direction of force (7). The program multiplies the computed crush energy by a so-called "correction factor" (of up to 2) which is a function of the angle between PDOF and a perpendicular to the deformed surface of the vehicle. In many vehicle-vehicle side impacts the struck vehicle has significant velocity and thus the PDOF's in many of these collisions differ significantly from the perpendicular. The "correction factor" is therefore very large in many instances. Compounding this is the reality that in these instances the PDOF angle is typically very difficult for even expert accident reconstructionists to estimate. Further, there has been no adequate justification given for the particular formulation for this factor, which seems to assume (contrary to experience) that vehicle structures are stiffer rather than more compliant when loaded angularly.

The field procedures used by the NASS teams to measure vehicle deformation are also in need of careful re-evaluation. For side impacts, the crush depth is measured at the maximum crush unless there is also sill crush, in which case the maximum crush and sill crush are numerically averaged. Therefore, for two identical vehicles with identical maximum crush, the one which has sill crush in addition to the maximum crush (and hence which has logically absorbed more energy) will actually be computed as having absorbed less energy.

Finally there is the problem of the "missing vehicle" algorithm. This "missing vehicle" technique is presumably the result of NHTSA's attempts to reduce the missing data problem (8). A study of the NCSS file indicates that a substantial number of cases do not have a calculated delta-V because one of the vehicles was not available for inspection by the investigating team. An algorithm was developed to estimate the energy absorbed by the missing car by calculating the apparent inter-vehicle forces from the crush and stiffness coefficients for the known car. Delta-V computations made by this method are thus subject to greater errors, particularly in view of the problems with the frontal and side stiffness coefficients noted earlier.

In conclusion, studies that attempt to address the relationships between injury and accident severity using NCSS and NASS data must be viewed with some skepticism in light of the many-faceted problems involved in the estimation of delta-V's, as delineated above.

SIDE VERSUS FRONTAL COLLISIONS

The vehicle safety community has made significant progress in the past 30 to 40 years in crashworthiness design improvements. Most of that effort, for appropriate reasons, has been concentrated on frontal collisions. Frontal and side collisions, however, are dramatically different for several traditional reasons: 1) the amount of crush space and structure present on the side of a vehicle are substantially less than that on the front of a vehicle, 2) as a result, in frontal impacts, intrusion is not a factor in producing injury in all but the most severe collisions, whereas in side impacts there is almost <u>always</u> intrusions into occupant seating areas, 3) ingress and egress of vehicles is through the side, and 4) sides have windows (also referred to as "glazing") that open and close. Since the automobile must operate within a worldwide system of streets, highways, garages, and parking facilities, it is unlikely to see these constraints altered.

LIMITED SIDE STRUCTURE AND CRUSH SPACE - The general approach to reducing injury exposure is to reduce the deceleration experienced as the occupant changes speed to match that of the vehicle during the collision. This is achieved by increasing the distance over which the occupant undergoes the speed change. In most frontal collisions, this is accomplished by linking the occupant to the occupant compartment by means of a restraint system. If the occupant is so restrained, both the substantial frontal crush and the space forward of the occupant inside the compartment ("rattlespace") can be effectively used to diminish occupant loadings. Obviously, this approach cannot offer as much benefit in side impacts.

In a side impact, the near-side occupant is seated very close to the collision object, separated only by a structure which, for practical reasons, cannot be sufficiently strong and stiff to keep the inner panel from moving inward in any but the lightest impacts, Figure 1. The problem is further compounded by architectural incompatibilities (e.g. in car-to-car crashes by the height mismatch between the bumpers of striking vehicles and the sill and floor structures of struck vehicles, and by the broad space between pillars in fixed-object collisions). Thus, in this type of collision situation, intrusion is almost always an issue. The near-side occupant is usually contacted by an accelerating inner door panel, resulting in a higher speed change for the occupant than for the center of gravity of the vehicle. In this situation, as shown in Figure 2, the presence of rattlespace can increase injury potential by allowing the interior surface to get a "running start" at the occupant (9). Intrusion above the struck vehicle beltline and into the window opening can also be an issue in side impacts, as structure from the striking vehicle may approach or penetrate the window opening, presenting a hard, generally blunt impact surface for the occupant's head.

On the other hand, intrusion is rarely a problem in frontal crashes because of the extensive structure in front of the passenger compartment and the relatively large amount of space available forward of the occupant. Frontal intrusion is seen mainly in severe high-speed crashes, unusual underride and override or narrow object situations, or oblique sideswipes in which structure may actually be peeled away from the passenger compartment.

INGRESS AND EGRESS - Since vehicles must have doors on the sides, the design challenge is significantly more complicated for side as opposed to frontal structures. Instead of continuous structural members running down the length of the side, there must be separate and distinct structures tied together at hinge and lock locations to accommodate the door. The force concentrations at these connections can be very great in collisions. Design options are limited by these connection points and their associated structure.

GLAZING - The upper portion of the side structure of almost all passenger cars is limited in terms of occupant protection. Side window safety glass, even if in the "up" position at impact, will generally disintegrate early in a side impact collision, usually before any occupant contact. The resulting openings are certainly a factor in partial and full ejections and therefore, head and neck injuries. There has been some effort in developing retractable side windows with an anchored inner plastic layer (10, 11). At the present time, applications of fixed side or retained membrane side glazing have yet to be developed to a marketable stage.

PADDING - Most efforts at improving side-impact occupant protection have focused on some combination of padding and stiffened side structure. Well-designed padding can reduce injury exposure in two ways: 1) by increasing somewhat the effective acceleration distance of the occupant, thereby reducing contact loads and 2) by distributing forces over larger areas, thereby reducing localized occupant loadings. The effectiveness of padding in side impacts is limited, however, because of the limited space available and the apparent negative reaction from consumers to a reduction in "elbow room".

The potential for reducing intrusion velocity by increasing door stiffness is also severely limited. The forces involved in moderate to severe side impacts are simply too great to allow practical side structures to prevent intrusion in most instances; intrusion velocity will always be of concern for near-side occupants in the crush zone. Consistent with this view is NHTSA's evaluation of the present FMVSS 214 (which essentially requires a door beam), which evaluation shows that the standard is effective only when the impact

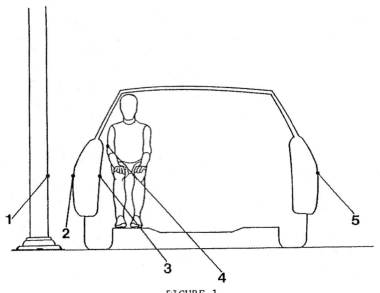

FIGURE 1
KEY POINTS OF INTEREST IN A FIXED-OBJECT SIDE IMPACT

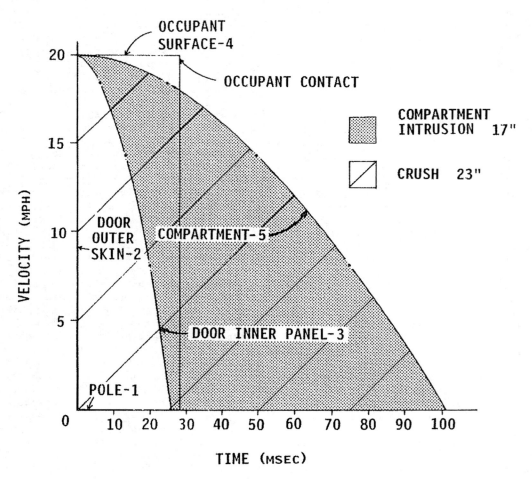

FIGURE 2
VELOCITY-TIME GRAPHS FOR A SUBCOMPACT VEHICLE IN A
20 mph FIXED POLE SIDE IMPACT

forces are primarily frontal, rear or non-horizontal (12).

Very high collision forces combine with physical constraints to greatly reduce the potential for significantly reducing intrusion velocities with reasonable structural reinforcements. Repeated research attempts to achieve significant benefit from stiffer structures have led several researchers to the conclusion that once proper attention is paid to door attachments and direct load paths, additional resources are better spent in padding (9). The most effective side structure may turn out to be the one that furnishes the best backup for the padding it supports, thus helping the padding to contact the occupant early, spread the intrusion contact and restraint forces somewhat, and provide the gentlest possible acceleration or "ride-up" to the intruder's velocity, while moving the occupant as carefully as practicable away from the space consumed by the intruder.

NHTSA'S PROPOSAL FOR THORACIC AND PELVIC INJURY

On January 27, 1988, the NHTSA issued its NPRM to revise the existing Federal Standard (FMVSS 214) on side-impact protection (1). This proposal involved the substitution of a full-scale dynamic-vehicle crash test to replace the current static side structure strength and stiffness requirements. In this proposed test, the striking vehicle is to be the newly-developed NHTSA Moving Deformable Barrier (MDB), originally intended to be a representation of an intermediate-sized vehicle (13). Compliance with the proposed revised standard would be on the basis of the Thoracic Trauma Index (TTI), an acceleration-based injury criterion using thoracic accelerometer data provided in a specialized Side-Impact Dummy (SID).

The NHTSA proposal is an interesting attempt at progress toward better side-impact protection. On the positive side, most safety researchers would agree that the proposed dynamic test can be more realistic than the present static crush test required by FMVSS 214, given appropriate dummy performance and injury criteria. Dummy-injury measures seem to be at least a potentially more rational method of judging a side-impact design as opposed to structural strength and door exterior deflection measures. On the other hand, NHTSA has proposed a relatively complex compliance test. The design tasks necessary to ensure compliance with the proposed standard will also be complex, especially since serious questions remain about the benefits of NHTSA's proposal, given the spectrum of real-world side impacts. Such benefits will depend on the nature of the final rule and the efficacy of the resulting designs in mitigating injury.

It is hoped the NHTSA notices will have the effect of rallying societal effort to identify and pursue rational objectives for evolving improvement in side impact. For this to happen, the research and development community must achieve agreement on what constitutes rational objectives. We believe there is basis for re-examination of some facets of the NHTSA proposal.

HONEYCOMB BARRIER FACE - Although the concept of a standardized crushing surface is conceptually appealing as a means of simulating the deformation of most side-impact partners, the NHTSA proposed aluminum honeycomb face for the MDB falls short of its reasonable performance standardization goals on several counts.

First, the honeycomb face itself does not demonstrate standardized or repeatable performance. Its manufacturing specifications do not properly regulate its crush characteristics, nor is it entirely reasonable to expect standardized crush performance from the honeycomb in the oblique buckling mode introduced by the crabbed-barrier configuration. An energy-absorbing material with reduced directional crush sensitivity is probably necessary if test variability is to be minimized. Second, the current honeycomb specification is admittedly too stiff to represent the frontal crush of virtually all passenger cars, though it is thought by some to represent light trucks reasonably well (13). Third, the honeycomb material is costly, and in short supply, introducing significant logistical and financial burdens to testing and research programs.

In summary, the non-standard, too-stiff, too-expensive, aluminum honeycomb barrier face does not add realism or effectiveness to the test, but does add cost, not only in material and logistical senses, but also in invalid test results, wasted time, and decreased test repeatability. It should be eliminated from the test requirement. In its place, a contoured rigid moving barrier should be used at an appropriately reduced test speed. If this approach should need refinement, a subsequent NPRM could be issued to upgrade the performance test to include an improved deformable barrier face when a device with appropriate performance, cost, availability, and repeatability has been developed and proven.

THE SID DUMMY - It is clear that the anthropometric test device (ATD) chest requires special treatment for human biofidelity in lateral impact, and that existing ATD thoraxes designed for frontal biofidelity have proven inadequate for the task (14). The proposed SID dummy simulates upper-arm inertia, introducing what may be an artifact on padding designs. It is not clear that its use will reduce real-world injuries unless it can be shown that (a) the majority of seriously-injured side-impact occupants are loaded through the upper arm in its anatomical unextended position and (b) that appropriate thoracic and abdominal dynamics are represented by the cadaver test data used to develop the SID. Recent studies call these two points into question (15).

THORACIC TRAUMA INDEX (TTI) INJURY CRITERION - The NPRM assumes a reduction of side-impact injury as a result of reductions in the TTI, based on a body of cadaveric-tolerance data. While the statistical correlation may be a good representation of the cadaver test data, it is difficult to apply to the design process until a reliable and economical computer simulation is available. The lumped-mass model presented in Reference 13 may be a good start toward the evolution of such a model. As it now stands, a designer is faced with performing multiple tests to evaluate padding-design changes, a costly and time-consuming procedure. A simpler injury index, more reflective of the physics of the injury process, would be preferred by the authors and by the vast majority of safety researchers with whom this topic has been discussed.

Given the broad spectrum of masses, stiffnesses, shapes and angles of trees, poles, posts, rails, car and truck fronts and corners and motorcycle components which may try to penetrate a car door, the task of improving occupant protection begins to look formidable indeed. But while a simple solution capable of resolving all side-impact issues may not be found at once, no progress can be made unless and until the basic vehicle-occupant kinematics in side impact are first understood.

NEAR-SIDE OCCUPANT/VEHICLE KINEMATICS

Figure 1 illustrates a typical fixed-object impact situation with a near-side occupant and identifies key points in the object/vehicle/occupant system (9). Figure 2 is a velocity-versus-time plot of the motion of these points in a 20 mph lateral test of a baseline 1972 Ford Pinto into a 14" diameter rigid pole (16). The velocity curves in Figure 2 are approximations of the velocities of the key points as identified in Figure 1. Since acceleration data for these points were not available, accelerations were estimated by using deformation measurements together with high-speed films.

The outer door surface (point 2) adjacent to the impact location comes immediately to rest upon impact. In contrast, the occupant compartment (point 5, represented by a point in the car side opposite the impact) comes to rest more gradually, in this case over a period of about 100 msec. The hatched area between these two curves (essentially the area under the curve for point 5) represents the vehicle crush in the plane of the collision, about 23 inches. The door inner panel (point 3) comes to rest much more quickly (in about 27 msec) than does the occupant compartment. The shaded area between the door inner and outer panel velocity curves (points 2 and 3) represents the door crush, about 6 inches, and the area between the door inner panel (point 3) and the occupant compartment (point 5) represents the intrusion into the compartment in the plane of the

collision, about 17 inches.

The motion of the outboard surface of a near-side occupant in the plane of the collision is illustrated by the curve for point 4 in Figure 2, starting about 4 inches away from the door inner panel. Under this baseline condition, occupant contact with the door inner panel (hip or torso) is estimated to occur at about 30 msec. By this time, the door inner panel (point 3) is at rest so that no ride-down benefit is realized. Additional padding of the door interior and inner panel could cause occupant contact to occur earlier, allowing some ridedown and peak-shaving benefits. Padding could also reduce the level of occupant deceleration by contributing a percentage of the additional padding distance as "stopping distance." An inflatable system in the door might also cause occupant deceleration to begin earlier and create a mechanism for increasing stopping distance. For any benefit to be realized, such a system would have to begin imparting significant occupant deceleration within the first 20-25 msec of the collision event.

Figure 3 identifies the key points of interest in an intersection-type car-to-car collision. Figure 4 is a time plot of the velocity of these points in a test collision involving full-size Fords (17). In this test, the struck vehicle was stationary and the striking vehicle moved at 40 mph. Figure 4 plots the velocities of the striking car's firewall (point 0) and bumper (point 1), the struck-side door inner panel (point 3), the far-side occupant compartment (point 5), and a dummy occupant (point 4) positioned on the struck side adjacent to the intruding door. As seen in Figure 4, the onset of change in occupant velocity is delayed until about 25-30 msec while the occupant "waits" until door intrusion advances through the rattlespace. As in the case of the fixed-object collision, padding and inflatable systems offer a potential of producing earlier occupant acceleration and increased acceleration distance. To be beneficial in this type of collision, an inflatable restraint would have to begin imparting occupant acceleration within 20 msec after initial vehicular contact.

Figure 5 is a velocity-time plot of a test impact in which a prototype of the recently proposed NHTSA moving deformable barrier (MDB) struck the side of a Chevrolet Citation at an angle of 60 degrees just behind the A-pillar. In accordance with the proposed test procedure, the MDB was "crabbed" at an angle and struck the stationary Citation at 33 mph to simulate a collision in which a striking vehicle travels at 30 mph and a struck vehicle at 15 mph (18). The severity of this configuration resulted in the door inner panel velocity actually exceeding the struck vehicle final velocity of about 22 feet per second (9). Occupant contact in this case was initiated at about 30-35 msec, the worst possible time, since the door inner panel was at or near its peak velocity of

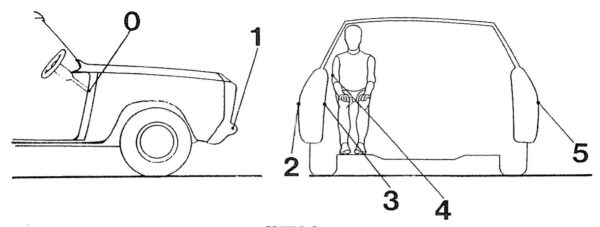

FIGURE 3
KEY POINTS OF INTEREST IN A
CAR-TO-CAR LATERAL COLLISION

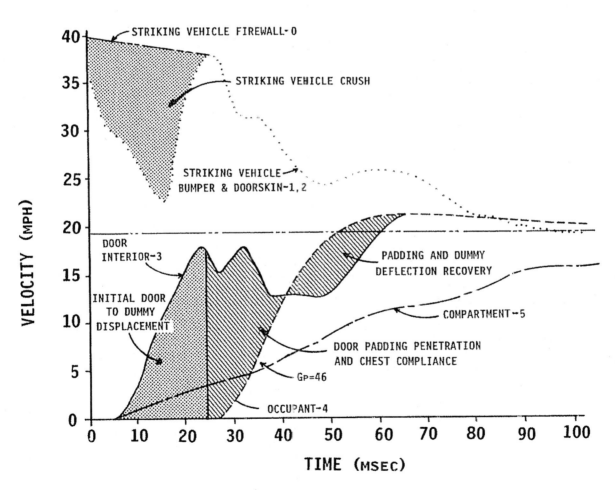

FIGURE 4
VEHICLE AND OCCUPANT KINEMATICS IN A
40 mph CAR-TO-CAR LATERAL IMPACT
(FIGURE REPRODUCED FROM REFERENCE 17)

nearly 40 feet per second. In this case, padding or an inflatable system could have similar benefits to the fixed-object case above, assuming occupant accelerations could be initiated before 20 msec.

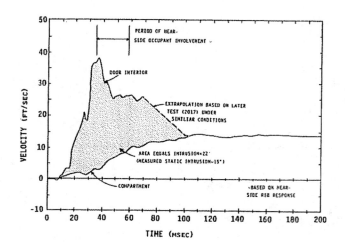

FIGURE 5
VEHICLE KINEMATICS IN A SIMULATED
MOVING-MOVING CAR-TO-CAR LATERAL IMPACT
(FIGURE REPRODUCED FROM REFERENCE 18)

In summary, at the theoretical level, an inflatable cushion in the door panel appears promising. It could potentially increase the effective occupant stopping distance by expanding into the "rattlespace" and could effectively distribute loads over large surface areas of the occupant. The major theoretical concern is whether the bag can inflate fast enough to provide a benefit without posing a significant deployment threat.

THE DOORBAG CONCEPT

Collision Safety Engineering has developed a prototype of a deployable door-mounted inflatable air cushion and pad system that offers the potential for significantly improving the side-impact injury-reduction capability of the vehicle interior, as compared to the performance of baseline vehicles.

Conceptually, the doorbag system incorporates mechanical performance features that address the significant parameters of the side-impact crash-protection problem. In the present configuration, padding is employed to provide load distribution and limit occupant accelerations. The padding is propelled toward the occupant by the deploying doorbag, utilizing available interior rattlespace and providing early occupant acceleration away from the intruding surfaces. Much of the normally empty space in the door interior is filled with foam to help provide a better load path for early load application. The doorbag is designed to be triggered very early in the event by a positive contact switch just inside

the outer door skin. The prototype configuration also employs an upward-deploying head bag which is interposed very quickly between the occupant's head and surfaces likely to cause head injuries.

DEVELOPMENT OF A TEST PROTOTYPE

The sensor, inflator, air cushion envelope, interior padding, and polyethelene foam were all integrated into a production-configured 1980 Chevrolet Citation driver's door in such a way that the window and door mechanisms could be operated normally. The door interior was provided with a two-inch thick layer of 20 psi polyethelene foam padding, enclosing the air cushion system. The system was designed so as to maximize the probability that an eventual production-engineered system could endure anticipated preimpact storage and function over the anticipated car life, without conflict with the normal door operation. The different features of the prototype system are briefly described as follows (Figure 6).

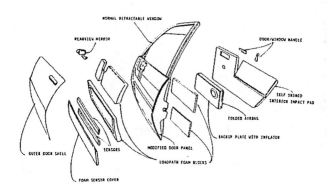

FIGURE 6
EXPLODED VIEW OF PROTOTYPE DOORBAG SYSTEM

INFLATOR - The inflator was made by Thiokol and was similar to those used in the Mercedes driver-system airbags, except for upscaled gas flow and pressure output. The inflator was located near the upper rear shoulder of the driver to minimize inflation time. Other locations were potentially slower and more complicated, requiring ductwork and diffusers to preserve window functions.

AIRBAG ENVELOPE - A single-chamber bag incorporating accordian folding was developed. A bag/inflator support structure was fabricated to support and to facilitate assembly of the system into the cutaway baseline door inner panel. High-density (30 psi) polystyrene foam was hand cut to fit into and occupy the door voids, without interference with window- and door-operating hardware, while providing a secure mounting for the prototype stripswitch sensor. Bag volume for the prototype was determined by comparing anticipated needs with volumes of successful driver systems. A 60 liter (2.1 cubic foot) bag volume was chosen.

INTERIOR PADDING - The interior upholstery was moved inward approximately 2 inches by installing a layer of 20 psi polyethelene foam on the interior door panel. Polyethelene foam was chosen for this research application because of its extremely good stiffness-to-weight ratio, low cost, and ease of manufacture. For production applications, the padding foam would require a durable skin with molded-in arm rest and handle recesses and would need to be subjected to appropriate performance and environmental engineering and testing.

It is anticipated that the interior padding will contribute to injury reduction for the arm and torso in both lateral and angled frontal lateral impacts. The protection capability of this foam layer was provided both by its geometry and deformational properties. The geometry contributed protection by providing an early force application, enchancing ride-up capabilities. The foam also provided the traditional padding functions of force limitation and enchanced force distribution. The two inches of interior space used by the interior foam pad may be small enough to be acceptable by consumers. The pad also has the potential advantage of furnishing some benefit with or without a deploying airbag (Figure 7).

STRUCTURAL CHANGES - The only structural change was removal of a portion of the interior door panel near the upper rear corner in the prototype to allow for a recessed mounting of the doorbag module. This was accomplished by cutting away the interior door panel and slightly rerouting the interior door- and lock-control rods. In a production system, the inner door panel stamping could probably be designed to accommodate the module with only slight change from its present configuration.

LOAD-PATH FOAM ELEMENTS - In the prototype system, a 30-psi polystyrene load path foam essentially filled the thickness of the Citation door between inner and outer sheet-metal panels. The foam was hand carved to fit between those panels and to sandwich the side-guard door beam. In a production model, the foam could probably be installed during the overall door assembly process before the inner and outer door panels are welded together. Provision would also be needed for a slot into which the window glass could be recessed, perhaps by using pre-cast coated styrofoam halves glued into place inside and outside a window slot.

SENSOR - The prototype sensor consisted of a series of corrugated contact switch elements mounted about one inch inside the outer door skin near the beltline stiffener fold, and embedded in the load path foam. In production, the outer section of load path foam could be preassembled to encapsulate the sensor system with wiring harness, backup capacitors, and diagnostic circuitry, before being installed

into the door structure (Figure 8).

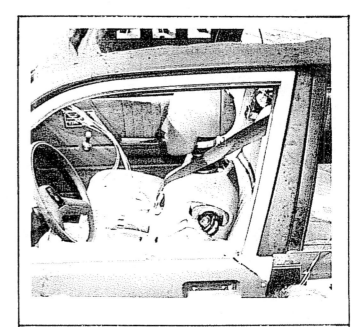

FIGURE 7
PRE-TEST PHOTOGRAPHS OF THE PROTOTYPE DOORBAG SYSTEM

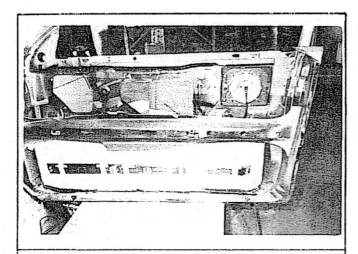

FIGURE 8
PHOTOGRAPHS OF THE PROTOTYPE DOORBAG SYSTEM

IMPACT TEST

A simulated vehicle-to-vehicle side-impact test was conducted at the CSE crash test facility. In this test, a rigid moving barrier (RMB) weighing 3003 lbs struck the left side of a stationary 2650 lb Citation automobile at a speed of 25.3 mph. An instrumented Hybrid 3 dummy occupied the driver seat of the 4-door Citation. Figure 9 shows the impact configuration, vehicle dimensions, etc.

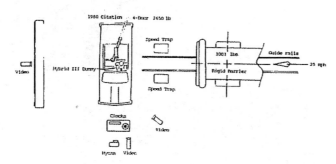

FIGURE 9
TEST CONFIGURATION

Test instrumentation consisted of a pressure transducer connected to the bag volume, a coordinated flash and electronic signal for inflator firing, high-speed and video-speed photography of the event with 600 rpm synchronous clocks in view, and three normal-speed video cameras. Stripswitch sensors were used to measure barrier-to-door contact and airbag/pad-to-dummy shoulder contact. Head, chest, and pelvis lateral accelerometers were placed in the dummy. The vehicle bench seat was adjusted to the center fore-aft position, and the instrument panel, windshield, driver's door glass, and a section of the roof panel and headliner were removed to facilitate photography.

The RMB impact speed was measured by dual break-wire time traps positioned just before impact. The brakes on all four wheels of the Citation were locked during the full impact event, and the brakes on the RMB were locked just subsequent to passing over the time traps and remained locked throughout the impact. Time plots of dummy head and chest data for this test are shown in Figures 10 and 11.

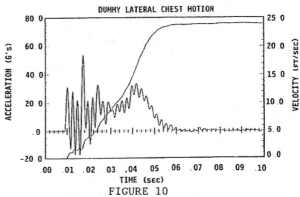

FIGURE 10
DUMMY LATERAL CHEST MOTION

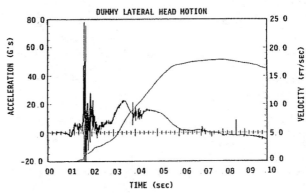

FIGURE 11
DUMMY LATERAL HEAD MOTION

The Citation moved about 8 feet post-impact and separated from the moving barrier. The overall residual vehicle crush was about 10.7 inches deep at height of deepest penetration. Analysis of the high-speed movies (Table 1) showed that the sensor ignition signal followed initial door contact by 1.5

TABLE 1
ANALYSIS OF HIGH SPEED FILM RECORD

The camera was running at an average speed of about 980 frames per second, as determined by the clock. Each frame is therefore about 1.02 milliseconds.

Frame	Event
0	Strobe fires, indicating first contact between barrier face and outer sheet metal of door.
3	Strobe flash dies
4	First visible pad movement
6	First contact between bag and dummy
10	First perceptible vehicle motion relative to stadia pole at base of windshield
12	Bag begins to rise above shoulder level
16	First motion of steering wheel
16-18	Bag contacts head
17	Steering wheel moves across to right
19	Vehicle starts to move at windshield header
20	First chest motion observed
23	Baffle stitching tears
24	Bag fills window area
26	Bag pushes outside window area
27	First motion of undamaged area of car relative to ground (front grill area, windshield header)
28	Dummy head moves, relative to vehicle, away from impact
30-36	Bag appears to be fully inflated
30	Steering wheel rim crosses centerline
36	First head motion
56	Steering wheel hub gets to center of vehicle and begins to rebound
60	Vehicles reach common speed
75	Barrier rebounding from door: vehicle speed about 21 ft/sec., barrier speed about 13 ft/sec.
78	Head rotated and moving ahead of vehicle

Speed of vehicle at separation was about 21 ft/sec.

Note: The dummy pelvis was not visible on film.

msec and that dummy contact force began to build about 6 msec after impact. Rapid pressure rise within the bag and full-bag deployment was achieved about 30-35 msec after impact. Time sequence photographs taken from the video camera are shown in Appendix 1.

Figure 12 is a speed-versus-time history illustrating important events and processes that transpired during the moving- barrier

test. The Citation and RMB centers of gravity reached a common speed at about 60 msec after impact. Based on the velocity change and impact duration, the average Citation center of gravity acceleration was about 10g. (Peaks of 15-25g in the vehicle pulse would reasonably be expected.) Dummy measurements showed modest accelerations, considering impact severity. The head experienced acceleration peaks of around 27g and an average of about 13g. Much more could be learned with the more elaborate photographic coverage and more complete electronic instrumentation planned for future testing.

The dummy chest lateral acceleration peaked at about 27g at 40 msec, while the chest velocity matched the vehicle velocity at about 55 msec. The estimated inner door panel velocity at chest level is based on crush measurements taken from the test vehicle. The potential chest contact velocity with the inner panel, which, without the inflatable system would have been reached at about 15 msec after contact, is about 35 ft/sec (see Figure 12). Thus, the chest velocity change was substantially reduced by the doorbag system, and the potential for injury was likewise reduced.

There was a tendency of the airbag to contact the left frontal lateral aspect of the dummy face, causing a relatively rapid rotational impulse of the head in the coronal plane. Although no quantitative definition of this impulse could be made using the existing instrumentation and photographic records, it will require adequate treatment in future doorbag design efforts. Whether such impulses are potentially injurious is presently unknown. The fidelity of the Hybrid III dummy to respond in this way is likewise unknown.

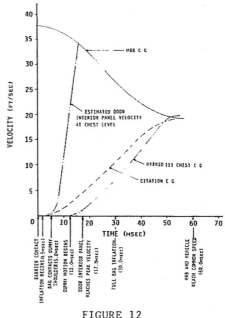

FIGURE 12
TEST VELOCITY HISTORY

DISCUSSION

The early preliminary testing produced encouraging results and suggests that a door-mounted airbag could provide meaningful head and chest protection in some side impacts. As with any single-injury countermeasure, its performance, however effective subsequent testing might show it to be, will probably not be adequate for all crash loading conditions or for all occupants. In the side impact, as in the frontal impact mode, the inflatable restraint offers the potential advantages of load distribution by mass and body contour matching and a force-limitation that can address a broadened impact-severity range. With side impacts in particular, it could potentially provide for early loading and utilization of the minimal available stroke distance for broad area load distribution and ride-up. If shown by further development to be a viable concept, its use could effectively double the average restraint-stroke distance, thereby reducing peak restraint forces and greatly reducing ATD-injury measures.

Common to other inflatable systems, its practicality and benefits will be affected by non-deployment, inadvertent deployment and reliability considerations. Of course resulting benefits will be had only by payment of increased consumer product cost. Whether this increased cost can be justified by injury reduction in the overall spectrum of real accidents is a matter requiring a good deal more testing and analysis than that accomplished thus far, and is, at any rate, a question of public policy, beyond the scope of this paper. A fuller understanding will require comparing the performance of a more highly developed doorbag prototype with alternative structure and padding systems.

CONCLUSIONS

Accident statistics, while incomplete in important areas, demonstrate that it is appropriate to allocate significant effort to understanding side impacts and addressing occupant-protection problems associated with them.

Recent NHTSA efforts to address these issues are interesting. Much work is needed, however, before an adequate understanding of side-impact injury mechanisms and associated occupant-protection issues are achieved. It may well be that some aspects of the NHTSA's proposals to establish a dynamic side-impact standard are premature and should be re-examined in some detail. Certainly, the lack of concensus about the proposed standard's repeatability, fidelity to real-world accidents, deformable aluminum honeycomb barrier face, side impact dummy performance, usefulness of the thoracic trauma index (TTI) and cost must be addressed.

The side-impact prototype doorbag system described in this paper represents an example of the kinds of injury-reduction methods that should be explored. Preliminary doorbag development work has indicated the following with respect to the prototype driver-doorbag padding system:

(1) it has the potential to be constructed and installed with state-of-the-art hardware;

(2) it can potentially be designed to deploy rapidly enough to fill the occupant rattlespace at shoulder and head levels in side impacts with wide collision partners;

(3) it has the potential for increasing the efficiency of padding systems for chest injury control by providing earlier contact and rideup;

(4) it may potentially prevent head egress through the window aperature in many moderately severe side impacts, and reduce head impact injury due to contact with interior and exterior surfaces; and

(5) it may broaden the options available to designers by maximizing the interior-occupant space for a given exterior envelope and level of occupant protection.

REFERENCES

1. Notice of Proposed Rulemaking to Revise Federal Motor Vehicle Safety Standard 214 - Side Impact Protection. Federal Register Volume 53, No. 17, January 27, 1988.

2. Advanced Notice of Proposed Rulemaking to Revise Federal Motor Vehicle Safety Standard 214 - Side Impact Protection. Federal Register Volume 53, No. 161, August 19, 1988.

3. Rouhana, S. and Foster, M., "Lateral Impact - An Analysis of the Statistics in the NCSS", SAE 851727, May, 1985.

4. Hackney, J., Hampton, C. Gabler, Kanianthra, J., and Cohen, D., "Update of the NHTSA Research Activity in Thoracic Side Impact Protection for the Front Seat Occupant", SAE 872207.

5. Partyka, S. and Rezabek, S., "Occupant Injury Patterns in Side Impacts - a Coordinated Industry/Government Accident Data Analysis", SAE 830459.

6. Ricci, L., "NCSS Statistics: Passenger Cars", NHTSA Contract No. DOT-HS-8-01944, Special Report, June, 1980.

7. Woolley, R., Warner, C., and Tagg, M., "Inaccuracies in the CRASH3 Program", SAE 850255.

8. Segal, D., McGrath, M. and Balasubramanian, N., "Supplemental National Crash Severity Study Accident Reconstruction", Prepared for U.S. DOT under Contract No. DTNH22-80-C-07065, Final Report, September, 1980.

9. Strother, C., Smith, G., James, M., and Warner, C., "Injury and Intrusion in Side Impacts and Rollovers", SAE 840403, February, 1984.

10. Clark, C. and Sursi, P., "The Ejection Reduction Possibilities of Glass-Plastic Glazing", SAE 840390, February, 1984.

11. Clark, C. and Sursi, P., "Car Crash Tests of Ejection Reduction by Glass-Plastic Side Glazing", SAE 851203, May, 1985.

12. Kahane, Charles J., "Evaluation of Side Structure Improvements in Response to Federal Motor Vehicle Safety Standard 214" NHTSA Report DOT-HS-806-314, November, 1982.

13. NHTSA Office of Regulatory Analysis, Plans, and Policies, "Preliminary Regulatory Impact Analysis - Side Impact, FMVSS 214", January, 1988.

14. Melvin, J., Robbins, H. and Stalnaker, R., "Side Impact Response and Injury", Sixth International Technical Conference of Experimental Safety Vehicles, U.S. Dept. of Transportation, p. 681, October, 1976.

15. Viano, D., "Evaluation of the SID Dummy and TTI Injury Criteria for Side Impact Testing", SAE 872208, November, 1987.

16. Tanner, R., "Crashworthiness of the Subcompact Vehicle", Final Report Under NHTSA Contract DOT-HS-113-3-746, November, 1975.

17. Greene, J., "Occupant Survivability in Lateral Collisions", Volumes I and II, Report Numbers DOT-HS-801-801 and DOT-HS-801-802, January, 1976

18. Hannenan, N., Schwarz, R., Struble, D., Syson, S., Wallace, G. and Forrest, S., "Design and Development of a Modified Production Vehicle for Enhanced Crashworthiness and Fuel Economy", Phase I Final Report Under NHTSA Contract DTNH22-81-C-07089, October, 1982.

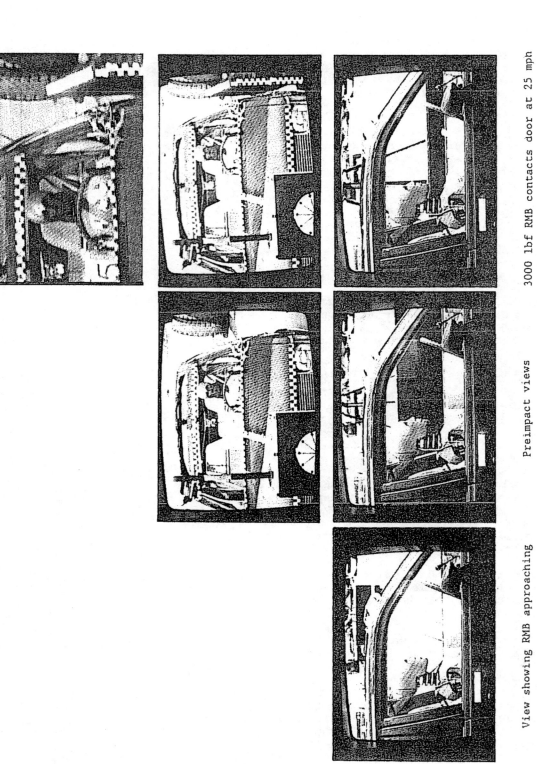

3000 lbf RMB contacts door at 25 mph
(note strobe)   0 msec

Preimpact views
−33 msec

View showing RMB approaching
−67 msec

APPENDIX 1:   TIME-SEQUENCE PHOTOS OF TEST EVENTS TAKEN FROM REAL-
              TIME VIDEOTAPE

About 100 msec

About 67 msec

About 33 msec -BAG FULLY INFLATED

1205

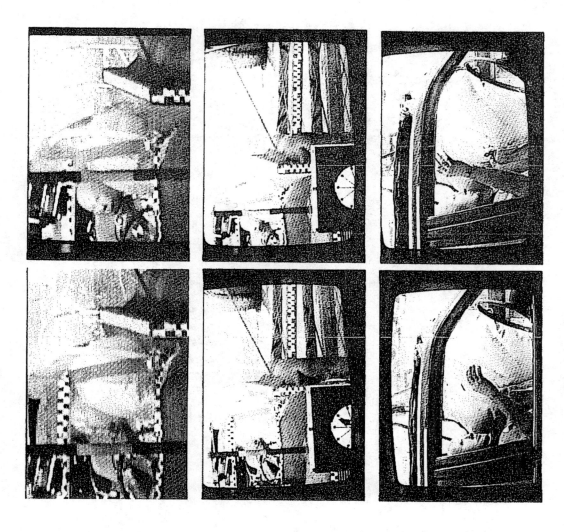

About 267 msec

About 133 msec

About 200 msec

About 233 msec

910901

Effectiveness of Safety Belts and Airbags in Preventing Fatal Injury

David C. Viano
Biomedical Science Dept
General Motors Research Laboratories
Warren, MI

ABSTRACT

Airbags and safety belts are now viewed as complements for occupant protection in a crash. There is also a view that no single solution exists to ensure safety and that a system of protective technologies is needed to maximize safety in the wide variety of real automotive crashes. This paper compares the fatality prevention effectiveness and biomechanical principles of occupant restraint systems. It focuses on the effectiveness of various systems in preventing fatal injury assuming the restraint is available and used. While lap-shoulder belts provide the greatest safety, airbags protect both belted and unbelted occupants.

RESTRAINT EFFECTIVENESS

Estimates of Effectiveness: Laboratory tests involving dummy injury assessment provide an objective evaluation of safety systems, but not an accurate estimate of restraint effectiveness in saving lives and preventing injury. In part, this is due to the limited type of crash testing conducted in relation to the wide range of real world crashes and the evolution of test dummies and injury criteria in simulating the responses of real occupants. While significant improvements have been made in the biofidelity of dummies, understanding of biomechanics, and validity of injury criteria that enable laboratory tests to better predict restraint effectiveness and better related to real-world safety performance, the most objective assessment of the effectiveness of occupant restraints is by analysis of real-world crashes.

Early epidemiologic studies of motor vehicle injury dealt with fleet evaluations of interior safety features, including the energy absorbing steering system, high penetration resistant windshields, and side-guard door beams. The first comprehensive study of lap-shoulder belts was conducted by Bohlin (1967) in Sweden and showed impressive injury and fatality prevention. Subsequent studies in various other countries have substantiated the earlier levels of safety belt effectiveness. However, variability in the underlying data and analysis approaches has led to a relatively wide range in the estimation of effectiveness in preventing fatality and serious injury.

The 1980's saw the introduction of sophisticated new statistical methods for epidemiologic analyses of field accident data. Evans (1986a) developed the double pair comparison procedure to isolate the effectiveness of belt use from other confounding factors in automotive crashes. Variations in the method have been used by Partyka, Kahane and others to determine the safety of lap-shoulder belts, rear lap belts and child safety seats. Evans (1991) has extended the approach to investigate the effects of alcohol use, occupant age, seating position, and direction of crash on fatality risks. The methodology has enabled accurate quantification of factors influencing crash injuries.

Statistical Analysis of Restraints in Fatal Crashes: Evans (1986a) developed the double paired comparison method to determine the effectiveness of occupant restraint as a function of seating position and crash direction. The method compares the number of fatalities to either of two occupants under two conditions, such as restrained or unrestrained, driver or passenger, and ejected or non-ejected. One of the occupants serves a normalizing or exposure estimating role for the frequency of fatality of the other under a particular crash situation. This forms the basis for the estimation of restraint effectiveness.

For example, the effectiveness of safety belt use by the right-front passenger (RFP) is determined by the double pair comparison using two sets of fatal crashes. The first set consists of a restrained RFP and an unrestrained driver, at least one of whom is killed. From the numbers of RFP and driver fatalities, a restrained RFP to unrestrained driver fatality ratio is calculated. From the second set of data on unrestrained RFPs and unrestrained drivers, an unrestrained RFP to unrestrained driver fatality ratio is similarly estimated. When the first fatality ratio is divided by the second, it gives the probability that a restrained RFP is killed compared to that of an unrestrained RFP in actual traffic accidents. This is the estimate of restraint system effectiveness.

Restraint Effectiveness: Determining the fatality prevention effectiveness of lap-shoulder belts was one of the first applications of the double pair comparison (Evans 1986b). Lap-shoulder belts were shown to be (41 ± 4)% effective in preventing fatality for front-seated occupants. They are (42 ± 4)% effective for drivers and (39 ± 4)% for right-front passengers (Table 1). The overall effectiveness was estimated at 43% by combining the average from Evans with those of NHTSA and others estimating 40-50% effectiveness.

Subsequently, the effectiveness of lap-shoulder belts was determined as a function of the direction of impact (Evans 1988c), including contributions from reducing ejection (Evans and Frick 1989). This type of analysis (Figure 1) helps differentiate two essential components of occupant protection with safety belt use. One is protection against ejection and is primarily due to the lap portion of the belt system. The other is mitigation of interior impact and is largely contributed by upper body restraint from the shoulder harness.

The relative safety contribution by reducing ejection with belt use significantly depends on the type of crash and is highest in primarily rollover accidents. Safety belt use is most effective in preventing driver fatality in crashes where rollover is the first harmful event and where occupant containment is a key feature of safety performance. The lap-shoulder belt system is least effective in preventing driver fatality in left-side impacts as the principal point of vehicle impact and deformation of the occupant compartment are critical factors in increasing fatality risk for an occupant.

The difference between overall lap-shoulder belt effectiveness and ejection reduction was used by Evans (1988c) to determine the safety benefit of a driver and passenger airbag system. The analysis inferred the fatality reduction effectiveness of the airbag only system as a component of impact mitigation and estimated an (18 ± 4)% effectiveness for the unbelted driver and (13 ± 4)% for the unbelted right-front passenger. This level compares favorably with the results of an expert judgment of fatality prevention potential of airbag restraints (Wilson and Savage 1973).

In a further study, Evans (1988a) determined an overall (18 ± 9)% effectiveness of lap-belt use by rear-seated occupants. Most of the safety benefit is from anti-ejection since this level compares favorably with the 17-19% effectiveness of lap-shoulder belts in preventing ejection. A major component of lap-belt effectiveness is occupant containment in the vehicle. For primarily frontal crashes, the effectiveness of lap-belt use by rear seated occupants has a larger variability and is relatively low for impact mitigation.

Lap-shoulder belts for rear outboard occupants are available in most new passenger vehicles and will increase in the vehicle fleet. However, there isn't sufficient crash injury data to conduct a statistical analysis of effectiveness. It is possible to use the understandings of restraint effectiveness in other seating positions to make a judgment about the level of impact mitigation and ejection prevention. Lap-shoulder belts are estimated to be 27% effective in preventing fatal injury in rear seats. This is essentially due to impact mitigation improvements.

Table 2 summarizes the available estimates of belt restraint and airbag effectiveness in preventing occupant fatalities. The results show the substantial effectiveness of lap-shoulder belts. They are clearly the principal safety feature in protecting occupants in severe crashes. Since airbags and other interior components do not hold the occupant in a seating position, they have lower overall effectiveness because of a much lower effectiveness in reducing ejection. Their contribution to overall crash protection is essentially limited to impact mitigation which is about half the overall benefit of lap-shoulder belt use.

The overall safety benefit of the combination of lap-shoulder belt use and airbag has not been determined from field accident data. However, it is possible to estimate the effectiveness by considering the current safety effectiveness studies and the frequency of unsurvivable crashes. Based on analysis of fatal crashes to unbelted occupants, Huelke, Sherman and Murphy (1979) estimated that approximately 50% of the crash fatalities are unpreventable by currently available occupant restraints. The limit, in part, reflects the

Table 1

Effectiveness of Occupant Protection Systems
(Adapted from Evans with additional estimates added)

Driver

Safety System	Impact Mitigation	Ejection Prevention	Overall Effectiveness
Lap-Shoulder Belt	(23 ± 4)%	(19 ± 1)%	(42 ± 4)%
Airbag	(18 ± 4)%	-	(18 ± 4)%
EA Steering System	6%	-	(6 ± 3)%
Lap-Shoulder Belt (Plus Air Bag)	[27%]	[19%]	[46%]*

Right Front Passenger

Safety System	Impact Mitigation	Ejection Prevention	Overall Effectiveness
Lap-Shoulder Belt	(22 ± 4)%	(17 ± 1)%	(39 ± 4)%
Airbag	(13 ± 4)%	-	(13 ± 4)%
Friendly Interior	[<6%]	-	[<6%]
Lap-Shoulder Belt (Plus Airbag)	[26%]	[17%]	[43%]

Rear-Seat Passenger

Safety System	Impact Mitigation	Ejection Prevention	Overall Effectiveness
Unbelted Rear Versus Front-Seat Position	-	-	(26 ± 2)%
Lap-Belt	(1 ± 9)%	(17 ± 1)%	(18 ± 9)%
Lap-Shoulder Belt	[10%]	[17%]	[27%]

*[] New estimates of restraint effectiveness based on judgement.

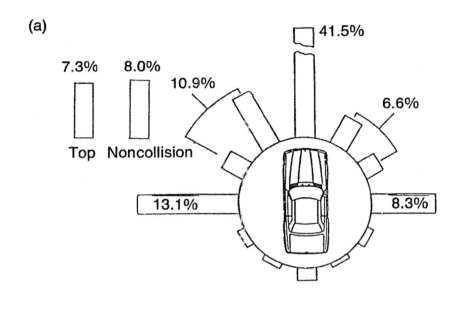

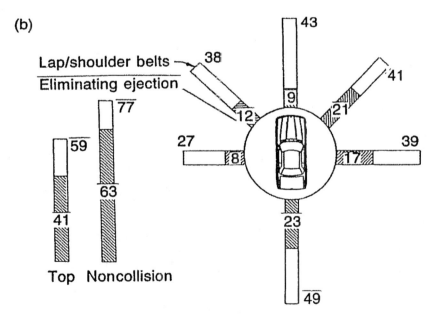

Figure 1: (a) Distribution of driver deaths by principal impact point and (b) effectiveness of lap/shoulder belts in preventing driver fatalities. The fraction of fatalities prevented by eliminating ejection is shown in the hashed portion of the bar according to impact direction. For example, in frontal (12 o'clock) crashes, lap/shoulder belts prevent 43% of driver fatalities; 9% of this is due to eliminating ejection, so that 34% is due to interior impact reduction (from Evans 1988c with permission).

Table 2

Effectiveness of Occupant Restraints

	Driver	Right Front Passenger	Rear Passenger
Lap-Shoulder Belts	(42 ± 4)%	(39 ± 4)%	[27%]
Airbag Only	(18 ± 4)%	(13 ± 4)%	--
Belts and Airbag	[46%]	[43%]	--
Lap Belt	--	--	(18 ± 9)%

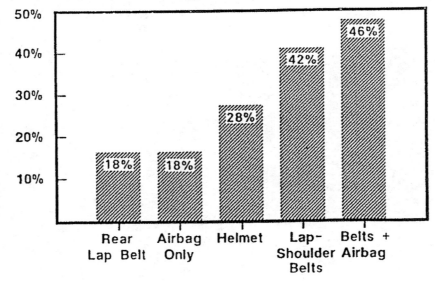

Figure 2: Effectiveness of safety devices in preventing fatal driver injury in severe motor vehicle crashes and motorcyclist injury with helmet use (from Viano 1990 with permission).

severity of many fatal crashes which may involve extreme vehicle damage and forces on the passenger compartment, unusual crash configurations and causes of death, and unique situations associated with particular seating positions and crash dynamics.

By comparing the level of unpreventable fatalities with the overall protective effect of lap-shoulder belts, Viano (1988a) found a maximum of 7% additional fatality prevention may occur with a combination of restraints or additional supplemental features. Based on a similar review of fatalities to lap-shoulder belted front-seated occupants, a potential 3-5% additional fatality prevention is estimated with a combination of lap-shoulder belt and airbag system.

Using a 4% additional benefit to safety belt wearers from airbags, there is an overall effectiveness of 46% for the belted driver with supplemental airbag (Figure 2) and 43% for the right-front passenger. However, this estimate is based on expert judgment and not statistical methods applied to crash injury data.

The 4% increment in effectiveness with the airbag supplementing lap-shoulder belt use is consistent with a 5% additional benefit determined by NHTSA (1984). The fact that a lap-shoulder belt and airbag are only 43% to 46% effective in preventing fatality underscores that absolute protection is not achievable by occupant restraints and that injury and fatality will continue to occur to belt wearers despite a significant overall net safety gain by restraint usage, even if augmented by airbags. This is because of accident severity, configuration, and limits of human tolerance.

Fatality risk depends on the particular occupant seating position, the principal impact point and the proximity of passenger compartment crush. Evans and Frick (1988) have shown (Figure 3) more than a 7 to 1 increase in fatality risk for a right-front passenger in right-side (nearside) versus a left-side (farside) impact.

A better appreciation of restraint effectiveness estimates may be possible by a fuller understanding of the biomechanics of restraint systems in reducing impact forces and controlling occupant kinematics over the wide variety of real-world crash. The following section expands on a recent review by Viano (1988b) and addresses key features of occupant restraint performance in frontal and rollover accidents.

RESTRAINT BIOMECHANICS AND PERFORMANCE

Driver Airbag: During the development of crash protection systems for automobile occupants in the early 1960's, many concepts for energy absorption and load distribution were conceived and evaluated. A driver airbag took advantage of rapid filling of a concealed bag to provide a cushion in front of an occupant in a crash (see review of SAE publications in Viano 1988c). This provided a large area to gradually decelerate the driver. However, it was necessary to provide lower torso restraint through either a lap-belt or knee bolster to prevent the driver from submarining the airbag (Figure 4a and b) and experiencing higher forces on interior impact in a frontal crash.

Although compressed gas in a cylinder was one of the early concepts, the eventual production system took advantage of a relatively small mass (70-100 g) of sodium azide. The chemical can rapidly inflate a 80-100 l driver airbag by ignition and conversion to harmless nitrogen gas when sensors detected a severe frontal crash. The airbag is stored inconspicuously in the interior and deployed only during a severe crash, sensors are located in the engine compartment to detect rapid decelerations of the front-end during a crash and an electronic system is used to monitor and initiate deployment of the airbag.

Production airbag systems were available from General Motors in 1974. However, the customer demand never developed for the safety option so the program was cancelled in 1976 after selling only 10,321 vehicles. Mertz (1988) recently published an analysis of the field performance of the airbag based on the work of Pursel et al (1978) using comparable non-airbag crashes. The driver airbag system was found to be 21% effective in preventing AIS 3+ injuries (16% for the passenger system), but was -34% ineffective for AIS 2+ passenger injury. However, the matched crash analysis only allows the determination of effectiveness in deployment accidents. The overall effectiveness of a driver airbag in typical crashes would be approximately 9% in preventing AIS 3+ driver injuries, assuming deployment accidents represent 45% of all severe crashes.

Many of the airbag inflation injuries occurred in moderate severity crashes without distortion of the occupant compartment. They were considered "unexpected" and caused by the occupant being near the cushion at the time of deployment. Horsch and Culver (1979) and more recently Horsch et al (1990) have observed significant forces adjacent to an airbag if the normal path of inflation is blocked by an occupant. Thus, the deployment, per se, of an airbag has the potential to seriously injury or kill, independent of crash severity.

The use of lap-shoulder belts helps minimize the risk of airbag injury, but there is a potential consequence of public misperception of driver airbag protection. Some car occupants may not recognize that the steering wheel airbag is only a supplement, and not an alternative, to the primary occupant protection system in the vehicle, the lap-shoulder belts.

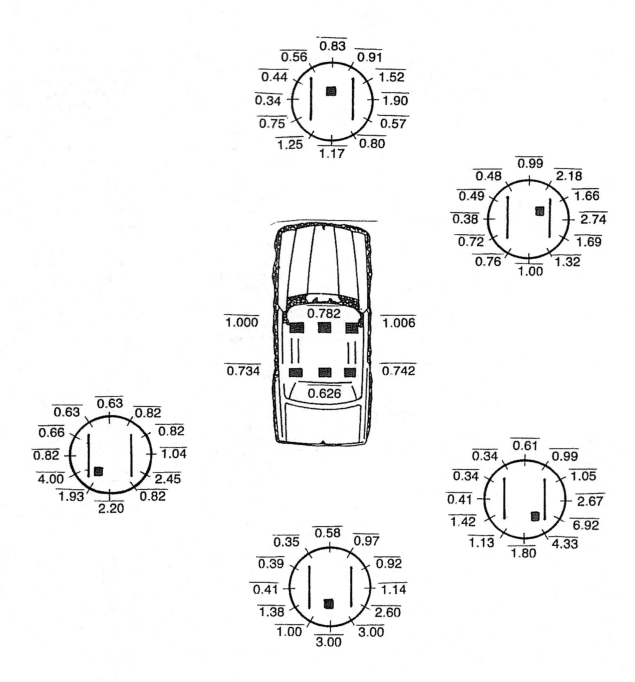

Figure 3: Relative fatality risk to passengers in different seating positions in relation to that of the driver set at 1.000 and fatality risk for passengers in various car seating positions relative to the principal direction of impact (based on Evans and Frick 1988 with permission).

If drivers fail to buckle up, they significantly reduce their driving safety, particularly in side impact and rollover crashes where the lack of belt restraint in the seat may lead to interior impact or ejection. Evans (1989) calculated a 41% increase in fatality risk by drivers ceasing to wear the lap-shoulder belt because they have a supplemental bag system. Fortunately, current belt use surveys are not finding a lower rate of belt use in airbag equipped vehicles.

Lap-belts: An early use of the lap-belt was to complement crash protection of the chest by a driver airbag or the energy absorbing (EA) steering system, another 1960's interior safety concept. The belt controlled forward excursion of the lower torso (Figure 4c) by restraining the pelvis and maintained an upright posture of the occupant (Horsch, Peterson and Viano, 1982).

In the absence of pelvic restraint, loads would be applied through the knees and seat pan to restrain the occupant and control the upright posture of the driver. These loads are necessary to take advantage of the cushioning affect of the airbag or EA steering column on the upper body. The lap-belt also complemented early designs for passenger protection by a padded dash board and found their way into rear seating positions to supplement padded seat backs (Figure 4d).

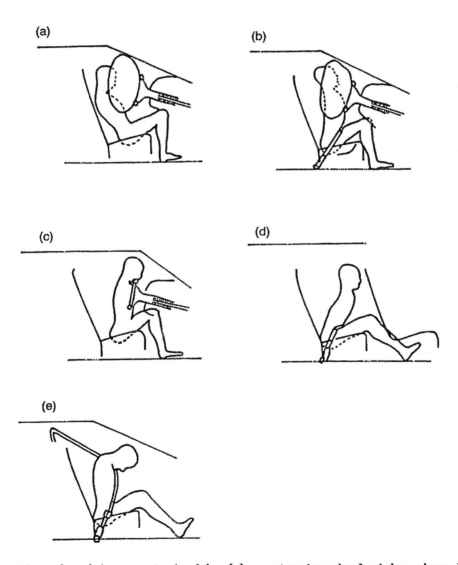

Figure 4: Kinematics of a driver restrained by (a) a steering wheel airbag where loads act on the upper body but not on the lower extremity, (b) the combination of lap belt and airbag giving pelvic and upper body restraint, (c) the energy absorbing steering system and knee bolster, and passenger restrained by (d) a lap belt in the rear seat and (e) a lap-shoulder belt in the right-front seating position.

Although much of the development work on restraints has involved frontal barrier and sled testing, safety engineers recognized the importance of the lap-belt in preventing ejection, which had been identified as the leading cause of fatality in the mid-1950's by crash investigations. Thus, occupant containment within the vehicle during a crash was a principal benefit of lap-belt usage. Since lap-belt restraint was only one part of the systems engineering for occupant protection by friendly interior components, development work focused on assuring that the various safety components worked together as a system in a crash.

During testing, it also became apparent that lap-belt use also reduced the potential for rear-seat occupant loading on the front-seat back. This is particularly important for belt restrained front-seat occupants, since Park (1987) has shown that unrestrained rear-seat occupants increase the fatality risk of belted front-seat occupants by $(4 \pm 2)\%$ because of the additional load applied on the restrained occupant. This decreases the safety effectiveness for a belted front-seat driver from 42% to 40% and a belted right-front passenger from 39% to 37%, a significant increment in crash injury risk.

With the advent of safety belt use in passenger cars in the late 1950's and early 60's, physicians started reporting on the injury patterns of belt restrained victims (Kulowski and Rost, 1956). The typical pattern of upper body injury to unrestrained occupants was replaced by belt related injury to abdominal organs and tissues. These injuries led to the phrase "seat belt syndrome" (Garrett and Braunstein 1962, Fish and Wright, 1965) as the new injury patterns in motor vehicle crashes received attention due to concentrated forces on the lower abdominal region for lap-belt wearers. In many cases improper belt wearing was identified as a cause of abdominal loading and injury, although other factors may play a role. Since belt use modifies injury patterns and it is not possible to prevent all injury to occupants in crashes, some in the public are suspicious about the safety effectiveness of belts, in spite of now over-whelming evidence of benefit.

Lap-Shoulder Belts: As safety engineers pursued continued improvements in crash protection, the combination of a lap and diagonal shoulder belt gained rapid acceptance (Bohlin 1967). Lap-shoulder belts provide occupant restraint during a crash by routing safety belts over the boney structures of the pelvis and shoulder. This takes advantage of a relatively high tolerance to impact forces for these regions of the skeleton and avoids concentrating load on the more compliant abdominal and thoracic regions. Control of occupant kinematics helps insure maximum protection by belt restraints.

The fundamentals of a high quality belt restraint system involve kinematic controls (Adomeit and Heger 1975, Adomeit 1977, 1979, Viano and Arepally, 1990) which maintain the lap portion of the belt low on the pelvis through adequate seat cushion support. This minimizes pelvic rotation and reduces the tendency for the lap-belt to slide off the illium and directly load the abdomen. Forward rotation of the upper torso of slightly greater than 90° upright posture (Figure 4e) directs a major portion of the upper torso restraint into the shoulder and upper chest region. This reduces loads on the more compliant areas of the lower rib cage. Each of the kinematic controls helps direct forces onto the skeletal structures above and below the center of gravity of the torso thus balancing the restraint and keeping loads away from more compliant body regions vulnerable to injury.

There is another component of quality restraint. Biomechanical responses related to occupant protection assessment need to be evaluated and kept below human tolerance for the severity of the crash test (Horsch 1987). Although there is a rich history in the use of chest acceleration as a measure of injury risk and the current standards require less than 60 g's for 3 ms duration, chest acceleration is an inadequate measure of injury risk.

Recent evaluations by Groesch et al (1986) have demonstrated that chest acceleration is not a logical measure of restraint effectiveness in real-world crashes or an adequate indicator of fundamentally different restraint configurations. The current evidence points to deformation of the body and its organs as the cause of injury, and the tolerance to deformation depends on the velocity of loading. The faster the deformation, the lower the tolerance to compression (normalized deformation). This type of injury is related to the Viscous response which is a cause of soft tissue injury and a measure of energy dissipated during rapid compression (Viano and Lau 1988).

With a significant increase in safety belt use after state passage of mandatory wearing laws and a greater health consciousness in America, Orsay et al (1988) has seen reductions in hospital admissions and injury severities for lap-shoulder belted occupants, even as higher relative numbers of belted victims are being treated.

Injury patterns associated with safety belt wearing are becoming better understood (Denis et al 1983, Arajarvi, Santavirta and Tolonen, 1987, Banerjee 1989). The potential for improper use has also continued with the advent of lap-shoulder belt systems, particularly placement of the shoulder harness under the arm (States et al 1987) and wearing of the lap-belt high on the abdomen with poor seating posture.

An improvement in passenger safety has been made by adding a shoulder belt to the out-board rear-seat lap-belt. This further improves the safety of rear-seat occupants, which is otherwise similar to belted front-seat occupants because of the inherently greater safety of rear seating positions (Evans 1988a). That is, the restraint effectiveness of lap-shoulder belt use by front-seat occupants is similar to lap-belt use by rear-seat occupants (39% effectiveness for lap-belted rear-seat occupants versus 43% for a lap-shoulder belted driver and 39% for the right-front passenger). Thus, combining the shoulder harness to the rear-seat lap-belt should increase effectiveness by mitigating interior contacts of the upper body. This should increase the safety of belted rear-seat occupants to an estimated level of 27% with an overall gain in safety over those belted in the front seat.

An analysis of rollover crashes by Huelke, Compton and Studer (1985) determined that safety belt use virtually eliminated the risk of paralyzing cervical injury and ejection. This observation is consistent with Evans' (1986b) finding that belt restraints are most effective in rollover crashes, particularly where rollover is the first harm event. In such crashes, lap-shoulder belts are (82 ± 5)% effective in preventing driver fatalities. Sixty-four percent (64%) of the overall effectiveness is by preventing ejection with the remaining 18% effectiveness by reducing forces due to interior impacts.

In contrast, unrestrained occupants in rollovers are subjected to a series of impacts and potential ejection during a rollover (Viano 1990). In a simulation of experimental rollovers by Robbins and Viano (1984), the complex kinematics of an unbelted occupant were studied using several crash scenarios, including complete occupant containment, and subsequent driver or passenger door opening (Figure 5). The potential for serious injury by interior impacts or ejection were apparent in the rollover sequences for the unrestrained driver.

PERCEPTIONS OF OCCUPANT RESTRAINTS

Interestingly, the overall fatality prevention effectiveness of a driver airbag without safety belt use and lap-belt use in the rear is similar. Both systems provide an 18% reduction in fatal crash injury risk. This level of effectiveness is far greater than built-in safety devices, such as the EA steering column at 6% effectiveness, and high penetration resistant windshields, side door-beams, and interior padding which have lower effectiveness levels. Therefore, airbags and rear-seat lap-belts are effective automotive safety features and are second only to front-seat lap-shoulder belts which have a 41% safety effectiveness. However, this is the effectiveness when the

airbag is available and the lap belt is used. It also does not consider the effects of occupancy rates on overall injury prevention.

Driver airbags and rear-seat lap-belts provide protection in different crash types. Airbag effectiveness is essentially due to interior impact mitigation in primarily frontal crashes. It has minimal effectiveness in lateral or rollover crashes. In contrast, rear-seat lap-belt effectiveness is essentially due to ejection prevention in primarily non-frontal crashes such as rollovers, since it is less effective in reducing impact forces in frontal crashes. Rear-seat lap-belts are currently available in virtually all passenger cars and only require buckling by an occupant to be effective. On the other hand, driver airbags are available in much fewer vehicles but provide crash protection independent of any action by the occupant.

Both systems have the possibility of modifying injury in particular situations. The high energy release of an airbag may injure an occupant against the system at the instant of deployment. Blocking the path of deployment increases pressures in the cushion during gas generation and develops high forces on the occupant. Since the force occurs with high velocity, there is a risk of injury by a Viscous mechanism.

Placing the lap-belt on the abdomen as compared to correctly on the pelvis does not direct restraining loads through the skeleton in severe frontal crashes. Abdominal loading and deformation occurs as forces in the lap-belt restrain forward excursion of the lower body. This situation, and another in which the lap-belt slips above the pelvis during a crash, may result in abdominal organ injury and hemorrhage by submarining pelvic restraint.

Since there is a general lack of information on the technical aspects, performance, and efficacy of safety systems, the public has developed an exaggerated perception of the safety effectiveness of a driver airbag and insecurities about rear-seat lap-belt use. Many falsely believe that airbags are safer than safety belts. The facts about safety systems need to be accurately covered in the news and conveyed to the public to gain understanding.

In addition, there should be a better awareness that safety devices cannot provide full or complete protection, function in a specific range of crash types, and may be associated with injury in particular crash situations. Misleading information or rumor about either safety benefits of crash protection components or injury risks, which may be infrequent in comparison to overall safety benefit, may cause occupants to reduce their overall

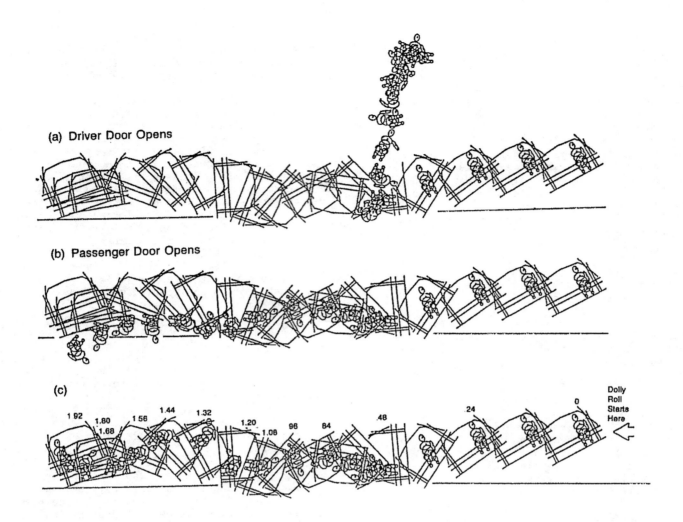

(a) Driver Door Opens

(b) Passenger Door Opens

(c)

1.92 1.80 1.56 1.44 1.32 1.20 .96 .84 .48 .24 0
1.68 1.08

Dolly
Roll
Starts
Here

Figure 5: Kinematics of an unrestrained driver in a rollover crash in which (a) the driver door opens and the occupant ejects and is thrown upward, (b) the passenger door opens, the driver ejects and is crushed between the rolling car and the road, and (c) the occupant is contained in the vehicle (from Robbins and Viano 1984 with permission).

driving safety by failing to properly use inherently safe technologies to protect themselves and their families.

REFERENCES

1. Adomeit, D. and Heger, A., "Motion Sequence Criteria and Design Proposals for Restraint Devices in Order to Avoid Unfavorable Biomechanic Conditions and Submarining." In Proceedings of the 19th Stapp Car Crash Conference, 139-166, SAE Technical Paper #751146, Society of Automotive Engineers, Warrendale, PA, 1975.

2. Adomeit, D., "Evaluation Methods for the Biomechanical Quality of Restraint Systems During Frontal Impact." In Proceedings of the 21th Stapp Car Crash Conference, 911-932, SAE Technical Paper #770936, Society of Automotive Engineers, Warrendale, PA, 1977.

3. Adomeit, D., "Seat Design — A Significant Factor for Safety Belt Effectiveness." In Proceedings of the 23rd Stapp Car Crash Conference, 39-68, SAE Technical Paper #791004, Society of Automotive Engineers, Warrendale, PA, 1979.

4. Arajarvi, E., Santavirta, S. and Tolonen, J., "Abdominal Injuries Sustained in Severe Traffic Accidents by Seat Belt Wearers." Journal of Trauma 27(4):393-397, 1987.

5. Banerjee, A., "Seat Belts and Injury Patterns: Evolution and Present Perspectives." Postgraduate Medical Journal, 65:199-204, 1989.

6. Bohlin, N.I., "A Statistical Analysis of 28,000 Accident Cases with Emphasis on Occupant Restraint Value." In Proceedings of the 11th Annual Stapp Car Crash Confer-

ence, University of California, Los Angeles, CA, October 10-11, 1967. SAE Technical Paper #670925, Warrendale, PA, Society of Automotive Engineers, 455-478, 1967.

7. Denis, R., Allard, M., Atlas, H. and Farkouh, E., "Changing Trends with Abdominal Injury in Seat Belt Wearers." _Journal of Trauma_ 23(11):1007-1008, 1983.

8. Evans, L., "The Effectiveness of Safety Belts in Preventing Fatalities." _Accident Analysis and Prevention_ 18:229:241, 1986b.

9. Evans, L., "Double Pair Comparison -- A New Method to Determine How Occupant Characteristics Affect Fatality in Risk in Traffic Crashes." _Accident Analysis and Prevention_ 18:217-227, 1986a.

10. Evans, L., "Passive Compared to Active Approaches to Reducing Occupant Fatalities." GM Research Publication GMR-6596, Experimental Safety Vehicles paper #ESV 89-5B-0-005, Goteborg, Sweden, May, 1989.

11. Evans, L. and Frick, M.C., "Potential Fatality Reductions Through Eliminating Occupant Ejection from Cars." _Accident Analysis and Prevention_, 22(2):169-182, 1989.

12. Evans, L., "Rear Seat Restraint System Effectiveness in Preventing Fatalities." _Accident Analysis and Prevention_ 20:129-136, 1988a.

13. Evans, L., "Occupant Protection Device Effectiveness in Preventing Fatalities." In _Proceedings of the 11th International Technical Conference on Experimental Safety Vehicles_, Washington, DC, May 12-15, 1987, U.S. Department of Transportation, National Highway Traffic Safety Administration, DOT HS 807 233, pp. 220-227, 1988b.

14. Evans, L., "Restraint Effectiveness Occupant Ejection from Cars, and Fatality Reductions." General Motors Research Report #GMR-6398, General Motors Research Laboratories, Warren, MI, 1988c.

15. Evans, L. and Frick, M., "Seating Position in Cars and Fatality Risk." _American Journal of Public Health_, 78:1456-1458, 1988.

16. Evans, L., _Traffic Safety and the Driver._ Van Nostrad Reinhold, January, 1991.

17. Fish, J. and Wright, R.H., "The Seat Belt Syndrome -- Does It Exist?" _Journal of Trauma_, 5:746-750, 1965.

18. Garrett, J.W. and Braunstein, P.W., "The Seat Belt Syndrome." _Journal of Trauma_ 2:220-238, 1962.

19. Groesch, L., Katz, E., Marwitz, H., Kassing, L., "New Measurement Methods to Assess the Improved Injury Protection of Airbag Systems." In _Proceedings of the 30th Annual Conference of the American Association for Automotive Medicine_, Association for the Advancement of Automotive Medicine; Des Plains, IL, 235-246, 1986.

20. Horsch, J.D. and Culver, C.C., "A Study of Driver Interactions with an Inflating Air Cushion." In _Proceedings of the 23rd Stapp Car Crash Conference_, SAE Technical Paper #791029, Society of Automotive Engineers, Warrendale, PA, 1979.

21. Horsch, J.D., Petersen, K.R. and Viano, D.C., "Laboratory Study of Factors Influencing the Performance of Energy Absorbing Steering Systems." _SAE Transactions_, Vol. 91, 1982, SAE Technical Paper #820475, in SAE Special Publication 507, _Occupant Interaction with the Energy Absorbing Steering System_, pp. 51-63, 1982.

22. Horsch, J.D., "Evaluation of Occupant Protection From Responses Measured in Laboratory Tests." SAE International Congress and Exposition, Detroit, MI, February 23-27, 1987. SAE Paper #870222. Society of Automotive Engineers, Warrendale, PA, 1987.

23. Horsch, J., Lau, I., Andrzejak, D., Viano, D., Melvin, J., Pearson, J., Cok, D., Miller, G., "Assessment of Airbag Deployment Loads." In the _Proceedings of the 34th Stapp Car Crash Conference_, pp. 276-288, SAE Technical Paper #902324, November 1990.

24. Huelke, D.F., Sherman, H.W., Murphy, M.J., "Effectiveness of Current and Future Restraint Systems in Fatal and Serious Injury Automobile Crashes." SAE Technical Paper #790323. Warrendale, PA, Society of Automotive Engineers, 1979.

25. Huelke, D.F., Compton, C. and Studer, R., "Injury Severity, Ejection, and Occupant Contacts in Passenger Car Rollover Crashes." SAE Paper #850336, Society of Automotive Engineers, Warrendale, PA, 1985.

26. Kulowski, J. and Rost, W.B., "Intra-Abdominal Injuries from Safety Belt in Auto Accident." _Archives of Surgery_, 73:970-971, 1956.

27. Mertz, H.J., "Restraint Performance of the 1973-76 GM Air Cushion Restraint System." SAE Technical Paper #880400, 1988:61-72, Society of Automotive Engineers, Warrendale, PA.

28. National Highway Traffic Safety Administration, "FMVSS 208 Regulatory Impact Analysis." Department of Transportation, 1984.

29. Orsay, E.M., Turnbull, T.L., Dunne, M., Barrett, J., Langenberg, P., and Orsay, C.P., "Prospective Study of the Effect of Safety Belts on Morbidity and Health Care Costs in Motor-Vehicle Accidents." _Journal of American Medical Association_ 260(24):3598-3603, 1988.

30. Park, S., "The Influence of Rear-Seat Occupants on Front-Seat Occupant Fatalities: The Unbelted Case." General Motors Research Laboratories Research Publication GMR-5664, January 8, 1987.

31. Pursel, H.D., Bryant, R.W., Scheel, J.W., and Yanik, A.J., "Matching Case Methodology for Measuring Restraint Effectiveness." SAE Technical Paper #780415, Society of Automotive Engineers, Warrendale, PA, February, 1978.

32. Robbins, D.H., and Viano, D.C., "MVMA-2D Modeling of Occupant Kinematics in Rollovers." SAE Transactions, Vol. 93, 1984. In Mathematical Simulation of Occupant and Vehicle Kinematics, (P-146), 65-77, SAE Technical Paper #840860, Society of Automotive Engineers, Warrendale, PA, 1984.

33. States, J.D., Huelke, D.F., Dance, M. and Green, R.N., "Fatal Injuries Caused by Underarm Use of Shoulder Belts." Journal of Trauma 27(7):740-, 1987.

34. Viano, D.C., "Limits and Challenges of Crash Protection." Accident Analysis and Prevention, 20(6):421-429, 1988a.

35. Viano, D.C., "Cause and Control of Automotive Trauma." Bulletin of the New York Academy of Medicine, Second Series, 64(5):376-421, 1988b.

36. Viano, D.C., (editor), SAE Passenger Car Inflatable Restraint Systems: A Compendium of Published Safety Research, SAE Progress in Technology PT-31, Society of Automotive Engineers, Warrendale, PA, 1988c.

37. Viano, D.C. and Lau, I.V., "A Viscous Tolerance Criterion for Soft Tissue Injury Assessment." Journal of Biomechanics, 21(5):387-399, 1988.

38. Viano, D.C., Arepally, S., "Assessing the Safety Performance of Occupant Restraint Systems." In the Proceedings of the 34th Stapp Car Crash Conference, pp. 301-328, SAE Technical Paper #902328, November, 1990.

39. Viano, D.C., "Cause and Control of Spinal Cord Injury in Automotive Crashes." G. Heiner Sell Lecture. Submitted to Paraplegia, 1990 and available until publication as GMR-6885, 12/7/89.

40. Viano, D.C., "Testimony Before the United States Senate Committee on Environment and Public Works Subcommittee on Water Resources, Transportation, and Infrastructure Concerning Senate Bill S.1007 The National Highway Fatality and Injury Reduction Act of 1989." October 17, 1989.

41. Wilson, R.A. and Savage, C.M., "Restraint System Effectiveness - A Study of Fatal Accidents." In the Proceedings of Automotive Safety Engineering Seminar, Warren, MI, General Motors Corporation, Automotive Safety Engineering, Environmental Activities Staff; 27-39. 1973.

Air Bag System for Side Impact Occupant Protection

Toru Kiuchi, Kenji Ogata
Toyota Motor Corporation
Charles Y. Warner
Collision Safety Engineering
John Jay Gordon
GMH Engineering

Abstract

Pilot and prototype designs of a door-mounted air bag system for occupant protection in side impact have been assembled and tested. The primary goal of the designs was to take advantage of the improved space utilization offered by the air bag when combined with the padding and structural benefits that are contemplated for torso injury. Another important goal of the project was the demonstration of the head-protection potential of such a system, attempting to interpose a pad between the head and side structures and intruding objects likely to cause impact injury.

The pilot design was subjected to a test program, providing a preliminary evaluation of a system which incorporates both head and torso protection in a single air bag system. The pilot design showed sufficient promise that a preliminary prototype design program was undertaken.

Full-scale crash tests of recent production 4-door sedans were conducted to establish baseline performance over a range of side-impact conditions. Design objectives were analyzed and subsystem performance goals were established and proven by component testing. The prototype system incorporated two kinds of sensor switches, a production steering wheel air bag inflator module, a large, flat, tethered air bag, and a fabric air bag cover, all mounted in a modified production door. The complete prototype system was evaluated in laboratory tests and full-scale crash tests, including FMVSS 214 crabbed moving deformable barrier (CMDB) tests employing the DOT/SID side-impact dummy. A very satisfactory performance was achieved, as demonstrated by comparison of dummy indices measured in baseline and air bag-equipped vehicles in full-scale crash tests. This paper outlines the designs and system configurations and discusses the results of the pilot and preliminary design test series.

Introduction

Fatalities and Injuries in Side Impact Accidents

Many head injuries occur in side impacts, due to contacts with interior structures or exterior intruding objects. According to accident data collected by the National Crash Severity Survey (NCSS) and the National Accident Sampling System (NASS), side impact accidents cause about 30 percent of all traffic accident occupant fatalities. Head injuries account for 40 percent of all fatalities to near-side front-seat occupants in side impacts, with chest injuries at 32 percent. Objects exterior to the vehicle are involved in about 40 percent of head injuries, presumably due to partial ejection or intrusion, while impacts with A-pillar (19%) and roof side rail (17%) structures make an almost equal contribution [DOT/NHTSA, 1990-1; Viano, 1987-1,2,3; Strother, 1990].

NHTSA Safety Rulemaking Activities

The most recent "Final Rule" of revised FMVSS 214 was issued in October, 1990, aimed at reducing thorax and pelvis injury indices, as measured in CMDB crashes. The test procedure includes the use of the DOT/Side Impact Dummy (DOT/SID) and the Thoracic Trauma Index (TTI(d)). The rule currently applies to passenger cars produced after September, 1993 for sale in the U.S. [DOT/NHTSA, 1990-2]. NHTSA is rightly considering means to reduce not only thorax and pelvis injuries but also head injuries in side impacts. The Advanced Notice of Proposed Rulemaking (ANPRM) issued in August of 1988 suggested head protection by the use of pads on pillars and rails, glass-plastic glazing in compartment sides, strengthened door hardware, etc. [DOT/NHTSA, 1988].

Upgraded Occupant Protection in Side Impact

It is extremely difficult to provide adequate stroke for the absorption of occupant second collision energy in side impacts [Warner, 1989; 1990-2]. Attempts to upgrade occupant protection in this crash mode have generally involved modifications of vehicle body side structure and imposition of paddings between occupant and interior for thorax, abdomen, and pelvis protection. The improvement of body side structure and interior padding configurations appears to be somewhat effective for reduction of chest injuries [Warner, 1990-1; Lau, 1989; Viano, 1989-2; Ridella, 1990]. It may also be partially effective for reduction of head injuries, but it is unresponsive to injuries caused by partial ejection or foreign body intrusion through the side glass. Our studies were directed at the feasibility of using air bags as supplemental side impact protection, with particular emphasis on head protection.

Toyota/CSE Pilot Study of Side Air Bags

A joint pilot study was initiated by Toyota and CSE in 1989, to investigate the potential of improved side impact occupant protection by use of supplemental air bag systems [Warner, 1988], with demonstration tests in Toyota production vehicles, and injury index comparisons with baseline FMVSS 214 tests and DOT/SID dummy.

Configuration of Pilot Side Air Bag System

The pilot system shown in Figures 1 and 2 was configured as an extension of findings from the early preliminary testing in 1980 Citation automobiles [Warner, 1988, 1990-3]. The bag and inflator were attached to a backup plate in the door inner panel, and covered with an energy-absorbing inner foam pad meant to reduce thoracic injury. The inflator was selected from among those already available for steering wheel air bags. The bag was designed to deploy inward and upward rapidly, in order to accommodate the limited distance available. It included distributed venting and tethers to control lateral expansion and encourage vertical deployment over the window space, with a width of about 100 mm and a volume of about 60 liters (Figures 3 and 4). Mechanical intrusion sensors were mounted at two levels near the door outer skin. Switch closures resulting from small deflections of the door provided the inflation signal for the bag system (Figure 5). Loadpath foams were included to support reaction forces and help with energy absorption. For the pilot study, no conventional interior trim was included on the door.

Performance Evaluation of Pilot System

The pilot system was evaluated in full-scale side impact test, using the FMVSS 214 procedure with the addition of a Hybrid-III dummy neck and head to the DOT/SID; results from the pilot system were compared with baseline tests from the same vehicle and procedures. Figure 6 shows ratios of the comparative dummy index values. While the TTI(d) reduction of 5% is not very significant, other body areas show impressive reductions ranging as high as 65%.

Prototype Air Bag System Performance Requirements

The potential of the air bag system for the reduction of occupant injuries in side impacts and the technology for the bag upward deployment were confirmed in the

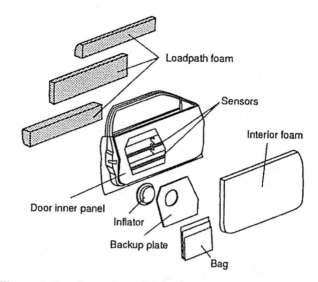

Figure 1. Configuration of the Pilot Air Bag System

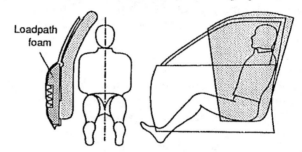

Figure 2. Frontal and Lateral Views of the Deployed Pilot Air Bag System

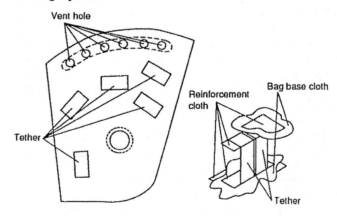

Figure 3. Shape of the Pilot Air Bag and Tether Configuration

pilot system study. Since the top priority of the pilot study was given to questions of overall technical feasibility, some of the basic door functions were ignored, making necessary further prototype design studies regarding the application of this type of system in mass production vehicles.

Baseline Full-Scale Impact Tests

Various types of side impacts, in terms of speed, angle and impact position may be observed in actual traffic accidents. Development of improved occupant protection

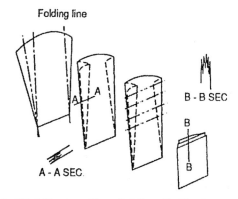

Figure 4. Pilot System Bag Folding Method

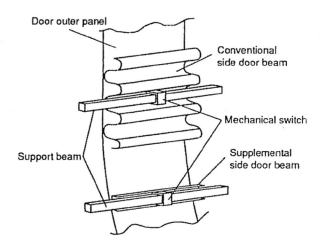

Figure 5. Structure of the Pilot System Sensor

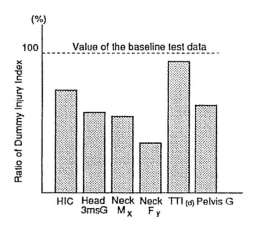

Figure 6. Dummy Injury Indices from Full-Scale Side Impact Test

suggests the need for consideration and evaluation of dummy occupant behavior over a wide range of side impact accident conditions. A matrix of eight typical side impacts representing some prominent types of injurious side impacts was selected for comparative evaluation of the prototype side impact air bag system, and full scale impact tests were carried out to evaluate them. Impact angles of 60 degrees and 90 degrees were selected for seven vehicle-to-vehicle tests, with striking and struck

vehicle speeds of 30 mph and 15 mph, respectively. Three relative positions along the struck vehicle side were evaluated, including the front and rear corner "L" configurations and the compartment "T" position specified in FMVSS 214. As in the pilot program, comparisons were directed at that baseline condition, and tests were set up to conform with its dummy, CMDB concepts, and injury indices for the six tests representing car-to-car exposures. For the last two tests, the CMDB was replaced with a light truck and a fixed pole, respectively, with the pole impact carried out with the car moving laterally into the pole at 20 mph.

Table 1 depicts the test configurations and presents important results, expressed as ratios of the result obtained in the baseline FMVSS 214 test shown in the first column of data. Dummy kinematics and secondary impacts between dummy and interior were studied in detail by high-speed cinematography. Partial head ejections out the window area were observed in some cases, as noted below. Observe that head injury indices are greater than the NHTSA baseline in six of the seven alternate tests, and that all indices are greater in the 60 degree compartment impact and the perpendicular light truck impact. Head injury indices are particularly high in the concentrated pole impact, resulting from direct contact of the head against the pole.

Table 1. Conditions and Results from Baseline Series of Full-Scale Crash Tests

		Test condition						Side impact against truck	Side impact against pole
		Side impact against MDB							
Dummy Injury Index	HIC	Ratio 1.0	0.43	1.56	1.90	4.60	2.07	2.84	16.79
	Head 3 ms G	1.0	0.71	1.34	1.20	2.69	1.56	1.45	1.60
	TTI(d)	1.0	0.42	0.48	0.39	2.34	0.42	1.28	0.90
	Pelvis G	1.0	0.16	0.17	0.17	1.87	0.09	1.14	0.86
Ejection		YES	YES	NO	NO	NO	NO	YES	YES
Interior contact		NO	NO	YES	YES	YES	YES	NO	NO
Exterior contact		NO	NO	NO	NO	NO	NO	YES	YES

Target Air Bag Performance

During the pilot study, bag occupant protection performance was given top priority at the expense of the normal door functions. Since the presence of an arm rest is considered essential, and since it has proven to be very difficult to deploy the bag both above and below the arm rest, it was decided to deploy the bag only above the arm rest. The pelvis portion of the occupant was thus excluded from the intended direct coverage area of the bag. Relationships among door intrusion, dummy motion, and elapsed time were determined by analysis of high speed films. The most stringent intrusion rates among the six baseline vehicle tests was found to occur under the FMVSS 214 test conditions, so that test was chosen for comparative evaluation of design goals. The relationship among bag thickness, deployment time and impact sensing can be approximated as shown in Figure 7, where door intrusion D (t) and bag thickness W (t) are plotted against time. The intrusion of the deploying bag surface B (t) can be expressed by Equation 1.

$$B(t) = D(t) + W(t) \qquad (1)$$

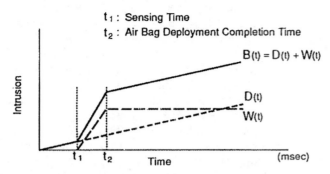

Figure 7. Intrusion of Air Bag Surface

It is of course desirable that contact between the air bag and the occupant occur at or after completion of air bag deployment, so that bag thickness may be used for energy management and its area may be used for force distribution over the occupant. As the initial distance between the door inner panel of the vehicle used in this study and the normally-seated DOT/SID occupant is 130 mm, Equation 1 becomes:

$$130 \leq D(t2) + W(t2) \qquad (2)$$

Figure 8 plots Equation 2, using the actual door intrusion determined by the baseline FMVSS 214 test, and demonstrates that bag thickness and deployment completion time must be traded off against one another. The bag should be as thick as possible in order to use it as effectively for occupant energy absorption stroke, so the bag thickness at chest height was given top priority. A target deployed thickness of 100 mm was selected to accommodate door intrusion, and tether length was set accordingly. Figure 9 shows the configuration of the bag at complete deployment, giving coverage from the arm rest to the roof side rail for head protection, as suggested by observed dummy kinematics and expected seating positions for a range of occupants.

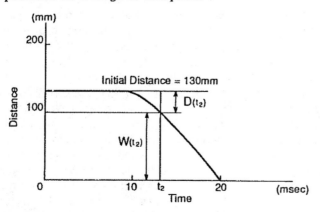

Figure 8. Relationship Between Door Intrusion and Air Bag Thickness

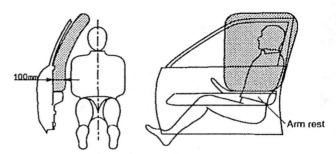

Figure 9. Frontal and Lateral Views of the Deployed Bag

The amount of energy to be absorbed in the secondary impact speed between the dummy chest and the door was calculated as 1000 J from the baseline FMVSS 214 test. Figure 10 shows the conceptional load-displacement characteristics of the door with and without the bag. A target energy absorption value of 500 J (50% of the total energy) was selected, assuming that the reduced door deformation by the bag will suppress the maximum load somewhat. With the bag thickness set at 100 mm, the bag must complete its deployment within 13 msec.

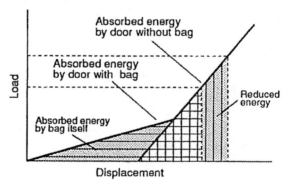

Figure 10. Load-Displacement Characteristics of Door Without and With Bag

The total time to 100 mm deployment was arbitrarily divided into 2 to 3 msec. for sensing and 10 to 11 msec. for bag deployment. Bag actuation duration to give adequate protection was targeted at 100 msec., based on dummy behavior in the tests, so that inflator characteristics and venting behavior were balanced around that goal.

The most important requirement for the sensors is to discriminate the need for bag deployment for occupant protection in side impact conditions, while avoiding inadvertent deployment in more normal conditions. The targeted 2-3 msec. sensing time is part of the 13 msec. deployment time, suggesting that G-sensors are impractical for this system. Another type of sensor capable of more rapid sensing is required.

Design of Prototype Side Air Bag System

Individual components that constitute the side air bag system were designed to be assembled into a front door, as shown in figure 11.

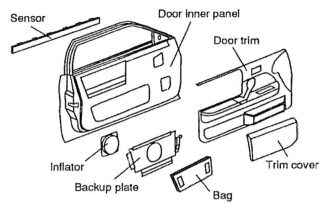

Figure 11. Configuration of the Air Bag System

Air Bag, Inflator and Backup Plate

The system consists of bag and inflator combined with backup plate, sensor, door panel and door trim, including air bag cover. Individual components were carefully designed to be compatible with all normal door functions. The bag was designed to meet the thickness and coverage ranges decided, with tether length set at 100 mm, as shown in Figure 12. Bag volume is about 40 liters.

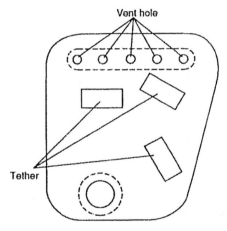

Figure 12. Shape of the Bag and Locations of the Tether

Development of a custom-designed inflator would require a much longer time than available under the prototype project, so it was decided to use an existing or modified inflator designed for frontal air bag systems.

Estimates of bag contact area with the occupant (500 cm²), bag thickness (100mm) and the energy (500 J), suggest a dynamic internal pressure requirement of about 10 N/cm². An optimal inflator was selected from among several available inflators by means of a series of inflation tests. The backup plate to which the bag and the inflator were assembled was designed so that the bag and the inflator were positioned within the limited inner space of the door without sacrificing normal door and window functions.

Door Structure and Trim Modifications

A large opening was provided within the door inner panel for the installation of the backup plate, with a compensating window sill reinforcement. This allowed the system to function without major changes in overall door deformation characteristics. The door trim was similar to that of the baseline production vehicle, but a newly-designed trim cover for the air bag module was added.

The upper portion of the trim cover was intended to break away inward, encouraging upward bag deployment.

Sensor Design

In place of an ordinary G-sensor, contact switch sensor system to be installed to the outermost location of the door was considered to sense impacts as quickly as possible. According to the results of the preliminary study regarding the characteristics of the contact switches, it was found that the activation pressure of such switch should be set rather high level to avoid an inadvertent air bag deployment when, for example, the door was opened and hit against a pole, tree, etc. where the relatively high load was concentrated around the narrow contact area. And it was also predicted that such a switch with higher activation pressure might not activate even when an ordinary vehicle collided into the side door. Therefore, it was decided to develop a contact switch sensor system consisting of two switch systems—the primary switch system to sense the impact with some other vehicle, and the secondary switch system that would not be activated in the impact on the pole when opening the door, but to be activated by a more severe side impact against a pole.

The primary switch system consists of plural contact switches installed at given intervals, and each one of them turns on easily by relatively low activation pressure. The primary switch system itself turns on only where two or more contact switches are turned on simultaneously, but it remains off where only one of the contact switches is activated by a mischievous action, etc.

The secondary switch system was installed in the longitudinal direction of the door, which would not be turned on when the door hit a pole or a tree, even if the door was accidentally and strongly opened, but turns on in a side impact against a pole, tree, etc., where the air bag deployment is required to reduce the occupant injury.

The total contact switch sensor system was so constructed that it would turn on if either the primary or secondary switch system described above turned on. Figure 13 shows the sensor configuration and the installation. The sensors were installed immediately inside the door outer panel along the pipe-wise side door beam so that they sense impacts as quickly as possible.

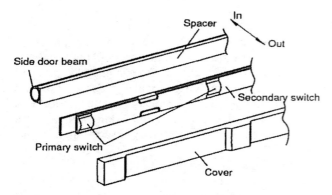

Figure 13. Configuration of the Sensor

Evaluation of Bag Performance in Static Bench Testing

Deployment tests were carried out to evaluate folding techniques, internal pressure, venting, and overall bag performance, leading to design modifications in an effort to optimize performance within the criteria outlined above. Figure 14 shows the final configuration of the prototype bag. Figure 15 presents the dynamic bag thickness at occupant chest height, taken from test films. Local bulging is obvious, extending locally beyond 100 mm, which gives the bag surface a mattress-like surface. This is fully acceptable, so long as the 13 msec. deployment time is achieved. Figure 16 shows the bag internal pressure vs. time, as recorded in a static deployment. The average pressure is lower in this static test than the 10 N/cm² dynamic pressure target, but pressure will be higher due to occupant contact forces in a dynamic deployment.

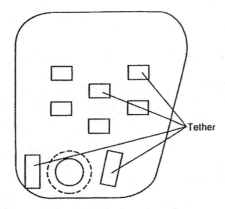

Figure 14. Final Configuration of the Bag and Tether Locations

Subsystem Deployment Tests on Vehicle

The prototype bag system was installed in a vehicle equipped with DOT/SID for a series of static deployments. Figure 17 represents the developing configuration during static deployment at various times after ignition; Figure 18 is a photograph taken at 25 msec. No unfavorable interactions with dummy or seat were noted, except that the motion of the trim cover induced a 70 g rib acceleration in the dummy. A subsequent test without the trim cover showed a reduction to 30 g. The trim cover was removed for the remainder of testing in the proto-

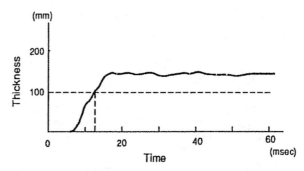

Figure 15. Bag Thickness at Occupant Chest Height

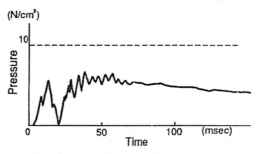

Figure 16. Bag Internal Pressure vs. Time

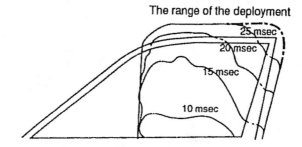

Figure 17. Process of Bag Deployment

Figure 18. Condition of the Bag at 25 msec

type project, with the idea that minor trim cover redesign efforts could resolve this dummy rib overload.

Impact tests were carried out to study the load-displacement characteristics of the air bag system as installed in the door. The door was supported against a rigid barrier face while the inflating bag was struck at full thickness from the inside by an impactor which simulated the occupant. The result was the load-displacement curve depicted in Figure 19. Target values of air bag energy and door energy were achieved.

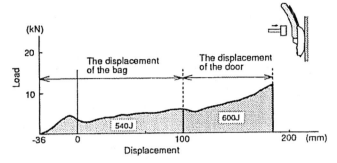

Figure 19. Load-displacement Characteristics of the Door With the Side Air Bag

Evaluation of Sensor Performance

The primary system used to sense striking vehicle side impact into the compartment must activated within 2 to 3 msec. Therefore it was decided to use switches which would be activated by a specified small activating stroke and load as the primary switch system.

Primary sensor performance was evaluated to demonstrate abilities to achieve sensing within 2-3 msec, by impacting a switch system subassembly attached to the rigid barrier by a FMVSS 214 moving barrier face at a speed of 3 m/s, with a resulting sensing time of 9 msec. Assuming an approximate inverse relationship of sensing time with speed, the required sensor performance was indicated. It was also verified that the primary switch system would not be activated when only one switch was turned on. Tests of the secondary sensor subsystem verified that it would not activate when impacted by a 160 mm diameter pole at maximum foreseeable occupant door opening force, while other tests verified that the secondary sensor system will activate when struck by a barrier-mounted 305 mm pole system having the same mass as the vehicle and a speed of 2.8 m/s, yielding a sensing time less than the selected target of 7 msec. This speed was selected as an appropriate threshold for deployment in pole side impact.

Full-Scale Side Impact Tests of the Prototype System

Full-scale FMVSS 214 side impact tests were carried out to evaluate the overall performance of the prototype system. The sensor system activated within 2 msec., and the bag deployed properly, with the dummy acceleration results shown in Figure 20. The most significant difference in the upper rib acceleration was that the first peak occurred earlier than in the baseline test, coinciding with the contact between dummy and air bag. The second peak occurred when the dummy contacted the door inner panel through the bag, the third by bottoming of available door structure deformation. The highest peak was reduced by 20 percent, compared with the baseline test. Accelerations of the lower spine do not show much

differences from the baseline, except for timing. This is attributable to the reduction of the rib acceleration and the reduction of the pelvis acceleration which will be described later. Overall thoracic trauma should have been reduced by the air bag, as suggested by the ten percent reduction in TTI(d) provided. Notably, pelvis acceleration was decreased by 24 percent compared to baseline test results, despite the fact that the bag did not contact the pelvis directly, probably due to pelvis motion in response to transmission of thorax accelerations through the spine, moving the pelvis away from its later door contact. Figure 21 shows reductions achieved by the prototype air bag system in several dummy injury indices.

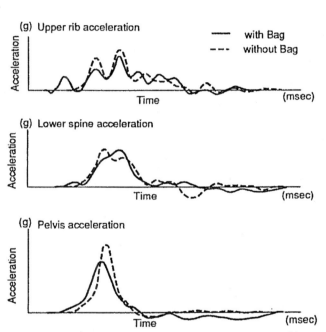

Figure 20. Dummy Acceleration from Full-Scale Side Impact Test With the Side Air Bag

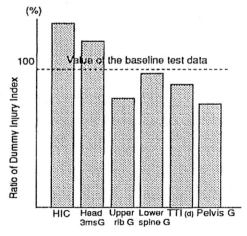

Figure 21. Dummy Injury Indices from Full-Scale Test: Prototype Side Impact Side Air Bag Compared to Baseline

Although head injury indices in the prototype system test are slightly higher than baseline, they are well below injury levels of concern. No head contact occurred with anything in the baseline test, nor with anything other than the air bag in the prototype system, so this comparison is somewhat mute regarding head injury effectiveness. Figure 22 compares head position at 80 msec., with and without the bag. Note the risk suggested by the partial ejection of the head in the baseline test and the protection provided by the bag. As a comparison, Figure 23 shows similar views from the baseline tests at the 30/15 mph condition with a light truck and with the pole at 20 mph. In both of these test cases, the dummy head contacted objects from outside the vehicle with serious injuries indicated.

Figure 22. Comparison of Dummy Head Behavior With Bag (Top) and Without Bag (Bottom)

The dynamic tests confirmed that the air bag system can be effective in overall occupant protection in side impact. Although the head protection potential has not yet been fully evaluated in a numerical sense, it is clear that the prototype air bag system can provide substantial and meaningful protection from this important injury source as well.

Conclusions

1. A pilot design and a more sophisticated preliminary prototype design for air bag systems for side impact

Figure 23. Impact of Dummy Head Against Harmful Objects: Against Light Truck (Top), Against Pole (Bottom)

protection from torso and head injuries have been assembled and evaluated in side impact testing conducted with the DOT/SID dummy and FMVSS 214 CMDB procedures and criteria.
- Folding methods and tether locations for deployment of the head-protection aspects of the bag have been clarified.
- Two kinds of sensor switches have been incorporated into a successful side impact intrusion sensing system.

2. A preliminary prototype system was developed by careful refinement of the design principles embodied in the pilot design. As compared to baseline vehicles, this improved design demonstrated substantial improvements in dummy injury measures in various vehicle-to-car and car-to-barrier crash testing.
- Thoracic injury risk is reduced by ten percent as measured by the DOT/SID and the TTI(d) criteria.
- While not specifically addressed as a design goal of the prototype air bag system, reductions in pelvic injury seem to be indicated by the test results.
- The test results have provided a clear demonstration of the potential effectiveness of the prototype system in the prevention of head injury in side impacts due to contact with vehicle interior surfaces and objects near or protruding inward through the side glass.

Future Research Objectives

The prototype side impact air bag system tested and studied in this project has demonstrated promise for further research in the following areas:

- The effect of interposition of an inflating bag on head injuries deserves further study in various impact modes. This may well prove to be the greatest potential benefit of the side impact air bag system.
- A complete study of the trade-offs in injury presented by other effects is called for, including: potential hearing damage due to the near proximity of the inflating bag, potential arm damage for the reported minority of occupants who may lean an arm on the window sill [Viano, 1989], potential arm or chest damage due to rapid opening of the air bag cover, etc. These concepts must be considered carefully for both intended and inadvertent deployment situations.
- Sensing strategies and hardware must be refined somewhat before proceeding to production. This will require careful study of the various impact situations and great effort to shorten the sensing time.
- Inevitable weight and cost penalties of incorporation of the side impact air bag system will need careful study before introduction of such systems in production vehicles.

References

DOT/NHTSA, 1990-1: Final Regulatory Impact Analysis: New Requirements for Passenger Cars to Meet a Dynamic Side Impact Test, FMVSS 214, U. S. Department of Transportation, Washington, D.C., August 1990.

DOT/NHTSA, 1990-2: CFR 49 Part 571. Docket No. 88-06, Notice 8, Federal Motor Vehicle Safety Standards: Side Impact Protection. U. S. Department of Transportation, Washington, D.C., October 1990.

DOT/NHTSA, 1988: CFR 49 Part 571. Docket No. 88-06, Notice 3, ANPRM FMVSS 214 Amendment to Side Impact Protection - Passenger Cars, U. S. Department of Transportation, Washington, D.C., August 1988.

Lau, 1989: "Design of a Modified Chest for EUROSID Providing Biofidelity and Injury Assessment" Ian V. Lau, David C. Viano, Clyde C. Culver and Edward Jedrzejczak, GMRL, [SAE Paper # 890881] SAE International Congress and Exposition, Detroit March, 1989.

Ridella, 1990: "Determining Tolerance to Compression and Viscous Injury in Frontal and Lateral Impacts" Stephen A. Ridella and David C. Viano, [SAE Paper # 902330] Thirty Fourth Stapp Car Crash Conference Proceedings Orlando, Fla. November 4-7, 1990.

Strother, 1990: "Reconstruction and Side Impact Societal Benefit" Charles E. Strother and Charles Y. Warner, SAE International Congress and Exposition, Detroit [SAE Paper # 900379], March, 1990.

Viano, 1987-1: "Evaluation of the SID Dummy and TTI Injury Criterion for Side Impact Testing" David C. Viano, GMRL [SAE Paper # 872208] Thirty First Stapp Car Crash Conference Proceedings New Orleans La. November 9-11, 1987

Viano, 1987-2: "Evaluation of the Benefit of Energy Absorbing Material in Side Impact Protection: Part I" David C. Viano, GMRL [SAE Paper # 872212] Thirty First Stapp Car Crash Conference Proceedings New Orleans La. November 9-11, 1987

Viano, 1987-3: "Evaluation of the Benefit of Energy Absorbing Material in Side Impact Protection: Part II" David C. Viano, GMRL [SAE Paper # 872213] Thirty First Stapp Car Crash Conference Proceedings New Orleans La. November 9-11, 1987

Viano, 1989-1: "Patterns of Arm Position During Normal Driving" Research Note: Human Factors, 1989, 31(6).

Viano, 1989-2: "Biomechanical Responses and Injuries in Blunt Lateral Impact" David C. Viano, GMRL [SAE Paper # 892432] Thirty Third Stapp Car Crash Conference Proceedings Washington, D.C. October 4-6, 1989.

Warner, 1988: "Inflatable Structures for Side Impact Crash Protection," Final Report, SBIR Phase I, Contract # DTRS-57-86-C-00089, U.S. Department of Transportation, Washington, D.C., April, 1988.

Warner, 1989: "Crash Protection in Near-Side Impact: Advantages of a Supplemental Inflatable Restraint," Charles Y. Warner, Charles E. Strother, Michael B. James, Donald E. Struble, and Timothy P. Egbert, SAE International Congress and Exposition, Detroit [SAE Paper # 890602], March, 1989.

Warner, 1990-1: "A Perspective on Side Impact Occupant Protection" Charles Y. Warner, SAE International Congress and Exposition, Detroit [SAE Paper # 900373], March, 1990.

Warner, 1990-2: "Application of Kinematic Concepts to Side Impact Injury Analysis," Charles Y. Warner and Charles E. Strother, SAE International Congress and Exposition, Detroit [SAE Paper # 900375], March, 1990.

Warner, 1990-3: U.S. Patent No. 4,966,388, "Inflatable Structures for Side Impact Protection," Charles Y. Warner, Charles E. Strother, Donald E. Struble & Milton G. Wille, U. S. Patent Office, Washington, D.C., October 30, 1990.

950865

Regulatory History of Automatic Crash Protection in FMVSS 208

Stephen R. Kratzke
National Highway Traffic Safety Administration

ABSTRACT

This paper summarizes the regulatory history of the automatic crash protection requirements in Federal Motor Vehicle Safety Standard 208. It is intended to give the reader an overview of the regulatory history involved in Standard 208, from its beginning in 1968 as a requirement that passenger cars be equipped with seat belts to its present requirement that, as of 1998, all passenger cars and light trucks must be equipped with air bags. It also discusses and summarizes the various court cases that have challenged different aspects of the automatic crash protection requirements.

INTRODUCTION

This paper traces the evolution of the occupant protection requirements in Federal Motor Vehicle Safety Standard 208 from its beginnings as a requirement for seat belts to be installed in passenger cars to its current requirements that each passenger car provide an air bag and a manual lap/shoulder belt for the driver and right front passenger position beginning in the 1998 model year (September 1, 1997) and that light trucks and vans provide an air bag and a manual lap/shoulder belt for the driver and right front passenger beginning in the 1999 model year (September 1, 1998). The purpose of this paper is to give the reader an understanding of how these requirements evolved into their present form and what purpose they are intended to serve.

THE FIRST OCCUPANT PROTECTION REQUIREMENTS IN STANDARD 208

ORIGINAL OCCUPANT PROTECTION REQUIREMENTS - Standard 208 was one of the 19 original Federal motor vehicle safety standards. It required that passenger cars provide a seat belt at every forward-facing designated seating position. This requirement took effect on January 1, 1968. There were no crash testing requirements to evaluate the protection afforded to vehicle occupants.

INITIAL AUTOMATIC PROTECTION REQUIREMENTS - It was not long, however, before the National Highway Traffic

Safety Administration (NHTSA) began to explore the possibility of requiring automatic crash protection in motor vehicles. Vehicles that provide automatic crash protection protect their occupants by means that require no action by the vehicle occupants. The two types of automatic crash protection that have been offered for sale on production vehicles are automatic seat belts and air bags. The effectiveness of a vehicle's automatic crash protection is assessed through crash testing. A vehicle must comply with specified injury criteria, as measured on a test dummy, when tested in a 30 mph crash test.

On July 2, 1969, NHTSA published an advance notice of proposed rulemaking (ANPRM) requesting comments on the merits of crash protection systems that protect vehicle occupants by means that require no action on the part of the occupants.[1] This notice specifically mentions the possibility of meeting such a requirement by means of air bags. After evaluating these comments, NHTSA published a notice proposing to require automatic crash protection for all passenger cars beginning January 1, 1972 and for all light trucks and vans beginning January 1, 1974.[2]

On November 3, 1970, NHTSA published a final rule that required automatic crash protection for all passenger cars as of July 1, 1973, and for most light trucks and vans as of July 1, 1974. Compliance would have been determined by a crash test with test dummies in the front outboard seats.[3] In the preamble to this rule, the agency made clear its position that automatic protection would supplant the need for vehicle occupants to use manual seat belts. NHTSA said: "Several comments recommended that the requirement for seat belts be retained, citing the benefits of keeping the driver in his seat during violent maneuvers and the possibility of failure of a passive system. It is the [NHTSA's] position that the possible benefits of required seat belts would not justify the costs to the manufacturers and to the public. Only a small percentage of the public uses the upper torso restraints that are presently furnished with passenger

---

[1] 34 FR 11148

[2] 35 FR 7187; May 7, 1970

[3] 35 FR 16927

cars."[4] In other words, passenger cars, light trucks, and vans would no longer be required to provide manual seat belts at each designated seating position once the automatic protection requirements took effect.

NHTSA received many petitions for reconsideration of this rule. In response to those petitions, NHTSA published a notice that postponed the effective date of the automatic protection requirements for passenger cars from July 1, 1973 until August 15, 1973, and granted a similar 45 day postponement for light trucks and vans, to correspond more precisely to the manufacturers' changeovers for a new model year's production.[5] This notice also repeated NHTSA's previous conclusion that the existing test dummy specifications published by the Society of Automotive Engineers were "the best available." The notice went on to state that, "NHTSA is sponsoring further research and examining all available data, however, with a view to issuance of further specifications for these devices."[6]

As indicated above, the automatic protection requirements did not require the use of any particular technology to achieve the desired result. Instead, manufacturers were free to use any means they chose so long as their vehicles provided the specified level of protection with no action by vehicle occupants. Up to this point, however, the regulatory notices dealing with automatic protection and the public comments responding to those notices had focused almost exclusively on air bag systems. In a July 8, 1971 final rule,[7] NHTSA added explicit language to Standard 208 to acknowledge that automatic belts could be used to meet the automatic protection requirements and to state the applicability of various requirements in the Standard to automatic belts (called "passive belts" at that time). On that same day, the agency published a proposal that automatic protection systems have a means of emergency release.[8] NHTSA suggested that automatic belts could use a spool-out mechanism and air bags would meet this requirement by deflating.

In response to petitions for reconsideration of the July 8, 1971 final rule on automatic belts, NHTSA excluded automatic belts from the assembly performance and webbing requirements. The agency explained that this change was made to allow manufacturers as much freedom in the design of automatic belt systems to fit the particular crash pulse of each car as they have in the design of other types of automatic protection systems.[9]

In addition to broadening the focus of its automatic protection rulemaking to recognize means other than air bags, the agency also introduced into its automatic protection rulemaking the concept of an ignition interlock system. Several

vehicle manufacturers asked NHTSA to delay the date by which automatic protection had to be installed in passenger cars because of unresolved technical problems with automatic protection systems. On October 1, 1971,[10] NHTSA proposed to postpone the effective date for mandatory automatic protection from August 15, 1973 until August 15, 1975. However, if a car produced between those dates did not provide automatic protection, it had to be equipped with an interlock system that would prevent the engine from starting if any front seat occupants did not have their manual seat belts fastened. Front outboard seating positions would also be subject to a crash test with a test dummy in each such seating position and the manual belt system fastened around the test dummy. The agency explained its proposal as follows: "It is intended by this option to provide a high level of seat belt usage, and to increase the life- and injury-saving effectiveness of installed belt systems, in the interim period before [automatic] systems are required." The ignition interlock option was adopted as proposed in a rule published February 24, 1972.[11]

THE COURT DECISION IN CHRYSLER V. DOT

Shortly after the March 10, 1971 rule was published requiring automatic protection in new vehicles, Chrysler, Jeep, American Motors, Ford, and the Automobile Importers of America filed lawsuits in the U.S. Court of Appeals for the Sixth Circuit challenging the automatic protection requirements. These lawsuits raised three primary arguments:

1. The automatic protection requirements were not "practicable," as required by NHTSA's authorizing legislation, because the technology needed to comply with automatic protection was not sufficiently developed as of that time. The manufacturers argued that NHTSA had no authority to establish a safety standard that required the industry to improve upon the existing technology. Under this view, automatic protection should not be required until devices to meet the requirement were sufficiently developed as of the date of the rulemaking so as to permit ready installation.

2. The automatic protection requirements did not "meet the need for motor vehicle safety," as required by NHTSA's authorizing legislation, because seat belts offered better occupant protection than automatic protection.

3. The automatic protection requirements were not "objective," as required by NHTSA's authorizing legislation, because the existing SAE Recommended Practice did not adequately specify sufficient details for the construction of the test dummy.

The Sixth Circuit announced its decision on these lawsuits on Dec 5, 1972 in an opinion titled Chrysler v. DOT, 472 F.2d 659 (6th Cir. 1972). The court ruled in favor of NHTSA on the first argument, stating that "the agency is empowered to issue safety standards which require improvements in existing technology or which require the development of new technology, and it is not limited to issuing standards based solely

---

[4] 35 FR 16928

[5] 36 FR 4600; March 10, 1971

[6] 36 FR 4602

[7] 36 FR 12858

[8] 36 FR 12866; July 8, 1971

[9] 36 FR 23725; December 14, 1971

---

[10] 36 FR 19266

[11] 37 FR 3911

on devices already fully developed."[12] The court also upheld the agency on the second point raised by the manufacturers, ruling that the question of whether to require automatic protection was delegated to NHTSA, that there was substantial support in the record for the decision to mandate automatic protection, and so the court had no basis for substituting its judgment for that of the agency.[13] However, the court found in favor of the manufacturers on the third argument. The court concluded that the SAE J963 test dummy incorporated in Standard 208 was not defined with sufficient specificity to meet the statutory requirement for "objectivity." Because of this shortcoming, this issue was remanded to the agency with instructions to delay the automatic crash protection requirement until a reasonable time after objective dummy specifications had been issued.[14]

The Sixth Circuit was petitioned by Ford to clarify the effect of its Chrysler decision on the requirement in Standard 208 for crash testing of the manual belts at front outboard seating positions in cars to be equipped with an ignition interlock. The Sixth Circuit announced its decision on this petition on February 2, 1973 in an opinion titled Ford v. NHTSA, 473 F.2d 1241 (6th Cir. 1973). The court ruled that its conclusion that the test dummy was not objective was equally applicable to this crash testing, and ordered the agency to delay the crash testing requirements for manual belt systems at front outboard seating positions in cars equipped with an ignition interlock.

The Ford v. NHTSA opinion was particularly problematic, because it overturned on February 2 an option that was scheduled to take effect on August 15 of that same year. All car manufacturers except General Motors had intended to choose the ignition interlock option, but it had now been declared invalid. On April 20, 1973, NHTSA addressed this problem by proposing to delete the requirement for crash testing of manual belts at the front outboard seats in cars equipped with an ignition interlock system.[15] NHTSA adopted this proposal in a June 20, 1973 final rule.[16] That June 20 final rule also announced NHTSA's position that the decision in the Chrysler case invalidated the automatic protection requirements, regardless of whether the language mandating automatic protection remained in the text of the standard. The agency also announced in the June 20 rule that additional rulemaking would be needed to reestablish an effective date for automatic protection requirements.

NHTSA proposed to adopt much more detailed test dummy specifications in a notice published April 2, 1973.[17] In fact, these proposed specifications were the existing specifications General Motors used for its Hybrid II test dummy. The Hybrid II test dummy was adopted as the test dummy to be used in Standard 208 compliance testing in a final rule published on August 1, 1973.[18] That August 1 rule also repeated the agency's previous announcement that it would give the public further notice and opportunity for comment before making a final decision on whether to reinstate the automatic protection requirements.

THE IGNITION INTERLOCK EXPERIENCE

As of August 15, 1973, all new cars had to be equipped either with automatic protection or an ignition interlock for both front outboard seating positions. General Motors sold a few thousand of its 1974 model year cars equipped with air bags that met the automatic protection requirement. Every other 1974 model year car sold in the United States came with an ignition interlock, which prevented the engine from operating if either the driver or front seat outboard passenger failed to fasten their manual seat belt.

In a March 19, 1974 notice, NHTSA described the public reaction to the ignition interlock as follows: "Public resistance to the belt-starter interlock system currently required (except on vehicles providing [automatic] protection) has been substantial, with current tallies of proper lap-shoulder belt usage on 1974 models running at or below the 60% level. Even that figure is probably optimistic as a measure of results to be achieved, in light of the likelihood that as time passes the awareness that the forcing systems can be disabled, and the means for doing so will become more widely disseminated, ..."[19] There were also speeches on the floor of both houses of Congress expressing the public's anger at the interlock requirement. On October 27, 1974, President Ford signed into law a bill that prohibited any Federal motor vehicle safety standard from requiring or permitting the use of any seat belt interlock system. In response to this change in the law, NHTSA published a final rule on October 31, 1974 that deleted the interlock option from Standard 208 effective immediately.[20]

While the interlock option was still in place, NHTSA also addressed the subject of automatic belts in response to a petition from Volkswagen. That company was going to introduce in its 1975 model year Rabbit models an automatic belt system that consisted of an upper torso restraint and knee bolsters, and asked for some changes and clarifications to the requirements of Standard 208 to ensure that this new belt system would comply with the standard. NHTSA proposed to require a manual single point emergency release mechanism on all automatic belts to allow vehicle occupants postcrash egress from the vehicle.[21] In the final rule adopting this proposal[22], NHTSA noted that one commenter had objected to the emergency release mechanism for automatic belts because the mechanism could be used to

---

[12] 472 F.2d 673

[13] 472 F.2d 674-675

[14] 472 F.2d 681

[15] 38 FR 9830

[16] 38 FR 16072

[17] 38 FR 8455

---

[18] 38 FR 20449

[19] 39 FR 10272

[20] 39 FR 38380

[21] 39 FR 3834; January 30, 1974

[22] 39 FR 14593; April 25, 1974

disconnect the belts in non-emergency situations. NHTSA responded that the advantages of having an emergency release mechanism outweighed the disadvantages of possible abuse.[23] Another commenter suggested that a lever or pushbutton that allowed the belt to spool out from the retractor instead of separating would be a more appropriate emergency release mechanism. NHTSA responded by stating that it believed the uniformity of having push-button action to release all motor vehicle seat belt assemblies was more compelling than the advantages suggested by the commenter of permitting a spool-out release.[24] The agency also provided more information about how it would determine whether an automatic belt system provided the necessary protection with "no action by vehicle occupants," as specified in Standard 208. Volkswagen did in fact introduce automatic belts as an optional piece of equipment on its 1975 Rabbit models.

THE COLEMAN DECISION

Standard 208 had never allowed the ignition interlock option to be anything more than an interim measure on the way to full automatic protection. The interlock was scheduled to expire as a permissible option on August 15, 1975 in any event. Thus, the change in the law eliminated the ignition interlock 10 months before it would have expired anyway. The more significant question at this point was whether rulemaking would be initiated to reinstate the automatic protection requirements and when any such requirements would take effect.

In a March 19, 1974 notice,[25] when the interlock option was still in place, NHTSA proposed to reinstate the automatic protection requirements for front outboard seating positions in passenger cars as of September 1, 1976. A little more than one year later, NHTSA again proposed to extend until September 1, 1976 the period during which manufacturers could comply with Standard 208 simply by installing manual belts at front seating positions.[26] However, this later notice did not propose any specific date for reinstating automatic protection requirements. In fact, this April 1975 notice announced that "A decision has not yet been made on the long-term requirements for occupant crash protection." The proposed extension of existing requirements until September 1, 1976 was made final in an August 13, 1975 rule.[27] The August 1975 rule stated: "While the NHTSA recognizes that the present crash protection options will in all likelihood be in effect for some period after August 31, 1976, the agency has not proposed more than the 1-year extension. ... The NHTSA intends to propose the long-term requirements for occupant crash protection, both for passenger cars and for light trucks and MPV's, as soon as possible."

The Secretary of Transportation, William T. Coleman, Jr., published a proposal on June 14, 1976 that had several noteworthy aspects.[28] This notice announced that Secretary Coleman was personally taking responsibility for the decision on automatic crash protection requirements in Standard 208. All previous decisions had been left to the NHTSA Administrator, subject to the review and approval of the Secretary. This notice proposed five options for dealing with occupant protection in frontal crashes. These were:

1. Continue existing manual belt requirements and continue further research to identify effective means of automatic protection.

2. Continue existing manual belt requirements and try to encourage States to pass mandatory belt use laws, which would substantially increase use of manual belts.

3. Continue existing manual belt requirements and conduct a Federally sponsored field test of automatic protection in vehicles used on the public roads.

4. Reinstate automatic protection requirements.

5. Require manufacturers to provide consumers with the option of ordering automatic protection in some or all of their models.

In accordance with this decision, NHTSA published a proposal[29] and final rule[30] extending the period during which manufacturers had the option of complying with Standard 208's occupant protection requirements by simply installing manual seat belts at all designated seating positions. That option was now extended until August 31, 1977, to allow time for the Secretary to announce his final decision on the automatic protection requirements.

On December 6, 1976, Secretary Coleman announced his decision on automatic protection. This decision was not published in the Federal Register, but written copies were placed in the public docket at the Department of Transportation. Secretary Coleman called on auto manufacturers to join with the Federal government in a demonstration program so that approximately 500,000 cars with passive restraint systems would be offered for sale at a reasonable cost to consumers in the 1979 and 1980 model years (i.e., the period beginning Sept. 1, 1978 and ending August 31, 1980). Secretary Coleman also announced that the Department of Transportation would make additional efforts to promote seat belt use during this period. At the end of this demonstration project, the Secretary concluded that the Department could make a more informed choice about the need for automatic protection requirements, based upon the real world experience gained from the 500,000 production vehicles that would be equipped with automatic protection and updated estimates of likely manual belt use in the future.

A notice appeared in the January 27, 1977 Federal Register[31] announcing Secretary Coleman's December 6, 1976 decision and "incorporating it by reference" in the Federal Register

---

[23] 39 FR 14593-14594

[24] 39 FR 14594

[25] 39 FR 10271

[26] 40 FR 16217; April 10, 1975

[27] 40 FR 33977

---

[28] 41 FR 24070

[29] 41 FR 29715; July 19, 1976

[30] 41 FR 36494; August 30, 1976

[31] 42 FR 5071; January 27, 1977

notice. The January 27 notice also extended indefinitely the option for manufacturers to comply with Standard 208 by installing belts at all designated seating positions in the vehicle.

THE ADAMS DECISION

Less than two months after the publication of the Federal Register notice announcing Secretary Coleman's final decision on the automatic protection requirements in Standard 208, a notice was published announcing that the new Secretary of Transportation was reexamining that decision. Thus, on March 24, 1977, a notice was published in the Federal Register[32] announcing that the new Secretary of Transportation, Brock Adams, was conducting a reexamination of Secretary Coleman's decision on automatic protection. This notice asked for public comments on the following three alternatives:

      1. Leave the Coleman decision in place;

      2. Reinstate the automatic protection requirements; or

      3. Seek to raise the usage level for manual belts by encouraging the States to pass belt use laws.

Before a final decision was announced on the automatic protection requirements for cars, NHTSA published a notice extending indefinitely the existing occupant protection requirements for light trucks and vans. This notice was published on June 2, 1977[33], and allowed manufacturers to either install automatic protection for the front outboard seating positions or to install manual seat belts at those seating positions. NHTSA explained in this notice that the agency had originally intended that manufacturers would have had the benefit of the experience with installing automatic protection in passenger cars before automatic protection systems were required in light trucks and vans. Since the Secretary of Transportation was in the process of deciding whether automatic protection systems should be required in passenger cars, it seemed premature to require those systems in light trucks and vans. Hence, NHTSA announced an indefinite extension of the option for light trucks and vans to meet the occupant protection requirements simply by installing manual seat belts at the front outboard seating positions. The agency noted that this indefinite extension did not preclude future rulemaking to modify the occupant protection requirements for light trucks and vans, but promised that notice and opportunity for comment would be provided prior to any modifications.[34]

On July 5, 1977, a final rule reinstating automatic protection requirements for passenger cars was published in the Federal Register.[35] The Department of Transportation concluded that automatic protection was necessary even though manual seat belts were "highly effective" at preventing injury and ejection,

because so few vehicle occupants used their manual belts.[36] The Department rejected the option of seeking mandatory seat belt use laws in the various States, because the prospects for success looked so poor.[37] The Department concluded that the demonstration programs called for in the Coleman decision were unnecessary, because they would have further delayed the mandatory introduction of occupant protection systems the Department had already found to be technologically feasible, practicable, and capable of offering substantial life-saving potential at reasonable costs.[38]

Accordingly, the July 5, 1977 rule required that all 1982 model year cars with a wheelbase over 114 inches be equipped with automatic protection. All 1983 model year cars with a wheelbase over 100 inches would be required to offer automatic protection, and all 1984 model year cars would be required to offer automatic protection. This gradual phased-in approach was intended to give vehicle manufacturers additional leadtime to overcome the greater difficulties of installing air bags in smaller cars and to give those efforts the benefit of the manufacturers' experience with installing air bags in larger cars.

Six petitions for reconsideration of this rule were filed. In addition, a group called the Pacific Legal Foundation filed a petition for a review of the rule in the U.S. Court of Appeals for the District of Columbia Circuit and asked the Department to stay the effectiveness of the automatic protection requirements for a period of time equal to the length of the judicial review. The vehicle manufacturers that filed petitions for reconsideration questioned the Department's analyses of the effectiveness of automatic protection systems. The Pacific Legal Foundation charged that the Department had failed to consider the public reaction to automatic protection and had ignored potential hazards posed by automatic protection systems. Ralph Nader and the Center for Auto Safety charged that the Department had improperly delayed implementation of the automatic protection requirements and that the Department had no authority to phase-in the automatic protection requirements gradually rather than requiring full compliance by the effective date. These petitions were denied in all significant respects by a December 5, 1977 notice in the Federal Register.[39]

THE COURT DECISION IN PACIFIC LEGAL FOUNDATION v. DOT

As indicated above, a group called Pacific Legal Foundation filed a lawsuit challenging the Adams decision to reinstate automatic protection requirements even before the Department had responded to the petitions for reconsideration of the Adams decision. After the Department denied their petitions, Ralph Nader and the Center for Auto Safety filed their own lawsuit challenging the Adams decision in the same court.

---

[32] 42 FR 15935

[33] 42 FR 28135

[34] 42 FR 28136; June 2, 1977

[35] 42 FR 34289

---

[36] 42 FR 34290; July 5, 1977

[37] 42 FR 34291-34292; July 5, 1977

[38] 42 FR 34291

[39] 42 FR 61466

The Pacific Legal Foundation argued that the Adams decision should be overturned because of three major shortcomings. First, the group argued that the data did not support the Secretary's findings on the effectiveness of air bags. Second, Pacific Legal Foundation argued that the Adams decision was unlawful because it failed to consider public reaction to the automatic protection requirements. Third, the group charged that the rule ignored collateral dangers to public safety posed by air bags. Ralph Nader and the Center for Auto Safety also challenged the rule, arguing that the Department had no authority to delay the effective date for the automatic protection requirements until the 1982 model year or to phase-in the requirements over three successive model years.

The D. C. Circuit announced its decision on this matter on Feb 1, 1979 in an opinion titled Pacific Legal Foundation v. DOT, 593 F.2d 1338 (D.C. Cir. 1979). The court upheld the Adams decision on all grounds. This court decision contains two especially significant findings. First, the court agreed with Pacific Legal Foundation that the Department must consider the likely public reaction to mandates in the safety standards in order to fulfill its responsibility to ensure that the standard is "practicable" and "meets the need for safety."[40] In this case, the court found that the Department had adequately considered the anticipated public reaction to the automatic protection requirements, notwithstanding the Department's claim that it was not required to consider the public reaction when promulgating safety standards. Second, the court expressly found that the Department could use a phase-in schedule when needed "to tailor safety standards to engineering reality."[41]

THE PECK DECISION

There had been only minor rulemaking notices from NHTSA dealing with the subject of automatic protection after the December 5, 1977 response to petitions for reconsideration of the Adams decision. Perhaps the most noteworthy rulemaking on automatic protection between 1978-1980 had to do with the emergency release mechanism on automatic belts. On May 22, 1978,[42] NHTSA published a notice proposing to allow automatic belts to use an emergency release mechanism other than the push-button, single-point release that had been required since the April 1974 final rule. This proposal was in response to a GM petition to allow a "spool release" design as the emergency release mechanism for automatic belts. The agency followed up this proposal with a November 13, 1978 final rule[43] that allowed automatic belts to use any emergency release mechanism that is a single-point release and that is accessible to a seated occupant. The agency explained its action thus: "This amendment will allow manufacturers to experiment with various [automatic] belt designs before the effective date of the [automatic protection] requirements and determine which

designs are the most effective and at the same time acceptable to the public."[44]

However, NHTSA's actions in 1981 ended this period of relative stability regarding the regulatory requirements for automatic protection. On February 12, 1981,[45] a notice signed by Drew Lewis, the new Secretary of Transportation, was published in the Federal Register. This notice proposed to delay the start of the phase-in for cars to be equipped with automatic protection for one year (from model year 1982 to model year 1983). The proposal to delay the phase-in was based upon the economic difficulties then confronting the automobile industry.

On April 9, 1981, two notices signed by Secretary Lewis were published in the Federal Register. The first was a final rule delaying for one additional year the start of the phase-in of automatic protection.[46] The second was a notice proposing three alternative courses of action regarding the future status of automatic protection requirements.[47] The three alternatives on which the Department sought comment were:

1. Retain the new phase-in schedule, but reverse the sequence of vehicles. In other words, small cars would be required to provide automatic protection first (in model year 1983), then mid-size cars (in model year 1984), and finally large cars (in model year 1985).

2. Allow one additional year of leadtime, but eliminate the phase-in. In other words, all cars would have to provide automatic protection beginning in model year 1984.

3. Rescind the requirements for automatic protection.

A final rule signed by NHTSA Administrator Raymond Peck that rescinded the automatic protection requirements for cars was published on October 29, 1981.[48] This rule indicated that there was significant uncertainty about the public acceptability of automatic protection. Vehicle manufacturers had overwhelmingly indicated that they would comply with the automatic protection requirements by installing detachable automatic belts in their new cars. The agency found substantial uncertainty about the likely use rates of these detachable automatic belts and announced that it could not reliably predict that detachable automatic belts would produce even a 5 percentage point increase in belt use over the existing belt use rates for manual belts.[49] The uncertain benefits combined with the relatively substantial costs for automatic protection led NHTSA to conclude that the automatic restraint requirement was no longer reasonable or practicable.

LEGAL BATTLE CULMINATING WITH SUPREME COURT DECISION ON AUTOMATIC PROTECTION

---

[40] 593 F.2d 1345-1346

[41] 593 F.2d 1348

[42] 43 FR 21912

[43] 43 FR 52493

[44] 43 FR 52494

[45] 46 FR 12033

[46] 46 FR 21172

[47] 46 FR 21205

[48] 46 FR 53419

[49] 46 FR 53423

Throughout the course of this standard, groups have looked to the Federal courts to overturn decisions on the automatic protection requirements with which the group is dissatisfied. This rescission of the automatic protection requirements proved no different. On November 23, 1981, State Farm Mutual Automobile Insurance Company, joined by several other petitioners, filed a lawsuit in the U.S. Court of Appeals for the District of Columbia Circuit challenging the rescission of the automatic protection requirements.

That court announced its decision on June 1, 1982 in an opinion titled State Farm v. DOT, 680 F.2d 206 (D.C. Cir. 1982). The court unanimously reversed NHTSA's rescission of the automatic protection requirements. The court found that NHTSA had failed to consider or analyze obvious alternatives to rescission and had offered no evidence to show that the 1977 conclusion that seat belt usage would increase with automatic belts was no longer true.[50]

On September 8, 1982, NHTSA filed a petition with the U.S. Supreme Court asking the Supreme Court to review the D.C. Circuit Court's decision. The Supreme Court granted that petition on November 8, 1982.[51] The Supreme Court announced its decision on this matter on June 24, 1983 in an opinion titled Motor Vehicle Manufacturers Association v. State Farm, 463 U.S. 29 (1983). The Supreme Court unanimously ruled that the rescission of the automatic protection requirements was unlawful, or "arbitrary and capricious," because the agency had failed to consider obvious alternatives to rescission and explain why alternatives short of rescission were not chosen.[52] The Court noted that air bags, nondetachable automatic belts, and detachable automatic belts were three existing technologies that could be used to comply with the automatic protection requirement. The Supreme Court said that, even if NHTSA were correct that detachable automatic belts would yield few benefits, that fact alone would not justify rescission. Instead, the Court said that fact would justify only the modification of the automatic protection requirements to prohibit compliance by means of detachable automatic belts.[53] The Court reasoned that the necessary next step to justify rescission was for NHTSA to adequately explain why it no longer believed that compliance

---

[50] 680 F.2d 230

[51] 459 U.S. 987 (1982)

[52] All nine Justices joined in the opinion finding the rescission of the automatic protection requirements was arbitrary and capricious because it failed to consider alternatives to rescission, such as permitting compliance only by means of air bags or nondetachable automatic belts, and explain why those alternatives were not adopted. However, in a separate opinion for four of the nine Justices, Justice Rehnquist expressed the view that the rescission of requirements permitting compliance with automatic protection by means of detachable automatic belts was satisfactorily justified and would not be considered "arbitrary and capricious." 463 U.S. 57-59 (1983).

[53] 463 U.S. 47 (1983)

with the automatic protection requirements by means of either air bags or nondetachable automatic belts would be an effective and cost-beneficial way of saving lives and preventing injuries.[54] Since NHTSA had not offered such an explanation, its rescission of the automatic protection requirements was unlawful. Accordingly, the Department was ordered to conduct further rulemaking on the status of the automatic protection requirements.

THE DOLE RULE AND THE RESULTING LEGAL CHALLENGE

Rulemaking on the automatic protection requirements was begun again on October 19, 1983, when the Department of Transportation published a notice seeking comments on a wide range of alternative actions the Department might take with respect to automatic protection in response to the Supreme Court's decision.[55] This notice sought comments on three broad possibilities - retain the automatic protection requirements in their existing form and establish a new compliance date, amend the automatic protection requirements, e.g., to preclude detachable automatic belts, or rescind the automatic protection requirements. In addition, this notice asked for comments on three other supplementary actions that could be taken in conjunction with any of these three broad possibilities. The supplementary actions were: (1) conduct a demonstration program for automatic protection, along the lines of the Coleman decision, (2) seek legislation to encourage States to pass mandatory seat belt use laws, and (3) seek legislation to require vehicle manufacturers to provide consumers with an option to select air bags or automatic belts, instead of manual belts, in their new cars.[56]

The Department received more than 6,000 comments on this notice. After reviewing the comments, the Department published a supplemental notice on May 14, 1984.[57] This notice asked for further comment on alternatives being considered regarding State laws mandating seat belt use, mandatory demonstration programs for automatic crash protection, and the possibility of requiring driver's side air bags on all small cars.

A final rule reinstituting automatic crash protection for cars was signed by Elizabeth Dole, the Secretary of Transportation and published on July 17, 1984.[58] This rule provides for a phase-in of automatic protection in cars beginning September 1, 1986 (the 1987 model year). All cars manufactured after September 1, 1989 (the 1990 model year) are required to provide automatic protection. During the phase-in of automatic protection, the rule encourages manufacturers to choose to install air bags instead of automatic belts by providing a 1.5 car credit for cars with driver air bags and any type of automatic protection

---

[54] 463 U.S. 49-51 (1983)

[55] 48 FR 48622

[56] 48 FR 48632

[57] 49 FR 20460

[58] 49 FR 28962

for the passenger. This rule also encouraged States to pass mandatory seat belt use laws. It did so by including a provision that the automatic protection requirements might be eliminated if the Secretary of Transportation made a determination by April 15, 1989 that enough States had enacted mandatory seat belt use laws that met criteria specified in Standard 208.

There were 16 petitions for reconsideration of this rule. NHTSA published a notice responding to those petitions on August 30, 1985.[59] This notice denied requests to change the phase-in schedule and to eliminate the possibility that automatic protection requirements might be rescinded if enough States enacted mandatory belt use laws. This notice also expanded NHTSA's efforts to encourage manufacturers to install air bags instead of automatic belts by adding a new one car credit. Under this new one car credit provision, manufacturers could comply with the automatic protection requirement by installing an air bag for the driver and manual lap/shoulder belt for the passenger. The one car credit provision was scheduled to remain in effect until the phase-in for automatic protection ended on August 31, 1989.

Shortly after the publication of this response to the petitions for reconsideration, a lawsuit challenging the July 1984 final rule was filed in the U.S. Court of Appeals for D.C. Circuit by State Farm Insurance Company and the State of New York. State Farm's primary argument was that the rule was unlawful because it would revoke automatic crash protection if enough States enacted mandatory belt use laws. New York argued that the rule was unlawful because it failed to require air bags or nondetachable automatic belts as the only permissible means of automatic protection.

The court announced its decision on September 18, 1986 in an opinion titled State Farm v. Dole, 802 F.2d 474 (D.C. Cir. 1986). The court upheld the Secretary's 1984 rule in all respects. The court indicated that State Farm's objection was based upon that insurer's speculation about what the Secretary might do if States passed laws and if those law were determined to comply with certain criteria. The court said that State Farm was not entitled to any legal relief until these potential events and determinations had actually occurred. In response to New York's argument that NHTSA should not have allowed detachable belts to be permitted as automatic protection, the court said: "The Safety Act does not require the Secretary to adopt the technological alternative providing the greatest degree of safety. The Act expressly permits the Secretary to consider such factors as reasonableness and practicality in addition to safety features. Both the Supreme Court and this court, moreover, have recognized the Secretary's authority to consider such factors as cost and public acceptance."[60]

THREE SIGNIFICANT RULES THAT GREW OUT OF THE DOLE RULE ON AUTOMATIC PROTECTION

USE OF A MORE ADVANCED TEST DUMMY - While the rulemaking that led to Secretary Dole's decision was underway, General Motors filed a petition for rulemaking asking

that a new test dummy it had developed, called the Hybrid III dummy, be permitted to be used in testing compliance with the provisions of Standard 208, including the automatic protection requirements. NHTSA granted this petition on July 20, 1984. On April 12, 1985,[61] NHTSA published a notice proposing to incorporate the Hybrid III test dummy as a permissible alternative for Standard 208 compliance testing. NHTSA explained its proposal as follows: "Based on its review of the available test data, the agency recognizes that the Hybrid III test dummy represents an appreciable advancement in the state-of-the-art of human simulation and that its use would be beneficial for continued improvement in vehicle safety."[62] This notice proposed to take advantage of the enhanced capabilities of the Hybrid III by adding new injury criteria for the neck, lower leg, facial laceration, and chest deflection. This notice also proposed to require the use of the Hybrid III test dummy for all Standard 208 compliance testing for cars manufactured on or after September 1, 1991.

NHTSA published a final rule in this area on July 25, 1986.[63] This rule adopted the Hybrid III test dummy for Standard 208 compliance testing and adopted the proposal that the Hybrid III be the exclusive dummy for Standard 208 compliance testing as of September 1, 1991. However, the July 1986 rule adopted only one additional injury criterion for use with the Hybrid III test dummy, the proposed one for chest deflection. In response to petitions for reconsideration of this July 1986 rule, NHTSA postponed the date for mandatory use of the Hybrid III test dummy to allow more time to examine technical issues that might arise from the use of this new test dummy.[64]

After completing its technical examination of these issues, NHTSA published a notice on December 10, 1992[65] proposing that the Hybrid III test dummy would be the only dummy used in Standard 208 compliance testing beginning September 1, 1996. After reviewing the comments on this proposal, NHTSA published a final rule giving one additional year of leadtime before requiring exclusive use of the Hybrid III test dummy.[66] The Hybrid III test dummy is now the only dummy that will be used for Standard 208 compliance testing of vehicles manufactured on or after September 1, 1997 (the 1998 model year).

CRASH TESTING OF MANUAL BELTS IN LIGHT TRUCKS AND VANS - On April 12, 1985, NHTSA published a notice that proposed, among other things, that the occupant protection afforded by manual seat belts installed in front outboard seating positions of light trucks and vans be evaluated according to the same crash test used to evaluate automatic

---

[59] 50 FR 35233

[60] 802 F.2d 474, at fn.23, 486-87 (1986)

[61] 50 FR 14602

[62] 50 FR 14603

[63] 51 FR 26688

[64] 53 FR 8755; March 17, 1988

[65] 57 FR 58437

[66] 58 FR 59189; November 8, 1993

protection.[67] NHTSA adopted this proposal in a final rule published November 23, 1987.[68] The November 1987 rule required light trucks and vans manufactured on or after September 1, 1991 (the 1992 model year) to be certified as complying with this crash testing requirement.

"ONE CAR CREDIT" FOR DRIVER AIR BAGS EXTENDED UNTIL AUGUST 31, 1993 - On June 11, 1986, Ford Motor Company filed a petition asking NHTSA to permit the production of cars with driver air bags and no automatic protection for the front seat passenger after September 1, 1989, the date on which this "one car credit" provision was scheduled to expire. In its petition, Ford said that, if this request were granted, Ford would "in all likelihood" install driver air bags in the majority of its North American-designed cars. NHTSA published a notice proposing to grant Ford's petition on November 25, 1986.[69] In that notice, NHTSA indicated that the proposed extension of the one car credit "would encourage the orderly development and production of passenger cars with full-front air bag systems."[70]

Comments supporting the proposed extension of the one car credit were submitted by air bag suppliers, insurance companies and their trade associations, vehicle manufacturers and their trade associations, and researchers and other organizations involved in highway safety issues. The only comments opposing this extension were submitted by the Center for Auto Safety and Robert Phelps, a private citizen. After considering these comments, NHTSA published a final rule extending the one car credit until August 31, 1993.[71] The agency explained that the extension of the one car credit would promote the widespread introduction of air bags. In addition, the agency concluded that there were a number of technical issues that still needed to be resolved before widespread introduction of passenger air bags would occur.

One petition for reconsideration of this rule was filed by Public Citizen, a group that had not previously participated in this rulemaking. NHTSA denied this petition in a notice published November 5, 1987.[72] In the denial, NHTSA again explained that it was seeking to encourage manufacturers to install air bags, instead of automatic belts, and that all available evidence indicated that the extension of the one car credit had increased the likelihood of widespread use of air bags.

Public Citizen filed a lawsuit in the U.S. Court of Appeals for the D.C. Circuit challenging NHTSA's extension of the one car credit until 1993. The court announced its decision on July 15, 1988 in an opinion titled Public Citizen v. Steed, 851 F.2d 444 (D.C. Cir. 1988). The court unanimously upheld the extension of the one car credit provision. The court specifically

found that NHTSA had extended the one car credit to encourage greater installation of air bags, because the agency believed th[at] air bags would offer long-term overall safety gains for vehicle occupants. The court said that, even if it accepted Public Citizen's assertion that the one car credit extension would perm[it] a near-term reduction in front seat occupant protection, "[i]t [is] within NHTSA's province to balance estimated long-term safety benefits against the possibility of a marginal short-term reduction in safety."[73]

AUTOMATIC PROTECTION IN LIGHT TRUCKS AND VANS

NHTSA had not taken up the question of automatic protection in light trucks and vans again since its June 197[7] notice indefinitely suspending the requirements for automatic protection in those vehicles. However, the agency's rulemaking on crash testing of manual belt systems in light trucks and vans had raised the issue of occupant protection in those vehicles. On January 9, 1990, NHTSA published a notice proposing to require automatic protection to be phased in for light trucks and vans in a manner that closely paralleled the recently completed phase-in of automatic protection for passenger cars.[74] A final rule requiring automatic protection in light trucks and vans was published on March 26, 1991.[75] This rule requires that 2[0] percent of each manufacturer's model year 1995 production of light trucks and vans provide automatic protection, 50 percent of model year 1996 production provide automatic protection, 9[0] percent of model year 1997 production provide automatic protection, and all 1998 light trucks and vans must provide automatic protection.

AIR BAGS REQUIRED AS THE MEANS OF AUTOMATIC PROTECTION

On December 18, 1991, then President Bush signed into law the Intermodal Surface Transportation Efficiency Act (ISTEA). Among other things, ISTEA requires that all passenger cars manufactured on or after September 1, 1997 (the 1998 model year) and light trucks and vans manufactured on or after September 1, 1998 (the 1999 model year) provide air bags at the driver and right front passenger positions. In response to this mandate, NHTSA published a notice on December 14, 1992.[76] This notice proposed to require that passenger cars and light trucks and vans comply with the automatic protection requirements by installing air bags and manual lap/shoulder seat belts at the driver and right front passenger positions. This NPRM also proposed to require that these vehicles have a label on the sun visor providing occupants with important safety information about the air bags and advising occupants that they must always wear their safety belts for maximum safety

---

[67] 50 FR 14589; April 12, 1985

[68] 52 FR 44898

[69] 51 FR 42598

[70] 51 FR 42599

[71] 52 FR 10096; March 30, 1987

[72] 52 FR 42440

[73] 851 F.2d 444, at 449 (1988).

[74] 55 FR 747

[75] 56 FR 12472

[76] 57 FR 59043

protection in all types of crashes. This proposal was adopted as a final rule in a notice published September 2, 1993.[77]

SUMMARY

The regulatory history of the automatic protection requirements in Standard 208 is obviously a long and complicated one that begins in the late 1960's and will continue into the late 1990's. However, there are eight major events that have been especially significant in that history. These are:

1. The 1972 court decision in Chrysler v. DOT. In this case, the court overturned the automatic protection requirements, although the court found that NHTSA had authority to promulgate such requirements and that the automatic protection requirements that were then established met the need for motor vehicle safety. However, the requirements were invalid because the specifications for a test dummy were inadequate.

2. The 1973 ignition interlock option and the 1974 Congressional disapproval of such an option. NHTSA thought it had an alternative to automatic protection that would allow it to achieve roughly the same benefits as automatic protection without getting into the technical and cost issues associated with automatic protection. However, the ignition interlock was so unpopular that Congress amended Federal law to provide that NHTSA could not require or even permit manufacturers to comply with a safety standard by means of an interlock.

3. The 1976 decision by Secretary Coleman to implement a demonstration program. This program was intended to resolve any lingering concerns about the public acceptability and real world effectiveness of automatic protection.

4. The 1977 decision by Secretary Adams to reimpose the automatic protection requirements on a phased-in schedule. This would have required passenger cars to provide automatic protection beginning in the 1982 model year.

5. The 1981 decisions by Secretary Lewis and Administrator Peck to rescind the automatic protection requirements. This rescission was based on the changed economic circumstances and the likely insignificance of the safety benefits that would result if vehicles provided automatic protection by means of detachable automatic belts.

6. The 1983 decision by the Supreme Court declaring the 1981 rescission of the automatic protection requirements unlawful. This decision guided the Department's subsequent consideration of these requirements.

7. The 1984 decision by Secretary Dole to reinstate the automatic protection requirements on a phased in schedule. This decision became the first requirement for automatic protection that actually went into effect.

8. The 1991 Federal law requiring that air bags, supplemented by manual lap/shoulder seat belts, be the means of automatic protection offered in all new cars by the 1998 model year and in all new light trucks and vans by the 1999 model year.

Most of the comments on the NHTSA proposal to implement the 1991 Federal law mandating air bags were directed toward the agency's proposed language for labels to be required on sun visors and on the proposed exemption procedures, with almost nothing said about air bags. This probably reflects the fact that all the commenters knew that Federal law mandated an air bag requirement, regardless of the comments. It is nevertheless ironic that 24 years after an automatic protection requirement was first proposed, the requirement to provide air bags in all passenger cars and light trucks and vans was adopted with so little comment. One would not have predicted this after all the regulatory notices, lawsuits, and other high profile actions that have been associated with the automatic protection requirements.

---

[77] 58 FR 46551

THE APPROVAL OF AIR-BAGS ETC.-
THE NEED FOR A STANDARD

Alan F. Charles
ACE Consultancy
United Kingdom
Paper Number 96-S1-W-24

ABSTRACT

This paper discusses the need for an international standard which may be used by National Competant Authorities (NCAs) to approve or authorise the acquisition of air-bags and seat-belt tensioners.

INTRODUCTION

Recent discussions in USA between the National Highway Traffic Safety Administration (NHTSA) and others (eg motor and insurance industry) on ways of reducing unwanted side-effects of air-bags has focussed on four main points:-
(a) changes in crash test requirements
(b) enforcement of seat-belt use regulation
(c) education of the public and
(d) the need for "smart" restraint systems.

All of these topics will be featured at 15th ESV but this paper will only consider one aspect of (d) ie it discusses the problems experienced by companies resulting from delays in obtaining approvals or authorisations for the components of smart systems and suggests that such delays could be avoided if national authorities based their approval or authorisation procedures on an international standard.

Approval or Authorisation

The requirement for approval or authorisation arises from the fact that the gas generators or inflators contain components which are considered as explosives by many national authorities and are thus subject to national regulations which consider both safety and security.

The process of approval or authorisation differs from nation to nation but in most cases the authority imposes a charge on the applicant and invariably causes a delay to the applicant.

Thus, the process of approval often creates barriers to the importation of foreign products.

Approval or authorisation should not be confused with the "classification" of the product as packaged for transport or with the whole vehicle "type approval" of the vehicle fitted with the devices.

(Details of "classification", including packaging and testing to the "UNITED NATIONS RECOMMENDATIONS ON THE TRANSPORT OF DANGEROUS GOODS, EDITION 9", may be obtained from the report, "REGULATIONS ON AIRBAGS AND SEAT-BELT TENSIONERS", by this author. Please contact ACE Consultancy on FAX +44 1606 861440)

The process of approval or authorisation is now further complicated by the introduction of "hybrid" devices in which the gas generator or inflator may contain both explosives and compressed gas and the approval of the device may be the responsibility of more than one authority (eg in Germany, BAM approves the explosive, TuV approves the compressed gas cylinder and gas.) In general, the NCA in the country of manufacture bases the approval on the results of tests and/or detailed inspection under national or domestic regulations. eg in USA, approval is based on Section 49 of the Code of Federal Regulations (CFR-49), whereas in countries such as Belgium, France, Germany, Italy, Spain, Switzerland and UK, approval or authorisation is under domestic explosives regulations, many of which were promulgated long before air-bags were even considered (eg UK 1875 Explosives Act).

Approval or Authorisation of Other Explosive Devices

International standards have already been developed for some explosive devices (eg CIP and SAAMI for sporting ammunition and SOLAS for marine pyrotechnics) and this has resulted in minimum delays and costs in obtaining approvals for importation of these items.

Since the publication of British Standard 7114 (on fireworks) in 1988, the procedures for importing fireworks into UK have been simplified and are now far less stringent than those for importing car air-bags or seat-belt tensioners!

EEC Directives.

Unfortunately, there is as yet no specific EEC Directive on air-bag inflators or seat-belt tensioners. Instead, each EU nation interprets non-specific Directives and applies them to their national legislation.

Because air-bag inflators may be considered as ammunition in one country (eg UK) and pyrotechnic articles in another (eg Germany), the devices may be not be subject to the same EEC Directives in each EU country (Subsidiarity?).
Also, because some EEC Directives are based on UN Recommendations on the Transport of Dangerous Goods, their applicability to articles outside their packaging is questionable.

ISO and GRSP

The standards being developed by WG8 of ISO TC22 and GRSP of UN-ECE WP29 may eventually be used to specify the quality, performance or efficacy of air-bags in vehicles but may not necessarily be appropriate for use by NCAs to approve or authorise air-bags for acquisition prior to fitting into the vehicles.

Proposed Standard.

It is proposed that a simple standard, based essentially on tests currently carried out by BAM (Germany) and US-DOT, (see figure 1) should be developed by representatives of "industry" and submitted to representatives of the national testing "authorities".

"Industry" could be represented by an organisation such as the Automotive Occupant Restraints Council (AORC) (formerly the American Seat-Belt Council), since the majority of air-bag and seat-belt manufacturers are members of AORC, and the Explosives, Propellant and Pyrotechnic (EPP) Group of the OECD-IGUS (Organisation for Economic Cooperation and Development - International Group on Unstable Substances) could represent the national testing authorities and/or NCAs.

The next meeting of EPP of OECD-IGUS is scheduled for Orlando, USA, from September 23rd to September 26th or 27th, '96, and representatives from the following countries will be invited to attend:-

BELGIUM, CANADA, CHINA, DENMARK, GERMANY, FINLAND, FRANCE, IRELAND, ITALY, JAPAN, NETHERLANDS, NORWAY,

POLAND, PORTUGAL, SPAIN, SWEDEN, SWITZERLAND, TAIWAN, UNITED KINGDOM and UNITED STATES OF AMERICA,

The same standard could be used by DGIII of EEC as the basis of an EEC Directive on vehicle air-bags and seat-belt tensioners.

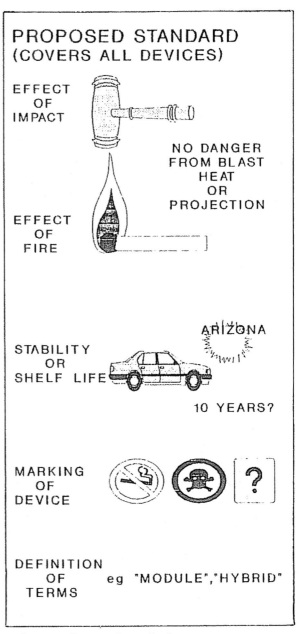

Figure 1. Proposed standard.

Airbag Technology: What it is and How it Came to Be

Donald E. Struble
Collision Safety Engineering

ABSTRACT

Since air bags emerged as an occupant protection concept in the early '70s, their development into a widely-available product has been lengthy, arduous, and the subject of an intense national debate. That debate is well documented and will not be repeated here. Rather, operating principles and design considerations are discussed, using systems and components from the developmental history of airbags as examples.

Design alternatives, crash test requirements, and performance limits are discussed. Sources of restraint system forces, and their connection with occupant size and position, are identified. Various types of inflators, and some of the considerations involved in "smart" systems, are presented. Sensor designs, and issues that influence the architecture of the sensor system, are discussed.

INTRODUCTION

When a vehicle crashes, it is acted upon by collision forces that tend to change its velocity. In a direction opposite to these forces, everything in the vehicle that can move does so. This includes the occupants, with their various articulated segments. A restraint system has the purpose of intercepting these motions and managing or eliminating the "second collisions" of an occupant's parts with potential contact surfaces, so that injuries can be mitigated to the extent possible.

An occupant restraint is a system, and works in harmony with the vehicle structure. The restraint system may include an air bag, which itself is a collection of components designed to work with each other and in cooperation with other parts of the vehicle.

RESTRAINT SYSTEMS IN FRONTAL CRASHES

In a crash, potential contact surfaces in the vehicle experience changes in their velocities, and to some extent, their directions of travel. Figure 1 shows the velocity-time history, in an actual crash, for the instrument panel. In contrast, a free particle would keep moving as before the crash. This behavior would be represented in Figure 1 as a horizontal line at 35 mph, in comparison to the

descending lines associated with a belted occupant. As time proceeds, the velocity differential between the occupant and the vehicle builds up. One might think of this as an accrual of a "velocity debt," which has to be paid back in order for the vehicle and its occupants to achieve a final common velocity (perhaps zero). In Figure 1, this debt would be the vertical distance between lines representing a potential contact surface, such as the instrument panel, and those representing the occupant. The debt is due and payable when contacts are made between the occupant and one or more contact surfaces. Figure 1 shows that debt management starts earlier for a belted occupant, and that the debt itself is much reduced.

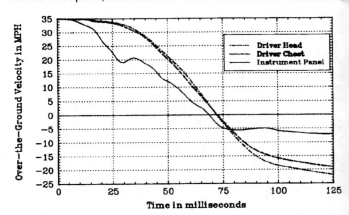

Figure 1. Velocity Histories in 35 mph Barrier Crash.

In this particular vehicle, there was 10.71 inches of space between the sternum and the steering wheel, and 23.18 inches between the head and the windshield, at the beginning of the crash. Figure 2 shows that a free particle would use up these distances in 54 and 75.6 milliseconds, respectively. If an unrestrained occupant behaves similarly, we would expect debt repayment to occur in earnest at about 76 milliseconds, as the head reaches the windshield, and the debt amounts to about the final velocity change, or delta-V (ΔV), of the vehicle.

Of course, these numbers will vary, depending on the crash, the initial position of the occupant, etc. Figures 1 and 2 show estimated occupant velocities and displacements, based on accelerometer data. These indicate that debt management began at about 25 milliseconds

straint process would have to begin by 50 milliseconds the latest.

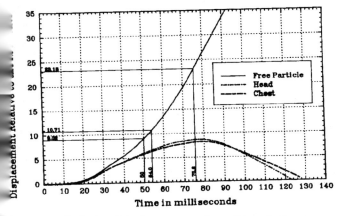

Figure 2. Relative Displacements in a 35 mph Barrier Crash.

The purpose of the restraint system is to intervene early, before the velocity debt becomes unmanageable, and to initiate a repayment program in the form of occupant accelerations - or decelerations (negative accelerations), as the case may be. An early start reduces the debt that has to be paid back, and at the same time, a good restraint system will tend to lengthen the payback period (to perhaps 150 milliseconds or more). If the crash is not excessively severe, the restraint system can keep the debt payments down to a level the human body can tolerate. The action of the restraint system in meting out accelerations to the occupant while the vehicle itself is still accelerating is known as "ride down."

The earliest restraints tended to be lap belts, which were thought to promote occupant retention within the vehicle, but which obviously could not provide much restraint for the upper body during severe frontal crashes. In the U.S., the usage of such systems was disappointingly low, and the prospects for shoulder belt usage were even more discouraging [DOT 78]. Thus, in the late '60s, air bags were seen as forcible intervention (debt management) for the vast majority occupants who would otherwise be unrestrained. This forcible intervention became known as "passive protection," and airbag requirements were promulgated by the NHTSA with unrestrained occupants in mind. The U.S. thus embarked on a technological odyssey in which the occupant protection standard, FMVSS 208, did not require dynamic (crash) testing of the most effective restraint system then and now available - seat belts. Meanwhile, the airbag designer was saddled with the task of protecting those who failed to use their available seat belts, and who would presently be in violation of the law in 49 states.

DEVELOPMENT OF THE AIRBAG CONCEPT

In this historical context, it is easy to see how the concept of air bags came into being: the forcible intervention would come in the form of an inflatable bag which would leap into the gap between the occupant and the vehicle interior. Once there, the bag's internal pressure would be uniformly applied to the occupant over a large area, not only meting out accelerations in controlled doses, but minimizing the stress on the occupant by distributing restraint the forces over as much area as possible. The result would be a more gradual pay down of the velocity debt, ideally with a minimum interest charge (in other words, rebound velocity). Unfortunately, the question of how to accomplish all this would not be a trivial one.

Of course, bags could not be pre-inflated because of the possibility of actually causing an accident. Since inflation had to begin after the crash started, the first step was to determine whether a bag could be inflated quickly enough. Figure 3 shows the time budget used by General Motors in their airbag vehicles of the mid-70s. These were very large sedans, so the time budget for today's vehicles would tend to be shorter, depending on such variables as the vehicle size, object struck, structural engagement, etc. In actual fact, the air bag must be inflated in a considerably shorter time, typically 25 to 30 milliseconds, because at the beginning of the event, the crash detection system requires a certain amount of time to do its work.

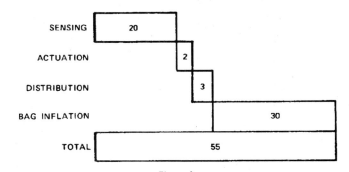

Figure 3. Typical Time Budget for Large Cars -GM

In addition to filling the air bag quickly, the inflation gas itself should not present hazards to the occupants or any one else involved with the vehicle from production to salvage. High-pressure air, stored in a tank and released with a quick-action valve, probably seemed an obvious choice to early airbag developers.

DRIVER-SIDE AIRBAG SYSTEMS

Design of a driver-side airbag system is heavily influenced by the available volume in the steering wheel hub. This volume is entirely insufficient to house a high-pressure tank, stowed air bag, and other hardware. Consequently, the only way to implement a stored-gas system on the driver side was to locate the tank elsewhere, and pipe the gas up the steering column. A prototype system, in which a single bottle inflated both driver and passenger systems, was developed by Volkswagen for its Type I Beetle, but without satisfactory results [Seiffert 72]. At Ford, after a long development effort to achieve ade-

with the automotive supplier Eaton, Yale, and Towne to develop a system for the right front passenger only [Frey 70]. The justification for this approach was that the driver was already protected by the combination of compressible steering column, padded hub, and yielding rim; it was the right front passenger who had the greatest need of an air bag. (On the other hand, there is almost always a driver present, whereas this seat is occupied only about 40 percent of the time, reducing the available benefit there.) In any case, a fleet of 831 1972 Mercury Montereys was fitted with passenger-side air bags for field testing, of which 126 were delivered to the U.S. Government.

SOLID PYROTECHNIC INFLATORS – In the meantime, General Motors and Chrysler were trying to find a gas source that could be fitted in the steering wheel hub. Chrysler's efforts utilized smokeless powder, but did not result in hardware being integrated into vehicles. GM, on the other hand, developed a driver system using the combustion of sodium azide to produce inflation gas. Sodium azide is a highly toxic and unstable compound and thus requires special care in production and disposal, but it has chemical reaction characteristics suitable for inflating air bags, and its primary combustion product is harmless nitrogen gas. Sodium azide has thus been the main ingredient of gas generants for driver's systems from the earliest times until the present, although it seems destined to be replaced by more environment-friendly materials.

Gas generant comes in solid form and is like a solid rocket propellant, in that the chemical reaction rate at any instant depends on the exposed surface area and the temperatures created by the reaction itself. These are controlled by the size, shape, and sodium azide content of the "grain" -- a rocket propellant term. Typically, the grain is about 60 per cent sodium azide, it is packaged in pellet form, and there is 75 to 100 grams of it in a driver's system.

The sodium azide pellets are part of a unit known as the inflator. The inflator housing is hermetically sealed to keep the pellets isolated and protected from vandalism until they are needed. Holes around the circumference of the housing allow the nitrogen gas to escape and provide axisymmetric bag filling, without creating any net thrust (for safety during handling). The inflator also contains a labyrinth of screens and perhaps baffles, to filter out particulates and to cool the hot nitrogen gas. Finally, the inflator contains a squib with electrical connections, so that the ignition of the pellets may be started. Figure 4 shows an inflator produced by Morton Thiokol.

AIRBAG MODULE – The inflator is part of a larger assembly called the airbag module, or simply module, which is the unit that actually gets installed in the vehicle. In addition to the inflator, the module contains the stowed air bag, which is securely fastened to the side of the module away from the occupant, so as to contain the pressure of the nitrogen gas when inflated. The exterior

surfaces of the module include the bag cover, which is typically plastic. The cover has molded-in lines where the material is weaker, allowing it to be split by the pressure of the inflating bag. The cover opens up, typically like petals of a flower, allowing the air bag to unfurl. A typical airbag module, installed in an airbag steering wheel, appears in Figure 5.

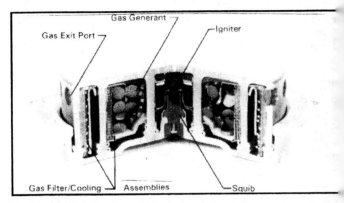

Figure 4. Typical Driver System Gas Generator - Morton Thiokol

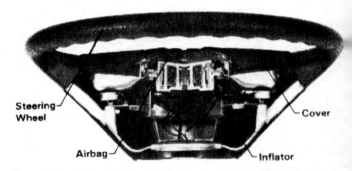

Figure 5. Driver Airbag Inflator and Module Assembly.

Clearly, the airbag module occupies precious real estate in the steering wheel hub, and significantly increases its mass. The volume must be kept to a minimum so as to avoid blocking the driver's view of the instrument cluster, or being in the way during emergency steering maneuvers. The added mass adds to the rotational inertia of the steering wheel; since this could adversely affect steering system returnability, the module mass must also be minimized. These factors have fostered the development of lighter and smaller inflators, and the use of thinner airbag material (so that the stowed air bag can be smaller and lighter).

The airbag cover must be durable, since it is often contacted by the driver. Since it is one of the most prominent features of the driver's station, it is important to the visual appeal of the vehicle, so the bag cover must also be attractive. It must protect the air bag from moisture, spilled liquids, etc. Most importantly, it must not impede the unfurling of the air bag, nor pose a hazard to vehicle occupants during deployment.

ELECTRICAL CONNECTIONS – Where the steering

otion between the airbag module electrical leads and e stationary wiring harness in the column. In the first-eneration GM system, electrical contact throughout the eering wheel motion was maintained via a special slip ng assembly. Redundant electrical contacts were used handle threats to reliability posed by electrical noise ue to friction. In the 1985 Ford Tempo/Topaz, and in any later designs, a spiral wire (like the main spring in a ock), having enough travel to accommodate steering heel turns from lock to lock, was used. A down side of e clock spring design is that repair personnel must be ure it is properly "rewound" when a steering wheel is eplaced.

TEERING WHEEL – The steering wheel is not only the aunch pad" for the air bag; it (and to a lesser extent, the indshield) also serves as the reaction surface. This ter-ninology means that restraint forces developed in the air ag itself have to go somewhere, and the primary load ath is through the steering wheel and into the column. onsequently, the design of an airbag wheel goes eyond merely accommodating the volume, mass, and lectrical connections of the module. The strength and tiffness of the spokes and rim must be sufficient to pro-ide a stable reaction surface, but as yielding as possible n non-deployment accidents involving unbelted occu-ants. The General Motors first-generation airbag wheel /as the result of a considerable development effort, and eems to have served as the point of departure for sub-equent airbag steering wheel designs.

TEERING COLUMN – In general, steering columns are esigned to limit rearward displacement (relative to the ompartment) in frontal crashes, as regulated by FMVSS 04, and they are also designed to limit the forces due to ccupant contact, by being able to stroke forward in the vent of such a contact. In this sense, the steering col-mn is part of the restraint system, even in a non-airbag ehicle.

n an airbag vehicle, however, contact is deliberately nade with the driver. If the driver is unbelted (as was assumed to be the case in the early 18

70s), virtually all of the upper body restraint forces pass hrough the column. A conventional column might stroke orward under these conditions, but at such a low force hat the stroking element reaches the end of its available ravel and abruptly "bottoms out" against some mechani-cal limit. An airbag column, on the other hand, is designed to move forward in a controlled fashion, which means that its inertia, and static friction in the column, must be dealt with carefully. Typically, this is done by starting the static stroke vs. force characteristic at a reduced level (perhaps by "pre-stroking" the energy-absorption unit), and then increasing the static force as stroke progresses. This generally means a redesign of the stroking element, and a strengthening of the cowl structure and bracketry through which the column passes

the loads on to the rest of the vehicle. GM's structural modifications are shown in Figure 6.

The legacy of research done at GM in the mid- to late-'60s on energy-absorbing steering columns is clearly seen in the design philosophy of their first-generation air-bag systems. The steering column was the primary energy-absorbing element; the air bag's function was to leap into the gap between the driver and the steering wheel, couple the driver to the mechanical energy-absorbing elements in the column, and provide a more uniform application of restraint loads on the occupant's upper body.

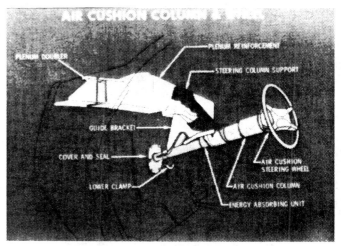

Figure 6. Structural Modifications Required for Driver System - GM

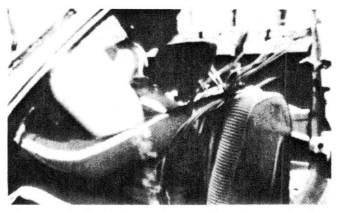

Figure 7. Belted Occupant Kinematics - Mercedes-Benz System

The column is aided and abetted in performing this duty if the torso is perpendicular to it. If a driver has a lap belt, the lap belt will limit the forward stroke of the pelvis and cause the upper body to pivot forward, thus helping to achieve the desired alignment. See Figure 7. For unbelted drivers, however, this does not occur; the whole body tends to translate forward with little articulation until interior contacts are made (typically, with the knees first). In this situation, we may well have a mostly erect torso moving straight forward toward an inclined steering wheel. As seen in Figure 8, a GM illustration of its first-generation airbag system, this geometry can cause the air bag to develop a wedge shape in side view. More importantly, it can cause a significant upward force component to be applied to the end of the column. See Figure 9.

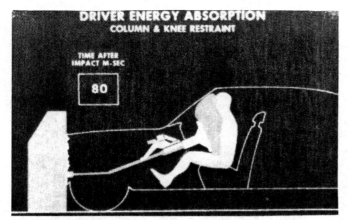

Figure 8. Unbelted Driver Kinematics - GM System

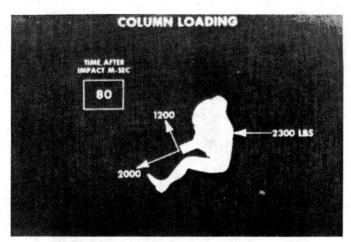

Figure 9. Column Loading - First-Generation GM System

These (off-axis) up-loads considerably increase the force levels required to stroke the column. The situation is rather like trying to close a chest of drawers by pushing on one side of the drawer. It may "jam," and resist closing altogether. To overcome such difficulties, one tends to design the stroking element for greater lateral stiffness, and less sensitivity to the minor misalignments that result from off-axis loads. An example of such a design is the steering column found in the Minicars Research Safety Vehicle (RSV), shown in Figure 10 [Struble 79]. Despite the lack of front-seat belt restraints, this system achieved dummy injury measures well below the FMVSS 208 criteria set for 30 mph, at delta-Vs in excess of 50 mph. At these speeds, one needed to use all the interior occupant stroking space that one could get. Reducing the articulation of the body segments (recall the lack of a lap belt) tended to keep the head away from the windshield and header, and was helpful in increasing the available stroking distance.

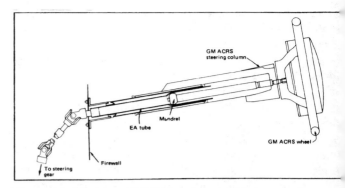

Figure 10. Minicars RSV Wheel and Column

Another approach to dealing with column up-loads is to reduce the angle between the torso and the steering wheel. For belted occupants, this is achieved by the lap belt, as described above. For unbelted occupants, particularly in the extreme crash conditions addressed in the RSV program, the need to reduce the angle resulted in a steering column rather more horizontal than most. Of course, such architecture causes the steering shaft to penetrate the dash panel at a water line much higher than the steering gear, so an intermediate shaft with double U-joints was required to make the connection. It is unknown how most drivers would have liked the more vertical steering wheel, or the closer positioning to the sternum.

Air bag – A third approach to up-load difficulties is to reduce the dependence on column stroking, which brings us to the design of the air bag itself. The first-generation GM air bag was about 22 inches in diameter when deployed, and extended about 10 inches rearward of the wheel rim [Campbell 72]. It was pressurized to approximately 3 psi, absent occupant loading, and was about 2.8 cubic feet in volume. This was larger than most current designs, such as the Mercedes-Benz bag shown in Figure 11, which tend to run about two cubic feet, or 60 liters, in volume.

11. Inflated Mercedes-Benz Air Bag

1246

he GM bag material itself was neoprene-coated nylon. he neoprene coating served to reduce the porosity of e bag, thus extending the duration of bag inflation. It as discovered, however, that the bag tended to act as a neumatic spring, storing energy as the occupant stroked rward into it, but then returning the energy to the occu- nt later [Klove 72]. This resulted in undesirable mounts of occupant rebound into the seat back and ead restraint. To counteract this tendency, the neoprene oating was subsequently removed; forcing gas through e fabric pores increased the energy dissipated (as pposed to energy being stored and subsequently eturned).

ubsequent bag designs have included actual holes, or nts, which took on a larger share of the task of absorb- g occupant energy. The vents are generally placed on e back side of the bag, away from the occupant, and re so located as to avoid blockage (by the steering heel spokes, instrument cluster brow, windshield, etc.). igure 12 is typical. The total vent area is determined uring the development process so as to provide the best ompromise among the disparate demands being made n the restraint system.

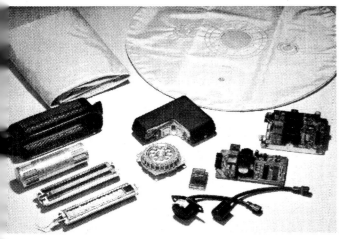

Figure 12. Typical Components, Including Vented Air Bag - Takata

As production volumes have increased, the bag has become perhaps the most labor-intensive component to manufacture and inspect. Typically, it has at least two major sections, which have to be sewn together in the tra- ditional manner, by a worker at a sewing machine. In addition, reinforcements have to be added at the vent noles, the edge or edges where it is attached to the mod- ule, and at the tethers (if any). More recent develop- ments indicate that bags may be woven, by machine, in a single piece.

Tethers are straps that connect the front and back sur- faces of the bag. They have become more widespread in recent years, and their purpose is to limit the travel of the bag front during deployment. For small occupants seated close to the steering wheel, tethers can reduce the occupant accelerations generated when the bag front contacts the sternum. Tethers also increase the bag's

aspect ratio (diameter divided by depth), improving the ability of the bag to protect occupants who load one side of the bag more than the other (due to angularity of impact or being out of position, for example).

The air bag used in the Minicars RSV had two chambers, as indicated in Figure 13 [Struble 79]. The inner cham- ber was connected directly to the inflator and filled first. Its relatively small size (28 l, or 1.0 cubic feet) provided a quick fill time, so it could jump into the gap between the steering wheel and the chest as soon as possible. It was aimed directly at the sternum, and this geometry was nstrumental in establishing the column angle. The chest oag was vented to the outer bag (48 l, or 1.7 cubic feet), which filled more slowly and provided restraint to the head. The timing of the chest and head bag inflation could be tuned somewhat by the size of the vent in the chest bag. Once the gas had been re-used in the head bag, it was vented to the atmosphere.

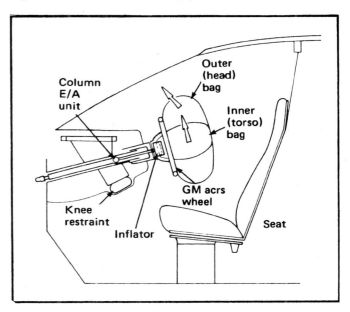

Figure 13. Minicars RSV Driver Airbag System

While the air bag inflates faster than the blink of an eye, a look at high-speed films will reveal that it is hardly instan- taneous. Indeed, there is an inflation sequence, and a review of a bag pressure time history, as seen in Figure 14, will reveal some distinct phases. First, there is a rela- tively high pressure spike of relatively short duration, associated with inflation gas being pumped into an extremely confined space behind the folded air bag. The bag is pressed against the bag cover with sufficient force to split the cover seams, and the cover opens. In very short order, there is a volume increase behind the air bag - a high percentage increase because the volume was so low to start with. Consequently, the bag pressure drops, typically to zero. This is known as the punch-out phase, because the bag is punching out through the airbag cover [Lau 93]. In the test from which Figure 14 was derived, the occupant was not close enough to interfere with bag deployment, and the punch-out phase lasted

out of the way to achieve fill, both the peak pressure and the duration of the punch-out phase can increase.

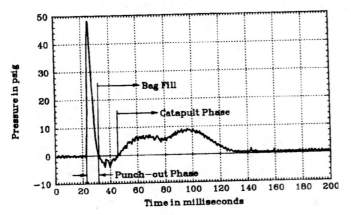

Figure 14. Airbag Pressure-Time Curve

The end of the punch-out phase marks the beginning of the bag fill phase. Now the airbag material has some velocity, and hence some momentum, as the cover doors swing open. Thus the fabric keeps moving, and as a result a negative gage pressure is created in the bag. During this time, the bag is seen in high speed films to be unfolding rapidly, with numerous sharp creases in the fabric. Bag unfurling motions are highly complex and three-dimensional, and the pattern depends on the folding process. The lateral portions of the bag may even appear to be sucked in. Of course, the inflator is continuing to produce gas throughout this time.

Finally, the bag material reaches the geometric limits of its travel, and the pressure climbs back up through zero. This is the earliest point at which bag pressure is available as an occupant restraint mechanism (just over 45 msec in Figure 14). In very short order, the creases come out of the material, and the bag assumes its inflated shape. The actual time budget used in the design for the system of Figure 14 is unknown, but it appears that 50 msec would have been an attainable goal.

AIRBAG PERFORMANCE – Airbag design is driven by a number of considerations, among which are the usual villains of cost, weight, and size. Occupant protection performance is dictated by the requirements of FMVSS 208, which has specified that dummy injury measures for the head, chest, and femurs be within certain limits for 30 mph frontal barrier crashes at any horizontal angle up to plus or minus 30 degrees from vehicle center line. FMVSS 208 has required that tests be run with the dummy occupants, representing 50th percentile males, unbelted. Generally, the angled barrier crashes involve longer stopping distances and softer crash pulses (vehicle acceleration versus time) and in that sense are less severe, but they do pose potential difficulties associated with the occupant moving into the bag at an angle. Another crash condition is the 35 mph frontal barrier test used in NHTSA's New Car Assessment Program (NCAP), in which the dummy occupants are belted. This

test is not required by the safety standards, but it has nevertheless become a de facto design requirement. These dichotomous test conditions mean that a single bullet (i.e., one system design) has to be fired at two targets. It is thus not surprising to find, in the earlier airbag designs NCAP-tested at 35 mph, that the FMVSS 208 injury criteria were exceeded. In fact, while some vehicles equipped with belts only have been meeting all the 208 criteria since the beginning of the 35-mph NCAP tests, it was not until 1988 that an airbag-equipped vehicle did so.

LOWER BODY RESTRAINT – The lack of a lap belt in the 30 mph test means that lower body restraint has to be provided by other means, typically by resisting the forward movement of the knees. The hardware involved is variously known as a knee bolster, knee restraint, or knee blocker. In any case, the design concept involves the knees engaging a deformable structure which limits knee movement to some extent, while maintaining femur loads within acceptable limits. These femur loads are transmitted to the pelvis, providing pelvic restraint. Attention has to be paid to the possibility of knee contact with the steering column, particularly during angular impacts.

Needless to say, the action of the knee restraint, and its effect on occupant kinematics, depends on the initial spacing between the knees and the restraint. Generally, the designer would like the spacing to be small, but care must be taken not to interfere with the operation of the foot pedals -- particularly the brake pedal. On the other hand, the need to reach the pedals tends to cause the driver to position the seat so that the knees are placed at a fairly uniform distance from the knee bolster, regardless of occupant size.

In a 35 mph NCAP test, pelvic restraint provided by the knee bolster, when combined with a lap belt, could be excessive. One option for dealing with this is to sew in one or more loops in the lap belt, which can pull out at a force level sufficient to provide occupant retention in non-frontal accidents, but allow enough pelvic motion to avoid unacceptable occupant kinematics. Webbing material with varying stretch characteristics can also be chosen. It's a bit of a balancing act.

UPPER BODY RESTRAINT – Similarly, the non-use of the torso belt in the FMVSS 208 test results in all the upper-body restraint being provided, in that test, by the air bag. The safety standard therefore effectively establishes the force and stroke requirements for the air bag, which may not be optimal for smaller occupants seated closer to the bag (see below). Even for a normally-seated 50th percentile male, the addition of a shoulder belt in the 35 mph NCAP test may mean an excessive amount of restraint force in that test, and indeed this could be the cause of the chest accelerations exceeding 60 Gs in NCAP tests of the earlier airbag cars. One approach to dealing with this situation is to adjust the timing of restraint forces from the air bag and the belt so that

n insight into this timing is provided in Figure 15, which ows the timing of chest accelerations and torso belt ads. We see that the belt loads reach their peak at out 50 msec, which closely corresponds to time bud- eted for airbag inflation that was mentioned earlier. ter 50 msec, we see the torso belt loads falling off while e chest accelerations continue more or less level until out 80 msec. The likely explanation is that the timing the belts and airbag inflation has been adjusted so that e air bag picks up where the torso belt leaves off, in rms of providing upper torso restraint.

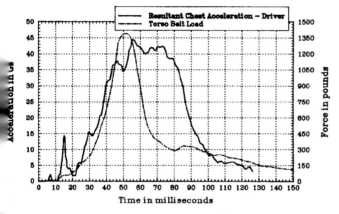

Figure 15. Belt Load and Chest Acceleration in 35 mph Barrier Test - 1994 Volvo 850

ARIABLES AFFECTING PERFORMANCE LIMITS – here are other variables not addressed by government standards or tests. Factors considered by the manufac- urers probably vary, but some that come to mind are occupant size (5th percentile female through 95th per- centile male), distance between the airbag cover and the sternum, and object struck (e.g., pole, offset barrier, etc.). Generally, protection of larger occupants is stroke-limited, which is to say that as crash severity is increased, some specified injury criteria limit is reached when the occu- pant stroke becomes excessive, resulting in bottoming out of the air bag or steering column, or contact with the interior. Protection of smaller occupants tends to be acceleration-limited. This is because restraint systems have to generate enough force to protect the many larger occupants in the population. In the same crash, these same force levels applied to smaller occupants will result in larger occupant accelerations. Smaller occupants may thus reach acceleration limit values at lower crash severi- ties than larger occupants will.

When the object struck is not a barrier, the crash pulse may be softer, but compartment intrusion may also be greater than in a barrier crash at the same speed, due to the concentration of crash forces on only a part of the structure. This could be reflected in displacement or rotation of the steering column, which could affect the "aim" of the airbag restraint forces, and the resulting occupant kinematics. It could also affect the sensing time, possibly producing a later deployment. This is a

function of the sensor system design, to be discussed later.

If the driver is sitting closer to the air bag than in the design condition, contact between the deploying bag front and the occupant can occur earlier, and at a higher contact velocity. This could increase the accelerations experienced by the occupant, particularly if he or she is small.

DESIGN PARAMETERS – To deal with all these require- ments, some of them in conflict with one another, the designer and developer of airbag systems has a number of parameters to work with. The crash pulse is an impor- tant one, albeit one that the airbag engineer may have lit- tle control of. Another one is the inflator characteristic, which is generally expressed in terms of the time histo- ries of pressure when the inflator is discharged into a fixed, closed volume (the so-called "tank test"). Typical tank test results are shown in Figure 16. Other variables include the column stroke characteristic (force vs. dis- tance), steering wheel location, airbag vent area, bag vol- ume, bag diameter, tether length, seat belt anchorage points, webbing stretch characteristics, and seat cushion stiffness.

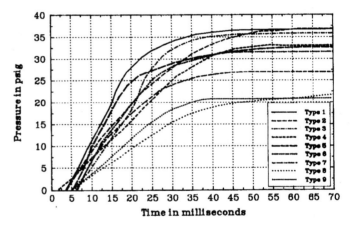

Figure 16. Tank Test Curves for Morton-Thiokol Inflators - 60 l, 22°C

PASSENGER-SIDE AIRBAG SYSTEMS

Perhaps the feature that most consistently distinguishes passenger-side systems from their driver-side counter- parts is the location of the launch pad: driver systems are mounted on the steering wheel, and passenger systems are mounted on, and part of, the instrument panel. This distinction causes a significant difference in the longitudi- nal distance between the airbag mount and the occu- pant's sternum. Moreover, the passenger side lacks a stroking element, like the steering column, that could be used for absorbing energy.

One approach to transferring driver's side airbag technol- ogy to the passenger's side would be to blur these dis- tinctions. In other words, move the launch pad aft and mount it on a stroking element. This concept led directly a design using a "bag-bolster" positioned rearward about even with the steering wheel. A bag-bolster system was sled-tested at Calspan [Romeo 75] at 75 kph (47 mph), but concerns remained regarding public acceptance of the appearance and ease of ingress and egress of such a design. Subsequent concepts for a passenger-side air bag have avoided significant changes in the instrument panel location.

In the Minicars RSV program, the objectives for the right front passenger system included the ability to undergo a crash at 50 mph delta-V, and still provide occupant protection within the 30 mph FMVSS 208 criteria. Originally, it was thought that a stroking dash (albeit conventionally-positioned) would be required to meet such ambitious goals. However, it turned out that the performance goals could be achieved, with room to spare, using venting alone. All other passenger-side airbag systems, as far as is known, have similarly relied upon venting for energy absorption.

BAG GEOMETRY – In the development process, it may be tempting to extrapolate from a driver side system by starting with a deeper (in the longitudinal direction) version of a driver bag, mounted on the instrument panel. This approach would not be valid because the aspect ratio of such a design would not be nearly high enough; the bag would buckle or be pushed aside by the occupant's motion. The bag needs to be wider (laterally) and taller (vertically) for two reasons: (1) to avoid instability (buckling) when loaded in compression, and (2) to handle the greater lateral deviation in an occupant's path during an angular collision, due to the occupant being farther from the bag when the crash event starts. This is true even when there is not a designated center seated position in the front seat. If there is a center position, and air-bag protection is provided for that occupant, the bag must be wider still.

These considerations lead to a considerably larger air bag. The first-generation GM bag, designed to protect both center and right front occupants, had a volume of about 14 cubic feet. Most present bags are much smaller. The inflated shape is typically a lateral cylinder with vertical ends, but the cross-section is not necessarily circular. See Figure 17.

The Minicars RSV passenger-side air bag was dual-chambered, for reasons that can be understood by comparing it to its counterpart on the driver's side. Refer to Figure 18. In this design, the inflator emptied directly into the lower, or torso bag, where earlier application of

restraint forces was needed. This bag had a volume of about 2.75 cubic feet (78 l). The torso bag was vented to the upper, or head bag, to pick up the head somewhat later. The volume of the head bag was about 3.0 cubic feet (85 l). Gas from the head bag was then vented to the atmosphere.

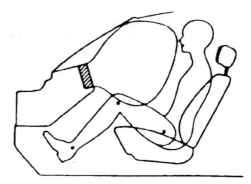

Figure 17. Mid-Mount Airbag Configuration

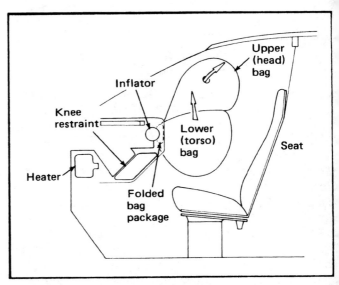

Figure 18. Minicars RSV Dual-Chambered Passenger Airbag System

The first-generation GM system also had a dual-chambered bag, but for entirely different reasons. One chamber was intended specifically for knee restraint. Since knee restraint loads are more concentrated than, say, the loads applied to the chest or the head, the knee bag operated at a much higher pressure. Due to its smaller size, it filled first. The other chamber, being larger, filled relatively more slowly and provided restraint to the torso and head. See Figure 19.

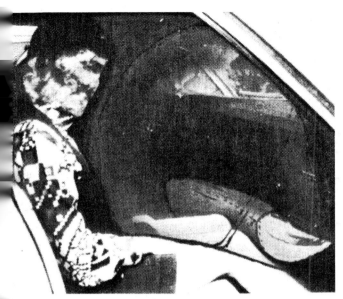

Figure 19. First-Generation GM Low-Mount Airbag with Knee Bag

TYPES OF AIRBAG MOUNTS – Compared to driver systems, passenger air bags have many more design options regarding their integration with the vehicle. The location and orientation of the module gives rise to some terminology regarding the type of mount. At the time it was designed, the Minicars RSV air bag was called a high-mount system, but as indicated in Figure 17, it would be called a mid-mount configuration in today's parlance. The airbag module is located on the aft face of the instrument panel, about where a glove box might traditionally be located. In this position, the air bag cannot, and is not intended to, provide lower-body restraint. The Minicars system was typical of many in that it was designed to be tested without the use of lap belts. Thus the lower-body restraint had to be provided by other means, such as a knee bolster. The module and the knee bolster make it very difficult to provide a traditional glove box in the instrument panel. Obviously, there is also a significant impact on the location and routing of heating, ventilation, air conditioning, and other components in the instrument panel. The bolster itself is similar in concept to those used on the driver's side, except that the designer does not have to deal with potentially hard contact surfaces presented by the steering column.

In another contrast to the driver's side, there is no requirement for the passenger to reach the pedals. For bench-seat vehicles, the passenger seat position may be determined by the driver. With bucket seats, the position may be a function of who is seated in the right rear seat, and how big they are. Variability in the size of the right front passenger provides (literally) another dimension. In any event, one can expect considerably more variation in the knee-to-knee-bolster distance at the beginning of the crash. A small occupant, seated far from the bag, could tend to submarine under the bag.

General Motors, in the design of their mid-1970s system, used an inflatable knee restraint to provide added toler-

ance to such variations. (Recall the very large size of the vehicles involved.) To effect the design, the airbag module was located lower on the instrument panel in what has come to be known as a low-mount configuration, illustrated in Figure 20.

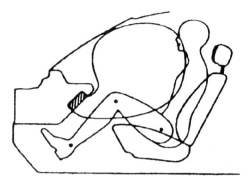

Figure 20. Low-Mount Airbag Configuration

A third design is known as a top-mount (or dash-top) configuration. In this system, the deploying bag is not directed aft at the chest or down at the knees, but rather upward toward the windshield. The airbag cover is typically on the top of the instrument panel, as seen in Figure 21. Top-mount designs were proposed in the early '70s, but were not used in either the GM or Minicars systems. More recently, this configuration has become widely used, for reasons to be discussed below.

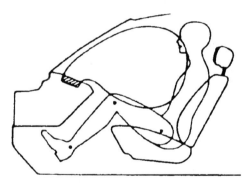

Figure 21. Top-Mount or Top-Dash Configuration

HOW RESTRAINT FORCES ARE DEVELOPED – When the occupant presses against an air bag, the pressure in the bag is transmitted directly across the layer of fabric, onto the occupant's body. Restraint forces are thus generated, but it would be a mistake to attribute all such forces to this mechanism (gas pressure).

Consider, for example, that with a 50th percentile male dummy and the seat in its middle position, there may be 560 mm (22 in) between the chest and the instrument panel. If the air bag is to fill this gap in 25 msec (say), the deploying air bag surface must move rearward at an average speed of at least 50 mph. Of course, the instantaneous speed varies, so one would expect the peak to be much higher. Indeed, film analysis of the deployment of various driver-side systems showed peak speeds ranging from about 100 mph to over 200 mph [Kossar 92]. Speeds at the time of facial contact were lower, of course,

When the air bag contacts the occupant, there is a momentum transfer between the two that depends on the portion of the air bag's mass brought to rest against the occupant, its velocity at contact, and the portion of the occupant's mass involved in the contact. This phenomenon is called bag slap, and can generate restraint forces when the bag pressure is low or even negative. See Figure 22. If multiple layers of fabric are involved (as when the bag is still partly folded, for example) and the brunt of the impact is taken by just the sternum or the head, the increased effective mass of the bag and the reduced effective mass of the occupant will combine to produce higher occupant accelerations. Obviously, if the occupant is initially positioned closer to the air bag, contact occurs sooner, earlier in the bag unfolding sequence, and possibly at a higher contact velocity.

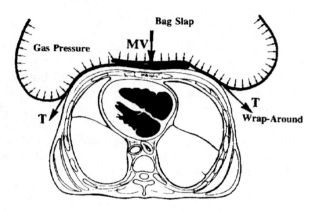

Figure 22. Sources of Restraint Forces

Clearly, bag slap is a greater concern for occupants having smaller mass, seated closer to the bag, and so positioned that the deployment forces are directed higher on the body.

Once momentum has been transferred to the occupant, the bag slap phase of the restraint process gives way to the catapult phase, wherein the occupant and aft bag material move together. As the occupant strokes into the bag, the pressure builds and generates restraint forces, as mentioned before. At the same time, membrane tension is created in the bag fabric, which is partially wrapped around the occupant. Because of occupant penetration, the bag tension has a rearward component at the locations where the fabric and the occupant cease to be in contact. These rearward components, or wrap-around forces, contribute significantly to restraint action during the catapult phase. This is particularly true on the passenger side, because the bag tends to be wider and deeper.

EFFECTS OF OCCUPANT SIZE AND POSITION – Since passengers don't operate foot pedals, they can choose to sit closer to the instrument panel (or farther away, of course) than they might on the driver side. In fact, the position of a bench seat may be controlled by the driver; with bucket seats, the position might depend more on the presence and size of a rear occupant than any-

thing else. At the same time, the front passenger doesn't have to be of driving age. One finds, therefore, a much wider range of occupant sizes, weights, ages, and seat positions on the passenger side.

If the occupant is short in stature, the bag may be directed more at the head and neck, and less at the torso. In this way, the relatively small effective mass of a child, for example, could be reduced still further. If the membrane tension is sufficiently high, coupled with a small effective mass, the bag fabric may actually snap taut, propelling the head rearward.

If restraint performance is optimized for a normally seated 50th percentile male with the seat in mid-position, one must expect some degradation of performance when any, or some combination of, these conditions is varied. Clearly, a system acting as if an unbelted 160-pound occupant is present is going to be much too powerful to be optimal for a child weighing a fifth as much. While the Minicars RSV system may have presented the most potential for occupant protection at the highest severities, at the same time it may also have posed a higher risk to small and out-of-position occupants. Short of tailoring the restraint system for the conditions (see below), the best one can do is make the greatest accommodation for the occupants most at risk (e.g., children, small adults, the elderly), while keeping performance for the "nominal" occupant and seat position within the limits imposed by Government testing.

Within a given airbag configuration, one can vary such parameters as the inflator charge, the angle and position at which the inflator is mounted, the fabric weight and vent area, and the folding pattern. Obviously, lighter fabric weight tends to reduce bag slap and stowage volume. Typically, the folding pattern has been found to be very important in addressing the needs of small or out-of-position occupants.

Another approach tried in the '70s was the so-called aspirated system [Katter 75]. This design concept stemmed, at least in part, from early concerns regarding overpressure in the compartment due to bag deployment. (Consider, for example, that the U.S. ESV family sedan designs of the early '70s had air bags in both front and rear seats [Alexander 74].) It was thought that overpressure might be alleviated if a supersonic ejector concept could be adapted to pump compartment air into the air bag. Such devices worked through viscous mixing, in a diffuser, of the air streams from the primary source (the inflator) and the secondary source (the compartment). The secondary flow continued until the bag pressure became high enough to stall the diffuser, at which time a check valve in the secondary air stream must close. Since the occurrence of stall depended on what the bag contacted, the system would naturally adjust the degree and rapidity of fill if the occupant were closer to the bag. Another advantage was the reduced inflation requirements for the inflator.

s it turned out, bag folding technique had a much larger influence than aspiration on the results for out-of-position children [Romeo 78]. As of this writing, there has been no known further development of aspirated systems.

TYPES OF INFLATORS – The volume of a passenger-side air bag is much larger than that of the driver-side bag, so one might expect a passenger-side inflator to need proportionately more gas generant. Of course, the packaging constraints are altogether different, meaning that there is little incentive to make a passenger-side inflator look much its driver-side counterpart. Rather, it is typically a circular cylinder, mounted so that its long axis lateral to the vehicle. As with a driver-side inflator, it typically has a hermetically-sealed metal housing, for the same reasons. Generally, nitrogen gas passes through a complex of screens and perhaps baffles, and exits through a series of holes or slots. These openings are evenly distributed from one end of the cylinder to the other, so as to provide an even fill across the width of the air bag.

Of course, other inflator configurations are possible. In the late '70s, when the inflator business was at its nadir, Calspan Corporation developed a design using two ganged driver inflators to fill a passenger bag [Romeo 78]. The primary motivation was the lack of a production passenger inflator, but utilizing two driver units with a common manifold has certain other advantages, discussed below.

Finally, we return to the type of inflator thought of first -- stored gas. Packaging constraints permit the storage bottle to be sufficiently close to the air bag to allow rapid inflation, but the same constraints cause the pressure in the bottle to be on the order of 3000 pounds per square inch. (The smaller the bottle, the higher the pressure required to store an amount of gas sufficient to inflate the bag.) Such pressures raise concerns about leakage (and thus reliability), when one considers that the system must remain absolutely leak-proof for perhaps 20 years. Nevertheless, the mid-70s passenger side systems by General Motors utilized stored-gas inflators, and successful deployments have occurred in cars of relatively advanced age.

The passenger-side module, shown in Figure 23, employed a membrane that was pierced to start the inflation process, and a manifold to distribute the inflation gas to the air bag. A characteristic of stored-gas systems is that the pressure in the tank is at its highest at time zero; as gas escapes from the tank (at the speed of sound), the remaining volume expands adiabatically (without heat loss), and as it does, the temperature drops. At the same time, the tank pressure drops.

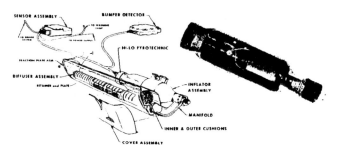

Figure 23. First-Generation GM Passenger Air Bag, with Inflator

The higher the storage pressure and the smaller the tank volume, the more steeply the pressure declines. At the same time, the peak noise level increases [Jones 71]. Questions have been raised regarding hearing loss, but the concerns seem to have subsided with the advent of pyrotechnic inflators on the passenger side. This is because pyrotechnic systems provide a more even gas flow, which reduces to some extent the concern about inflation noise. Because the gas generant grain can be modified to some extent without changing the inflator housing, the gas flow is more readily tailored or adjusted than with a stored-gas system.

At the same time, the combustion temperature in a pyrotechnic inflator is much higher than ambient, such that ambient temperature variations do not have much effect on gas generation. Stored-gas systems are thought to be more sensitive in this regard.

Since the inflation gas expands as it passes through the opening and into the air bag, it cools (and in fact gets very cold). If the gas pressure within the bag were insufficient, it could be increased by heating the gas. This thought gives rise to the concept of an "augmented" or "hybrid" inflator, in which heat is applied to the gas on its way to the air bag.

The GM passenger-side systems of the mid-1970 utilized this concept. A small charge of pyrotechnic materials could be ignited, not for the purpose of materially increasing the amount of inflation gas, but rather to add energy to the inflation gas by raising its temperature, and causing it to expand more. The increased temperature reduced the sensitivity to variations in ambient temperature [Seiffert 72].

DUAL-LEVEL INFLATORS – The engineers at Calspan were not the only ones to design a passenger system using two driver inflators. A 1979 paper described a Mercedes-Benz design in which "incremental deployment" of one or both inflators could advantageous in low-speed impacts, or for out-of-position occupants or children [Reidelbach 79]. The wording of the paper suggests that such a feature was not implemented, nor was it at Calspan. With GM's design, however, the choice could be made between a "low-level" deployment in which just the stored gas was released, or a "high-level" deployment which also involved the ignition of the augmented inflator charge.

Of course, to make a choice there must be some logic employed, and there must be sensors to provide the inputs. With the GM system, the choice involved the nature of the crash pulse. GM's design employed two impulse detectors on the bumper near the frame attachments, plus sensors on the dash panel. Generally, the trigger level was lower for the bumper units than for the dash panel sensors. Lower-speed crashes, underrides, center pole impacts, etc. might cause the lower-level units to trigger, but not necessitate a maximum-level inflation (as would be achieved by igniting the augmented pyrotechnic charge). Therefore, the logic was this: to release the stored gas if the lower-level sensors triggered, but to ignite the augmented charge only if the higher-level sensor triggered as well. It is noteworthy that the logic involved only the nature of the crash as experienced by the vehicle; no decisions were made on the basis of conditions inside the compartment.

FROM DUAL-LEVEL INFLATORS TO "SMART" SYSTEMS – The decision whether to have a high- or a low-level deployment could be made on the basis of factors other than the crash characteristics seen by the vehicle. For example, Is the seat occupied? If the seat is occupied, is the belt being worn? In this electronic age, we could envision a so-called "smart" system, in which the deployment logic could be based on detecting the presence of a child seat (particularly a rear-facing one), discriminating between humans and various objects in the seat, and/or detecting the size of the occupant, his or her proximity to the instrument panel, etc. Again, it is noteworthy that all considerations involve the occupant and conditions inside the passenger compartment, in contrast to the logic utilized in the first-generation General Motors air cushion system of the mid-70s.

The term "smart" has more recently acquired an official definition, thanks to the NHTSA. In a rule issued in late 1996 [NHTSA 96], an airbag system is considered "smart" if:

- It does not deploy if the mass on the seat is 30 kg or less.
- It does not deploy if a rear facing child seat or out-of-position occupant is present.
- It does deploy if a properly belted child is present and there is no risk of injury.

Obviously, the ability to adjust the air bag's deployment to these and other factors is highly dependent on sensor technology, which is still in a state of intensive development at this writing.

Among the competing sensor technologies are the following:

- Sensing occupant weight by detecting the seat cushion deflection. Such a device could be fooled by heavy objects, would not be sensitive to occupant position, and would require extensive cushion redesign to incorporate the sensor.
- Infrared sensors to measure body heat. Here, the challenge is to distinguish the body heat "signal" (about 37°C) from the high "noise" levels due to variables such as compartment temperature (which can vary from -20°C to +70°C), heated seats, and heat absorbing clothing.
- Detecting changes in capacitance due to an occupant. Such technology could detect occupant position, and sensors could be located in a variety of places, but at the same time they could be fooled by conductive materials, including moisture.
- Ultrasound could be used, as during pregnancy, and sensors could be located in a variety of places, but would be sensitive to temperature and humidity. Large objects could block sound waves.
- A semiconductor-based "seeing eye" could discriminate between the visual appearance of an occupied seat and an empty one, and thus could detect occupant presence and position. However, high resolution plus pattern recognition equals high cost.

All of these systems are "passive" in the sense that occupant does not have to (and indeed, cannot be expected to) wear a reflector or transponder to "talk back" to the sensor or sensors. An "active" system, by way of contrast, includes such a device, which makes the sensor's task easier and improves the quality of communication between sensor and object. This concept would be practical for specific hardware designed to fit on an automobile seat, such as a child seat. A rear-facing child seat could have a reflector or transponder which could cause the air bag to be depowered or deactivated if the child seat is placed in the front seat, despite warnings to the contrary. Alternatively, a reflector or transponder on a forward-facing child seat could provide quality information on the child seat's presence and distance from the airbag module. The problem, of course, is the installed base of millions of child seats not so equipped.

CRASH SENSORS

At the instant of contact, the vehicle will not yet "know" that it is in a crash. The best that can be done is to have sensors and associated electronics continually on "sentry duty," keeping track of velocities and accelerations, and looking for the telltale signs that a crash, and not just a hard bump, has begun. The sensor system will also have to distinguish the direction of the crash, and whether the crash severity warrants a deployment. This is a tall order and takes some time, called sensing time -- perhaps on the order of 20 to 25 milliseconds. What is left of the first 50 milliseconds can be devoted to filling the air bag.

DEPLOYMENT THRESHOLD – Clearly, there are circumstances in which airbag deployment is undesirable. In non-frontal impacts (e.g., side, rear, and rollovers), the injury hazards may not be amenable to reduction by an air bag. In any case, airbag deployment necessitates replacement, which may add considerably to the repair bill. In property-damage-only accidents, airbag replacement cost could represent a significant part of the total. If airbag deployment were perceived as unnecessary, and if the replacement cost were high, negative reaction could hinder the public acceptance of air bags.

Similarly, the injury potential may be low in low speed non-deployment accidents, particularly if the belts are being worn; at the same time, the energy of airbag deployment could actually increase the probability of injury. Therefore, for frontal impacts a deployment threshold is created. Below the threshold the air bag should never deploy, and above the threshold it always should. Of course, every crash has a different signature, so for design purposes the threshold is described in terms of barrier impacts. At one time, the Government was proposing a 15 mph barrier impact as the minimum to initiate airbag deployment [NHSB 70], probably with an eye to the 15 mph interior impact requirements of FMVSS 201 [FMVSS 201]. In any event, a requirement for deployment threshold was never promulgated, but similar considerations for protecting unrestrained occupants has tended to cause thresholds to be set at about 12 mph in frontal barrier impacts.

Of course, variability in crashes means that the threshold has to have some thickness. In other words, there has to be a "gray area" within which deployment may or may not occur. The specification for this threshold can, and probably does, vary with vehicle design, but generally the specification does reflect a gray area. The low end of the gray area, below which the bag should never deploy, may be set to 8 mph in a barrier crash, and the upper end, above which it should always deploy, may be 14 mph, again in a barrier crash. The sensor system is designed to respond only to the frontal component of an accident, so deployment occurs only if the frontal component exceeds the threshold.

The setting of the deployment threshold can depend on many factors, including the usage of seat belts. Some systems have the ability to adjust the threshold up or down automatically, based on whether the belts are being worn, since for seat belt users the potential benefits of air bags are reduced at low speeds. Such decision-making capability is one aspect of a "smart" restraint design.

RELIABILITY CONSIDERATIONS – Deployment threshold is related to another topic - reliability. Air bags are unique among automotive systems. They are different from brakes, for example, which can be disassembled for inspection or maintenance, and which can give can give cues regarding their condition whenever they are used. On the other hand, air bags may remain unused for long periods -- perhaps 20 years -- but they must remain fully ready to perform when needed, and not deploy when not needed. Therefore, an airbag system must have a built-in readiness tester that can immediately inform the operator or the repair technician when there is a problem, and a diagnostic system that will indicate where the problem is. Even so, the consequences of an error (especially a failure to deploy when needed) can be so serious that reliability targets have been set at levels comparable to man-rated space missions [Jones 70]. These levels are achieved by quality assurance inspection and testing at a 100 percent level, as opposed to a statistical sampling process. In other words, every step in the manufacture and assembly of every airbag system entails inspection, testing if appropriate, and documentation, and every vehicle is subjected to a complete diagnostic test procedure before it is offered for sale. Thereafter, every system is electronically tested every time the engine is started.

ELECTRICAL CONNECTIONS – If the airbag system wires were manipulated by service personnel, there would be increased odds of a wire being cut, incompletely re-connected, or not re-connected at all. Therefore, the airbag wiring is contained in a separate and independent harness, wrapped with a material of a distinctive color (typically yellow), and routed inconspicuously. Harness routing is also chosen so as to minimize the likelihood of being pinched or crushed during the crash.

The connections to this harness are crucial to system reliability. Typically, each connector employs dual contacts, gold plated to prevent any compromise of electrical continuity due to corrosion. As shown in Figure 24, part of a bar code may be printed on each of the mating connector housings, so a bar-code reader can verify and record that connector is correctly assembled [Kobayashi 87].

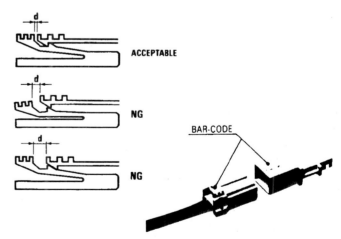

Figure 24. Machine-Checkable Airbag Harness Connector - Honda

THE NEED FOR FIELD TESTING – Because of the extremely stringent reliability requirements, and the relative scarcity of accidents warranting an airbag deploy-

prior to vehicle introduction, which would give adequate insights regarding reliability issues [Jones 71]. Thus we find manufacturers being very cautious, with the early airbag designs being introduced in limited quantities, in fleet settings. A notable example is GM's 1000-car "green fleet" in 1973 [Smith 73].

Another example is the NHTSA-sponsored airbag retrofit program for 539 police cars in 1983-85 [Romeo 84].The intent of this program was to design, test, and evaluate a driver airbag retrofit system, and using such production hardware as existed at the time, to manufacture and install retrofit kits into Highway Patrol vehicles in various states. Sensors and diagnostic systems were supplied by the Technar Corporation, the gas generator came from Bayern Chemie, and the steering wheel and airbag module were made by Takata Corporation. Important lessons were learned, but plans to offer a retrofit kit on a wider scale were thwarted by the inability to obtain adequate products liability insurance [DeLorenzo 86].

Other important field test fleets in the U.S. were the 5300 Ford Tempos for the U.S. Government in 1985 [Maugh 85], further described below, and Chrysler's introductory fleet in 1988 [Edwards 91]. Deployment accidents in all these fleets were investigated in detail [Mertz 88], and valuable insights were obtained from this field experience.

HOW RELIABILITY REQUIREMENTS AFFECT SENSOR SYSTEM ARCHITECTURE – Basically, reliability entails the avoidance of two kinds of errors: false negatives and false positives. In other words, a sensor should avoid a failure to trigger when it is supposed to, and avoid triggering when it is not supposed to. The probability of a false negative can be reduced by placing more sentries on duty, and empowering any one of them to sound the alarm. In engineering terms, one would have multiple sensors connected in parallel. However, the probability of a false positive (i.e., an unwanted deployment) increases with the number of sensors in parallel, since the probability of a false positive for the system as a whole is a combination of the individual false positive probabilities.

On the other hand, the system-wide probability of a false positive can be reduced by placing two or more sensors in series. In this system, any one sentry can squelch the alarm. Consequently, the probability of a false negative increases with the number of sensors in series.

If several sensors are employed, many combinations of series and parallel connections are possible. In the General Motors system of the 1970s, the driver and low-level passenger circuits were connected through sensors on both the positive and the negative (ground) side; i.e., in series. On each side, there was a low-level G switch and a bumper switch in parallel. The same parallel combination was used on the negative side of the high-level passenger circuit, but the positive side was connected through a separate high-level G switch [Louckes 73].

See Figure 25. The G switches, actuation circuits, backup energy source, diagnostics, and crash monitoring and recording devices were included in a single component mounted in the passenger compartment.

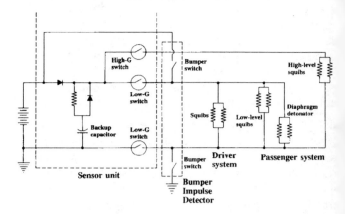

Figure 25. Sensor Circuit, First-Generation GM System

The result of this logic was that a low-level crash would trigger a driver airbag inflation and a low-level passenger system inflation. A high-level passenger system inflation would occur only if the low-level circuit were completed on the negative side, and the high-level sensor closed (on the positive side).

The switch used in GM's Bumper Impulse Detector was a prototype of sensor designs to come, in that it consisted of an inertial mass and a spring; for the switch to close, the acceleration would have to be strong enough to overcome the resistance of the spring and would have to last long enough for the mass to travel the requisite distance to reach an electrical contact. See Figure 26. Consequently, this device was a mechanical integrator of accelerations, in which switch closure depended, for the most part, on the velocity change. It was calibrated to close in an 11 mph barrier crash, and its purpose was to achieve early detection of the crash. (Recall that the G switches, in contrast, were well aft, in the passenger compartment.) The Bumper Impulse Detector employed dual contacts and springs in series.

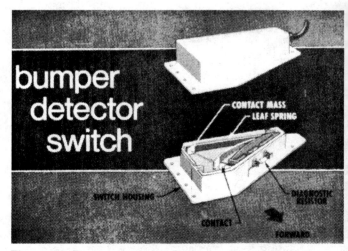

Figure 26. Bumper Impulse Detector, First-Generation GM System

The G switches consisted of a mass on a wire, which formed a pendulum, as shown in Figure 27. The mass was held in its aft most position by a magnet; if the vehicle acceleration were strong enough, the mass could break free of the magnet and move forward in a wedge-shaped groove which extended to 37 degrees either side of straight ahead. The low-level G switch was calibrated (by means of the magnet strength) to close at an acceleration level corresponding to an 11 mph barrier crash. Obviously, the acceleration level would vary with the vehicle and the location of the switch, but in the GM design the level was 18 Gs. The high-level G switch was reportedly calibrated to close at 30 Gs [Louckes 73].

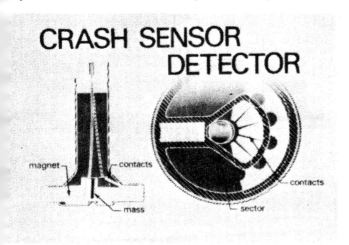

Figure 27. G-Switch Sensor, First-Generation GM

The bumper presented a very harsh environment for any electrical device, including the Bumper Impulse Detector. Designers of subsequent systems have refrained from placing sensors in such a location.

The 1985 Ford Tempo/Topaz system (driver side only) employed five sensors, which were functionally divided into two groups. One group, the secondary or "safing" sensors, were intended to minimize the probability of false positives (inadvertent deployments), and were therefore, as a group, wired in series with the group of primary or "discriminating" sensors. To avoid raising false negatives (failure to deploy), the safing sensors were set to trigger at lower crash severities than the discriminating sensors were.

The safing sensors were wired in parallel with each other and were located on vehicle center line at two locations: the top of the radiator support and on the dash panel. The discriminating sensors were connected in series with the safing sensors, which could thus switch Aon@ frequently without necessarily resulting in a deployment. See Figure 28. The discriminating sensors, wired in parallel with each other, were located on the left and right

front fender aprons, and at the top center of the radiator support, as indicated conceptually in Figure 29.

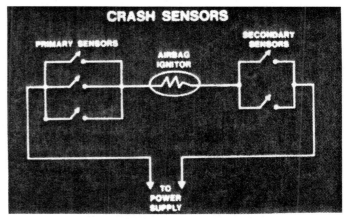

Figure 28. 1985 Ford Tempo/Topaz Sensor Circuit

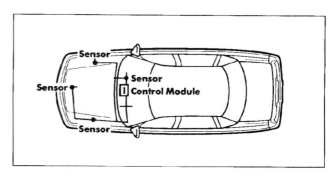

Figure 29. Schematic of Distributed-Sensor System

The sensors were manufactured by the Breed Corporation, and employed a ball that, at a specified deceleration, would pull away from a bias magnet and move through a tube toward a set of electrical contacts, which would have to be bridged to activate the inflator. A small clearance gap between the ball and the tube damped the response to higher-frequency accelerations. This concept is shown in Figure 30. It was simple and reliable, but sensitivity to cross-axis accelerations and temperature variations made its performance more difficult to predict in low speed impacts.

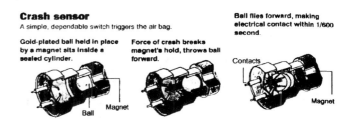

Figure 30. Ball-in-Tube Sensor Function

SENSOR CALIBRATION – Beyond meeting the specifications for deployment threshold, there is a second requirement that influences how sensors are calibrated: sensing time, or time-to-fire. Recall that the sensing time budget is typically 25 or 30 milliseconds. In this time frame, the discriminating sensors must make their decision. Figure 31 shows the time history of ΔV accumulation for four hypothetical situations involving different speeds and objects struck. The upper curve describes an obvious deployment accident, and it has distinguished itself from the others by 15 msec or so. The remaining three curves include one other deployment event and two for which deployment is probably not warranted; yet in the first 30 msec of the crash it is difficult to tell which is which.

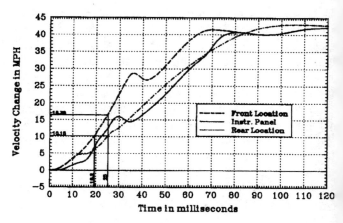

Figure 32. Typical Structural Responses for 35 mph Barrier Crash

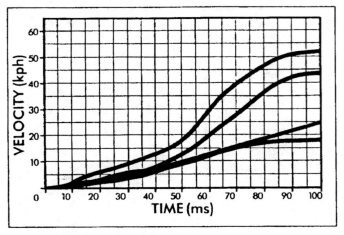

Figure 31. Typical Deployment and Non-Deployment Events - NEC

Recognizing that the crash pulse, and hence the ΔV accumulation, depend on where the sensor is placed, one can attempt to sort out the curves of Figure 31 in a shorter time by placing a the discriminating sensors at a variety of locations on the vehicle. Note, however, that the ΔV required to trigger the sensors will be well below the ΔV ultimately experienced in the crash.

We can illustrate the considerations involved by assuming that a sensor is a ΔV switch; i.e., that its trigger decision is made solely on the basis of the ΔV seen at the sensor location. (As indicated previously, this is not a bad assumption for traditional mechanical sensors.) Figure 32 shows the ΔV accumulation in a typical 35 mph barrier crash. If one needed a 25 msec sensing time for such conditions, and if one had a sensor at a rear location (such as near the B-pillar), one would need about a 10 mph trigger level. If that level proved to be too low (for deployment threshold considerations, say), then one could choose to move the sensor to a forward location; there, it could achieve a 25 msec sensing time with a trigger level of just over 16 mph. Alternatively, the same trigger level, in the forward location, would reduce the sensing time to just over 19 msec.

The results of these considerations are seen in the design of the Ford Tempo/Topaz system. Calibration variables for the sensors included the distance between the ball's rest position and the contacts, its clearance between the guide tube inner walls, and the strength of the bias magnet. These variables were tuned by subjecting them to haversine acceleration pulses of various amplitudes and durations, worked out by Breed to meet Ford's requirements regarding deployment threshold and sensing time. For the discriminating sensors, the "must deploy" threshold in 14 mph barrier crashes was tested with "soft," "medium," and "hard" haversine pulses at 10 mph, and the "no deploy" threshold in 8 mph barrier crashes was tested with "soft," "medium," and "hard" pulses at 6.5 mph. Safing sensor calibration pulses were at 3.5 mph for the forward sensor and 1.1 mph for the aft one. At Ford, each and every sensor was tested twice: once during the manufacturing process and again after final sensor assembly.

This distinction between safing and discriminating sensors, and the use of haversine pulses to test their calibration, has become typical of sensor system design to the present.

TYPES OF SENSOR DESIGNS – Airbag sensor concepts since the '70s have shown their origins in the GM Bumper Impulse Detector: an inertial mass being required to overcome a bias force (from a magnet, for example) and travel a certain distance before electrical contact is made. One implementation of this concept was the ball-in-tube device described previously. Another was the widely used "rolamite" sensor developed by the Technar Corporation (now a part of TRW). See Figure 33. A roller was wrapped with a spring band, which tended to keep the roller pushed against a stop. In a crash, the roller moved forward without slipping (or damping), unrolling the spring band in so doing, and covering the distance to the electrical contacts (again, gold-plated) if the crash was severe enough. The rolling action reduced friction to a minimum; the surface on which the lling occurred was slightly arched. The calibration level as determined by the spring band and the mass of the ller. For a particular application, each sensor would be librated via a set screw, adjusting the position of the ller at rest.

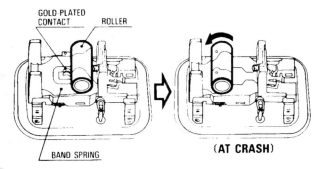

Figure 33. Rolamite Sensor Action - Honda

nother electro-mechanical sensor is the gas-damped iaphragm type, shown in Figure 34. A diaphragm, to which the inertial mass is mounted, provides both the as force and segregation of the gas volumes. Gas is ermitted to pass from one chamber to another through ne or more orifices, creating a damped response.

s noted previously, such systems are electrical switches hat respond to velocity change, by perform a mechanical ntegration of acceleration. Another approach would be o measure the strain in the bias spring, by making that pring a piezoresistive or piezoelectric element in a solid tate electrical circuit. This results in a very small device n which the only motion is beam bending. The acceleraion can be integrated digitally, or processed in other vays. For example, acceleration and velocity could be ised together in some mathematical formula or algorithm NEC 95].

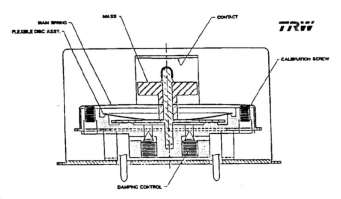

Figure 34. Gas-Damped Sensor - TRW

Electronic sensor systems could be tuned to a particular vehicle by adjusting the parameters or code stored in firmware. This approach opens the door to a great deal of design flexibility, and does so with reduced weight, cost, and complexity (and thus increased reliability). At this writing, such systems are common, and there is every reason to expect them to predominate in the future.

FROM MULTI-POINT TO SINGLE-POINT SYSTEMS –
Both the early GM and Ford systems involved multiple sensors at multiple locations, and we have seen how such a strategy results in different calibrations for different sensors. It also results in a considerable wiring harness, which, as we have seen before, entails special care to maintain reliability.

It is desirable, from both cost and reliability perspectives, to decrease the amount of wiring in the air bag system. A Breed Corporation concept carried this objective to its logical conclusion: a purely mechanical system, located at the inflator, that both sensed the crash and initiated the bag inflation [Breed 85]. In this system, a spring-loaded firing pin was held in place by a lever, which itself was held in place by a bias spring. See Figure 35. The lever could be moved by a sensing mass, and if the motion were sufficient, the firing pin would be released, and propelled into a stab primer.

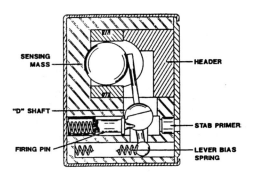

Figure 35. All Mechanical Sensor - Breed

Having no sensors any closer to the crush zone than the steering wheel hub caused some concern regarding sensing time. However, the reduction of axial play in the column produced sensing times on the order of 30 milliseconds in 30 mph frontal barrier tests. The system was intended to be retro-fitted into Chevrolet Impala police vehicles.

The general trend, however, has been to stay with electrical sensing elements, so that the increasing computational power, cost-effectiveness, space efficency, and reduced weight of digital electronics can be used to fullest advantage. This approach allows the integration of sensors, diagnostics, backup power supply, and the logic elements into a single sealed unit. These are known as single-point systems, although separate safing and discriminating sensors are generally employed. Since the electronic environment is much less hostile in the occupant compartment than in the engine compartment, the air bag electronic module is often placed on the tunnel, near or at the dash panel. Because no information comes directly from the crush zone, increased emphasis is placed on signal processing and digital logc.

CONCLUSION

Among all safety system concepts that have ever found their way into production automobiles, the air bag arguably represents the most dramatic departure ever taken from traditional automotive technology. It started in a time of technological optimism when Americans were heading to the Moon, yet refusing to fasten their seat belts. At that time, the air bag seemed like a way to solve a behavioral problem by technological means.

Since then, many nations have shown the way in achieving widespread belt use, and American belt use, while still far behind, has risen to levels not imagined in the '70s or early '80s. Belt systems have thus garnered much of the safety benefit that was assumed to be available to air bags. At the same time, the operating airbag concept has changed from being the only restraint widely used, to being strictly supplementary to the belts. We find that air bags are playing a more limited role, and are contending for a smaller portion of the available safety benefit.

Notwithstanding these developments, air bags were greeted enthusiastically by the public (with perhaps some help from the various passive belt designs). They became almost a litmus test of automotive safety. The development of these devices, and their availability in large numbers, represent astounding technological achievements, and are tributes to the many thousands of individuals, beyond the few cited in this paper, who made it all happen. Air bags are now busily saving lives and reducing injuries, though not necessarily in the numbers originally envisioned.

However, there have also been continued warnings of technological problems and potential adverse side effects, and some of these have come to pass as vast numbers of airbag-equipped vehicles have taken to the road. We are now engaged in redoubled efforts to address these, and the near future will contain many new developments. The work is not yet finished.

REFERENCES

1. [Alexander 74] *An Evaluation of the U.S. Family Sedan Experimental Safety Vehicle (ESV) Project*, GH Alexander, RD Vergara, JT Herridge, W Millicovsky, and MR Neale, Final Report, Contract DOT-HS-322-3-621-1, October 1974.
2. [Breed 85] "The Breed All-Mechanical Airbag Module," A Breed, SAE Paper 856014, *Proceedings, Tenth International Technical Conference on Experimental Safety Vehicles*, 1985.
3. [Campbell 72] "Air Cushion Restraint Systems Development and Vehicle Application," DD Campbell, SAE Paper 720407, May 1972.
4. [DeLorenzo 86] "Supplier cuts off air bag retrofitter - Lack of adequate liability insurance cited," M DeLorenzo, Automotive News, 10 October 1986.
5. [DOT 78] "New DOT Study Finds Only 14% of Drivers Use Auto Safety Belts," Press Release, U.S. Department of Transportation, 15 December 1978.
6. [Edwards 91] "A Preliminary Field Analysis of Chrysler Air bag Effectiveness," WR Edwards, *Proceedings, Thirteenth International Technical Conference on Experimental Safety Vehicles*, November 1991.
7. [FMVSS 201] National Highway Traffic Safety Administration, FMVSS 201, *Interior Impact Protection*.
8. [Frey 70] "History of Air Bag Development," SM Frey, *Proceedings, International Conference on Passive Restraints*, May 1970.
9. [Jones 70] "Inflatable Passive Air Restraint System Crash Sensors," TO Jones, *Proceedings, International Conference on Passive Restraints*, May 1970.
10. [Jones 71] "Crash Sensor Development," TO Jones and OT McCarter, SAE Paper 710016, January 1971.
11. [Katter 75] *Development of Improved Inflation Techniques*, LB Katter, Final Report, Contract DOT-HS-344-3-690, September 1975.
12. [Klove 72] "Special Problems and Considerations in the Development of Air Cushion Restraint Systems," EH Klove Jr. and RN Oglesby, SAE Paper 720411, May 1972.
13. [Kobayashi 87] "Reliability Considerations in the Design of an Air Bag System," S Kobayashi, K Honda, and K Shitanoki, *Proceedings, Eleventh International Technical Conference on Experimental Safety Vehicles*, May 1987.
14. [Kossar 92] "Air Bag Deployment Characteristics," LK Sullivan and JM Kossar, Final Report No. DOT HS 807 869, February 1992.
15. [Lau 93] "Mechanism of Injury from Air Bag Deployment Loads," IV Lau, JD Horsch, DC Viano, and DV Andrzejak, *Accident Analysis and Prevention*, Pergamon Press, Vol. 25, No. 1, February 1993.
16. [Loukes 73] "General Motors Driver Air Cushion Restraint System," TN Louckes, RJ Slifka, TC Powell, and SG Dunford, SAE Paper 730605, May 1973.
17. [Maugh 85] "Supplemental Driver Airbag System - Ford Motor Company Tempo and Topaz Vehicles," RE Maugh, SAE Paper 856015, July 1985.
18. [Mertz 88] "Restraint Performance of the 1973-76 GM Air Cushion Restraint System," HJ Mertz, SAE Paper 880400, February 1988.
19. [NEC 95] Advertising material from NEC Technologies, Inc., 1995.
20. [NHTSA 96] *Final Rule on Labels*, National Highway Traffic Safety Administration, 27 November 1996.
21. [NHSB 70] National Highway Safety Bureau: Proposed Amendment to Motor Vehicle Safety Standard 208, *Occupant Crash Protection*, 3 November 1970.
22. [Reidelbach 79] "Advanced Restraint System Concepts," W Reidelbach and H Scholz, SAE Paper 790321, February 1979.
23. [Romeo 75] "Front Passenger Passive Restraint for Small Car, High Speed, Frontal Impacts," DJ Romeo, SAE Paper 751170, November 1975.
24. [Romeo 78] *Front Passenger Aspirator Air Bag System for Small Cars*, DJ Romeo, Final Technical Report, Phase II, Contract DOT-HS-5-01254, March 1978.
25. [Romeo 84] "Driver Air Bag Police Fleet Demonstration Program - A 15-Month Progress Report," DJ Romeo and JB Morris, SAE Paper 841216, October 1984.
26. [Seiffert 72] "Development Problems with Inflatable Restraints in Small Passenger Vehicles," UW Seiffert and GH Borenius, SAE Paper 720409, May 1972.

7. [Smith 73] "The 1,000 Car Air Cushion Field Trial Program," GR Smith and MR Bennett, *Proceedings, Automotive Safety Engineering Seminar*, June 1973.

8. [Struble 79] "Status Report of Minicars= Research Safety Vehicle," DE Struble, *Proceedings, Seventh International Technical Conference on Experimental Safety Vehicles*, June 1979.

CONTACT

Donald E. Struble, Ph.D.
Collision Safety Engineering, Inc.
320 West Peoria Avenue, Suite B-145
Phoenix, AZ 85023   Phone: (602) 395-1011

THE COMBINATION OF A NEW AIR BAG TECHNOLOGY WITH A BELT LOAD LIMITER

Farid Bendjellal
Gilbert Walfisch
Christian Steyer
Jean-Yves Forêt Bruno
Xavier Trosseille
Renault
France
Paper Number 98-S5-O-14

986101

ABSTRACT

This study deals with the development of a restraint system in order to improve occupant protection in frontal impact. In frontal collisions where vehicle intrusion is minor, the main lesions caused to occupants are thoracic, mainly rib fractures resulting from the seat-belt. In collisions where intrusion is substantial, the lower members are particularly vulnerable. In the coming years, we will see developments which include more solidly-built cars, as offset crash test procedures are widely used to evaluate the passive safety of production vehicles. If this trend will continue, restraint forces from the belt will increase and as a consequence more thoracic injuries will occur in frontal collisions.

In order to address this risk, it has become necessary to work on an optimized limitation of the restraining forces, while taking account of the broadest possible population, especially elderly people. A first step in this reduction was taken in 1995 with the introduction of the first-generation Programmed Restraint System (PRS), with a seat-belt force threshold of 6 kN combined with a belt pretensioner. Thirty seven frontal accident cases involving this type of restraint were investigated.

Analysis of these data combined with findings from the University of Heidelberg / NHTSA study, shows that it is necessary to go a step further by reducing the shoulder belt force to 4 kN. As this objective cannot be achieved with a standard restraint system, it was necessary to redesign the airbag and its operating mode that is, a new seat-belt + airbag combination called PRS II.

This paper summarizes the data obtained with the 6 kN load limiter restraint in real-world collisions. A description of the new system is given and its performance in offset crash configurations with respect to a European standard belt + air bag system is discussed.

INTRODUCTION

IMPORTANCE OF FRONTAL COLLISIONS. Detailed analyses of all fatal accident reports in France in 1990 and of the accidentology file of the PSA/Renault Laboratory enabled to determine the distribution of fatalities and seriously injured occupants with respect to collision configurations. The percentages related to frontal impact are respectively 50°/0 and 70%, as shown in Figure 1 ; illustrating the predominant role of this crash configuration on occupant injuries In order to assess the distribution of lesions in frontal collisions as regards the main body segments, an analysis was conducted on 100 belted front seat occupants taking into consideration serious injuries. Figure 2 presents the distribution of AIS 3+ injuries for the head, the thorax, the abdomen and the lower members. It can be observed that the thoracic risk is highest for the passenger, and secondly, for the driver For the latter, injuries to the lower member's constitute the most frequent risk. Since 1992, improvements have been noted in Europe in cars as regards the resistance of the passenger compartment, especially the reduction in intrusion In addition the majority of cars are today equipped with belt pretensioners. The combination of these improvements would suggest a certain benefit in reducing the severity of injuries to the occupant. To assess this hypothesis two accident files, including belted drivers involved in frontal collisions, were selected. The first file (A) comprises 2000 vehicles manufactured before 1991 and with no belt pretensioners in the restraint system. The second file (B)

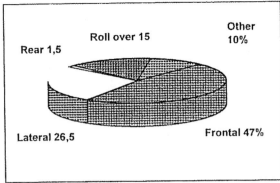

Figure 1a : Distribution of fatalities per collision type LAB PSA/Renault accident database.

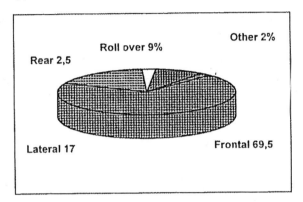

Figure 1b : Distribution of severely injured occupants per collision type LAB PSA/Renault accident database.

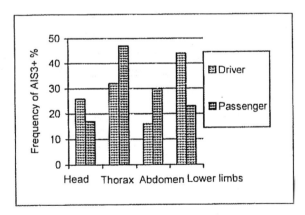

Figure 2 : Distribution of severe injuries per body regions for 100 seriously injured occupants (MAIS 3+) in frontal collisions Driver and front seat passengers (belted).

includes 160 vehicles, manufactured since 1992, all equipped with belt pretensioners and structural reinforcements. The two files were compared taking into account the frequency of moderate to serious injuries, AIS 2+ , corresponding to the main body segments, as shown in Figure 3.

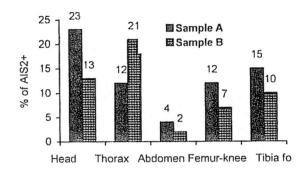

Figure 3 : Risk of AIS 2+ injuries in frontal collisions involving belted drivers. Comparison of 2 accident samples with cars manufactured before 1991 (A) and cars manufactured since 1992 (B).

When comparing files A and B, a tendency in the reduction of injury frequency is observed for the head, the abdomen, the lower limbs. For the thoracic segment an opposite trend appears with an increase of risk. As this tendency to reduce intrusion will continue and, as airbags will become more widespread in Europe, one may expect gains as regards the risk of injuries to the head and lower members, and abdominal risks will be maintained. For the thorax, there will be increased risk since rigidifying the structure will result in a direct increase in restraining forces on the occupant. A study presented by Bendjellal, 1997 (1), showed accident cases in frontal collisions with cars manufactured after 1992, where front seat occupants, restrained with a combination of a belt pretensioner and an air bag, sustained severe thoracic lesions.

The study presented in this paper was initiated in order to address this rising risk of chest injuries.

THORACIC RISK LINKED TO SEAT-BELT

BELT INDUCED INJURIES AND OCCUPANT AGE - The 3-point seat-belt was designed to protect the occupant as regards contact with the passenger compartment and to avoid ejection from the vehicle. In order to provide this protection, the seat-belt exerts substantial and localized forces on the thoracic cavity. These forces, which may reach 10 kN, generate broken ribs which may or may not be combined with internal lesions of the thorax. The first relationship between seat-belt tension and the associated thoracic risk level was established by J.Y. Forêt Bruno in 1978 (2) based on an analysis of 90 accident cases. The vehicles in question, sold in France in the

1970's, were equipped with 3-point static seat-belts in the front seats with a load limiter located in the belt webbing between the occupant's shoulder and the upper anchorage point. The load limitation was obtained by tearing of the stitching which was used to sew loops in the webbing. In case of impact the stitching tore, thus allowing more webbing from the loop: as a consequence the torso can move relative to the vehicle at a controlled load level. A view of such a load limiter before and after impact is shown in Figure 4a and its force-time response in dynamic test is illustrated in Figure 4b.

a) Load limiter before and after impact

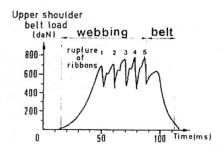

FORCE LIMITER type B

b) Response of load limiter in dynamic test

Figure 4 : A load limiter installed in cars sold in France between 1970 and 1977

In 1989, other cases were added to this investigation, bringing the total of this database up to 290 accidents (3). The key point of this unique database is the possibility of showing a relationship between the seat-belt tension exerted on the occupant, his age and the type of resulting lesions: this relationship, reproduced from Forêt Bruno study (3), is given in Figure 5. This data clearly shows that thoracic risk among occupants restrained by seat-belts increases with age and that a shoulder belt force of 8 to 9 kN may induce a high risk for the chest.

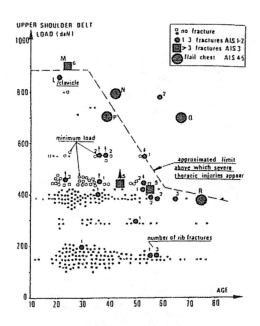

Figure 5: Relationship between shoulder belt tension, age of occupant and injury severity to the chest. Reproduced from (3).

LIMITATION OF THORACIC RISK LINKED TO SEAT-BELT

The data discussed in the previous sections and these accident cases show the necessity of reducing seat-belt tension forces in frontal crashes. An initial stage, consisting of limiting this force to 6 kN, was carried out in 1995 on Renault vehicles with the introduction of the PRS system (Programmed Restraint System). This system is comprised of a pretensioner pyrotechnic buckle, a retractor webbing clamp and a steel part, fastened between the retractor and the seat-belt anchoring point as shown in Figure 6. This part, designed to deform at a given level of force, acts like a force limiter. The system's operating method includes 3 phases: at the beginning of the impact (15 milliseconds), the buckle pretensioner triggers in order to take up the seat-belt/occupant slack. The occupant's coupling is increased in this phase with the action of the strap blocking mechanism in the retractor (17 ms). This combination enables one to substantially reduce the occupant's initial displacement. In Phase 2, restraining forces are gradually applied. When the belt tension level reaches 6 kN (70 milliseconds) the force limiter comes into play, authorizing controlled displacement of the retractor in the B-pillar, upwards. Movement of the retractor will enable a displacement of the torso under controlled load, thus allowing the rib cage to be relieved of seat-belt stresses. Complete operation of this device is shown in Figure 7

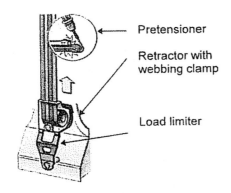

Figure 6: The Programmed Restraint System installed in Renault cars since 1995 (6 kN shoulder belt load limiter)

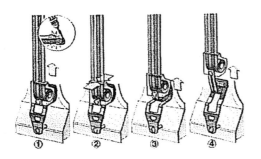

Figure 7. The PRS operating mode - Phase 1 Initial part of the crash and belt pretension activation, Phase 2 Action of the webbing clamp, Phase 3 Load limiter activation, Phase 4 End of impact

Behaviour of the Programmed Restraint System in Real - World Accidents - To date, 80 accident cases related to frontal collisions with cars equipped with this device have been investigated since 1995. Thirty seven cases are discussed in this paper. The main parameters of this sample are summarized in Figure A1 in the appendix. Age distribution of occupants ranges from 17 years to 72 years with 11 cases (30%) with age < 25 years, 7 cases (19%) with age ranging from 26 to 35 years, 3 cases (8%) between 36 and 45 years, 8 cases (21.5%) between 46 and 55 years, and 8 cases (21.5%) with age > 56 years. The severity of the collisions, expressed in terms of EES, ranges from 35 km/h to 75 km/h. Nearly half of this sample (48.6%) corresponds to a severity which is superior to EES of 55 km/h.

Regarding the injury severity for the thorax, only 2 cases(5.4%) are related to an AIS level of 3. In the first case the driver a 72 old male sustained 3 right rib fractures and lung contusion. The car was involved in an offset collision to the left, with an overlap of 85% and with an EES of 50- 55 km/h. Except fractures

to the left metatarsi, no other injuries were found. In the second case two front seat occupant were involved; a 58 years old male in the driver position and a 60 years old female in the passenger position. The driver sustained a fracture to the sternum (AIS 2) and the passenger had 4 left rib fractures (AIS 3). In both accident the shoulder belt load. estimated from the PRS deployment, was 6 kN. The other thoracic AIS levels observed for the rest of the sample are AIS 2 with 7 cases (19%) , AIS 1 with 13 cases (35%) and AIS 0 with 15 cases (40%).

Regarding the overlap distribution among these accident cases , half of the sample corresponds to offset configuration with an overlap below 74%° and the other half is close to a full barrier test. An illustration of one accident, case No . 12041, is given in Figure 8 with photographs of the car deformation and the PRS deployment.

8a   8b

Figure 8 : Illustration of car deformation in a frontal collision Case No 12041 (9a) and the PRS actual deployment (9b).

Belt limitation threshold - The accident cases presented in the previous section, are encouraging, but they show that a threshold of 6 kN for belt load limitation is not sufficient to prevent a risk of serious injury to the thorax as 2 cases with occupants having sustained an AIS 3 level were found. This observation is consistent with the data from Forêt Bruno (3) published in 1989. It is therefore necessary to go a step further in the reduction of shoulder belt load. As this reduction will result in an increase in excursions of the head and thorax it is therefore essential that with this kind of seat belt it is necessary to combine the pretensioner and quite obviously the airbag. The combination of an airbag and a 3 point belt restraint is discussed in various publications among them are the paper from Kompass in 1994 (4), the study of Kallieris et al in 1995 (5) and Mertz et al investigation in 1995 (6). According to the data discussed in Kallieris paper (5) and the mathematical simulation investigated by NHTSA (5) for a variety of crash conditions (frontal and rollover) crash severities and occupant sizes (5, 50 and 95 percentiles) a threshold of 4 kN for the shoulder belt load limitation appears to be suitable for reducing the risk for thoracic injury without negative consequences on other injury measurements. Therefore a 4 kN load limitation threshold is chosen for the belt system. Whilst working in the same stopping distance for the thorax, i.e. a distance from thorax to steering wheel of 300 to 350 mm, it is necessary for the airbag to play an important role by taking part of the thoracic restraint. The question is : which type of air bag has to be chosen for this occupant protection approach ?

Air bag accident data in the USA and in France - When the FMVSS 208 was introduced in the USA in the beginning of the 1980's, according to investigations carried out by NHTSA, most people did not use seat belts. The percentage of people wearing seat belts at that time was on the order of 15%; this suggested the necessity of protecting the majority of unbelted occupants, by means of a restraint system independent of the seat belt The physics of a vehicle, impacting a rigid barrier at 50 km/h and with 50th percentile dummies not restrained by a seat belt, imposed de facto paddings or knee plates for the protection of the femurs and knees and the airbag for protection of the upper part of the body. The performance of such a restraint system combined with the seat belt is quite positive with more than 1500 lives saved (7) during the 1990 1996 period. However cases of fatal accidents have been noted involving either adults not restrained by seat belts or else children in rearfacing seats or even children without any restraint system whatsoever. This problem stems mainly from the energy parameters of the airbag dimensioned in order to absorb energy on the order of 3000 J In comparison, a Eurobag or « facebag ». designed to protect the head of a 50° percentile restrained by seat belts has an energy potential of 200 J. If one wants to design a seat belt airbag restraint system which takes account of OOP situations, it is therefore necessary to explore other possibilities.

Current situation in France - Out of the total number of automobiles in France - some 25 million - only 2 to 3% of vehicles are equipped with driver airbags. We lack data on airbag efficiency in Europe since the target survey files remain statistically low in comparison with the USA, only 100 cases have been studied in France by the Laboratoire d Accidenlologie et

de Biomécanique Peugeot Renault; 75 cases involved frontal collisions with belted drivers.

In Figure 9 a risk comparison for the head, with and without airbags, is given. For the 25 to 45 km/h speed range, one notes moderate lesions (11%) for cases with no air bag as opposed to 0% for cases with air bag. For the 46 to 65 km/h speed range, the frequency of AIS > 2 is 40% without air bag and only 14°/o with air bag. No facial fractures were observed with air bag, whereas half of the sample without air bag represents facial fractures. The tendency of air bag to improve head protection is confirmed.

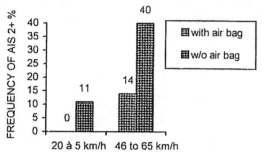

Figure 9: Accident survey with frontal collisions involving occupants with 3 points belt + Air bag restraint system. Frequency of AIS 2+ injuries to the head. All cases with Eurobag type of air bag.

SPECIFICATIONS FOR AN OPTIMIZED SEAT-BELT + AIR BAG RESTRAINT SYSTEM - The basic principle is that occupant restraint energy must be managed, whilst complying with human tolerance limits. In this context, the thoracic cavity is more tolerant to distributed pressure (air bag) than to a very localized pressure (belt). With the same stopping distance for the occupant in the vehicle, it is possible for the airbag to take a part of the seat-belt forces.

Once the basic elements of the seat-belt, that is, pyrotechnic pretensioner and force-limitation, have been determined, it is now necessary to define the airbag characteristics. The corresponding specification is based on 2 separate parts: to contribute actively to restraining the occupant, and to control the aggressiveness of the deployment of the airbag. This results in the 3 following main functions:

1. The airbag must inflate very early on in the impact and "wait for" the occupant's contact; this is the anticipation function, analogous to the effect of a pretensioner on the PRS seat-belt.
2. Having a law of force which is as constant as possible. This is equivalent to controlling

the pressure in the airbag and the force exerted on the occupant. This is similar to the action of the force-limiter in the PRS seat-belt.

3. These two functions result in an increase in the generator power in relation to the Eurobag. In order to control the bag aggressiveness in OOP situations, it is necessary to compensate this through a more elaborate airbag-folding strategy, in order to reduce the punch out transmitted to the occupant. This objective results in a deployment mode distributed in 3 directions: first downwards and sideways and then toward the occupant.

Based on these elements, a new airbag has been developed in the frame of the new system called the PRS II.

DEVELOPMENT OF THE PRS-II DESCRIPTION AND VALIDATION

The system comprises 3 main components. These are the pretensioner, the belt load limiter and the air bag.

The pretensioner - This is a device which enables the seat-belt strap to be drawn taut very quickly at the initial moment of impact. For the PRS-II system, and given the experience acquired on Renault vehicles since 1992, a pyrotechnic buckle pretensioner has again been selected, especially for its efficiency with respect to submarining. In 4 milliseconds, it enables to take up the seat-belt slack and secure the occupant to the seat.

The seat-belt force-limiter - The force limitation function is located at the core of the retractor with a torsion bar whose plastic deformation comes into play as soon as the seat-belt force at the shoulder reaches 4 kN. For this function an another option is to use the deformable steel plate, as in the PRS-1 generation, providing a sufficient space in the B-pillar packaging.

The airbag - The airbag is a 60 liters bag with a pressure limitation function and a folding which allows a deployment from top to bottom and to the sides. As opposite to Eurobag, this bag is defined to protect the head and the thorax.

There are different ways to control the pressure of the air bag; the system described here refers to a set of vents in a row, contained in a meltable seam. After an impact, the air bag deploys to its full volume, while the vent is still closed. Ones At a given pressure of the gas inside the bag, the seam tears and the vents open successively. The restraint force acting on the occupant from the air bag is thus controlled.

Development of the PRS II - After a computer simulation phase, the opening pressure of the airbag vents has been validated during tests using a free fall pendulum system. At the same time, the seat-belt force limiter was developed. Then, sled tests were conducted in order to fine-tune the system's characteristics. The validation program also included static tests in OOP , according to ISO recommendations (8), and crash tests with vehicles.

Figure 10 provides a description of PRS-II components. The operating phases of the system, as obtained in a 50% offset rigid barrier test, are illustrated in the same figure where the 4 upper sequences indicate the air bag work and the 4 lower sequences relate to the belt actions. Sequence 1 in Figure 10 represents the firing of the belt pretensioner at 12 ms followed by the start of air bag deployment at 15 ms. Note that once the air bag deployment is achieved (sequence 2) , the vents are still closed; the air bag is waiting for the occupant. When the thorax contacts the air bag, in sequence 3, the seam covering the vents starts to tear, thus liberating the first vent. The bag pressure is now under control; in sequence 4 the belt load limiter function starts to work in conjunction with the opening of the remaining vents in the air bag: with this last sequence the thoracic restraint loads are controlled thorough the impact duration.

COMPARISON OF PRS II WITH A CONVENTIONAL RESTRAINT SYSTEM - Various mathematical simulations and sled tests were conducted in order to assess the PRS-I I performances in frontal collisions. In addition two crash tests with the same vehicle model (mass of the vehicle 1200 kg) were performed; the test configuration corresponds

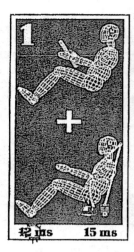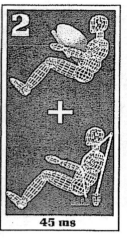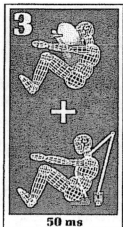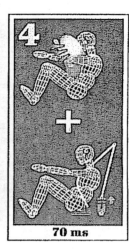

Figure 10: PRS II principle - 1- Pretensioner action and air bag deployment ; 2- air bag full deployment ; 3- Opening of the first vent of the bag ; 4- Combination of belt load limiter action and air bag pressure control (opening of the other air bag vents).

to a 50 % offset rigid barrier test at 56 km/h One of the vehicle was equipped with a conventional belt + air bag system; the belt included a pyrotechnic buckle pretensioner and the air bag was of Eurobag type(volume of 45 liters) The other vehicle had the PRS-II system. In the front seats of both vehicles instrumented Hybrid III 50° dummies were installed. The results from both tests are illustrated in Table 1 and time-histories for the head acceleration, the chest acceleration and the shoulder belt load are provided in Figure A2 in the appendix. With the PRS-II the head HIC and 3ms acceleration are reduced, respectively 75% and 55% : the neck shearing force is also reduced respectively 60% for -Fx and 57% for +Fx. The neck extension moment is increased with the PRS II with a maximum of 35 Nm as opposed to 11 Nm with the conventional system. The shoulder belt load reduction with the PRS-II is significant -55%, as a direct result of the combined work of the belt and the airbag. The thoracic acceleration is also reduced but the amount of reduction (24%) is smaller than those observed with the other criteria. This last result shows that 1) the occupant stopping distance is the same for the 2 systems we are comparing and 2) the energy distribution on the thorax is spread differently with the PRS-II. Chest injury parameters, such as the chest deflection and the VC, cannot be compared as the data corresponding to the conventional system (with the same vehicle) are not available. The maximum chest deflection and VC with the PRS II are 25 mm and 0.09 m/s. Compared to the conventional system the PRS II allowed an increased x-displacement of the chest (+60 mm).

Results of PRS-II validation in vehicles tests - Vehicles from the same model whose front seats were equipped with PRS 11 were tested according to 3 impact configurations . 1 Rigid obstacle, 15° barrier, 50°70 offset and a speed of 56 km/h according to AMS procedure (9), 2 Deformable barrier at 0°, 40°70 offset and a speed of 56 km/h. This configuration reproduces the future European regulatory test, ECE 94 (10), 3. Full rigid barrier, wall at 0°, and a speed of 56 km/h. This test is the representation of the New Car Assessment Program (NCAP) as used by NHTSA in the USA. The interest of such a test matrix is to combine demanding conditions for the restraint system - the case of the US NCAP test - and for the structure of the vehicle with the other two offset crashes. The first offset test condition allows to assess both the structure of the vehicle and the restraint system. The test according to the procedure defined by the EEVC (ECE 94) is a special case, since this configuration enables to simulate a car to car collision and also to judge the quality of the triggering system for the belt restraint and the airbag restraint, in particular as the first part of the crash is soft compared to the two other test configurations.

The results of these tests are documented in Table 2, which includes the resulting accelerations of the head and thorax, the Head Injury Criterion (HIC 36 ms) the upper neck shear force, the upper neck extension moment, and the shoulder belt tension, the chest acceleration, the chest deflection and VC. All the maximum values refer to measurements obtained from Hybrid III 50° percentile dummy, for both the driver and passenger.

Table 1: 50% offset rigid barrier test at 56 km/h. Comparison of PRS II responses with those of a conventional belt + air bag system - Same vehicle used in both tests, driver data.

	Measurements & injury criteria with a Hybrid III 50° percentile dummy	50% Offset rigid barrier test, 56 km/h with a conventional restraint system	50% Offset rigid barrier test, 56 km/h with the PRS II
Restraint system	Buckle pretensioner activation time (ms)	18	16
	Belt pretension (mm)	49	49
	Initiation of belt load limitation (ms)	None	70
	Duration of belt load limitation (ms)	None	40
	Air bag type	Eurobag 45 liters	PRS II 60 liters
	Time of actiation of air bag pressure limitation (ms)	None	72
Body segments Head	HIC 36 ms	763	186
	3 ms acceleration (G)	74	33
Neck	Shear Force -Fx (kN)	0 5	0 2
	Shear Force +Fx (kN)	0 7	0 3
	Extension moment (Nm)	11	35
Thorax	Shoulder Belt Load (kN)	9 7	4 3
	3 ms acceleration (G)	53	40
	Chest deflection (mm)	na	25
	VC (m/s)	na	0 09
	Thoracic X-displacement measured at shoulder level (mm)	290	350

Table 2 : Summary of crash test results with a production vehicle (mass 1200 kg) equipped with PRS II in offset and full barrier tests. Driver and passenger injury criteria and measurements.

Body segments	Measurements & injury criteria with a Hybrid III 50° percentile dummy	50% Offset rigid barrier test, 56 km/h		100% rigid barrier test, 56 kmh US NCAP		40% Offset deformable barrier test, 56 km/h EEVC procedure	
		Driver	Passenger	Driver	Passenger	Driver	Passenger
Head	HIC 36 ms	186	257	347	519	74	111
Neck	3 ms acceleration (G)	33	37	45	53	24	28
	Shear Force -Fx (kN)	0.2	0.03	0 5	1.2	0 009	0 02
	Shear Force +Fx (kN)	0 3	0 6	0 4	0.2	0 5	0 3
	Extension moment (Nm)	35	28	29	na	11	12
Thorax	Shoulder Belt Load (kN)	4 3	4 5	4 6	4 8	3 9	4 1
	3 ms acceleration (G)	40	36	42	45	23	24
	Chest deflection (mm)	25	27	40	40	23	15
	VC (m/s)	0 09	0.22	0.64	0.64	0.01	0 03

The PRS-II system behaved well in all 3 configurations, both belt load limiter and air bag pressure limiter worked. The shoulder belt tension was between 3.9 kN and 4.8 kN. The lowest value was recorded for this parameter was obtained in the EEVC test (for the driver) and the highest value in the US NCAP test (for the passenger). This difference is due to the friction in the D-ring. As this friction is directly related to the dummy forward displacement, its effect on the shoulder peak load is more pronounced for the passenger. Chest accelerations were all below 46 G; this result indicates no chest to steering wheel contact. Neither head to steering wheel contact was observed as illustrated by the low values recorded with the HIC (between 74 and 519) and with the head 3ms acceleration - between 24 G and 53 G. Chest deflections ranges from 15 mm to 40 mm; the lowest value was obtained in the EEVC test for the passenger and the highest in the NCAP test for both the driver and the passenger.

VC values were between 0.03 m/s and 0.64 m/s. Both head and chest accelerations and also chest deflections and VC's ensure that the use of belt load limitation, in the test conditions described here, combined with air bag pressure control has no negative effects on injury measures.

Consideration of neck secondary risk in OOP - An evaluation of the new airbag was performed in static deployment tests using the Hybrid III 50° dummy, in order to measure the risk for the neck region. The results indicate that none of the IARV (11) levels was exceeded. Detailed results of these tests, as well as a biomechanical evaluation of this airbag can be found in Trosseille paper (12).

SUMMARY AND CONCLUSION

This study was initiated to address the rising risk of belt induced chest injuries in frontal impact. The starting point was the analysis of 290 frontal accident cases with vehicles that were equipped in France in the 1970's with a belt load limiter in front seats. The load limiter was based on a tear-webbing principle and was located near the upper belt anchorage point. This database shows that older people ($\geq$ 50 years) may sustain severe chest injuries. Based on this experience a program was initiated at Renault with a view to reduce the shoulder belt load. In 1995, a belt restraint system called PRS was introduced; it comprises a combination of a pyrotechnic pretensioner located at the buckle, a clamp retractor and a steel part attached to the retractor and to the belt anchorage point. This

steel part designed to deform at a given load, acts as a load limiter. This allowed to control the shoulder belt load at 6 kN level. Accident cases involving this type of restraint were collected and analyzed; in particular the behavior of the belt load limiter was investigated in relation with occupant injuries. The data from 37 cases with belted front seat occupants, are reported in this paper. Crash severities ranged from 35 km/h to 75 km/h. A significant part of this sample, 27% of occupants with age > 50 years, sustained minor to moderate chest injuries. The combination of belt pretension and a 6 kN belt load limitation appears to have benefits in reducing thoracic loads from the belt for this population; the 6 kN level is however not sufficient to cover the whole population. Thus, a further step in reducing the shoulder belt load is necessary. As this reduction will involve increased excursions of the head and the thorax, the belt load limitation has to be combined with an air bag.

The combination of an air bag and a 3-point belt restraint was discussed in various publications among them are the paper from Kompass in 1994 (4), the study of Kallieris et al. in 1995 (5) and Mertz et al. investigation in 1995 (6). According to the data discussed in Kallieris paper (5) and the mathematical simulation investigated by NHTSA (5) for a variety of crash conditions (frontal and rollover), crash severities and occupant sizes (5°, 50° and 95° percentiles) a threshold of 4 kN for the shoulder belt load limitation appears to be suitable for reducing the risk for thoracic injury, without negative consequences on other injury measurements.

From the experience acquired with the PRS a new approach in the occupant restraint system was developed. The PRS-11 combines a pyrotechnic buckle pretensioner with a 4 kN belt load limiter and an air bag specially designed with respect to 2 key factors: a deployment to the sides and from top to bottom in order to reduce the risk in OOP situations and a pressure control which operates when a certain load is applied by the thorax. One the major concern with the belt load limitation was the possibility to increase the injury risk for the head and for the thorax. A comparison of PRS II with a conventional restraint system was performed, for the driver, on the basis of offset frontal collisions involving the same car model. The data with PRS II show substantial reductions for the head and chest acceleration, HIC values and neck shear forces. Neck extension moment is increased with the PRS II but the value, 35 Nm, remains below the 57

Nm suggested IARV (11). Maximum chest deflection and VC obtained with PRS II were 25 mm and 0 09 m/s These data were not compared to those of conventional system, as the corresponding data were not available for the same car model

The PRS II was also evaluated in 3 frontal collisions: offset rigid barrier test at 56 km/h, offset deformable barrier test at 56 km/h (EEVC frontal impact test procedure), and in full rigid barrier test at 56 km/h (NHTSA frontal NCAP test) In the test conditions described here, the combination of a 4 kN belt load limitation with the pretensioner and air bag pressure control has no negative effects on injury measures

ACKNOWLEDGMENTS

The PRS 1I principle was initiated by the Renault Safety Eng Dept Many people were involved in its development in particular vehicle platforms teams and advanced research groups The authors would like to thank Michel Kozireff and his team from Autoliv France for their important contribution This study was made possible thanks to the sled and crash tests performed at the Renault Lardy Test Center Views and opinions expressed here are those of the authors and not necessarily those of Renault.

REFERENCES

1- F Bendjellal, G Walfisch, C Steyer, P Ventre, J.Y Forêt-Bruno, X Trosseille, J.P Lassau : « The Programmed Restraint System - A Lesson from Accidentology ». SAE Paper No.973333, Proceedings of Stapp Car Crash Conference, Orlando, USA, 1997

2- J.Y Forêt-Bruno et all : « Correlation Between Thoracic Lesions and Force Values Measured at Shoulder of 92 Belted Occupants Involved in Real Accidents ». SAE Paper No.780892, Proceedings of Stapp Car Crash Conference, USA, 1978.

3- J.Y Forêt-Bruno, F. Brun-Cassan, C. Brigout, C Tarrière : « Thoracic Deflection of Hybrid III: Dummy Responses for Simulation of Real Accidents » In proceedings of the 12th International Technical Conference on Experimental Safety Vehicles Goteborg, Sweden, May 1989 -

4- K Kompass: « Opportunities and Limits of an Air bag Optimization Based on the Passive Requirements of Standard 208 » Paper No

94-S4-O-08, 14th ESV Conference Munich , Germany

5- D Kallieris, A Rizzeti, R Mattern, R Morgan, R Eppinger and L Keenan : « On the Synergism of the Driver Air bag and the 3-point Belt in Frontal Collisions ».SAE Paper No.952700, Proceedings of Stapp Car Crash Conference, USA, 1995.

6- H.J Mertz, J E Williamson and D.A Lugt : « The Effect of Limiting Shoulder Belt Load with Air Bag Restraint ». SAE Paper No.9508861 International Congress and Exposition Detroit, USA, 1995.

7- ISO/TC22/SC10/WG3 : « NHTSA Communication on Air Bag Related Accidents » Meeting in Delft, The Nederlands, May, 1997

8- ISO/DTR 10982 - Road Vehicles - Test Procedures for Evaluating Out of Position Vehicle Occupant Interactions with deploying Air Bags. November, 1995.

9- Auto Motor und Sport - May 1997 Report - Germany

10- « Uniform Provisions Concerning the Approval of Vehicles with Regard to the Protection of the Occupants in the Event of frontal Collision » Official Journal of European Communities, Brussels, July 1996

11- H.J Mertz : « Anthropomorphic Test Devices « Accidental Injury - Biomechanics and Prevention, A M Nahum and J.M Melvin, eds, Springer - Verlag 1 New York, 1993

12- X Trosseille, et all : « Evaluation of Secondary Risk with a New Programmed Restraining System » Paper No 98-S5-W-24 Written paper, 16th International Technical Conference on the Enhanced Safety of Vehicles (ESV), Windsor, Canada, June 1998.

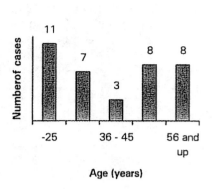

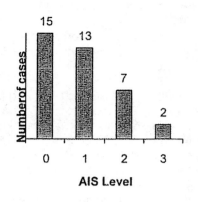

a) : Age distribution

c): Maximum thoracic AIS

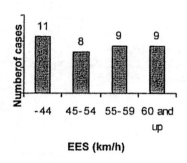

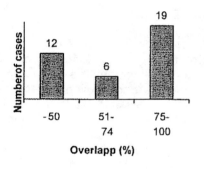

b). Collisions' severity

d): Range of overlap

Figure A1 : Summary of data from accident investigations with frontal collisions involving the PRS

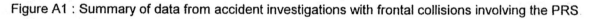

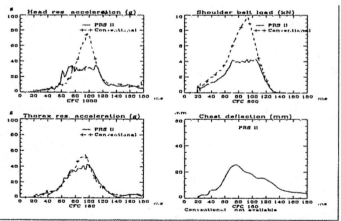

Figure A2 : Comparison of PRS II responses with those of a conventional belt + air bag system. Driver data from a 56 km/h offset rigid barrier test.

A HALF CENTURY OF ATTEMPTS TO RE-SOLVE VEHICLE OCCUPANT SAFETY: UNDERSTANDING SEATBELT AND AIRBAG TECHNOLOGY

Wendy Waters
Michael J. Macnabb
New Directions Road Safety Institute
Betty Brown
Insurance Corporation of British Columbia
Canada
Paper number 98-S6-W-24

ABSTRACT

In road safety, a common perception exists that technology and/or regulation can solve problems, and does so in a sequential and progressive manner. This is not always the case. Technology is no panacea and government interventions can do as much harm as good. Using historical methodologies, this paper explores the multiple attempts and failures of manufacturers, governments, and other groups to solve the rather simple safety concept of crash harm reduction through properly restrained vehicle occupants. This historical-methodology approach is suggested as an effective evaluation tool to measure other road safety interventions.

INTRODUCTION

Seatbelts save lives. No responsible road safety professional today would dispute this fact. They have been in use for approximately forty years and evidence of their effectiveness is abundant. Yet usage rates in the United States today remain shockingly low (around 60 percent), especially when contrasted with Canada, Australia, and Western Europe with rates approaching or exceeding 90 percent [1]. Comparing the experience in the US with that of other countries offers insights into the nature of seatbelt use and how road safety interventions work (or do not).

The availability of the technologies of seatbelts and passive restraints have failed to solve the problem of injuries and deaths in the United States caused by the occupants colliding with the interior of the vehicle or being ejected, after the vehicle has hit another object. Yet, the technology of seatbelts has allowed other countries to solve this problem to a large extent. The US problem then, is not with insufficient technology, but with the failure of drivers and passengers to use it. A reason for this behavior rests in the history of the relationship between US society and seatbelts, including the politics involved.

This paper explores the successive cycles of government intervention in the United States, each one an attempt to solve the problem of the human collision [2] Using a comparative-world methodology, we contrast the case of the United States with that of Canada (especially British Columbia) and to a lesser extent with Australia and Europe. This approach illuminates the extent to which seatbelt usage has been cultural and political and demonstrates the need to consider social and human factors when evaluating or designing road safety initiatives. The political history of seatbelts in the US and society's interaction with both the belts and the politics, contributed to widespread apathy and even antipathy toward them, which has been a factor in the continued problem of deaths and injuries to unbelted Americans.

The First Attempt (to solve the problem): Government Regulated Seatbelts, 1966-1970

Initially, in the late 1950s automobile manufacturers introduced seatbelts to solve the problem of keeping the driver in his or her seat following a minor collision such that control of the vehicle could be maintained. They became an option on new vehicles—albeit not a popular one. In the mid 1960s legislators and activists (Ralph Nader being the most memorable of them) re-defined the problem to which seatbelts were the solution—they argued that seatbelts could prevent thousands of accident-related injuries and deaths by reducing the severity of the "second collision" between the occupant and the interior of the vehicle or from the occupant being ejected during an accident (the first collision being between the automobile

---

[1] For Canadian statistics see Transport Canada Road Safety, Leaflet CL 9709 (E)

[2] That is, preventing injury to people after the vehicle has hit something. We acknowledge that road safety involves much more than seatbelt usage, but this paper is only about the problem of occupant protection

and another object)[3] Reducing second collision injuries and fatalities has remained a problem in the US for the rest of the century Despite claims by individuals such as Ralph Nader that having seatbelts in every vehicle would solve the problem of preventing the so-called secondary collision, this failed to happen because people did not wear them

From the early 1960s seatbelts were available as options on most American-made cars In 1963 only 9 percent of cars had belts, yet usage rates ranged in those vehicles from 47 percent always using them on local trips to 74 percent on longer trips[1] Approximately 30 percent of vehicles on the road in 1966 had them, although a National Safety Council survey found that full-time usage rates among people who chose option seatbelts was 44 percent (67 percent said they used them on longer journeys exceeding 25 miles)[5] Given a choice, automobile makers and consumers did not often opt for seatbelts (but, it's worth noting that those whose chose them as an option— who were actively involved in obtaining them—tended to use them)

For the those concerned with national public safety-- such as health officials, certain governors, senators, and congressmen, and consumer advocates including Ralph Nader— something had to be done about the thousands being killed each year (43,400 in 1963[6] and approximately 50,000 by 1966[7]) Their solution was to legislate seatbelt installation along with a range of safety guidelines to make the interior of the vehicle less dangerous[8] In 1966 the US government created a separate Department of Transportation with a mandate to set standards and to put in place mechanisms to monitor them (soon the National Highway Traffic Safety Administration [NHTSA] would be

created for this purpose)[9]

The first motor vehicle safety standards went into effect in 1968 These safety standards and the creation of NHTSA were large steps forward in making motor vehicle travel safer But the introduction of seatbelts as standard equipment on vehicles failed to make Americans buckle up and injury rates remained high (Usage and accident rates at this time were similar in Canada where the majority of vehicles were produced by US manufacturers)

The automobile manufactures (Chrysler, American Motors, General Motors, and Ford) predicted as much Prior to the Motor Vehicle Safety Standards they argued that the public would not wear seatbelts, and that making them mandatory would ruin the styling of their vehicles and reduce sales[10] Auto makers further argued that Americans were not ready for seatbelts and would resent having something they did not want, and the costs for it, imposed upon them[11] The manufacturers claimed to have an interest in safety, but insisted that it could be better achieved through improved highways and driver education—not federally imposed standards[12] While it is indisputable that the auto manufacturers' main motivation in making these arguments was their complete hostility to any government regulation of their industry, hindsight shows they had some valid points[13]

Over thirty years later, it is worth examining their

---

[9]This was several years after the United States government mandated that any vehicle purchased for government use through the General Services Administration be equipped with seatbelts and other safety-related equipment The opposition to this government stance on the part of the automobile companies is written up in NYT, 31 August 1964, p 27 The state of New York had also already ordered lap belts on all vehicles sold in the state NYT, 18 September 1964, p 34

[10]NYT, 19 February 1965, p 37

[11]NYT, 18 September 1964, p 34; 24 February 1965, p 81

[12]Despite claims to be concerned about safety, General Motors under tight questioning from Senator Robert F Kennedy during government hearings on this issue admitted to making $1 7 BILLION in profits during the previous year, and spending only $1 2 million on safety research and initiatives Other manufacturers showed similar records Newsweek, 26 July 1965, pp 67-68

[13]It should be noted that government involvement in the industry has always been huge—through constructing highways the US government has given an enormous subsidy to the industry

---

[3]Ralph Nader, Unsafe at Any Speed: The Designed in Dangers of the American Automobile (New York: Grossman Publishers, 1965, 1972), especially chapter three

[4]New York Times (hereafter NYT), 12 April 1967, section XIV, p 31

[5]NYT, 10 April 1966, section XII, p 9 Another survey, this one carried out by the Auto Industries Highway Safety Committee, found that 38 percent of drivers with seatbelts "sometimes" used them on shorter trips and 25 percent sometimes used them on longer trips

[6]NYT, 7 April 1964, p 34

[7]Business Week, 11 June 1966, p 179

[8]Removing or re-designing dangerous protruding objects such as the metal "cookie cutter" ring on the steering wheel, were among the changes to design mandated by this legislation

arguments Legislating seatbelt installation did not solve occupant restraint problems, but it also did not cause a reduction in sales nor make people fear automobile use To their pleasant surprise. auto makers did not experience a decline in sales as a result of this legislation If anything, the increased attention to safety on the newer cars became selling features as a result of a new public interest in the issue Regardless of whether people wanted to wear a seatbelt all the time. many wanted them there along with the other new safety features of collapsible steering wheel, dual braking systems, a padded dash board, and safety door latches [14] People seem to believe Nader who had said that people may cause accidents, but cars causes injuries [15] Arguably, since this time a culture of conspicuous consumption of safety features from airbags to anti-lock brakes and four-wheel drive has emerged. making them emblems of wealth or class status as much as safety devices

As the auto makers predicted, people did not like or wear seat belts A historical perspective suggests, however, that the automobile manufacturers in the United States themselves played a large role in making their own prophesy come true The evidence presented below indicates that through making seatbelts especially ugly and uncomfortable. publicly raising concerns about price increases. and desperately arguing the (minute) potential dangers of belts, they made the arrival of the seatbelt era in America more cumbersome, controversial, and difficult than it needed to be

For example, take the engineering and styling of the belts By the early 1960s. seatbelts in Europe had already evolved into an early version of the self-adjusting, three-point, fully-retractable harnesses that are in common use today [16] The European models were readily available as examples on the thousands of imported automobiles solid in the US each year US manufacturers chose instead to install manual-adjusting, especially large, belts that restricted movement, and installed shoulder belts separate from lap belts, making it necessary for the user to do up

two separate buckles [17] Moreover because these shoulder belts were not self adjusting, drivers wearing them often could not reach components on or near the dash board Ralph Nader became especially critical of the manufacturer's tactic, suggesting that the deliberately engineering belts for "human irritation "[18]

The manufacturers complained loudly to the public and in the press about these belts Executives publicly bemoaned the ugliness of the belts and how they detracted from the car's appearance One likened them to "spaghetti" while another to the "vines" in "Tarzan's cave "[19] While Volkswagon and Volvo promoted the safety features (including belts) on their vehicles in their advertising and public relations, the US auto makers complained that seatbelts ruined the car's aesthetic appeal and raised prices [20] A Chrysler executive commented that "We can't think of a better way of doing it " Yet, the European example was right in front of them This executive further commented that the inconvenience of the belt design does not increase the chance that riders will wear them, thereby publicly encouraging people not to do so [21]

The motivation for the auto makers' tactic was their resentment of government regulation The dialog, as reported in the newspapers, between them and the US government (and Ralph Nader) suggests a war for public support on the question of regulating the automobile industry The manufacturers chose to make seatbelts the focus of their objections to the new regulations—even though these rules also included many other safety features In the press manufacturers told Americans that no conclusive evidence existed on the benefits of safety belts and that adding them and other design modifications to automobiles would raise prices significantly

Manufacturers also called attention to the minor injuries that seatbelts cause (neglecting to mention that this was while saving one's life), and asserted that insufficient data existed to warrant their widespread use They especially attacked shoulder harnesses for the abrasions they left on the necks of people in accidents (again ignoring the lifesaving that went on in the process) If people wanted an excuse for not taking the trouble to buckle their seatbelts, the manufacturers gave it to them A 1967 New York Times reporter even commented that the controversy raised over shoulder harnesses probably degraded the strap

---

[14] NYT, 19 August 1965, p 13, discusses Dodge stressing 12 new safety features on it's higher priced vehicles

[15] Nader views discussed in Business Week, 11 June 1966, p 179

[16] Business Week, 11 June 1966, p 192 discussed this safety belt and an article on 23 April 1966, pp 52-54, discussed seatbelts and safety features on Volvo and SAAB vehicles, imported into the United States The existence of European superiority on safety belts was also discussed in NYT, 18 September, 1964, p 34

[17] NYT, 2 April 1967, section XIV, p 28

[18] Ralph Nader writing in NYT, 21 March 1968, p 12A

[19] NYT, 2 April 1967, section XIV, p 28

[20] NYT, 22 August 1967, p 41

[21] NYT, 31 March 1968, p 12A

so much that Americans would not ever use it even if the belt were improved or subsequent research negated the significance of the abrasions (both of which did occur)[22] The combination of manufacturers negative attitudes toward seatbelts and unsubstantiated concerns about their safety could not have made seatbelts appealing to the average American

Ralph Nader along with several senators fought back hard, especially on the subject of costs To General Motors executives who complained of the costs, they countered that it was the world's most profitable corporation and therefore could absorb a few dollars for safety[23] Subsequently Senators Warren G Magnuson (Democrat from Washington) and Walter Mondale (Democrat from Minnesota) found evidence that the manufacturers had grossly inflated the costs of seatbelts in their propaganda The senators' own research suggested that the costs of the new seatbelt was approximately $3, while the manufacturers stated the costs to range from $23 to $34[24] It appears the automobile companies hoped to convince people to write their political representatives and ask for the repudiation of the motor vehicle safety standards through which, in their view, the government forced people to buy things, like seatbelts, that the did not want[25] But the auto companies failed to understand the situation: not even the most right-wing Republicans on the government's safety committees took up the position of the automobile manufacturers Supporting safety standards was politically popular as most Americans supported the idea generally[26] It took the auto manufacturers a few years to recognize the new reality

The end result of these seatbelt-focused exchanges was not public opinion against the regulations; people believed that making cars safer was a good idea Instead this dialog contributed to negative opinions towards seatbelts specifically and helped instill the view that they were something being imposed on Americans by a "big brother" government that was growing Nearly fifteen years later, in letters to the editor the public continued to echo these same sentiments[27] The result of people not wearing belts was thousands needlessly dying, which in turn brought more government intervention in the industry and in Americans' lives—not less

The Second Attempt: Additional Seatbelt Paraphernalia, 1971-1976

Because the imposition of safety standards failed to solve the problem of carnage caused by the second collision by the early 1970s, advocates for public safety decided they needed to undertake greater measures The secretary of transportation believed that he had five choices (retain the present rules, conduct a five-year field test of air bags, require air bags as an option on all new cars, make seatbelt use mandatory, or mandate passive restraints on all cars starting with the 1980 model year)[28] The choice in the US, where the government now had some control over the automobiles on the market (a luxury that Canadian or Australian governments did not have) was to turn to new technology (while Canada and Australia turned to mandatory [seatbelt] usage laws [MULs])

In the early 1970s the United States had recently been to the moon, proving its technological capacity to be unmatched in the world A faith in technology permeated US culture It became the prescription for the country's ill of motor vehicle accident casualties

The US government regulators took three steps in the early 1970s aimed at increasing the amount of technology on vehicles The first was to convince manufacturers to experiment with the relatively new airbag technology with a goal of introducing it within a few years[29] The second move was to attempt to increase belt usage through reminder systems that buzzed when the seatbelt was not fastened All cars manufactured for the 1971 model year (and subsequent years) had this feature But, usage rates remained low, bringing yet another cycle of legislation— insisting that all new vehicles for the 1974 model year would have an interlock system installed which would prevent the vehicle from being started unless the seatbelt were fastened Opposition to interlock technology was

---

[22]NYT, 27 August 1967, section IV, p 13 Dr Haddon, director of the National Highway Safety Bureau, undertook investigation and reported his findings in January 1968, noting that in Sweden a study of 28,000 crashes that involved lap and shoulder seatbelts saw no one killed at speeds under 60 miles per hour NYT, January 1, 1968

[23]NYT, 7 January 1968, p 54

[24]NYT, 8 January 1968, p 47

[25]NYT, 15 September 1968, p 46

[26]Elizabeth Brenner Drew, "The Politics of Auto Safety," Atlantic Monthly (October 1966): 95-102

[27]NYT, 6 November 1981, p 30; and 31 December 1984, p A26

[28]NYT, 2 August 1976, p 24

[29]Popular Mechanics (February 1971), pp 64-65 Experiments with airbags began in the late 1960s with Ford forming a partnership with Eaton Yale & Towne, Inc to develop the airbag They were tested by the Air Force using baboons Newsweek, 1 January 1968

widespread and probably created as much resentment toward seatbelts and government demands that people wear them as it did converts to the wearing of them. It should be noted that the Canadian government did not demand the interlock system and most manufacturers either left it off of vehicles being shipped to Canada or gave Canadians a bypass switch.[30]

The Third Attempt: Airbags and other Passive Restraints, 1976-1983

With interlock devices not working, US legislators and certain lobbyists proceeded to their third choice of airbags or other passive restraints (such as the automatic seatbelt developed first by Volkswagon).[31] The public widely seems to have embraced the idea of airbags as it gave an excuse for not becoming accustomed to seatbelt wearing and it fit with the western (and especially American) cultural tendency to see technology as a panacea thereby absolving individuals and society of taking responsibility for their own behavior.[32] Industry at first balked at the idea of airbags. They cited their excessive costs and the stressed the dangers that they believed inherent in airbags—especially to children.[33] The industry had "cried wolf" when it protested seatbelts on the basis of their safety, which made their calls of dangers with airbags much less credible at the time (nevertheless, the history of their use in the 1990s has born out these concerns to be real issues with airbags).

In 1976 airbags seemed like the only technology that might save Americans from themselves and NHTSA sought to make them mandatory. Manufacturers protested adamantly. The auto makers received a slight compromise from the transportation secretary William T Coleman in 1976. who seems to have listened to the safety concerns. The auto companies agreed to make 250,000 vehicles with airbags each year, that would be sold to consumers and monitored by NHTSA to gather information about them.[34] Soon, in 1977. a new transportation secretary (Brock Adams) ordered that airbags or automatic lap and shoulder restrains be installed in all standard and luxury automobiles by 1982, and in all smaller cars by the 1984 model year.[35] General Motors protested this in 1979. still arguing that airbags might injure small children.[36]

In the 1980s the Reagan administration reversed pending legislation that would mandate passive restraints (airbags or automatic belts) in all vehicles. This move belonged to a general policy of deregulating American industries. The President called for improved driver training as the solution, rather than vehicle regulation.[37]

Consumer advocate groups and automobile insurance companies took the government to court over the reversal of this bill, and won.[38] Reagan's Transportation Secretary Elizabeth Dole was told that the problem of occupant restraint had to be solved to save lives, and she was given a year to draft new, replacement legislation or the old bill passed in 1977 would be re-instated. In 1983 she introduced a compromise that included phasing in airbags (on 25 percent of new vehicles after September 1 1987. 40 percent after September 1, 1988, and 100 percent by September 1, 1989. But, she also legislated that this requirement would be removed if enough individual states passed mandatory usage laws (MULs) that taken together covered at least 2/3 of the American population.[39] Automobile companies suddenly became huge proponents of seatbelts and MULs.[40]

The Fourth Attempt: Mandatory Usage Laws, 1984-1990s

More than a decade after parts of Australia made seatbelt usage mandatory. eight years after Canada began doing so, and after thirty-two other countries had adopted MULs, the US states began to look at the issue.[41] On

---

[30]Vancouver Province, 17 July 1973, p. 5. It should be noted that most vehicles in Canada were US made, or made for the US market. 1973 probably marked the first year that the standards would be different

[31]Popular Science (March 1974), p. 93

[32]Business Week, 4 July 1977, p. 20 Discusses this as a problem with promoting airbags to a large extent

[33]NYT, 3 July 1977, section IV p. 6

[34]NYT. 12 December 1976, p. 6

[35]NYT, 1 July 1977, p. 1

[36]NYT, 2 October 1979, p. A17

[37]Motor Trend (April 1981), p. 32. NYT. 24 October 1981, p. 1

[38]NYT, 25 June 1983, section one, pp. 1.8; This decision to insist that the government return to the passive restraint technology approach could be interpreted as a legal statement supporting the notion of technology as a panacea for a major social or behavioral problem—that of people refusing to buckle a seatbelt

[39]NYT, 12 July 1984, p. A18

[40]On auto company involvement see NYT, 6 December 1983, p. 31. 25 April 1984, p. 22. 12 July 1984, p. A19

[41]NYT, 6 June 1984, p. D25. 32 countries, 7 Canadian

January 1, 1985 the law went into effect in New York state and over the next few years other states passed similar legislation. Unlike in Canada or Australia, where the passage of such laws have contributed to long-term substantial decreases in the number and severity of injuries caused by the second collision, their effect in the United States has been more limited. This indicates that neither seatbelts, nor the law, alone or combined, contain the entire solution to the problem (if it did, US rates would resemble more closely those of other countries) [42]

Several likely reasons exist for the failure of seatbelts and MULs to save Americans, which will be explored here. One possible explanation for the law's failure to raise US usage rates to the levels seen elsewhere is that the law has often had limited enforceability. In some states (although not New York) it was a secondary enforcement law; police officers could not pull a vehicle over solely for the infraction of not wearing a seatbelt—there had to be another reason and the seatbelt would become an additional, discretionary ticket. This weakened regulation decreased the seriousness of the issue in people's minds. Although New York kept it a primary offense, it did not experience the same long-term levels of compliance as Canada, likely because the police themselves did not take enforcement of the MUL as seriously [43]

Perhaps a bigger explanation for why MULs in the US have been less effective than elsewhere has been the lack of accompanying awareness of the need for seatbelts on the part of the US public. Compare the arguments for and against MULs in the US (especially NY state) and Canada (taking British Columbia [BC], which enacted an MUL in 1977, as the main source of data). In BC, newspaper editorials, letters to the editor and newspaper reports stressed the importance to the BC economy of passing such a law. With government-run medical insurance and motor vehicle insurance, the costs of unbelted drivers to the provincial economy became clear to most voters. For example, the BC Medical Association (Physicians) in 1976 argued that injuries cost on average $4000 a piece, and deaths $150,000 and that 115 of the 717 people who died in automobile crashes in the province

the previous year would have survived had they been wearing seatbelts (thus a needless cost of $17.25 million dollars) [44] Because medical insurance came from tax revenue and vehicle insurance was run by the government (and thus considered like a tax) people could understand that taxes would go up if claims from injuries and deaths did not go down.

In the United States the costs to society were less clear for the average person than they were in BC. With hundreds of auto insurance companies and medical coverage companies to chose from in the US, and with a large population, the effect on society of the unbelted driver was less evident to the average person—although known to federal government agencies such as NHTSA and the insurance companies.

Arguments for and against the MULs given in the newspapers, by interests groups, and everyday citizens, differed between New York and BC. The argument against an MUL made frequently in the US—that the unbelted driver only endangers him- or herself—was quickly negated in the British Columbia campaign. Not only did an unbelted driver cost society, according to reporters and letters to the editor, but the unbelted driver could also lose control of the vehicle following a first collision and would be unable to avoid hitting another vehicle or pedestrian. British Columbians stressed the need to protect society in general ahead of any arguments about individual rights. Whereas in New York and the US, citizens stressed that individual rights should come before measures to protect society at large--even if society paid for the medical and vehicle losses through higher insurance rates. Fears of an Orwellian "Big Brother" government were often repeated by politicians and citizens in New York as the reason to oppose MULs [45] Meanwhile, the state of Virginia refused to go along with the federal push for MULs on the principle that the state had a proud history of opposing the federal government—safety, monetary losses, and lives lost were subordinated to a political and cultural principle [46]

Along with automobile companies, insurance companies became prominent proponents of MULs in the US [47] While in general this is similar to BC where the one automobile insurance company, ICBC (the Insurance Corporation of British Columbia) [48] actively supported the

---

provinces and the US territory of Puerto Rico had passed such legislation. The countries that had done so included Japan, Britain, France and the Soviet Union.

[42] While one could argue that this is because Australians or Canadians are more law abiding generally than the average American, no solid evidence exists to support this.

[43] NYT, 28 February 1985, p. B5; A police Chief named Margeson is quoted as saying only a few tickets in his jurisdiction had been issued, that "It's not a priority."

---

[44] Vancouver Province, 9 April 1976, p. 33

[45] NYT, 12 October 1984, p. C1. That this debate happened in 1984 contributed to Orwellian interpretations.

[46] NYT, 28 February 1985, p. B5

[47] NYT, 25 June 1983, p. 8

[48] All motorists in the province must insure their vehicles

concept of an MUL, there is also a substantial qualitative difference. Whatever else British Columbians felt about ICBC, they could recognize it as an exclusively BC entity, designed to serve residents of the province. American insurance companies generally transcend state and regional boundaries, and Americans may not have seen them as having local and community interests foremost in their minds (ahead of their own financial statements). Thus, many citizens may have written off MULs as the imposition of a powerful insurance lobby in Washington DC and the state capitals, and not something as emerging from society.

In British Columbia the MUL did not result from federal government initiative, but from Provincial concerns and studies, and citizen interest. In 1975 ICBC sponsored a safety conference that examined the impressive results in the Australian state of Victoria (20 percent decline in fatalities and a 50 percent decline in hospital admissions from car accident injuries[49]) and the need for reducing government pay outs to injured individuals through government-run medical insurance and car insurance.[50] Citizens came to support the idea of an MUL and even push politicians who acquired cold feet, concerned about public reaction to a perception of an imposition on civil liberties.[51] Politicians debated the MUL for two years before finally passing the legislation.

This political waffling ironically may have made the laws more popular as people were able to fault the government for inaction on a proposal that would save lives and money.[52] Of course some people opposed the concept of legislating seatbelt use, but the majority seemed to accept its necessity as everyone paid for the costs of injuries and fatalities.[53] Education also played a large role in

fomenting public support in British Columbia. In BC the MUL was combined with an intense education campaign—before and after the passage of the law—in the schools, at fairs, and in the media stressing why one should wear a belt.[54]

When the BC MUL finally passed in 1977 (with only one legislator opposing the bill[55]), approximately 65 percent of citizens supported it. It went into effect on October 1, 1977 and statistics (73 percent usage in the Vancouver and Victoria areas in March 1978) suggest that the majority of those who opposed it, wore their belts anyway. Prior to the MUL, only 28 percent in these areas used safety belts.[56] (By contrast in the United States in 1978 metropolitan-area usage was 14 percent).[57]

To contrast these facts with those from New York reveals striking differences. Governor Cuomo of New York approximated that during the time the state legislature debated the MUL, correspondence received from state residents was about "18,000-to-1 against".[58] Moreover, politicians in New York and other states were far from unanimous in their votes for the law. Most laws that did

---

through ICBC, a government-owned and regulated company, created in 1972

[49]Vancouver Province, 26 March 1975, p 8

[50]Vancouver Sun, 24 January 1976, p 5 and 12 March 1976, p 5. Vancouver Province, 26 March 1975, p 8

[51]Vancouver Province, 29 October 1975, p 4 an editorial notes that the highways minister said that any government with guts should pass a seat belt law, and that he supported one, but that the public should expect him or his government to initiate such a law. New Democratic Party (a semi-socialist party) Premier Dave Barrett echoed these remarks in the Vancouver Sun, 19 November 1975, p 53

[52]Vancouver Sun, 12 April 1975, p 43 and 24 January 1976, p 5 ; Vancouver Province 21 June 1975, p 5;

[53]Examples of opposition in letters to the editor in the Vancouver Sun , such as 20 September 1975, p 5. Article

stating that the majority polled favored the law in Vancouver Sun, 21 October 1975, p 15

[54]Advertisements promoting the MUL included instructions on how to wear a seatbelt and why one should wear one. For example, see Vancouver Sun, 24 September 1977, p 27. ICBC also toured the Seat Belt Convincer, a seat mounted on a ramp that people could sit in, belted in. An attendant would pull a trigger sending the seat sliding down a 12 foot incline, coming to an abrupt stop at 9.6 km/h, producing a jolt sufficient to demonstrate the utility of the seatbelt: The Colonist (Victoria BC), 19 November 1977, p 11

[55]Vancouver Sun 26 March 1977, p 16. By contrast when the state of Washington passed seat-belt legislation nearly a decade later, the vote was 33 to 15. Vancouver Sun, 8 March 1986

[56]Vancouver Sun, 9 January 1978, p A12. Vancouver Sun, 13 May 1978, p A8. Within a year, these numbers dropped significantly , to only approximately 55 percent of drivers buckling up by December 1980. The Colonist (Victoria BC), 19 December 1980, p 6. Subsequent studies following the implementation of MULs elsewhere in Canada reveal a pattern of high initial compliance, followed by a lessening of usage rates.

[57]NYT, 17 December 1978, p 34

[58]NYT, 1 February 1985, p B2

pass, did so with a bare majority [59] The one similarity between New York and BC was that, at least initially after the MUL went into effect, the majority of people (over 70 percent) buckled up regardless of their opinion of the law [60] Subsequently, in both places rates dropped (to 40 percent in NY after just three months, and in BC it gradually fell to a low of 55 percent over the next few years) [61] But since that time BC's rate has steadily increased reaching near 90 percent while that of NY has grown much more slowly

This BC increase happened because of direct interventions on the part of ICBC Following the decline to 55 percent, ICBC established its Traffic Safety Division with a mandate to promote seatbelt usage and other safer driving behaviors ICBC created a three pronged approach to safety initiatives that has proven successful on many campaigns to this day The first prong is police involvement through road checks and other enforcement programs (known as STEP—Selective Traffic Enforcement Program) The second aspect is a corresponding education campaign on radio and in newspapers (and more recently, television) that promotes the reason for the initiative and informs people that the police are actively looking for violators The third prong involves making use of local traffic safety committees (usually comprised of representatives from town government, educational institutions, related businesses, and citizens groups) to promote the initiative locally through such means as fairs, contests, or banners in key locations; this last aspect gave communities partial "ownership" of the problem and the solution process This approach was first used successfully to reduce drinking and driving (and the program remains in place today, providing consistent and sustained pressure)

In 1983 ICBC applied the approach to achieve compliance with the seatbelt MUL, and by maintaining the program through the years has helped bring the steady increase in seatbelt usage rates In British Columbia these campaigns have included a particular focus on children and youth, with remarkable success [62] At a time in their lives

when they are supposed to be risk-takers, BC's youth has a high rate of seatbelt usage (and also thanks to these education programs, the lowest rate of drinking-and-driving incidents) [63] Given the contrast with the US, BC's long-term commitment to road-safety education has likely played a significant role in reaching a 90th percentile usage rate

In the US some efforts at education occurred at the federal and state levels, but the quality and commitment appears to have been much lower [64] In New York, legislators intended that the law itself would be the educator (and not an intimidator) But without accompanying education, people viewed the law as a nuisance, and not a real reason to buckle up [65] Indeed, through the 1980s a large percentage of Americans continued to believe that it is better to be thrown free of the vehicle in an accident [66] All of this suggests a need for increased educational efforts

Overall, the contrast in US and Canadian MUL experience demonstrates the necessity of public involvement in creating the legislation whether directly through lobbying or indirectly through interacting with education programs or media reports that convince people of the need for a new regulation or a certain behavior Having a comprehensive, multi-faceted education program in place before, during, and after the discussion of the MUL, contributed to favorable public interest in seatbelts

---

but also in car accidents) or leaving the driving population, thereby increasing the percentage of people wearing seatbelts through natural aging of the population Furthermore, psychologists have identified that people become more cautious as they enter their 30s and 40s, or have children The US baby boomer population themselves, by moving into this more conservative age group helped to raise usage rates These generational factors alone do not, however, account for the increased rate of usage into the 90 percentile range in Australia and Canada

[59] In New York in the lower house it was 82-60, only 6 votes more than the needed number for passage NYT, 22 June 1984, p B3 In the NY senate the vote was 37 to 22 (NYT, 26 June 1984)

[60] NYT, 1 February 1985, p B2

[61] NYT, 9 May 1985, p A13

[62] All regions have seen a gradual increase in seatbelt usage as the population has aged Moreover, younger generations have generally had higher usage rates than older ones This suggests that those most opposed and most unaccustomed to wearing belts are gradually dying off (mostly of old age,

[63] Canadian Medical Association Journal 157, no 12, 15 December 1997, pp 1661-1662

[64] Whether this is due to the nature of the education campaigns or the amount of money spent on them is not known but is a question worthy of study

[65] NYT, 28 February 1985, p B5, shows examples of these opinions

[66] NYT 26 September 1984, p C1 This report on education declared it a failure as fewer than 15 percent of Americans wear seat belts This article also reported the persistence of a myth that it is better to be thrown free of a vehicle during an accident

usage Yet BC has some preconditions that New York does not In Canada there has generally been a culture of accepting government regulation and direction and being angry when the government is not perceived as protecting its citizens Moreover, with government-run medical and vehicle insurance, it was much easier for British Columbians to understand the cost to them personally of a society that does not buckle up [67]

The Fifth Attempt: Airbags Revisited, 1990s

With US MULs still failing to reduce second collision casualties sufficiently, the US government returned to airbags and passive restraints in the 1990s All passenger cars produced today for the US market must have airbags (in addition to seatbelts and buzzers, and MULs in most states) Yet, airbags are not as neat and simple a solution as seatbelts when the latter are used Airbags only inflate once, are useless the occupants during any subsequent collisions or roll-overs, and cannot help them in incidents that do not involve a front-end collision Used in conjunction with seatbelts, airbags provide approximately 5 percent more protection in frontal crashes Yet they also have inherent dangers

Until 1998 airbags exploded at such a high velocity as to be potentially dangerous The airbag was designed to prevent serious injury to an unbelted 50th percentile male crashing at 50 kilometers and hour; but the power required to do this has proved deadly to smaller occupants (especially women and children) Making them mandatory on all vehicles meant that those people willing to wear belts faced unnecessary dangers Because the majority of Americans did not buckle up, law makers and engineers began opting for a technology that was not necessarily more effective than a properly buckled three-point harness in a frontal collision

This illustrates an industrial, one-size-fits-all mentality No discussion has emerged until this past year of offering different types of technologies to suit individual needs (and still meet a federal occupant protection criteria) Recent Canadian regulators have demanded that airbags on vehicles destined for Canada be depowered (because the majority of Canadians wear their belts and do not require such a powerful bang for adequate supplementary restraint)—a first step toward a more flexible view of safety technology [68] Law makers and manufacturers in the US

seem unwilling to acknowledge that safety might require a more flexible approach than the industrial paradigm Solving the problem of occupant restraint may require acknowledging this human factor--everyone is not created equal nor uses safety technology in the same way (As GM safety engineer Paul Skeels said in 1966, designing an automobile interior for safety would be different for a belted versus an unbelted occupant [69])

CONCLUSIONS

The Problem Persists

Although US efforts to protect people from the second collision have been less successful than those in other places, there have been some gains US Injury and fatality rates did begin to fall in the 1980s, for the first time in history [70] Recent surveys undertaken by NHTSA as part of President Clinton's new seatbelt usage drive suggest that over half of Americans favor a primary enforcement MUL Although this same survey suggests that, at best, 66 percent of Americans buckle up every time they enter a vehicle, it also indicates that more people are starting to recognize the importance of seatbelt use [71]

The history of the interaction of seatbelt technology with the US public suggests that there are large obstacles for society and safety advocates to overcome in order to see widespread usage and a resolution to the problem of preventing second collision injuries so long desired Efforts at improving safety have often created greater resentment toward seatbelts and government safety measures among large sectors of US society Negative memories of seatbelts and government intervention can be passed to the next generation, and continued low usage rates suggests

---

Canadian Press Newswire, 1 November 1996

[69]Business Week, 11 June 1966, p 184

[70]NYT, 5 February 1984, p 22 In 1983 43,028 people were killed on US highways, the lowest level in 20 years according to Transportation Secretary Elizabeth Dole, or 2 6 deaths per 100 million vehicle miles—the lowest level ever recorded She attributed the drop to seatbelt use and anti-drinking-and-driving campaigns In 1980 the death rate had reached 51,091

[71]From the NHTSA website (http://www nhtsa dot gov/people/injury/buckleplan/presbel t2/) This is inferred from the fact that 76 percent of drivers said they wear a seatbelt "all the time" when driving but over 10 percent of this group also stated that at least once in the previous week they had not worn the belt That they lied suggests they know that they should be wearing it

---

[67]Canadian settlers followed the Mounted Police and the Hudson's Bay Company—the government--, while in the US the procedure was the reverse

[68]Financial Post Daily, 14 November 1997, p 30 and

that they have been Seatbelt education at the elementary school level might be necessary to counter parental influence Ultimately to improve US usage, road safety promoters need to understand the failures and design programs with these in mind The solution will likely be one that brings everyday citizens into the process and that allows them to understand the need for belts Sustained, region-based education and enforcement programs, perhaps based on the BC model (but adapted to local, US conditions) is one possible way to take control of the situation, rather than waiting for a technological solution

This exploration of seatbelts and airbags demonstrates that technology alone can not solve problems Further experiments with technology are not the answers to the US problem of excessive injuries and deaths caused by second collision injuries Through this comparative-regional methodology this paper illustrates the problem in fact rests with the political, cultural, and historical context of that technology This paper also shows that the failure to address the cultural and political context of seatbelt technology in the US has resulted in five unsuccessful and different attempts to decrease the severity of second collisions through government legislation and additional technology (but without much public or community involvement) [72] Each attempt has been complicated by, and has further exacerbated, the culture and politics of seatbelt usage

Comments on the methodologies

Road safety concerns have been with society since the onset of the automotive era The issue of traffic safety generally, like that of occupant protection specifically, has a complex history and one that has not developed in isolation from the people that use it

One way to understand what has helped to prevent deaths and injuries on the road (and why) or what did not help significantly (and why not) requires analyzing the situation in such a way as to establish variables and constants Because history cannot be repeated in a lab, "virtual" variables and constants can be established through comparisons and contrasts with other regions that had the same problems, but achieved different outcomes from interventions That is what we have done here One of our constants is the MUL, while the multiple variables include: the BC combination of community involvement, police action, and sustained education on the issue of seatbelt wearing; the US style of federally-led usage initiatives; and

the contrast in the initial conditions in each place such as cultural attitudes toward government regulation and the economic context of socialized medical and vehicle insurance [73] With so many variables, determining the crucial ones is not an exact science, but it is one of the best means available to understand the social mechanisms involved

Another way to understand the reasons for success and failure is to compare one historical era with another (put another way, through historical reflection) We compared attitudes toward technology in the 1960s and 1970s with more present day perspectives, and noticed the extent to which politicians, manufacturers, and the public considered technology as a panacea—and the more, the better

ACKNOWLEDGMENTS

We are grateful to the Insurance Corporation of British Columbia (ICBC) for helping to fund the research and writing of this paper and would also like to thank Zoë Bennett for research assistance

---

[72]By contrast BC went from the first attempt (seatbelts in all vehicles) to the forth (MULs), without the steps in between

---

[73]A good analysis of comparative methodology can be found in Theda Skocpol and Margaret Somers, "The Uses of Comparative History in Macrosocial Inquiry," Comparative Studies in Society and History 22 (1980): 174-195

1999-01-1066

Rear Impact Air Bag Protection System

Gerald J. Keller and H. John Miller
Breed Technologies, Inc.

ABSTRACT

At least 160 reports and documents have been written on rear impact crashes and occupant protection since 1964. Reviewing automotive crash safety research has identified the importance of reducing relative motion between the head, neck, and torso. Although rear impact accidents produce relatively few severe injuries, National Automotive Sampling System (NASS) Crashworthiness Data System (CDS) data base analysis has determined that rear impact accidents are second only to frontal impacts in terms of cost to society.

This paper discusses a new system level approach for rear impact occupant protection. After reviewing automotive safety research, it is proposed that optimized rear impact protection is offered by supporting the vehicle occupant's head, neck, and torso in rear impact accidents. An unfurled air bag has been developed that provides energy absorption for the torso by deploying inside the seat back cushion, and extending vertically to support the occupant's head. Preliminary dynamic testing indicates that the Rear Impact Air Bag allows the occupant to ride down the cushion onto the seat back structure reducing 50% Hybrid III Male ATD injury scores.

INTRODUCTION

Rear impact accidents may be associated with a variety of soft tissue cervical spine injuries and complaints. These are sometimes termed Whiplash Associated Disorder Syndrome (WADS) [23]. Despite detailed field and laboratory research no acceptable WADS threshold parameters have been agreed upon by the automotive safety community. It has been reported that some physical symptoms of rear impact injuries do not occur until after the event, and can subsequently cause long term disability [19]. Pioneering volunteer research conducted by Mertz and Patrick [15] found that controlling relative motion between the head, neck, and torso can reduce injuries.

LITERATURE REVIEW

Detailed research into rear impact injury mechanisms and mitigation has been conducted for over thirty years. Several theories on injury mitigation have been consistently researched and quantified:

a. Reduction of relative motion between the head, neck, and torso.

b. Minimization of hyperflexion and hyperextension motion of the cervical spine and head.

c. Minimization of the occupant ramping effect on the seat back surface.

d. Minimization of occupant rebound.

Figure 1 illustrates a summarized concept of rear impact occupant kinematics as depicted by McConnel et al [13], as well as Thompson et al [36]. The vehicle occupant initially translates rearward into the vehicle seat and may ramp, or move vertically up the seat back. As the occupant's head and neck move rise above the headrest support, the head and neck rotate rearward (extension motion). After rearward motion has stopped, the occupant may rebound or move forward off the seat back surface.

The pelvis and torso have an important role in cervical spine and head kinematics. Dynamic testing by Svennson et al [32], indicated that the pelvic/torso interaction with the automotive seat back influences the initial and final position of the head and neck for the Hybrid III Anthropomorphic Test Dummy (ATD). The orientation of the torso can influence head and neck kinematics by changing the orientation of the Cervical Spine at the base of the neck. Image #4 in Figure 1 below illustrates the effect of forward motion of the pelvis changing the orientation of the torso, and subsequently the orientation of the Cervical Spine.

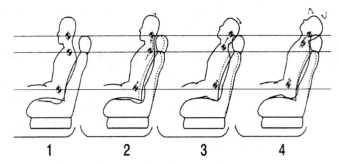

Figure 1. Conceptual Rear Impact Occupant Kinematics and Effect of Pelvis / Torso Angle on orientation of T1 and cervical spine.

The benefits of minimizing relative motion between the head, neck, and torso have been documented in volunteer research conducted by Mertz / Patrick in 1967 [15]. Volunteers were estimated to be capable of tolerating rear impact simulations up to 70.8 kph (44 mph) with mild discomfort by initially locating their heads on a rigid headrest during dynamic sled testing. A 1994 Volvo Car Corporation internal safety report [8] outlined field crash investigations indicating that when an occupant's head was initially in contact the headrest during a rear impact accident no injuries occurred.

Chalmers University of Technology indicated that rapid extension-flexion motion of the cervical spine might cause soft tissue injuries without exceeding voluntary range of motion [33]. Live pigs were subjected to rapid extension-flexion motion of the cervical spine. Spinal fluid pressure was measured at various locations along the cervical spine. Post-test examinations revealed injury to the nerve-root regions of the cervical and upper thoracic spine. Researchers theorized that the rapid pressure change of the spinal fluid had stressed the surrounding soft tissues. Assessment of this injury threshold for humans could not be extrapolated from the data.

Yang et al [42], proposed the risk for soft tissue injuries in rear impact accidents is increased by combination of compression and shearing motion of the cervical spine. Volunteer testing conducted by McConnell et al [13], documented vertical occupant acceleration relative to seat back surface in low speed rear impact testing. Compressive force load is applied to the cervical spine due to both the vertical acceleration of the thoracic spine and the inertial resistance of the head. Testing found that axial compression of the cervical spine increased the risk of soft tissue injury when encountering shearing forces typically associated with relative motion between the head, neck and torso in rear impact accidents.

Automotive seat back characteristics influence occupant vertical acceleration relative to the seat back surface, or ramping phenomenon during rear impact accidents. Viano [37], reported the energy absorption performance of the seat back, the initial inclination, and shape of the automotive seat back have been identified as factors in occupant ramping and rebound.

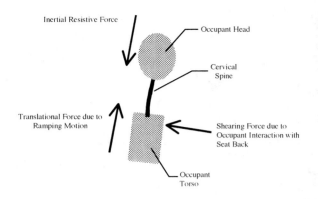

Figure 2. Free Body Diagram of compressive loading of the cervical spine.

Occupant rebound effect can be aggravated by an elastic response of the seat back. States et al [29], proposed the elastic rebound of the seat back could cause the torso to move forward while the occupant's head is still translating rearward. This effect would increase the relative velocity between the head, neck and torso augmenting the risk for neck injury.

In 1993 Svennson et al, addressed the potential of seat back and headrest designs to reduce neck injuries in rear impact accidents [31] & [32]. The distance between the occupant's head and headrest was found to have greatest impact on head and neck motion. A stiffened seat back in conjunction with optimized upper and lower seat back energy absorption properties, were found to be an effective method to reduce head and torso motion. Additionally, the distance between the head and headrest was reduced by allowing the occupant to compress the upper portion of the seat back early in the event before significant relative motion between the head and torso occurred.

Mechanical energy absorbing seat back and articulating headrest concepts have been developed that help control occupant kinematics and reduce injuries in rear impact accidents. Many current products are offered: Delphi Interior and Lighting Systems Catcher's Mitt™ (Wiklund et al. [41]), Autoliv Anti-Whiplash System™, Volvo Motor Corporation WHIPS System (Lundell et al. [12]), and an unnamed Toyota Motor Corporation seat design (Sekizuka [26]).

All four injury factors initially listed above were addressed in a study completed in 1971 by the University of Michigan Transportation Research Institute (UMTRI) on deployable headrest concepts [14]. Prototypes were constructed and tested in car to car rear impact crash tests at 32, 96, 128 kph (20, 60, and 80 mph) impact velocities. The inflatable headrest design utilized a sealed air bag cushion and a pyrotechnic-hybrid style airbag inflator. It was discovered that an optimized relationship between the head restraint and seat back structure is essential for meeting the performance goals noted above. The UMTRI research study recommended using energy absorbing materials in the seat back in contion with a deployable headrest to maintain torso/ad geometry during rear impact accidents.

ASS DATABASE ANALYSIS

order to identify opportunities for improved occupant otection in rear impact accidents NASS and CDS accient databases were analyzed from 1988 – 1996. The st to society of rear impact accidents has been estiated using frequency of rear impact accidents and erage cost of whiplash injuries. More detailed NASS-DS data base analysis has shown trends in accident npact angle and type of accident. Occupant interaction ith the vehicle seat and causes for seat failures in rear npact accidents were also examined.

ccident data analysis has indicated the effect on society f rear impact accidents despite low frequency of severe juries. 9.9% of all passenger cars involved in tow away ccidents from 1988-1996 in the NASS Database were volved in rear impact accidents. More significantly 23% f the total number injuries reported for all crash types rontal, side, rear), is attributed to rear impact accidents. verall rear impact ranks second only to frontal impact in rms of cost to society. Using 1988-1994 NASS Data, e National Highway Traffic Safety Administration NHTSA) estimated 742,340 whiplash injuries occurred nnually in passenger cars, light trucks, and vans. In rms of 1996 average societal costs (excluding property amage) the projected annual cost rear impact accidents as $4.5 billion per year [19].

able 1 describes the relationship between impact angle nd types of rear impact crashes. The majority of rear npacts involve two vehicles traveling in the same direcion (6:00 impact angle). A 'rear-end impact' describes a ear impact occurring from behind the target vehicle. A sideswipe impact' describes a rear impact occurring at n oblique angle from behind or adjacent to the target ehicle. A 'turning impact' describes a rear impact when a striking vehicle turns into the path of the target vehicle. Any situation not addressed by rear-end, sideswipe, or urning impact is coded as 'Other'.

Table 1. Impact Angle vs. Rear Impact Accident Type for all passenger cars involved in tow away accidents as reported in 1988-1996 NASS-CDS Database.

	Other*	Rear End Impact	Sideswipe Impact	Turning Impact
5:00	0.69%	1.10%	1.15%	1.91%
6:00	7.11%	**66.01%**	2.28%	1.25%
7:00	0.58%	1.68%	1.96%	2.44%
Other*	3.00%	7.65%	0.26%	0.93%
Total	11.37%	**76.44%**	5.65%	6.53%

The NASS database was researched with regard to the driver seating position when a seat failure occurs. Despite the narrow focus of this investigation, Table 2 demonstrates the high incidence of occupant interaction with the seat back frame during rear impact accidents.

Table 2. NASS-CDS 1990-1996 Rear Impact Crashes, Frequency of Driver Seat Damage for Rear-End Impacts for Rear-End Impact Style crashes. Data represents incidents when seat failure occurs.

Adjusters Fail	3.6%
Seat Back Fail	**27.3%**
Tracks/Anchors Fail	8.3%
Deformed by Occupant	**53.1%**
Deformed by Compartment Intrusion	1.2%
Combination	5.0%
Other	1.5%

REAR IMPACT AIR BAG SYSTEM

Breed Technologies has been researching a new type of inflatable countermeasure for rear impact protection that has the potential to provide improvements in all the aforementioned injury modes. The Rear Impact Air Bag System consists of an unfurled air bag system in the seat back structure with a deployable head restraint section. The inflator is located away from the occupant's head near the base of the seat. The headrest portion of the cushion is folded and packaged adjacent to the head rest, deploying through a tear seam in the seat back cushion trim material

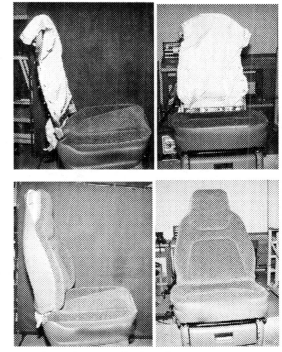

Figure 3. Rear Impact Air Bag System

This technology is intended to address several criteria outlined in the literature review:

a. Reducing relative motion between the head, neck, and torso by providing overall support to the entire body.

b. Reducing shear loading and reduce occupant rebound by allowing the occupant to ride down the cushion into the seat back frame.

c. Reduce offset distance between the occupant's head and the headrest.

Figure 4 illustrates the Breed Rear Impact Air Bag concept with respect to rear impact as depicted in literature.

Baseline Condition

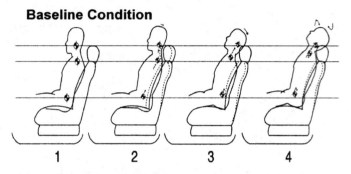

1   2   3   4

Rear Impact Air Bag

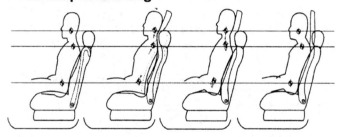

Figure 4.   Conceptual Effect of Rear Impact Air Bag System

TEST SETUP

26 dynamic sled tests, as well as 12 static deployment tests were conducted to investigate this concept. The data presented in this paper was generated by HYGE dynamic sled testing at 32 kph velocity using the following sled test pulse. Static air bag evaluation testing data proved to be inconsistent with dynamic test data due to the lack of occupant interaction with the seat back surface in a normally seated position.

Preliminary evaluation testing was conducted with a belted 50% Male Hybrid III ATD. Prasad et al [21], evaluated Hybrid III ATD instrumentation options for rear impact testing and found the standard Hybrid III ATD suitable for rear impact testing. The 50% ATD was centered in the seat with the torso resting against the seat back surface in normally seated position. A gap of 100 millimeters was used to locate the 50 % ATD's head from the headrest surface.

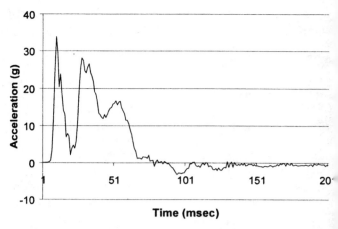

Figure 5.   32 kph Rear Impact Sled Test Pulse

A fixed headrest seat was chosen for preliminary evaluation testing to eliminate the variable of headrest position. Automotive seating is known to have a wide range of dynamic rear impact performance: Benson et al [3], Prasad et al. [22], Saczalski et al. [25], and Svensson et al. [32]. In order to normalize dynamic rear impact performance for development testing, the rear impact air bag system was evaluated using a stiffened seat frame with an energy absorbing recliner. Figure 5 shows the seat recliner mechanism replaced with a steel bar with a notch location to induce controlled bending failure. Notched steel bars were welded across both recliners to simulate a double side recliner arrangement. The perimeter of the seat frame back frame and base was also reinforced. The notched steel bar recliner arrangement was similar to the design used by Stroller, et al [30].

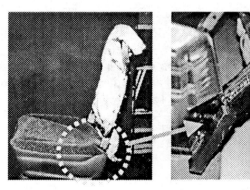

Figure 6.   Seat Recliner Modification Shown in Post Test Condition

Finite element analysis to predict the bending moment of the seat recliner mechanism. Simulated as ASTM 1010 Steel the reinforcement bars was modeled using ABAQUS/Explicit Finite Element Analysis Software using an applied torque load of 500 N-m. The analysis predicted a rotation of 30 degrees for with estimated recliner strength of 1129 N-m installed on both sides of the seat.

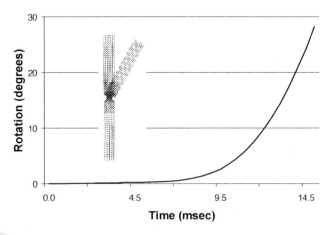

Figure 7. Finite Element Analysis of Notched Steel Seat Recliner Reinforcement. Inset image illustrates finite element bending initial and final positions.

TEST DATA

The following figures illustrate the dynamic performance of the rear impact air bag system in comparison with a baseline seat.

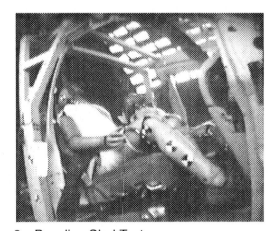

Figure 8. Baseline Sled Test

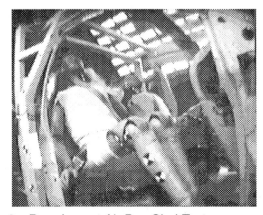

Figure 9. Rear Impact Air Bag Sled Test

Figures 10 & 11 illustrate that no appreciable ramping or lateral motion of the crash test dummy was observed in either the baseline or air bag tests. In both cases the

seat back recliner rotated approximately 30 degrees (see Figure 6), indicating that the modified seat system was capable of controlling ATD kinematics in the test.

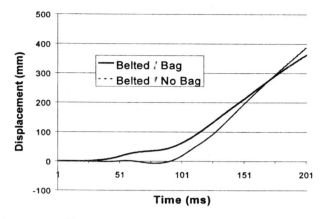

Figure 10. Chest Vertical Displacement

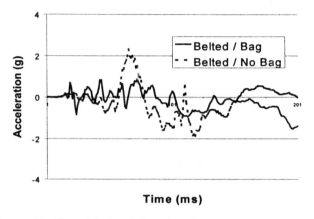

Figure 11. Chest Lateral Acceleration

Calculated Occupant Free Flight Velocity is an indication of potential occupant motion with respect to the test environment. Figures 12 indicates that occupant motion into the seat back was finished by approximately 60 milliseconds into the test.

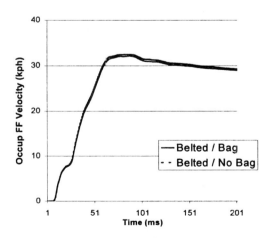

Figure 12. Calculated Occupant Free Flight Velocity

Figures 13-18 illustrate ATD interaction with the air bag system. Injury scores at 40-60 milliseconds were comparable, indicating no significant change to overall occupant kinematics due to interaction with the air bag system. Improvements in rebound are shown in the reduction of peak acceleration, shear force, and neck tension are observed at approximately 100 milliseconds into the test.

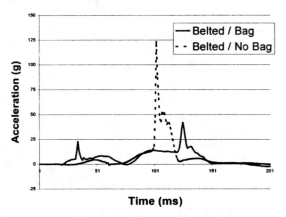

Figure 13. Head – X Acceleration

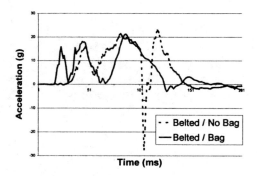

Figure 14. Chest X-Acceleration

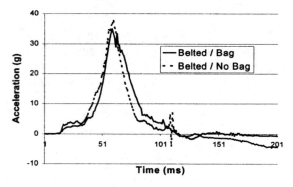

Figure 15. Pelvic X Acceleration

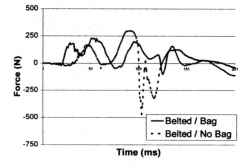

Figure 16. Upper Neck Shear Force

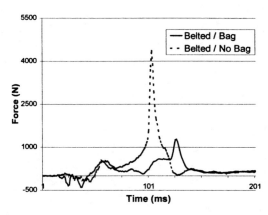

Figure 17. Neck Tension Force

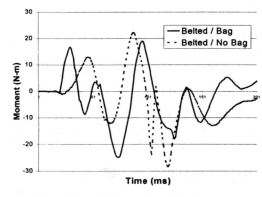

Figure 18. Upper Neck Extension Moment

DISCUSSION

The Hybrid III upper torso and lumbar are known to be too stiff to accurately predict the biomechanical interaction of the occupant with the seat back surface [32]. However, until more detailed biomechanical math modeling or sled simulation can be completed, the test data shown in Figures 12-19 provides a preliminary evaluation of the technology.

Lower neck extension moment data was not available for this test. The value of the lower neck load cell for evaluating seat performance, and the risk of neck injury in rear impact simulations was established by Prasad et al. [21]. Future development work will utilize Lower Neck

CONCLUSION

A review of automotive safety research has indicated the importance of addressing rear impact protection as a system level solution supporting the head, neck, and torso. A primary enabler for reducing rear impact injuries is minimizing relative motion between the head, neck and torso. Newer research has indicated that soft tissue injuries in the neck can occur in rear impact accidents without exceeding normal range of motion for vehicle occupants.

order to further understand the initiative for improved rear impact protection, NASS CDS data was used to quantify the cost to society and factors associated with rear impact accidents. NHTSA has estimated the social cost of Rear Impact at 4.5 million dollar per year. NASS CDS data base research has also indicated high frequency of 6:00 Impact Angle Rear Impact Accidents, and occupant deformation of the seat back leading to seat failure.

A deployable rear impact air bag system has been developed that provides rear impact protection by supporting the torso, neck, and head as a system. Preliminary testing has been conducted using a 50% Male Hybrid III ATD with a modified seat system with a 32 kph impact velocity. Analysis of the test data indicated that no appreciable jumping or lateral motion of the ATD occured. The modified seating system proved to be capable of controlling the occupant's rearward motion in both test conditions with 30 degrees of deformation. Preliminary test data indicates that the deployment of the air bag system did not significantly affect overall occupant kinematics. Reductions in injury scores late in the test show that the ATD was able to ride down the cushion into the seat back frame reducing the rebound effect off the seat back surface.

Initial test results are encouraging and development work continues on this technology.

ACKNOWLEDGMENTS

The authors would like to express sincerest thanks to John Burdock, Lynette Curtiss, and Susan Richards for their contributions to this paper.

The authors would also like to acknowledge Dr. Rose Ray and Edmund Lau of Exponent for their contributions to this paper.

REFERENCES

1. Anderson R., Welcher J., Szabo T., Eubanks J., Haight W.; "Effect of Braking on Human Occupant and Vehicle Kinematics in Low Speed Rear End Collisions"; SAE International Conference and Exposition; Society of Automotive Engineers, Inc., Warrendale PA USA, February 1998 – 983158.

2. Begeman P., Visarious H., Noite L., Prasad P.; "Viscoelastic Shear Responses of the Cadaver and Hybrid III Lumbar Spine"; 38th Stapp Car Crash Conference; Society of Automotive Engineers, Inc., Warrendale PA USA, November 1994 – 942205.

3. Benson B., Smith G., Kent R., Monson C.; "Effect of Seat Stiffness in Out of Position Occupant Response in Rear End Collisions"; 40th Stapp Car Crash Conference; Society of Automotive Engineers, Inc., Warrendale PA USA, November 1996 – 962434.

4. Digges K., Morris J., Malliaris A.; "Safety Performance of Motor Vehicle Seats"; SAE International Conference and Exposition; Society of Automotive Engineers, Inc., Warrendale PA USA, March 1993 – 930348.

5. Foret-Bruno J., Dauvilliers F., Tarriere C., Mack P.; "Influence of Seat and Head Rest Stiffness on the Risk of Cervical Injuries in Rear Impact"; 13th International Technical Conference on Experimental Safety Vehicles; National Highway Traffic Safety Administration, Washington, DC, USA, November 1991- 916126.

6. Foret-Bruno J., Dauvilliers F., Tarriere C., Mack P.; "Influence of the Seat and Head Rest Stiffness on the Risk of Cervical Injuries";

7. Hilyard J., Melvin J., McElhaney J.; "Deployable Head Restraints"; Department of Transportation Report; DOT-HS-031-2-281, January 31, 1973.

8. Jakobsson L., Norin H., Jernstrom C., Svensson S., Johnson P., Hellman I., Svesson M.; "Analysis of Head and Neck Responses in rear end impacts - a new human-like model"; 1994 Volvo Corporation Internal Safety Report.

9. James M., Strother C., Warner C., Decker R., Perl R.; "Occupant Protection in Rear-end Collisions: " Safety Priorities and Seat Belt Effectiveness"; 35th Stapp Car Crash Conference; Society of Automotive Engineers, Inc., Warrendale, PA, USA, November 1991 – 912913.

10. Kroonenberg A., Phillippens M., Cappon H., Wismans J., Hell W., Langwieder K.; "Human Head-Neck Response During Low-Speed Rear End Impacts"; SAE International Congress and Exposition; Society of Automotive Engineers, Inc., Warrendale, PA, USA, February 1998 – 980298.

11. Lau, E., Exuzides A.; "High Speed Rear Impact Crashes, Analysis of Field Accident Experience"; High Speed Rear Impact TOPTEC; Society of Automotive Engineers, Inc., Warrendale, PA, USA, October 27-28, 1997.

12. Lundell B., Jakobsson L., Alfredsson B.; "The WHIPS Seat – A Car Seat for Improved Protection Against Neck Injuries in Rear End Impacts"; 18th International Technical Conference on the Enhanced Safety of Vehicles; National Highway Traffic Safety Administration, Windsor, Ontario Canada, May 1998 – 98-S7-O-08.

13. McConnel W., Howard R., Guzman H., Bomar J., Raddin J., Raddin J., Benedict J., Smith H., Hatsell C.; "Analysis of Human Test Subject Kinematic Responses to Low Velocity Rear End Impacts"; SAE International Congress and Exposition; Society of Automotive Engineers, Inc., Warrendale, PA, USA, March 1993 – 930889.

14. Melvin J., McElhaney J., Roberts V., Portnoy H.; "Deployable Head Restraints – A Feasibility Study"; 15th Stapp Car Crash Conference; Society of Automotive Engineers, Inc., Warrendale, PA, USA, November 1971 – 710853.

15. Mertz H., Patrick L.; "Investigation of the Kinematics and Kinetics of Whiplash", 11th Stapp Car Crash Conference; Society of Automotive Engineers, Inc., Warrendale, PA, USA; October 1967 – 670919.

16. Mertz H., Prasad P., Irwin A.; "Injury Risk Curves for Children and Adults in Frontal and Rear Collisions"; 41st Stapp Car Crash Conference; Society of Automotive Engineers, Inc., Warrendale, PA, USA, November 1997- 973318.

17. NHTSA / University of Virginia; "Analytical Modeling of Occupant Seating/Restraint Systems"; Department of Transportation Report; DTRS-57-90-C-0092 February 1994.

18. NHTSA / University of Virginia; "Simulation of Occupant and Seat Responses in Rear Impacts"; Department of Transportation Report; DTRS-57-93-C-00105, September 1995.

19. Office of Crashworthiness Standards, National Highway Traffic Safety Administration; "Head Restraints – Identification of Issues Relevant to Regulation, Design, and Effectiveness"; www.nhtsa.dot.gov:80/cars/rules/CrashWorthy/status9.html; updated 12/98

20. Parkin S., Mackay G., Hassan A., Graham R.; "Rear End Collisions and Seat Performance – to yield or not to yield"; 39th AAAM Conference Proceedings, page 231.

21. Prasad P., Kim A., Weerappuli D.; "Biofidelity of Anthropomorphic Test Devices for Rear Impact"; 41st Annual Stapp Car Crash Conference; Society of Automotive Engineers, Inc., Warrendale, PA, USA, November 1997 – 973342.

22. Prasad P., Kim A., Weerappuli D., Roberts V., Schneider D.; "Relationships Between Passenger Car Seat Back Strength and Occupant Injury Severity in Rear End Collisions: Field and Laboratory Studies"; 41st Annual Stapp Car Crash Conference; Society of Automotive Engineers, Inc., Warrendale, PA, USA, November 1997 – 973343.

23. Quebec Task Force on Whiplash-Associated Disorders; Whiplash associated disorders (WAD) Redefining '"whiplash" and its management; Quebec City, QC Societe de l'assurance automobile du Quebec; January 1995

24. Reeves G., Bowerman N.; "The potential for an Improved Rear Impact Test Procedure"; 14th International Technical Conference on the Enhanced Safety of Vehicles; National Highway Traffic Safety Administration, Washington, D.C., USA, May 1994 – 946174.

25. Saczalski K., Syson S., Hille R., Pozzi M.; "Field Accident Evaluations and Experimental Study of Seat Back Performance Relative to Rear-Impact Occupant Protection"; SAE International Congress and Exposition; Society of Automotive Engineers, Inc., Warrendale, PA, USA, March 1993 – 930346.

26. Sekizuka M.; "Seat Designs for Whiplash Injury Lessening"; 18th International Technical Conference on the Enhanced Safety of Vehicles; National Highway Traffic Safety Administration, Windsor, Ontario Canada, May 1998 – 98-S7-O-06.

27. Siegmund G., King D., Lawerence J., Wheeler J., Brault J., Smith T.; "Head Kinetic Response of Human Subjects in Low Speed Rear-End Collisions" 41st Annual Stapp Car Crash Conference; Society of Automotive Engineers, Inc., Warrendale, PA, USA November 1997 – 973341.

28. States, J., Korn, M., Masengill, J; American Association for Automotive Medicine, 13th annual conference; October 1969 - 1969-12-0004.

29. States J., Balcerak J., Williams J., Morris A., Babcock W., Polvino R., Riger P., Dawley R.; "Injury Frequency and Head Restraint Effectiveness in Rear-End Impact Accidents"; 16th Stapp Car Crash Conference; Society of Automotive Engineers, Inc., Warrendale, PA, USA, November 1972 – 720967.

30. Strother C., James M., Gordon J.; "Response of Out-of-Position Dummies in Rear Impact"; SAE International Congress and Exposition; Society of Automotive Engineers, Inc., Warrendale, PA, USA, February 1994 – 941055.

31. Svensson M., Lovsund P., Haland Y., Larsson S. "The Influence of Seat-Back and Head-Restraint Properties on the Head-Neck Motion During Rear Impact"; 1993 ICROMBI Conference; paper number 1993-13-0028.

32. Svensson M., Lovsund P., Haland Y., Larsson S. "Rear-End Collisions – A Study of the Influence of Backrest Properties using a New Dummy Neck" SAE International Congress and Exposition; Society of Automotive Engineers, Inc., Warrendale, PA, USA March 1993 – 930343.

33. Svensson M., Alumna B., Lovsund P., Hanson H. Seeman T., Sunesson A., Ortengren T.; "Pressure Effects in the Spinal Canal during Whiplash Extension Motion - Possible Cause of Injury to the Cervical Spinal Ganglia"; 1993 ICROMBI Conference; paper number 1993-13-0013.

34. Szabo T., Welcher J.; "Human Subject Kinematics and Electromygraphic Activity During Low Speed Rear Impacts"; 40th Stapp Car Crash Conference; Society of Automotive Engineers, Inc., Warrendale, PA, USA, November 1996 – 962432.

35. Szabo T., Welcher J., Anderson J., Rice M., Ward P., Paulo L., Carpenter N.; "Human Occupant Kinematic Response to Low Speed Rear-End Impacts"; SAE International Congress and Exposition; Society of Automotive Engineers, Inc., Warrendale, PA, February 1994 940532.

36. Thomson R., Romilly D., Navin F., Macnabb M.; "Dynamic Requirements of Automobile Seatbacks"; SAE International Congress and Exposition; Society of Automotive Engineers, Inc., Warrendale, PA, USA, March 1993 – 930349.

37. Viano D.; "Influence of Seat Back Angle on Occupant Kinematics in Simulated Rear-End Impacts"; 36th Stapp Car Crash Conference; Society of Automotive Engineers, Inc., Warrendale, PA, USA, November 1992 – 922521.

8. Viano D.; "Restraint of Belted or Unbelted Occupant by Seat in Rear-End Impacts"; 36th Stapp Car Crash Conference; Society of Automotive Engineers, Inc., Warrendale, PA, USA, November 1992 – 922522.

9. Viano D., Gargan M.; "Headrest position during normal driving; implications to neck injury risks in rear crashes"'; 39th AAAM Conference Proceedings, page 215.

10. Warner C., Stother C., James M., Decker R.; "Occupant Protection in Rear-End Collisions: II The Role of Seat Back Deformation in Injury Reduction"; 35th Stapp Car Crash Conference; Society of Automotive Engineers, Inc., Warrendale, PA, USA, November 1991 – 912914.

11. Wiklund K., Larsson H.; "SAAB active head restraint (SAHR) – Seat design to reduce the risk of neck injuries in rear impacts; "; SAE International Congress and Exposition; Society of Automotive Engineers, Inc., Warrendale, PA, USA, February 1998 – 980297.

12. Yang K., Begeman P., Muser M., Niederer P., Walz F.; "On the Role of Cervical Facet Joints in Rear End Impact Neck Injury Mechanisms"; SAE International Congress and Exposition; Society of Automotive Engineers, Inc., Warrendale, PA, USA, February 1997 – 970497.

Advanced Air Bag Systems and Occupant Protection Recent Modifications to FMVSS 208

John E. Hinger and Harold E. Clyde
Exponent® Failure Analysis Associates, Inc

ABSTRACT

Because of a rising number of air bag related injuries and specific Congressional instructions, FMVSS 208 was revised in March 1997. At that time, the changes allowed manufacturers to quickly implement redesigned air bags that were less powerful with the goal of reducing air bag related injuries. The legislature has since mandated additional revisions to FMVSS 208 to ensure use of new technologies for the protection of occupants of varying stature. This paper presents an overview of some of the more significant modifications to FMVSS 208 and discusses some of the challenges for advanced air bag systems.

INTRODUCTION

Frontal crash protection for vehicle occupants has been extensively debated for many years. Frontal crashes are the leading cause of fatalities to front seat occupants, particularly to unrestrained occupants. On May 12, 2000, the National Highway Traffic Safety Administration (NHTSA) published an amended version of Federal Motor Vehicle Safety Standard (FMVSS) 208 for occupant safety in motor vehicles. Changes in FMVSS 208 were made pursuant to Congressional instruction to NHTSA through the Transportation Equity Act for the 21st Century which required NHTSA to "improve occupant protection for occupants of different sizes, belted and unbelted … while minimizing the risk to infants, children, and other occupants from injuries and deaths caused by air bags, by means that include advanced air bags."[1] The new requirements will help ensure that advanced air bag technologies, meant to improve the injury reducing capabilities of air bag systems, will be installed in future motor vehicles. The key modifications, included in this standard, were designed to reduce the risk of injury associated with the passive occupant protection provided by an air bag, including that to small children and out-of-position occupants, and to increase new vehicle testing requirements that are to include the use of additional sizes of test dummies.

Previous SAE papers have presented the regulator history of air bag systems through the 1990s. This paper presents recent regulatory changes and how these have modified FMVSS 208. These topics include the phased-in adaptation of advanced air bag technologies through 2010; details of the new testing requirements, injury criteria, and use of additional sizes of crash test dummies; and a discussion of the current issues regarding these technologies. Implementation of the new requirements will occur in two stages with a phase-in period for each stage. The first stage is intended to minimize the risk of air bags to vehicle occupants, especially women drivers and children in the front passenger seat. The second stage is designed to improve protection for belted occupants and represents a fundamental change in occupant protection philosophy since there will be higher test speed requirements for restrained dummies than for unrestrained dummies, causing designers to focus on each test condition independent of the other.

FRONTAL INJURY

Air bag systems have been effective in reducing injuries and death to front seat occupants during collisions. Air bags have reduced driver fatality risk by 31% in pure frontal crashes and by 11% in all crashes [2], and reduced fatalities for right-front occupants, age 13 and older, by 32% in pure frontal crashes.[2] It is estimated that the combination of seat belts and air bags are 75 percent effective in preventing serious head injuries and 66 percent effective in preventing serious chest injuries.[3] Since the incorporation of air bags into passenger vehicles in 1986 and through August 1, 2000, it is estimated that air bags have saved the lives of 5,899 front seat occupants.[4] NHTSA estimates that air bags may save up to 3,200 lives annually when all passenger cars and light trucks are equipped with air bags. The benefits of air bags have not been without some occupant safety cost. As of August 1, 2000, there have been 167 confirmed air bag-related fatalities.[4] Of these fatalities, 99 were children, 62 were drivers and 6 were passengers. The one common fact to each of these air bag fatalities is that the occupant was very close to the air bag when it started to deploy.[1] During May, 2000,

NTSA issued a final rule amending FMVSS 208 to improve the frontal crash protection of air bags for all occupants and to reduce the risk of air bag induced injuries to occupants, particularly small women and children.

PREVIOUS RULEMAKING

Previous rulemaking has attempted to address some of the issues surrounding the risk of injuries to small women and children. This rulemaking has regulated the air bag warning labels provided in vehicles, the availability of on-off switches for air bags, the option of a sled test to certify vehicles with a "depowered" air bag design and the requirements for deactivation of air bags.[5]

TEA 21

The Transportation Equity Act for the 21st Century (TEA 21) was enacted in June 1998, by Congress requiring NHTSA to issue a rule amending FMVSS 208 to improve occupant protection for occupants of different sizes, belted and unbelted, under Federal Motor Vehicle Safety Standard No. 208, while minimizing the risk to infants, children, and other occupants from injuries and deaths caused by air bags, by means that include advanced air bags."[1] To achieve these goals, NHTSA is requiring vehicles to meet broader test requirements with an assortment of new dummies to ensure that occupants are properly protected under a wider variety of crash conditions.

FMVSS 208

The purpose of FMVSS 208 is to reduce the number of deaths of vehicle occupants, and severity of injuries, by specifying vehicle crashworthiness requirements and specifying equipment requirements for active and passive restraint systems.[1] Originally, the standard specified the types of restraints required and was later amended to specify performance requirements for test dummies seated in the front outboard seating positions of passenger cars. Previously, FMVSS 208 required all passenger cars manufactured after September 1, 1997, and light trucks manufactured after September 1, 1998 to be equipped with driver and passenger air bags, along with manual lap and shoulder belts. Specific sun visor warning labels have also been required. Currently, for unbelted occupants, manufacturers have the option of certifying vehicles with a 48 km/h (30 mph) frontal barrier crash test or with an approximately 48 km/h delta-V generic pulse sled test. For belted occupants, the vehicles are to be certified with 48 km/h frontal and oblique barrier tests. To comply with FMVSS 208, instrumented test dummies are positioned in the front outboard seating positions for each test and must meet specific injury criteria established in the standard.[6]

MODIFICATIONS TO FMVSS 208

The recent modifications to FMVSS 208 will ensure that advanced air bag technologies will be adapted to make vehicles more effective in protecting occupants and reduce the risk of air bag induced injuries. These modifications specifically address small stature drivers, child occupants and the average male. The standard requires new dynamic and static testing, including changing the way head injury risk is measured, adding new neck injury criteria and reducing the allowable chest deflection. New warning labels have also been required for vehicles with advance air bag systems.

SMALL STATURE OCCUPANTS

TEA 21 specifically targeted the adaptation of advanced air bag technologies to improve the occupant protection of small stature drivers. Specific language in the law requires that the protection of small drivers be taken into account for all occupant protection requirements.

The 5th percentile female dummy has been added to most of the new crash test requirements. It will be used in both belted and unbelted vehicle crash tests and for out-of-position occupant testing.

CHILDREN

Manufacturers have the option of selecting one of two alternative test methods to reduce the risk of air bag-induced injuries to infants in child restraint systems. The manufactures' options include either suppressing the air bag deployment or deploying the passenger air bag in a low risk manner in the presence of a 12-month-old Child Restraint Air Bag Interaction (CRABI) dummy in a rear facing child safety seat (RFCSS) or a convertible child restraint in the rear-facing mode.

To reduce the risk of air bag-induced injuries to small children in the front seat, manufacturers will conduct tests using 3-year-old and 6-year-old child dummies. The manufacturer has the option of either suppressing the air bag deployment if a child is present, deploying it in a low risk manner if a 3-year-old and 6-year old child dummy is out-of-position or suppressing the air bag deployment when an occupant is out-of-position.

Additional injury parameters were developed with limits that vary between infant, child, and adult dummies. The required warning labels will still state that children are safest in the back seat.

PHASE-IN REQUIREMENTS

The new requirements of FMVSS 208 are scheduled to take effect over a seven-year period. This period is divided into two phases. The first phase occurs over

4 years and addresses the reduction of injury due to occupant air bag interactions. The second phase occurs over 3 years and addresses increased occupant protection of restrained occupants.

FIRST PHASE – AIR BAG INJURY REDUCTION

The first phase-in period adds the requirements of 5th percentile dummy barrier testing (frontal and offset), air bag suppression or multistage inflation in the presence of rear facing child seats, 3 and 6 year-old child dummies or out-of-position occupants, and 5th percentile out-of-position driver occupants.

The first phase-in requires that a percentage of each manufacturers light vehicle production meet the requirements of the standard by a specified date. If a manufacturer exceeds the percentage required early in the phase-in, they receive credit for future years, but must be 100% compliant by September 1, 2006. The first phase-in is follows:

35% of production beginning September 1, 2003

65% of production beginning September 1, 2004

100% of production beginning September 1, 2005

SECOND PHASE – RESTRAINED OCCUPANTS

The second phase-in requires that the speed of frontal barrier testing with belted 50th percentile male dummies be raised from 48 km/h to 56 km/h (30 mph to 35 mph). The second phase-in also allows manufacturers to receive future credit for meeting the requirements earlier. All vehicles must 100% compliant by September 1, 2010. The second phase-in is follows:

35% of production beginning September 1, 2007

65% of production beginning September 1, 2008

100% of production beginning September 1, 2009.

DYNAMIC TESTING REQUIREMENTS

The dynamic test matrix for FMVSS has been drastically modified by the new requirements. The previous crash test requirements of FMVSS 208 required only the use of a belted and unbelted 50th percentile male dummy in a 48 km/h rigid and oblique barrier test. This could be supplemented with the 48 km/h sled test option for air bag certification.

The new requirements include testing with belted and unbelted 5th percentile female and 50th percentile male dummies into a frontal rigid barrier. Additionally, the unbelted 50th percentile male is used in an oblique rigid barrier test and a belted 5th percentile female dummy is used in a deformable offset barrier test. Chart 1

summarizes the test dummy size, restraint use, barrier type, impact speed and angle for the new requirements.

Chart 1. FMVSS 208 required crash test matrix

ATD	Seat Belt	Barrier Type	Crash Type	Speed (mph)	Angle
50th	Yes	Rigid	Frontal	0-30*	90
5th	Yes	Rigid	Frontal	0-30	90
50th	No	Rigid	Frontal	20-25	90
5th	No	Rigid	Frontal	20-25	90
50th	No	Rigid	Frontal	20-25	30
5th	Yes	Deformable	Offset Frontal	0-25	90

* Raised to 35 mph starting in the 2008 model year

The 40% offset deformable barrier test, with a 5th percentile female dummy, is believed to simulate real world crashes and has been designed to test air bag systems for deployment and driver interaction with the air bag system.

NEW STATIC TEST REQUIREMENTS

Additional static tests of air bag systems are now required to determine the systems interaction with small adults, children, and infant restraint devices. These tests include static testing of the air bag system with out-of-position occupants. The system must either inflate in a low risk manner or suppress the deployment of the air bag if an out of position driver were detected.

Upon the detection of a small right front seat occupant, the system must either inflate at a low-speed impact inflation rate, or suppress the air bag deployment, or suppresses the air bag deployment if the child moves close to the air bag during an impact.

With the presence of a child safety seat in the right front position, the system must either inflate at a low-speed impact inflation rate, or suppress the inflation of the air bag.

INJURY CRITERIA

In addition to Head Injury Criteria (HIC), femur load, chest deflection and chest acceleration, new standards are established for neck injury and chest deflection.

The calculation formula and upper acceptable limit of HIC has been modified. The new formula calculates HIC based on maximum head accelerations over a 15 millisecond time frame, instead of the previous 36 millisecond duration. The maximum acceptable limit for the HIC_{15} calculation is not to exceed the value of 700 for the 50th percentile male and 5th percentile female dummies.

Chest acceleration is not to exceed 60 G's for the 50th percentile male and 5th percentile female dummies. hile chest deflection shall not exceed 52 mm for the percentile female and 63 mm for the 50th percentile male.

Femur loads shall not exceed 6805 N for the percentile female and 10,000 N for the 50th percentile male.

The neck injury combines neck load and moments into an injury standard. This new value is called Nij. Rather than evaluating each neck load and moment separately, Nij combines axial neck load (Fz) and bending moment (Mocy) into a single cumulative score. The Nij is defined as:

$$Nij = \left(\frac{Fz}{Fzc} \right) + \left(\frac{Mocy}{Myc} \right) \qquad (1)$$

Where Fzc is the critical axial load of 2800 N in both tension and compression. Myc is the critical bending moment about the occipital condyle and is 93 Nm in flexion and 37 Nm in extension.

The Nij cannot exceed 1 at any measured load condition during the test event. Additionally the peak axial force (Fz) cannot exceed 1490 N in tension or 1820 N in compression.

OTHER ISSUES

The 1997 revisions to FMVSS 208 for certification allowed manufacturers to use depowered air bags with the use of the sled test option. The sled test option for certification will be eliminated because of the variability in the structural characteristics of vehicles within the US fleet and the direct consequences on test conditions and results due to these variations. Sled testing did not account for variations in crash pulse to individual vehicle designs since a single pulse was used for all vehicles regardless of vehicle weight, stiffness, or size class. Therefore, NHTSA has determined that only full-scale crash tests would be allowed for certification.

ADAPTATION OF ADVANCED AIR BAG TECHNOLOGY

Under the requirements of TEA 21, FMVSS 208 has been modified to both allow the use of advanced technologies, and provide a means by which a manufacturer can certify new technologies for use in production vehicles. The regulations within 208 are specifically vague in the requirements of which advanced technologies will be allowed. This vagueness in requirements was designed so that new technologies (currently under development or yet to be invented) would not be stifled in their development by the regulations.

One area under development is the use of multi-level inflation systems. These systems could provide differing levels of protection for low-speed and high-s

collisions or be tailored to reduce the injury potential for out of positions occupants. Vehicles equipped with a multi-level inflation system will be required to meet the injury criteria with a low risk (level) deployment in a rigid barrier test with unbelted 5th percentile dummies in both outboard seating positions at 26 km/h (16 mph). This test, combined with the unbelted 32 km/h (20 mph) rigid barrier test requirements, provides the flexibility to develop a multi-level inflation system that can protect occupants in low speed crashes and reduce the injury to out-of-position occupants when a high-level deployment occurs. High-level deployments provide additional protection to occupants in severe crashes.

Another area under development is a weight sensor that can detect the presence of a child (infant to 6-year-old) and suppress the deployment of the passenger air bag. Currently, there is no suitable test dummy to test suppression systems for children ages 7 to 12.

AIR BAG WARNING LABELS

Since 1995 model year, NHSTA has specified the content of vehicle warning labels for supplemental restraint systems. These original warning label requirements were subsequently revised by NHTSA for 1997 model year vehicles. For vehicles equipped with advance air bag systems a new sun visor-warning label will be used. Also, if manufacturers deem it necessary to supplement the new required label with additional information on a separate label they will now permitted to do this. The new required label should be visible with the sun visor in the down position. The new required label is shown in Figure 1.

Additionally, a new removable label should be placed on the right side of the occupant compartment near the supplemental restraint system prior to its sale. This new label is shown in Figure 2.

Figure 1. Sun visor warning label

Label Outline, Vertical and Horizontal Lines Black

Bottom Text and Artwork Black with White Background

Top Text Black with Yellow Background

⚠ WARNING

EVEN WITH ADVANCED AIR BAGS
- Children can be killed or seriously injured by the air bag
- The back seat is the safest place for children
- Always use seat belts and child restraints
- See owner's manual for more information about air bags

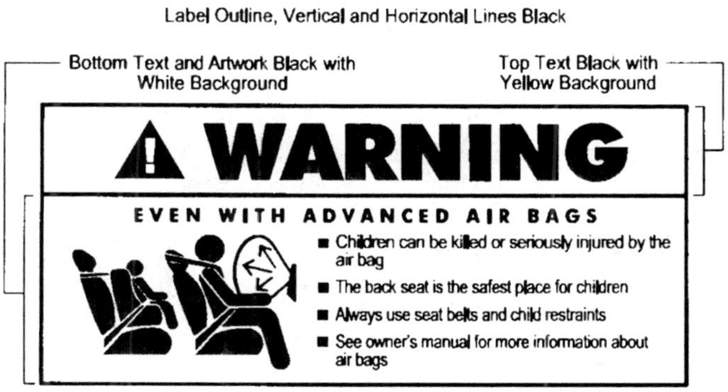

Figure 2. Removable dash label

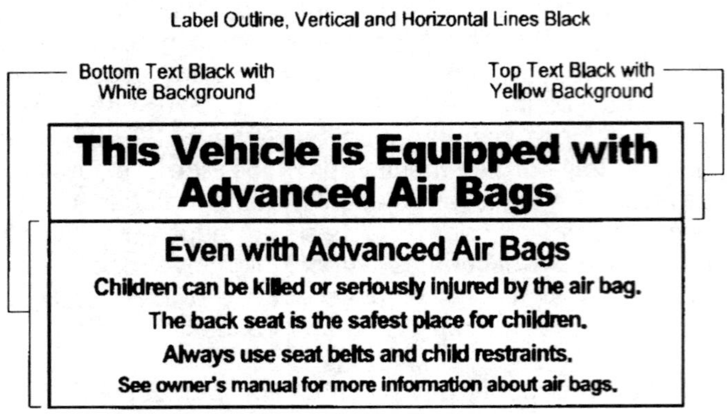

Label Outline, Vertical and Horizontal Lines Black

Bottom Text Black with
White Background

Top Text Black with
Yellow Background

This Vehicle is Equipped with Advanced Air Bags

Even with Advanced Air Bags

Children can be killed or seriously injured by the air bag.

The back seat is the safest place for children.

Always use seat belts and child restraints.

See owner's manual for more information about air bags.

FUNDAMENTAL CHANGE IN OCCUPANT PROTECTION PHILOSOPHY

All previous standards only addressed injury criteria with a 50th percentile male dummy. No regulatory consideration was given to out of position occupants or occupants of varying statures. Additionally, the requirements for belted and unbelted occupants remained identical. By separating the belted occupant requirements from the unbelted occupant, NHSTA has made a philosophical change in the viewing of injuries for occupants.

This change allows for better protection of smaller stature occupants. The physics of slowing an unbelted occupant and a belted occupant vary greatly. A belted occupant will rely primarily on the belt system to provide ride-down time and limit deceleration forces and the unbelted occupant will rely on the air bag system to reduce contact forces with the vehicle interior. For the belted occupant, the air bag system is truly a supplemental system to the seat belt system, while the unbelted occupant relies on the air bag system as their primary protection.

To date, air bags have been designed for the benefit of belted and unbelted 50th percentile male occupant in a 30 mph barrier collision due to certification requirements. Under the recent modifications to FMVSS 208, air bag systems in the future will be designed to benefit occupants of various sizes, regardless of belt usage.

The law in 49 of the 50 states currently requires the use of seat belts. However, the national seat belt usage rate in the United States is only about 69%.[3] The recent modifications to FMVSS will allow air bag system designers the flexibility to provide protection to belted occupants, while reducing the injury potential to unbelted occupants, particularly to occupants that have historically had some increased risk of injuries from deploying air bags.

CURRENT ISSUES

Prior to modifying FMVSS 208, NHTSA and NASA agreed to cooperative effort that leveraged NASA's expertise in advance technologies to understand the parameters affecting air bag systems, to assess air bag technology state-of-the-art and to identify new concepts for air bag systems. The Jet Propulsion Laboratories (JPL) was selected by NASA to undertake this investigation. Due the volume of information contained in JPL's report, only highlights of the findings will be presented and readers interested in more details should read the report.[7]

"Air bag systems are a significant engineering design challenge because they deploy rapidly and with great force toward an approaching occupant. Their deployments are based on predictions of crash severity early in the event, …"[7]

Advanced restraint systems can improve the safety of air bag systems by providing more information about the type and severity of crash and tailoring the deployment characteristics to individual occupants. "Improving air bag safety is an incremental process, and implementation of advance technology will be evolutionary."[7]

It was anticipated that advanced technologies such as improved crash sensors, belt use sensors, seat position sensors, automatic suppression, two-stage inflators, compartmented air bags would be available by model year 2001, and by model year 2003, more sophisticated occupant sensing systems could be incorporated to suppress inflation of the air bag system when it has a high likelihood of injuring a front seat occupant. Along with these technologies, there comes a predicted risk of air-bag-induced injuries from the unreliability of the advanced systems. The development of advanced restraint systems is influenced by government regulatory requirements and industrial costs. Chart 2 identifies advanced air bag technologies that are currently under development or could be developed in the future and the expected readiness date of these technologies as determined by JPL.[7]

Advanced air bag technologies can make air bag systems safer, but only if the technologies are deemed reliable. Without reliable technologies, an air bag system may deploy when not needed, may not deploy when needed or may deploy in a manner not optimal for the occupant. It will remain the goal of air bag designer to only implement new technologies when they have been determined to be reliable.

Chart 2. Advance Air Bag Technologies

Technology	Description	Readiness Date
Pre-crash sensing	Remote sensing for early crash severity determination	Could be available by MY2001
Belt use sensor	Determines whether or not a seat belt is being used	Could be available by MY2000
Belt spool-out sensor	Aid in determining occupant size	Could be available by MY2001
Seat position sensor	Used to estimate driver size and proximity to air bag	Could be available by MY2001
Occupant classification sensor	Measure occupant weight and presence	Could be available by MY2000
Occupant proximity sensor	Provide range information between occupants and interior	Could be available by MY2000/2001
Inflatable seat belts	A portion of seatbelt is inflated to augment the belt function	Could be available by MY2001

ADOPTED ADVANCE AIR BAG TECHNOLOGIES

Currently, some manufacturers have already begun to incorporate advanced air bag technologies into their newer vehicles. As of MY2000, advanced air bag technologies that have been adopted in some vehicles include: dual-stage passenger air bags, advanced crash severity sensors, belt use sensors, dual-threshold deployments for driver and passenger air bags, and driver seat position sensors.

CONCLUSIONS

The modifications to FMVSS 208 will provide greater protection to small stature drivers, front seat child passengers and front seat infant passengers. New air bag designs that are developed to meet the new requirements will create less risk of serious air bag induced injuries to occupants than current air bag designs. It is expected that new air bag system designs will incorporate advanced air bag technologies.

In addition to using a larger family of test dummies to evaluate air bag system performance, new designs will be required to meet more stringent injury criteria.

The new requirements of FMVSS 208 will be phased-in over a seven-year period. The first phase, covering occupant air bag injury reduction, is to begin with model year 2003, and the second phase, covering increased protection for restrained occupants, will be completed by model year 2010.

Seat belts still offer the most effective protection to occupants and will remain the primary restraint system for all occupants. The continuous development and manufacture of improved occupant crash protection systems in motor vehicles remains the goal of FMVSS 208.

This paper provides a summary the most recent modifications to FMVSS 208. Although, the new requirements have been adopted, some aspects of the new requirements are still being addressed through the rulemaking process and could lead to further changes in the standard prior to its implementation in 2003.

REFERENCES

1. Department of Transportation, NHTSA, 49CFR Parts 552, 571 and 595, [Docket No. NHTSA 007013; Notice 1], May 2000.
2. Fourth Report to Congress, Effectiveness of Occupant Protection Systems and their Use, NHTSA May 1999
3. NHTSA Safety Fact Sheet, 11/2/99.
4. NHTSA Special Crash Investigation report, 8/1/00.
5. Recent Regulatory History of Air Bags, SAE 980650.
6. Code of Federal Regulations, Transportation, part 571.208, October 1, 1997.
7. Advance Air Bag Technology Assessment, JPL, April 1998.

CONTACT

John E. Hinger is a managing engineer with Exponent Failure Analysis Associates. He has previously worked in the automotive field for several automotive manufacturers. He has a B.S. in mechanical engineering from the University of Illinois. He may be contacted at jhinger@exponent.com.

Harold E. Clyde, P.E. is a senior engineer with Exponent Failure Analysis Associates. He has a M.S. in mechanical engineering from Brigham Young University. He may be contacted at hclyde@exponent.com. Either author may be contacted through Exponent at (650) 688-7282.

Theoretical Evaluation of the Requirements of the 1999 Advanced Airbag SNPRM

Part One: Design Space Constraint Analysi

Tony R. Laituri, N. Sriram, Brian P. Kachnowsk
Brion R. Scheidel and Priya Prasa
Ford Motor Co

ABSTRACT

In the 1999 Supplemental Notice for Proposed Rulemaking (SNPRM) for Advanced Airbags, the National Highway Traffic Safety Administration (NHTSA) sought comments on the maximum speed at which the high-speed, unbelted occupant test suite will be conducted, i.e., 48 kph vs. 40 kph. To help address this question, an analysis of constraints was performed via extensive mathematical modeling of a theoretical restraint system. First, math models (correlated with several existing physical tests) were used to predict the occupant responses associated with 336 different theoretical dual-stage driver airbag designs subjected to six specific Regulated and non-Regulated tests. Second, the pertinent, predicted occupant responses for all 336 designs were compared with a set of generic acceptance criteria for the six distinct performance constraints (where two of the six represented the aforementioned "high-speed" unbelted occupant test suite and where "high-speed" was set equal to either 48 or 40 kph). Finally, statistics were generated to help evaluate the stringency of the various performance constraints.

Results from the assessment for a modeled, prototype, mid-sized passenger car included the following: (1) <u>None of the 336</u> theoretical dual-stage driver airbag designs satisfied the generic acceptance criteria set when the unbelted rigid fixed barrier testing constraints were run at 48 kph, (2) <u>21 of the 336</u> satisfied the generic acceptance criteria set when the unbelted rigid fixed barrier testing constraints were run at 40 kph, and (3) When considering the discarded designs, <u>nearly all</u> of them were predicted to not comply due to (at least one of) the <u>unbelted</u> occupant performance constraints of the generic acceptance criteria set.

1.0 BACKGROUND

The Federal Motor Vehicle Safety Standard No. 208 (FMVSS 208) was amended in 1997 to include an alternative to the 48 kph, full-vehicle, rigid fixed barrier test inv

ing unbelted occupants. This amendment involved wha came to be known as the "generic sled test," i.e., ar unbelted, mid-sized male instrumented test dummy sub jected to a 48 kph ÷V, 17.2 G, 125 ms half-sine wave "generic" crash pulse on a hydraulically-controlled sled This sled test alternative was made available by NHTSA ir order to expedite the introduction of depowered airbags -- a means to reduce further the already low real-world risk o injury related to airbag inflation.

The dual-stage airbag inflator is one of the possible design elements of an advanced restraint system that is anticipated to help reduce even further the risk of airbag inflation-related injury. The function of the dual-stage inflator is to provide two separate injections of gas into the airbag. If dual-stage inflators are activated by new, advanced-technology crash-sensing systems, occupant restraint systems could potentially be more adaptable and yield airbag energies more commensurate with vehicle crash severities, i.e., lower-energy inflations for moderate-severity crashes and higher-energy inflations for the more severe crashes.

In the midst of significant manufacturer and supplier efforts to voluntarily develop and implement these 1st-generation advanced restraint systems, NHTSA issued a Notice for Proposed Rulemaking (NPRM) (NHTSA, 1998) and a Supplemental Notice for Proposed Rulemaking (NHTSA, 1999). Therein, NHTSA proposed broad-ranging new elements to Regulated testing -- four of present interest: (1) An end to the aforementioned 1997 amendment to FMVSS 208, thereby reverting to the high-speed rigid fixed barrier vehicle testing with an unbelted, mid-sized male Hybrid III test dummy (HIII50), (2) The introduction of high-speed rigid fixed barrier vehicle testing with an unbelted, small-sized female Hybrid III test dummy (HIII05) in addition to the aforementioned unbelted HIII50 high-speed rigid fixed barrier vehicle test, (3) The introduction of a set of static, out-of-position occupant (OOPO) tests, and (4) The introduction of new neck injury criteria for both instrumented

he present study provided a theoretical, mathematical odel-based analysis of the predicted effects of Regulated nd non-Regulated performance constraints on available esign space, i.e., the designs that satisfied the related set generic acceptance criteria.

0 MATHEMATICAL MODEL DEVELOPMENT

1 MODELS FOR DYNAMIC PERFORMANCE TESTS

ne first step toward attempting to predict the effects of ternative frontal impact Regulations on airbag designs volved development of occupant response models of xisting physical tests. The modeling software chosen was ladymo3D developed by TNO. The physical tests involved non-production, mid-sized passenger car equipped with a rototype advanced restraint system. The airbag system cluded a dual-stage inflator. The seat belt system cluded a load-limited retractor and a pyrotechnic preten- ioner at the belt buckle. The validation cases consisted of ariations on crash type, speed, severity, restraint level, ccupant size, and occupant seating position (see Figure 1 r one example, i.e., the USA New Car Assessment Pro- ram (NCAP) test).

ix validation cases were studied in order to attempt to cor- :late math model occupant responses with those of physi- al experiments while applying consistent modeling ractices in all of the cases (see Appendix 1). During the alidation process, special attention was given to six occu- ›ant responses: head acceleration, upper neck fore and aft hear force, upper neck tensile axial force, corrected upper »eck extension moment, chest acceleration, and chest leflection. Additionally, pelvis displacement and femur ›ad comparisons between test and simulation were made vith intent to gain acceptable model correlation with the

time histories of the lower-body responses observed in the tests. Other comparisons included column stroke and seat belt forces (when applicable).

The results of the validation effort and detailed comments on the applied modeling procedures are given in Appendix 1.

Given acknowledged test-to-test variability of occupant responses, the correlations between the simulated occu- pant responses and the actual, physical test case responses (for all of the validation cases shown in Appen- dix 1) were deemed acceptable.

2.2 MODELS FOR STATIC PERFORMANCE TESTS

In an effort to reduce even further the airbag inflation- related risks to occupants in contact with the airbag module at the time of airbag deployment, static, dummy chest-on- module testing is also a proposed Regulated test condition.

Previous research (Laituri, et al., 1999) was used to relate various inflator predictor variables to a pertinent occupant response for the static, HIII50 chest-on-module test condi- tion (ISO, 1998). Regression analysis was used to show good agreement when the predictor variable was a tank- test derived parameter (designated as the "10ms-win- dowed inflator thrust variable") and the response variable was the peak viscous criterion, V^*C_{max} (for the HIII50 in the aforementioned test condition of Figure 2). However, ster- nal velocity has also been considered as a measure of the thoracic injury risk. Accordingly, after revisiting the data used to generate the aforementioned thrust variable- V^*C_{max} regression equation, the following linear regression equation (with a correlation coefficient, R^2, equal to 0.87 for 18 data points) was derived:

$$SV_{HIII50} = 0.67 + 0.0122 \left. \dot{m}_{max} \right|_{10ms} \sqrt{RT_{inf}} \qquad (1)$$

where SV_{HIII50} is sternum velocity in m/s, $\left. \dot{m}_{max} \right|_{10ms}$ is the peak 10-ms windowed inflator mass flow rate in kg/s, $\div$ is the inflator gas specific heat ratio, and T_{inf} is the modeled inflator exit gas temperature in degrees Kelvin. The term in parentheses in Eq. (1) is the 10ms-windowed inflator thrust variable in N.

Moreover, the chest-on-module test condition with the small female instrumented test dummy needed to be con- sidered (since it is one of the proposed Regulated tests). Physical tests in which nominally identical airbag systems interacted with both the HIII50 and HIII05 in the chest-on- module test condition were analyzed. For a small sample (N=3), the following approximation resulted:

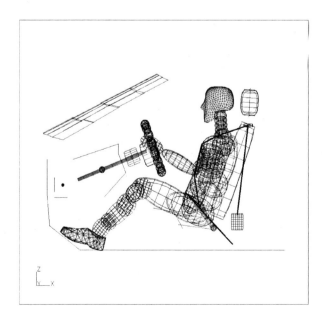

Figure 1: Example of Madymo3D Occupant Model

$$SV_{HIII05} \div 1.1(SV_{HIII50}) \qquad (2)$$

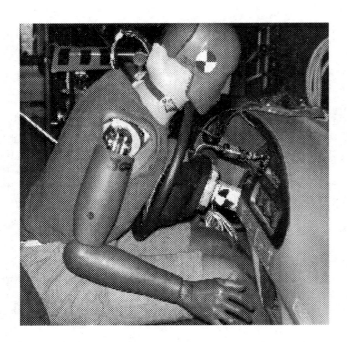

Figure 2: Chest-on-Module Test Condition (HIII50)

These mathematical models served as the foundation for an extensive parametric study to help better understand the effect of alternative frontal impact Regulations on potential airbag designs.

3.0 DESIGN SPACE SURVEY TECHNIQUE

3.1 OBJECTIVES

In order to attempt to predict the effects of the aforementioned performance constraints on available airbag design space, four major topics were addressed: (1) Specification of the sampled design space for seat belts and airbags, (2) The selection of performance tests to which the resulting studied designs were subjected, (3) The means by which the related occupant responses were estimated, and (4) The generic criteria by which suitable (acceptable) designs were identified.

3.2 DESIGN SPACE SAMPLING: SEAT BELTS

Recent research has indicated that retractors with a load limit set at approximately 4 kN can provide an effective balance between the required restraining belt load and rib fracture risk potential in real-world crash events (Foret-Bruno, et al., 1998). Additionally, occupants in real-world crashes do not always have their seat belts on as tightly as test dummies in controlled laboratory tests (Bauberger, et al., 1996). Therefore, when a belt was called for in this study, the modeled seat belt was assumed to be a design constant that consisted of both a 4 kN load-limited retractor and a pyrotechnic, buckle pretensioner.

3.3 DESIGN SPACE SAMPLING: AIRBAGS

The airbag design variables examined in this study included airbag venting, airbag size, and inflator characteristics.

Airbag venting is typically designated as the "number of holes" x the "vent hole diameter." A range of vent sizes typical of both full-powered and depowered airbag designs was considered in this study, viz., 2x10 mm, 2x15 mm, 2x20 mm, 2x25 mm, 2x30 mm, 2x35 mm, and 2x40 mm.

Driver airbag size is typically specified by its unfolded diameter. A range of airbag sizes typical of present depowered airbags was considered in this study, viz., 610 mm, 648 mm, and 673 mm.

An internal tether was assumed present, but primarily for purposes of design space sampling simplification, its length was assumed fixed at 254 mm.

Prior experimental research on the potential relationship between responses for in-position, HIII50s and out-of-position, HIII50s helped identify the related design challenge associated with inflator selection (Prasad, et al., 1996). Accordingly, a rather fine resolution of the design space associated with the dual-stage inflator was warranted for this study.

Nineteen theoretical inflators and one experimental, prototype dual-stage driver inflator were studied. The prototype inflator served as a baseline from which the other theoretical inflators were derived. The nineteen variations were derived by scaling the mass flow rates of the baseline Stage I and Stage II contributions of the inflator by +/- 25% increments from the baseline (while keeping the gas constituents constant) as shown in Figure 3. More details concerning these theoretical inflators are found in Table 1.

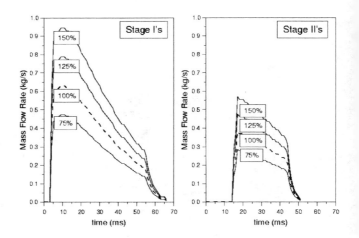

Figure 3: Theoretical Dual Stage Inflator Constructs

Stage I Scale	Stage II Scale	Peak Tank Pressure	Peak 0 ms-based Tank Pressure Rise Rate	Peak 10 ms-based Tank Pressure Rise Rate	Generated Gas Mass
(% of Ref)	(% of Ref)	(kPa g)	(kPa/ms)	(kPa/ms)	(kg)
75	0	210.76	6.49	6.28	0.0167
100	0	277.20	8.62	8.30	0.0223
125	0	342.75	10.74	10.29	0.0279
150	0	407.68	12.85	12.26	0.0335
75	75	259.43	8.26	7.66	0.0238
100	75	325.58	10.17	9.48	0.0294
125	75	391.01	12.05	11.30	0.0349
150	75	455.91	13.90	13.18	0.0405
75	100	275.69	8.98	8.28	0.0261
100	100	341.76	10.88	10.08	0.0317
125	100	407.15	12.75	11.87	0.0373
150	100	472.04	14.60	13.65	0.0429
75	125	291.96	9.69	8.90	0.0285
100	125	357.95	11.58	10.69	0.0340
125	125	423.31	13.45	12.47	0.0396
150	125	488.19	15.29	14.23	0.0452
75	150	308.22	10.40	9.53	0.0308
100	150	374.15	12.29	11.30	0.0364
125	150	439.48	14.15	13.07	0.0420
150	150	504.34	15.99	14.82	0.0476

3.4 CONSIDERED PERFORMANCE TESTS

Tests used to help assess vehicle crashworthiness may be broken into two groups: Regulated and non-Regulated. For purposes of this study, six performance tests (and the related performance constraints) were considered:

Three "Regulated" tests:

• The high-speed, unbelted occupant SNPRM test suite consisting of rigid fixed barrier, full-vehicle testing with both HIII50 and HIII05 unbelted drivers. This two-test suite of tests was studied with "high-speed" set to equal either 48 or 40 kph (hereafter, designated as "RFB48" and "RFB40," respectively.)

• The static, out-of-position occupant test involving

HIII05 in the chest-on-module condition (hereafter, designated as "OOPO HIII05.")

Three "non-Regulated" tests:

• The USA New Car Assessment Program test for frontal impact involving rigid fixed barrier, full-vehicle testing at 56 kph with a belted HIII50 driver (hereafter designated "NCAP.").

• The lower-speed, airbag deployment threshold test involving deformable barrier, full-vehicle testing with a fully-forward belted HIII05 driver. The test chosen as a surrogate for this crash environment was the belted HIII05 subjected to the generic sled test pulse (hereafter, designated as "Canadian test surrogate" or, for short, "Canadian.")

• The static, out-of-position occupant test involving the HIII50 in the chest-on-module condition (ISO Technical Report 10982, 1998) (hereafter, designated as "OOPO HIII50.")

It should be noted that these tests, while numerous, should be considered non-comprehensive with respect to the actual, total number of actual performance constraints that will be placed upon future airbag designs. Further discussion of this topic is left for a later section.

3.5 OCCUPANT RESPONSE PREDICTION PROCEDURES

3.5.1 DYNAMIC PERFORMANCE TESTS

The occupant responses associated with performance tests involving dummies in motion when the airbags are deployed were estimated via the Madymo3D math modeling methods discussed in Section 2.1 and Appendix 1. See Table 2 for more details regarding the monitored, dynamic, in-position occupant responses.

3.5.2 STATIC OOPO PERFORMANCE TESTS

The occupant responses associated with performance tests involving dummies at rest when the airbags are deployed were estimated via the regression and body-size sternum velocity approximations discussed in Section 2.2. See Table 2 for more details regarding the monitored, static, out-of-position occupant responses.

Note that, since the studied inflators were theoretical, the variables on the right-hand side of Eq. (1) needed to be determined for each the inflators.

The $\dot{m}_{max}|_{10ms}$ term was derived from a Madymo3D mathematical model of a tank test involving a studied inflator. The T_{inf} term was modeled as a constant (Wang, et al., 1988) and estimated from the TNO Madymo Tank Analysis code (Madymo Utilities Manual, 1999). Both the gas constant, R, and the constant-pressure gas specific heat, $c_{p,}$

were derived from Amagat's rule for mixtures (Van Wylen, et al., 1978). The ideal gas equation was then used to derive both the constant-volume specific heat, c_v, and $\div$ for the gas mixture. The results of these calculations are shown in Table 3. With the 10ms-windowed inflator thrust variable calculated for each of the theoretical inflators, SV_{HIII50} and SV_{HIII05} were estimated via Eqs. (1) and (2), respectively.

3.6 GENERIC ACCEPTANCE CRITERIA

In order to assess compliance or non-compliance of the studied designs subjected to the aforementioned Regulated and non-Regulated performance tests, generic acceptance criteria were derived. Specifically, pertinent occupant responses associated with each of the performance tests were compared with both published injury assessment reference values (IARVs) and proposed injury criteria performance limits (ICPLs) with applied, generic safety factors, defined as margins to increase confidence in compliance. Further details follow.

3.6.1 PUBLISHED IARVS

The published occupant response IARVs of Table 4 were essential in the generation of the set of generic acceptance criteria (designated, "SGAC").

Table 2:
Summary of Performance Tests and
the Related Pertinent Occupant Responses
and Prediction Methods

Performance Test	Dummy	Pertinent Occupant Responses	Source of Predicted Occupant Response
High-Speed Unbelted Test	HIII50	HIC15, Upper Neck Axial Force, Upper Neck Shear Force, Upper Neck Moment, Upper Neck N_{TE}, Chest G's (3ms cumdur), Chest Deflection	Madymo3D Model
High-Speed Unbelted Test	HIII05	HIC15, Upper Neck Axial Force, Upper Neck Shear Force, Upper Neck Moment, Upper Neck N_{TE}, Chest G's (3ms cumdur), Chest Deflection	Madymo3D Model
NCAP	HIII50	CPI	Madymo3D Model
Canadian	HIII05	Upper Neck Moment	Madymo3D Model
OOPO HIII50	HIII50	Sternum Velocity	Regression Model
OOPO HIII05	HIII05	Sternum Velocity	Regression Model

3.6.2 PROPOSED INJURY CRITERIA PERFORMANCE LIMITS

NHTSA's NPRM called for a new Regulated response, viz the upper neck combined loading criterion, N_{ij}. The ij indices represent various combinations of tension, compression, extension, and flexion. Since combined tension-extension loading is generally considered the most prevalent loading condition associated with airbags in frontal impacts, only N_{TE} was considered in this study. Additionally, only NPRM-related estimates of N_{TE} were made in this study since the SNPRM had not been published when the mathematical model-based simulations were being first

Table 3:
Inflator Characteristics for Theoretical Inflators derived from Tank Test Models and Amagat's Rule of Mixtures

Stage I Scale	Stage II Scale	Peak 10ms-based Mass Flow Rate	Inflator Gas Temp	$m_{max}\big|_{10ms}\sqrt{\div RT_{inf}}$
(% of Ref)	(% of Ref)	(kg/s)	(K)	(N)
75	0	0.462	672	264
100	0	0.616	671	351
125	0	0.771	671	439
150	0	0.925	670	526
75	75	0.705	594	378
100	75	0.853	608	462
125	75	1.001	619	548
150	75	1.150	626	633
75	100	0.794	578	420
100	100	0.940	594	504
125	100	1.088	605	588
150	100	1.236	614	673
75	125	0.882	563	460
100	125	1.029	582	546
125	125	1.175	594	630
150	125	1.322	604	715
75	150	0.971	552	502
100	150	1.117	570	587
125	150	1.264	583	671
150	150	1.410	593	755

Body Part	Response	HIII05	HIII50	Source
Head	HIC15	779	700	(AAMA, 1998)
Upper Neck	Ext. Moment	39 N-m	77 N-m	(AAMA, 1998)
Upper Neck	Neck Tensile Force	2070 N	3290 N	(AAMA, 1998)
Upper Neck	Neck Shear Force	2068 N	3100 N	(AAMA, 1998)
Chest	Chest Acceleration	73 G's (3ms cumdur)	60 G's (3ms cumdur)	(AAMA, 1998)
Chest	Chest Acceleration	60 G's (3ms cumdur)	60 G's (3ms cumdur)	(NHTSA, SNPRM, 1999)
Chest	Chest Deflection	53 mm	65 mm	(AAMA, 1998)
Chest	Sternum Velocity	8.2 m/s	8.2 m/s	(AAMA, 1998)
Upper Neck	Combined Tension-Extension Response	1.0	1.0	(NHTSA, SNPRM, 1999)

onducted. The equation for N_{TE} is

$$N_{TE} = \max\left[\frac{+F_z(t)}{+F_{z_1}} + \frac{M_y(t)}{M_{y_1}}\right] \quad (3)$$

where F_z is the upper neck tensile force and M_y is the corrected upper neck extension moment. F_{z_1} was defined as 3200 N for HIII05 and 3600 N for HIII50, and M_{y_1} was defined as 60 N-m for HIII05 and 125 N-m for HIII50. It should be noted that all upper neck moments reported herein were "corrected," i.e., neck moments at the upper neck load cell were transferred to be about the occipital condyles. NHTSA's proposed ICPL for N_{TE} of 1.0 was used to generate the related generic acceptance criterion.

3.6.3 NON-REGULATED TEST METRICS

3.6.3.1 NCAP

The NCAP test score is based on the combined probability of injury, CPI, defined as

$$CPI = P_{head} + P_{chest} - (P_{head} + P_{chest}) \quad (4)$$

where P_i represents the probability of life-threatening injury attributed to either the head or chest.

The equations used to calculate the NCAP score were

$$P_{head} = \frac{1}{(1 + e^{(5.02 - 0.00351 + HIC36)})} \quad (5)$$

and

$$P_{chest} = \frac{1}{(1 + e^{(5.55 - 0.0693 + ChestGs)})} \quad (6)$$

subject to the following scoring system:

5 star: $0.0 < CPI + 0.10$

4 star: $0.10 < CPI + 0.20$ (7)

3 star: $0.20 < CPI + 0.35$

Since NCAP is not a Regulated crash test, the CPI had to be chosen in order to quantify the related generic acceptance criterion. Accordingly, a CPI score of 0.20 was chosen.

3.6.3.2 CANADIAN SURROGATE TEST

This test, involving a belted HIII05, was previously described in Section 3.4. The HIII05 upper neck extension moment is typically the occupant response of interest in this performance test. However, since this is not a regulated test, the magnitude of moment had to be chosen in order to quantify the related generic acceptance criterion. Accordingly, the published IARV, 39 N-m of Table 4, was chosen.

3.6.4 OOPO HIII50 AND OOPO HIII05

The OOPO HIII50 and OOPO HIII05 (chest-on-module) performance tests were previously discussed in Section 2.2. For this study, these tests were assessed by the AAMA IARVs of Table 4, i.e., sternum velocities of 8.2 m/s were used to generate the related generic acceptance criteria.

Note that the HIII05 is the targeted Regulation device for this mode of testing (NHTSA, 1998), but for the purposes of this study, the OOPO HIII50 test was considered as a non-Regulated performance test.

Moreover, whereas the chest-on-module performance tests primarily help assess thoracic injury risks, another static OOPO test was proposed to better help assess neck injury risks (NHTSA, 1998). By not accounting for this test, the studied OOPO performance constraints were deemed non-comprehensive.

3.6.5 SAFETY FACTORS

Safety factors, defined as margins to ensure confidence in compliance (to compensate for system variability), were used in this study. The generic safety factors applied to ere

$$SF_{reg} \div \frac{IARV \ or \ ICPL}{response} = 1.25 \qquad (8)$$

and those applied to non-Regulated responses were

$$SF_{nonreg} \div \frac{IARV \ or \ Target}{response} = 1.11 \qquad (9)$$

Eqs. (8) and (9) could also be considered as generic compliance margins of 20 and 10%, respectively. It is accordingly noted that, for the purposes of this study, greater compliance margins was applied to Regulated responses than to non-Regulated responses cf., Table 5. Additionally, it should be noted that, since discussion continued as to which occupant response (chest acceleration, chest deflection, sternal velocity, or the viscous criterion) better predicts OOPO thoracic injury potential, less stringent weighting was applied to the estimated OOPO-related occupant responses. Again, it should be recalled that generic safety factors were applied in this theoretical study; Precise statistical compliance factors and specific corporate emphasis were beyond the scope of this study.

The collective results of Sections 3.4 through 3.6.5 were used to derive the set of generic acceptance criteria for all studied performance tests, e.g., for the NCAP performance test, the related generic acceptance criterion was equal to 0.18 (i.e., $SF_{nonreg} \times Target_{NCAP} = 0.9 \times 0.2$).

With both a means to estimate the pertinent occupant occupant responses for the performance tests for a wide variety of designs and a means to assess the stringency of the accompanying performance constraints, the design space study was then conducted.

4.0 DESIGN SPACE SCREENING EXERCISE (DSSE)

In order to predict the effects that alternative frontal impact test Regulations would have on the number of candidate dual-stage driver airbag designs (when assessed subject to a set of generic acceptance criteria established for the six considered performance tests), the following computational exercise was executed:

Step 1:
All of the studied airbag design variable combinations were subjected to mathematical simulations of each of the potential, unbelted FMVSS 208 tests.

Step 2:
All of the studied airbag design variable combinations were subjected to mathematical simulations of both the NCAP and Canadian tests.

Step 3:
All of the studied airbag design variable combinations were subjected to the mathematical model representations of the HIII50 and HIII05 chest-on-module tests.

Table 5:
Safety Factor Schedule for Generic Acceptance Criteria

Performance Test	Dummy	Pertinent Occupant Responses	Safety Factor for Constraint-Related Occupant Responses
High-Speed Unbelted Test	HIII50	HIC15, Upper Neck Axial Force, Upper Neck Shear Force, Upper Neck Moment, Upper Neck N_{TE}, Chest G's (3ms cumdur), Chest Deflection	1.25
High-Speed Unbelted Test	HIII05	HIC15, Upper Neck Axial Force, Upper Neck Shear Force, Upper Neck Moment, Upper Neck N_{TE}, Chest G's (3ms cumdur), Chest Deflection	1.25
NCAP	HIII50	CPI	1.11
Canadian	HIII05	Upper Neck Moment	1.11
OOPO HIII50	HIII50	Sternum Velocity	1.11
OOPO HIII05	HIII05	Sternum Velocity	1.11

Step 4:
Each of the predicted occupant responses from Steps 1-3 were compared with the set of generic acceptance criteria (with $SF_{reg} = 1.25$ and $SF_{nonreg} = 1.11$).

For the purposes of this study, designs predicted to satisfy the set of generic acceptance criteria were binned as "acceptable" while designs that did not satisfy were binned as "unacceptable."

Step 5:
Statistics such as "Evaluated Designs Deemed Unacceptable (%)" were calculated for each performance constraint, e.g., Evaluated Designs Deemed Unacceptable (%) = 92% when considering the HIII50 occupant Chest G's in the rigid fixed barrier case at 48 kph (i.e., 310 designs of the studied 336 did not have the predicted 3ms cumdur Chest G's to be less than 48 G's (= 60/1.25)).

This 5-step exercise, designated the "Design Space Screening Exercise" (or "DSSE"), can be expressed in mathematical notation as follows:

$$\begin{Bmatrix} vent \\ bagsize \\ stage1scale \\ stage2scale \end{Bmatrix} = \begin{Bmatrix} 2 \times 10, \ 2 \times 15, \ 2 \times 20, \ 2 \times 25, \ 2 \times 30, \ 2 \times 35, \ 2 \times 40mm \\ 610, \ 648, \ 673mm \\ 0.75, \ 1.00, \ 1.25, \ 1.50 \\ 0.0, \ 0.75, \ 1.00, \ 1.25, \ 1.50 \end{Bmatrix} \quad vs \quad \{constraints\} \ (10)$$

...here the studied performance constraints were

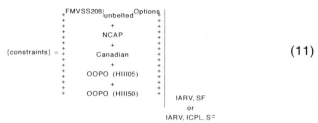

$$\{constraints\} = \begin{array}{c} +FMVSS208|_{unbelted}^{Options} \\ + \\ NCAP \\ + \\ Canadian \\ + \\ OOPO\ (HIII05) \\ + \\ OOPO\ (HIII50) \end{array} \quad \begin{array}{c} \\ \\ \\ \\ \\ \\ \\ \\ IARV,\ SF \\ or \\ IARV,\ ICPL,\ S^= \end{array} \qquad (11)$$

...nd recall

$$FMVSS208|_{unbelted}^{Options} = \begin{array}{ccc} +RFB48-HIII50 & + & RFB48-HIII05 \\ +RFB40-HIII50 & + & RFB40-HIII05 \end{array} \quad (12)$$

...he related crash pulses associated with this specific study ...f a prototype mid-sized, passenger car are shown in Figure .... Note that, due to rebound effects in the barrier tests, in...egration of the crash pulse deceleration does not exactly ...ield the speeds in the aforementioned case designations.

...0 RESULTS OF DESIGN SPACE
SAMPLING EXERCISE

...One theoretical airbag deployment strategy was assessed ...ia the DSSE, specifically, a "Non-Uniform Deployment

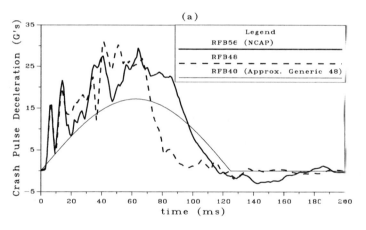

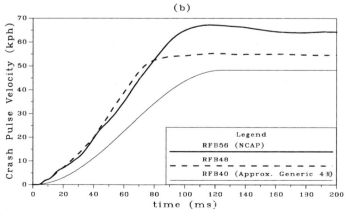

Figure 4: Crash Pulse Information

Schedule" consisting of:

• Stage I-only deployments for all studied performance tests involving the HIII05 (irrespective of restraint level and crash speed/severity), and

• Stage I+II deployments for the studied rigid fixed barrier crash tests, i.e., RFB40, RFB48, and NCAP, and

• Stage I+II deployments possible in the OOPO HIII50 performance test.

5.1 RESULTS OF DESIGN SPACE
SCREENING EXERCISE - HIII50

Given the application of the set of generic acceptance criteria, Figure 5 shows the screening (thrifting) effects on design space observed for the HIII50 (where the studied airbag deployment schedule called for deployment of Stage I+II airbags). Note the following:

For the dynamic performance constraints:

- For the constraint set that included the RFB48 test suite, 92% of the 336 studied designs were discarded due to Chest G considerations, and

- For the constraint set that included the RFB40 test suite, 0% of the 336 studied designs were discarded due to Chest G considerations.

For the static, OOPO performance constraints:

- For the constraint set that included the RFB48 test suite, 50% of the 336 studied designs were discarded due to sternal velocity considerations, and

- For the constraint set that included the RFB40 test suite, 50% of the 336 studied designs were discarded due to sternal velocity considerations.

5.2 RESULTS OF DESIGN SPACE
SCREENING EXERCISE - HIII05

Similarly, given the application of the set of generic acceptance criteria, Figure 6 shows the screening (thrifting) effects on design space observed for the HIII05 (where the studied airbag deployment schedule called for deployment of only Stage I airbags). Note the following:

For the dynamic performance constraints:

- For the constraint set that included the RFB48 test suite, 76% of the 84 studied designs were discarded due to N_{TE} considerations, and

- For the constraint set that included the RFB40 test suite, 51% of the 84 studied designs were discarded due to N_{TE} ...ations, and

- For the constraint set that included the RFB48 test suite, 58% of the 84 studied designs were discarded due to chest deflection considerations, and

- For the constraint set that included the RFB40 test suite, 32% of the 84 studied designs were discarded due to chest deflection considerations.

For the static, OOPO performance constraints:

- For the constraint set that included the RFB48 test suite, 25% of the 84 studied designs were discarded due to sternum velocity considerations, and

- For the constraint set that included the RFB40 test suite, 25% of the 84 studied designs were discarded due to sternum velocity considerations.

5.3 RESULTS OF DESIGN SPACE SCREENING EXERCISE - RESTRAINT LEVEL CONSIDERATION

Another way to analyze the results of this study was to consider non-acceptance from the standpoint of restraint level. Table 6 illustrates that <u>nearly all</u> of the non-acceptable studied designs that were predicted to not satisfy the set of generic acceptance criteria were discarded due to a lack of compliance with one (or more) of the performance constraints associated with the unbelted occupant tests.

5.4 RESULTS OF DESIGN SPACE SCREENING EXERCISE - AGGREGATE

When the results of Sections 5.2 through 5.3 were considered collectively, the screening exercise yielded the results shown in Table 7.

Observations from Figures 5 and 6 and Tables 6 and 7 indicated a significant reduction in the available design space when contrasting the 40 vs. 48 kph high-speed unbelted occupant-influenced performance constraints of the SGAC. Specifically, when the generic acceptance criteria set contained the unbelted rigid fixed barrier testing constraints run at 48 kph, not a single design (in the design space studied) satisfied the considered set of generic acceptance criteria. This reduction was almost entirely attributable to the <u>unbelted</u> occupant portion of the performance constraints.

6.0 DESIGN SELECTION PROCEDURE

A procedure was developed to determine the extent to which a design satisfied the constraint set. Accordingly, this procedure was used to help select designs that best attempted to satisfy the chosen constraints.

The "Design Selection Procedure" was a natural extension of the "Design Space Screening Exercise."

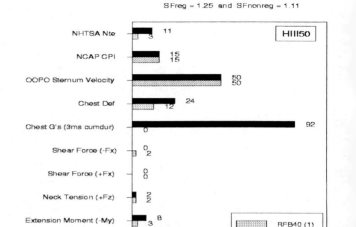

Figure 5: Results of Design Space Screening Exercise for Two High-Speed Unbelted Occupant Test Suites: HIII50 Test Dummy Considerations

(1) Assumption, from NHTSA 1999 SNPRM, that RFB40 approx. same severity as Generic48

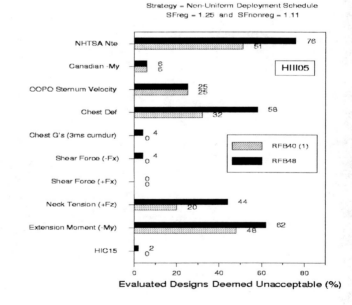

Figure 6: Results of Design Space Screening Exercise for Two High-Speed Unbelted Occupant Test Suites: HIII05 Test Dummy Considerations

(1) Assumption, from NHTSA 1999 SNPRM, that RFB40 approx. same severity as Generic48

this procedure to identify the design that best satisfied the constraints was established by defining a "Selected Airbag Design Acceptance Factor," i.e.,

Selected Airbag Design Acceptance Factor" = minimum [maximum (all normalized Regulated and non-Regulated occupant responses)]

over the domain of 336 studied designs, where

normalized occupant response = predicted occupant response / related component of the SGAC.

The results of the design selection procedure (as before, applied with intent to contrast the effect of varied high-speed, unbelted constraints) are shown in Table 8.

Note that, for continuity purposes, the results of Table 7 are presented in the second column of Table 8.

The third column of Table 8 contains the design specifications for the airbag that was chosen (out of the 336 studied designs) subject to the aforementioned procedure.

The fourth column of Table 8 contains that design's "Selected Airbag Design Acceptance Factor" -- a derived number which can be interpreted as the level of compliance for the design that best satisfied the most-restricting performance constraints. Accordingly, from inspection of Table 8, no design was predicted to have a normalized occupant response lower than 0.85. Moreover, the 0.85 value implies both SGAC compliance and that approximately 15% of the design space remained available for purposes of design selection after the design space screening exercise was completed. This available design space could be possibly used to help address other constraints not considered in this study, e.g., the aforementioned OOPO performance tests for the neck, non-Regulated frontal impact performance tests such as the high-speed, deformable offset barrier test conducted by the Insurance Institute for Highway Safety, 5-star USA NCAP performance, etc.

Table 6:
Results of the Design Space Screening Exercise
with Respect to the Level of Restraint
(Non-Uniform Deployment Schedule)
with SF_{reg} = 1.25 and SF_{nonreg} = 1.11

Scenario	Studied Designs	% Non-Acceptable (present IARVs + proposed ICPL) with	
		all constraints	unbelted only constraints
"High-Speed" Unbelted Occupant Component of SGAC			
RFB40[1]	336 (HIII50), 84 (HIII05)	94	93
RFB48	336 (HIII50), 84 (HIII05)	100	100

Table 7:
Results of the Design Space Screening Exercise
with Aggregate Considerations
(Non-Uniform Deployment Schedule)
with SF_{reg} = 1.25 and SF_{nonreg} = 1.11

"High-Speed" Unbelted Occupant Component of SGAC	Studied Designs	Accepted Designs (present IARVs + proposed ICPL)
RFB40[1]	336 (HIII50), 84 (HIII05)	21
RFB48	336 (HIII50), 84 (HIII05)	0

Table 8:
Results of the
Design Space Screening Exercise
and the
Design Selection Procedure

"High-Speed" Unbelted Occupant Component of SGAC	Number of Accepted Designs [2] (out of 336)	Selected Airbag Details [3]	Selected Airbag Design Acceptance Factor	Limiting Factor
RFB40 (approx by Generic48 [1])	n = 21 (6%)	Inflator Stage I = 100% Inflator Stage II = 75% Airbag=648mm 2x20 mm vents	0.85	Sternum Vel (OOPO-HIII50)
RFB48	n = 0 (0%)	Inflator Stage I = 125% Inflator Stage II = 100% Airbag=610mm 2x25 mm vents	1.07	Chest G's (in position - unbelted HIII50)

(1) Assumption, from NHTSA 1999 SNPRM, that RFB40 approx. same severity as Generic48

(1) NHTSA 1999 SNPRM
(2) via "Design Space Screening Exercise"

The RFB48 case, in contradistinction to the RFB40 case, was estimated to have a 1.07 value. This prediction indicated that, of the domain of 336 studied designs, none were found to satisfy the SGAC, i.e., an absence of available design space where the design that best satisfied was 7% higher than dictated by the SGAC.

Finally, it should be noted from inspection of Table 8 and Table 3 that for the designs selected to best satisfy the studied performance constraints, the RFB48-based design was predicted to require higher mass flow rates for both Stage I and Stage II than those required for the RFB40-based design.

7.0 CONCLUSIONS

Advanced airbags with dual-stage inflators have been identified as a potential means to further enhance occupant protection when used in conjunction with advanced-technology, crash-sensing systems. With intent to comment on the maximum speed at which the high-speed, unbelted occupant test suite of the 1999 SNPRM for Advanced Airbags will be conducted (48 vs. 40 kph), an extensive parametric study involving 336 different theoretical dual-stage driver airbag systems in a prototype mid-sized passenger car was conducted.

The two candidates for the high-speed, unbelted occupant performance test suite were considered in conjunction with four other performance tests, viz., NCAP, Canadian, OOPO HIII50, and OOPO HIII05 via mathematical modeling. By comparing the pertinent, predicted occupant responses with a set of generic occupant response acceptance criteria derived for the studied performance tests (which involved safety factor-weighted IARVs/ICPLs and chosen non-Regulated targets), available design space was estimated, i.e., the number of theoretical dual-stage driver airbag designs that satisfied the set of generic occupant response acceptance criteria.

The conclusions drawn from this study included:

• The introduction of both the HIII05 and the HIII50 into the proposed, future Regulation was predicted to significantly reduce the available design space (considered in this study).

• None of the 336 studied theoretical dual-stage driver airbag designs satisfied the generic acceptance criteria set when the unbelted rigid fixed barrier testing constraints were run at 48 kph.

• 21 out of the 336 satisfied the generic acceptance criteria set when the unbelted rigid fixed barrier testing constraints were run at 40 kph.

• When considering the discarded designs, nearly all of them were predicted to not comply due to (at least one of) the unbelted occupant performance constraints of the generic acceptance criteria set.

• For the designs selected to best satisfy the studied, non comprehensive performance constraints, the RFB48-based design was predicted to require higher mass flow rates for both Stage I and Stage II than those required for the RFB40 based design. Accordingly, the anticipated positive real world benefits of dual-stage driver inflators may be reduced. The estimated, projected significance of this conclusion i beyond the scope of the present study and was left for the second part of this research.

• Given the above conclusions (and the attendant assumptions), the 40 kph high-speed unbelted occupant test suite was considered to be significantly more amenable to the application of dual-stage driver airbags than its 48 kph counterpart.

It should be noted that, implicit to these results, the static OOPO requirement was only partially addressed; No HIII05 neck tests were considered. Also, other non-Regulated frontal impact performance tests such as the high-speed deformable offset barrier test conducted by the Insurance Institute for Highway Safety were not considered. Accordingly, by the addition of those performance constraints, the available design space may be even more limited.

ACKNOWLEDGMENTS

The authors would like to thank the following contributors to this study: Kris Warmann for providing much of the information regarding the prototype hardware; Linda Rink for providing baseline inflator characteristics predictions; T.C. Weng for discussions regarding many of the occupant model inputs; Scott Schmidt and Dave Clark for providing the AAMA reference material; Jeff Nadeau, Stacy Nadeau, and Phil Przybylo for reviewing this document.

REFERENCES

American Automobile Manufacturers Association (AAMA), "Proposal for Dummy Response Limits for FMVSS 208 Compliance Testing," Docket No. NHTSA 98-4405, Notice 1, AAMA S98-13, Attachment C, December 1998.

Bauberger, Alfred and Dieter Schaper, "Belt Pretensioning and Standardized 'Slack' Dummy," 96-S10-W28, 15th International Technical Conference on Enhanced Safety of Vehicles, Melbourne, Australia, May 1996.

Foret-Bruno, J-Y, et al., "Thoracic Injury Risk in Frontal Car Crashes with Occupant Restrained with Belt Load Limiter, 42nd Stapp Car Crash Conference Proceedings, Paper 983166, November 1998.

ISO Technical Report 10982, "Road Vehicles-Test Procedures for Evaluating Out-of-Position Vehicle Occupant Interactions with Deploying Air Bags, March 15, 1998.

Laituri, Tony R. and Priya Prasad, "Correlation of Driver Inflator Predictor Variables with the Viscous Criterion for the id-Sized Male, Instrumented Test Dummy in the Chest-n-Module Condition, SAE Paper 1999-01-0763.

adymo Database Manual, Version 5.3, TNO Road-Vehies Research Institute, 1997.

adymo Utilities Manual, Version 5.4, TNO Road-Vehies Research Institute, 1999.

ational Highway Traffic Safety Administration, Preliminary Economic Assessment, FMVSS No. 208, Advanced irbags, Office of Regulatory Analysis & Evaluation Plans nd Policy, August 1998.

ational Highway Traffic Safety Administration, Supplemental Notice for Proposed Rulemaking, FMVSS No. 208, dvanced Airbags, September 1999.

rasad, Priya, and Tony R. Laituri, "Consideration of Belt-d FMVSS 208 Testing," 96-S3-O-03, 15th International echnical Conference on Enhanced Safety of Vehicles, Melbourne, Australia, May 1996.

AE J211, Instrumentation for Impact Test - SAE J211 AE Recommended Practice, October 1988.

an Wylen, Gordon J., and Richard E. Sontag, Funda-nentals of Classical Thermodynamics, John Wiley & ons, 1978.

Vang, J.T. and Donald J. Nefske, "A New CAL3D Airbag nflation Model," SAE Paper 880654, 1988.

DEFINITIONS, ACRONYMS, ABBREVIATIONS

ACRONYMS

AMA	American Automobile Manufacturers Association
CPI	Combined Probability of Injury
DSSE	Design Space Screening Exercise
FMVSS	Federal Motor Vehicle Safety Standard
HIC	Head Injury Criteria
HIII05	Hybrid III, Small-female test dummy
HIII50	Hybrid III, Mid-sized male test dummy
IARV	Injury Assessment Reference Value
ICPL	Injury Criteria Performance Limit
ISO	International Standards Organization
NCAP	New Car Assessment Program
NHTSA	National Highway Traffic Safety Administration
NPRM	Notice for Proposed Rulemaking
OOPO	Out-of-Position Occupant (chest-on-module)
SF	Generic Safety Factor
SGAC	Set of Generic Acceptance Criteria
SNPRM	Supplementary Notice for Proposed Rulemaking
SV	Sternum Velocity

APPENDIX 1

The validation cases (and their accompanying hardware) are outlined in Table A1. Note that "Generic" refers to "generic pulse" and "Rigid" refers to "rigid fixed barrier." Hereafter, the case numbers of Table A1 will serve as designations for the various test type/speed/severity/restraint level/occupant size/seating position variables.

A comparison of the aforementioned six modeled and physical occupant responses for the "NCAP" case, i.e., Case 1, is shown in Figure A1 where all occupant responses (for both model and test) were filtered subject to SAE J211 protocols (SAE J211, 1988). Unless otherwise noted on the model/test comparison plots, the ordinate range for each of the occupant responses in Figure A1 (and the validation plots thereafter) was determined from recently-published response limits known as injury criteria performance limits (ICPLs) (NHTSA, 1998). It should be noted that, in general, the response limits are dummy-size dependent. Displaying the results in this manner helped identify responses near proposed ICPLs. The model-test comparisons for Cases 1 through 6 are found in Figures A1 through A6, respectively.

Table A1:
Validation Cases for Madymo3D Modeling (Driver)

Case Number	Type	Speed (kph)	Test Count	Severity	Restraint Level	Occ	Seat Position	Inflator
NCAP (Case 1)	Vehicle	56	1	Rigid	Belt+Bag	HIII50	Mid	Stages I + II
Canadian (Case 2)	Sled	48	1	Generic	Belt+Bag	HIII05	Full-Forward	Stage I
FMVSS 208 (Case 3)	Sled	48	1	Generic	Bag	HIII50	Mid	Stage I
FMVSS 208 (Case 4)	Vehicle	48	2	Rigid	Bag	HIII50	Mid	Stages I + II
FMVSS 208 (Case 5)	Vehicle	48	1	Rigid	Bag	HIII05	Full-Forward	Stage I
FMVSS 208 (Case 6)	Vehicle	48	1	Rigid	Bag	HIII05	Full-Forward	Stages I + II

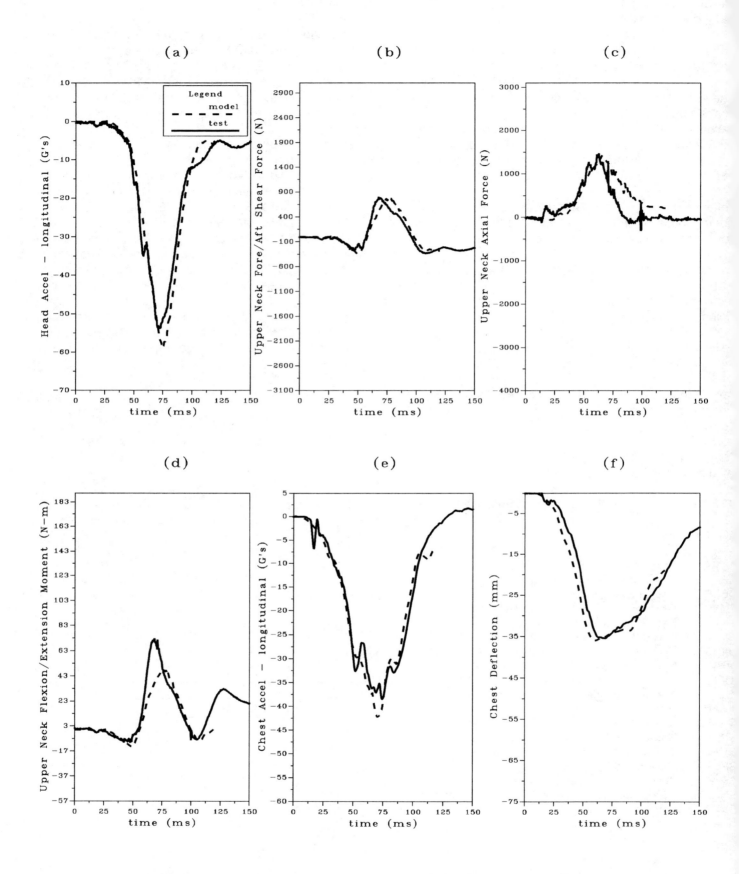

Figure A1: Mathematical Model Validation (Occupant Responses for Case 1)

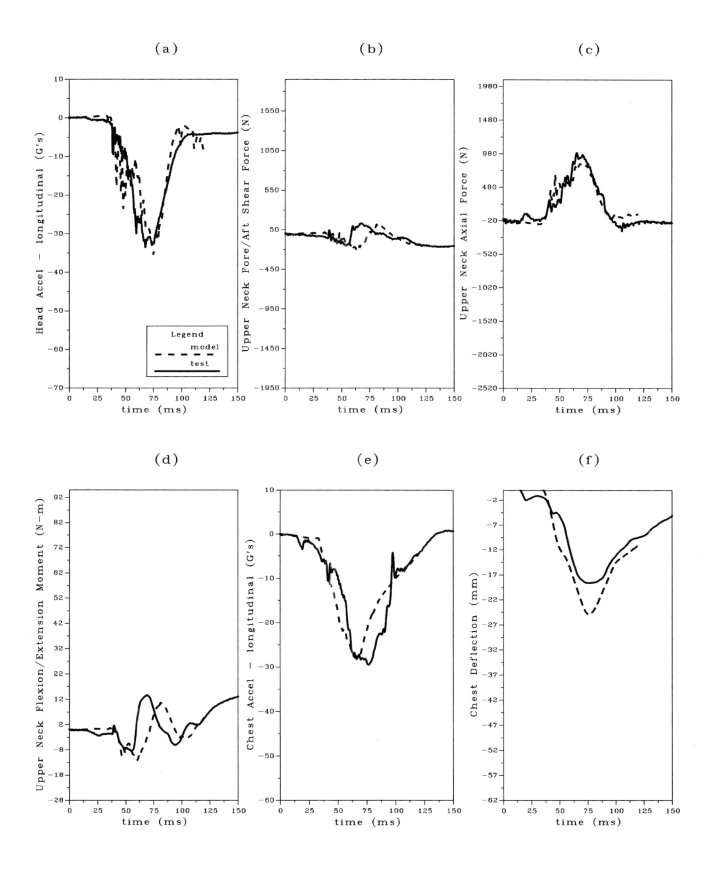

Figure A2: Mathematical Model Validation (Occupant Responses for Case 2)

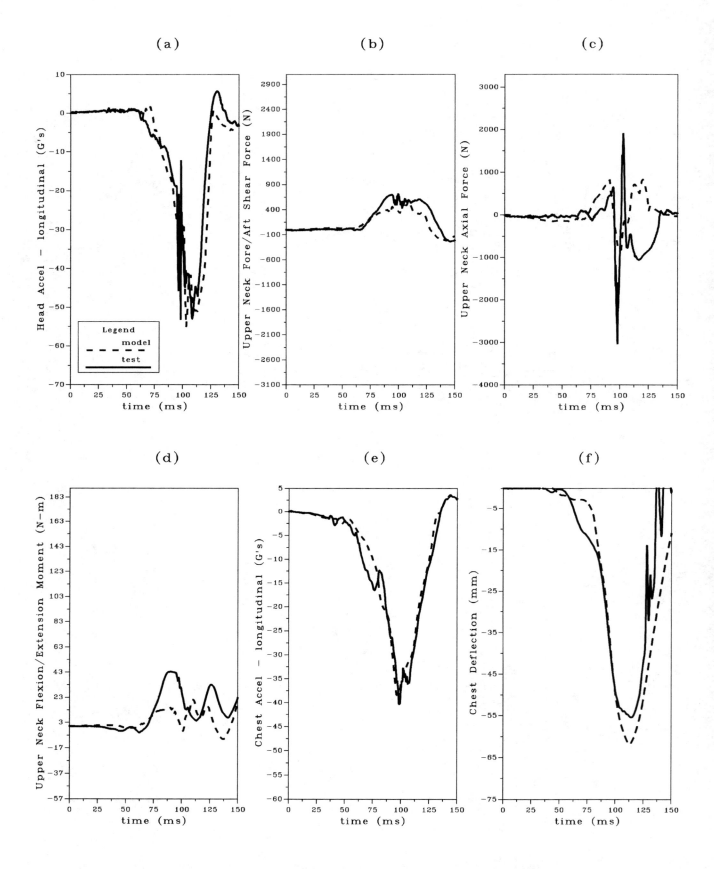

Figure A3: Mathematical Model Validation (Occupant Responses for Case 3)

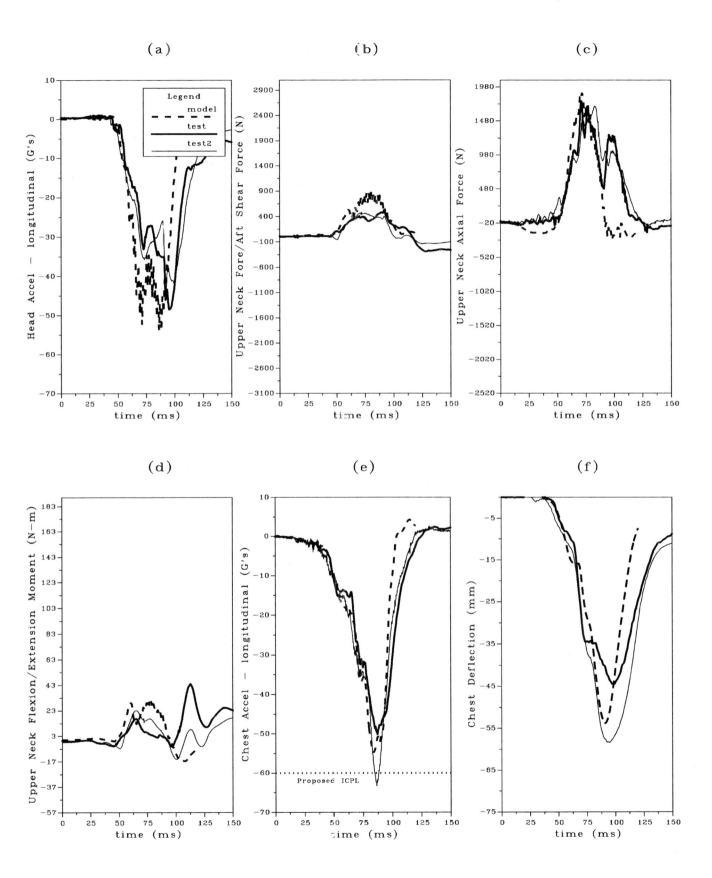

Figure A4:   Mathematical Model Validation (Occupant Responses for Case 4)

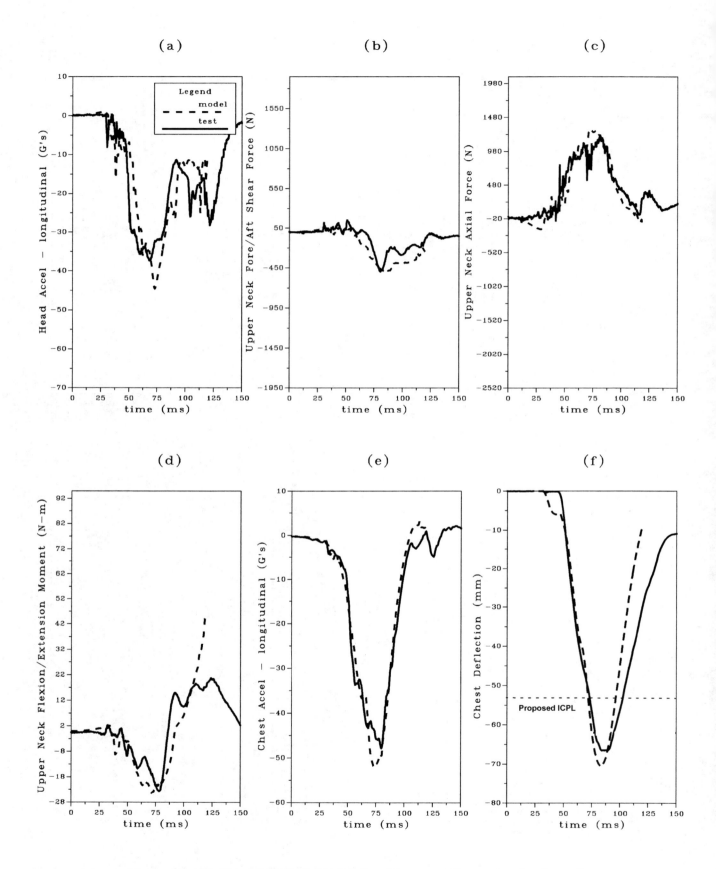

Figure A5: Mathematical Model Validation (Occupant Responses for Case 5)

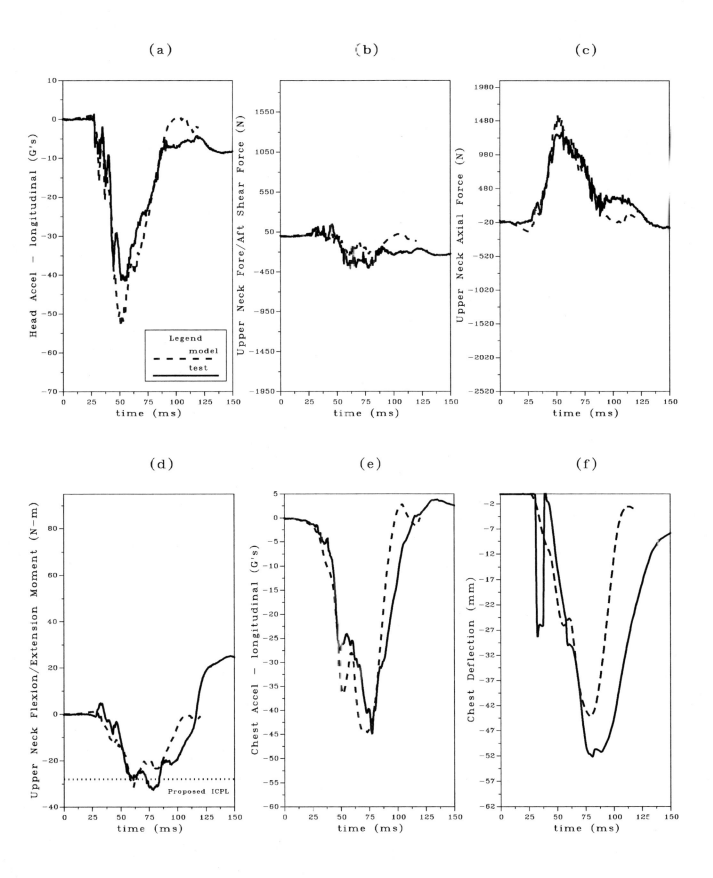

Figure A6: Mathematical Model Validation (Occupant Responses for Case 6)

Pertinent details of the model(s) included:

Simulation code: Madymo3D version 5.3

Dummies: Rigid-link representations from TNO database v. 5.3.1 with later-added facetted head(s)

Airbags: Finite element representations

Airbag unfolding: Actual initial size via initial metric approach (with no self contact)

Airbag inflator characteristics: Time-dependent mass flow rate and temperature from supplier

Airbag inflation: 6-jet array to better represent observed airbag trajectory

Airbag tethers: 9-strap array to represent a circular ring

Airbag material: Density, thickness, and isotropic modulus of elasticity estimated from testing

Airbag leakage: Discrete vent and seam losses modeled via discharge coefficient (Note: A seam leakage estimate was based on airbag diameter and was introduced via a modified vent discharge coefficient)

Airbag permeability: Estimated from free inflation testing

Airbag trigger times: In accordance with specific events or estimated from thumb rules

Airbag energy losses due to module: Modeled as "rigid" mesh with a time lag from the airbag trigger time that is directly proportional to the inflator's peak tank pressure rise rate

Seat belt retractor: Estimated force-deflection characteristic from supplier testing

Seat belt pyrotechnic pretensioner: Modeled as an initially-locked, pretensioned spring

Seat belt pyrotechnic pretensioner trigger times: Set equal to airbag trigger times

Seat belt slack: Slight amounts present (approximated)

Dummy position: In accordance with specific event (and instrumented dummy size)

Crash pulse: Estimated from average of B-pillar left and right rocker decelerations

Intrusions: Time-dependent functions from vehicle structural models and post-crash inspections

Column stroke: Predicted via a combination of the column force-deflection characteristic and interaction with intr

components which have prescribed motion

Steering Wheel: Compliant

Occupant chest/steering wheel contacts: Conducted with "Evaluation" feature of Madymo3D

2002-01-0186

Performance of Depowered Air Bags in Real World Crashes

J. Augenstein, E. Perdeck and J. Stratton
William Lehman Injury Research Center, University of Miami School of Medicine

K. Digges and J. Steps
The National Crash Analysis Center, George Washington University

ABSTRACT

During the period 1992 through 2000, the William Lehman Injury Research Center collected crash and injury data on 141 drivers and 41 right front passengers in frontal crashes with air bag deployment. Among these cases were twenty-eight cases with depowered air bags. The paper compares the crash characteristics for injured occupants in vehicles with 1st generation and depowered air bags.

The population with 1st generation air bags contains unexpected fatalities among as well as fatalities at low delta-V's. To date, these populations are absent among the fatally injured occupants of vehicles with depowered air bags. The depowered cases include both belted and unbelted survivors at crash severities above 40 mph delta-V. The maximum injury in these severe crashes was AIS 3 with no evidence of unsatisfactory air bag performance. However, serious internal chest injuries were observed in two cases with unrestrained drivers at crash severities of 19 and 24 mph.

INTRODUCTION

The Lehman Injury Research Center air bag data represents a near census of the seriously and fatally injured air bag protected occupants in Miami and Southern Florida. The database of frontal crashes involving 1st generation air bags contains 154 occupants, 45 of whom were fatally injured.

In the database of 1st generation air bags, 13 of the 45 fatalities were seniors (over 65) and 4 were small children. Eight of the fatalities were at speeds below 15 mph.

The depowered air bag database contains 28 occupants, 4 of whom were fatally injured. In the depowered cases, there have been no child fatalities and no fatalities as speeds below 20 mph. Three of the four fatalities were in severe crashes with delta-V greater than 40 mph and with massive intrusion of the occupant compartment.

Among the depowered cases there were four survivors of extremely severe crashes – greater than 40 mph. Two occupants were belt restrained and two were unrestrained. The chest injuries were of remarkably low severity and no head injuries were sustained to either occupant. These cases suggest that air bags as currently depowered provide protection in high severity crashes to both restrained and unrestrained occupants. These cases are summarized in the sections to follow.

There were two cases of severe injuries to unrestrained occupants at moderate crash severities – 19 and 24 mph. In these cases, the occupant may have been close to the deploying bag. The case summaries follow.

COMPARISON OF FATAL CRASHES

The distribution of crash severity for the fatally injured drivers with 1st generation and depowered air bags is shown in Table 1. The table shows the number of fatalities in each delta-V (mph) increment. The 1st generation air bag fatalities are fairly uniformly distributed. There are a relatively large number of fatalities in the lower speed ranges for the 1st generation air bags.

Table 1. Distribution of Driver Fatalities by Delta-V

Delta-V	1st Generation	Depowered
0-15	5	0
16-20	3	0
21-25	5	1
26-30	5	0
31-35	3	0
36-40	5	0
40+	6	1
Total	32	2

Table 2. Distribution of Driver Fatalities by Age

Age	1st Generation	Depowered
15-20	3	0
21-30	5	0
31-40	3	1
41-50	6	1
51-60	5	0
61-70	5	0
71+	5	0
Total	32	2

The age distribution of the fatally injured drivers with 1st generation and depowered air bags is shown in Table 2. There are a relatively large number of fatally injured older occupants among the fatally injured group with 1st generation air bags.

The distribution of crash severity for the fatally injured right front passengers with 1st generation and depowered air bags is shown in

Table 3. The 1st generation air bag fatalities are fairly uniformly distributed. There are a relatively large number of fatalities in the lowest speed range (0-10 mph) for the 1st generation air bags.

Table 3. Distribution of Passenger Fatalities by Delta-V

Delta-V	1st Generation	Depowered
0-10	3	0
11-20	1	0
21-30	4	0
31-40	2	0
40+	3	2
Total	13	2

The age distribution of the fatally injured right front passengers with 1st generation and depowered air bags is shown in Table 4. There are a relatively large number of fatally injured children among the fatally injured group with 1st generation air bags. To date no injuries to children have been observed in the depowered Lehman Center cases.

Table 4. Distribution of Passenger Fatalities by Age

Age	1st Generation	Depowered
0-3	4	0
21-30	2	0
31-40	0	1
41-50	1	0
51-60	2	0
61-70	1	0
71+	3	1
Total	13	2

SEVERE CRASHES WITH FAVORABLE AIR BAG PERFORMANCE

Four extremely severe crashes of drivers with depowered air bags are in the Lehman Center database. These cases had a delta-V of 40 mph or greater. Two of the drivers were restrained and two were unrestrained. None had any significant head injuries.

ase D015-99

he case vehicle was a 1999 Mitsubishi
Mirage, involved in a frontal offset crash with a
992 Honda Accord. The maximum crush was
9" and the delta-V was 45 mph. There was
xtensive intrusion in the driver location – 17"
t the toepan, and 9" at the dashboard. The
river was a 34 year old male, 6'1" tall,
veighing 185 lbs., restrained by a lap and
houlder belt and a depowered air bag. The
ase vehicle is shown in Figure 1. The injuries
vere as follows:

aceration, Ant. Right Seer, Right Lobe
    AIS-2
Fracture, Anterior Right Rib Cage, Multiple
    AIS-3
Fracture, Post. Left Lumbar Vertebra, Multiple
    AIS-2
Fracture, Anterior Shaft of Femur, Left
    AIS-3
2 minor Lacerations and Contusions
    AIS-1

Figure 1. D01599 Case Vehicle

CASE D023-00

The case vehicle was a 1998 Toyota Tacoma
Pickup that impacted an embankment at 12
o'clock with a delta-V of 40 mph. The driver
was a 33 year old male, 68" tall, 176 lbs. He
was restrained by a 3-point seatbelt, and the
air bag deployed. The case vehicle is shown in
Figure 2. The injuries were as follows:

Fracture of Shaft, Right Femur   AIS-3
Fracture of Intertrochanteric Crest of Right
Femur   AIS-3

There were no head or chest injuries.

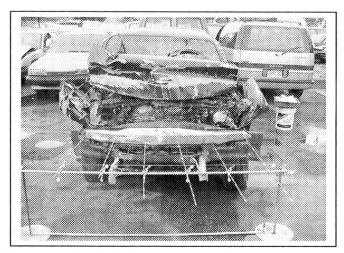

Figure 2. D023-00 Case Vehicle

Case U001-99

The case vehicle was a 1999 Dodge Ram 1500
Pickup, involved in a frontal offset crash with
an unyielding pole. The delta-V was 43 mph
and the pole intruded into the right occupant
compartment. The right front passenger was
an unrestrained 29 year old female. The air
bag on the passenger side had been
deactivated by the on-off switch and it did not
deploy. The passenger was fatally injured from
multiple AIS 6 level head and chest injuries.
The driver was an unrestrained 29 year old
male protected by the driver air bag. He
sustained no head injury or chest injures
greater than AIS 1. He did sustain ulna and
femur fractures. The case vehicle is shown in
Figure 3.

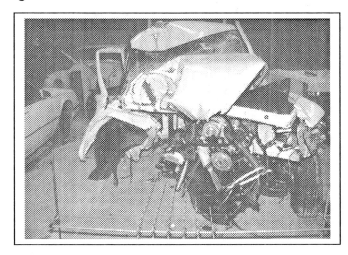

Figure 3. Case Vehicle, U001-99

Case D022-00

The case vehicle was a 2000 Dodge Ram 1500 Pickup that impacted an unyielding pole at 1 o'clock with a delta-V of 41 mph. The unrestrained driver was a 28 year old male, 71" tall, 189 lbs. The vehicle sustained a max of 59" of crush. The vehicle is shown in Figure 4. The only injury was an AIS-3 laceration of the spleen.

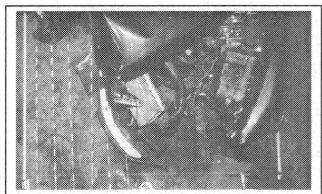

Figure 4. Case Vehicle D023-00

MODERATE SEVERITY CRASHES WITH UNFAVORABLE AIR BAG PERFORMANCE

There were two cases at moderate crash severity in which severe or fatal injuries resulted.

Case D018-99

Figure 5. D01899 Case Vehicle

The case vehicle was a 1999 Pontiac that

and the delta-V was 24 mph. There was 2" of dashboard intrusion. The driver was a 44 year old male, 5'8" tall, weighing 148 lbs. He was unrestrained and the depowered air bag deployed. The case vehicle is shown in Figure 5. The injuries were as follows:

Sternum Fracture	AIS-2
Tear of the Pericardium	AIS-2
Heart Valve Laceration	AIS-4
Ventricle Laceration	AIS-6

CASE D010-99

The case vehicle was a 1998 Honda Accord that impacted the right side of a 1982 Buick Park Avenue at the rear wheel. The delta-V was 19 mph. There was no intrusion. The driver was an unrestrained 18 year old female, 5'4" tall, weighing 130 lbs. The case vehicle is shown in Figure 6. The injuries were as follows:

Laceration, Anterior, Right Medial Segment of Liver, Left Lobe	AIS-5
Avulsion, Anterior Knee, Right	AIS-1
Multiple Contusions, Upper Chest	AIS-1

Figure 6. D01099 Case Vehicle

SEVERE FATAL CASES

The depowered air bag cases with fatalities and crash severities above 40 mph involved massive intrusion of the occupant compartment. The vehicles in these cases were a 1998 Chevrolet Cavalier that impacted a van with a 55 mph delta-V and a 1998 Pontiac Bonneville that impacted a tractor-trailer with a delta-V of 40 mph. These vehicles are shown in Figures 7 and 8.

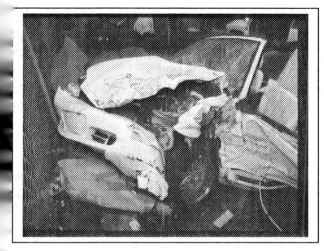

Figure 7. Fatal Crash with Delta-V 55 mph.

complex occupant loading. One such crash is non-fatal case D016-99. This crash involved a frontal underride of a trailer followed by a severe rear impact. The air bag deployed in the frontal impact. However, the occupant compartment intrusion resulted from the rear impact that produced 39 inches of crush. The restrained 22 year old female driver suffered AIS2 rib fractures and an AIS 3 liver laceration. These were attributed to occupant compartment intrusion from the second impact. The damage from the rear impact is shown in Figure 9. The occupant compartment intrusion is shown in Figure 10.

Figure 9. D01699 Case Vehicle Rear Impact

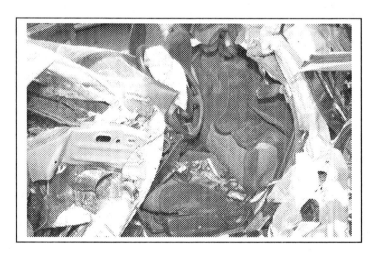

Figure 10. D01699 Case Vehicle Intrusion

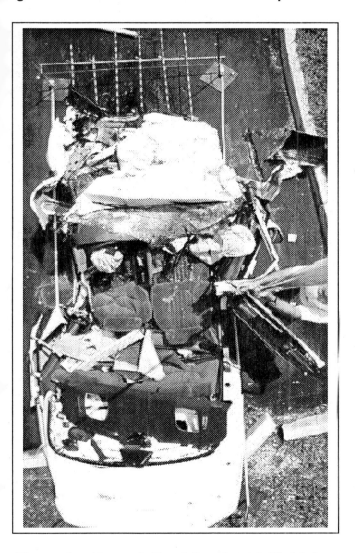

Figure 8. 40+ mph Fatal Crash with Multiple Impacts

MULTIPLE IMPACT CASE

Crashes involving airbag deployment

DISCUSSION

At present, there is insufficient data on depowered air bags to make statistically significant comparisons with 1st generation air bags. However, to date the data from the William Lehman Injury Research Center contains no fatalities at low crash speeds when depowered air bags were present. There have been no reported fatalities to children.

The two of the cases with depowered air bags produced unexpected injuries to unrestrained drivers. These cases were D018-99 and D010-99, summarized earlier. In all other cases, the depowered air bags performed as expected. In all other cases with crash severity less than 40 mph there were no AIS 3+ head and only one AIS 3 chest injury. The AIS 3 chest injury involved intrusion from a rear impact (Case D016-99). In two cases with delta-V greater than 40 mph the depowered air bags performed exceptionally well. In all cases above 40 mph with fatalities, the occupant compartment suffered extensive intrusion.

CONCLUSION

In the limited number of depowered cases investigated by the William Lehman Injury Research Center, the performance of depowered air bags has been very good. High speed protection at crash severities greater than 40 mph has been observed for both restrained and unrestrained occupants. The database of depowered air bags contains no significant injuries in very low speed crashes and no injuries to children.

Three of four fatalities occurred in crashes that were so severe that the occupant compartment was destroyed.

However, serious internal chest injuries were observed in two cases with unrestrained drivers at crash severities of 19 and 24 mph. One of these crashes produced in a fatal heart injury and the other an AIS 5 liver injury. These cases contained the only unexpected injuries among the population protected by depowered air bags.

ACKNOWLEDGEMENT

The authors would like to express appreciation to the Alliance of Automobile Manufacturers for sponsoring this research.

About the Editor

Daniel J. Holt holds a Masters of Science degree in Mechanical Engineering and a Masters of Science degree in Aerospace Engineering. He is currently the Editor-at-Large for SAE's Automotive Engineering International magazine. For 18 years Mr. Holt was the Editor-in-Chief of the SAE Magazines Division where he was responsible for the editorial content of Automotive Engineering International, Aerospace Engineering, Off-Highway Engineering, and other SAE magazines.

He has written numerous articles in the area of safety, crash testing, and new vehicle technology.

Prior to joining SAE Mr. Holt was a biomedical engineer working with the Orthopedic Surgery Group at West Virginia University. He was responsible for developing devices to aid orthopedic surgeons and presented a number of papers on crash testing and fracture healing.

Mr. Holt is a member of Sigma Gamma Tau and a charter member of West Virginia University's Academy of Distinguished Alumni in Aerospace Engineering. He is also a member of SAE.